Advanced Mathematics
Precalculus with Discrete Mathematics and Data Analysis

To lay the groundwork for further study of mathematics at the college level, **all standard precalculus topics** are presented, as well as substantial new material on **discrete mathematics** and **data analysis,** pp. x–xv (contents).

Numerous **applications** lessons, examples, and exercises establish the importance of mathematics to everyday life and a variety of scientific and technical fields, pp. 43–48, 698–700.

Integration of technology throughout the lesson presentations, examples, and exercises fosters effective learning and prepares students for participation in a technological society, pp. 75, 131, 690.

The wide variety of types and multiple levels of **exercises** meet many different teaching and learning needs. New communication exercises, including **Reading, Writing, Discussion,** and **Visual Thinking** exercises, are labeled for easy reference, as are **application exercises,** pp. 107, 129, 154.

Numerous worked-out examples, important results in tinted boxes, Class Exercises, Chapter Summaries, Chapter Tests, and **Cumulative Reviews** make the text accessible and easy to use, pp. 126, 172, 176.

Activities in the lesson presentations, **Investigation and Research exercises,** and end-of-chapter **Projects** are especially useful in promoting active learning, pp. 19, 156, 210.

The flexible outline, incorporation of technology, applications orientation, and provisions for active student learning meet **contemporary standards** (see pp. xvi–xxv).

Careers in Mathematics pages, Biographies, and other text references, establish the **multicultural nature of mathematics** and its importance to people of varied interests and background today, pp. xxxiv, 366, 537.

A section on **College Entrance Examinations** gives students an excellent opportunity to prepare for these important examinations, pp. 790–799.

Teacher's Resources File includes **Tests, Alternative Assessment, Activities Book, Student Resource Guide, Warm-Up Exercises Transparencies,** and **Using Technology** blackline masters, as well as **Precalculus Plotter Plus software.** A separate **Solution Key** is also available. See pp. T12–T21.

Advanced Matnatics

Precalculus with Discrete Mathematics and Data Analysis

Richard G. Br

Teacher's Edition

Editorial Adviser

Andrew M. Gleason

Teacher Consultants

Martha A. Brown
Dane R. Camp
Maria F. G. Fierro
Wallis Green
Linda Hunter
Carolyn Kennedy

McDougal Litell Inc.

A Houghton Mifflin Company
Evanston, Illinois   Boston   Dallas   Phoenix

THE AUTHOR

Richard G. Brown, Mathematics Teac Academy, Exeter, New Hampshire. A teacher and author, Mr. Brown has t of mathematics courses for both students and teachers at several schoo. His affiliations have included the Newton (Massachusetts) High Schoo f New Hampshire, Arizona State University, and the North Carolina Schoo Mathematics during the school year beginning in 1983. In 1989 he was a dential Award for Excellence in Mathematics Teaching. Currently a memb Consortium Council, he is an active participant in professional mathem ns and the author of mathematics texts and journal articles.

EDITORIAL ADVISER

Andrew M. Gleason, Hollis Professor of Math atural Philosophy, Harvard University. Professor Gleason is a well known nematician and a member of the National Academy of Sciences. He has ser nt of the American Mathematical Society.

ACKNOWLEDGMENTS

The author wishes to thank the following teacher ontributions to this teacher's edition: Celia Lazarski, Glenbard North High Scho ream, Illinois; Dane R. Camp, Downer's Grove South High School, Downer's Grov onathan Choate, Groton School, Groton, Massachusetts; Spruill Kilgore, Phi r Academy, Exeter, New Hampshire; Dennis McCowan, Weston High School, lassachusetts; Lois A. Martin, Downingtown High School, Downingtown, Pennsylva ony V. Piccolino, Irondequoit High School, Rochester, New York; James Sconyers, E on High School, Terra Alta, West Virginia; Larry Urbaniak, York High School, Elm nois.

ISBN: 0-395-42169-1

3456789-D-97 96 95

Contents

The program

The Advanced Mathematics advantages — comprehensive content, technology, and applications.

DISCRETE MATHEMATICS and DATA ANALYSIS are presented through practical applications.

Advanced Mathematics
Contents

COMPREHENSIVE CONTENT lets you tailor a course to fit the needs of you and your students.

TECHNOLOGY, integrated appropriately, facilitates the teaching and learning of mathematics.

TEXAS INSTRUMENTS TI-81

VARIED APPLICATIONS establish the importance of mathematics in everyday life.

Applications

Varied applications establish the importance of mathematics in everyday life.

10-3 Double-Angle and Half-Angle Formulas

Objective *To derive and apply double-angle and half-angle formulas.*

Trigonometric functions are used in science and engineering to study light and sound waves. An important application is the wave pattern of a vibrating string. Consider the wave of a note sounded by a violin. Not only does the string vibrate as a whole, producing a fundamental tone, but it also vibrates in halves, thirds, and progressively smaller segments, producing overtones (called harmonics). An equation of this type of wave involves sums of sines and cosines of x, $2x$, $3x$, and greater multiples of x. The computer-generated graphs below illustrate this.

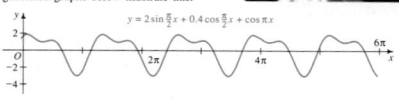
$$y = 2\sin\tfrac{\pi}{2}x + 0.4\cos\tfrac{\pi}{2}x + \cos\pi x$$

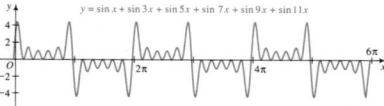
$$y = \sin x + \sin 3x + \sin 5x + \sin 7x + \sin 9x + \sin 11x$$

If you know the value of $\sin\alpha$, you do *not* double it to find $\sin 2\alpha$. Nor do you halve it to find $\sin\tfrac{1}{2}\alpha$. To see that this is true, complete the following activity.

Activity

1. On a single set of axes, graph $y = \sin 2x$ and $y = 2\sin x$. Do the graphs coincide? What does this tell you about the values of $\sin 2x$ and $2\sin x$?
2. On a single set of axes, graph $y = \sin\tfrac{1}{2}x$ and $y = \tfrac{1}{2}\sin x$. Do the graphs coincide? What does this tell you about the values of $\sin\tfrac{1}{2}x$ and $\tfrac{1}{2}\sin x$?

380 Chapter Ten

14-1 Matrix Addition and Scalar Multiplication

Objective *To find the sum, difference, or scalar multiples of matrices.*

An automobile dealer sells four different models whose fuel economy is shown below.

	Sports car	Sedan	Station wagon	Van
Miles per gallon for city driving	17	22	17	16
Miles per gallon for highway driving	23	30	24	19

The information in this table can be displayed as a rectangular array of numbers enclosed by brackets, as shown below. Such an array is called a **matrix** (plural, *matrices*). A matrix is usually named by a capital letter. In this example, E is used to name the matrix.

$$E = \begin{bmatrix} 17 & 22 & 17 & 16 \\ 23 & 30 & 24 & 19 \end{bmatrix}$$

Each number in a matrix is an **element** (or **entry**) of the matrix. The **dimensions** of a matrix are the number of rows and columns. Matrix E has two rows and four columns, so we say that E is a 2×4 (read "two by four") matrix, denoted by $E_{2\times 4}$. (The number of rows is first and the number of columns is second.)

Sometimes it is helpful to label the rows and columns to remind us of what each represents. In the matrix below, the row labels remind us of the city and highway driving conditions and the column labels remind us of the model. We say that the fuel economy matrix, E, is a "driving-condition by model" matrix.

$$\begin{array}{c} \\ c \\ h \end{array}\begin{array}{cccc} sp & se & sw & v \end{array} \\ \begin{bmatrix} 17 & 22 & 17 & 16 \\ 23 & 30 & 24 & 19 \end{bmatrix} = E$$

Of course, the same information could be given by interchanging the rows and columns of E. Matrix F, shown on the next page, is called the **transpose** of matrix E, and is denoted by E^t.

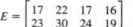 Computer-aided design allows us to model three-dimensional objects on a two-dimensional screen. We can adjust the size, shape, and orientation of parts without the time and expense of constructing physical prototypes.

Matrices **517**

Use the followi[ng]
38, you will ne[ed]

Suppose x_1 and x_2 ar[e]
increasing in its dom[ain]
its domain if $f(x_2) < f(x_1)$ whenever $x_2 > x_1$.

37. **a.** For what value(s) of m is the linear function $f(x) = mx$ an increasing function? a decreasing function? neither?
    b. *Visual Thinking* Describe what the definitions of increasing and decreasing functions imply about the graphs of the functions.

38. Graph each function using a computer or a graphing calculator. Then use the graph to tell whether the function is increasing or decreasing in its domain.
    a. $f(x) = x^3 + x - 1$ **b.** $f(x) = \sqrt[5]{1 - x^3}$ **c.** $f(x) = \dfrac{10}{1 + 2^x}$

39. *Visual Thinking* Suppose the graph of an increasing function is reflected in (a) the x-axis, (b) the y-axis, and (c) the line $y = x$. In each case, tell whether the reflected graph represents an increasing or a decreasing function.

40. On one set of axes, graph $y = |x - 2|$, $y = |x - 2| - 2$, and $y = ||x - 2| - 2|$.

41. Use symmetry to sketch the graph of $x^{2/3} + y^{2/3} = 1$.

4-4 Periodic Functions; Stretching and Translating Graphs

Objective *To determine periodicity and amplitude from graphs, to stretch and shrink graphs both vertically and horizontally, and to translate graphs.*

Periodic Functions

The world is full of periodic phenomena. The tides come in and go out again and again, each *cycle* or *period* lasting about 12.4 h. The amount of daylight increases and decreases with a period of one year. The functions that describe periodic behavior are called *periodic* functions.

A function f is **periodic** if there is a positive number p, called a **period** of f, such that

$$f(x + p) = f(x)$$

for all x in the domain of f. The smallest period of a periodic function is called the **fundamental period** of the function.

138 Chapter Four

SECTION INTRODUCTIONS involve students through relevant applications.

T6

Applications of Probability

16-5 Working with Conditional Probability

Objective *To solve problems involving conditional probability.*

In this section we will consider the probability of a certain cause when a certain effect is observed. For example, a flu can cause symptoms such as high fever and sore throat, but these can be symptoms of other disorders besides flu. Example 1 examines the probability that a person who has a fever (an effect) also has the flu (a possible cause of the fever).

Example During a flu epidemic, 35% of a school's students have the flu. Of those with the flu, 90% have high temperatures. However, a high temperature is also possible for people without the flu; in fact, the school nurse estimates that 12% of those without the flu have high temperatures.
a. Incorporate the facts given above into a tree diagram.
b. About what percent of the student body have a high temperature?
c. If a student has a high temperature, what is the probability that the student has the flu?

Solution **a.** In the tree diagram at the right, F and F' represent the events "has the flu" and "does not have the flu," and T and T' represent "has a high temperature" and "does not have a high temperature." The probabilities in red come directly from the description in the example. The probabilities in blue are deduced from the fact that all the branches from any given point of a tree must have probabilities that total 1.

0.35 0.65
F F'
0.90 0.10 0.12 0.88
T T' T T'

b. To find $P(T)$, the probability that a student has a high temperature, we add the probabilities of the two paths leading to a T:
$$P(T) = P(F \text{ and } T) + P(F' \text{ and } T)$$
$$= 0.35 \times 0.90 + 0.65 \times 0.12$$
$$= 0.315 + 0.078 = 0.393$$

Thus, 39.3% of the student body have a high temperature.

c. Since a high temperature already exists, we consider only the *portion* of the students who have high temperatures. Part (b) has shown that this portion, 0.393, is the sum of 0.315 (those with high temperatures and the flu) and 0.078 (those with high temperatures and no flu). Thus:
$$P(F \mid T) = \frac{P(F \text{ and } T)}{P(T)} = \frac{0.315}{0.393} \approx 0.802$$

624 *Chapter Sixteen*

Example 5 A house bought five years ago for $100,000 was just sold for $135,000. To the nearest tenth of a percent what was the annual growth rate?

Solution
$$A(t) = A_0(1 + r)^t$$
Since $A_0 = 100{,}000$
and $A(5) = 135{,}000$,
$$135{,}000 = 100{,}000\,(1 + r)^5.$$
Therefore:
$$1.35 = (1 + r)^5$$
$$(1.35)^{1/5} = (1 + r)$$
$$r \approx 0.0619$$

To the nearest tenth of a percent, the growth rate was 6.2%.

CLASS EXERCISES

Simplify each expression.

1. a. $4^{1/2}$ b. $4^{-1/2}$ 2. a. $4^{3/2}$ b. $4^{-3/2}$
3. a. $-9^{1/2}$ b. $-9^{-1/2}$ 4. a. $(3^{1/2} \cdot 5^{1/2})^2$ b. $(3^{1/2} + 5^{1/2})^2$
5. $\left(\frac{49}{25}\right)^{-1/2}$ 6. $\left(\frac{4}{9}\right)^{3/2}$ 7. $(8^{-1/6})^{-2}$ 8. $8^{3/2} \cdot 2^{3/2}$
9. $(2x^{-1/3})^3$ 10. $\left(\frac{125}{x^6}\right)^{1/3}$ 11. $\frac{x^{1/3}}{2x^{-2/3}}$ 12. $2x^{3/2} \cdot 4x^{-1/2}$

13. **Consumer Economics** The cost in dollars of a new pair of running shoes t years from now is $C(t) = 62(1.05)^t$.
 a. What is the cost now?
 b. To find the cost in 2.5 years, use $t = \underline{\ ?\ }$.
 c. To find the cost 9 months ago, use $t = \underline{\ ?\ }$.

Solve.

14. $3^{2x} = 3^{12}$ 15. $9^x = 3^5$
16. $x^{2/3} = 9$ 17. $x^{-1/2} = 4$

18. **Discussion** If the exponent is an integer n, then the laws of exponents are true for both positive and negative bases. When the definition of exponent is extended to rational numbers, the base must be restricted to positive numbers. Discuss why this restriction must be made.

Exponents and Logarithms **177**

---

11. $|y| > 1$ 12. $|x| \le 2$ 13. $|x - 3| < 2$
14. $2 < |x + 4| < 3$ 15. $y > x^3 - 9x$ 16. $y < x^4 - 5x^2 + 4$

In Exercises 17–30, graph the solution set of the given system of inequalities.

17. $x \ge 0$ 18. $y \le 0$ 19. $x < 0$
 $x + 2y \le 4$ $2x + y \le 4$ $3x - 2y \le -6$
20. $y < 0$ 21. $y \ge x^2 - 2$ 22. $y \le 6 - x^2$
 $x - y > -1$ $y < x$ $2x - y \le -3$
B 23. $y \le x^3 + 4$ 24. $y \le x^2 + x - 2$
 $y > x^2 + x - 6$ $y \le -x^2 + x + 12$
25. $0 \le x \le 3$ 26. $-1 \le x \le 4$
 $0 \le y \le 2$ $-2 \le y \le 5$
 $y \le 2 - x$ $5y \ge x + 11$
27. $|x| < 3$ 28. $|x| \ge 2$
 $|y| < 1$ $|y| \le 4$
29. $1 \le |x - 4| \le 3$ 30. $1 \le |x + 3| \le 3$
 $1 \le |y - 4| \le 3$ $1 \le |y + 4| \le 4$

31. **Geology** Over the years, Yellowstone National Park rangers have compiled data on the eruption of the Old Faithful geyser. The graph below shows the results of the data. For example, if the eruption lasts 3 min, the next eruption will most likely occur sometime between 60 min and 80 min later. Eruptions lasting less than 1.5 min or more than 5.5 min are rare. Describe the shaded region in the graph by a set of inequalities.

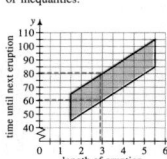

Inequalities **107**

EXAMPLES walk students through the mathematics needed to solve real-world problems.

EXERCISES provide a range of applications and connections in and out of the classroom.

Technology

Technology, integrated appropriately, facilitates the teaching and learning of mathematics.

10-4 Solving Trigonometric Equations

Objective   *To use identities to solve trigonometric equations.*

The acceleration of a body falling toward the Earth's surface is called acceleration due to gravity, often denoted by g. In theoretical physics, g is usually considered constant. However, g is not actually constant but varies slightly with latitude. A good approximation to the value of g can be found by using the following formula, which expresses g in terms of θ, the latitude in degrees.

Gravity acting on water creates this waterfall.

$$g = 9.78049(1 + 0.005288 \sin^2 \theta - 0.000006 \sin^2 2\theta)$$

For example, if you live in Chicago, which has a latitude of 42°N, $\sin \theta \approx 0.6691$ and $\sin 2\theta \approx 0.9945$. Therefore, $g \approx 9.8036$ m/s².

As you can see from this example, some problems involve trigonometric equations that have multiples of angles or numbers. The following suggests two methods that may be helpful in solving such equations. The first method gives a graphical method using a graphing calculator or computer, and the second method gives ways for solving the equation algebraically.

Methods for Solving the Trigonometric Equation $f(x) = g(x)$

Method 1

Use a graphing calculator or computer to graph $y = f(x)$ and $y = g(x)$ on the same set of axes. Use the zoom or trace feature to find the x-coordinates of any intersection points of the two graphs.

Method 2

Use the following guidelines.
a. It may be helpful to draw a quick sketch of $y = f(x)$ and $y = g(x)$ to see roughly where the solutions are.
b. If the equation involves functions of $2x$ and x, transform the functions of $2x$ into functions of x only by using identities.
c. If the equation involves functions of $2x$ only, it is usually better to solve for $2x$ directly and then solve for x.
d. Be careful not to lose roots when you divide both sides of an equation by a function of the variable. Review the discussion about losing roots on pages 32–33.

Example 1   Solve $\cos 2x = 1 - \sin x$ for $0 \le x < 2\pi$.

Solution   **Method 1**   The diagram below shows the graphs of $y = \cos 2x$ and $y = 1 - \sin x$ on the same set of axes. There are four solutions in the interval $0 \le x < 2\pi$. From the diagram, you can see that 0 and π are solutions. Using a zoom feature, you can find that $0.52 \approx \dfrac{\pi}{6}$ and $2.62 \approx \dfrac{5\pi}{6}$ are also solutions.

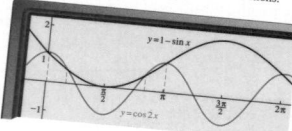

Method 2

$$\cos 2x = 1 - \sin x$$
$$1 - 2\sin^2 x = 1 - \sin x$$
$$2\sin^2 x = \sin x$$

$\sin x = 0$	$2 \sin x = 1$
$x = 0, \ \pi$	$\sin x = \dfrac{1}{2}$
	$x = \dfrac{\pi}{6}, \ \dfrac{5\pi}{6}$

Graphs are useful for solving not only trigonometric equations but also trigonometric inequalities. For example, to solve the inequality $\cos 2x < 1 - \sin x$, look at the graph shown in Example 1. You can see that the graph of $y = \cos 2x$ is *below* the graph of $y = 1 - \sin x$ when $\dfrac{\pi}{6} < x < \dfrac{5\pi}{6}$ and when $\pi < x < 2\pi$.

Example 2   Solve $3 \cos 2x + \cos x = 2$ for $0 \le x < 2\pi$.

Solution

$$3 \cos 2x + \cos x = 2$$
$$3(2 \cos^2 x - 1) + \cos x = 2 \leftarrow \cos 2x = 2 \cos^2 x - 1$$
$$6 \cos^2 x + \cos x - 5 = 0$$
$$(6 \cos x - 5)(\cos x + 1) = 0$$

$$\cos x = \frac{5}{6} \quad \text{or} \quad \cos x = -1$$
$$x \approx 0.59, \ 5.70 \qquad\qquad x = \pi$$
$$\qquad\qquad\qquad\qquad x \approx 3.14$$

1. The two programs below evaluate $P(x) = 2x^4 - 5x^3 + 7x^2 - 9x + 11$ for integral values of x from 1 to 100. In the first program $P(x)$ is in its given form, while in the second program $P(x)$ is in the form used for synthetic substitution: $x(x(x(2x - 5) + 7) - 9) + 11$. Determine the running time for each program. Which one is faster?

```
10 FOR X = 1 TO 100
20 LET P = 2 * X^4 - 5 * X^3 + 7 * X^2 - 9 * X + 11
30 NEXT X
40 PRINT ''DONE''
50 END
```

```
10 FOR X = 1 TO 100
20 LET P = X * (X * (X * (2 * X - 5) + 7) - 9) + 11
30 NEXT X
40 PRINT ''DONE''
50 END
```

2. Consider a third program, just like those in Exercise 1 except for line 20:

```
20 LET P = 2*X*X*X*X - 5*X*X*X + 7*X*X - 9 * X + 11
```

How does the running time for this program compare with the running times for the programs in Exercise 1?

2-2 Synthetic Division; The Remainder and Factor Theorems

Objective To use synthetic division and to apply the remainder and factor theorems.

In earlier algebra courses, you learned how to add, subtract, multiply, and divide polynomials. What follows is an example of finding the quotient and remainder when $2x^4 - 15x^2 - 10x + 5$ is divided by $x - 3$ using long division.

$$
\begin{array}{r}
2x^3 + 6x^2 + 3x - 1 \quad \longleftarrow \text{ Quotient} \\
\text{Divisor} \longrightarrow x - 3 \overline{)2x^4 + 0x^3 - 15x^2 - 10x + 5} \quad \longleftarrow \text{ Dividend} \\
\underline{2x^4 - 6x^3} \\
6x^3 - 15x^2 \\
\underline{6x^3 - 18x^2} \\
3x^2 - 10x \\
\underline{3x^2 - 9x} \\
-1x + 5 \\
\underline{-1x + 3} \\
2 \quad \longleftarrow \text{ Remainder}
\end{array}
$$

The diagram on the previous page illustrates the convention that the pole is placed at the origin with the polar axis coinciding with the positive x-axis and with positive angles measured in the counterclockwise direction. The given equations can be used as formulas for converting from one coordinate system to another. Many scientific and graphing calculators have built-in conversion operations.

Example 2 **a.** Give polar coordinates for the point $(3, 4)$.
b. Give the rectangular coordinates for the point $(3, 30°)$.

Solution **a.** $r = \pm\sqrt{x^2 + y^2} = \pm\sqrt{3^2 + 4^2} = \pm 5$
$\tan \theta = \dfrac{y}{x} = \dfrac{4}{3}$. Thus, $\theta \approx 53.1°$ is one possible value of the angle. $(3, 4)$ is in Quadrant I. If $\theta = 53.1°$, then the point is on the terminal ray, and so $r = 5$. Therefore, polar coordinates are $(5, 53.1°)$. Another possible pair of polar coordinates of the same point is $(-5, 233.1°)$.
b. $x = r \cos 30° = 3 \cos 30°$ and $y = r \sin 30° = 3 \sin 30°$. Thus, rectangular coordinates of the point are $\left(\dfrac{3\sqrt{3}}{2}, \dfrac{3}{2}\right)$.

The graph of a polar equation can be drawn by a graphing calculator, as described below, provided the calculator has a polar mode or a parametric mode. Many calculators use t instead of θ.

If the calculator has polar mode, enter the equation: $r = f(t)$
If the calculator has parametric mode, enter the equations: $x = r \cos t = f(t) \cos t$
$y = r \sin t = f(t) \sin t$

The two equations for x and y are called *parametric equations*. We will study them in greater detail in Chapter 12, where t often represents time in a practical problem.

Example 3 Sketch the polar graph of $r = 2 \sin 2\theta$.

Solution With a graphing calculator
In polar mode, enter:
$r = 2 \sin 2t$
In parametric mode, enter:
$x = 2 \sin 2t \cos t$
$y = 2 \sin 2t \sin t$
The following values were used for the graph of $r = 2 \sin 2t$ shown:
$0 \le t \le 2\pi$, step: $\dfrac{\pi}{30}$
$-2 \le x \le 2$, scale: 1
$\le y \le 2$, scale: 1
(continues on the next page.)

For Exercises 23–27, first write a matrix equation, and then solve the equation. You will need to use computer software or a calculator that computes matrix operations in order to solve the equation.

23. $3x + 5y - 9z = 26$
$4x + 7y + 2z = -7$
$6x - 9y - 8z = 3$

24. $3a + 4b - 5c - 11d = 5$
$5a - 2b - 7c + 8d = 6$
$2a - 4b - 8c + 16d = -3$
$3b + 2c + 8d = 4$

25. Nutrition A dietician wants to combine four foods (I, II, III, and IV) to make a meal having 78 units of vitamin A, 67 units of vitamin B, 146 units of vitamin C, and 153 units of vitamin D. The matrix below gives the vitamin content per ounce of each food. How many ounces of each food should be included in the meal? Begin by writing the unknowns in a 4×1 matrix. Set up a matrix equation and then solve, giving answers to the nearest ounce.

$$
\begin{array}{c}
\quad\quad\text{I}\;\;\text{II}\;\;\text{III}\;\;\text{IV} \\
\begin{array}{c} A \\ B \\ C \\ D \end{array}
\begin{bmatrix} 3 & 2 & 2 & 6 \\ 2 & 3 & 5 & 0 \\ 8 & 6 & 4 & 7 \\ 5 & 5 & 8 & 6 \end{bmatrix}
\end{array}
$$

26. a. Manufacturing A company makes oak tables, chairs, and desks. Each item requires labor time in minutes, as given in the matrix below. The amount of time available for labor each week is 20,250 min for carpentry, 12,070 min for assembly, and 17,000 min for finishing. If the production manager wants to use all of the available labor, how many tables, chairs, and desks should the manager schedule for production each week?

$$
\begin{array}{c}
\quad\quad\text{tables}\;\;\text{chairs}\;\;\text{desks} \\
\begin{array}{c} \text{carpentry} \\ \text{assembly} \\ \text{finishing} \end{array}
\begin{bmatrix} 120 & 105 & 125 \\ 40 & 65 & 110 \\ 80 & 90 & 125 \end{bmatrix}
\end{array}
$$

b. Suppose that because of vacation schedules the amount of labor available is less for the coming week. The amount of labor available is 14,960 min for carpentry, 8,970 min for assembly, and 12,590 min for finishing. How many tables, chairs, and desks should the manager schedule for this week?

27. Suppose the points $(4, 11)$, $(-6, -9)$, and $(8, 61)$ are known to lie on the parabola $y = ax^2 + bx + c$. Write three equations in terms of a, b, and c. Express this system as a single matrix equation, and then solve to find an equation of the parabola.

Students EXPLORE and STUDY polynomial functions, trigonometry, polar graphs, matrix applications, statistics, curve fitting, mathematical modeling, and programming techniques.

Discrete mathematics and data analysis

The mathematics needed in a changing society is presented through practical applications.

39. Physics From a point 30 yd directly in front of the goal posts, a football is kicked at angle of elevation θ with initial velocity v. If a coordinate system is set up with the ball at $(0, 0)$ as shown, then the position (x, y) of the ball t seconds after it is kicked is given by the parametric equations

$$x = (v \cos \theta)t \quad \text{and} \quad y = (v \sin \theta)t - 16t^2.$$

a. Suppose the initial velocity of the ball is 60 ft/s and the angle of elevation is $30°$. Find an equation in x and y that describes the path of the ball.

b. The goal post crossbar is 10 ft above the ground. Under the conditions stated in part (a), will the ball pass over the goal post crossbar?

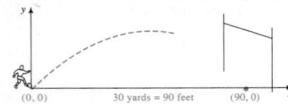

(0, 0)   30 yards = 90 feet   (90, 0)

c. To simulate the motion of the football, use a computer or a graphing calculator to graph the parametric equations of the ball's path when $v = 60$ ft/s and $\theta = 30°$. What parametric equations can you use to represent the goal post with a vertical line segment 10 units long? (For the most effective simulation, the goal post should be drawn first.)

d. Keeping $v = 60$ ft/s, vary the value of θ. To the nearest degree, find the minimum value of θ that will allow the ball to pass over the crossbar.

40. Physics A projectile is fired from a cannon whose angle of elevation is θ and whose muzzle velocity is v. If the muzzle is at the origin of a coordinate system, the position $P(x, y)$ of the particle t seconds later is given by the parametric equations $x = (v \cos \theta)t$ and $y = (v \sin \theta)t - 5t^2$.

a. What do these parametric equations give when $\theta = 90°$?

b. Show that the projectile will hit the ground $(y = 0)$ when $t = \dfrac{v \sin \theta}{5}$. Then show that the x-value when the projectile hits the ground is $x = \dfrac{v^2 \sin 2\theta}{10}$.

c. For what value of θ will the cannon fire the longest horizontal distance?

d. Graph the projectile's path for $v = 100$ m/s and $\theta = 15°$, $45°$, and $60°$.

e. Find a Cartesian equation of the projectile's path. Is the path parabolic?

Vectors and Determinants **439**

Chapter 12
VECTORS AND DETERMINANTS
Aviation, Navigation, Physics, Sports

T10

Chapter 13
SEQUENCES AND SERIES
Physics, Consumer Economics, Geography, Chemistry, Finance, Medicine, Ecology

26. Visual Thinking Let r_n represent the number of are drawn in a plane such that no two lines concurrent. The diagrams below illustrate r_1, equation for r_n.

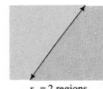

$r_1 = 2$ regions

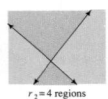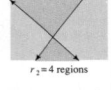

$r_2 = 4$ regions

27. Visual Thinking Let p_n represent the number when n lines are drawn in a plane such that no three lines are concurrent. The diagrams in E $p_2 = 1$, and $p_3 = 3$.

a. Find p_4 and p_5.

b. Give a recursion equation for p_n.

28. Writing For what kind of problem would you be more likely to use a recursive definition of a sequence than an explicit definition? Why? For what type of problem would an explicit definition be more useful?

For the parts of the following exercises designated in green, you will need to use a computer or a programmable calculator.

29. Finance On the birth of their daughter, Mr. and Mrs. Swift began saving for her college education by investing $5000 in an annuity account paying 10% interest per year. Each year on their daughter's birthday they invested $2000 more in the account.

a. Let A_n represent the amount in the account on their daughter's nth birthday. Give a recursive definition for A_n.

b. Find the amount that will be in the account on her 18th birthday.

30. Finance When Mr. Tallchief reaches his retirement age of 65, he expects to have a retirement account worth about $400,000. One month after he retires, and every month thereafter, he intends to withdraw $4000 from the account. The balance will be invested at 9% annual interest compounded monthly.

a. Let A_n represent the amount in the account n months after Mr. Tallchief's retirement. Give a recursive definition for A_n.

b. When will there be no money left in the account?

Sequences and Series **483**

14-6 Transformation Matrices

Objective To find the images of points under different types of transformations using matrices.

Many movies and TV programs have introductions in which computer graphics are used to enlarge or rotate an image or printed message. Often the image or message seems to be transformed into three dimensions even though you are looking at a two-dimensional screen. Geometric transformations in two dimensions can be represented by 2×2 matrices and three-dimensional transformations can be represented by 3×3 matrices.

In Chapter 4 you learned that a function is a correspondence or rule that assigns to every element in one set of objects (the domain) exactly one element in a second set of objects (the range). For most of the functions studied in this book, the domain and range have been sets of numbers. If the domain and range are sets of points, then the function is usually called a **transformation of the plane**.

Consider the two-dimensional transformation T that maps each point $P(x, y)$ to its image point $P'(x + 2y, 3x + y)$. Under this *linear transformation*, line segments are mapped onto line segments, so the effect of T on an n-sided figure is easily found by finding just the images of the vertices of the figure. For example, consider $\triangle ABC$ shown below. To find its image, $\triangle A'B'C'$, we find the images of $A(0, 2)$, $B(3, 1)$, and $C(2, -3)$ using $T: (x, y) \to (x + 2y, 3x + y)$ as follows:

Since $A = (0, 2)$, $T: (0, 2) \to (0 + 2(2), 3(0) + 2)$,   so $A' = (4, 2)$.
Since $B = (3, 1)$, $T: (3, 1) \to (3 + 2(1), 3(3) + 1)$,   so $B' = (5, 10)$.
Since $C = (2, -3)$, $T: (2, -3) \to (2 + 2(-3), 3(2) + -3)$,   so $C' = (-4, 3)$.

Plot $A'(4, 2)$, $B'(5, 10)$, and $C'(-4, 3)$ and draw $\triangle A'B'C'$. Notice that $\triangle A'B'C'$ has the reverse orientation to that of $\triangle ABC$: reading the vertices in alphabetical order, one must move clockwise around $\triangle ABC$ and counterclockwise around $\triangle A'B'C'$.

In the transformation T above, if the image of $P(x, y)$ is $P'(x', y')$, then:

$$x + 2y = x'$$
$$3x + y = y'$$

These equations can be simplified by writing a single matrix equation.

$$\begin{bmatrix} 1 & 2 \\ 3 & 1 \end{bmatrix} \begin{bmatrix} x \\ y \end{bmatrix} = \begin{bmatrix} x' \\ y' \end{bmatrix}$$

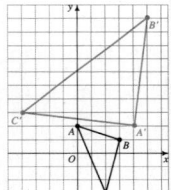

...matrices **551**

Chapter 14
MATRICES
Business, Manufacturing, Animal Science, Industrial Design, Nutrition, Ecology, Geography, Transportation, Psychology, Statistics, Biology

...ay need to consult an
...ch set.)
...South America,
...ies bordering
...Sea,
...ies bordering
...h the Amazon
...senting A, P, and R
... eight regions of yo...
...ed by that region.

protein called the Rhesus factor on its surface. If the protein is present, t... blood is said to be ''Rh positive''; if not, the blood is ''Rh negative.'' Abo... 85% of the American population is ''Rh positive.'' Using this informatio... redraw the first Venn diagram on page 565 and introduce a third set, Rh+, th... overlaps sets A and B. Then indicate the approximate number of people w... belong to each of the eight regions of the new Venn diagram.

19. In a parking lot containing 85 cars, there are 45 cars with automatic transmi... sions, 43 cars with rear-wheel drive, and 46 cars with four-cylinder engine... Of the cars with automatic transmissions, 26 also have rear-wheel drive. ... the cars with rear-wheel drive, 29 also have four-cylinder engines. Of the ca... with four-cylinder engines, 27 also have automatic transmissions. There are ... cars with all three features.
   a. How many cars do not have automatic transmissions and rear-wheel dri... but do have four-cylinder engines?
   b. How many cars do not have any of the three features?

20. Of the 415 girls at Gorham High School last year, 100 played fall sports, 98 played winter sports, and 96 played spring sports. Twenty-two girls played sports all three seasons while 40 played only in fall, 47 only in win- ter, and 33 only in spring.
   a. How many girls played fall and winter sports but not a spring sport?
   b. How many girls did not play sports in any of the three seasons?

570 *Chapter Fifteen*

---

30. **Sports** Refer to Exercise 29 and suppose that the *home* team has a probability p of winning each game and a probability $q = 1 - p$ of losing.
   a. If team X is the home team for games 1, 2, 6, and 7, what is the probability that team X wins the championship? Give the answer in terms of p only.
   b. Use a computer or graphing calculator to find the approximate value of p that makes team X's probability of winning the championship 0.5.

16-3 The Binomial Probability Theorem

Objective To use the binomial probability theorem to find the probability of a given outcome on repeated independent trials of a binomial experiment and to approximate the probability when the trials are not independent.

Consider the experiment in which a die is rolled three times. On each roll, the probability of getting a six (S) is $p = \frac{1}{6}$ and the probability of getting a non-six (N) is $1 - p = \frac{5}{6}$. These probabilities are shown in the tree diagram at the right. We can use the tree diagram to create the table below, which summarizes the results of the experiment.

	3 sixes	2 sixes			1 six			0 sixes
Outcome	SSS	SSN	SNS	NSS	SNN	NSN	NNS	NNN
Probability	$\frac{1}{6}\cdot\frac{1}{6}\cdot\frac{1}{6}$	$\frac{1}{6}\cdot\frac{1}{6}\cdot\frac{5}{6}$	$\frac{1}{6}\cdot\frac{5}{6}\cdot\frac{1}{6}$	$\frac{5}{6}\cdot\frac{1}{6}\cdot\frac{1}{6}$	$\frac{1}{6}\cdot\frac{5}{6}\cdot\frac{5}{6}$	$\frac{5}{6}\cdot\frac{1}{6}\cdot\frac{5}{6}$	$\frac{5}{6}\cdot\frac{5}{6}\cdot\frac{1}{6}$	$\frac{5}{6}\cdot\frac{5}{6}\cdot\frac{5}{6}$
	$\left(\frac{1}{6}\right)^3$	$3\left(\frac{1}{6}\right)^2\left(\frac{5}{6}\right)$			$3\left(\frac{1}{6}\right)\left(\frac{5}{6}\right)^2$			$\left(\frac{5}{6}\right)^3$

Now compare:
$$[p + (1 - p)]^3 = p^3 + \qquad 3p^2(1 - \ldots$$

If the die were rolled 4 times... distribution of probabilities:

Outcome	4 sixes	3 sixes
Probability	$\left(\frac{1}{6}\right)^4$	$4\left(\frac{1}{6}\right)^3\left(\frac{5}{6}\right.$

$$\ldots - p)]^4 = p^4 + 4p^3(1 - p$$

---

   a. Use the graph at the right on the preceding page to find a linear equation for Y in terms of X.
   b. Use your answer to part (a) to find an equation giving y in terms of x.
   c. Suppose that some time in the future a new planet is discovered and its average distance from the sun is estimated at 7000 million kilometers. Use your equation in part (b) to estimate its time of orbit around the sun.

17. **Astronomy** In 1610, Galileo discovered how the time, T, required for each of Jupiter's satellites to revolve about Jupiter is related to the average distance, a, of the satellite from Jupiter. By plotting log T versus log a, see if you can discover the relationship.

Satellite	a (kilometers)	T (hours)
Io	422,000	42.5
Europa	671,000	85.2
Ganymede	1,070,000	171.7
Callisto	1,883,000	400.5

18. **Sports** The winning times (in seconds) for various men's races (in meters) in the 1988 Olympics are given below.

x = distance (m)	100	200	400	800	1500	5000	10,000
y = time (s)	9.92	19.75	43.87	103.45	215.96	791.70	1641.46

   a. Plot log y versus log x. Find an equation of the resulting line.
   b. Find an equation that gives y in terms of x.
   c. If there had been a 1000-meter race in these Olympics, what would you predict the winning time would have been?

19. **Forestry** Forest managers keep records giving the number of board feet B (in hundreds of board feet) of lumber produced by harvested white pine trees with diameter d (in inches). Consider the following data for nine white pine.

			17	18	19	20	21	22	23
			27.61	33.00	39.06	45.84	53.38	61.72	70.90

...s log d and show that you get a line with slope about 3.
...n why part (a) shows that volume varies approximately ... f the diameter.
...n why the result of part (b) is not surprising if you think of ... cylinder-like object whose height h is proportional to the

Curve Fitting and Models **703**

...normal distribution is extremely important in that it occurs in a wide variety of data, including heights and weights of people, dimensions of manufactured goods, test scores, blood cholesterol levels, and times for marathon races.

Consider the following experiment: A psychologist asks 100 children to solve a certain puzzle and records their solution times. The results are shown in figure (c) below. This histogram gives the percent of children who completed the puzzle in the indicated time interval. Each interval includes its left endpoint but not its right endpoint. Thus, 1% of the children solved the puzzle in a time greater than or equal to 25 s, but less than 35 s.

precipitation in cm
0 1 2 3 4 5 6
(b)

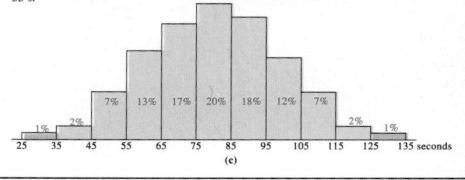

1% 2% 7% 13% 17% 20% 18% 12% 7% 2% 1%
25 35 45 55 65 75 85 95 105 115 125 135 seconds
(c)

Statistics **661**

Teaching support

Additional resources provide flexible teaching options and address the needs of a wide range of students.

TEACHER'S EDITION
- Student pages with overprinted answers
- Convenient side-column format
- Alternate chapter tests
- Professional articles
- Assignment Guide
- Wealth of teaching suggestions

SOLUTION KEY
- Worked-out solutions to all exercises
- Contains all necessary diagrams

OVERHEAD VISUALS
- Color transparencies
- Presentation scripts

TEACHER'S RESOURCES FILE
Convenient five-pocket portfolio includes:

TESTS
- Chapter Tests
- Cumulative Tests
- Answers

ALTERNATIVE ASSESSMENT
- Portfolio Assessment
- Nonroutine Problems
- College Entrance Examinations
- Spiral Review

USING TECHNOLOGY
- Graphing Calculator Activities
- Computer Activities

ACTIVITIES BOOK
- Applications
- Enrichment
- Creative Problem Solving

USING PRECALCULUS PLOTTER PLUS
- Enrichment Topics
- Activities
- Precalculus Plotter Plus Software

STUDENT RESOURCE GUIDE
- Sections for study and review
- Chapter Reviews
- English/Spanish Glossary

WARM-UP EXERCISES
- Class-starter Transparencies

Teacher's Edition

Convenient format for ease of teaching — front of book pages, interleaved pages, side-column material, and overprinted answers

FRONT OF BOOK PAGES
- Overview of the Program
- Professional Articles on Technology, Communication, Connections, and Discrete Mathematics
- Assignment Guide
- Alternate Chapter Tests and Cumulative Reviews

INTERLEAVED PAGES
- Chapter Overview
- Chapter Objectives
- Supplementary Resources List
- Software
- Pacing Guide
- Teaching Notes
  - Presenting the Section
  - Using Technology
  - Applications
  - Cooperative Learning
  - Communication
  - Review
  - Assessment

T14

Chapter 1 Linear and Quadratic Functions

Overview
This chapter reviews the coordinate geometry of lines and the various methods of finding the real and imaginary roots of quadratic equations. Students graph linear and quadratic functions and find the coordinates of any points of intersection of two lines or of a line and a parabola. Using slope along with the distance and midpoint formulas, students verify properties of special quadrilaterals and find equations of medians, altitudes, and perpendicular bisectors of sides of various triangles. Students also solve a variety of real-world problems for which linear or quadratic functions are models.

Objectives
- **1-1** To find the intersection of two lines and to find the length and midpoint of a segment.
- **1-2** To find the slope of a line and to determine whether two lines are parallel, perpendicular, or neither.
- **1-3** To find an equation of a line given certain geometric properties of the line.
- **1-4** To model real-world situations by means of linear functions.
- **1-5** To add, subtract, multiply, and divide complex numbers.
- **1-6** To solve quadratic equations using different methods.
- **1-7** To define and graph quadratic functions.
- **1-8** To model real-world situations using quadratic functions.

Supplementary Resources
Tests, pp. 1–2
Alt. Assess., pp. 1–5, 63
Activities, pp. 1–2
Using Tech., pp. 6–7, 8, 9–10
Student Res. Guide, pp. 1–11

Software
Precalculus Plotter Plus
Activities, pp. 23, 24, 25
Equation Solver, Function Plotter, Conics Quiz

Pacing Guide

Section	1-1	1-2	1-3	1-4	1-5	1-6	1-7	1-8	Review	Test	Total
Standard (days)	1	1	2	1	1	1	1	2	1	1	12
	1	1	1	1	1	1	1	1	1	1	10

Teaching Notes

1-1 pages 1–7

Presenting the Section
You might want to emphasize which pairs of simultaneous linear equations have no solutions and which have infinitely many solutions. Graphically, two parallel lines imply there is no solution, and coincident lines imply there are infinitely many solutions. Algebraically, solving the system would lead to a false statement (like $0 = 2$) if there is no solution, and a statement which is always true (like $0 = 0$) if there are infinitely many solutions. Show examples of each and have the students graph them.

Using Technology
Encourage students to use calculators to check their estimates of solution points to linear equations or to systems of linear equations. Calculators with graphing capabilities and computer graphing programs are also excellent motivational tools for graphing linear equations. Students enter a linear equation in y-form (y is written in terms of x) and the points are plotted automatically. A table of values can be displayed, if needed. Students can also graph more than one line at a time, observe the intersection of the lines, guess its coordinates and then zoom in on the solution to check their guesses. Students may not be able to find the exact solution by zooming in, but can find values as close as necessary. Emphasize the importance of using algebra to determine the exact solution to a system of linear equations.

Cooperative Learning
Learning can be enhanced with group arrangements of three or four students each. Topics include going over more difficult exercises and review of (1) terminology; (2) how to find the solution points of the equation of a line; (3) the various methods of solving a system of linear equations simultaneously; (4) applications of the distance formula, including area problems, and of systems of linear equations; and (5) various approaches to sketching the graph of a line. Encourage students to ask and answer questions within their groups and offer alternate solutions if possible.

Applications
If 3 pencils and 2 pads of paper cost $1.30 while 5 pencils and one pad of paper cost $1.00, what would be the cost of each pencil and each pad of paper? This is a mixture problem, a common application of linear systems of equations. If a is the cost of each pencil and b is the cost of each pad of paper, then

$$3a + 2b = 1.30$$

and

$$5a + b = 1.00,$$

or each pencil costs $.10 and each pad of paper costs $.50. Linear systems of equations also apply to some uniform motion problems, investment problems, other mixture problems, and number problems commonly studied in algebra. Some business applications include finding the break-even point of the cost of production. In all of these problems, students are asked to identify two independent linear relationships and generate two related equations with the same two variables.

Communication
You may wish to summarize and discuss the general form of a linear equation by using the following chart. Ask students to explain why the descriptions following the general form are true and to give additional examples of each description.

Equation	Description	Example
$Ax + By = C$	General form	$2x + 3y = 6$
$Ax + By = 0$	Contains $(0, 0)$	$2x + 3y = 0$
$Ax = C$	Vertical	$2x = 6$
$By = C$	Horizontal	$3y = 6$

Review
You might need to review the Pythagorean theorem before assigning Exercise 34, page 7. If (x_1, y_1) and (x_2, y_2) are the coordinates of the endpoints of a line segment, then $|y_2 - y_1|$ is the vertical distance and $|x_2 - x_1|$ is the horizontal distance to a third point (x_2, y_1). This third point is a vertex of a right triangle with the given line segment as its hypotenuse. The length of the line segment is then

$$\sqrt{(x_2 - x_1)^2 + (y_2 - y_1)^2}.$$

45. Discussion According to legend Manhattan Island was purchased in 1626 for trinkets worth about $24. If the $24 had been invested instead at a rate of 6% interest per year, what would be the value of the money in 1996? Compare this with a recent total of $34,000,000,000 in assessed values for Manhattan Island. about $55,000,000,000

C 46. Solve $2^x + 8 \cdot 2^{-x} = 9$.

47. Solve $2^x + 2^{-x} = \frac{5}{2}$.

48. Solve $2^{2x} - 3 \cdot 2^{x+1} + 8 = 0$.

49. Solve $3^{2x+1} - 10 \cdot 3^x + 3 = 0$.

46. 0,3 47. 1,−1 48. 2,1 49. −1,1

5-2 Growth and Decay: Rational Exponents

Objective   To define and apply rational exponents.

In Section 5-1, we considered a 9% annual growth in the cost of a hamburger. We saw that if a hamburger costs $4 now (time $t = 0$), then its cost t years from now will be $C(t) = 4(1.09)^t$. To find the cost one-half year from now, we must evaluate an expression with a fractional exponent:

$$C\left(\frac{1}{2}\right) = 4(1.09)^{1/2}$$

Although a scientific calculator easily gives an approximate value of $4.18, there remains the question of what a fractional exponent means.

The definition of a rational exponent given below is made in such a way that the laws for integral exponents on page 170 will continue to hold. However, the base b must now be a positive real number other than 1.

Definition of $b^{1/2}$:   If exponent law 7 is to hold for rational exponents, then

$$(b^{1/2})^2 = b^{(1/2)(2)} = b^1.$$

Since we know $(\sqrt{b})^2 = b$, we define $b^{1/2}$ to be $\sqrt{b}$.

Definition of $b^{3/2}$: $b^{3/2} = (b^{1/2})^3 = (\sqrt{b})^3$ and $b^{3/2} = (b^3)^{1/2} = \sqrt{b^3}$.

Either $(\sqrt{b})^3$ or $\sqrt{b^3}$ can be used as a definition of $b^{3/2}$.

Definition of $b^{p/q}$:   Using reasoning similar to that above, we make these definitions:

$$b^{1/q} = \sqrt[q]{b} \quad \text{and} \quad b^{p/q} = (\sqrt[q]{b})^p \text{ or } \sqrt[q]{b^p}$$

Exponents and Logarithms   **175**

Exercise Note

Students should rewrite Exercises 46–49 as quadratic equations before solving. For example, by substituting y for 2^x in Exercise 46, the equation becomes $y + 8 \cdot \frac{1}{y} = 9$, or $y^2 + 8 = 9y$.

Teaching Notes, p. 168B

Warm-Up Exercises

Simplify each expression.

1. $-\sqrt[4]{(-3)^4}$ −3

2. $\sqrt[3]{10^6}$ 100

3. $\left(\sqrt{\frac{a^4}{4}}\right)^3$ $\frac{a^6}{8}$

4. $\sqrt{32} \cdot \sqrt{18}$ 24

5. $\frac{x}{\sqrt[3]{x}} \cdot \sqrt[3]{x^2}$

6. $\sqrt[3]{-\frac{16}{125}}$ $-\frac{2}{5}\sqrt[3]{2}$

Motivating the Section

In this section, students will solve problems like the following: Suppose college tuition increased from $4000 to $10,000 over 10 years. What is the annual rate of increase? (About 9.6%)

Assessment Note

After discussing the definitions on this page, ask students questions like the following:

1. What is $16^{3/2}$? (64)

2. What is $6^{3/2}$? ($\approx$14.7)

3. What is $-6^{3/2}$? ($\approx$−14.7)

4. What is $(-6)^{3/2}$? (Undefined)

SIDE-COLUMN MATERIAL

Warm-Up Exercises
These exercises review previously taught skills or prerequisite skills necessary for the new content.

Motivating the Section
A few sentences that describe each section's purpose and significance, often with a real-world application.

Communication Note
Suggestions for enhancing students' abilities in and appreciation of this aspect of mathematics.

Making Connections
These notes connect current topics to other branches of mathematics.

Error Analysis
These notes point out common mistakes and misconceptions that students may have, and suggest other pitfalls you should caution students about.

In addition to those sections listed above, the side-column notes may include:

- Supplementary Materials ● Problem Solving
- Using Technology ● Activity Note ● Review Note
- Cooperative Learning ● Additional Answers
- Suggested Assignments ● Application ● Exercise Note
- Example Note ● Assessment Note ● Mathematical Note
- Additional Examples

T15

Using technology

Students gain a deeper understanding of mathematics by using calculators and computers.

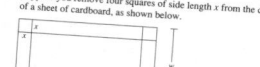

Graphing Calculator: Maximums and Minimums
For use with Section 2-4

Sketch the graph of each of the following polynomial functions for $-10 \leq x \leq 10$ and $-10 \leq y \leq 10$. Use the trace and zoom features to determine the maximum or minimum values and zeros of each function.

1. $f(x) = 2x^2 + 4x$
2. $h(x) = -x^2 + 6x - 7$
3. $g(x) = x^2 - 4x + 4$
4. $f(x) = x^2(x - 2)(x + 1)$
5. $k(x) = |-x^3 + 5x|$
6. $g(x) = -2x^2 + 4x - 7$

7. Suppose you remove four squares of side length x from the corners of a sheet of cardboard, as shown below.

The volume of the open box formed by this piece of cardboard is $V(x) = (l - 2x)(w - 2x)x$.

a. Write the volume as a function of x if the cardboard is 16 cm long and 12 cm wide.

b. Set the dimensions of the screen at $0 \leq x \leq 10$ and $-50 \leq y \leq 250$, x-scale: 1, y-scale: 50. Graph $V(x)$ and sketch the graph.

c. For which values of x is $V(x)$ positive? For which values of x is $V(x)$ negative? Which values of x should be excluded as solutions to this problem? Why?

d. Use the trace and zoom features to find the approximate maximum value of $V(x)$. What value of x gives the maximum volume? What is the approximate maximum volume?

e. Find the maximum volume of each open box formed from a rectangular piece of cardboard with the dimensions below by graphing each function and using the trace and zoom features.

Length	Width	Maximum
15 in.	10 in.	
10 ft	8 ft	

Activities using graphing, spreadsheet, or programming capabilities of a graphing calculator or software help students explore math topics.

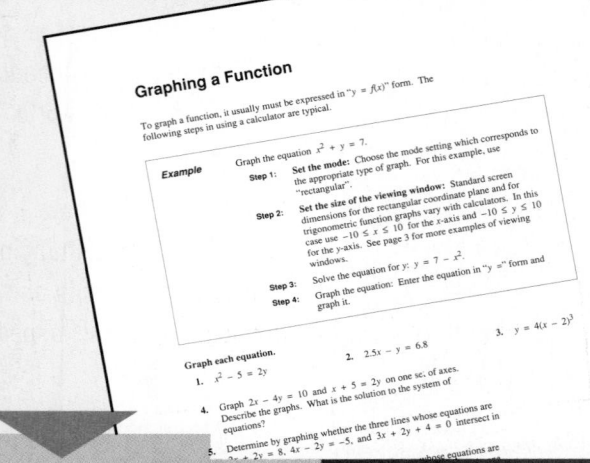

Graphing a Function

To graph a function, it usually must be expressed in "$y = f(x)$" form. The following steps in using a calculator are typical.

Example — Graph the equation $x^2 + y = 7$.

Step 1: **Set the mode:** Choose the mode setting which corresponds to the appropriate type of graph. For this example, use "rectangular".

Step 2: **Set the size of the viewing window:** Standard screen dimensions for the rectangular coordinate plane and for trigonometric function graphs vary with calculators. In this case use $-10 \leq x \leq 10$ for the x-axis and $-10 \leq y \leq 10$ for the y-axis. See page 3 for more examples of viewing windows.

Step 3: Solve the equation for y: $y = 7 - x^2$.

Step 4: Graph the equation: Enter the equation in "$y =$" form and graph it.

Graph each equation.
1. $x^2 - 5 = 2y$
2. $2.5x - y = 6.8$
3. $y = 4(x - 2)^3$

4. Graph $2x - 4y = 10$ and $x + 5 = 2y$ on one sc. of axes. Describe the graphs. What is the solution to the system of equations?

5. Determine by graphing whether the three lines whose equations are ... intersect in ...

Students are introduced to the basic procedures required to use a graphing calculator successfully.

Using Technology
by Celia Lazarski

Calculator and Computer Activities

Advanced Mathematics
Precalculus with Discrete Mathematics and Data Analysis

Houghton Mifflin

Tips for Sta...

TEXAS INSTRUMENT...
To enter statistical data:

a. Press [2nd] [STAT] and select DATA from the menu. Press [2] (ClrStat) and [ENTER] to clear all old data from the memory.

b. Press [2nd] [STAT] and select DATA again. Press [1] (Edit). Enter each value; press [ENTER] after each entry.

To graph data:
a. Store the data as described above.

b. Press [2nd] [DRAW] and select [1] (ClrDraw) to clear the screen. Press [ENTER]

c. Press [RANGE] and enter values appropriate for the data.

d. Press [2nd] [STAT] and select DRAW. Select [1] (Histogram), [2] (Scatter plot), or [3] (xy Line). Press [ENTER]

To fit an exponential model to a set of data:
a. Store the data as described above.

b. Press [2nd] [STAT] and select CALC to choose a model. Press [4] (ExpReg) [ENTER] to choose an exponential regression curve.

c. Press [VARS] and select LR. Press [4] (RegEQ) to display the regression equation.

CASIO fx-7700...

Tips on certain graphing calculator models help students in a broad range of applications.

NAME _____ DATE _____

[Precalculus Plotter Plus]

Matrix Transformations of Plane Vectors

Use the Matrix Calculator program. You will also need graph paper.

1. Polygon P has six vertices: (1, 1), (1, 4), (2, 4), (2, 2), (3, 2), (3, 1).
   a. Graph polygon P on a piece of graph paper. What is its area? _____
   b. Matrix P in the chart below represents polygon P. Enter matrix P by selecting DEFINE and choosing "New matrix." Enter new variable P. The size is 2 rows and 6 columns. Use the ⟨Tab⟩ key to enter the elements of the matrix. Press ⟨Esc⟩ when you are done.
   c. Matrix T in the chart below represents a matrix that will transform each point of polygon P. Enter new matrix T. Press ⟨Esc⟩.
   d. Choose "New equation." Define a new matrix I with the equation $I = TP$. Matrix I represents the image polygon (the result of a transformation on polygon P). Press ⟨Esc⟩.
   e. Define the scalar value $D = \det(T)$ similarly. This places the value of the determinant of matrix T into scalar variable D. Press ⟨Esc⟩ twice.
   f. Select RESULTS and use the arrow keys to view P, T, I, and D. Record the results in the first row of the chart below.

For exercises 2–6, edit matrix T by selecting DEFINE and highlighting "T." Press ⟨Return⟩ or ⟨Enter⟩ three times until the first matrix element is highlighted. Edit the elements of the matrix as shown in the chart, then press ⟨Esc⟩. Select RESULTS and record the new values for I and D. Then draw each image polygon (matrix I) on graph paper and calculate its area.

	T	P	$I = T \cdot P$	$D = \det(T)$	Area
1.	$\begin{bmatrix} 1 & 0 \\ 0 & 1 \end{bmatrix}$	$\begin{bmatrix} 1 & 1 & 2 & 2 & 3 & 3 \\ 1 & 4 & 4 & 2 & 2 & 1 \end{bmatrix}$			
2.	$\begin{bmatrix} -2 & 0 \\ 0 & -2 \end{bmatrix}$	$\begin{bmatrix} 1 & 1 & 2 & 2 & 3 & 3 \\ 1 & 4 & 4 & 2 & 2 & 1 \end{bmatrix}$			
3.	$\begin{bmatrix} 2 & 1 \\ 0 & 1 \end{bmatrix}$	$\begin{bmatrix} 1 & 1 & 2 & 2 & 3 & 3 \\ 1 & 4 & 4 & 2 & 2 & 1 \end{bmatrix}$			
4.	$\begin{bmatrix} -1 & 0 \\ 1 & 2 \end{bmatrix}$	$\begin{bmatrix} 1 & 1 & 2 & 2 & 3 & 3 \\ 1 & 4 & 4 & 2 & 2 & 1 \end{bmatrix}$			
5.	$\begin{bmatrix} -1 & 0 \\ 0 & 1 \end{bmatrix}$	$\begin{bmatrix} 1 & 1 & 2 & 2 & 3 & 3 \\ 1 & 4 & 4 & 2 & 2 & 1 \end{bmatrix}$			
6.	$\begin{bmatrix} -1 & 1 \\ 1 & -1 \end{bmatrix}$	$\begin{bmatrix} 1 & 1 & 2 & 2 & 3 & 3 \\ 1 & 4 & 4 & 2 & 2 & 1 \end{bmatrix}$			

7. Describe how the absolute value of the determinant of the transformation matrix is related to the area of the resulting image polygon, and what effect the sign of the determinant has on the polygon.

Enrichment Topics

Solving Simple Radical Equations by Graphing

One method for solving an equation of the form $f(x) = g(x)$ involves graphing the equations $y = f(x)$ and $y = g(x)$ on the same set of axes and then determining the x-coordinate of each point where the two curves intersect. Students can use the Function Plotter program and its ZOOM feature to do this, or the Equation Solver program and its TRACE feature.

For example, have the students solve the radical equation $\sqrt{x+2} = 4 - x$ by graphing $y = \sqrt{x+2}$ and $y = 4 - x$. (Note that the expression $\sqrt{x+2}$ is typically entered as sqr(x + 2).) Ask the students for the x-coordinate of the point of intersection of the two curves. [2] Then have them substitute this value into the equation $\sqrt{x+2} = 4 - x$ to confirm that it is a solution.

To provide more of a challenge, ask the students to solve $\sqrt{x+2} = 5 - x$ by graphing. In this case the x-coordinate of the point of intersection is irrational $\left(\frac{11 - \sqrt{29}}{2} \right)$, so the students must ZOOM or TRACE to get a reasonable approximation. [2.8]

An alternate approach to solving an equation of the form $f(x) = g(x)$ involves rewriting the equation as $f(x) - g(x) = 0$. When $y = f(x) - g(x)$ is graphed, any x-intercepts will be solutions of the equation. For example, the students can solve $\sqrt{x+2} = 4 - x$ by graphing $y = \sqrt{x+2} - (4 - x)$ and determining the x-coordinates of the point(s) where the graph crosses the x-axis.

Graphing to Find nth Roots of Real Numbers

You can have students approximate the nth root of a number a by using the Equation Solver program to graph $y = x^n - a$. (Use Function Plotter on the Macintosh.) For example, have the students use the Equation Solver program to graph $y = x^2 - 2$. By using TRACE, students can find the x-coordinates of the points where the parabola crosses the x-axis, which are $\pm\sqrt{2}$.

Warn students that they may have to press ⟨ ⟩ and then move the trace box again several times to get a good approximation of the roots. Roughly one decimal place of accuracy is added for each new approximation. For example, a typical series of approximations of $\sqrt{2}$ using this method is: first approximation: 1.42857143; second approximation: 1.41285714; third approximation: 1.4142083.

You might ask students to explain why this method works. [On the x-axis, $y = 0$, so the equation becomes $0 = x^2 - a$. The roots of this equation are $\pm\sqrt{a}$.]

Likewise, the students can approximate $\sqrt[3]{-4}$ by using TRACE to find where the graph of $y = x^3 + 4$ crosses the x-axis. A typical series of approximations is: first approximation: −1.61904762; second approximation: −1.58857143; third approximation: −1.58741667.

Overview of the Disk

The *Precalculus Plotter Plus* disk can be used independently by students or by the teacher as a classroom presentation device. Refer to the User's Manual: Introduction and Worked Examples (pages 3–17) for detailed instruction on all the options available with the *Precalculus Plotter Plus* disk. The Worked Examples may also be used to present and motivate lesson material, and to introduce students to the *Precalculus Plotter Plus* disk.

What can *Precalculus Plotter Plus* do?

Plotters: These programs allow you to enter equations, decide the scales of the axes, and plot the graphs.

Quizzes: These programs check understanding by presenting graphs of a given type, and asking the student to determine equations for them.

Utilities: **Data Analyzer:** a utility program for presenting histograms, scatterplots, and regression lines of spreadsheet data;
Sequence Experiment (Apple and IBM only): a program that generates sequences defined explicitly or recursively, and calculates partial sums;
Matrix Calculator: a utility program that allows students to operate on matrices;
Sampling Experiment (Apple and IBM only): a simulation that chooses samples from a population with specified statistical characteristics.

Plotter Plus User's Manual: *Macintosh*

Follow the instruction sheet that comes with the disk. Once you have inserted the disk, click on the *Plotter Plus* icon. You will see the title screen. Click OK or press the ⟨Enter⟩ key. The menu bar will show File, Edit, and **Programs**. Go to **Programs** and select the program you want from the list of ten (shown at the right). Another menu bar will appear. A typical plotter menu bar shows File, Edit, **Programs**, **Equations**, and **Graph**. To enter the equations, go to **Equations** and select Define **Equations**. You can jump from one text box to another to enter coefficients or constants by using the Tab key or the mouse, and move inside a text box by using arrow keys or the mouse. Under **Graph**, you can select Scale at any time. After you have drawn a graph, you can Zoom In or Zoom Out. Help can be obtained by choosing it from the menu bar or by pressing ⌘-⟨?⟩. You can choose another program by going to **Programs** in the menu bar and selecting a new program. The Macintosh version ... similar to trace on many graphing calculators, that ... includes ...

Line Plotter
Line Quiz
Parabola Plotter
Parabola Quiz
Function Plotter
Inequality Plotter
Conics Plotter
Polar Graph Plotter
Matrix Calculator
Data Analyzer

Using Precalculus Plotter Plus
by David L. Myers

Includes: User's Manual · Worked Examples · Enrichment Topics · Activity Sheets · Answers · Correlations to *Advanced Mathematics, Introductory Analysis,* and *Trigonometry with Applications*

Advanced Mathematics
Precalculus with Discrete Mathematics and Data Analysis

Houghton Mifflin

T17

Student Resource Guide

The guide provides vocabulary review, worked-out examples, and additional problem sets.

Student Resource Guide
for Study and Review

Includes: Key Terms and Main Ideas ·
Worked Examples · Practice Exercises ·
Chapter and Cumulative Reviews · Spanish Glossary

Advanced Mathematics
Precalculus with Discrete Mathematics and Data Analysis

Houghton Mifflin

T18

Exponential Functions

Sections 5-3 and 5-4

OVERVIEW Section 5-3 presents irrational exponents a
Section 5-4 presents the number *e* and the value
Both sections focus on applications such as
compound interest.

KEY TERMS

Exponential function with base *b* (p. 181)
a function of the form $f(x) = ab^x$ where *a* and *b* are positive numbers and $b \neq 1$

e (p. 186)
the value that $\left(1 + \frac{1}{n}\right)^n$ approaches as *n* becomes very large
a number that is approximately equal to 2.7183

Natural exponential function (p. 187)
the function $y = e^x$

UNDERSTANDING THE MAIN IDEAS

Exponential functions
- The graph of an exponential function is a smooth, unbroken curve. This means that an exponent can be a rational or an irrational number. The easiest way to evaluate an expression with an irrational exponent, such as $\pi^{\sqrt{7}}$, is to use your calculator: $[\pi][y^x][7][\sqrt{x}][=][20.669736]$.
- You can use given information to specify an exponential function. (See Example 1 on text page 181.)
- If your calculator does not have an "*e*" button, use the $[INV]$, and $[\ln x]$ keys to find and then store a value for *e*, or simply store 2.7182818 in memory to use as an approximation

Applications
- When an exponential function is used to model exponential growth or decay, its equation is generally written in one of the following forms rather than in the form $f(t) = ab^t$:

Copyright © by Houghton Mifflin Company. All

Sections 5-3 and 5-4

1. $A(t) = A_0(1 + r)^t \leftarrow \begin{array}{l} A_0 = \text{amount at time } t = 0 \\ r = \text{growth/decay rate written as a decimal} \end{array}$

2. $A(t) = A_0 b^{t/k} \leftarrow \begin{array}{l} A_0 = \text{amount at time } t = 0 \\ k = \text{time needed to multiply } A_0 \text{ by } b \end{array}$

   Example: $A(t) = A_0\left(\frac{1}{2}\right)^{t/h} \leftarrow h = \text{half-life of a substance}$

- Use the "Rule of 72" to estimate how long it takes for a quantity to double:
  If a quantity is growing at *r*% per year (or other time unit), then the doubling time is about $(72 \div r)$ years (or other time unit).
- If P_0 dollars are invested at an annual rate *r* (expressed as a decimal) and compounded *n* times a year, then the amount in the account after *t* years is given by $P(t) = P_0\left(1 + \frac{r}{n}\right)^{nt}$. If $1 grows to $(1 + x)$ dollars, then the *effective annual yield* is x (expressed as a percent.)
- If P_0 is an initial amount (dollars, population, and so on) that is compounded continuously at an annual rate *r*, then the amount at a future time *t* is given by $P(t) = P_0 e^{rt}$.

CHECKING THE MAIN IDEAS

For Exercises 1–3, match each situation with the appropriate model.
1. A cell population of P_0 doubles every 8 hours. What will the population be in *t* hours?

   A. $P_0 \cdot e^{0.08t}$
2. P_0 dollars are invested at 8% annual interest compounded continuously. What would the investment be worth in *t* years?

   B. $P_0 \cdot 8^{t/2}$
3. The value of a machine worth P_0 dollars now decreases in value by 8% per year. What will the value of the car be in *t* years?

   C. $P_0 \cdot 2^{t/8}$

   D. $P_0(0.92)^t$
4. Evaluate $\sqrt{3}^{\sqrt{2}}$ to the nearest tenth.

   E. $P_0(1.08)^t$
5. Find an exponential function *f* such that $f(0) = 40$ and $f(2) = 90$. (See Example 1 on text page 181.)
6. If $1000 is invested at 8% per year, in about how many years will the investment be worth $2000?
7. **Critical Thinking** Explain why the value of an exponential function with positive base *b* must be a positive number.
8. Evaluate e^3 and 3^e to the nearest tenth. Which value is greater?
9. Suppose that $1000 is invested at 8% annual interest compounded monthly.
   a. Find the value of the investment after one year.
   b. Find the effective annual yield.
10. A cell population *t* days from now is given by $A(t) = 8e$
    a. What is the cell population
    b. Find the cell population 4 da

50 *ADVANCED MATHEMATICS Student*

Chapter 5

...IEW

...and Logarithms

...exercises before trying the Practice Test fo
...ulty with a particular problem, review the in
...ch expression. *(Section 5-1)*

b. $-(3a^2)(2a^3)^2$ c. $\left(\frac{a}{b^3}\right)^{-}$

2. Suppose the value of a U.S. dollar decreases 5% per year.
value of a U.S. dollar, to the nearest cent, in 4 years. *(Sect...*

3. Simplify **(a)** $\left(\frac{1}{1000}\right)^{-1/3}$ and **(b)** $(81^{-3/2})^{-1/2}$. *(Section 5-2)*

4. Solve $16 = 2^{x+1}$. *(Section 5-2)*

5. Find an exponential function *f* such that $f(0) = 8$ and $f(2)$ *(Section 5-3)*

6. Write the formula for continuous compounding. Then use value of $1 invested at 5% annual interest compounded c... for one year. *(Section 5-4)*

7. Write the equation $\log_2 8 = 3$ in exponential form. *(Secti...*

8. Find the value of $\log \frac{1}{10}$. *(Section 5-5)*

9. Simplify **(a)** $\frac{1}{2}\log_2 36 - \log_2 3$ and **(b)** $\ln \sqrt[3]{e}$. *(Section 5-...*

10. Use a calculator to solve $3^x = 40$ for *x* to the nearest hund... *(Section 5-7)*

PRACTICE TEST
Chapter 5

Simplify each expression.
1. $\sqrt[4]{\frac{16^{k-1}}{16^{k-6}}}$ 2. $\frac{(3c^{-5/3})^2(2c^{-2/3})}{3(c^{-7})^3}$ 3. $\log_b b^{...}$

4. $(2^{-2} - 4^{-2})^{-1}$ 5. $(2^{-2} \cdot 4^{-2})^{-1}$ 6. $3^{(\log_3 5...}$

7. $\ln e^3$ 8. $e^{\ln 5}$

Express each of the following in terms of $\log A$ and $\log B$.

10. $\log \frac{10A}{B^2}$ 11. $\log_{AB} 10$

The temperature *T* (in °F) of a glass of lemonade left in a r...
t minutes is $T(t) = 95 - 60e^{-0.15t}$.
12. What is the temperature after 5 minutes?
13. How soon will the temperature be 80°F?

Ten thousand dollars is invested at 10% compounded quar...
14. What is its value of the investment after one year?
15. How long does it take for the investment to double in value?

Copyright © by Houghton Mifflin Company. All rights reserved.   *ADVANCED MATHEMATICS Student Resource Guide* **55**

Use a calculator to find the value of *x* to the nea...
16. $e^x = 12$ 17. $(1.02)^x = 2$
19. Express *y* in terms of *x* if $\log y = 4\log x - ...$
Solve. Express *x* as a logarithm if necessary.
20. $e^{2x} - 5e^x + 4 = 0$ 21. $4^{2x} - ...$

MIXED REVIEW
Chapters 1–5

1. Solve $|2x - 5| \leq 4$ and graph the solution on...
Let $f(x) = 2x^2 - x^4$.
2. Sketch the graph of $y \leq P(x)$.
3. Find the domain, range, and zeros of *f*.
4. Describe the graph of $x = f(y)$. Tell whether
would represent a function.

Solve each equation or system of equations.
5. $25^{x-3} = 125^{4/3}$ 6. $\begin{array}{l} y = 2x^2 - 5x - \\ 3x + y = 1 \end{array}$

Simplify.
8. $3 - (1 - 2i)^2$ 9. $\log 0.001 + 10^{2\log 3}$ 10. $(a^{-1} - b^{-1})(a + b)^{-1}$
$\triangle ABC$ has vertices $A(0, 0)$, $B(10, 0)$ and $C(2, 4)$.
11. Find an equation of the line through *C* that contains the median to $\overline{AB}$.
12. Find the perimeter of $\triangle ABC$.
13. The sum of the roots of a cubic equation is -2 and the product is 12. If one root is -1, find the other two roots.
14. Tell whether the graph of the equation $|x| - |y| = 2$ has symmetry in (i) the *x*-axis, (ii) the *y*-axis, (iii) the line $y = x$, and/or (iv) the origin.
15. Suppose $1000 is invested at 6% per year. How much more will the investment be worth after 2 years if the interest is compounded continuously rather than monthly?
16. Solve $e^x = 100$ to the nearest hundredth.
17. Let $f(x) = \sqrt{x}$ and $g(x) = 2x - 1$. Find **(a)** $(f + g)(x)$, **(b)** $(f$ and **(c)** $(g \circ f)(x)$.
18. Express the area of an equilateral triangle in terms of its height
19. Sketch the graph of a periodic function $f(x)$ with amplitude 3 a... period 4. Then sketch the graph of $f(\frac{1}{2}x)$ on the same set of axe...
20. The sector of a circle shown at the right has radius *r* and perim... 50 cm. Let $A(r)$ be the area function of the sector.
    a. If $A(1) = 24$, $A(5) = 100$, and $A(10) = 150$, find a quadra... function for $A(r)$.
    b. Find the maximum possible area of the sector.

56 *ADVANCED MATHEMATICS Student Resource Guide*   Copyright © by Houghton Miffl...

SECTIONS FOR STUDY AND REVIEW
- Overview
- Key terms
- Understanding the main idea
- Checking the main idea
- Using the main idea

CHAPTER REVIEW
- Quick check
- Practice test
- Mixed review

Enrichment develops understanding of concepts.

Activities

The Activities Book offers a wide range of topics that appeal to students.

Applications get students involved.

Creative problem solving motivates students.

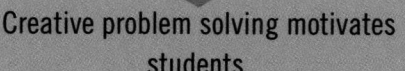

Assessment

Your opportunity to monitor students' understanding

Alternative assessment suggestions are proposed for each section of the textbook.

- Group Activities
- Journal Topics
- Nonroutine Problems
- Open-ended Questions
- Performance Tasks
- Portfolio Projects
- Presentations
- Research Projects
- Teaching Questions

Alternative Assessment

Portfolio Assessment
Nonroutine Problems
College Entrance Examinations
Spiral Review

Advanced Mathematics
Precalculus with Discrete Mathematics and Data Analysis

Tests
by Dane R. Camp

Chapter Tests
Cumulative Tests

Blackline Masters with Answer Key

Advanced Mathematics
Precalculus with Discrete Mathematics and Data Analysis

Houghton Mifflin

The *Tests* booklet contains a Chapter Test for each of the twenty chapters in the textbook plus nine Cumulative Tests.

Advanced **Mathematics**
Precalculus with Discrete Mathematics and Data Analysis

Overhead Visuals

VISUAL 13 • **The Normal Curve**

...tion represented by a normal curve, and to use ...ine probabilities.

...h Lesson 15-3, pages 719–722, and Lesson

...dard normal distribution. The graph is called ...of its shape.

...etry? Yes; it is symmetric to the y-axis.
...ptotes? Yes; it is asymptotic to the x-axis.
...urve and the x-axis is 1.
...urve, the x-axis, the y-axis, and the line

...eft of $x = 1.5$? $0.5 + a$

Add Sheet 3 to Sheets 1 and 2.
5. The area between $x = 0$ and $x = 2$ in a standard normal distribution is about 0.477.
   a. What percent of the data is within two standard deviations of the mean? About 95%
   b. What percent of the data is more than two standard deviations from the mean? About 5%
   c. What is the area under the curve between $x = -2$ and $x = 1$? About 0.817

Add Sheet 4 to the other sheets.
6. The area of one of the green regions is about 0.02.
   a. What percent of the data is within three standard deviations of the mean? About 99%
   b. What percent of the data is more than three standard deviations from the mean? About 1%

Leave all the sheets on the projector to answer Question 7.
7. Scores on a fitness test are normally distributed with mean 200 and

Warm-Up Exercises for Section 3-2

The zero(s) of each polynomial function are given. Use the zeros as boundary values to find the intervals in which
(a) $P(x) \geq 0$ and (b) $P(x) < 0$.

Warm-Up Exercises for Section 3-3

Consider the points $(3, -4)$, $(-3, 4)$, and $(-3, -4)$. For each of the following inequalities, two of the points lie in the graph of the inequality and the third ... lie in the

Warm-Up Exercises in the Teacher's Edition as prepared transparencies
(From Teacher's Resources File)

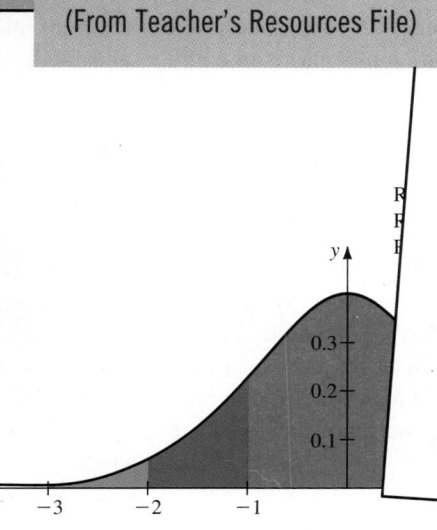

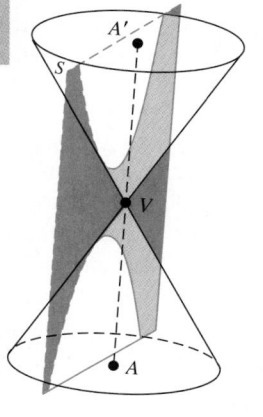

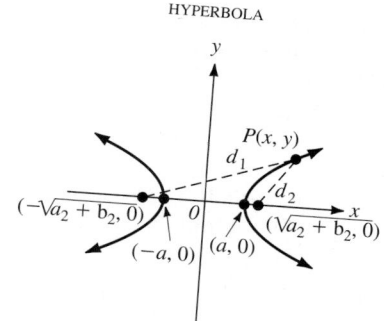

HYPERBOLA

Conditions: $|d_1 - d_2| = 2a$.
Equation: $\dfrac{x^2}{a^2} - \dfrac{y^2}{b^2} = 1$.

Overview of the Program

Goals

The following goals have guided the development of this textbook program:

- Provide a strong foundation of precalculus concepts, techniques, and applications to prepare students for more advanced work.
- Place appropriate emphasis on discrete mathematics and data analysis—subjects that provide the mathematical framework for many important contemporary applications.
- Show how technology can be used as a tool to facilitate learning and doing mathematics.
- Present topics in a way that encourages students to become actively involved and accommodates different learning and teaching styles.
- Develop students' quantitative reasoning and problem-solving skills.
- Develop students' abilities to understand and communicate mathematical ideas effectively.
- Increase students' appreciation of mathematics through seeing a wide range of mathematical applications and learning about the contributions of men and women from a number of different backgrounds to the development of the mathematical sciences.
- Offer comprehensive teaching support in the Teacher's Edition and supplementary publications, especially for new topics and approaches in the curriculum.

Flexibility

This textbook has been designed to meet the needs of a variety of course options. Because of its comprehensive topic coverage and extensive and varied exercise sets, it provides plenty of material from which to tailor a course to fit your students' needs, your curriculum requirements, and your teaching preferences. The *Assignment Guide*, pages T33–T41, offers some suggestions for structuring and pacing your course.

Sequencing of Topics

The twenty chapters of this textbook have been divided into the following four units:

1. Functions, Graphs, and Applications (Chs. 1–6)
2. Trigonometry (Chs. 7–11)
3. Discrete Mathematics and Data Analysis (Chs. 12–18)
4. Limits and Introduction to Calculus (Chs. 19–20)

Chapters 1–6 cover the core precalculus topics. The first five chapters should be covered in order. (With an accelerated class, you may wish to cover them quickly, or even begin with Chapter 4.) Chapter 6 can be taught at any time during the course. Chapters 7–11 provide a mini-course in trigonometry, which can be omitted if students have already studied trigonometry. If you wish to emphasize discrete mathematics and data analysis in your course, you can teach Chapters 13–18 directly following Chapter 5. (Chapter 12 requires some familiarity with trigonometry.)

Technology

Because calculators and computers can greatly enhance the teaching and learning of precalculus topics, suggestions for using these technological tools have been integrated throughout. The article *Technology* on page T24 explains how the textbook and supplementary publications encourage and facilitate the use of technology.

Meeting Contemporary Standards

The recommendations of the *Curriculum and Evaluation Standards* of the National Council of Teachers of Mathematics, and those of other organizations committed to excellence in mathematics education, have been addressed in a number of ways throughout the course. See pages xvi–xxiii for examples of how the text promotes problem solving, communication, applications, and reasoning, as discussed in the *Standards*. See also Index entries for these topics.

Connections: Modeling, Integration

The modeling connections between problem situations and their mathematical representations are discussed throughout the course in a variety of contexts. Students are given practice in modeling real-life situations using polynomial, exponential, logarithmic, and trigonometric functions. Connections among mathematical topics, especially algebra and geometry, are discussed and used to solve problems throughout. See the article *Making Connections*, pages T26 and T27, for a more detailed discussion.

Learning and Teaching Styles

Advanced Mathematics supports a variety of learning and teaching styles. The clear, readable text, the worked-out examples, and the charts, illustrations, and highlighting of important results help students develop their abilities to understand and use symbolic and visual representations of mathematical ideas.

The *Activities* and *Projects* offer interesting material for varying the mode of classroom instruction, with many of these being suitable for group work. (See, for example, pages 62–64, 131, 187, and 210.) The wide variety of exercises provides a balance of practice with concepts, skills, and applications.

In addition, the supplementary *Student Resource Guide* contains study and review materials that meet the needs of a wide range of students.

Active Learning

Recent research into the way in which students learn has revealed that learning is not a process of passively absorbing isolated bits of information but is rather a process of actively constructing meaning by interpreting new information in light of what is already known. Successful learners are able to analyze and interpret given information and then make conjectures that go beyond what is presented explicitly.

This book encourages students to become active learners and to expand their thinking skills. For example, *Activities* within lesson presentations, communication exercises, *Investigation* exercises, *Research* exercises, and *Projects* are all useful for helping students develop their abilities to analyze, interpret, and make conjectures. See the Index entry *Reasoning* and the discussion of *journal writing* in the article on *Communication*, page T25.

Multicultural Considerations

Studies of future employment needs indicate that many jobs in the 21st century will require greater mathematical skills than previously. Thus, it is important to encourage all students in our multicultural society to study as much mathematics as possible in order to prepare for their future careers. To help students appreciate the variety of career opportunities that are open to people with a background in mathematics, several special sections on careers that use mathematics have been included in the textbook. See *Careers in Mathematics and Science*, pages xxxiv–xxxvii, *Architecture and Urban Planning*, page 117, *The People of Mathematics*, pages 366–367, and *Genetics and Statistics*, page 713. In these sections, men and women from a variety of cultural backgrounds are highlighted. The *Biographies* also feature the contributions of mathematicians and scientists from a variety of backgrounds (see pages 375, 499, 537, and 736, for instance).

In addition, in the supplementary *Activities Book*, there are some activities that focus on multicultural aspects of mathematics.

Assessment

An effective assessment program is consistent with the goals of the course that it accompanies. Since the goals of *Advanced Mathematics* include helping students develop thinking and problem-solving skills as well as master mathematical content, a varied and comprehensive assessment program is needed. For this reason, in addition to the chapter tests and cumulative reviews in this Teacher's Edition, two separate booklets of supplementary assessment materials are provided. The *Tests* include chapter tests and cumulative reviews to evaluate students' mastery of course content. The *Alternative Assessment* booklet includes suggestions for portfolio assessment and nonroutine problems to challenge students' thinking and problem-solving skills, as well as questions in college-entrance-examination format and spiral reviews.

Technology

Using Technology with the Course

The availability of powerful calculators and computers for classroom use has significantly expanded the types of problems and the range of topics that can be studied in a precalculus course. (See, for example, the discussion of *iterated functions* and *fractals* in Chapter 19.) With the aid of technology, students can concentrate on exploring, understanding, and applying mathematics, without becoming bogged down in calculations or tedious plotting of points. Technology is particularly useful in the work with polynomial functions, trigonometry, polar graphs, matrix applications, statistics, curve fitting, and mathematical modeling. Since technology has such a great potential for facilitating the teaching and learning of precalculus, suggestions for using it have been integrated throughout the textbook. The technology logo (see page xxiv) is used to indicate places in the course where technology is desirable or necessary.

The introductory section *Using a Graphing Calculator* on pages xxvi–xxxiii demonstrates the basic features and uses of a graphing calculator. You may wish to have students read the section and work the exercises at the beginning of the year.

Throughout the text, examples and exercises that use a graphing calculator can also be done with computer graphing software such as Houghton Mifflin's *Precalculus Plotter Plus*. This software (shown on page xxv) is accompanied by a booklet that includes a user's guide with worked examples, enrichment topics, and 47 activities in worksheet format.

The supplementary publication *Using Technology* contains calculator and computer activities that offer additional ways of using technology with the course.

The Role of Technology in the Course

Although technology has the power to enhance *Advanced Mathematics*, technology does not drive the course. The focus of the course is the *mathematics*. Technology is a tool that supports and extends—but does not dominate—the teaching and learning of precalculus topics.

Special emphasis has been given in *Advanced Mathematics* to using calculators and computers to explore mathematical concepts. Activities and exercises with an exploratory flavor are included throughout the book. The Activity on reflection of graphs on page 131, the Activity on the derivation of the number *e* on page 187, and the Calculator Exercises on pages 498–499 are a few examples.

We strongly encourage you to use technology with your *Advanced Mathematics* classes whenever appropriate. At the same time, we recognize that not all classes have access to calculators or computers on a regular basis. Therefore, this book has been designed so that it can be used without technology, although some parts of the text and the exercises will have to be omitted. These parts are clearly highlighted by the technology logo. Tables of squares and square roots, natural logarithms, common logarithms, trigonometric functions in degrees and minutes, trigonometric functions in decimal degrees, trigonometric functions in radians, and values of e^x and e^{-x} are provided at the back of the book for those who do not have access to calculators. In addition, alternative solutions, with and without the use of technology, are included in some cases. See, for example, the alternative methods of solutions shown on pages 298, 303, 325–326, and 387.

Reading List

Kelman, Peter, et al. *Computers in Teaching Mathematics*. Reading, MA: Addison-Wesley, 1983.

Masalski, William J. *How to Use the Spreadsheet as a Tool in the Secondary Mathematics Classroom*. Reston, VA: NCTM, 1989.

Oldknow, A. J. *Microcomputers in Geometry*. New York: Halsted Press, 1987.

Smith, D. V., and A. J. Oldknow. *Learning Mathematics with Micros*. New York: Halsted Press, 1983.

The documentation and support material accompanying software packages and calculators can also be quite helpful.

Communication

To learn and make use of mathematics, it is important for students to become comfortable with communicating mathematical ideas to others.

Reading

Students will find many new words in this book. Students who don't remember the mathematical meaning of a word should look it up in the *Glossary*, pages 860–865, or the *Index*, pages 866–878. In addition to specialized vocabulary, the mathematics in this textbook is expressed using concise notation that involves symbols and abbreviations. A complete list of symbols is provided on pages 856–857. Activities focusing on the use of symbols and vocabulary can be found on pages 246 and 623.

Using Diagrams and Models

Encourage students to study the drawings that accompany the text in this book. Frequently the diagrams summarize a lot of material that would be inconvenient to write out explicitly. (A graph of a function, for example, provides a record of infinitely many ordered pairs.) Point out to students that drawing a sketch may help them understand the conditions that define a problem. A careful drawing may even provide an approximate answer that will lead to a complete solution. Drawing diagrams and building models are nonverbal ways of communicating mathematical ideas to other people. Such diagrams or models can be used as a framework to guide subsequent discussion, and may help overcome language barriers between people of different backgrounds. An activity focusing on the use of drawings is found on page 763.

Listening and Speaking

Since a great deal of students' time in math class is spent in listening to the teacher and to other students, it makes sense to listen carefully and critically. Encourage questions and comments, but make sure that students understand the importance of allowing a speaker to finish before they speak. This is especially important in cooperative or small-group situations. Students should come to realize that the different perspectives of group members may be key elements in the solution of a problem. Activities focused on developing listening and speaking skills in small group discussions appear on page 379.

Journal Writing

Keeping a mathematics journal is one of the ways in which students can improve their understanding and enjoyment of mathematics. Suggest that they write a few paragraphs each day, setting down their thoughts as they occur without worrying too much about grammar and spelling.

In a mathematics journal, students can express some of the feelings they experience as they learn mathematics: feelings of frustration, accomplishment, confusion, or insight. They can explore connections among different topics, evaluate progress, list questions that remain unanswered, develop conjectures, or just draw sketches.

You may want to use some of the suggestions listed below as your students approach a journal-writing session.

1. Select a homework problem that has given you trouble and write about the different approaches you tried, including ones that did not work. Describe the difficulties you encountered. Did any of your "false starts" either lead you to understand the problem better or suggest another approach?
2. Write a history of your life as a mathematics student. What was the best experience you ever had in math class? What was the worst experience?
3. Imagine that a friend has been absent from math class for exactly two weeks. Write out the essentials that your friend should know in order to catch up.
4. Formulate a mathematical question that is an extension of a problem that you have already worked on.
5. Give advice to a student who will be taking this math course next year.

Making Connections

As students enter this course, you can expect a level of maturity not typical of the less advanced student. They have attained higher levels of mathematical maturity and sophistication, by virtue of the number and scope of mathematics courses they have already mastered. Students such as these have the potential to benefit greatly from more advanced courses. They may, too, be more keenly aware of the world around them, and more consciously wonder about the reasons for studying a given mathematical topic. These are the students who can appreciate connections in mathematics.

Connections Within Mathematics

The precalculus course includes material from a number of branches of mathematics, thereby enabling students to experience connections among them. These are the interrelationships among topics and processes, some perhaps encountered previously as the student spiraled upward through the mathematics curriculum. A student may study the same concept expressed in a number of manifestations. Students at the level represented by this course are ready to see these connections, to understand them, and even to appreciate them. A short but not exhaustive list follows.

- *Functions:* The function concept is used to progress from the familiar linear and quadratic functions, building to polynomial, exponential, and trigonometric functions.
- *Functions and Inverses:* Function techniques are used to define important new types of functions. These include exponential, logarithmic, trigonometric, and inverse trigonometric functions.
- *Algebra and Geometry:* Connections between algebra and geometry are exploited throughout Chapter 2, where polynomial functions and equations are examined. The parallel development here is between the algebraically stated function or equation and the geometric representation as a graph. Characteristics of polynomials, such as

zeros and maxima or minima, are translated directly into geometric terms. The graph and its behavior are used to answer important questions about the function or equation. Graphing technology has enabled this connection to be exploited very fully.

- *Experimental and Theoretical Mathematics:* Newton's Law of Cooling provides an ideal setting for math students to simulate the role of experimental scientist. In the experimental phase, they perform an experiment, collect data, and graph the data. To analyze the results, they then develop a particular exponential function as the model of the situation.
- *Circular Functions and Series:* Circular and trigonometric functions are introduced early, with the usual definitions. Students study series in Chapter 13, and in Section 19-4, they are presented with a whole new way of looking at sines and cosines. This approach to circular functions as power series is also quite applicable in helping students understand some computational aspects of the circular functions.
- *Coordinate Systems and Complex Numbers:* Representing points in a plane takes on a new dimension as students are introduced to polar coordinates as an alternative to the rectangular coordinates they've used for years already. Complex numbers can then be studied from two points of view, using both rectangular and polar forms of a given complex number.
- *Parametric Equations:* Students have graphed equations for years. In this course, they have extensive experience with graphing, moving on to graphs of new classes of equations such as trigonometric and exponential equations. In the context of vectors, the concept of parametric equations is introduced. At this point, students have an altogether new way of viewing the graphing process, which can be used to picture a number of equations that they have studied only in explicit form before. The variables x and y are seen as

determined by a third variable, the parameter, rather than as dependent on each other.

- *Recursive and Closed-Form Definitions:* Sequences are approached in two ways. Closed-form definitions are the initial introduction. They are followed by recursive definitions, in which a given term of a sequence is related to the preceding term. The two forms are seen to be equivalent, with each especially suited to its own purposes or applications.
- *Matrices and Linear Systems:* Students have learned to solve systems through algebraic transformations. Now they are shown that matrices can be used to represent systems efficiently and compactly, and that matrix techniques can be used to produce an actual solution of a system. This method is shown to be quite powerful because of its generality and adaptability to computer applications.

Connections Through Modeling

Besides connections within mathematics, there are the connections between mathematics and those disciplines and applications for which mathematics provides the model. These are the connections that are likely to become the most important in later life.

Although it is not feasible to discuss all of the connections presented through modeling, it is possible to highlight a few important ones.

- *Compound Interest:* Like many topics, this one is first introduced at a very basic level, illustrating the idea of a function and what the graph of a function shows. Later this topic is modeled by an exponential function, and deeper analysis is possible. Finally, in an example of both of the kinds of connections mentioned earlier, geometric sequences provide an apt model for compound interest and hint at a connection to exponential functions.
- *Half-Life:* This high-interest topic is dealt with from two major points of view. The first introduction is in connection with exponential functions. Later the phenomenon is modeled through geometric sequences. Both treatments allow numerous meaningful application problems.

This sampling of topics gives an idea of the extent to which mathematical connections are incorporated in *Advanced Mathematics*. Other areas of connection are evident throughout the book, especially among applications. Understanding and appreciating these connections provide a fitting culmination to the high school student's years of study of mathematics.

Discrete Mathematics

During the past few decades, a number of related mathematical areas of study have come to the fore under the heading *discrete mathematics*. This field bears a fruitful relation to computer science, contributing to improved computer design and benefiting from computer solutions to certain of its problems. A related field greatly enriched by current technology is that of *data analysis*. Discrete mathematics is central to the mathematical analysis of countless economic and social problems; data analysis is particularly relevant when large amounts of data are involved in a decision. It is not hard to see why study within these fields is becoming recommended as a part of the high school curriculum.

Discrete Mathematics

Discrete mathematics is not easily defined. The definition given in the *Curriculum and Evaluation Standards* of the National Council of Teachers of Mathematics is "the study of mathematical properties of sets and systems that have only a finite number of elements." Common discrete mathematics topics include the development and analysis of algorithms, combinatorics, probability, graph theory, iteration and recursion, linear programming, and matrix operations.

The significance of discrete mathematics lies in its great variety of practical applications. Discrete mathematics comes into play in the fields of business and finance, industrial design, operations management, manufacturing, and physiology, to name just a few. It is sometimes described as the mathematics needed in order to be a responsible citizen in our society. A responsible citizen should be able to:

- collect and interpret data and identify and analyze trends.
- build mathematical models that can be used to study problems.
- construct algorithms to solve problems.

Personal finance provides an everyday application of discrete mathematics. Every citizen needs to be familiar with the mathematics of finance in order to understand how savings accounts, annuities, mortgages, and credit cards work. Discrete mathematics helps students view financial problems from a new perspective. For example, a savings account illustrates what is called a *dynamical system*. A dynamical system consists of a starting value for the system (the initial deposit in a savings account) and a rule for computing the next value of the system, often called a *recurrence relation* or a *difference equation* (the next balance is equal to the current balance plus the interest accumulated on that balance). The study of a dynamical system shows what happens when the recurrence relation is applied over and over again.

Example: Compound Interest

Suppose you invest \$100 in an account that pays 10% interest annually. If you let B_n be the value of the account after n years, then you can describe the account as the system that starts with $B_0 = 100$ and for which

$$B_n = B_{n-1} + (0.1)B_{n-1} = (1.1)B_{n-1}.$$

In other words, you start with $B_0 = 100$ and then *iterate*, or apply over and over again, the formula $B_n = (1.1)B_{n-1}$. The resulting sequence of numbers gives the value for the account in succeeding years. If you now deposit \$125 each year, you have an annuity with payments of \$125. To model this, you could apply the formula

$$B_n = (1.1)B_{n-1} + 125.$$

Example: Mortgage Loan

It is very easy to model a mortgage. Let B_n be the amount owed to the bank after n payment periods. Assume the buyer of a home has borrowed \$20,000 at 9% interest per year (or 0.75% interest per month) and is paying \$325 per month in payments. The dynamical system with $B_0 = 20,000$ and the recurrence relation $B_n = (1.0075)B_{n-1} - 325$ can be used to model the value of the mortgage's current balance.

The dynamics of financial mathematics is not difficult to understand: the same process is repeated over and over again.

Current Developments

Why hasn't this approach been used in the curriculum before? Because it is computation intensive: to get the next value, one has to know the preceding value. The process can be extremely repetitive and time consuming. To get around this, mathematicians had to discover very complicated formulas for calculating such values.

Now that students have the computational power of computers, this easy-to-understand approach can become part of the secondary curriculum. It should also be pointed out that this approach works very well in conjunction with spreadsheet analysis. Spreadsheets are used in many fields to expedite the performance of complex calculations on vast amounts of data.

Chaos and Fractals

Another application of the field of dynamical systems is in understanding two very popular new mathematical fields: *chaos theory* and *fractal geometry*. The study of chaos theory stems from the discovery that even very simple systems that depend on only one variable can behave in random, unpredictable ways. Scientists hope that through studying the behavior of simple dynamical systems, they will be able to understand complicated systems, such as weather patterns or economic systems.

Fractal geometry is a theory of geometric forms that model naturally occurring shapes such as coastlines, mountains, clouds, and snowflakes. Fractals are formed by the iteration of certain geometric constructions, thereby producing dynamical systems. The inherent beauty of computer-generated fractals, and their relation to dynamical systems, have made them popular subjects for study in high schools.

Data Analysis

Data analysis is another field that has become far more accessible with the advent of computers and hand-held graphing calculators. One has only to observe the amount of data encountered by most people day after day in order to see the need for students to be able to interpret and analyze data in order to make informed decisions. Students must understand how to collect and organize data and must then be able to draw conclusions or make predictions from them. They must, for example, understand the difference between data that show linear growth and data that show exponential growth: In the long run, there is a big difference between being given a raise of $500 per year and receiving a 10% raise per year.

To be able to analyze data, students will need to understand the basics of probability theory and combinatorics, the mathematics of counting. They will also need to understand the basics of descriptive statistics and exploratory data analysis. They will need the ability to apply this knowledge to a wide range of topics, such as biology, medicine, economics, manufacturing, and sports.

The following list of references will help you learn more about discrete mathematics and data analysis.

Discrete Mathematics

Biggs, Norman L. *Discrete Mathematics*, revised ed. New York: Oxford University Press, 1990.

Chartrand, Gary. *Introductory Graph Theory*. New York: Dover Publications, 1985.

Dossey, John A., et al. *Discrete Mathematics*. Glenview, IL: Scott, Foresman, 1987.

Gerstein, Larry. *Discrete Mathematics and Algebraic Structures*. New York: W. H. Freeman, 1987.

Kalmanson, K. *An Introduction to Discrete Mathematics*. Reading, MA: Addison-Wesley, 1986.

Data Analysis

Brook, Richard J.; Gregory C. Arnold; Thomas H. Hassard; and Robert M. Pringle, eds. *The Fascination of Statistics*. New York: Marcel Dekker, 1986.

Moore, David S. *Statistics: Concepts and Controversies*, 3rd ed. New York: W. H. Freeman, 1991.

Moore, David S., and George P. McCabe. *Introduction to the Practice of Statistics*. New York: W. H. Freeman, 1985.

North Carolina School of Science and Mathematics. *New Topics for Secondary School Mathematics: Data Analysis*. Reston, VA: NCTM, 1988.

Reading List

Applications

Crutchfield, James P.; J. Doyne Farmer; Norman H. Packard; and Robert S. Shaw. "Chaos." *Scientific American*, December 1986, pp. 46–57.

Egsgard, John; Gary Flewelling; Craig Newell; and Wendy Warburton. *Making Connections with Mathematics*. Providence, RI: Janson Publications, 1988.

Kastner, Bernice. *Space Mathematics: A Resource for Secondary School Teachers*. Palo Alto, CA: Dale Seymour Publications, 1988.

Kumar, Surya. *HiMap, Module 14, Decision Making and Math Models*. Arlington, MA: Consortium for Mathematics and Its Applications, 1989.

Assessment

Charles, Randall I., and Edward A. Silver, eds. *The Teaching and Assessing of Mathematical Problem Solving*. Reston, VA: NCTM, 1988.

Charles, Randall; Frank Lester; and Phares O'Daffer. *How to Evaluate Progress in Problem Solving*. Reston, VA: NCTM, 1987.

Clarke, David J. *MCTP Professional Development Package: Assessment Alternatives in Mathematics*. Canberra, Australia: Curriculum Development Centre, 1989.

Stenmark, Jean K. *Assessment Alternatives in Mathematics: An Overview of Assessment Techniques that Promote Learning*. Berkeley, CA: Lawrence Hall of Science, 1989.

Communication

Connolly, Paul, and Teresa Vilardi, eds. *Writing to Learn Mathematics and Science*. New York: Teachers College Press, 1989.

Pimm, David. *Speaking Mathematically: Communication in Mathematics Classrooms*. New York: Routledge, 1987.

Silver, Edward; Jeremy Kilpatrick; and Beth Schlesinger. *Thinking Through Mathematics: Fostering Inquiry and Communication in Mathematics Classrooms*. New York: College Board Publications, 1990.

Curriculum and Learning Theory

Marzano, Robert J., et al. *Dimensions of Thinking: A Framework for Curriculum and Instruction*. Alexandria, VA: Association for Supervision and Curriculum Development, 1988.

Mathematical Sciences Education Board of the National Research Council. *Reshaping School Mathematics: A Philosophy and Framework for Curriculum*. Washington, D.C.: National Academy Press, 1990.

Resnick, Lauren B. *Education and Learning to Think*. Washington, D.C.: National Academy Press, 1987.

Stewart, Ian. "Change," in Steen, Lynn Arthur. *On the Shoulders of Giants: New Approaches to Numeracy*. Washington, D.C.: National Academy Press, 1990, pp. 183–217.

Data Analysis

Brook, Richard J.; Gregory C. Arnold; Thomas H. Hassard; and Robert M. Pringle, eds. *The Fascination of Statistics*. New York: Marcel Dekker, 1986.

Moore, David S. *Statistics: Concepts and Controversies,* 3rd ed. New York: W. H. Freeman, 1991.

Moore, David S., and George P. McCabe. *Introduction to the Practice of Statistics*. New York: W. H. Freeman, 1985.

North Carolina School of Science and Mathematics. *New Topics for Secondary School Mathematics: Data Analysis*. Reston, VA: NCTM, 1988.

Discrete Mathematics

Biggs, Norman L. *Discrete Mathematics*, revised ed. New York: Oxford University Press, 1990.

Chartrand, Gary. *Introductory Graph Theory*. New York: Dover Publications, 1985.

Devaney, Robert L. *Chaos, Fractals, and Dynamics: Computer Experiments in Mathematics*. Reading, MA: Addison-Wesley, 1990.

Dossey, John A., et al. *Discrete Mathematics*. Glenview, IL: Scott, Foresman, 1987.

Finkbeiner, Daniel T., II, and Wendell D. Lindstrom. *A Primer of Discrete Mathematics*. New York: W. H. Freeman, 1987.

Gerstein, Larry. *Discrete Mathematics and Algebraic Structures*. New York: W. H. Freeman, 1987.

Kalmanson, K. *An Introduction to Discrete Mathematics*. Reading, MA: Addison-Wesley, 1986.

North Carolina School of Science and Mathematics. *New Topics for Secondary School Mathematics: Matrices*. Reston, VA: NCTM, 1988.

Roberts, Fred S. *Discrete Mathematical Models with Applications to Social, Biological, and Environmental Problems*. New York: Prentice-Hall, 1976.

Mathematics

Consortium for Mathematics and Its Applications. *For All Practical Purposes: Introduction to Contemporary Mathematics*. New York: W. H. Freeman, 1988.

Dunham, William. *Journey Through Genius: The Great Theorems of Mathematics*. New York: John Wiley & Sons, 1990.

Steen, Lynn Arthur, ed. *On the Shoulders of Giants: New Approaches to Numeracy*. Washington, D.C.: National Academy Press, 1990.

Problem Solving

Krulik, Stephen, and Jesse A. Rudnik. *Problem Solving: A Handbook for Senior High School Teachers*. Needham, MA: Allyn and Bacon, 1988.

Krulik, Stephen, ed. *1980 NCTM Yearbook: Problem Solving in School Mathematics*. Reston, VA: NCTM, 1980.

London, Robert. *Nonroutine Problems: Doing Mathematics*. Providence, RI: Janson Publications, 1989.

Rachlin, Sid, ed. *Problem Solving in the Mathematics Classroom*. Reston, VA: NCTM, 1982.

Technology

Barrett, Gloria, and John Goebbel. "The Impact of Graphing Calculators on the Teaching and Learning of Mathematics," in *1990 NCTM Yearbook: Teaching and Learning Mathematics in the 1990s*. Reston, VA: NCTM, 1990, pp. 205–211.

Churchhouse, Robert F., et al., eds. *International Commission on Mathematical Instruction Study Series: The Influence of Computers and Informatics on Mathematics and Its Teaching*. Cambridge, England: Cambridge University Press, 1986.

Demana, Franklin, and Bert K. Waits. "Enhancing Mathematics Teaching and Learning through Technology," in *1990 NCTM Yearbook: Teaching and Learning Mathematics in the 1990s*. Reston, VA: NCTM, 1990, pp. 212–222.

Fey, James T., ed. *Computing and Mathematics: The Impact on Secondary School Curricula*. Reston, VA: NCTM, 1984.

Heid, M. K. "Resequencing Skills and Concepts in Applied Calculus Using the Computer as a Tool." *Journal for Research in Mathematics Education*, January 1988, pp. 3–25.

Assignment Guide

This guide presents suggested assignments for three year-long courses, which are among the many courses that may be taught using this book. They are intended to serve as guides to help you plan the course that best meets the needs of your students and curriculum.

The Chart

Daily suggested assignments for the three year-long courses described below appear in the chart on the following pages. Side columns for Chapters 1–13 show assignments for the standard and comprehensive courses. Side columns for Chapters 14–19 show the assignments for the discrete mathematics course.

In Chapters 9 and 15, the sets of Mixed Exercises, which cover material in a group of lessons, appear in the time schedule with the last lesson they cover.

In general, two days are left for testing. The first day assigns the Chapter Test in the book as a review. The second day is for administering a test for the chapter from the Teacher's Edition, from an ancillary publication, or of your own.

Year Courses

A 160-day schedule is shown for each course with each schedule divided into (a) two approximately equal semesters and (b) three roughly equivalent trimesters. Since each section in the book contains more exercises than students would normally be expected to complete within a single assignment, the suggested daily assignments include only a portion of the exercises. Most concepts are covered by several exercises, and thus an assignment of selected exercises still provides sufficient work for mastery of the objectives of the lesson.

Class Exercises are not mentioned in any suggested assignment. These exercises continue the development of the concepts introduced in the lesson and sometimes prepare students for the Written Exercises. Therefore, we expect that most teachers will use these exercises in class. If it suits your method of teaching, however, the Class Exercises may be assigned as homework.

Standard Course

The standard course includes all of Chapters 1–11, the material on vectors in two dimensions from Chapter 12, and the finite sequences and series material from Chapter 13.

Comprehensive Course

The comprehensive course covers all of Chapters 1–13 and moves through the earlier chapters at a quicker pace than the standard course.

Discrete Mathematics Course

This course is designed for students with strong mathematics backgrounds. This course includes Chapters 1, 2, 4, 5, and 12–18. It also includes the linear programming material from Chapter 3, the first three lessons from Chapter 8 so that students can do the exercises with trigonometric modeling, and the material on limits in Chapter 19. Three projects (see pages 210–211, 470–471, and 714–715) are also included. Only one day is allotted for review and/or testing in Chapters 3, 8, and 19 because only a few sections are covered.

Semester and Quarter Courses

A variety of quarter and semester courses may also be taught. A few of the possibilities include the following:

- quarter course in trigonometry
- quarter course in discrete mathematics
- quarter course in data analysis
- semester course in elementary functions
- semester course in analytic geometry and trigonometry

Suggested Schedule for Standard Course

	Semester 1						Semester 2						
Chapter	1	2	3	4	5	6	7	8	9	10	11	12	13
Days	12	13	10	15	16	17	14	12	12	10	12	8	9

Trimester 1 — Trimester 2 — Trimester 3

Suggested Schedule for Comprehensive Course

	Semester 1							Semester 2					
Chapter	1	2	3	4	5	6	7	8	9	10	11	12	13
Days	10	11	7	15	14	15	11	12	11	9	11	17	17

Trimester 1 — Trimester 2 — Trimester 3

Suggested Schedule for Discrete Mathematics Course

	Semester 1						Semester 2						
Chapter	1	2	3	4	5	8	12	13	14	15	16	17	18	19
Days	11	11	6	12	15	7	17	14	11	15	12	11	12	6

Trimester 1 — Trimester 2 — Trimester 3

Daily Assignments

Day	Standard Course		Comprehensive Course		Discrete Math Course	
1	1-1	5–6/Exs. 3, 7, 10, 11, 17, 19, 20, 21, 22, 25, 27, 32	1-1	5–7/Exs. 10, 11, 17, 19, 20, 21, 22, 25, 27, 29, 32, 33, 35	1-1	5–6/Exs. 3, 7, 10, 11, 17, 19, 20, 21, 22, 25, 27, 32
2	1-2	11–13/1–27 odd, Calc. Exs. 1, 2	1-2	11–13/13–33 odd, Calc. Exs. 1, 2	1-2	11–13/1–27 odd, Calc. Exs. 1, 2
3	1-3	16–17/1–19 odd	1-3	16–18/1, 5, 9, 13, 15, 21, 22, 23, 27, 28, 29	1-3	16–18/21, 22, 23, 25, 27, 28
4	1-3	17–18/21, 22, 23, 25, 27, 28	1-4	23–25/13–24 all	1-4	22–25/1–23 odd
5	1-4	22–25/1–23 odd	1-5	28–29/1–45 odd	1-5	28–29/1–39 odd
6	1-5	28–29/1–39 odd	1-6	35–36/17–45 odd	1-6	35–36/3–39 every 3rd
7	1-6	35–36/3–39 every 3rd	1-7	41–42/3–33 every 3rd, 35–42 all	1-7	41–42/3–33 every 3rd, 35, 36

Day	Standard Course	Comprehensive Course	Discrete Math Course
8	**1-7** 41–42/3–33 every 3rd, 35, 36	**1-8** 45–48/1, 5, 8, 11–17 all	**1-8** 45–47/1–10 all
9	**1-8** 45–47/1–10 all	**Chapter Test** 50–51	**1-8** 47–48/11–17 all
10	**1-8** 47–48/11–17 all	**Chapter Test** T42–T43	**Chapter Test** 50–51
11	**Chapter Test** 50–51	**2-1** 56–57/1–31 odd, 34, 35	**Chapter Test** T42–T43
12	**Chapter Test** T42–T43	**2-2** 61/1–25 odd	**2-1** 56–57/1–31 odd
13	**2-1** 56–57/1–31 odd	**2-3** 66–68/1–39 every 3rd	**2-2** 61/1–25 odd
14	**2-2** 61/1–25 odd	**2-4** 71–72/quadratic: 1–13 odd	**2-2** 61/2–26 even
15	**2-2** 61/2–26 even	**2-4** 72–74/cubic: 1–9 odd, Calc. Ex. 1–2	**2-3** 66/1–19 odd
16	**2-3** 66/1–19 odd	**2-5** 78/1–17 odd	**2-3** 66–68/21–39 odd
17	**2-3** 66–68/21–39 odd	**2-6** 83–84/1–21 odd	**2-4** 71–72/quadratic: 1–11 odd
18	**2-4** 71–72/quadratic: 1–11 odd	**2-6** 84–85/23–41 odd	**2-4** 72–74/cubic: 1–7 odd
19	**2-4** 72–74/cubic: 1–7 odd	**2-7** 89–91/1, 5, 9, 13, 17, 21, 25, 27, 29, 31–39 all	**2-5** 78/1–15 odd
20	**2-5** 78/1–15 odd	**Chapter Test** 93	**2-6** 83–85/1–37 every 4th
21	**2-6** 83–84/1–21 odd	**Chapter Test** T44	**2-7** 89–91/1–27 odd
22	**2-6** 84/23–37 odd	**3-1** 98–99/1–30 every 3rd, 33–37	**Chapter Test** 93
23	**2-7** 89–91/1–27 odd	**3-2** 103–104/25–44 all	**Chapter Test** T44
24	**Chapter Test** 93	**3-3** 106–108/3–33 every 3rd, 35, 36	**3-3** 106–107/1–19 odd
25	**Chapter Test** T44	**3-4** 112–114/1–13 odd	**3-3** 107–108/21–35 odd
26	**3-1** 98–99/1–31 odd	**Chapter Test** 115	**3-4** 112/1–6 all
27	**3-2** 103/1–19 odd	**Chapter Test** T45	**3-4** 113–114/7–12 all
28	**3-2** 103–104/21–41 odd	**Cumulative Review** 116/1–17 all	**Chapter Test** 115/3–8 all and/or T45/3–8 all
29	**3-3** 106–107/1–19 odd	**4-1** 122–124/1–27 odd, 22	**Cumulative Review** 116/1–17 all
30	**3-3** 107–108/21–35 odd	**4-2** 128–129/1–23 odd	**4-1** 122–124/1–21 odd

Day	Standard Course	Comprehensive Course	Discrete Math Course
31	3-4 112/1–6 all	4-2 129–131/27–39 odd	4-2 128–129/1–17 odd
32	3-4 113–114/7–12 all	4-3 136–137/1–25 odd	4-2 129–130/19–37 odd
33	Chapter Test 115	4-3 137–138/27–35 odd, 36–41 all	4-3 136–137/1, 5, 9, 13, 17, 21, 25, 27, 29, 31, 33, 35
34	Chapter Test T45	4-4 143–144/1–8 all	4-4 143–146/1–19 odd
35	Cumulative Review 116/1–17 all	4-4 144–146/9–19 all	4-5 149/1–10 all
36	4-1 122–124/1–21 odd	4-5 149–150/1–19 odd	4-5 150/11–27 odd, Calc. Exs. 1, 2, 3
37	4-2 128–129/1–17 odd	4-5 150/21–31 all, Calc. Exs. 1, 2, 3	4-6 154–157/1–19 odd
38	4-2 129–130/19–37 odd	4-6 154–156/1–10 all	4-7 161–162/1–13 odd
39	4-3 136–137/1–19 odd	4-6 156–157/11–21 odd	4-7 162–165/15–27 odd
40	4-3 137–138/21–39 odd	4-7 161–163/1–17 odd	Chapter Test 166–167
41	4-4 143–144/1–8 all	4-7 163–165/19–31 odd	Chapter Test T46–T47
42	4-4 144–145/9–15 all	Chapter Test 166–167	5-1 173–175/3–45 every 3rd
43	4-5 149/1–10 all	Chapter Test T46–T47	5-2 178–179/3–36 every 3rd
44	4-5 150/12–26 even, Calc. Exs. 1, 2, 3	5-1 173–175/1–45 every 3rd, 46–49 all	5-2 179–180/37–57 odd
45	4-6 154–156/1–10 all	5-2 178/3–36 every 3rd	5-3 183–184/1–13 all
46	4-6 156–157/11–19 odd	5-2 179–180/37–57 odd	5-3 184–186/15–27 odd
47	4-7 161–162/1–13 odd	5-3 183–184/1–15 all	5-4 189–190/1–23 odd, 22
48	4-7 162–164/15–27 odd	5-3 183–186/17–29 odd	5-5 194–195/1–19 odd
49	Chapter Test 166–167	5-4 189–190/1–23 odd, 22	5-5 195–197/21–48 every 3rd
50	Chapter Test T46–T47	5-5 194–195/1–19 odd	5-6 200/3–33 every 3rd
51	5-1 173/1–21 odd	5-5 195–197/21–51 every 3rd	5-6 201–203/35–51 odd
52	5-1 173–175/23–45 odd	5-6 200–201/3–39 every 3rd	5-7 205–206/1–19 odd
53	5-2 178/3–36 every 3rd	5-6 202–203/41–51 odd, 52–54 all	5-7 206–207/23–43 odd
54	5-2 179–180/37–57 odd	5-7 205–206/1–23 odd	Chapter Test 209
55	5-3 183–184/1–13 all	5-7 206–207/25–43 odd, 44, 45	Chapter Test T48

Day	Standard Course	Comprehensive Course	Discrete Math Course
56	**5-3** 184–186/15–27 odd	**Chapter Test** 209	**Project** 210–211
57	**5-4** 189–190/1–13 all	**Chapter Test** T48	**8-1** 299/1–21 odd
58	**5-4** 189–190/14–23 all	**6-1** 218/1–13 odd	**8-1** 299–301/23–35 odd, 45
59	**5-5** 194–195/1–19 odd	**6-1** 219/14–18 all	**8-2** 305–306/1, 5, 9, 11, 15, 17, 19
60	**5-5** 195–197/21–48 every 3rd	**6-2** 222–223/1–35 odd	**8-2** 306–308/21–33 odd
61	**5-6** 200/3–33 every 3rd	**6-2** 223–224/37–59 odd, 48	**8-3** 313/1–13 odd
62	**5-6** 201–202/35–51 odd	**6-3** 228–229/1–23 odd	**8-3** 314–316/15–23 odd
63	**5-7** 205–206/1–19 odd	**6-3** 229–230/25–37 odd, 38–42 all	**Chapter Test** 328/1–7 all and/or T51/1–7 all
64	**5-7** 206–207/23–43 odd	**6-4** 235–236/1–23 odd, 24	**12-1** 423–426/1, 3, 9, 11, 12, 19, 23, 25
65	**Chapter Test** 209	**6-4** 236–237/25–45 odd, 26	**12-2** 429–430/1, 3, 5, 7, 9, 13, 15, 17
66	**Chapter Test** T48	**6-5** 240–241/1–31 odd	**12-2** 430–432/19, 21, 23, 25, 29, 31, 33, 39
67	**6-1** 218/1–9 odd	**6-6** 244–246/1–31 odd	**12-3** 435–436/1–19 odd
68	**6-1** 218–219/11, 15, 16, 17	**6-7** 250–251/1–21 odd	**12-3** 437–440/23, 25, 27, 29, 33, 38, 39
69	**6-2** 222–223/1–35 odd	**Chapter Test** 253	**12-4** 444–446/1, 3, 9, 13, 15, 17, 21, 23, 25
70	**6-2** 223–224/37, 41, 43–53 odd, 48	**Chapter Test** T49	**12-5** 450–451/1, 5, 9, 11, 13, 15, 17, 21, 23, 25
71	**6-3** 228–229/1–23 odd	**Cumulative Review** 254–255/1–12 all	**12-5** 451–452/25, 27, 31, 33, 35, 39, 41
72	**6-3** 229–330/25–41 odd	**Cumulative Review** 255/13–24 all	**12-6** 455–456/1, 9, 11, 13, 15, 19, 21
73	**6-4** 235–236/1–23 odd	**7-1** 261–262/1–33 odd	**12-6** 456–458/23, 27, 31, 33, 35, 39
74	**6-4** 236–237/25–43 odd	**7-2** 264–267/1, 5, 9, 13, 15, 16, 17, 21, 23, 25	**12-7** 460–461/1, 4, 5, 7, 11, 13
75	**6-5** 240–241/1–15 odd	**7-3** 272–274/1–23 odd	**12-8** 463–464/1, 3, 7, 9, 11
76	**6-5** 241/17–31 odd, 20	**7-3** 273–274/29–43 all	**12-8** 464/13, 15, 17, 21, 22
77	**6-6** 244–245/1–15 odd	**7-4** 279–280/1–19 odd	**12-9** 467/1, 3, 5, 9, 13, 15

Day	Standard Course	Comprehensive Course	Discrete Math Course
78	**6-6** 245–246/17–29 odd	**7-4** 280–282/21–33 odd, 24, 34	**Chapter Test** 469
79	**6-7** 250–251/1–15 odd	**7-5** 285–286/3, 6, 9, 11, 15, 17, 20, 21, 23, 25, 27	**Chapter Test** T55
80	**Chapter Test** 253	**7-6** 289–290/1–21 odd	**Project** 470–471
81	**Chapter Test** T49	**7-6** 291/22–30 all	**13-1** 476–477/1, 5, 9, 11, 13, 17, 21, 25, 29, 33, 37, 41
82	**Cumulative Review** 254–255/1–12 all	**Chapter Test** 293	**13-1** 477–479/43, 47, 49, 53, 55, 59, 61
83	**Cumulative Review** 254–255/13–24 all	**Chapter Test** T50	**13-2** 481–482/1, 5, 9, 11, 13, 15, 17, 19, 21, 22
84	**7-1** 261/1–15 odd	**8-1** 299–300/1–29 odd	**13-2** 482–485/23, 25, 28, 31, 33
85	**7-1** 262/17–33 odd, 32	**8-1** 300–301/31–45 odd	**13-3** 489–490/1, 3, 7, 9, 11, 13, 17, 27, 29, 31
86	**7-2** 264–265/1–15 odd	**8-2** 305–306/1, 5, 9, 11, 15, 17, 19, 23	**13-3** 490–492/33, 37, 38, 40, 42, 43
87	**7-2** 266–267/16–23 all	**8-2** 306–308/25, 27, 29–36 all	**13-4** 496–498/1–33 odd
88	**7-3** 272/1–15 odd	**8-3** 313–314/1–15 odd	**13-5** 502–503/1, 3, 7, 9, 11, 13, 15, 19, 21, 23
89	**7-3** 272–274/17–27 odd, 31–41 odd	**8-3** 314–316/17, 19, 21, 24–26 all	**13-5** 503–506/29, 31, 33, 35, 41, 43, Calc. Ex.
90	**7-4** 279–280/1–17 odd	**8-4** 321/1–23 odd	**13-6** 508/1–21 odd
91	**7-4** 280–281/19, 24, 27, 29, 31, 33	**8-4** 321–322/28–39 all, 42	**13-6** 509–510/23–37 odd
92	**7-5** 285–286/1–15 odd	**8-5** 326/1–15 odd	**13-7** 513/1–11 odd
93	**7-5** 286/17–27 odd	**8-5** 326–327/17–35 odd	**13-7** 513–514/13, 16, 19, 21, 25
94	**7-6** 289–290/1–13 odd	**Chapter Test** 328–329	**Chapter Test** 515
95	**7-6** 290–291/15–21 all, 27, 29	**Chapter Test** T51	**Chapter Test** T56
96	**Chapter Test** 293	**9-1** 334–336/1, 7, 13–25 odd	**14-1** 520–522/1, 3, 9, 12, 15, 16, 17, 22
97	**Chapter Test** T50	**9-1** 337–338/31, 33, 35, 37, 39, 40	**14-2** 526–529/1–19 odd

Day	Standard Course	Comprehensive Course	Discrete Math Course
98	**8-1** 299/1–21 odd	**9-2** 342–344/1, 7, 9, 15, 19, 27, 31, 33, 35	**14-3** 534–535/1, 9, 13, 15, 20, 21
99	**8-1** 299–300/23–35 odd, 45	**9-3** 347–349/3, 7, 11, 15, 19, 21–25 all	**14-3** 536–537/23, 25, 26, 27
100	**8-2** 305–306/1, 5, 9, 11, 15, 17, 19	**9-4** 352–354/1, 7, 9, 13, 15, 17, 18, 21, 23	**14-4** 540–543/1, 3, 5, 7, 9, 12, 13
101	**8-2** 306–307/21–33 odd	**Mixed Exercises** 355–356/1–21 odd	**14-5** 546–548/1–7 odd
102	**8-3** 313/1–13 odd	**Mixed Exercises** 356–358/23, 24, 25, 29, 36, 40, 43	**14-5** 548–550/9, 11, 13, 14
103	**8-3** 314–316/15–23 odd	**9-5** 362/1–9 all	**14-6** 557–559/1, 3, 5, 13, 15, 19, 25
104	**8-4** 321/1–15 odd	**9-5** 363/11–18 all	**Chapter Test** 560–561
105	**8-4** 321–322/17–23 odd, 27–35 odd	**Chapter Test** 365	**Chapter Test** T57–T58
106	**8-5** 326/1–15 odd	**Chapter Test** T52	**Cumulative Review** 562–563/1–25 odd
107	**8-5** 326–327/17–31 odd	**10-1** 373/1–27 odd	**15-1** 568–571/1, 7, 9, 13, 15, 17, 19
108	**Chapter Test** 328–329	**10-1** 373–374/29–43 odd	**15-2** 575–576/1–15 odd
109	**Chapter Test** T51	**10-2** 377–379/1, 5, 7, 9, 11, 13, 15, 17, 19, 23, 29	**15-2** 576–577/17–31 odd
110	**9-1** 334–336/1–7 odd, 13–21 odd	**10-3** 383–384/1–25 odd	**15-3** 580–581/1–15 odd
111	**9-1** 336–338/23, 25, 27, 31, 33, 34, 36, 37	**10-3** 384–385/31–49 odd	**15-3** 581–582/17–23 all, 25, 29
112	**9-2** 342/1–15 odd	**10-4** 389–390/1–21 odd	**15-4** 585–587/3, 5, 9, 11, 13, 16, 17
113	**9-2** 343/19–27 odd	**10-4** 390–391/23–31 odd, 33–41 all, 43	**Mixed Exercises** 587–588/1–13 all
114	**9-3** 347–348/3, 7, 11, 15, 19, 21, 22	**Chapter Test** 393	**Mixed Exercises** 588–589/14–24 all
115	**9-4** 352–354/1, 7, 9, 13, 15, 17, 21	**Chapter Test** T53	**15-5** 592/1–13 odd
116	**Mixed Exercises** 355–356/1–19 odd	**11-1** 400–401/1, 5, 9, 11, 13, 17, 19, 21	**15-5** 592–594/17–23 all
117	**Mixed Exercises** 356–357/21–29 odd	**11-1** 401–402/25, 27; 29–34 all, 39	**Chapter Test** 595/1–8 all

Day	Standard Course	Comprehensive Course	Discrete Math Course
118	**9-5** 362/1–9 all	**11-2** 406/1–19 odd	**Chapter Test** T59
119	**9-5** 363/11–17 odd	**11-2** 406–407/23–29 all	**16-1** 603–605/1, 5, 7, 9, 15, 17, 25, 27, 29
120	**Chapter Test** 365	**11-3** 410–411/3, 5, 9, 11, 13, 14, 16	**16-2** 609–611/1, 5, 7, 11, 13–18 all
121	**Chapter Test** T52	**11-4** 413–414/1, 3, 6, 7, 9, 11	**16-2** 612/21–25 all, 27, 29
122	**10-1** 373/1–21 odd	**11-4** 414/12–17	**16-3** 616–618/7–19 odd, 20
123	**10-1** 373–374/23–37 odd	**Chapter Test** 415	**16-4** 621–622/1–12 all
124	**10-2** 377–378/1–10 all	**Chapter Test** T54	**16-4** 622–623/13–20 all
125	**10-2** 378/11, 13, 14, 17, 19, 23, 25	**Cumulative Review** 416/1–13 odd	**16-5** 626–627/1–9 odd
126	**10-3** 383–384/1–23 odd	**Cumulative Review** 417/14, 15, 19, 21, 22, 25–31 odd	**16-5** 628–629/11–17 odd
127	**10-3** 384–385/31–47 odd	**12-1** 423–426/1, 3, 9, 11, 12, 19, 23, 25	**16-6** 633–634/1–13 odd
128	**10-4** 389/1–15 odd	**12-2** 429–430/1, 3, 5, 7, 9, 13, 15, 17	**16-6** 634–635/14, 15–23 odd
129	**10-4** 390/17–33 odd	**12-2** 430–432/19, 21, 23, 25, 29, 31, 33, 39	**Chapter Test** 636–637
130	**Chapter Test** 393	**12-3** 435–436/1–19 odd	**Chapter Test** T60
131	**Chapter Test** T53	**12-3** 436–439/23, 25, 27, 29, 33, 38, 39	**17-1** 643–645/1, 5, 7, 9, 11, 13
132	**11-1** 400–401/1, 5, 9, 13, 17, 19, 21	**12-4** 444–445/1, 3, 9, 13, 15, 17, 21, 23, 25	**17-1** 645–648/15, 17, 18, 23, 24, 25
133	**11-1** 401/25, 27, 29, 33	**12-5** 450–451/1, 5, 9, 11, 13, 15, 19, 21, 23	**17-2** 651–653/1, 3, 5, 7
134	**11-2** 406/1–15 odd	**12-5** 451–452/25, 27, 31, 33, 35, 39, 41	**17-3** 658/1–8 all
135	**11-2** 406/17, 19, 21, 23, 25–28 all	**12-6** 455–456/1, 9, 11, 13, 15, 19, 21	**17-3** 659–660/9–15 all
136	**11-3** 410/1, 3, 5, 7	**12-6** 456–458/23, 27, 29, 31, 33, 37, 39	**17-4** 667–668/1–11 odd
137	**11-3** 410/9, 11, 13, 14	**12-7** 460–461/1, 4, 5, 7, 11, 13	**17-4** 668–669/12–16 all
138	**11-4** 413/1, 3, 6, 9	**12-8** 463–464/1, 3, 7, 9, 11	**17-5** 673–674/1–11 odd
139	**11-4** 414/11, 12, 13, 15	**12-8** 464/13, 15, 17, 21, 22	**17-6** 678/1–15 odd

Day	Standard Course	Comprehensive Course	Discrete Math Course
140	Chapter Test 415	12-9 467/1, 3, 5, 7, 9	17-6 679–680/17, 19, 21–23 all
141	Chapter Test T54	12-9 467/11, 13, 14, 15	Chapter Test 681
142	Cumulative Review 416/1–13 odd	Chapter Test 469	Chapter Test T61
143	Cumulative Review 417/14, 15, 19, 21, 22, 25–31 odd	Chapter Test T55	18-1 688–689/1–13 odd
144	12-1 423–426/1, 3, 9, 11, 12, 19, 25	13-1 476–477/1, 5, 9, 13, 17, 21, 25, 29, 33, 37, 41	18-1 690–691/18–21 all
145	12-2 429–430/1, 3, 5, 7, 9, 13, 15, 17	13-1 477–478/43, 47, 49, 53, 55, 59, 61	18-2 694–695/1, 3, 5, 7, 9, 10
146	12-2 430–431/19, 21, 23, 25, 29, 33	13-2 481–482/1, 5, 9, 11, 13, 15, 17, 19, 21, 22	18-2 695–697/13–17 all
147	12-3 435–436/1–17 odd	13-2 482–485/23, 25, 28, 31, 33	18-3 701/1–15 odd
148	12-3 436–439/19, 21, 25, 27, 33, 39	13-3 489–490/1, 3, 7, 9, 11, 13, 17, 27, 29, 31	18-3 703–704/17, 19, 20, 21
149	12-4 444/1, 3, 9, 13, 15, 17, 21	13-3 490–491/33, 37, 38, 40, 42, 43	18-4 705–706/1–5 all
150	Chapter Test 469/1–7 all	13-4 496–497/1–33 odd	18-4 707–709/6, 9, 10, 11
151	Chapter Test T55/1–7 all	13-5 502–503/1, 3, 7, 9, 11, 13, 15, 19, 21, 23	Chapter Test 710
152	13-1 476–477/1, 5, 9, 13, 17, 21, 25, 29, 33, 37, 41	13-5 503–506/29, 31, 33, 35, 41, 43, Calc. Ex.	Chapter Test T62
153	13-2 481–485/1–19 odd	13-6 508/1–21 odd	Cumulative Review 711–712/1–22 all
154	13-2 481–485/21–27 odd	13-6 508–509/23–37 odd	Project 714–715
155	13-3 489–492/1, 3, 7, 9, 11, 13, 17, 27	13-7 513/1–11 odd	19-1 724–725/1–27 odd
156	13-4 496–497/1–25 odd	13-7 513–514/13, 16, 19, 21, 25	19-1 725/29–47 odd
157	Chapter Test 515/1–9 all	Chapter Test 515	19-2 728/1–11 odd
158	Chapter Test T56/1–9 all	Chapter Test T56	19-2 728–729/4, 12–17 all
159	Reviews T65–T73/selection	Reviews T65–T73/selection	Chapter Test 749/1–2 and/or T63/1–2
160	Reviews T65–T73/selection	Reviews T65–T73/selection	Reviews T65–T73/selection

Chapter Tests

1. Let $A = (-4, 7)$ and $B = (4, -5)$.
   a. Find the length of $\overline{AB}$.
   b. Find the coordinates of the midpoint of $\overline{AB}$.

2. Find the value of a if it is known that the point $(-3, 7)$ lies on the line $2x + ay = 26$.

3. Solve the equations $3x - 2y = 3$ and $5x + 4y = 16$ simultaneously. Then sketch the graphs of the lines and label the intersection point with its coordinates.

4. Find the slope and the y-intercept of the line $3x + 2y = -1$.

5. Tell which of the following equations have parallel line graphs and which have perpendicular line graphs.

   a. $3x - 2y = 12$ b. $y = \frac{3}{2}x + 1$ c. $4x + 6y + 20 = 0$

6. Write an equation of the line through the points $(6, -2)$ and $(3, 7)$.

7. Write an equation of the line through the point $(2, 5)$ and parallel to the line $4x - 3y = -50$.

8. Write an equation of the line with x-intercept 2 and y-intercept -3.

9. Write an equation of the vertical line through the point $(5, 1)$.

10. a. *Writing* Describe the steps for finding an equation of the median from vertex A in $\triangle ABC$.
    b. Write an equation of the median from A if $A(2, 1)$, $B(4, 8)$, and $C(8, -2)$ are the vertices of $\triangle ABC$.

11. Twyla buys a 10-ride pass so that she can use public transportation to go to and from work. The pass costs \$15 and is worth \$0 after the tenth ride.
    a. Give an equation of the linear function that models the value of the pass as a function of the number of rides completed.
    b. Use the equation to find the value of the pass after the sixth ride.

Express each complex number in the form $a + bi$.

12. $\sqrt{-12} - \sqrt{-48}$

13. $(3 - 2i)^2$

14. $\dfrac{1}{4 - 7i}$

15. $\dfrac{\sqrt{3} + 2i}{\sqrt{3} - 2i}$

16. $4(3 + 4i) - 5i(1 + i)$

17. i^{23}

18. Solve for x.
    a. $12x^2 + 12 = 25x$ b. $x^2 + 4x = 7$ c. $x^2 - 6x + 18 = 0$

(Continues on next page)

19. Find the discriminant of $3x^2 - 2x - 5 = 0$ and then solve by whichever method seems easiest.

20. Sketch the graph of each parabola. Label the vertex, the axis of symmetry, and the intercepts.
    a. $y = -4 + 2(x + 1)^2$ **b.** $y = -x^2 - 4x + 2$

21. Sketch the graphs of the line $x + 2y = 1$ and the parabola $y = x^2 - 4x - 4$. Find the coordinates of any points of intersection.

22. At a photo store, a particular style of frame with nonreflective glass is priced according to the size of the frame. A 3 × 5 frame sells for $5.75, a 4 × 6 frame sells for $8.00, and a 5 × 7 frame sells for $10.75.
    a. Find a quadratic function of the form $f(x) = ax^2 + bx + c$ that models the frame-with-glass price (in dollars) in terms of the smaller dimension of the frame.
    b. Using this model, find the price of a 7 × 9 frame.

1. If $P(x) = 4x^3 - 5x^2 + 1$, use synthetic substitution to find:

   a. $P(2)$ **b.** $P\left(\dfrac{1}{2}\right)$ **c.** $P(i)$

2. If -2 is a zero of the polynomial $P(x) = 2x^2 + x + k$, find the value of k.

3. A polynomial $P(x)$ is divided by $x - 2$. The quotient is $2x^2 + x - 2$ and the remainder is -1. Find $P(x)$.

4. Two roots of the equation

   $$x^4 + x^3 - 5x^2 + x - 6 = 0$$

   are $x = 2$ and $x = -3$. Find the remaining roots.

5. Sketch the graph of

   $$y = -x^2(x - 1)(x + 2).$$

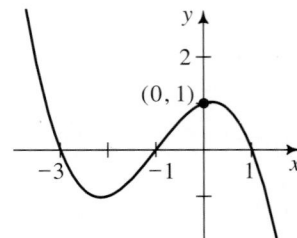

6. Give an equation of the cubic graph shown at the right.

Ex. 6

7. A rectangular enclosure, subdivided into three congruent pens as shown, is to be made using a barn as one side and 120 m of fencing for the rest of the enclosure. Find the value of x that gives the maximum area for the enclosure.

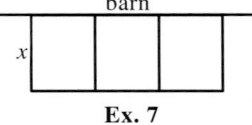

Ex. 7

 Part (b) of Exercise 8 requires the use of a computer or graphing calculator.

8. **a.** If $P(x) = 2x^4 - 2x - 1$, explain how you know that the equation $P(x) = 0$ has at least one real root between $x = 1$ and $x = 2$.

   b. Use a computer or graphing calculator to find each root of $P(x) = 0$ to the nearest tenth.

9. Solve $2x^3 + 2x^2 = 4x + 4$.

10. Find all real and imaginary roots of $2x^4 + 3x^3 + 7x^2 - 7x - 5 = 0$.

11. Find the sum and product of the roots of $3x^3 - 2x^2 + x - 6 = 0$.

12. Find a cubic equation with integral coefficients having $1 + i$ and 3 as roots.

13. *Writing* Suppose you are given a cubic equation with integral coefficients. Write a paragraph in which you discuss the number and the nature of the equation's possible roots. (Consider such questions as: How many distinct roots are possible? How many must be real? How many can be imaginary?)

Solve and graph each inequality or equation on a number line.

1. **a.** $-2(4x - 3) < 12 - 6x$ **b.** $\frac{2}{3}(2x + 1) \leq \frac{2x - 6}{5}$

2. **a.** $|x + 4| < 3$ **b.** $|x - 2| > 5$
    c. $|2x - 5| = 5$ **d.** $|3 - 4x| \leq 7$

3. Solve each inequality.
    a. $(x - 2)(x + 3)^3(2x - 1)^2 \geq 0$ **b.** $3x^2 - 5x + 5 < 0$

    c. $2x^3 - x \leq -x^2$ **d.** $\dfrac{(2x + 1)(x - 5)}{(x - 3)^3} \leq 0$

4. *Writing* Describe the set of points in the coordinate plane that satisfies the inequality $2x + 3y \geq 12$.

5. Sketch the graph of each inequality in the coordinate plane.
    a. $5 - 2y \leq 4x$ **b.** $|y| > 1$

6. Graph the solution set of each system.
    a. $x > 0$ **b.** $y \leq -x^2 + 4x$
       $2x - y \leq -1$ $y \geq 2x - 3$

7. Give a set of inequalities that defines the shaded region shown at the right.

8. A manufacturer of metal widgets makes two types, A and B. Each type A widget requires 0.5 h of cutting, 1.5 h of welding, and 1 h of finishing. Each type B widget requires 0.5 h of cutting, 1 h of welding, and 1.5 h of finishing. Each day, 15 h of cutting time, 36 h of welding time, and 42 h of finishing time are available. If the profits on a type A widget and on a type B widget are $3 and $4, respectively, how many of each type of widget should be made each day to maximize the total profits?

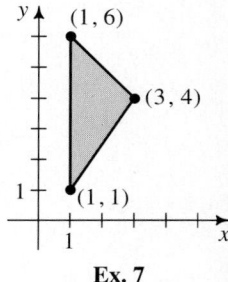

Ex. 7

1. Give the domain, range, and zeros of the function $f(x) = \sqrt{x^2 - 16}$.

2. **a.** Graph the function $g(x) = \begin{cases} x + 2 & \text{if } x < -1 \\ x^2 & \text{if } -1 \le x \le 1 \\ 1 & \text{if } x > 1 \end{cases}$

    b. Use the graph in part (a) to find the range and zeros of g.

3. Let $f(x) = x^2 - 3x + 2$ and $g(x) = x - 2$. Find:

    a. $(f + g)(x)$ **b.** $(f - g)(x)$ **c.** $(f \cdot g)(x)$ **d.** $\left(\dfrac{f}{g}\right)(x)$

4. Using the functions f and g in Exercise 3, find:
    a. $(f \circ g)(x)$ **b.** $(g \circ f)(x)$

5. **Writing** Describe, in general terms, how to find the domain of the composite function $(f \circ g)(x)$. Then illustrate with a specific example using the functions

    $f(x) = \dfrac{1}{x^2 - 1}$ and $g(x) = \sqrt{x + 1}$.

6. Determine whether the graph of $x^2 - xy + y^2 = 6$ has symmetry in: (i) the x-axis, (ii) the y-axis, (iii) the line $y = x$, and (iv) the origin.

7. Given the graph of $y = f(x)$ shown at the right, sketch the graph of each of the following.

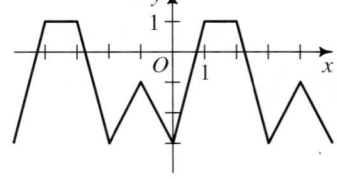

(1, 5)

$(-3, 1)$ $y = f(x)$

$(-1, -1)$

Ex. 7

    a. $y = \dfrac{1}{2}f(x)$ **b.** $y = |f(x)|$

    c. $y = -f(x)$ **d.** $y = f(x - 2)$

8. The graph of a periodic function $y = g(x)$ is shown at the right.
    a. What is the fundamental period of g?
    b. What are the maximum and minimum values of g?
    c. What is the amplitude of g?

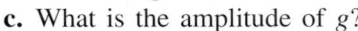

Ex. 8

9. **a.** Which one of the two functions $f(x) = \sqrt{9 - x^2}$ and $g(x) = 2x - 3$ has an inverse? Find a rule for the inverse.

    b. Explain why the other function does not have an inverse.

10. The length c of the hypotenuse of a right triangle is a function of the lengths a and b of the legs.
    a. State a rule for the function $c(a, b)$.
    b. Find $c(3, 4)$ and $c(12, 5)$.
    c. Draw the curve of constant length $c(a, b) = 2$ in an ab-plane.

(*Continues on next page*)

11. The water in a cylindrical tank 4 ft in diameter empties through a hole in the bottom. Assuming that the water has a depth of 16 ft at time $t = 0$ and empties at the rate of 5 ft^3/s, express the depth of the water as a function of time.

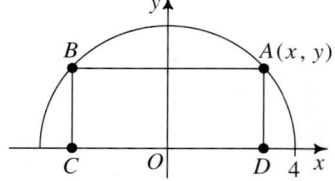 **Part (c) of Exercise 12 requires the use of a computer or graphing calculator.**

12. Rectangle $ABCD$ has two vertices on the semicircle $y = \sqrt{16 - x^2}$ and two vertices on the x-axis, as shown at the right.

a. Express the area of the rectangle as a function of the x-coordinate of A.

b. What is the domain of the area function?

c. Use a computer or graphing calculator to find the maximum area.

1. Give a general expression for the value of a $100,000 investment t years from now if its value increases at the rate of 12% per year.

2. Evaluate each expression.

    a. $\dfrac{3^{-5} \cdot 3^{10}}{3^2}$

    b. $(4^{-2} + 4^0)^{-1}$

    c. $\sqrt{\dfrac{9^5}{3^{-2}}}$

    d. $\dfrac{2^{-4} + 4^{-3}}{2^{-3}}$

3. Solve each equation.

    a. $3^{8-x} = 27^{2-x}$

    b. $4\sqrt{32} = 2^{3x}$

4. A gallon of gasoline cost $1.03 two years ago. Now it costs $1.59. To the nearest percent, what has been the annual rate of increase in the cost?

5. Graph $y = 3^x$ and $y = \left(\dfrac{1}{3}\right)^x$ on the same set of axes. How are the graphs related?

6. **Writing** Explain what it means for the half-life of a radioactive isotope to be 2 years. Also give the mathematical model for the exponential decay of 5 g of the isotope.

7. Suppose that $1200 is invested at an interest rate of 9.6%. How much is the investment worth after 2 years if interest is compounded **(a)** monthly? **(b)** continuously?

8. Given that $\log 40 \approx 1.6021$, find the value of:

    a. $\log 4$
    b. $\log 400$
    c. $\log 0.25$

9. Find the exact value of:

    a. $\log_2 16$
    b. $\log_{27} 3$
    c. $5^{\log_4 64}$

10. Express y as a function of x.

    a. $\log_3 y = 4 \log_3 (2x)$
    b. $\log_3 y = 2 + \log_3 x$

11. Express the following in terms of $\log_b M$ and $\log_b N$.

    a. $\log_b \sqrt[3]{\dfrac{M^3}{N^2}}$

    b. $\log_b M^3 N^2$

12. Solve $\log_2 (x - 2) + \log_2 (x + 5) = 1$.

13. a. Between what two consecutive integers must $\log_4 67$ lie?

    b. Use the change-of-base formula to express $\log_4 67$ in terms of common logarithms and then approximate it to the nearest thousandth.

14. Solve $3^x = 31$ for x to the nearest hundredth.

1. In parallelogram $ABCD$ shown at the right, X and Y are the respective midpoints of $\overline{BC}$ and $\overline{AD}$. Prove that $AXCY$ is a parallelogram.

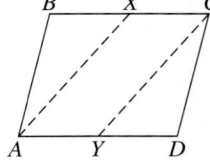

Ex. 1

2. Find the coordinates of any points where the line $-x + y = 3$ and the circle $x^2 + y^2 = 9$ intersect.

3. Find an equation of the circle having $(4, 8)$ and $(6, -2)$ as endpoints of a diameter.

4. Find the coordinates of the vertices and the foci of the ellipse with equation $16x^2 + 9y^2 = 144$. Sketch the ellipse.

5. Find the coordinates of any points where the line $x - 3y = 6$ and the ellipse $x^2 + 9y^2 = 36$ intersect.

6. Find an equation for the ellipse that has $(-3, 0)$ and $(3, 0)$ as endpoints of its minor axis and $(0, -5)$ and $(0, 5)$ as endpoints of its major axis.

7. Give an equation of the hyperbola centered at the origin, having a vertex at $(4, 0)$, and having a focus at $(2\sqrt{7}, 0)$.

8. Graph the hyperbola $xy + 10 = 0$ and give equations for its asymptotes.

9. Find an equation for the hyperbola that has asymptotes with equations $y = \pm 3x$ and a vertex $(0, 1)$.

10. Find the coordinates of the vertex and focus, and the equation of the directrix, of the parabola $y = 2x^2$.

11. Find an equation of the parabola whose directrix is $y = 2$ and whose focus is $(0, -4)$.

12. On the same set of axes, graph $x^2 + y^2 = 16$ and $x^2 + 6y^2 = 96$. Solve these equations simultaneously.

13. Given that the following equations are not degenerate conics, tell whether the graph is a circle, an ellipse, a hyperbola, or a parabola.
    a. $x^2 + y^2 + xy + x - y - 15 = 0$
    b. $2x^2 + 4xy - y^2 + 12 = 0$
    c. $x^2 + 4xy + 4y^2 + 2x - 6y + 24 = 0$

14. Find and simplify an equation for the set of points $P(x, y)$ that are half as far from the point $F(1, 0)$ as they are from the line $x = 4$.

15. **Writing** Using only the *signs* of the coefficients A, B, and C in the equation $Ax^2 + By^2 + C = 0$ (where $A \neq 0$, $B \neq 0$, and $C \neq 0$), describe the possibilities for the graph of the equation.

1. **a.** Convert $-90°$ to radians. Give the answer in terms of π.
   b. Convert $212°$ to radians. Give the answer to the nearest hundredth.

2. **a.** Convert $\dfrac{7\pi}{6}$ radians to degrees.

   b. Convert 3.5 radians to degrees. Give the answer to the nearest ten minutes or tenth of a degree.

3. Find two angles, one positive and one negative, that are coterminal with each given angle.

   a. $-100°$ **b.** $320.3°$ **c.** $\dfrac{7\pi}{4}$ **d.** $220°40'$

4. A sector of a circle has radius 5 cm and central angle $137°$. Find its approximate arc length and area.

5. The sun is about 9×10^7 mi from Earth, and its apparent size is about 0.0043 radians. What is the sun's approximate diameter?

6. Find $\sin \theta$ and $\cos \theta$.

   a. 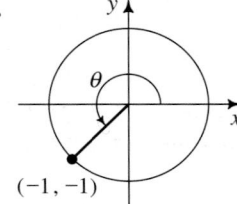   **b.** 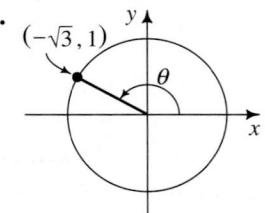   **c.**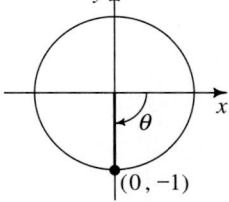

7. Complete each statement with one of the symbols $<$, $>$, or $=$.
   a. $\sin 70°$ __?__ $\sin 65°$ **b.** $\cos 70°$ __?__ $\cos 65°$
   c. $\cos 40°$ __?__ $\cos 320°$ **d.** $\sin 313°$ __?__ $\sin 314°$

8. Give the exact value of each expression in simplest radical form.

   a. $\sin \dfrac{5\pi}{6}$ **b.** $\cos (-180°)$ **c.** $\sin 210°$ **d.** $\cos \dfrac{5\pi}{6}$

9. Give the exact value of each expression or state that the value is undefined.

   a. $\csc 135°$ **b.** $\sec \dfrac{2\pi}{3}$ **c.** $\cot (-60°)$ **d.** $\tan (-\pi)$

10. If $\cot x = -\dfrac{1}{3}$ where $\dfrac{\pi}{2} < x < \pi$, find the values of the other five trigonometric functions.

11. Give the exact value of each expression.

    a. $\text{Tan}^{-1}\left(-\dfrac{\sqrt{3}}{3}\right)$ **b.** $\sec\left(\text{Sin}^{-1}\dfrac{1}{2}\right)$ **c.** $\csc\left(\text{Cos}^{-1}\left(-\dfrac{3}{5}\right)\right)$

12. *Writing* To obtain the inverse cosine function, we restrict the domain of $f(x) = \cos x$ to $0 \le x \le \pi$. Explain why this restriction is necessary.

1. Find the angle of inclination of the line $2x + 3y = 9$ to the nearest degree.

2. Solve $5 \sin \theta = -1$ for $0 \le \theta < 360°$. Give answers to the nearest tenth of a degree.

3. Solve $3 - 2 \csc x = 17$ for $0 \le x < 2\pi$. Give answers to the nearest hundredth of a radian.

4. Solve $2 \cos 4x = \sqrt{3}$ for $0 \le x < 2\pi$.

5. Graph $y = 2 \cos \frac{x}{3}$, and give its period and amplitude.

6. Find an equation of the trigonometric function whose graph is shown at the right.

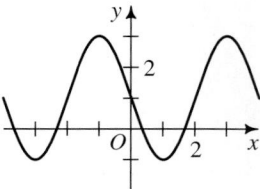

Ex. 6

7. At Ocean Tide Dock the first high tide today occurs at 2:00 A.M. with depth 6 m, and the first low tide occurs at 8:30 A.M. with depth 2 m.
   a. Sketch and label a graph showing the depth of the water at the dock as a function of time after midnight.
   b. *Writing* Suppose a tanker that requires at least 4 m of water depth is planning to dock after 9:00 A.M. Write a paragraph in which you describe how you would use the graph in part (a) to determine the earliest possible time that the tanker can dock and the longest period of time that the tanker can be docked before the water level becomes too low.

8. Simplify each expression.
   a. $\tan A \ (\csc A - \sin A)$
   b. $\dfrac{\tan \theta}{\cos (90° - \theta)}$
   c. $(\csc x + \cot x)(1 - \cos x)$
   d. $\dfrac{\tan \theta + \cot \theta}{\sec^2 \theta}$

9. Solve each equation for $0 \le x < 2\pi$. Give answers to the nearest hundredth of a radian when necessary.
   a. $\cos x = 2 \sin x$
   b. $\tan x - \cot x = 0$

10. Solve each equation for $0° \le \theta < 360°$. Give answers to the nearest tenth of a degree.
    a. $2 \sin^2 \theta = 3 \cos \theta + 3$
    b. $\sin \theta \tan \theta = 2 \sin \theta$

Where appropriate, give angle measures to the nearest tenth of a degree and lengths of sides in simplest radical form or to three significant digits.

1. The sides of an isosceles triangle have lengths 7, 10, and 10. What are the measures of its angles?

2. At a distance of 200 m, the angle of elevation to the top of a building is 70°. About how tall is the building?

3. A regular octagon is inscribed in a circle of radius 6 in. Find the area of the octagon.

4. Find the area of the quadrilateral shown at the right.

5. How many different triangles PQR can be constructed using the given information?
   a. $p = 5$, $q = 5$, $\angle Q = 74°$
   b. $p = 7$, $q = 8$, $\angle P = 23°$

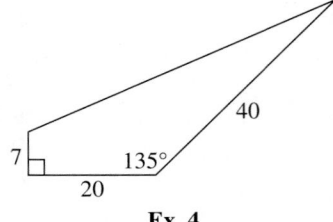

Ex. 4

6. Observers at points A and B, 50 km apart, sight an airplane at angles of elevation of 40° and 80°, respectively, as shown in the diagram. How far is the plane from each observer?

7. A triangle has sides of length 5, 9, and 10. What is the measure of its smallest angle?

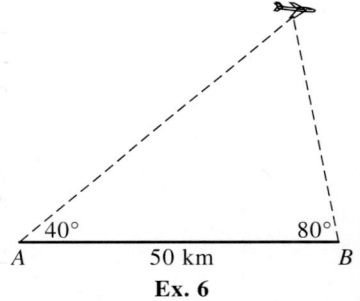

Ex. 6

8. Two ships leave a port on courses that differ by 70° and each travels at 25 knots (nautical miles per hour). In terms of nautical miles, how far apart are the ships after 1 h?

9. *Writing* Suppose you know the lengths of all three sides of a triangle as well as the measure of one of the angles. To find the measure of one of the other two angles, would you use the law of sines or the law of cosines? Write a paragraph in which you explain your choice.

10. After leaving an airport, a plane flies for 1.75 h at a speed of 200 km/h on a course of 100°. The plane then flies for 2 h at a speed of 250 km/h on a course of 40°. At this time, how far from the airport is the plane?

1. Simplify the given expression.
   a. $\cos 75° \cos 15° - \sin 75° \sin 15°$
   b. $\sin 75° \cos 15° - \cos 75° \sin 15°$
   c. $\sin (30° + x) + \sin (30° - x)$
   d. $\cos (45° - x) - \cos (45° + x)$

2. Find the exact value of $\cos 75°$.

3. Find $\tan \left(\dfrac{7\pi}{4} - \theta \right)$ when $\tan \theta = \dfrac{1}{3}$.

4. If $\tan \alpha = \dfrac{1}{3}$ and $\tan \beta = \dfrac{1}{2}$, show that $\tan (\alpha + \beta) = 1$.

5. *Writing* Explain why the formula for $\tan (\alpha + \beta)$ cannot be used to simplify the expression $\tan \left(\dfrac{\pi}{2} + \theta \right)$. How else could this expression be simplified? Try your idea and give the result.

6. Suppose $\angle A$ is acute and $\cos A = \dfrac{5}{13}$. Find each of the following.
   a. $\sin A$ b. $\cos 2A$
   c. $\sin 2A$ d. $\sin 4A$

7. Simplify the given expression.
   a. $\dfrac{1 + \cos 2x}{\sin 2x}$ b. $(1 + \cot^2 y)(\cos 2y + 1)$

   c. $\dfrac{\tan t}{\sec t - 1}$ d. $\cos^2 \dfrac{x}{4} - \sin^2 \dfrac{x}{4}$

8. Evaluate the given expression.
   a. $\sin \dfrac{\pi}{12} \cos \dfrac{\pi}{12}$ b. $1 - 2 \sin^2 \dfrac{5\pi}{12}$

9. Prove that the given equation is an identity.
   a. $(1 + \tan^2 x)(1 + \cos 2x) = 2$
   b. $\dfrac{\cos \theta \csc \theta}{\tan \theta + \cot \theta} = \sin^2 \theta + \cos 2\theta$

10. a. On the same set of axes, sketch the graphs of $y = \sin 2\theta$ and $y = \cos \theta$ for $0° \le \theta < 360°$.
    b. Determine where the graphs intersect by solving $\sin 2\theta = \cos \theta$.

11. Solve the equation $\cos 2x = \sin x - 2$ either graphically or algebraically for $0 \le x < 2\pi$.

1. Give polar coordinates or rectangular coordinates for each point, as indicated.
   a. $(3, -3)$, polar   **b.** $(6, -90°)$, rect.   **c.** $(0, 2)$, polar
   d. $(8, -\pi)$, rect.   **e.** $(2, 60°)$, rect.   **f.** $(1, -\sqrt{3})$, polar

2. **a.** Sketch the polar graph of $r = 2 \sin 2\theta$.
   b. Give a rectangular equation of this graph.

3. Sketch the polar graph of each equation.
   a. $r = 5$ **b.** $\theta = -1$ **c.** $r = 1 + \sin \theta$

4. Let $z_1 = -\sqrt{3} + i$ and $z_2 = 4 + 4i$.
   a. Express z_1, z_2, and $z_1 z_2$ in polar form.
   b. Show z_1, z_2, and $z_1 z_2$ in an Argand diagram.

5. Let $z = 3 \text{ cis } 150°$.
   a. Find z^2 in polar form and in rectangular form.
   b. Show that z^2 in polar form agrees with z^2 in rectangular form.

6. **a.** Express $z = 1 - i$ in polar form.
   b. Show z, z^2, z^3, and z^4 in an Argand diagram.
   c. Find z^{10}.

7. ***Writing*** Write a paragraph giving a geometric interpretation of the n nth roots of a complex number.

8. Show that $\sqrt[3]{2} \text{ cis } 130°$ is a cube root of $\sqrt{3} + i$, and find the other two cube roots.

1. $\mathbf{F}_1$ is a force of 8 N pulling an object due south, and $\mathbf{F}_2$ is a force of 15 N pulling the object due east. Find the direction (measured clockwise from north) and magnitude of the resultant force $\mathbf{F}_3$. Make a vector diagram.

2. In the diagram at the right, $AT:TC = 2:3$. If $\overrightarrow{BA} = \mathbf{u}$ and $\overrightarrow{BC} = \mathbf{v}$, express each of the following in the form $r\mathbf{u} + s\mathbf{v}$.

   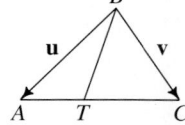

   a. $\overrightarrow{AB}$ **b.** $\overrightarrow{CA}$
   c. $\overrightarrow{CT}$ **d.** $\overrightarrow{BT}$

3. A plane is on a course of $200°$ at a speed of 520 mi/h. What are the north-south and east-west components of the plane's velocity vector?

4. Given the points $A(2, 4)$, $B(0, -5)$, and $C(7, 0)$, find the coordinates of a point $D(x, y)$ such that quadrilateral $ABCD$ is a parallelogram.

5. An object moves with constant velocity so that its position at time t is $(x, y) = (1, 3) + t(2, -3)$.
   a. Find the object's velocity and speed.
   b. Find a pair of parametric equations of the line along which the object moves.
   c. When and where does the object cross the parabola $y = 2x^2 - 6x$?

6. **Writing** Suppose $\mathbf{u}$ and $\mathbf{v}$ are nonzero vectors such that $\mathbf{u} \cdot \mathbf{v} = |\mathbf{u}||\mathbf{v}|$. Describe the relationship between $\mathbf{u}$ and $\mathbf{v}$, and explain your reasoning.

7. Given the points $A(1, 3)$, $B(-2, 7)$, and $C(0, -6)$, find the measure of $\angle B$ to the nearest tenth of a degree.

8. If $A = (2, -4, -5)$ and $B = (-1, 0, 7)$, find **(a)** the length of $\overline{AB}$ and **(b)** the midpoint of $\overline{AB}$.

9. Given the line L with equation $(x, y, z) = (0, -1, 6) + t(4, 1, 3)$.
   a. Write a vector equation for the line through $(2, 1, 7)$ parallel to L.
   b. Where does line L intersect the xz-plane?

10. A plane is tangent to the sphere $x^2 + (y - 1)^2 + (z + 3)^2 = 54$ at the point $(1, 3, 4)$. Find an equation of the plane.

11. Evaluate: **a.** $\begin{vmatrix} 1 & 2 & 0 \\ 3 & 4 & 1 \\ 2 & -1 & 0 \end{vmatrix}$ **b.** $\begin{vmatrix} 0 & 0 & 0 & 5 \\ 4 & 2 & 0 & 0 \\ 2 & 1 & 2 & 0 \\ 1 & 0 & 0 & -1 \end{vmatrix}$

12. Use Cramer's rule to solve: $4x + 5y = -12$
    $6x - 11y = 19$

13. **a.** Find a vector perpendicular to the plane determined by the points $A(2, 0, 1)$, $B(-4, 0, 3)$, and $C(4, 5, 2)$.
    b. Find the area of $\triangle ABC$.

1. State whether each sequence is arithmetic, geometric or neither, and find a formula for t_n in terms of n.

   a. 17, 12, 7, 2, . . .   **b.** 3, 8, 15, 24, 35, . . .   **c.** -81, 27, -9, 3, . . .

2. In an arithmetic sequence, $t_3 = 23$ and $t_6 = 50$. Find t_{24}.

3. In a geometric sequence, $t_3 = 4$ and $t_6 = \dfrac{4}{27}$. Find t_{10}.

4. List the first six terms of the sequence defined recursively by $t_1 = 1$, $t_2 = 3$, and $t_n = 3t_{n-1} + t_{n-2}$.

5. Give a recursive definition for the sequence 1, 4, 13, 40, ...

6. Consider the sequence of positive 3-digit integers ending in 6.

   a. Find t_1, t_2, t_3, and an explicit formula for t_n.

   b. Find S_1, S_2, S_3, and an explicit formula for S_n.

7. Find the sum of the first 8 terms of the following geometric series:

$$1 - \frac{1}{3} + \frac{1}{9} - \frac{1}{27} + \cdots$$

8. *Writing* Explain the difference between a sequence that has no limit and a sequence whose limit is infinity. Give an example of each.

9. Find the value of each limit if it exists. If the limit is infinity, say so.

   a. $\displaystyle\lim_{n \to \infty} \frac{5n^2 + n}{4n^2 - 2n + 5}$ **b.** $\displaystyle\lim_{n \to \infty} n \sin n\pi$ **c.** $\displaystyle\lim_{n \to \infty} (1.01)^n$

10. Express 0.232323 . . . as an infinite series and as a rational number.

11. Find **(a)** the interval of convergence and **(b)** the sum, expressed in terms of x, for the series $1 + \dfrac{2x}{3} + \dfrac{4x^2}{9} + \cdots$.

12. Express each of the following series using sigma notation.

    a. $-\dfrac{1}{4} + \dfrac{1}{16} - \dfrac{1}{36} + \dfrac{1}{64} - \dfrac{1}{100}$

    b. $8 + 5 + 2 - 1 - 4 - 7 - 10 - 13$

13. Evaluate $\displaystyle\sum_{k=1}^{20} 2k(k-1)$.

14. Prove by mathematical induction:

$$3 + 7 + 11 + \cdots + (4n - 1) = n(2n + 1)$$

1. A children's toy store sells Mama, Papa, and Baby Bears, each in small, medium, and large sizes. During one week, the store sold 10 small-sized, 8 medium-sized, and 20 large-sized Mama Bears. It also sold 6 small-sized, 10 medium-sized, and 8 large-sized Papa Bears and 14 small-sized, 20 medium-sized, and 14 large-sized Baby Bears.

   a. Express this information in a sales matrix S.

   b. Because this was a special sale week, the store expects to sell 50% fewer of each type of bear next week. How many of each medium-sized bear would it expect to sell? In terms of S, what matrix gives the amount of expected sales for each type of bear? Write this matrix.

2. If $A = \begin{bmatrix} 1 & 2 & 0 \\ -4 & 1 & 5 \end{bmatrix}$ and $B = \begin{bmatrix} -3 & 4 & 6 \\ -2 & -1 & 2 \end{bmatrix}$, find $3A - 2B$.

3. **Writing** Suppose A is a 2×2 nonzero matrix. Write a paragraph in which you state whether A must have an additive inverse and whether A must have a multiplicative inverse. In each case, give a convincing argument to support your conclusion.

4. Buddy's Nut Shop sells two varieties of mixed nuts. The mixes contain different amounts of peanuts (p), hazelnuts (h), cashews (c), and walnuts (w). Matrix P below gives the portion of the total weight of each mix for each type of nut. The owner plans to change the mixes soon, and so decides to combine what is left of the two mixes and sell this combined mixture. Matrix N gives the current amounts for each mix.

$$\begin{array}{c} \\ \text{regular} \\ \text{deluxe mix} \end{array} \begin{array}{cccc} \text{p} & \text{h} & \text{c} & \text{w} \end{array} \\ \begin{bmatrix} 45\% & 20\% & 15\% & 20\% \\ 20\% & 30\% & 25\% & 25\% \end{bmatrix} = P$$

$$\begin{array}{c} \\ \text{weight in lbs} \end{array} \begin{array}{cc} \text{regular} & \text{deluxe} \end{array} \\ \begin{bmatrix} 16 & 4 \end{bmatrix} = N$$

   a. What is the meaning of matrix NP? Find this matrix.

   b. What is the amount of hazelnuts in the combined mixture?

5. Solve this system using matrices: $\begin{aligned} 2x - 3y &= 13 \\ 2x - 7y &= 25 \end{aligned}$

(Continues on next page)

6. The diagram below indicates how various sites in a park are connected by roads.

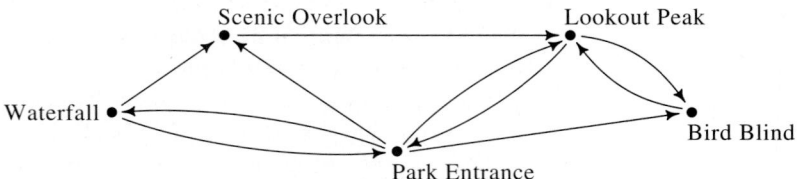

Scenic Overlook
Lookout Peak
Waterfall
Bird Blind
Park Entrance

a. Write the matrix M that models this network. Label the rows and columns in alphabetical order.
b. Find M^2. Which site has the greatest number of two-step paths to the other sites?

7. a. A survey done by a certain city has found that from one year to the next, 6% of those living in the city will move to its suburbs while 2% of those living in its suburbs will move into the city. Write a transition matrix T that describes this situation.
b. If 70% of the population is now in the city and 30% is in the suburbs, write the current population matrix P_0.
c. Find the percentages of the population living in the city and in the suburbs one year from now.

8. Consider the transformation $T:(x, y) \rightarrow (3x + y, x - 2y)$.
a. Draw $\triangle PQR$ with vertices $P(0, -1)$, $Q(1, 3)$, and $R(-2, 2)$.
b. Draw $\triangle P'Q'R'$, the image of $\triangle PQR$ under the transformation T.
c. Find the transformation matrix T and its determinant. Explain what this value tells you about the relationship between $\triangle PQR$ and $\triangle P'Q'R'$.

1. Of the 360 children attending a matinee at a movie theater, 210 buy popcorn, 245 buy something to drink, and 140 buy both. How many children buy neither popcorn nor something to drink?

2. A football team has 3 guards, 4 tackles, 6 ends, and 2 centers on its roster. In how many ways can a line consisting of 1 left guard, 1 right guard, 1 left tackle, 1 right tackle, 1 left end, 1 right end, and 1 center be formed?

3. A bank has each of its customers select a password consisting of 5 different letters or 5 different digits. How many different passwords are possible?

4. A bookshelf has space for 6 books. If 8 different books are available, how many different arrangements can be made on the shelf?

5. Ten teams are in an athletic conference. How many games have been played after each team has played every other team once?

6. A boat has 4 red, 4 blue, and 2 yellow flags with which to signal other boats. All 10 flags are flown in various sequences to denote different messages. How many such sequences are possible?

7. *Writing* Write a paragraph in which you compare the number of selections of six items from a collection of ten and the number of arrangements of ten items where six are alike and the other four are alike.

8. In the expansion of $(x - 3)^7$, find the coefficients of the terms containing x^5 and x^4.

1. **Writing** Suppose the probability of an event can be determined both theoretically and empirically. Explain why the theoretical and empirical probabilities may not be the same.

2. An experiment consists of randomly drawing a card from a standard deck and rolling a die. Find the probability that:
   a. the card is a face card and the die shows a 5
   b. the card is *not* a face card and the die shows a 5
   c. the card is *not* a face card and the die does *not* show a 5

3. A fish bowl contains slips of paper numbered 1 through 9. Two slips are drawn, one after the other and without replacement. Find the probability of each of the following events.
   a. Both numbers are even.
   b. The second number drawn is 3.
   c. The first number drawn is greater than 7 and the second number drawn is less than 3.

4. For a certain brand of marigold seeds, the seeds sprout on average 9 out of 10 times. If 4 seeds are selected at random and planted, find the probability that:
   a. all 4 seeds sprout
   b. at least 3 of the 4 seeds sprout

5. A Central School PTA committee is to consist of 3 teachers and 3 parents, who are to be chosen at random from the 20 teachers and 100 parents involved in the PTA. If half of the teachers and 80 parents are women, find the probability that the committee has:
   a. only female members
   b. only 1 parent and only 1 teacher who are women

6. Ninety-five percent of the sneakers manufactured by a shoe company have no defects. In order to find the 5% that do have defects, inspectors carefully look over every pair of sneakers. Still, the inspectors sometimes make mistakes because 4% of the defective pairs pass inspection and 2% of the good pairs fail the inspection test.
   a. Incorporate the facts given above into a tree diagram.
   b. What percent of the pairs of sneakers do *not* pass inspection?
   c. If a pair of sneakers does *not* pass inspection, what is the probability that it does *not* have a defect?

7. Decide if the following is a fair game; if it is not a fair game, state which player has the advantage: Two dice are rolled. If the sum is less than 7, player A wins $5 from player B. Otherwise B wins $4 from A.

1. **Writing** Write a paragraph in which you discuss what is meant by a *measure of central tendency* and a *measure of dispersion*. Also explain why using only a measure of central tendency to characterize a set of data is inadequate. Give an example to support your explanation.

2. The table below gives a month's sales data for a real estate office.

Price (× $1000)	70–89.9	90–109.9	110–129.9	130–149.9
Number of houses sold	2	7	9	12

   a. Draw a histogram for the data.
   b. Using the middle price (80, 100, and so on) as the representative price for each class, find the mean, median, and mode.

3. The stem-and-leaf plot at the right gives the number of points that a basketball player scored in 24 games.

   $$
   \begin{array}{c|l}
   1 & 8\ 9\ 9 \\
   2 & 0\ 1\ 2\ 2\ 3\ 3\ 5\ 5\ 6\ 7\ 8\ 9\ 9\ 9\ 9 \\
   3 & 1\ 1\ 2\ 4\ 4\ 8
   \end{array}
   $$

   a. Find the median, lower quartile, and upper quartile.
   b. Find the range and interquartile range.
   c. Draw a box-and-whisker plot for the data.

4. Manuel took an English test and a mathematics test on the same day. He scored 80 on both tests. The English class had a mean score of 70 and a standard deviation of 10 on its test, while the mathematics class had a mean score of 74 with a standard deviation of 5 on its test. For which test did Manuel rank higher with respect to his class?

5. Calculate the mean, variance, and standard deviation of the data 14, 17, 20, 21, 22, 24, 25, 31, 42.

6. At a large university, the mathematics-plus-verbal standardized test scores of incoming students are normally distributed with a mean score of 1100 and a standard deviation of 100. Find the percent of incoming students with scores **(a)** greater than 1300 and **(b)** less than 1000. Also find **(c)** the 90th percentile and **(d)** the third quartile of the test scores.

7. At a certain college, 75% of the students live on campus and the rest live off campus. In a survey of 50 randomly selected students from each of these two groups, 29 on-campus students and 17 off-campus students favored a change in housing rules. Estimate the percent of the entire student population favoring the change.

8. In a survey of 900 high school students, 750 said that their favorite type of pizza topping was pepperoni. Find both a 95% and a 99% confidence interval for p, the true proportion of high school students who prefer pepperoni pizza.

1. The accountant for a small business correlated the weekly advertising budget and weekly sales figures for one year. Letting x represent advertising dollars and y represent sales dollars, the accountant found the following: $\bar{x} = 1200$; $\bar{y} = 8400$; $\overline{xy} = 10,100,000$; $s_x = 108$; and $s_y = 246$.
   a. Calculate the correlation coefficient.
   b. Find an equation of the least-squares line.
   c. *Writing*   Evaluate the usefulness of the least-squares line in predicting the weekly sales figures for a given advertising budget.

2. Let x represent a basketball player's scoring average, and let y represent that player's height. If x and y have a high correlation, can you infer that being tall causes a high scoring average? Explain.

3. Express y in terms of x if $\log y = 2x - 1$.

4. The table gives car owners' average annual expenditure y, in dollars, for gasoline for various years, x.

x	1935	1945	1955	1965	1975	1985
y	35	57	93	152	247	400

   a. Fit an exponential curve to the data.
   b. Predict the annual expenditure for gasoline in 2000.

5. Express y in terms of x if $\log y = 2.5 \log x + 2$.

6. *Writing*   Let x represent the circumference of an orange, and let y represent the weight of the orange. The data (x, y) can be modeled by a power function $y = ax^m$. Using what you know about geometry, conjecture what the value of m is. Then write a few sentences justifying your answer.

7. The table below gives the mean orbital velocity v, in miles per second, for three planets, and the mean distance d, in millions of miles, of these planets from the sun.

	$d\ (\times 10^6\ \text{mi})$	$v\ (\text{in mi/s})$
Mercury	36	29.8
Venus	67.25	21.8
Mars	141.7	15.0

   a. Find an equation that gives v in terms of d.
   b. Use your equation to predict the mean orbital velocity of Earth, given that Earth is 93 million miles from the sun.

1. Evaluate each limit.

   a. $\displaystyle\lim_{x\to\infty} \frac{2x^2 - x}{x^2 + 1}$

   b. $\displaystyle\lim_{x\to 2} \frac{x^2 - 2x}{x^2 - 4}$

   c. $\displaystyle\lim_{x\to 2} \frac{x^3 - 8}{x - 2}$

2. Explain why $\displaystyle\lim_{x\to\infty} \sin x$ does not exist.

3. Sketch the graph of each rational function. Show vertical and horizontal asymptotes and x-intercepts.

   a. $y = \dfrac{x}{x + 4}$

   b. $y = \dfrac{x^2 - 1}{x^2 + 1}$

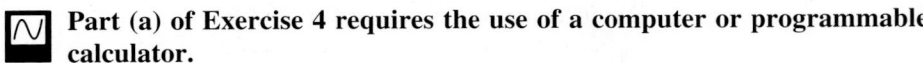 **Part (a) of Exercise 4 requires the use of a computer or programmable calculator.**

4. **a.** Using rectangles of width 0.01, modify and run the program on page 730 to approximate the area under the curve $y = 9 - x^2$ from $x = -3$ to $x = 3$.
   b. Using part (a) and the symmetry of the curve $y = 9 - x^2$, give an approximation of the area under the curve from $x = 0$ to $x = 3$.

5. Use series (3) on page 733 to find an infinite series for $\sin(-1)$.

6. Use series (5) on page 733 to find an infinite series for $\ln(1 + x^3)$.

7. If $f(x) = e^{-x} + 2$, use a calculator to approximate $\displaystyle\lim_{n\to\infty} f^n(1)$.

8. Analyze all orbits for iterations of the function $f(x) = x^3 - x^2$.

9. *Writing* Write a paragraph in which you discuss the reasoning behind the logistic function $f(x) = cx(1 - x)$. (That is, tell what assumption this population-growth model makes about the relationship between one generation and the next.) Also describe the types of population growth that arise from iterating the logistic function for various values of c.

1. Find the derivative of each function. Express the derivative without using fractional or negative exponents.

   a. $f(x) = 4x^3 + 5x^2 - 2x + 3$

   b. $f(x) = \dfrac{4}{x^2} + \dfrac{3}{x} + 4x + 2$

   c. $f(x) = \sqrt[4]{5x^3} - \sqrt[4]{\dfrac{1}{x}}$

   d. $f(x) = \dfrac{2}{x^5} + 3x^5$

2. *Writing* Write a paragraph in which you state the definition of the derivative of a function $f(x)$ and geometrically relate the derivative to the graph of f.

3. Graph the function $f(x) = x^3 + x^2 - x + 1$. Identify any local maximum and local minimum points.

4. Given a function $f(x)$, if $f'(2) = 0$, $f'(x) < 0$ for $x < 2$, and $f'(x) > 0$ for $x > 2$, what can you say about $f(2)$?

5. **a.** A rectangular solid has a base that is twice as long as it is wide, as shown. Express its surface area A as a function of x and h.

   b. If the volume of the solid is 1125 cubic units, express h in terms of x.

   c. Use parts (a) and (b) to express the surface area A as a function of x alone.

   d. Find the dimensions that minimize the surface area of the solid. What is the surface area of this solid?

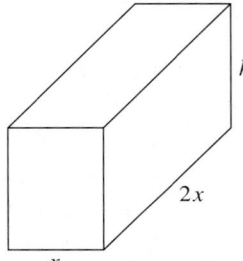

6. A particle is moving along a number line so that its position to the right of the origin at time t is $s(t) = 4t^2 - 5t + 4$. Find the position, velocity, and acceleration of the particle at time $t = 4$.

7. An arrow is shot upward from the ground next to a tower that is 48 ft high. If the initial velocity of the arrow is 96 ft/s, then $s(t) = 96t - 16t^2$ gives the arrow's height, in feet, above the ground t seconds after the arrow is shot.

   a. In about how many seconds does the arrow pass the top of the tower on its upward trip?

   b. How high does the arrow travel before it starts to descend?

   c. If it lands on the top of the tower, about how long was the arrow in the air?

   d. At about what velocity does the arrow hit the top of the tower?

Cumulative Reviews

Review of Functions, Graphs, and Applications (Chapters 1–6)

1. Consider points $A(2, -3)$, $B(6, 5)$, and $C(0, 5)$.

  a. Sketch $\triangle ABC$ in a coordinate plane.

  b. Find the length of $\overline{AB}$.

  c. Find the coordinates of the midpoint of $\overline{AB}$.

  d. Find the area of $\triangle ABC$.

  e. Find an equation of $\overleftrightarrow{AB}$.

  f. Find an equation of the perpendicular bisector of $\overline{AB}$.

  g. Find an equation of the perpendicular bisector of $\overline{BC}$.

  h. Use the results of parts (f) and (g) to find the center and radius of the circle containing points A, B, and C.

  i. Find an equation of the circle in part (h).

  j. Graph the circle in part (h) on the same set of axes used to sketch $\triangle ABC$ in part (a).

  k. Find the x- and y-intercepts of the circle in part (h).

  l. Find the coefficients a, b, and c in the equation $y = ax^2 + bx + c$ so that the graph of the equation contains points A, B, and C.

  m. Graph the equation in part (1) on the same set of axes used to graph the circle in part (j).

  n. Identify the curve in part (m).

  o. For the curve in part (m), find the coordinates of the vertex and the focus, and the equations of the axis of symmetry and the directrix.

  p. Find the x- and y-intercepts of the curve in part (m).

  q. The curve in part (m) and the circle in part (j) intersect at points A, B, and C. Find the coordinates of the fourth point of intersection.

  r. Find the coordinates of any points where the curve in part (m) and the line in part (f) intersect.

2. Consider the polynomial

$$P(x) = 2x^5 - 3x^4 - 2x^3 + 7x^2 - 2x - 6.$$

  a. *Writing* Explain how you know that the equation $P(x) = 0$ has at least one real root between $x = 1$ and $x = 2$.

  b. Find all real and imaginary roots of $P(x) = 0$. Identify any multiple roots.

  c. Solve $P(x) \geq 0$.

  d. Graph $y = P(x)$.

3. A quartic polynomial $P(x)$ with real coefficients has zeros $2 + i$ and $3 - 2i$.

  a. Find the other zeros.

  b. *Writing* Explain why knowing only the zeros of $P(x)$ is not enough to find a rule for $P(x)$.

  c. If $P(0) = 13$, find a rule for $P(x)$.

  d. Using your answer to part (c), find $P(-1)$.

> **Exercise 4 requires the use of a calculator or computer.**

4. Use a calculator or computer to approximate to the nearest hundredth the two real roots of the equation $x^4 - x^3 - 2x^2 - 3x - 1 = 0$.

5. From a platform 35 m above the ground, a ball is thrown upward with an initial speed of 30 m/s. The approximate height of the ball above the ground t seconds later is given by $h(t) = 35 + 30t - 5t^2$.

  a. After how many seconds does the ball hit the ground?

  b. What is the domain of h?

  c. What is the range of h?

  d. After how many seconds does the ball reach its maximum height above the ground?

6. Find the sum and product of the roots of the equation $3x^3 + 6x^2 - 2x + 9 = 0$.

7. For the equation $3x^3 + ax^2 - 9x + 6 = 0$, the sum of the roots is twice the product of the roots. Find a.

8. Solve $2|x - 3| + 5 = 15$.

9. Solve $|3x - 5| \geq 2$ and graph the solution on a number line.

10. Let R be the solution set of the system of inequalities $y \geq x^2 - 2x - 8$ and $x + y \leq -2$.

   a. Graph R.

   b. **Writing** Explain how you could approximate the area of R.

11. Suppose (x, y) is a point in the shaded region shown at the right.

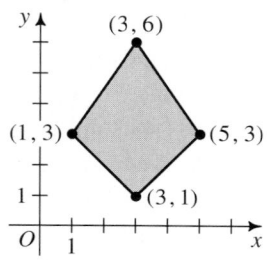

   a. Give the maximum and minimum values of $2x + 3y$.

   b. Give the maximum and minimum values of $5x + y$.

12. Consider the functions
$$f(x) = x^2, \quad g(x) = x - 1, \quad \text{and} \quad h(x) = \sqrt{x}.$$

   a. Give the domain and range of each function.

   b. For each function that has an inverse, find the inverse.

   c. Two of the six composite functions that can be formed using all three functions are $r(x) = \sqrt{x^2 - 1}$ and $s(x) = (\sqrt{x} - 1)^2$. Express r and s in terms of f, g, and h.

   d. Give simplified rules for the other four composite functions from part (c).

   e. Give the domain and range of each composite function in part (d).

   f. **Writing** Explain why $g(x)$ and the four composite functions in part (c) are all different functions.

13. The graph of $y = f(x)$ is shown at the right. Graph each of the following.

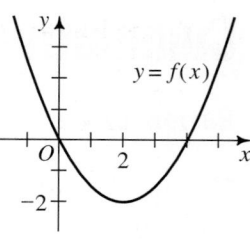

   a. $y = |f(x)|$

   b. $y = -f(x)$

   c. $y = f(-x)$

   d. $y = f(x + 1) + 1$

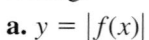 **Part (e) of Exercise 14 requires the use of a graphing calculator or computer.**

14. An open-top box is to be made so that its width is 4 ft and its volume is 40 ft^3. The base of the box costs $4/ft^2 and the sides cost $2/ft^2.

   a. Express the cost of the box as a function of its length l and height h.

   b. Find a relationship between l and h.

   c. Express the cost as a function of h only.

   d. Give the domain of the cost function.

   e. Use a graphing calculator or computer to approximate the dimensions of the box having least cost.

15. Without using a calculator, evaluate each of the following.

   a. $e^{2 \ln 3}$

   b. $\dfrac{2^3 \cdot 2^{-5}}{2^{-7}}$

   c. $\log_4 64$

   d. $\log_4 32$

   e. $\log_2 \sqrt[4]{\dfrac{1}{8}}$

   f. $(3^{-2} + 3^0)^{-2}$

   g. $\log 25 + \log 4$

16. Use a calculator to solve or evaluate each of the following.

   a. $\log_5 27$

   b. $3^{x-2} = 30$

   c. $(x - 2)^3 = 30$

d. The half-life of a radioactive substance is 2 days. If you start with 20 g of the substance, how much remains after 4 days? after 5 days?

e. What is the value of a $100 investment after 20 years if the annual interest rate is 10% and interest is compounded monthly? continuously?

17. Without using a calculator, solve each of the following.

a. $\log_4 (x + 3) + \log_4 (x - 3) = 2$

b. $\sqrt{\dfrac{4^{x+3}}{16^x}} = 32$

c. $\log_9 x = 1.5$

18. For the exponential function $f(x) = ab^x$, suppose $f(2) = 2$ and $f(4) = 18$.

a. Find a and b.

b. Find $f^{-1}(54)$.

19. Consider the hyperbola

$$9x^2 - 16y^2 - 72x - 64y - 64 = 0.$$

a. Graph the hyperbola.

b. Give the coordinates of the center, foci, and vertices.

c. Give the equations of the asymptotes.

d. Find an equation of an ellipse that has vertices at the foci of the hyperbola and foci at the vertices of the hyperbola.

e. Find a function $f(x)$ whose graph is a parabola that has its vertex at the left focus of the hyperbola and that contains the origin.

f. For the parabola in part (e), give the coordinates of the focus, and equations of the directrix and axis of symmetry.

g. Find the eccentricity of the hyperbola, the ellipse, and the parabola.

Review of Trigonometry (Chapters 7–11)

1. Given that $\sin \alpha = -\dfrac{3}{5}$ and $\cos \beta = \dfrac{7}{25}$, where $\pi < \alpha < \dfrac{3\pi}{2} < \beta < 2\pi$, find each of the following.

a. $\cos \alpha$ **b.** $\sin \beta$

c. $\sin (\alpha + \beta)$ **d.** $\cos (\alpha - \beta)$

e. $\tan \dfrac{1}{2} \alpha$ **f.** $\sin 2\beta$

2. Evaluate.

a. $\sin (\pi - x) + \cos \left(\dfrac{3\pi}{2} - x \right)$

b. $\sec \dfrac{3\pi}{4} \csc \dfrac{5\pi}{4} + \tan \dfrac{4\pi}{3} \cot \dfrac{11\pi}{6}$

c. $\text{Tan}^{-1} (-\sqrt{3})$

d. $\sin (\text{Cos}^{-1} (0.6))$

3. Simplify.

a. $\dfrac{\tan x - \sin x \cos x}{\sin^2 x}$

b. $\dfrac{(\sec x + 1)(1 - \cos x)}{\sin x}$

c. $\dfrac{1}{\sec x - \tan x} - \dfrac{1}{\sec x + \tan x}$

d. $\text{Tan}^{-1} \dfrac{1}{3} + \text{Tan}^{-1} \dfrac{1}{2}$

e. $\text{Sin}^{-1} (\sin x)$, $\dfrac{\pi}{2} \le x \le \pi$

4. Show that $y = \sin x$ is a root of the equation $y^2 + y \cot x \cos x - 1 = 0$.

5. Find the solutions of each equation in the indicated interval.

a. $(2 \cos x - \sqrt{3})(\tan x + \sqrt{3}) = 0$, $-\pi \le x \le \pi$

b. $\tan x \sin x + \tan x = 0$, $-2\pi \le x \le 0$

c. $3 \sin x + \sqrt{3} \cos x = 0$, $90° \le x \le 450°$

d. $\sin 2x + \cos 2x = \sec 2x$, $-90° \le x \le 90°$

6. A sector of a circle of radius 5 in. has a central angle of 72°.

   a. Find the arc length of the sector.

   b. Find the area of the sector.

   c. Suppose the area of the sector is doubled by changing the radius of the circle while keeping the central angle fixed. Find the radius of the new circle and the arc length of the sector.

   d. Suppose the area of the sector is doubled by changing the central angle while keeping the radius fixed. Find the measure of the new central angle and the arc length of the sector.

   e. If we define the perimeter of a sector as $P = s + 2r$ where s is the arc length of the sector and r is the radius of the circle, then which of the two techniques described in parts (c) and (d) gives a smaller perimeter of the sector?

7. Graph $y = \sin x$ for $-2\pi \le x \le 2\pi$.

8. Consider the function $f(x) = 4 \sin\left(\frac{\pi}{2}x\right) + 2$.

   a. What is the amplitude of f?

   b. What is the fundamental period of f?

   c. Graph f.

   d. Find the first three positive values of x for which $f(x) = 5$.

   e. Find a function in terms of cosine that is equivalent to f.

9. **a.** Find the angle of inclination of the line $2x + 3y = 7$.

   b. Find the angle of inclination of the line $x - 3y = -1$.

   c. Find the acute angle of intersection of the lines in parts (a) and (b).

   d. Find an equation of the line containing the point $(2, 1)$ and having an angle of inclination equal to the angle found in part (c).

   e. Graph all three lines from parts (a), (b), and (d) on the same set of axes.

10. Two cars leave the same town at the same time. One heads east at 30 mi/h, and the other in a direction of N30°W at 40 mi/h. How far apart are the cars in 1 h?

11. The lengths of two sides of a parallelogram are 6 in. and 8 in., and the measure of the included angle is 72°18′. Find the length of the longer diagonal to the nearest hundredth of an inch.

12. A car rounds a curve of radius 528 ft at the rate of 60 mi/h. Through how many radians does the car turn in 9 s?

13. An island is near two coastal towns, A and B. Town B is 15 mi south of town A. The island's bearing from A is S57.1°E, and its bearing from B is N12.6°E. How far is the island from the closer town?

14. A tourist is traveling by car due east on a road through a forest. From point A on the road, a path heads due northeast to an observation tower. The tower is 4 mi from A, and a ranger in the tower can see for a distance of 3.5 mi. For what length of road can the ranger see the tourist's car?

15. Find the area of a triangle with sides of length 4, 5, and 6.

16. Find y in the diagram below if $\tan x = \frac{1}{3}$.

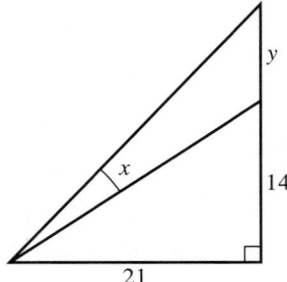

17. Consider the complex number $z = -8 + 8i\sqrt{3}$.

   a. Show z in an Argand diagram.

   b. Find $|z|$.

   c. Express z in polar form.

d. Find the four fourth roots of z in rectangular form.

e. Graph the roots from part (d) in the complex plane.

f. What polygon do the four points in part (e) form?

18. Graph each polar equation.

   a. $r = \cos 2\theta$

   b. $r = 1 + 2 \sin \theta$

   c. $r = \theta,\ 0 \le \theta \le 4\pi$

19. Consider the polar equation $r = 2 \sin \theta$.

   a. Graph the equation.

   b. Give a rectangular equation for the equation.

   c. *Writing* Explain why the graph in part (a) is consistent with the equation in part (b).

20. Express each product in polar and rectangular form.

   a. $(3 \text{ cis } 60°)(2 \text{ cis } 75°)$

   b. $(\sqrt{3} - i)(-2 + 2i\sqrt{3})$

Review of Discrete Mathematics (Chapters 12–14)

1. Use parallelogram $ABCD$ to express $\overrightarrow{DB}$ in terms of:

   a. $\overrightarrow{DC}$ and $\overrightarrow{CB}$

   b. $\overrightarrow{DC}$ and $\overrightarrow{DA}$

   c. $\overrightarrow{AB}$ and $\overrightarrow{BC}$

   d. $\overrightarrow{BA}$ and $\overrightarrow{BC}$

2. Consider vectors $\mathbf{u} = (-3, 5)$, $\mathbf{v} = (3, -6)$, and $\mathbf{w} = (2, 1)$.

   a. Find $\mathbf{v} \cdot \mathbf{w}$.

   b. What can you conclude about $\mathbf{v}$ and $\mathbf{w}$ from part (a)?

   c. Find the angle between $\mathbf{u}$ and $\mathbf{v}$.

   d. Find a unit vector in the same direction as $\mathbf{v}$.

3. Find the angle between the vectors $(8, -10, 6)$ and $(5, 0, 4)$.

4. A plane flying at 600 mi/h on a course of 75° is blown off course by a 50 mi/h easterly wind.

   a. Illustrate with a vector diagram.

   b. Find the direction angle and resultant speed of the plane.

5. a. Find an equation of the sphere that has a diameter with endpoints $A(2, -2, 2)$ and $B(8, 6, -22)$.

   b. Find an equation of the plane tangent to the sphere at A.

6. Which of the following planes are parallel and which are perpendicular?

   a. $2x + y - 3z = 12$

   b. $4x + 2y - 6z = 2$

   c. $3x - 9y - z = 1$

7. Solve using Cramer's rule:

$$\begin{aligned} x + y + z &= 6 \\ x - 2y + 3z &= 6 \\ 2x + y - 2z &= -2 \end{aligned}$$

8. Find the area of the parallelogram with sides determined by the vectors $\mathbf{u} = (1, 5)$ and $\mathbf{v} = (3, 8)$.

9. Evaluate:

$$\begin{vmatrix} 2 & 0 & 1 & 0 \\ 0 & 0 & 1 & 0 \\ -2 & 0 & 1 & 4 \\ 1 & 1 & 0 & -1 \end{vmatrix}$$

10. Consider the sequence 3, 6, 11, 18, 27, ...

   a. State whether the sequence is arithmetic, geometric, or neither.

   b. Find t_6.

   c. Give a recursive definition for the nth term, t_n.

   d. Give an explicit definition for t_n.

   e. Find t_{100}.

f. Writing State whether you used part (c) or part (d) to complete part (e). Explain why.

11. Consider the sequence 4, 11, 18, 25, ...

    a. State whether the sequence is arithmetic, geometric, or neither.

    b. Find t_6.

    c. Give a recursive definition for the nth term, t_n.

    d. Give an explicit definition for t_n.

    e. Find t_{100}.

    f. If you were to graph the sequence, you would obtain a set of discrete points that lie on some curve. Identify this curve.

    g. Find the sum of the first 20 terms of the sequence.

12. Consider the sequence 5, 10, 20, 40, ...

    a. State whether the sequence is arithmetic, geometric, or neither.

    b. Find t_6.

    c. Give a recursive definition for the nth term, t_n.

    d. Give an explicit definition for t_n.

    e. Find t_{100}.

    f. If you were to graph the sequence, you would obtain a set of discrete points that lie on some curve. Identify this curve.

    g. Find the sum of the first 20 terms of the sequence.

13. Consider the series $\sum\limits_{n=1}^{\infty} x^{-n}$.

    a. What is the interval of convergence?

    b. In terms of x, find the sum of the series.

14. Consider the series $1 + \sin^2 x + \sin^4 x + \cdots$.

    a. What is the interval of convergence?

    b. In terms of x, find the sum of the series.

    c. Use part (b) to simplify $\sum\limits_{n=0}^{\infty} \sin 2x \sin^{2n} x$.

15. Evaluate.

    a. $\lim\limits_{n \to \infty} \dfrac{n-1}{n+1}$

    b. $\lim\limits_{n \to \infty} \dfrac{5n^2 - 7n}{4n^2 + 1000n}$

    c. $\lim\limits_{n \to \infty} \text{Tan}^{-1} n$

    d. $\lim\limits_{n \to \infty} \left(1 - \dfrac{1}{n}\right)^n$

16. Consider the series

    $$2 \cdot 3 + 4 \cdot 5 + 6 \cdot 7 + \cdots + 100 \cdot 101.$$

    a. Express the series using sigma notation.

    b. Find the sum of the series.

17. Use mathematical induction to prove that $7^n - 1$ is a multiple of 6 for all positive integers n.

18. Consider the matrices

    $$A = \begin{bmatrix} 2 & 1 & 3 \\ 4 & -2 & 5 \end{bmatrix} \text{ and } B = \begin{bmatrix} 3 & 7 & 9 \\ -8 & 5 & 4 \end{bmatrix}.$$

    a. Find $A + 2B$.

    b. Find A^t.

19. Consider the matrices

    $$A = \begin{bmatrix} 2 & 1 \\ -4 & 7 \end{bmatrix} \text{ and } B = \begin{bmatrix} 3 & 4 & 7 \\ 2 & -4 & 1 \end{bmatrix}.$$

    a. Find whichever product, AB or BA, exists.

    b. Find A^{-1}.

    c. Find B^{-1}.

    d. Write a matrix equation for the system of equations $2x + y = 9$ and $-4x + 7y = 26$, and use A^{-1} to find the solution.

20. Consider the communication network shown.

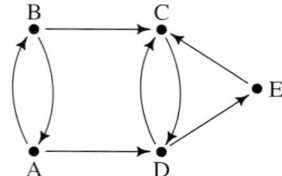

    a. Write a communication matrix M for the network.

    b. Find M^2.

    c. Find $M + M^2$ and interpret the result.

21. A survey of commuters who work in a large city revealed that 60% of the commuters currently do not use public transportation. However, 8% plan to switch to public transportation next year, while 6% of those using public transportation plan not to use it next year.

  a. Give the transition matrix T.

  b. What percent of the commuters will use public transportation next year?

  c. *Writing* Explain what will happen if the rates of change remain constant from one year to the next. Also explain how you could obtain this result mathematically.

22. Consider the transformation

$$T:(x, y) \rightarrow (3x + 2y, 2x - y).$$

  a. Write T as a matrix, and find the determinant of this matrix.

  b. In a coordinate plane, plot the points $A(3, 1)$, $B(5, 4)$, and $C(6, -2)$. Then find and plot their images, A', B', and C', under T.

  c. Find the ratio of the area of $\triangle A'B'C'$ to the area of $\triangle ABC$.

Review of Data Analysis (Chapters 15–18)

1. Five couples are having dinner together.

  a. In how many ways can the ten people line up at a buffet table?

  b. In how many of the arrangements of part (a) are the people arranged alternately by sex?

  c. In how many of the arrangements of part (a) are the people arranged both by couple and alternately by sex?

  d. In how many ways can the people be seated around a circular dinner table?

2. A committee of four is to be randomly selected from a group of five men and seven women.

  a. Find the probability that the committee is composed only of women.

  b. Find the probability that the committee is composed only of men.

  c. Find the probability that there is at least one man on the committee.

3. The probability that Rajiv answers any test question correctly is $\frac{3}{4}$, while the probability that Reba answers any question correctly is $\frac{4}{5}$. For a given question, find the probability that:

  a. both answer it correctly

  b. neither answers it correctly

  c. Rajiv answers it correctly and Reba doesn't

  d. Reba answers it correctly and Rajiv doesn't

  e. at least one of them answers it correctly

  f. Rajiv answers 5 out of 7 questions correctly

4. All canoes made by a sporting goods company are inspected before shipping, and 97% of the canoes have no defects. The inspectors correctly identify 80% of the defective canoes and 100% of the good canoes.

  a. What percent of the canoes pass inspection?

  b. If a canoe passes inspection, what is the probability it is defective?

5. The heights, in inches, for fifteen players on a basketball team are as follows: 74, 79, 78, 79, 82, 80, 82, 83, 86, 82, 80, 81, 84, 83, 81.

  a. Make a stem-and-leaf plot of the data.

  b. Draw a histogram of the data.

  c. Find the median, lower quartile, and upper quartile.

  d. Find the range and interquartile range.

  e. Draw a box-and-whisker plot of the data.

  f. Find the mean and standard deviation.

g. What percent of the heights are within one standard deviation of the mean? two standard deviations?

h. What percent of the data are within one and two standard deviations of the mean for a *normal distribution?*

i. Writing Using the results of parts (g) and (h), comment on how well a normal curve would fit the data.

6. The table below gives the scores of ten students on a midyear exam (x) and a final exam (y).

x	67	50	71	71	77	50	72	81	96	99
y	68	66	78	76	82	66	74	78	99	99

a. Make a scatter plot of the data.

b. Determine the correlation coefficient and interpret the result.

c. Find an equation of the least-squares line.

d. If a student's score on the midyear exam is 74, use the least-squares equation to estimate the student's score on the final exam.

7. A coffee machine is supposed to dispense 6 ounces per cup. The amount of coffee actually dispensed is normally distributed with a standard deviation of 0.5 ounce.

a. What percent of the cups will have more than 6.7 ounces?

b. What is the probability that a cup contains between 5.8 ounces and 6.2 ounces?

c. If 7-ounce cups are used, how many cups out of 1000 are likely to overflow?

8. In a random sample of 1000 Chicago families that have television sets, it was found that 825 also have a VCR. Find 95% and 99% confidence intervals for the actual proportion of families in Chicago with a VCR.

9. The table below gives the tuition costs, in dollars, at a local college for various years from 1955 to 1985.

1955	1960	1965	1970	1975	1980	1985
1000	1470	2160	3175	4660	6850	10,050

a. Fit an exponential curve to the data.

b. Estimate the college's tuition cost in 1995.

c. Estimate the tuition cost in 2000.

10. The table below gives the cost y, in dollars, to build a storage tank of capacity x, in gallons.

x	100	500	1000	2000
y	200	580	920	1460

a. Fit a power curve to the data.

b. Estimate the cost of a 30,000-gallon tank.

Review of Limits and Introduction to Calculus (Chapters 19–20)

1. Evaluate each limit or state that it does not exist.

a. $\lim\limits_{x \to 0} \dfrac{x^3 + 2x}{x^2 - x}$

b. $\lim\limits_{x \to 1^+} \dfrac{x - 1}{\sqrt{x - 1}}$

c. $\lim\limits_{x \to \infty} \dfrac{2x^3 + 5x}{5x^3 - 7x^2 + 1}$

d. $\lim\limits_{x \to -\infty} \dfrac{|x^3 + 6x|}{4x^3 + 3x^2}$

e. $\lim\limits_{h \to 0} \dfrac{e^{1+h} - e}{h}$ (Use a calculator.)

f. What familiar number is the limit in part (e)?

2. Graph each function. Show vertical and horizontal asymptotes and x-intercepts.

a. $y = \dfrac{x + 1}{x - 1}$

b. $y = \dfrac{x^2 + 1}{x^2 - 4}$

c. $y = \dfrac{x^2 - 1}{x - 1}$

d. $y = \dfrac{2x^2}{x^2 - 4x - 5}$

3. a. Write a series expansion for sin (0.3).

b. Use two terms of the series in part (a) to approximate sin (0.3).

c. Use a calculator to evaluate sin (0.3).

d. *Writing* Comment on the results of parts (b) and (c).

4. Find the first four terms in the orbit of $x_0 = 1$ for iterations of the function $f(x) = \dfrac{1}{e^x + 1}$.

5. Consider the function $f(x) = 2 + \dfrac{1}{x}$.

a. Use a web diagram to analyze the orbit of $x_0 = 1$ for iterations of f.

b. Use a calculator to find $\lim\limits_{n \to \infty} f^n(1)$ to the nearest hundredth.

c. Find the exact value of the limit in part (b).

6. A savings account has an annual interest rate of 8% and an opening balance of $1000. No withdrawals or deposits are made.

a. Use a recursive formula to describe the sequence of yearly account balances.

b. Use an iterated function to describe the sequence of yearly account balances.

c. What is the balance after 15 years?

7. Consider the function $f(x) = \sqrt[3]{x}$.

a. Express the derivative of f without negative or fractional exponents.

b. Let $L(x)$ be the linear function whose graph is tangent to the graph of $f(x)$ at $x = 64$. Find $L(x)$.

c. Find $L(65)$.

d. Find the error if you use $L(65)$ as an approximation for $f(65)$.

8. Consider the function
$$f(x) = x^3 - 4x^2 - 3x + 12.$$

a. Find all the zeros of f.

b. Find the intervals on which f is increasing.

c. Find the intervals on which f is decreasing.

d. Find the coordinates of any relative maximum points on the graph of f.

e. Find the coordinates of any relative minimum points on the graph of f.

f. Use the results of parts (a)–(e) to sketch the graph of f.

9. The position at time t of a particle moving along the x-axis is given by $x = t^3 - 9t^2 + 24t$.

a. In what direction is the particle moving at time $t = 0$?

b. Find all values of t for which the particle is moving to the left.

c. For what values of t does the particle change direction?

d. What is the position of the particle at time $t = 3$?

e. How far does the particle travel from time $t = 0$ to time $t = 3$?

10. A rectangle is formed so that one side is on the x-axis, and the opposite side has endpoints on the part of the parabola $y = 4 - 4x^2$ above the x-axis. Find the maximum area of the rectangle.

Answers

Chapter 1 Test

1.a. $4\sqrt{13}$ **b.** $(0, 1)$ **2.** $\dfrac{32}{7}$

3.

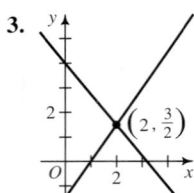

4. slope: $-\dfrac{3}{2}$; y-int.: $-\dfrac{1}{2}$

5. $\mathbf{a} \parallel \mathbf{b}$; $\mathbf{a} \perp \mathbf{c}$; $\mathbf{b} \perp \mathbf{c}$

6. $3x + y = 16$

7. $4x - 3y = -7$

8. $3x - 2y = 6$

9. $x = 5$

10.a. Use the coordinates of B and C to find the coordinates of M, the midpoint of $\overline{BC}$. Then use the coordinates of A and M to find an equation of the line containing the median $\overline{AM}$.   **b.** $y = \dfrac{1}{2}x$

11.a. $V(r) = 15 - 1.5r$ **b.** \$6   **12.** $-2i\sqrt{3}$

13. $5 - 12i$ **14.** $\dfrac{4}{65} + \dfrac{7}{65}i$ **15.** $-\dfrac{1}{7} + \dfrac{4\sqrt{3}}{7}i$

16. $17 + 11i$

17. $-i$

18.a. $\dfrac{3}{4}, \dfrac{4}{3}$

   b. $-2 \pm \sqrt{11}$

   c. $3 \pm 3i$

19. 64; $-1, \dfrac{5}{3}$

20a.

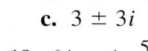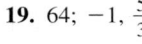

   b.

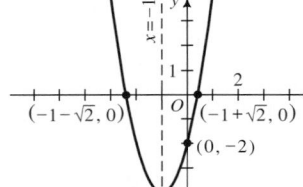

21.

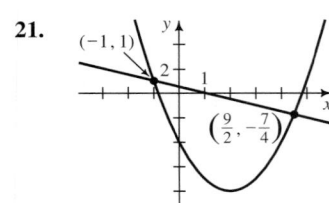

22.a. $f(x) = \dfrac{1}{4}x^2 + \dfrac{1}{2}x + 2$ **b.** \$17.75

Chapter 2 Test

1.a. 13 **b.** $\dfrac{1}{4}$ **c.** $6 - 4i$ **5.**

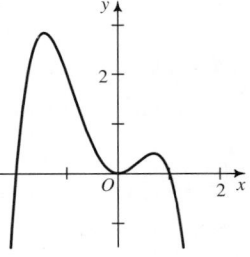

2. -6

3. $2x^3 - 3x^2 - 4x + 3$

4. $\pm i$

6. $y = -\dfrac{1}{3}(x + 3)(x + 1)(x - 1)$ **7.** 15

8.a. $P(1) = -1 < 0$; $P(2) = 27 > 0$; by the location principle, $P(x) = 0$ must have at least one real root between $x = 1$ and $x = 2$.   **b.** $-0.5, 1.1$ **9.** $-1, \pm\sqrt{2}$

10. $-\dfrac{1}{2}, 1, -1 \pm 2i$ **11.** sum: $\dfrac{2}{3}$; product: 2

12. Answers may vary, but they should be of the form $a(x^3 - 5x^2 + 8x - 6) = 0$ where a is an integer.   **13.** A cubic equation with integral coefficients can have 1, 2, or 3 distinct roots. There will be exactly 3 roots if multiple roots are counted separately. One of the roots must always be real. The other two roots must both be real or both be imaginary. If imaginary, the other two roots must be conjugates.

Chapter 3 Test

1.a. $x > -3$

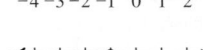

   b. $x \le -2$

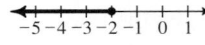

2.a. $-7 < x < -1$

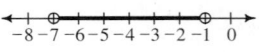

   b. $x < -3$ or $x > 7$

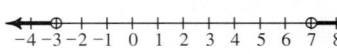

   c. $x = 0$ or $x = 5$

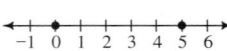

   d. $-1 \le x \le \dfrac{5}{2}$

3.a. $x \le -3$ or $x = \dfrac{1}{2}$ or $x \ge 2$ **b.** No solution

   c. $x \le -1$ or $0 \le x \le \dfrac{1}{2}$ **d.** $x \le -\dfrac{1}{2}$ or $3 < x \le 5$

4. The set consists of points in the coordinate plane that are on or above the line $2x + 3y = 12$.

5. a. 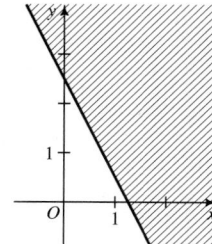 **b.**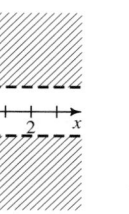

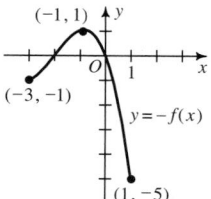

d.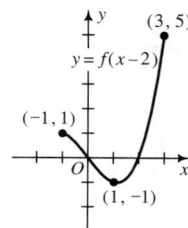

6. a. 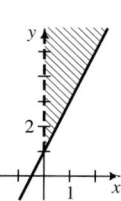 **b.**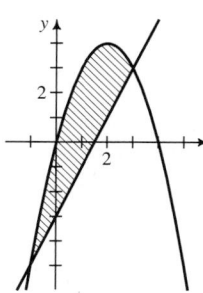

7. $x \geq 1$; $3x - 2y \leq 1$; $x + y \leq 7$
8. 6 type A widgets and 24 type B widgets

8. a. 5 **b.** max.: 1; min.: -3 **c.** 2
9. a. g; $g^{-1}(x) = \dfrac{x + 3}{2}$ **b.** f is not a one-to-one function.

10. a. $c(a, b) = \sqrt{a^2 + b^2}$ **b.** 5; 13 **c.**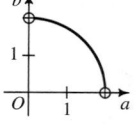

11. $d(t) = 16 - \dfrac{5}{4\pi}t$
12. a. $A(x) = 2x\sqrt{16 - x^2}$ **b.** $0 \leq x \leq 4$ **c.** 16

Chapter 4 Test

1. domain: $x \leq -4$ or $x \geq 4$; range: $f(x) \geq 0$; zeros: ± 4

2. a. 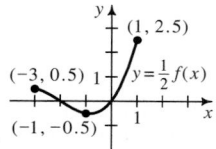 **b.** range: $g(x) \leq 1$; zeros: -2, 0

3. a. $x^2 - 2x$ **b.** $x^2 - 4x + 4$ **c.** $x^3 - 5x^2 + 8x - 4$
d. $x - 1$, $x \neq 2$ **4. a.** $x^2 - 7x + 12$ **b.** $x^2 - 3x$
5. For a number x to be in the domain of $(f \circ g)(x)$, the number must be in the domain of g, and $g(x)$ must be in the domain of f. For example, consider the functions $f(x) = \dfrac{1}{x^2 - 1}$ and $g(x) = \sqrt{x + 1}$. Since the domain of g is $\{x | x \geq -1\}$, the domain of $f \circ g$ is $\{x | x \geq -1 \text{ and } g(x) \neq \pm 1\}$. Since $g(x) \neq -1$ for any x, and since $g(x) = 1$ if $x = 0$, the domain of $f \circ g$ is $\{x | x \geq -1, x \neq 0\}$. **6.** iii, iv

7. a. 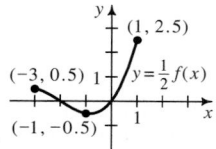 **b.**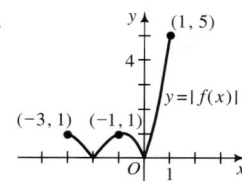

Chapter 5 Test

1. $V(t) = 100{,}000(1.12)^t$ **2. a.** 27 **b.** $\dfrac{16}{17}$ **c.** 729 **d.** $\dfrac{5}{8}$

3. a. -1 **b.** $\dfrac{3}{2}$ **4.** 24%
5. The graphs are reflections of each other in the y-axis.

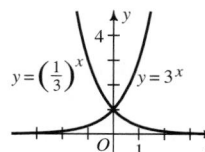

6. If the half-life of a radioactive isotope is 2 years, then during any 2-year period, half of the radioactive isotope decays and half remains. If we begin with 5 g of the radioactive isotope, then the amount remaining (in grams) after t years is given by $A(t) = 5\left(\dfrac{1}{2}\right)^{t/2}$.

7. a. \$1452.89 **b.** \$1454.00
8. a. 0.6021 **b.** 2.6021 **c.** -0.6021
9. a. 4 **b.** $\dfrac{1}{3}$ **c.** 125
10. a. $y = 16x^4$ **b.** $y = 9x$
11. a. $\log_b M - \dfrac{2}{3}\log_b N$ **b.** $3\log_b M + 2\log_b N$
12. $\dfrac{-3 + \sqrt{57}}{2}$
13. a. 3 and 4 **b.** 3.033 **14.** 3.13

Chapter 6 Test

1. Refer to the diagram.
$AX = \sqrt{[(x+a) - 0]^2 + (y-0)^2} = \sqrt{(x+a)^2 + y^2}$;
$CY = \sqrt{[(2x+a) - x]^2 + (y-0)^2} = \sqrt{(x+a)^2 + y^2}$;
so $AX = CY$. $AY = x$; $CX = (2x+a) - (x+a) = x$; so
$AY = CX$. Since opposite sides are congruent, $AXCY$ is a
parallelogram.

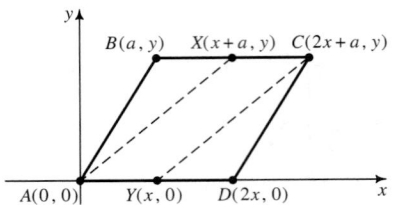

2. $(0, 3)$, $(-3, 0)$ **3.** $(x-5)^2 + (y-3)^2 = 26$
4. vertices: $(0, 4)$, $(0, -4)$; foci: $(0, \sqrt{7})$, $(0, -\sqrt{7})$
5. $(6, 0)$, $(0, -2)$ **6.** $\dfrac{x^2}{9} + \dfrac{y^2}{25} = 1$ **7.** $\dfrac{x^2}{16} - \dfrac{y^2}{12} = 1$

8.

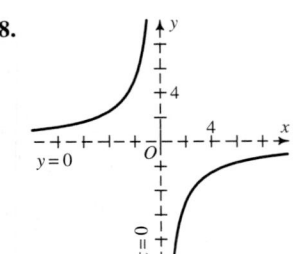

9. $y^2 - 9x^2 = 1$
10. vertex: $(0, 0)$;
 focus: $\left(0, \dfrac{1}{8}\right)$;
 directrix: $y = -\dfrac{1}{8}$
11. $y = -\dfrac{1}{12}x^2 - 1$

12.

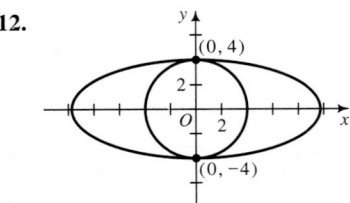

13. a. ellipse
 b. hyperbola
 c. parabola
14. $\dfrac{x^2}{4} + \dfrac{y^2}{3} = 1$

15. If $A = B$ and $AC < 0$, then the graph of $Ax^2 + By^2 + C = 0$ is a circle. If $A \neq B$, $AB > 0$, and $AC < 0$, then the graph is an ellipse. If $AB > 0$ and $AC > 0$, then there is no graph. If $AB < 0$, then the graph is a hyperbola.

Chapter 7 Test

1. a. $-\dfrac{\pi}{2}$ **b.** 3.70   **2. a.** 210°   **b.** 200.5°   **3.** Answers
may vary. Examples are given.   **a.** 260°, −460°
b. 680.3°, −39.7°   **c.** $\dfrac{15\pi}{4}$, $-\dfrac{\pi}{4}$ **d.** 580°40′, −139°20′

4. 11.96 cm; 29.89 cm² **5.** 3.9×10^5 mi
6. a. $\sin \theta = -\dfrac{\sqrt{2}}{2}$; $\cos \theta = -\dfrac{\sqrt{2}}{2}$
 b. $\sin \theta = \dfrac{1}{2}$; $\cos \theta = -\dfrac{\sqrt{3}}{2}$ **c.** $\sin \theta = -1$; $\cos \theta = 0$
7. a. >   **b.** <   **c.** =   **d.** <   **8. a.** $\dfrac{1}{2}$ **b.** −1   **c.** $-\dfrac{1}{2}$
d. $-\dfrac{\sqrt{3}}{2}$ **9. a.** $\sqrt{2}$ **b.** −2   **c.** $-\dfrac{\sqrt{3}}{3}$ **d.** 0
10. $\sin x = \dfrac{3\sqrt{10}}{10}$; $\cos x = -\dfrac{\sqrt{10}}{10}$; $\tan x = -3$;
$\sec x = -\sqrt{10}$; $\csc x = \dfrac{\sqrt{10}}{3}$ **11. a.** $-\dfrac{\pi}{6}$ **b.** $\dfrac{2\sqrt{3}}{3}$ **c.** $\dfrac{5}{4}$
12. The function $f(x) = \cos x$ is not one-to-one when the
domain consists of all real x. If we restrict the domain to
$0 \leq x \leq \pi$, however, the function is one-to-one and
therefore has an inverse.

Chapter 8 Test

1. 146°   **2.** 191.5°, 348.5°   **3.** 3.28, 6.14
4. $\dfrac{\pi}{24}$, $\dfrac{11\pi}{24}$,   **5.** period: 6π; amp.: 2
 $\dfrac{13\pi}{24}$, $\dfrac{23\pi}{24}$,
 $\dfrac{25\pi}{24}$, $\dfrac{35\pi}{24}$,
 $\dfrac{37\pi}{24}$, $\dfrac{47\pi}{24}$

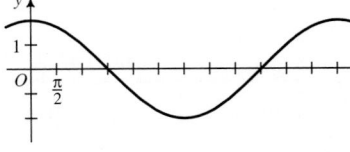

6. $y = 2\cos \dfrac{\pi}{2}(x+1) + 1$, or $y = -2\sin \dfrac{\pi}{2}x + 1$

7. a.

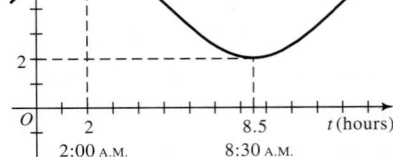

b. To find the earliest possible time that the tanker can
dock, examine the graph in part (a) to see where the curve
first crosses from below to above the line $D = 4$ after
$t = 9$; this occurs at time $t = 11.75$ (11:45 A.M.). To find
the longest period of time that the tanker can be docked
before the water level becomes too low, again examine the
graph and note that the curve crosses from above to below
the line $D = 4$ at time $t = 18.25$ (6:15 P.M.); thus, the
tanker can be docked for $18.25 - 11.75 = 6.5$ h.
8. a. $\cos A$ **b.** $\sec \theta$ **c.** $\sin x$ **d.** $\cot \theta$

9.a. 0.46, 3.61   **b.** $\dfrac{\pi}{4}, \dfrac{3\pi}{4}, \dfrac{5\pi}{4}, \dfrac{7\pi}{4}$

10.a. 120°, 180°, 240°   **b.** 0°, 63.4°, 180°, 243.4°

Chapter 9 Test

1. 69.5°, 69.5°, 41.0°   **2.** 549 m   **3.** $72\sqrt{2}$ in^2 **4.** 452
5.a. 1   **b.** 2   **6.** 56.9 km from A, 37.1 km from B
7. 29.9°   **8.** 28.7 nautical miles   **9.** Use the law of
cosines, because it (unlike the law of sines) can
distinguish between obtuse and acute angles.   **10.** 740 km

Chapter 10 Test

1.a. 0   **b.** $\dfrac{\sqrt{3}}{2}$ **c.** $\cos x$ **d.** $\sqrt{2}\sin x$ **2.** $\dfrac{\sqrt{6}-\sqrt{2}}{4}$

3. -2 **4.** $\tan(\alpha+\beta)=\dfrac{\tan\alpha+\tan\beta}{1-\tan\alpha\tan\beta}=$

$\dfrac{\frac{1}{3}+\frac{1}{2}}{1-\frac{1}{3}\cdot\frac{1}{2}}=\dfrac{\frac{5}{6}}{\frac{5}{6}}=1$ **5.** Since $\tan\dfrac{\pi}{2}$ is undefined, you

cannot simplify $\tan\left(\dfrac{\pi}{2}+\theta\right)=\dfrac{\tan\dfrac{\pi}{2}+\tan\theta}{1-\tan\dfrac{\pi}{2}\tan\theta}$. Instead, let

$\tan\left(\dfrac{\pi}{2}+\theta\right)=\dfrac{\sin\left(\dfrac{\pi}{2}+\theta\right)}{\cos\left(\dfrac{\pi}{2}+\theta\right)}$. Then $\tan\left(\dfrac{\pi}{2}+\theta\right)=$

$\dfrac{\sin\dfrac{\pi}{2}\cos\theta+\cos\dfrac{\pi}{2}\sin\theta}{\cos\dfrac{\pi}{2}\cos\theta-\sin\dfrac{\pi}{2}\sin\theta}=\dfrac{1\cdot\cos\theta+0\cdot\sin\theta}{0\cdot\cos\theta-1\cdot\sin\theta}=\dfrac{\cos\theta}{-\sin\theta}=$

$-\cot\theta$.   **6.a.** $\dfrac{12}{13}$ **b.** $-\dfrac{119}{169}$ **c.** $\dfrac{120}{169}$ **d.** $-\dfrac{28,560}{28,561}$

7.a. $\cot x$ **b.** $2\cot^2 y$ **c.** $\cot\dfrac{1}{2}t$ **d.** $\cos\dfrac{1}{2}x$

8.a. $\dfrac{1}{4}$ **b.** $-\dfrac{\sqrt{3}}{2}$ **9.a.** $(1+\tan^2 x)(1+\cos 2x)=$

$(\sec^2 x)[1+(2\cos^2 x-1)]=\left(\dfrac{1}{\cos^2 x}\right)\left(2\cos^2 x\right)=2$

b. $\dfrac{\cos\theta\csc\theta}{\tan\theta+\cot\theta}=\dfrac{\cot\theta}{\tan\theta+\cot\theta}\cdot\dfrac{\tan\theta}{\tan\theta}=\dfrac{1}{\tan^2\theta+1}=$

$\dfrac{1}{\sec^2\theta}=\cos^2\theta=(1-\cos^2\theta)+(2\cos^2\theta-1)=$

$\sin^2\theta+\cos 2\theta$

10.a. See graph.
  b. 30°, 90°,
    150°, 270°

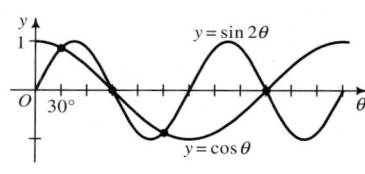

11. $\dfrac{\pi}{2}$

Chapter 11 Test

1.a. $(3\sqrt{2},-45°)$ **b.** $(0,-6)$
   c. $(2,90°)$ **d.** $(-8,0)$
   e. $(1,\sqrt{3})$ **f.** $(2,300°)$
2.a. See graph at right.
   b. $(x^2+y^2)^3=16x^2y^2$

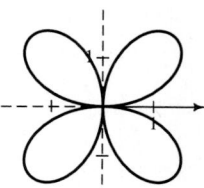

3.a.

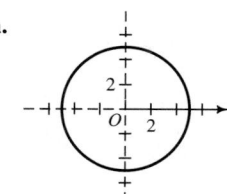

b.

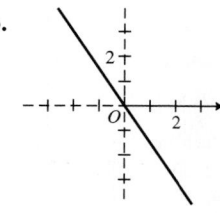

c.

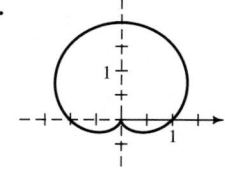

4.a. $z_1=2\,\text{cis}\,150°$;   **b.**
   $z_2=4\sqrt{2}\,\text{cis}\,45°$;
   $z_1z_2=8\sqrt{2}\,\text{cis}\,195°$

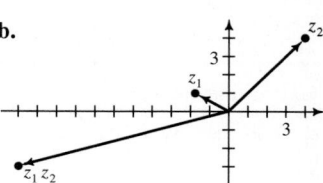

5.a. $z^2=9\,\text{cis}\,300°=\dfrac{9}{2}-\dfrac{9\sqrt{3}}{2}i$ **b.** In rectangular

form, $z=-\dfrac{3\sqrt{3}}{2}+\dfrac{3}{2}i$. Thus, $z^2=\left(-\dfrac{3\sqrt{3}}{2}+\dfrac{3}{2}i\right)^2=$

$\dfrac{27}{4}-\dfrac{9\sqrt{3}}{2}i+\dfrac{9}{4}i^2=\dfrac{9}{2}-\dfrac{9\sqrt{3}}{2}i$, which agrees with (a).

6.a. $\sqrt{2}\,\text{cis}(-45°)$
  b. See graph at right.
  c. $-32i$
7. When the n nth roots of a
complex number are graphed,
they are the vertices of a
regular polygon of n sides.

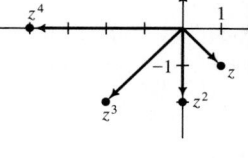

8. $(\sqrt[3]{2}\,\text{cis}\,130°)^3=2\,\text{cis}\,390°=2\,\text{cis}\,30°=$

$2\left(\dfrac{\sqrt{3}}{2}+\dfrac{1}{2}i\right)=\sqrt{3}+i$, so $\sqrt[3]{2}\,\text{cis}\,130°$ is a cube root of

$\sqrt{3}+i$. The other two cube roots are $\sqrt[3]{2}\,\text{cis}\,10°$ and
$\sqrt[3]{2}\,\text{cis}\,250°$.

Chapter 12 Test

1. direction: $118°$; magnitude: 17 N

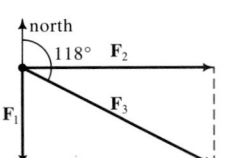

2.a. $-\mathbf{u}$
  b. $\mathbf{u} - \mathbf{v}$
  c. $\dfrac{3}{5}\mathbf{u} - \dfrac{3}{5}\mathbf{v}$
  d. $\dfrac{3}{5}\mathbf{u} + \dfrac{2}{5}\mathbf{v}$

3. 488.6 mi/h south, 177.9 mi/h west **4.** $D(9, 9)$
5.a. $\mathbf{v} = (2, -3)$, $|\mathbf{v}| = \sqrt{13}$ **b.** $x = 1 + 2t$; $y = 3 - 3t$
c. At time $t = -\dfrac{7}{8}$, the object is at point $\left(-\dfrac{3}{4}, \dfrac{45}{8}\right)$ on the parabola. Also, at time $t = 1$, the object is at point $(3, 0)$ on the parabola.
6. If θ is the angle between $\mathbf{u}$ and $\mathbf{v}$, then $\mathbf{u} \cdot \mathbf{v} = |\mathbf{u}||\mathbf{v}|$ implies that $\cos \theta = 1$, or $\theta = 0°$. Thus, $\mathbf{u}$ and $\mathbf{v}$ have the same direction. **7.** $28.1°$ **8.a.** 13 **b.** $\left(\dfrac{1}{2}, -2, 1\right)$
9.a. $(x, y, z) = (2, 1, 7) + t(4, 1, 3)$ **b.** $(4, 0, 9)$
10. $x + 2y + 7z = 35$ **11.a.** 5 **b.** -20 **12.** $\left(-\dfrac{1}{2}, -2\right)$
13.a. $(10, 10, 30)$ **b.** $5\sqrt{11}$

Chapter 13 Test

1.a. arithmetic; $t_n = 22 - 5n$ **b.** neither; $t_n = (n + 1)^2 - 1$
c. geometric; $t_n = 243\left(-\dfrac{1}{3}\right)^n$ **2.** 212 **3.** $\dfrac{4}{2187}$
4. $1, 3, 10, 33, 109, 360$ **5.** $t_1 = 1$; $t_n = t_{n-1} + 3^{n-1}$
6.a. $t_1 = 106$; $t_2 = 116$; $t_3 = 126$; $t_n = 96 + 10n$
b. $S_1 = 106$; $S_2 = 222$; $S_3 = 348$; $S_n = 101n + 5n^2$
7. $\dfrac{1640}{2187}$ **8.** If the terms of a sequence increase without bound, the limit of the sequence is infinity. For example, the limit of the sequence $1, 4, 9, 16, \ldots$ is infinity, and we write $\lim\limits_{n \to \infty} n^2 = \infty$. On the other hand, if the terms of a sequence do not increase (or decrease) without bound and do not approach a single finite number, then the sequence has no limit. For example, the sequence defined by $t_n = \dfrac{n(-1)^n}{n + 1}$ has no limit because the terms alternately approach 1 (for even n) and -1 (for odd n). **9.a.** $\dfrac{5}{4}$
b. 0 **c.** ∞ **10.** $\sum\limits_{n=1}^{\infty} 23\left(\dfrac{1}{100}\right)^n = \dfrac{23}{99}$ **11.a.** $|x| < \dfrac{3}{2}$
b. $\dfrac{3}{3 - 2x}$ **12.a.** $\sum\limits_{n=1}^{5} \dfrac{(-1)^n}{4n^2}$ **b.** $\sum\limits_{n=1}^{8} (11 - 3n)$
13. 5320

14. (Step 1) $n = 1$: $n(2n + 1) = 1(2 \cdot 1 + 1) = 3$. (Step 2) Assume true for $n = k$: $3 + 7 + 11 + \cdots + (4k - 1) = k(2k + 1)$. Prove true for $n = k + 1$: $3 + 7 + 11 + \cdots + (4k - 1) + [4(k + 1) - 1] = k(2k + 1) + 4k + 3 = 2k^2 + 5k + 3 = (k + 1)(2k + 3) = (k + 1)[2(k + 1) + 1]$.

Chapter 14 Test

1.a.
$$\begin{array}{c} \\ S \\ M \\ L \end{array} \begin{array}{ccc} M & P & B \\ \left[\begin{array}{ccc} 10 & 6 & 14 \\ 8 & 10 & 20 \\ 20 & 8 & 14 \end{array}\right] \end{array} = S$$

b. 4 medium-sized Mama Bears, 5 medium-sized Papa Bears, and 10 medium-sized Baby Bears;
$$0.5S = \begin{bmatrix} 5 & 3 & 7 \\ 4 & 5 & 10 \\ 10 & 4 & 7 \end{bmatrix}$$

2. $\begin{bmatrix} 9 & -2 & -12 \\ -8 & 5 & 11 \end{bmatrix}$

3. A 2×2 nonzero matrix A always has an additive identity, $-A$. To prove this, let $A = \begin{bmatrix} a & b \\ c & d \end{bmatrix}$; then $-A = \begin{bmatrix} -a & -b \\ -c & -d \end{bmatrix}$ and $A + (-A) = \begin{bmatrix} a & b \\ c & d \end{bmatrix} + \begin{bmatrix} -a & -b \\ -c & -d \end{bmatrix} = \begin{bmatrix} 0 & 0 \\ 0 & 0 \end{bmatrix} = 0$ (likewise for $(-A) + A$). On the other hand, a 2×2 nonzero matrix A may not have a multiplicative inverse. For example, consider $A = \begin{bmatrix} 1 & 0 \\ 1 & 0 \end{bmatrix}$. If we assume that $B = \begin{bmatrix} a & b \\ c & d \end{bmatrix}$ is the multiplicative inverse of A, then $AB = I$, or $\begin{bmatrix} 1 & 0 \\ 1 & 0 \end{bmatrix}\begin{bmatrix} a & b \\ c & d \end{bmatrix} = \begin{bmatrix} a & b \\ a & b \end{bmatrix} = \begin{bmatrix} 1 & 0 \\ 0 & 1 \end{bmatrix}$, but this implies that $a = 1$ and $a = 0$ as well as $b = 0$ and $b = 1$, an obvious impossibility. Thus, A has no multiplicative inverse. **4.** NP gives the weight, in pounds, of each type of nut in the mixture. $NP = [8 \quad 4.4 \quad 3.4 \quad 4.2]$ **b.** 4.4 lb
5. $(2, -3)$

6.a. From
$$\begin{array}{c} \\ BB \\ LP \\ PE \\ SO \\ W \end{array} \begin{array}{ccccc} & & \text{To} & & \\ BB & LP & PE & SO & W \\ \left[\begin{array}{ccccc} 0 & 1 & 0 & 0 & 0 \\ 1 & 0 & 1 & 0 & 0 \\ 1 & 1 & 0 & 1 & 1 \\ 0 & 1 & 0 & 0 & 0 \\ 0 & 0 & 1 & 1 & 0 \end{array}\right] \end{array} = M$$

b. $M^2 = \begin{bmatrix} 1 & 0 & 1 & 0 & 0 \\ 1 & 2 & 0 & 1 & 1 \\ 1 & 2 & 2 & 1 & 0 \\ 1 & 0 & 1 & 0 & 0 \\ 1 & 2 & 0 & 1 & 1 \end{bmatrix}$; the park entrance, with 6 two-step paths.

7.a. From $\begin{array}{c}\\C\\S\end{array}\begin{array}{cc}C & S\\\end{array}\begin{bmatrix}0.94 & 0.06\\0.02 & 0.98\end{bmatrix} = T$ **b.** $\begin{array}{cc}C & S\end{array}\begin{bmatrix}0.70 & 0.30\end{bmatrix} = P_0$

c. $P_0T = [0.664 \quad 0.336]$;
66.4% in the city, and 33.6% in the suburbs

8.a,b.

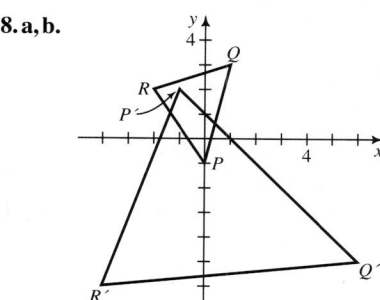

c. $T = \begin{bmatrix}3 & 1\\1 & -2\end{bmatrix}$; $|T| = -7$. The determinant -7 means that the area of $\triangle P'Q'R'$ is 7 times the area of $\triangle PQR$ and that the orientation of the triangle is reversed.

Chapter 15 Test

1. 45 **2.** 540 **3.** 7,923,840 **4.** 20,160 **5.** 45
6. 3150 **7.** The number of selections of 6 items from a collection of 10 is $_{10}C_6 = \dfrac{10!}{4! \, 6!}$. The number of arrangements of 10 items where 6 are alike and the remaining 4 are alike is $\dfrac{10!}{6! \, 4!}$. Therefore, there are 210 possibilities in each case. **8.** $189x^5$; $-945x^4$

Chapter 16 Test

1. Consider flipping a fair coin. The theoretical probability of getting "heads" is always $\dfrac{1}{2}$. An empirical probability, however, can vary. For example, suppose the coin is flipped by two people 20 times each. If the first person gets 9 "heads," then that person assigns an empirical probability of $\dfrac{9}{20}$ to getting "heads." If the second person gets 11 "heads," then that person assigns an empirical probability of $\dfrac{11}{20}$ to getting "heads." Because these empirical probabilities are *approximations* of the theoretical probability, we cannot expect them to equal the theoretical probability or each other.

2.a. $\dfrac{1}{26}$ **b.** $\dfrac{5}{39}$ **c.** $\dfrac{25}{39}$ **3.a.** $\dfrac{1}{6}$ **b.** $\dfrac{1}{9}$ **c.** $\dfrac{1}{18}$
4.a. ≈ 0.656 **b.** ≈ 0.948 **5.a.** ≈ 0.053 **b.** ≈ 0.037

6.a. 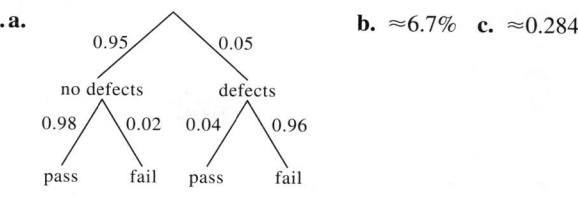 **b.** $\approx 6.7\%$ **c.** ≈ 0.284

7. The game is not fair; B has the advantage.

Chapter 17 Test

1. As its name implies, a *measure of central tendency* indicates some "center" of the data. (For example, the median is the halfway point in a set of ordered data.) A *measure of dispersion* indicates the spread of data about some center. Using only a measure of central tendency to characterize a set of data is inadequate, because we then have no knowledge of the variability of the data. For example, consider the two sets {7, 8, 8, 9} and {1, 4, 12, 15}. The mean in each case is 8, but the standard deviations are about 0.7 and 5.7, respectively, which indicate that the data in the second set are much more dispersed.

2.a.

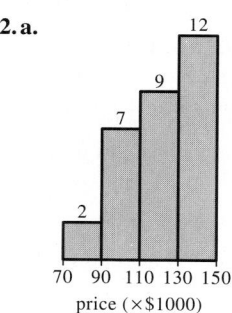

price (×$1000)

b. mean: $120,667;
median: $120,000;
mode: $140,000

3.a. median: 26.5;
lower quartile: 22;
upper quartile: 30
b. range: 20;
interquartile range: 8

c.
```
    18  22  26  30  34  38
```

4. the mathematics test **5.** $\bar{x} = 24$; $s^2 \approx 61.3$; $s \approx 7.83$
6.a. 2.28% **b.** 15.87% **c.** 1230 **d.** 1170 **7.** 52%
8. 95% confidence interval: $0.808 < p < 0.858$;
99% confidence interval: $0.796 < p < 0.871$

Chapter 18 Test

1.a. ≈ 0.753 **b.** $y = 1.715x + 6342$ **c.** Since the correlation coefficient in part (a) is fairly close to 1, the least-squares line should give reasonable predictions of weekly sales figures for given advertising budgets. However, once the business' ads have saturated the local advertising market, an increase in the advertising budget will probably not bring about an increase in sales as predicted by the least-squares line.
2. No, a high correlation can occur without there being a cause-and-effect relationship. Factors other than height, such as skill and aggressiveness, contribute to a high scoring average. **3.** $y = 0.1(100)^x$ **4.a.** $y = 6.36(1.05)^x$, where x is the number of years since 1900 **b.** \$836
5. $y = 100x^{2.5}$
6. $m = 3$. If we assume that the weight y of an orange is proportional to its volume V, then $y = bV$ for some constant b. Since an orange is approximately a sphere, $V = \frac{4}{3}\pi r^3$ where r is the orange's radius. If the circumference of the orange is x, then $x = 2\pi r$, so $r = \frac{x}{2\pi}$ and $V = \frac{4}{3}\pi\left(\frac{x}{2\pi}\right)^3 = \frac{1}{6\pi^2}x^3$. Thus, $y = b\left(\frac{1}{6\pi^2}x^3\right) = ax^3$ for some constant a. **7.a.** $v = 179.4d^{-0.5}$ **b.** 18.6 m/s

Chapter 19 Test

1.a. 2 **b.** $\frac{1}{2}$ **c.** 12 **2.** As x increases without bound, $\sin x$ oscillates between -1 and 1; that is, the value of the sine function neither approaches some finite number nor increases (or decreases) without bound. Thus, $\lim_{x\to\infty} \sin x$ does not exist.

3.a.

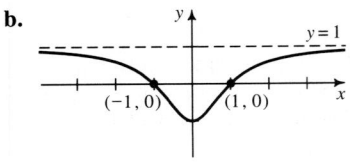

b.

Chapter 20 Test

4.a. 36 **b.** 18 **5.** $-1 + \frac{1}{3!} - \frac{1}{5!} + \frac{1}{7!} - \frac{1}{9!} + \cdots$
6. $x^3 - \frac{x^6}{2} + \frac{x^9}{3} - \frac{x^{12}}{4} + \cdots$, $-1 < x \le 1$ **7.** 2.12
8. The fixed points are $\frac{1-\sqrt{5}}{2}$, 0, and $\frac{1+\sqrt{5}}{2}$. For $\frac{1-\sqrt{5}}{2} \le x_0 \le \frac{1+\sqrt{5}}{2}$, $\lim_{n\to\infty} f^n(x_0) = 0$.
For $x_0 < \frac{1-\sqrt{5}}{2}$, $\lim_{n\to\infty} f^n(x_0) = -\infty$. For $x_0 > \frac{1+\sqrt{5}}{2}$, $\lim_{n\to\infty} f^n(x_0) = \infty$.
9. The logistic function $f(x) = cx(1 - x)$ is based on the assumption that the next generation, $f(x)$, is jointly proportional to the current generation, x, and the "room available for growth," $1 - x$. When this function is iterated for various values for c, three basic types of population growth occur: the population approaches an equilibrium point, the population repeatedly cycles through some sequence of values, or the population behaves chaotically (that is, with no apparent predictability).

Chapter 20 Test

1.a. $f'(x) = 12x^2 + 10x - 2$ **b.** $f'(x) = -\frac{8}{x^3} - \frac{3}{x^2} + 4$
c. $f'(x) = \frac{3}{4}\sqrt[4]{\frac{5}{x}} + \frac{1}{4}\sqrt[4]{\frac{1}{x^5}}$ **d.** $f'(x) = -\frac{10}{x^6} + 15x^4$
2. The derivative of a function $f(x)$ is defined as $f'(x) = \lim_{h\to 0} \frac{f(x+h) - f(x)}{h}$. The ratio $\frac{f(x+h) - f(x)}{h}$ gives the slope of the line through the points $(x, f(x))$ and $(x + h, f(x + h))$ on the graph of f. As h approaches 0, $\frac{f(x+h) - f(x)}{h}$ approaches the slope of the line tangent to the graph of f at $(x, f(x))$. Thus, $f'(x)$ gives the slope of the curve (which is defined to be the slope of the tangent) at $(x, f(x))$.

3. $(-1, 2)$ **4.** $f(2)$ is a local minimum.

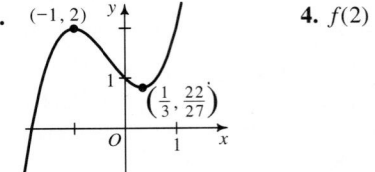

5.a. $A(x, h) = 4x^2 + 6xh$ **b.** $h = \frac{1125}{2x^2}$
c. $A(x) = 4x^2 + 3375x^{-1}$ **d.** $x = \frac{15}{2}$ and $h = 10$;
$A = 675$ sq. units **6.** $s(4) = 48$; $v(4) = 27$; $a(4) = 8$
7.a. 0.55 s **b.** 144 ft **c.** 5.45 s **d.** -78.4 ft/s

Cumulative Review for Chapters 1–6

1. a.

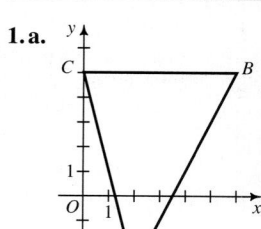

b. $4\sqrt{5}$
c. $(4, 1)$
d. 24
e. $2x - y = 7$
f. $x + 2y = 6$
g. $x = 3$
h. $\left(3, \dfrac{3}{2}\right)$; $r = \dfrac{\sqrt{85}}{2}$
i. $(x - 3)^2 + \left(y - \dfrac{3}{2}\right)^2 = \dfrac{85}{4}$

j.

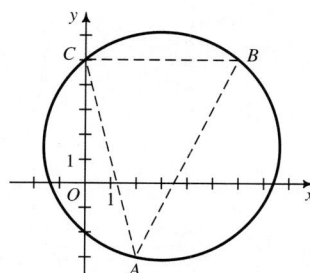

k. x-int.: $3 \pm \sqrt{19}$;
  y-int.: $-2, 5$
l. $a = 1$;
  $b = -6$;
  $c = 5$

m.

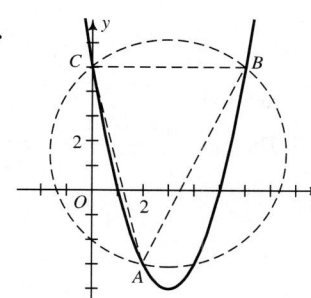

n. parabola
o. $V(3, -4)$;
  $F\left(3, -\dfrac{15}{4}\right)$;
  axis: $x = 3$;
  directrix: $y = -\dfrac{17}{4}$
p. x-int.: $1, 5$;
  y-int.: 5
q. $(4, -3)$

r. $\left(\dfrac{11 - \sqrt{89}}{4}, \dfrac{13 + \sqrt{89}}{8}\right)$; $\left(\dfrac{11 + \sqrt{89}}{4}, \dfrac{13 - \sqrt{89}}{8}\right)$

2. a. $P(1) = -4 < 0$;
  $P(2) = 18 > 0$; by the
  location principle,
  $P(x) = 0$ must have at
  least one real root
  between $x = 1$ and $x = 2$

b. -1 (a double root), $\dfrac{3}{2}$,
  $1 \pm i$

c. $x = -1$ or $x \geq \dfrac{3}{2}$

d.

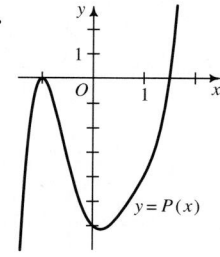

3. a. $2 - i$, $3 + 2i$ **b.** $P(x)$ must have the form
$a[x - (2 + i)][x - (2 - i)][x - (3 - 2i)][x - (3 + 2i)]$
where a is any nonzero real number. Choosing different
values for a gives different rules for $P(x)$.

c. $P(x) = \dfrac{1}{5}(x^4 - 10x^3 + 42x^2 - 82x + 65)$ **d.** 40

4. $-0.41, 2.41$ **5. a.** 7 s   **b.** $0 \leq t \leq 7$ **c.** $0 \leq h(t) \leq 80$
d. 3 s   **6.** sum: -2; product: -3 **7.** 12 **8.** $-2, 8$

9. $x \leq 1$ or $x \geq \dfrac{7}{3}$

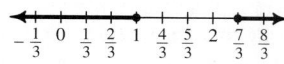

10. a.

b. Answers may vary.
Example: Find the area
of the triangle with
vertices $(-2, 0)$, $(3, -5)$,
and $(1, -9)$.

11. a. max.: 24; min.: 9 **b.** max.: 28; min.: 8

12. a.

Function	Domain	Range
f	all reals	$f(x) \geq 0$
g	all reals	all reals
h	$x \geq 0$	$h(x) \geq 0$

b. $g^{-1}(x) = x + 1$; $h^{-1}(x) = x^2$ for $x \geq 0$
c. $r = h \circ g \circ f$; $s = f \circ g \circ h$
d. $f(h(g(x))) = (\sqrt{x - 1})^2 = x - 1$;
$g(f(h(x))) = (\sqrt{x})^2 - 1 = x - 1$;
$g(h(f(x))) = \sqrt{x^2} - 1 = |x| - 1$;
$h(f(g(x))) = \sqrt{(x - 1)^2} = |x - 1|$

e.

Function	Domain	Range
$f \circ h \circ g$	$x \geq 1$	$f(h(g(x))) \geq 0$
$g \circ f \circ h$	$x \geq 0$	$g(f(h(x))) \geq -1$
$g \circ h \circ f$	all reals	$g(h(f(x))) \geq -1$
$h \circ f \circ g$	all reals	$h(f(g(x))) \geq 0$

f. The function g has the set of all real numbers as both
its domain and range. The composite functions in part (d)
have restricted ranges and, in the case of $f \circ h \circ g$ and
$g \circ f \circ h$, restricted domains.

13. a. 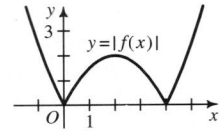   **b.**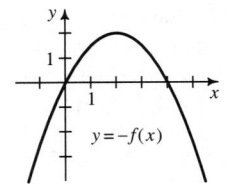

c. 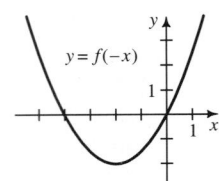   **d.**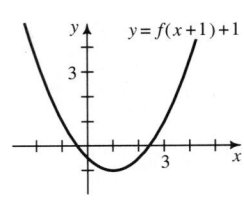

14. a. $C(l, h) = 16l + 4lh + 16h$ **b.** $lh = 10$, or $l = \dfrac{10}{h}$

c. $C(h) = \dfrac{160}{h} + 40 + 16h$ **d.** $h > 0$ **e.** 3.16 ft long,

4 ft wide, 3.16 ft high   **15. a.** 9   **b.** 32   **c.** 3   **d.** 2.5

e. $-\dfrac{3}{4}$ **f.** $\dfrac{81}{100}$ **g.** 2   **16. a.** 2.048   **b.** 5.096   **c.** 5.107

d. 5 g; 3.54 g   **e.** \$732.81; \$738.91   **17. a.** 5   **b.** -2

c. 27   **18. a.** $a = \dfrac{2}{9}$; $b = 3$ **b.** 5

19. a.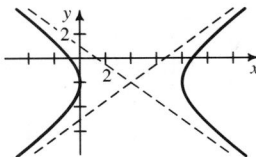

b. center: $(4, -2)$; vertices: $(0, -2)$, $(8, -2)$;
foci: $(-1, -2)$, $(9, -2)$ **c.** $3x + 4y = 4$; $3x - 4y = 20$

d. $\dfrac{(x - 4)^2}{25} + \dfrac{(y + 2)^2}{9} = 1$ **e.** $f(x) = 2(x + 1)^2 - 2$

f. focus: $\left(-1, -\dfrac{15}{8}\right)$; directrix: $y = -\dfrac{17}{8}$; axis: $x = -1$

g. hyperbola: $e = \dfrac{5}{4}$; ellipse: $e = \dfrac{4}{5}$; parabola: $e = 1$

Cumulative Review for Chapters 7–11

1. a. $-\dfrac{4}{5}$ **b.** $-\dfrac{24}{25}$ **c.** $\dfrac{3}{5}$ **d.** $\dfrac{44}{125}$ **e.** -3 **f.** $-\dfrac{336}{625}$

2. a. 0   **b.** -1 **c.** $-\dfrac{\pi}{3}$ **d.** 0.8

3. a. $\tan x$ **b.** $\tan x$ **c.** $2 \tan x$ **d.** $\dfrac{\pi}{4}$ **e.** $\pi - x$

4. Substituting $\sin x$ for y in $y^2 + y \cot x \cos x - 1$,
we have $(\sin x)^2 + \sin x \cot x \cos x - 1 =$
$\sin^2 x + \cos^2 x - 1 = 1 - 1 = 0$.

5. a. $-\dfrac{\pi}{3}, -\dfrac{\pi}{6}, \dfrac{\pi}{6}, \dfrac{2\pi}{3}$ **b.** $-2\pi, -\pi, 0$ **c.** $150°, 330°$

d. $-90°, -67.5°, 0°, 22.5°, 90°$ **6. a.** 2π in.   **b.** 5π in^2

c. $r = 5\sqrt{2}$ in.; $s = 2\pi\sqrt{2}$ in.   **d.** $\theta = 144°$; $s = 4\pi$ in.

e. the technique described in part (d)

7.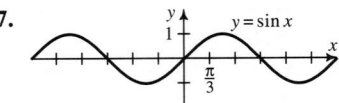

8. a. 4
b. 4
c. See graph at right.
d. 0.54, 1.46, 4.54

e. $g(x) = 4\cos\left(\dfrac{\pi}{2}(x - 1)\right) + 2$

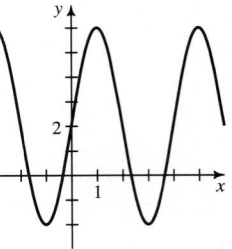

9. a. 146.3°
b. 18.4°
c. 52.1°
d. $9x - 7y = 11$
e. See graph at right.
10. 60.83 mi
11. 11.37 in.

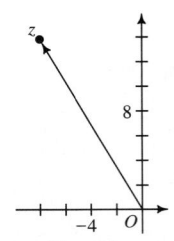

12. 1.5 radians   **13.** 3.49 mi from A   **14.** 4.12 mi
15. 9.92   **16.** $y = 13$ **17. a.** See graph below.   **b.** 16
c. 16 cis 120°   **d.** $\sqrt{3} + i, -1 + i\sqrt{3}, -\sqrt{3} - i, 1 - i\sqrt{3}$
e. See graph below.   **f.** a square

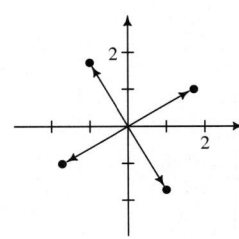

Ex. 17.a.   Ex. 17.e.

18. a. 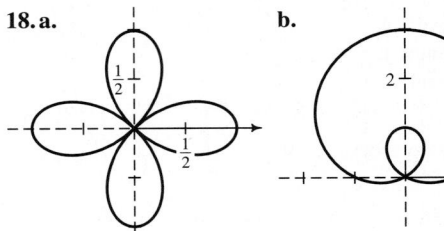   **b.**

c.

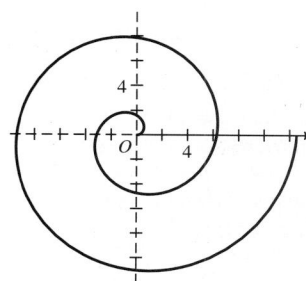

19. a.

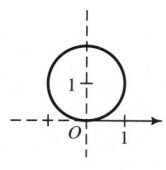

b. $x^2 + y^2 = 2y$

c. Writing the equation in part (b) as $x^2 + (y - 1)^2 = 1$, we know that the graph is a circle with center $(0, 1)$ and radius 1. This description matches the polar graph in part (a).

20. a. 6 cis 135°; $-3\sqrt{2} + 3i\sqrt{2}$ **b.** 8 cis 90°; $8i$

Cumulative Review for Chapters 12–14

1. a. $\overrightarrow{DC} + \overrightarrow{CB}$ **b.** $\overrightarrow{DC} + \overrightarrow{DA}$ **c.** $\overrightarrow{AB} - \overrightarrow{BC}$
d. $-\overrightarrow{BC} - \overrightarrow{BA}$ **2. a.** 0   **b.** $\mathbf{v} \perp \mathbf{w}$ **c.** 175.6°
d. $\left(\dfrac{\sqrt{5}}{5}, -\dfrac{2\sqrt{5}}{5}\right)$ **3.** 45.03°

4. a.

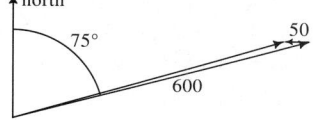

b. 73.66°; 551.9 mi/h

5. a. $(x - 5)^2 + (y - 2)^2 + (z + 10)^2 = 169$
b. $3x + 4y - 12z = -26$ **6.** $\mathbf{a} \parallel \mathbf{b}$; $\mathbf{a} \perp \mathbf{c}$; $\mathbf{b} \perp \mathbf{c}$
7. $(1, 2, 3)$ **8.** 7   **9.** 8   **10. a.** neither   **b.** 38
c. $t_1 = 3$; $t_n = t_{n-1} + (2n - 1)$ **d.** $t_n = n^2 + 2$
e. 10,002   **f.** Use part (d), because it allows you to find t_{100} directly. Part (c) requires that you know all the terms preceding t_{100} before you can find t_{100}.
11. a. arithmetic   **b.** 39   **c.** $t_1 = 4$; $t_n = t_{n-1} + 7$
d. $t_n = 7n - 3$ **e.** 697   **f.** the line $y = 7x - 3$
g. 1410   **12. a.** geometric   **b.** 160   **c.** $t_1 = 5$;
$t_n = 2t_{n-1}$ **d.** $t_n = 5 \cdot 2^{n-1}$ **e.** $5 \cdot 2^{99} \approx 3 \times 10^{30}$
f. the exponential curve $y = 5 \cdot 2^{x-1}$ **g.** 5,242,875
13. a. $|x| > 1$ **b.** $\dfrac{1}{x - 1}$ **14. a.** all real x except odd

multiples of $\dfrac{\pi}{2}$ **b.** $\sec^2 x$ **c.** $2 \tan x$ **15. a.** 1   **b.** $\dfrac{5}{4}$

c. $\dfrac{\pi}{2}$ **d.** $\dfrac{1}{e}$ **16. a.** $\displaystyle\sum_{n=1}^{50} 2n(2n + 1)$ **b.** 174,250

17. (Step 1) $n = 1$: $7^n - 1 = 7^1 - 1 = 6$, a multiple of 6. (Step 2) Assume true for $n = k$: $7^k - 1 = 6m$ for some integer m. Prove true for $n = k + 1$: $7^{k+1} - 1 = 7^k \cdot 7 - 1 = 7^k(6 + 1) - 1 = 6 \cdot 7^k + 7^k - 1 = 6 \cdot 7^k + 6m = 6(7^k + m)$. Since k is an integer, 7^k is an integer and so is $7^k + m$. Thus, $7^{k+1} - 1$ is a multiple of 6.

18. a. $\begin{bmatrix} 8 & 15 & 21 \\ -12 & 8 & 13 \end{bmatrix}$ **b.** $\begin{bmatrix} 2 & 4 \\ 1 & -2 \\ 3 & 5 \end{bmatrix}$

19. a. $AB = \begin{bmatrix} 8 & 4 & 15 \\ 2 & -44 & -21 \end{bmatrix}$ **b.** $\begin{bmatrix} \dfrac{7}{18} & -\dfrac{1}{18} \\ \dfrac{2}{9} & \dfrac{1}{9} \end{bmatrix}$

c. B^{-1} does not exist.

d. $\begin{bmatrix} 2 & 1 \\ -4 & 7 \end{bmatrix} \begin{bmatrix} x \\ y \end{bmatrix} = \begin{bmatrix} 9 \\ 26 \end{bmatrix}$; $x = \dfrac{37}{18}$; $y = \dfrac{44}{9}$

20. a. From

	To				
	A	B	C	D	E
A	0	1	0	1	0
B	1	0	1	0	0
C	0	0	0	1	0
D	0	0	1	0	1
E	0	0	1	0	0

b. $\begin{bmatrix} 1 & 0 & 2 & 0 & 1 \\ 0 & 1 & 0 & 2 & 0 \\ 0 & 0 & 1 & 0 & 1 \\ 0 & 0 & 1 & 1 & 0 \\ 0 & 0 & 0 & 1 & 0 \end{bmatrix}$

c. $\begin{bmatrix} 1 & 1 & 2 & 1 & 1 \\ 1 & 1 & 1 & 2 & 0 \\ 0 & 0 & 1 & 1 & 1 \\ 0 & 0 & 2 & 1 & 1 \\ 0 & 0 & 1 & 1 & 0 \end{bmatrix}$; this matrix represents the total number of 1-step and 2-step paths between pairs of points in the communication network.

21. a. From $\begin{array}{c} \text{use} \\ \text{don't} \end{array} \begin{bmatrix} 0.94 & 0.06 \\ 0.08 & 0.92 \end{bmatrix}$ (To use/don't use)   **b.** 42.4%

c. The percentages will level off to about 57.1% using public transportation and 42.9% not using public transportation. To find these limiting percentages, let $P_0 = [0.4 \quad 0.6]$ and consider $P_0 T^n$ as n, the number of years from now, increases.

22. a. $T = \begin{bmatrix} 3 & 2 \\ 2 & -1 \end{bmatrix}$; $|T| = -7$

b. $A'(11, 5)$,
$B'(23, 6)$,
$C'(14, 14)$

c. 7:1

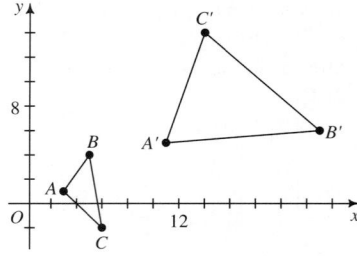

b. $r \approx 0.943$. The correlation between midyear exam scores and final exam scores is positive and strong.
c. $y = 0.697x + 27.46$ **d.** 79
7. a. 8.1% **b.** ≈ 0.31 **c.** ≈ 23 cups
8. 95% confidence interval: $0.801 < p < 0.849$;
99% confidence interval: $0.789 < p < 0.861$
9. a. $C(t) = 1000(1.08)^t$ where t is the number of years since 1955 **b.** \$21,725 **c.** \$31,920
10. a. $y = 9.41x^{0.663}$ **b.** \$8750

Cumulative Review for Chapters 15–18

1. a. 3,628,800 **b.** 28,800 **c.** 240 **d.** 362,880

2. a. $\dfrac{7}{99}$ **b.** $\dfrac{1}{99}$ **c.** $\dfrac{92}{99}$ **3. a.** $\dfrac{3}{5}$ **b.** $\dfrac{1}{20}$ **c.** $\dfrac{3}{20}$ **d.** $\dfrac{1}{5}$

e. $\dfrac{19}{20}$ **f.** ≈ 0.31 **4. a.** 97.6% **b.** ≈ 0.0061

5. a.
```
7 | 4
· | 8 9 9
8 | 0 0 1 1 2 2 2 3 3 4
· | 6
```
7|4 represents 74 in.

b.

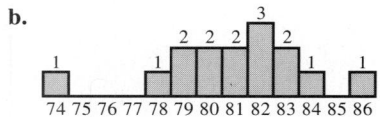

c. median: 81; lower quartile: 79; upper quartile: 83
d. range: 12; interquartile range: 4

e.

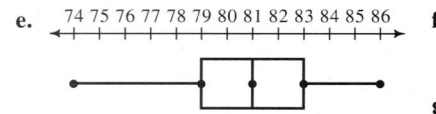

f. mean: ≈ 80.9;
std. dev.:
≈ 2.7
g. 73%; 93%
h. 68%; 95%

i. Based on the results of parts (g) and (h), we might conclude that a normal curve fits the height data fairly well.

6. a.

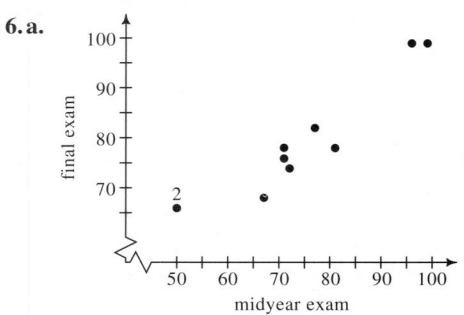

Cumulative Review for Chapters 19–20

1. a. -2 **b.** 0 **c.** $\dfrac{2}{5}$ **d.** $-\dfrac{1}{4}$ **e.** ≈ 2.718 **f.** e

2. a.

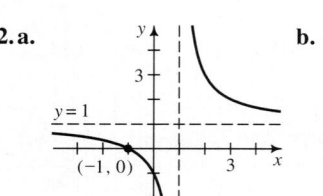

b.

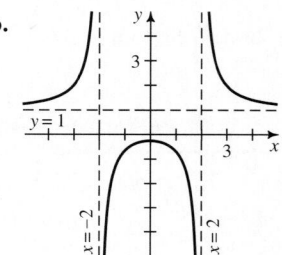

c.

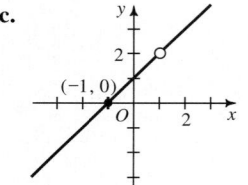

d.

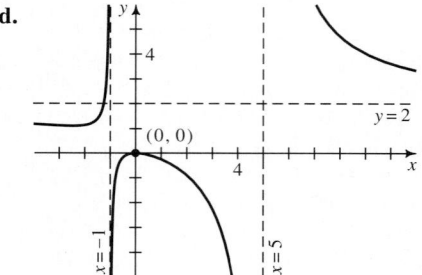

3. a. $0.3 - \dfrac{(0.3)^3}{3!} + \dfrac{(0.3)^5}{5!} - \dfrac{(0.3)^7}{7!} + \cdots$

b. 0.2955 **c.** 0.2955202 **d.** The results of parts (b) and (c) are in remarkable agreement.
4. 1, 0.269, 0.433, 0.393

5. a.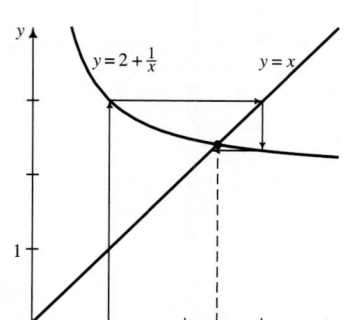

$y = 2 + \frac{1}{x}$ $y = x$

$x_0 = 1$

b. 2.41

c. $1 + \sqrt{2}$

6. a. $x_0 = 1000; x_n = 1.08x_{n-1}$ **b.** Iterate $f(x) = 1.08x$ using the seed $x_0 = 1000.$ **c.** \$3172.17

7. a. $f'(x) = \frac{\sqrt[3]{x}}{3x}$ **b.** $L(x) = \frac{1}{48}x + \frac{8}{3}$ **c.** $\frac{193}{48} \approx 4.0208$

d. ≈ 0.0001

8. a. $\pm\sqrt{3}, 4$ **b.** $x < -\frac{1}{3}; x > 3$ **c.** $-\frac{1}{3} < x < 3$

d. $\left(-\frac{1}{3}, \frac{338}{27}\right)$ **e.** $(3, -6)$ **f.**

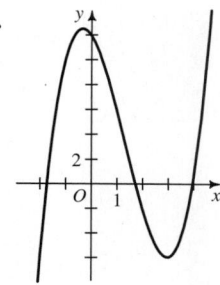

9. a. to the right   **b.** $2 < t < 4$ **c.** $t = 2; t = 4$
d. $x = 18$ **e.** 22

10. $\frac{16\sqrt{3}}{9}$

Advanced Mathematics

Precalculus
with Discrete Mathematics and Data Analysis

Richard G. Brown

Editorial Adviser

Andrew M. Gleason

Teacher Consultants

Martha A. Brown
Dane R. Camp
Maria F. G. Fierro
Wallis Green
Linda Hunter
Carolyn Kennedy

McDougal Littell Inc.

A Houghton Mifflin Company

Evanston, Illinois   Boston   Dallas   Phoenix

AUTHOR

Richard G. Brown, Mathematics Teacher, Phillips Exeter Academy, Exeter, New Hampshire. *A teacher and author, Mr. Brown has taught a wide range of mathematics courses for both students and teachers at several schools and universities. His affiliations have included the Newton (Massachusetts) High School, the University of New Hampshire, Arizona State University, and the North Carolina School for Science and Mathematics during the school year beginning in 1983. Currently a member of the COMAP Consortium Council, he is an active participant in professional mathematics organizations and the author of mathematics texts and journal articles.*

EDITORIAL ADVISER

Andrew M. Gleason, Hollis Professor of Mathematics and Natural Philosophy, Harvard University. *Professor Gleason is a well-known research mathematician and a member of the National Academy of Sciences. He has served as President of the American Mathematical Society.*

TEACHER CONSULTANTS

Martha A. Brown, Supervisor of Mathematics,
Prince George's County Public Schools, Prince George's County, Maryland
Dane R. Camp, Mathematics Teacher,
Downer's Grove High School, Downer's Grove, Illinois
Maria F. G. Fierro, Mathematics Department Chairperson,
Cerritos High School, Cerritos, California
Wallis Green, Mathematics Teacher,
C. E. Jordan High School, Durham, North Carolina
Linda Hunter, Mathematics Department Chairperson,
Douglas MacArthur High School, San Antonio, Texas
Carolyn Kennedy, Mathematics Teacher,
West High School, Columbus, Ohio

TECHNOLOGY CONSULTANT

Wade Ellis, Jr., Mathematics Instructor,
West Valley College, Saratoga, California

ACKNOWLEDGEMENTS

The author wishes to thank **Jonathan Choate,** Mathematics Teacher, Groton School, Groton, Massachusetts, for contributing the material on dynamical systems in Chapter 19. The author also wishes to thank **Donna DiFranco,** Professor of Mathematics, Bentley College, Waltham, Massachusetts, for contributing the projects that follow Chapters 5, 9, 12, 18, and 19.

ISBN: 0-395-42168-3

123456789 - D - 97 96 95 94 93

Introduction

Dear Student,

Welcome to this course in *Advanced Mathematics*! This year you will explore many new topics as well as extend those you have studied in previous courses.

I have written this book with the goal of making mathematics clear, interesting, and relevant. As a result, you will see many real-world applications of the topics you study. For example:

- Exponential functions model population growth, decline of natural resources, and cost of a college education.

- Logarithmic functions measure the intensity of earthquakes, the loudness of music, and the brightness of stars.

- Trigonometric functions describe AM/FM radio waves, the pattern of the tides, and the daily change in the time of sunset.

- Discrete mathematics provides techniques for calculating the return on a financial investment, deciding which mix of products to manufacture, and predicting the course of a flu epidemic.

- Probability theory predicts potential gains and losses with car insurance, business decisions, and even multiple choice tests.

- Statistics helps organize and analyze many types of numerical information, such as percentile ranks, sports data, and advertising claims, that bombard us daily.

As you see the wide range of fields that use mathematics, think about how mathematics will play a role in your own future. Take full advantage of this course to prepare for the many opportunities ahead.

I'd like to offer you the advice that I offer my own students:

> Math is not a spectator sport!

Don't just watch *other* people do mathematics! Stay actively involved by doing the activities, participating in classroom discussions and group work, reading the text and examples, and working on the exercises and projects.

I hope you find the course useful, stimulating, and enjoyable.

Sincerely,

Richard G. Brown

Richard G. Brown

Contents

Applications

Calculate a swimmer's rate of speed or see how the State Conservation Department uses estimation skills to approximate the number of deer in a mountainous area.

Problem Solving

Take a closer look at the motion of the tide through trigonometric modeling.

Reasoning

Create your own argument proving something is either right or wrong.

Communication

Increase your ability to explain a math concept clearly and concisely.

Technology

Move into the wonderful world of mathematics through calculators and computers.

Functions, Graphs, and Applications

Functions, Graphs, and Applications

Trigonometry

Trigonometry

Discrete Mathematics and Data Analysis

Discrete Mathematics and Data Analysis

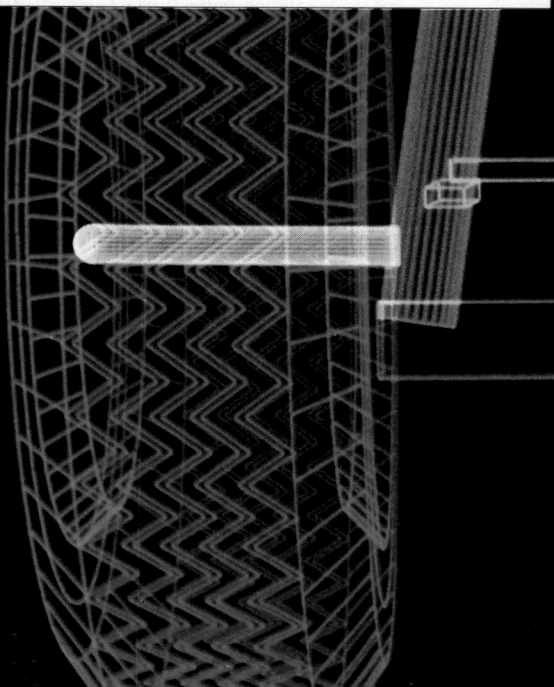

Discrete Mathematics and Data Analysis

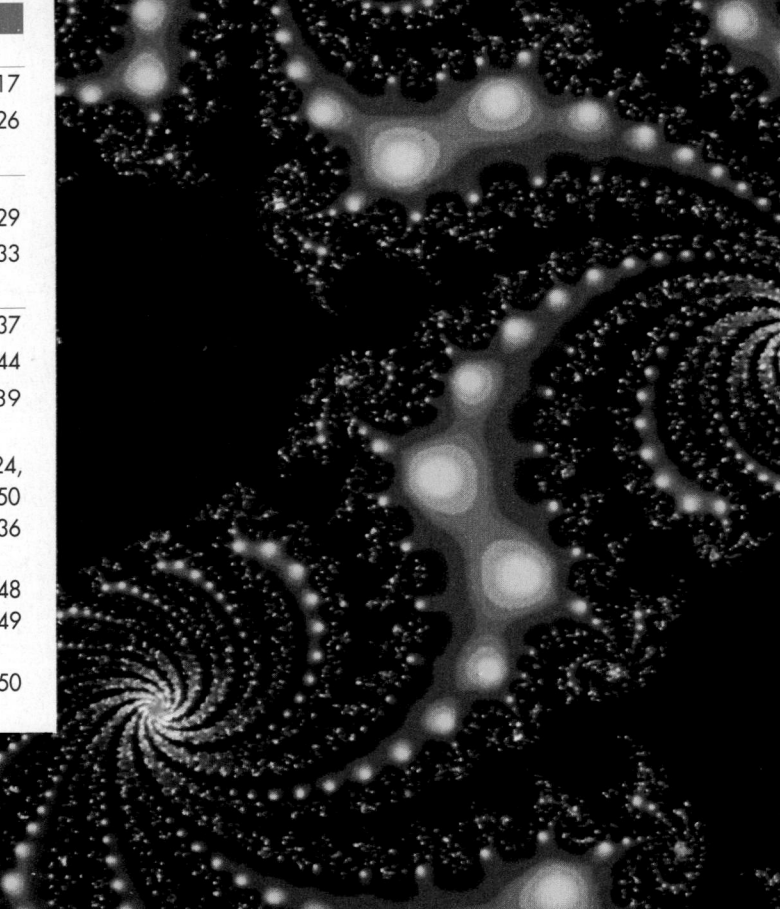

Limits and Introduction to Calculus

Applications

"When am I ever going to use this?"

Business

Business
Communications
Consumer Economics
Economics
Finance
Industrial Design
Linear Programming
Manufacturing
Operations Research
Statistics

Social Science

Archaeology
Education
Polling
Psychology
Transportation

Daily Life

Carpentry
Forestry
Horticulture
Landscaping
Music
Plumbing
Recreation
Sports

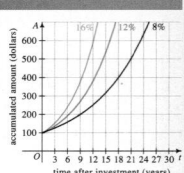

9. **Discussion** The amount that a person pays for auto insurance is a function of many variables. Discuss what some of these variables are.

10. **a.** If $f(x, y) = \sqrt{x^2 + y^2}$, find $f(3, 4)$, $f(-4, 3)$, and $f(0, 5)$.
   b. Sketch some constant curves of the function f.

11. Name several pairs (a, b) not in the domain of $f(a, b) = \dfrac{a + b}{a - b}$.

WRITTEN EXERCISES

A 1. **Consumer Economics** If \$100 is invested at interest rate r compounded annually, then the accumulated amount t years later is $A = 100(1 + r)^t$.
   a. The formula shows that A is a function of ? and ?.
   b. The graphs of A versus t for several constant values of r are shown at the right. The equation of the rightmost curve is $A = 100(1.08)^t$. What are the equations of the other two curves?
   c. About how many years does it take to double your money at 8% interest compounded annually? at 12% interest? at 16% interest?

2. **Consumer Economics** The graph at the right below shows the fuel efficiency at various speeds for 2400 lb, 3000 lb, and 3600 lb cars. The fuel efficiency is given in miles per gallon (mi/gal).
   a. Fuel efficiency is a function of ? and ?.
   b. At what other speed will a 3000 lb car have approximately the same fuel efficiency as it has when traveling at 25 mi/h?
   c. At what speeds does a 3600 lb car have approximately the same fuel efficiency as a 2400 lb car traveling at 55 mi/h?
   d. Regardless of what a car weighs, at approximately what speed is maximum fuel efficiency reached?

154 Chapter Four

B 37. **Finance** A house bought for \$50,000 in 1980 was sold ... To the nearest percent, what was the annual rate of ap... the value of the house?

38. **Consumer Economics** The price of firewood four years ago was \$140 per cord. Today, a cord of wood costs \$182. To the nearest percent, what has been the annual rate of increase in the cost?

39. **Economics** The *consumer price index* (CPI) is a measure of the average cost of goods and services. The United States government set the index at 100 for the period 1982-1984. In 1988, the index was 118.3. What was the average annual rate of increase (to the nearest tenth of a percent) from 1984 to 1988?

40. **Research** Look in an almanac to find the current consumer price index. Then determine the average annual rate of increase (to the nearest tenth of a percent) for this index since 1984. (See Exercise 39.)

41. **Education** Yearly expenses at a state university have increased from \$14,000 to \$18,500 in the last 4 years. What has been the average annual growth rate in expenses? If this growth rate continues, what will the expenses be 4 years from now?

42. **Research** Find the present cost of something that interests you and its cost several years ago. Find the average annual growth rate.

Solve.

43. **a.** $(4x)^3 = 9^6$ **b.** $3^{4x} = 9^6$

44. **a.** $(4 - x)^{1/2}$ ...

45. $\dfrac{2^{t^2}}{2^t} = 64$

46. $\dfrac{5^{t^2}}{(5^t)^2} = 125$

47. $\sqrt{\dfrac{9^{t+3}}{27^t}} = 81$

48. $\sqrt[3]{\dfrac{8^{t+1}}{16^t}} = 3$...

Factor. (In Exercise 49(a), factor out $a^{1/2}b^{1/2}$.)

49. **a.** $a^{3/2}b^{1/2} - a^{1/2}b^{3/2}$ **b.** $a^{1/2}b^{-1/2} - ...$

50. **a.** $(x - 1)^{1/2} - x(x - 1)^{-1/2}$ **b.** $(x + 1)^{3/2} - ...$

51. **a.** $(x^2 + 1)^{3/2} - x^2(x^2 + 1)^{1/2}$ **b.** $(x^2 + 2)^{1/2}$ ...

52. **a.** $(2x + 1)^{2/3} - 4(2x + 1)^{-1/3}$ **b.** $(1 + x^2)^{-3}$ ...

Exponents ...

---

15. $|x - 4| < 3$ 16. $|x - 7| > 3$

17. $|x + 7| \geq 3$ 18. $|x + 2| < -1$

19. $|x - 8| = 4$ 20. $|x + 3| = 7$

21. $|2x - 4| \leq 5$ 22. $|3x - 9| \geq 9$

23. $|4x + 8| \leq 9$ 24. $|6 - 3x| < 12$

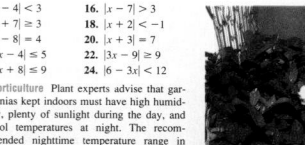

B 25. **Horticulture** Plant experts advise that gardenias kept indoors must have high humidity, plenty of sunlight during the day, and cool temperatures at night. The recommended nighttime temperature range in degrees Fahrenheit is $60° \leq F \leq 65°$. Given that $C = \dfrac{5}{9}(F - 32)$, express the corresponding temperature range in degrees Celsius.

26. **a.** **Writing** Use the geometric definition of $|x|$ and the definition of a combined inequality to write a sentence that gives the meaning of the expression $2 < |x| < 4$.
   b. Solve the inequality $2 < |x| < 4$ and graph its solution.
   c. Solve the inequality $1 \leq |x| < 5$ and graph its solution.

In Exercises 27–32, solve the given inequality and graph its solution.

27. $1 \leq |x - 4| \leq 3$ 28. $2 < |x - 6| \leq 5$

29. $0 \leq |x - 7| < 2$ 30. $0 < |x + 3| < 1$

31. $\dfrac{1}{x} < 2$ (*Hint:* Consider two cases.)   32. $\dfrac{1}{x - 1} > 4$

Use the following definition of $|x|$ to complete Exercises 33–37:

$$|x| = \begin{cases} x & \text{if } x \geq 0. \\ -x & \text{if } x < 0. \end{cases}$$

Solve.

C 33. $|x| + |x - 2| = 2$ 34. $|x| + |x - 2| > 5$

35. Show that $|ab| = |a| \cdot |b|$ when (1) $a > 0$ and $b > 0$; (2) $a < 0$ and $b > 0$; (3) $a > 0$ and $b < 0$; and (4) $a < 0$ and $b < 0$.

36. **a.** Give three examples illustrating the triangle inequality:
   $$|a + b| \leq |a| + |b|$$
   b. **Investigation** In Chapter 11 we will define the absolute value of a complex number as follows: $|x + yi| = \sqrt{x^2 + y^2}$. Decide whether the triangle inequality holds if a and b are complex numbers.

37. Use the triangle inequality stated in Exercise 36(a) to prove:
   a. $|a - b| \leq |a| + |b|$ **b.** $|a| - |b| \leq |a - b|$

Inequalities **99**

Applications

Earth Science

- Conservation
- Geography
- Geology
- Meteorology
- Oceanography

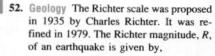

C **52. Geology** The Richter scale was proposed in 1935 by Charles Richter. It was refined in 1979. The Richter magnitude, R, of an earthquake is given by,

$$R = 0.67 \log (0.37E) + 1.46,$$

where E is the energy in kW · h released by the earthquake.
a. Show that $E = 2.7 \cdot 10^{(R - 1.46)/0.67}$.
b. Show that if R increases by 1 unit, E increases by a factor of about 31.

53. If $9^{(9^9)}$ is multiplied out and is typed on a strip of paper, 3 digits per centimeter, about how many kilometers long would the paper be? (*Hint:* The common logarithm of a number can tell you how many digits the number has.)

The San Andreas fault

54. Suppose that all you know about a function f is that $f(ab) = f(a) + f(b)$ for all positive numbers a and b.
a. Find $f(1)$.
b. Prove that $f(a^2) = 2f(a)$ and $f(a^3) = 3f(a)$. What generalization does this suggest?
c. Prove that $f(\sqrt{a}) = \frac{1}{2}f(a)$ and $f(\sqrt[3]{a}) = \frac{1}{3}f(a)$. What generalization does this suggest?
d. Prove that $f\left(\frac{1}{b}\right) = -f(b)$.
e. Prove that $f\left(\frac{a}{b}\right) = f(a) - f(b)$.
f. Try to find a function f that satisfies the original equation.
g. If $f(10) = 1$, find the values of x for which $f(x) = 2$ and $f(x) = 3$.

5-7 Exponential Equations; Changing Bases

Objective To solve exponential equations and to change logarithms from one base to another.

An **exponential equation** is an equation that contains a variable in the exponent. Here are exponential equations you can solve from Section 5-2.

$$2^{t-3} = 8 \qquad 9^{2t} = 3\sqrt[3]{3}$$

These exponential equations are special because both sides of each equation can easily be expressed as powers of the same number. Usually exponential equations cannot be solved this way. In this section, you will see how to use logarithms to solve exponential equations involving a variable such as time, t.

Exponents and Logarithms **203**

WRITTEN EXERCISES

In Exercises 1–4, state any errors that you think might occur in the sampling situation discussed.

A **1.** A senator explains a vote in favor of a new bill by stating that 60% of the mail received favored the bill.

2. A city newspaper asks readers to mail in a form on which they can choose one of three ways to finance improvements to the zoo. Based on the responses, the newspaper reports that financing the improvements by raising the admission charge is the first choice of the majority of the citizens.

3. A radio talk show host invites listeners to telephone the station and talk about their feelings on a proposed highway to be built in their county.

4. A newspaper reporter randomly stops people going in a grocery store and asks, "Does your family use newspaper coupons?"

5. In Farmington High School, there are 360 ninth- or tenth-grade students and 320 eleventh- or twelfth-grade students. A poll shows that 12 out of 40 ninth- or tenth-grade students intend to vote for Lahey as student council president. It also shows that 24 out of 40 eleventh- or twelfth-grade students intend to vote for Lahey. Estimate the percent of the student body favoring Lahey.

6. Twenty-five percent of a city's employees live in the city and 75% live in the surrounding suburbs. The mayor conducts a small survey asking 10 employees who live in the city and 10 who live in the suburbs if they would prefer increases in pay or increases in benefits. Eight of the 10 city dwellers and four of the 10 suburbanites preferred an increase in pay. Estimate the percent of city employees preferring an increase in pay to an increase in benefits.

B **7.** Estimate the total number of years of service of 200 factory workers based on a sample of size $n = 5$ where the years of service reported are 15, 8, 20, 5, 12.

8. Manufacturing In one hour, a factory packaged 150 boxes of light bulbs. Estimate the total number of defective light bulbs packaged that hour based on a sample of 8 boxes containing the following numbers of defective bulbs: 5, 3, 5, 1, 2, 9, 4, 3.

9. Ecology The State Conservation Department wants to estimate the number of deer in a mountainous area. It captures and tags 80 deer and then releases them. Later it captures 156 deer and finds that 12 are tagged. Estimate the number of deer in the area.

Statistics **6**

Life Science

- Agriculture
- Animal Science
- Biology
- Ecology
- Medicine
- Nutrition
- Physiology

---

8. Investigation Draw two vectors **a** and **b**. Sketch $\mathbf{a} - \mathbf{b}$, $3(\mathbf{a} - \mathbf{b})$, $3\mathbf{a} - 3\mathbf{b}$. What do you notice?

9. Navigation A ship travels 200 km west from port and then 240 km due south before it is disabled. Illustrate this in a vector diagram. Use trigonometry to find the course that a rescue ship must take from port in order to reach the disabled ship.

10. Physics On graph paper, make a diagram that illustrates a force of 8 N and a force of 6 N west acting on a body. Illustrate the resultant sum of the two forces and estimate its strength. Then, using trigonometry, determine the approximate direction (as a number of degrees west of north) of this force.

11. Aviation On graph paper, make a diagram that illustrates the velocity of an airplane heading east at 400 knots. Illustrate a wind velocity of 50 knots blowing toward the northeast. If the airplane encounters this wind, illustrate its resultant velocity. Estimate the resultant speed and direction of the airplane. (The direction is the angle the resultant makes with due north measured clockwise from north.)

12. a. Sports A swimmer leaves point A swimming south across a river at 2 km/h. The river is 4 km wide and flows east at the rate of 1 km/h. Make a vector diagram showing her resultant velocity.
b. Calculate her resultant speed.
c. How long will it take her to swim across the river? How far east does she swim?

13. Aviation On graph paper, make a vector diagram showing an airplane heading southwest at 600 knots and encountering a wind blowing from the west. Show the plane's resultant velocity when the wind blows at **(a)** 30 knots, **(b)** 60 knots, and **(c)** 90 knots.

14. Navigation On graph paper, make a vector diagram showing a motorboat heading east at 10 knots. Add to your diagram a vector representing a current moving southeast. Show the boat's resultant velocity if the current moves at **(a)** 2 knots, **(b)** 3 knots, and **(c)** 4 knots.

15. a. Navigation Make a diagram showing the result of sailing a ship 3 mi on a course of $040°$ followed by sailing it 8 mi on a course of $100°$.
b. From your diagram, estimate the distance of the ship from its starting point.
c. Find the exact distance of the ship from its starting point by using the law of cosines.

424 *Chapter Twelve*

Physical Science

- Astronomy
- Aviation
- Chemistry
- Civil Engineering
- Electronics
- Engineering
- Mechanics
- Navigation
- Optics
- Physics
- Space Science
- Surveying
- Telecommunications
- Thermodynamics

Problem Solving

"Did you ever try to solve a problem but didn't know where to begin?"

1-7 Quadratic Functions and Their Graphs

Objective To define and graph quadratic functions.

The graph of the **quadratic function** $f(x) = ax^2 + bx + c$, where $a \neq 0$, is the set of points (x, y) that satisfy the equation $y = ax^2 + bx + c$. This graph is a *parabola*, a curve that can be seen, for example, in the cables of a suspension bridge and in the path of a thrown ball. Parabolas can also be defined geometrically, as you will see in Chapter 6.

 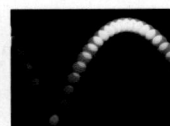

If a graph has an *axis of symmetry*, then when you fold the graph along this axis, the two halves of the graph coincide. The graph of a quadratic function has a vertical axis of symmetry, or **axis**. The **vertex** of the parabola is the point where the axis of symmetry intersects the parabola. If $a > 0$, the parabola opens upward, and the function has a minimum value. If $a < 0$, the parabola opens downward, and the function has a maximum value. The bigger $|a|$ is, the narrower the parabola is. In the figure at the far right below, the graph of $y = 3x^2$ is narrower than the graph of $y = x^2$.

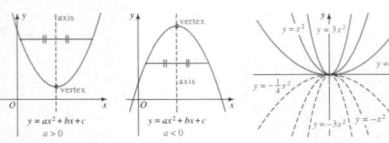

The y-intercept of a parabola with equation $y = ax^2 + bx + c$ is c. The x-intercepts are the real roots of $ax^2 + bx + c = 0$. Since the quadratic equation may have two, one, or no ... $b^2 - 4ac$, we have ... shown in the diagra...

Sections

include mathematical modeling to help you describe real-world situations.

Activity

Fasten the ends of a string to a piece of cardboard with thumbtacks. Make sure the string has some slack. Keeping the string taut, draw a curve on the cardboard as shown. Describe the curve traced by the pencil point P. Repeat the experiment by moving the tacks farther apart or closer together.

If $F_1(c, 0)$ and $F_2(-c, 0)$ are two fixed points in the plane and a is a constant, $0 < c < a$, then the set of all points P in the plane such that
$$PF_1 + PF_2 = 2a$$
is an **ellipse**. This is the *geometric definition* of an ... F_1 ... F_2 are called the **foci** of the ... of *focus*.) In the Activ... cks are the foci and the

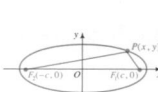

... on for the ellipse with foci $F_1(c, 0)$ and $F_2(-c, 0)$, we let ... the ellipse, then express PF_1 and PF_2 in terms of x, y, and ... $2a$, we have:

$$\sqrt{\cdots} + \sqrt{(x - (-c))^2 + y^2} = 2a$$
$$\sqrt{(x + c)^2 + y^2} = 2a - \sqrt{(x - c)^2 + y^2}$$
$$\cdots y^2 = 4a^2 - 4a\sqrt{(x - c)^2 + y^2} + [(x - c)^2 + y^2]$$
$$\cdots y^2 = 4a^2 - 4a\sqrt{(x - c)^2 + y^2} + x^2 - 2cx + c^2 + y^2$$
$$\cdots cx = 4a^2 - 4a\sqrt{(x - c)^2 + y^2}$$

... ranging terms:
$$cx - a^2 = -a\sqrt{(x - c)^2 + y^2}$$
$$c^2x^2 - 2ca^2x + a^4 = a^2[(x - c)^2 + y^2]$$
$$= a^2x^2 - 2ca^2x + a^2c^2 + a^2y^2$$
$$c^2x^2 + a^4 = a^2x^2 + a^2c^2 + a^2y^2$$
$$a^4 - a^2c^2 = a^2x^2 - c^2x^2 + a^2y^2$$
$$a^2(a^2 - c^2) = (a^2 - c^2)x^2 + a^2y^2$$
... r than c, substitute $b^2 = a^2 - c^2$:
$$a^2b^2 = b^2x^2 + a^2y^2$$
... have: $\dfrac{x^2}{a^2} + \dfrac{y^2}{b^2} = 1$, where $b^2 = a^2 - c^2$

Activities

get you involved and help build understanding.

A hands-on activity where you construct an ellipse.

10-4 Solving Trigonometric Equations

Objective To use identities to solve trigonometric equations.

The acceleration of a body falling toward the Earth's surface is called acceleration due to gravity, often denoted by g. In theoretical physics, g is usually considered constant. However, g is not actually constant but varies slightly with latitude. A good approximation to the value of g can be found by using the following formula, which expresses g in terms of θ, the latitude in degrees.

Gravity acting on water creates this waterfall.

$$g \approx 9.78049(1 + 0.005288 \sin^2 \theta - 0.000006 \sin^2 2\theta)$$

For example, if you live in Chicago, which has a latitude of 42°N, $\sin \theta \approx 0.6691$ and $\sin 2\theta \approx 0.9945$. Therefore, $g \approx 9.8036$ m/s².

As you can see from this example, some problems involve trigonometric equations that have multiples of angles or numbers. The following suggests two methods that may be helpful in solving such equations. The first method gives a graphical method using a graphing calculator or computer, and the second method gives ways for solving the equation algebraically.

Methods for Solving the Trigonometric Equation $f(x) = g(x)$

Method 1

Use a graphing calculator or computer to graph $y = f(x)$ and $y = g(x)$ on the same set of axes. Use the zoom or trace feature to find the x-coordinates of any intersection points of the two graphs.

Method 2

Use the following guidelines.

a. It may be helpful to draw a quick sketch of $y = f(x)$ and $y = g(x)$ to see roughly where the solutions are.

b. If the equation involves functions of $2x$ and x, transform the functions of $2x$ into functions of x by using identities.

c. If the equation involves functions of $2x$ only, it is usually better to solve for $2x$ directly and then solve for x.

d. Be careful not to lose roots when you divide both sides of an equation by a function of the variable. Review the discussion about losing roots on pages 32–33.

Sections

present alternative methods of solving problems.

B **23.** If 4 is a zero of $f(x) = 3x^3 + kx - 2$, find the value of k.

24. If $2i$ is a zero of $f(x) = x^4 + x^2 + a$, find the value of a.

25. A quadratic polynomial P has leading coefficient -2, a constant term of 6, and no linear term. Find the zeros of P.

26. The leading coefficient of a cubic polynomial P is 2, and the coefficient of the linear term is -5. If $P(0) = 7$ and $P(2) = 21$, find $P(3)$.

27. a. What are the zeros of the quadratic function $f(x) = 2(x - 1)(x - 4)$?
   b. Find a quadratic function with zeros 2 and 3.
   c. Find a cubic function with zeros 0, 2, and 3.

28. a. What are the zeros of the quadratic function $f(x) = 3(x + 1)(x - 2)$?
   b. Find a quadratic function with zeros 3 and -4.
   c. Find a quartic function with zeros ± 3 and ± 4.

29. If $f(x) = 7x + 2$, find: **a.** $f(9.2) - f(8.2)$ **b.** $f(x + 1) - f(x)$

30. If $g(x) = 3 - 8x$, find: **a.** $g(6.25) - g(4.25)$ **b.** $g(x + 2) - g(x)$

31. If $f(x) = mx + k$, show that the value of $\dfrac{f(x + h) - f(x)}{h}$, where $h \neq 0$, does not depend on x or h. Interpret this result graphically.

32. If $f(x) = x^2$ and $h \neq 0$, find the value of $\dfrac{f(x + h) - f(x)}{h}$. Is the value of this expression independent of the values of x and h, as it was for the function of Exercise 31?

33. *Investigation* Multiply several pairs of nonzero polynomials. What is the relationship between the degree of the product and the degrees of the factors? How can you use this relationship to justify saying that the polynomial 0 has no degree, even though all other constant polynomials have degree 0?

C **34. a.** Consider the following table of values for the quadratic function $f(x) = x^2 + 2x + 3$.

x	0	1	2	3	4	5
$f(x)$	3	6	11	18	27	38
differences		3				

   What pattern do you observe in t
   b. Make a difference table similar
   $g(x) = 2x^2 - 3x - 1$. What patt
   c. Do the differences in the values
   like those you observed in parts
   difference table.

35. *Investigation* Experiment with diff
   (See Exercise 34.) Can you detect a

Investigations

help you discover relationships and connections in mathematics.

PROJECT

Chaos in the Complex Plane

If you've explored some of the dynamical systems presented in Sections 19-5 and 19-6, then you're probably well aware that iterating a function—no matter how simple its rule—can produce unexpectedly complicated orbits for certain seeds. The results are even more unusual and exciting when a function's domain and range, which until now were sets of real numbers, are allowed to be sets of complex numbers instead. In fact, iterating a function with a complex domain and range can produce exceptionally intricate and beautiful graphic images, like the one shown.

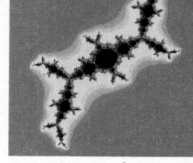

Filled-in Julia set of $f(z) = z^2 - 0.112 + 0.861i$

Materials:
a computer or programmable calculator with graphics mode (preferably color graphics)   a printer that can print a graphics screen

Complex Orbits

Suppose we have a function $f(z)$ where z is a complex number. Iterating $f(z)$ using some complex seed z_0 gives the orbit of z_0:

$$z_0$$
$$z_1 = f(z_0)$$
$$z_2 = f(z_1) = f^2(z_0)$$
$$z_3 = f(z_2) = f^3(z_0)$$
$$\vdots$$

For example, the orbit of $z_0 = 1 +$
$f(z) = z^2$ is:

$$z_0 = 1 + 2i$$
$$z_1 = (1 + 2i)^2 = -3 + 4i$$
$$z_2 = (-3 + 4i)^2 = -7 - 24i$$
$$z_3 = (-7 - 24i)^2 = -527 +$$
$$\vdots$$

The calculation of such orbits is best
tor. At the top of the next page is a B
$f(z) = z^2$ for any seed z_0.

Projects

provide you with opportunities to explore interesting topics through experiments.
Discover how patterns can produce exceptionally intricate and beautiful graphic images.

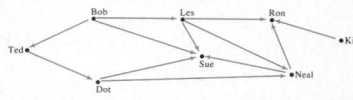 For some parts of the following exercises, you may wish to use computer software or a calculator that performs matrix operations.

7. a. Ship A sends messages to Ship B. Ship B sends and receives messages from Ships C and E. Ship D sends and receives messages from Ship C. Ship E receives messages from Ships A and C. Draw this network.
   b. Write the matrix M that models this communication network, labeling rows and columns alphabetically.
   c. Find M^2. Explain what the element in the fifth row, second column means.
   d. Find the matrix that represents the number of ways messages can be sent from one ship to another using at most one relay.
   e. Reason from the network diagram what the last element in row 4 of M^3 is. (You do not need to calculate M^3.)

8. a. Suppose radios of varying quality are used by forest rangers to communicate with each other in a large national park. Make a diagram showing five rangers, some pairs having two-way communication, some pairs having one-way communication, and some having no communication.
   b. Write a matrix M to model the communication between the forest rangers. Find $M + M^2$. What information does this matrix give you?

B **9.** The diagram below models a rumor network.

Bob — Les — Ron
Ted
Sue — Kim
Dot — Neal

If a person has only outgoing arrows, then that person is a transmitter of the rumor. A person having only incoming arrows is a receiver of the rumor. A person having both outgoing and incoming arrows is a relay point for the spreading of the rumor.
   a. Identify the transmitters, receivers, and relays for this network.
   b. How can you identify a transmitter by looking at the corresponding rumor matrix? How can you identify a receiver?

10. Let M be a matrix that illustrates communication among several Ham radio operators. If M^2 contains no zeros, what can you conclude?

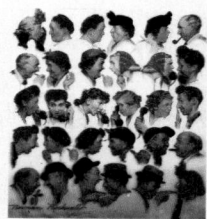

Current technology

is used to solve complex real-world problems.

Reasoning

"What do you think?"
"Prove it to me!"
"You need to decide."

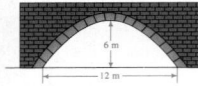
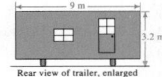

42. **Transportation** A hauling company needs to determine whether a large house trailer can be moved along a highway that passes under a bridge with an opening in the shape of a parabolic arch, 12 m wide at the base and 6 m high in the center. If the trailer is 9 m wide and 3.2 m tall (measured from the ground to the top of the trailer), will it fit under the bridge?

6 m

12 m

9 m

3.2 m

Rear view of trailer, enlarged

8 Quadratic Models

Objective To model real-world situations using quadratic functions.

...entists are fascinated by the ability of birds to travel great distances without ...pping. Using wind tunnels to monitor parakeets' oxygen consumption and car-... dioxide production in flight, they have investigated the rate at which parakeets ...end energy in level flight. The data in the table below show the number of ...ries burned per gram hour (the number of calories needed to move each gram ...body weight for one hour) for three different flight speeds.

Speed (in mi/h)	12	22	30
Calories burned	152	105	165

If we want to predict the bird's energy ...at some other flight speed, we will want ...ind a good mathematical model based on ...se data. The information in the table is ...tted on the graph in figure (a). Note that ...the flight speed increases, the energy ...ended decreases and then increases. ...arly, a linear model would not be suita-... to use. We can find many kinds of ...ves that pass through these data points. ...is process is called *curve-fitting*, which ...will discuss in more detail in Chapter ...) One possible curve is a parabola with ...ation $C(x) = ax^2 + bx + c$. To find the ...ues of a, b, and c, substitute the data ...m the graph into this equation, as shown ...the next page.

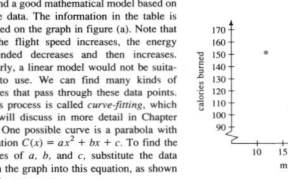

(a)

Linear and Quadratic Functions **43**

---

17. **Writing** In each of Exercises 15 and 16 you made a prediction. Which of these predictions do you think has a better chance of being correct? Write a few sentences explaining your reasoning.

The table shows statistics for the ten players of a college basketball team in its opening game of the season. Use the data to complete Exercises 18–20.

minutes played	32	30	25	21	21	18	16	15	12	10
points scored	22	15	11	8	6	4	8	10	4	6
fouls committed	4	3	5	2	3	2	2	1	0	1

Use computer software to make a scatter plot, draw the least-squares line, find its equation, and find the correlation coefficient for each set of data.

	x	y
18.	minutes	points
19.	minutes	fouls
20.	fouls	points

21. **Writing** Refer to the results of Exercise 20. Does the positive correlation between fouls committed and points scored suggest a cause-and-effect relationship between fouls and points? As a coach, would you advise playing more defensively and thus committing more fouls as a way of increasing your players' points scored? Write a short paragraph explaining your reasoning.

22. The diagram at the right shows a scatter plot of a set of data and a line with slope m that contains the point $(\bar{x}, \bar{y})$.

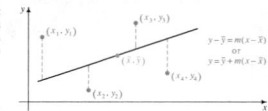

$y - \bar{y} = m(x - \bar{x})$
or
$y = \bar{y} + m(x - \bar{x})$

a. For each point (x_i, y_i), let d_i represent the vertical distance between the point and the line, that is, $d_i = |y_i - y|$. Show that:

$$d_i = |(y_i - \bar{y}) - m(x_i - \bar{x})|$$

690 *Chapter Eighteen*

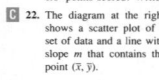

Decision making

Discuss the validity of a conjecture using math

---

13. In the *random-number table* shown below, the probability that each digit, 0–9, occurs in any given position is 0.1. We can use the table to perform *Monte Carlo simulations* in order to find certain probabilities empirically.

```
49487 52802  28667 62058  87822 14704  18519 17889  45869 14454
29480 91539  46317 84803  86056 62812  33584 70391  77749 64906
25252 97738  23901 11106  86864 55808  22557 23214  15021 54268
02431 42193  96960 19620  29188 05863  92900 06836  13433 21709
69414 89353  70724 67893  23218 72452  03095 68333  13751 37260
```

For example, to use Monte Carlo simulations to find the probability of getting 0 "heads" in 3 tosses of a coin, first associate the digits 0–4 with the outcome "heads" and the digits 5–9 with the outcome "tails." Then the first row of the random-number table represents 16 simulations of 3 tosses of a coin.

```
494 87  5 280 2  28 667  620 58  8 782 2  14 704  185 19  1 788 9  45 869  144 54
HTH TT   T HTH TH   HTHT T   THHTT   H THH TT   HTTH T   HHT HH   H TTTT T   HHTT HT   THH TH
```

a. Of the 16 simulations, 5 produce 0 "heads." Thus, based on 16 simulations, $P(0$ "heads"$) = 0.3125$. Compare this probability to the theoretical probability of getting 0 "heads" in 3 tosses of a coin (see Exercise 1(d)).

b. Begin with the last 2 digits of the first row and perform 14 more simulations. Using all 30 simulations, find the probability of getting 0 "heads."

c. How many simulations can be done using the whole table? Based on these simulations, what is the probability of getting 0 "heads"?

d. **Discussion** When you use Monte Carlo simulations to approximate the theoretical probability of an event, will increasing the number of simulations always improve the accuracy of the approximation? Explain.

14. a. **Discussion** In order to use Monte Carlo simulations (see Exercise 13) to find the probabilities asked for in Exercises 2–12, how should the digits 0–9 be assigned to the possible outcomes for each exercise? (*Hint:* It may be helpful to eliminate some digits. Cross off all occurrences of those digits in the table before actually doing a simulation.)

b. Perform Monte Carlo simulations for Exercise 2 and Exercise 7. Compare the probabilities you find to the theoretical probabilities found using the binomial probability theorem.

15. a. **Sports** A basketball player's free-throw percent is .750. What is the probability that she scores on exactly 4 of her next 5 free throws?

b. **Discussion** Using the binomial probability theorem to complete part (a) assumes that the probability of a successful free throw is always .750. Is this assumption valid? Explain.

Probability **617**

Follow a logical argument

Suppose that a small TV manufacturing company produces console and portable TV's using three different machines, A, B, and C. The table below shows how many hours are required on each machine per day in order to produce a console TV or a portable TV.

Machine	Console	Portable	Hours available
A	1 h	2 h	16
B	1 h	1 h	9
C	4 h	1 h	24

These requirements for the different machines can be described by the following inequalities, where x represents the number of console TV's and y represents the number of portable TV's:

a. $x \geq 0$
b. $y \geq 0$ $\left\{$ The number of TV's cannot be negative.

c. $x + 2y \leq 16$ $\left\{$ Machine A needs 1 hour for each console TV and 2 hours for each portable. Thus, for x console TV's and y portable TV's Machine A needs $x + 2y$ hours. Since this machine is available for at most 16 hours a day, $x + 2y \leq 16$.

d. $x + y \leq 9$
e. $4x + y \leq 24$ $\left\{$ The last two inequalities are similar to the one above. Machines B and C are available for at most 9 hours and 24 hours respectively.

The possible solutions to any problem subject to these constraints would have to satisfy all five inequalities. The easiest way to find these values is to find the region that is formed by the graphs of the system of inequalities. A step-by-step procedure is shown below.

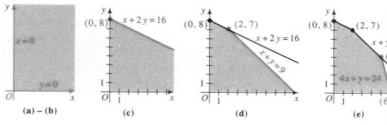

(a) – (b)   **(c)**   **(d)**   **(e)**

Any point (x, y) in graph (e) is called a *feasible solution*. The set of these points is called the *feasible region*.

Inequalities

Make and test conjectures

You may find it helpful to have a graphing calculator to complete Exercises 35 and 36.

B 35. a. Predict how the graph of $y = \log_b\left(\frac{1}{x}\right)$ is related to the graph of $y = \log_b x$.
  b. Check your prediction by graphing $y = \log x$ and $y = \log\left(\frac{1}{x}\right)$ on a single set of axes.

36. a. Predict how the graph of $y = \log_b x^2$ is related to the graph of $y = \log_b x$.
  b. Check your prediction by graphing $y = \log x$ and $y = \log x^2$ on a single set of axes.

37. a. If $f(x) = \log_2 x$, show that $f(2x) = f(x) + 1$.
  b. Part (a) shows that horizontally shrinking the graph of f by a factor of 2 is equivalent to __?__.

38. a. If $f(x) = \log_3 x$, show that $f\left(\frac{x}{3}\right) = f(x) - 1$.
  b. Make a statement similar to that in part (b) of Exercise 37.

39. Suppose that $\log y = ax + b$, where a and b are real numbers and a is nonzero. Express y in terms of x. What kind of expression defines $\log y$ in terms of x? What kind of expression defines y in terms of x?

40. Conservation The expiration time T of a natural resource is the time remaining until it is all used. If one assumes that the current growth rate of consumption remains constant, then the expiration time in years is given by

$$T = \frac{1}{r} \ln\left(\frac{rR}{C} + 1\right),$$

where C = current consumption, r = current growth rate of consumption, and R = resource size. Suppose that the world's consumption of oil is growing at the rate of 7% per year ($r = 0.07$) and current consumption is approximately 17×10^9 barrels per year. Find the expiration time for the following estimates of R.

10^9 barrels of remaining

10^9 barrels of remaining (shale oil)

---

9. Biology A colony of bacteria decays so that the population t days from now is given by $A(t) = 1000\left(\frac{1}{2}\right)^{t/4}$.
  a. What is the amount present when $t = 0$?
  b. How much will be present in 4 days?
  c. What is the half-life?

10. Discussion Can the data below be described as exponential growth?

x	0	1	2	3	4
y	4	4.4	4.8	5.3	5.8

Justify your answer by using a graph, an equation, or a logical argument. Can you think of a situation that the data might describe?

WRITTEN EXERCISES

Evaluate each expression with a calculator.

A 1. 6^π and π^6 **2.** $3.6^{\sqrt{2}}$ and $\sqrt{2}^{3.6}$

Find an exponential function having the given values.

3. $f(0) = 3$, $f(1) = 15$ **4.** $f(0) = 5$, $f(3) = 40$
5. $f(0) = 64$, $f(2) = 4$ **6.** $f(0) = 80$, $f(4) = 5$

7. Physics The half-life of a radioactive isotope is 4 days. If 3.2 kg are present now, how much will be present after:
  a. 4 days?   **b.** 8 days?   **c.** 20 days?   **d.** t days?

8. Physics The half-life of radium is about 1600 years. If 1 kg is present now, how much will be present after:
  a. 3200 years?   **b.** 16,000 years?   **c.** 800 years?   **d.** t years?

9. Physics The table shows the amount $A(t)$ in grams of a radioactive element present after t days. Suppose that $A(t)$ decays exponentially.

t (days)	0	2	4	6	8	10
$A(t)$	320	226	160	115	80	57

  a. What is the half-life of the element?
  b. About how much will be present after 16 days?
  c. Find an equation for $A(t)$.

Construct and present valid arguments

Reasoning

Communication

"Have you ever had trouble explaining something because you didn't fully understand it?"

Discussion

Further your understanding of math by discussing it with others.

Reading

Learn about a variety of careers that use mathematics.

Research

Investigate interesting facts and applications of mathematics.

CLASS EXERCISES

1. As in Example 1, suppose a die is rolled. Find each probability.
   a. P(perfect square) **b.** P(factor of 60) **c.** P(negative number)

2. If a card is drawn at random from a standard deck of 52 cards, what is the probability of getting:
   a. the queen of hearts? **b.** a heart? **c.** a queen?
   d. a red card? **e.** a face card? **f.** a red face card?

3. If two dice are rolled, what is the probability that both show the same number?

4. If two dice are rolled, find the probability of getting:
   a. a sum of 3
   b. a sum of 4
   c. a sum of 3 or 4

5. **Meteorology** If the probability of rain tomorrow is 40%, what is the probability of no rain tomorrow?

6. **Occupational Safety** If the probability of no accidents in a manufacturing plant during one month is 0.82, what is the probability of at least one accident?

7. *Discussion* In the solution of Example 3, another possible sample space is the set of the 11 possible sums for the two dice:

   $$S = \{2, 3, 4, 5, 6, 7, 8, 9, 10, 11, 12\}$$

   Since two of these 11 possibilities correspond to a sum of 9 or 10, some people might be tempted to say that the probability of a sum of 9 or 10 is $\frac{2}{11}$. Discuss why this reasoning is incorrect.

8. A penny, a nickel, and a dime are each tossed.
   a. Does the set $S = \{0, 1, 2, 3\}$, which gives the four different numbers of "heads" that could come up, satisfy the definition of a sample space?
   b. What is wrong with reasoning that since one of the four sample points in part (a) corresponds to 2 "heads," then $P(2 \text{ "heads"}) = \frac{1}{4}$?

9. A card is picked at random from a standard deck. Explain why the set $S = \{\text{club, spade, red card, face card}\}$ is *not* a sample space.

10. Comment on the following reasoning:
    There are 3 states whose names begin with the letter C (California, Colorado, and Connecticut); call them "C-states." Likewise, there are 3 "O-states" (Ohio, Oklahoma, and Oregon). Thus, if a person is chosen at random from the U.S. population, that person has the same probability of being from a "C-state" as from an "O-state."

Careers in Genetics and Statistics

The twentieth century has seen unprecedented progress in our ability to understand, control, and cure disease. Many fields of study contribute to medical research, and the methods of data analysis play an important role.

GENETICS

Genetics, the study of heredity, encompasses a number of areas of research. Examples include population genetics, which uses statistical analysis to find patterns in gene distribution and the occurrence of genetic diseases; microbial [genetics], which explores the process of mutation by studying rapidly multiplying microorganisms; and cytogenetics, which looks at hereditary activity at the cellular level.

Agnes Stroud-Lee, a University of Chicago Ph.D., has done research in cytogenetics at the Los Alamos National Laboratory. By observing the effects of radiation and toxic chemicals on chromosomes, she sought clues to abnormalities in genetic material. Her results have improved our understanding of birth defects and cancer. "I have made inroads," she says, ". . . other scientists can use my work to further their pursuit of science." Ms. Stroud-Lee retired in 1979.

Agnes Stroud-Lee

STATISTICS

Joe Fred Gonzalez, Jr.

Statisticians are experts in the design of experiments, surveys, and questionnaires used to generate accurate and useful data, and in the meaningful analysis of data. Their skills are indispensable to medical and health research.

The National Center for Health Statistics is an important source of information for researchers and policymakers. NCHS analysis of data reveals connections between such varied health factors as socioeconomic status, type of health care used, nutrition, exposure to risks, and age.

Joe Fred Gonzalez, Jr., a mathematical statistician for NCHS whose work includes developing estimation procedures and national sample survey designs, always found math "challenging and intellectually stimulating." His desire to understand applications of mathematics led him to an M.S. degree in statistics from the George Washington University. Believing in the value of education, Mr. Gonzalez is also a part-time university and college professor.

Careers in Genetics and Statistics   **713**

Find two angles, one positive and one negative, that are coterminal with each given angle.

17. **a.** $500°$ **b.** $-60°$ **c.** $\frac{\pi}{4}$ **d.** $-\frac{2\pi}{3}$
18. **a.** $1000°$ **b.** $-100°$ **c.** $\frac{4\pi}{3}$ **d.** $-\frac{\pi}{6}$
19. **a.** $28.5°$ **b.** $116.3°$ **c.** $-60.4°$ **d.** $-315.3°$
20. **a.** $38.4°$ **b.** $127.6°$ **c.** $-50.8°$ **d.** $-320.7°$
21. **a.** $360°30'$ **b.** $-90°40'$ **c.** $3°21'$ **d.** $115°15'$
22. **a.** $180°20'$ **b.** $-270°30'$ **c.** $11°44'$ **d.** $172°11'$

23. Give an expression in terms of the integer n for the measure of all angles that are coterminal with an angle of $29.7°$.

24. Give an expression in terms of the integer n for the measure of all angles that are coterminal with an angle of $-116°10'$.

Each of Exercises 25–30 gives the speed of a revolving gear. Find (a) the number of degrees per minute through which each gear turns and (b) the number of radians per minute. Give answers to the nearest hundredth.

25. 35 rpm 26. 27 rpm 27. 2.5 rpm
28. 6.5 rpm 29. 14.6 rpm 30. 19.8 rpm

31. *Reading* On page 257, you were told that when a car with wheels of radius 14 in. is driven at 35 mi/h, the wheels turn at an approximate rate of 420 rpm. Show how to obtain this rate of turn.

32. **Recreation** Suppose you can ride a bicycle a distance of 5 mi in 15 min. If you ride at a constant speed and if the bicycle's wheels have diameter 27 in., find the wheels' approximate rate of turn (in rpm).

33. **Research** Consult an encyclopedia or an atlas to see how points on a world map are located by using *latitude* and *longitude* coordinates given in degrees, minutes, and seconds.
    a. If you travel south from a given point on Earth, about how many miles do you have to go to traverse an angle of 1°?
    b. Explain why your answer to part (a) might be different if you travel west instead of south.

34. **Research** Consult a book of astronomy or a star atlas to see how stars on a celestial map are located by using angles of *right ascension* and *declination*. Describe how each of these angles is measured, and give examples.

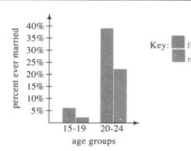

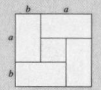

22. Geography The table below gives data on the percent of the U.S. population ever married.

Age group	15–19	20–24	25–29	30–34	35–44	45–54
Percent of females ever married	6	39	71	84	92	95
Percent of males ever married	2	22	57	75	89	94

a. At the right is part of a comparative histogram showing two side-by-side bars for each of the age groups. Complete the histogram.

b. *Writing* Write a paragraph explaining what the data show.

Key: female, male

percent ever married / age groups (15-19, 20-24)

23. Nutrition The table below gives data on the recommended energy intake, in kilocalories (Cal), for average females and males in various age groups.

Age group	11–14	15–18	19–22	23–50	51–75	≥76
Energy needs (in Cal) for females	2200	2100	2100	2000	1800	1600
Energy needs (in Cal) for males	2700	2800	2900	2700	2400	2050

a. Draw a comparative histogram showing two side-by-side bars for each of the age groups. (See Exercise 22 for an example of such a histogram.)

b. *Writing* Write a paragraph explaining what the data show.

24. For a group of 10 teenagers, the mean age is 17.1, the median is 16.5, and the mode is 16. If a 21-year-old joins the group, give the mean, median, and mode for the ages of the 11 people.

Statistics **647**

Writing

Bring together your math skills and your writing skills. Express, in your own words, the meaning and understanding of concepts.

29. a. If $f(x) = x^{2/3}$, find $f'(x)$.

b. For what values of x is $f'(x)$ undefined?

c. Sketch the graph of $f(x)$. (If you use a computer or graphing calculator, you may need to enter the rule for the function as $(x^2)^{1/3}$ to obtain the complete graph.) Explain how the graph of $f(x)$ supports the result of part (b).

30. Repeat Exercise 29 using the function $f(x) = x^{1/3}$.

In Exercises 31–36, find a function that has the given derivative.

31. $f'(x) = 4x^3$ **32.** $f'(x) = 5x^4$ **33.** $g'(x) = 3x^5$

34. $g'(x) = x^7$ **35.** $h'(x) = 3\sqrt{x}$ **36.** $h'(x) = \sqrt[3]{x}$

C 37. a. Use the binomial theorem (page 591) to write the first three terms and the last term in the expansion of $(x + h)^n$.

b. Use part (a) and the definition of $f'(x)$ to prove Theorem 2 on page 760 for positive integral values of n.
(*Note:* The binomial theorem can be generalized to apply to any nonzero real number n. This form of the binomial theorem can then be used to give a general proof of Theorem 2.)

38. Prove Theorem 3 on page 760. (*Hint:* Use the definition of $f'(x)$.)

39. Prove Theorem 4 on page 760. (*Hint:* Use the definition of $f'(x)$.)

COMMUNICATION: *Visual Thinking*

The saying "a picture is worth a thousand words" is often true in mathematics. One well-drawn diagram may be a very convincing argument on its own, without being a formal proof. Diagrams and sketches can offer evidence that a conjecture may be true, and they can offer clues that lead to interesting discoveries.

For example, consider the diagram at the right, where the lengths a and b are unequal positive numbers.

1. Explain how the areas of the squares and rectangles in the diagram allow you to conclude that $(a + b)^2 = (a - b)^2 + 4ab$.

2. Explain how the saying "the whole is greater than its parts" allows you to conclude from the diagram that $(a + b)^2 > 4ab$.

3. By taking the square root of each side of the inequality above, what relationship do you discover between the arithmetic mean, $\frac{a + b}{2}$, and the geometric mean, $\sqrt{ab}$, of a and b?

An Introduction to Calculus

Exercises

Combine your knowledge of math and other subjects to explain interesting problems.

WRITTEN EXERCISES

For Exercises 1–4, suppose a card is drawn from a well-shuffled standard deck of 52 cards. Find the probability of drawing each of the following.

A 1. a. a black card   **b.** a spade   **c.** not a spade

2. a. a black face card   **b.** a black jack   **c.** not a black ja

3. a. a red diamond   **b.** a black diamond   **c.** not a black di

4. a. a jack   **b.** a jack or king   **c.** neither jack n

5. Mr. and Mrs. Smith each bought 10 raffle tickets. Each of their three children bought 4 tickets. If 4280 tickets were sold in all, what is the probability that the grand prize winner is:
a. Mr. or Mrs. Smith?   **b.** one of the 5 Smiths?   **c.** none of the S

6. One of the integers between 11 and 20, inclusive, is picked at random. What is the probability that the integer is:
a. even?   **b.** divisible by 3?   **c.** a prime?

7. New York City is divided into five boroughs: Manhattan, Queens, the Bronx, Brooklyn, and Staten Island. Suppose that a New York City telephone number is randomly chosen.
a. Explain why the probability that it is a Manhattan telephone number is *not* $\frac{1}{5}$.
b. What do you need to know in order to find the correct probability in part (a)?

8. Suppose that a member of the U.S. Senate and a member of the U.S. House of Representatives are randomly chosen to be photographed with the President. Explain why $\frac{1}{50}$ is the probability that the senator is from Iowa, and why $\frac{1}{50}$ is *not* the probability that the representative is from Iowa.

In Exercises 9–12, use the table on page 601, which gives the 36 equally likely outcomes when two dice are rolled. Find the probability of each event.

9. a. Sum is 6.   **b.** Sum is 7.   **c.** Sum is 8.

10. a. Sum is even.   **b.** Sum is 12.   **c.** Sum is less than 12.

11. The two dice show different numbers.

12. The red die shows a greater number than the white die.

Probability **603**

Visual Thinking

Learn to communicate visually through a diagram or picture.

Communication

Technology

"Technology enhances understanding, encourages exploration, and opens the door to many new and fascinating applications."

Technology Logos
– – – – – – – – –

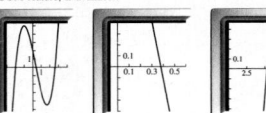 indicates advanced technology suggested to get the full benefit of studying the lesson or exercise.

b. indicates specific exercises for which advanced technology is necessary.

■ Polynomial Equations

2-5 Using Technology to Approximate Roots of Polynomial Equations

Objective *To use technology to approximate the real roots of a polynomial equation.*

You already know how to solve linear and quadratic equations. In this section we will discuss how to solve a polynomial equation of higher degree, such as

$$x^3 - x^2 - 5x + 2 = 0,$$

by using a computer, or a graphing or programmable calculator. In the next section we will see that some polynomial equations can be solved by factoring.

Example 1 Solve $x^3 - x^2 - 5x + 2 = 0$ by using a computer or graphing calculator.

Solution Draw the graph of the polynomial function

$$y = x^3 - x^2 - 5x + 2$$

with a computer or graphing calculator. The x-intercepts of the graph are the roots of the equation. If estimating the x-intercepts is difficult because the graph is not labeled, you could use the TRACE feature, if available. To improve the accuracy of the approximations, you could rescale or use the ZOOM feature, if available.

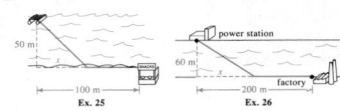

Graph of function for $-3 \le x \le 4$ Enlargement of graph for $0 \le x \le 0.6$ Enlargement of graph for $2.4 \le x \le 2.8$

From the graphs of $y = x^3 - x^2 - 5x + 2$ shown above, we estimate that the roots of $x^3 - x^2 - 5x + 2 = 0$ are $x = -2$, $x = 0.4$, and $x = 2.6$.

In the next example, we will solve the equation $x^3 - x^2 - 5x + 2 = 0$ without drawing a graph. We will instead use the *location principle*, stated at the top of the next page.

Polynomial Functions **75**

---

24. Manufacturing A sheet of metal is 60 cm wide and 10 m long. It is bent along its width to form a gutter with a cross section that is an isosceles trapezoid with 120° angles, as shown at the right.
a. Express the volume V of the gutter as a function of x, the length in centimeters of one of the equal sides. (*Hint*: Volume = area of trapezoid × length of gutter)
b. For what value of x is the volume of the gutter a maximum

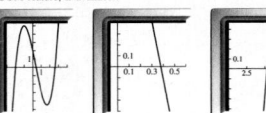 **Part (b) of Exercises 25–29 requires the use of a computer calculator.**

25. From a raft 50 m offshore, a lifeguard wants to swim to shore snack bar 100 m down the beach, as shown at the left below.
a. If the lifeguard swims at 1 m/s and runs at 3 m/s, express th ming and running time t as a function of the distance x s diagram.
b. Use a computer or graphing calculator to find the minimum time.

50 m 60 m power station factory 200 m 100 m
Ex. 25 **Ex. 26**

26. Engineering A power station and a factory are on opposite sides of a river 60 m wide, as shown at the right above. A cable must be run from the power station to the factory. It costs $25 per meter to run the cable in the river and $20 per meter on land.
a. Express the total cost C as a function of x, the distance downstream from the power station to the point where the cable touches the land.
b. Use a computer or graphing calculator to find the minimum cost.

27. Landscaping A rectangular area of 60 m² has a wall as one of its sides, as shown. The sides perpendicular to the wall are made of fencing that costs $6 per meter. The side parallel to the wall is made of decorative fencing that costs $8 per meter.
a. Express the total cost C of the fencing as a function of the length x of a side perpendicular to the wall.
b. Use a computer or graphing calculator to find the minimum cost.

C 28. A cylinder is inscribed in a sphere of radius 1.
a. Express the volume V of the cylinder as a function of the base radius r.
b. Use a computer or graphing calculator to find the radius and height of the cylinder having maximum volume.

164 *Chapter Four*

Functions
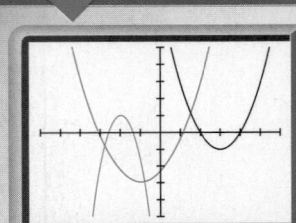

Statistics
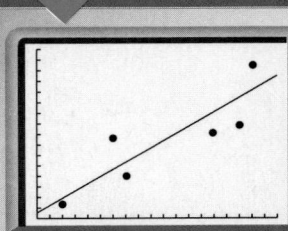

Polar Graphs
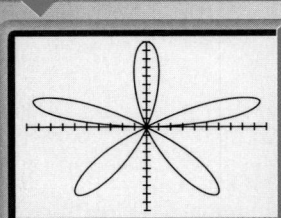

Technology

Using Precalculus Plotter Plus
by David L. Myers

Includes: User's Manual · Worked Examples · Enrichment Topics · Activity Sheets · Answers · Correlations to Advanced Mathematics, Introductory Analysis, and Trigonometry with Applications

Advanced Mathematics
Precalculus with Discrete Mathematics and Data Analysis

Houghton Mifflin

Precalculus Plotter Plus 2-07168
Graphing · Quizzes · Explorations
David L. Myers · Macintosh Computers
(1 MB, System 6.0.5 or later)

© 1993 by Houghton Mifflin Company. All rights reserved. Copies of this software may be made for purposes of execution of this software program provided each copy shows the copyright notice. Macintosh is a registered trademark of Apple Computer, Inc.

TO START: Double-click on Plotter Plus icon.

The diagram on the previous page illustrates the convention [...] placed at the origin with the polar axis coinciding with the positive [...] positive angles measured in the counterclockwise direction. The giv[...] can be used as formulas for converting from one coordinate system [...] Many scientific and graphing calculators have built-in conversion o[...]

Example 2   **a.** Give polar coordinates for the point (3, 4).
b. Give the rectangular coordinates for the point (3, [...]

Solution   **a.** $r = \pm\sqrt{x^2 + y^2} = \pm\sqrt{3^2 + 4^2} = \pm 5$

$\tan\theta = \dfrac{y}{x} = \dfrac{4}{3}$. Thus, $\theta \approx 53.1°$ is one possible val[...]

(3, 4) is in Quadrant I. If $\theta = 53.1°$, then the point is [...]
ray, and so $r = 5$. Therefore, polar coordinates are (5, [...]
possible pair of polar coordinates of the same point is [...]
b. $x = r\cos 30° = 3\cos 30°$ and $y = r\sin 30° = 3\sin 30°$[...]

Thus, rectangular coordinates of the point are $\left(\dfrac{3\sqrt{3}}{2}, \dfrac{3}{2}\right)$

The graph of a polar equation can be drawn by a graphing calculator, [...]
described below, provided the calculator has a polar mode or a paramet[...]
mode. Many calculators use t instead of θ.

If the calculator has polar mode, enter the equation: $r = f(t)$
If the calculator has parametric mode, enter the equations: $x = r\cos t = f(t)$
$y = r\sin t = f(t)$

The two equations for x and y are called *parametric equations*. We will study them [...]
in greater detail in Chapter 12, where t often represents time in a practical problem.

Example 3   Sketch the polar graph of $r = 2\sin 2\theta$.

Solution   **With a graphing calculator**
In polar mode, enter:
$r = 2\sin 2t$
In parametric mode, enter:
$x = 2\sin 2t\cos t$
$y = 2\sin 2t\sin t$
The following values
were used for the graph
of $r = 2\sin 2t$ shown:
$0 \leq t \leq 2\pi$, step: $\dfrac{\pi}{30}$
$-2 \leq x \leq 2$, scale: 1
$-2 \leq y \leq 2$, scale: 1

(Solution continues on the next page.)

Polar Coordinates and Complex Numbers **397**

Precalculus Plotter Plus

Graphing software that lets you

- Enter equations and plot the graphs.

- Use data to create histograms, scatter plots, and regression lines of spreadsheet data.

- Test statistical hypotheses through sampling experiments.

- Explore sequences and series.

- Perform matrix operations.

- Check your understanding by determining equations for displayed graphs.

Inequalities

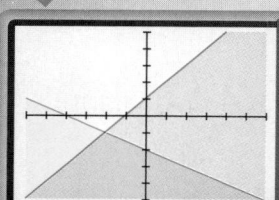

Matrices

$$T = \begin{bmatrix} 1 & 1 & 2 & -5 \\ 3 & 1 & -1 & -3 \\ 2 & 2 & 3 & 6 \end{bmatrix}$$

Trigonometry

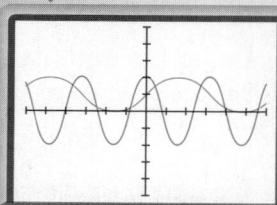

These pages will introduce you to the basic features of most graphing calculators. Because of the variety of graphing calculators available, specific keystrokes are not given. Refer to your calculator's instruction manual for details.

Setting the RANGE Variables

When using a graphing calculator to display graphs, think of the screen as a "viewing window" that shows a portion of the coordinate plane. On many calculators, the "standard" viewing window uses values from −10 to 10 on both axes. You adjust the viewing window by entering values for the *RANGE variables*, which appear on the screen when you press the RANGE key.

The x-axis will be shown for −2 ≤ x ≤ 4.

The y-axis will be shown for −3 ≤ y ≤ 3.

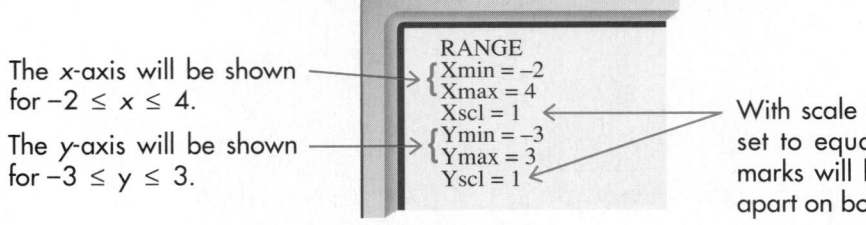

```
RANGE
Xmin = −2
Xmax = 4
Xscl = 1
Ymin = −3
Ymax = 3
Yscl = 1
```

With scale variables set to equal 1, tick marks will be 1 unit apart on both axes.

Graphing a Function

To graph a function, enter its equation and set the RANGE variables for an appropriate viewing window. (You may need to experiment to find the best viewing window.) Here, the graph of the cubic equation $y = x^3 - 3x^2 + 2$ is shown using the viewing window described above. Note that the scale labels shown here, and on similar diagrams throughout the book, do not actually appear on a calculator display.

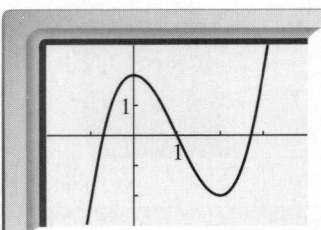

Function form

An equation to be graphed must be entered in the form $y = \ldots$, that is, y must be expressed as a *function* of x. For example, before graphing $x^2y + y = 4$, first solve the equation for y. Enter the equivalent equation $y = \dfrac{4}{x^2 + 1}$.

Using parentheses

Be careful when you enter an equation like the one discussed above.

$$\text{Enter} \quad Y = 4 \div (X^2 + 1), \quad \text{not} \quad Y = 4 \div X^2 + 1.$$

The second equation will be interpreted as $y = \dfrac{4}{x^2} + 1$, not as $y = \dfrac{4}{x^2 + 1}$.

Try This

1. Enter and graph each equation, using an appropriate viewing window. You may need to solve for y first.

   a. $y = x^2 + x - 5$ **b.** $6 + 2y = x$ **c.** $|x| - y = -8$

Appearance of Graphs

Graphs displayed on a graphing calculator often look distorted. Adjusting some of the settings on your calculator may improve a graph's appearance.

Squaring the screen

A *square screen* is a viewing window with equal unit spacing on the two axes.

Standard viewing window   **Square screen window**

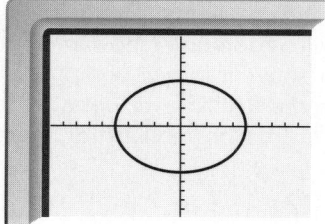   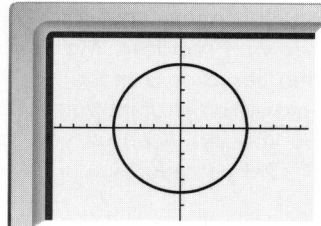

Here, the circle $x^2 + y^2 = 25$ appears stretched horizontally.
    Displayed on a square screen, the same graph is undistorted.

On many graphing calculators, the ratio of the screen's height to its width is about 2 to 3. For a square screen, choose values for the RANGE variables that make the "length" of the y-axis about two-thirds the "length" of the x-axis:

$$(\text{Ymax} - \text{Ymin}) \approx \frac{2}{3}(\text{Xmax} - \text{Xmin})$$

Connected mode; graphs with asymptotes

When your calculator is in *connected mode*, the individual plotted points on a graph are joined by line segments. As a result, graphs often look jagged. Also, the separate pieces of a graph with vertical asymptotes may appear to be connected. If you take your calculator out of connected mode, only points *on* the graph will be plotted. This may give you a better sense of the true shape of the graph, although there can be large gaps between the points.

ZOOM

How many x-intercepts does the graph of $y = x^3 - 6.2x^2 + 9.6x + 0.05$ (shown below) have between $x = 2$ and $x = 4$? To answer this question, use the calculator's *ZOOM feature* to enlarge the section of the graph near the point (3, 0). On many calculators, you can do this by creating a "ZOOM BOX" around the point of interest. The contents of this box can then be drawn at full-screen size.

Standard viewing window

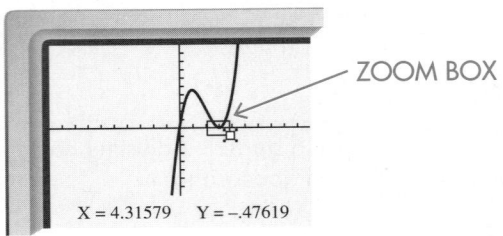

X = 4.31579   Y = −.47619

ZOOM-BOX window

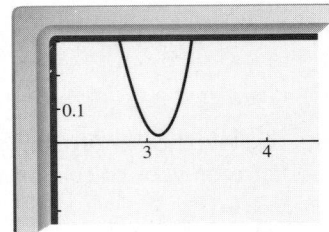

The coordinates of one corner of the ZOOM BOX are displayed.

Now you can see that the graph has no x-intercepts near $x = 3$.

On most graphing calculators, the ZOOM feature offers several ways to adjust the viewing window. Many calculators will "zoom-in" (show a smaller portion of the coordinate plane), or "zoom-out" (show a larger portion of the coordinate plane) on a point you select, changing the RANGE variables by factors that you specify. Consult your calculator's instruction manual for specific details on ZOOM procedures.

Try This

2. Use the fourth-degree equation $y = 30x^4 + 122x^3 - 3x^2 - 492x$.

a. Graph the equation using a viewing window with $-4 \le x \le 4$ and $-500 \le y \le 1000$.

b. Use a ZOOM BOX to enlarge the graph's "flat" section. For an even more detailed view of this portion of the graph, set Ymin = 450 and Ymax = 500. Describe the shape of the "flat" section of the graph. A sideways S

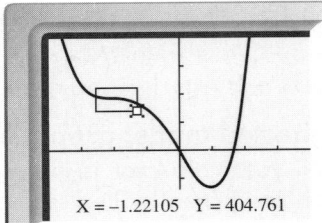

X = −1.22105   Y = 404.761

TRACE

After a graph is displayed, you can use the calculator's *TRACE feature*. When you press the TRACE key, a flashing cursor appears on the graph. The x- and y-coordinates of the cursor's location are shown at the bottom of the screen. Press the left- and right-arrow keys to move the TRACE cursor along the graph.

Finding a point of interest

You can use the TRACE and ZOOM features to find the coordinates of a point of interest on a graph, such as an x-intercept or a high or low point. Consider the graph shown in Exercise 2, which has an x-intercept between 1 and 2. To find the coordinates of this x-intercept, begin by pressing the TRACE key.

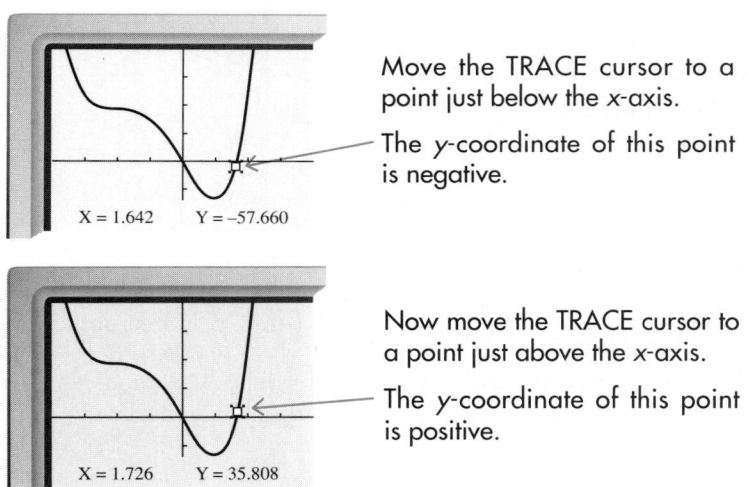

Move the TRACE cursor to a point just below the x-axis.

The y-coordinate of this point is negative.

X = 1.642   Y = −57.660

Now move the TRACE cursor to a point just above the x-axis.

The y-coordinate of this point is positive.

X = 1.726   Y = 35.808

Somewhere between these two points, the graph must cross the x-axis (at a point where $y = 0$). Therefore, the x-intercept is between 1.64 and 1.73.

"Zoom in" on a point near the graph's x-intercept. Move the TRACE cursor along the graph until it is just below the x-axis when $x \approx 1.695$ and then just above the x-axis when $x \approx 1.697$.

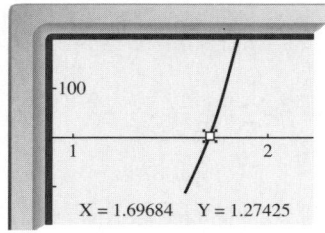

X = 1.69684   Y = 1.27425

When rounded to the nearest hundredth, the two x-values mentioned above are the same. Thus, to the nearest hundredth, the x-intercept is $x = 1.70$. If you wish, you can increase the accuracy of this approximation by repeating this process.

Finding an intersection point of two graphs

You can use a similar process to find the coordinates of an intersection point of two graphs. When the TRACE cursor is on one graph near the intersection point, note the value of the x-coordinate. Then press the up-arrow key. This moves the TRACE cursor to the point on the other graph that has the *same x-coordinate*. Compare the y-coordinates of the two points. You can "zoom-in" and repeat this process until the y-coordinates are the same to the desired degree of accuracy.

Try This

3. a. Graph $y = 2^{-x} - 5$, as shown at the right. Find the x-intercept of the graph to the nearest tenth. -2.3

b. Add the graph of $y = 2x$ to the same set of axes. Find the coordinates of the intersection point of the two graphs to the nearest hundredth. $(-1.28, -2.57)$

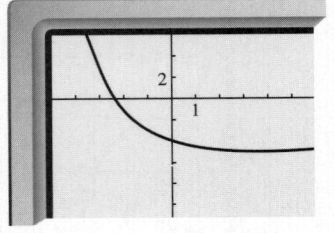

Solving Equations with a Graphing Calculator

Finding approximate solutions to equations that are difficult or impossible to solve algebraically is a powerful and important use of a graphing calculator.

Using an x-intercept to solve an equation

To solve the fourth-degree equation $x^4 - 6x + 4 = 0$, use the graph of the related fourth-degree function $y = x^4 - 6x + 4$.

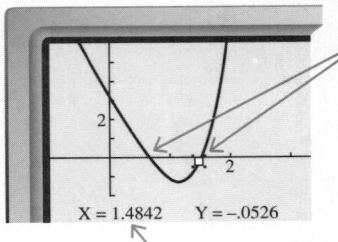

X = 1.4842   Y = −.0526

For any point on the x-axis, $y = 0$. Thus, each x-intercept of the function

$$y = x^4 - 6x + 4$$

is a solution of the equation

$$x^4 - 6x + 4 = 0.$$

To the nearest tenth, one solution of $x^4 - 6x + 4 = 0$ is $x = 1.5$.

You can use the ZOOM and TRACE features to find both real solutions of the equation $x^4 - 6x + 4 = 0$ to any desired degree of accuracy.

Using an intersection point to solve an equation

To solve the radical equation $\sqrt{x} + 3 = 7 - x$, use the graphs of the two functions $y = \sqrt{x} + 3$ and $y = 7 - x$, drawn on the same set of axes.

The coordinates of the intersection point (a, b) must satisfy both equations:

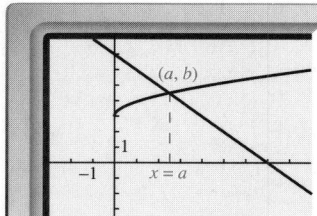

$$b = \sqrt{a} + 3 \text{ and } b = 7 - a$$

Thus, $\sqrt{a} + 3 = 7 - a$.

Therefore, $x = a$ is a solution of the equation $\sqrt{x} + 3 = 7 - x$.

Using ZOOM and TRACE, you can find that the solution is $x \approx 2.44$.

Try This

4. For each equation, find all real solutions to the nearest tenth.

   a. $0 = x^5 - 3x^2 + 3$ −0.9 **b.** $\sqrt{x + 5} = |x|$ −1.8, 2.8

5. Use a graph to determine the *number* of real solutions of this equation.

$$x + 4 = -x^4 + 3x^3 - 1.5x + 5 \quad \text{2}$$

Graphing Parametric Equations

In the seconds after a baseball is hit, the ball has moved both horizontally and vertically, as shown in the diagram.

Instead of using one equation to describe the path of the ball, you can use *two* equations, one to express x in terms of t (the time in seconds) and one to express y in terms of t:

$$x = 100t \quad \text{and} \quad y = -16t^2 + 40t + 3$$

These two equations, used to express two variables *(x and y)* in terms of a third variable *(t)* are called *parametric equations*. The variable t is the *parameter* used to define the variables x and y.

Parametric mode

If your graphing calculator has a built-in parametric mode, you can enter and graph parametric equations that express the variables x and y in terms of the variable t. The RANGE-variables screen will have three additional quantities for you to specify—Tmin, Tmax, and Tstep (called "pitch" on some calculators).

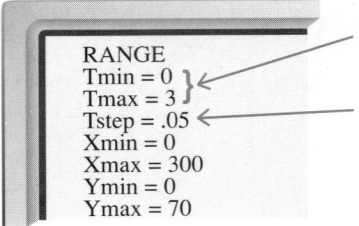

```
RANGE
Tmin = 0
Tmax = 3
Tstep = .05
Xmin = 0
Xmax = 300
Ymin = 0
Ymax = 70
```

The calculator will use t-values from $t = 0$ to $t = 3$.

The difference between successive t-values will be 0.05.

Note: The x- and y-scale variables, omitted here, will also appear.

Each t-value is substituted in both the equation for x and the equation for y; then the point (x, y) is plotted on the calculator screen.

You will learn more about parametric equations in Sections 11-1 and 12-3.

Try This

6. Graph the parametric equations given on page xxxi for the path of the ball. Set the RANGE variables as shown on the calculator screen above. Then use the TRACE feature to determine how many seconds it takes the ball to reach its maximum height. 1.25 s

Other Capabilities of a Graphing Calculator

In addition to its ability to display the graphs of functions, your graphing calculator may have other capabilities that will be useful to you in this course.

Statistical graphs

Many graphing calculators can display histograms, line graphs, or scatter plots of data you have entered. For example, the histogram at the right displays the data from the list that appears on page 639 of fifty scores on a standardized mathematics achievement test. The histogram shows the number of students whose scores fell in each 100-point interval. You can obtain statistics

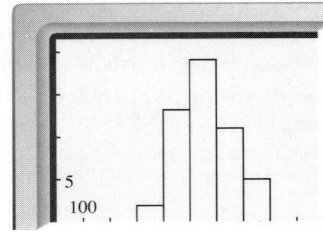

about the data, such as the mean and standard deviation (see Chapter 17), from a graphing calculator just as you would from a scientific calculator.

Curve fitting

Curve fitting (see Chapter 18) is the process of finding an equation that describes a set of ordered pairs. Often, the first step is to graph the paired data in a scatter plot. For example, the chart shown below gives the winning times in the men's Olympic 400 m freestyle swimming race. The data can be entered in a graphing calculator and then displayed as plotted points (y, s), where y is the number of years since 1900 and s is the time in seconds.

Year	Winning time (seconds)
1972	240.27
1976	231.93
1980	231.31
1984	231.23
1988	226.95
1992	225.00

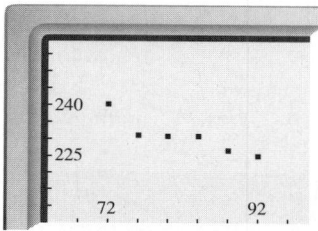

The relationship between y and s can be approximated by a line. A graphing calculator will give an equation of the "line of best fit," $s = -0.653y + 285$, and a correlation coefficient that reflects how well the equation models the data.

Matrices

Many graphing calculators allow you to enter numerical information in *matrix form* (see Chapter 14) and can then perform a variety of *matrix operations*.

For example, the matrices A and B shown below display the quiz and homework averages of 3 precalculus students for each of two units.

Unit 1   Quiz HW

$$\begin{matrix}\text{Mary} \\ \text{José} \\ \text{Kasha}\end{matrix} \begin{bmatrix} 68 & 75 \\ 89 & 77 \\ 92 & 95 \end{bmatrix} = A$$

Unit 2   Quiz HW

$$\begin{matrix}\text{Mary} \\ \text{José} \\ \text{Kasha}\end{matrix} \begin{bmatrix} 76 & 80 \\ 81 & 81 \\ 85 & 90 \end{bmatrix} = B$$

Unit 1 will be worth 40% and Unit 2 will be worth 60% of each student's midterm quiz and homework grades. A graphing calculator with matrix capabilities can calculate $0.4a + 0.6b$ for each pair of corresponding elements in matrices A and B, and display the results in a new matrix that gives each student's quiz and homework averages.

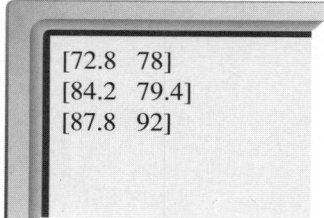

And more

Throughout the year, you are sure to find other topics that can be explored with a graphing calculator, and you will discover new methods and uses for this valuable tool. Be sure to share your discoveries with your classmates and your teacher.

Careers in Mathematics and Science

Today's society is changing so rapidly that you cannot foresee the career opportunities that may be available to you within a few years. You can be sure, however, that many of the most exciting careers will involve the use of mathematics, science, and technology. Recent studies have shown that even now there is a shortage of scientists and engineers. By choosing to complete four years of high school mathematics you have already made one decision that will help you keep your options open. Continuing to study mathematics and related fields in college will prepare you to take advantage of a wide variety of career choices, including jobs in the social sciences and the arts, as well as engineering, medicine, and scientific research.

Now meet seven people whose careers use the tools and methods of mathematics, the language of science and technology.

Martha C. Zuniga

IMMUNOLOGY

An immunologist investigates the body's immune response, the process by which the body identifies, reacts to, and fights off toxic or disease-causing agents. Research in immunology contributes to the medical profession's ability to enhance the body's capacity to fight cancer and other life-threatening diseases.

Martha C. Zuniga, who holds a Ph.D. in biology from Yale University, conducts research on the immune system's ability to distinguish alien virus and tumor cells from the body's healthy tissues. In 1989 the National Science Foundation honored her with its Presidential Young Investigator Award, a grant to provide her with funding to continue her research.

As a recipient of the NSF award, Dr. Zuniga maintains an active role in undergraduate education, helping to promote the importance of studying science. At University of California, Santa Cruz, where she teaches immunology, Dr. Zuniga encourages her students to enjoy the challenges of difficult academic work. "People get excited about heroism, and I think there can be heroism in intellectual pursuits," says Dr. Zuniga, who also works with students through the Society for the Advancement of Chicanos and Native Americans in Science.

COMPUTER GRAPHIC ARTS

Computer graphics have expanded the world of many artists. Computers help artists who work in traditional media to make decisions about composition, design, and color. For other artists, the end product of the artistic process is an image stored in the computer's memory and then reproduced on film, printers, or video.

Midori Kitagawa De Leon, a Ph.D. student at the Visualization Laboratory at Texas A & M University, became interested in computer graphics while majoring in oil painting at the Women's College of Fine Arts, in Tokyo. She went on to earn an M.A. in Computer Graphics and Animation from Ohio State University. There she wrote programs to generate three-dimensional "branching objects," such as trees and other plant-like forms, and to simulate the growth of the plants using genetically determined growth patterns as well as information about their environment.

Midori Kitagawa De Leon

Ms. Kitagawa De Leon's artistic work, which includes film animations of surrealistic plant life, has been featured in computer graphics art shows and magazines. In addition to its artistic value, her work will help landscape architects, who will use the programs to project the future appearance of their designs.

Guion Bluford, Jr.

SPACE EXPLORATION

Any United States citizen holding a degree in mathematics, science, or engineering can apply to be an astronaut. Once selected, an astronaut candidate goes through a one-year training program, and then may spend many more years working on the ground before getting an opportunity to fly.

Guion Bluford, Jr. was a teenager when the space age began in 1957. He was fascinated by flying objects, from the model airplanes he built to the newspapers he tossed each day on his paper route, and he dreamed of entering the relatively new field of aerospace engineering. He graduated from Pennsylvania State University and the Air Force Institute of Technology, earning his Ph.D. in aerospace engineering in 1978. In the same year, NASA accepted him into its astronaut training program.

In August of 1983 Colonel Bluford flew on NASA's eighth shuttle mission, becoming the first African-American to travel into space. Two years later, his second space flight saw one Dutch, two West German, and five American astronauts fly together on a mission run by West Germany. In 1991 Colonel Bluford completed a third mission.

Robert K. Whitman

ENGINEERING

Aerospace, biomedical, chemical, environmental, industrial—these are a few of the branches of the vast field of engineering. Engineers apply mathematics and science to the solution of practical problems. They design airplanes, buildings, highways, artificial limbs, lasers, and computers, among other things.

Robert K. Whitman, deputy director of The American Indian Science and Engineering Society, remembers his early interest in engineering. ''I . . . took apart my parents' radio, much to their dismay. I wanted to see . . . what made it pick up sounds. . . . I read . . . that engineering requires a lot of training in math and science. In high school, I took all the math and science courses I could get.'' Bob Whitman went on to study electrical engineering at the University of New Mexico and at Colorado State University. He received a scholarship from NASA and an Outstanding Achievement Award from the Navajo Nation (1978). He has worked for IBM, where his projects included developing printed circuit boards used in electronic equipment, and designing computer software.

COMPUTER PROGRAMMING

Computers help doctors make diagnoses, assist architects with their designs, and regulate the functioning of all kinds of machines from heart pacemakers to rocket engines. To carry out these and other tasks, computers require explicit instructions; thus, the power of computers to change and improve our lives can be realized only through the imagination and ingenuity of the people who program them.

Martine Kempf was still in her twenties when she succeeded in programming a computer to recognize and respond to the human voice. Using her program she invented the Katalavox, a small black box containing a voice-activated microcomputer. Installed on a wheelchair, the Katalavox gives quadriplegics unprecedented control of their lives. In addition, the Katalavox has become indispensable in the operating room, where surgeons' hands are free to work while surgical microscopes are guided by voice.

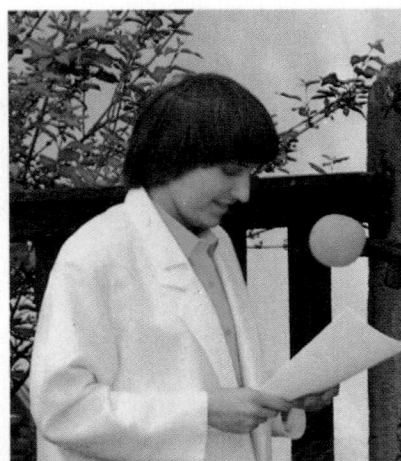

Martine Kempf

Ms. Kempf, a native of France, was studying astronomy at the University of Bonn in 1982 when she first wrote the Katalavox program. She later moved to the United States to run her own business marketing her invention.

PSYCHOLOGY

Psychologists study human and animal behavior, focusing on the mental functions involved in the emotional, intellectual, and physical development of individuals. Trained in the methods of scientific research and mathematical analysis of data, they formulate and test theories to explain and predict behavior. They also study the influence of the mind and body on each other.

Patricia Cowings, a psychophysiologist for the Space Life Sciences Division of NASA, studies the body's response to the weightlessness experienced in space travel. She teaches astronauts to control, through mental techniques, such physical functions as blood pressure and heart rate, allowing them to counteract some of the negative effects of weightlessness.

Dr. Cowings first studied psychology at the State University of New York at Stony Brook, and then pursued graduate degrees at the University of California at Davis. While still a graduate student she began working for NASA at the Ames Research Center. She earned a Ph.D. in psychology in 1973.

Patricia Cowings

Samuel Chao Chung Ting

PHYSICS

Physicists seek to understand all aspects of matter and energy, the fundamental components of our world. They pursue knowledge about subjects as varied as electromagnetism, optics, thermodynamics, acoustics, and quantum theory. All branches of physics require a thorough background in higher mathematics.

Samuel Chao Chung Ting, a researcher in the field of high-energy particle physics, studies the composition and behavior of subatomic particles. In 1974, while working at Brookhaven National Laboratory, Dr. Ting demonstrated the existence of the J particle, a discovery heralded as an important breakthrough in twentieth-century understanding of atomic structure. Also known as *psi*, the particle was independently observed by Dr. Burton Richter of the Stanford Linear Accelerator Center. In 1976 Dr. Ting and Dr. Richter jointly received the Nobel Prize for their discovery.

American by birth, Dr. Ting grew up in China, returning to the United States to attend the University of Michigan, where he earned a Ph.D. in physics. In 1967 he joined the faculty of the Massachusetts Institute of Technology.

Chapter 1 Linear and Quadratic Functions

Overview

This chapter reviews the coordinate geometry of lines and the various methods of finding the real and imaginary roots of quadratic equations. Students graph linear and quadratic functions and find the coordinates of any points of intersection of two lines or of a line and a parabola. Using slope along with the distance and midpoint formulas, students verify properties of special quadrilaterals and find equations of medians, altitudes, and perpendicular bisectors of sides of various triangles. Students also solve a variety of real-world problems for which linear or quadratic functions are models.

Objectives

1-1 To find the intersection of two lines and to find the length and midpoint of a segment.
1-2 To find the slope of a line and to determine whether two lines are parallel, perpendicular, or neither.
1-3 To find an equation of a line given certain geometric properties of the line.
1-4 To model real-world situations by means of linear functions.
1-5 To add, subtract, multiply, and divide complex numbers.
1-6 To solve quadratic equations using different methods.
1-7 To define and graph quadratic functions.
1-8 To model real-world situations using quadratic functions.

Supplementary Resources

Tests, pp. 1–2
Alt. Assess., pp. 1–5, 63
Activities, pp. 1–2
Using Tech., pp. 6–7, 8, 9–10
Student Res. Guide, pp. 1–11

Software

Precalculus Plotter Plus
Activities, pp. 23, 24, 25
Equation Solver, Function Plotter,
    Conics Quiz

Pacing Guide

Section	1-1	1-2	1-3	1-4	1-5	1-6	1-7	1-8	Review	Test	Total
Standard (days)	1	1	2	1	1	1	1	2	1	1	12
Comprehensive (days)	1	1	1	1	1	1	1	1	1	1	10

Teaching Notes

Presenting the Section

You might want to emphasize which pairs of simultaneous linear equations have no solutions and which have infinitely many solutions. Graphically, two parallel lines imply there is no solution, and coincident lines imply there are infinitely many solutions. Algebraically, solving the system would lead to a false statement (like $0 = 2$) if there is no solution, and a statement which is always true (like $0 = 0$) if there are infinitely many solutions. Show examples of each and have the students graph them.

Using Technology

Encourage students to use calculators to check their estimates of solution points to linear equations or to systems of linear equations. Calculators with graphing capabilities and computer graphing programs are also excellent motivational tools for graphing linear equations. Students enter a linear equation in y-form (y is written in terms of x) and the points are plotted automatically. A table of values can be displayed, if needed. Students can also graph more than one line at a time, observe the intersection of the lines, guess its coordinates and then zoom in on the solution to check their guesses. Students may not be able to find the exact solution by zooming in, but can find values as close as necessary. Emphasize the importance of using algebra to determine the exact solution to a system of linear equations.

Cooperative Learning

Learning can be enhanced with group arrangements of three or four students each. Topics include going over more difficult exercises and review of (1) terminology; (2) how to find the solution points of the equation of a line; (3) the various methods of solving a system of linear equations simultaneously; (4) applications of the distance formula, including area problems, and of systems of linear equations; and (5) various approaches to sketching the graph of a line. Encourage students to ask and answer questions within their groups and offer alternate solutions if possible.

Applications

If 3 pencils and 2 pads of paper cost $1.30 while 5 pencils and one pad of paper cost $1.00, what would be the cost of each pencil and each pad of paper? This is a mixture problem, a common application of linear systems of equations. If a is the cost of each pencil and b is the cost of each pad of paper, then

$$3a + 2b = 1.30$$

and

$$5a + b = 1.00,$$

or each pencil costs $.10 and each pad of paper costs $.50. Linear systems of equations also apply to some uniform motion problems, investment problems, other mixture problems, and number problems commonly studied in algebra. Some business applications include finding the break-even point of the cost of production. In all of these problems, students are asked to identify two independent linear relationships and generate two related equations with the same two variables.

Communication

You may wish to summarize and discuss the general form of a linear equation by using the following chart. Ask students to explain why the descriptions following the general form are true and to give additional examples of each description.

Equation	Description	Example
$Ax + By = C$	General form	$2x + 3y = 6$
$Ax + By = 0$	Contains $(0, 0)$	$2x + 3y = 0$
$Ax = C$	Vertical	$2x = 6$
$By = C$	Horizontal	$3y = 6$

Review

You might need to review the Pythagorean theorem before assigning Exercise 34, page 7. If (x_1, y_1) and (x_2, y_2) are the coordinates of the endpoints of a line segment, then $|y_2 - y_1|$ is the vertical distance and $|x_2 - x_1|$ is the horizontal distance to a third point (x_2, y_1). This third point is a vertex of a right triangle with the given line segment as its hypotenuse. The length of the line segment is then

$$\sqrt{(x_2 - x_1)^2 + (y_2 - y_1)^2}.$$

Presenting the Section

An algebraic way to introduce slope is to draw the line $y = x$. The line will make a 45° angle with the x-axis and pass through the origin. Then draw other lines through the origin with positive slopes. Those lines with first-quadrant points above the line $y = x$ will have slopes greater than 1; those with first-quadrant points below the line $y = x$ will have slopes less than 1. Students can also see that the slopes approach 0 as the lines get closer to the x-axis, and the slopes get infinitely greater as the lines get closer to the y-axis.

To avoid students' confusing the slope of a line with its associated equation, it is important to emphasize that the slope of a line is a measure of its steepness, not a way to determine the line uniquely. You can illustrate this by laying several yardsticks against a desk so that they all have the same slope. The diagrams on page 9 also illustrate this idea.

It is important for students to understand the difference between a line with no slope (no number) and a line with zero slope (zero is a number). They should note that when calculating the slope of a line given the coordinates of two of its points, the order of subtraction is very important to avoid getting an answer that has the opposite sign.

Using Technology

You might want to prepare students for the theorem stated on page 9 by having them first graph $y = mx + k$ for various values of k. A graphing calculator or computer will work nicely for this activity. Students will be able to graph a whole family of lines very quickly by using this technology, and they will not have any difficulty making the generalizations that changing k but not m gives parallel lines and that changing m but not k gives concurrent lines.

Geometric drawing software provides an excellent way to verify the theorem on page 10. Have students graph lines with opposite reciprocal slopes, label them, and then use the angle measuring capabilities of the drawing software to confirm that the angle between the lines is 90 degrees.

A computer also can be used to analyze data having a linear relationship. Data points can be entered from the terminal, the computer can determine the ''line of best fit,'' and then an equation can be written for that line. The equation of the line can then be used to describe trends in the data and to determine other data points.

Applications

Students may enjoy thinking about and discussing some real-world applications with positive, zero, or negative ''slope.'' Encourage them to come up with examples that use the word slope (like ''ski slope'') and examples that apply the idea of a constant increase or decrease (pitch of a roof, grade of a road or hill, constant increases in population, and so on). Linear models for sales predictions are easy to find in newspapers, and most students should understand that offering a 20% discount on all items in a clothing store would yield a linear equation with sales price and list price as variables. You may wish to have a few students prepare a short report and present it to the class.

Presenting the Section

You may want to discuss the difference between an equation and a graph. An equation expresses a specific algebraic relationship among variables and numbers. A graph is a set of points used to visualize the solutions to an equation. Emphasize that while a linear equation can be written in any one of several different forms, as listed on page 14, recognizing the easiest form to use will be helpful to students in finding an equation of a line given certain geometric properties of the line. Stress that the slope-intercept form is most useful if an equation is to be graphed, and that it arises naturally from applications involving a constant increase or decrease.

Equations of lines written in general form arise naturally from applications such as the following: How many pounds of apples costing 90¢/lb and how many pounds of oranges costing 50¢/lb need to be added together to total $5.00? This problem is a linear combination of the costs of the apples and oranges and can be described by a linear equation in general form.

The point-slope form of the equation of a line is used often to find the equation of a line through two given points on a graph.

Using Technology

Geometric drawing software can be used to demonstrate Example 4 on page 15. First, the points A and B are plotted and a line segment is drawn connecting them. The computer can then be instructed to locate the midpoint of $\overline{AB}$, label it, and draw a segment perpendicular to $\overline{AB}$. Alternately, you can test to see if a point is on the perpendicular bisector of the segment by first connecting it to the midpoint and then using the angle measuring feature of the software to determine if that segment forms a 90° angle with $\overline{AB}$.

1-4 | pages 19–25

Presenting the Section

The objective of this section is to help students make connections between a problem situation and the use of a function as a model. You might start the discussion by having students generate a list of real-world examples of dependent and independent variables like, ''As the cost of an item goes up, the related sales tax goes up. Tax depends on cost.'' If some of the examples students come up with are not linear, it would be an appropriate time to discuss what makes a function linear and what does not. Refer to Sections 1-7 or 1-8 for examples of nonlinear functions.

A ''function machine'' can also be used to introduce the concept of function. Draw the following diagram on the chalkboard.

| Input value | → | Function rule | → | Output value |

Have students give examples of the ''function rule'' and input values. See if they can predict the output values. Linear and quadratic examples (see Section 1-7) from mathematics can be used as well as real-life examples such as sales tax = tax rate × sales price.

Cooperative Learning

You might want students to work in groups of three or four to discuss the attributes of a function and to come up with examples of equations that are not functions. Include activities that help students decide when relationships are linear and when they are not. Students should be able to distinguish between constant and nonconstant functions and identify the dependent and independent variables.

You might also want to use the group environment to assign a class project to create a mathematical model for a real-world situation. Let each group choose an area of interest such as business or sports. Students can use almanacs, do research, or use newspapers to gather data for plotting by hand or by the use of a computer to get the ''line of best fit.'' Other tasks for individual members can be to predict other data points or describe trends in the data. In any case, have the groups share their model linear functions with the class.

Applications

In physics, a quantitative description of the uniform motion of an automobile is the linear function

$$s = v_0 t,$$

where v_0 is the constant velocity, t is the time, and s is the distance traveled. Have students plot data values for this function and predict the distance traveled for various times. You can also use this example to lead into a discussion of the possibility of nonlinear functions. If an automobile is accelerating uniformly, the distance it will travel is given by

$$s = v_0 t + \frac{1}{2}at^2,$$

where v_0 is the initial velocity, a is the acceleration, t is the time, and s is the distance. This is a quadratic function (Section 1-8).

A good example of a step function is the rate structure of the U.S. Postal Service. Ask students to find out what the rates are and plot them. See if they can determine an associated function.

Communication

Straight-line graphs are good choices to help students understand the concept of function. The data values are easy to calculate, and there are many real-life models from which to choose. Encourage students to discuss any examples they are aware of as well as the examples in the section itself. Make sure they try to express clearly the dependent relationships between variables. Stress that functions are extremely important in all of mathematics, and that they can be described with graphs, equations, words, or lists of data values (sets of points).

Presenting the Section

Complex numbers are introduced in this chapter so that a quadratic equation can be solved completely. You might want to review the algebra of adding and multiplying binomials before assigning the exercises.

You can summarize the hierarchy within the complex number system by using the following chart.

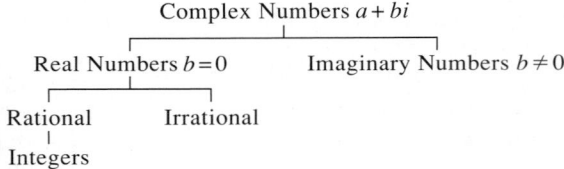

Ask students to give several examples for each number group above and to express the real numbers in complex form as well (for example, $6 = 6 + 0i$).

You can summarize the operations on complex numbers as follows:

1. Addition: add corresponding real parts and imaginary parts separately.
2. Subtraction: subtract corresponding real parts and imaginary parts separately.
3. Multiplication: use FOIL and replace i^2 with -1.
4. Division: multiply numerator and denominator by the conjugate of the denominator.

Cooperative Learning

You might want to use small groups to go over the answers to the exercises in this section. Monitor the groups and provide assistance as needed. Encourage students to comment on the solutions of other students. Have them discuss their proofs of Exercises 42–46, page 29.

Applications

Complex numbers have extensive applications in science and engineering, especially electrical engineering and the theories of alternating currents and of fluid dynamics. Have interested students do brief written or oral reports on some uses of $\sqrt{-1}$.

Presenting the Section

Students typically have problems simplifying radicals when using the quadratic formula. You might need to review radicals before starting this section.

A simple chart summarizing the meaning of the discriminant follows.

Value of discriminant	Number of solutions	Type of solution
negative	2	complex conjugates
zero	1	real double root
positive	2	real

Before assigning Exercises 25–28, page 35, stress the importance of checking for extraneous roots. Students need to keep the restrictions on the variable in mind as they solve these exercises to be sure that no solution gives a zero denominator. For Exercises 31 and 32 you might need to review some similarity concepts from geometry.

Using Technology

Students who need more practice can use technology for solving quadratic equations. A function plotter with a zoom feature can be used to get a decimal approximation for the zeros. Some computer-based graphing utilities automatically give the zeros of the function.

Using a calculator to simplify the process of solving quadratic equations is invaluable. Some calculators are programmable to the extent that the student only needs to enter the coefficients of the terms. Because a calculator solution is expressed in decimal form, some students may think the solution is irrational. Using calculator solutions can be helpful in reviewing terminating, repeating, and nonterminating decimals. Quadratics with complex solutions cannot be solved directly on many calculators because they will not calculate the square root of a negative number. In those cases, students should work with the real and imaginary parts of the solutions separately.

Assessment

Have students give examples of quadratics and the methods they would choose to solve them. For example, for $3x^2 + 2x - 1 = 0$, the coefficients are all integers and $b^2 - 4ac$ is a perfect square (16); therefore, the quadratic can be solved by factoring. The factors are $(3x - 1)$ and $(x + 1)$.

1-7 | pages 37–43

Presenting the Section

Using technology is a good way to demonstrate the effect of changing the coefficient a in the quadratic function $y = ax^2 + bx + c$. As the value of $|a|$ increases, the graph gets "thinner." As it decreases, the graph "flattens out." Point out that all parabolas have the same basic shape.

Suggest that students always graph the axis of symmetry and take advantage of the symmetry of a parabola when graphing by hand. For quadratic functions, each point (except the vertex) has a corresponding point on the other side of the axis with the same y-coordinate.

Be sure to point out that there are parabolas that are *not* quadratic functions. One such group of parabolas have the form $x = a(y - h)^2 + k$.

Using Technology

Function plotting programs are quite useful because of their rapid graphing of quadratic functions. While graphing parabolas by hand can be very time consuming, especially to get a smooth curve, graphing them on a computer or graphing calculator is almost instantaneous. Zooming is also a useful feature, if available, to locate both the vertex and the zeros of the function. These graphing programs can also be used in Exercises 21–24, page 41, to zoom in on the coordinates of the intersection of a line and a parabola.

Cooperative Learning

Exercises 39–40, page 42, would be good ones for students to go over in groups. Have them discuss their proofs with each other, and offer examples to show that they understand how each of the statements is applied.

Applications

There are many real-world applications of quadratic functions; some are given in the next section. For many applications a quadratic function fits the data, and the function then becomes a model for that type of real-world situation. As a model, it can be used to find maximum or minimum points. Finding the maximum or minimum value of a function is an important topic in calculus.

1-8 | pages 43–48

Presenting the Section

The examples and exercises in this section should demonstrate that real-life situations for which quadratic models can be constructed are quite diverse. Be sure students understand that at least three data pairs are needed to determine the coefficients a, b, and c of the quadratic function $f(x) = ax^2 + bx + c$. Of course, the three data pairs, when graphed, should appear to lie on a parabola to suggest a quadratic model in the first place.

Cooperative Learning

If time permits, students can work in groups to construct their own quadratic models. They should generate or find some experimental data and try to fit a quadratic curve to the data. There are computer programs available that can help students determine the equation of a curve that fits the data. Depending on what the data are, students can then discuss projected values the curve implies and possible maximum and minimum values. If time permits, have the groups present their models to the class for discussion. Other follow-up work could be for groups to compare their experimental data with published data on their topic.

1 Linear and Quadratic Functions

■ Linear Functions

1-1 Points and Lines

Objective *To find the intersection of two lines and to find the length and the coordinates of the midpoint of a segment.*

Each point in the plane can be associated with an ordered pair of numbers, called the **coordinates** of the point. Also, each ordered pair of numbers can be associated with a point in the plane. The association of points and ordered pairs is the basis of *coordinate geometry*, a branch of mathematics that connects geometric and algebraic ideas.

To set up a coordinate system, we can choose two perpendicular lines, one horizontal and the other vertical, as the **x-axis** and the **y-axis** and designate their point of intersection as the **origin**. Using a convenient unit of measure, we mark off the axes as number lines with zero located at the origin. The axes divide the plane into four **quadrants**.

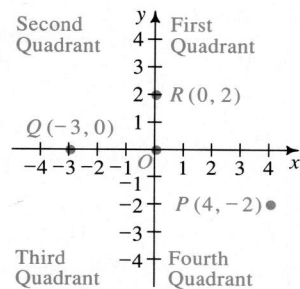

The diagram shows that P has **x-coordinate** 4 and **y-coordinate** -2. We write $P(4, -2)$. Points with x-coordinate 0, such as R, lie on the y-axis. Points with y-coordinate 0, such as Q, lie on the x-axis. The coordinates of the origin, O, are $(0, 0)$.

Linear Equations

A **solution** of the equation $2x - 3y = 12$ is an ordered pair of numbers that makes the equation true. For example, $(0, -4)$ is a solution because $2(0) - 3(-4) = 12$. Several solutions are shown in the diagram. The set of all points in the plane corresponding to solutions of an equation is called the **graph** of the equation. The graph of $2x - 3y = 12$ is the line shown in the diagram. We call -4 the **y-intercept** of the graph because the line intersects the y-axis at $(0, -4)$. We call 6 the **x-intercept** of the graph because the line intersects the x-axis at $(6, 0)$.

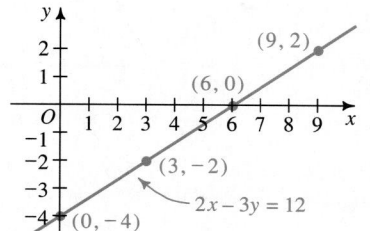

Any equation of the form $Ax + By = C$, where A and B are not both 0, is called a **linear equation** because its graph is a line. Conversely, any line in the plane is the graph of a linear equation. The graph is often referred to as "the line $Ax + By = C$." We call $Ax + By = C$ the **general form** of a linear equation.

◀ In this view of Chicago from the air, the regular pattern of streets intersecting at right angles suggests a coordinate grid.

Linear and Quadratic Functions **1**

Teaching Notes, p. 1B

Warm-Up Exercises

Find $\frac{a+b}{2}$ and $(a-b)^2$ for the given values of a and b.

1. $a = 3$, $b = 3$ 3; 0

2. $a = 2$, $b = -1$ 0.5; 9

3. $a = -5$, $b = 1$ -2; 36

4. $a = 0$, $b = -7$ -3.5; 49

5. $a = \frac{1}{2}$, $b = -1$ $-\frac{1}{4}$; $\frac{9}{4}$

Motivating the Section

In economics, a "break-even analysis" might include plotting linear relationships such as business costs and revenues on the same set of axes and then determining where their graphs intersect. The point where costs equal revenues is known as the "break-even point." In this section, students will find the intersection point of two lines.

Problem Solving

Encourage students to prepare charts that organize solution points of the equation of a line. You can have them fill in a table of values and discuss the relationship between the table, the graph, and the equation. Using tables will help prepare students for the concept of *function* in Chapter 4.

Have students discuss why A and B cannot both be 0 in the general form of a linear equation. (If they are, the equation reduces to $0 = C$, a false statement unless C is also 0.)

Example 1 Sketch the graph of $3x + 2y = 18$.

Solution One way to sketch the graph is to find the intercepts.

Step 1 To find the *y*-intercept, let $x = 0$.
$$3(0) + 2y = 18$$
$$y = 9$$

The line passes through $(0, 9)$, so the *y*-intercept is 9.

To find the *x*-intercept, let $y = 0$.
$$3x + 2(0) = 18$$
$$x = 6$$

The line passes through $(6, 0)$, so the *x*-intercept is 6.

Step 2 Plot $(0, 9)$ and $(6, 0)$. Draw a straight line through them. It is always a good idea to check that you have drawn the line correctly. Select a different point on the line and determine whether its coordinates satisfy the equation. In this case, $(4, 3)$ does check since $3(4) + 2(3) = 18$.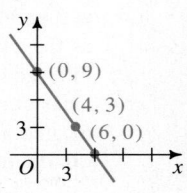

When one of the constants *A*, *B*, or *C* in $Ax + By = C$ is 0, you can draw certain conclusions about the graph. In figure (a), $C = 0$ and the line contains the origin. In figure (b), $A = 0$ and the line is horizontal. In figure (c), $B = 0$ and the line is vertical.

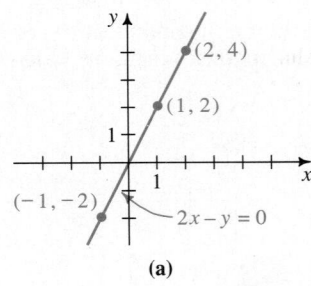

(a)

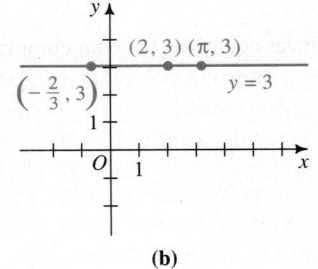

(b)

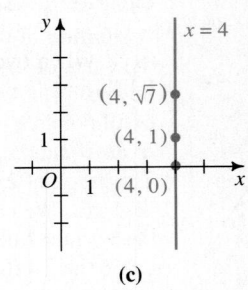

(c)

Intersection of Lines

You can determine where two lines intersect by drawing their graphs or by solving their equations simultaneously. Consider the following pair of linear equations:

$$2x + 5y = 10 \qquad (1)$$
$$3x + 4y = 12 \qquad (2)$$

You can make hand-drawn sketches or you can use a graphing calculator or computer to obtain the graphs shown at the right. (If you use a calculator or computer, be sure to write y in terms of x.) From the figure it seems that at the point of intersection, x is a little less than 3 and y is a little less than 1. (With a calculator's trace feature or a computer's zoom feature, you can get better approximations to x and y.) Be aware that solutions found by graphing are not always exact. An algebraic solution yields the exact values.

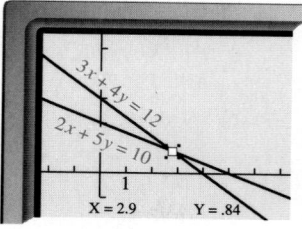

To solve the equations simultaneously, you can multiply both sides of equation (1) by 3 and both sides of equation (2) by 2. Then subtract the second equation from the first equation.

$$3(2x + 5y) = 3(10) \longrightarrow 6x + 15y = 30$$
$$2(3x + 4y) = 2(12) \longrightarrow \underline{6x + 8y = 24}$$
$$7y = 6$$
$$y = \frac{6}{7}$$

Now substitute $\frac{6}{7}$ into equation (1) and solve for x: $\quad 2x + 5\left(\frac{6}{7}\right) = 10$
$$x = \frac{20}{7}$$

Thus $\left(\frac{20}{7}, \frac{6}{7}\right)$ is the common solution of the two equations (or the intersection point of their graphs). Notice that the graphical estimate $x \approx 3$ and $y \approx 1$ is close to the exact answer. You can use a graphical estimate as a check on an algebraic solution.

When two linear equations have no common solution, their graphs are parallel lines. When two linear equations have infinitely many common solutions, the equations have the same graph.

no common solution:
$$6x + 4y = 8$$
$$3x + 2y = 1$$

infinitely many common solutions:
$$6x + 4y = 8$$
$$3x + 2y = 4$$

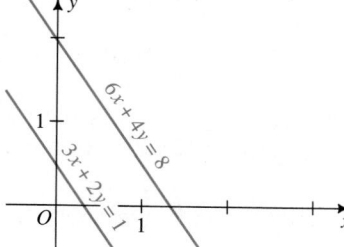

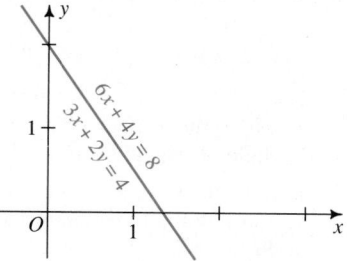

For further details on using a graphing calculator to approximate the intersection point of two graphs (or to find other points of interest), refer to the sections on the ZOOM and TRACE features in the article *Using a Graphing Calculator* (see pages xxviii–xxx).

Additional Examples

1. Solve: $3x - y = 9 \qquad (1)$
$\qquad\qquad 7x - 5y = 25 \qquad (2)$
Then graph the equations.

Solution 1: Substitution method. From eq. (1), $y = 3x - 9$. From eq. (2), $7x - 5y = 25$.
$7x - 5(3x - 9) = 25$
$-8x = -20; \ x = 2.5$
$y = 3(2.5) - 9 = -1.5$

Solution 2: Multiply with the addition or subtraction method.
$5(3x - y) = 5(9)$
$15x - 5y = 45$
$\underline{7x - 5y = 25}$
$\qquad\quad 8x = 20; \ x = 2.5$
$3(2.5) - y = 9; \ y = -1.5$
Thus, the solution is $(2.5, -1.5)$.

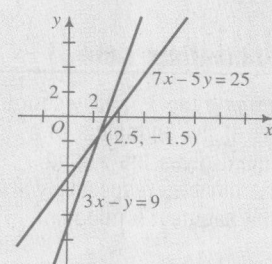

(*Continues on next page*)

2. Use $A(2, 0)$, $B(0, 8)$, $C(-4, 7)$, and $D(-2, -1)$.
   a. Show that $\overline{AC}$ and $\overline{BD}$ bisect each other.
   b. Show that $AC = BD$.
   c. What kind of figure is $ABCD$?

   a. For $\overline{AC}$ and $\overline{BD}$ to bisect each other, they must have the same midpoint, M. Midpoint of $AC =$
   $\left(\frac{2-4}{2}, \frac{0+7}{2}\right) = \left(-1, \frac{7}{2}\right)$
   Midpoint of $\overline{BD} =$
   $\left(\frac{0-2}{2}, \frac{8-1}{2}\right) = \left(-1, \frac{7}{2}\right)$
   Thus, $\overline{AC}$ and $\overline{BD}$ bisect each other.

   b. $AC =$
   $\sqrt{(-4-2)^2 + (7-0)^2}$
   $= \sqrt{85}$
   $BD =$
   $\sqrt{(-2-0)^2 + (-1-8)^2}$
   $= \sqrt{85}$
   $AC = BD$

   c. Since the diagonals bisect each other, $ABCD$ is a parallelogram. Since the diagonals of this parallelogram have the same length, $ABCD$ must be a rectangle.

Mathematical Note

Averaging the x- and y-coordinates of the endpoints of a segment gives the x- and y-coordinates, respectively, of the segment's midpoint.

⋈ Using Technology

Encourage students with programmable calculators to write programs that calculate the distance between two points or the midpoint of the segment connecting the points.

We denote the line segment with endpoints A and B as $\overline{AB}$ and its length as AB. You can use the formulas below to find AB and the midpoint of $\overline{AB}$. You are asked to derive these formulas in Written Exercises 34 and 35.

The Distance and Midpoint Formulas

Let $A = (x_1, y_1)$, $B = (x_2, y_2)$, and M be the midpoint of $\overline{AB}$. Then:

$$AB = \sqrt{(x_2 - x_1)^2 + (y_2 - y_1)^2} \qquad \text{(distance formula)}$$

$$M = \left(\frac{x_1 + x_2}{2}, \frac{y_1 + y_2}{2}\right) \qquad \text{(midpoint formula)}$$

Example 2 If $A = (-1, 9)$ and $B = (4, -3)$, find:
   a. the length of $\overline{AB}$
   b. the coordinates of the midpoint of $\overline{AB}$

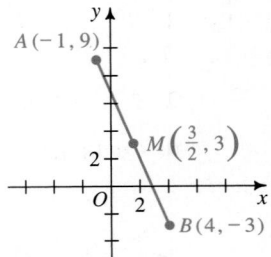

Solution a. Find the distance between A and B.
   $AB = \sqrt{(4 - (-1))^2 + (-3 - 9)^2}$
   $= \sqrt{25 + 144}$
   $= 13$

   b. $M = \left(\frac{-1 + 4}{2}, \frac{9 + (-3)}{2}\right) = \left(\frac{3}{2}, 3\right)$

CLASS EXERCISES

Find the length and the coordinates of the midpoint of $\overline{CD}$.

1. $C(0, 0)$, $D(8, 6)$ 10; (4, 3)
2. $C(4, 2)$, $D(6, 6)$ $2\sqrt{5}$; (5, 4)
3. $C(-3, 4)$, $D(3, -2)$ $6\sqrt{2}$; (0, 1)
4. $C(7, -9)$, $D(7, -1)$ 8; (7, -5)
5. Let $A = (2, 3)$, $B = (6, 7)$, and M be the midpoint of $\overline{AB}$. Find the coordinates of the midpoints of $\overline{AM}$ and $\overline{MB}$. (3, 4); (5, 6)
6. Which of the following points are on the line $2x + 3y = 15$? a, b, d
   a. $(3, 3)$ b. $(9, -1)$
   c. $(2.5, 3.5)$ d. $(-10.5, 12)$
7. The point $(8, 4)$ is on the horizontal line h.
   a. Name three other points on h. See below.
   b. State an equation of h. $y = 4$
8. The point $(8, 4)$ is on the vertical line v.
   a. Name three other points on v. See below.
   b. State an equation of v. $x = 8$

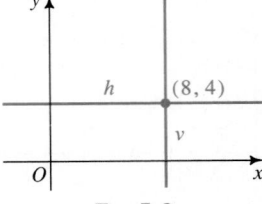

Exs. 7, 8

7. a. For example, $(0, 4)$, $(1, 4)$, $(2, 4)$

8. a. For example, $(8, -2)$, $(8, 0)$, $(8, 8)$

9. Find the coordinates of the points where the line $4x - 3y = 18$ intersects the axes. (0, −6), (4.5, 0)

10. The diagram at the right shows the graphs of the equations $x + y = 5$ and $2x - y = 1$.
    a. Estimate the coordinates of the point of intersection. (2, 3)
    b. Find the exact coordinates of the point of intersection by solving the equations simultaneously. (2, 3)
    c. Compare the solution with your estimate.

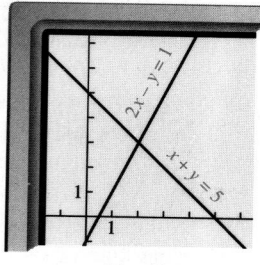

Ex. 10

WRITTEN EXERCISES

Find the length of $\overline{CD}$ and the coordinates of the midpoint of $\overline{CD}$.

A
1. $C(1, 0)$, $D(7, 8)$ 10; (4, 4)
2. $C(3, 3)$, $D(15, 12)$ 15; (9, 7.5)
3. $C(-8, -3)$, $D(7, 5)$ 17; $\left(-\frac{1}{2}, 1\right)$
4. $C(-2, -1)$, $D(4, 9)$ $2\sqrt{34}$; (1, 4)
5. $C\left(\frac{1}{2}, \frac{9}{2}\right)$, $D\left(-2, -\frac{3}{2}\right)$ $\frac{13}{2}$; $\left(-\frac{3}{4}, \frac{3}{2}\right)$
6. $C\left(\frac{7}{2}, -1\right)$, $D\left(-\frac{5}{2}, \frac{7}{2}\right)$ $\frac{15}{2}$; $\left(\frac{1}{2}, \frac{5}{4}\right)$
7. $C(4.8, 2.2)$, $D(4.8, -2.8)$ 5; (4.8, −0.3)
8. $C(1.7, 5.7)$, $D(-2.3, 5.7)$ 4; (−0.3, 5.7)

9. Which of the following points are on the graph of $3x - 2y = 15$? a, c
   a. (9, 6)　　**b.** (8, 4)　　**c.** $\left(-\frac{4}{3}, -\frac{19}{2}\right)$　　**d.** (3.4, −3.2)　　**e.** (−9, −22)

10. Which of the following points are on the graph of $-5x + 4y = 18$? a, d, e
    a. (−1.2, 3.0)　　**b.** $\left(3, -\frac{3}{4}\right)$　　**c.** (−18, 24)　　**d.** (−6, −3)　　**e.** (3.6, 9)

In Exercises 11 and 12, graph each equation. Label the origin and the x- and y-intercepts as O, P, and Q, respectively. Find the area of $\triangle OPQ$.

11. $3x - 2y = 6$ 3
12. $4x + 3y = 24$ 24

13. On a single set of axes, sketch the horizontal line through (4, 3) and the vertical line through (5, −2). What is the intersection of these lines? What are the equations of these lines? (5, 3); $y = 3$, $x = 5$

14. Repeat Exercise 13 for the horizontal line through (−2, −1) and the vertical line through (−2, 3). (−2, −1); $y = -1$, $x = -2$

In Exercises 15–18, solve the given pair of equations simultaneously. Then sketch the graphs of the equations and label the intersection point.

15. $3x - 5y = 9$
    $x + y = 3$ (3, 0)
16. $2x + 3y = 15$
    $4x - 9y = 3$ $\left(\frac{24}{5}, \frac{9}{5}\right)$
17. $x - 3y = 4$
    $5x + y = -8$ $\left(-\frac{5}{4}, -\frac{7}{4}\right)$
18. $-2x - 6y = 18$
    $x - 3y = 6$ $\left(-\frac{3}{2}, -\frac{5}{2}\right)$

Using Technology

A graphing calculator can be used in Class Exercise 10 to estimate the point of intersection. Use the zoom feature to refine the estimate.

Review Note

For Written Exercises 11 and 12, remind students that the area of a right triangle is one-half the product of the lengths of the legs.

19. Plot $A(1, 7)$, $B(3, 5)$, $C(4, -1)$, and $D(2, 1)$. Use the distance formula to show that the opposite sides of quadrilateral $ABCD$ are equal in length. What kind of figure is $ABCD$? Parallelogram

20. Plot $A(-6, 3)$, $B(-1, 6)$, $C(2, 1)$, and $D(-3, -2)$. Use the distance formula to show that quadrilateral $ABCD$ is a square. (*Hint:* Show that the four sides are equal in length and that the two diagonals are equal in length.)

21. Plot $A(5, 1)$, $B(7, -1)$, $C(1, -3)$, and $D(-1, -1)$. Use the midpoint formula to show that the diagonals of quadrilateral $ABCD$ have the same midpoint. What kind of quadrilateral is $ABCD$? Parallelogram

22. Plot $A(2, 0)$, $B(4, -6)$, $C(9, 1)$, and $D(7, 7)$. Show that $\overline{AC}$ and $\overline{BD}$ bisect each other. What kind of quadrilateral is $ABCD$? Parallelogram

23. Given $A(-3, 3)$, $B(1, 11)$, and $C(3, 15)$, show that B is on $\overline{AC}$ by showing that $AB + BC = AC$.   **27.** $(9, 0)$, $(-15, 0)$

24. Repeat Exercise 23 for $A(-3, 7)$, $B(-1, 4)$, and $C(3, -2)$.   **28.** $(0, 3)$, $(0, -1)$

B 25. **a.** Show that $P(4, 2)$ is equidistant from $A(9, 2)$ and $B(1, 6)$.
   b. If $(2, k)$ is equidistant from A and B, find the value of k. -2

26. **a.** Show that $P(1, 4)$ is equidistant from $A(-5, -3)$ and $B(-1, -5)$.
   b. If $(3, k)$ is equidistant from A and B, find the value of k. 8

27. P is a point on the x-axis 13 units from the point $(-3, 5)$. Find all the possible coordinates for P.

28. Q is a point on the y-axis $2\sqrt{10}$ units from the point $(6, 1)$. Find all the possible coordinates of Q.

Parallel and intersecting lines create visual interest in the Pyramid at the Louvre Museum, Paris.

29. Show that the three lines whose equations are $x + 3y = 19$, $2x - 5y = 5$, and $x - 2y = 4$ intersect in one point. $(10, 3)$

30. Determine whether the three lines whose equations are $3x + 2y = 4$, $5x - 2y = 0$, and $4x + 3y = 3$ intersect in one point. No

31. **a.** Plot the points $A(-6, 7)$, $B(6, 3)$, and $C(-2, -1)$. Then show that $(BC)^2 + (AC)^2 = (AB)^2$. What can you conclude about $\angle C$? $\angle C$ is a rt $\angle$.
   b. Give the coordinates of the midpoint, M, of $\overline{AB}$. Verify these coordinates by showing that $CM = \frac{1}{2}AB$. $(0, 5)$

32. The area of a triangle with sides a, b, and c units long can be found using *Hero's (or Heron's) formula*:

$$\text{Area} = \sqrt{s(s - a)(s - b)(s - c)} \text{ where } s = \frac{a + b + c}{2}$$

Find to the nearest tenth the area of the triangle with vertices $A(-13, 2)$, $B(5, 17)$, and $C(22, -4)$. 316.5

C **33.** In $\triangle ABC$, $D(7, 3)$ is the midpoint of $\overline{AB}$, $E(10, 9)$ is the midpoint of $\overline{BC}$, and $A(2, -1)$, $F(5, 5)$ is the midpoint of $\overline{AC}$. Find the coordinates of A, B, and C. $B(12, 7)$, $C(8, 11)$

34. In this proof, you may assume the following: The distance between two points on the same vertical line is the absolute value of the difference in y-coordinates; the distance between two points on the same horizontal line is the absolute value of the difference in x-coordinates. Note that the first quadrant is used for convenience.

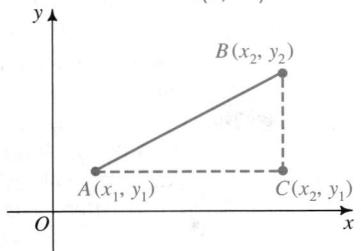

Given: $A = (x_1, y_1)$, $B = (x_2, y_2)$

Prove: $AB = \sqrt{(x_2 - x_1)^2 + (y_2 - y_1)^2}$

35. Given: $A = (x_1, y_1)$, $B = (x_2, y_2)$
M is the midpoint of $\overline{AB}$.
P is the midpoint of $\overline{BC}$.
Q is the midpoint of $\overline{AC}$.

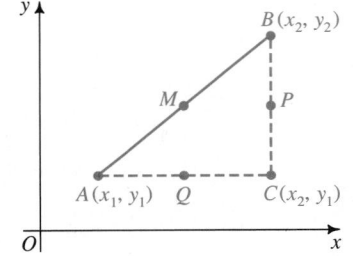

Prove: **a.** $P = \left(x_2, \dfrac{y_1 + y_2}{2}\right)$

b. $Q = \left(\dfrac{x_1 + x_2}{2}, y_1\right)$

c. Points M and P have the same y-coordinate.

d. Points M and Q have the same x-coordinate.

e. $M = \left(\dfrac{x_1 + x_2}{2}, \dfrac{y_1 + y_2}{2}\right)$

36. Three vertices of a parallelogram have coordinates $(-3, 1)$, $(1, 4)$, and $(4, 3)$. Find the coordinates of the fourth vertex. How many possible answers are there? $(0, 0)$ or $(-6, 2)$ or $(8, 6)$; 3

1-2 Slopes of Lines

Objective *To find the slope of a line and to determine whether two lines are parallel, perpendicular, or neither.*

The slope of a nonvertical line is a number measuring the steepness of the line relative to the x-axis. Let (x_1, y_1) and (x_2, y_2) be any two points on a line. The difference in the y values gives the *rise*, and the difference in the x values gives the *run*. The slope is the ratio of the rise to the run. That is, the slope m is defined by:

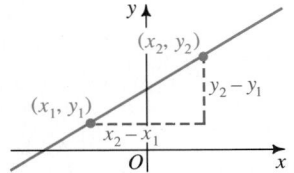

$$m = \frac{y_2 - y_1}{x_2 - x_1}$$

Linear and Quadratic Functions   **7**

Problem Solving

Encourage students to solve Exercise 36 visually by plotting the points first, completing the parallelogram with one point, and then completing a triangle with the three given points as midpoints of the sides. This method will help them locate the other two points (for a total of three possible solutions).

Teaching Notes, p. 1C

Warm-Up Exercises

Graph the given pair of equations. Then tell whether the graphed lines intersect, are parallel, or coincide.

1. $3x + 2y = 6$, $3x + 2y = -6$ parallel

2. $x - y = 4$, $x + y = 4$ intersect

3. $2x - y = 1$, $y - 2x = -1$ coincide

4. $x + 3y = 2$, $3x + y = 2$ intersect

5. $-4x + 3y = 12$, $4x - 3y = 12$ parallel

Motivating the Section

A good way to motivate a discussion of slope is to ask students if train tracks are ever sloped. Most students will say "yes" but do not know that the slope must be very gradual (usually with a rise of less than 3%). Finding the slope of a line is an objective of this section.

Some important facts about the slope of a line follow.

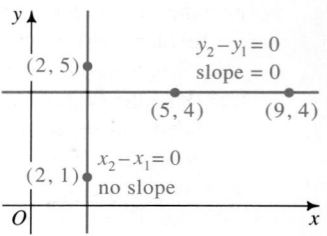

1. a. Horizontal lines have a slope of 0, because $y_2 - y_1 = 0$ for all *y*.

   b. Vertical lines have no slope, because $x_2 - x_1 = 0$ for all *x*.

   It should be apparent that having *no slope* is quite different from having *zero* slope.

2. The figures below show lines with *positive slope*. Lines with positive slope rise to the right as you look at points with greater and greater *x*-coordinates. The greater the slope, the more steeply the line rises.

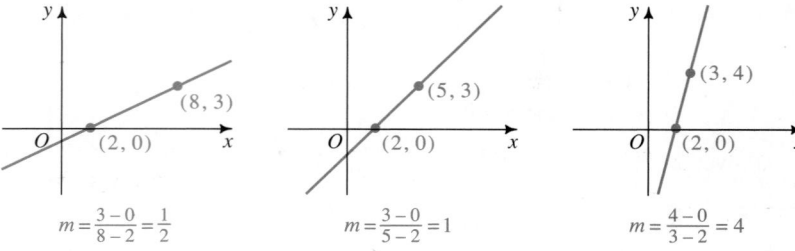

3. The figures below show lines with *negative slope*. Lines with negative slope fall to the right as you look at points with greater and greater *x*-coordinates. The greater the absolute value of the slope, the more steeply the line falls.

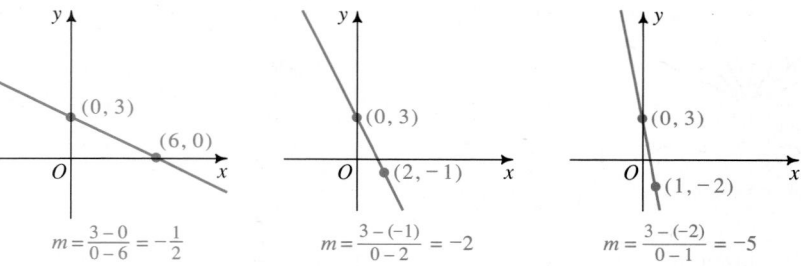

The following theorem provides an easy way to find the slope of a line from its equation.

Slope of a Line

The slope of $y = mx + k$ is *m*.
Since *m* is a constant value, the slope of a line is constant.

8   *Chapter One*

The fact that the slope is constant tells us that any two points on a line will always give the same value for the slope. See Written Exercise 27 for an algebraic proof of the theorem on the previous page.

The numbers m and k in $y = mx + k$ provide us with a mental picture of the line. We know that m is the slope of the line. The number k is the y-intercept, since $y = m(0) + k = k$. For these reasons, we refer to the form $y = mx + k$ as the **slope-intercept form**.

Example 1 What are the slope and y-intercept of the line $7x + 13y = 26$?

Solution Write the equation in slope-intercept form.

$$7x + 13y = 26$$
$$13y = -7x + 26$$
$$y = -\frac{7}{13}x + 2$$

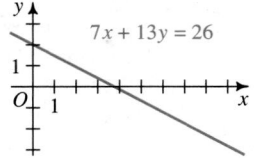

The slope is $-\frac{7}{13}$ and the y-intercept is 2.

The diagrams below illustrate the effect of m and k on the graph of an equation written in slope-intercept form. Notice that different lines with the same slope are parallel.

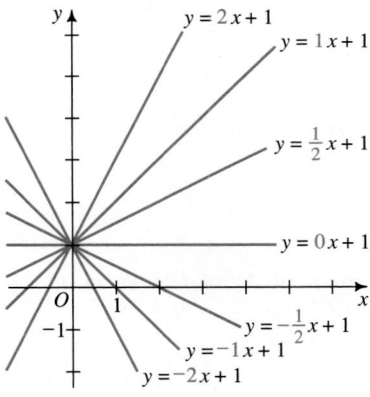
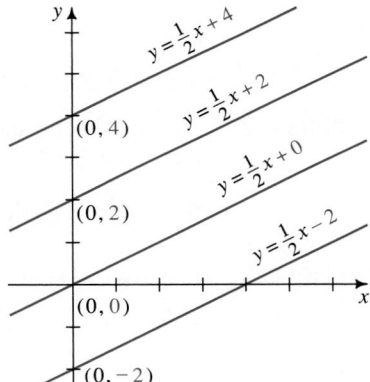

The two theorems that follow show how the concept of slope is related to the geometric ideas of parallel and perpendicular lines. See Written Exercises 28 and 29 for an algebraic proof and a geometric proof of the theorem below.

Slopes of Parallel Lines

Two nonvertical lines are parallel if and only if they have the same slope.

Linear and Quadratic Functions **9**

Point out that changing m but not k gives a whole family of lines that intersect at the same point on the y-axis. As the absolute value of m gets larger, the line gets steeper. Changing k but not m gives a whole family of lines parallel to each other.

Making Connections

You might want to introduce the idea of *translation* by showing that if the line $y = mx$ is moved up or down $|k|$ units, then $y = mx + k$ is the equation of the new line.

◩ Using Technology

A graphing calculator or computer can be used to demonstrate the effect of changing m or k on the graph of a line $y = mx + k$.

1. A line *l* has equation $x + 2y = 5$. What is the slope of a line **(a)** parallel to *l*? **(b)** perpendicular to *l*?

Write the equation in slope-intercept form, $y = mx + k$:

$2y = -x + 5$;

$y = -\frac{1}{2}x + \frac{5}{2}$

a. Line *l* has slope $-\frac{1}{2}$.
Any line parallel to *l* has slope $-\frac{1}{2}$.

b. Any line perpendicular to *l* has slope 2, since

$2 \cdot -\frac{1}{2} = -1$.

2. Find the value of *k* if the line joining $(2, k)$ and $(4, 5)$ and the line $y = 3x + 1$ are **(a)** parallel, **(b)** perpendicular.

a. The line with equation $y = 3x + 1$ has slope 3. The line joining $(2, k)$ and $(4, 5)$ has slope $\frac{k-5}{2-4} = \frac{k-5}{-2}$. Parallel lines have equal slopes, so $\frac{k-5}{-2} = 3$. Then $k - 5 = -6$ and $k = -1$.

b. The slope of the line joining $(2, k)$ and $(4, 5)$ must be $-\frac{1}{3}$. Thus,

$\frac{k-5}{-2} = \frac{-1}{3}$;

$3k - 15 = 2$; $k = \frac{17}{3}$.

As with the slopes of parallel lines, there is a special relationship between the slopes of perpendicular lines. The following activity, which leads us to the next theorem, demonstrates this relationship for line segments of equal length.

Activity

A right triangle is placed along the *x*-axis. An identical triangle is placed along the *y*-axis.

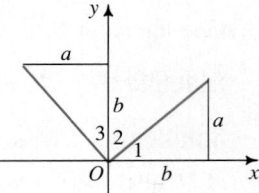

a. How are $\angle 1$ and $\angle 2$ related? Complementary
How are $\angle 1$ and $\angle 3$ related? Congruent

b. Deduce how $\angle 2$ and $\angle 3$ are related. Complementary

c. Deduce the relationship between the red and blue segments. Perpendicular

d. What is the slope of the red segment?
What is the slope of the blue segment? $\frac{a}{b}$; $-\frac{b}{a}$

Slopes of Perpendicular Lines

Two lines are perpendicular if and only if their slopes are *negative reciprocals* of each other. That is, if the slopes are m_1 and m_2, then

$$m_1 = -\frac{1}{m_2}, \text{ or } m_1 \cdot m_2 = -1.$$

See Written Exercises 30 and 31 for a geometric proof and an algebraic proof.

Example 2 The equations of three lines are given. Which lines are parallel and which are perpendicular?

$$l_1: y = \frac{3}{4}x - 7 \qquad l_2: 4x + 3y = 10 \qquad l_3: 3x - 4y = 11$$

Solution Find the slopes of the lines by rewriting their equations in slope-intercept form.

$$l_1: y = \frac{3}{4}x - 7 \qquad\qquad m_1 = \frac{3}{4}$$

$$l_2: y = -\frac{4}{3}x + \frac{10}{3} \qquad m_2 = -\frac{4}{3}$$

$$l_3: y = \frac{3}{4}x - \frac{11}{4} \qquad\qquad m_3 = \frac{3}{4}$$

Since both m_1 and m_3 are equal to $\frac{3}{4}$, lines l_1 and l_3 are parallel. Since

$$m_2 \cdot m_1 = m_2 \cdot m_3 = -\frac{4}{3} \cdot \frac{3}{4} = -1, \text{ line } l_2 \text{ is perpendicular to lines}$$

l_1 and l_3.

CLASS EXERCISES

Find the slope of the line joining the given points.

1. $(0, 0)$, $(5, 7)$ $\frac{7}{5}$ **2.** $(2, 0)$, $(3, 6)$ 6   **3.** $(-2, 3)$, $(1, 0)$ -1

4. $(2, -6)$, $(8, -3)$ $\frac{1}{2}$ **5.** $(-8, 2)$, $(8, 2)$ 0   **6.** $(5, -3)$, $(5, -7)$ No slope

Find the slope and *y*-intercept of the line whose equation is given.

7. $y = 3x + 4$ 3; 4   **8.** $y = \frac{3}{5}x - 3$ $\frac{3}{5}$; -3 **9.** $4x + 3y = 9$ $-\frac{4}{3}$; 3   **10.** $y = -2$

0; -2

11. Explain why a vertical line has no slope.

12. A line *l* has slope -2. What is the slope of a line perpendicular to *l*? $\frac{1}{2}$

13. A line *l* has equation $3x + 2y = 10$. What is the slope of a line **(a)** parallel to *l* and **(b)** perpendicular to *l*? See below.

14. a. Show that opposite sides of quadrilateral *ABCD* are parallel.

b. Show that adjacent sides of quadrilateral *ABCD* are perpendicular.

13. a. $-\frac{3}{2}$ **b.** $\frac{2}{3}$

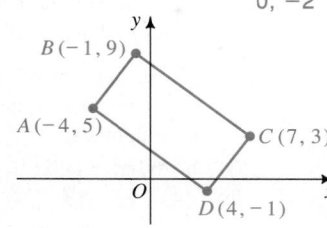

WRITTEN EXERCISES

Find the slope of the line joining the points whose coordinates are given.

1. $(4, 2)$, $(9, 5)$ $\frac{3}{5}$ **2.** $(0, 4)$, $(12, 0)$ $-\frac{1}{3}$

3. $(-4, -2)$, $(2, -6)$ $-\frac{2}{3}$ **4.** $(-2, 6)$, $(2, -2)$ -2

5. $(8, 5)$, $(-7, 5)$ 0   **6.** $(-3, 8)$, $(-3, -2)$ No slope

7. $\left(-\frac{1}{3}, 2\right)$, $\left(\frac{1}{4}, -\frac{1}{3}\right)$ -4 **8.** $(0.25, 1.5)$, $(0.5, 1)$ -2

9. (a, b), (b, a) -1 **10.** $\left(a, \frac{a}{b}\right)$, $\left(b, \frac{b}{a}\right)$ $\frac{a+b}{ab}$

Find the slope and the *y*-intercept of the line whose equation is given.

11. $y = 3x + 5$ 3; 5   **12.** $y = 4 - 5x$ -5; 4   **13.** $4x - 2y = 8$ 2; -4

14. $3x + 9y = 7$ $-\frac{1}{3}$; $\frac{7}{9}$ **15.** $3y = 11x$ $\frac{11}{3}$; 0   **16.** $y = 5$ 0; 5

Tell which of the given equations have parallel line graphs and which have perpendicular line graphs.

17. a. $y = \frac{5}{2}x - 8$ **b.** $-15x + 6y - 10 = 0$ **c.** $4x + 10y = 15$

See below.

18. a. $3y = 5x - 5$ **b.** $y = -\frac{3}{5}x + 4$ **c.** $10y = -6x - 7$

17. a $\parallel$ b, a $\perp$ c, b $\perp$ c   **18.** b $\parallel$ c, a $\perp$ b, a $\perp$ c

Linear and Quadratic Functions   **11**

Additional Answers
Class Exercises

11. The *x*-coordinates of any two points on a vertical line are equal. Substituting any two points (x_1, y_1) and (x_1, y_2) in the slope formula gives $\frac{y_2 - y_1}{0}$, which is undefined.

Using Technology

You might have students write short programs to find the slope of a line given two points on the line (see Written Exercises 1–8).

Exercise Note

In Exercises 17b and 17c, students may incorrectly assume that the slope is $-A$ (of the form $Ax + By = C$). Point out that the correct slope is $-A/B$.

Suggested Assignments

Standard
11/1–27 odd; Calc. Exs. 1, 2
Comprehensive
11/13–33 odd; Calc. Exs. 1, 2

Supplementary Materials

Alternative Assessment, 2
Precalculus Plotter Plus, 24

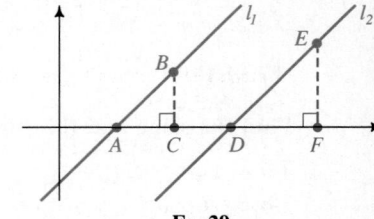

Cooperative Learning

You might want students to go over their solutions to Exercises 23–26 and 29–30 in small groups. Have them review the appropriate theorems from geometry.

Additional Answers
Written Exercises

19. Product of slopes =

$$\frac{2-(-3)}{7-2} \cdot \frac{2-7}{2-(-3)} =$$

$$\frac{5}{5} \cdot \frac{-5}{5} = -1$$

20. Rewrite the equation of line 2 in slope-intercept form: $y = \frac{3}{5}x - 3$.

Product of slopes =

$$\frac{-2-3}{5-2} \cdot \frac{3}{5} = \frac{-5}{3} \cdot \frac{3}{5} = -1$$

23. a. Slope of $\overline{AB}$ = slope of $\overline{CD} = \frac{5}{3}$; slope of $\overline{BC} =$

slope of $\overline{AD} = \frac{1}{3}$; so

$\overline{AB} \parallel \overline{CD}$, $\overline{BC} \parallel \overline{AD}$.

b. Midpoint of $\overline{AC}$ =

$$\left(\frac{-4+8}{2}, \frac{-6+6}{2}\right) =$$

$(2, 0) =$

$$\left(\frac{2+2}{2}, \frac{4+(-4)}{2}\right) =$$

midpoint of $\overline{BD}$.

---

19. Show that the line through $(2, -3)$ and $(7, 2)$ is perpendicular to the line through $(-3, 7)$ and $(2, 2)$.

20. Show that the line through $(2, 3)$ and $(5, -2)$ is perpendicular to the line $3x - 5y = 15$.

21. Find the value of k if the line joining $(4, k)$ and $(6, 8)$ and the line joining $(-1, 4)$ and $(0, 8)$ are **(a)** parallel, **(b)** perpendicular. **a.** 0 **b.** $\frac{17}{2}$

22. Find the value of h if the line joining $(3, h)$ and $(5, 10)$ and the line $y = 3x + 4$ are **(a)** parallel, **(b)** perpendicular. **a.** 4 **b.** $\frac{32}{3}$

B **23.** Given the points $A(-4, -6)$, $B(2, 4)$, $C(8, 6)$, and $D(2, -4)$:
  a. Show by using slopes that quadrilateral $ABCD$ is a parallelogram.
  b. Verify that both diagonals have the same midpoint.

24. Given the points $A(-4, 1)$, $B(2, 3)$, $C(4, 9)$, and $D(-2, 7)$:
  a. Show that quadrilateral $ABCD$ is a parallelogram with perpendicular diagonals.
  b. What special name is given to a quadrilateral like $ABCD$? Rhombus

25. Show that the points $(-4, -5)$, $(-3, 0)$, $(0, 2)$, and $(5, 1)$ are the vertices of an isosceles trapezoid.

26. Show that $(-1, -1)$, $(9, 4)$, $(20, 6)$, and $(10, 1)$ are the vertices of a rhombus, and then find the area of this rhombus. Area = 35

27. Let (x_1, y_1) and (x_2, y_2) be any two points on the line $y = Mx + k$.

Show that $M = \dfrac{y_2 - y_1}{x_2 - x_1}$.

28. a. Suppose lines l_1 and l_2 have equations $y = m_1x + b_1$ and $y = m_2x + b_2$, respectively. By solving these equations simultaneously, show that the

lines intersect when $x = \dfrac{b_2 - b_1}{m_1 - m_2}$.

  b. What can you say about this value of x when $m_1 = m_2$? What can you say about the intersection of the lines when $m_1 = m_2$?

29. a. Express the slopes of lines l_1 and l_2 in terms of the lengths of the sides of the triangles shown.
  b. Prove that if the slopes of lines l_1 and l_2 are equal, then:
    (1) $\triangle ABC \sim \triangle DEF$
    (2) $\angle BAC \cong \angle EDF$
    (3) $l_1 \parallel l_2$
  c. *The converse of part (b).*
    Prove that if $l_1 \parallel l_2$, then:
    (1) $\angle BAC \cong \angle EDF$
    (2) $\triangle ABC \sim \triangle DEF$
    (3) slope of l_1 = slope of l_2

Ex. 29

29. a. Slope of $l_1 = \dfrac{BC}{AC}$,

slope of $l_2 = \dfrac{EF}{DF}$

---

12   *Chapter One*

30. a. Express the slopes of lines l_1 and l_2 in terms of the lengths of the sides of the triangles shown.
  b. Prove that if the slopes of lines l_1 and l_2 are negative reciprocals of each other, then:
  (1) $\triangle OAB \sim \triangle OCD$
  (2) $\angle 1 \cong \angle 2$
  (3) $l_1 \perp l_2$
  c. *The converse of part (b).*
  Prove that if $l_1 \perp l_2$ then:
  (1) $\angle 1 \cong \angle 2$
  (2) $\triangle OAB \sim \triangle OCD$
  (3) slope of $l_1 \cdot$ slope of $l_2 = -1$

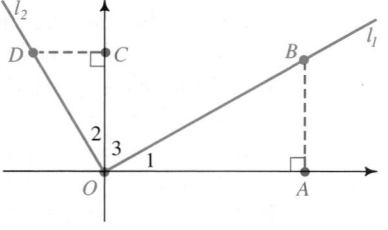

Ex. 30

30. a. Slope of $l_1 = \dfrac{BA}{OA}$,

slope of $l_2 = -\dfrac{OC}{DC}$

C **31.** The diagram at the right shows lines l_1 and l_2, chosen arbitrarily to intersect at $O(0, 0)$. These lines, whose equations are $y = m_1x$ and $y = m_2x$, intersect the vertical line $x = 1$ at $A(1, m_1)$ and $B(1, m_2)$.
  a. Use the distance formula to find OA, OB, and AB.
  b. Assume that $l_1 \perp l_2$. Use part (a) to show that if
$$(OA)^2 + (OB)^2 = (AB)^2,$$
  then $m_1 \cdot m_2 = -1$.
  c. Prove the converse to the statement in part (b).

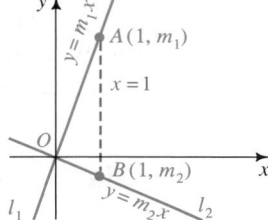

Ex. 31

32. $ABCD$ is a square with vertices $A(0, 0)$ and $B(6, 8)$. Give the coordinates of C and D. (Two answers are possible.) $C(-2, 14)$, $D(-8, 6)$ or $C(14, 2)$, $D(8, -6)$

33. $PQRS$ is a rectangle with vertices $P(-4, -1)$ and $Q(-6, 5)$ and $PQ = 2(QR)$. Find the coordinates of R and S. $R(-3, 6)$, $S(-1, 0)$ or $R(-9, 4)$, $S(-7, -2)$

▮▮▮▮ CALCULATOR EXERCISES

Consider the points $A(-1.8, 2.3)$, $B(-0.4, 4.4)$, and $C(2.4, -0.5)$.

1. Verify that $\triangle ABC$ is a right triangle and find its area. See below.

2. Show that the midpoint of the hypotenuse is equidistant from the three vertices.

1. slope $\overline{AB} \cdot$ slope $\overline{AC} = (1.5)(-0.\overline{6}) = -1$, so $\overline{AB} \perp \overline{AC}$; 6.37

2. Midpoint M of $\overline{BC}$ is $(1.0, 1.95)$. $AM = BM = CM = \sqrt{7.9625}$

Linear and Quadratic Functions **13**

Warm-Up Exercises

Find the slope, x-intercept, and y-intercept of the line whose equation is given.

1. $2x - 3y = 9$ $\frac{2}{3}; \frac{9}{2}; -3$

2. $x + 2y = 1$ $-\frac{1}{2}; 1; \frac{1}{2}$

3. $-4x + y = 2$ $4; -\frac{1}{2}; 2$

4. $x = -1$
no slope; -1; no y-int.

Motivating the Section

Finding the equation of a line is important in many applications. For example, if data points representing consumer spending trends over a period of time were plotted and found to have a "line of best fit," you could use the properties of the line to determine other data points that satisfy the equation of the line.

Problem Solving

You might want to ask your more capable students to show how the different forms of the equation of a line can be derived from the general form. Have them give some examples that show when certain forms are not applicable (for example, slope-intercept form is not applicable for vertical lines).

Review Note

Before doing Example 4, you might want to review the geometric development of the perpendicular bisector of a segment.

1-3 Finding Equations of Lines

Objective   *To find an equation of a line given certain geometric properties of the line.*

A linear equation in x and y can be written in several different forms, four of which are shown below.

The **general** form	$Ax + By = C$	
The **slope-intercept** form	$y = mx + k$	Line has slope m and y-intercept k.
The **point-slope** form*	$\frac{y - y_1}{x - x_1} = m$	Line has slope m and contains (x_1, y_1).
The **intercept** form	$\frac{x}{a} + \frac{y}{b} = 1$	Line has x-intercept a and y-intercept b.

When finding an equation of a line, sometimes it is easier to use one form rather than another. The following examples illustrate this.

Example 1   Find an equation of the line with x-intercept 8 and y-intercept 4.

Solution   Use the intercept form with $a = 8$ and $b = 4$.

$$\frac{x}{8} + \frac{y}{4} = 1$$

The equation in general form is $x + 2y = 8$.

Example 2   Find an equation of the line through $(1, 4)$ and $(5, -2)$.

Solution   **Step 1**   First we find the slope of the line.

$$m = \frac{4 - (-2)}{1 - 5} = -\frac{3}{2}$$

Step 2   Use the point-slope form with slope $-\frac{3}{2}$ and a point on the line, choosing either $(1, 4)$ or $(5, -2)$.

Using $(1, 4)$:   Using $(5, -2)$:

$$\frac{y - 4}{x - 1} = -\frac{3}{2} \qquad\qquad \frac{y + 2}{x - 5} = -\frac{3}{2}$$

Each equation can be written in general form as $3x + 2y = 11$.

*The point-slope equation is sometimes written in the form $y - y_1 = m(x - x_1)$ to make it formally apparent that (x_1, y_1) satisfies the equation.

Example 3 Find an equation of the line with y-intercept -3 and parallel to the line $2x + 5y = 8$.

Solution Rewriting the equation $2x + 5y = 8$ in slope-intercept form you get $y = -\frac{2}{5}x + \frac{8}{5}$. Thus, the slope of this line and the slope of the parallel line are both $-\frac{2}{5}$. Now use the slope-intercept form.

$$y = mx + k \longrightarrow y = -\frac{2}{5}x - 3$$

The equation in general form is $2x + 5y = -15$.

Example 4 illustrates two different ways of finding the perpendicular bisector of a line segment.

Example 4 Write an equation of the perpendicular bisector of the segment joining $A(-2, 3)$ and $B(4, -5)$.

Solution 1

Step 1 The slope of $\overline{AB}$ is

$$\frac{-5 - 3}{4 - (-2)} = -\frac{8}{6} = -\frac{4}{3}.$$

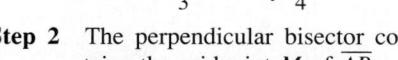

The slope of the perpendicular bisector is the *negative reciprocal* of $-\frac{4}{3}$, namely $\frac{3}{4}$.

Step 2 The perpendicular bisector contains the midpoint M of $\overline{AB}$:

$$M = \left(\frac{-2 + 4}{2}, \frac{3 + (-5)}{2}\right) = (1, -1)$$

Step 3 The equation of the line through $(1, -1)$ with slope $\frac{3}{4}$ is

$$\frac{y - (-1)}{x - 1} = \frac{3}{4},$$

which can be written in general form as $3x - 4y = 7$.

Solution 2 This method uses the fact that $P(x, y)$ is on the perpendicular bisector of $\overline{AB}$ if and only if $PA = PB$. Using the distance formula, proceed as follows:

$$PA = PB$$
$$\sqrt{(x + 2)^2 + (y - 3)^2} = \sqrt{(x - 4)^2 + (y + 5)^2}$$
$$(x^2 + 4x + 4) + (y^2 - 6y + 9) = (x^2 - 8x + 16) + (y^2 + 10y + 25)$$
$$4x - 6y + 13 = -8x + 10y + 41$$
$$12x - 16y = 28$$
$$3x - 4y = 7$$

Linear and Quadratic Functions **15**

1. Find an equation of the line with slope $-\frac{1}{4}$ and x-intercept -6.

The line contains $(-6, 0)$ and has slope $-\frac{1}{4}$.

Point-slope form:
$$\frac{y - 0}{x - (-6)} = -\frac{1}{4}$$
$$\frac{y}{x + 6} = \frac{-1}{4}$$
$$-x - 6 = 4y$$
$$\therefore x + 4y = -6$$

Slope-intercept form:
$$y = -\frac{1}{4}x + k$$
$$0 = -\frac{1}{4}(-6) + k$$
$$-\frac{3}{2} = k$$
$$y = -\frac{1}{4}x - \frac{3}{2}$$
$$\therefore x + 4y = -6$$

2. Find an equation of the line through $(1, -4)$ and perpendicular to the line $2x - y = 4$.

The line $2x - y = 4$, or $y = 2x - 4$, has slope 2. Thus, any line perpendicular to it has slope $-\frac{1}{2}$.

Point-slope form:
$$\frac{y + 4}{x - 1} = -\frac{1}{2}$$
$$2y + 8 = -x + 1$$
$$\therefore x + 2y = -7$$

Slope-intercept form:
$$y = -\frac{1}{2}x + k$$
$$-4 = -\frac{1}{2}(1) + k$$
$$-\frac{7}{2} = k$$
$$y = -\frac{1}{2}x - \frac{7}{2}$$
$$\therefore x + 2y = -7$$

CLASS EXERCISES

Find an equation in a convenient form of the line described.

1. The line with slope $\frac{5}{3}$ and y-intercept -2 $y = \frac{5}{3}x - 2$

2. The line through $(-4, 6)$ and parallel to the line $y = 3x + 8$ $\frac{y - 6}{x + 4} = 3$

3. The line through $(1, 2)$ and perpendicular to the line $y = 4x - 3$ $\frac{y - 2}{x - 1} = -\frac{1}{4}$

4. The line with x-intercept -1 and y-intercept 6 $-x + \frac{y}{6} = 1$

5. The line through $(8, 3)$ and $(2, -1)$ $\frac{y + 1}{x - 2} = \frac{2}{3}$

6. In the special case of a nonvertical line passing through the origin, the general form of its equation, $Ax + By = C$, becomes $Ax + By = 0$. Write **(a)** the slope-intercept form, **(b)** the point-slope form, and **(c)** the intercept form of a nonvertical line passing through the origin.
a. $y = mx$ **b.** $\frac{y}{x} = m$ **c.** No intercept form

WRITTEN EXERCISES

Write an equation of the line described. Answers are given in general form.

A **1.** The line with slope -2 and y-intercept 8 $2x + y = 8$

2. The line with slope $\frac{3}{5}$ and passing through the origin $3x - 5y = 0$

3. The line with y-intercept 4 and x-intercept -2 $2x - y = -4$

4. The line with y-intercept -6 and parallel to the line $5x + 4y = 1$ $5x + 4y = -24$

5. The line through $(-1, 4)$ and $(5, 8)$ **6.** The line through $(0, 5)$ and $(6, 1)$

7. The horizontal line through $(5, -7)$ **8.** The vertical line through $(5, -7)$

9. The line through $(2, -7)$ and $(2, 3)$ **10.** The line through $(5, -3)$ and $(2, -3)$

11. The line parallel to the line $0.3x - 1.2y = 6.4$ and with y-intercept 1.8 $0.25x - y = -1.8$

12. The line through $(-2, 4)$ parallel to the line through $(1, 1)$ and $(5, 7)$

13. The line through $(8, -2)$ perpendicular to the line $y = 7 - 2x$

14. The line through the origin perpendicular to the line $x - 3y = 9$

15. The perpendicular bisector of the segment joining $(0, 3)$ and $(-4, 5)$

16. The perpendicular bisector of the segment joining $(2, 4)$ and $(4, -4)$

The rows of crops suggest parallel lines.

12. $3x - 2y = -14$ **13.** $x - 2y = 12$
14. $3x + y = 0$ **15.** $2x - y = -8$ **16.** $x - 4y = 3$

16   *Chapter One*

B **17.** Given $A(2, 0)$ and $B(8, 4)$, show that $P(3, 5)$ is on the perpendicular bisector of $\overline{AB}$ by these two methods.

Method 1. Show $\overline{AB} \perp \overline{PM}$ where M is the midpoint of $\overline{AB}$.
Method 2. Show $PA = PB$.

18. Repeat Exercise 17 using $A(0, 7)$, $B(2, -1)$, and $P(5, 4)$.

19. $\triangle PQR$ has vertices $P(4, -1)$, $Q(-2, 7)$, and $R(9, 9)$.
  a. Find an equation of the median from R. $3x - 4y = -9$
  b. Find an equation of the altitude from R. $3x - 4y = -9$
  c. Are your answers to (a) and (b) the same? Is $\triangle PQR$ isosceles? Yes; yes

20. $\triangle DEF$ has vertices $D(-2, 5)$, $E(6, -1)$, and $F(5, 6)$.
  a. Verify that $\triangle DEF$ is isosceles. $DF = EF = 5\sqrt{2}$
  b. Write an equation of the bisector of $\angle F$. $4x - 3y = 2$

21. $\triangle ABC$ has vertices $A(-2, -1)$, $B(0, 7)$, and $C(8, 3)$.
  a. Write equations for the three medians of the triangle. $y = 3, y = x + 1, y = -2x + 7$
  b. Show that the three medians intersect at a single point G, called the *centroid*, or center of gravity, of the triangle. $G(2, 3)$

22. $\triangle PQR$ has vertices $P(0, 4)$, $Q(8, -2)$ and $R(7, 5)$.
  a. Write equations for the perpendicular bisectors of the three sides of the triangle. $4x - 3y = 13, 7x + y = 29, x - 7y = -3$
  b. Show that the perpendicular bisectors intersect at a single point C, called the *circumcenter* of the triangle. $C(4, 1)$
  c. Note that C is the midpoint of $\overline{PQ}$. What kind of triangle is $\triangle PQR$? isos. right $\triangle$

23. $\triangle KLM$ has vertices $K(0, 0)$, $L(18, 0)$, and $M(6, 12)$. $x = 6, x - y = 0,$
  a. Write equations for the altitudes to the three sides of the triangle. $x + 2y = 18$
  b. Show that the altitudes intersect at a single point O, called the *orthocenter* of the triangle. $O(6, 6)$ **24. d.** slope $\overline{GC} = -\frac{3}{4} = $ slope $\overline{CO}$

24. $\triangle RST$ has vertices $R(1, 2)$, $S(25, 2)$, and $T(10, 20)$.
  a. Find the centroid G of $\triangle RST$ using the method of Exercise 21. $(12, 8)$
  b. Find the circumcenter C of $\triangle RST$ using the method of Exercise 22. $(13, 7.25)$
  c. Find the orthocenter O of $\triangle RST$ using the method of Exercise 23. $(10, 9.5)$
  d. Show that G, C, and O lie on the same line, which is called *Euler's line*. See above.

25. Each of the lines shown below passes through $(2, 1)$ and forms a triangle with the axes. Which of these three triangles has the least area?

least area

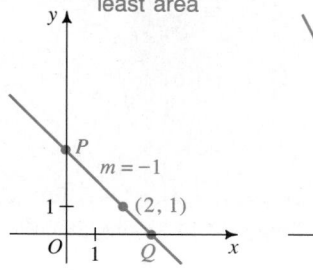
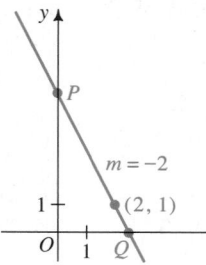
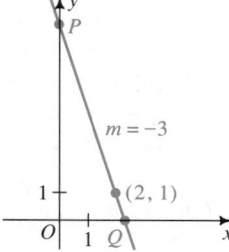

Linear and Quadratic Functions   **17**

Review Note

You might need to review the geometric meaning of median, altitude, centroid, circumcenter, and orthocenter of a triangle before assigning Exercises 19–24.

🔊 **Using Technology**

You can use geometric drawing software to help students visualize the triangles and related segments created in Exercises 19–24.

Mathematical Note

Exercise 29 develops a formula for the distance from a point P to a line $ax + by = c$. This formula can be generalized to give the distance from the point (x_1, y_1, z_1) to the plane $ax + by + cz = d$ as:

$$\frac{|ax_1 + by_1 + cz_1 - d|}{\sqrt{a^2 + b^2 + c^2}}$$

26. Consider the set of all lines through $(2, 6)$ with negative slope. Let P and Q be the points where the line intersects the y- and x-axes, respectively. Determine by experimentation the slope of the line that gives $\triangle OPQ$ the least area. -3

27. The vertices of $\triangle ABC$ are $A(8, 5)$, $B(0, 1)$, and $C(9, -2)$.
    a. Find the length and an equation of $\overline{BC}$. $3\sqrt{10}$; $x + 3y = 3$
    b. Find an equation of the altitude from A to $\overline{BC}$. $3x - y = 19$
    c. Find the point where the altitude from A intersects $\overline{BC}$. $(6, -1)$
    d. Find the length of the altitude from A to $\overline{BC}$. $2\sqrt{10}$
    e. Find the area of $\triangle ABC$. 30

28. Find the distance from the point $(9, 5)$ to the line $4x - 3y = -4$. 5

C 29. Given the point $P(x_1, y_1)$, and the line l with equation $ax + by = c$:
    a. Write an equation of the line j that passes through P and is perpendicular to l. $bx - ay = bx_1 - ay_1$
    b. Show that l and j intersect at the point

    $$Q\left(\frac{b^2x_1 - aby_1 + ac}{a^2 + b^2}, \frac{a^2y_1 - abx_1 + bc}{a^2 + b^2}\right).$$

    c. PQ is the distance between the point P and the line l. Show that

    $$PQ = \frac{|ax_1 + by_1 - c|}{\sqrt{a^2 + b^2}}.$$

    d. Use the formula in part (c) to solve Exercise 28.

30. a. Use the formula derived in Exercise 29 to verify that the point $(7, 9)$ is 3 units from the lines $12x - 5y = 0$ and $3x - 4y = 0$.
    b. Tell why $P(7, 9)$ is on the bisector of $\angle DOE$. Write an equation of this angle bisector and note that its slope is *not* the average of the slopes of lines OD and OE.

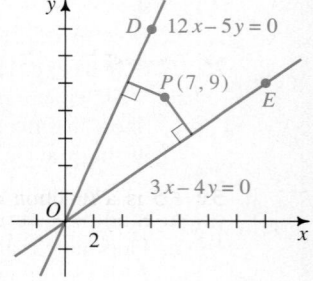

31. a. Suppose $P(x, y)$ is a point in the interior of $\angle DOE$, as shown in the diagram. Explain why P is on the bisector of $\angle DOE$ if

    $$\frac{|a_1x + b_1y - c_1|}{\sqrt{a_1^2 + b_1^2}} = \frac{|a_2x + b_2y - c_2|}{\sqrt{a_2^2 + b_2^2}}.$$

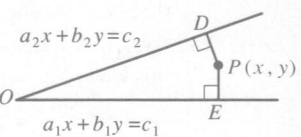

    b. Lines l_1 and l_2 are defined by the equations below.

    $$l_1: 4x - 3y = -6$$
    $$l_2: 3x - 4y = -4$$

    Use the equation of part (a) to write equations of the two lines that bisect the angles formed by l_1 and l_2. $x + y = -2$, $7x - 7y = -10$

1-4 Linear Functions and Models

Objective   *To model real-world situations by means of linear functions.*

A rail on a railroad track expands with heat and contracts with cold. Since its length depends on the temperature, we say that its length is a *function* of the temperature. If L stands for length in meters and T stands for temperature in degrees Celsius, then for a 10 m rail, L and T are related by this formula:

$$L = 0.0001T + 10$$

Activity

a. Complete the table of values below for $L = 0.0001T + 10$.

T (in °C)	0°	10°	20°	30°	40°
L (in m)	10	?	?	?	?
		10.001	10.002	10.003	10.004

b. Graph the values found in part (a) using T as the horizontal axis and L as the vertical axis in a coordinate plane.

c. The graph in part (b) will look very much like a horizontal line. Why?
_____   Slope is close to 0.

A *function* describes a dependent relationship between quantities. For example, the value of the expression $3x - 5$ depends on the value of x, so we say that $3x - 5$ is a function of x and write $f(x) = 3x - 5$. The function notation $f(x)$ is read "f of x." It represents the value of the function f evaluated at x. Thus, if $f(x) = 3x - 5$, then $f(2) = 1$. Notice that $f\!\left(\dfrac{5}{3}\right) = 0$, so we say that $\dfrac{5}{3}$ is a *zero* of the function f. In general, if $f(a) = 0$, then a is called a **zero** of the function f.

Some examples of linear functions are given below.

$f(x) = 3x - 5$	f is a linear function of x.
$L(T) = 0.0001T + 10$	L is a linear function of T.
$g(s) = -1.2s + 4.7$	g is a linear function of s.
$h(t) = 3$	h is a linear function of t.

Linear functions have the form $f(x) = mx + k$. The function h above is linear because it can be written as $h(t) = 0t + 3$. This special kind of linear function is called a *constant function*. The graphs of linear functions are straight lines or points on a line, as Example 1 shows on the next page.

Linear and Quadratic Functions   **19**

Warm-Up Exercises

Write an equation of the line.

1. the line through $(-2, 1)$ and $(0, 5)$ $y = 2x + 5$

2. the line with slope -1 and passing through $(1, 3)$
$y = -x + 4$

3. the line parallel to the line $2x + y = 3$ and with y-intercept 1 $y = -2x + 1$

4. the line perpendicular to the line $x - 3y = 4$ and passing through $(2, -4)$
$y = -3x + 2$

Motivating the Section

A dependent relationship between quantities in real life often is modeled by a mathematical function. For example, the cost of automobile insurance is a function of the age of the car. Ask students for other examples.

Communication Note

Make sure students do not think that the use of parentheses in function notation implies multiplication.

Problem Solving

It may be helpful for students to make a table of values for many of the functions in this section. If a function is linear, only two points are needed to determine its graph. A third point may be used as a check. Have students give an example of a linear equation that is not a function. (Answers will vary. Example: $x = 2$)

Example 1 The senior class has paid $200 to rent a roller skating rink for a fund raising party. Tickets for the party are $5 each.

   a. Express the net income as a function of the number of tickets sold.
   b. Graph the function. Identify the point at which the class begins to make a profit.

Solution **a.** Let $I(n) =$ the net income when n tickets are sold.

Net income is the amount of money remaining after operating expenses are deducted from the amount of money earned.

$I(n) = \$5 \times$ the number of tickets sold $- \$200$ rental fee
$I(n) = 5n - 200$

b. The graph of I is shown in red below. Since n must be a nonnegative integer, the graph consists of discrete (isolated) points on a line. However, the line that contains the discrete points is often given as a sketch of the function. (See the graph in blue.) From these graphs you can see that $n = 40$ is a zero of the income function. Therefore, the class must sell more than 40 tickets to make a profit.

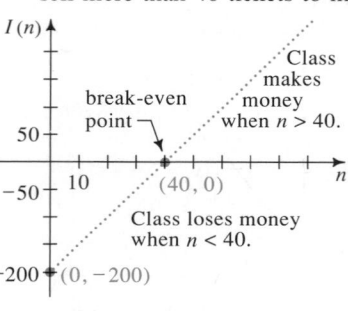

Graph of function I as discrete points on a line

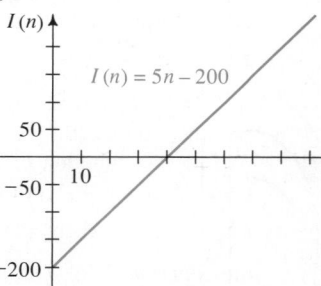

Graph of function I as a continuous portion of a line

The *domain* of a function is the set of values for which the function is defined. In the function $I(n) = 5n - 200$ in Example 1, n represents the number of tickets sold, so the domain is the set of nonnegative integers. Substituting $n = -3$ or $n = 4.5$ does not make sense in this case. You can think of the domain of a function as the set of input values. The set of output values is called the *range* of the function. When you substitute the domain values $n = 0, 1, 2, 3, \ldots$, you get the range values $I(n) = -200, -195, -190, -185, \ldots$.

The function $I(n) = 5n - 200$ is said to *model* income. A **mathematical model** is one or more functions, graphs, tables, equations, or inequalities that describe a real-world situation. Example 2 illustrates another model.

20 *Chapter One*

Example 2   Suppose that it costs 50 cents for the first minute of a long distance telephone call and 20 cents for each additional minute or fraction thereof. Give a graphical model of the cost of a call lasting t minutes.

Solution   It is important to realize that the cost of a call lasting 2 min and 12 seconds (s) is the same as a 3 min call. Likewise, a $3\frac{1}{2}$ min call costs the same as a 4 min call. The table below lists some of the input and output values for this cost function.

Input (time in minutes) t	Output (cost in cents) $C(t)$
$0 < t$ and $t \le 1$	50
$1 < t$ and $t \le 2$	$50 + 20(1) = 70$
$2 < t$ and $t \le 3$	$50 + 20(2) = 90$
$3 < t$ and $t \le 4$	$50 + 20(3) = 110$

The function that models this cost is a *step function*, so named because its graph has steps. See figure (a) below. Because the linear function $f(x) = 50 + 20x$ gives good approximations to the costs, it can also be used to model the cost of long distance calls. See figure (b). Be aware that if you use the linear function, you will get overestimates of the cost, as you can see in figure (c), which shows the graphs of both functions on the same set of axes.

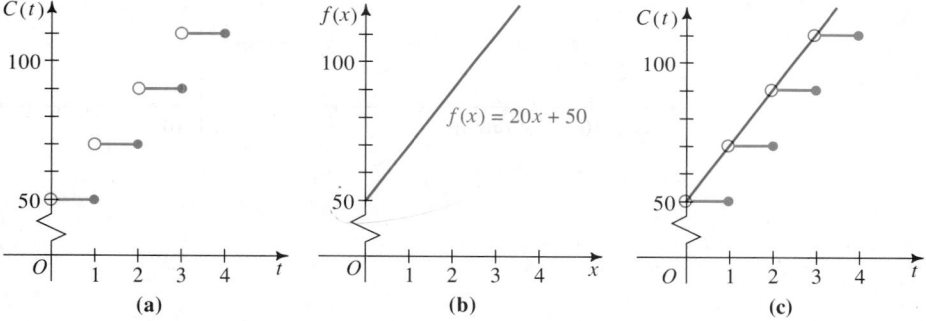

(a)   (b)   (c)

CLASS EXERCISES

1. **a.** If $f(x) = 5x - 10$, find $f(3)$ and $f(-3)$. 5; −25
   b. Find the zero of f. 2
   c. Graph the function. Where does the graph of f intersect the vertical axis? the horizontal axis? (0, −10); (2, 0)
   d. Where does the zero of f appear on the graph of f?
   At the intersection with the horizontal axis

Linear and Quadratic Functions   **21**

Additional Examples cont.

2. Let g be a linear function such that $g(1) = 2$ and $g(5) = 4$.
   a. Sketch the graph of g.
   b. Find an equation for $g(x)$.
   c. Find $g(-1)$.

   a. Draw the line through (1, 2) and (5, 4).

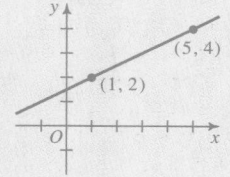

   b. Since $g(x) = mx + k$,
   $2 = m \cdot 1 + k$
   $\underline{4 = m \cdot 5 + k}$
   $2 = 4m;\ m = \frac{1}{2}$

   $2 = \frac{1}{2} \cdot 1 + k;\ k = \frac{3}{2}$

   $\therefore\ g(x) = \frac{1}{2}x + \frac{3}{2}$

   c. From the graph, $g(-1) = 1$. Or, from the equation, $g(-1) = \frac{1}{2}(-1) + \frac{3}{2} = 1$.

Suggested Assignments

Standard
22/1–23 odd
Comprehensive
23/13–24

Supplementary Materials

Alternative Assessment, 3
Using Technology, 6–7
Student Resource Guide, 1–4

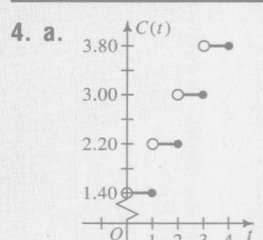
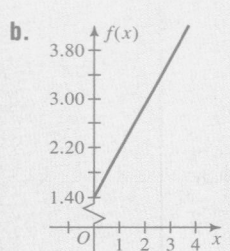
2. Which of the following are linear functions? a, b
   a. $f(t) = \dfrac{8 - 9t}{3}$ b. $g(x) = \dfrac{x}{4} - 3$ c. $h(x) = \dfrac{4}{x} - 3$

3. Using the equation $s + 2t = 8$, express t as a function of s. $t(s) = -\dfrac{s}{2} + 4$

4. It costs \$1.40 for the first minute of a phone call to Paris, France, and \$.80 for each additional minute or fraction thereof.
   a. Draw the graph of a step function that models this cost.
   b. Sketch the graph of a linear function that approximates this cost.

5. **Writing** Suppose that $f(x) = mx + k$. Write an expression for $f(a) - f(b)$ in terms of a, b, and m only. In your own words, explain how this formula can be used to evaluate mentally $f(79.6) - f(69.6)$ for the function $f(x) = 50x + 387$. Use the formula to find the value of $f(12.8) - f(12.3)$. $f(a) - f(b) = m(a - b)$; 25

WRITTEN EXERCISES

[A]

1. a. If $f(x) = \dfrac{3}{4}x - \dfrac{1}{2}$, find $f(2)$ and $f(-2)$. 1; -2 b. Find the zero of f. $\dfrac{2}{3}$

2. a. If $C(n) = 20 - \dfrac{5}{8}n$, find $C(0)$ and $C(16)$. 20; 10   b. Find the zero of C. 32

3. Let $f(x) = 3x - 7$. Decide whether $f(2) + f(6) = f(8)$. No

4. Let $h(t) = \dfrac{9 - 4t}{2}$. Decide whether $h(4.5) - h(3.5) = h(1)$. No

5. Consider the constant functions $g(x) = 2$ and $h(x) = -1$.
   a. Graph the two functions. What are the slopes of their graphs? 0
   b. Write the functions g and h in the form $f(x) = mx + k$. $g(x) = 0x + 2$, $h(x) = 0x - 1$

6. Consider the constant function $P(x) = -0.5$.
   a. Find $P(1269.35)$. -0.5
   b. Does the function P have any zeros? Explain. No; $0 = -0.5$ has no sols.

7. a. What is the slope of the graph of $f(x) = 1.5x - 2$? 1.5
   b. What is the zero of the function? $\dfrac{4}{3}$
   c. What are the intercepts of the graph? x-int. $\dfrac{4}{3}$, y-int. -2

8. a. What is the slope of the graph of $C(t) = 80t + 5.2$? 80
   b. Where does the graph of the function C intersect the vertical axis? (0, 5.2)

9. The graph of the linear function f has slope -2 and intersects the n-axis at $n = 6$. Find an equation for $f(n)$. $f(n) = -2n + 12$

10. The zero of the linear function S is 3. The graph of this function intersects the vertical axis at $S(x) = -2$. Find an equation for $S(x)$. $S(x) = \dfrac{2}{3}x - 2$

22   *Chapter One*

11. Let f be a linear function such that $f(1) = 5$ and $f(3) = 9$.
    a. Sketch the graph of f.
    b. Find an equation for $f(x)$. $f(x) = 2x + 3$

12. Let g be a linear function such that $g(-1) = -3$ and $g(-4) = 12$.
    a. Sketch the graph of g.
    b. Find an equation for $g(x)$. $g(x) = -5x - 8$

13. Bill bikes 4 km to school. After 5 min he is 3.2 km from the school. The graph of the function that models Bill's distance to school is shown in red. Because of hills and traffic conditions along the way, Bill's speed varies. Nevertheless, a linear function, whose graph is shown in blue, can also be used to model his distance as a function of his time spent bicycling.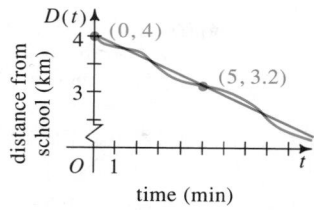
    a. Give an equation of the linear function.
    b. Use the equation to find approximately how long it takes Bill to bicycle to school.   **a.** $D(t) = -0.16t + 4$ **b.** About 25 min

14. Different round-trip airfares from Boston to various cities are shown in the table below.

From Boston to:	Buffalo	Chicago	Houston	Los Angeles	Hawaii
distance (mi)	400	850	1600	2600	5100
cost (dollars)	118	198	298	350	400

    a. Using this data, sketch a graph of the cost as a function of distance.
    b. Does it appear that a linear function would give good approximations to this cost function? Explain.

B **15.** Maria Correia's new car costs \$280 per month for car payments and insurance. She estimates that gas and maintenance cost \$0.15 per mile.
    a. Express her total monthly cost as a function of the miles driven during the month. $C(m) = 0.15m + 280$
    b. What is the slope of the graph of the cost function? 0.15

16. **Business** A recording studio invests \$24,000 to produce a master tape of a singing group. It costs \$1.50 to make each copy of the master and cover the operating costs.
    a. Express the cost of producing t tapes as a function $C(t)$. $C(t) = 1.5t + 24,000$
    b. If each tape is sold for \$6.50, express the revenue (the total amount received from the sale) as a function $R(t)$. $R(t) = 6.5t$
    c. Sketch the graphs of the functions C and R.
    d. Find the coordinates of the break-even point (that is, where the graphs intersect). (4800, 31,200)

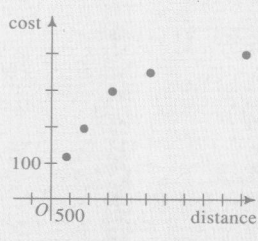
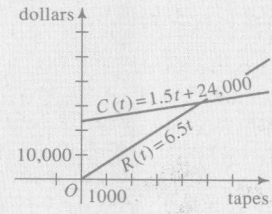

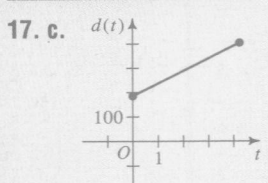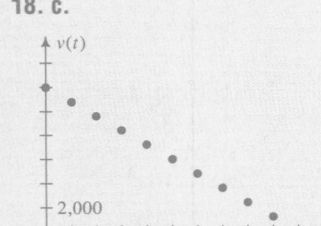
17. A salesman whose home is in Dallas, Texas, leaves Shreveport, Louisiana, heading east for Jackson, Mississippi. He drives the 220 mi trip at a steady speed, arriving in Jackson exactly $4\frac{1}{4}$ h after leaving Shreveport. Let t be the traveling time in hours after the salesman leaves Shreveport. The salesman's distance from Dallas, $d(t)$, is a linear function of t.

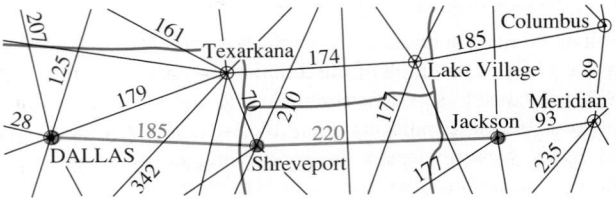

  a. Find to the nearest tenth the salesman's average driving speed after leaving Shreveport, and then find an equation for $d(t)$. 51.8 mi/h; $d(t) = 185 + 51.8t$
  b. What is the domain of the function d? Set of reals from 0 to 4.25 incl.
  c. Sketch the graph of d and give its slope. Slope is 51.8.
  d. What is the relationship between the salesman's driving speed and the slope of the graph of the function? They are equal.

18. **Accounting** A businesswoman buys a $12,000 computer system. The value of the system is considered to depreciate by a fixed amount, which is 10% of the purchase price each year over a ten-year period. (This method of determining the value is called *straight-line depreciation*.)
  a. Express the value of the computer system as a function of the number of years since its purchase. $v(t) = 12,000 - 1200t$
  b. What is the domain of the function? Set of integers from 0 to 10 incl.
  c. Sketch the graph of the function and give its slope. Slope is −1200.
  d. What is the relationship between the amount of yearly depreciation of the computer system and the slope of the graph of the function? They are equal.

19. *Writing* Write a sentence that uses the common meaning of the word slope, and then write a sentence that uses the mathematical meaning of the word slope. Write a paragraph that compares and contrasts the common and mathematical meanings.

20. At a city garage, it costs $4 to park for the first hour and $2 for each additional hour or fraction thereof. The fee is a function of the time parked.
  a. Sketch the graph of a function that models this parking fee.
  b. Sketch the graph of a linear function that approximates the parking fee. Find an equation for this linear function. Answers may vary: $C(t) = 2t + 4$

21. A taxicab driver charges $2.00 for the first half mile and $.40 for each additional quarter mile. The fare is a function of the distance traveled.
  a. Sketch the graph of a function that models this fare.
  b. Sketch the graph of a linear function that approximates the fare. Find an equation for this linear function. Answers may vary: $C(m) = 1.6m + 1.2$

22. The table below shows that the measure of each angle of a regular polygon is a function of the number of sides.

number of sides	3	4	5	6	. . .	n
angle measure	60°	90°	108°	120°	. . .	$A(n)$

   a. Explain how the data in the table tell us that the function is not linear.
   b. What is the domain of the function? The set of integers ≥ 3
   c. Use your knowledge of geometry to write an equation for $A(n)$. $A(n) = 180 - \dfrac{360}{n}$

23. The last test that Mr. Clements gave was so hard that he decided to scale the grades upward. He decided to raise the lowest score of 47 to a 65 and the highest score of 78 to a 90. Find a linear function that would give a fair way to convert the other test scores. $f(x) = \dfrac{25}{31}x + \dfrac{840}{31}$

C **24.** When David Arnold started out on his trip, his odometer read **45973** and his fuel gauge read $\frac{7}{8}$ full. Exactly 1 h 48 min later, the odometer read **46081** and the fuel gauge read $\frac{1}{2}$ full.

   a. Under what conditions does it seem reasonable that the odometer reading would be a linear function of the fuel gauge reading? Driving at a fixed speed
   b. Find an equation for the linear function, and sketch its graph. See below.
   c. What is the real-world significance of the point where the graph of the function crosses the vertical axis? Odometer reading when the fuel tank is empty
   d. When David sees the $\frac{1}{8}$ full reading on his fuel gauge, he wonders if he has enough gas to get to Hartford, 40 mi away, driving at the same speed. Can he reach Hartford without refueling? No
   e. Express the odometer reading as a function of the time spent driving.

   b. $f(x) = -288x + 46225$ **e.** $f(t) = t + 45973$, t in minutes

■ Quadratic Functions

1-5 The Complex Numbers

Objective   To add, subtract, multiply, and divide complex numbers.

Throughout the history of mathematics, new kinds of numbers have been invented to fill deficiencies in the existing number system. In earliest times, there were only the *counting numbers*, 1, 2, 3, ... The Egyptians and the ancient Greeks invented the *rational numbers*, so named because they are ratios of integers, to represent fractional parts of quantities. The Greeks also discovered that some numbers were not rational. For example, the ratio of the length of a diagonal of a square to the length of a side cannot be represented as the quotient of two integers. We know that this ratio is $\sqrt{2}$, an *irrational number*.

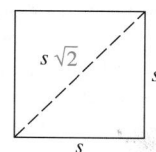

Warm-Up Exercises

Carry out the indicated operation and simplify.

1. $(5 - 2x) + (-3 + x)$
   $2 - x$

2. $(-1 + 3x) - (1 + 2x)$
   $-2 + x$

3. $(2 + 5x)(1 - x)$
   $2 + 3x - 5x^2$

4. $(-3 + 4x)^2$
   $9 - 24x + 16x^2$

5. $(2 - 7x)(2 + 7x)$ $4 - 49x^2$

Motivating the Section

Complex numbers, the topic introduced in this section, have applications in electronics. Electrical engineers often express current, voltage, and impedance (total effective resistance) in a simplified electrical circuit as complex numbers.

Assessment Note

Ask students for examples of counting numbers, rational numbers, and irrational numbers to see if students classify the numbers correctly.

Communication Note

Emphasize that i is a numeral that stands for a specific number, namely $\sqrt{-1}$. It is not a variable.

Mathematical Note

Point out that complex numbers satisfy the associative, commutative, and distributive properties of addition and multiplication, and they have additive and multiplicative inverses. Also, the number i can be defined as a solution of $x^2 + 1 = 0$, an equation that does not have real solutions.

Review Note

Before doing Examples 1 and 2, you might want to review how to simplify radical expressions.

Error Analysis

Students commonly mistake $\sqrt{-4} \cdot \sqrt{-9}$ for $\sqrt{36}$, or 6. In fact, $\sqrt{-4} \cdot \sqrt{-9} = -6$, not 6. The product property of radicals cannot be used for complex numbers. See Class Exercise 11, page 28.

Additional Examples

1. Simplify:
   a. $\sqrt{-3} \cdot \sqrt{-6}$
   b. $\dfrac{1}{1-i}$
   c. i^{23}
   a. $\sqrt{-3} \cdot \sqrt{-6} = $
   $i\sqrt{3} \cdot i\sqrt{6} = i^2\sqrt{18} = $
   $-1(3\sqrt{2}) = -3\sqrt{2}$
   (Note: $\sqrt{a} \cdot \sqrt{b} \neq \sqrt{ab}$
   if $a < 0$ and $b < 0$.)

Real Numbers

Zero was originally used by the Hindus to represent an empty column in a counting board similar to an abacus. The concept was brought to Europe by way of Arabia in the 9th century. The negative numbers were invented very much later in Renaissance Europe. One story tells that the plus and minus signs were first used in German warehouses to indicate whether crates of goods weighed more or less than some standard amount.

Numbers that are used for measurement and that can be represented on a continuous number line belong to the *real number system*. The real numbers consist of zero and all positive and negative integers, rational numbers, and irrational numbers. One of the basic properties of real numbers is that their squares are never negative.

Complex Numbers

In the 16th century, a few mathematicians began to work with numbers whose squares *are* negative. The word ''imaginary'' gradually came to be used to describe such numbers as $\sqrt{-1}$ and $\sqrt{-15}$. The use of the word imaginary reflects some of the original uneasiness many mathematicians had with nonreal numbers. Today, the phrase ''imaginary numbers'' seems a little unfortunate since these numbers are firmly established in mathematics. They are routinely used in advanced mathematics, electrical AC circuits, map making, and quantum mechanics, to name just a few fields.

We define the **imaginary unit** i with the following properties:

$$i = \sqrt{-1} \text{ and } i^2 = -1.$$

We then define the square root of any negative number as follows:

$$\text{If } a > 0, \sqrt{-a} = i\sqrt{a}.$$

Even though there is now a definition for the square root of a negative number, we cannot assume that all the square root properties of positive numbers will also be true for negative real numbers. For example, the property

$$\sqrt{a} \cdot \sqrt{b} = \sqrt{ab}$$

is true when a and b are both positive real numbers, but not when a and b are both negative real numbers. (See Class Exercise 11.)

Example 1 Simplify: **a.** $\sqrt{-25}$ **b.** $\sqrt{-7}$

Solution **a.** $\sqrt{-25} = i\sqrt{25} = 5i$ **b.** $\sqrt{-7} = i\sqrt{7}$

Example 2 Simplify $\sqrt{-9} - 2\sqrt{-25}$.

Solution $\sqrt{-9} - 2\sqrt{-25} = i\sqrt{9} - 2i\sqrt{25}$
$$= 3i - 2 \cdot 5i$$
$$= -7i$$

Any number of the form $a + bi$, where a and b are real numbers and i is the imaginary unit, is called a **complex number**. For example, $5i$, $3 + 4i$, $5.2 - i\sqrt{7}$, 6, and 0 are all complex numbers. In $a + bi$, a is called the **real part** and b is called the **imaginary part** of the complex number. If $b \neq 0$, the number is called an **imaginary number**.

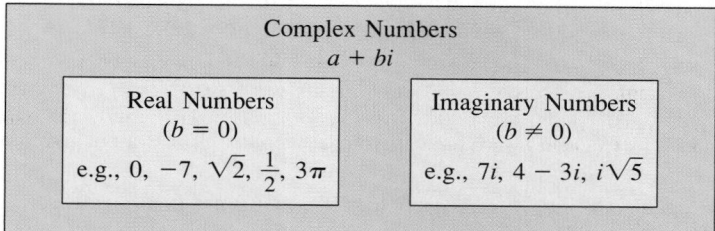

As the Venn diagram illustrates, you can think of the complex numbers as consisting of two non-overlapping sets, the real numbers and the imaginary numbers. Those imaginary numbers in which $a = 0$, such as $3i$, $-i$, and $i\sqrt{7}$, are called **pure imaginary** numbers.

Two complex numbers $a + bi$ and $c + di$ are **equal** if and only if $a = c$ and $b = d$. The examples below show that you can add or multiply two complex numbers simply by treating i as if it were a variable and using the distributive laws.

Example 3 $(2 + 3i) + (4 + 5i) = 6 + 8i$

Example 4 $(2 + 3i)(4 + 5i) = 8 + 10i + 12i + 15i^2$
$$= 8 + 22i + (15)(-1) \longleftarrow i^2 = -1$$
$$= -7 + 22i$$

The complex numbers $a + bi$ and $a - bi$ are called **complex conjugates**. Their sum is a real number, and their product is a nonnegative real number. (See Written Exercises 35 and 36). The conjugate of the complex number $z = a + bi$ is often denoted by $\bar{z} = a - bi$. In Example 5, we use the fact that the product of z and $\bar{z}$ is a real number.

Example 5 Express $\dfrac{5 - 2i}{4 + 3i}$ in the form $a + bi$.

Solution Multiply the numerator and the denominator by the conjugate of the denominator $4 + 3i$.
$$\frac{5 - 2i}{4 + 3i} = \frac{5 - 2i}{4 + 3i} \cdot \frac{4 - 3i}{4 - 3i} = \frac{20 - 15i - 8i + 6i^2}{16 - 9i^2}$$
$$= \frac{20 - 23i - 6}{16 + 9}$$
$$= \frac{14 - 23i}{25} = \frac{14}{25} - \frac{23}{25}i$$

b. $\dfrac{1}{1 - i} = \dfrac{1}{1 - i} \cdot \dfrac{1 + i}{1 + i} =$

$\dfrac{1 + i}{1 - (-1)} = \dfrac{1}{2} + \dfrac{1}{2}i$

(*Note:* Since $\dfrac{1}{1 - i} =$

$\dfrac{1}{2} + \dfrac{1}{2}i$, $\dfrac{1}{2} + \dfrac{1}{2}i$ is the reciprocal of $1 - i$.)

c. $i^{23} = i^{20} \cdot i^3 =$
$(i^4)^5 \cdot (i^2 \cdot i) =$
$[(i^2)^2]^5 \cdot (-1 \cdot i) =$
$[(-1)^2]^5 \cdot (-i) =$
$1^5 \cdot (-i) = -i$

2. Show that $2 + 3i$ is a square root of $-5 + 12i$.

We need to show that $(2 + 3i)^2 = -5 + 12i$.
$(2 + 3i)(2 + 3i) =$
$4 + 12i + 9i^2 =$
$(4 - 9) + 12i = -5 + 12i$
Thus, $2 + 3i$ is a square root of $-5 + 12i$.

Assessment Note

You can evaluate students' understanding of the definition of a complex number by drawing a Venn diagram of the complex numbers on the chalkboard and having students decide how to categorize the following numbers: $2i - 2i$, $3 - 3i$, $9i^6$, $(2i)(-2)$, $(3 - 3i)(9i^6)$. (Answers: real, complex, real, pure imaginary, complex.) You can also have students give their own examples of real, pure imaginary, and complex numbers and include them in the diagram.

Example Note

Students can use FOIL in Example 4 to help them multiply the binomials.

Problem Solving

Ask students to generalize Class Exercise 10 by simplifying i^n when n is 1, 2, or 3 more than a multiple of 4. (Consecutive integral powers of i repeatedly cycle through the sequence $1, i, -1, -i$.)

Making Connections

Written Exercises 19–24 are simplified in a way that is similar to the way expressions containing a radical in the denominator are simplified. You first find the conjugate of the denominator, then multiply the numerator and denominator by that conjugate.

Assessment Note

Ask students to give an example of an irrational number that is the product of a complex number and its conjugate. (Example: $(\sqrt[3]{3} - 2i) \cdot (\sqrt[3]{3} + 2i) = \sqrt[3]{9} + 4$)

Suggested Assignments

Standard
28/1–39 odd
Comprehensive
28/1–45 odd

Supplementary Materials

Alternative Assessment, 4

CLASS EXERCISES

Simplify.

1. $(2 + 6i) + (5 - 4i)$ $7 + 2i$
2. $(5 - 2i) - (6 + 3i)$ $-1 - 5i$
3. $(3 + i)(3 - i)$ 10
4. $(9 + i)(3 - i)$ $28 - 6i$
5. $(4 + i)(4 - i)$ 17
6. $(3 + 5i)(3 - 5i)$ 34
7. $(\sqrt{2} + i)(\sqrt{2} - i)$ 3
8. $(a + bi)(a - bi)$ $a^2 + b^2$
9. Find real numbers x and y such that $4 - 5i = 2x + yi$. $x = 2, y = -5$
10. Find i^4, i^8, i^{12}, and i^{16}. What can you say about i^n if n is a positive multiple of 4? $1; 1; 1; 1;$ equal to 1
11. The statement $\sqrt{a}\sqrt{b} = \sqrt{ab}$ is true if a and b are both positive.
    a. Verify that $\sqrt{4} \cdot \sqrt{9} = \sqrt{4 \cdot 9}$. $\sqrt{4} \cdot \sqrt{9} = 2 \cdot 3 = 6 = \sqrt{36} = \sqrt{4 \cdot 9}$
    b. Verify that $\sqrt{-4} \cdot \sqrt{9} = \sqrt{-4 \cdot 9}$. $\sqrt{-4} \cdot \sqrt{9} = 2i \cdot 3 = 6i = \sqrt{-36} = \sqrt{-4 \cdot 9}$
    c. Compare the value of $\sqrt{-4} \cdot \sqrt{-9}$ with the value of $\sqrt{-4 \cdot (-9)}$. $-6 \neq 6$
    d. Is the statement $\sqrt{a}\sqrt{b} = \sqrt{ab}$ true if both a and b are negative? No

WRITTEN EXERCISES

Simplify each expression.

A
1. $\sqrt{-4} + \sqrt{-16} + \sqrt{-1}$ $7i$
2. $\sqrt{-49} - \sqrt{-9} + \sqrt{-36}$ $10i$
3. $\sqrt{-1}\sqrt{-9}$ -3
4. $\sqrt{-2}\sqrt{-5}$ $-\sqrt{10}$
5. $\dfrac{\sqrt{-12}}{\sqrt{-3}}$ 2
6. $\dfrac{\sqrt{-25}}{\sqrt{-50}}$ $\dfrac{\sqrt{2}}{2}$
7. $(4 - 3i) + (-6 + 8i)$ $-2 + 5i$
8. $(7 - 8i) - (6 + 2i)$ $1 - 10i$
9. $4(3 + 5i) - 2(2 - 6i)$ $8 + 32i$
10. $\dfrac{1}{6}(7 - 2i) + \dfrac{2}{3}(5 - 5i)$ $\dfrac{9}{2} - \dfrac{11}{3}i$
11. $(6 - i)(6 + i)$ 37
12. $(7 + 3i)(7 - 3i)$ 58
13. $(5 + i\sqrt{5})(5 - i\sqrt{5})$ 30
14. $(\sqrt{3} + 4i\sqrt{2})(\sqrt{3} - 4i\sqrt{2})$ 35
15. $(8 + 3i)(2 - 5i)$ $31 - 34i$
16. $(5 - 2i)(-1 + 3i)$ $1 + 17i$
17. $(4 - 5i)^2$ $-9 - 40i$
18. $(4 + 7i)^2$ $-33 + 56i$

Write each expression in the form $a + bi$.

19. $\dfrac{1}{2 + 5i}$ $\dfrac{2}{29} - \dfrac{5}{29}i$
20. $\dfrac{1}{4 - 3i}$ $\dfrac{4}{25} + \dfrac{3}{25}i$
21. $\dfrac{5 + i}{5 - i}$ $\dfrac{12}{13} + \dfrac{5}{13}i$
22. $\dfrac{3 - 2i}{3 + 2i}$ $\dfrac{5}{13} - \dfrac{12}{13}i$
23. $\dfrac{3 + i\sqrt{2}}{7 - i\sqrt{2}}$ $\dfrac{19}{51} + \dfrac{10\sqrt{2}}{51}i$
24. $\dfrac{2 + i\sqrt{5}}{3 - i\sqrt{5}}$ $\dfrac{1}{14} + \dfrac{5\sqrt{5}}{14}i$

28   *Chapter One*

B 25. $\dfrac{5}{i}$ $-5i$

26. $\dfrac{i^2 + 2i^3}{i}$ $-2 + i$

27. $i + i^2 + i^3 + i^4 + i^5$ i

28. $i^{46} + i^{47}$ $-1 - i$

29. i^{-3} i

30. i^{-6} -1

31. i^{-35} i

32. $(i^n)^4$, where n is any integer 1

33. Find real numbers x and y such that $(2x + y) + (3 - 5x)i = 1 - 7i$. $x = 2, y = -3$

34. Find real numbers x and y such that $(3x - 4y) + (6x + 2y)i = 5i$. $x = \dfrac{2}{3}, y = \dfrac{1}{2}$

35. Show that the sum of $a + bi$ and its conjugate is a real number. See below.

36. Show that the product of $a + bi$ and its conjugate is a nonnegative real number. $(a + bi)(a - bi) = a^2 + abi - abi - b^2i^2 = a^2 + b^2 \geq 0$

37. **a.** How could you show that 79 is a square root of 6241 without using a calculator? Calculate 79^2 and compare with 6241.

   b. How could you show that $3 - i$ is a square root of $8 - 6i$? Calculate $(3 - i)^2$.

38. Show that $4 - 3i$ is a square root of $7 - 24i$.

39. Show that $\dfrac{\sqrt{2}}{2}(1 + i)$ is a square root of i.

40. Find the square roots of $3 + 4i$. $2 + i, -2 - i$

C 41. If $z = \bar{z}$, what special kind of number is z? real number

42. Show that $\overline{z_1 + z_2} = \overline{z_1} + \overline{z_2}$.

43. Show that $\overline{z_1 z_2} = \overline{z_1} \cdot \overline{z_2}$.

44. Show that $\overline{z_1/z_2} = \overline{z_1}/\overline{z_2}$ if $z_2 \neq 0$.

45. Use Exercise 43 to show that $\overline{z^2} = (\bar{z})^2$.

46. **a.** Use Exercises 43 and 45 to show that $\overline{z^3} = (\bar{z})^3$.

   b. Make a generalization of Exercise 45 and part (a) above.

These windmills produce electricity. Complex numbers are used in the study of electricity.

▮▮▮▮ CALCULATOR EXERCISE

Find $(6 + 7i)^8$. Use the fact that $(x + yi)^2 = (x^2 - y^2) + (2xy)i$.

42,660,913 + 30,082,416i

▨▨▨ COMPUTER EXERCISE

Write a program that will compute $(a + bi)^n$ when you input the positive integer n and the real numbers a and b.

35. $(a + bi) + (a - bi) = (a + a) + (b - b)i = 2a$

Review Note

You might need to review the laws of exponents before assigning the exercises on this page. Note that i^{-n} is defined to be $\dfrac{1}{i^n}$.

Communication Note

Have students verbalize the results of Exercises 42–44 as theorems.

Making Connections

Exercises 37–40 can help students make connections with finding roots of polynomial equations in Sections 2-6 and 2-7. Roots of quadratic equations are found in Section 1-8.

Additional Answers Written Exercises

42. $\overline{z_1 + z_2} =$
$\overline{(a + bi) + (c + di)} =$
$\overline{(a + c) + (b + d)i} =$
$(a + c) - (b + d)i =$
$(a - bi) + (c - di) =$
$\overline{a + bi} + \overline{c + di} = \overline{z_1} + \overline{z_2}$

43. $\overline{z_1 z_2} = \overline{(a + bi)(c + di)} =$
$\overline{(ac - bd) + (ad + bc)i} =$
$(ac - bd) - (ad + bc)i =$
$ac - bci - adi - bd =$
$ac - bci - adi + bdi^2 =$
$(a - bi)(c - di) =$
$\overline{(a + bi)} \cdot \overline{(c + di)} =$
$\overline{z_1} \cdot \overline{z_2}$

Warm-Up Exercises

Factor each expression.

1. $x^2 - 3x$ $x(x - 3)$

2. $x^2 - 4x - 5$
  $(x - 5)(x + 1)$

3. $6x^2 + 5x - 6$
  $(2x + 3)(3x - 2)$

4. $12 + 5x - 2x^2$
  $(4 - x)(3 + 2x)$

5. $x^2 - 8x + 16$ $(x - 4)^2$

Motivating the Section

Quadratic functions are used in physics to model the path of an object thrown in the air. For example, if a missile is fired straight up into the air, its height above the ground is given by a quadratic function, and the amount of time it takes to return to the ground can be determined by solving a quadratic equation.

Assessment Note

See if students can summarize how to solve a quadratic equation by using the method of completing the square.

Making Connections

The method of completing the square is important for working with conic sections (Chapter 6).

1-6 Solving Quadratic Equations

Objective   *To solve quadratic equations using different methods.*

Any equation that can be written in the form

$$ax^2 + bx + c = 0,$$

where $a \neq 0$, is called a **quadratic equation**. A **root**, or **solution**, of a quadratic equation is a value of the variable that satisfies the equation. Three methods for solving quadratic equations are:

     (1) factoring   (2) completing the square   (3) the quadratic formula

Factoring

Whenever the product of two factors is zero, at least one of the factors must be zero. For example, if $(3x - 2)(x + 4) = 0$, then $3x - 2 = 0$ or $x + 4 = 0$. A quadratic equation must be written in the **standard form** $ax^2 + bx + c = 0$ before it can be solved by factoring.

Example 1   Solve $(3x - 2)(x + 4) = -11$.

Solution

$$3x^2 + 10x - 8 = -11$$
$$3x^2 + 10x + 3 = 0$$
$$(3x + 1)(x + 3) = 0$$

Therefore, $3x + 1 = 0$ or $x + 3 = 0$

$$x = -\frac{1}{3} \text{ or } x = -3$$

The solutions are -3 and $-\frac{1}{3}$.

Completing the Square

The method of transforming a quadratic equation so that one side is a perfect square trinomial is called **completing the square**.

Example 2   Solve $2x^2 - 12x - 7 = 0$.

Solution

Step 1   Divide both sides by the coefficient of x^2 so that x^2 will have a coefficient of 1. $x^2 - 6x - \frac{7}{2} = 0$

Step 2   Subtract the constant term from both sides. $x^2 - 6x = \frac{7}{2}$

Step 3 Complete the square. Add the square of one half the coefficient of x to both sides.

$$x^2 - 6x + (-3)^2 = \frac{7}{2} + (-3)^2$$

$$(x - 3)^2 = \frac{25}{2}$$

Step 4 Take the square root of both sides and solve for x.

$$x - 3 = \pm \frac{5\sqrt{2}}{2}$$

$$x = 3 \pm \frac{5\sqrt{2}}{2}$$

The solutions are $3 + \frac{5\sqrt{2}}{2}$ and $3 - \frac{5\sqrt{2}}{2}$.

The Quadratic Formula

The *quadratic formula*, shown below, is derived by completing the square. (See Written Exercise 44.)

The Quadratic Formula

The roots of the quadratic equation $ax^2 + bx + c = 0$ are given by:

$$x = \frac{-b \pm \sqrt{b^2 - 4ac}}{2a} \qquad (a \neq 0)$$

Example 3 Solve $2x^2 + 7 = 4x$.

Solution Rewrite the equation in standard form:

$$2x^2 - 4x + 7 = 0$$

Substitute $a = 2$, $b = -4$, and $c = 7$ into the quadratic formula.

$$x = \frac{-(-4) \pm \sqrt{(-4)^2 - 4 \cdot 2 \cdot 7}}{2 \cdot 2}$$

$$= \frac{4 \pm \sqrt{-40}}{4} = \frac{4 \pm 2i\sqrt{10}}{4}$$

$$= 1 \pm \frac{i\sqrt{10}}{2}$$

The solutions are $1 - \frac{i\sqrt{10}}{2}$ and $1 + \frac{i\sqrt{10}}{2}$.

The Discriminant $b^2 - 4ac$

The quantity that appears beneath the radical sign in the quadratic formula, $b^2 - 4ac$, can tell you whether the roots of a quadratic equation are real or imaginary, as described on the next page. Because of this "discriminating ability," $b^2 - 4ac$ is called the **discriminant**.

Mathematical Note

Stress that the quadratic formula will give imaginary roots as well as real roots. Remind students that "$-b$" is "the opposite of b" and can be a positive number (when b is negative).

Error Analysis

Students may try to use the quadratic formula without expressing the equation in standard form first. Caution students to write the equation with 0 on one side before using the quadratic formula.

◻ Using Technology

Encourage students with programmable calculators to write programs to solve quadratic equations by using the quadratic formula. The values of a, b, and c must be entered, and decimal approximations for the roots of an equation should be returned.

Making Connections

If a quadratic function is graphed (Section 1-7), the number of times the graph crosses the x-axis equals the number of real roots of the related quadratic equation. If the graph does not cross the x-axis, then the roots are imaginary. If the graph intersects the x-axis once, then there is one real (double) root.

Error Analysis

Students sometimes divide both sides of an equation by a common factor and consequently lose a root. Caution students to solve equations by either of the two methods shown at the top of page 33.

The Nature of the Discriminant

Given the quadratic equation $ax^2 + bx + c = 0$, where a, b, and c are real numbers:

If $b^2 - 4ac < 0$, there are two conjugate imaginary roots.
If $b^2 - 4ac = 0$, there is one real root (called a *double root*).
If $b^2 - 4ac > 0$, there are two different real roots.

In the special case where the equation $ax^2 + bx + c = 0$ has integral coefficients and $b^2 - 4ac$ is the square of an integer, the equation has rational roots. That is, if $b^2 - 4ac$ is the square of an integer, then $ax^2 + bx + c$ has factors with integral coefficients.

Choosing a Method of Solution

Although you can use the quadratic formula to solve *any* quadratic equation, it is often much easier to factor or complete the square. The list below suggests when to use which method.

Situation	Method to use
a, b, and c are integers and $b^2 - 4ac$ is a perfect square.	factoring
The equation has the form $x^2 + (\text{even number})x + \text{constant} = 0$.	completing the square

In the remaining situations the quadratic formula is usually the easiest to use, especially when solving an equation like $px^2 + qx + r = 0$, where the coefficients are letters. Also, the quadratic formula is the most efficient method to use when approximating real roots of a quadratic equation using a computer or calculator.

Losing or Gaining a Root

If an equation contains variables on both sides or variables in a denominator, then you must carefully organize your method for solving in order not to lose a root or gain a root.

Losing a Root. It is possible to lose a root by dividing both sides of an equation by a common factor.

Incorrect method

$$4x(x - 1) = 3(x - 1)^2 \longleftarrow \text{Divide both sides by } (x - 1).$$
$$4x = 3(x - 1)$$
$$4x = 3x - 3$$
$$x = -3$$

Actually, as shown below, the roots of $4x(x - 1) = 3(x - 1)^2$ are 1 and -3. When we divided both sides of the equation by $x - 1$, we lost a root. Two ways to avoid this mistake are shown below. Both methods are correct.

Method 1

If there is a factor common to both sides of the equation, remember to include as roots all values that make this factor zero.

$$4x(x - 1) = 3(x - 1)^2$$

If $x - 1 = 0$, then $x = 1$ and both sides of the equation equal zero. Thus, 1 is a root.	If $x - 1 \neq 0$, we can divide both sides by $x - 1$, getting $4x = 3(x - 1)$. Thus, -3 is a root.

Method 2

Bring all terms to one side of the equation and then solve.

$$4x(x - 1) = 3(x - 1)^2$$
$$4x(x - 1) - 3(x - 1)^2 = 0$$
$$4x^2 - 4x - 3x^2 + 6x - 3 = 0$$
$$x^2 + 2x - 3 = 0$$
$$(x - 1)(x + 3) = 0$$
$$x = 1 \text{ or } x = -3$$

Gaining a Root. It is possible to gain a root by squaring both sides of an equation. (See Class Exercise 19.) Another possible way to gain a root is by multiplying both sides of an equation by an expression, as in the following example. Any gained root, called an **extraneous root**, satisfies the transformed equation but not the original equation. Always check your solutions in the *original equation* in order to eliminate extraneous roots.

Example 4 Solve $\dfrac{x + 2}{x - 2} + \dfrac{x - 2}{x + 2} = \dfrac{8 - 4x}{x^2 - 4}$.

Solution Multiply both sides of the equation by $(x + 2)(x - 2)$.
$$(x + 2)^2 + (x - 2)^2 = 8 - 4x$$
$$(x^2 + 4x + 4) + (x^2 - 4x + 4) = 8 - 4x$$
$$2x^2 + 8 = 8 - 4x$$
$$2x^2 + 4x = 0$$
$$2x(x + 2) = 0$$
$$x = 0, -2$$

Check:

$x = 0$	$x = -2$
$\dfrac{0 + 2}{0 - 2} + \dfrac{0 - 2}{0 + 2} \stackrel{?}{=} \dfrac{8 - 0}{0^2 - 4}$	$\dfrac{-2 + 2}{-2 - 2} + \dfrac{-2 - 2}{-2 + 2} \stackrel{?}{=} \dfrac{8 - 4(-2)}{(-2)^2 - 4}$
$-1 + (-1) = -2$	Since two denominators are zero, the equation is meaningless. Thus, -2 is *not* a root of the original equation.
Thus, 0 is a root.	

Therefore, the solution is $x = 0$.

CLASS EXERCISES

What must be added to the following expressions to complete the square?

1. $x^2 + 8x +$ ___?___ 16 **2.** $x^2 - 10x +$ ___?___ 25 **3.** $x^2 + 7x +$ ___?___ $\frac{49}{4}$ **4.** $x^2 + 2ax +$ $\frac{?}{a^2}$

Does factoring, completing the square, or using the quadratic formula seem to you the easiest method for solving the following equations?

5. $x^2 + 14x = 374$ Complete square **6.** $4x^2 - 5x = 0$ Factor **7.** $3x^2 + 11x - 4 = 0$ Factor

8. $px^2 + qx + r = 0$ Quadratic formula **9.** $4x^2 - x - 7 = 0$ Quadratic formula **10.** $x^2 - 14x - 736 = 0$ Complete square

Exercises 11–13 refer to the quadratic equation $ax^2 + bx + c = 0$, where a, b, and c are real numbers.

11. Explain why the equation has imaginary roots when $b^2 - 4ac < 0$.

12. Explain why the equation has only one root, and why this root is real, when $b^2 - 4ac = 0$. Does it seem reasonable to you to call this sort of root a *double root*?

13. If a, b, and c are integers and $b^2 - 4ac$ is the square of an integer, explain why the equation has rational roots.

14. The discriminant of the equation $x^2 - \sqrt{5}x + 1 = 0$ is the square of an integer, but the roots of the equation are not rational. Does the equation offer a contradiction of the result of Exercise 13?

15. Can a quadratic equation with irrational coefficients be solved by using the quadratic formula? Explain your answer.

16. Comment on the following method for solving the equation $9x^2 + 36x = x + 4$.

Dividing by $x + 4$ causes the root -4 to be lost.

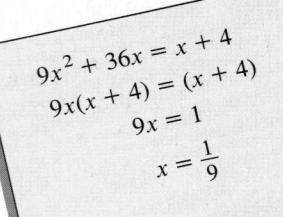

$9x^2 + 36x = x + 4$
$9x(x + 4) = (x + 4)$
$9x = 1$
$x = \frac{1}{9}$

17. Solve $(2x - 5)(x + 6) = 7(x + 6)$. ± 6

18. Solve $(3x - 5)(4x - 1) = (4x - 1)$. $2, \frac{1}{4}$

19. Comment on the following method for solving the equation $\sqrt{x + 6} = x$.

Squaring introduces the extraneous root -2, which does not satisfy the original equation.

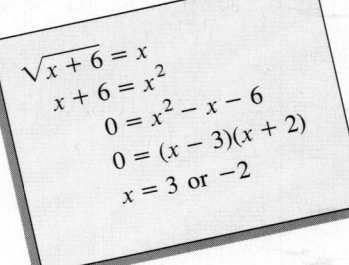

$\sqrt{x + 6} = x$
$x + 6 = x^2$
$0 = x^2 - x - 6$
$0 = (x - 3)(x + 2)$
$x = 3$ or -2

34   *Chapter One*

WRITTEN EXERCISES

Solve by factoring.

A

1. $3x^2 - 4x - 7 = 0$ $\ -1, \dfrac{7}{3}$

2. $4x^2 - 8x - 32 = 0$ $\ 4, -2$

3. $(2x - 3)(x + 4) = 6$ $\ 2, -\dfrac{9}{2}$

4. $(3y - 2)(y + 4) = 24$ $\ 2, -\dfrac{16}{3}$

Solve by completing the square. Give both real and imaginary roots.

5. $x^2 - 10x = 1575$ $\ 45, -35$

6. $x^2 - 6x = 391$ $\ 23, -17$

7. $2z^2 - 16z - 1768 = 0$ $\ 34, -26$

8. $x^2 - 8x - 20 = 0$ $\ 10, -2$

9. $x^2 + 6x + 10 = 0$ $\ -3 \pm i$

10. $y^2 + 10y + 35 = 0$ $\ -5 \pm i\sqrt{10}$

Solve by using the quadratic formula. Give your answers in simplest radical form. Give both real and imaginary roots.

11. $5x^2 + 2x - 1 = 0$ $\ \dfrac{-1 \pm \sqrt{6}}{5}$

12. $4x^2 - 4x - 17 = 0$ $\ \dfrac{1 \pm 3\sqrt{2}}{2}$

13. $3t^2 = 12t - 15$ $\ 2 \pm i$

14. $5u^2 + 2 = 5u$ $\ \dfrac{5 \pm i\sqrt{15}}{10}$

15. $\dfrac{4}{v} = \dfrac{v - 6}{v - 4}$ $\ 2, 8$

16. $\dfrac{4}{z} = \dfrac{3z}{z - 3}$ $\ \dfrac{2 \pm 4i\sqrt{2}}{3}$

Solve by whichever method seems easiest. Give both real and imaginary roots. Be sure not to lose or gain roots.

17. $8x^2 = 7 - 10x$ $\ \dfrac{1}{2}, -\dfrac{7}{4}$

18. $4t = 1 + 15t^2$ $\ \dfrac{2 \pm i\sqrt{11}}{15}$

19. $(3x - 2)^2 = 121$ $\ -3, \dfrac{13}{3}$

20. $(4y + 4)^2 = -16$ $\ -1 \pm i$

21. $(4x + 7)(x - 1) = 2(x - 1)$ $\ 1, -\dfrac{5}{4}$

22. $(2x + 1)(4x - 3) = 3(4x - 3)^2$ $\ 1, \dfrac{3}{4}$

23. $2w(4w - 1) = w(1 - 4w)$ $\ 0, \dfrac{1}{4}$

24. $3(2x - 3)^2 = 4x(3 - 2x)$ $\ \dfrac{3}{2}, \dfrac{9}{10}$

25. $\dfrac{x + 3}{x - 3} + \dfrac{x - 3}{x + 3} = \dfrac{18 - 6x}{x^2 - 9}$ $\ 0$

26. $\dfrac{r}{r - 1} - \dfrac{r}{r + 1} = \dfrac{2}{r^2 - 1}$ $\ $ No sols.

27. $\dfrac{t^2 + 1}{t + 2} = \dfrac{t}{3} + \dfrac{5}{t + 2}$ $\ 3$

28. $\dfrac{x + 2}{x^2 - x - 6} = 3 - \dfrac{4}{x - 3}$ $\ \dfrac{14}{3}$

29. $2\sqrt{x} = x - 8$ $\ 16$

30. $\sqrt{2x + 5} = x + 1$ $\ 2$

In Exercises 31 and 32, $\overline{DE}$ is parallel to $\overline{BC}$. Find the value of x.

B 31.
6

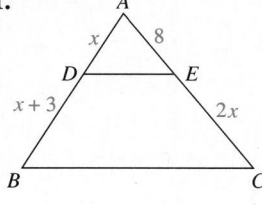

32.
2

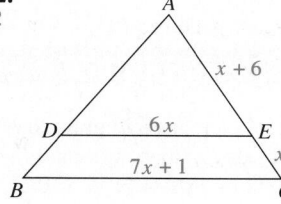

Linear and Quadratic Functions   **35**

Assessment Note

Before assigning the exercises, you might want to quiz students orally about the values of a, b, and c when applied to a quadratic equation such as $6x^2 = 7 + 2x$. Ask students to explain why $b \neq 2$ and $c \neq 7$.

Review Note

For Written Exercises 31 and 32, you might need to review the proportions that can be derived from similar triangles.

Suggested Assignments

Standard
35/3–39 every 3rd
Comprehensive
35/17–45 odd

Supplementary Materials

Alternative Assessment, 4

Review Note

For Written Exercises 33 and 34, you might need to review the relevant theorems from geometry.

◩ Using Technology

For Exercise 43, refer students to the discussion of solving equations with a graphing calculator in the article *Using a Graphing Calculator* (see pages xxx–xxxi).

Additional Answers
Written Exercises

36. **a.** $64 - 20k$

   b. $k = \frac{16}{5}$

   c. $k < \frac{16}{5}$

   d. $k > \frac{16}{5}$

   e. For example, $k = 3$, $k = -4$, $k = 0$

43. (41) 0.22;
   (42) No solution
   No extraneous solutions are encountered with this method.

In Exercises 33 and 34, $\overleftrightarrow{PB}$ intersects the circle at points A and B, and $\overleftrightarrow{PD}$ intersects the circle at points C and D. By theorems from geometry, $PA \cdot PB = PC \cdot PD$ in each case. Find the value of x.

33.

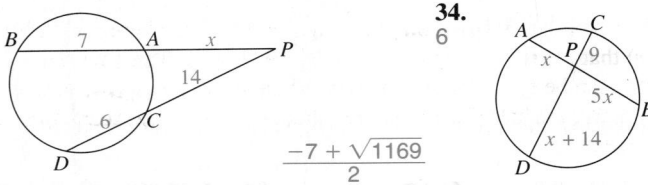

$$\frac{-7 + \sqrt{1169}}{2}$$

34.
6

35. a. What is the discriminant of the equation $4x^2 + 8x + k = 0$? $64 - 16k$
   b. For what value of k will the equation have a double root? $k = 4$
   c. For what values of k will the equation have two real roots? $k < 4$
   d. For what values of k will the equation have imaginary roots? $k > 4$
   e. Name three values of k for which the given equation has rational roots.
   See below.
36. Repeat Exercise 35 for the equation $5x^2 + 8x + k = 0$.

35. e. For example, $k = 0$, $k = -12$, $k = -5$

In Exercises 37–40, use the quadratic formula to solve each equation.

37. $\sqrt{2}x^2 - 5x + \sqrt{8} = 0$ $2\sqrt{2}, \frac{\sqrt{2}}{2}$

38. $4x^2 - 2\sqrt{5}x - 1 = 0$ $\frac{\sqrt{5} \pm 3}{4}$

39. $x^2 - 6ix - 9 = 0$ $3i$

40. $ix^2 - 3x - 2i = 0$ $-i, -2i$

C Solve the equations in Exercises 41 and 42. You will need to square twice.

41. $\sqrt{2x + 5} = 2\sqrt{2x} + 1$ $\frac{2}{9}$

42. $\sqrt{y - 3} = 1 - \sqrt{2y - 4}$ No sol.

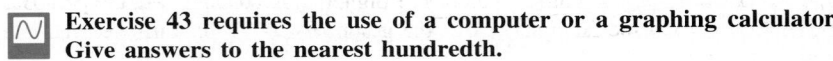 Exercise 43 requires the use of a computer or a graphing calculator. Give answers to the nearest hundredth.

43. Use a computer or a graphing calculator to solve the equations in Exercises 41 and 42. For example, for Exercise 41, you could graph $y = \sqrt{2x + 5}$ and $y = 2\sqrt{2x} + 1$ on the same set of axes and then use the ZOOM feature to obtain approximations of the x-coordinates of any points of intersection. Do you encounter any extraneous roots using this method?

44. Derive the quadratic formula. (*Hint*: Solve the general quadratic equation $ax^2 + bx + c = 0$, $a \neq 0$, by completing the square.)

45. Let a be a positive integer such that $ax^2 + x - 6$ can be factored. Find:
   a. The five smallest values of a **b.** Two values of a greater than 100
   1, 2, 5, 7, 12   For example, 155, 222

////COMPUTER EXERCISE

Write a computer program that will print the roots of $ax^2 + bx + c = 0$ when you input real numbers a, b, and c. Have the program print REAL ROOTS, DOUBLE ROOT, or IMAGINARY ROOTS as appropriate.

36   *Chapter One*

1-7 Quadratic Functions and Their Graphs

| Objective | To define and graph quadratic functions. |

The graph of the **quadratic function** $f(x) = ax^2 + bx + c$, where $a \neq 0$, is the set of points (x, y) that satisfy the equation $y = ax^2 + bx + c$. This graph is a *parabola*, a curve that can be seen, for example, in the cables of a suspension bridge and in the path of a thrown ball. Parabolas can also be defined geometrically, as you will see in Chapter 6.

 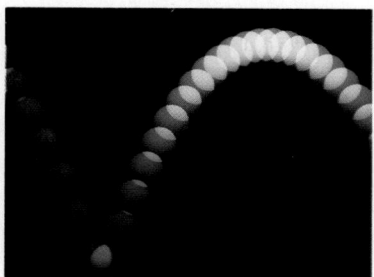

If a graph has an *axis of symmetry*, then when you fold the graph along this axis, the two halves of the graph coincide. The graph of a quadratic function has a vertical axis of symmetry, or **axis**. The **vertex** of the parabola is the point where the axis of symmetry intersects the parabola. If $a > 0$, the parabola opens upward, and the function has a minimum value. If $a < 0$, the parabola opens downward, and the function has a maximum value. The bigger $|a|$ is, the narrower the parabola is. In the figure at the far right below, the graph of $y = 3x^2$ is narrower than the graph of $y = x^2$.

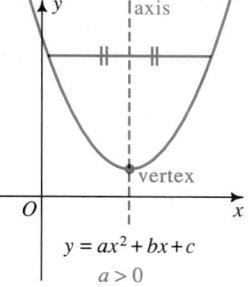 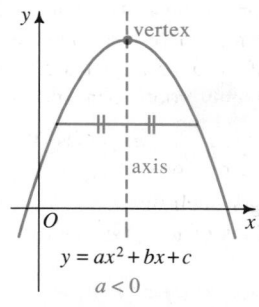 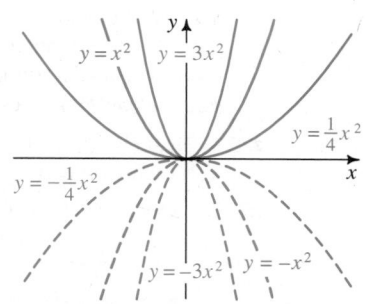

The *y*-intercept of a parabola with equation $y = ax^2 + bx + c$ is c. The *x*-intercepts are the real roots of $ax^2 + bx + c = 0$. Since the quadratic equation may have two, one, or no real roots, depending on the value of the discriminant $b^2 - 4ac$, we have three possibilities regarding the *x*-intercepts of a parabola, as shown in the diagrams on the next page.

Linear and Quadratic Functions **37**

Solve each equation.

1. $x^2 + 6x = 0$ $-6; 0$
2. $x^2 - 6x = 0$ $0; 6$
3. $x^2 + 6x + 9 = 0$ -3
4. $x^2 - 6x + 9 = 0$ 3
5. $x^2 + 6x + 5 = 0$ $-5; -1$

Motivating the Section

A surface produced by rotating a parabola (defined by a quadratic equation) is used in making telescopes and mirrors. Some antennas that are used for television reception from satellites are also parabolic. A parabola is the graph of a quadratic function, the topic of this section.

Using Technology

A graphing calculator is an excellent resource to help students see how the value of a in the equation $y = ax^2$ affects the graph. Have students observe the effect of changing a by graphing several parabolas on the same set of axes. Also have students verify that when $a > 0$, the parabola opens upward and the related function has a minimum value; when $a < 0$, the parabola opens downward and the related function has a maximum value.

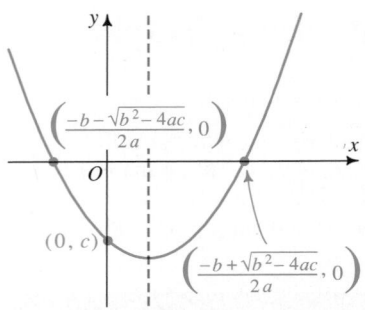
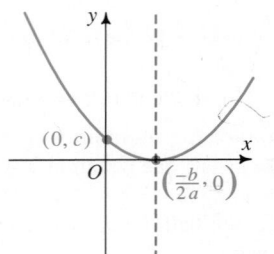
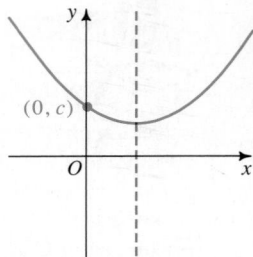

If $b^2 - 4ac > 0$, there are two *x*-intercepts.

If $b^2 - 4ac = 0$, there is one *x*-intercept (at a point where the parabola and *x*-axis are tangent).

If $b^2 - 4ac < 0$, there are no *x*-intercepts.

The axis of symmetry is a vertical line midway between the *x*-intercepts. The equation of the axis is $x = -\frac{b}{2a}$. (See Exercises 39 and 40.) In fact, this is the equation of the axis even if there are fewer than two *x*-intercepts.

Example 1 Sketch the parabola $y = 2x^2 - 8x + 5$. Label the intercepts, axis of symmetry, and vertex.

Solution

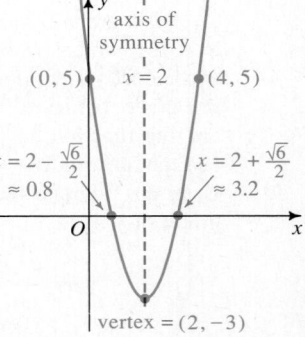

Step 1 To find the *y*-intercept, you let $x = 0$. The *y*-intercept is 5.

Step 2 To find the *x*-intercepts, solve
$$2x^2 - 8x + 5 = 0.$$
$$x = 2 \pm \frac{\sqrt{6}}{2}$$

Step 3 The axis of symmetry is the line
$$x = -\frac{b}{2a} = -\frac{-8}{2 \cdot 2} = 2.$$

Step 4 Since the vertex is on the axis of symmetry, its *x*-coordinate is 2. To find the *y*-coordinate of the vertex, substitute $x = 2$ into the equation of the parabola.
$$y = 2(2)^2 - 8(2) + 5 = -3$$
Thus, the vertex is $(2, -3)$.

If the equation of a parabola is written in the form
$$y = a(x - h)^2 + k,$$
then the vertex is (h, k). For example:

The vertex of the parabola $y = 2(x - 3)^2 + 7$ is $(3, 7)$.
The vertex of the parabola $y = -4(x + 9)^2 + 2$ is $(-9, 2)$.

Example 2 a. Find the vertex of the parabola $y = -2x^2 + 12x + 4$ by completing the square.
b. Find the x- and y-intercepts.

Solution a.
$$y = -2x^2 + 12x + 4$$
$$= -2(x^2 - 6x \quad\quad) + 4$$
$$= -2(x^2 - 6x + 9) + 4 - (-2)(9)$$
$$y = -2(x - 3)^2 + 22$$

The vertex is (3, 22).

b. When $x = 0$, $y = 4$. Thus, the y-intercept is 4, which is the constant term of the original equation.

To find the x-intercepts, let $y = 0$. Using the completed square form of the given equation:

$$-2(x - 3)^2 + 22 = 0$$
$$-2(x - 3)^2 = -22$$
$$(x - 3)^2 = 11$$
$$x = 3 \pm \sqrt{11}$$

Sketching the Graph of $y = ax^2 + bx + c$, $a \neq 0$

1. Get a quick mental picture of the graph as follows:
   a. If $a > 0$, the graph opens upward.
   If $a < 0$, the graph opens downward.
   b. If $b^2 - 4ac > 0$, the graph has two x-intercepts.
   If $b^2 - 4ac = 0$, the graph has one x-intercept.
   If $b^2 - 4ac < 0$, the graph has no x-intercepts.
2. Find the axis and vertex of the parabola as follows:

   Method 1. The equation of the axis is $x = -\frac{b}{2a}$. Substitute this value of x into $y = ax^2 + bx + c$ to find the y-coordinate of the vertex.

   Method 2. Rewrite the equation in the form
   $$y = a(x - h)^2 + k.$$
   The vertex is (h, k) and the axis is $x = h$.
3. Find the x- and y-intercepts.
   a. Using $y = ax^2 + bx + c$, the y-intercept is always c.
   b. To find the x-intercepts, solve $ax^2 + bx + c = 0$ (from using Method 1) or solve $a(x - h)^2 + k = 0$ (from using Method 2).

Additional Examples cont.

2. Find an equation of the quadratic function f with $f(-1) = -7$ and maximum value $f(2) = -1$. Show that the function has no x-intercepts.
   Use the form $f(x) = a(x - h)^2 + k$. Since the vertex is $(2, -1)$, $f(x) = a(x - 2)^2 - 1$. Since $(-1, -7)$ lies on the graph, $-7 = a(-1 - 2)^2 - 1 = 9a - 1$; $a = -\frac{2}{3}$. Thus, $f(x) = -\frac{2}{3}(x - 2)^2 - 1$.
   The parabola opens downward from its vertex $(2, -1)$, which is below the x-axis, so the parabola has no x-intercepts. (Or, $-\frac{2}{3}(x - 2)^2 - 1 = 0$; $(x - 2)^2 = -\frac{3}{2}$, which has no real solutions.)

Mathematical Note

If a quadratic function is given in the form $y = a(x - h)^2 + k$, the following method can be used to determine its graph:

1. graph $y = ax^2$;
2. move (translate) the graph $|h|$ units to the right ($h > 0$) or to the left ($h < 0$);
3. move (translate) the graph $|k|$ units up ($k > 0$) or down ($k < 0$).

Point out that the maximum or minimum value of a quadratic function is the y-coordinate of the vertex.

Example 3 presents two methods for finding the intersection points of a line and a parabola. The second method uses a computer or a graphing calculator.

Example 3 Where does the line $y = 2x + 5$ intersect the parabola $y = 8 - x^2$?

Solution

Method 1 Set $2x + 5$ equal to $8 - x^2$ and solve for x.

$$2x + 5 = 8 - x^2$$
$$x^2 + 2x - 3 = 0$$
$$(x + 3)(x - 1) = 0$$
$$x = -3 \text{ or } x = 1$$

Substitute $x = -3$ and $x = 1$ into the equation $y = 2x + 5$ to get $y = -1$ and $y = 7$, respectively. The intersection points are $(-3, -1)$ and $(1, 7)$.

Method 2 Use a computer or a graphing calculator to graph the equations $y = 2x + 5$ and $y = 8 - x^2$ on a single set of axes. Using the TRACE feature, you can find the coordinates of the intersection points to be $(-3, -1)$ and $(1, 7)$.

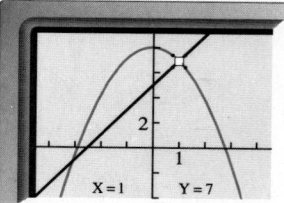

Example 4 Find an equation of the function whose graph is a parabola with x-intercepts 1 and 4 and y-intercept -8.

Solution Any parabola with x-intercepts 1 and 4 has an equation of the form $y = a(x - 1)(x - 4)$ for some constant a. We use the fact that the parabola contains $(0, -8)$ to find the value of a.

$$-8 = a(0 - 1)(0 - 4)$$
$$-2 = a$$

Then the equation is $y = -2(x - 1)(x - 4)$, or $f(x) = -2x^2 + 10x - 8$.

CLASS EXERCISES

For each quadratic function, tell whether its graph (a) opens upward or downward, and (b) intersects the x-axis in two, one, or no points.

1. $f(x) = x^2 - 2x + 3$
Up; none

2. $f(x) = -2x^2 - 12x - 18$
Down; one

3. $f(x) = 9 - x^2$
Down; two

Without calculating, find the vertex of each parabola.

4. $y = (x - 5)^2 + 4$ $(5, 4)$

5. $y = (x + 2)^2 - 3$ $(-2, -3)$

6. $y = -x^2 - 10$
$(0, -10)$

7. Explain in your own words the relationship between the zeros of a quadratic function and the x-intercepts of its graph.

40   *Chapter One*

WRITTEN EXERCISES

In Exercises 1–20, sketch each parabola. Label the vertex, the axis of symmetry, and the *x*- and *y*-intercepts.

In Exercises 1–8, use the method of Example 1.

A
1. **a.** $y = x^2 - 6x$ **b.** $y = x^2 - 6x + 9$ **c.** $y = x^2 - 6x + 10$
2. **a.** $y = -x^2 + 4x$ **b.** $y = -x^2 + 4x - 4$ **c.** $y = -x^2 + 4x - 8$
3. $y = (x + 5)(x + 3)$ **4.** $y = (x - 4)(x + 2)$ **5.** $y = 9 - x^2$
6. $y = x^2 - 16$ **7.** $y = x^2 - 2x - 15$ **8.** $y = x^2 + 3x - 10$

In Exercises 9–14, use the method of Example 2.

9. $y = x^2 - 2x - 7$ **10.** $y = x^2 + 4x + 9$ **11.** $y = 4x^2 - 8x + 2$
12. $y = -3x^2 - 12x - 3$ **13.** $y = \frac{1}{2}x^2 + 4x + 8$ **14.** $y = -\frac{1}{3}x^2 + 2x + 1$

In Exercises 15–20, use the method that seems most appropriate to you.

15. $y = 4x^2 - 16x + 15$ **16.** $y = -3(x + 3)^2 + 12$ **17.** $y = x^2 + 2x + 17$
18. $y = 2(x + 3)^2 + 6$ **19.** $y = (2x - 7)(2x - 1)$ **20.** $y = -\frac{1}{4}x^2 + 6$

 In Exercises 21–24, sketch the graphs of the line and the parabola on a single set of axes. Find the coordinates of any points of intersection by solving algebraically or by using a computer or a graphing calculator.

21. $y = 4 - 2x$, $y = x^2 - 6x + 8$ (2, 0)   **22.** $y = x + 3$, $y = 4x - x^2$ None
23. $y + x = -6$, $y = x^2 + 6x$ (−1, −5),   **24.** $2x + y = 10$, $y = 9 - x^2$ (1, 8)
                         (−6, 0)

 You may find it helpful to have a computer or a graphing calculator to complete Exercises 25 and 26.

B
25. On a single set of axes, graph $y = x^2 - 4$, $y = 2x - 4$, $y = 2x - 5$, and $y = 2x - 6$. Which line appears to be tangent to the parabola? Show algebraically that this line intersects the parabola in only one point. $y = 2x - 5$
26. On a single set of axes, graph $y = x^2 + 5x$ and $y = x + k$ for various values of k. By experimenting, find the value of k that appears to make the line tangent to the parabola. Show algebraically that your answer is correct. −4

In Exercises 27–34, find an equation of the quadratic function described.
                                        $f(x) = -3x^2 + 3x + 6$
27. Its graph is a parabola with *x*-intercepts 2 and −1 and *y*-intercept 6.
28. The function f has zeros 5 and 1 and $f(0) = 1$. $f(x) = \frac{1}{5}x^2 - \frac{6}{5}x + 1$
29. Its graph is a parabola with vertex (4, 8) and passing through the origin. (*Hint:* start with the form of the equation at the bottom of page 38.) $f(x) = -\frac{1}{2}x^2 + 4x$

Linear and Quadratic Functions   **41**

---

30. $f(x) = \frac{8}{9}x^2 - \frac{16}{3}x$

31. $f(x) = \frac{7}{4}(x - 3)^2 - 5$

32. $g(x) = -\frac{1}{2}(x + 1)^2 + 6$

33. $f(x) = \frac{3}{8}(x - 4)^2$

34. $f(x) = -2x^2 + 4x + 6$

35.

  a. $y = -\frac{29}{3600}(x - 60)^2 + 30$

  b. Yes

37. a. $f(1 + k) = 2(1 + k)^2 - 4(1 + k) + 7 = 2k^2 + 5;$
  $f(1 - k) = 2(1 - k)^2 - 4(1 - k) + 7 = 2k^2 + 5$

38. a. $h(3 + k) = 2(3 + k)^2 - 12(3 + k) = 2k^2 - 18 = 2k^2 + h(3) \geqslant h(3)$
  (since $k^2 \geqslant 0$). So the y-value at any point on the parabola is greater than or equal to $h(3)$. Thus $(3, h(3))$ is the minimum point of the parabola.

  b. $f(4 + k) = 9 + 8(4 + k) - (4 + k)^2 = 25 - k^2 = f(4) - k^2 \leqslant f(4)$ (since $k^2 \geqslant 0$). Thus $(4, f(4))$ is the maximum point on the parabola.

39. $\frac{1}{2}\left(\frac{-b + \sqrt{b^2 - 4ac}}{2a} + \frac{-b - \sqrt{b^2 - 4ac}}{2a}\right) = \frac{1}{2}\left(\frac{-2b}{2a}\right) = \frac{-b}{2a}$

---

30. The graph is a parabola with vertex $(3, -8)$ and passing through the origin.

31. The minimum value of f is $f(3) = -5$, and $f(1) = 2$.

32. The maximum value of g is $g(-1) = 6$, and $g(-3) = 4$.

33. Its graph is a parabola tangent to the x-axis at $(4, 0)$ with y-intercept 6.

34. The function f has zeros -1 and 3 and a maximum value of 8.

35. Sports A baseball player tries to hit a ball over an outfield fence that is 4 m high and 110 m from home plate. The ball is hit 1 m above home plate and reaches its highest point 30 m above a point on the ground that is 60 m from home plate.
  a. Make a sketch showing the path of the baseball. If home plate is at the origin of a coordinate system, find an equation of the parabolic path of the baseball.
  b. Will the ball go over the outfield fence?

36. Sports From the center of the 20 yd (60 ft) line, a football player attempts to make a field goal by kicking the ball directly toward the goal posts, which are 90 ft away. The goal-post crossbar is 10 ft above the ground. The ball reaches its highest altitude of 32 ft at a point 48 ft from where it was kicked.
  a. Make a sketch showing the path of the football. If the point from which the ball is kicked is the origin of a coordinate system, find an equation of the parabolic path of the football.
  b. Will the kicker make the field goal?   **a.** $y = -\frac{1}{72}(x - 48)^2 + 32$ **b.** No

37. a. If $f(x) = 2x^2 - 4x + 7$, show that $f(1 + k) = f(1 - k)$ for all real k.
  b. How does part (a) show that the axis of symmetry is the line $x = 1$?

38. a. If $h(x) = 2x^2 - 12x$, show that $h(3) \leqslant h(3 + k)$ for all real k. What does this tell you about the graph of $h(x)$?
  b. If $f(x) = 9 + 8x - x^2$, show that $f(4) \geqslant f(4 + k)$ for all real k. What does this tell you about the graph of $f(x)$?

39. Show that the average of the roots of $ax^2 + bx + c = 0$ is $-\frac{b}{2a}$.

40. If $ax^2 + bx + c = a(x - h)^2 + k$, prove $h = -\frac{b}{2a}$ and $k = \frac{4ac - b^2}{4a}$.

 **For part (a) of Exercise 41, you will need to use a computer or a graphing calculator.**

C **41. a.** Graph the parabola $y = x^2$. Using the ZOOM feature, enlarge a small section of the parabola that contains the point $(1, 1)$ until the graph on the screen looks nearly linear. Use two points on this portion of the parabola to estimate the slope of a line tangent to the parabola at the point $(1, 1)$. 2
  b. Using your estimate from part (a), find an equation of a line tangent to the parabola $y = x^2$ at the point $(1, 1)$. Show algebraically that this line intersects the parabola only at the point $(1, 1)$. $y = 2x - 1$

42   *Chapter One*

42. Transportation A hauling company needs to determine whether a large house trailer can be moved along a highway that passes under a bridge with an opening in the shape of a parabolic arch, 12 m wide at the base and 6 m high in the center. If the trailer is 9 m wide and 3.2 m tall (measured from the ground to the top of the trailer), will it fit under the bridge? No

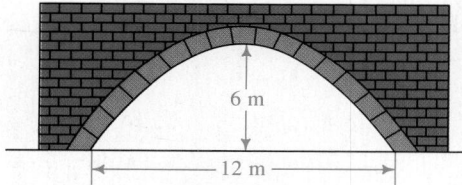

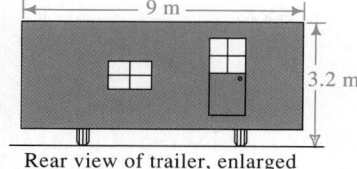

Rear view of trailer, enlarged

1-8 Quadratic Models

Objective *To model real-world situations using quadratic functions.*

Scientists are fascinated by the ability of birds to travel great distances without stopping. Using wind tunnels to monitor parakeets' oxygen consumption and carbon dioxide production in flight, they have investigated the rate at which parakeets expend energy in level flight. The data in the table below show the number of calories burned per gram hour (the number of calories needed to move each gram of body weight for one hour) for three different flight speeds.

Speed (in mi/h)	12	22	30
Calories burned	152	105	165

If we want to predict the bird's energy use at some other flight speed, we will want to find a good mathematical model based on these data. The information in the table is plotted on the graph in figure (a). Note that as the flight speed increases, the energy expended decreases and then increases. Clearly, a linear model would not be suitable to use. We can find many kinds of curves that pass through these data points. (This process is called *curve-fitting*, which we will discuss in more detail in Chapter 18.) One possible curve is a parabola with equation $C(x) = ax^2 + bx + c$. To find the values of a, b, and c, substitute the data from the graph into this equation, as shown on the next page.

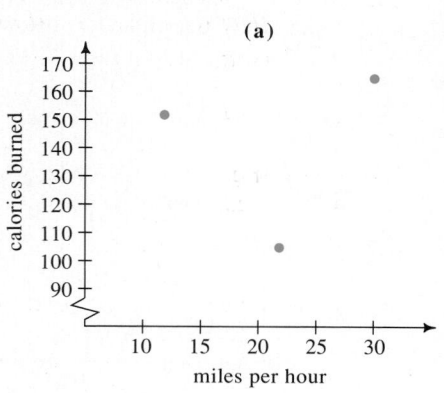

Teaching Notes, p. 1F

Warm-Up Exercises

Find the vertex, *x*-intercept(s), and *y*-intercept of the parabola whose equation is given.

1. $y = (x - 1)^2 + 2$
(1, 2); none; 3

2. $y = x^2 + 10x + 25$
(−5, 0); −5; 25

3. $y = 4 - x^2$ (0, 4); ±2; 4

4. $y = x^2 - 3x - 4$
(1.5, −6.25); −1, 4; −4

5. $y = (x - 1)(x + 3)$
(−1, −4); 1, −3; −3

Motivating the Section

The use of functions to model real-world situations is common in physics, mathematics, and engineering. With the increased use of graphing calculators and computers, students can plot data and go through a curve fitting process to find a function that fits the data. One type of function model, the quadratic model, is discussed in this section.

Mathematical Note

Linear models have the form $C(x) = bx + c$. Quadratic models have the form $C(x) = ax^2 + bx + c$.

Some advanced scientific calculators have matrix algebra
features that can be used to
solve a system of linear equations (see Example 1). The
coefficients of the system are
entered as elements of a matrix, and row operations are
performed on it to obtain the
reduced matrix. See Section
14-3 for a discussion of how
to solve a system of linear
equations by matrix operations.

Cooperative Learning

An important part of using a
model effectively is the interpretation of the results. Have
students in small groups discuss the features listed on
this page. Have them try to
come up with other situations
that lead to quadratic models.

Suggested Assignments

Standard
Day 1: 45/1–10
Day 2: 47/11–17
Comprehensive
45/1, 5, 8, 11–17

Supplementary Materials

Alternative Assessment, 5
Using Technology, 9–10
Student Resource Guide, 5–9

Example 1   Find a quadratic function in the form $C(x) = ax^2 + bx + c$ given that $C(12) = 152$, $C(22) = 105$, and $C(30) = 165$.

Solution

$C(12) = a \cdot 12^2 + b \cdot 12 + c$ $144a + 12b + c = 152$ (1)
$C(22) = a \cdot 22^2 + b \cdot 22 + c$ $484a + 22b + c = 105$ (2)
$C(30) = a \cdot 30^2 + b \cdot 30 + c$ $900a + 30b + c = 165$ (3)

Subtract (1) from (2): $340a + 10b = -47$ (4)

Subtract (2) from (3): $416a + 8b = 60$ (5)

Multiply (5) by $\frac{5}{4}$: $520a + 10b = 75$ (6)

Subtract (4) from (6): $180a \qquad = 122$
$a \qquad \approx 0.678$

Substituting $a = 0.678$ into equation (4) gives $b = -27.752$.

Substituting $a = 0.678$ and $b = -27.752$ into equation (3) gives $c \approx 387.360$. Thus,

$$C(x) = 0.678x^2 - 27.752x + 387.360.$$

We say that $C(x)$ is a *quadratic model* of how the energy expenditure of parakeets depends on flight speed.

Generally speaking, quadratic models are often used to model real-world situations that have certain features, some of which are:

1. Values decrease and then increase, as in the example of the energy expenditure of parakeets in flight. See figure (a) on the previous page.
2. Values increase and then decrease. For example, the number of cars parked at a large office complex on a typical weekday is a quadratic function of the time of day. See figure (b) below.
3. Values depend on surface area. For example, the area A of a circular region is a quadratic function of the radius r, $A = \pi r^2$. In particular, the cost of a plain pizza is a function of area and therefore a quadratic function of the radius (or diameter). See figure (c) below.

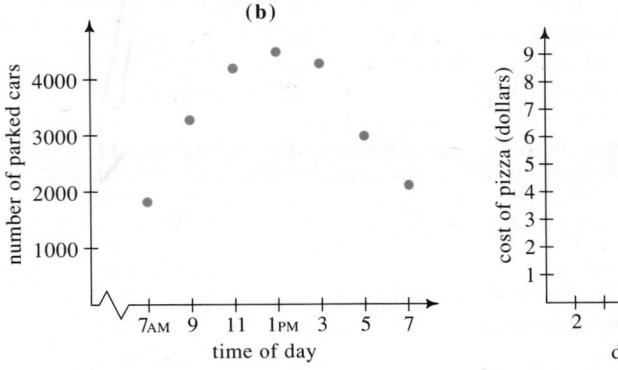

Once we have a model, we can determine its reliability by testing how well other experimental values satisfy the model. If after several experiments we find that the actual data agree with the model, then we can use the model to predict other data values with confidence.

Example 2 Using the given model, determine the number of calories burned per gram hour by a parakeet flying level at 26 mi/h. Compare your answer with the actual laboratory result of 123.5 calories per gram hour.

Solution Substitute the speed of 26 mi/h into the quadratic equation:

$$C(x) = 0.678x^2 - 27.752x + 387.360$$
$$C(26) = 0.678(26)^2 - 27.752(26) + 387.360 = 124.136$$

Thus, the energy expenditure of a parakeet flying at 26 mi/h is about 124 calories per gram hour. Since this result is approximately within one unit of the actual energy expenditure of 123.5 calories per gram hour, we can assume that our model is probably a good one.

CLASS EXERCISES

Discuss whether the relationship described can be modeled by (1) a linear function, (2) a quadratic function, or (3) some other kind of function.

1. The cost of paving a square lot depends on the length of a side. Quadratic
2. The weekly pay of a salesperson who earns $200 plus 10% of sales depends on total sales. Linear
3. The altitude of a ball hit into the air depends on the time it is in flight. Quadratic
4. A hot cup of coffee is poured and allowed to sit. Its temperature depends on the time since it was poured. Other
5. *Visual Thinking* What can you conclude about the flight efficiency of parakeets from the graph on page 43?

WRITTEN EXERCISES

In Exercises 1–4, use the given values to find an equation of the form $f(x) = ax^2 + bx + c$.

A
1. $f(0) = 5$, $f(1) = 10$, $f(2) = 19$ $f(x) = 2x^2 + 3x + 5$
2. $f(1) = 4$, $f(2) = 12$, $f(4) = 46$ $f(x) = 3x^2 - x + 2$
3. $f(0) = 6$, $f(2) = 18$, $f(4) = 34$ $f(x) = \frac{1}{2}x^2 + 5x + 6$
4. $f(1) = 10.5$, $f(2) = 13$, $f(5) = 32.5$ $f(x) = x^2 - 0.5x + 10$

5. As the flight speed of parakeets increases from 12 mi/h, the energy consumption decreases, and the birds achieve peak efficiency at approximately 22 mi/h. As speed increases from 22 mi/h, efficiency decreases. Parakeets appear to fly most efficiently in the middle of their speed range.

Additional Examples

1. In an electric circuit, the available power P in watts when a current of I amperes is flowing is given by $P = 110I - 11I^2$.
   a. If the current is increased from 2 amperes to 3 amperes, by how much will the power increase?
   b. Find the maximum power that can be produced by the circuit.
   a. $P(2) = 110(2) - 11(2^2) = 176$; $P(3) = 110(3) - 11(3^2) = 231$; the power increases by $231 - 176$, or 55 watts.
   b. The maximum occurs at $I = \frac{-b}{2a} = \frac{-110}{2(-11)} = 5$; $P(5) = 110(5) - 11(5^2) = 275$; the maximum power is 275 watts.

(*Continues on next page*)

2. The table shows the height in feet, $h(t)$, t seconds after a ball is thrown upward.

t (seconds)	0	1	3
$h(t)$ (feet)	72	112	96

a. Explain why a quadratic model is appropriate and find a rule for $h(t)$.
b. Find the maximum height attained by the ball.
c. When does the ball hit the ground?

a. The height of the ball first increases and then decreases.
Let $h(t) = at^2 + bt + c$.
Since $h(0) = 72$:
$72 = a \cdot 0^2 + b \cdot 0 + c$;
$c = 72$, so
$h(t) = at^2 + bt + 72$.
Since $h(1) = 112$:
$112 = a \cdot 1^2 + b \cdot 1 + 72$; $a + b = 40$. Also,
$96 = a \cdot 3^2 + b \cdot 3 + 72$;
so $9a + 3b = 24$, or
$3a + b = 8$.
$\underline{\begin{array}{r} a + b = 40 \\ 3a + b = 8 \end{array}}$
$2a = -32$; $a = -16$
$-16 + b = 40$; $b = 56$
$h(t) = -16t^2 + 56t + 72$
b. The maximum height occurs when
$t = \frac{-b}{2a} = \frac{-56}{-32} = 1.75$;
$h(1.75) = -16(1.75)^2 + 56(1.75) + 72 = 121$.
Max. height is 121 ft.
c. Find t such that $h(t) = 0$;
$-16t^2 + 56t + 72 = 0$
$-8(2t^2 - 7t - 9) = 0$
$-8(2t - 9)(t + 1) = 0$
$t = 4.5$ or $t = -1$ (reject).
Thus, the ball hits the ground in 4.5 s.

---

5. Consumer Economics The table below shows the price in dollars, $P(x)$, for pizzas with diameter x inches. The quadratic function that fits the data is $P(x) = 0.00625x^2 + 0.2625x + 2$.

x	8	10	16
$P(x)$	4.50	5.25	7.80

a. Verify that the data fit the function.
b. Does the model give a reasonable price for an 18 in. pizza? $8.75, yes   **c.** No
c. Assuming that the pizza maker is willing to make 4 in. pizza snacks, does the model give a price you would be willing to pay?

6. Physics If you drive at x miles per hour and apply your brakes, your stopping distance in feet is approximately $f(x) = x + \frac{x^2}{20}$.

a. By how much will your stopping distance increase if you increase your speed from 20 mi/h to 30 mi/h? 35 ft
b. By how much will your stopping distance increase if you increase your speed from 50 mi/h to 60 mi/h? 65 ft
c. An important driving guideline is that you leave one car length between your car and the car in front of you for every 10 mi/h of speed. Thus, you should leave three car lengths if you are driving at 30 mi/h and 5 car lengths if you are driving at 50 mi/h. Is this guideline consistent with the model $f(x)$? No, it assumes a linear function.

7. Consumer Economics Suppose your car contains just one gallon of gas. Driving at 20 mi/h you can go 26 mi. Likewise, you can go 34 mi driving at 40 mi/h and 32 mi driving at 50 mi/h.

a. Find a quadratic function that models this data. $D(s) = -0.02s^2 + 1.6s + 2$
b. How far could you go if you drove at 65 mi/h? 21.5 mi
c. The nearest gas station is 16 mi away. If the speed limit is 55 mi/h, at what maximum speed could you drive and still reach it? 55 mi/h

8. Agriculture A farmer planting wheat has found that the yield per acre in bushels depends on the amount of seed sown. By experimenting, he has collected the data given in the following table.

number of bags of seed	3	4	5
number of bushels per acre	70	85	80

a. Give reasons why the yield could decrease when the number of bags increases from 4 bags to 5 bags. For example, increased crowding of seeds
b. Find a quadratic function that models the information in the table.
c. What number of bags will maximize the yield per acre?
b. $f(x) = -10x^2 + 85x - 95$ **c.** $4\frac{1}{4}$ bags

9. Consumer Economics A bakery sells cheesecakes having diameters 6 in., 8 in., and 10 in., all of the same height. Suppose that they cost $8, $12, and $20, respectively.

  a. Which size gives you the most cheesecake per dollar? 8 in.

  b. Find a quadratic function that models how the cost of a cheesecake varies with diameter. $f(x) = \frac{1}{2}x^2 - 5x + 20$

10. Electronics In an electric circuit, the available power in watts is $P(I)$, where I measures the current in amperes. In a 120-volt circuit with a resistance of 16 ohms, electric currents of 2, 4, and 5 amperes produce 176, 224, and 200 watts of power, respectively.

  a. Find a quadratic function that models how the power in the circuit varies with current. $P(I) = -16I^2 + 120I$

  b. Suppose you can set the current in the circuit to 1, 2, 3, 4, or 5 amperes. According to your model, which setting yields the most power? Which setting yields the most watts of power per ampere? 4 amperes; 1 ampere

For exercises 11–14, use the following: Suppose that an object is thrown into the air with an initial upward velocity of v_0 meters per second from a height h_0 meters above the ground. Then, t seconds later, its height $h(t)$ meters above the ground is modeled by the function $h(t) = -4.9t^2 + v_0t + h_0$. (This model doesn't account for air resistance.)

B **11. Physics** A stone is thrown with an upward velocity of 14 m/s from a cliff 30 m high.

  a. Find its height above the ground t seconds later. $h(t) = -4.9t^2 + 14t + 30$

  b. When will the stone reach its highest elevation? **b.** $\frac{10}{7}$ s later **c.** $\frac{30}{7}$ s later

  c. When will the stone hit the ground?

12. Physics A ball is dropped out of a window of a tall building. If it hits the ground 3.5 s later, how high above the ground is the window? About 60 m

13. Physics One half second after springing from a high diving board, a diver reaches her highest point above the water, 4.225 m. If the diving board is 3 m above the water, how long is the diver in the air? $\frac{10}{7}$ s

14. Physics A stone is dropped into a deep, dark well. Although you don't see the stone hit the water, you hear it splash 4.0 s after it is dropped. What is the distance from the top of the well to the surface of the water? (*Hint:* Sound travels at about 343 m/s.) About 71 m

15. Business A manufacturer charges $24 for stereo headphones and has been selling about 1000 of them a week. He estimates that for every $1 price reduction, 100 more headphones can be sold per week. (For example, he could sell 1100 headphones at $23 each and 1200 headphones at $22 each.)

  a. Let $24 - x$ be the reduced price per set of headphones. Write a quadratic function that gives the total revenue received by the manufacturer in a week. $R(x) = -100x^2 + 1400x + 24{,}000$

  b. What price will maximize the total revenue? $17

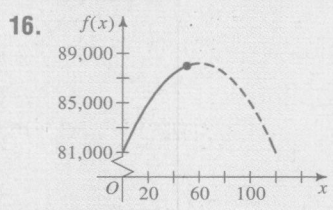
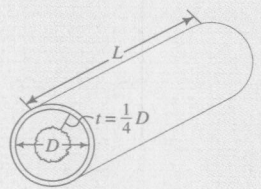

16. Business An airplane with a 200-passenger capacity is to be chartered for a transatlantic flight. The airplane cannot be chartered unless there are at least 150 passengers. The cost of the trip is $540 per passenger, except that the airline company agrees to reduce everyone's ticket price by $2 for *each* ticket sold in excess of 150. Let x be the number of tickets sold in excess of 150 and let $f(x)$ be the total ticket income that the company will receive. Draw a graph of $f(x)$ and specify the domain, range, and maximum value of $f(x)$.

17. Plumbing A water pipe of fixed length L has a carrying capacity that depends on the inner diameter of the pipe. The pipe initially has inner diameter D, but over many years, as mineral deposits accumulate inside the pipe, its carrying capacity is reduced.

a. Give a quadratic function $f(t)$ that models the carrying capacity of the pipe as a function of the thickness t of mineral deposits. Your function will be in terms of the constants D and L and the variable t. (*Hint:* The volume of a cylinder is $\pi r^2 h$.) See below.

b. Show that $f(\frac{1}{4}D) = \frac{1}{4} \cdot f(0)$, and then make a diagram illustrating this fact. Explain in your own words what your diagram says about the carrying capacity of the pipe. **17. a.** $f(t) = \pi L \left(\frac{D}{2} - t\right)^2$

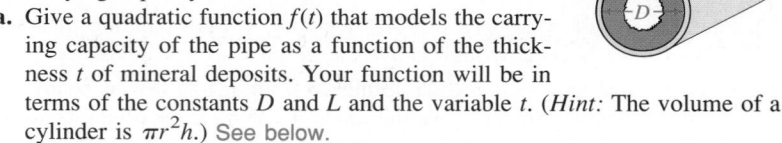

Chapter Summary

1. Given any two points $A(x_1, y_1)$ and $B(x_2, y_2)$:

a. The distance $AB = \sqrt{(x_2 - x_1)^2 + (y_2 - y_1)^2}$.

b. The *midpoint* of $\overline{AB}$ is $\left(\dfrac{x_1 + x_2}{2}, \dfrac{y_1 + y_2}{2}\right)$.

c. The slope of line AB is $\dfrac{y_2 - y_1}{x_2 - x_1}$ if $x_1 \neq x_2$.

d. If $x_1 = x_2$, then line AB is a vertical line and has no slope.

2. The graph of a linear equation $Ax + By = C$, where A and B are not both zero, is a line. If the equation is written in the form $y = mx + k$, the slope is m and the y-intercept is k. The equation of a vertical line has the form $x = C$; vertical lines have no slope.

3. Two different nonvertical lines are parallel if and only if they have the same slope. Two lines with slopes m_1 and m_2 are perpendicular if and only if

$$m_1 = -\frac{1}{m_2}, \text{ or } m_1 \cdot m_2 = -1.$$

4. **a.** A linear equation can be written in several different forms, as shown in the chart below.

The **general** form	$Ax + By = C$	
The **slope-intercept** form	$y = mx + k$	Line has slope m and y-intercept k.
The **point-slope** form	$\dfrac{y - y_1}{x - x_1} = m$	Line has slope m and contains (x_1, y_1).
The **intercept** form	$\dfrac{x}{a} + \dfrac{y}{b} = 1$	Line has x-intercept a and y-intercept b.

   b. To find an equation of a line, use whatever form is the most appropriate. See Examples 1 through 3 on pages 14–15.

5. A *function* describes a dependent relationship between quantities. A *linear function* has the form $f(x) = mx + k$. If $f(a) = 0$, then a is a *zero* of the function f. Certain real-world situations can be effectively modeled by linear functions.

6. **a.** The *imaginary unit i* is defined with the following properties:

$$i = \sqrt{-1} \text{ and } i^2 = -1.$$

   If $a > 0$, then $\sqrt{-a} = i\sqrt{a}$.

   b. A complex number is a number of the form $a + bi$ where a and b are real numbers. The *complex conjugate* of a complex number $z = a + bi$ is $\bar{z} = a - bi$.

   c. The real numbers and the imaginary numbers are subsets of the complex numbers. See the Venn diagram on page 27.

7. A quadratic equation, $ax^2 + bx + c = 0$, can be solved by factoring, by completing the square, or by using the quadratic formula:

$$x = \frac{-b \pm \sqrt{b^2 - 4ac}}{2a}$$

If the *discriminant* $b^2 - 4ac > 0$, there are two real roots.
If $b^2 - 4ac = 0$, there is one real root (called a double root).
If $b^2 - 4ac < 0$, there are two complex conjugate roots.

Linear and Quadratic Functions   **49**

Communication Note

The point-slope form mentioned in item 4 of the summary can also be written as $y - y_1 = m(x - x_1)$.

Supplementary Materials

Tests, 1–2
Alternative Assessment, 63
Student Resource Guide, 10–11

10. a. Find the slope m and the midpoint M of the segment. Write the equation of the line with slope $-\dfrac{1}{m}$ and containing M.

8. a. A *quadratic function* has the form $f(x) = ax^2 + bx + c$. Its graph is a parabola whose *axis of symmetry* is the line $x = -\dfrac{b}{2a}$, and whose *vertex* is the point where the parabola intersects this line.

b. If the equation of a parabola is written in the form $y = a(x - h)^2 + k$, then the vertex of the parabola is the point (h, k).

c. The procedure for graphing a quadratic function is outlined on page 39.

9. Quadratic functions can be used effectively to model real-world situations, especially situations in which (1) values decrease and then increase, (2) values increase and then decrease, or (3) values depend on surface area.

Key vocabulary and ideas

linear equation (p. 1)	complex number (p. 27)
distance formula, midpoint formula (p. 4)	complex conjugate (p. 27)
slope of a line (pp. 7–8)	quadratic equation (p. 30)
slope-intercept form (p. 9)	completing the square (pp. 30–31)
parallel and perpendicular lines (pp. 9–10)	quadratic formula (p. 31)
function, zero of a function (p. 19)	discriminant (pp. 31–32)
linear function (p. 19)	quadratic function, parabola (p. 37)
imaginary unit i (p. 26)	axis of symmetry, vertex (p. 37)

Chapter Test

1. Let $A = (-2, -6)$ and $B = (-4, 2)$. Find: **a.** $2\sqrt{17}$ **b.** $(-3, -2)$ 1-1
   a. The length of $\overline{AB}$ **b.** The coordinates of the midpoint of $\overline{AB}$

2. Find the value of a if the point $(4, -2)$ lies on the line $2x + ay = 14$. -3

3. Solve the equations $2x + 3y = 2$ and $6x - y = -4$ simultaneously. Then sketch the graphs of the lines and label the intersection point. $\left(-\dfrac{1}{2}, 1\right)$

4. Find the slope and the y-intercept of the line $4x - 2y = 7$. $2; -\dfrac{7}{2}$ 1-2

5. Tell which of the following equations have parallel line graphs and which have perpendicular line graphs. $b \parallel c$, $a \perp b$, $a \perp c$

   a. $2x + 3y = 1$ **b.** $y = \dfrac{3}{2}x + 3$ **c.** $6x - 4y - 10 = 0$

6. Write an equation of the line through $(6, -2)$ and $(-3, 1)$. $x + 3y = 0$ 1-3

7. Write an equation of the line through $(5, 5)$ and parallel to the line $4x + 3y = -2$. $4x + 3y = 35$

8. Write an equation of the line with x-intercept -3 and y-intercept -5. $5x + 3y = -15$

9. Write an equation of the vertical line through $(4, -2)$. $x = 4$

10. a. *Writing* Describe the steps of finding an equation of the perpendicular bisector of a line segment.

   b. Write an equation of the perpendicular bisector of the segment joining the points $(7, 0)$ and $(1, 8)$. $3x - 4y = -4$

11. A teacher must grade 23 exam papers. Working for 63 min, she is able to grade exactly seven exams. The function that models the number of exams graded is shown in red at the right. Because some exams take longer to grade than others, the teacher does not work at a constant rate. Nevertheless, her rate is close enough to being constant that a linear function (whose graph is shown in blue) can be used effectively to model her progress.

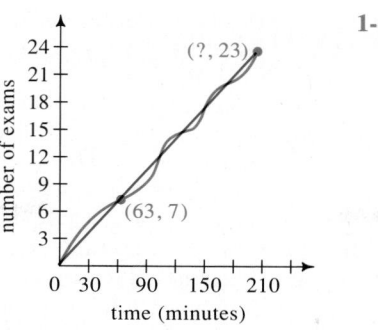

1-4

a. Give an equation of the linear function that models the number of exams as a function of time. See below.

b. Use the equation to find approximately how long it will take to grade all the exams.   **a.** $f(t) = \frac{1}{9}t$ **b.** 207 min = 3 h 27 min

Express each complex number in the form $a + bi$.

12. $\sqrt{-50} - \sqrt{-8}$ $3i\sqrt{2}$

13. $(2 + 3i)^2$ $-5 + 12i$ 1-5

14. $\dfrac{1}{2 + 3i}$ $\dfrac{2}{13} - \dfrac{3}{13}i$

15. $\dfrac{\sqrt{3} + i}{\sqrt{3} - i}$ $\dfrac{1}{2} + \dfrac{\sqrt{3}}{2}i$

16. $4(3 + 2i) - 5(1 - i)$ $7 + 13i$ **17.** i^{17} i

18. Solve for x.   1-6
  a. $7x^2 - 2 = 5x$ $1, -\frac{2}{7}$ **b.** $x^2 - 4x = 9$ $\dfrac{2 \pm \sqrt{13}}{}$ **c.** $x^2 + 2x + 2 = 0$ $-1 \pm i$

19. Find the discriminant of $3x^2 - 2x + 2 = 0$ and then solve by whichever method seems easiest. $-20;\ \dfrac{1 \pm i\sqrt{5}}{3}$

20. Sketch the graph of each parabola. Label the vertex, the axis of symmetry, and the intercepts.   1-7
  a. $y = 8 - 2(x - 1)^2$ **b.** $y = x^2 - 6x + 5$

21. Sketch the graphs of the line $2x - y = -2$ and the parabola $y = -x^2 + 4x + 1$. Find the coordinates of any points of intersection. (1, 4)

22. A photo processing lab lists the following prices for reprints: 29¢ for a   1-8
3 × 5 print; 39¢ for a 4 × 6 print; and 55¢ for a 5 × 7 print.
  a. Find a quadratic function of the form $f(x) = ax^2 + bx + c$ that models the cost (in cents) per print in terms of the smaller dimension of the print size. $f(x) = 3x^2 - 11x + 35$
  b. Using this model, find the approximate price of a 7 × 9 print. $1.05

20. a.

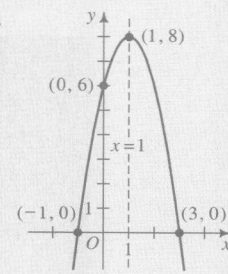

b.

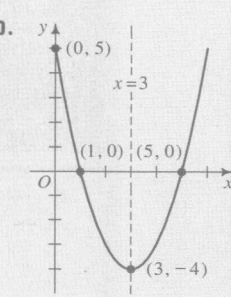

21.

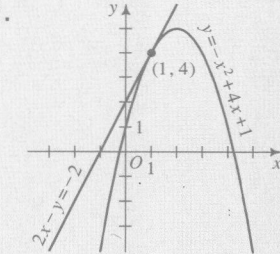

Linear and Quadratic Functions   **51**

Chapter 2 Polynomial Functions

Overview

In this chapter, students solve polynomial equations, both with and without the use of technology, and draw the graphs of polynomial functions. Students also find maximum and minimum points on quadratic and cubic graphs and then apply this knowledge to extreme-value problems from physics, business, and manufacturing. The chapter concludes with a few of the classical theorems about polynomial equations and a brief history of mathematicians' attempts to solve such equations.

Objectives

2-1 To identify a polynomial function, to evaluate it using synthetic substitution, and to determine its zeros.

2-2 To use synthetic division and to apply the remainder and factor theorems.

2-3 To graph a polynomial function and to determine an equation for a polynomial graph.

2-4 To write a polynomial function for a given situation and to find the maximum or minimum value of the function.

2-5 To use technology to approximate the real roots of a polynomial equation.

2-6 To solve polynomial equations by various methods of factoring, including the use of the rational root theorem.

2-7 To apply general theorems about polynomial equations.

Supplementary Resources

Tests, pp. 3–4
Alt. Assess., pp. 6–8, 63
Activities, pp. 3–5
Using Tech., pp. 11, 12, 39–40, 41
Student Res. Guide, pp. 12–23

Software

Precalculus Plotter Plus
Activities, pp. 26, 27
Function Plotter, Equation Solver

Pacing Guide

Section	2-1	2-2	2-3	2-4	2-5	2-6	2-7	Review	Test	Total
Standard (days)	1	2	2	2	1	2	1	1	1	13
Comprehensive (days)	1	1	1	2	1	2	1	1	1	11

Teaching Notes

2-1 | pages 53–58

Presenting the Section

This section introduces students to the vocabulary of polynomials. You should emphasize the different roles of the coefficients (which are constants) and the variables. For example, 7^2x is a linear polynomial but $7x^2$ is quadratic; $\left(\frac{\pi}{1+\pi}\right)x^3$ is a polynomial but $5^3\left(\frac{x}{1+x}\right)$ is not. The distinction between the words *zero* and *root* is also subtle; the former refers to a function, the latter to an equation. For example, 2 is a zero of the polynomial $P(x) = x^2 - 4$; $x = 2$ is a root of the equation $x^2 - 4 = 0$. Compare both to the x-intercept of the graph of a polynomial.

If the coefficients of a polynomial are integers, a table of values may be made easily by using synthetic substitution. Be sure students remember to use 0 as a coefficient whenever their polynomial has no term of a given power.

Using Technology

You may want to make use of a computer to produce a table of values for polynomials. Students can write their own programs; the first program on page 58 is a good model. Function plotting software often includes the option of a table of values. Ask your students under what conditions they should use synthetic substitution and when using a computer is preferable.

2-2 | pages 58–61

Presenting the Section

In this section students learn why knowing a root of a polynomial equation helps in factoring the polynomial. Point out that if you know one root of a cubic equation, you can find all three, since the quotient in the synthetic division process will be quadratic, and then the quadratic formula will give the other two roots.

You should make sure your students avoid the pitfalls of synthetic division: (1) getting the sign of the divisor backward (to divide by $x + 2$, use -2 rather than 2 as the divisor); (2) leaving out coefficients that are zero; and (3) forgetting whether to add or subtract at each step.

As students learn to divide out given factors, they will wonder how to decide what factors to try. You may answer that this is a difficult question that will be considered later in the chapter (Section 2-6). You can also justify your choices by an analogy with factoring a quadratic expression, as in the examples in Section 2-6.

Students may ask about using synthetic division to divide by $2x + 1$ or $x^2 - 4$. Reply that synthetic division may be used to solve such problems, but that it cannot be used directly. For example, to divide $2x^2 - 5x - 3$ by $2x + 1$, note that $(2x + 1)(x - 3)$ is the same as $\left(x + \frac{1}{2}\right)(2x - 6)$; to divide by $x^2 - 4$, divide by $x + 2$ and then by $x - 2$.

Cooperative Learning

Students have difficulty realizing when they should use simple substitution in an exercise stated as a division problem; this is what the factor and remainder theorems are all about. If you ask students in small groups to find the remainder when $x^{36} + 4x^{27} + 7$ is divided by $x + 1$, someone is likely to think of the easy way while the others are writing down zeros laboriously. The longer students have struggled, the more clearly they will understand and remember the point: the desired answer is $(-1)^{36} + 4(-1)^{27} + 7 = 4$.

2-3 | pages 62–68

Presenting the Section

Students have little intuitive idea of what polynomial graphs should look like. When graphing with pencil and paper, students first plot the x-intercepts of a graph and then analyze the sign of the function between and beyond the x-intercepts. There are two reasons for the sign analysis: (1) to determine whether the graph crosses the x-axis at an x-intercept or whether the graph is tangent there, and (2) to distinguish between the true graph and its reflection in the x-axis (which has the same x-intercepts). (In the latter case, an appeal can be made to the sign of the leading coefficient.)

Again, generalizing from parabolas, students will wonder how to "find the vertex," by which they mean how to find the coordinates of the local high or low points. You can tell them that the problem is more difficult if the polynomial is not quadratic, and that they must wait until the next section to answer the question. (Make it clear that only quadratic polynomials have vertices.)

Using Technology

In this section, students can explore the shapes of cubic and quartic functions by using a computer or graphing calculator. This can be done by having students use the technology to either draw the graphs or check their hand-drawn graphs.

Have students begin by graphing polynomials that are factored. Ask students to characterize the way squared or cubed factors affect the graph of the polynomial. Give them practice in graphing a polynomial with adjacent zeros associated with squared and cubed factors. Continue until students can predict what the graph will look like.

Students usually expect local maximums and minimums to occur halfway between x-intercepts (based on their experience of the symmetry of quadratic functions). The graph of $y = (x + 2)^2(x - 2)$ should convince them otherwise.

2-4   pages 68–74

Presenting the Section

In this section, we turn to the problem of finding the extreme values of polynomials, a problem so difficult that calculus was used to solve it until recently. Since about 1975 computers have made these problems accessible to precalculus students.

Applications

In many cases, the costs or profits for a business can be modeled with a polynomial. The objective is to minimize costs and maximize profits. In manufacturing, farming, weather forecasting, space travel, and many other areas, the problem of finding the maximum or minimum values of a function is important.

The method of solution introduced in this section for polynomials is also used for other functions. Unless the polynomial is quadratic, a computer, graphing calculator, or calculus usually is necessary.

Using Technology

If you are going to find maximum and minimum values by using a BASIC program, you may wish to use the one on page 70. A more elementary technique is to print out a table of values. Point out that if the table has (for example) the following entries, it is possible for the maximum value to occur anywhere between 0.2 and 0.4, so the next refinement of the FOR-NEXT loop should be FOR X = .2 TO .4 STEP .01.

X	Y
.2	5.8
.3	6.4
.4	5.2
.5	5.0

Results obtained from computers or graphing calculators may be given in scientific notation, which requires explanation. Many computers and calculators use E for "exponent," so 3.14159E–01 means 3.14159×10^{-1} or 0.314159.

Allow students sufficient time to use a computer or calculator in solving a problem. Even if students understand the method, their work must be free of errors to be successful, and correcting errors is time consuming.

Assessment

To evaluate students' mastery of the ideas discussed above, you can ask them to write a computer program on paper, or give them a printout of a table of values of a polynomial and ask them to use it to find the minimum or maximum value (or the zeros). If you have access to enough computers, you can direct students to use whatever method seems appropriate. If students must work at adjacent computers and you want them to analyze different polynomials, you might give them equivalent polynomials obtained either by using the negative of a basic polynomial or by translating the basic polynomial to the left or to the right.

Presenting the Section

When computers and calculators are used to approximate the roots of polynomial equations, students must do several steps correctly before they get any correct results. To make error-correction easier, have students work in groups until each student is familiar with all the necessary procedures.

Using Technology

Before you ask students to use technology, be sure to try all of the procedures yourself. In particular, try the TRACE, ZOOM, and RESCALE features to see how they work.

Students sometimes try to zoom in on more than one root at a time. Although this may seem to be efficient, it usually results in very distorted enlargements of the graph. Students should instead zoom in and estimate only one root at a time; zooming out to see the "big picture" will be necessary between zoom-ins on individual roots.

When solving equations using the table-of-values method in Example 2, students may make the step size too small and get a long column of numbers that goes by on the screen too fast to be read. See if your computer has a "hold screen" feature that freezes the output temporarily.

Sometimes, when students try to get more accuracy in their table of values, they copy down the wrong interval, and their results will be very confusing. They may also make mistakes with negative numbers and write "FOR X = −1.1 TO −1.2" when they mean "FOR X = −1.2 TO −1.1." Another common mistake is forgetting to use the * sign for multiplication.

Assessment

You will probably need different methods for assessing students' progress, depending on whether they are writing their own BASIC programs. If they are writing their own programs, have them write the programs on paper, then use the computer and write down the answer. Students can include their first and last versions of line 20 in Example 2. If they are using graphing software or a programmable calculator, you may have to observe them during the process.

Presenting the Section

Students sometimes have difficulty factoring expressions containing four terms because they think in terms of "grouping" rather than "looking for a common term." Hence, they may believe that the equation $x^2(x + 3) + 4 = 0$ is equivalent to the equation $(x^2 + 4)(x + 3) = 0$ because the terms have "just been regrouped." Remind students of the distributive property, $ab + ac = a(b + c)$, or use the rational root theorem to avoid the difficulty altogether. You might point out that some equations can be solved easily by factoring but cannot be solved at all by the rational root theorem. An example is Exercise 1, which has no rational roots.

Assessment

Ask students to write down the conditions under which they would use technology for graphing and under which they would use paper and pencil. Then ask students if Exercise 27 on page 84 can be done more easily by graphing in order to find the factors. Also, ask students how they can use technology to answer questions about tangency, such as those in Exercises 31–36 on page 84.

Presenting the Section

This section summarizes some of the classical theorems in mathematics for polynomial equations. Some of the proofs are not accessible, however, to students at this level. For example, the proof of the fundamental theorem of algebra requires the use of either complex analysis or abstract algebra.

Notice that the coefficients in Theorem 3 must indeed be rational instead of real; otherwise, the polynomial $P(x) = (x - (1 - \sqrt{2}))(x - 3)$ would be a counterexample.

Communication

After students have studied the theorems in this section, you might ask selected students to explain the meaning of each theorem, using mathematical language correctly.

2 *Polynomial Functions*

Zeros and Factors of Polynomial Functions

2-1 Polynomials

Objective   *To identify a polynomial function, to evaluate it using synthetic substitution, and to determine its zeros.*

In Chapter 1 we discussed linear functions like $f(x) = 2x - 1$ and quadratic functions like $g(x) = x^2 - 3x + 5$. Linear and quadratic functions belong to a family of functions known as *polynomial functions*.

A **polynomial** in x is an expression that can be written in the form

$$a_n x^n + a_{n-1} x^{n-1} + \cdots + a_2 x^2 + a_1 x + a_0$$

where n is a nonnegative integer. The expressions $a_n x^n$, $a_{n-1} x^{n-1}$, ..., $a_2 x^2$, $a_1 x$, a_0 are called the **terms** of the polynomial, and the numbers a_n, a_{n-1}, ..., a_2, a_1, a_0 are called the **coefficients** of the polynomial. In this book, the coefficients are real numbers, but the values of x may be real or imaginary numbers. (In more advanced courses, even the coefficients may be imaginary.)

Although the terms of a polynomial can be written in any order, we usually write them in descending powers of x. The term containing the highest power of x is called the **leading term**. The coefficient of the leading term is called the **leading coefficient**, and the power of x contained in the leading term is called the **degree** of the polynomial. The polynomial 0 has no degree. Polynomials of the first few degrees have special names, as indicated below.

Degree	Name	Example
0	constant	5
1	linear	$3x + 2$
2	quadratic	$x^2 - 4$
3	cubic	$x^3 + 2x + 1$
4	quartic	$-3x^4 + x$
5	quintic	$x^5 + \pi x^4 - 3.1x^3 + 11$

To identify a particular term of a polynomial, we use the name associated with the power of x contained in the term. For example, the polynomial $x^2 - 4$ has a *quadratic* term of x^2, no *linear* term, and a *constant* term of -4. Note that the leading coefficient of $x^2 - 4$ is understood to be 1.

Every polynomial defines a function, often called P. Any value of x for which $P(x) = 0$ is a **root** of the equation and a **zero** of the function.

◀ The shape of this roller coaster, with its peaks and valleys, suggests the graph of a polynomial equation.

Polynomial Functions   **53**

Example Note

Students should notice that even though Example 1b is not a polynomial, a zero still can be found. Discuss why $x = 1$ cannot be a solution of $g(x) = 0$. (The fraction would be undefined because the denominator would equal 0.)

Note that in Example 2, $P(2) = 0$. This means that 2 is a zero of the polynomial. It also means that $(x - 2)$ is a factor of $P(x)$ (see Section 2-2).

Additional Examples

1. Let $f(x) = \dfrac{2x^3 - 18x}{x + 3}$.

   a. Tell whether f is a polynomial function.
   b. Give each value of x for which f is undefined.
   c. Give the zeros of f.

   a. No, f is a quotient of two polynomial functions, but f is not a polynomial function itself.
   b. When $x = -3$, the denominator has the value 0. Thus, f is undefined for $x = -3$.
   c. $\dfrac{2x^3 - 18x}{x + 3} = \dfrac{2x(x^2 - 9)}{x + 3} =$
   $\dfrac{2x(x + 3)(x - 3)}{x + 3} =$
   $2x(x - 3)$ if $x \neq -3$.
   Thus, the zeros of f are 0 and 3.

Example 1 State whether each function is a polynomial function. Give the zeros of each function, if they exist.

   a. $f(x) = 2x^3 - 32x$ 　　　　　　**b.** $g(x) = \dfrac{x + 1}{x - 1}$

Solution **a.** The function f is a polynomial function. To find its zeros, we solve $f(x) = 0$:

$$2x^3 - 32x = 0$$
$$2x(x^2 - 16) = 0$$
$$2x(x + 4)(x - 4) = 0$$
$$x = 0 \text{ or } x = -4 \text{ or } x = 4$$

The zeros of f are 0, -4, and 4.

   b. The function g is *not* a polynomial function. To find its zeros, we solve $g(x) = 0$:

$$\frac{x + 1}{x - 1} = 0 \longleftarrow \begin{cases} \text{The fraction equals } 0 \\ \text{when its numerator equals } 0. \end{cases}$$

$$x + 1 = 0$$
$$x = -1$$

The only zero of g is -1.

Given a polynomial $P(x)$, we can substitute not only numbers but also variable expressions for x, as shown in the following example.

Example 2 If $P(x) = 3x^4 - 7x^3 - 5x^2 + 9x + 10$, find:
   a. $P(2)$ 　　　　　　　　　　**b.** $P(-3n)$

Solution **a.** $P(2) = 3 \cdot 2^4 - 7 \cdot 2^3 - 5 \cdot 2^2 + 9 \cdot 2 + 10$
   $= 48 - 56 - 20 + 18 + 10$
   $= 0$

   b. $P(-3n) = 3(-3n)^4 - 7(-3n)^3 - 5(-3n)^2 + 9(-3n) + 10$
   $= 3(81n^4) - 7(-27n^3) - 5(9n^2) + 9(-3n) + 10$
   $= 243n^4 + 189n^3 - 45n^2 - 27n + 10$

In Example 2, $P(2)$ can be evaluated more simply using *synthetic substitution*. The basis for this technique involves rewriting the polynomial as follows:

$$P(x) = 3x^4 - 7x^3 - 5x^2 + 9x + 10$$
$$= x(3x^3 - 7x^2 - 5x + 9) + 10$$
$$= x(x(3x^2 - 7x - 5) + 9) + 10$$
$$= x(x(x(3x - 7) - 5) + 9) + 10$$

Since $P(x) = x(x(x(3x - 7) - 5) + 9) + 10$, we can evaluate $P(x)$ for any value of x by writing the coefficients in order,

$$3, -7, -5, 9, \text{ and } 10,$$

and then applying the following algorithm, called **synthetic substitution**:

1. Start with the leading coefficient.
2. Repeat the following until the last coefficient is reached:
   a. Multiply by the value of x.
   b. Add the next coefficient.

The following example illustrates this algorithm.

Example 3 **a.** If $P(x) = 3x^4 - 7x^3 - 5x^2 + 9x + 10$, find $P(2)$.
 b. If $S(x) = 3x^4 - 5x^2 + 9x + 10$, find $S(-2)$.

Solution **a.**

$$
\underline{2\,\rvert} \quad\; 3 \qquad -7 \qquad -5 \qquad 9 \qquad 10
$$
$$
\qquad\qquad \nearrow 6 \;\nearrow -2 \;\nearrow -14 \;\nearrow -10
$$
$$
\qquad 3 \qquad -1 \qquad -7 \qquad -5 \quad \underline{\;0\;} \;\longleftarrow\; P(2) = 0
$$

b. Use 0 as the coefficient of the missing cubic term.

$$
\underline{-2\,\rvert} \quad\; 3 \qquad\; 0 \qquad -5 \qquad 9 \qquad 10
$$
$$
\qquad\qquad -6 \;\nearrow 12 \;\nearrow -14 \;\nearrow 10
$$
$$
\qquad 3 \qquad -6 \qquad 7 \qquad -5 \quad \underline{\;20\;} \;\longleftarrow\; S(-2) = 20
$$

When doing synthetic substitution yourself, you do not need to write the red marks shown in the solution of Example 3.

CLASS EXERCISES

Classify each polynomial function as linear, quadratic, cubic, quartic, or quintic. Also give its leading term, leading coefficient, and degree.

1. $f(x) = 2 + 3x + 5x^2$ Quadratic; $5x^2$; 5; 2 **2.** $g(x) = 5x^4 - x^2$ Quartic; $5x^4$; 5; 4
3. $h(x) = 7 - 4x^3$ Cubic; $-4x^3$; -4; 3 **4.** $P(x) = x^4 + 9x^2 - 1 - 2x^5$
Quintic; $-2x^5$; -2; 5

If $f(x) = 2x^2 + 5$, find each of the following. (Remember: $i = \sqrt{-1}$.)

5. $f(3)$ 23 **6.** $f(3i)$ -13 **7.** $f(3n)$ $18n^2 + 5$ **8.** $f(n + 3)$
$2n^2 + 12n + 23$

In Exercises 9–12, give the zeros of each function.

9. $f(x) = (x + 5)(3x - 9)$ $-5, 3$ **10.** $g(x) = (x^2 - 16)(x^2 + 25)$ $\pm 4, \pm 5i$
11. $P(x) = x^2 - 6x + 9$ 3 **12.** $h(x) = ax^2 + bx + c$ See below.

13. If $P(x) = 2x^3 - 9x^2 + 27$, use synthetic substitution to find $P(3)$ and $P(-3)$. 0; -108

12. $\dfrac{-b \pm \sqrt{b^2 - 4ac}}{2a}$

Polynomial Functions **55**

2. If $g(x) = 2x^2 - 3x$, find $g(3i)$ and $g(-2x)$ by substitution. Then use synthetic substitution to verify your first answer.
$g(3i) = 2(3i)^2 - 3(3i) = 2(9i^2) - 9i = -18 - 9i$
$g(-2x) = 2(-2x)^2 - 3(-2x) = 2(4x^2) + 6x = 8x^2 + 6x$

$$
\underline{3i\,\rvert} \quad 2 \qquad -3 \qquad\qquad 0
$$
$$
\qquad\qquad\quad 6i \qquad -18 - 9i
$$
$$
\qquad 2 \;\; -3 + 6i \;\;\underline{\,\rvert\,-18 - 9i}
$$
$\therefore g(3i) = -18 - 9i$

Error Analysis

When doing synthetic substitution, students may forget to include zeros for the missing terms in the polynomials (those terms with coefficients equal to 0). Caution students that if the polynomial is of degree n, then $n + 1$ coefficients are needed to use the synthetic substitution algorithm.

Exercise Note

For Class Exercise 8, point out that if $n + 3$ replaces x in a polynomial, students will be finding a power of $(n + 3)$ for any term with a degree higher than one.

Review Note

You might need to review how to find the complex roots of $x^2 + 25 = 0$. (See Class Exercise 10.)

WRITTEN EXERCISES

In Exercises 1–12, state whether the function is a polynomial function. Give the zeros of each function, if they exist.

A 1. $f(x) = 17 - 3x$ Yes; $\frac{17}{3}$ 2. $g(x) = x^2 - 6x + 8$ Yes; 2, 4   3. $h(x) = 9$ Yes; none

4. $k(x) = x - \dfrac{1}{x}$ No; -1, 1   5. $m(x) = \dfrac{x^2 - 3x - 4}{x^2 + 1}$ No; -1, 4   6. $n(x) = \dfrac{x^2 + 2}{2}$ Yes; $\pm i\sqrt{2}$

7. $p(x) = x^3 - 9x$ Yes; -3, 0, 3   8. $q(x) = 1 - 5x + 6x^2$ See below.

9. $f(x) = px^2 + r + qx$ See below.   10. $r(x) = (x - 7)^2(x^2 + 7)$ Yes; 7, $\pm i\sqrt{7}$

11. $v(x) = x^3 + 2x^2 + x$ Yes; -1, 0   12. $w(x) = 2x^4 - x^3 - x^2$ Yes; $-\frac{1}{2}$, 0, 1

13. a. Show that the expression $\dfrac{x^3 + 2x^2 - x - 2}{x + 2}$ can be simplified to $x^2 - 1$.

  b. Explain why the first expression is not a polynomial but the second is.

  c. Give the values of x, if any, for which each expression is undefined.

14. a. For what values of x is $f(x) = \dfrac{2x^3 - 3x^2 - 8x + 12}{x - 2}$ undefined? 2

  b. Give the zeros of f. -2, $\frac{3}{2}$ 8. Yes; $\frac{1}{3}$, $\frac{1}{2}$ 9. Yes; $\dfrac{-q \pm \sqrt{q^2 - 4pr}}{2p}$

In Exercises 15–18, find the values of the function. (Remember: $i = \sqrt{-1}$.) Simplify the answers.

15. $h(x) = 2x^2 - 5x + 6$
  a. $h(-1)$ 13   b. $h(2i)$ $-2 - 10i$ c. $h(1 + i)$ $1 - i$ d. $h(3a)$ $18a^2 - 15a + 6$

16. $P(x) = 8x - 4x^2$ 16. d. $\dfrac{16}{x} - \dfrac{16}{x^2}$
  a. $P(2\sqrt{3})$ $16\sqrt{3} - 48$ b. $P(1 - \sqrt{2})$ -4 c. $P(1 + 2i)$ 20   d. $P\left(\dfrac{2}{x}\right)$

17. $f(x) = x^3 - 9x$
  a. $f\left(-\dfrac{\sqrt{2}}{3}\right)$ $\dfrac{79\sqrt{2}}{27}$ b. $f(i\sqrt{3})$ $-12i\sqrt{3}$ c. $f\left(\dfrac{x}{3}\right)$ $\dfrac{x^3}{27} - 3x$ d. $f(x - 3)$ $x^3 - 9x^2 + 18x$

18. $k(x) = x^2(x^2 + 16)$ 18. a. $740 - 416\sqrt{2}$
  a. $k(4 - \sqrt{2})$ b. $k(1 + i)$ $-4 + 32i$ c. $k\left(\dfrac{p}{q}\right)$ See below.   d. $k(x^2)$ $x^8 + 16x^4$

19. If $P(x) = 4x^3 - 5x^2 + 7x - 9$, use synthetic substitution to find:
  a. $P(2)$ 17   b. $P(3)$ 75   c. $P(-2)$ -75 d. $P(-3)$ -183

20. If $P(x) = x^4 - 3x^2 + 7x + 8$, use synthetic substitution to find:
  a. $P(3)$ 83   b. $P(-3)$ 41   c. $P(-1)$ -1 d. $P(-2)$ -2

21. If $P(x) = 3x^3 - 7x^2 + 2x + 3$, use synthetic substitution to find:
  a. $P\left(\dfrac{1}{3}\right)$ 3   b. $P\left(-\dfrac{2}{3}\right)$ $-\dfrac{7}{3}$

22. If $P(x) = 4x^4 - 3x^3 + 7x - 2$, use synthetic substitution to find:
  a. $P\left(-\dfrac{1}{4}\right)$ $-\dfrac{59}{16}$ b. $P\left(\dfrac{3}{4}\right)$ $\dfrac{13}{4}$ 18. c. $\dfrac{p^4}{q^4} + \dfrac{16p^2}{q^2}$

56   *Chapter Two*

27. b. $f(x) = a(x - 2)(x - 3)$ for any real $a \neq 0$ **c.** $g(x) = ax(x - 2)(x - 3)$ for any real $a \neq 0$

B **23.** If 4 is a zero of $f(x) = 3x^3 + kx - 2$, find the value of k. -47.5

24. If $2i$ is a zero of $f(x) = x^4 + x^2 + a$, find the value of a. -12

25. A quadratic polynomial P has leading coefficient -2, a constant term of 6, and no linear term. Find the zeros of P. $\pm\sqrt{3}$

26. The leading coefficient of a cubic polynomial P is 2, and the coefficient of the linear term is -5. If $P(0) = 7$ and $P(2) = 21$, find $P(3)$. 64

27. a. What are the zeros of the quadratic function $f(x) = 2(x - 1)(x - 4)$? 1, 4
b. Find a quadratic function with zeros 2 and 3. See above.
c. Find a cubic function with zeros 0, 2, and 3.

28. a. What are the zeros of the quadratic function $f(x) = 3(x + 1)(x - 2)$? $-1, 2$
b. Find a quadratic function with zeros 3 and -4. See below.
c. Find a quartic function with zeros ± 3 and ± 4.

29. If $f(x) = 7x + 2$, find: **a.** $f(9.2) - f(8.2)$ 7   **b.** $f(x + 1) - f(x)$ 7

30. If $g(x) = 3 - 8x$, find: **a.** $g(6.25) - g(4.25)$ -16 **b.** $g(x + 2) - g(x)$ -16

31. If $f(x) = mx + k$, show that the value of $\dfrac{f(x + h) - f(x)}{h}$, where $h \neq 0$, does not depend on x or h. Interpret this result graphically.

32. If $f(x) = x^2$ and $h \neq 0$, find the value of $\dfrac{f(x + h) - f(x)}{h}$. Is the value of this expression independent of the values of x and h, as it was for the function of Exercise 31? $2x + h$; no

33. *Investigation* Multiply several pairs of nonzero polynomials. What is the relationship between the degree of the product and the degrees of the factors? How can you use this relationship to justify saying that the polynomial 0 has no degree, even though all other constant polynomials have degree 0?

C **34. a.** Consider the following table of values for the quadratic function $f(x) = x^2 + 2x + 3$.

x	0	1	2	3	4	5
$f(x)$	3	6	11	18	27	38
differences		3	5	7	9	11

What pattern do you observe in the differences?
b. Make a difference table similar to the one in part (a) for the function $g(x) = 2x^2 - 3x - 1$. What pattern do you observe in the differences?
c. Do the differences in the values of *all* quadratic functions form a pattern like those you observed in parts (a) and (b)? Justify your answer with a difference table.

35. *Investigation* Experiment with differences in the values of cubic functions. (See Exercise 34.) Can you detect any patterns?

28. b. $f(x) = a(x - 3)(x + 4)$ for any real $a \neq 0$
c. $g(x) = a(x - 3)(x + 3)(x - 4)(x + 4)$ for any real $a \neq 0$

Polynomial Functions **57**

Warm-Up Exercises

Find the required polynomial.

1. $\dfrac{x^2 - 4x - 5}{x + 1} = \underline{\quad?\quad}$ $x - 5$

2. $(3x^2 - 7)(x - 4) + 2 = \underline{\quad?\quad}$
   $3x^3 - 12x^2 - 7x + 30$

3. $\dfrac{5x^3 + 18x^2 - 6x + 1}{x + 4} =$
   $5x^2 - 2x + 2 + \dfrac{?}{x + 4}$ -7

4. dividend $= 3x^4 - 2x^2 - 11$,
   divisor $= x^2 - 3$, remainder $= 10$, quotient $= \underline{\quad?\quad}$
   $3x^2 + 7$

5. divisor $= x + 5$, quotient $= 2x^3 - x + 1$, remainder $= -4$, dividend $= \underline{\quad?\quad}$
   $2x^4 + 10x^3 - x^2 - 4x + 1$

Motivating the Section

Suppose we want to know if $x - 1$ is a factor of the polynomial $x^{1000} + x^{500} + 1$. A computer can be used to solve this problem, but it is much easier to solve if synthetic substitution and the factor theorem, both topics discussed in this section, are used. (The answer is $P(1) = 1^{1000} + 1^{500} + 1 = 3$. By the factor theorem, $x - 1$ is *not* a factor of the polynomial.)

//// COMPUTER EXERCISES

1. The two programs below evaluate $P(x) = 2x^4 - 5x^3 + 7x^2 - 9x + 11$ for integral values of x from 1 to 100. In the first program $P(x)$ is in its given form, while in the second program $P(x)$ is in the form used for synthetic substitution: $x(x(x(2x - 5) + 7) - 9) + 11$. Determine the running time for each program. Which one is faster? Times may vary. The second program is faster.

```
10   FOR X = 1 TO 100
20   LET P = 2 * X^4 - 5 * X^3 + 7 * X^2 - 9 * X + 11
30   NEXT X
40   PRINT ''DONE''
50   END
```

```
10   FOR X = 1 TO 100
20   LET P = X * (X * (X * (2 * X - 5) + 7) - 9) + 11
30   NEXT X
40   PRINT ''DONE''
50   END
```

2. Consider a third program, just like those in Exercise 1 except for line 20:

```
20 LET P = 2*X*X*X*X - 5*X*X*X + 7*X*X - 9 * X + 11
```

How does the running time for this program compare with the running times for the programs in Exercise 1? This program is faster than the first, but slower than the second.

2-2 Synthetic Division; The Remainder and Factor Theorems

| Objective | To use synthetic division and to apply the remainder and factor theorems.

In earlier algebra courses, you learned how to add, subtract, multiply, and divide polynomials. What follows is an example of finding the quotient and remainder when $2x^4 - 15x^2 - 10x + 5$ is divided by $x - 3$ using long division.

$$
\begin{array}{r}
2x^3 + 6x^2 + 3x - 1 \quad \longleftarrow \text{ Quotient}\\
x - 3\overline{\smash{\big)}\,2x^4 + 0x^3 - 15x^2 - 10x + 5} \quad \longleftarrow \text{ Dividend}\\
\underline{2x^4 - 6x^3}\\
6x^3 - 15x^2\\
\underline{6x^3 - 18x^2}\\
3x^2 - 10x\\
\underline{3x^2 - 9x}\\
-1x + 5\\
\underline{-1x + 3}\\
2 \quad \longleftarrow \text{ Remainder}
\end{array}
$$

Now compare dividing $P(x) = 2x^4 - 15x^2 - 10x + 5$ by $x - 3$ with evaluating $P(3)$ by synthetic substitution:

$$
\begin{array}{r|rrrrr}
3 & 2 & 0 & -15 & -10 & 5 \\
& & 6 & 18 & 9 & -3 \\
\hline
& 2 & 6 & 3 & -1 & \underline{|\,2\,} \quad \leftarrow P(3) = 2
\end{array}
$$

Notice that $P(3)$, shown in red, is the remainder when $P(x)$ is divided by $x - 3$. Because this is true for any polynomial $P(x)$ and any divisor $x - a$, we have the following theorem.

The Remainder Theorem

When a polynomial $P(x)$ is divided by $x - a$, the remainder is $P(a)$.

Proof   When a polynomial $P(x)$ is divided by $x - a$, the quotient $Q(x)$ is a polynomial in x and the remainder R is a constant. These polynomials are related as follows:

$$\text{Dividend} = \text{Divisor} \times \text{Quotient} + \text{Remainder}$$

$$P(x) = (x - a) \cdot Q(x) + R$$

Substituting a for x in this equation, we get:

$$
\begin{aligned}
P(a) &= (a - a) \cdot Q(a) + R \\
&= 0 \cdot Q(a) + R \\
&= R
\end{aligned}
$$

Look again at the synthetic substitution shown above. Notice that the numbers shown in blue are the coefficients of the quotient when $P(x)$ is divided by $x - 3$. Because synthetic substitution gives not only the remainder but also the quotient, we can use it in place of long division. When we do, we often refer to it as *synthetic division*.

Example 1   Divide $P(x) = x^3 + 5x^2 + 5x - 2$ by $x + 2$.

Solution   After rewriting $x + 2$ as $x - (-2)$, use synthetic division:

$$
\begin{array}{r|rrrr}
-2 & 1 & 5 & 5 & -2 \\
& & -2 & -6 & 2 \\
\hline
& 1 & 3 & -1 & \underline{|\,0\,} \\
& \downarrow & \downarrow & \downarrow & \downarrow \\
& 1x^2 + & 3x & -1 &
\end{array}
$$

The quotient is: $1x^2 + 3x - 1$

The remainder is: 0

1. Find the quotient and the remainder when $x^3 + x^2 - 10x + 8$ is divided by $x + 4$.

$$
\begin{array}{r|rrrr}
-4 & 1 & 1 & -10 & 8 \\
& & -4 & 12 & -8 \\
\hline
& 1 & -3 & 2 & \underline{|\,0\,} \\
& \downarrow & \downarrow & \downarrow & \\
& 1x^2 & -\ 3x & +\ 2 &
\end{array}
$$

The quotient is $x^2 - 3x + 2$ and the remainder is 0.

2. If $x = 2$ is a root of $2x^3 + 5x^2 - 23x + 10 = 0$, find the remaining roots.

If $x = 2$ is a root of the polynomial equation, then $x - 2$ is a factor of $2x^3 + 5x^2 - 23x + 10$.

$$
\begin{array}{r|rrrr}
2 & 2 & 5 & -23 & 10 \\
& & 4 & 18 & -10 \\
\hline
& 2 & 9 & -5 & 0
\end{array}
$$

Thus, $(x - 2)(2x^2 + 9x - 5) = 0$. The other two roots are obtained by factoring:

$$
\begin{aligned}
2x^2 + 9x - 5 &= 0 \\
(2x - 1)(x + 5) &= 0 \\
2x - 1 = 0 \text{ or } x + 5 &= 0 \\
x = \tfrac{1}{2} \text{ or } x &= -5
\end{aligned}
$$

The other roots are $x = \dfrac{1}{2}$ and $x = -5$.

Assessment Note

See if students can use the remainder theorem to find $P(x)$ if $P(x)$ divided by $(x - 1)$ has a remainder of -2 and a quotient of $x^3 + x^2 - x - 1$. $(P(x) = x^4 - 2x^2 - 1)$

Since the remainder in Example 1 is 0, we have:

$$P(x) = x^3 + 5x^2 + 5x - 2 = (x + 2)(x^2 + 3x - 1)$$

We see that $P(x)$ is the product of two *factors*, one of which is $x + 2$. This leads us to another important theorem.

The Factor Theorem

For a polynomial $P(x)$, $x - a$ is a factor if and only if $P(a) = 0$.

Example 2 If $P(x) = 2x^4 + 5x^3 - 8x^2 - 17x - 6$, determine whether each of the following is a factor of $P(x)$: **a.** $x - 1$ **b.** $x - 2$

Solution **a.** Use direct substitution to find $P(1)$:

$$P(1) = 2 \cdot 1^4 + 5 \cdot 1^3 - 8 \cdot 1^2 - 17 \cdot 1 - 6$$
$$= 2 + 5 - 8 - 17 - 6 = -24$$

Since $P(1) \neq 0$, $x - 1$ is not a factor of $P(x)$.

b. Use synthetic substitution to find $P(2)$:

$$
\begin{array}{r|rrrrr}
2 & 2 & 5 & -8 & -17 & -6 \\
  &   & 4 & 18 & 20 & 6 \\
\hline
  & 2 & 9 & 10 & 3 & \underline{|0\,} \leftarrow P(2) = 0
\end{array}
$$

Since $P(2) = 0$, $x - 2$ is a factor of $P(x)$.

3. a. $2x^3 - 5x^2 + 8x - 4$; $x - \frac{1}{2}$ **b.** $2x^2 - 4x + 6$; -1 **c.** -1

CLASS EXERCISES

1. a. What is the remainder when $P(x) = x^{15} + 3x^{10} + 2$ is divided by $x - 1$? 6
 b. What is the remainder when $P(x)$ is divided by $x + 1$? 4

2. At the right is a division problem
 that has been done synthetically.
 a. What is the dividend? What is
 the divisor? $x^4 - 8x^2 + 5x - 1$; $x + 3$
 b. What is the quotient? What is the remainder? $x^3 - 3x^2 + x + 2$; -7
 c. What is the value of the polynomial when $x = -3$? -7

$$
\begin{array}{r|rrrrr}
-3 & 1 & 0 & -8 & 5 & -1 \\
   &   & -3 & 9 & -3 & -6 \\
\hline
   & 1 & -3 & 1 & 2 & \underline{|-7\,}
\end{array}
$$

3. Repeat parts (a) and (b) of Exercise 2 for
 the synthetic division problem shown at
 the right. Then give the value of the polynomial when $x = \frac{1}{2}$. See above.

$$
\begin{array}{r|rrrr}
\frac{1}{2} & 2 & -5 & 8 & -4 \\
   &   & 1 & -2 & 3 \\
\hline
   & 2 & -4 & 6 & \underline{|-1\,}
\end{array}
$$

4. Is $x + 1$ a factor of $P(x) = x^3 + 3x^2 + x - 1$? If so, how would you find the other factors of $P(x)$?

5. Explain why the factor theorem is a special case of the remainder theorem.

Cooperative Learning

Have students discuss Class Exercise 5 in groups to make sure they understand how the factor theorem is a special case of the remainder theorem.

Mathematical Note

The remainder in the synthetic division process must always be a constant, since the degree of the divisor is 1 and the remainder must have a lower degree. The synthetic division process also can be used for a polynomial with complex coefficients.

The factor theorem is a *biconditional*, that is, it is true "both ways." Therefore, when testing a number a, if $P(a) \neq 0$, then a is not a root.

Additional Answers
Class Exercises

4. Yes; use the quadratic formula on the quotient to obtain the quotient's zeros, say a and b, so that the other factors of $P(x)$ are $(x - a)$ and $(x - b)$.

5. $x - a$ is a factor of $P(x)$ if and only if the division of $P(x)$ by $x - a$ results in a remainder of 0. Since the remainder theorem tells us that the remainder is $P(a)$, $x - a$ is a factor of $P(x)$ if and only if $P(a) = 0$.

Suggested Assignments

Standard
Day 1: 61/1–25 odd
Day 2: 61/2–26 even
Comprehensive
61/1–25 odd

A 1. Find the remainder when $x^5 - 2x^3 + x^2 - 4$ is divided by:
   a. $x - 1$ -4 **b.** $x + 1$ -2 **c.** $x - 2$ 16   **d.** $x + 2$ -16

2. Find the remainder when $x^3 - 3x^2 + 5$ is divided by:
   a. $x - 2$ 1   **b.** $x + 2$ -15 **c.** $x - 3$ 5   **d.** $x + 3$ -49

In Exercises 3–10, find the quotient and the remainder when the first polynomial is divided by the second. 3–10. See below.

3. $x^3 - 2x^2 + 5x + 1; \ x - 1$ 4. $2x^3 + x^2 + 3x + 7; \ x + 2$

5. $x^4 - 2x^3 + 5x + 2; \ x + 1$ 6. $2x^4 - 3x^3 + 4x^2 - 5x + 2; \ x - 1$

7. $x^5 + x^3 + x; \ x - 3$ 8. $x^2 - 3x^4; \ x + 2$

9. $3x^4 - 2x^3 + 5x^2 + x + 1; \ x^2 + 2x$ 10. $x^5 + 3x^2 + 4; \ x^2 + 2x + 1$

11. Determine whether $x - 1$ or $x + 1$ is a factor of $x^{100} - 4x^{99} + 3$. Yes; no

12. Determine whether $x - 2$ or $x + 2$ is a factor of $x^{20} - 4x^{18} + 3x - 6$. Yes; no

13. Which of the following are factors of $P(x) = x^3 - 5x^2 + 3x + 9$?
    a. $x - 1$ No   **b.** $x + 3$ No   **c.** $x - 3$ Yes

14. Which of the following are factors of $P(x) = x^4 - 3x^3 + 5x - 2$?
    a. $x + 2$ No   **b.** $x - 2$ Yes   **c.** $x + 4$ No

15. Show that $x - a$ is a factor of $x^n - a^n$ for any positive integer n.

16. Show that $x + a$ is a factor of $x^n + a^n$ for any positive odd integer n.

17. When a polynomial $P(x)$ is divided by $2x + 1$, the quotient is $x^2 - x + 4$ and the remainder is 3. Find $P(x)$. $2x^3 - x^2 + 7x + 7$

18. When a polynomial $P(x)$ is divided by $3x - 4$, the quotient is $x^3 + 2x + 2$ and the remainder is -1. Find $P(x)$. $3x^4 - 4x^3 + 6x^2 - 2x - 9$

In Exercises 19–24, you are given a polynomial equation and one or more of its roots. Find the remaining roots.

19. $2x^3 - 5x^2 - 4x + 3 = 0$; root: $x = 3$ $-1, \frac{1}{2}$

20. $6x^3 + 11x^2 - 4x - 4 = 0$; root: $x = -2$ $-\frac{1}{2}, \frac{2}{3}$

B 21. $2x^4 - 9x^3 + 2x^2 + 9x - 4 = 0$; roots: $x = -1, \ x = 1$ $\frac{1}{2}, 4$

22. $4x^4 - 4x^3 - 25x^2 + x + 6 = 0$; roots: $x = -2, \ x = 3$ $\pm\frac{1}{2}$

23. $x^4 + 3x^3 - 3x^2 + 3x - 4 = 0$; roots: $x = -4, \ x = 1$ $\pm i$

24. $x^4 - 2x^3 + x^2 - 4 = 0$; roots: $x = -1, \ x = 2$ $\frac{1 \pm i\sqrt{7}}{2}$

25. Use the factor theorem to show that $x - a$ is a factor of
    $$x^2(a - b) + a^2(b - x) + b^2(x - a).$$

26. Use the factor theorem to show that $x - c$ is a factor of
    $$(x - b)^3 + (b - c)^3 + (c - x)^3.$$

3. $x^2 - x + 4$; 5

4. $2x^2 - 3x + 9$; -11

5. $x^3 - 3x^2 + 3x + 2$; 0

6. $2x^3 - x^2 + 3x - 2$; 0

7. $x^4 + 3x^3 + 10x^2 + 30x + 91$; 273

8. $-3x^3 + 6x^2 - 11x + 22$; -44

9. $3x^2 - 8x + 21$; $-41x + 1$

10. $x^3 - 2x^2 + 3x - 1$; $-x + 5$

Supplementary Materials

Alternative Assessment, 6
Student Resource Guide, 12–15

Mathematical Note

For Exercises 21–24, in which more than one root is given, students should divide by the first root and then divide the resulting quotient polynomial (later called the "depressed" polynomial) by the second root, since it will have to be one of its factors.

Additional Answers
Written Exercises

15. If $P(x) = x^n - a^n$, then $P(a) = a^n - a^n = 0$. By the factor theorem, $x - a$ is a factor of $P(x)$.

16. If $P(x) = x^n + a^n$ (n odd), then $P(-a) = (-a)^n + a^n = -a^n + a^n = 0$. By the factor theorem, $x + a$ is a factor of $P(x)$.

25. If $P(x) = x^2(a - b) + a^2(b - x) + b^2(x - a)$, then $P(a) = a^2(a - b) + a^2(b - a) + b^2(a - a) = a^3 - a^2b + a^2b - a^3 + b^2(0) = 0$, so $x - a$ is a factor of $P(x)$.

26. If $P(x) = (x - b)^3 + (b - c)^3 + (c - x)^3$, then $P(c) = (c - b)^3 + (b - c)^3 + (c - c)^3 = -(b - c)^3 + (b - c)^3 + 0 = 0$, so $x - c$ is a factor of $P(x)$.

Give the real zeros of each function.

1. $P(x) = (x - 7)^2(2x + 1)$

$7, -\dfrac{1}{2}$

2. $f(x) = x^3 + 2x^2 + x$ $0, -1$

3. $g(x) = 6x^2 - 7x - 20$

$\dfrac{5}{2}, -\dfrac{4}{3}$

4. $h(x) = x^3 - 5x^2 + x - 5$

5

5. $k(x) = 4x^4 - 17x^2 + 18$

$\pm\dfrac{3}{2}, \pm\sqrt{2}$

Motivating the Section

In many real-world situations, functions that are of degree 3 or higher are used as models. In this section, students will graph polynomials of degree 3 or higher.

Communication Note

The graphs given on this page show the *general* shape of graphs of cubic functions. When students draw their own graphs, they should show tick marks and scales on both axes.

Mathematical Note

The graph of a polynomial function is a continuous, unbroken curve with only smooth, rounded turns. The points where the graph turns are called "local minimums" or "local maximums."

■ Graphs; Maximums and Minimums

2-3 Graphing Polynomial Functions

Objective   *To graph a polynomial function and to determine an equation for a polynomial graph.*

Graphs of Cubic and Quartic Functions

Activity 1   **a.** As x increases without bound, $f(x)$ increases without bound; as x decreases without bound, $f(x)$ decreases without bound.

 In this activity, you will use a computer or graphing calculator to explore the shape of the graphs of cubic functions.

a. Graph $f(x) = 2x^3 - 7x^2 + 4x + 20$. Adjust the viewing window until you think the complete shape of the graph is visible. Then re-adjust the window, or use the TRACE feature, to determine what happens to the value of $f(x)$ as the value of x increases or decreases without bound. See above.

b. Repeat part (a) using the function $f(x) = -x^3 - 5x^2 - 12$. See below.

c. Repeat part (a) using other cubic functions, some with positive leading coefficients and some with negative leading coefficients.

d. Based on your work, describe the general shape of the graph of a cubic function. How does the sign of the leading coefficient affect the graph?

b. As x increases without bound, $f(x)$ decreases without bound; as x decreases without bound, $f(x)$ increases without bound.

Generally speaking, the graph of $y = ax^3 + bx^2 + cx + d$ is shaped like an unbroken "sideways S." (Unbroken curves are said to be *continuous,* a concept that will be discussed in Section 19-1.) Although the graph has some bends in it, overall, the curve rises from left to right if $a > 0$ and the curve falls from left to right if $a < 0$. The "generic" cubic graphs shown below illustrate these facts.

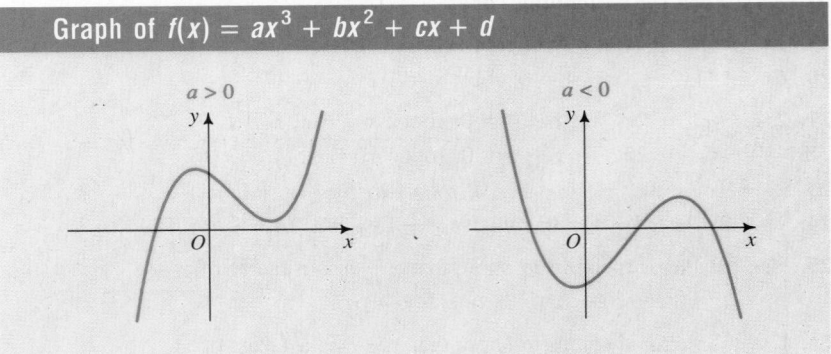

Graph of $f(x) = ax^3 + bx^2 + cx + d$

You can quickly sketch the graph of a cubic function by hand if the function is in factored form, as in the example on the next page.

Example 1 Sketch the graph of the factored cubic function $f(x) = (x + 1)(x - 1)(x - 2)$.

Solution

Step 1 Find and plot the zeros of the function:

$$(x + 1)(x - 1)(x - 2) = 0$$
$$x = -1, x = 1, x = 2$$

The zeros of f are -1, 1, and 2.

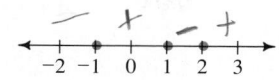

Step 2 Perform a sign analysis of $f(x)$ by testing one value of x from each of the intervals determined by the zeros.

Interval: $x > 2$ Test: $x = 3$
$$f(3) = \underbrace{(3 + 1)(3 - 1)(3 - 2)}_{\substack{+ \quad\quad + \quad\quad + \\ \text{signs of} \\ \text{factors}}} \underbrace{= +}_{\substack{\text{sign of} \\ \text{product}}}$$

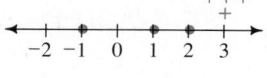

Interval: $1 < x < 2$ Test: $x = 1.5$
$$f(1.5) = (1.5 + 1)(1.5 - 1)(1.5 - 2) = -$$

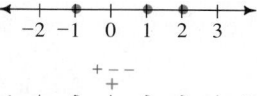

Interval: $-1 < x < 1$ Test: $x = 0$
$$f(0) = (0 + 1)(0 - 1)(0 - 2) = +$$

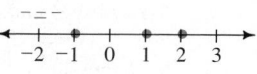

Interval: $x < -1$ Test: $x = -2$
$$f(-2) = (-2 + 1)(-2 - 1)(-2 - 2) = -$$

Step 3 Sketch the graph: Step 1 gives you the x-intercepts of the graph, and Step 2 tells you where the graph is above or below the x-axis.

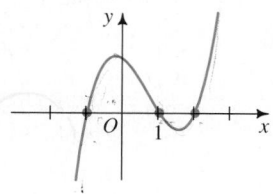

Activity 2

 Use the method of Example 1 to sketch the graphs of the following polynomial functions. Use a computer or graphing calculator to check your graphs.

a. $y = x(x + 2)(x - 1)$ **b.** $y = (x + 3)(x + 1)(x - 1)(x - 2)$
c. $y = -x(x + 2)(x - 1)$ **d.** $y = -(x + 3)(x + 1)(x - 1)(x - 2)$

The graphs of parts (b) and (d) of Activity 2 illustrate that the graph of a quartic function generally has a "W-shape" or "M-shape," as shown on the next page.

Error Analysis

Caution students to put the intervals determined by the zeros in increasing (or decreasing) order of the zeros, since not doing so may cause a misinterpretation of where the graph is above or below the x-axis.

Example Note

In Step 3 of Example 1, point out that no scale on the y-axis is shown because exactly how high/low the graph rises/falls above/below the x-axis is not our concern here.

Using Technology

In Activity 1, students used a graphing calculator to explore the characteristics of the graphs of cubic functions. Students should recognize, however, that the goal of the section is to quickly sketch the graphs of factored polynomial functions *by hand*. In Activity 2, technology should be used only to *check* sketches.

Activity Note

Ask students to generalize about the shape and direction of the graph of a quartic function based on their graphs in parts (b) and (d).

Suppose the polynomial function $P(x)$ contains the factor $(x - a)^k$, which yields k repeated zeros. If k is odd, then the graph of $P(x)$ will *cross* the x-axis at $x = a$; and if k is even, then the graph will be *tangent* to the x-axis at $x = a$.

Additional Examples

1. Factor $f(x) = x^3 - 2x^2 - 4x + 8$ and sketch its graph.

Use factoring by grouping:
$x^3 - 2x^2 - 4x + 8 =$
$x^2(x - 2) - 4(x - 2) =$
$(x^2 - 4)(x - 2) =$
$(x + 2)(x - 2)^2$
(Note: Synthetic division can also be used to determine the factors.) The function has zeros -2 and 2, with 2 a double zero. Test a value in each interval to obtain this sign graph:

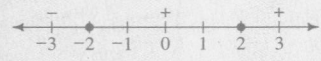

Use the sign graph to sketch the graph of $f(x)$:

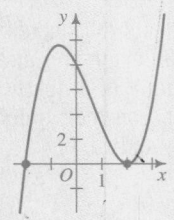

Graph of $f(x) = ax^4 + bx^3 + cx^2 + dx + e$

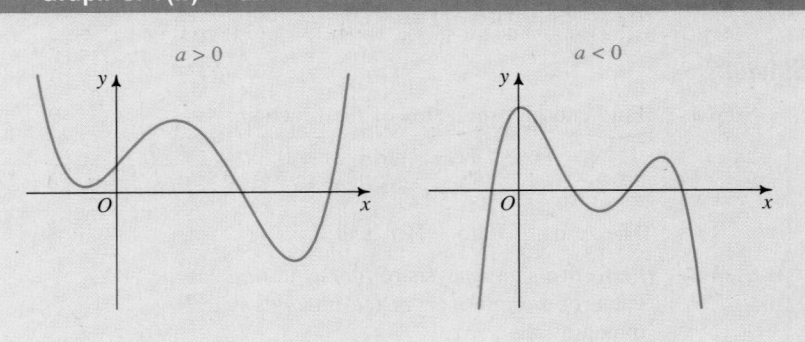

$a > 0$ $a < 0$

Activity 3

In part (a) of this activity, you may wish to use a computer or a graphing calculator either to draw the graphs or to check your hand-drawn graphs.

a. In Example 1 we considered the graph of $y = (x + 1)(x - 1)(x - 2)$. Quickly sketch the three related graphs of

$$y = (x + 1)(x - 1)(x - k)$$

for $k = 1.5$, $k = 1.3$, and $k = 1.1$. What happens to the graphs as the value of k approaches 1? The distance between the x-intercepts k and 1 approaches 0.

b. In general, what effect does a squared factor, as in $y = (x + 1)(x - 1)^2$, have on the graph of a polynomial function? Use a sign analysis to explain your answer. The graph is tangent to the x-axis at the root associated with the squared factor. For example, a sign analysis of $y = (x + 1)(x - 1)^2$ shows that $y > 0$ for both $x > 1$ and $-1 < x < 1$. Thus, the graph cannot cross the x-axis at the x-intercept $x = 1$.

Effect of a Squared or Cubed Factor

If a polynomial $P(x)$ has a squared factor such as $(x - c)^2$, then $x = c$ is a **double root** of $P(x) = 0$. In this case, the graph of $y = P(x)$ is tangent to the x-axis at $x = c$, as shown in the graphs below.

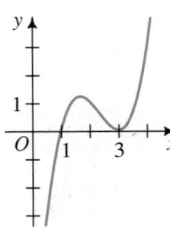

Cubic
$y = (x-1)(x-3)^2$
(1)

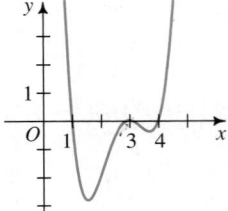

Quartic
$y = (x-1)(x-3)^2(x-4)$
(2)

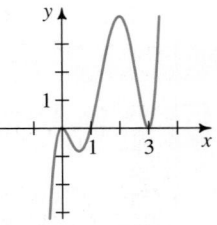

Quintic (5th degree)
$y = x^2(x-1)(x-3)^2$
(3)

64   *Chapter Two*

If a polynomial $P(x)$ has a cubed factor such as $(x - c)^3$, then $x = c$ is a **triple root** of $P(x) = 0$. In this case, the graph of $y = P(x)$ flattens out around $(c, 0)$ and crosses the x-axis at this point, as shown in the graphs below.

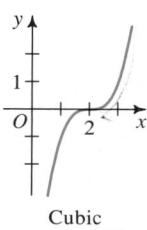

Cubic
$y = (x - 2)^3$
(4)

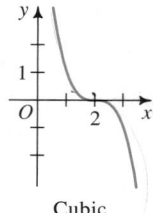

Cubic
$y = -(x - 2)^3$
(5)

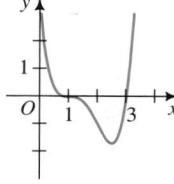

Quartic
$y = (x - 1)^3(x - 3)$
(6)

CLASS EXERCISES

1. *Discussion* The general shape and direction of the graph of a polynomial function $y = P(x)$ are determined by the "end-behavior" of the function, that is, the values $P(x)$ takes on as the value of x increases or decreases without bound. Discuss how both the degree of $P(x)$ and the sign of the leading coefficient affect the end-behavior and the shape of the graph of $y = P(x)$.

2. *Discussion* When the value of x is "close to" c, the value of $(x - c)^3$ is close to 0. Use this fact to help explain why the graph of a polynomial with a cubed factor $(x - c)^3$ flattens out around the x-intercept $x = c$.

3. Study the graphs in figures (1)–(6) on this and the preceding page. For each polynomial function $y = P(x)$, tell how many distinct real roots the equation $P(x) = 0$ will have. Indicate which real roots are double roots or triple roots.

5. **a.** $y = (x - 1)(x - 4)^2$ **b.** $y = -x(x + 1)(x - 3)$ **c.** $y = -(x - 1)^2$ **d.** $y = x(x - 2)^3$

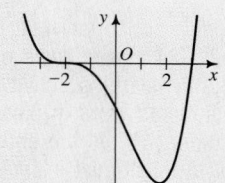 **In Exercises 4 and 5, you may wish to use a computer or a graphing calculator to check your answers.**

4. Sketch the graph of each equation.
   a. $y = (x + 2)(x - 2)(x - 4)$ **b.** $y = x(x + 2)(1 - x)$ **c.** $y = -x(x + 3)^2$
   d. $y = -x^2(x + 4)(x - 4)$ **e.** $y = -x^3(x - 2)$ **f.** $y = x^3(2 + x)$

5. In Exercise 4, you sketched the graph of each equation. Now, reverse the process and give an equation for each of the following polynomial graphs. Since the y-axis has no indicated scale, more than one equation can be correct.
See above.

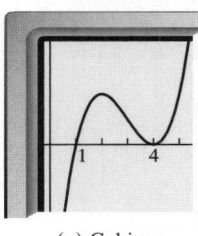

(a) Cubic

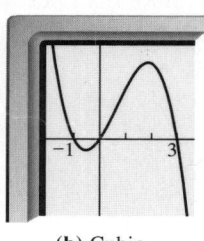

(b) Cubic

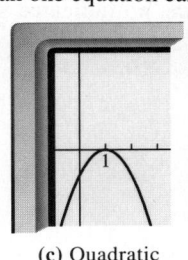

(c) Quadratic

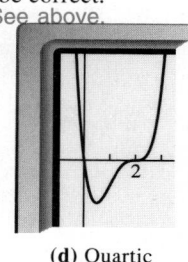

(d) Quartic

WRITTEN EXERCISES

In Exercises 1–12, sketch the graph of each equation.

A 1. $y = (x + 1)(x - 2)(x - 4)$
2. $y = -(x + 3)(x + 2)(x - 1)$
3. $y = -x(x + 5)(x + 3)$
4. $y = x(x - 1)(x - 4)$
5. $y = x^2(x + 2)$
6. $y = (x - 1)^3$
7. $y = x(1 - x)(1 + x)(2 + x)$
8. $y = x(x + 2)(x - 2)(x - 1)$
9. $y = (x + 1)^3(x - 2)$
10. $y = -x^2(2 - x)^2$
11. $y = x^2(x + 2)(x - 1)(x + 1)$
12. $y = x^2(1 - x)^2(2 + x)$
17. $(x + 1)(x - 1)^3$
18. $(2x + 1)(2x - 1)(x - 3)^2$

In Exercises 13–18, factor each polynomial function and sketch its graph.

13. $f(x) = x^3 - 4x$ $x(x + 2)(x - 2)$
14. $f(x) = x^3 - 4x^2 - 5x$ $x(x + 1)(x - 5)$
15. $f(x) = x^4 - x^2$ $x^2(x + 1)(x - 1)$
16. $f(x) = x^4 - 2x^3 + x^2$ $x^2(x - 1)^2$
17. $f(x) = x^4 - 2x^3 + 2x - 1$ See above.
(*Hint:* $x = 1$ is a triple root.)
18. $f(x) = 4x^4 - 24x^3 + 35x^2 + 6x - 9$
(*Hint:* $x = 3$ is a double root.) See above.

19. The graphs of $y = x^2$ and $y = x^4$ are shown at the right.
   a. Copy these graphs and then add the graph of $y = x^6$. (You may wish to use a computer or graphing calculator to check your graph.)
   b. What three points are common to all three graphs? $(-1, 1)$, $(0, 0)$, $(1, 1)$

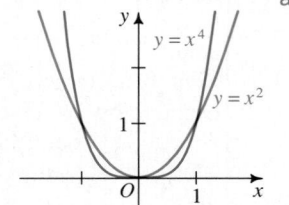

20. The graphs of $y = x^3$ and $y = x^5$ are shown at the right.
   a. Copy these graphs and then add the graph of $y = x^7$. (You may wish to use a computer or graphing calculator to check your graph.)
   b. What three points are common to all three graphs? $(-1, -1)$, $(0, 0)$, $(1, 1)$

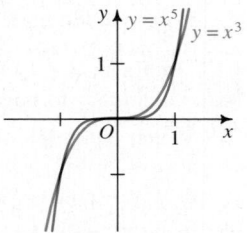

In Exercises 21–24, give an equation for each polynomial graph shown. Since the y-axis has no indicated scale, more than one answer is possible.

21.

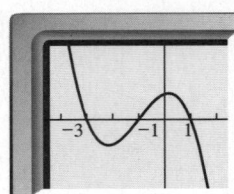

Cubic
$y = -(x + 3)(x + 1)(x - 1)$

22.

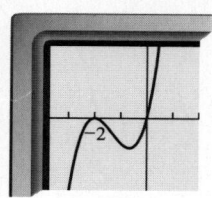

Cubic
$y = x(x + 2)^2$

66   *Chapter Two*

23.

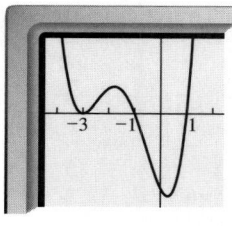

Quartic

$y = (x + 3)^2(x + 1)(x - 1)$

24.

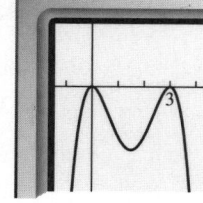

Quartic

$y = -x^2(x - 3)^2$

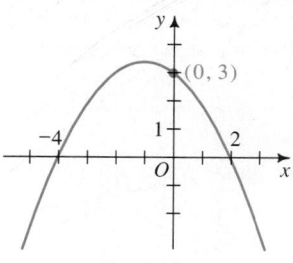 **Use a computer or graphing calculator to complete part (b) of Exercises 25 and 26.**

25. a. Sketch the graph of $P(x) = 16x - x^3$.
   b. Evaluate $P(2)$ and $P(2.1)$ and conclude that the highest point of the graph in the interval $0 \le x \le 4$ does *not* occur midway between the roots $x = 0$ and $x = 4$. Find to the nearest tenth the x-coordinate of the highest point.
   c. Show that $P(-x) = -P(x)$. (Because of this property, the origin is called a *point of symmetry* of the graph. This idea will be discussed in Section 4-3.)

26. a. Sketch the graph of $P(x) = x^4 - 4x^2$.
   b. Evaluate $P(1)$ and $P(1.1)$ and conclude that the lowest point of the graph in the interval $0 \le x \le 2$ does *not* occur midway between the roots $x = 0$ and $x = 2$. Find to the nearest tenth the x-coordinate of the lowest point.
   c. Show that $P(-x) = P(x)$. (Because of this property, the y-axis is called a *line of symmetry* of the graph. This idea will be discussed in Section 4-3.)

Sketch the graphs of the given equations on a single set of axes. Then determine the coordinates of any points of intersection. Although you may wish to use a computer or graphing calculator, the intersection points can be found just as easily by solving the system algebraically.

B **27.** $y = x^3 - 4x$ $(0, 0)$, $(1, -3)$, $(-1, 3)$
     $y = -3x$

28. $y = -x(x - 2)^2$
    $y = -x$
    $(0, 0)$, $(1, -1)$, $(3, -3)$

In Exercises 29–32, give an equation for each polynomial graph shown. Unlike Exercises 21–24, a scale on the y-axis is given.

29.

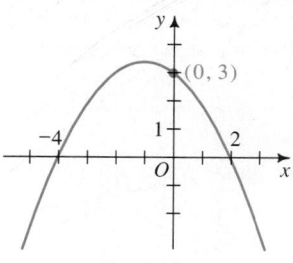

Quadratic

$y = -\dfrac{3}{8}(x + 4)(x - 2)$

30.

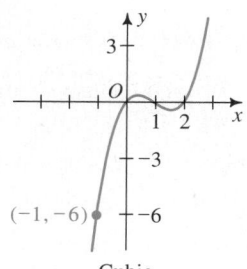

Cubic

$y = x(x - 1)(x - 2)$

Review Note

Review how to solve a system of equations before assigning Exercises 27 and 28.

Exercise Note

Students should recognize that each of Exercises 21–24 determines a *family* of polynomial functions because no points (other than the x-intercepts) are indicated on the graph. Thus, for Exercise 21, any answer of the form $y = a(x + 3)(x + 1)(x - 1)$ where $a < 0$ is acceptable. (Have students use technology to graph this function for various negative values of a.)

Once a particular point (other than the x-intercepts) is given, however, only one member of the family will have the given graph. This is the case in Exercises 29–32.

Additional Answers
Written Exercises

25. b. $P(2) = 24$; $P(2.1) = 24.339$. Since $P(2.1) > P(2)$, a local maximum does *not* occur at $x = 2$. A local maximum occurs at $x \approx 2.3$.
   c. $P(-x) = 16(-x) - (-x)^3 = -16x + x^3 = -(16x - x^3) = -P(x)$

26. b. $P(1) = -3$; $P(1.1) = -3.3759$. Since $P(1.1) < P(1)$, a local minimum does *not* occur at $x = 1$. A local minimum instead occurs at $x \approx 1.4$.
   c. $P(-x) = (-x)^4 - 4(-x)^2 = x^4 - 4x^2 = P(x)$

31.

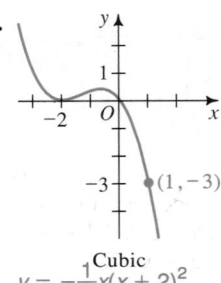

Cubic
$y = -\frac{1}{3}x(x + 2)^2$

32.

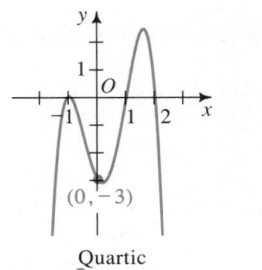

Quartic
$y = -\frac{3}{2}(x + 1)^2(x - 1)(x - 2)$

In Exercises 33 and 34, assume that the zeros of each polynomial function described are real and distinct. Sketch the graph of the function. If such a graph is impossible, say so.

33. A cubic function with **(a)** 3 zeros, **(b)** 2 zeros, **(c)** 1 zero, **(d)** no zeros.

34. A quartic function with **(a)** 4 zeros, **(b)** 3 zeros, **(c)** 2 zeros, **(d)** 1 zero, **(e)** no zeros.

35. A sixth degree polynomial function has r distinct real zeros. What are the possible values of r? 0, 1, 2, 3, 4, 5, 6

36. A fifth degree polynomial function has r distinct real zeros. What are the possible values of r? 1, 2, 3, 4, 5

37. Find an equation of the cubic function whose graph passes through the points $(3, 0)$ and $(1, 4)$, and is tangent to the x-axis at the origin. $y = -2x^2(x - 3)$

38. Find an equation of the quartic function whose graph passes through $(0, -2)$ and is tangent to the x-axis at $(-1, 0)$ and $(2, 0)$. $y = -\frac{1}{2}(x + 1)^2(x - 2)^2$

39. If $P(x)$ is a cubic polynomial such that $P(-3) = P(-1) = P(2) = 0$ and $P(0) = 6$, find $P(x)$. $-(x + 3)(x + 1)(x - 2)$

40. If $P(x)$ is a cubic polynomial such that $P(0) = 0$, $P(2) = -4$, and $P(x)$ is positive only when $x > 4$, find $P(x)$. $\frac{1}{2}x^2(x - 4)$

2-4 Finding Maximums and Minimums of Polynomial Functions

Objective   *To write a polynomial function for a given situation and to find the maximum or minimum value of the function.*

Quadratic Functions

In Section 1-7, we saw that the maximum or minimum value of the quadratic function $f(x) = ax^2 + bx + c$ occurs at $x = -\frac{b}{2a}$. We use this fact in Example 1 at the top of the next page.

Warm-Up Exercises

Tell whether the parabola defined by the given equation opens upward or downward. Then find the vertex of the parabola.

1. $y = x^2 + 4x$ up; $(-2, -4)$

2. $y = -x^2 + 5x + 3$ down; $\left(\frac{5}{2}, \frac{37}{4}\right)$

3. $y = 5 + 8x - 4x^2$ down; $(1, 9)$

4. $y = 2x^2 + 1$ up; $(0, 1)$

5. $y = 25x^2 + 150x + 128$ up; $(-3, -97)$

6. $y = -1.2x^2 - 3x + 8$ down; $(-1.25, 9.875)$

Motivating the Section

Suppose an architect wants to design a physical fitness area in the shape of a rectangle with a semicircle at each end. If a 1000-meter running track is to be built around this shape, what dimensions will give the maximum area of the rectangle? Problems such as this one can be solved by using the techniques developed in this section.

Example 1

A rectangular dog pen is constructed using a barn wall as one side and 60 m of fencing for the other three sides. Find the dimensions of the pen that give the greatest area.

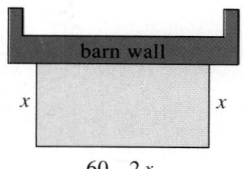

Solution

Let x = the length in meters of each side touching the barn. Then $60 - 2x$ = the length in meters of the side parallel to the barn. If the lengths of these sides are to be positive, then x must be between 0 and 30. Letting $A(x)$ represent the area of the pen, we have:

$$A(x) = x(60 - 2x)$$
$$= -2x^2 + 60x$$

The maximum value of $A(x)$ occurs at $x = -\dfrac{60}{2(-2)} = 15$. Thus, the area will be a maximum when the dimensions are 15 m by 30 m, with the longer side being parallel to the barn. The maximum area is $A(15) = 450$ m^2.

Cubic Functions

When the graph of a cubic function has a "peak" and a "valley" as shown at the right, we say that the function has a *local maximum* at the highest point of the peak (point H) and a *local minimum* at the lowest point of the valley (point L).

 In the following example we use a computer or graphing calculator to approximate the local maximum of a cubic function.

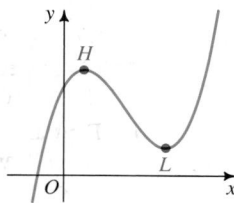

Example 2

Squares with sides of length x are cut from the corners of a rectangular piece of sheet metal with dimensions of 6 in. and 10 in. The metal is then folded to make an open-top box. What is the maximum volume of such a box?

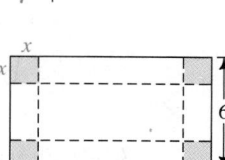

Solution

The dimensions of the box are:

$$\text{height} = x$$
$$\text{length} = 10 - 2x$$
$$\text{width} = 6 - 2x$$

If these dimensions are to be positive, then x must be between 0 and 3.

(Solution continues on the next page.)

Communication Note

Discuss why the word "local" must be used to describe the peaks and valleys in the graphs. (The graphs are unbounded, so they do not have maximum or minimum points except within a given interval.)

Additional Examples

1. Two numbers have a difference of 6. Find their minimum possible product.

   Let x and $x + 6$ represent the numbers. Then $P(x) = x(x + 6) = x^2 + 6x$. The minimum value occurs when $x = -\dfrac{6}{2(1)} = -3$. $P(-3) = -3(-3 + 6) = -9$. The minimum product, -9, occurs for the numbers -3 and 3.

2. An open box is to be formed by cutting squares from a square sheet of metal 10 cm on a side and then folding up the sides, as shown below.

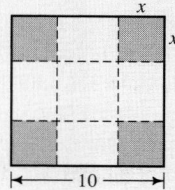

   a. Show that the volume of the box is given by $V(x) = x(10 - 2x)^2$.
   b. Find the domain of V.
   c. Find the approximate value of x that maximizes the volume. Then give the approximate maximum volume.

(Continues on next page)

a. The base of the box is a square with sides of length $10 - 2x$. The box has height x. Thus,
$V(x) = lwh = (10 - 2x)(10 - 2x)x = x(10 - 2x)^2$.

b. Since $x > 0$ and $10 - 2x > 0$, $0 < x < 5$.

c. Use either a computer or graphing calculator to obtain the graph of $V(x) = x(10 - 2x)^2$ on the interval $0 < x < 5$ and then zoom in on the peak of the graph, or modify the program given in Example 2 as follows:

20 FOR X = 0 TO 5 STEP 0.1
30 LET V = X * (10 − 2 * X)^2

Whichever method is used, you should find that the maximum volume, about 74.1 cm^3, is obtained when $x \approx 1.7$.

Review Note

For Class Exercises 1–4, you may want to remind students that the quadratic function $f(x) = ax^2 + bx + c$ has a minimum value if $a > 0$ and a maximum value if $a < 0$.

Letting $V(x)$ represent the volume of the box, we have:

$$V(x) = x(10 - 2x)(6 - 2x)$$

One way to find the maximum value of V is to use a computer or graphing calculator to draw the graph of $y = V(x)$, as shown at the right. Then you can either trace the curve to its highest point between 0 and 3 or enlarge a region containing the highest point. Either way you can find that the highest point is approximately (1.2, 32.8). Thus, the maximum volume of the box is approximately 32.8 in.^{3}

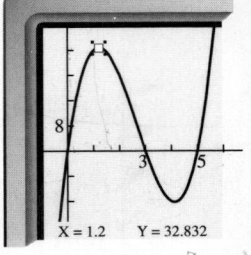

X = 1.2   Y = 32.832

Another way to find the maximum value of $V(x)$ is to use a simple program like the one below on a computer or programmable calculator. The FOR-NEXT loop in lines 20–70 evaluates the volume V for values of X between 0 and 3. The variable M, introduced in line 10, will ultimately be the maximum value of V, but initially it is zero. Each time that a volume V is greater than the previous maximum M, line 50 replaces the previous value of M with the value of V, and line 60 replaces $X1$ with the corresponding value of X.

```
10 LET M = 0
20 FOR X = 0 TO 3 STEP 0.01
30 LET V = X * (6 - 2 * X) * (10 - 2 * X)
40 IF V <= M THEN 70
50 LET M = V
60 LET X1 = X
70 NEXT X
80 PRINT ''MAXIMUM VOLUME IS APPROXIMATELY ''; M;
   '' AT X = ''; X1
90 END
```

The program produces this output:

```
MAXIMUM VOLUME IS APPROXIMATELY 32.835 AT X = 1.21
```

CLASS EXERCISES

In Exercises 1–4, state (a) whether the quadratic function has a maximum or a minimum value and (b) the value of x at which the maximum or minimum occurs.

1. $f(x) = (x - 1)(x - 7)$ **a.** min. **b.** 4
2. $g(x) = 8 - (x - 2)^2$ **a.** max. **b.** 2
3. $h(x) = 2x^2 - 6x + 9$ **a.** min. **b.** $\dfrac{3}{2}$
4. $k(x) = 1 - 4x - 3x^2$ **a.** max. **b.** $-\dfrac{2}{3}$
5. Does every cubic function have a local maximum and a local minimum? Explain. No; for example, $f(x) = x^3$.

70   *Chapter Two*

---

6. Reading Why do you think the adjective "local" is used when describing a maximum or minimum value of a cubic function but not of a quadratic function?

7. Note that the graph of $y = V(x)$ shown in the solution of Example 2 has a local minimum when x is between 3 and 5. Describe how you would modify the computer program to find this local minimum. Ignore the fact that $V(x)$ does not represent a volume outside the interval $0 < x < 3$.

WRITTEN EXERCISES

Quadratic functions

A

1. Farming A farmer wants to make a rectangular enclosure using a wall as one side and 120 m of fencing for the other three sides.

    a. Express the area in terms of x and state the domain of the area function. $x(120 - 2x)$; $0 < x < 60$
    b. Find the value of x that gives the greatest area. 30

2. A rectangle has a perimeter of 80 cm. If its width is x, express its length and its area in terms of x. What is the maximum area of the rectangle? See below.

3. Suppose you have 102 m of fencing to make two side-by-side rectangular enclosures, as shown. What is the maximum area that you can enclose? 433.5 m²

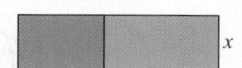

4. Suppose you have to use exactly 200 m of fencing to make either one square enclosure or two separate square enclosures of any sizes you wish. What plan gives you the least area? the greatest area? See below.

5. Suppose a scientist makes two measurements x_1 and x_2 of a quantity whose true measure x is not known. The errors are then $|x - x_1|$ and $|x - x_2|$. What number can be assigned to x so that the sum of the squares of the errors is a minimum?

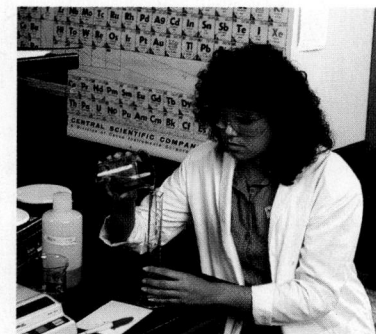

6. Repeat Exercise 5 if three measurements x_1, x_2, and x_3 are made of a quantity whose true measure x is not known.

B

7. Generalize the results of Exercises 5 and 6.

8. Show that of all rectangles having perimeter P, the square has the greatest area.

9. Physics If a ball is thrown vertically upward at 30 m/s, then its approximate height in meters t seconds later is given by $h(t) = 30t - 5t^2$.
    a. After how many seconds does the ball hit the ground? 6 s
    b. What is the domain of h? $0 \le t \le 6$
    c. How high does the ball go? 45 m   **2.** $40 - x$; $x(40 - x)$; 400 cm²

4. Two squares with 25 m sides; one square with 50 m sides

7. Change line 20 to:
20 FOR X = 3 TO 5 STEP 0.01
Change line 40 to:
40 IF V > = M THEN 70
Change line 80 to:
80 PRINT "MINIMUM VALUE IS APPROXIMATELY "; M; " AT X = "; X1

Making Connections

Exercise 8 shows that the limit of the area of a rectangle with a given perimeter is the area of the square with that perimeter.

Additional Answers Written Exercises

5. $\dfrac{x_1 + x_2}{2}$

6. $\dfrac{x_1 + x_2 + x_3}{3}$

7. If n measurements x_1, x_2, ..., x_n are made of a quantity whose true measure x is not known, then the sum of the squares of the errors is a minimum if the value $\dfrac{x_1 + x_2 + \cdots + x_n}{n}$ (that is, the arithmetic average of the n measurements) is assigned to x.

Suggested Assignments

Standard
Day 1: 71/1–11 odd
Day 2: 72/1–7 odd
Comprehensive
Day 1: 71/1–13 odd
Day 2: 72/1–9 odd; Calc. Exs. 1, 2

10. **Physics** If a ball is thrown upward from a building 30 m tall and the ball has a vertical velocity of 25 m/s, then its approximate height above the ground t seconds later is given by $h(t) = 30 + 25t - 5t^2$.
   a. After how many seconds does the ball hit the ground? 6 s
   b. What is the domain of h? $0 \le t \le 6$
   c. How high does the ball go? 61.25 m

11. **Business** The publisher of a magazine that has a circulation of 80,000 and sells for $1.60 a copy decides to raise the price of the magazine because of increased production and distribution costs. By surveying the readers of the magazine, the publisher finds that the magazine will lose 10,000 readers for each $.40 increase in price. What price per copy maximizes the income? $2.40

12. **Business** An orange grower has 400 crates of fruit ready for market and will have 20 more for each day the grower waits. The present price is $60 per crate and will drop an estimated $2 per day for each day waited. In how many days should the grower ship the crop to maximize his income? 5 days

C 13. In the isosceles triangle shown at the right, two rectangles are inscribed. Many others could also be inscribed. What are the dimensions of the inscribed rectangle having the largest possible area? Width = 9; height = 6

14. Given the points $A(0, 5)$, $B(3, 7)$, and $C(6, 2)$, find the point P on the x-axis that minimizes $(PA)^2 + (PB)^2 + (PC)^2$. (3, 0)

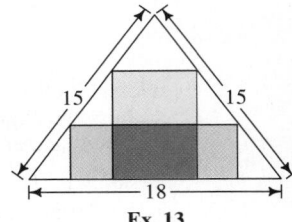

Ex. 13

Cubic functions

In Exercises 1–9, you will need to use a computer or calculator as shown in Example 2. Give approximate values to the nearest hundredth unless otherwise instructed.

A 1. **Manufacturing** As the diagram indicates, a manufacturer cuts squares from the corners of an 8 cm by 14 cm piece of sheet metal and then folds the metal to make an open-top box.
   a. Show that the volume of the box is: See below.
      $$V(x) = x(8 - 2x)(14 - 2x)$$
   b. What is the domain of V? $0 < x < 4$
   c. Find the approximate value of x that maximizes the volume. Then give the approximate maximum volume. 1.64; 82.98 cm³
      1. a. Volume = (length)(width)(height) = $(14 - 2x)(8 - 2x)(x)$

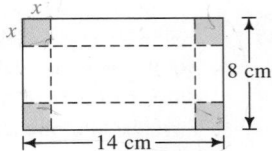

2. **Manufacturing** A 10 cm by 20 cm piece of sheet metal is cut and folded as indicated in the diagram to make a box with a top.

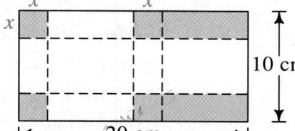

a. Show that the volume of the box is:

$$V(x) = x(10 - x)(10 - 2x)$$

b. What is the domain of V? $0 < x < 5$

c. Find the approximate value of x that maximizes the volume. Then give the approximate maximum volume. 2.11; 96.22 cm^3

3. **Manufacturing** In a rectangular piece of cardboard with perimeter 20 ft, three parallel and equally spaced creases are made, as shown at the left below. The cardboard is then folded to make a rectangular box with open square ends.

a. Show that the volume of the box is $V(x) = x^2(10 - 4x)$.

b. What is the domain of V? $0 < x < 2.5$

c. Find the approximate value of x that maximizes the volume. Then give the approximate maximum volume. 1.67; 9.26 ft^3

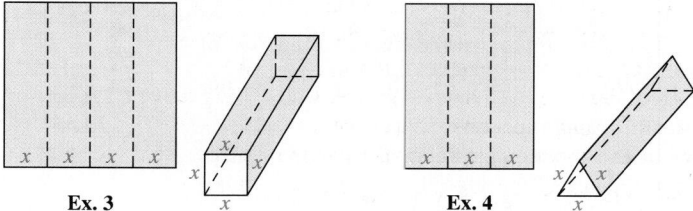

Ex. 3   Ex. 4

4. **Manufacturing** In a rectangular piece of cardboard with perimeter 30 in., two parallel and equally spaced creases are made, as shown at the right above. The cardboard is then folded to make a prism with open ends that are equilateral triangles.

a. Show that the volume of the prism is $V(x) = \left(\frac{\sqrt{3}}{4}x^2\right)(15 - 3x)$.

b. What is the domain of V? $0 < x < 5$

c. Find the approximate value of x that maximizes the volume. Then give the approximate maximum volume. 3.33; 24.06 in.^{3}

B **5.** **Manufacturing** A rectangular piece of sheet metal with perimeter 50 cm is rolled into a cylinder with open ends, as shown at the right.

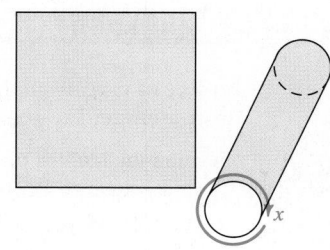

a. Express the volume of the cylinder as a function of x. Then give the domain of this function. See below.

b. Find the approximate value of x that maximizes the volume. Then give the approximate maximum volume. 16.67; 184.21 cm^3

5. a. 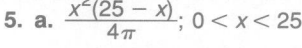 $\frac{x^2(25 - x)}{4\pi}$; $0 < x < 25$

Polynomial Functions   **73**

Additional Answers Written Exercises

2. a. $V = lwh =$
$\left(\frac{20 - 2x}{2}\right)(10 - 2x)(x) =$
$x(10 - x)(10 - 2x)$

3. a. $V = lwh =$
$\left(\frac{20 - 8x}{2}\right)(x)(x) =$
$x^2(10 - 4x)$

4. a. $V = Bh =$
$\left[\frac{1}{2}x\left(\frac{\sqrt{3}}{2}x\right)\right]\left(\frac{30 - 6x}{2}\right) =$
$\left(\frac{\sqrt{3}}{4}x^2\right)(15 - 3x)$

Review Note

You might want to review the volume formulas from geometry for prisms and cylinders. In each case, the volume is found by multiplying the area of the base and the height. Also review that the volume of a cone is $\frac{1}{3}\pi r^2 h$ and the volume of a sphere is $\frac{4}{3}\pi r^3$.

6. A cylinder is generated by rotating a rectangle with perimeter 12 in. about one of its sides, as shown at the right. $\pi x^2(6-x); 0 < x < 6$
   a. Express the volume of the cylinder as a function of x. Then give the domain of this function.
   b. Find the approximate value of x that maximizes the volume. Then give the approximate maximum volume. 4.00; 100.53 in.^{3}

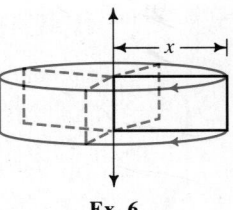

Ex. 6

7. A cylinder is inscribed in a sphere with radius 5, as shown at the right. $2\pi x(25 - x^2); 0 < x < 5$
   a. Express the volume of the cylinder as a function of x. Then give the domain of this function.
   b. Find the approximate value of x that maximizes the volume. Then give the approximate maximum volume. 2.89; 302.30

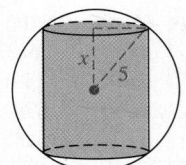

Ex. 7

8. A cone is inscribed in a sphere of radius 6, as shown at the right.
   a. Express the volume of the cone as a function of x. $\left(Hint: V = \frac{1}{3}\pi r^2 h\right)$ Then give the domain of this function. See below.
   b. Find the approximate value of x that maximizes the volume. Then give the approximate maximum volume. 2.00; 268.08

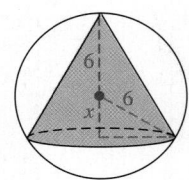

Ex. 8

C 9. A cylinder is inscribed in a cone with height 10 and a base of radius 5, as shown at the right. Find the approximate values of r and h for which the volume of the cylinder is a maximum. Then give the approximate maximum volume. 3.33; 3.33; 116.36

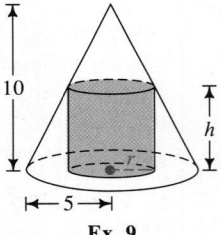

Ex. 9

▮▮▮▮ CALCULATOR EXERCISES

A piece of wire 40 cm long is to be cut into two pieces. One piece will be bent to form a circle; the other will be bent to form a square.

1. Find the lengths of the two pieces that cause the sum of the areas of the circle and square to be a minimum. Use 17.6 cm for circle, 22.4 cm for square.

2. How could you make the total area of the circle and the square a maximum? Use all of the wire for the circle.

$$8. \ \frac{1}{3}\pi(36 - x^2)(6 + x); 0 < x < 6$$

74   *Chapter Two*

■ Polynomial Equations

2-5 Using Technology to Approximate Roots of Polynomial Equations

Objective *To use technology to approximate the real roots of a polynomial equation.*

You already know how to solve linear and quadratic equations. In this section we will discuss how to solve a polynomial equation of higher degree, such as

$$x^3 - x^2 - 5x + 2 = 0,$$

by using a computer, or a graphing or programmable calculator. In the next section we will see that some polynomial equations can be solved by factoring.

Example 1 Solve $x^3 - x^2 - 5x + 2 = 0$ by using a computer or graphing calculator.

Solution Draw the graph of the polynomial function

$$y = x^3 - x^2 - 5x + 2$$

with a computer or graphing calculator. The x-intercepts of the graph are the roots of the equation. If estimating the x-intercepts is difficult because the graph is not labeled, you could use the TRACE feature, if available. To improve the accuracy of the approximations, you could rescale or use the ZOOM feature, if available.

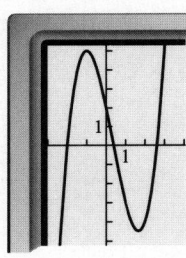 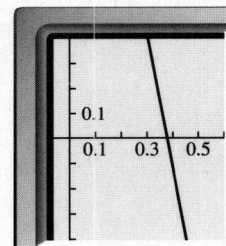 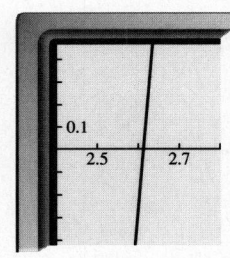

Graph of function for $-3 \le x \le 4$ Enlargement of graph for $0 \le x \le 0.6$ Enlargement of graph for $2.4 \le x \le 2.8$

From the graphs of $y = x^3 - x^2 - 5x + 2$ shown above, we estimate that the roots of $x^3 - x^2 - 5x + 2 = 0$ are $x = -2$, $x = 0.4$, and $x = 2.6$.

In the next example, we will solve the equation $x^3 - x^2 - 5x + 2 = 0$ without drawing a graph. We will instead use the *location principle*, stated at the top of the next page.

Teaching Notes, p. 52D

Warm-Up Exercises

Graph the related polynomial function. Then tell the *number* of real roots of the polynomial equation.

1. $x^3 - 2x + 3 = 0$ one

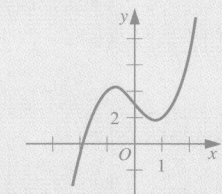

2. $x^3 - 4x - 1 = 0$ three

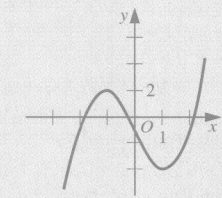

3. $x^4 - x^3 + 2 = 0$ none

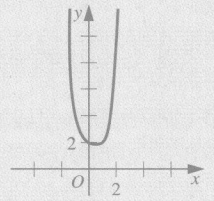

Motivating the Section

In real-life situations, calculators and computers are used routinely to approximate the roots of a polynomial equation.

◩ Using Technology

Students may want to refer to the discussion of finding a point of interest in the article *Using a Graphing Calculator* (see page xxix).

1. Use a computer or a graphing calculator to estimate the real roots of $x^3 - 3x + 1 = 0$.

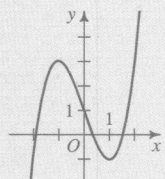

The graph of $y = x^3 - 3x + 1$ above shows that there are roots between -2 and -1, 0 and 1, and 1 and 2. By rescaling or using the ZOOM feature, the roots can be estimated more precisely as -1.9, 0.3, and 1.5.

2. Show that $x^4 - 2x^3 - 7x^2 + 10x + 10 = 0$ has at least two real roots between 2 and 3.

Method 1: Run this program on a computer or programmable calculator:
```
10 PRINT "X", "P(X)"
20 FOR X = 2 TO 3 STEP 0.1
30 PRINT X, X^4 - 2 *
   X^3 - 7 * X^2 + 10 *
   X + 10
40 NEXT X
50 END
```
The resulting table shows that $P(x) \approx 0$ when $x \approx 2.24$ and $x \approx 2.73$.

Method 2: A simpler method is to use synthetic substitution to show that $P(2) = 2$, $P(2.5) = -\frac{15}{16}$, and $P(3) = 4$. Since there are two changes of sign in the interval $2 < x < 3$, there are at least two roots in that interval.

The Location Principle

If $P(x)$ is a polynomial with real coefficients and a and b are real numbers such that $P(a)$ and $P(b)$ have opposite signs, then between a and b there is at least one real root r of the equation $P(x) = 0$.

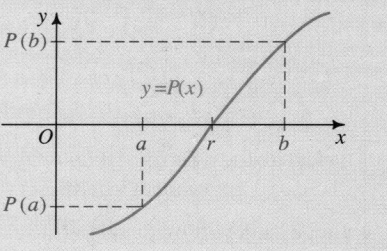

Thus, if we want to find the roots of the polynomial equation $P(x) = 0$, the location principle tells us to look for changes of sign in a table of values for $P(x)$.

Example 2 Solve $x^3 - x^2 - 5x + 2 = 0$ without drawing a graph.

Solution By using a computer or programmable calculator, we can obtain a table of values for the polynomial $P(x) = x^3 - x^2 - 5x + 2$ from a simple program like the one below. Although the interval $-4 \le x \le 4$ used in the program is wide enough for locating all the real roots of the given equation, wider intervals may be needed for other equations.

```
10 PRINT ''X'',''P(X)''
20 FOR X = -4 TO 4
30 PRINT X, X^3 - X^2 - 5 * X + 2
40 NEXT X
50 END
```

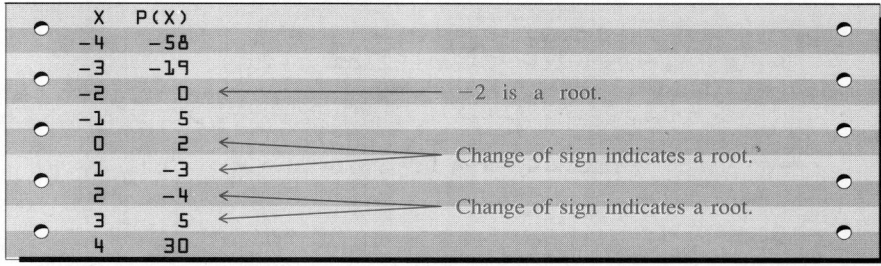

X	P(X)	
-4	-58	
-3	-19	
-2	0	← -2 is a root.
-1	5	
0	2	← Change of sign indicates a root.
1	-3	←
2	-4	← Change of sign indicates a root.
3	5	←
4	30	

The table shows that -2 is a root, that another root is between 0 and 1, and that still another root is between 2 and 3. To find more accurately the root between 0 and 1, merely change line 20 to:

```
20  FOR X = 0 TO 1 STEP 0.1
```

The table at the top of the next page shows the results of this change in the program.

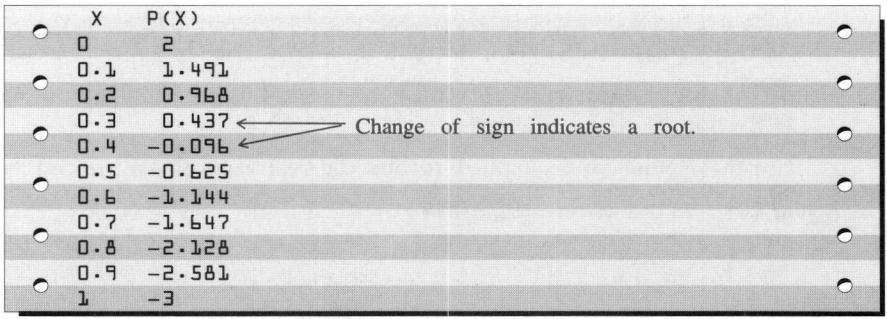

```
     X    P(X)
     0    2
     0.1  1.491
     0.2  0.968
     0.3  0.437  ←
     0.4  -0.096 ←——— Change of sign indicates a root.
     0.5  -0.625
     0.6  -1.144
     0.7  -1.647
     0.8  -2.128
     0.9  -2.581
     1    -3
```

You can obtain greater accuracy by changing line 20 once again.

```
20   FOR X = 0.3 TO 0.4 STEP 0.01
```

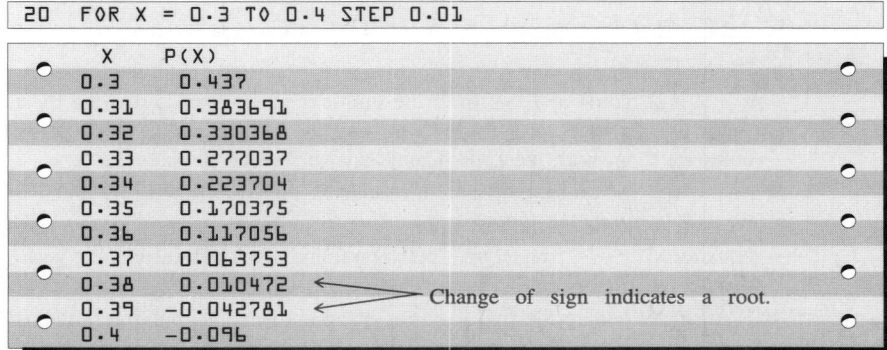

```
     X     P(X)
     0.3   0.437
     0.31  0.383691
     0.32  0.330368
     0.33  0.277037
     0.34  0.223704
     0.35  0.170375
     0.36  0.117056
     0.37  0.063753
     0.38  0.010472 ←
     0.39  -0.042781 ←——— Change of sign indicates a root.
     0.4   -0.096
```

Of the two values 0.010472 and −0.042781, the first is closer to 0. We therefore take $x = 0.38$ rather than $x = 0.39$ as the better approximation of the root between 0 and 1.

5. Answers may vary: $x \approx -2.2$, $x \approx 0.4$, and $x \approx 2.8$.

CLASS EXERCISES

In Exercises 1–4, mentally locate between two consecutive integers a real root of each equation.

1. $x^3 + x - 5 = 0$ 1, 2
2. $2x^3 - x - 3 = 0$
3. $x^4 - 2x^3 - 3x - 3 = 0$ −1, 0; 2, 3
4. $x^3 + 5x^2 - 3 = 0$

5. Using the table of values below, give the approximate roots of $P(x) = 0$.

x	−3	−2	−1	0	1	2	3	4
$P(x)$	4	−1	−5	−2	3	10	−3	−16

6. Explain why you cannot use the location principle to find a double root of a polynomial equation. No change in sign at a double root.

Example Note

Different computers and programmable calculators may give values of $P(x)$ that are different from the ones shown in Example 2. In most cases, the differences will be so slight as *not* to affect the approximations of the roots.

WRITTEN EXERCISES

 Many of these Written Exercises require the use of a computer or calculator.

Use the method of Example 1 to find the real roots to the nearest tenth.

A

1. $2x^3 + x^2 + 4x - 15 = 0$ 1.5
2. $3x^3 + x^2 + 2x + 8 = 0$ −1.3
3. $x^3 - x^2 - 4x - 2 = 0$ −1, −0.7, 2.7
4. $x^3 - 2x^2 - 7x + 2 = 0$ −2, 0.3, 3.7
5. $x^4 - 3x^2 = 9$ ±2.2
6. $x^4 - 5x^3 + x^2 - 10x - 2 = 0$ −0.2, 5.2

Use the method of Example 2 to find the real roots to the nearest tenth.

7. $x^3 - 5x^2 - 3x - 7 = 0$ 5.7
8. $x^3 + 2 = 3x^2$ −0.7, 1, 2.7
9. $1.2x^4 - 0.7x^2 = 3.6x$ 0, 1.6
10. $0.23x^4 - 0.5x^3 = 4.7$ −1.7, 3.0

B **11.** Example 2 shows that $x = -2$ is a zero of $P(x) = x^3 - x^2 - 5x + 2$ so that $x + 2$ is a factor of $P(x)$. Find the quotient when $P(x)$ is divided by $x + 2$. Then use the quadratic formula to find the two zeros of the quotient and compare them with the approximate answers given in Example 1.

12. Show that $9x^4 - 8x^2 + 1 = 0$ has at least two real roots between 0 and 1.

13. Explain why the method of Example 2 does not reveal any real roots of the equation $42x^2 - 13x + 1 = 0$ between 0 and 1 when the step size is 0.1, but does reveal real roots between 0.1 and 0.2 when the step size is 0.01.

14. *Writing* Compare the two methods of approximating roots presented in Examples 1 and 2. Write a paragraph or two in which you discuss any limitations that each method may have. Also identify which method you prefer to use and explain why.

15. A box has length 4, width 2, and height 1. If another box with twice the volume has dimensions $4 + x$, $2 + x$, and $1 + x$, find the value of x to the nearest hundredth.   0.46

16. Suppose each face of a 3 cm × 5 cm × 4 cm block of wood is shaved (or planed) by x cm. If the shaved block has half the original volume, find the value of x to the nearest hundredth.   0.39

C **17.** Given a polynomial function P such that $P(a) < 0$ and $P(b) > 0$, we can use a method called *linear interpolation* to estimate the zero between a and b. As its name implies, this method assumes that the graph of P is approximately linear between a and b. For example, in the figure at the right, compare the dotted line and the graph of P between points A and B. The zero of P is approximately the x-intercept of the dotted line. Show that this x-intercept is:

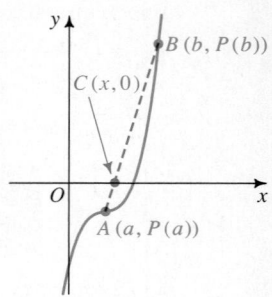

$$x = a - \frac{P(a) \cdot (b - a)}{P(b) - P(a)}$$

18. Two buildings are d units apart. A ladder 20 units long has its foot resting against building 1 and its top against the side of building 2. A second ladder 15 units long has its foot against building 2 and its top against the side of building 1. The ladders touch each other at a point c units above the ground.

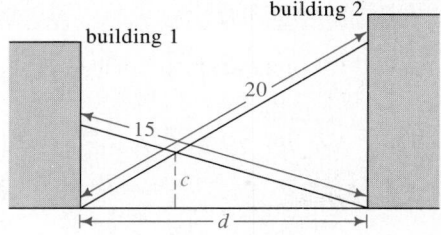

a. Show that

$$\frac{1}{\sqrt{400 - d^2}} + \frac{1}{\sqrt{225 - d^2}} = \frac{1}{c}.$$ (*Hint:* Use similar triangles.)

b. If $d = 12$, use part (a) to find c. 5.76

c. If $c = 8$, find d. Since it is difficult to solve for d directly, use a computer or calculator to approximate d to the nearest hundredth by finding the value of d for which the expression below changes sign. 5.95

$$\frac{1}{\sqrt{400 - d^2}} + \frac{1}{\sqrt{225 - d^2}} - \frac{1}{8}$$

////COMPUTER EXERCISE

The *bracket-and-halving method* described below is particularly well suited to a computer. We illustrate the method with the equation:

$$P(x) = x^3 + x - 1 = 0$$

1. First, we mentally locate a root between two integers or have the computer do this. In this case, there is a root between $x = 0$ and $x = 1$ because $P(0)$ is negative and $P(1)$ is positive.

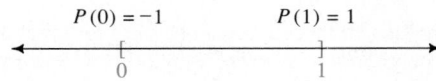

2. Now we divide the interval from 0 to 1 in half and determine in which half the root lies. Since $P\left(\frac{1}{2}\right) = \frac{1}{8} + \frac{1}{2} - 1 < 0$ and $P(1) > 0$, the root is between $\frac{1}{2}$ and 1.

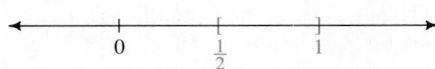

(Computer Exercise continues on the next page.)

**Additional Answers
Computer Exercise**

10 PRINT "ENTER THE DEGREE OF THE POLYNOMIAL P(X)."
20 INPUT N
30 DIM P(N)
40 FOR D = N TO 0 STEP −1
50 PRINT "ENTER THE COEFFICIENT OF THE TERM HAVING DEGREE ";D;"."
60 INPUT P(D)
70 NEXT D
80 PRINT "ENTER TWO NUMBERS A AND B SUCH THAT P(A) AND P(B) HAVE OPPOSITE SIGNS."
90 INPUT A,B
100 LET M = (A + B) / 2
110 LET PA = P(N): LET PB = P(N): LET PM = P(N)
120 FOR D = N − 1 TO 0 STEP −1
130 LET PA = PA * A + P(D): LET PB = PB * B + P(D): LET PM = PM * M + P(D)
140 NEXT D
150 IF ABS (PM) < 0.01 THEN 200
160 IF PA * PM < 0 THEN LET B = M
170 IF PB * PM < 0 THEN LET A = M
180 PRINT "A ZERO OF P(X) IS BETWEEN ";A; " AND ";B;"."
190 GOTO 100
200 PRINT "THE ZERO IS APPROXIMATELY "; M;"."
210 END
To get greater accuracy in the approximation of the root, change the "0.01" in line 150 to a smaller positive number.

3. Next, we divide the interval from $\frac{1}{2}$ to 1 in half and again find the half containing the root. Since $P\left(\frac{3}{4}\right) = \frac{27}{64} + \frac{3}{4} - 1 > 0$ and $P\left(\frac{1}{2}\right) < 0$, the root is between $\frac{1}{2}$ and $\frac{3}{4}$.

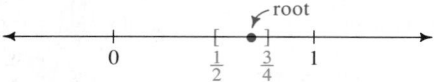

4. We continue to find intervals that bracket the root and then halve these intervals until we have the accuracy we desire.

Write a computer program that uses the bracket-and-halving-method, also called a *binary search* method, to approximate a zero of a polynomial function P. The following algorithm may be helpful.

1. Input the degree of P.
2. Input the coefficients of P.
3. Input numbers a and b for which $P(a)$ and $P(b)$ have opposite signs.
4. Calculate $m = \frac{a+b}{2}$. Then calculate $P(m)$, $P(a)$, and $P(b)$.
5. If $P(m)$ is sufficiently close to 0, then print that m is an approximation of the zero and stop. Otherwise go to step 6.
6. If $P(a)$ and $P(m)$ have opposite signs, let $b = m$, print that the zero is between a and b, and return to step 4. Otherwise let $a = m$, print that the zero is between a and b, and return to step 4.

2-6 Solving Polynomial Equations by Factoring

Objective   *To solve polynomial equations by various methods of factoring, including the use of the rational root theorem.*

We have seen how a computer or calculator can be used to search for and approximate the real roots of a polynomial equation. Very often these roots are irrational, but if any are rational they can be found by various methods of factoring. The first example illustrates a method that involves grouping terms.

Example 1   *Solving a polynomial equation by factoring after grouping terms*
Solve: $x^3 + 5x^2 - 4x - 20 = 0$

Solution
$$(x^3 + 5x^2) - (4x + 20) = 0$$
$$x^2(x + 5) - 4(x + 5) = 0$$
$$(x^2 - 4)(x + 5) = 0$$
$$(x + 2)(x - 2)(x + 5) = 0$$
$$x = -2, \; x = 2, \; x = -5$$

Teaching Notes, p. 52D

Warm-Up Exercises

Give all the factors of each integer.

1. 15 $\pm 1, \pm 3, \pm 5, \pm 15$
2. 9 $\pm 1, \pm 3, \pm 9$
3. 11 $\pm 1, \pm 11$
4. 20 $\pm 1, \pm 2, \pm 4, \pm 5, \pm 10, \pm 20$
5. 1 ± 1
6. 8 $\pm 1, \pm 2, \pm 4, \pm 8$

Motivating the Section

The technology-based equation-solving techniques studied in Section 2-5 are powerful, yet they have their limitations. For example, they usually give only *approximate* roots, and they cannot locate *imaginary* roots. The factoring techniques of this section, while having their own limitations, can overcome the limitations of the technology-based techniques.

Another method of solving higher-degree polynomial equations involves recognizing polynomials that have a *quadratic form*. For example, $3x^4 + 2x^2 - 4$ has a quadratic form, because if we let $y = x^2$, we have

$$3x^4 + 2x^2 - 4 = 3(x^2)^2 + 2x^2 - 4$$
$$= 3y^2 + 2y - 4,$$

which is quadratic in y.

Example 2   *Solving a polynomial equation that has a quadratic form*
Solve: $2x^4 - x^2 - 3 = 0$

Solution   Let $y = x^2$ and proceed as follows:

$$2x^4 - x^2 - 3 = 0$$
$$2(x^2)^2 - x^2 - 3 = 0 \longleftarrow \text{This is quadratic in } x^2.$$
$$2y^2 - y - 3 = 0 \longleftarrow \text{This is quadratic in } y.$$
$$(2y - 3)(y + 1) = 0$$

$$y = \frac{3}{2} \qquad\qquad\qquad \text{or} \qquad\qquad y = -1$$
$$x^2 = \frac{3}{2} \qquad\qquad\qquad\qquad\qquad\qquad x^2 = -1$$
$$x = \pm\sqrt{\frac{3}{2}} = \pm\frac{\sqrt{6}}{2} \qquad\qquad\qquad\qquad x = \pm i$$

A third method of solving higher-degree polynomial equations is based on generalizing the method of factoring. Consider a specific equation like $12x^2 - 8x - 15 = 0$. If you were using factoring to find the rational roots of this equation, you might begin by writing:

$$12x^2 - 8x - 15 = 0$$
$$(\quad)(\quad) = 0$$

For your first factor you might try $12x - 5$ or $6x - 5$ or many other possibilities. But you would not try $7x - 5$ because 7 does not divide the leading coefficient 12. Nor would you try $3x - 4$ because 4 does not divide the constant term -15.

Now the same reasoning holds if you were going to use factoring to solve the general polynomial equation:

$$a_n x^n + a_{n-1} x^{n-1} + \cdots + a_0 = 0$$
$$(qx - p)(\qquad\qquad\qquad) = 0$$

If $qx - p$ is a factor $\left(\text{so that } x = \frac{p}{q} \text{ is a root}\right)$, then q must divide a_n, and p must divide a_0. This result, known as the *rational root theorem*, is formally stated at the top of the next page.

Additional Examples cont.

To find the other roots, use the depressed equation $x^3 + 2x^2 + x + 2 = 0$.
Apply the rational root theorem or factor by grouping:

$x^3 + 2x^2 + x + 2 = 0$

$x^2(x + 2) + 1(x + 2) = 0$

$(x^2 + 1)(x + 2) = 0$

$x^2 = -1$ or $x = -2$

$x = \pm i$ or $x = -2$

Thus, the roots are -2, 2, i, and $-i$.

2. Solve $3x^3 + 8x^2 = 9x - 2$.

Write the equation in standard form:

$3x^3 + 8x^2 - 9x + 2 = 0$

If $x = \dfrac{p}{q}$ is a rational root, then $p = \pm 1$ or ± 2, and $q = \pm 1$ or ± 3. Therefore, $\dfrac{p}{q}$ can be ± 1, ± 2, $\pm\dfrac{1}{3}$, or $\pm\dfrac{2}{3}$. Use standard synthetic substitution or the short form below to test each possible root:

	3	8	-9	2
1	3	11	2	4
-1	3	5	-14	16
2	3	14	19	40
-2	3	2	-13	28
$\frac{1}{3}$	3	9	-6	0

Thus, $\dfrac{1}{3}$ is a root. The resulting depressed equation, $3x^2 + 9x - 6 = 0$, is quadratic. Use the quadratic formula:

$3(x^2 + 3x - 2) = 0$

$x = \dfrac{-3 \pm \sqrt{17}}{2}$

$\therefore\ x = \dfrac{1}{3}$ or $x = \dfrac{-3 \pm \sqrt{17}}{2}$

The Rational Root Theorem

Let $P(x)$ be a polynomial of degree n with integral coefficients and a nonzero constant term:

$$P(x) = a_n x^n + a_{n-1}x^{n-1} + \cdots + a_0,\ \text{where } a_0 \neq 0$$

If one of the roots of the equation $P(x) = 0$ is $x = \dfrac{p}{q}$ where p and q are nonzero integers with no common factor other than 1, then p must be a factor of a_0, and q must be a factor of a_n.

Example 3
 a. According to the rational root theorem, what are the possible rational roots of $P(x) = 3x^4 + 13x^3 + 15x^2 - 4 = 0$?
 b. Determine whether any of the possible rational roots really *are* roots. Then find all other roots, real or imaginary.

Solution
 a. $x = \dfrac{p}{q}$ is a possible rational root if p divides -4 and q divides 3. Thus p could equal ± 1, ± 2, or ± 4, and q could equal ± 1 or ± 3. Therefore, $\dfrac{p}{q}$ could equal ± 1, ± 2, ± 4, $\pm\dfrac{1}{3}$, $\pm\dfrac{2}{3}$, or $\pm\dfrac{4}{3}$.

 b. We use an abbreviated form of synthetic substitution to check the integral possibilities first.

	3	13	15	0	-4	$\leftarrow$ coefficients of $P(x)$
1	3	16	31	31	27	$\leftarrow P(1) = 27$
-1	3	10	5	-5	1	$\leftarrow P(-1) = 1$
2	3	19	53	106	208	$\leftarrow P(2) = 208$
-2	3	7	1	-2	0	$\leftarrow P(-2) = 0$

The last line shows that $x = -2$ is a root, and we have:

$P(x) = 3x^4 + 13x^3 + 15x^2 - 4$

$ = (x + 2)(3x^3 + 7x^2 + x - 2)$

To find the other roots of $P(x) = 0$, we now solve the associated *depressed equation*:

$$3x^3 + 7x^2 + x - 2 = 0$$

Picking up where we left off in our list of possible rational roots, we check $x = -2$ again because it may be a double root:

	3	7	1	-2	
-2	3	1	-1	0	$\leftarrow P(-2) = 0$

So $x = -2$ is indeed a double root, and the resulting depressed equation is quadratic, which we can solve using the quadratic formula:

$$3x^2 + x - 1 = 0$$

$$x = \frac{-1 \pm \sqrt{13}}{6}$$

The roots are therefore $x = -2$ (a double root) and $x = \frac{-1 \pm \sqrt{13}}{6}$.

CLASS EXERCISES

1. Solve $x^3 + 4x^2 - 9x - 36 = 0$ by grouping terms. $-4, \pm 3$

2. **a.** Explain why the equation $x^4 - 5x^2 + 4 = 0$ has a quadratic form.
   b. How would you solve this equation?

3. Which of the following equations have a quadratic form?
   a. $(2x - 1)^2 - 5(2x - 1) + 4 = 0$ Yes   **b.** $x^6 - 5x^3 + 4 = 0$ Yes

4. **a.** Describe how you would solve $x^3 + 2x^2 - 3x = 0$.
   b. In general, for what cubic equations can you use the method of part (a)?

5. According to the rational root theorem, what are the possible rational roots of each equation? See below.
   a. $2x^3 - 3x^2 + 9x - 4 = 0$ $\pm 4, \pm 2, \pm 1, \pm \frac{1}{2}$ **b.** $6x^4 - 2x^2 + 9x + 5 = 0$

6. If $P(x)$ is a polynomial such that $P(1) = -1$ and $P(2) = 3$, must $P(x) = 0$ have a rational root between $x = 1$ and $x = 2$? Justify your answer.

7. **a.** If $P(x) = x^3 + 2x^2 + x + 1$, explain why the equation $P(x) = 0$ has a real root between $x = -2$ and $x = -1$. $P(-2) = -1 < 0$; $P(-1) = 1 > 0$
   b. Determine whether this real root is rational or irrational. Irrational
   c. Explain why this equation has no positive roots. $P(x) > 1$ for all $x > 0$

5. **b.** $\pm 5, \pm 1, \pm \frac{5}{2}, \pm \frac{1}{2}, \pm \frac{5}{3}, \pm \frac{1}{3}, \pm \frac{5}{6}, \pm \frac{1}{6}$

WRITTEN EXERCISES

For Exercises 1–10, tell whether each equation is: (a) a polynomial equation that can be solved by grouping terms, or (b) a polynomial equation that has a quadratic form. Then solve the equation.

A

1. $x^4 - 4x^2 - 12 = 0$ b; $\pm\sqrt{6}, \pm i\sqrt{2}$ 2. $x^3 + 6x^2 - 4x - 24 = 0$ a; $-6, \pm 2$
3. $3x^3 - 16x^2 - 12x + 64 = 0$ a; $\pm 2, \frac{16}{3}$ 4. $x^4 - 7x^2 - 8 = 0$ b; $\pm 2\sqrt{2}, \pm i$
5. $2x^4 = -7x^2 + 15$ b; $\pm\frac{\sqrt{6}}{2}, \pm i\sqrt{5}$ 6. $2x^3 - x^2 - 2x + 1 = 0$ a; $\pm 1, \frac{1}{2}$
7. $x^3 + 2x^2 - 6x = 12$ a; $-2, \pm\sqrt{6}$ 8. $2x^3 - 3x^2 = 12 - 8x$ a; $\frac{3}{2}, \pm 2i$
9. $10x^3 + 5x = 6x^2 + 3$ a; $\frac{3}{5}, \pm\frac{\sqrt{2}}{2}i$ 10. $2x^4 + 3x^2 - 20 = 0$ b; $\pm\frac{\sqrt{10}}{2}, \pm 2i$

11. Show that $x^6 - 7x^3 - 8 = 0$ has a quadratic form. Then find the two real roots and the four imaginary roots of this equation. $-1, 2, -1 \pm i\sqrt{3}, \frac{1}{2} \pm \frac{\sqrt{3}}{2}i$

Polynomial Functions **83**

After discussing the rational root theorem, ask students for the possible rational roots of $6x^5 - 13x^2 + 4x - 2 = 0$. ($p$ could equal $\pm 2, \pm 1$, and q could equal $\pm 6, \pm 3, \pm 2, \pm 1$; therefore, the roots could be $\frac{p}{q}$, or $\pm\frac{1}{6}, \pm\frac{1}{3}, \pm\frac{1}{2}, \pm\frac{2}{3}, \pm 1, \pm 2$.)

Additional Answers
Class Exercises

2. **a.** Let $y = x^2$. Then $x^4 - 5x^2 + 4 = 0$ becomes $y^2 - 5y + 4 = 0$, which is quadratic in y.
   b. Solve $y^2 - 5y + 4 = 0$ for y by factoring. Then substitute x^2 for y and solve for x.

4. **a.** First factor out x. Then factor (or use the quadratic formula on) the quadratic that remains.
   b. The method of part (a) can be used only on cubic equations with no constant term.

6. No; while there must be a *real* root between $x = 1$ and $x = 2$, the root may be irrational.

Suggested Assignments

Standard
Day 1: 83/1–21 odd
Day 2: 84/23–37 odd
Comprehensive
Day 1: 83/1–21 odd
Day 2: 84/23–41 odd

Supplementary Materials

Alternative Assessment, 8

12. **a.** Let $y = x^2$. Write the equation $x^4 - x^2 - 1 = 0$ in terms of y and then solve for y, expressing solutions as decimals rounded to the nearest hundredth. $y^2 - y - 1 = 0$; 1.62, −0.62

   b. Use your solutions from part (a) to approximate the two real roots and the two imaginary roots of the equation $x^4 - x^2 - 1 = 0$. Give each real root to the nearest hundredth, and for each imaginary root bi, give b to the nearest hundredth. ±1.27, ±0.79i

Use the rational root theorem to solve each equation, giving all real and imaginary roots. 18. $-1, -\frac{2}{3}, 2$ 22. $-\frac{1}{2}, 2, \frac{-1 \pm \sqrt{5}}{2}$

13. $x^3 - x^2 - x + 1 = 0$ ±1

14. $x^3 + 2x^2 - x - 2 = 0$ −2, ±1

15. $x^4 - 10x^2 + 9 = 0$ ±1, ±3

16. $2x^3 - 9x^2 + 3x + 4 = 0$ $-\frac{1}{2}, 1, 4$

17. $3x^3 - 4x^2 - 5x + 2 = 0$ $-1, \frac{1}{3}, 2$

18. $3x^4 + 2x^3 - 9x^2 - 12x - 4 = 0$

19. $x^4 + 2x^3 - 2x^2 - 6x - 3 = 0$ $-1, \pm\sqrt{3}$

20. $3x^3 - x^2 - 36x + 12 = 0$ $\frac{1}{3}, \pm 2\sqrt{3}$

21. $2x^3 - 7x + 2 = 0$ $-2, \frac{2 \pm \sqrt{2}}{2}$

22. $2x^4 - x^3 - 7x^2 + x + 2 = 0$

In Exercises 23–28, factor each polynomial function and sketch its graph.

23. $f(x) = x^3 + 2x^2 - 9x - 18$

24. $g(x) = 4x^3 + x^2 - 18x$

25. $h(x) = x^4 - 11x^2 + 28$

26. $k(x) = -x^4 + 10x^2 - 24$

27. $m(x) = 4x^3 - 13x - 6$

28. $n(x) = 1 - 3x^2 - 2x^3$

29. Show that the equation $x^3 + x^2 - 3 = 0$ has no rational roots, but that it does have an irrational root between $x = 1$ and $x = 2$.

30. Show that the equation $3x^3 - 4x^2 + 5x - 2 = 0$ has no rational roots, but that it does have an irrational root between $x = 0$ and $x = 1$.

36. $(3, 0)$, $(\sqrt[3]{2}, 2\sqrt[3]{2} - 6)$

Sketch the graphs of the two given equations on a single set of axes. Then determine algebraically where the graphs intersect or are tangent. $(\sqrt{2}, 4)$, $(-\sqrt{2}, 4)$

B 31. $y = x^3 - 3x$
   $y = 2$ (−1, 2), (2, 2)

32. $y = x^3 - x$
   $y = 3x$ (−2, −6), (0, 0), (2, 6)

33. $y = 4x^2 - x^4$
   $y = 4$

34. $y = x^4 - 6x^2$
   $y = -9$ $(-\sqrt{3}, -9)$, $(\sqrt{3}, -9)$

35. $y = x^3 + 4x^2$
   $y = 3x + 18$ (−3, 9), (2, 24)

36. $y = x^4 - 3x^3$
   $y = 2x - 6$
   See above.

37. Show that the positive real cube root of 12 is irrational. (*Hint*: What equation must the cube root of 12 satisfy?)

38. Show that the positive fifth root of 100 is irrational.

39. **a.** The slant height of a cone is 3 and its height is h. Show that its volume is:
   $$V(h) = \frac{1}{3}\pi(9 - h^2)h$$

   b. Find the two values of h for which $V(h) = \frac{10}{3}\pi$. 2, $-1 + \sqrt{6}$

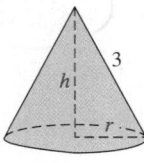

40. a. A cylinder is inscribed in a sphere of radius 4. Show that the volume of the cylinder is:

$$V(x) = 2\pi x(16 - x^2)$$

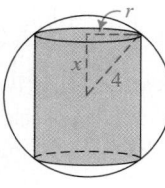

b. Find the two values of x for which $V(x) = 42\pi.$ 3, $\dfrac{-3 + \sqrt{37}}{2}$

C **41.** A wooden block is in the shape of a rectangular prism with dimensions n cm, $(n + 3)$ cm, and $(n + 9)$ cm, for some integer n. The surface of the block is painted and the block is then cut into 1 cm cubes by cuts parallel to the faces. If exactly half of these cubes have no paint on them, find the dimensions of the original block. 7 cm × 10 cm × 16 cm

2-7 General Results for Polynomial Equations

Objective   *To apply general theorems about polynomial equations.*

In this section, we will state five general theorems about polynomial equations, some of which you may have already discovered yourself as you were studying this chapter. The first theorem, known as the *fundamental theorem of algebra*, is a cornerstone of much advanced work in mathematics. Its proof is beyond the scope of this book, but proofs of the other theorems in this section will be suggested in the exercises.

Theorem 1. The Fundamental Theorem of Algebra

In the complex number system consisting of all real and imaginary numbers, if $P(x)$ is a polynomial of degree n ($n > 0$) with complex coefficients, then the equation $P(x) = 0$ has exactly n roots (provided a double root is counted as 2 roots, a triple root is counted as 3 roots, and so on).

By the fundamental theorem of algebra, we can expect a cubic equation like $2ix^3 + \sqrt{5}x^2 + (3 + 2i)x + 7 = 0$ to have 3 roots and a quartic equation like $3x^4 - 11x^3 + 19x^2 + 25x - 36 = 0$ to have 4 roots. Sometimes these roots come in pairs, as the next two theorems indicate.

Theorem 2. Complex Conjugates Theorem

If $P(x)$ is a polynomial with real coefficients, and $a + bi$ is an imaginary root of the equation $P(x) = 0$, then $a - bi$ is also a root.

Teaching Notes, p. 52D

Warm-Up Exercises

Find **(a)** the sum and **(b)** the product of each pair of numbers.

1. $\sqrt{3}, -\sqrt{3}$ **a.** 0, **b.** -3

2. $-2 - \sqrt{5}, -2 + \sqrt{5}$
a. -4, **b.** -1

3. $6 - \dfrac{\sqrt{3}}{2}, 6 + \dfrac{\sqrt{3}}{2}$

a. 12, **b.** $35\dfrac{1}{4}$

4. $i\sqrt{7}, -i\sqrt{7}$ **a.** 0, **b.** 7

5. $4 + 2i, 4 - 2i$ **a.** 8, **b.** 20

6. $\dfrac{2 - i\sqrt{2}}{6}, \dfrac{2 + i\sqrt{2}}{6}$

a. $\dfrac{2}{3}$, **b.** $\dfrac{1}{6}$

Motivating the Section

You might introduce this section by discussing the historical development presented on page 88.

Assessment Note

After discussing Theorems 1 and 2, you might ask students questions like the following:

1. How many roots (counting multiple roots separately) would a polynomial equation of degree 10 have? (10)

2. If $1 + i$ is a root of a polynomial equation with real coefficients, what is another root? $(1 - i)$

After discussing Theorems 3–5, you might ask students questions like the following:

1. If a 6th-degree polynomial equation with rational coefficients has $2 + \sqrt{5}$ as a root, what is another root? $(2 - \sqrt{5})$

2. What is the sum of the roots of $-2x^3 + 3x^2 + 1 = 0$? $\left(\frac{3}{2}\right)$ What is the product of the roots? $\left(\frac{1}{2}\right)$

Theorem 3

Suppose $P(x)$ is a polynomial with rational coefficients, and a and b are rational numbers such that $\sqrt{b}$ is irrational. If $a + \sqrt{b}$ is a root of the equation $P(x) = 0$, then $a - \sqrt{b}$ is also a root.

By the complex conjugates theorem, if we know that $1 + i\sqrt{2}$ is a root of the equation $x^3 - x^2 + x + 3 = 0$, then we also know that $1 - i\sqrt{2}$ is a root. Similarly, by Theorem 3, if we know that $\frac{3}{2} + \sqrt{5}$ is a root of the equation $4x^3 - 16x^2 + x + 11 = 0$, then we also know that $\frac{3}{2} - \sqrt{5}$ is a root.

If we apply the fundamental theorem of algebra and the complex conjugates theorem to a cubic polynomial with real coefficients, we can say that the polynomial has either three real roots or one real root and a pair of imaginary roots. In any event, the polynomial must have at least one real root. The next theorem generalizes this result.

Theorem 4

If $P(x)$ is a polynomial of odd degree with real coefficients, then the equation $P(x) = 0$ has at least one real root.

The next theorem states an interesting relationship between the roots of a polynomial equation and the coefficients of the polynomial.

Theorem 5

For the equation $a_n x^n + a_{n-1} x^{n-1} + a_{n-2} x^{n-2} + \cdots + a_0 = 0$, with $a_n \neq 0$:

the sum of the roots is $-\dfrac{a_{n-1}}{a_n}$;

the product of the roots is $\begin{cases} \dfrac{a_0}{a_n} & \text{if } n \text{ is even.} \\[2mm] -\dfrac{a_0}{a_n} & \text{if } n \text{ is odd.} \end{cases}$

By Theorem 5, we know that for the equation $2x^3 - 5x^2 - 3x + 9 = 0$, the sum of the roots is $\frac{5}{2}$ and the product of the roots is $-\frac{9}{2}$. Similarly, for the equation $x^4 + x^2 + x + 3 = 0$, the sum of the roots is 0 (because there is no cubic term) and the product of the roots is 3.

Applying Theorem 5 to the general quadratic equation

$$ax^2 + bx + c = 0,$$

we know that the sum of the roots is $-\dfrac{b}{a}$ and the product of the roots is $\dfrac{c}{a}$. Thus, after writing the general quadratic equation as

$$x^2 - \left(-\dfrac{b}{a}\right)x + \dfrac{c}{a} = 0,$$

we see that the equation has the form

$$x^2 - (\text{sum of the roots})x + (\text{product of the roots}) = 0.$$

This form gives us a pattern for obtaining a quadratic equation from its roots.

Example 1   Find a quadratic equation with roots $2 \pm 3i$.

Solution   Find the sum of the roots: $(2 + 3i) + (2 - 3i) = 4$
Find the product of the roots: $(2 + 3i)(2 - 3i) = 13$
Write an equation: $x^2 - (\text{sum})x + \text{product} = 0$
$$x^2 - 4x + 13 = 0$$

Example 2   Find a cubic equation with integral coefficients that has no quadratic term and $3 + i\sqrt{2}$ as one of the roots.

Solution   If a cubic equation of the form $a_3x^3 + a_2x^2 + a_1x + a_0 = 0$ has no quadratic term, then $a_2 = 0$ and the sum of the roots is $-\dfrac{a_2}{a_3} = 0$. Since $3 + i\sqrt{2}$ is a root, $3 - i\sqrt{2}$ must be another root. Therefore, if r is the third root, we have:

$$(3 + i\sqrt{2}) + (3 - i\sqrt{2}) + r = 0$$
$$r = -6$$

Now that we know all the roots, one way to obtain a cubic equation is to write a quadratic equation with roots $3 \pm i\sqrt{2}$ first:

$$x^2 - (\text{sum})x + \text{product} = 0$$
$$x^2 - 6x + 11 = 0$$

Then a cubic equation with roots -6 and $3 \pm i\sqrt{2}$ is:

$$(x + 6)(x^2 - 6x + 11) = 0$$
$$x^3 - 25x + 66 = 0$$

Another way to find such an equation is to write a product of factors in the form $(x - \text{root})$, with one factor for each root:

$$[x - (3 + i\sqrt{2})][x - (3 - i\sqrt{2})][x - (-6)] = 0$$

Polynomial Functions   **87**

1. Find a cubic equation with integral coefficients and roots $1 - \sqrt{6}$ and $-\dfrac{3}{2}$.

   By the fundamental theorem of algebra, the equation has 3 roots. By Theorem 3, the third root must be $1 + \sqrt{6}$. The roots $1 + \sqrt{6}$ and $1 - \sqrt{6}$ have the sum 2 and the product $1 - 6 = -5$. Thus, using the form $x^2 - (\text{sum})x + \text{product} = 0$, we have $x^2 - 2x - 5 = 0$. A cubic equation with the required roots is:

   $$\left(x + \dfrac{3}{2}\right)(x^2 - 2x - 5) = 0$$
   $$(2x + 3)(x^2 - 2x - 5) = 0$$
   $$2x^3 - x^2 - 16x - 15 = 0$$

2. Find integers b and c such that the equation $x^3 + bx^2 + cx - 10 = 0$ has $-2 + i$ as a root.

   By the complex conjugates theorem, $-2 - i$ is also a root. By Theorem 5, the sum of the roots is $-\dfrac{b}{1} = -b$ and the product of the roots is $-\dfrac{-10}{1} = 10$. If r is the third root, then $(-2 + i)(-2 - i)r = 10$; $5r = 10$; $r = 2$. Thus, $(-2 + i) + (-2 - i) + 2 = -b$; $-2 = -b$; $b = 2$. Since 2 is a root of $x^3 + 2x^2 + cx - 10 = 0$, $2^3 + 2 \cdot 2^2 + 2c - 10 = 0$; $2c = -6$; $c = -3$.

Historical Development

The five theorems in this section have been proved in the last 300 years or so, but the study of polynomial equations has a long history, going back to the time of the ancient Greeks.

Before the Renaissance   The Greeks were able to solve quadratic equations by a geometric method, but an algebraic method leading to the quadratic formula was not discovered until centuries later. The methods of equation solving of the Hindu Brahmagupta (about A.D. 628) or the Persian Omar Khayyam (about 1100) look complicated to us today because the notation used was cumbersome. Even as late as the 1500's, mathematicians were using abbreviated Latin words in their equations. For example,

$$4 \text{ Se.} - 5 \text{ Pri.} - 7 \text{ N. aequatur } 0$$

was used instead of $4x^2 - 5x - 7 = 0$.

The Renaissance   In a work published in 1545, the Italian mathematician Girolamo Cardano (1501–1576) stated a "cubic formula" and a "quartic formula." These formulas gave the solutions of cubic and quartic equations using only radicals and the coefficients of the equations. Cardano's presentation of the formulas largely depended on the work of others. For example, Cardano solved cubic equations by first putting them in the form $x^3 + mx^2 = n$ and then applying a method of solution that he obtained from the Italian mathematician Tartaglia (about 1499–1557). Similarly, Cardano solved quartic equations by a method that he learned from his own student Lodovico Ferrari (1522–1565).

The Modern Age   The next goal of mathematicians was to discover a "quintic formula." In 1824, however, a young Norwegian, Niels Henrik Abel (1802–1829), proved that such a formula does not exist. That is, he proved that it is impossible, in general, to express the roots of a fifth-degree equation in terms of radicals and coefficients.

The theory of polynomial equations has been used to answer a number of questions in geometry. For example, since the days of ancient Greece, people have tried to find a way to trisect an arbitrary angle with a straightedge and compass. Finally, in the 1800's, the theory of equations was used to prove that no such general method exists.

As another example, in 1796 the German mathematician *Carl Friedrich Gauss* (1777–1855), at the age of 19, used the equation $x^n = 1$ to determine which n-sided regular polygons could be constructed with straightedge and compass. In short, the study of polynomial equations has helped link together algebra and geometry.

Carl Friedrich Gauss

CLASS EXERCISES

1. A cubic equation with real coefficients has roots -1 and $\sqrt{3} + 2i$. What is the third root? $\sqrt{3} - 2i$

2. A quartic equation with integral coefficients has roots $-3 + \sqrt{2}$ and $-2i$. What are the other roots? $-3 - \sqrt{2},\ 2i$

3. State the sum and product of the roots of each equation. **3. a.** $-\frac{5}{2};\ -\frac{3}{2}$
   a. $2x^2 + 5x - 3 = 0$ **b.** $4x^3 - 2x^2 + 5x - 6 = 0\ \frac{1}{2};\ \frac{3}{2}$
   c. $x^4 + 2x^3 + 3x^2 = 0\ -2;\ 0$ **d.** $x^5 + 32 = 0\ 0;\ -32$

4. *Visual Thinking* Use a graph to explain why a cubic equation with real coefficients must have at least one real root. Graph must cross x-axis at least once.

5. *Visual Thinking* Use a graph to explain why a quartic equation with real coefficients does not necessarily have any real roots. Graph may lie entirely above or below x-axis.

WRITTEN EXERCISES

4. T; for example, $(x + 3)(x - 4)\left(x - (1 - \sqrt{2})\right)(x^2 + 1) = 0$

Tell whether the statements in Exercises 1–8 are true or false. Justify your answers. (All equations mentioned have real coefficients.)

A

1. Some cubic equations have no real roots. F; Theorem 4

2. Every polynomial equation has at least one real root. F; for example, $x^2 + 1 = 0$

3. The roots of a certain quartic equation are $\pm\frac{1}{2}$, 0, and $1 + i$. F; $1 - i$ must also be a root

4. The roots of a certain fifth-degree equation are $-3, 4, 1 - \sqrt{2}$, and $\pm i$. See above.

5. It is possible for the graph of a cubic function to be tangent to the x-axis at $x = -2$, $x = 1$, and $x = 6$. F; 3 points of tangency imply degree ≥ 6

6. No polynomial equation can have an odd number of imaginary roots. T; Theorem 2

7. Suppose $P(x)$ is a polynomial with rational coefficients, and b is rational but $\sqrt{b}$ is irrational. If $\sqrt{b}$ is a root of the equation $P(x) = 0$, then $-\sqrt{b}$ is also a root. T; Theorem 3 with $a = 0$

8. If $a + bi$, $b \neq 0$, is a root of the polynomial equation $P(x) = 0$, then the equation must have an even number of roots. F; for example, $x^3 + 1 = 0$ has 2 imaginary roots and 1 real root

In Exercises 9–12, find the sum and product of the roots of the given equation.

9. $4x^2 - 3x + 6 = 0\ \frac{3}{4};\ \frac{3}{2}$ 10. $6x^3 - 9x^2 + x = 0\ \frac{3}{2};\ 0$
11. $3x^3 + 5x^2 - x - 2 = 0\ -\frac{5}{3};\ \frac{2}{3}$ 12. $x^4 - 4x^2 = 5\ 0;\ -5$

In Exercises 13–16, find a quadratic equation with integral coefficients that has the given roots. Answers may vary. Examples are given. $3x^2 - 2x + 1 = 0$

13. $1 \pm i$ 14. $4 \pm \sqrt{3}$ 15. $3 \pm \sqrt{2}$ 16. $\dfrac{1 \pm i\sqrt{2}}{3}$
   $x^2 - 2x + 2 = 0$ $x^2 - 8x + 13 = 0$ $x^2 - 6x + 7 = 0$

Cooperative Learning

You might want students to discuss Written Exercises 1–8 in groups. Suggest that they offer counterexamples for the false statements.

Suggested Assignments

Standard
89/1–27 odd
Comprehensive
89/1, 5, 9, 13, 17, 21, 25, 27, 29, 31–39

Supplementary Materials

Alternative Assessment, 8
Using Technology, 12
Student Resource Guide, 19–21

29. a. Let $P(x) = x^3 + \frac{b}{a}x^2 +$

$\frac{c}{a}x + \frac{d}{a}$. If $P(x) = 0$ has

roots r_1, r_2, and r_3, then
the factors of $P(x)$ are
$x - r_1$, $x - r_2$, and
$x - r_3$, and $P(x) =$
$k(x - r_1)(x - r_2)(x - r_3)$.
Since the leading
coefficient of $P(x)$ is 1,
$k = 1$.

30. Suppose $ax^4 + bx^3 +$
$cx^2 + dx + e = 0$ has
roots r_1, r_2, r_3, and r_4.

Then $x^4 + \frac{b}{a}x^3 + \frac{c}{a}x^2 +$

$\frac{d}{a}x + \frac{e}{a} = 0$ has the same

roots and $x^4 + \frac{b}{a}x^3 +$

$\frac{c}{a}x^2 + \frac{d}{a}x + \frac{e}{a} = (x - r_1) \cdot$

$(x - r_2)(x - r_3)(x - r_4) =$
$x^4 -$
$(r_1 + r_2 + r_3 + r_4)x^3 +$
$(r_1r_2 + r_1r_3 + r_1r_4 +$
$r_2r_3 + r_2r_4 + r_3r_4)x^2 -$
$(r_1r_2r_3 + r_1r_2r_4 + r_1r_3r_4 +$
$r_2r_3r_4)x + r_1r_2r_3r_4$. Equat-
ing the coefficients of the
cubic and constant terms

gives $-\frac{b}{a} = r_1 + r_2 +$

$r_3 + r_4$ and $\frac{e}{a} = r_1r_2r_3r_4$.

---

23. $x^4 - 10x^3 + 29x^2 - 10x + 28 = 0$

17. A cubic equation with integral coefficients has no quadratic term. If one root is
$2 + i\sqrt{5}$, what are the other roots? $-4,\ 2 - i\sqrt{5}$

18. A quartic equation with integral coefficients has no cubic term and no constant
term. If one root is $3 - i\sqrt{7}$, what are the other roots? $0,\ -6,\ 3 + i\sqrt{7}$

**In Exercises 19–22, find a cubic equation with integral coefficients that has
the given roots.** Answers may vary.

19. 2 and $4 + i$ $x^3 - 10x^2 + 33x - 34 = 0$ **20.** 3 and $7 - i$ $x^3 - 17x^2 + 92x - 150 = 0$

21. $\frac{4 + i\sqrt{3}}{2}$ and -1 $4x^3 - 12x^2 + 3x + 19 = 0$ **22.** $i\sqrt{2}$ and 5 $x^3 - 5x^2 + 2x - 10 = 0$
See above.

23. Find a quartic equation with integral coefficients that has roots $5 - i\sqrt{3}$ and i.

24. Find a quartic equation with integral coefficients that has roots $1 + i\sqrt{7}$ and
$2 + i$. $x^4 - 6x^3 + 21x^2 - 42x + 40 = 0$

25. Find integers c and d such that the equation $x^3 + cx + d = 0$ has $1 + \sqrt{3}$ as
one of its roots. $c = -6,\ d = -4$

26. Find integers b and c such that the equation $x^3 + bx^2 + cx + 8 = 0$ has $2i$ as
one of its roots. $b = 2,\ c = 4$

B   **27.** Find an integer c such that the equation $4x^3 + cx - 27 = 0$ has a double root. -27

28. Find an integer d such that the equation $x^3 + 4x^2 - 9x + d = 0$ has two roots
that are additive inverses of each other. -36

29. Suppose $ax^3 + bx^2 + cx + d = 0$ has roots r_1, r_2, and r_3. Then

$$x^3 + \frac{b}{a}x^2 + \frac{c}{a}x + \frac{d}{a} = 0$$

has these same roots.

a. Explain why $x^3 + \frac{b}{a}x^2 + \frac{c}{a}x + \frac{d}{a} = (x - r_1)(x - r_2)(x - r_3)$.

b. By carrying out the multiplication of factors on the right side of the equation in part (a), show that:

$$-\frac{b}{a} = r_1 + r_2 + r_3 \text{ and } -\frac{d}{a} = r_1r_2r_3$$

c. Conclude from part (b) that the average of the roots of the cubic equation

is $-\frac{b}{3a}$. $\frac{1}{3}(r_1 + r_2 + r_3) = \frac{1}{3}\left(-\frac{b}{a}\right) = -\frac{b}{3a}$

30. Use the technique of Exercise 29 to show that if

$$ax^4 + bx^3 + cx^2 + dx + e = 0$$

has roots r_1, r_2, r_3, and r_4, then:

$$-\frac{b}{a} = r_1 + r_2 + r_3 + r_4 \text{ and } \frac{e}{a} = r_1r_2r_3r_4$$

The purpose of Exercises 31–38 is to prove the complex conjugates theorem, page 85, for a cubic polynomial. (These exercises should be done sequentially.) Let $x = r + si$ and $z = u + vi$. The conjugates are then $\bar{x} = r - si$ and $\bar{z} = u - vi$.

31. Prove that $\overline{x + z} = \bar{x} + \bar{z}$. (The conjugate of a sum is the sum of the conjugates.)

32. Prove that $\overline{xz} = \bar{x} \cdot \bar{z}$. (The conjugate of a product is the product of the conjugates.)

33. Prove that $\overline{x^2} = (\bar{x})^2$. (The conjugate of a square is the square of the conjugate.)
$\overline{x^2} = \overline{x \cdot x} = \bar{x} \cdot \bar{x} = (\bar{x})^2$

34. Prove that $\overline{x^3} = (\bar{x})^3$. (The conjugate of a cube is the cube of the conjugate.)
$\overline{x^3} = \overline{x^2 \cdot x} = \overline{x^2} \cdot \bar{x} = (\bar{x})^2 \cdot \bar{x} = (\bar{x})^3$

35. a. Prove that the conjugate of a real number a is a. $\bar{a} = \overline{a + 0i} = a - 0i = a$
  b. Prove that $\overline{ax^3} = a(\bar{x})^3$. $\overline{ax^3} = \bar{a} \cdot \overline{x^3} = a(\bar{x})^3$

36. Prove that $\overline{bx^2} = b(\bar{x})^2$ and $\overline{cx} = c\bar{x}$. $\overline{bx^2} = \bar{b} \cdot \overline{x^2} = b(\bar{x})^2$; $\overline{cx} = \bar{c} \cdot \bar{x} = c\bar{x}$

37. Prove that $\overline{ax^3 + bx^2 + cx + d} = a(\bar{x})^3 + b(\bar{x})^2 + c(\bar{x}) + d$.

38. Prove that if $P(x) = ax^3 + bx^2 + cx + d$ and $P(x) = 0$, then $P(\bar{x}) = 0$.

Note: The proof just given can be generalized by noting that $\overline{ax^n} = a(\bar{x})^n$. Moreover, a proof of Theorem 3, page 86, is similar to that just given if you interpret $\overline{a + \sqrt{b}}$ to mean $a - \sqrt{b}$, where a and b are rational numbers and $\sqrt{b}$ is an irrational number.

39. If $P(x) = x^3 - x$, the result of Exercise 37 tells you that $\overline{P(1 + i)} = P(1 - i)$. Verify this result by expressing $\overline{P(1 + i)}$ and $P(1 - i)$ in the form $a + bi$. $-3 - i$

40. *Reading* Consider the following statement:

Every polynomial equation with complex (that is, real or imaginary) coefficients and degree n ($n > 0$) has at least one complex root.

Explain why this statement is equivalent to the fundamental theorem of algebra stated on page 85.

Chapter Summary

1. A *polynomial* in x is an expression that can be written in the form

$$a_n x^n + a_{n-1} x^{n-1} + \cdots + a_2 x^2 + a_1 x + a_0,$$

where n is a nonnegative integer. *Synthetic substitution* can be used to evaluate a polynomial. If the value of a polynomial is 0 for some number, then the number is called a *zero* of the related polynomial function.

Polynomial Functions **91**

Cooperative Learning

You might want students to discuss Written Exercises 31–38 in small groups. Suggest that they offer explanations for their proof steps.

Additional Answers
Written Exercises

31. $\overline{x + z} =$
$\overline{(r + si) + (u + vi)} =$
$\overline{(r + u) + (s + v)i} =$
$(r + u) - (s + v)i =$
$(r - si) + (u - vi) =$
$\bar{x} + \bar{z}$

32. $\overline{xz} = \overline{(r + si)(u + vi)} =$
$\overline{(ru - sv) + (su + rv)i} =$
$(ru - sv) - (su + rv)i$;
$\bar{x} \cdot \bar{z} = \overline{r + si} \cdot \overline{u + vi} =$
$(r - si)(u - vi) =$
$ru - sui - rvi + svi^2 =$
$(ru - sv) - (su + rv)i$;
so $\overline{xz} = \bar{x} \cdot \bar{z}$.

37. $\overline{ax^3 + bx^2 + cx + d} =$
$\overline{ax^3} + \overline{bx^2} + \overline{cx} + \overline{d}$
(by Ex. 31) $=$
$a(\bar{x})^3 + b(\bar{x})^2 + c(\bar{x}) + d$
(by Exs. 35 and 36)

38. If $P(x) = 0$, then $P(\bar{x}) =$
$a(\bar{x})^3 + b(\bar{x})^2 + c(\bar{x}) + d$
$= \overline{ax^3 + bx^2 + cx + d}$
(by Ex. 37) $=$
$\overline{P(x)} = \bar{0} = 0$.

Supplementary Materials

Tests, 3–4
Alternative Assessment, 63
Student Resource Guide, 22–23

2. If you divide a polynomial $P(x)$ by $x - a$, you obtain a quotient $Q(x)$ and a remainder R where:

$$P(x) = (x - a) \cdot Q(x) + R$$

From this equation we have the *remainder theorem*:

When $P(x)$ is divided by $x - a$, the remainder $R = P(a)$.

We also have the *factor theorem*:

$x - a$ is a factor of $P(x)$ if and only if $P(a) = 0$.

3. Because synthetic substitution gives the quotient and remainder when a polynomial $P(x)$ is divided by $x - a$, it is often called *synthetic division*.

4. **a.** The general shapes of the graphs of cubic and quartic functions are shown on pages 62 and 64.
   b. If a polynomial $P(x)$ has a squared factor $(x - c)^2$, then the graph of $y = P(x)$ is tangent to the x-axis at $x = c$. (See figures (1)–(3) on page 64.) If it has a cubed factor $(x - c)^3$, then the graph of $y = P(x)$ flattens out at $x = c$ and crosses the x-axis at this point. (See figures (4)–(6) on page 65.)
   c. Example 1 on page 63 demonstrates a useful technique for sketching the graph of a factored polynomial function.

5. The quadratic function $f(x) = ax^2 + bx + c$ has a maximum or minimum value at $x = -\dfrac{b}{2a}$. A cubic function can have a local maximum and a local minimum, which can be approximated using a computer or calculator.

6. Sections 2-5 and 2-6 present several methods of solving polynomial equations.
   a. Technology is used in Section 2-5 to approximate the real roots of a polynomial equation. One approach is based on the *location principle*:

   If $P(x)$ is a polynomial with real coefficients and a and b are real numbers such that $P(a)$ and $P(b)$ have opposite signs, then between a and b there is at least one real root r of the equation $P(x) = 0$.

   b. Various methods of factoring are used in Section 2-6 to solve polynomial equations. One method is based on the *rational root theorem*:

   If $P(x) = a_n x^n + \cdots + a_0$ has integral coefficients with $a_0 \neq 0$ and if one root of $P(x) = 0$ is $x = \dfrac{p}{q}$ where p and q are nonzero integers with no common factor other than 1, then p must be a factor of a_0, and q must be a factor of a_n.

7. Section 2-7 presents five theorems about polynomial equations and their roots. The most important of these is the *fundamental theorem of algebra*:

   In the complex number system consisting of all real and imaginary numbers, if $P(x)$ is a polynomial of degree n ($n > 0$) with complex coefficients, then the equation $P(x) = 0$ has exactly n roots (provided multiple roots are counted individually).

92   *Chapter Two*

Key vocabulary and ideas

polynomial (p. 53)
synthetic substitution (p. 55)
remainder theorem (p. 59)
synthetic division (p. 59)
factor theorem (p. 60)
double root (p. 64)
triple root (p. 65)

local maximum, local minimum (p. 69)
location principle (p. 76)
rational root theorem (p. 82)
fundamental theorem of algebra (p. 85)
complex conjugates theorem (p. 85)
sum and product of the roots of a
    polynomial equation (p. 86)

Chapter Test

1. If $P(x) = 4x^3 - 5x - 2$, use synthetic substitution to find: **2-1**
   a. $P(2)$ 20   **b.** $P(-0.5)$ 0   **c.** $P(i)$ $-2 - 9i$

2. If 2 is a zero of $P(x) = 3x^2 + kx - 8$, find the value of k. -2

3. When a polynomial $P(x)$ is divided by $x + 3$, the quotient is   **2-2**
   $x^2 - x + 5$ and the remainder is 2. Find $P(x)$. $x^3 + 2x^2 + 2x + 17$

4. Two roots of the equation $x^4 - 3x^3 - 14x^2 + 12x + 40 = 0$ are
   $x = 2$ and $x = 5$. Find the remaining roots. -2 (a double root)

5. Sketch the graph of $y = x(x + 1)(x - 2)^2$.   **2-3**

6. Give an equation of the cubic graph shown
   at the right. $y = \frac{1}{2}(x + 2)(x - 1)(x - 3)$

7. A rectangular enclosure is to be made using   **2-4**
   a barn as one side and 80 m of fencing to
   form the other three sides. What is the maximum area of such an enclosure? 800 m^2

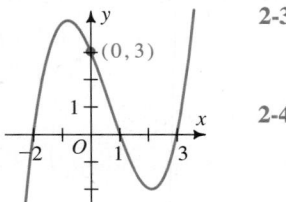
(0, 3)

Part (b) of Exercise 8 requires the use of a computer or graphing
    calculator.

8. **a.** Use the location principle: $P(1) = -2 < 0$ and $P(2) = 7 > 0$.

8. **a.** If $P(x) = 2x^4 - 3x^3 - 1$, explain how you know that the equation   **2-5**
       $P(x) = 0$ has at least one real root between $x = 1$ and $x = 2$.
   b. Find each real root of $P(x) = 0$ to the nearest tenth. $-0.6, 1.6$

9. Solve $2x^3 - 3x^2 = 6x - 9$. $\frac{3}{2}, \pm\sqrt{3}$ **2-6**

10. Find all real and imaginary roots of the polynomial equation
    $3x^4 - x^3 + 4x^2 - 2x - 4 = 0$. $-\frac{2}{3}, 1, \pm i\sqrt{2}$

11. **Writing** Write a paragraph or two in which you compare and con-
    trast the two basic methods for solving polynomial equations pre-
    sented in this chapter: using technology and factoring. $\frac{5}{2}; -4$

12. Find the sum and product of the roots of $2x^3 - 5x^2 - x + 8 = 0$.   **2-7**

13. Find a cubic equation with integral coefficients and roots $3i$ and -5. See below.

14. $P(x)$ is a cubic polynomial with real coefficients. Explain why an
    imaginary number cannot be a double root of the equation $P(x) = 0$.

13. For example, $x^3 + 5x^2 + 9x + 45 = 0$.   *Polynomial Functions*   **93**

Chapter 3 Inequalities

Overview

In this chapter, students solve and graph linear and polynomial inequalities in one or two variables. They also solve and graph absolute-value inequalities and graph the solution sets of systems of inequalities. Students conclude their work with inequalities by setting up and solving linear programming models, in which a function's maximum or minimum value on a feasible region is found using the corner-point principle.

Objectives

3-1 To solve and graph linear inequalities in one variable.
3-2 To solve and graph polynomial inequalities in one variable.
3-3 To graph polynomial inequalities in two variables and to graph the solution set of a system of inequalities.
3-4 To solve certain applied problems using linear programming.

Supplementary Resources

Tests, pp. 5–6, 7–8
Alt. Assess., pp. 9–10, 64
Activities, pp. 6–7
Using Tech., p. 13
Student Res. Guide, pp. 24–32

Software

Precalculus Plotter Plus
Activities, pp. 28, 29–30
Function Plotter, Linear Programming

Pacing Guide

Section	3-1	3-2	3-3	3-4	Review	Test	C. Rev.	Total
Standard (days)	1	2	2	2	1	1	1	10
Comprehensive (days)	1	1	1	1	1	1	1	7

Teaching Notes

Presenting the Section

You might want to discuss the properties of real numbers used to solve Example 1 on page 95. Emphasize that not reversing the inequality sign is a common error to avoid when multiplying or dividing both sides of an inequality by a negative number. You might want to point out that when graphing the solution, if the variable is on the left, the direction of the shading will match the inequality sign. Encourage students to check a point on the line, such as 0, to make sure the graph is correct.

Overlays used on an overhead projector might be helpful to demonstrate combination graphs, especially for graphs of linear inequalities that involve absolute value. Stress that solutions to sentences such as $|x| \le c$ and $|x| < c$ are "and" sentences, and solutions to sentences such as $|x| \ge c$ and $|x| > c$ are "or" sentences.

The graphs of inequalities such as $|x - a| < b$ should be emphasized because of future work with limits. For example, the definition of $\lim_{x \to c} f(x) = L$ presented on page 723 involves the inequalities $0 < |x - c| < \delta$ and $|f(x) - L| < \epsilon$.

Using Technology

Trial and error is a good problem-solving technique that can be enhanced by being able to check numbers quickly. Encourage students to use calculators or computers to check or find some solutions to inequalities. Programmable calculators or computers also can be used as good motivational tools to find some solutions to absolute-value sentences. For example, students can find 11 solutions to $|x + 2| \le 5$ by running the following simple BASIC program:

```
10 FOR X = -10 TO 10
20 IF ABS(X + 2) < = 5 THEN PRINT X
30 NEXT X
40 END
```

Using software that graphs number-line inequalities is ideal for checking solutions to one-variable inequalities.

Cooperative Learning

Learning can be enhanced when students work in groups of three or four. Groups can review how to solve absolute-value sentences both geometrically and algebraically, as organized in the charts on pages 96 and 97, and they can discuss when absolute-value sentences (such as $|2x - 9| < -6$) have no solutions. They can also go over Exercises 35–37, page 99, to help explain why $|ab| = |a| \cdot |b|$ and to review the triangle inequality and other absolute-value statements that it implies. Encourage students to ask and answer questions and to compare graphs for accuracy. Groups can also compile their own lists of applications of combined inequalities, such as those discussed in Applications below, and present them to the class.

Applications

Examples of combined linear inequalities can be used as a motivator for this section. For example, most students have no difficulty in telling you what minimum score they need on the next test to get a certain average. If their scores on four tests are 82, 63, 78, and 90, and they want their average to be between 80 and 90 after the fifth test, then their fifth score x must satisfy this inequality:

$$80 < \frac{82 + 63 + 78 + 90 + x}{5} < 90$$

Many real-world questions can be expressed as linear inequalities as well. Some examples are as follows:

1. What are the possible legal speeds on a highway that posts a minimum speed of 45 mi/h and a maximum speed of 55 mi/h?

2. What is the tax on an income of at least $6000 but less than $6050?

3. How should $10,000 be split into investments earning 8% and 6% annual interest so that a minimum of $750 is earned at 8%?

4. If adults buy three times as many tickets as students, at most how many student tickets at $4 and adult tickets at $7 can be bought with $80?

Encourage students to discuss why these examples are linear inequalities and how to express them algebraically.

Communication

Point out that "<" in a sentence is "less than," while "less than" in an expression like "6 less than x" implies subtraction.

Students may be confused by the difference between "and" and "or" in everyday use versus "and" and "or" in mathematical use. Point out that in mathematics, two statements joined by "and" must both be true, while only one of two statements joined by "or" needs to be true. Therefore, "and" graphs are intersection graphs while "or" graphs are union graphs. Have students discuss their understandings of the terms *and*, *or*, *intersection*, and *union*.

3-2   pages 100–104

Presenting the Section

This section extends the work with polynomials in Chapter 2. Stress that in order to use Method 1, page 100 (using a sign graph of $P(x)$), the inequalities need to be expressed with 0 on one side of the inequality sign. Method 1 works best if $P(x)$ is factorable.

Inequalities involving rational expressions also can be solved by using Method 1, as shown in Example 3, page 102. The zeros of the numerator and denominator are plotted and the signs of the intervals checked. Notice that in Exercise 29, page 103, both the numerator and denominator have no real roots (use the discriminant to check that they are imaginary roots). Thus, $\dfrac{2x^2 + 7x + 8}{x^2 + 1}$ is always positive and the solution set is {all real numbers}.

Method 2, page 100, involves analyzing a graph of $P(x)$. Students learned how to graph $P(x)$ in Section 2-3 and should now feel comfortable with this method.

Using Technology

Graphing calculators and computers are excellent motivational tools for graphing polynomial inequalities. Refer to the discussion in Section 2-5 on how to use technology to approximate the roots of polynomial equations.

Interested students can be encouraged to write their own computer programs to print solutions to quadratic inequalities.

Cooperative Learning

You might have students work in groups of three or four to do the Activity and Example 2 on page 101. You can also use groups to go over the calculator/computer solutions to Exercises 31–34, page 104. Encourage students to ask and answer questions within their groups, including how to use technology effectively.

Applications

If a projectile is fired straight up from the ground, its height above the ground is modeled by the graph of a quadratic function. If the initial velocity is known, the time it takes to return to ground level can be determined. By using a quadratic inequality, you can determine the time period during which the projectile exceeds a certain height (or is less than a certain height). (See, for example, Exercise 7 on page 782. Students should be able to answer all but part (d) of this exercise.)

Polynomial inequalities can also be applied to compound interest. For example, an investment of P dollars at $r\%$ interest compounded annually for two years increases to $P(1 + 0.01\ r)^2$ dollars. If an investment of $1000 is to increase to at least $1100, you can find the minimum value for the interest rate.

3-3   pages 104–108

Presenting the Section

Graphs of inequalities in two variables consist of regions in the plane. The associated equation is plotted first and then points in each region are checked to see if the region is part of the solution. If the inequalities are written in the form $y < P(x)$, solution points are below the graph of $P(x)$. Similarly, if $y > P(x)$, then solution points are above the graph of $P(x)$.

The use of an overhead projector is an effective way to develop the concepts in this section. You can show the difference between a number-line graph and a coordinate-plane graph, as shown in the Problems at the top of page 106. You can also graph the solution sets of systems of inequalities, as in

Exercises 17–30, page 107. These kinds of exercises are important in linear programming applications (Section 3-4) and in applications presented later in the textbook.

The concept of *tolerance* in manufacturing is exemplified by Exercise 32, page 108. Manufactured items cannot be truly congruent, so a tolerance level is set (sometimes by government standards) that specifies an acceptable range of sizes for an item to work properly.

Using Technology

A computer or graphing calculator is useful when graphing polynomial inequalities in two variables. Of course, only boundary curves can (usually) be obtained; students must decide for themselves whether the boundaries are solid or dashed and where the shading should occur.

Cooperative Learning

Exercise 31, page 107, and Exercises 32–36, page 108, are good preliminary exercises for Section 3-4. Have students go over their solutions in groups so that each exercise can be discussed thoroughly. Encourage students to discuss both their methods of solution and their answers with each other.

Applications

Mathematical applications of systems of linear inequalities include the linear programming models presented in Section 3-4. Systems of linear inequalities also are used in physics, statistics, engineering, and economics.

3-4   pages 108–114

Presenting the Section

This section illustrates the use of inequalities in decision making by using linear programming models. Emphasize that the models are realistic and that businesses rely on computers to solve even more complex problems (with more variables and constraints).

Note that for n linear inequalities, there are $\frac{n(n-1)}{2}$ pairs of simultaneous linear equations to solve to determine the vertices of the feasible region.

Encourage students to organize their data in charts, as shown on page 109 and in the exercises. The values in the charts can be used to answer other questions, such as "What do points along the x-axis signify?" and "Which points in the region correspond to no profit?" Have students make as many generalizations about the data and feasible region as possible. This will help them better understand the problem and the solution process.

The example on page 111 illustrates that sometimes the feasible region can be unbounded, yet a minimum value can still be obtained.

Using Technology

The use of a graphing calculator or a computer should be encouraged for solving linear programming problems. The TRACE and ZOOM features of some calculators and computer graphing software can be used to determine the coordinates of the vertices of a feasible region.

There is software available (such as Houghton Mifflin's *Precalculus Plotter Plus*) that specifically graphs the feasible regions of linear programming problems. Such software includes an EVALUATE feature to obtain the value of the function to be maximized or minimized.

Applications

Important applications of linear programming are maximizing profit or income in manufacturing, agriculture, or other production businesses, and minimizing the costs of producing certain items. Have students compile a list of as many applications as they can find.

Assessment

You may wish to have students summarize verbally or in writing the linear programming procedure:
1. define the variables;
2. write a system of linear inequalities (constraints) to fit the problem;
3. graph the solution set of each inequality;
4. find the coordinates of the vertices of the feasible region;
5. evaluate the linear expressions at the vertices;
6. determine the maximum or minimum value of whatever function is under consideration (cost, profit, and so on).

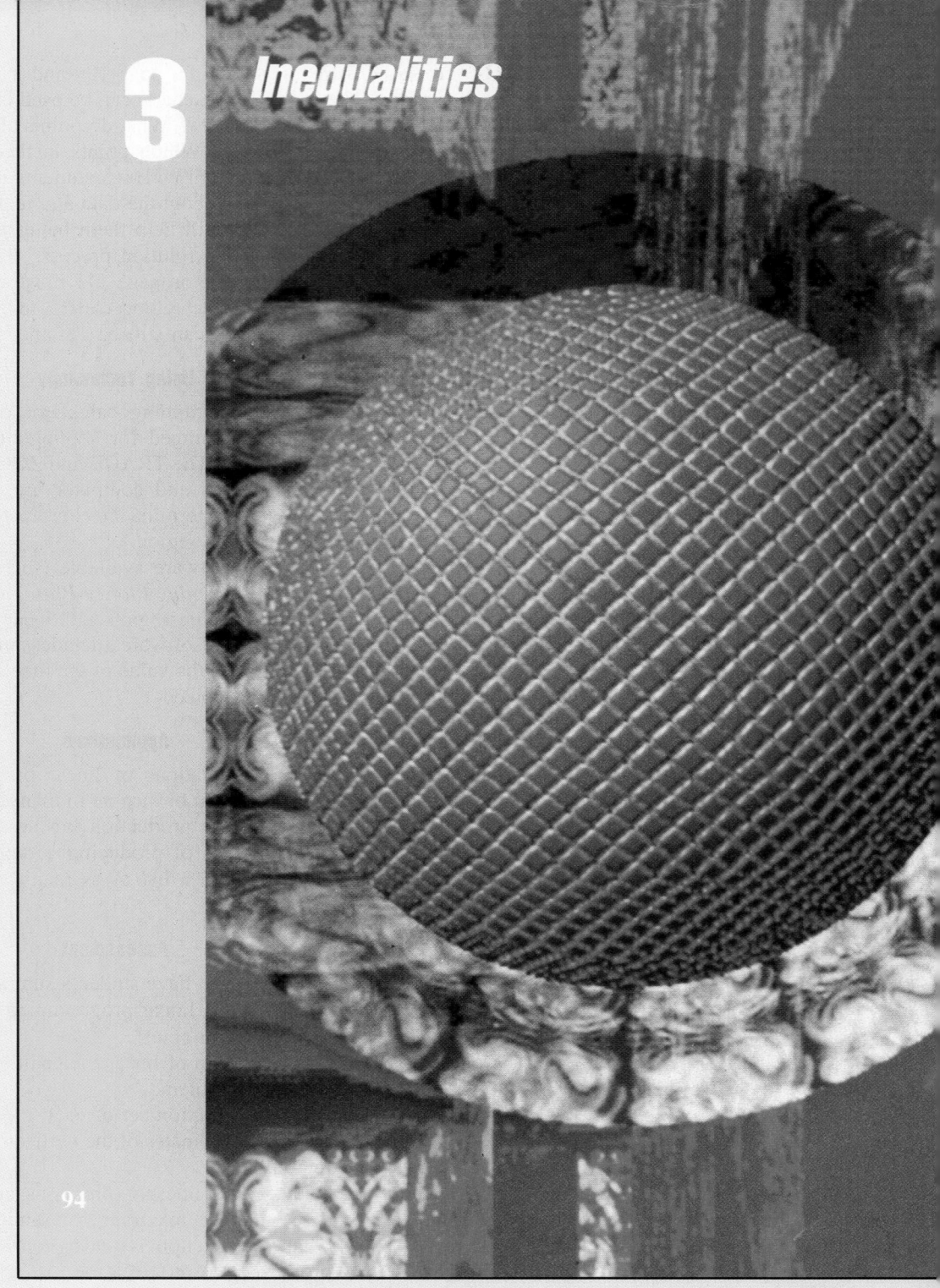

3 Inequalities

94

Inequalities in One Variable

3-1 Linear Inequalities; Absolute Value

Objective *To solve and graph linear inequalities in one variable.*

In this chapter you will study linear and polynomial inequalities and their graphs. Inequalities can be written in one or more variables.

Linear Inequalities: $|3x - 1| < 4$ $2x + 3y > 6$
Polynomial Inequalities: $x^3 - x > 0$ $y < 3x^2 - 4x + 1$

The properties of real numbers used in solving linear inequalities are similar to the properties used in solving linear equations.

1. You can add the same number to (or subtract the same number from) both sides of an inequality. That is, if $a < b$, then $a + c < b + c$.
2. You can multiply (or divide) both sides of an inequality by the same positive number. That is, if $a < b$ and $c > 0$, then $ac < bc$.
3. You can multiply (or divide) both sides of an inequality by the same negative number *if you reverse the inequality sign*. That is, if $a < b$ and $c < 0$, then $ac > bc$.

Example 1   **a.** $3x - 4 \le 10 + x$ **b.** $\dfrac{8 - 2x}{5} > 6$
$\qquad\qquad\qquad 3x \le 14 + x \qquad\qquad\qquad 8 - 2x > 30$
$\qquad\qquad\qquad 2x \le 14 \qquad\qquad\qquad\qquad -2x > 22$
$\qquad\qquad\qquad\ x \le 7 \qquad\qquad\qquad\qquad\ \ x < -11$

In the solution to part (b), the inequality sign *is reversed* because we divided both sides of the inequality $-2x > 22$ by -2.

The graphs of several linear inequalities are shown below. If an endpoint is to be included in the graph, a solid dot (●) is used. Otherwise, an open dot (○) indicates that the endpoint is not to be included in the graph. The last inequality means $-3 < x$ and $x \le 4$. The graph of this *combined inequality* is the intersection of the first two graphs.

$x \le 4$

$x > -3$

$-3 < x \le 4$

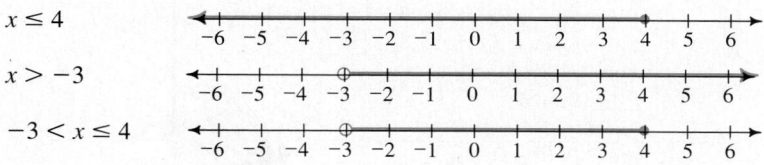

Many contemporary works of art owe their existence to mathematics. "Redbal," pictured at the left, was created entirely with the computer by twentieth-century artist David Em.

Inequalities   **95**

1. Solve and graph the solution.

   a. $-7x - 1 > 13$

   b. $\dfrac{2x + 3}{9} \leq \dfrac{x - 1}{6}$

   a. $-7x - 1 > 13$
      $-7x > 14$
      $\dfrac{-7x}{-7} < \dfrac{14}{-7}$
      $x < -2$

      $\xleftarrow{\quad} \overset{-4\ -3\ -2\ -1\ \ 0\ \ 1}{+\!+\!+\!\circ\!+\!+\!+\!+} \xrightarrow{\quad}$

   b. $18\left(\dfrac{2x + 3}{9}\right) \leq 18\left(\dfrac{x - 1}{6}\right)$
      $4x + 6 \leq 3x - 3$
      $4x \leq 3x - 9$
      $x \leq -9$

      $\xleftarrow{\quad} \overset{-11\ -10\ -9\ \ -8\ \ -7}{+\!+\!\bullet\!+\!+\!+\!+} \xrightarrow{\quad}$

2. Solve by the algebraic method and the geometric method.

   a. $|2x + 3| = 1$
   b. $|x - 2| \geq 3$

   a. Algebraic Method:
      $2x + 3 = \pm 1$
      $2x = -2$ or $2x = -4$
      $x = -1$ or $x = -2$

   Geometric Method:
      $|2x + 3| = 1$
      $2\left|x + \dfrac{3}{2}\right| = 1$
      $\left|x - \left(-\dfrac{3}{2}\right)\right| = \dfrac{1}{2}$

   The distance from x to $-\dfrac{3}{2}$ is equal to $\dfrac{1}{2}$.

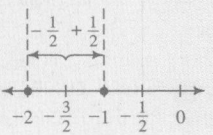

   Each method gives $x = -2$ or $x = -1$.

Warning. Care must be taken when inequalities are combined. For example, the statement $6 < x < 1$ says that $6 < x$ *and* $x < 1$. Since no real number x satisfies both these inequalities, the combined inequality represents the empty set.

Absolute Value

The absolute value of a number x, denoted $|x|$, can be interpreted geometrically as the distance from x to zero in either direction on the number line. Thinking in this way, you can solve linear equations and inequalities involving $|x|$. In all three cases below, c represents a distance, so $c \geq 0$.

Sentences Involving $|x|$			
Sentence	**Meaning**	**Graph**	**Solution**
$|x| = c$	The distance from x to 0 is *exactly* c units.	$\xleftarrow{\ \bullet\ \ \ \ \ \ \bullet\ } \atop {-c\ \ 0\ \ c}$	$x = c$ or $x = -c$
$|x| < c$	The distance from x to 0 is *less than* c units.	$\xleftarrow{\ \circ\ \ \ \ \ \ \circ\ } \atop {-c\ \ 0\ \ c}$	$-c < x < c$
$|x| > c$	The distance from x to 0 is *greater than* c units.	$\xleftarrow{\ \circ\ \ \ \ \ \ \circ\ } \atop {-c\ \ 0\ \ c}$	$x < -c$ or $x > c$

Equations and inequalities involving $|x - k|$, where k is a constant, can be solved by interpreting $|x - k|$ as the distance from x to k on the number line.

Sentences Involving $|x - k|$			
Sentence	**Meaning**	**Graph**	**Solution**
$|x - 5| = 3$	The distance from x to 5 is 3 units.	$\underset{2\ 3\ 4\ 5\ 6\ 7\ 8}{\overset{-3\ \ \ \ +3}{}}$	$x = 2$ or $x = 8$
$|x - 1| < 2$	The distance from x to 1 is less than 2 units.	$\underset{-1\ 0\ 1\ 2\ 3}{\overset{-2\ \ +2}{}}$	$-1 < x < 3$
$|x + 3| > 2$ or $|x - (-3)| > 2$	The distance from x to -3 is greater than 2 units.	$\underset{-5\ -4\ -3\ -2\ -1\ \ 0}{\overset{-2\ \ +2}{}}$	$x < -5$ or $x > -1$

Absolute-value sentences can be solved not only by the geometric method shown on the previous page but also algebraically by using the three sets of equivalent sentences listed below. The equivalent sentences are obtained by replacing x by $ax + b$ in the first chart given on page 96. In all three cases, $c \geq 0$.

Sentence	Equivalent Sentence
$\lvert ax + b \rvert = c$	$ax + b = \pm c$
$\lvert ax + b \rvert < c$	$-c < ax + b < c$
$\lvert ax + b \rvert > c$	$ax + b < -c$ or $ax + b > c$

Example 2 Solve: **a.** $\lvert 3x - 9 \rvert > 4$ **b.** $\lvert 2x + 5 \rvert \leq 7$

Solution   **Algebraic Method**

a. $\lvert 3x - 9 \rvert > 4$ means $3x - 9 < -4$ or $3x - 9 > 4$.

$$3x < 5 \quad or \quad 3x > 13$$

$$\text{Thus, } x < \frac{5}{3} \quad or \quad x > \frac{13}{3}.$$

b. $\lvert 2x + 5 \rvert \leq 7$ means $-7 \leq 2x + 5 \leq 7$.

$$-7 \leq 2x + 5 \quad and \quad 2x + 5 \leq 7$$

$$-12 \leq 2x \qquad and \quad 2x \leq 2$$

$$-6 \leq x \qquad and \quad x \leq 1$$

$$\text{Thus, } -6 \leq x \leq 1.$$

Geometric Method: A geometric solution can also be given to these inequalities. To do so, we must use the fact that

$$\lvert ab \rvert = \lvert a \rvert \cdot \lvert b \rvert.$$

(See Exercise 35 on page 99.) This allows us to conclude, for example, that $\lvert 3x - 9 \rvert = \lvert 3 \rvert \cdot \lvert x - 3 \rvert = 3 \lvert x - 3 \rvert$.

a. $\lvert 3x - 9 \rvert > 4$

$3 \lvert x - 3 \rvert > 4$

$\lvert x - 3 \rvert > \dfrac{4}{3}$

This means the distance from x to 3 is greater than $\frac{4}{3}$, which gives the picture above. Thus, $x < \frac{5}{3}$ or $x > \frac{13}{3}$.

(Solution continues on the next page.)

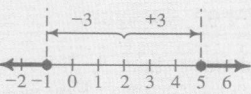

Problem Solving

Ask students to give an example of an absolute-value sentence that is always true, for example $|2x - 4| > -1$. Ask students to give an example of an absolute-value sentence that is never true, for example $|2x - 4| < -1$.

Error Analysis

Students tend to do exercises such as Written Exercise 13 quickly without paying attention to the real nature of the problem. The absolute value of a number is never less than -8 (or any other negative number). By comparison, for Exercise 14, since the absolute value of any number is always at least 0, the solution is "all numbers." Caution students to read problems carefully.

Suggested Assignments

Standard
98/1–31 odd
Comprehensive
98/1–30 every 3rd, 33–37

Supplementary Materials

Alternative Assessment, 9
Activities Book, 6–7
Precalculus Plotter Plus, 28

b.
$$|2x + 5| \le 7$$
$$2\left|x + \left(\frac{5}{2}\right)\right| \le 7$$
$$\left|x - \left(-\frac{5}{2}\right)\right| \le \frac{7}{2}$$

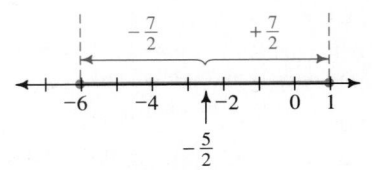

This means the distance from x to $-\frac{5}{2}$ is less than or equal to $\frac{7}{2}$, as illustrated above. Thus, $-6 \le x \le 1$.

CLASS EXERCISES

Solve each inequality.

1. $3x > -12$ $x > -4$

2. $-3x > -12$ $x < 4$

3. $1 - 2x < 11$ $x > -5$

4. $|x| < 2$ $-2 < x < 2$

5. $|x - 5| < 2$ $3 < x < 7$

6. $|x + 3| > 1$
$x < -4$ or $x > -2$

In Exercises 7–12, match the inequalities with graphs a–f.

7. $1 < x \le 4$ f

8. $4 \le x$ a

9. $x < 1$ e

10. $x < 1$ or $x \ge 4$ c

11. $6 \ge x \ge 4$ b

12. $6 \le x \le 4$ d

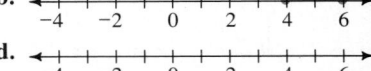

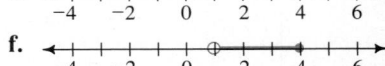

WRITTEN EXERCISES

In Exercises 1–24, solve the given equation or inequality and graph its solution. If there is no solution, say so.

A

1. $7x - 12 < 9$ $x < 3$

2. $8x + 6 > 30$ $x > 3$

3. $\dfrac{15 - 6x}{3} > 5$ $x < 0$

4. $\dfrac{8 - 11x}{4} \le 13$ $x \ge -4$

5. $\dfrac{1}{4}(x - 1) \le \dfrac{x + 4}{6}$ $x \le 11$

6. $\dfrac{2 - x}{3} < \dfrac{3 - 2x}{5}$ $x < -1$

7. $8x + 3(x + 1) > 5x - (9x - 6)$

8. $2x(6x - 1) \ge (3x - 2)(4x + 3)$

9. $\dfrac{x + 2}{4} - \dfrac{2 - x}{3} + \dfrac{4x - 5}{6} < 4$

10. $\dfrac{4}{3}\left(x - \dfrac{1}{2}\right) + \dfrac{1}{2}x \ge \dfrac{2}{3}\left(2x - \dfrac{5}{2}\right)$

11. $|x| < 3$ $-3 < x < 3$

12. $|x| \ge 5$ $x \le -5$ or $x \ge 5$

13. $|x| < -8$ No solution

14. $|x| > -2$ All real numbers

7. $x > \dfrac{1}{5}$ **8.** $x \le 2$ **9.** $x < 4$ **10.** $x \ge -2$

15. $|x - 4| < 3$ **16.** $|x - 7| > 3$
17. $|x + 7| \geq 3$ **18.** $|x + 2| < -1$
19. $|x - 8| = 4$ **20.** $|x + 3| = 7$
21. $|2x - 4| \leq 5$ **22.** $|3x - 9| \geq 9$
23. $|4x + 8| \leq 9$ **24.** $|6 - 3x| < 12$

B **25. Horticulture** Plant experts advise that gardenias kept indoors must have high humidity, plenty of sunlight during the day, and cool temperatures at night. The recommended nighttime temperature range in degrees Fahrenheit is $60° \leq F \leq 65°$. Given that $C = \frac{5}{9}(F - 32)$, express the corresponding temperature range in degrees Celsius. $15\frac{5}{9}° \leq C \leq 18\frac{1}{3}°$

26. a. Writing Use the geometric definition of $|x|$ and the definition of a combined inequality to write a sentence that gives the meaning of the expression $2 < |x| < 4$. See below.
  b. Solve the inequality $2 < |x| < 4$ and graph its solution. $-4 < x < -2$ or $2 < x < 4$
  c. Solve the inequality $1 \leq |x| < 5$ and graph its solution. $-5 < x \leq -1$ or $1 \leq x < 5$

In Exercises 27–32, solve the given inequality and graph its solution.

27. $1 \leq |x - 4| \leq 3$ $1 \leq x \leq 3$ or $5 \leq x \leq 7$ **28.** $2 < |x - 6| \leq 5$ $1 \leq x < 4$ or $8 < x \leq 11$
29. $0 \leq |x - 7| < 2$ $5 < x < 9$ **30.** $0 < |x + 3| < 1$ See below.
31. $\frac{1}{x} < 2$ (*Hint:* Consider two cases.) **32.** $\frac{1}{x - 1} > 4$ $1 < x < \frac{5}{4}$
31. $x < 0$ or $x > \frac{1}{2}$ **30.** $-4 < x < -3$ or $-3 < x < -2$

Use the following definition of $|x|$ to complete Exercises 33–37:

$$|x| = \begin{cases} x \text{ if } x \geq 0. \\ -x \text{ if } x < 0. \end{cases}$$

26. a. The distance from x to 0 is greater than 2 units, and the distance from x to 0 is less than 4 units.

Solve.

C **33.** $|x| + |x - 2| = 2$ $0 \leq x \leq 2$ **34.** $|x| + |x - 2| > 5$ $x < -\frac{3}{2}$ or $x > \frac{7}{2}$

35. Show that $|ab| = |a| \cdot |b|$ when (1) $a > 0$ and $b > 0$; (2) $a < 0$ and $b > 0$; (3) $a > 0$ and $b < 0$; and (4) $a < 0$ and $b < 0$.

36. a. Give three examples illustrating the triangle inequality: Answers will vary.
$$|a + b| \leq |a| + |b|$$
  b. Investigation In Chapter 11 we will define the absolute value of a complex number as follows: $|x + yi| = \sqrt{x^2 + y^2}$. Decide whether the triangle inequality holds if a and b are complex numbers. Yes

37. Use the triangle inequality stated in Exercise 36(a) to prove:
  a. $|a - b| \leq |a| + |b|$ **b.** $|a| - |b| \leq |a - b|$

Inequalities **99**

Teaching Notes, p. 94C

Warm-Up Exercises

The zeros of each polynomial function are given. Use the zeros as boundary values to find the intervals in which (a) $P(x) \geq 0$ and (b) $P(x) < 0$.

1. $P(x) = (x + 3)(x - 3)$; zeros: 3, −3
  a. $x \leq -3$; $x \geq 3$
  b. $-3 < x < 3$

2. $P(x) = (4 - 3x)(5 + x)$; zeros: $-5, \frac{4}{3}$

  a. $-5 \leq x \leq \frac{4}{3}$

  b. $x < -5$ or $x > \frac{4}{3}$

3. $P(x) = (x - 7)(x + 3) \cdot (x - 4)$; zeros: −3, 4, 7
  a. $-3 \leq x \leq 4$; $x \geq 7$
  b. $x < -3$; $4 < x < 7$

4. $P(x) = (x - 4)(2x + 1) \cdot (x - 2)^2$; zeros: $-\frac{1}{2}$, 2, 4

  a. $x \leq -\frac{1}{2}$; $x \geq 4$; $x = 2$

  b. $-\frac{1}{2} < x < 2$; $2 < x < 4$

5. $P(x) = x^2 - 4x + 4$; zero: 2 **a.** all real nos., **b.** none

Motivating the Section

Polynomial inequalities have real-life applications such as the following: Determine the width of a uniform cement strip around a rectangular swimming pool of known dimensions, say 30 ft by 40 ft, if enough material is purchased to cover 500 ft^2 of area. The width w of a strip must satisfy the inequality $(30 + 2w)(40 + 2w) - (30)(40) \leq 500$, or $4w^2 + 140w - 500 \leq 0$.

3-2 Polynomial Inequalities in One Variable

Objective   *To solve and graph polynomial inequalities in one variable.*

Let $P(x)$ be any polynomial. Then $P(x) < 0$ and $P(x) > 0$ are called **polynomial inequalities**. There are two ways to solve a polynomial inequality.

Method 1 Use a *sign graph* of $P(x)$.

This method was first introduced in Section 2-3 and is illustrated below in Examples 1–3.

Method 2 Analyze a graph of $P(x)$.

Note that $P(x) > 0$ when the graph is *above* the x-axis, and $P(x) < 0$ when the graph is *below* the x-axis.

Using a sign graph is an easy way to solve a polynomial inequality if the polynomial is factorable. Recall that to perform a sign analysis of a polynomial $P(x)$, you test one value of x from each of the intervals determined by the zeros of $P(x)$. Then you determine the sign of $P(x)$ in each of these intervals.

Example 1   Solve $x^3 - 2x^2 - 3x < 0$ by using a sign graph.

Solution

Step 1   Find zeros of the polynomial: $P(x) = x^3 - 2x^2 - 3x$
$$P(x) = x(x^2 - 2x - 3)$$
$$P(x) = x(x + 1)(x - 3)$$

The zeros are −1, 0, and 3.

Plot the zeros on a number line. Use open dots since $P(x)$ is *strictly less* than zero.

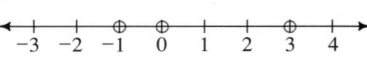

Step 2   Test $x = -2$, $x = -\frac{1}{2}$, $x = 1$, and $x = 4$, for example, to find the sign of $P(x)$ in each interval.

$$P(x) = x(x + 1)(x - 3)$$
$$P(-2) = -2(-2 + 1)(-2 - 3) = (-)(-)(-) = -$$
$$P\left(-\frac{1}{2}\right) = -\frac{1}{2}\left(-\frac{1}{2} + 1\right)\left(-\frac{1}{2} - 3\right) = (-)(+)(-) = +$$
$$P(1) = 1(1 + 1)(1 - 3) = (+)(+)(-) = -$$
$$P(4) = 4(4 + 1)(4 - 3) = (+)(+)(+) = +$$

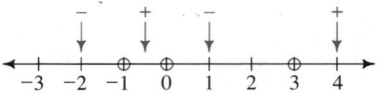

The solution of $P(x) < 0$ is $x < -1$ or $0 < x < 3$.

Notice that the polynomial in Example 1 changes signs at each of its zeros. Is this fact true for all polynomials? The following activity explores this question.

Activity

a. What is the sign of $(x - 1)^2$ for $x < 1$? positive for $x > 1$? positive The square of a nonzero

b. Explain why $(x - 1)^2$ does not change sign at $x = 1$. number is always positive.

c. What is the sign of the product $-2(x + 3)(x - 1)^2$ for $-3 < x < 1$? negative
What is the sign of the product $-2(x + 3)(x - 1)^2$ for $x > 1$? negative
Explain why this product should not change sign at $x = 1$. $(x - 1)$ is squared, and multiplication by a positive number does not change the sign of the product.

Not all polynomials change sign at a zero. A polynomial $P(x)$ will not change sign at a zero c if c corresponds to the squared factor $(x - c)^2$. Example 2 illustrates such a polynomial.

Example 2 Solve $(x^2 - 1)(x - 4)^2 \geq 0$.

Solution

Step 1 Find the zeros of the polynomial: $P(x) = (x^2 - 1)(x - 4)^2$
$$P(x) = (x + 1)(x - 1)(x - 4)^2$$
The zeros are -1, 1, and 4.

Since $P(x)$ is greater than or equal to zero, use solid dots to plot the zeros.

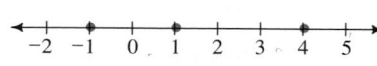

Step 2 Test $x = -2$, $x = 0$, $x = 2$, and $x = 5$, for example, to find the sign of $P(x)$ in each interval.
$$P(x) = (x + 1)(x - 1)(x - 4)^2$$
$$P(-2) = (-)(-)(-)^2 = +$$
$$P(0) = (+)(-)(-)^2 = -$$
$$P(2) = (+)(+)(-)^2 = +$$
$$P(5) = (+)(+)(+)^2 = +$$

The solution of $P(x) \geq 0$ is $x \leq -1$ or $x \geq 1$.

The technique used in Example 2 can be used to solve inequalities involving functions of the form $\dfrac{P(x)}{Q(x)}$ where $P(x)$ and $Q(x)$ are polynomials or products of polynomials.

Example Note

For Example 2, a chart like the one below can be used to analyze the sign of $P(x)$. In the chart, columns A, B, and C give the signs of the factors $x + 1$, $x - 1$, and $(x - 4)^2$, respectively.

	A	B	C	$P(x)$
$x < -1$	−	−	+	+
$-1 < x < 1$	+	−	+	−
$1 < x < 4$	+	+	+	+
$x > 4$	+	+	+	+

Additional Examples

1. Solve: $2x^2 + 3x - 5 < 0$.

Let $P(x) = 2x^2 + 3x - 5$.
Method 1: Factor and use a sign graph.
The zeros of $P(x) =$
$(2x + 5)(x - 1)$ are $-\dfrac{5}{2}$
and 1. Use open dots at
$-\dfrac{5}{2}$ and 1 since $P(x)$ is
strictly less than 0.

Test $x = -3$, $x = 0$, and
$x = 2$:
$$P(-3) = (-)(-) = +$$
$$P(0) = (+)(-) = -$$
$$P(2) = (+)(+) = +$$

$P(x) < 0$ if $-\dfrac{5}{2} < x < 1$.
(*Continues on next page*)

Method 2: Graph the parabola.

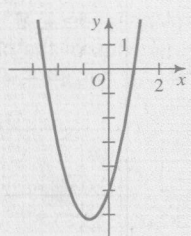

The x-intercepts are $-\frac{5}{2}$ and 1. $P(x) < 0$ when the graph is below the x-axis. The solution is $-\frac{5}{2} < x < 1$.

2. Solve: $x^4 - 4x^2 \geq 0$.

$x^2(x^2 - 4) \geq 0$

$x^2(x + 2)(x - 2) \geq 0$

When $x = 1$, $x^4 - 4x^2 = -3 < 0$. The polynomial changes sign at 2 and -2, but not at 0, since 0 is a double zero of $x^2(x + 2)(x - 2)$. Thus:

The solution is $x \leq -2$, $x = 0$, or $x \geq 2$.

Example 3 Solve $\dfrac{(x + 2)(x - 5)^2}{x - 4} \leq 0$.

Solution

Step 1 Plot the zeros of all linear factors occurring in a numerator or a denominator. In this case there are zeros at -2, 4, and 5. Since $f(x)$ can equal 0, use solid dots for -2 and 5. Since the denominator can never be zero, use an open dot for 4 to remind yourself to omit it from the solution.

Step 2 Do a sign analysis of $f(x) = \dfrac{(x + 2)(x - 5)^2}{x - 4}$ by testing convenient values in each of the intervals determined by the zeros above. For example, $f(0) = \dfrac{(+)(-)^2}{(-)} = -$.

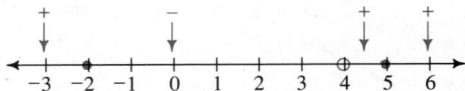

The solution of $f(x) \leq 0$ is $-2 \leq x < 4$ or $x = 5$.

 A quicker method for solving a polynomial inequality is to use a computer or a graphing calculator to graph the polynomial, and then analyze the graph, as shown in Example 4.

Example 4 Solve $2x^3 + x^2 - 8x + 3 > 0$ by using a computer or graphing calculator to sketch a graph.

Solution A graph of $P(x) = 2x^3 + x^2 - 8x + 3$ is shown below. Note that the zeros of P are about -2.4, 0.4, and 1.5. (Zeros can be found more accurately using a zoom feature.) The exact zeros, which can be found by using the techniques of Chapter 2, are 1.5 and $-1 \pm \sqrt{2}$.

$P(x) > 0$ when the graph is above the x-axis. Looking at the graph, we can see that $P(x) > 0$ when $-2.4 < x < 0.4$ or when $x > 1.5$.

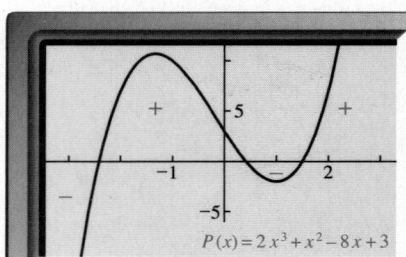

$P(x) = 2x^3 + x^2 - 8x + 3$

The solution is $-2.4 < x < 0.4$ or $x > 1.5$.

CLASS EXERCISES

In Exercises 1 and 2, use the given graph to solve each inequality.

1. $x^3 - 4x^2 - 4x + 16 > 0$

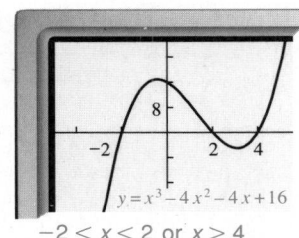

$y = x^3 - 4x^2 - 4x + 16$

$-2 < x < 2$ or $x > 4$

2. $x^3 + 5x^2 + 3x - 9 \leq 0$

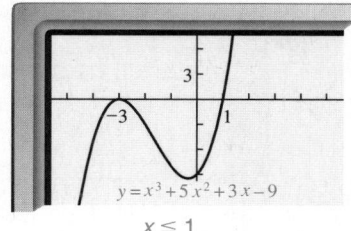

$y = x^3 + 5x^2 + 3x - 9$

$x \leq 1$

Use a sign graph to solve each inequality.

3. $(x - 1)(x - 2)(x - 3)^2(x - 4) > 0$ $\quad$ $1 < x < 2$ or $x > 4$

4. $\dfrac{(x - 2)(x - 7)}{x - 4} < 0$ $\quad x < 2$ or $4 < x < 7$

5. $(x^2 + 1)(x - 5) \leq 0$ $\quad x \leq 5$

6. $|x^2 - 2x| < 0$ $\quad$ No solution

3. $1 < x < 2$ or $x > 4$ $\quad$ **4.** $x < 1$ or $3 < x < 5$ $\quad$ **5.** $x < 4$, $x \neq 2$, $x \neq 3$ $\quad$ **6.** $-3 < x < -2$

WRITTEN EXERCISES

Solve each inequality. $\quad$ **14.** $x \leq -3$ or $0 \leq x \leq \dfrac{2}{3}$ $\quad$ **19.** $x < -2$ or $\dfrac{1}{2} < x < 1$

A

1. $(x - 3)(x + 4) > 0$ $\quad x < -4$ or $x > 3$

2. $(x + 7)(x + 9) > 0$ $\quad x < -9$ or $x > -7$

3. $(x - 1)(x - 2)(x - 4) > 0$ $\quad$ See above.

4. $(1 - x)(x - 3)(x - 5) > 0$

5. $(x - 4)(x - 2)^2(x - 3)^2 < 0$

6. $(2x - 5)^2(x + 3)(x + 2) < 0$

7. $x^2 - 2x - 15 < 0$ $\quad -3 < x < 5$

8. $x^2 + 3x - 18 > 0$ $\quad x < -6$ or $x > 3$

9. $2x^2 - x - 3 \geq 0$ $\quad x \leq -1$ or $x \geq 1.5$

10. $1 - 2x - 3x^2 < 0$ $\quad x < -1$ or $x > \dfrac{1}{3}$

11. $2x^2 + 5x - 7 \leq 0$ $\quad -3.5 \leq x \leq 1$

12. $x^2 - 8x + 16 \leq 0$ $\quad x = 4$

13. $x^4 - 3x^2 - 10 > 0$ $\quad x < -\sqrt{5}$ or $x > \sqrt{5}$

14. $3x^3 + 7x^2 - 6x \leq 0$ $\quad$ See above.

15. $a^3 + 2a^2 - 4a - 8 > 0$ $\quad a > 2$

16. $b^4 - 16 < 0$ $\quad -2 < b < 2$

17. $n^3 - 7n + 6 < 0$ $\quad n < -3$ or $1 < n < 2$

18. $2y^3 + 3y^2 - 1 \geq 0$ $\quad y = -\big\{$ or $y \geq 0.5$

19. $2x^3 + x^2 - 5x < -2$ $\quad$ See above.

20. $r^3 - 9r > 8r^2$ $\quad -1 < r < 0$ or $r > 9$

21. $y^4 + y^3 < 4(y + 4)$ $\quad -2 < y < 2$

22. $x^4 + 6 < 5x^2$ $\quad$ See below.

23. $4x^4 - 4x^3 - 3x^2 + 4x - 1 > 0$ $\quad x < -1$ or $x > 1$

24. $2y^3 + y^2 - 12y + 9 < 0$ $\quad y < -3$ or $1 < y < 1.5$

B

25. $\dfrac{(x - 3)(x - 4)}{(x - 5)(x - 6)^2} < 0$ $\quad x < 3$ or $4 < x < 5$

26. $\dfrac{(x + 1)(x - 3)^2}{(x - 5)^2} > 0$ $\quad$ See below.

27. $\dfrac{(2x - 5)^3}{x^2 - 3x - 28} \geq 0$ $\quad$ See below.

28. $\dfrac{(3n - 12)^2}{3n^2 - 12} \leq 0$ $\quad -2 < n < 2$ or $n = 4$

29. $\dfrac{2x^2 + 7x + 8}{x^2 + 1} > 0$ $\quad$ All real nos.

30. $\dfrac{n^2 + 4n + 4}{n^2 + 4n} > 0$ $\quad n < -4$ or $n > 0$

22. $-\sqrt{3} < x < -\sqrt{2}$ or $\sqrt{2} < x < \sqrt{3}$

27. $-4 < x \leq \dfrac{5}{2}$ or $x > 7$

26. $-1 < x < 3$, $3 < x < 5$, or $x > 5$

Inequalities $\quad$ **103**

Exercise Notes

In Class Exercise 6, students might find it helpful to sketch the graph of $y = |x^2 - 2x|$ in order to see that no part of the graph is below the x-axis.

The polynomials in Written Exercises 7–24 and 27–30 need to be factored first. In Written Exercises 19–22, students must rewrite the inequalities so that 0 is on one side before factoring.

Suggested Assignments

Standard
Day 1: 103/1–19 odd
Day 2: 103/21–41 odd
Comprehensive
103/25–44

Supplementary Materials

Alternative Assessment, 9
Using Technology, 13
Student Resource Guide, 24–27

Additional Answers
Written Exercises

31. $-2 < x < -\sqrt{3}$ or $x > \sqrt{3}$
$-2 < x < -1.73$ or $x > 1.73$

32. $x \leq -1$ or $-\dfrac{1}{4} \leq x \leq 2$
$x \leq -1$ or $-0.25 \leq x \leq 2$

33. $1 - \sqrt{2} \leq x \leq \dfrac{1}{2}$ or $x \geq 1 + \sqrt{2}$
$-0.41 \leq x \leq 0.50$ or $x \geq 2.41$

34. $-\sqrt{5} < x < -\dfrac{1}{2}$ or $\dfrac{2}{3} < x < \sqrt{5}$
$-2.24 < x < -0.50$ or $0.67 < x < 2.24$

31. $x^3 + 2x^2 - 3x - 6 > 0$

32. $4x^3 - 3x^2 - 9x - 2 \le 0$

33. $2x^3 - 5x^2 + 1 \ge 0$

34. $6x^4 - x^3 - 32x^2 + 5x + 10 < 0$

For Exercises 35 and 36, use the graphs to solve the inequality $P(x) > Q(x)$.

35.

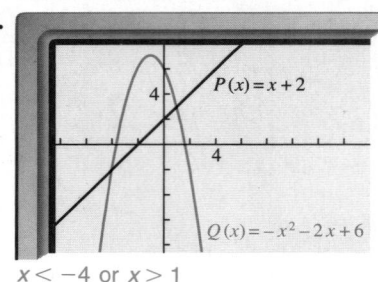

$x < -4$ or $x > 1$

36.

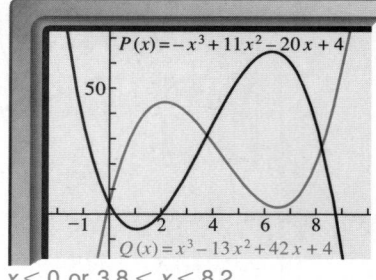

$x < 0$ or $3.8 < x < 8.2$

For Exercises 37–42, use a computer or a graphing calculator, and the method suggested by Exercises 35 and 36, to solve each inequality.

37. $x^2 + x + 14 > -x + 10$ All real nos.

38. $x^3 - 2x^2 - 2x - 3 \le |x - 1|$ $x \le 3.15$

39. $2 - x \le \sqrt{x + 1}$ $x \ge 0.70$

40. $x^2 < \sqrt{4x^2 + 5}$ $-2.24 < x < 2.24$

41. $\sqrt{x^2 - 9} \le \sqrt{25 - x^2}$
$-4.12 \le x \le -3$ or $3 \le x \le 4.12$

42. $\sqrt{3x + 7} > 3\sqrt{x} - 1$ $0 \le x < 2.62$

Solve each inequality.

C **43.** $\dfrac{|2x - 5|}{x^2 + 4} < 0$ No solution

44. $\dfrac{6x^2 + 13x - 5}{|5x^3 - 10|} < 0$ $-\dfrac{5}{2} < x < \dfrac{1}{3}$

■ Inequalities in Two Variables

3-3 Polynomial Inequalities in Two Variables

| *Objective* | To graph polynomial inequalities in two variables and to graph the solution set of a system of inequalities. |

Like solutions to equations in two variables, solutions to inequalities in two variables are ordered pairs of numbers. Thus, graphs of inequalities in two variables consist of points in the plane.

The graphs of linear and quadratic inequalities in two variables are closely related to the graphs of lines and parabolas, as illustrated on the next page. If a line has equation $y = mx + k$, any point (x, y) *above* the line satisfies the inequality $y > mx + k$. Any point (x, y) *below* the line $y = mx + k$ satisfies the inequality

---

Warm-Up Exercises

Consider the points $(3, -4)$, $(-3, 4)$, and $(-3, -4)$. For each of the following inequalities, two of the points lie in the graph of the inequality and the third does not. Which two points lie in the graph?

1. $y < -x$ $(3, -4)$; $(-3, -4)$

2. $y \ge 2(x + 1)$ $(-3, 4)$; $(-3, -4)$

3. $y > x^2 - x - 12$ $(3, -4)$; $(-3, 4)$

4. $y \le |x|$ $(3, -4)$; $(-3, -4)$

5. $2x + 5y \le -10$ $(3, -4)$; $(-3, -4)$

6. $y < 2x^2 - 7x - 4$ $(-3, 4)$; $(-3, -4)$

Motivating the Section

Imagine an ant traveling along the line $y = mx + k$ in the coordinate plane. As long as the ant stays on the line, the coordinates of its position satisfy the equation $y = mx + k$. If the ant moves off the line, however, the coordinates of its position will be such that $y \ne mx + k$. In this case, either $y > mx + k$ or $y < mx + k$, depending on which side of the line the ant is. Graphing such inequalities in two variables is the objective of this section.

$y < mx + k$. A similar statement can be made for points (x, y) above and below a parabola.

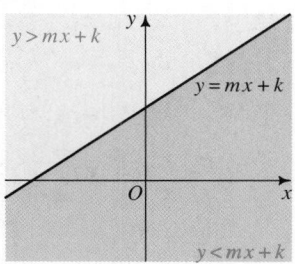

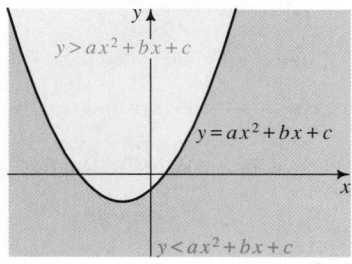

Example 1   Sketch the graph of $y > x^2 - 2x - 8$.

Solution

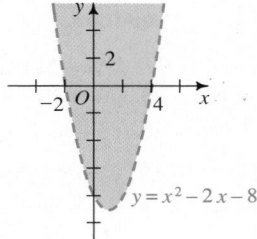

Method 1   Follow the approach suggested above. If a parabola has equation $y = x^2 - 2x - 8$, then any point (x, y) above the parabola satisfies the inequality. We show the parabola as a dashed curve since points on the parabola are not included in the solution.

Method 2   Draw the graph of $y = x^2 - 2x - 8$, which separates the plane into two regions. Choose a point in one region and check to see whether it satisfies the inequality. For example, the point $(0, 0)$ satisfies the inequality. Thus the region containing $(0, 0)$ is shaded as the graph of the inequality.

If you want to graph the solution of a system of inequalities, you find the intersection of the graphs of the individual inequalities.

Example 2   Graph the solution set of the inequalities $y \le 4 - x^2$ and $y > x + 2$.

Solution

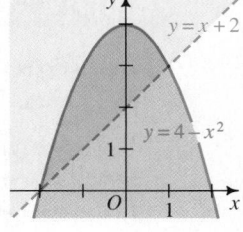

Step 1   The graph of $y \le 4 - x^2$ consists of points on or below the parabola $y = 4 - x^2$. This region is shaded in red.

Step 2   The graph of $y > x + 2$, shaded in blue, consists of points above the line $y = x + 2$. Because the points of the line are not included in the graph, you use a dashed line.

Step 3   The intersection of the blue and red regions is the graph of the solution set of the system of inequalities.

Inequalities   **105**

Assessment Note

To check students' understanding of where to shade when graphing a linear inequality, ask students the following questions:

1. Is the graph of $y \ge 2x - 1$ above or below the line $y = 2x - 1$? (above)

2. Is the graph of $2x - 1 \ge y$ above or below the line $y = 2x - 1$? (below)

Additional Examples

1. Sketch the graph of:
   a. $|x + 1| > 3$
   b. $y \le x^3 + 2x^2$

   a. $|x + 1| > 3$ means $x + 1 < -3$ or $x + 1 > 3$; $x < -4$ or $x > 2$. Graph $x = -4$ and $x = 2$ as dashed lines and shade the two appropriate regions.

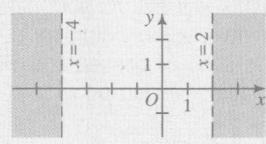

   b. Begin by graphing $y = x^3 + 2x^2 = x^2(x + 2)$. Note that $x^2(x + 2)$ has zeros -2 and 0 (a double zero). Then shade the region below the graph since $y \le x^3 + 2x^2$.

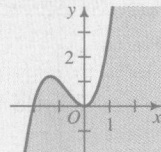

(Continues on next page)

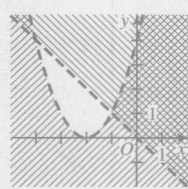
If an inequality contains just one variable, you must be aware of whether you are working just on the number line or in the whole coordinate plane. The two possibilities for the inequality $x \geq 3$ are shown below.

Problem   Find all x such that $x \geq 3$.   **Problem**   Find all (x, y) such that $x \geq 3$.

Solution 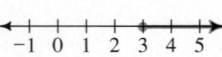   **Solution**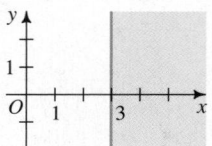

8. $x \geq 0$ **9.** $0 \leq y \leq 3$ **10.** $y \geq x^2$
$\quad\ y \geq 0$ $\quad\ x \geq 0$ $\quad\ \ y < x + 2$
$\quad\ y \leq -x + 2$ $\quad\ y \leq 4 - 2x$

CLASS EXERCISES

In each of the following exercises, the coordinates of a point and an equation are given. Tell whether the point is above, on, or below the graph of the equation.

1. $(3, 4)$; $y = x$ Above   **2.** $(3, 6)$; $y = 2x + 1$ Below
3. $(1, -1)$; $y = 3x - 4$ On   **4.** $(0, 9)$; $y = x^2 - 8x + 9$ On
5. $(-1, 12)$; $y = 6 - 5x - x^2$ Above   **6.** $(2, 10)$; $y = x^3 + x^2 + x + 1$ Below

7. a. If a point (x, y) is in the first quadrant, what inequalities must be satisfied?
   b. What if (x, y) is in the second quadrant? third quadrant? fourth quadrant?
      a. $x > 0$, $y > 0$ **b.** $x < 0$, $y > 0$; $x < 0$, $y < 0$; $x > 0$, $y < 0$

Give a set of inequalities that defines the shaded regions. See above.

8. 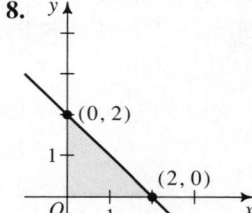   **9.** 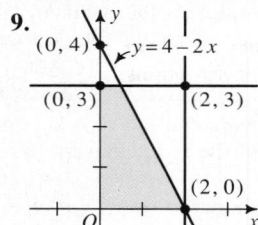   **10.**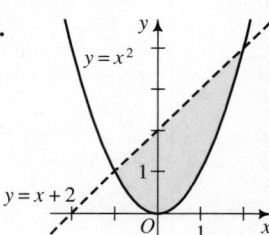

WRITTEN EXERCISES

In Exercises 1–16, sketch the graph of the given inequality.

A   **1.** $y \geq x$ **2.** $y \leq 2x - 1$
    3. $3x + 4y < 12$ **4.** $4 - 2x < 3y$
    5. $y \leq x^2$ **6.** $y \geq x^2 + 4x + 8$
    7. $y < 2x^2 - 4x + 1$ **8.** $y > 3x^2 + 6x - 2$
    9. $0 \leq x \leq 2$ **10.** $-1 \leq y \leq 3$

106   *Chapter Three*

11. $|y| > 1$ **12.** $|x| \le 2$ **13.** $|x - 3| < 2$

14. $2 < |x + 4| < 3$ **15.** $y > x^3 - 9x$ **16.** $y < x^4 - 5x^2 + 4$

In Exercises 17–30, graph the solution set of the given system of inequalities.

17. $x \ge 0$
  $x + 2y \le 4$

18. $y \le 0$
  $2x + y \le 4$

19. $x < 0$
  $3x - 2y \le -6$

20. $y < 0$
  $x - y > -1$

21. $y \ge x^2 - 2$
  $y < x$

22. $y \le 6 - x^2$
  $2x - y \le -3$

B **23.** $y \le x^3 + 4$
  $y > x^2 + x - 6$

24. $y \le x^2 + x - 2$
  $y \le -x^2 + x + 12$

25. $0 \le x \le 3$
  $0 \le y \le 2$
  $y \le 2 - x$

26. $-1 \le x \le 4$
  $-2 \le y \le 5$
  $5y \ge x + 11$

27. $|x| < 3$
  $|y| < 1$

28. $|x| \ge 2$
  $|y| \le 4$

29. $1 \le |x - 4| \le 3$
  $1 \le |y - 4| \le 3$

30. $1 \le |x + 3| \le 3$
  $1 \le |y + 4| \le 4$

31. Geology Over the years, Yellowstone National Park rangers have compiled data on the eruption of the Old Faithful geyser. The graph below shows the results of the data. For example, if the eruption lasts 3 min, the next eruption will most likely occur sometime between 60 min and 80 min later. Eruptions lasting less than 1.5 min or more than 5.5 min are rare. Describe the shaded region in the graph by a set of inequalities.

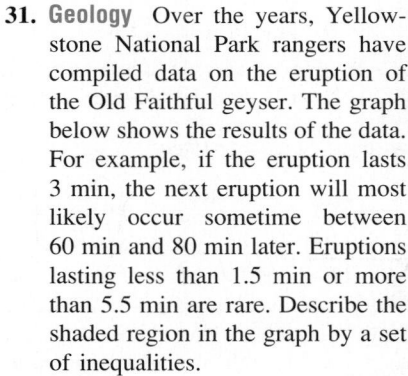

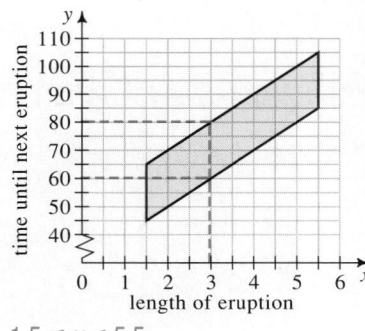

$1.5 \le x \le 5.5$
$10x + 30 \le y \le 10x + 50$

Students should recognize that many of the inequalities given in the Written Exercises are *combined* inequalities, which will require graphing the intersection or union of the associated simple inequalities. For example, Exercise 10 involves graphing the intersection of $y \ge -1$ and $y \le 3$, while Exercise 11 involves graphing the union of $y < -1$ and $y > 1$.

Making Connections

Labeling the corner points for the intersection of a set of inequalities (see Exercises 33 and 34) is important in linear programming applications (see Section 3-4).

32. Manufacturing A manufacturer makes rods and washers for machines. The rods have an outside diameter of $x = 2.00 \pm 0.10$ cm. The washers have an inside diameter of $y = 2.08 \pm 0.05$ cm.

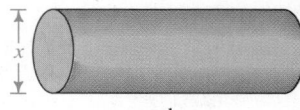

rod

a. Make a graph showing the points (x, y) that give the possible dimensions for x and y.

b. In order for a washer to fit on a rod, y must be larger than x. Shade a subset of your graph in part (a) to show the points (x, y) that satisfy this condition.

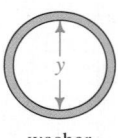

washer

33. On one set of axes, graph each of the following inequalities. Then shade the region that satisfies all the inequalities. Label each corner point of this region with its coordinates.

(1) $y \geq x$ (2) $x \geq 3$ (3) $y \leq 8$ (4) $x + 2y \geq 16$

34. Repeat Exercise 33 for the following inequalities:

(1) $x \geq 0$ (2) $y \geq 3$ (3) $x + y \leq 12$ (4) $3y - x \leq 12$

Sketch the graph of each equation or inequality.

C 35. $|x| + |y| = 2$ **36.** $|2x - 3y| < 6$

3-4 Linear Programming

Objective   *To solve certain applied problems using linear programming.*

Running a profitable business requires a careful balancing of resources (for example, people's time, materials, and machine availability). A manager must choose the best use of these resources. Often the range of possible choices can be described by a set of linear inequalities called *constraints*. In most situations the number of alternative solutions to the constraints is so great that it is hard to find the best one. *Linear programming*, a relatively new branch of mathematics, provides mathematical methods for finding the best solution.

Linear programming was developed by mathematician George Dantzig during World War II to improve large-scale military planning. Around 1950, industry began using these methods for practical planning and scheduling. Today, by using linear programming methods on computers, many otherwise impossible decision-making problems can be solved, such as the routing of millions of long-distance calls over an immense network of telephone lines.

Teaching Notes, p. 94D

Warm-Up Exercises

Consider the points $A(0, 3)$, $B(2, 2)$, $C(3, 1)$, and $D(4, 0)$. Find the value of the given expression at each of the given points.

1. $3x + 4y$ 12; 14; 13; 12

2. $5x + 4y$ 12; 18; 19; 20

3. $50x + 200y$ 600; 500; 350; 200

4. $6.5x + 2y$ 6; 17; 21.5; 26

Motivating the Section

Linear programming is a fairly new branch of mathematics that has many applications in the business world. It is used as a method for determining the maximum profit or minimum cost under a variety of (linear) constraints. Students will get a good introduction to applications of linear programming in this section.

Suppose that a small TV manufacturing company produces console and portable TV's using three different machines, A, B, and C. The table below shows how many hours are required on each machine per day in order to produce a console TV or a portable TV.

Machine	Console	Portable	Hours available
A	1 h	2 h	16
B	1 h	1 h	9
C	4 h	1 h	24

These requirements for the different machines can be described by the following inequalities, where x represents the number of console TV's and y represents the number of portable TV's:

a. $x \geq 0$
b. $y \geq 0$ } The number of TV's cannot be negative.

c. $x + 2y \leq 16$ { Machine A needs 1 hour for each console TV and 2 hours for each portable. Thus, for x console TV's and y portable TV's, Machine A needs $x + 2y$ hours. Since this machine is available for at most 16 hours a day, $x + 2y \leq 16$.

d. $x + y \leq 9$
e. $4x + y \leq 24$ } The last two inequalities are similar to the one above. Machines B and C are available for at most 9 hours and 24 hours, respectively.

The possible solutions to any problem subject to these constraints would have to satisfy all five inequalities. The easiest way to find these values is to find the region that is formed by the graphs of the system of inequalities. A step-by-step procedure is shown below.

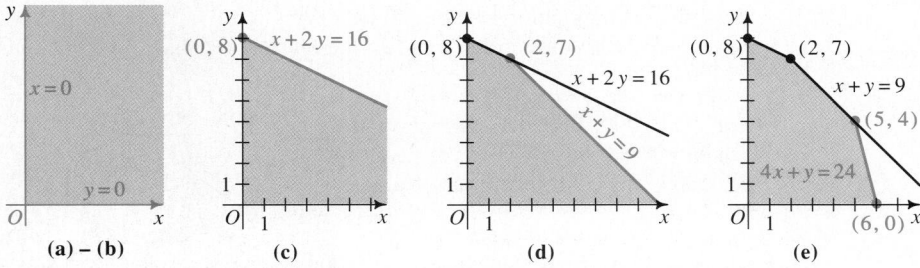

Any point (x, y) in graph (e) is called a *feasible solution*. The set of these points is called the *feasible region*.

Communication Note

Have students use a dictionary to find the meaning of any unfamiliar words in this section, such as *constraint* and *feasible*.

Problem Solving

Encourage students to organize data into charts, like the one shown on this page.

Review Note

You may need to review finding slopes and y-intercepts of lines before discussing "profit lines" on this page. Since the equation $P = 60x + 40y$ can be written as $y = -\dfrac{3}{2}x + \dfrac{P}{40}$, the profit equation determines a family of parallel lines whose y-intercepts depend on P.

▨ Using Technology

There is software available that graphs linear programming models. Such software can be used to find the value of $P = Ax + By$ at any point in a feasible region.

Communication Note

Have a student explain the corner-point principle to the class. Be sure students understand that any point of the feasible region satisfies the constraints, and that the maximum or minimum values of $P = Ax + By$ occur at the vertices of the region.

Maximizing a Profit. **Suppose that the manufacturing company described on the previous page makes a $60 profit on each console TV and a $40 profit on each portable TV. How many console TV's and how many portable TV's should be produced each day to maximize the profit?** The profit for x console TV's and y portable TV's is:

$$P = 60x + 40y \text{ (dollars)}$$

Since there are so many feasible solutions, it would be undesirable to evaluate $P = 60x + 40y$ at all feasible solutions in order to determine which one gives the greatest profit. However, we can narrow the search by observing what happens when we plot several *profit lines* on the graph of the feasible region, as shown at the right. We can see that as the amount of profit increases, the corresponding profit lines move farther away from the origin. We have to exclude any profit value greater than 460 since these profit lines contain no points that lie in the feasible region. Thus, the "greatest-profit line" containing points in the feasible region is the line $60x + 40y = 460$. The only feasible solution on this line occurs at $(5, 4)$, which is a vertex, or *corner point*, of the feasible region. Therefore, the company will make the maximum profit of $460 when 5 console TV's and 4 portable TV's are produced each day.

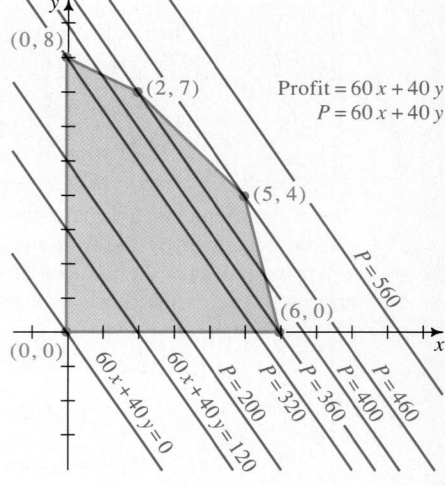

Notice that the "least-profit line" goes through the corner point $(0, 0)$. In general, the greatest or least value of a linear expression (profit or cost, to name two) will always occur at a corner point of the feasible region.

The Corner-Point Principle

A maximum or minimum value of a linear expression $P = Ax + By$, if it exists, will occur at a corner point of the feasible region.

The corner-point principle applies only to *convex* polygonal regions. A polygonal region is convex if every line segment joining two points of the region is contained in the region. Regions (a) and (b) are convex, but region (c) is not.

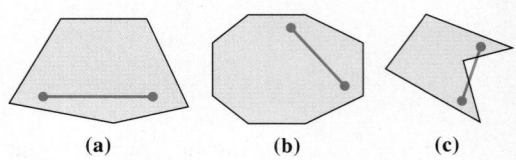

(a)   (b)   (c)

Because of the corner-point principle, we need to evaluate a specific expression only at the corner points of the feasible region in order to find an optimal value, as the example below shows.

Example   *Minimizing a Cost.* Every day Rhonda Miller needs a dietary supplement of 4 mg of vitamin A, 11 mg of vitamin B, and 100 mg of vitamin C. Either of two brands of vitamin pills can be used: Brand X at 6¢ a pill or Brand Y at 8¢ a pill. The chart below shows that a Brand X pill supplies 2 mg of vitamin A, 3 mg of vitamin B, and 25 mg of vitamin C. Likewise, a Brand Y pill supplies 1, 4, and 50 mg of vitamins A, B, and C, respectively. How many pills of each brand should she take each day in order to satisfy the minimum daily need most economically?

	Brand X	Brand Y	Minimum daily need
Vitamin A	2 mg	1 mg	4 mg
Vitamin B	3 mg	4 mg	11 mg
Vitamin C	25 mg	50 mg	100 mg
Cost per pill	6¢	8¢	

Solution   Let x = number of Brand X pills, and y = number of Brand Y pills.

Step 1   Set up a system of inequalities that describe the constraints.

$$x \geq 0$$
$$y \geq 0$$
$$2x + y \geq 4$$
$$3x + 4y \geq 11$$
$$x + 2y \geq 4$$

Step 2   Determine the feasible region by graphing the system of inequalities, as shown at the right above.

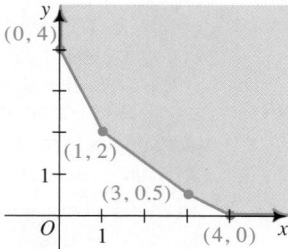

Step 3   The cost (in cents) can be expressed by the equation $C = 6x + 8y$. Evaluate the cost at each corner point.

Corner point	Value of $6x + 8y$
(0, 4)	$6(0) + 8(4) = 32¢$
(1, 2)	$6(1) + 8(2) = 22¢$ } Minimum
(3, 0.5)	$6(3) + 8(0.5) = 22¢$ } cost
(4, 0)	$6(4) + 8(0) = 24¢$

The minimum cost occurs at both (1, 2) and (3, 0.5). It is, however, inconvenient to take half a pill. Therefore, the best choice is 1 Brand X pill and 2 Brand Y pills.

Inequalities   **111**

Mathematical Note

The example on this page illustrates that sometimes the feasible region can be unbounded, yet a minimum value can still be obtained.

Additional Examples

1. A convex polygonal region has vertices (0, 8), (2, 5), (4, 3), and (6, 0). Minimize $C = 4x + 3y$.

Point	$4x + 3y$
(0, 8)	24
(2, 5)	23
(4, 3)	25
(6, 0)	24

C is minimized when $x = 2$ and $y = 5$.

2. A factory produces short-sleeved and long-sleeved shirts. A short-sleeved shirt requires 30 minutes of labor, a long-sleeved shirt requires 45 minutes of labor, and 240 hours of labor are available per day. The maximum number of shirts that can be packaged in a day is 400, so no more than 400 shirts should be produced.
   a. If the profits on a short-sleeved shirt and a long-sleeved shirt are $11 and $16, respectively, find the maximum possible daily profit.
   b. Explain why the factory manager might decide to produce different numbers of shirts than the ones giving maximum profit.

(*Continues on next page*)

a. Graph the following system:

$x \geq 0; \ y \geq 0;$

$\frac{1}{2}x + \frac{3}{4}y \leq 240;$

$x + y \leq 400$

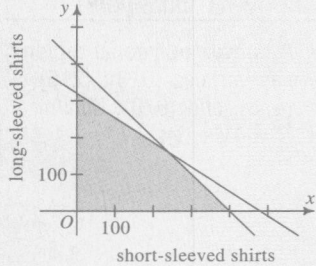

short-sleeved shirts

$P = 11x + 16y$
$(400, 0): P = \$4400$
$(240, 160): P = \$5200$
$(0, 320): P = \$5120$
The maximum daily profit is $5200.

b. The method used assumes that all the shirts produced can be sold. If this is not true, for example, if long-sleeved shirts become unpopular, the manager might decide to increase production of short-sleeved shirts and decrease production of long-sleeved shirts.

Suggested Assignments

Standard
Day 1: 112/1–6
Day 2: 113/7–12
Comprehensive
112/1–13 odd

WRITTEN EXERCISES

A 1. **a.** Refer to the example on page 110 and suppose each console TV manufactured produces an $80 profit and each portable TV manufactured produces a $55 profit. How many of each kind of TV should be manufactured to maximize the profit? 5 consoles, 4 portables

b. Suppose the profit on a console TV is $90 but is only $20 on a portable TV. Show that the maximum profit is achieved by producing only console TV's.

2. **a.** *Writing* Refer to the example on page 111. For each inequality listed in Step 1 of the solution, write a sentence that explains in your own words what the inequality means.

b. Suppose each Brand X pill costs 3¢ and each Brand Y pill costs 9¢. What combination of pills should Rhonda take to minimize the cost? 4 Brand X

c. Suppose each Brand X pill costs 10¢ and each Brand Y pill costs 4¢. Show that the cost is minimized by taking only Brand Y pills.

3. Suppose the data of Exercise 1 are changed as shown below. How many of each kind of TV should be produced to maximize the profit?
5 consoles and 3 portables

	Console	Portable	Total hours available per day
Machine A	1	3	18
Machine B	1	1	8
Machine C	3	1	18
Profit per TV	$70	$40	

4. Suppose the data of Exercise 2 are changed as shown below. How many of each type of pill should be taken daily to keep the cost at a minimum?
3 Brand X and 1 Brand Y

	Brand X	Brand Y	Minimum daily need
Vitamin A	4 mg	2 mg	10 mg
Vitamin B	6 mg	6 mg	24 mg
Vitamin C	25 mg	50 mg	125 mg
Cost per pill	12¢	15¢	

In Exercises 5–8, find the point or points (x, y) that satisfy the given constraints and give the stated objective.

5. Given: $x + 2y \leq 4$
   $x + y \leq 3$
   $x \geq 0, \ y \geq 0$
   Maximize $P = x + 3y$. (0, 2)

6. Given: $x + 2y \geq 6$
   $x + 3y \geq 7$
   $x \geq 0, \ y \geq 0$
   Minimize $C = 3x + 4y$. (0, 3)

7. Given: $2x + y \geq 12$
$x + y \geq 10$
$x + 3y \geq 16$
$x \geq 0,\ y \geq 0$

Minimize: **a.** $C = 5x + 6y$ (7, 3)
           b. $C = 5x + 5y$ (2, 8), (7, 3)

8. Given: $3x + 4y \leq 36$
$3x + 2y \leq 24$
$x \leq 6$
$x \geq 0,\ y \geq 0$

Maximize: **a.** $P = 3x + 3y$ (4, 6)
           b. $P = 3x + 4y$ (0, 9), (4, 6)

B **9. Agriculture** A farmer can make a profit of $250 for every ton of alfalfa harvested and $350 for every ton of corn harvested. Corn requires 3 h per ton to harvest while alfalfa requires only 2 h per ton. Each crop requires 1 h per ton for planting. The planting time available is 500 h and the harvesting time available is 1200 h. Organize the given information in a table.

	Alfalfa	Corn	Hours available
Planting time (hours/ton)	? 1	? 1	? 500
Harvest time (hours/ton)	? 2	? 3	? 1200
Profit (dollars/ton)	?250	? 350	

Let x and y represent the tons of alfalfa and corn to be grown, and write the inequalities that must be satisfied. How much of each crop should the farmer grow in order to maximize the profit? 300 tons alfalfa, 200 tons corn

10. Manufacturing A manufacturer of stereo speakers makes two kinds of speakers, an economy model that sells for $50 and a deluxe model that sells for $200. The deluxe model uses 1 woofer (a low frequency range speaker), 2 tweeters (high frequency range speakers) and 1 mid-frequency range speaker. The economy model uses 1 tweeter, 1 mid-frequency range speaker, and no woofers. The manufacturer's current inventory consists of 20 woofers, 45 tweeters, and 35 mid-frequency range speakers. Organize the given information in a table.

	Economy	Deluxe	Number available
Number of woofers	? 0	? 1	? 20
Number of tweeters	? 1	? 2	? 45
Number of mid-ranges	? 1	? 1	? 35
Income per model	? $50	? $200	

Let x and y represent the number of economy and deluxe models to be produced, and write the inequalities that must be satisfied. How many of each model should be manufactured to maximize the income from their sale?
5 economy, 20 deluxe

Supplementary Materials

Alternative Assessment, 10
Precalculus Plotter Plus, 29–30
Student Resource Guide, 28–30

Additional Answers
Written Exercises

1. b. $P = 90x + 20y$; at (0, 0), $P = 0$; at (0, 8), $P = 160$; at (2, 7), $P = 320$; at (5, 4), $P = 530$; at (6, 0), $P = 540$. The maximum profit, $540, occurs with 6 consoles and 0 portables.

2. a. The number of pills of either brand cannot be negative. The combined amount of vitamin A must be greater than or equal to 4 mg. The combined amount of vitamin B must be greater than or equal to 11 mg. The combined amount of vitamin C must be greater than or equal to 100 mg.

  c. $C = 10x + 4y$; at (0, 4), $C = 16$; at (1, 2), $C = 18$; at (3, 0.5), $C = 32$; at (4, 0), $C = 40$. The minimum cost, 16 cents, occurs with 0 Brand X pills and 4 Brand Y pills.

Cooperative Learning

Students generally find the B-level exercises difficult to do by themselves. Consider giving them class time to work in groups so that they can cooperatively determine the function to be maximized or minimized, the feasible region, and the solution of the problem.

**Additional Answers
Written Exercises**

11. b. A maximum profit of
$960 occurs for any of
the following: 12 racing
and 30 free-style, 14
racing and 27 free-style,
16 racing and 24 free-
style, 18 racing and 21
free-style, 20 racing and
18 free-style.

Supplementary Materials

Tests, 5–6, 7–8
Alternative Assessment, 64
Student Resource Guide,
31–32

11. a. Manufacturing A sporting goods company manufactures two types of skis: a racing model and a
free-style model. Each pair of racing skis requires
3 h of labor, and the company can produce at
most 20 pairs of racing skis per day. Each pair of
free-style skis requires 2 h of labor, and the company can produce at most 30 pairs of free-style
skis per day. The maximum number of hours of
labor available for ski production is 96. The profit
on each pair of racing skis is $30 and the profit on
each pair of free-style skis is $40. Find the num-
ber of each that should be manufactured in order
to maximize profits.

   b. Suppose the profit on each pair of free-style skis
is $20. Show that there is more than one way to
maximize profits.

12. Manufacturing A game company is developing two new games, a board game
and a mechanical game. Each board game requires $\frac{1}{2}$ h to manufacture, $\frac{1}{2}$ h to
assemble, and $\frac{1}{4}$ h to inspect and package. Each mechanical game requires 1 h
to manufacture, $\frac{1}{2}$ h to assemble, and $\frac{1}{2}$ h to inspect and package. In a given
week there are 40 h available for manufacturing, 32 h for assembly, and 18 h
for inspection and packaging. Suppose the profit on each board game is $10
and the profit on each mechanical game is $15. How many of each game
should be produced each week for maximum profit?   56 board games,
8 mechanical games

C **13.** Let x and y be positive integers such that $x + y \le 13$ and $3x + y \le 24$. Find
the maximum value of: **a.** $x + y$ 13   **b.** $4x + y$ 32   **c.** $2x + y$ 18

Chapter Summary

1. The rules for solving *linear inequalities* are similar to the rules for solving
linear equations and are summarized on page 95. The graph of a linear inequality in one variable consists of points on a number line.

2. Equations and inequalities involving $|x - k|$ can be solved in two ways:
  a. By interpreting $|x - k|$ as the distance from x to k on the number line. (See
the examples given in the chart on page 96.)
  b. By using an equivalent sentence and then solving it algebraically. (See the
set of equivalent sentences given in the chart on page 97.)

3. The solution to a *polynomial inequality* $P(x) > 0$ (or $P(x) < 0$) consists of
those values of x for which the graph of the polynomial $y = P(x)$ is above (or
below) the x-axis. As an alternative to analyzing the graph of the polynomial
$P(x)$, you can analyze a sign graph of the factors of $P(x)$ to solve the polynomial inequality. These methods are illustrated in the examples of Section 3–2.

114 *Chapter Three*

4. Solutions to *inequalities in two variables* are ordered pairs of real numbers. Their graphs consist of points in the coordinate plane. (See Example 1 on page 105.) The graph of the solution of a system of inequalities is the intersection of the graphs of the individual inequalities. (See Example 2 on page 105.)

5. *Linear programming* is a method for solving certain decision-making problems where a linear expression is to be maximized or minimized. This method uses the *corner-point principle*, which is stated on page 110.

Key vocabulary and ideas

linear inequality (p. 95)
absolute value (p. 96)
polynomial inequality (p. 100)
sign graph (p. 100)

inequality in two variables (p. 104)
system of inequalities (p. 105)
linear programming (p. 108)
feasible region (p. 109)

Chapter Test

Solve and graph each inequality on a number line.

1. **a.** $3(8 - 4x) < 9 - 7x$ $x > 3$ **b.** $\frac{1}{3}(x - 3) \leq \frac{x - 2}{4}$ $x \leq 6$ **3-1**

2. **a.** $|x - 3| < 5$ $-2 < x < 8$ **b.** $|x + 6| > 4$ $x < -10$ or $x > -2$
   c. $|3x + 4| = 4$ $x = 0$ or $x = -\frac{8}{3}$ **d.** $|7 - 5x| \leq 2$ $1 \leq x \leq \frac{9}{5}$

3. Solve each inequality. $x \leq -4$ or $x \geq \frac{1}{2}$ **3-2**
   a. $(2x - 1)(x + 4)(x - 3)^2 \geq 0$ **b.** $4x^2 + 5x - 6 < 0$ $-2 < x < \frac{3}{4}$
   c. $2x^3 - x^2 > x$ $-\frac{1}{2} < x < 0$ or $x > 1$ **d.** $\frac{(x - 6)(x + 1)}{(x - 3)^2} \leq 0$ See below.

4. **Writing** Describe the set of points in the coordinate plane that satisfies the inequality $y < 2x - 5$. The points below the line $y = 2x - 5$ **3-3**

5. Graph each inequality in the coordinate plane.   **3. d.** $-1 \leq x < 3$ or $3 < x \leq 6$
   a. $2 - 3y \geq 3x$ **b.** $|x| < 3$

6. Graph the solution set of each system.
   a. $x > 0$ **b.** $y \geq x^2 - 2x$
   $\quad 3x + 4y < 2$ $\quad y \leq 6 - x$

7. Give a set of inequalities that defines the shaded region shown at the right.

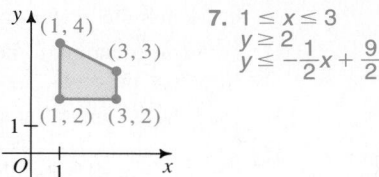

7. $1 \leq x \leq 3$
$y \geq 2$
$y \leq -\frac{1}{2}x + \frac{9}{2}$

8. Cars and trucks are made in a factory that is divided into two shops.   **3-4**
   Shop 1 performs the basic assembly operation, working 6 person-days on each truck and only 3 person-days on each car. Shop 2 performs finishing operations, working 4 person-days on each car or truck that it produces. Shop 1 has 150 person-days per week available, while Shop 2 has 120 person-days per week. The manufacturer makes a profit of $500 on each truck and $350 on each car. How many of each should be produced each week to maximize the profit? 20 trucks, 10 cars

1. a.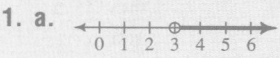

b.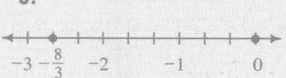

2. a.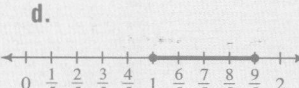

b.

c.

d.

6. a.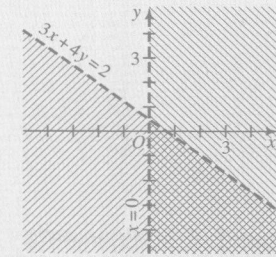

b.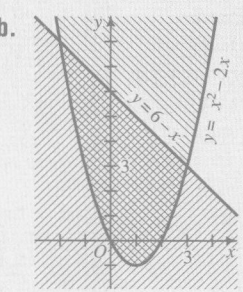

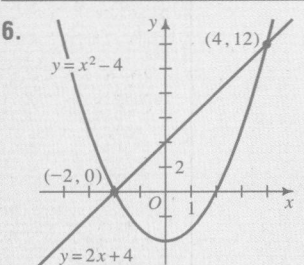
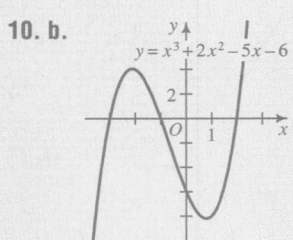
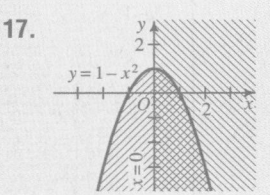
Cumulative Review

1. Let $A = (-4, 5)$ and $B = (-7, -2)$. Find the length of $\overline{AB}$ and the coordinates of the midpoint of $\overline{AB}$. $\sqrt{58}; \left(-\frac{11}{2}, \frac{3}{2}\right)$ **5. b.** $\frac{3 \pm \sqrt{2}}{2}$

2. Write an equation of the line described.
   a. The line through $(4, -2)$ and $(2, -4)$ $x - y = 6$
   b. The line with y-intercept 5 and parallel to $2x - 3y = 3$ $2x - 3y = -15$

3. Each mailing sent out by a charitable organization costs $1027. The average donation received as a result of the mailing is $12.50.
   a. Express the organization's profit as a function of the number of donations received. $P(d) = 12.5d - 1027$
   b. How many donations are needed to realize a profit of $500? 123

4. Simplify each expression. $_{-2i}$ $^{-12\sqrt{3}}$
   a. $\sqrt{-1} + \sqrt{-25} - \sqrt{-64}$ b. $\sqrt{-12}\sqrt{-36}$ c. $\frac{3-i}{6+2i}$ $\frac{2}{5} - \frac{3}{10}i$

5. Solve each equation. Give both real and imaginary roots.
   a. $x^2 - 6x = -18$ b. $4x^2 - 12x + 7 = 0$ c. $\sqrt{2x+1} = \sqrt{x-3} + 2$
   $3 \pm 3i$ See above.   4, 12

6. Sketch the graph of the line $y = 2x + 4$ and the parabola $y = x^2 - 4$. Find the coordinates of any points of intersection. $(-2, 0), (4, 12)$

7. Write an equation of the quadratic function f whose graph has x-intercepts 3 and 7 and $f(5) = 8$. $f(x) = -2x^2 + 20x - 42$

8. Write an equation of the form $f(x) = ax^2 + bx + c$ in **8.** $f(x) = \frac{1}{4}x^2 + x - 6$ which $f(0) = -6$, $f(2) = -3$, and $f(4) = 2$.

9. If $P(x) = 2x^3 - x^2 + 3$, use synthetic substitution to find: $4 - 2i$
   a. $P(2)$ 15   b. $P(-1)$ 0   c. $P\left(-\frac{3}{2}\right)$ -6 d. $P(i)$

10. a. One root of $P(x) = x^3 + 2x^2 - 5x - 6$ is $x = -1$. Find the other roots. 2, -3
    b. Sketch the graph of $y = x^3 + 2x^2 - 5x - 6$.

11. Write an equation of the cubic function whose graph is shown at the right. $y = \frac{1}{6}(x + 2)^2(x - 3)$

Ex. 11

12. Find the maximum area of a rectangle whose perimeter is 100cm. 625 cm^2

13. Use a computer or a graphing calculator to find the real roots of $x^3 - 3x - 1 = 0$ to the nearest tenth. $-1.5, -0.3, 1.9$

14. Find all real and imaginary roots of $x^3 - 3x^2 + 4x - 12 = 0$. $\pm 2i, 3$

15. Find a cubic equation with integral coefficients and roots $-\frac{1}{3}$ and $2 + 2i$. $3x^3 - 11x^2 + 20x + 8 = 0$

16. Solve and graph each inequality on a number line.
    a. $|x - 5| > 12$ $x < -7$ or $x > 17$ b. $(x - 2)^2(x + 3)(x - 1) \geq 0$ $x \leq -3$ or $x \geq 1$

17. Graph the solution set of the system of inequalities $x \geq 0$ and $y \leq 1 - x^2$.

Careers in Architecture and Urban Planning

A city is a complex and endlessly evolving world. People who design the buildings and plan the layout of communities shape the character of individual cities, and have a significant influence on the quality of urban life.

ARCHITECTURE

An architect must have a unique blend of talents, a strong sense of visual aesthetics, and the technical expertise needed to create well-constructed, functional buildings. With a strong background in mathematics an architect is able to design beautiful and efficient uses of three-dimensional space.

Minoru Yamasaki (1912–1986) began to think of a career in architecture when he saw his uncle's plans for the United States Embassy in Tokyo. Twenty-five years later, Yamasaki also designed a building for Japan. His commission for the American Consulate in Kobe helped build his growing reputation as an innovative architect.

Yamasaki, famous for designing New York City's World Trade Center, moved away from the starkness that characterized much of modern architecture, believing that good buildings have "a joyful quality," and "provide the delight of change and surprise." Many of his buildings won honors and awards.

Minoru Yamasaki

URBAN PLANNING

Architects often collaborate with urban planners who decide the types and locations of new buildings in a city. Urban planners, whose jobs may also include designing traffic control systems or helping businesses with management and marketing issues, base their recommendations on extensive analysis of population growth, economic trends, environmental impact, and transportation requirements.

Yvonne Goldsberry holds master's degrees in Urban Planning and Public Health from Columbia University. In 1987, using her experience as a planner of health care for the elderly, Ms. Goldsberry founded Aureus, Inc., a company designed to provide the skills of an urban planner to a variety of health and human service organizations, including hospitals, mental health centers, adoption agencies, nursing homes, and child care facilities.

Yvonne Goldsberry

Chapter 4 Functions

Overview

The concept of function is one of the most important ideas in mathematics and in the fields to which mathematics is applied. For example, the use of functions has been crucial in physics, where an understanding of velocity and acceleration as functions of time has resulted in an ability to determine the trajectories of objects—whether man-made or natural—in space.

In this chapter, students build on their knowledge of polynomial functions from Chapters 1 and 2. After learning a general definition of function, students combine functions in a variety of ways and explore the relationships between the graph of a function and an algebraic rule for the function. Students also consider functions of more than one variable, and they use functions to solve real-world problems.

Objectives

4-1 To identify a function, to determine the domain, range, and zeros of a function, and to graph a function.

4-2 To perform operations on functions and to determine the domains of the resulting functions.

4-3 To reflect graphs and to use symmetry to sketch graphs.

4-4 To determine periodicity and amplitude from graphs, to stretch and shrink graphs both vertically and horizontally, and to translate graphs.

4-5 To find the inverse of a function, if the inverse exists.

4-6 To graph functions of two variables in a two-dimensional coordinate system and to read such graphs.

4-7 To form a function of one variable from a verbal description and, when appropriate, to determine the minimum or maximum value of the function.

Supplementary Resources

Tests, pp. 9–10
Alt. Assess., pp. 10–14, 64
Activities, pp. 8–9
Using Tech., pp. 2–4, 14, 15–16
Student Res. Guide, pp. 33–45

Software

Precalculus Plotter Plus
Activities, pp. 31–32
Inverse Plotter

Pacing Guide

Section	4-1	4-2	4-3	4-4	4-5	4-6	4-7	Review	Test	Total
Standard (days)	1	2	2	2	2	2	2	1	1	15
Comprehensive (days)	1	2	2	2	2	2	2	1	1	15

Teaching Notes

4-1 | **pages 119–124**

Presenting the Section

This section introduces students to the vocabulary of functions. Point out the following facts: (1) the use of function notation helps to specify the independent variable (for example, $f(x) = mx + k$ gives more information than $y = mx + k$); (2) the letters used are immaterial (instead of $f(x) = mx + k$, we could have written $g(u) = mu + k$); and (3) the graph of a function is understood to have the independent variable on the horizontal axis and the dependent variable on the vertical axis.

Students need to learn to think both in terms of the graph of the function and the algebraic definition or rule for the function. The range of a function usually is determined by looking at its graph, whereas the domain is determined by the algebraic definition. Students are uneasy about specifying domains unless they are assured that problems arise only when a function has a denominator (which must not be zero) or an even root (the radicand must not be negative).

When students are asked to determine whether a given equation in two variables specifies a function, they may refer either to the graph (the vertical line test) or to the equation (whether there is ever more than one y-value for some x-value).

You may need to remind students that the symbol $\sqrt{x}$ refers to the positive number whose square is x. You may challenge them to define the function whose graph is the top (or bottom) half of the graph of $x = y^2$. (This skill is also needed when graphing such equations on graphing calculators.)

Using Technology

Exercise 10b, page 123, may be your students' first encounter with $\frac{0}{0}$, since substituting -2 for t gives zero in both the numerator and the denominator. Ask students if the functions f and g of Exercise 10 are the same function and how their graphs are related. See if your graphing calculator or computer has the ability to distinguish the two graphs: the graph of g is the same as the graph of f except that the graph of g does not contain $(-2, 1)$.

Exercises 17 and 18 may be students' first encounter with a function whose rule varies depending on the domain. After students have graphed these functions with pencil and paper, challenge them to graph the functions using a computer or graphing calculator.

4-2 | **pages 124–131**

Presenting the Section

Your students' mathematical experience so far has been limited to the four operations of addition, subtraction, multiplication, and division, as they are performed on numbers. In this section, students learn that the four arithmetic operations also can be performed on functions, and that there is an entirely new operation for functions called *composition*. Some confusion may arise because of the parentheses in the expression $f(g(x))$. Although students know that the parentheses do not mean multiplication in $f(x)$, they may revert to multiplication when they try to compute $f(g(x))$.

Continue to emphasize that composition of functions is a form of substitution. It is a big conceptual leap to go from finding $f(2)$ if $f(x) = x^2 + 1$ by substituting 2 for x, to finding $f(a)$ by substituting a for x, to finding $f(x - 3)$ by substituting $x - 3$ for x. Students have not had to substitute an expression in x for x before, and they will need a great deal of practice before they are comfortable with the process. You may find a difference in the speed with which students comprehend this new operation, and those who catch on quickly may become impatient with those who do not. More examples will help those students having difficulty with the concept of composition of functions.

Another new concept in this section is the addition and subtraction of graphs. Emphasize that for a given value of x, you are to add or subtract the corresponding y-values of the two functions.

Using Technology

If you are using a graphing calculator, familiarize your students with its capacity for adding, subtracting, multiplying, and dividing functions. Does it provide for composing functions? What are the capabilities of your computer software?

Applications

Exercises 27 and 28, pages 129 and 130, are an introduction to finding the composite of two functions when only their graphs are given. Other applied problems in the section require the substitution of variables represented by different letters and thus are easier.

4-3 | pages 131–138

Presenting the Section

The material in this section is introduced in the Activities, page 131. By working through a sufficient number of examples, your students can discover the properties for themselves.

Cooperative Learning

Students seem to enjoy drawing graphs on a chalkboard or overhead transparency sheet, using different colors. An effective way to introduce the symmetries of graphs is to create groups of two or three students, asking each group to graph pairs of functions using different colors. Some students may check the various results with a graphing calculator. Begin these activities using quadratic functions. Absolute-value graphs are a challenge even when the original function is linear. Ask students to explain the patterns they see.

To show reflections in the y-axis, begin with linear functions such as $y = 3x + 1$ and $y = -3x + 1$. When students recognize the pattern, see if they can guess the equation of the reflection of the graph of $y = x^2 + 4x + 3$ in the y-axis.

To show the interchange of variables, you may have students graph simple quadratic equations in which a y-term is squared but not an x-term, such as $x = y^2$. Again, have groups graph this equation and $y = x^2$ in different colors. See if students can draw in a line of symmetry. You should progress to graphs of equations like $x = 4 + y^2$ and $x = 6y - y^2$.

As you discuss the point of symmetry of the graph of a cubic, you may wish to introduce the terms *concavity* and *inflection point*. A graph is concave up if it ''holds water'' and concave down if it ''sheds water,'' and it has an inflection point where it changes concavity.

Notice that Exercises 31–36, page 137, explore the properties of *odd* and *even* functions and that Exercise 37, page 138, introduces *increasing* and *decreasing* functions.

4-4 | pages 138–146

Presenting the Section

The concepts of *period* and *amplitude* are introduced in this section. Even though students may not have encountered an equation for a periodic function, they will understand the concept. Be sure students understand the difference between a period of a function and the fundamental period. Ask them if 8 or 12 is a period of the function in Example 1 (yes). Also ask if $f(x) = 1$ is periodic (yes) and what its fundamental period is (it has none).

You can introduce translations, stretches, and shrinks by building on your students' knowledge of the graphs of quadratic functions. To emphasize the difference between $y = cf(x)$ and $y = f(cx)$, have students compare the graphs of the following functions:

$$y = x^2, \ y = \frac{x^2}{2}, \ y = \left(\frac{x}{2}\right)^2, \ y = 2x^2, \text{ and } y = (2x)^2.$$

Likewise, have students observe translations by comparing the graphs of $y = x^2$, $y = x^2 - 1$, $y = (x - 1)^2$, $y = x^2 + 1$, and $y = (x + 1)^2$.

Using Technology

You may use graphing calculators or computers to introduce this section. Give students pairs of functions to graph and ask them to identify the algebraic and geometric patterns they see and to relate them to each other.

4-5 | pages 146–150

Presenting the Section

In Section 4-3, students became familiar with the idea of reflecting a graph through the line $y = x$, and they learned that the equation of the new graph can be found by interchanging x and y in the equation for the first graph. In this section, students learn that if both graphs represent functions, the two functions are called *inverses* of each other.

Cooperative Learning

You may wish to have students prove that if the point (a, b) is on the graph of a function and the graph is reflected in the line $y = x$, (b, a) is the corresponding point on the reflected graph. Have two or three students work together. You may want to tell them that both algebraic and geometric proofs are possible. When the groups think they have a proof, ask for one person to present a group's solution. When that proof is clear to everyone, ask for a presentation of an alternate proof.

The following is an algebraic proof (it assumes that the function has an inverse). Let (a, b) be on the graph of $y = f(x)$. Then $b = f(a)$ and $f^{-1}(b) = f^{-1}(f(a))$. Since $f^{-1}(b) = a$, then (b, a) is on the graph of $y = f^{-1}(x)$.

A geometric proof uses coordinate and transformational geometry. As shown below, $\triangle PRT$ is reflected in the line $y = x$ to get $\triangle QRT$. Since reflection is an isometry, $\triangle PRT \cong \triangle QRT$. Using the fact that $\overline{PR}$ is parallel to the y-axis, you can show that $\angle QRT \cong \angle 1$. Therefore $\overline{RQ}$ is parallel to the x-axis, and the y-coordinate of Q is a. Since $RP = RQ$, $\sqrt{(a - a)^2 + (b - a)^2} = \sqrt{(x - a)^2 + (a - a)^2}$, so the x-coordinate of Q is b. Thus, if $P(a, b)$ is on the graph of $y = f(x)$, then $Q(b, a)$ is on the graph of $y = f^{-1}(x)$.

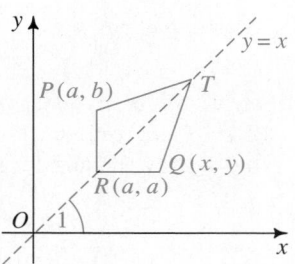

4-6 | pages 151–157

Presenting the Section

In this introduction to functions of two or more variables, students will concentrate on learning concepts rather than skills. They will learn how to read sophisticated graphs (such as those in the first four exercises), but will draw only elementary ones. The exercises may take less time than usual because there are few computations.

One potentially confusing concept is notational: the notation $P(x, y)$ for a point in the xy-plane is the same as $P(l, w)$ for a function (perimeter) of two variables (length and width). However, the meaning of the notation should be clear from the context.

In the Example, ask your students if the curves of constant height are linear or quadratic functions of the radius (quadratic); ask also about the curves of constant radius (linear).

Applications

The graphs presented in this section are the types one might find in a scientific report in a newspaper or in a science magazine. Students should find it quite satisfying to understand such graphs. Additional examples can be found in a textbook on multivariable calculus. You may also wish to use a topographic map of your area as an example. Such a map may be available in a local bookstore.

4-7 | pages 157–165

Presenting the Section

In this section, students form functions that model various applications. Instruct students to write down what is given and what function is desired, to identify the variables, and to draw any appropriate diagrams. The exercises are almost entirely conceptual. Since there are very few calculations in the exercises, students who do not see how to write a given function will have very little to show for their efforts and may become frustrated. Thus, it may be a good idea to assign one or two of these problems each day over a period of time, or to have students discuss the exercises in groups. You can reassure those students having difficulty by pointing out that problems such as the ones in this section are some of the most difficult problems in the study of precalculus.

Applications

Modeling real-world situations is the first step in any application of mathematics. This section is an excellent introduction to the difficulties involved in using mathematics to solve applied problems.

4 Functions

Properties of Functions

4-1 Functions

Objective   *To identify a function, to determine the domain, range, and zeros of a function, and to graph a function.*

You have already studied linear functions, quadratic functions, and polynomial functions. The purpose of this chapter is to consider functions in general. You will see how functions can be combined to produce new functions and how simple changes in a function's rule will change its graph. These ideas will be used later in the book when you study exponential functions, logarithmic functions, and trigonometric functions.

Definition of a Function

A **function** is a correspondence or rule that assigns to every element in a set D exactly one element in a set R. The set D is called the **domain** of the function, and the set R is called the **range**.

The diagram at the right shows a function f mapping, or pairing, a domain element x to a range element $f(x)$, read "the value of f at x" or "f of x." Although f names the function and $f(x)$ gives its value at x, we sometimes refer to the function $f(x)$, thereby indicating both the function f and the variable x of its domain.

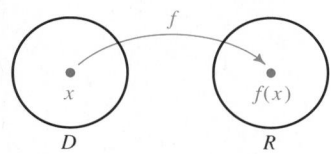

In this book the domain and range of a function are typically sets of numbers, but this need not be the case, as the following illustrations show.

1. Birthday function B maps a person to his or her birthday. For example,

$$B(\text{George Washington}) = \text{February 22}$$

  and $B(\text{Martin Luther King, Jr.}) = \text{January 15.}$

2. Area function A maps a geometric figure to its area. For example, if $PQRS$ is a square with sides of length 5, then $A(PQRS) = 5^2$.

We can treat a function f as a set of ordered pairs (x, y) such that x is an element of the domain of f and y is the corresponding element of the range. This is written formally as $\{(x, y) \mid y = f(x)\}$ or more simply as

$$y = f(x).$$

Although the letters f, x, and y are commonly used in general discussions of functions, other letters can be used. For example, $v = g(u)$ is a function g that assigns a domain element u to a range element v.

◀ Weather maps often involve the visual representation of functions of more than one variable. Each colored band in the satellite photo at the left represents a particular range of wind velocity plotted as a function of latitude and longitude.

Functions   **119**

Warm-Up Exercises

Find all values of x for which the given function has the value 0.

1. $f(x) = 9 - x^2$ ± 3

2. $g(x) = x^2 + 3x - 10$ $-5, 2$

3. $h(x) = |x + 8|$ -8

4. $k(x) = \sqrt{x + 2}$ -2

5. $r(x) = (2x - 3)^3$ $\frac{3}{2}$

6. $s(x) = \dfrac{2}{x - 1}$ none

Motivating the Section

If an automobile travels at a uniform speed of 50 mi/h, how far does it travel in 0.5 h? This simple problem is an example of distance expressed as a function of time. In real life, functions are very important in physics, engineering, economics, and other areas. An introduction to functions is the subject of this section.

Problem Solving

The mapping shown on this page and the analysis shown on the next page are two ways to display the correspondence of functions. Ask students to think of another way. (Draw two parallel number lines with one representing x and the other representing $f(x)$, and then draw lines from the domain value to the range value.)

Assessment Note

After discussing Example 1, ask students to give the domain for each of the following functions.

1. $g(x) = \dfrac{1}{x^2 + 1}$

$\{x \mid x \text{ is real}\}$

2. $g(x) = \dfrac{1}{\sqrt{x^2 - 1}}$

$\{x \mid x < -1 \text{ or } x > 1\}$

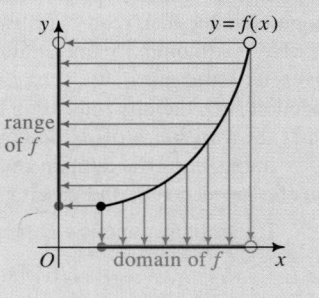 Using Technology

For Example 1, you can have students graph the functions using a graphing calculator or computer. They can then analyze the domains of the functions as indicated in the chart on this page.

Problem Solving

Have students give examples of functions that are not defined for certain values, for example $f(x) = \dfrac{1}{x}$.

Communication Note

The *graph* of a function is the set of all points in the plane determined by the function. Members of the range are called *values* of the function.

A function is frequently given in terms of a rule and a domain. If the domain of a function is not specified, then it is understood to consist of those real numbers for which the function produces real values.

Example 1   Give the domain of each function.

a. $g(x) = \dfrac{1}{x - 7}$
b. $h(r) = \sqrt{1 - r^2}$

Solution   **a.** The domain of g is the set of all real numbers except 7, since $\dfrac{1}{x - 7}$ is not defined when the denominator is 0.

b. $\sqrt{1 - r^2}$ is a real number only if $1 - r^2 \geq 0$. Therefore, the domain of h is $\{r \mid -1 \leq r \leq 1\}$.

The Graph of a Function

The graph of a function $y = f(x)$ consists of all points $(x, f(x))$ in an xy-plane. We can obtain the domain, range, and zeros of a function from its graph, as indicated below.

Analyzing the Graph of a Function

1. The domain of $y = f(x)$ is the set of x-coordinates of points on the graph of f. Projecting the graph of f on the x-axis gives a graph of the domain.

2. The range of $y = f(x)$ is the set of y-coordinates of points on the graph of f. Projecting the graph of f on the y-axis gives a graph of the range.

3. The zeros of $y = f(x)$ are the x-intercepts of the graph.

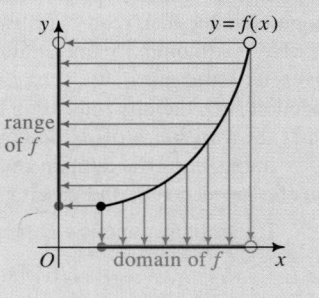

Example 2   Use the graph to give the domain, range, and zeros of each function.

a.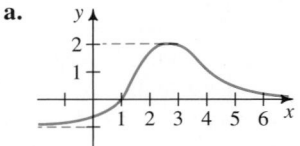

b.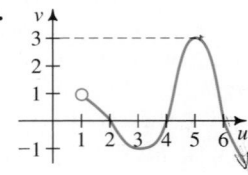

Solution   **a.** Domain = {all real numbers}
Range = $\{y \mid -1 < y \leq 2\}$
Zero: 1

b. Domain = $\{u \mid u > 1\}$
Range = $\{v \mid v \leq 3\}$
Zeros: 2, 4, and 6

120   *Chapter Four*

In applied settings, we often work with equations relating two or more measured quantities. For example, if gasoline is priced at $1.25 per gallon at a certain gas station, then a pump registers the cost C of g gallons of gas by using the equation

$$C = 1.25g.$$

Alternatively, the equation

$$C(g) = 1.25g$$

emphasizes that C is a function of g. Because the cost depends on the number of gallons, C and g are called **dependent** and **independent variables**, respectively.

Functions are a subset of the more general class of correspondences called **relations**. A **relation** is *any* correspondence or rule that pairs the members of two sets (the domain and the range). In general, a relation expressed as an equation in two variables, say x and y, defines y as a function of x provided there is exactly one y-value for each x-value. For example, the relation described by the equation $x^2y = 4x - y$ defines y as a function of x, but the relation described by the equation $y^2 = 4x^2$ does not. One way to determine this is to solve each equation for y, if possible. The first equation becomes $y = \dfrac{4x}{x^2 + 1}$. Because there is only one y-value for each x-value, this relation is a function. The second equation becomes $y = \pm 2x$. Because there are two y-values for each nonzero x-value, this relation is not a function. The graphs of these two equations are shown below.

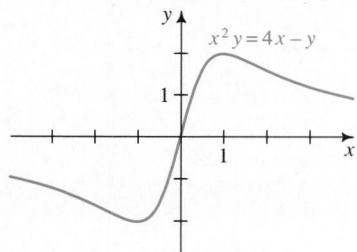

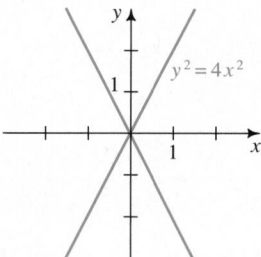

Another way to determine whether an equation in x and y defines y as a function of x is to apply the *vertical-line test* to the graph of the equation.

The Vertical-Line Test

If no vertical line intersects a given graph in more than one point, then the graph is the graph of a function.

Functions　**121**

Mathematical Note

In general, the dependent variable is associated with the vertical axis (the *y*-axis), and the independent variable is associated with the horizontal axis (the *x*-axis).

◩ Using Technology

Since graphing calculators graph only functions, try this experiment with students. Suppose you want to determine if $x = y^2$ is a function. Have students try to graph it. Since the format for graphing with a calculator is always "$y = $," the equation must be solved for y. You get $y = \sqrt{x}$ and $y = -\sqrt{x}$, which must be graphed separately to get the entire graph of $x = y^2$. This is not a function because there are two values of y for each value of x.

Additional Examples

1. Sketch the graph of each set. Tell if the graph is the graph of a function. If it is, give the domain, the range, and the zeros.
   a. $\{(x, y) \mid y = 4 - x^2\}$
   b. $\{(x, y) \mid x = |y|\}$

   a.

   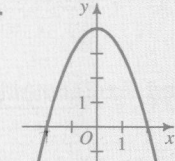

   The graph is a function. The domain is {real numbers}. The range is $\{y \mid y \leq 4\}$. Zeros: ± 2

(Continues on next page)

Additional Examples cont.

b.

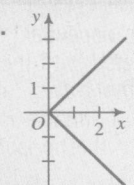

Not a function. For every positive x, there are two values of y.

2. Sketch the graph of each function. Use the graph to find the range and zeros of the function.

a. $f(x) = x^2 - x - 6$

b. $g(u) = \begin{cases} 1 & \text{if } u > 0 \\ 0 & \text{if } u = 0 \\ -1 & \text{if } u < 0 \end{cases}$

a. $f(x) = (x - 3)(x + 2)$

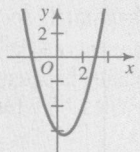

Range $= \{y | y \geq -6.25\}$
Zeros: -2, 3

b.

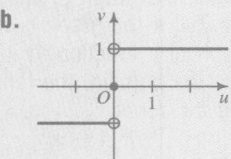

Range $= \{-1, 0, 1\}$
Zero: 0

Suggested Assignments

Standard
122/1–21 odd
Comprehensive
122/1–27 odd, 22

Supplementary Materials

Alternative Assessment, 10
Using Technology, 2–4, 14

122   Chapter 4

VLT = Vertical-line test

CLASS EXERCISES

1. a. What are the domain and range of the cost-of-gas function described on the preceding page? $g \geq 0$; $C \geq 0$
   b. Sketch the graph of the cost-of-gas function in a gC-plane.

2. Give the domain of each function.

   a. $f(x) = \dfrac{4}{x - 2}$ $x \neq 2$ **b.** $g(t) = \sqrt{t}$ $t \geq 0$ **c.** $h(s) = \sqrt{s - 4}$ $s \geq 4$

3. a. Show that 2 is a zero of the function $f(x) = 3x^2 - 12$. $f(2) = 3 \cdot 2^2 - 12 = 0$
   b. What is another zero? -2
   c. Find the range of the function. $f(x) \geq -12$

4. Give the domain, range, and zeros of the function whose graph is shown at the right.

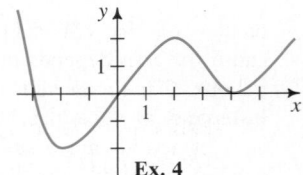

Ex. 4

5. Apply the vertical-line test to the graphs of $x^2y = 4x - y$ and $y^2 = 4x^2$ shown on the preceding page. Passes VLT; fails VLT

6. Which of the relations described by the two equations $x^2 - y = 1$ and $y^2 - x = 1$ does *not* define y as a function of x? Explain.
$y^2 - x = 1 \Rightarrow y = \pm\sqrt{x + 1}$, so there are two y-values for each $x > -1$

Tell whether the relation described by each equation defines y as a function of x. If it does, sketch the graph.

7. $y = 2x + 5$ Yes   **8.** $y = |x|$ Yes   **9.** $x = |y|$ No   **10.** $y^2 = x^2$ No

Tell whether the graph of each relation is the graph of a function. Explain your answer.

11.

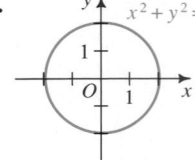

No; fails VLT

12.

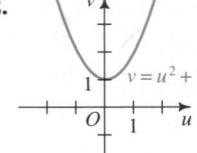

Yes; passes VLT

13.

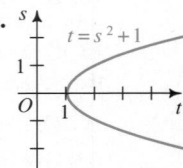

No; fails VLT

WRITTEN EXERCISES

In Exercises 1–6, tell whether the graph of each relation is the graph of a function. If it is, give the domain and range of the function.

A **1.**

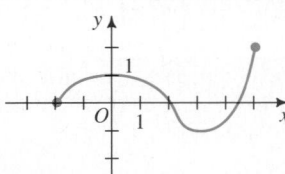

Yes; $-2 \leq x \leq 5$; $-1 \leq y \leq 2$

2.

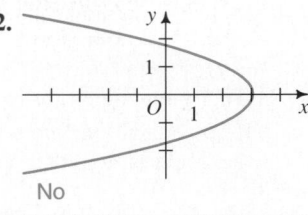

No

122   *Chapter Four*

3.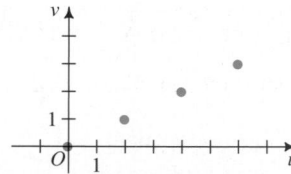

Yes; 0, 2, 4, 6; 0, 1, 2, 3

4.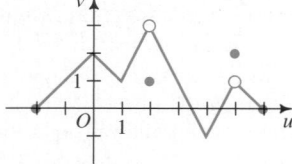

Yes; $-2 \le u \le 6$; $-1 \le v < 3$

5.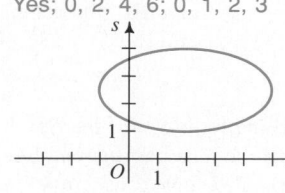

No

6.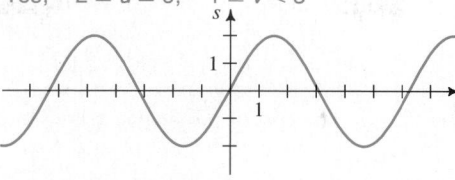

Yes; all real numbers; $-2 \le s \le 2$

7. Explain why the equation $x^2 + y^2 = 1$ does not define y as a function of x.

8. Explain why the equation $x^3 + y^3 = 1$ defines y as a function of x.

Give the domain of each function.

9. a. $f(x) = \dfrac{1}{x} \quad x \ne 0$
   b. $g(x) = \dfrac{1}{x - 9} \quad x \ne 9$
   c. $h(x) = \dfrac{3x}{x^2 - 4} \quad x \ne \pm 2$

10. a. $f(t) = \dfrac{1}{t + 3} \quad t \ne -3$
   b. $g(t) = \dfrac{t + 2}{t^2 + 5t + 6} \quad t \ne -3, -2$
   c. $h(t) = \dfrac{2t^2 - 3}{t^3 - 9t} \quad t \ne 0, \pm 3$

Give the domain, range, and zeros of each function.

11. a. $f(x) = |x|$
   b. $g(x) = |x - 2|$
   c. $h(x) = |x| - 2$

12. a. $f(t) = \sqrt{t}$
   b. $g(t) = \sqrt{9 - t}$
   c. $h(t) = \sqrt{9 - t^2}$

In Exercises 13–18, sketch the graph of each function. Use the graph to find the range and zeros of the function.

13. $f(x) = x^2 - 6x + 8$
14. $g(x) = 4 - (x - 3)^2$

15. $f(t) = (t - 2)^3$
16. $g(t) = t^3 + 4t^2 - t - 4$

17. $h(u) = \begin{cases} u^2 & \text{if } -2 \le u < 1 \\ 2 - u & \text{if } 1 \le u < 4 \end{cases}$

18. $g(u) = \begin{cases} u - 1 & \text{if } u < 0 \\ u^2 - 2u - 3 & \text{if } 0 \le u \le 3 \\ 0 & \text{if } u > 3 \end{cases}$

19. a. Let V be the function that assigns to each solid its volume. If C is a cylinder with radius 3 and height 4, find $V(C)$. 36π
   b. Give the domain and range of V. All solids; nonnegative real numbers

20. a. A formula from geometry states that $S = (n - 2)180°$. Give the meaning of this formula. $S =$ sum of the measures of the interior angles of an n-gon
   b. Is S a function of n? If so, give the domain and range of this function. Yes; integers greater than 2; positive multiples of 180°

Mathematical Note

Exercise 3 is an example of a *discrete* function. Point out that many real-life functions are also sets of discrete data values.

△ Using Technology

Have students use a graphing calculator or computer to check the solutions to Exercises 11–16. Some advanced graphing calculators and computer software can even handle the *piecewise* functions in Exercises 17 and 18. Check the instruction manuals to see how.

Additional Answers
Written Exercises

8. Write as $y = \sqrt[3]{1 - x^3}$. There is only one y-value for each x-value.

12. a. $D = \{t | t \ge 0\}$; $R = \{f(t) | f(t) \ge 0\}$; zero: 0
   b. $D = \{t | t \le 9\}$; $R = \{g(t) | g(t) \ge 0\}$; zero: 9
   c. $D = \{t | -3 \le t \le 3\}$; $R = \{h(t) | 0 \le h(t) \le 3\}$; zeros: ± 3

14. $R = \{g(x) | g(x) \le 4\}$; zeros: 1, 5

16. $R = \{\text{all real numbers}\}$; zeros: $-4, -1, 1$

18. $R = \{g(u) | g(u) \le 0\}$; zeros: all $u \ge 3$

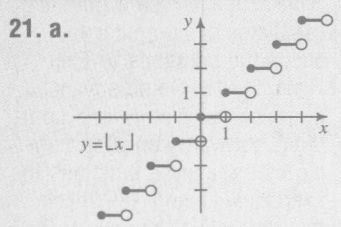
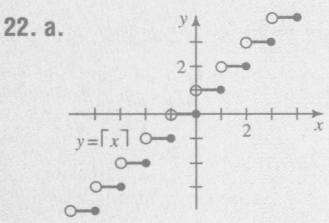
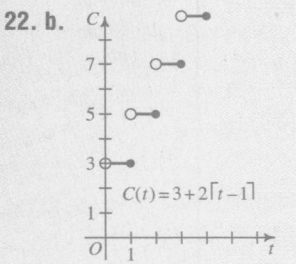
B 21. The **greatest integer function** assigns to each number the greatest integer less than or equal to the number. If we denote the greatest integer in x by $\lfloor x \rfloor$, then we have $\lfloor 5.28 \rfloor = 5$, $\lfloor 5 \rfloor = 5$, $\lfloor \pi \rfloor = 3$, and $\lfloor -1.7 \rfloor = -2$.
   a. Sketch the graph of $y = \lfloor x \rfloor$.
   b. Give the domain and range of the greatest integer function.

22. The greatest integer function $f(x) = \lfloor x \rfloor$, described in Exercise 21, is sometimes called "the floor of x." By contrast, $c(x) = \lceil x \rceil$ is called "the ceiling of x" and is the least integer greater than or equal to x. Thus $\lceil 5.28 \rceil = 6$, $\lceil 5 \rceil = 5$, $\lceil \pi \rceil = 4$, and $\lceil -1.7 \rceil = -1$.
   a. Sketch the graph of $y = \lceil x \rceil$.
   b. The cost of parking a car in a municipal parking lot is \$3 for the first hour or any part thereof, plus \$2 for each additional hour or part thereof. Sketch the graph of this cost function and find a rule for the cost C as a function of time t. Your rule should use the ceiling function. $C(t) = 3 + 2\lceil t - 1 \rceil$

23. *Writing* Think about what it means for two functions to be *equal*. Would you say that the functions $f(x) = |x|$ and $g(x) = \sqrt{x^2}$ are equal? Are the functions $f(x) = |x|$ and $h(x) = (\sqrt{x})^2$ equal? Write a brief defense of your conclusions.

24. *Research* Use a mathematics dictionary to find the meaning of the phrase *implicit function*. Then determine what implicit functions, if any, are defined by each of the following equations. $y = \pm x$
   a. $x = y^2$ $y = \pm\sqrt{x}$ **b.** $x^2 + y^2 = 1$ $y = \pm\sqrt{1 - x^2}$ **c.** $x^2 - y^2 = 0$

25. For which of the following functions does $f(a + b) = f(a) + f(b)$? d
   a. $f(x) = x^2$ **b.** $f(x) = \frac{1}{x}$ **c.** $f(x) = 4x + 1$ **d.** $f(x) = 4x$

26. For which of the functions in Exercise 25 does $f(ab) = f(a) \cdot f(b)$? a, b

27. If $f(a + b) = f(a) + f(b)$ for some function f, prove that $f(0) = 0$.

28. If $f(ab) = f(a) \cdot f(b)$ for some function f, prove that $f(1) = 1$.

4-2 Operations on Functions

Objective   *To perform operations on functions and to determine the domains of the resulting functions.*

Suppose a company manufactures and sells a certain product. If the cost of manufacturing x items of the product is given by the function $C(x)$ and the revenue generated by the sale of the x items is given by the function $R(x)$, then the company's profit is given by the function $P(x)$ where

$$P(x) = R(x) - C(x).$$

That is, the profit function is the *difference* between the revenue and cost functions. As this example from economics suggests, it is possible to combine two given functions to produce a new function.

The Sum, Difference, Product, and Quotient of Functions

Each function listed below is defined for all x in the domains of both f and g.

1. **Sum** of f and g: $(f + g)(x) = f(x) + g(x)$
2. **Difference** of f and g: $(f - g)(x) = f(x) - g(x)$
3. **Product** of f and g: $(f \cdot g)(x) = f(x) \cdot g(x)$
4. **Quotient** of f and g: $\left(\dfrac{f}{g}\right)(x) = \dfrac{f(x)}{g(x)}$, provided $g(x) \neq 0$

Example 1   Let $f(x) = x + 1$ and $g(x) = x^2 - 1$. Find a rule for each of the following functions.

    a. $(f + g)(x)$ **b.** $\left(\dfrac{f}{g}\right)(x)$

Solution   **a.** $(f + g)(x) = f(x) + g(x) = (x + 1) + (x^2 - 1)$
$$= x^2 + x$$

    b. $\left(\dfrac{f}{g}\right)(x) = \dfrac{f(x)}{g(x)} = \dfrac{x + 1}{x^2 - 1}$
$$= \dfrac{x + 1}{(x + 1)(x - 1)}$$
$$= \dfrac{1}{x - 1}, \text{ provided } x \neq \pm 1$$

The graph of the sum function, $f + g$, can be obtained directly from the graphs of f and g. As the diagram at the left below shows, vertical arrows from the x-axis to the graph of g are "added" to the graph of f to obtain the graph of $f + g$.

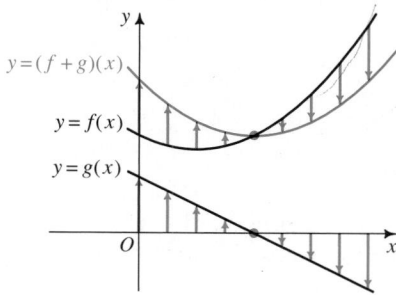   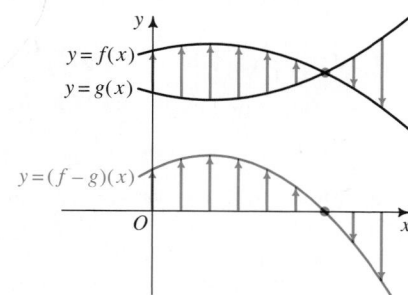

The diagram at the right above shows that the graph of the difference function, $f - g$, gives the vertical distance from the graph of g to the graph of f. Wherever the graph of f lies above the graph of g, $f - g$ is positive, and wherever the graph of f lies below the graph of g, $f - g$ is negative.

Warm-Up Exercises cont.

1. $f\left(-\dfrac{1}{2}\right)$ $\dfrac{3}{4}$

2. $f(1 - \sqrt{3})$ $3 - \sqrt{3}$

3. $f(x + 1)$ $x^2 + x$

4. $g(-1.2)$ 1.25

5. $g(-\sqrt{3})$ $2 + \sqrt{3}$

6. $g(x^2 - 2)$ $\dfrac{1}{x^2}$

Motivating the Section

The example from economics presented on page 124 shows that performing operations on functions sometimes leads to new functions, the topic of this section.

Problem Solving

Suppose $f(x) = x$ and $g(x) = x^2$. Have students use the rules for the arithmetic of functions given on this page to determine if, in general:

1. $(f + g)(x) = (g + f)(x)$ (yes)
2. $(f - g)(x) = (g - f)(x)$ (no)
3. $(f \cdot g)(x) = (g \cdot f)(x)$ (yes)
4. $\left(\dfrac{f}{g}\right)(x) = \left(\dfrac{g}{f}\right)(x)$ (no)

Making Connections

The difference function $(f - g)(x)$ can be used to solve the equation $f(x) = g(x)$. See Exercise 13, page 129.

A fifth way of combining functions can be illustrated by the sport of cycling. When an 18-speed touring bicycle is in sixth gear, the gear ratio is 3 : 2, which means that the wheels of the bicycle revolve 3 times for every 2 revolutions of the pedals. This relationship can be expressed as

$$w = \frac{3}{2}p$$

where w and p represent wheel and pedal revolutions, respectively. Since the wheels of a touring bicycle have a diameter of 27 in., w revolutions of the wheels move the bicycle a distance d, in inches, given by:

$$d = 27\pi w$$

Notice that $d = 27\pi w$ gives distance as a function of wheel revolutions and that $w = \frac{3}{2}p$ gives wheel revolutions as a function of pedal revolutions. By substituting $\frac{3}{2}p$ for w in $d = 27\pi w$, we get

$$d = 27\pi\left(\frac{3}{2}p\right) = 40.5\pi p,$$

which gives distance as a function of pedal revolutions. This function $d(p)$ is said to be the *composite* of the functions $d(w)$ and $w(p)$.

The Composite of Functions

The **composite** of f and g, denoted $f \circ g$, is defined by two conditions:

1. $(f \circ g)(x) = f(g(x))$, which is read "f circle g of x equals f of g of x";

2. x is in the domain of g and $g(x)$ is in the domain of f.

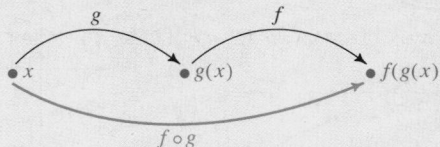

The domain of $f \circ g$ is the set of x satisfying condition (2) above. The operation that combines f and g to produce their composite is called the **composition** of functions.

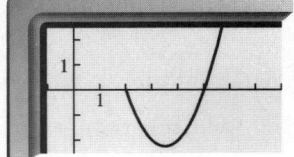

As Example 2 shows, using a computer or graphing calculator to graph a composite function helps you see that its domain may need to be restricted.

Example 2 Let $f(x) = x^4 - 3x^2$ and $g(x) = \sqrt{x - 2}$. Find a rule for $(f \circ g)(x)$ and give the domain of the composite function. Confirm the domain by using a computer or graphing calculator to graph $y = (f \circ g)(x)$.

Solution

$$(f \circ g)(x) = f(g(x)) = f(\sqrt{x - 2}) \longleftarrow g(x) \text{ is defined for } x \geq 2.$$
$$= (\sqrt{x - 2})^4 - 3(\sqrt{x - 2})^2 \longleftarrow f(g(x)) \text{ is defined}$$
$$= x^2 - 7x + 10 \qquad\qquad\text{for } x \geq 2.$$

The domain of the composite function $f \circ g$ is $\{x \mid x \geq 2\}$ even though the expression $x^2 - 7x + 10$ is also defined for $x < 2$. To use a graphing calculator to confirm the domain, enter the equation of $f \circ g$ in its unsimplified form:

$$y = (\sqrt{x - 2})^4 - 3(\sqrt{x - 2})^2$$

Then the graph of $y = (f \circ g)(x)$ is the portion of the parabola $y = x^2 - 7x + 10$ for which $x \geq 2$.

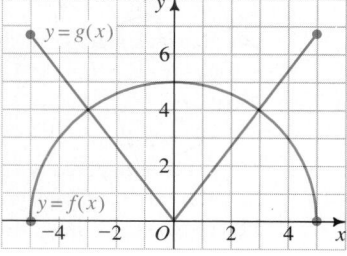

Example 3 Let $f(x) = \dfrac{1}{x}$ and $g(x) = x + 1$. Find rules for $(f \circ g)(x)$ and $(g \circ f)(x)$ and give the domain of each composite function.

Solution

$$(f \circ g)(x) = f(g(x)) = f(x + 1) = \frac{1}{x + 1} \quad \text{Domain} = \{x \mid x \neq -1\}$$

$$(g \circ f)(x) = g(f(x)) = g\left(\frac{1}{x}\right) = \frac{1}{x} + 1 \quad \text{Domain} = \{x \mid x \neq 0\}$$

Notice in Example 3 that $(f \circ g)(x) \neq (g \circ f)(x)$. Since composition is not necessarily commutative, the *order* of two functions being composed is important.

CLASS EXERCISES

For Exercises 1–4, use the graphs of f and g shown at the right.

1. Find each of the following.
   a. $(f + g)(0)$ 5   b. $(f + g)(3)$ 8
2. For what values of x is: $-3 < x < 3$
   a. $(f - g)(x) = 0$? ± 3 b. $(f - g)(x) > 0$?
3. Sketch the graph of each of the following.
   a. $y = (f + g)(x)$ b. $y = (f - g)(x)$
4. Sketch the graph of each of the following.
   a. $y = f(x) + 1$ b. $y = g(x) - 1$

Functions **127**

In Example 2, instead of entering the equation of $f \circ g$ in its unsimplified form, you may want to enter the equations of g and $f \circ g$ as follows:

$$y_1 = \sqrt{x - 2}$$
$$y_2 = (y_1)^4 - 3(y_1)^2$$

Additional Examples cont.

2. Let $f(x) = x^3 + 1$, $g(x) = \sqrt[3]{x - 1}$, and $h(x) = \sqrt[3]{x + 1}$.
   a. Show that $f(g(x)) = g(f(x))$ for all x.
   b. Show that $f(h(-1)) = h(f(-1))$.
   c. Show that $f(h(x)) \neq h(f(x))$ if $x \neq -1$.

   a. $f(g(x)) = f(\sqrt[3]{x - 1}) =$
   $(\sqrt[3]{x - 1})^3 + 1 =$
   $(x - 1) + 1 = x$
   $g(f(x)) = g(x^3 + 1) =$
   $\sqrt[3]{(x^3 + 1) - 1} =$
   $\sqrt[3]{x^3} = x$
   $\therefore f(g(x)) = g(f(x))$
   for all x.

   b. $h(-1) = \sqrt[3]{-1 + 1} = 0$
   $f(h(-1)) = f(0) =$
   $0^3 + 1 = 1$
   $f(-1) = (-1)^3 + 1 = 0$
   $h(f(-1)) = h(0) =$
   $\sqrt[3]{0 + 1} = 1$
   $\therefore f(h(-1)) = 1 =$
   $h(f(-1))$

   c. $f(h(x)) = f(\sqrt[3]{x + 1}) =$
   $(\sqrt[3]{x + 1})^3 + 1 =$
   $(x + 1) + 1 = x + 2$
   $h(f(x)) = h(x^3 + 1) =$
   $\sqrt[3]{(x^3 + 1) + 1} =$
   $\sqrt[3]{x^3 + 2}$

(Continues on next page)

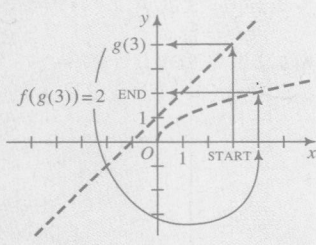
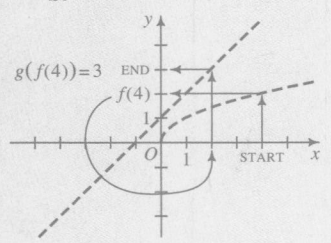
In Exercises 5–10, let $f(x) = x^2 + x$ and $g(x) = x + 1$. Find each of the following.

5. $(f + g)(x)$ $x^2 + 2x + 1$

6. $(f - g)(x)$ $x^2 - 1$

7. $(f \cdot g)(x)$ $x^3 + 2x^2 + x$

8. $\left(\dfrac{f}{g}\right)(x)$ x, for $x \neq -1$

9. a. $f(g(2))$ 12   **b.** $(f \circ g)(x)$ $x^2 + 3x + 2$

10. a. $g(f(2))$ 7   **b.** $(g \circ f)(x)$ $x^2 + x + 1$

11. *Visual Thinking* On a single set of axes, sketch the graphs of $f(x) = \sqrt{x}$ and $g(x) = x + 1$. Demonstrate how the graphs of f and g can be used to find the value of **(a)** $f(g(3))$ and **(b)** $g(f(4))$.

12. If $F(X)$ is the father of X and $M(X)$ is the mother of X, what expression represents the maternal grandfather of X? the paternal grandmother of X? $F(M(X))$; $M(F(X))$

13. If B is the birthday function defined on page 119 and if F is the father function defined in Exercise 12, which of the composite functions $B \circ F$ and $F \circ B$ is defined? $B \circ F$

WRITTEN EXERCISES

A

1. Copy the graph of $y = f(x)$ shown at the right. On a single set of axes, draw the graphs of $y = f(x) + 2$ and $y = f(x) - 3$.

2. On a single set of axes, graph $y = |x|$, $y = |x| + 5$, and $y = |x| - 4$.

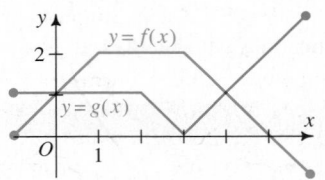

Ex. 1

3. The graphs of $y = f(x)$ and $y = g(x)$ are shown at the left below.
   a. Find $f(1) - g(1)$. 1
   b. For what values of x is $f(x) - g(x)$ positive? negative? zero? $0 < x < 4$; $-1 \leq x < 0$ or $4 < x \leq 6$; $0, 4$
   c. What is the maximum value of $f(x) - g(x)$? 2

Ex. 3

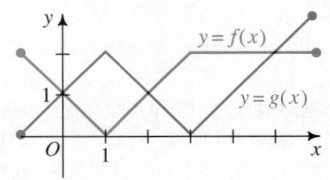

Ex. 4

4. Repeat Exercise 3 using the graphs of $y = f(x)$ and $y = g(x)$ shown at the right above. **a.** -2 **b.** $-1 \leq x < 0$ or $2 < x < 5$; $0 < x < 2$ or $5 < x \leq 6$; $0, 2, 5$ **c.** 2

5–10. Let $f(x) = x^3 - 1$ and $g(x) = x - 1$. Evaluate the expressions in Class Exercises 5–10. **5.** $x^3 + x - 2$ **6.** $x^3 - x$ **7.** $x^4 - x^3 - x + 1$ See below.

11. On a single set of axes, graph $f(x) = x$ in one color, graph $g(x) = |x|$ in a second color, and graph $f + g$ in a third color.

12. Using the graphs given in Exercise 4, graph $f + g$.
   8. $x^2 + x + 1$ for $x \neq 1$ **9. a.** 0 **b.** $x^3 - 3x^2 + 3x - 2$ **10. a.** 6 **b.** $x^3 - 2$

13. **Visual Thinking** Given two functions f and g, one way to obtain the real solutions of the equation $f(x) = g(x)$ is to graph the equations $y = f(x)$ and $y = g(x)$ in an xy-plane and then find the x-coordinates of any points of intersection. Describe another way to solve $f(x) = g(x)$ that also involves graphing in an xy-plane but that is based on the difference function $f - g$.
Graph $y = (f - g)(x)$ and find the x-intercepts.

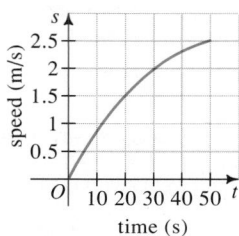 **Use a computer or graphing calculator and one of the methods from Exercise 13 to find the real solutions of each of the following equations. Give answers to the nearest hundredth.**

± 0.71

14. $x^3 = x + 1$ 1.32 15. $\sqrt{x + 1} = 2x$ 0.64 16. $\sqrt{1 - x^2} = |x|$

B 17. Let $f(x) = 2x - 3$, $g(x) = \dfrac{x + 3}{2}$, and $h(x) = 3x + 2$.

  a. Show that $f(g(x)) = g(f(x))$ for all x.
  b. Show that $f(h(x)) \neq h(f(x))$ for any x.

18. Let $f(x) = x^3$, $g(x) = \sqrt[3]{x}$, and $h(x) = 3x$.
  a. Show that $f(g(x)) = g(f(x))$ for all x.
  b. Show that $f(h(x)) = h(f(x))$ for only one value of x.

Let $f(x) = \sqrt{x}$, $g(x) = 6x - 3$, and $h(x) = \dfrac{x}{3}$. Find each of the following.

$\sqrt{2x - 3}$ $2\sqrt{x} - 1$

19. a. $f(g(h(6)))$ 3   b. $f(g(h(x)))$ 20. a. $h(g(f(4)))$ 3   b. $h(g(f(x)))$

21. a. $h\!\left(f\!\left(g\!\left(\tfrac{1}{2}\right)\right)\right)$ 0   b. $h(f(g(x)))$ $\dfrac{\sqrt{6x - 3}}{3}$ 22. a. $g(h(f(9)))$ 3   b. $g(h(f(x)))$

$2\sqrt{x} - 3$

Let $f(x) = x^3$, $g(x) = \sqrt{x}$, $h(x) = x - 4$, and $j(x) = 2x$. Express each function k as a composite of three of these four functions.

23. $k(x) = 2(x - 4)^3$ $j(f(h(x)))$ 24. $k(x) = \sqrt{(x - 4)^3}$ $g(f(h(x)))$
25. $k(x) = (2x - 8)^3$ $f(j(h(x)))$ 26. $k(x) = \sqrt{x^3 - 4}$ $g(h(f(x)))$

27. **Physiology** The graph at the left below shows a swimmer's speed s as a function of time t. The graph at the right below shows the swimmer's oxygen consumption c as a function of s. Time is measured in seconds, speed in meters per second, and oxygen consumption in liters per minute.
  a. What are the speed and oxygen consumption after 20 s of swimming? 1.5 m/s; 9 L/min
  b. How many seconds have elapsed if the swimmer's oxygen consumption is 15 L/min? 30 s

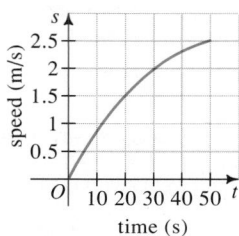

time (s)

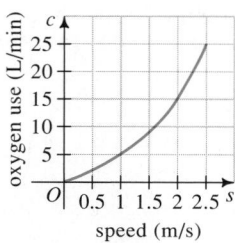

speed (m/s)

Functions **129**

Using Technology

For Exercises 14–16, refer students to the discussion of solving equations with a graphing calculator in the article *Using a Graphing Calculator* (see pages xxx–xxxi).

Error Analysis

Students often evaluate the composition of functions incorrectly because they are not careful about order. (See Exercises 19–22.) Emphasize that the innermost function is evaluated first. Evaluation then proceeds outward.

**Additional Answers
Written Exercises**

17. a. $f(g(x)) = f\!\left(\dfrac{x + 3}{2}\right) =$
    $2\!\left(\dfrac{x + 3}{2}\right) - 3 = x$
    $g(f(x)) = g(2x - 3) =$
    $\dfrac{(2x - 3) + 3}{2} = x$
    $\therefore f(g(x)) = g(f(x))$ for all x.
  b. $f(h(x)) = f(3x + 2) =$
    $2(3x + 2) - 3 = 6x + 1$
    $h(f(x)) = h(2x - 3) =$
    $3(2x - 3) + 2 = 6x - 7$
    $\therefore f(h(x)) \neq h(f(x))$ for all x.

18. a. $f(g(x)) = f\!\left(\sqrt[3]{x}\right) =$
    $\left(\sqrt[3]{x}\right)^3 = x$
    $g(f(x)) = g\!\left(x^3\right) =$
    $\sqrt[3]{x^3} = x$
    $\therefore f(g(x)) = g(f(x))$ for all x.
  b. $f(h(x)) = f(3x) =$
    $(3x)^3 = 27x^3$
    $h(f(x)) = h\!\left(x^3\right) = 3x^3$
    $\therefore f(h(x)) = h(f(x))$ only for $x = 0$.

28. Consumer Economics The graph at the left below shows a car's fuel economy e as a function of the speed s at which the car is driven. The graph at the right below shows the per-mile fuel cost c as a function of e. Fuel economy is measured in miles per gallon, speed in miles per hour, and fuel cost in cents per mile.

   a. If the car is driven at 55 mi/h, what is the fuel cost? 4 cents/mile
   b. If the fuel cost is to be kept at or below 4 cents per mile, at what speeds should the car be driven? Between 25 mi/h and 55 mi/h

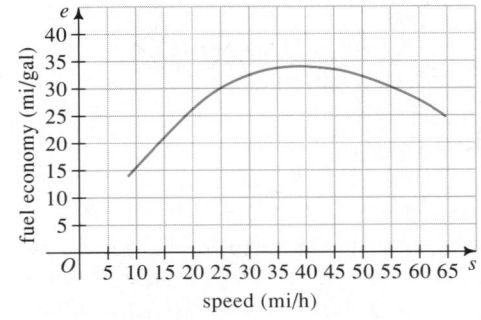
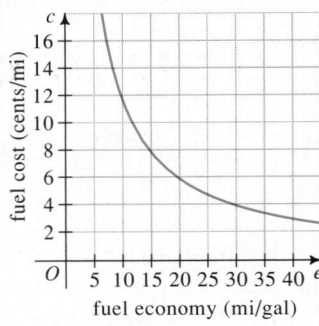

speed (mi/h)   fuel economy (mi/gal)

29. a. Express the radius r of a circle as a function of the circumference C.
   b. Express the area A of the circle as a function of C.

30. a. Express the area A and perimeter P of a semicircular region in terms of the radius r.
   b. Express A as a function of P.

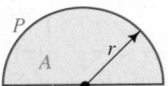

31. Physics The speed s of sound in air is given by the formula $s = 331 + 0.6C$ where s is measured in meters per second and C is the Celsius temperature. If $C = \dfrac{5}{9}(F - 32)$, express s as a function of F, the Fahrenheit temperature.

32. The surface area and volume of a sphere are given in terms of the radius by the following formulas: $A = 4\pi r^2$ and $V = \dfrac{4}{3}\pi r^3$.

   a. Express r as a function of A. $r(A) = \dfrac{\sqrt{A}}{2\sqrt{\pi}}$ **b.** Express V as a function of A. $V(A) = \dfrac{A\sqrt{A}}{6\sqrt{\pi}}$

◰ In Exercises 33–36, find rules for $(f \circ g)(x)$ and $(g \circ f)(x)$ and give the domain of each composite function. You may wish to use a computer or graphing calculator to confirm your answer for the domain.

33. $f(x) = 2x$, $g(x) = \sqrt{16 - x^2}$ **34.** $f(x) = \sqrt{x}$, $g(x) = \dfrac{1}{x - 4}$

35. $f(x) = x^2$, $g(x) = \sqrt{1 - x}$ **36.** $f(x) = x^2$, $g(x) = \sqrt{16 - x^2}$

37. If $g(x) = \dfrac{x + 3}{2}$, find $g(g(1))$, $g(g(g(1)))$, and $g(g(g(g(1))))$. 2.5, 2.75, 2.875

38. If $f(x) = 2x - 1$, show that $f(f(x)) = 4x - 3$. Find $f(f(f(x)))$.

   38. $f(f(x)) = 2(2x - 1) - 1 = 4x - 3$; $f(f(f(x))) = 8x - 7$

C 39. **Physics** The luminous intensity I, measured in candela (cd), of a 100 watt light bulb is 130 cd. The law of illumination states that $E = \dfrac{I}{d^2}$, where E is the illumination and d is the distance in meters to the light bulb. Suppose you hold a book 1 m away from a 100 watt bulb and begin walking away from the bulb at a rate of 1 m/s.

  a. Express E in terms of the time t in seconds after you begin walking.
  b. When will the illumination on the book be 1% of its original value?

  a. $E = \dfrac{130}{(t+1)^2}$ **b.** After 9 s

■ Graphs and Inverses of Functions

4-3 Reflecting Graphs; Symmetry

Objective   *To reflect graphs and to use symmetry to sketch graphs.*

In this section and the next we will see how the graph of an equation is transformed when the equation is altered. This will allow us to graph a simple equation and—by reflecting it, stretching or shrinking it, or sliding it—to obtain the graph of a related, more complicated equation.

   We begin by considering the *reflection* of a graph in a line. The *line of reflection* acts like a mirror and is located halfway between a point and its reflection. (If the point being reflected is on the line, then the point is its own reflection.)

Activities

 By completing the following activities, you will see how a change in an equation results in a reflection of its graph in some line. For parts (a) and (b) of each activity, graph each pair of equations on a single set of axes. You may wish to use a computer or graphing calculator to help you with your graphs.

  1. a. Graph $y = x^2$ and $y = -x^2$.
     b. Graph $y = x^3 + 2x^2$ and $y = -(x^3 + 2x^2)$.
     c. In general, how are the graphs of $y = f(x)$ and $y = -f(x)$ related?
  2. a. Graph $y = x^2 - 1$ and $y = |x^2 - 1|$.
     b. Graph $y = x(x - 1)(x - 3)$ and $y = |x(x - 1)(x - 3)|$.
     c. In general, how are the graphs of $y = f(x)$ and $y = |f(x)|$ related?
  3. a. Graph $y = 2x - 1$ and $y = 2(-x) - 1$.
     b. Graph $y = \sqrt{x}$ and $y = \sqrt{-x}$.
     c. In general, how are the graphs of $y = f(x)$ and $y = f(-x)$ related?
  4. a. Graph $y = 2x + 1$ and $x = 2y + 1$.
     b. Graph $y = x^2$ and $x = y^2$.
     c. In general, how is the graph of an equation affected when you interchange the variables in the equation?

Functions   **131**

Use the graph to tell the new location of the given point after it is reflected, or flipped, over the specified line.

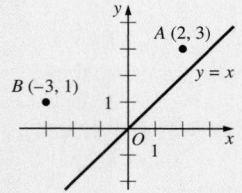

  1. A; x-axis $(2, -3)$
  2. A; y-axis $(-2, 3)$
  3. B; x-axis $(-3, -1)$
  4. B; y-axis $(3, 1)$
  5. A; line $y = x$ $(3, 2)$
  6. B; line $y = x$ $(1, -3)$

Motivating the Section

Symmetry in nature is apparent everywhere. Ask students for some examples of "bilateral symmetry" (animals have it) and other symmetry in nature. This section shows how to use line or point symmetry to sketch a graph.

Using Technology

Encourage students to use graphing calculators to complete parts (a) and (b) of the activities, then make a conjecture about how the graphs are related (part (c)), and then use the graphing calculator again to test the conjecture (by graphing more examples). Remind students that to use a graphing calculator, you must solve the equation for y.

As the activities on the preceding page show, simple changes in an equation can produce reflections of its graph in the x-axis, the y-axis, and the line $y = x$.

Reflection in the *x*-axis

The graph of $y = -f(x)$ is obtained by reflecting the graph of $y = f(x)$ in the x-axis. In the graphs below, notice that each point (x, y) on the original (red) graph becomes the point $(x, -y)$ on the reflected (blue) graph.

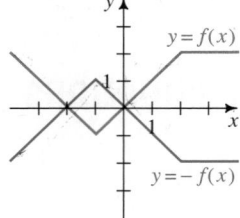 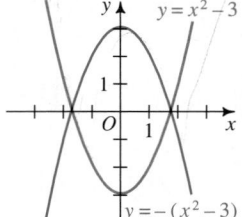

The graph of $y = |f(x)|$ is identical to the graph of $y = f(x)$ when $f(x) \geq 0$ and is identical to the graph of $y = -f(x)$ when $f(x) < 0$. This principle is applied to the graphs shown above to produce the following graphs. Notice that the graphs do not dip below the x-axis.

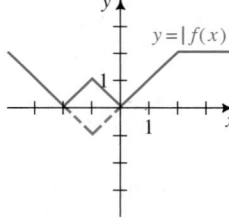 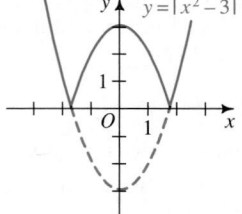

Reflection in the *y*-axis

The graph of $y = f(-x)$ is obtained by reflecting the graph of $y = f(x)$ in the y-axis. In the graphs below, notice that each point (x, y) on the original (red) graph becomes the point $(-x, y)$ on the reflected (blue) graph.

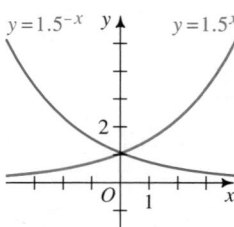 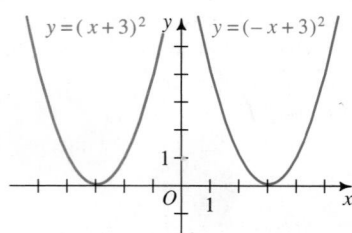

Reflection in the Line $y = x$

Reflecting the graph of an equation in the line $y = x$ is equivalent to interchanging x and y in the equation.

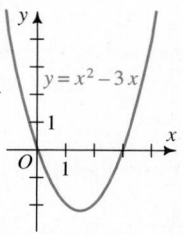

Original graph and equation

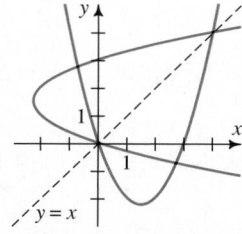

Reflection in $y = x$

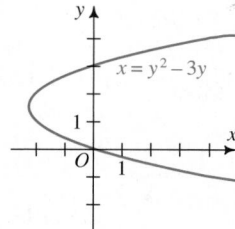

Reflected graph and altered equation

Symmetry

A line l is called an **axis of symmetry** of a graph if it is possible to pair the points of the graph in such a way that l is the perpendicular bisector of the segment joining each pair. (See the figure at the left below.)

A point O is called a **point of symmetry** of a graph if it is possible to pair the points of the graph in such a way that O is the midpoint of the segment joining each pair. (See the figure at the right below.)

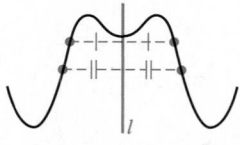

l = axis of symmetry

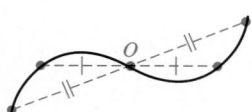

O = point of symmetry

As you know, the graph of $y = ax^2 + bx + c$ has an axis of symmetry with equation $x = -\dfrac{b}{2a}$. In Exercise 16 on page 145, we will show that the graph of $y = ax^3 + bx^2 + cx + d$ has a point of symmetry at $x = -\dfrac{b}{3a}$.

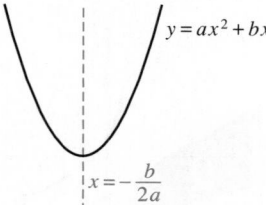

The graph of every quadratic function has a line of symmetry.

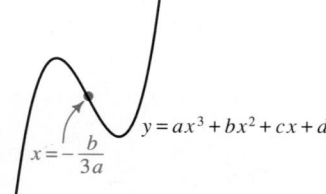

The graph of every cubic function has a point of symmetry.

Making Connections

Reflections in the line $y = x$ will be important in the study of inverse functions (see Section 4-5).

Mathematical Note

A figure has an axis of symmetry if the two "halves" of the figure coincide when the figure is folded along the axis. A figure has a point of symmetry if a rotation of 180 degrees about the point produces the same figure.

Error Analysis

Encourage students who have difficulty reflecting in the line $y = x$ to interchange x with y in the original equation to form a new equation. They can then graph the new equation.

Error Analysis

The symmetry rules on this page will be difficult for some students to understand. Have them test other examples so they will understand the types of symmetry better. A graphing calculator or a computer might be helpful.

Additional Examples

1. Let $f(x) = 2x - 3$. Sketch each graph.
   a. $y = -f(x)$
   b. $y = |f(x)|$
   c. $y = f(-x)$

a.

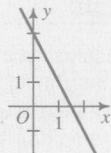

b.

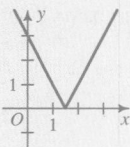

c.

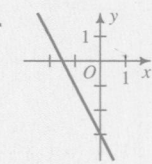

Special Tests for the Symmetry of a Graph

Type of Symmetry	Example
Symmetry in the x-axis $f(\ ) \to -f(\ )$ *Meaning:* $(x, -y)$ is on the graph whenever (x, y) is. *Testing an equation of a graph:* In the equation, leave x alone and substitute $-y$ for y. Does an equivalent equation result?	$y^2x = 1$ equivalent $(-y)^2x = 1$ 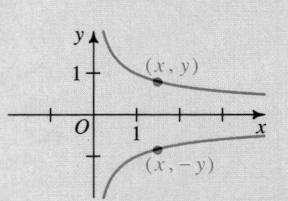
Symmetry in the y-axis $f(\ ) \to f(-\ )$ *Meaning:* $(-x, y)$ is on the graph whenever (x, y) is. *Testing an equation of a graph:* In the equation, substitute $-x$ for x and leave y alone. Does an equivalent equation result?	$y = x^2$ equivalent $y = (-x)^2$ 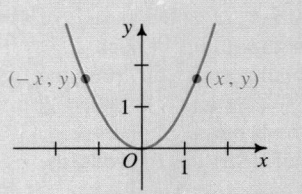 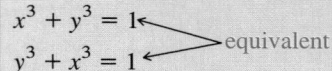
Symmetry in the line $y = x$ *Meaning:* (y, x) is on the graph whenever (x, y) is. *Testing an equation of a graph:* In the equation, interchange x and y. Does an equivalent equation result?	$x^3 + y^3 = 1$ equivalent $y^3 + x^3 = 1$ 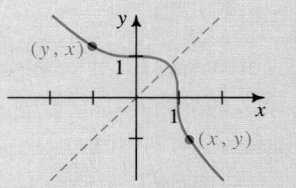
Symmetry in the origin *Meaning:* $(-x, -y)$ is on the graph whenever (x, y) is. *Testing an equation of a graph:* In the equation, substitute $-x$ for x and $-y$ for y. Does an equivalent equation result?	$y = x^3$ equivalent $-y = (-x)^3$ 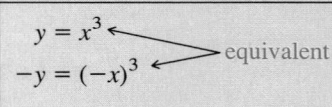 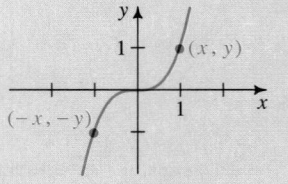

134 *Chapter Four*

Example

Use symmetry to sketch the graph of $y^4 = x + 1$.

Solution

The equation $y^4 = x + 1$ can be rewritten as $y = \pm\sqrt[4]{x + 1}$. You can see from either of these equations that the graph has symmetry in the x-axis. Therefore, you only need to graph $y = \sqrt[4]{x + 1}$ (by plotting a few points or by using a computer or graphing calculator). You can then reflect the graph in the x-axis to obtain the graph of $y = -\sqrt[4]{x + 1}$. The two pieces together comprise the complete graph of $y = \pm\sqrt[4]{x + 1}$.

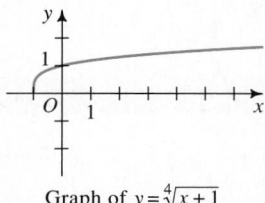

Graph of $y = \sqrt[4]{x+1}$

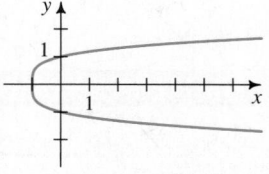

Graph of $y = \pm\sqrt[4]{x+1}$
or $y^4 = x + 1$

CLASS EXERCISES

1. The graph of $y = f(x)$ is shown at the right. Sketch the graph of each of the following equations.
   a. $y = -f(x)$
   b. $y = |f(x)|$
   c. $y = f(-x)$

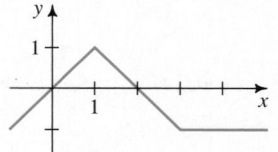

2. For each graph, give the equation(s) of any line(s) of symmetry and give the coordinates of any point of symmetry.

   a.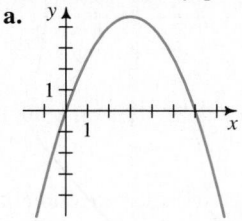

   b.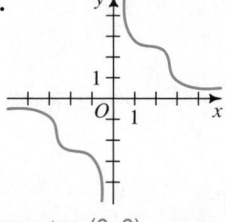

   c.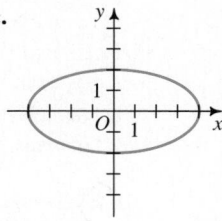

   $x = 3$ $y = \pm x$; (0, 0) $x = 0, y = 0$; (0, 0)

3. Tell whether the graph of each equation has symmetry in:
   (i) the x-axis, (ii) the y-axis, (iii) the line $y = x$, and (iv) the origin.
   a. $x^4 + y^4 = 1$ i, ii, iii, iv   b. $xy^3 = 1$ iv   c. $x(x + y) = 1$ iv

4. *Visual Thinking* If a graph has symmetry in both the x- and y-axes, what other symmetry must it have? Explain.

2. Use symmetry to sketch the graph of each equation.
   a. $|x + y| = 2$
   b. $x|y| = 4$

   a. The graph does not have symmetry in the x-axis or y-axis, but does have symmetry in the line $y = x$ and in the origin.

   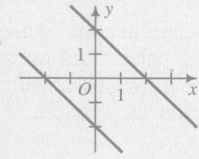

   Note that $|x + y| = 2$ is equivalent to $x + y = \pm 2$. This verifies the graph above, obtained by symmetry.

   b. The graph has symmetry in the x-axis and no other line. Notice that x cannot be negative.

   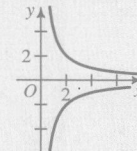

Example Note

Doing a seemingly complicated sketch of a graph can be aided by the use of symmetry tests, as the example on this page shows. The graph of the given equation has symmetry in the x-axis, so only a few points need to be plotted. The rest can be done by reflection.

5. **Visual Thinking** Can a graph that has symmetry in the x-axis be the graph of a function? Explain.

6. **Visual Thinking** Describe how to obtain the graph of $y = \sqrt{|x|}$ from the graph of $y = \sqrt{x}$.

7. Give the equation of the axis of symmetry for the graph of each quadratic function.
   a. $f(x) = x^2 - 8x - 7$ $x = 4$
   b. $g(x) = 8x - 4x^2$ $x = 1$
   c. $h(x) = x^2 + 3$ $x = 0$

8. Give the coordinates of the point of symmetry for the graph of each cubic function.
   a. $f(x) = x^3 - 6x^2 + 5x + 7$ (2, 1)
   b. $g(x) = 9x + 6x^2 + 2x^3$ (−1, −5)
   c. $h(x) = 3x^3 - 3x + 7$ (0, 7)

WRITTEN EXERCISES

In Exercises 1–4, the graph of $y = f(x)$ is given. Sketch the graphs of:
a. $y = -f(x)$ b. $y = |f(x)|$ c. $y = f(-x)$

A 1.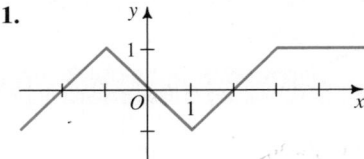

2.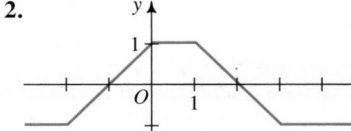

3.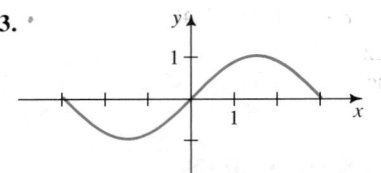

4.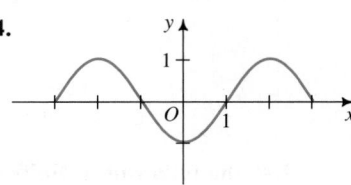

5. Sketch the graphs of $y = x^2 - 9$, $y = 9 - x^2$, and $y = |9 - x^2|$ on a single set of axes.

6. Sketch the graphs of $y = |x| - 2$, $y = 2 - |x|$, and $y = |2 - |x||$ on a single set of axes.

In Exercises 7–14, sketch the graph of each equation and the reflection of the graph in the line $y = x$. Then give an equation of the reflected graph.

7. $y = 3x - 4$ 8. $y = \frac{1}{2}x + 1$ 9. $y = x^2 - 2x$ 10. $y = x^2 + 3x$

11. $y = x^3$ 12. $y = \sqrt{x}$ 13. $y = |x| + 2$ 14. $y = |x| - 3$

15. Test each equation to see if its graph has symmetry in:
    (i) the x-axis, (ii) the y-axis, (iii) the line $y = x$, and (iv) the origin.
    a. $y^2 - xy = 2$ iv   b. $x^2 + y^2 = 1$ i, ii, iii, iv   c. $y = x|x|$ iv

16. Repeat Exercise 15 for each of the following equations.

    a. $x^2 + xy = 4$ iv   **b.** $|x| + |y| = 1$ **c.** $y = \dfrac{x}{|x|}$ iv

                                     i, ii, iii, iv

Use symmetry to sketch the graph of each equation.

B **17.** $|x| + |y| = 2$ **18.** $|x|^{1/2} + |y|^{1/2} = 2$

    19. $x^2 y = 1$ **20.** $|xy| = 12$

In Exercises 21–26, graph each parabola, showing the vertex with its coordinates and the axis of symmetry with its equation. The pairs of graphs in Exercises 21–23 should be done on a single set of axes.

21. a. $y = (x - 3)^2 + 5$ **22. a.** $y = 2(x + 1)^2 + 3$ **23. a.** $y = 3 - (x - 4)^2$
    b. $x = (y - 3)^2 + 5$ **b.** $x = 2(y + 1)^2 + 3$ **b.** $x = 3 - (y - 4)^2$

24. $x = 2(y + 1)^2 + 4$ **25.** $x = y^2 + 6y + 8$ **26.** $x = y^2 + 2y - 3$

27. The graph of a cubic function has a local minimum at $(5, -3)$ and a point of symmetry at $(0, 4)$. At what point does a local maximum occur? $(-5, 11)$

28. a. Find the point of symmetry of the graph of the cubic function $f(x) = -x^3 + 15x^2 - 48x + 45$. $(5, 55)$

    b. The function has a local minimum at $(2, 1)$. At what point does a local maximum occur? $(8, 109)$

29. a. Graph $y = 3x^2 - x^3$. At what point does a local minimum occur? $(0, 0)$

    b. Find the point of symmetry of the graph and then deduce the coordinates of the point where a local maximum occurs. $(1, 2); (2, 4)$

30. a. Graph $y = -x^3 - 6x^2 - 9x$. At what point does a local minimum occur? $(-3, 0)$

    b. Find the point of symmetry and then deduce the coordinates of the point where a local maximum occurs. $(-2, 2); (-1, 4)$

Use the following definitions to complete Exercises 31–36.

                    f is an **even** function if $f(-x) = f(x)$.   _y-axis_

                    f is an **odd** function if $f(-x) = -f(x)$.   _origin_

31. Classify each function as even, odd, or neither.   _Neither_

    a. $f(x) = x^2$ Even   **b.** $f(x) = x^3$ Odd   **c.** $f(x) = x^2 - x$

    d. $f(x) = x^4 + 2x^2$ Even   **e.** $f(x) = x^3 + 3x^2$ Neither   **f.** $f(x) = x^5 - 4x^3$
                                                                           Odd

32. Use the results of Exercise 31 to guess the reasons for using the terms "even" and "odd" as they are applied to polynomial functions.

33. a. What kind of symmetry does the graph of an even function have? In the y-axis
    b. What kind of symmetry does the graph of an odd function have? In the origin

34. Study the graphs shown in Exercises 3 and 4. Then tell whether each function graphed is even or odd. **3.** Odd   **4.** Even

35. If f and g are both odd functions and $h(x) = f(x) \cdot g(x)$, prove that h is even.

36. If f is an even function and g is an odd function, prove that $h(x) = f(x) \cdot g(x)$ is odd.

Functions   **137**

Review Note

You might need to review the general standard form for a parabola, $y = a(x - h)^2 + k$, with vertex (h, k). You also may want to review the ideas of "local minimum" and "local maximum."

Making Connections

In Chapter 7, students will study the sine function (an odd function) and the cosine function (an even function). (See Exercises 31–36.)

Additional Answers
Written Exercises

21. a. $(3, 5)$; $x = 3$
   b. $(5, 3)$; $y = 3$

22. a. $(-1, 3)$; $x = -1$
   b. $(3, -1)$; $y = -1$

23. a. $(4, 3)$; $x = 4$
   b. $(3, 4)$; $y = 4$

24. $(4, -1)$; $y = -1$

25. $(-1, -3)$; $y = -3$

26. $(-4, -1)$; $y = -1$

32. A polynomial function must contain only even powers of the variable to be an even function and only odd powers of the variable to be an odd function.

35. $h(-x) = f(-x) \cdot g(-x)$
          $= [-f(x)][-g(x)]$
          $= f(x) \cdot g(x)$
          $= h(x)$
  $\therefore$ h is even.

36. $h(-x) = f(-x) \cdot g(-x)$
          $= f(x)[-g(x)]$
          $= -[f(x) \cdot g(x)]$
          $= -h(x)$
  $\therefore$ h is odd.

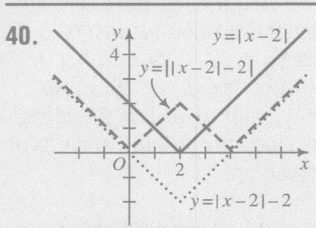
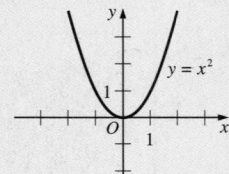
∿ **Use the following definition to complete Exercises 37–39. For Exercise 38, you will need to use a computer or graphing calculator.**

Suppose x_1 and x_2 are any two domain elements of a function f. We say that f is **increasing** in its domain if $f(x_2) > f(x_1)$ whenever $x_2 > x_1$, and f is **decreasing** in its domain if $f(x_2) < f(x_1)$ whenever $x_2 > x_1$. **37. b.** Increasing: graph rises from left to right; decreasing: graph falls from left to right.

37. a. For what value(s) of m is the linear function $f(x) = mx$ an increasing function? a decreasing function? neither? $m > 0$; $m < 0$; $m = 0$

  b. *Visual Thinking* Describe what the definitions of increasing and decreasing functions imply about the graphs of the functions. See above.

38. Graph each function using a computer or a graphing calculator. Then use the graph to tell whether the function is increasing or decreasing in its domain.

  a. $f(x) = x^3 + x - 1$ inc.   **b.** $f(x) = \sqrt[3]{1 - x^3}$ dec.   **c.** $f(x) = \dfrac{10}{1 + 2^x}$ dec.

39. *Visual Thinking* Suppose the graph of an increasing function is reflected in **(a)** the x-axis, **(b)** the y-axis, and **(c)** the line $y = x$. In each case, tell whether the reflected graph represents an increasing or a decreasing function.
a. dec. **b.** dec. **c.** inc.

40. On one set of axes, graph $y = |x - 2|$, $y = |x - 2| - 2$, and $y = ||x - 2| - 2|$.

C **41.** Use symmetry to sketch the graph of $x^{2/3} + y^{2/3} = 1$.

4-4 Periodic Functions; Stretching and Translating Graphs

| **Objective** | To determine periodicity and amplitude from graphs, to stretch and shrink graphs both vertically and horizontally, and to translate graphs. |

Periodic Functions

The world is full of periodic phenomena. The tides come in and go out again and again, each *cycle* or *period* lasting about 12.4 h. The amount of daylight increases and decreases with a period of one year. The functions that describe periodic behavior are called *periodic* functions.

A function f is **periodic** if there is a positive number p, called a **period** of f, such that

$$f(x + p) = f(x)$$

for all x in the domain of f. The smallest period of a periodic function is called the **fundamental period** of the function.

The definition of a periodic function implies that if f is a periodic function with period p, then $f(x) = f(x + mp)$ for all x and any integer m.

Example 1   The graph of a periodic function f is shown below. Find:
   a. the fundamental period of f **b.** $f(99)$

Divide by period
= f(rem) = ANS.

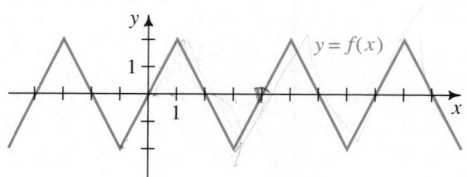

Solution   **a.** If we start at the origin and follow the graph to the right, the graph takes 4 units to complete one up-and-down cycle; another such cycle then begins. Thus, f is a periodic function with fundamental period 4.
   b. Since the fundamental period of f is 4 and $99 \div 4 = 24$ with remainder 3, we have:

$$f(99) = f(99 - 24 \cdot 4) = f(3) = -2$$

If a periodic function has a maximum value M and a minimum value m, then the **amplitude** A of the function is given by:

$$A = \frac{M - m}{2}$$

Example 2   Find the amplitude of the function f described in Example 1.

Solution   The function's maximum value is $M = 2$; its minimum value is $m = -2$. The amplitude is half the difference between M and m:

$$A = \frac{2 - (-2)}{2} = 2$$

Stretching and Shrinking Graphs

If you go into a house of mirrors at a circus or carnival, you may see your reflection distorted by some of the mirrors. For example, one mirror might make you look tall and thin, while another might make you look short and broad. Just as mirrors can stretch or shrink your reflection both vertically and horizontally, it is possible to stretch or shrink the graph of an equation both vertically and horizontally.

Functions   **139**

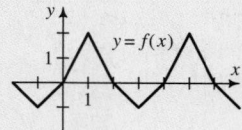

c.

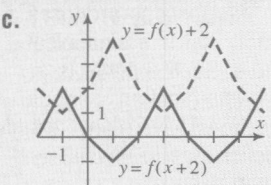

d.

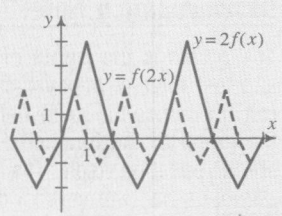

2. Use the graph of $y = \log_3 x$ shown below to describe each graph and name three points on it.

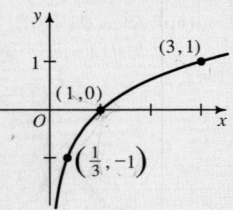

a. $y - 2 = \log_3(x + 1)$
b. $y = \log_3(-x)$
c. $y = |\log_3 x|$
d. $y = -\log_3 x$
e. $x = \log_3 y$
f. $y = \frac{1}{3}\log_3 x$
g. $y = \log_3(2x)$

The graph of $y = cf(x)$ where c is positive (and not equal to 1) is obtained by *vertically* stretching or shrinking the graph of $y = f(x)$. For example, in the graphs below, notice that points on the x-axis remain fixed, while all other points move away from the x-axis for $c > 1$ (a vertical stretch) or toward the x-axis for $0 < c < 1$ (a vertical shrink).

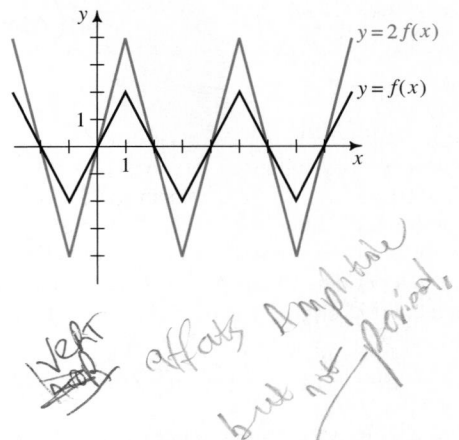

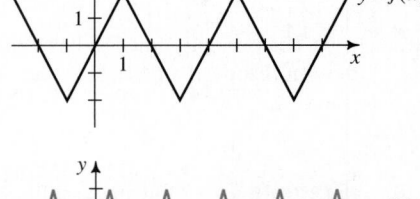

The graph of $y = f(cx)$ where c is positive (and not equal to 1) is obtained by *horizontally* stretching or shrinking the graph of $y = f(x)$. For example, in the graphs at the right, notice that points on the y-axis remain fixed, while all other points move toward the y-axis for $c > 1$ (a horizontal shrink) or away from the y-axis for $0 < c < 1$ (a horizontal stretch).

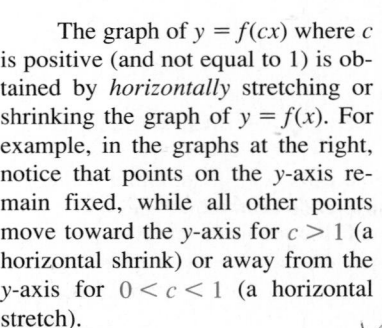

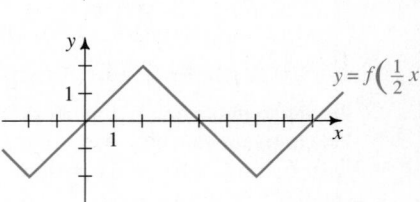

The graphs shown above are all based on a periodic function f with fundamental period 4 and amplitude 2. Notice that a vertical stretching or shrinking of the graph of f affects the amplitude but not the period, and a horizontal stretching or shrinking of the graph affects the period but not the amplitude. A more formal statement of these results is given at the top of the next page.

Changing the Period and Amplitude of a Periodic Function

If a periodic function f has period p and amplitude A, then:

$y = cf(x)$ has period p and amplitude cA, and

$y = f(cx)$ has period $\frac{p}{c}$ and amplitude A.

[handwritten: Vert stretching] [handwritten: hor. stretching] [handwritten: mult y value by c] [handwritten: mult x value by $\frac{1}{c}$]

Translating Graphs

The graph of $y - k = f(x - h)$ is obtained by translating the graph of $y = f(x)$ horizontally h units and vertically k units. For example, as shown at the right below, the graph of $y - 1 = (x - 2)^2$ is the graph of $y = x^2$ translated 2 units horizontally and 1 unit vertically.

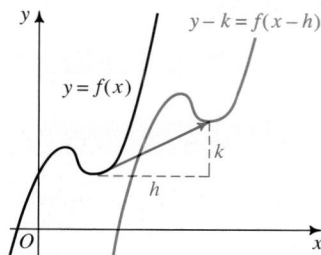
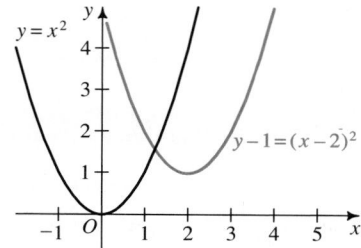

Example 3 Sketch the graphs of the following equations:

$$y = |x|, \; y - 2 = |x - 3|, \; \text{and} \; y = |x + 5|$$

Solution Once we have graphed $y = |x|$, we can slide the graph 3 units horizontally and 2 units vertically to obtain the graph of $y - 2 = |x - 3|$. Since the equation $y = |x + 5|$ is equivalent to $y = |x - (-5)|$, we slide the original graph -5 units horizontally and 0 units vertically to obtain the graph of $y = |x + 5|$.

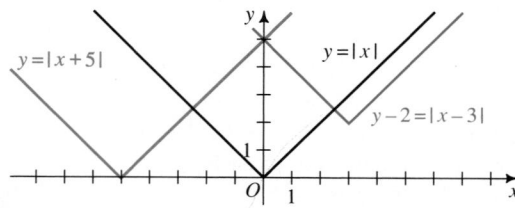

In this chapter we have seen how certain simple changes in the equation of a curve can stretch or shrink the curve, reflect it, or translate it. These results are summarized at the top of the next page.

Functions **141**

a. The graph of $y = \log_3 x$ is shifted 1 unit to the left and 2 units up; $\left(-\frac{2}{3}, 1\right)$, $(0, 2)$, $(2, 3)$.

b. Reflect the graph in the y-axis; $\left(-\frac{1}{3}, -1\right)$, $(-1, 0)$, $(-3, 1)$. *[handwritten: x → -x]*

c. The portion of the graph in the interval $0 < x < 1$ is reflected in the x-axis; $\left(\frac{1}{3}, 1\right)$, $(1, 0)$, $(3, 1)$.

d. The graph is reflected in the x-axis; $\left(\frac{1}{3}, 1\right)$, $(1, 0)$, $(3, -1)$. *[handwritten: y → -y]*

e. The graph is reflected in the line $y = x$; $\left(-1, \frac{1}{3}\right)$, $(0, 1)$, $(1, 3)$.

f. The graph is shrunk vertically; $\left(\frac{1}{3}, -\frac{1}{3}\right)$, $(1, 0)$, $\left(3, \frac{1}{3}\right)$.

g. The graph is shrunk horizontally; $\left(\frac{1}{6}, -1\right)$, $\left(\frac{1}{2}, 0\right)$, $\left(\frac{3}{2}, 1\right)$.

Cooperative Learning

Using $c = 2$ or $c = \frac{1}{2}$ and $f(x) = x^2$, have students work in groups to discuss the various changes to the graph of $y = f(x)$ listed in the chart on this page. Also have students discuss Class Exercises 5 and 6 in groups to see if they understand the concepts.

Additional Answers
Class Exercises

5. Let $(a, 0)$ be a point on the graph of $y = f(x)$. If the graph is vertically stretched or shrunk, the equation of the new graph is $y = cf(x)$. At $x = a$, $y = cf(a) = c \cdot 0 = 0$. Thus, $(a, 0)$ is also on the new graph.

6. Let $(0, b)$ be a point on the graph of $y = f(x)$. If the graph is horizontally stretched or shrunk, the equation of the new graph is $y = f(cx)$. At $x = 0$, $y = f(c \cdot 0) = f(0) = b$. Thus, $(0, b)$ is also on the new graph.

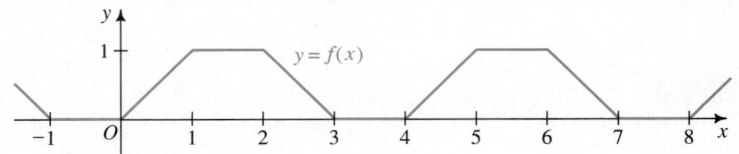

If the equation $y = f(x)$ is changed to:	Then the graph of $y = f(x)$ is:
$y = -f(x)$	reflected in the x-axis.
$y = \|f(x)\|$	unchanged when $f(x) \geq 0$ and reflected in the x-axis when $f(x) < 0$.
$y = f(-x)$	reflected in the y-axis.
$x = f(y)$	reflected in the line $y = x$.
$y = cf(x), c > 1$	stretched vertically.
$y = cf(x), 0 < c < 1$	shrunk vertically.
$y = f(cx), c > 1$	shrunk horizontally.
$y = f(cx), 0 < c < 1$	stretched horizontally.
$y - k = f(x - h)$	translated h units horizontally and k units vertically.

(handwritten annotations: Y→-Y; Y→Y; X→-X; multiply y value by c, effect AMPLITUDE/RANGE ÷ by c; ÷x val by c, effects period $\frac{p}{c}$; per AMP, RANGE STAY SAME)

CLASS EXERCISES

For Exercises 1–4, refer to the graph of a function $y = f(x)$ shown below.

(graph of $y = f(x)$, a periodic trapezoidal wave with axes labeled from -1 to 8 on the x-axis and 1 on the y-axis)

1. **a.** What is the fundamental period of f? 4
   b. What is the amplitude? $\frac{1}{2}$
   c. Find $f(25)$ and $f(-25)$. 1; 0

2. Sketch the graphs of $y = 2f(x)$ and $y = \frac{1}{2}f(x)$. Give the fundamental period and amplitude of each. $p = 4$, $A = 1$; $p = 4$, $A = \frac{1}{4}$

3. Sketch the graphs of $y = f(2x)$ and $y = f\left(\frac{1}{2}x\right)$. Give the fundamental period and amplitude of each. $p = 2$, $A = \frac{1}{2}$; $p = 8$, $A = \frac{1}{2}$

4. Sketch the graphs of $y = f(x) - 2$ and $y = f(x - 2)$. Give the fundamental period and amplitude of each. $p = 4$, $A = \frac{1}{2}$; $p = 4$, $A = \frac{1}{2}$

5. If a graph is vertically stretched or shrunk, explain why points on the x-axis remain fixed.

6. If a graph is horizontally stretched or shrunk, explain why points on the y-axis remain fixed.

142   *Chapter Four*

WRITTEN EXERCISES

☒ Using Technology

Students should do Exercises 6 and 7 by hand but check their work with graphing calculators.

In Exercises 1–4, the graph of a function f is given. Tell whether f appears to be periodic. If so, give its fundamental period and its amplitude, and then find $f(1000)$ and $f(-1000)$.

Yes; $p = 0.8$; $A = 0.5$; 0, 0

A **1.**

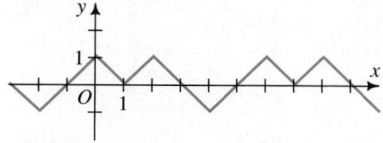

Yes; $p = 6$, $A = 1$; −1, 1

2.

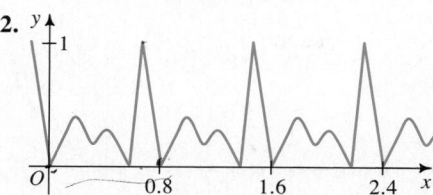

3.

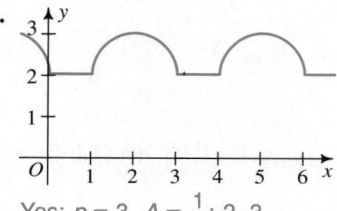

Yes; $p = 3$, $A = \frac{1}{2}$; 2, 3

4. No

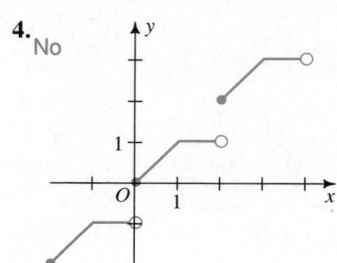

Error Analysis

If students have difficulty starting Exercises 6 or 7, refer them to Example 3 on page 141. They should start by graphing the basic function, then stretch, shrink, or translate it.

Suggested Assignments

Standard
Day 1: 143/1–8
Day 2: 144/9–15
Comprehensive
Day 1: 143/1–8
Day 2: 144/9–19

Supplementary Materials

Alternative Assessment, 12

5. Use the graph of $y = f(x)$, shown at the right, to sketch the graph of each of the following.

  a. $y = 2f(x)$ **b.** $y = -\frac{1}{2}f(x)$

  c. $y = f(-2x)$ **d.** $y = f\left(\frac{1}{2}x\right)$

  e. $y = f\left(x - \frac{1}{2}\right)$ **f.** $y = f(-x) + 1$

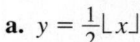

6. The greatest integer function $y = \lfloor x \rfloor$ gives the greatest integer less than or equal to x. Thus, $\lfloor 2.1 \rfloor = 2$ and $\lfloor -3.1 \rfloor = -4$. Use the graph of this function, shown at the right, to sketch the graph of each of the following.

  a. $y = \frac{1}{2}\lfloor x \rfloor$ **b.** $y = -2\lfloor x \rfloor$

  c. $y = \left\lfloor -\frac{1}{2}x \right\rfloor$ **d.** $y = \lfloor 2x \rfloor$

  e. $y = \lfloor x - 1 \rfloor$ **f.** $y = 2\lfloor x \rfloor + 1$

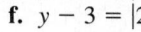
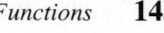

7. Sketch the graph of each of the following.
  a. $y + 2 = |x|$ **b.** $y = |x - 3|$ **c.** $y - 4 = |x + 5|$
  d. $y = 2|x + 1|$ **e.** $y + 1 = -|x|$ **f.** $y - 3 = |2x|$

8. Sketch the graph of each of the following.
   a. $y - 1 = \sqrt{x}$
   b. $y = \sqrt{x + 4}$
   c. $y + 2 = \sqrt{x - 5}$
   d. $y = 2\sqrt{x - 3}$
   e. $y - 2 = \sqrt{-x}$
   f. $y - 4 = \sqrt{4x}$

B **9.** Use the graph of $y = 2^x$, shown at the left below, to sketch the graph of each of the following.
   a. $y = 2^{-x}$
   b. $y = 2^{x-1}$
   c. $y = 3 - 2^x$
   d. $x = 2^y$

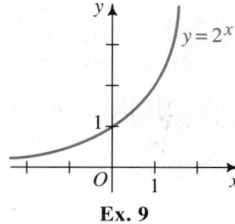

Ex. 9

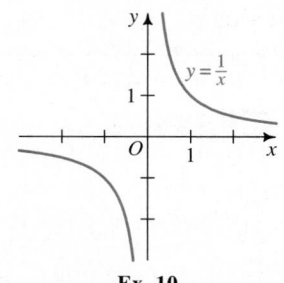

Ex. 10

10. Use the graph of $y = \dfrac{1}{x}$, shown at the right above, to sketch the graph of each of the following.
   a. $y = -\dfrac{1}{x}$
   b. $y = \dfrac{1}{x - 2}$
   c. $y = 1 + \dfrac{1}{x}$
   d. $x = \dfrac{1}{y}$

11. Given that the equation of a circle with radius 3 and center $(0, 0)$ is $x^2 + y^2 = 9$ (graph at right), deduce the equation of the circle if it is translated so that its center is $(8, 4)$. $(x - 8)^2 + (y - 4)^2 = 9$

12. a. Refer to the circle with equation $x^2 + y^2 = 9$ in Exercise 11. Sketch the graph of $\left(\dfrac{x}{2}\right)^2 + y^2 = 9$.

   b. In Exercise 11, the area of the circular region is $\pi r^2 = \pi \cdot 3^2 = 9\pi$. Make a conjecture about the area of the region enclosed by the graph in part (a).

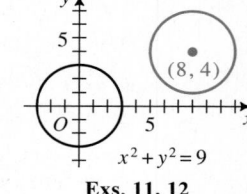

Exs. 11, 12

13. The graph of $y = f(x)$, shown at the right, has x-intercepts at 0 and 6 and a local maximum at $(4, 32)$.
   a. Where do the x-intercepts and local maximum occur on the graph of $y = f(2x)$? 0, 3; (2, 32)
   b. Where do the x-intercepts and local maximum occur on the graph of $y = 2f(x)$? 0, 6; (4, 64)
   c. Where do the x-intercepts and local maximum occur on the graph of $y = f(x - 2)$? 2, 8; (6, 32)
   d. Where do the x-intercepts and local maximum occur on the graph of $y = f(x + 2)$? −2, 4; (2, 32)
   e. If f is a cubic polynomial, find a rule for $f(x)$. $-x^2(x - 6)$

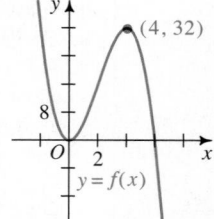

14. a. Sketch the graph of $y = x^3 - 3x^2 + 2x$ and label all intercepts.

b. Sketch the graph of $y = \left(\frac{1}{2}x\right)^3 - 3\left(\frac{1}{2}x\right)^2 + 2\left(\frac{1}{2}x\right)$ by using the graph of part (a). Label all intercepts.

 The first figure in Exercise 15 shows the calculator-drawn graph of $y = |x|$. **Changes in the equation $y = |x|$ cause its graph to be reflected, stretched, shrunk, and/or translated to produce the graphs shown in parts (a)–(e). Give an equation for each graph. You may wish to use a computer or a graphing calculator to confirm your answers.**

15.

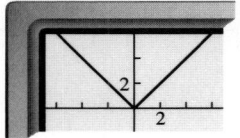

a. $y = |x - 4| + 2$

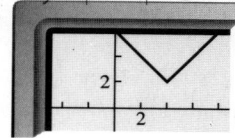

b. $y = -|x| + 4$

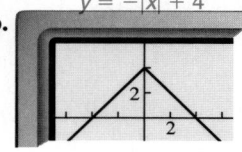

c.

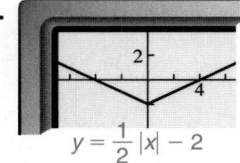

$y = \frac{1}{2}|x| - 2$

d.

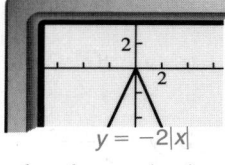

$y = -2|x|$

e.

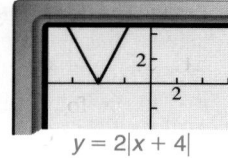

$y = 2|x + 4|$

16. In this exercise, you will show that the graph of

$$y = ax^3 + bx^2 + cx + d$$

has a point of symmetry at $x = -\dfrac{b}{3a}$.

a. If $y = f(x) = ax^3 + px$, show that $f(-x) = -f(x)$.

b. Explain how part (a) shows that the origin is a symmetry point of the graph of $y = ax^3 + px$.

c. Explain why $(0, q)$ is a symmetry point for the graph of $y = ax^3 + px + q$.

d. Explain why (h, q) is a symmetry point for the graph of $y = a(x - h)^3 + p(x - h) + q$.

e. Suppose the equation

$$y = ax^3 + bx^2 + cx + d$$

is rewritten in the equivalent form

$$y = a(x - h)^3 + p(x - h) + q.$$

By comparing the coefficients of the x^2 terms, show that $h = -\dfrac{b}{3a}$. Then use the results of part (d) to conclude that $y = ax^3 + bx^2 + cx + d$ has a point of symmetry at $x = -\dfrac{b}{3a}$.

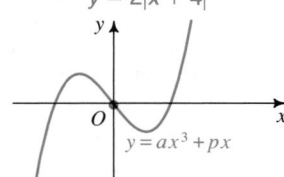

$y = ax^3 + px$

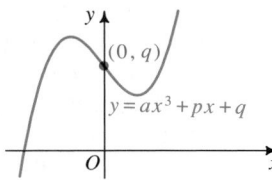

$y = ax^3 + px + q$

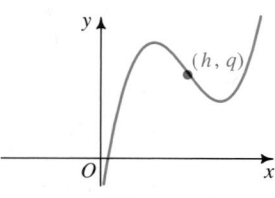

$y = a(x - h)^3 + p(x - h) + q$

Solve each equation for *y*.

1. $x = 2y - 7$ $y = \dfrac{x+7}{2}$

2. $x = \dfrac{1}{3}y - \dfrac{3}{2}$ $y = 3x + \dfrac{9}{2}$

3. $x = \dfrac{1}{y-1}$ $y = \dfrac{1+x}{x}$

4. $x = y^3 + 8$ $y = \sqrt[3]{x - 8}$

Motivating the Section

Conversion formulas can be expressed as inverse pairs, the subject of this section. For example, $f = \dfrac{1}{12}i$ and $i = 12f$ are formulas for converting inches to feet and feet to inches, respectively. Ask students to name other such pairs of conversion formulas.

Mathematical Note

If two functions are inverses of each other, their composition gives the *identity* function, $f(x) = x$. For the functions *F* and *C* given on this page, one formula "undoes" the other one.

---

17. If the periodic functions *f* and *g* both have fundamental period *p* and $h(x) = (f + g)(x)$, show that $h(x + p) = h(x)$ for all *x* in the domain of *h*.

18. If *f* has fundamental period 2 and *g* has fundamental period 3, what is the fundamental period of $f + g$? 6

C **19.** Is a constant function like $f(x) = 5$ periodic? If so, does it have a fundamental period? Explain.

20. A function *f* is defined for all real numbers as follows: $f(x) = 1$ if *x* is rational, and $f(x) = 0$ if *x* is irrational. Show that if *p* is a rational number, then $f(x + p) = f(x)$ for all *x*. Does *f* have a fundamental period? Explain.

4-5 Inverse Functions

Objective   *To find the inverse of a function, if the inverse exists.*

Conversion formulas, such as those used in converting between U.S. customary and metric measurements, come in pairs. For example, consider the conversion formulas for the Fahrenheit and Celsius temperature scales:

$$F = \frac{9}{5}C + 32 \text{ and } C = \frac{5}{9}(F - 32)$$

The first formula gives a Fahrenheit temperature *F* as a function of a Celsius temperature *C*, while the second formula gives *C* as a function of *F*.

Notice that in the first formula, $F = 32$ when $C = 0$, and in the second formula, $C = 0$ when $F = 32$. Because each formula undoes what the other one does, the formulas are examples of *inverses*.

Two functions *f* and *g* are called **inverse functions** if the following statements are true:

1. $g(f(x)) = x$ for all *x* in the domain of *f*;
2. $f(g(x)) = x$ for all *x* in the domain of *g*.

Example 1   If $f(x) = \dfrac{x-1}{2}$ and $g(x) = 2x + 1$, show that *f* and *g* are inverses of each other.

Solution   We must check that conditions (1) and (2) stated in the definition above are satisfied:

1. $g(f(x)) = g\left(\dfrac{x-1}{2}\right) = 2\left(\dfrac{x-1}{2}\right) + 1 = x$ for all *x*

2. $f(g(x)) = f(2x + 1) = \dfrac{(2x+1)-1}{2} = x$ for all *x*

We denote the inverse of a function f by f^{-1}, read "f inverse." The symbol $f^{-1}(x)$, read "f inverse of x," is the value of f^{-1} at x. Note that $f^{-1}(x)$ does *not* mean $\dfrac{1}{f(x)}$.

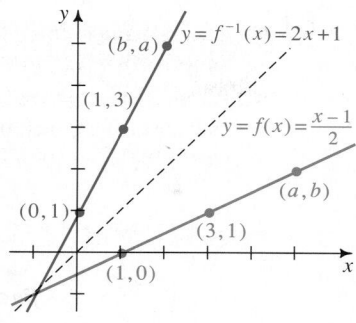

The graphs of f and f^{-1} in Example 1 are shown at the right. Notice that for every point (a, b) on the graph of $y = f(x)$, the point (b, a) is on the graph of $y = f^{-1}(x)$. This means that the graph of f^{-1} is the reflection of the graph of f in the line $y = x$.

When a function f has an inverse, the graph of f^{-1} can be obtained from the graph of f by changing every point (x, y) on the graph to the point (y, x). Similarly, the rule for $f^{-1}(x)$ can be obtained by interchanging x and y in the equation $y = f(x)$, as shown in the following example.

Example 2 Let $f(x) = 4 - x^2$ for $x \geq 0$.
  a. Sketch the graph of $y = f^{-1}(x)$.
  b. Find a rule for $f^{-1}(x)$.

Solution
  a. Since the domain of $f(x) = 4 - x^2$ is $x \geq 0$, the graph of $y = f(x)$ is half of a parabola, as shown in red at the right. By reflecting the graph of $y = f(x)$ in the line $y = x$, we obtain the graph of $y = f^{-1}(x)$, as shown in blue at the right.

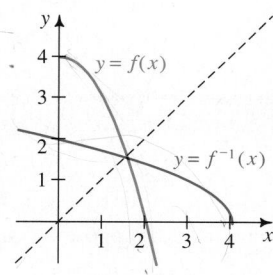

  b. 1. Set $y = f(x)$. $y = 4 - x^2$, $x \geq 0$ ⟵ describes $y = f(x)$

   2. Interchange x and y. $x = 4 - y^2$, $y \geq 0$ ⟵ describes $y = f^{-1}(x)$

   3. Solve for y. $y = \pm\sqrt{4 - x}$, $y \geq 0$
      $y = \sqrt{4 - x}$

   We therefore have the rule $f^{-1}(x) = \sqrt{4 - x}$. The graph shows that the domain of f^{-1} is $\{x \mid x \leq 4\}$.

Note that in Example 2, if the domain of $f(x) = 4 - x^2$ were the set of all real numbers instead of nonnegative real numbers, the reflection of the graph of $y = f(x)$ in the line $y = x$ would not be the graph of a function. Thus, not all functions have inverses.

1. State whether each function has an inverse. If f^{-1} exists, find a rule for $f^{-1}(x)$ and show that $f(f^{-1}(x)) = f^{-1}(f(x)) = x$.

a. $f(x) = |x + 1|$

b. $f(x) = \dfrac{1}{x+1}$, $x \neq -1$

a. No inverse since the function is not one-to-one.

b. There is an inverse, since the function is one-to-one. Let $x = \dfrac{1}{y+1}$; $x(y+1) = 1$;

$y + 1 = \dfrac{1}{x}$; $y = \dfrac{1-x}{x}$;

$\therefore f^{-1}(x) = \dfrac{1-x}{x}$, $x \neq 0$.

$f(f^{-1}(x)) = f\left(\dfrac{1-x}{x}\right) =$

$\dfrac{1}{\dfrac{1-x}{x} + 1} =$

$\dfrac{x}{(1-x)+x} = x.$

$f^{-1}(f(x)) =$

$f^{-1}\left(\dfrac{1}{x+1}\right) =$

$\dfrac{1 - \dfrac{1}{x+1}}{\dfrac{1}{x+1}} \cdot \dfrac{x+1}{x+1} =$

$\dfrac{(x+1) - 1}{1} = x.$

2. Show that $h^{-1}(x) = h(x)$ if $h(x) = -\dfrac{2}{x}$.

Interchange x and y in

$y = -\dfrac{2}{x}$ to obtain $x = -\dfrac{2}{y}$. Then $xy = -2$ and $y = -\dfrac{2}{x}$. Thus, $h^{-1}(x) = -\dfrac{2}{x} = h(x)$.

A function $y = f(x)$ that has an inverse is called a **one-to-one function**, because not only does each x-value correspond to exactly one y-value, but also each y-value corresponds to exactly one x-value. We can determine whether a function is one-to-one by applying the *horizontal-line test* to its graph.

The Horizontal-Line Test

If the graph of the function $y = f(x)$ is such that no horizontal line intersects the graph in more than one point, then f is one-to-one and has an inverse.

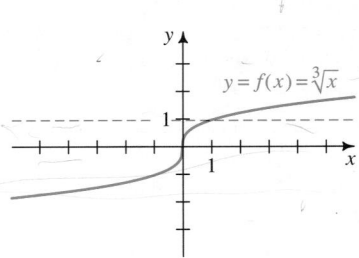

f is one-to-one and has an inverse.

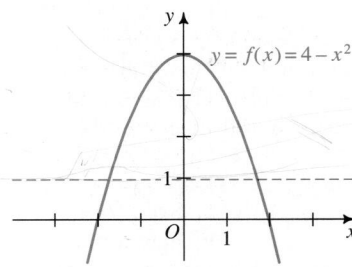

f is not one-to-one and has no inverse. (Compare with f in Example 2.)

3. a. $\dfrac{1}{4}x$ **b.** $\dfrac{x-2}{3}$ **c.** $\dfrac{x+1}{2}$ **d.** $\dfrac{4-x}{5}$

CLASS EXERCISES

1. Consider the birthday function described on page 119. Is this function one-to-one? Does it have an inverse? No; no

2. Suppose a function f has an inverse. If $f(2) = 3$, find each of the following.
   a. $f^{-1}(3)$ 2   **b.** $f(f^{-1}(3))$ 3   **c.** $f^{-1}(f(2))$ 2

3. Find a rule for $g^{-1}(x)$ if: See above.
   a. $g(x) = 4x$ **b.** $g(x) = 3x + 2$
   c. $g(x) = 2x - 1$ **d.** $g(x) = 4 - 5x$

4. The graph of $f(x) = x^3$ is shown at the right.
   a. Name several points on the graph of $y = f^{-1}(x)$ and then sketch the graph. $(8, 2)$, $(1, 1)$, $(-1, -1)$, $(-8, -2)$
   b. Find a rule for $f^{-1}(x)$. $\sqrt[3]{x}$

5. a. Is the function $f(x) = x^3 - 1$ a one-to-one function? Explain. Yes
   b. Is the function $g(x) = x^3 - x$ a one-to-one function? Explain. No; $g(-1) = g(1) = 0$

6. Consider the two temperature conversion formulas given on page 146. Explain how one formula can be obtained from the other.
   Solve for the desired variable.

Ex. 4

7. The graphs of f, g, and h are shown below. Which functions are one-to-one? Which functions have inverses? g; g

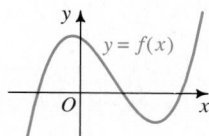

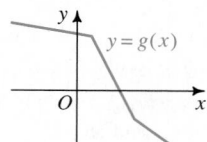

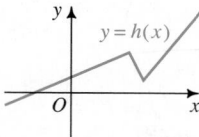

8. Which of the following functions have inverses? b, d
  a. $f(x) = |x|$
  b. $f(x) = x^3$
  c. $f(x) = x^4$
  d. $f(x) = x^4$, $x \leq 0$

9. On the dial or the buttons of a telephone, a telephone function T pairs letters of the alphabet with the digits 2–9. For example, $T(A) = 2$ and $T(D) = 3$. Does T have an inverse? Explain. No. $T(A) = T(B) = 2$, so T is not one-to-one.

10. Explain how the vertical-line test (given on page 121) can be used to justify the horizontal-line test (given on page 148).

11. Explain why the domain of a one-to-one function f is the range of f^{-1} and why the domain of f^{-1} is the range of f.

WRITTEN EXERCISES

A **1.** Suppose a function f has an inverse. If $f(2) = 6$ and $f(3) = 7$, find:
  a. $f^{-1}(6)$ 2
  b. $f^{-1}(f(3))$ 3
  c. $f(f^{-1}(7))$ 7

2. Suppose a function f has an inverse. If $f(0) = -1$ and $f(-1) = 2$, find:
  a. $f^{-1}(-1)$ 0
  b. $f^{-1}(f(0))$ 0
  c. $f(f^{-1}(2))$ 2

3. If $g(3) = 5$ and $g(-1) = 5$, explain why g has no inverse. g is not one-to-one.

4. Explain why $f(x) = x^3 + x^2$ has no inverse. f is not one-to-one: $f(-1) = f(0) = 0$.

5. Let $h(x) = 4x - 3$.
  a. Sketch the graphs of h and h^{-1}.
  b. Find a rule for $h^{-1}(x)$. $\dfrac{x+3}{4}$

6. Let $L(x) = \frac{1}{2}x - 4$.
  a. Sketch the graphs of L and L^{-1}.
  b. Find a rule for $L^{-1}(x)$. $2x + 8$

In Exercises 7–10, the graph of a function is given. State whether the function has an inverse.

7.
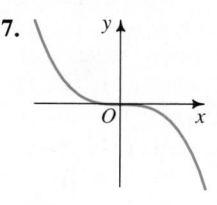

Yes

8.
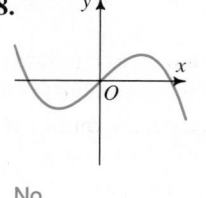

No

9.
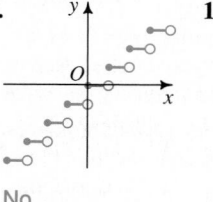

No

10.
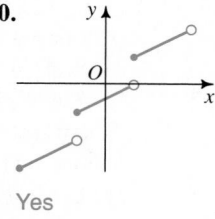

Yes

Additional Answers
Class Exercises

10. To obtain the graph of the inverse of a function f, we reflect the graph of f in the line $y = x$. This reflection transforms horizontal lines to vertical lines. Thus, if the graph of f passes the horizontal-line test, then the inverse graph passes the vertical-line test and is the graph of a function.

Additional Answers
Written Exercises

5. a.
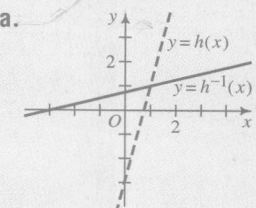

6. a.
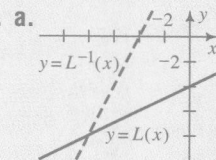

Suggested Assignments

Standard
Day 1: 149/1–10
Day 2: 150/12–26 even; Calc.
Exs. 1–3
Comprehensive
Day 1: 149/1–19 odd
Day 2: 150/21–31; Calc.
Exs. 1–3

Supplementary Materials

Alternative Assessment, 12
Precalculus Plotter Plus, 31–32
Student Resource Guide, 36–40

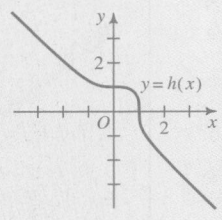
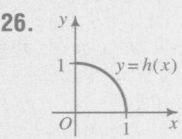
State whether the function f has an inverse. If f^{-1} exists, find a rule for $f^{-1}(x)$ and show that $f(f^{-1}(x)) = f^{-1}(f(x)) = x$.

11. $f(x) = 3x - 5$ Yes; $\frac{x+5}{3}$ **12.** $f(x) = |x| - 2$ No **13.** $f(x) = \sqrt[4]{x}$ Yes; x^4, $x \geq 0$

14. $f(x) = \frac{1}{x}$ Yes; $\frac{1}{x}$, $x \neq 0$ **15.** $f(x) = \frac{1}{x^2}$ No **16.** $f(x) = \sqrt{5 - x}$
Yes; $5 - x^2$, $x \geq 0$

17. $f(x) = \sqrt{4 - x^2}$ No **18.** $f(x) = \sqrt{5 - x^2}$ No **19.** $f(x) = \sqrt[3]{1 + x^3}$
Yes; $\sqrt[3]{x^3 - 1}$

Sketch the graphs of g and g^{-1}. Then find a rule for $g^{-1}(x)$.
$-\sqrt{9 - x}$, $x \leq 9$

B **20.** $g(x) = x^2 + 2$, $x \geq 0$ $\sqrt{x - 2}$, $x \geq 2$ **21.** $g(x) = 9 - x^2$, $x \leq 0$

22. $g(x) = (x - 1)^2 + 1$, $x \leq 1$ **23.** $g(x) = (x - 4)^2 - 1$, $x \geq 4$
$1 - \sqrt{x - 1}$, $x \geq 1$ $4 + \sqrt{x + 1}$, $x \geq -1$

In Exercises 24–26, show that $h^{-1}(x) = h(x)$. Then sketch the graph of h. You may wish to use a computer or graphing calculator.

24. $h(x) = \sqrt[3]{1 - x^3}$ **25.** $h(x) = \frac{x}{x - 1}$ **26.** $h(x) = \sqrt{1 - x^2}$, $x \geq 0$

27. a. Using the results of Exercises 24–26, state how the graph of h is related to the line $y = x$ when $h^{-1}(x) = h(x)$. Graph of h is symmetric in the line $y = x$.
   b. Find a function h, different from those in Exercises 24–26, such that $h^{-1}(x) = h(x)$. For example: $h(x) = -x$

28. Refer to the definition of an increasing function given on page 138.
   a. Explain why an increasing function must have an inverse. Inc. function is one-to-one
   b. Suppose f is an increasing function. Is f^{-1} also an increasing function? Explain your answer and support it with at least two examples. Yes

C **29.** Which statement below is true? Prove it. ii
   (i) $(f \circ g)^{-1} = f^{-1} \circ g^{-1}$ (ii) $(f \circ g)^{-1} = g^{-1} \circ f^{-1}$

30. If f is a linear function such that $f(x + 2) - f(x) = 6$, find the value of $f^{-1}(x + 2) - f^{-1}(x)$. $\frac{2}{3}$

31. Suppose a, b, and c are constants such that $a \neq 0$. Let $P(x) = ax^2 + bx + c$ for $x \leq -\frac{b}{2a}$. Find a rule for $P^{-1}(x)$. $\dfrac{-b - \sqrt{b^2 - 4a(c - x)}}{2a}$, $b^2 - 4a(c - x) \geq 0$

▮▮▮▮ CALCULATOR EXERCISES

In the next chapter we will define the functions $f(x) = e^x$ and $g(x) = \ln x$. You can use a calculator to learn something about these functions.
f and g are inverse functions.

1. Enter any number. Press the $\boxed{e^x}$ and $\boxed{\ln x}$ keys alternately several times. What do you notice? Repeat this process for several other numbers. How would you describe the relationship between $f(x) = e^x$ and $g(x) = \ln x$?

2. By entering various numbers, determine whether $f(x) = e^x$ is defined for all real numbers. It is.

3. By experimenting, determine the domain of $g(x) = \ln x$. $x > 0$

■ Applications of Functions

4-6 Functions of Two Variables

Objective | *To graph functions of two variables in a two-dimensional coordinate system and to read such graphs.*

The formula for the perimeter of a rectangle, $P = 2l + 2w$, tells us that the perimeter P depends on both the length l and the width w of the rectangle. In other words, P is a function of two variables, l and w, and we write

$$P(l, w) = 2l + 2w$$

to emphasize the functional relationship.

Graphing a function of two variables requires a three-dimensional coordinate system as shown at the right. Notice that there are three axes, each perpendicular to the other two. We will study such coordinate systems in Chapter 12, but for now let us consider two ways to describe P in a two-dimensional coordinate system.

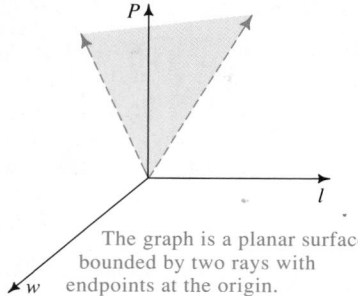

The graph is a planar surface bounded by two rays with endpoints at the origin.

Graph of $P(l, w) = 2l + 2w$
(where $l > 0$ and $w > 0$)
in a 3-dimensional coordinate system

One way is to draw curves for various constant values of either w or l. For example, by using constant values of w, we get such equations as $P(l, 1) = 2l + 2$, $P(l, 2) = 2l + 4$, and $P(l, 3) = 2l + 6$. The graphs of these equations, called curves of constant width, are shown in the lP-plane at the left below.

Another way to describe P in a two-dimensional plane is to draw curves along which the value of the perimeter function is constant. For example, $P(l, w) = 8$ when $(l, w) = (1, 3), (2, 2),$ and $(3, 1)$. These points and others for which $P(l, w) = 8$ are shown on the red curve in the lw-plane at the right below. Other curves of constant perimeter are also shown.

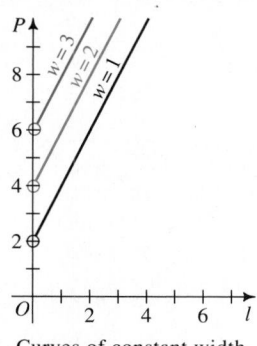

Curves of constant width

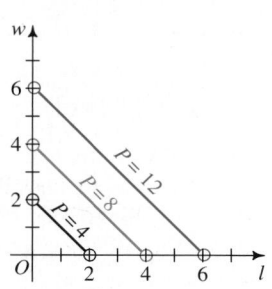

Curves of constant perimeter

Functions **151**

1. The total surface area of a cone is given by $A = \pi r^2 + \pi r l$ where r is the base radius and l is the slant height.
   a. The total area is a function of what two variables?
   b. Draw a graph of l versus r for $A = 24\pi$, $A = 48\pi$, and $A = 96\pi$ (curves of constant area).
   c. Show that $A(2r, 2l) = 4 \cdot A(r, l)$.
   d. Explain the statement in part (c).

   a. the radius and slant height
   b.

   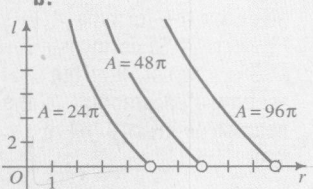

   c. $A(2r, 2l) =$
      $\pi(2r)^2 + \pi(2r)(2l) =$
      $4(\pi r^2 + \pi r l) =$
      $4 \cdot A(r, l)$
   d. If the base radius and slant height of a cone are both doubled, then the total surface area is quadrupled.

2. Describe the domain of
   $f(x, y) = \dfrac{x^2 + y^2}{x^2 - xy - 2y^2}.$

   The function is not defined when $x^2 - xy - 2y^2 = 0$, that is, when $(x - 2y) \cdot (x + y) = 0$. Thus the domain is $\{(x, y)\,|\,x \neq 2y$ and $x \neq -y\}$.

Example

The volume of a cylinder is given by the formula $V = \pi r^2 h$.
a. Sketch the graphs of V versus r for $h = 1$, $h = 2$, and $h = 3$. (These graphs are called curves of constant height.)
b. Sketch the graphs of h versus r for $V = 36\pi$ and $V = 72\pi$. (These graphs are called curves of constant volume.)
c. If h is held constant, what happens to V when r is doubled?

solve for h in terms of r

Solution

a. Since $V(r, h) = \pi r^2 h$, the three curves to be graphed have equations:

   $$V(r, 1) = \pi r^2,$$
   $$V(r, 2) = 2\pi r^2,$$
   and $$V(r, 3) = 3\pi r^2.$$

   With both r and V being positive quantities, we use only the first quadrant when graphing the equations. Each curve is half of a parabola, as shown.

b. If $V = 36\pi$, we substitute $\pi r^2 h$ for V and solve for h:

   $$\pi r^2 h = 36\pi$$
   $$h = \frac{36}{r^2}$$

   We use the last equation to make a table of values:

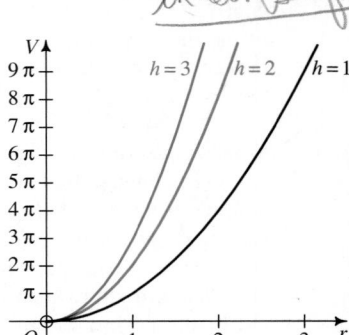

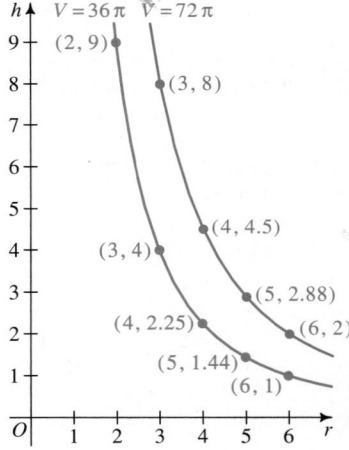

r	1	2	3	4	5	6
$h = \dfrac{36}{r^2}$	36	9	4	2.25	1.44	1

   We then plot the pairs (r, h) to obtain the curve for $V = 36\pi$. In the same way, we sketch the curve for $V = 72\pi$. Both curves are shown at the right.

c. Holding h constant and doubling r, we have:

   $$V(2r, h) = \pi(2r)^2 h = 4 \cdot \pi r^2 h = 4 \cdot V(r, h)$$

   Thus, when h is held constant and r is doubled, V is quadrupled.

Real-world functions of more than one variable are often graphed in two dimensions. For example, meteorologists produce maps like the one below that show curves of constant barometric pressure.

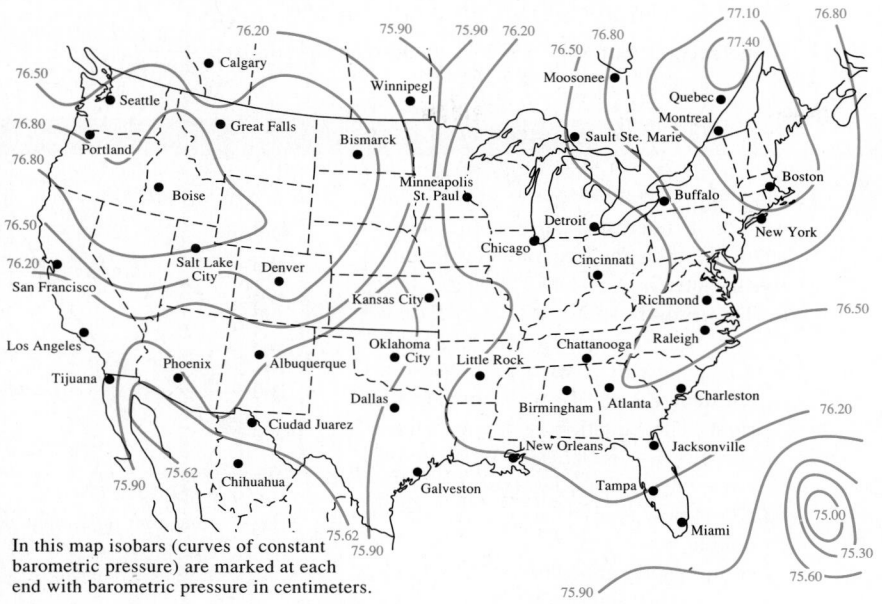

In this map isobars (curves of constant barometric pressure) are marked at each end with barometric pressure in centimeters.

CLASS EXERCISES

Each of the following formulas from mathematics and science expresses some quantity as a function of two variables. Describe this functional relationship in words.

1. $A = lw$ **2.** $d = rt$ **3.** $V = \frac{1}{3}\pi r^2 h$

4. $F = ma$ **5.** $D = \frac{m}{V}$ **6.** $A = 2\pi rh + 2\pi r^2$

7. First-quadrant rays with vertices at the origin

7. Visual Thinking Given the formula *density = mass ÷ volume* (see Exercise 5), what do curves of constant density look like in an *mV*-plane?

8. Refer to the barometric pressure graph shown above.
   a. Name two other cities with approximately the same barometric pressure as New Orleans. Tampa, San Francisco
   b. Bad weather usually accompanies low pressure. In which cities is the weather apt to be poorest? Chihuahua, Ciudad Juarez, Phoenix
   c. Good weather usually accompanies high pressure. In which cities is the weather apt to be best? Quebec, Montreal, Boston

Answers may vary. Examples are given.

1. The area *A* of a rectangle is a function of the length *l* and width *w*.

2. Distance *d* is a function of rate *r* and time *t*.

3. The volume *V* of a cone is a function of the base radius *r* and the height *h*.

4. Force *F* is a function of mass *m* and acceleration *a*.

5. Density *D* is a function of mass *m* and volume *V*.

6. The surface area *A* of a cylinder is a function of the base radius *r* and the height *h*.

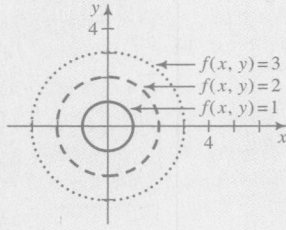
9. For example: driver's age, where driver lives, driving record

9. **Discussion** The amount that a person pays for auto insurance is a function of many variables. Discuss what some of these variables are.

10. **a.** If $f(x, y) = \sqrt{x^2 + y^2}$, find $f(3, 4)$, $f(-4, 3)$, and $f(0, 5)$. 5; 5; 5
   b. Sketch some constant curves of the function f.

11. Name several pairs (a, b) not in the domain of $f(a, b) = \dfrac{a + b}{a - b}$. For example: $(-1, -1)$, $(0, 0)$, $(2, 2)$

WRITTEN EXERCISES

1. b. $A = 100(1.12)^t$, $A = 100(1.16)^t$

A

1. **Consumer Economics** If \$100 is invested at interest rate r compounded annually, then the accumulated amount t years later is $A = 100(1 + r)^t$.
   a. The formula shows that A is a function of __?__ and __?__. r, t
   b. The graphs of A versus t for several constant values of r are shown at the right. The equation of the rightmost curve is $A = 100(1.08)^t$. What are the equations of the other two curves?
   c. About how many years does it take to double your money at 8% interest compounded annually? at 12% interest? at 16% interest? 9 yr; 6 yr; 5 yr

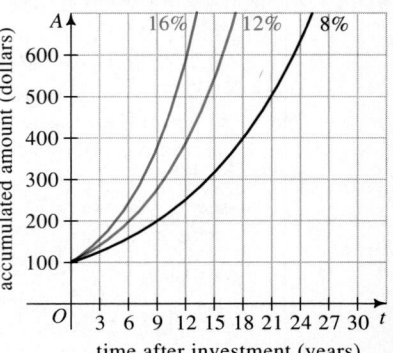

2. **Consumer Economics** The graph at the right below shows the fuel efficiency at various speeds for 2400 lb, 3000 lb, and 3600 lb cars. The fuel efficiency is given in miles per gallon (mi/gal).
   a. Fuel efficiency is a function of __?__ and __?__.
   b. At what other speed will a 3000 lb car have approximately the same fuel efficiency as it has when traveling at 25 mi/h?
   c. At what speeds does a 3600 lb car have approximately the same fuel efficiency as a 2400 lb car traveling at 55 mi/h?
   d. Regardless of what a car weighs, at approximately what speed is maximum fuel efficiency reached?

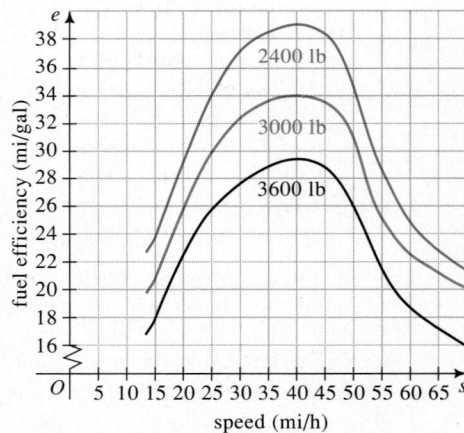

2. a. speed, weight   **b.** 50 mi/h   **c.** 32 mi/h, 47 mi/h   **d.** 40 mi/h

154   *Chapter Four*

3. **Meteorology** The graph shows the wind-chill equivalent temperatures for recorded temperatures of 30°F, 20°F, and 10°F.
   a. Describe the functional relationship that the graph depicts.
   b. If the recorded temperature is 10°F, approximately what wind speed will produce a wind-chill equivalent temperature of −20°F? 16 mi/h
   c. Would you rather be in a place where the recorded temperature is 30°F with a 30 mi/h wind or 20°F with a 10 mi/h wind? Why? The latter *feels* warmer.
   d. Regardless of the recorded temperature, is there a greater change in the wind-chill equivalent temperature when the wind speed is between 5 mi/h and 10 mi/h or when it is between 35 mi/h and 40 mi/h? Between 5 mi/h and 10 mi/h

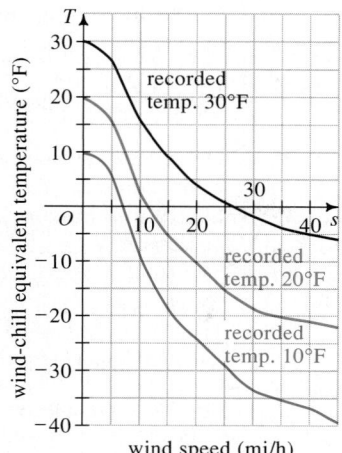

4. **Physiology** The graph shows the number of calories burned each hour when 100 lb, 140 lb, and 180 lb people walk at various speeds.
   a. Describe the functional relationship that the graph depicts.
   b. At approximately what speed must a 140 lb person walk in order to burn 450 Cal/h? 4.6 mi/h
   c. At approximately what speed must a 100 lb person walk in order to burn as many calories per hour as a 180 lb person walking at 2 mi/h? 4 mi/h
   d. Regardless of what a person weighs, is there a greater change in the number of calories burned per hour when the person increases his or her walking speed from 3 mi/h to 3.5 mi/h or from 4 mi/h to 4.5 mi/h? From 4 mi/h to 4.5 mi/h

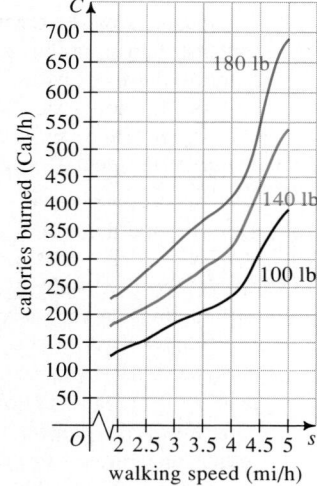

5. **a.** If $A(b, h) = \frac{1}{2}bh$, find $A(8, 3)$ and $A(16, 6)$. 12; 48
   b. Give a geometric interpretation of the function A.
   c. Give several pairs (b, h) for which $A(b, h) = 5$. For example: (2, 5), (1, 10), (20, 0.5)

6. **a.** If $V(r, h) = \frac{1}{3}\pi r^2 h$, find $V(2, 6)$ and $V(4, 12)$. 8π; 64π
   b. Give a geometric interpretation of the function V.
   c. Give several pairs (r, h) for which $V(r, h) = 75\pi$. For example: (5, 9), (3, 25), $\left(10, \frac{9}{4}\right)$

Functions   **155**

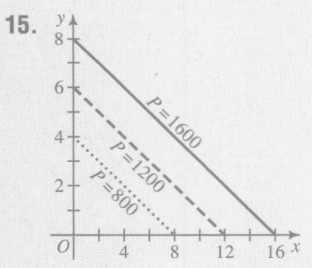
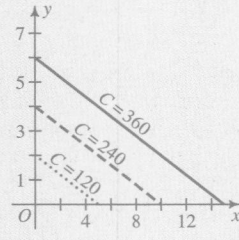
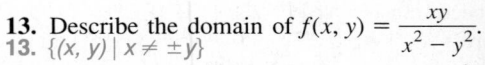
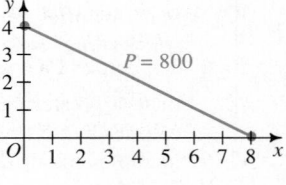
7. a. If $A(b, h) = \frac{1}{2}bh$, show that $A(3b, 3h) = 9 \cdot A(b, h)$.

b. Give a verbal description of the equation $A(3b, 3h) = 9 \cdot A(b, h)$.

8. a. If $V(r, h) = \pi r^2 h$, show that $V(2r, 2h) = 8 \cdot V(r, h)$.

b. Give a verbal description of the equation $V(2r, 2h) = 8 \cdot V(r, h)$.

9. a. If you travel at a constant rate r for t hours, the distance d that you travel is a function of r and t. Give a rule for this function. $d(r, t) = rt$

b. Sketch the constant distance curves $d(r, t) = 200$ and $d(r, t) = 400$ in an rt-plane. $V(s, h) = s^2h$

10. a. The volume V of a square prism is a function of the length s of a side of the square base and of the height h of the prism. Give a rule for this function.

b. Sketch two curves of constant volume in an sh-plane.

B **11.** **Research** Look up "contour map" in a dictionary or encyclopedia. Draw a contour map of the butte shown at the right.

12. **Research** Most daily newspapers show a weather map on which curves of constant temperature are drawn. Obtain such a map and from it read the approximate forecast temperatures in Los Angeles, Seattle, Chicago, Dallas, New York City, and Miami.

13. Describe the domain of $f(x, y) = \dfrac{xy}{x^2 - y^2}$.
13. $\{(x, y) \mid x \neq \pm y\}$

14. Describe the domain of $g(x, y) = \dfrac{4 + y^2}{4 - x^2}$.
14. $\{(x, y) \mid x \neq \pm 2\}$

15. **Retailing** The profit in selling x black-and-white television sets and y color television sets is $P(x, y) = 100x + 200y$ dollars, where $x \geq 0$ and $y \geq 0$. All points (x, y) on the line segment labeled $P = 800$ will give a profit of \$800. This line is a *constant profit line*. Copy this graph and draw the constant profit lines $P(x, y) = 1200$ and $P(x, y) = 1600$.

16. **Manufacturing** The cost of producing x wooden tennis rackets and y graphite tennis rackets is $C(x, y) = 24x + 60y$. Draw several lines of constant cost in an xy-plane.

17. a. If $A(l, w, h) = 2lw + 2wh + 2lh$, find $A(4, 3, 5)$ and $A(6, 4, 7)$. 94; 188

b. Give a geometric interpretation of the function A. Surface area of rect. prism

18. a. If $A(b_1, b_2, h) = \frac{1}{2}(b_1 + b_2)h$, find $A(3, 4, 2)$ and $A(5, 7, 3)$. 7; 18

b. Give a geometric interpretation of the function A. Area of trapezoid

For Exercises 19 and 20, refer to textbooks in the given fields to find several examples of functional relationships involving three or more variables. Be prepared to discuss these relationships in class.

19. **Research** Physics, chemistry, or biology

20. **Research** Medicine, psychology, or economics

C 21. **Sports** For some international sailing competitions, the rating R of a yacht is a function of several variables:

$$R(A, L, V) = 0.9\left(\frac{L\sqrt{A}}{12\sqrt[3]{V}} + \frac{L + \sqrt{A}}{4}\right)$$

where A = surface area of the sails in square meters, L = length of the yacht in meters, and V = volume of water yacht displaces in cubic meters. To be in the R-5.5 class, a yacht must have a rating less than 5.5. Does a yacht that has 37 m^2 of sail, is 10 m long, and displaces 8.5 m^3 of water qualify?

No; $R(37, 10, 8.5) \approx 5.85 > 5.5$

4-7 Forming Functions from Verbal Descriptions

Objective *To form a function of one variable from a verbal description and, when appropriate, to determine the minimum or maximum value of the function.*

An important concern of mathematics is finding the minimum or maximum value of a function. You have already seen quadratic and cubic examples that involve minimizing costs and maximizing profits. Other such applications use mathematics to minimize the structural stress on a girder or to maximize the volume of a container made from a given amount of material.

Minimum and maximum values are often referred to as *extreme values*. Approximate extreme values of a function can be found using a computer or graphing calculator. Exact extreme values are most often found using calculus.

Whether technology or calculus is used, we almost always need to write a rule for the function to be minimized or maximized. If the rule depends on two or more variables, then we also need to find a relationship among the variables so that the function can be written in terms of only one variable. Developing these skills is the goal of this section.

Teaching Notes, p. 118D

Warm-Up Exercises

Use the figure to complete Exercises 1–4.

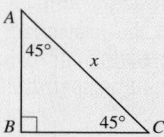

1. Express AB in terms of x.
$\dfrac{x\sqrt{2}}{2}$

2. Express BC in terms of x.
$\dfrac{x\sqrt{2}}{2}$

3. Express the area of $\triangle ABC$ in terms of x. $\dfrac{1}{4}x^2$

4. Express the perimeter of $\triangle ABC$ in terms of x.
$(1 + \sqrt{2})x$

Motivating the Section

In the business world, efficiency and profitability are two important concerns. Efficiency usually involves minimizing the time it takes to get something done, while profitability usually involves maximizing revenue (or minimizing costs). Finding the minimum or maximum value of a function is one of the objectives of this section.

☒ Using Technology

Have students use graphing calculators to obtain the graphs at the bottom of this page. Then have them zoom in on the high point of the graph in the first quadrant to obtain a better approximation of the volume than that given in the text.

Example 1

An open-top box with a square base is to be constructed from sheet metal in such a way that the completed box is made of 2 m² of sheet metal. Express the volume of the box as a function of the base width.

Solution

1. Sketch the box. Let h be the height of the box, and let w be the width. The volume of the box is a function of h and w:

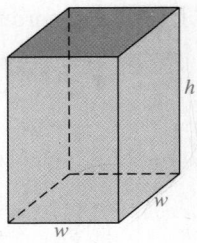

$$V(w, h) = w^2 h$$

2. The equation above gives V in terms of w and h. To get V in terms of w alone, we must replace h in the above equation with a function of w. We use the fact that the box is made of 2 m² of sheet metal.

$$\text{Area of sheeting used} = 2$$
$$\text{Area of the base} + 4 \cdot (\text{Area of a side}) = 2$$
$$w^2 \qquad + \qquad 4wh \qquad = 2$$
$$h = \frac{2 - w^2}{4w}$$

3. To find V in terms of w alone, substitute $\frac{2 - w^2}{4w}$ for h.

$$V(w, h) = w^2 h$$
$$V(w) = w^2\left(\frac{2 - w^2}{4w}\right) = \frac{2w - w^3}{4}$$

In Example 1, the volume of the box was expressed as a function of its width: $V(w) = \frac{2w - w^3}{4}$. To find the maximum volume, you can use a computer or graphing calculator to obtain the graph at the left below. Notice that this graph includes some values of w not in the domain of the function. By zooming in on the high point of the graph in the first quadrant, you can obtain the "blowup" shown at the right below. This shows that the maximum volume is approximately $V(0.8) = 0.3 \text{ m}^3$.

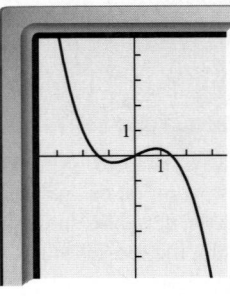

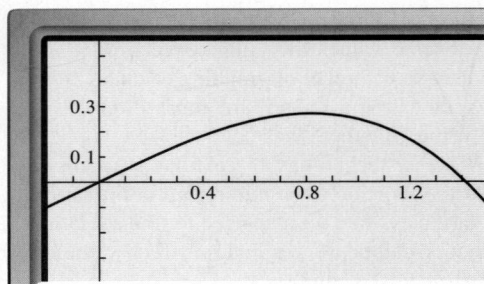

Example 2   A north-south bridle path intersects an east-west river at point O. At noon, a horse and rider leave O traveling north at 12 km/h. At the same time, a boat is 25 km east of O traveling west at 16 km/h. Express the distance d between the horse and the boat as a function of the time t in hours after noon.

Solution   1. Make a sketch showing the horse and boat at some time t. Let h be the horse's distance from O, and let b be the boat's distance east of O. Since the horse is traveling from O at 12 km/h,

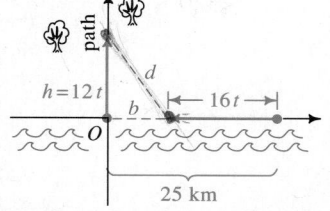

$$h = 12t.$$

Since the boat is 25 km from O and traveling toward O at 16 km/h,

$$b = 25 - 16t.$$

2. By the Pythagorean theorem,

$$d(h, b) = \sqrt{h^2 + b^2}.$$

3. To find d in terms of t, substitute $12t$ for h and $25 - 16t$ for b.

$$d(t) = \sqrt{(12t)^2 + (25 - 16t)^2}$$
$$= \sqrt{144t^2 + 625 - 800t + 256t^2}$$
$$= \sqrt{400t^2 - 800t + 625}$$

In Example 2, it is possible to determine when the horse and boat are closest to each other (that is, when the distance between them is a minimum) and what the minimum distance is. Since $d(t)$ is the square root of the function

$$f(t) = 400t^2 - 800t + 625,$$

$d(t)$ will be a minimum when $f(t)$ is a minimum. Since the minimum or maximum value of $f(x) = ax^2 + bx + c$ occurs at $x = -\dfrac{b}{2a}$, the minimum value of $f(t) = 400t^2 - 800t + 625$ occurs at:

$$t = -\frac{-800}{2(400)} = 1$$

Therefore, the horse and boat are closest 1 h after noon. The minimum distance between them is $d(1)$.

$$d(1) = \sqrt{400 \cdot 1^2 - 800 \cdot 1 + 625}$$
$$= \sqrt{225} = 15$$

Thus, the minimum distance between the horse and boat is 15 km.

Functions   **159**

Review Note

Students may need to be re-minded of the Pythagorean theorem, which they will need for many of the exercises.

Example 3 Water flows into a conical tank 100 cm wide and 250 cm deep at a rate of $40 \text{ cm}^3/\text{s}$. Find the volume V of the water in the tank as a function of the height h of the water. Represent h as a function of the time t that the water has been flowing into the empty tank.

Solution

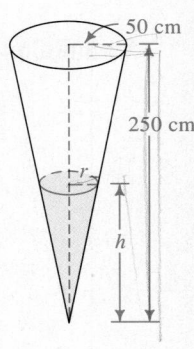

1. Make a sketch. Let r be the radius of the water surface, and let h be the height of the water in the tank. The volume of water in the tank is a function of r and h:

$$V(r, h) = \frac{1}{3}\pi r^2 h$$

2. Find an expression for r in terms of h by using the similar triangles from a cross section of the tank.

$$\frac{r}{h} = \frac{50}{250}$$

$$r = \frac{h}{5}$$

3. To find V in terms of h alone, substitute $\frac{h}{5}$ for r.

$$V(r, h) = \frac{1}{3}\pi r^2 h$$

$$V(h) = \frac{1}{3}\pi \left(\frac{h}{5}\right)^2 h = \frac{\pi}{75}h^3$$

4. To represent h as a function of the time t, note that the volume of water in the tank at t seconds is $V = 40t$. Therefore, substitute $40t$ for V and solve for h:

$$40t = \frac{\pi}{75}h^3$$

$$h = \sqrt[3]{\frac{3000t}{\pi}}$$

Thus, $h(t) = \sqrt[3]{\frac{3000t}{\pi}}$.

1. b. $s(d) = \dfrac{d\sqrt{2}}{2}$

CLASS EXERCISES

1. A square has a side of length s and a diagonal of length d, as shown.
   a. Express d as a function of s. $d(s) = s\sqrt{2}$
   b. Express s as a function of d. See above.
   c. Express the area A of the square as a function of d. $A(d) = \dfrac{d^2}{2}$

2. a. Express the volume V of a cube as a function of the length e of an edge.
   b. The length of a diagonal of the cube is $d = e\sqrt{3}$. Express V as a function of d. 2. a. $V(e) = e^3$ b. $V(d) = \dfrac{d^3\sqrt{3}}{9}$

160 *Chapter Four*

3. In the figure at the left below, point A is 4 km north of point C, point B is 8 km east of C, and P is a point on $\overline{BC}$ at a distance of x km from C.
   a. Express $AP + PB$ as a function of x. $\sqrt{16 + x^2} + 8 - x$
   b. What is the domain of this function? $0 \le x \le 8$

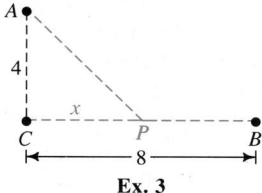

Ex. 3

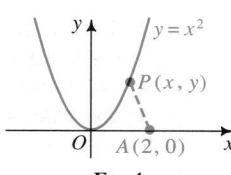

Ex. 4

$d(x, y) = \sqrt{(x-2)^2 + y^2}$

4. $P(x, y)$ is a point on the parabola $y = x^2$, as shown at the right above.
   a. Express the distance from P to $A(2, 0)$ as a function of x and y.
   b. Express the distance as a function of x alone. $d(x) = \sqrt{(x-2)^2 + x^4}$
   c. How can you find the minimum value of the function in part (b)?

5. A runner starts north from point O at 6 m/s. At the same time, a second runner sprints east from O at 8 m/s. (See the figure at the right.) Find the distance d between the runners t seconds later. $d = 10t$

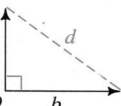

WRITTEN EXERCISES

A 1. Express the area A of a 30°-60°-90° triangle as a function of the length h of the hypotenuse. $A(h) = \dfrac{h^2\sqrt{3}}{8}$

2. Express the area A of an equilateral triangle as a function of the perimeter P. $A(P) = \dfrac{P^2\sqrt{3}}{36}$

3. A tourist walks n km at 4 km/h and then travels $2n$ km at 36 km/h by bus. Express the total traveling time t (in hours) as a function of n. $t(n) = \dfrac{11n}{36}$

4. A student holds a ball of string attached to a kite, as shown at the right. The string is held 1 m above the ground and rises at a 30° angle to the horizontal. If the student lets the string out at a rate of 2 m/s, express the kite's height h (in meters) as a function of the time t (in seconds) after the kite begins to fly. $h(t) = t + 1$

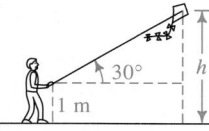

5. A store owner bought n dozen toy boats at a cost of $3.00 per dozen, and sold them at $.75 apiece. Express the profit P (in dollars) as a function of n. $P(n) = 6n$

6. The cost of renting a large boat is 30 dollars per hour plus a usage fee roughly equivalent to x^3 cents per hour when the boat is operated at a speed of x km/h. Express the cost C (in cents per kilometer) as a function of x. $C(x) = \dfrac{3000 + x^3}{x}$

7. The height of a cylinder is twice the diameter. Express the total surface area A as a function of the height h. $A(h) = \dfrac{5}{8}\pi h^2$

8. A pile of sand is in the shape of a cone with a diameter that is twice the height. Express the volume V of sand as a function of the height h. $V(h) = \dfrac{1}{3}\pi h^3$

Functions **161**

9. A light 3 m above the ground causes a boy 1.8 m tall to cast a shadow s meters long measured along the ground, as shown at the right. Express s as a function of d, the boy's distance in meters from the light. $s(d) = 1.5d$

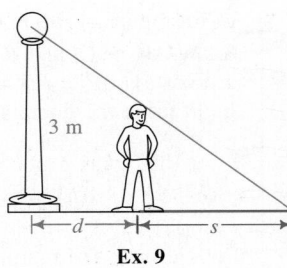
Ex. 9

10. When a girl 1.75 m tall stands between a wall and a light on the ground 15 m away, she casts a shadow h meters high on the wall, as shown at the right. Express h as a function of d, the girl's distance in meters from the light. $h(d) = \dfrac{105}{4d}$

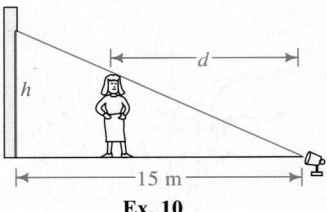
Ex. 10

11. A box with a square base has a surface area (including the top) of 3 m^2. Express the volume V of the box as a function of the width w of the base. $V(w) = \dfrac{3w - 2w^3}{4}$

12. A box with a square base and no top has a volume of 6 m^3. Express the total surface area A of the box as a function of the width w of the base. $A(w) = \dfrac{w^3 + 24}{w}$

13. A stone is thrown into a lake, and t seconds after the splash the diameter of the circle of ripples is t meters.
   a. Express the circumference C of this circle as a function of t. $C(t) = \pi t$
   b. Express the area A of this circle as a function of t. $A(t) = \dfrac{\pi}{4}t^2$

14. A balloon is inflated in such a way that its volume increases at a rate of 20 cm^3/s.
   a. If the volume of the balloon was 100 cm^3 when the process of inflation began, what will the volume be after t seconds of inflation? $100 + 20t$
   b. Assuming that the balloon is spherical while it is being inflated, express the radius r of the balloon as a function of t. $r(t) = \sqrt[3]{\dfrac{75 + 15t}{\pi}}$

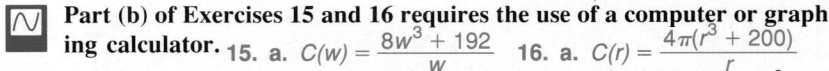

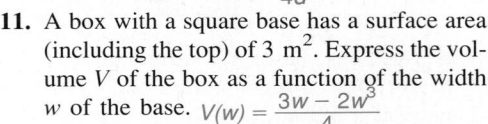 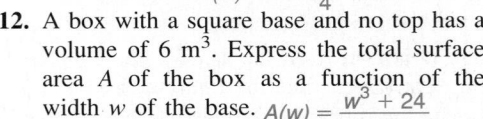

Part (b) of Exercises 15 and 16 requires the use of a computer or graphing calculator. **15. a.** $C(w) = \dfrac{8w^3 + 192}{w}$ **16. a.** $C(r) = \dfrac{4\pi(r^3 + 200)}{r}$

B 15. **Manufacturing** A box with a square base and no top has volume 8 m^3. The material for the base costs $8 per square meter, and the material for the sides costs $6 per square meter.
   a. Express the cost C of the materials used to make the box as a function of the width w of the base. See above.
   b. Use a computer or graphing calculator to find the minimum cost. About $126

16. **Manufacturing** A cylindrical can has a volume of 400π cm^3. The material for the top and bottom costs 2¢ per square centimeter. The material for the vertical surface costs 1¢ per square centimeter.
   a. Express the cost C of the materials used to make the can as a function of the radius r. See above.
   b. Use a computer or graphing calculator to find the minimum cost. About $8.12

17. At 2:00 P.M. bike A is 4 km north of point C and traveling south at 16 km/h. At the same time, bike B is 2 km east of C and traveling east at 12 km/h.
  a. Show that t hours after 2:00 P.M. the distance between the bikes is:
$$\sqrt{400t^2 - 80t + 20}$$
  b. At what time is the distance between the bikes the least? 2:06 P.M.
  c. What is the distance between the bikes when they are closest? 4 km

18. A car leaves Oak Corners at 11:33 A.M. traveling south at 70 km/h. At the same time, another car is 65 km west of Oak Corners traveling east at 90 km/h.
  a. Express the distance d between the cars as a function of the time t after the first car left Oak Corners. See below.
  b. Show that the cars are closest to each other at noon.

19. Engineering Water is flowing at a rate of 5 m³/s into the conical tank shown at the right.
  a. Find the volume V of the water as a function of the water level h. $V(h) = \dfrac{\pi}{48}h^3$
  b. Find h as a function of the time t during which water has been flowing into the tank. See below.

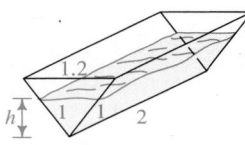

Ex. 19

20. Engineering A trough is 2 m long, and its ends are triangles with sides of length 1 m, 1 m, and 1.2 m as shown at the right.
  a. Find the volume V of the water in the trough as a function of the water level h. $V(h) = 1.5h^2$
  b. If water is pumped into the empty trough at the rate of 6 L/min, find the water level h as a function of the time t after the pumping begins. (1 m³ = 1000 L) $h(t) = \sqrt{0.004t}$

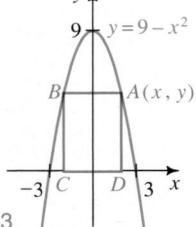

Ex. 20

21. $P(x, y)$ is an arbitrary point on the line $2x + y = 10$.
  a. Express the distance d from the origin to P as a function of the x-coordinate of P. $d(x) = \sqrt{5x^2 - 40x + 100}$
  b. What are the domain and range of this function? All real numbers; $d(x) \geq 2\sqrt{5}$

22. $P(x, y)$ is an arbitrary point on the parabola $y = x^2$.
  a. Express the distance d from P to the point $A(0, 1)$ as a function of the y-coordinate of P. $d(y) = \sqrt{y^2 - y + 1}$
  b. What is the minimum distance d? $\dfrac{\sqrt{3}}{2}$

23. As shown at the right, rectangle $ABCD$ has vertices C and D on the x-axis and vertices A and B on the part of the parabola $y = 9 - x^2$ that is above the x-axis.
  a. Express the perimeter P of the rectangle as a function of the x-coordinate of A. $P(x) = -2x^2 + 4x + 18$
  b. What is the domain of the perimeter function? $0 < x < 3$
  c. For what value of x is the perimeter a maximum? 1

Ex. 23

18. a. $d(t) = 5\sqrt{520t^2 - 468t + 169}$ **19. b.** $h(t) = \sqrt[3]{\dfrac{240t}{\pi}}$

**Additional Answers
Written Exercises**

17. a. $d(t) =$
  $\sqrt{(AC)^2 + (BC)^2} =$
  $\sqrt{400t^2 - 80t + 20}$

18. b. The minimum value of $d(t)$ occurs at the minimum value of $f(t) = 520t^2 - 468t + 169$, which is when $t = -\dfrac{-468}{2(520)} = 0.45$. Since 0.45 h = 27 min, the cars are closest at 11:33 + :27, or 12 noon.

24. **Manufacturing** A sheet of metal is 60 cm wide and 10 m long. It is bent along its width to form a gutter with a cross section that is an isosceles trapezoid with 120° angles, as shown at the right.

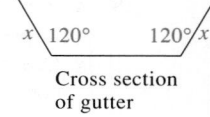

Cross section of gutter

   a. Express the volume V of the gutter as a function of x, the length in centimeters of one of the equal sides. (*Hint*: Volume = area of trapezoid × length of gutter) $V(x) = 750\sqrt{3}(40x - x^2)$
   b. For what value of x is the volume of the gutter a maximum? 20

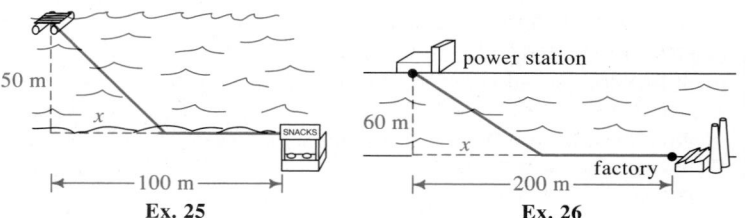 **Part (b) of Exercises 25–29 requires the use of a computer or graphing calculator.**

27. a. $C(x) = \dfrac{12x^2 + 480}{x}$

25. From a raft 50 m offshore, a lifeguard wants to swim to shore and run to a snack bar 100 m down the beach, as shown at the left below.
   a. If the lifeguard swims at 1 m/s and runs at 3 m/s, express the total swimming and running time t as a function of the distance x shown in the diagram. $t(x) = \sqrt{2500 + x^2} + \dfrac{1}{3}(100 - x)$
   b. Use a computer or graphing calculator to find the minimum time. About 80.5 s

50 m

x

SNACKS

|← 100 m →|

Ex. 25

power station

60 m

x

factory

|← 200 m →|

Ex. 26

26. **Engineering** A power station and a factory are on opposite sides of a river 60 m wide, as shown at the right above. A cable must be run from the power station to the factory. It costs $25 per meter to run the cable in the river and $20 per meter on land. $C(x) = 25\sqrt{3600 + x^2} + 20(200 - x)$
   a. Express the total cost C as a function of x, the distance downstream from the power station to the point where the cable touches the land.
   b. Use a computer or graphing calculator to find the minimum cost. $4900

27. **Landscaping** A rectangular area of 60 m^2 has a wall as one of its sides, as shown. The sides perpendicular to the wall are made of fencing that costs $6 per meter. The side parallel to the wall is made of decorative fencing that costs $8 per meter.

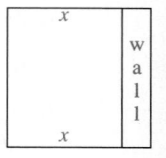

   a. Express the total cost C of the fencing as a function of the length x of a side perpendicular to the wall. See above.
   b. Use a computer or graphing calculator to find the minimum cost. About $152

C 28. A cylinder is inscribed in a sphere of radius 1. $V(r) = 2\pi r^2\sqrt{1 - r^2}$
   a. Express the volume V of the cylinder as a function of the base radius r.
   b. Use a computer or graphing calculator to find the radius and height of the cylinder having maximum volume. $r \approx 0.82$, $h \approx 1.14$

29. A cone circumscribes a sphere with radius 1. $\quad V(r) = \dfrac{2\pi r^4}{3(r^2 - 1)}$
   a. Express the volume V of the cone as a function of the base radius r.
   b. Use a computer or graphing calculator to find the radius and height of the cone having minimum volume. $r \approx 1.41$, $h \approx 4.02$

30. **Sports** A baseball diamond is a square with 90 ft sides. A runner (R in the diagram at the right) has taken a 9 ft lead from first base. At the moment the ball is pitched, the runner runs toward second base at 27 ft/s.

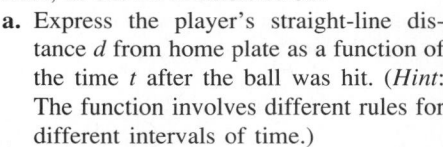

   a. Express the runner's straight-line distance d from home plate as a function of the time t after the ball is thrown.
   b. What are the domain and range of this function? See below.

31. **Sports** A baseball player hits a ball into the farthest corner of the outfield and tries for an inside-the-park home run; that is, the player tries to run the bases and make it to home plate safely. Suppose that the player runs at 30 ft/s and stays strictly on the base lines, as shown in Exercise 30.

   a. Express the player's straight-line distance d from home plate as a function of the time t after the ball was hit. (*Hint*: The function involves different rules for different intervals of time.)
   b. Sketch the graph of the function in part (a).

30. **a.** $d(t) = 9\sqrt{9t^2 + 6t + 101}$ **b.** $0 \le t \le 3$; $9\sqrt{101} \le d(t) \le 90\sqrt{2}$

 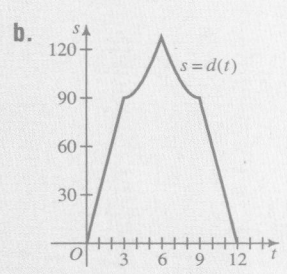
Chapter Summary

1. A *function* consists of a set of real numbers, called the *domain* of the function, and a rule that assigns to each element in the domain exactly one real number. The set of real numbers assigned by the rule is called the *range* of the function. The *graph* of a function is the set of points corresponding to the ordered pairs that satisfy the functional rule.

2. Functions can be *added*, *subtracted*, *multiplied*, or *divided*:

$$(f + g)(x) = f(x) + g(x) \qquad (f - g)(x) = f(x) - g(x)$$

$$(f \cdot g)(x) = f(x) \cdot g(x) \qquad \left(\dfrac{f}{g}\right)(x) = \dfrac{f(x)}{g(x)} \text{ provided } g(x) \ne 0$$

3. The *composite* of f and g, denoted $f \circ g$, is defined by $(f \circ g)(x) = f(g(x))$ such that x is in the domain of g and $g(x)$ is in the domain of f.

4. Simple changes in an equation of a graph can cause the graph to be reflected, translated, stretched, or shrunk. The chart on page 142 summarizes these changes.

Functions **165**

Additional Answers
Chapter Test

2.

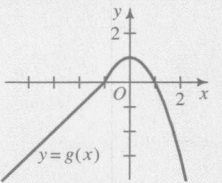

6. a.

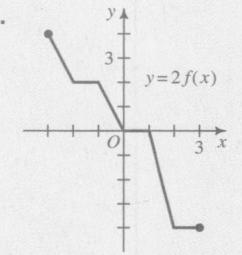

b.

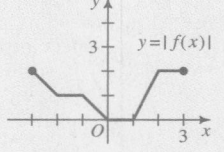

c.

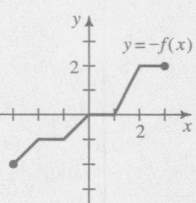

d.

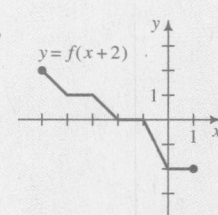

---

5. Methods for testing an equation or its graph for symmetry in the x-axis, in the y-axis, in the origin, or in the line $y = x$ are summarized on page 134.

6. Properties of functions can be used to help graph and analyze functions:

   f is an *even* function if $f(-x) = f(x)$ for all x in the domain of f; the graph of an even function has symmetry in the y-axis.

   f is an *odd* function if $f(-x) = -f(x)$ for all x in the domain of f; the graph of an odd function has symmetry in the origin.

7. A function $y = f(x)$ is *periodic* if there is a positive number p, called a *period* of f, such that $f(x + p) = f(x)$ for all x in the domain of f. The smallest period of f is called the *fundamental period*. If f has maximum value M and minimum value m, then the *amplitude* of f is defined to be $\dfrac{M - m}{2}$.

8. If f is periodic with period p and amplitude A, then $y = cf(x)$ has period p and amplitude cA, and $y = f(cx)$ has period $\dfrac{p}{c}$ and amplitude A.

9. If $f(g(x)) = x$ for all x in the domain of g and $g(f(x)) = x$ for all x in the domain of f, then f and g are *inverses*. The inverse of f is denoted f^{-1} and its graph is the reflection of the graph of f in the line $y = x$. f^{-1} exists if f is *one-to-one*, that is, every horizontal line intersects the graph of f in at most one point.

10. In many situations, a problem is modeled by a function of more than one variable. By expressing one variable in terms of the other, a function of two variables can often be transformed into a function of one variable.

Key vocabulary and ideas

function, domain, range (p. 119)	periodic function, period (p. 138)
relation (p. 121)	amplitude (p. 139)
vertical-line test (p. 121)	vertical stretch or shrink (p. 140)
sum and difference of f and g (p. 125)	horizontal stretch or shrink (p. 140)
product and quotient of f and g (p. 125)	translation (p. 141)
composite function (p. 126)	inverse functions (p. 146)
reflection (pp. 131–133)	one-to-one function (p. 148)
axis of symmetry (p. 133)	horizontal-line test (p. 148)
point of symmetry (p. 133)	function of two variables (p. 151)

Chapter Test

$-3 \le x \le 3;\ 0 \le f(x) \le 3;\ \pm 3$

1. Give the domain, range, and zeros of the function $f(x) = \sqrt{9 - x^2}$.   **4-1**

2. Graph $g(x) = \begin{cases} x + 1 & \text{if } x < -1 \\ 1 - x^2 & \text{if } x \ge -1 \end{cases}$. Find the range and zeros of g. $g(x) \le 1;\ \pm 1$

3. Let $f(x) = x^2 + 2x$ and $g(x) = x + 2$. Find: **4-2**

   a. $(f + g)(x)$ **b.** $(f - g)(x)$ **c.** $(f \cdot g)(x)$ **d.** $\left(\dfrac{f}{g}\right)(x)$

---

4. Using the functions f and g in Exercise 3, find:

a. $(f \circ g)(x)$ $x^2 + 6x + 8$ **b.** $(g \circ f)(x)$ $x^2 + 2x + 2$

5. Determine whether the graph of $x^2 - xy = 4$ has symmetry in: (i) the x-axis, (ii) the y-axis, (iii) the line $y = x$, and (iv) the origin. iv   4-3

6. Given the graph of $y = f(x)$ shown at the right, sketch the graph of each of the following.   4-4

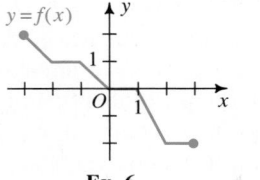

$y = f(x)$

a. $y = 2f(x)$ **b.** $y = |f(x)|$
c. $y = -f(x)$ **d.** $y = f(x + 2)$

Ex. 6

7. The graph of a periodic function $y = g(x)$ is shown at the right.

a. What is the fundamental period of g? 4

b. What are the maximum and minimum values of g? 4; 1

c. What is the amplitude of g? $\frac{3}{2}$

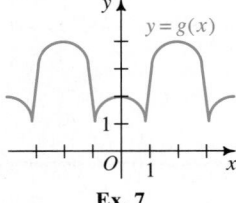

$y = g(x)$

Ex. 7

8. *Writing* Describe and illustrate what happens when the graph of a periodic function with period p is horizontally translated p units. Then use the definition of a periodic function to explain why this happens.

9. a. Which one of the two functions $f(x) = \sqrt{3 + x^2}$ and $g(x) = 3 + x$ has an inverse? Find a rule for the inverse. g; $g^{-1}(x) = x - 3$ 4-5

b. Explain why the other function does not have an inverse.

10. The area A of a triangle is a function of the base b and height h.   4-6

a. Express A as a function of b and h. $A(b, h) = \frac{1}{2}bh$

b. Find $A(3, 4)$ and $A(6, 5)$. 6; 15

c. Draw the curve of constant area $A(b, h) = 3$ in a bh-plane.

11. A cylindrical tank 4 ft in diameter fills with water at the rate of 10 ft³/s. Express the depth of the water in the tank as a function of the time t in seconds. Assume the tank is empty at time $t = 0$. $h(t) = \frac{5t}{2\pi}$ 4-7

Part (c) of Exercise 12 requires the use of a computer or graphing calculator.

12. As shown at the right, triangle OAB is an isosceles triangle with vertex O at the origin and vertices A and B on the part of the parabola $y = 9 - x^2$ that is above the x-axis.

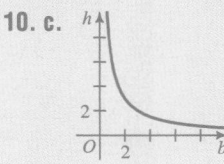

$y = 9 - x^2$

$A(x, y)$

a. Express the area of the triangle as a function of the x-coordinate of A. $A(x) = x(9 - x^2)$

b. What is the domain of the area function? $0 < x \le 3$

c. Use a computer or graphing calculator to find the maximum area. ≈ 10.4

8. If the graph of a periodic function with period p is horizontally translated p units, the translated graph coincides with the original graph. The definition of a periodic function says that $f(x + p) = f(x)$ for all x and some positive number p. Thus, the graphs of $y = f(x + p)$ and $y = f(x)$ are identical (that is, the graph of f translated p units to the *left* coincides with the graph of f). Since replacing x with $x - p$ in the definition gives $f(x) = f(x - p)$ for all x, the graphs of $y = f(x)$ and $y = f(x - p)$ are also identical (that is, the graph of f translated p units to the *right* coincides with the graph of f).

9. b. $f(x) = \sqrt{3 + x^2}$ does not have an inverse because f is not one-to-one: $f(-1) = f(1) = 2$, for example.

10. c.

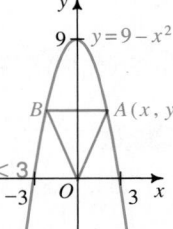

Chapter 5   Exponents and Logarithms

Overview

In this chapter, students study the laws of exponents and logarithms, and then apply them to real-life situations. The first three sections introduce integral, rational, and real exponents. In these sections, students simplify exponential expressions and investigate exponential growth and decay models. The fourth section introduces the number e and the natural exponential function e^x, which students apply to compound-interest problems. The last three sections introduce logarithms. In these sections, students simplify logarithmic expressions, examine models based on logarithms (such as the Richter scale), and solve exponential equations using logarithms.

Objectives

5-1   To define and apply integral exponents.
5-2   To define and apply rational exponents.
5-3   To define and use exponential functions.
5-4   To define and apply the natural exponential function.
5-5   To define and apply logarithms.
5-6   To prove and apply laws of logarithms.
5-7   To solve exponential equations and to change logarithms from one base to another.

Supplementary Resources

Tests, pp. 11–12
Alt. Assess., pp. 15–18, 65
Activities, pp. 10–11
Using Tech., pp. 17–18, 19–20
Student Res. Guide, pp. 46–56

Software

Precalculus Plotter Plus
Activities, pp. 33–34, 35–36
Inverse Plotter, Function Plotter

Pacing Guide

Section	5-1	5-2	5-3	5-4	5-5	5-6	5-7	Review	Test	Total
Standard (days)	2	2	2	2	2	2	2	1	1	16
Comprehensive (days)	1	2	2	1	2	2	2	1	1	14

Teaching Notes

5-1   pages 169–175

Presenting the Section

While the laws of exponents are review for most students, the applications in this lesson are not. The growth and decay models are good motivators for introducing exponents.

Tables are used in this chapter (such as the one on page 169) because later concepts arise naturally from limiting values in tables. Tables also make the use of exponents in functions more obvious. Tables suggest functions, and if the variables occur as exponents, they are exponential functions. Encourage students to extend the tables, if necessary, to see the patterns in the tables.

The laws of exponents are divided into laws for multiplying numbers with the same base and laws for multiplying numbers with the same exponent. This means that law 1 does not apply to problems such as $4^3 \cdot 3^2$ or to Exercise 37, page 174 (unless the numbers are first converted to the same base). Also, emphasize that an exponential expression is not considered simplified unless it contains no negative exponents and no powers of powers.

Example 4, page 171, illustrates that you cannot simplify a power of a sum or difference by distributing the exponent over each individual term. This is a common error that students make. Students need to be reminded that the laws of exponents are for products of numbers and expressions, not for sums. Emphasize the definition of negative exponents. For example, stress that $\left(-\frac{1}{3}\right)^{-3} = -27$, not $\frac{1}{27}$.

Using Technology

Scientific calculators are time savers for many of the problems in this section. You might need to show students how to use the power key (x^y or y^x), and you can anticipate some confusion about simplifying expressions like $(-3)^4$ and -3^4. Using parentheses and the change-of-sign key ($+/-$) is critical for many problems. Students should also be careful about calculating the power of a negative number, which some calculators cannot do directly. Caution students that some calculators will indicate an error when there are too many digits in an answer, while others use scientific notation to display answers with many digits.

Have students use the laws of exponents by substituting numbers and evaluating expressions on their calculators, including b^0, b^{-x}, and $(a^2 b^2)^{-1}$. Have them evaluate 0^0, 2^{100}, and 10^{20} to see how the calculator responds. It might be fun for students to estimate how high a stack of paper 0.01 in. thick will become if it is folded in half 20 times, and then check the result on a calculator. (About 10,486 in.)

Cooperative Learning

Groups of several students each can review the laws of exponents or review how to simplify expressions containing zero or negative exponents. Have them go over the definitions on page 171 by discussing examples.

Applications

Applications of exponents in growth and decay models are numerous, including the business and economics examples presented in this section. This content leads to a discussion of exponential functions in Section 5-3. Students should realize that negative exponents are used extensively to represent the size or mass of very small objects, while large powers of 10 are used to describe astronomically large distances and objects. Encourage students to think of formulas and applications that also use exponents, like Avogadro's number (6.02×10^{23}) in chemistry. Ask students if they can define "googol" (10^{100}) and "googolplex" (a googol of googols). Googols are used often to count the number of stars in the universe.

5-2   pages 175–180

Presenting the Section

This section defines rational exponents and shows how they are used in growth and decay models. The laws of exponents still hold with the base b now a positive real number other than 1. That means that although $16^{1/4}$ is defined, $(-16)^{1/4}$ is not, because it is not real (see Example 1, page 176). Emphasize that simplified expressions are ones with no negative exponents, no fractional exponents in the denominator of a fraction, and no complex fractions (a fraction divided by a fraction), and with the power as small as possible (this is equivalent to an index as small as possible if the expression is expressed in radical form).

Using Technology

Scientific calculators are recommended to help simplify the work in this section, although answers in decimal form will result. Caution students to make sure they use parentheses around an exponent that is of the form p/q so that the order of operations will be done correctly. Programmable calculators can help students to do compound interest problems like Example 5, page 177, and Exercises 37–41, page 179.

Applications

Applications of rational exponents occur in exponential growth or decay problems as well as in business, economics, consumer economics, finance, education, and music applications. Have students bring in additional examples from newspapers or magazines.

Physics applications are numerous. For example, the period t of a pendulum varies directly with the square root of its length and is given by

$$t = 2\pi\sqrt{\frac{l}{g}},$$

where l is the length and g is a constant. The frequency of a violin string is a function of the inverse of its length, and a function of the square root of the tension on the string.

Cooperative Learning

You might want to have groups of students go over the definitions on page 175, giving examples and discussing why the base must be a positive real number other than 1 (Class Exercise 18, page 177). Students should ask questions within their groups, including how to find rational powers of numbers effectively with calculators, how to express numbers with radical signs and negative exponents, and how to do problems like those given in Exercises 17–28, page 178.

Groups could do Exercises 40 and 42, page 179. Each member of a group could find the present cost

of an item in a category from the table below and then determine the cost of the item several years ago by using the percent change in the category. The group can then discuss the results and make corrections or additions as necessary. Extended work, which leads into Section 5-3, could include trying to fit exponential curves to the data.

Consumer Price Indexes					
Year	All	Energy	Food	Shelter	Transport.
*1984	103.9	100.9	103.2	104.0	103.7
*1985	107.6	101.6	105.6	109.8	106.4
*1986	109.6	88.2	109.0	115.8	102.3
*1987	113.6	88.6	113.5	121.3	105.4
*1988	118.3	89.3	118.2	127.1	108.7
**1989	124.0	94.3	124.9	132.8	114.1

* U.S. Bureau of the Census, *Statistical Abstract of the United States: 1990*, Table 762.
** U.S. Bureau of Labor Statistics, *Monthly Labor Review*, January 1990, Table 1.

5-3 | pages 180–186

Presenting the Section

Emphasize the three forms of an exponential function: (1) $f(x) = ab^x$, (2) $A(t) = A_0(1 + r)^t$ (see Example 2, page 181), and (3) $A(t) = A_0 b^{t/k}$ (see Example 3, page 182). Note that the range is the set of positive real numbers, which is why the graphs shown on page 181 are all above the x-axis. Students will have to decide which of the three forms fits each exercise.

Exponential form (2) above applies to the growth of an investment earning compound interest. It might be helpful to show it in chart form with replacements for r and A_0.

Time	Balance
0	$A(0) = A_0$
1	$A(1) = A_0(1 + r)$
2	$A(2) = A(1)(1 + r) = A_0(1 + r)(1 + r)$
	$= A_0(1 + r)^2$
3	$A(3) = A(2)(1 + r) = A_0(1 + r)^2(1 + r)$
	$= A_0(1 + r)^3$
t	$A(t) = A(t - 1)(1 + r) = A_0(1 + r)^t$

Using Technology

Computers or graphing calculators can be used to help students see patterns in the graphs of exponential functions, as shown on page 181. Have students substitute several values of $b > 1$ in $f(x) = ab^x$ while keeping a constant. All the graphs rise to the right, and as b gets larger, the graphs get steeper. Moreover, all the graphs have the point $(0, a)$ in common. To see graphs that fall to the right, substitute values of b between 0 and 1 and do the same experiment. Then have students predict the answers to Exercises 19–22, page 185, before graphing them. Students should check their answers with graphing calculators or computers.

Cooperative Learning

Exercise 25, page 186, is an excellent exercise to assign as a group project. Have students meet to discuss the requirements of the investigation and then collect the information. Students can meet again to discuss their data and answers. Encourage them to discuss how the money in an account would grow through compounding. Interested students should be encouraged to graph their data with curve-fitting software or by hand. You might want to give extra credit to students who display their results and discuss them with the class.

Applications

There are many applications of exponential growth. These include natural growth of bacteria in a culture, radioactive decay, carbon dating, chemical decomposition, compound interest, inflation, depreciation, population growth, decay of the current in an electrical circuit, and human cell growth. When exponential functions are used to describe growth and decay, the variable t is often used to represent time.

Newton's law of cooling is a physics application of exponential growth. It states that the rate of change in the temperature of a warm body is proportional to the difference between its temperature and the temperature of its surroundings; over time, the temperature change is exponential. See the Project on pages 210 and 211.

An application from biology is to formulate a "logistics curve," which may apply to populations with an initial rapid growth rate followed by a declining growth rate, as shown in the graph below.

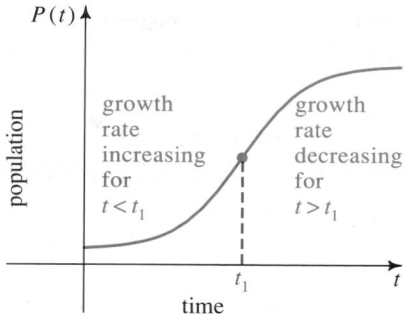

Notice that the graph of the increasing growth rate looks exponential. Bacteria growth sometimes follows a logistics curve.

5-4   pages 186–190

Presenting the Section

The natural exponential function e^x is introduced in this section both as an extension of the compound interest formula with continuous compounding, $A(t) = A_0(1 + r)^t$, and as a use of the irrational number e defined as the limit of $\left(1 + \dfrac{1}{n}\right)^n$ as n approaches infinity (see page 186). If P dollars are invested at an annual interest rate of r and compounded continuously, the investment formula becomes $P(t) = P_0 e^{rt}$, where $P(t)$ is the final balance, P_0 is the initial investment, r is a constant (annual interest rate), and t is time (in years). The same formula is used extensively in growth and decay problems (see Section 5-3), where $P(t)$ is the final amount, P_0 is the initial amount, r is a constant (positive for growth and negative for decay), and t is time.

Using Technology

Scientific calculators with a natural exponential key (e^x) are recommended to help students estimate values they will be working with in this section. (Some calculators may not have a natural exponential key but do have a power key $(x^y$ or $y^x)$. In such cases, students might want to store the value of e (2.71828) in memory and use it as the base for the power key.)

Interested students can be encouraged to write computer programs that calculate the value of an investment using the interest formula for continuous compounding. Point out that the logic of their programs is the same as that used by banks and other investment businesses. A sample BASIC program is given below for continuous compounding.

```
10 INPUT "ENTER THE INITIAL INVEST-
   MENT: "; P0
20 INPUT "ENTER THE RATE: "; R
30 INPUT "ENTER THE TIME: "; T
40 LET PT = P0 * 2.71828^(R * T)
50 PRINT "AMOUNT AFTER ";T;" YEARS
   IS "; PT
60 END
```

Cooperative Learning

Encourage students to work in pairs to do the Class Exercises, pages 188 and 189. Have partners experiment with estimating and checking more values, and then discuss what they think "e" is and how it is used in the Exercises for this section. Have them also discuss the concept of "limit."

You can also use groups to go over the Activity on page 187 and Exercises 19–23, page 190 (these are problems which will occur again in calculus). Have students share their methods of solution as well as their answers.

Applications

Many of the applications in Sections 5-1 to 5-3 also can be expressed as functions of the number e. Compound interest is one of the most important of these, along with growth-and-decay and statistics applications. A general equation for a normal curve is $y = ne^{-kx^2}$, where k is a positive constant. Have students evaluate this function for several values of n and k.

5-5   pages 191–197

Presenting the Section

The use of scientific calculators with logarithm keys is essential in this section. Have students experiment with evaluating logarithms on their calculators until they understand the definition of logarithm as an exponent. For example, have them evaluate the common logarithms of the powers of 10 listed in chart form on page 191. Students should also evaluate nonintegral powers of 10, and verify that the logarithm of each one of them is the power of 10 they used. Point out that an important use of logarithms is to solve exponential equations with variable exponents (see Section 5-7).

You might want to use an overhead projector or a function plotting program with a projection system to demonstrate some of the properties of the graph of the base b logarithmic function.

Using Technology

The use of scientific calculators with a common logarithm key ($\log x$) and a natural logarithm key ($\ln x$) will speed up the calculations in this section. Point out to students that they are less likely to make errors with calculators than with tables (remember that logarithms were invented to speed up calculations).

Function plotting programs can be used to explore logarithmic functions. Have students graph the equations $y = \log x$, $y = \log 2x$, and $y = 2 \log x$ on the same set of axes to study the effect of the constant 2. Also, have students "zoom in" to see the asymptotic behavior of logarithmic graphs near the y-axis. Finally, have students graph the line $y = x$ and various exponential functions to see the inverse relationship with logarithmic functions.

Cooperative Learning

Consider using small groups to go over Class Exercises 1–9 and Written Exercises 1–8 and 11–18, pages 194 and 195. These exercises help students learn the definitions of $\log_b x = a$ and $\ln x = k$. The group environment will allow students to explain incorrect answers to each other and to become better calculator users. Also, encourage students to find logarithms without calculators whenever possible and to discuss the patterns they see in part (d) of Written Exercises 12, 13, and 14, page 194.

You might also want students to go over their proofs that log 2 is irrational (Exercise 52, page 197). Have the best proofs written on the chalkboard for discussion.

Applications

The Weber-Fechner law describes how people react to stimuli and differentiate phenomena. It states that the perceived intensity of a stimulus is given by

$$S = C \ln \left(\frac{I}{I_0} \right)$$

where I is the real intensity of the stimulus, I_0 is the smallest observed value of the stimulus, and C is a constant related to the stimulus. This law can be applied to loudness of sounds, brightness of light, pitch of musical tones, and other physical phenomena.

Point out that logarithms are used in advanced mathematics, engineering, and many of the sciences, including chemistry to find the pH of a solution (see Exercise 45, page 196) and seismology to measure the magnitude of earthquakes.

5-6 | pages 197–203

Presenting the Section

Point out that the laws of logarithms are important in science and engineering applications as well as in advanced mathematics because they are used to convert complicated expressions into simpler ones.

You might need to do several examples before assigning Exercises 25–28, page 200. Use the inverse properties of logarithmic and exponential functions to show that

$$\ln e^x = e^{\ln x} = x \text{ or } \log_b b^x = b^{\log_b x} = x.$$

Using Technology

Scientific calculators are recommended both for doing the logarithm calculations and for helping students see the patterns in some of the exercises.

Cooperative Learning

You might want students to discuss among themselves Class Exercise 21, page 199, and Exercises 35–39 and 53–54, pages 201–203. This can be done in a group setting so that all students have an opportunity to get their questions answered and can compare methods for proof questions. Monitor the student discussions and offer assistance when necessary.

Applications

Applications of logarithmic equations include the Richter scale system of measuring the severity of earthquakes discussed in Exercises 51 and 52, pages 202 and 203, and the conservation example on page 201. Logarithmic equations also are used to measure the intensity of sound waves in decibels, the pH level of a chemical solution, and the brightness of light (see Section 5-5).

5-7 | pages 203–207

Presenting the Section

This section can be used to bring closure to the chapter because students use common and natural logarithms to solve exponential equations like those given in Sections 5-1 to 5-4. Students first isolate the exponential expression, take the common or natural logarithm of both sides (the natural logarithm is used if e is involved), and then use the definition and the properties of logarithms to solve for the variable exponent. This method is based on the inverse properties of logarithmic and exponential functions. The change of base formula, $\log_b c = \frac{\log_a c}{\log_a b}$, also is introduced to enable students to find the logarithm of a number in terms of the logarithm of another known base (usually 10). Have students try to find $\log_3 8$ with their calculators to motivate the need for this formula.

Using Technology

Scientific calculators are recommended for doing the logarithm calculations, and graphing calculators or computers can be used for extended work with Exercises 13–19, pages 205 and 206.

Assessment

You might want to have students make up a summary list of all the applications presented in this chapter, give the equations which are used to model the applications, and show how to solve each of the equations. Encourage students to discuss the methods of solutions, while you monitor the discussions to evaluate students' knowledge of the material.

5 *Exponents and Logarithms*

168

■ Exponents

5-1 Growth and Decay: Integral Exponents

| Objective | To define and apply integral exponents. |

Suppose that the cost of a hamburger has been increasing at the rate of 9% per year. Then, each year the cost is 1.09 times the cost in the previous year. Suppose that the cost now is $4. Some projected future costs are given in the table below.

Time (years from now)	0	1	2	3	t
Cost (dollars)	4	4(1.09)	$4(1.09)^2$	$4(1.09)^3$	$4(1.09)^t$

$\times 1.09 \quad \times 1.09 \quad \times 1.09 \quad \times 1.09$

The table suggests that the cost is a function of time t. Since the variable t occurs as an exponent, the cost is said to be an exponential function of time:

$$C(t) = 4(1.09)^t$$

When $t > 0$ the function gives future costs, and when $t < 0$ it gives costs in the past. Example 1 illustrates this.

Example 1   Use the cost function, $C(t) = 4(1.09)^t$, to find the cost of a hamburger (a) 5 years from now and (b) 5 years ago.

Solution   Values of $C(t)$ are most easily found with a scientific calculator. (Use the exponent key.)
   a. $C(5) = 4(1.09)^5 \approx 6.15$. The cost will be about $6.15.
   b. $C(-5) = 4(1.09)^{-5} \approx 2.60$. The cost was about $2.60.

We will soon give exact meaning to expressions that involve negative exponents. However, first let us contrast the cost of a hamburger that has been increasing at 9% per year with the cost of a graphing calculator that has been *decreasing* at 9% per year. Each year the calculator cost is $100\% - 9\%$, or 91%, of what it was the year before.

Hamburger cost: $4 now Cost increasing at 9% per year $C(t) = 4(1.09)^t$	Calculator cost: $70 now Cost decreasing at 9% per year $C(t) = 70(0.91)^t$

◀ Ancient Egyptian astronomers used geometry to predict the future position of stars and planets. Current archeologists use mathematics to explore the Egyptian past. *Exponential functions,* for instance, are used in the carbon dating of ancient artifacts, such as this ceremonial boat dating from the reign of King Cheops, around 2400 B.C.

Exponents and Logarithms   **169**

As one would expect, a graph of exponential growth is increasing, while a graph of exponential decay is decreasing. In real-life situations, a certain population may experience exponential growth for a period of time, but this usually cannot be sustained (due to environmental factors); the population eventually "levels off" (that is, reaches a *steady state*).

Problem Solving

Have students apply the idea of exponential growth to practical problems such as the following: If grocery prices increased by 1% per day for a year, how much would $100 in groceries cost at the end of one year? (About $3778) How much of an annual increase is 1% per month? (About 12.7%)

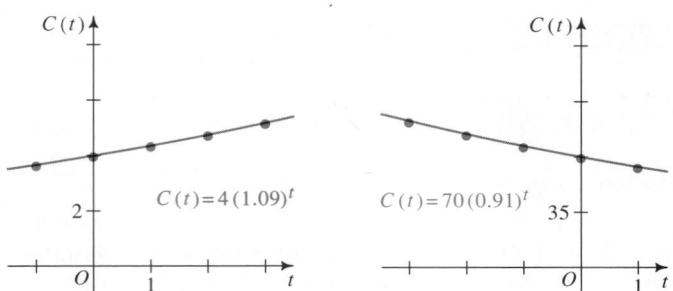

The graph at the left above shows exponential growth. The one at the right above shows exponential decay. Growth and decay can be modeled by:

$$A(t) = A_0(1 + r)^t,$$

where A_0 is the initial amount, the amount at time $t = 0$, and r is the growth rate. If $r > 0$, then the initial amount grows exponentially. If $-1 < r < 0$, then the initial amount decays exponentially.

Example 2 Suppose that a radioactive isotope decays so that the radioactivity present decreases by 15% per day. If 40 kg are present now, find the amount present (**a**) 6 days from now and (**b**) 6 days ago.

Solution $A(t) = A_0(1 + r)^t = 40(1 - 0.15)^t = 40(0.85)^t$

 a. $A(6) = 40(0.85)^6 \approx 15.1$ **b.** $A(-6) = 40(0.85)^{-6} \approx 106.1$
  There will be about 15.1 kg. There was about 106.1 kg.

Although we have used a calculator to evaluate expressions involving negative exponents, we must still define them. To do this, we will first review the laws of exponents for positive integers.

Laws of Exponents

Same bases

 1. $b^x \cdot b^y = b^{x+y}$ **2.** $\dfrac{b^x}{b^y} = b^{x-y}$ $(b \neq 0)$

 3. If $b \neq 0$, 1, or -1, then $b^x = b^y$ if and only if $x = y$.

Same exponents

 4. $(ab)^x = a^x b^x$ **5.** $\left(\dfrac{a}{b}\right)^x = \dfrac{a^x}{b^x}$ $(b \neq 0)$

 6. If $x \neq 0$, $a > 0$, and $b > 0$, then $a^x = b^x$ if and only if $a = b$.

Power of a power
 7. $(b^x)^y = b^{xy}$

If the laws on the preceding page are to make sense for the zero exponent and negative integral exponents, then such exponents must be defined as follows.

Definition of b^0: If law 1 is to hold for $y = 0$, then we have
$$b^x \cdot b^0 = b^{x+0} = b^x.$$

Since b^0 behaves like the number 1, we define it to be 1:
$$b^0 = 1 \ (b \neq 0)$$

Definition of b^{-x}: If law 1 is to hold for $y = -x$ and $b \neq 0$, then we have
$$b^x \cdot b^{-x} = b^{x+(-x)} = b^0 = 1.$$

Since b^x and b^{-x} have a product of 1, they are reciprocals of each other. Therefore, we define:
$$b^{-x} = \frac{1}{b^x} \ (x > 0 \text{ and } b \neq 0)$$

An expression is simplified when it contains neither negative exponents nor powers of powers. This is illustrated in the examples below.

Example 3 Simplify $\left(\dfrac{b^2}{a}\right)^{-2}\left(\dfrac{a^2}{b}\right)^{-3}$, where $a \neq 0$ and $b \neq 0$.

Solution $\left(\dfrac{b^2}{a}\right)^{-2}\left(\dfrac{a^2}{b}\right)^{-3} = \dfrac{(b^2)^{-2}}{a^{-2}} \cdot \dfrac{(a^2)^{-3}}{b^{-3}} = \dfrac{b^{-4}a^{-6}}{a^{-2}b^{-3}} = \dfrac{1}{a^4 b}$

Example 4 Simplify $(a^{-2} + b^{-2})^{-1}$, where $a \neq 0$ and $b \neq 0$.

Solution $(a^{-2} + b^{-2})^{-1} = \left(\dfrac{1}{a^2} + \dfrac{1}{b^2}\right)^{-1} = \left(\dfrac{b^2 + a^2}{a^2 b^2}\right)^{-1} = \dfrac{a^2 b^2}{b^2 + a^2}$

As Example 4 shows, you do not simplify a power of a sum or difference by distributing the exponent over the individual terms. Laws 4 and 5 apply only to a power of a product or quotient:
$$(a^{-2}b^{-2})^{-1} = a^2 b^2 \quad \text{but} \quad (a^{-2} + b^{-2})^{-1} \neq a^2 + b^2$$

Example 5 Simplify **(a)** $\dfrac{x^5 + x^{-2}}{x^{-3}}$ and **(b)** $\dfrac{x^5 \cdot x^{-2}}{x^{-3}}$, where $x \neq 0$.

Solution
a. $\dfrac{x^5 + x^{-2}}{x^{-3}} = \dfrac{x^5}{x^{-3}} + \dfrac{x^{-2}}{x^{-3}}$
$= x^8 + x$

b. $\dfrac{x^5 \cdot x^{-2}}{x^{-3}} = \dfrac{x^3}{x^{-3}}$
$= x^6$

We shall assume throughout the rest of this book that variables are restricted so that there are no denominators of zero. In this section only, we assume that variables appearing as exponents represent integers.

Exponents and Logarithms   **171**

After discussing the definitions of b^0 and b^{-x}, ask students questions like the following:

1. What is 1000^0? (1)

2. What is 1000^{-1} (0.001)

3. What is $(1000^{-1})^0$? (1)

Error Analysis

For Example 4, some students may want to "distribute" the power to each of the terms in the parentheses. Caution them that powers do not distribute over addition or subtraction. To help them see this, have them evaluate $(a^{-2} + b^{-2})^{-1}$ and $a^2 + b^2$ for given values of a and b, such as $a = 1$ and $b = 2$. $((1^{-2} + 2^{-2})^{-1} = 0.8$; $1^2 + 2^2 = 5$; so $(1^{-2} + 2^{-2})^{-1} \neq 1^2 + 2^2$.)

Additional Examples

1. In a certain city, the value of a house is increasing at a rate of 16% annually.
  a. Find the value of a $100,000 house in 4 yr.
  b. In how many years will the value of the house be approximately double what it is now?

  Let $C(t) = C_0(1 + r)^t = 100,000(1 + 0.16)^t = 100,000(1.16)^t$
  a. $C(4) = 100,000(1.16)^4 \approx \$181,000$
  b. Substitute consecutive integral values of t on a calculator: $C(5) \approx \$210,000$. Thus, the house will double in value in about 5 years.

(*Continues on next page*)

2. Simplify by using powers of the same base.

a. $\dfrac{2^8 \cdot 4^{-5}}{16^{-1}}$

b. $\sqrt{\dfrac{25^3 \cdot 5^n}{125^{2-n}}}$

a. $\dfrac{2^8 \cdot 4^{-5}}{16^{-1}} =$

$\dfrac{2^8 \cdot (2^2)^{-5}}{(2^4)^{-1}} = \dfrac{2^8 \cdot 2^{-10}}{2^{-4}} =$

$\dfrac{2^{-2}}{2^{-4}} = 2^{-2-(-4)} =$

$2^2 = 4$

b. $\sqrt{\dfrac{25^3 \cdot 5^n}{125^{2-n}}} =$

$\sqrt{\dfrac{25^3 \cdot 5^n}{125^2 \cdot 125^{-n}}} =$

$\sqrt{\dfrac{(5^2)^3 \cdot 5^n}{(5^3)^2 \cdot (5^3)^{-n}}} =$

$\sqrt{\dfrac{5^6}{5^6} \cdot \dfrac{5^n}{5^{-3n}}} = \sqrt{5^{4n}} =$

$\sqrt{5^{2n} \cdot 5^{2n}} = 5^{2n}$

Error Analysis

Class Exercise 10 will be difficult for some students because they will apply the laws of exponents incorrectly. They need to recognize that the *exponents* (and not the *bases*) are the same here, so laws 4 and 5 (rather than laws 1 and 2) apply.

CLASS EXERCISES

Copy and complete each table.

1.

a.	If A_0 increases by	3%	15%	4.6%	120%	?
b.	then multiply A_0 by	?	?	?	?	1.105
		1.03	1.15	1.046	2.2	10.5%

2.

a.	If A_0 decreases by	12%	7.5%	80%	?	100%
b.	then multiply A_0 by	?	?	?	0.68	?
		0.88	0.925	0.20	32%	0

3.

	Item	Annual rate of increase	Cost now	Cost in t years	
a.	Bike	5%	$200.00	?	$200(1.05)^t$
b.	Jeans	8%	$20.00	?	$20(1.08)^t$
c.	Loaf of bread	?	$1.25	$1.25(1.06)^t$	6%

4.

	Item	Annual rate of decrease	Value now	Value in t years	
a.	Car	20%	$9800	?	$9800(0.8)^t$
b.	Boat	15%	$2200	?	$2200(0.85)^t$
c.	Skates	?	$100	$100(0.75)^t$	25%

Simplify.

5. a. 8^{-1} $\dfrac{1}{8}$ **b.** 8^{-2} $\dfrac{1}{64}$ **6. a.** $\left(\dfrac{2}{3}\right)^{-1}$ $\dfrac{3}{2}$ **b.** $\left(\dfrac{2}{3}\right)^{-2}$ $\dfrac{9}{4}$

7. a. $4 \cdot 3^{-2}$ $\dfrac{4}{9}$ **b.** $(4 \cdot 3)^{-2}$ $\dfrac{1}{144}$ **8. a.** $(2^{-1} \cdot 4^{-1})^{-1}$ 8   **b.** $(2^{-1} + 4^{-1})^{-1}$ $\dfrac{4}{3}$

9. $\dfrac{12^3}{6^3}$ 8   **10.** $\dfrac{8^n \cdot 3^n}{4^n}$ 6^n **11.** $\dfrac{(-2n)^2}{-2n^2}$ -2 **12.** $x^{-3}(x^5 + x^3)$ $x^2 + 1$

13. $\dfrac{3a^3 \cdot 6a^6}{a^{-1}}$ $18a^{10}$ **14.** $\dfrac{3a^3 - 6a^6}{a^{-1}}$ $3a^4 - 6a^7$ **15.** $\dfrac{5b^3 + 10b^6}{b^{-2}}$ $5b^5 + 10b^8$ **16.** $\dfrac{5a^3 \cdot 10b^6}{b^{-2}}$ $50a^3b^8$

WRITTEN EXERCISES

Simplify each expression. **3. a.** $\frac{5}{8}$ **b.** $\frac{1}{1000}$ **4. a.** 128 **b.** 8 **6. a.** $\frac{x^{12}}{16}$ **b.** $\frac{2}{x^7}$

A

1. a. $(-4)^{-2}$ $\frac{1}{16}$ **b.** -4^{-2} $-\frac{1}{16}$ **2. a.** $(-3)^{-4}$ $\frac{1}{81}$ **b.** -3^{-4} $-\frac{1}{81}$

3. a. $5 \cdot 2^{-3}$ **b.** $(5 \cdot 2)^{-3}$ **4. a.** $2 \div 4^{-3}$ **b.** $(2 \div 4)^{-3}$

5. a. $(x^{-2})^{-1}$ x^2 **b.** $((x^{-2})^{-1})^0$ 1 **6. a.** $(2x^{-3})^{-4}$ **b.** $2x^{-3} \cdot x^{-4}$

7. a. $(3a^{-1})^{-1}$ **b.** $(3 + a^{-1})^{-1}$ **8. a.** $(5x^2 \cdot x^{-2})^2$ **b.** $(5x^2 + x^{-2})^2$

9. a. $(2^{-2} + 2^{-3})^{-1}$ **b.** $(2^{-2} \cdot 2^{-3})^{-1}$ **10. a.** $(4^{-1} - 2^{-1})^2$ **b.** $(4^{-1} \div 2^{-1})^2$

11. a. $(a^{-1} - b^{-1})^{-1}$ **b.** $(a^{-1} \cdot b^{-1})^{-1}$ **12. a.** $(2 + x^{-2})^{-2}$ **b.** $(2 \cdot x^{-2})^{-2}$

Copy and complete the table. The cost of each item grows exponentially.

	Item	Annual rate of increase	Cost now	Cost in 10 years	Cost in 20 years	
13.	Airplane ticket	15%	$300	?	?	$1214; $4910
14.	Swim suit	8%	$35	?	?	$76; $163
15.	Jar of mustard	7%	$1	?	?	$1.97; $3.87
16.	College tuition	10%	$12,000	?	?	$31,125; $80,730

Copy and complete the table. The value of each item decays exponentially.

	Item	Annual rate of decrease	Value now	Value in 3 years	Value in 6 years	
17.	Farm tractor	25%	$65,000	?	?	$27,422; $11,569
18.	Industrial equipment	10%	$200,000	?	?	See below.
19.	Value of the dollar	6%	$1	?	?	$0.83; $0.69
20.	Value of the dollar	8%	$1	?	?	$0.78; $0.61

18. $145,800; $106,288

Simplify each expression.

21. $(3a^{-2})^3 \cdot 3a^5$ $\frac{81}{a}$ **22.** $(-4x^3)^2 \cdot 3x^{-2}$ $48x^4$ **23.** $(3n^2)^{-1} (3n^2)^7$ $3^6 n^{12}$

24. $(2r^{-1})^4 (4r^2)^{-2}$ $\frac{1}{r^8}$ **25.** $\frac{(2a^{-1})^2}{(2a^{-1})^{-2}}$ $\frac{16}{a^4}$ **26.** $\frac{(-3n^{-3})^2}{-9n^{-4}}$ $-\frac{1}{n^2}$

27. $\left(\frac{a}{b^2}\right)^{-2} \left(\frac{a}{b^2}\right)^{-3}$ $\frac{b^{10}}{a^5}$ **28.** $\frac{(-2r)^4}{(-2r)^{-2}}$ $64r^6$ **29.** $2x^{-3}(x^5 - 2x^3)$ $2x^2 - 4$

30. $xy^{-2}(xy^2 - 3y^3)$ $x^2 - 3xy$ **31.** $\frac{6a^{-2} + 9a^2}{3a^{-2}}$ $2 + 3a^4$ **32.** $\frac{8n^4 - 4n^{-2}}{2n^{-2}}$ $4n^6 - 2$

Exponents and Logarithms **173**

Additional Answers
Written Exercises

7. a. $\frac{a}{3}$

b. $\frac{a}{3a + 1}$

8. a. 25

b. $\frac{(5x^4 + 1)^2}{x^4}$

9. a. $\frac{8}{3}$

b. 32

10. a. $\frac{1}{16}$

b. $\frac{1}{4}$

11. a. $\frac{ab}{b - a}$

b. ab

12. a. $\frac{x^4}{(2x^2 + 1)^2}$

b. $\frac{x^4}{4}$

Suggested Assignments

Standard
Day 1: 173/1–21 odd
Day 2: 173/23–45 odd
Comprehensive
173/1–45 every 3rd, 46–49

Supplementary Materials

Alternative Assessment, 15
Precalculus Plotter Plus,
35–36

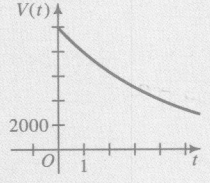
33. **Business** The value of a new car decreases 20% each year. Complete the table. The value $V(t)$ of the car is in dollars and its age t is in years. Give each value to the nearest hundred. Using the values in your table, make a graph to show the relationship between $V(t)$ and t.

t	0	1	2	3	4	5
$V(t)$	10,000	?	?	?	?	?

34. **Business** The value in dollars of a car t years from now is $V(t) = 12,500(0.85)^t$. **(a)** What is the annual rate of depreciation, the rate at which the car loses value? **(b)** In how many years will the value of the car be approximately half what it is now? **a.** 15% **b.** 4 years

35. **Consumer Economics** If grocery prices increase 1% per month for a whole year, how much would groceries that cost $100 at the beginning of the year cost at the end of the year? $112.68

36. **Consumer Economics** The cost of goods and services in an urban area increased 1.5% last month. At this rate, what will be the annual rate of increase? 19.6%

Simplify by using powers of the same base.

37. **a.** $\dfrac{3^5 \cdot 9^4}{27^4}$ 3   **b.** $\dfrac{125^{-3} \cdot 25}{5^{-8}}$ 5   **c.** $\sqrt{\dfrac{8^n \cdot 2^7}{4^{-n}}}$ $\sqrt{2^{5n+7}}$

38. **a.** $\dfrac{4^9 \cdot 8^{-4}}{16^3}$ $\dfrac{1}{64}$ **b.** $\dfrac{3^7 \cdot 9^5}{\sqrt{27^{12}}}$ $\dfrac{1}{3}$ **c.** $\sqrt[3]{\dfrac{125^n \cdot 5^{4n}}{25^{-n}5^{3n}}}$

Write as a power of b.

39. **a.** $\dfrac{(b^n)^3}{b^n \cdot b^n}$ b^n **b.** $\dfrac{(b^n)^2}{b^n \cdot b^{n+2}}$ b^{-2} **c.** $\dfrac{b \cdot b^n}{(b^3)^n}$ b^{1-2n}

40. **a.** $\sqrt{\dfrac{b^{2n}}{b^{-2n}}}$ b^{2n} **b.** $\dfrac{(b \cdot b^n)^2}{(b^2)^n}$ b^2 **c.** $\sqrt{\dfrac{b^{1-n}}{b^{n-1}}}$ b^{1-n}

Simplify. (*Hint:* **In Exercise 41(a), multiply the numerator and the denominator by 2^3.**)

B 41. **a.** $\dfrac{2^{-1}}{2^{-2} + 2^{-3}}$ $\dfrac{4}{3}$ **b.** $\dfrac{4^{-5}}{4^{-2} + 4^{-3}}$ $\dfrac{1}{80}$ 42. **a.** $\dfrac{3^{-2}}{3^{-3} + 3^{-2}}$ $\dfrac{3}{4}$ **b.** $\dfrac{2^{-1} - 2^{-2}}{2^{-1} + 2^{-2}}$ $\dfrac{1}{3}$

43. **a.** $\dfrac{x^{-2} - y^{-2}}{x^{-1} - y^{-1}}$ $\dfrac{y+x}{xy}$ **b.** $\dfrac{1 - y^{-1}}{y - y^{-1}}$ $\dfrac{1}{y+1}$ 44. **a.** $\dfrac{x^{-1}}{x - x^{-1}}$ $\dfrac{1}{x^2 - 1}$ **b.** $\dfrac{4 - x^{-4}}{2 - x^{-2}}$ $\dfrac{2x^2 + 1}{x^2}$

45. *Discussion* According to legend Manhattan Island was purchased in 1626 for trinkets worth about $24. If the $24 had been invested instead at a rate of 6% interest per year, what would be the value of the money in 1996? Compare this with a recent total of $34,000,000,000 in assessed values for Manhattan Island. about $55,000,000,000

C **46.** Solve $2^x + 8 \cdot 2^{-x} = 9$.

47. Solve $2^x + 2^{-x} = \frac{5}{2}$.

48. Solve $2^{2x} - 3 \cdot 2^{x+1} + 8 = 0$.

49. Solve $3^{2x+1} - 10 \cdot 3^x + 3 = 0$.

46. 0,3   **47.** 1,−1   **48.** 2,1   **49.** −1,1

5-2 Growth and Decay: Rational Exponents

Objective   *To define and apply rational exponents.*

In Section 5-1, we considered a 9% annual growth in the cost of a hamburger. We saw that if a hamburger costs $4 now (time $t = 0$), then its cost t years from now will be $C(t) = 4(1.09)^t$. To find the cost one-half year from now, we must evaluate an expression with a fractional exponent:

$$C\left(\frac{1}{2}\right) = 4(1.09)^{1/2}$$

Although a scientific calculator easily gives an approximate value of $4.18, there remains the question of what a fractional exponent means.

The definition of a rational exponent given below is made in such a way that the laws for integral exponents on page 170 will continue to hold. However, the base b must now be a positive real number other than 1.

Definition of $b^{1/2}$:   If exponent law 7 is to hold for rational exponents, then

$$(b^{1/2})^2 = b^{(1/2)(2)} = b^1.$$

Since we know $(\sqrt{b})^2 = b$, we define $b^{1/2}$ to be $\sqrt{b}$.

Definition of $b^{3/2}$: $b^{3/2} = (b^{1/2})^3 = (\sqrt{b})^3$ and $b^{3/2} = (b^3)^{1/2} = \sqrt{b^3}$

Either $(\sqrt{b})^3$ or $\sqrt{b^3}$ can be used as a definition of $b^{3/2}$.

Definition of $b^{p/q}$:   Using reasoning similar to that above, we make these definitions:

$$b^{1/q} = \sqrt[q]{b} \quad \text{and} \quad b^{p/q} = (\sqrt[q]{b})^p \text{ or } \sqrt[q]{b^p}$$

Exponents and Logarithms   **175**

Error Analysis

Students commonly make mistakes when they use calculators to find rational powers of numbers. To guarantee the correct order of operations, students may need to use parentheses when entering the exponent in an expression like $6^{2/3}$ on their calculators.

Example Note

In Example 1, when rewriting an expression with a rational exponent, emphasize that the denominator represents the root and the numerator represents the power.

In Example 3, the laws of exponents apply because the expressions on either side of the equals sign have been rewritten so that they have the same base. Students should be aware that this is necessary. In Example 4, the laws of exponents apply because the expressions on either side of the equals sign have been raised to the same power.

Making Connections

Examples 3 and 4 demonstrate how the laws of exponents can be used to solve equations with exponents. Later in this chapter, logarithms will be used to solve equations with exponents.

Example 1 Simplify: **a.** $16^{1/4}$ **b.** $16^{-1/4}$ **c.** $8^{2/3}$ **d.** $8^{-2/3}$

Solution **a.** $16^{1/4} = \sqrt[4]{16} = 2$ **b.** $16^{-1/4} = (16^{1/4})^{-1} = 2^{-1} = \frac{1}{2}$

c. $8^{2/3} = (\sqrt[3]{8})^2 = 4$ **d.** $8^{-2/3} = (8^{2/3})^{-1} = 4^{-1} = \frac{1}{4}$

You can use a calculator to evaluate expressions that contain rational powers. For example, $5^{2/3} \approx 2.9240$.

Example 2 Suppose a car presently worth \$8200 depreciates 20% per year. About how much will it be worth 2 years and 3 months from now?

Solution Express 2 years and 3 months as 2.25 years. Then:

$$C(t) = 8200(1 - 0.20)^t = 8200(0.80)^t$$
$$C(2.25) = 8200(0.80)^{2.25} \approx 4963$$

It will be worth about \$4963.

Examples 3 and 4 show how the laws of exponents can be used to solve two kinds of equations containing exponents.

Example 3 Solve **(a)** $2^x = \frac{1}{8}$ and **(b)** $9^{x+1} = \sqrt{27}$.

Solution Express both sides of each equation as powers of the same base. Then apply law 3.

a.
$$2^x = \frac{1}{8}$$
$$2^x = 2^{-3}$$
$$x = -3$$

b.
$$9^{x+1} = \sqrt{27}$$
$$(3^2)^{x+1} = \sqrt{3^3}$$
$$3^{2x+2} = 3^{3/2}$$
$$2x + 2 = \frac{3}{2}$$
$$x = -\frac{1}{4}$$

Example 4 Solve **(a)** $4x^{3/2} = 32$ and **(b)** $(x - 1)^{-1/4} - 2 = 0$.

Solution **a.**
$$4x^{3/2} = 32$$
$$x^{3/2} = 8$$
$$(x^{3/2})^{2/3} = 8^{2/3}$$
$$x = 4$$

b.
$$(x - 1)^{-1/4} - 2 = 0$$
$$(x - 1)^{-1/4} = 2$$
$$((x - 1)^{-1/4})^{-4} = 2^{-4}$$
$$x - 1 = \frac{1}{16}$$
$$x = \frac{17}{16}$$

Example 5   A house bought five years ago for $100,000 was just sold for $135,000. To the nearest tenth of a percent what was the annual growth rate?

Solution
$$A(t) = A_0(1 + r)^t$$
Since $A_0 = 100,000$
and $A(5) = 135,000$,
$$135,000 = 100,000(1 + r)^5.$$
Therefore:
$$1.35 = (1 + r)^5$$
$$(1.35)^{1/5} = (1 + r)$$
$$r \approx 0.0619$$

To the nearest tenth of a percent, the growth rate was 6.2%.

CLASS EXERCISES

Simplify each expression.

4. b. $8 + 2\sqrt{15}$

1. a. $4^{1/2}$ 2   b. $4^{-1/2}$ $\frac{1}{2}$ 2. a. $4^{3/2}$ 8   b. $4^{-3/2}$ $\frac{1}{8}$

3. a. $-9^{1/2}$ -3 b. $-9^{-1/2}$ $-\frac{1}{3}$ 4. a. $(3^{1/2} \cdot 5^{1/2})^2$ 15   b. $(3^{1/2} + 5^{1/2})^2$

5. $\left(\frac{49}{25}\right)^{-1/2}$ $\frac{5}{7}$ 6. $\left(\frac{4}{9}\right)^{3/2}$ $\frac{8}{27}$ 7. $(8^{-1/6})^{-2}$ 2   8. $8^{3/2} \cdot 2^{3/2}$ 64

9. $(2x^{-1/3})^3$ $\frac{8}{x}$ 10. $\left(\frac{125}{x^6}\right)^{1/3}$ $\frac{5}{x^2}$ 11. $\frac{x^{1/3}}{2x^{-2/3}}$ $\frac{x}{2}$ 12. $2x^{3/2} \cdot 4x^{-1/2}$ 8x

13. **Consumer Economics**   The cost in dollars of a new pair of running shoes t years from now is $C(t) = 62(1.05)^t$.
   a. What is the cost now? $62
   b. To find the cost in 2.5 years, use $t = $ __?__. 2.5
   c. To find the cost 9 months ago, use $t = $ __?__. -0.75

Solve.

14. $3^{2x} = 3^{12}$ 6   15. $9^x = 3^5$ $\frac{5}{2}$

16. $x^{2/3} = 9$ 27   17. $x^{-1/2} = 4$ $\frac{1}{16}$

18. **Discussion**   If the exponent is an integer n, then the laws of exponents are true for both positive and negative bases. When the definition of exponent is extended to rational numbers, the base must be restricted to positive numbers. Discuss why this restriction must be made.
   $b^{p/q} = (\sqrt[q]{b})^p$ and $\sqrt[q]{b}$ is not a real number when q is even and b is negative.

1. Write each expression using positive rational exponents and no radical signs.
   a. $\left(\sqrt[6]{27a^{-1}b^2}\right)^4$
   b. $\sqrt[3]{x} \cdot \sqrt[6]{x} \div \sqrt[5]{x}$

   a. $\left(\sqrt[6]{27a^{-1}b^2}\right)^4 =$
   $\left(27a^{-1}b^2\right)^{4/6} =$
   $\left(\frac{27b^2}{a}\right)^{2/3} =$
   $\frac{27^{2/3} \cdot \left(b^2\right)^{2/3}}{a^{2/3}} =$
   $\frac{\left(\sqrt[3]{27}\right)^2 \cdot b^{4/3}}{a^{2/3}} =$
   $\frac{9b^{4/3}}{a^{2/3}}$

   b. $\sqrt[3]{x} \cdot \sqrt[6]{x} \div \sqrt[5]{x} =$
   $\frac{x^{1/3} \cdot x^{1/6}}{x^{1/5}} =$
   $\frac{x^{(2/6)+(1/6)}}{x^{1/5}} =$
   $x^{(1/2)-(1/5)} = x^{3/10}$

2. Simplify.
   a. $3x^{-1/3}\left(x^{1/3} - 4x^{4/3}\right)$
   b. $\frac{9a^{1/2}b^{-1/2} - 6a^2b^{1/2}}{12ab^{-3/2}}$

   a. $3x^{-1/3}\left(x^{1/3} - 4x^{4/3}\right) =$
   $3x^0 - 12x^{3/3} =$
   $3 \cdot 1 - 12x = 3 - 12x$
   b. $\frac{9a^{1/2}b^{-1/2} - 6a^2b^{1/2}}{12ab^{-3/2}} =$
   $\frac{9a^{1/2}b^{-1/2}}{12ab^{-3/2}} - \frac{6a^2b^{1/2}}{12ab^{-3/2}} =$
   $\frac{3b^{2/2}}{4a^{1/2}} - \frac{ab^{4/2}}{2} =$
   $\frac{3b}{4a^{1/2}} - \frac{ab^2}{2}$

Mathematical Note

An expression is not considered simplified unless there are no negative exponents or complex fractions (fractions over fractions) and unless powers (or indices) are as small as possible.

Suggested Assignments

Standard
Day 1: 178/3–36 every 3rd
Day 2: 179/37–57 odd
Comprehensive
Day 1: 178/3–36 every 3rd
Day 2: 179/37–57 odd

Supplementary Materials

Alternative Assessent, 15
Student Resource Guide, 46–48

WRITTEN EXERCISES

Write each expression using a radical sign and no negative exponents.

A **1. a.** $x^{2/3}$ $\sqrt[3]{x^2}$ **b.** $x^{3/2}$ $\sqrt{x^3}$ **c.** $5^{1/2} \cdot x^{-1/2}$ $\sqrt{\dfrac{5}{x}}$ **d.** $6^{1/3} \cdot x^{2/3}$ $\sqrt[3]{6x^2}$

2. a. $3y^{2/5}$ $3\sqrt[5]{y^2}$ **b.** $(3y)^{2/5}$ $\sqrt[5]{9y^2}$ **c.** $a^{4/7} \cdot b^{-4/7}$ $\sqrt[7]{\dfrac{a^4}{b^4}}$ **d.** $a^{1/10} \cdot b^{-1/5}$ $\sqrt[10]{\dfrac{a}{b^2}}$

Write each expression using positive rational exponents.

3. a. $\sqrt{x^5}$ $x^{5/2}$ **b.** $\sqrt[3]{y^2}$ $y^{2/3}$ **c.** $(\sqrt[6]{2a})^5$ $(2a)^{5/6}$ **d.** $\sqrt{x} \cdot \sqrt[3]{x} \cdot \sqrt[6]{x}$ x

4. a. $\sqrt[3]{8x^7}$ $2x^{7/3}$ **b.** $(\sqrt[4]{16x})^3$ $8x^{3/4}$ **c.** $\sqrt[3]{27x^{-6}y^2}$ $\dfrac{3y^{2/3}}{x^2}$ **d.** $\sqrt[4]{x} \cdot \sqrt[3]{x} \div \sqrt[6]{x}$ $x^{5/12}$

Simplify. **11.** $\dfrac{a^4}{4}$ **12.** $\dfrac{n^{15/2}}{27}$ **13.** $2x^2$ **14.** $4a^3$

5. a. $\left(\dfrac{9}{25}\right)^{1/2}$ $\dfrac{3}{5}$ **b.** $\left(\dfrac{9}{25}\right)^{-1/2}$ $\dfrac{5}{3}$ **c.** $\left(\dfrac{9}{25}\right)^{5/2}$ $\dfrac{243}{3125}$ **d.** $\left(\dfrac{9}{25}\right)^{-1.5}$ $\dfrac{125}{27}$

6. a. $\left(\dfrac{27}{8}\right)^{1/3}$ $\dfrac{3}{2}$ **b.** $\left(\dfrac{27}{8}\right)^{2/3}$ $\dfrac{9}{4}$ **c.** $\left(\dfrac{27}{8}\right)^{-2/3}$ $\dfrac{4}{9}$ **d.** $\left(\dfrac{27}{8}\right)^{0}$ 1

7. $(16^{-3/5})^{5/4}$ $\dfrac{1}{8}$ **8.** $(25^{-1/3})^{-3/2}$ 5 **9.** $(81^{1/2} - 9^{1/2})^2$ 36 **10.** $(3^{-2} + 4^{-2})^{-1/2}$ $\dfrac{12}{5}$

11. $(8a^{-6})^{-2/3}$ **12.** $(9n^{-5})^{-3/2}$ **13.** $(4x^{-3})^{-1/2} \cdot 4x^{1/2}$ **14.** $(4a^3)^{1/3} \div (4a^3)^{-2/3}$

15. Consumer Economics The cost of a certain brand of camera has been increasing at 8% per year. If a camera now costs $150, find the cost:
a. 2 years and 6 months from now $182 **b.** 4 years and 3 months ago $108

16. Business The value of a computer depreciates at the rate of 25% per year. If a computer is now worth $2400, find its approximate value:
a. 3 years and 6 months from now $877 **b.** 20 months ago $3877

Simplify. **17.** $a^2 - 2a$ **18.** $2n + 2$ **19.** $x^2 - 2x$ **20.** $2n^2 - 6n$ **24.** $2x - 1$

17. $a^{1/2}(a^{3/2} - 2a^{1/2})$ **18.** $2n^{1/3}(n^{2/3} + n^{-1/3})$ **19.** $x^{-1/2}(x^{5/2} - 2x^{3/2})$

20. $2n^{-2/3}(n^{8/3} - 3n^{5/3})$ **21.** $\dfrac{x^{1/2} - 2x^{-1/2}}{x^{-1/2}}$ $x - 2$ **22.** $\dfrac{y^{-1/3} - 3y^{2/3}}{y^{-4/3}}$ $y - 3y^2$

23. $\dfrac{2n^{1/3} - 4n^{-2/3}}{2n^{-2/3}}$ $n - 2$ **24.** $\dfrac{x^{-1/2}(2x^{1/2} - x^{-1/2})}{x^{-1}}$ **25.** $\dfrac{2n^{1/3}(3n^{1/3} - 4n^{4/3})}{2n^{-1/3}}$ $3n - 4n^2$

26. $\dfrac{4ab^{-1/2} - 2ab^{1/2}}{(a^2b)^{-1/2}}$ $4a^2 - 2a^2b$ **27.** $\dfrac{(\sqrt[3]{4a})^2}{\sqrt[6]{4a}}$ $2\sqrt{a}$ **28.** $\dfrac{(\sqrt{2x})^5}{(\sqrt{2x})^9}$ $\dfrac{1}{4x^2}$

Solve.

29. $8^x = 2^6$ 2 **30.** $9^{4x} = 81$ $\dfrac{1}{2}$ **31.** $8^{x-1} = 2^{x+1}$ 2

32. $9^x = 3^{10}$ 5 **33.** $8^x = 2^7 \cdot 4^9$ $\dfrac{25}{3}$ **34.** $27^{1-x} = \left(\dfrac{1}{9}\right)^{2-x}$ $\dfrac{7}{5}$

35. a. $(8x)^{-3} = 64$ $\dfrac{1}{32}$ **b.** $8x^{-3} = 64$ $\dfrac{1}{2}$ **c.** $(8 + x)^{-3} = 64$ $-\dfrac{31}{4}$

36. a. $(2x)^{-2} = 16$ $\pm\dfrac{1}{8}$ **b.** $2x^{-2} = 16$ $\pm\dfrac{\sqrt{2}}{4}$ **c.** $4(x - 2)^{-2} = 16$ $\dfrac{5}{2}$ or $\dfrac{3}{2}$

B **37. Finance** A house bought for $50,000 in 1980 was sold for $150,000 in 1990. To the nearest percent, what was the annual rate of appreciation (increase) in the value of the house? 12%

38. Consumer Economics The price of firewood four years ago was $140 per cord. Today, a cord of wood costs $182. To the nearest percent, what has been the annual rate of increase in the cost? 7%

39. Economics The *consumer price index* (CPI) is a measure of the average cost of goods and services. The United States government set the index at 100 for the period 1982-1984. In 1988, the index was 118.3. What was the average annual rate of increase (to the nearest tenth of a percent) from 1984 to 1988? 4.3%

40. Research Look in an almanac to find the current consumer price index. Then determine the average annual rate of increase (to the nearest tenth of a percent) for this index since 1984. (See Exercise 39.)

41. Education Yearly expenses at a state university have increased from $14,000 to $18,500 in the last 4 years. What has been the average annual growth rate in expenses? If this growth rate continues, what will the expenses be 4 years from now?

42. Research Find the present cost of something that interests you and its cost several years ago. Find the average annual growth rate.

41. approx. 7%; $24,400

Solve.

43. a. $(4x)^3 = 9^{\frac{81}{4}}$ **b.** $3^{4x} = 9^{\frac{3}{6}}$ **44. a.** $(4 - x)^{1/2} = 8^{\frac{-60}{}}$ **b.** $\left(\dfrac{1}{2}\right)^{4 - x} = 8^{\frac{7}{8}}$

45. $\dfrac{2^{x^2}}{2^x} = 64$ $-2, 3$ **46.** $\dfrac{5^{x^2}}{(5^x)^2} = 125$ $-1, 3$

47. $\sqrt{\dfrac{9^{x + 3}}{27^x}} = 81$ -2 **48.** $\sqrt[3]{\dfrac{8^{x + 1}}{16^x}} = 32$ -12

50. b. $(x + 1)^{1/2}(x - 3)$

Factor. (In Exercise 49(a), factor out $a^{1/2}b^{1/2}$.) **51. b.** $2(x^2 + 2)^{-1/2}$

49. a. $a^{3/2}b^{1/2} - a^{1/2}b^{3/2}$ $a^{1/2}b^{1/2}(a - b)$ **b.** $a^{1/2}b^{-1/2} - a^{3/2}b^{1/2}$ $a^{1/2}b^{-1/2}(1 - ab)$

50. a. $(x - 1)^{1/2} - x(x - 1)^{-1/2}$ $-(x - 1)^{-1/2}$ **b.** $(x + 1)^{3/2} - 4(x + 1)^{1/2}$

51. a. $(x^2 + 1)^{3/2} - x^2(x^2 + 1)^{1/2}$ $(x^2 + 1)^{1/2}$ **b.** $(x^2 + 2)^{1/2} - x^2(x^2 + 2)^{-1/2}$

52. a. $(2x + 1)^{2/3} - 4(2x + 1)^{-1/3}$ **b.** $(1 + x^2)^{-3/2} - (1 + x^2)^{-1/2}$
           $(2x + 1)^{-1/3}(2x - 3)$ $-x^2(1 + x^2)^{-3/2}$

Exercise Note

For Exercises 49–52, students simplify expressions with rational exponents. In each case, all the terms have a common factor which is itself an exponential expression. In Exercise 50a, $(x - 1)^{-1/2}$ divides each term and is therefore a common factor. Note that
$$\frac{(x - 1)^{1/2}}{(x - 1)^{-1/2}} =$$
$$(x - 1)^{(1/2)-(-1/2)} = x - 1.$$

Warm-Up Exercises

Find the value of x.

1. $x = 4 \cdot 2^3$ 32

2. $x = 8 \cdot \left(\frac{1}{2}\right)^4$ $\frac{1}{2}$

3. $409{,}600 = 100 \cdot 4^x$ 6

4. $9 = 81 \cdot \left(\frac{1}{3}\right)^x$ 2

5. $5203.02 = x \cdot (1.01)^4$
5000

6. $655.36 = x \cdot (0.8)^5$ 2000

Motivating the Section

Many real-life populations grow exponentially. For example, if an initial population of 100 bacteria doubles every hour, in 24 hours the number of bacteria will be 1,677,721,600, or 100×2^{24}. The number of bacteria after a specified length of time is an exponential function, the topic of this section.

If you play the A below middle C on a piano, the piano string vibrates with a frequency of 220 vibrations per second, or 220 hertz (Hz). Frequencies for some notes above A are as follows:

Note	A$^\#$ (A sharp)	B	Middle C	C$^\#$
Frequency	$220 \cdot 2^{1/12}$	$220 \cdot 2^{2/12}$	$220 \cdot 2^{3/12}$	$220 \cdot 2^{4/12}$

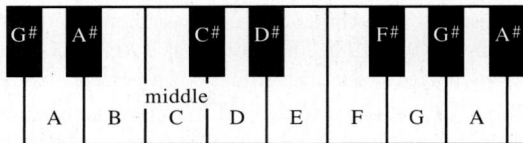

53. Music Give the frequencies of the notes D through A above middle C.

54. a. Music The frequency of G$^\#$ below middle C is $220 \cdot 2^{-1/12}$ Hz and the frequency of G$^\#$ above middle C is $220 \cdot 2^{11/12}$ Hz. Find the ratio of these frequencies. (*Note*: The interval between these two notes is called an "octave.") 2:1

b. If a song is played in the key of C, the interval from C to the G above is called a "fifth." Show that the ratio of these two frequencies is approximately 3 to 2. $220 \cdot 2^{10/12} : 220 \cdot 2^{3/12} \approx 3{:}2$

Solve for x by rewriting the equation in quadratic form.

55. $x^{2/3} - 7x^{1/3} + 12 = 0$ (*Hint:* $x^{2/3} = (x^{1/3})^2$)

56. $x^{4/3} - 6x^{2/3} + 8 = 0$ $2\sqrt{2}$, 8

57. $9^{2x} - 2 \cdot 9^x - 3 = 0$ $\frac{1}{2}$

58. $4^{2x} - 10 \cdot 4^x + 16 = 0$

55. 27, 64

58. $\frac{1}{2}$, $\frac{3}{2}$

5-3 Exponential Functions

Objective *To define and use exponential functions.*

We have defined rational exponents, but not irrational ones. For example, how is $3^{\sqrt{2}}$ defined? To find out, complete the following activity.

Activity

The first row in the table shows a sequence of decimal approximations for $\sqrt{2}$. Copy the table. Complete it by using the power key on a calculator. (Write all the digits displayed.) What do you notice about 3^x?

x	1.4	1.41	1.414	1.4142	1.41421
3^x	?	?	?	?	?

4.6555367, 4.706965, 4.727695, 4.7287339, 4.7287859

As you discovered, the sequence $3^{1.4}, 3^{1.41}, 3^{1.414}, \ldots$ seems to approach a fixed number. It can be proved that if any sequence has $\sqrt{2}$ as its limit, then the corresponding sequence of powers of 3 also has a limit, defined to be $3^{\sqrt{2}}$.

Any function of the form $f(x) = ab^x$, where $a > 0$, $b > 0$, and $b \neq 1$ is called an **exponential function with base b**. Its domain is the set of all real numbers. The range is the set of positive real numbers. See the graphs below.

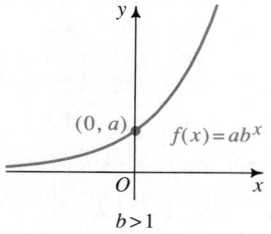
$b > 1$

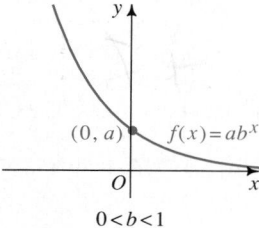
$0 < b < 1$

Example 1 If f is an exponential function, $f(0) = 3$, and $f(2) = 12$, find $f(-2)$.

Solution Since f is an exponential function, $f(x) = ab^x$ for some constants a and b. We use the given values of f to find a and b.

Since $f(0) = 3$, then $ab^0 = 3$. So $a = 3$.

Since $a = 3$ and $f(2) = 12$, then $3b^2 = 12$. So $b = 2$.

Therefore, $f(x) = 3 \cdot 2^x$. Thus, $f(-2) = 3 \cdot 2^{-2} = \dfrac{3}{4}$.

When exponential functions are used to describe exponential growth and decay, the variable t is often used to represent time. Although these functions can always be written as $f(t) = ab^t$, they are often written in other forms such as the two listed below.

(1) $A(t) = A_0(1 + r)^t$ $\quad A_0 =$ amount at time $t = 0$ and $r =$ growth rate

(2) $A(t) = A_0 b^{t/k}$ $\qquad k =$ time needed to multiply A_0 by b

Example 2 A bank advertises that if you open a savings account, you can double your money in 12 years. Express $A(t)$, the amount of money after t years, in each of the two forms listed above.

Solution Since 12 years is the time needed to multiply A_0 by 2, form (2) gives:

$$A(t) = A_0 \cdot 2^{t/12}$$

Notice that $A(12) = A_0 \cdot 2^{12/12} = 2A_0$.

To express $A(t)$ in form (1), reason as follows:

$$A(t) = A_0 \cdot 2^{t/12} = A_0 \cdot (2^{1/12})^t$$
$$\approx A_0(1.059)^t = A_0(1 + 0.059)^t$$

Assessment Note

After discussing the rule of 72, ask students about how long it will take for a savings account of $1000 to grow to $2000 if it earns a 6% annual rate of interest. ($72 \div 6 = 12$ years)

Making Connections

Some students will know the half-life of certain radioactive substances from their study of chemistry. Ask them to share this information with the class when discussing Example 3.

Additional Examples

1. The table shows the number of bacteria in a nutrient solution at various times. The bacteria population is growing exponentially.

t = time in hours	$P(t)$ = bacteria population
3	260
6	450
9	780

a. How did the population change from time $t = $ 3 h to time $t = $ 9 h?
b. Find an equation for $P(t)$.
c. What will the population be after 1 day?

a. When t increases from 3 to 9, the population triples. Thus, $P(t) = P_0 \cdot 3^{t/6}$.
b. Since $P(6) = 450$, $450 = P_0 \cdot 3^{6/6}$; $P_0 = 150$; $P(t) = 150 \cdot 3^{t/6}$.
c. $P(24) = 150 \cdot 3^{24/6} = 12{,}150$

The so-called rule of 72 stated below provides an estimate of the time it takes for a quantity to double.

The Rule of 72

If a quantity is growing at $r\%$ per year (or month), then the **doubling time** is approximately $(72 \div r)$ years (or months).

For example, if a quantity grows at 8% per month, then its doubling time will be about $72 \div 8 = 9$ months. If a population grows exponentially at 2% per year, then it will double in about $72 \div 2 = 36$ years.

Example 3 A radioactive isotope has a **half-life** of 5 days. This means that half the substance decays in 5 days. At what rate does the substance decay each day?

Solution We use the form: $A(t) = A_0 b^{t/k} = A_0\left(\dfrac{1}{2}\right)^{t/5}$

(Notice that $A(5) = A_0 \cdot \dfrac{1}{2}$, which agrees with the half-life being 5 days.)

To find the daily decay rate, we rewrite $A(t)$ as follows.

$$A(t) = A_0\left(\left(\frac{1}{2}\right)^{1/5}\right)^t \approx A_0(0.87)^t = A_0(1 - 0.13)^t$$

Thus, the daily decay rate is approximately 13%.

CLASS EXERCISES

1. The graph of $y = ab^x$ has y-intercept 7. Find the value of a. 7
2. If $h(x) = ab^x$, $h(0) = 5$, and $h(1) = 15$, find the values of a and b. $a = 5$; $b = 3$
3. For what values of b is $g(x) = b^x$ an increasing function? a decreasing function? (See the definitions of increasing and decreasing on page 138.) $b > 1$; $0 < b < 1$
4. State the domain and range of $f(x) = 3 \cdot 4^x$. Dom: real numbers; Range: $\{y | y > 0\}$
5. Give an integer estimate for 5^π. Find 5^π on a calculator. What is the difference between your estimate and the calculator value? 125; 156.9925; about 32
6. **Finance** According to the rule of 72, how long does it take to double an investment that grows at the rate of 9% per year? 6% per month? 8 years; 1 year
7. **Finance** Suppose that today you invest some money that grows to the amount $A(t) = 1000 \cdot 2^{t/10}$ in t years.
   a. How much money did you invest? $1000
   b. How long does it take to double your money? 10 years
8. **Biology** Suppose that t hours from now the population of a bacteria colony is given by $P(t) = 90(100)^{t/8}$.
   a. What is the population when $t = 0$? 90
   b. How long does it take for the population to be multiplied by 100? 8 hours

182   *Chapter Five*

9. a. 1000   **9. b.** 500   **9. c.** 4 days

9. Biology A colony of bacteria decays so that the population t days from now is given by $A(t) = 1000\left(\dfrac{1}{2}\right)^{t/4}$.

a. What is the amount present when $t = 0$?

b. How much will be present in 4 days?

c. What is the half-life?

10. *Discussion* Can the data below be described as exponential growth? Yes

x	0	1	2	3	4
y	4	4.4	4.8	5.3	5.8

Justify your answer by using a graph, an equation, or a logical argument. Can you think of a situation that the data might describe?
For example: $4 invested at 10%

WRITTEN EXERCISES

Evaluate each expression with a calculator.

A **1.** 6^{π} and π^{6} 278.4; 961.4

2. $3.6^{\sqrt{2}}$ and $\sqrt{2}^{3.6}$ 6.12; 3.48

Find an exponential function having the given values.

3. $f(0) = 3$, $f(1) = 15$ $f(x) = 3 \cdot 5^{x}$

4. $f(0) = 5$, $f(3) = 40$ $f(x) = 5 \cdot 2^{x}$

5. $f(0) = 64$, $f(2) = 4$ $f(x) = 64\left(\dfrac{1}{4}\right)^{x}$

6. $f(0) = 80$, $f(4) = 5$ $f(x) = 80\left(\dfrac{1}{2}\right)^{x}$

7. Physics The half-life of a radioactive isotope is 4 days. If 3.2 kg are present now, how much will be present after: $3.2\left(\dfrac{1}{2}\right)^{t/4}$ kg

a. 4 days? 1.6 kg   **b.** 8 days? 0.8 kg   **c.** 20 days? 0.1 kg   **d.** t days?

8. Physics The half-life of radium is about 1600 years. If 1 kg is present now, how much will be present after: $\left(\dfrac{1}{2}\right)^{t/1600}$ kg

a. 3200 years? 250 g   **b.** 16,000 years? 0.98 g   **c.** 800 years? 707 g   **d.** t years?

9. Physics The table shows the amount $A(t)$ in grams of a radioactive element present after t days. Suppose that $A(t)$ decays exponentially.

t(days)	0	2	4	6	8	10
$A(t)$	320	226	160	115	80	57

a. What is the half-life of the element? 4 days

b. About how much will be present after 16 days? 20 g

c. Find an equation for $A(t)$. $A(t) = 320\left(\dfrac{1}{2}\right)^{t/4}$

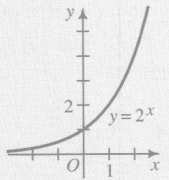
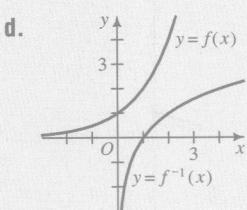

Ⓝ **Using Technology**

A curve-fitting program can be used to plot the data in Exercise 10. Have students tell whether the graph looks exponential and have the program estimate the function.

Review Note

You may want to review how to determine if a function has an inverse before assigning Exercise 18.

---

10. Geography The table shows the population $P(t)$ (in thousands) for a small mythical nation at various times.

t (year)	1825	1850	1875	1900	1925	1950	1975
$P(t)$	200	252	318	401	504	635	800

a. Does it appear that this population is growing exponentially? Yes
b. About how long does it take for the population to double? 75 years
c. Find an equation for $P(t)$. (*Hint:* The exponent contains $t - 1825$.)

11. Business The value of a car t years from now is given by $V(t) = 4000(0.85)^t$.
a. What is the annual rate of depreciation? 15%
b. In how many years will the value of the car be about half what it is now? 4 years

12. a. Geography Suppose the population of a nation grows at 3% per year. If the population was 30,000,000 people in 1990, what will be the population, to the nearest million, in the year 2000? 40 million people
b. According to the rule of 72, how long does it take for the population to double? 24 years

13. a. Finance If $1000 is invested so that it grows at the rate of 10% per year, what will the investment be worth in 20 years? $6727
b. According to the rule of 72, in approximately how many years will the investment double in value? 7 years

14. Biology A bacteria colony triples every 4 days. The population is P_0 bacteria. What will the population $P(t)$ be t days later? $P(t) = P_0(3)^{t/4}$

15. Consumer Economics If the price of sneakers increases 6% per year, about how long will it take for the price to double? 12 yr

16. Medicine When a certain medicine enters the blood stream, it gradually dilutes, decreasing exponentially with a half-life of 3 days. The initial amount of the medicine in the blood stream is A_0 milliliters. What will the amount be 30 days later? $A_0(1/2)^{10}$

17. Medicine An amount A_0 of radioactive iodine has a half-life of 8.1 days. In terms of A_0, how much is present after 5 days? (Radioactive iodine is used to evaluate the health of the thyroid gland.)

18. a. Let $f(x) = 2^x$. Complete the table.

x	-2	-1	0	1	2
$f(x)$	?	?	?	?	?

$\dfrac{1}{4}; \dfrac{1}{2}; 1; 2; 4$

b. Graph the function by plotting points.

c. Explain why f has an inverse. It is one-to-one.

d. Graph f^{-1} on the same set of axes as f.

e. State the domain and range of f and of f^{-1}.

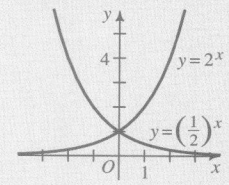 **You may find it helpful to have a graphing calculator to complete Exercises 19–21. You need a graphing calculator for Exercise 22.**

B **19. a.** Graph the functions $y = 2^x$ and $y = \left(\frac{1}{2}\right)^x$ on a single set of axes.

  b. How are the graphs related to each other?

20. a. Graph the functions $y = 4^x$ and $y = 4^{x/2}$ on a single set of axes.

  b. How are the graphs related to each other?

21. a. Graph the functions $y = 2^x$ and $y = 2^{x-1}$ on a single set of axes.

  b. How are these graphs related to each other?

22. a. Graph the functions $y = x^2$ and $y = 2^x$ on a single set of axes.

  b. Use a graphing calculator to solve $x^2 = 2^x$.

23. a. **Geography** The bar graph below gives the population of the United States for each census from 1800 to 1980. The tops of the bars lie approximately on the curve $y = ab^x$, where x is the number of years since 1800. Find the values of a and b to the nearest thousandth. Ans. may vary: 18.924, 1.014

  b. Use part (a) to predict the population in the years 2000 and 2050. In 2000, 305.2 million; In 2050, 611.6 million

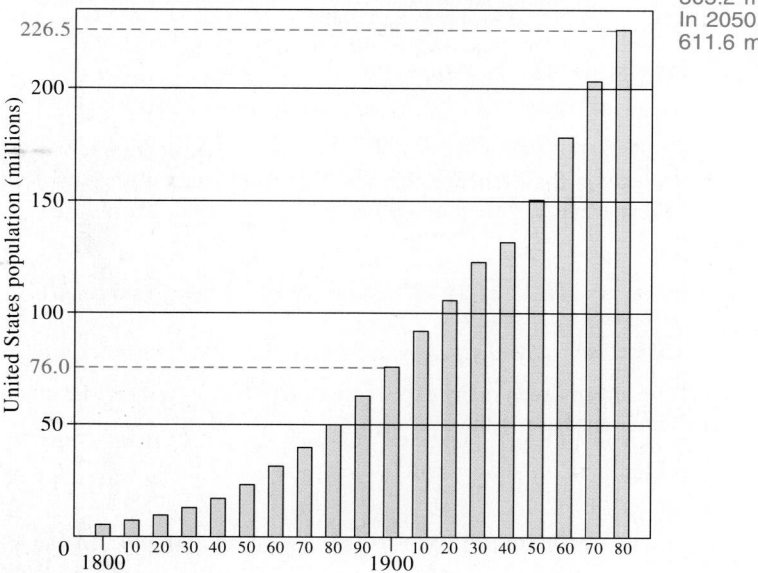

24. a. If $f(x) = 2^x$, show that $f(x + 1) = 2f(x)$.

  b. Translating the graph of f 1 unit (left or right?) is equivalent to (vertically or horizontally?) stretching the graph of f by ? units. left; vertically; 2

19. a.

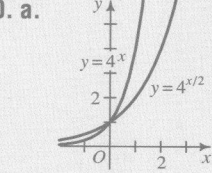

b. The graphs are reflections of each other in the y-axis.

20. a.

b. The graph of $y = 4^{x/2}$ is the graph of $y = 4^x$ stretched horizontally by a factor of 2.

21. a.

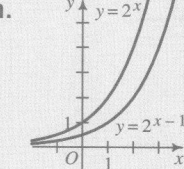

b. The graph of $y = 2^{x-1}$ is the graph of $y = 2^x$ shifted one unit to the right.

22. a.

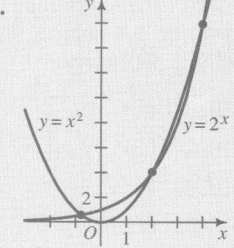

b. -0.76666, 2, 4

24. a. $f(x + 1) = 2^{x+1} = 2^x \cdot 2^1 = 2 \cdot 2^x = 2f(x)$

26. c. For example, the graph could describe a population of animals which decreases due to predators. When the number of predators decreases, the population increases again.

27. $f(x + 2) - f(x) =$
$10^{2(x+2)+1} - 10^{2x+1} =$
$10^{2x+5} - 10^{2x+1} =$
$10^{2x} \cdot 10^5 - 10^{2x} \cdot 10 =$
$10^{2x}(10^5 - 10) =$
$10^{2x}(99,990)$

29. $f(0) = f(0 + 0) =$
$f(0) \cdot f(0)$. Since $f(x) > 0$,
$f(x) \neq 0$; hence
$f(0) = \dfrac{f(0)}{f(0)} = 1$.
$f(2x) = f(x + x) =$
$f(x) \cdot f(x) = [f(x)]^2$.
$f(3x) = f(2x + x) =$
$f(2x) \cdot f(x) =$
$[f(x)]^2 \cdot f(x) = [f(x)]^3$.

Teaching Notes, p. 168D

Warm-Up Exercises

Evaluate each expression. Give answers to the nearest hundredth, if necessary.

1. $1000(1.06)$ 1060

2. $1000(1.03)^2$ 1060.9

3. $1000(1.015)^4$ 1061.36

4. $1000(1.005)^{12}$ 1061.68

5. $1000\left(1 + \dfrac{0.06}{360}\right)^{360}$
1061.83

6. $1000(2.7183)^{0.06}$ 1061.84

25. *Investigation* Go to a bank in your area.
  a. What kinds of savings accounts does the bank offer? List the characteristics of each type of account. Write down the interest rate, any conditions placed on opening the account, any penalties placed on withdrawals, and so forth.
  b. Suppose that you have $1000 to invest. Which of the plans available would you choose?
  c. Make a table showing how the money in the account would grow.
  d. Discuss your investigation with the class.

26. *Visual Thinking* In the diagram, the red and blue curves are exponential in behavior.
  a. Describe the behavior of $P(t)$ on the interval $0 \leq t \leq r$. Decreasing
  b. Describe the behavior of $P(t)$ on the interval $r < t \leq s$. Increasing
  c. Describe a real-world situation for which the diagram could be a model. Ans. may vary.

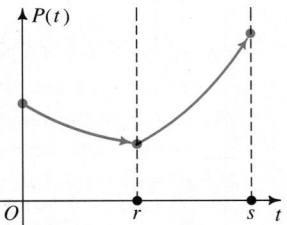

27. If $f(x) = 10^{2x+1}$ and x is a nonnegative integer, show that $f(x + 2) - f(x)$ is divisible by 99.

28. Over the past 60 years, the population of a country increased from 100 million people to 200 million people.
  a. Find an equation that gives the population t years from now.
  b. Use a calculator and trial and error to find approximately when the population will reach 300 million people.

28. a. $P(t) = (2 \cdot 10^8)(2)^{t/60}$
  b. 35 years

C **29.** Suppose that $f(x)$ is positive for all real x and

$$f(x + y) = f(x) \cdot f(y)$$

for all x and y. Prove that $f(0) = 1$, $f(2x) = [f(x)]^2$, and $f(3x) = [f(x)]^3$. Give an equation for a function that has the property that $f(x + y) = f(x) \cdot f(y)$ for all real x and y. Ans. may vary. For example, $f(x) = 2^x$.

5-4 The Number *e* and the Function *e*^{x}

Objective   *To define and apply the natural exponential function.*

You have already seen many exponential functions. In advanced mathematics, the most important base is the irrational number e, defined as

$$\lim_{n \to \infty} \left(1 + \frac{1}{n}\right)^n.$$

This is read "the limit of $\left(1 + \dfrac{1}{n}\right)^n$ as n approaches infinity." Although you will not study limits until Chapter 13, you can get an idea of the value of e by completing the Activity on the next page.

1. Copy and complete the table by using a calculator. What do you notice about the successive values of $\left(1 + \frac{1}{n}\right)^n$?

2. Using a graphing calculator, graph $y = \left(1 + \frac{1}{x}\right)^x$. What happens to y when x becomes large?

n	$\left(1 + \frac{1}{n}\right)^n$	
10	?	2.59374246
100	?	2.70481383
1000	?	2.71692393
10,000	?	2.71814593
100,000	?	2.71826824

You can see that as n increases, $\left(1 + \frac{1}{n}\right)^n$ appears to get closer and closer to 2.718 . . . . The Swiss mathematician Leonhard Euler (1707–1783) proved this, and the limit is called e in his honor. Values of e^x can be obtained using a calculator or table (page 821). The function e^x is called the **natural exponential function**. In Exercise 13, you will be asked to draw its graph and the graph of its inverse. The number e is extremely important in advanced mathematics. It appears in unexpected places including statistics and physics.

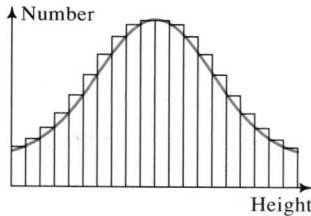

Statistics: Distribution of heights of 18-year-olds

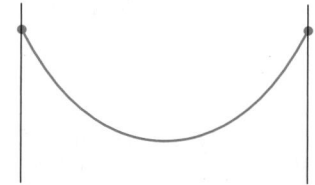

Physics: Hanging rope

The equations of these two graphs contain e^x.

Compound Interest and the Number *e*

Suppose you invest P dollars (the *principal*) at 12% interest compounded semiannually. Each half year, your money grows by 6%. At the end of the year, you have $P(1.06)^2 = 1.1236P$ dollars. If the interest had been compounded quarterly (4 times per year), then it would grow by 3% each quarter. At the end of the year, you would have $P(1.03)^4 = 1.1255P$ dollars. See the table at the top of the next page. The table also shows the results when the 12% annual interest rate is compounded monthly and daily. For simplicity, the investment is $1.

Motivating the Section

Students of psychology may have heard of the "normal curve" as it relates to human behavior and to testing situations. An equation of the normal curve involves the irrational number *e*, which students study in this section.

 Using Technology

Encourage students with scientific calculators to extend the table of values in the Activity to see how many table entries it takes before the differences in the successive values of $\left(1 + \frac{1}{n}\right)^n$ become very small. This means that the value of $\left(1 + \frac{1}{n}\right)^n$ is approaching some number, which we call *e*. When *n* is 1,000,000, the value of *e* and the value of $\left(1 + \frac{1}{n}\right)^n$ are equal to 5 decimal places.

Making Connections

Emphasize that the natural exponential function is used extensively in further study of mathematics, especially calculus.

Interest period	% growth each period	Growth factor during period	Amount
Annually	12%	$1 + \frac{0.12}{1}$	$1.12^1 = 1.12$
Semiannually	6%	$1 + \frac{0.12}{2}$	$1.06^2 = 1.1236$
Quarterly	3%	$1 + \frac{0.12}{4}$	$1.03^4 \approx 1.1255$
Monthly	1%	$1 + \frac{0.12}{12}$	$1.01^{12} \approx 1.1268$
Daily (360 days)	$\frac{12}{360}\%$	$1 + \frac{0.12}{360}$	$\left(1 + \frac{0.12}{360}\right)^{360} \approx 1.1275$
k times per year	$\frac{12}{k}\%$	$1 + \frac{0.12}{k}$	$\left(1 + \frac{0.12}{k}\right)^k$

The numbers in the right-hand column of the table suggest that the amount approaches a fixed value. To find this number, note that

$$\left(1 + \frac{0.12}{k}\right)^k = \left(1 + \frac{1}{\frac{k}{0.12}}\right)^k = \left[\left(1 + \frac{1}{\frac{k}{0.12}}\right)^{k/0.12}\right]^{0.12} = \left[\left(1 + \frac{1}{n}\right)^n\right]^{0.12}$$

where $n = \frac{k}{0.12}$. Since the limit of $\left(1 + \frac{1}{n}\right)^n$ as n increases is e, the limit of the expression above is $e^{0.12}$, approximately 1.1275. Since $1 grows to $1.1275 in one year, a bank that compounds daily will advertise that its 12% annual interest is equivalent to a 12.75% *effective annual yield*.

In general, if you invest P dollars at an annual rate r (expressed as a decimal) compounded continuously, then t years later your investment will be worth Pe^{rt} dollars. (See Written Exercises 17 and 18.)

The same principle applies to any quantity, such as population, where compounding takes place "all the time." If P_0 is the initial amount, then the amount at any future time t is $P(t) = P_0 e^{rt}$.

CLASS EXERCISES

Use a calculator to evaluate the following.

1. e^2 7.3891 **2.** $e^{3.2}$ 24.5325 **3.** e^{-4} 0.0183 **4.** $e^{\sqrt{2}}$ 4.1133 **5.** e^1 2.7183

6. Which is larger, e^π or π^e? Try to estimate before using your calculator. e^π

7. a. *Reading* From your reading, how is e defined? $\displaystyle\lim_{n \to \infty}\left(1 + \frac{1}{n}\right)^n = e$

b. Estimate the value of $\left(1 + \frac{1}{1,000,000}\right)^{1,000,000}$. 2.718

188 *Chapter Five*

8. Finance If money is invested at 8% compounded semiannually, then each year the investment is multiplied by 1.04^2. What is the investment multiplied by if interest is compounded: **b.** $(1.0067)^{12}$ **c.** $e^{0.08}$
a. quarterly? $(1.02)^4$ **b.** 12 times a year?   **c.** continuously?

9. Finance A bank advertises that its 5% annual interest rate compounded daily is equivalent to a 5.13% effective annual yield. What does this mean?
Daily compounded at 5% yields same interest as 5.13% annually.

WRITTEN EXERCISES

A **1. a.** Evaluate $\left(1 + \dfrac{1}{n}\right)^n$ when $n = 5000$, and $n = 5,000,000$. 2.71801; 2.718282   **6. a.** \$1.0824 **b.** \$1.0830 **c.** \$1.0833

  b. Compare your answers in part (a) with an approximation for e. approx. equal

2. a. Evaluate $\left(1 - \dfrac{1}{n}\right)^n$ for $n = 100$, $n = 10,000$, and $n = 1,000,000$. 0.36603; 0.36786; 0.36788

  b. Compare your answers in part (a) with an approximation for e^{-1}. 0.36788; approx. equal

  c. What appears to be $\displaystyle\lim_{n\to\infty}\left(1 - \dfrac{1}{n}\right)^n$? e^{-1}

3. Which is larger, $e^{\sqrt{2}}$ or $\sqrt{2}^e$? Try to estimate before using your calculator. $e^{\sqrt{2}}$

4. Evaluate: **a.** $e^{0.08}$ 1.0833   **b.** $e^{-0.08}$ 0.9231   **c.** $e^{4/3}$ 3.7937

5. Finance Suppose you invest \$1.00 at 6% annual interest. Calculate the amount that you would have after one year if the interest is compounded **(a)** quarterly, **(b)** monthly, **(c)** continuously. **a.** \$1.0614 **b.** \$1.0617 **c.** \$1.0618

6. Finance Repeat Exercise 5 if the annual rate is 8%. See above.

7. Finance One hundred dollars deposited in a bank that compounds interest quarterly yields \$107.50 over 1 year. Find the effective annual yield. 7.5%

8. Finance After a year during which interest is compounded quarterly, an investment of \$800 is worth \$851. What is the effective annual yield? 6.375%

9. With which plan would an investor earn more, Plan A or B? Plan A
Plan A: A 6% annual rate compounded annually over a 10-year period
Plan B: A 5.5% annual rate compounded quarterly over a 10-year period

10. With which plan would an investor earn more, Plan A or B? Plan A
Plan A: An 8% annual rate compounded quarterly for 5 years
Plan B: A 7.5% annual rate compounded daily for 5 years

11. Finance Suppose that \$1000 is invested at 7% interest compounded continuously. How much money would be in the bank after 5 years? \$1419.07

12. Biology A population of ladybugs rapidly multiplies so that the population t days from now is given by $A(t) = 3000e^{0.01t}$.
a. How many ladybugs are present now?
b. How many will there be after a week?
a. 3000 **b.** 3218

---

2. Use the graph of $y = e^x$ to graph **(a)** $y = e^{x+1}$ and **(b)** $y = e^x + 1$.

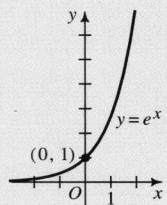

a.

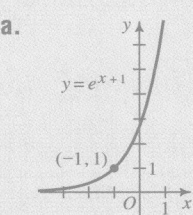

b.

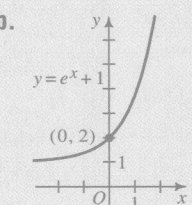

⬜ Using Technology

Calculators are necessary for most of the Class and Written Exercises.

Suggested Assignments

Standard
Day 1: 189/1–13
Day 2: 190/14–23
Comprehensive
189/1–23 odd, 22

Supplementary Materials

Alternative Assessment, 16
Student Resource Guide, 49–51

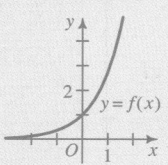
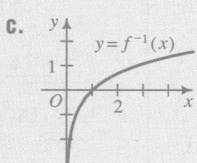
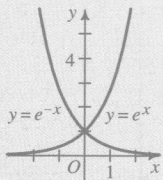
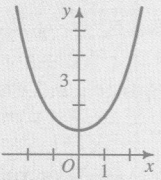
190 Chapter 5

---

14. They are reflections of each other in the *y*-axis.

You might find it helpful to have a graphing calculator to complete Exercises 13–16.

13. a. Use a calculator to sketch the graph of $f(x) = e^x$.
  b. Explain why this function has an inverse. Function is one-to-one.
  c. Without using a calculator, graph the inverse.

14. On a single set of axes, graph $y = e^x$ and $y = e^{-x}$. How are the graphs related?
  See above.

15. Graph $y = \dfrac{e^x + e^{-x}}{2}$. (See the hanging rope on page 187.)

16. Sketch the graph of $y = e^{-x^2}$. (See the "bell-shaped curve" on page 187.)

B **17. a.** **Finance** Suppose that P dollars is invested at an annual interest rate r (r a decimal) with interest compounded k times a year. Explain why the value of the investment at the end of the year is $P\left(1 + \dfrac{r}{k}\right)^k$.
  b. Show that the expression in part (a) is approximately Pe^r if k is large.

18. a. **Finance** Suppose that \$10,000 is invested at an annual rate of 9% and that interest is compounded every second for 365 days. Find the value of this investment at the end of one year by using the expression in part (a) of Exercise 17. (First find k, the number of seconds in one year.)
  b. Compare your answer in part (a) with the value of $10{,}000e^{0.09}$.
  a. Ans. may vary: about \$10,941.74 **b.** 10,941.74

C **19.** It can be proved that $e = 1 + \dfrac{1}{1!} + \dfrac{1}{2!} + \dfrac{1}{3!} + \cdots + \dfrac{1}{n!} + \cdots$.

  Approximate e by using the first five terms. (*Note:* $n!$, read "n factorial," denotes $n(n - 1)(n - 2) \cdots 2 \cdot 1$. For example, $3! = 3 \cdot 2 \cdot 1 = 6$.) 2.7083

20. The points $A(0, f(0))$ and $B(h, f(h))$ are on the graph of $f(x) = e^x$. Find the slope of $\overline{AB}$ if h is **(a)** 1, **(b)** 0.1, and **(c)** 0.01.

21. It can be proved that the line $y = 1 + x$ is tangent to the graph of $y = e^x$ at $(0, 1)$. Thus, $e^x \approx 1 + x$ when $|x|$ is small. Show that $(1 + x)^{1/x}$ is approximately e.

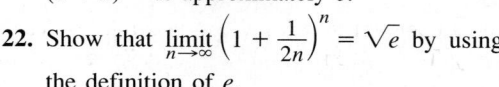

22. Show that $\underset{n\to\infty}{\text{limit}} \left(1 + \dfrac{1}{2n}\right)^n = \sqrt{e}$ by using the definition of e.

23. Determine the value of $\underset{n\to\infty}{\text{limit}} \left(1 + \dfrac{2}{n}\right)^n$.

  (See Exercise 22.) e^2
  20. a. 1.7183 **b.** 1.0517 **c.** 1.0050

////COMPUTER EXERCISE

Write a program that prints a table giving the value of a \$1000 investment compounded quarterly at 4.5%, 5%, 5.5%, 6%, and 6.5% annual interest rates after each of ten successive years.

Logarithms

5-5 Logarithmic Functions

Objective *To define and apply logarithms.*

Now we will explore *logarithms*, numbers used to measure the severity of earthquakes, the loudness of sounds, and the brightness of stars.

We define the *common logarithm* of an integral power of 10 to be its exponent, as shown in the chart below.

×10 ×10 ×10 ×10

Number	10^0	10^1	10^2	10^3	10^4	$\cdots$	10^k
Common logarithm	0	1	2	3	4	$\cdots$	k

+1 +1 +1 +1

We denote the common logarithm of x by $\log x$. For example,

$$\log 10^3 = 3,\ \log 10^4 = 4,\ \text{and } \log 10^k = k.$$

In general, the **common logarithm** of any positive real number x is defined to be the exponent you get when you write x as a power of 10.

$$\log x = a \text{ if and only if } 10^a = x.$$

For example:

$$\log 6.3 \approx 0.8 \text{ because } 6.3 \approx 10^{0.8}$$

You can verify this by evaluating $10^{0.8}$ on a calculator.

You can also use a scientific calculator to find common logarithms directly. On most calculators, you enter the number whose common logarithm you wish to find and then press the "log" key.

The table above illustrates that common logarithms increase arithmetically when numbers increase exponentially. This fact enables us to make a linear scale to measure quantities that increase exponentially.

Common logarithms are useful in applications involving the perception of sound. Every sound has an intensity level due to the power of the sound wave. In the table on the next page, I_0 represents the intensity of a sound barely audible. The intensity level I of any other sound is measured in terms of I_0. (See column 2 of the table on the next page.) The human ear perceives a sound as soft or loud. The unit for measuring the loudness of a sound is the decibel (dB) and it is related to the intensity of a sound by:

$$\text{decibel level of } I = 10 \log \frac{I}{I_0}$$

Teaching Notes, p. 168E

Warm-Up Exercises

Find the value of x.

1. $10^x = 100{,}000$ 5

2. $2^x = \frac{1}{64}$ −6

3. $e^x = e^{-x}$ 0

4. $5^x = 5^{1/x}$ ±1

5. $4^x = -1$ no solution

6. $4^x = 2048$ 5.5

Motivating the Section

Some students may have learned in chemistry about the pH measure of a solution. Ask if they know what the pH level of pure water is. (7) The pH scale is used to describe the concentration of hydrogen ions in a solution (the scale ranges from 0 to 14). This is really a logarithmic scale. In this section, students will be introduced to logarithmic functions.

Using Technology

Most scientific calculators have a common logarithm key. Have students familiarize themselves with it by verifying the common logarithms for the table of values given on this page.

Error Analysis

Students who do not understand the decibel table given on this page should be encouraged to verify each of the entries in the decibel level column by performing a calculation similar to that given above the table.

Assessment Note

Ask students to use the chart on this page to determine the ratio of the sound intensity level of leaves in a breeze to that of soft recorded music.

$\left(\dfrac{10^2 I_0}{10^4 I_0} = \dfrac{1}{10^2} = \dfrac{1}{100} \right)$

Cooperative Learning

Have students discuss Example 1 in small groups so that they can verify algebraically how the numbers are generated. Have students verify the calculations with calculators.

(See column 3 of the table below.) For example, a loud stereo set has an intensity level of $10^8 I_0$ and is perceived to have a decibel level of:

$$10 \log \frac{10^8 I_0}{I_0} = 10 \log 10^8$$
$$= 10 \times 8 = 80$$

Sound	I	Decibel level
Barely audible	I_0	0
Whisper	$10 I_0$	10
Leaves in a breeze	$10^2 I_0$	20
Soft recorded music	$10^4 I_0$	40
Two-person conversation	$10^6 I_0$	60
Loud stereo set	$10^8 I_0$	80
Subway train	$10^{10} I_0$	100
Jet at takeoff	$10^{12} I_0$	120
Pain in eardrum	$10^{13} I_0$	130

Many people think that when the intensity of a sound is doubled, the decibel level is doubled also. The following example shows that this is not so.

Example 1 Two loud stereos are playing the same music simultaneously at 80 dB each. What is the decibel level of the combined sound? By how many decibels is the decibel level of the two stereos greater than the decibel level of one stereo?

Solution Since one stereo at 80 dB has an intensity $10^8 I_0$, two stereos will have an intensity that is twice that amount.

$$2(10^8 I_0)$$

With a calculator, we find that the decibel level corresponding to the two stereos is

$$10 \log\left(\frac{2 \times 10^8 I_0}{I_0} \right) \approx 83.$$

Since the decibel level of one stereo is 80 dB, there is only about a 3 dB increase in the decibel level when the two stereos are played at the same time.

The decibel scale is an example of a *logarithmic scale*. Such a scale is also used to measure acidity and brightness. (See Exercises 45 and 46.)

Common logarithms have base 10. Logarithms to other bases are sometimes used. The *logarithm to base b* of a positive number x, denoted by $\log_b x$, is defined to be the exponent a that you get when you write x as a power of b. (*Note*: $b > 0$, $b \ne 1$)

$$\log_b x = a \text{ if and only if } x = b^a.$$

Example 2
$\log_5 25 = 2$ because $5^2 = 25$.
$\log_5 125 = 3$ because $5^3 = 125$.
$\log_2 \dfrac{1}{8} = -3$ because $2^{-3} = \dfrac{1}{8}$.

The base b logarithmic function, whose graph is shown in blue below, is the inverse of the base b exponential function, whose graph is shown in red. Notice the domain and range of each function.

Base b exponential function:
$f(x) = b^x$
Domain: All reals
Range: Positive reals

Base b logarithmic function:
$f^{-1}(x) = \log_b x$
Domain: Positive reals
Range: All reals

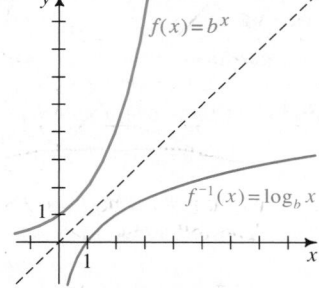

The most important logarithmic function in advanced mathematics and statistics has the number e as its base. This function is called the *natural logarithm function*. The natural logarithm of x is usually denoted $\ln x$ although sometimes it is written $\log_e x$. The value of $\ln x$ can be found by using a scientific calculator.

$$\ln x = k \quad \text{if and only if} \quad e^k = x.$$

For example,

$$\ln 5 \approx 1.6 \text{ because } e^{1.6} \approx 5.$$

Example 3   Find the value of x to the nearest hundredth.
    a. $10^x = 75$ **b.** $e^x = 75$

Solution   **a.** By definition, x is the common logarithm of 75.
         Thus, $x = \log 75 \approx 1.88$.
    b. By definition, x is the natural logarithm of 75.
         Thus, $x = \ln 75 \approx 4.32$.

Exponents and Logarithms   **193**

Additional Examples cont.

b. Let $\ln \dfrac{1}{e^3} = x$.

Then $e^x = \dfrac{1}{e^3} = e^{-3}$.

Thus, $x = -3$.

c. Let $\log \dfrac{1}{10,000} = x$.

Then $10^x = \dfrac{1}{10,000} = 10^{-4}$. Thus, $x = -4$.

d. Let $\log_5 1 = x$.

Then $5^x = 1$.

Since $5^0 = 1$, $x = 0$.

2. Solve.
a. $\log x = 4$
b. $\ln x = \dfrac{1}{2}$
c. $\log x = -1.2$

a. $x = 10^4 = 10,000$
b. $x = e^{1/2} = \sqrt{e} \approx 1.65$
c. $x = 10^{-1.2} \approx 0.063$

Additional Answers
Class Exercises

2. You cannot find the logarithm of a negative number. The domain of a logarithmic function is positive reals.

3. a. 1.9031 **b.** 2.9031
c. 4.9031 **d.** −0.0969
e. −1.0969

Suggested Assignments

Standard:
Day 1: 194/1–19 odd
Day 2: 195/21–48 every 3rd
Comprehensive
Day 1: 194/1–19 odd
Day 2: 195/21–51 every 3rd

Supplementary Materials

Alternative Assessment, 17
Using Technology, 17–18

CLASS EXERCISES

a. 1.4771 **b.** 2.7686 **c.** 3.8325 **d.** −0.5229 **e.** −3.5229
1. Use a calculator to approximate each logarithm to four decimal places.
  a. $\log 30$ **b.** $\log 587$ **c.** $\log 6800$ **d.** $\log 0.3$ **e.** $\log 0.0003$

2. Try to find $\log(-7)$ on a calculator. What happens? Why? Error message

3. Suppose you are told that $\log 8 \approx 0.9031$. Find each of the following without using a calculator. Then check your answers with a calculator.
  a. $\log 80$ **b.** $\log 800$ **c.** $\log 80,000$ **d.** $\log 0.8$ **e.** $\log 0.08$

4. a. The statement $\log_2 16 = 4$ means that $16 = \underline{\ ?\ }$. 2^4
  b. The statement $\log 31 \approx 1.49$ means that $31 \approx \underline{\ ?\ }$. $10^{1.49}$
  c. The statement $\log 61 \approx 1.79$ means that $61 \approx \underline{\ ?\ }$. $10^{1.79}$

5. Find each logarithm without using a calculator.
  a. $\log_7 49$ 2 **b.** $\log_2 16$ 4 **c.** $\log_2 \dfrac{1}{8}$ −3 **d.** $\log_5 \dfrac{1}{5}$ −1 **e.** $\log_5 \sqrt{5}$ $\dfrac{1}{2}$

6. Use a calculator to approximate each logarithm to four decimal places.
  a. $\ln 2$ 0.6931 **b.** $\ln 3$ 1.0986 **c.** $\ln 2.7$ 0.9933 **d.** $\ln 2.8$ 1.0296 **e.** $\ln e$ 1

7. Find the value of x to the nearest hundredth: **(a)** $10^x = 50$ **(b)** $e^x = 50$
  1.70 3.91

Solve each equation. (Do not use a calculator.)

8. a. $\log_5 x = 2$ 25 **b.** $\log_6 x = 2$ 36 **c.** $\log x = 2$ 100 **d.** $\ln x = 2$ e^2

9. a. $\log_x 121 = 2$ 11 **b.** $\log_x 64 = 3$ 4 **c.** $\log_x \left(\dfrac{1}{2}\right) = -1$ 2 **d.** $\log_x \sqrt{6} = \dfrac{1}{2}$ 6

WRITTEN EXERCISES

Write each equation in exponential form.

$4^2 = 16$ ⟍ $4^3 = 64$ ⟍ $6^{-2} = \dfrac{1}{36}$ ⟍ $4^{1.5} = 8$

A **1.** $\log_4 16 = 2$ **2.** $\log_4 64 = 3$ **3.** $\log_6 \left(\dfrac{1}{36}\right) = -2$ **4.** $\log_4 8 = 1.5$

$10^3 = 1000$ ⟍ $10^{1.6} \approx 40$ ⟍ ⟍ $e^{-1.6} \approx 0.2$

5. $\log 1000 = 3$ **6.** $\log 40 \approx 1.6$ **7.** $\ln 8 \approx 2.1$ $e^{2.1} \approx 8$ **8.** $\ln 0.2 \approx -1.6$

9. a. What does it mean to say that x is the common logarithm of N? $10^x = N$
  b. Solve **(1)** $10^x = 7$ and **(2)** $10^x = 0.562$ for x to the nearest hundredth. See below
  c. What does it mean to say that x is the natural logarithm of N? $e^x = N$
  d. Solve **(1)** $e^x = 12$ and **(2)** $e^x = 0.06$ for x to the nearest hundredth. See below.

10. Find the value of x to the nearest hundredth: **(a)** $10^x = 170$ **(b)** $e^x = 500$
  9. b. 0.85; −0.25 **d.** 2.48; −2.81 **10. a.** 2.23 **b.** 6.21

Find each logarithm. (Do not use a calculator.)

11. a. $\log 100$ 2 **b.** $\log 10,000$ 4 **c.** $\log 0.01$ −2 **d.** $\log 0.0001$ −4

12. a. $\log_2 4$ 2 **b.** $\log_2 32$ 5 **c.** $\log_2 64$ 6 **d.** $\log_2 2^{10}$ 10

13. a. $\log_3 9$ 2 **b.** $\log_3 27$ 3 **c.** $\log_3 243$ 5 **d.** $\log_3 3^8$ 8

14. a. $\log_5 0.2$ −1 **b.** $\log_5 \dfrac{1}{125}$ −3 **c.** $\log_5 \sqrt[3]{5}$ $\dfrac{1}{3}$ **d.** $\log_5 1$ 0

15. **a.** $\log_4 64$ 3   **b.** $\log_4 \frac{1}{64}$ -3 **c.** $\log_4 \sqrt[4]{4}$ $\frac{1}{4}$ **d.** $\log_4 1$ 0

16. **a.** $\log_6 36$ 2   **b.** $\log_{36} 6$ $\frac{1}{2}$ **c.** $\log_6 6\sqrt{6}$ $\frac{3}{2}$ **d.** $\log_6 \sqrt[3]{\frac{1}{6}}$ $-\frac{1}{3}$

17. **a.** $\ln e$ 1   **b.** $\ln e^2$ 2   **c.** $\ln \frac{1}{e}$ -1 **d.** $\ln \sqrt{e}$ $\frac{1}{2}$

18. **a.** $\log 10^8$ 8   **b.** $\log_2 2^8$ 8   **c.** $\log_5 5^8$ 8   **d.** $\ln e^8$ 8

19. Given $\log 4.17 \approx 0.6201$, find: **a.** $\log 417$ **b.** $\log 0.417$ **c.** $\log 0.0417$

20. Given $\log 6.92 \approx 0.8401$, find: **a.** $\log 692$ **b.** $\log 0.692$ **c.** $\log 0.00692$

21. Given $\ln 10 \approx 2.3026$, find: **a.** $\ln 0.1$ **b.** $\ln 0.01$ **c.** $\ln 100$

22. Given $\ln 5 \approx 1.6094$, find: **a.** $\ln 0.2$ **b.** $\ln 25$ **c.** $\ln 0.04$

B 23. **Physics** Find the decibel level for each sound with the given intensity I.
    a. Average car at 70 km/h, $I = 10^{6.8}I_0$ 68 dB   **b.** Whisper, $I = 10^{1.5}I_0$ 15 dB

24. **Physics** Find the decibel level for each sound with the given intensity I.   75 dB
    a. Softly played flute, $I = 10^{4.1}I_0$ 41 dB   **b.** Vacuum cleaner, $I = 10^{7.5}I_0$

25. **a.** **Physics** Find the decibel level of two stereos, playing the same music simultaneously at 62 dB. 65 dB
    b. Find the decibel level if three stereos play instead of two. 67 dB

26. **Physics** The decibel level of one car accelerating from rest to 50 km/h is 80 dB. Find the decibel level of four similar cars accelerating at once. 86 dB

27. Graph $f(x) = 2^x$ and $f^{-1}(x) = \log_2 x$ on a single set of axes. Give the domain and range of each function.

28. **a.** Find $2^{\log_2 8}$, $5^{\log_5 25}$, and $3^{\log_3 x}$.   8, 25, x
    b. If $f(x) = 3^x$, what is $f^{-1}(x)$? What is $(f \circ f^{-1})(x)$? $\log_3 x$; x

    29. **a.** $(f \circ g)(x) = e^{\ln x} = x$, $x > 0$; $(g \circ f)(x) = \ln(e^x) = x$, x is any real no.

 Part (b) of Exercises 29 and 30 requires the use of a computer or a graphing calculator. You may also use a computer or a graphing calculator to confirm your answers to part (b) of Exercises 31 and 32.

29. **a.** Consider the functions $f(x) = e^x$ and $g(x) = \ln x$. Find rules for $(f \circ g)(x)$ and $(g \circ f)(x)$ and give the domain of each composite function. See above.
    b. To confirm the domains in part (a), graph the composite functions using a computer or a graphing calculator. Enter the equations of the composite functions in their *unsimplified* form. (See Example 2 on page 127.)

30. Repeat Exercise 29 for the functions $f(x) = e^{-x}$ and $g(x) = -\ln x$. See below.

31. **a.** On separate sets of axes, graph $y = \log |x|$ and $y = |\log x|$.
    b. Give the domain and range of each function.

32. Repeat Exercise 31 for the functions $y = -\ln x$ and $y = \ln(-x)$.

33. Give the domain, range, and zeros of $y = \log x + 3$ and $y = \log(x + 3)$.

34. Give the domain, range, and zeros of $y = \log_2(x - 2)$ and $y = \log_2 x - 2$.

30. **a.** $(f \circ g)(x) = e^{-(-\ln x)} = x$, $x > 0$; $(g \circ f)(x) = -\ln(e^{-x}) = x$, x is any real no.

Exponents and Logarithms   **195**

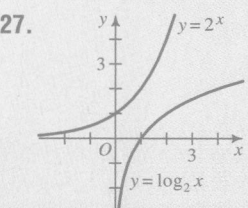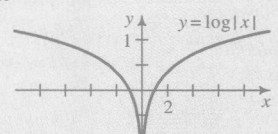
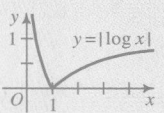

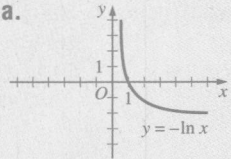
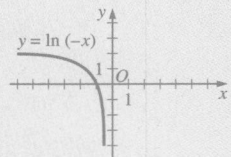
Solve for x without using a calculator. You may leave answers in terms of e if necessary.

35. a. $\log x = 3$ 1000   b. $\log |x| = 3$ ± 1000 c. $\log |x - 1| = 3$ 1001, -999

36. a. $\log_6 x = 2$ 36   b. $\ln x = 2$ e^2 c. $\ln |x| = 2$ $\pm e^2$

37. a. $\log_4 x = 1.5$ 8   b. $\ln x = 1.5$ $e^{3/2}$ c. $\ln x = 0$ 1

38. a. $\log(x^2 - 1) = 2$ $\pm\sqrt{101}$ b. $\ln(x^2 - 1) = 2$ $\pm\sqrt{e^2 + 1}$ c. $\ln |x| = 1$ $\pm e$

39. a. $\log_5(\log_3 x) = 0$ 3   b. $\log(\log x) = 1$ 10^{10} c. $\ln(\ln x) = 1$ e^e

40. a. $\log_6(\log_2 x) = 1$ 64   b. $\ln(x - 2) = 1$ $e + 2$ c. $(\log x)^2 = 4$ 0.01, 100

Solve for x using a calculator. Give answers to the nearest hundredth.

41. a. $\log x = 0.7$ 5.01   b. $\log x = 3.7$ 5011.87   c. $\log x = -0.3$ 0.50

42. a. $\log x = 1.4$ 25.12   b. $\log x = 0.4$ 2.51   c. $\log x = -0.6$ 0.25

43. a. $\ln x = 4.2$ 66.69   b. $\ln x = -1.5$ 0.22   c. $e^x = 5$ 1.61

44. a. $\ln x = 1.73$ 5.64   b. $\ln x = -0.52$ 0.59   c. $e^x = 16$ 2.77

45. **Chemistry** The pH of a solution is a measure of how acidic or alkaline the solution is. The pH of a solution is given by:

$$\text{pH} = -\log_{10} \text{(hydrogen ion concentration)}$$

Pure water, which has a pH of 7, is considered neutral. A solution with pH less than 7 is acidic; a solution with pH greater than 7 is alkaline. Find the pH of the following solutions and classify them as acidic, neutral, or alkaline.

Solution	Hydrogen ion concentration (moles per liter)	
Human gastric juices	10^{-2}	2; acidic
Acid rain	3×10^{-5}	4.52; acidic
Pure water	10^{-7}	7; neutral
Good soil for vegetables	5×10^{-7}	6.30; acidic
Sea water	10^{-8}	8; alkaline

46. **Astronomy** The observed brightness of stars is classified by magnitude. Two stars can be compared by giving their magnitude difference d or their brightness ratio r. The numbers d and r are related by the equation $d = 2.5 \log r$. Comparing a first magnitude star with a sixth magnitude star, we have that $d = 6 - 1 = 5$. Find the value of r. What can you say about the relative brightness of the two stars? $r = 100$; One star is 100 times brighter than the other.

47. a. Compare $\log_4 16$ and $\log_{16} 4$.
  b. Compare $\log_9 27$ and $\log_{27} 9$.
  c. State and prove a generalization based on parts (a) and (b).

48. b. $\dfrac{1}{2} + \dfrac{3}{2} = 2$

$2 + 3 = 5$

48. a. Show that $\log_2 4 + \log_2 8 = \log_2 32$ by finding the three logarithms.
  b. Verify that $\log_9 3 + \log_9 27 = \log_9 81$. See above.
  c. State and prove a generalization based on parts (a) and (b). $\log_b x + \log_b y = \log_b xy$

49. a. If $\log y = 1.5x - 2$, show that $y \approx 0.01(31.6)^x$.
  b. If $\log y = 0.5x + 1$, express y in terms of x. $y \approx 10(3.16)^x$

50. a. If $\ln y = 4x + 2$, show that $y \approx 7.4(54.6)^x$.
  b. If $\ln y = 1 - 0.1x$, express y in terms of x. $y \approx 2.72(0.905)^x$

C **51.** In advanced mathematics, it can be proved that the number of prime numbers less than a positive integer n is approximately $\dfrac{n}{\ln n}$.
  a. About how many primes are less than **(1)** 1000 and **(2)** 1,000,000? 145; 72,382
  b. There are four prime numbers less than 10. We say that the density of the primes in the interval from 1 to 10 is $\dfrac{4}{10}$. Use your answers to part (a) to find the approximate density of primes in the intervals from 1 to 1000 and from 1 to 1,000,000. 0.145; 0.072
  c. What happens to the density of primes less than n as n increases? Approaches zero

52. Prove that $\log 2$ is irrational. (*Hint:* Assume that $\log 2 = \dfrac{p}{q}$, where p and q are integers and $\dfrac{p}{q}$ is in lowest terms.)

5-6 Laws of Logarithms

Objective *To prove and apply laws of logarithms.*

Since the logarithmic function $y = \log_b x$ is the inverse of the exponential function $y = b^x$, it is not surprising that the laws of logarithms are very closely related to the laws of exponents on page 170.

Laws of Logarithms

If M and N are positive real numbers and b is a positive number other than 1, then:

1. $\log_b MN = \log_b M + \log_b N$

2. $\log_b \dfrac{M}{N} = \log_b M - \log_b N$

3. $\log_b M = \log_b N$ if and only if $M = N$

4. $\log_b M^k = k \log_b M$, for any real number k

Exponents and Logarithms **197**

**Additional Answers
Written Exercises**

47. a. 2; $\dfrac{1}{2}$ **b.** $\dfrac{3}{2}$, $\dfrac{2}{3}$

  c. $\log_a b = \dfrac{1}{\log_b a}$

48. c. Proof: Let $c = \log_b x$ and $d = \log_b y$. Then $b^c = x$ and $b^d = y$. Hence $xy = b^c b^d = b^{c + d}$. Therefore $\log_b xy = c + d$, and so $\log_b xy = \log_b x + \log_b y$.

49. a. If $\log y = 1.5x - 2$, then $y = 10^{1.5x - 2} = 10^{-2} 10^{1.5x} = 10^{-2}(10^{1.5})^x \approx 0.01(31.6)^x$.

50. a. If $\ln y = 4x + 2$, then $y = e^{4x + 2} = e^2(e^4)^x \approx 7.4(54.6)^x$.

52. Assume $\log 2 = \dfrac{p}{q}$, where p and q are integers and $\dfrac{p}{q}$ is in lowest terms. Then $2 = 10^{p/q}$ and $2^q = 10^p$. Thus $2^q = 2^p \cdot 5^p$ and $2^{q - p} = 5^p$, which is impossible since no integral power of 2 is equal to an integral power of 5. Therefore $\log 2 \neq \dfrac{p}{q}$, and so $\log 2$ is irrational.

Teaching Notes, p. 168F

Warm-Up Exercises

1. Explain why $\sqrt[3]{x^3} = (\sqrt[3]{x})^3$.
  The cube root and cube functions are inverses, so $\sqrt[3]{x^3} = x = (\sqrt[3]{x})^3$.

(*Continues on next page*)

2. Write the equation $\log_3\left(\frac{1}{27}\right) = -3$ in exponential form.

$3^{-3} = \frac{1}{27}$

3. Write the equation $128^{1/7} = 2$ in logarithmic form. $\log_{128} 2 = \frac{1}{7}$

4. Simplify.

a. $5^{\log_5 25}$ 25

b. $\log 10^9$ 9

Motivating the Section

The laws of logarithms were more important before the days of calculators. In calculus and statistics, the use of logarithms makes many calculations and formulas manageable.

Additional Examples

1. Simplify.

a. $\frac{1}{3} \log 64 + 2 \log 5$

b. $e^{\ln (1/2)x}$

c. $\log 3 - \log 6 - \log 5$

a. $\frac{1}{3} \log 64 + 2 \log 5 =$

$\log 64^{1/3} + \log 5^2 =$

$\log (4^3)^{1/3} + \log 25 =$

$\log (4 \cdot 25) =$

$\log 10^2 = 2$

b. Since e^x and $\ln x$ are inverses, $e^{\ln (1/2)x} = \frac{1}{2}x$.

c. $\log 3 - \log 6 - \log 5 =$

$\log 3 - (\log 6 + \log 5) =$

$\log 3 - \log 30 =$

$\log \frac{3}{30} =$

$\log 10^{-1} = -1$

To prove law 1, let $\log_b M = x$ and $\log_b N = y$. Then $M = b^x$ and $N = b^y$.

$$MN = b^x \cdot b^y = b^{x+y}$$

Therefore:

$$\log_b MN = x + y$$
$$\log_b MN = \log_b M + \log_b N$$

Laws 2 and 4 are proved in a similar fashion. Law 3 is a restatement, in terms of logarithms, of the third law of exponents on page 170.

If you know the logarithms of M and N, then you can use the laws of logarithms to find the logarithm of a more complicated expression in M and N.

Example 1   Express $\log_b MN^2$ in terms of $\log_b M$ and $\log_b N$.

Solution $\log_b MN^2 = \log_b M + \log_b N^2$ (law 1)

$\qquad\qquad\quad = \log_b M + 2 \log_b N$ (law 4)

Example 2   Express $\log_b \sqrt{\frac{M^3}{N}}$ in terms of $\log_b M$ and $\log_b N$.

Solution $\log_b \sqrt{\frac{M^3}{N}} = \log_b \left(\frac{M^3}{N}\right)^{1/2} = \frac{1}{2} \log_b \left(\frac{M^3}{N}\right)$ (law 4)

$\qquad\qquad\qquad\qquad\quad = \frac{1}{2}(\log_b M^3 - \log_b N)$ (law 2)

$\qquad\qquad\qquad\qquad\quad = \frac{1}{2}(3 \log_b M - \log_b N)$ (law 4)

In Examples 1 and 2, the logarithm of an expression was written in terms of separate logarithms. In Examples 3 and 4, separate logarithms are combined into a single logarithm.

Example 3   Simplify $\log 45 - 2 \log 3$.

Solution $\log 45 - 2 \log 3 = \log 45 - \log 3^2$ (law 4)

$\qquad\qquad\qquad\quad = \log \frac{45}{3^2}$ (law 2)

$\qquad\qquad\qquad\quad = \log 5$

Example 4   Express y in terms of x if $\ln y = \frac{1}{3}\ln x + \ln 4$.

Solution $\ln y = \frac{1}{3}\ln x + \ln 4 = \ln x^{1/3} + \ln 4$ (law 4)

$\qquad\qquad\qquad \ln y = \ln 4x^{1/3}$ (law 1)

$\qquad\qquad\qquad\quad y = 4x^{1/3}$ (law 3)

Our final example shows how properties of logarithms can be used to solve certain equations.

Example 5   Solve $\log_2 x + \log_2 (x - 2) = 3$.

Solution
$$\log_2 x + \log_2 (x - 2) = 3$$
$$\log_2 x(x - 2) = 3$$
$$x(x - 2) = 8 \longleftarrow 2^3 = 8$$
$$x^2 - 2x - 8 = 0$$
$$(x - 4)(x + 2) = 0$$
$$x = 4 \text{ or } -2$$

Since $\log x$ is not defined for negative x, -2 is *not* a solution. The only solution is $x = 4$.

CLASS EXERCISES

Express the common logarithm of each of the following in terms of $\log M$ and $\log N$.

$2 \log M - \log N$ $\qquad$ $\frac{1}{3}(\log M + \log N)$ $\qquad$ $2 \log M - 3 \log N$

1. $M^2 N$ $\qquad$ **2.** $\dfrac{M^2}{N}$ $\qquad$ **3.** $\sqrt{\dfrac{M}{N}}$ $\qquad$ **4.** $\sqrt[3]{MN}$ $\qquad$ **5.** $M\sqrt{N}$ $\qquad$ **6.** $\dfrac{M^2}{N^3}$

$2 \log M + \log N$ $\qquad$ $\frac{1}{2}(\log M - \log N)$ $\qquad$ $\log M + \frac{1}{2}\log N$

Use the laws of logarithms to express each of the following as a single logarithm.

7. $\log_5 2 + \log_5 3$ $\log_5 6$ $\qquad\qquad$ **8.** $\log_3 5 + \log_3 4$ $\log_3 20$

9. $\log 12 - \log 3$ $\log 4$ $\qquad\qquad$ **10.** $\log 3 + \log 6 - \log 2$ $\log 9$

11. $\ln 4 + 2 \ln 3$ $\ln 36$ $\qquad\qquad$ **12.** $\frac{1}{2}\ln 25 - \ln 2$ $\ln \frac{5}{2}$

13. $\log M + 2 \log N$ $\log MN^2$ $\qquad$ **14.** $2 \log P - \log Q$ $\log \left(\dfrac{P^2}{Q}\right)$

15. $\log_b M + \log_b N + \log_b P$ $\log_b MNP$ $\qquad$ **16.** $\log_b M + \log_b N - 3 \log_b P$ $\log_b \dfrac{MN}{P^3}$

17. $\frac{1}{2}\ln a - \frac{1}{2}\ln b$ $\ln \left(\dfrac{a}{b}\right)^{1/2}$ $\qquad\qquad$ **18.** $\ln c + \frac{1}{3}\ln d$ $\ln cd^{1/3}$

In Exercises 19 and 20, give an example to show that in general each statement is false.

19. $\log_b (M + N) = \log_b M + \log_b N$ $\qquad$ **20.** $\log_b \left(\dfrac{M}{N}\right) = \dfrac{\log_b M}{\log_b N}$

21. *Discussion* What is wrong with the following argument?

Since $\frac{1}{8} < \frac{1}{4}$, $\log \frac{1}{8} < \log \frac{1}{4}$. Therefore, $\log \left(\frac{1}{2}\right)^3 < \log \left(\frac{1}{2}\right)^2$. This means that $3 \log \left(\frac{1}{2}\right) < 2 \log \left(\frac{1}{2}\right)$. Thus, $3 < 2$. $\log\left(\frac{1}{2}\right) < 0$, and dividing by a neg. number reverses the order of the inequality.

Additional Examples cont.

2. If $\log_6 2 = x$ and $\log_6 5 = y$, express each logarithm in terms of x and y.
a. $\log_6 40$
b. $\log_6 3$
c. $\log_6 15$

a. $\log_6 (8 \cdot 5) =$
$\log_6 8 + \log_6 5 =$
$\log_6 2^3 + \log_6 5 =$
$3 \log_6 2 + \log_6 5 =$
$3x + y$

b. $\log_6 3 = \log_6 \dfrac{6}{2} =$
$\log_6 6 - \log_6 2 =$
$1 - x$

c. $\log_6 15 = \log_6 \dfrac{6 \cdot 5}{2} =$
$\log_6 6 + \log_6 5 -$
$\log_6 2 = 1 + y - x$

Error Analysis

Class Exercises 19 and 20 point out two common misperceptions about logarithms among students. Be sure to discuss these exercises with your students.

Additional Answers
Class Exercises

19. For example,
$\log 10 + \log 1 = 1 + 0 = 1 \neq \log 11$.

20. For example, $\dfrac{\log 1}{\log 10} = \dfrac{0}{1} =$
$0 \neq \log \left(\dfrac{1}{10}\right)$.

WRITTEN EXERCISES

In Exercises 1–6, write each expression in terms of log M and log N.

A   **1.** $\log (MN)^2$ **2.** $\log \dfrac{M}{N^2}$ **3.** $\log \sqrt[3]{\dfrac{M}{N}}$ **4.** $\log M\sqrt[4]{N}$ **5.** $\log M^2\sqrt{N}$ **6.** $\log \dfrac{1}{M}$

Write each expression as a rational number or as a single logarithm.

7. $\log 2 + \log 3 + \log 4$ $\log 24$

8. $\log 8 + \log 5 - \log 4$ 1

9. $\frac{1}{2}\log_6 9 + \log_6 5$ $\log_6 15$

10. $\log_2 48 - \frac{1}{3}\log_2 27$ 4

11. $2 \ln 6 - \ln 3$ $\ln 12$

12. $\frac{1}{2}\ln 5 + 3 \ln 2$ $\ln 8\sqrt{5}$

13. $\log M - 3 \log N$ $\log \dfrac{M}{N^3}$

14. $4 \log M + \frac{1}{2}\log N$ $\log M^4\sqrt{N}$

15. $\log A + 2 \log B - 3 \log C$ $\log \dfrac{AB^2}{C^3}$

16. $\frac{1}{2}(\log_b M + \log_b N - \log_b P)$ $\log_b \sqrt{\dfrac{MN}{P}}$

17. $\frac{1}{3}(2 \log_b M - \log_b N - \log_b P)$ See below.

18. $5(\log_b A + \log_b B) - 2 \log_b C$ See below.

19. $\log \pi + 2 \log r$ $\log \pi r^2$

20. $\log 4 - \log 3 + \log \pi + 3 \log r$ $\log \dfrac{4}{3}\pi r^3$

21. $\ln 2 + \ln 6 - \frac{1}{2}\ln 9$ $\ln 4$

22. $\ln 10 - \ln 5 - \frac{1}{3}\ln 8$ 0

Simplify each expression.   **17.** $\log_b \sqrt[3]{\dfrac{M^2}{NP}}$ **18.** $\log_b \dfrac{(AB)^5}{C^2}$

23. a. $\ln e^2$ 2 **b.** $\ln e^3$ 3 **c.** $\ln \dfrac{1}{e}$ -1 **d.** $\ln \sqrt{e}$ $\dfrac{1}{2}$

24. a. $\ln e^4$ 4 **b.** $\ln \dfrac{1}{e^3}$ -3 **c.** $\ln \sqrt[3]{e}$ $\dfrac{1}{3}$ **d.** $\ln 1$ 0

25. a. $\ln e^x$ x **b.** $e^{\ln x}$ x **c.** $e^{2 \ln x}$ x^2 **d.** $e^{-\ln x}$ $\dfrac{1}{x}$

26. a. $\ln e^{3x}$ $3x$ **b.** $e^{3 \ln x}$ x^3 **c.** $e^{\ln \sqrt{x}}$ $\sqrt{x}$ **d.** $e^{(-1/2) \ln x}$ $x^{-1/2}$

27. a. $10^{\log 6}$ 6 **b.** $10^{2 \log 6}$ 36 **c.** $10^{3 + \log 4}$ 4000 **d.** $e^{3 + \ln 4}$ $4e^3$

28. a. $10^{3 \log 5}$ 125 **b.** $e^{3 \ln 5}$ 125 **c.** $10^{1 + 2 \log x}$ $10x^2$ **d.** $e^{1 + 2 \ln x}$ ex^2

Express y in terms of x.

29. a. $\log y = 2 \log x$ $y = x^2$ **b.** $\log y = 3 \log x + \log 5$ $y = 5x^3$

30. a. $\ln y - \ln x = 2 \ln 7$ $y = 49x$ **b.** $\ln y = 2 \ln x - \ln 4$ $y = 0.25x^2$

31. a. $\log y = -\log x$ $y = \dfrac{1}{x}$ **b.** $\log y = 2 \log x + \log 2$ $y = 2x^2$

32. a. $\log y + \frac{1}{2}\log x = \log 3$ $y = \dfrac{3}{\sqrt{x}}$ **b.** $\ln y = \frac{1}{3}(\ln 4 + \ln x)$ $y = \sqrt[3]{4x}$

33. a. $\log y = 1.2x - 1$ $y = 0.1(10^{1.2})^x$ **b.** $\ln y = 1.2x - 1$ $y = \dfrac{1}{e}(e^{1.2})^x$

34. a. $\log y = 3 - 0.5x$ $y = 1000(10^{-0.5})^x$ **b.** $\ln y = 3 - 0.5x$ $y = e^3(e^{-0.5})^x$

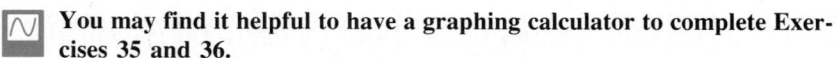 You may find it helpful to have a graphing calculator to complete Exercises 35 and 36.

B **35. a.** Predict how the graph of $y = \log_b\left(\frac{1}{x}\right)$ is related to the graph of $y = \log_b x$. $\log_b \frac{1}{x} = -\log_b x$

b. Check your prediction by graphing $y = \log x$ and $y = \log\left(\frac{1}{x}\right)$ on a single set of axes.
$\log_b x^2 = 2\log_b x, \; x > 0$

36. a. Predict how the graph of $y = \log_b x^2$ is related to the graph of $y = \log_b x$.

b. Check your prediction by graphing $y = \log x$ and $y = \log x^2$ on a single set of axes.

37. a. If $f(x) = \log_2 x$, show that $f(2x) = f(x) + 1$.

b. Part (a) shows that horizontally shrinking the graph of f by a factor of 2 is equivalent to __?__. Shifting it up one unit

38. a. If $f(x) = \log_3 x$, show that $f\left(\frac{x}{3}\right) = f(x) - 1$.

b. Make a statement similar to that in part (b) of Exercise 37.

39. Suppose that $\log y = ax + b$, where a and b are real numbers and a is nonzero. Express y in terms of x. What kind of expression defines $\log y$ in terms of x? What kind of expression defines y in terms of x? $y = 10^b(10^a)^x$; linear; exponential

40. Conservation The expiration time T of a natural resource is the time remaining until it is all used. If one assumes that the current growth rate of consumption remains constant, then the expiration time in years is given by

$$T = \frac{1}{r} \ln\left(\frac{rR}{C} + 1\right),$$

where C = current consumption, r = current growth rate of consumption, and R = resource size. Suppose that the world's consumption of oil is growing at the rate of 7% per year $(r = 0.07)$ and current consumption is approximately 17×10^9 barrels per year. Find the expiration time for the following estimates of R.

a. $R \approx 1691 \times 10^9$ barrels (estimate of remaining crude oil) 29.6 years

b. $R \approx 1881 \times 10^9$ barrels (estimate of remaining crude plus shale oil) 31 years

Exponents and Logarithms **201**

 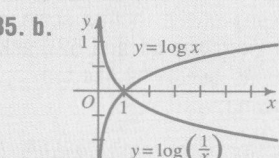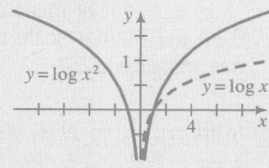

Exercise Note

To check students' understanding of the Richter scale (see Exercises 51 and 52), ask students questions like the following:

1. In terms of the *amplitude* of the seismic wave, how many times more intense is an earthquake measuring 9 on the Richter scale than one measuring 7? ($10^2 = 100$)

2. In terms of the *energy* released, how many times more intense is an earthquake measuring 9 than one measuring 7? ($31^2 = 961$)

Error Analysis

For Exercises 43–50, make sure students check their solutions. Remind them that logarithms are not defined for negative numbers.

Additional Answers Written Exercises

52. a. $R = 0.67 \log (0.37E) + 1.46$, $\frac{R - 1.46}{0.67} = \log (0.37E)$, $0.37E = 10^{(R-1.46)/0.67}$, $E = 2.7 \cdot 10^{(R-1.46)/0.67}$

b. Let $E(R) = 2.7 \cdot 10^{(R-1.46)/0.67}$. Then $E(R + 1) = 2.7 \cdot 10^{(R+1-1.46)/0.67} = 2.7 \cdot 10^{1/0.67} \cdot 10^{(R-1.46)/0.67} \approx 31 \cdot E(R)$.

---

41. If $\log_8 3 = r$ and $\log_8 5 = s$, express each logarithm in terms of r and s. **d.** $r - 2$
    a. $\log_8 75$ $r + 2s$ **b.** $\log_8 225$ $2r + 2s$ **c.** $\log_8 0.12$ $r - 2s$ **d.** $\log_8 \frac{3}{64}$

42. If $\log_9 5 = x$ and $\log_9 4 = y$, express each logarithm in terms of x and y.
    a. $\log_9 100$ $2x + y$ **b.** $\log_9 36$ $1 + y$ **c.** $\log_9 \left(6\frac{1}{4}\right)$ $2x - y$ **d.** $\log_9 3.2$
    d. $2y - x$

In Exercises 43–46, solve the given equation.

43. **a.** $\log_2 (x + 2) + \log_2 5 = 4$ $\frac{6}{5}$ **b.** $\log_4 (2x + 1) - \log_4 (x - 2) = 1$ $\frac{9}{2}$

44. **a.** $\log_6 (x + 1) + \log_6 x = 1$ 2 **b.** $\log_3 x + \log_3 (x - 2) = 1$ 3

45. **a.** $\log_4 (x - 4) + \log_4 x = \log_4 5$ 5 **b.** $\log_2 (x^2 + 8) = \log_2 x + \log_2 6$ $2, 4$

46. **a.** $\ln (x^2) = 16$ $\pm e^8$ **b.** $\ln \frac{1}{x} = -5$ e^5

47. **a.** For what values of M is $\log_2 M < 0$? $0 < M < 1$
    b. Use your answer to part (a) to solve the inequality $\log_2 \left(\frac{x-1}{2}\right) < 0$. $1 < x < 3$
    c. For what values of M is $\log_3 M > 2$? $M > 9$ **d.** $x < -2$ or $x > 2$
    d. Use your answer to part (c) to solve the inequality $\log_3 (x^2 + 5) > 2$.
    e. Use the fact that $\log_b M > \log_b N$ only if $M > N$ to solve the inequality $\log_6 5x > 2 \log_6 x$. $0 < x < 5$

In Exercises 48–50, solve the given inequality.

48. **a.** $\ln (x - 4) + \ln 3 \le 0$ $4 < x \le \frac{13}{3}$ **b.** $\log (5 - x) - \log 7 > 0$ $x < -2$

49. **a.** $2 \log x < \log (2x - 1)$ No sol.   **b.** $\ln (x + 1) - \ln 2 > 3$ $x > 2e^3 - 1$

50. **a.** $\log_2 (x + 5) + \log_2 (x - 2) \ge 3$ $x \ge 3$ **b.** $\log_4 (x - 1) + \log_4 (x + 1) < \log_4 6$
    $1 < x < \sqrt{7}$

The *Richter scale* is a system for rating the severity of an earthquake. The severity can be measured either by the amplitude of the seismic wave or by the energy released by the earthquake. A one-point increase in the Richter scale number corresponds to a ten-fold increase in the *amplitude* of the seismic wave and to a thirty-one-fold increase in the *energy* released. Exercises 51 and 52 give some details.

51. **Geology** Richter scale numbers for some earthquakes that have occurred in this century are given in the table at the right. To find the ratio of the seismic wave amplitudes for the quakes in Japan and Alaska, we write:

$$\frac{10^{8.9}}{10^{8.4}} = 10^{0.5} \approx 3.16$$

≈ 16

a. Find the ratio of wave amplitudes for the 1906 and 1989 California earthquakes.

b. Find the ratio of wave amplitudes for the earthquakes in (1) Iran (1968) and Yugoslavia, and (2) Iran (1990) and Yugoslavia.

$\approx 25, \approx 50$

Year	Place	Richter scale
1906	California	8.3
1933	Japan	8.9
1963	Yugoslavia	6.0
1964	Alaska	8.4
1968	Iran	7.4
1989	California	7.1
1990	Iran	7.7

C **52. Geology** The Richter scale was proposed in 1935 by Charles Richter. It was refined in 1979. The Richter magnitude, R, of an earthquake is given by,

$$R = 0.67 \log (0.37E) + 1.46,$$

where E is the energy in kW · h released by the earthquake.
  a. Show that $E = 2.7 \cdot 10^{(R - 1.46)/0.67}$.
  b. Show that if R increases by 1 unit, E increases by a factor of about 31.

The San Andreas fault

53. If $9^{(9^9)}$ is multiplied out and is typed on a strip of paper, 3 digits per centimeter, about how many kilometers long would the paper be? (*Hint:* The common logarithm of a number can tell you how many digits the number has.) ≈1230 km

54. Suppose that all you know about a function f is that $f(ab) = f(a) + f(b)$ for all positive numbers a and b.
  a. Find $f(1)$. 0
  b. Prove that $f(a^2) = 2f(a)$ and $f(a^3) = 3f(a)$. What generalization does this suggest? $f(a^n) = nf(a)$, for n a pos. int.
  c. Prove that $f(\sqrt{a}) = \frac{1}{2}f(a)$ and $f(\sqrt[3]{a}) = \frac{1}{3}f(a)$. What generalization does this suggest? $f(\sqrt[n]{a}) = \frac{1}{n}f(a)$, for n a pos. int.
  d. Prove that $f\left(\frac{1}{b}\right) = -f(b)$.
  e. Prove that $f\left(\frac{a}{b}\right) = f(a) - f(b)$.
    $f(x) = \log_b x, \ b > 0, \ b \neq 1$
  f. Try to find a function f that satisfies the original equation.
  g. If $f(10) = 1$, find the values of x for which $f(x) = 2$ and $f(x) = 3$. 100, 1000

5-7 Exponential Equations; Changing Bases

Objective *To solve exponential equations and to change logarithms from one base to another.*

An **exponential equation** is an equation that contains a variable in the exponent. Here are exponential equations you can solve from Section 5-2.

$$2^{t-3} = 8 \qquad 9^{2t} = 3\sqrt[3]{3}$$

These exponential equations are special because both sides of each equation can easily be expressed as powers of the same number. Usually exponential equations cannot be solved this way. In this section, you will see how to use logarithms to solve exponential equations involving a variable such as time, t.

Exponents and Logarithms **203**

Teaching Notes, p. 168F

Warm-Up Exercises

Solve each equation.

1. $7^x = 2401$ 4
2. $4^x = 8^{x-1}$ 3

Between what two integers does the solution lie?

3. $10^x = 123,456$ 5 and 6
4. $4^x = 400$ 4 and 5
5. $9^x = 6$ 0 and 1
6. $10^x = 0.7$ 0 and −1

Motivating the Section

In this section, students solve exponential equations by using logarithms. Therefore, exponential growth-and-decay applications can now be expressed with variable exponents, and the exponent can be determined without resorting to trial-and-error methods. For example, the problem of how long it takes for a $100 savings account to grow to $1000 if the interest rate is 6% compounded annually can be expressed as the equation
$1000 = 100(1 + 0.06)^t$
where t is time. (In this case, $t \approx 39.5$ years.)

1. Solve.
   a. $3^{-x} = 0.7$
   b. $3^x = 9\sqrt[3]{3}$
   c. $(1.1)^x = 2$
   d. $x = \log_7 2$

   a. $\log 3^{-x} = \log 0.7$
   $-x \log 3 = \log 0.7$
   $x = -\dfrac{\log 0.7}{\log 3} \approx 0.325$

   b. $3^x = 3^2 \cdot 3^{1/3} = 3^{7/3}$
   $x = \dfrac{7}{3}$

   c. $\log (1.1)^x = \log 2$
   $x \log 1.1 = \log 2$
   $x = \dfrac{\log 2}{\log 1.1} \approx 7.27$

   d. $x = \dfrac{\log 2}{\log 7} \approx 0.356$

2. The half-life of carbon-11 is 20 minutes. How long will it take for 800 g of carbon-11 to decay to:
   a. 640 g   b. 40 g

   a. $A(t) = A_0\left(\dfrac{1}{2}\right)^{t/20}$

   $640 = 800\left(\dfrac{1}{2}\right)^{t/20}$

   $0.8 = \left(\dfrac{1}{2}\right)^{t/20}$

   $\log 0.8 = \log \left(\dfrac{1}{2}\right)^{t/20}$

   $\log 0.8 = \dfrac{t}{20}\left(\log \dfrac{1}{2}\right)$

   $\dfrac{t}{20} = \dfrac{\log 0.8}{\log 0.5}$

   $t \approx 6.4$; about 6.4 minutes

   b. $40 = 800\left(\dfrac{1}{2}\right)^{t/20}$

   $0.05 = (0.5)^{t/20}$

   $\log 0.05 = \log (0.5)^{t/20}$

   $\log 0.05 = \dfrac{t}{20} \log 0.5$

   $t = 20\left(\dfrac{\log 0.05}{\log 0.5}\right)$

   $t \approx 86.4$; about 86.4 minutes

Example 1

In 1990, there were about 5.4 billion people in the world. If the population has been growing at 1.95% per year, estimate the year when the population will be 8 billion people.

Solution

$$A_0 (1 + r)^t = A(t)$$
$$5.4(1 + 0.0195)^t = 8$$
$$1.0195^t = 1.4815$$
$$\log (1.0195)^t = \log 1.4815$$
$$t \log (1.0195) = \log 1.4815$$

Thus: $t = \dfrac{\log 1.4815}{\log 1.0195} \approx 20.35$

The population will reach 8 billion people around the year 2010.

Example 2

Suppose you invest P dollars at an annual rate of 6% compounded daily. How long does it take (a) to increase your investment by 50%? (b) to double your money?

Solution

a. Since the interest is compounded daily, we can reliably use the formula for continuous compounding, $P(t) = Pe^{rt}$. We need to find the value of t for which $P(t) = P + 50\%P = 1.5P$.

$$P(t) = Pe^{rt} = 1.5P$$
$$Pe^{0.06t} = 1.5P$$
$$e^{0.06t} = 1.5$$

The most convenient logarithm to use is the natural logarithm.

$$\ln e^{0.06t} = \ln 1.5$$
$$0.06t \cdot \ln e = \ln 1.5$$
$$(0.06t)1 \approx 0.4055$$

Thus: $t \approx 6.76$
It will take about 6.76 years, or 6 years and 9 months.

b. Use the technique shown in part (a).

$$Pe^{0.06t} = 2P$$
$$e^{0.06t} = 2$$
$$0.06t = \ln 2$$

Thus: $t = \dfrac{\ln 2}{0.06} \approx \dfrac{0.693}{0.06} = 11.55$

It will take about 11.55 years, or 11 years and 6 months. (Note that the rule of 72 gives 12 years as an estimate.)

The **change-of-base formula** enables us to find the logarithm of a number in one base if logarithms in another base are known. The formula is stated below, and you are asked to derive it in Exercise 21.

$$\log_b c = \frac{\log_a c}{\log_a b}$$

For example, $\log_5 8 = \dfrac{\log 8}{\log 5} \approx \dfrac{0.903}{0.700} = 1.2900.$

CLASS EXERCISES

Solve each equation. Leave answers in radical form when possible.

1. $x^3 = 81$ $\sqrt[3]{81}$ **2.** $3^x = 81$ 4   **3.** $4^x = 81$ 3.17   **4.** $x^4 = 8$ $\pm\sqrt[4]{8}$

Use a calculator to find the value of x to the nearest hundredth.

5. $10^x = 3$ 0.48   **6.** $10^x = 8.1$ 0.91   **7.** $10^x = 256$ 2.41   **8.** $100^x = 302$ 1.24

 You may find it helpful to have a graphing calculator to complete Exercise 9.

9. *Discussion* Explain how the change-of-base formula enables you to use a graphing calculator or computer to graph $y = \log_2 x$. Graph $y = \dfrac{1}{\log 2} \log x$.

WRITTEN EXERCISES

Use a calculator to find the value of x to the nearest hundredth.

A
1. $3^x = 12$ 2.26   **2.** $2^x = 100$ 6.64   **3.** $(1.06)^x = 3$ 18.85   **4.** $(0.98)^x = 0.5$ 34.31
5. $e^x = 18$ 2.89   **6.** $e^{-x} = 0.01$ 4.61   **7.** $\sqrt{e^x} = 50$ 7.82   **8.** $(e^x)^3 = 200$ 1.77

For each pair of equations, solve one of them by using powers of the same number. To the nearest hundredth, solve the other by using logarithms.

9. a. $4^x = 16\sqrt{2}$ $\frac{9}{4}$ **b.** $4^x = 20$ 2.16   **10. a.** $9^x = \dfrac{3}{3^x}$ $\frac{1}{3}$ **b.** $9^x = 4$ 0.63

11. a. $25^x = \sqrt[5]{5^x}$ 0   **b.** $25^x = 2$ 0.22   **12. a.** $8^x = \sqrt[3]{\dfrac{2}{4^x}}$ $\frac{1}{11}$ **b.** $8^x = \sqrt[3]{5}$ 0.26

13. Geography The population of Kenya reached 25,000,000 people in 1990. When will it reach 50,000,000 people? Assume an annual rate of increase of 4.1%. 2007

14. Finance An investment is made at 7% annual interest compounded daily. How long does it take to triple the investment? 15.7 years

Exponents and Logarithms **205**

Supplementary Materials

Alternative Assessment, 18
Precalculus Plotter Plus,
33–34
Student Resource Guide,
52–54

Additional Answers
Written Exercises

16. $P(t) = Pe^{0.01rt}$, where r is the interest rate expressed as a percent. The amount invested is doubled when $Pe^{0.01rt} = 2P$, or $e^{0.01rt} = 2$. Then $0.01rt = \ln 2$, and $t = \dfrac{\ln 2}{0.01r} \approx \dfrac{0.693}{0.01r} = \dfrac{69.3}{r}$.

17. The amount invested is tripled when $Pe^{0.01rt} = 3P$, or $e^{0.01rt} = 3$. Then $0.01rt = \ln 3$, and $t = \dfrac{\ln 3}{0.01r} \approx \dfrac{1.10}{0.01r} = \dfrac{110}{r}$.

21. Let $x = \log_b c$. Then $b^x = c$. Taking the base a logarithm of both sides gives $\log_a b^x = \log_a c$, or $x \log_a b = \log_a c$. Thus $x = \dfrac{\log_a c}{\log_a b}$ and therefore $\log_b c = \dfrac{\log_a c}{\log_a b}$.

23. a. If $x = 2^x$, then $\log x = x \log 2$. It is impossible to isolate x.
b. Graph $y = x$ and $y = 2^x$ on the same set of axes and look for the intersection points.
c. Graph $y = x + 1$ and $y = 2^x$ on the same set of axes and look for the intersection points. Graph $y = x + 2$ and $y = 2^x$ on the same set of axes and look for the intersection points.

---

15. Finance A $10,000 certificate of deposit at a certain bank will double in value in 9 years.
 a. Give a formula for the accumulated amount t years after the investment is made. $A(t) = 10{,}000(1.08)^t$
 b. How long does it take for the money to triple in value? 14.3 years

B **16. Finance** According to the rule of 72, an investment at $r\%$ interest compounded continuously will double in approximately $\dfrac{72}{r}$ years. Show that a more accurate doubling time is $\dfrac{69.3}{r}$ years. (Note that the usual formula, $P(t) = Pe^{rt}$, where r is a decimal, must be rewritten as $P(t) = Pe^{0.01rt}$ where r is a percent.)

17. Finance Prove that the time needed to triple an investment at $r\%$ interest compounded continuously is approximately $\dfrac{110}{r}$.

18. Finance Tell how long it takes for $100 to become $1000 if it is invested at 8% interest compounded:
 a. annually 29.9 years **b.** quarterly 29.1 years **c.** daily 28.8 years

19. Physics A radioactive isotope has a half-life of 9.6 h.
 a. If there is 1 kg of the isotope now, how much will there be in 24 h? 177 g
 b. How long does it take for the isotope to decay to 1 g? 95.7 h

20. a. Find $\log_6 88$ by using the change-of-base formula. 2.5
 b. Find $\log_6 88$ by solving $6^x = 88$. 2.5

21. Derive the change-of-base formula.

22. Find the value of x to the nearest tenth.
 a. $x^5 = 98$ 2.5 **b.** $5^x = 98$ 2.8

> You may find it helpful to have a graphing calculator to complete Exercises 23 and 24.

23. Discussion Suppose that you wished to solve $x = 2^x$.
 a. What happens when you try to solve $x = 2^x$ by taking the logarithm of each side of the equation?
 b. Give another approach to solve $x = 2^x$. How many solutions exist? None
 c. Discuss the approach you would take to solve $x + 1 = 2^x$ and to solve $x + 2 = 2^x$. How many solutions are there in each case? 2; 2
 d. Based on your answers to parts (a)–(c), how many solutions are there to $x + c = 2^x$, where c is a whole number? 0 or 2

24. Given: $x^x = \pi$. Find the value of x to the nearest hundredth. 1.85

Solve. Express x as a logarithm if necessary.

25. $2^{2x} - 2^x - 6 = 0$ (*Hint:* $2^{2x} = (2^x)^2$) $\log_2 3$
26. $3^{2x} - 5 \cdot 3^x + 4 = 0$ 0, $\log_3 4$
27. $e^{2x} - 5e^x + 6 = 0$ $\ln 2$, $\ln 3$
28. $e^{2x} - e^x - 6 = 0$ $\ln 3$
29. $3^{2x+1} - 7 \cdot 3^x + 2 = 0$ $\log_3 2$, -1
30. $e^x + e^{-x} = 4$ $\ln(2 - \sqrt{3})$, $\ln(2 + \sqrt{3})$

32. a. $x > 3.5$ **b.** $x < 0$ **c.** $x < 0$ **d.** No solution

31. a. If $b^m > b^n$ and $b > 1$, what can you say about m and n? $m > n$

   b. If $b^m > b^n$ and $0 < b < 1$, what can you say about m and n? $m < n$

32. Solve: **a.** $8^x > 8^{7-x}$ **b.** $0.6^{5x} > 0.6^{x/2}$ **c.** $e^{3x} < e^x$ **d.** $\left(\dfrac{1}{2}\right)^x > 2^{6-x}$

33. Archaeology All living organisms contain a small amount of carbon 14, denoted C^{14}, a radio-active isotope. When an organism dies, the amount of C^{14} present decays exponentially. By measuring the radioactivity $N(t)$ of, say, an ancient skeleton of an animal and by comparing that radioactivity with the radioactivity N_0 of living animals, archaeologists can tell approximately when the animal died.

   a. Given that the half-life of C^{14} is about 5700 years, write an equation relating $N(t)$, N_0, and the time t since the animal's death. See below.

   b. Suppose it is found that, for a certain animal, $N(t) = \dfrac{1}{10}N_0$. To the nearest 100 years, how long ago did the animal die? 18,900 years

34. Archaeology An archaeologist unearths a piece of wood that may have come from the Hanging Gardens of Babylon, about 600 B.C. The amount of radioactive C^{14} in the wood is $N(t) = 0.8N_0$. Is it possible that the wood could be from the Hanging Gardens? (The half-life of C^{14} is about 5700 years.) No.

35. Prove: $\log_b c = \dfrac{1}{\log_c b}$ **33. a.** $N(t) = N_0\left(\dfrac{1}{2}\right)^{t/5700}$ **36.** Prove: $(\log_a b)(\log_b c) = \log_a c$

Evaluate each expression. (Use the results of Exercises 35 and 36.)

37. $\log_3 2 \cdot \log_2 27$ 3 **38.** $\log_{25} 8 \cdot \log_8 5$ $\dfrac{1}{2}$ **39.** $\dfrac{1}{\log_2 6} + \dfrac{1}{\log_3 6}$ 1 **40.** $\dfrac{1}{\log_4 6} + \dfrac{1}{\log_9 6}$ 2

42. a. $-3.98 < x < 0.54$ **b.** No solution **c.** $x > 0.32$ **d.** $0 < x \le 1$

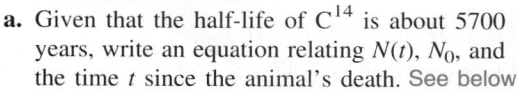 **For Exercises 41 and 42, use a computer or a graphing calculator to solve each inequality. Give answers to the nearest hundredth.**

41. On a single set of axes, graph $y = \log_2 (x - 1)$ and $y = \log_3 x$. (See Class Exercise 9 on page 205.) Use your graph to solve $\log_2 (x - 1) > \log_3 x$. $x > 3$

42. Solve each inequality using the method suggested in Exercise 41. See above.

   a. $e^x < \ln (x + 5)$ **b.** $2^x \le \log_5 x$ **c.** $\log 20x > 2^{-x}$ **d.** $\log x \ge \log_4 x^2$

43. Oceanography After passing through a material t centimeters thick, the intensity $I(t)$ of a light beam is given by $I(t) = (4^{-ct})I_0$, where I_0 is the initial intensity and c is a constant called the absorption factor. Ocean water absorbs light with an absorption factor of $c = 0.0101$. At what depth will a beam of light be reduced to 50% of its initial intensity? 2% of its initial intensity? 49.5 cm; 279.4 cm

C **44.** Prove: $a^{\log b} = b^{\log a}$ **45.** Prove: $\dfrac{1}{\log_a ab} + \dfrac{1}{\log_b ab} = 1$

Cooperative Learning

Have students go over the proofs in Exercises 35, 36, 44, and 45 in groups. They should discuss each step and offer suggestions and alternate solutions.

**Additional Answers
Written Exercises**

35. Let $x = \log_c b$. Then $c^x = b$ and $c = b^{1/x}$. Thus $\log_b c = \dfrac{1}{x}$, and so $\log_b c = \dfrac{1}{\log_c b}$.

36. Let $x = \log_a b$ and $y = \log_b c$. Then $a^x = b$ and $b^y = c$. Substituting for b gives $(a^x)^y = c$. Thus $a^{xy} = c$ and $\log_a c = xy = (\log_a b)(\log_b c)$.

44. Let $x = a^{\log b}$. Then $\log x = \log (a^{\log b}) = \log b \log a$. Let $y = b^{\log a}$. Then $\log y = \log (b^{\log a}) = \log a \log b$. Since $\log x = \log y$, $x = y$. Thus $a^{\log b} = b^{\log a}$.

45. $\dfrac{1}{\log_a ab} + \dfrac{1}{\log_b ab} = $
$\dfrac{1}{\log_a a + \log_a b} + \dfrac{1}{\log_b a + \log_b b} = $
$\dfrac{1}{1 + \log_a b} + \dfrac{1}{\dfrac{1}{\log_a b} + 1} = $
$\dfrac{1}{1 + \log_a b} + \dfrac{\log_a b}{1 + \log_a b} = $
$\dfrac{1 + \log_a b}{1 + \log_a b} = 1.$

Chapter Summary

1. An *exponential function* has the form $f(x) = ab^x$, where $a > 0$, $b > 0$, and $b \neq 1$. Such functions are commonly used for calculating such things as the value of an investment deposited at a certain rate of compound interest, how much of a substance remains after a period of radioactive decay, or the size of a population growing at a certain rate.

2. The *laws of exponents* are given on page 170. They are used to define the zero exponent and negative integral exponents:

$$b^0 = 1$$

$$b^{-x} = \frac{1}{b^x}$$

Integral exponents are used to define rational exponents:

$$b^{p/q} = (\sqrt[q]{b})^p = \sqrt[q]{b^p},$$

where $b > 0$, p and q are integers, and $q \neq 0$.

3. The *rule of 72* provides an approximation of the doubling time for exponential growth. If a quantity is growing at $r\%$ per year, then

$$\text{doubling time} \approx 72 \div r.$$

4. The number e is defined as

$$\lim_{n \to \infty} \left(1 + \frac{1}{n}\right)^n.$$

The number e is the base for the *natural exponential function* and its inverse, the *natural logarithm function*.

5. The *logarithm* of x to the base b ($b > 0$, $b \neq 1$) is the exponent a such that $x = b^a$. Thus,

$$\log_b x = a \text{ if and only if } x = b^a.$$

Common logarithms, logarithms to base 10, are usually denoted $\log x$, while natural logarithms, those to base e, are usually denoted $\ln x$.

6. Since logarithms are exponents, the laws of logarithms, found on page 197, are closely related to the laws of exponents stated on page 170.

7. *Exponential equations* can be solved by writing both sides of the equation in terms of a common base (Section 5-2) or taking logarithms of both sides (Section 5-7).

8. The *change-of-base formula* enables you to write logarithms in any given base in terms of logarithms in any other base.

$$\log_b c = \frac{\log_a c}{\log_a b}$$

exponential function (p. 169, p. 181) natural exponential function (p. 187)
exponential growth and decay (p. 170) common logarithm (p. 191)
zero exponent (p. 171) logarithm to base b (p. 193)
negative exponent (p. 171) logarithmic function (p. 193)
rational exponent (p. 175) natural logarithm function (p. 193)
the rule of 72 (p. 182) exponential equation (p. 203)
the number e (p. 186) change-of-base formula (p. 205)

Chapter Test

1. Give a general expression for the value of a piece of property   5-1
   t years from now if its current value is \$150,000 and property values
   are increasing at the rate of 9% per year. $v(t) = 150{,}000(1.09)^t$

2. Evaluate the following.
   a. $\dfrac{2^5 \cdot 2^{-4}}{2^{-2}}$ 8   **b.** $(5^{-2} + 5^0)^{-1}$ $\dfrac{25}{26}$ **c.** $\sqrt{\dfrac{4^6}{2^{-4}}}$ 256   **d.** $\dfrac{3^{-3} + 9^{-2}}{3^{-3}}$ $\dfrac{4}{3}$

3. Solve each equation: **a.** $2^{6-x} = 4^{2+x}$ $\dfrac{2}{3}$ **b.** $3\sqrt{27} = 9^{2x}$ $\dfrac{5}{8}$ 5-2

4. A gallon of milk cost \$1.99 two years ago. Now it costs \$2.19. To the
   nearest percent, what has been the annual rate of increase in the cost? 5%

5. Graph $y = 2^x$ and $y = 2^{-x}$ on a single set of axes. How are the graphs   5-3
   related? Reflections in the y-axis

6. **Writing** Explain how $P(t) = P_0(1 + r)^t$ can be used to model exponential growth and decay. Give an example of each.

7. Suppose that \$1500 is invested at an interest rate of 8.5%. How much   5-4
   is the investment worth after 18 months if interest is compounded
   (a) quarterly? **(b)** continuously? \$1701.70; \$1703.98

8. Given that $\log 25 \approx 1.3979$, find the value of: 5-5
   a. $\log 2.5$ 0.3979   **b.** $\log 2500$ 3.3979   **c.** $\log 0.04$ -1.3979

9. Find the exact value of: **a.** $\log_2 8$ 3   **b.** $\log_8 2$ $\dfrac{1}{3}$ **c.** $2^{\log_4 64}$ 8

10. Express y as a function of x.   5-6
    a. $\log_2 y = 2 \log_2 (2x)$ $y = 4x^2$ **b.** $\log_2 y = 2 + \log_2 x$ $y = 4x$

11. Express each of the following in terms of $\log_b M$ and $\log_b N$.
    a. $\log_b \sqrt[3]{\dfrac{M^2}{N}}$ $\dfrac{2}{3}\log_b M - \dfrac{1}{3}\log_b N$ **b.** $\log_b M^2N^3$ $2\log_b M + 3\log_b N$

12. Solve $\log_5 x + \log_5 (x - 4) = 1$. 5

13. **a.** Between what two consecutive integers must $\log_5 21$ lie? 1 and 2   5-7
    b. Use the change-of-base formula to express $\log_5 21$ in terms of
    common logarithms and then evaluate it to the nearest thousandth. $\dfrac{\log 21}{\log 5} \approx 1.892$

14. Solve $5^x = 8$ for x to the nearest hundredth. $x \approx 1.29$

5.

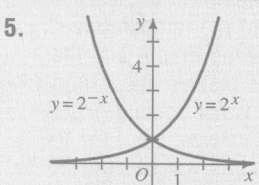

6. Answers will vary. If the
   initial amount P_0 is growing at a rate of r%, then
   the amount present at time
   t is $P(t) = P_0\left(1 + \dfrac{r}{100}\right)^t$. If
   P_0 is decaying at a rate of
   r%, then $P(t) =$
   $P_0\left(1 - \dfrac{r}{100}\right)^t$. For example,
   \$1000 invested at an an-
   nual rate of 5% will give
   $1000(1 + 0.05)^2$ dollars at
   the end of two years. A car
   worth \$10,000 depreciating
   at an annual rate of 7%
   will be worth
   $10{,}000(1 - 0.07)^2$ dollars
   at the end of two years.

Project Note

Many variables affect the shape of the temperature-versus-time graph, and therefore the mathematical model, for the water-cooling experiment. For example, how the water cools will be affected by the composition of the cup (that is, whether the cup is styrofoam, glass, ceramic, and so on), the surface area of the water, and whether the water has impurities. Thus, even if students all carefully control both the room temperature and the initial temperature of the water, they are still likely to obtain slightly different results because of the other uncontrolled variables. You might have students investigate and report on the effects of these other variables.

PROJECT

Newton's Law of Cooling

The best way to get a real feeling for mathematics is to see it in action. You've already seen functions that describe various processes. Did you ever wonder how you might develop a function to describe a process yourself?

Since this might be your first attempt, let's consider a relatively simple occurrence, such as cooling. Intuitively, you know that a cup of hot water set in a cool room will cool to room temperature. There must be some relationship between the temperature of the water over time and the temperature of the room.

In this project, you will see how to gather data for a cooling experiment, model the data, report your findings, and explore the experiment further.

Materials:

a standard lab thermometer   a graphing calculator or software
a cup of hot water   a refrigerator

Gather the Data.

■ Record the temperature, T_r, of the room.
■ Next, fill a cup with some hot tap water. Place the thermometer in the cup. (Make certain that the water is not too hot for the thermometer scale.)
■ Record the initial water temperature. Every five minutes thereafter, record the elapsed time t and the water temperature $T(t)$ until the temperature stops changing.
■ Using a software package or a programmable graphing calculator, graph the ordered pairs $(t, T(t))$. Typically, your plot should resemble the graph below.

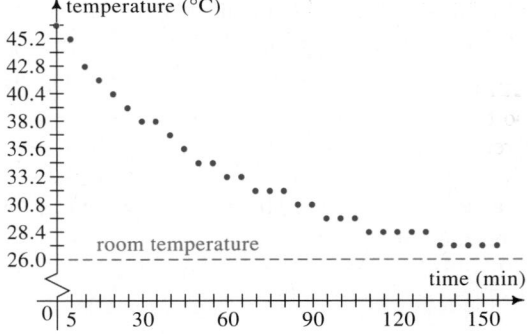

The graph probably confirms your expectations. The water temperature decreases, and the water cools more rapidly at first and then more slowly.

Model the Data.

In this part of the project, you will look for a function that reasonably represents the table of ordered pairs and the graph obtained. Which of the types of functions that you have studied so far might be a good model?

- Let's consider the exponential function $T(t) = Ce^{-kt}$, where C and k are positive constants. Why would there be a negative sign in front of k? (For a hint, see Exercise 14 in Section 5-4.)

    As t increases, what does Ce^{-kt} approach? (Use your calculator.) Now think about this. After a long period of time, what would you expect the temperature of the water to approach? If the temperature of the water approaches room temperature, then our exponential should approach T_r, or 26.0, instead of 0. Let's adjust our formula for $T(t)$ accordingly and write:

$$T(t) = Ce^{-kt} + T_r$$

    Notice that our function $T(t)$ is the exponential $Ce^{-k}t$ translated T_r units vertically (Section 4-4).

- Next, find the value of $T(0)$. What does this tell us about the value of C? More specifically, then:

$$T(t) = (T(0) - T_r)e^{-kt} + T_r$$

- How can we find a value for k? If the function fits all of the data, then any one of the ordered pairs should satisfy the function. For instance, we can use the ordered pair (30, 38.0) from the given data to find that $k \approx 0.02$. Then:

$$T(t) = 20.4e^{-0.02t} + 26.0$$

    Try it! Find a value of k and the corresponding function by using one of the other ordered pairs.

- How well do you think the function above will fit the ordered pairs? To find out how well, try graphing the data points and the function on a single set of axes. Because of the variation in measurements and conditions, chances are high that there will be places where the data points do not lie exactly on the graph of the function. The value of k, after all, is just an estimate. A more accurate method of fitting an exponential function to a set of data involves all of the data points. You will see the method in Chapter 18.

Report Your Results.

A good written report should be logically organized and clearly written. It should enable someone to reproduce the experiment successfully. Be sure to state conditions, such as room temperature and initial water temperature, a listing of your ordered pairs of data, and your graph. Specify your calculations for finding the values of C and k, and your resulting function for modeling the data. In your report, analyze how well your function seems to fit your data. Explain why you do or do not think that cooling can be described by an exponential function.

Extend the Project.

Let's go one step further. What do you think would happen if instead of letting the cup of hot water cool to room temperature, you placed it in the refrigerator? To get an accurate comparison, heat the water to the same initial temperature and record your data over the same time intervals. Graph both your new and old ordered pairs on a single set of axes. How has the colder environment affected the cooling process?

Chapter 6 Analytic Geometry

Overview

Analytic geometry was introduced in Chapter 1 with the study of the distance formula and the equations of lines. In this chapter, students use analytic methods to prove some familiar theorems from geometry and to find the equations of circles, ellipses, hyperbolas, and parabolas. After examining these conic sections separately, students find the points of intersection of pairs of conics by using both algebraic and geometric methods. Finally, students consider a common definition of the conics based on eccentricity, and they identify the graphs of second-degree equations using the coefficients in the equations.

Objectives

6-1 To prove theorems from geometry by using coordinates.

6-2 To find equations of circles and to find the coordinates of any points where circles and lines meet.

6-3 To find equations of ellipses and to graph them.

6-4 To find equations of hyperbolas and to graph them.

6-5 To find equations of parabolas and to graph them.

6-6 To solve systems of second-degree equations.

6-7 To define conic sections in terms of eccentricity and to classify the graph of a second-degree equation by examining the coefficients in the equation.

Supplementary Resources

Tests, pp. 13–15, 16–18
Alt. Assess., pp. 18–22, 65
Activities, pp. 12–14, 15–16
Using Tech., pp. 5, 21–22
Student Res. Guide, pp. 57–65

Software

Precalculus Plotter Plus
Activities, pp. 63, 64, 65
Conics Plotter, Conics Quiz,
    Function Plotter

Pacing Guide

Section	6-1	6-2	6-3	6-4	6-5	6-6	6-7	Review	Test	C. Rev.	Total
Standard (days)	2	2	2	2	2	2	1	1	1	2	17
Comprehensive (days)	2	2	2	2	1	1	1	1	1	2	15

Teaching Notes

6-1 | **pages 214–219**

Presenting the Section

You can demonstrate convenient placement of axes for several geometric figures by using an overhead transparency of a coordinate grid and precut clear geometric figures. Rotate a set of coordinate axes in various positions around the figures, explaining to the students why one placement would be easier to work with than another placement.

Cooperative Learning

Have students work in groups to go over the proofs in this section. Each member of the group can be assigned one of the following tasks:

1. Read the problem and identify the hypothesis and conclusion.
2. Make a coordinate diagram of the polygon in the problem.
3. Reword what is to be proved in algebraic terms.
4. Write the proof.
5. Discuss possible alternate methods of proof.

6-2 | **pages 219–225**

Presenting the Section

You might want to introduce this lesson by first reviewing the distance formula for the length of a segment. Draw a circle with $(0, 0)$ as its center and radius 4. Label the points of the circle that are obvious, for example, $(4, 0)$, $(-4, 0)$, $(0, 4)$, and $(0, -4)$. Then show students how using the distance formula with the coordinates of a point of the circle and the center yields the equation $x^2 + y^2 = 16$. Generalize this method using a circle with center $(0, 0)$ and radius r to derive the equation $x^2 + y^2 = r^2$.

Next, translate the circle with center $(0, 0)$ and radius 4 so that its center moves to $(3, -4)$. Emphasize that this circle is congruent to the first circle, but every point is shifted three units to the right and four units down. Use the distance formula again to show that the equation of this circle is

$$(x - 3)^2 + (y + 4)^2 = 16.$$

You can then generalize this method using an arbitrary circle with center (h, k) and radius r to derive the equation $(x - h)^2 + (y - k)^2 = r^2$.

Using Technology

Demonstrate to students that if the equation of a circle is rewritten as two equations, then a graphing calculator can graph the two equations on the same set of axes. Point out that graphs may not be perfect circles because the screens are not square. Adjust the range parameters as necessary. Another problem that can arise is that some points of the circle may appear to be missing because the width of each pixel of the screen is "wider than a point." Several successive points may not be graphed because they are in a direct vertical line with this "wide" pixel. Ask students to use their imaginations when viewing these graphs.

Cooperative Learning

You might want students to work in groups to write an algorithm for solving Written Exercises 50–53 on page 225. If the tasks are shared, have each student present and teach his or her results to the other members of the group. There might be some students who are interested in translating the algorithm into a computer program that the class can use.

6-3 | **pages 225–231**

Presenting the Section

You might want to introduce this section by continuing the discussion on circles from Section 6-2. Use a circle made with string or a rubber band stretched on a geoboard. Show how stretching the band can make the circle look like an ellipse. Designate the major axis and minor axis for the ellipse on the geoboard, and then measure them. When the lengths of the major and minor axes are equal, the ellipse is a circle. Discuss how stretching the circle can be described algebraically, which is the objective of this section. You might also want to get students involved by having them make their own models. They can use thumbtacks and string to make homemade models. Have them experiment to see that when the foci and center become the same point, the ellipse becomes a circle.

Using Technology

There are numerous software packages available (for example, *Precalculus Plotter Plus, Graph Wiz, Master Grapher*) that allow students to enter the equation of a conic and have it graphed. Some of these packages allow students to experiment with the graph before printing it by adjusting the range parameters to see a larger (or smaller) version of the graph and by tracing and zooming in on important points such as the vertices.

Applications

The orbits of the planets and of some comets are elliptical. For example, the orbit of Halley's comet is an ellipse with minor axis approximately 9.1 astronomical units (AU) and with major axis about 36.2 AU. (An astronomical unit is the distance between the Earth and the sun, which is about 93 million miles.) Have students try to graph this orbit with the sun at one focus so that they can see the relative distance traveled by the comet. The time it takes to complete one orbit, which is its period, is 76 years. (By comparison, the period of the comet Encke is 3.3 years.)

Cooperative Learning

The following research project may be fun for your students. Divide your students into groups, and assign to each group a comet to research. Each group should gather statistics about that comet (including its period), try to determine the length of the major and minor axis of the elliptical path of the comet, graph the path of the comet to show an ellipse (which can be done on poster board), and discuss their findings with the rest of the class.

6-4 **pages 231–237**

Presenting the Section

You should relate the derivation of the equation of a hyperbola to the derivation of the equation of an ellipse, shown on page 226. To shorten the work when deriving the equation of a hyperbola, you could make substitutions where appropriate in the derivation given on page 226. Emphasize that the derivation makes use of the distance formula.

When showing students how to graph a hyperbola, emphasize the following steps:

1. Graph the asymptotes.
2. Draw a rectangle with dimensions $2a$ and $2b$.
3. Plot the vertices.
4. Sketch the curve through the vertices.

Cooperative Learning

Have students work in groups to do the following activity. Ask students to draw two points 9 units apart and label the points A and B. Then have students use a compass to draw 14 concentric circles with center A, increasing the radius of each subsequent circle by 1 unit. Repeat these steps using point B as the center of the concentric circles. (See the figure below.) Tell students they have created "conic graph paper." Then ask students to plot all points that satisfy this condition: the difference in the distance between the point and A and the distance between the point and B is 7 units. For example, students should plot the point where the second circle with center A intersects the ninth circle with center B. When all possible points have been plotted, the students can draw a hyperbola. The students can use this conic graph paper to draw other hyperbolas.

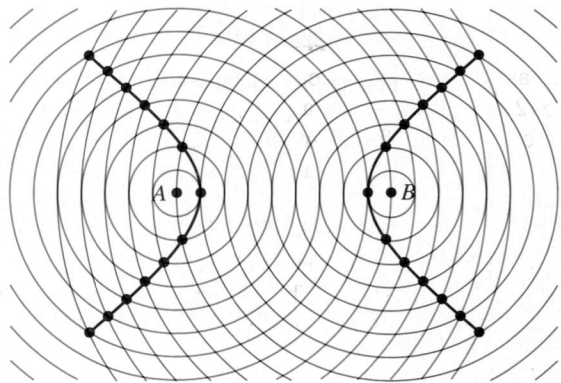

Applications

The reflection property of the hyperbola can be found in the construction of a reflecting telescope that has a larger parabolic mirror and a smaller hyperbolic mirror. Both mirrors share a common focus. An example of this type of telescope is the 200-inch telescope on Mount Palomar in California.

The LORAN (Long Range Navigational) system used to locate ships at sea is based on the intersection of three hyperbolas.

6-5 | pages 238–241

Presenting the Section

You might want to review the discussion of parabolas given in Section 1-7. Then tell students that parabolas can be defined another way and show how the equation is derived. Point out that all parabolas defined by equations (1)–(4) on page 238 have the same basic shape. Shifting the parabola around the coordinate plane just shows a different orientation or location of it.

Using Technology

Encourage students to use graphing calculators or available software packages to graph parabolas. When using graphing calculators, students should remember that most calculators require functions. Tell students that to graph the equation $x = \frac{1}{4p}y^2$ they need to enter the equations $y = \sqrt{4px}$ and $y = -\sqrt{4px}$.

Applications

It has been known since the time of Galileo (1564–1642) that the path of a projectile is a parabola. Parabolas can be found in many places. Communication systems such as satellite dishes, electronic surveillance systems, and telescopes all use parabolic reflectors. Parabolic reflectors can be used to receive parallel light rays or sound waves, which are then reflected to the focus. The rays or waves, which are concentrated at the focus, are then recorded.

6-6 | pages 242–246

Presenting the Section

You might want to introduce systems of second-degree equations with chalkboard drawings of the different number of geometric solutions possible. You can use the following drawings as examples of 0, 1, 2, 3, 4, or infinitely many solutions.

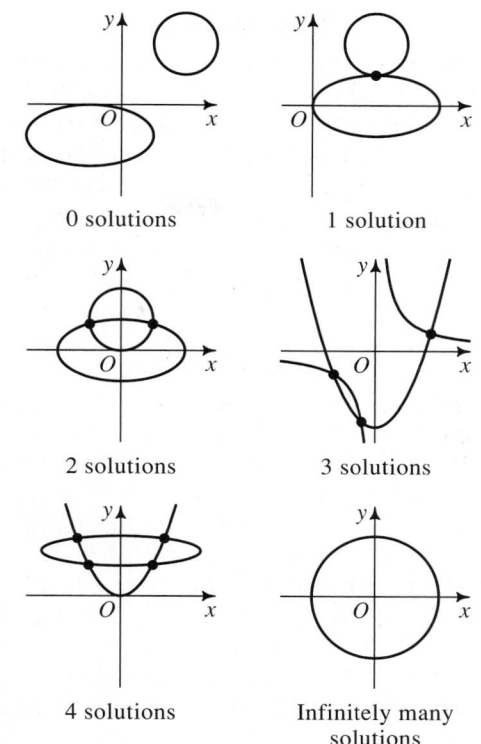

0 solutions

1 solution

2 solutions

3 solutions

4 solutions

Infinitely many solutions

Point out how the algebraic, geometric, and technological methods support each other in this section. For example, after using an algebraic method, solutions can be checked graphically by using either technology or graph paper.

Using Technology

If students use a graphing calculator or a computer to solve systems of second-degree equations, caution them to zoom in as close as possible to the intersection points, because two graphs may appear to intersect when, in fact, they do not.

6-7 | pages 247–251

Presenting the Section

Begin by listing the different conics. Ask your students to compare the conics and suggest what properties these conics might have in common. Then discuss with students how the eccentricity of conic sections gives a common definition for conics.

6 *Analytic Geometry*

212

Introduction

Analytic geometry is the study of geometric problems by means of analytic (or algebraic) methods. You saw some analytic geometry when you studied the distance formula, midpoint formula, and equations of lines in Sections 1-1 to 1-3. You saw more analytic geometry in Section 1-7 when you studied quadratic functions and their graphs, parabolas.

Imagine that the double cones shown below are extended indefinitely up and down. When these cones are sliced by a plane tilted at various angles, the resulting cross sections are called **conic sections**. As shown below, a circle, an ellipse, a hyperbola, and a parabola are conic sections.

Problem Solving

Ask your students if there are any other shapes other than a double cone such that a circle or an ellipse is formed when the shape is sliced by a plane. (Answers may vary. For example, a cylinder sliced by a plane at certain angles.)

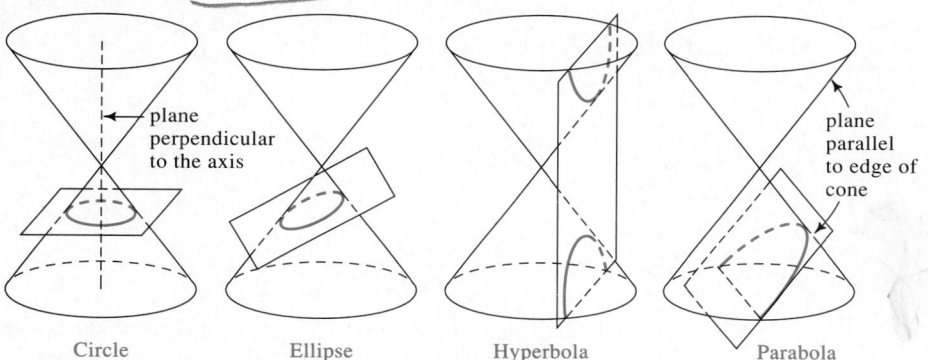

| Circle | Ellipse | Hyperbola | Parabola |

plane perpendicular to the axis

plane parallel to edge of cone

It is also possible to slice the double cone to obtain a single point, a line, or a pair of lines. Do you see how? These extreme cases are called *degenerate* conic sections.

Using analytic geometry, we can find equations for these curves. Each equation is a special case of the **general second-degree equation** in two variables:

$$Ax^2 + Bxy + Cy^2 + Dx + Ey + F = 0$$

For example,

$$x^2 + y^2 = 9$$
$$x^2 + 4y^2 = 9$$
$$x^2 - y^2 = 9$$
$$\frac{1}{4}x^2 - y = 0$$

are equations of a *circle*, an *ellipse*, a *hyperbola*, and a *parabola*, respectively.

Before we study the conic sections, we will show how to use analytic geometry to prove theorems.

◀ In this Cassegrain reflecting telescope at Arizona's Kitt Peak Observatory, a large parabolic mirror reflects light to its focus, which is also one of the foci of a smaller hyperbolic mirror. The mirror reflects the light to the eyepiece, located at the second hyperbolic focus.

Analytic Geometry **213**

Warm-Up Exercises

A triangle has vertices $R(a, b)$, $S(a + c, b)$ and $T(a + c, b + c)$, where a, b, and c represent positive numbers.

1. Show that $\overline{RS} \perp \overline{ST}$.
$\overline{RS}$ has slope $\frac{b - b}{(a + c) - a} =$ 0, so $\overline{RS}$ is a horizontal segment. S and T have the same x-coordinate, so $\overline{ST}$ is a vertical segment. Hence, $\overline{RS} \perp \overline{ST}$.

2. What kind of triangle is $\triangle RST$?
Since $\overline{RS} \perp \overline{ST}$, $\triangle RST$ is a right triangle.

3. Find RS, ST, and RT. What do you notice?
$RS = c$, $ST = c$, and $RT = \sqrt{2c^2} = c\sqrt{2}$. Since $RS = ST$, $\triangle RST$ is isosceles.

4. Use your answers from Exercises 2 and 3 to find the measure of $\angle RTS$.
Since $\triangle RST$ is an isosceles right triangle, $m\angle RTS = m\angle TRS = 45°$.

Motivating the Section

Some theorems from geometry are easier to prove by using coordinate methods. This section demonstrates coordinate proofs.

6-1 Coordinate Proofs

Objective   *To prove theorems from geometry by using coordinates.*

In this section, you will see how to use analytic methods to prove theorems from geometry.

If a theorem is about a right triangle, there are several ways that we can place the coordinate axes on the triangle and then assign coordinates to the vertices of the triangle. For example, figures (1) and (2) below illustrate a right triangle with legs a and b units long. Most people prefer to work with figure (2). Because more of the coordinates are zero, the work is easier. Figure (3) shows that even if the right triangle is "tilted," we can choose the coordinate axes in such a way that the vertices of the triangle have several zero coordinates.

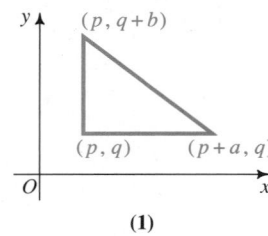
(1)

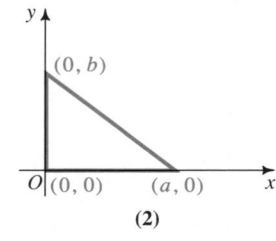
(2)

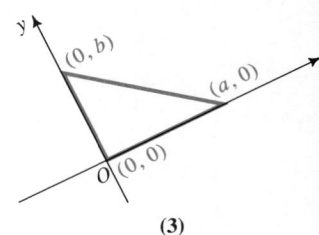
(3)

Similarly, if we wish to prove a theorem about a trapezoid or a parallelogram, we can always choose the axes in such a way that one of the vertices of the figure is at the origin and one of its parallel sides lies on the x-axis.

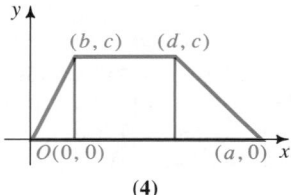
(4)

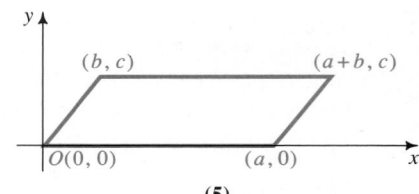
(5)

Example 1   Prove that the midpoint of the hypotenuse of a right triangle is equidistant from the three vertices.

Solution   **Step 1**   *First, we make a coordinate diagram of the triangle and note what we are given and what we must prove.*

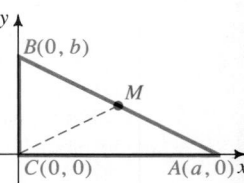

    Given: $\angle C$ is a right angle.
            M is the midpoint of $\overline{AB}$.
    Prove: $MC = MA$
            (We already know that $MB = MA$.)

Step 2 *Next, we use what is given to add information to the diagram or to express algebraically any given fact not shown in the original diagram.* In this example, we use the given fact that M is the midpoint of $\overline{AB}$ to find the coordinates of M.

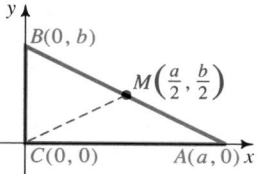

Step 3 *Finally, we reword what we are trying to prove in algebraic terms.* To prove $MC = MA$:

$$MC = \sqrt{\left(\frac{a}{2} - 0\right)^2 + \left(\frac{b}{2} - 0\right)^2} = \sqrt{\left(\frac{a}{2}\right)^2 + \left(\frac{b}{2}\right)^2}$$

$$= \sqrt{\frac{a^2}{4} + \frac{b^2}{4}}$$

$$MA = \sqrt{\left(\frac{a}{2} - a\right)^2 + \left(\frac{b}{2} - 0\right)^2} = \sqrt{\left(-\frac{a}{2}\right)^2 + \left(\frac{b}{2}\right)^2}$$

$$= \sqrt{\frac{a^2}{4} + \frac{b^2}{4}}$$

Therefore, $MC = MA$. Since $MA = MB$, we have that $MA = MB = MC$.

Example 2 Prove that the median of a trapezoid is parallel to the bases and has length equal to the average of the lengths of the bases.

Solution **Step 1** *We show a diagram, and the "Given" and "Prove."* Place the x-axis along the longer base of the trapezoid, with the origin at the endpoint of the longer base. Since the bases of a trapezoid are parallel and base $\overline{OP}$ has been chosen to be horizontal, base $\overline{RQ}$ is also horizontal. Thus R and Q have the same y-coordinate.

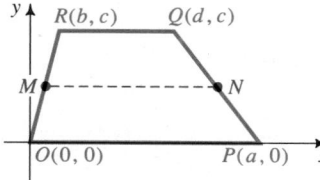

Given: Figure $OPQR$ is a trapezoid. Points M and N are midpoints of $\overline{OR}$ and $\overline{PQ}$, respectively.

Prove: (1) $\overline{MN} \parallel \overline{OP}$ and (2) $MN = \dfrac{OP + RQ}{2}$.

(Solution continues on the next page.)

Analytic Geometry **215**

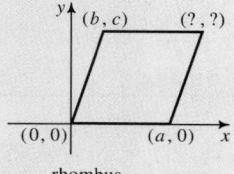

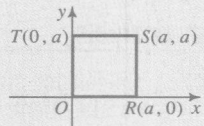
Step 2 *Next, we use what is given to add information to the diagram. In this example, we use the fact that M and N are midpoints to find their coordinates.*

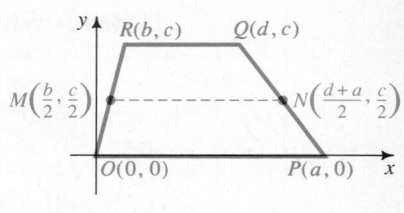

Step 3 *We reword what we are to prove in algebraic terms.*

(1) To prove $\overline{MN} \parallel \overline{OP}$, we must show that $\overline{MN}$ and $\overline{OP}$ have the same slope. A quick check shows that both slopes are zero, so this part of the proof is done.

(2) Lastly, we use algebra to show that $MN = \frac{1}{2}(OP + RQ)$.

$$MN = \frac{d+a}{2} - \frac{b}{2}$$

$$\frac{1}{2}(OP + RQ) = \frac{1}{2}(a + (d-b)) = \frac{d+a}{2} - \frac{b}{2}$$

Therefore, $MN = \frac{1}{2}(OP + RQ)$.

Example 3 Prove that the altitudes of a triangle meet in one point, that is, they are *concurrent.*

Solution **Step 1** *We show a diagram and the "Given" and "Prove."*

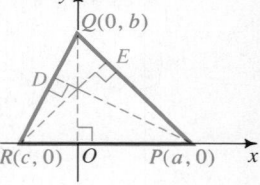

Given: $\triangle PQR$ with altitudes $\overline{PD}$, $\overline{QO}$, and $\overline{RE}$.

Prove: Lines PD, QO, and RE have a point in common. Notice that the axes are placed in such a way that one of the altitudes lies on the y-axis.

Step 2 *We use the given information* to express algebraically the fact that $\overline{PD}$, $\overline{QO}$, and $\overline{RE}$ are altitudes.

a. To find the slope of line PD, we note that the slope of line QR is $\dfrac{b-0}{0-c} = -\dfrac{b}{c}$, so that the slope of line PD is $\dfrac{c}{b}$.

Since line PD contains the point $(a, 0)$, its equation is

$$\frac{y-0}{x-a} = \frac{c}{b}, \text{ or } cx - by = ca.$$

b. Likewise, an equation of line RE is

$$\frac{y-0}{x-c} = \frac{a}{b}, \text{ or } ax - by = ca.$$

c. An equation of the vertical line QO is $x = 0$.

Step 3 *We reword what we are to prove in algebraic terms.* To prove that lines *PD*, *QO*, and *RE* have a point in common, we must show that their equations have a common solution. Using subtraction to solve

$$cx - by = ca$$
$$ax - by = ca,$$

we get $$cx - ax = 0.$$

Therefore, $x(c - a) = 0$. Since $c \neq a$, $x = 0$. Substituting 0 for x in the equation $cx - by = ca$, we get $y = -\frac{ca}{b}$. Thus, the lines *PD* and *RE* intersect at $\left(0, -\frac{ca}{b}\right)$, a point on the y-axis, that is, on altitude *QO*, so we are done. (The point of concurrency of the altitudes is called the *orthocenter* of the triangle.)

Summary of Methods Commonly Used in Coordinate Proofs

1. To prove line segments equal, use the distance formula to show that they have the same length.

2. To prove nonvertical lines parallel, show that they have the same slope.

3. To prove lines perpendicular, show that the product of their slopes is -1.

4. To prove that two line segments bisect each other, use the midpoint formula to show that each segment has the same midpoint.

5. To show that lines are concurrent, show that their equations have a common solution.

CLASS EXERCISES

1. Study the coordinates of the vertices in the following diagrams and tell which figures represent isosceles triangles. a, c

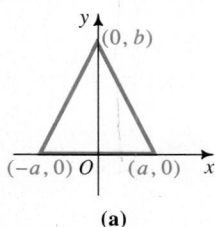

(a)

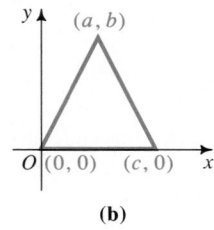

(b)

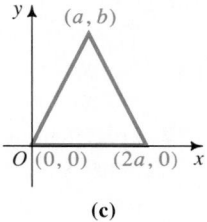

(c)

Additional Examples cont.

2. Use figure (a) of Class Exercise 1 to prove that the line segments joining the midpoints of successive sides of any isosceles triangle form an isosceles triangle.

Let $L = (-a, 0)$, $M = (a, 0)$, and $N = (0, b)$. Then $\overline{LM}$ has midpoint $O(0, 0)$, $\overline{MN}$ has midpoint $P\left(\frac{a}{2}, \frac{b}{2}\right)$, and $\overline{LN}$ has midpoint $Q\left(-\frac{a}{2}, \frac{b}{2}\right)$.

$OP = \sqrt{\left(\frac{a}{2}\right)^2 + \left(\frac{b}{2}\right)^2} = \frac{1}{2}\sqrt{a^2 + b^2}$ and $OQ = \sqrt{\left(-\frac{a}{2}\right)^2 + \left(\frac{b}{2}\right)^2} = \frac{1}{2}\sqrt{a^2 + b^2}$. Thus, $OP = OQ$ and $\triangle OPQ$ is isosceles.

Suggested Assignments

Standard
Day 1: 218/1–9 odd
Day 2: 218/11, 15–17
Comprehensive
Day 1: 218/1–13 odd
Day 2: 219/14–18

Supplementary Materials

Alternative Assessment, 18

Review Note

You might want to review altitudes, medians, diagonals, midpoints, and the distance formula before assigning these exercises.

2. Which of the following diagrams represent parallelograms? a, b, c

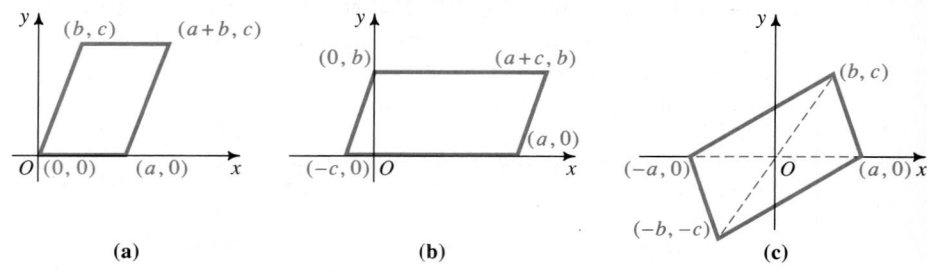

(a)   (b)   (c)

WRITTEN EXERCISES

In these exercises, do not use any theorems from geometry other than the Pythagorean Theorem. You may use the formulas and theorems in Sections 1-1 and 1-2, and results proved in earlier exercises of this section.

A **1.** Use figure (a) or (c) of Class Exercise 1 to prove that the medians to the legs of an isosceles triangle are equal in length.

2. Prove that if a triangle has two congruent medians, then it is isosceles.

3. Prove that the line segment joining the midpoints of two sides of a triangle is parallel to the third side and has length half that of the third side.

4. Use either figure (a) or (b) of Class Exercise 2 to prove that the diagonals of a parallelogram bisect each other.

5. Prove that if the diagonals of a quadrilateral bisect each other, then the quadrilateral is a parallelogram. How does this result show that any parallelogram can be represented by figure (c) of Class Exercise 2?

6. Prove that the lengths of the diagonals of a rectangle are equal.

7. Prove that the lengths of the diagonals of an isosceles trapezoid are equal.

8. Prove that if the diagonals of a trapezoid are congruent, then the trapezoid is isosceles.

9. Prove that the line segments joining the midpoints of successive sides of any quadrilateral form a parallelogram.

10. Prove that the line segments joining the midpoints of successive sides of any rectangle form a rhombus.

B **11.** Prove that if the diagonals of a parallelogram are perpendicular, then the parallelogram is a rhombus.

12. Prove that the diagonals of a rhombus are perpendicular.

13. Suppose that Q is any point in the plane of rectangle $RSTU$. Prove that $(QR)^2 + (QT)^2 = (QS)^2 + (QU)^2$.

14. Use the diagram at the right to prove that P is on the perpendicular bisector of $\overline{AB}$ if $PA = PB$.

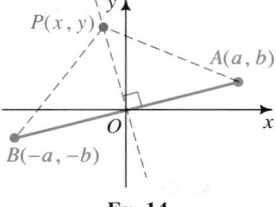

Ex. 14

15. **a.** Use a figure similar to the one in Example 3 to prove that the medians of $\triangle PQR$ meet in a point G. (This point is called the *centroid* of the triangle.)
    b. Show that the x-coordinate of the centroid is the average of the x-coordinates of P, Q, and R, and that the y-coordinate of the centroid is the average of the y-coordinates of P, Q, and R.

16. Using Exercise 15, prove that the centroid G divides each median in a 2:1 ratio.

17. **a.** Use a figure similar to the one in Example 3 to prove that the perpendicular bisectors of the sides of $\triangle PQR$ meet in a point C. This point is called the *circumcenter* of the triangle.
    b. Prove that C is equidistant from P, Q, and R.

C 18. Let G be the intersection point of the medians of a triangle, let C be the intersection of the perpendicular bisectors of the sides, and let H be the intersection of the altitudes. Prove that G, C, and H are collinear and that $GH = 2GC$. (*Hint:* Use the results of Example 3 and Exercises 15 and 17.)

6-2 Equations of Circles

Objective To find equations of circles and to find the coordinates of any points where circles and lines meet.

The set of all points $P(x, y)$ in the plane that are 5 units from the point $C(2, 4)$ is a *circle*. To find an equation of this circle, we use the distance formula:

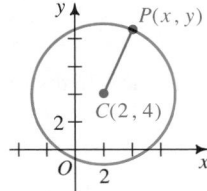

$$PC = 5$$
$$\sqrt{(x - 2)^2 + (y - 4)^2} = 5$$
$$(x - 2)^2 + (y - 4)^2 = 25$$

In general, if $P(x, y)$ is on the circle with center $C(h, k)$ and radius r, then:

$$PC = r$$
$$\sqrt{(x - h)^2 + (y - k)^2} = r$$
$$(x - h)^2 + (y - k)^2 = r^2$$

The steps leading to the last equation can be reversed to show that any point $P(x, y)$ satisfying the equation is on the circle with center (h, k) and radius r.

Making Connections

Exercises 15–18 illustrate to students the strong connection between algebra and geometry.

Teaching Notes, p. 212B

Warm-Up Exercises

1. Complete each square.
   a. $x^2 + 18x +$ _?_ 81
   b. $y^2 - 7y +$ _?_ $\frac{49}{4}$

2. Describe a general method for completing the square if the coefficient of the quadratic term is 1.
   Add the square of half the coefficient of the linear term.

3. Solve the system.
   $y = 2x$
   $x^2 + y^2 = 20$
   $(2, 4), (-2, -4)$

4. Describe a general method for solving the system.
   $y = mx + b$
   $x^2 + y^2 = r^2$
   Substitute $mx + b$ for y in the second equation and then solve for x.

Motivating the Section

Tell students that circles, the topic of this section, are used often in many areas of construction. The tunnels from England to France under the English Channel have circular cross-sections with a diameter of 25 feet. Ask students to give examples of uses of circles they have observed.

If the center of the circle is the origin, the equation becomes:

$$x^2 + y^2 = r^2$$

For brevity, we often refer to a circle with equation $x^2 + y^2 = r^2$, for example, as "the circle $x^2 + y^2 = r^2$."

Example 1   Find the center and radius of each circle.
a. $(x - 3)^2 + (y + 7)^2 = 19$
b. $x^2 + y^2 - 6x + 4y - 12 = 0$

Solution   a. Since $(x - 3)^2 + (y + 7)^2 = (x - 3)^2 + (y - (-7))^2$, the center of the circle is $(3, -7)$. The radius is $\sqrt{19}$.

b. We rewrite the equation in *center-radius form* by completing the squares in x and y.

Original equation: $x^2 + y^2 - 6x + 4y - 12 = 0$

$$(x^2 - 6x \quad) + (y^2 + 4y \quad) = 12$$
$$(x^2 - 6x + 9) + (y^2 + 4y + 4) = 12 + 9 + 4$$

Center-radius form: $\quad (x - 3)^2 + (y + 2)^2 = 25$

Thus, the center of the circle is $(3, -2)$ and the radius is 5.

 In Example 2, you will see how to use a graphing calculator (or computer) to graph an equation that represents a circle.

Example 2   Use a graphing calculator to graph the equation in Example 1(b).

Solution
$$x^2 + y^2 - 6x + 4y - 12 = 0$$
$$(x - 3)^2 + (y + 2)^2 = 25 \longleftarrow \text{center-radius form}$$

Solve for y in terms of x.
$$(y + 2)^2 = 25 - (x - 3)^2$$
$$y_1 = -2 + \sqrt{25 - (x - 3)^2} \quad (1)$$
$$y_2 = -2 - \sqrt{25 - (x - 3)^2} \quad (2)$$

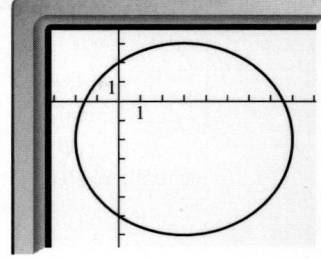

The figure at the right was obtained by graphing each of the *two* equations above on the same set of axes. Equation (1) represents the top half of the circle; equation (2) represents the bottom half.

Notice that the graph does not appear to be a perfect circle. You should consult your calculator manual to determine how to adjust the display to make the graph appear as a circle.

Example 3 Find the coordinates of the points where the line $y = 2x - 2$ and the circle $x^2 + y^2 = 25$ intersect.

Solution **Step 1** To find the coordinates of their intersection points, A and B, we solve these two equations simultaneously:

$$y = 2x - 2 \quad (1)$$
$$x^2 + y^2 = 25 \quad (2)$$

Step 2 Substituting for y in equation (2), we get:

$$x^2 + (2x - 2)^2 = 25$$
$$x^2 + 4x^2 - 8x + 4 = 25$$
$$5x^2 - 8x - 21 = 0$$
$$(5x + 7)(x - 3) = 0$$

Therefore, $x = -\dfrac{7}{5}$ or $x = 3$.

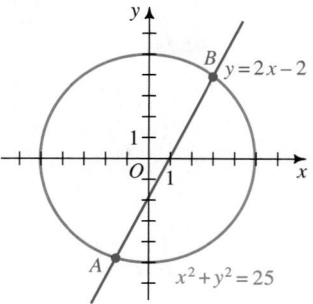

Step 3 Substituting these values for x in equation (1), we get:

$$y = 2\left(-\frac{7}{5}\right) - 2 = -\frac{24}{5} \quad \text{and} \quad y = 2(3) - 2 = 4$$

Thus, $A = \left(-\dfrac{7}{5}, -\dfrac{24}{5}\right)$ and $B = (3, 4)$. Check this result by substituting the coordinates of A and B in equations (1) and (2).

The solution to Example 3 involves an algebraic method, which is summarized below. You can also find the solution by simultaneously graphing the line and the circle with a graphing calculator or computer software.

> ## To Find the Intersection of a Line and a Circle Algebraically:
>
> 1. Solve the linear equation for y in terms of x (or x in terms of y).
> 2. Substitute this expression for y (or x) in the equation of the circle. Then solve the resulting quadratic equation.
> 3. Substitute each real x-solution from Step 2 in the *linear* equation to get the corresponding value of y (or vice versa). Each point (x, y) is an intersection point.
> 4. You can check your result by substituting the coordinates of the intersection points in the two original equations.

When a line and a circle do not intersect, the quadratic equation in Step 2 will have a negative discriminant and the equation will have only imaginary roots. If the discriminant is zero, then there is only one real root. This indicates that the line intersects the circle in a single point and is tangent to the circle.

Additional Examples

1. Find an equation of the circle that passes through $(-7, 4)$ and has y-intercepts 11 and -13.

   Substitute $(0, 11)$ and $(0, -13)$ in the equation $(x - h)^2 + (y - k)^2 = r^2$:
   $h^2 + (11 - k)^2 = r^2$
   $h^2 + (-13 - k)^2 = r^2$
   Subtracting gives
   $(121 - 22k + k^2) - (169 + 26k + k^2) = 0;$
   $-48 - 48k = 0;\ k = -1.$
   Substitute $(-7, 4)$ and $(0, 11)$ in the equation $(x - h)^2 + (y + 1)^2 = r^2$:
   $(-7 - h)^2 + (4 + 1)^2 = r^2$
   $(0 - h)^2 + (11 + 1)^2 = r^2$
   Subtracting gives
   $(-7 - h)^2 - h^2 + 25 - 144 = 0;\ -70 + 14h = 0;$
   $h = 5.$
   Substitute $(0, 11)$ in $(x - 5)^2 + (y + 1)^2 = r^2$:
   $(0 - 5)^2 + (11 + 1)^2 = r^2;$
   $169 = r^2.$ Equation:
   $(x - 5)^2 + (y + 1)^2 = 169.$

2. Find an equation of the line tangent to the circle $x^2 + y^2 + 4x - 46 = 0$ at $P(3, 5)$.

   Write the equation in center-radius form:
   $(x^2 + 4x + ?) + y^2 = 46$
   $(x^2 + 4x + 4) + y^2 = 50$
   $(x + 2)^2 + y^2 = 50$
   P is on the circle because $(3 + 2)^2 + 5^2 = 50.$

   (*Continues on next page*)

The circle has center $O(-2, 0)$. Radius $\overline{OP}$ has slope $\frac{5 - 0}{3 - (-2)} = 1$. Since the tangent is $\perp$ to $\overline{OP}$, the tangent has slope -1. The tangent line contains $P(3, 5)$. Thus, it has equation $y - 5 = -1(x - 3)$, or $x + y = 8$.

Additional Answers
Class Exercises

10. a. All points less than 5 units from the origin.
b. All points more than 5 units from the origin.

11. Solve $3y + x = 6$ for x in terms of y. Substitute this expression for x in the equation $x^2 + y^2 = 10$. Solve the resulting quadratic equation. Substitute each solution for y in the linear equation to get the corresponding value of x.

13. Answers may vary. Alter the scale, condensing the x-axis (or stretching the y-axis) to compensate for the distortion.

Suggested Assignments

Standard
Day 1: 222/1–35 odd
Day 2: 223/37, 41–53 odd, 48
Comprehensive
Day 1: 222/1–35 odd
Day 2: 223/37–59 odd, 48

Supplementary Materials

Alternative Assessment, 19
Using Technology, 5

CLASS EXERCISES

Find the center and radius of each circle whose equation is given.

1. $x^2 + y^2 = 16$ $C(0, 0)$; $r = 4$
2. $(x - 2)^2 + (y - 7)^2 = 36$ $C(2, 7)$; $r = 6$
3. $(x - 4)^2 + (y + 7)^2 = 7$ $C(4, -7)$; $r = \sqrt{7}$
4. $x^2 + y^2 + 12y = 0$ $C(0, -6)$; $r = 6$
5. $x^2 - 2x + y^2 - 6y = 9$ $C(1, 3)$; $r = \sqrt{19}$
6. $4x^2 + 4y^2 = 36$ $C(0, 0)$; $r = 3$

In Exercises 7–9, find an equation of the circle described.

7. The circle with center $(7, 3)$ and radius 6. $(x - 7)^2 + (y - 3)^2 = 36$
8. The circle with center $(-5, 4)$ and radius $\sqrt{2}$. $(x + 5)^2 + (y - 4)^2 = 2$
9. The circle with center $(0, 0)$ that passes through $(-5, 12)$. $x^2 + y^2 = 169$
10. The graph of $x^2 + y^2 = 25$ consists of all points in the plane that are 5 units from the origin. Describe the graphs of:
a. $x^2 + y^2 < 25$ **b.** $x^2 + y^2 > 25$
11. *Discussion* Suppose that you wish to find where the line $3y + x = 6$ intersects the circle $x^2 + y^2 = 10$. Describe what you would do.
12. *Discussion* To find the intersection of the line $y = x + 8$ and the circle $x^2 + y^2 = 16$, Janice solved the equations simultaneously and found that $x = -4 \pm 2i\sqrt{2}$. What do the imaginary roots tell her? See below.
13. *Discussion* Describe how to use a graphing calculator to graph $x^2 + y^2 = 9$ so that the circle does not appear distorted or flattened.
12. The line and the circle do not intersect.

WRITTEN EXERCISES

2. $(x - 5)^2 + (y + 6)^2 = 49$ **4.** $(x - a)^2 + (y - b)^2 = f^2$
In Exercises 1–12, write an equation of the circle described. **6.** $(x + 4)^2 + (y - 2)^2 = 7$

A
1. $C(4, 3)$, $r = 2$ $(x - 4)^2 + (y - 3)^2 = 4$ **2.** $C(5, -6)$, $r = 7$ See above.
3. $C(-4, -9)$, $r = 3$ $(x + 4)^2 + (y + 9)^2 = 9$ **4.** $C(a, b)$, $r = f$
5. $C(6, 0)$, $r = \sqrt{15}$ $(x - 6)^2 + y^2 = 15$ **6.** $C(-4, 2)$, $r = \sqrt{7}$
7. The center is $(2, 3)$; the circle passes through $(5, 6)$. $(x - 2)^2 + (y - 3)^2 = 18$
8. The points $(8, 0)$ and $(0, 6)$ are endpoints of a diameter. $(x - 4)^2 + (y - 3)^2 = 25$
9. The center is $(5, -4)$ and the circle is tangent to the x-axis. $(x - 5)^2 + (y + 4)^2 = 16$
10. The center is $(-3, 1)$ and the circle is tangent to the line $x = 4$. See below.
11. The circle is tangent to the x-axis at $(4, 0)$ and has y-intercepts -2 and -8. See below.
12. The circle contains $(-2, 16)$ and has x-intercepts -2 and -32. See below.
10. $(x + 3)^2 + (y - 1)^2 = 49$ **11.** $(x - 4)^2 + (y + 5)^2 = 25$ **12.** $(x + 17)^2 + (y - 8)^2 = 289$
Write each equation in center-radius form. Give the center and radius.

13. $x^2 + y^2 - 2x - 8y + 16 = 0$ **14.** $x^2 + y^2 - 4x + 6y + 4 = 0$
15. $x^2 + y^2 - 12y + 25 = 0$ **16.** $x^2 + y^2 + 14x = 0$
17. $2x^2 + 2y^2 - 10x - 18y = 1$ **18.** $2x^2 + 2y^2 - 5x + y = 0$

19. Show that the line $y = 2x + 8$ contains the center of the circle $x^2 + y^2 + 6x - 4y + 8 = 0$. $C(-3, 2)$; $2 = 2(-3) + 8$

20. Determine whether the line $3x + 2y = 6$ contains the center of the circle $x^2 + y^2 + 4x - 12y + 24 = 0$. Yes, $3x + 2y = 6$ contains $C(-2, 6)$.

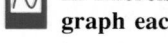 **In Exercises 21–23, use a graphing calculator or computer software to graph each equation.**

21. $x^2 + y^2 = 50$ 22. $(x - 3)^2 + y^2 = 36$ 23. $x^2 + y^2 - 6y = 40$

In Exercises 24–26, on a single set of axes, sketch the graph of each semicircle whose equation is given.

24. **a.** $y = \sqrt{9 - x^2}$ **b.** $y = -\sqrt{9 - x^2}$

25. **a.** $x = \sqrt{9 - y^2}$ **b.** $x = -\sqrt{9 - y^2}$

26. **a.** $y = \sqrt{16 - (x - 5)^2}$ **b.** $x = 5 - \sqrt{16 - y^2}$

In Exercises 27–34, graph the equations. Solve the equations simultaneously to find the coordinates of any intersection points of their graphs. If the graphs are tangent or fail to intersect, say so.

27. $x + y = 23$, $x^2 + y^2 = 289$ 28. $9y - 8x = 10$, $x^2 + y^2 = 100$

29. $2x - y = 7$, $x^2 + y^2 = 7$ 30. $x + 2y = 10$, $x^2 + y^2 = 20$

31. $5x + 2y = -1$, $x^2 + y^2 = 169$ 32. $y = 5$, $x^2 + y^2 - 4x - 6y = -9$

33. $y = \sqrt{3}x$, $x^2 + (y - 4)^2 = 16$ 34. $x - y = 3$, $x^2 + y^2 - 10x + 4y = -13$

35. *Writing* Write a description of the graphs of $x^2 + y^2 < 1$ and $x^2 + y^2 > 1$.

36. Sketch the graph of $(x - 3)^2 + (y - 4)^2 \le 25$.

B 37. The line $x - 2y = 15$ intersects the circle $x^2 + y^2 = 50$ in points A and B. Show that the line joining the center of the circle to the midpoint of $\overline{AB}$ is perpendicular to $\overline{AB}$.

38. Find the length of a tangent line segment from $(10, 5)$ to the circle $x^2 + y^2 = 25$. 10

39. $P(2, 3)$ is on the circle with center $O(0, 0)$.
     a. Write an equation of the circle. See below.
     b. Write an equation for the tangent l to the circle at P. (*Hint: l is perpendicular to $\overline{OP}$.*)

40. Show that $P(4, 2)$ is on the circle with equation $(x - 3)^2 + (y - 4)^2 = 5$. Find an equation of the tangent to the circle at P. (*Hint: See the hint for Exercise 39.*) $y = \dfrac{x}{2}$

41. A circle with center $C(2, 4)$ has radius 13.
     a. Verify that $A(14, 9)$ and $B(7, 16)$ are points on this circle.
     b. If M is the midpoint of $\overline{AB}$, show that $\overline{CM} \perp \overline{AB}$.

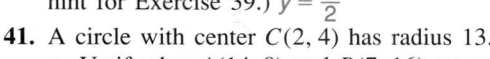

39. **a.** $x^2 + y^2 = 13$ **b.** $2x + 3y = 13$ *Analytic Geometry*   **223**

42. A circle with center $C(-4, 0)$ has radius 15.
 a. Verify that $A(8, 9)$ and $B(-13, 12)$ are points on this circle.
 b. Write an equation of the perpendicular bisector of $\overline{AB}$ and show that the coordinates of point C satisfy the equation.
 c. What theorem from geometry does this exercise illustrate?

43. A diameter of a circle has endpoints $A(13, 0)$ and $B(-13, 0)$.
 a. Show that $P(-5, 12)$ is a point on this circle.
 b. Show that $\overline{PA}$ and $\overline{PB}$ are perpendicular.

44. **a.** Find the coordinates of A and of B if $\overline{AB}$ is a horizontal diameter of the circle $x^2 + y^2 - 34x = 0$. $A(0, 0)$, $B(34, 0)$
 b. Show that $P(2, 8)$ is a point on this circle and that $\overline{PA} \perp \overline{PB}$.

45. Given $O(0, 0)$ and $N(12, 0)$, find an equation in terms of x and y for all points $P(x, y)$ such that $\overline{PO} \perp \overline{PN}$. Simplify this equation and show that P is on a circle. What are the center and radius of the circle? $(x - 6)^2 + y^2 = 36$; $C(6, 0)$, $r = 6$

46. **Discussion** Given $A(6, 8)$ and $B(-6, -8)$, write an equation in terms of x and y for all points $P(x, y)$ such that $\overline{PA} \perp \overline{PB}$. Simplify the equation and interpret your answer. $x^2 + y^2 = 100$

 You may find it helpful to have a graphing calculator to complete Exercises 47–49.

47. A triangle is inscribed in a semicircle as shown at the right.
 a. Find an equation for the semicircle.
 b. Write a function $A(x)$ for the area of the triangle.
 c. What is the domain of the function from part (b)?
 d. Graph $y = A(x)$.
 e. Use the graph from part (d) to find the value of x that maximizes $A(x)$.
 f. What is the maximum value of $A(x)$?

48. **a.** A rectangle is inscribed in a circle of radius 4 as shown at the left below. Write a function $A(x)$ for the area of the rectangle.
 b. Graph $y = A(x)$. Use the graph from part (a) to find the value of x that maximizes $A(x)$. What is the maximum value of $A(x)$?

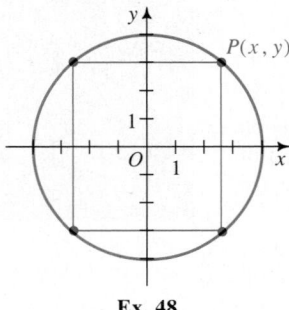

Ex. 48

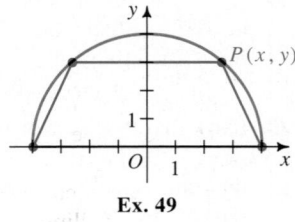

Ex. 49

49. a. An isosceles trapezoid is inscribed in a semicircle of radius 4 as shown at the bottom of page 224. Write a function $A(x)$ for the area of the trapezoid.
   b. Find the value of x that maximizes $A(x)$. **a.** $A(x) = (x + 4)\sqrt{16 - x^2}$ **b.** 2

50. $(x - 1)^2 + (y - 1)^2 = 2$ **51.** $(x - 3)^2 + (y - 4)^2 = 25$

Find an equation of the circle that contains the given points.

50. $A(0, 0)$, $B(2, 0)$, and $C(2, 2)$ See above.   **51.** $P(0, 0)$, $Q(6, 0)$, and $R(0, 8)$

52. $L(8, 2)$, $M(1, 9)$, and $N(1, 1)$
   $(x - 4)^2 + (y - 5)^2 = 25$

53. $D(7, 5)$, $E(1, -7)$, and $F(9, -1)$
   $(x - 2)^2 + y^2 = 50$

In Exercises 54–57, describe the set of points satisfying each equation.

54. $x^2 + y^2 + 2x + 2y + 2 = 0$

55. $x^2 + y^2 - 6x + 8y + 26 = 0$

C **56.** $(x^2 + y^2 - 1)(x^2 + y^2 - 4) = 0$

57. $x^3y + xy^3 - xy = 0$

58. a. A point (x, y) lies inside the circle $x^2 + y^2 = 2$ and above the line $y = 1$. Give two inequalities that must be satisfied. $x^2 + y^2 < 2$, $y > 1$
   b. Sketch the region in which the point lies and find the area of the region. $A = \dfrac{\pi}{2} - 1$

59. a. Sketch the set of points that satisfies $|x| \geq 1$ and $x^2 + y^2 \leq 4$.
   b. Find the area of the region. $A = \dfrac{8\pi}{3} - 2\sqrt{3}$

60. Suppose that point $P(a, b)$ is any point on the circle with center $O(0, 0)$ and radius r. Suppose that line l is perpendicular to $\overline{OP}$ at P. Prove that l is tangent to the circle as follows:
   a. Show that the equation of l can be written $ax + by = r^2$.
   b. Solve the equations of l and the circle simultaneously. Show that there is only one solution.

6-3 Ellipses

| **Objective** | To find equations of ellipses and to graph them. |

If you stand a few feet away from a wall and shine a flashlight against it, you can make a lighted area in the form of an oval, or *ellipse*. The ellipse appears to be an elongated circle. Try it.

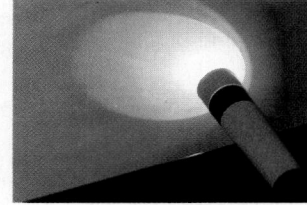

   Ellipses are found in many applications. Light or sound waves from one *focus* F_1 are reflected from the ellipse to the other focus F_2. This reflection property is used to make a whispering chamber, where a person whispering at F_1 can be heard at F_2. The United States Capitol Building has such a chamber.

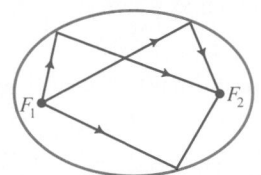

   To sketch an ellipse, complete the Activity on the next page.

Analytic Geometry   **225**

For Exercises 1–3, consider the equation $\dfrac{x^2}{9} + \dfrac{y^2}{25} = 1$.

1. Find the x-intercepts of the graph of the equation. ± 3

2. Find the y-intercepts of the graph of the equation. ± 5

3. How is the graph of $\dfrac{(x - 2)^2}{9} + \dfrac{(y + 1)^2}{25} = 1$ related to the graph of $\dfrac{x^2}{9} + \dfrac{y^2}{25} = 1$?

The graph of $\dfrac{(x - 2)^2}{9} + \dfrac{(y + 1)^2}{25} = 1$ is obtained by translating the graph of $\dfrac{x^2}{9} + \dfrac{y^2}{25} = 1$ 2 units to the right and 1 unit down.

Motivating the Section

Tell your students that they can find many examples of ellipses in everyday life. For example, a running track suggests an ellipse. In Australia, there is a type of football game played on a field that suggests an ellipse.

Communication Note

You might want to assign a research project on "whispering galleries." Ask students to write a report giving the famous galleries and discussing why locating the foci of ellipses is important in these galleries.

Problem Solving

Tell students to consider the equation $\frac{x^2}{a^2} + \frac{y^2}{b^2} = 1$, and ask students the following questions. What happens to the equation of an ellipse if $a = b$? Since $b^2 = a^2 - c^2$, what happens geometrically to the ellipse? (If $a = b$, then the equation becomes $x^2 + y^2 = a^2$, which is a circle. If $a = b$ in the equation $b^2 = a^2 - c^2$, then $c = 0$, and the two foci become one point, namely the center of a circle.)

Fasten the ends of a string to a piece of cardboard with thumbtacks. Make sure the string has some slack. Keeping the string taut, draw a curve on the cardboard as shown. Describe the curve traced by the pencil point P. Repeat the experiment by moving the tacks farther apart or closer together. The curve is an ellipse. As the tacks move farther apart, the ellipse becomes more elongated.

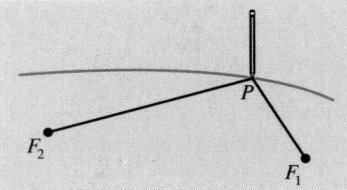

If $F_1(c, 0)$ and $F_2(-c, 0)$ are two fixed points in the plane and a is a constant, $0 < c < a$, then the set of all points P in the plane such that

$$PF_1 + PF_2 = 2a$$

is an **ellipse**. This is the *geometric definition* of an ellipse. Points F_1 and F_2 are called the **foci** of the ellipse. (*Foci* is the plural of *focus*.) In the Activity above, the thumbtacks are the foci and the string has length $2a$.

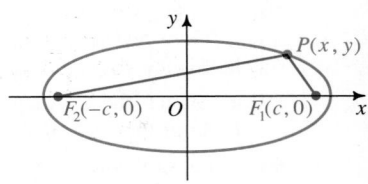

To find an equation for the ellipse with foci $F_1(c, 0)$ and $F_2(-c, 0)$, we let $P(x, y)$ be any point on the ellipse, then express PF_1 and PF_2 in terms of x, y, and c. Since $PF_1 + PF_2 = 2a$, we have:

$$\sqrt{(x - c)^2 + y^2} + \sqrt{(x - (-c))^2 + y^2} = 2a$$
$$\sqrt{(x + c)^2 + y^2} = 2a - \sqrt{(x - c)^2 + y^2}$$

Squaring: $(x + c)^2 + y^2 = 4a^2 - 4a\sqrt{(x - c)^2 + y^2} + [(x - c)^2 + y^2]$

$$x^2 + 2cx + c^2 + y^2 = 4a^2 - 4a\sqrt{(x - c)^2 + y^2} + x^2 - 2cx + c^2 + y^2$$

Simplifying: $4cx = 4a^2 - 4a\sqrt{(x - c)^2 + y^2}$

Dividing by 4 and rearranging terms:

$$cx - a^2 = -a\sqrt{(x - c)^2 + y^2}$$

Squaring again: $c^2x^2 - 2ca^2x + a^4 = a^2[(x - c)^2 + y^2]$
$$= a^2x^2 - 2ca^2x + a^2c^2 + a^2y^2$$

Simplifying: $c^2x^2 + a^4 = a^2x^2 + a^2c^2 + a^2y^2$

Rearranging terms: $a^4 - a^2c^2 = a^2x^2 - c^2x^2 + a^2y^2$
$$a^2(a^2 - c^2) = (a^2 - c^2)x^2 + a^2y^2$$

Since a must be greater than c, substitute $b^2 = a^2 - c^2$:

$$a^2b^2 = b^2x^2 + a^2y^2$$

Dividing by a^2b^2, we have: $\dfrac{x^2}{a^2} + \dfrac{y^2}{b^2} = 1$, where $b^2 = a^2 - c^2$

226   *Chapter Six*

The equation $\dfrac{x^2}{a^2} + \dfrac{y^2}{b^2} = 1$ is an *algebraic definition* of an ellipse with center at the origin and foci on the x-axis. Sometimes it is convenient to consider an ellipse with foci $(0, c)$ and $(0, -c)$. In this case, we can derive an equation using similar reasoning and obtain:

$$\dfrac{y^2}{a^2} + \dfrac{x^2}{b^2} = 1, \text{ where } b^2 = a^2 - c^2$$

The steps in the derivation of these equations can be reversed to show that any point satisfying the equations must also satisfy the condition $PF_1 + PF_2 = 2a$. Thus the algebraic and geometric definitions are equivalent.

All ellipses have two axes of symmetry. In the figures below, the axes of symmetry are the x- and y-axes. The portions of the axes of symmetry that lie on or within the ellipse are called the **major axis** and **minor axis** of the ellipse. The endpoints of the major axis are called the **vertices** of the ellipse. The midpoint of the major axis is the **center** of the ellipse.

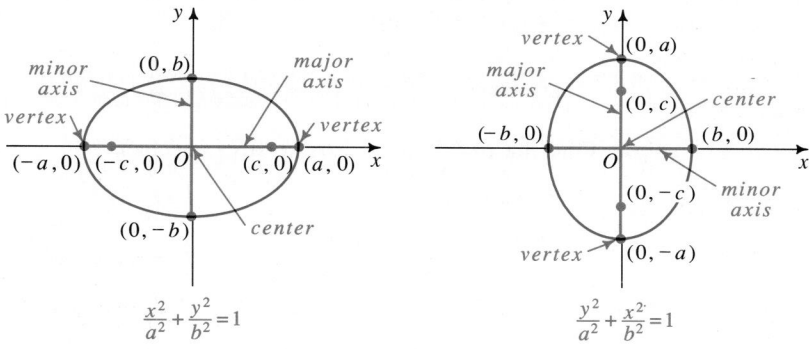

$$\dfrac{x^2}{a^2} + \dfrac{y^2}{b^2} = 1 \qquad\qquad \dfrac{y^2}{a^2} + \dfrac{x^2}{b^2} = 1$$

Example Find an equation of the ellipse with center at the origin, one vertex at $(0, 5)$, and one focus at $(0, 2)$. Sketch the ellipse, and label the vertices and the endpoints of the minor axis.

Solution Since one focus is $(0, 2)$, we have that $c = 2$. One vertex is at $(0, 5)$, so $a = 5$ and:

$$b^2 = a^2 - c^2 = 5^2 - 2^2 = 21$$

We substitute to write an equation of the ellipse:

$$\dfrac{y^2}{a^2} + \dfrac{x^2}{b^2} = 1$$

$$\dfrac{y^2}{25} + \dfrac{x^2}{21} = 1$$

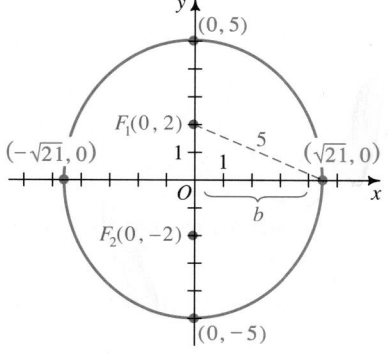

1. Find an equation of the ellipse with center at the origin, vertex at $(0, 4)$, and minor axis 4 units long. Find the coordinates of its foci.

   Since the vertex is on the y-axis, the equation has the form $\dfrac{y^2}{a^2} + \dfrac{x^2}{b^2} = 1$. Also, $a = 4$ and since $2b = 4$, $b = 2$. Thus, $\dfrac{y^2}{16} + \dfrac{x^2}{4} = 1$ is an equation. $c^2 = a^2 - b^2 = 16 - 4 = 12$. Therefore, the foci are $(0, \pm c)$, or $(0, \pm 2\sqrt{3})$.

2. Sketch the graphs of $4x^2 + y^2 = 64$ and $x + y = 4$ on the same set of axes. Then solve the system algebraically.

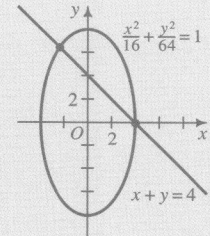

   $y = 4 - x$, so
   $4x^2 + (4 - x)^2 = 64$;
   $5x^2 - 8x - 48 = 0$;
   $(5x + 12)(x - 4) = 0$;
   $x = -2.4$ or $x = 4$;
   $(-2.4, 6.4)$, $(4, 0)$.

Mathematical Note

The area of an ellipse with major and minor axes of length $2a$ and $2b$ is πab. This formula becomes the formula for the area of a circle (πr^2) when the major and minor axes of the ellipse are equal.

CLASS EXERCISES

1. Study the ellipse shown and state: **1. d.** $\frac{x^2}{100} + \frac{y^2}{64} = 1$
   a. whether its major axis is horizontal or vertical. Horiz.
   b. its vertices and foci. $V(\pm 10, 0)$, $F(\pm 6, 0)$
   c. the constant value of $PF_1 + PF_2$. 20
       (*Hint:* Suppose P is at a vertex.)
   d. an equation of the ellipse. See above.

2. An ellipse has equation $\frac{x^2}{25} + \frac{y^2}{169} = 1$.

   a. Is its major axis horizontal or vertical? Vert.
   b. Find the coordinates of its vertices and foci. $V(0, \pm 13)$, $F(0, \pm 12)$

3. Describe the graphs of $\frac{x^2}{9} + y^2 < 1$ and $\frac{x^2}{9} + y^2 > 1$.

4. *Discussion* If F_1 and F_2 are fixed points in space, describe the set of points P such that (**a**) $PF_1 + PF_2 = 8$ and (**b**) $PF_1 + PF_2 < 8$.

WRITTEN EXERCISES

Sketch each ellipse. Find the coordinates of its vertices and foci.

A

1. $\frac{x^2}{36} + \frac{y^2}{16} = 1$ 　　　　**2.** $\frac{x^2}{4} + \frac{y^2}{9} = 1$ 　　　　**3.** $\frac{x^2}{16} + \frac{y^2}{25} = 1$

4. $4x^2 + 25y^2 = 100$ 　　　**5.** $9x^2 + 25y^2 = 225$ 　　　**6.** $6.25x^2 + 4y^2 = 25$

7. Sketch the graph of each inequality: (**a**) $\frac{x^2}{25} + \frac{y^2}{9} \le 1$ 　(**b**) $\frac{x^2}{4} + \frac{y^2}{16} \ge 1$

Find the domain and range of each function. Then graph the function. You may find a graphing calculator helpful.

8. $y = 3\sqrt{1 - x^2}$ 　　　　　　　　　　**9.** $y = -\frac{1}{3}\sqrt{36 - x^2}$

10. a. On a single set of axes, graph $x^2 + y^2 = 1$ and $\left(\frac{x}{3}\right)^2 + y^2 = 1$.

   b. Give the area of the circle and guess the area of the ellipse.

11. a. On a single set of axes, graph $x^2 + y^2 = 1$ and $x^2 + \left(\frac{y}{2}\right)^2 = 1$.

   b. Give the area of the circle and guess the area of the ellipse.

Each ellipse has its center at the origin. Find an equation of the ellipse.

12. Vertex $(7, 0)$; minor axis 2 units long　**13.** Vertex $(0, -9)$; minor axis 6 units long

14. Vertex $(0, -13)$; focus $(0, -5)$　　　**15.** Vertex $(17, 0)$; focus $(8, 0)$

12. $\frac{x^2}{49} + y^2 = 1$ 　**13.** $\frac{x^2}{9} + \frac{y^2}{81} = 1$ 　**14.** $\frac{x^2}{144} + \frac{y^2}{169} = 1$ 　**15.** $\frac{x^2}{289} + \frac{y^2}{225} = 1$

228 　*Chapter Six*

16. a. *Investigation* Use the method described in the Activity to draw a large ellipse filling most of a full page of paper. Label the thumbtack holes F_2 and F_1. Then pick any point P on the ellipse. Draw $\overline{F_2P}$, $\overline{F_1P}$, and the tangent to the ellipse at P. Use a protractor to show that the angle between $\overline{F_2P}$ and the tangent is congruent to the angle between $\overline{F_1P}$ and the tangent. Repeat the experiment for another point P.

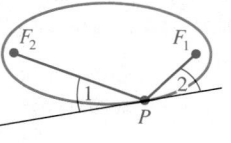

b. What property of the ellipse does part (a) illustrate?

17. *Investigation* Refer to the Activity on page 226. What happens to an ellipse as its foci F_1 and F_2 move toward each other? if F_1 and F_2 coincide? The ellipse becomes more circular. When $F_1 = F_2$, the figure is a circle.

Sketch the graphs of the given equations on a single set of axes. Then determine algebraically where the graphs intersect.

18. $9x^2 + 2y^2 = 18$ $(0, -3),$ $\left(-\frac{4}{3}, 1\right)$
    $3x + y = -3$

19. $x^2 + 4y^2 = 400$ $(12, -8),$ $(16, -6)$
    $x - 2y = 28$

20. $2x^2 + y^2 = 9$ $(-2, 1)$
    $y - 4x = 9$

21. $x^2 + 4y^2 = 16$ $(2, \sqrt{3}),$ $(-2, \sqrt{3})$
    $|x| = 2$ $(-2, -\sqrt{3}),$ $(2, -\sqrt{3})$

22. a. What happens to the graph of an equation of an ellipse when x is replaced by $x - 5$? The graph is shifted 5 units to the right.

    b. Graph $\dfrac{x^2}{4} + y^2 = 1$ and $\dfrac{(x - 5)^2}{4} + y^2 = 1$ on a single set of axes.

23. a. What happens to the graph of an equation of an ellipse when y is replaced by $y + 6$? The graph is shifted 6 units down.

    b. Graph $\dfrac{x^2}{4} + \dfrac{y^2}{9} = 1$ and $\dfrac{x^2}{4} + \dfrac{(y + 6)^2}{9} = 1$ on a single set of axes.

Sketch each ellipse. Label the center and vertices.

 24. $\dfrac{(x - 3)^2}{4} + \dfrac{(y - 6)^2}{25} = 1$

25. $\dfrac{(x + 5)^2}{25} + \dfrac{(y - 4)^2}{16} = 1$

26. $(x + 7)^2 + \dfrac{(y - 5)^2}{9} = 1$

27. $4(x + 2)^2 + (y - 5)^2 = 4$

 You may find it helpful to have a graphing calculator to complete Exercise 28(b).

28. In the figure at the right, rectangle $PQRS$ is inscribed in the ellipse $\dfrac{x^2}{16} + \dfrac{y^2}{4} = 1$.

    a. Show that the area of the rectangle is
$$A(x) = 4x\sqrt{4 - \dfrac{x^2}{4}}.$$

    b. Approximate the maximum area to the nearest tenth of a square unit. 16.0

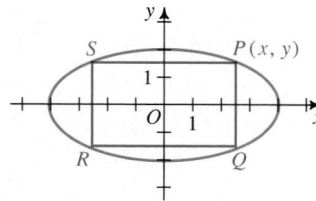

Exercise Note

The reflected-wave property of ellipses illustrated in Exercise 16 is the basis for the whispering-chamber effect discussed on page 225.

◩ Using Technology

A graphing calculator or a computer can be used to check the graphs in Exercises 22–27. Students may need to solve for y.

For $\dfrac{(x - h)^2}{a^2} + \dfrac{(y - k)^2}{b^2} = 1$,

use

$$y = k \pm \sqrt{\dfrac{a^2b^2 - b^2(x - h)^2}{a^2}}.$$

29. a. If the ellipse is a circle,
$a = b = r$ and
$V = \frac{4}{3}\pi r^3$, the formula
for volume of a sphere.

38. $\dfrac{(x-3)^2}{25} + \dfrac{(y-7)^2}{16} = 1$

39. $\dfrac{4(x-4)^2}{39} + \dfrac{(y+1)^2}{16} = 1$

40. $\dfrac{(x-5)^2}{12} + \dfrac{(y-5)^2}{16} = 1$

41. $\dfrac{(x-5)^2}{25} + \dfrac{(y-6)^2}{36} = 1$

42. a. $\sqrt{(x-3)^2 + y^2} +$
$\sqrt{(x+3)^2 + y^2} = 10$

b. $\dfrac{x^2}{25} + \dfrac{y^2}{16} = 1$

---

29. When the ellipse $\dfrac{x^2}{a^2} + \dfrac{y^2}{b^2} = 1$ is rotated about either of its axes, an *ellipsoid* is formed.

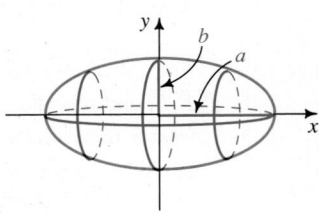

a. The volume of the ellipsoid shown is $V = \frac{4}{3}\pi b^2 a$. Interpret this volume if the original ellipse is a circle.

b. Sketch the ellipsoid formed by rotating the given ellipse about its minor axis and guess its volume. $\frac{4}{3}\pi a^2 b$

Sketch each ellipse. Find the coordinates of its vertices and foci.

30. $\dfrac{(x-5)^2}{25} + \dfrac{(y+3)^2}{9} = 1$ V: $(0, -3)$, $(10, -3)$; F: $(1, -3)$, $(9, -3)$

31. $\dfrac{(x+6)^2}{12} + \dfrac{(y-4)^2}{16} = 1$ V: $(-6, 0)$, $(-6, 8)$; F: $(-6, 2)$, $(-6, 6)$

32. $9(x-3)^2 + 4(y+5)^2 = 36$ V: $(3, -2)$, $(3, -8)$; F: $(3, -5 + \sqrt{5})$, $(3, -5 - \sqrt{5})$

33. $(x+1)^2 + 4(y+3)^2 = 9$ V: $(-4, -3)$, $(2, -3)$; F: $\left(-1 + \dfrac{3\sqrt{3}}{2}, -3\right)$, $\left(-1 - \dfrac{3\sqrt{3}}{2}, -3\right)$

34. $x^2 + 25y^2 - 6x - 100y + 84 = 0$
(*Hint*: Complete the squares in x and in y. Begin by rewriting the equation in this form: $(x^2 - 6x + \quad) + 25(y^2 - 4y + \quad) = -84$.) See below.

35. $9x^2 + y^2 + 18x - 6y + 9 = 0$ V: $(-1, 6)$, $(-1, 0)$; F: $(-1, 3 + 2\sqrt{2})$, $(-1, 3 - 2\sqrt{2})$

36. $9x^2 + 16y^2 - 18x - 64y - 71 = 0$ V: $(-3, 2)$, $(5, 2)$; F: $(1 + \sqrt{7}, 2)$, $(1 - \sqrt{7}, 2)$

34. V: $(-2, 2)$, $(8, 2)$; F: $(3 + 2\sqrt{6}, 2)$, $(3 - 2\sqrt{6}, 2)$

> You may find a graphing calculator or graphing software helpful to complete Exercise 37.

37. A graphing calculator or computer software can be used to sketch the graph of an equation that represents a function.

a. Solve $\dfrac{x^2}{36} + \dfrac{y^2}{16} = 1$ for y. The result involves two equations. $y = \pm\frac{2}{3}\sqrt{36 - x^2}$

b. Graph the ellipse in part (a). Does the result agree with the graph in Exercise 1? Yes

In Exercises 38–41, find an equation of the ellipse described.

38. Center is $(3, 7)$; one focus is $(6, 7)$; one vertex is $(8, 7)$.

39. Center is $(4, -1)$; one vertex is $(4, -5)$; one focus is $(4, -3.5)$.

40. Vertices are $(5, 9)$ and $(5, 1)$; one focus is $(5, 7)$.

41. Center is $(5, 6)$; the ellipse is tangent to both axes.

42. a. Suppose that $F_1 = (3, 0)$, $F_2 = (-3, 0)$, and $P = (x, y)$. Using the distance formula, write an equation that expresses the fact that $PF_1 + PF_2 = 10$.

b. Simplify your equation to one of the form $\dfrac{x^2}{a^2} + \dfrac{y^2}{b^2} = 1$.

230 *Chapter Six*

1. **a.** To approximate the area of the ellipse $\dfrac{x^2}{25} + \dfrac{y^2}{9} = 1$, shown at the right, you can add the areas of the ten rectangles in the first quadrant and multiply the sum by 4. The area of the leftmost rectangle is $\dfrac{1}{2} \cdot 3$; the area of the next is $\dfrac{1}{2}y_2$; and the area of the next is $\dfrac{1}{2}y_3$. How do you find y_2 and y_3?

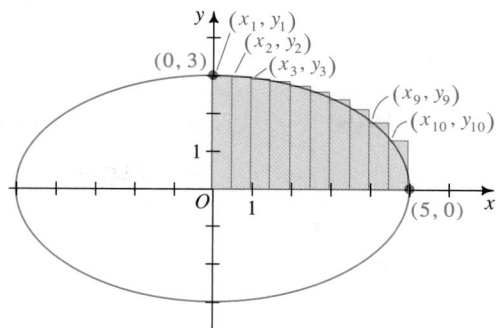

   b. Write a program to print the approximate area of the ellipse.

2. Modify the program to print a more accurate approximation of the area.

6-4 Hyperbolas

Objective   *To find equations of hyperbolas and to graph them.*

The equation of a *hyperbola* with center at the origin and horizontal and vertical axes of symmetry is very similar to the equation of an ellipse. Moreover, the terminology for ellipses and hyperbolas is very similar.

Suppose that $F_1(c, 0)$ and $F_2(-c, 0)$ are fixed points in the plane, and that a is a constant, $a < c$. The *geometric definition* of a **hyperbola** states that a hyperbola is the set of all points $P(x, y)$ in the plane such that
$$|PF_1 - PF_2| = 2a.$$

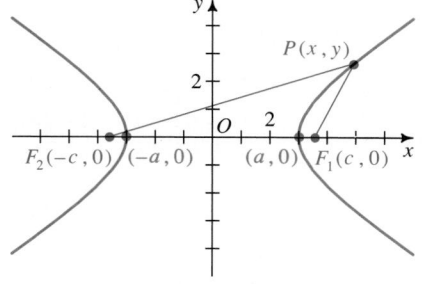

To derive an equation of a hyperbola with foci $F_1(c, 0)$ and $F_2(-c, 0)$, we begin as follows:
$$|PF_1 - PF_2| = 2a$$
$$\sqrt{(x - c)^2 + y^2} - \sqrt{(x + c)^2 + y^2} = \pm 2a$$

The rest of the derivation is similar to that for the ellipse on page 226, except that we let $b^2 = c^2 - a^2$ rather than $b^2 = a^2 - c^2$, since $c > a$. The equation that results is $\dfrac{x^2}{a^2} - \dfrac{y^2}{b^2} = 1$.

Analytic Geometry   **231**

Making Connections

Approximating the area under a curve by adding areas of rectangles is important for work in calculus.

Teaching Notes, p. 212C

Warm-Up Exercises

For Exercises 1-4, consider the equation $y^2 - x^2 = 9$.

1. Find the x- and y- intercepts of the graph of the equation.
   No x-intercepts;
   y-intercepts, ± 3.

2. Use the chart on page 134 to determine the symmetries of the graph of the equation. Symmetry in the x-axis, in the y-axis, and in the origin.

3. Solve the equation for y.
   $y = \pm\sqrt{9 + x^2}$

4. Write the equation that would translate the graph of $y^2 - x^2 = 9$ 2 units to the left and 3 units up.
   $(y - 3)^2 - (x + 2)^2 = 9$

Motivating the Section

Tell your students that hyperbolas are important as graphs of equations found in chemistry (models of atoms with hyperbolic orbits for the positively charged particles), physics (the theory of relativity's use of hyperbolic space), economics (demand curves), and other sciences.

Mathematical Note

Tell students that the line segment connecting the vertices of a hyperbola is known as the "transverse axis." The foci always lie on the line containing this axis.

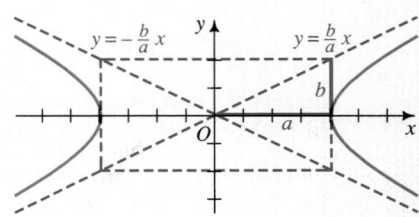

Ⓝ Using Technology

Emphasize to students that when graphing hyperbolas using a graphing calculator, they may need to express the equation as a function. Use

$$y = \pm\sqrt{\frac{b^2 x^2}{a^2} - b^2} \text{ or}$$

$$y = \pm\sqrt{\frac{a^2 x^2}{b^2} - a^2},$$

depending on the form of the hyperbola's equation.

Additional Examples

1. Find an equation of the hyperbola with center at the origin, a vertex at $(0, 4)$, and an asymptote with equation $y = \frac{2}{3}x$.

The vertex is on the y-axis, so the equation has the form $\frac{y^2}{a^2} - \frac{x^2}{b^2} = 1$.

An asymptote is $y = \frac{2}{3}x = \frac{a}{b}x$ and $a = 4$, so $\frac{2}{3} = \frac{4}{b}$ and $b = 6$.

$\frac{y^2}{4^2} - \frac{x^2}{6^2} = 1$; $\frac{y^2}{16} - \frac{x^2}{36} = 1$

Its foci are $(0, \pm c)$, where $c^2 = a^2 + b^2 = 16 + 36 = 52$. The foci are $(0, \pm 2\sqrt{13})$.

Asymptotes

When we solve $\dfrac{x^2}{a^2} - \dfrac{y^2}{b^2} = 1$ for y, we obtain:

$$y = \pm\frac{b}{a}\sqrt{x^2 - a^2}$$

If $|x|$ is very large, $\sqrt{x^2 - a^2}$ is approximately the same as $\sqrt{x^2} = |x|$. Therefore, when $|x|$ is very large,

$$y = \pm\frac{b}{a}\sqrt{x^2 - a^2} \approx \pm\frac{b}{a}x.$$

The lines $y = \pm\frac{b}{a}x$ are called the *asymptotes* of the hyperbola. An **asymptote** of a curve is a line that the curve approaches more and more closely, that is, the distance between the curve and its asymptote becomes less and less as $|x|$ becomes large. See the following figure and table.

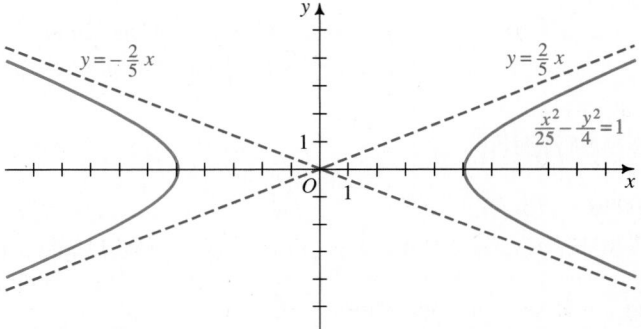

x	10	20	100	1000
y (on asymptote) $y = \frac{2}{5}x$	4	8	40	400
y (on hyperbola) $y = \frac{2}{5}\sqrt{x^2 - 25}$	3.46	7.75	39.95	399.995
vertical distance (difference in y's)	0.54	0.25	0.05	0.005

The asymptotes for $\dfrac{x^2}{a^2} - \dfrac{y^2}{b^2} = 1$ contain the diagonals of a rectangle with dimensions $2a$ and $2b$. The asymptotes and the rectangle are shown in blue. Drawing the rectangle and its diagonals is a good first step in graphing the equation.

232 *Chapter Six*

Example 1   Graph the hyperbola $\dfrac{x^2}{36} - \dfrac{y^2}{9} = 1$. Find its foci.

Solution

1. Since the hyperbola has x-intercepts ± 6, the hyperbola contains $(6, 0)$ and $(-6, 0)$. Draw the rectangle centered at the origin having length $2 \cdot 6 = 12$ and width $2 \cdot 3 = 6$. Sketch the diagonals of the rectangle.

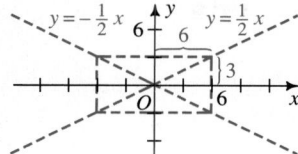 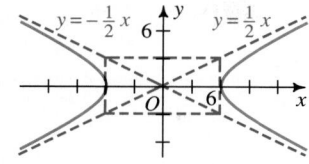

2. Sketch the curves through the vertices, $(\pm 6, 0)$, extending them towards the asymptotes as shown at the right above.
3. To find the foci, note that $c^2 = a^2 + b^2 = 36 + 9 = 45$. Thus, $c = \pm\sqrt{45} = \pm 3\sqrt{5}$. So the foci are $(\pm 3\sqrt{5}, 0)$.

The graph of

$$\frac{y^2}{a^2} - \frac{x^2}{b^2} = 1$$

is also a hyperbola, but with a vertical major axis. Its asymptotes have equations $y = \pm\dfrac{a}{b}x$. Contrast the hyperbola in Example 2 with the one in Example 1.

Example 2   Graph the hyperbola $\dfrac{y^2}{9} - \dfrac{x^2}{36} = 1$. Find its foci.

Solution

1. Since the hyperbola has y-intercepts ± 3, the hyperbola contains $(0, -3)$ and $(0, 3)$. Draw the rectangle centered at the origin having length $2 \cdot 6 = 12$ and width $2 \cdot 3 = 6$. Sketch the diagonals of the rectangle.

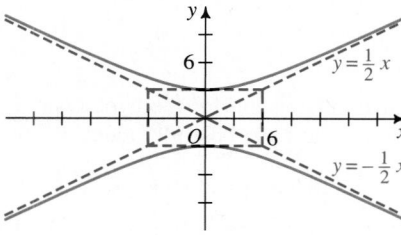

2. Sketch the curves through the vertices, $(0, \pm 3)$, extending them towards the asymptotes as shown above.
3. To find the foci, note that $c^2 = a^2 + b^2 = 9 + 36 = 45$ and $c = \pm 3\sqrt{5}$. So the foci are $(0, \pm 3\sqrt{5})$.

Analytic Geometry   **233**

Additional Examples cont.

2. Sketch the hyperbola $x^2 - 9y^2 + 2x + 36y - 44 = 0$. Find the coordinates of its vertices and foci and the equations of its asymptotes.

$(x^2 + 2x + ?) - (9y^2 - 36y + ?) = 44$
$(x^2 + 2x + 1) - 9(y^2 - 4y + 4) = 44 + 1 - 36$
$(x + 1)^2 - 9(y - 2)^2 = 9$
$\dfrac{(x + 1)^2}{9} - \dfrac{(y - 2)^2}{1} = 1$

The graph is obtained by shifting the graph of $\dfrac{x^2}{9} - \dfrac{y^2}{1} = 1$ to the left 1 unit and up 2 units.

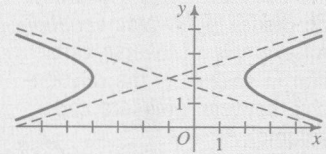

Vertices: Shifting $(\pm 3, 0)$ gives $(-4, 2)$ and $(2, 2)$.
Foci: Shifting $(\pm\sqrt{10}, 0)$ gives $(-1 \pm \sqrt{10}, 2)$.
Asymptotes: Shifting $y = \pm\dfrac{1}{3}x$ gives

$y - 2 = \pm\dfrac{1}{3}(x + 1)$, or

$x - 3y = -7$ and $x + 3y = 5$.

The diagrams below show hyperbolas with simple equations, that is, equations of the form $xy = k$ where k is a nonzero constant. The coordinate axes are the asymptotes for the hyperbolas.

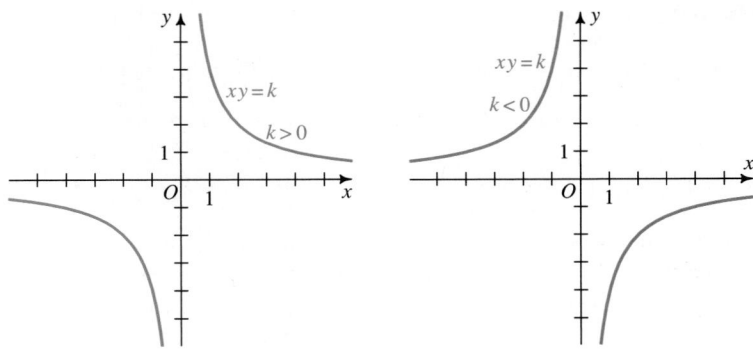

CLASS EXERCISES

1. **a.** Give the vertices and foci for the hyperbola at the left below. $V: (\pm 2, 0), F: (\pm\sqrt{13}, 0)$
   b. Is the major axis horizontal or vertical? Horizontal
   c. Find an equation for the hyperbola and equations of its asymptotes. $\dfrac{x^2}{4} - \dfrac{y^2}{9} = 1$, $y = \pm\dfrac{3}{2}x$

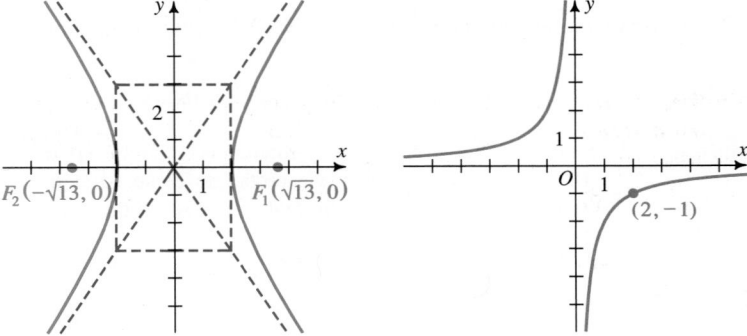

2. **a.** Give an equation for the hyperbola at the right above. $y = -\dfrac{2}{x}$
   b. Rotate the hyperbola 90° about the origin. What is its new equation? $y = \dfrac{2}{x}$

3. A hyperbola has equation $\dfrac{y^2}{25} - \dfrac{x^2}{1} = 1$.

   a. Is its major axis horizontal or vertical? Explain. Vertical
   b. What are its vertices and foci? $V: (0, \pm 5), F: (0, \pm\sqrt{26})$
   c. What are the equations of its asymptotes? $y = \pm 5x$
   d. If the hyperbola were translated 6 units to the right and 5 units down, what would be its new equation? $\dfrac{(y + 5)^2}{25} - \dfrac{(x - 6)^2}{1} = 1$

4. Describe the graphs of $x^2 - y^2 < 4$ and $x^2 - y^2 > 4$.

234   *Chapter Six*

In Exercises 1–8, sketch each hyperbola and its asymptotes. Give equations for the asymptotes.

 1. $\dfrac{x^2}{9} - \dfrac{y^2}{4} = 1$ **2.** $\dfrac{y^2}{16} - \dfrac{x^2}{4} = 1$

3. $x^2 - y^2 = 1$ **4.** $y^2 - x^2 = 1$

5. $4y^2 - x^2 = 4$ **6.** $25x^2 - 16y^2 = 400$

7. $xy = 4$ **8.** $xy = -4$

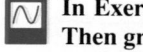 In Exercises 9 and 10, give the domain and the range of each function. Then graph the function. You may use a graphing calculator if you wish.

9. $y = \sqrt{x^2 - 4}$ **10.** $y = -\sqrt{9 + x^2}$

In Exercises 11 and 12, sketch each hyperbola and its asymptotes. Give equations for the asymptotes.

11. a. $xy = 12$ **b.** $xy = -12$ **c.** $(x + 3)(y + 4) = 12$

12. a. $xy = 8$ **b.** $xy = -8$ **c.** $(x - 2)(y + 4) = -8$

13. On a single set of axes, graph the hyperbolas with equations $x^2 - y^2 = k$, where $k = 4, 1, -1,$ and -4.

14. On a single set of axes, graph the hyperbolas with equations $xy = k$, where $k = 12, 8, 4,$ and 1.

In Exercises 15–18, sketch the graphs of the given equations on a single set of axes. Then determine algebraically the coordinates of any points where the graphs intersect.

15. $x^2 - y^2 = 16$ (4.1, 0.9)
$x + y = 5$

16. $y^2 - x^2 = 94$ $\left(-\dfrac{85}{6}, -\dfrac{103}{6}\right)$
$x - y = 3$

17. $x + 2y = 6$ (−4, 5), (10, −2)
$xy = -20$

18. $2x + y = -14$ (−4, −6), (−3, −8)
$xy = 24$

In Exercises 19–22, find an equation of the hyperbola, with center at the origin, that satisfies the given conditions.

19. A vertex at $(6, 0)$ and a focus at $(10, 0)$ $\dfrac{x^2}{36} - \dfrac{y^2}{64} = 1$

20. A vertex at $(0, -12)$ and a focus at $(0, -13)$

21. A vertex at $(0, -2)$ and an asymptote with equation $y = -x$

22. A vertex at $(8, 0)$ and an asymptote with equation $y = \dfrac{1}{2}x$

20. $\dfrac{y^2}{144} - \dfrac{x^2}{25} = 1$ **21.** $\dfrac{y^2}{4} - \dfrac{x^2}{4} = 1$ **22.** $\dfrac{x^2}{64} - \dfrac{y^2}{16} = 1$

Error Analysis

Some students may be confused when finding the asymptotes for Exercises 1–6. Stress that if the x^2-term is positive, then the asymptotes are of the form $y = \pm\dfrac{b}{a}x$. If the y^2-term is positive, the asymptotes are of the form $y = \pm\dfrac{a}{b}x$.

In Exercise 22 some students may assume that $b = 1$ and $a = 2$ because the equation of the asymptotes is $y = \pm\dfrac{1}{2}x$. Stress that the number $\dfrac{1}{2}$ gives the *ratio* of b to a.

Additional Answers
Written Exercises

1. $y = \pm\dfrac{2}{3}x$

2. $y = \pm 2x$

3. $y = \pm x$

4. $y = \pm x$

5. $y = \pm\dfrac{1}{2}x$

6. $y = \pm\dfrac{5}{4}x$

7. $x = 0, y = 0$

8. $x = 0, y = 0$

9. Domain = $\{x \mid |x| \geq 2\}$; range = $\{y \mid y \geq 0\}$

10. Domain = {all reals}; range = $\{y \mid y \leq -3\}$

Suggested Assignments

Standard
Day 1: 235/1–23 odd
Day 2: 236/25–43 odd
Comprehensive
Day 1: 235/1–23 odd, 24
Day 2: 236/25–45 odd, 26

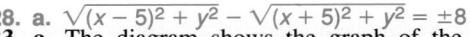

28. **a.** $\sqrt{(x-5)^2 + y^2} - \sqrt{(x+5)^2 + y^2} = \pm 8$

23. **a.** The diagram shows the graph of the hyperbola $\dfrac{x^2}{4} - y^2 = 1$. Show that this equation can be rewritten as $y = \pm\dfrac{1}{2}\sqrt{x^2 - 4}$.

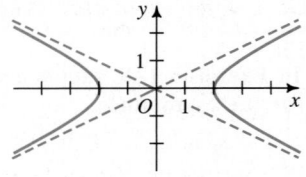

   b. Find the vertical distance between the hyperbola and its asymptote $y = \dfrac{1}{2}x$ when $x = 5$, $x = 10$, and $x = 100$. 0.209, 0.101, 0.010

24. Graph $y^2 - x^2 = 9$ and its asymptotes. Find a value of x for which the vertical distance between the hyperbola and an asymptote is less than 0.001. $|x| > 4499.9995$

B 25. *Visual Thinking* When the hyperbola with equation $\dfrac{x^2}{a^2} - \dfrac{y^2}{b^2} = 1$ is rotated about its vertical axis of symmetry, a *hyperboloid of one sheet* is formed. (See the figure at the right.) When the hyperbola is rotated about its horizontal axis of symmetry, a *hyperboloid of two sheets* is formed. Sketch this hyperboloid.

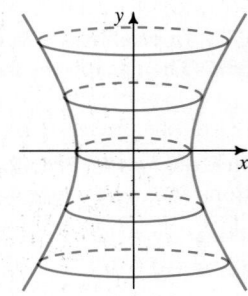

26. **a.** *Visual Thinking* The region enclosed by the hyperbola $y^2 - x^2 = 16$ and the lines $x = \pm 3$ is rotated about the x-axis. Sketch the solid formed.
   b. Think of the solid as inscribed in a cylinder. Find the volume of the cylinder by using $V = \pi r^2 h$. 150π

27. *Investigation* This exercise illustrates a reflection property of a hyperbola: The lines drawn from a point on a hyperbola to the foci form equal angles with the tangent at that point.

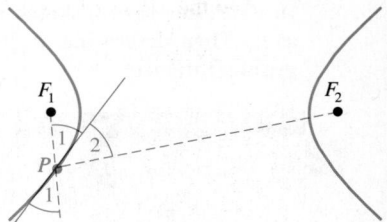

   a. Carefully trace the hyperbola and the foci of the hyperbola at the right.
   b. On one of the branches, choose a point P different from the one shown. Carefully sketch a line tangent to the hyperbola at P. Draw lines PF_1 and PF_2.
   c. Measure $\angle 1$ and $\angle 2$. What is the relationship between these angles? Equal

28. **a.** Suppose that $F_1 = (5, 0)$, $F_2 = (-5, 0)$, and $P = (x, y)$. Write an equation that expresses the fact that $|PF_1 - PF_2| = 8$. See above.
   b. Simplify your equation to the form $\dfrac{x^2}{a^2} - \dfrac{y^2}{b^2} = 1$. $\dfrac{x^2}{16} - \dfrac{y^2}{9} = 1$

29. **a.** Suppose that $F_1 = (0, 10)$, $F_2 = (0, -10)$, and $P = (x, y)$. Write an equation that expresses the fact that $|PF_1 - PF_2| = 12$. See below.
   b. Simplify your equation to the form $\dfrac{y^2}{a^2} - \dfrac{x^2}{b^2} = 1$. $\dfrac{y^2}{36} - \dfrac{x^2}{64} = 1$

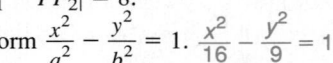

236   *Chapter Six*   29. **a.** $\sqrt{x^2 + (y-10)^2} - \sqrt{x^2 + (y+10)^2} = \pm 12$

On a single set of axes, sketch the graphs of the given equations.

30. $x^2 - y^2 = 1$
   $(x - 4)^2 - (y - 3)^2 = 1$

31. $y^2 - 4x^2 = 4$
   $(y + 5)^2 - 4x^2 = 4$

32. Suppose the hyperbola in Example 1 is translated 8 units right and 6 units up. Give the equation of the new hyperbola and the equations of its asymptotes.

33. Suppose the hyperbola in Example 1 is rotated 90° about the origin. Give the equation of the new hyperbola and the equations of its asymptotes.

Sketch the graph of each inequality.

34. **a.** $xy \geq 8$ **b.** $xy \geq 1$ **c.** $xy \geq 0$
35. **a.** $xy \leq -4$ **b.** $xy \leq -1$ **c.** $xy \leq 0$
36. $x^2 - 4y^2 < 4$ 37. $y^2 - x^2 > 1$ 38. $(x - 5)^2 - (y - 4)^2 \geq 1$

Sketch the hyperbola whose equation is given. Find the coordinates of the vertices and the foci, and equations for the asymptotes.

39. $\dfrac{(x - 6)^2}{36} - \dfrac{(y - 8)^2}{64} = 1$ 40. $\dfrac{(y + 5)^2}{16} - \dfrac{x^2}{9} = 1$

41. $y^2 - x^2 - 2y + 4x - 4 = 0$ 42. $x^2 - 4y^2 - 2x + 16y - 19 = 0$

Find an equation of the hyperbola described.

43. Center is $(5, 0)$; one vertex is $(9, 0)$; one focus is $(10, 0)$. $\dfrac{(x - 5)^2}{16} - \dfrac{y^2}{9} = 1$

44. Vertices are $(4, 0)$ and $(4, 8)$; asymptotes have slopes ± 1. See below.

C 45. Find the values of a and b that make the hyperbola $\dfrac{x^2}{a^2} - \dfrac{y^2}{b^2} = 1$ congruent to the hyperbola $xy = 1$. $a = b = \sqrt{2}$

44. $\dfrac{(y - 4)^2}{16} - \dfrac{(x - 4)^2}{16} = 1$

Apollonius (262–200 B.C.)

Apollonius of Perga was an astronomer and geometer noted for his eight-volume work entitled *On Conic Sections*. In it, he derived all the conic sections from a plane intersecting a right double cone. He was the first to use the terms *ellipse*, *parabola*, and *hyperbola* in reference to conic sections.

The problem of constructing a circle tangent to three given circles in the plane bears his name. It is called the problem of Apollonius.

Additional Answers Written Exercises

32. $\dfrac{(x - 8)^2}{36} - \dfrac{(y - 6)^2}{9} = 1$;
   $-x + 2y = 4$,
   $x + 2y = 20$

33. $\dfrac{y^2}{36} - \dfrac{x^2}{9} = 1$; $y = \pm 2x$

39. Vertices: $(12, 8)$, $(0, 8)$; foci: $(16, 8)$, $(-4, 8)$; asymptotes: $-4x + 3y = 0$, $4x + 3y = 48$

40. Vertices: $(0, -1)$, $(0, -9)$; foci: $(0, 0)$, $(0, -10)$; asymptotes: $y = \dfrac{4}{3}x - 5$, $y = -\dfrac{4}{3}x - 5$

41. Vertices: $(2, 2)$, $(2, 0)$; foci: $(2, 1 + \sqrt{2})$, $(2, 1 - \sqrt{2})$; asymptotes: $y = x - 1$, $y = -x + 3$

42. Vertices: $(3, 2)$, $(-1, 2)$; foci: $(1 + \sqrt{5}, 2)$, $(1 - \sqrt{5}, 2)$, asymptotes: $x - 2y = -3$, $x + 2y = 5$

For Exercises 1–4, consider the equation $y = 2x^2 + 8x - 3$.

1. Complete: The graph of the equation is called a __?__.
   parabola

2. Find the x- and y-intercepts of the graph of the equation.
   x-intercepts: $\dfrac{-4 \pm \sqrt{22}}{2}$;
   y-intercept: -3

3. Find the x-coordinate of the vertex of the graph of the equation. Then find the y-coordinate of the vertex.
   -2; -11

4. Determine the symmetries of the graph of the equation.
   symmetric in the line $x = -2$

5. Find the domain and range of the related function.
   domain: all real numbers, range: $\{y | y \geq -11\}$

Motivating the Section

The shape formed by rotating a parabola about its axis is known as a paraboloid. It is a common shape used for reflectors in searchlights and automobile headlights. The light emanating from a bulb placed at a focus is then reflected parallel to the parabola's axis.

6-5 Parabolas

| **Objective** | *To find equations of parabolas and to graph them.* |

When we discussed quadratic functions in Section 1-7, we stated that their graphs are parabolas. In this section, we will state the geometric and algebraic definitions of a parabola and examine some of the properties of a parabola.

A **parabola** is the set of all points P in the plane that are equidistant from a fixed point F, the **focus**, and a fixed line d (not containing F), the **directrix**. That is, if F and d are fixed, then the set of all points P such that $PF = PN$ is a parabola.

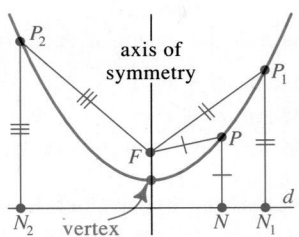

The line through the focus F and perpendicular to the directrix d is the line of symmetry for the parabola. It intersects the parabola at its *vertex*.

If the line $y = -p$, $p \neq 0$, is the directrix of a parabola with focus $F(0, p)$, then by the geometric definition above:

$$PF = PN$$
$$\sqrt{(x - 0)^2 + (y - p)^2} = |y + p|$$

Squaring and simplifying we get:

$$x^2 + (y - p)^2 = (y + p)^2$$
$$x^2 + y^2 - 2yp + p^2 = y^2 + 2yp + p^2$$
$$x^2 = 4py$$
$$(1) \quad y = \frac{1}{4p}x^2$$

$$(1) \quad y = \frac{1}{4p}x^2$$

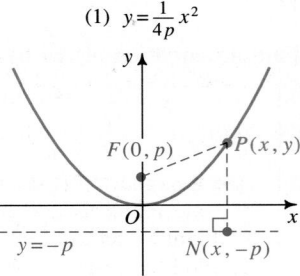

The figures below show other parabolas with vertex $(0, 0)$. The derivation of each equation is analogous to that shown above.

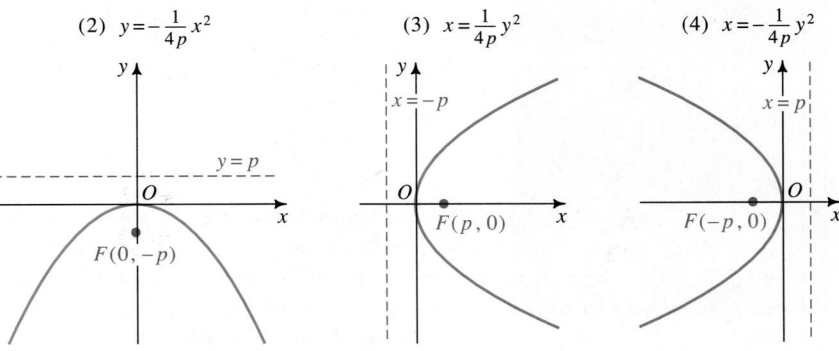

$$(2) \quad y = -\frac{1}{4p}x^2 \qquad (3) \quad x = \frac{1}{4p}y^2 \qquad (4) \quad x = -\frac{1}{4p}y^2$$

Example 1　Find the focus and directrix of each parabola whose equation is given.

　　　a. $y = 2x^2$　　　　　　　　　**b.** $x = \frac{1}{20}y^2$

Solution　**a.** Comparing $y = 2x^2$ with $y = \frac{1}{4p}x^2$, we see that $p = \frac{1}{8}$. Thus, the focus

　　　　　is $F\left(0, \frac{1}{8}\right)$ and the directrix is $y = -\frac{1}{8}$.

　　　　b. Comparing $x = \frac{1}{20}y^2$ with $x = \frac{1}{4p}y^2$, we see that $p = 5$. Thus, the

　　　　　focus is $F(5, 0)$ and the directrix is $x = -5$.

Example 2　Find an equation of the parabola with vertex $(0, 0)$ and directrix $x = 2$.

Solution　A sketch shows that the parabola has an equation of type (4), $x = -\frac{1}{4p}y^2$. Since p is the distance from the vertex to the directrix, $p = 2$. Thus, $x = -\frac{1}{8}y^2$.

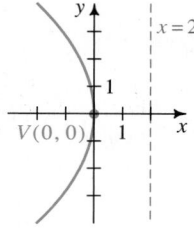

CLASS EXERCISES

1. Points A, B, C, and P are on the parabola $y = \frac{1}{4}x^2 - 1$, whose focus is $F(0, 0)$ and whose directrix d is the line $y = -2$. (See the figure at the left below.)

　　a. Find the distances from A to F and A to d. 1, 1
　　b. Find the distances from B to F and B to d. 2, 2
　　c. Find the distances from C to F and C to d. 5, 5
　　d. Find the distances from P to F and P to d. $\sqrt{x^2 + y^2}$, $y + 2$

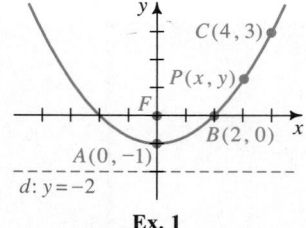

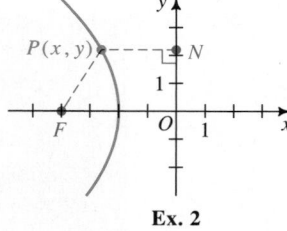

　　　　　　Ex. 1　　　　　　　　　　　　**Ex. 2**

2. The point $P(x, y)$ is on the parabola whose focus F is $(-4, 0)$ and whose directrix is the y-axis. (See the figure at the right above.) (1) $\sqrt{(x + 4)^2 + y^2}$; (2) $|x|$
　　a. Express (1) the distance PF and (2) the distance PN in terms of x and y.
　　b. Explain how to *derive* an equation of the parabola. Express PF and PN in terms of x and y. Set the expressions equal to one another and simplify.

Analytic Geometry　**239**

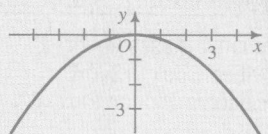

3. **Writing** In a few sentences, describe how to tell whether a parabola opens up, down, left, or right when you are given an equation for the parabola.

Give the focus and the directrix of each parabola. Tell whether the parabola opens up, down, left, or right.

4. $y = \frac{1}{4}x^2$ 5. $y = -\frac{1}{4}x^2$ 6. $x = y^2$ 7. $x = -y^2$

8. **Discussion** You know that a parabola can open up, down, left, or right. Is it possible for a parabola to open in some other direction? Explain.

9. **Discussion** Explain how to use the graph of the equation $y = x^2$ to graph the equation $y - k = (x - h)^2$. Shift $y = x^2$ to the right h and up k.

WRITTEN EXERCISES

For each parabola give the coordinates of its vertex and focus and the equation of its directrix.

 1. a. $y = \frac{1}{8}x^2$ b. $x = \frac{1}{8}y^2$ 2. a. $y = -\frac{1}{12}x^2$ b. $x = -\frac{1}{12}y^2$

3. a. $y = -2x^2$ b. $x = -2y^2$ 4. a. $y = x^2 + 1$ b. $x = y^2 + 1$

In Exercises 5–8, for each parabola give the coordinates of its vertex and focus and the equation of its directrix. (Use translations.)

5. $y - 1 = \frac{1}{4}(x - 2)^2$ 6. $y + 3 = \frac{1}{8}(x - 5)^2$

7. $x - 4 = (y - 7)^2$ 8. $x + 2 = -2(y - 3)^2$

9. **Telecommunications** A satellite dish used to receive television signals has a parabolic cross-section as shown at the left below. Incoming signals that are parallel to the axis of the parabola are reflected to its focus. Thus, the incoming signal is magnified. Find an equation of the parabola. $x = \frac{1}{16}y^2$

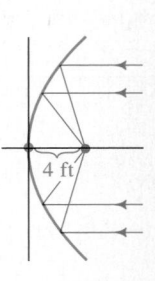

10. *Investigation* On a sheet of graph paper, carefully draw a *large* graph of the equation $x = \frac{1}{16}y^2$. Draw a tangent line to the parabola at any point P. Mark two points A and B on the tangent line. Join P to the focus F (which you must first determine). Draw a ray through P parallel to the axis of the parabola. Label a point C on the ray. With a protractor, measure $\angle CPB$ and $\angle FPA$. What do you observe? Repeat the experiment for another point P on the parabola. The angles are equal.

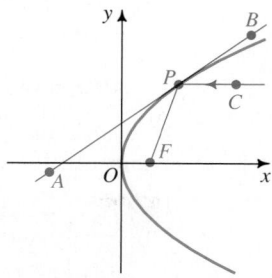

Find an equation of each parabola. Sketch its graph.

11. Focus, $(-1, 0)$; directrix, $x = 1$ **12.** Focus, $(0, 0.25)$; directrix, $y = -0.25$

13. Vertex, $(0, 0)$; focus, $(0, -0.25)$ **14.** Vertex, $(0, 0)$; focus, $(5, 0)$

15. Vertex, $(0, 0)$; directrix, $x = 4$ **16.** Vertex, $(0, 0)$; directrix, $y = -2$

17. Focus, $(0, 2)$; directrix, $y = 0$ **18.** Focus, $(3, 0)$; directrix, $x = -3$

Sketch the graph of each "half-parabola." You may use a graphing calculator if you wish.

19. **a.** $y = \sqrt{x}$ **b.** $y = -\sqrt{x}$ **c.** $y = -\sqrt{x - 3}$ **d.** $y = \sqrt{x + 2} - 1$

20. **a.** $x = \sqrt{y}$ **b.** $x = -\sqrt{y}$ **c.** $x = \sqrt{y - 4}$ **d.** $x = -\sqrt{y + 3} + 1$

21. The diagram at the right shows many points $P(x, y)$ that are equidistant from $F(0, 6)$ and the x-axis.

  a. Write an equation that states that $PF = PN$. $\sqrt{x^2 + (y - 6)^2} = y$

  b. Rewrite the equation in the form

$$y - k = \frac{1}{4p}(x - h)^2.$$
$$y - 3 = \frac{1}{12}(x - 0)^2$$

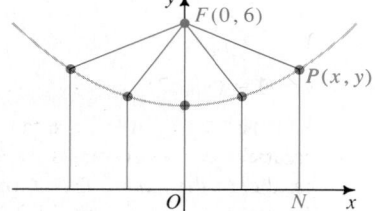

The coordinates of F and an equation of line d are given. Write and simplify an equation that specifies the set of points $P(x, y)$ that are equidistant from F and d.

22. $F(0, 2)$ **23.** $F(0, -3)$ **24.** $F(-1, 3)$ **25.** $F(-7, -5)$
    d: $y = -2$ d: $x = -2$ d: $x = 0$ d: $y = -7$
                                                              See below.

Find the vertex, focus, and directrix of each parabola. Graph the equation. Check your graph with a graphing calculator.

26. $y = 2x^2 - 8x + 3$ **27.** $y = 5 - 6x - 3x^2$

28. $6x - x^2 = 8y + 1$ **29.** $4y = x^2 - 8x + 12$

30. $x = y^2 - 2y - 5$ **31.** $y^2 - 6y + 16x + 25 = 0$

22. $y = \frac{1}{8}x^2$ **23.** $x = \frac{1}{4}y^2 + \frac{3}{2}y + \frac{5}{4}$ **24.** $x = -\frac{1}{2}y^2 + 3y - 5$ **25.** $y = \frac{1}{4}x^2 + \frac{7}{2}x + \frac{25}{4}$

Analytic Geometry   **241**

1. Solve the system graphically: $xy = 3$
$$3x - 2y = 3$$

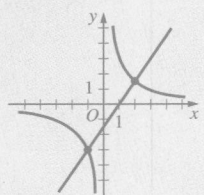

$(-1, -3), \left(2, \dfrac{3}{2}\right)$

2. Solve the system by substitution:
$$y = x + 1$$
$$y = x^2 - 2x - 3$$
$(4, 5)(-1, 0)$

3. Solve the system by using linear combination:
$$x^2 + y^2 = 4$$
$$4x^2 - y^2 = 1$$
$(1, \sqrt{3}), (-1, \sqrt{3}),$
$(-1, -\sqrt{3}), (1, -\sqrt{3})$

4. Solve the system by using linear combination:
$$x^2 + y^2 = 4$$
$$4x^2 + 9y^2 = 36$$
$(0, 2), (0, -2)$

Motivating the Section

If a system of second-degree equations is used to model the elliptical orbits of the Earth and a comet, we can solve the system to determine where the orbits cross.

Communication Note

Be sure to stress that "solving a system" means finding all *real* solutions. Students should reject solutions that are not real, as illustrated in Example 3 on page 244.

6-6 Systems of Second-Degree Equations

Objective   *To solve systems of second-degree equations.*

In this section, we will investigate three methods of solving a system of second-degree equations in two variables.

> ### Methods for Solving Systems of Second-Degree Equations
>
> **Method 1**   Algebraic approaches, such as substitution or elimination
> **Method 2**   A graphing calculator or a computer graphing utility
> **Method 3**   A combination of algebraic and graphing methods

 In the first solution to Example 1, you can see how to use technology to solve a system of second-degree equations.

Example 1   Solve the system: $xy = 6$ and $x^2 + 4y^2 = 64$

Solution 1   To solve the system graphically, you must first solve each equation for y.

$$y = \frac{6}{x} \qquad\qquad \text{hyperbola}$$

$$y = \sqrt{16 - \frac{x^2}{4}} \qquad \text{upper half of ellipse}$$

$$y = -\sqrt{16 - \frac{x^2}{4}} \qquad \text{lower half of ellipse}$$

The figure shows a calculator or computer display of the graphs.

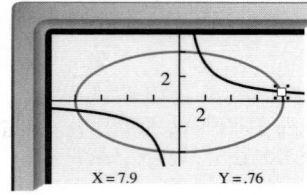

As you can see, there are four solutions. The figure suggests that the solutions in the third quadrant are reflections in the origin of the two solutions in the first quadrant. (You can verify this: If (a, b) satisfies each equation, show that $(-a, -b)$ satisfies each equation.) Using a calculator or computer, you can find that approximate solutions are $(1.5, 3.9)$, $(7.9, 0.76)$, $(-1.5, -3.9)$, and $(-7.9, -0.76)$.

Alternatively, the system in Example 1 can be solved algebraically by using substitution.

Solution 2

Step 1 *Solve one equation for x or y.* Solve $xy = 6$ for x: $x = \dfrac{6}{y}$

Step 2 *Substitute in the other equation and solve.* Substitute $\dfrac{6}{y}$ for x in the second equation.

$$\left(\frac{6}{y}\right)^2 + 4y^2 = \frac{36}{y^2} + 4y^2 = 64$$
$$4y^4 - 64y^2 + 36 = 0$$
$$y^4 - 16y^2 + 9 = 0$$

Using the quadratic formula, we get:

$$y^2 = \frac{16 \pm \sqrt{16^2 - 4 \cdot 9}}{2} = 8 \pm \sqrt{55} \approx 15.4 \text{ or } 0.584$$
$$y \approx \pm 3.9 \text{ or } \pm 0.76$$

Step 3 *Find the corresponding values of x.* When $y = 3.9$, for example, $x = 6 \div 3.9 \approx 1.5$. The four approximate solutions are $(1.5, 3.9)$, $(7.9, 0.76)$, $(-1.5, -3.9)$, and $(-7.9, -0.76)$.

Example 2 The figure shows two circles with equations $x^2 + y^2 = 20$ and $(x - 5)^2 + (y - 5)^2 = 10$. Find the coordinates of their points of intersection, points A and B.

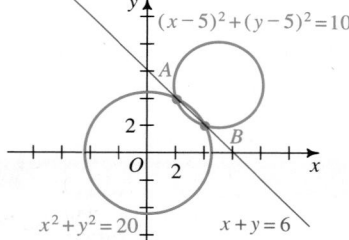

Solution Write both equations in expanded form and then subtract.

$$\begin{aligned} x^2 + y^2 &= 20 \quad (1)\\ x^2 + y^2 - 10x - 10y &= -40 \quad (2)\\ \hline 10x + 10y &= 60\\ x + y &= 6 \quad (3) \end{aligned}$$

Equation (3) is the line containing A and B. From (3), $y = 6 - x$. Then substituting in (1) we get:

$$x^2 + (6 - x)^2 = 20$$
$$x^2 - 6x + 8 = 0$$
$$(x - 2)(x - 4) = 0$$
$$x = 2 \text{ or } 4$$

When $x = 2$, $y = 6 - x = 4$. When $x = 4$, $y = 6 - x = 2$. Thus, the intersection points are $A(2, 4)$ and $B(4, 2)$.

Analytic Geometry **243**

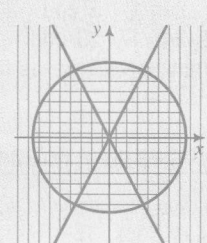

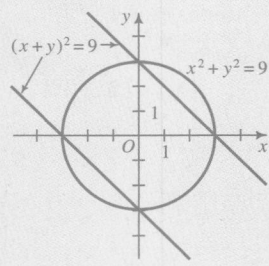
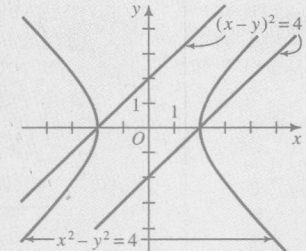
Example 3   Solve $2x^2 + y^2 = 4$ and $x^2 - 2y^2 = 12$ simultaneously.

Solution 1   Multiply the first equation by 2. Then add.

$$
\begin{array}{r}
4x^2 + 2y^2 = 8 \\
x^2 - 2y^2 = 12 \\
\hline
5x^2 = 20
\end{array}
$$

Therefore, $x = \pm 2$. Substitute $x = \pm 2$ in either equation to obtain $y = \pm 2i$. Since there is no solution (x, y) where x and y are real numbers, $2x^2 + y^2 = 4$ and $x^2 - 2y^2 = 12$ have no common solution.

Alternatively, you can solve the system by graphing the upper half of the ellipse and the hyperbola and then using symmetry.

Solution 2   Solving $2x^2 + y^2 = 4$ and $x^2 - 2y^2 = 12$ for y, we get $y = \sqrt{4 - 2x^2}$ for the upper half of the ellipse and $y = \sqrt{\dfrac{x^2}{2} - 6}$ for the upper half of the hyperbola. By symmetry, we can see that there are no solutions.

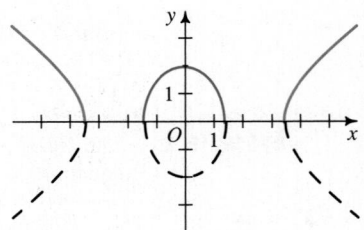

CLASS EXERCISES

1. **Visual Thinking** Two quadratic equations in two variables may have 4, 3, 2, 1, or 0 real solutions. Illustrate each of these cases with a circle and a parabola.
2. Compare the graphs of $x^2 + y^2 = 9$ and $(x + y)^2 = 9$.
3. Compare the graphs of $x^2 - y^2 = 4$ and $(x - y)^2 = 4$.

WRITTEN EXERCISES

Sketch the graphs of each pair of circles to determine the number of points of intersection. If the circles are tangent or fail to intersect, say so. Then solve the system.

A
1. $x^2 + y^2 = 16$, $(x - 4)^2 + y^2 = 16$ $(2, 2\sqrt{3})$, $(2, -2\sqrt{3})$
2. $x^2 + y^2 = 4$, $x^2 + (y - 6)^2 = 25$ **2.** $\left(-\dfrac{\sqrt{39}}{4}, \dfrac{5}{4}\right)$, $\left(\dfrac{\sqrt{39}}{4}, \dfrac{5}{4}\right)$
3. $x^2 + y^2 = 20$, $(x - 2)^2 + (y + 1)^2 = 13$ $(4, 2)$, $(0.8, -4.4)$
4. $x^2 + y^2 - 4y = 0$, $x^2 + y^2 - 2x = 4$ $(2, 2)$, $(-1.2, 0.4)$
5. $x^2 + y^2 = 5$, $x^2 + y^2 - 12x + 6y = -25$ Tangent at $(2, -1)$
6. $x^2 + y^2 = 4$, $x^2 + y^2 - 6x - 6y = -14$ No point of intersection

7. b. $9x^2 + 9y^2 = 162$; a circle

7. *Visual Thinking* Consider the ellipses $4x^2 + 5y^2 = 81$ and $5x^2 + 4y^2 = 81$.

    a. In how many points do they intersect? Give a convincing argument to justify your answer. Illustrate your answer graphically. 4

    b. Add the given equations. What is the graph of the resulting equation? See above.

    c. Why do the intersection points of the ellipses lie on the graph from part (b)?

8. Two ellipses have equations $5x^2 + 3y^2 = 64$ and $3x^2 + 5y^2 = 64$. Show algebraically that their points of intersection are on the circle $x^2 + y^2 = 16$.

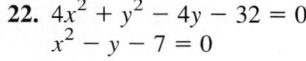 **Solve each system algebraically or with a graphing calculator or computer software. (Note: For many of these systems, an algebraic approach may be more efficient.)**

9. $x^2 + 4y^2 = 16$
  $x^2 + y^2 = 4$

10. $4x^2 + y^2 = 16$
  $x^2 - y^2 = -4$

11. $x^2 + 9y^2 = 36$
  $x^2 - 2y = 4$

12. $4x^2 + 4y^2 = 25$
  $2x + y^2 = 1$

13. $x^2 + y^2 = 25$
  $xy = -12$

14. $y^2 - x^2 = 64$
  $xy = 24$

15. $x^2 - y^2 = 1$
  $y = -1 - x^2$

16. $9x^2 + 4y^2 = 36$
  $x^2 + y^2 = 16$

17. $xy = 4$
  $y = -4x^2$

18. $9x^2 - 16y^2 = 144$
  $x + y^2 = -4$

19. $(x + 3)^2 + y^2 = 1$
  $x = -y^2$

20. $x^2 + 6y^2 = 9$
  $5x^2 + y^2 = 16$

B **21.** $x^2 + y^2 = 9$
  $x^2 + y^2 + 8x + 7 = 0$

22. $4x^2 + y^2 - 4y - 32 = 0$
  $x^2 - y - 7 = 0$

23. a. The circles $x^2 + y^2 = 20$ and $(x - 6)^2 + (y - 3)^2 = 5$ are tangent. Find the coordinates of the point of tangency and then find an equation of the common internal tangent shown. (*Hint*: How is this line related to the line containing the centers?) See below.

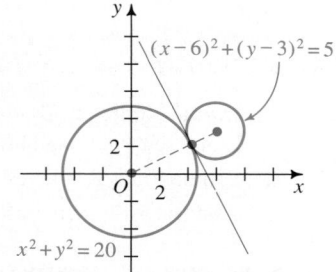

    b. If you subtract the equations of the two circles from each other, you get a linear equation. Find this equation. Compare it with the equation found in part (a).

24. a. Find the radii and the distance between the centers of the circles with equations $x^2 + y^2 = 225$ and $(x - 6)^2 + (y - 8)^2 = 25$. See below.

    b. Use your answers to part (a) to conclude that the circles must be internally tangent. Make a sketch.

    c. Find an equation of the common tangent of the two circles. $3x + 4y = 75$

24. a. Radii: 15, 5; distance: 10

Graph the solution set of each inequality or system of inequalities.

25. $x^2 - 9y^2 \le 0$

26. $y^2 - 2xy \ge 0$

27. $x^2 + 4y^2 - 10x + 24y + 61 \le 0$

28. $x^2 - y^2 + 2x - 2y \ge 0$

23. a. Tangent at (4, 2); $y = -2x + 10$ **b.** $y = -2x + 10$

**Additional Answers
Written Exercises**

7. c. If a point (a, b) is on both ellipses, then $4a^2 + 5b^2 = 81$ and $5a^2 + 4b^2 = 81$, so $9a^2 + 9b^2 = 162$.

8. If a point (a, b) is on both ellipses, $5a^2 + 3b^2 = 64$ and $3a^2 + 5b^2 = 64$, so $8a^2 + 8b^2 = 128$, or $a^2 + b^2 = 16$. Thus, (a, b) is on the circle $x^2 + y^2 = 16$.

9. $(0, 2)$, $(0, -2)$

10. $\left(\dfrac{2\sqrt{15}}{5}, \dfrac{4\sqrt{10}}{5}\right)$, $\left(\dfrac{2\sqrt{15}}{5}, -\dfrac{4\sqrt{10}}{5}\right)$, $\left(-\dfrac{2\sqrt{15}}{5}, \dfrac{4\sqrt{10}}{5}\right)$, $\left(-\dfrac{2\sqrt{15}}{5}, -\dfrac{4\sqrt{10}}{5}\right)$

11. $(0, -2)$, $\left(\dfrac{2\sqrt{17}}{3}, \dfrac{16}{9}\right)$, $\left(-\dfrac{2\sqrt{17}}{3}, \dfrac{16}{9}\right)$

12. $\left(-\dfrac{3}{2}, 2\right)$, $\left(-\dfrac{3}{2}, -2\right)$

13. $(4, -3)$, $(-4, 3)$, $(3, -4)$, $(-3, 4)$

14. $(2\sqrt{2}, 6\sqrt{2})$, $(-2\sqrt{2}, -6\sqrt{2})$

15. no point of intersection

16. no point of intersection

17. $(-1, -4)$

18. $(-4, 0)$

19. no point of intersection

20. $(\sqrt{3}, 1)$, $(\sqrt{3}, -1)$, $(-\sqrt{3}, -1)$, $(-\sqrt{3}, 1)$

21. $(-2, -\sqrt{5})$, $(-2, \sqrt{5})$

22. $(-3, 2)$, $(3, 2)$, $(\sqrt{5}, -2)$, $(-\sqrt{5}, -2)$

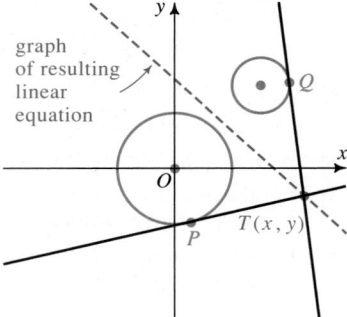

Using Technology

Graphing calculators may be used for Exercises 25–30.

Error Analysis

Students may shade the wrong part of the plane when solving Exercises 25–30. Remind students to check a point in the shaded region to make sure it satisfies each inequality.

29. $9x^2 - y^2 \geq 0$
   $x^2 + y^2 \leq 10$

30. $x^2 + 4y^2 \leq 16$
   $x^2 - 2xy \geq 0$

C 31. Carpentry A board 90 cm long just fits inside the bottom of a box 80 cm long and 60 cm wide. How wide is the board? About 10.6 cm

32. **a.** Given equations of two non-intersecting circles, show that the graph of the difference of the equations consists of those points $T(x, y)$ for which tangents $\overline{TP}$ and $\overline{TQ}$ are congruent.

b. Does the result in part (a) hold if the circles intersect? Yes

graph of resulting linear equation

COMMUNICATION: Reading

René Descartes was the first mathematician to use letters from the end of the alphabet to represent variables and letters from the beginning of the alphabet to represent constants, much as we do today. In the equation $Ax^2 + Bxy + Cy^2 + Dx + Ey + F = 0$, the letters A, B, C, D, E, and F represent constants, and the letters x and y represent variables.

The *American Heritage Dictionary* entry for the letter *c* is given below. Pick a letter of the alphabet. Write as many mathematical or scientific uses of that letter as you can without using a dictionary. Use a dictionary to expand your list.

c, C (sē) *n., pl.* **c's** or *rare* **cs**. **C's** or **Cs. 1.** The third letter of the modern English alphabet. See **alphabet. 2.** Any of the speech sounds represented by this letter. **c, C, c., C. Note:** As an abbreviation or symbol, *c* may be a small or a capital letter, with or without a period. Established forms or those generally preferred precede the definition. When no form is given, all four forms are in general use in that sense. **1. c** *Physics.* candle. **2. C** *Electricity.* capacitance. **3. c., C.** capacity. **4. c., C.** cape. **5. c** carat. **6. C** The symbol for the element carbon. **7. c., C.** carton. **8. c., C.** case. **9. c., C.** *Baseball.* catcher. **10. C.** Catholic. **11. C** Celsius. **12. C.** Celtic. **13. c., C.** cent. **14. c** centi-. **15. C** centigrade. **16. c., C.** centime. **17. c., C.** century. **18. C.** chancellor. **19. c., C.** chapter. **20. C** *Physics.* charge conjugation. **21. C.** chief. **22. c., C.** church. **23. c., C.** circa. **24. C.** city. **25. c.** cloudy. **26. C.** companion. **27. c., C.** congius. **28. C.** Congress. **29. C.** Conservative. **30. c, C** *Mathematics.* constant. **31. c., C.** consul. **32. c., C.** copy. **33. c., C.** copyright. **34. c., C.** corps. **35. C** coulomb. **36. C.** court. **37. c** cubic. **38. c.** cup. **39. C** The Roman numeral for 100 (Latin *centum*). **40.** The third in a series. **41. C** The third highest in quality or rank: *a mark of C on a term paper.* **42. C** *Music.* **a.** The first tone in the scale of C major, or the third tone in the relative minor scale. **b.** The key or a scale in which C is the tonic. **c.** A written or printed note representing this tone. **d.** A string, key, or pipe tuned to the pitch of this tone.

6-7 A New Look at Conic Sections

Objective *To define conic sections in terms of eccentricity and to classify the graph of a second-degree equation by examining the coefficients in the equation.*

Although the various conic sections look quite different from one another, they do have some common properties.
1. The conic sections result from slicing a double cone with a plane. (See page 213.)
2. The conic sections can all be obtained from the single definition given below.
3. The conic sections all have second-degree equations. (See the discussion and the theorem on the next page.)

Common Definition of Conic Sections

Let F (the **focus**) be a fixed point not on a fixed line d (the **directrix**). Let P be a point in the plane of d and F, and let PD be the perpendicular distance from P to d. Consider the set of points for which the ratio $PF:PD$ is the constant e. The number e is the **eccentricity** of the conic section.

This set of points is:
(1) an ellipse if $0 < e < 1$;
(2) a parabola if $e = 1$;
(3) a hyperbola if $e > 1$.

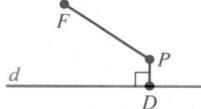

The diagram below suggests that once line d and point F are specified, the parabola (shown in blue) with directrix d and focus F separates the set of ellipses and the set of hyperbolas having this same directrix and focus. Notice that as the value of e approaches 0, the ellipse becomes more circular. The eccentricity of a circle is 0.

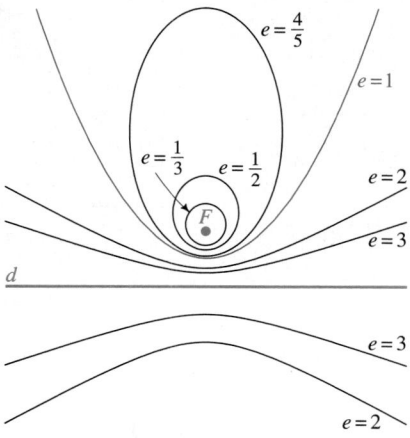

Warm-Up Exercises

Describe the graph of each equation.

1. $x^2 + y^2 = 0$ the point $(0, 0)$

2. $x^2 + y^2 = -1$ no graph

3. $x^2 - y^2 = 0$ the lines $y = x$ and $y = -x$

4. $x^2 - y^2 = -1$ hyperbola with a vertical major axis

Motivating the Section

The eccentricity of the orbit of the comet Kohoutek is so close to 1 that its orbit is very elongated and its period is very long. The concept of eccentricity and the single definition of the conic sections presented in this section help students realize that conics do have common properties and that algebraically they can all be represented with second-degree equations.

Mathematical Note

The angle of the plane that intersects a cone is determined by the value of $B^2 - 4AC$. If t is the measure of the acute angle between the axis of a cone and the edge of the cone, and θ is the smallest angle between the axis and the plane, $t < \theta < 90°$ for an ellipse, $\theta = t$ for a parabola, and $\theta < t$ for a hyperbola.

1. In the figure shown below, point F has coordinates $(0, 9)$ and line d has equation $y = 4$. If the distance from $P(x, y)$ to F is $\frac{3}{2}$ the distance from P to d, show that P lies on a hyperbola.

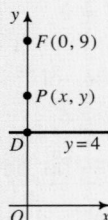

The sketch shows that P is a point such that $PF = \frac{3}{2}PD$.

If $\frac{PF}{PD} = \frac{3}{2}$, $2PF = 3PD$.

$2\sqrt{(x - 0)^2 + (y - 9)^2} = 3\sqrt{(x - x)^2 + (y - 4)^2}$;

$4x^2 + 4(y - 9)^2 = 9(y - 4)^2$;

$4x^2 - 5y^2 = -180$;

$\frac{y^2}{36} - \frac{x^2}{45} = 1$

This is an equation of a hyperbola.

2. The graph of each given equation is not a degenerate conic. Use the theorem in this section to identify the graph of the equation.
   a. $4x^2 + 4xy - y^2 = 16$
   b. $x^2 + \sqrt{3}xy + 2y^2 = 1$

   a. $B^2 - 4AC =$
      $4^2 - 4(4)(-1) = 32 > 0$.
      Thus, the graph is a hyperbola.
   b. $B^2 - 4AC =$
      $(\sqrt{3})^2 - 4(1)(2) = -5 < 0$. Thus, the graph is an ellipse.

Example 1 Let the focus F be $(0, 3)$ and let the directrix d have equation $y = 12$. Find the equation of the set of points P for which $\frac{PF}{PD} = \frac{1}{2}$, and identify the graph.

Solution

$$2(PF) = PD$$
$$2\sqrt{(x - 0)^2 + (y - 3)^2} = |12 - y|$$
$$4[x^2 + (y - 3)^2] = (12 - y)^2$$
$$4x^2 + 4y^2 - 24y + 36 = 144 - 24y + y^2$$
$$4x^2 + 3y^2 = 108$$
$$\frac{x^2}{27} + \frac{y^2}{36} = 1$$

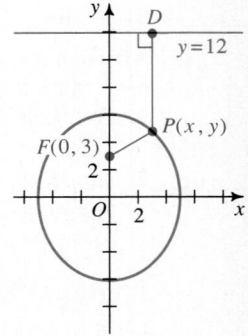

The graph is an ellipse with vertices $(0, \pm6)$ and foci $(0, \pm3)$.

Second-Degree Equations in Two Variables

A conic section can be considered the graph of a *second-degree equation*

$$Ax^2 + Bxy + Cy^2 + Dx + Ey + F = 0,$$

where A, B, and C are not all 0. For example, if $A = 4$, $B = 0$, $C = 3$, $D = 0$, $E = 0$, and $F = -108$, then we have $4x^2 + 3y^2 - 108 = 0$. In the example above, we saw that the graph of this equation is an ellipse.

So far we have considered primarily those second-degree equations in which $B = 0$. The one exception has been the hyperbola of the form $xy = k$. The theorem below enables us to classify a conic section by looking at its equation.

Classifying Second-Degree Equations

Consider the equation $Ax^2 + Bxy + Cy^2 + Dx + Ey + F = 0$. If A, B, and C are not all 0 and if the graph is not degenerate, then:
(1) the graph is a circle or an ellipse if $B^2 - 4AC < 0$; (In a circle, $B = 0$ and $A = C$.)
(2) the graph is a parabola if $B^2 - 4AC = 0$;
(3) the graph is a hyperbola if $B^2 - 4AC > 0$.

Example 2 Identify the graph of the equation $x^2 - 2xy + 3y^2 - 1 = 0$.

Solution
$$B^2 - 4AC = (-2)^2 - 4(1)(3)$$
$$= -8 < 0$$
Since $B \neq 0$, the graph is an ellipse.

1. b. 8, 10, 5, $|y + 6|$

Use the diagram for Exercises 1 and 2.

1. Find the distances from A, B, C, and P to:
   4, 6, 8, $|x - 3|$
   a. line d_1 with equation $x = 3$.
   b. line d_2 with equation $y = -6$.

2. State an equation that expresses the fact that:
   a. $PO = 2PA$
   b. $PO = \frac{1}{2}PB$
   c. the distance between P and C is equal to the distance between P and d_2.
   d. the distance between P and B is two times the distance between P and d_2.
   e. the distance between P and A is $\frac{1}{2}$ the distance between P and d_2.

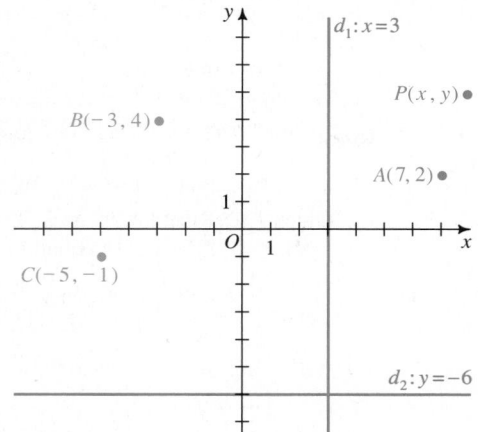

3. **Aviation** The shock wave of a supersonic jet flying parallel to the ground is a cone. Points on the intersection of the cone and the ground receive a sonic boom at the same time. What conic section is the intersection? Hyperbola

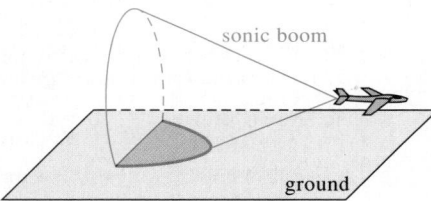

sonic boom

ground

4. Given that the following equations are not degenerate conics, tell whether the graph of each is a circle, an ellipse, a parabola, or a hyperbola.
   a. $x^2 - 3xy + 2y^2 + 2x - y + 6 = 0$ Hyperbola
   b. $3x^2 - 4xy + 2y^2 - 3y = 0$ Ellipse
   c. $x^2 - 6xy + 9y^2 + x - y - 1 = 0$ Parabola
   d. $144x^2 + 144y^2 - 216x + 96y - 47 = 0$ Circle

6. Two lines: $x - 3y = 0$ and $x + 3y = 0$

Describe the graph of each degenerate conic. If there is no graph, say so. No graph

5. $x^2 + 9y^2 = 0$ (0,0)   6. $x^2 - 9y^2 = 0$ See above. 7. $x^2 + 9y^2 + 1 = 0$

8. **Discussion** Describe the graph of $Ax^2 + Cy^2 = 0$ if A and C are not both 0.

9. **Investigation** Shine a flashlight against a wall so that the edge of the lighted area is (a) a circle, (b) a parabola, and (c) a hyperbola. Relate the results of your investigation to the diagrams of the conic sections on page 213.

Analytic Geometry **249**

Chapter 6   **249**

WRITTEN EXERCISES

A **1. a.** Find an equation for the ellipse.

  b. Find the coordinates of F, the focus on the positive x-axis.

  c. The distance from $P(x, y)$ to F is 0.5 times the distance from P to the line $d: x = 4$. Write and simplify an equation to show that P lies on the ellipse shown.

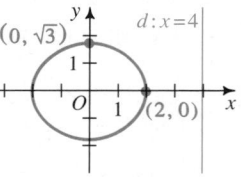

2. Given $F(0, 8)$ and $d: y = \frac{25}{2}$. The distance from $P(x, y)$ to F is $\frac{4}{5}$ the distance from P to d. Write an equation to show that P lies on an ellipse.

3. Given $F(-6, 0)$ and $d: x = -\frac{3}{2}$. The distance from $P(x, y)$ to F is twice the distance from P to d. Write an equation to show that P lies on a hyperbola.

The following are not degenerate conics. Identify the graph of each.

4. $x^2 - 2xy - y^2 = 4$ Hyperbola

5. $x^2 - xy + 2y^2 = 2$ Ellipse

6. $y = x - \frac{1}{x}$ Hyperbola

7. $x^2 - 3xy + y^2 = 5$ Hyperbola

8. $x^2 + xy + y^2 - 4x\sqrt{2} - 4y\sqrt{2} = 0$ Ellipse

9. $x^2 + xy + y^2 - x\sqrt{2} - y\sqrt{2} = 0$ Ellipse

Describe the graph of each equation. If there is no graph, say so.

10. $4x^2 - y^2 = 0$

11. $4x^2 + y^2 = -1$ No graph

12. $y^2 - 3yx = 0$

10. Two lines: $2x - y = 0$, $2x + y = 0$

12. Two lines: $y = 0$, $y = 3x$

When an equation of an ellipse is given in the form shown on page 226, it can be shown that $e = \frac{c}{a}$. Use this fact in Exercises 13 and 14.

B **13. Astronomy** Earth's orbit is an ellipse with the sun almost at one focus. During January, Earth is closest to the sun, a distance of 9.14×10^7 mi. During July, Earth is farthest from the sun, a distance of 9.44×10^7 mi. Find the eccentricity of Earth's orbit. 0.016

14. Astronomy Halley's comet has an elliptical orbit with the sun at a focus. When it is closest to the sun, it passes within 8.8×10^7 km of the sun. Its greatest distance from the sun is about 5.282×10^9 km. Sketch the orbit and give its eccentricity. 0.967

Solve. 16. $\left(\dfrac{\sqrt{6}}{3}, \dfrac{2\sqrt{6}}{3}\right), \left(-\dfrac{\sqrt{6}}{3}, -\dfrac{2\sqrt{6}}{3}\right), (2, -1), (-2, 1)$

15. $x^2 + y^2 = 9 \ (\pm 3, 0), (0, \pm 3)$
$(x + y)^2 = 9$

16. $2x^2 + 3xy - 2y^2 = 0$
$x^2 + 2y^2 = 6$

17. $x^2 + y^2 = 40 \ (2\sqrt{5}, 2\sqrt{5}), (-2\sqrt{5}, -2\sqrt{5})$
$x^2 + 2xy - 3y^2 = 0 \ (6, -2), (-6, 2)$

18. $3x^2 - 4xy - 4y^2 = 0$
$x - 2y^2 + 4 = 0$

C 19. $x^2 + xy + y^2 = 3 \ (\pm\sqrt{3}, 0), (2, -1),$
$x^2 - y^2 = 3 \ (-2, 1)$

20. $x^2 + 3y^2 = 3$
$3x^2 - xy = 6$

21. The diagram at the left below shows an ellipse formed by a plane cutting through a cone. It also shows two spheres tangent to the plane of the ellipse at A and B. They are also tangent to the cone, touching the cone in the parallel circles C_1 and C_2. Let P be any point on the ellipse. (For simplicity, P is shown where it is.)
a. Explain why $PA = PM$ and $PB = PN$.
b. Prove $PA + PB$ is a constant, thus proving that A and B are the foci of an ellipse. (The spheres are called **Dandelin spheres** after the Belgian mathematician Germinal Pierre Dandelin (1794–1847) who discovered them in 1822.)

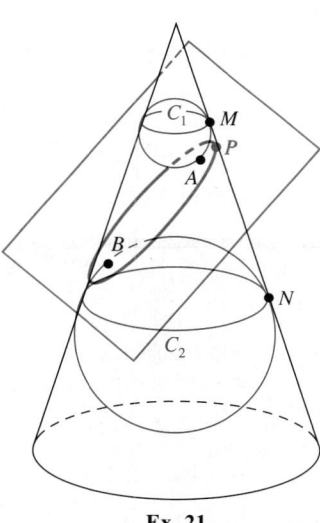

Ex. 21

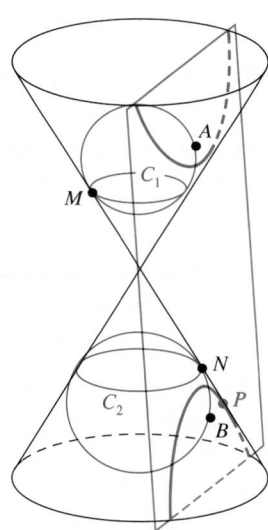

Ex. 22

22. The diagram at the right above shows a hyperbola formed by a plane cutting a double cone. Two spheres are tangent to the plane of the hyperbola at A and B. They are also tangent to the cone, touching the cone in the parallel circles C_1 and C_2. Let P be any point on the hyperbola. Show that $PA - PB$ is a constant, thus proving that A and B are the foci of a hyperbola.

Additional Answers
Chapter Test

1. Let the square have sides of length a. Let the four vertices be the origin, $A(a, 0)$, $B(a, a)$ and $C(0, a)$. Then, slope $\overline{AC} = \frac{a-0}{0-a} = -1$ and

   slope $\overline{OB} = \frac{0-a}{0-a} = 1$.

   (slope $\overline{AC}$)(slope $\overline{OB}$) = -1. Thus $\overline{AC} \perp \overline{OB}$. $AC = \sqrt{(a-0)^2 + (0-a)^2} = a\sqrt{2}$ and $OB = \sqrt{(a-0)^2 + (a-0)^2} = a\sqrt{2}$. Thus, $AC = OB$. Therefore, the diagonals are perpendicular and congruent.

6. $\frac{x^2}{9} + \frac{y^2}{16} = 1$

7. a. $\frac{y^2}{9} - \frac{x^2}{4} = 1$

   b. $\frac{x^2}{4} - \frac{y^2}{16} = 1$

8. Asymptotes: $x = 0$, $y = 0$

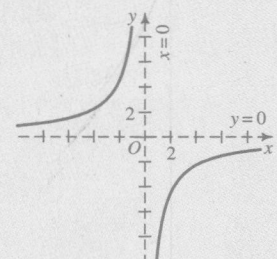

10. $x = \frac{1}{12}y^2$

Chapter Summary

1. To prove a theorem from geometry by using coordinates, introduce a coordinate system in which the figure involved has as many zero coordinates as possible. Choose one or more of the methods listed on page 217.

2. A *circle* with radius r and center $C(h, k)$ has an equation of the form
$$(x - h)^2 + (y - k)^2 = r^2.$$
To find the intersection of a line and circle, follow the procedure on page 221.

3. If F_1 and F_2 are two fixed points (the *foci*) in the plane and a is a positive real number, then the set of all points $P(x, y)$ in the plane such that
$$PF_1 + PF_2 = 2a$$
is an *ellipse*. An ellipse centered at the origin and having horizontal or vertical major axis has an equation of the form
$$\frac{x^2}{a^2} + \frac{y^2}{b^2} = 1 \text{ or } \frac{y^2}{a^2} + \frac{x^2}{b^2} = 1.$$

4. If F_1 and F_2 are two fixed points (the *foci*) in the plane and a is a positive real number, then the set of all points $P(x, y)$ in the plane such that
$$|PF_1 - PF_2| = 2a$$
is a *hyperbola*. A hyperbola centered at the origin and opening either horizontally or vertically has an equation of the form
$$\frac{x^2}{a^2} - \frac{y^2}{b^2} = 1 \text{ or } \frac{y^2}{a^2} - \frac{x^2}{b^2} = 1.$$

5. If F is a fixed point (the *focus*) in the plane and d (the *directrix*) is a fixed line not containing F, then the set of points equidistant from F and d is a *parabola*. A parabola whose vertex is at the origin and whose directrix is either horizontal or vertical has an equation of the form
$$y = \pm\frac{1}{4p}x^2 \text{ or } x = \pm\frac{1}{4p}y^2.$$

6. A system of two second-degree equations in two variables can be solved by algebraic methods, graphical methods, or a combination of algebraic and graphical methods.

7. Let F (the *focus*) be a fixed point not on a fixed line d (the *directrix*) and let e (the *eccentricity*) be positive. Let P be any point in the plane such that
$$\frac{PF}{PD} = e, \text{ where } PD \text{ is the distance from } P \text{ to } d.$$
If $0 < e < 1$, the set of points is an *ellipse*.
If $e = 1$, the set of points is a *parabola*.
If $e > 1$, the set of points is a *hyperbola*.
    Conic sections are graphs of second-degree equations. A conic section can be identified by examining the coefficients of its equation.

Chapter Test

1. Prove that the diagonals of a square are perpendicular and congruent. **6-1**

2. Find the coordinates of any points where the line $x - y = 2$ and the circle $x^2 + y^2 = 4$ intersect. (0, −2), (2, 0) **6-2**

3. Find an equation of the circle having $(2, 5)$ and $(-2, -1)$ as endpoints of a diameter. $x^2 + (y - 2)^2 = 13$

4. Find the coordinates of the vertices and the foci of the ellipse with equation $9x^2 + 5y^2 = 45$. Sketch the ellipse. $V(0, \pm 3)$, $F(0, \pm 2)$ **6-3**

5. Find the coordinates of any points where the line $2x + 3y = 6$ and the ellipse $4x^2 + 9y^2 = 36$ intersect. (3, 0), (0, 2)

6. Find an equation for the ellipse that has $(0, -4)$ and $(0, 4)$ as vertices and $(-3, 0)$ and $(3, 0)$ as endpoints of its minor axis.

7. Find an equation for the hyperbola that satisfies the given conditions. **6-4**
   a. Center at $(0, 0)$, a vertex at $(0, -3)$, and a focus at $(0, -\sqrt{13})$.
   b. A vertex at $(2, 0)$ and asymptotes with equations $y = \pm 2x$.

8. Graph the hyperbola $xy = -9$ and give equations for its asymptotes.

9. Find the coordinates of the vertex and focus, and the equation of the **6-5**
   directrix, of the parabola $y = \frac{1}{6}x^2$. $V(0, 0)$, $F\left(0, \frac{3}{2}\right)$, $y = -\frac{3}{2}$

10. Find an equation for the parabola whose directrix is $x = -3$ and whose focus is $(3, 0)$.

11. On a single set of axes, graph $x^2 + y^2 = 25$ and $x^2 + 10y^2 = 169$. **6-6**
    Solve these equations simultaneously. (3, ±4), (−3, ±4)

12. Given that the following equations are not degenerate conics, tell **6-7**
    whether the graph is a circle, an ellipse, a hyperbola, or a parabola.
    a. $4x^2 + 4y^2 - 8x + 24y - 15 = 0$ Circle
    b. $-x^2 + 2xy - 3y^2 + 12 = 0$ Ellipse
    c. $2x^2 - 12xy + 18y^2 + 2x - 6y + 24 = 0$ Parabola

13. Find and simplify an equation that expresses that $P(x, y)$ is equidistant from the point $F(1, 3)$ and the line $y = -4$. See below.

14. *Writing* In a few sentences, describe the possibilities for the graph of the equation $Ax^2 + Bxy + Cy^2 = 0$, where $AC = 0$.

13. $\sqrt{(x - 1)^2 + (y - 3)^2} = |y + 4|$ simplifies to $y + \frac{1}{2} = \frac{1}{14}(x - 1)^2$.

Analytic Geometry **253**

Additional Answers
Chapter Test

11.

14. If $A = B = C = 0$, the graph is the entire xy-plane. If $A = B = 0$ and $C \neq 0$, the graph is the x-axis; if $B = C = 0$ and $A \neq 0$, the graph is the y-axis; if $A = C = 0$ and $B \neq 0$, the graph is the x- and y-axes. If only $A = 0$, the graph is two lines: the x-axis and $Bx + Cy = 0$. If only $C = 0$, the graph is two lines: the y-axis and $Ax + By = 0$.

Suggested Assignments

Standard
Day 1: 254/1–12
Day 2: 255/13–24
Comprehensive
Day 1: 254/1–12
Day 2: 255/13–24

Additional Answers
Cumulative Review

1.

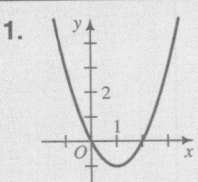

5.a.

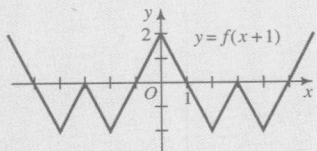

b.

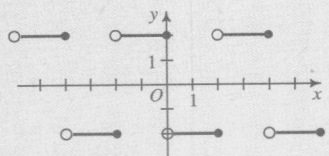

6.a. $f^{-1}(x) = \dfrac{8 - x}{2}$

b. No inverse

c. $f^{-1}(x) = \dfrac{1 + 2x}{x}$

7. c.

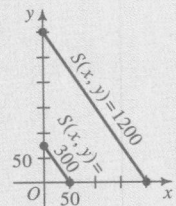

Cumulative Review

1. Sketch the graph of the function $f(x) = x^2 - 2x$. Then use the graph to find the range and zeros of f. Range $= \{y \mid y \ge -1\}$; zeros: 0, 2

2. If $f(x) = x - 1$ and $g(x) = 1 - x$, find:
  a. $(f + g)(x)$ 0   **b.** $(f - g)(x)$ $2x - 2$ **c.** $(f \cdot g)(x)$ $-x^2 + 2x - 1$
  d. $\left(\dfrac{f}{g}\right)(x)$ $-1,\ x \ne 1$ **e.** $(f \circ g)(x)$ $-x$ **f.** $(g \circ f)(x)$ $2 - x$

3. Tell whether the graph of each equation has symmetry in (i) the x-axis, (ii) the y-axis, (iii) the line $y = x$, and (iv) the origin.
  a. $xy^2 + x^2y = 4$ iii   **b.** $y = x^4 + 2x^2 + 1$ ii   **c.** $y^2 = |x| + 1$ i, ii, iv

4. Give the fundamental period and amplitude of each periodic function whose graph is shown. **a.** 6; 2   **b.** 4; 2

  a.

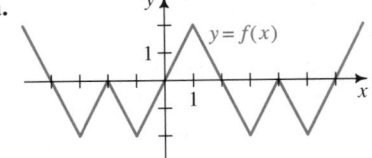

  b.

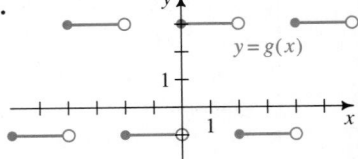

5. a. Using the graph of $y = f(x)$ in part (a) of Exercise 4, sketch the graph of $y = f(x + 1)$.
  b. Using the graph of $y = g(x)$ in part (b) of Exercise 4, sketch the graph of $y = g(-x) - 1$.

6. Tell whether each function f has an inverse. If f^{-1} exists, find a rule for $f^{-1}(x)$ and show that $f(f^{-1}(x)) = f^{-1}(f(x)) = x$.
  a. $f(x) = 8 - 2x$ **b.** $f(x) = x^3 - x^2$ **c.** $f(x) = \dfrac{1}{x - 2}$

7. Tickets to a show cost $6.00 for adults and $4.00 for children. Let x and y represent the numbers of tickets sold to adults and children, respectively.
  a. Find a rule for the total sales function $S(x, y)$. $S(x, y) = 6x + 4y$
  b. Find $S(40, 25)$ and $S(32, 48)$. 340; 384
  c. Sketch the graphs of the constant sales curves $S(x, y) = 300$ and $S(x, y) = 1200$ in an xy-plane.

8. Simplify.
  a. $(4^{-1} + 2^{-1})^2$ $\dfrac{9}{16}$ **b.** $(8y^4)(2y^{-3})^{-2}$ $2y^{10}$ **c.** $\dfrac{(3x^2)^{-1}}{6x^{-3}}$ $\dfrac{x}{18}$
  d. $\left(\dfrac{81}{64}\right)^{-1/2}$ $\dfrac{8}{9}$ **e.** $(27^{-2})^{-1/3}$ 9   **f.** $a^{3/4}(a^{-3/4} - a^{1/4})$ $1 - a$

9. Solve.
  a. $3^{4x+1} = 81$ $\dfrac{3}{4}$ **b.** $\dfrac{1}{5}x^{2/3} = 20$ 1000   **c.** $(x - 1)^{-2} = 25$ $\dfrac{4}{5}$ or $\dfrac{6}{5}$

10. Find an exponential function f such that $f(0) = \dfrac{1}{2}$ and $f(4) = \dfrac{9}{2}$. $f(x) = \dfrac{1}{2}(\sqrt{3})^x$

---

254   *Cumulative Review*

11. The table shows the number $N(t)$ of bacteria in a colony at various times t (in hours).

t	0	1	2	3	4	5
$N(t)$	150	189	238	300	378	476

    a. What is the doubling time for this colony of bacteria? 3 h
    b. Find an equation for $N(t)$. $N(t) = 150 \cdot 2^{t/3}$
    c. How many bacteria will there be after one full day? 38,400

12. Suppose you invest \$100 at 7% annual interest. Calculate the amount that you would have after one year if interest is compounded:
    a. quarterly \$107.19   **b.** monthly \$107.23   **c.** continuously   \$107.25

13. Which plan yields a greater return on an investment? Plan A

        Plan A: 8.1% annual rate compounded quarterly
        Plan B: 8% annual rate compounded continuously

14. Simplify.
    a. $\log \dfrac{1}{100}$ -2 **b.** $\log_2 16$ 4   **c.** $\log_{125} 5$ $\dfrac{1}{3}$
    d. $\ln \dfrac{1}{e^5}$ -5 **e.** $e^{2 \ln 3}$ 9   **f.** $10^{1 + \log 5}$ 50

15. Write each expression in terms of $\log M$ and $\log N$.
    a. $\log M^2 N^3$ **b.** $\log \sqrt{\dfrac{M^3}{N}}$ **c.** $\log 100 M \sqrt{N}$
  $2 \log M + 3 \log N$ $\dfrac{3}{2} \log M - \dfrac{1}{2} \log N$ $2 + \log M + \dfrac{1}{2} \log N$

16. Express y in terms of x.
    a. $\log y = \log x + 2$ $100x$ **b.** $\ln y = \ln 2 - 3 \ln x$ $\dfrac{2}{x^3}$ **c.** $\log y = 2.5x + 1$
                                                          $y = 10 \cdot 10^{2.5x}$

17. To the nearest hundredth, solve each equation.
    a. $(e^x)^2 = 64$ 2.08   **b.** $2^{x + 1} = 50$ 4.64   **c.** $(1.04)^x = 2$ 17.67

18. Prove that the line segments joining the midpoints of successive sides of any rhombus form a rectangle.

19. Find the center and radius of the circle $x^2 + y^2 - 4x + 14y + 28 = 0$. $(2, -7)$; 5

20. Sketch the ellipse $\dfrac{x^2}{9} + \dfrac{y^2}{16} = 1$. Find the coordinates of its center, vertices, and foci. C: $(0, 0)$; V: $(0, \pm 4)$; F: $(0, \pm \sqrt{7})$

21. Find an equation of the hyperbola centered at the origin that has a focus at $(\sqrt{5}, 0)$ and a vertex at $(2, 0)$. $\dfrac{x^2}{4} - y^2 = 1$

22. Sketch the parabola $x + 2 = -\dfrac{1}{4}(y - 1)^2$, and give its vertex, focus, and directrix. V: $(-2, 1)$; F: $(-3, 1)$; d: $x = -1$

23. Solve the system: $x^2 + y^2 = 25$ $(\pm 3, 4)$
                   $2x^2 - 3y = 6$

24. Given that $x^2 - 2xy + 4y^2 = 4$ is not a degenerate conic, identify the graph of the equation. Ellipse

18. *PQRS* is a rhombus. $A = (a, b)$; $B = (-a, b)$; $C = (-a, -b)$; $D = (a, -b)$. Slope of $\overline{AB} = 0$; slope of $\overline{BC}$ is undefined; therefore $\overline{AB} \perp \overline{BC}$. $AB = DC = 2a$ and $BC = AD = 2b$, so *ABCD* is a parallelogram. A parallelogram with one right angle is a rectangle, so *ABCD* is a rectangle.

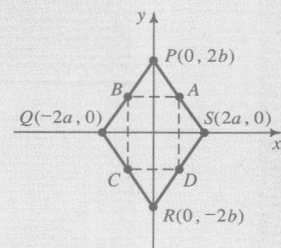

20.

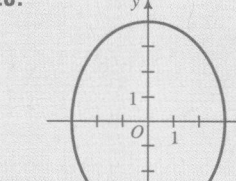

22.

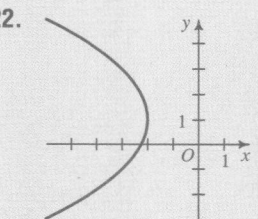

Chapter 7 Trigonometric Functions

Overview

Although students may be familiar with trigonometric *ratios* from previous mathematics courses, this chapter uses trigonometric *functions* to begin the study of trigonometry. In the first two sections, students convert between degree and radian measures of angles, and find arc lengths and areas of sectors of circles. The next three sections introduce students to the six trigonometric functions using the coordinates of the point where the terminal ray of an angle in standard position intersects a circle centered at the origin; students then evaluate and graph the trigonometric functions. The last section introduces the inverse trigonometric functions, which students also evaluate.

Objectives

7-1 To find the measure of an angle in either degrees or radians and to find coterminal angles.

7-2 To find the arc length and area of a sector of a circle and to solve problems involving apparent size.

7-3 To use the definitions of sine and cosine to find values of these functions and to solve simple trigonometric equations.

7-4 To use reference angles, calculators or tables, and special angles to find values of the sine and cosine functions and to sketch the graphs of these functions.

7-5 To find values of the tangent, cotangent, secant, and coseant functions and to sketch the functions' graphs.

7-6 To find values of the inverse trigonometric functions.

Supplementary Resources

Tests, pp. 19–20
Alt. Assess., pp. 22–24, 66
Activities, pp. 17–18
Using Tech., p. 23
Student Res. Guide, pp. 66–78

Software

Precalculus Plotter Plus
Activities, pp. 37–38, 39–40, 41–42
Function Plotter, Circular Function Quiz

Pacing Guide

Section	7-1	7-2	7-3	7-4	7-5	7-6	Review	Test	Total
Standard (days)	2	2	2	2	2	2	1	1	14
Comprehensive (days)	1	1	2	2	1	2	1	1	11

Teaching Notes

Presenting the Section

In this section, an angle is represented as a rotation about a point and is measured in degrees or radians. A large rotation is usually measured in revolutions.

You might need to emphasize the basic relationship between radians and degrees as described on page 258. Stress that to convert from radians to degrees you multiply by $\frac{180}{\pi}$, and to convert from degrees to radians you multiply by $\frac{\pi}{180}$. Alternatively, you can use the following proportion:

$$\frac{\text{radian measure}}{2\pi} = \frac{\text{degree measure}}{360}$$

You might want to encourage students to learn the degree and radian measures of frequently used angles, as shown on page 259.

Using Technology

The sine and inverse sine keys on scientific calculators can be used to help students learn how to convert between degrees and radians. For example, starting in degree mode, have students find sin 30°. (0.5) Then have them switch to radian mode and find Sin^{-1} 0.5. (0.5235987756) Therefore, 30° ≈ 0.524 radians. Note that this method of conversion is limited to angles that are between −90° and 90°, because that is the range of the inverse sine function. Discussing this method will give you an opportunity to introduce the concept of an inverse trigonometric function.

Many scientific calculators have a key to convert an angle expressed in degrees-minutes-seconds to decimal form, as well as a key to convert decimal angles to degrees-minutes-seconds form.

Cooperative Learning

Groups of several students each can go over Exercises 33 and 34 on page 262. Have students do their own research and discuss their answers and examples within each group. Have one group explain its answers to the class.

Presenting the Section

You might want to introduce this section by first reviewing "arc" and "intercept" and then reviewing the familiar circumference formula ($C = 2\pi r$) and area formula ($A = \pi r^2$) for circles. You might also include the geometric formula for the area of a sector of a circle, $\frac{m\widehat{AB}}{360}\pi r^2$, in which $m\widehat{AB}$ is the degree measure of an arc of a circle. You can then show how the formulas given on page 263 are derived from these familiar formulas by using θ to represent the measure of the central angle associated with $\widehat{AB}$. Point out how the formulas become simpler when the angles are expressed in radians.

Using Technology

The use of scientific calculators is important in this section to simplify some of the computations. You may need to review scientific notation; also make sure students know how to enter on their calculators a number written in scientific notation.

Students with programmable calculators can be encouraged to write programs that use the formulas on page 263 and then perform the calculations automatically. Have them include conversion factors between radians and degrees if appropriate.

Geometry software that calculates the area of a sector and the length of an arc can be used for extra practice and review.

Cooperative Learning

Here is an activity for which students can work in pairs. Each pair of students needs a ruler, a nickel, a dime, and a quarter. Give the following instructions:
1. Place the quarter on your desk. Then, with one eye closed and the other looking directly down on the quarter, position your head so that your open eye is 12 in. above the quarter.
2. Place the dime between your open eye and the quarter, and move the dime back and forth until it just "eclipses" the quarter (that is, until the dime and quarter have the same apparent size). Your partner should then measure the distance from your open eye to the dime.

3. Measure the diameters of the nickel, dime, and quarter. Using these measurements as well as the measurements from steps (1) and (2), you and your partner should try to predict how far from your open eye the nickel must be placed so that it "eclipses" the quarter.
4. Repeat step (2) using the nickel instead of the dime to check the prediction that you and your partner made in step (3).

Applications

The *linear speed* and the *angular speed* of an object moving uniformly along a circular path can be derived from the formula $s = r\theta$ for the length of a circular arc of radius r and with θ as its central angle. If the object moves a distance s and through an angle θ in time t, then the object's linear and angular speeds are $\frac{s}{t}$ and $\frac{\theta}{t}$, respectively. (Angular speed is usually given in revolutions per minute or radians per second.)

7-3 pages 268–274

Presenting the Section

The sine and cosine functions are defined in terms of the coordinates of a point on a circle. Because the sine and cosine are ratios, they do not depend on the actual radius of the circle and therefore can be defined using any circle, although the unit circle (radius = 1) is most convenient.

The wrapping function described in Exercises 43 and 44, page 274, can be a good way of introducing the sine and cosine functions. Physical models of the wrapping process will help students see how each real number t can be associated with an x-coordinate (in the case of cosine) or a y-coordinate (in the case of sine) of a point on the unit circle.

Using Technology

Students can use scientific calculators to verify the sine and cosine values of any angle. Have students focus on angles in a particular quadrant and then evaluate the sine and cosine of the angles (both in degree mode and in radian mode). This should help students see the signs of the values and even the patterns (increasing/decreasing) that emerge for the values. Extend the use of the calculator to include angles greater than 360° (or 2π radians) to verify the periodicity of the sine and cosine functions.

7-4 pages 275–282

Presenting the Section

The use of reference angles is important to help students find the exact sine and cosine values of angles that are multiples of 30° or 45°. Reference angles also are used in solving equations like $\cos x = -0.5$: In degree mode, calculate $\mathrm{Cos}^{-1}(-0.5)$. This gives one solution, $x = 120°$, but you cannot get other solutions, such as $x = 240°$, without reference angles.

Point out to students that reference angles are positive acute angles measured from the x-axis. Encourage students to make sketches to help them determine the correct reference angle. Emphasize that if a table is used, reference angles must be used to find $\sin \theta$ and $\cos \theta$ if θ is not in the first quadrant.

You might want to explore the periodicity of the sine and cosine functions by using an overhead projector and transparencies of the graphs of sine and cosine. Superimpose copies of one period of the sine curve to show how it repeats itself. Transparencies also can be used to show how the cosine graph has the same shape and period as the sine graph. (In fact, it can be treated as a horizontal translation of the sine graph.)

Using Technology

Students can use graphing calculators to obtain the graphs of the sine and cosine functions. They can then use the trace feature to confirm the entries in the table on page 277 (although they will need decimal equivalents for the irrational entries) and to see that the values of the sine and cosine functions repeat (with a fundamental period of 360° or 2π radians).

Applications

The pure tone produced by a vibrating tuning fork can be represented mathematically by a *sinusoid*, a graph representing a sine or cosine curve. An oscilloscope produces an *oscillogram*, which may also depict a sine wave.

Assessment

A good way to evaluate students' understanding of the values of the sine and cosine functions is to have them draw a large circle and label all angles that are multiples of 30° and 45° with the corresponding exact values of the sine and cosine of the angles.

7-5 | pages 282–286

Presenting the Section

The definitions of the other trigonometric functions involve restrictions on either the x- or y-coordinate of a point on the terminal ray of an angle θ. That is, if $x = 0$ (when the value of the cosine function is 0), the tangent and secant functions are undefined; if $y = 0$ (when the value of the sine function is 0), the cotangent and cosecant functions are undefined.

As stated in Section 7-3 for the sine and cosine functions, the trigonometric function of an angle θ expressed in radians is related to the trigonometric function of the real number θ.

Emphasize the sign chart on page 282, and have students verify that the entries are correct.

Using Technology

Students without cotangent, cosecant, and secant keys should use the reciprocal key $(1/x)$ on their calculators. For example, to find csc 30°, they would first find sin 30° and then press the reciprocal key.

A graphing calculator is very helpful to show the tangent and secant graphs. Students can trace along the tangent graph, for example, to verify that $|\tan \theta|$ gets very large as θ approaches an odd multiple of $\frac{\pi}{2}$. This method helps students see that the tangent graph has vertical asymptotes.

Cooperative Learning

Students can work in small groups to go over their answers to Exercise 20, page 286. Encourage students to look for accuracy in group answers and to discuss restrictions on the domain or range of each function. Have students discuss their approaches to sketching the graphs of the functions.

Communication

You may wish to have students summarize in writing how to evaluate a trigonometric function of an angle:

1. Find the associated reference angle.
2. Determine the value of the trigonometric function of the reference angle.
3. Affix the sign based on the quadrant containing the terminal ray of the given angle.

Ask a few students to read their summaries to the class in order to check their understanding.

7-6 | pages 286–291

Presenting the Section

Use the horizontal line test (see page 148) to show students the need for restricting the domains of the sine, cosine, and tangent functions so that their inverses exist. Point out that other restricted domains (such as $\left\{ x \mid \frac{\pi}{2} < x < \frac{3\pi}{2} \right\}$ for tan x) would also result in one-to-one functions, but that the restricted domains used here are the most convenient. You might want to mention that the trigonometric functions with restricted domains are sometimes called the *principal* trigonometric functions. (For example, the principal tangent function is Tan x.) The principal trigonometric functions take on all the values of their unrestricted counterparts, but they do not *repeat* these values periodically.

By using transparencies on an overhead projector, you can easily show why the domains of the trigonometric functions must be restricted, how to restrict them to produce the principal trigonometric functions, and how to obtain the graphs of the inverse trigonometric functions by reflection in the line $y = x$.

Applications

The angle of elevation of the sun is the angle through which an observer must raise his or her line of sight above the horizontal in order to see the sun (see Chapter 9). The inverse trigonometric functions can be used to express the angle of elevation as a function of time.

7 *Trigonometric Functions*

256

■ Angles, Arcs, and Sectors

7-1 Measurement of Angles

Objective *To find the measure of an angle in either degrees or radians and to find coterminal angles.*

The word *trigonometry* comes from two Greek words, *trigonon* and *metron*, meaning "triangle measurement." The earliest use of trigonometry may have been for surveying land in ancient Egypt after the Nile River's annual flooding washed away property boundaries. In Chapter 9 we will discuss this use of trigonometry in greater detail. In Chapter 8 we will discuss more modern applications of trigonometry, such as the analysis of radio waves. The foundation for these applications is laid in this chapter, where we discuss the definitions and properties of the trigonometric functions.

In trigonometry, an angle often represents a rotation about a point. Thus, the angle θ shown is the result of rotating its *initial ray* to its *terminal ray*.

A common unit for measuring very large angles is the **revolution**, a complete circular motion. For example, when a car with wheels of radius 14 in. is driven at 35 mi/h, the wheels turn at an approximate rate of 420 revolutions per minute (abbreviated rpm).

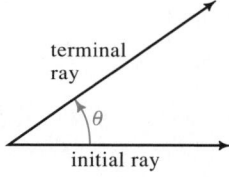

A common unit for measuring smaller angles is the **degree**, of which there are 360 in one revolution. For example, when a door is opened, the doorknob is usually turned $\frac{1}{4}$ revolution, or 90 degrees.

The convention of having 360 degrees in 1 revolution can be traced to the fact that the Babylonian numeration system was based on the number 60. One theory suggests that Babylonian mathematicians subdivided the angles of an equilateral triangle into 60 equal parts (eventually called degrees). Since six equilateral triangles can be arranged within a circle, 1 revolution contained $6 \times 60 = 360$ degrees.

Angles can be measured more precisely by dividing 1 degree into 60 **minutes**, and by dividing 1 minute into 60 **seconds**. For example, an angle of 25 degrees, 20 minutes, and 6 seconds is written $25°20'6''$.

Angles can also be measured in decimal degrees. To convert between decimal degrees and degrees, minutes, and seconds, you can reason as follows:

$$12.3° = 12° + 0.3(60)' = 12°18'$$

$$25°20'6'' = 25° + \left(\frac{20}{60}\right)° + \left(\frac{6}{3600}\right)° = 25.335°$$

◀ Swiftly falling water propels this water wheel. Can you see how the radius of the wheel and the speed of the water determine the speed of revolution of the shaft of the water wheel?

Trigonometric Functions **257**

Teaching Notes, p. 256B

Warm-Up Exercises

Simplify each expression.

1. $135 \times \frac{\pi}{180}$ $\frac{3\pi}{4}$

2. $-285 \times \frac{\pi}{180}$ $-\frac{19\pi}{12}$

3. $\frac{7\pi}{15} \times \frac{180}{\pi}$ 84

4. $-\frac{16\pi}{9} \times \frac{180}{\pi}$ -320

5. Evaluate $75.2 \times \frac{\pi}{180}$ to the nearest hundredth. 1.31

Motivating the Section

Trigonometry has been used for centuries in navigation and surveying. These and other applications of trigonometry involve measuring angles, the topic of this section.

Review Note

You might want to review the geometric definition of an angle as the union of two rays with a common endpoint.

Mathematical Note

The symbol θ is a Greek letter (theta) often used to name an angle or to represent the measure of an angle.

◫ Using Technology

Calculators with degree keys can be used to convert between decimal degrees and degrees, minutes, and seconds.

Relatively recently in mathematical history, another unit of angle measurement, the *radian*, has come into widespread use. When an arc of a circle has the same length as the radius of the circle, as shown at the left below, the measure of the central angle, $\angle AOB$, is by definition 1 radian. Likewise, a central angle has a measure of 1.5 radians when the length of the intercepted arc is 1.5 times the radius, as shown at the right below.

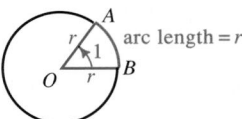
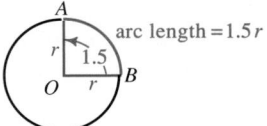

In general, the **radian measure** of the central angle, $\angle AOB$, is the number of *radius* units in the length of arc AB. This accounts for the name radian. In the diagram at the right, the measure θ (Greek theta) of the central angle is:

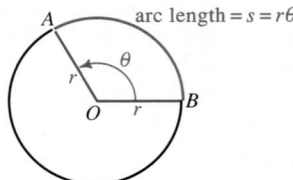

$$\theta = \frac{s}{r}$$

Let us use this equation to see how many radians correspond to 1 revolution. Since the arc length of 1 revolution is the circumference of the circle, $2\pi r$, we have

$$\theta = \frac{s}{r} = \frac{2\pi r}{r} = 2\pi.$$

Thus, 1 revolution measured in radians is 2π and measured in degrees is 360. We therefore have 2π radians = 360 degrees, or

$$\pi \text{ radians} = 180 \text{ degrees.}$$

This gives us the following two conversion formulas:

$$1 \text{ radian} = \frac{180}{\pi} \text{ degrees} \approx 57.2958 \text{ degrees}$$

$$1 \text{ degree} = \frac{\pi}{180} \text{ radians} \approx 0.0174533 \text{ radians}$$

Example 1   **a.** Convert 196° to radians (to the nearest hundredth).
    b. Convert 1.35 radians to decimal degrees (to the nearest tenth) and to degrees and minutes (to the nearest ten minutes).

Solution   Use a calculator and the conversion formulas above. Note that some calculators have the conversion formulas already built in; consult the instruction manual for your calculator.

    a. $196° = 196 \times \dfrac{\pi}{180} \approx 3.42$ radians

    b. 1.35 radians $= 1.35 \times \dfrac{180}{\pi} \approx 77.3° \approx 77°20'$

Angle measures that can be expressed evenly in degrees cannot be expressed evenly in radians, and vice versa. That is why angles measured in radians are frequently given as fractional multiples of π. Angles whose measures are multiples of $\frac{\pi}{4}$, $\frac{\pi}{3}$, and $\frac{\pi}{6}$ appear often in trigonometry. The diagrams below will help you keep the degree conversions for these special angles in mind. Note that a degree measure, such as 45°, is usually written with the degree symbol (°), while a radian measure, such as $\frac{\pi}{4}$, is usually written without any symbol.

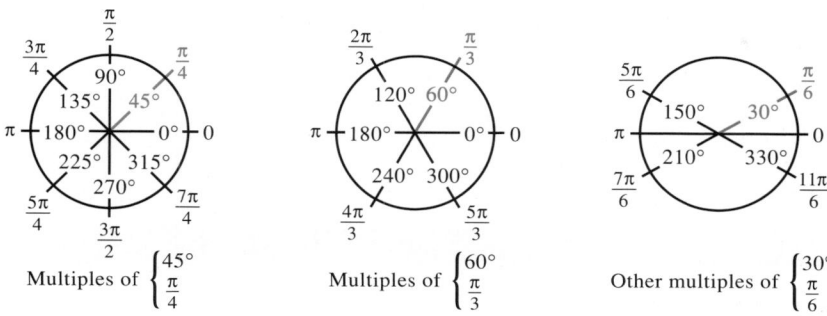

When an angle is shown in a coordinate plane, it usually appears in **standard position**, with its vertex at the origin and its initial ray along the positive x-axis. Moreover, we consider a counterclockwise rotation to be positive and a clockwise rotation to be negative. The diagrams below give examples of positive and negative angles. (In this book we often do not distinguish between an angle and its measure. Thus, by "positive and negative angles" we mean angles with positive and negative measures.)

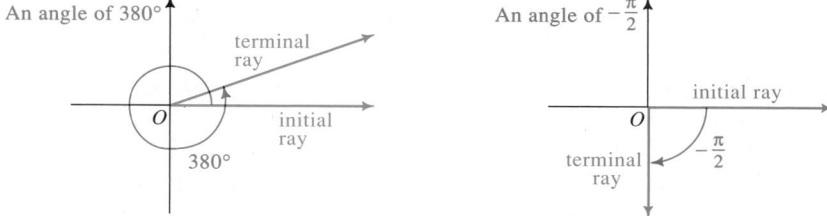

If the terminal ray of an angle in standard position lies in the first quadrant, as shown at the left above, the angle is said to be a first-quadrant angle. Second-, third-, and fourth-quadrant angles are similarly defined. If the terminal ray of an angle in standard position lies along an axis, as shown at the right above, the angle is called a **quadrantal angle**. The measure of a quadrantal angle is always a multiple of 90°, or $\frac{\pi}{2}$.

Error Analysis

Students sometimes have difficulty finding a coterminal angle for an angle θ when $|\theta| > 2\pi$. If $\theta > 2\pi$, students will need to subtract 2π from θ *more than once* to obtain a negative coterminal angle. Likewise, if $\theta < -2\pi$, students will need to add 2π to θ *more than once* to obtain a positive coterminal angle.

Mathematical Note

For an angle θ expressed in degrees, all coterminal angles have the form $\theta + 360n°$ where n is an integer. For an angle θ expressed in radians, all coterminal angles have the form $\theta + 2n\pi$ radians.

Additional Answers
Class Exercises

1. a. π

  b. $\dfrac{\pi}{2}$

  c. $\dfrac{7\pi}{4}$

  d. $\dfrac{\pi}{3}$

  e. $\dfrac{2\pi}{3}$

  f. $\dfrac{4\pi}{3}$

  g. $\dfrac{\pi}{6}$

  h. $\dfrac{\pi}{180}$

Two angles in standard position are called **coterminal** angles if they have the same terminal ray. For any given angle there are infinitely many coterminal angles.

Example 2 Find two angles, one positive and one negative, that are coterminal with the angle $\dfrac{\pi}{4}$. Sketch all three angles.

Solution A positive angle coterminal with $\dfrac{\pi}{4}$ is:

$$\frac{\pi}{4} + 2\pi = \frac{9\pi}{4}$$

A negative angle coterminal with $\dfrac{\pi}{4}$ is:

$$\frac{\pi}{4} - 2\pi = -\frac{7\pi}{4}$$

The three angles are shown at the right.

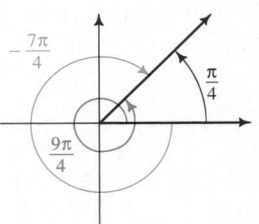

CLASS EXERCISES

1. Convert each degree measure to radians. Leave answers in terms of π.
   a. $180°$ b. $90°$ c. $315°$ d. $60°$
   e. $120°$ f. $240°$ g. $30°$ h. $1°$

2. Convert each radian measure to degrees.
   a. 2π $360°$ b. π $180°$ c. $\dfrac{\pi}{2}$ $90°$ d. $\dfrac{\pi}{4}$ $45°$

   e. $\dfrac{3\pi}{4}$ $135°$ f. $\dfrac{5\pi}{3}$ $300°$ g. $\dfrac{11\pi}{6}$ $330°$ h. $\dfrac{5\pi}{6}$ $150°$

3. Find two angles, one positive and one negative, that are coterminal with each given angle. Answers may vary.
   a. $10°$ $370°$, $-350°$ b. $100°$ $460°$, $-260°$ c. $-5°$ $355°$, $-365°$ d. $400°$ $40°$, $-320°$

   e. π 3π, $-\pi$ f. $\dfrac{\pi}{2}$ $\dfrac{5\pi}{2}$, $-\dfrac{3\pi}{2}$ g. $-\dfrac{\pi}{3}$ $\dfrac{5\pi}{3}$, $-\dfrac{7\pi}{3}$ h. 4π $\pm 2\pi$

4. a. The equation $\theta = (60 + 360n)°$, where n is an integer, represents all angles θ coterminal with an angle of __?__°. $60°$
   b. What would be the equivalent equation in radians? $\theta = \dfrac{\pi}{3} + 2n\pi$

5. Give the radian measure of θ in each of the following diagrams. a. 1 b. 2 c. 0.75

   a. 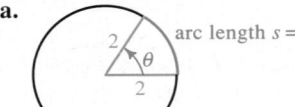 arc length $s = 2$

   b. 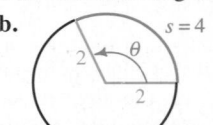 $s = 4$

   c. 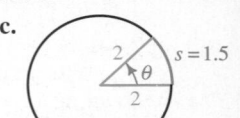 $s = 1.5$

6. Find the degree measure of an angle formed by each rotation described.
   a. $1\dfrac{2}{3}$ revolutions counterclockwise $600°$ b. $2\dfrac{3}{4}$ revolutions clockwise $-990°$

WRITTEN EXERCISES

Convert each degree measure to radians. Leave answers in terms of π.

A
1. a. 315°　　**b.** 225°　　**c.** 15°　　**d.** −45°
2. a. −90°　　**b.** 135°　　**c.** −180°　　**d.** −225°
3. a. −120°　　**b.** −240°　　**c.** 300°　　**d.** 360°
4. a. 210°　　**b.** −135°　　**c.** −210°　　**d.** −315°

Convert each radian measure to degrees.

5. a. $-\frac{\pi}{2}$ −90°　　**b.** $\frac{4\pi}{3}$ 240°　　**c.** $-\frac{3\pi}{4}$ −135°　　**d.** $-\frac{\pi}{6}$ −30°

6. a. $-\frac{5\pi}{6}$ −150°　　**b.** -2π −360°　　**c.** $\frac{5\pi}{4}$ 225°　　**d.** $-\frac{\pi}{3}$ −60°

7. a. π 180°　　**b.** $-\frac{3\pi}{2}$ −270°　　**c.** $\frac{2\pi}{3}$ 120°　　**d.** $\frac{7\pi}{6}$ 210°

8. a. $-\frac{\pi}{4}$ −45°　　**b.** $\frac{7\pi}{4}$ 315°　　**c.** 4π 720°　　**d.** $\frac{11\pi}{6}$ 330°

9. Give the radian measure of θ if:
  a. $r = 5$ and $s = 6$ 1.2
  b. $r = 8$ and $s = 6$ 0.75

10. Give the radian measure of θ if:
  a. $r = 4$ and $s = 5$ 1.25
  b. $r = 6$ and $s = 15$ 2.5

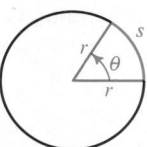

Exs. 9, 10

Convert each degree measure to radians. Give answers to the nearest hundredth of a radian.

11. a. 95° 1.66　　**b.** 110° 1.92　　**c.** 95°10′ 1.66　　**d.** 119.2° 2.08
12. a. 212° 3.70　　**b.** 365° 6.37　　**c.** 200°40′ 3.50　　**d.** 240.8° 4.20

Convert each radian measure to degrees. Give answers to the nearest ten minutes or tenth of a degree.

75°40′, 75.6°
13. a. 1.6 91°40′, 91.7°　**b.** 1.7 97°20′, 97.4°　**c.** 1.21 69°20′, 69.3°　**d.** 1.32
14. a. 2.2 126°, 126.1°　**b.** 3.7 212°, 212.0°　**c.** 2.82 161°30′, 161.6°　**d.** 3.41 195°20′, 195.4°

Visual Thinking Estimate (by sight) the size in radians of each angle shown below. Then measure each angle with a protractor and convert from degrees to radians to find its actual size.

15. 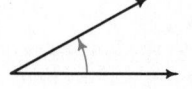 0.5　　　　**16.**  1.75

Additional Answers
Written Exercises

1. a. $\frac{7\pi}{4}$　**b.** $\frac{5\pi}{4}$
　c. $\frac{\pi}{12}$　**d.** $-\frac{\pi}{4}$

2. a. $-\frac{\pi}{2}$　**b.** $\frac{3\pi}{4}$
　c. $-\pi$　**d.** $-\frac{5\pi}{4}$

3. a. $-\frac{2\pi}{3}$　**b.** $-\frac{4\pi}{3}$
　c. $\frac{5\pi}{3}$　**d.** 2π

4. a. $\frac{7\pi}{6}$　**b.** $-\frac{3\pi}{4}$
　c. $-\frac{7\pi}{6}$　**d.** $-\frac{7\pi}{4}$

Suggested Assignments

Standard
Day 1: 261/1–15 odd
Day 2: 262/17–33 odd, 32
Comprehensive
261/1–33 odd

Supplementary Materials

Alternative Assessment, 22

Additional Answers
Written Exercises

19. a. 388.5°, −331.5°
  b. 476.3°, −243.7°
  c. 299.6°, −420.4°
  d. 44.7°, −675.3°

20. a. 398.4°, −321.6°
  b. 487.6°, −232.4°
  c. 309.2°, −410.8°
  d. 39.3°, −680.7°

21. a. 0°30′, −359°30′
  b. 269°20′, −450°40′
  c. 363°21′, −356°39′
  d. 475°15′, −244°45′

22. a. 540°20′, −179°40′
  b. 89°30′, −630°30′
  c. 371°44′, −348°16′
  d. 532°11′, −187°49′

31. $35 \text{ mi/h} = \frac{35}{60} \text{ mi/min} =$
$\frac{35}{60}(5280)(12)$ in./min, or
36,960 in./min;
1 revolution = $r\theta =$
$14(2\pi)$ in., or 28π in.; rate
of turn $= \frac{36{,}960 \text{ in./min}}{28\pi \text{ in./rev}} \approx$
420 rpm

Find two angles, one positive and one negative, that are coterminal with each given angle. Answers may vary.

17. a. 500° 140°, −220° **b.** −60° 300°, −420° **c.** $\frac{\pi}{4}$ $\frac{9\pi}{4}$, $-\frac{7\pi}{4}$ **d.** $-\frac{2\pi}{3}$ $\frac{4\pi}{3}$, $-\frac{8\pi}{3}$

18. a. 1000° 280°, −80° **b.** −100° 260°, −460° **c.** $\frac{4\pi}{3}$ $\frac{10\pi}{3}$, $-\frac{2\pi}{3}$ **d.** $-\frac{\pi}{6}$ See below.

B **19. a.** 28.5° **b.** 116.3° **c.** −60.4° **d.** −315.3°
20. a. 38.4° **b.** 127.6° **c.** −50.8° **d.** −320.7°
21. a. 360°30′ **b.** −90°40′ **c.** 3°21′ **d.** 115°15′
22. a. 180°20′ **b.** −270°30′ **c.** 11°44′ **d.** 172°11′

23. Give an expression in terms of the integer n for the measure of all angles that are coterminal with an angle of 29.7°. $29.7° + n \cdot 360°$

24. Give an expression in terms of the integer n for the measure of all angles that are coterminal with an angle of −116°10′. $−116°10′ + n \cdot 360°$ **18. d.** $\frac{11\pi}{6}$, $-\frac{13\pi}{6}$

Each of Exercises 25–30 gives the speed of a revolving gear. Find (a) the number of degrees per minute through which each gear turns and (b) the number of radians per minute. Give answers to the nearest hundredth.

25. 35 rpm 12,600°; 219.91 **26.** 27 rpm 9720°; 169.65 **27.** 2.5 rpm 15.71 900°;

28. 6.5 rpm 2340°; 40.84 **29.** 14.6 rpm 5256°; 91.73 **30.** 19.8 rpm 7128°; 124.41

31. Reading On page 257, you were told that when a car with wheels of radius 14 in. is driven at 35 mi/h, the wheels turn at an approximate rate of 420 rpm. Show how to obtain this rate of turn.

32. Recreation Suppose you can ride a bicycle a distance of 5 mi in 15 min. If you ride at a constant speed and if the bicycle's wheels have diameter 27 in., find the wheels' approximate rate of turn (in rpm). about 249 rpm

33. Research Consult an encyclopedia or an atlas to see how points on a world map are located by using *latitude* and *longitude* coordinates given in degrees, minutes, and seconds.
  a. If you travel south from a given point on Earth, about how many miles do you have to go to traverse an angle of 1°?
  b. Explain why your answer to part (a) might be different if you travel west instead of south.

34. Research Consult a book of astronomy or a star atlas to see how stars on a celestial map are located by using angles of *right ascension* and *declination*. Describe how each of these angles is measured, and give examples.

7-2 Sectors of Circles

Objective *To find the arc length and area of a sector of a circle and to solve problems involving apparent size.*

A **sector** of a circle, shaded in red at the right below, is the region bounded by a central angle and the intercepted arc. Your geometrical intuition should tell you that the length s of the arc is some fraction of the circumference of the circle and that the area K of the sector is the same fraction of the area of the circle.

For example, suppose the central angle of a sector is 60° and the radius is 12. Then the arc length of the sector is $\frac{60}{360} = \frac{1}{6}$ of the whole circumference, or $\frac{1}{6}(2\pi r) = \frac{1}{6}(2\pi \cdot 12) = 4\pi$. Simi-

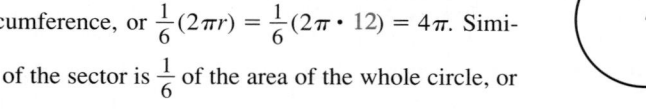

larly, the area of the sector is $\frac{1}{6}$ of the area of the whole circle, or

$$\frac{1}{6}\pi r^2 = \frac{1}{6}\pi \cdot 12^2 = 24\pi.$$

In general, we have the following formulas for the arc length s and area K of a sector with central angle θ.

If θ is in degrees, then: (1) $s = \frac{\theta}{360} \cdot 2\pi r$ (2) $K = \frac{\theta}{360} \cdot \pi r^2$

If θ is in radians, then: (1a) $s = r\theta$ (2a) $K = \frac{1}{2}r^2\theta$

Notice that formulas (1a) and (2a) are more straightforward than formulas (1) and (2). In fact, one reason for using radian measure is that many formulas in calculus are expressed more simply in radians than in degrees.

By combining formulas (1a) and (2a), we can obtain a third area formula:

$$K = \frac{1}{2}r^2\theta = \frac{1}{2}r(r\theta) = \frac{1}{2}rs$$

     (2b) $K = \frac{1}{2}rs$

Example 1   A sector of a circle has arc length 6 cm and area 75 cm². Find its radius and the measure of its central angle.

Solution   Using formula (2b), we have: $75 = \frac{1}{2}r(6)$

$$r = 25$$

Then, using formula (1a), we have: $6 = 25\theta$

$$\theta = \frac{6}{25} = 0.24$$

Thus, the radius is 25 cm and the central angle is 0.24 radians $\approx 14°$.

Trigonometric Functions   **263**

Warm-Up Exercises

Evaluate each expression for $r = 2.4$ and $\theta = 90°$. Give the answer in terms of π.

1. $\frac{\theta}{360} \cdot 2\pi r$ 1.2π

2. $\frac{\theta}{360} \cdot \pi r^2$ 1.44π

Evaluate each expression for $r = 12$ and $\theta = 1.5$.

3. $r\theta$ 18   **4.** $\frac{1}{2}r^2\theta$ 108

5. If $s = 18$, $\frac{1}{2}rs = 24$, and $s = r\theta$, find θ.   6.75

Motivating the Section

You can introduce this section by discussing great circle routes that are traveled by airplanes. The lengths of these routes are the lengths of arcs.

Error Analysis

Remind students to check the mode settings on their calculators before using the formulas on this page. Formulas (1) and (2) require degree mode, while formulas (1a), (2a), and (2b) require radian mode.

Communication Note

Have students read the formulas on this page aloud. They should replace variables with descriptive word phrases. For example, formula (1a) states, "The arc length of a sector is the product of the radius and the central angle of the sector."

Mathematical Note

Emphasize that an object's apparent size is the measure of the angle that it subtends at our eyes. The distance of an object from our eyes is the radius of a large circle with our eyes at the center. The diameter of the object is given by $s = r\theta$, with r being the object's distance from our eyes, and θ being the apparent size of the object.

Additional Examples

1. A sector has perimeter 16 cm and area 15 cm^2. Find its radius r and arc length s.

$P = 2r + s = 16$;
$s = 16 - 2r$
$\frac{1}{2}rs = 15$
$rs = 30$
$r(16 - 2r) = 30$
$r^2 - 8r + 15 = 0$
$(r - 5)(r - 3) = 0$
$r = 5$ and
$s = 16 - 2(5) = 6$, or
$r = 3$ and
$s = 16 - 2(3) = 10$.

2. A phonograph record with diameter 12 in. turns at $33\frac{1}{3}$ rpm. Find the distance that a point on the rim travels in one minute.

Each minute the point travels through $\theta = 33\frac{1}{3}(2\pi) = \frac{200\pi}{3}$ radians.

$s = r\theta = 6\left(\frac{200\pi}{3}\right) = 400\pi$ in. ≈ 1257 in. ≈ 105 ft.

Apparent Size

When there is nothing in our field of vision against which to judge the size of an object, we perceive the object to be smaller when it is farther away. For example, the sun is much larger than the moon, but we perceive the sun to be about the same size as the moon because the sun is so much farther from Earth. Thus, how big an object looks depends not only on its size but also on the angle that it subtends at our eyes. The measure of this angle is called the object's **apparent size**.

Example 2 Jupiter has an apparent size of 0.01° when it is 8×10^8 km from Earth. Find the approximate diameter of Jupiter.

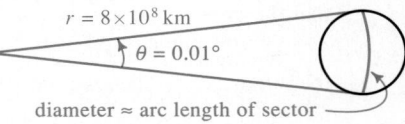

$r = 8 \times 10^8$ km
$\theta = 0.01°$
diameter ≈ arc length of sector

Solution As the exaggerated diagram above indicates, the diameter of Jupiter is approximately the same as the arc length of a sector with central angle 0.01° and radius 8×10^8 km. Using formula (1), we have:

$$\text{diameter} \approx s \approx \frac{0.01}{360}(2\pi)(8 \times 10^8) \approx 140{,}000 \text{ km}$$

CLASS EXERCISES

Find the arc length and area of each sector. 1. π; 2π 2. 4π; 12π 3. 8; 8

1.

$r = 4$
45°

2.

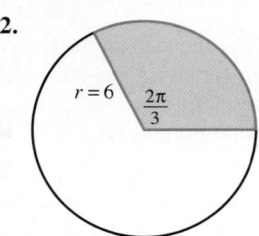

$r = 6$
$\frac{2\pi}{3}$

3.

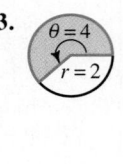

$\theta = 4$
$r = 2$

4. **a.** The apparent size of a tall building 2 km away is 0.05 radians. What is the building's approximate height? 100 m
   b. Explain why you can apply an arc length formula to $\triangle ABC$ and get a good approximation of BC.
   Because in a large circle, a small arc is very nearly a straight line.

A 0.05 2 km B C

WRITTEN EXERCISES

A 1. A sector of a circle has radius 6 cm and central angle 0.5 radians. Find its arc length and area. 3 cm; 9 cm^2

2. A sector of a circle has radius 5 cm and central angle 3 radians. Find its arc length and area. 15 cm; 37.5 cm^2

3. A sector of a circle has arc length 11 cm and central angle 2.2 radians. Find its radius and area. 5 cm; 27.5 cm^2

4. A sector of a circle has arc length 2 cm and central angle 0.4 radians. Find its radius and area. 5 cm; 5 cm^2

5. A sector of a circle has area 25 cm^2 and central angle 0.5 radians. Find its radius and arc length. 10 cm; 5 cm

6. A sector of a circle has area 90 cm^2 and central angle 0.2 radians. Find its radius and arc length. 30 cm; 6 cm

7. A sector of a circle has central angle 30° and arc length 3.5 cm. Find its area to the nearest square centimeter. 12 cm^2

8. A sector of a circle has central angle 24° and arc length 8.4 cm. Find its area to the nearest square centimeter. 84 cm^2

9. A sector of a circle has perimeter 7 cm and area 3 cm^2. Find all possible radii. 1.5 cm or 2 cm

10. A sector of a circle has perimeter 12 cm and area 8 cm^2. Find all possible radii. 2 cm or 4 cm

11. **Astronomy** The diameter of the moon is about 3500 km. Its apparent size is about 0.0087 radians. About how far is it from Earth? 402,000 km

12. **Astronomy** At its closest approach, Mars is about 5.6×10^7 km from Earth and its apparent size is about 0.00012 radians. What is the approximate diameter of Mars? 6720 km

13. **Physics** A compact disc player uses a laser to read music from a disc. The player varies the rotational speed of the disc depending on the position of the laser. When the laser is at the outer edge of the disc, the player spins the disc at the slowest speed, 200 rpm. **a.** 72,000°; 1257 **b.** 7477 cm **c.** About 125 cm/s
   a. At the slowest speed, through how many degrees does the disc turn in a minute? Through how many radians does it turn in a minute?
   b. If the diameter of the disc is 11.9 cm, find the approximate distance that a point on the outer edge travels at the slowest speed in 1 min.
   c. Use part (b) to give the speed in cm/s.

14. **Physics** To make a clay vase, an artist uses a potter's wheel that has a diameter of 13 in. and spins at 120 rpm. Find the approximate distance traveled in 1 min by a point on the outer edge of the wheel. 4901 in.

B 15. **Astronomy** The moon and the sun have approximately the same apparent size for viewers on Earth. The distances from Earth to the moon and to the sun are about 4×10^5 km and 1.5×10^8 km, respectively. The diameter of the moon is about 3500 km. What is the approximate diameter of the sun? 1.3×10^6 km

N **Using Technology**

Students will need scientific calculators for many of the Written Exercises.

Suggested Assignments

Standard
Day 1: 264/1–15 odd
Day 2: 266/16–23
Comprehensive
264/1, 5, 9, 13, 15–17, 21, 23, 25

Supplementary Materials

Alternative Assessment, 23
Activities Book, 17–18
Student Resource Guide, 66–68

16. **Astronomy** This exercise will show how the Greek mathematician and astronomer Eratosthenes (about 276 B.C.–194 B.C.) determined the circumference of Earth. It was reported to him that at noon on the first day of summer the sun was directly overhead in the city of Syene because there was no shadow in a deep well. Eratosthenes observed at this same time in the city of Alexandria that the sun's rays made an angle $\theta = 7.2°$ with a vertical pole.

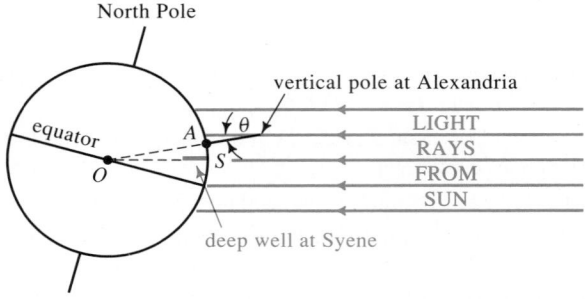

a. How did Eratosthenes conclude that the measure of $\angle AOS = \theta = 7.2°$?
b. If Alexandria was known to be 5000 stadia due north of Syene, show how Eratosthenes could conclude that the circumference of Earth was about 250,000 stadia. (1 stadium ≈ 0.168 km)
c. Given that

$$\text{percent difference} = \frac{\text{old value} - \text{modern value}}{\text{modern value}} \times 100,$$

what is the percent difference between Eratosthenes' value for the circumference of Earth and the modern value of 40,067 km? 4.8%

17. **Farming** A cow at C is tethered to a post alongside a barn 10 m wide and 30 m long. If the post is 10 m from a corner of the barn and if the rope is 30 m long, find the cow's total grazing area to the nearest square meter. 1885 m²

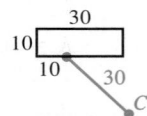

18. **Optics** What is the apparent size of an object 1 cm long held 80 cm from your eyes?

19. **Optics** You are traveling in a car toward a certain mountain at a speed of 80 km/h. The apparent size of the mountain is 0.5°. Fifteen minutes later the same mountain has an apparent size of 1°. About how tall is the mountain? 350 m

20. **Optics** A ship is approaching a lighthouse known to be 20 m high. The apparent size of the lighthouse is 0.005 radians. Ten minutes later the lighthouse has an apparent size of 0.010 radians. What is the approximate speed of the ship (in km/h)? 12 km/h

21. Astronomy Some stars are so far away that their positions appear fixed as Earth orbits the sun. Other stars, however, appear over time to shift their positions relative to the background of ''fixed'' stars. Suppose that the star shown below appears to shift through an arc of $\theta = 0°0'1.5''$ when viewed on the first day of winter and the first day of summer. If the distance from Earth to the sun is about 1.5×10^8 km, find the approximate distance from Earth to the star. 4.1×10^{13} km

Earth Dec. 22

θ — star

Earth June 22

22. Astronomy Give the distance found in Exercise 21 in light years. (A *light year* is the distance light travels in one year. Use the fact that light travels 3.00×10^8 m/s.) **4.4 light years**

23. *Writing* A point traveling along a circle has both a *linear* speed, defined as the length of arc traversed per unit of time, and an *angular* speed, defined as the measure of angle moved through per unit of time. Write a paragraph in which you compare the linear and angular speeds of two points on a rotating line (for example, the tip of a clock's hand versus a point on the hand closer to the center of rotation). Then discuss the implication this has for ice skaters who form a rotating line by interlocking their arms and skating in a circle.

24. The sector shown at the right has perimeter 20 cm.

  a. Show that $\theta = \dfrac{20}{r} - 2$ and that the area of the sector is $K = 10r - r^2$.

  b. What value of r gives the maximum possible area of a sector of perimeter 20 cm? (*Hint: K* is a quadratic function of *r*.) **5 cm**

  c. What is the measure of the central angle of the sector of maximum area? **2**

25. The purpose of this exercise is to derive the formula for the area of a circle by first deriving the formula for the area of a sector. Consider the sector with radius r and arc length s shown in the diagram at the right. Inscribed in the sector are n congruent isosceles triangles, each with height h and base b.

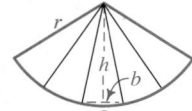

  a. Show that the total area of the inscribed triangles is $\dfrac{1}{2}nbh$.

  b. As n increases, h gets closer and closer to __?__, and nb gets closer and closer to __?__.

  c. Use parts (a) and (b) to derive the formula $K = \dfrac{1}{2}rs$.

  d. Derive the formula for the area of a circle from $K = \dfrac{1}{2}rs$.

Application

Students may be interested in knowing more about how to measure the distance from Earth to a star or how to find distances between stars. Have them research the phenomenon of parallax and present their findings to the class.

Making Connections

You can use Exercise 25 to introduce the idea of approaching a limit (see Chapter 19).

Additional Answers Written Exercises

23. The angular speed is the same for any two points on a rotating line. The linear speed varies, however: the greater the distance from the center, the greater the linear speed. Thus, the farther a skater is from the center, the faster he or she must skate in order to keep the line straight.

24. a. $20 = 2r + s = 2r + r\theta$;
$\theta = \dfrac{20 - 2r}{r} = \dfrac{20}{r} - 2$.
$K = \dfrac{1}{2}r^2\theta =$
$\dfrac{1}{2}r^2\left(\dfrac{20}{r} - 2\right) =$
$10r - r^2$

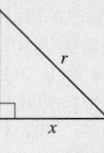
■ The Trigonometric Functions

7-3 The Sine and Cosine Functions

Objective *To use the definitions of sine and cosine to find values of these functions and to solve simple trigonometric equations.*

If you have ever ridden on a Ferris wheel, you may have wondered how to find your height above the ground at any given moment. Suppose a Ferris wheel has a radius of 20 ft and revolves at 5 rpm. If the bottom of the Ferris wheel sits 4 ft off the ground, then t seconds after the ride begins, a rider's height h above the ground is given in feet by:

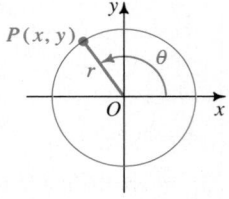

$$h = 24 + 20\sin(30t - 90)°$$

The "sin" that appears in the equation above is an abbreviation of the *sine* function, one of the two trigonometric functions that we will discuss in this section.

Suppose $P(x, y)$ is a point on the circle $x^2 + y^2 = r^2$ and θ is an angle in standard position with terminal ray OP, as shown at the right. We define the **sine** of θ, denoted $\sin\theta$, by:

$$\sin\theta = \frac{y}{r}$$

and we define the **cosine** of θ, denoted $\cos\theta$, by:

$$\cos\theta = \frac{x}{r}$$

Example 1 If the terminal ray of an angle θ in standard position passes through $(-3, 2)$, find $\sin\theta$ and $\cos\theta$.

Solution Make a sketch as shown. To find the radius r of the circle, use the equation $x^2 + y^2 = r^2$ with $x = -3$ and $y = 2$:

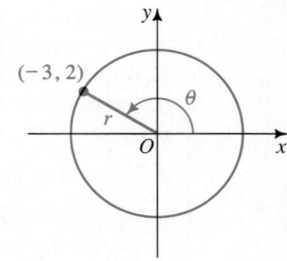

$$(-3)^2 + 2^2 = 13 = r^2$$
$$\sqrt{13} = r$$

Thus: $\sin\theta = \dfrac{y}{r} = \dfrac{2}{\sqrt{13}} = \dfrac{2\sqrt{13}}{13}$

and $\cos\theta = \dfrac{x}{r} = \dfrac{-3}{\sqrt{13}} = -\dfrac{3\sqrt{13}}{13}$

Example 2 If θ is a fourth-quadrant angle and $\sin \theta = -\dfrac{5}{13}$, find $\cos \theta$.

Solution Make a sketch of a circle with radius 13 as shown. Since $\sin \theta = \dfrac{y}{r} = -\dfrac{5}{13}$ and r is always positive, $y = -5$. To find x, use the circle's equation, $x^2 + y^2 = r^2$:

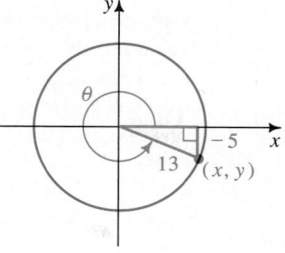

$$x^2 + (-5)^2 = 13^2$$
$$x^2 + 25 = 169$$
$$x^2 = 144$$
$$x = \pm 12$$

Since θ is a fourth-quadrant angle, $x = 12$. Thus, $\cos \theta = \dfrac{x}{r} = \dfrac{12}{13}$.

Although the definitions of $\sin \theta$ and $\cos \theta$ involve the radius r of a circle, the values of $\sin \theta$ and $\cos \theta$ depend only on θ, as the following activity shows.

Activity

You will need graph paper, a ruler, a compass, a protractor, and a calculator.
a. Using graph paper, draw an xy-plane and an acute angle θ in standard position.
b. Draw three concentric circles centered at the origin, and mark the points P_1, P_2, and P_3 where the circles intersect the terminal ray of θ.
c. Carefully measure the radii r_1, r_2, and r_3 of the three circles as well as the vertical distances y_1, y_2, and y_3 between P_1, P_2, and P_3 and the x-axis.
d. Use a calculator to compute $\dfrac{y_1}{r_1}$, $\dfrac{y_2}{r_2}$, and $\dfrac{y_3}{r_3}$ to the nearest hundredth. Each ratio is an approximation of the sine of θ. What do you observe about the ratios? *They are equal.*
e. Use your knowledge of geometry to support your observation from part (d). *Corresponding sides of similar triangles are proportional.*

The circle $x^2 + y^2 = 1$ has radius 1 and is therefore called the **unit circle**. This circle is the easiest one with which to work because, as the diagram shows, $\sin \theta$ and $\cos \theta$ are simply the y- and x-coordinates of the point where the terminal ray of θ intersects the circle.

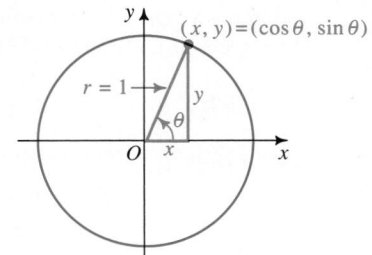

$$\sin \theta = \frac{y}{r} = \frac{y}{1} = y$$

$$\cos \theta = \frac{x}{r} = \frac{x}{1} = x$$

When a circle is used to define the trigonometric functions, they are sometimes called *circular functions*. (See Exercise 44 for another way to use the unit circle to define the trigonometric, or circular, functions.)

Trigonometric Functions **269**

2. Complete each statement using $<$, $>$, or $=$.
   a. $\sin 30°$ _?_ $\sin (-30°)$
   b. $\cos 30°$ _?_ $\cos (-30°)$
   c. $\cos 300°$ _?_ $\cos 330°$

   a.

   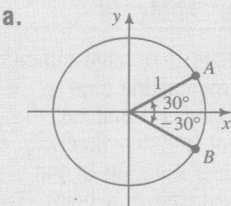

   $\sin 30° = y$-coordinate of $A > 0$. $\sin (-30°) = y$-coordinate of $B < 0$. Since $\sin 30° > 0$ and $\sin (-30°) < 0$, $\sin 30° > \sin (-30°)$.

   b. $\cos 30° = x$-coordinate of A. $\cos (-30°) = x$-coordinate of B. Since A and B lie on the same vertical line, $\cos 30° = \cos (-30°)$.

   c.

   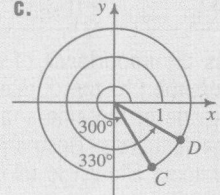

   $\cos 300° = x$-coordinate of C. $\cos 330° = x$-coordinate of D. From the diagram, the x-coordinate of C is less than the x-coordinate of D. Therefore, $\cos 300° < \cos 330°$.

From the definitions and diagram at the bottom of the preceding page, we can see that the domain of both the sine and cosine functions is the set of all real numbers, since sin θ and cos θ are defined for any angle θ. Also, the range of both functions is the set of all real numbers between -1 and 1 inclusive, since sin θ and cos θ are the coordinates of points on the unit circle.

The diagrams below indicate where the sine and cosine functions have positive and negative values. For example, if θ is a second-quadrant angle, sin θ is positive and cos θ is negative.

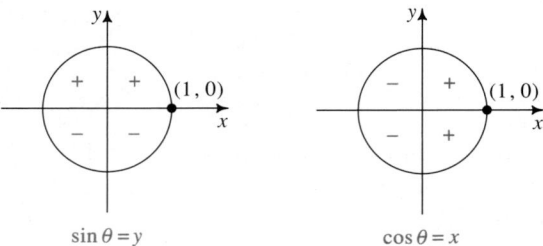

Example 3   Find: **a.** sin 90°   **b.** sin 450°   **c.** cos $(-\pi)$

Solution

a. 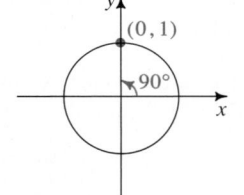   **b.** 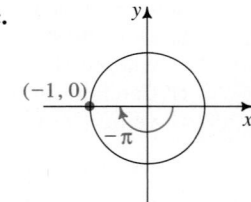   **c.**

sin 90° = y-coordinate = 1   sin 450° = y-coordinate = 1   cos$(-\pi)$ = x-coordinate = -1

As figures (a) and (b) in Example 3 show, $\theta = 90°$ and $\theta = 450°$ are two solutions of the trigonometric equation sin $\theta = 1$. The following example shows that there are infinitely many solutions of this equation.

Example 4   Solve sin $\theta = 1$ for θ in degrees.

Solution   You already know that $\theta = 90°$ is one solution of the equation sin $\theta = 1$. Since any angle coterminal with 90° also has 1 as its sine value,

$$\theta = 90°, \ 90° \pm 360°, \ 90° \pm 2 \cdot 360°, \ 90° \pm 3 \cdot 360°, \ldots$$

are all solutions of the equation. They can be written more conveniently as $\theta = 90° + n \cdot 360°$, where n is an integer. (In radians, the solutions would be written as $\theta = \frac{\pi}{2} + n \cdot 2\pi$ or $\theta = \frac{\pi}{2} + 2n\pi$.)

270   *Chapter Seven*

From Example 4 and the definitions of $\sin \theta$ and $\cos \theta$, you can see that the sine and cosine functions repeat their values every 360° or 2π radians. Formally this means that for all θ:

$$\sin (\theta + 360°) = \sin \theta$$
$$\cos (\theta + 360°) = \cos \theta$$

$$\sin (\theta + 2\pi) = \sin \theta$$
$$\cos (\theta + 2\pi) = \cos \theta$$

We summarize these facts by saying that the sine and cosine functions are *periodic* and that they have a *fundamental period* of 360°, or 2π radians. It is the periodic nature of these functions that makes them useful in describing many repetitive phenomena such as tides, sound waves, and the orbital paths of satellites.

CLASS EXERCISES

Find $\sin \theta$ and $\cos \theta$.

1.
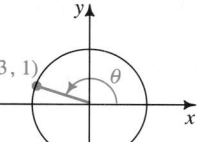
$\dfrac{\sqrt{10}}{10}, -\dfrac{3\sqrt{10}}{10}$

2. $-\dfrac{2\sqrt{5}}{5}, \dfrac{\sqrt{5}}{5}$
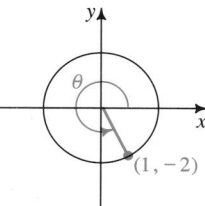

3. $-\dfrac{4\sqrt{41}}{41}, -\dfrac{5\sqrt{41}}{41}$
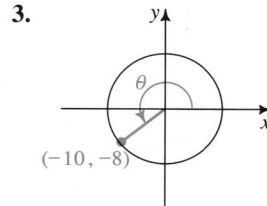

4. State whether each expression is positive or negative.
 a. $\sin 165°$ Pos.
 b. $\sin 265°$ Neg.
 c. $\cos 210°$ Neg.
 d. $\cos 310°$ Pos.
 e. $\sin \dfrac{5\pi}{6}$ Pos.
 f. $\cos \dfrac{5\pi}{6}$ Neg.
 g. $\sin \dfrac{4\pi}{3}$ Neg.
 h. $\cos \dfrac{5\pi}{3}$ Pos.
 i. $\sin 2$ Pos.
 j. $\cos 2$ Neg.
 k. $\sin 4$ Neg.
 l. $\cos 4$ Neg.

5. Does $\cos \theta$ increase or decrease as:
 a. θ increases from 0° to 90°? Dec.
 b. θ increases from 90° to 180°? Dec.
 c. θ increases from 180° to 270°? Inc.
 d. θ increases from 270° to 360°? Inc.

6. Answer Exercise 5 for $\sin \theta$. **a.** Inc. **b.** Dec. **c.** Dec. **d.** Inc.

7. Use the unit circle to justify the fact that for all θ:

$$(\cos \theta)^2 + (\sin \theta)^2 = 1$$

8. There are infinitely many values of θ for which $\cos \theta = 0$. Name several. $\pm 90°, \pm 270°$

9. a. Explain the meaning of $\theta = 45° + n \cdot 360°$, where n is an integer. See below.
 b. What is the equivalent statement if θ is expressed in radians? $\theta = \dfrac{\pi}{4} + 2n\pi$
 a. θ is coterminal with an angle of 45°.

Trigonometric Functions **271**

Review Note

You may wish to have students review the concept of periodicity by having students reread the beginning of Section 4-4.

Exercise Note

For Class Exercise 4, you might need to remind students that angles not labeled in degrees are taken to be in radians.

Problem Solving

You might have students organize their answers to Class Exercises 5 and 6 in a chart. The chart can be used to make generalizations about the patterns in the values of the sine and cosine functions as the terminal side of θ passes through the four quadrants. Scientific calculators may help students to see the patterns for $\cos \theta$ and for $\sin \theta$.

◪ Using Technology

Calculators would be useful to check the identity given in Class Exercise 7. Have students do this by substituting several values for θ.

Additional Answers
Class Exercises

7. Suppose the terminal ray of θ passes through the point (x, y) on the unit circle. Then $\cos \theta = x$ and $\sin \theta = y$. Since an equation of the unit circle is $x^2 + y^2 = 1$, we have $(\cos \theta)^2 + (\sin \theta)^2 = 1$.

WRITTEN EXERCISES

Find the value of each expression without using a calculator or table.

A 1. **a.** $\sin 180°$ 0 **b.** $\cos 180°$ −1 **c.** $\sin 270°$ −1 **d.** $\cos 270°$ 0

2. **a.** $\sin (-90°)$ −1 **b.** $\cos (-90°)$ 0 **c.** $\sin 360°$ 0 **d.** $\cos 360°$ 1

3. **a.** $\sin (-\pi)$ 0 **b.** $\cos \pi$ −1 **c.** $\sin \frac{3\pi}{2}$ −1 **d.** $\cos \frac{\pi}{2}$ 0

4. **a.** $\cos 2\pi$ 1 **b.** $\sin \left(-\frac{\pi}{2}\right)$ −1 **c.** $\sin 3\pi$ 0 **d.** $\cos \left(-\frac{3\pi}{2}\right)$ 0

Name each quadrant described.

5. **a.** $\sin \theta > 0$ and $\cos \theta < 0$ II **b.** $\sin \theta < 0$ and $\cos \theta < 0$ III

6. **a.** $\sin \theta < 0$ and $\cos \theta > 0$ IV **b.** $\sin \theta > 0$ and $\sin (90° + \theta) > 0$ I

Without using a calculator or table, solve each equation for *all* θ in radians.

7. **a.** $\sin \theta = 1$ $\frac{\pi}{2} + 2n\pi$ **b.** $\cos \theta = -1$ $\pi + 2n\pi$ **c.** $\sin \theta = 0$ $n\pi$ **d.** $\sin \theta = 2$ No sol.

8. **a.** $\cos \theta = 1$ $2n\pi$ **b.** $\sin \theta = -1$ $\frac{3\pi}{2} + 2n\pi$ **c.** $\cos \theta = 0$ $\frac{\pi}{2} + n\pi$ **d.** $\cos \theta = -3$ No sol.

Without using a calculator or table, state whether each expression is positive, negative, or zero. P = positive; N = negative

9. **a.** $\sin 4\pi$ 0 **b.** $\cos \frac{7\pi}{6}$ N **c.** $\sin \left(-\frac{\pi}{4}\right)$ N **d.** $\cos \frac{3\pi}{4}$ N

10. **a.** $\cos 3\pi$ N **b.** $\sin \frac{2\pi}{3}$ P **c.** $\sin \frac{11\pi}{6}$ N **d.** $\cos \left(-\frac{\pi}{2}\right)$ 0

11. **a.** $\sin 60°$ P **b.** $\cos (-120°)$ N **c.** $\cos 300°$ P **d.** $\sin (-210°)$ P

12. **a.** $\cos 45°$ P **b.** $\sin 135°$ P **c.** $\cos (-225°)$ N **d.** $\sin (-315°)$ P

13. **a.** $\sin \frac{7\pi}{4}$ N **b.** $\sin \left(-\frac{\pi}{6}\right)$ N **c.** $\cos \frac{3\pi}{2}$ 0 **d.** $\cos \frac{\pi}{3}$ P

14. **a.** $\cos \left(-\frac{\pi}{3}\right)$ P **b.** $\sin \frac{\pi}{6}$ P **c.** $\sin \frac{5\pi}{4}$ N **d.** $\cos \frac{7\pi}{4}$ P

15. **a.** $\cos 89°$ P **b.** $\cos 91°$ N **c.** $\sin 720°$ 0 **d.** $\sin (-270°)$ P

16. **a.** $\sin 1°$ P **b.** $\sin (-1°)$ N **c.** $\cos 90°$ 0 **d.** $\cos 540°$ N

Find $\sin \theta$ and $\cos \theta$.

17. $\frac{4}{5}; \frac{3}{5}$ 18. $-\frac{\sqrt{10}}{10}; -\frac{3\sqrt{10}}{10}$ 19. $\frac{12}{13}; -\frac{5}{13}$ 20. $-\frac{\sqrt{5}}{5}; \frac{2\sqrt{5}}{5}$

17.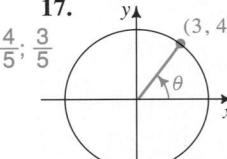

18.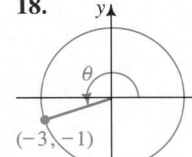

19.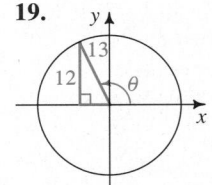

20.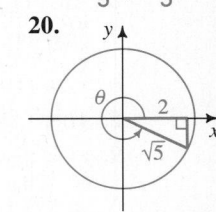

Complete the table. (A sketch like the one in Example 2 may be helpful.)

21. 22. 23. 24. 25. 26. 27. 28.

Quadrant	I	II	III	IV	II	III	IV	II
$\sin \theta$	$\frac{3}{5}$	$\frac{5}{13}$	?	?	$\frac{1}{5}$	$-\frac{3}{7}$	?	$\frac{1}{9}$
$\cos \theta$	?	?	$-\frac{24}{25}$	$\frac{15}{17}$	?	?	$\frac{3}{4}$	?

B **29. a.** What are the coordinates of points P and Q where the line $y = \frac{1}{2}$ intersects the unit circle? (Refer to the diagram at the left below.)

   b. Explain how part (a) shows that if $\sin \theta = \frac{1}{2}$, then $\cos \theta = \pm\frac{\sqrt{3}}{2}$.

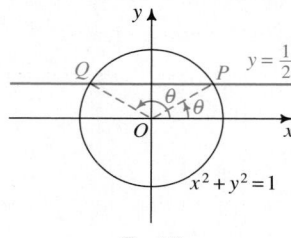

Ex. 29

Ex. 30

30. a. What are the coordinates of points P and Q where the line $x = -\frac{1}{2}$ intersects the unit circle? (Refer to the diagram at the right above.)

   b. Explain how part (a) shows that if $\cos \theta = -\frac{1}{2}$, then $\sin \theta = \pm\frac{\sqrt{3}}{2}$.

31. *Investigation* In the diagram of the unit circle at the right, z is measured in radians.

   a. Show that the length of arc PQ is z.
   b. Show that the length of $\overline{PA}$ is $\sin z$.
   c. What do parts (a) and (b) imply about the relationship between $\sin z$ and z for a small angle z? Confirm this relationship by using a calculator to compare $\sin z$ and z when z is a very small number of radians.

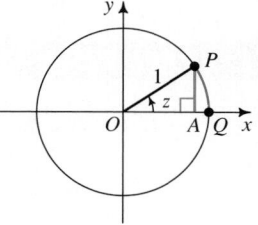

32. *Investigation* Refer to the diagram for Exercise 31.

   a. Show that the length of $\overline{OA}$ is $\cos z$.
   b. Use the results of part (c) of Exercise 31 and part (a) of this exercise to find an algebraic expression involving z (measure of the angle in radians) that approximates $\cos z$ for a small angle z.
   c. Use a calculator to check the accuracy of the approximation in part (b) when z is a very small number of radians.

Trigonometric Functions **273**

Mathematical Note

Use of the wrapping function (see Exercise 44) guarantees that there is a one-to-one correspondence between the set of real numbers and the unit circle. Thus, a real number can be related to the radian measure of an angle.

Making Connections

Using calculus, it can be shown that sin x and cos x can be approximated by the respective infinite sums used in the Computer Exercises. Values of x must be given in radians.

Additional Answers
Written Exercises

43. a. $W(2) = (0, 1)$; $W(3) = (0, 0)$; $W(4) = (1, 0)$; $W(5) = (1, 1)$

b. W is a periodic function with fundamental period 4 because $W(t + 4) = W(t)$ for all t.

c. $c(t + 4) = x$-coordinate of $W(t + 4) = x$-coordinate of $W(t) = c(t)$; thus, c is periodic. $s(t + 4) = y$-coordinate of $W(t + 4) = y$-coordinate of $W(t) = s(t)$; thus, s is periodic.

d.
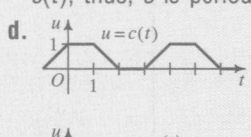
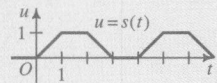

Without using a calculator or table, complete each statement with one of the symbols <, >, or =.

33. sin 40° __?__ sin 30° >

34. cos 40° __?__ cos 30° <

35. sin 172° __?__ sin 8° =

36. sin 310° __?__ sin 230° =

37. sin 130° __?__ sin 50° =

38. cos 50° __?__ cos (−50°) =

39. cos 214° __?__ cos 213° >

40. sin 169° __?__ sin 168° <

41. List in order of increasing size: sin 1, sin 2, sin 3, sin 4 sin 4, sin 3, sin 1, sin 2

42. List in order of increasing size: cos 1, cos 2, cos 3, cos 4 cos 3, cos 4, cos 2, cos 1

C **43.** Consider a special type of function called a *wrapping function*. This function, denoted by W, wraps a vertical number line whose origin is at $R(1, 0)$ around a unit square, as shown at the right. With each real number t on the vertical number line, W associates a point $P(x, y)$ on the square. For example, $W(1) = (1, 1)$ and $W(-1) = (0, 0)$. From W we can define two simpler functions:

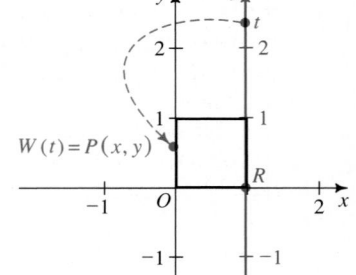

$$c(t) = x\text{-coordinate of } P,$$

and $\quad s(t) = y\text{-coordinate of } P.$

a. Find $W(2)$, $W(3)$, $W(4)$, and $W(5)$.
b. Explain why W is a periodic function and give its fundamental period.
c. Explain how the periodicity of W guarantees the periodicity of c and s.
d. Sketch the graphs of $u = c(t)$ and $u = s(t)$ in separate *tu*-planes.

44. *Writing* Suppose the unit square in Exercise 43 is replaced with the unit circle. Write a paragraph in which you describe how the wrapping function can now be used to define the circular functions sine and cosine.

////COMPUTER EXERCISES

1. Use a computer to obtain the approximate value (to five decimal places) of

$$x - \frac{x^3}{3!} + \frac{x^5}{5!} - \frac{x^7}{7!} + \frac{x^9}{9!} - \cdots$$

when $x = 1$, $x = 2$, and $x = \frac{\pi}{2} \approx 1.5708$. Compare the results with the values of SIN(1), SIN(2), and SIN(1.5708) given directly by the computer.

2. Use a computer to obtain the approximate value (to five decimal places) of

$$1 - \frac{x^2}{2!} + \frac{x^4}{4!} - \frac{x^6}{6!} + \frac{x^8}{8!} - \cdots$$

when $x = 1$, $x = 2$, and $x = \pi \approx 3.1416$. Compare the results with the values of COS(1), COS(2), and COS(3.1416) given directly by the computer.

7-4 Evaluating and Graphing Sine and Cosine

Objective | *To use reference angles, calculators or tables, and special angles to find values of the sine and cosine functions and to sketch the graphs of these functions.*

Reference Angles

Let α (Greek alpha) be an acute angle in standard position. Suppose, for example, that $\alpha = 20°$. Notice that the terminal ray of $\alpha = 20°$ and the terminal ray of $180° - \alpha = 160°$ are symmetric in the y-axis. If the sine and cosine of $\alpha = 20°$ are known, then the sine and cosine of $160°$ can be deduced, as shown in the diagram at the right.

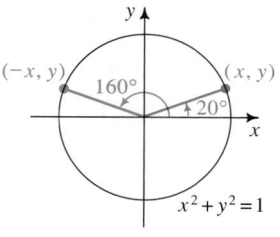

$$\sin 160° = y = \sin 20°$$
$$\cos 160° = -x = -\cos 20°$$

The angle $\alpha = 20°$ is called the *reference angle* for the $160°$ angle. It is also the reference angle for the $200°$ and $340°$ angles shown below.

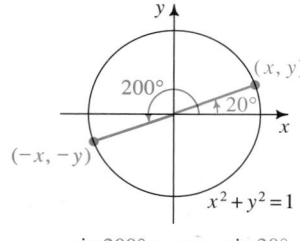

$$\sin 200° = -y = -\sin 20°$$
$$\cos 200° = -x = -\cos 20°$$

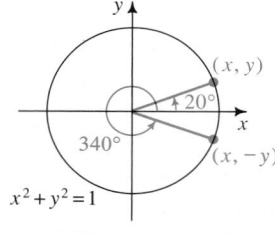

$$\sin 340° = -y = -\sin 20°$$
$$\cos 340° = x = \cos 20°$$

In general, the acute angle α is the **reference angle** for the angles $180° - \alpha$, $180° + \alpha$, and $360° - \alpha$ as well as all coterminal angles. In other words, the reference angle for any angle θ is the acute positive angle α formed by the terminal ray of θ and the x-axis.

Example 1 | Express $\sin 695°$ in terms of a reference angle.

Solution | An angle between $0°$ and $360°$ that is coterminal with a $695°$ angle is:

$$695° - 360° = 335°$$

The reference angle for $335°$ is:

$$360° - 335° = 25°$$

(See the diagram at the right.) Since $695°$ is a fourth-quadrant angle, $\sin 695° < 0$. Thus:

$$\sin 695° = -\sin 25°$$

Teaching Notes, p. 256C

Warm-Up Exercises

Copy and complete the chart.

	Quad	Sign of:			
		x	y	$\cos\theta$	$\sin\theta$
1.	I	$+$	$+$	$+$	$+$
2.	II	$-$	$+$	$-$	$+$
3.	III	$-$	$-$	$-$	$-$
4.	IV	$+$	$-$	$+$	$-$

Motivating the Section

When an oscilloscope is used to analyze a pure tone, such as that produced by striking a tuning fork, a sine wave results. Graphing the sine and cosine functions is one of the objectives of this section.

Mathematical Note

Point out that the sine or cosine of the reference angle for a given angle gives the absolute value of the sine or cosine of the given angle. The quadrant in which the terminal ray of the given angle lies determines the sign of the sine or cosine of the given angle.

Additional Examples

1. Find the exact value of each expression.

   a. $\sin \dfrac{7\pi}{6}$ **b.** $\cos \dfrac{7\pi}{6}$

   c. $\log_4 \sin 150°$

   a. $\sin \dfrac{7\pi}{6} = -\sin \dfrac{\pi}{6}$ since

   $\dfrac{7\pi}{6}$ is a third-quadrant

   angle. $\sin \dfrac{7\pi}{6} = -\dfrac{1}{2}$.

   b. $\cos \dfrac{7\pi}{6} = -\cos \dfrac{\pi}{6}$

   since $\dfrac{7\pi}{6}$ is a third-

   quadrant angle.

   $\cos \dfrac{7\pi}{6} = -\dfrac{\sqrt{3}}{2}$.

   c. $\sin 150° = \sin 30° = \dfrac{1}{2}$

   $\log_4 \sin 150° = \log_4 \dfrac{1}{2}$

   Let $\log_4 \dfrac{1}{2} = x$.

   Then: $4^x = \dfrac{1}{2}$

   $(2^2)^x = 2^{-1}$

   $2x = -1$

   $x = -\dfrac{1}{2} = \log_4 \sin 150°$

2. Sketch the graphs of

   $y = -\sin x$ and $y = \left|\dfrac{x}{3}\right|$

   on the same set of axes. How many solutions does

   the equation $-\sin x = \left|\dfrac{x}{3}\right|$

   have?

   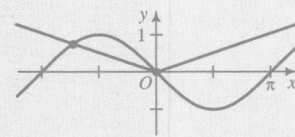

   The graph shows that there are two solutions, one at the origin and the other in the third quadrant.

Using Calculators or Tables

The easiest way to find the sine or cosine of most angles is to use a calculator that has the sine and cosine functions. Always be sure to check whether the calculator is in degree or radian mode.

If you do not have access to a calculator, there are tables at the back of the book that evaluate $\sin \theta$ and $\cos \theta$ for first-quadrant values of θ. Instructions for using trigonometric tables are on page 800.

Example 2 Find the value of each expression to four decimal places.
a. $\sin 122°$ **b.** $\cos 237°$ **c.** $\cos 5$ **d.** $\sin (-2)$

Solution Note that in parts (a) and (b) the angles are given in degrees, while in parts (c) and (d) the angles are given in radians. Use your calculator and compare your results with those below.
a. $\sin 122° \approx 0.8480$ **b.** $\cos 237° \approx -0.5446$
c. $\cos 5 \approx 0.2837$ **d.** $\sin (-2) \approx -0.9093$

Finding Sines and Cosines of Special Angles

Because angles that are multiples of 30° and 45° occur often in mathematics, it can be useful to know their sine and cosine values without resorting to a calculator or table. To do this, you need the following facts.

1. In a 30°-60°-90° triangle, the sides are in the ratio $1:\sqrt{3}:2$. (Note that in this ratio, 1 corresponds to the side opposite the 30° angle, $\sqrt{3}$ to the side opposite the 60° angle, and 2 to the side opposite the 90° angle.)

2. In a 45°-45°-90° triangle, the sides are in the ratio $1:1:\sqrt{2}$, or $\sqrt{2}:\sqrt{2}:2$.

These facts are used in the diagrams below to obtain the values of $\sin \theta$ and $\cos \theta$ for $\theta = 30°$, $\theta = 45°$, and $\theta = 60°$.

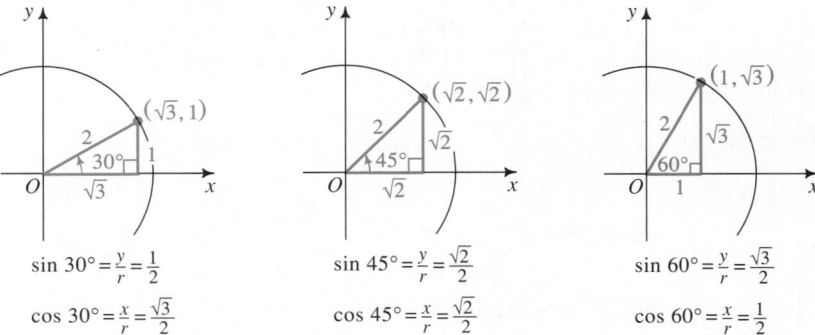

$$\sin 30° = \frac{y}{r} = \frac{1}{2}$$
$$\cos 30° = \frac{x}{r} = \frac{\sqrt{3}}{2}$$

$$\sin 45° = \frac{y}{r} = \frac{\sqrt{2}}{2}$$
$$\cos 45° = \frac{x}{r} = \frac{\sqrt{2}}{2}$$

$$\sin 60° = \frac{y}{r} = \frac{\sqrt{3}}{2}$$
$$\cos 60° = \frac{x}{r} = \frac{1}{2}$$

The information obtained from the diagrams above is summarized in the table at the top of the next page.

Although the table at the right only gives the sine and cosine values of special angles from 0° to 90°, reference angles can be used to find other multiples of 30° and 45°. For example:

$$\sin 210° = -\sin 30° = -\frac{1}{2}$$

$$\cos 315° = \cos 45° = \frac{\sqrt{2}}{2}$$

θ (degrees)	θ (radians)	$\sin \theta$	$\cos \theta$
0°	0	0	1
30°	$\frac{\pi}{6}$	$\frac{1}{2}$	$\frac{\sqrt{3}}{2}$
45°	$\frac{\pi}{4}$	$\frac{\sqrt{2}}{2}$	$\frac{\sqrt{2}}{2}$
60°	$\frac{\pi}{3}$	$\frac{\sqrt{3}}{2}$	$\frac{1}{2}$
90°	$\frac{\pi}{2}$	1	0

As the table suggests, $\sin \theta$ and $\cos \theta$ are both one-to-one functions for first-quadrant θ, a fact that we use in the next example to solve a geography problem.

Example 3

The *latitude* of a point on Earth is the degree measure of the shortest arc from that point to the equator. For example, the latitude of point P in the diagram equals the degree measure of arc PE. At what latitude is the circumference of the circle of latitude at P half the distance around the equator?

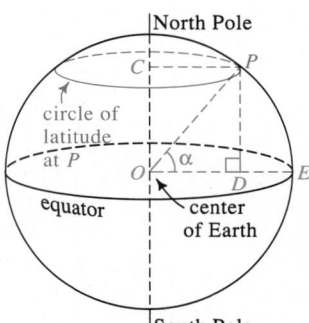

Solution

Let R be the radius of Earth, and let r be the radius of the circle of latitude at P. Then:

circumference of circle of latitude at $P = \frac{1}{2}$(circumference of Earth)

$$2\pi r = \frac{1}{2}(2\pi R)$$

$$r = \frac{1}{2}R$$

As shown in the diagram, point D is the intersection of the perpendicular from P to $\overline{OE}$. Since planes containing circles of latitude are perpendicular to the north-south axis of Earth, quadrilateral $OCPD$ is a rectangle, so that $OD = CP = r$ and

$$\cos \alpha = \frac{OD}{OP} = \frac{r}{R} = \frac{\frac{1}{2}R}{R} = \frac{1}{2}.$$

Since α is an acute angle, its measure must be 60° (see the table above). Thus, the latitude of point P is 60°N (north of the equator).

Trigonometric Functions **277**

Review Note

You might need to review 30°-60°-90° and 45°-45°-90° right-triangle relationships before finding the sine and cosine of angles that are multiples of 30° and 45°.

Error Analysis

Since students may have difficulty remembering the special values of $\sin \theta$ and $\cos \theta$, you may wish to show students that the $\sin \theta$ and $\cos \theta$ columns in the table on this page can be written in the following form.

$\sin \theta$	$\cos \theta$
$\frac{\sqrt{0}}{2}$	$\frac{\sqrt{4}}{2}$
$\frac{\sqrt{1}}{2}$	$\frac{\sqrt{3}}{2}$
$\frac{\sqrt{2}}{2}$	$\frac{\sqrt{2}}{2}$
$\frac{\sqrt{3}}{2}$	$\frac{\sqrt{1}}{2}$
$\frac{\sqrt{4}}{2}$	$\frac{\sqrt{0}}{2}$

Communication Note

Point out that all points on the equator have a latitude of 0°. The North Pole has a latitude of 90°N, and the South Pole has a latitude of 90°S. Have students verify that a 1° separation in circles of latitude is about 69 miles (or 110 km).

Mathematical Note

The calculation done in Example 3 is independent of the actual radius of Earth (3963 mi) because cosine is a ratio, and evaluating it does not depend on the radius.

Graphs of Sine and Cosine

To graph the sine function, imagine a particle on the unit circle that starts at $(1, 0)$ and rotates counterclockwise around the origin. Every position (x, y) of the particle corresponds to an angle θ, where $\sin \theta = y$ by definition. As the particle rotates through the four quadrants, we get the four pieces of the sine graph shown below.

   I From 0° to 90°, the y-coordinate increases from 0 to 1.

   II From 90° to 180°, the y-coordinate decreases from 1 to 0.

   III From 180° to 270°, the y-coordinate decreases from 0 to -1.

   IV From 270° to 360°, the y-coordinate increases from -1 to 0.

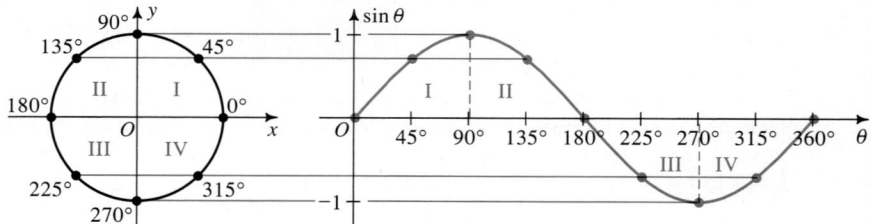

Since the sine function is periodic with a fundamental period of 360°, the graph above can be extended left and right as shown below.

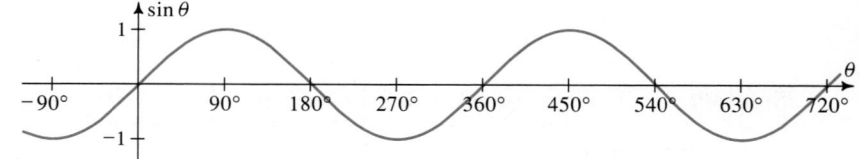

To graph the cosine function, we analyze the x-coordinate of the rotating particle in a similar manner. The cosine graph is shown below.

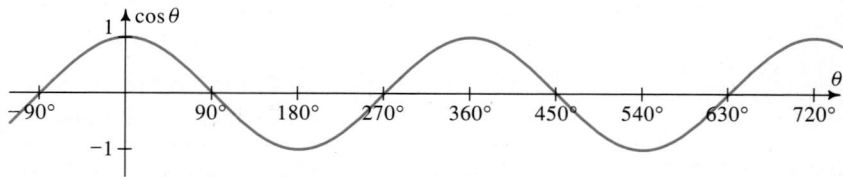

CLASS EXERCISES

1. Find the reference angle for θ. $\pi - 3 \approx 0.14$
     a. $\theta = 170° \ 10°$ **b.** $\theta = 310° \ 50°$ **c.** $\theta = 205.1° \ 25.1°$ **d.** $\theta = 3$
2. Name another angle between 0° and 360° with the same sine as 70°. 110°

3. Name another angle between 0° and 360° with the same cosine as 40°. 320°

4. Give each of the following in terms of the sine of a reference angle. sin 40°
   a. sin 170° sin 10°   **b.** sin 330° −sin 30°   **c.** sin (−15°) −sin 15° **d.** sin 400°

5. Give each of the following in terms of the cosine of a reference angle.
   a. cos 160°   **b.** cos 182° −cos 2°   **c.** cos (−100°)   **d.** cos 365°
   −cos 20°   −cos 80°   cos 5°

Use a calculator or table to find the value of each expression to four decimal places. 7. c. −0.8161 d. 0.8471 8. d. −0.4161

−0.9397   −0.6428
6. a. sin 188° −0.1392 **b.** sin (−110°)   **c.** cos 350° 0.9848   **d.** cos (−230°)

7. a. cos 10.2° 0.9842 **b.** sin 28.6° 0.4787   **c.** sin (−54.7°)   **d.** cos (−32.1°)

8. a. sin 3 0.1411   **b.** cos 4 −0.6536   **c.** sin (−1) −0.8415 **d.** cos (−2)

9. a. cos 2.5 −0.8011 **b.** cos (−0.73) 0.7452 **c.** sin (−3.4) 0.2555 **d.** sin 0.39
11. a. 1/2 **b.** −1/2 **c.** −1/2 **d.** 1/2   0.3802

Study the sine and cosine values of 30°, 45°, and 60°. Then give the exact value of each expression in simplest radical form. 11. a.-d., See above.

10. a. $\sin 45°\ \dfrac{\sqrt{2}}{2}$ **b.** $\sin 135°\ \dfrac{\sqrt{2}}{2}$ **c.** $\sin 225°\ -\dfrac{\sqrt{2}}{2}$ **d.** $\sin 315°\ -\dfrac{\sqrt{2}}{2}$

11. a. cos 60°   **b.** cos 120°   **c.** cos 240°   **d.** cos 300°

12. a. $\sin 30°\ \dfrac{1}{2}$ **b.** $\sin (-30°)\ -\dfrac{1}{2}$ **c.** $\cos 30°\ \dfrac{\sqrt{3}}{2}$ **d.** $\cos (-30°)\ \dfrac{\sqrt{3}}{2}$

13. a. $\sin 330°\ -\dfrac{1}{2}$ **b.** $\cos 330°\ \dfrac{\sqrt{3}}{2}$ **c.** $\sin \dfrac{7\pi}{6}\ -\dfrac{1}{2}$ **d.** $\cos \dfrac{7\pi}{6}\ -\dfrac{\sqrt{3}}{2}$

14. a. $\cos \dfrac{\pi}{4}\ \dfrac{\sqrt{2}}{2}$ **b.** $\sin \left(-\dfrac{\pi}{3}\right)\ -\dfrac{\sqrt{3}}{2}$ **c.** $\cos \dfrac{5\pi}{6}\ -\dfrac{\sqrt{3}}{2}$ **d.** sin (−300°)
                                                                                                                    $\dfrac{\sqrt{3}}{2}$

15. For the graphs of the sine and cosine functions shown on the preceding page, the θ-axis is labeled in degrees. Redraw each graph, labeling the θ-axis in radians this time. (Your labels should be in terms of π.)

16. *Visual Thinking* Explain how translating the cosine graph can be used to justify the fact that for all θ:
$$\cos (\theta - 90°) = \sin \theta$$

17. a. What symmetry does the graph of the sine function have? Symmetry in the origin
   b. What symmetry does the graph of the cosine function have? Symmetry in the y-axis

18. Use Exercise 16 and part (b) of Exercise 17 to justify the fact that for all θ:
$$\cos (90° - \theta) = \sin \theta$$

3. a. −cos 44.5° **d.** sin 21°   **4. a.** −cos 72.1° **b.** −sin 88.7° **d.** −cos 48°

WRITTEN EXERCISES

2. d. −sin 80°

Express each of the following in terms of a reference angle.
                                                                        −cos 40°
A **1. a.** sin 128° sin 52°   **b.** cos 128° −cos 52° **c.** sin (−37°) −sin 37° **d.** cos 500°

   2. a. sin 310° −sin 50°   **b.** cos 310° cos 50°   **c.** cos (−53°) cos 53°   **d.** sin 1000°

   3. a. cos 224.5°   **b.** cos 658° cos 62°   **c.** sin 145.7° sin 34.3° **d.** sin (−201°)

   4. a. cos 107.9°   **b.** sin 271.3°   **c.** sin 834° sin 66°   **d.** cos (−132°)

Trigonometric Functions **279**

Problem Solving

You might want students to justify the identity
sin (θ − 90°) = −cos θ
(see Class Exercise 18).

Additional Answers
Class Exercises

11. a. $\dfrac{1}{2}$ **b.** $-\dfrac{1}{2}$

   c. $-\dfrac{1}{2}$ **d.** $\dfrac{1}{2}$

16. If the cosine graph is translated 90° to the right, it coincides with the sine graph. Thus,
cos (θ − 90°) = sin θ
for all θ.

18. Since the cosine graph has symmetry in the y-axis, cos (−θ) = cos θ for all θ. Thus,
cos (90° − θ) =
cos (−(θ − 90°)) =
cos (θ − 90°). Since
cos (θ − 90°) = sin θ, we also have cos (90° − θ) = sin θ for all θ.

Suggested Assignments

Standard
Day 1: 279/1–17 odd
Day 2: 280/19, 24, 27, 29, 31, 33
Comprehensive
Day 1: 279/1–19 odd
Day 2: 280/21–33 odd, 24, 34

Supplementary Materials

Alternative Assessment, 24
Using Technology, 23
Precalculus Plotter Plus, 37–38, 39–40
Student Resource Guide, 69–72

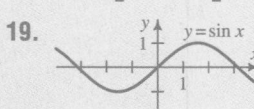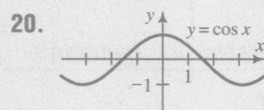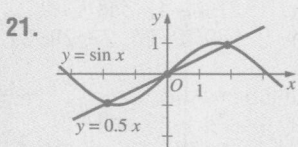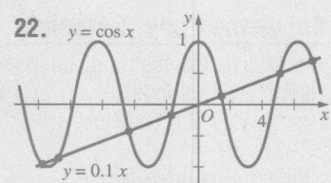
Use a calculator or table to find the value of each expression to four decimal places. 6. d. 0.8957 7. d. −0.4586 8. d. −0.6704 9. d. 0.2837

5. **a.** sin 28° 0.4695 **b.** sin 46° 0.7193 **c.** cos 65.3° 0.4179 **d.** sin 0.49 0.4706
6. **a.** sin 81° 0.9877 **b.** cos 0.3° 1.0000 **c.** cos 1.17 0.3902 **d.** sin 1.11
7. **a.** sin 160° 0.3420 **b.** cos 205.3° −0.9041 **c.** cos 302.1° 0.5314 **d.** sin 207°18′
8. **a.** cos 238° −0.5299 **b.** sin 403° 0.6820 **c.** sin 285.7° −0.9627 **d.** cos 132°6′
9. **a.** sin 1.2 0.9320 **b.** sin 2.2 0.8085 **c.** cos 3.5 −0.9365 **d.** cos 5
10. **a.** sin 3.54 −0.3880 **b.** cos 6 0.9602 **c.** sin (−2) −0.9093 **d.** cos 3 −0.9900

Study the sine and cosine values of 30°, 45°, and 60°. Then give the exact value of each expression in simplest radical form. 14. d. $\dfrac{\sqrt{2}}{2}$ 16. d. $-\dfrac{\sqrt{3}}{2}$

11. **a.** cos 45° **b.** cos 135° **c.** sin 210° **d.** cos 150°
12. **a.** sin (−45°) **b.** cos 315° **c.** cos 225° **d.** sin (−225°)
13. **a.** sin 150° **b.** cos (−240°) **c.** sin (−135°) **d.** cos (−30°)
14. **a.** cos 210° $-\dfrac{\sqrt{3}}{2}$ **b.** cos 90° 0 **c.** sin (−120°) $-\dfrac{\sqrt{3}}{2}$ **d.** sin (−315°)
15. **a.** $\cos\dfrac{\pi}{6}$ $\dfrac{\sqrt{3}}{2}$ **b.** $\sin\dfrac{\pi}{3}$ $\dfrac{\sqrt{3}}{2}$ **c.** $\cos\dfrac{2\pi}{3}$ $-\dfrac{1}{2}$ **d.** $\sin\dfrac{3\pi}{4}$ $\dfrac{\sqrt{2}}{2}$
16. **a.** $\cos\dfrac{\pi}{4}$ $\dfrac{\sqrt{2}}{2}$ **b.** $\sin\left(-\dfrac{\pi}{4}\right)$ $-\dfrac{\sqrt{2}}{2}$ **c.** $\sin\dfrac{5\pi}{3}$ $-\dfrac{\sqrt{3}}{2}$ **d.** $\cos\left(-\dfrac{7\pi}{6}\right)$
17. **a.** $\cos 2\pi$ 1 **b.** $\sin\dfrac{11\pi}{6}$ $-\dfrac{1}{2}$ **c.** $\cos\left(-\dfrac{5\pi}{6}\right)$ $-\dfrac{\sqrt{3}}{2}$ **d.** $\cos\dfrac{3\pi}{4}$ $-\dfrac{\sqrt{2}}{2}$
18. **a.** $\cos\left(-\dfrac{\pi}{3}\right)$ $\dfrac{1}{2}$ **b.** $\sin\pi$ 0 **c.** $\sin\dfrac{5\pi}{4}$ $-\dfrac{\sqrt{2}}{2}$ **d.** $\sin\left(-\dfrac{\pi}{6}\right)$ $-\dfrac{1}{2}$

19. The "natural" way to graph $y = \sin x$ or $y = \cos x$ is to measure x in radians and use the same real-number scale on both axes. Use this method to sketch the graph of $y = \sin x$. (Note that π is a little more than 3 units on the x-axis.)

20. Sketch the graph of $y = \cos x$ using the method described in Exercise 19.

◭ **For part (b) of Exercises 21–23, you will need to use a computer or a graphing calculator. Be sure to use radian measure for x.**

21. **a.** On a single set of axes, graph $y = \sin x$ and $y = 0.5x$. How many solutions does the equation $\sin x = 0.5x$ have? 3
    b. Find each solution of the equation to the nearest hundredth. −1.90, 0, 1.90
22. **a.** On a single set of axes, graph $y = \cos x$ and $y = 0.1x$. How many solutions does the equation $\cos x = 0.1x$ have? 7
    b. Find the smallest positive solution of the equation to the nearest tenth. 1.4
23. **a.** On a single set of axes, graph $y = |\cos x|$ and $y = x^2$. How many solutions does the equation $|\cos x| = x^2$ have? 2
    b. Find each solution of the equation to the nearest hundredth. −0.82, 0.82

24. Sketch the graph of $y = |\sin x|$. Then give the domain, range, and fundamental period of the functions $y = \sin x$ and $y = |\sin x|$.

280 *Chapter Seven*

25. **Visual Thinking** Imagine a particle starting at (1, 0) and making one counter-clockwise revolution on the unit circle. Let θ be the angle in standard position that corresponds to the particle's position.
   a. At how many points along the path of the particle are the x- and y-coordinates equal? 2
   b. What values of θ correspond to the points in part (a)? 45°, 225°
   c. On a single set of axes, sketch the graphs of the sine and cosine functions for $0° \le \theta \le 360°$. Use the graphs to show where $\sin \theta = \cos \theta$ in order to verify your answers in part (b).

For Exercises 26–33, use 3963 mi for the radius of Earth.

B 26. **Geography** The latitude of Durham, North Carolina, is 36°N. About how far from Durham is the North Pole? 3735 mi

27. **Geography** The latitude of Lima, Peru, is 12°S. About how far from Lima is the South Pole? 5395 mi

28. **Geography** Beijing, China, is due north of Perth, Australia. The latitude of Beijing is 39°55′N and the latitude of Perth is 31°58′S. About how far apart are the two cities? 4972 mi

29. **Geography** Memphis, Tennessee, is due north of New Orleans, Louisiana. The latitude of Memphis is 35°6′N and the latitude of New Orleans is 30°N. About how far apart are the two cities? 353 mi

30. **Physics** Earth's rotational speed at the equator is found by dividing the circumference of the equator by 24 hours:
$$24{,}902 \text{ mi} \div 24 \text{ h} \approx 1038 \text{ mi/h}$$
What is Earth's rotational speed at (**a**) Bangor, Maine (latitude 45°N) and (**b**) Esquina, Argentina (latitude 30°S)? **a.** 734 mi/h   **b.** 899 mi/h

31. **Physics** What is Earth's rotational speed at (**a**) Anchorage, Alaska (latitude 61°N) and (**b**) Rio de Janeiro, Brazil (latitude 23°S)? **a.** 503 mi/h   **b.** 955 mi/h

32. **a.** **Physics** Show that at latitude L, Earth's rotational speed in miles per hour is approximately equal to 1038 cos L.
   b. **Physics** Find Earth's rotational speed at the North Pole. 0 mi/h

33. **Physics** Rome, Italy, and Boston, Massachusetts, have approximately the same latitude (42°N). A plane flying from Rome due west to Boston is able to "stay with the sun," leaving Rome with the sun overhead and landing in Boston with the sun overhead. How fast is the plane flying? (*Hint*: See part (a) of Exercise 32.) 771 mi/h

Trigonometric Functions **281**

23.
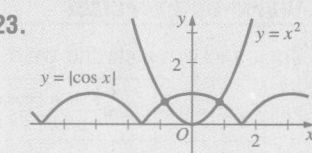

24.
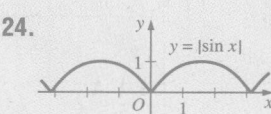

For $y = \sin x$, the domain is {all real numbers}, the range is $\{y \mid -1 \le y \le 1\}$, and the period is 2π. For $y = |\sin x|$, the domain is {all real numbers}, the range is $\{y \mid 0 \le y \le 1\}$, and the period is π.

25. c.
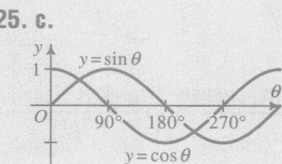

32. a. The circle at latitude L has radius $r = 3963 \cos L$. The circumference of this circle is $C = 2\pi r = 2\pi(3963 \cos L) = 7926\pi \cos L$. Thus, the rotational speed $= \dfrac{7926\pi \cos L}{24} \approx 1038 \cos L$ mi/h.

Copy and complete the chart.

	Quad	$\sin \theta$	$\cos \theta$	$\dfrac{\sin \theta}{\cos \theta}$	$\dfrac{\cos \theta}{\sin \theta}$
1.	I	$\dfrac{15}{17}$	?	?	?
2.	II	$\dfrac{\sqrt{3}}{2}$	?	?	?
3.	III	?	$-\dfrac{7}{25}$	?	?
4.	IV	?	$\dfrac{\sqrt{2}}{2}$	?	?

1. $\dfrac{8}{17}; \dfrac{15}{8}; \dfrac{8}{15}$

2. $-\dfrac{1}{2}; -\sqrt{3}; -\dfrac{\sqrt{3}}{3}$

3. $-\dfrac{24}{25}; \dfrac{24}{7}; \dfrac{7}{24}$

4. $-\dfrac{\sqrt{2}}{2}; -1; -1$

Motivating the Section

Introducing the other trigonometric functions lays the groundwork for solving triangles (see Chapter 9), a topic having many applications in navigation and surveying. Defining the other trigonometric functions is the objective of this section.

C **34. Mechanics** A piston rod $\overline{PQ}$, 4 units long, is connected to the rim of a wheel at point P, and to a piston at point Q. As P moves counterclockwise around the wheel at 1 radian per second, Q slides left and right in the piston. What are the coordinates of P and Q in terms of time t in seconds? Assume that P is at $(1, 0)$ when $t = 0$. $P(\cos t, \sin t)$, $Q(\cos t + \sqrt{16 - \sin^2 t}, 0)$

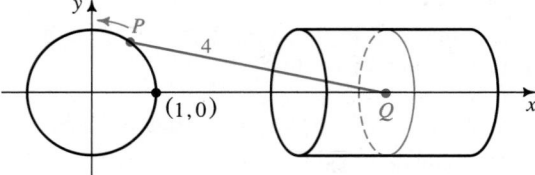

7-5 The Other Trigonometric Functions

> **Objective** To find values of the tangent, cotangent, secant, and cosecant functions and to sketch the functions' graphs.

We can define four other trigonometric functions of an angle θ in terms of the x- and y-coordinates of a point on the terminal ray of θ.

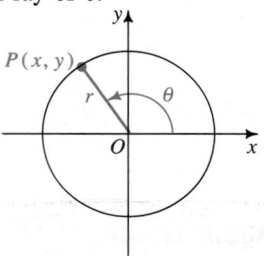

tangent of θ: $\tan \theta = \dfrac{y}{x}, \; x \neq 0$

cotangent of θ: $\cot \theta = \dfrac{x}{y}, \; y \neq 0$

secant of θ: $\sec \theta = \dfrac{r}{x}, \; x \neq 0$

cosecant of θ: $\csc \theta = \dfrac{r}{y}, \; y \neq 0$

Since $\cos \theta = \dfrac{x}{r}$ and $\sin \theta = \dfrac{y}{r}$, we can write these four new functions in terms of $\cos \theta$ and $\sin \theta$:

$$\tan \theta = \frac{\sin \theta}{\cos \theta} \qquad \cot \theta = \frac{\cos \theta}{\sin \theta} \qquad \sec \theta = \frac{1}{\cos \theta} \qquad \csc \theta = \frac{1}{\sin \theta}$$

Notice that $\sec \theta$ and $\cos \theta$ are reciprocals, as are $\csc \theta$ and $\sin \theta$. This is also true of $\cot \theta$ and $\tan \theta$:

$$\cot \theta = \frac{1}{\tan \theta}$$

The signs of these functions in the various quadrants are summarized in the table at the right.

	I	II	III	IV
$\sin \theta$ and $\csc \theta$	+	+	−	−
$\cos \theta$ and $\sec \theta$	+	−	−	+
$\tan \theta$ and $\cot \theta$	+	−	+	−

---

282 *Chapter Seven*

Example 1 Find the value of each expression to four significant digits.
 a. tan 203°   **b.** cot 165°   **c.** csc (−1)   **d.** sec 11

Solution Use a calculator to check each answer given below. If your calculator does not have the cotangent, secant, and cosecant as built-in functions, you must use the reciprocal relationships noted at the bottom of the preceding page. Be sure your calculator is in degree mode for parts (a) and (b), and in radian mode for parts (c) and (d).

a. $\tan 203° \approx 0.4245$ **b.** $\cot 165° = \dfrac{1}{\tan 165°} \approx -3.732$

c. $\csc (-1) = \dfrac{1}{\sin (-1)} \approx -1.188$ **d.** $\sec 11 = \dfrac{1}{\cos 11} \approx 226.0$

Example 2 If θ is a second-quadrant angle and $\tan \theta = -\dfrac{3}{4}$, find the values of the other five trigonometric functions.

Solution Since θ is a second-quadrant angle and $\tan \theta = \dfrac{y}{x} = -\dfrac{3}{4}$, we can draw a diagram as shown at the right. Then:

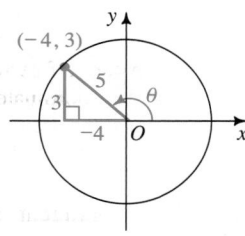

$$\cos \theta = \frac{x}{r} = -\frac{4}{5} \qquad \sec \theta = \frac{1}{\cos \theta} = -\frac{5}{4}$$

$$\sin \theta = \frac{y}{r} = \frac{3}{5} \qquad \csc \theta = \frac{1}{\sin \theta} = \frac{5}{3}$$

$$\cot \theta = \frac{1}{\tan \theta} = -\frac{4}{3}$$

The Tangent Graph

Imagine on the unit circle a particle P that starts at $(1, 0)$ and rotates counterclockwise around the origin. Every position (x, y) of the particle corresponds to an angle θ, where $\tan \theta = \dfrac{y}{x}$ by definition.

First consider what happens as the particle moves through the first quadrant. When P is at $(1, 0)$, $\theta = 0$ and $\tan \theta = \dfrac{0}{1} = 0$. As P moves toward $(0, 1)$, y increases and x decreases, so that $\tan \theta = \dfrac{y}{x}$ gets larger. When P reaches $(0, 1)$, $\theta = \dfrac{\pi}{2}$. Although the definition of tangent would suggest that $\tan \dfrac{\pi}{2}$ is $\dfrac{1}{0}$, this expression is undefined, so we say that $\tan \dfrac{\pi}{2}$ is also undefined.

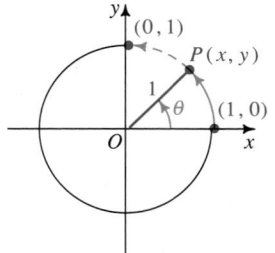

You may need to tell students to use the reciprocal key (1/*x*) if their calculators do not have built-in cotangent, secant, and cosecant keys.

Example Note

The Pythagorean theorem is used implicitly in Example 2.

Additional Example

1. If $\csc \theta = -\dfrac{17}{15}$ and $-90° < \theta < 90°$, find the values of the other five trigonometric functions.

Since $\csc \theta = -\dfrac{17}{15}$,

$\sin \theta = \dfrac{1}{\csc \theta} = -\dfrac{15}{17} = \dfrac{y}{r}$.

Since $\sin \theta < 0$ and $-90° < \theta < 90°$, θ is a fourth-quadrant angle. Thus, $x > 0$.

$x = \sqrt{r^2 - y^2} = \sqrt{17^2 - (-15)^2} = 8$.

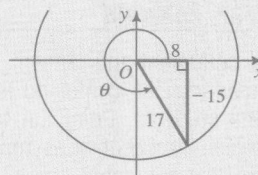

$\cos \theta = \dfrac{x}{r} = \dfrac{8}{17}$;

$\sec \theta = \dfrac{1}{\cos \theta} = \dfrac{17}{8}$;

$\tan \theta = \dfrac{y}{x} = -\dfrac{15}{8}$;

$\cot \theta = \dfrac{1}{\tan \theta} = -\dfrac{8}{15}$

As the particle continues to move around the unit circle, we can analyze the other values of tan θ and obtain the tangent graph shown below. Notice that the graph has vertical asymptotes at odd multiples of $\frac{\pi}{2}$. Also notice that the tangent function is periodic with fundamental period π (or 180°). The graph of the cotangent function is similar and is left for you to sketch (see Written Exercise 9).

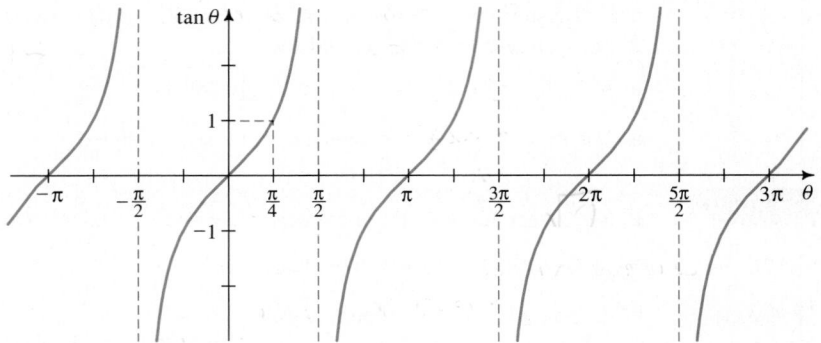

The Secant Graph

Since the secant function is the reciprocal of the cosine function, we can obtain the secant graph (shown in red below) using the cosine graph (black) and these facts:

1. sec $\theta = 1$ when cos $\theta = 1$: at $\theta = 0$, $\pm 2\pi$, $\pm 4\pi$, . . .

2. sec $\theta = -1$ when cos $\theta = -1$: at $\theta = \pm\pi$, $\pm 3\pi$, $\pm 5\pi$, . . .

3. sec θ is undefined when cos $\theta = 0$: at $\theta = \pm \frac{\pi}{2}$, $\pm\frac{3\pi}{2}$, $\pm\frac{5\pi}{2}$, . . .

4. |sec θ| gets larger as |cos θ| gets smaller.

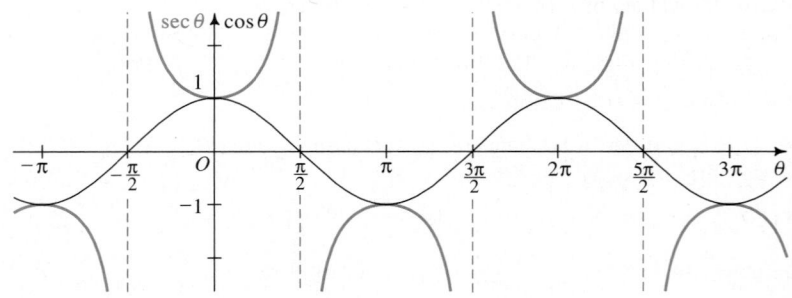

Notice that the graph of the secant function has vertical asymptotes at odd multiples of $\frac{\pi}{2}$. Also notice that the secant function, like the cosine function, is periodic with fundamental period 2π (or 360°). The cosecant graph is like the secant graph and can be found by a similar analysis (see Written Exercise 10).

284   *Chapter Seven*

1. a. $90° + n \cdot 180°$ **b.** $90° + n \cdot 360°$ **c.** $135° + n \cdot 180°$ **d.** $n \cdot 180°$

CLASS EXERCISES

1. Give the values of θ (in degrees) for which: See above.
   a. $\cot \theta = 0$ **b.** $\csc \theta = 1$ **c.** $\cot \theta = -1$ **d.** $\csc \theta$ is undefined

2. If $\sec 15° \approx 1.035$, give the approximate value of: **a.** 1.035 **b.** −1.035 **c.** −1.035
   a. $\sec(-15°)$ **b.** $\sec 165°$ **c.** $\sec 195°$ **d.** $\sec 345°$ **d.** 1.035

3. Find the value of each expression to four significant digits. See below.
   a. $\tan 2$ **b.** $\cot 185°$ **c.** $\csc 3$ **d.** $\sec(-22°)$

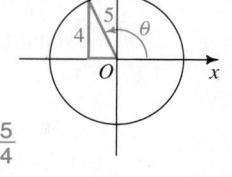

4. In what quadrant is θ if $\csc \theta < 0$ and $\tan \theta > 0$? III

5. The diagram shows an angle θ with $\sin \theta = \dfrac{4}{5}$. Find:
   a. $\cos \theta\ -\dfrac{3}{5}$ **b.** $\tan \theta\ -\dfrac{4}{3}$ **c.** $\cot \theta\ -\dfrac{3}{4}$ **d.** $\sec \theta\ -\dfrac{5}{3}$ **e.** $\csc \theta\ \dfrac{5}{4}$

   3. a. −2.185   **b.** 11.43   **c.** 7.086   **d.** 1.079

WRITTEN EXERCISES

Find the value of each expression to four significant digits.

A

1. **a.** $\tan 100°$ −5.671   **b.** $\cot 276°$ −0.1051   **c.** $\csc 5$ −1.043   **d.** $\sec 2.14$ −1.855
2. **a.** $\sec(-11°)$ 1.019   **b.** $\csc 233°$ −1.252   **c.** $\tan 3$ −0.1425   **d.** $\cot 7.28$ 0.6466

Express each of the following in terms of a reference angle.

3. **a.** $\sin 195°$ −sin 15°   **b.** $\sec 280°$ sec 80°   **c.** $\tan(-140°)$ tan 40°   **d.** $\sec 2$ −sec 1.14
4. **a.** $\cot 285°$ −cot 75°   **b.** $\sec(-105°)$ −sec 75°   **c.** $\csc 600°$ −csc 60°   **d.** $\tan 3$ −tan 0.14
5. **a.** $\tan 820°$ −tan 80°   **b.** $\sec 290°$ sec 70°   **c.** $\cot 185°$ cot 5°   **d.** $\csc 4$ −csc 0.86
6. **a.** $\tan 160°$ −tan 20°   **b.** $\csc 115°$ csc 65°   **c.** $\sec 235°$ −sec 55°   **d.** $\cot 5$ −cot 1.28

7. Give the values of x (in radians) for which $\csc x$ is:
   a. undefined $n\pi$ **b.** 0 None   **c.** 1 $\dfrac{\pi}{2} + 2n\pi$ **d.** −1 $\dfrac{3\pi}{2} + 2n\pi$

8. Give the values of x (in radians) for which $\cot x$ is:
   a. undefined $n\pi$ **b.** 0 $\dfrac{\pi}{2} + n\pi$ **c.** 1 $\dfrac{\pi}{4} + n\pi$ **d.** −1 $\dfrac{3\pi}{4} + n\pi$

9. Sketch a graph of $\cot \theta$ versus θ for $-\pi \le \theta \le 3\pi$. Be sure to show where any vertical asymptotes occur.

You may find it helpful to have a computer or a graphing calculator to complete Exercise 10.

10. **Writing** On a single set of axes, graph $y = \sin x$ and $y = \csc x$, showing at least two full periods of each function. Using the reciprocal relationship between sine and cosecant, explain the features (x-intercepts, vertical asymptotes, periodicity, and so on) of the cosecant graph.

11. On a single set of axes, sketch the graphs of $y = \tan x$ and $y = 2x$. How many solutions does the equation $\tan x = 2x$ have? Infinitely many

12. On a single set of axes, sketch the graphs of $y = \sin x$ and $y = \sec x$. How many solutions does the equation $\sin x = \sec x$ have? None

Trigonometric Functions   **285**

△ **Using Technology**

Written Exercises 1 and 2 may be done with a calculator. A graphing calculator can be used for Written Exercises 11 and 12.

Error Analysis

Students who have difficulty locating asymptotes should be encouraged to draw circles with the quadrantal angles marked, and to evaluate the six trigonometric functions for the quadrantal angles. This will help them see when the tangent, cotangent, secant, and cosecant functions are undefined.

Suggested Assignments

Standard
Day 1: 285/1–15 odd
Day 2: 286/17–27 odd
Comprehensive
285/3, 6, 9, 11, 15, 17, 20, 21–27 odd

Supplementary Materials

Alternative Assessment, 24
Precalculus Plotter Plus, 41–42

14. $\sin x = -\dfrac{7}{25}$; $\tan x =$

$-\dfrac{7}{24}$; $\csc x = -\dfrac{25}{7}$;

$\sec x = \dfrac{25}{24}$; $\cot x = -\dfrac{24}{7}$

16. $\sin x = \dfrac{5}{13}$; $\cos x = -\dfrac{12}{13}$;

$\tan x = -\dfrac{5}{12}$; $\sec x =$

$-\dfrac{13}{12}$; $\csc x = \dfrac{13}{5}$

18. $\sin x = -\dfrac{1}{5}$; $\cos x =$

$-\dfrac{2\sqrt{6}}{5}$; $\tan x = \dfrac{\sqrt{6}}{12}$;

$\sec x = -\dfrac{5\sqrt{6}}{12}$;

$\cot x = 2\sqrt{6}$

Warm-Up Exercises

Solve each equation for θ if $0° \le \theta < 360°$.

1. $\sin \theta = 1$ 90°

2. $\tan \theta = -\sqrt{3}$ 120°, 300°

3. $\cos \theta = 2$ no solution

4. $\tan \theta = 0$ 0°, 180°

5. $\sin \theta = -\dfrac{\sqrt{3}}{2}$ 240°, 300°

6. $\cos \theta = -\dfrac{\sqrt{3}}{2}$ 150°, 210°

In Exercises 13–18, find the values of the other five trigonometric functions.

13. $\sin x = \dfrac{5}{13}$, $\dfrac{\pi}{2} < x < \pi$

14. $\cos x = \dfrac{24}{25}$, $-\dfrac{\pi}{2} < x < 0$

15. $\tan x = \dfrac{3}{4}$, $\pi < x < 2\pi$

16. $\cot x = -\dfrac{12}{5}$, $0 < x < \pi$

17. $\sec x = -3$, $0 < x < \pi$

18. $\csc x = -5$, $\dfrac{\pi}{2} < x < \dfrac{3\pi}{2}$

19. Explain why the cotangent graph has vertical asymptotes at multiples of π.

20. **Reading** Prepare a summary table for the six trigonometric functions introduced in this section and the preceding one. The table should give the definition, the domain, the range, the fundamental period, and a sketch of the graph of each function.

B 21. **a.** Verify that $1 + \tan^2 \dfrac{\pi}{3} = \sec^2 \dfrac{\pi}{3}$. $\left[\textit{Note: } \tan^2 \dfrac{\pi}{3} \text{ means } \left(\tan \dfrac{\pi}{3}\right)^2.\right]$

    b. Can you find any other values of x for which $1 + \tan^2 x = \sec^2 x$?

22. **a.** Evaluate $1 + \cot^2 x$ and $\csc^2 x$ for $x = \dfrac{\pi}{2}$, $x = \dfrac{3\pi}{4}$, and $x = \dfrac{7\pi}{6}$.

    b. Make a conjecture about the relationship between $1 + \cot^2 x$ and $\csc^2 x$. Prove your conjecture using Exercise 7 on page 271 and the definitions of cotangent and cosecant. 23. **a.** $-\dfrac{1}{2}$ **b.** -2 **c.** $\dfrac{\sqrt{3}}{2}$ **d.** $-\sqrt{3}$

Find the exact value of each expression or state that the value is undefined.
See above.
23. **a.** $\cos 120°$ **b.** $\sec 120°$ **c.** $\sin 120°$ **d.** $\tan 120°$

24. **a.** $\sin 225°$ $-\dfrac{\sqrt{2}}{2}$ **b.** $\csc 225°$ $-\sqrt{2}$ **c.** $\tan 225°$ 1 **d.** $\sec 225°$ $-\sqrt{2}$

25. **a.** $\csc 90°$ 1 **b.** $\sec 180°$ -1 **c.** $\tan 240°$ $\sqrt{3}$ **d.** $\cot 0°$ Undef.

26. **a.** $\csc 150°$ 2 **b.** $\csc 0°$ Undef.   **c.** $\tan 315°$ -1 **d.** $\sec 315°$ $\sqrt{2}$

27. **a.** $\csc \pi$ Undef.   **b.** $\tan \dfrac{2\pi}{3}$ $-\sqrt{3}$ **c.** $\cot \dfrac{\pi}{2}$ 0 **d.** $\sec \dfrac{5\pi}{6}$ $-\dfrac{2\sqrt{3}}{3}$

28. **a.** $\tan \dfrac{\pi}{2}$ Undef.   **b.** $\cot \dfrac{7\pi}{4}$ -1 **c.** $\sec(-3\pi)$ -1 **d.** $\csc \dfrac{7\pi}{6}$ -2

7-6 The Inverse Trigonometric Functions

Objective   *To find values of the inverse trigonometric functions.*

From the graph of $f(x) = \tan x$ shown on the left at the top of the next page, we can see that the tangent function is not one-to-one and thus has no inverse. However, if we restrict x to the interval $-\dfrac{\pi}{2} < x < \dfrac{\pi}{2}$, the restricted function, which we denote $F(x) = \text{Tan } x$, is one-to-one. Its inverse is denoted $\text{Tan}^{-1} x$ and is read "the inverse tangent of x." Notice that $\text{Tan}^{-1} x = y$ means that $\tan y = x$ and $-\dfrac{\pi}{2} < y < \dfrac{\pi}{2}$.

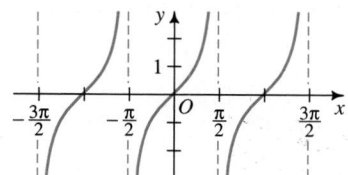

$f(x) = \tan x$ has no inverse.

Domain: $\{x \mid x \neq \frac{\pi}{2} + n\pi\}$

Range: Real numbers

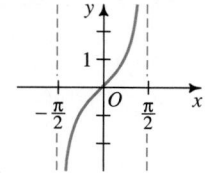

$F(x) = \text{Tan } x$ has an inverse.

Domain: $\{x \mid -\frac{\pi}{2} < x < \frac{\pi}{2}\}$

Range: Real numbers

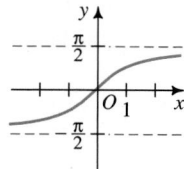

$F^{-1}(x) = \text{Tan}^{-1} x$

Domain: Real numbers

Range: $\{y \mid -\frac{\pi}{2} < y < \frac{\pi}{2}\}$

Example 1 Find $\text{Tan}^{-1} 2$ with a calculator.

Solution On most calculators, the inverse tangent function is symbolized by $\tan^{-1}$, INV tan, or ARC tan. Moreover, the calculator will give you answers in degrees or radians.

With the calculator in degree mode, $\text{Tan}^{-1} 2 \approx 63.4°$.

With the calculator in radian mode, $\text{Tan}^{-1} 2 \approx 1.11$.

Example 2 Find $\text{Tan}^{-1}(-1)$ without a calculator.

Solution $\text{Tan}^{-1}(-1) = x$ means that $\tan x = -1$ and $-\frac{\pi}{2} < x < \frac{\pi}{2}$.

Thus, $\text{Tan}^{-1}(-1) = -\frac{\pi}{4}$.

By considering only the solid portion of the graph of $g(x) = \sin x$ shown at the left below, we obtain a new function $G(x) = \text{Sin } x$ with domain $-\frac{\pi}{2} \leq x \leq \frac{\pi}{2}$. This function has an inverse, $G^{-1}(x) = \text{Sin}^{-1} x$, whose graph is shown at the right below. Note that $\text{Sin}^{-1} x = y$ means that $\sin y = x$ and $-\frac{\pi}{2} \leq y \leq \frac{\pi}{2}$.

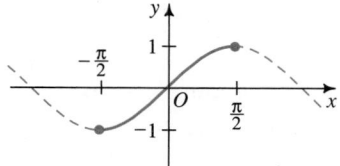

$G(x) = \text{Sin } x$ has an inverse.

Domain: $\{x \mid -\frac{\pi}{2} \leq x \leq \frac{\pi}{2}\}$

Range: $\{y \mid -1 \leq y \leq 1\}$

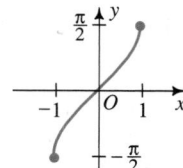

$G^{-1}(x) = \text{Sin}^{-1} x$

Domain: $\{x \mid -1 \leq x \leq 1\}$

Range: $\{y \mid -\frac{\pi}{2} \leq y \leq \frac{\pi}{2}\}$

Trigonometric Functions **287**

Motivating the Section

The angle of banking on a race track or roadway depends on the speed v and curve radius r and is given by the formula $\tan \theta = \frac{v^2}{rg}$ where g is the gravitational constant. Solving this formula for θ is an application from physics of the inverse trigonometric functions, the topic of this section.

Additional Examples

1. Evaluate each expression without using a calculator.
   a. $\text{Tan}^{-1}\sqrt{3}$
   b. $\text{Sin}^{-1}(-1)$
   c. $\text{Cos}^{-1}\left(-\frac{\sqrt{3}}{2}\right)$

   a. $\text{Tan}^{-1}\sqrt{3}$ is the number between $-\frac{\pi}{2}$ and $\frac{\pi}{2}$ whose tangent is $\sqrt{3}$. Since $\tan \frac{\pi}{3} = \sqrt{3}$, $\text{Tan}^{-1}\sqrt{3} = \frac{\pi}{3}$.

   b. If $\text{Sin}^{-1}(-1) = x$, then $\sin x = -1$ and $-\frac{\pi}{2} \leq x \leq \frac{\pi}{2}$. Since $\sin\left(-\frac{\pi}{2}\right) = -1$, $\text{Sin}^{-1}(-1) = -\frac{\pi}{2}$.

   c. If $\text{Cos}^{-1}\left(-\frac{\sqrt{3}}{2}\right) = x$, then $\cos x = -\frac{\sqrt{3}}{2}$ and $0 \leq x \leq \pi$. Since $\cos \frac{5\pi}{6} = -\frac{\sqrt{3}}{2}$, $\text{Cos}^{-1}\left(-\frac{\sqrt{3}}{2}\right) = \frac{5\pi}{6}$.

(*Continues on next page*)

2. Find an approximate value and the exact value of $\csc\left(\text{Cos}^{-1}(-0.4)\right)$.

Using a calculator in radian mode, you find that $\text{Cos}^{-1}(-0.4) \approx 1.9823132$ and $\csc 1.9823132 \approx 1.09$. To find the exact value, let $\theta = \text{Cos}^{-1}(-0.4)$. Then $\cos\theta = -0.4$ and since $0° \le \theta \le 180°$, θ is a second-quadrant angle. Draw one of the diagrams below:

(a)

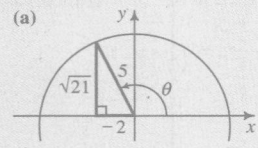

(b)

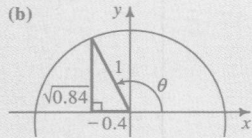

From diagram (a),
$$\csc\left(\text{Cos}^{-1}(-0.4)\right) =$$
$$\csc\theta = \frac{1}{\sin\theta} =$$
$$\frac{r}{y} = \frac{5}{\sqrt{21}} = \frac{5\sqrt{21}}{21}.$$

From diagram (b),
$$\csc\left(\text{Cos}^{-1}(-0.4)\right) =$$
$$\csc\theta = \frac{1}{\sin\theta} = \frac{r}{y} =$$
$$\frac{1}{\sqrt{0.84}} = \sqrt{\frac{100}{84}} = \frac{5}{\sqrt{21}} =$$
$$\frac{5\sqrt{21}}{21}.$$

Error Analysis

Students will have difficulties with problems like part (b) of Example 4. You might need to do several more examples.

Similarly, by considering only the solid portion of the graph of $h(x) = \cos x$ at the left below, we obtain a new function $H(x) = \text{Cos } x$ with domain $0 \le x \le \pi$. This function has an inverse, $H^{-1}(x) = \text{Cos}^{-1} x$, whose graph is shown at the right below. Note that $\text{Cos}^{-1} x = y$ means that $\cos y = x$ and $0 \le y \le \pi$.

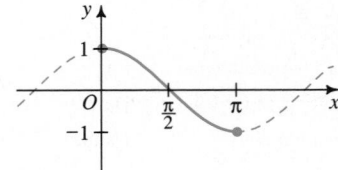

$H(x) = \text{Cos } x$ has an inverse.
Domain: $\{x \mid 0 \le x \le \pi\}$
Range: $\{y \mid -1 \le y \le 1\}$

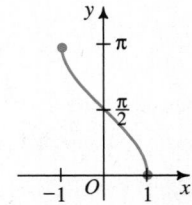

$H^{-1}(x) = \text{Cos}^{-1} x$
Domain: $\{x \mid -1 \le x \le 1\}$
Range: $\{y \mid 0 \le y \le \pi\}$

Example 3   **a.** Find $\text{Sin}^{-1}(-0.8)$ with a calculator.
b. Find $\text{Cos}^{-1}(-0.5)$ without a calculator.

Solution   **a.** With the calculator in radian mode, $\text{Sin}^{-1}(-0.8) \approx -0.93$.
b. $\text{Cos}^{-1}(-0.5) = x$ means that $\cos x = -0.5$ and $0 \le x \le \pi$.
Thus, $\text{Cos}^{-1}(-0.5) = \dfrac{2\pi}{3}$.

Example 4   **a.** Find $\cos\left(\text{Tan}^{-1}\left(-\dfrac{2}{3}\right)\right)$ with a calculator.

b. Find $\cos\left(\text{Tan}^{-1}\left(-\dfrac{2}{3}\right)\right)$ without a calculator.

Solution   **a.** With the calculator in either degree or radian mode:
$$\cos\left(\text{Tan}^{-1}\left(-\frac{2}{3}\right)\right) \approx 0.83$$

b. Let $\theta = \text{Tan}^{-1}\left(-\dfrac{2}{3}\right)$, so that θ is a fourth-quadrant angle such that $\tan\theta = -\dfrac{2}{3}$. After sketching a right triangle as in the diagram at the right, we see:

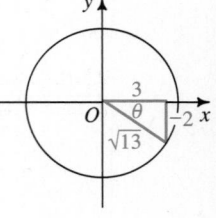

$$\cos\left(\text{Tan}^{-1}\left(-\frac{2}{3}\right)\right) = \cos\theta$$

$$= \frac{3}{\sqrt{13}} = \frac{3\sqrt{13}}{13}$$

(Use a calculator to compare this result with part (a).)

Find the value of each expression to the nearest tenth of a degree.

1. Tan^{-1} 1.2 50.2°
2. Sin^{-1} (−0.3) −17.5°
3. Cos^{-1} (−0.425) 115.2°

Find the value of each expression to the nearest hundredth of a radian.

4. Tan^{-1} (−2.9) −1.24
5. Sin^{-1} 0.75 0.85
6. Cos^{-1} 0.058 1.51

7. What happens when you try to evaluate Sin^{-1} 1.7 with a calculator? Explain why this happens. See below.

8. Use the diagram at the right to find the value of each expression.

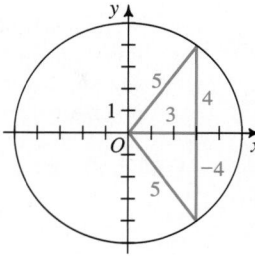

a. $\cos\left(\text{Sin}^{-1}\dfrac{4}{5}\right)$ $\dfrac{3}{5}$

b. $\tan\left(\text{Sin}^{-1}\dfrac{4}{5}\right)$ $\dfrac{4}{3}$

c. $\cot\left(\text{Sin}^{-1}\dfrac{4}{5}\right)$ $\dfrac{3}{4}$

7. An error message appears, because 1.7 is not in the domain of $y = \text{Sin}^{-1} x$.

WRITTEN EXERCISES

Use a calculator or table to find the value of each expression to the nearest tenth of a degree.

A **1. a.** Sin^{-1} 0.9 64.2° **b.** Sin^{-1} (−0.9) −64.2° **c.** $\text{Cos}^{-1}\dfrac{3}{4}$ 41.4° **d.** $\text{Cos}^{-1}\left(-\dfrac{3}{4}\right)$ 138.6°

2. a. $\text{Tan}^{-1}\dfrac{7}{3}$ 66.8° **b.** $\text{Tan}^{-1}\left(-\dfrac{7}{3}\right)$ −66.8° **c.** Cos^{-1} 0.4 66.4° **d.** Cos^{-1} (−0.4) 113.6°

Use a calculator or table to find the value of each expression to the nearest hundredth of a radian.

3. a. Tan^{-1} 0.23 0.23 **b.** Tan^{-1} (−0.23) −0.23 **c.** Cos^{-1} 0.345 1.22 **d.** Cos^{-1} (−0.345) 1.92

4. a. $\text{Sin}^{-1}\dfrac{3}{8}$ 0.38 **b.** $\text{Sin}^{-1}\left(-\dfrac{3}{8}\right)$ −0.38 **c.** $\text{Cos}^{-1}\dfrac{5}{6}$ 0.59 **d.** $\text{Cos}^{-1}\left(-\dfrac{5}{6}\right)$ 2.56

Without using a calculator or table, find the value of each expression in radians. Many answers can be given in terms of π.

5. a. Sin^{-1} 0 0 **b.** Cos^{-1} 0 $\dfrac{\pi}{2}$ **c.** Tan^{-1} 1 $\dfrac{\pi}{4}$ **d.** Tan^{-1} (−1) $-\dfrac{\pi}{4}$

6. a. Sin^{-1} 1 $\dfrac{\pi}{2}$ **b.** Sin^{-1} (−1) $-\dfrac{\pi}{2}$ **c.** Cos^{-1} 1 0 **d.** Cos^{-1} (−1) π

7. a. $\text{Sin}^{-1}\dfrac{1}{2}$ $\dfrac{\pi}{6}$ **b.** $\text{Sin}^{-1}\left(-\dfrac{1}{2}\right)$ $-\dfrac{\pi}{6}$ **c.** $\text{Cos}^{-1}\dfrac{1}{2}$ $\dfrac{\pi}{3}$ **d.** $\text{Cos}^{-1}\left(-\dfrac{1}{2}\right)$ $\dfrac{2\pi}{3}$

8. a. $\text{Sin}^{-1}\dfrac{\sqrt{2}}{2}$ $\dfrac{\pi}{4}$ **b.** $\text{Sin}^{-1}\left(-\dfrac{\sqrt{2}}{2}\right)$ $-\dfrac{\pi}{4}$ **c.** $\text{Cos}^{-1}\dfrac{\sqrt{2}}{2}$ $\dfrac{\pi}{4}$ **d.** $\text{Cos}^{-1}\left(-\dfrac{\sqrt{2}}{2}\right)$ $\dfrac{3\pi}{4}$

Trigonometric Functions **289**

Exercise Note

For Written Exercises 5–8, make sure students give answers within the range of $\text{Sin}^{-1} x$, $\text{Cos}^{-1} x$, and $\text{Tan}^{-1} x$.

Suggested Assignments

Standard
Day 1: 289/1–13 odd
Day 2: 290/15–21, 27, 29
Comprehensive
Day 1: 289/1–21 odd
Day 2: 291/22–30

Supplementary Materials

Alternative Assessment, 24
Student Resource Guide, 73–76

10. For any x between -1 and 1 inclusive, we can define the inverse cosine function, denoted Arccos (read "arc cosine"), as follows: Arccos $x = s$ provided $\cos s = x$ where s is the length of the arc with endpoints $R(1, 0)$ and $P(x, y)$ on the unit circle. (Note that P must be on the upper half of the circle.) Since the definition describes the range of the inverse cosine function as a set of arc lengths, the function's denotation (Arccos) is appropriate.

14. a. $0.98;\ \dfrac{5\sqrt{26}}{26}$

  b. $-0.20;\ -\dfrac{\sqrt{26}}{26}$

9. What is wrong with the expression $\text{Cos}^{-1} 3$? What happens when you try to evaluate it with a calculator?

10. *Writing* Each of the trigonometric functions can be defined in terms of arc lengths instead of angle measures. In the diagram, angle θ (in standard position) intercepts arc PR on the unit circle. If arc PR has length s, then we can define the cosine of s as:

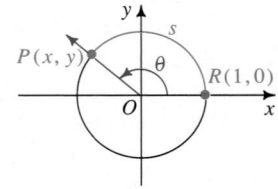

$$\cos s = \cos \theta = x\text{-coordinate of } P$$

Write a similar definition for the inverse cosine function and explain why the inverse cosine function is sometimes denoted Arccos (read "arc cosine"). (By similar reasoning, the inverse sine function is sometimes denoted Arcsin, the inverse tangent function is sometimes denoted Arctan, and so on.)

Find both the approximate value and the exact value of each expression.

11. a. $\tan\left(\text{Cos}^{-1} \dfrac{12}{13}\right)$ $0.42;\ \dfrac{5}{12}$ **b.** $\tan\left(\text{Cos}^{-1}\left(-\dfrac{12}{13}\right)\right)$ $-0.42;\ -\dfrac{5}{12}$

12. a. $\sin\left(\text{Cos}^{-1} \dfrac{1}{5}\right)$ $0.98;\ \dfrac{2\sqrt{6}}{5}$ **b.** $\sin\left(\text{Cos}^{-1}\left(-\dfrac{1}{5}\right)\right)$ $0.98;\ \dfrac{2\sqrt{6}}{5}$

13. a. $\csc\left(\text{Tan}^{-1} 1.05\right)$ $1.38;\ \dfrac{29}{21}$ **b.** $\sec\left(\text{Sin}^{-1}(-0.5)\right)$ $1.15;\ \dfrac{2\sqrt{3}}{3}$

14. a. $\cos\left(\text{Tan}^{-1}(-0.2)\right)$ **b.** $\sin\left(\text{Tan}^{-1}(-0.2)\right)$

> ◪ **You may find it helpful to have a computer or a graphing calculator to complete Exercises 15–18. Be sure to use radian measure for x.**

B **15. a.** Graph $y = \sin(\text{Sin}^{-1} x)$. Give the domain, range, and a simplified rule for the function $y = \sin(\text{Sin}^{-1} x)$. $-1 \le x \le 1;\ -1 \le y \le 1;\ y = x,\text{ for } -1 \le x \le 1$
  b. Graph $y = \text{Sin}^{-1}(\sin x)$.
  c. Does your graph from part (b) contradict the property of inverse functions which states that $f^{-1}(f(x)) = x$ for all x in the domain of f? Explain. No

16. a. Graph $y = \text{Cos}^{-1}(\cos x)$. Give the coordinates of the high and low points of the graph. High: $(\pi + 2n\pi, \pi)$; low: $(2n\pi, 0)$
  b. For what values of x is $\text{Cos}^{-1}(\cos x) = x$? $0 \le x \le \pi$
  c. Explain why $\text{Cos}^{-1}(\cos x) = x$ is not true for all values of x.

17. *Writing* In a paragraph, compare the graphs of $y = \tan(\text{Tan}^{-1} x)$ and $y = \text{Tan}^{-1}(\tan x)$.

18. Graph $f(x) = \text{Cos}^{-1} x + \text{Cos}^{-1}(-x)$. Give the domain and range of f. $-1 \le x \le 1;\ y = \pi$

Determine whether each equation is true for all x for which both sides of the equation are defined. If it is not true, give a counterexample.

19. a. $\tan(\text{Tan}^{-1} x) = x$ True   **b.** $\text{Tan}^{-1}(\tan x) = x$ Not true

20. a. $\text{Sin}^{-1}(-x) = -\text{Sin}^{-1} x$ True   **b.** $\text{Cos}^{-1}(-x) = -\text{Cos}^{-1} x$ Not true

21. a. $\text{Cos}^{-1} x = \dfrac{\pi}{2} - \text{Sin}^{-1} x$ True   **b.** $\text{Tan}^{-1} \dfrac{1}{x} = \dfrac{\pi}{2} - \text{Tan}^{-1} x$ Not true

22. Let $y = \text{Cot } x$ be a function identical to the cotangent function except that its domain is $0 < x < \pi$. State the domain and range of $y = \text{Cot}^{-1} x$ and sketch its graph. All real x; $0 < y < \pi$

23. Let $y = \text{Sec } x$ be a function identical to the secant function except that its domain is $0 \le x \le \pi$, $x \ne \dfrac{\pi}{2}$. State the domain and range of $y = \text{Sec}^{-1} x$ and sketch its graph. $x \le -1$ or $x \ge 1$; $0 \le y < \dfrac{\pi}{2}$ or $\dfrac{\pi}{2} < y \le \pi$

24. Let $y = \text{Csc } x$ be a function identical to the cosecant function except that its domain is $-\dfrac{\pi}{2} \le x \le \dfrac{\pi}{2}$, $x \ne 0$. State the domain and range of $y = \text{Csc}^{-1} x$ and sketch its graph. $x \le -1$ or $x \ge 1$; $-\dfrac{\pi}{2} \le y < 0$ or $0 < y \le \dfrac{\pi}{2}$

25. **a.** Compare the values of $\text{Sec}^{-1} 2$ and $\text{Cos}^{-1} \dfrac{1}{2}$. $\text{Sec}^{-1} 2 = \text{Cos}^{-1} \dfrac{1}{2} = \dfrac{\pi}{3}$
   b. Define $\text{Sec}^{-1} x$ in terms of the inverse cosine function. See below.

26. Define $\text{Csc}^{-1} x$ in terms of the inverse sine function.

25. **b.** $\text{Sec}^{-1} x = \text{Cos}^{-1} \dfrac{1}{x}$ 26. $\text{Csc}^{-1} x = \text{Sin}^{-1} \dfrac{1}{x}$

In Exercises 27 and 28, the given expression always has the same value for all x between -1 and 1, inclusive. Find this value and explain why the expression has a constant value.

27. $\text{Cos}^{-1} x + \text{Cos}^{-1} (-x)$ π

28. $\text{Sin}^{-1} x + \text{Sin}^{-1} (-x)$ 0

29. Using calculus, one can prove that
$$\text{Tan}^{-1} x = x - \frac{x^3}{3} + \frac{x^5}{5} - \frac{x^7}{7} + \cdots, \text{ for } |x| \le 1.$$

Use this relationship to show that $\pi = 4\left(1 - \dfrac{1}{3} + \dfrac{1}{5} - \dfrac{1}{7} + \cdots\right)$.

C 30. If $\text{Cos}^{-1} x + \text{Cos}^{-1} y = \dfrac{\pi}{2}$, show that $x^2 + y^2 = 1$. (*Hint*: Use Exercise 18 on page 279 and Exercise 7 on page 271.)

Chapter Summary

1. In trigonometry, an angle is often formed by rotating an *initial ray* to a *terminal ray*. When an angle's vertex is at the origin and its initial ray is along the positive x-axis, the angle is in *standard position*. Two angles in standard position are *coterminal angles* if they have the same terminal ray.

2. Angles can be measured in *revolutions*, *degrees*, or *radians*. Conversion formulas between degrees and radians are given on page 258.

3. The radian measure of $\angle AOB$, shown at the right, is given by $\theta = \dfrac{s}{r}$. Thus, the length of arc AB is $s = r\theta$, and the area K of sector AOB is $K = \dfrac{1}{2}r^2\theta$ (θ in radians) or $K = \dfrac{1}{2}rs$.

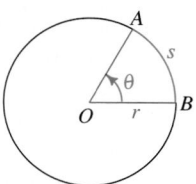

Cooperative Learning

You might want students to go over their proofs for Exercise 30 in small groups. Encourage the students in each group to discuss, correct, and modify their steps, if necessary, and to offer alternate proofs.

Additional Answers
Written Exercises

15. c. The function $y = \sin x$ is not one-to-one and is not the inverse of $y = \text{Sin}^{-1}$; $y = \text{Sin}^{-1}$ has inverse $y = \text{Sin } x$, defined on page 287.

16. c. The function $y = \cos x$ is not one-to-one and is not the inverse of $y = \text{Cos}^{-1}$. $\text{Cos}^{-1} (\cos x) = x$ is true for $0 \le x \le \pi$, the values of x in the domain of $y = \text{Cos } x$ (defined on page 288), which *is* the inverse of $y = \text{Cos}^{-1}$.

22.

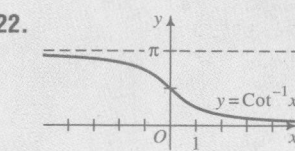

23.

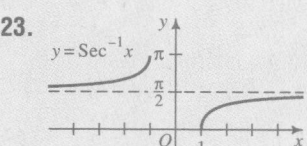

24.

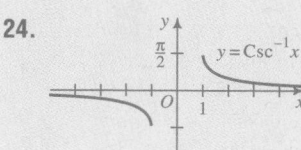

4. If $P(x, y)$ is a point on the unit circle $x^2 + y^2 = 1$, and if θ is an angle in standard position with terminal ray OP, then the six *trigonometric functions* are defined as shown in the table below. Graphs of the sine, cosine, tangent, and secant functions are shown on pages 278 and 284.

Function	$\sin \theta = y$	$\cos \theta = x$	$\tan \theta =$ $\dfrac{\sin \theta}{\cos \theta}$	$\cot \theta =$ $\dfrac{\cos \theta}{\sin \theta}$	$\sec \theta =$ $\dfrac{1}{\cos \theta}$	$\csc \theta =$ $\dfrac{1}{\sin \theta}$
Domain	all θ	all θ	$\theta \neq \dfrac{\pi}{2} + n\pi$	$\theta \neq n\pi$	$\theta \neq \dfrac{\pi}{2} + n\pi$	$\theta \neq n\pi$
Range	$|\sin \theta| \leq 1$	$|\cos \theta| \leq 1$	all reals	all reals	$|\sec \theta| \geq 1$	$|\csc \theta| \geq 1$
Period	2π	2π	π	π	2π	2π

5. Values of the sine and cosine functions for the angles 30°, 45°, and 60° are given in the table on page 277. To find the corresponding values of the other four trigonometric functions, use the definitions shown in the table above.

6. The acute angle α is the *reference angle* of $\theta = 180° \pm \alpha$, $\theta = 360° \pm \alpha$, and all coterminal angles. To find a trigonometric function of θ, first evaluate the function at α, the positive acute angle that θ makes with the x-axis. Use a plus or a minus sign depending on the quadrant involved. The diagram shows which functions are positive in each quadrant.

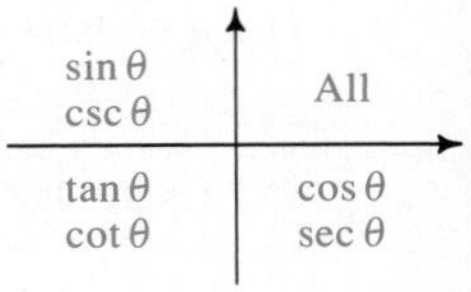

$\sin \theta$ $\csc \theta$	All
$\tan \theta$ $\cot \theta$	$\cos \theta$ $\sec \theta$

7. a. By restricting the domains of the trigonometric functions, we can define new one-to-one functions that have inverses. The domains and ranges of these *inverse trigonometric functions* are as follows:

Function	$y = \text{Sin}^{-1} x$	$y = \text{Cos}^{-1} x$	$y = \text{Tan}^{-1} x$
Domain	$-1 \leq x \leq 1$	$-1 \leq x \leq 1$	all real x
Range	$-\dfrac{\pi}{2} \leq y \leq \dfrac{\pi}{2}$	$0 \leq y \leq \pi$	$-\dfrac{\pi}{2} < y < \dfrac{\pi}{2}$

b. The equation $y = \text{Sin}^{-1} x$ is read "y is the inverse sine of x" and means that y is the value for which $\sin y = x$ where $-\dfrac{\pi}{2} \leq y \leq \dfrac{\pi}{2}$.

c. Graphs of three inverse trigonometric functions are on pages 287 and 288.

Key vocabulary and ideas

angle, initial ray, terminal ray (p. 257)	sine and cosine functions (p. 268)
degree, minute, second (p. 257)	unit circle, circular function (p. 269)
radian (p. 258)	reference angle (p. 275)
standard position (p. 259)	tangent and cotangent functions (p. 282)
coterminal angles (p. 260)	secant and cosecant functions (p. 282)
length of arc, area of a sector (p. 263)	Tan^{-1}, Sin^{-1}, Cos^{-1} (pp. 286–288)

1. a. Convert 270° to radians. Give the answer in terms of π. $\frac{3\pi}{2}$ 7-1
   b. Convert 192° to radians. Give the answer to the nearest hundredth. 3.35

2. a. Convert $\frac{5\pi}{3}$ radians to degrees. 300°

   b. Convert 2.5 radians to degrees. Give the answer to the nearest ten minutes or tenth of a degree. 143°10' or 143.2°

3. Find two angles, one positive and one negative, that are coterminal with each given angle.   Ans. may vary.
   a. −200°   **b.** 313.2°   **c.** $\frac{5\pi}{6}$ **d.** 142°10'
   160°, −560°   673.2°, −46.8° $\frac{17\pi}{6}$, $-\frac{7\pi}{6}$ 502°10', −217°50'

4. A sector of a circle has radius 5 cm and central angle 48°. Find its approximate arc length and area. 4.2 cm; 10.5 cm²   7-2

5. The moon is about 4×10^5 km from Earth, and its apparent size is about 0.0087 radians. What is the moon's approximate diameter? 3480 km

6. Find $\sin \theta$ and $\cos \theta$.   7-3

   a.   **b.**   **c.**

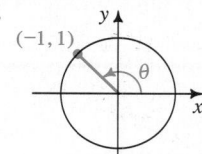 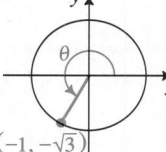 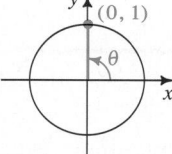

7. Complete each statement with one of the symbols <, >, or = .
   a. $\sin 60°$ __?__ $\sin 65°$ <   **b.** $\cos 60°$ __?__ $\cos 65°$ >
   c. $\sin 20°$ __?__ $\sin 160°$ =   **d.** $\cos 184°$ __?__ $\cos 185°$ <

8. Give the exact value of each expression in simplest radical form.   7-4
   a. $\sin \frac{5\pi}{4}$ $-\frac{\sqrt{2}}{2}$ **b.** $\cos(-90°)$ 0   **c.** $\sin 150°$ $\frac{1}{2}$ **d.** $\cos \frac{11\pi}{6}$ $\frac{\sqrt{3}}{2}$

9. Give the exact value of each expression or state that the value is undefined.   7-5
   a. $\tan 135°$ −1   **b.** $\cot \frac{2\pi}{3}$ $-\frac{\sqrt{3}}{3}$ **c.** $\sec(-60°)$ 2   **d.** $\csc(-\pi)$ Undef.

10. If $\tan x = -\frac{1}{3}$ where $\frac{\pi}{2} < x < \pi$, find the values of the other five trigonometric functions.

11. Give the exact value of each expression.   7-6
   a. $\text{Tan}^{-1} \sqrt{3}$ $\frac{\pi}{3}$ **b.** $\cot\left(\text{Sin}^{-1} \frac{1}{2}\right)$ $\sqrt{3}$ **c.** $\sec\left(\text{Cos}^{-1}\left(-\frac{3}{5}\right)\right)$ $-\frac{5}{3}$

12. *Writing* To obtain the inverse sine function, we restrict the domain of $f(x) = \sin x$ to $-\frac{\pi}{2} \le x \le \frac{\pi}{2}$. Write a short explanation in which you discuss why this restriction is necessary.

Trigonometric Functions   **293**

**Additional Answers
Chapter Test**

6. a. $\sin \theta = \frac{\sqrt{2}}{2}$;

    $\cos \theta = -\frac{\sqrt{2}}{2}$

  b. $\sin \theta = -\frac{\sqrt{3}}{2}$;

    $\cos \theta = -\frac{1}{2}$

  c. $\sin \theta = 1$; $\cos \theta = 0$

10. $\sin x = \frac{\sqrt{10}}{10}$;

    $\cos x = -\frac{3\sqrt{10}}{10}$;

    $\csc x = \sqrt{10}$;

    $\sec x = -\frac{\sqrt{10}}{3}$;

    $\cot x = -3$

12. Restricting the domain of $f(x) = \sin x$ to the interval $-\frac{\pi}{2} \le x \le \frac{\pi}{2}$ is necessary to make the sine function one-to-one.

Chapter 8 Trigonometric Equations and Applications

Overview

This chapter focuses on the solution of trigonometric equations and the applications of periodic functions to a wide variety of fields. In the first section, students solve simple trigonometric equations and apply them to problems from analytic geometry, such as finding the inclination of a line and the direction angle of a conic. In the next two sections, students explore changes in the sine and cosine curves, including period and amplitude changes and horizontal and vertical translations; students then use such changes to develop models of real-world periodic phenomena from physics, astronomy, meteorology, and oceanography. In the last two sections, students prove a variety of trigonometric identities and use them to simplify and solve more difficult trigonometric expressions and equations.

Objectives

8-1 To solve simple trigonometric equations and to apply them.
8-2 To find equations of different sine and cosine curves and to apply these equations.
8-3 To use trigonometric functions to model periodic behavior.
8-4 To simplify trigonometric expressions and to prove trigonometric identities.
8-5 To use trigonometric identities or technology to solve more difficult trigonometric equations.

Supplementary Resources

Tests, pp. 21–22
Alt. Assess., pp. 25–26, 66
Activities, pp. 19–20
Using Tech., p. 24
Student Res. Guide, pp. 79–86

Software

Precalculus Plotter Plus
Activities, pp. 43, 44–45, 46–47, 48–49
Function Plotter

Pacing Guide

Section	8-1	8-2	8-3	8-4	8-5	Review	Test	Total
Standard (days)	2	2	2	2	2	1	1	12
Comprehensive (days)	2	2	2	2	2	1	1	12

Teaching Notes

8-1 | pages 295–301

Presenting the Section

It is important to distinguish between the general solution of a trigonometric equation and particular solutions. If an interval, such as $0 \leq x < 2\pi$, is specified, then we are looking for particular solutions. When writing the general solution of a trigonometric equation, be sure students avoid mixing degree measure with radian measure, such as $\frac{\pi}{6} + n \cdot 360°$.

In solving trigonometric equations, encourage the use of scientific calculators rather than trigonometric tables. Remind students to check that the calculator is in the proper mode (degrees when the specified interval is $0° \leq \theta < 360°$, and radians when the specified interval is $0 \leq x < 2\pi$). Include at least one example whose solution set is empty, such as $2 \cos x + 3 = 8$.

Finding the inclination of a nonvertical line is equivalent to solving the trigonometric equation $\tan \alpha = m$, where $0° \leq \alpha < 180°$ and $\alpha \neq 90°$. Stress the need for first writing the equation of the line in slope-intercept form.

The section concludes with a discussion of sketching a rotated conic (that is, a conic having one of its axes inclined at an angle α, $0° < \alpha < 90°$, to the x-axis). Both methods given on page 298 for sketching such conics are interesting. Each method beautifully integrates algebraic techniques into the geometric setting.

Using Technology

A graphing calculator or computer is a useful tool in sketching rotated conics. Demonstrating an example similar to the one on page 298 helps students think about Class Exercise 15, page 299, regarding the necessity for entering two equations to obtain the graph of a conic on a graphing calculator.

8-2 | pages 301–308

Presenting the Section

The primary objective in this section is to have students quickly sketch sine and cosine curves by analyzing the constants A and B in the equations $y = A \sin Bx$ and $y = A \cos Bx$. Constantly reinforce the notion that the amplitude involves a vertical stretching or shrinking, the period involves a horizontal stretching or shrinking, and the presence of a negative sign, as in $-\cos x$, causes a reflection in the x-axis. If we allow B to be negative, then the period is given by $\frac{2\pi}{|B|}$. (Note that a negative value of B has no effect on a cosine curve, since cosine is an even function, but a negative value of B does result in a reflection of a sine curve in the x-axis, since sine is an odd function.) Discourage students from "point plotting" to sketch sine and cosine curves.

Students sometimes confuse expressions like $\sin 3x$ and $3 \sin x$. One simple way to correct this error is to use a specific value of x, say $x = \frac{\pi}{6}$, in comparing $\sin 3x$ and $3 \sin x$. $\left(\sin 3\left(\frac{\pi}{6}\right) = 1 \text{ and } 3 \sin \frac{\pi}{6} = \frac{3}{2} \right)$ Another way to show that these two expressions are not equal is to utilize the concepts of the section and compare the graphs of $y = 3 \sin x$ and $y = \sin 3x$. Students should see that, in general, $3 \sin x \neq \sin 3x$.

Using Technology

As a follow-up to the above discussion, have students use a graphing calculator to determine for which value(s) of x on the interval $0 \leq x < 2\pi$ it is true that $3 \sin x = \sin 3x$.

A graphing calculator also vividly displays for students the relationships among several curves and their points of intersection, such as those in Exercises 1–4, page 305.

Cooperative Learning

You might wish to divide the class into groups and have each group work on graphing equations such as:

$$y = \tan 2x, \ y = -\frac{1}{2} \tan 3x, \text{ and } y = 3 \tan \frac{x}{2}$$

From their results, have students determine a formula for finding the period of the tangent function.

Presenting the Section

Emphasize that $y = \sin\left(x - \frac{\pi}{2}\right)$ represents a horizontal shifting (translation) of the basic sine graph $\frac{\pi}{2}$ units to the right. We say that the *phase shift* is $\frac{\pi}{2}$. Students sometimes think that the presence of the subtraction sign in the above equation signifies a translation to the left. Show them that the point where $x = 0$ on the graph of $y = \sin x$ corresponds to the point where $x - \frac{\pi}{2} = 0$, or $x = \frac{\pi}{2}$, on the graph of $y = \sin\left(x - \frac{\pi}{2}\right)$; thus, the point $(0, 0)$ on the graph of $y = \sin x$ has been shifted to $\left(\frac{\pi}{2}, 0\right)$.

Students sometimes forget that amplitude is defined to be half the difference between the maximum and minimum values of the function, that is, as $\frac{M - m}{2}$. They erroneously define amplitude to be the maximum displacement of the graph from the x-axis instead of defining it as the maximum displacement from the axis of the curve, which is the horizontal line $y = k$ for a sine or cosine curve that has been translated vertically k units.

Make certain that students can determine what transformations of the graph of $y = \sin x$ are represented by the equation $y = k + A \sin B(x - h)$. For example, in the equation $y = 3 - 4 \sin\left(x + \frac{\pi}{2}\right)$, the graph of $y = \sin x$ is translated $\frac{\pi}{2}$ units to the left, translated upward 3 units, stretched vertically by a factor of 4, and reflected in its axis, $y = 3$.

Review

You may want to spend some time making certain that students are able to work comfortably with expressions such as $\sin 2$ or $\cos 3.1$. Students get so accustomed to working with radian measures involving π that they subsequently experience difficulty in estimating the value of $\sin 2$ or $\cos 3.1$ or in determining whether the given expression is positive or negative. A good review exercise is to ask students to complete a table like the following one:

	Sign	Estimated absolute value
$\sin 2$	$+$	Between $\frac{\sqrt{3}}{2}$ and 1
$\cos 3.1$	$-$	Between $\frac{\sqrt{3}}{2}$ and 1
$\sin(-2.6)$	$-$	Between $\frac{1}{2}$ and $\frac{\sqrt{3}}{2}$
$\cos 6.2$	$+$	Between $\frac{\sqrt{3}}{2}$ and 1
$\cos(-1)$	$+$	Between $\frac{1}{2}$ and $\frac{\sqrt{3}}{2}$

Assessment

Present an equation like $y = 2 - 4 \cos\left(x - \frac{\pi}{6}\right)$ and ask students to write a paragraph describing specifically what the various constants and signs do to the graph of $y = \cos x$. This is an opportunity for students to write about mathematical concepts and to explain their understanding of these concepts.

Presenting the Section

Although the material in this section begins with familiar relationships among the trigonometric functions, the emphasis should be on student discovery of these relationships via graphing calculator or computer and via transformations of trigonometric graphs. When discussing the reciprocal relationships, be sure students avoid writing $\sin^{-1}\theta$ for $(\sin \theta)^{-1}$. The former represents the inverse sine function, while the latter represents the reciprocal of the sine function, which is the cosecant.

Emphasize that an identity is true for all values of the variable for which the statement is meaningful. Thus, in the cofunction relationship $\cot \theta = \tan(90° - \theta)$, since the value $\theta = 0°$ does not make the statement meaningful, it is incorrect to write $\cot 0° = \tan(90° - 0°) = \tan 90°$.

Review

Students should be aware that there are three types of statements in mathematics: identities, conditional statements, and false statements. Identities are statements that are always true for all values of the variable for which the statement is meaningful; conditional statements are true for some (but not all) values of the variable; and false statements are true for no values of the variable. Examples of these three types follow:

(1) Identities: $x^2 - 9 = (x + 3)(x - 3)$
$x^2 \geq 0$
$|5 - x| = |x - 5|$

(2) Conditional statements: $2x + 3 = 9$
$|x - 6| = 8$
$3x - 1 < 14$

(3) False statements: $x^2 < 0$
$|2x - 1| = -5$
$\sqrt{x - 5} + 5 = 0$

The second part of Section 8-4, which is devoted to the use of trigonometric identities to prove other identities and to simplify expressions, is intended to help students develop facility in working with trigonometric relationships. It is not intended that students should be asked to memorize these identities.

Assessment

Refer students to the six negative relationships given on page 317 and ask them to determine whether each function is even, odd, or neither (see Exercises 31–36, page 137). (Since $\sin(-x) = -\sin x$, $\csc(-x) = -\csc x$, $\tan(-x) = -\tan x$, and $\cot(-x) = -\cot x$, these four functions are odd and therefore have graphs that are symmetric in the origin. Since $\cos(-x) = \cos x$ and $\sec(-x) = \sec x$, these two functions are even and therefore have graphs that are symmetric in the y-axis.)

Communication

Assigning Exercise 39 on page 322 gives students an opportunity to write about a mathematical situation and to focus their attention on the "why." This particular exercise is valuable even for students who do not have a graphing calculator.

Presenting the Section

Students traditionally have difficulty with solving trigonometric equations containing more than one trigonometric function or having a quadratic form. Before showing examples whose solutions require the use of a calculator, illustrate various examples each with a slightly different twist and whose solutions are special angles. Ask for particular solutions in some examples and general solutions in others. Caution students about dividing both sides of an equation by expressions such as $\cos x$, which may result in the loss of roots.

Often students are told that for equations containing more than one trigonometric function, they should rewrite the equation using only one trigonometric function. This is not always the case. For instance, in the equation $2 \sin x \cos x = 3 \sin x$, retaining both trigonometric functions is much easier than attempting to rewrite the equation strictly in terms of sine or cosine. You may wish to use the following examples to discuss with students the various approaches to solving trigonometric equations.

(a) $2 \sin x \cos x = 3 \sin x$ for $0 \leq x < 2\pi$
(b) $2 \cos^2 x + \cos x - 1 = 0$ for $0 \leq x < 2\pi$
(c) $4 \cos^2 x - 1 = 0$
(d) $\sin x + \sqrt{3} \cos x = 0$

This section is an excellent one for making mathematical connections. For example, stress that knowing how to solve the equation $3 \sin^2 x + 2 \sin x - 1 = 0$ corresponds to solving the quadratic equation $3y^2 + 2y - 1 = 0$. Also, remind students that since the range of the sine and cosine functions is the set of real numbers between -1 and 1, inclusive, an equation like $\cos x = 2$ has no solution.

Using Technology

Since students are familiar with methods for solving nonfactorable quadratic equations, it is appropriate to include trigonometric equations (quadratic in form) whose solutions are best estimated either by using inverse keys on a scientific calculator or by observing the points where the graphs cross the x-axis on a graphing calculator.

8 Trigonometric Equations and Applications

294

Equations and Applications of Sine Waves

8-1 Simple Trigonometric Equations

Objective   *To solve simple trigonometric equations and to apply them.*

The sine graph at the right illustrates that there are many solutions to the trigonometric equation $\sin x = 0.5$. We know that $x = \dfrac{\pi}{6}$ and $x = \dfrac{5\pi}{6}$ are two *particular* solutions. Since the period of $\sin x$ is 2π, we can add integral multiples of 2π to get the other solutions.

$$x = \frac{\pi}{6} + 2n\pi \quad \text{and} \quad x = \frac{5\pi}{6} + 2n\pi,$$

where n is any integer, are the *general solutions* of $\sin x = 0.5$.

When solving an equation such as $\sin x = 0.6$, we can use a calculator or a table of values.

Example 1   Find the values of x between 0 and 2π for which $\sin x = 0.6$.

Solution

Step 1 Set the calculator in radian mode and use the inverse sine key. Thus,

$$x = \operatorname{Sin}^{-1} 0.6 \approx 0.6435.$$

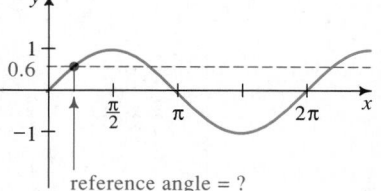

reference angle = ?

Step 2 0.6435 is the reference angle for other solutions. Since $\sin x$ is positive, a Quadrant II angle also satisfies the equation.

$x = \pi - 0.6435 \approx 2.4981$. Therefore, $x \approx 0.6435,\ 2.4981$.

Note that if you had been asked to find *all* values of x for which $\sin x = 0.6$, then your answer would be

$x \approx 0.6435 + 2\pi n$ and $x \approx 2.4981 + 2\pi n$, for any integer n.

To solve an equation involving a single trigonometric function, we first transform the equation so that the function is alone on one side of the equals sign. Then we follow the same procedure used in Example 1.

◀ The hour of sunset varies with the time of year; tides advance and recede in a pattern that is determined by the moon. Studied since ancient times, these natural cycles can be modeled by sine and cosine curves.

Trigonometric Equations and Applications   **295**

If a line's angle of inclination α is between 0° and 90°, $\tan \alpha > 0$ and hence the slope of the line is positive so that the line "rises" from left to right. If α is between 90° and 180°, $\tan \alpha < 0$, and the slope of the line is negative so that the line "falls" from left to right.

Example 2 To the nearest tenth of a degree, solve $3 \cos \theta + 9 = 7$ for $0° \le \theta < 360°$.

Solution

$$3 \cos \theta + 9 = 7$$
$$3 \cos \theta = -2$$
$$\cos \theta = -\frac{2}{3}$$

Since $\cos \theta < 0$, the solutions are between 90° and 180°, and 180° and 270°. (See graph.) One way to find these solutions is to first find the reference angle by ignoring the negative sign.

The reference angle is:

$$\text{Cos}^{-1}\left(\frac{2}{3}\right) \approx 48.2°$$

The first solution is:

$$\theta \approx 180° - 48.2° = 131.8°$$

The second solution is:

$$\theta \approx 180° + 48.2° = 228.2°$$

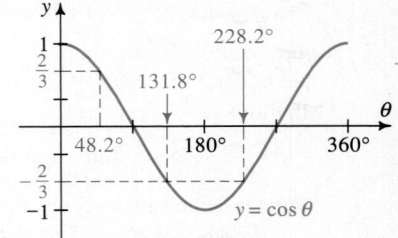

Inclination and Slope

The **inclination** of a line is the angle α, where $0° \le \alpha < 180°$, that is measured from the positive x-axis to the line. The line at the left below has inclination 35°. The line at the right below has inclination 155°. The theorem that follows states that the slope of a nonvertical line is the tangent of its inclination.

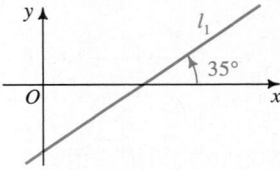

slope of $l_1 = \tan 35° \approx 0.7002$

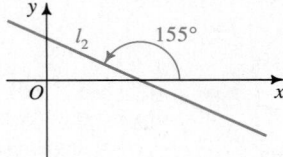

slope of $l_2 = \tan 155° \approx -0.4663$

Theorem

For any line with slope m and inclination α,

$$m = \tan \alpha \qquad \text{if } \alpha \ne 90°.$$

If $\alpha = 90°$, then the line has no slope. (The line is vertical.)

Proof: Refer to the diagram on the next page. Suppose that l_1 passes through the origin and has inclination α. Let P be the point where l_1 intersects the unit circle.

296 *Chapter Eight*

Then $P = (\cos \alpha, \sin \alpha)$. The slope of l_1 is

$$m = \frac{\sin \alpha - 0}{\cos \alpha - 0} = \tan \alpha.$$

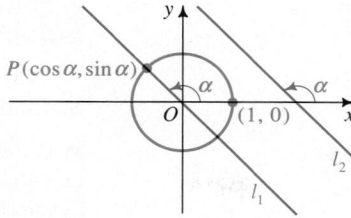

If the line l_2 has inclination α and does not contain the origin, it is parallel to l_1. Therefore, the slope of l_2 also equals $\tan \alpha$.

Example 3 To the nearest degree, find the inclination of the line $2x + 5y = 15$.

Solution Rewrite the equation as $y = -\frac{2}{5}x + 3$.

$$\text{slope} = -\frac{2}{5} = \tan \alpha$$

$$\alpha = \text{Tan}^{-1}\left(-\frac{2}{5}\right) \approx -21.8° \text{ (The reference angle is 21.8°.)}$$

Since $\tan \alpha$ is negative and α is a positive angle, $90° < \alpha < 180°$. The inclination is $180° - 21.8° \approx 158.2°$.

In Section 6-7, you learned to graph conic sections whose equations have no xy-term. That is, equations of the form

$$Ax^2 + Bxy + Cy^2 + Dx + Ey + F = 0,$$

where $B = 0$. The graph at the right shows a conic section with center at the origin whose equation has an xy-term ($B \neq 0$). Conics like this have one of their two axes inclined at an angle α to the x-axis. To find this *direction angle* α, use the formula below.

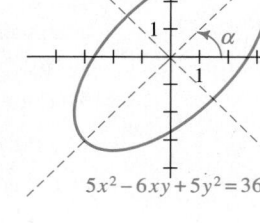

$$\alpha = \frac{\pi}{4} \qquad \text{if } A = C$$

$$\tan 2\alpha = \frac{B}{A - C} \qquad \text{if } A \neq C, \text{ and } 0 < 2\alpha < \pi$$

The direction angle α is useful in finding the equations of the axes of these conic sections. This is shown in Method 1 of Example 4 on the next page.

Example 4 Identify the graph of the equation, find the direction angle α, and sketch the curve $x^2 - 2xy + 3y^2 = 1$.

Solution $A \neq C$ and $B^2 - 4AC = (-2)^2 - 4(1)(3) = -8 < 0$, so the graph is an ellipse.

$$\tan 2\alpha = \frac{B}{A - C} = \frac{-2}{1 - 3} = 1 \qquad \text{Thus, } 2\alpha = \frac{\pi}{4} \text{ and } \alpha = \frac{\pi}{8}.$$

(Solution continues on the next page.)

Review Note

Briefly review with students that if not all of A, B, and C are zero, and if the graph of $Ax^2 + Bxy + Cy^2 + Dx + Ey + F = 0$ is not degenerate, then the graph is:

(1) an ellipse if $B^2 - 4AC < 0$;

(2) a parabola if $B^2 - 4AC = 0$;

(3) a hyperbola if $B^2 - 4AC > 0$.

Trigonometric Equations and Applications **297**

In the solution to Example 4, solving the equation
$x^2 - 2xy + 3y^2 = 1$ for y
requires that the equation be written in standard form,
$3y^2 + (-2x)y + (x^2 - 1) = 0$.
Emphasize that the coefficients are 3, $-2x$, and $x^2 - 1$.

Additional Examples

1. Solve $\cot^2 \theta = 4$ for $0° \leq \theta < 360°$. Give solutions to the nearest tenth of a degree.

   $\cot \theta = \pm 2$
   Notice that there are four solutions between 0° and 360°, one in each quadrant. Use a calculator in degree mode to find the reference angle, 26.6°.
   Thus, the solutions are 26.6°,
   $180° - 26.6° = 153.4°$,
   $180° + 26.6° = 206.6°$,
   and $360° - 26.6° = 333.4°$.

2. Find the inclination of the line joining the points $(-3, -5)$ and $(3, -3)$.

   $\text{slope} = \dfrac{-3 - (-5)}{3 - (-3)} = \dfrac{1}{3}$

   $\tan \theta = \dfrac{1}{3}$

   $\theta = \text{Tan}^{-1} \dfrac{1}{3} \approx 18.4°$

 To sketch the conic, we can use a graphing calculator or a computer, as shown in Method 1, or we can use substitution and our knowledge of the particular conic, as shown in Method 2.

Method 1

With a graphing calculator, it is often necessary to rewrite the equation so that it is quadratic in y. Then the equation can be solved for y by using the quadratic formula.

$$x^2 - 2xy + 3y^2 = 1$$
$$3y^2 + (-2x)y + (x^2 - 1) = 0$$
$$y = \frac{-(-2x) \pm \sqrt{(-2x)^2 - 4 \cdot 3(x^2 - 1)}}{2 \cdot 3}$$
$$y = \frac{2x \pm \sqrt{12 - 8x^2}}{6}$$

Rewrite the equation above as two equations:

$$y_1 = \frac{2x + \sqrt{12 - 8x^2}}{6} \quad \text{and}$$

$$y_2 = \frac{2x - \sqrt{12 - 8x^2}}{6}$$

Entering these two functions into a graphing calculator or computer gives the graph shown at the right.

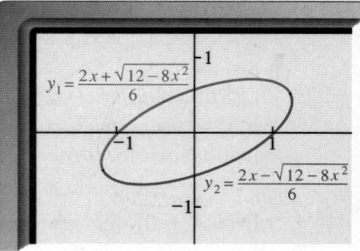

Method 2

Use the direction angle found on page 297 to find equations of the axes of the ellipse. Since $\alpha = \dfrac{\pi}{8}$, the slope of one axis is $\tan \dfrac{\pi}{8} \approx 0.41$, and the slope of the other axis is $-\dfrac{1}{0.41} \approx -2.4$. Thus, equations of the two axes are $y = 0.41x$ and $y = -2.4x$. By substituting for y into $x^2 - 2xy + 3y^2 = 1$, we find that these axes intersect the ellipse at points $(1.2, 0.50)$, $(-1.2, -0.50)$, $(0.21, -0.50)$, and $(-0.21, 0.50)$. Also, by substituting $y = 0$ and $x = 0$, we find that the graph intersects the x-axis at $(\pm 1, 0)$ and intersects the y-axis at $\left(0, \pm \dfrac{\sqrt{3}}{3}\right)$. We can use these points to sketch the graph of the ellipse, as shown at the right.

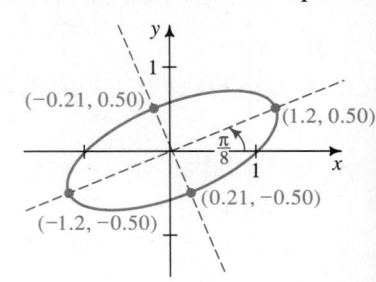

CLASS EXERCISES

Solve for $0° \le \theta < 360°$ without using tables or a calculator.

1. $\cos \theta = \frac{1}{2}$ 60°, 300°   **2.** $\sin \theta = -4$ No sol.   **3.** $\csc \theta = 2$ 30°, 150°   **4.** $\cot \theta = -1$ 135°, 315°

Solve for $0 \le x < 2\pi$ without using tables or a calculator.

5. $\cos x = -\frac{\sqrt{3}}{2}$ $\frac{5\pi}{6}, \frac{7\pi}{6}$ **6.** $\cot x = 1$ $\frac{\pi}{4}, \frac{5\pi}{4}$ **7.** $\tan x = -\sqrt{3}$ $\frac{2\pi}{3}, \frac{5\pi}{3}$ **8.** $\sec x = \frac{1}{2}$ No sol.

Solve for θ in degrees, giving *all* solutions.

9. $\cos \theta = -1$ **10.** $\sin \theta = -\frac{\sqrt{2}}{2}$ **11.** $\tan \theta = 1$

9. $(180 + 360n)°$
10. $(225 + 360n)°$, $(315 + 360n)°$
11. $(45 + 180n)°$

12. The inclination of a line is 140°. Find its slope. -0.8391

13. The slope of a line is $\frac{3}{5}$. Find its inclination. 31°

14. Find to the nearest degree a direction angle α for the conic whose equation is $5x^2 + 2xy - y^2 = 6$. 9°

15. *Discussion* Comment on why you think it may be necessary to enter two equations when graphing a conic using a graphing calculator, as mentioned in Method 1 of Example 4.

3. 50.2°, 230.2° **6.** 4.1°, 175.9° **9.** 146.4°, 213.6° **12.** 38.7°, 218.7°

WRITTEN EXERCISES

A **Solve for $0° \le \theta < 360°$. Give answers to the nearest tenth of a degree.**

1. $\sin \theta = -0.7$ 224.4°, 315.6° **2.** $\cos \theta = 0.42$ 65.2°, 294.8° **3.** $\tan \theta = 1.2$ See above.
4. $\cot \theta = -0.3$ 106.7°, 286.7° **5.** $\sec \theta = -5$ 101.5°, 258.5° **6.** $\csc \theta = 14$
7. $3 \cos \theta = 1$ 70.5°, 289.5° **8.** $4 \sin \theta = 3$ 48.6°, 131.4° **9.** $5 \sec \theta + 6 = 0$
10. $2 \tan \theta + 1 = 0$ 153.4°, 333.4°   **11.** $6 \csc \theta - 9 = 0$ 41.8°, 138.2°   **12.** $4 \cot \theta - 5 = 0$

Solve for $0 \le x < 2\pi$. Give answers to the nearest hundredth of a radian.

13. $\tan x = -1.5$ 2.16, 5.30   **14.** $\sec x = 2.5$ 1.16, 5.12   **15.** $\csc x = -1.4$ See below.
16. $\cos x = -0.8$ 2.50, 3.79   **17.** $\cot x = 6$ 0.17, 3.31   **18.** $3 \sin x + 2 = 4$
19. $8 = 9 \cos x + 2$ 0.84, 5.44   **20.** $\frac{5 \csc x}{3} = \frac{9}{4}$ 0.83, 2.31   **21.** $\frac{3 \cot x}{4} + 1 = 0$

Find the slope and equation of each line described. Sketch the line.

22. inclination = 45°
y-intercept = 4
1; $y = x + 4$

23. inclination = 120°
contains (2, 3)
$-\sqrt{3}$; $y = -\sqrt{3}x + 3 + 2\sqrt{3}$

24. inclination = 158°
contains (−3, 5)
-0.40; $y = -0.40x + 3.80$

15. 3.94, 5.49 **18.** 0.73, 2.41 **21.** 2.50, 5.64

Additional Answers Class Exercises

15. You can only graph a function on most graphing calculators. Since a conic is a relation, solving a conic's equation for y in terms of x usually results in two equations (involving a positive and a negative square root).

Additional Answers Written Exercises

3. 50.2°, 230.2°
6. 4.1°, 175.9°
9. 146.4°, 213.6°
12. 38.7°, 218.7°
15. 3.94, 5.49
18. 0.73, 2.41
21. 2.50, 5.64

Suggested Assignments

Standard
Day 1: 299/1–21 odd
Day 2: 299/23–35 odd, 45
Comprehensive
Day 1: 299/1–29 odd
Day 2: 300/31–45 odd

Supplementary Materials

Alternative Assessment, 25
Using Technology, 24

Find the inclination of each line. Give your answers to the nearest degree.

25. The line $3x + 5y = 8$ 149°

26. The line $x - 4y = 7$ 14°

27. The line joining $(-1, 2)$ and $(4, 1)$ 169°

28. The line joining $(-1, 1)$ and $(4, 2)$ 11°

29. A line perpendicular to $4x + 3y = 12$ 37°

30. A line parallel to $2x + 3y = -6$ 146°

Solve for $0 \le x < 2\pi$ without using tables or a calculator.

B 31. $|\csc x| = 1$ $\dfrac{\pi}{2}, \dfrac{3\pi}{2}$

32. $|\sec x| = \sqrt{2}$ $\dfrac{\pi}{4}, \dfrac{3\pi}{4}, \dfrac{5\pi}{4}, \dfrac{7\pi}{4}$

33. $\log_2 (\sin x) = 0$ $\dfrac{\pi}{2}$

34. $\log_2 (\cos x) = -1$ $\dfrac{\pi}{3}, \dfrac{5\pi}{3}$

35. $\log_3 (\tan x) = \dfrac{1}{2}$ $\dfrac{\pi}{3}, \dfrac{4\pi}{3}$

36. $\log_{\sqrt{3}} (\cot x) = 1$ $\dfrac{\pi}{6}, \dfrac{7\pi}{6}$

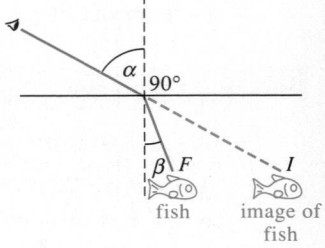 **Identify the graph of each equation, find its direction angle α, and sketch the curve. You may wish to use a graphing calculator or computer.**

37. $x^2 + xy + y^2 = 1$

38. $x^2 - 2xy - y^2 = 4$

39. $x^2 - xy + y^2 = 1$

40. $x^2 - xy + 2y^2 = 2$

41. $x^2 - 3xy + y^2 = -6$

42. $y = x - \dfrac{1}{x}$

43. $x^2 - 2xy + y^2 - 4\sqrt{2}x - 4\sqrt{2}y = 0$

44. $x^2 - 2xy + y^2 - \sqrt{2}x - \sqrt{2}y = 0$

You may find it helpful to have a computer or a graphing calculator to complete part (b) of Exercise 45. Use degree measure for α and β.

45. **Optics** The rays of light that are reflected from a fish in a pond to your eyes travel at different speeds through water and through air. Consequently, these rays bend at the water's surface. Your mind, which expects light to travel in a straight line, perceives the fish to be at location I when the fish is actually at location F. According to Snell's law,

$$\frac{\text{speed of light in air}}{\text{speed of light in water}} = \frac{\sin \alpha}{\sin \beta}.$$

a. $0° \le \alpha \le 90°$, $0° \le \beta \le 48.6°$

a. Using the fact that the speed of light in air is 3.00×10^8 m/s and the speed of light in water is 2.25×10^8 m/s, show that $\beta = \text{Sin}^{-1}(0.75 \sin \alpha)$. Give an appropriate domain and range for this function.

b. Use the graphs of $y = \beta$ and $y = 0.5\alpha$ to decide if there is a positive angle α for which $\beta = \dfrac{1}{2}\alpha$. No

46. Navigation A *great circle* on Earth's surface is a circle whose center is at Earth's center. The shortest distance measured in degrees between points A and B on Earth's surface is along the arc of a great circle. To find this shortest distance, we can work with the spherical triangle ABN shown in the diagram, whose sides are arcs of great circles. The important measurements in this triangle are:

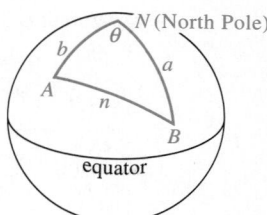

    great circle arc $AB = n°$
    great circle arc $AN = b° = 90° -$ latitude of A
    great circle arc $BN = a° = 90° -$ latitude of B
    $\theta =$ difference in longitudes of A and B

Use the formula $\cos n = \cos a \cos b + \sin a \sin b \cos \theta$ and the following data to find the great circle distance measured in degrees between Rome and Boston. Give your answer to the nearest tenth of a degree. 59.4°

	Latitude	Longitude
Rome	41.53°N	12.30°E
Boston	42.20°N	71.05°W

8-2 Sine and Cosine Curves

Objective	To find equations of different sine and cosine curves and to apply these equations.

Recall from Section 4-4 that the graph of $y = cf(x)$ can be obtained by vertically stretching or shrinking the graph of $y = f(x)$. This is illustrated by the sine and cosine curves below.

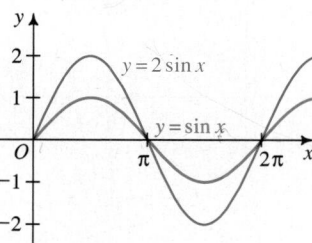
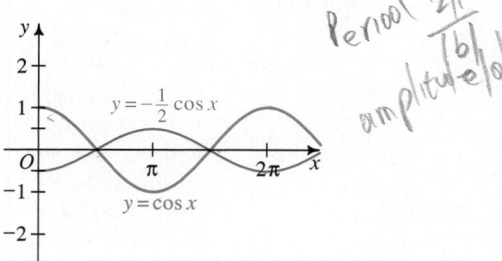

(handwritten: Period $\frac{2\pi}{|b|}$ amplitude $|a|$)

Notice that $y = -\frac{1}{2}\cos x$ is the graph of $y = \frac{1}{2}\cos x$ that has been reflected in the x-axis. You can see that the graph of $y = 2\sin x$ has amplitude 2 and that the graph of $y = -\frac{1}{2}\cos x$ has amplitude $\frac{1}{2}$.

Teaching Notes, p. 294B

Warm-Up Exercises

1. Sketch the graph of $y = \sin x$ for $-2\pi \le x \le 2\pi$. Name the amplitude and period of the function. (See graph below.); 1; 2π

2. On your graph from Exercise 1, add the graph of $y = -\frac{1}{2}$. Name the x-coordinates of the points where the graphs intersect in the interval $-2\pi \le x \le 2\pi$. (See graph below.); $-\frac{\pi}{6}, -\frac{5\pi}{6}, \frac{7\pi}{6}, \frac{11\pi}{6}$

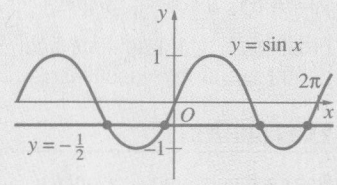

3. Sketch the graph of $y = \cos x$ for $-2\pi \le x \le 2\pi$. Name the amplitude and period of the function. (See graph below.); 1; 2π

4. On your graph from Exercise 3, add the graph of $y = 1$. Name the x-coordinates of the intersection points in the interval $-2\pi \le x \le 2\pi$. (See graph below.); -2π; 0; 2π

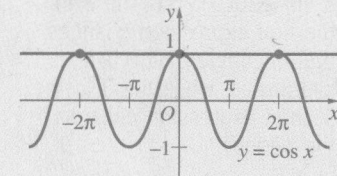

Also in Section 4-4, you learned that the graph of $y = f(cx)$ is obtained by horizontally stretching or shrinking the graph of $y = f(x)$. This is illustrated below.

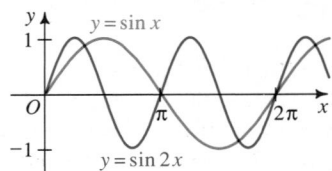 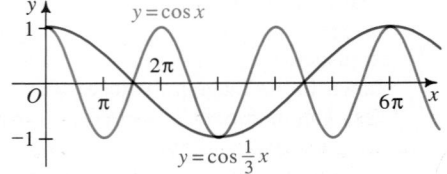

The period of $y = \sin x$ and $y = \cos x$ is 2π. (In this chapter, when we refer to the period p of a function, we mean the *fundamental period* of the function.) You can see from above that the period of $y = \sin 2x$ is $\frac{2\pi}{2}$, or π, and that the period of $y = \cos \frac{1}{3}x$ is $\frac{2\pi}{\frac{1}{3}}$, or 6π.

In general, we can determine useful information about the graphs of $y = A \sin Bx$ and $y = A \cos Bx$ by analyzing the factors A and B.

Period and Amplitude of Sine and Cosine Curves

For functions $y = A \sin Bx$ and $y = A \cos Bx$ ($A \neq 0$ and $B > 0$):

$$\text{amplitude} = |A| \qquad \text{period} = \frac{2\pi}{B}$$

Example 1   Give the amplitude and period of the function $y = -4 \sin 3x$. Then sketch at least one cycle of its graph.

Solution   amplitude $= |-4| = 4$

period $= \dfrac{2\pi}{B} = \dfrac{2\pi}{3}$

Reflect the graph of $y = 4 \sin 3x$ in the x-axis to get the graph of $y = -4 \sin 3x$.

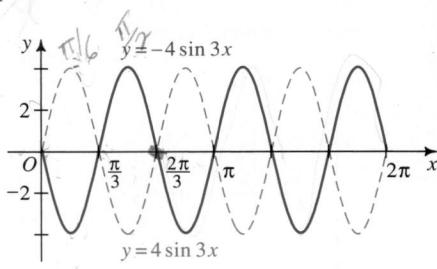

Example 2   Give the amplitude, period, and an equation of the curve shown at the right.

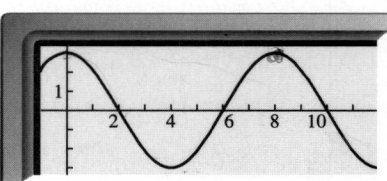

Solution   Use the cosine curve, $y = A \cos Bx$, with amplitude 3 and period 8. Thus, $8 = \dfrac{2\pi}{B}$, which gives $B = \dfrac{\pi}{4}$. The equation is

$$y = A \cos Bx = 3 \cos \dfrac{\pi}{4}x.$$

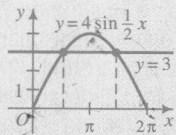 Example 3 shows two methods of solving a trigonometric equation. Method 1 is an algebraic solution, as developed in Section 8-1. Method 2 is a graphical solution in which a graphing calculator or a computer can be used.

Example 3   Solve the equation $6 \sin 2x = 5$ for $0 \le x < 2\pi$. Give answers to the nearest hundredth of a radian.

Solution   **Method 1** Transform the equation as follows:

$$6 \sin 2x = 5$$

$$\sin 2x = \dfrac{5}{6}$$

$$2x \approx 0.99, 2.16, 7.27, 8.44 \quad \longleftarrow \text{ If } 0 \le x < 2\pi, \text{ then}$$
$$x \approx 0.49, 1.08, 3.63, 4.22 \qquad\qquad 0 \le 2x < 4\pi.$$

Method 2 Use a graphing calculator or computer to sketch the graphs of $y = 5$ and $y = 6 \sin 2x$ on the same set of axes. Then find the x-coordinates of the intersection points over the interval $0 \le x < 2\pi$.

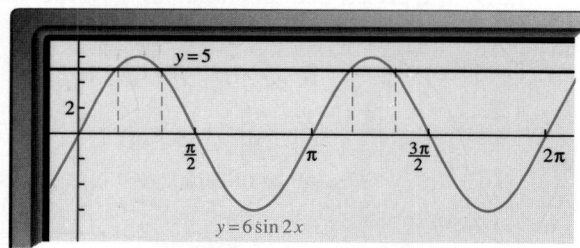

Using the zoom feature to get more accuracy, you will be able to see that the line $y = 5$ intersects the curve $y = 6 \sin 2x$ at $x \approx 0.49, 1.08, 3.63,$ and 4.22.

Applications to Electricity

Most household circuits in the United States are 60-cycle alternating current (AC) circuits. This means that the voltage oscillates like the sine curve at a *frequency* of 60 cycles per second. In other words, 1 cycle is completed every $\dfrac{1}{60}$ of a second, which is the period. The period of a sine or cosine curve is always the reciprocal of the frequency.

Trigonometric Equations and Applications   **303**

Chapter 8   **303**

Additional Examples

1. Give the amplitude and period of $y = -3 \cos \dfrac{\pi}{3}x$. Then sketch its graph.
   Amplitude $= |-3| = 3$
   Period $= 2\pi \div \dfrac{\pi}{3} =$
   $2\pi \cdot \dfrac{3}{\pi} = 6$

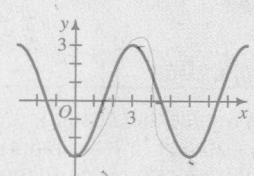

2. Graph $y = 4 \sin \dfrac{1}{2}x$ and $y = 3$ on the same set of axes to determine the number of solutions that the equation $4 \sin \dfrac{1}{2}x = 3$ has in the interval $0 \le x < 2\pi$. Approximate each solution to the nearest hundredth.

   The graph shows that there are two solutions in the interval $0 \le x < 2\pi$.
   $4 \sin \dfrac{1}{2}x = 3$
   $\sin \dfrac{1}{2}x = 0.75$
   $\dfrac{1}{2}x = \text{Sin}^{-1}\, 0.75 \approx 0.848;$
   $x \approx 1.70$
   Also, $\dfrac{1}{2}x \approx \pi - 0.848 \approx$
   $2.294;\ x \approx 4.59.$
   (*Note:* A zoom feature on a graphing calculator can be used to approximate or verify solutions.)

The frequency f of a function of the form $g(t) = A \cos Bt$, where t represents time, is the number of cycles completed per unit of time. The period p, on the other hand, is the amount of time needed to complete one cycle. Thus, $f = \frac{1}{p}$, and since $p = \frac{2\pi}{B}$, $f = \frac{B}{2\pi}$.

Application

Have students consult a physics textbook to learn more about the role of the trigonometric functions in describing the phenomenon of waves in electricity.

Cooperative Learning

You may want to use Class Exercises 5 and 6 as the basis of a cooperative learning activity. Have students work in pairs. Beginning with the equation $y = \sin x$ or $y = \cos x$, one student changes the equation so that its period and/or amplitude are altered. After the student graphs the equation, either with a graphing calculator or by hand, the other student is asked to give an equation of the graph. The students can then switch roles.

In the following diagram, the frequency is 60, and the period is $\frac{-1}{60}$. Thus, $\frac{1}{60} = \frac{2\pi}{B}$, and so $B = 120\pi$. Most household circuits are called 110-volt circuits because they deliver energy at the same rate as a direct current of 110 volts. In actuality, maximum voltage in a 110-volt circuit is $110\sqrt{2}$ volts, and so $A = 110\sqrt{2}$. The voltage equation is:

$$V = A \sin Bx$$
$$V = 110\sqrt{2} \sin 120\pi t \quad (t \text{ in seconds})$$

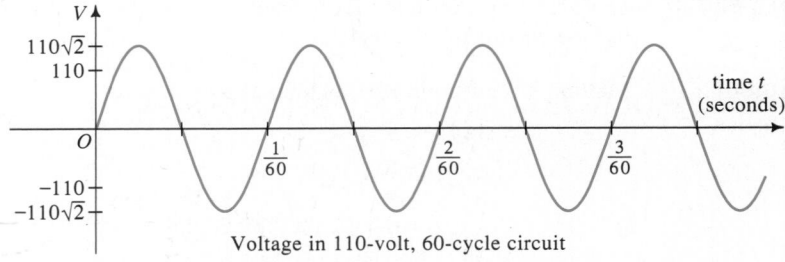

Voltage in 110-volt, 60-cycle circuit

CLASS EXERCISES

Give the period and amplitude of each function.

1. $y = 4 \cos 2x$ $\pi,\ 4$ **2.** $y = 3 \sin \frac{1}{2}x$ $4\pi,\ 3$

3. $y = 5 \sin \frac{2\pi}{7}x$ $7,\ 5$ **4.** $y = 6 \cos \frac{2\pi}{3}t$ $3,\ 6$

Give the period, the amplitude, and an equation for each curve.

5.   **6.**

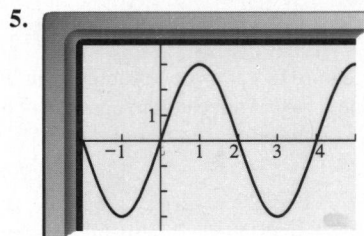

$4,\ 3,\ y = 3 \sin \frac{\pi}{2}x$ $4\pi,\ 2,\ y = -2 \cos \frac{1}{2}x$

7. *Visual Thinking* Draw a quick sketch to tell how many solutions each equation has between 0 and 2π.

    a. $\sin x = 1$ 1   **b.** $\sin 2x = \frac{1}{2}$ 4   **c.** $\sin 3x = -1$ 3

304   *Chapter Eight*

A For Exercises 1–4, sketch and label the graphs given on a single set of axes.

1. $y = \cos x$, $y = 3 \cos x$, $y = \frac{1}{3} \cos x$

2. $y = \sin x$, $y = 4 \sin x$, $y = -4 \sin x$

3. $y = \sin x$, $y = \sin \frac{1}{2}x$

4. $y = \cos x$, $y = \cos 3x$

Give the amplitude and period of each function. Then sketch its graph.

5. $y = 2 \sin 3x$ $2, \frac{2\pi}{3}$

6. $y = 4 \cos 2x$ $4, \pi$

7. $y = -2 \cos 2t$ $2, \pi$

8. $y = -4 \sin \frac{t}{3}$ $4, 6\pi$

9. $y = \frac{1}{2} \cos 2\pi t$ $\frac{1}{2}, 1$

10. $y = 1.5 \sin \frac{\pi}{2}x$ $1.5, 4$

Give the amplitude, period, and an equation for each curve.

11.

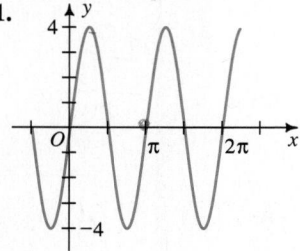

$4, \pi, y = 4 \sin 2x$

12.

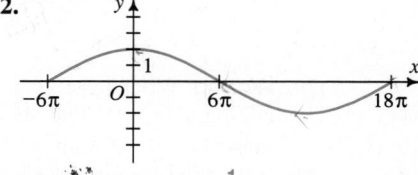

$2, 24\pi, y = 2 \cos \frac{1}{12}x$

13.

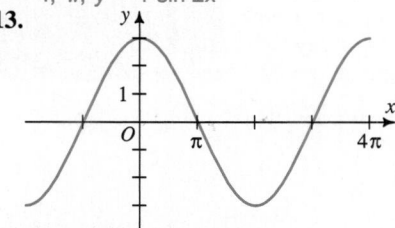

$3, 4\pi, y = 3 \cos \frac{1}{2}x$

14.

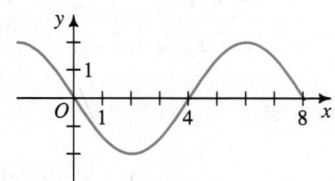

$2, 8, y = -2 \sin \frac{\pi}{4}x$

15.

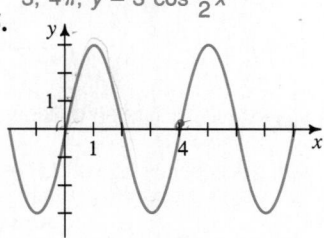

$3, 4, y = 3 \sin \frac{\pi}{2}x$

16.

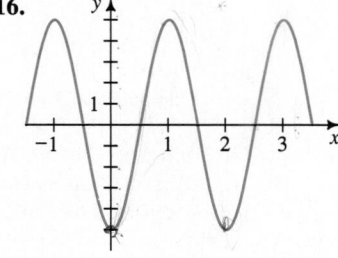

$5, 2, y = -5 \cos \pi x$

Trigonometric Equations and Applications   **305**

Additional Answers
Written Exercises

1.
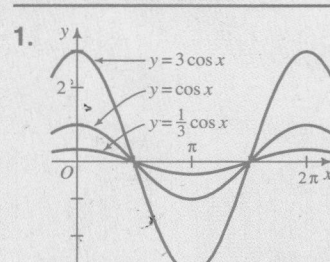

2.
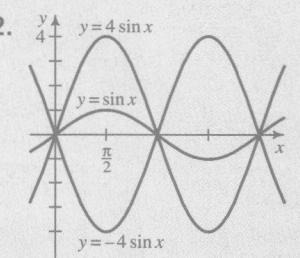

3.
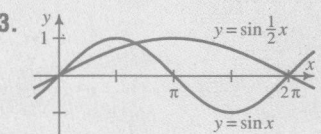

4.
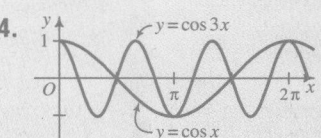

Suggested Assignments

Standard
Day 1: 305/1, 5, 9, 11, 15, 17, 19
Day 2: 306/21–33 odd
Comprehensive
Day 1: 305/1, 5, 9, 11, 15, 17, 19, 23
Day 2: 306/25, 27, 29–36

Supplementary Materials

Alternative Assessment, 25
Activities Book, 19–20

17. A sine curve varies between 4 and −4 with period 12. Find its equation.

18. A cosine curve varies between −9 and 9 with period 5. Find its equation.

   17. $y = \pm 4 \sin \frac{\pi}{6}x$ **18.** $y = \pm 9 \cos \frac{2\pi}{5}x$

Solve each equation for $0 \le x < 2\pi$ either algebraically or graphically using a computer or graphing calculator. Give answers to the nearest hundredth of a radian.

19. a. $\cos x = -1$ **b.** $\cos 2x = -1$ **c.** $\cos 3x = -1$

20. a. $2 \sin x = 1$ **b.** $2 \sin 2x = 1$ **c.** $2 \sin \frac{x}{2} = 1$

B   **21.** $8 \cos 2x = 1$ **22.** $5 \sin 3x = -2$ **23.** $3 \sin \frac{x}{2} = -1$

24. $1.5 \cos \frac{x}{2} = \frac{1}{2}$ **25.** $4 \sin \frac{\pi}{2}x = 1$ **26.** $-3 \cos \frac{\pi}{4}x = 1$

27. Sketch the graphs of $y = \tan 2x$ and $y = \tan \frac{1}{2}x$.

28. Sketch the graphs of $y = \sec 2x$ and $y = \sec \frac{1}{2}x$.

29. *Writing* Write a paragraph to explain why it does not make sense to talk about the amplitude of the graph of $y = \tan x$ and of the graph of $y = \sec x$.

30. **Electronics** An AM radio wave means that the *amplitude* is *modulated*. Thus, instead of a constant amplitude A, there is a varying amplitude $A(t)$ as shown below. Each AM radio station in the United States is assigned a frequency between 540 kHz and 1700 kHz. (A hertz (Hz) is one cycle per second, so 1 kHz is 1000 cycles per second.)

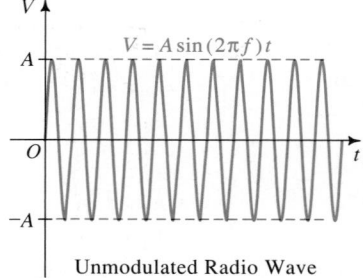

Unmodulated Radio Wave

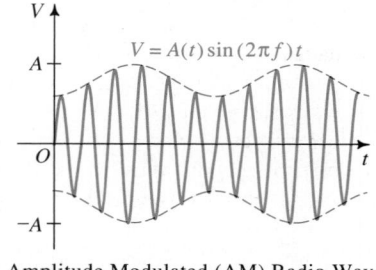

Amplitude Modulated (AM) Radio Wave

a. Give the period of an AM radio wave with a frequency of 800 kHz. 1.25×10^{-6} s

b. An AM radio wave has equation $V(t) = A(t) \sin 1,850,000\pi t$. What is its frequency in kHz?   925 kHz

31. Electronics The initials FM stand for *frequency modulation*. In FM broadcasting, information is communicated by varying the frequency; that is, the constant frequency f is replaced by a variable frequency $f(t)$, which is a function of time. An exaggerated graph of this situation appears at the right. A complication of FM broadcasting is that $f(t)$ must remain near the radio station's assigned frequency. FM frequencies are given in MHz (millions of cycles per second).

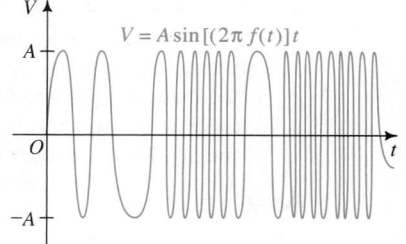

$V = A \sin[(2\pi f(t)]t$

Frequency Modulated (FM) Radio Wave

a. Suppose that an FM radio wave has an equation of the following form: $V(t) = A \sin [2\pi(200{,}000{,}000 + 10{,}000 \sin 500\pi t)]t$. Thus, the varying frequency is $f(t) = 200{,}000{,}000 + 10{,}000 \sin 500\pi t$. What is the assigned frequency in MHz? 200 MHz

b. Give the maximum and minimum frequencies of this FM radio wave over a long period of time. 200.01 MHz, 199.99 MHz

32. Research In a reference book find information about two of the following:

a. the range of assigned frequencies of the FM broadcasting channels in the United States

b. the frequency ranges for VHF (very high frequency) and UHF (ultra high frequency) television

c. the frequencies of ultraviolet light and infrared light

d. the frequency of the musical pitch called "middle C"

33. Physics Suppose a weight with mass m grams hangs on a spring. If you pull the weight A centimeters downward and let go of it, the weight will oscillate according to the formula below. Let d represent the displacement in centimeters t seconds after the initial displacement. Let k be a constant measuring the spring's stiffness.

$$d = -A \cos\left[\left(\sqrt{\frac{k}{m}}\right)t\right]$$

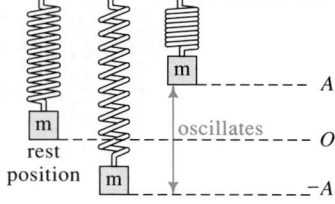

$$p = \frac{2\pi}{\sqrt{\frac{k}{m}}} = 2\pi\sqrt{\frac{m}{k}}$$

a. Show that the period of the motion is $2\pi\sqrt{\frac{m}{k}}$.

b. Suppose you put a weight with known mass 100 g on the spring and you time its period to be $t = 1.1$ s. Find the value of the spring constant k. $k \approx 3263$

c. Suppose you put a weight with unknown mass m on the spring and time its period to be $t = 1.4$ s. Find m. (Use the value of k found in part (b).) $m \approx 162$ g

Trigonometric Equations and Applications   **307**

34. Physics When a pendulum swings back and forth through a small arc, its horizontal displacement is given by the formula below. Let D be the horizontal displacement in centimeters of the pendulum t seconds after passing through O, let A represent the maximum displacement, and let l represent the length of the pendulum.

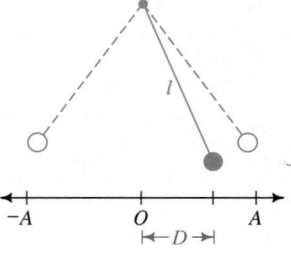

$$D \approx A \sin \left[\left(\sqrt{\frac{980}{l}} \right) t \right]$$

a. If the length l of the pendulum is 100 cm, find the earliest time t for which the displacement is maximum. 0.5018 s

b. How long is a clock pendulum that has a period of 1 second? 24.8 cm

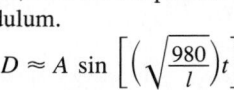

35. On the same set of axes, sketch the graphs of $y = 2^{-x}$, $y = -2^{-x}$, and $y = 2^{-x} \sin x$ for $x > 0$.

36. Music When the note called *concert* A is sounded on a piano, the piano string vibrates with a frequency of 440 Hz. The equation that gives the displacement of a point on the vibrating piano string is

$$D = B(2^{-t}) \sin 880\pi t,$$

where t is the number of seconds after the string is struck and B is a constant that depends on how hard the string is hit and the point's position on the string. (This formula applies only for the first few seconds.)

a. What part of the equation models the gradual dying out of the vibrations? 2^{-t}

b. What is the earliest time t for which the displacement is maximum? 5.68×10^{-4} s

c. The piano string for the A an octave lower than concert A has a frequency of 220 Hz. Write an equation for its displacement. $D = B(2^{-t}) \sin 440\pi t$

8-3 Modeling Periodic Behavior

Objective   *To use trigonometric functions to model periodic behavior.*

When a graph is translated h units horizontally and k units vertically, then x and y must be replaced by $(x - h)$ and $(y - k)$, respectively. This idea was introduced in Section 4-4 and is illustrated below and on the next page.

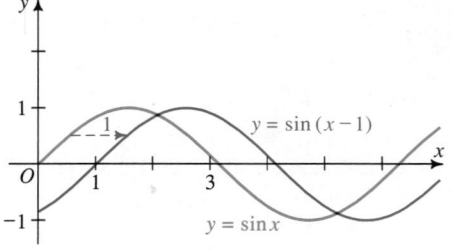

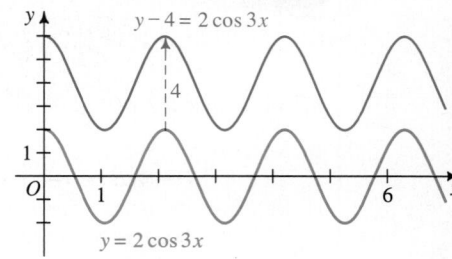

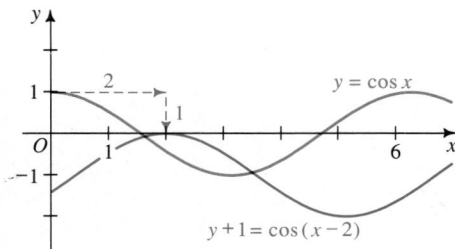

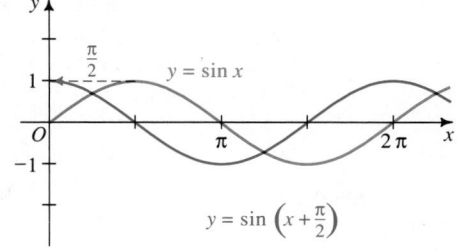

$$y + 1 = \cos(x - 2)$$

$$y = \sin\left(x + \frac{\pi}{2}\right)$$

Notice that the last graph, whose equation is $y = \sin(x + \frac{\pi}{2})$, is also the graph of $y = \cos x$. Thus, the cosine curve is the sine curve that has been shifted $\frac{\pi}{2}$ units to the left. Since the cosine curve is congruent to the sine curve, we refer to sine and cosine curves as *sine waves*.

General Sine Waves

If the graphs of $y = A \sin Bx$ and $y = A \cos Bx$ are translated horizontally h units and vertically k units, then the resulting graphs have equations

$$y - k = A \sin B(x - h) \quad \text{and} \quad y - k = A \cos B(x - h).$$

Example 1 Give an equation of the sine wave shown at the right.

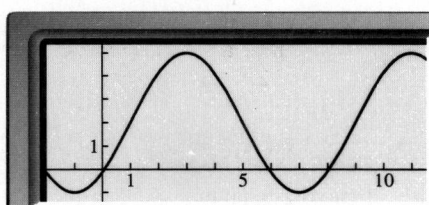

Solution The graph suggests a translation of $y = A \sin Bx$ or $y = A \cos Bx$. To find A and B, reason as follows:

amplitude: $\quad A = \dfrac{\text{Max} - \text{min}}{2} = \dfrac{5 - (-1)}{2} = 3$

period: $\quad p = $ horizontal distance between successive maximums
$\quad\quad = 11 - 3 = 8$

Since $8 = \dfrac{2\pi}{B}$, $B = \dfrac{\pi}{4}$.

The given sine wave is a translation of $y = 3 \sin \frac{\pi}{4}x$ or of $y = 3 \cos \frac{\pi}{4}x$.

(Solution continues on the next page.)

Trigonometric Equations and Applications **309**

Mathematical Note

Emphasize that the cosine and sine curves are out of phase by $\frac{\pi}{2}$ units; that is, a translation of the sine curve $\frac{\pi}{2}$ units to the left yields the cosine curve.

Example Note

In the solution of Example 1, the "max" and "min" referred to in the calculation of the amplitude are the y-coordinates of the highest and lowest points on the given graph.

1. Sketch the graph of

$y = 2 \sin \left(x + \frac{\pi}{3}\right) + 3.$

Begin by shifting the graph of $y = 2 \sin x$ to the left by $\frac{\pi}{3}$ units. Then raise that graph 3 units up. The resulting graph has maximum value $2(1) + 3 = 5$ and minimum value $2(-1) + 3 = 1.$

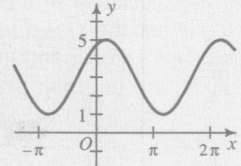

2. At a certain spot the water level is 6 m at low tide and 14 m at high tide. If the tide has period 12.4 h, find an equation for the water level.

The amplitude is $\frac{1}{2}(14 - 6) = 4.$

The average water level is $\frac{1}{2}(6 + 14) = 10.$

Since $\frac{2\pi}{B} = 12.4,$

$B \approx 0.507.$ Thus,
$L = 4 \cos 0.507t + 10$ or
$L = 4 \sin 0.507t + 10$ is an equation for the water level.

To find the amounts of the translation, first find the *axis of the wave*, which is the horizontal line midway between the maximum and minimum points of the curve.

$$\text{axis of wave:} \quad y = \frac{\text{Max} + \text{min}}{2} = \frac{5 + (-1)}{2} = 2$$

(1) If the equation is to be in terms of cosine, select a highest point on the curve. Determine the translation amounts in moving from the point $(0, A)$ to this point. See the diagram at the left below.

(2) If the equation is to be in terms of sine, select a point where the curve intersects its axis. Determine the translation amounts in moving from the point $(0, 0)$ to this point. See the diagram at the right below.

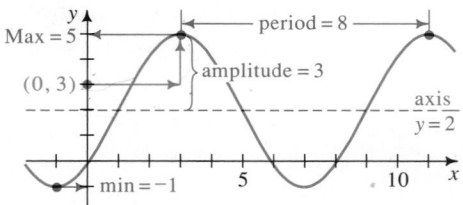

The graph of $y = 3 \cos \frac{\pi}{4}x$ is translated 3 units right and 2 units up.

$$y - 2 = 3 \cos \frac{\pi}{4}(x - 3)$$

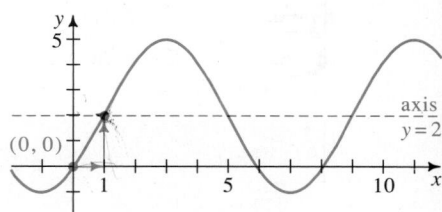

The graph of $y = 3 \sin \frac{\pi}{4}x$ is translated 1 unit right and 2 units up.

$$y - 2 = 3 \sin \frac{\pi}{4}(x - 1)$$

Thus, an equation of the given graph is

$$y - 2 = 3 \cos \frac{\pi}{4}(x - 3) \quad \text{or} \quad y - 2 = 3 \sin \frac{\pi}{4}(x - 1).$$

Trigonometric Models

Trigonometric functions are useful in solving many problems that involve periodic behavior, such as the motion of the tide. In many applications, the variable might represent something other than angles. For example, an application might involve $\sin t$ or $\cos t$, where t represents time. In this case, you evaluate $\sin t$ or $\cos t$ as if t were in radians.

 As the example on the next page shows, a sketch of a graph showing given information is very helpful in analyzing a problem in order to find an equation of the trigonometric model. Then you could use a graphing calculator or a computer to draw quickly the functions involved in the equation, and then zoom in on the intersection points to find the time values.

Example 2 The depth of water at the end of a pier varies with the tides throughout the day. Today the high tide occurs at 4:15 A.M. with a depth of 5.2 m. The low tide occurs at 10:27 A.M. with a depth of 2.0 m.

    a. Find a trigonometric equation that models the depth of the water t hours after midnight.

    b. Find the depth of the water at noon.

    c. A large boat needs at least 3 m of water to moor at the end of the pier. During what time period after noon can it safely moor?

Application

The example on this page provides an excellent real-world application. You should discuss this example thoroughly with your class before assigning Exercises 15 and 16 on page 314.

Solution   **a.** First make a sketch, noting the given information.

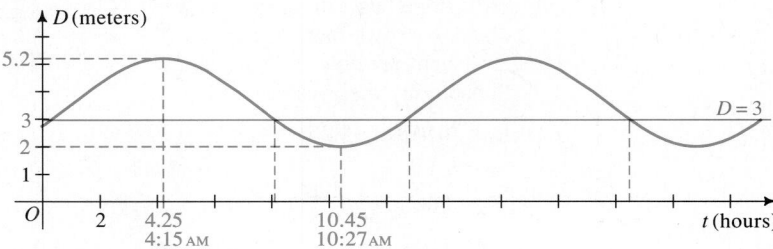

The amplitude is $A = \dfrac{5.2 - 2.0}{2} = 1.6$

The period is $p = 2 \cdot (\text{time of low tide} - \text{time of high tide})$

                        $= 2(10.45 - 4.25)$

                        $= 12.4$

              Since $12.4 = \dfrac{2\pi}{B}$, $B \approx 0.507$.

The axis of the wave has the equation $y = \dfrac{5.2 + 2.0}{2} = 3.6$.

Thus, an equation that models the depth D at time t hours after midnight is:

$$D - 3.6 = 1.6 \cos [0.507(t - 4.25)], \text{ or}$$
$$D = 1.6 \cos [0.507(t - 4.25)] + 3.6$$

    b. Substitute $t = 12$ into the equation above to find the depth at noon.

$$D = 1.6 \cos [0.507(12 - 4.25)] + 3.6$$
$$D \approx 2.47 \text{ m}$$

    c. Using a graphing calculator or a computer, first enter the equation $D - 3.6 = 1.6 \cos [0.507(t - 4.25)]$ to get the curve shown above. Then graph the line $D = 3$ and find that it intersects the curve at $t = 8.1$, $t = 12.8$, and $t = 20.5$ (the times when the depth is 3 m).

       Without a graphing calculator, you can find these values of t algebraically, as shown on the next page.

Trigonometric Equations and Applications   **311**

Suggested Assignments

Standard
Day 1: 313/1–13 odd
Day 2: 314/15–23 odd
Comprehensive
Day 1: 313/1–15 odd
Day 2: 314/17, 19, 21, 24–26

Supplementary Materials

Alternative Assessment, 26
Precalculus Plotter Plus,
48–49
Student Resource Guide,
79–81

$$D = 1.6 \cos [0.507(t - 4.25)] + 3.6$$
$$3 = 1.6 \cos [0.507(t - 4.25)] + 3.6$$
$$-0.6 = 1.6 \cos [0.507(t - 4.25)]$$
$$-0.375 = \cos [0.507(t - 4.25)]$$
$$\text{Cos}^{-1}(-0.375) = 0.507(t - 4.25)$$

Use a scientific calculator to find that $\text{Cos}^{-1}(-0.375) \approx 1.96$. Any radian value coterminal with 1.96 or with -1.96 would also have -0.375 as its cosine value. The diagram shows that 4.32 and 8.24 are such possible radian values. Substituting these values gives:

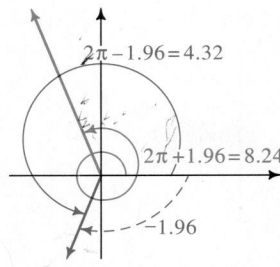

$1.96 = 0.507(t - 4.25)$ $\quad$ $4.32 = 0.507(t - 4.25)$ $\quad$ $8.24 = 0.507(t - 4.25)$

$\quad t \approx 8.1$ h $\qquad\qquad\quad t \approx 12.8$ h $\qquad\qquad\quad t \approx 20.5$ h

$\quad t = 8:06$ A.M. $\qquad\qquad t = 12:48$ P.M. $\qquad\qquad t = 8:30$ P.M.

You should locate these three times on the graph shown on page 311. Looking at the graph, you can see that the curve is above the line $D = 3$ when $12.8 < t < 20.5$. Therefore, the boat can safely moor between 12:48 P.M. and 8:30 P.M.

CLASS EXERCISES

For Exercises 1–5, refer to the sine wave shown.

1. What is the amplitude? 2
2. What is the period? 4
3. What is the axis of the wave? $y = 5$
4. **a.** If the wave is considered to be the translation of the graph of $y = A \sin Bx$, what are the horizontal and vertical translation amounts? horiz. 0, ver. 5
   b. What is this equation? $y = 5 + 2 \sin \frac{\pi}{2}x$
5. **a.** If the wave is considered to be the translation of the graph of $y = A \cos Bx$, what are the translation amounts? horiz. 1, ver. 5
   b. What is this equation? $y = 5 + 2 \cos \frac{\pi}{2}(x - 1)$

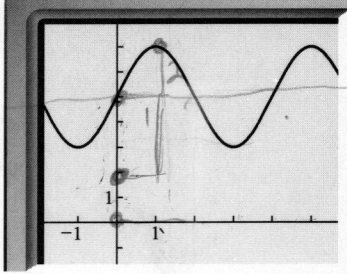

6. The equation of a sine wave is $y - 6 = 4 \sin 3(x + 5)$. What do the numbers 3, 4, 5, and 6 tell you about the wave? 3, period $= 2\pi \div 3$; 4, amplitude $= 4$; 5, Horiz. shift is 5 to the left; 6, Ver. shift is 6 up

312 *Chapter Eight*

Give an equation for each trigonometric graph.

A 1.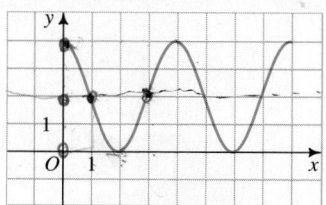

2.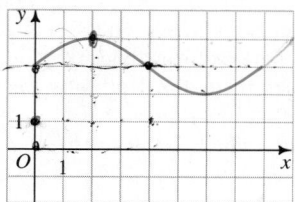

3.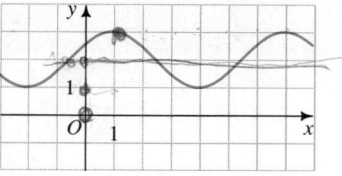

4.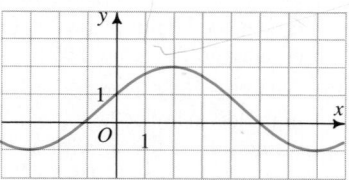

5.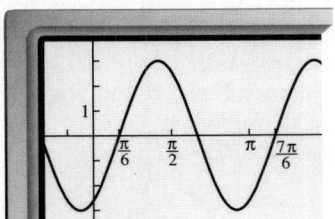

6.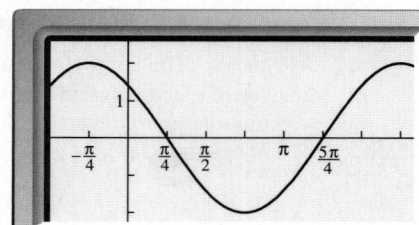

7.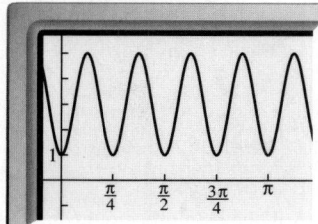

8.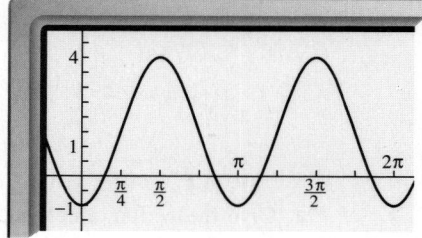

Sketch the graph of each equation.

9. $y = 3 + 5 \sin 2x$

10. $y = -2 + 2 \cos \frac{1}{2}x$

11. $y = -3 \sin\left(x - \frac{\pi}{6}\right)$

12. $y = 4 - 4 \cos 2(x - \pi)$

13. $y - 3 = 2 \cos \frac{\pi}{2}(x - 1)$

14. $y + 3 = 6 \sin \frac{\pi}{4}(x + 2)$

Trigonometric Equations and Applications **313**

Cooperative Learning

You may want to use Written Exercises 1–6 as the basis of a cooperative learning activity. Have students work in pairs. One student writes the equation of a general sine wave. After the student graphs the equation, either with a graphing calculator or by hand, the other student is asked to give an equation of the graph. The students can then switch roles.

Additional Answers Written Exercises

1. $y = 2 + 2 \cos \frac{\pi}{2}x$ or

   $y = 2 + 2 \sin \frac{\pi}{2}(x - 3)$

2. $y = 3 + \cos \frac{\pi}{4}(x - 2)$ or

   $y = 3 + \sin \frac{\pi}{4}x$

3. $y = 2 + \cos \frac{\pi}{3}(x - 1)$ or

   $y = 2 + \sin \frac{\pi}{3}\left(x + \frac{1}{2}\right)$

4. $y = \frac{1}{2} + \frac{3}{2} \cos \frac{\pi}{5}(x - 2)$ or

   $y = \frac{1}{2} + \frac{3}{2} \sin \frac{\pi}{5}\left(x + \frac{1}{2}\right)$

5. $y = 3 \cos 2\left(x - \frac{5\pi}{12}\right)$ or

   $y = 3 \sin 2\left(x - \frac{\pi}{6}\right)$

6. $y = 2 \cos \left(x + \frac{\pi}{4}\right)$ or

   $y = 2 \sin \left(x + \frac{3\pi}{4}\right)$

7. $y = 3 + 2 \cos 8\left(x - \frac{\pi}{8}\right)$ or

   $y = 3 + 2 \sin 8\left(x - \frac{\pi}{16}\right)$

8. $y = \frac{3}{2} + \frac{5}{2} \cos 2\left(x - \frac{\pi}{2}\right)$

   or $y = \frac{3}{2} + \frac{5}{2} \sin 2\left(x - \frac{\pi}{4}\right)$

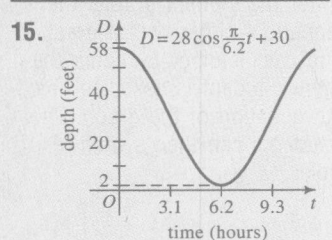
15. Oceanography The Bay of Fundy is an inlet of the Atlantic Ocean bounded by Maine and New Brunswick on the north and Nova Scotia on the south. It is famous for its high tides. At a dock there, the depth of water is 2 ft at low tide and 58 ft at high tide, which occurs 6 h 12 min after low tide. Draw a graph showing the depth of water at the dock as a function of the time since high tide occurs. Find an equation of your graph.

16. Oceanography Rework the tide problem in Example 2 assuming that the first high tide today occurs at 3:00 A.M. with a depth of 4.0 m, and the first low tide occurs at 9:24 A.M. with a depth of 1.8 m.
16. a. $D = 2.9 + 1.1 \cos[0.491(t - 3)]$ **b.** 2.58 m **c.** 12:47 P.M. − 6:49 P.M.

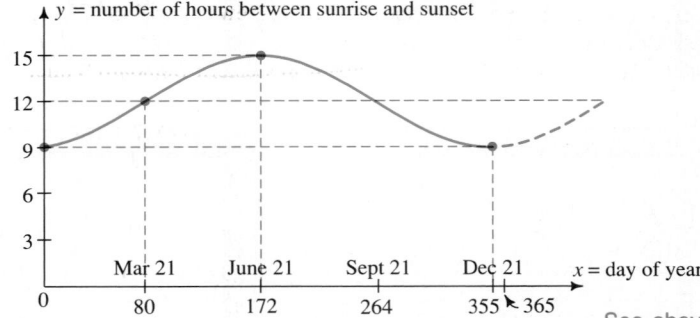 **For some parts of the following exercises, you may wish to use a graphing calculator or a computer. Be sure to use radian measure.**
17. a. 365, 3, $y = 12 + 3 \sin \frac{2\pi}{365}(x - 80$

B **17. Astronomy** The approximate number of hours between sunrise and sunset in Denver, Colorado, varies throughout the year as shown in the graph. This variation is approximated by the sine wave shown below.

See above.
a. Give the period, amplitude, and equation of the daylight-hours graph.
b. Find the amount of daylight in Denver on January 1 and on July 4. 9.1 h, 14.9 h
c. Over the course of a year, during what period of time is the amount of daylight in Denver at least 14 hours? The 123rd day (May 3) through the 220th day
d. If you were to draw a daylight-hours graph for Seattle, Washington, which   (Aug. 8) is north of Denver, do you think its amplitude would be less than or greater than that for the Denver curve? Greater

18. Astronomy The graph given in Exercise 17 applies to Denver and to all other locations at latitude 39°44′N. Modify the graph for a location at 39°44′S. Give an equation of your modified graph. $y = 12 - 3 \sin \frac{2\pi}{365}(x - 80)$

314 *Chapter Eight*

19. Meteorology Average monthly temperatures for New Orleans are plotted for the middle of each month. These points are connected to give a smooth curve, shown below, that gives an approximation of the average daily temperatures.

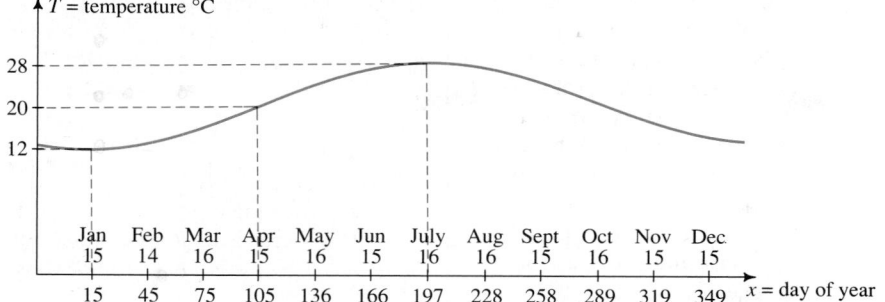

a. Find an equation of the average-daily-temperature graph.
b. Over the course of a year, during what periods of time is the average daily temperature in New Orleans no more than 15°C?

20. Meteorology The average maximum and minimum temperatures of two cities are given below. For each location, sketch a temperature sine wave like that in Exercise 19. Then give an equation of the curve.
a. Winnipeg, Canada: 26°C (July 16) and −14°C (January 15)
b. Rio de Janeiro, Brazil: 28°C (January 15) and 22°C (July 16)

21. Astronomy The graph at the left below shows the time of sunsets occurring every other day during September in Exeter, New Hampshire. The graph at the right shows the time of sunsets on either the 21st or 22nd day of each month for the entire year. All times are given in Eastern Standard Time.

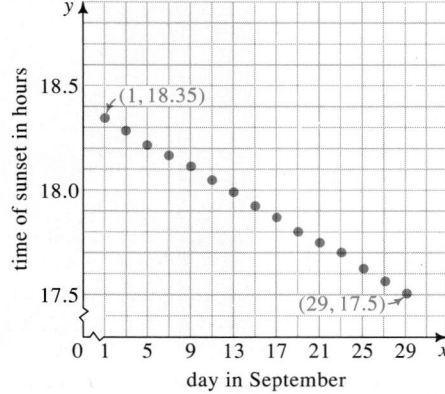

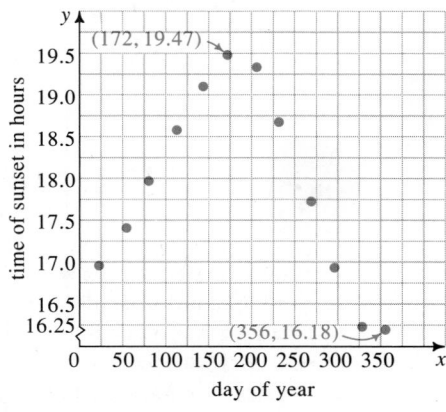

a. Find a linear function that models the data in the graph at the left.
b. Find a trigonometric function that models the data in the graph at the right.
c. Use the function in part (b) to find the period of time, over the course of a year, during which the sun sets after 7:00 P.M. EST.

Additional Answers
Written Exercises

19. a. $T =$
$20 + 8 \cos \frac{2\pi}{365}(x - 197)$
b. The 1st day (Jan. 1) through the 66th day (Mar. 7), the 328th day (Nov. 24) through the 365th day (Dec. 31)

20. a. $T =$
$6 + 20 \sin \frac{2\pi}{365}(x - 105)$
b. $T =$
$25 + 3 \cos \frac{2\pi}{365}(x - 15)$

21. a. $y = -0.03x + 18.38$
b. $y = 17.83 +$
$1.65 \cos \frac{2\pi}{365}(x - 172)$
c. The 127th day (May 7) through the 217th day (Aug. 5)

22. **Research** Refer to Exercise 20. Consult a local almanac or newspaper to find the average maximum and minimum temperatures for the area where you live. Sketch a temperature sine wave and give an equation of the curve.

23. **Research** Refer to Exercise 21. Consult a local almanac or newspaper to find the date and time of the latest sunrise and of the earliest sunrise in your area. (For simplicity, use standard time, not daylight-saving time.)
    a. Draw a graph that approximates the times of sunrises for the whole year and give its equation.
    b. What is the time of sunrise on your birthday?

    26. $h = 35 + 20 \cos \frac{500}{63} t$

24. **Thermodynamics** On a cold winter day, a house is heated until it is warm enough for the thermostat to turn off the heat. Then the house cools until it is cool enough for the thermostat to turn on the heat again. Assume that this periodic change in temperature can be modeled by a trigonometric function.
    a. Find an equation of this function given the following data: See below.

    1. Thermostat turns heat on at 10:15 A.M. when house temperature is 18°C.

    2. Thermostat turns heat off at 10:30 A.M. when house temperature is 20°C.

    b. **Writing** Write a paragraph explaining why this trigonometric function might not be a good model. Think about how long it takes for the heating and cooling parts of the cycle. Think about the effects of temperature changes outdoors. 24. a. $y = 19 + \cos 4\pi(x - 10.5)$

C 25. **Physics** Suppose that after you are loaded into a Ferris wheel car, the wheel begins turning at 4 rpm. The wheel has diameter 12 m and the bottom seat of the wheel is 1 m above the ground. (See the diagram below.) Express the height h of the seat above the ground as a function of time t seconds after it begins turning. (*Suggestion*: Begin your solution by sketching the graph of height versus time. Label the times at which you reach the lowest and highest points.)

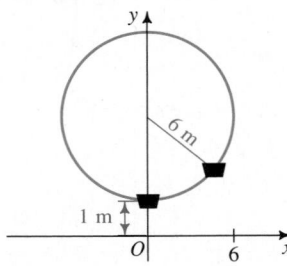

25. $h = 7 + 6 \cos \frac{2\pi}{15}(t - 7.5)$

26. **Physics** A reflector is fastened to the front wheel of a bicycle 20 cm from the center of the wheel. The diameter of the wheel and inflated tire is 70 cm. If the bike is traveling at 10 km/h, express the height of the reflector above the ground as a function of time. Assume that at time $t = 0$ seconds, the reflector is at its highest point. (*Hint*: The hardest part is determining the period.) See above.

Identities and Equations

8-4 Relationships Among the Functions

Objective *To simplify trigonometric expressions and to prove trigonometric identities.*

In this section, we will investigate some of the relationships among the trigonometric functions. Some of these relationships you have seen before. In Section 7-5, you saw that certain functions are reciprocals of others.

Reciprocal Relationships

$$\csc \theta = \frac{1}{\sin \theta} \qquad \sec \theta = \frac{1}{\cos \theta} \qquad \cot \theta = \frac{1}{\tan \theta}$$

Recall that if P is a point on the unit circle as shown, then $x = \cos \theta$ and $y = \sin \theta$. From the symmetry of a circle, the following relationships are true.

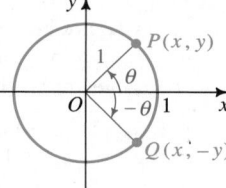

$$\sin (-\theta) = -y = -\sin \theta$$
$$\cos (-\theta) = x = \cos \theta$$

Using the reciprocal relationships, we can find negative relationships for the other functions.

Relationships with Negatives

$\sin (-\theta) = -\sin \theta$	and	$\cos (-\theta) = \cos \theta$
$\csc (-\theta) = -\csc \theta$	and	$\sec (-\theta) = \sec \theta$
$\tan (-\theta) = -\tan \theta$	and	$\cot (-\theta) = -\cot \theta$

We can find other relationships using the unit circle. Again, refer to the diagram above. Since $P(\cos \theta, \sin \theta)$ is on the unit circle,

$$(\sin \theta)^2 + (\cos \theta)^2 = x^2 + y^2 = 1.$$

This is the first of the three Pythagorean relationships that follow. You are asked to explore the other two relationships in Activity 2 and in Exercises 25 and 26.

Pythagorean Relationships

$$\sin^2 \theta + \cos^2 \theta = 1 \qquad 1 + \tan^2 \theta = \sec^2 \theta \qquad 1 + \cot^2 \theta = \csc^2 \theta$$

Trigonometric Equations and Applications **317**

Warm-Up Exercises

1. Write each function in terms of sin θ and/or cos θ.
   a. tan θ $\frac{\sin \theta}{\cos \theta}$
   b. sec θ $\frac{1}{\cos \theta}$
   c. csc θ $\frac{1}{\sin \theta}$
   d. cot θ $\frac{\cos \theta}{\sin \theta}$

2. Name the reciprocal function of the given function. (Hint: Use your answers from Exercise 1.)
   a. sin θ csc θ
   b. cos θ sec θ
   c. tan θ cot θ
   d. sec θ cos θ

3. Simplify: $\frac{1}{\sin x \csc x}$ 1

4. Simplify.
   a. $\frac{t - \frac{1}{t}}{t + 1}$ $\frac{t - 1}{t}$
   b. $\frac{\sin \theta - \frac{1}{\sin \theta}}{\sin \theta + 1}$ $\frac{\sin \theta - 1}{\sin \theta}$

Motivating the Section

Ask students to simplify the algebraic expression $\frac{2x^2 - 1}{1 + x\sqrt{2}}$. $(x\sqrt{2} - 1)$ Then explain to students that the relationships developed in this section will enable them to simplify complicated trigonometric expressions such as

$$\frac{1}{\cos \theta + \frac{\sin^2 \theta}{\cos \theta}}.$$

The sine and cosine are called **cofunctions**, as are the tangent and cotangent, and the secant and cosecant. There is a special relationship between a function and its cofunction, as you will discover in the following activity.

Activity 1

1. Use a calculator to evaluate the following.
   a. sin 50°, cos 40° 0.776, 0.776   b. sin 25°, cos 65° 0.423, 0.423
   c. cos 11°, sin 79° 0.982, 0.982   d. sin 83°, cos 7° 0.993, 0.993
2. Complete each of the following.
   a. sin 18° = cos (?°) 72   b. cos 89° = sin (?°) 1
   c. sin θ = cos (?) 90° − θ d. cos θ = sin (?) 90° − θ

The reason for the cofunction relationships can be seen from the diagram at the right. If the sum of the measures of $\angle POA$ and $\angle P'OA$ is 90°, then P and P' are symmetric with respect to the line $y = x$. Hence, if $P = (a, b)$, then $P' = (b, a)$. Consequently,

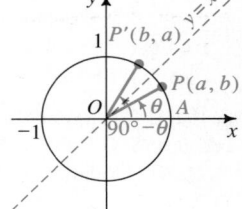

$$\sin \theta = y\text{-coordinate of } P \qquad \cos \theta = x\text{-coordinate of } P$$
$$= x\text{-coordinate of } P' \qquad\qquad = y\text{-coordinate of } P'$$
$$= \cos (90° - \theta) \qquad\qquad = \sin (90° - \theta)$$

You should convince yourself that this argument remains valid if the diagram is changed so that θ is not in Quadrant I. In general,

function of θ = cofunction of the complement of θ.

The cofunction relationships are summarized below.

Cofunction Relationships		
sin θ = cos (90° − θ)	and	cos θ = sin (90° − θ)
tan θ = cot (90° − θ)	and	cot θ = tan (90° − θ)
sec θ = csc (90° − θ)	and	csc θ = sec (90° − θ)

Identities

Each of the trigonometric relationships given is true for all values of the variable for which *each side of the equation is defined*. Such relationships are called **trigonometric identities**, just as $(a + b)^2 = a^2 + 2ab + b^2$ is called an *algebraic identity*. The following activity will help you understand identities.

Activity 2

 Use a graphing calculator or computer to answer each question. Be sure you use radian measure. If your calculator does not have a particular function, use its reciprocal function. For example, you can use $\dfrac{1}{\cos x}$ instead of sec x.

1. **a.** On the same set of axes, graph $y = 1 + \tan^2 x$ and $y = \sec^2 x$.
   b. Compare the two graphs. Would you say that the equation $1 + \tan^2 x = \sec^2 x$ is an identity? Explain. Yes.
2. **a.** On the same set of axes, graph $y = \sin 2x$ and $y = 2 \sin x$.
   b. Compare the two graphs. Would you say that the equation $\sin 2x = 2 \sin x$ is an identity? Explain. No.
3. **a.** Graph $y = \sec x - \sin x \tan x$. What is the domain of the sine function? of the secant function? of the tangent function? See below.
   b. On the same set of axes, graph $y = \cos x$. What is the domain of the cosine function? All real numbers
   c. If you eliminate those values of x for which any of these functions are undefined, what can you say about the comparison of the two graphs for the remaining values of x? Does the screen display suggest an identity? They coincide. Yes, $\cos x = \sec x - \sin x \tan x$.

   3. a. All real numbers; all real numbers except $\dfrac{n\pi}{2}$; all real numbers except $\dfrac{n\pi}{2}$ ($n \in$ odd integers)

In the following examples, we will use trigonometric identities to simplify expressions and to prove other identities.

Example 1   Simplify $\sec x - \sin x \tan x$.

Solution   Express each function in terms of $\sin x$ and $\cos x$.

$$\sec x - \sin x \tan x = \frac{1}{\cos x} - \sin x \cdot \frac{\sin x}{\cos x}$$

$$= \frac{1 - \sin^2 x}{\cos x}$$

$$= \frac{\cos^2 x}{\cos x}$$

$$= \cos x$$

Example 2   Prove: $\dfrac{\cot A (1 + \tan^2 A)}{\tan A} = \csc^2 A$

Solution   Use identities to simplify the expression on either the left or right side of the equals sign. Usually, it's a good idea to choose the more complicated expression. Here, start simplifying the left side by replacing the expression $1 + \tan^2 A$ with $\sec^2 A$, as shown on the next page.

Trigonometric Equations and Applications   **319**

Mathematical Note

In proving a trigonometric identity, stress that the left and right sides should be "manipulated" independently. Warn students against adding the same quantity to both sides or multiplying both sides by the same quantity, since doing so *assumes* equality; it is equality that we are trying to establish.

Additional Examples

1. Simplify.
   a. $\sin \theta \sec \theta \cot \theta$
   b. $1 + \tan^2 (x - 90°)$
   c. $\csc x (\cos^3 x \tan x - \sin x)$

   a. $\sin \theta \cdot \dfrac{1}{\cos \theta} \cdot \dfrac{\cos \theta}{\sin \theta} = 1$
   b. $1 + \tan^2 [-(90° - x)] =$
   $1 + [-\tan (90° - x)]^2 =$
   $1 + \cot^2 x = \csc^2 x$
   c. $\dfrac{1}{\sin x} \left(\cos^3 x \cdot \dfrac{\sin x}{\cos x} - \right.$
   $\left. \sin x \right) = \cos^2 x - 1 =$
   $-(1 - \cos^2 x) =$
   $-\sin^2 x$

2. Prove:
   $$\frac{1}{1 - \sec A} + \frac{1}{1 + \sec A} = -2 \cot^2 A$$
   $$\frac{1}{1 - \sec A} + \frac{1}{1 + \sec A} =$$
   $$\frac{1 + \sec A}{1 - \sec^2 A} + \frac{1 - \sec A}{1 - \sec^2 A} =$$
   $$\frac{2}{-\tan^2 A} = -2 \cot^2 A$$

Exercise Note

Class Exercises 1–3 emphasize alternate ways of writing the Pythagorean relationships which are useful for proving other trigonometric identities. Class Exercise 7 reinforces the idea that the identities are valid only if the arguments are equal. Be sure that students know that, in general, $\sin^2 A + \cos^2 B \neq 1$.

Additional Answers
Class Exercises

8. a. To simplify first expression: multiply numerator and denominator by t. To simplify second expression: multiply numerator and denominator by $\tan A$.
b. To simplify first expression: multiply numerator and denominator of first fraction by a and of second fraction by b. To simplify second expression: multiply numerator and denominator of first fraction by $\sec \theta$ and of second fraction by $\tan \theta$.
c. To simplify first expression: multiply numerator and denominator by xy. To simplify second expression: multiply numerator and denominator by $\cos \theta \sin \theta$.
d. To simplify first expression: multiply numerator and denominator by x. To simplify second expression: multiply numerator and denominator by $\cos \theta$.

$$\frac{\cot A \,(1 + \tan^2 A)}{\tan A} = \frac{\cot A \cdot \sec^2 A}{\tan A} \quad \longleftarrow \quad 1 + \tan^2 A = \sec^2 A$$

$$= \cot^2 A \cdot \sec^2 A \quad \longleftarrow \quad \frac{1}{\tan A} = \cot A$$

$$= \frac{\cos^2 A}{\sin^2 A} \cdot \frac{1}{\cos^2 A} \quad \longleftarrow \quad \cot A = \frac{\cos A}{\sin A} \text{ and}$$

$$= \frac{1}{\sin^2 A} \qquad\qquad\qquad \sec A = \frac{1}{\cos A}$$

$$= \csc^2 A$$

Since the result gives the expression on the other side of the equals sign, the given equation is an identity.

Sometimes when simplifying an expression, you may not see which identity applies. If this happens, try expressing the functions involved in terms of sine and cosine only. Usually this method takes longer, but it can be effective if all else fails.

CLASS EXERCISES

1. Since $\sin^2 \theta + \cos^2 \theta = 1$:　**a.** $1 - \sin^2 \theta = \underline{\ ?\ }$　**b.** $1 - \cos^2 \theta = \underline{\ ?\ }$ $\cos^2\theta$ ... $\sin^2\theta$
2. Since $1 + \tan^2 \theta = \sec^2 \theta$:　**a.** $\sec^2 \theta - 1 = \underline{\ ?\ }$　**b.** $\sec^2 \theta - \tan^2 \theta = \underline{\ ?\ }$ 1
3. Since $1 + \cot^2 \theta = \csc^2 \theta$:　**a.** $\csc^2 \theta - 1 = \underline{\ ?\ }$　**b.** $\csc^2 \theta - \cot^2 \theta = \underline{\ ?\ }$ 1
2. a. $\tan^2 \theta$　$\cot^2 \theta$

Simplify each expression.

4. a. $\tan \theta \cdot \cos \theta$ $\sin \theta$　**b.** $\tan (90° - A) \cot A$　**c.** $\cos\left(\dfrac{\pi}{2} - x\right) \sin x$ $\tan^2 x$

5. a. $(1 - \sin x)(1 + \sin x)$ $\cos^2 x$　**b.** $\sin^2 x - 1$ $-\cos^2 x$　**c.** $(\sec x - 1)(\sec x + 1)$

6. a. $\tan A \cdot \cot A$ 1　**b.** $\cot y \cdot \sin y$ $\cos y$　**c.** $\cot^2 x - \csc^2 x$ -1

7. Evaluate each expression.

a. $\sin^2 \dfrac{5\pi}{6} + \cos^2 \dfrac{5\pi}{6}$ 1　**b.** $\sec^2 \pi - \tan^2 \pi$ 1　**c.** $\csc^2 135° - \cot^2 135°$ 1

8. *Discussion* Tell how you would simplify each complex fraction. The result of the first fraction given should help you simplify the second fraction.

a. $\dfrac{t + \dfrac{1}{t}}{t}$　$\dfrac{\tan A + \dfrac{1}{\tan A}}{\tan A}$　**b.** $\dfrac{\dfrac{a}{1} - \dfrac{b}{1}}{\dfrac{1}{a} \ \ \dfrac{1}{b}}$　$\dfrac{\sec \theta}{\cos \theta} - \dfrac{\tan \theta}{\cot \theta}$

c. $\dfrac{\dfrac{y}{x} + \dfrac{x}{y}}{\dfrac{1}{xy}}$　$\dfrac{\dfrac{\sin \theta}{\cos \theta} + \dfrac{\cos \theta}{\sin \theta}}{\dfrac{1}{\cos \theta \sin \theta}}$　**d.** $\dfrac{1}{x + \dfrac{y^2}{x}}$　$\dfrac{1}{\cos \theta + \dfrac{\sin^2 \theta}{\cos \theta}}$

WRITTEN EXERCISES

Simplify.

A **1. a.** $\cos^2 \theta + \sin^2 \theta$ 1   **b.** $(1 - \cos \theta)(1 + \cos \theta)$ $\sin^2 \theta$ **c.** $(\sin \theta - 1)(\sin \theta + 1)$ $-\cos^2 \theta$

2. a. $1 + \tan^2 \theta$ $\sec^2 \theta$ **b.** $(\sec x - 1)(\sec x + 1)$ $\tan^2 x$ **c.** $\tan^2 x - \sec^2 x$ -1

3. a. $1 + \cot^2 A$ $\csc^2 A$ **b.** $(\csc A - 1)(\csc A + 1)$ $\cot^2 A$ **c.** $\dfrac{1}{\sin^2 A} - \dfrac{1}{\tan^2 A}$ 1

4. a. $\dfrac{1}{\cos(90° - \theta)}$ $\csc \theta$ **b.** $1 - \dfrac{\sin^2 \theta}{\tan^2 \theta}$ $\sin^2 \theta$ **c.** $\dfrac{1}{\cos^2 \theta} - \dfrac{1}{\cot^2 \theta}$ 1

5. a. $\cos \theta \cot(90° - \theta)$ $\sin \theta$ **b.** $\csc^2 x (1 - \cos^2 x)$ 1   **c.** $\cos \theta(\sec \theta - \cos \theta)$ $\sin^2 \theta$

6. a. $\cot A \sec A \sin A$ 1   **b.** $\cos^2 A (\sec^2 A - 1)$ $\sin^2 A$ **c.** $\sin \theta(\csc \theta - \sin \theta)$ $\cos^2 \theta$

7. $\sin A \tan A + \sin(90° - A) \sec A$ **8.** $\csc A - \cos A \cot A$ $\sin A$

9. $(\sec B - \tan B)(\sec B + \tan B)$ 1   **10.** $(1 - \cos B)(\csc B + \cot B)$ $\sin B$

11. $(\csc x - \cot x)(\sec x + 1)$ $\tan x$ **12.** $(1 - \cos x)(1 + \sec x)$ $\cos x$ $\sin^2 x$

Simplify each expression.

13. $\dfrac{\sin x \cos x}{1 - \cos^2 x}$ $\cot x$ **14.** $\dfrac{\tan x + \cot x}{\sec^2 x}$ $\cot x$

15. $(\sin x + \cos x)^2 + (\sin x - \cos x)^2$ 2   **16.** $(\sec^2 \theta - 1)(\csc^2 \theta - 1)$ 1

17. $\dfrac{\cot^2 \theta}{1 + \csc \theta} + \sin \theta \csc \theta$ $\csc \theta$ **18.** $\dfrac{\tan^2 \theta}{\sec \theta + 1} + 1$ $\sec \theta$

19. $\cos^3 y + \cos y \sin^2 y$ $\cos y$ **20.** $\dfrac{\sec y + \csc y}{1 + \tan y}$ $\csc y$

B **21.** $\dfrac{\cos \theta}{1 + \sin \theta} + \dfrac{1 + \sin \theta}{\cos \theta}$ 2 $\sec \theta$ **22.** $\dfrac{\sin \theta \cot \theta + \cos \theta}{2 \tan(90° - \theta)}$ $\sin \theta$

23. $\dfrac{\sin^4 \theta - \cos^4 \theta}{\sin^2 \theta - \cos^2 \theta}$ 1   **24.** $\dfrac{\sin^2 \theta}{1 + \cos \theta}$ (*Hint:* See Exercise 1(b).) $1 - \cos \theta$

25. Use the equation $\sin^2 \theta + \cos^2 \theta = 1$ to prove that $\tan^2 \theta + 1 = \sec^2 \theta$.

26. Use the equation $\sin^2 \theta + \cos^2 \theta = 1$ to prove that $\cot^2 \theta + 1 = \csc^2 \theta$.

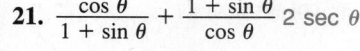 **In Exercises 27 and 28, use a graphing calculator or a computer to graph the given functions.**

27. Graph the function $y = \sin^4 x + 2 \sin^2 x \cos^2 x + \cos^4 x$. What is the domain of this function? What other function could have this same graph? Use trigonometric relationships to verify the suggested identity. All real numbers; $y = 1$

28. Graph the function $y = (\sin x \div \cos x) \div \tan x$. What is the domain of this function? What other function could have this same graph? Use trigonometric relationships to verify the suggested identity. $x \neq \dfrac{n\pi}{2}$, $n \in$ integers; $y = 1$, $x \neq \dfrac{n\pi}{2}$, $n \in$ integers

Additional Answers
Written Exercises

25. $\sin^2 \theta + \cos^2 \theta = 1$;
$\dfrac{\sin^2 \theta}{\cos^2 \theta} + \dfrac{\cos^2 \theta}{\cos^2 \theta} = \dfrac{1}{\cos^2 \theta}$;
$\tan^2 \theta + 1 = \sec^2 \theta$

26. $\sin^2 \theta + \cos^2 \theta = 1$;
$\dfrac{\sin^2 \theta}{\sin^2 \theta} + \dfrac{\cos^2 \theta}{\sin^2 \theta} = \dfrac{1}{\sin^2 \theta}$;
$1 + \cot^2 \theta = \csc^2 \theta$

Suggested Assignments

Standard
Day 1: 321/1–15 odd
Day 2: 321/17–23 odd,
27–35 odd
Comprehensive
Day 1: 321/1–23 odd
Day 2: 321/28–39, 42

Supplementary Materials

Alternative Assessment, 26
Precalculus Plotter Plus, 43

37. c. $\dfrac{PQ}{OQ} = \dfrac{AB}{OB}, \dfrac{\sin\theta}{\cos\theta} = \dfrac{AB}{1},$
$\tan\theta = AB; \dfrac{OP}{OQ} = \dfrac{AO}{BO},$
$\dfrac{1}{\cos\theta} = \dfrac{AO}{1},$ sec $\theta = AO$

e. $(AB)^2 + (BO)^2 = (AO)^2;$
since $AB = \tan\theta$ and
$AO = \sec\theta$ from
part (c),
$(\tan\theta)^2 + 1^2 = (\sec\theta)^2$
or $\sec^2\theta = 1 + \tan^2\theta.$

f. $AP = AO - OP =$
$\sec\theta - 1; AC = AO +$
$OC = \sec\theta + 1;$ since
$(AB)^2 = AP \cdot AC,$
$\tan^2\theta =$
$(\sec\theta - 1)(\sec\theta + 1)$

38. Draw $\overline{PQ} \perp \overline{OD}$ at point Q.
By AA Similarity, $\triangle OQP \sim$
$\triangle ODC,$ so $\dfrac{PQ}{CD} = \dfrac{OQ}{OD}$ and
$\dfrac{PO}{CO} = \dfrac{QO}{DO},$ thus, $\dfrac{PQ}{OQ} = \dfrac{CD}{OD}$
and $\dfrac{PO}{QO} = \dfrac{CO}{DO}; PQ =$
$\cos\theta, OQ = \sin\theta,$ and
$DO = PO = 1; CD =$
$OD \cdot \dfrac{PQ}{OQ} = 1 \cdot \dfrac{\cos\theta}{\sin\theta} =$
$\cot\theta$ and $CO =$
$DO \cdot \dfrac{PO}{QO} = 1 \cdot \dfrac{1}{\sin\theta} =$
$\csc\theta.$

39. $\sec^2 x = 1 + \tan^2 x,$ so
$\sec x = \pm\sqrt{1 + \tan^2 x}.$
Graphing $y = \sqrt{1 + \tan^2 x}$
gives the graph of $y =$
$|\sec x|,$ not $y = \sec x.$

40. $\tan\theta = \pm\dfrac{\sqrt{1 - \cos^2\theta}}{\cos\theta}$

41. $\sec\theta = \pm\dfrac{1}{\sqrt{1 - \sin^2\theta}}$

In Exercises 29–36, prove the given identity.

29. $\cot^2\theta + \cos^2\theta + \sin^2\theta = \csc^2\theta$

30. $\dfrac{\cot\theta - \tan\theta}{\sin\theta \cos\theta} = \csc^2\theta - \sec^2\theta$

31. $\dfrac{\sin\theta}{\csc\theta} + \dfrac{\cos\theta}{\sec\theta} = \sin\theta \csc\theta$

32. $\dfrac{1 - \sin^2\theta}{1 + \cot^2\theta} = \sin^2\theta \cos^2\theta$

33. $\tan^2 x - \sin^2 x = \tan^2 x \sin^2 x$

34. $\dfrac{\tan^2 x}{1 + \tan^2 x} = \sin^2 x$

35. $\dfrac{\sin\theta}{\sin\theta + \cos\theta} = \dfrac{\tan\theta}{1 + \tan\theta}$

36. $\dfrac{\tan x}{1 + \sec x} + \dfrac{1 + \sec x}{\tan x} = 2\csc x$

37. $\overline{AB}$ is tangent to the unit circle at $B(1, 0)$.
a. Why is $\triangle OPQ \sim \triangle OAB$? AA Similarity
b. Use part (a) to explain why

$$\dfrac{PQ}{OQ} = \dfrac{AB}{OB} \text{ and } \dfrac{OP}{OQ} = \dfrac{AO}{BO}.$$

Corr. sides of sim. △ are in proportion.
c. Use part (b) to show that $AB = \tan\theta$ and $AO = \sec\theta.$
d. *Visual Thinking* Use the diagram to explain why the
name *tangent* is given to the expression $\dfrac{\sin x}{\cos x}$ and the
name *secant* is given to the expression $\dfrac{1}{\cos x}.$ See below.
e. Use right triangle AOB to prove that
$\sec^2\theta = 1 + \tan^2\theta.$
f. Extend $\overline{AO}$ to intersect the circle at C. A theorem
from geometry states that $(AB)^2 = AP \cdot AC.$ Use
this fact to prove $\tan^2\theta = (\sec\theta - 1)(\sec\theta + 1).$

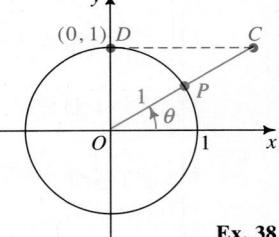
Ex. 37

38. $\overline{CD}$ is tangent to the unit circle at $D(0, 1)$. Show that
$$CD = \cot\theta \text{ and } CO = \csc\theta.$$
(*Hint:* See Exercise 37.)

Ex. 38

39. *Writing* Jon expected that the graph of $y = \sqrt{1 + \tan^2 x}$ would be the same
as the graph of $y = \sec x,$ but his graphing calculator showed that this was not
the case. Write a paragraph explaining why this happened.

40. Express $\tan\theta$ in terms of $\cos\theta$ only.
41. Express $\sec\theta$ in terms of $\sin\theta$ only.

C **42.** Prove $\sqrt{\dfrac{1 - \sin x}{1 + \sin x}} = |\sec x - \tan x|.$ For what values of x is this identity true?

37. d. $\overline{AB}$ is tangent to the circle, $\overline{AO}$ is contained in a secant of the circle.

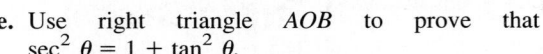

///// COMPUTER EXERCISE

Imagine that your computer can calculate only the sine function. Write a program
for which the input is any $x,$ where $0 \le x \le \dfrac{\pi}{2},$ and the outputs are the six trigonometric functions of $x.$ (*Hint:* See Exercise 41.)

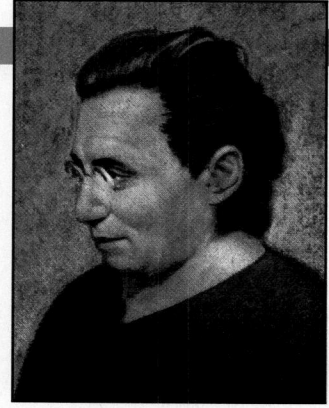

Amalie Emmy Noether (1882–1935)

Emmy Noether was born and educated in Erlangen, Germany. She lectured at the University of Göttingen from 1915 until 1933 when she and other Jewish mathematicians were denied the right to teach. She emigrated to the United States, where she became a visiting professor at Bryn Mawr College and lectured at the Institute of Advanced Study at Princeton.

Noether's contributions centered on invariants and on noncommutative algebras. Her work on invariants culminated in a theorem, known to physicists as Noether's theorem, which is basic to the general theory of relativity. Her impact, however, extended far beyond her own work. Her insight, advice, and encouragement affected the research and publications of many colleagues and students.

Additional Answers
Computer Exercise

```
10   PRINT "X = ";
20   INPUT X
30   PRINT "SIN X = ";SIN(X)
40   PRINT "COS X = ";
     (1 − (SIN(X))^2)^.5
50   PRINT "TAN X = ";
     SIN(X)/((1 −
     (SIN(X))^2)^.5)
60   PRINT "CSC X = ";
     1/SIN(X)
70   PRINT "SEC X = ";
     1/((1 − (SIN(X))^2)^.5)
80   PRINT "COT X = ";
     ((1 − (SIN(X))^2)^.5)/
     SIN(X)
90   END
```

8-5 Solving More Difficult Trigonometric Equations

Objective *To use trigonometric identities or technology to solve more difficult trigonometric equations.*

Many trigonometric equations can be solved in the same way that algebraic equations are solved.

Example 1 Solve $2 \sin^2 \theta - 1 = 0$ for $0° \le \theta < 360°$.

Solution First solve for $\sin^2 \theta$.

$$2 \sin^2 \theta - 1 = 0$$
$$2 \sin^2 \theta = 1$$
$$\sin^2 \theta = \frac{1}{2}$$
$$\sin \theta = \pm \frac{\sqrt{2}}{2} \longleftarrow \sin^2 \theta = \frac{1}{2} \text{ implies that } \sin \theta = \pm \sqrt{\frac{1}{2}}.$$
$$\theta = 45°, 135°, 225°, \text{ and } 315°$$

Notice that there are four solutions because $\sin \theta$ can be either positive or negative, and so we need to look at all four quadrants.

Teaching Notes, p. 294D

Warm-Up Exercises

Solve each equation.

1. $2x^2 - x - 3 = 0$ $\frac{3}{2}, -1$

2. $3x^2 + 5x + 1 = 0$ $\frac{-5 \pm \sqrt{13}}{6}$

3. $\sqrt{1 - x^2} = x - 1$ 1

4. $2\sqrt{1 - x^2} = x + 2$ $0, -\frac{4}{5}$

5. $(2x - 8)(2x - 6) = 0$ 4, 3

Motivating the Section

Ask students to describe the procedure they would use to solve a quadratic equation like $x^2 + 5x - 6 = 0$. (Factor: $(x + 6)(x - 1) = 0$. Use the zero-product rule: $x + 6 = 0$ or $x - 1 = 0$. Solve the linear equations: $x = -6$ or $x = 1$.) Then explain that this same procedure can be applied in solving more complicated trigonometric equations.

Trigonometric Equations and Applications **323**

Making Connections

As a follow-up to Example 1, discuss a trigonometric equation that is quadratic in form but does not factor, such as $\cos^2 \theta - 3 \cos \theta + 1 = 0$ for $0° \leq \theta < 360°$. Use the quadratic formula to solve for $\cos \theta$, and find approximate solutions using a calculator.

Example Note

In discussing Example 3, warn students against dividing both sides of the given equation by $\sin x$, since the roots $x = 0$ and $x = \pi$ would then be lost.

Additional Examples

1. Solve for $0° \leq \theta < 360°$:
$2 \sin \theta - \csc \theta = 1$

$2 \sin \theta - \dfrac{1}{\sin \theta} = 1$

$2 \sin^2 \theta - \sin \theta - 1 = 0$
$(2 \sin \theta + 1)(\sin \theta - 1) = 0$

$\sin \theta = -\dfrac{1}{2}$ or $\sin \theta = 1$

$\theta = 210°, 330°$ or $\theta = 90°$

2. Solve for $0 \leq x < 2\pi$:
$2 \cot^2 x + 3 \csc x = 1$

Method 1: Algebraic
$2(\csc^2 x - 1) +$
$3 \csc x - 1 = 0$
$2 \csc^2 x + 3 \csc x - 3 = 0$
Let $y = \csc x$. Then:
$2y^2 + 3y - 3 = 0$

$y = \dfrac{-3 \pm \sqrt{3^2 - 4(2)(-3)}}{2(2)}$

$y = \dfrac{-3 \pm \sqrt{33}}{4}$

$y \approx 0.6861$ (reject, since $|\csc x| \geq 1$ for all x) or
$y \approx -2.1861$
$\csc x \approx -2.1861$

$\sin x \approx \dfrac{1}{-2.1861} \approx -0.4574$

$x \approx \pi + 0.475 \approx 3.62$ or
$x \approx 2\pi - 0.475 \approx 5.81$

Compare the solving of the quadratic equation $x^2 - 3x - 4 = 0$ with that of $\cos^2 \theta - 3 \cos \theta - 4 = 0$:

$$x^2 - 3x - 4 = 0 \qquad\qquad \cos^2 \theta - 3 \cos \theta - 4 = 0$$
$$(x + 1)(x - 4) = 0 \qquad (\cos \theta + 1)(\cos \theta - 4) = 0$$
$$x + 1 = 0 \ \text{ or } \ x - 4 = 0 \qquad \cos \theta = -1 \ \text{ or } \ \cos \theta = 4$$
$$x = -1 \ \text{ or } \ x = 4 \qquad\qquad \theta = 180° + n \cdot 360°$$

(Since the range of the cosine function is all real numbers between -1 and 1, $\cos \theta = 4$ has no solution.)

Some trigonometric equations that are not quadratic can be transformed into equations that have the quadratic form.

Example 2   Solve $\sin^2 x - \sin x = \cos^2 x$ for $0 \leq x < 2\pi$.

Solution   To get an equation involving only $\sin x$, substitute $1 - \sin^2 x$ for $\cos^2 x$.

$$\sin^2 x - \sin x = \cos^2 x$$
$$\sin^2 x - \sin x = 1 - \sin^2 x$$
$$2 \sin^2 x - \sin x - 1 = 0$$
$$(2 \sin x + 1)(\sin x - 1) = 0$$
$$\sin x = -\frac{1}{2} \text{ or } \sin x = 1$$
$$x = \frac{7\pi}{6}, \frac{11\pi}{6}, \frac{\pi}{2}$$

Example 3   Solve $\sin x \tan x = 3 \sin x$ for $0 \leq x < 2\pi$. Give answers to the nearest hundredth of a radian.

Solution
$$\sin x \tan x = 3 \sin x$$
$$\sin x \tan x - 3 \sin x = 0$$
$$\sin x (\tan x - 3) = 0$$
$$\sin x = 0 \qquad\text{or}\quad \tan x = 3$$
$$x \approx 0, 3.14 \qquad\qquad x \approx 1.25, 4.39$$

The solutions are 0, 1.25, 3.14, and 4.39.

Notice that in Example 3 we did not divide both sides of the equation by the factor $\sin x$. Doing so would have caused us to lose a root. (See the discussion concerning losing roots on page 32.) In the next example, there is no common factor for both sides of the equation. In this case, there is no difficulty in dividing both sides by $\sin \theta$, as long as $\sin \theta \neq 0$, of course. We do not lose a root since values of θ for which $\sin \theta = 0$ are clearly not solutions to the original equation.

324   *Chapter Eight*

Example 4   Solve $2 \sin \theta = \cos \theta$ for $0° \le \theta < 360°$.

Solution

$2 \sin \theta = \cos \theta$

$2 = \dfrac{\cos \theta}{\sin \theta}$ ⟵ Divide both sides by $\sin \theta$.

$2 = \cot \theta$ ⟵ You can use $\tan \theta = \dfrac{1}{2}$ instead.

$\theta \approx 26.6°, 206.6°$

 The next example shows two strategies for solving a trigonometric equation. One method is to write a simpler equation by using identities and then solve the equation algebraically. Another method is to use a graphing calculator or computer to draw graphs and then zoom in on the intersection points.

Example 5   Solve $2 \sin \theta = \cos \theta + 1$ for $0° \le \theta < 360°$.

Solution

Method 1

Try to rewrite the equation so that there is only one function. You can replace $\sin \theta$ by $\pm \sqrt{1 - \cos^2 \theta}$ since $\sin^2 \theta + \cos^2 \theta = 1$. Then the equation is in terms of $\cos \theta$ only.

$$2 \sin \theta = \cos \theta + 1$$
$$2(\pm\sqrt{1 - \cos^2 \theta}) = \cos \theta + 1$$
$$[2(\pm\sqrt{1 - \cos^2 \theta})]^2 = (\cos \theta + 1)^2$$
$$4(1 - \cos^2 \theta) = \cos^2 \theta + 2 \cos \theta + 1$$
$$5 \cos^2 \theta + 2 \cos \theta - 3 = 0$$
$$(5 \cos \theta - 3)(\cos \theta + 1) = 0$$
$$\cos \theta = 0.6 \qquad \text{or} \qquad \cos \theta = -1$$
$$\theta \approx 53.1°, 306.9° \qquad\qquad \theta = 180°$$

Since we squared the original equation, it is possible that we may have gained a root. Therefore, we must check each of these solutions in the *original equation*.

$2 \sin 53.1° \overset{?}{=} \cos 53.1° + 1$ $\qquad\qquad$ $2 \sin 306.9° \overset{?}{=} \cos 306.9° + 1$

$1.6 = 1.6$ $\qquad\qquad\qquad\qquad\qquad$ $-1.6 \ne 1.6$

$2 \sin 180° \overset{?}{=} \cos 180° + 1$

$0 = 0$

Thus, 306.9° is an extraneous root and must be rejected.

The solutions are 53.1° and 180°.

(Solution continues on the next page.)

Additional Examples cont.

Method 2: Graphical
Graph $y = 2 \div (\tan x)^2 + 3 \div \sin x - 1$ for $0 \le x < 6.5$.
A quick check shows that the two solutions in the domain $0 \le x < 2\pi$ occur in the intervals $3 < x < 4$ and $5 < x < 6$. By using the zoom or trace features, you find that the approximate solutions are 3.62 and 5.81.

Example Note

In discussing Example 4, note that dividing both sides of the given equation by $\sin \theta$ is acceptable, because the factor does not "disappear." That is, division by $\sin \theta$ merely moves the factor from the left to the right side of the equation. (This is not the case in Example 3, however, where division by $\sin x$ cancels the factor from both sides and results in a loss of roots.)

If students do not seem to understand when it is and is not appropriate to divide by a variable expression, you can show them how to solve the equation in Example 4 by squaring both sides instead (see Method 1 of Example 5).

Review Note

Remind students that when solving a radical equation, there is the possible presence of extraneous roots. Therefore, they must check each solution in the original equation. Example 5 highlights this point.

Method 2

Use a computer or graphing calculator to graph the equations

$$y = 2 \sin x \text{ and } y = \cos x + 1.$$

(If you can't use your computer in the degree mode, use the radian mode and then convert your answers to degrees.) For the graphs, choose the intervals

$$0° \le x < 360° \text{ and } -2 \le y \le 2.$$

The graphs are shown at the right. Notice that for values of x between 0° and 360°, the graphs intersect in two points. By using a zoom (or trace) feature, you can find that the solutions are about 53.1° and 180°.

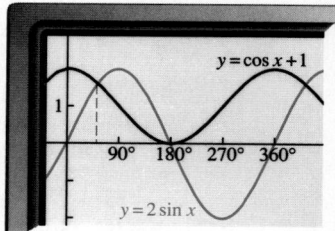

WRITTEN EXERCISES

Solve for $0° \le \theta < 360°$. Give answers to the nearest tenth of a degree.

 A

1. $\sec^2 \theta = 9$

2. $\tan^2 \theta = 1$

3. $1 - \csc^2 \theta = -3$

4. $8 \cos^2 \theta - 3 = 1$

5. $6 \sin^2 \theta - 7 \sin \theta + 2 = 0$

6. $2 \tan^2 \theta = 3 \tan \theta - 1$

7. $6 \sin^2 \theta = 7 - 5 \cos \theta$
  60°, 70.5°, 289.5°, 300°

8. $\cos^2 \theta - 3 \sin \theta = 3$ 270°

Solve for $0 \le x < 2\pi$. Give answers to the nearest hundredth of a radian when necessary.

9. $\cos x \tan x = \cos x$ $\frac{\pi}{4}, \frac{5\pi}{4}$

10. $\sec x \sin x = 2 \sin x$ $0, \frac{\pi}{3}, \pi, \frac{5\pi}{3}$

11. $\sin^2 x = \sin x$ $0, \frac{\pi}{2}, \pi$

12. $\tan^2 x = \tan x$ $0, \frac{\pi}{4}, \pi, \frac{5\pi}{4}$

13. $2 \cos^2 x = \cos x$ $\frac{\pi}{3}, \frac{\pi}{2}, \frac{3\pi}{2}, \frac{5\pi}{3}$

14. $3 \sin x = \cos x$ 0.32, 3.46

15. $\sin x + \cos x = 0$ $\frac{3\pi}{4}, \frac{7\pi}{4}$

16. $\sec x = 2 \csc x$
  1.11, 4.25

 Solve each equation for $0 \le x < 2\pi$ algebraically by using identities or graphically by using a graphing calculator or computer. Give answers to the nearest hundredth of a radian when necessary. $0, \frac{\pi}{4}, \frac{3\pi}{4}, \pi, \frac{5\pi}{4}, \frac{7\pi}{4}$

 B

17. $\tan^2 x = 2 \tan x \sin x$ $0, \frac{\pi}{3}, \pi, \frac{5\pi}{3}$

18. $2 \sin x \cos x = \tan x$

19. $2 \csc^2 x = 3 \cot^2 x - 1$

20. $2 \sec^2 x + \tan x = 5$

21. $\sin^2 x + \sin x - 1 = 0$

22. $\cos^2 x - 2 \cos x - 1 = 0$

23. $3 \cos x \cot x + 7 = 5 \csc x$

24. $2 \sin^3 x - \sin^2 x - 2 \sin x + 1 = 0$

25. $2 \cos^2 x - \cos x = 2 - \sec x$
  $0, \frac{\pi}{3}, \pi, \frac{5\pi}{3}$

26. $\csc^2 x - 2 \csc x = 2 - 4 \sin x$
  $\frac{\pi}{6}, \frac{\pi}{4}, \frac{3\pi}{4}, \frac{5\pi}{6}, \frac{5\pi}{4}, \frac{7\pi}{4}$

326   *Chapter Eight*

27. $2 \cos x = 2 + \sin x$ 0, 5.36   **28.** $1 - \sin x = 3 \cos x$ $\frac{\pi}{2}$, 5.36

 Using Technology

Exercise 35 provides a creative use of technology to solve a problem. You might consider assigning this exercise to all students.

 Solve each equation for $0 \le x < 2\pi$ by using a graphing calculator or computer. Give answers to the nearest hundredth of a radian.

29. $\cos x = x$ 0.74   **30.** $\tan x = x$ 0, 4.49   **31.** $\sin x = \frac{x}{2}$ 0, 1.90

32. $\sin x = 2x$ 0   **33.** $\cos x = 2x - 3$ 1.52   **34.** $\sin x = x^2$ 0, 0.88

C **35.** *Investigation* Consider the graphs of $y = \cos x$ and $y = x$ shown at the left below. Suppose you choose an arbitrary value of x between 0 and $\frac{\pi}{2}$. From this x-value move directly up to point A on the cosine curve. To get to point B, move horizontally to the line $y = x$ and from there move vertically back to the cosine curve. To get to point C, move horizontally and vertically again.

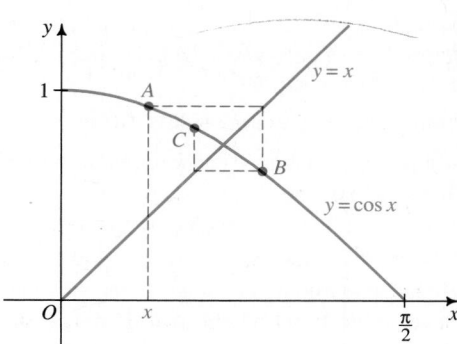

A set of mirrors iterates the basic image of two eggs to create a striking photograph.

a. The x-coordinate of point A is obviously x. Find the x-coordinates of points B and C in terms of x. B: $\cos x$, C: $\cos(\cos x)$

b. If you continue to move from point to point on the cosine curve as described above, what appears to happen? You approach the pt. of intersection.

c. Based on your answers to parts (a) and (b), what relationship exists between the x-coordinates of the points A, B, C, . . . and the solution of the equation $\cos x = x$? The x-coords. approach the sol.

d. Describe a method for approximating the solution of the equation $\cos x = x$ using a scientific calculator. See below.

e. Implement your method of part (d) to find the solution of the equation $\cos x = x$ to the nearest thousandth of a radian. 0.739

f. Determine whether the initial choice of x needs to be limited to the interval from 0 to $\frac{\pi}{2}$. No

d. Enter a guess. Press the cos key until the values agree to the desired number of places.

Trigonometric Equations and Applications   **327**

Chapter Summary

1. Techniques for solving simple *trigonometric equations* are illustrated in Examples 1 and 2 of Section 8-1.

2. If a line has *inclination* α and *slope m*, then $m = \tan \alpha$.

3. If $Ax^2 + Bxy + Cy^2 + Dx + Ey + F = 0$ is the equation of a non-degenerate conic section, one of its axes has *direction angle* α, where α is an angle formed by the positive *x*-axis and this axis of the conic. Also:

$$\alpha = \frac{\pi}{4} \text{ (if } A = C) \text{ and } \tan 2\alpha = \frac{B}{A - C} \text{ (if } A \neq C, 0 < 2\alpha < \pi)$$

Example 4 on pages 297–298 shows how to graph such a conic.

4. The functions $y = A \sin Bx$ and $y = A \cos Bx$ have amplitude $|A|$ and period $\frac{2\pi}{B}$, where $A \neq 0$ and $B > 0$.

5. Sine curves and cosine curves are both called *sine waves*. If the graphs of $y = A \sin Bx$ and $y = A \cos Bx$ are translated horizontally h units and vertically k units, then the resulting graphs have equations

$$y - k = A \sin B(x - h) \text{ and } y - k = A \cos B(x - h).$$

Information about how to find the translation amounts h and k for either a sine curve or a cosine curve is given on page 310.

6. The graphs of the sine waves have many applications, including the description of tides, radio waves, sunrises, musical tones, and the motion of springs.

7. A *trigonometric identity* is an equation that is true for all values of the variable for which both sides of the equation are defined. In proving such identities, the following relationships are helpful: the reciprocal relationships, the negative relationships, the Pythagorean relationships, and the cofunction relationships. These relationships are summarized on pages 317–318.

8. More advanced techniques for solving various trigonometric equations either algebraically or graphically are discussed in Examples 1–5 of Section 8-5.

Key vocabulary and ideas

inclination of a line, direction angle of a conic (pp. 296–297)
period and amplitude of sine and cosine curves (p. 302)
equations of general sine waves (p. 309)
trigonometric identity (p. 318)

Chapter Test

1. Find the inclination of the line $x - 3y = 9$ to the nearest degree. 18° **8-1**

2. Solve $5 \cos \theta = -1$ for $0° \leq \theta < 360°$. Give answers to the nearest tenth of a degree. 101.5°, 258.5°

3. Solve $3 - \csc x = 7$ for $0 \le x < 2\pi$. Give answers to the nearest hundredth of a radian. 3.39, 6.03

4. Solve $2 \sin 3x = \sqrt{2}$ for $0 \le x < 2\pi$. $\dfrac{\pi}{12}, \dfrac{\pi}{4}, \dfrac{3\pi}{4}, \dfrac{11\pi}{12}, \dfrac{17\pi}{12}, \dfrac{19\pi}{12}$ 8-2

5. Graph $y = 3 \sin \dfrac{x}{2}$, and give its period and amplitude. 4π, 3

6. Find an equation for the trigonometric function whose graph is shown   8-3
below. $y = 1 + 2 \sin \dfrac{\pi}{2}x$

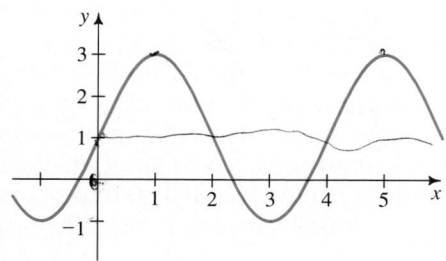

7. At Ocean Tide Dock the first high tide today occurs at 2:00 A.M. with depth 5 m, and the first low tide occurs at 8:30 A.M. with depth 2.2 m.
   a. Sketch and label a graph showing the depth of the water at the dock as a function of time after midnight.
   b. **Writing** Write a paragraph or two to describe how you would use the given information about the high and low tides to find a trigonometric equation of this function.

8. Simplify each expression: 8-4

   a. $\cot A (\sec A - \cos A) \sin A$

   b. $\dfrac{\cot \theta}{\sin (90° - \theta)} \csc \theta$

   c. $(\sec x + \tan x) (1 - \sin x) \cos x$

   d. $\dfrac{\cot \alpha + \tan \alpha}{\csc^2 \alpha} \tan \alpha$

9. Solve each equation for $0 \le x < 2\pi$. Give answers to the nearest   8-5
hundredth of a radian when necessary.
   a. $2 \cos x = \sin x$ 1.11, 4.25
   b. $\sin x = \csc x$ $\dfrac{\pi}{2}, \dfrac{3\pi}{2}$

10. Solve each equation for $0° \le \theta < 360°$. Give answers to the nearest tenth of a degree.
   a. $2 \cos^2 \theta + 3 \sin \theta - 3 = 0$
      30°, 90°, 150°
   b. $\cos \theta \cot \theta = 2 \cos \theta$
      26.6°, 90°, 206.6°, 270°

5.

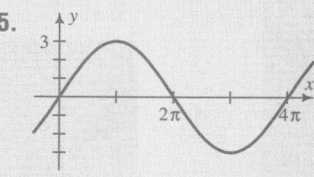

7. a.

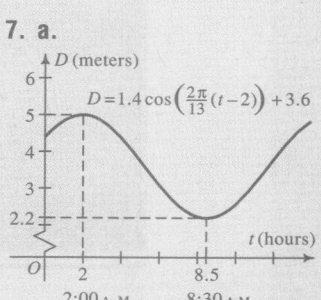

b. Answers may vary. The difference in time between high and low tides is half the period. The difference between the heights of the high and low tides is twice the amplitude. The average of the high and low tides is the axis of the function, or the vertical shift. The time of the first high tide after midnight would be the horizontal shift if the function is expressed in terms of cosine.

Chapter 9 Triangle Trigonometry

Overview

In this chapter, students use trigonometry to find the unknown sides or angles of a triangle. The first section defines the trigonometric functions of an acute angle of a right triangle as ratios of the lengths of the sides; students then use these definitions to find unknown parts of right triangles and to solve such application problems as finding angles of elevation and depression. The second section presents the formula for the area of a triangle in terms of the sine of one of the angles; students then find the areas of various figures, including quadrilaterals and sectors of circles. The next two sections present the law of sines and the law of cosines, which students use to solve triangles. In the last section, students use their knowledge of triangle trigonometry to solve problems from navigation and surveying.

Objectives

9-1 To use trigonometry to find unknown sides or angles of a right triangle.

9-2 To find the area of a triangle given the lengths of two sides and the measure of the included angle.

9-3 To use the law of sines to find unknown parts of a triangle.

9-4 To use the law of cosines to find unknown parts of a triangle.

9-5 To use trigonometry to solve navigation and surveying problems.

Supplementary Resources

Tests, pp. 23–24, 25–26
Alt. Assess., pp. 27–29, 67
Activities, pp. 21–22, 23
Using Tech., pp. 25–26
Student Res. Guide, pp. 87–94

Pacing Guide

Section	9-1	9-2	9-3	9-4	9-5	Review	Test	Total
Standard (days)	2	2	1	3	2	1	1	12
Comprehensive (days)	2	1	1	3	2	1	1	11

Teaching Notes

9-1 | pages 331–338

Presenting the Section

You might want to introduce this section by first re-viewing the Pythagorean theorem and how it is used to find a missing side of a right triangle given the other two sides. Also review that the sum of the meas-ures of the angles of a triangle is 180°, and that in a right triangle the acute angles are complementary.

Stress the need to analyze each problem care-fully in order to determine the correct trigonometric ratio to find the unknown angle or side. Students should draw diagrams as accurately as possible and label them carefully. Encourage finding exact values when possible, especially in the case of 30°-60°-90° and 45°-45°-90° triangles.

Emphasize that the definitions of the trigono-metric functions of the acute angles of a right trian-gle are special cases of the trigonometric functions defined in Chapter 7, where the angle is given in standard position. The values of sin θ, cos θ, and tan θ depend only on the size of the angle θ, not on the size of the triangle. In this section, since the trig-onometric functions are defined in terms of acute angles, the functions all have positive values.

Using Technology

The use of scientific calculators can speed up the computations in this section, including finding the values of the trigonometric functions. Tell students to use the reciprocal key to find the values of the reciprocal trigonometric functions, if necessary.

Applications

Many applications of trigonometry involve solving right triangles, that is, finding the unknown sides and angles. A surveyor uses a transit to determine the measures of the acute angles between each of two reference points and the point where the transit is located. Calculations involving right-triangle trigo-nometry may then be appropriate to find the distance between the reference points.

Assessment

You might consider evaluating students on their knowledge of how an angle of elevation can be used to find the height of an object by having them do an experiment similar to the one described in Written Exercise 15, page 335. (Rather than find the height of a flagpole as in the exercise, students can find the height of a tree, building, or other tall object.)

1. Determine the angle of elevation θ of the top of the object by using a protractor.
2. Measure the height, h, of the student's eyes above the ground.
3. Measure the actual distance, d, along the ground from the student to the object.
4. Calculate the height of the object according to the equation: height = $d \tan \theta + h$.
5. If possible, measure the actual height of the object.

Communication

You might want to have students discuss some gen-eral guidelines for solving trigonometric problems that involve right triangles.

1. Read the problem carefully and draw a diagram that matches the given information.
2. Label angles and sides with known measures.
3. Choose the correct trigonometric function that can be used to solve for the unknown side or angle.
4. Use the trigonometric function to write an equa-tion and solve it.

9-2 | pages 339–344

Presenting the Section

You may want to begin this section by reviewing the conditions for triangle congruence from geometry, including SAS, ASA, and SSS. These conditions state that a triangle is determined uniquely if certain facts about the triangle are known. As you discuss the area formulas on page 339, point out that they involve the SAS condition. Since a triangle is deter-mined uniquely if two sides and the included angle are known, the area of the triangle can be found.

Using Technology

You can help students gain insight into the area formulas if you have them keep the lengths of the sides constant, say 7 and 4 as in Example 1, page 339, and then have them use calculators or computers to generate a table of values for the area K of the triangle as the angle between the two sides increases. Their tables should suggest that the area K increases as the angle increases from $0°$ to $90°$ and then decreases as the angle increases from $90°$ to $180°$. Use the table to show that the maximum value of the area occurs when the included angle between the sides is $90°$.

Cooperative Learning

You might want to use "jigsawing" to go over difficult problems, such as Exercises 32–36, page 344. Divide students into groups and give each group one of the exercises to do. Have the students work out the solution to the exercise individually and then gather as a group to discuss it. Each group should discuss the exercise until all students understand and agree on the solution. All groups should then split up and redistribute themselves so that new groups are formed with at least one student from each of the original groups. Each solution can then be presented to the members of the new group. Have students share ideas about how they arrived at the complete solution to each exercise, and how calculators or computers were used.

Assessment

You will need to decide ahead of time if you will accept approximate answers for some of the exercises in this section or if you want students to express answers exactly. For example, the exact answer for Exercise 11, page 342, is $3200\sqrt{2}$ cm^2. Students using calculators may instead give an approximation of that answer. Consider accepting either answer.

9-3 | pages 345–349

Presenting the Section

Emphasize that the law of sines is applicable to all triangles when either (1) two angles and any side are known, or (2) two sides and the angle opposite one

of them are known (called the "ambiguous case"). The first case always yields a triangle, but as Activities 1 and 2 on page 346 demonstrate, for the second case there are times when it is not possible to construct a unique triangle—or any triangle at all—from the given information. This is equivalent to the SSA situation that students studied in geometry.

Point out that when the SSA situation applies, it may be possible to construct two triangles, one triangle, or no triangle. The three possibilities mean that you need other information to determine specifically which one it is, or that you need to apply the law of cosines (Section 9-4) first. Note that the law of sines cannot be used to find the angle measures of a triangle if three sides (SSS) and no angle are given. (Use the law of cosines, Section 9-4.)

You may want to rewrite the area formulas on page 339 in terms of the law of sines. For example, if

$$K = \tfrac{1}{2}ab \sin C, \quad \text{then} \quad K = \tfrac{1}{2}\left(\frac{\sin A \sin B}{\sin C}\right)c^2.$$

Have students verify this formula and see if they can also write the equivalent area formulas for the other formulas on page 339. See Written Exercise 25, page 349.

You might want to show how the law of sines, when applied to right triangles, is consistent with the definition of the sine of an acute angle given in Section 9-1.

Using Technology

Students with programmable calculators can be encouraged to write a program to use the law of sines automatically. For example, given side b and angles A and B, the program can calculate a using

$$a = \frac{b \sin A}{\sin B}.$$

Students also can be encouraged to write a program that determines the number of solutions of a triangle for which the lengths of two sides and the angle opposite one of them are known.

Geometry software that constructs triangles can be used to demonstrate the various situations to which the law of sines applies (that is, the AAS, ASA, and SSA cases). Have students input various lengths of sides and measures of angles to test each case. Have the computer construct the triangles, if possible, and give the other angle and side measurements for the triangles.

Cooperative Learning

You might want to use a group environment to complete Activities 1 and 2, page 346, and Written Exercise 1, page 347. Provide compasses, if necessary, and have students do the constructions together so that they can discuss the results, criticize solutions, and offer suggestions.

Applications

Triangulation is a technique involving the law of sines that is used to determine the location of an object. If an object is illuminated by a beam of light from two different sources, then the angle that each beam makes with the ground can be calculated and the law of sines can be applied to determine the distance to the object. This method also can be modified to calculate distances to objects that are very far away, such as objects in space.

Assessment

To evaluate students' ability to solve problems using the law of sines, give them a list of problems similar to Written Exercises 3–14, pages 347–348, and ask them to give an expression that will return the solution if it is entered in a calculator.

9-4 | pages 350–354

Presenting the Section

The law of cosines allows you to solve a triangle when you have the SAS situation and you need to find the length of the third side, or when you have the SSS situation and you need to find the measure of any angle. Point out that the law of cosines is not adequate when the given parts of the triangle involve the ASA, AAS, or SSA situations. The law of sines must be used in these cases.

Emphasize that the Pythagorean theorem is a special case of the law of cosines when finding the length of the hypotenuse of a right triangle. Since $\cos 90° = 0$, the cosine term in the law of cosines equals 0 and the law reduces to the Pythagorean theorem.

Using Technology

You might want to encourage students with programmable calculators to reduce the number of computations they do by writing a program that solves law-of-cosines problems. This program should handle the SAS and SSS situations separately.

Cooperative Learning

Groups of students can work to write their own joint computer program that can be used to solve a triangle, given some combination of sides and angles. See the Computer Exercise on page 355 for details.

Applications

The law of cosines can be used in many applications dealing with triangles, including applied problems using vectors (see Chapter 12) and finding the area of a triangle for which three sides are known. For example, if a triangular sail measures 10 ft × 15 ft × 12 ft, its area can be found by first using the law of cosines to determine an angle of the triangle and then substituting the sine of the angle into one of the area formulas given on page 339.

9-5 | pages 359–364

Presenting the Section

Make sure you stress how angles are measured in each type of application presented in this section. In navigation applications, north is 0° and the bearing of the ship or airplane is measured clockwise from the north line. In surveying applications, the angle measures are east or west of the north-south line and are therefore usually given as acute angles. Show students the actual measuring instruments if available, such as magnetic compasses. Historical instruments in a museum might also be of interest.

Cooperative Learning

Have students work in groups to go over the exercises in this section. Have them check each other's diagrams and discuss the concepts of ''course'' and ''bearing'' as they relate to the problems. You might want groups to be responsible for discussing only certain problems and then have students from each group explain the problems to the class.

9 Triangle Trigonometry

330

9-1 Solving Right Triangles

| Objective | To use trigonometry to find unknown sides or angles of a right triangle. |

In Chapter 7, we defined the trigonometric functions in terms of coordinates of points on a circle. In this chapter, our emphasis shifts from circles to triangles. When certain parts (sides and angles) of a triangle are known, you will see that trigonometric relationships can be used to find the unknown parts. This is called **solving a triangle**. For example, if you know the lengths of the sides of a triangle, then you can find the measures of its angles. In this section, we will consider how trigonometry can be applied to right triangles.

The right triangles shown in the diagrams at the right both have an acute angle of measure θ and are, therefore, similar. Consequently, the lengths of corresponding sides are proportional, and we have the equations shown below:

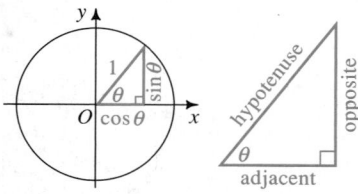

$$\frac{\sin \theta}{1} = \frac{\text{opposite}}{\text{hypotenuse}} \qquad \frac{\cos \theta}{1} = \frac{\text{adjacent}}{\text{hypotenuse}} \qquad \tan \theta = \frac{\sin \theta}{\cos \theta} = \frac{\text{opposite}}{\text{adjacent}}$$

By the reciprocal relationships (see page 282), we also have:

$$\csc \theta = \frac{\text{hypotenuse}}{\text{opposite}} \qquad \sec \theta = \frac{\text{hypotenuse}}{\text{adjacent}} \qquad \cot \theta = \frac{\text{adjacent}}{\text{opposite}}$$

Applications of these equations are given in the following examples. In Example 1, notice the convention of using a capital letter to denote an angle and the corresponding lower-case letter to denote the length of the side opposite that angle.

Example 1 For the right triangle ABC shown at the right, find the value of b to three significant digits.

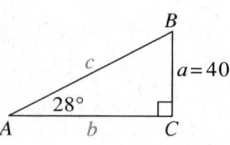

Solution To find the value of b, use either $\tan 28°$ or $\cot 28°$.

Using $\tan 28°$:

$$\tan 28° = \frac{\text{opposite}}{\text{adjacent}} = \frac{40}{b}$$

$$b = \frac{40}{\tan 28°} \approx 75.2$$

Using $\cot 28°$:

$$\cot 28° = \frac{\text{adjacent}}{\text{opposite}} = \frac{b}{40}$$

$$b = 40 \cot 28° \approx 75.2$$

◄ Triangle trigonometry is especially useful in navigation. The pilot of the ship shown can use it to find the height of the mountains around the ship, to measure the width of the channel ahead, and to plot the safest course along that channel.

Triangle Trigonometry **331**

Warm-Up Exercises

Find the value of x.

1.

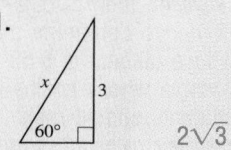

$2\sqrt{3}$

2.

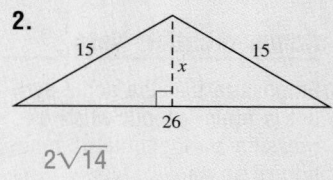

$2\sqrt{14}$

3. $x = \text{Sin}^{-1} 0.9 \approx 64.2°$

4. $x = \text{Tan}^{-1} 4.5 \approx 77.5°$

5. $\frac{x}{20} = \sin 30° \quad 10$

6. $\frac{x}{12} = \cos 75° \approx 3.11$

Motivating the Section

Being able to find the length of each side of a triangle and the measure of each angle is important in triangular bracing, the use of triangles to support construction in homes and buildings. Right-triangle trigonometry, the topic of this section, allows students to solve a right triangle completely.

◪ Using Technology

Students who do not have secant, cosecant, and cotangent keys on their calculators should use the reciprocal key along with the cosine, sine, and tangent keys.

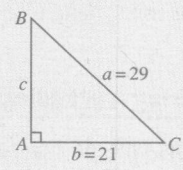
Example 2 The safety instructions for a 20 ft ladder indicate that the ladder should not be inclined at more than a 70° angle with the ground. Suppose the ladder is leaned against a house at this angle, as shown. Find **(a)** the distance x from the base of the house to the foot of the ladder and **(b)** the height y reached by the ladder.

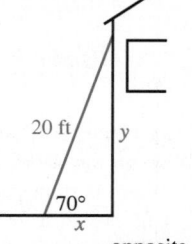

Solution

a. $\cos 70° = \dfrac{\text{adjacent}}{\text{hypotenuse}} = \dfrac{x}{20}$

$x = 20 \cos 70° \approx 6.84$

The foot of the ladder is about 6.84 ft from the base of the house.

b. $\sin 70° = \dfrac{\text{opposite}}{\text{hypotenuse}} = \dfrac{y}{20}$

$y = 20 \sin 70° \approx 18.8$

The ladder reaches about 18.8 ft above the ground.

Example 3 The highest tower in the world is in Toronto, Canada, and is 553 m high. An observer at point A, 100 m from the center of the tower's base, sights the top of the tower. The *angle of elevation* is $\angle A$. Find the measure of this angle to the nearest tenth of a degree.

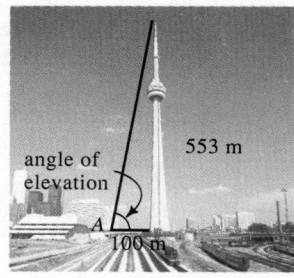

Solution

$\tan A = \dfrac{\text{opposite}}{\text{adjacent}} = \dfrac{553}{100} = 5.53$

$\angle A = \text{Tan}^{-1} \, 5.53 \approx 79.7°$

$\begin{cases} \tan A \text{ is an abbreviation} \\ \text{for the tangent of } \angle A. \end{cases}$

Because we can divide an isosceles triangle into two congruent right triangles, we can apply trigonometry to isosceles triangles, as shown in the following example.

Example 4 A triangle has sides of lengths 8, 8, and 4. Find the measures of the angles of the triangle to the nearest tenth of a degree.

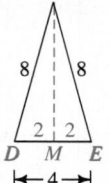

Solution By drawing the altitude to the base of isosceles triangle DEF, we get two congruent right triangles. In $\triangle DMF$, we have:

$\cos D = \dfrac{\text{adjacent}}{\text{hypotenuse}} = \dfrac{2}{8} = 0.25$

$\angle D = \text{Cos}^{-1} \, 0.25 \approx 75.5°$

Thus, $\angle E = \angle D \approx 75.5°$, and $\angle F \approx 180° - 2(75.5°) = 29.0°$.

When calculators or trigonometric tables are used, only approximate results are possible in most cases. Throughout this chapter, therefore, you should round angle measures to the nearest tenth of a degree and give lengths accurate to three significant digits.

CLASS EXERCISES

1. Match each element of row A with an element in row B.

A: sine cosine tangent cotangent secant cosecant

B: $\dfrac{\text{opposite}}{\text{adjacent}}$ $\dfrac{\text{opposite}}{\text{hypotenuse}}$ $\dfrac{\text{adjacent}}{\text{opposite}}$ $\dfrac{\text{hypotenuse}}{\text{opposite}}$ $\dfrac{\text{adjacent}}{\text{hypotenuse}}$ $\dfrac{\text{hypotenuse}}{\text{adjacent}}$

2. Refer to the diagram at the right.
   a. Express the sine, cosine, and tangent of α in terms of x, y, and z.
   b. Express the sine, cosine, and tangent of β in terms of x, y, and z.
   c. Tell whether each of the following statements is true or false.

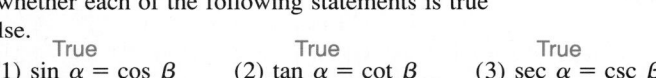

         True   True   True
      (1) $\sin \alpha = \cos \beta$ (2) $\tan \alpha = \cot \beta$ (3) $\sec \alpha = \csc \beta$

3. **Reading** In the solution of Example 1, suppose a calculator is not available. Would you prefer to find the value of b using $\tan 28°$ or $\cot 28°$? Explain.

4. a. For right triangle ABC in Example 1, use trigonometry to find the value of c to three significant digits. 85.2
   b. Knowing the values of a and b in Example 1, how could you find the value of c without using trigonometry? Use the Pythagorean theorem.

5. For each triangle, state two equations (using reciprocal trigonometric functions) that can be used to find the value of x.

   a. 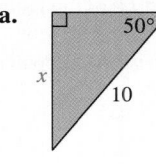   b. 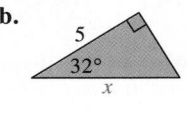   c. 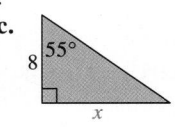   d.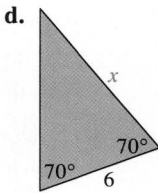

6. For each triangle, state two equations (using reciprocal trigonometric functions) that can be used to find the measure of θ.

   a. 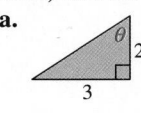   b. 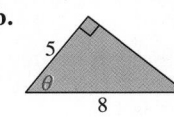   c. 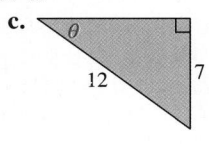   d.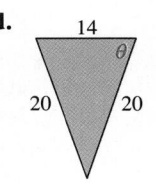

Triangle Trigonometry   **333**

2. A rhombus with perimeter 40 cm has a 70° angle. Find the lengths of its diagonals.

   Draw a sketch. Remember that the diagonals of a rhombus are perpendicular to and bisect each other. They also bisect the angles of the rhombus.

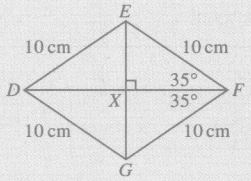

$$\sin 35° = \frac{EX}{10}$$
$$EX = 10 \sin 35°$$
$$EG = 20 \sin 35° \approx 11.5 \text{ cm}$$
$$\cos 35° = \frac{FX}{10}$$
$$FX = 10 \cos 35°$$
$$DF = 20 \cos 35° \approx 16.4 \text{ cm}$$

Mathematical Note

The paragraph at the top of this page addresses, in a broad way, the issue of accuracy of results. Generally speaking, results should be no more accurate than the given information. Be sure students understand that even though a calculator can display 8 or more digits, all these digits may not belong in the answer.

◺ Using Technology

Have students use their calculators to verify each of the statements in Class Exercise 2c for various values of α and β (where α and β are complementary angles).

Encourage students to memorize the exact values of the trigonometric functions for the special angles 30°, 45°, and 60°.

angle	30°	45°	60°
sine	$\dfrac{1}{2}$	$\dfrac{\sqrt{2}}{2}$	$\dfrac{\sqrt{3}}{2}$
cosine	$\dfrac{\sqrt{3}}{2}$	$\dfrac{\sqrt{2}}{2}$	$\dfrac{1}{2}$
tangent	$\dfrac{\sqrt{3}}{3}$	1	$\sqrt{3}$

Additional Answers
Written Exercises

5. a. $\dfrac{5}{13}$ **b.** $\dfrac{5}{13}$ **c.** $\dfrac{5}{12}$

d. $\dfrac{5}{12}$ **e.** $\dfrac{13}{12}$ **f.** $\dfrac{13}{12}$

10. a. 1, $\sqrt{3}$

b. $\dfrac{1}{2}$; $\dfrac{\sqrt{3}}{2}$; $\dfrac{\sqrt{3}}{3}$; $\sqrt{3}$

11. The length of each segment is the value of the tangent of the corresponding angle. For example, $\tan 10° = \dfrac{CB_1}{1} = CB_1$.

Suggested Assignments

Standard
Day 1: 334/1–7 odd,
13–21 odd
Day 2: 336/23, 25, 27, 31, 33, 34, 36, 37
Comprehensive
Day 1: 334/1, 7, 13–25 odd
Day 2: 337/31, 33, 35, 37, 39, 40

Supplementary Materials

Alternative Assessment, 27

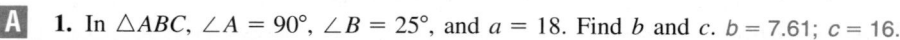

WRITTEN EXERCISES

Throughout the exercises, give angle measures to the nearest tenth of a degree and lengths to three significant digits.

A **1.** In $\triangle ABC$, $\angle A = 90°$, $\angle B = 25°$, and $a = 18$. Find b and c. $b = 7.61$; $c = 16.3$

2. In $\triangle PQR$, $\angle P = 90°$, $\angle Q = 64°$, and $p = 27$. Find q and r. $q = 24.3$; $r = 11.8$

3. In $\triangle DEF$, $\angle D = 90°$, $\angle E = 12°$, and $e = 9$. Find d and f. $d = 43.3$; $f = 42.3$

4. In $\triangle XYZ$, $\angle X = 90°$, $\angle Y = 37°$, and $z = 25$. Find x and y.

5. Use the diagram at the right to find: $x = 31.3$; $y = 18.8$

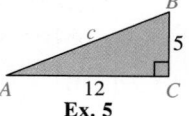

   a. $\sin A$ **b.** $\cos B$ **c.** $\tan A$
   d. $\cot B$ **e.** $\sec A$ **f.** $\csc B$

Ex. 5

6. Sketch $\triangle ABC$ with $\angle C = 90°$. What is the relationship between:
   a. $\sin A$ and $\cos B$? =   **b.** $\tan A$ and $\cot B$? =   **c.** $\sec A$ and $\csc B$? =

7. Find the measures of the acute angles of a 3-4-5 right triangle. 36.9°, 53.1°

8. Find the measures of the acute angles of a right triangle whose legs are 9 cm and 16 cm long. 29.4°, 60.6°

9. The legs of an isosceles right triangle are 1 unit long.
   a. Find the length of the hypotenuse in simplest radical form. $\sqrt{2}$
   b. Use part (a) to find the exact value of each of the following. 1; $\dfrac{\sqrt{2}}{2}$; $\dfrac{\sqrt{2}}{2}$
      (1) $\tan 45°$ (2) $\sin 45°$ (3) $\cos 45°$
   c. Convert the answers in part (b) to decimal form. Compare these with the values of $\tan 45°$, $\sin 45°$, and $\cos 45°$ obtained from the calculator.

10. The hypotenuse of a 30°-60°-90° triangle is 2 units long.
   a. Find the lengths of the legs in simplest radical form.
   b. Use part (a) to find the exact value of each of the following.
      (1) $\sin 30°$ (2) $\sin 60°$ (3) $\tan 30°$ (4) $\tan 60°$
   c. Convert the answers in part (b) to decimal form. Compare these with the values of $\sin 30°$, $\sin 60°$, $\tan 30°$, and $\tan 60°$ obtained from the calculator.

📺 **For Exercise 11, you may wish to use a computer with software that draws and measures geometric figures.**

11. Draw a horizontal segment AC of length 1 unit and a vertical ray CB, as shown. Then draw segments that make angles of 10°, 20°, . . . , 80° with $\overline{AC}$. Measure the lengths CB_1, CB_2, . . . , CB_8. What is the significance of these lengths?

12. In the diagram that you drew for Exercise 11, note that $\angle CAB_2$ is twice as large as $\angle CAB_1$. Is $\overline{CB_2}$ twice as long as $\overline{CB_1}$? Is $\overline{CB_4}$ four times as long as $\overline{CB_1}$? No; no

13. **Aviation** An airplane is at an elevation of 35,000 ft when it begins its approach to an airport. Its *angle of descent* is 6°.

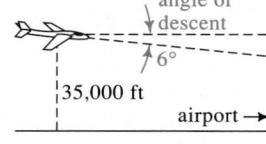

   a. What is the distance between the airport and the point on the ground directly below the airplane? 333,000 ft

   b. What is the approximate air distance between the plane and the airport? What assumptions did you make in finding this distance? 335,000 ft

14. **Navigation** A lighthouse keeper observes that there is a 3° *angle of depression* between the horizontal and the line of sight to a ship. If the keeper is 19 m above the water, how far is the ship from shore? 363 m

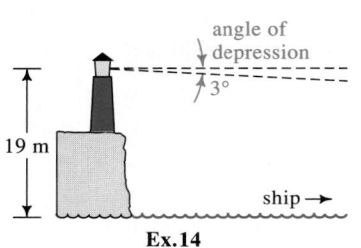

Ex. 14

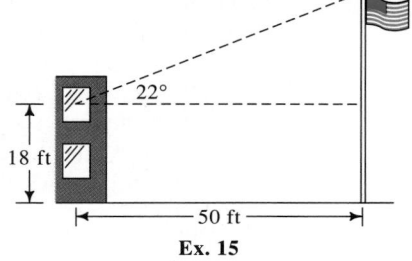

Ex. 15

15. A student looks out of a second-story school window and sees the top of the school flagpole at an angle of elevation of 22°. The student is 18 ft above the ground and 50 ft from the flagpole. Find the height of the flagpole. 38.2 ft

16. For an observer at point A, 250 m from a building, the angle of elevation of the top of the building is 5°. In Chapter 7, we said that $\triangle ABC$ is about the same as a sector with central angle A.

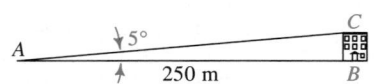

   a. Use the arc length formula $s = r\theta$ to approximate BC. (Remember to express θ in radians.) 21.8 m

   b. Use right-triangle trigonometry to find BC more accurately. Compare your answers. 21.9 m; approximately the same

17. Find the measures of the angles of an isosceles triangle whose sides are 6, 6, and 8. Also find the area of the triangle. 48.2°, 48.2°, 83.6°; area = 17.9

18. The legs of an isosceles triangle are each 21 cm long and the angle between them has measure 52°. What is the length of the third side? 18.4 cm

B 19. In the figure at the right, $\overline{PA}$ and $\overline{PB}$ are tangents to a circle with radius $OA = 6$. If the measure of $\angle APB$ is 42°, find PA and PB. $PA = PB = 15.6$

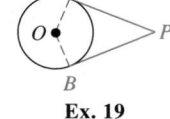

Ex. 19

20. Sketch the circle $(x - 6)^2 + (y - 8)^2 = 9$ and the two tangents to the circle from the origin. Find the measure of the angle between the tangents. 34.9°

Communication Note

An *angle of descent* and an *angle of depression* are determined in the same way. An angle of depression is the angle formed by the line of sight with the horizontal when looking down at an object.

Mathematical Note

If an observer at *A* is higher than an observer at *B*, then the angle of depression at *A* (when *B* is observed) and the angle of elevation at *B* (when *A* is observed) are congruent alternate interior angles (because horizontal lines at *A* and *B* are parallel).

Problem Solving

Encourage students to draw diagrams as accurately as possible and to label them carefully. Accurate diagrams help students keep track of given information so that they can apply the trigonometric relationships correctly.

Error Analysis

Drawing a diagram showing an angle of elevation or an angle of depression is often difficult for students. Encourage them to draw a horizontal line (which represents looking straight ahead) first and then draw either a line of sight with *positive* slope (that is, a *raised* line of sight) for an angle of elevation or a line of sight with *negative* slope (that is, a *lowered* line of sight) for an angle of depression.

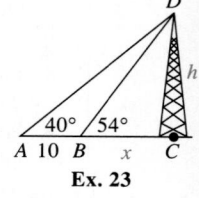

Exercise Note

For Exercise 27b, you might want students to find the other trigonometric functions of $\angle A$ in terms of x:

$$\csc A = \frac{1}{x}$$

$$\cos A = \sqrt{1 - x^2}$$

$$\sec A = \frac{1}{\sqrt{1 - x^2}}$$

$$\tan A = \frac{x}{\sqrt{1 - x^2}}$$

$$\cot A = \frac{\sqrt{1 - x^2}}{x}$$

Students can do the same for Exercise 28b, page 337.

---

21. An isosceles trapezoid has sides whose lengths are in the ratio 5 : 8 : 5 : 14. Find the measure of the angle between one of the legs and the shorter base. (*Hint:* Draw an altitude.) 126.9°

22. A rectangle is 14 cm wide and 48 cm long. Find the measure of the acute angle between its diagonals. 32.5°

23. From points A and B, 10 m apart, the angles of elevation of the top of a tower are 40° and 54°, respectively, as shown at the right. Find the tower's height. (*Hint:* Write two equations in the unknowns x and h.) 21.5 m

Ex. 23

24. Navigation From a ship off-shore, the angle of elevation of a hill is 1.1°. After the ship moves inland at 4.5 knots for 20 min, the angle of elevation is 1.4°. How high is the hill? (1 knot = 1 nautical mile per hour ≈ 6080 feet per hour) 817 ft

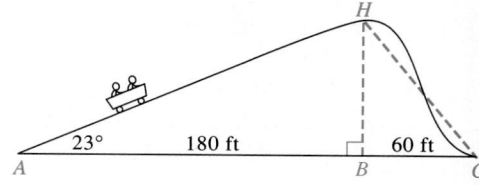

25. Physics The roller coaster car shown in the diagram takes 23.5 s to go up the 23° incline $\overline{AH}$ and only 2.8 s to go down the drop from H to C. The car covers horizontal distances of 180 ft on the incline and 60 ft on the drop.

a. How high is the roller coaster above point B? 76.4 ft
b. Find the distances AH and HC. AH = 196 ft; HC = 97.1 ft
c. How fast (in ft/s) does the car go up the incline? 8.3 ft/s
d. What is the approximate average speed of the car as it goes down the drop? (Assume that the car travels along $\overline{HC}$. Since the actual path is longer than $\overline{HC}$, is your approximate answer too big or too small?) 34.7 ft/s; too small

26. Space Science From a point 1.5 mi from a launch pad at Cape Canaveral, an ob-server sights a space shuttle at an angle of elevation of 10° moments after it is launched. After 10 s, the angle of elevation is 50°. **a.** 1.52 mi **b.** 0.152 mi/s

a. How far does the space shuttle travel vertically during the 10 s interval?
b. What is the average speed of the space shuttle during that time interval?

27. a. Draw $\triangle ABC$ with $\angle C = 90°$ and $\sin A = \frac{5}{13}$. Without using a calculator or table, find sec A and csc A. **a.** $\frac{13}{12}$; $\frac{13}{5}$ **b.** $\frac{1}{\sqrt{1 - x^2}}$
b. Given that $\sin A = x$, find sec A in terms of x.

336 *Chapter Nine*

28. a. Draw $\triangle PQR$ with $\angle R = 90°$ and $\tan P = \frac{1}{4}$. Without using a calculator or table, find $\sin P$ and $\cos Q$. **a.** $\frac{\sqrt{17}}{7}; \frac{\sqrt{17}}{7}$ **b.** $\frac{x}{\sqrt{1+x^2}}$

b. Given that $\tan P = x$, find $\sin P$ in terms of x.

29. A line tangent to the circle $x^2 + y^2 = 1$ at P intersects the axes at T and S, as shown in the diagram.

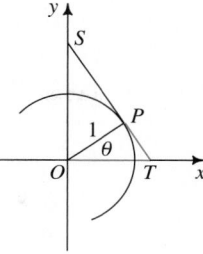

a. Show that $PT = \tan \theta$ and $OT = \sec \theta$. (This may explain the use of the names *tangent* and *secant*. Another interpretation is given in Exercise 37 on page 322.)

b. Show that $\sec^2 \theta = 1 + \tan^2 \theta$.

c. *Visual Thinking* Describe what happens to point T as θ increases from 0° to 90°, from 90° to 180°, from 180° to 270°, and from 270° to 360°. Your answers should agree with the graphs of $y = \tan \theta$ and $y = \sec \theta$.

30. a. Use the diagram in Exercise 29 to show that $PS = \cot \theta$ and $OS = \csc \theta$.

b. Show that $\csc^2 \theta = 1 + \cot^2 \theta$.

c. *Visual Thinking* Describe what happens to point S as θ increases from 0° to 90°, from 90° to 180°, from 180° to 270°, and from 270° to 360°. Your answers should agree with the graphs of $y = \cot \theta$ and $y = \csc \theta$.

31. Use the diagram at the left below. Express x and y in terms of α, β, and a. $x = a \sin \beta \cot \alpha$; $y = a \sin \beta \csc \alpha$

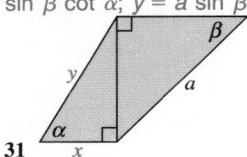

Ex. 31

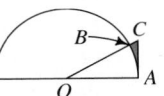

Ex. 32

32. Use the diagram at the right above. Show that:

a. $h = a \sin B = b \sin A$ **b.** area of $\triangle ABC = \frac{1}{2}ac \sin B = \frac{1}{2}bc \sin A$

33. In the diagram at the right, $\overline{AC}$ is tangent to a semicircle with radius $OA = 12$ cm. If the measure of $\angle AOB$ is 28°, find to the nearest square centimeter the area of the region that is inside $\triangle AOC$ and outside the circle. 3 cm^2

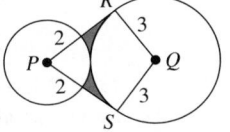

34. As shown, two circles with radii 2 and 3 and centers P and Q, respectively, are externally tangent. From P, tangents $\overline{PR}$ and $\overline{PS}$ are drawn to the larger circle.

a. Find the measures of $\angle RPS$ and $\angle RQS$ to the nearest hundredth of a radian. $\angle RPS = 1.29$; $\angle RQS = 1.85$

b. Find the area of the shaded region to the nearest hundredth of a square unit. 1.08

35. Each leg of an isosceles triangle is 8 cm long. Express the measure of the triangle's vertex angle as a function of the length of the triangle's base. Give an appropriate domain and range for this function.

36. The diagonals of a rectangle are 20 cm long. Express the measure of the acute angle formed by the diagonals as a function of the length of the rectangle's shorter sides. Give an appropriate domain and range for this function.

Have students use their calculators to verify the identities in Exercises 29b and 30b for various values of θ. Trigonometric identities will be discussed in Chapter 10.

Additional Answers
Written Exercises

32. a. Since $\sin A = \frac{h}{b}$, $h = b \sin A$. Also, since $\sin B = \frac{h}{a}$, $h = a \sin B$. Thus, $h = b \sin A = a \sin B$.

b. Area of $\triangle ABC = \frac{1}{2}$(base)(height) $= \frac{1}{2}ch$. Since $h = b \sin A = a \sin B$, area of $\triangle ABC = \frac{1}{2}ac \sin B = \frac{1}{2}bc \sin A$.

35. $\theta = 2 \operatorname{Sin}^{-1}\left(\frac{b}{16}\right)$, where b is the length in centimeters of the triangle's base and θ is the measure of the triangle vertex angle; $0 < b < 16$, $0° < \theta < 180°$

36. $\theta = 2 \operatorname{Sin}^{-1}\left(\frac{s}{20}\right)$, where s is the length in centimeters of the rectangle's shorter side and θ is the measure of the acute angle formed by the rectangle's diagonals; $0 < s < 10\sqrt{2}$, $0° < \theta < 90°$

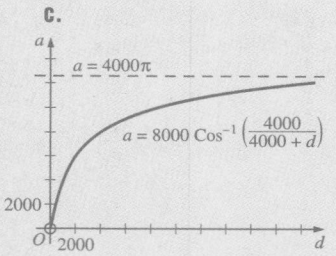
37. Aviation An airplane at an elevation of 30,000 ft begins to descend toward the runway on which it will land. Let d be the horizontal distance between the plane and the runway. Let θ be the plane's angle of descent (see Exercise 13).
  a. Express θ as a function of d. See below.
  b. Suppose that the horizontal distance between the airplane and the runway is 60 mi. Find to the nearest tenth the angle at which the airplane must descend in order to land on the runway. (Recall that 1 mi = 5280 ft.) 5.4°

37. a. Answers may vary. Example: $\theta = \text{Tan}^{-1} \left(\frac{30,000}{d} \right)$

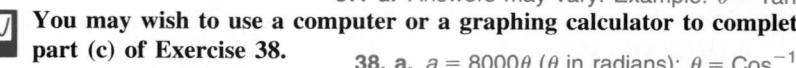 You may wish to use a computer or a graphing calculator to complete part (c) of Exercise 38.

38. a. $a = 8000\theta$ (θ in radians); $\theta = \text{Cos}^{-1} \left(\frac{4000}{4000 + d} \right)$

38. Space Science A telecommunications satellite S orbits above the earth. Suppose the shaded region, determined by tangents such as $\overline{SA}$ and $\overline{SB}$, is the portion of the earth's surface that can send and receive signals via the satellite. Assume point E to be the center of the earth and the radius of the earth to be 4000 mi.
  a. Arc APB is an arc of a great circle (see Exercise 46 on page 301). The length of arc APB is the maximum distance a between two cities that can communicate via the satellite. Express a in terms of the central angle θ, and express θ in terms of the satellite's distance d above the earth's surface. See above.
  b. Use your answers in part (a) to express a as a function of d.
  c. Give an appropriate domain and range for your function in part (b) and sketch its graph. $d > 0$, $0 < a < 4000\pi$
  d. Many satellites are in *geostationary* orbits about 22,300 mi above the earth's surface. At this distance, the satellite's speed matches that of the earth's rotation, causing the satellite to appear stationary to an observer on the earth. If satellite S is in a geostationary orbit, can it transmit signals from Lima, Peru, to Cairo, Egypt, which is about 7726 mi away? Yes

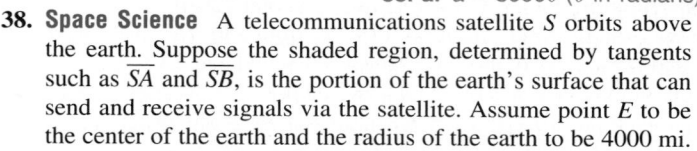

C **39.** Find the measure of the angle θ formed by a diagonal of a cube and a diagonal of one of the faces of the cube. 35.3°

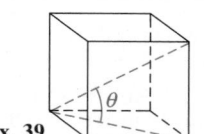
Ex. 39

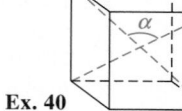
Ex. 40

40. Find the measure of the angle α formed by two diagonals of a cube. 109.5°

41. Derive a formula for the area A of a regular polygon of n sides inscribed in a circle of radius r. $A = nr^2 \sin\theta \cos\theta$

42. Derive a formula for the area K of a regular polygon of n sides circumscribed about a circle of radius r. $K = nr^2 \tan\theta$

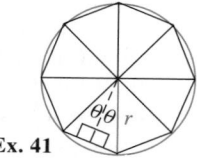
Ex. 41

Ex. 42

//// COMPUTER EXERCISE

Use the answers from Exercises 41 and 42 to write a program that prints the values of A, K, and $K - A$ for $r = 1$ and $n = 10, 20, 30, \ldots, 100$. Interpret the results.

338   *Chapter Nine*

9-2 The Area of a Triangle

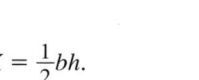 To find the area of a triangle given the lengths of two sides and the measure of the included angle.

When the lengths of two sides of a triangle and the measure of the included angle are known, the triangle is uniquely determined. This fact is a consequence of the side-angle-side (SAS) condition for congruence. We now consider the problem of expressing the area of the triangle in terms of these measurements.

Suppose that we are given the lengths a and b and the measure of $\angle C$ in $\triangle ABC$. If the length of the altitude from B is h, then the area of the triangle is given by

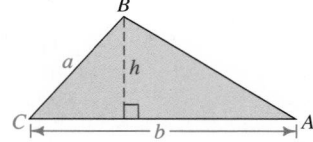

$$K = \frac{1}{2}bh.$$

By right-triangle trigonometry, we know that $\sin C = \frac{h}{a}$, or

$$h = a \sin C.$$

Substituting $a \sin C$ for h, we find that

$$K = \frac{1}{2}ab \sin C.$$

If some other pair of sides and the included angle of $\triangle ABC$ were known, we could repeat the procedure for finding the area and thereby obtain two other area formulas. All three formulas are stated below.

The Area of a Triangle

The area K of $\triangle ABC$ is given by:

$$K = \frac{1}{2}ab \sin C = \frac{1}{2}bc \sin A = \frac{1}{2}ac \sin B$$

Notice that the area formulas all have the basic pattern:

$$K = \frac{1}{2} \cdot (\text{one side}) \cdot (\text{another side}) \cdot (\text{sine of included angle})$$

Example 1   Two sides of a triangle have lengths 7 cm and 4 cm. The angle between the sides measures 73°. Find the area of the triangle.

Solution $K = \frac{1}{2} \cdot 7 \cdot 4 \cdot \sin 73° \approx 13.4$

Thus, the area is about 13.4 cm².

Triangle Trigonometry   **339**

Teaching Notes, p. 330B

Warm-Up Exercises

Evaluate each expression. Give answers in radical form.

1. $\frac{1}{2} \cdot 12 \cdot 7 \cdot \sin 135°$ $21\sqrt{2}$

2. $\frac{1}{2} \cdot 9 \cdot 6 \cdot \sin 60°$ $\frac{27\sqrt{3}}{2}$

Solve each equation for $0° \leq \theta < 180°$. Give answers to the nearest tenth of a degree.

3. $\frac{1}{2} \cdot 10 \cdot 8 \cdot \sin \theta = 30$
   48.6°, 131.4°

4. $\frac{1}{2} \cdot 6 \cdot 6 \cdot \sin \theta = 12$
   41.8°, 138.2°

Motivating the Section

Being able to find the area of a triangle as a function of two sides and the included angle is useful in certain engineering applications. In this section, the area of a triangle and engineering applications are presented.

Additional Examples

1. Find the exact area of a regular hexagon inscribed in a unit circle. Then approximate the area to three significant digits.

(*Continues on next page*)

Additional Examples cont.

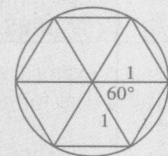

The diagonals divide the inscribed hexagon into six congruent triangles.

$K = 6\left(\frac{1}{2} \cdot 1 \cdot 1 \cdot \sin 60°\right) =$

$3 \cdot \frac{\sqrt{3}}{2} = \frac{3}{2}\sqrt{3}$ sq. units

$K \approx 2.60$ sq. units

2. Adjacent sides of a parallelogram have lengths 12.5 cm and 8 cm. The measure of the included angle is 40°. Find the area of the parallelogram to three significant digits.

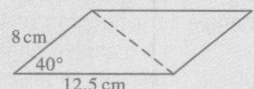

The two triangles are congruent. Thus:

$K =$

$2\left(\frac{1}{2} \cdot 8 \cdot 12.5 \cdot \sin 40°\right) \approx$

$100(0.6428) \approx 64.3$ cm^2

◪ Using Technology

For Example 3, you might want students to use the zoom and trace features of a computer or graphing calculator to verify that the graph of

$y = \alpha - \sin \alpha - \frac{\pi}{2}$ crosses

the α-axis at $\alpha \approx 2.3$. This represents an angle measure of 2.3 radians.

Example 2 The area of $\triangle PQR$ is 15. If $p = 5$ and $q = 10$, find all possible measures of $\angle R$.

Solution Since $K = \frac{1}{2}pq \sin R$,

$$15 = \frac{1}{2} \cdot 5 \cdot 10 \cdot \sin R = 25 \sin R.$$

Thus, $\sin R = \frac{15}{25} = 0.6$. Since $\angle R$ could be acute *or* obtuse,

$$\angle R = \text{Sin}^{-1} 0.6 \approx 36.9° \quad \text{or} \quad \angle R \approx 180° - 36.9° = 143.1°$$

The diagrams below show the two possibilities for $\triangle PQR$.

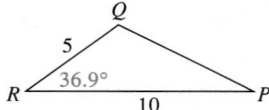   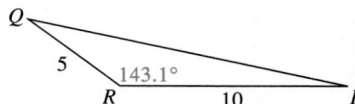

As shown at the right, a **segment** of a circle is the region bounded by an arc of the circle and the chord connecting the endpoints of the arc.

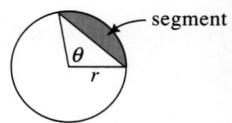

◪ Next we use a computer or graphing calculator to solve an equation based on the area of a segment.

Example 3 The diagram shows an end view of a cylindrical oil tank with radius 1 m. Through a hole in the top, a vertical rod is lowered to touch the bottom of the tank. When the rod is removed, the oil level in the tank can be read from the oil mark on the rod. Where on the rod should marks be put to show that the tank is $\frac{3}{4}$ full, $\frac{1}{2}$ full, and $\frac{1}{4}$ full?

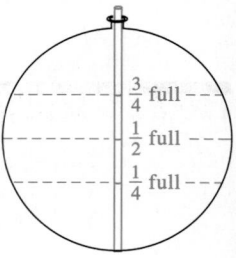

Solution Since the radius of the tank is $r = 1$ m, we can use the unit circle $x^2 + y^2 = 1$, shown at the right. To find where to put the "$\frac{3}{4}$ full" mark, we need to find the horizontal line $y = k$ that cuts off a segment having $\frac{1}{4}$ of the circle's area, or

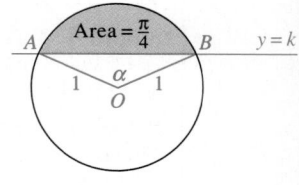

$$\frac{1}{4}\pi r^2 = \frac{1}{4}(\pi \cdot 1^2) = \frac{\pi}{4}.$$

340   *Chapter Nine*

Area of segment = area of sector AOB − area of $\triangle AOB$

$$\frac{\pi}{4} = \frac{1}{2} \cdot 1^2 \cdot \alpha - \frac{1}{2} \cdot 1^2 \cdot \sin \alpha$$

$$\frac{\pi}{4} = \frac{1}{2}(\alpha - \sin \alpha)$$

$$\frac{\pi}{2} = \alpha - \sin \alpha$$

$$0 = \alpha - \sin \alpha - \frac{\pi}{2}$$

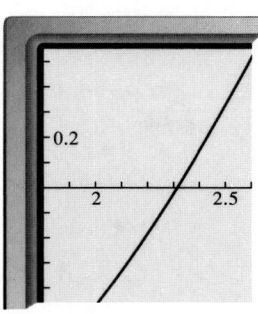

To solve the last equation, we graph $y = \alpha - \sin \alpha - \frac{\pi}{2}$ using a computer or graphing calculator. From the graph at the right, we see that $y = 0$ when $\alpha \approx 2.3$. Using this value of α and the diagram at the right, we have:

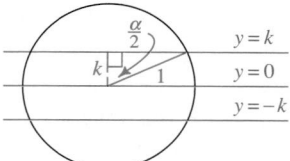

$$k = \cos \frac{\alpha}{2}$$

$$\approx \cos \frac{2.3}{2} \approx 0.4$$

From the symmetry of the circle we deduce that the horizontal lines dividing the circle into four parts with equal areas have equations $y \approx 0.4$, $y = 0$, and $y \approx -0.4$. Thus, if one end of a vertical rod is at the bottom of the tank, then marks should be placed 0.6 m, 1 m, and 1.4 m from that end to indicate when the tank is $\frac{1}{4}$ full, $\frac{1}{2}$ full, and $\frac{3}{4}$ full, respectively.

CLASS EXERCISES

1. Two adjacent sides of a triangle have lengths 5 cm and 8 cm.
   a. If these sides form a 30° angle, what is the area of the triangle? 10 cm²
   b. If these sides form a 150° angle, what is the area of the triangle? 10 cm²

Find the area of each triangle.

2.
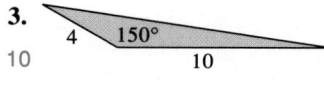

3.
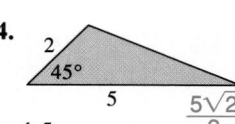

4.
$\frac{5\sqrt{2}}{2}$

5. A triangle with area 5 cm² has two sides of lengths 4 cm and 5 cm.
   a. Find the sine of the angle included between these sides. 0.5
   b. Find the two possible measures of the included angle. 30°, 150°

6. Find the area of a segment of a circle with radius 2 if the measure of the central angle of the segment is $\frac{\pi}{6}$. $\frac{\pi}{3} - 1 \approx 0.047$

Exercise Note

Encourage students to do Class Exercises 1–5 without the use of a calculator or table.

Mathematical Note

In Written Exercises 1–4, if two triangles have two pairs of corresponding sides equal and the corresponding included angles are supplementary, then their areas will be equal. This result will be used later when applying the law of sines to find the included angle of a triangle when the area is given.

Additional Answers
Written Exercises

15. Area $K = 6 \sin \theta$;
domain = $\{\theta | 0 < \theta < \pi\}$;
range = $\{K | 0 < K \le 6\}$

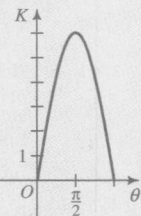

Suggested Assignments

Standard
Day 1: 342/1–15 odd
Day 2: 343/19–27 odd
Comprehensive
342/1, 7, 9, 15, 19, 27, 31, 33, 35

Supplementary Materials

Alternative Assessment, 27
Activities Book, 21–22
Student Resource Guide, 87–89

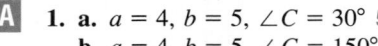

WRITTEN EXERCISES

Throughout the exercises, give areas in radical form or to three significant digits. Give lengths to three significant digits and angle measures to the nearest tenth of a degree. In Exercises 1–4, find the area of each $\triangle ABC$.

A **1. a.** $a = 4$, $b = 5$, $\angle C = 30°$ 5
  b. $a = 4$, $b = 5$, $\angle C = 150°$ 5

3. a. $a = 6$, $c = 2$, $\angle B = 45°$ $3\sqrt{2}$
  b. $a = 6$, $c = 2$, $\angle B = 135°$ $3\sqrt{2}$

2. a. $b = 3$, $c = 8$, $\angle A = 120°$
  b. $b = 3$, $c = 8$, $\angle A = 60°$

4. a. $a = 10$, $b = 20$, $\angle C = 70°$
  b. $a = 10$, $b = 20$, $\angle C = 110°$

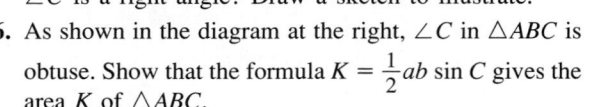
2. a. $6\sqrt{3}$ **b.** $6\sqrt{3}$ **4. a.** 94.0 **b.** 94.0

5. What does the formula $K = \frac{1}{2}ab \sin C$ become when $\angle C$ is a right angle? Draw a sketch to illustrate.

6. As shown in the diagram at the right, $\angle C$ in $\triangle ABC$ is obtuse. Show that the formula $K = \frac{1}{2}ab \sin C$ gives the area K of $\triangle ABC$.

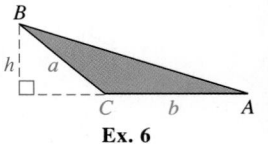
Ex. 6

7. Find the area of $\triangle XYZ$ if $x = 16$, $y = 25$, and $\angle Z = 52°$. 158

8. Find the area of $\triangle RST$ if $\angle S = 125°$, $r = 6$, and $t = 15$. 36.9

9. The area of $\triangle ABC$ is 15. If $a = 12$ and $b = 5$, find the measure(s) of $\angle C$. 30°, 150°

10. The area of $\triangle PQR$ is 9. If $q = 4$ and $r = 9$, find the measure(s) of $\angle P$. 30°, 150°

11. Find the area of a regular octagon inscribed in a circle of radius 40 cm. $3200\sqrt{2}$ cm^2

12. Find the area of a regular 12-sided polygon inscribed in a circle of radius 8 cm. 192 cm^2

13. Adjacent sides of a parallelogram have lengths 6 cm and 7 cm, and the measure of the included angle is 30°. Find the area of the parallelogram. 21 cm^2

14. Sketch a parallelogram with sides of lengths a and b and with an acute angle θ. Express the area of the parallelogram in terms of a, b, and θ. $A = ab \sin \theta$

15. Suppose a triangle has two sides of lengths 3 cm and 4 cm and an included angle θ. Express the area of the triangle as a function of θ. State the domain and range of the function and sketch its graph.

16. Suppose a triangle has two sides of lengths a and b. If the angle between these sides varies, what is the maximum possible area that the triangle can attain? What can you say about the minimum possible area? $\frac{1}{2}ab$; no min.

B **17. a.** Given $\triangle ABC$ with an inscribed circle as shown at the right, show that the radius r of the circle is given by:

$$r = \frac{2(\text{area of } \triangle ABC)}{\text{perimeter of } \triangle ABC}$$

(*Hint*: If I is the center of the inscribed circle, then: area of $\triangle ABI$ + area of $\triangle ACI$ + area of $\triangle BCI$ = area of $\triangle ABC$.)

b. Find the radius of the inscribed circle if $AB = AC = 10$ and $BC = 16$. $2\frac{2}{3}$

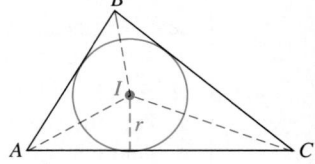

342 *Chapter Nine*

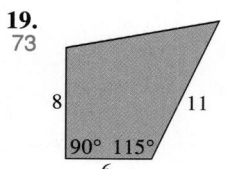 **Part (b) of Exercise 18 requires the use of a computer or graphing calculator. Give your answer to the nearest tenth.**

18. In isosceles triangle ABC, $AB = BC = 10$ and $AC = 2x$.
    a. Use Exercise 17 to show that the radius of the inscribed circle is:
    $$r = \frac{x\sqrt{100 - x^2}}{10 + x}$$
    b. Use a computer or graphing calculator to find the value of x that maximizes r. Interpret your answer. $x \approx 6.18$

Find the area of each quadrilateral to the nearest square unit.

19.
73
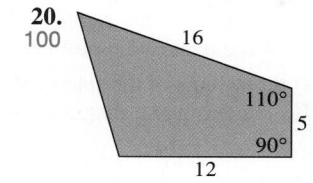

20.
100

21. Find the area of a segment of a circle of radius 5 if the measure of the central angle of the segment is 2 radians. 13.6

22. Find the area of the segment formed by a chord 24 cm long in a circle of radius 13 cm. 139 cm²

 Exercises 23 and 24 require the use of a computer or graphing calculator. Give answers to the nearest tenth.
2.5 radians
23. In a circle of radius 10, there is a segment with area 95. Use a computer or graphing calculator to find the measure of the central angle of the segment.

24. For the cylindrical oil tank described in Example 3 on pages 340 and 341, use a computer or graphing calculator to determine where on a measuring rod marks should be put to show that the tank is $\frac{1}{3}$ full and $\frac{2}{3}$ full. 0.7 m and 1.3 m

25. *Visual Thinking* In Example 3 on pages 340 and 341, suppose the cylindrical oil tank sits upright on one of its circular ends (with the opening for the measuring rod at the other end). Describe how the problem of measuring the amount of oil in the tank changes.

26. *Research* Find out and report on what the typical shape of a car's fuel tank is and how the amount of fuel in the tank is measured and indicated on the car's fuel gauge. Is the measuring instrument designed to give a truly accurate measurement of fuel? If not, why not?

Graph the region satisfying both inequalities and find its area.

27. $x^2 + y^2 \le 36$, $y \ge 3$ 22.1   28. $x^2 + y^2 \le 9$, $x \ge 1$ 8.25

C 29. $x^2 + y^2 \le 9$, $x^2 + y^2 - 10x + 9 \le 0$ 6.64   30. $x^2 + y^2 - 8y \le 0$, $x^2 + y^2 \le 16$ 19.7

Triangle Trigonometry   **343**

Mathematical Note

Exercises 21 and 22 can be done by using the formula

$$A = \frac{1}{2}r^2(\theta - \sin\theta)$$

to find the area of a shaded segment in terms of the central angle θ (in radians) and the radius r.

Using Technology

A computer or graphing calculator can be used to obtain the boundaries of the graphs for Written Exercises 27–30.

Additional Answers Written Exercises

18. **a.** Altitude from B to midpoint of $\overline{AC}$ has length $\sqrt{100 - x^2}$.
Area of $\triangle ABC =$
$\frac{1}{2}(2x)(\sqrt{100 - x^2}) =$
$x\sqrt{100 - x^2}$.
Using the formula of Exercise 17, we have
$r = \frac{2x\sqrt{100 - x^2}}{10 + 10 + 2x} = \frac{x\sqrt{100 - x^2}}{10 + x}$.
b. The largest inscribed circle does *not* occur when the triangle is equilateral.

25. Measuring the amount of oil is simpler. The volume of oil is directly proportional to the depth of oil. Thus, for example, the tank is $\frac{1}{4}$ full when the depth of the oil is one-quarter the height of the tank.

31. In the diagram at the right, $\alpha = 60°$ and $\beta = 60°$.
  a. Express the areas of $\triangle BCD$ and $\triangle ACD$ in terms of a, b, and x.
  b. Express the area of $\triangle ABC$ in terms of a and b.
  c. Show that $x = \frac{ab}{a + b}$.
  d. Does part (c) agree with what you know from geometry if $a = b$? Explain.

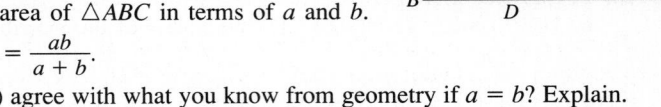

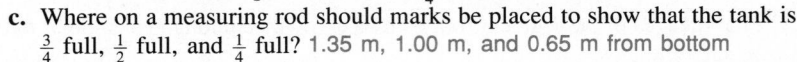 **Exercises 32–36 require the use of a computer or graphing calculator. Give answers to the nearest hundredth.**

32. Engineering Suppose an oil tank is a sphere with radius 1 unit, as shown at the right.
  a. Find the exact volume of a sphere with radius 1. $\frac{4}{3} \pi$
  b. If the volume of the spherical cap formed by an angle α (as shown in the diagram) is given by

$$\frac{2\pi}{3} - \pi \cos \frac{\alpha}{2} + \frac{\pi}{3} \cos^3 \frac{\alpha}{2},$$

  use a computer or graphing calculator to find the value of α for which the spherical tank is $\frac{3}{4}$ full. 2.43 radians
  c. Where on a measuring rod should marks be placed to show that the tank is $\frac{3}{4}$ full, $\frac{1}{2}$ full, and $\frac{1}{4}$ full? 1.35 m, 1.00 m, and 0.65 m from bottom

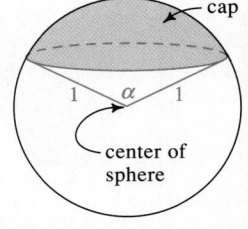

33. Engineering A fuel tank has a cross section whose shape is a 2 m × 2 m square capped at the top and bottom by semicircles, as shown at the left below. Use a computer or graphing calculator to determine how to mark a measuring rod to show that the tank is only 10% full. 0.56 m from bottom

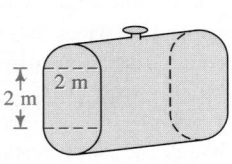

Ex. 33

Ex. 34

34. A goat is tethered to a stake at the edge of a circular field with radius 1 unit, as shown at the right above. Use a computer or graphing calculator to determine how long the rope should be so that the goat can graze over half the field. 1.16

35. $\overline{RP}$ is a tangent to a circle with radius 1, as shown at the right. If the area of the region shaded blue equals the area of the region shaded red, use a computer or graphing calculator to find θ in radians. 1.17

36. Use the diagram for Exercise 35 and a computer or graphing calculator to find the value of θ for which $\overline{QR}$ and arc QP have the same length. 1.07

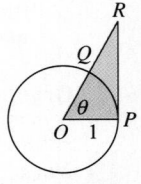

Exs. 35, 36

9-3 The Law of Sines

Objective *To use the law of sines to find unknown parts of a triangle.*

When there are several methods for solving a problem, a comparison of the solutions can lead to new and useful results. From Section 9-2, we know three ways to find the area K of $\triangle ABC$, depending on which pair of sides is known.

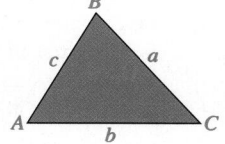

$$K = \frac{1}{2}bc \sin A = \frac{1}{2}ac \sin B = \frac{1}{2}ab \sin C$$

If each of these expressions is divided by $\frac{1}{2}abc$, we obtain the *law of sines*.

The Law of Sines

In $\triangle ABC$, $\dfrac{\sin A}{a} = \dfrac{\sin B}{b} = \dfrac{\sin C}{c}$.

If we know two angles and a side of a triangle, then we can use the law of sines to find the other sides, as shown in the following example.

Example 1 A civil engineer wants to determine the distances from points A and B to an inaccessible point C, as shown. From direct measurement the engineer knows that $AB = 25$ m, $\angle A = 110°$, and $\angle B = 20°$. Find AC and BC.

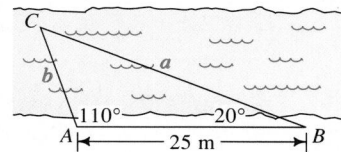

Solution First find the measure of $\angle C$:

$$\angle C = 180° - (110° + 20°) = 50°$$

Then, by the law of sines:

$$\frac{\sin 110°}{a} = \frac{\sin 20°}{b} = \frac{\sin 50°}{25}$$

Therefore:

$$a = \frac{25 \sin 110°}{\sin 50°} \approx 30.7 \quad \text{and} \quad b = \frac{25 \sin 20°}{\sin 50°} \approx 11.2$$

Thus, $BC \approx 30.7$ m and $AC \approx 11.2$ m.

In the activity at the top of the next page, you are given the lengths of two sides of a triangle and the measure of a nonincluded angle. You are then asked to investigate whether it is possible to construct the triangle and, if so, whether the triangle is unique.

Triangle Trigonometry **345**

Warm-Up Exercises

For Exercises 1-8, refer to the figure. Tell whether a triangle can be formed that satisfies the given information. If so, use your knowledge of geometry to decide if the triangle is uniquely determined.

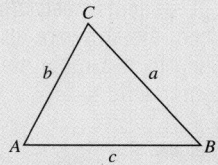

1. $\angle A = 40°$, $\angle B = 50°$ yes; no

2. $a = 8$, $b = 6$, $c = 7$ yes; yes

3. $\angle A = 30°$, $a = 6$, $b = 8$ yes; no

4. $\angle A = 50°$, $c = 10$, $\angle B = 42°$ yes; yes

5. $a = 7$, $b = 10$, $c = 2$ no

6. $\angle A = 85°$, $\angle C = 95°$, $a = 8$ no

7. $b = 12$, $\angle A = 100°$, $c = 14$ yes; yes

8. $b = 15$, $\angle A = 70°$, $\angle B = 20°$ yes; yes

Motivating the Section

Being able to solve any triangle completely is useful for many applications, including how to find the distance to an inaccessible object and how to find the area of any triangle. The law of sines, the topic of this section, allows the complete solution of certain triangles in which three parts are known.

 Using Technology

Geometry software can be used to complete Activities 1 and 2. Students can input the measurements given in the activities and have the computer try to construct the triangle. The measure feature of the software can then be used to solve the triangle, if it exists. If students need more practice, the software also can be used to test other sets of measurements.

Problem Solving

Encourage students to use diagrams to help them determine when triangles are possible for the SSA case.

Additional Examples

1. Solve $\triangle RST$ if $\angle S = 40°$, $r = 30$, and $s = 20$. Give angle measures to the nearest tenth of a degree and lengths to three significant digits.

$\frac{\sin R}{r} = \frac{\sin S}{s}$

$\frac{\sin R}{30} = \frac{\sin 40°}{20}$

$\sin R \approx 1.5(0.6428) = 0.9642$

$\angle R \approx 74.6°$ or $\angle R \approx 180° - 74.6° = 105.4°$

If $\angle R \approx 74.6°$, then $\angle T \approx 180° - 40° - 74.6° = 65.4°$

and $\frac{\sin S}{s} = \frac{\sin T}{t}$,

$\frac{\sin 40°}{20} = \frac{\sin 65.4°}{t}$,

$t \approx \frac{20(0.9092)}{0.6428} \approx 28.3$.

Thus, one solution is $\angle R = 74.6°$, $\angle T = 65.4°$, and $t = 28.3$.

346 Chapter 9

Activity 1

For this activity, use a ruler, compass, and protractor. Draw $\angle A$ with measure 30°. Along one ray of $\angle A$, locate point C 10 cm from point A. For each of the following compass settings, draw a large arc. Then tell whether the arc crosses the other ray of $\angle A$ and, if so, in how many points.
a. Compass at C and opened to 4 cm No
b. Compass at C and opened to 5 cm Yes; 1
c. Compass at C and opened to 6 cm Yes; 2

Activity 1 shows that when you are given the lengths of two sides of a triangle and the measure of a nonincluded angle (SSA), it may be possible to construct no triangle, one triangle, or two triangles. For this reason the SSA situation is called the *ambiguous case*.

Activity 2

Show that your answers to Activity 1 agree with what the law of sines would give in each of the following SSA situations. **a.** No solution
a. If $\angle A = 30°$, $b = 10$, and $a = 4$, find $\angle B$.
b. If $\angle A = 30°$, $b = 10$, and $a = 5$, find $\angle B$. 90°
c. If $\angle A = 30°$, $b = 10$, and $a = 6$, find $\angle B$. 56.4°, 123.6°

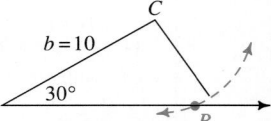

Example 2 In $\triangle RST$, $\angle S = 126°$, $s = 12$, and $t = 7$. Determine whether $\angle T$ exists. If so, find all possible measures of $\angle T$.

Solution Make a sketch, as shown. By the law of sines:

$$\frac{\sin T}{7} = \frac{\sin 126°}{12}$$

$$\sin T = \frac{7 \sin 126°}{12} \approx 0.4719$$

$$\angle T \approx 28.2° \quad \text{or} \quad \angle T \approx 151.8°$$

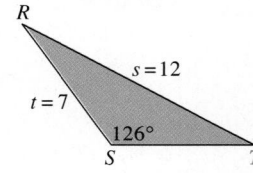

It seems that $\angle T$ exists and that there are two possible measures of $\angle T$. A triangle cannot have two obtuse angles, however, so we must reject $\angle T \approx 151.8°$. Thus, $\angle T \approx 28.2°$.

CLASS EXERCISES

1. *Reading* Rephrase the law of sines in words rather than symbols:

In any triangle,   ?   is constant.

346 *Chapter Nine*

2. The diagram at the right shows a 30°-60°-90° triangle with a hypotenuse of length 2. Find the values of the ratios $\frac{\sin A}{a}$, $\frac{\sin B}{b}$, and $\frac{\sin C}{c}$. Are the ratios equal?

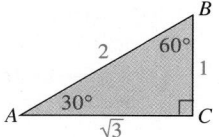

2. $\frac{1}{2}$, $\frac{1}{2}$, $\frac{1}{2}$; yes

In Exercises 3–5, consider $\triangle ABC$.

3. If $\angle A \geq 90°$, what can you conclude about the measure of $\angle B$? Explain. $\angle B < 90°$

4. If $\angle B$ has a greater measure than $\angle C$, what must be true of b and c? Why? $b > c$

5. If $a > b$, what must be true of $\angle A$ and $\angle B$? Why? $\angle A > \angle B$

In Exercises 6–8, state an equation that you can use to solve for x.

6.

7.

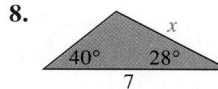

8.

9. If $a = 8$ and $b = 6$, draw a diagram to show that $\triangle ABC$ is uniquely determined for each of the given measures of $\angle A$: **a.** 45° **b.** 90° **c.** 120°

10. Use the law of sines to show that $\triangle ABC$ in part (a) of Exercise 9 is unique.

11. Use the law of sines to show that there is no $\triangle XYZ$ with $\angle X = 30°$, $x = 3$, and $y = 8$.

WRITTEN EXERCISES

A 1. The purpose of this exercise is to determine whether $\triangle ABC$ can be constructed when two lengths a and b and the measure of an acute angle A are given. As shown at the right, first $\angle A$ and then the side of length b are constructed; finally a circular arc is drawn with center C and radius a. Given that the distance from C to the side opposite C is $b \sin A$, determine for each of the following conditions whether 0, 1, or 2 triangles can be formed.

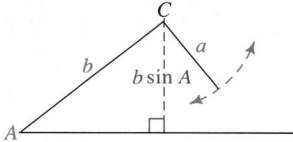

  a. $a < b \sin A$ 0   **b.** $a = b \sin A$ 1
  c. $b \sin A < a < b$ 2   **d.** $a \geq b$ 1

2. Use Exercise 1 or the law of sines to determine whether there are 0, 1, or 2 triangles possible for each of the following sets of measurements.

  a. $a = 2$, $b = 4$, $\angle A = 22°$ 2   **b.** $b = 3$, $c = 6$, $\angle B = 30°$ 1
  c. $a = 7$, $c = 5$, $\angle A = 68°$ 1   **d.** $b = 4$, $c = 3$, $\angle C = 76°$ 0

In Exercises 3–14, solve each $\triangle ABC$. Give angle measures to the nearest tenth of a degree and lengths in simplest radical form or to three significant digits. Be alert to problems with no solution or two solutions.

3. $\angle A = 45°$, $\angle B = 60°$, $a = 14$ **4.** $\angle B = 30°$, $\angle C = 45°$, $b = 9$

5. $\angle B = 30°$, $\angle A = 135°$, $b = 4$ **6.** $\angle A = 60°$, $\angle B = 75°$, $c = 10$

If $\angle R \approx 105.4°$, then $\angle T \approx 180° - 40° - 105.4° = 34.6°$

and $\frac{\sin S}{s} = \frac{\sin T}{t}$,

$\frac{\sin 40°}{20} = \frac{\sin 34.6°}{t}$,

$t \approx \frac{20(0.5678)}{0.6428} \approx 17.7$.

Thus, another solution is $\angle R = 105.4°$, $\angle T = 34.6°$, and $t = 17.7$.

2. In $\triangle ABC$, $\cos A = \frac{1}{2}$,

$\cos B = -\frac{1}{4}$, and $a = 6$.

Find the value of b in simplest radical form.

Recall from Section 8-4 that $\sin^2 \theta + \cos^2 \theta = 1$ for every angle θ. Also, if θ is the measure of an angle of a triangle, then $0° < \theta < 180°$, so $\sin \theta > 0$. Thus:

$\sin A = \sqrt{1 - \left(\frac{1}{2}\right)^2} = \frac{\sqrt{3}}{2}$

and $\sin B =$
$\sqrt{1 - \left(-\frac{1}{4}\right)^2} = \frac{\sqrt{15}}{4}$.

$\frac{\sin A}{a} = \frac{\sin B}{b}$

$\frac{\frac{\sqrt{3}}{2}}{6} = \frac{\frac{\sqrt{15}}{4}}{b}$

$\frac{\sqrt{3}}{2} b = \frac{3}{2}\sqrt{15}$

$b = \frac{3\sqrt{15}}{\sqrt{3}} = 3\sqrt{\frac{15}{3}} =$

$3\sqrt{5}$

Suggested Assignments

Standard
347/3, 7, 11, 15, 19, 21, 22
Comprehensive
347/3, 7, 11, 15, 19, 21–25

7. $\angle C = 25°$, $b = 3$, $c = 2$

8. $\angle B = 36°$, $a = 10$, $b = 8$

9. $\angle A = 76°$, $a = 12$, $b = 4$

10. $\angle B = 130°$, $b = 15$, $c = 11$

11. $\angle C = 88°$, $b = 7$, $c = 7$

12. $\angle A = 95°$, $a = 13$, $c = 10$

13. $\angle B = 40°$, $a = 12$, $b = 6$

14. $\angle C = 112°$, $c = 5$, $a = 7$

15, 16. See below.

15. In $\triangle RST$, $\angle R = 140°$ and $s = \frac{3}{4}r$. Find the measures of $\angle S$ and $\angle T$.

16. In $\triangle DEF$, $\angle F = 120°$ and $f = \frac{4}{3}e$. Find the measures of $\angle D$ and $\angle E$.

17. A fire tower at point A is 30 km north of a fire tower at point B. A fire at point F is observed from both towers. If $\angle FAB = 54°$ and $\angle ABF = 31°$, find AF. **15.5 km**

18. From lighthouses P and Q, 16 km apart, a disabled ship S is sighted. If $\angle SPQ = 44°$ and $\angle SQP = 66°$, find the distance from S to the nearer lighthouse. **11.8 km**

19. In $\triangle ABC$, $\tan A = \frac{3}{4}$, $\tan B = 1$, and $a = 10$. Find b in simplest radical form. $\dfrac{25\sqrt{2}}{3}$

20. In $\triangle ABC$, $\cos A = \frac{1}{2}$, $\cos B = -\frac{1}{4}$, and $a = 6$. Find b in simplest radical form. $3\sqrt{5}$

15. $\angle S = 28.8°$, $\angle T = 11.2°$ **16.** $\angle D = 19.5°$, $\angle E = 40.5°$

B **21. Navigation** A ship passes by buoy B which is known to be 3000 yd from peninsula P. The ship is steaming east along line BE and $\angle PBE$ is measured as 28°. After 10 min, the ship is at S and $\angle PSE$ is measured as 63°.

a. How far from the peninsula is the ship when it is at S? **1580 yd**

b. If the ship continues east, what is the closest it will get to the peninsula? **1410 yd**

c. How fast (in yd/min) is the ship traveling? **193 yd/min**

d. Ship speeds are often given in knots, where 1 knot = 1 nautical mile per hour ≈ 6080 feet per hour. Convert your answer in part (c) to knots. **5.71 knots**

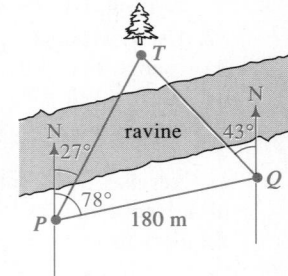

22. Surveying From points P and Q, 180 m apart, a tree at T is sighted on the opposite side of a deep ravine. From point P, a compass indicates that the angle between the north-south line and line of sight $\overline{PT}$ is 27° and that the angle between the north-south line and $\overline{PQ}$ is 78°. From point Q, the angle between the north-south line and $\overline{QT}$ is 43°.

a. How far from P is the tree? **164 m**

b. How far from P is the point on $\overline{PQ}$ that is closest to the tree? **103 m**

348 *Chapter Nine*

23. The purpose of this exercise is to show that in $\triangle ABC$ the three equal ratios $\dfrac{a}{\sin A}$, $\dfrac{b}{\sin B}$, and $\dfrac{c}{\sin C}$ are each equal to the diameter of the circumscribed circle of $\triangle ABC$.

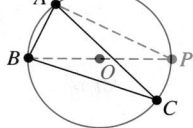

   a. Let the circle with center O be the circumscribed circle and let $\overline{BP}$ be the diameter through B. Show that $\angle P$ and $\angle C$ have the same measure.

   b. Show that $\dfrac{AB}{BP} = \sin P$.

   c. Use parts (a) and (b) to show that
$$\text{diameter} = \frac{c}{\sin C} = \frac{b}{\sin B} = \frac{a}{\sin A}.$$

24. A triangle with angles 50°, 60°, and 70° has all three vertices on a circle of radius 8 cm. Find the lengths of the three sides. (*Hint*: See Exercise 23.)
   12.3 cm, 13.9 cm, 15.0 cm

25. a. Prove that the area of $\triangle ABC$ is given by $K = \dfrac{1}{2}\left(\dfrac{\sin B\,\sin C}{\sin A}\right)a^2$.

   b. State two other formulas for K, one involving b and one involving c.

26. Use the results of Exercises 23 and 25 to show that the ratio of the area of $\triangle ABC$ to the area of its circumscribed circle is $\dfrac{2}{\pi}\sin A\,\sin B\,\sin C$.

27. The purpose of this exercise is to use the law of sines to prove that an angle bisector in a triangle divides the opposite side in the ratio of the two adjacent sides: that is, $\dfrac{x}{y} = \dfrac{a}{b}$ in the diagram at the right.

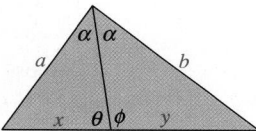

   a. Show that $\sin \theta = \sin \phi$. (ϕ is the Greek letter *phi*.)

   b. Show that $\dfrac{a}{x} = \dfrac{\sin \theta}{\sin \alpha}$ and $\dfrac{b}{y} = \dfrac{\sin \phi}{\sin \alpha}$.

   c. Prove the stated theorem.

C 28. In the diagram at the left below, the two circles with centers P and Q have radii 1 and 2, respectively. P is the midpoint of $\overline{OQ}$. An arbitrary line through O intersects the circles at A, B, C, and D.

   a. Use the law of sines to prove that A is the midpoint of $\overline{OC}$.

   b. What special property do you think B has?

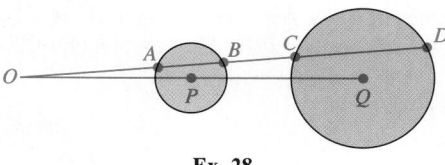

Ex. 28

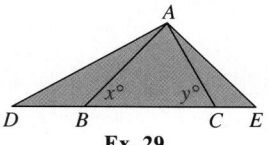

Ex. 29

29. In the diagram at the right above, $\angle DAC = \angle BAE = 90°$. Prove that $DE = BC \tan x \tan y$.

Cooperative Learning

You might want students to evaluate each other's proofs for Written Exercises 23 and 25–29. Have them work in small groups to do so.

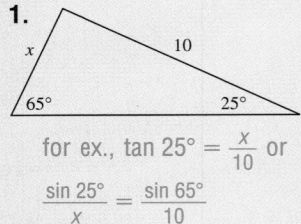
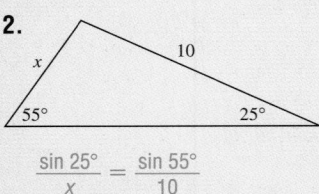
9-4 The Law of Cosines

| **Objective** | *To use the law of cosines to find unknown parts of a triangle.* |

In Section 9-2 we mentioned that by the SAS condition for congruence, a triangle is uniquely determined if the lengths of two sides and the measure of the included angle are known. By the side-side-side (SSS) condition for congruence, a triangle is also uniquely determined if the lengths of three sides are known. The *law of cosines* can be used to solve a triangle in either of these two cases.

The Law of Cosines

$$\text{In } \triangle ABC, \ c^2 = a^2 + b^2 - 2ab \cos C.$$

To derive this law, we introduce a coordinate system by placing the *x*-axis along $\overline{BC}$ of $\triangle ABC$ and the *y*-axis at *C* so that $C = (0, 0)$, $B = (a, 0)$, and $A = (b \cos C, b \sin C)$, as shown. Using the distance formula, we have:

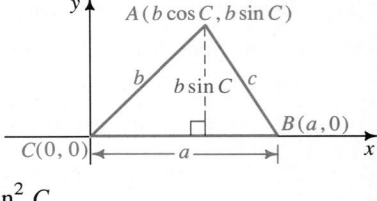

$$c^2 = (AB)^2 = (b \cos C - a)^2 + (b \sin C - 0)^2$$
$$= b^2 \cos^2 C - 2ab \cos C + a^2 + b^2 \sin^2 C$$
$$= b^2(\cos^2 C + \sin^2 C) - 2ab \cos C + a^2$$
$$= b^2 \cdot 1 - 2ab \cos C + a^2$$
$$= a^2 + b^2 - 2ab \cos C$$

To help you remember the law of cosines, notice that it has the basic pattern:

$$\begin{pmatrix} \text{side} \\ \text{opposite} \\ \text{angle} \end{pmatrix}^2 = \begin{pmatrix} \text{side} \\ \text{adjacent} \\ \text{to angle} \end{pmatrix}^2 + \begin{pmatrix} \text{other side} \\ \text{adjacent} \\ \text{to angle} \end{pmatrix}^2 - 2\begin{pmatrix} \text{one} \\ \text{adjacent} \\ \text{side} \end{pmatrix}\begin{pmatrix} \text{other} \\ \text{adjacent} \\ \text{side} \end{pmatrix} \cos (\text{angle})$$

Note that when $\angle C = 90°$ the law of cosines reduces to $a^2 + b^2 = c^2$. Therefore the law of cosines includes the Pythagorean theorem as a special case and is, consequently, more flexible and useful than the Pythagorean theorem. When $\angle C$ is acute, c^2 is less than $a^2 + b^2$ by the amount $2ab \cos C$; when $\angle C$ is obtuse, $\cos C$ is negative and so c^2 is greater than $a^2 + b^2$.

Example 1   Suppose that two sides of a triangle have lengths 3 cm and 7 cm and that the angle between them measures 130°. Find the length of the third side.

Solution   Make a sketch, as shown.

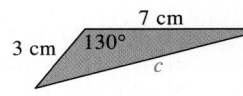

$$c^2 = 3^2 + 7^2 - 2 \cdot 3 \cdot 7 \cdot \cos 130°$$
$$\approx 85.0$$

Thus, the length of the third side is about $\sqrt{85.0}$ cm, or about 9.22 cm.

If we solve the law of cosines for cos C, we obtain:

$$\cos C = \frac{a^2 + b^2 - c^2}{2ab}$$

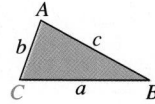

In this form, the law of cosines can be used to find the measures of the angles of a triangle when the lengths of the three sides are known. The basic pattern for this form of the law of cosines is:

$$\cos (\text{angle}) = \frac{(\text{adjacent})^2 + (\text{adjacent})^2 - (\text{opposite})^2}{2 \cdot (\text{adjacent}) \cdot (\text{adjacent})}$$

Example 2   The lengths of the sides of a triangle are 5, 10, and 12. Solve the triangle.

Solution   Make a sketch of the triangle, as shown.

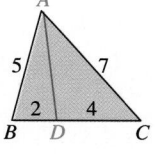

1. $\cos \alpha = \dfrac{5^2 + 10^2 - 12^2}{2 \cdot 5 \cdot 10} = -0.19$

   $\alpha \approx 101.0°$

2. $\cos \beta = \dfrac{12^2 + 10^2 - 5^2}{2 \cdot 12 \cdot 10} = 0.9125$

   $\beta \approx 24.1°$

3. $\theta \approx 180° - (101.0° + 24.1°) = 54.9°$

   Using the law of cosines, we can easily identify acute and obtuse angles. For instance, in Example 2, since cos α is negative and cos β is positive, we know that α is an obtuse angle and β is an acute angle. The law of sines does not distinguish between acute and obtuse angles, however, because both types of angle have positive sine values.

Example 3   In the diagram at the right, $AB = 5$, $BD = 2$, $DC = 4$, and $CA = 7$. Find AD.

Solution   First we apply the law of cosines to $\triangle ABC$:

$$\cos B = \frac{5^2 + 6^2 - 7^2}{2 \cdot 5 \cdot 6} = 0.2$$

Then we apply the law of cosines to $\triangle ABD$:

$$(AD)^2 = 5^2 + 2^2 - 2 \cdot 5 \cdot 2 \cdot 0.2 = 25$$

Thus, $AD = \sqrt{25} = 5$.

   In this section and the preceding section, we have seen various applications of the laws of sines and cosines. The situations in which these laws can be used are summarized at the top of the next page. The summary assumes that once the measures of two angles of a triangle are known, the measure of the third angle is found using the geometric fact that the sum of the measures of the angles is 180°.

Mathematical Note

Point out that there are *three* alternate forms of the law of cosines: In $\triangle ABC$,

$$\cos A = \frac{b^2 + c^2 - a^2}{2bc},$$

$$\cos B = \frac{a^2 + c^2 - b^2}{2ac},$$

$$\cos C = \frac{a^2 + b^2 - c^2}{2ab}.$$

Making Connections

Example 2 shows that the law of cosines can be used to find the measures of the angles of any triangle for which the sides are known. This is the SSS case from geometry.

Additional Examples

1. A triangle has sides of lengths 6, 12, and 15.
   a. Find the measure of the smallest angle.
   b. Find the length of the median to the longest side.

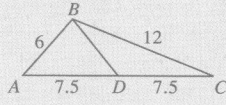

   a. The smallest angle of $\triangle ABC$ is opposite the shortest side, $\overline{AB}$.

   $\cos C = \dfrac{12^2 + 15^2 - 6^2}{2 \cdot 12 \cdot 15} =$ 0.925
   $\angle C = \text{Cos}^{-1} \, 0.925 \approx$ 22.3°
   b. In $\triangle BCD$,
   $(BD)^2 = 12^2 + (7.5)^2 - 2(12)(7.5)(0.925) =$ 33.75, and $BD \approx 5.81$

(*Continues on next page*)

Additional Examples cont.

2. A parallelogram has diagonals of lengths 20 cm and 12 cm. If the diagonals intersect to form a 60° angle, find the perimeter of the parallelogram.

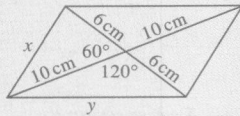

$$x^2 = 10^2 + 6^2 -$$
$$2 \cdot 10 \cdot 6 \cdot \cos 60° =$$
$$136 - 120 \cdot \frac{1}{2} = 76$$
$$x = \sqrt{76} = 2\sqrt{19} \text{ cm}$$
$$y^2 = 10^2 + 6^2 -$$
$$2 \cdot 10 \cdot 6 \cdot \cos 120° =$$
$$136 - 120\left(-\frac{1}{2}\right) = 196$$
$$y = 14 \text{ cm}$$
$$\text{perimeter} = 2x + 2y =$$
$$(4\sqrt{19} + 28) \text{ cm}$$

Exercise Note

Class Exercise 6 refers to the fact that the largest angle of a triangle is opposite the longest side, and the smallest angle is opposite the shortest side. Using these inequalities can help students find mistakes in their solutions.

Suggested Assignments

Standard
Day 1: 352/1, 7, 9, 13, 15, 17, 21
Day 2: 355/1–19 odd
Day 3: 356/21–29 odd
Comprehensive
Day 1: 352/1, 7, 9, 13, 15, 17, 18, 21, 23
Day 2: 355/1–21 odd
Day 3: 356/23–25, 29, 36, 40, 43

Given:	Use:	To find:
SAS	law of cosines	the third side and then one of the remaining angles.
SSS	law of cosines	any two angles.
ASA or AAS	law of sines	the remaining sides.
SSA	law of sines	an angle opposite a given side and then the third side. (Note that 0, 1, or 2 triangles are possible.)

When you use the law of sines, remember that every acute angle and its supplement have the same sine value. Class Exercises 4–7 show how you can tell which angle is correct for a given triangle.

CLASS EXERCISES

In Exercises 1–3, state an equation that you can use to solve for *x*.

1.

2.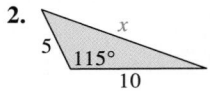

3.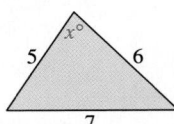

$x^2 = 5^2 + 6^2 - 2 \cdot 5 \cdot 6 \cos 35°$ $x^2 = 5^2 + 10^2 - 2 \cdot 5 \cdot 10 \cos 115°$ $\cos x° = \dfrac{5^2 + 6^2 - 7^2}{2 \cdot 5 \cdot 6}$

In Exercises 4–7, consider $\triangle XYZ$, where $x = 4$, $y = 8$, and $\angle Z = 50°$.

4. Use the law of cosines to find *z* to the nearest hundredth. 6.23

5. Use the law of sines to find the measure of $\angle Y$ to the nearest tenth of a degree. Then find the measure of $\angle X$. $\angle Y = 100.4°$; $\angle X = 29.6°$

6. Since $x < z < y$, what can you say about the measures of $\angle X$, $\angle Y$, and $\angle Z$?

7. Do your answers to Exercises 5 and 6 agree? If not, find your error.
   6. $\angle X < \angle Z < \angle Y$

WRITTEN EXERCISES

Solve each triangle. Give lengths to three significant digits and angle measures to the nearest tenth of a degree.

1. $a = 8$, $b = 5$, $\angle C = 60°$ **2.** $t = 16$, $s = 14$, $\angle R = 120°$

3. $x = 9$, $y = 40$, $z = 41$ **4.** $a = 6$, $b = 10$, $c = 7$

5. $p = 3$, $q = 8$, $\angle R = 50°$ **6.** $d = 5$, $e = 9$, $\angle F = 115°$

7. $a = 8$, $b = 7$, $c = 13$ **8.** $x = 10$, $y = 11$, $z = 12$

In Exercises 9 and 10, use the method of Example 3, page 351, to find AD in the diagram at the right.

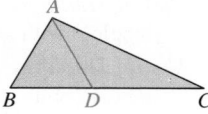

9. $AB = 8$, $BD = 7$, $DC = 5$, $AC = 10$ 7.07

10. $AB = 5$, $BD = 5$, $DC = 3$, $AC = 7$ 5

In Exercises 11 and 12, find the length of the median from A in the given $\triangle ABC$. Give your answers in simplest radical form.

11. $a = 8$, $b = 4$, $c = 6\sqrt{10}$

12. $a = 12$, $b = 13$, $c = 5\sqrt{61}$

13. A parallelogram has a 70° angle and sides 6 cm and 10 cm long. How long are its diagonals? 9.74 cm, 13.3 cm

14. An isosceles trapezoid has a height of 4 cm and bases 3 cm and 7 cm long. How long are its diagonals? 6.40 cm

Find the area of each quadrilateral to the nearest square unit.

B **15.**
146

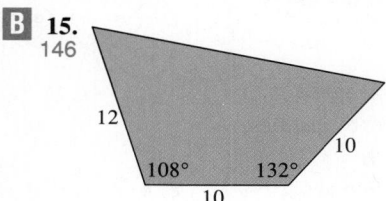

16.
70.7

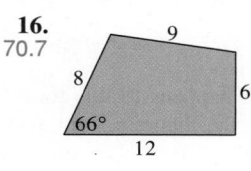

17. **Aviation** Two airplanes, at points A and B in the diagram at the left below, have elevations of 23,000 ft and 18,000 ft, respectively. Both are flying east toward an airport control tower at T. From T, the angle of elevation of the airplane at A is 4°, and the angle of elevation of the airplane at B is 2.5°. How far apart (in mi) are the airplanes? (5280 ft = 1 mi) 15.8 mi

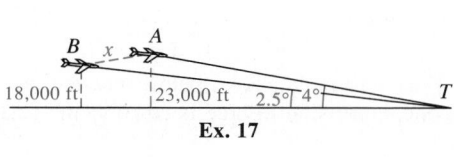

Ex. 17

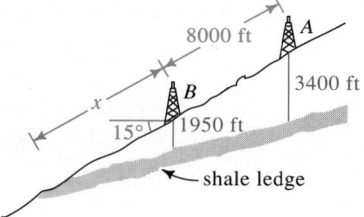

Ex. 18

18. **Geology** In the diagram at the right above, an oil well at A was drilled 3400 ft before it hit a ledge of shale. The same ledge was only 1950 ft deep when drilled from a well at B, which is 8000 ft directly downhill from A. The hill is inclined at 15° to the horizontal.
   a. If you assume that the ledge lies in a plane, how far down the hill from B would you expect the shale ledge to emerge? 10,800 ft
   b. What is the angle between the ledge and the hill? 10.4°

Additional Answers Written Exercises

1. $\angle A = 81.8°$, $\angle B = 38.2°$, $c = 7$

2. $\angle S = 27.8°$, $\angle T = 32.2°$, $r = 26$

3. $\angle X = 12.7°$, $\angle Y = 77.3°$, $\angle Z = 90°$

4. $\angle A = 36.2°$, $\angle B = 100.3°$, $\angle C = 43.5°$

5. $\angle P = 20.7°$, $\angle Q = 109.3°$, $r = 6.49$

6. $\angle D = 22.2°$, $\angle E = 42.8°$, $f = 12.0$

7. $\angle A = 32.2°$, $\angle B = 27.8°$, $\angle C = 120.0°$

8. $\angle X = 51.3°$, $\angle Y = 59.2°$, $\angle Z = 69.5°$

Supplementary Materials

Alternative Assessment, 28
Activities Book, 23
Using Technology, 25–26

19. **Visual Thinking** Give a geometric interpretation of the law of cosines for the "triangle" with two sides a and b and "included angle" $C = 0°$.

20. **Visual Thinking** Give a geometric interpretation of the law of cosines for the "triangle" with two sides a and b and "included angle" $C = 180°$.

21. In $\triangle ABC$, $AB = 20$, $BC = 15$, $AC = 10$, and $\overline{CD}$ bisects $\angle ACB$, as shown at the right.

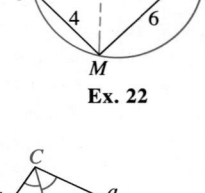

   a. Use the results of Exercise 27, page 349, to find AD and DB. *AD = 8, DB = 12*

   b. Use the method of Example 3, page 351, to find CD. $3\sqrt{6}$

22. Quadrilateral $JKLM$ is inscribed in a circle as shown. Find the length of each diagonal. (*Hint:* Opposite angles of an inscribed quadrilateral are supplementary. Express $\cos L$ in terms of $\cos J$ in order to find KM.) $KM = 7$, $JL = 7\frac{1}{7}$

23. **a.** Prove that in $\triangle ABC$ the length x of the median from C is given by:
$$x = \frac{1}{2}\sqrt{2a^2 + 2b^2 - c^2}$$

   b. What happens to this formula when $\angle C = 90°$?

   c. State the theorem from geometry that part (b) justifies.

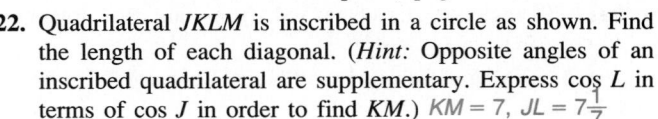

Ex. 22

24. In the diagram at the right, $\overline{CD}$ bisects $\angle ACB$.

   a. Use the method of part (a) of Exercise 21 to show:
$$p = \frac{bc}{a + b} \quad \text{and} \quad q = \frac{ac}{a + b}$$

   b. Prove that $x^2 = ab - pq$.

   c. Use part (b) to find CD in Exercise 21.

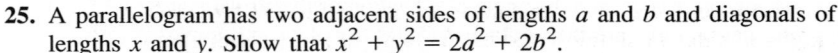

25. A parallelogram has two adjacent sides of lengths a and b and diagonals of lengths x and y. Show that $x^2 + y^2 = 2a^2 + 2b^2$.

26. In this exercise, you are to derive *Hero's formula* for the area K of $\triangle ABC$:
$$K = \sqrt{s(s - a)(s - b)(s - c)} \quad \text{where } s = \frac{a + b + c}{2}$$

   a. In the partial derivation of Hero's formula shown below, use the law of cosines to show that the second expression in blue is equal to the first.

$$\begin{aligned}
\sin^2 C &= 1 - \cos^2 C \\
&= (1 + \cos C)(1 - \cos C) \\
&= \frac{(a + b)^2 - c^2}{2ab} \cdot \frac{c^2 - (a - b)^2}{2ab} \\
&= \frac{(a + b + c)(a + b - c)(c + a - b)(c - a + b)}{4a^2b^2} \\
&= \frac{(2s)(2s - 2c)(2s - 2b)(2s - 2a)}{4a^2b^2}
\end{aligned}$$

   b. Use the formula $K = \frac{1}{2}ab \sin C$ to complete the derivation in part (a).

Write a program that accepts the input of the measures of any three parts of a triangle and then prints the measures of the remaining three parts. You might want to work in a group of five students and write a structured program that begins with a main routine like the one below (written in BASIC).

```
10 PRINT ''INDICATE THE PROBLEM TYPE.''
20 INPUT ''(ENTER SSS, SAS, ASA, AAS, OR SSA): ''; T$
30 IF T$ = ''SSS'' THEN GOSUB 100
40 IF T$ = ''SAS'' THEN GOSUB 200
50 IF T$ = ''ASA'' THEN GOSUB 300
60 IF T$ = ''AAS'' THEN GOSUB 400
70 IF T$ = ''SSA'' THEN GOSUB 500
80 END
```

Each member of the group can write one of the subroutines called by the main routine. Since many programming languages use radians instead of degrees, you may need to convert between the two types of angle measurement. Also, since many programming languages do not have inverse sine and inverse cosine as built-in functions but do have inverse tangent as a built-in function, you may need to use the following relationships, which are valid for $|x| < 1$:

$$\text{Sin}^{-1} x = \text{Tan}^{-1} \frac{x}{\sqrt{1 - x^2}} \quad \text{and} \quad \text{Cos}^{-1} x = \frac{\pi}{2} - \text{Tan}^{-1} \frac{x}{\sqrt{1 - x^2}}$$

MIXED TRIGONOMETRY EXERCISES

1. Area = 16.1; $p = 5.57$ **2.** Area = 26.8; $r = 15.5$

Where appropriate, give angle measures to the nearest tenth of a degree and lengths of sides in simplest radical form or to three significant digits.

A

1. Find the area of $\triangle PQR$ if $q = 6$, $r = 7$, and $\angle P = 50°$. Also find p.

2. Find the area of $\triangle PQR$ if $p = 7$, $q = 10$, and $\angle R = 130°$. Also find r.

3. Find the measure of the largest angle in a triangle with sides having lengths $3\sqrt{6}$, $6\sqrt{3}$, and $9\sqrt{2}$. 90°

4. Find the measure of the smallest angle in a triangle with sides having lengths 7, 12, and 13. 32.2°

5. In $\triangle RST$, $\angle R = 75°$, $\angle S = 45°$, and $t = 3$. Find r and s. $r \approx 3.35$; $s = \sqrt{6} \approx 2.45$

6. In $\triangle RST$, $\angle S = 100°$, $\angle T = 30°$, and $r = 8$. Find s and t. $s \approx 10.3$; $t \approx 5.22$

7. Three measurements in $\triangle ABC$ are given as $\angle A = 60°$, $a = 4$, and $b = 5$. Show that at least one of the measurements is incorrect. See below.

8. A regular polygon with 180 sides is inscribed in a circle with radius 1. Find its area. Compare your answer with π. Area ≈ 3.141; $\pi \approx 3.142$

9. In $\triangle ABC$, $\cos A = -0.6$. Find $\sin A$ and $\tan A$. $\sin A = 0.8$; $\tan A = -\frac{4}{3}$

10. In $\triangle ABC$, $a = 17$, $b = 10$, and $c = 21$. Find $\cos A$ and $\sin A$. $\cos A = 0.6$ $\sin A = 0.8$

7. $5 \sin 60°$ cannot equal $4 \sin B$.

Triangle Trigonometry **355**

You might want to have students who are working on the Computer Exercise discuss such side issues as the following:

1. Can any of the subroutines be written in such a way that they can be used for more than one type of problem?

2. If the law of sines applies to the given data, does that mean that the law of cosines cannot apply?

3. How were the given identities for $\text{Sin}^{-1} x$ and $\text{Cos}^{-1} x$ derived?

Exercise Note

For Mixed Trigonometry Exercises 3 and 4, remind students that the largest angle is opposite the longest side and the smallest angle is opposite the shortest side.

Problem Solving

Students might enjoy answering Exercise 11 for a variety of ramp angles, from 1° to 8°, with 8° giving the minimum horizontal length of the ramp.

Exercise Note

For Exercise 17, remind students that the ratio of the areas of two similar triangles with corresponding side lengths a and b is $a^2 : b^2$.

Assessment Note

Have students explain why there are two answers for Exercise 19.

17.8 ft

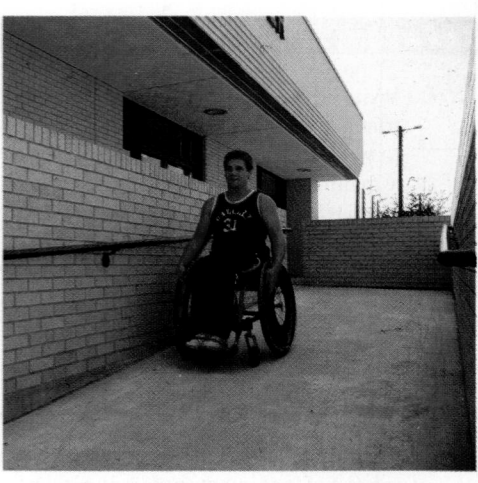

11. **Civil Engineering** A wheelchair ramp must rise 30 in. to meet the front door of a public library. If the ramp's angle of elevation is not to exceed 8°, what is the minimum horizontal length of the ramp?

12. **Navigation** A submarine dives at an angle of 16° with the horizontal. If it takes 4 min to dive from the surface to a depth of 300 ft, how fast does it move along its sloping path downward? Give your answer in feet per minute. Then convert it to nautical miles per hour. (*Note:* 1 nautical mile per hour ≈ 6080 feet per hour) 272 ft/min, 2.69 knots

13. In △XYZ, ∠X = 21.1°, x = 6, and y = 9. Find the measure(s) of ∠Y. 32.7°, 147.3°

14. In parallelogram ABCD, ∠A = 60°, AB = 5, and AD = 8.
    a. Find the area of ABCD. $20\sqrt{3}$ **b.** Find the lengths of both diagonals. 7, $\sqrt{129}$

15. A triangle has area 21 cm² and two of its sides are 9 cm and 14 cm long. Find the possible measures of the angle formed by these sides. 19.5°, 160.5°

16. In △DEF, ∠D = 36°, ∠E = 64°, and f = 8. Find d and e. d = 4.77, e = 7.30

17. In the diagram at the right, △ABC is similar to △DEF and ∠A = 120°. a = 14.0, e = 8.57
    a. Find the lengths a, e, and f. f = 14.3
    b. Find the ratio of the areas of the triangles. 49 : 100

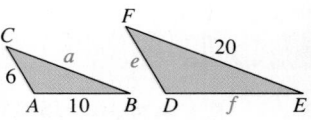

18. The diagonals of a parallelogram have lengths 8 and 14 and they meet at a 60° angle. Find the area and the perimeter of the parallelogram. See below.

19. An obtuse triangle with area 12 has two sides of lengths 4 and 10. Find the length of the third side. (There are two answers.) $6\sqrt{5} \approx 13.4$; $2\sqrt{13} \approx 7.21$

20. The perimeter of a regular decagon (10 sides) is 240. Find its area. ≈4430

21. If fencing costs $2.50 per foot, how much will it cost to buy fencing to go around the plot of land shown at the left below? $478.77

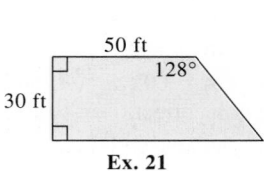

Ex. 21

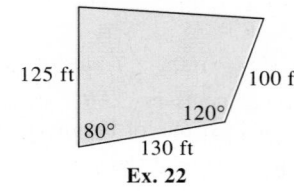

Ex. 22

22. In the township of Madison, rural undeveloped land is taxed at a rate of $115 per acre. Find the tax on the plot of land shown at the right above. (*Note:* 1 acre = 43,560 ft²) $41.63   **18.** A = $28\sqrt{3}$; p = $2\sqrt{37} + 2\sqrt{93}$

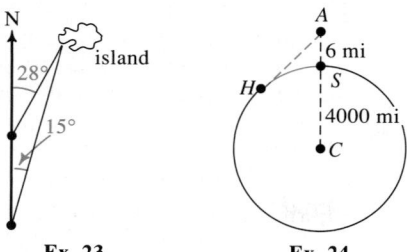

1.15 nautical mi ≈ 1.32 mi

23. **Navigation** A ship is steaming north at 6 knots (6 nautical miles per hour) when the captain sights a small island at an angle of 15° to the east of the ship's course, as shown at the right. After 10 min, the angle is 28°. How far away is the island at this moment?

24. **Aviation** An airplane at A is flying at a height of 6 mi above Earth's surface at S, as shown at the far right.

Ex. 23　　Ex. 24

　a. Find the distance to the nearest tenth of a mile from A to the horizon H. (The radius of Earth is about 4000 mi.) 219.2 mi

　b. Find the curved distance to the nearest tenth of a mile from S along Earth's surface to H. 219.0 mi

25. In $\triangle ABC$, $\tan A = 1$, $\tan B = \frac{3}{4}$, and $b = 18$. Find $\sin A$, $\sin B$, and a. See below.

26. In $\triangle DEF$, $\sec F = -\sqrt{2}$. Find the measure of $\angle F$ and $\tan F$. $\angle F = 135°$, $\tan F = -1$

27. If $180° < x < 360°$ and $\tan x = -\frac{1}{5}$, find $\sin x$ and $\cos x$. $\sin x = -\frac{\sqrt{26}}{26}$; $\cos x = \frac{5\sqrt{26}}{26}$

28. The area of $\triangle PQR$ is 84. If $r = 14$ and $q = 13$, find $\sin P$. Use a trigonometric identity to find two possible values of $\cos P$ and two possible values of p.

25. $\sin A = \frac{\sqrt{2}}{2}$; $\sin B = \frac{3}{5}$; $a = 15\sqrt{2}$　　28. $\sin P = \frac{12}{13}$; $\cos P = \pm\frac{5}{13}$; $p = 15$, $\sqrt{505}$

In geometry you can prove two triangles congruent by SSS. This means that when the lengths of the three sides of a triangle are given, its shape is completely determined. Exercises 29 and 30 below illustrate this principle. Exercises 31 and 32 illustrate the principle for SAS and ASA.

B 29. In $\triangle ABC$, $a = 5$, $b = 8$, and $c = 7$. (SSS)
　　a. Solve $\triangle ABC$.
　　b. Find the area of $\triangle ABC$. Use the formula $K = \frac{1}{2} ab \sin C$ or Hero's formula (Exercise 26, page 354).
　　c. Find the length of the altitude to $\overline{AC}$.
　　d. Find the length of the median from B to $\overline{AC}$ (Exercise 23, page 354).
　　e. Find the length of the angle bisector from B to $\overline{AC}$ (Exercise 24, page 354).
　　f. Find the radius R of the circumscribed circle (Exercise 23, page 349).
　　g. Find the radius r of the inscribed circle (Exercise 17, page 342).

30. Repeat Exercise 29 if $a = 13$, $b = 14$, and $c = 15$. (SSS)

31. Repeat Exercise 29 if $a = 9$, $b = 10$, and $\cos C = -\frac{3}{5}$. (SAS)

32. Repeat Exercise 29 if $\sin A = \frac{8}{17}$, $c = 21$, and $\sin B = \frac{\sqrt{2}}{2}$. (ASA) Assume that $\angle B$ is obtuse.

33. a. The consecutive sides of a quadrilateral inscribed in a circle have lengths 1, 4, 3, and 2. Find the length of each diagonal, using the fact that opposite angles are supplementary. Approximately 2.80, 3.92
　　b. Check your answer to part (a) by using Ptolemy's theorem: If $ABCD$ is inscribed in a circle, then $AC \cdot BD = AB \cdot DC + AD \cdot BC$. $(2.80)(3.92) = 10.976 \approx 11 = (1)(3) + (4)(2)$

Triangle Trigonometry **357**

Chapter 9　**357**

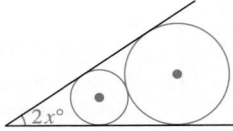

Using Technology

Have students use calculators to check the identity in Exercise 37.

34. $\triangle DEF$ is inscribed in a circle of radius 3. If $\angle D = 120°$ and $\angle E = 15°$, find the lengths d, e, and f. (*Hint:* Find the measures of arc DE and arc DF.)
$d = 3\sqrt{3} \approx 5.20$; $e \approx 1.55$; $f = 3\sqrt{2} \approx 4.24$

35. Prove that the altitude from A in $\triangle ABC$ has length $h = \dfrac{a}{\cot B + \cot C}$.

C 36. Two circles are externally tangent. Common tangents to the circles form an angle of measure $2x$. Prove that the ratio of the radii of the circles is $\dfrac{1 - \sin x}{1 + \sin x}$. (*Hint:* Express $\sin x$ in terms of the two radii.)

37. Prove that $\cot A + \cot B + \cot C = \dfrac{a^2 + b^2 + c^2}{4(\text{area of } \triangle ABC)}$ in any $\triangle ABC$.

38. Given $\triangle ABC$ with $c^2 = \dfrac{a^3 + b^3 + c^3}{a + b + c}$, find the measure of $\angle C$. 60°

39. As shown at the right, a ladder, $\overline{AC}$, leans against the side of a house at an angle α with the ground. Suppose that the foot of the ladder slides y units from A to B, the top of the ladder slides x units from C to D, and the ladder makes an angle β with the ground. Prove:

$$x = \dfrac{y(\sin \alpha - \sin \beta)}{\cos \beta - \cos \alpha}$$

40. A sailor at the seashore watches a ship with a smokestack 30 m above water level as the ship steams out to sea. The sailor's eye level is 4 m above water level. About how far is the ship from shore when the stack disappears from the sailor's view? (The radius of Earth is about 6400 km.) 26.8 km

41. In $\triangle XYZ$, $\overline{YR}$ and $\overline{ZS}$ are altitudes. Prove that $RS = x \cos X$.

42. **a.** As shown in the diagram at the left below, $\overline{AD}$ is a diameter of the circle and is tangent to line l at D. If $AD = 1$ and $\overline{BC} \perp l$, show that:

$$AB + BC = 1 + \cos \theta - \cos^2 \theta$$

b. What value of θ makes the sum $AB + BC$ a maximum? $\theta = 60°$

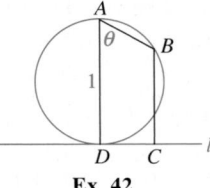

Ex. 42

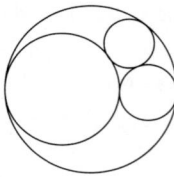

Ex. 43

43. Each circle in the diagram at the right above is tangent to the other three circles. The largest three circles have radii of 1, 2, and 3, respectively. Find the radius of the smallest circle. $\dfrac{6}{7}$

9-5 Applications of Trigonometry to Navigation and Surveying

Objective *To use trigonometry to solve navigation and surveying problems.*

As shown below, the *course* of a ship or plane is the angle, measured clockwise, from the north direction to the direction of the ship or plane.

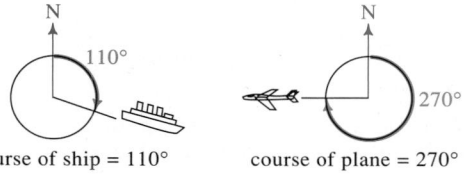

course of ship = 110°   course of plane = 270°

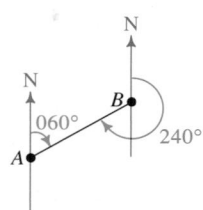

As shown at the right, the *compass bearing* of one location from another is measured in the same way. Note that compass bearings and courses are given with three digits, such as 060° rather than 60°.

bearing of B from A = 060°
bearing of A from B = 240°

Example 1   A ship proceeds on a course of 300° for 2 hours at a speed of 15 knots (1 knot = 1 nautical mile per hour). Then it changes course to 230°, continuing at 15 knots for 3 more hours. At that time, how far is the ship from its starting point?

Solution   Make a diagram, as shown below.

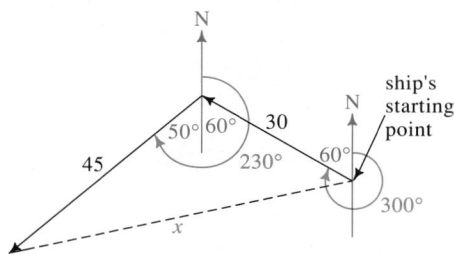

The ship travels first along a path of length $2 \cdot 15 = 30$ nautical miles and then along a path of length $3 \cdot 15 = 45$ nautical miles. The angle between the two paths is 110°. (You can find this angle by drawing north-south lines and using geometry.) To find x, the distance of the ship from its starting point, use the law of cosines:

$$x^2 = 30^2 + 45^2 - 2 \cdot 30 \cdot 45 \cdot \cos 110° \approx 3848$$

Thus, $x \approx \sqrt{3848} \approx 62.0$ nautical miles.

Triangle Trigonometry   **359**

Warm-Up Exercises

For Exercises 1–4, consider the following problem. Dennis and Robyn walked 6 mi due north from their cabin, turned 40° toward the east, and walked 4 mi more.

1. Make a scale drawing to illustrate the facts.

2. Using a ruler and your scale drawing, find the approximate distance that Dennis and Robyn were from their cabin after walking the 10 mi. Give your answer to the nearest mile.   9 mi

3. Describe how you could compute the distance to the cabin.   Note that the angle included between the sides of lengths 6 mi and 4 mi has measure 140°. Use the law of cosines.

4. Is there enough information given to completely solve the triangle?   yes

Motivating the Section

Ask students if they have ever seen a surveyor at work. Point out that the instruments used by a surveyor are for measuring angles and distances. Surveying applications (as well as navigation applications) are presented in this section.

Making Connections

You might have students use a town or state map to practice using compass readings.

Communication Note

A compass reading of N20°E is the direction 20° east of north. Likewise, a compass reading of S30°W is the direction 30° west of south.

Additional Examples

1. Town *T* is 8 km northeast of village *V*. City *C* is 4 km from *T* on a bearing of 150° from *T*. What is the bearing and distance of *C* from *V*?

   Make a diagram.

   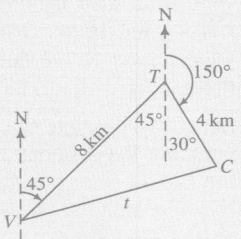

   Notice that the two angles marked 45° are alternate interior angles formed by two parallel lines and a transversal. $t^2 = 8^2 + 4^2 - 2 \cdot 8 \cdot 4 \cdot \cos 75° \approx 80 - 64(0.2588) \approx 63.44$
   $t \approx 7.96$ km
   $$\frac{\sin T}{t} = \frac{\sin V}{v}$$
   $$\frac{\sin 75°}{7.96} = \frac{\sin V}{4}$$
   $\sin V \approx \dfrac{4(0.9659)}{7.96} \approx 0.4854$
   $\angle V \approx 29.0°$
   Thus, *C* is about 7.96 km from *V* at a bearing of 45° + 29.0°, or 74°.

In surveying, a compass reading is usually given as an acute angle from the north-south line toward the east or west. A few examples are shown below.

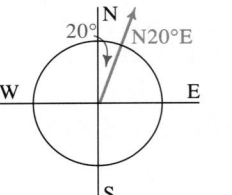
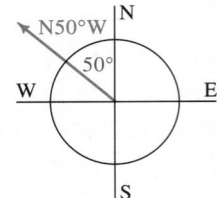
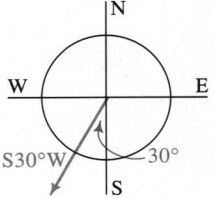

Example 2 Very often a plot of land is taxed according to its area. Sketch the plot of land described. Then find its area.

From a granite post, proceed 195 ft east along Tasker Hill Road, then along a bearing of S32°E for 260 ft, then along a bearing of S68°W for 385 ft, and finally along a line back to the granite post.

Solution We first sketch the plot of land, one side at a time and in the sequence described. (See the diagram at the left below.)

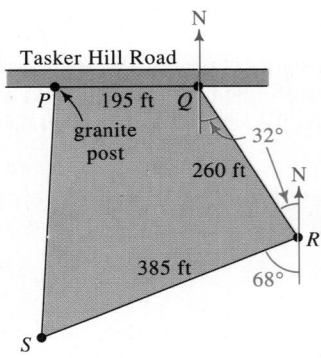
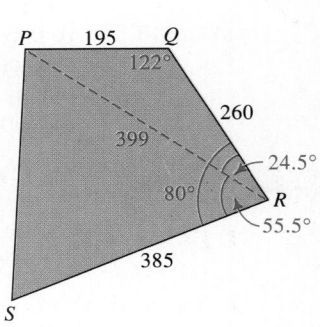

From the bearings given, we deduce that:

$$\angle PQR = 90° + 32° = 122°$$
$$\angle QRS = 180° - (32° + 68°) = 80°$$

To find the area of *PQRS*, we divide the quadrilateral into two triangles by introducing $\overline{PR}$. (See the diagram at the right above.) We can find the area of $\triangle PQR$ directly:

$$\text{Area of } \triangle PQR = \frac{1}{2} \cdot PQ \cdot QR \cdot \sin Q$$
$$= \frac{1}{2} \cdot 195 \cdot 260 \cdot \sin 122° \approx 21{,}500 \text{ ft}^2$$

To find the area of $\triangle PRS$, we must first find PR and $\angle PRS$.

To find PR, we use the law of cosines:

$$PR^2 = 195^2 + 260^2 - 2 \cdot 195 \cdot 260 \cdot \cos 122°$$
$$\approx 159{,}000$$

Therefore, $PR \approx \sqrt{159{,}000} \approx 399$ ft.

To find $\angle PRS$, we find $\angle PRQ$ by the law of sines:

$$\frac{\sin PRQ}{195} = \frac{\sin 122°}{399}$$

$$\sin PRQ = \frac{195 \sin 122°}{399} \approx 0.4145$$

$$\angle PRQ \approx 24.5°$$

Therefore, $\angle PRS = \angle QRS - \angle PRQ \approx 80° - 24.5° = 55.5°$.
Knowing that $PR \approx 399$ ft and $\angle PRS \approx 55.5°$, we have:

$$\text{Area of } \triangle PRS = \tfrac{1}{2} \cdot PR \cdot RS \cdot \sin PRS$$

$$\approx \tfrac{1}{2} \cdot 399 \cdot 385 \cdot \sin 55.5° \approx 63{,}300 \text{ ft}^2$$

Thus, we have:

$$\text{Area of quadrilateral } PQRS = \text{area of } \triangle PQR + \text{area of } \triangle PRS$$
$$\approx 21{,}500 + 63{,}300$$
$$= 84{,}800 \text{ ft}^2$$

CLASS EXERCISES

1. **Reading** In the solution of Example 1, justify the fact that the angle opposite the side of length x in the diagram has a measure of $110°$.

2. Suppose land is taxed at a rate of $75 per acre. Determine the approximate tax on the land in Example 2. (1 acre = 43,560 ft^2) $146

Visual Thinking In each diagram, a north-south line is given. If $\overrightarrow{OT}$ represents the path of a ship, estimate its course.

3.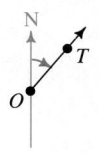

040°

4.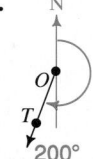

115°

5.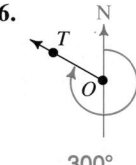

200°

6.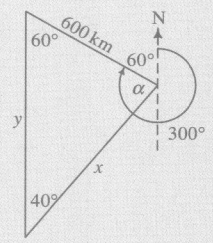

300°

7. **Visual Thinking** Suppose point X is directly east of point Y.
   a. Give the bearing of X from Y. 090°
   b. Give the bearing of Y from X. 270°

---

2. A plane flies 600 km on a course of 300°. It then flies south for a while and finally flies on a 40° course to return to its starting point. Find the total distance traveled.

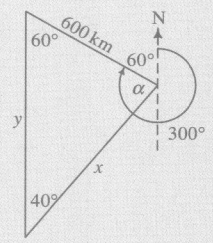

Notice that $\alpha = 80°$.

$$\frac{\sin 40°}{600} = \frac{\sin 60°}{x}$$

$$x \approx \frac{600(0.8660)}{0.6428} \approx 808 \text{ km}$$

$$\frac{\sin 40°}{600} = \frac{\sin 80°}{y}$$

$$y \approx \frac{600(0.9848)}{0.6428} \approx 919 \text{ km}$$

Thus, the total distance traveled is about $600 + 808 + 919 \approx 2327$ km.

Example Note

Example 2 has many intermediate steps before the final answer is found. For these intermediate steps, note that lengths and areas are given to three significant digits and that angle measures are given to the nearest tenth of a degree, in keeping with the statement at the top of page 333.

Problem Solving

Making sketches of the compass readings in Class Exercises 8–11 will help students visualize compass readings.

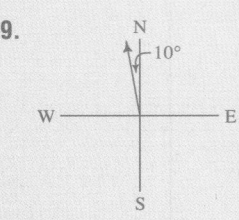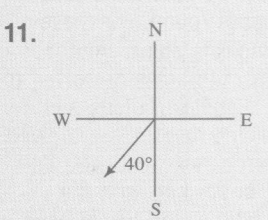
For Exercises 8–11, make a sketch of each compass reading.

8. N70°E
9. N10°W
10. S15°E
11. S40°W

12. Match each compass direction with a course.

Direction		Course
045°	northeast	135°
135°	southeast	225°
315°	northwest	315°
225°	southwest	045°

13. A famous Alfred Hitchcock movie is *North by Northwest*. This direction is midway between north and northwest. What course corresponds to this direction? 337.5°

14. From one corner of a triangular plot of land, a surveyor determines the directions to the other two corners to be N32°E and S76°E. What is the measure of the angle formed by the edges of the plot of land at the corner where the surveyor is? 72°

WRITTEN EXERCISES

In Exercises 1–4, draw a diagram like those in Class Exercises 3–6 to show the path of a ship proceeding on each given course.

A 1. 070°
2. 150°
3. 340°
4. 225°

5. **Navigation** Ship A sights ship B on a compass bearing of 080°. Make a sketch and give the compass bearing of ship A from ship B. 260°

6. **Navigation** Ship X sights ship Y on a bearing of 308°. What is the bearing of X from Y? 128°

7. **Aviation** An airplane flies on a course of 110° at a speed of 1200 km/h. How far east of its starting point is it after 2 h? 2255 km

8. A hunter walks east for 1 h and then north for $1\frac{1}{2}$ h. What course should the hunter take to return to his starting point? What assumptions do you make to answer the question?

9. Point B is 10 km north of point A, and point C is 10 km from B on a bearing of 060° from B. Find the bearing and distance of C from A. 030°; 17.3 km

10. 067.5°; 7.39 km   **11.** 18.6 nautical miles

10. Point *S* is 4 km west of point *R*, and point *T* is 4 km southwest of *S*. Find the bearing and distance of *R* from *T*.

11. Navigation Traveling at a speed of 10 knots, a ship proceeds south from its port for $1\frac{1}{2}$ h and then changes course to 130° for $\frac{1}{2}$ h. At this time, how far from port is the ship?

12. Navigation A sailboat leaves its dock and proceeds east for 2 mi. It then changes course to 205° until it is due south of its dock. How far south is this? 4.29 mi

13. Navigation Two ships, A and B, leave port at the same time. Ship A proceeds at 12 knots on a course of 040°, while ship B proceeds at 9 knots on a course of 115°. After 2 h, ship A loses power and radios for help. How far and on what course must ship B travel to reach ship A? See below.

14. Navigation A ship leaves port and sails northwest for 1 h and then northeast for 2 h. If it does not change speed, find what course the ship should take to return directly to port. Also find how long this return will take. 198.4°; 2.24 h

13. 26 nautical mi; 358°

In Exercises 15–18, sketch each plot of land described and find its area.

125,000 m²

B **15. Surveying** From an iron post, proceed 500 m northeast to the brook, then 300 m east along the brook to the old mill, then 200 m S15°E to a post on the edge of Wiggin's Road, and finally along Wiggin's Road back to the iron post.

16. Surveying From a cement marker, proceed 260 m southwest to the river, then 240 m south along the river to the bridge, then 280 m N40°E to a sign on the edge of Sycamore Lane, and finally along Sycamore Lane back to the cement marker. 40,500 m²

17. Surveying From the southeast corner of the cemetery on Burnham Road, proceed S78°W for 250 m along the southern boundary of the cemetery until a granite post is reached, then S15°E for 180 m to Allard Road, then N78°E along Allard Road until it intersects Burnham Road, and finally N30°E along Burnham Road back to the starting point. 29,600 m²

18. Surveying From the intersection of Simpson's Road and Mulberry Lane, proceed N32°W for 320 m along Simpson's Road, then S56°W for 280 m to the old oak tree, then S22°E until Mulberry Lane is reached, and finally N68°E along Mulberry Lane back to the starting point. 87,100 m²

Triangle Trigonometry   **363**

19. "True north" is the north *geographic* pole near the center of the Arctic Ocean where all lines of longitude meet. "Magnetic north" is the north *magnetic* pole, the point toward which a north-seeking compass needle points. Although the position of true north is fixed, the north magnetic pole continually shifts; in fact, it can move many miles in just a few years.

Compass variation is the angle between magnetic north and true north. When magnetic north pulls a compass needle to the left of true north, the variation is *west;* when magnetic north pulls a compass needle to the right of true north, the variation is *east.* (See the illustrations on the next page.) Because the north magnetic pole shifts over time, the world's dividing lines between east and west variation also shift. Generally speaking, however, west variation occurs in the eastern parts of North America, South America, and Asia, and in most of Europe and Africa; east variation occurs in the western parts of North America, South America, and Asia, and in most of Australia. (Consult an isogonic map for more details.)

(*Continues on next page*)

19. Research There is a difference between magnetic north, the direction in which a compass needle points, and true north. Research what this difference is and what the phrase ''compass variation'' means. In what parts of the world is the variation ''east'' and in what parts is it ''west''? Also explain what is meant by the mariner's rhyme:

Error west, compass best.
Error east, compass least.

Chapter Summary

1. If you know the lengths of two sides of a right triangle, or the length of one side and the measure of one acute angle, then you can find the measures of the remaining sides and angles using the trigonometric functions, whose definitions are given below. (The definitions are based on the diagram at the right.)

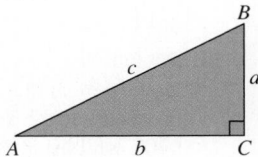

$$\sin A = \frac{\text{opposite}}{\text{hypotenuse}} = \frac{a}{c} \qquad \csc A = \frac{1}{\sin A} = \frac{\text{hypotenuse}}{\text{opposite}} = \frac{c}{a}$$

$$\cos A = \frac{\text{adjacent}}{\text{hypotenuse}} = \frac{b}{c} \qquad \sec A = \frac{1}{\cos A} = \frac{\text{hypotenuse}}{\text{adjacent}} = \frac{c}{b}$$

$$\tan A = \frac{\text{opposite}}{\text{adjacent}} = \frac{a}{b} \qquad \cot A = \frac{1}{\tan A} = \frac{\text{adjacent}}{\text{opposite}} = \frac{b}{a}$$

2. Knowing the lengths a and b of two sides of a triangle and the measure of the included angle C, you can obtain the area K of the triangle using the formula:

$$K = \frac{1}{2}ab \sin C$$

3. In any $\triangle ABC$, the following relationships hold:

Law of sines: $\quad \dfrac{\sin A}{a} = \dfrac{\sin B}{b} = \dfrac{\sin C}{c}$

Law of cosines: $\quad c^2 = a^2 + b^2 - 2ab \cos C$

Law of cosines (alternate form): $\quad \cos C = \dfrac{a^2 + b^2 - c^2}{2ab}$

4. You can use the law of sines and the law of cosines to solve triangles, as summarized on page 352. It is important to remember that the SSA (side-side-nonincluded angle) case may result in 0, 1, or 2 triangles.

5. The application of trigonometry, particularly the laws of sines and cosines, to navigation and surveying is discussed in Section 9-5. In navigation, north represents 0° and the *course* or *compass bearing* of a ship or plane is measured by a clockwise rotation from north. In surveying, angles are given acute measures east or west of a north-south line.

364 *Chapter Nine*

Chapter Test

Where appropriate, give angle measures to the nearest tenth of a degree and lengths of sides in simplest radical form or to three significant digits.

1. The sides of an isosceles triangle have lengths 5, 10, and 10. What are the measures of its angles? 75.5°, 75.5°, 29.0°   **9-1**

2. At a distance of 100 m, the angle of elevation to the top of a fir tree is 28°. About how tall is the tree? 53.2 m

3. A regular pentagon is inscribed in a circle of radius 4 in. Find the area of the pentagon. 38.0 in.^{2} **9-2**

4. Find the area of the quadrilateral shown at the right. 448

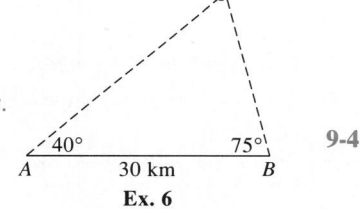

5. How many different triangles *PQR* can be constructed using the given information?   **9-3**
   a. $p = 5$, $q = 4$, $\angle Q = 74°$ 0
   b. $p = 9$, $q = 8$, $\angle P = 23°$ 1

Ex. 4

7.

6. Observers at points *A* and *B*, 30 km apart, sight an airplane at angles of elevation of 40° and 75°, respectively, as shown in the diagram. How far is the plane from each observer? See below.

7. A triangle has sides of length 5, 8, and 10. What is the measure of its largest angle? 97.9°   **9-4**

Ex. 6

8. Two hikers follow a trail that splits into two forks. Each hiker takes a different fork. The forks diverge at an angle of 67° and both hikers walk at a speed of 3.5 mi/h. How far apart are the hikers after 1 h? 3.86 mi

9. *Writing* When only three parts of a triangle are known, the law of sines and the law of cosines can be used to find the unknown parts. Discuss the specific circumstances for using each law. See below.

10. After leaving an airport, a plane flies for 1.5 h at a speed of 200 km/h on a course of 200°. Then, on a course of 340°, the plane flies for 2 h at a speed of 250 km/h. At this time, how far from the airport is the plane? 332 km   **9-5**

6. ≈32.0 km from *A*, ≈21.3 km from *B*   9. Refer to the chart at the top of page 352.

Triangle Trigonometry   **365**

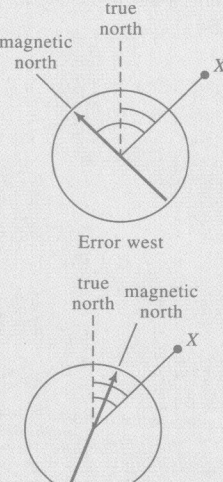

The People of Mathematics

Why do you think people choose to become mathematicians? What events in their lives steer them in the mathematical direction? How has mathematics changed over the centuries? Is mathematics discovered or created? Maybe you've thought about questions like these before. The answers, though, may actually surprise you. Here's a chance to do a little research in some areas of mathematics that perhaps you haven't yet formally explored.

Mathematics is such an extensive field, it's easy to get lost in all of the available information. You'll probably find it helpful, then, to narrow your study by choosing from the following list of suggested topics. Your goal will be to summarize your findings in a report of at least six pages in length.

Mathematical Essays and Autobiographies

- *Adventures of a Mathematician* by Stanislaw M. Ulam
- *I Want to Be a Mathematician* by Paul R. Halmos
- ''Mathematical Creation'' by Henri Poincaré
- ''The Mathematician'' by John von Neumann
- *A Mathematician's Apology* by G. H. Hardy
- ''Olga Taussky-Todd: An Autobiographical Essay'' by Olga Taussky-Todd

The Development of Numbers

- The mathematics of the Rhind papyrus
- The Mayan numeration system and the representation of zero
- The Babylonian (or Sumerian) numeration system
- The Hindu-Arabic numeration system
- Irrational numbers and the school of Pythagoras
- Calculators of π: Archimedes (Greece), Zŭ Chōngzhī and Zŭ Gĕng (China), Aryabhata (India), al-Kashi (Persia), Viète (France), Kanada (Japan), Chudnovsky brothers (United States)

Famous Problems and Paradoxes

- The Chinese remainder theorem
- The Cantor set
- Goldbach's conjecture
- Russell's paradox
- The Möbius strip
- Chaotic motion and the Lorenzian waterwheel

A Few Masters of Discovery

- Benjamin Banneker
- János Bolyai
- Leonhard Euler
- Sophie Germain
- Sofia Kovalevskaia (also known as Sonya Kovalevsky)
- George Polya
- Srinivasa Ramanujan

People and Computational Machines

- The abacus, including the Chinese *suan pan* and the Japanese *soroban*
- Napier's rods
- Charles Babbage's difference engine
- Ada Byron Lovelace and the art of programming
- Alan Turing and the Turing machine
- Early computers: Mark I, ENIAC, EDSAC, and UNIVAC I

Writing Your Report

A report on any mathematical topic should be more than just a list of facts. It should also convey the elements of discovery and excitement. Imagine what it must have felt like to be among the first to experience the development of a new idea or theory. Try to reconstruct each episode. Include worked-out examples and diagrams or photographs. Investigate the social and historical events surrounding your subjects. Also investigate the economic or scientific needs that might have promoted each discovery. Finally, convince your readers that the mathematicians you mention are or were real people. Describe their personalities and traits as well as their accomplishments. Use quotations from their writings whenever possible.

Some Resources

Albers, Donald J., and G. L. Alexanderson, eds. *Mathematical People: Profiles and Interviews.* Boston: Birkhäuser, 1985.

Bell, E. T. *Men of Mathematics.* New York: Simon and Schuster, 1937.

Boyer, Carl B. *A History of Mathematics.* New York: John Wiley and Sons, 1968.

Bunch, Bryan H. *Mathematical Fallacies and Paradoxes.* New York: Van Nostrand Reinhold, 1982.

Campbell, Douglas M., and John C. Higgins, eds. *Mathematics: People, Problems, Results.* Belmont, CA: Wadsworth, 1984.

Davis, Philip J., and Reuben Hersh. *The Mathematical Experience.* Boston: Houghton Mifflin, 1982.

Evans, Christopher. *The Making of the Micro: A History of the Computer.* New York: Van Nostrand Reinhold, 1981.

Gleick, James. *Chaos: Making a New Science.* New York: Penguin Books, 1988.

Haber, Louis. *Black Pioneers of Science and Invention.* New York: Harcourt, Brace, and World, 1970.

Hoffman, Paul. *Archimedes' Revenge: The Joys and Perils of Mathematics.* New York: W. W. Norton, 1988.

Ifrah, Georges. *From One to Zero: A Universal History of Numbers.* New York: Viking, 1985.

Newman, James R. *The World of Mathematics.* New York: Simon and Schuster, 1956.

Osen, Lynn. *Women in Mathematics.* Cambridge, MA: MIT Press, 1974.

Perl, Teri. *Math Equals: Biographies of Women Mathematicians + Related Activities.* Menlo Park, CA: Addison Wesley, 1978.

Peterson, Ivars. *Islands of Truth: A Mathematical Mystery Cruise.* New York: W. H. Freeman, 1990.

Yǎn, Lǐ, and Dù Shírán. *Chinese Mathematics: A Concise History.* Oxford: Clarendon Press, 1987.

A Mayan Pyramid

Chapter 10 Trigonometric Addition Formulas

Overview

This chapter focuses on the sum, difference, double-angle, and half-angle formulas for sine, cosine, and tangent. In the first section, the difference formula for cosine is derived analytically, and from this identity flow the formulas for $\cos(\alpha + \beta)$ and $\sin(\alpha \pm \beta)$. In the next two sections, formulas for $\tan(\alpha \pm \beta)$, $\sin 2\alpha$, $\cos 2\alpha$, $\tan 2\alpha$, $\sin \frac{\alpha}{2}$, $\cos \frac{\alpha}{2}$, and $\tan \frac{\alpha}{2}$ are derived. Throughout the first three sections, students use the given formulas to simplify expressions, to prove identities, and to find the exact values for the sine, cosine, and tangent of certain angles. In the last section, students solve trigonometric equations either algebraically (using the formulas presented in this chapter) or geometrically (using technology).

Objectives

10-1 To derive and apply formulas for $\cos(\alpha \pm \beta)$ and for $\sin(\alpha \pm \beta)$.
10-2 To derive and apply formulas for $\tan(\alpha \pm \beta)$.
10-3 To derive and apply double-angle and half-angle formulas.
10-4 To use identities to solve trigonometric equations.

Supplementary Resources

Tests, pp. 27–28
Alt. Assess., pp. 29–31, 67
Activities, pp. 24–25, 26–27
Student Res. Guide, pp. 95–99

Software

Precalculus Plotter Plus
Activities, pp. 52, 53–54
Function Plotter

Pacing Guide

Section	10-1	10-2	10-3	10-4	Review	Test	Total
Standard (days)	2	2	2	2	1	1	10
Comprehensive (days)	2	1	2	2	1	1	9

Teaching Notes

10-1 | **pages 369–374**

Presenting the Section

You might begin teaching this section by immediately dispelling the notion that $\sin(\alpha + \beta) = \sin\alpha + \sin\beta$ or that $\cos(\alpha + \beta) = \cos\alpha + \cos\beta$ for all α and β. Have students confirm that these are not identities by substituting specific values for α and β, say $\alpha = 60°$ and $\beta = 30°$.

After deriving the four formulas in this section, ask students for which angles they are now able to find exact values by using these formulas. (Multiples of 15°) Also elicit from students ways to write inequalities that place an angle in a particular quadrant. For example, to place angle α in the second quadrant, they would use $\frac{\pi}{2} < \alpha < \pi$.

Using Technology

Because the sum and difference formulas for sine and cosine involve two variables (α and β), you cannot use a computer or graphing calculator to graph each side of one of these identities and confirm that the graphs coincide. However, you *can* hold one of the variables constant and graph the two sides of the resulting one-variable identity. For example, the identity

$$\cos\left(x + \frac{\pi}{4}\right) = \frac{\sqrt{2}}{2}\cos x - \frac{\sqrt{2}}{2}\sin x$$

can be confirmed by graphing

$$y = \cos\left(x + \frac{\pi}{4}\right) \text{ and } y = \frac{\sqrt{2}}{2}\cos x - \frac{\sqrt{2}}{2}\sin x.$$

Cooperative Learning

Since students seem to have difficulty with exercises such as finding $\sin(\alpha + \beta)$ given that $\sin\alpha = \frac{3}{5}$, $\sin\beta = \frac{24}{25}$, and $0 < \alpha < \frac{\pi}{2} < \beta < \pi$, assign each of four groups of students one of Exercises 27–30 on page 373. Have each group do the problem as stated. Then change the location of the angles, changing the sign of the cosine, sine, or tangent of the angle as necessary, and have each group do the resulting problem.

Assign one of Exercises 39a, 39c, 40a, and 41a to each of four groups. One person in each group should write the group's solution on a transparency, and another person should present it to the class.

Communication

Since many students may still feel uncomfortable with the use of α and β to name angles, point out that they are the first two letters of the Greek alphabet, *alpha* and *beta*. It is from these two letters that we get the word *alphabet*.

Review

Before introducing the section, review the unit circle and the coordinates of points on it in terms of the cosine and sine of an angle. Also review the law of cosines from Chapter 9 and the distance formula from Chapter 1.

Assessment

Assign this problem: Find $\sin 60°$ in as many different ways as you can. Several of the methods students may think of are: to use a calculator; to use a graphing calculator with the trace feature or a graphing program on a computer; to use a table; to find $\cos 60°$ and then use the Pythagorean relationship $\sin^2 60° + \cos^2 60° = 1$; to find $\sin(30° + 30°)$; to find $\sin(90° - 30°)$; or to use a 30°-60°-90° triangle. Have students write down and then compare their methods.

10-2 | **pages 375-379**

Presenting the Section

Begin this section by pointing out the fact that the tangent of an angle equals the sine of the angle divided by the cosine of the angle. Also remind students that multiplying an expression by another expression equivalent to 1 (no matter how complicated it appears) does not change the value of the original expression.

You may want to write the sum and difference formulas for tangent as follows:

$$\tan(\alpha \pm \beta) = \frac{\tan\alpha \pm \tan\beta}{1 \mp \tan\alpha\tan\beta}$$

Challenge students to derive formulas for cot $(\alpha + \beta)$ and cot $(\alpha - \beta)$ in terms of cot α and cot β. $\left(\cot (\alpha \pm \beta) = \frac{\cot \alpha \cot \beta \mp 1}{\cot \alpha \pm \cot \beta} \right)$

When discussing the material on "Angles Between Two Lines," you will need to use the theorem from geometry that an exterior angle of a triangle equals the sum of the two remote interior angles. Also, point out to students that an angle of inclination is measured from the positive x-axis to the line in a counterclockwise direction.

Cooperative Learning

The Communication feature at the end of this section (page 379) gives students an opportunity to work in small groups.

10-3 | pages 380–385

Presenting the Section

Several of the formulas from the previous two sections are now used to derive the double-angle and half-angle formulas. Students still might be tempted to think that sin 2α can be written as $2 \sin \alpha$ and that sin $\frac{1}{2}\alpha$ can be written as $\frac{1}{2} \sin \alpha$. The activities presented at the beginning of the section, where the students graph $y = \sin 2x$ and $y = 2 \sin x$, $y = \sin \frac{1}{2}x$ and $y = \frac{1}{2} \sin x$, are important to help dispel these erroneous ideas.

Review the formula $\sin^2 \alpha + \cos^2 \alpha = 1$ and its other forms: $\sin^2 \alpha = 1 - \cos^2 \alpha$ and $\cos^2 \alpha = 1 - \sin^2 \alpha$. These formulas will be used in deriving two other formulas for cos $2x$ and eventually two other formulas for tan $\frac{x}{2}$.

Using Technology

This is a good time to use a graphing calculator and its trace feature to examine the graphs of $y = \sin x$, $y = 2 \sin x$, $y = \sin 2x$, $y = \frac{1}{2} \sin x$, and $y = \sin \frac{1}{2}x$. Relate these to the familiar graphs of $y = x^2$, $y = 2x^2$, $y = (2x)^2$, $y = \frac{1}{2}x^2$, and $y = \left(\frac{1}{2}x\right)^2$. The trace feature is helpful in finding the range and period of the variations in the sine function.

Cooperative Learning

Have students work in groups of four to find applications involving one of the formulas from Sections 10-1 through 10-3. Many physics problems on electricity, heat, dynamics, aviation, navigation, and other topics use these formulas. Have each group present its problems to the class.

Review

For Exercise 45, page 385, review (1) the relationship between an exterior angle and the two remote interior angles; (2) how you know when a triangle is isosceles and what you then know about its angles; and (3) how you can use right-triangle trigonometry to find the tangent of an angle.

Assessment

This is a good time to assess students' knowledge of all the formulas presented in this chapter. Ask students to explain how the formula for cos $(\alpha - \beta)$ was used to derive the others. The diagram below shows the "family tree" of derivations.

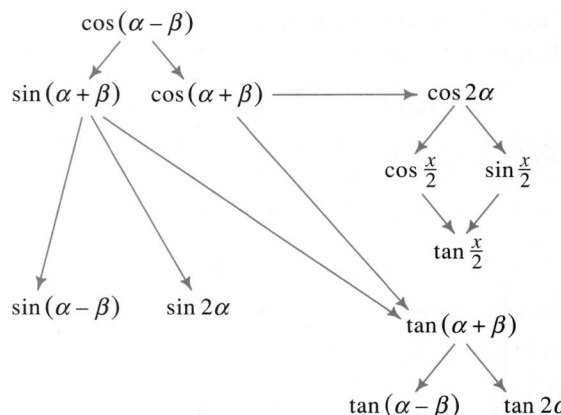

Have each student find the sine, cosine and tangent of a particular angle using the sum, difference, double-angle, and half-angle formulas. Instead of using angles that produce only "nice" answers, let the students use scientific calculators. Every student can be given a different angle with which to work.

Some examples of ways to find $\sin 26°$ are as follows:

$$\sin 26° = \sin (10° + 16°)$$
$$= \sin 10° \cos 16° + \cos 10° \sin 16°$$
$$\approx 0.4384$$
$$\sin 26° = \sin (30° - 4°)$$
$$\sin 26° = \sin \frac{52°}{2}$$
$$\sin 26° = \sin (2 \cdot 13°)$$

When students finish their problems, have them find $\cos 26°$ and $\tan 26°$ in the same manner.

10-4 | pages 386–391

Presenting the Section

Introduce the section by discussing ways of solving equations (including using a graphing calculator or a graphing program on a computer), and pitfalls to avoid (including dividing by a variable expression and losing roots). For many of the problems, even if a graphing calculator or computer is not used, sketching a graph by hand gives an idea not only of the number of roots, but also of some specific ones.

Be sure students understand that trigonometric equations generally have infinitely many solutions. The given restrictions on x (usually $0 \le x < 2\pi$ or $0° \le x < 360°$) result in a finite number of solutions.

Since none of the four examples in this section involves solving trigonometric inequalities (see the paragraph between Examples 1 and 2, however), you should present your own examples. For instance, you can show students how to solve the inequality $3 \cos 2x + \cos x > 2$ for $0 \le x < 2\pi$. (Note that this is a variation of Example 2.) Either perform a sign analysis (see Section 3-2) on the intervals determined by the roots of the related equation, or use a computer or graphing calculator. In the latter case, the two equations $y = 3 \cos 2x + \cos x$ and $y = 2$ can be graphed together and the intervals where the first graph is above the second can be noted, or the single equation $y = 3 \cos 2x + \cos x - 2$ can be graphed and the intervals where the graph is above the x-axis can be noted. (The solution of the inequality is $0 \le x < 0.59$ or $5.70 < x < 2\pi$.)

Using Technology

Graphing calculators and computers with graphing software are invaluable tools for solving trigonometric equations. Be sure students know how to use the zoom feature (or to rescale) in order to give solutions correct to the nearest hundredth. Once solutions are found, encourage students to check them by using scientific calculators after substituting the solutions back into the equations.

Cooperative Learning

Students can work in groups of four to write problems for which the answers are given. For example, if the answers are $x = 0°, 45°, 180°$, and $225°$, then the problem might be $\tan x \sin x - \sin x = 0$ or $\sin^2 x - \sin x \cos x = 0$. Have each person write a problem for a given set of answers. Problems should be discussed among the members of the group.

10 Trigonometric Addition Formulas

368

10-1 Formulas for cos (α ± β) and sin (α ± β)

Objective *To derive and apply formulas for cos (α ± β) and for sin (α ± β).*

The photograph at the right shows a monitor (an electrocardiograph) recording the heartbeats of a healthy athlete. The monitor shows a wave that depicts the occurrence of each heartbeat over a period of time. The equation of this wave involves sines and cosines of α, 2α, 3α, and larger multiples of α. Our goal in this chapter is to gain experience working with expressions like $\sin 2\alpha$ and $\cos 3\alpha$. We will derive formulas showing, for example, how $\cos 2\alpha$ is related to the cosine of α and how $\cos (\alpha + \beta)$ is related to the sine and cosine of α and β.

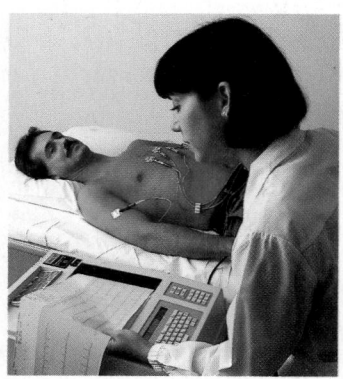

Formulas for cos (α ± β)

To find a formula for $\cos (\alpha - \beta)$, let A and B be points on the unit circle with coordinates as shown in the diagram at the right. Then the measure of $\angle AOB$ is $\alpha - \beta$. The distance AB can be found by using either the law of cosines or the distance formula.

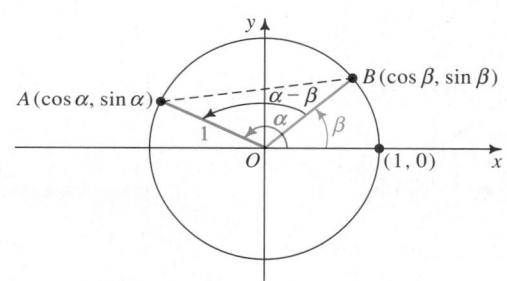

Using the law of cosines:

$$(AB)^2 = 1^2 + 1^2 - 2 \cdot 1 \cdot 1 \cdot \cos (\alpha - \beta) = 2 - 2 \cos (\alpha - \beta)$$

Using the distance formula:

$$
\begin{aligned}
(AB)^2 &= (\cos \alpha - \cos \beta)^2 + (\sin \alpha - \sin \beta)^2 \\
&= \cos^2 \alpha - 2 \cos \alpha \cos \beta + \cos^2 \beta + \sin^2 \alpha - 2 \sin \alpha \sin \beta + \sin^2 \beta \\
&= (\cos^2 \alpha + \sin^2 \alpha) + (\cos^2 \beta + \sin^2 \beta) - 2(\cos \alpha \cos \beta + \sin \alpha \sin \beta) \\
&= 2 - 2(\cos \alpha \cos \beta + \sin \alpha \sin \beta)
\end{aligned}
$$

Therefore,

$$2 - 2 \cos (\alpha - \beta) = 2 - 2(\cos \alpha \cos \beta + \sin \alpha \sin \beta).$$
$$\cos (\alpha - \beta) = \cos \alpha \cos \beta + \sin \alpha \sin \beta \qquad (1)$$

◀ The sound of every stringed instrument can be represented by the graph of a trigonometric function. The curve is determined by unique patterns of tone and overtone, which translate into sums of trigonometric functions.

Trigonometric Addition Formulas **369**

Warm-Up Exercises

Express each angle (a) as a sum and (b) as a difference of multiples of 30°, 45°, or 60°. Answers may vary. Examples are given.

1. 255° **a.** 225° + 30°, **b.** 315° − 60°

2. 195° **a.** 135° + 60°, **b.** 225° − 30°

3. 345° **a.** 315° + 30°, **b.** 390° − 45°

Express each angle (a) as a sum and (b) as a difference of multiples of $\frac{\pi}{6}$, $\frac{\pi}{4}$, or $\frac{\pi}{3}$. Answers may vary. Examples are given.

4. $\frac{19\pi}{12}$ **a.** $\frac{5\pi}{4} + \frac{\pi}{3}$, **b.** $\frac{11\pi}{6} - \frac{\pi}{4}$

5. $-\frac{\pi}{12}$ **a.** $-\frac{\pi}{3} + \frac{\pi}{4}$, **b.** $\frac{\pi}{6} - \frac{\pi}{4}$

Motivating the Section

Read aloud the introductory paragraph of this section in class. Point out that many scientific applications can be modeled by using trigonometry.

Assessment Note

You might have students give justifications for the steps taken in the derivations of the sum and difference formulas for sine and cosine.

Mathematical Note

The unit circle can be used to show visually that sin α and sin (−α) are opposites but that cos α and cos (−α) are equal.

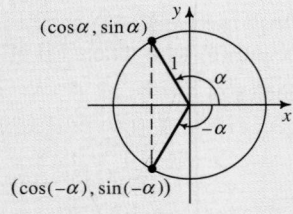

$(\cos\alpha, \sin\alpha)$

$(\cos(-\alpha), \sin(-\alpha))$

$\cos(-\alpha) = \cos\alpha$
$\sin(-\alpha) = -\sin\alpha$

Problem Solving

To show that there is more than one way to find sin 15°, have students work Example 1 using angles of 60° and 45°.

To obtain a formula for cos $(\alpha + \beta)$, we can use the formula for cos $(\alpha - \beta)$ and replace β with $-\beta$. Recall that cos $(-\beta)$ = cos β and sin $(-\beta)$ = $-\sin \beta$.

$$\cos (\alpha - (-\beta)) = \cos \alpha \cos (-\beta) + \sin \alpha \sin (-\beta)$$
$$\cos (\alpha + \beta) = \cos \alpha \cos \beta - \sin \alpha \sin \beta \qquad (2)$$

Formulas for sin $(\alpha \pm \beta)$

To find a formula for sin $(\alpha + \beta)$, we use the cofunction relationship

$$\sin \theta = \cos \left(\frac{\pi}{2} - \theta\right).$$

Let $\theta = \alpha + \beta$. Then:

$$\sin (\alpha + \beta) = \cos \left[\frac{\pi}{2} - (\alpha + \beta)\right]$$
$$= \cos \left[\left(\frac{\pi}{2} - \alpha\right) - \beta\right]$$
$$= \cos \left(\frac{\pi}{2} - \alpha\right) \cos \beta + \sin \left(\frac{\pi}{2} - \alpha\right) \sin \beta$$

Since $\cos \left(\frac{\pi}{2} - \alpha\right) = \sin \alpha$ and $\sin \left(\frac{\pi}{2} - \alpha\right) = \cos \alpha$:

$$\sin (\alpha + \beta) = \sin \alpha \cos \beta + \cos \alpha \sin \beta \qquad (3)$$

If we use the formula for sin $(\alpha + \beta)$ and replace β with $-\beta$, we get:

$$\sin (\alpha - \beta) = \sin \alpha \cos \beta - \cos \alpha \sin \beta \qquad (4)$$

Sum and Difference Formulas for Cosine and Sine

$$\cos (\alpha \pm \beta) = \cos \alpha \cos \beta \mp \sin \alpha \sin \beta$$
$$\sin (\alpha \pm \beta) = \sin \alpha \cos \beta \pm \cos \alpha \sin \beta$$

There are two main purposes for the addition formulas: finding *exact* values of trigonometric expressions and simplifying expressions to obtain other identities. The following examples illustrate how to use these formulas.

Example 1   Find the exact value of sin 15°.

Solution   Use formula (4) to get the exact value.

$$\sin 15° = \sin (45° - 30°)$$
$$= \sin 45° \cos 30° - \cos 45° \sin 30°$$
$$= \frac{\sqrt{2}}{2} \cdot \frac{\sqrt{3}}{2} - \frac{\sqrt{2}}{2} \cdot \frac{1}{2} = \frac{\sqrt{6} - \sqrt{2}}{4}$$

Example 2 Find the exact value of: **a.** $\cos 50° \cos 10° - \sin 50° \sin 10°$

b. $\sin \dfrac{5\pi}{12} \cos \dfrac{\pi}{12} + \cos \dfrac{5\pi}{12} \sin \dfrac{\pi}{12}$

Solution The two given expressions have the patterns shown in formulas (2) and (3), respectively.

a.
$$\cos \alpha \cos \beta - \sin \alpha \sin \beta = \cos (\alpha + \beta)$$
$$\cos 50° \cos 10° - \sin 50° \sin 10° = \cos (50° + 10°)$$
$$= \cos 60°$$
$$= \frac{1}{2}$$

b.
$$\sin \alpha \cos \beta + \cos \alpha \sin \beta = \sin (\alpha + \beta)$$
$$\sin \frac{5\pi}{12} \cos \frac{\pi}{12} + \cos \frac{5\pi}{12} \sin \frac{\pi}{12} = \sin \left(\frac{5\pi}{12} + \frac{\pi}{12} \right)$$
$$= \sin \frac{\pi}{2}$$
$$= 1$$

Example 3 Suppose that $\sin \alpha = \dfrac{4}{5}$ and $\sin \beta = \dfrac{5}{13}$, where $0 < \alpha < \dfrac{\pi}{2}$ and $\dfrac{\pi}{2} < \beta < \pi$. Find $\cos (\alpha + \beta)$.

Solution Sketch right-triangle diagrams to help find $\cos \alpha$ and $\cos \beta$. If $\sin \alpha = \dfrac{4}{5}$ and $0 < \alpha < \dfrac{\pi}{2}$, then $\cos \alpha = \dfrac{3}{5}$.

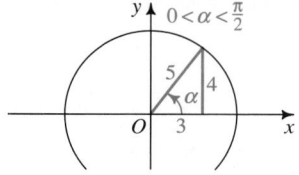 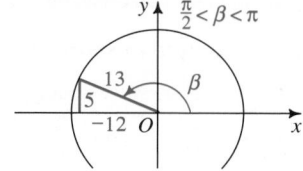

If $\sin \beta = \dfrac{5}{13}$ and $\dfrac{\pi}{2} < \beta < \pi$, then $\cos \beta = -\dfrac{12}{13}$.

Thus,
$$\cos (\alpha + \beta) = \cos \alpha \cos \beta - \sin \alpha \sin \beta$$
$$\cos (\alpha + \beta) = \left(\frac{3}{5} \right) \left(-\frac{12}{13} \right) - \left(\frac{4}{5} \right) \left(\frac{5}{13} \right) = -\frac{56}{65}$$

The sum or difference formulas can be used to verify many identities that we have seen, such as $\sin (90° - \theta) = \cos \theta$, and also to derive new identities.

Trigonometric Addition Formulas **371**

Additional Examples

1. Evaluate each expression without using a calculator or a table.
   a. sin 285° **b.** cos 285°

   Note that there are many ways to evaluate these expressions (for example, expressing 285° as 240° + 45°, or 315° − 30°, and so on).

   a. $\sin 285° =$
   $\sin (225° + 60°) =$
   $\sin 225° \cos 60° +$
   $\cos 225° \sin 60° =$
   $-\dfrac{\sqrt{2}}{2} \cdot \dfrac{1}{2} +$
   $\left(-\dfrac{\sqrt{2}}{2}\right)\dfrac{\sqrt{3}}{2} =$
   $-\dfrac{\sqrt{2} + \sqrt{6}}{4}$

   b. $\cos 285° =$
   $\cos (225° + 60°) =$
   $\cos 225° \cos 60° -$
   $\sin 225° \sin 60° =$
   $-\dfrac{\sqrt{2}}{2} \cdot \dfrac{1}{2} -$
   $\left(-\dfrac{\sqrt{2}}{2}\right)\dfrac{\sqrt{3}}{2} =$
   $\dfrac{\sqrt{6} - \sqrt{2}}{4}$

Example 4 Show that $\sin \left(\dfrac{3\pi}{2} - x\right) = -\cos x$.

Solution By formula (4),

$$\sin \left(\frac{3\pi}{2} - x\right) = \sin \left(\frac{3\pi}{2}\right) \cos x - \cos \left(\frac{3\pi}{2}\right) \sin x$$
$$= (-1) \cos x - (0) \sin x$$
$$= -\cos x$$

Sometimes a problem involving a sum can be more easily solved if the sum can be expressed as a product. The formulas given below are derived from the addition formulas in Exercises 39–41.

Rewriting a Sum or Difference as a Product

$$\sin x + \sin y = 2 \sin \frac{x + y}{2} \cos \frac{x - y}{2}$$

$$\sin x - \sin y = 2 \cos \frac{x + y}{2} \sin \frac{x - y}{2}$$

$$\cos x + \cos y = 2 \cos \frac{x + y}{2} \cos \frac{x - y}{2}$$

$$\cos x - \cos y = -2 \sin \frac{x + y}{2} \sin \frac{x - y}{2}$$

CLASS EXERCISES

Find an example to show that, in general, each statement is true.
Answers may vary.

1. $\sin (x + y) \neq \sin x + \sin y$
For example, let $x = 60°$, $y = 30°$.

2. $\cos (a - b) \neq \cos a - \cos b$
For example, let $a = 60°$, $b = 0°$.

Are there any values of a and b for which each statement is true? If so, give an example. Answers may vary.

3. $\sin (a - b) = \sin a - \sin b$
Yes. For example, let $b = 0°$.

4. $\sin (a + b) = \sin a + \sin b$
Yes. For example, let $b = 0°$.

Simplify each expression. Do not evaluate.

5. $\sin 1° \cos 2° + \cos 1° \sin 2°$ $\sin 3°$

6. $\sin 20° \cos 15° - \cos 20° \sin 15°$ $\sin 5°$

7. $\cos \dfrac{\pi}{4} \cos \dfrac{\pi}{4} - \sin \dfrac{\pi}{4} \sin \dfrac{\pi}{4}$ $\cos \dfrac{\pi}{2}$

8. $\cos 75° \cos 25° + \sin 75° \sin 25°$ $\cos 50°$

9. *Discussion* Explain how the identity $\sin (-\alpha) = -\sin \alpha$ is a special case of the difference formula for sine. $\sin (-\alpha) = \sin (0 - \alpha) =$
$\sin 0 \cos \alpha - \cos 0 \sin \alpha = 0 - \sin \alpha = -\sin \alpha$

372 *Chapter Ten*

Simplify the given expression.

A 1. $\sin 75° \cos 15° + \cos 75° \sin 15°$ 1

2. $\cos 105° \cos 15° + \sin 105° \sin 15°$ 0

3. $\cos \frac{5\pi}{12} \cos \frac{\pi}{12} - \sin \frac{5\pi}{12} \sin \frac{\pi}{12}$ 0

4. $\sin \frac{4\pi}{3} \cos \frac{\pi}{3} - \cos \frac{4\pi}{3} \sin \frac{\pi}{3}$ 0

5. $\sin 3x \cos 2x - \cos 3x \sin 2x$ $\sin x$

6. $\cos 2x \cos x + \sin 2x \sin x$ $\cos x$

In Exercises 7–10, prove that the given equation is an identity.

7. $\sin (x + \pi) = -\sin x$

8. $\cos (\pi + x) = -\cos x$

9. $\cos \left(x + \frac{\pi}{2}\right) = -\sin x$

10. $\cos \left(\frac{\pi}{2} - x\right) = \sin x$

11. *Visual Thinking* Translate the graph of $y = \sin x$ to the right by π units, and then reflect this curve about the x-axis. What graph results? $y = \sin x$

12. *Visual Thinking* Translate the graph of $y = \cos x$ to the left by $\frac{\pi}{2}$ units, and then reflect this curve about the x-axis. What graph results? $y = \sin x$

Find the exact value of each expression.

13. $\cos 75°$ $\frac{\sqrt{6} - \sqrt{2}}{4}$

14. $\cos 15°$ $\frac{\sqrt{6} + \sqrt{2}}{4}$

15. $\cos 105°$ $\frac{\sqrt{2} - \sqrt{6}}{4}$

16. $\sin 105°$ $\frac{\sqrt{6} + \sqrt{2}}{4}$

17. $\sin (-15°)$ $\frac{\sqrt{2} - \sqrt{6}}{4}$

18. $\cos (-165°)$ $\frac{-\sqrt{6} - \sqrt{2}}{4}$

19. $\sin \frac{7\pi}{12}$ $\frac{\sqrt{6} + \sqrt{2}}{4}$

20. $\sin \frac{11\pi}{12}$ $\frac{\sqrt{6} - \sqrt{2}}{4}$

Simplify the given expression.

21. $\sin (30° + \theta) + \sin (30° - \theta)$ $\cos \theta$

22. $\cos (30° + \theta) + \cos (30° - \theta)$ $\sqrt{3} \cos \theta$

23. $\cos \left(\frac{\pi}{3} + x\right) + \cos \left(\frac{\pi}{3} - x\right)$ $\cos x$

24. $\sin \left(\frac{\pi}{4} + x\right) + \sin \left(\frac{\pi}{4} - x\right)$ $\sqrt{2} \cos x$

25. $\cos \left(\frac{3\pi}{2} + x\right) + \cos \left(\frac{3\pi}{2} - x\right)$ 0

26. $\sin (\pi + x) + \sin (\pi - x)$ 0

B 27. Suppose that $\sin \alpha = \frac{3}{5}$ and $\sin \beta = \frac{24}{25}$, where $0 < \alpha < \frac{\pi}{2} < \beta < \pi$. Find $\sin (\alpha + \beta)$. $\frac{3}{5}$

28. Suppose that $\sin \alpha = \frac{4}{5}$ and $\sin \beta = \frac{1}{2}$, where $\frac{\pi}{2} < \beta < \alpha < \pi$. Find $\sin (\alpha - \beta)$. $\frac{3 - 4\sqrt{3}}{10}$

29. Suppose that $\tan \alpha = \frac{4}{3}$ and $\tan \beta = \frac{12}{5}$, where $0 < \alpha < \beta < \frac{\pi}{2}$. Find $\cos (\alpha - \beta)$. $\frac{63}{65}$

30. Suppose that $\sec \alpha = \frac{5}{4}$ and $\tan \beta = -1$, where $0 < \alpha < \frac{\pi}{2} < \beta < \pi$. Find $\cos (\alpha + \beta)$. $-\frac{7\sqrt{2}}{10}$

---

2. Prove that $\sec \left(\frac{\pi}{2} - x\right) = \csc x$.

$$\sec \left(\frac{\pi}{2} - x\right) =$$

$$\frac{1}{\cos \left(\frac{\pi}{2} - x\right)} =$$

$$\frac{1}{\cos \frac{\pi}{2} \cos x + \sin \frac{\pi}{2} \sin x} =$$

$$\frac{1}{0 + 1 \cdot \sin x} = \frac{1}{\sin x} = \csc x$$

Cooperative Learning

You might have students work in groups and think of ways that they can translate and reflect the graphs of the sine and cosine functions with the result being one of these graphs (see Exercises 11 and 12). They should then write identities based on these transformations, and they can check their answers on a graphing calculator.

Suggested Assignments

Standard
Day 1: 373/1–21 odd
Day 2: 373/23–37 odd
Comprehensive
Day 1: 373/1–27 odd
Day 2: 373/29–43 odd

Supplementary Materials

Alternative Assessment, 29–30
Activities Book, 24–25
Precalculus Plotter Plus, 53–54

Use a graphing calculator set in radian mode to sketch each graph. Explain how the graph is related to a familiar graph and why.

31. $y = \sin x \cos 1 + \cos x \sin 1$ **32.** $y = \cos x \cos 2 + \sin x \sin 2$

Simplify the given expression.

33. $\dfrac{\sin (\alpha + \beta) + \sin (\alpha - \beta)}{\cos \alpha \cos \beta}$ $2 \tan \alpha$

34. $\dfrac{\cos (\alpha + \beta) + \cos (\alpha - \beta)}{\sin \alpha \sin \beta}$ $\dfrac{2 \cot \alpha \cot \beta}{}$

35. $\cos x \cos y (\tan x + \tan y)$ $\sin (x + y)$

36. $\sin x \sin y (\cot x \cot y - 1)$ $\cos (x + y)$

37. $\sin (x + y) \sec x \sec y$ $\tan x + \tan y$

38. $\cos \left(x + \dfrac{\pi}{3}\right) + \sin \left(x - \dfrac{\pi}{6}\right)$ 0

39. a. Derive the formula $\sin x + \sin y = 2 \sin \dfrac{x + y}{2} \cos \dfrac{x - y}{2}$.

(*Hint:* Show that $\sin (\alpha + \beta) + \sin (\alpha - \beta) = 2 \sin \alpha \cos \beta$, and then substitute $\alpha = \dfrac{x + y}{2}$ and $\beta = \dfrac{x - y}{2}$.)

b. Use the formula in part (a) to show that $\sin 40° + \sin 20° = \cos 10°$.
c. Use the formula in part (a) to derive the formula for $\sin x - \sin y$.

40. a. Derive the formula $\cos x + \cos y = 2 \cos \dfrac{x + y}{2} \cos \dfrac{x - y}{2}$. (Simplify $\cos (\alpha + \beta) + \cos (\alpha - \beta)$. Then substitute $\alpha = \dfrac{x + y}{2}$ and $\beta = \dfrac{x - y}{2}$, as was done in Exercise 39.)

b. Find the exact value of $\cos 105° + \cos 15°$. $\dfrac{\sqrt{2}}{2}$

41. a. Derive the formula $\cos x - \cos y = -2 \sin \dfrac{x + y}{2} \sin \dfrac{x - y}{2}$. (*Hint:* See Exercise 39.)

b. Find the exact value of $\cos 75° - \cos 15°$. $-\dfrac{\sqrt{2}}{2}$

C **42.** Evaluate $\sin \left(\text{Tan}^{-1} \dfrac{1}{2} + \text{Tan}^{-1} \dfrac{1}{3}\right)$ without using a calculator or tables. $\dfrac{\sqrt{2}}{2}$

43. Evaluate $\cos \left(\text{Tan}^{-1} \dfrac{1}{2} + \text{Tan}^{-1} 2\right)$ without using a calculator or tables. 0

44. Ptolemy's theorem states that if *ABCD* is inscribed in a circle, then

$$AB \cdot CD + AD \cdot BC = AC \cdot BD.$$

Consider a special case of this theorem in which $\overline{BD}$ is a diameter with length 1.
a. Show that $AB = \sin \alpha$ and $AD = \cos \alpha$.
b. Find *BC* and *CD* in terms of β.
c. Show that $AC = \sin (\alpha + \beta)$.
d. Use parts (a), (b), and (c), and Ptolemy's theorem to derive the formula for $\sin (\alpha + \beta)$.

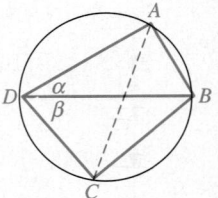

374 *Chapter Ten*

David Blackwell (1919–)

David Blackwell expected to become a school-teacher like his grandfather, but his interest in statistics led to a Ph.D. at the age of 22. After teaching at Howard University for ten years, he became a professor of statistics at the University of California at Berkeley in 1954.

Blackwell has made contributions to many fields of mathematics, including probability theory, game theory, set theory, and dynamic programming. In 1965, Blackwell was the first African American mathematician elected to the National Academy of Sciences.

10-2 Formulas for tan $(\alpha \pm \beta)$

Objective   *To derive and apply formulas for tan $(\alpha \pm \beta)$.*

To derive a formula expressing tan $(\alpha + \beta)$ in terms of tan α and tan β, we use the formulas for sin $(\alpha + \beta)$ and cos $(\alpha + \beta)$.

$$\tan (\alpha + \beta) = \frac{\sin (\alpha + \beta)}{\cos (\alpha + \beta)}$$
$$= \frac{\sin \alpha \cos \beta + \cos \alpha \sin \beta}{\cos \alpha \cos \beta - \sin \alpha \sin \beta}$$

Dividing the numerator and the denominator by cos α cos β, we obtain:

$$\tan (\alpha + \beta) = \frac{\dfrac{\sin \alpha \cos \beta}{\cos \alpha \cos \beta} + \dfrac{\cos \alpha \sin \beta}{\cos \alpha \cos \beta}}{\dfrac{\cos \alpha \cos \beta}{\cos \alpha \cos \beta} - \dfrac{\sin \alpha \sin \beta}{\cos \alpha \cos \beta}} = \frac{\dfrac{\sin \alpha}{\cos \alpha} + \dfrac{\sin \beta}{\cos \beta}}{1 - \dfrac{\sin \alpha}{\cos \alpha} \cdot \dfrac{\sin \beta}{\cos \beta}}$$

Therefore, we have the following formula.

Sum Formula for Tangent
$$\tan (\alpha + \beta) = \frac{\tan \alpha + \tan \beta}{1 - \tan \alpha \tan \beta} \qquad (5)$$

This identity is valid for all values of α and β for which tan α, tan β, and tan $(\alpha + \beta)$ are defined.

Trigonometric Addition Formulas   **375**

To derive a formula for $\tan(\alpha - \beta)$, simply replace β with $-\beta$ in formula (5) and use the fact that $\tan(-\beta) = -\tan \beta$. Thus, we have the following formula.

Difference Formula for Tangent

$$\tan(\alpha - \beta) = \frac{\tan \alpha - \tan \beta}{1 + \tan \alpha \tan \beta} \qquad (6)$$

Example 1 Suppose $\tan \alpha = \dfrac{1}{3}$ and $\tan \beta = \dfrac{1}{2}$.

a. Find $\tan(\alpha + \beta)$. **b.** Show that $\text{Tan}^{-1} \dfrac{1}{3} + \text{Tan}^{-1} \dfrac{1}{2} = \dfrac{\pi}{4}$.

Solution **a.** $\tan(\alpha + \beta) = \dfrac{\tan \alpha + \tan \beta}{1 - \tan \alpha \tan \beta} = \dfrac{\dfrac{1}{3} + \dfrac{1}{2}}{1 - \dfrac{1}{3} \cdot \dfrac{1}{2}} = \dfrac{\dfrac{5}{6}}{\dfrac{5}{6}} = 1$

b. Let $\alpha = \text{Tan}^{-1} \dfrac{1}{3}$ and $\beta = \text{Tan}^{-1} \dfrac{1}{2}$. Since α and β are both between 0 and $\dfrac{\pi}{2}$, $\alpha + \beta$ is between 0 and π. From part (a), we know that $\tan(\alpha + \beta) = 1$. Thus, since $\dfrac{\pi}{4}$ is the only angle between 0 and π whose tangent is 1,

$$\alpha + \beta = \frac{\pi}{4}.$$

That is,

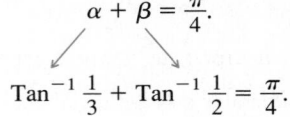

$$\text{Tan}^{-1} \frac{1}{3} + \text{Tan}^{-1} \frac{1}{2} = \frac{\pi}{4}.$$

Angles between Two Lines

The formula for $\tan(\alpha - \beta)$ can be used to find an angle between two intersecting lines. For example, consider the angle θ between lines l_1 and l_2 with slopes m_1 and m_2. Suppose the inclinations of these lines are α and β, respectively. Recall from Section 8-1 that the slope of a nonvertical line is the tangent of its inclination. Thus,

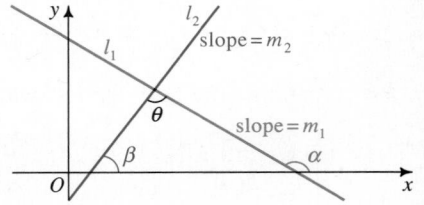

$$\tan \alpha = m_1 \text{ and } \tan \beta = m_2.$$

Since the measure of an exterior angle of a triangle equals the sum of the measures of the opposite interior angles, $\alpha = \beta + \theta$. Therefore,

$$\theta = \alpha - \beta \text{ and } \tan \theta = \tan(\alpha - \beta).$$

376 *Chapter Ten*

Substituting in $\tan(\alpha - \beta) = \dfrac{\tan \alpha - \tan \beta}{1 + \tan \alpha \tan \beta}$ gives the following formula:

$$\tan \theta = \frac{m_1 - m_2}{1 + m_1 m_2}.$$

Example 2 Find an angle between the lines $y = 3x + 1$ and $y = 5 - 2x$.

Solution The lines have slopes 3 and -2. Thus, we can either let $m_1 = 3$ and $m_2 = -2$ or let $m_1 = -2$ and $m_2 = 3$. These two possibilities give us the two supplementary angles, θ_1 and θ_2, that are formed by the lines.

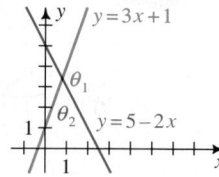

(1) The first case gives

$$\tan \theta_1 = \frac{3 - (-2)}{1 + 3(-2)} = -1. \text{ So, } \theta_1 = 135°.$$

(2) The second case gives

$$\tan \theta_2 = \frac{-2 - 3}{1 + (-2)3} = 1. \text{ So, } \theta_2 = 45°.$$

CLASS EXERCISES

1. Suppose $\tan \alpha = 2$ and $\tan \beta = 3$. Find **(a)** $\tan(\alpha + \beta)$ and **(b)** $\tan(\alpha - \beta)$. $-1; -\dfrac{1}{7}$

2. Find the exact value: **a.** $\dfrac{\tan 15° + \tan 30°}{1 - \tan 15° \tan 30°}$ **b.** $\dfrac{\tan 85° - \tan 25°}{1 + \tan 85° \tan 25°}$ $1; \sqrt{3}$

3. *Discussion* Interpret the formula for finding an angle between two lines when $1 + m_1 m_2 = 0$. The lines are perpendicular.

WRITTEN EXERCISES

In Exercises 1 and 2, find **(a)** $\tan(\alpha + \beta)$ and **(b)** $\tan(\alpha - \beta)$.

A 1. $\tan \alpha = \dfrac{2}{3}$ and $\tan \beta = \dfrac{1}{2}$ $\dfrac{7}{4}; \dfrac{1}{8}$ 2. $\tan \alpha = 2$ and $\tan \beta = -\dfrac{1}{3}$ $1; 7$

In Exercises 3–6, find the exact value of the given expression.

3. $\dfrac{\tan 75° - \tan 30°}{1 + \tan 75° \tan 30°}$ 1 4. $\dfrac{\tan 100° + \tan 50°}{1 - \tan 100° \tan 50°}$ $-\dfrac{\sqrt{3}}{3}$

5. $\dfrac{\tan \dfrac{2\pi}{3} + \tan \dfrac{\pi}{12}}{1 - \tan \dfrac{2\pi}{3} \tan \dfrac{\pi}{12}}$ -1 6. $\dfrac{\tan \dfrac{4\pi}{3} - \tan \dfrac{\pi}{12}}{1 + \tan \dfrac{4\pi}{3} \tan \dfrac{\pi}{12}}$ 1

Trigonometric Addition Formulas **377**

2. If $\dfrac{\pi}{2} < \alpha < \pi < \beta < \dfrac{3\pi}{2}$, $\cos \alpha = -\dfrac{15}{17}$, and $\tan \beta = \dfrac{7}{24}$, find $\sin(\alpha + \beta)$, $\cos(\alpha + \beta)$, $\tan(\alpha + \beta)$, $\sin(\alpha - \beta)$, $\cos(\alpha - \beta)$, and $\tan(\alpha - \beta)$.

Draw right-triangle diagrams.

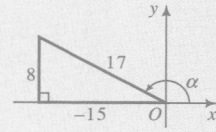

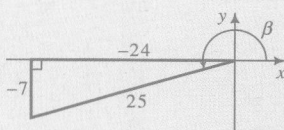

Using the diagrams, we see that:

$\sin \alpha = \dfrac{8}{17}$; $\tan \alpha = -\dfrac{8}{15}$;

$\sin \beta = -\dfrac{7}{25}$; and

$\cos \beta = -\dfrac{24}{25}$.

$\sin(\alpha + \beta) = \dfrac{8}{17}\left(-\dfrac{24}{25}\right) + \left(-\dfrac{15}{17}\right)\left(-\dfrac{7}{25}\right) = -\dfrac{87}{425}$.

$\sin(\alpha - \beta) = \dfrac{8}{17}\left(-\dfrac{24}{25}\right) - \left(-\dfrac{15}{17}\right)\left(-\dfrac{7}{25}\right) = -\dfrac{297}{425}$.

$\cos(\alpha + \beta) = \left(-\dfrac{15}{17}\right)\left(-\dfrac{24}{25}\right) - \dfrac{8}{17}\left(-\dfrac{7}{25}\right) = \dfrac{416}{425}$.

$\cos(\alpha - \beta) = \left(-\dfrac{15}{17}\right)\left(-\dfrac{24}{25}\right) + \dfrac{8}{17}\left(-\dfrac{7}{25}\right) = \dfrac{304}{425}$.

(*Continues on next page*)

$$\tan(\alpha + \beta) =$$

$$\frac{-\frac{8}{15} + \frac{7}{24}}{1 - \left(-\frac{8}{15}\right)\left(\frac{7}{24}\right)} =$$

$$\frac{\frac{-29}{120}}{\frac{52}{45}} = -\frac{87}{416}$$

$$\left(\text{or } \tan(\alpha + \beta) = \frac{\sin(\alpha + \beta)}{\cos(\alpha + \beta)} = -\frac{87}{416}\right).$$

$$\tan(\alpha - \beta) =$$

$$\frac{-\frac{8}{15} - \frac{7}{24}}{1 + \left(-\frac{8}{15}\right)\left(\frac{7}{24}\right)} =$$

$$\frac{-\frac{99}{120}}{\frac{38}{45}} = -\frac{297}{304}$$

$$\left(\text{or } \tan(\alpha - \beta) = \frac{\sin(\alpha - \beta)}{\cos(\alpha - \beta)} = -\frac{297}{304}\right).$$

◪ Using Technology

If students have programmable calculators, they might like to write a program for finding the angle between two lines.

Suggested Assignments

Standard
Day 1: 377/1–10
Day 2: 378/11, 13, 14, 17, 19, 23, 25
Comprehensive
377/1, 5–19 odd, 23, 29

Supplementary Materials

Alternative Assessment, 30

---

7. Evaluate $\tan\left(\frac{\pi}{4} + \theta\right)$ when $\tan\theta = \frac{1}{2}$. 3

8. Evaluate $\tan\left(\frac{3\pi}{4} - \theta\right)$ when $\tan\theta = \frac{1}{3}$. -2

9. Show that $\tan(-\alpha) = -\tan\alpha$.

10. **Visual Thinking** Simplify $\tan(x + \pi)$ and interpret your answer graphically. What does your answer illustrate about the period of the tangent function?

11. Evaluate $\tan 75°$ and $\tan 165°$ without using a calculator or tables. $2 + \sqrt{3}, -2 + \sqrt{3}$

12. Evaluate $\tan 15°$ and $\tan 105°$ without using a calculator or tables. $2 - \sqrt{3}, -2 - \sqrt{3}$

13. Find the two supplementary angles formed by the line $y = 3x - 5$ and the line $y = x + 4$. 26.6°, 153.4°

14. Find the two supplementary angles formed by the line $3x + 2y = 5$ and the line $4x - 3y = 1$. 70.6°, 109.4°

◪ **Use a graphing calculator set in radian mode to sketch each graph. Explain how the graph is related to the graph of $y = \tan x$ and why.**

15. $y = \dfrac{\tan x + \tan 1}{1 - \tan x \tan 1}$

16. $y = \dfrac{\tan x - \tan\frac{\pi}{8}}{1 + \tan x \tan\frac{\pi}{8}}$

17. Suppose $\tan\alpha = \frac{1}{4}$ and $\tan\beta = \frac{3}{5}$, where $0 < \alpha < \beta < \frac{\pi}{2}$. Find $\tan(\alpha + \beta)$.
Then show that $\text{Tan}^{-1}\frac{1}{4} + \text{Tan}^{-1}\frac{3}{5} = \frac{\pi}{4}$.

18. Suppose $\tan\alpha = 3$ and $\tan\beta = \frac{1}{2}$, where $0 < \beta < \alpha < \frac{\pi}{2}$. Find $\tan(\alpha - \beta)$.
Then show that $\text{Tan}^{-1}\,3 - \text{Tan}^{-1}\frac{1}{2} = \frac{\pi}{4}$.

19. Suppose $\alpha = \text{Tan}^{-1}\,2$ and $\beta = \text{Tan}^{-1}\,3$. Show that $\tan(\alpha + \beta) = -1$.

20. Suppose $\alpha = \text{Tan}^{-1}\,5$ and $\beta = \text{Tan}^{-1}\frac{2}{3}$. Show that $\alpha - \beta = \frac{\pi}{4}$.

21. **Investigation** For what values of α and β does $\tan(\alpha + \beta) = \tan\alpha + \tan\beta$? For what values of α and β does $\tan(\alpha - \beta) = \tan\alpha - \tan\beta$? See below.

22. Given $A(3, 1)$, $B(14, -1)$, and $C(5, 5)$, find the measure of $\angle BAC$ by using (a) the slopes of lines AB and AC and (b) the law of cosines. 73.7°; 73.7°

23. Suppose $\cot\alpha = 2$ and $\cot\beta = \frac{2}{3}$, where $0 < \alpha < \beta < \frac{\pi}{2}$. Find $\cot(\alpha + \beta)$. $\frac{1}{8}$

24. Suppose $\cot\alpha = \frac{3}{2}$ and $\cot\beta = \frac{1}{2}$, where $0 < \alpha < \beta < \frac{\pi}{2}$. Find $\cot(\alpha - \beta)$. $-\frac{7}{4}$

25. Suppose $\sin\alpha = \frac{3}{5}$ and $\cos\beta = \frac{5}{13}$, where $0 < \alpha < \beta < \frac{\pi}{2}$. Find:
   a. $\sin(\alpha + \beta)$ $\frac{63}{65}$ b. $\cos(\alpha + \beta)$ $-\frac{16}{65}$ c. $\tan(\alpha + \beta)$ $-\frac{63}{16}$

---

21. Both are true for α or $\beta = n\pi$ where n is an integer.

26. Suppose $\sin \alpha = \dfrac{4}{5}$ and $\tan \beta = -\dfrac{3}{4}$, where $\dfrac{\pi}{2} < \alpha < \beta < \pi$. Find:

    a. $\sin (\alpha + \beta)$ -1 **b.** $\cos (\alpha + \beta)$ 0 **c.** $\tan (\alpha + \beta)$ undefined

[C] 27. a. Line l bisects θ, the angle formed by lines l_1 and l_2. If the slopes of these three lines are m, m_1, and m_2, respectively, show that:

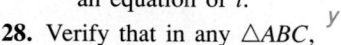

$$\frac{m_1 - m}{1 + m_1 m} = \frac{m - m_2}{1 + m_2 m}$$

    b. If l_1 and l_2 have equations $y = 2x$ and $y = x$, find an equation of l. $y = \dfrac{1 + \sqrt{10}}{3}\, x$

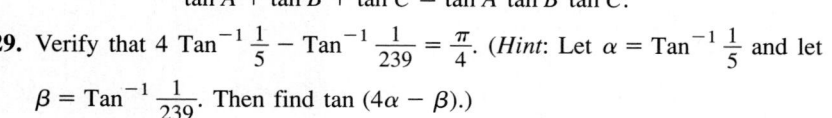

    l_1 (slope m_1)

    l_2 (slope m_2)

    l (slope m)

28. Verify that in any $\triangle ABC$,

$$\tan A + \tan B + \tan C = \tan A \tan B \tan C.$$

29. Verify that $4 \,\mathrm{Tan}^{-1}\dfrac{1}{5} - \mathrm{Tan}^{-1}\dfrac{1}{239} = \dfrac{\pi}{4}$. (*Hint*: Let $\alpha = \mathrm{Tan}^{-1}\dfrac{1}{5}$ and let $\beta = \mathrm{Tan}^{-1}\dfrac{1}{239}$. Then find $\tan(4\alpha - \beta)$.)

COMMUNICATION: Discussion

Small Group Discussion

Often when you are faced with a difficult problem, you can respond better if you have discussed it with others. Working cooperatively in a small group allows each of you to contribute your strengths and skills to solving the problem. Sometimes another person's perspective helps you better understand a problem. Likewise, your perspective may help someone else in the group gain insight. When working in a group, it is important that all members participate. This means that you must speak up when you have ideas and also listen carefully to the ideas of other persons in your group.

Working in groups of three or four, discuss one of the following questions. Then present your results to the entire class.

1. The decimal expansion of $\dfrac{1}{7}$, namely $0.\overline{142857}$, has a six-digit block of repeating digits. What is the maximum length of the block of repeating digits in the decimal expansion of $\dfrac{1}{n}$? Explain your answer.

2. Are there more rational numbers or irrational numbers?

3. Is the sum of two sine curves always another sine curve?

4. Why does it make sense that $b^0 = 1$ if $b \ne 0$? Why is 0^0 undefined?

Trigonometric Addition Formulas   **379**

Warm-Up Exercises

Using a calculator to evaluate each expression, tell whether the expression is equal to sin 50° or cos 50°.

1. $1 - 2\sin^2 25°$ cos 50°

2. $\cos^2 25° - \sin^2 25°$
   cos 50°

3. $2\sin 25° \cos 25°$ sin 50°

4. $2\cos^2 25° - 1$ cos 50°

5. $\sqrt{\dfrac{1 - \cos 100°}{2}}$ sin 50°

6. $\sqrt{\dfrac{1 + \cos 100°}{2}}$ cos 50°

Motivating the Section

Being able to work with functions of $2x$, $3x$, $\frac{1}{2}x$, and so on is important to accoustical engineers, as noted in the first paragraph of this section.

10-3 Double-Angle and Half-Angle Formulas

Objective   *To derive and apply double-angle and half-angle formulas.*

Trigonometric functions are used in science and engineering to study light and sound waves. An important application is the wave pattern of a vibrating string. Consider the wave of a note sounded by a violin. Not only does the string vibrate as a whole, producing a fundamental tone, but it also vibrates in halves, thirds, and progressively smaller segments, producing overtones (called harmonics). An equation of this type of wave involves sums of sines and cosines of x, $2x$, $3x$, and greater multiples of x. The computer-generated graphs below illustrate this.

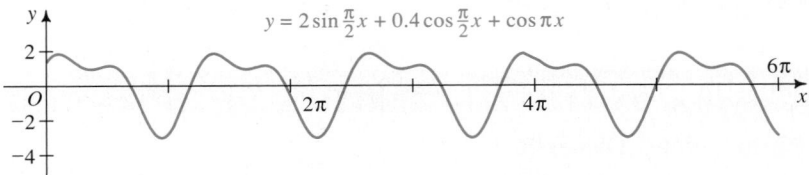

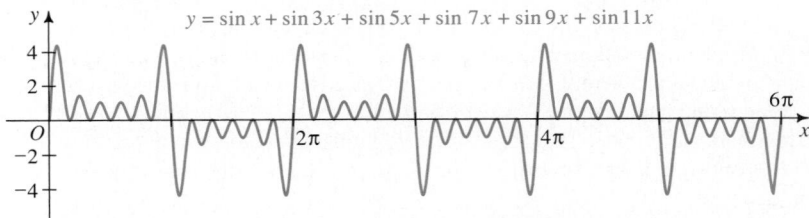

If you know the value of sin α, you do *not* double it to find sin 2α. Nor do you halve it to find sin $\frac{1}{2}\alpha$. To see that this is true, complete the following activity.
1. No. The values are not the same.   2. No. The values are not the same.

Activity

1. On a single set of axes, graph $y = \sin 2x$ and $y = 2\sin x$. Do the graphs coincide? What does this tell you about the values of sin $2x$ and $2\sin x$? See above.

2. On a single set of axes, graph $y = \sin \frac{1}{2}x$ and $y = \frac{1}{2}\sin x$. Do the graphs coincide? What does this tell you about the values of sin $\frac{1}{2}x$ and $\frac{1}{2}\sin x$? See above.

Double-Angle Formulas

The following double-angle formulas are special cases of the formulas for $\sin(\alpha + \beta)$, $\cos(\alpha + \beta)$, and $\tan(\alpha + \beta)$. If we let $\beta = \alpha$ in these formulas we obtain the following formulas.

$$\sin(\alpha + \beta) = \sin\alpha\cos\beta + \cos\alpha\sin\beta$$
$$\sin(\alpha + \alpha) = \sin\alpha\cos\alpha + \cos\alpha\sin\alpha$$
$$\sin 2\alpha = 2\sin\alpha\cos\alpha \tag{7}$$
$$\cos(\alpha + \beta) = \cos\alpha\cos\beta - \sin\alpha\sin\beta$$
$$\cos(\alpha + \alpha) = \cos\alpha\cos\alpha - \sin\alpha\sin\alpha$$
$$\cos 2\alpha = \cos^2\alpha - \sin^2\alpha \tag{8a}$$

Using the fact that $\sin^2\alpha + \cos^2\alpha = 1$, we can obtain alternative formulas for $\cos 2\alpha$:

$$\cos 2\alpha = 1 - 2\sin^2\alpha \tag{8b}$$
$$\cos 2\alpha = 2\cos^2\alpha - 1 \tag{8c}$$

To express $\tan 2\alpha$ in terms of $\tan\alpha$, we again let $\beta = \alpha$:

$$\tan(\alpha + \beta) = \frac{\tan\alpha + \tan\beta}{1 - \tan\alpha\tan\beta}$$

$$\tan(\alpha + \alpha) = \frac{\tan\alpha + \tan\alpha}{1 - \tan\alpha\tan\alpha}$$

$$\tan 2\alpha = \frac{2\tan\alpha}{1 - \tan^2\alpha} \tag{9}$$

Example 1 If $\sin\alpha = \frac{4}{5}$ and $0 < \alpha < \frac{\pi}{2}$, find $\sin 2\alpha$, $\cos 2\alpha$, and $\tan 2\alpha$.

Solution From the given information and the diagram, we know that $\cos\alpha = \frac{3}{5}$ and $\tan\alpha = \frac{4}{3}$.

$$\sin 2\alpha = 2\sin\alpha\cos\alpha$$
$$= 2\left(\frac{4}{5}\right)\left(\frac{3}{5}\right) = \frac{24}{25}$$
$$\cos 2\alpha = \cos^2\alpha - \sin^2\alpha$$
$$= \left(\frac{3}{5}\right)^2 - \left(\frac{4}{5}\right)^2 = -\frac{7}{25}$$

$$\tan 2\alpha = \frac{2\tan\alpha}{1 - \tan^2\alpha} = \frac{2\left(\frac{4}{3}\right)}{1 - \left(\frac{4}{3}\right)^2} = -\frac{24}{7}$$

1. Simplify.

   a. $\sin^2\left(\frac{\alpha}{2}\right) - \cos^2\left(\frac{\alpha}{2}\right)$

   b. $\dfrac{2\tan 157.5°}{1 - \tan^2 157.5°}$

   a. Let $\theta = \frac{\alpha}{2}$.
   $$\sin^2\theta - \cos^2\theta =$$
   $$-(\cos^2\theta - \sin^2\theta) =$$
   $$-\cos 2\theta =$$
   $$-\cos 2\left(\frac{\alpha}{2}\right) =$$
   $$-\cos\alpha$$

   b. $\dfrac{2\tan 157.5°}{1 - \tan^2 157.5°} =$
   $$\tan 2(157.5°) =$$
   $$\tan 315° = -1$$

2. Use a half-angle formula to evaluate each expression.

   a. $\sin(-67.5°)$

   b. $\cos\dfrac{\pi}{4}$

   a. $\sin(-67.5°) =$
   $$\sin\left(-\frac{135°}{2}\right) =$$
   $$-\sqrt{\frac{1 - \cos(-135°)}{2}} =$$
   $$-\sqrt{\frac{1 - \left(-\frac{\sqrt{2}}{2}\right)}{2}} =$$
   $$-\sqrt{\frac{2 + \sqrt{2}}{4}} =$$
   $$-\frac{1}{2}\sqrt{2 + \sqrt{2}}$$

   b. $\cos\dfrac{\pi}{4} =$
   $$\sqrt{\frac{1 + \cos\frac{\pi}{2}}{2}} =$$
   $$\sqrt{\frac{1 + 0}{2}} = \sqrt{\frac{1}{2}} = \frac{\sqrt{2}}{2}$$

Trigonometric Addition Formulas **381**

Example 2 Derive a formula for $\sin 4x$ in terms of functions of x.

Solution

$$\sin 4x = \sin [2(2x)] = 2 \sin 2x \cos 2x$$
$$= 2(2 \sin x \cos x)(\cos^2 x - \sin^2 x)$$
$$= 4 \sin x \cos^3 x - 4 \sin^3 x \cos x$$

Using another form for $\cos 2x$ will yield different but equivalent results.

Half-Angle Formulas

To obtain the sine and cosine half-angle formulas, we use formulas (8b) and (8c), replacing α with $\frac{x}{2}$.

$$\cos 2\alpha = 1 - 2 \sin^2 \alpha$$

$$\cos 2\left(\frac{x}{2}\right) = 1 - 2 \sin^2 \frac{x}{2}$$

$$\cos x = 1 - 2 \sin^2 \frac{x}{2}$$

$$2 \sin^2 \frac{x}{2} = 1 - \cos x$$

$$\sin \frac{x}{2} = \pm \sqrt{\frac{1 - \cos x}{2}} \qquad (10)$$

$$\cos 2\alpha = 2 \cos^2 \alpha - 1$$

$$\cos 2\left(\frac{x}{2}\right) = 2 \cos^2 \frac{x}{2} - 1$$

$$\cos x = 2 \cos^2 \frac{x}{2} - 1$$

$$2 \cos^2 \frac{x}{2} = 1 + \cos x$$

$$\cos \frac{x}{2} = \pm \sqrt{\frac{1 + \cos x}{2}} \qquad (11)$$

When you use the half-angle formulas, choose $+$ or $-$ depending on the quadrant in which $\frac{x}{2}$ lies.

Example 3 Find the exact value of $\cos \frac{5\pi}{8}$.

Solution Because $\frac{5\pi}{8}$ is in the second quadrant, $\cos \frac{5\pi}{8}$ is negative. Since $\frac{5\pi}{8} = \frac{1}{2}\left(\frac{5\pi}{4}\right)$, we can let $x = \frac{5\pi}{4}$ in formula (11).

$$\cos \frac{5\pi}{8} = -\sqrt{\frac{1 + \cos \frac{5\pi}{4}}{2}} = -\sqrt{\frac{1 + \left(-\frac{\sqrt{2}}{2}\right)}{2}}$$

$$= -\sqrt{\frac{2 - \sqrt{2}}{4}} = -\frac{\sqrt{2 - \sqrt{2}}}{2}$$

To derive a formula for $\tan \frac{x}{2}$, divide equation (10) by equation (11):

$$\tan \frac{x}{2} = \pm \sqrt{\frac{1 - \cos x}{1 + \cos x}} \qquad (12)$$

However, the following formulas, which can be derived by simplifying the radical expression in formula (12), may be more useful.

$$\tan \frac{x}{2} = \frac{\sin x}{1 + \cos x} \quad \text{(12a)} \qquad \tan \frac{x}{2} = \frac{1 - \cos x}{\sin x} \quad \text{(12b)}$$

Notice that these formulas don't need the ambiguous sign $\pm$. (Why?) You are asked to prove these formulas in Exercises 37 and 38.

The following table summarizes the double-angle and half-angle formulas.

Double-Angle and Half-Angle Formulas

$$\sin 2\alpha = 2 \sin \alpha \cos \alpha$$

$$\cos 2\alpha = \cos^2 \alpha - \sin^2 \alpha = 1 - 2 \sin^2 \alpha = 2 \cos^2 \alpha - 1$$

$$\tan 2\alpha = \frac{2 \tan \alpha}{1 - \tan^2 \alpha}$$

$$\sin \frac{\alpha}{2} = \pm \sqrt{\frac{1 - \cos \alpha}{2}} \qquad \cos \frac{\alpha}{2} = \pm \sqrt{\frac{1 + \cos \alpha}{2}}$$

$$\tan \frac{\alpha}{2} = \pm \sqrt{\frac{1 - \cos \alpha}{1 + \cos \alpha}} = \frac{\sin \alpha}{1 + \cos \alpha} = \frac{1 - \cos \alpha}{\sin \alpha}$$

CLASS EXERCISES

In Exercises 1–10, simplify the given expression.

1. $2 \sin 10° \cos 10° \sin 20°$

2. $\cos^2 15° - \sin^2 15° \cos 30° = \frac{\sqrt{3}}{2}$

3. $1 - 2 \sin^2 35° \cos 70°$

4. $2 \cos^2 25° - 1 \cos 50°$

5. $\dfrac{2 \tan 50°}{1 - \tan^2 50°} \tan 100°$

6. $\dfrac{2 \tan 40°}{1 - \tan^2 40°} \tan 80°$

7. $1 - \sin^2 x \cos^2 x$

8. $1 - 2 \sin^2 x \cos 2x$

9. $2 \sin 3\alpha \cos 3\alpha \sin 6\alpha$

10. $\cos^2 5\theta - \sin^2 5\theta \cos 10\theta$

11. Discussion Given that $\cos 70° \approx 0.3420$, explain how you can find $\cos 35°$.
Use the formula for $\cos \frac{\alpha}{2}$ with $\alpha = 70°$.

WRITTEN EXERCISES

In Exercises 1–12, simplify the given expression.

A

1. $2 \cos^2 10° - 1 \cos 20°$

2. $2 \sin \frac{\alpha}{2} \cos \frac{\alpha}{2} \sin \alpha$

3. $\dfrac{4 \tan \beta}{1 - \tan^2 \beta} 2 \tan 2\beta$

4. $1 - 2 \sin^2 20° \cos 40°$

5. $2 \sin 35° \cos 35° \sin 70°$

6. $\cos^2 4A - \sin^2 4A \cos 8A$

Trigonometric Addition Formulas **383**

Mathematical Note

Since formulas (12a) and (12b) at the top of this page give the correct sign directly, they do not need the $\pm$ sign and are therefore less confusing to use than the other formulas for $\tan \frac{x}{2}$.

Exercise Note

You might have students actually carry out the calculation in Class Exercise 11.

Suggested Assignments

Standard
Day 1: 383/1–23 odd
Day 2: 384/31–47 odd
Comprehensive
Day 1: 383/1–25 odd
Day 2: 384/31–49 odd

Supplementary Materials

Alternative Assessment, 31
Activities Book, 26–27
Precalculus Plotter Plus, 52

Since students may be tempted to treat an identity as an equation (see Exercises 31–38), suggest that they think of a double yellow line between the two sides. Tell them not to cross the line when verifying the identity. (For example, they should not add something to both sides.) Instead, they should modify each side independently.

**Additional Answers
Written Exercises**

25. a. $\dfrac{\sqrt{2} - \sqrt{6}}{4}$

  b. $-\dfrac{\sqrt{2 - \sqrt{3}}}{2}$

26. a. $\dfrac{\sqrt{6} + \sqrt{2}}{4}$

  b. $\dfrac{\sqrt{2 + \sqrt{3}}}{2}$

27. range: $-2 \le y \le 1.13$,
 period: 2π

28. range: $-2.78 \le y \le 2.78$,
 period: 2π

29. range: $-6 \le y \le 4.13$,
 period: π

30. range: $-5 \le y \le 4.62$,
 period: 4π

7. $\dfrac{2 \tan 25°}{1 - \tan^2 25°}$ $\tan 50°$

8. $2 \cos^2 3\alpha - 1$ $\cos 6\alpha$

9. $1 - 2 \sin^2 \dfrac{x}{2}$ $\cos x$

10. $\dfrac{\cos^2 40° - \sin^2 40°}{\cos 80°}$

11. $\dfrac{\sqrt{\dfrac{1 - \cos 80°}{2}}}{\sin 40°}$

12. $\dfrac{\sqrt{\dfrac{1 + \cos 70°}{2}}}{\cos 35°}$

In Exercises 13–18, find the exact value of the given expression.

13. $2 \cos^2 \dfrac{\pi}{8} - 1$ $\dfrac{\sqrt{2}}{2}$

14. $\dfrac{2 \tan \dfrac{\pi}{8}}{1 - \tan^2 \dfrac{\pi}{8}}$ 1

15. $\cos^2 \dfrac{\pi}{12} - \sin^2 \dfrac{\pi}{12}$ $\dfrac{\sqrt{3}}{2}$

16. $1 - 2 \sin^2 \dfrac{7\pi}{12}$ $-\dfrac{\sqrt{3}}{2}$

17. $\sin 15° \cos 15°$ $\dfrac{1}{4}$

18. $4 \sin \dfrac{2\pi}{3} \cos \dfrac{2\pi}{3}$ $-\sqrt{3}$

In Exercises 19–24, $\angle A$ is acute. **19.** $\dfrac{120}{169}, \dfrac{119}{169}$ **20.** $\dfrac{3}{5}, \dfrac{4}{3}$ **21.** $\dfrac{24}{25}, \dfrac{336}{625}$
22–24. See below.

19. If $\sin A = \dfrac{5}{13}$, find $\sin 2A$ and $\cos 2A$.

20. If $\tan A = \dfrac{1}{2}$, find $\cos 2A$ and $\tan 2A$.

21. If $\sin A = \dfrac{3}{5}$, find $\sin 2A$ and $\sin 4A$.

22. If $\cos A = \dfrac{1}{3}$, find $\cos 2A$ and $\cos 4A$.

23. If $\cos A = \dfrac{1}{5}$, find $\cos 2A$ and $\cos \dfrac{A}{2}$.

24. If $\cos A = \dfrac{1}{4}$, find $\sin 2A$ and $\sin \dfrac{A}{2}$.

25. Find $\cos 105°$ using **(a)** an addition formula and **(b)** a half-angle formula.

26. Find $\sin 75°$ using **(a)** an addition formula and **(b)** a half-angle formula.

Use a graphing calculator or computer to sketch the graph of each function. Then give the range and period of the function.

B **27.** $y = \sin x + \cos 2x$

28. $y = \sin 3x + 2 \cos x$

29. $y = 2 \sin 2x + 4 \cos 4x$

30. $y = 3 \cos \dfrac{x}{2} - 2 \cos 3x$

In Exercises 31–38, prove that the given equation is an identity.

31. $\dfrac{\sin 2A}{1 - \cos 2A} = \cot A$

32. $\dfrac{1 - \cos 2A}{1 + \cos 2A} = \tan^2 A$

33. $\left(\sin \dfrac{x}{2} + \cos \dfrac{x}{2}\right)^2 = 1 + \sin x$

34. $\sin 4x = 4 \sin x \cos x \cos 2x$

35. $\dfrac{1 - \tan^2 x}{1 + \tan^2 x} = \cos 2x$

36. $\dfrac{1 + \sin A - \cos 2A}{\cos A + \sin 2A} = \tan A$

37. $\tan \dfrac{x}{2} = \dfrac{\sin x}{1 + \cos x}$

38. $\tan \dfrac{x}{2} = \dfrac{1 - \cos x}{\sin x}$

Simplify the given expression. **22.** $-\dfrac{7}{9}, \dfrac{17}{81}$ **23.** $-\dfrac{23}{25}, \dfrac{\sqrt{15}}{5}$ **24.** $\dfrac{\sqrt{15}}{8}, \dfrac{\sqrt{6}}{4}$

39. $\dfrac{1 + \cos 2x}{\cot x}$ $\sin 2x$

40. $\dfrac{(1 + \tan^2 x)(1 - \cos 2x)}{2}$ $\tan^2 x$

384 *Chapter Ten*

41. $(1 - \sin^2 x)(1 - \tan^2 x) \cos 2x$

42. $\sin x \tan x + \cos 2x \sec x \cos x$

43. $\dfrac{\sin 3x}{\sin x} - \dfrac{\cos 3x}{\cos x}$ 2

44. $\cos^2\left(\dfrac{\pi}{4} - \dfrac{x}{2}\right) - \sin^2\left(\dfrac{\pi}{4} - \dfrac{x}{2}\right) \sin x$

45. *Visual Thinking* The diagram at the right shows a unit circle, with the measure of $\angle BOC = \theta$.

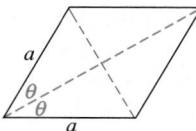

   a. Explain why the measure of $\angle BAO = \dfrac{\theta}{2}$.

   b. Explain why $\tan \dfrac{\theta}{2} = \dfrac{\sin \theta}{1 + \cos \theta}$.

46. Use the rhombus at the right and the steps below to derive the formula for $\sin 2\theta$ for $0° < \theta < 90°$.

   a. Show that the area of the rhombus is $a^2 \sin 2\theta$.

   b. Use the fact that the diagonals of a rhombus are perpendicular to show that its area is
   $$2a^2 \sin \theta \cos \theta.$$

   c. Use your answers from parts (a) and (b) to obtain a formula for $\sin 2\theta$.

47. Find $\log_2 2 + \log_2 (\sin x) + \log_2 (\cos x)$ when $x = \dfrac{\pi}{12}$. -1

48. Find $\dfrac{4^{2\cos^2 \theta}}{4}$ when $\theta = \dfrac{\pi}{3}$. $\dfrac{1}{2}$

Prove that the given equation is an identity.

49. $\sin 3x = 3 \sin x - 4 \sin^3 x$

50. $\cos 3x = 4 \cos^3 x - 3 \cos x$

 51. Given: $QR = QS = 1$, $\angle Q = 36°$, and $\overline{RT}$ bisects $\angle R$.

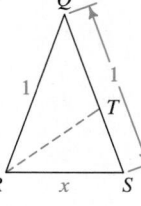

   a. Prove that $\triangle QRS$ is similar to $\triangle RST$.

   b. Use similar triangles to show that $\dfrac{x}{1 - x} = \dfrac{1}{x}$.

   c. Show that $x = \dfrac{\sqrt{5} - 1}{2}$.

   d. Draw the bisector of $\angle Q$. Show that
   $$\sin 18° = \dfrac{\sqrt{5} - 1}{4} = \cos 72°.$$

52. In $\triangle ABC$, the measure of $\angle B$ is twice the measure of $\angle C$.

   a. Use the law of sines to show that $b = 2c \cos C$.

   b. Use the law of cosines to show that $b^2 = c(a + c)$.

53. The lengths of the sides of a triangle are consecutive integers n, $n + 1$, and $n + 2$, and the largest angle is twice the smallest angle θ.

   a. Use the law of sines to show that $\cos \theta = \dfrac{n + 2}{2n}$.

   b. Use the law of cosines to show that $\cos \theta = \dfrac{n + 5}{2(n + 2)}$.

   c. Use parts (a) and (b) to find n. 4

Warm-Up Exercises

Use a calculator to solve each equation for $0° \leq x < 360°$.

1. $\sin x = \frac{1}{2}$ 30°, 150°

2. $\cos x = -1$ 180°

3. $\tan x = 0$ 0°, 180°

Solve each equation.

4. $3x^2 - 11x - 4 = 0$
  $-\frac{1}{3}, 4$

5. $6x - x^2 = 1 + 2x + 2x^2$
  $\frac{1}{3}, 1$

Motivating the Section

Remind students that they have already solved trigonometric equations in Chapter 8. In this section, students will use the identities of this chapter in solving other, more complicated trigonometric equations.

10-4 Solving Trigonometric Equations

Objective *To use identities to solve trigonometric equations.*

The acceleration of a body falling toward the Earth's surface is called acceleration due to gravity, often denoted by g. In theoretical physics, g is usually considered constant. However, g is not actually constant but varies slightly with latitude. A good approximation to the value of g can be found by using the following formula, which expresses g in terms of θ, the latitude in degrees.

Gravity acting on water creates this waterfall.

$$g \approx 9.78049(1 + 0.005288 \sin^2 \theta - 0.000006 \sin^2 2\theta)$$

For example, if you live in Chicago, which has a latitude of 42°N, $\sin \theta \approx 0.6691$ and $\sin 2\theta \approx 0.9945$. Therefore, $g \approx 9.8036$ m/s^2.

 As you can see from this example, some problems involve trigonometric equations that have multiples of angles or numbers. The following suggests two methods that may be helpful in solving such equations. The first method gives a graphical method using a graphing calculator or computer, and the second method gives ways for solving the equation algebraically.

Methods for Solving the Trigonometric Equation $f(x) = g(x)$

Method 1

Use a graphing calculator or computer to graph $y = f(x)$ and $y = g(x)$ on the same set of axes. Use the zoom or trace feature to find the x-coordinates of any intersection points of the two graphs.

Method 2

Use the following guidelines.

a. It may be helpful to draw a quick sketch of $y = f(x)$ and $y = g(x)$ to see roughly where the solutions are.

b. If the equation involves functions of $2x$ and x, transform the functions of $2x$ into functions of x by using identities.

c. If the equation involves functions of $2x$ only, it is usually better to solve for $2x$ directly and then solve for x.

d. Be careful not to lose roots when you divide both sides of an equation by a function of the variable. Review the discussion about losing roots on pages 32–33.

Example 1 Solve $\cos 2x = 1 - \sin x$ for $0 \le x < 2\pi$.

Solution **Method 1** The diagram below shows the graphs of $y = \cos 2x$ and $y = 1 - \sin x$ on the same set of axes. There are four solutions in the interval $0 \le x < 2\pi$. From the diagram, you can see that 0 and π are solutions. Using a zoom feature, you can find that $0.52 \approx \dfrac{\pi}{6}$ and $2.62 \approx \dfrac{5\pi}{6}$ are also solutions.

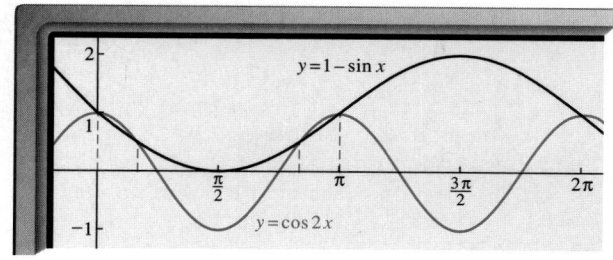

Method 2
$$\cos 2x = 1 - \sin x$$
$$1 - 2 \sin^2 x = 1 - \sin x$$
$$2 \sin^2 x = \sin x$$

$\sin x = 0$	$2 \sin x = 1$
$x = 0, \pi$	$\sin x = \dfrac{1}{2}$
	$x = \dfrac{\pi}{6}, \dfrac{5\pi}{6}$

Graphs are useful for solving not only trigonometric equations but also trigonometric inequalities. For example, to solve the inequality $\cos 2x < 1 - \sin x$, look at the graph shown in Example 1. You can see that the graph of $y = \cos 2x$ is *below* the graph of $y = 1 - \sin x$ when $\dfrac{\pi}{6} < x < \dfrac{5\pi}{6}$ and when $\pi < x < 2\pi$.

Example 2 Solve $3 \cos 2x + \cos x = 2$ for $0 \le x < 2\pi$.

Solution
$$3 \cos 2x + \cos x = 2$$
$$3(2 \cos^2 x - 1) + \cos x = 2 \leftarrow \cos 2x = 2 \cos^2 x - 1$$
$$6 \cos^2 x + \cos x - 5 = 0$$
$$(6 \cos x - 5)(\cos x + 1) = 0$$

$$\cos x = \frac{5}{6} \qquad \text{or} \qquad \cos x = -1$$
$$x \approx 0.59,\ 5.70 \qquad\qquad x = \pi$$
$$\qquad\qquad\qquad\qquad x \approx 3.14$$

Example Note

When discussing Example 1, remind students that trigonometric equations have infinitely many solutions since trigonometric functions are periodic. However, since any solution is an angle coterminal with one in the domain $0° \le x < 360°$ or $0 \le x < 2\pi$, this is the domain to which the solutions are restricted. (In order to draw students' attention to finding all solutions, you might want to change the domain once in a while.)

In method 2 of Example 1, you could also make one side equal 0 and factor the other side:
$$2 \sin^2 x - \sin x = 0$$
$$\sin x (2 \sin x - 1) = 0$$
$$\sin x = 0 \text{ or } 2 \sin x - 1 = 0$$
$$\sin x = 0 \text{ or } \sin x = \frac{1}{2}$$
$$x = 0, \pi \text{ or } x = \frac{\pi}{6}, \frac{5\pi}{6}$$

Using Technology

Note that the use of a scientific calculator is assumed in the solution of Example 2. A calculator gives only $x \approx 0.59$ as a solution of $\cos x = \dfrac{5}{6}$, however. Students must recognize for themselves that $x \approx 5.70$ is also a solution.

Trigonometric Addition Formulas **387**

1. Solve $\cos 2x - \sin x > 0$ for $0 \le x < 2\pi$ by using identities.

$\cos 2x - \sin x > 0$

$1 - 2\sin^2 x - \sin x > 0$

$2\sin^2 x + \sin x - 1 < 0$

$(2\sin x - 1)(\sin x + 1) < 0$

The zeros of the related equation are at $\sin x = \frac{1}{2}$ and $\sin x = -1$, that is, $x = \frac{\pi}{6}, \frac{5\pi}{6}, \frac{3\pi}{2}$. Do a sign analysis as in Section 3-2.

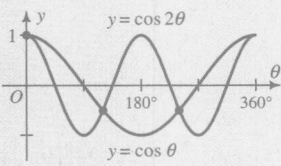

Solution: $0 \le x < \frac{\pi}{6}$, or

$\frac{5\pi}{6} < x < \frac{3\pi}{2}$, or

$\frac{3\pi}{2} < x < 2\pi$.

2. Solve $\cos 2\theta = \cos \theta$ for $0° \le \theta < 360°$.

Method 1: Algebraic

$\cos 2\theta = \cos \theta$

$2\cos^2 \theta - 1 = \cos \theta$

$2\cos^2 \theta - \cos \theta - 1 = 0$

$(\cos \theta - 1)(2\cos \theta + 1) = 0$

$\cos \theta = 1$ or $\cos \theta = -\frac{1}{2}$

$\theta = 0°$ or $\theta = 120°, 240°$

Method 2: Graphical

Graph $y = \cos 2\theta$ and $y = \cos \theta$ together.

The graph shows three points of intersection: at $\theta = 0°, 120°, 240°$.

Example 3 Solve $2 \sin 2x = 1$ for $0° \le x < 360°$.

Solution The graphs of $y = 2 \sin 2x$ and $y = 1$ over the interval $0° \le x < 360°$ are shown below. You can see that there are four possible solutions.

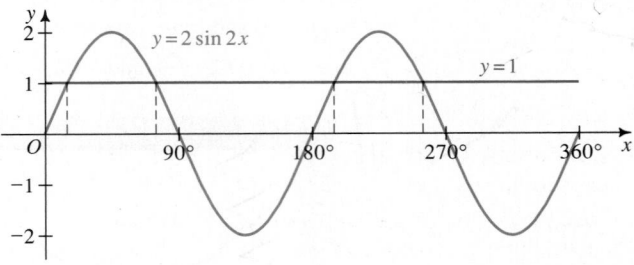

The most efficient way to solve this equation algebraically is to solve directly for $2x$.

$$\sin 2x = \frac{1}{2}$$

$$2x = 30°, 150°, 390°, 510° \quad \longleftarrow \quad 0° \le 2x < 720°$$

$$x = 15°, 75°, 195°, 255°$$

Examples 1 through 3 show that some trigonometric equations can be solved either algebraically or graphically. Example 4 shows that for some equations the best method is to use a graphing calculator or computer.

Example 4 Find the smallest positive root of $\sin 2.8x = \cos x$.

Solution You can make a sketch of the graphs of $y = \sin 2.8x$ and $y = \cos x$ on the same set of axes to see approximately where the solutions are, but then finding an algebraic solution would be very difficult. In this case, the most efficient method is to use a graphing calculator or a computer. The diagram below indicates that there are six roots between 0 and 2π. You are interested in the smallest one.

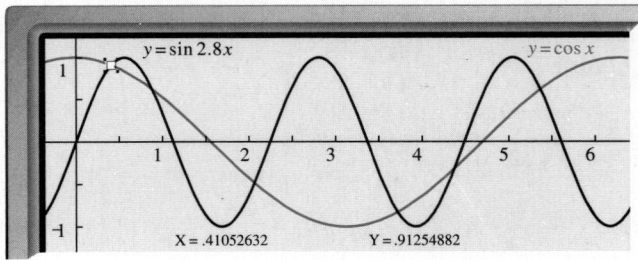

Using the zoom or trace feature, you will find that $x \approx 0.41$.

CLASS EXERCISES

Describe the method you would use to solve the following equations for $0° \leq x < 360°$. Some of these equations can be solved by methods discussed in Chapter 8.

1. $\cos 2x = \cos x$

2. $\sin^2 x = \sin x$

3. $\sin x = \cos x$

4. $\sin 2x = \cos 2x$

5. $\sin 3x = \cos 3x$

6. $\tan (x - 10°) = 1$

7. *Discussion* When solving $\sin 4x = \sin 2x$, would you begin by writing the expressions in terms of functions of x or $2x$? Why? *2x*

8. *Discussion* When solving $\cos 4x = 1 - 3 \cos 2x$, would you begin by writing the expressions in terms of functions of x or $2x$? Why? *2x*

9. a. On a single set of axes, sketch the graphs of $y = \sin x$ and $y = \cos x$ over the interval $0 \leq x \leq \dfrac{\pi}{2}$.

b. Find the x-coordinates of any points of intersection. $\frac{\pi}{4}$

c. Over what interval(s) is $\sin x > \cos x$?

d. Over what interval(s) is $\sin x < \cos x$? **c.** $\frac{\pi}{4} < x \leq \frac{\pi}{2}$ **d.** $0 \leq x < \frac{\pi}{4}$

WRITTEN EXERCISES

A **1–8.** Solve the equations in Class Exercises 1–8 for $0° \leq x < 360°$.

9. a. On a single set of axes, sketch the graphs of $y = \sin 2x$ and $y = \tan x$ for $0 \leq x < 2\pi$.

b. Find the x-coordinates of the points where the graphs intersect by solving $\sin 2x = \tan x$.

c. Solve $\sin 2x > \tan x$ for $0 \leq x < 2\pi$.

10. a. On a single set of axes, sketch the graphs of $y = \cos 2x$ and $y = 3 \cos x$ for $0 \leq x < 2\pi$.

b. Find the x-coordinates of the points where the graphs intersect by solving $\cos 2x = 3 \cos x$. 1.86, 4.43

c. Solve $\cos 2x < 3 \cos x$ for $0 \leq x < 2\pi$. $0 \leq x < 1.86$, $4.43 < x < 2\pi$

 In Exercises 11–16, solve each inequality for $0 \leq x < 2\pi$. Give answers to the nearest hundredth. You may wish to use a graphing calculator or computer. **11.** $0 \leq x < 1.11$, $4.25 < x \leq 6.28$

12. $0 < x < 1.05$, $1.57 < x < 3.14$, $4.71 < x < 5.24$

11. $\cos x > \dfrac{1}{2} \sin x$

12. $\tan x < 2 \sin x$

$1.11 \leq x \leq 4.25$

13. $\sin \left(3x - \dfrac{\pi}{2}\right) > 0$

14. $2 \cos x \leq \cos \left(x - \dfrac{\pi}{2}\right)$

15. $\cos x \leq \sin 2x$

16. $\cos 2x \geq 5 - \cos x$ No solution

$0.52 \leq x \leq 1.57$, $2.62 \leq x \leq 4.71$

13. $0.52 < x < 1.57$, $2.62 < x < 3.67$, $4.71 < x < 5.76$

Trigonometric Addition Formulas **389**

Additional Answers
Class Exercises

1. Substitute $2 \cos^2 x - 1$ for $\cos 2x$ and solve algebraically for x.

2. Solve algebraically for x.

3. Rewrite as $\tan x = 1$, $\cos x \neq 0$, and solve algebraically for x.

4. Rewrite as $\tan 2x = 1$, $\cos 2x \neq 0$, and solve algebraically for $2x$.

5. Rewrite as $\tan 3x = 1$, $\cos 3x \neq 0$, and solve algebraically for $3x$.

6. Solve algebraically for $x - 10°$.

Additional Answers
Written Exercises

1. $0°$, $120°$, $240°$

2. $0°$, $90°$, $180°$ **3.** $45°$, $225°$

4. $22.5°$, $112.5°$, $202.5°$, $292.5°$

5. $15°$, $75°$, $135°$, $195°$, $255°$, $315°$

6. $55°$, $235°$

7. $0°$, $30°$, $90°$, $150°$, $180°$, $210°$, $270°$, $330°$

8. $30°$, $150°$, $210°$, $330°$

Suggested Assignments

Standard
Day 1: 389/1–15 odd
Day 2: 390/17–33 odd
Comprehensive
Day 1: 389/1–21 odd
Day 2: 390/23–31 odd, 33–41, 43

Supplementary Materials

Alternative Assessment, 31
Student Resource Guide, 95–97

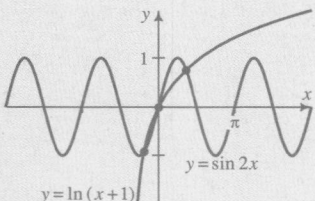
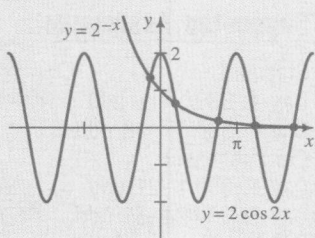
In Exercises 17–22, solve each equation for $0° \le x < 360°$ by using trigonometric identities. Give answers to the nearest tenth when necessary. You may wish to check your answers using a graphing calculator or computer.

B 17. $2 \cos (x + 45°) = 1$ 15°, 255°   18. $\cot (x - 20°) = 1$ 65°, 245°

19. $\sin (60° - x) = 2 \sin x$ 19.1°, 199.1°   20. $2 \sin (30° + x) = 3 \cos x$

21. $\sin x = \sin 2x$ 0°, 60°, 180°, 300°   22. $\tan^2 2x - 1 = 0$ See below.

20. 49.1°, 229.1°   22. 22.5°, 67.5°, 112.5°, 157.5°, 202.5°, 247.5°, 292.5°, 337.5°

In Exercises 23–32, solve each equation for $0 \le x < 2\pi$ by using trigonometric identities. Give answers to the nearest hundredth when necessary. You may wish to check your answers using a graphing calculator or computer.

23. $\sin x \cos x = \dfrac{1}{2}$ $\dfrac{\pi}{4}, \dfrac{5\pi}{4}$ 24. $\cos (3x + \pi) = \dfrac{\sqrt{2}}{2}$

25. $\tan 2x = 3 \tan x$ 26. $\tan 2x + \tan x = 0$

27. $\cos 2x = 5 \sin^2 x - \cos^2 x$ 28. $\sin 2x \sec x + 2 \cos x = 0$

29. $3 \sin x = 1 + \cos 2x$ $\dfrac{\pi}{6}, \dfrac{5\pi}{6}$ 30. $\sin 2x = 5 \cos^2 x$ $\dfrac{\pi}{2}, \dfrac{3\pi}{2}$, 1.19, 4.33

31. $\cos 2x = \sec x$ 0   32. $\sin x \cos 2x = 1$ $\dfrac{3\pi}{2}$

33. **a.** Is it possible to use trigonometric identities to solve $\sin 2x = x$? No
    b. On a single set of axes, sketch the graphs of $y = \sin 2x$ and $y = x$.
    c. How many solutions are there to the equation $\sin 2x = x$? 3
    d. Use a scientific calculator and trial and error to approximate the positive solution(s) to the nearest tenth. 0.9

34. **Visual Thinking** On the same set of axes, sketch the graphs of $y = \sin 2x$ and $y = \ln (x + 1)$ for $-2\pi < x < 2\pi$. Then tell *how many* solutions the equation $\sin 2x = \ln (x + 1)$ has in the interval $-2\pi < x < 2\pi$. 3

35. **Visual Thinking** On the same set of axes, sketch the graphs of $y = 2 \cos 2x$ and $y = 2^{-x}$ for $-2\pi < x < 2\pi$. Then tell *how many* solutions the equation $2 \cos 2x = 2^{-x}$ has in the interval $-2\pi < x < 2\pi$. 5

Dolphins locate objects by means of ultrasonic pulses.

Solve each equation graphically for $-2\pi \leq x < 2\pi$ by using a graphing calculator or computer. Give answers to the nearest hundredth.

-4.49, -1.79, 1.05, 5.24   ± 5.41, ±3.80, ±2.14, 0

36. $\sin 1.5x = 2 \cos x$ **37.** $\tan 2x = x$

38. $\sin 2x = \ln (x + 1)$ -0.61, 0, 1.14   **39.** $2 \cos 2x = 2^{-x}$ -0.42, 0.62,
                                                          2.40, 3.91, 5.50

Solve each equation for $0 \leq x < 2\pi$.

C **40.** $\sin 3x = \sin 5x + \sin x$ (*Hint:* Use one of the formulas given on page 372.)

41. $\cos 3x + \cos x = \cos 2x$ **42.** $\sin 3x - \sin x = 2 \cos 2x$

Solve.

43. $\text{Cos}^{-1} 2x = \text{Sin}^{-1} x \dfrac{\sqrt{5}}{5} \approx 0.447$ **44.** $\text{Tan}^{-1} 2x = \text{Sin}^{-1} x$ $0, \pm\dfrac{\sqrt{3}}{2}$

////COMPUTER EXERCISES

Write a computer program that will print a table of sines and cosines for $1°, 2°, \ldots,$ $90°$ given only that $\sin 1° = 0.017452406$. *Hint:* Here is how you could begin:

1. Since $\sin^2 1° + \cos^2 1° = 1$,

$$\cos 1° = \sqrt{1 - \sin^2 1°} = \sqrt{1 - (0.017452406)^2},$$

which the computer may evaluate as 0.999847695.

2. $\sin 2° = \sin (1° + 1°) = \sin 1° \cos 1° + \cos 1° \sin 1° = 0.034899495.$

$\cos 2° = \sqrt{1 - \sin^2 2°} = 0.999390827.$

3. $\sin 3° = \sin (2° + 1°) = \sin 2° \cos 1° + \cos 2° \sin 1°,$ and so on.

▮ Chapter Summary

1. The *sum and difference formulas* for sine, cosine, and tangent are as follows:

$$\sin (\alpha + \beta) = \sin \alpha \cos \beta + \cos \alpha \sin \beta$$

$$\sin (\alpha - \beta) = \sin \alpha \cos \beta - \cos \alpha \sin \beta$$

$$\cos (\alpha + \beta) = \cos \alpha \cos \beta - \sin \alpha \sin \beta$$

$$\cos (\alpha - \beta) = \cos \alpha \cos \beta + \sin \alpha \sin \beta$$

$$\tan (\alpha + \beta) = \frac{\tan \alpha + \tan \beta}{1 - \tan \alpha \tan \beta}$$

$$\tan (\alpha - \beta) = \frac{\tan \alpha - \tan \beta}{1 + \tan \alpha \tan \beta}$$

Supplementary Materials

Tests, 27–28
Alternative Assessment, 67
Student Resource Guide, 98–99

2. The formulas for rewriting a trigonometric sum or difference as a product are as follows:

$$\sin x + \sin y = 2 \sin \frac{x+y}{2} \cos \frac{x-y}{2}$$

$$\sin x - \sin y = 2 \cos \frac{x+y}{2} \sin \frac{x-y}{2}$$

$$\cos x + \cos y = 2 \cos \frac{x+y}{2} \cos \frac{x-y}{2}$$

$$\cos x - \cos y = -2 \sin \frac{x+y}{2} \sin \frac{x-y}{2}$$

3. An angle θ formed by intersecting lines l_1 and l_2 with slopes m_1 and m_2 can be found by using the following formula:

$$\tan \theta = \frac{m_1 - m_2}{1 + m_1 m_2}$$

4. The *double-angle formulas* shown below are derived from the formulas for $\sin(\alpha + \beta)$, $\cos(\alpha + \beta)$, and $\tan(\alpha + \beta)$.

$$\sin 2\alpha = 2 \sin \alpha \cos \alpha$$

$$\cos 2\alpha = \cos^2 \alpha - \sin^2 \alpha = 1 - 2 \sin^2 \alpha = 2 \cos^2 \alpha - 1$$

$$\tan 2\alpha = \frac{2 \tan \alpha}{1 - \tan^2 \alpha}$$

5. The *half-angle formulas* shown below are derived from two of the double-angle formulas for cosine.

$$\sin \frac{\alpha}{2} = \pm\sqrt{\frac{1 - \cos \alpha}{2}}$$

$$\cos \frac{\alpha}{2} = \pm\sqrt{\frac{1 + \cos \alpha}{2}}$$

$$\tan \frac{\alpha}{2} = \pm\sqrt{\frac{1 - \cos \alpha}{1 + \cos a}} = \frac{\sin \alpha}{1 + \cos \alpha} = \frac{1 - \cos \alpha}{\sin \alpha}$$

6. Suggested methods for solving a trigonometric equation either graphically or algebraically are given on page 386.

Key vocabulary and ideas

sum and difference formulas for cosine and sine (p. 370)
formulas for rewriting a sum or difference as a product (p. 372)
sum and difference formulas for tangent (pp. 375–376)
angle formed by intersecting lines (pp. 376–377)
double-angle formulas (p. 381, p. 383)
half-angle formulas (pp. 382–383)

1. Simplify the given expression.   10-1
   a. $\cos 75° \cos 15° + \sin 75° \sin 15°$ $\frac{1}{2}$
   b. $\sin 75° \cos 15° + \cos 75° \sin 15°$ 1
   c. $\cos (30° + x) + \cos (30° - x)$ $\sqrt{3} \cos x$
   d. $\sin (45° - x) - \sin (45° + x)$ $-\sqrt{2} \sin x$

2. Find the exact value of $\cos 15°$.   2. $\dfrac{\sqrt{6} + \sqrt{2}}{4}$

3. Find $\tan \left(\dfrac{5\pi}{4} - \theta \right)$ when $\tan \theta = -\dfrac{1}{3}$.   2   10-2

4. If $\tan \alpha = \dfrac{4}{3}$ and $\tan \beta = -\dfrac{1}{2}$, show that $\tan (\alpha + \beta) = \tan (\pi - \beta)$. See below.

5. **Writing** Consider the lines $y = 2x + 1$ and $y = 4 - 3x$. Write a paragraph explaining how it is possible to find two different angles that are formed by the intersection of these lines. What is the relationship between these angles? They are supplementary.

6. Suppose $\angle A$ is acute and $\cos A = \dfrac{4}{5}$. Find each of the following: 10-3

   a. $\sin A$ a. $\dfrac{3}{5}$ c. $\dfrac{24}{25}$ b. $\cos 2A$ b. $\dfrac{7}{25}$ d. $\dfrac{336}{625}$
   c. $\sin 2A$ d. $\sin 4A$

7. Simplify the given expression.

   a. $\dfrac{\sin 2x}{1 - \cos 2x} \cot x$ b. $(1 + \tan^2 y)(\cos 2y - 1)$ $-2 \tan^2 y$

   c. $\dfrac{\tan t}{\sec t + 1} \tan \dfrac{t}{2}$ d. $\cos^2 \dfrac{x}{2} - \sin^2 \dfrac{x}{2}$ $\cos x$

8. Evaluate the given expression.

   a. $2 \cos^2 \dfrac{\pi}{12} - 1$ $\dfrac{\sqrt{3}}{2}$ b. $4 \sin \dfrac{\pi}{6} \cos \dfrac{\pi}{6}$ $\sqrt{3}$

9. Prove that the given equation is an identity.
   a. $(1 + \cot^2 x)(1 - \cos 2x) = 2$
   b. $\dfrac{\sin \theta \sec \theta}{\tan \theta + \cot \theta} = \cos^2 \theta - \cos 2\theta$

10. a. On the same set of axes, sketch the graphs of $y = \cos 2\theta$ and   10-4
    $y = \sin \theta$ for $0° \le \theta < 360°$.
    b. Determine where the graphs intersect by solving $\cos 2\theta = \sin \theta$.   30°, 150°, 270°

11. Solve the equation $\cos 2x = \cos x + 2$ either graphically or algebraically for $0 \le x < 2\pi$. π

4. $\tan (\alpha + \beta) = \dfrac{\tan \alpha + \tan \beta}{1 - \tan \alpha \tan \beta} = \dfrac{\dfrac{4}{3} + \left(-\dfrac{1}{2} \right)}{1 - \left(\dfrac{4}{3} \right)\left(-\dfrac{1}{2} \right)} = \dfrac{1}{2}$

$\tan (\pi - \beta) = \dfrac{\tan \pi - \tan \beta}{1 + \tan \pi \tan \beta} = \dfrac{0 - \left(-\dfrac{1}{2} \right)}{1 + (0)\left(-\dfrac{1}{2} \right)} = \dfrac{1}{2}$

**Additional Answers
Chapter Test**

5. If one angle is θ_1 and the other is θ_2, we can use the slopes of the two lines to find the angles.

$\tan \theta_1 = \dfrac{m_1 - m_2}{1 + m_1 m_2}$ and

$\tan \theta_2 = \dfrac{m_2 - m_1}{1 + m_2 m_1}$. Alternatively, we could find just one angle using the slopes and find the second using the fact that $2\theta_1 + 2\theta_2 = 360°$.

9. a. $(1 + \cot^2 x) \cdot$
   $(1 - \cos 2x) =$
   $(\csc^2 x) \cdot$
   $(1 - (1 - 2 \sin^2 x)) =$
   $(\csc^2 x)(2 \sin^2 x) =$
   2

   b. $\dfrac{\sin \theta \sec \theta}{\tan \theta + \cot \theta} = \dfrac{\sin \theta}{\cos \theta} \div$
   $\left(\dfrac{\sin \theta}{\cos \theta} + \dfrac{\cos \theta}{\sin \theta} \right) =$
   $\dfrac{\sin \theta}{\cos \theta} \div$
   $\left(\dfrac{\sin^2 \theta + \cos^2 \theta}{\cos \theta \sin \theta} \right) =$
   $\dfrac{\sin \theta}{\cos \theta} \cdot \dfrac{\cos \theta \sin \theta}{1} =$
   $\sin^2 \theta = 1 - \cos^2 \theta =$
   $\cos^2 \theta + \sin^2 \theta -$
   $\cos^2 \theta =$
   $\cos^2 \theta - \cos 2\theta$

10. a.

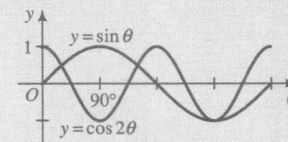

Chapter 11 Polar Coordinates and Complex Numbers

Overview

This chapter introduces students to an entirely new way of locating points in a plane. A point P is still represented by an ordered pair, but the components are polar coordinates, where the first component measures the directed distance of P from the pole, O, of the polar axis, and the second component measures the angle formed by ray OP and the polar axis. This representation is well suited for describing the motion of an object with respect to some fixed point, which is a central concept in many applications of physics, such as radar.

In the first two sections of the chapter, students learn how the same point may be represented in both polar and rectangular coordinates, how to graph equations in the new system, and how to transform equations from one system to the other. In the last two sections, students use polar coordinates to study powers and roots of complex numbers.

Objectives

11-1 To graph polar equations.
11-2 To write complex numbers in polar form and to find products in polar form.
11-3 To use De Moivre's theorem to find powers of complex numbers.
11-4 To find roots of complex numbers.

Supplementary Resources

Tests, pp. 29–30, 31–32, 33–39
Alt. Assess., pp. 32–34, 68
Activities, pp. 28–29
Using Tech., pp. 27–28
Student Res. Guide, pp. 100–105

Software

Precalculus Plotter Plus
Activities, pp. 55–56, 57–58, 59–60, 61–62
Polar Graph Plotter, Conics Plotter

Pacing Guide

Section	11-1	11-2	11-3	11-4	Review	Test	C. Rev.	Total
Standard (days)	2	2	2	2	1	1	2	12
Comprehensive (days)	2	2	1	2	1	1	2	11

Teaching Notes

Presenting the Section

Before you introduce polar coordinates, you may want to review the definitions of sine and cosine from Section 7–3. In that section, we began with a circle of radius r and an angle θ and noted that the x- and y-coordinates of a point on the circle were given by $x = r \cos \theta$ and $y = r \sin \theta$. Now the process is reversed: given a point P with coordinates (x, y), we find the circle (by finding its radius) and the angle. Tell students that they will study a new system that is different from the rectangular coordinate system with which they are so familiar.

Graphs of functions given in polar form are investigated immediately. You may wish to have your students compare graphs of trigonometric functions using rectangular coordinates with the corresponding graphs using polar coordinates.

Using Technology

As is pointed out in the text, many scientific and graphing calculators have a built-in function for converting between rectangular and polar coordinates. Parametric equations are introduced in the context of graphing polar equations using a graphing calculator that has a parametric mode.

Presenting the Section

Complex numbers are represented on a graph, called an Argand diagram, using the fact that the complex number $a + bi$ corresponds uniquely to the ordered pair (a, b). In representing complex numbers in the plane, the x- and y-axes become the real and imaginary axes. Then the complex number $a + bi$ is related to its polar form, which is (r, θ). To conveniently represent a number in polar form rather than rectangular form, the notation r cis θ is used instead of $a + bi$.

The remainder of the section introduces the polar representation of the product of two complex numbers. The efficiency of polar notation is made evident by comparing the multiplication of complex numbers expressed in polar form to the multiplication of complex numbers expressed in rectangular form.

Cooperative Learning

Once your students have mastered the conversion between rectangular and polar forms, they should be able to discover the theorem on page 404 for themselves. You may want to have your students do the following exercise by working in groups of two or three. Let $z_1 = 6 + 6i\sqrt{3}$ and $z_2 = -2 - 3i\sqrt{3}$. Ask students to compute $z_1 z_2$, and then change z_1, z_2, and $z_1 z_2$ to polar form. Have students graph z_1, z_2, and $z_1 z_2$, and ask if they see a pattern emerging. If they don't see the pattern, repeat the exercise with $z_1 = -4\sqrt{3} + 4i$ and $z_2 = 2i$.

Presenting the Section

De Moivre's theorem and the Argand diagram together reveal beautiful patterns in powers of complex numbers. If the absolute value of a complex number is greater than 1, the successive powers of that number, when plotted on an Argand diagram, form a counterclockwise spiral out from the origin. If the absolute value is less than 1, the powers spiral into the origin. If the absolute value equals 1, the powers remain on the unit circle. In this case, the powers eventually repeat themselves.

Presenting the Section

This section explores the use of De Moivre's theorem to find all of the nth roots of a complex number. There are n such roots. In order to prepare your students to find all the possible roots, you may need to review the procedure for finding all the solutions of a trigonometric equation such as $\cos 3x = 0$, or, equivalently, $\cos 3x = \cos 90°$.

Review

Before assigning the exercises in this section, you may want to review algebraic procedures discussed in Chapter 2 for solving equations like $z^3 + 8 = 0$.

11 Polar Coordinates and Complex Numbers

394

11-1 Polar Coordinates and Graphs

Objective   *To graph polar equations.*

The position of a point P in the plane can be described by giving its directed distance r from a fixed point O, called the **pole**, and the measure of an angle formed by ray OP and a reference ray, called the **polar axis**.

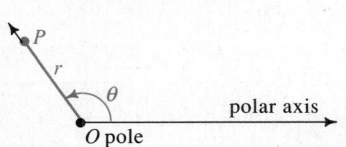

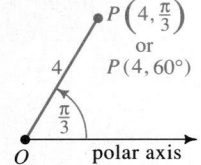

The number r and the angle θ are called **polar coordinates** of P, and we write $P = (r, \theta)$. The angle θ can be measured either in degrees or radians. Figures (a) and (b) show that the same point can be represented by more than one pair of polar coordinates. To find a point when r is negative, simply find the ray that forms the angle θ with the polar axis, and then go r units in the opposite direction from the ray, as shown in figure (c).

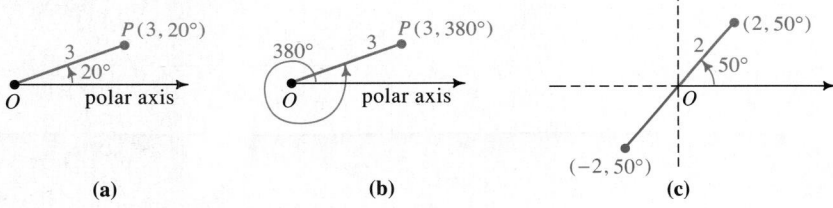

(a)   (b)   (c)

Another way to find a point when $r < 0$ is to plot the point $(|r|, \theta + \pi)$ or the point $(|r|, \theta + 180°)$. This is illustrated below.

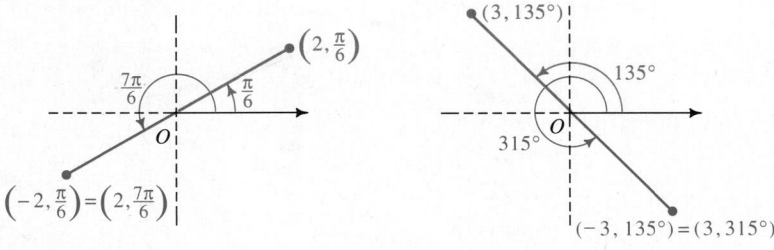

◁ The graceful shape of this modern staircase echoes the lines of the chambered nautilus. The underlying mathematical curve is the *equiangular spiral*, a shape that can be described very simply with polar coordinates.

Polar Coordinates and Complex Numbers   **395**

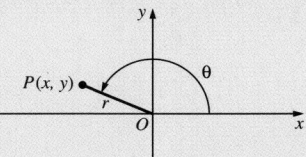

Assessment Note

After discussing polar coordinates, you may want to check students' understanding by having them plot the points $\left(\frac{1}{2}, 225°\right)$ and $\left(-1, \frac{2\pi}{3}\right)$.

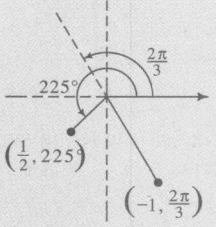

Review Note

Before discussing the conversion between polar and rectangular coordinates, you might have students review the right-triangle definitions of sine, cosine, and tangent given on page 331.

An equation in r and θ is called a **polar equation**. To sketch its graph, called a **polar graph,** you plot in a *polar coordinate system* several points (r, θ) that satisfy the equation, and then draw a smooth curve through them.

Example 1   Sketch the polar graph of $r = 2 \cos \theta$.

Solution   Make a table of values to find values of r for selected values of θ.

θ	0°	30°	45°	60°	90°	120°	135°	150°	180°
r	2	$\sqrt{3}$	$\sqrt{2}$	1	0	-1	$-\sqrt{2}$	$-\sqrt{3}$	-2

θ	210°	225°	240°	270°	300°	315°	330°	360°
r	$-\sqrt{3}$	$-\sqrt{2}$	-1	0	1	$\sqrt{2}$	$\sqrt{3}$	2

From the table, you can see that the graph is complete for values of θ between 0° and 180°. For values of θ between 180° and 360°, the same graph is traced again. The figure below suggests that the graph is a circle.

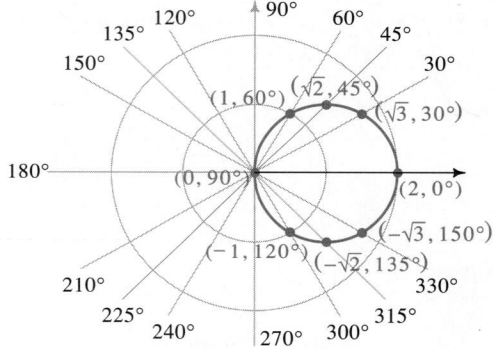

The relationship between polar coordinates and Cartesian (or rectangular) coordinates of any point is given by the diagram and the equations below.

Polar and Rectangular Conversions

Converting from polar to rectangular coordinates:

$x = r \cos \theta \qquad y = r \sin \theta$

Converting from rectangular to polar coordinates:

$r = \pm\sqrt{x^2 + y^2} \qquad \tan \theta = \dfrac{y}{x}$

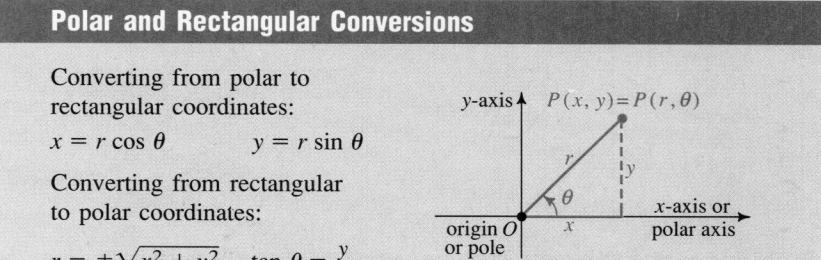

396   *Chapter Eleven*

The diagram on the previous page illustrates the convention that the pole is placed at the origin with the polar axis coinciding with the positive x-axis and with positive angles measured in the counterclockwise direction. The given equations can be used as formulas for converting from one coordinate system to another. Many scientific and graphing calculators have built-in conversion operations.

Example 2 **a.** Give polar coordinates for the point $(3, 4)$.
b. Give the rectangular coordinates for the point $(3, 30°)$.

Solution **a.** $r = \pm\sqrt{x^2 + y^2} = \pm\sqrt{3^2 + 4^2} = \pm 5$

$\tan \theta = \dfrac{y}{x} = \dfrac{4}{3}$. Thus, $\theta \approx 53.1°$ is one possible value of the angle. $(3, 4)$ is in Quadrant I. If $\theta = 53.1°$, then the point is on the terminal ray, and so $r = 5$. Therefore, polar coordinates are $(5, 53.1°)$. Another possible pair of polar coordinates of the same point is $(-5, 233.1°)$.

b. $x = r \cos 30° = 3 \cos 30°$ and $y = r \sin 30° = 3 \sin 30°$.

Thus, rectangular coordinates of the point are $\left(\dfrac{3\sqrt{3}}{2}, \dfrac{3}{2}\right)$.

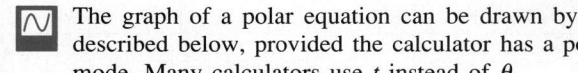 The graph of a polar equation can be drawn by a graphing calculator, as described below, provided the calculator has a polar mode or a parametric mode. Many calculators use t instead of θ.

If the calculator has polar mode, enter the equation: $r = f(t)$
If the calculator has parametric mode, enter the equations: $x = r \cos t = f(t) \cos t$
$y = r \sin t = f(t) \sin t$

The two equations for x and y are called *parametric equations*. We will study them in greater detail in Chapter 12, where t often represents time in a practical problem.

Example 3 Sketch the polar graph of $r = 2 \sin 2\theta$.

Solution **With a graphing calculator**
In polar mode, enter:
$r = 2 \sin 2t$

In parametric mode, enter:
$x = 2 \sin 2t \cos t$
$y = 2 \sin 2t \sin t$

The following values were used for the graph of $r = 2 \sin 2t$ shown:

$0 \le t \le 2\pi$, step: $\dfrac{\pi}{30}$
$-2 \le x \le 2$, scale: 1
$-2 \le y \le 2$, scale: 1

(Solution continues on the next page.

Polar Coordinates and Complex Numbers **397**

Mathematical Note

Example 4 shows that polar equations are useful in certain situations because such equations may be much simpler than the corresponding rectangular equations. Thus, the point of the example is not to become proficient at changing from one form of an equation to the other, but to appreciate the efficiency of the new notation.

Additional Examples

1. Give two polar coordinates for each point. When appropriate, give θ to the nearest tenth of a degree.
   a. $(0, -2)$
   b. $(6\sqrt{3}, -6)$
   c. $(-5, 12)$

   a. $r = \pm\sqrt{0^2 + (-2)^2} = \pm 2$. If $r = 2$:

   $\cos\theta = \dfrac{x}{r} = \dfrac{0}{2} = 0$

   $\sin\theta = \dfrac{y}{r} = \dfrac{-2}{2} = -1$

   Examples: $(2, 270°)$ and $(-2, 90°)$.

   b. $r = \pm\sqrt{(6\sqrt{3})^2 + (-6)^2} = \pm 12$. If $r = 12$:

   $\cos\theta = \dfrac{x}{r} = \dfrac{6\sqrt{3}}{12} = \dfrac{\sqrt{3}}{2}$

   $\sin\theta = \dfrac{y}{r} = \dfrac{-6}{12} = -\dfrac{1}{2}$

   $(6\sqrt{3}, -6)$ is in Quadrant IV.
   Examples: $(12, 330°)$ and $(-12, 150°)$.

   c. $r = \pm\sqrt{(-5)^2 + 12^2} = \pm 13$

   $\tan\theta = \dfrac{y}{x} = \dfrac{12}{-5} = -2.4$

   $(-5, 12)$ is in Quad. II.
   Examples: $(13, 112.6°)$ and $(-13, 292.6°)$.

Without a graphing calculator

Make a table of values and plot the points. The four polar plots below show how the graph evolves as the value of θ increases. The completed graph, the last graph shown at the right, is called a *four-leaved rose*.

θ	0°	30°	45°	60°	90°
r	0	$\sqrt{3}$	2	$\sqrt{3}$	0

θ	120°	135°	150°	180°
r	$-\sqrt{3}$	-2	$-\sqrt{3}$	0

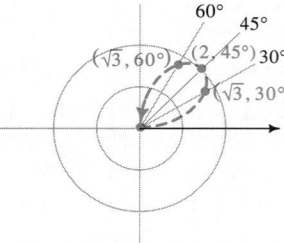

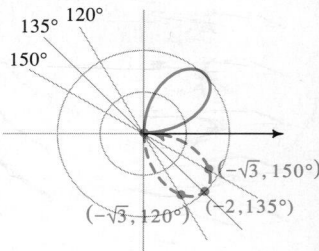

θ	210°	225°	240°	270°
r	$\sqrt{3}$	2	$\sqrt{3}$	0

θ	300°	315°	330°	360°
r	$-\sqrt{3}$	-2	$-\sqrt{3}$	0

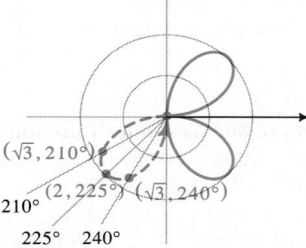

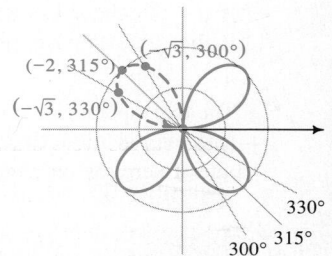

Example 4 Find a rectangular equation for $r = 2\sin 2\theta$, whose graph was shown in Example 3.

Solution Use the conversion formulas given on page 396.

$$r = 2\sin 2\theta$$
$$r = 4\sin\theta\cos\theta$$
$$r = 4 \cdot \frac{y}{r} \cdot \frac{x}{r}$$
$$r^3 = 4xy$$
$$(\pm\sqrt{x^2 + y^2})^3 = 4xy$$
$$(x^2 + y^2)^3 = 16x^2y^2$$

Example 4 illustrates that the polar equation of a curve can be much simpler than its rectangular equation. Sometimes the rectangular equation can be the simpler one. The polar equations of the following curves are all rather simple, but the corresponding rectangular equations are extremely complicated.

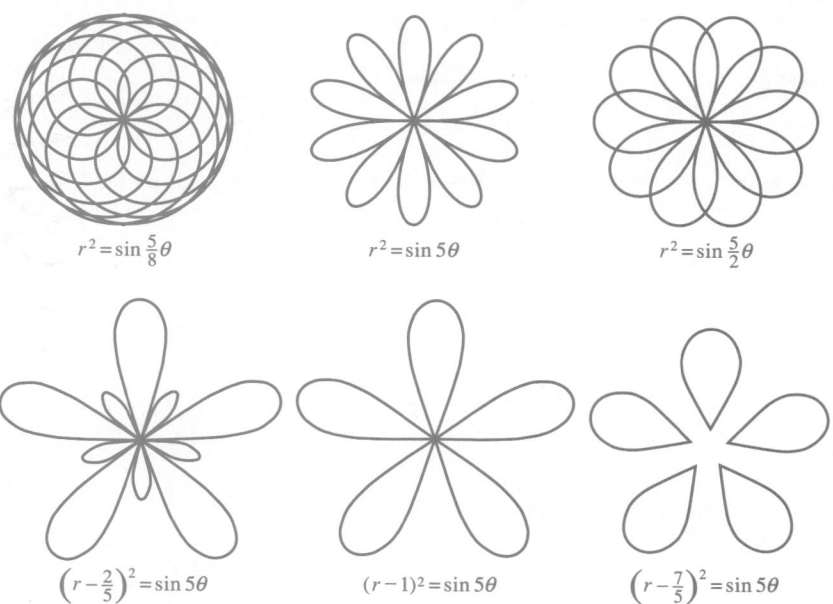

$r^2 = \sin \frac{5}{8}\theta$ $\qquad$ $r^2 = \sin 5\theta$ $\qquad$ $r^2 = \sin \frac{5}{2}\theta$

$\left(r - \frac{2}{5}\right)^2 = \sin 5\theta$ $\qquad$ $(r-1)^2 = \sin 5\theta$ $\qquad$ $\left(r - \frac{7}{5}\right)^2 = \sin 5\theta$

These curves were drawn using the *Precalculus Plotter Plus* software. The Computer Exercises on page 402 will ask you to experiment with polar graphs.

CLASS EXERCISES

1. Plot each point whose polar coordinates are given. Then give two other pairs of polar coordinates for the same point.
   a. (6, 50°) (6, 410°) (−6, −130°) b. (−7, 120°) (7, 300°) (7, −60°) c. $\left(3, \frac{\pi}{4}\right)$ d. $\left(-4, \frac{5\pi}{6}\right)\left(4, \frac{11\pi}{6}\right)$

2. Plot each point whose polar coordinates are given. Then give two other pairs of polar coordinates for the same point. $\left(4, -\frac{\pi}{6}\right)$
   a. (5, −45°) (−5, 135°) (5, 315°) b. (−2, −60°) (2, 120°) (−2, 300°) c. $\left(1, -\frac{\pi}{2}\right)$ d. $\left(-\frac{3}{2}, -3\pi\right) \left(\frac{3}{2}, 0\right)$
   $\left(-\frac{3}{2}, \pi\right)$

3. Give polar coordinates for the point (1, −1). ($\sqrt{2}$, 315°)

4. Give the rectangular coordinates for the point (−8, 60°). (−4, −4$\sqrt{3}$)

5. **Discussion** Find the rectangular equation for $r = 2 \cos \theta$, whose polar graph is shown on page 396. What is the graph of the rectangular equation? Does the rectangular equation confirm the polar graph? Explain. $(x − 1)^2 + y^2 = 1$; a circle with radius 1 and center (1, 0); yes; both graphs are the same.

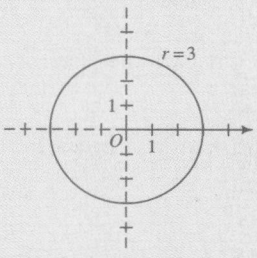
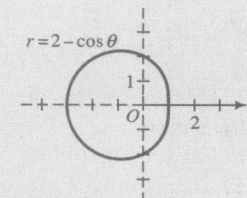

WRITTEN EXERCISES

A Plot each point whose polar coordinates are given. Then give two other pairs of polar coordinates for the point.

1.   (2, 400°), (−2, 220°)   (3, 260°), (−3, 80°)   See below.   (4, 190°), (−4, 370°)
    a. $A(2, 40°)$ **b.** $B(3, -100°)$ **c.** $C\left(5, \dfrac{3\pi}{2}\right)$ **d.** $D(-4, 10°)$

2.   (1, −280°), (−1, 260°)   See below.   (2, 130°), (−2, 310°)   (3, 290°), (3, −70°)
    a. $A(1, 80°)$ **b.** $B\left(6, -\dfrac{\pi}{6}\right)$ **c.** $C(2, 490°)$ **d.** $D(-3, 110°)$

3. a. $(2\sqrt{2}, 135°)$

Give polar coordinates (r, θ), where θ is in degrees, for each point.

3.   See above.   (2, 150°)
    a. $(-2, 2)$ **b.** (5, 0) $(5, 0°)$ **c.** $(\sqrt{2}, -\sqrt{2})$ **d.** $(-\sqrt{3}, 1)$

4.   (2, −45°)
    a. $(-4, -4)$ **b.** $(3, -3\sqrt{3})$ **c.** (0, 6) $(6, 90°)$ **d.** (0, 0)
     $(4\sqrt{2}, 225°)$ $(6, -60°)$ (0, 0°)

Give polar coordinates (r, θ), where θ is in radians, for each point.

5. a. $(-1, -1)$ $\left(\sqrt{2}, \dfrac{5\pi}{4}\right)$ **b.** (0, 12) $\left(12, \dfrac{\pi}{2}\right)$ **c.** $\left(\dfrac{1}{2}, \dfrac{\sqrt{3}}{2}\right)$ $\left(1, \dfrac{\pi}{3}\right)$ **d.** $(-2, 0)$ $(2, \pi)$

6. a. $\left(\dfrac{\sqrt{2}}{2}, \dfrac{\sqrt{2}}{2}\right)\left(1, \dfrac{\pi}{4}\right)$ **b.** $(-\sqrt{3}, 3)$ **c.** $(0, -4)$ $\left(4, \dfrac{3\pi}{2}\right)$ **d.** $(e, -e)$
                                 $\left(2\sqrt{3}, \dfrac{2\pi}{3}\right)$ $\left(e\sqrt{2}, -\dfrac{\pi}{4}\right)$

Give the rectangular coordinates for each point.

      $(-2, 2\sqrt{3})$ (0, −3) $(-\sqrt{2}, \sqrt{2})$

7. a. $(4, 120°)$ **b.** $(-3, 90°)$ **c.** $\left(1, \dfrac{5\pi}{6}\right)\left(-\dfrac{\sqrt{3}}{2}, \dfrac{1}{2}\right)$ **d.** $\left(2, \dfrac{3\pi}{4}\right)$

8. a. $(2, 225°)$ **b.** $(6, -30°)$ **c.** $\left(10, -\dfrac{3\pi}{2}\right)$ (0, 10) **d.** $\left(-4, \dfrac{\pi}{3}\right)$
    $(-\sqrt{2}, -\sqrt{2})$ $(3\sqrt{3}, -3)$ $(-2, -2\sqrt{3})$

Find the rectangular coordinates for each point to three decimal places.

9. a. $(1, 20°)$ **b.** $(2, 20°)$ **c.** (1, 2)   **d.** $(1, -2)$

10. a. $(1, 65°)$ **b.** $(4, 65°)$ **c.** (2, 3)   **d.** $(2, -3)$

11. Find polar coordinates for each point. Give r to the nearest tenth and θ to the nearest tenth of a degree.
                                                         (7.2, −56.3°)
    a. (3, 4) (5, 53.1°)   **b.** (1, 2) (2.2, 63.4°)   **c.** $(-2, 3)$(3.6, 123.7°)**d.** $(4, -6)$

12. Find polar coordinates for each point. Give r to the nearest tenth and θ to the nearest tenth of a radian.
                                                         (10, 2.5)
    a. (5, 2)   **b.** $(8, -6)$ **c.** $(-1, -4)$ **d.** $(-8, 6)$
    (5.4, 0.4)   (10, −0.6)   (4.1, −1.8)

Ⓝ In Exercises 13–22, sketch the polar graph of each equation. Also, give a rectangular equation of each graph. You may wish to use a computer or a graphing calculator to sketch these graphs.

                                                      2. b. $\left(6, \dfrac{11\pi}{6}\right), \left(-6, \dfrac{5\pi}{6}\right)$

13. $r = \sin \theta$ **1. c.** $\left(5, -\dfrac{\pi}{2}\right), \left(-5, \dfrac{\pi}{2}\right)$ **14.** $r = \cos \theta$

15. $r = -3 \cos \theta$ **16.** $r = -5 \sin \theta$

17. $r = 1 - \sin \theta$ (cardioid)   **18.** $r = 2 + 2 \cos \theta$ (cardioid)

400   *Chapter Eleven*

19. $r = 1 + 2 \sin \theta$ (limaçon)   **20.** $r = 1 - 2 \cos \theta$ (limaçon)

21. $r = \sec \theta$ **22.** $r = 2 \csc \theta$

23. a. Show that the formulas for converting from rectangular to polar coordinates given at the bottom of page 396 come from the definitions of the sine and cosine functions given on page 268.

   b. Use the equations in part (a) and the Pythagorean theorem to derive the formulas for converting from polar to rectangular coordinates.

24. a. The rectangular equations $x = 4$ and $y = 2$ have very simple graphs. Sketch them in a rectangular coordinate system.

   b. Do the polar equations $r = 4$ and $\theta = 2$ (radians) have simple graphs? Sketch them in a polar coordinate system.

   c. Show that the line with rectangular equation $x = 3$ has polar equation $r = 3 \sec \theta$.

   d. Find a polar equation of the line with rectangular equation $y = 3$.

 In Exercises 25–34, sketch the polar graph of each equation. Use radian measure for θ. You may wish to use a computer or a graphing calculator to sketch these graphs.

B **25.** $r = \cos 2\theta$ (four-leaved rose)

26. $r = \cos 4\theta$ (eight-leaved rose)

27. $r = \sin 3\theta$ (three-leaved rose)

28. $r = \sin 5\theta$ (five-leaved rose)

29. $r = \theta$ (spiral of Archimedes)

30. $r = \dfrac{1}{\theta}$ (hyperbolic spiral)

31. $r = e^{\theta}$ (logarithmic spiral)

32. $r^2 \theta = 1$ (lituus)

$\left(Hint: r = \dfrac{1}{\sqrt{\theta}} \text{ and } r = -\dfrac{1}{\sqrt{\theta}} \right).$

33. $r^2 = 4 \sin 2\theta$ (lemniscate)

34. $r = \cos \dfrac{1}{2}\theta$

35. *Reading* Look up the definitions of *cardioid* and *limaçon* in a mathematics dictionary. How are these two curves related, and why do you think the name cardioid is given to its curve? Study the equations in Exercises 17–20. How can you identify the type of curve from its equation?

36. *Research* The limaçon was first studied and named by Étienne Pascal (1588?–1651), and so this curve is usually called *Pascal's limaçon*. The lemniscate was first studied by Jacques Bernoulli (1654–1705), and so this curve is frequently called the *lemniscate of Bernoulli*. Write a brief report on these two mathematicians, describing their background and some of their important contributions to mathematics.

Polar Coordinates and Complex Numbers   **401**

Additional Answers
Written Exercises

19. $(x^2 + y^2 - 2y)^2 = x^2 + y^2$

20. $(x^2 + y^2 + 2x)^2 = x^2 + y^2$

21. $x = 1$

22. $y = 2$

24. d. $r = 3 \csc \theta$

29.

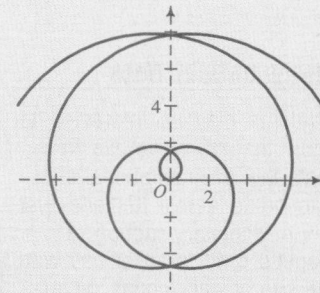

35. The cardioid is a special case of the limaçon, and it is so named because it is a heart-shaped curve.

C 37. *Investigation* In Chapter 4 we discussed how a rectangular equation can be tested to determine if its graph has symmetry. (See symmetry tests on page 134.) Discover some symmetry tests for polar equations. Write your results so that each test is clearly stated and an example of the test is given.

38. A conic section with eccentricity e has its focus at the origin. Its directrix is the line $y = -p$, where $p > 0$. Use the eccentricity definition of a conic given on page 247 to show that the polar equation of the conic is:

$$r = \frac{ep}{1 - e \sin \theta}$$

39. a. Find polar coordinates for the point of intersection of the graphs of $r = 1 + \sin \theta$ and $r = 2 \sin \theta$.

b. Now sketch the two graphs and note that they have another common point whose polar coordinates do not satisfy both equations. Explain why this happens.

40. Suppose $P(r, \theta)$ is any point on the circle with center $C = (a, \alpha)$ and radius R. Use the law of cosines to show that the polar equation of the circle is:

$$r^2 - 2ar \cos (\theta - \alpha) + a^2 = R^2$$

///// COMPUTER EXERCISES

You will need to use either computer graphing software or a graphing calculator that graphs polar equations.

1. a. Graph $r = \sin kt$ or $r = \cos kt$ for $k = 1, 3$, and 5.
   b. Predict what the graph of $r = \sin kt$ or $r = \cos kt$ will look like for larger odd values of k. Then check your prediction.
   c. Graph $r = \sin kt$ or $r = \cos kt$ for $k = 2, 4$, and 6.
   d. Predict what the graph of $r = \sin kt$ or $r = \cos kt$ will look like for larger even values of k. Then check your prediction.
   e. Make a generalization about the graph of an equation of the form $r = a \sin nt$ or $r = a \cos nt$, where n is a positive integer.

2. Graph $r^2 = k^2 \sin 2t$ and $r^2 = k^2 \cos 2t$ for $k = 1, 2$, and 3. Make a generalization about these graphs, which are all lemniscates.

3. Graph $r = \sin \frac{a}{b}t$, where a and b are both odd. Describe in a few sentences how the shape of the graph depends on the values of a and b.

4. The six equations whose graphs are shown on page 399 are all special cases of the equation $(r - c)^2 = \sin \frac{a}{b}t$. Experiment with values of b and c to see if you can discover any patterns. You might consider letting $c = 0$ while gradually varying b. Then try letting $b = 1$ while gradually varying c. Finally, let a vary while keeping b and c fixed. Write a paragraph or two describing your results and conclusions.

402   *Chapter Eleven*

11-2 Geometric Representation of Complex Numbers

Objective   *To write complex numbers in polar form and to find products in polar form.*

In 1806, Jean Robert Argand, a Parisian bookkeeper, published an essay in which complex numbers were represented geometrically. The diagram at the right, known as an **Argand diagram**, shows that the complex number $3 + 4i$ can be represented by the point $(3, 4)$ or by an arrow from the origin to $(3, 4)$. Similarly, the number $5 - 2i$ can be represented by a specific point or arrow. These different representations will be discussed again in

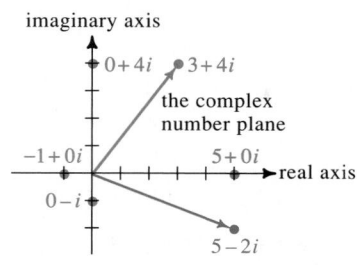

Chapter 12. When the plane is used to represent complex numbers, the plane is called the **complex number plane**, or simply the **complex plane**. An important subset of this plane is the real number line. A typical point on this line has coordinates $(a, 0)$, and represents the complex number $a + 0i$. A point on the vertical axis, or imaginary axis, has coordinates $(0, b)$ and represents $0 + bi$, a pure imaginary number.

The point representing the complex number $z = a + bi$ can be given either in rectangular coordinates (a, b) or in polar coordinates (r, θ). (See the diagram below.) We can specify a complex number in two ways:

rectangular form: $z = a + bi$

polar form: $z = r \cos \theta + (r \sin \theta)i$
$z = r(\cos \theta + i \sin \theta)$
$z = r \operatorname{cis} \theta$

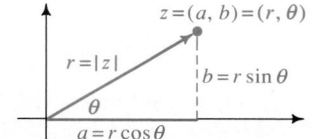

The polar form $r(\cos \theta + i \sin \theta)$ is often abbreviated as $r \operatorname{cis} \theta$ (pronounced "r siss theta"). For example, we write $2 \operatorname{cis} 30°$ as the abbreviation for $2(\cos 30° + i \sin 30°)$.

The length of the arrow representing z is called the **absolute value** of z, and is defined to be $|z| = \sqrt{a^2 + b^2}$. Since $r = \sqrt{a^2 + b^2} = |z|$, which is nonnegative, we will see only positive values of r in polar coordinates of a complex number z. The angle θ is called the *polar angle* of z. Throughout the rest of this book, we will give final answers using positive radii and positive polar angles.

Many scientific calculators have keys that enable us to convert from rectangular form to polar form and vice versa. These conversion keys allow us to express very easily a complex number in either polar or rectangular form, as discussed in Examples 1 and 2.

Warm-Up Exercises

Give polar coordinates (r, θ), where θ is in degrees, for each point. If necessary, give θ to the nearest tenth of a degree.

1. $(-4, 4)$ $(4\sqrt{2}, 135°)$

2. $(-5, -12)$ $(13, 247.4°)$

Multiply.

3. $(1 - i)(4 + i)$ $5 - 3i$

4. $(1 + i\sqrt{3})(1 - i\sqrt{3})$ 4

Motivating the Section

This section extends the study of polar coordinates to their use in representing complex numbers. As motivation for using the new notation, ask students to find the product of two complex numbers in rectangular form. For example, ask students to find the product $(\sqrt{2} + 3i)(1 - 2i)$. $((6 + \sqrt{2}) + (3 - 2\sqrt{2})i)$ Explain that using the polar form of complex numbers makes finding products much easier.

Additional Examples

1. Express each product or quotient in polar and rectangular form.
   a. $(3 \text{ cis } 165°)(4 \text{ cis } 45°)$
   b. $(3 \text{ cis } 165°) \div (4 \text{ cis } 45°)$

   a. $(3 \cdot 4)(\text{cis } (165° + 45°))$
   $= 12 \text{ cis } 210°$
   In rectangular form,
   $12\left(-\dfrac{\sqrt{3}}{2} - \dfrac{1}{2}i\right) =$
   $-6\sqrt{3} - 6i.$

   b. If $z_1 = r \text{ cis } \alpha$ and $z_2 = r \text{ cis } \beta$, where $z_2 \neq 0 + 0i$, then
   $\dfrac{z_1}{z_2} = \dfrac{r}{s} \text{ cis } (\alpha - \beta).$
   (Students are asked to prove this in Exercise 31.)
   $\dfrac{3 \text{ cis } 165°}{4 \text{ cis } 45°} =$
   $\dfrac{3}{4} \text{ cis } (165° - 45°) =$
   $\dfrac{3}{4} \text{ cis } 120°$
   In rectangular form,
   $\dfrac{3}{4}\left(-\dfrac{1}{2} + \dfrac{\sqrt{3}}{2}i\right) =$
   $-\dfrac{3}{8} + \dfrac{3\sqrt{3}}{8}i.$

Example 1 Express $2 \text{ cis } 50°$ in rectangular form.

Solution **Method 1** Using the polar to rectangular conversion key on a calculator, we find that $(2, 50°) \approx (1.29, 1.53)$. Thus, the rectangular form of $2 \text{ cis } 50°$ is $1.29 + 1.53i$.

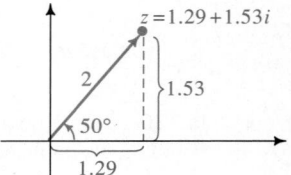

Method 2 Use the fact that $a = r \cos \theta$ and $b = r \sin \theta$, as follows:
$2 \text{ cis } 50° = 2(\cos 50° + i \sin 50°)$
$= (2 \cos 50°) + (2 \sin 50°)i \approx 1.29 + 1.53i$

Example 2 Express $-1 - 2i$ in polar form.

Solution **Method 1** Using the rectangular to polar conversion key on a calculator, we find that $(-1, -2) \approx (2.24, -117°)$, or $(2.24, 243°)$. Thus, the polar form of $-1 - 2i$ is $2.24 \text{ cis } 243°$.

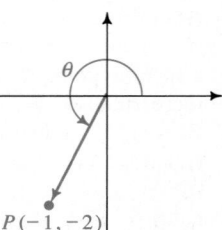

Method 2 $r = |-1 - 2i| = \sqrt{(-1)^2 + (-2)^2} = \sqrt{5} \approx 2.24$
$\text{Tan}^{-1}\left(\dfrac{-2}{-1}\right) \approx 63°; \ \theta \approx 63° + 180° = 243°$

Thus, the polar form of $-1 - 2i$ is $\sqrt{5} \text{ cis } 243°$, or $2.24 \text{ cis } 243°$.

The theorem below shows how to multiply complex numbers in polar form.

Product of Two Complex Numbers in Polar Form

To multiply two complex numbers in polar form:

1. Multiply their absolute values.
2. Add their polar angles.

In other words, if $z_1 = r \text{ cis } \alpha$ and $z_2 = s \text{ cis } \beta$, then:

$z_1 z_2 = (r \text{ cis } \alpha) \cdot (s \text{ cis } \beta)$
$= rs \text{ cis}(\alpha + \beta)$

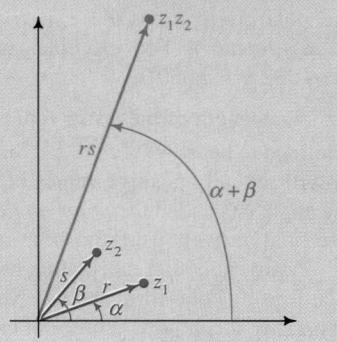

404   *Chapter Eleven*

Proof:

$$r \text{ cis } \alpha \cdot s \text{ cis } \beta = r(\cos \alpha + i \sin \alpha) \cdot s(\cos \beta + i \sin \beta)$$
$$= rs[(\cos \alpha \cos \beta - \sin \alpha \sin \beta) +$$
$$i(\sin \alpha \cos \beta + \cos \alpha \sin \beta)]$$
$$= rs[\cos(\alpha + \beta) + i \sin(\alpha + \beta)]$$
$$= rs \text{ cis}(\alpha + \beta)$$

Example 3   Express $z_1 = 3 + 3i$, $z_2 = 1 + i\sqrt{3}$, and $z_1 z_2$ in polar form. Graph each complex number.

Solution   Using either method of Example 2:

$$z_1 = 3\sqrt{2} \text{ cis } 45° \text{ and}$$
$$z_2 = 2 \text{ cis } 60°$$

By the previous theorem:

$$z_1 z_2 = 3\sqrt{2} \cdot 2 \text{ cis}(45° + 60°)$$
$$= 6\sqrt{2} \text{ cis } 105°$$

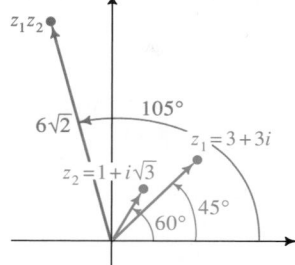

To see the usefulness of the previous theorem, let's compare the multiplication of two complex numbers in polar form with the multiplication of the same numbers in rectangular form. Consider the numbers:

$$z_1 = 2 \text{ cis } 50° \approx 1.286 + 1.532i \quad \text{and} \quad z_2 = 3 \text{ cis } 20° \approx 2.819 + 1.026i$$

Polar Method: $(2 \text{ cis } 50°)(3 \text{ cis } 20°) = 6 \text{ cis } 70°$

Rectangular Method: $(1.286 + 1.532i)(2.819 + 1.026i) \approx$
$$3.625 + 1.319i + 4.319i + 1.572i^2 \approx 2.05 + 5.64i$$

You can see that the two answers agree by expressing $6 \text{ cis } 70°$ in rectangular form.

$$6 \text{ cis } 70° = (6 \cos 70°) + (6 \sin 70°)i \approx 2.05 + 5.64i$$

CLASS EXERCISES

Express each complex number in polar form.

1. $1 + i$ $\sqrt{2} \text{ cis } 45°$ **2.** i $\text{ cis } 90°$ **3.** -3 $3 \text{ cis } 180°$ **4.** $\sqrt{3} - i$ $2 \text{ cis } 330°$

5. Reading What does the expression $2 \text{ cis } \frac{\pi}{4}$ mean? $2\left(\cos \frac{\pi}{4} + i \sin \frac{\pi}{4}\right)$

Express each complex number in rectangular form.

6. $5 \text{ cis } 90°$ $5i$ **7.** $3 \text{ cis } \pi$ -3 **8.** $4 \text{ cis } 45°$ $2\sqrt{2} + 2i\sqrt{2}$ **9.** $6 \text{ cis } 30°$ $3\sqrt{3} + 3i$

Express each product in polar form.

10. $(4 \text{ cis } 25°)(6 \text{ cis } 35°)$ $24 \text{ cis } 60°$ **11.** $\left(5 \text{ cis } \frac{\pi}{4}\right)\left(2 \text{ cis } \frac{3\pi}{4}\right)$ $10 \text{ cis } \pi$

2. Let $z_1 = 2 - 2i$ and $z_2 = 3i$.

a. Find $z_1 z_2$ in rectangular form.

b. Find z_1, z_2, and $z_1 z_2$ in polar form.

c. Convert $z_1 z_2$ in part (b) to rectangular form. Does your answer agree with the one in part (a)?

a. $(2 - 2i)3i = 6i - 6i^2 = 6 + 6i$

b. $r_1 = \sqrt{2^2 + (-2)^2} = 2\sqrt{2}$

$\tan \theta_1 = \frac{-2}{2} = -1$

and θ_1 is in Quadrant IV.

Thus, $\theta_1 = -\frac{\pi}{4}$.

$z_1 = 2\sqrt{2} \text{ cis}\left(-\frac{\pi}{4}\right)$

$r_2 = \sqrt{0^2 + 3^2} = 3$

$\theta_2 = \frac{\pi}{2}$

$z_2 = 3 \text{ cis } \frac{\pi}{2}$

$z_1 z_2 =$
$\left(2\sqrt{2} \text{ cis}\left(-\frac{\pi}{4}\right)\right) \cdot$
$\left(3 \text{ cis } \frac{\pi}{2}\right) = 6\sqrt{2} \text{ cis } \frac{\pi}{4}$

c. $6\sqrt{2}\left(\cos \frac{\pi}{4} + i \sin \frac{\pi}{4}\right) =$

$6\sqrt{2}\left(\frac{\sqrt{2}}{2} + \frac{\sqrt{2}}{2}i\right) =$

$6 + 6i$; yes

WRITTEN EXERCISES

Express each complex number in polar form. Give angle measures to the nearest degree when necessary.

$2\sqrt{2}$ cis $225°$

A **1.** $-1 + i$ $\sqrt{2}$ cis $135°$ **2.** $-3i$ 3 cis $270°$ **3.** $1 + i\sqrt{3}$ 2 cis $60°$ **4.** $-2 - 2i$

5. -7 7 cis $180°$ **6.** $2\sqrt{3} + 2i$ **7.** $3 - 4i$ **8.** $5 + 12i$
  4 cis $30°$ 5 cis $307°$ 13 cis $67°$

Express each complex number in rectangular form.

9. 6 cis $100°$ **10.** 8 cis $230°$ **11.** 9 cis $\dfrac{4\pi}{3}$ **12.** 2 cis $\dfrac{3\pi}{4}$
  $-1.04 + 5.91i$ $-5.14 - 6.13i$ $-\sqrt{2} + i\sqrt{2}$

Express each product in polar and rectangular form. $-\dfrac{9}{2} - \dfrac{9\sqrt{3}}{2}i$

13. $(5$ cis $30°)(2$ cis $60°)$ 10 cis $90°$; $10i$ **14.** $(2$ cis $115°)(3$ cis $65°)$ 6 cis $180°$; -6

15. $\left(8 \text{ cis } \dfrac{\pi}{3}\right)\left(\dfrac{1}{2} \text{ cis}\left(-\dfrac{2\pi}{3}\right)\right)$ 4 cis $\dfrac{5\pi}{3}$;   **16.** $\left(4 \text{ cis } \dfrac{\pi}{4}\right)\left(3 \text{ cis } \dfrac{\pi}{2}\right)$ 12 cis $\dfrac{3\pi}{4}$;
  $2 - 2i\sqrt{3}$ $-6\sqrt{2} + 6i\sqrt{2}$

For Exercises 17–22:
(a) Find z_1z_2 in rectangular form by multiplying z_1 and z_2.
(b) Find z_1, z_2, and z_1z_2 in polar form. Show that z_1z_2 in polar form agrees with z_1z_2 in rectangular form.
(c) Show z_1, z_2, and z_1z_2 in an Argand diagram.

17. $z_1 = 2 + 2i\sqrt{3}$, $z_2 = \sqrt{3} - i$ **18.** $z_1 = 3 + 3i$, $z_2 = -2i$

19. $z_1 = 2 + 2i$, $z_2 = 2 - 2i$ **20.** $z_1 = -1 + i\sqrt{3}$, $z_2 = -1 - i\sqrt{3}$

21. $z_1 = 6 - 4i$, $z_2 = -5 + 2i$ **22.** $z_1 = -3 - 5i$, $z_2 = 4 + 7i$

23. *Visual Thinking* Suppose that $a + bi$ is multiplied by $1 + i$. By how many degrees must the arrow from $(0, 0)$ to (a, b) be rotated to coincide with the arrow from $(0, 0)$ to the product? $45°$

24. *Visual Thinking* Suppose that $a + bi$ is multiplied by $\sqrt{3} + i$. By how many degrees must the arrow from $(0, 0)$ to (a, b) be rotated to coincide with the arrow from $(0, 0)$ to the product? $30°$

Exercises 25–28 should be worked in order because they preview the next section.

25. Restate the theorem on page 404 for the special case when $z_2 = z_1$.

26. Find $(\cos \theta + i \sin \theta)^2$. Show that your result equals $\cos 2\theta + i \sin 2\theta$.

27. From Exercise 26, $(\cos \theta + i \sin \theta)^2 = \cos 2\theta + i \sin 2\theta$. Make a conjecture about $(\cos \theta + i \sin \theta)^3$. See below.

28. The diagram at the right shows the complex number $z = \cos 30° + i \sin 30°$. Copy this diagram, and on it show z^2, z^3, and z^4. (*Hint*: Use the results of Exercises 26 and 27.) **27.** $(\cos \theta + i \sin \theta)^3 = \cos 3\theta + i \sin 3\theta$

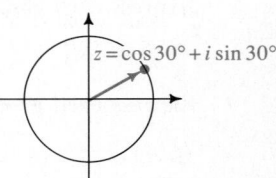

$z = \cos 30° + i \sin 30°$

B **29. a.** If $z = r(\cos \theta + i \sin \theta)$, then the conjugate of z is $\bar{z} = r(\cos \theta - i \sin \theta)$. Represent z and $\bar{z}$ in an Argand diagram.

b. *Visual Thinking* Show that $\cos(-\theta) + i \sin(-\theta) = \cos \theta - i \sin \theta$. Use this fact and your diagram in part (a) to explain why $\bar{z} = r \text{ cis } (-\theta)$ if $z = r \text{ cis } \theta$.

30. If $z = r \text{ cis } \theta = r(\cos \theta + i \sin \theta)$, show that $\dfrac{1}{z} = \dfrac{1}{r} \text{cis}(-\theta)$.

31. This exercise shows that to divide two complex numbers, you divide their absolute values and subtract their polar angles. Let $z_1 = r \text{ cis } \alpha$ and $z_2 = s \text{ cis } \beta$, where $z_2 \neq 0 + 0i$. Prove that

$$\frac{z_1}{z_2} = \frac{r}{s} \text{cis}(\alpha - \beta).$$

(*Hint:* Use the results of Exercise 30.)

C **32.** (An alternate proof of the theorem on page 404.)

a. Let $z_1 = a + bi$ and $z_2 = c + di$. Show z_1 and z_2 in an Argand diagram. Prove that the product is given by $z_1z_2 = (ac - bd) + (bc + ad)i$.

b. On your diagram, let α, β, and θ be the polar angles for z_1, z_2, and z_1z_2, respectively. Then $\tan \alpha = \underline{\ ?\ }$, $\tan \beta = \underline{\ ?\ }$, and $\tan \theta = \underline{\ ?\ }$.

c. Use the values of $\tan \alpha$ and $\tan \beta$ in part (b) and the formula for $\tan (\alpha + \beta)$ to show that $\tan (\alpha + \beta) = \tan \theta$. Then you can conclude that $\alpha + \beta = \theta$.

d. By definition, $|z_1| = \sqrt{a^2 + b^2}$, $|z_2| = \underline{\ ?\ }$, and $|z_1z_2| = \underline{\ ?\ }$. Show that $|z_1z_2| = |z_1| \cdot |z_2|$.

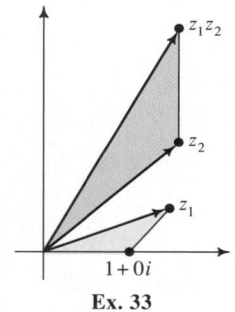

Ex. 33

33. *Writing* Write a paragraph explaining why the two triangles in the diagram at the right are similar.

11–3 Powers of Complex Numbers

Objective *To use De Moivre's theorem to find powers of complex numbers.*

The polar form of a complex number makes it easy to find powers of a complex number. To find the square of a complex number, we use the theorem on page 404.

$$(r \text{ cis } \alpha)^2 = (r \text{ cis } \alpha)(r \text{ cis } \alpha) = r \cdot r \text{ cis } (\alpha + \alpha) = r^2 \text{ cis } 2\alpha$$

To find the cube of $r \text{ cis } \alpha$, we use the square of $r \text{ cis } \alpha$ and the same theorem.

$$\begin{aligned}
(r \text{ cis } \alpha)^3 &= (r \text{ cis } \alpha)^2(r \text{ cis } \alpha) \\
&= (r^2 \text{ cis } 2\alpha)(r \text{ cis } \alpha) \\
&= r^2 \cdot r \text{ cis } (2\alpha + \alpha) \\
&= r^3 \text{ cis } 3\alpha
\end{aligned}$$

Polar Coordinates and Complex Numbers   **407**

Supplementary Materials

Alternative Assessment, 33

Teaching Notes, p. 394B

Warm-Up Exercises

Express each complex number in polar form.

1. $\sqrt{3} - i$ $2 \text{ cis } 330°$

2. $-3 - 3i$ $3\sqrt{2} \text{ cis } 225°$

3. -5 $5 \text{ cis } 180°$

4. $-\dfrac{3}{2}i$ $\dfrac{3}{2} \text{ cis } 270°$

Write each expression in rectangular form.

5. $(\sqrt{2})^5 \text{ cis } (5 \cdot 27°)$
  $-4 + 4i$

6. $3^6 \text{ cis } (6 \cdot 15°)$ $729i$

Motivating the Section

Tell students that representing complex numbers in polar form makes it easy to explore the powers and roots of complex numbers. You might want to review Written Exercises 25–28 from Section 11-2 as an introduction to De Moivre's theorem.

Additional Examples

1. Compute $(\sqrt{3} - i)^4$ by the following two methods:
   a. Multiply out $(\sqrt{3} - i)^4$ and simplify.
   b. Apply De Moivre's theorem. Then express your answer in rectangular form.
   a. $(\sqrt{3} - i)^2 = 2 - 2i\sqrt{3}$
   $(\sqrt{3} - i)^4 =$
   $(2 - 2i\sqrt{3})^2 =$
   $-8 - 8i\sqrt{3}$
   b. $\sqrt{3} - i = 2 \operatorname{cis}(-30°)$
   $(2 \operatorname{cis}(-30°))^4 =$
   $2^4 \operatorname{cis}(4(-30°)) =$
   $16 \operatorname{cis}(-120°);$
   $16(\cos(-120°) +$
   $i\sin(-120°)) =$
   $16\left(-\dfrac{1}{2} + i\left(-\dfrac{\sqrt{3}}{2}\right)\right) =$
   $-8 - 8i\sqrt{3}$

2. Let $z = -1$. Use De Moivre's theorem to show that the positive powers of z are alternately equal to ± 1.
   $z = 1 \operatorname{cis} 180°$. If n is a positive odd integer, then
   $z^n = 1^n \operatorname{cis}(n \cdot 180°) =$
   $\cos(n \cdot 180°) +$
   $i\sin(n \cdot 180°) =$
   $-1 + 0i = -1$. If n is a positive even integer, then
   $z^n = \cos(n \cdot 180°) +$
   $i\sin(n \cdot 180°) =$
   $1 + 0i = 1$.

The pattern just shown can be generalized for any positive integer n:

$$(r \operatorname{cis} \alpha)^n = r^n \operatorname{cis} n\alpha$$

In fact, this equation holds for *any* integer n and even for fractional values of n, as we will see in the next section. This result is known as De Moivre's Theorem.

De Moivre's Theorem

If $z = r \operatorname{cis} \theta$, then $z^n = r^n \operatorname{cis} n\theta$.

Example If $z = \dfrac{1}{2} + \dfrac{\sqrt{3}}{2}i$, find z^2, z^3, z^4, z^5, and z^6. Plot these on an Argand diagram.

Solution First we put z into polar form, and then use De Moivre's theorem.

$$z = \frac{1}{2} + \frac{\sqrt{3}}{2}i = 1 \operatorname{cis} 60°$$

$$z^2 = 1^2 \operatorname{cis}(2 \cdot 60°) = 1 \operatorname{cis} 120° = -\frac{1}{2} + \frac{\sqrt{3}}{2}i$$

$$z^3 = 1^3 \operatorname{cis}(3 \cdot 60°) = 1 \operatorname{cis} 180° = -1$$

$$z^4 = 1^4 \operatorname{cis}(4 \cdot 60°) = 1 \operatorname{cis} 240° = -\frac{1}{2} - \frac{\sqrt{3}}{2}i$$

$$z^5 = 1^5 \operatorname{cis}(5 \cdot 60°) = 1 \operatorname{cis} 300° = \frac{1}{2} - \frac{\sqrt{3}}{2}i$$

$$z^6 = 1^6 \operatorname{cis}(6 \cdot 60°) = 1 \operatorname{cis} 360° = 1$$

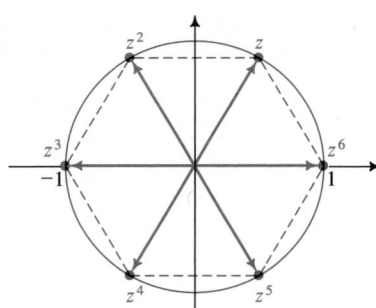

The Argand diagram above shows that these six powers of z are closely related to the construction of a regular hexagon inscribed in a circle of radius 1.

The path taken by the points corresponding to the powers of a given complex number z depends greatly on z. To see a few of the possible paths, complete the activity given on the next page.

408 *Chapter Eleven*

Activity

Express each complex number z in polar form. Then graph z^n for $n = 1, 2, \ldots, 6$. Describe the path taken by the points graphed.

1. $z = 1 + i$
$\sqrt{2}$ cis 45°

2. $z = \dfrac{1}{2} + \dfrac{1}{2}i$
$\dfrac{\sqrt{2}}{2}$ cis 45°

3. $z = \sqrt{3} + i$
2 cis 30°

The paths discovered in the activity are called *equiangular spirals* and can be found throughout nature. For example, the horns and claws of some animals, the seeds of a sunflower, and the chambers of a nautilus shell form equiangular spirals.

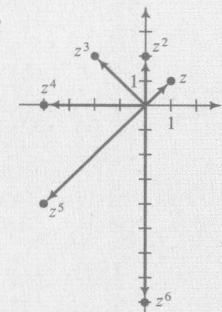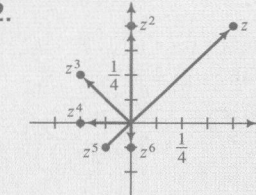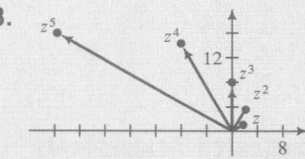
CLASS EXERCISES

Give the polar form of each of the following.

1. $(2 \text{ cis } 45°)^2$ 4 cis 90°

2. $(2 \text{ cis } 45°)^3$ 8 cis 135°

3. $(\sqrt{2} \text{ cis } (-18°))^4$ 4 cis (−72°)

4. $(1 \text{ cis } 36°)^{10}$ 1 cis 360°

5. $\left(4 \text{ cis } \dfrac{\pi}{6}\right)^3$ 64 cis $\dfrac{\pi}{2}$

6. $\left(\sqrt{3} \text{ cis } \dfrac{5\pi}{6}\right)^6$ 27 cis 5π

7. Let $z = -1 + i$.
  a. Express z and z^6 in polar form. $\sqrt{2}$ cis 135°, 8 cis 90°
  b. Express z^6 in rectangular form. 8i

8. Let $z = 1 - i$.
  a. Find z^{10} by writing z as $\sqrt{2}$ cis 315°, applying De Moivre's theorem, and simplifying the result. 32 cis 3150° = 32 cis 270° = −32i
  b. Find z^{10} by writing z as $\sqrt{2}$ cis(−45°), applying De Moivre's theorem, and simplifying the result. 32 cis (−450°) = 32 cis 270° = −32i
  c. *Discussion* Which calculation was easier? Explain why. Part (b);
  −450° is easier to simplify than 3150° since 3150° involves a larger multiple of 360°.

◩ Using Technology

For Class Exercise 8c, you may wish to give students the following algorithm for simplifying 3150° to 270°: Step 1 is to divide 3150 by 360, getting 8.75. Step 2 is to multiply 360 by 8, getting 2880. Step 3 is to subtract 2880 from 3150, getting 270.

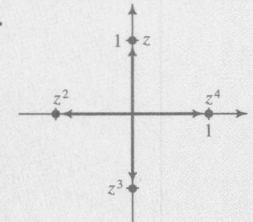
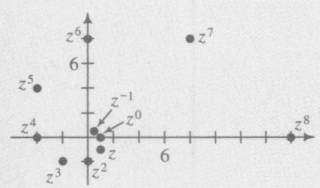
WRITTEN EXERCISES

A **1.** If $z = \dfrac{\sqrt{3}}{2} + \dfrac{1}{2}i$, express z in polar form. Calculate z^2 and z^3 by using De Moivre's theorem. Show z, z^2, and z^3 in an Argand diagram. 1 cis 30°; 1 cis 60°; 1 cis 90°

2. If $z = i$, show z, z^2, z^3, and z^4 in an Argand diagram.

3. a. Express $z = 1 - i$ in polar form. Show z^{-1}, z^0, z, z^2, z^3, z^4, z^5, z^6, z^7, and z^8 in an Argand diagram. $\sqrt{2}$ cis 315°

    b. Show that $z^{12} = -64$. $(\sqrt{2})^{12}$ cis $[12(-45°)] = 64$ cis$(-540°) = -64$

4. a. Express $z = 1 + i\sqrt{3}$ in polar form. Show z^{-2}, z^{-1}, z^0, z, z^2, z^3, and z^4 in an Argand diagram. 2 cis 60°

    b. Show that $z^{18} = 2^{18}$. $z^{18} = 2^{18}$ cis $(18 \cdot 60°) = 2^{18}$ cis 1080° = 2^{18}

5. Compute $(1 - i\sqrt{3})^3$ by two methods.

    a. Simplify $(1 - i\sqrt{3})(1 - i\sqrt{3})(1 - i\sqrt{3})$. -8

    b. Apply De Moivre's theorem and express your answer in rectangular form. -8

6. Compute $(-1 - i)^3$ by two methods.

    a. Simplify $(-1 - i)(-1 - i)(-1 - i)$. $2 - 2i$

    b. Apply De Moivre's theorem and express your answer in rectangular form. $2 - 2i$

7. Refer to the diagram on page 408. Do you see that z^7 simplifies to z? Yes Simplify z^{14}, z^{40}, z^{-1}, and z^{-11}. z^2, z^4, z^5, z

8. Let $z = \cos \theta + i \sin \theta$. Evaluate $|z^{100}|$. 1

B **9. a.** Verify that De Moivre's theorem is true when $n = 0$.

    b. In Exercise 30 on page 407, you showed that for $z \ne 0 + 0i$,

$$\frac{1}{z} = \frac{1}{r} \text{ cis}(-\theta).$$

    Use this fact to prove De Moivre's theorem for the case when $n = -1$ by writing $z^{-1} = \dfrac{1}{z}$ and then simplifying.

10. In Exercise 9, you proved De Moivre's theorem for $n = -1$. Now prove it for $n = -2$ by writing $z^{-2} = (z^{-1})^2$. (A similar proof of De Moivre's theorem can be made for any negative integer n.)

11. Make an Argand diagram showing the number -1. On the same diagram, show a number z whose square is -1.

12. Make an Argand diagram showing the number $-i$. On the same diagram, show a number z whose cube is $-i$.

In Exercises 13–16, use the following identity: $re^{i\theta} = r$ cis θ.

13. Show that $e^{i\pi} = -1$. $e^{i\pi} = 1$ cis $\pi = \cos \pi + i \sin \pi = -1$

14. Is i^i a real number or an imaginary number? (*Hint:* Let $r = 1$ and $\theta = \dfrac{\pi}{2}$ in the identity $re^{i\theta} = r(\cos \theta + i \sin \theta)$, simplify the result, and then raise both sides of the equation to the power i.) Find i^i. Real; 0.208

410   *Chapter Eleven*

15. Prove De Moivre's theorem for all real numbers n. Assume that the laws of exponents hold for imaginary exponents.

16. Use the laws of exponents to simplify the product $(re^{i\alpha})(se^{i\beta})$. Interpret your answer by using the identity $re^{i\theta} = r \operatorname{cis} \theta$.

C 17. The curve through the powers of a complex number z is called an **equiangular spiral** because it crosses all lines through the origin at the same angle. You can get a rough idea why this happens by studying the diagram below.

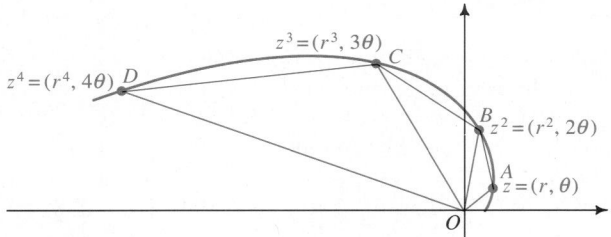

a. Prove that $\triangle OAB$, $\triangle OBC$, and $\triangle OCD$ are all similar. (It is easiest to use the SAS Similarity theorem. This proof can be extended to include all other triangles formed by the lines through the origin and the points representing the powers of z.)

b. Why are $\angle OAB$, $\angle OBC$, and $\angle OCD$ all congruent? (The fact that these angles are congruent suggests that the lines through the origin meet the curve at the same angle.)

18. Draw the polar graph of $\ln r = \theta$. (You may wish to use a computer or graphing calculator.) Your result should look like the equiangular spiral of Exercise 17. Because the polar equation contains $\ln r$, the spiral is also called a **logarithmic spiral.**

René Descartes (1596–1650)

The philosopher René Descartes was educated in France, and pursued military service in the Netherlands. It is thought that his army work with maps may have led to his notation known today as *Cartesian coordinates*, which is the foundation of analytic geometry.

Descartes modernized algebraic notation. He was the first to use the last letters of the alphabet to represent unknowns and the beginning letters to represent constants. He also introduced the terms *function*, *real*, and *imaginary*.

Polar Coordinates and Complex Numbers **411**

Warm-Up Exercises

Evaluate each expression for the specified values of k. Give answers in rectangular form.

1. $\operatorname{cis}\left(\dfrac{90°}{2} + \dfrac{k \cdot 360°}{2}\right)$ for
   $k = 0, 1$
   $\dfrac{\sqrt{2}}{2} + \dfrac{\sqrt{2}}{2}i, \ -\dfrac{\sqrt{2}}{2} - \dfrac{\sqrt{2}}{2}i$

2. $\operatorname{cis}\left(\dfrac{180°}{3} + \dfrac{k \cdot 360°}{3}\right)$ for
   $k = 0, 1, 2$
   $\dfrac{1}{2} + \dfrac{\sqrt{3}}{2}i, \ -1, \ \dfrac{1}{2} - \dfrac{\sqrt{3}}{2}i$

3. $8^{1/3} \cdot \operatorname{cis}\left(\dfrac{270°}{3} + \dfrac{k \cdot 360°}{3}\right)$
   for $k = 0, 1, 2$
   $2i, \ -\sqrt{3} - i, \ \sqrt{3} - i$

4. $4^{1/4} \cdot \operatorname{cis}\left(\dfrac{180°}{4} + \dfrac{k \cdot 360°}{4}\right)$
   for $k = 0, 1, 2, 3$ $1 + i,$
   $-1 + i, \ -1 - i, \ 1 - i$

Motivating the Section

The number 1 has three cube roots, one real and two imaginary, which can be found by solving the equation $x^3 - 1 = 0$. De Moivre's theorem provides another method for finding the cube roots of 1.

Additional Examples

1. Find the three cube roots of -1.
   $z = -1 + 0i = 1 \operatorname{cis} 180°$. The three cube roots of z are $\sqrt[3]{1} \operatorname{cis} \alpha$, where $\alpha = \dfrac{180° + k \cdot 360°}{3}$, $k = 0, 1, 2$.
   $z_1 = \operatorname{cis} 60° = \dfrac{1}{2} + \dfrac{\sqrt{3}}{2}i,$
   $z_2 = \operatorname{cis} 180° = -1$, and
   $z_3 = \operatorname{cis} 300° = \dfrac{1}{2} - \dfrac{\sqrt{3}}{2}i.$

11–4 Roots of Complex Numbers

Objective   *To find roots of complex numbers.*

In this section, we use DeMoivre's Theorem to find roots of complex numbers. In general, a nonzero complex number will have two square roots, three cube roots, four fourth roots, and k kth roots. Here are two examples:

Example 1   Find the cube roots of $8i$.

Solution   This problem can be solved by thinking of $8i$ and its cube root, $z = r \operatorname{cis} \alpha$, as points in the plane. Since z is a cube root of $8i$,

$$z^3 = 8i$$
$$(r \operatorname{cis} \alpha)^3 = 8 \operatorname{cis} 90°$$
$$r^3 \operatorname{cis} 3\alpha = 8 \operatorname{cis} 90°$$

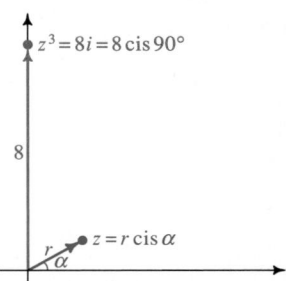

Thus, $\quad r^3 = 8 \quad$ and $\quad 3\alpha = 90°$ or $(90° +$ a multiple of $360°)$
$\qquad\quad r = 2 \qquad\qquad\quad \alpha = 30°$ or $(30° +$ a multiple of $120°)$
$\qquad\qquad\qquad\qquad\qquad\quad \alpha = 30°, 30° + 120°, 30° + 240°$
$\qquad\qquad\qquad\qquad\qquad\quad \alpha = 30°, 150°, 270°$

Therefore, the three cube roots of $8i$ are:

$$z_1 = 2 \operatorname{cis} 30° = \sqrt{3} + i$$
$$z_2 = 2 \operatorname{cis} 150° = -\sqrt{3} + i$$
$$z_3 = 2 \operatorname{cis} 270° = -2i$$

As a check, you can use De Moivre's theorem to cube z_1, z_2, z_3 and see that the result in each case is $8i$.

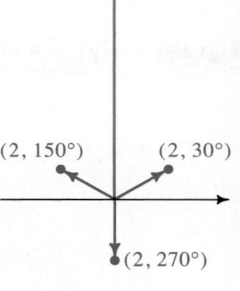

From the analysis above, we see that if $z = r \operatorname{cis} \theta$, then:

$$\sqrt[3]{z} = z^{1/3} = r^{1/3} \operatorname{cis}\left(\dfrac{\theta}{3} + \dfrac{k \cdot 360°}{3}\right) \text{ for } k = 0, 1, 2$$

Example 2   Find the four fourth roots of -16.

Solution   The complex number $-16 + 0i$ can be expressed in polar form as $16 \operatorname{cis} \pi$. If $z = r \operatorname{cis} \alpha$ is a fourth root of -16, then

$$z^4 = -16$$
$$(r \operatorname{cis} \alpha)^4 = 16 \operatorname{cis} \pi$$
$$r^4 \operatorname{cis} 4\alpha = 16 \operatorname{cis} \pi$$

Thus, $r^4 = 16$ and $4\alpha = \pi$ or $(\pi + \text{multiples of } 2\pi)$

$\qquad r = 16^{1/4} \qquad \alpha = \dfrac{\pi}{4}$ or $\left(\dfrac{\pi}{4} + \text{multiples of } \dfrac{\pi}{2}\right)$

$\qquad r = 2 \qquad\qquad \alpha = \dfrac{\pi}{4}, \dfrac{3\pi}{4}, \dfrac{5\pi}{4}, \dfrac{7\pi}{4}$

The four fourth roots of -16 are:

$z_1 = 2 \text{ cis } \dfrac{\pi}{4} = \sqrt{2} + i\sqrt{2}$

$z_2 = 2 \text{ cis } \dfrac{3\pi}{4} = -\sqrt{2} + i\sqrt{2}$

$z_3 = 2 \text{ cis } \dfrac{5\pi}{4} = -\sqrt{2} - i\sqrt{2}$

$z_4 = 2 \text{ cis } \dfrac{7\pi}{4} = \sqrt{2} - i\sqrt{2}$

The method given in these examples can be generalized as follows:

The *n* th Roots of a Complex Number

The n nth roots of $z = r \text{ cis } \theta$ are:

$$\sqrt[n]{z} = z^{1/n} = r^{1/n} \text{ cis } \left(\dfrac{\theta}{n} + \dfrac{k \cdot 360°}{n}\right) \text{ for } k = 0, 1, 2, \ldots, n - 1$$

WRITTEN EXERCISES

Find the indicated roots of each number.

A **1.** cube roots of i **2.** cube roots of $-i$ **3.** cube roots of 8

4. cube roots of -8 **5.** fourth roots of 16 **6.** fourth roots of $-\dfrac{1}{2} - \dfrac{\sqrt{3}}{2}i$

7. a. Refer to the answers in Example 1. Show that $z_1 + z_2 + z_3 = 0$ and that $z_1 z_2 z_3 = 8i$.
   b. Any cube root of $8i$ must satisfy the equation $z^3 = 8i$. Show that the results in part (a) verify Theorem 5 on page 86.

8. a. Refer to the answers in Example 2. Show that $z_1 + z_2 + z_3 + z_4 = 0$ and that $z_1 z_2 z_3 z_4 = 16$.
   b. Any fourth root of -16 must satisfy $z^4 = -16$. Show that the results in part (a) verify Theorem 5 on page 86.

9. The three cube roots of 8 must satisfy the equation $z^3 - 8 = 0$. Solve this equation. (Choose one of the methods given in Chapter 2.) Your answers should agree with those of Exercise 3. $2, -1 + i\sqrt{3}, -1 - i\sqrt{3}$

2. a. Find the square roots of $-2 + 2i\sqrt{3}$.
   b. Plot $-2 + 2i\sqrt{3}$ and its square roots on an Argand diagram.

   a. The square roots of $-2 + i\sqrt{3} = 4 \text{ cis } 120°$ are $z_1 = \sqrt{4} \text{ cis } \dfrac{120°}{2} = 2 \text{ cis } 60°$ and $z_2 = \sqrt{4} \text{ cis } \dfrac{120° + 360°}{2} = 2 \text{ cis } 240°$.

   b.

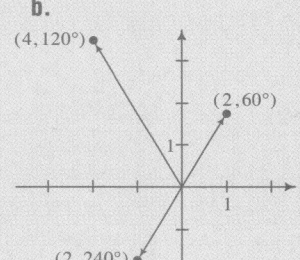

Additional Answers
Written Exercises

1. $\dfrac{\sqrt{3}}{2} + \dfrac{1}{2}i, -\dfrac{\sqrt{3}}{2} + \dfrac{1}{2}i, -i$

2. $\dfrac{\sqrt{3}}{2} - \dfrac{1}{2}i, -\dfrac{\sqrt{3}}{2} - \dfrac{1}{2}i, i$

3. $2, -1 + i\sqrt{3}, -1 - i\sqrt{3}$

4. $-2, 1 + i\sqrt{3}, 1 - i\sqrt{3}$

5. $2, -2, 2i, -2i$

6. $\dfrac{1}{2} + \dfrac{\sqrt{3}}{2}i, -\dfrac{\sqrt{3}}{2} + \dfrac{1}{2}i,$
   $-\dfrac{1}{2} - \dfrac{\sqrt{3}}{2}i, \dfrac{\sqrt{3}}{2} - \dfrac{1}{2}i$

Suggested Assignments

Standard
Day 1: 413/1, 3, 6, 9
Day 2: 414/11–13, 15
Comprehensive
Day 1: 413/1, 3, 6, 7, 9, 11
Day 2: 414/12–17

15. $1.7895 + 0.3155i, -1.1680 + 1.3920i, -0.6215 - 1.7075i$

10. The three cube roots of -8 must satisfy the equation $z^3 + 8 = 0$. Solve this equation. (Choose one of the methods given in Chapter 2.) Your answers should agree with those of Exercise 4. $-2, 1 - i\sqrt{3}, 1 + i\sqrt{3}$

B **11. a.** Factor $z^4 + 4$ by writing it as $(z^4 + 4z^2 + 4) - 4z^2$, which is the difference of two squares. $(z^2 + 2z + 2)(z^2 - 2z + 2)$

 b. Use part (a) to solve $z^4 = -4$, and plot the roots of the equation on an Argand diagram.

 c. Use De Moivre's theorem to verify your answer to part (b).

12. *Visual Thinking* Study the diagrams for Examples 1 and 2, and describe how the roots of complex numbers are related to regular polygons.

13. a. *Visual Thinking* Describe the regular polygon whose vertices are roots of the equation $z^5 = 100{,}000$.

 b. *Visual Thinking* Describe the regular polygon whose vertices are roots of the equation $z^9 = -1$.

See above.

14. Find the fifth roots of 32. See below. **15.** Find the cube roots of $3\sqrt{3} + 3i$.

16. A prime number of the form $2^{2^n} + 1$ is called a *Fermat prime*. 3, 5,

 a. Find the first four Fermat primes by substituting $n = 0, 1, 2,$ and 3. 17, 257

 b. Carl Friedrich Gauss (1777–1855), considered one of the greatest mathematicians of all time, used complex numbers to help determine which regular polygons could be constructed with straightedge and compass. His discovery: A regular polygon of N sides can be constructed if and only if $N = 2^j$ ($j \geq 2$) or $N = $ a product of different Fermat primes $\times 2^k$ ($k \geq 0$). For example, a polygon with $3 \times 5 \times 2^3 = 120$ sides is constructible. Decide whether a polygon is constructible if it has 32 sides, 17 sides, 7 sides, 9 sides, 10 sides, 255 sides. Yes, yes, no, no, yes, yes

 c. Explain what the solutions to the equation $z^{17} = 1$ have to do with a regular polygon of 17 sides. They form the vertices of the polygon when graphed.

C **17.** The roots of $z^n = 1$ are vertices of a regular polygon. Give the perimeter of an n-sided regular polygon as a trigonometric function of n. $P(n) = 2n \sin\left(\dfrac{180°}{n}\right)$

14. $2, 0.6180 + 1.9021i, -1.6180 + 1.1756i, -1.6180 - 1.1756i, 0.6180 - 1.9021i$

▰ Chapter Summary

1. The point $P(r, \theta)$ is described by its *polar coordinates*, r and θ, where r is the directed distance from the *pole* to P and θ is the *polar angle* measured from the polar axis to the ray OP. Although a point has only one pair of rectangular coordinates, it has many pairs of polar coordinates. For example, $(2, 20°) = (-2, 200°) = (-2, -160°)$ all represent the same point.

2. The formulas for converting from polar to rectangular coordinates and from rectangular to polar coordinates are given on page 396.

3. The *complex plane* can be represented by an *Argand diagram*. In this diagram, the complex number $a + bi$ is represented by the point (a, b) or by an arrow from the origin to (a, b).

4. The *absolute value* of a complex number $z = a + bi$ is $|z| = \sqrt{a^2 + b^2}$. Thus, $|z| = r$. The diagram at the right illustrates two ways of expressing z:

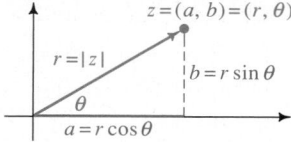

$$z = (a, b) = (r, \theta)$$
$$r = |z| \qquad b = r \sin \theta$$
$$a = r \cos \theta$$

rectangular form: $z = a + bi$
polar form: $z = r \operatorname{cis} \theta$

5. If $z_1 = r \operatorname{cis} \alpha$ and $z_2 = s \operatorname{cis} \beta$, then

$$z_1 z_2 = (r \operatorname{cis} \alpha)(s \operatorname{cis} \beta) = rs \operatorname{cis} (\alpha + \beta).$$

Powers of complex numbers can be found using De Moivre's theorem:

If $z = r \operatorname{cis} \theta$, then $z^n = r^n \operatorname{cis} n\theta$.

6. The *n* *n*th roots of $z = r \operatorname{cis} \theta$ are:

$$\sqrt[n]{z} = z^{1/n} = r^{1/n} \operatorname{cis}\!\left(\frac{\theta}{n} + \frac{k \cdot 360°}{n}\right) \text{ for } k = 0, 1, 2, \ldots, n - 1$$

Key vocabulary and ideas

polar coordinates (p. 395)
polar equation, polar graph (p. 396)
Argand diagram (p. 403)
complex plane (p. 403)
rectangular form (p. 403)

polar form (p. 403)
absolute value of $z = a + bi$ (p. 403)
product of two complex numbers (p. 404)
De Moivre's theorem (p. 408)
*n*th roots of a complex number (p. 413)

Chapter Test

        a. $(3\sqrt{2}, 45°)$ **e.** $(-1, -\sqrt{3})$

1. Give polar or rectangular coordinates for each point, as indicated.   **11-1**
   a. $(3, 3)$, polar   **b.** $(6, 90°)$, rect. $(0, 6)$ **c.** $(0, -2)$, polar $(2, 270°)$
   d. $(8, \pi)$, rect. $(-8, 0)$ **e.** $(-2, 60°)$, rect.   **f.** $(-1, \sqrt{3})$, polar $(2, 120°)$

2. a. Sketch the polar graph of $r = 1 - 2 \cos \theta$.
   b. Give a rectangular equation of this graph. $(x^2 + y^2 + 2x)^2 = x^2 + y^2$

3. Sketch the polar graph of: **a.** $r = 2$ **b.** $\theta = 2$ **c.** $r = \theta, \theta \geq 0°$

4. a. Express $z_1 = \sqrt{3} - i$, $z_2 = 4 + 4i$, and $z_1 z_2$ in polar form.   **11-2**
   b. Show z_1, z_2, and $z_1 z_2$ in an Argand diagram. **a.** $2 \operatorname{cis} 330°$, $4\sqrt{2} \operatorname{cis} 45°$, $8\sqrt{2} \operatorname{cis} 15°$

5. a. Let $z = 2 \operatorname{cis} 120°$. Find z^2 in polar form and in rectangular form. See below.
   b. Show that z^2 in polar form agrees with z^2 in rectangular form.

6. a. Express $z = -1 - i$ in polar form. $\sqrt{2} \operatorname{cis} 225°$ **11-3**
   b. Show z, z^2, z^3, and z^4 in an Argand diagram.
   c. Find z^{10}. $32i$ **5. a.** $4 \operatorname{cis} 240°$, $-2 - 2i\sqrt{3}$

7. *Writing* Describe how to find the *n* *n*th roots of a complex number.   **11-4**
   Explain why this method is a special case of De Moivre's theorem.

8. Show that $\sqrt[6]{2} \operatorname{cis} 135°$ is a cube root of $1 + i$, and find the other two cube roots. The other roots are $\sqrt[6]{2} \operatorname{cis} 15°$ and $\sqrt[6]{2} \operatorname{cis} 255°$.

Polar Coordinates and Complex Numbers   **415**

Additional Answers
Chapter Test

2. a.

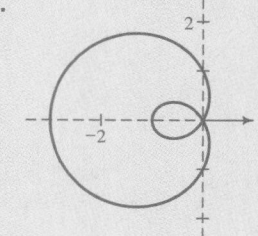

3. a–c.

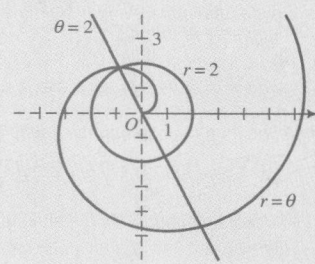

4. b.

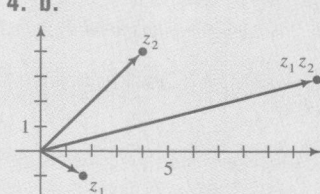

6. b.

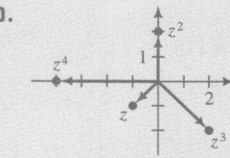

7. a. $\sin \frac{\pi}{6} = \frac{1}{2}$;

$\cos \frac{\pi}{6} = \frac{\sqrt{3}}{2}$;

$\tan \frac{\pi}{6} = \frac{\sqrt{3}}{3}$;

$\csc \frac{\pi}{6} = 2$;

$\sec \frac{\pi}{6} = \frac{2\sqrt{3}}{3}$;

$\cot \frac{\pi}{6} = \sqrt{3}$

b. $\sin 225° = -\frac{\sqrt{2}}{2}$;

$\cos 225° = -\frac{\sqrt{2}}{2}$;

$\tan 225° = 1$;
$\csc 225° = -\sqrt{2}$;
$\sec 225° = -\sqrt{2}$;
$\cot 225° = 1$

c. $\sin \left(-\frac{\pi}{2}\right) = -1$;

$\cos \left(-\frac{\pi}{2}\right) = 0$;

$\tan \left(-\frac{\pi}{2}\right)$ is undefined;

$\csc \left(-\frac{\pi}{2}\right) = -1$;

$\sec \left(-\frac{\pi}{2}\right)$ is undefined;

$\cot \left(-\frac{\pi}{2}\right) = 0$

8. $\sin x = -\frac{12}{13}$;

$\cos x = -\frac{5}{13}$;

$\tan x = \frac{12}{5}$;

$\csc x = -\frac{13}{12}$;

$\sec x = -\frac{13}{5}$

---

1. Find two angles, one positive and one negative, that are coterminal with the given angle.

   a. $-35°$ 325°, $-395°$ **b.** $\frac{5\pi}{6}$ $\frac{17\pi}{6}$, $-\frac{7\pi}{6}$

2. a. Convert 165° to radians. Give the answer in terms of π. $\frac{11\pi}{12}$
   b. Convert 208° to radians. Give the answer to the nearest hundredth of a radian.   3.63

3. a. Convert $-\frac{7\pi}{6}$ to degrees. $-210°$

   b. Convert 1.8 radians to degrees. Give the answer to the nearest ten minutes or tenth of a degree.   103°10′ or 103.1°

4. A sector of a circle has area 8.1π cm² and central angle 36°. Find its radius and arc length. $r = 9$ cm; $s = 1.8\pi$ cm

5. If θ is a third-quadrant angle and $\cos \theta = -\frac{2}{3}$, find $\sin \theta$. $-\frac{\sqrt{5}}{3}$

6. Express each of the following in terms of a reference angle. $-\sin 62°$
   a. $\cos 236°$ $-\cos 56°$ **b.** $\sin 485°$ $\sin 55°$ **c.** $\sin (-62°)$

7. Give the exact value of the six trigonometric functions for each angle.

   a. $\frac{\pi}{6}$ **b.** 225°   **c.** $-\frac{\pi}{2}$

8. Let $\cot x = \frac{5}{12}$, where $\pi < x < \frac{3\pi}{2}$. Find the values of the other five trigonometric functions.

9. a. Sketch the graph of $y = \tan x$ for $-\pi \le x \le 2\pi$.
   b. Give the domain, range, and period of the tangent function.

10. Without using a calculator or a table, find the value of each expression. Leave your answers in terms of π whenever appropriate.

   a. $\mathrm{Tan}^{-1} 1$ $\frac{\pi}{4}$ **b.** $\mathrm{Sin}^{-1} \frac{\sqrt{3}}{2}$ $\frac{\pi}{3}$

   c. $\mathrm{Cos}^{-1} \left(-\frac{\sqrt{3}}{2}\right)$ $\frac{5\pi}{6}$ **d.** $\mathrm{Tan}^{-1} \sqrt{3}$ $\frac{\pi}{3}$

   e. $\mathrm{Sin}^{-1} 0$ 0   **f.** $\sin \left(\mathrm{Cos}^{-1} \frac{3}{5}\right)$ $\frac{4}{5}$

   g. $\sec \left(\mathrm{Tan}^{-1} \frac{5}{12}\right)$ $\frac{13}{12}$ **h.** $\cos (\mathrm{Sin}^{-1} 1)$ 0

11. Solve the equation $2 \sec x - 5 = 0$ for $0 \le x < 2\pi$. Give your answer to the nearest hundredth of a radian.   1.16, 5.12

12. Find the inclination of the line $3x - 4y = 6$ to the nearest degree.   37°

13. Sketch the graph of each equation.

   a. $y = -2 \sin \frac{x}{2}$ **b.** $y + 1 = 3 \cos \frac{\pi}{4}(x - 2)$

---

416   *Cumulative Review*

14. Simplify $\dfrac{1 - \cos\theta}{\sin^2\theta} + \tan^2\theta - \sec^2\theta.\ -\dfrac{\cos\theta}{1 + \cos\theta}$ **15.** $0, \dfrac{\pi}{4}, \pi, \dfrac{5\pi}{4}$

15. Solve the equation $\cos x \sin x = \sin^2 x$ for $0 \le x < 2\pi$. See above.

16. For the right triangle shown at the right, find the missing lengths to three significant digits. $a \approx 7.42,\ b \approx 11.9$

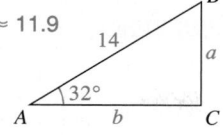

17. A parallelogram has sides of lengths 4 cm and 9 cm with an included angle of 40°. Find the area of the parallelogram to the nearest tenth of a square centimeter. 23.1 cm^2

18. Two forest rangers at observation posts A and B, 10 mi apart, spot a fire at point F. The forest rangers know that $\angle FAB = 110°$ and $\angle FBA = 25°$. To the nearest tenth of a mile, how far is the fire from the nearer observation post? 6.0 mi

19. Suppose that two sides of a triangle have lengths 8.4 and 7.6 with an included angle of 82°. Find the length of the third side to three significant digits. 10.5

20. Two people are standing 100 m apart on the bank of a river that flows due east. If a rock on the opposite bank is along a bearing of N25°E from one person and along a bearing of N15°W from the other person, what is the width of the river? 136 m

21. Let α and β be acute angles with $\sin\alpha = \dfrac{4}{5}$ and $\sin\beta = \dfrac{5}{13}$. Find $\cos(\alpha - \beta)$. $\dfrac{56}{65}$

22. Find the exact value of $\sin 75°$. $\dfrac{\sqrt{2} + \sqrt{6}}{4}$

23. Find $\tan\left(\dfrac{3\pi}{4} + \theta\right)$ when $\tan\theta = -\dfrac{1}{2}$. -3

24. Suppose α is an acute angle and $\cos\alpha = \dfrac{1}{3}$. Find $\sin 2\alpha$ and $\cos\dfrac{1}{2}\alpha$. $\dfrac{4\sqrt{2}}{9}; \dfrac{\sqrt{6}}{3}$

25. Give polar coordinates for the point $(-1, \sqrt{3})$. Then give two other pairs of polar coordinates for the same point. (2, 120°), (2, −240°), (2, 480°)

26. Give the rectangular coordinates for the point $(-6, 30°)$. $(-3\sqrt{3}, -3)$

27. **a.** Sketch the polar graph of $r = 4\cos\theta$.
    b. Give a rectangular equation of this graph.

28. Sketch the polar graph of $r = 1 - \sin\theta$. Then identify the graph.

29. Let $z_1 = 1 - i\sqrt{3}$ and $z_2 = -1 + i\sqrt{3}$.
    a. Express $z_1 z_2$ in rectangular form. $2 + 2i\sqrt{3}$
    b. Express z_1, z_2, and $z_1 z_2$ in polar form. 2 cis 300°; 2 cis 120°; 4 cis 420°
    c. Show that your answers in parts (a) and (b) agree. 4 cis 420° = 2 + 2i√3

30. Let $z = -2 - 2i$. Use DeMoivre's theorem to find z^6. $-512i$

31. Find the cube roots of $1 - i$. Write your answers in polar form.
    $\sqrt[6]{2}$ cis 105°, $\sqrt[6]{2}$ cis 225°, $\sqrt[6]{2}$ cis 345°

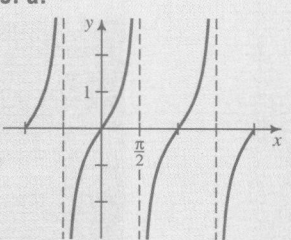
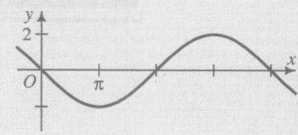
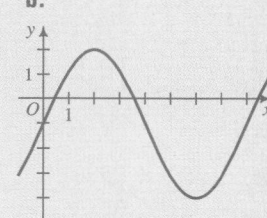
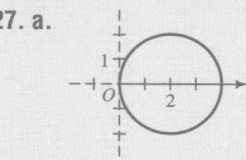
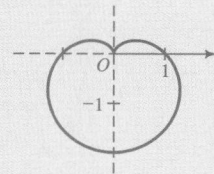

Chapter 12 Vectors and Determinants

Overview

In this chapter, vectors in two and three dimensions are represented geometrically and algebraically. Students learn how to perform basic operations on vectors, including addition, scalar multiplication, finding the dot product, and finding the cross product. They also learn how to convert vector equations into parametric equations and to draw the graphs of such equations. Students also solve physics and navigation problems involving vectors in two dimensions.

Determinants are introduced to help find the intersection of lines in a plane or lines in space, and Cramer's rule is used to solve a system of equations. Students also find vector and Cartesian equations of planes, the angle between a line and a plane, and the distance between a point and a plane.

Objectives

12-1 To perform basic operations on vectors.
12-2 To use coordinates to perform vector operations.
12-3 To use vector and parametric equations to describe motion in the plane.
12-4 To define and apply the dot product.
12-5 To extend vectors to three dimensions and to apply them.
12-6 To sketch planes and to find equations of planes.
12-7 To define and evaluate determinants.
12-8 To use determinants to solve algebraic and geometric problems.
12-9 To define and apply the cross product.

Supplementary Resources

Tests, pp. 40–41
Alt. Assess., pp. 34–38, 68
Activities, pp. 30, 31–35
Using Tech., pp. 29–30, 31–32
Student Res. Guide, pp. 106–120

Software

Precalculus Plotter Plus
Activities, p. 66
Polar Graph Plotter, Conics Plotter

Pacing Guide

Section	12-1	12-2	12-3	12-4	12-5	12-6	12-7	12-8	12-9	Review	Test	Total
Standard (days)	1	2	2	1	—	—	—	—	—	1	1	8
Comprehensive (days)	1	2	2	1	2	2	1	2	2	1	1	17

Teaching Notes

12-1 | **pages 419–426**

Presenting the Section

You might want to introduce this section by discussing *vector* and *scalar quantities* and giving some examples of each. A vector quantity has both direction and magnitude, while a scalar quantity has only magnitude. For example, the velocity of an automobile is a vector quantity, but the speed of an automobile (which is registered by the speedometer) is a scalar; it does not imply the direction in which the automobile is traveling. Other vector quantities are force, acceleration, momentum, and torque. Some scalar quantities are distance, time, volume, work, and mass. Students who are studying physics should be familiar with these examples.

Be sure students understand how to move vectors when using the parallelogram method of addition. Stress the importance of drawing accurate scale diagrams. (You might suggest that students use graph paper.) Point out that multiplication of a vector, as presented in this section, is *scalar* multiplication: It is multiplication of the vector by a real number and can be thought of as "repeated addition" of the vector.

You might want to summarize some of the properties of vector addition and scalar multiplication by using the vectors **u**, **v**, and **w** and scalars k and m:

1. Vector addition is commutative:
$$\mathbf{u} + \mathbf{v} = \mathbf{v} + \mathbf{u}$$
2. Vector addition is associative:
$$(\mathbf{u} + \mathbf{v}) + \mathbf{w} = \mathbf{u} + (\mathbf{v} + \mathbf{w})$$
3. Scalar multiplication is associative:
$$k(m\mathbf{u}) = (km)\mathbf{u}$$
4. Scalar multiplication is distributive:
$$(k + m)\mathbf{u} = k\mathbf{u} + m\mathbf{u}$$
$$k(\mathbf{u} + \mathbf{v}) = k\mathbf{u} + k\mathbf{v}$$
5. The length of $k\mathbf{u}$ is $|k|$ times the length of **u**:
$$|k\mathbf{u}| = |k||\mathbf{u}|$$

Using Technology

Accurate scale drawings of vectors can be done with certain computer drawing programs, sometimes called "drafting" programs. Students can use the drawing mode of these programs to perform arithmetic operations with the vectors, including addition, subtraction, and scalar multiplication.

Cooperative Learning

Use a small-group environment to review the navigation problems from Chapter 9 and to go over the application exercises on page 424. Instruct students to look for the common errors that can be made in drawing scale diagrams of vectors, including using the wrong directions for the vectors, measuring the lengths incorrectly, and not forming the parallelogram correctly to show the resultant. In the navigation problems, a vector's direction is the angle that the resultant makes with due north as measured clockwise from north.

Applications

Many applications of vectors are from physics. Physicists use vectors to represent two or more forces acting on a body at the same time. The combined effect of these forces (the resultant) can be represented geometrically by the parallelogram rule.

12-2 | **pages 426–432**

Presenting the Section

The operations of addition and scalar multiplication of vectors are not difficult for students to understand when the vectors are given in component form. You might want to use the properties of vector addition and scalar multiplication given in Presenting the Section for Section 12-1 to show how each property applies to vectors expressed in component form. Also, discuss how to convert components from rectangular form to polar form and vice versa (see Chapter 11).

Cooperative Learning

You can motivate the discussion of vectors by having groups of students perform the simple physics experiment presented in Class Exercise 6 on page 429. Have each group set up its own experiment as it is described. Even though the magnitude of the force is too difficult to measure, students still should be able to verify the results qualitatively. Students with a knowledge of physics can provide more detailed explanations.

Applications

Vectors in component form are closely related to complex numbers, as noted on page 427. To make this relationship clear, you might want to review operations on, and geometric representations of, complex numbers (see Sections 1-5 and 11-2) and compare them with vectors.

12-3 **pages 432–440**

Presenting the Section

Parametric equations will be a new topic for many students, so you might want to spend time showing them (1) how to find the parametric equations from a vector equation and (2) how to find a rectangular equation from the parametric equations. For (1), see page 433. For (2), eliminate the parameter t from the parametric equations as follows:

Since $x = 5 - 3t$, $t = -\frac{x-5}{3}$.

Since $y = 3 + 4t$, $y = 3 + 4\left(-\frac{x-5}{3}\right)$,

$y = 3 - \frac{4}{3}x + \frac{20}{3}$,

or $y = -\frac{4}{3}x + \frac{29}{3}$.

Emphasize that a line has infinitely many vector and parametric equations.

Use the discussion at the top of page 435 to show how to find the rectangular equation of a curve from its parametric equations.

You might want to present the following summary of parametric equations for conics.

Parametric equations	Conic
$x = r \cos t$ $y = r \sin t$	Circle
$x = a \cos t$ $y = b \sin t$	Ellipse
$x = a \sec t$ $\left(t \neq \frac{(2k+1)\pi}{2}\right)$ $y = b \tan t$	Hyperbola

Using Technology

You may want students to obtain graphs of parametric equations using advanced graphing calculators. Parametric equations can be used to describe the motion of a point along a planar curve. The calculators need to be set in parametric mode rather than in function mode to enter the equations. After graphing problems such as Example 2, page 433, students can use the calculators' trace feature to generate a table of values for the graph if desired. Students also can use the trace feature to find the point where the graph of an object moving at a constant velocity intersects some other graph (see Exercises 19–22, page 436, for example).

Applications

Applications of vectors include air and water navigation problems. For example, an airplane navigator is concerned with the velocity vector representing the airplane's motion and the velocity vector representing the wind's motion.

Cooperative Learning

You may want to use small groups to go over Exercises 28, 29, 30a, and 31a, page 438. Have students discuss the similarities and differences in the parametric equations for the circle, ellipse, and hyperbola (see Presenting the Section) and how they relate to the exercises. If possible, have students use graphing calculators or computers to observe the differences in the sets of equations as they are being graphed. Have each group come up with a generalization about the form of the parametric equations for each of these graphs and also how translating the center of the conic changes its parametric equations.

12-4 **pages 441–446**

Presenting the Section

You might want to introduce the dot product of two vectors by first reviewing how to find the slope of a line connecting two points and how to determine if two lines are perpendicular. Then show how the coordinate methods can be applied to vectors.

Once the dot product is defined, it can be used to determine: (1) if two vectors are parallel or perpendicular, (2) the angle between two vectors, and (3) the solution of some problems from physics.

Communication

Have students state the meaning of each property of the dot product of two vectors given on page 442. For example, students should recognize that the first property states that the dot product is commutative.

12-5 | pages 446–452

Presenting the Section

You might want to introduce the three-dimensional coordinate system using two adjacent walls and the floor of the classroom. These "planes" intersect in lines that can be thought of as the x-, y-, and z-axes, and the axes meet in a point (a corner of the room) that is the origin of the coordinate system. Point out that if the walls and floor were extended infinitely far, they would divide three-dimensional space into eight regions, called octants. Have students practice locating points in space by means of ordered triples. For example, an apple on your desk is so many feet from one wall, so many feet from the other wall, and so many feet above the floor. The numbers x, y, and z in the ordered triple (x, y, z) become the x-, y-, and z-components of the vector from the origin to the point (x, y, z).

Show students how the distance and midpoint formulas in two dimensions generalize to the formulas in three dimensions. Review how to find the equation of a circle given the endpoints of a diameter and show how a similar argument can be used to find the equation of a sphere.

You might want to review the vector operations in two dimensions (page 427) and then introduce the operations in three dimensions. Give examples of each property.

Using Technology

Three-dimensional graphing can be done with several software packages, including *Mathematica* and *Master Grapher*. Encourage students to use such software to explore perspective by looking at a three-dimensional drawing from a variety of angles.

Applications

The geometric representation of the solution of a system of equations in three unknowns is a common use of vectors in three dimensions. When the solution to such a system is a line, using the vector or parametric forms of the line's equation may be helpful in determining particular solution points.

Assessment

You might want to ask students to summarize the basic vector properties for two and three dimensions and to give an example of each.

If (x_1, y_1) and (x_2, y_2) are vectors in a plane, and (x_1, y_1, z_1), (x_2, y_2, z_2), and (x_3, y_3, z_3) are vectors in space, and k is a scalar, then:

Vector addition:
$$(x_1, y_1) + (x_2, y_2) = (x_1 + x_2, y_1 + y_2)$$
$$(x_1, y_1, z_1) + (x_2, y_2, z_2) =$$
$$(x_1 + x_2, y_1 + y_2, z_1 + z_2)$$

Scalar multiplication:
$$k(x_1, y_1) = (kx_1, ky_1)$$
$$k(x_1, y_1, z_1) = (kx_1, ky_1, kz_1)$$

Magnitude:
$$|(x_1, y_1)| = \sqrt{x_1{}^2 + y_1{}^2}$$
$$|(x_1, y_1, z_1)| = \sqrt{x_1{}^2 + y_1{}^2 + z_1{}^2}$$

Dot product:
$$(x_1, y_1) \cdot (x_2, y_2) = x_1 x_2 + y_1 y_2$$
$$(x_1, y_1, z_1) \cdot (x_2, y_2, z_2) = x_1 x_2 + y_1 y_2 + z_1 z_2$$

Cooperative Learning

Have students discuss, in small groups, Class Exercises 5, 7, and 8, page 449.

You may also wish to point out that the vectors (a, b) and $(-b, a)$ are perpendicular vectors since $(a, b) \cdot (-b, a) = -ab + ab = 0$. Challenge the groups to find, in terms of a, b, and c, a vector perpendicular to (a, b, c). (Example: $(b - c, c - a, a - b)$)

12-6 | pages 452–458

Presenting the Section

You might want to review how to set up a three-dimensional coordinate system. Use models to demonstrate planes and their equations.

Show how the Cartesian equation of a plane is derived from the vector equation.

Applications

In this section, students will find the distance from a point to a plane (see Exercise 38, page 458). The formula is important because it can be used to find the distance from the vertex of a solid, such as a tetrahedron, to the face opposite the vertex.

12-7 | pages 458–461

Presenting the Section

When discussing how to evaluate a 3×3 determinant by expanding by the minors of the elements in the first row, show students the following equation:

$$\begin{vmatrix} a_1 & a_2 & a_3 \\ b_1 & b_2 & b_3 \\ c_1 & c_2 & c_3 \end{vmatrix} =$$

$$a_1 \begin{vmatrix} b_2 & b_3 \\ c_2 & c_3 \end{vmatrix} - a_2 \begin{vmatrix} b_1 & b_3 \\ c_1 & c_3 \end{vmatrix} + a_3 \begin{vmatrix} b_1 & b_2 \\ c_1 & c_2 \end{vmatrix}$$

Then ask students to write a similar equation for expanding by the minors of the elements in the first column.

The following is an additional method for finding the solution for Example 2, page 459. It applies only to a 3×3 determinant.

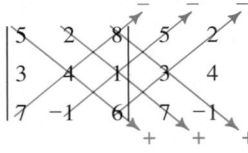

Adjoin the first two columns to the right of the determinant. Calculate the six products shown by the arrows. Use the signs as designated for each product. The sum of the products is the determinant. Thus, the determinant's value is $(5)(4)(6) + (2)(1)(7) + (8)(3)(-1) - (7)(4)(8) - (-1)(1)(5) - (6)(3)(2) = -145$.

Using Technology

Determinants can be evaluated easily on certain advanced scientific calculators or with many mathe-matical software packages for computers. Have students use technology to check their answers to Written Exercises 7–10, 12, and 13, page 461.

Cooperative Learning

Students can learn a great deal about the properties of determinants if they evaluate the same determinant in different ways. Have each row of the class expand a determinant by a different row or by a different column and compare results. Then have each row explore one of the options listed below and compare results again.

1. The value of a determinant is 0 if any row (column) of the determinant is 0.
2. Multiplying a row (column) of a determinant by any constant has the same effect as multiplying the value of the determinant by the constant.
3. In a determinant, you can subtract a row (column) or a multiple of a row (column) from another row (column) without changing the value of the determinant.
4. Interchanging two rows or columns of a determinant changes the sign of the value of the determinant.

Applications

Determinants have applications in solving algebraic and geometric problems, including finding the solution to a system of n linear equations in n variables.

Assessment

Ask students to explain how to expand a determinant by minors.

1. Multiply each element in a row or column by its minor.
2. Find the corresponding position of the element in the "checkerboard pattern" of plus and minus signs. If the product in step 1 should be multiplied by -1, do that.
3. Add all the products. The sum is the value of the determinant.

Also ask students to explain why the properties listed in Cooperative Learning (above) are true.

Presenting the Section

Emphasize that the equations in a linear system should be in standard form before applying Cramer's rule. Also, be sure students understand that D is the determinant of the coefficients of the variables and that D_x is formed from D by replacing the column of x-coefficients with the column of constants (similarly for D_y and D_z).

Using Technology

The determinants needed to apply Cramer's rule can be evaluated easily on certain advanced scientific calculators or with many mathematical software packages for computers. Have students use technology to check their answers to Written Exercises 17–19, page 464.

Cooperative Learning

You might want to design an exercise in which each member of a group of four students is asked to solve a system of three equations in three variables by Cramer's rule. Have each member solve a different type of system: one in which the determinant in the denominator is 0; one in which the determinant in a numerator is 0, but the determinant in the denominator is not 0; one in which the determinant in a numerator and in the denominator is 0; and one in which the determinant in a numerator and in the denominator is not 0. Ask students collectively to draw some conclusions about the applicability of Cramer's rule to these systems and to give a geometric interpretation of their conclusions.

1. If the determinant in each denominator is not 0, then the system has a unique solution.
2. If the determinant in each numerator and in each denominator is 0, then the system has infinitely many solutions.
3. If the determinant in each numerator is not 0 and the determinant in each denominator is 0, then the system has no solution.

When three variables are involved, the graph of each equation in a system is a plane, so the three statements above can be interpreted geometrically as listed at the top of the next column.

1. The three planes intersect in a unique point.
2. The three planes intersect in a line.
3. The three planes are parallel, or they intersect two at a time (no common line or point of intersection).

Presenting the Section

Before presenting this section, review the algebraic representation of vectors from Section 12-2 and vector equations from Section 12-3. Also, review the fact that the associative and commutative laws hold for the addition of vectors and that scalar multiplication is distributive over addition of vectors.

You might want to point out that the dot product of two vectors associates a number (scalar) with each ordered pair of vectors, while the cross product associates a vector with each ordered pair of vectors.

The properties of the cross product, page 465, imply the following:
1. If either $\mathbf{u}$ or $\mathbf{v}$ is a zero vector, then $\mathbf{u} \times \mathbf{v} = \mathbf{0}$.
2. If $\mathbf{u}$ and $\mathbf{v}$ are parallel vectors, then $\mathbf{u} \times \mathbf{v} = \mathbf{0}$.
3. If $\mathbf{u} \times \mathbf{v} \neq \mathbf{0}$ and $\mathbf{u}$ and $\mathbf{v}$ are not parallel vectors, then $\mathbf{u} \times \mathbf{v}$ will be perpendicular to both $\mathbf{u}$ and $\mathbf{v}$, and the magnitude of $\mathbf{u} \times \mathbf{v}$ is the area of the parallelogram formed by $\mathbf{u}$ and $\mathbf{v}$.

Cooperative Learning

You might want students to go over Exercise 15, page 467, in small groups. First have them do the exercise individually and then discuss the different methods of finding the area of a triangle as a group. Ask students when each method applies and what similarities and differences they notice.

Assessment

Given three points $P(a_1, b_1, c_1)$, $Q(a_2, b_2, c_2)$, and $R(a_3, b_3, c_3)$, ask students to summarize how to find the equation of the plane containing the points.
1. Find a nonzero vector perpendicular to the plane containing P, Q, and R.
2. Find the cross product $\overrightarrow{PQ} \times \overrightarrow{PR}$.
3. Substitute the coordinates of P, Q, or R into the equation $\overrightarrow{PQ} \times \overrightarrow{PR} = d$ to find d and thereby obtain the equation of the plane.

12

Vectors and Determinants

418

■ Properties and Basic Operations

12-1 Geometric Representation of Vectors

| **Objective** | *To perform basic operations on vectors.* |

Vectors are quantities that are described by a *direction* and a *magnitude* (size). A force, for example, is a vector quantity because to describe a force, you must specify the direction in which it acts and its strength. Another example of a vector is velocity. The velocity of an airplane is described by its direction and speed.

The velocities of two airplanes each heading northeast at 700 knots are represented by the arrows **u** and **v** in the diagram below. We write **u** = **v** to indicate that both planes have the same velocity even though the two arrows are different. In general, any two arrows with the same length and the same direction represent the same vector. The diagram below also shows a third airplane with speed 700 knots, but because it is heading in a different direction, its velocity vector **w** does not equal either **u** or **v**.

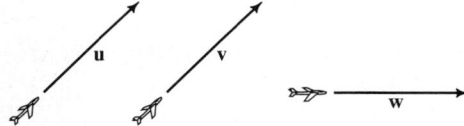

The *magnitude* of a vector **v** (also called the *absolute value* of **v**) is denoted $|\mathbf{v}|$. In the diagram above, $|\mathbf{v}| = 700$. Since $|\mathbf{w}|$ also equals 700, we have that $|\mathbf{w}| = |\mathbf{v}|$ even though $\mathbf{w} \neq \mathbf{v}$.

Addition of Vectors

If a vector **v** is pictured by an arrow from point A to point B, then it is customary to write $\mathbf{v} = \overrightarrow{AB}$. In the diagram, $\overrightarrow{AB}$ represents the motion of an object 15 units south from A to B. If the object is then moved 10 units southeast from B to C, that motion is represented by $\overrightarrow{BC}$. Since the result of moving an object first from A to B and then from B to C is the same as moving the object directly from A to C, it is natural to write

$$\overrightarrow{AB} + \overrightarrow{BC} = \overrightarrow{AC}.$$

We say that $\overrightarrow{AC}$ is the *vector sum* of $\overrightarrow{AB}$ and $\overrightarrow{BC}$.

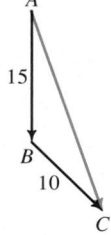

◀ A hang glider gains altitude when the upward vectors of force produced by rising warm air masses, or thermals, exceed the downward vectors of force produced by gravity.

Vectors and Determinants **419**

Teaching Notes, p. 418B

Warm-Up Exercises

Solve each triangle.

1.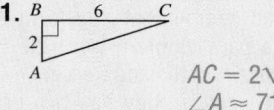

$AC = 2\sqrt{10}$;
$\angle A \approx 71.6°$;
$\angle C \approx 18.4°$

2.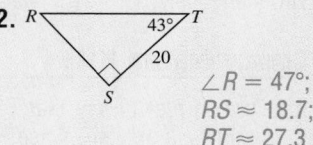

$\angle R = 47°$;
$RS \approx 18.7$;
$RT \approx 27.3$

3.

$EF \approx 5.74$;
$\angle F \approx 53.1°$;
$\angle E \approx 91.9°$

Motivating the Section

In physics, vectors are used to describe quantities that have a magnitude and a direction. For example, the motion of a person swimming across a river or of a person walking on a moving train can be modeled by vectors.

Communication Note

Students should notice that a vector named with a single lowercase letter (such as **v**) is represented in boldface print in this text. This distinguishes a vector from the ordinary variable *v*. Absolute value bars are used to represent the magnitude of a vector. A vector always has a nonnegative length.

The addition of two vectors is a commutative operation. In other words, the order in which the vectors are added does not make any difference. You can see this in the vector diagrams below where the red arrows denoting **a** + **b** and **b** + **a** have the same length and direction.

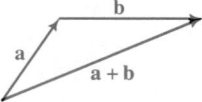 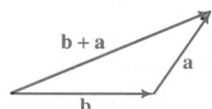

If the two diagrams above are moved together, a parallelogram is formed. This suggests that another way to add **a** and **b** is to draw a parallelogram $OACB$ with sides $\overrightarrow{OA} = $ **a** and $\overrightarrow{OB} = $ **b**. The diagonal $\overrightarrow{OC}$ of the parallelogram is the sum. This method is frequently used in physics problems involving forces that are combined. The vector sum of the forces is called the *resultant* of the forces. These ideas are illustrated in Example 1 in which the unit of force used is the *newton*. One **newton** (N) is the force needed to bring a 1 kg mass to a speed of 1 meter per second in 1 second.

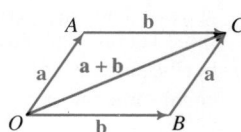

Example 1   **a.** Make a scale drawing showing a force of 20 N pulling an object east and another force of 10 N pulling the object in the compass direction 150°. Draw the resultant force vector and use your drawing to approximate the magnitude and direction of the resultant force.

b. Use trigonometry to find the magnitude (to the nearest hundredth) and the direction (to the nearest tenth of a degree) of the resultant.

Solution   **a.** The given forces are $\overrightarrow{OA}$ and $\overrightarrow{OB}$ and the resultant force is $\overrightarrow{OC}$. Note that $\angle AOB = 150° - 90° = 60°$.

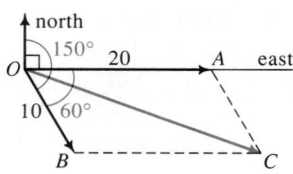

By measuring with a ruler and protractor, you can determine that the magnitude of $\overrightarrow{OC}$ is about 26 and that the direction of $\overrightarrow{OC}$ is about 110°. Therefore, $\overrightarrow{OC}$ represents a force of 26 N pulling the object in the compass direction 110°.

420   *Chapter Twelve*

b. Use the law of cosines in $\triangle OBC$. Since $\angle B$ is a supplement of $\angle AOB$, $\angle OBC = 120°$. Therefore:

$$|\overrightarrow{OC}|^2 = 10^2 + 20^2 - 2 \cdot 10 \cdot 20 \cos 120° = 700$$

Therefore, $|\overrightarrow{OC}| \approx 26.46$.

To find the direction of the resultant, use the law of sines.

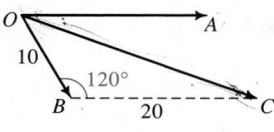

$$\frac{\sin C}{10} = \frac{\sin 120°}{26.46}$$

$$\sin C = 0.3273$$

$$\angle C \approx 19.1°$$

Since $\angle C \approx 19.1°$, $\angle AOC \approx 19.1°$ also. Therefore, the direction of $\overrightarrow{OC}$ is about $90° + 19.1° = 109.1°$.

Vector Subtraction

The *negative* of a vector **v**, denoted $-\mathbf{v}$, has the same length as **v** but the opposite direction. The sum of **v** and $-\mathbf{v}$ is the zero vector **0**. It is best thought of as a point.

Vectors can be subtracted as well as added.

$\mathbf{v} - \mathbf{w}$ means $\mathbf{v} + (-\mathbf{w})$.

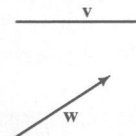
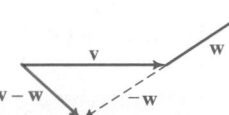

Multiples of a Vector

The vector sum $\mathbf{v} + \mathbf{v}$ is abbreviated as $2\mathbf{v}$. Likewise, $\mathbf{v} + \mathbf{v} + \mathbf{v} = 3\mathbf{v}$. The diagram below shows that the arrows representing $2\mathbf{v}$ and $3\mathbf{v}$ have the same direction as the arrow representing **v**, but that they are two and three times as long.

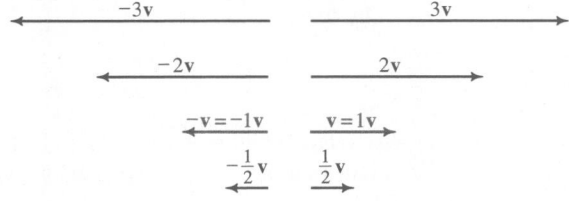

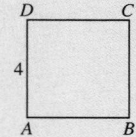

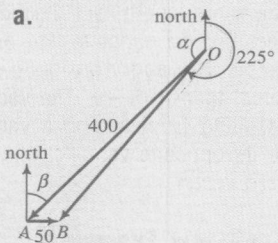
In general, if k is a positive real number, then $k\mathbf{v}$ is the vector with the same direction as $\mathbf{v}$ but with an absolute value k times as large. If $k < 0$, then $k\mathbf{v}$ has the same direction as $-\mathbf{v}$ and has an absolute value $|k|$ times as large. If $k \neq 0$, then $\frac{\mathbf{v}}{k}$ is defined to be equal to the vector $\frac{1}{k}\mathbf{v}$.

Example 2 Vectors $\mathbf{u}$ and $\mathbf{v}$ are shown at the right. Use these vectors to sketch the following:
a. $2\mathbf{u} + 3\mathbf{v}$
b. $\mathbf{u} - 2\mathbf{v}$

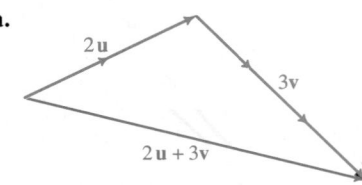

Solution **a.** 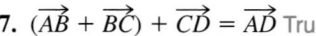 **b.**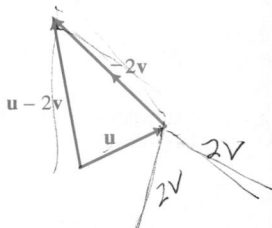

When working with vectors, it is customary to refer to real numbers as **scalars**. When this is done, the operation of multiplying a vector $\mathbf{v}$ by a scalar k is called **scalar multiplication**. This operation has the following properties. If $\mathbf{v}$ and $\mathbf{w}$ are vectors and k and m are scalars, then:

$$\left. \begin{array}{l} k(\mathbf{v} + \mathbf{w}) = k\mathbf{v} + k\mathbf{w} \\ (k + m)\mathbf{v} = k\mathbf{v} + m\mathbf{v} \end{array} \right\} \quad \text{Distributive laws}$$

$$k(m\mathbf{v}) = (km)\mathbf{v} \quad \text{Associative law}$$

CLASS EXERCISES

Quadrilateral *ABCD* is the parallelogram shown below. Tell whether each of the following is true or false.

1. $\overrightarrow{BC} + \overrightarrow{BA} = \overrightarrow{BD}$ True
2. $|\overrightarrow{BC}| + |\overrightarrow{BA}| = |\overrightarrow{BD}|$ False
3. $\overrightarrow{AO} = \overrightarrow{AC}$ False
4. $\overrightarrow{AB} + \overrightarrow{CD} = \mathbf{0}$ True
5. $\overrightarrow{AO} = \overrightarrow{OC}$ True
6. $\overrightarrow{AO} = \frac{1}{2}\overrightarrow{AC}$ True

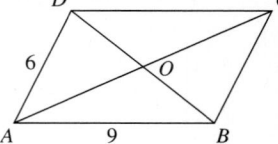

7. $(\overrightarrow{AB} + \overrightarrow{BC}) + \overrightarrow{CD} = \overrightarrow{AD}$ True
8. $\overrightarrow{AB} + (\overrightarrow{BC} + \overrightarrow{CD}) = \overrightarrow{AD}$ True

9. Exercises 7 and 8 show that vector addition is associative. Is addition of real numbers associative? Is subtraction of real numbers associative? yes; no

Complete the following statements. The parallelogram *ABCD* is shown at the bottom of the preceding page. 16. $\overrightarrow{OD}$; $\overrightarrow{AD}$ 17. $\overrightarrow{DB}$; $\overrightarrow{DC}$

10. $\overrightarrow{AD} = \underline{\ ?\ }\ \overrightarrow{BC}$

11. $|\overrightarrow{AD}| = \underline{\ ?\ }\ 6$

12. $\frac{1}{2}\overrightarrow{BD} = \underline{\ ?\ }\ \overrightarrow{BO}$ or $\overrightarrow{OD}$

13. $2\overrightarrow{AO} = \underline{\ ?\ }\ \overrightarrow{AC}$

14. $\overrightarrow{AB} + \overrightarrow{AD} = \underline{\ ?\ }\ \overrightarrow{AC}$

15. $\overrightarrow{AD} + \overrightarrow{DC} + \overrightarrow{CB} = \underline{\ ?\ }\ \overrightarrow{AB}$

16. $\overrightarrow{AO} - \overrightarrow{DO} = \overrightarrow{AO} + \underline{\ ?\ } = \underline{\ ?\ }$

17. $\overrightarrow{BC} - \overrightarrow{BD} = \overrightarrow{BC} + \underline{\ ?\ } = \underline{\ ?\ }$

18. **Reading** Explain the two different uses of the plus sign in Exercises 1 and 2 on page 422.

19. **Reading** Explain the two different uses of the plus sign in the distributive law $(k + m)\mathbf{v} = k\mathbf{v} + m\mathbf{v}$.

20. **Reading** In the associative law, $k(m\mathbf{v}) = (km)\mathbf{v}$, two different kinds of multiplication are used on the right side of the equation. Explain.

21. **Discussion** Does the statement $k + \mathbf{v} = \mathbf{v} + k$ make sense? Explain.
No; cannot add a scalar and a vector

WRITTEN EXERCISES

 1. **a.** Sketch an arrow representing a velocity **v** of 100 mi/h to the northwest.
   b. Sketch 2**v** and tell what velocity it represents. 200 mi/h to the northwest
   c. Sketch −**v** and tell what velocity it represents. 100 mi/h to the southeast

2. **a.** Sketch an arrow representing a ship's trip **w** of 200 mi on a course of 200°.
   b. Sketch 3**w** and tell what trip it represents. 600 mi on course of 200°
   c. Sketch −1.5**w** and tell what trip it represents. 300 mi on course of 20°

3. Draw rectangle *PQRS* with *PQ* = 4 and *QR* = 3. Complete.
   a. $\overrightarrow{PQ} + \overrightarrow{QR} = \underline{\ ?\ }\ \overrightarrow{PR}$ **b.** $|\overrightarrow{PQ} + \overrightarrow{QR}| = \underline{\ ?\ }\ 5$ **c.** $\overrightarrow{PQ} + \overrightarrow{RS} = \underline{\ ?\ }\ 0$

4. Draw parallelogram *ABCD* with *AB* = 10, *BC* = 6, and ∠*A* = 60°. Complete.
   a. $\overrightarrow{AB} + \overrightarrow{BC} = \underline{\ ?\ }\ \overrightarrow{AC}$ **b.** $|\overrightarrow{AB} + \overrightarrow{BC}| = \underline{\ ?\ }\ 14$ **c.** $|\overrightarrow{AB} + \overrightarrow{BC} + \overrightarrow{CD}| = \underline{\ ?\ }\ 6$

In Exercises 5 and 6, copy vectors **u** and **v**. Then sketch **u** + **v**, **u** + 2**v**, **u** + 3**v**, and **u** − **v**.

5.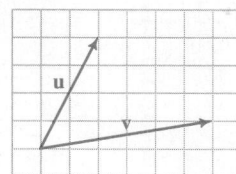

6.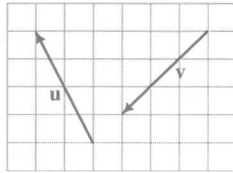

7. **Investigation** Draw two vectors **u** and **v**. Sketch **u** + **v**, 2(**u** + **v**), and 2**u** + 2**v**. What do you notice? 2(**u** + **v**) = 2**u** + 2**v**

Additional Answers
Class Exercises

18. Exercise 1 is addition of vectors. Exercise 2 is addition of real numbers.

19. The plus sign on the left side of $(k + m)\mathbf{v} = k\mathbf{v} + m\mathbf{v}$ indicates addition of real numbers. On the right, it indicates addition of vectors.

20. *km* is multiplication of two real numbers. That product multiplied by **v** is scalar multiplication.

Additional Answers
Written Exercises

5.

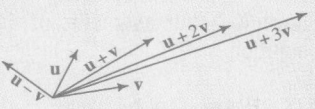

6.

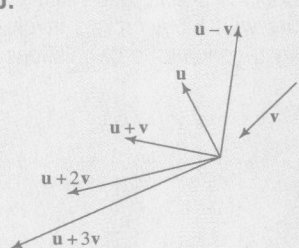

Suggested Assignments

Standard
423/1, 3, 9, 11, 12, 19, 25
Comprehensive
423/1, 3, 9, 11, 12, 19, 23, 25

Supplementary Materials

Alternative Assessment, 34

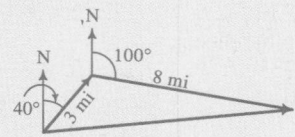
10. 10 N; 36.9° west of north

8. **Investigation** Draw two vectors **a** and **b**. Sketch **a** − **b**, 3(**a** − **b**), and 3**a** − 3**b**. What do you notice? 3(**a** − **b**) = 3**a** − 3**b**

9. **Navigation** A ship travels 200 km west from port and then 240 km due south before it is disabled. Illustrate this in a vector diagram. Use trigonometry to find the course that a rescue ship must take from port in order to reach the disabled ship. 219.8°

10. **Physics** On graph paper, make a diagram that illustrates a force of 8 N north and a force of 6 N west acting on a body. Illustrate the resultant sum of these two forces and estimate its strength. Then, using trigonometry, determine the approximate direction (as a number of degrees west of north) of this force. See above

11. **Aviation** On graph paper, make a diagram that illustrates the velocity of an airplane heading east at 400 knots. Illustrate a wind velocity of 50 knots blowing toward the northeast. If the airplane encounters this wind, illustrate its resultant velocity. Estimate the resultant speed and direction of the airplane. (The direction is the angle the resultant makes with due north measured clockwise from north.) 436.8 knots; 85.4°

12. a. **Sports** A swimmer leaves point A swimming south across a river at 2 km/h. The river is 4 km wide and flows east at the rate of 1 km/h. Make a vector diagram showing her resultant velocity.
    b. Calculate her resultant speed. 2.24 km/h
    c. How long will it take her to swim across the river? How far east does she swim? 2 h; 2 km

13. **Aviation** On graph paper, make a vector diagram showing an airplane heading southwest at 600 knots and encountering a wind blowing from the west. Show the plane's resultant velocity when the wind blows at (a) 30 knots, (b) 60 knots, and (c) 90 knots.

14. **Navigation** On graph paper, make a vector diagram showing a motorboat heading east at 10 knots. Add to your diagram a vector representing a current moving southeast. Show the boat's resultant velocity if the current moves at (a) 2 knots, (b) 3 knots, and (c) 4 knots.

15. a. **Navigation** Make a diagram showing the result of sailing a ship 3 mi on a course of 040° followed by sailing it 8 mi on a course of 100°.
    b. From your diagram, estimate the distance of the ship from its starting point.
    c. Find the exact distance of the ship from its starting point by using the law of cosines. $\sqrt{97}$ mi

424   *Chapter Twelve*

16. a. Physics F_1 is a force of 3 N pulling an object north and F_2 is a force of 5 N pulling the object in the compass direction 060°. Sketch both force vectors.

  b. Sketch the resultant force vector $F_1 + F_2$ and use the law of cosines and the law of sines to find its magnitude and direction. 7; 038°

  c. Give the magnitude and direction of the force F_3 such that $F_1 + F_2 + F_3 = 0$. 7; 218°

B **17.** In the diagram, M and N are the midpoints of $\overline{PQ}$ and $\overline{PR}$.

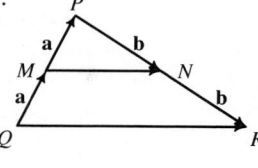

  a. Express $\overrightarrow{MN}$ in terms of **a** and **b**. a + b

  b. Express $\overrightarrow{QR}$ in terms of **a** and **b**. 2(a + b) = 2a + 2b

  c. Your answers to parts (a) and (b) show that $\overrightarrow{QR} = 2\overrightarrow{MN}$. What theorem about the segment joining the midpoints of two sides of a triangle does this equation suggest?

18. a. Complete the following equations given that $\overline{MN}$ is the median of trapezoid $ABCD$.

  (1) $\overrightarrow{AM} + \overrightarrow{MN} + \overrightarrow{ND} = $? $\overrightarrow{AD}$

  (2) $\overrightarrow{BM} + \overrightarrow{MN} + \overrightarrow{NC} = $? $\overrightarrow{BC}$

  (3) $\overrightarrow{AM} + \overrightarrow{BM} = $? 0

  (4) $\overrightarrow{ND} + \overrightarrow{NC} = $? 0

  b. Add equations (1) and (2), and by using (3) and (4), simplify the resulting equation to $2\overrightarrow{MN} = \overrightarrow{AD} + \overrightarrow{BC}$. What theorem about the median of a trapezoid does this equation suggest?

19. Point Q divides $\overline{MN}$ in the ratio $2:1$, that is, $\overrightarrow{MQ} = 2\overrightarrow{QN}$. If $v = \overrightarrow{MN}$, express each vector in terms of **v**.

  a. $\overrightarrow{MQ}$ $\frac{2}{3}$v **b.** $\overrightarrow{QN}$ $\frac{1}{3}$v **c.** $\overrightarrow{NQ}$ $-\frac{1}{3}$v

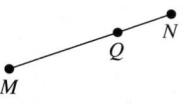

Ex. 19

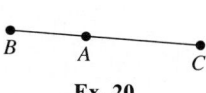

Ex. 20

20. Point A divides $\overline{BC}$ in the ratio $2:3$, that is, $\dfrac{BA}{AC} = \dfrac{2}{3}$. If $v = \overrightarrow{BA}$, express each vector in terms of **v**.

  a. $\overrightarrow{CA}$ $-\frac{3}{2}$v **b.** $\overrightarrow{BC}$ $\frac{5}{2}$v **c.** $\overrightarrow{AB}$ $-$v

21. In the diagram, $u = \overrightarrow{AB}$ and $v = \overrightarrow{BD}$. The midpoint of $\overline{AD}$ is E and $\dfrac{BD}{DC} = \dfrac{1}{3}$. Express each of the following vectors in the form $r\mathbf{u} + s\mathbf{v}$, where r and s are real numbers. For example, $\overrightarrow{AC} = \overrightarrow{AB} + \overrightarrow{BC} = \mathbf{u} + 4\mathbf{v}$.

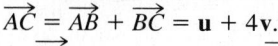

  a. $\overrightarrow{AD}$ u + v **b.** $\overrightarrow{AE}$ $\frac{1}{2}$u + $\frac{1}{2}$v **c.** $\overrightarrow{BE}$ $\frac{1}{2}$v − $\frac{1}{2}$u **d.** $\overrightarrow{EC}$ $\frac{1}{2}$u + $\frac{7}{2}$v

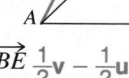

Problem Solving

As an extension of Exercises 17 and 18, ask students to try to prove some other theorems from geometry. One such theorem might be: The length of the median to the hypotenuse of a right triangle is half the length of the hypotenuse.

Additional Answers
Written Exercises

17. c. The segment joining the midpoints of two sides of a triangle is parallel to the third side and is half as long as the third side.

18. b. The median of a trapezoid is parallel to its bases, and the length of the median is the mean of the lengths of the bases.

Vectors and Determinants **425**

22. In the diagram, $ABCD$ is a parallelogram. If $\overrightarrow{AB} = \mathbf{x}$ and $\overrightarrow{AD} = \mathbf{y}$, express each vector in terms of **x** and **y**.
    a. $\overrightarrow{BC}$ b. $\overrightarrow{CD}$ c. $\overrightarrow{AC}$ d. $\overrightarrow{AO}$ e. $\overrightarrow{BO}$
    y $-\mathbf{x}$ $\mathbf{x} + \mathbf{y}$ $\frac{1}{2}(\mathbf{x} + \mathbf{y})$ $\frac{1}{2}(\mathbf{y} - \mathbf{x})$

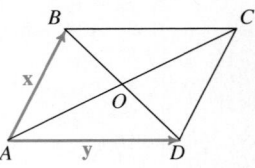

23. In the diagram, $DH:HF = 1:2$ and $EG:GH = 2:1$. If $\mathbf{v} = \overrightarrow{ED}$ and $\mathbf{w} = \overrightarrow{EF}$, express each vector in terms of **v** and **w**.
    a. $\overrightarrow{DF}$ b. $\overrightarrow{DH}$ c. $\overrightarrow{EH}$ d. $\overrightarrow{EG}$ e. $\overrightarrow{DG}$
    $\mathbf{w} - \mathbf{v}$ $\frac{1}{3}(\mathbf{w} - \mathbf{v})$ $\frac{2}{3}\mathbf{v} + \frac{1}{3}\mathbf{w}$ $\frac{4}{9}\mathbf{v} + \frac{2}{9}\mathbf{w}$ $\frac{2}{9}\mathbf{w} - \frac{5}{9}\mathbf{v}$

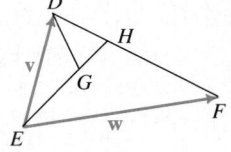

24. **Navigation** Ship A is 50 nautical miles west of ship B which is heading south at 8 knots (8 nautical miles per hour). If A is to rendezvous with B in 3 h, what should be the course and the speed of ship A? Solve this problem in two ways: **(a)** by making an accurate vector diagram and estimating the answers by measuring, and **(b)** by using right triangle trigonometry. 18.5 knots at 115.6°

25. **Aviation** Jenny Wu wants to fly from Chicago to Atlanta as quickly as possible. If there were no wind, her course would be 150° and she would fly at her plane's top speed of 450 knots. What course should she plan if there is a 75 knot wind blowing toward the east? Solve this problem in two ways. First, make an accurate vector diagram and measure her course with a protractor. Second, use the law of sines. 158.9°

12-2 Algebraic Representation of Vectors

Objective   *To use coordinates to perform vector operations.*

In the coordinate system shown at the right, the vector **v** consists of a 2 unit change in the x-direction and a -3 unit change in the y-direction. The numbers 2 and -3 are called the components of **v**. When we write $\mathbf{v} = (2, -3)$, we are expressing **v** in **component form**.

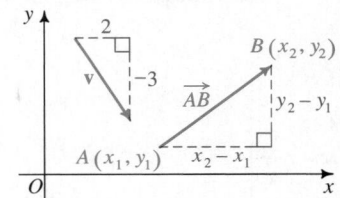

   The diagram also shows that the components of $\overrightarrow{AB}$ can be found by subtracting the coordinates of points A and B as follows:

$$\overrightarrow{AB} = (x_2 - x_1, y_2 - y_1)$$

The magnitude of $\overrightarrow{AB}$ is then found by using the distance formula.

$$|\overrightarrow{AB}| = \sqrt{(x_2 - x_1)^2 + (y_2 - y_1)^2}$$

Example 1   Given $A(4, 2)$ and $B(9, -1)$, express $\overrightarrow{AB}$ in component form. Find $|\overrightarrow{AB}|$.

Solution $\overrightarrow{AB} = (9 - 4, -1 - 2) = (5, -3)$
$|\overrightarrow{AB}| = \sqrt{5^2 + (-3)^2} = \sqrt{34}$

The two diagrams below show that the rectangular coordinates, a and b, and the polar coordinates, r and θ, of a point closely tie together three important ideas of mathematics: (1) point, (2) vector, and (3) complex number. Although (a, b) can represent both the point P and the vector $\overrightarrow{OP}$, you can always tell which meaning is intended from the context.

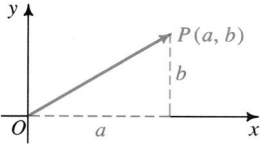

Point: $P(a, b)$
Vector: $\overrightarrow{OP} = (a, b)$
Complex number: $z = a + bi$

Point: $P(r, \theta)$
Vector: $\overrightarrow{OP} = (r \cos \theta, r \sin \theta)$
Complex number: $z = r(\cos \theta + i \sin \theta) = r \operatorname{cis} \theta$

Example 2   A force **F** of 10 N acts at an angle of 130° with the positive x-axis. Find **F** in component form.

Solution $|\mathbf{F}| = 10$, so $\mathbf{F} = (10 \cos 130°, 10 \sin 130°) \approx (-6.43, 7.66)$.

In terms of coordinates, vector operations are defined as below.

Vector Operations with Coordinates

Vector Addition: $\mathbf{v} + \mathbf{u} = (a, b) + (c, d) = (a + c, b + d)$ Figure (a)
Vector Subtraction: $\mathbf{v} - \mathbf{u} = (a, b) - (c, d) = (a - c, b - d)$
Scalar Multiplication: $k\mathbf{v} = k(a, b) = (ka, kb)$ Figure (b)

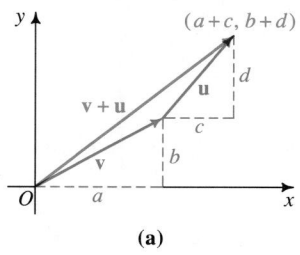

(a)

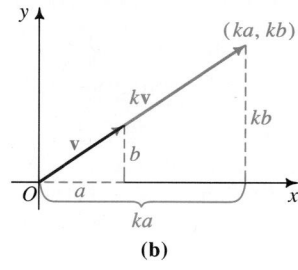

(b)

Vectors and Determinants   **427**

Error Analysis

When students express $\overrightarrow{AB}$ in component form, as in Example 1, make sure they combine the components in the correct order; that is, each component of A must be subtracted from the corresponding component of B. The order is not reversible.

Assessment Note

After discussing Example 1, ask students to do the following:

1. Given $A(-2, 4)$ and $B(1, 1)$, find $\overrightarrow{AB}$ in component form. $(3, -3)$

2. Find $|\overrightarrow{AB}|$. $(3\sqrt{2})$

3. Give polar coordinates for $\overrightarrow{AB}$. $(3\sqrt{2}, 315°)$

Communication Note

Have students express the vector operations defined on this page in words. For example: "To add two vectors, add their corresponding components."

Additional Examples

1. Let $\mathbf{u} = (-5, 2)$ and $\mathbf{v} = (-3, -6)$. Find:

a. $2\mathbf{u} + \dfrac{1}{3}\mathbf{v}$

b. $\left|2\mathbf{u} + \dfrac{1}{3}\mathbf{v}\right|$

c. $\left|2\mathbf{u}\right| + \left|\dfrac{1}{3}\mathbf{v}\right|$

d. $\dfrac{\mathbf{u} - \mathbf{v}}{|\mathbf{u} - \mathbf{v}|}$

e. the polar form of $\mathbf{v}$

a. $2(-5, 2) +$
$\dfrac{1}{3}(-3, -6) =$
$(-10 - 1, 4 - 2) =$
$(-11, 2)$

b. $\sqrt{(-11)^2 + 2^2} =$
$\sqrt{125} = 5\sqrt{5}$

c. $2\mathbf{u} = (-10, 4)$
$\dfrac{1}{3}\mathbf{v} = (-1, -2)$
$\left|2\mathbf{u}\right| + \left|\dfrac{1}{3}\mathbf{v}\right| =$
$\sqrt{(-10)^2 + 4^2} +$
$\sqrt{(-1)^2 + (-2)^2} =$
$2\sqrt{29} + \sqrt{5}$

d. $\mathbf{u} - \mathbf{v} =$
$(-5, 2) - (-3, -6) =$
$(-5 + 3, 2 + 6) =$
$(-2, 8);$
$|\mathbf{u} - \mathbf{v}| =$
$\sqrt{(-2)^2 + 8^2} = 2\sqrt{17};$
$\dfrac{\mathbf{u} - \mathbf{v}}{|\mathbf{u} - \mathbf{v}|} =$
$\left(\dfrac{-2}{2\sqrt{17}}, \dfrac{8}{2\sqrt{17}}\right) =$
$\left(-\dfrac{\sqrt{17}}{17}, \dfrac{4\sqrt{17}}{17}\right)$

e. $|\mathbf{v}| =$
$\sqrt{(-3)^2 + (-6)^2} =$
$3\sqrt{5}$
Let θ be the direction of
$\mathbf{v}$. Since θ is a third-
quadrant angle with
$\tan \theta = \dfrac{-6}{-3} = 2,$
$\theta \approx 243.4°$. Thus, $\mathbf{v} \approx$
$(3\sqrt{5}, 243.4°).$

Example 3 If $\mathbf{u} = (1, -3)$ and $\mathbf{v} = (2, 5)$, find:

a. $\mathbf{u} + \mathbf{v}$ **b.** $\mathbf{u} - \mathbf{v}$ **c.** $2\mathbf{u} - 3\mathbf{v}$

Solution **a.** $\mathbf{u} + \mathbf{v} = (1, -3) + (2, 5)$
$= (1 + 2, -3 + 5) = (3, 2)$

b. $\mathbf{u} - \mathbf{v} = (1, -3) - (2, 5)$
$= (1 - 2, -3 - 5) = (-1, -8)$

c. $2\mathbf{u} - 3\mathbf{v} = 2(1, -3) - 3(2, 5)$
$= (2, -6) - (6, 15) = (-4, -21)$

Example 4 If $A = (0, 4)$ and $B = (6, 1)$, find **(a)** the coordinates of point P that is $\dfrac{2}{3}$ of the way from A to B, and **(b)** the coordinates of point Q that is $\dfrac{5}{6}$ of the way from A to B.

Solution **a.** Refer to the diagram. Notice that:

$\overrightarrow{AB} = (6 - 0, 1 - 4) = (6, -3)$
$\overrightarrow{OP} = \overrightarrow{OA} + \overrightarrow{AP}$
$= \overrightarrow{OA} + \dfrac{2}{3}\overrightarrow{AB}$
$= (0, 4) + \dfrac{2}{3}(6, -3)$
$= (4, 2)$

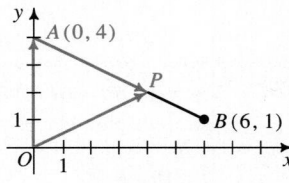

The coordinates of P are $(4, 2)$.

b. Similarly, $\overrightarrow{OQ} = \overrightarrow{OA} + \dfrac{5}{6}\overrightarrow{AB}$. The coordinates of Q are $(5, 1.5)$.

CLASS EXERCISES

$(1, 4); (4, -4)$

1. a. What are the components of $\overrightarrow{AB}$? of $\overrightarrow{CD}$?

b. Find $|\overrightarrow{AB}|$ and $|\overrightarrow{CD}|$. $\sqrt{17}; 4\sqrt{2}$

c. Find the coordinates of the point $\dfrac{1}{4}$ of the way from A to B. $\left(\dfrac{17}{4}, 1\right)$

d. Find the coordinates of the point $\dfrac{3}{4}$ of the way from C to D. $(0, -2)$

2. If $\mathbf{v} = (1, 2)$ and $\mathbf{u} = (3, 0)$, find:

a. $3\mathbf{v}$ $(3, 6)$ **b.** $\mathbf{v} + \mathbf{u}$ $(4, 2)$

c. $\mathbf{v} - \mathbf{u}$ $(-2, 2)$ **d.** $2\mathbf{v} + 3\mathbf{u}$ $(11, 4)$

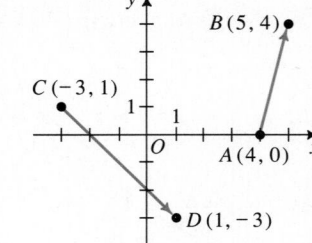

4. $2\mathbf{v} = (6\cos 40°, 6\sin 40°)$; $-\mathbf{v} = (-3\cos 40°, -3\sin 40°)$

3. Physics A force $\mathbf{F}$ of 2 N is represented in the diagram at the right.
  a. Express $\mathbf{F}$ in component form. $(-\sqrt{3}, 1)$
  b. What is $|\mathbf{F}|$? 2

4. If $\mathbf{v} = (3\cos 40°, 3\sin 40°)$, write $2\mathbf{v}$ and $-\mathbf{v}$ in component form. See above.

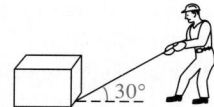
Ex. 3

5. a. If $\overrightarrow{AB} = (3, 2)$ and $A = (4, 0)$, find the coordinates of point B. $(7, 2)$
  b. If $\overrightarrow{CD} = (4, -1)$ and $D = (8, 8)$, find the coordinates of point C. $(4, 9)$

6. Physics Suppose you were moving a heavy box across a floor by pulling on a rope attached near the bottom of the front of the box. If you pull with a force of 80 lb at an angle of 30°, which component (horizontal or vertical) will move the box across the floor? What is the magnitude of this component? Would it be easier to move the box if the angle were smaller?
Horizontal; $40\sqrt{3}$ lb ≈ 69.3 lb; yes

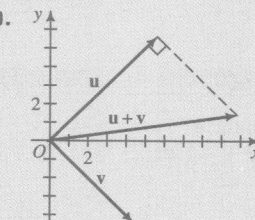

WRITTEN EXERCISES

Plot points A and B. Give the component form of $\overrightarrow{AB}$ and find $|\overrightarrow{AB}|$.

A
  1. $A(1, -2)$, $B(3, -2)$ $(2, 0)$; 2
  2. $A(4, 2)$, $B(0, -1)$ $(-4, -3)$; 5
  3. $A(-3, -5)$, $B(-5, 1)$ $(-2, 6)$; $2\sqrt{10}$
  4. $A(7, -2)$, $B(3, -2)$ $(-4, 0)$; 4

Polar coordinates of point P are given and O is the origin. Draw vector $\overrightarrow{OP}$ and give its component form.
  (1.85, 5.71)   (-6.13, 5.14)   (-1, -1.73)   (5.66, -5.66)
  5. $P(6, 72°)$
  6. $P(8, 140°)$
  7. $P\left(2, \dfrac{4\pi}{3}\right)$
  8. $P\left(8, -\dfrac{\pi}{4}\right)$

Let $\mathbf{u} = (3, 1)$, $\mathbf{v} = (-8, 4)$, and $\mathbf{w} = (-6, -2)$. Calculate each expression.

  9. a. $\mathbf{u} + \mathbf{v}$ $(-5, 5)$ **b.** $\mathbf{u} - \mathbf{v}$ $(11, -3)$ **c.** $3\mathbf{u} + \mathbf{w}$ $(3, 1)$ **d.** $|3\mathbf{u} + \mathbf{w}|$ $\sqrt{10}$
  10. a. $\mathbf{v} + \mathbf{w}$ $(-14, 2)$ **b.** $2(\mathbf{v} + \mathbf{w})$ $(-28, 4)$ **c.** $2\mathbf{v} + 2\mathbf{w}$ $(-28, 4)$ **d.** $|2\mathbf{v} + 2\mathbf{w}|$ $20\sqrt{2}$

  11. a. $\mathbf{u} + \frac{1}{2}\mathbf{w}$ $\mathbf{0} = (0, 0)$ **b.** $\left|\mathbf{u} + \frac{1}{2}\mathbf{w}\right|$ $|\mathbf{0}| = 0$ **c.** $|\mathbf{u}| + \left|\frac{1}{2}\mathbf{w}\right|$ $2\sqrt{10}$

  12. a. $\frac{7}{6}\mathbf{v} - \frac{2}{3}\mathbf{v}$ $\frac{1}{2}\mathbf{v} = (-4, 2)$ **b.** $\left(\frac{7}{6} - \frac{2}{3}\right)\mathbf{v}$ $\frac{1}{2}\mathbf{v} = (-4, 2)$ **c.** $\dfrac{\mathbf{v}}{|\mathbf{v}|}$ $\left(-\dfrac{2\sqrt{5}}{5}, \dfrac{\sqrt{5}}{5}\right)$

In Exercises 13–18, find the coordinates of the point P described.

  13. $A(0, 0)$, $B(6, 3)$; $\frac{1}{2}$ of the way from A to B. $\left(3, \dfrac{3}{2}\right)$

  14. $A(1, 4)$, $B(5, -4)$; $\frac{1}{4}$ of the way from A to B. $(2, 2)$

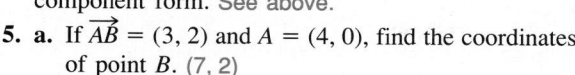

15. $A(7, -2)$, $B(2, 8)$; $\frac{4}{5}$ of the way from A to B. $(3, 6)$

16. $A(-3, -4)$, $B(-6, 1)$; $\frac{1}{6}$ of the way from A to B. $\left(-\frac{7}{2}, -\frac{19}{6}\right)$

17. $A(-7, -4)$, $B(-1, -1)$; $\frac{3}{5}$ of the way from A to B. $\left(-\frac{17}{5}, -\frac{11}{5}\right)$

18. $A(3, -2)$, $B(5, 1)$; $\frac{2}{3}$ of the way from A to B. $\left(\frac{13}{3}, 0\right)$

19. *Writing* Write a few sentences in which you compare the absolute value of a complex number with the absolute value of a vector.

20. *Visual Thinking* Make a sketch showing the complex number $z = 1 + i$ represented by the vector $(1, 1)$. On your sketch, draw vectors to represent the complex numbers z^2, z^3, and z^4. Find the components of each vector.

21. Physics Suppose that you pull a child in a wagon by pulling a rope that makes a 60° angle with the ground.

   a. If the pulling force **F** is 40 lb in the direction of the rope, give the horizontal and vertical components of the force. $(20, 20\sqrt{3})$

   b. Which component, horizontal or vertical, moves the wagon along the ground? Horizontal

   c. *Visual Thinking* If the rope made a 50° angle with the ground rather than a 60° angle, would the wagon move more easily or less easily? Why?

22. Physics Suppose that a 50 lb box sits on a ramp that makes a 20° angle with the horizontal.

   a. Find the component of the 50 lb force that is parallel to the ramp. 17.1 lb

   b. Find the component of the 50 lb force that is perpendicular to the ramp. 47.0 lb

   c. The component of force parallel to the ramp tends to push the box down the ramp. What force is needed to counteract that component? 17.1 lb up the ramp

   d. If the ramp makes a 50° angle with the horizontal, will the box have a greater tendency to slide down the ramp? What force is needed to keep the box from sliding down the ramp? Yes; 38.3 lb up the ramp

23. Navigation A ship is on a course of 108° at a speed of 15 knots. What is the north-south component of its velocity vector? What is the east-west component? 4.6 knots south; 14.3 knots east

24. Aviation An airplane is heading on a course of 240° at 700 knots.

   a. What is the east-west component of its velocity? 606.2 knots west

   b. If the plane encounters a 50 knot wind blowing from the west, what will be the east-west component of the plane's resultant velocity? 556.2 knots west

35. Circle: radius = 2; center at A

25. A bug crawls at a constant speed across the plane so that its position at time t is $(x, y) = (3 + t, 1 + 2t)$. Thus:

at time $t = 0$ $(x, y) = (3 + 0, 1 + 2 \cdot 0) = (3, 1)$
at time $t = 4$ $(x, y) = (3 + 4, 1 + 2 \cdot 4) = (7, 9)$
at time $t = 8$ $(x, y) = (3 + 8, 1 + 2 \cdot 8) = (11, 17)$

Draw a coordinate system and show the bug's position for integral times t between 0 and 8 inclusive.

B **26.** *Writing* Given: A and B are fixed points. In a few sentences, describe the set of points P such that $\overrightarrow{AP} = t\overrightarrow{AB}$. Consider the following three cases:

(1) $0 \leq t \leq 1$ **(2)** $t \geq 1$ **(3)** $t \leq 0$

27. a. Make a sketch illustrating that $|-2\mathbf{v}| = 2|\mathbf{v}|$.
   b. Prove that $|k\mathbf{v}| = |k||\mathbf{v}|$. (*Hint:* Let $\mathbf{v} = (a, b)$.)

28. a. Make a sketch showing two vectors $\mathbf{u}$ and $\mathbf{v}$ such that $|\mathbf{u} + \mathbf{v}| \neq |\mathbf{u}| + |\mathbf{v}|$.
   b. Make a sketch showing two vectors $\mathbf{a}$ and $\mathbf{b}$ such that $|\mathbf{a} + \mathbf{b}| = |\mathbf{a}| + |\mathbf{b}|$.

29. Quadrilateral $ABCD$ is a parallelogram with vertices $A = (1, 2)$, $B = (3, 8)$, $C = (9, 10)$, and $D = (x, y)$. Find the values of x and y. (*Hint:* $\overrightarrow{AB} = \overrightarrow{DC}$.) $(7, 4)$

30. Repeat Exercise 29 if $A = (-2, 1)$, $B = (-5, -3)$, and $C = (3, -2)$. $(6, 2)$

31. Refer to the diagram at the right.
   a. Find real numbers r and s such that $r(1, 2) + s(3, 0) = (9, 6)$. (*Hint:* Solve two simultaneous equations in r and s.) $r = 3, s = 2$
   b. Relate your answer to the diagram.

32. Find real numbers r and s such that $r(1, 2) + s(3, 0) = (-1, 4)$. Illustrate your solution with a diagram similar to the one in Exercise 31. $r = 2, s = -1$

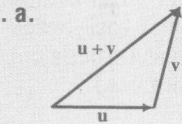

Ex. 31

33. Find a scalar r such that $|r(3, 4)| = 1$. $\pm\dfrac{1}{5}$

34. Find a vector of length 1 in the same direction as $(-4, 3)$. $\left(-\dfrac{4}{5}, \dfrac{3}{5}\right)$

35. If $A = (4, 0)$, describe the set of points P such that $|\overrightarrow{AP}| = 2$. See above.

36. If $A = (4, 0)$ and $B = (-4, 0)$, describe the set of points P in the plane such that: **a.** $|\overrightarrow{AP}| + |\overrightarrow{BP}| = 10$ See below.   **b.** $|\overrightarrow{AP} + \overrightarrow{BP}| = 10$ See below.

37. Let $A = (x_1, y_1)$, $B = (x_2, y_2)$, and M be the midpoint of $\overline{AB}$. Use the method in Example 4 to prove the midpoint formula in Section 1-1.

38. The diagonals of quadrilateral $PQRS$ have the same midpoint M. Use Exercise 37 to prove that $\overrightarrow{SP} = \overrightarrow{RQ}$ and $\overrightarrow{SR} = \overrightarrow{PQ}$. What geometry theorem does this prove?

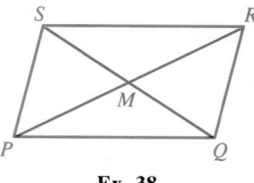

Ex. 38

36. a. Ellipse: foci at A and B **b.** Circle: radius = 5; center at midpoint of $\overline{AB}$

Cooperative Learning

You may wish to have students work in groups to go over the proofs in Exercises 37 and 38.

**Additional Answers
Written Exercises**

26. (1) The set of points P is the line segment $\overline{AB}$.
   (2) The points P on $\overleftrightarrow{AB}$ such that B is between A and P.
   (3) The points P on $\overleftrightarrow{AB}$ such that A is between B and P.

27. a.

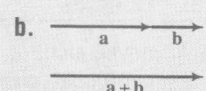

   b. Let $\mathbf{v} = (a, b)$. Then $k\mathbf{v} = (ka, kb)$, and
   $|\mathbf{v}| = \sqrt{a^2 + b^2}$. $|k\mathbf{v}| = \sqrt{k^2a^2 + k^2b^2} = |k|\sqrt{a^2 + b^2} = |k||\mathbf{v}|$.

28. a.

b.

31. b. The sides of the parallelogram shown in the diagram have lengths of $3|(1, 2)|$ and $2|(3, 0)|$.

39. Given: the points with polar coordinates $P(10, 30°)$ and $Q(6, -30°)$
  a. Make a vector diagram that shows $\overrightarrow{OP}$, $\overrightarrow{OQ}$, and $\overrightarrow{OP} + \overrightarrow{OQ}$.
  b. Find $|\overrightarrow{OP} + \overrightarrow{OQ}|$. 14

C **40.** A theorem from geometry states that the centroid G of $\triangle ABC$ is the point $\frac{2}{3}$ of the way from each vertex to the midpoint of the opposite side.
  a. Find a formula for the coordinates of G in terms of the coordinates of A, B, and C. $G = (\frac{1}{3}(a_1 + b_1 + c_1), \frac{1}{3}(a_2 + b_2 + c_2))$
  b. Given the points $A(-2, 5)$, $B(3, 7)$ and $C(5, -3)$, find the centroid of $\triangle ABC$. (2, 3)

12-3 Vector and Parametric Equations: Motion in a Plane

Objective   *To use vector and parametric equations to describe motion in the plane.*

Vector and Parametric Equations of Lines

Suppose that P is any point on line AB. Then as the diagram at the right suggests

$$\overrightarrow{OP} = \overrightarrow{OA} + t\overrightarrow{AB}$$

for some real number t. This equation is called a **vector equation** of line AB and $\overrightarrow{AB}$ is called a **direction vector** of the line.

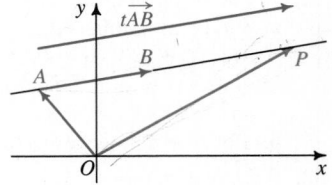

Example 1   Find a vector equation of the line through points $A(3, 4)$ and $B(5, 5)$.

Solution   First, find a direction vector of line AB.

$$\overrightarrow{AB} = (5 - 3, 5 - 4) = (2, 1)$$

Then, let $P = (x, y)$ and substitute into the general vector equation.

$$\overrightarrow{OP} = \overrightarrow{OA} + t\overrightarrow{AB}$$
$$(x, y) = (3, 4) + t(2, 1)$$

    In Example 1, the vector equation of line AB contains the variable t. If t is interpreted as time, then you can think of P as moving along the line, taking various positions at various times t. The diagram at the right shows this. Notice that during one unit of time, P moves 2 units right and 1 unit up. In other words, P moves with constant velocity $\mathbf{v} = (2, 1)$. Its speed is $|\mathbf{v}| = \sqrt{2^2 + 1^2} = \sqrt{5}$.

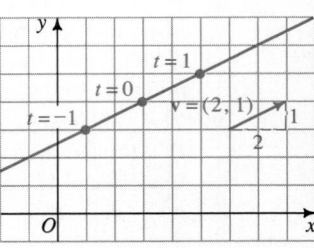

It is instructive to compare the diagram at the bottom of the preceding page with the vector equation of the line.

$$(x, y) = (3, 4) + t(2, 1)$$

position at time $t = 0$ constant velocity

Vector Equation for an Object Moving with Constant Velocity

$$(x, y) = (x_0, y_0) + t(a, b)$$

position at time $t = 0$ constant velocity

Example 2 Suppose an object which is moving with constant velocity is at point $A(5, 3)$ when the time $t = 0$ seconds and at point $B(-4, 15)$ when the time $t = 3$ seconds.
a. Find the velocity and speed of the object.
b. Find a vector equation that describes the motion of the object.

Solution **a.** $\overrightarrow{AB} = (-4, 15) - (5, 3) = (-9, 12)$
Since in 3 s, the object moves $(-9, 12)$, then in 1 s it moves

$$\frac{1}{3}(-9, 12) = (-3, 4).$$

Thus, its velocity is
$$\mathbf{v} = (-3, 4)$$
and its speed is
$$|\mathbf{v}| = \sqrt{(-3)^2 + 4^2} = 5.$$

b. $(x, y) = (x_0, y_0) + t(a, b)$
$(x, y) = (5, 3) + t(-3, 4)$

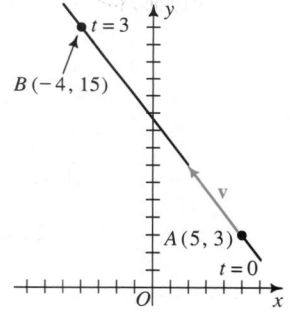

From the vector equation of the line in Example 2, we can derive two equations that give the x- and y-coordinates of the moving object at time t. These equations are called the **parametric equations** of the line, and t is called the **parameter**. They are derived as follows:

Vector equation: $(x, y) = (5, 3) + t(-3, 4)$
$$= (5 - 3t, 3 + 4t)$$

Parametric equations: $x = 5 - 3t$ and $y = 3 + 4t$

Parametric equations of a line are useful in finding the times and positions at which an object moving with constant velocity crosses a curve whose equation is known. This situation is illustrated in the next example.

Vectors and Determinants **433**

Review Note

You might need to review how to solve quadratic equations, including factoring, before doing Example 3.

⌐N Using Technology

A computer or graphing calculator that graphs parametric equations is an excellent tool for simulating motion. In the Activity on this page and in many of the exercises on following pages, students will enjoy using technology to explore the concept of motion in a plane.

Mathematical Note

Polar graphs and parametric equations are related as follows: If $r = f(\theta)$, then this curve is defined parametrically to be $x(t) = f(t) \cos t$ and $y(t) = f(t) \sin t$. These equations allow you to use parametric graphing utilities to graph polar equations.

Activity Note

You may want to extend the Activity to include questions like those in Exercise 26 on page 437.

Example 3 An object moves along a line in such a way that its x- and y-coordinates at time t are $x = 1 - t$ and $y = 1 + 2t$. When and where does the object cross the circle $x^2 + (y - 1)^2 = 25$?

Solution Substitute the parametric equations in the equation of the circle and solve for t.

$$x^2 + (y - 1)^2 = 25$$
$$(1 - t)^2 + (1 + 2t - 1)^2 = 25$$
$$1 - 2t + t^2 + 4t^2 = 25$$
$$5t^2 - 2t - 24 = 0$$
$$(5t - 12)(t + 2) = 0$$

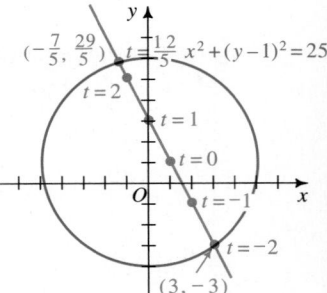

Thus, the times of crossing are $t = \frac{12}{5}$ and $t = -2$.

To find the points of crossing, substitute t in the parametric equations:

When $t = \frac{12}{5}$, we get $x = 1 - \frac{12}{5} = -\frac{7}{5}$ and $y = 1 + 2\left(\frac{12}{5}\right) = \frac{29}{5}$.

When $t = -2$, we get $x = 1 - (-2) = 3$ and $y = 1 + 2(-2) = -3$.

Thus, the points of crossing are $\left(-\frac{7}{5}, \frac{29}{5}\right)$ and $(3, -3)$.

t = when

(x,y) = where

It is easy to convert the parametric equations of the line in Example 3 to a single Cartesian equation in x and y. We eliminate t, using the following steps:

Solve the equation $x = 1 - t$ for t: $t = 1 - x$
Substitute for t in the equation for y: $y = 1 + 2t = 1 + 2(1 - x) = 3 - 2x$

Therefore, a Cartesian equation of the line is $y = -2x + 3$.

Parametric Equations of Curves

Objects don't always move along a line. In the following Activity, you will use a computer or a graphing calculator to explore the motion of a point along a curve.

Activity   **a.** Beginning and ending at (4, 0), the cursor travels around the circle of radius centered at the origin, completing one counterclockwise revolution.

⌐N Use a computer or a graphing calculator that graphs parametric equations. Be sure to use radian measure for t, letting the value of t vary from 0 to 6.28.

a. Using a square screen (see page xxvii), graph the parametric equations $x = 4 \cos t$ and $y = 4 \sin t$. Observe the motion of the TRACE cursor as the value of t varies from 0 to 6.28. Describe the motion. Where does the cursor begin and end, and in what direction does it move? See above.

b. Change the equations to $x = 4 \cos(-t)$ and $y = 4 \sin(-t)$ and again use the TRACE feature to observe the motion of the cursor for $0 \le t \le 6.28$. Describe how the change in the equation affects the motion. The direction is clockwise.

c. Repeat part (b) using the equations $x = 4 \cos 3t$ and $y = 4 \sin 3t$. The cursor completes three revolutions (counterclockwise).

Parametric equations can be used to describe the motion of a point along a variety of curves. In the Activity on the preceding page, you observed the motion of a point as it traveled around the path shown in the diagram at the left below. At the right below, a point moves along a *Lissajous curve*. In each case, as the value of *t* varies from 0 to 2π, the point makes one complete trip around the curve.

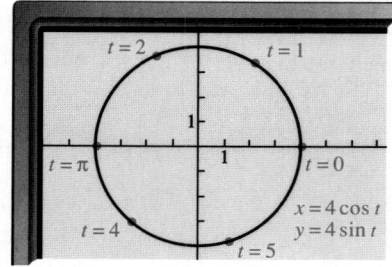
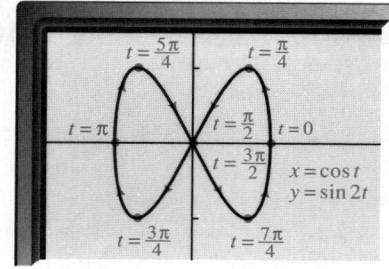

To convert a pair of parametric equations involving trigonometric functions to a single Cartesian equation in x and y, we often use the Pythagorean relationship $\sin^2 t + \cos^2 t = 1$. For example, to show that the curve at the left above is a circle, use the equations $x = 4 \cos t$ and $y = 4 \sin t$ to find the value of $x^2 + y^2$:

$$x^2 + y^2 = (4 \cos t)^2 + (4 \sin t)^2 = 16(\cos^2 t + \sin^2 t) = 16$$

Therefore the parametric equations $x = 4 \cos t$ and $y = 4 \sin t$ satisfy the Cartesian equation $x^2 + y^2 = 16$, which is the equation of a circle.

CLASS EXERCISES

2. $(x, y) = (-3, 1) + t(4, 3)$

1. A line has vector equation $(x, y) = (2, -5) + t(1, 3)$. **a.** $(2, -5), (3, -2), (4, 1)$
   a. Name three points on the line.   **b.** Find a direction vector of the line. $(1, 3)$
   c. Give a pair of parametric equations of the line. $x = 2 + t, y = -5 + 3t$

2. An object moves with constant velocity along a line from $A(-3, 1)$ at time $t = 0$ through $B(5, 7)$ at time $t = 2$. Give a vector equation of line AB. See above.

3. Show that the parametric equations in part (c) of the Activity on page 434 describe the circle with Cartesian equation $x^2 + y^2 = 16$.
   $x^2 + y^2 = (4 \cos 3t)^2 + (4 \sin 3t)^2 = 16(\cos^2 3t + \sin^2 3t) = 16$

WRITTEN EXERCISES

Find vector and parametric equations for each specified line.
   $(x, y) = (1, 5) + t(2, -1); x = 1 + 2t, y = 5 - t$

A **1.** The line through $(1, 5)$ with direction vector $(2, -1)$

2. The line with x-intercept 4 and y-intercept -3 $(x, y) = (4, 0) + t(4, 3); x = 4 + 4t, y = 3t$

3. The line through $(1, 0)$ and $(3, -4)$ **4.** The line through $(3, 1)$ and $(-4, -4)$

5. The line through $(-2, 3)$ and $(5, 1)$ **6.** The line through $(7, 5)$ with inclination 45°

7. The horizontal line through (π, e) **8.** The vertical line through $(\sqrt{2}, \sqrt{3})$
$(x, y) = (\pi, e) + t(1, 0); x = \pi + t, y = e$ $(x, y) = (\sqrt{2}, \sqrt{3}) + t(0, 1); x = \sqrt{2}, y = \sqrt{3} + t$

Vectors and Determinants **435**

Suggested Assignments

Standard
Day 1: 435/1–17 odd
Day 2: 436/19, 21, 25, 27, 33, 39
Comprehensive
Day 1: 435/1–19 odd
Day 2: 437/23, 25, 27, 29, 33, 38, 39

Supplementary Materials

Alternative Assessment, 35
Using Technology, 29–30, 31–32

Additional Answers
Written Exercises

3. $(x, y) = (1, 0) + t(2, -4);$
   $x = 1 + 2t, y = -4t$

4. $(x, y) = (3, 1) +$
   $t(-7, -5); x = 3 - 7t,$
   $y = 1 - 5t$

5. $(x, y) = (-2, 3) +$
   $t(7, -2); x = -2 + 7t,$
   $y = 3 - 2t$

6. $(x, y) = (7, 5) + t(1, 1);$
   $x = 7 + t, y = 5 + t$

9. a.

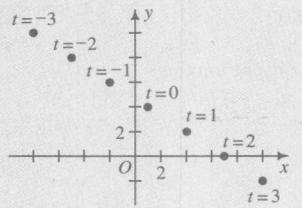

10. a.

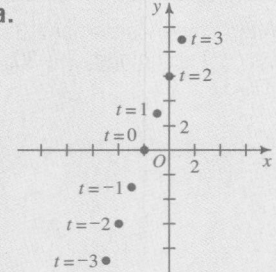

21.

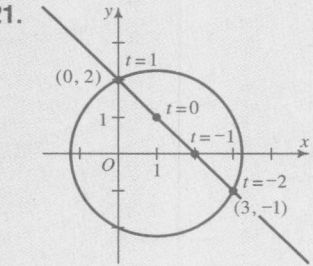

22.

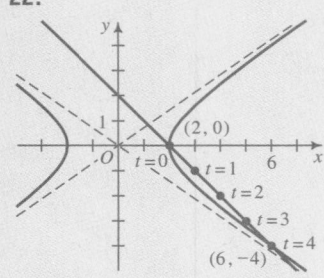

9. b. $\mathbf{v} = (3, -2)$; $|\mathbf{v}| = \sqrt{13}$

In Exercises 9 and 10, a point moves in the plane so that its position $P(x, y)$ at time t is given by the specified vector equation.
a. Graph the point's position at the times $t = 0, 1, 2, 3, -1, -2,$ and -3.
b. Find the velocity and speed of the moving point.
c. Find the parametric equations of the moving point.

9. $(x, y) = (1, 4) + t(3, -2)$ See above.
   c. $x = 1 + 3t, y = 4 - 2t$

b. $\mathbf{v} = (1, 3)$, $|\mathbf{v}| = \sqrt{10}$
10. $(x, y) = (-2, 0) + t(1, 3)$
   c. $x = -2 + t, y = 3t$

In Exercises 11 and 12, find vector and parametric equations of the moving object described. 11. $(x, y) = (2, 3) + t(3, -1)$; $x = 2 + 3t, y = 3 - t$
                12. $(x, y) = (1, -5) + t(1, -1)$; $x = 1 + t, y = -5 - t$

11. Velocity $= (3, -1)$ and position at time $t = 0$ is $(2, 3)$.

12. Velocity $= (1, -1)$ and position at time $t = 0$ is $(1, -5)$.

13. A line has vector equation $(x, y) = (3, 2) + t(2, 4)$. Give a pair of parametric equations and a Cartesian equation of the line. $x = 3 + 2t, y = 2 + 4t; 2x - y = 4$

14. A line has parametric equations $x = 5 - t$ and $y = 4 + 2t$. Give a vector equation and a Cartesian equation of the line. $(x, y) = (5, 4) + t(-1, 2); 2x + y = 14$

15. a. *Writing* Describe the line having parametric equations $x = 2$ and $y = t$.
   b. Give a direction vector of the line. $(0, 1)$ Vert. line through $(2, 0)$
   c. What can you say about the slope of the line? Slope is undefined.

16. a. *Writing* Describe the line having parametric equations $x = t$ and $y = 3$.
   b. Give a direction vector of the line. $(1, 0)$ Hor. line through $(0, 3)$
   c. What is the slope of the line? 0

17. a. A line has direction vector $(2, 3)$. What is the slope of the line? 1.5
   b. A line has direction vector $(4, 6)$. What is the slope of the line? 1.5
   c. Explain why the following lines are parallel:
        $(x, y) = (8, 1) + r(2, 3)$ and $(x, y) = (2, 5) + s(4, 6)$ Slopes are $=$.
   d. Find a vector equation of the line through $(7, 9)$ and parallel to these lines. See below.

18. Find a vector equation of the line through $(2, 1)$ and parallel to the line $(x, y) = (-2, 7) + t(3, 5)$. $(x, y) = (2, 1) + t(3, 5)$ 17. d. $(x, y) = (7, 9) + t(2, 3)$

19. At time t, the position of an object moving with constant velocity is given by the parametric equations $x = 2 - 3t$ and $y = -1 + 2t$.
   a. What are the velocity and speed of the object? $\mathbf{v} = (-3, 2)$; $|\mathbf{v}| = \sqrt{13}$
   b. When and where does it cross the line $x + y = 2$? $t = -1$; $(5, -3)$

20. At time t, the position of an object moving with constant velocity is given by the parametric equations $x = 1 + 3t$ and $y = 2 - 4t$.
   a. What are the velocity and speed of the object? $\mathbf{v} = (3, -4)$; $|\mathbf{v}| = 5$
   b. When and where does it cross the x-axis? $t = \frac{1}{2}$; $\left(\frac{5}{2}, 0\right)$

B 21. An object moves with constant velocity so that its position at time t is $(x, y) = (1, 1) + t(-1, 1)$. When and where does the object cross the circle $(x - 1)^2 + y^2 = 5$? Illustrate with a sketch. $t = 1, (0, 2)$; $t = -2, (3, -1)$

22. An object moves with constant velocity so that its position at time t is $(x, y) = (2, 0) + t(1, -1)$. When and where does the object cross the hyperbola $x^2 - 2y^2 = 4$? Illustrate with a sketch. $t = 0, (2, 0)$; $t = 4, (6, -4)$

In part (b) of Exercises 23 and 24, use a computer or a graphing calculator that will graph two sets of parametric equations simultaneously.

23. A spider and a fly crawl so that their positions at time t (in seconds) are:

   spider: $(x, y) = (-2, 5) + t(1, -2)$ fly: $(x, y) = (1, 1) + t(-1, 1)$

   a. Make a sketch showing the position of the spider and the position of the fly at various times. Do their lines of travel meet? Yes, at $(-1, 3)$

   b. You can use a computer or a graphing calculator to simulate the movement of the spider and the fly. Graph the parametric equations of their paths simultaneously. Do the spider and the fly appear to meet? If so, use the TRACE feature to determine when and where they meet. No

   c. It is possible to determine algebraically whether the spider meets the fly. Just equate the x-coordinates of the positions of the bugs and also the y-coordinates. This gives the following equations:

   $$x = -2 + t = 1 - t$$
   $$y = 5 - 2t = 1 + t$$

   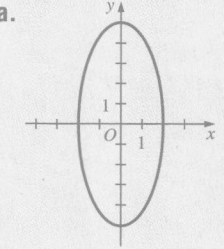

   The spider will meet the fly if and only if there is a single value of t satisfying both equations. Is there such a value of t? No

24. Repeat Exercise 23 for these equations:

   spider: $(x, y) = (3, -2) + t(2, 1)$
   fly: $(x, y) = (-1, 6) + t(4, -3)$

   24. a. Yes, at $(7, 0)$ **b.** Yes, at $(7, 0)$ when $t = 2$ **c.** Yes, $t = 2$

25. Without graphing, describe the curve with parametric equations $x = r \cos t$ and $y = r \sin t$. (*Hint*: What is the value of $x^2 + y^2$?) Circle: radius $= r$; center $(0, 0)$

26. Answers may vary. **a.** $x = 8 \cos 2t$, $y = 8 \sin 2t$ **b.** $x = 8 \cos(-t)$, $y = 8 \sin(-t)$

 In Exercises 26–27, you may wish to confirm your answers by using a computer or a graphing calculator that graphs parametric equations.

   c. $x = -8 \cos t$, $y = -8 \sin t$ **d.** $x = -8 \sin 2t$, $y = -8 \cos 2t$

26. If an object moves so that its position at time t is given by the parametric equations $x = 8 \cos t$ and $y = 8 \sin t$, then the object travels counterclockwise around the circle of radius 8 centered at the origin. The object begins at the point $(8, 0)$ and completes one revolution as t varies from 0 to 2π. Give a pair of parametric equations that will cause the object's motion to occur in each of the following ways as t varies from 0 to 2π. See above.

   a. The object begins at $(8, 0)$ and completes 2 counterclockwise revolutions.

   b. The object begins at $(8, 0)$ and completes 1 clockwise revolution.

   c. The object begins at $(-8, 0)$ and completes 1 counterclockwise revolution.

   d. The object begins at $(0, -8)$ and completes 2 clockwise revolutions.

27. **a.** Before graphing, describe what you think the curve with parametric equations $x = 2 \cos t$ and $y = 5 \sin t$ looks like. Then graph the equations.

   b. Find a Cartesian equation for the parametric equations given in part (a).

   c. An ellipse has Cartesian equation $4x^2 + y^2 = 36$. What do you think the parametric equations of the ellipse are? Check your answer by graphing.

   27. a. Ellipse: vertices $(0, \pm 5)$

   b. $25x^2 + 4y^2 = 100$ **c.** $x = 3 \cos t$, $y = 6 \sin t$ *Vectors and Determinants*   **437**

After students use graphing calculators to complete Exercises 27 and 28, you may want to comment on how much easier it is to enter equations of circles and ellipses in parametric form, rather than in rectangular form as done in Chapter 6. The same is true for hyperbolas (see Exercise 31).

Additional Answers
Written Exercises

27. **a.**

c.

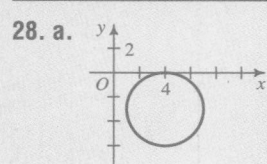
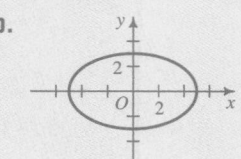
29. b. Before: $x = 6 \cos t$, $y = 6 \sin t$; after: $x = 3 + 6 \cos t$, $y = 5 + 6 \sin t$

 In Exercise 28(a), use a computer or a graphing calculator to graph the given parametric equations. Use a square screen (see page xxvii).

28. a. Graph $x = 4 + 3 \cos t$ and $y = -3 + 3 \sin t$.
   b. Describe the graph in a sentence and give its Cartesian equation. See below.

29. a. After the circle $x^2 + y^2 = 36$ is translated 3 units right and 5 units up, what does its Cartesian equation become? $(x - 3)^2 + (y - 5)^2 = 36$
   b. Give parametric equations of the circle before and after it is translated. See above

30. a. Show that the parametric equations $x = a \cos t$, $y = b \sin t$ satisfy the Cartesian equation of the ellipse $\dfrac{x^2}{a^2} + \dfrac{y^2}{b^2} = 1$.

   b. Sketch the ellipse with parametric equations $x = 5 \cos t$, $y = 3 \sin t$.

31. a. Show that the parametric equations $x = a \sec t$, $y = b \tan t$ satisfy the Cartesian equation of the hyperbola $\dfrac{x^2}{a^2} - \dfrac{y^2}{b^2} = 1$.

   b. A particle moves so that its position (x, y) at time t is given by $x = \sec t$, $y = \tan t$. Sketch its path and tell what quadrant the particle is located in during each of the following time periods:

   (1) $0 < t < \dfrac{\pi}{2}$ I   (2) $\dfrac{\pi}{2} < t < \pi$ III   (3) $\pi < t < \dfrac{3\pi}{2}$ II   (4) $\dfrac{3\pi}{2} < t < 2\pi$ IV

32. Discussion If $x = \sin t$ and $y = \sin t$, then $y = x$. However, the graph of the parametric equations is only *part* of the line $y = x$. Explain.

33. Discussion The graph of $x = \tan t$, $y = \tan t$ is different from the graph of $x = \sec t$, $y = \sec t$ even though both sets of parametric equations satisfy the relationship $y = x$. Explain.

28. b. Circle: $C = (4, -3)$, $r = 3$; $(x - 4)^2 + (y + 3)^2 = 9$

 In Exercises 34–37, use a computer or a graphing calculator to draw the graphs of the given parametric equations.

34. $x = \cos 3t$, $y = \sin 5t$ **35.** $x = \cos t$, $y = \sin 4t$

36. $x = \cos t$, $y = \cos 2t$ **37.** $x = \cos 2t$, $y = \sin t$

 In some parts of the following exercises, you will need to use a computer or a graphing calculator that graphs parametric equations.

38. Physics The diagram at the right shows a 45 m tower whose top is represented by the point $(0, 45)$. If a ball is thrown horizontally from the top of the tower at 30 m/s, then its position (x, y), t seconds later, is given by

$x = 30t$ and $y = 45 - 4.9t^2$.   **a.** 90.9 m   **c.** $y = 45 - \dfrac{49}{9000} x^2$

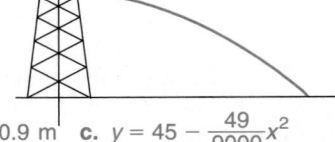

   a. How far from the base of the tower does the ball land?
   b. Use a computer or a graphing calculator to graph the parametric equations given above. Use the TRACE feature to confirm your answer in part (a).
   c. Find a single equation in x and y that describes the path of the ball.

438   *Chapter Twelve*

39. Physics From a point 30 yd directly in front of the goal posts, a football is kicked at angle of elevation θ with initial velocity v. If a coordinate system is set up with the ball at $(0, 0)$ as shown, then the position (x, y) of the ball t seconds after it is kicked is given by the parametric equations

$$x = (v \cos \theta)t \text{ and } y = (v \sin \theta)t - 16t^2. \quad y = \frac{\sqrt{3}}{3}x - \frac{4}{675}x^2$$

a. Suppose the initial velocity of the ball is 60 ft/s and the angle of elevation is 30°. Find an equation in x and y that describes the path of the ball.

b. The goal post crossbar is 10 ft above the ground. Under the conditions stated in part (a), will the ball pass over the goal post crossbar? No

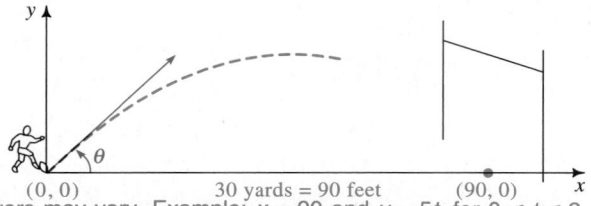

(0, 0)   30 yards = 90 feet   (90, 0)

c. Answers may vary. Example: $x = 90$ and $y = 5t$, for $0 \le t \le 2$

c. To simulate the motion of the football, use a computer or a graphing calculator to graph the parametric equations of the ball's path when $v = 60$ ft/s and $\theta = 30°$. What parametric equations can you use to represent the goal post with a vertical line segment 10 units long? (For the most effective simulation, the goal post should be drawn first.) See above.

d. Keeping $v = 60$ ft/s, vary the value of θ. To the nearest degree, find the minimum value of θ that will allow the ball to pass over the crossbar. 36°

40. Physics A projectile is fired from a cannon whose angle of elevation is θ and whose muzzle velocity is v. If the muzzle is at the origin of a coordinate system, the position $P(x, y)$ of the particle t seconds later is given by the parametric equations $x = (v \cos \theta)t$ and $y = (v \sin \theta)t - 5t^2$.

a. What do these parametric equations give when $\theta = 90°$?

b. Show that the projectile will hit the ground $(y = 0)$ when $t = \frac{v \sin \theta}{5}$. Then show that the x-value when the projectile hits the ground is $x = \frac{v^2 \sin 2\theta}{10}$.

c. For what value of θ will the cannon fire the longest horizontal distance?

d. Graph the projectile's path for $v = 100$ m/s and $\theta = 15°$, $45°$, and $60°$.

e. Find a Cartesian equation of the projectile's path. Is the path parabolic?

Vectors and Determinants **439**

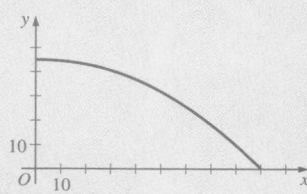

41. A circle with radius 1 rolls along the x-axis at one unit per second. At $t = 0$, P is at the origin. The path traced by P as the circle rolls is called a **cycloid**.

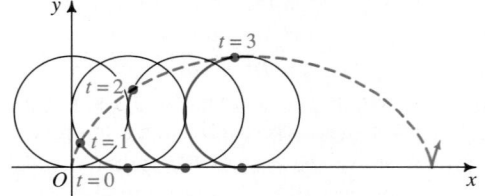 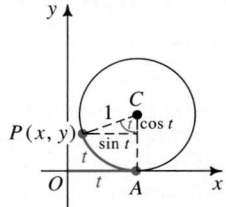

a. Refer to the diagram at the right above and explain why at time t, $OA = t$, arc $AP = t$, and the measure of $\angle ACP$ is t radians.

b. Use $\overrightarrow{OP} = \overrightarrow{OA} + \overrightarrow{AC} + \overrightarrow{CP}$ to show that

$$(x, y) = (t - \sin t, 1 - \cos t).$$

c. Give a pair of parametric equations for the cycloid.

Note: Three of the cycloid's properties are:

1. The length of one arch is 8 times the radius of the rolling circle.
2. The area under one arch is three times the area of the rolling circle.
3. Of all the paths from A to B that a sliding particle acting under gravity can take, the fastest is along an inverted cycloid arch.

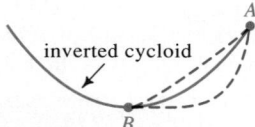

inverted cycloid

//// COMPUTER EXERCISES

1. To approximate the length of one arch of a cycloid having parametric equations $x = t - \sin t$ and $y = 1 - \cos t$, subdivide the time interval into 100 subintervals each of length $L = \frac{2\pi}{100}$, evaluate x and y for $t = 0, L, 2L, 3L, \ldots,$ $100L$, and add the lengths of the line segments joining (x_0, y_0), (x_1, y_1), $(x_2, y_2), \ldots, (x_{100}, y_{100})$. Write a computer program to do this.

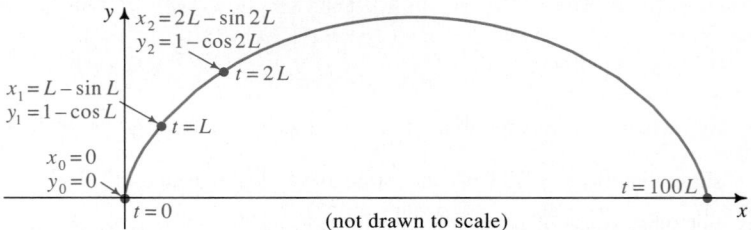

(not drawn to scale)

2. Use a computer to approximate the area under one arch of the cycloid.

12-4 Parallel and Perpendicular Vectors; Dot Product

Objective *To define and apply the dot product.*

Geometrically, two vectors are *parallel* if the lines that contain them are parallel. Algebraically, we say that if $v_2 = kv_1$ for some real number k, then v_1 and v_2 are **parallel**. Similarly, two vectors are *perpendicular* if the lines that contain them are perpendicular. Refer to the diagram below where $v_1 = (x_1, y_1)$ and $v_2 = (x_2, y_2)$. If the lines that contain v_1 and v_2 are perpendicular, then the product of their slopes is -1. That is:

$$\left(\frac{y_1}{x_1}\right)\left(\frac{y_2}{x_2}\right) = -1$$
$$y_1 y_2 = -x_1 x_2$$
$$x_1 x_2 + y_1 y_2 = 0$$

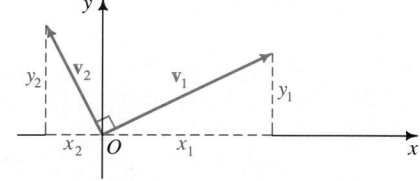

Note that if one vector is horizontal and the other is vertical, then we still have $x_1 x_2 + y_1 y_2 = 0$.

The quantity $x_1 x_2 + y_1 y_2$ occurs often enough in vector work that it is given a special name.

The Dot Product

If $v_1 = (x_1, y_1)$ and $v_2 = (x_2, y_2)$, then the **dot product** of the vectors v_1 and v_2, denoted by $v_1 \cdot v_2$, is defined by

$$v_1 \cdot v_2 = x_1 x_2 + y_1 y_2.$$

The dot product is also called the *scalar product*, because $v_1 \cdot v_2$ is a scalar quantity.

Remarks
1. Perpendicular vectors are sometimes called **orthogonal** vectors.
2. If $v_1 = kv_2$, it is customary to call v_1 parallel to v_2 even though the arrows representing v_1 and v_2 may be collinear.
3. The zero vector is both parallel and perpendicular to all vectors.

Example 1 If $u = (3, -6)$, $v = (4, 2)$, and $w = (-12, -6)$. Find $u \cdot v$ and $v \cdot w$. Show that u and v are perpendicular and that v and w are parallel.

Solution
$u \cdot v = (3, -6) \cdot (4, 2) = 3(4) + (-6)(2) = 0$
$v \cdot w = (4, 2) \cdot (-12, -6) = 4(-12) + 2(-6) = -60$

Since $u \cdot v = 0$, u and v are perpendicular.
Since $w = (-12, -6) = -3(4, 2) = -3v$, v and w are parallel.

Review Note

You might want to review the law of cosines before discussing the angle between two vectors.

Additional Examples

1. Let $\mathbf{u} = (8, -4)$ and $\mathbf{v} = (2, 1)$.

  a. Show that $\mathbf{u} \cdot \mathbf{v} = \mathbf{v} \cdot \mathbf{u}$.

  b. Find the angle θ between $\mathbf{u}$ and $\mathbf{v}$ to the nearest tenth of a degree.

  c. Find a vector that is parallel to $\mathbf{u}$.

  d. Find a vector that is perpendicular to $\mathbf{v}$.

  a. $\mathbf{u} \cdot \mathbf{v} = (8, -4) \cdot (2, 1) =$
$8(2) + (-4)(1) = 12$;
$\mathbf{v} \cdot \mathbf{u} = (2, 1) \cdot (8, -4) =$
$2(8) + 1(-4) = 12$

  b. $|\mathbf{u}| = \sqrt{8^2 + (-4)^2} =$
$4\sqrt{5}$

  $|\mathbf{v}| = \sqrt{2^2 + 1^2} = \sqrt{5}$

  $\cos \theta = \dfrac{\mathbf{u} \cdot \mathbf{v}}{|\mathbf{u}||\mathbf{v}|} =$

  $\dfrac{12}{4\sqrt{5} \cdot \sqrt{5}} = 0.6$

  $\theta \approx 53.1°$

  c. Find (a, b) such that $(a, b) = k(8, -4)$.

  Example: If $k = \dfrac{1}{2}$,

  $(a, b) = (4, -2)$.

  d. Find (r, s) such that $(r, s) \cdot (2, 1) = 0$;
$2r + s = 0$. Example:
If $s = -2$, $r = 1$;
$(r, s) = (1, -2)$.

The dot product has many properties that follow from its definition. Proofs of the properties listed below are left as exercises.

Properties of the Dot Product

1. $\mathbf{u} \cdot \mathbf{v} = \mathbf{v} \cdot \mathbf{u}$
2. $\mathbf{u} \cdot \mathbf{u} = |\mathbf{u}|^2$
3. $k(\mathbf{u} \cdot \mathbf{v}) = (k\mathbf{u}) \cdot \mathbf{v}$
4. $\mathbf{u} \cdot (\mathbf{v} + \mathbf{w}) = \mathbf{u} \cdot \mathbf{v} + \mathbf{u} \cdot \mathbf{w}$

The following formula provides an easy way to find the measure of the angle between two nonzero vectors when we know their magnitudes and their dot product.

The Angle Between Two Vectors

$$\cos \theta = \frac{\mathbf{u} \cdot \mathbf{v}}{|\mathbf{u}||\mathbf{v}|}, \text{ where } 0° \le \theta \le 180°$$

To prove this formula, we sketch the vector diagram shown at the right. In the diagram, $\mathbf{u} = (x_1, y_1)$, $\mathbf{v} = (x_2, y_2)$, and θ is the angle between these vectors. Note that the sides of the triangle have lengths:

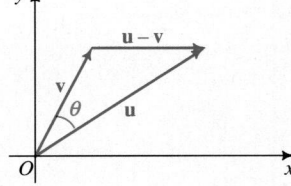

$$|\mathbf{u}| = \sqrt{x_1^2 + y_1^2}$$
$$|\mathbf{v}| = \sqrt{x_2^2 + y_2^2}$$
$$|\mathbf{u} - \mathbf{v}| = \sqrt{(x_1 - x_2)^2 + (y_1 - y_2)^2}$$

From the law of cosines,

$$\cos \theta = \frac{|\mathbf{u}|^2 + |\mathbf{v}|^2 - |\mathbf{u}-\mathbf{v}|^2}{2|\mathbf{u}||\mathbf{v}|},$$

we get:

$$\cos \theta = \frac{(x_1^2 + y_1^2) + (x_2^2 + y_2^2) - [(x_1 - x_2)^2 + (y_1 - y_2)^2]}{2|\mathbf{u}||\mathbf{v}|}$$

$$= \frac{2x_1x_2 + 2y_1y_2}{2|\mathbf{u}||\mathbf{v}|}$$

$$= \frac{x_1x_2 + y_1y_2}{|\mathbf{u}||\mathbf{v}|}$$

Therefore, $\cos \theta = \dfrac{\mathbf{u} \cdot \mathbf{v}}{|\mathbf{u}||\mathbf{v}|}$.

Example 2 To the nearest degree, find the measure of θ, the angle between the vectors $(1, 2)$ and $(-3, 1)$.

Solution

$$\cos \theta = \frac{(1, 2) \cdot (-3, 1)}{|(1, 2)||(-3, 1)|}$$

$$= \frac{-1}{\sqrt{5}\sqrt{10}} \approx -0.1414$$

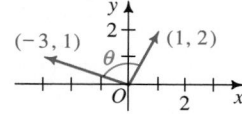

Therefore, $\theta \approx 98°$.

Example 3 In $\triangle PQR$, $P = (2, 1)$, $Q = (4, 7)$, and $R = (-2, 4)$. Find the measure of $\angle P$ to the nearest tenth of a degree.

Solution First note that $\angle P$ is formed by $\overrightarrow{PQ}$ and $\overrightarrow{PR}$.

$$\cos P = \frac{\overrightarrow{PQ} \cdot \overrightarrow{PR}}{|\overrightarrow{PQ}||\overrightarrow{PR}|}$$

Next find the component form of $\overrightarrow{PQ}$ and $\overrightarrow{PR}$.

$$\overrightarrow{PQ} = (4 - 2, 7 - 1) = (2, 6)$$
$$\overrightarrow{PR} = (-2 - 2, 4 - 1) = (-4, 3)$$

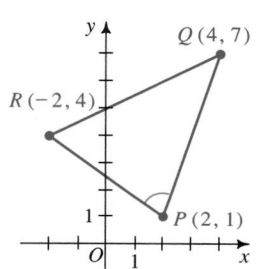

Therefore, $\cos P = \dfrac{(2, 6) \cdot (-4, 3)}{|(2, 6)||(-4, 3)|}$

$$= \frac{10}{\sqrt{40}\sqrt{25}} \approx 0.3162.$$

Therefore, $\angle P \approx 71.6°$.

CLASS EXERCISES

1. Which of the following three vectors are parallel and which are perpendicular?
   $\mathbf{u} = (4, -6)$ $\mathbf{v} = (-2, 3)$ $\mathbf{w} = (9, 6)$ $\mathbf{u} \parallel \mathbf{v}; \mathbf{v} \perp \mathbf{w}; \mathbf{u} \perp \mathbf{w}$

2. If $\mathbf{u} = (3, 4)$ and $\mathbf{v} = (-2, 2)$, find:
   a. $\mathbf{u} \cdot \mathbf{v}$ 2 **b.** $2(\mathbf{u} \cdot \mathbf{v})$ 4 **c.** $(2\mathbf{u}) \cdot \mathbf{v}$ 4 **d.** $\mathbf{u} \cdot (2\mathbf{v})$ 4

3. If $\mathbf{u} = (3, 4)$, find: **a.** $|\mathbf{u}|$ 5 **b.** $|\mathbf{u}|^2$ 25 **c.** $\mathbf{u} \cdot \mathbf{u}$ 25

4. If θ is the angle between $\mathbf{u} = (7, 1)$ and $\mathbf{v} = (5, 5)$, find:
   a. $|\mathbf{u}|$ $5\sqrt{2}$ **b.** $|\mathbf{v}|$ $5\sqrt{2}$
   c. $\mathbf{u} \cdot \mathbf{v}$ 40 **d.** θ to the nearest degree

5. For the figure shown, find $\cos \theta$ **(a)** by using the dot product and **(b)** by using right-triangle trigonometry. (Your answers should agree.) 0.8

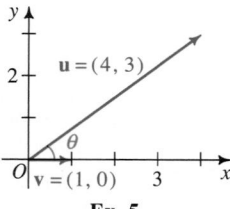

Ex. 5

2. Let $\mathbf{u} = (x_1, y_1)$ and $\mathbf{v} = (x_2, y_2)$. Show that if $\mathbf{u}$ and $\mathbf{v}$ are perpendicular, then $|\mathbf{u} + \mathbf{v}| = |\mathbf{u} - \mathbf{v}|$.

   Since $\mathbf{u}$ and $\mathbf{v}$ are perpendicular, $\mathbf{u} \cdot \mathbf{v} = x_1 x_2 + y_1 y_2 = 0$.
   $\mathbf{u} + \mathbf{v} = (x_1 + x_2, y_1 + y_2)$
   $|\mathbf{u} + \mathbf{v}| =$
   $\sqrt{(x_1 + x_2)^2 + (y_1 + y_2)^2}$
   $= \sqrt{x_1^2 + x_2^2 + y_1^2 + y_2^2}$
   since $x_1 x_2 + y_1 y_2 = 0$.
   $\mathbf{u} - \mathbf{v} = (x_1 - x_2, y_1 - y_2)$
   $|\mathbf{u} - \mathbf{v}| =$
   $\sqrt{(x_1 - x_2)^2 + (y_1 - y_2)^2}$
   $= \sqrt{x_1^2 + x_2^2 + y_1^2 + y_2^2}$
   Thus, $|\mathbf{u} + \mathbf{v}| = |\mathbf{u} - \mathbf{v}|$.

🔲 **Using Technology**

A scientific calculator can be used to find the inverse cosine in Examples 2 and 3.

Suggested Assignments

Standard
444/1, 3, 9, 13, 15, 17, 21
Comprehensive
444/1, 3, 9, 13, 15, 17, 21, 23, 25

Supplementary Materials

Alternative Assessment, 36
Student Resource Guide, 109–111

6. $\mathbf{u} \cdot \mathbf{u} = (a, b) \cdot (a, b) =$
$a^2 + b^2$; $|\mathbf{u}|^2 =$
$(\sqrt{a^2 + b^2})^2 = a^2 + b^2 =$
$\mathbf{u} \cdot \mathbf{u}$

8. a. $\mathbf{u} \cdot \mathbf{v} = 1(-4) + 3(2) =$
$-4 + 6 = 2$; $\mathbf{v} \cdot \mathbf{u} =$
$(-4)1 + 2(3) =$
$-4 + 6 = 2$
b. $3(\mathbf{u} \cdot \mathbf{v}) = 3(2) = 6$;
$\mathbf{u} \cdot (3\mathbf{v}) =$
$(1, 3) \cdot (-12, 6) =$
$-12 + 18 = 6$

10. $\mathbf{u} \cdot (\mathbf{v} + \mathbf{w}) = (1, -4) \cdot$
$(-1, 3) = -1 + (-12) =$
-13; $\mathbf{u} \cdot \mathbf{v} + \mathbf{u} \cdot \mathbf{w} =$
$(1, -4) \cdot (-2, -2) +$
$(1, -4) \cdot (1, 5) =$
$6 - 19 = -13$

12. $\cos^{-1}\left(\dfrac{\mathbf{u} \cdot \mathbf{v}}{|\mathbf{u}||\mathbf{v}|}\right) =$
$\cos^{-1}\left(\dfrac{2 - 15}{\sqrt{13}\sqrt{26}}\right) =$
$\cos^{-1}\left(\dfrac{-13}{13\sqrt{2}}\right) =$
$\cos^{-1}\left(-\dfrac{\sqrt{2}}{2}\right) = 135°$

17. $\overrightarrow{PQ} = (2, 1)$; $\overrightarrow{PR} = (3, 4)$;
$\cos P = \dfrac{\overrightarrow{PQ} \cdot \overrightarrow{PR}}{|\overrightarrow{PQ}||\overrightarrow{PR}|} =$
$\dfrac{(2, 1) \cdot (3, 4)}{(\sqrt{5})(5)} = \dfrac{10}{5\sqrt{5}} = \dfrac{2}{\sqrt{5}}$

18. $\overrightarrow{RS} = (2, 6)$; $\overrightarrow{RT} = (7, 1)$;
$\cos R = \dfrac{\overrightarrow{RS} \cdot \overrightarrow{RT}}{|\overrightarrow{RS}||\overrightarrow{RT}|} =$
$\dfrac{(2, 6) \cdot (7, 1)}{\sqrt{40}\sqrt{50}} = \dfrac{20}{\sqrt{2000}} =$
$\dfrac{20}{20\sqrt{5}} = \dfrac{1}{\sqrt{5}} = \dfrac{\sqrt{5}}{5}$

WRITTEN EXERCISES

A   **1.** Find: **a.** $(2, 3) \cdot (4, -5)$ -7 **b.** $(3, -5) \cdot (7, 4)$ 1

2. Find: **a.** $(-3, 0) \cdot (5, 7)$ -15 **b.** $\left(\dfrac{3}{5}, \dfrac{4}{5}\right) \cdot \left(\dfrac{1}{2}, -\dfrac{3}{2}\right)$ $-\dfrac{9}{10}$

3. Find the value of a if the vectors $(4, 6)$ and $(a, 3)$ are **(a)** parallel, **(b)** perpendicular.   2; -4.5

4. Find the value of a if the vectors $(6, -8)$ and $(4, a)$ are **(a)** parallel, **(b)** perpendicular. $-\dfrac{16}{3}$; 3

5. If $\mathbf{u} = (-2, 3)$, find **(a)** $\mathbf{u} \cdot \mathbf{u}$ and **(b)** $|\mathbf{u}|^2$.   13; 13

6. Let $\mathbf{u} = (a, b)$. Show that $\mathbf{u} \cdot \mathbf{u} = |\mathbf{u}|^2$.

7. If $\mathbf{u} = (5, -3)$ and $\mathbf{v} = (3, 7)$, verify that:
   a. $\mathbf{u} \cdot \mathbf{v} = \mathbf{v} \cdot \mathbf{u}$ **b.** $2(\mathbf{u} \cdot \mathbf{v}) = (2\mathbf{u}) \cdot \mathbf{v}$

8. If $\mathbf{u} = (1, 3)$ and $\mathbf{v} = (-4, 2)$, show that:
   a. $\mathbf{u} \cdot \mathbf{v} = \mathbf{v} \cdot \mathbf{u}$ **b.** $3(\mathbf{u} \cdot \mathbf{v}) = \mathbf{u} \cdot (3\mathbf{v})$

In Exercises 9 and 10, verify that $\mathbf{u} \cdot (\mathbf{v} + \mathbf{w}) = \mathbf{u} \cdot \mathbf{v} + \mathbf{u} \cdot \mathbf{w}$ for the given vectors u, v, and w.

9. $\mathbf{u} = (-2, 5)$, $\mathbf{v} = (1, 3)$, and $\mathbf{w} = (-1, 2)$

10. $\mathbf{u} = (1, -4)$, $\mathbf{v} = (-2, -2)$, and $\mathbf{w} = (1, 5)$

11. Verify that the angle between $\mathbf{u} = (1, 3)$ and $\mathbf{v} = (2, 1)$ is $45°$.

12. Verify that the angle between $\mathbf{u} = (2, 3)$ and $\mathbf{v} = (1, -5)$ is $135°$.

In Exercises 13–16, give angle measures to the nearest tenth of a degree.

13. Find the measure of the angle between $\mathbf{u} = (3, -4)$ and $\mathbf{v} = (3, 4)$. $106.3°$

14. Find the measure of the angle between $\mathbf{u} = (1, 3)$ and $\mathbf{v} = (-8, 5)$. $76.4°$

15. Given $A(1, 5)$, $B(4, 6)$, and $C(2, 8)$, find the measure of $\angle A$. $53.1°$

16. Given $A(-5, 1)$, $B(-3, 3)$, and $C(2, 2)$, find the measure of $\angle A$. $36.9°$

17. Given $P(0, 3)$, $Q(2, 4)$, and $R(3, 7)$, verify that $\cos P = \dfrac{2}{\sqrt{5}}$.

18. Given $R(0, -3)$, $S(2, 3)$, and $T(7, -2)$, verify that $\cos R = \dfrac{1}{\sqrt{5}}$.

19. a. Given $A(1, -3)$, $B(-1, 3)$, and $C(6, 2)$, find $\cos C$ and $\sin C$. $\cos C = 0.6$, $\sin C = 0.8$
   b. Use the formula Area $= \dfrac{1}{2}ab \sin C$ to find the area of $\triangle ABC$.   20

20. Repeat Exercise 19 for $A(5, 3)$, $B(3, 0)$, and $C(2, 2)$.   **a.** $\dfrac{\sqrt{2}}{10}$, $\dfrac{7\sqrt{2}}{10}$ **b.** 3.5

21. Given $A(3, 1)$, $B(14, -1)$, and $C(5, 5)$, find the measure of $\angle BAC$ **(a)** by using the dot product and **(b)** by using angles of inclination. (See page 296 for the inclination formula.) $73.7°$

22. Given $A(0, 0)$, $B(2, 3)$, and $C(1, -5)$, find the measure of $\angle BAC$ **(a)** by using the dot product and **(b)** by using angles of inclination. (See page 296 for the inclination formula.) 135°

B 23. Let $\mathbf{u} = (x_1, y_1)$, $\mathbf{v} = (x_2, y_2)$, and $\mathbf{w} = (x_3, y_3)$. Show that
$$\mathbf{u} \cdot (\mathbf{v} + \mathbf{w}) = \mathbf{u} \cdot \mathbf{v} + \mathbf{u} \cdot \mathbf{w}$$
by expressing both sides of the equation in terms of components.

24. Let $\mathbf{u} = (x_1, y_1)$, $\mathbf{v} = (x_2, y_2)$, and let k be a scalar. Show that
$$\mathbf{u} \cdot \mathbf{v} = \mathbf{v} \cdot \mathbf{u} \qquad \text{and} \qquad (k\mathbf{u}) \cdot \mathbf{v} = k(\mathbf{u} \cdot \mathbf{v}).$$

25. **Physics** The vector $\mathbf{s}$ represents the displacement of a sled that is pulled with force $\mathbf{F}$. The *work* done in moving the sled in the direction of $\mathbf{s}$ is defined to be:

component of $\mathbf{F}$ in the direction of $\mathbf{s} \times$ distance sled moves

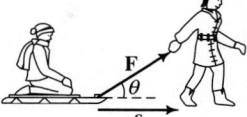

a. Show that this definition can be written as:
$$\text{Work} = |\mathbf{F}| \cos \theta \times |\mathbf{s}|$$

b. Use part (a) and the formula for the angle between two vectors to show that:
$$\text{Work} = \mathbf{F} \cdot \mathbf{s}$$

c. Find the work done if $\mathbf{F} = (10, 3)$ and $\mathbf{s} = (35, 0)$.

26. **Physics** Refer to the diagram in Exercise 25. Write a few sentences to explain what happens to the work done in moving the sled in the direction of $\mathbf{s}$ if $|\mathbf{F}|$ remains the same and θ increases from 0° to 90°.

27. The purpose of this exercise is to prove the Pythagorean theorem using the dot product. In the right triangle, let $\mathbf{u} = \overrightarrow{AB}$ and $\mathbf{v} = \overrightarrow{BC}$.

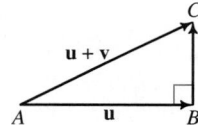

a. Note that $\overrightarrow{AC} = \mathbf{u} + \mathbf{v}$, so that
$$|\overrightarrow{AC}|^2 = |\mathbf{u} + \mathbf{v}|^2 = (\mathbf{u} + \mathbf{v}) \cdot (\mathbf{u} + \mathbf{v}).$$
Show that this last equation simplifies to
$$|\mathbf{u} + \mathbf{v}|^2 = |\mathbf{u}|^2 + 2(\mathbf{u} \cdot \mathbf{v}) + |\mathbf{v}|^2.$$

b. Explain why $\mathbf{u} \cdot \mathbf{v} = 0$.

c. Use part (a) and part (b) to prove that $|\overrightarrow{AC}|^2 = |\overrightarrow{AB}|^2 + |\overrightarrow{BC}|^2$.

28. Use dot products to show that if $ABCD$ is a parallelogram, then
$$|\overrightarrow{AC}|^2 + |\overrightarrow{BD}|^2 = 2|\overrightarrow{AB}|^2 + 2|\overrightarrow{AD}|^2.$$
(*Hint*: Let $\mathbf{u} = \overrightarrow{AB}$ and $\mathbf{v} = \overrightarrow{AD}$. Then use dot products to express each side of the equation in terms of $\mathbf{u}$ and $\mathbf{v}$.)

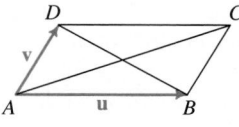

29. **a.** Show that $(\mathbf{u} + \mathbf{v}) \cdot (\mathbf{u} - \mathbf{v}) = 0$ if and only if $|\mathbf{u}| = |\mathbf{v}|$.
b. Use part (a) and the figure for Exercise 28 to prove that the diagonals of a parallelogram are perpendicular if and only if the parallelogram is a rhombus.

Warm-Up Exercises

1. Show that the vectors (4, 6) and (3, −2) are perpendicular.
$4 \cdot 3 + 6(−2) = 0$

2. Find the midpoint of $\overline{RS}$, given $R(0, −7)$ and $S(−6, 3)$. $(−3, −2)$

3. Find a vector equation and the corresponding parametric equations for the line containing $A(5, −3)$ and $B(3, 1)$. For ex., $(x, y) = (5, −3) + t(−2, 4)$; $x = 5 − 2t$, $y = −3 + 4t$

4. To the nearest tenth of a degree, find the measure of the angle between the vectors (1, −1) and (−1, −3). $63.4°$

5. Find the center and radius of the circle whose equation is $x^2 + 2x + y^2 − 8y = 8$. $(−1, 4)$; 5

6. Where does the line with vector equation $(x, y) = (1, 1) + t(−1, 0)$ intersect the circle whose equation is given in Exercise 5? $(−5, 1)$ and $(3, 1)$

Motivating the Section

The location of an airplane in flight requires three coordinates: latitude, longitude, and altitude. We can therefore define a three-dimensional position vector for the plane. Working with vectors in three-dimensional space is the objective of this section.

C **30.** The purpose of this exercise is to use vectors to prove the theorem that the lines containing the altitudes of a triangle are concurrent.
Given: $\overline{AD}$ and $\overline{BE}$ are altitudes of △ABC.

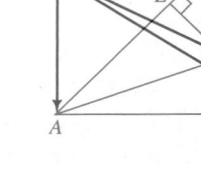

(1) Show that if line AP is perpendicular to $\overline{BC}$, then:
$$(\overrightarrow{OP} − \overrightarrow{OA}) \cdot (\overrightarrow{OC} − \overrightarrow{OB}) = 0$$

(2) Multiply out the expression on the left side.

(3) Prove: If line BP is perpendicular to $\overline{CA}$, then:
$$(\overrightarrow{OP} − \overrightarrow{OB}) \cdot (\overrightarrow{OA} − \overrightarrow{OC}) = 0$$

(4) Multiply out the expression on the left side.

(5) Add the equations in (2) and (4) and factor the result.

(6) Show how your answer in (5) proves that line CP is perpendicular to $\overline{AB}$. Is the theorem proved?

31. Use $\cos \theta = \dfrac{\mathbf{u} \cdot \mathbf{v}}{|\mathbf{u}||\mathbf{v}|}$ to prove that $|\mathbf{u} \cdot \mathbf{v}| \le |\mathbf{u}||\mathbf{v}|$. Then use this result and the result of Exercise 27(a) to prove that $|\mathbf{u} + \mathbf{v}| \le |\mathbf{u}| + |\mathbf{v}|$.

■ Three Dimensions

12-5 Vectors in Three Dimensions

Objective   *To extend vectors to three dimensions and to apply them.*

To start our study of vectors in three dimensions, we begin by setting up a three-dimensional coordinate system. In this system, we have three *axes* meeting at the *origin O* as shown at the right. The plane containing the *x*- and *y*-axes is called the **xy-coordinate plane**. The **xz-coordinate plane** and the **yz-coordinate plane** are similarly defined. These planes divide space into eight regions called **octants**. We locate points in space by means of ordered triples. In the figure, *P* is represented by (2, 3, 4). The numbers 2, 3,

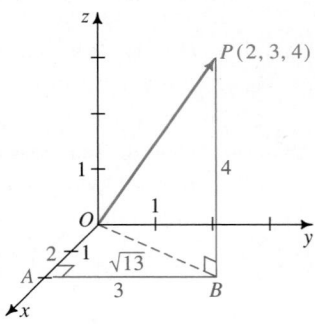

and 4 are not only the *x*-, *y*-, and *z*-coordinates of point *P* but they are also the *x*-, *y*-, and *z*-components of the vector $\overrightarrow{OP}$, which is also written (2, 3, 4).

To find $|\overrightarrow{OP}|$, refer to the figure at the bottom of the preceding page. Note that in right $\triangle OAB$,

$$|\overrightarrow{OB}| = \sqrt{2^2 + 3^2} = \sqrt{13},$$

and in right $\triangle OBP$,

$$|\overrightarrow{OP}| = \sqrt{(\sqrt{13})^2 + 4^2} = \sqrt{29}.$$

In general, if $A(x_1, y_1, z_1)$ and $B(x_2, y_2, z_2)$ are two points in three-dimensional space, then

$$\overrightarrow{AB} = (x_2 - x_1, y_2 - y_1, z_2 - z_1).$$

The diagram to the right can be used to derive the following distance and midpoint formulas. (See Written Exercises 29 and 30.) Notice that the distance and midpoint formulas for a three-dimensional coordinate system are generalizations of the two-dimensional formulas.

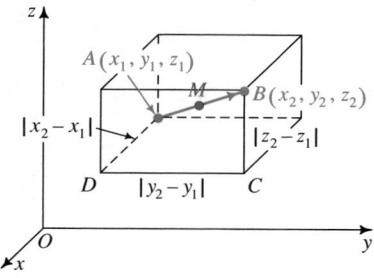

The Distance and Midpoint Formulas

$$|\overrightarrow{AB}| = \sqrt{(x_2 - x_1)^2 + (y_2 - y_1)^2 + (z_2 - z_1)^2}$$
$$= \text{the distance between } A \text{ and } B$$

The midpoint of $\overline{AB} = \left(\dfrac{x_1 + x_2}{2}, \dfrac{y_1 + y_2}{2}, \dfrac{z_1 + z_2}{2}\right).$

Example 1   A sphere has points $A(8, -2, 3)$ and $B(4, 0, 7)$ as endpoints of a diameter.
  a. Find the center C and radius r of the sphere.
  b. Find an equation of the sphere.

Solution   **a.** $C = $ midpoint of $\overline{AB}$
$$= \left(\frac{8 + 4}{2}, \frac{-2 + 0}{2}, \frac{3 + 7}{2}\right) = (6, -1, 5)$$

Use the distance formula to find r.
$$r = CB = \sqrt{(4 - 6)^2 + (0 - (-1))^2 + (7 - 5)^2} = 3$$

  b. The distance between $C(6, -1, 5)$ and $P(x, y, z)$, an arbitrary point on the sphere, equals the radius, 3. Thus:
$$PC = \sqrt{(x - 6)^2 + (y + 1)^2 + (z - 5)^2} = 3$$

Equivalently:
$$(x - 6)^2 + (y + 1)^2 + (z - 5)^2 = 9$$

Vectors and Determinants   **447**

The result of Example 1 can be generalized to give the following. Notice the similarity between the equation of a sphere and the equation of a circle.

Additional Examples cont.

2. Given line L with equation $(x, y, z) = (7, 3, -1) + t(2, 0, -1)$.
   a. Explain why the line is parallel to the xz-plane.
   b. Show that $P(1, 3, 2)$ lies on L.
   c. Where does L intersect the xy-plane?
   d. Where does L intersect the sphere with center $(3, 0, -9)$ and radius 13?

   a. Every point on L has y-coordinate 3. Thus, the line is three units to the right of the xz-plane and parallel to it.
   b. When $t = -3$, $(x, y, z) = (7 + (-3)2, 3 + (-3)0, -1 + (-3)(-1)) = (1, 3, 2)$.
   c. L intersects the xy-plane when $z = 0$; that is, $-1 - t = 0$; $t = -1$; $(x, y, z) = (7, 3, -1) + (-1)(2, 0, -1) = (5, 3, 0)$.
   d. Equation of sphere: $(x - 3)^2 + y^2 + (z + 9)^2 = 13^2$; parametric equations: $x = 7 + 2t, y = 3, z = -1 - t$; thus, $(7 + 2t - 3)^2 + (3^2) + (-1 - t + 9)^2 = 169$; simplifying yields $5t^2 + 89 = 169$; $t^2 = 16$; $t = \pm 4$ $(x, y, z) = (-1, 3, 3)$ when $t = -4$ and $(x, y, z) = (15, 3, -5)$ when $t = 4$.

Equation of the Sphere with Center (x_0, y_0, z_0) and Radius r

$$(x - x_0)^2 + (y - y_0)^2 + (z - z_0)^2 = r^2$$

Vector and parametric equations of a line in three dimensions are similar to those of a line in two dimensions.

The Line Containing (x_0, y_0, z_0) with Direction (a, b, c)

Vector equation: $(x, y, z) = (x_0, y_0, z_0) + t(a, b, c)$
Parametric equations: $x = x_0 + at \qquad y = y_0 + bt \qquad z = z_0 + ct$

Example 2   Find vector and parametric equations of the line containing $A(2, 3, 1)$ and $B(5, 4, 6)$.

Solution   A direction vector of the line is $\overrightarrow{AB} = (5 - 2, 4 - 3, 6 - 1) = (3, 1, 5)$.

A corresponding vector equation is $(x, y, z) = (2, 3, 1) + t(3, 1, 5)$.

Parametric equations of the line are:
$$x = 2 + 3t \quad y = 3 + t \quad z = 1 + 5t$$

The preceding discussion and examples show that vector work in three dimensions closely parallels that in two dimensions. The table below summarizes properties of vectors in three dimensions.

Basic Vector Properties

If $\mathbf{u} = (x_1, y_1, z_1)$ and $\mathbf{v} = (x_2, y_2, z_2)$, then we have the following:
1. Vector addition: $\mathbf{u} + \mathbf{v} = (x_1, y_1, z_1) + (x_2, y_2, z_2)$
   $= (x_1 + x_2, y_1 + y_2, z_1 + z_2)$
2. Scalar multiplication: $k\mathbf{u} = k(x_1, y_1, z_1) = (kx_1, ky_1, kz_1)$
3. The magnitude of $\mathbf{u}$: $|\mathbf{u}| = \sqrt{x_1^2 + y_1^2 + z_1^2}$
4. Dot product: $\mathbf{u} \cdot \mathbf{v} = x_1x_2 + y_1y_2 + z_1z_2$
5. The angle θ between two vectors: $\cos \theta = \dfrac{\mathbf{u} \cdot \mathbf{v}}{|\mathbf{u}||\mathbf{v}|}$

Note that perpendicular vectors need not intersect. Also recall that the zero vector is parallel and perpendicular to all vectors.

Example 3 Find a vector equation of the line through $(1, 5, -2)$ and parallel to the line L with equation $(x, y, z) = (8, 0, 1) + t(4, 3, 2)$.

Solution From the equation for L, $(4, 3, 2)$ is a direction vector of L. Thus, $(4, 3, 2)$ is a direction vector for any line parallel to L. Hence, an equation of the line containing $(1, 5, -2)$ and having direction vector $(4, 3, 2)$ is $(x, y, z) = (1, 5, -2) + t(4, 3, 2)$.

Example 4 Find the measure of θ, the angle between the vectors $(4, -5, 3)$ and $(7, 0, -1)$.

Solution $\cos \theta = \dfrac{(4, -5, 3) \cdot (7, 0, -1)}{|(4, -5, 3)||(7, 0, -1)|} = \dfrac{28 + 0 - 3}{\sqrt{50}\sqrt{50}} = \dfrac{25}{50} = \dfrac{1}{2}$

Thus, $\theta = 60°$.

Mathematical Note

The parametric equations for the line in Example 3 are $x = 1 + 4t$, $y = 5 + 3t$, and $z = -2 + 2t$. The so-called *symmetric* equations of the line can be derived by solving for t in each of the parametric equations and then setting the results equal to each other to get:

$$\frac{x - 1}{4} = \frac{y - 5}{3} = \frac{z + 2}{2}$$

CLASS EXERCISES

1. Find the length and midpoint of $\overline{AB}$. **1. a.** $\sqrt{14}$; $\left(\frac{1}{2}, -1, \frac{3}{2}\right)$ **b.** $2\sqrt{14}$; $(4, 2, -1)$
   a. $A = (0, 0, 0)$ and $B = (1, -2, 3)$
   b. $A = (3, 0, -4)$ and $B = (5, 4, 2)$

2. The figure at the right shows a rectangular box with one vertex at the origin O of a three-dimensional coordinate system and three edges along the axes as shown.
   a. If point $G = (4, 5, 3)$, give the coordinates of points A, B, C, D, E, and F.
   b. Find $|\overrightarrow{OG}|$. $5\sqrt{2}$

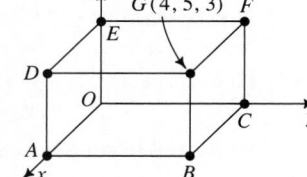

3. Find an equation of the sphere with radius 5 and center at **(a)** $(0, 0, 0)$ and at **(b)** $(1, 2, 3)$. $x^2 + y^2 + z^2 = 25$; $(x - 1)^2 + (y - 2)^2 + (z - 3)^2 = 25$

4. Simplify: **a.** $(3, 5, -2) + 2(1, 2, 3)$ **b.** $(3, 8, 1) \cdot (4, -1, 4)$ 8 **a.** $(5, 9, 4)$

5. *Discussion* How can you tell if two vectors are perpendicular? How can you tell if they are parallel?

6. Line L has vector equation $(x, y, z) = (2, 5, 1) + t(6, 7, 8)$. Answers may vary.
   a. Name two points on L. $(2, 5, 1)$, $(8, 12, 9)$
   b. Find three parametric equations of L. $x = 2 + 6t$, $y = 5 + 7t$, $z = 1 + 8t$
   c. Find a direction vector of L. $(6, 7, 8)$
   d. Explain why L is parallel to line $(x, y, z) = (8, -2, 2) + s(6, 7, 8)$. See below.
   e. Explain why L is perpendicular to line $(x, y, z) = (2, 5, 1) + s(4, 0, -3)$. See below.

7. *Discussion* Explain how to find where the line **6. d.** Dir. vectors are =.
   $$(x, y, z) = (-6, 0, 3) + s(4, 1, -1)$$
   intersects the sphere $x^2 + y^2 + z^2 = 9$. **6. e.** Dot product of dir. vectors = 0.

8. *Discussion* Suppose that $\cos \theta = \pm 1$, where θ is the angle between two vectors. What can you conclude about the vectors? The vectors are parallel.

Problem Solving

Ask students to determine when a vector in three dimensions is parallel to a plane containing any two of the three coordinate axes. (When one of the vector's components is 0.)

Additional Answers Class Exercises

2. **a.** $A = (4, 0, 0)$,
   $B = (4, 5, 0)$,
   $C = (0, 5, 0)$,
   $D = (4, 0, 3)$,
   $E = (0, 0, 3)$,
   $F = (0, 5, 3)$

5. If the dot product equals zero, the two vectors are perpendicular. If one vector is a scalar multiple of the other, the two vectors are parallel.

7. Substitute parametric equations obtained from the vector equation of the line into equation of sphere.

Vectors and Determinants **449**

Additional Answers
Written Exercises

5. $A = (5, 0, 0)$,
   $B = (5, 6, 0)$,
   $C = (0, 6, 0)$,
   $D = (5, 0, 4)$,
   $E = (0, 0, 4)$,
   $F = (0, 6, 4)$;
   $|\overrightarrow{OG}| = \sqrt{77}$

6. $A = (6, 0, 0)$,
   $B = (6, 7, 0)$,
   $C = (0, 7, 0)$,
   $D = (6, 0, 3)$,
   $E = (0, 0, 3)$,
   $F = (0, 7, 3)$;
   $|\overrightarrow{OG}| = \sqrt{94}$

12. $(7 - 1)^2 + (7 - 5)^2 +$
    $(6 - 3)^2 = 6^2 + 2^2 +$
    $3^2 = 36 + 4 + 9 = 49$

19. a. $\cos A = \dfrac{\sqrt{3}}{2}$, $\sin A = \dfrac{1}{2}$

    b. Area $= \dfrac{\sqrt{3}}{2}$

20. a. $\cos A = -\dfrac{1}{3}$,

    $\sin A = \dfrac{2\sqrt{2}}{3}$

    b. $5\sqrt{2}$

Suggested Assignments

Comprehensive
Day 1: 450/1, 5, 9, 11, 13, 15, 19, 21, 23
Day 2: 451/25, 27, 31, 33, 35, 39, 41

Supplementary Materials

Alternative Assessment, 36
Activities Book, 31–35

WRITTEN EXERCISES

1. $2\sqrt{6}$; $(1, 4, -1)$ 3. $2\sqrt{14}$; $(1, -2, 1)$ 4. $2\sqrt{21}$; $(0, 0, 0)$

In Exercises 1–4, find the length and midpoint of $\overrightarrow{AB}$.

   $2\sqrt{17}$; $(2, 4, -4)$

A 1. $A = (2, 5, -3)$ and $B = (0, 3, 1)$ See   2. $A = (2, 8, -5)$ and $B = (2, 0, -3)$

   3. $A = (3, -5, 0)$ and $B = (-1, 1, 2)$ above.   4. $A = (4, 1, 2)$ and $B = (-4, -1, -2)$

5. Refer to Class Exercise 2 and suppose that $G = (5, 6, 4)$. What are the coordinates of points A, B, C, D, E, and F? Find $|\overrightarrow{OG}|$.

6. Refer to Class Exercise 2 and suppose that $G = (6, 7, 3)$. What are the coordinates of points A, B, C, D, E, and F? Find $|\overrightarrow{OG}|$.

In Exercises 7 and 8, simplify each expression. 8. a. $(7, 5, -4)$

7. a. $(3, 8, -2) + 2(4, -1, 2)$ $(11, 6, 2)$ b. $(1, -8, 6) \cdot (5, 2, 1)$ -5 c. $|(3, 5, 1)|$ $\sqrt{35}$

8. a. $(8, 7, 4) - (2, 0, 9) + (1, -2, 1)$ b. $(1, 4, 2) \cdot (3, 1, 2)$ 11   c. $|(6, 2, 3)|$ 7

9. Are vectors $(3, -7, 1)$ and $(6, 3, 3)$ perpendicular? Yes

10. Find the value of k if $(2, k, -3)$ and $(4, 2, 6)$ are perpendicular. 5

11. Find an equation of the sphere with radius 2 and center at **(a)** the origin and **(b)** $(3, -1, 2)$. $x^2 + y^2 + z^2 = 4$; $(x - 3)^2 + (y + 1)^2 + (z - 2)^2 = 4$

12. Find an equation of the sphere with radius 7 and center $(1, 5, 3)$. Show that the point $(7, 7, 6)$ is on the sphere. $(x - 1)^2 + (y - 5)^2 + (z - 3)^2 = 49$

13. Find the center and radius of the sphere with equation
    $$x^2 + y^2 + z^2 + 2x - 4y - 6z = 11.\ (-1, 2, 3);\ 5$$
    (*Hint:* Complete the squares in x, y, and z.)

14. Find the center and radius of the sphere with equation
    $$x^2 + y^2 + z^2 - 6x + 10y + 2z = 65.\ (3, -5, -1);\ 10$$

15. Find the angle between $(8, 6, 0)$ and $(2, -1, 2)$ to the nearest tenth of a degree. 70.5°

16. Find the angle between $(2, 2, 1)$ and $(3, 6, -2)$ to the nearest tenth of a degree. 40.4°

17. Let $A = (1, 3, 4)$, $B = (3, -1, 0)$, and $C = (3, 2, 6)$.
    a. Show that $\overrightarrow{AB}$ and $\overrightarrow{AC}$ are perpendicular. $\overrightarrow{AB} \cdot \overrightarrow{AC} = (2, -4, -4) \cdot (2, -1, 2) = 0$
    b. Find the area of right triangle ABC. 9

18. Repeat Exercise 17 if $A = (3, 7, -5)$, $B = (5, 9, -4)$, and $C = (7, 5, -9)$. See below.

19. a. Given $A(0, 0, 0)$, $B(1, 1, -2)$, and $C(0, 1, -1)$, find $\cos A$ and $\sin A$.
    b. Find the area of $\triangle ABC$.

20. Repeat Exercise 19 given the points $A(1, 2, 3)$, $B(1, 5, 7)$, and $C(3, 3, 1)$.

21. Line L has vector equation $(x, y, z) = (-2, 0, 1) + t(4, -1, 1)$.
    a. Find three parametric equations of L. $x = -2 + 4t$, $y = -t$, $z = 1 + t$
    b. Name two points on L. $(-2, 0, 1)$, $(2, -1, 2)$ $(-10, 2, -1)$
    c. Which of the points $(-10, 2, -1)$, $(18, -2, 6)$, and $(14, -5, 4)$ are on L?
    d. Write a vector equation of the line containing $(1, 2, 3)$ and parallel to L.
       $$(x, y, z) = (1, 2, 3) + t(4, -1, 1)$$

18. a. $\overrightarrow{AB} \cdot \overrightarrow{AC} = (2, 2, 1) \cdot (4, -2, -4) = 0$ b. 9

---

450   *Chapter Twelve*

22. Repeat Exercise 21 if the vector equation of L is
$$(x, y, z) = (6, -1, 0) + t(2, -1, 1).$$

23. Write vector and parametric equations for the line containing $A(4, 2, -1)$ and $B(6, 3, 2)$. $(x, y, z) = (4, 2, -1) + t(2, 1, 3);$ $x = 4 + 2t,$ $y = 2 + t,$ $z = -1 + 3t$

24. Write vector and parametric equations for the line containing $A(2, 3, 1)$ and $B(4, -1, 3)$. $(x, y, z) = (2, 3, 1) + t(2, -4, 2);$ $x = 2 + 2t,$ $y = 3 - 4t,$ $z = 1 + 2t$

B **25.** Explain why the line $(x, y, z) = (4, -5, 2) + t(2, 1, 0)$ is parallel to the xy-plane. Direction vector is parallel because z component is zero.

26. Describe the line $(x, y, z) = (1, 2, 3) + t(0, 0, 1)$. Where does it intersect the xy-plane? $(1, 2, 0)$

27. Given line L with equation:
$(x, y, z) = (5, 3, 0) + t(1, 4, -6)$
  a. Find vector and parametric equations of the line through $(1, 2, 3)$ and parallel to L.
  b. Find vector and parametric equations of a line through $(5, 3, 0)$ and perpendicular to L.

28. Given line L with equation:
$(x, y, z) = (1, 3, -2) + t(2, 1, 1)$
  a. Find vector and parametric equations of the line through $(4, 5, 6)$ and parallel to L.
  b. Find vector and parametric equations of a line through $(1, 3, -2)$ and perpendicular to L.

Parallel and perpendicular lines dominate the design of this office building.

29. Refer to the diagram on page 447. Let $A = (x_1, y_1, z_1)$ and $B = (x_2, y_2, z_2)$. Prove the distance formula.

30. Refer to the diagram on page 447. Let $M(x, y, z)$ be the midpoint of $\overline{AB}$. Prove the midpoint formula.

37. Answers may vary. **a.** $4x + 6y + 12z = 49$ **b.** $4 \cdot 1 + 6 \cdot \frac{3}{2} + 12 \cdot 3 = 49$

In Exercises 31–36, O is the origin, A and B are fixed points, and P is an arbitrary point. Describe the set of points P that satisfy the given equation in (a) a two-dimensional plane and (b) three-dimensional space.

31. $|\overrightarrow{PA}| = |\overrightarrow{PB}|$ **32.** $\overrightarrow{PA} \cdot \overrightarrow{AB} = 0$ **33.** $\overrightarrow{PA} \cdot \overrightarrow{PB} = 0$

34. $|\overrightarrow{PA}| = 10$ **35.** $|\overrightarrow{PA}| + |\overrightarrow{PB}| = 10$ **36.** $\overrightarrow{OP} = \overrightarrow{OA} + t\overrightarrow{AB}, |t| \leq 1$

37. a. If $P(x, y, z)$ is equidistant from $A(0, 0, 0)$ and $B(2, 3, 6)$, find and simplify an equation that must be satisfied by x, y, and z. See above.
  b. Verify that the coordinates of the midpoint of $\overline{AB}$ satisfy the equation.

38. a. Give an equation in x, y, and z that satisfies $\overrightarrow{PA} \cdot \overrightarrow{AB} = 0$ if $P = (x, y, z)$, $A = (3, 5, 2)$, and $B(6, 6, 4)$. $3x + y + 2z = 18$
  b. Compare your answer with Exercise 32. Both are planes containing A.

22. Answers may vary. Examples are given.
  a. $x = 6 + 2t,$ $y = -1 - t,$ $z = t$
  b. $(6, -1, 0),$ $(8, -2, 1)$
  c. $(14, -5, 4)$
  d. $(x, y, z) = (1, 2, 3) + t(2, -1, 1)$

26. The line is parallel to the z-axis, perpendicular to the xy-plane, and goes through the point $(1, 2, 3)$.

27. Answers may vary. Examples are given.
  a. $(x, y, z) = (1, 2, 3) + t(1, 4, -6);$ $x = 1 + t,$ $y = 2 + 4t,$ $z = 3 - 6t$
  b. $(x, y, z) = (5, 3, 0) + t(6, -3, -1);$ $x = 5 + 6t,$ $y = 3 - 3t,$ $z = -t$

28. Answers may vary. Examples are given.
  a. $(x, y, z) = (4, 5, 6) + t(2, 1, 1)$
  $x = 4 + 2t,$ $y = 5 + t,$ $z = 6 + t$
  b. $(x, y, z) = (1, 3, -2) + t(1, -1, -1)$
  $x = 1 + t,$ $y = 3 - t,$ $z = -2 - t$

39. a. Give an equation in x, y, and z that satisfies $\overrightarrow{PA} \cdot \overrightarrow{PB} = 0$ if $P = (x, y, z)$, $A = (2, 2, 1)$, and $B = (-2, -2, -1)$. $x^2 + y^2 + z^2 = 9$

  b. Compare your answer with Exercise 33. Both are spheres.

40. Where does the line $(x, y, z) = (6, -5, 4) + t(3, -2, 4)$ intersect **(a)** the xy-plane, **(b)** the yz-plane, and **(c)** the xz-plane? See below.

41. Describe the set of points S in the xy-plane that are also on the sphere whose equation is $(x - 1)^2 + (y - 2)^2 + (z - 3)^2 = 25$. Give an equation of S. See above.

42. Where does the line $(x, y, z) = (3, 4, -2) + t(2, -1, 2)$ and the sphere whose equation is $(x - 3)^2 + (y - 4)^2 + (z + 2)^2 = 36$ intersect? (7, 2, 2), (−1, 6, −6)

C 43. a. Show that the lines with equations $(x, y, z) = (3, 0, -4) + t(1, 2, 3)$ and $(x, y, z) = (1, 5, 2) + s(2, 1, 4)$ have no points in common.

  b. Part (a) shows that the lines do not intersect. Are they parallel? No

44. a. Show that the lines with equations $(x, y, z) = (-1, 5, 0) + t(1, 2, -1)$ and $(x, y, z) = (0, 1, 3) + s(1, -1, 1)$ intersect.

  b. Find the coordinates of their point of intersection. (−2, 3, 1)

45. a. Explain why the following lines are perpendicular:

L_1: $(x, y, z) = (3, -1, 5) + t(2, 3, 4)$ and L_2: $(x, y, z) = (3, -1, 5) + s(2, 0, -1)$

  b. Find an equation of a line perpendicular to both L_1 and L_2.
  Answers may vary. Example: $(x, y, z) = (3, -1, 5) + r(3, -10, 6)$

12-6 Vectors and Planes
40. (3, −3, 0); (0, −1, −4); $\left(-\frac{3}{2}, 0, -6\right)$

Objective   *To sketch planes and to find equations of planes.*

In two-dimensional space, the graph of $ax + by = c$ is a line provided that a and b are not both 0. In three-dimensional space, the graph of $ax + by + cz = d$ is a *plane* provided that a, b, and c are not all 0. For example, the graph of $2x + 3y + 4z = 12$ is a plane. To sketch it, we locate the points where the plane intersects the axes.

Setting y and $z = 0$ gives $2x = 12$. Thus, $x = 6$ and 6 is the x-intercept.
Setting x and $z = 0$ gives $3y = 12$. Thus, $y = 4$ and 4 is the y-intercept.
Setting x and $y = 0$ gives $4z = 12$. Thus, $z = 3$ and 3 is the z-intercept.

The diagram below shows that portion of the plane in the *first octant*, the region where all coordinates are positive.

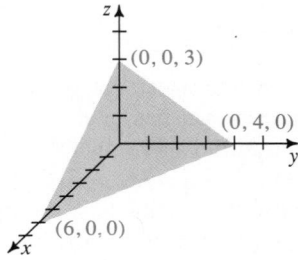

Parts of some other planes are shown below along with their equations.

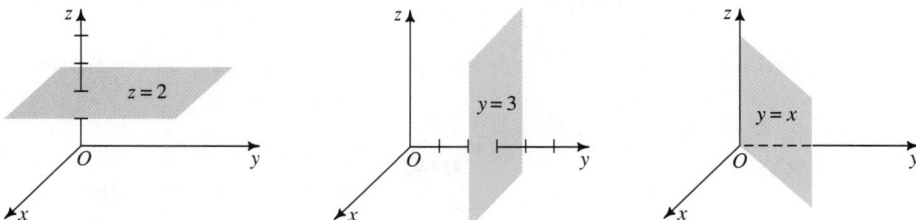

If a vector $\overrightarrow{AN}$ is perpendicular to a plane containing A, B, and C, as shown, then it is perpendicular to $\overrightarrow{AB}$, $\overrightarrow{AC}$, and all other vectors in the plane. Thus, if P is any point in the plane, then

$$\overrightarrow{AN} \cdot \overrightarrow{AP} = 0.$$

The equation above is called a *vector equation of the plane*.

To derive a Cartesian equation of the plane, let $A(x_0, y_0, z_0)$ be a fixed point in the plane and let $P(x, y, z)$ be any point in the plane. Then

$$\overrightarrow{AP} = (x - x_0, y - y_0, z - z_0).$$

Since $\overrightarrow{AN} = (a, b, c)$ is perpendicular to the plane, $\overrightarrow{AN}$ is perpendicular to $\overrightarrow{AP}$. Therefore:

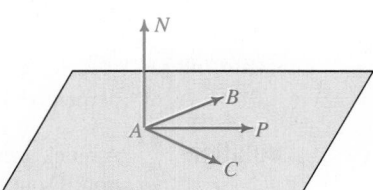

$$\overrightarrow{AN} \cdot \overrightarrow{AP} = 0$$
$$(a, b, c) \cdot (x - x_0, y - y_0, z - z_0) = 0$$
$$a(x - x_0) + b(y - y_0) + c(z - z_0) = 0$$
$$ax + by + cz = ax_0 + by_0 + cz_0$$
$$ax + by + cz = d, \text{ where } d = ax_0 + by_0 + cz_0$$

This result is summarized below.

The Cartesian Equation of a Plane

If (a, b, c) is a nonzero vector perpendicular to a plane at the point (x_0, y_0, z_0), then an equation of the plane is:

$$ax + by + cz = d, \text{ where } d = ax_0 + by_0 + cz_0.$$

Because the steps of the derivation above are reversible, the converse is also true.

Vectors and Determinants   **453**

Assessment Note

After discussing the Cartesian equation of a plane, ask students to give an equation for each of the planes containing a pair of coordinate axes. (*xy*-plane: $z = 0$; *xz*-plane: $y = 0$; *yz*-plane: $x = 0$)

Additional Examples

1. Find an equation of the plane tangent to the sphere $(x - 7)^2 + (y - 5)^2 + (z + 1)^2 = 121$ at the point $P(-2, -1, 1)$.

   Notice that $(-2 - 7)^2 + (-1 - 5)^2 + (1 + 1)^2 = 121$, so P lies on the sphere. The sphere has center $C(7, 5, -1)$; $\overrightarrow{CP} = (-2 - 7, -1 - 5, 1 - (-1)) = (-9, -6, 2)$; since $\overrightarrow{CP}$ is perpendicular to the tangent plane, the plane has an equation of the form $-9x - 6y + 2z = d$. Since P lies on the plane, $-9(-2) - 6(-1) + 2(1) = d$; $d = 26$; $-9x - 6y + 2z = 26$, or $9x + 6y - 2z = -26$.

(*Continues on next page*)

2. Let nonzero vectors **u** and **v** be perpendicular, respectively, to planes P and Q. Then P and Q are perpendicular if **u** and **v** are perpendicular, and P and Q are parallel if **u** and **v** are parallel.
   a. Show that the planes with equations
   $12x - 6y - 3z = 7$ and
   $x + y + 2z = 5$ are perpendicular.
   b. Show that the planes with equations
   $12x - 6y - 3z = 7$ and
   $-8x + 4y + 2z = 5$ are parallel.

   a. $\mathbf{u} = (12, -6, -3)$ and $\mathbf{v} = (1, 1, 2);\ \mathbf{u} \cdot \mathbf{v} = 12(1) + (-6)(1) + (-3)(2) = 0$, so **u** and **v** are perpendicular. Therefore, the planes are perpendicular.
   b. $\mathbf{u} = (12, -6, -3)$ and $\mathbf{v} = (-8, 4, 2)$. Since $\mathbf{v} = -\frac{2}{3}\mathbf{u}$, **u** and **v** are parallel. Therefore, the planes are parallel.

Example 1   Vector $(3, 4, -2)$ is perpendicular to a plane that contains $A(0, 1, 2)$. Find an equation of the plane.

Solution   Since $(3, 4, -2)$ is perpendicular to the plane, an equation of the plane is
$$3x + 4y - 2z = d.$$
To find d, substitute the coordinates of A into the equation.
$$3(0) + 4(1) - 2(2) = 0 = d$$
An equation of the plane is $3x + 4y - 2z = 0$.
(Notice that this plane contains the origin, since $(0, 0, 0)$ satisfies the equation.)

Example 2   If $A = (1, 0, 2)$ and $B = (3, -4, 6)$, find a Cartesian equation of the plane perpendicular to $\overline{AB}$ at its midpoint.

Solution   A quick sketch helps to begin the solution. From the sketch at the right, the plane is perpendicular to $\overline{AB}$ at its midpoint M:

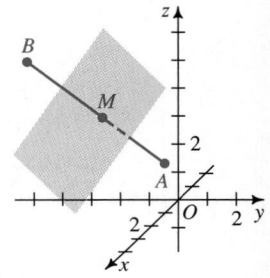

$$M = \left(\frac{1+3}{2}, \frac{0+(-4)}{2}, \frac{2+6}{2}\right)$$
$$= (2, -2, 4)$$
$$\overrightarrow{MB} = (3 - 2, -4 - (-2), 6 - 4) = (1, -2, 2)$$

Since $\overline{AB}$ is perpendicular to the plane, $\overrightarrow{MB}$ is perpendicular to the plane. Therefore, a Cartesian equation of the plane is $1x - 2y + 2z = d$.

To find d, substitute the coordinates of M into this equation. Therefore, an equation of the plane is:
$$x - 2y + 2z = 14$$

CLASS EXERCISES

x-int. $= 2$, y-int. $= 6$, z-int. $= 3$
1. Find the x-, y-, and z-intercepts of the plane $3x + y + 2z = 6$.

2. A rectangular box is shown. Find an equation of the plane that contains:
   a. the top of the box $z = 4$
   b. the bottom of the box $z = 0$
   c. the front of the box $x = 3$
   d. the right side of the box $y = 6$

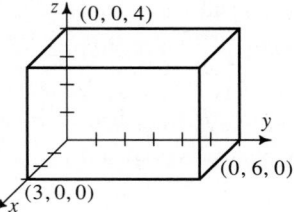

3. Name a vector perpendicular to each plane.　(3, 0, −4)
   a. $2x + 3y + 4z = 10$ (2, 3, 4) **b.** $3x − 4z = 12$　　**c.** $z = 5$ (0, 0, 1)

4. The vector $(1, 1, 1)$ is perpendicular to a plane that contains $(3, 4, 5)$. What is a Cartesian equation of the plane? $x + y + z = 12$

5. The vector $(6, 7, 8)$ is perpendicular to a plane that contains $(0, 0, 0)$. What is a Cartesian equation of the plane? $6x + 7y + 8z = 0$

WRITTEN EXERCISES

Sketch each plane whose equation is given.

A 1. $2x + 3y + 6z = 12$ 　　　　　　　　　 2. $3x + y + 2z = 6$
　 3. $5x − 2y + 2z = 10$ 　　　　　　　　　 4. $2x + 4y − z = 8$
　 5. $z = 2$ 　　　　　　　　　　　　　　　　 6. $y = 4$

7. Let $E(4, 6, 1)$ be a vertex of the rectangular box shown. Find a Cartesian equation of each plane.
   a. plane $DEFG$ $z = 1$ **b.** plane $ABED$ $x = 4$
   c. plane $BCFE$ $y = 6$ **d.** plane $OADG$ $y = 0$

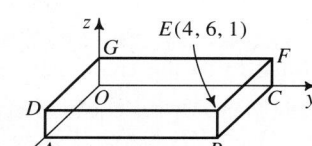

8. Repeat Exercise 7 with $E(5, 7, 2)$.
   a. $z = 2$ **b.** $x = 5$ **c.** $y = 7$ **d.** $y = 0$

Find a vector perpendicular to the plane whose equation is given. Answers may vary.

9. $3x + 4y + 6z = 12$ (3, 4, 6) 　　　　 10. $3x − 5y + 4z = 0$ (3, −5, 4)
11. $x + y = 4$ (1, 1, 0) 　　　　　　　　　 12. $x − z = 0$ (1, 0, −1)
13. $z = 1$ (0, 0, 1) 　　　　　　　　　　　 14. $y = −3$ (0, 1, 0)
16. $x − 4y + 2z = 7$

In Exercises 15–18, find a Cartesian equation of the plane described.
　　　　　　　　　　　　　　　　　　　 $2x + 3y + 5z = 44$
15. Vector $(2, 3, 5)$ is perpendicular to the plane that contains point $A(3, 1, 7)$.

16. Vector $(1, −4, 2)$ is perpendicular to the plane that contains point $A(3, 0, 2)$. See above.

17. Vector $(0, 0, 1)$ is perpendicular to the plane that contains point $A(1, 4, 5)$. $z = 5$

18. Vector $(2, 0, 3)$ is perpendicular to the plane that contains point $A(3, 8, −2)$. $2x + 3z = 0$

19. Consider the points $A(2, 2, 2)$ and $B(4, 6, 8)$.
   a. Find a Cartesian equation of the plane that is perpendicular to $\overline{AB}$ at its midpoint M. $x + 2y + 3z = 26$
   b. Show that the point $P(2, 0, 8)$ satisfies your answer to part (a). $2 + 2(0) + 3(8) = 26$
   c. Show that $PA = PB$. See below.

20. Consider the points $A(−2, 4, 1)$ and $B(0, 2, −1)$.
   a. Find a Cartesian equation of the plane that is perpendicular to $\overline{AB}$ at its midpoint M. $x − y − z = −4$
   b. Find two points in this plane and show that each of these points is equidistant from A and B.

19. c. $PA = \sqrt{(2 − 2)^2 + (2 − 0)^2 + (2 − 8)^2} = 2\sqrt{10} = \sqrt{(4 − 2)^2 + (6 − 0)^2 + (8 − 8)^2} = PB$

Error Analysis

Students may have difficulty sketching the planes in Written Exercises 1–6. Have them first determine the intercepts and then sketch the plane containing them.

Suggested Assignments

Comprehensive
Day 1: 455/1, 9, 11, 13, 15, 19, 21
Day 2: 456/23, 27, 29, 31, 33, 37, 39

Supplementary Materials

Alternative Assessment, 37
Student Resource Guide, 112–114

23.

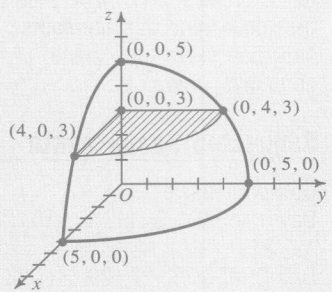

25. a. The angle opposite α in the quadrilateral is supplementary to α (since the other two angles add to 180°). This angle is also supplementary to θ. Hence $\alpha = \theta$ since they are each supplementary to the same angle.
b. This is the formula for the cosine of the angle between two vectors.

$$2^2 + 1^2 + 2^2 = 9$$

21. a. Show that the point $A(2, 1, 2)$ is on the sphere $x^2 + y^2 + z^2 = 9$.
b. Write a Cartesian equation of the plane tangent to the sphere at $A(2, 1, 2)$. (*Hint:* $\overrightarrow{OA} \perp$ the plane.) $2x + y + 2z = 9$

22. Find an equation of the plane tangent to the sphere
$$(x - 1)^2 + (y - 1)^2 + (z - 1)^2 = 49$$
at the point $(7, -1, 4)$. $6x - 2y + 3z = 56$

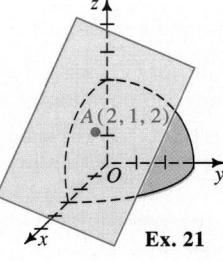

Ex. 21

B 23. The plane $z = 3$ intersects the sphere $x^2 + y^2 + z^2 = 25$ in a circle. Illustrate this with a sketch and find the area of the circle. 16π

24. The plane $y = 5$ intersects the sphere $x^2 + y^2 + z^2 = 169$ in a circle. Illustrate this with a sketch and find the area of the circle. 144π

Exercises 25–32 deal with the angle θ between two planes. This angle is defined to be the angle between vectors perpendicular to the planes.

25. Plane M_1 has equation $a_1x + b_1y + c_1z = d_1$. Plane M_2 has equation $a_2x + b_2y + c_2z = d_2$.
a. Explain why, in the diagram, $\theta = \alpha$.
b. Explain why
$$\cos \theta = \cos \alpha = \frac{(a_1, b_1, c_1) \cdot (a_2, b_2, c_2)}{|(a_1, b_1, c_1)||(a_2, b_2, c_2)|}.$$

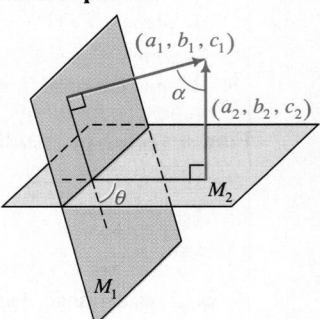

26. To the nearest tenth of a degree, find the measure of the angle between the planes $2x + 2y - z = 3$ and $x + 2y + z = 5$. Use the formula in Exercise 25(b). 47.1°

27. To the nearest tenth of a degree, find the measure of the angle between the planes $2x + 2y - z = 3$ and $4x - 3y + 2z = 5$. Use the formula in Exercise 25(b). 90°

In Exercises 28–33, you will need to know that two planes are perpendicular (parallel) if vectors perpendicular to the planes are perpendicular (parallel).

28. a. Are the planes $3x + 4y + 2z = 5$ and $2x - y - z = 3$ perpendicular? Yes
b. Are the planes $4x - 5y + 6z = 0$ and $3x - 2z = 7$ perpendicular? Yes

29. a. Name a vector perpendicular to the plane $2x + 3y - z = 6$. $(2, 3, -1)$
b. Is the vector in your answer to part (a) perpendicular to the plane $2x + 3y - z = 12$? Yes
c. Sketch both planes on a single set of axes. Are the two planes parallel? Yes
d. Is the plane $4x + 6y - 2z = 4$ parallel to the first plane? Yes

30. Which of the following planes are perpendicular and which are parallel?

M_1: $3x + 2y - z = 6$ $M_1 \parallel M_2$

M_2: $6x + 4y - 2z = 8$ $M_1 \perp M_3$

M_3: $4x - 2y + 8z = 7$ $M_2 \perp M_3$

31. Given plane M with equation $4x + y - 3z = 10$, find a Cartesian equation of a plane **(a)** parallel to M and **(b)** perpendicular to M.

32. Repeat Exercise 31 if plane M has equation $5x - 2y + z = 2$.

33. a. What is a direction vector of the line $(x, y, z) = (3, 1, 4) + t(4, -5, 2)$?

b. Explain why this line is parallel to the plane $2x + 2y + z = 7$.

Note the clever use of parallel and perpendicular planes in this building.

Exercises 34–36 deal with the angle α between a line and a plane.

34. The angle α between a line l and a plane M is the acute angle between the line and the projection of the line onto the plane.

a. Suppose that the line has direction vector (p, q, r) and that (a, b, c) is perpendicular to the plane. Show that

$$\sin \alpha = \frac{(a, b, c) \cdot (p, q, r)}{|(a, b, c)||(p, q, r)|}.$$

b. Find the angle between the line $(x, y, z) = (2, 1, 1) + t(0, 1, -1)$ and the plane $x + 2y - 2z = 6$ to the nearest tenth of a degree. 70.5°

35. a. Explain why the line $(x, y, z) = (8, 9, 10) + t(3, 4, 5)$ is perpendicular to the plane $3x + 4y + 5z = 10$.

b. Where does the line intersect the plane? (2, 1, 0)

36. a. Explain why the line $(x, y, z) = (4, 2, 1) + t(2, 3, 4)$ is perpendicular to the plane $2x + 3y + 4z = 76$.

b. Where does the line intersect the plane? (8, 8, 9)

Exercises 37–39 deal with the distance between a point and a plane.

37. The purpose of this exercise is to find the distance between the point $(3, 1, 5)$ and the plane $2x + 2y + z = 4$.

a. Find a vector equation of the line that contains the point $(3, 1, 5)$ and is perpendicular to the plane $2x + 2y + z = 4$. $(x, y, z) = (3, 1, 5) + t(2, 2, 1)$

b. Where does the line intersect the plane? (1, -1, 4)

c. Find the distance between the point $(3, 1, 5)$ and your answer to part (b). 3

Vectors and Determinants   **457**

Additional Answers
Written Exercises

31–33. a. Answers may vary. Examples are given.

31. a. $4x + y - 3z = 0$
b. $3x + 4z = 0$

32. a. $5x - 2y + z = 1$
b. $2x + 5y = 1$

33. a. $(4, -5, 2)$
b. $(4, -5, 2) \cdot (2, 2, 1) = 0$. Therefore the direction vector of the line is perpendicular to $(2, 2, 1)$. Since $(2, 2, 1)$ is also perpendicular to the plane, the line and the plane are parallel.

Let $\begin{vmatrix} a & b \\ c & d \end{vmatrix}$ represent the expression $ad - bc$. That is, evaluate such an expression by taking the difference of the products of the diagonals in the indicated order.

1. Evaluate $\begin{vmatrix} 3 & 2 \\ 5 & 7 \end{vmatrix}$. 11

2. Evaluate $\begin{vmatrix} 5 & 7 \\ 3 & 2 \end{vmatrix}$. Compare your answer to that for Exercise 1. What seems to happen to the value of the expression when the rows are reversed? -11; the value is multiplied by -1.

3. Can the value of $\begin{vmatrix} a & b \\ c & d \end{vmatrix}$ be zero? Explain your answer. Yes; if $ad = bc$. For ex., $\begin{vmatrix} 1 & 2 \\ 3 & 6 \end{vmatrix} = 0$.

4. Show that if the rows are identical, then the value of the expression is zero. $\begin{vmatrix} a & b \\ a & b \end{vmatrix} = ab - ba = 0$

Motivating the Section

Point out that students have solved systems of equations since their elementary algebra courses. Determinants may help simplify finding the solution to a system of n equations in n unknowns. Defining and evaluating determinants is the topic of this section.

C 38. Use the method of Exercise 37 to prove that the distance between the point (x_0, y_0, z_0) and the plane $ax + by + cz + d = 0$ is given by
$$\frac{|ax_0 + by_0 + cz_0 + d|}{\sqrt{a^2 + b^2 + c^2}}.$$

39. In a plane, you are given point (x_0, y_0) and the line $ax + by + c = 0$. Conjecture a formula for the distance between the point and the line. (*Hint:* The formula is similar to that in Exercise 38.) $\dfrac{|ax_0 + by_0 + c|}{\sqrt{a^2 + b^2}}$

■ Determinants and Their Applications

12-7 Determinants

Objective   *To define and evaluate determinants.*

The expression $\begin{vmatrix} a_1 & a_2 \\ b_1 & b_2 \end{vmatrix}$ is called a 2×2 **determinant** (2 rows and 2 columns). To find the value of this determinant:

take the product of these numbers ⟶
then subtract the product of these numbers. ⟶

Thus, the value of a 2×2 determinant is $a_1b_2 - a_2b_1$.

Example 1 $\begin{vmatrix} 3 & 4 \\ 2 & 7 \end{vmatrix} = (3)(7) - (2)(4) = 13$

The general 3×3 determinant is shown at the right. Each *element*, or number, in this determinant is associated with a 2×2 determinant called its *minor*. The **minor** of an element is the determinant that remains when you cross out the row and column containing that element.

$\begin{vmatrix} a_1 & a_2 & a_3 \\ b_1 & b_2 & b_3 \\ c_1 & c_2 & c_3 \end{vmatrix}$

The minor of element a_2 is $\begin{vmatrix} a_1 & a_2 & a_3 \\ b_1 & b_2 & b_3 \\ c_1 & c_2 & c_3 \end{vmatrix} \longrightarrow \begin{vmatrix} b_1 & b_3 \\ c_1 & c_3 \end{vmatrix}$.

You can evaluate a 3×3 determinant by expanding by the minors of the elements in any row or column. Consider the position of each element in the original determinant and find its corresponding position on the checkerboard pattern at the right. This pattern is used in adding and subtracting the product of the row (column) elements with their corresponding minors.

$\begin{vmatrix} + & - & + \\ - & + & - \\ + & - & + \end{vmatrix}$

Example 2 Evaluate:
$$\begin{vmatrix} 5 & 2 & 8 \\ 3 & 4 & 1 \\ 7 & -1 & 6 \end{vmatrix}$$

Solution 1 Expand by minors of the first row.

$$\begin{vmatrix} 5 & 2 & 8 \\ 3 & 4 & 1 \\ 7 & -1 & 6 \end{vmatrix} = 5 \begin{vmatrix} 4 & 1 \\ -1 & 6 \end{vmatrix} - 2 \begin{vmatrix} 3 & 1 \\ 7 & 6 \end{vmatrix} + 8 \begin{vmatrix} 3 & 4 \\ 7 & -1 \end{vmatrix}$$

$$= 5(25) - 2(11) + 8(-31)$$

$$= -145$$

Solution 2 Alternatively, you can expand by minors of the second column.

$$\begin{vmatrix} 5 & 2 & 8 \\ 3 & 4 & 1 \\ 7 & -1 & 6 \end{vmatrix} = -2 \begin{vmatrix} 3 & 1 \\ 7 & 6 \end{vmatrix} + 4 \begin{vmatrix} 5 & 8 \\ 7 & 6 \end{vmatrix} - (-1) \begin{vmatrix} 5 & 8 \\ 3 & 1 \end{vmatrix}$$

$$= -2(11) + 4(-26) + 1(-19)$$

$$= -145$$

Activity

The determinant in Example 2 was evaluated (1) by minors of the first row and (2) by minors of the second column.
a. Expand this determinant by minors of the second row.
b. Expand this determinant by minors of the third column.

Examples 1 and 2 have shown how to evaluate 2×2 and 3×3 determinants with paper and pencil. For $n \times n$ determinants with $n \geq 4$, it is easier to use certain graphing calculators or computer software. Without such technology, pencil and paper calculations can be made easier by transforming the determinant into an equivalent one in which a row (column) has several zero elements. To do this, multiply any row (column) by a constant and add the resulting products to the elements of another row (column).

$$\begin{vmatrix} 2 & 5 & 3 & 7 \\ 4 & 1 & 0 & 2 \\ 3 & 2 & 2 & 1 \\ -2 & 8 & 1 & 0 \end{vmatrix}$$
— Add $(-3 \times$ row 4$)$ to row 1. $\longrightarrow$
— Add $(-2 \times$ row 4$)$ to row 3. $\longrightarrow$
$$\begin{vmatrix} 8 & -19 & 0 & 7 \\ 4 & 1 & 0 & 2 \\ 7 & -14 & 0 & 1 \\ -2 & 8 & 1 & 0 \end{vmatrix}$$

Once you have a row (column) with several zeros, you can use a 4×4 checkerboard pattern of $(+)$ and $(-)$ signs. Since column 3 in the determinant at the right above now contains three zeros, we will expand by its minors as shown at the top of the following page.

1. Show that:
$$\begin{vmatrix} ka & kb \\ kc & kd \end{vmatrix} = k^2 \begin{vmatrix} a & b \\ c & d \end{vmatrix}$$

$$\begin{vmatrix} ka & kb \\ kc & kd \end{vmatrix} =$$
$$ka(kd) - kc(kb) =$$
$$k^2(ad - bc) =$$
$$k^2 \begin{vmatrix} a & b \\ c & d \end{vmatrix}.$$

2. Show that:
$$\begin{vmatrix} a & b & c \\ d & e & f \\ g & h & i \end{vmatrix} = - \begin{vmatrix} d & e & f \\ a & b & c \\ g & h & i \end{vmatrix}$$

$$\begin{vmatrix} a & b & c \\ d & e & f \\ g & h & i \end{vmatrix} =$$
$$a \begin{vmatrix} e & f \\ h & i \end{vmatrix} - b \begin{vmatrix} d & f \\ g & i \end{vmatrix} +$$
$$c \begin{vmatrix} d & e \\ g & h \end{vmatrix} =$$
$$a(ei - fh) - b(di - fg) + c(dh - eg);$$

expanding $- \begin{vmatrix} d & e & f \\ a & b & c \\ g & h & i \end{vmatrix}$

by minors of the second row gives

$$-\left[-a \begin{vmatrix} e & f \\ h & i \end{vmatrix} + b \begin{vmatrix} d & f \\ g & i \end{vmatrix} - \right.$$
$$\left. c \begin{vmatrix} d & e \\ g & h \end{vmatrix} \right] = a \begin{vmatrix} e & f \\ h & i \end{vmatrix} -$$
$$b \begin{vmatrix} d & f \\ g & i \end{vmatrix} + c \begin{vmatrix} d & e \\ g & h \end{vmatrix} =$$
$$a(ei - fh) - b(di - fg) + c(dh - eg).$$

$$+0\begin{vmatrix}4&1&2\\7&-14&1\\-2&8&0\end{vmatrix}-0\begin{vmatrix}8&-19&7\\7&-14&1\\-2&8&0\end{vmatrix}+0\begin{vmatrix}8&-19&7\\4&1&2\\-2&8&0\end{vmatrix}-1\begin{vmatrix}8&-19&7\\4&1&2\\7&-14&1\end{vmatrix}$$

The first three products above are 0, and the 3×3 determinant in the fourth product can be evaluated directly as in Example 2. Alternatively, that determinant can be transformed so that it has a row or column with two zeros.

CLASS EXERCISES

Evaluate each determinant.

1. $\begin{vmatrix}2&3\\4&5\end{vmatrix}$ -2 **2.** $\begin{vmatrix}8&5\\4&4\end{vmatrix}$ 12 **3.** $\begin{vmatrix}3&-5\\7&10\end{vmatrix}$ 65 **4.** $\begin{vmatrix}-5&4\\-4&-3\end{vmatrix}$ 31

5. Complete, expanding by minors of the top row.

$$\begin{vmatrix}3&2&4\\5&7&1\\8&6&9\end{vmatrix}=\frac{3}{?}\begin{vmatrix}7&1\\6&9\end{vmatrix}-\frac{2}{?}\begin{vmatrix}5&1\\8&9\end{vmatrix}+\frac{4}{?}\begin{vmatrix}5&7\\8&6\end{vmatrix}$$

6. Complete, expanding by minors of the third column.

$$\begin{vmatrix}3&2&4\\5&7&1\\8&6&9\end{vmatrix}=4\begin{vmatrix}5&7\\?&?\\?&?\\8&6\end{vmatrix}-1\begin{vmatrix}3&2\\?&?\\?&?\\8&6\end{vmatrix}+9\begin{vmatrix}3&2\\?&?\\?&?\\5&7\end{vmatrix}$$

7. Evaluate mentally: $\begin{vmatrix}1&2&3\\4&5&6\\0&0&1\end{vmatrix}$ -3

WRITTEN EXERCISES

Evaluate each determinant.

A **1.** $\begin{vmatrix}2&7\\8&4\end{vmatrix}$ -48 **2.** $\begin{vmatrix}-3&5\\-7&2\end{vmatrix}$ 29 **3.** $\begin{vmatrix}25&125\\75&250\end{vmatrix}$ -3125

4. Show that $k \cdot \begin{vmatrix}a&b\\c&d\end{vmatrix}=\begin{vmatrix}ka&kb\\c&d\end{vmatrix}=\begin{vmatrix}a&b\\kc&kd\end{vmatrix}$.

5. Show that $\begin{vmatrix}a&b\\c&d\end{vmatrix}=\begin{vmatrix}a-c&b-d\\c&d\end{vmatrix}$. Then evaluate $\begin{vmatrix}387&411\\385&410\end{vmatrix}$. 435

6. True or false? $\begin{vmatrix}a&b\\c&d\end{vmatrix}=\begin{vmatrix}a-kc&b-kd\\c&d\end{vmatrix}$ for every real number k. True

460 *Chapter Twelve*

Problem Solving

Ask students what properties of determinants are implied by Written Exercises 5 and 6. (You can subtract a row or a multiple of a row from another row without changing the value of the determinant.)

Mathematical Note

The properties of determinants given in Written Exercises 5 and 6 can be used to help reduce a determinant to one in which there are several zeros in a row or column so that the determinant can be evaluated more easily.

Additional Answers Written Exercises

4. $k\begin{vmatrix}a&b\\c&d\end{vmatrix}=k(ad-bc)$;

$\begin{vmatrix}ka&kb\\c&d\end{vmatrix}=kad-kbc=$

$k(ad-bc);\begin{vmatrix}a&b\\kc&kd\end{vmatrix}=$

$a(kd)-(kc)b=$

$k(ad-bc)$

5. $\begin{vmatrix}a&b\\c&d\end{vmatrix}=ad-bc$;

$\begin{vmatrix}a-c&b-d\\c&d\end{vmatrix}=$

$(a-c)d-c(b-d)=$

$ad-cd-bc+cd=$

$ad-bc=\begin{vmatrix}a&b\\c&d\end{vmatrix}$

Suggested Assignments

Comprehensive
460/1, 4, 5, 7, 11, 13

Supplementary Materials

Alternative Assessment, 37
Precalculus Plotter Plus, 66

Evaluate.

7. $\begin{vmatrix} 4 & -7 & 3 \\ 2 & 0 & 0 \\ 5 & 1 & 6 \end{vmatrix}$ $\overset{90}{}$

8. $\begin{vmatrix} -1 & 3 & 2 \\ 4 & 0 & 1 \\ 1 & 5 & 0 \end{vmatrix}$ $\overset{48}{}$

9. $\begin{vmatrix} 1 & -3 & 4 \\ 0 & 1 & 1 \\ 5 & -2 & 3 \end{vmatrix}$ $\overset{-30}{}$

10. $\begin{vmatrix} 1 & 2 & 3 \\ 2 & 4 & 6 \\ 17 & 18 & 19 \end{vmatrix}$ $\overset{0}{}$

B   11. *Discussion*   In Exercise 10, one row is a multiple of another row and the value of the determinant is zero. Is this a coincidence? Justify your answer. No

12. Evaluate: $\begin{vmatrix} 2 & -4 & 7 & 3 \\ 0 & 5 & 1 & -2 \\ -2 & 1 & 0 & -3 \\ 0 & -6 & 4 & 2 \end{vmatrix}$ 32

13. Evaluate: $\begin{vmatrix} -2 & 1 & 5 & 0 \\ 3 & 4 & -2 & 1 \\ 0 & 0 & 1 & 2 \\ 1 & 0 & 0 & -3 \end{vmatrix}$ 78

C   14. Consider the linear equations $a_1x + b_1y = c_1$, and $a_2x + b_2y = c_2$.

   a. Solve these equations for x and y in terms of the constants.

   b. Show that your answer can be written as

$$x = \frac{\begin{vmatrix} c_1 & b_1 \\ c_2 & b_2 \end{vmatrix}}{\begin{vmatrix} a_1 & b_1 \\ a_2 & b_2 \end{vmatrix}} \quad \text{and} \quad y = \frac{\begin{vmatrix} a_1 & c_1 \\ a_2 & c_2 \end{vmatrix}}{\begin{vmatrix} a_1 & b_1 \\ a_2 & b_2 \end{vmatrix}}.$$

12-8 Applications of Determinants

Objective   *To use determinants to solve algebraic and geometric problems.*

Determinants can be used to solve n equations in n variables. The method used is called *Cramer's rule* and is illustrated in Examples 1 and 2. If $n \geq 4$, this method involves laborious paper and pencil calculations. As a result, a special calculator or computer is a better tool when $n \geq 4$.

Example 1   Use Cramer's rule to solve: $3x + 5y = 7$
                                      $4x + 9y = 11$

Solution   **Step 1**   Calculate three determinants.

$$D = \begin{vmatrix} 3 & 5 \\ 4 & 9 \end{vmatrix} = 7 \qquad D \text{ consists of the coefficients of } x \text{ and } y.$$

D_x is formed from D by replacing the first column of D by the constant terms. D_y is formed from D by replacing the second column of D by the constant terms.

$$D_x = \begin{vmatrix} 7 & 5 \\ 11 & 9 \end{vmatrix} = 8 \qquad D_y = \begin{vmatrix} 3 & 7 \\ 4 & 11 \end{vmatrix} = 5$$

      Step 2 $x = \dfrac{D_x}{D} = \dfrac{8}{7}$ and $y = \dfrac{Dy}{D} = \dfrac{5}{7}$.

Vectors and Determinants   **461**

Using Technology

You may wish to have students use computers or advanced scientific calculators to check their answers when evaluating determinants, as in Exercises 7–10, 12, and 13.

Problem Solving

Ask students how interchanging two rows of a determinant affects the value of the determinant. (The value of the changed determinant is the opposite of the value of the unchanged determinant.)

Teaching Notes, p. 418F

Warm-Up Exercises

Evaluate each determinant.

1. $\begin{vmatrix} 4 & -3 \\ -1 & -5 \end{vmatrix}$ -23 2. $\begin{vmatrix} 3 & -2 \\ 1 & 2 \end{vmatrix}$ 8

3. $\begin{vmatrix} a & a \\ 1 & a \end{vmatrix}$ $a^2 - a$

4. $\begin{vmatrix} 5 & 0 & -1 \\ 0 & 3 & -2 \\ 4 & -1 & 1 \end{vmatrix}$ 17

5. $\begin{vmatrix} 2 & -1 & 3 \\ -3 & -2 & 1 \\ 1 & 3 & 2 \end{vmatrix}$ -42

Motivating the Section

Point out that determinants have a number of uses in mathematics, such as solving a system of linear equations and finding the areas of geometric figures.

Additional Examples

1. Solve: $ax + by = 0$
$bx + ay = b - a$

$D = \begin{vmatrix} a & b \\ b & a \end{vmatrix} = a^2 - b^2$

$D_x = \begin{vmatrix} 0 & b \\ b-a & a \end{vmatrix} =$
$-b(b-a)$

$D_y = \begin{vmatrix} a & 0 \\ b & b-a \end{vmatrix} =$
$a(b-a)$

$x = \dfrac{-b(b-a)}{a^2 - b^2} =$
$\dfrac{-b(b-a)}{(a-b)(a+b)} = \dfrac{b}{a+b}$

$y = \dfrac{a(b-a)}{a^2-b^2} = -\dfrac{a}{a+b}$

2. A right rectangular solid has edges $\overline{OA}$, $\overline{OB}$, and $\overline{OC}$, where $A = (2, -1, -2)$, $B = (0, 2, -1)$, and $C = (5, 2, 4)$.
  a. Find the volume by using the formula $V = lwh$.
  b. Find the volume by using determinants.

a. $l = OA =$
$\sqrt{2^2 + (-1)^2 + (-2)^2} =$
$3; w = OB = \sqrt{5};$
$h = OC = \sqrt{45} = 3\sqrt{5};$
$V = 3 \cdot \sqrt{5} \cdot 3\sqrt{5} =$
45 cubic units

b. $V = \begin{vmatrix} 2 & -1 & -2 \\ 0 & 2 & -1 \\ 5 & 2 & 4 \end{vmatrix} =$
$2(8 + 2) + 5(1 + 4) =$
45 cubic units

Example 2   Solve the system of equations: $3x - y + 2z = 4$
$2x + 3y - z = 14$
$7x - 4y + 3z = -4$

Solution   **Step 1**   Write and evaluate four determinants.

$D = \begin{vmatrix} 3 & -1 & 2 \\ 2 & 3 & -1 \\ 7 & -4 & 3 \end{vmatrix}$ $D_x = \begin{vmatrix} 4 & -1 & 2 \\ 14 & 3 & -1 \\ -4 & -4 & 3 \end{vmatrix}$

$D_y = \begin{vmatrix} 3 & 4 & 2 \\ 2 & 14 & -1 \\ 7 & -4 & 3 \end{vmatrix}$ $D_z = \begin{vmatrix} 3 & -1 & 4 \\ 2 & 3 & 14 \\ 7 & -4 & -4 \end{vmatrix}$

Notice that the numbers in D are the coefficients of x, y, and z in the three equations. To form D_x, the first column of D is replaced by the constants on the right sides of the equations. Similarly, the second and third columns of D are replaced by the constants to get D_y and D_z, respectively. Using the methods of Section 12-7 to expand D, D_x, D_y, and D_z, we get:

$$D = -30, \quad D_x = -30, \quad D_y = -150, \quad D_z = -90$$

Step 2   Therefore,

$$x = \frac{D_x}{D} = \frac{-30}{-30} = 1, \; y = \frac{D_y}{D} = \frac{-150}{-30} = 5, \; z = \frac{D_z}{D} = \frac{-90}{-30} = 3$$

Thus, the solution is $(x, y, z) = (1, 5, 3)$.

Determinants are useful in geometry as well as in algebra. We can use determinants to find area and volume, as shown below.

Area

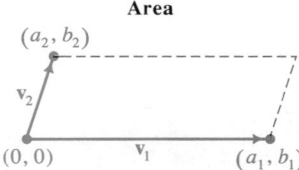

Volume

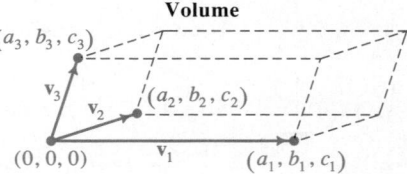

The area of a parallelogram with sides determined by $\mathbf{v}_1$ and $\mathbf{v}_2$ is the absolute value of:

$\begin{vmatrix} a_1 & b_1 \\ a_2 & b_2 \end{vmatrix}$

The volume of a parallelepiped with edges determined by $\mathbf{v}_1$, $\mathbf{v}_2$, and $\mathbf{v}_3$ is the absolute value of:

$\begin{vmatrix} a_1 & b_1 & c_1 \\ a_2 & b_2 & c_2 \\ a_3 & b_3 & c_3 \end{vmatrix}$

Example 3 Find the area of the triangle with vertices $P(1, 2)$, $Q(3, 6)$, and $R(6, 1)$.

Solution $\overrightarrow{PQ}$ and $\overrightarrow{PR}$ determine the sides of triangle PQR and parallelogram $PQSR$.

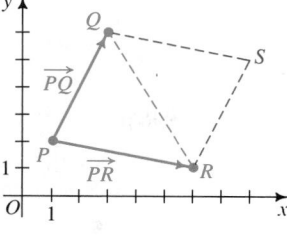

$$\overrightarrow{PQ} = (3, 6) - (1, 2) = (2, 4)$$
$$\overrightarrow{PR} = (6, 1) - (1, 2) = (5, -1)$$

Area of $\triangle PQR = \dfrac{1}{2} \cdot$ area of $\square PQSR$

Area $= \dfrac{1}{2} \cdot$ absolute value of $\begin{vmatrix} 2 & 4 \\ 5 & -1 \end{vmatrix}$

$\qquad = \dfrac{1}{2} |-22| = 11$

The minus sign in the next-to-last step indicates that P, Q, and R have a negative (clockwise) orientation.

CLASS EXERCISES

1. When determinants are used to solve the system of equations $\begin{matrix} 3x + 4y = 7 \\ 5x + 6y = 8 \end{matrix}$,

then $x = \dfrac{|?|}{|?|}$ and $y = \dfrac{|?|}{|?|}$.

2. When determinants are used to solve the system of equations
$3x + y - 2z = 4$
$2x - y + 4z = 1$, then $x = \dfrac{|?|}{|?|}$, $y = \dfrac{|?|}{|?|}$, and $z = \dfrac{|?|}{|?|}$.
$x + 2y + 7z = 3$

3. What happens when you try to use Cramer's rule to find the solutions for the
system of equations $\begin{matrix} 2x + y = 3 \\ 2x + y = 5 \end{matrix}$? The denominator $= 0$. There is no sol. because the lines are parallel.

4. What is the area of the parallelogram with sides determined by vectors
$\mathbf{v}_1 = (1, 3)$ and $\mathbf{v}_2 = (2, 4)$? 2

WRITTEN EXERCISES

**In Exercises 1–4, solve each system of equations by using Cramer's rule.
Then, sketch the graphs of the two equations and label their intersection with
the common solution of the equations.**

A **1.** $5x - 4y = 1$ $\quad$(1, 1)
$\quad$ $3x + 2y = 5$

2. $5x - 2y = 11$ $\quad$(3, 2)
$\quad$ $x + 3y = 9$

3. $3x + 2y = -1$ $\quad$(1, -2)
$\quad$ $2x - y = 4$

4. $7x + y = 7$ $\quad$(0, 7)
$\quad$ $-x + 2y = 14$

Vectors and Determinants **463**

**Additional Answers
Class Exercises**

1. $x = \dfrac{\begin{vmatrix} 7 & 4 \\ 8 & 6 \end{vmatrix}}{\begin{vmatrix} 3 & 4 \\ 5 & 6 \end{vmatrix}}$,

$y = \dfrac{\begin{vmatrix} 3 & 7 \\ 5 & 8 \end{vmatrix}}{\begin{vmatrix} 3 & 4 \\ 5 & 6 \end{vmatrix}}$

2. $x = \dfrac{\begin{vmatrix} 4 & 1 & -2 \\ 1 & -1 & 4 \\ 3 & 2 & 7 \end{vmatrix}}{\begin{vmatrix} 3 & 1 & -2 \\ 2 & -1 & 4 \\ 1 & 2 & 7 \end{vmatrix}}$,

$y = \dfrac{\begin{vmatrix} 3 & 4 & -2 \\ 2 & 1 & 4 \\ 1 & 3 & 7 \end{vmatrix}}{\begin{vmatrix} 3 & 1 & -2 \\ 2 & -1 & 4 \\ 1 & 2 & 7 \end{vmatrix}}$,

$z = \dfrac{\begin{vmatrix} 3 & 1 & 4 \\ 2 & -1 & 1 \\ 1 & 2 & 3 \end{vmatrix}}{\begin{vmatrix} 3 & 1 & -2 \\ 2 & -1 & 4 \\ 1 & 2 & 7 \end{vmatrix}}$,

Suggested Assignments

Comprehensive
Day 1: 463/1, 3, 7, 9, 11
Day 2: 464/13, 15, 17, 21, 22

Supplementary Materials

Alternative Assessment, 38

Solve each system of equations by using Cramer's rule.

5. $ax + by = 1$ $\left(\dfrac{1}{a+b}, \dfrac{1}{a+b}\right)$
   $bx + ay = 1$

6. $ax + by = c$ $\left(\dfrac{6c}{5a}, -\dfrac{c}{5b}\right)$
   $3ax - 2by = 4c$

7. **a.** Draw the graphs of $9x - 6y = 3$ and $6x - 4y = 10$. Then solve the system of equations by using determinants. No sol.

   b. If $ax + by = c$ and $dx + ey = f$ have parallel line graphs, what determinant must be equal to zero? $\begin{vmatrix} a & b \\ d & e \end{vmatrix}$

8. **a.** Draw the graphs of $12x + 8y = -4$ and $6x + 4y = -2$. Then solve the system of equations by using determinants. Sol. is line $3x + 2y = -1$

   b. If $ax + by = c$ and $dx + ey = f$ have the same graph, what determinants must be equal to zero? $D = D_x = D_y = 0$

In Exercises 9–14, find the area of each figure, given points $P(4, 3)$, $Q(7, -1)$, $R(2, 3)$, $S(-3, 6)$, $T(-5, 4)$, and $V(-2, -5)$.

9. $\triangle PQR$ 4

10. $\triangle PQS$ $\dfrac{19}{2}$

11. $\triangle RST$ 8

12. $\triangle PSV$ 37

13. $\square$ with sides $\overline{PR}$, $\overline{PS}$ 6

14. $\square$ with sides $\overline{PS}$, $\overline{PT}$ 20

15. **Discussion** If the "area" of $\triangle LMN$ is zero, what can you conclude about points L, M, and N? They are collinear.

16. **Discussion** If a parallelepiped with sides $\overline{DE}$, $\overline{DF}$, and $\overline{DG}$ has "zero volume," what can you say about points D, E, F, and G? They are coplanar.

In Exercises 17–19, solve each system of equations by using Cramer's rule.

B 17. $x - 2y + 3z = 2$ (3, 2, 1)
   $2x - 3y + z = 1$
   $3x - y + 2z = 9$

18. $4x - 2y + 3z = 2$ (1, 1, 0)
   $5x - 6y + 2z = -1$
   $3x + 4y - 5z = 7$

19. $3x - 2y + z = 7$ (2, 0, 1)
   $2x + y - 3z = 1$
   $x + 2y + 2z = 4$

In Exercises 20–22, find the volume of each solid, given points $D(5, 1, 0)$, $E(3, 1, 4)$, $F(0, 2, -1)$, $G(5, 2, 0)$, and $H(3, 1, 3)$.

20. Parallelepiped with edges $\overline{DE}$, $\overline{DF}$, and $\overline{DG}$ 22

21. Parallelepiped with edges $\overline{EF}$, $\overline{EG}$, and $\overline{EH}$ 5

C 22. Pyramid with edges $\overline{DF}$, $\overline{DG}$, and $\overline{DH}$ (*Hint:* The volume of a pyramid is one-sixth the volume of a parallelepiped.) $\dfrac{17}{6}$

12-9 Determinants and Vectors in Three Dimensions

Objective   *To define and apply the cross product.*

The unit vectors in the positive x, y, and z directions are often denoted by the symbols $\mathbf{i}$, $\mathbf{j}$, and $\mathbf{k}$. This means that if point $P = (3, 5, 6)$, then

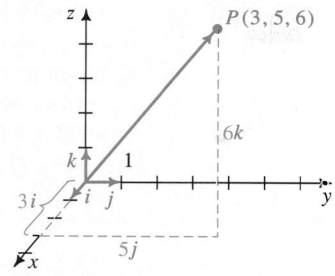

$$\overrightarrow{OP} = (3, 5, 6)$$
$$= 3\mathbf{i} + 5\mathbf{j} + 6\mathbf{k}.$$

This representation of a vector is useful in defining a new vector operation, the *cross product* of two vectors.

If $\mathbf{v}_1 = (a_1, b_1, c_1)$ and $\mathbf{v}_2 = (a_2, b_2, c_2)$, then the **cross product** of $\mathbf{v}_1$ and $\mathbf{v}_2$, denoted by $\mathbf{v}_1 \times \mathbf{v}_2$, is found by evaluating:

$$\begin{vmatrix} \mathbf{i} & \mathbf{j} & \mathbf{k} \\ a_1 & b_1 & c_1 \\ a_2 & b_2 & c_2 \end{vmatrix}$$

For example, if $\mathbf{v}_1 = (2, 3, 4)$ and $\mathbf{v}_2 = (1, 0, 5)$, then:

$$\mathbf{v}_1 \times \mathbf{v}_2 = \begin{vmatrix} \mathbf{i} & \mathbf{j} & \mathbf{k} \\ 2 & 3 & 4 \\ 1 & 0 & 5 \end{vmatrix} = \begin{vmatrix} 3 & 4 \\ 0 & 5 \end{vmatrix}\mathbf{i} - \begin{vmatrix} 2 & 4 \\ 1 & 5 \end{vmatrix}\mathbf{j} + \begin{vmatrix} 2 & 3 \\ 1 & 0 \end{vmatrix}\mathbf{k}$$
$$= 15\mathbf{i} - 6\mathbf{j} - 3\mathbf{k}, \text{ or } (15, -6, -3)$$

The following list gives some of the major properties of the cross product. The figure at the right below illustrates properties 1–3.

Properties of the Cross Product

1. $\mathbf{u} \times \mathbf{v}$ is perpendicular to $\mathbf{u}$ and to $\mathbf{v}$.
2. $\mathbf{v} \times \mathbf{u} = -(\mathbf{u} \times \mathbf{v})$, that is, $\mathbf{v} \times \mathbf{u}$ and $\mathbf{u} \times \mathbf{v}$ have opposite directions.
3. $|\mathbf{u} \times \mathbf{v}| = |\mathbf{u}||\mathbf{v}| \sin \theta$, where θ is the angle between $\mathbf{u}$ and $\mathbf{v}$. Geometrically, the magnitude of $\mathbf{u} \times \mathbf{v}$ is the area of the parallelogram formed by $\mathbf{u}$ and $\mathbf{v}$.
4. $\mathbf{u} \times (\mathbf{v} + \mathbf{w}) = (\mathbf{u} \times \mathbf{v}) + (\mathbf{u} \times \mathbf{w})$
5. $\mathbf{u}$ is parallel to $\mathbf{v}$ if and only if $\mathbf{u} \times \mathbf{v} = \mathbf{0}$.

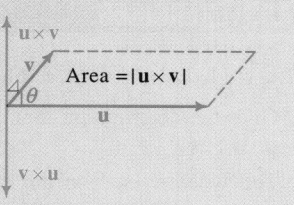

Area $= |\mathbf{u} \times \mathbf{v}|$

Teaching Notes, p. 418F

Warm-Up Exercises

1. Find a Cartesian equation of the plane that contains the point $P(1, -1, 2)$ and to which the vector $(2, 3, 1)$ is perpendicular.
   $2x + 3y + z = 1$

2. Show that the point $Q(-5, 4, -1)$ lies on the plane described in Ex. 1.
   $2(-5) + 3(4) - 1 = 1$

   Evaluate each determinant. Leave answers in terms of $\mathbf{i}$, $\mathbf{j}$, and $\mathbf{k}$.

3. $\begin{vmatrix} \mathbf{i} & \mathbf{j} & \mathbf{k} \\ 1 & 1 & 0 \\ 2 & 0 & -1 \end{vmatrix}$ 4. $\begin{vmatrix} \mathbf{i} & \mathbf{j} & \mathbf{k} \\ 4 & 3 & -1 \\ 3 & -2 & 5 \end{vmatrix}$

   $-\mathbf{i} + \mathbf{j} - 2\mathbf{k}$ $13\mathbf{i} - 23\mathbf{j} - 17\mathbf{k}$

Motivating the Section

The cross product of two vectors is useful in determining the equation of a plane and the area of geometric figures.

Additional Examples

1. Let $\mathbf{u} = (1, 0, -2)$ and $\mathbf{v} = (-5, 2, 4)$.
   a. Use the dot product to verify that $\mathbf{u} \times \mathbf{v}$ and $\mathbf{v} \times \mathbf{u}$ are perpendicular to both $\mathbf{u}$ and $\mathbf{v}$.
   b. Find $\sin \theta$ where θ is the angle between $\mathbf{u}$ and $\mathbf{v}$.

   a. $(1, 0, -2) \cdot (4, 6, 2) = 4 + 0 - 4 = 0$
   $(1, 0, -2) \cdot (-4, -6, -2) = -4 + 0 + 4 = 0$
   $(-5, 2, 4) \cdot (4, 6, 2) = -20 + 12 + 8 = 0$
   $(-5, 2, 4) \cdot (-4, -6, -2) = 20 - 12 - 8 = 0$

(Continues on next page)

Additional Examples cont.

b. $\sin \theta = \dfrac{|u \times v|}{|u||v|} =$

$\dfrac{|(4, 6, 2)|}{\sqrt{5} \cdot \sqrt{45}} = \dfrac{2\sqrt{14}}{15}$

2. Given points $A(0, 1, -3)$, $B(2, -2, -9)$, and $C(-2, 0, -1)$, find the area of $\triangle ABC$ **(a)** by using a cross product and **(b)** by using a dot product.

a. $\overrightarrow{AB} = (2, -3, -6)$ and $\overrightarrow{AC} = (-2, -1, 2)$.

Area $= \frac{1}{2}|\overrightarrow{AB} \times \overrightarrow{AC}|$

$\overrightarrow{AB} \times \overrightarrow{AC} =$

$\begin{vmatrix} i & j & k \\ 2 & -3 & -6 \\ -2 & -1 & 2 \end{vmatrix} =$

$-12i + 8j - 8k$

$\frac{1}{2}|\overrightarrow{AB} \times \overrightarrow{AC}| =$

$\frac{1}{2}\sqrt{144 + 64 + 64} =$

$\frac{1}{2} \cdot 4\sqrt{17} = 2\sqrt{17}$

b. $\cos A = \dfrac{\overrightarrow{AB} \cdot \overrightarrow{AC}}{|\overrightarrow{AB}||\overrightarrow{AC}|} =$

$\dfrac{(2, -3, -6) \cdot (-2, -1, 2)}{\sqrt{49} \cdot \sqrt{9}} =$

$-\dfrac{13}{21}$

$\sin A = \sqrt{1 - \left(-\dfrac{13}{21}\right)^2} =$

$\dfrac{4\sqrt{17}}{21}$

Area $= \frac{1}{2}|\overrightarrow{AB}||\overrightarrow{AC}| \sin A =$

$\frac{1}{2}\sqrt{49}\sqrt{9} \cdot \dfrac{4\sqrt{17}}{21} =$

$2\sqrt{17}$

Example

a. Find a nonzero vector perpendicular to the plane containing the points $P(1, 0, 3)$, $Q(2, 5, 0)$, and $R(3, 1, 4)$.

b. Find an equation of the plane containing P, Q, and R.

c. Find the area of the parallelogram with sides formed by $\overrightarrow{PQ}$ and $\overrightarrow{PR}$.

Solution

Without actually plotting the points, make a quick sketch.

a. By property 1, $\overrightarrow{PQ} \times \overrightarrow{PR}$ is a vector perpendicular to the plane containing P, Q, and R. Since $\overrightarrow{PQ} = (1, 5, -3)$ and $\overrightarrow{PR} = (2, 1, 1)$

$\overrightarrow{PQ} \times \overrightarrow{PR} = \begin{vmatrix} i & j & k \\ 1 & 5 & -3 \\ 2 & 1 & 1 \end{vmatrix} = 8i - 7j - 9k$, or $(8, -7, -9)$.

b. Since $(8, -7, -9)$ is perpendicular to the plane, an equation of the plane is $8x - 7y - 9z = d$ (by the theorem in Section 12-6). To find d, substitute $P(1, 0, 3)$ (or Q or R) in the equation:

$$8 \cdot 1 - 7 \cdot 0 - 9 \cdot 3 = d$$
$$d = -19$$

Thus, $8x - 7y - 9z = -19$ is an equation of the plane. You can check this by substituting the coordinates of P, Q, and R into the equation.

c. By property 3, the area of the parallelogram is given by:

$$|\overrightarrow{PQ} \times \overrightarrow{PR}| = |(8, -7, -9)| = \sqrt{8^2 + (-7)^2 + (-9)^2} = \sqrt{194}$$

The area is $\sqrt{194}$ square units, or about 13.9 square units.

If a parallelogram lies in the xy-plane, its area can be found by using 2×2 determinants, as in Section 12-8. Otherwise, the method shown in Example 2(c) above can be used.

CLASS EXERCISES

Let $u = (1, 2, 3)$, $v = (4, 5, 6)$, and $w = (7, 8, 9)$.

1. Give the determinant that can be used to evaluate each cross product.
  a. $u \times v$ **b.** $v \times w$

2. Tell whether each of the following is a vector or a scalar.
  a. $u \times v$ Vector   **b.** $u \cdot v$ Scalar   **c.** $u + v$ Vector
  d. $u \cdot (u + v)$ Scalar   **e.** $u \times (u + v)$ Vector   **f.** $u \cdot (u \times v)$ Scalar

3. *Discussion* Why does the following equation make no sense? ($\times$ denotes the cross product) $u \cdot (v \times w) = (u \cdot v) \times (u \cdot w)$ Right side shows the cross product of two real numbers, which makes no sense.

---

466   *Chapter Twelve*

WRITTEN EXERCISES

1. $(-1, -5, 4); (1, 5, -4);$ yes

In Exercises 1–8, let $\mathbf{u} = (4, 0, 1)$, $\mathbf{v} = (5, -1, 0)$, and $\mathbf{w} = (-3, 1, -2)$.

A

1. Calculate $\mathbf{v} \times \mathbf{u}$ and $\mathbf{u} \times \mathbf{v}$. Do your results agree with property 2? See above.

2. Calculate $\mathbf{v} \times \mathbf{w}$. Use your answer to find $\mathbf{w} \times \mathbf{v}$ immediately. $(2, 10, 2); (-2, -10, -2)$

3. Verify that $\mathbf{u} \times \mathbf{v}$ is perpendicular to $\mathbf{u}$ and to $\mathbf{v}$.

4. a. Use a cross product to find a vector perpendicular to the plane of $\mathbf{u}$ and $\mathbf{w}$.

    b. Use a dot product to find a vector perpendicular to $\mathbf{w}$.

5. Find the area of the parallelogram determined by $\mathbf{u}$ and $\mathbf{v}$. $\sqrt{42}$

6. Find the area of the triangle determined by $\mathbf{u}$ and $\mathbf{w}$. $\frac{1}{2}\sqrt{42}$

7. Verify that $\mathbf{u} \times (\mathbf{v} + \mathbf{w}) = (\mathbf{u} \times \mathbf{v}) + (\mathbf{u} \times \mathbf{w})$.

8. Verify that $\mathbf{u} \cdot (\mathbf{v} + \mathbf{w}) = \mathbf{u} \cdot \mathbf{v} + \mathbf{u} \cdot \mathbf{w}$.

In Exercises 9–12, (a) find a vector perpendicular to the plane determined by P, Q, and R, (b) find a Cartesian equation of the plane determined by P, Q, and R, and (c) find the area of $\triangle PQR$.

9. $P(1, 1, 0)$, $Q(-1, 0, 2)$, $R(2, 1, 1)$ **10.** $P(0, 0, 0)$, $Q(3, -1, 2)$, $R(4, 5, -2)$

11. $P(2, 3, -4)$, $Q(2, 1, 0)$, $R(3, 3, -2)$ **12.** $P(0, 2, 3)$, $Q(1, 1, 1)$, $R(2, 1, 3)$

B

13. Angle θ is between vectors $\mathbf{u} = (1, 2, 2)$ and $\mathbf{v} = (4, 3, 0)$. **a.** $\frac{\sqrt{5}}{3}$ **b.** $\frac{2}{3}$

    a. Find $\sin \theta$ by using property 3 of the cross product.

    b. Find $\cos \theta$ by using property 5 of the dot product (page 448).

    c. Verify that $\sin^2 \theta + \cos^2 \theta = 1$.   **13. c.** $\left(\frac{\sqrt{5}}{3}\right)^2 + \left(\frac{2}{3}\right)^2 = 1$

14. Repeat Exercise 13 for the vectors $\mathbf{u} = (1, 2, 1)$ and $\mathbf{v} = (1, 1, 2)$. See below.

15. Given $A(11, 17, 0)$, $B(-8, 12, 0)$, and $C(25, 13, 0)$, find the area of $\triangle ABC$ by:

    a. using determinants. (Since all the points lie in the xy-plane, use the method of Example 3 in Section 12-8.)

    b. using cross products.   **14. a.** $\frac{\sqrt{11}}{6}$ **b.** $\frac{5}{6}$ **c.** $\left(\frac{\sqrt{11}}{6}\right)^2 + \left(\frac{5}{6}\right)^2 = 1$

    c. using Hero's formula. (See page 6.)

    d. using the dot product and the formula: Area $= \frac{1}{2}ab \sin \theta$.

    Do your answers in parts (a), (b), (c), and (d) agree?

C

16. Let $\mathbf{u} = (a_1, b_1, c_1)$, $\mathbf{v} = (a_2, b_2, c_2)$, and $\mathbf{w} = (a_3, b_3, c_3)$.

    a. Prove that $|\mathbf{u} \times \mathbf{v}|^2 = |\mathbf{u}|^2|\mathbf{v}|^2 - (\mathbf{u} \cdot \mathbf{v})^2$.

    b. Prove the five properties of the cross product listed on page 465.

////COMPUTER EXERCISE

Write a computer program that will print a Cartesian equation of the plane determined by three noncollinear points P, Q, and R, when you input the coordinates of P, Q, and R. (*Hint:* Use the cross product.)

Chapter Summary

1. A *vector* is a quantity that has both magnitude and direction. Vectors can be represented geometrically by arrows or algebraically by ordered pairs of numbers. The ordered pairs can be in *polar form* or in *component form*. Rules for vector operations in components are given on page 427.

2. **a.** If P is any point on line AB, then
$$\overrightarrow{OP} = \overrightarrow{OA} + t\overrightarrow{AB}$$
is a *vector equation* of line AB, and $\overrightarrow{AB}$ is a *direction vector* of the line.

   b. A *vector equation* of an object moving with constant velocity along the line is:

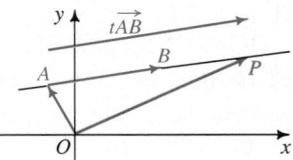

$$\text{new position} = \text{initial position} + t(\text{constant velocity})$$
$$(x, y) = (x_0, y_0) + t(a, b)$$

   c. From the vector equation of the line, we can obtain the two parametric equations $x = x_0 + ta$ and $y = y_0 + tb$. These give the x- and y- coordinates of points on the line in terms of the parameter t (t is often considered to represent time).

3. **a.** If $\mathbf{u} = (x_1, y_1)$ and $\mathbf{v} = (x_2, y_2)$, then the *dot product* of $\mathbf{u}$ and $\mathbf{v}$ is:
$$\mathbf{u} \cdot \mathbf{v} = x_1 x_2 + y_1 y_2$$
Some properties of the dot product are listed on page 442.

   b. The angle θ between two vectors $\mathbf{u}$ and $\mathbf{v}$ is given by the formula
$$\cos \theta = \frac{\mathbf{u} \cdot \mathbf{v}}{|\mathbf{u}| \, |\mathbf{v}|}.$$

   c. Vectors $\mathbf{u}$ and $\mathbf{v}$ are *perpendicular* if and only if $\mathbf{u} \cdot \mathbf{v} = 0$.

   d. Nonzero vectors $\mathbf{u}$ and $\mathbf{v}$ are *parallel* if and only if $\mathbf{u} = k\mathbf{v}$.

4. Vectors in three-dimensional space are ordered triples of numbers. The basic properties of these vectors, summarized in the box on page 448, are analogous to those of vectors in the plane.

5. The graph of $ax + by + cz = d$ is a plane (provided a, b and c are not all equal to zero). The nonzero vector (a, b, c) is perpendicular to this plane at the point (x_0, y_0, z_0), where $d = ax_0 + by_0 + cz_0$.

6. A *determinant* is a square array of numbers evaluated as shown on page 458. The value of an n by n determinant can be computed by expanding by the minors of one of its rows or columns.

7. *Cramer's Rule* involves using determinants to solve a system of n linear equations in n variables. Determinants can also be used to find areas and volumes.

8. The *cross product* of the vectors $\mathbf{v}_1 = (a_1, b_1, c_1)$ and $\mathbf{v}_2 = (a_2, b_2, c_2)$ is defined on page 465. Properties of the cross product are given in the box on page 465. The cross product is useful in finding an equation of a plane, as shown in the example on page 466.

468 *Chapter Twelve*

vector, direction, magnitude (p. 419)   dot product (p. 441)
component form (p. 426)   angle between two vectors (p. 442)
vector equation, direction vector (p. 432)   Cartesian equation of a plane (p. 453)
parametric equations (p. 433)   Cramer's rule (pp. 461–462)

Chapter Test

Additional Answers
Chapter Test

2. a. $-\mathbf{u}$
   b. $\mathbf{v} - \mathbf{u}$
   c. $\frac{1}{3}\mathbf{v} - \frac{1}{3}\mathbf{u}$
   d. $\frac{2}{3}\mathbf{u} + \frac{1}{3}\mathbf{v}$

5. a. $\mathbf{v} = (1, 4);\ |\mathbf{v}| = \sqrt{17}$
   b. $x = -2 + t,$
       $y = -2 + 4t$
   c. $t = \frac{3}{2},\ \left(-\frac{1}{2}, 4\right);\ t = 5,$
       $(3, 18)$

6. $\mathbf{u}$ and $\mathbf{v}$ are opposite in direction. Since $\cos\theta = \frac{\mathbf{u}\cdot\mathbf{v}}{|\mathbf{u}||\mathbf{v}|}$, $\mathbf{u}\cdot\mathbf{v} = |\mathbf{u}||\mathbf{v}|\cos\theta$. If $\mathbf{u}\cdot\mathbf{v} = -|\mathbf{u}||\mathbf{v}|$, then $\cos\theta = -1$. Therefore, $\theta = 180°$ and $\mathbf{u}$ and $\mathbf{v}$ are opposite in direction.

1. An object is pulled due south by a force $\mathbf{F}_1$ of 5 N, and due east by a   12-1
force $\mathbf{F}_2$ of 12 N. Find the direction and magnitude of $\mathbf{F}_3 = \mathbf{F}_1 + \mathbf{F}_2$. 112.6°; 13 N

2. In the diagram, $AT:TC = 1:2$. Express each of
the following in the form $r\mathbf{u} + s\mathbf{v}$:
a. $\overrightarrow{AB}$ **b.** $\overrightarrow{AC}$ **c.** $\overrightarrow{AT}$ **d.** $\overrightarrow{BT}$

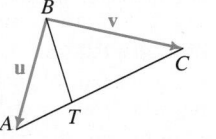

3. A plane is on a course of 320° at a speed of   12-2
480 mi/h. What are the north-south and east-
west components of its velocity vector? 367.7 mi/h N; 308.5 mi/h W

4. Given the points $A(1, 3)$, $B(0, -3)$, and $C(5, 0)$, find the coordinates
of a point $D(x, y)$ such that quadrilateral $ABCD$ is a parallelogram. (6, 6)

5. a. Find the velocity and speed of an object that moves with constant   12-3
velocity so that its position at time t is $(x, y) = (-2, -2) + t(1, 4)$.
b. Find a pair of parametric equations of the path of the object.
c. When and where does it intersect the parabola $y = 4x^2 - 6x$?

6. Writing Let $\mathbf{u}$ and $\mathbf{v}$ be nonzero vectors such that $\mathbf{u}\cdot\mathbf{v} = -|\mathbf{u}||\mathbf{v}|$.   12-4
Describe the relationship between $\mathbf{u}$ and $\mathbf{v}$. Explain your reasoning.

7. Given the points $A(1, 3)$, $B(-2, 7)$, and $C(0, -6)$, find the measure of
$\angle A$ to the nearest tenth of a degree. 136.8°

8. If $A = (4, 0, -5)$ and $B = (-6, -2, 7)$, find **(a)** the length of $\overline{AB}$ and   12-5
(b) the midpoint of $\overline{AB}$. $2\sqrt{62}$; $(-1, -1, 1)$ **9. a.** $(x, y, z) = (2, 1, 7) + t(4, -1, -3)$

9. Line L has equation $(x, y, z) = (0, -1, 6) + t(4, -1, -3)$.
a. Write a vector equation for the line through $(2, 1, 7)$ parallel to L. See above.
b. Where does line L intersect the xz-plane? $(-4, 0, 9)$

10. Find an equation of the plane that is tangent to the sphere with equa-   12-6
tion $x^2 + (y + 1)^2 + (z - 3)^2 = 33$ at the point $(2, -3, -2)$. $2x - 2y - 5z = 20$

11. Evaluate: **a.** $\begin{vmatrix} 2 & 0 & 3 \\ 4 & 2 & -2 \\ -3 & 0 & 1 \end{vmatrix}$ 22   **b.** $\begin{vmatrix} 0 & 0 & 0 & 5 \\ -1 & 2 & 0 & 0 \\ 3 & 1 & 2 & 0 \\ -4 & 0 & 0 & -1 \end{vmatrix}$ 80   **13. b.** $\frac{\sqrt{2948}}{2} \approx 27.1$ 12-7

12. Use Cramer's Rule to solve: $\begin{aligned} 7x + 4y &= 19 \\ 3x - 10y &= 14 \end{aligned}$ $\left(3, -\frac{1}{2}\right)$ 12-8
13. a. Example: $(-16, 46, -24)$

13. $A(0, 1, -3)$, $B(-6, 1, 1)$, and $C(4, 5, 2)$ determine a plane. Find: 12-9
a. A vector perpendicular to the plane   **b.** The area of $\triangle ABC$ See above.

Vectors and Determinants   **469**

Uniform Circular Motion

Did you ever wonder why clothes end up pressed against the outside of the washer when the cycle is completed? Did you ever watch a line of skaters make a whip by whirling in a circle? Why does the person on the outside move so much faster than the others? The answer lies in the concepts of force and uniform circular motion. Most likely, you're already familiar with force and motion along a straight line. Here are a few experiments to get a better sense of force and motion in a circle.

Materials:

a piece of string 1 m in length   two pairs of heavy socks
a circular paper plate   a marble or a small ball

Make Investigations.

- First, tie one pair of socks securely to one end of the string. Hold the other end of the string and *carefully* swing the socks slowly around your head in a circle. Be sure to note the pull of the weight on your arm.
- Now, roll up the string a little and try to swing the socks so that they stay up in a circle again. Does using a shorter length of string seem to increase or decrease the speed needed?
- Repeat the experiment using both pairs of socks at once. Why do you think the pull on your arm is greater?
- What do you think would happen if you let go of the string? *Don't try it!* To get the same effect, try this experiment instead. Take a paper plate and roll a marble or a small ball around the inside of the rim. Next, cut out a circular sector equal to one-fourth of the circle, and roll the ball again. In which direction does the ball roll when it reaches the cut edge?

Do a Vector Analysis.

As you know from the examples and exercises concerning force and velocity in Chapter 12, you can use vectors to help explain the results of your investigations.

- Begin by drawing a circle with radius r to depict the path of your object. Think of the radius as the string that holds the object in the path. Let's use polar coordinates (r, θ) with $\theta = \beta t$, where t = elapsed time in seconds and β = the angular speed in radians per second. For simplicity, let $r = 1$. Also, choose a smaller value for β, such as 0.5, and a value for t. Show the position of your object in a diagram by drawing the vector

$$\mathbf{r} = (r \cos \beta t, r \sin \beta t).$$

- Motion, of course, involves a change in position. Suppose that you wanted to find the rate of change of the object's position exactly at time t. This rate, called the *instantaneous velocity* of the object at time t (which will be discussed in Chapter 20) can be represented by the vector

$$\mathbf{v} = (-\beta r \sin \beta t, \beta r \cos \beta t).$$

Before locating **v**, recall the result of your paper plate investigation. We hope you found that at the instant the marble or small ball reached the cut edge, it rolled ahead in a direction that was perpendicular to the radius of the plate. Therefore, **v** must be perpendicular to **r**. Why? Draw **v**, placing its initial point at the terminal point of **r**. Use the dot product to show that **v** and **r** are perpendicular.

■ The *speed* of an object is defined as the magnitude of the velocity. Show that the magnitude of **v** is a constant. Does this result surprise you?

■ In Chapter 20, you'll also learn about acceleration. *Acceleration* is the rate of change of an object's velocity. For example, you can actually feel acceleration when a car in which you're riding suddenly slows down. The instantaneous acceleration of your object can be represented on your diagram by the vector

$$\mathbf{a} = (-\beta^2 r \cos \beta t, -\beta^2 r \sin \beta t) = -\beta^2 \mathbf{r}.$$

Draw **a** and notice its direction. The vector **a** is called *centripetal acceleration* since it always points towards the center of the circular motion.

■ One of the most useful formulas in physics is

$$\mathbf{F} = m\mathbf{a},$$

where **F** represents force, **a** represents acceleration, and the scalar m is the mass of the object. Let $m = 1.5$, and locate **F** on your diagram. The vector **F** is called the *centripetal force*. Why? It corresponds, for example, to the force you exert on the object by holding it in place with the string. Now, draw $-\mathbf{F}$, which is equal and opposite to **F**. The vector $-\mathbf{F}$ is called the *centrifugal force*. Why? This force corresponds to the pull that you feel on your arm. Apply the formulas to show that using a heavier weight or a faster angular speed results in a greater pull. Explain why centrifugal force sends the ketchup to the top of the bottle when swung in a big circle.

■ Here's an extra challenge: Use concentric circles and the given formulas to illustrate why using a shorter string at the same angular speed decreases the pull on your arm.

Report Your Results.

In this project there are a lot of little experiments and related analyses to perform, so be sure to try them all and answer each of the questions asked. As a suggestion, you might organize your report so that each investigation is followed by a corresponding vector analysis. Remember to state your choices for β, t, and m, and to clearly label each diagram. Include enough diagrams to provide a careful illustration of each idea. Don't forget to try the extra challenge!

Chapter 13 Sequences and Series

Overview

This chapter introduces finite and infinite sequences and series. With regard to sequences, students learn to identify arithmetic and geometric sequences and to define them explicitly (Section 13-1) and recursively (Section 13-2); students also find the limit, if it exists, of an infinite sequence (Section 13-4). With regard to series, students find the sums of finite series (Section 13-3) and infinite series (Section 13-5); students also use sigma notation to represent a series (Section 13-6). The chapter concludes with the topic of mathematical induction, which students use to prove statements for all positive integers.

Objectives

13-1 To identify an arithmetic or geometric sequence and find a formula for its nth term.
13-2 To use sequences defined recursively to solve problems.
13-3 To find the sum of the first n terms of arithmetic or geometric series.
13-4 To find or estimate the limit of an infinite sequence or to determine that the limit does not exist.
13-5 To find the sum of an infinite geometric series.
13-6 To represent series using sigma notation.
13-7 To use mathematical induction to prove that a statement is true.

Supplementary Resources

Tests, pp. 42–43
Alt. Assess., pp. 39–43, 69
Activities, pp. 36–37
Using Tech., pp. 33–34
Student Res. Guide, pp. 121–129

Software

Precalculus Plotter Plus
Activities, pp. 68–69, 70, 71–72, 73–74, 75
Sequence Experiment, Function Plotter

Pacing Guide

Section	13-1	13-2	13-3	13-4	13-5	13-6	13-7	Review	Test	Total
Standard (days)	1	2	1	1	—	—	—	1	1	7
Comprehensive (days)	2	2	2	1	2	2	2	1	1	15

Teaching Notes

13-1 | **pages 473–479**

Presenting the Section

For most students the notion of a sequence as a set of numbers arranged in a definite order is intuitively clear. However, you might want to introduce this section by giving some everyday examples of sequences, such as student rank and grade-point average, finishing position and time in a race, and so on. Ask students for other examples and then move to more specifically mathematical ones; for example, the sequence of positive even integers. See if students can give you mathematical examples before they turn to the ones in the book.

Emphasize that the common difference for an arithmetic sequence can be found by subtracting any term from its following term, while the common ratio for a geometric sequence can be found by dividing any term by its preceding term.

Using Technology

If students are having trouble seeing the relationship between the common difference of an arithmetic sequence and the slope of the line containing its graph, visualization may be aided by using a graphing calculator.

13-2 | **pages 479–485**

Presenting the Section

This section introduces recursive definitions for sequences. Explicit definitions from the preceding section and recursive definitions from this section are then contrasted and related. Students need to become comfortable with both forms of definitions and should become good at the skill of changing from one type of definition to the other.

You may wish to begin by giving many examples of both explicit and recursive definitions for various sequences. Make certain that students understand the difference and that they are comfortable with the notation.

Next, you might have the students practice finding recursive definitions from explicit ones and explicit definitions from recursive ones. Examples:

Explicit Definition	Recursive Definition
$O_n = 2n - 1$	$O_1 = 1$ $O_{n+1} = O_n + 2$
$L_n = n - 1$	$L_1 = 0$ $L_{n+1} = L_n + 1$
$S_n = n^2$	$S_1 = 1$ $S_{n+1} = S_n + (2n + 1)$

Using Technology

A computer program is used to solve the problem in the example on pages 480 and 481. Note the use of the greatest integer function, INT, in the program. This is needed because the actual value of P contains a decimal portion that must be truncated for the population to be a whole number.

Review

Solving systems of equations can be reviewed when finding explicit definitions of sequences for recursive ones. For example, given the recursive definition $Z_1 = 2$ and $Z_{n+1} = Z_n + (2n + 1)$, you can find an explicit definition as follows:

1. Begin by writing down the terms of the sequence and then subtracting each term from the following term. This gives a sequence of *first-order* differences:

   Given sequence: 2, 5, 10, 17, 26, 37, . . .
   1st-order differences: 3, 5, 7, 9, 11, . . .

2. Use the sequence of first-order differences to obtain the sequence of second-order differences. if this is a constant sequence, then stop; if not, continue finding sequences of higher-order differences until a constant sequence is obtained. For the given sequence we have:

   2nd-order differences: 2, 2, 2, 2, . . .

3. The order of the constant sequence gives the highest power of the variable in the explicit definition (which is assumed to be a polynomial). For the given sequence, the explicit definition is quadratic: $Z_n = an^2 + bn + c$.

4. Set up and solve a system of equations to obtain the explicit definition. Since the first three ordered pairs for $Z_n = an^2 + bn + c$ are $(n, Z_n) = (1, 2), (2, 5),$ and $(3, 10)$, we have:

$$n = 1: \quad a(1) + b(1) + c = 2$$
$$n = 2: \quad a(4) + b(2) + c = 5$$
$$n = 3: \quad a(9) + b(3) + c = 10$$

Solving the system, we find $a = 1$, $b = 0$, and $c = 1$. Thus, the explicit definition for the given sequence is $Z_n = n^2 + 1$.

13-3 | pages 486–492

Presenting the Section

The main idea for students to grasp in this section is what a series is and what its relationship to a sequence is. You may wish to begin by giving a recursive definition for a particular sequence. For example, $F_1 = 3$ and $F_{n+1} = F_n + 4$. Then list the terms of this sequence:

1st term: $F_1 = 3$
2nd term: $F_2 = 7$
3rd term: $F_3 = 11$
4th term: $F_4 = 15$

You might ask the students to give the explicit definition. ($F_n = 4n - 1$) Then list the sums of the terms.

1 term: $\quad 3$
2 terms: $\quad 3 + 7 = 10$
3 terms: $\quad 3 + 7 + 11 = 21$
4 terms: $\quad 3 + 7 + 11 + 15 = 36$

These are the series and sums for the related finite sequences. You might ask for the next series and sum. ($3 + 7 + 11 + 15 + 19 = 55$)

Communication

You might ask students how they would formally define a series, based on the above example. (Using the term ''sum'' is the key to a good definition.) You might also want to note the difference between a series (which is an *indicated* sum) and the actual sum of the series. Then have students read page 486 and discuss the ideas presented.

Using Technology

Discuss the BASIC program given for Example 1 on pages 486 and 487. You might point out that the variable S in the program is a *running sum*; that is, the value of S is found not by adding together all 20 terms of the series at one time, but rather by adding the terms one at a time to an accumulated sum.

13-4 | pages 493–499

Presenting the Section

This section introduces the notion of a limit, which is a principal concept of calculus. Students need to acquire an intuitive understanding of this concept. Therefore, you may wish to begin by considering numerous examples without much emphasis on rigor. Drawing graphs should greatly help students visualize limits.

Assessment

You might want to review this material by giving students statements like the following ones to classify as true or false.

1. If the limit of a convergent sequence is a positive number, then all terms of the sequence are positive numbers. (False)
2. A term of a convergent sequence can be equal to the limit of the sequence. (True)
3. If the terms of a sequence are alternately positive and negative numbers, then the sequence cannot be convergent. (False)

In discussing statements such as these, students should give reasons for their answers. The process of verbalizing their thinking should help them master the concepts and make it easier for you to assess their level of understanding.

Using Technology

You might want to have students guess the limits of various sequences. Then have them use calculators to compute quickly the terms corresponding to large values of n. By doing this, they should easily be able to ''see'' where a sequence is heading.

Presenting the Section

The primary focus of this section is the concept of an infinite geometric series. Students need to know what it is, how to determine its sum, and how to apply this information to various types of problems. You might want to start with an infinite geometric sequence and build a corresponding series from it. It is important to point out that because it is impossible to sum the infinitely many terms of an infinite series, we look at the finite partial sums instead; if the sequence of partial sums has a limit, then we *define* the sum of the infinite series to be this limit.

Review

Example 3 on page 502 is a good illustration of the use of a geometric series, but students should also be familiar with a simpler approach (learned in algebra) to finding the fraction for a repeating decimal: In Example 3, $n = 0.\overline{45}$, so:

$$\begin{array}{r} 100n = 45.\overline{45} \\ - \quad 1n = 0.\overline{45} \\ \hline 99n = 45 \\ n = \frac{45}{99} = \frac{5}{11} \end{array}$$

Cooperative Learning

Ask each group of "paired partners" to consider questions like the ones below. Have them mutually decide upon their answers and write down reasons. Then match each pair of students with another pair and have the foursome share what they have written. The group of four should decide which reason is best for each question and submit a copy of their final four selections to you.

1. If two infinite geometric series have the same finite sum, are they the same series? (Not necessarily)
2. Can the sum of a convergent infinite geometric series be less than its first term? (Yes)
3. Can the sum of an infinite geometric series in which $a > 0$ and $|r| < 1$ be negative? (No)
4. Does an infinite geometric series exist for which $a = 6$ and $S = \frac{2}{3}$? (No)

Presenting the Section

You might want to begin this section by introducing sigma notation and having students practice just "plugging in" the index values and expanding the notation. Emphasize that the index of summation should match the variable in the expression. For example $\sum\limits_{k=1}^{3} 2j$ means $2j + 2j + 2j$, since there is no k in the summand for which to substitute. Also note that $\sum\limits_{k=1}^{3} 7 = 7 + 7 + 7 = 21$.

You might want to show how to change the upper and lower limits of summation in sigma notation without changing the series that the sigma notation represents. To do so, add a constant to both the lower and upper limits of summation and then replace the variable in the expression by the sum of the variable and the opposite of the constant. Here are two examples:

$$\sum_{j=4}^{6} \frac{j}{j+2} = \sum_{j=0}^{2} \frac{j+4}{j+6} \qquad \sum_{k=3}^{7} \frac{k}{k+1} = \sum_{k=1}^{5} \frac{k+2}{k+3}$$

Presenting the Section

Students will need time to be convinced that mathematical induction is a reliable technique and to become at ease with this new method of proof. You will probably need to show many examples of mathematical-induction proofs. Emphasize the two parts of the proofs. Be specific about how you want the proofs to be written. It is best not to be too formal, but do give the students some structure.

Cooperative Learning

Because students generally find proofs by mathematical induction difficult to do, have students work in small groups to discuss and, if necessary, correct their proofs. You might challenge the groups to find alternate proofs (that is, proofs that do not involve mathematical induction) for some of the statements.

13 *Sequences and Series*

Finite Sequences and Series

13-1 Arithmetic and Geometric Sequences

Objective *To identify an arithmetic or geometric sequence and find a formula for its nth term.*

A sequence is a set of numbers, called terms, arranged in a particular order. We begin with two of the simplest types of sequences: *arithmetic* and *geometric*.

Arithmetic Sequences

A sequence of numbers is called an **arithmetic sequence** if the difference of any two consecutive terms is constant. This difference is called the *common difference*. The following sequences are all arithmetic.

$$2, 6, 10, 14, 18, \ldots \text{ difference} = 4$$
$$17, 10, 3, -4, -11, -18, \ldots \text{ difference} = -7$$
$$a, a + d, a + 2d, a + 3d, a + 4d, \ldots \text{ difference} = d$$

Geometric Sequences

A sequence of numbers is called a **geometric sequence** if the ratio of any two consecutive terms is constant. This ratio is called the *common ratio*. The following sequences are all geometric.

$$1, 3, 9, 27, 81, \ldots \text{ ratio} = 3$$
$$64, -32, 16, -8, 4, \ldots \text{ ratio} = -\tfrac{1}{2}$$
$$a, ar, ar^2, ar^3, ar^4, \ldots \text{ ratio} = r$$

The same notation can be used for all types of sequences. The first term of a sequence is often denoted by t_1, the second and third terms by t_2 and t_3, and so on. The nth term of the sequence is then denoted by t_n. Some sequences can be defined by particular rules or formulas. If you have a formula for t_n in terms of n, you can find the value of any term of the sequence. For example, suppose a sequence has the formula

$$t_n = n^2 + 1.$$

Then:

$$t_1 = 1^2 + 1 = 2$$
$$t_2 = 2^2 + 1 = 5$$
$$t_3 = 3^2 + 1 = 10$$
$$t_4 = 4^2 + 1 = 17, \text{ and so on}$$

Note that this sequence is neither arithmetic nor geometric.

◀ In the building pictured at the left, the architect Harry Wolf was inspired by proportions derived from the Fibonacci sequence, producing a harmonious effect. This sequence runs: 1, 1, 2, 3, 5, 8, 13, . . . Can you find how Wolf has used these values?

Teaching Notes, p. 472B

Warm-Up Exercises

Look for a pattern and predict the next number or expression in each list.

1. 1000, 500, 250, 125, ? 62.5
2. 1, 2, 4, 7, 11, 16, ? 22
3. 1, −3, 9, −27, ? 81
4. 8, 3, −2, −7, ? −12
5. $2, 2\sqrt{2}, 4, 4\sqrt{2}$, ? 8
6. $7a + 4b, 6a + 5b, 5a + 6b, 4a + 7b$, ? $3a + 8b$

Motivating the Section

Ask students to list some sequences with which they are already familiar, such as the positive odd integers, the integral powers of 2 that are greater than 1, and the positive multiples of 3. Ask students if they can give an expression for the nth term of each sequence. (For the examples given above, the nth terms are $2n - 1$, 2^n, and $3n$.) Then point out that classifying and working with such sequences are the objectives of this section.

Assessment Note

Ask students to identify the type of sequence and the common difference or ratio for each of the following:

1. 5, 10, 20, 40, . . . (Geometric; ratio = 2)
2. 6, 1, −4, −9, . . . (Arithmetic; diff. = −5)

Formulas for the nth term of general arithmetic and geometric sequences are given below. Notice how similar these formulas are.

Arithmetic Sequence:

$$t_n = t_1 + (n - 1)d$$

To get the nth term,   start with the first term   and add the difference $n - 1$ times.

Geometric Sequence:

$$t_n = t_1 \cdot r^{(n-1)}$$

To get the nth term,   start with the first term   and multiply by the ratio $n - 1$ times.

Definition of a Sequence

By now you probably have a good intuitive idea of what a sequence is, but can you give a precise definition of a sequence? In mathematics, a **sequence** is usually defined to be a function whose domain is the set of positive integers. For example, the sequence with nth term

$$t_n = 4n - 2$$

can be thought of as the function

$$t(n) = 4n - 2, \text{ where } n \text{ is a positive integer.}$$

Since a sequence is a function, to graph a sequence such as $1, 4, 7, 10, \ldots,$ you plot the points $(1, 1), (2, 4), (3, 7), (4, 10)$. You can see how an arithmetic sequence is related to its graph by completing the following activity.

Activity

For each sequence, **(a) graph the sequence** and **(b) find the slope of the line containing the graph of the sequence.**

1. $0, 1, 2, 3, \ldots$

Slope = 1

2. $18, 14, 10, \ldots$

Slope = -4

3. $5.3, 6, 6.7, \ldots$

Slope = 0.7

The activity above demonstrates that the graph of any arithmetic sequence consists of discrete points on a line whose slope is the same as the common difference of the sequence.

Example 1   **a.** Find a formula for the nth term of the arithmetic sequence

$$3, 5, 7, \ldots.$$

b. Sketch the graph of the sequence.

Solution

a. Use the formula $t_n = t_1 + (n - 1)d$. For this sequence, t_1 is 3 and the common difference d is 2.

$$t_n = 3 + (n - 1) \cdot 2$$
$$= 2n + 1$$

b. The equation $t_n = 2n + 1$ describes a line, but t_n is defined only when n is a positive integer. The graph of the sequence consists of discrete points on the line $t = 2n + 1$.

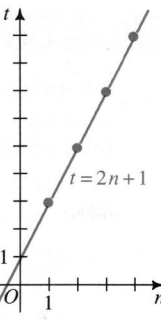

Example 2

a. Find a formula for the nth term of the geometric sequence

$$3, 4.5, 6.75, \ldots .$$

b. Sketch the graph of the sequence.

Solution

a. Use the formula $t_n = t_1 \cdot r^{(n-1)}$. In this case, t_1 is 3 and the common ratio r is 1.5.

$$t_n = 3 \cdot 1.5^{(n-1)}$$
$$= 3 \cdot \frac{1.5^n}{1.5^1}$$
$$= \frac{3}{1.5} \cdot (1.5)^n = 2 \cdot (1.5)^n$$

b. The equation $t_n = 2 \cdot (1.5)^n$ is an exponential equation whose graph has an intercept on the t-axis at 2. The graph of the sequence consists of discrete points on the exponential curve $t = 2(1.5)^n$.

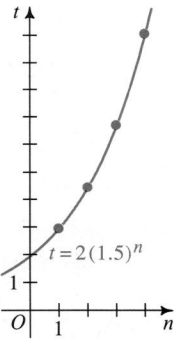

The graphs of all geometric sequences are similar to the one in Example 2. The points lie along an exponential curve whose base is the same as the common ratio of the sequence.

The graph of a sequence or the formula for its nth term can give you information about all of the terms of the sequence. Example 3 shows how you can find any term of a geometric sequence if you know two of its terms.

Example 3

In a geometric sequence, $t_3 = 12$ and $t_6 = 96$. Find t_{11}.

Solution

Substituting in the formula $t_n = t_1 \cdot r^{(n-1)}$ gives

$$t_3 = t_1 \cdot r^2 = 12 \text{ and } t_6 = t_1 \cdot r^5 = 96.$$

Thus, $t_1 = \frac{12}{r^2} = \frac{96}{r^5}$, which gives $r^3 = 8$ and $r = 2$.

Since $t_1 = 3$, $t_{11} = t_1 \cdot r^{10} = 3 \cdot (2)^{10} = 3072$.

Sequences and Series **475**

Additional Examples cont.

2. In a certain sequence, $t_2 = 2$ and $t_5 = 16$. Find t_{10} if the sequence is **(a)** arithmetic and **(b)** geometric.

a. $t_2 = 2 = t_1 + (2 - 1)d$;
$2 = t_1 + d$
$t_5 = 16 = t_1 + (5 - 1)d$;
$16 = t_1 + 4d$

$14 = 3d; \ d = \frac{14}{3}$

$t_1 = 2 - \frac{14}{3} = -\frac{8}{3}$

$t_{10} = t_1 + (10 - 1)d =$
$-\frac{8}{3} + 9\left(\frac{14}{3}\right) = \frac{118}{3}$

b. $t_2 = 2 = t_1 \cdot r^{2-1}$;
$2 = t_1 \cdot r$
$t_5 = 16 = t_1 \cdot r^{5-1}$;
$16 = t_1 \cdot r^4$

$t_1 = \frac{2}{r}; \ t_1 = \frac{16}{r^4}$

$\frac{2}{r} = \frac{16}{r^4}$

$r^3 = 8; \ r = 2$

$t_1 = \frac{2}{r} = \frac{2}{2} = 1$

$t_{10} = t_1 \cdot r^{10-1} =$
$1 \cdot 2^9 = 512$

CLASS EXERCISES

For Exercises 1–6, state whether the given sequence is arithmetic, geometric, or neither. If arithmetic, give the common difference, and if geometric, give the common ratio.

1. 3, 8, 13, 18, . . . Arithmetic; 5   **2.** 4, 8, 16, 32, . . . Geometric; 2

3. 2, 5, 10, 17, . . . Neither   **4.** 23, 17, 11, 5, . . . Arithmetic; -6

5. 27, -18, 12, -8, . . . Geometric; $-\frac{2}{3}$ **6.** 1, -3, 5, -7, . . . Neither

For Exercises 7–10, state the first four terms of the specified sequence. Then tell whether the sequence is arithmetic, geometric, or neither.

7. $t_n = 5n + 2$ 7, 12, 17, 22; arithmetic

8. $t_n = \dfrac{n+1}{n+2}$ $\frac{2}{3}$, $\frac{3}{4}$, $\frac{4}{5}$, $\frac{5}{6}$; neither

9. $t_n = 3^n$ 3, 9, 27, 81; geometric

10. $t_n = n^3$ 1, 8, 27, 64; neither

11. Discussion Does the sequence 2, 2, 2, . . . satisfy the definition of:
  a. an arithmetic sequence? Yes; $d = 0$ **b.** a geometric sequence? Yes; $r = 1$

12. Visual Thinking How does the graph of the sequence with nth term $t_n = 3n - 5$ differ from the graph of the line $y = 3x - 5$? The graph of the sequence is a set of discrete points; the graph of the line is continuous.

WRITTEN EXERCISES

Find the first four terms of the given sequence and state whether the sequence is arithmetic, geometric, or neither.

A

1. $t_n = 2n + 3$ **2.** $t_n = n^3 + 1$

3. $t_n = 3 \cdot 2^n$ **4.** $t_n = 3 - 7n$

5. $t_n = n + \dfrac{1}{n}$ **6.** $t_n = (-2)^n$

7. $t_n = (-1)^n \cdot n$ **8.** $t_n = 16 \cdot 2^{2n}$

9. $t_n = \sin \dfrac{n\pi}{2}$ **10.** $t_n = \cos n\pi$

Find a formula for t_n and sketch the graph of each arithmetic or geometric sequence.

11. 1, 4, 7, 10, . . . $-2 + 3n$

12. 8, 6, 4, 2, . . . $10 - 2n$

13. 8, 4, 2, 1, . . . $16(0.5)^n$

14. 0.3, 0.9, 2.7, 8.1, . . . $(0.1)(3^n)$

15. 30, 26, 22, 18, . . . $34 - 4n$

16. 24, -12, 6, -3, . . . $-48\left(-\dfrac{1}{2}\right)^n$

The sequence of numbers in each column of a calendar is arithmetic.

State whether the given sequence is arithmetic, geometric, or neither. Find a formula for t_n, the nth term of each sequence.

17. 17, 21, 25, 29, . . . Arith.; $13 + 4n$

18. 15, 7, -1, -9, . . . Arith.; $23 - 8n$

19. 8, 12, 18, 27, . . . Geom.; $\dfrac{16}{3}\left(\dfrac{3}{2}\right)^n$

20. 100, -50, 25, -12.5, . . . See below.

21. 1, 4, 9, 16, . . . Neither; n^2

22. $\dfrac{1}{2}, \dfrac{2}{3}, \dfrac{3}{4}, \dfrac{4}{5}, \cdots$ Neither; $\dfrac{n}{n+1}$

23. 11, 101, 1001, 10001, . . . Neither; $10^n + 1$

24. $\dfrac{2}{1}, \dfrac{3}{4}, \dfrac{4}{9}, \dfrac{5}{16}, \cdots$ Neither; $\dfrac{n+1}{n^2}$

25. $2a - 2b$, $3a - b$, $4a$, $5a + b$, . . . Arith.; $(a - 3b) + n(a + b)$

26. $\dfrac{a}{9}, \dfrac{a^2}{18}, \dfrac{a^3}{36}, \dfrac{a^4}{72}, \cdots$ Geom.; $\dfrac{2}{9}\left(\dfrac{a}{2}\right)^n$

27. $2^{2/3}, 2^{5/3}, 2^{8/3}, 2^{11/3}, \ldots$ Geom.; $2^{-1/3}(2^n)$

28. $\sqrt{2}, \sqrt[3]{2}, \sqrt[4]{2}, \sqrt[5]{2}, \ldots$ Neither; $2^{1/(n+1)}$

Find the indicated term of each arithmetic sequence.

29. $t_1 = 15$, $t_2 = 21$, $t_{20} = ?$ 129

30. $t_1 = 76$, $t_3 = 70$, $t_{101} = ?$ -224

31. $t_3 = 8$, $t_5 = 14$, $t_{50} = ?$ 149

32. $t_8 = 25$, $t_{20} = 61$, $t_1 = ?$ 4

Find the indicated term of each geometric sequence.   **20.** Geom.; $-200\left(-\dfrac{1}{2}\right)^n$

33. $t_1 = 2^{-4}$, $t_2 = 2^{-3}$, $t_{12} = ?$ 128

34. $t_1 = 2$, $t_2 = 2^{3/2}$, $t_{13} = ?$ 128

35. $t_2 = 64$, $t_5 = -8$, $t_9 = ?$ $-\dfrac{1}{2}$

36. $t_1 = 81$, $t_4 = 24$, $t_7 = ?$ $\dfrac{64}{9}$

37. How many terms are in the arithmetic sequence 18, 24, . . . , 336?   54

38. How many terms are in the arithmetic sequence 178, 170, . . . , 2?   23

B   **39.** Find the number of multiples of 7 between 30 and 300.   38

40. Find the number of multiples of 6 between 28 and 280.   42

41. How many 3-digit numbers are divisible by 4 and 6?   75

42. How many 4-digit numbers are *not* divisible by 11?   8181

43. True or false? If the sequence a, b, c is arithmetic, so is the sequence $\sin a$, $\sin b$, $\sin c$.   False

44. True or false? If the sequence a, b, c is arithmetic, then the sequence 2^a, 2^b, 2^c is geometric.   True

45. Explain why the logarithms of the terms of the geometric sequence a, ar, ar^2, ar^3, . . . form an arithmetic sequence. $t_n = \log a + (n - 1)\log r$

46. **Writing** One of the principal ideas in the work of the English economist Thomas Malthus (1766–1834) is often expressed as "food supply rises arithmetically, but population increases geometrically." Write a paragraph or two restating this idea in terms of sequences. If Malthus's idea is correct, what kind of consequences can be expected? Explain your answer.

47. Find x if the sequence 2, 8, $3x + 5$ is (a) arithmetic and (b) geometric.   3; 9

48. Find x if the sequence 4, x, $\dfrac{3}{2}x$ is (a) arithmetic and (b) geometric.   8; 6

Exercise Note

For Exercises 17–28, you might want to point out that we *assume* that the pattern in each sequence continues. For example, in Exercise 17, we assume that the next term after 29 is 33, even though the next term could be any number.

Additional Answers Written Exercises

46. Answers will vary. Example: $F_n = F_1 + (n - 1)d$ and $P_n = P_1(r)^{n-1}$. In words, this idea means that population tends to increase more rapidly than food supplies. If this is true, either population growth must be slowed to ensure adequate food supplies, or there will be an inadequate supply of food for the existing population level.

53. For example: 1, 43, 85, 127, 169, . . .

49. Find x and y if the sequence y, $2x + y$, $7y$, 20, . . . is arithmetic. $x = 3$, $y = 2$

50. Find x and y if the sequence $2y$, $2xy$, 2, $\dfrac{xy}{2}$, . . . is geometric. $x = \pm\dfrac{1}{2}$; $y = 4$

51. Physics If the half-life of an element is 2 days, what fractional amount of the element remains after 2 days? 4 days? 10 days? d days? See below.

52. Consumer Economics If P dollars is invested at 8% interest compounded quarterly, what is the value of the investment after 1 quarter? 2 quarters? q quarters? n years? $1.02P$; $(1.02)^2P$; $(1.02)^qP$; $(1.02)^{4n}P$

53. Find an arithmetic sequence none of whose terms is divisible by 2, 3, or 7. See above

54. Find an arithmetic sequence all of whose terms are multiples of 2 and 3 but not multiples of 4 or 5. For example: 6, 66, 126, 186, . . .

55. Consider the two arithmetic sequences: **51.** $\dfrac{1}{2}$; $\dfrac{1}{4}$; $\dfrac{1}{32}$; $\dfrac{1}{2^{d/2}}$

A: 3, 14, 25, . . .   B: 2, 9, 16, . . . 58, 135, 212, 289, 366

Write the first five terms of sequence A that are also terms of sequence B.

56. a. Suppose you are given the terms $t_2 = 18$ and $t_4 = 8$. Find two *different* $r = \pm\dfrac{2}{3}$; possible common ratios. Then find the two sequences having these terms. $\dfrac{81}{2}\left(\pm\dfrac{2}{3}\right)$

b. Given two terms of a geometric sequence, under what conditions will there be two different common ratios that could be used to find two sequences that have the given terms? Both term numbers odd or both term numbers even

57. Find all right triangles having sides with integral lengths that form an arithmetic sequence. Triangles with sides $3n$, $4n$, $5n$

58. Prove that there are no right triangles having sides with integral lengths that form a geometric sequence.

For Exercises 59–64, use the following definitions as needed:

The **arithmetic mean** of the numbers a and b is $\dfrac{a + b}{2}$.

The **geometric mean** of the positive numbers a and b is $\sqrt{ab}$.

59. Find the arithmetic mean and the geometric mean of:
a. 4 and 9 6.5; 6   **b.** 5 and 10 7.5; $5\sqrt{2}$

60. Assuming that a and b are positive, find x if the sequence a, x, b is: **a.** $\dfrac{a + b}{2}$
a. arithmetic   **b.** geometric $\pm\sqrt{ab}$

61. $\overline{CD}$ is the altitude to the hypotenuse of the right triangle ABC.
a. Prove that $\triangle ADC \sim \triangle ACB$ and, therefore, that $\dfrac{x}{b} = \dfrac{b}{c}$; that is, b is the geometric mean of x and c.
b. Likewise it can be shown that a is the geometric mean of _?_ and _?_ .
c. Use the result of parts (a) and (b) to prove that $a^2 + b^2 = c^2$.

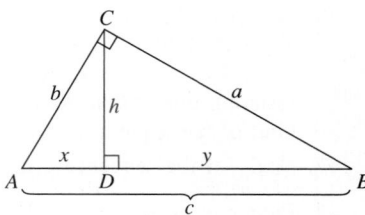

62. Refer to the figure for Exercise 61. Use similar triangles to prove that h is the geometric mean of x and y.

C 63. Prove that the arithmetic mean of two positive numbers is never less than their geometric mean.

64. Consider a circle with radius 6. Find A, the area of the inscribed equilateral triangle, and B, the area of the circumscribed equilateral triangle. Show that the geometric mean of A and B is the area of the inscribed regular hexagon.

13-2 Recursive Definitions

Objective   *To use sequences defined recursively to solve problems.*

Sometimes a sequence is defined by giving the value of t_n in terms of the preceding term, t_{n-1}. For example, consider the sequence defined by the following formulas:

$$t_1 = 3$$
$$t_n = 2t_{n-1} + 1$$

The second formula above states that the nth term is one more than twice the $(n-1)$st term. Knowing that the sequence begins with $t_1 = 3$, we can determine the first few terms of the sequence as follows:

$$t_n = 2t_{n-1} + 1 \longrightarrow \quad t_2 = 2t_1 + 1 = 2(3) + 1 = 7$$
$$t_3 = 2t_2 + 1 = 2(7) + 1 = 15$$
$$t_4 = 2t_3 + 1 = 2(15) + 1 = 31$$

The formulas $t_1 = 3$ and $t_n = 2t_{n-1} + 1$ give a *recursive definition* for the sequence 3, 7, 15, 31, ... A recursive definition consists of two parts:

1. An *initial condition* that tells where the sequence starts.
2. A *recursion equation* (or *recursion formula*) that tells how any term in the sequence is related to the preceding term.

In Section 13-1, you were given *explicit definitions* for sequences. To see the contrast between a recursive definition and an explicit definition, consider the arithmetic sequence

$$23, 20, 17, 14, \ldots .$$

Recursive definition: $t_1 = 23$ Initial condition.

$t_n = t_{n-1} - 3$ Recursion equation says that each term is 3 less than the preceding term.

Explicit definition: $t_n = 26 - 3n$ t_n is given explicitly in terms of n.

Sometimes it is possible to find an explicit definition for a sequence when you are given its recursive definition, as the activity on the next page shows.

**Additional Answers
Written Exercises**

62. Since $\angle ACD$ and $\angle B$ are both complements of $\angle DCB$, $\angle B \cong \angle ACD$. Also, $\angle ADC \cong \angle CDB$. Thus $\triangle ADC \sim \triangle CDB$ by AA Similarity and $\frac{AD}{CD} = \frac{DC}{DB}$, $\frac{x}{h} = \frac{h}{y}$, $h = \sqrt{xy}$. Therefore h is the geometric mean of x and y.

64. $A = 27\sqrt{3}$; $B = 108\sqrt{3}$; the area of the inscribed regular hexagon is $54\sqrt{3}$. $\sqrt{AB} = \sqrt{27 \cdot 108 \cdot 3} = 54\sqrt{3}$.

Teaching Notes, p. 472B

Warm-Up Exercises

Examine each sequence and tell how you could compute the twentieth term if you were given the nineteenth term.

1. 6, 15, 24, 33, $\cdots$ Add 9 to t_{19}.

2. 5, 10, 20, 40, $\cdots$ Multiply t_{19} by 2.

3. 1, 2, 6, 24, 120, $\cdots$ Multiply t_{19} by 20.

4. $-1, -3, -5, -7, \cdots$ Subtract 2 from t_{19}.

5. 1, 3, 7, 15, 31, $\cdots$ Multiply t_{19} by 2 and then add 1.

6. 1, 2, 8, 64, 1024, $\cdots$ Multiply t_{19} by 2^{19}.

Chapter 13   **479**

Present students with a sequence like 3, 7, 11, 15, ... Ask them what the term after 15 is. (19) Then ask them how they knew that the next term is 19. Chances are, they will say that the sequence is arithmetic with common difference 4, so the term after 15 is 15 + 4 = 19. Point out that this approach is based on the *recursion* formula $t_n = t_{n-1} + 4$ (instead of the *explicit* formula $t_n = 3 + (n-1)4$) for the given sequence.

Additional Examples

1. Find the second, third, fourth, and fifth terms of each sequence. Classify each sequence as arithmetic, geometric, or neither. Then find t_{10}.
   a. $t_1 = 2$; $t_n = 3 \cdot t_{n-1}$
   b. $t_1 = 1$; $t_n = t_{n-1} + n$
   c. $t_1 = 20$; $t_n = t_{n-1} - 3$

   a. $t_2 = 3 \cdot 2 = 6$; $t_3 = 3 \cdot 6 = 18$; $t_4 = 3 \cdot 18 = 54$; $t_5 = 3 \cdot 54 = 162$; geometric with $r = 3$; $t_{10} = 2 \cdot 3^{10-1} = 39,366$
   b. $t_2 = 1 + 2 = 3$; $t_3 = 3 + 3 = 6$; $t_4 = 6 + 4 = 10$; $t_5 = 10 + 5 = 15$; neither; $t_{10} = 15 + 6 + 7 + 8 + 9 + 10 = 55$
   c. $t_2 = 20 - 3 = 17$; $t_3 = 17 - 3 = 14$; $t_4 = 14 - 3 = 11$; $t_5 = 11 - 3 = 8$; arithmetic with $d = -3$; $t_{10} = 20 + (10 - 1)(-3) = -7$

Activity

1. $t_1 = 3$; $t_n = t_{n-1} + 4$ 3, 7, 11, 15, 19; arithmetic
   a. Give the first five terms of this sequence. What kind of sequence is it?
   b. Find a formula for t_n in terms of n. $t_n = 4n - 1$

2. $t_1 = 64$; $t_n = \frac{1}{2} t_{n-1}$ 64, 32, 16, 8, 4; geometric
   a. Give the first five terms of this sequence. What kind of sequence is it?
   b. Find a formula for t_n in terms of n. $t_n = 128\left(\frac{1}{2}\right)^n$

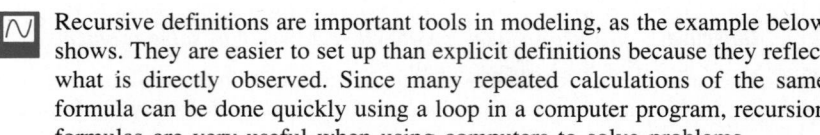

Recursive definitions are important tools in modeling, as the example below shows. They are easier to set up than explicit definitions because they reflect what is directly observed. Since many repeated calculations of the same formula can be done quickly using a loop in a computer program, recursion formulas are very useful when using computers to solve problems.

Example   The population of a certain country grows as a result of two conditions:

1. The annual population growth is 1% of those already in the country. This growth rate equals the birth rate minus the death rate.
2. 20,000 people immigrate into the country each year.

If the population now is 5,000,000 people, what will the population be in 20 years?

Solution   If you ignore the immigration for a moment and consider only the 1% growth rate, then $P_n = 1.01P_{n-1}$. Since immigration adds another 20,000 persons each year, the recursion equation is

$$P_n = 1.01P_{n-1} + 20,000.$$

In many practical situations, it is convenient to use a zero subscript to indicate an initial observation. If we let $P_0 = 5,000,000$ represent the population now, then P_1 is the population at the end of one year, and P_n is the population at the end of n years. To calculate the population for the first few years, you could use a calculator and the recursion equation.

$$P_0 = 5,000,000$$
$$P_1 = 1.01 \cdot 5,000,000 + 20,000 = 5,070,000$$
$$P_2 = 1.01 \cdot 5,070,000 + 20,000 = 5,140,700$$
$$P_3 = 1.01 \cdot 5,140,700 + 20,000 = 5,212,107$$

However, to calculate the population for each of the next 20 years, it is easier to write a simple computer program or use a spreadsheet program. The BASIC program given on the next page uses the recursion equation above to find the population.

```
10 LET P = 5000000 ←————————— This is the initial population P₀.
20 FOR N = 1 TO 20
30 LET P = 1.01*P+20000 ←———— Pₙ = 1.01 · Pₙ₋₁ + 20,000
40 NEXT N
50 PRINT ''THE POPULATION IN '';N-1;'' YEARS IS '';INT(P)
60 END
```

The program produces this output:

```
○    THE POPULATION IN 20 YEARS IS 6541330    ○
```

CLASS EXERCISES

Give the first four terms of each sequence.

10, 12, 15, 19

1. $t_1 = 5;\ t_n = t_{n-1} + 3$ 5, 8, 11, 14

2. $t_1 = 10;\ t_n = t_{n-1} + n$

3. $t_1 = 3;\ t_n = 2t_{n-1}$ 3, 6, 12, 24

4. $t_1 = 4;\ t_n = 2t_{n-1} - 1$ 4, 7, 13, 25

5. Sometimes a recursive definition will tell how the $(n + 1)$st term is related to the nth term. Give the first four terms of the sequence defined recursively by $t_1 = 5$ and $t_{n+1} = 2t_n + n$. 5, 11, 24, 51

6. Sometimes a recursive definition gives t_n in terms of more than one preceding term. Give the first four terms of the sequence defined recursively by $t_1 = 4$, $t_2 = 16$, and $t_n = t_{n-1} + t_{n-2}$. 4, 16, 20, 36

WRITTEN EXERCISES

Find the third, fourth, and fifth terms of each sequence.
4. 26, 666, 443,546
8. 8, 32, 256

 1. $t_1 = 6;\ t_n = t_{n-1} + 4$ 14, 18, 22

2. $t_1 = 9;\ t_n = \frac{1}{3}t_{n-1}$ 1, $\frac{1}{3}$, $\frac{1}{9}$

3. $t_1 = 1;\ t_n = 3t_{n-1} - 1$ 5, 14, 41

4. $t_1 = 4;\ t_n = (t_{n-1})^2 - 10$ See above.

5. $t_1 = 1;\ t_n = t_{n-1} + 2n - 1$ 9, 16, 25

6. $t_1 = \frac{1}{2};\ t_n = \frac{n}{n+1}(t_{n-1} + 1)$ $\frac{3}{2}$, 2, $\frac{5}{2}$

7. $t_1 = 2;\ t_2 = 4;\ t_n = t_{n-1} + t_{n-2}$ 6, 10, 16

8. $t_1 = 2;\ t_2 = 4;\ t_n = t_{n-1} \cdot t_{n-2}$ See above.

9. $t_1 = 5;\ t_2 = 8;\ t_n = (t_{n-1} - t_{n-2})^2$ 9, 1, 64

10. $t_1 = 7;\ t_2 = 3;\ t_n = t_{n-1} - 2t_{n-2}$ −11, −17, 5

11. Find an explicit definition for the sequence in Exercise 1. $t_n = 2 + 4n$

12. Find an explicit definition for the sequence in Exercise 2. $t_n = 27\left(\frac{1}{3}\right)^n$

Give a recursive definition for each sequence.

13. 9, 13, 17, 21, . . .

14. 81, 27, 9, 3, . . .

15. 1, 3, 7, 15, 31, 63, . . .

16. 1, 3, 7, 13, 21, 31, . . .

17. 1, 3, 6, 10, 15, 21, . . .

18. 1, 2, 6, 24, 120, 720, . . .

---

2. Evaluate 1^2, $(1 + 2)^2$, $(1 + 2 + 3)^2$, and $(1 + 2 + 3 + 4)^2$. Find a recursive definition for this sequence.

$t_1 = 1^2 = 1$
$t_2 = (1 + 2)^2 = 9$
$t_3 = (1 + 2 + 3)^2 = 36$
$t_4 = (1 + 2 + 3 + 4)^2 = 100$

Notice that $t_2 - t_1 = 8 = 2^3$, $t_3 - t_2 = 27 = 3^3$, and $t_4 - t_3 = 64 = 4^3$. Thus, $t_1 = 1$ and $t_n = t_{n-1} + n^3$ is a recursive definition for the sequence.

▧ Using Technology

In the program given on this page, truncating the population is left until the last step (that is, until year 20). You might have students use a computer or programmable calculator to see what happens if truncating is done at *each* step instead. (Change line 30 to: 30 LET P = INT(1.01 ∗ P + 20000)) You could also have students *round* at each step. (Change line 30 to: 30 LET P = INT(1.01 ∗ P + 20000.5))

Additional Answers Written Exercises

13. $t_1 = 9;\ t_n = t_{n-1} + 4$

14. $t_1 = 81;\ t_n = \frac{1}{3}t_{n-1}$

15. $t_1 = 1;\ t_n = t_{n-1} + 2^{n-1}$

16. $t_1 = 1;\ t_n = t_{n-1} + 2(n - 1)$

17. $t_1 = 1;\ t_n = t_{n-1} + n$

18. $t_1 = 1;\ t_n = n \cdot t_{n-1}$

19. **a.** Give the first eight terms of the sequence defined recursively by $t_1 = 3$, $t_2 = 5$, and $t_n = t_{n-1} - t_{n-2}$. 3, 5, 2, −3, −5, −2, 3, 5
    b. Observing the pattern you get in part (a), tell what the 1000th term of the sequence will be. −3

20. **a.** Give the first eight terms of the sequence defined recursively by $t_1 = 4$, $t_2 = 8$, and $t_n = \dfrac{t_{n-1}}{t_{n-2}}$. 4, 8, 2, $\frac{1}{4}$, $\frac{1}{8}$, $\frac{1}{2}$, 4, 8
    b. Observing the pattern you get in part (a), tell what the 1000th term of the sequence will be. $\frac{1}{4}$

21. **Geography** Refer to the example on page 480. Suppose the population of those living in the country grows 2% per year, and that an additional 50,000 people immigrate into the country every year. $P_0 = 5,000,000$ and $P_n =$
    a. Give a recursion equation for P_n, the population in n years. $(1.02)P_{n-1} + 50,000$
    b. If the population now is 8,500,000, what will the population be in 5 years? 9,644,889

22. **Chemistry** Each day 8% of a quantity of radioactive iodine will decay.
    a. Express this fact with a recursion equation. $Q_0 = Q$; $Q_n = 0.92\,Q_{n-1}$
    b. Choosing a value for the initial amount of iodine, find the approximate half-life of the iodine. About 8 days

B 23. **Visual Thinking** Let S_n represent the number of dots in an n by n square array. Pretend you have forgotten that $S_n = n^2$. Give a recursion equation that tells how S_{n+1} is related to S_n by reasoning how many extra dots are needed to form the $(n + 1)$st square array from the previous nth square array. Illustrate your answer with a diagram of dots. $S_{n+1} = S_n + 2n + 1$

24. Suppose that everyone in a room shakes hands with everyone else exactly once. Let H_n represent the number of handshakes if there are n people ($n \geq 2$) in the room. Give a recursion equation that tells how H_n is related to H_{n-1}. (*Hint:* Suppose you know H_{n-1}. If another person enters the room, how many additional handshakes will there be?) $H_n = H_{n-1} + n - 1$

25. **Visual Thinking** Let d_n represent the number of diagonals that can be drawn in an n-sided polygon. The diagram at the left below shows that a hexagon has 9 diagonals, and so $d_6 = 9$.

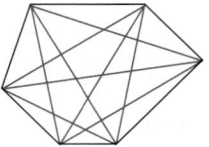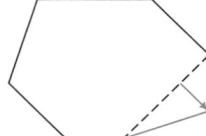

    a. Imagine pushing out one side of the hexagon so that a polygon of 7 sides is formed. (See the diagram at the right above.) How many additional diagonals can be drawn? 5
    b. Imagine pushing out one side of a polygon with $n - 1$ sides so that an n-sided polygon is formed. Tell how many additional diagonals can be drawn. Then write a recursion equation for d_n. $n - 2$ additional diagonals; $d_4 = 2$; $d_n = d_{n-1} + n - 2$

26. *Visual Thinking* Let r_n represent the number of regions formed when n lines are drawn in a plane such that no two lines are parallel and no three are concurrent. The diagrams below illustrate r_1, r_2, and r_3. Give the recursion equation for r_n. $r_1 = 2$; $r_n = r_{n-1} + n$

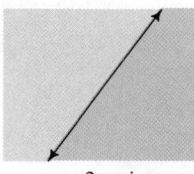
$r_1 = 2$ regions

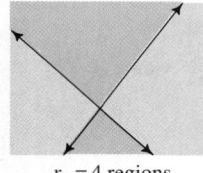
$r_2 = 4$ regions

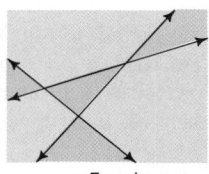
$r_3 = 7$ regions

27. *Visual Thinking* Let p_n represent the number of intersection points created when n lines are drawn in a plane such that no two lines are parallel and no three lines are concurrent. The diagrams in Exercise 26 show that $p_1 = 0$, $p_2 = 1$, and $p_3 = 3$.
  a. Find p_4 and p_5. 6, 10
  b. Give a recursion equation for p_n. $p_1 = 0$; $p_n = p_{n-1} + n - 1$

28. *Writing* For what kind of problem would you be more likely to use a recursive definition of a sequence than an explicit definition? Why? For what type of problem would an explicit definition be more useful?

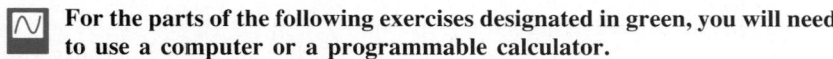 **For the parts of the following exercises designated in green, you will need to use a computer or a programmable calculator.**

29. **Finance** On the birth of their daughter, Mr. and Mrs. Swift began saving for her college education by investing $5000 in an annuity account paying 10% interest per year. Each year on their daughter's birthday they invested $2000 more in the account.
  a. Let A_n represent the amount in the account on their daughter's nth birthday. Give a recursive definition for A_n. $A_0 = 5000$; $A_n = 1.1A_{n-1} + 2000$
  b. Find the amount that will be in the account on her 18th birthday. $118,997.93

30. **Finance** When Mr. Tallchief reaches his retirement age of 65, he expects to have a retirement account worth about $400,000. One month after he retires, and every month thereafter, he intends to withdraw $4000 from the account. The balance will be invested at 9% annual interest compounded monthly.

  a. Let A_n represent the amount in the account n months after Mr. Tallchief's retirement. Give a recursive definition for A_n. See below.
  b. When will there be no money left in the account? after 185 months
  a. $A_0 = 400,000$; $A_n = A_{n-1}(1.0075) - 4000$

◹ Using Technology

The program used in the example on page 480 can be changed appropriately for Exercise 21.

Exercise Note

Students are often interested in problems like those in Exercises 29 and 30. You might want to pose additional problems by giving variations on these or even combining ideas. For example: If your father followed Mr. Swift's plan until you were 18 and then put the $2000 each year toward his own retirement, how close to Mr. Tallchief's $400,000 would he come?

Additional Answers Written Exercises

28. Answers may vary. Example: You would be more likely to use a recursive definition when you need to find the term following a given list of terms in a sequence. You would be more likely to use an explicit definition when you need to find a term t_n for which n is large.

Additional Answers
Written Exercises

31. b. To transfer n disks from A to B, first transfer the top $(n - 1)$ disks from A to C. (This takes M_{n-1} moves.) Then move the nth disk to B (one move). Finally, transfer the $(n - 1)$ disks from C to B (M_{n-1} moves). Thus, $M_n = M_{n-1} + 1 + M_{n-1} = 2M_{n-1} + 1$.

c. $M_1 = 1$; $M_2 = 2M_1 + 1 = 2(1) + 1 = 3$; $M_3 = 2M_2 + 1 = 2 \cdot 3 + 1 = 7$.

31. The Tower of Hanoi puzzle consists of a block of wood with three posts, A, B, and C. On post A there are eight disks of diminishing size from bottom to top. The task is to transfer all eight disks from post A to one of the other two posts given that:

1. only one disk can be moved at a time;
2. no disk can be placed on top of a smaller disk.

a. Let M_n represent the minimum number of moves needed to move n disks from post A to one of the other posts. Obviously, $M_1 = 1$. What are M_2 and M_3? (To try this puzzle, you could use coins of different sizes and three locations A, B, and C on a sheet of paper to represent the three posts.) 3, 7

b. Suppose you know how to move $(n - 1)$ disks or coins from post A to another post, and that to do so requires M_{n-1} moves. Explain why $M_n = 2M_{n-1} + 1$.

c. Use your answer to part (b) to check your values for M_2 and M_3. Then find M_4, M_5, M_6, M_7, and M_8. 15, 31, 63, 127, 255

d. Find an explicit definition for M_n. $M_n = 2^n - 1$

e. According to legend, in the great Temple of Benares, there is an altar with three diamond needles. At the beginning of time, 64 gold rings of decreasing radius from bottom to top were placed on one of these needles. Day and night, priests sit before the altar transferring one gold ring per second in accordance with the two rules given above. The legend also says that when all 64 rings have been transferred to one of the other diamond needles, the world will come to an end. How long will it take the priests to transfer all the rings? (You can use your answer to part (d), or you can find a computer solution using the recursion equation given in part (b).) 5.8×10^{11} yr

32. Here is a famous problem posed in the thirteenth century by Leonardo de Pisa, better known as Fibonacci: Suppose we have one pair of newborn rabbits of both genders. We assume that the following conditions are true.

1. It takes a newborn rabbit one month to become an adult.
2. A pair of adult rabbits of both genders will produce one pair of newborn rabbits of both genders each month, beginning one month after becoming adults.
3. The rabbits do not die.

How many rabbits will there be one year later? (To help with your solution to this problem, write a recursion equation in which r_n, the number of pairs of rabbits n months from now, is expressed in terms of r_{n-1} and r_{n-2}. Remember that $r_0 = 1$.) $r_n = r_{n-1} + r_{n-2}$; 233 pairs or 466 rabbits

33. Physics *Newton's Law of Cooling* states that over equal time periods (for example, one minute) the change in an object's temperature is proportional to the difference between the temperature of the object at the beginning of the time period and the room temperature. Thus, if t_0 is the temperature of the object when it is placed in a room whose temperature is R, then t_n, the object's temperature n minutes later, is given by the recursion equation

$$t_n - t_{n-1} = k(t_{n-1} - R),$$

a. -0.05; negative sign indicates that the coffee is cooling.

where k is a proportionality constant.
  a. Suppose that a cup of coffee at $98°$ C is placed in a room whose temperature is $18°$ C, and that 1 min later the coffee has cooled to $94°$ C. Find the proportionality constant k, and explain the significance of its negative sign.
  b. Show that the recursion equation given above can be rewritten as $t_n = 0.95t_{n-1} + 0.9$.
  c. Use a computer to print out a table of values for t_n. When will the cup of coffee cool to less than $75°$ C? After about 7 min

34. Medicine Suppose that in a closed community with population P, a flu epidemic begins and that the number of people *newly* exposed to the flu on a given day is proportional to the number not yet exposed on the previous day.
  a. If f_n represents the number of people exposed to the flu n days after it begins, explain how the description above leads to the recursion equation $f_n - f_{n-1} = k(P - f_{n-1})$, where k is a proportionality constant.
  b. Suppose in a college community of 2500 students, the flu begins with 100 students exposed to the flu; that is, $f_0 = 100$. On the next day, $f_1 = 220$. Find the value of k, and then show that $f_n = 0.95f_{n-1} + 125$. $k = 0.05$
  c. Use a computer to print out a table of values for f_n. About how long does it take before the flu spreads through the whole student body? 166 days

C 35. a. Ecology Suppose that a certain population grows from one generation to the next in such a way that the population *increase* is proportional to the population of the previous generation. Explain how this growth can be modeled by the equation $P_n - P_{n-1} = kP_{n-1}$. In your explanation, tell what P_n and P_{n-1} represent.
  b. Suppose that a population has a carrying capacity C. This is the maximum size beyond which the population cannot grow. As the population gets closer to C, overcrowding tends to reduce population increases. Sketch a graph that models this type of population growth with overcrowding.
  c. The population growth described in part (b) can be modeled by modifying the equation in part (a) so that the new equation is

$$P_n - P_{n-1} = kP_{n-1}(C - P_{n-1}).$$

Write a sentence or two explaining what this recursion equation says.
  d. For a population of fish in an aquarium, let $C = 500$ and $k = 0.002$. Choose a value for the initial population, and use a computer to print the values given by the recursion equation in part (c). How well do the computer results agree with your graph in part (b)?

Sequences and Series **485**

Use the formulas $S = \frac{n(f + l)}{2}$ and $l = f + (n - 1)d$ and the given values to find the values of the unknown variables.

1. $n = 50$, $f = 1$, $l = 50$; $S = \underline{\ ?\ }$, $d = \underline{\ ?\ }$ 1275; 1

2. $n = 15$, $f = 8$, $d = 3$; $l = \underline{\ ?\ }$, $S = \underline{\ ?\ }$ 50; 435

3. $l = -12$, $f = 100$, $d = -2$; $n = \underline{\ ?\ }$, $S = \underline{\ ?\ }$ 57; 2508

Use the formulas $S = \frac{a(1 - r^n)}{1 - r}$ and $l = ar^{n-1}$ and the given values to find the values of the unknown variables.

4. $a = 16$, $r = \frac{1}{2}$, $n = 6$;

$l = \underline{\ ?\ }$, $S = \underline{\ ?\ }$ $\frac{1}{2}$; $31\frac{1}{2}$

5. $a = 16$, $r = -\frac{1}{2}$, $l = -2$;

$n = \underline{\ ?\ }$, $S = \underline{\ ?\ }$ 4; 10

Motivating the Section

There is a piece of mathematical lore to the effect that when a very young Gauss was asked to sum the first 100 counting numbers, he was able to do this in a matter of seconds—without a calculator or computer! As an introduction to this section, help your students discover how Gauss was able to do this. (Sum the first and last terms (1 + 100 = 101), the second and next-to-last terms (2 + 99 = 101), and so on to get 50 sums of 101; thus, the sum of the first 100 counting numbers is 50 · 101 = 5050.)

13-3 Arithmetic and Geometric Series and Their Sums

Objective *To find the sum of the first n terms of arithmetic or geometric series.*

The words *sequence* and *series* are often used interchangeably in everyday conversation. For example, a person may refer to a *sequence of events* or to a *series of events*. In mathematics, however, a distinction is made between a sequence of numbers and a series of numbers. This distinction between a sequence and a series can best be made by considering some examples.

Finite sequence:	2, 6, 10, 14
Finite series:	2 + 6 + 10 + 14
Infinite sequence:	$\frac{1}{2}, \frac{1}{4}, \frac{1}{8}, \frac{1}{16}, \cdots$
Infinite series:	$\frac{1}{2} + \frac{1}{4} + \frac{1}{8} + \frac{1}{16} + \cdots$

As you can see from the examples above, a **series** is an indicated sum of the terms of a sequence. The *sum* of the finite series $2 + 6 + 10 + 14$ is 32. In Section 13-5 we will see that the sum of the infinite series $\frac{1}{2} + \frac{1}{4} + \frac{1}{8} + \frac{1}{16} + \cdots$ is 1. However, not all infinite series have sums. In this section we will consider only finite series and their sums.

If S_n represents the sum of n terms of a series, then S_n can be expressed either explicitly or recursively as follows:

Explicit definition of S_n: $S_n = t_1 + t_2 + t_3 + \cdots + t_n$

Recursive definition of S_n: $S_0 = 0$
   $S_n = S_{n-1} + t_n$ for $n \geq 1$

 The recursive definition can be used as the basis for a computer program that will find the sum of a series. Example 1 illustrates how this is done.

Example 1   Find the sum of the cubes of the first twenty positive integers.

Solution   We want $S_{20} = 1^3 + 2^3 + 3^3 + \cdots + 20^3$.

First find a formula for t_n in terms of n. The nth term is $t_n = n^3$, and is represented by T in line 30 of the computer program given on the next page. The recursive definition of the sum S_n, which is represented by S, occurs in lines 10 and 40.

```
10 LET S = 0 ←——————————— Initial condition $S_0 = 0$
20 FOR N = 1 TO 20
30 LET T = N^3
40 LET S = S + T ←——————— Recursion equation $S_n = S_{n-1} + t_n$
50 NEXT N
60 PRINT ''THE SUM OF THE FIRST ''; N-1; '' TERMS IS ''; S
70 END
```

When the program is run, the computer gives this output:

```
⊙  THE SUM OF THE FIRST 20 TERMS IS 44100                    ⊙
```

You can use the method of Example 1 to find the sum of any finite series. However, if the series is arithmetic or geometric, you can also use the formulas given in the following two theorems.

Sum of a Finite Arithmetic Series

The sum of the first n terms of an arithmetic series is

$$S_n = \frac{n(t_1 + t_n)}{2}.$$

Proof: Write the series for S_n twice, the second time with the order of the terms reversed. Then add the two equations, term by term:

$$S_n = \quad t_1 \quad + (t_1 + d) + (t_1 + 2d) + \cdots + (t_n - d) + t_n$$
$$\underline{S_n = \quad t_n \quad + (t_n - d) + (t_n - 2d) + \cdots + (t_1 + d) + t_1}$$
$$2S_n = (t_1 + t_n) + (t_1 + t_n) + (t_1 + t_n) + \cdots + (t_1 + t_n) + (t_1 + t_n)$$

Since there are n of these $(t_1 + t_n)$ terms,

$$2S_n = n(t_1 + t_n)$$
$$S_n = \frac{n(t_1 + t_n)}{2}$$

Example 2 Find the sum of the first 25 terms of the arithmetic series

$$11 + 14 + 17 + 20 + \cdots.$$

Solution **Step 1:** First find the 25th term.

$$t_{25} = t_1 + (n - 1)d = 11 + (25 - 1)3 = 83$$

Step 2: $S_{25} = \frac{n(t_1 + t_{25})}{2} = \frac{25(11 + 83)}{2} = 1175$

Assessment Note

After defining a series, ask students to give the series that corresponds to each of these sequences:

1. 3, 7, 11, 15
   $(3 + 7 + 11 + 15)$

2. 1, 3, 5, ..., $2n - 1$
   $(1 + 3 + 5 + \cdots + 2n - 1)$

3. $a_1, a_2, a_3, ..., a_n$
   $(a_1 + a_2 + a_3 + \cdots + a_n)$

Using Technology

Depending on the computer used, the program on this page may give a value of S that is very close to, but not exactly, an integer. This error is due to rounding because of the base-2 system used. The rounding can be avoided by using N ∗ N ∗ N instead of N^3 in line 30.

Mathematical Note

For the sum of a finite arithmetic series, if the first term t_1 and the common difference d are known, then $S_n = \frac{n[2t_1 + (n-1)d]}{2}$.

Communication Note

Students may better remember the formula for the sum of the first n terms of an arithmetic series if you point out that the sum equals the product of the number of terms and the average of the first and last terms.

For a finite geometric series, $(t_1)r^{n-1} = t_n$, so $(t_1)r^n = (t_n)r$. Therefore, the sum of a finite geometric series can be written as $S_n = \dfrac{t_1 - (t_n)r}{1 - r}$.

Additional Examples

1. Find the sum of each series.
   a. $3 + 9 + 27 + \cdots + 6561$
   b. The sum of the first n positive even integers in terms of n

a. The series is geometric with $t_1 = 3$ and $r = 3$.
$t_n = 6561 = 3 \cdot 3^{n-1} = 3^n$, so $n = 8$.
$S_8 = \dfrac{t_1(1 - r^8)}{1 - r} =$
$\dfrac{3(1 - 3^8)}{1 - 3} = 9840$

b. The series $2 + 4 + 6 + 8 + \cdots + 2n$ is arithmetic with $d = 2$.

Thus, $S_n = \dfrac{n(t_1 + t_n)}{2} =$
$\dfrac{n(2 + 2n)}{2} = \dfrac{2n(n + 1)}{2} =$
$n(n + 1)$.
(Alternatively, $S_1 = 2$, $S_2 = 6$, $S_3 = 12$, $S_4 = 20$, and so on. Examining differences gives $S_n = S_{n-1} + 2n$ with $S_1 = 2$.)

2. Find the sum of all positive 3-digit integers that are divisible by 8.

We consider 100, 101, 102, ..., 999 and see that $t_1 = 104$ and $t_n = 992$.
$992 = 104 + (n - 1)8$, so $n = 112$.
$S_{112} = \dfrac{112(104 + 992)}{2} = 61{,}376$

Sum of a Finite Geometric Series

The sum of the first n terms of a geometric series is

$$S_n = \frac{t_1(1 - r^n)}{1 - r},$$

where r is the common ratio and $r \neq 1$.

Proof: Multiply the series for S_n by the common ratio and then subtract the resulting series from the original one, as shown below.

$$S_n = t_1 + t_1r + t_1r^2 + \cdots + t_1r^{n-2} + t_1r^{n-1}$$
$$rS_n = \qquad t_1r + t_1r^2 + \cdots + t_1r^{n-2} + t_1r^{n-1} + t_1r^n$$

$$S_n - rS_n = t_1 + 0 + 0 + \cdots + \quad 0 \quad + \quad 0 \quad - t_1r^n$$
$$S_n - rS_n = t_1 - t_1r^n$$
$$S_n(1 - r) = t_1(1 - r^n)$$
$$S_n = \frac{t_1(1 - r^n)}{1 - r}, \; r \neq 1$$

This formula is not defined for $r = 1$. If r does equal 1, however, the geometric series is simply a series of repeated numbers, such as $a + a + \cdots + a$, whose sum is na.

Example 3 Find the sum of the first 10 terms of the geometric series

$$2 - 6 + 18 - 54 + \cdots.$$

Solution $S_{10} = \dfrac{t_1(1 - r^{10})}{1 - r}$

$= \dfrac{2(1 - (-3)^{10})}{1 - (-3)} = \dfrac{2(1 - 3^{10})}{4} = -29{,}524$

CLASS EXERCISES

Answers may vary. Example: 1, 3, 5, ... ; $1 + 3 + 5 + \cdots$
1. Give an example of an arithmetic sequence and of an arithmetic series.
2. Which theorem can be used to find the sum of the finite series $1 + 2 + 4 + 8 + 16 + \cdots + 1024$? What is this sum? See below.
3. a. Is the series $3 + 5 + 9 + 17 + \cdots + 32{,}769$ arithmetic, geometric, or neither? Neither $\qquad S_0 = 0, \; S_n = S_{n-1} + t_n$, where
   b. Give a recursive definition of the series. $t_n = 2^n + 1 \; (n \geq 1)$
   c. How would you modify the computer program in Example 1 to find the sum of this series? Change lines 20 and 30 as follows:
   20 FOR N = 1 TO 15   30 LET T = 2^N + 1
2. Sum of Finite Geometric Series; 2047

For each of the arithmetic series in Exercises 1–8, find the specified sum.

A

1. S_{10}: $t_1 = 3$, $t_{10} = 39$ 210

2. S_{200}: $t_1 = 18$, $t_{200} = 472$ 49,000

3. S_{50}: $5 + 10 + 15 + \cdots$ 6375

4. S_{25}: $17 + 25 + 33 + \cdots$ 2825

5. S_{12}: $t_n = 5 + 3n$ 294

6. S_{40}: $t_1 = 5$, $t_3 = 11$ 2540

7. $1 + 2 + 3 + \cdots + 1000$ 500,500

8. $3 + 7 + 11 + \cdots + 99$ 1275

9. Find S_8 for the geometric series with $t_1 = 8$ and **(a)** $r = \frac{1}{2}$, **(b)** $r = -\frac{1}{2}$. $\frac{255}{16}$; $\frac{85}{16}$

10. Show that the sum of the first 10 terms of the geometric series

$$1 + \frac{1}{3} + \frac{1}{9} + \frac{1}{27} + \cdots$$

is twice the sum of the first 10 terms of the series

$$1 - \frac{1}{3} + \frac{1}{9} - \frac{1}{27} + \cdots.$$

11. Show that $1 + 2 + 4 + \cdots + 2^{n-1} = 2^n - 1$.

12. Show that $9 + 90 + 900 + \cdots + 9 \cdot 10^{n-1} = 10^n - 1$.

13. Show that $\sqrt{2} + 2 + 2\sqrt{2} + \cdots + 64 = \frac{63\sqrt{2}}{\sqrt{2} - 1}$.

14. For the series $1 + \sqrt[5]{3} + \sqrt[5]{3^2} + \sqrt[5]{3^3} + \cdots$, show that

$$S_{15} = \frac{26}{3^{1/5} - 1}.$$

15. Show that the sum of the first n positive integers is $\frac{n(n + 1)}{2}$.

16. Let S_n be the sum of the first n positive odd integers.
  a. Evaluate S_1, S_2, S_3, and S_4. 1, 4, 9, 16
  b. Suggest and prove a formula for S_n. $S_n = n^2$

$1 + 3 + 5 + 7 = 4^2$

Ex. 16

Find a formula for t_n, the nth term of the series. Then give a recursive definition for S_n, the sum of n terms of the series. $t_n = 3^{n-1}$; $S_0 = 0$, $S_n = S_{n-1} + 3^{n-1}$

17. $2 + 4 + 6 + 8 + \cdots$

18. $1 + 3 + 9 + 27 + \cdots$

19. $8 + 12 + 18 + 27 + \cdots$ See below.

20. $50 + 47 + 44 + 41 + \cdots$
$t_n = 53 - 3n$; $S_0 = 0$, $S_n = S_{n-1} + 53 - 3n$

 For Exercises 21–26, you will need to use a computer.

Write a computer program, like the one given in Example 1, that will print the sum of each series. Then run the program to find the sum. $4.49317984 \times 10^{17}$

21. $1^4 + 2^4 + 3^4 + \cdots + 10^4$ 25,333

22. $1^1 + 2^2 + 3^3 + \cdots + 15^{15}$

23. $10^2 + 20^2 + 30^2 + \cdots + 1000^2$ 33,835,000

24. $\sqrt{5} + \sqrt{10} + \sqrt{15} + \cdots + \sqrt{200}$ 383.744568

17. $t_n = 2n$; $S_0 = 0$, $S_n = S_{n-1} + 2n$

19. $t_n = 8\left(\frac{3}{2}\right)^{n-1}$; $S_0 = 0$, $S_n = S_{n-1} + 8\left(\frac{3}{2}\right)^{n-1}$

Additional Answers
Written Exercises

10. In $1 + \frac{1}{3} + \cdots$, $r = \frac{1}{3}$, so

$$S_{10} = \frac{1\left(1 - \left(\frac{1}{3}\right)^{10}\right)}{1 - \frac{1}{3}} =$$

$$\frac{3}{2}\left(1 - \left(\frac{1}{3}\right)^{10}\right).$$

In $1 - \frac{1}{3} + \cdots$, $r = -\frac{1}{3}$,

so $S_{10} =$

$$\frac{1\left(1 - \left(-\frac{1}{3}\right)^{10}\right)}{1 + \frac{1}{3}} =$$

$$\frac{3}{4}\left(1 - \left(\frac{1}{3}\right)^{10}\right). \text{ Thus,}$$

$$2 \cdot \frac{3}{4}\left(1 - \left(\frac{1}{3}\right)^{10}\right) =$$

$$\frac{3}{2}\left(1 - \left(\frac{1}{3}\right)^{10}\right).$$

11. $t_1 = 1$ and $r = 2$, so $S_n = \frac{1(1 - 2^n)}{1 - 2} = 2^n - 1$.

12. $t_1 = 9$ and $r = 10$, so $S_n = \frac{9(1 - 10^n)}{1 - 10} = -1(1 - 10^n) = 10^n - 1$.

13. $64 = \sqrt{2} \cdot (\sqrt{2})^{n-1} = (\sqrt{2})^n$; $2^6 = 2^{(1/2)n}$; $6 = \frac{1}{2}n$; $n = 12$;

$$S_{12} = \frac{\sqrt{2}(1 - (\sqrt{2})^{12})}{1 - \sqrt{2}} = \frac{\sqrt{2}(1 - 64)}{1 - \sqrt{2}} = \frac{63\sqrt{2}}{\sqrt{2} - 1}$$

14. $S_{15} = \frac{1(1 - (3^{1/5})^{15})}{1 - 3^{1/5}} = \frac{1 - 3^3}{1 - 3^{1/5}} = \frac{26}{3^{1/5} - 1}$

15. $t_1 = 1$, $t_n = n$, and $d = 1$; $S_n = \frac{n(1 + n)}{2} = \frac{n(n + 1)}{2}$

Additional Answers
Written Exercises

25. b. 20 FOR N = 1 TO 25
      30 LET T = 2^(−N)

26. a. 30 LET T = 1/3^N

Suggested Assignments

Standard
489/1, 3, 7, 9, 11, 13, 17, 27
Comprehensive
Day 1: 489/1, 3, 7, 9, 11, 13,
17, 27, 29, 31
Day 2: 490/33, 37, 38, 40, 42,
43

Supplementary Materials

Alternative Assessment, 40
Using Technology, 33–34
Precalculus Plotter Plus,
68–69, 70, 71–72
Student Resource Guide,
121–123

B 25. a. Refer to the computer program given in Example 1. Delete line 60 and insert the following line.

$$\text{45 PRINT N,S}$$

A table is printed giving the sum of the first n terms for each number n from 1 to 20.

Explain what happens to the computer's output.

b. Change lines 20 and 30 in the program so that the computer will print the sums $S_1, S_2, S_3, \ldots, S_{25}$ for the series

$$2^{-1} + 2^{-2} + 2^{-3} + \cdots + 2^{-25}.$$

c. Find S_{25} using the formula given on page 488. (Your answer should be approximately the same as the last number that the computer prints.) 0.99999997

d. Guess the sum of the *infinite* series

$$2^{-1} + 2^{-2} + 2^{-3} + \cdots + 2^{-n} + \cdots. \quad 1$$

26. a. Modify the computer program in Exercise 25 and run it for the series

$$\frac{1}{3} + \frac{1}{3^2} + \frac{1}{3^3} + \cdots + \frac{1}{3^{25}}.$$

b. Find S_{25} using the formula given on page 488. $\frac{1}{2}$

c. Guess the sum of the *infinite* series

$$\frac{1}{3} + \frac{1}{3^2} + \frac{1}{3^3} + \cdots + \frac{1}{3^n} + \cdots. \quad \frac{1}{2}$$

27. The sum of the first n terms of a series is $S_n = n^2 + 4n$. Find t_1, t_2, and t_3. 5, 7, 9

28. The sum of the first n terms of a series is $S_n = 2n^2$.
   a. Find t_1, t_2, and t_3. 2, 6, 10
   b. Find $S_n − S_{n-1}$. $4n − 2$

29. Find the sum of all multiples of 3 between 1 and 1000. 166,833

30. Find the sum of all positive 3-digit numbers divisible by 6. 82,350

31. Find the sum of all positive 3-digit numbers whose last digit is 3. 49,320

32. Find the sum of all positive odd numbers less than 400 that are divisible by 5. 8000

33. Find the sum of the series $1 − 3 + 5 − 7 + 9 − 11 + \cdots + 1001$. 501

34. Find the sum of the series $1 + 2 + 4 + 5 + 7 + 8 + 10 + 11 + \cdots + 299$, which is the sum of the integers except for multiples of 3. 30,000

35. The originator of a chain letter writes 5 letters instructing each recipient to write 5 similar letters to additional people. Then these people each send 5 similar letters to other people. Determine the number of people who should receive letters if the chain continues unbroken for 12 steps. Explain why the process always fails. (There are laws forbidding chain letters that request money.) 305,175,780; see below.

36. Value Appliance Store has radios that can be purchased on a daily installment plan. For a particular type of radio, you pay only 1 cent the first day, 2 cents the next day, 4 cents the next day, 8 cents the next day, and so on, for 14 days. How much does the radio cost? $163.83

35. The process always fails because the number of people who have not yet received letters is eventually exhausted.

37. a. If you go back through ten generations, how many ancestors do you have? Count your parents as the first generation back, your four grandparents as the second generation, and so on. (Assume there are no duplications.) 2046

b. How many generations back must you go in order to have more than one million ancestors? 19

38. a. Finance Suppose a doctor earns $40,000 during the first year of practice. Suppose also that each succeeding year the salary increases 10%. What is the total of the doctor's salaries over the first 10 years? $637,496.98

b. How many years must the doctor work if the salary total is to exceed a million dollars? 14 yr

39. The number $T_n = 1 + 2 + 3 + \cdots + n = \dfrac{n(n+1)}{2}$ is sometimes called a **triangular number** because it is possible to represent the number by a triangular array of dots, as shown.

$T_4 = 1 + 2 + 3 + 4 = 10$

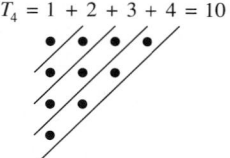

a. Evaluate T_1, T_2, T_3, T_4, and T_5. 1, 3, 6, 10, 15

b. Add any two consecutive triangular numbers. Then make a conjecture and prove it. $T_n + T_{n+1} = (n+1)^2$

40. Consider the series $1^3 + 2^3 + 3^3 + \cdots + n^3$. Evaluate S_1, S_2, S_3, and S_4. 1 9 36 100 Then suggest a formula for S_n. (*Hint:* See Exercise 39.) $S_n = \left(\dfrac{n(n+1)}{2}\right)^2$

41. The sum of every row, column, and diagonal of a magic square equals the magic number M. For the 3-by-3 and 4-by-4 magic squares shown below, M is 15 and 34, respectively. Find the value of M for an n-by-n magic square that contains consecutive integers starting with 1. (*Hint:* How many numbers are in an n-by-n square? What is their sum? Deduce the row sum M.) Check that your formula gives $M(3) = 15$ and $M(4) = 34$. $M(n) = \dfrac{n(n^2+1)}{2}$

$M = 15$

8	1	6
3	5	7
4	9	2

$M = 34$

1	15	14	4
12	6	7	9
8	10	11	5
13	3	2	16

42. Finance If you invest $1000 per year every year for 10 years and if your money is compounded annually at 12%, how much money will you have at the time you make your tenth investment? $17,548.74

43. Finance If you invest P dollars at an interest rate r compounded annually, after n years you will have $P(1 + r)^n$ dollars. Suppose you invest P dollars every year for n years. Show that at the time you make your nth investment, you will have $\dfrac{P[(1 + r)^n - 1]}{r}$ dollars.

**Additional Answers
Written Exercises**

39. b. $T_n + T_{n+1} = \dfrac{n(n+1)}{2} +$

$\dfrac{(n+1)(n+2)}{2} =$

$\dfrac{(n+1)[n+(n+2)]}{2} =$

$\dfrac{(n+1)(2n+2)}{2} =$

$(n+1)^2$

43. At the time of the nth investment, the first investment of P dollars yields $P(1 + r)^{n-1}$ dollars, the second investment yields $P(1 + r)^{n-2}$ dollars, and so on. The series

$P(1 + r)^{n-1} +$
$P(1 + r)^{n-2} + \cdots + P$

is geometric with common ratio $(1 + r)^{-1}$. The sum of the series is
$S_n = P(1 + r)^{n-1} \cdot$

$\dfrac{1 - [(1+r)^{-1}]^n}{1 - (1+r)^{-1}} =$

$\dfrac{P(1+r)^{n-1}[1 - (1+r)^{-n}]}{1 - (1+r)^{-1}} =$

$\dfrac{P\left[\dfrac{(1+r)^n - 1}{1+r}\right]}{\dfrac{(1+r) - 1}{1+r}} =$

$\dfrac{P[(1+r)^n - 1]}{r}$ dollars.

44. a. $A_1 = (1 + r)A_0 - P = (1 + r)A - P$; $A_2 = (1 + r)A_1 - P = (1 + r)[(1 + r)A - P] - P = (1 + r)^2 A - [(1 + r) + 1]P$; $A_3 = (1 + r)A_2 - P = (1 + r)[(1 + r)^2 A - [(1 + r) + 1]P] - P = (1 + r)^3 A - [(1 + r)^2 + (1 + r) + 1]P$

b. Rearrange the terms in the bracketed series as $1 + (1 + r) + \cdots + (1 + r)^{n-1}$. This is a geometric series with common ratio $1 + r$;
$$S_n = \frac{1[1 - (1 + r)^n]}{1 - (1 + r)} = \frac{1 - (1 + r)^n}{-r} = \frac{(1 + r)^n - 1}{r}.$$ Thus,
$$A_n = (1 + r)^n A - \left[\frac{(1 + r)^n - 1}{r}\right]P.$$

c. After paying n installments, the loan is paid, and so $A_n = 0$. Thus,
$$0 = (1 + r)^n A - \left[\frac{(1 + r)^n - 1}{r}\right]P;$$
$$P = (1 + r)^n A \div \frac{(1 + r)^n - 1}{r} = (1 + r)^n A \cdot \frac{r}{(1 + r)^n - 1} = \frac{Ar(1 + r)^n}{(1 + r)^n - 1}.$$

44. Finance The purpose of this exercise is to develop a formula for the monthly payment P that is required to repay a loan of A dollars in n monthly installments, with the interest on the unpaid balance equal to a monthly rate r. Let A_k = amount still owed after paying k installments. Then interest for the $(k + 1)$st month is rA_k, and the principal paid off in the $(k + 1)$st payment is $P - rA_k$. Thus,

$$A_{k+1} = A_k - (P - rA_k) = (1 + r)A_k - P.$$

a. Use the equation above and the fact that $A_0 = A$ to find A_1, A_2, and A_3 in terms of A, r, and P.

b. Generalizing from A_1, A_2, and A_3, you get
$$A_n = (1 + r)^n A - [(1 + r)^{n-1} + (1 + r)^{n-2} + \cdots + (1 + r) + 1]P.$$
Since the bracketed quantity above is a geometric series, show that the equation can be rewritten
$$A_n = (1 + r)^n A - \left[\frac{(1 + r)^n - 1}{r}\right]P.$$

c. Since $A_n = 0$ (why?), use the last equation in part (b) to show that
$$P = \frac{Ar(1 + r)^n}{(1 + r)^n - 1}.$$

C **45. Finance** A *direct reduction loan* is often used for buying cars and for mortgages on homes. You pay interest only on that portion of the loan that you have not repaid (that is, the unpaid balance). Therefore the monthly payment you make has two parts: I, which is the interest on the unpaid balance, and R, which is used to reduce the balance.

Suppose you take out a car loan for $10,000 at 12% annual interest, calculated monthly, to be repaid in 5 years. Your monthly payment will be P.

a. Use the formula in Exercise 44(c) to show that $P = \$222.44$. Remember that r = monthly rate = $\dfrac{12\%}{12} = 0.01$.

b. Study the table below. Then calculate the entries in the last row of the table.

n = payment number	A_n = amount outstanding after payment	I_n = interest on A_n	R_n = loan reduction = \$222.44 − I_n
0	10,000.00	100.00	122.44
1	9,877.56	98.78	123.66
2	?	?	?
	9753.90	97.54	124.90

c. Express A_n in terms of A_{n-1}. $A_n = 1.01 A_{n-1} - 222.44$

d. Show that the values of R_n form a geometric sequence with ratio 1.01, which is (1 + monthly rate).

e. (Optional) Write a computer program that will print out the table above for all 60 payments. Label the column headings N, A, I, and R.

492 *Chapter Thirteen*

Infinite Sequences and Series

13-4 Limits of Infinite Sequences

Objective — *To find or estimate the limit of an infinite sequence or to determine that the limit does not exist.*

A sequence that does not have a last term is called *infinite*. Consider the infinite geometric sequence

$$\frac{1}{2}, \frac{1}{4}, \frac{1}{8}, \frac{1}{16}, \ldots, \left(\frac{1}{2}\right)^n, \ldots$$

The terms of this sequence are surely getting smaller, but how small do they get? With a calculator or logarithms, we can calculate that $t_{10} = \left(\frac{1}{2}\right)^{10} \approx 0.001$ and $t_{100} = \left(\frac{1}{2}\right)^{100} \approx 0.0000000000000000000000000000001$. When you substitute larger and larger values of n, $t_n = \left(\frac{1}{2}\right)^n$ becomes a smaller and smaller positive number. It never becomes zero, but we can make t_n come as close to zero as we like just by finding a large enough value for n.

The graph at the right illustrates this idea. No matter how small a positive number k we choose, we can always make t_n be within k units of zero just by going far enough to the right on the graph.

The preceding discussion can be summarized by the following equation:

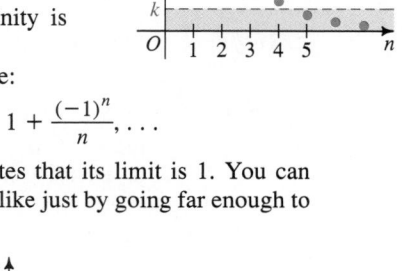

$$\lim_{n \to \infty} \left(\frac{1}{2}\right)^n = 0$$

This is read "the limit of $\left(\frac{1}{2}\right)^n$ as n goes to infinity is zero."

As another illustration consider this sequence:

$$1 - \frac{1}{1}, 1 + \frac{1}{2}, 1 - \frac{1}{3}, 1 + \frac{1}{4}, \ldots, 1 + \frac{(-1)^n}{n}, \ldots$$

The graph of this sequence, shown below, illustrates that its limit is 1. You can make the terms of the sequence as close to 1 as you like just by going far enough to the right on the graph.

$$\lim_{n \to \infty} \left(1 + \frac{(-1)^n}{n}\right) = 1$$

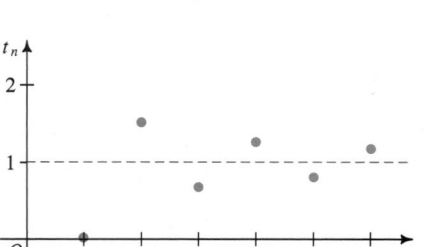

In everyday usage, the word "limit" sometimes suggests a barrier, but in mathematical usage it is better to think of a limit as a target. Thus, the limit 1 is a target approached more and more closely by $1 + \frac{(-1)^n}{n}$ as n gets larger and larger.

Sequences and Series **493**

Teaching Notes, p. 472C

Warm-Up Exercises

Use a calculator to evaluate each expression. Copy the display shown on your calculator and explain what it means. Answers may vary. An example is given.

1. $(0.9)^{1000}$ 1.7479 − 46; the value is about 1.7479×10^{-46}, which ≈ 0.

2. $(1.1)^{1000}$ 2.4699 41; the value is about 2.4699×10^{41}, a very large number.

3. $\left(1 + \frac{1}{10,000}\right)^{10,000}$ 2.7181459, which $\approx e$

4. $\frac{3(10,000) + 2}{4(10,000)^3 - 1}$ 7.5005 − 09; the value is about 7.5005×10^{-9}, a very small number.

Write each expression in simplest form using positive rational exponents.

5. $\frac{\sqrt{n}}{n^4}$ $\frac{1}{n^{7/2}}$ 6. $\frac{\sqrt[3]{n^2}}{n^2}$ $\frac{1}{n^{4/3}}$

Motivating the Section

A fund-raising telethon has a goal of $500,000. Suppose that $\frac{1}{2}$ of the goal is reached during the first 8 hours, $\frac{1}{4}$ of the goal is reached during the second 8 hours, $\frac{1}{8}$ of the goal is reached during the third 8 hours, and so on. Does it seem that the goal will never be reached? It is theoretically possible to reach the goal, but it will take an infinite amount of time.

Additional Examples

1. Write an expression for t_n, the nth term of each sequence. Then find the limit, or state that the limit does not exist.

   a. $\dfrac{1}{2}, \dfrac{3}{4}, \dfrac{7}{8}, \dfrac{15}{16}, \dfrac{31}{32}, \dfrac{63}{64}, \ldots$

   b. $\dfrac{1}{2}, \dfrac{4}{3}, \dfrac{9}{4}, \dfrac{16}{5}, \ldots$

   a. $t_n = \dfrac{2^n - 1}{2^n}$

   Dividing the numerator and the denominator by 2^n gives $t_n = \dfrac{1 - \dfrac{1}{2^n}}{1}$. As $n \to \infty$, $\dfrac{1}{2^n} \to 0$. Thus, as $n \to \infty$, $t_n \to \dfrac{1 - 0}{1} = 1$. The limit is 1.

   b. $t_n = \dfrac{n^2}{n + 1}$

   Dividing the numerator and denominator by n gives $t_n = \dfrac{n}{1 + \dfrac{1}{n}}$. As $n \to \infty$, $\dfrac{1}{n} \to 0$. Thus, as $n \to \infty$, $t_n \to \dfrac{n}{1 + 0} = n$. But $n \to \infty$, so $t_n \to \infty$.

You can often estimate the limit of an infinite sequence by substituting large values of n in the formula for the nth term, as illustrated in the next example.

Example 1 Find: **a.** $\displaystyle\lim_{n \to \infty} \sin\left(\dfrac{1}{n}\right)$ **b.** $\displaystyle\lim_{n \to \infty} (0.99)^n$

Solution

a. When $n = 100$, $\sin\left(\dfrac{1}{n}\right) = \sin\left(\dfrac{1}{100}\right) \approx 0.01$. As n gets larger, $\dfrac{1}{n}$ approaches 0 and so does $\sin\dfrac{1}{n}$. Thus, it seems that $\displaystyle\lim_{n \to \infty} \sin\left(\dfrac{1}{n}\right) = 0$.

b. We can evaluate $(0.99)^n$ for large n with logarithms or a calculator. For example,
$$(0.99)^{1000} \approx 4.3 \times 10^{-5} \text{ and } (0.99)^{10,000} \approx 2.2 \times 10^{-44}.$$
Thus, it seems that $\displaystyle\lim_{n \to \infty} (0.99)^n = 0$.

We have just estimated that $\displaystyle\lim_{n \to \infty} (0.99)^n$ is 0, and earlier we saw that $\displaystyle\lim_{n \to \infty} \left(\dfrac{1}{2}\right)^n = 0$. These two examples are special cases of the following theorem.

Theorem
If $|r| < 1$, then $\displaystyle\lim_{n \to \infty} r^n = 0$.

In Example 1 we estimated limits by considering t_n for large values of n. In the next example, we show how to change the form of t_n to assist in finding a limit.

Example 2 Find: **a.** $\displaystyle\lim_{n \to \infty} \dfrac{n^2 + 1}{2n^2 - 3n}$ **b.** $\displaystyle\lim_{n \to \infty} \dfrac{5n^2 + \sqrt{n}}{3n^3 + 7}$

Solution In both parts (a) and (b), we divide numerator and denominator by the highest power of n that occurs in the denominator.

a. Dividing numerator and denominator by n^2, we have:
$$\dfrac{n^2 + 1}{2n^2 - 3n} = \dfrac{1 + \dfrac{1}{n^2}}{2 - \dfrac{3}{n}}$$

Notice that when n is very large, $\dfrac{1}{n^2}$ and $\dfrac{3}{n}$ are very near 0. Therefore,
$$\dfrac{n^2 + 1}{2n^2 - 3n} \approx \dfrac{1}{2}$$
when n is very large. For this reason, we say:
$$\lim_{n \to \infty} \dfrac{n^2 + 1}{2n^2 - 3n} = \dfrac{1}{2}$$

b. Dividing numerator and denominator by n^3, we have:

$$\frac{5n^2 + \sqrt{n}}{3n^3 + 7} = \frac{\dfrac{5}{n} + \dfrac{1}{n^{5/2}}}{3 + \dfrac{7}{n^3}}$$

Notice that when n is very large, $\dfrac{5}{n}$, $\dfrac{1}{n^{5/2}}$, and $\dfrac{7}{n^3}$ are very near 0.

Therefore,
$$\frac{5n^2 + \sqrt{n}}{3n^3 + 7} \approx \frac{0 + 0}{3 + 0} = 0$$

when n is very large. For this reason, we say:

$$\lim_{n \to \infty} \frac{5n^2 + \sqrt{n}}{3n^3 + 7} = 0$$

Situations in Which a Sequence Has No Limit

Not all sequences have limits. If the terms of a sequence do not "home in" on a single value, we say that the sequence has no limit or that the limit of the sequence does not exist. For example, the following sequence has no limit:

$$\frac{1}{2}, \; -\frac{2}{3}, \; \frac{3}{4}, \; -\frac{4}{5}, \ldots, \; \frac{(-1)^{n+1} \cdot n}{n + 1}, \ldots$$

The diagram shows the graph of this sequence. Notice that the odd-numbered terms form a sequence with limit 1. Similarly, the even-numbered terms form a sequence with limit -1. Nevertheless, there is no single limiting number for all the terms of the sequence. Thus, we say that this sequence has no limit.

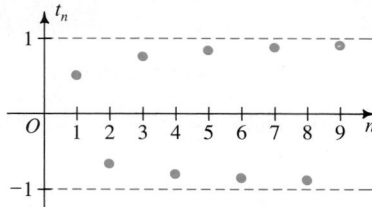

Infinite Limits

Sometimes the terms of a sequence increase or decrease without bound. Here are two examples:

(a) $3, 7, 11, 15, \ldots, 4n - 1, \ldots$ (b) $-10, -100, -1000, \ldots, -10^n, \ldots$

The terms in sequence (a) continue to increase as n becomes larger, but they do not approach a fixed number as a target. No matter what fixed number is selected, the terms of the sequence will eventually exceed it. For this reason, we say that the limit of $4n - 1$ as n increases without bound is infinity. We write this as

$$\lim_{n \to \infty} (4n - 1) = \infty.$$

Similarly, the terms in sequence (b) decrease without bound; that is, they are farther and farther to the left of zero on the number line. We say that the limit of -10^n as n increases without bound is negative infinity and write

$$\lim_{n \to \infty} (-10^n) = -\infty.$$

Sequences and Series   **495**

2. Find the limit or state that the limit does not exist.

a. $\lim\limits_{n \to \infty} e^n$

b. $\lim\limits_{n \to \infty} \dfrac{\tan n\pi}{n}$

a. The terms of the sequence e, e^2, e^3, e^4, ... become arbitrarily large as $n \to \infty$, since $e \approx 2.7$; thus, the limit is infinity.

b. For all n, $\tan n\pi = 0$.
Also, $\lim\limits_{n \to \infty} \dfrac{1}{n} = 0$. Thus,
$$\lim_{n \to \infty} \frac{\tan n\pi}{n} = 0 \cdot 0 = 0.$$

Review Note

A quick review of fractional exponents might be helpful for part (b) of Example 2:

$$\frac{\sqrt{n}}{n^3} = \frac{n^{1/2}}{n^3} = \frac{1}{n^{3-1/2}} = \frac{1}{n^{5/2}}$$

Mathematical Note

A sequence that has a limit is called a *convergent* sequence. A sequence that does not have a limit is called a *divergent* sequence.

Communication Note

A class discussion about the concept of infinity may be very instructive and helpful. Many students tend to think of infinity as a number. Connecting infinity to the idea of "increasing without bound" and negative infinity to the idea of "decreasing without bound" is important.

Cooperative Learning

Class Exercises 1–10 provide a good opportunity for using "paired partners." Have one of the partners work the odd-numbered exercises and the other work the even-numbered ones. Then have the partners share results and explain their answers to one another.

Suggested Assignments

Standard
496/1–25 odd
Comprehensive
496/1–33 odd

Supplementary Materials

Alternative Assessment, 40–41
Precalculus Plotter Plus,
73–74

It is important to realize that infinity is neither a number nor a place. When we say that "n goes to infinity," we mean that n increases without bound. When we say that "the limit is infinity," we mean that the terms are increasing without bound, not that they are approaching some number.

Example 3   Show that $\lim\limits_{n\to\infty}\dfrac{7n^3}{4n^2-5}=\infty$.

Solution   Dividing numerator and denominator by n^2, the highest power of n in the denominator, we get:

$$\frac{7n^3}{4n^2-5}=\frac{7n}{4-\dfrac{5}{n^2}}$$

When n is very large, $\dfrac{5}{n^2}$ is very near 0. Therefore, $\dfrac{7n^3}{4n^2-5}\approx\dfrac{7n}{4}$ when n is very large. Thus, since $\dfrac{7n}{4}$ increases without bound as n does,

$$\lim_{n\to\infty}\frac{7n}{4}=\infty \quad\text{and}\quad \lim_{n\to\infty}\frac{7n^3}{4n^2-5}=\infty.$$

CLASS EXERCISES

Find the following limits.

1. $\lim\limits_{n\to\infty}\dfrac{n}{n+1}$ 1

2. $\lim\limits_{n\to\infty}\dfrac{n^2-1}{n^2}$ 1

3. $\lim\limits_{n\to\infty}\dfrac{2n+1}{3n+1}$ $\dfrac{2}{3}$

4. $\lim\limits_{n\to\infty}\dfrac{8n^2-3n}{5n^2+7}$ $\dfrac{8}{5}$

5. $\lim\limits_{n\to\infty}\cos\left(\dfrac{1}{n}\right)$ 1

6. $\lim\limits_{n\to\infty}\log\left[\cos\left(\dfrac{1}{n}\right)\right]$ 0

7. $\lim\limits_{n\to\infty}(0.999)^n$ 0

8. $\lim\limits_{n\to\infty}(1.001)^n$ ∞

9. $\lim\limits_{n\to\infty}\dfrac{n^4}{2n+1}$ ∞

10. $\lim\limits_{n\to\infty}\dfrac{n^2+9{,}999{,}999}{n^3}$ 0

11. **Discussion**   Do you think the sequence 1, 0, 1, 0, 1, 0, . . . has a limit? No. Why? The terms do not "home in" on a single value.

WRITTEN EXERCISES

In Exercises 1–12, find the given limit.

A

1. $\lim\limits_{n\to\infty}\dfrac{n+5}{n}$ 1

2. $\lim\limits_{n\to\infty}\dfrac{n^2+1}{n^2}$ 1

3. $\lim\limits_{n\to\infty}\left[1+\dfrac{(-1)^n}{n}\right]$ 1

4. $\lim\limits_{n\to\infty}\dfrac{4n-3}{2n+1}$ 2

5. $\lim\limits_{n\to\infty}\dfrac{3n^2+5n}{8n^2}$ $\dfrac{3}{8}$

6. $\lim\limits_{n\to\infty}\dfrac{2n^4}{6n^5+7}$ 0

496   *Chapter Thirteen*

7. $\lim_{n \to \infty} \tan\left(\dfrac{1}{n}\right)$ 0

8. $\lim_{n \to \infty} \sec\left(\dfrac{1}{n}\right)$ 1

9. $\lim_{n \to \infty} \dfrac{\sqrt{n}}{n+1}$ 0

10. $\lim_{n \to \infty} \dfrac{5n^{2/3} - 8n}{6n - 1}$ $-\dfrac{4}{3}$

11. $\lim_{n \to \infty} \log\left(\dfrac{n+1}{n}\right)$ 0

12. $\lim_{n \to \infty} \log \sqrt[n]{10}$ 0

In Exercises 13–18, find the limit of the specified sequence or state that the limit does not exist.

13. $\dfrac{1}{3}, -\dfrac{1}{9}, \dfrac{1}{27}, -\dfrac{1}{81}, \dfrac{1}{243}, \ldots$ 0

14. $1, -4, 9, -16, 25, -36, \ldots$ Does not exist

15. $\dfrac{3}{2}, -\dfrac{4}{3}, \dfrac{5}{4}, -\dfrac{6}{5}, \dfrac{7}{6}, -\dfrac{8}{7}, \cdots$ Does not exist

16. $\dfrac{1}{10}, -\dfrac{2}{10^2}, \dfrac{3}{10^3}, -\dfrac{4}{10^4}, \dfrac{5}{10^5}, \ldots$ 0

17. $t_n = \cos\left(\dfrac{n\pi}{2}\right)$ Does not exist

18. $t_n = \sin(n\pi)$ 0

In Exercises 19–30, evaluate the given limit or state that the limit does not exist. If the sequence approaches ∞ or $-\infty$, so state.

19. $\lim_{n \to \infty} \dfrac{5n^{5/2} - 7n}{n^2 + 10n}$ ∞

20. $\lim_{n \to \infty} \dfrac{5n}{n^{1/2} - 3}$ ∞

21. $\lim_{n \to \infty} \log\left(\dfrac{1}{n}\right)$ $-\infty$

22. $\lim_{n \to \infty} \dfrac{\sin n}{n}$ 0

23. $\lim_{n \to \infty} \dfrac{\cos(n\pi)}{n}$ 0

24. $\lim_{n \to \infty} e^{-n}$ 0

25. $\lim_{n \to \infty} \tan\left(\dfrac{\pi}{4} + n\pi\right)$ 1

26. $\lim_{n \to \infty} 2^n$ ∞

B **27.** $\lim_{n \to \infty} \log\left[\sec\left(\dfrac{1}{n}\right)\right]$ 0

28. $\lim_{n \to \infty} [(-1)^n - 1]$ Does not exist

29. $\lim_{n \to \infty} \dfrac{\sqrt{n+1}}{\sqrt{n-1}}$ 1

30. $\lim_{n \to \infty} \dfrac{\sqrt[3]{8n^2 - 5n + 1}}{\sqrt[3]{n^2 + 7n - 3}}$ 2

31. a. If the nth term of a geometric series is $t_n = \left(\dfrac{1}{3}\right)^{n-1}$, show that
$$S_n = \dfrac{3}{2}\left[1 - \left(\dfrac{1}{3}\right)^n\right].$$

b. Find $\lim_{n \to \infty} S_n$. $\dfrac{3}{2}$

32. Consider the formula for the sum of the series $t_1 + t_1 r + \cdots + t_1 r^{n-1}$ and then suggest a formula for the sum of the following infinite series:
$$t_1 + t_1 r + t_1 r^2 + \cdots, \text{ where } |r| < 1 \quad \dfrac{t_1}{1-r}$$

33. Recall that $e = \lim_{n \to \infty} \left(1 + \dfrac{1}{n}\right)^n$. Find $\lim_{n \to \infty} \left(1 + \dfrac{2}{n}\right)^n$ by noting that
$$\left(1 + \dfrac{2}{n}\right)^n = \left(1 + \dfrac{1}{\frac{n}{2}}\right)^n = \left[\left(1 + \dfrac{1}{\frac{n}{2}}\right)^{n/2}\right]^2. \quad e^2$$

34. a. Show that $\lim\limits_{n\to\infty}\left(1 + \dfrac{3}{n}\right)^n = e^3$.   **b.** Evaluate $\lim\limits_{n\to\infty}\left(1 + \dfrac{1}{2n}\right)^n$. $e^{1/2}$

(*Hint:* See Exercise 33.)

35. The area A under the curve $y = x^3$ between $x = 0$ and $x = 1$ can be approximated by adding the areas of n rectangles as shown.

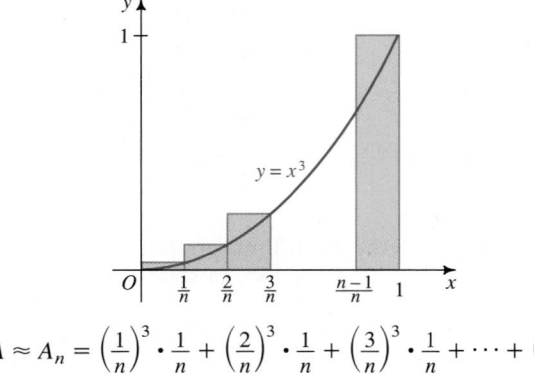

$$A \approx A_n = \left(\dfrac{1}{n}\right)^3 \cdot \dfrac{1}{n} + \left(\dfrac{2}{n}\right)^3 \cdot \dfrac{1}{n} + \left(\dfrac{3}{n}\right)^3 \cdot \dfrac{1}{n} + \cdots + \left(\dfrac{n}{n}\right)^3 \cdot \dfrac{1}{n}$$

$$= \dfrac{1}{n^4}(1^3 + 2^3 + 3^3 + \cdots + n^3)$$

a. According to the result of Exercise 40 in Section 13–3,

$$1^3 + 2^3 + 3^3 + \cdots + n^3 = \left[\dfrac{n(n+1)}{2}\right]^2.$$

Use this formula to show that $A_n = \dfrac{n^2 + 2n + 1}{4n^2}$.

b. As n becomes very large, what value is A_n approaching? $\dfrac{1}{4}$

c. Find A by evaluating $\lim\limits_{n\to\infty} A_n$. $\dfrac{1}{4}$

C **36.** Use the procedure in Exercise 35 to find the area under the curve $y = x^2$ between 0 and 1. You will need to know that

$$1^2 + 2^2 + 3^2 + \cdots + n^2 = \dfrac{n(n+1)(2n+1)}{6}. \quad \dfrac{1}{3}$$

IIII CALCULATOR EXERCISES

1. a. Guess the value of $\lim\limits_{n\to\infty}\left(\sqrt{n+1} - \sqrt{n}\right)$ by evaluating $\sqrt{n+1} - \sqrt{n}$ for several large values of n. 0

b. Multiply $\sqrt{n+1} - \sqrt{n}$ by $\dfrac{\sqrt{n+1} + \sqrt{n}}{\sqrt{n+1} + \sqrt{n}}$. $\dfrac{1}{\sqrt{n+1} + \sqrt{n}}$

c. Determine what happens to the expression in part (b) when n becomes very large. Does your answer agree with your answer to part (a)? The value of the expression approaches zero. Yes

2. A sequence is defined recursively by the equations

$$t_1 = 1, \text{ and } t_n = \frac{t_{n-1}}{2} + \frac{1}{t_{n-1}}.$$

  a. Find decimal approximations for the first five terms of the sequence. 1, 1.5, 1.4166667, 1.4142157, 1.4142136

  b. Suggest a limit for this sequence. $\sqrt{2}$

3. a. Evaluate $\sqrt{1 + \sqrt{1 + \sqrt{1 + \sqrt{1 + \cdots}}}}$ by considering its value to be the limit of the following sequence as n approaches ∞:

$$\sqrt{1 + \sqrt{1}}, \ \sqrt{1 + \sqrt{1 + \sqrt{1}}}, \ \sqrt{1 + \sqrt{1 + \sqrt{1 + \sqrt{1}}}}, \text{ and so on.} \quad 1.618034$$

  b. Compare your answer with the golden ratio, $R = \dfrac{1 + \sqrt{5}}{2}$. Same

//// COMPUTER EXERCISE

A student leaves home to go to the movies. Halfway there, the student remembers some uncompleted homework and heads back home. Halfway back home the student has a change of mind and heads back to the movies. You guessed it! Halfway back to the movies, the student, overcome by an attack of conscience, heads back to complete the homework. Suppose the student continues in this fashion. Write a computer program to calculate how far from home the student is after each of the first twenty changes of mind. (If you think of the student as moving along a number line with home at 0 and the movies at 1, does the student appear to be approaching a limiting point on the number line?)

Manuel Sandoval Vallarta (1899–1977)

Manuel Sandoval Vallarta, a mathematician and physicist, was a professor at the Massachusetts Institute of Technology. He studied relativity with Albert Einstein and electromagnetic theory with Max Planck. In 1946 he returned to his native Mexico, where he founded the Instituto Nacional de Energía Nuclear.

A pioneer in the field of cosmic-ray physics, and an advocate for the peaceful uses of atomic energy, Sandoval Vallarta was president of the Latin American Council on Cosmic Rays, and represented Mexico in conferences of the United Nations Atomic Energy Agency.

Sequences and Series   **499**

```
10 LET M = 1
20 LET P = 0
30 FOR N = 1 TO 20 STEP 2
40 LET D = .5 * (M − P)
50 LET P = P + D
60 PRINT P
70 LET P = .5 * P
80 PRINT P
90 NEXT N
100 END
```

After twenty changes of mind, the student will be approximately $\frac{1}{3}$ of the distance from home; however, after 21 changes of mind, the student will be approximately $\frac{2}{3}$ of the distance from home. The student does not approach a limiting point.

Warm-Up Exercises

Use your calculator to find S_1, S_2, S_3, S_4, and S_5. Then state whether the infinite series seems to have a limit or not. If so, predict the limit.

1. $1 - \dfrac{1}{3} + \dfrac{1}{9} - \dfrac{1}{27} + \dfrac{1}{81} - \cdots$ 1; $0.\overline{6}$; $0.\overline{7}$; $0.\overline{740}$; about 0.753; yes; about 0.75

2. $1 + 2 + 4 + 8 + 16 + \cdots$ 1; 3; 7; 15; 31; no

3. $0.3 + 0.03 + 0.003 + 0.0003 + \cdots$ 0.3; 0.33; 0.333; 0.3333; 0.33333; yes; $\dfrac{1}{3}$

4. $1 + 0.5 + (0.5)^2 + (0.5)^3 + (0.5)^4 + \cdots$ 1; 1.5; 1.75; 1.875; 1.9375; yes; 2

Motivating the Section

Have students discuss the following, somewhat fanciful, situation and find out what rule the National Football League might have that would relate to it: In a football game, the team on offense has the ball inside the 10-yard line of the defensive team. The defensive team commits a series of rule infractions for which it is penalized. Each penalty amounts to one half the remaining distance to the goal line. Will the football ever cross the goal line for a touchdown as a result of penalties alone? If so, after how many penalties will a touchdown be declared?

13-5 Sums of Infinite Series

| Objective | To find the sum of an infinite geometric series. |

The sum of an infinite series is very closely connected to the limit of an infinite sequence. To see this, consider the infinite geometric series

$$\frac{1}{2} + \frac{1}{4} + \frac{1}{8} + \frac{1}{16} + \cdots + \left(\frac{1}{2}\right)^n + \cdots.$$

Associated with this series is the following *sequence of partial sums:*

$$S_1 = \frac{1}{2}$$

$$S_2 = \frac{1}{2} + \frac{1}{4} = \frac{3}{4}$$

$$S_3 = \frac{1}{2} + \frac{1}{4} + \frac{1}{8} = \frac{7}{8}$$

$$\vdots$$

$$S_n = \frac{1}{2} + \frac{1}{4} + \frac{1}{8} + \frac{1}{16} + \cdots + \left(\frac{1}{2}\right)^n$$

$$= \frac{\frac{1}{2}\left(1 - \left(\frac{1}{2}\right)^n\right)}{1 - \frac{1}{2}} \longleftarrow \text{ using the formula } S_n = \frac{t_1(1 - r^n)}{1 - r}$$

$$= 1 - \left(\frac{1}{2}\right)^n$$

Since the sequence of partial sums $\frac{1}{2}, \frac{3}{4}, \frac{7}{8}, \ldots, 1 - (\frac{1}{2})^n$ has limit 1, we say that the infinite series has **limit** 1 or has the **sum** 1.

In general, for any infinite series $t_1 + t_2 + \cdots + t_n + \cdots$,

$$S_n = t_1 + t_2 + \cdots + t_n$$

is called the **nth partial sum**. If the **sequence of partial sums** $S_1, S_2, \ldots, S_n, \ldots$ has a finite limit S, then the infinite series is said to **converge** to the sum S. If the sequence of partial sums approaches infinity or has no finite limit, the infinite series is said to **diverge.**

Since we already have a formula for the nth partial sum of a geometric series, we can prove a theorem that tells when such series converge.

Sum of an Infinite Geometric Series

If $|r| < 1$, the infinite geometric series

$$t_1 + t_1 r + t_1 r^2 + \cdots + t_1 r^n + \cdots$$

converges to the sum $S = \dfrac{t_1}{1 - r}$.

If $|r| \geq 1$ and $t_1 \neq 0$, then the series diverges.

---

500 *Chapter Thirteen*

Proof: The nth partial sum of the geometric series is

$$S_n = \frac{t_1(1 - r^n)}{1 - r}.$$

1. Therefore, if $|r| < 1$:

$$\lim_{n \to \infty} S_n = \lim_{n \to \infty} \frac{t_1(1 - r^n)}{1 - r}$$

$$= \frac{t_1(1 - 0)}{1 - r} \text{ since } \lim_{n \to \infty} r^n = 0 \text{ when } |r| < 1$$

$$= \frac{t_1}{1 - r}$$

2. However, if $|r| > 1$, r^n increases or decreases without bound as n approaches infinity. Thus, S_n becomes infinite and the series diverges.

3. If $r = 1$, the series becomes the divergent series

$$t_1 + t_1 + t_1 + \cdots$$

4. If $r = -1$, the series becomes the divergent series

$$t_1 - t_1 + t_1 - t_1 + \cdots$$

Example 1 Find the sum of the infinite geometric series

$$9 - 6 + 4 - \cdots.$$

Solution Since $t_1 = 9$ and $r = -\dfrac{2}{3}$,

$$S = \frac{t_1}{1 - r} = \frac{9}{1 - \left(-\dfrac{2}{3}\right)} = \frac{27}{5}.$$

Example 2 For what values of x does the following infinite series converge?

$$1 + (x - 2) + (x - 2)^2 + (x - 2)^3 + \cdots$$

Solution This is an infinite geometric series with $r = x - 2$. By the theorem on page 500, the series converges when $|r| < 1$; that is, when $|x - 2| < 1$, or

$$1 < x < 3.$$

This interval $1 < x < 3$ for which the series converges is called the *interval of convergence* for the series.

Our final example on the next page illustrates two important facts about repeating decimals. First, they can be written as infinite geometric series, and second, they represent rational numbers.

Mathematical Note

For $|r| > 1$, an infinite geometric series grows without bound. For example, the series $1 + 2 + 4 + 8 + \cdots$ has common ratio $r = 2$. If we use the given theorem, the sum of the series is $S = \dfrac{1}{1 - 2} = -1$. It is obvious that this is not a correct sum.

Additional Examples

1. Find the first three terms of an infinite geometric sequence with sum 16 and common ratio $-\dfrac{1}{2}$.

$$S = \frac{t_1}{1 - r}; \quad 16 = \frac{t_1}{1 - \left(-\dfrac{1}{2}\right)}$$

$$t_1 = 24$$

$$t_2 = 24\left(-\frac{1}{2}\right) = -12$$

$$t_3 = (-12)\left(-\frac{1}{2}\right) = 6$$

2. Show that the series $1 + \cos^2 x + \cos^4 x + \cos^6 x + \cdots$ is geometric and converges to $\csc^2 x$ if $x \neq n\pi$, where n is an integer.

The common ratio is $\cos^2 x$ and $t_1 = 1$; the infinite geometric series has sum $\dfrac{t_1}{1 - r} =$

$$\frac{1}{1 - \cos^2 x} = \frac{1}{\sin^2 x} =$$

$\csc^2 x$ provided $|r| < 1$; $|\cos^2 x| < 1$; $-1 < \cos^2 x < 1$; $x \neq n\pi$, where n is an integer.

7. For an arithmetic series with $t_1 \neq 0$ and $d \neq 0$, $S_n = \frac{n(t_1 + t_n)}{2}$; as $n \to \infty$, $\left| \frac{n(t_1 + t_n)}{2} \right| \to \infty$, so an infinite arithmetic series diverges.

8. The series diverges; each term of the series is greater than or equal to $\frac{1}{2}$, so the sequence of partial sums approaches infinity.

Suggested Assignments

Comprehensive
Day 1: 502/1, 3, 7, 9, 11, 13, 15, 19, 21, 23
Day 2: 503/29, 31, 33, 35, 41, 43; Calc. Ex.

Supplementary Materials

Alternative Assessment, 41
Precalculus Plotter Plus, 75

Example 3   The infinite, repeating decimal 0.454545 . . . can be written as the infinite series

$$0.45 + 0.0045 + 0.000045 + \cdots.$$

What is the sum of this series?

Solution   This is a geometric series with $t_1 = 0.45$ and $r = 0.01$. Therefore,

$$S = \frac{t_1}{1 - r} = \frac{0.45}{1 - 0.01} = \frac{0.45}{0.99} = \frac{5}{11}.$$

CLASS EXERCISES

For each infinite geometric series, find S_1, S_2, S_3, and S_4. Also, find the sum of the series if it converges. $\frac{1}{2}, \frac{1}{4}, \frac{3}{8}, \frac{5}{16}; \frac{1}{3}$

1. $1 + \frac{1}{3} + \frac{1}{9} + \frac{1}{27} + \cdots$ 1, $\frac{4}{3}, \frac{13}{9}, \frac{40}{27}; \frac{3}{2}$ 2. $\frac{1}{2} - \frac{1}{4} + \frac{1}{8} - \frac{1}{16} + \cdots$

3. $1 + 3 + 9 + 27 + \cdots$ 1, 4, 13, 40; diverges   4. $1 + 0.1 + 0.01 + 0.001 + \cdots$ 1, 1.1, 1.11, 1.111; $\frac{10}{9}$

5. Find the interval of convergence for each series.
   a. $1 + x + x^2 + x^3 + \cdots$ $-1 < x < 1$ **b.** $1 + 2x + 4x^2 + 8x^3 + \cdots$ $-\frac{1}{2} < x < \frac{1}{2}$

6. Express 0.3333 . . . as an infinite geometric series. For this series, determine **(a)** t_1, **(b)** r, and **(c)** the sum. 0.3; 0.1; $\frac{1}{3}$

7. *Discussion* Consider any infinite arithmetic series for which $t_1 \neq 0$ and $d \neq 0$. Explain why this series diverges.

8. *Discussion* Do you think the series $\frac{1}{2} + \frac{2}{3} + \frac{3}{4} + \frac{4}{5} + \cdots + \frac{n}{n + 1} + \cdots$ converges or diverges? Tell why. Diverges

WRITTEN EXERCISES

In Exercises 1–8, find the sum of the given infinite geometric series.

A 1. $1 + \frac{1}{2} + \frac{1}{4} + \frac{1}{8} + \cdots$ 2   2. $1 - \frac{1}{3} + \frac{1}{9} - \frac{1}{27} + \cdots$ $\frac{3}{4}$

3. $24 - 12 + 6 - 3 + \cdots$ 16   4. $\frac{1}{4} + \frac{1}{16} + \frac{1}{64} + \cdots$ $\frac{1}{3}$

5. $5 + 5^{-1} + 5^{-3} + \cdots$ $\frac{125}{24}$ 6. $\sqrt{27} + \sqrt{9} + \sqrt{3} + \cdots$ $\frac{1}{2}(9 + 9\sqrt{3})$

7. $t_1 + t_2 + \cdots + t_n + \cdots$, where $t_n = 8(5)^{-n}$ 2   8. $t_1 + t_2 + \cdots + t_n + \cdots$, where $t_n = (-2)^{1-n}$ $\frac{2}{3}$

9. Find the common ratio of an infinite geometric series with sum 8 and first term 4. $\frac{1}{2}$

10. Find the first three terms of an infinite geometric series with sum 81 and common ratio $\frac{1}{3}$. 54, 18, 6

---

502   *Chapter Thirteen*

11. What is the value of x if the series $1 + 2x + 4x^2 + \cdots$ converges to $\frac{3}{5}$? $-\frac{1}{3}$

12. What is the value of x if the series $x^2 - x^3 + x^4 - \cdots$ converges to $\frac{x}{5}$? 0 or $\frac{1}{4}$

For each infinite geometric series, find (a) the interval of convergence and (b) the sum, expressed in terms of x.

13. $1 + x^2 + x^4 + x^6 + \cdots$ $-1 < x < 1$; $\frac{1}{1 - x^2}$
14. $1 + 3x + 9x^2 + \cdots$ $-\frac{1}{3} < x < \frac{1}{3}$; $\frac{1}{1 - 3x}$

15. $1 + (x - 3) + (x - 3)^2 + \cdots$ **15–18,** See below.
16. $1 - (x - 1) + (x - 1)^2 - \cdots$

17. $1 - \frac{2}{x} + \frac{4}{x^2} - \frac{8}{x^3} + \cdots$
18. $\frac{x^2}{3} - \frac{x^4}{6} + \frac{x^6}{12} - \cdots$

19. Show that the series $\sin^2 x + \sin^4 x + \sin^6 x + \cdots$ converges to $\tan^2 x$ if $x \ne \frac{\pi}{2} + n\pi$.

20. **a.** Show that the series $\tan^2 x - \tan^4 x + \tan^6 x - \cdots$ converges to $\sin^2 x$ if $-\frac{\pi}{4} < x < \frac{\pi}{4}$.

   15. $2 < x < 4$; $\frac{1}{4 - x}$ **16.** $0 < x < 2$; $\frac{1}{x}$ **17.** $x < -2$ or $x > 2$; $\frac{x}{x + 2}$

   b. Are there other values of x for which the series converges?

21. **Writing** Explain why there is no infinite geometric series with first term 10 and sum 4.

22. **Writing** Explain why the sum of an infinite geometric series is positive if and only if the first term is positive.

   18. $-\sqrt{2} < x < \sqrt{2}$; $\frac{2x^2}{6 + 3x^2}$

In Exercises 23–28, use the method of Example 3 to express the given repeating decimal as a rational number.

23. $0.777\ldots$ $\frac{7}{9}$
24. $0.636363\ldots$ $\frac{7}{11}$
25. $44.444\ldots$ $\frac{400}{9}$
26. $5.363636\ldots$ $\frac{59}{11}$
27. $0.142857142857\ldots$ $\frac{1}{7}$
28. $0.0123123123\ldots$ $\frac{41}{3330}$

For the series in Exercises 29–32, find the first four partial sums, S_1, S_2, S_3, S_4, and suggest a formula for S_n. Then find the sum of the infinite series by evaluating $\lim\limits_{n \to \infty} S_n$. Note that since these series are not geometric, you cannot use the sum formula given on page 500.

B

29. $\frac{1}{1 \cdot 2} + \frac{1}{2 \cdot 3} + \frac{1}{3 \cdot 4} + \cdots + \frac{1}{n(n + 1)} + \cdots$ $\frac{1}{2}, \frac{2}{3}, \frac{3}{4}, \frac{4}{5}; \frac{n}{n + 1}; 1$

30. $\frac{1}{1 \cdot 3} + \frac{1}{3 \cdot 5} + \frac{1}{5 \cdot 7} + \cdots + \frac{1}{(2n - 1)(2n + 1)} + \cdots$ $\frac{1}{3}, \frac{2}{5}, \frac{3}{7}, \frac{4}{9}; \frac{n}{2n + 1}; \frac{1}{2}$

31. $\frac{1}{1 \cdot 4} + \frac{1}{4 \cdot 7} + \frac{1}{7 \cdot 10} + \cdots + \frac{1}{(3n - 2)(3n + 1)} + \cdots$ $\frac{1}{4}, \frac{2}{7}, \frac{3}{10}, \frac{4}{13}; \frac{n}{3n + 1}; \frac{1}{3}$

32. $\frac{3}{1 \cdot 4} + \frac{5}{4 \cdot 9} + \frac{7}{9 \cdot 16} + \cdots + \frac{2n + 1}{n^2(n + 1)^2} + \cdots$ $\frac{3}{4}, \frac{8}{9}, \frac{15}{16}, \frac{24}{25}; \frac{n^2 + 2n}{(n + 1)^2}; 1$

Exercise Note

For Exercises 23–28, see the teaching notes for this section for another method of deriving the rational numbers. You might want students to practice both methods.

Additional Answers Written Exercises

19. $t_1 = \sin^2 x$ and $r = \sin^2 x$;
   $S = \dfrac{\sin^2 x}{1 - \sin^2 x} = \dfrac{\sin^2 x}{\cos^2 x} = \tan^2 x$ provided $|r| < 1$;
   $|\sin^2 x| < 1$;
   $-1 < \sin^2 x < 1$;
   $x \ne \frac{\pi}{2} + n\pi$, n an integer.

20. **a.** $t_1 = \tan^2 x$ and $r = -\tan^2 x$; if $-\frac{\pi}{4} < x < \frac{\pi}{4}$, then $-1 < \tan x < 1$ and $|-\tan^2 x| < 1$, so the series converges.
   $S = \dfrac{\tan^2 x}{1 - (-\tan^2 x)} = \dfrac{\tan^2 x}{\sec^2 x} = \dfrac{\sin^2 x}{\cos^2 x} \cdot \cos^2 x = \sin^2 x$

   b. $|r| < 1$ when $|-\tan^2 x| < 1$, or $-1 < \tan x < 1$; this is true, for example, when $\frac{3\pi}{4} < x < \frac{5\pi}{4}$; and in general, when $-\frac{\pi}{4} + n\pi < x < \frac{\pi}{4} + n\pi$, n an integer.

22. If an infinite geometric series has a sum S, then $S = \dfrac{t_1}{1 - r}$ and $|r| < 1$; thus $1 - r > 0$; S is positive if and only if t_1 is positive.

33. **Physics** A ball is dropped from a height of 8 m. Each time it hits the ground, it rebounds $\frac{3}{4}$ of the distance it has fallen. In theory, how far will the ball travel before coming to rest? 56 m   **35. a.** 288   **b.** $96 + 48\sqrt{2}$

34. Repeat Exercise 33 for a ball dropped from a height of 10 m if it rebounds 95% of the distance it falls each time. 390 m

35. Each side of a square has length 12. The midpoints of the sides of the square are joined to form another square, and the midpoints of this square are joined to form still another square. If this process is continued indefinitely, find **(a)** the sum of the areas of all the squares and **(b)** the sum of the perimeters. See above

36. Each side of an equilateral triangle has length 12. The midpoints of the sides of the triangle are joined to form another equilateral triangle, and the midpoints of this triangle are joined to form still another triangle. If this process is continued indefinitely, find **(a)** the sum of the areas of all the triangles and **(b)** the sum of the perimeters. $48\sqrt{3}$; 72

37. S_n is the nth partial sum and S is the limit of the geometric series

$$1 + \frac{1}{2} + \frac{1}{4} + \frac{1}{8} + \cdots.$$

What is the smallest value of n for which $S - S_n < 0.0001$? 15

38. Repeat Exercise 37 for the geometric series $1 + \frac{2}{3} + \frac{4}{9} + \cdots.$ 26

39. Here is an old paradox: Achilles races a turtle who has a 100-meter head start. If Achilles runs 10 m/s and the turtle only 1 m/s, when will Achilles overtake the turtle?

Erroneous Solution When Achilles covers the 100-meter head start, the turtle has moved 10 m ahead. And when Achilles covers this 10 m, the turtle has moved 1 m ahead. Every time Achilles runs to where the turtle was, the turtle has moved ahead. Thus, Achilles can *never* catch the turtle.

Correct Solution: Let $t_1 = $ the time for Achilles to cover the 100-meter head start; let $t_2 = $ the time to cover the next 10 m; let $t_3 = $ the time for the next 1 m, and so on. Find the first three terms and the sum of the infinite series $t_1 + t_2 + t_3 + \cdots.$ 10, 1, 0.1; $\frac{100}{9}$ s

40. **Writing** Comment on the following paradox: You can *never* leave the room in which you are sitting because in order to do so, you must first walk halfway to the door. Then you must walk half the remaining distance to the door, and then half the next remaining distance. Since you must continue to cover the halves of these remaining distances an infinite number of times, you can never leave the room.

Exercises 41–44 deal with sequences and series of complex numbers and should be done sequentially.

41. In an arithmetic sequence of complex numbers, $t_1 = 1 + 3i$ and $t_2 = 3 + 4i$. Find the next three terms. Also find t_{25} and S_{25}. $5 + 5i$, $7 + 6i$, $9 + 7i$; $49 + 27i$; $625 + 375i$

42. In a geometric sequence of complex numbers, $t_1 = i$ and the common ratio is $r = 2i$. Find the next four terms of the sequence and represent all five terms graphically in an Argand diagram. Find the sum of these terms. $-2, -4i, 8, 16i;$ $6 + 13i$

43. The formula for the sum of an infinite geometric series given on page 500 holds for complex numbers as well as for real numbers. Use this formula to find the sum of the following series.

 a. $1 + \dfrac{i}{2} + \dfrac{i^2}{4} + \dfrac{i^3}{8} + \dfrac{i^4}{16} + \cdots$ $\dfrac{4}{5} + \dfrac{2}{5}i$ **b.** $27 - 9i + 3i^2 - i^3 + \cdots$ $\dfrac{243}{10} - \dfrac{81}{10}i$

44. A bug leaves the origin and crawls 1 unit east, $\frac{1}{2}$ unit north, $\frac{1}{4}$ unit west, $\frac{1}{8}$ unit south, and so forth, as shown. Each segment of its journey can be considered as a complex number or vector. Hence, the bug's ultimate destination can be considered as the sum

$$S = t_1 + t_2 + t_3 + t_4 + \cdots$$

of an infinite geometric series. $\frac{1}{2}i$

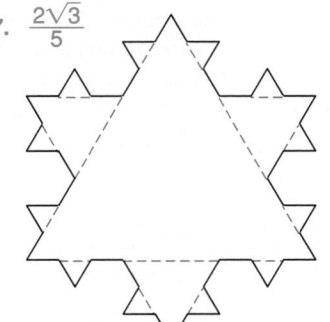

$t_3 = -\frac{1}{4} + 0i$

$t_4 = 0 - \frac{1}{8}i$

$t_2 = 0 + \frac{1}{2}i$

origin

$t_1 = 1 + 0i$

 a. Find r for this series.

 b. Use the formula $S = \dfrac{t_1}{1 - r}$ to show that $S = \dfrac{4}{5} + \dfrac{2}{5}i$. Thus, the ultimate destination point is $\left(\dfrac{4}{5}, \dfrac{2}{5}\right)$.

 c. How far must the bug crawl to reach its ultimate destination? 2 units

C   **45.** Consider an infinite geometric series of positive terms that converges to S. What fractional part of S is the sum of the odd-numbered terms? the sum of the even-numbered terms? (Give answers in terms of r, the common ratio of the series.) $\dfrac{1}{1 + r}; \dfrac{r}{1 + r}$

46. a. Prove that every rational number $\dfrac{a}{b}$ can be expressed as a repeating decimal. (For example, $\dfrac{2}{11} = 0.181818\ldots$ and $\dfrac{1}{4} = 0.25000\ldots$ or $0.24999\ldots.$)

 b. Prove that every repeating decimal represents a rational number.

Exercises 47 and 48 refer to the "snowflake curve" defined as follows: The sides of an equilateral triangle are trisected. A new equilateral triangle is placed on the middle third of each trisection. The sides common to the previous figure and the new triangles are then removed. This process continues indefinitely using the sides of the last figure obtained.

47. $\dfrac{2\sqrt{3}}{5}$

47. Find the area enclosed by the snowflake curve if each side of the initial equilateral triangle is one unit in length. See above.

48. Show that the limit of the sequence of perimeters of the snowflake curve is infinite.

Review Note

In connection with Exercise 46, remind students that non-repeating decimals represent real numbers that are irrational.

Additional Answers Written Exercises

44. b. $t_1 = 1$, $t_2 = \dfrac{1}{2}i$, $t_3 = -\dfrac{1}{4}$, and so on; $S =$

$$\dfrac{1}{1 - \dfrac{1}{2}i} = \dfrac{2}{2 - i} \cdot \dfrac{2 + i}{2 + i} =$$

$$\dfrac{4 + 2i}{5} = \dfrac{4}{5} + \dfrac{2}{5}i, \text{ or}$$

$$\left(\dfrac{4}{5}, \dfrac{2}{5}\right).$$

48. Let each side of the original triangle be 1 unit long. The sequence of perimeters is 3, 4, $\dfrac{16}{3}$, $\dfrac{64}{9}$, $\ldots$, which is an infinite geometric sequence whose common ratio is $\dfrac{4}{3}$; since $\dfrac{4}{3} > 1$, the series diverges and the perimeter is infinite.

Warm-Up Exercises

Find a formula for t_n (the nth term) and give the values of n for the first and last terms of the given series.

1. $1 + 6 + 11 + 16 + \cdots +$ 91 $\quad t_n = 1 + (n - 1)5$ or $t_n = 5n - 4$; 1; 19

2. $1 + 3 + 9 + 27 + \cdots +$ 6561 $\quad t_n = 3^{n-1}$; 1; 9

3. $1 + 8 + 27 + 64 + \cdots +$ 1000 $\quad t_n = n^3$; 1; 10

4. $\frac{1}{2} - \frac{2}{3} + \frac{3}{4} - \frac{4}{5} + \cdots +$ $\frac{99}{100}$ $\quad t_n = (-1)^{n+1} \cdot \frac{n}{n+1}$; 1; 99

5. $64 - 32 + 16 - 8 + \cdots -$ $\frac{1}{8}$ $\quad t_n = 64\left(-\frac{1}{2}\right)^{n-1}$; 1; 10

6. $10 + 3 - 4 - 11 - \cdots -$ 200 $t_n = 10 + (n - 1)(-7)$ or $t_n = 17 - 7n$; 1; 31

Motivating the Section

Ask students to write down expressions for the sum of the first n even integers $(2 + 4 + 6 + \cdots + 2n)$ and for the first n odd integers $(1 + 3 + 5 + \cdots + 2n - 1)$. Then explain that in this section, students will learn to use a shorter, more convenient notation for series like these. Introduce the sigma notation for these two series.

$\left(\sum_{k=1}^{n} 2k \text{ and } \sum_{k=1}^{n} (2k - 1) \right)$

▌▌▌▌ CALCULATOR EXERCISE

The infinite series $\frac{1}{5} + \frac{2}{25} + \frac{4}{125} + \frac{8}{625} + \cdots$ converges to $\frac{1}{3} = 0.3333 \ldots$.

Determine how many terms of the series must be added to make the sum:
a. correct in at least the first 2 decimal places, that is 0.33. 6
b. correct in at least the first 3 decimal places, that is 0.333. 8
c. correct in at least the first 4 decimal places, that is 0.3333. 11

13-6 Sigma Notation

Objective   *To represent series using sigma notation.*

The Greek letter Σ (sigma) is often used in mathematics to express a series or its sum in abbreviated form. For example, $\sum_{k=1}^{100} k^2$ represents the series whose terms are obtained by evaluating k^2 first for $k = 1$, then for $k = 2$, then for $k = 3$, and so on until $k = 100$. That is,

$$\sum_{k=1}^{100} k^2 = 1^2 + 2^2 + 3^2 + \cdots + 100^2.$$

The symbol on the left above may be read as "the sum of k^2 for values of k from 1 to 100." Similarly, the symbol $\sum_{k=5}^{10} 3k$ may read as "the sum of $3k$ for values of k from 5 to 10." This symbol represents the series whose terms are obtained by evaluating $3k$ first for $k = 5$, then for $k = 6$, and so on, to $k = 10$. That is,

$$\sum_{k=5}^{10} 3k = 3 \cdot 5 + 3 \cdot 6 + 3 \cdot 7 + 3 \cdot 8 + 3 \cdot 9 + 3 \cdot 10 = 135.$$

The symbol $3k$ is called the **summand,** the numbers 5 and 10 are called the **limits of summation,** and the symbol k is called the **index.** Any letter can be used for the index. For example:

$$\sum_{n=1}^{5} \left(-\tfrac{1}{2}\right)^n = \left(-\tfrac{1}{2}\right)^1 + \left(-\tfrac{1}{2}\right)^2 + \left(-\tfrac{1}{2}\right)^3 + \left(-\tfrac{1}{2}\right)^4 + \left(-\tfrac{1}{2}\right)^5$$
$$= -\tfrac{1}{2} + \tfrac{1}{4} - \tfrac{1}{8} + \tfrac{1}{16} - \tfrac{1}{32} = -\tfrac{11}{32}$$

Sigma notation can also be used to represent an infinite series and its sum. For example:

$$\sum_{j=0}^{\infty} \left(\tfrac{1}{2}\right)^j = \left(\tfrac{1}{2}\right)^0 + \left(\tfrac{1}{2}\right)^1 + \left(\tfrac{1}{2}\right)^2 + \left(\tfrac{1}{2}\right)^3 + \cdots$$
$$= 1 + \tfrac{1}{2} + \tfrac{1}{4} + \tfrac{1}{8} + \cdots = 2$$

The symbol $\sum_{j=0}^{\infty} \left(\frac{1}{2}\right)^j$ is read "the sum of $\left(\frac{1}{2}\right)^j$ for values of j from 0 to infinity"; it represents both the infinite geometric series at the bottom of page 506 and its sum, 2.

When you get used to using sigma notation, you will find it much easier than manipulating series that are written in expanded form. The following properties are consequences of the commutative, associative, and distributive properties of the real and imaginary numbers. These properties are proved in Exercises 29 and 30.

Properties of Finite Sums

1. $\sum_{i=1}^{n} (a_i + b_i) = \sum_{i=1}^{n} a_i + \sum_{i=1}^{n} b_i$
    2. $\sum_{i=1}^{n} ca_i = c \sum_{i=1}^{n} a_i$

The properties above can be used together with previously derived sums to derive the sums of many other series. Example 1 illustrates how these properties may be used with the following known sums.

sum of integers: $\sum_{k=1}^{n} k = \frac{n(n+1)}{2}$ (Proved in Section 13-3)

sum of squares: $\sum_{k=1}^{n} k^2 = \frac{n(n+1)(2n+1)}{6}$ (To be proved in Section 13-7)

sum of cubes: $\sum_{k=1}^{n} k^3 = \left[\frac{n(n+1)}{2}\right]^2$ (To be proved in Section 13-7)

Example Express $1 \cdot 2 + 3 \cdot 4 + 5 \cdot 6 + \cdots + 199 \cdot 200$ in sigma notation and evaluate.

Solution $1 \cdot 2 + 3 \cdot 4 + \cdots + 199 \cdot 200 = \sum_{k=1}^{100} (2k-1)(2k)$

$$= \sum_{k=1}^{100} (4k^2 - 2k)$$

$$= 4 \sum_{k=1}^{100} k^2 - 2 \sum_{k=1}^{100} k$$

$$= 4 \left[\frac{100 \cdot 101 \cdot 201}{6}\right] - 2 \left[\frac{100 \cdot 101}{2}\right]$$

$$= 1{,}353{,}400 - 10{,}100 = 1{,}343{,}300$$

Additional Examples

1. Express each series using sigma notation and find the sum.
   a. $2 + 9 + 16 + \cdots + 100$
   b. $1 + \frac{1}{2} + \frac{1}{4} + \frac{1}{8} + \cdots$
   a. The series is arithmetic with $t_1 = 2$ and $t_n = 100 = 2 + (n-1)7$; $n = 15$;
   $$\sum_{k=1}^{15} [2 + (k-1)7] =$$
   $$\sum_{k=1}^{15} (7k - 5) =$$
   $$\frac{15(2 + 100)}{2} = 765$$
   b. The series is geometric with $t_1 = 1$ and $r = \frac{1}{2}$.
   $$\sum_{n=1}^{\infty} 2^{-(n-1)} \text{ or } \sum_{n=1}^{\infty} \frac{1}{2^{n-1}}$$
   $$S = \frac{1}{1 - \frac{1}{2}} = 2$$

2. Show that
   $$\sum_{k=1}^{n} (2k-1) = n^2.$$
   $$\sum_{k=1}^{n} (2k-1) = 1 + 3 + 5 + \cdots + (2n-1) =$$
   $$\frac{n[1 + (2n-1)]}{2} = \frac{n(2n)}{2} = n^2$$

Supplementary Materials

Alternative Assessment, 41–42

Additional Answers
Written Exercises

1. $2 + 3 + 4 + 5 + 6$

2. $5 + 10 + 15 + 20 + \cdots + 50$

3. $1 + \frac{1}{2} + \frac{1}{3} + \frac{1}{4} + \frac{1}{5}$

4. $5 + 9 + 13 + 17 + 21$

5. $3 + 1 + \frac{1}{3} + \frac{1}{9} + \cdots$

6. $2 - 3 + 4 - 5 + 6 - \cdots$

7. $\frac{1}{16} + \frac{1}{4} + 1 + 4 + 16$

8. $13 + 10 + 7 + 4 + 1 + 2 + 5$

17. $\sum_{t=1}^{4} \log t = \log 1 + \log 2 + \log 3 + \log 4 = \log (1 \cdot 2 \cdot 3 \cdot 4) = \log 24$

18. $\sum_{k=1}^{4} k \log 2 = 1 \log 2 + 2 \log 2 + 3 \log 2 + 4 \log 2 = 10 \log 2 = \log 2^{10}$

CLASS EXERCISES

Give each series in expanded form.

9 + 16 + 25 + 36

$1 + \frac{1}{2} + \frac{1}{3} + \frac{1}{4} + \cdots$

1. $\sum_{k=1}^{4} 5k$ **2.** $\sum_{k=3}^{6} k^2$ **3.** $\sum_{j=2}^{8} (-1)^j$ **4.** $\sum_{n=1}^{\infty} \frac{1}{n}$

5 + 10 + 15 + 20

1 − 1 + 1 − 1 + 1 − 1 + 1

Express each of the following series using sigma notation.

5. $4 + 9 + 16 + 25 + 36$ $\sum_{k=2}^{6} k^2$ **6.** $\frac{1}{2} + \frac{2}{3} + \frac{3}{4} + \frac{4}{5}$ $\sum_{k=1}^{4} \frac{k}{k+1}$

7. The arithmetic series: $3 + 6 + 9 + \cdots + 300$ $\sum_{k=1}^{100} 3k$

8. The infinite geometric series: $\frac{1}{3} + \frac{1}{9} + \frac{1}{27} + \frac{1}{81} + \cdots$ $\sum_{k=1}^{\infty} 3^{-k}$

WRITTEN EXERCISES

In Exercises 1–8, write the given series in expanded form.

A **1.** $\sum_{k=2}^{6} k$ **2.** $\sum_{k=1}^{10} 5k$ **3.** $\sum_{k=1}^{5} \frac{1}{k}$ **4.** $\sum_{n=3}^{7} (4n - 7)$

5. $\sum_{n=0}^{\infty} 3^{1-n}$ **6.** $\sum_{j=2}^{\infty} j(-1)^j$ **7.** $\sum_{t=-2}^{2} 4^t$ **8.** $\sum_{s=0}^{6} |13 - 3s|$

In Exercises 9–16, express the given series using sigma notation.

9. $4 + 8 + 12 + 16 + 20$ See below. **10.** $1 + 2 + 4 + 8 + 16 + 32$ $\sum_{k=0}^{5} 2^k$

11. $5 + 9 + 13 + \cdots + 101$ See below. **12.** $2 + 4 + 6 + 8 + \cdots + 200$ See below.

13. $1 + \frac{1}{4} + \frac{1}{9} + \frac{1}{16} + \cdots$ $\sum_{k=1}^{\infty} \frac{1}{k^2}$ **14.** $\frac{1}{2} + \frac{1}{4} + \frac{1}{6} + \frac{1}{8} + \cdots$ $\sum_{k=1}^{\infty} \frac{1}{2k}$

15. $\sin x + \sin 2x + \sin 3x + \cdots$ $\sum_{k=1}^{\infty} \sin kx$ **16.** $48 + 24 + 12 + 6 + \cdots$ $\sum_{k=0}^{\infty} 48\left(\frac{1}{2}\right)^k$

17. Show that $\sum_{t=1}^{4} \log t = \log 24$. **18.** Show that $\sum_{k=1}^{4} k \log 2 = \log 2^{10}$.

19. Evaluate: $\sum_{k=1}^{100} \cos k\pi$ 0 **20.** Evaluate: $\sum_{k=1}^{50} \sin \left(k \cdot \frac{\pi}{2}\right)$ 1

B **21.** Evaluate: **a.** $\sum_{n=1}^{8} \left(\frac{\sqrt{2}}{2} + \frac{\sqrt{2}}{2}i\right)^n$ 0 **b.** $\sum_{n=1}^{8} \left|\left(\frac{\sqrt{2}}{2} + \frac{\sqrt{2}}{2}i\right)^n\right|$ 8

22. Evaluate: $\sum_{n=1}^{\infty} \left(\frac{1}{2} + \frac{1}{2}i\right)^n$ (*Hint:* This is a geometric series with $|r| < 1$.) i

9. $\sum_{k=1}^{5} 4k$ **11.** $\sum_{k=1}^{25} 4k + 1$ **12.** $\sum_{k=1}^{100} 2k$

508 *Chapter Thirteen*

23. Show that $\displaystyle\sum_{k=1}^{8} (-1)^k \log k = \log\left(\dfrac{2\cdot 4\cdot 6\cdot 8}{1\cdot 3\cdot 5\cdot 7}\right)$ **24.** Evaluate $\displaystyle\sum_{k=1}^{100} \lfloor\sqrt{k}\rfloor$, where 625

26. a. $\displaystyle\sum_{k=0}^{5} 27\left(-\dfrac{1}{3}\right)^k$ **26. b.** $\displaystyle\sum_{k=0}^{5} (-27)\left(-\dfrac{1}{3}\right)^k$ $\lfloor k\rfloor$ is the greatest integer of k.

In Exercises 25–28, express the given series using sigma notation. $\displaystyle\sum_{k=0}^{5} -1\left(-\dfrac{1}{2}\right)^k$

25. a. $1 - \dfrac{1}{2} + \dfrac{1}{4} - \dfrac{1}{8} + \dfrac{1}{16} - \dfrac{1}{32}$ $\displaystyle\sum_{k=0}^{5}\left(-\dfrac{1}{2}\right)^k$ **b.** $-1 + \dfrac{1}{2} - \dfrac{1}{4} + \dfrac{1}{8} - \dfrac{1}{16} + \dfrac{1}{32}$

26. a. $27 - 9 + 3 - 1 + \dfrac{1}{3} - \dfrac{1}{9}$ See above. **b.** $-27 + 9 - 3 + 1 - \dfrac{1}{3} + \dfrac{1}{9}$ See above.

27. $1 - 3 + 5 - 7 + \cdots - 99$ See below. **28.** $-2 + 4 - 6 + \cdots + 100$ See below.

Show how the commutative, associative, and distributive properties are used to prove each of the following.
Answers may vary.

27. $\displaystyle\sum_{k=0}^{49} (-1)^k(2k+1)$ **28.** $\displaystyle\sum_{k=1}^{50} (-1)^k 2k$

29. $\displaystyle\sum_{i=1}^{n} (a_i + b_i) = \sum_{i=1}^{n} a_i + \sum_{i=1}^{n} b_i$ **30.** $\displaystyle\sum_{i=1}^{n} ca_i = c\sum_{i=1}^{n} a_i$

Use the properties of finite sums to prove each of the following.

31. $\displaystyle\sum_{i=1}^{n} (a_i + b_i)^2 = \sum_{i=1}^{n} a_i^2 + 2\sum_{i=1}^{n} a_i b_i + \sum_{i=1}^{n} b_i^2$

32. $\displaystyle\sum_{x=1}^{n} (ax^2 + bx + c) = a\sum_{x=1}^{n} x^2 + b\sum_{x=1}^{n} x + cn$

Evaluate the following series using the method shown in the example.

33. $1\cdot 2 + 2\cdot 3 + 3\cdot 4 + \cdots + 100\cdot 101$ 343,400

34. $1\cdot 3 + 3\cdot 5 + 5\cdot 7 + \cdots + 99\cdot 101$ 171,650

35. $1\cdot 2\cdot 3 + 2\cdot 3\cdot 4 + 3\cdot 4\cdot 5 + \cdots + 20\cdot 21\cdot 22$ 53,130

36. $2\cdot 4 + 6\cdot 8 + 10\cdot 12 + \cdots + 398\cdot 400$ 5,373,200

C **37. a.** An 8-by-8 checkerboard has 64 little squares and many other squares of various sizes. How many squares does it have in all?
  b. Find the total number of squares in an n-by-n checkerboard.

23. $\displaystyle\sum_{k=1}^{8} (-1)^k \log k =$
$-1 \log 1 + 1 \log 2 - 1 \log 3 + \cdots + 1 \log 8 =$
$(\log 2 + \log 4 + \log 6 + \log 8) - (\log 1 + \log 3 + \log 5 + \log 7) =$
$\log (2\cdot 4\cdot 6\cdot 8) - \log (1\cdot 3\cdot 5\cdot 7) =$
$\log\left(\dfrac{2\cdot 4\cdot 6\cdot 8}{1\cdot 3\cdot 5\cdot 7}\right)$

29. $\displaystyle\sum_{i=1}^{n} (a_i + b_i) = (a_1 + b_1) + (a_2 + b_2) + \cdots + (a_n + b_n) =$
$(a_1 + a_2 + \cdots + a_n) + (b_1 + b_2 + \cdots + b_n) =$
$\displaystyle\sum_{i=1}^{n} a_i + \sum_{i=1}^{n} b_i$

30. $\displaystyle\sum_{i=1}^{n} ca_i =$
$ca_1 + ca_2 + \cdots + ca_n =$
$c(a_1 + a_2 + \cdots + a_n) =$
$c\displaystyle\sum_{i=1}^{n} a_i$

37. a. An 8-by-8 checkerboard contains
1 8-by-8 square,
4 7-by-7 squares,
9 6-by-6 squares,
16 5-by-5 squares,
25 4-by-4 squares,
36 3-by-3 squares,
49 2-by-2 squares, and
64 1-by-1 squares,
or $\displaystyle\sum_{k=1}^{8} k^2 = \dfrac{8(9)(17)}{6} = 204$ squares in all.
  b. $\displaystyle\sum_{k=1}^{n} k^2 = \dfrac{n(n+1)(2n+1)}{6}$

38. a. The 3-by-3-by-3 cube shown has 27 little cubes but also many other cubes of various sizes. How many cubes does it have in all? 36

b. Find the total number of cubes in an *n*-by-*n*-by-*n* cube. See below.

39. A stack of oranges is compactly arranged so the bottom layer consists of oranges in an equilateral triangle with *n* oranges on a side. The layer next to the bottom consists of an equilateral triangle of oranges with $n - 1$ oranges on a side. This pattern continues upward with one orange on the top. How many oranges are there? See below.

Fourth layer

//// **COMPUTER EXERCISE**

The series $\displaystyle\sum_{n=1}^{\infty} \frac{1}{n}$ diverges, but very slowly. Find how many terms are needed to make the partial sum $1 + \frac{1}{2} + \frac{1}{3} + \cdots + \frac{1}{n}$ greater than: (a) 3; (b) 4; (c) 10.
(The answer is surprising!)

$\quad$ 11 $\quad$ 31 $\quad$ 12,367

13-7 Mathematical Induction

Objective *To use mathematical induction to prove that a statement is true.*

Observe the pattern in the following statements:

$$\frac{1}{1 \cdot 2} = \frac{1}{2}$$

$$\frac{1}{1 \cdot 2} + \frac{1}{2 \cdot 3} = \frac{2}{3}$$

$$\frac{1}{1 \cdot 2} + \frac{1}{2 \cdot 3} + \frac{1}{3 \cdot 4} = \frac{3}{4}$$

It *appears* that

$$\frac{1}{1 \cdot 2} + \frac{1}{2 \cdot 3} + \frac{1}{3 \cdot 4} + \cdots + \frac{1}{n(n + 1)} = \frac{n}{n + 1}.$$

Nevertheless, this appearance does not constitute a *proof* that the statement is true for all positive integers *n*. One way to prove the statement true is to use a method called **mathematical induction.**

38. b. $\displaystyle\sum_{k=1}^{n} k^3$, or $\left(\frac{n(n + 1)}{2}\right)^2$

39. $\frac{1}{6}(n(n + 1)(n + 2))$

Proof by Mathematical Induction

Let S be a statement in terms of a positive integer n.

Step 1. Show that S is true for $n = 1$.

Step 2. Assume that S is true for $n = k$, where k is a positive integer, and then prove that S must be true for $n = k + 1$.

If you can do both steps (1) and (2) above, then you can conclude that S is true for all positive integers. The reason is that once you know that S is true for $n = 1$, step (2) tells you it is true for $n = 1 + 1 = 2$. Applying step (2) again, S must be true for $n = 2 + 1 = 3$, then for $n = 3 + 1 = 4$, and so on.

You might think of mathematical induction as something like setting up dominoes so they will all fall down. Set up the dominoes so that if any one domino falls, the next domino will fall. Then knock over the first domino, which will knock over the second, which will knock over the third, and so on. Thus, you can conclude that all the dominoes will fall down.

Example 1   Prove that $\dfrac{1}{1 \cdot 2} + \dfrac{1}{2 \cdot 3} + \dfrac{1}{3 \cdot 4} + \cdots + \dfrac{1}{n(n + 1)} = \dfrac{n}{n + 1}$ for all positive integers n.

Solution

Step 1   Show that the statement is true for $n = 1$.

$\dfrac{1}{1(1 + 1)} = \dfrac{1}{1 + 1}$. Yes, it is true.

Step 2   Assume that the statement is true for $n = k$, and then prove that it must be true for $n = k + 1$.

Assume: $\dfrac{1}{1 \cdot 2} + \dfrac{1}{2 \cdot 3} + \cdots + \dfrac{1}{k(k + 1)} = \dfrac{k}{k + 1}$

Prove:

$\dfrac{1}{1 \cdot 2} + \dfrac{1}{2 \cdot 3} + \cdots + \dfrac{1}{k(k + 1)} + \dfrac{1}{(k + 1)[(k + 1) + 1]} = \dfrac{k + 1}{(k + 1) + 1}$, or

$\dfrac{1}{1 \cdot 2} + \dfrac{1}{2 \cdot 3} + \cdots + \dfrac{1}{k(k + 1)} + \dfrac{1}{(k + 1)(k + 2)} = \dfrac{k + 1}{k + 2}$

Basic strategy at this stage of proof: Show that the left side of the ''Prove'' statement is equal to the right side. To do this, take the left side and try to simplify it by using the commutative, associative, and distributive properties and the ''Assume'' statement. This is shown in the proof on the next page.

1. Use mathematical induction to prove that the following statement is true for all positive integers n.
$1 + 4 + 7 + \cdots + (3n - 2) = \dfrac{3n^2 - n}{2}$

Step 1: Show that the statement is true for $n = 1$.
$3 \cdot 1 - 2 = 1$; $\dfrac{3(1)^2 - 1}{2} = 1$; so the statement is true for $n = 1$.

Step 2: Assume the statement true for $n = k$:
$1 + 4 + 7 + \cdots + (3k - 2) = \dfrac{3k^2 - k}{2}$

Prove the statement true for $n = k + 1$:
$1 + 4 + \cdots + (3k - 2) + (3(k + 1) - 2) = \dfrac{3(k + 1)^2 - (k + 1)}{2}$

The right side of the equation simplifies to $\dfrac{3k^2 + 5k + 2}{2}$. By assumption, $1 + 4 + 7 + \cdots + (3k - 2) = \dfrac{3k^2 - k}{2}$, so the left side of the equation simplifies to
$\dfrac{3k^2 - k}{2} + (3k + 1) = \dfrac{3k^2 - k + 6k + 2}{2} = \dfrac{3k^2 + 5k + 2}{2}$. Since we have shown that the two sides are the same, the proof is complete.

(*Continues on next page*)

2. Write out the first few terms of the sequence defined recursively by $a_1 = \frac{1}{2}$ and $a_n = a_{n-1} + \frac{1}{n(n+1)}$. Then suggest and prove a formula in terms of n for the nth term, a_n.

Sequence: $\frac{1}{2}, \frac{2}{3}, \frac{3}{4}, \frac{4}{5}, \cdots$

Formula: $a_n = \frac{n}{n+1}$

Proof: If $a_k = \frac{k}{k+1}$, then

$$a_{k+1} = \frac{k}{k+1} + \frac{1}{(k+1)[(k+1)+1]} =$$

$$\frac{k}{k+1} + \frac{1}{(k+1)(k+2)} =$$

$$\frac{k(k+2)+1}{(k+1)(k+2)} =$$

$$\frac{k^2+2k+1}{(k+1)(k+2)} =$$

$$\frac{(k+1)^2}{(k+1)(k+2)} = \frac{k+1}{k+2} =$$

$$\frac{k+1}{(k+1)+1}.$$

Error Analysis

A student's inability to accurately perform algebraic manipulations is going to show up very quickly in Step 2 of many of these proofs.

Review Note

Review adding fractions with unlike denominators.

Proof:

$$\frac{1}{1\cdot 2} + \frac{1}{2\cdot 3} + \cdots + \frac{1}{k(k+1)} + \frac{1}{(k+1)(k+2)}$$

$$= \underbrace{\frac{1}{1\cdot 2} + \frac{1}{2\cdot 3} + \cdots + \frac{1}{k(k+1)}}_{\text{Use the "Assume" statement.}} + \frac{1}{(k+1)(k+2)}$$

$$= \frac{k}{k+1} + \frac{1}{(k+1)(k+2)}$$

$$= \frac{k(k+2)}{(k+1)(k+2)} + \frac{1}{(k+1)(k+2)}$$

$$= \frac{k^2+2k+1}{(k+1)(k+2)}$$

$$= \frac{(k+1)^2}{(k+1)(k+2)}$$

$$= \frac{k+1}{k+2}$$

Example 2 Prove that $n^3 + 2n$ is a multiple of 3 for all positive integers n.

Solution

Step 1 Show that the statement is true for $n = 1$.

$1^3 + 2\cdot 1 = 3$ is a multiple of 3. Yes, it is true.

Step 2 Assume that the statement is true for $n = k$, and then prove that it must be true for $n = k + 1$.

Assume: $k^3 + 2k$ is a multiple of 3.
Prove: $(k+1)^3 + 2(k+1)$ is a multiple of 3.

Proof: Use the same basic strategy as in Example 1.

$$(k+1)^3 + 2(k+1) = k^3 + 3k^2 + 3k + 1 + 2k + 2$$

$$= \underbrace{(k^3 + 2k)}_{\substack{\text{Use the} \\ \text{assumed} \\ \text{statement.}}} + 3(k^2 + k + 1)$$

$$= \text{a multiple of } 3 + \text{a multiple of } 3$$

$$= \text{a multiple of } 3$$

In the last line of the proof, we have used the fact that the sum of two multiples of 3, say $3i$ and $3j$, is another multiple of 3, namely $3(i + j)$.

WRITTEN EXERCISES

Use mathematical induction to prove the statements in these exercises. *Note that some of these statements can also be proved by other methods.*

A **1.** $1 + 2 + \cdots + n = \dfrac{n(n + 1)}{2}$

2. $1 + 3 + \cdots + (2n - 1) = n^2$

3. $\displaystyle\sum_{i=1}^{n} 2i = n^2 + n$ **19.** $\dfrac{1}{2}, \dfrac{1}{3}, \dfrac{1}{4}, \cdots, \dfrac{1}{n+1}$

4. $\displaystyle\sum_{i=1}^{n} 2^{i-1} = 2^n - 1$

5. $\displaystyle\sum_{i=1}^{n} i^2 = \dfrac{n(n + 1)(2n + 1)}{6}$

6. $\displaystyle\sum_{i=1}^{n} i(i + 1) = \dfrac{n(n + 1)(n + 2)}{3}$

7. $\displaystyle\sum_{i=1}^{n} (2i - 1)^2 = \dfrac{n(2n - 1)(2n + 1)}{3}$

8. $(1 + x)^n \geq 1 + nx$, where $x > -1$

9. $\displaystyle\sum_{i=1}^{n} (i \cdot 2^{i-1}) = 1 + (n - 1) \cdot 2^n$

10. $18^n - 1$ is a multiple of 17.

11. $11^n - 4^n$ is a multiple of 7.

12. $n(n^2 + 5)$ is a multiple of 6.

For Exercises 13 and 14, decide which values of n will make the inequality true and then prove your answer is correct.

B **13.** $2^n > 2n$ $n \geq 3$

14. $n! > 2^n$ $n \geq 4$

15. If, in a room with n people ($n \geq 2$), every person shakes hands once with every other person, prove that there are $\dfrac{n^2 - n}{2}$ handshakes.

16. Prove that the sum of the cubes of any three consecutive positive integers is a multiple of 9.

17. Use the triangle inequality (see Exercise 36 on page 99) to prove that
$$|a_1 + a_2 + \cdots + a_n| \leq |a_1| + |a_2| + \cdots |a_n|.$$

18. Prove De Moivre's theorem (page 408) for every positive integer n.

In Exercises 19–21, write out the first few terms of each sequence. Then suggest and prove a formula in terms of n for the nth term, a_n.

19. $a_1 = \dfrac{1}{2},\ a_n = a_{n-1} - \dfrac{a_{n-1}}{n + 1}$ See above.

20. $a_1 = 1,\ a_n = 2a_{n-1} + 1$ $1, 3, 7, 15, \ldots, 2^n - 1$

21. $a_1 = \dfrac{1}{4},\ a_n = a_{n-1} + \dfrac{1}{(3n - 2)(3n + 1)}$ $\dfrac{1}{4}, \dfrac{2}{7}, \dfrac{3}{10}, \dfrac{4}{13}, \cdots, \dfrac{n}{3n + 1}$

Sequences and Series **513**

Cooperative Learning

For Exercises 19–21, you might divide the class into groups of three and give each group a different recursive definition. Each group should find and prove the resulting explicit definition.

Then pair the groups. Have the paired groups exchange papers. Each group should write a critique of the work of the other group.

Suggested Assignments

Comprehensive
Day 1: 513/1–11 odd
Day 2: 513/13, 16, 19, 21 25

Supplementary Materials

Alternative Assessment, 42–43
Student Resource Guide, 124–127

In Exercises 22–24, write out the first few terms of each sequence. Then suggest and prove a formula in terms of n for the nth partial sum, S_n.

C **22.** $S_n = \sum_{i=1}^{n} i^3$ 1, 9, 36, 100, $\ldots$; $\left(\dfrac{n(n+1)}{2}\right)^2$

23. 1, 5, 23, 119, $\ldots$; $(n+1)! - 1$

23. $S_n = \sum_{i=1}^{n} (i \cdot i!)$ *(Hint:* Look at the sequence of factorials of integers.)

24. 6, 30, 90, 210, $\ldots$; $\dfrac{n(n+1)(n+2)(n+3)}{4}$

24. $S_n = \sum_{i=1}^{n} i(i+1)(i+2)$ *(Hint:* See Exercises 1 and 6.)

25. Prove that a convex polygon with n sides $(n \geq 3)$ has $\dfrac{n^2 - 3n}{2}$ diagonals.

26. Prove that it is possible to pay any debt of \$4, \$5, \$6, \$7, $\ldots$, \$$n$, and so on, by using only \$2 bills and \$5 bills. For example, a debt of \$11 can be paid with three \$2 bills and one \$5 bill, or \$11 = 3 $\cdot$ \$2 + 1 $\cdot$ \$5.

Chapter Summary

1. A *sequence* is a function whose domain is the set of positive integers. A sequence can be defined explicitly by a formula such as $t_n = 3n + 6$ or recursively by a pair of formulas, such as $t_1 = 6$ and $t_n = 7 - t_{n-1}$.

2. A sequence is *arithmetic* if the difference d of any two consecutive terms is constant. A sequence is *geometric* if the ratio r of any two consecutive terms is constant.

3. A series is an indicated sum of the terms of a sequence. The formulas on pages 487 and 488 can be used to find the sums of finite arithmetic series and finite geometric series, respectively.

4. If the nth term of a sequence becomes arbitrarily close to a number L as n gets larger and larger, we write $\lim_{n \to \infty} t_n = L$. If there is no single limiting number L for all terms of the sequence, we say that the sequence has no limit. A sequence whose terms increase (or decrease) without bound is said to have a limit of ∞ (or $-\infty$).

5. Associated with any infinite series $t_1 + t_2 + t_3 + \cdots$, there is a *sequence of partial sums:*
$$S_1 = t_1, \; S_2 = t_1 + t_2, \; S_3 = t_1 + t_2 + t_3, \text{ and so on.}$$
If this sequence has a limit L, we say that the series *converges* to the sum L. If the sequence of partial sums has no limit, the series *diverges*.

6. The infinite geometric series $t_1 + t_1 r + t_1 r^2 + \cdots$ converges to the sum
$$S = \frac{t_1}{1 - r} \text{ if } |r| < 1. \text{ The series diverges if } |r| \geq 1 \text{ and } t_1 \neq 0.$$

7. Any infinite repeating decimal can be written as an infinite geometric series whose sum is a rational number.

8. The Greek letter Σ (sigma) is often used to express a series or its sum. Some properties of finite sums are given on page 507.

9. *Mathematical induction,* used to prove that a statement S is true for all positive integers, involves two steps: (1) Show that S is true for $n = 1$, and (2) prove that S is true for $n = k + 1$ whenever it is true for $n = k$.

Key vocabulary and ideas

sequence (p. 473, p. 474)
arithmetic and geometric sequences (p. 473)
recursive and explicit definitions (p. 479)
series, sum of a series (p. 486)
limit of a sequence (p. 493)

sequence of partial sums (p. 500)
sum of an infinite series (p. 500)
converge, diverge (p. 500)
sigma notation (p. 506)
mathematical induction (p. 510)

1. **a.** arithmetic; $t_n = 14 - 4n$ **b.** neither; $t_n = n^2 + 2$ **c.** geometric; $t_n = \frac{2}{3}(-3)^n$

Chapter Test

1. State whether each sequence is arithmetic, geometric, or neither, and   **13-1**
   find a formula for t_n in terms of n. See above.
   a. $10, 6, 2, -2, \ldots$ **b.** $3, 6, 11, 18, 27, \ldots$ **c.** $-2, 6, -18, 54, \ldots$

2. In an arithmetic sequence, $t_2 = 6$ and $t_6 = 16$. Find t_{21}. 53.5

3. In a geometric sequence, $t_3 = 9$ and $t_6 = \frac{9}{8}$. Find t_{12}. $\frac{9}{512}$

4. In a sequence, $t_1 = 1$, $t_2 = 4$, and $t_n = 2t_{n-1} + t_{n-2}$. Find t_6. 128   **13-2**

5. Give a recursive definition for the sequence $1, 5, 21, 85, \ldots$. $t_1 = 1$, $t_n = 4t_{n-1} + 1$

6. For the sequence of positive 3-digit integers ending in 4, find an   **13-3**
   explicit formula for **(a)** t_n and **(b)** S_n. **a.** $t_n = 94 + 10n$ **b.** $S_n = 99n + 5n^2$

7. Find the sum of the first 6 terms of the series $27 - 9 + 3 - 1 + \cdots$. $\frac{182}{9}$

8. *Writing* Briefly tell what the word "limit" means in mathematics.   **13-4**

9. Find: **a.** $\lim_{n \to \infty} \dfrac{3n^2 + 1}{4n^2 - 2n + 5}$ $\frac{3}{4}$ **b.** $\lim_{n \to \infty} n \cos n\pi$ Does not exist **c.** $\lim_{n \to \infty} (0.59)^n$ 0

10. Express $0.131313\ldots$ as an infinite series and as a rational number. $\frac{13}{99}$ **13-5**

11. Find **(a)** the interval of convergence and **(b)** the sum, expressed in
    terms of x, for the series $1 + \dfrac{3x}{2} + \dfrac{9x^2}{4} + \cdots$. **a.** $-\frac{2}{3} < x < \frac{2}{3}$ **b.** $\dfrac{2}{2 - 3x}$

12. Express each of the following series using sigma notation.   **13-6**
    See below.
    a. $1 - \dfrac{1}{9} + \dfrac{1}{25} - \dfrac{1}{49} + \dfrac{1}{81}$ **b.** $7 + 3 - 1 - 5 - 9 - 13 - 17$

13. Evaluate $\displaystyle\sum_{k=1}^{20} 3k(k + 2)$. 9870   **12. a.** $\displaystyle\sum_{n=0}^{4} \dfrac{(-1)^n}{(2n + 1)^2}$ **b.** $\displaystyle\sum_{n=1}^{7} (11 - 4n)$

14. Prove by mathematical induction: **13-7**

$$2 + 7 + 12 + \cdots + (5n - 3) = \frac{n(5n - 1)}{2}$$

8. Answers will vary. Example: An infinite sequence has a limit L if the nth term gets arbitrarily close to L as n becomes larger.

10. $0.131313\ldots =$
$$\sum_{n=1}^{\infty} 13(10)^{-2n} = \frac{13}{99}$$

14. Step 1: $\dfrac{1(5 \cdot 1 - 1)}{2} = 2$;
the statement is true for $n = 1$.
Step 2: Assume that the statement is true for $n = k$: $2 + 7 + 12 + \cdots + (5k - 3) = \dfrac{k(5k - 1)}{2}$. Then $2 + 7 + 12 + \cdots + (5k - 3) + [5(k + 1) - 3] = \dfrac{k(5k - 1)}{2} + (5k + 2) = \dfrac{5k^2 - k}{2} + \dfrac{10k + 4}{2} = \dfrac{5k^2 + 9k + 4}{2} = \dfrac{(k + 1)(5k + 4)}{2} = \dfrac{(k + 1)[5(k + 1) - 1]}{2}$; the statement is true for $n = k + 1$.

Chapter 14 Matrices

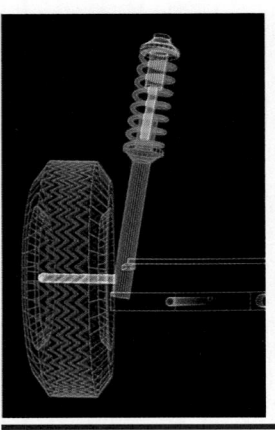

Overview

This chapter introduces the fundamentals of matrix arithmetic and the application of matrices to a wide variety of fields. In fact, applications are often used to motivate the rules for matrix arithmetic.

In the first two sections, students learn to perform matrix addition, scalar multiplication, and matrix multiplication. In the next section, students use matrices to solve systems of linear equations. In the last three sections, students set up communication matrices (which students use to solve communication-network problems), transition matrices (which students use to make predictions about the future states of various systems), and transformation matrices (which students use to analyze transformations of the coordinate plane).

Objectives

14-1 To find the sum, difference, or scalar multiples of matrices.
14-2 To find the product of two matrices.
14-3 To find the inverse of a 2×2 matrix and to solve linear systems using matrices.
14-4 To solve communication network problems using matrices.
14-5 To make predictions using powers of matrices.
14-6 To find the images of points under different types of transformations using matrices.

Supplementary Resources

Tests, pp. 44–45, 46–47
Alt. Assess., pp. 43–46, 69
Activities, pp. 38–40
Using Tech., pp. 35–36, 44–45
Student Res. Guide, pp. 130–139

Software

Precalculus Plotter Plus
Activities, pp. 67, 93–94, 95–96, 97–98
Matrix Calculator

Pacing Guide

Section	14-1	14-2	14-3	14-4	14-5	14-6	Review	Test	C. Rev.	Total
Discrete Math (days)	1	1	2	1	2	1	1	1	1	11

Teaching Notes

Presenting the Section

This section focuses on the basic definitions and skills upon which further study of matrices depends. It is important that students master these definitions and skills: matrix, dimensions, scalar, matrix addition, scalar multiplication, and matrix equality. Some adjustment in thinking will be necessary as students learn to deal with relationships and operations defined in terms of elements of matrices. Make sure students realize that the elements of a matrix are real numbers but that matrices themselves constitute a new system of mathematical entities with its own arithmetic.

Applications

The chapter's emphasis on applications begins with the very first example. Applications like those in Exercises 15 and 16 help reinforce the element-by-element nature of matrix addition and subtraction. Addition or subtraction of corresponding elements is the only way these operations could be meaningful in the problem settings.

Communication

Turn the tables on the usual "story problem" procedure. Have each student write a matrix and exchange it with another student. Then have students write a scenario that could be represented by the given matrix. Alternatively, give a matrix and ask students to write the "story" described by that matrix. The diversity of student answers will be interesting.

Review

This section provides a good opportunity to review the field properties in the new context of matrix addition and scalar multiplication. Once these operations have been defined, students might be encouraged to investigate whether the operations are commutative and associative, what identities there are, and whether $k(A + B) = kA + kB$ or whether $(m + n)A = mA + nA$, where k, m, and n are scalars and A and B are matrices.

Presenting the Section

Use the example on page 523 to introduce matrix multiplication. Then define the operation and demonstrate how the product is found using Example 1 and other pairs of matrices. Before going on to consider applications or properties of matrix multiplication, give students the chance to practice multiplication until they have grasped the "row-by-column" algorithm. Applications like those in Example 2 and Exercises 15–22 will clarify why matrices must be compatible if we are to multiply them. That is, if A is $m \times n$, then B must be $n \times p$ in order to multiply A and B.

Assessment

To determine whether students recognize when matrices are compatible for multiplication and to assess students' depth of understanding of the meaningfulness of the product of two matrices, give students the following matrices after discussing Example 2.

Sandwich quantities (month by type):

$$\begin{array}{c} \\ \text{May} \\ \text{June} \end{array} \begin{array}{ccc} \text{chicken} & \text{ham} & \text{beef} \\ \left[\begin{array}{ccc} 5500 & 1800 & 2300 \\ 5800 & 1600 & 2700 \end{array}\right. & & \left.\vphantom{\begin{array}{c}5500\\5800\end{array}}\right] \end{array} = A$$

Sandwich prices (type by month):

$$\begin{array}{c} \\ \text{chicken} \\ \text{ham} \\ \text{beef} \end{array} \begin{array}{cc} \text{May} & \text{June} \\ \left[\begin{array}{cc} 1.50 & 1.80 \\ 1.75 & 2.00 \\ 1.75 & 1.75 \end{array}\right] \end{array} = B$$

Observe class discussion, or have students write descriptions of the information contained in A and B. Ask them to explain the difference between AB and BA in terms of the information represented by the matrix products. Note that both products are defined, and can be considered meaningful.

Using Technology

Students often find matrix multiplication to be tedious and time-consuming, with many opportunities to make annoying errors in the numerous operations of arithmetic required. They may be eager to learn about the software packages that may be available for use on either school or home computers. Some students may have the interest and ability to write their own programs to find matrix products. Matrix arithmetic programs can also be very helpful later in verifying inverses of matrices.

The TI–81 programmable graphing calculator has a matrix feature that permits students to define matrices (up to 6×6) and to perform all basic operations on matrices. Caution students that care must be taken in entering the matrices, but then the tedious and time-consuming calculations can be performed quickly by the calculator.

14-3   pages 530–537

Presenting the Section

The discussion of matrix arithmetic continues, culminating with inverses and their use in solving systems of linear equations. Point out that identity matrices and multiplicative inverses for matrices are by definition and of necessity square matrices.

Point out that $|A|$ is always a real number, so $\frac{1}{|A|}$ is a real number if $|A| \neq 0$. Therefore the inverse matrix, $A^{-1} = \frac{1}{|A|} \begin{bmatrix} d & -b \\ -c & a \end{bmatrix}$, when defined, is the product of a scalar and a matrix. The scalar can be written outside the matrix or multiplied by each term inside the matrix, depending on the situation. Although derivation of the formula for the inverse of a 2×2 matrix (see Written Exercise 29) may be too challenging for some students, all students will be able to understand a verification of the formula:

Let $A = \begin{bmatrix} a & b \\ c & d \end{bmatrix}$ where $ad - bc \neq 0$, and let

$$B = \frac{1}{|A|} \begin{bmatrix} d & -b \\ -c & a \end{bmatrix} = \begin{bmatrix} \dfrac{d}{ad-bc} & -\dfrac{b}{ad-bc} \\ -\dfrac{c}{ad-bc} & \dfrac{a}{ad-bc} \end{bmatrix}.$$

Evaluate AB and BA to show that each product is $I_{2\times2} = \begin{bmatrix} 1 & 0 \\ 0 & 1 \end{bmatrix}$, and so $A = B^{-1}$ and $B = A^{-1}$.

Using Technology

The goal here is to focus on concepts and applications, not the details of long computational processes. Encourage students to use programmable

calculators or computer software when inverses of matrices larger than 2×2 matrices are needed.

Review

Before discussing properties of matrix operations, review the properties of real numbers. These will include the additive and multiplicative identities and inverses, as well as the associative, commutative, and distributive properties of addition and multiplication.

Students may be interested to learn that matrices do not have the familiar zero-product property: For real numbers x and y, if $xy = 0$ then $x = 0$ or $y = 0$. (See Written Exercise 28.) This is not the case for matrices, however. For example, let $A = \begin{bmatrix} 1 & 0 \\ 1 & 0 \end{bmatrix}$ and $B = \begin{bmatrix} 0 & 0 \\ 1 & -1 \end{bmatrix}$. Thus, $A \neq O_{2\times2}$ and $B \neq O_{2\times2}$, yet $AB = O_{2\times2}$ and $BA = O_{2\times2}$.

14-4   pages 537–543

Presenting the Section

Beginning with this section, applications of matrices are highlighted. Students will be interested to see the direct translation of a communication network to a matrix. Emphasize that a communication matrix consists only of 1's and 0's. Powers of these matrices, however, can contain other nonnegative integers. Ask students to interpret the extreme cases of communication matrices containing all zeros (no communication is possible) or all ones (every position can communicate with every other, including itself).

Clarify the meaning of a power of a communication matrix M. Make sure students understand the difference between n-step relays, represented by M^n, and all possible paths using n or fewer steps, represented by the matrix $M + M^2 + \cdots + M^n$.

Communication

Communication matrices provide an excellent opportunity for students to practice visualization skills. Given a communication matrix, students can turn it directly into an equivalent graphic image, the network diagram. This gives the rather abstract matrix a visual concreteness.

Presenting the Section

Transition matrices build on students' experience in the last section. Here, as there, students repeatedly multiply by a special matrix to produce data for a new stage in the analysis of a problem. A major difference, of course, is that a transition matrix does not necessarily contain only ones and zeros. In fact, once you feel students have understood the basic idea of a transition matrix, you might ask them to devise and explain a transition matrix with only ones and zeros. A few examples are shown below.

$$\begin{array}{c} \quad A\,B\,C \\ \begin{array}{c} A \\ B \\ C \end{array}\!\!\left[\begin{array}{ccc} 1 & 0 & 0 \\ 0 & 1 & 0 \\ 0 & 0 & 1 \end{array}\right] \end{array}$$ (Nobody changes brands.)

$$\begin{array}{c} \quad A\,B\,C \\ \begin{array}{c} A \\ B \\ C \end{array}\!\!\left[\begin{array}{ccc} 1 & 0 & 0 \\ 0 & 1 & 0 \\ 0 & 1 & 0 \end{array}\right] \end{array}$$ (Brand C loses all its market shares to brand B.)

$$\begin{array}{c} \quad A\,B\,C \\ \begin{array}{c} A \\ B \\ C \end{array}\!\!\left[\begin{array}{ccc} 0 & 1 & 0 \\ 0 & 0 & 1 \\ 1 & 0 & 0 \end{array}\right] \end{array}$$ (Everybody changes brands.)

The discussion following Example 2 should be analyzed carefully. This will enable students to see why a given entry of $P_0 T^n$ shows the market share of a given brand n months from now.

The idea of a steady state deserves careful development. In the discussion after Example 2, students need to see that a limiting condition exists. Further computation of P_n for very large values of n, say $n = 50$ or $n = 100$, produces negligible change in the resulting market-share matrix. This stability of market shares is what is meant by a steady state.

Assessment

To check students' understanding of transition matrices, assign a problem like Class Exercise 3 after discussing Example 2. Have students respond either orally or in writing to the question of how they would decide if this situation leads to a steady state. If necessary, ask which values of P_n they would compute and why. Ask how they would use computed values of P_n to decide whether a steady state is reached.

Presenting the Section

Students have already had experience with transformations, both earlier in this course and possibly in previous courses. Matrices provide a new way to treat transformations, one that is very precise and compact.

Students will appreciate being able to find images of any number of points in the xy-plane by finding the product of a 2×2 and a $2 \times n$ matrix. In addition, Theorem 2 provides a very practical technique for writing the matrix for a particular transformation.

Theorem 3 and its discussion give students another way to relate transformations as functions and transformation matrices.

Cooperative Learning

Have students work in groups of 4 or 5 to play a variation on the old game of "Guess My Function." Choose a member of each group to be the Image Maker (IM). This player secretly writes the matrix for a transformation, and the other players try to guess the transformation. The players give the IM an ordered pair, and the IM will give the image of that point under the chosen transformation. By observing points and their images, keeping a record of guesses and answers, and graphing the results, players can eventually determine the transformation. If the game becomes too easy, you may want to impose a special rule: Nobody can give $(1, 0)$ or $(0, 1)$ for an input point.

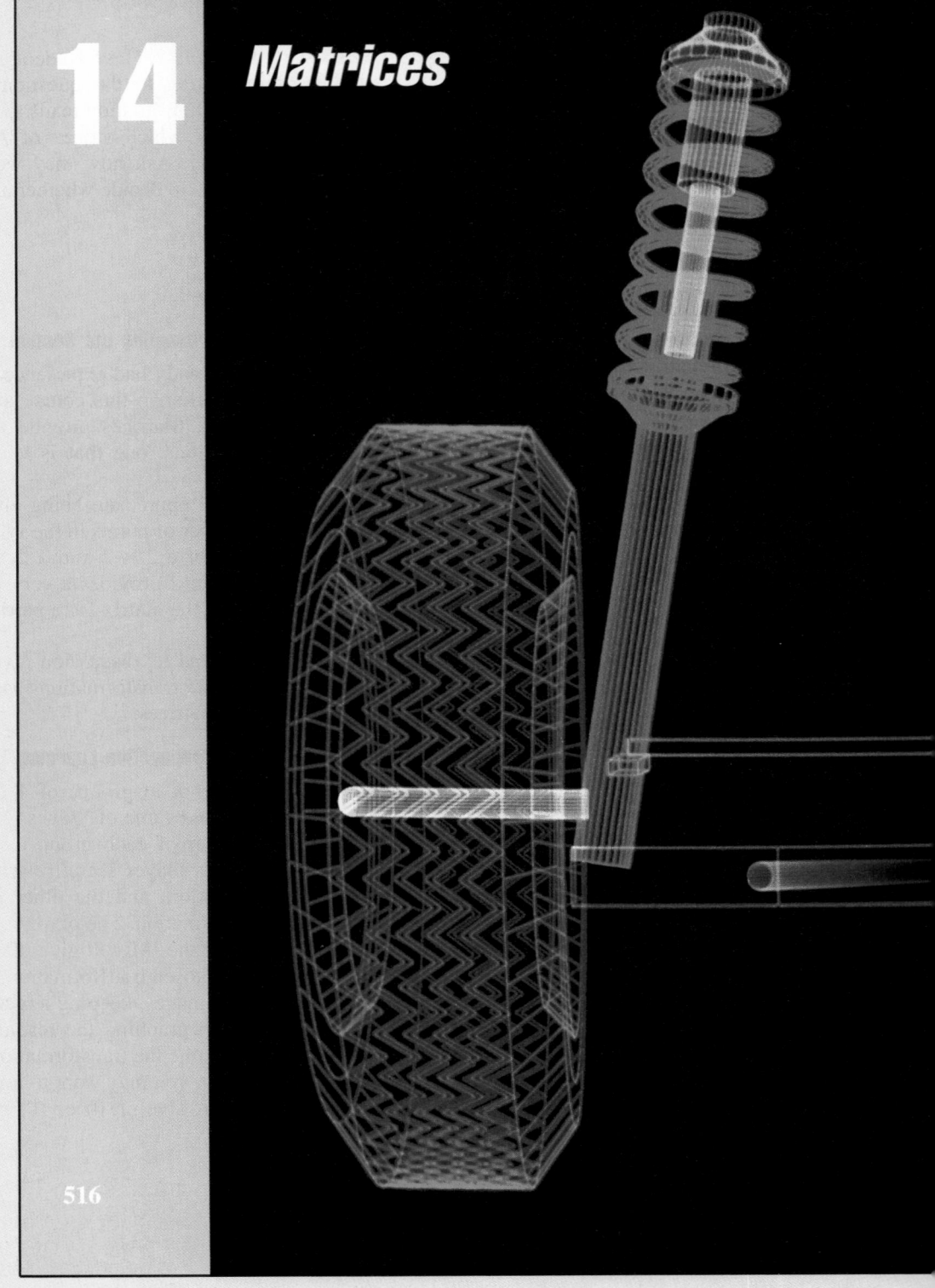

14

Matrices

516

14-1 Matrix Addition and Scalar Multiplication

Objective *To find the sum, difference, or scalar multiples of matrices.*

An automobile dealer sells four different models whose fuel economy is shown below.

	Sports car	Sedan	Station wagon	Van
Miles per gallon for city driving	17	22	17	16
Miles per gallon for highway driving	23	30	24	19

The information in this table can be displayed as a rectangular array of numbers enclosed by brackets, as shown below. Such an array is called a **matrix** (plural, *matrices*). A matrix is usually named by a capital letter. In this example, E is used to name the matrix.

$$E = \begin{bmatrix} 17 & 22 & 17 & 16 \\ 23 & 30 & 24 & 19 \end{bmatrix}$$

Each number in a matrix is an **element** (or **entry**) of the matrix. The **dimensions** of a matrix are the number of rows and columns. Matrix E has two rows and four columns, so we say that E is a 2×4 (read "two by four") matrix, denoted by $E_{2 \times 4}$. (The number of rows is first and the number of columns is second.)

Sometimes it is helpful to label the rows and columns to remind us of what each represents. In the matrix below, the row labels remind us of the city and highway driving conditions and the column labels remind us of the model. We say that the fuel economy matrix, E, is a "driving-condition by model" matrix.

$$\begin{array}{c} \\ c \\ h \end{array} \begin{array}{cccc} \text{sp} & \text{se} & \text{sw} & \text{v} \\ \begin{bmatrix} 17 & 22 & 17 & 16 \\ 23 & 30 & 24 & 19 \end{bmatrix} \end{array} = E$$

Of course, the same information could be given by interchanging the rows and columns of E. Matrix F, shown on the next page, is called the **transpose** of matrix E, and is denoted by E^t.

◀ Computer-aided design allows us to model three-dimensional objects on a two-dimensional screen. We can adjust the size, shape, and orientation of parts without the time and expense of constructing physical prototypes.

Additional Examples

1. Let $A = \begin{bmatrix} 2 & -1 \\ 4 & 0 \\ 0 & -8 \end{bmatrix}$

and $B = \begin{bmatrix} -6 & 3 & 5 \\ 0 & 7 & -4 \end{bmatrix}$.

a. Find $A^t + B$ and $A + B^t$. Are the resulting matrices equal?

b. Find $A^t - 2B$.

a. $A^t + B =$
$\begin{bmatrix} 2 & 4 & 0 \\ -1 & 0 & -8 \end{bmatrix} + \begin{bmatrix} -6 & 3 & 5 \\ 0 & 7 & -4 \end{bmatrix} = \begin{bmatrix} -4 & 7 & 5 \\ -1 & 7 & -12 \end{bmatrix}$

$A + B^t = \begin{bmatrix} 2 & -1 \\ 4 & 0 \\ 0 & -8 \end{bmatrix} + \begin{bmatrix} -6 & 0 \\ 3 & 7 \\ 5 & -4 \end{bmatrix} = \begin{bmatrix} -4 & -1 \\ 7 & 7 \\ 5 & -12 \end{bmatrix}$; no

b. $A^t - 2B =$
$\begin{bmatrix} 2 & 4 & 0 \\ -1 & 0 & -8 \end{bmatrix} - \begin{bmatrix} -12 & 6 & 10 \\ 0 & 14 & -8 \end{bmatrix} = \begin{bmatrix} 14 & -2 & -10 \\ -1 & -14 & 0 \end{bmatrix}$

2. Find the values of the variables for which the given statement is true.
$\begin{bmatrix} a & b \\ c & d \end{bmatrix} - \begin{bmatrix} 2 & -3 \\ 5 & -1 \end{bmatrix} = \begin{bmatrix} 7 & 2.5 \\ -1 & 0 \end{bmatrix}$

$\begin{bmatrix} a-2 & b+3 \\ c-5 & d+1 \end{bmatrix} = \begin{bmatrix} 7 & 2.5 \\ -1 & 0 \end{bmatrix}$

$a - 2 = 7; a = 9$
$b + 3 = 2.5; b = -0.5$
$c - 5 = -1; c = 4$
$d + 1 = 0; d = -1$

$$\begin{array}{c} \begin{array}{cc} c & h \end{array} \\ \begin{array}{c} sp \\ se \\ sw \\ v \end{array} \begin{bmatrix} 17 & 23 \\ 22 & 30 \\ 17 & 24 \\ 16 & 19 \end{bmatrix} \end{array} = F = E^t$$

Notice that the rows of F are the same as the columns of E. The dimensions of F are 4×2 (a "model by driving-condition" matrix).

Suppose that the Environmental Protection Agency (EPA) mandates that all of these fuel performance figures must increase 10% by the year 1998. This means that every element of matrix E must be multiplied by 1.10. The resulting matrix is denoted by $1.1E$.

$$1.1E = 1.1\begin{bmatrix} 17 & 22 & 17 & 16 \\ 23 & 30 & 24 & 19 \end{bmatrix} = \begin{bmatrix} (1.1)17 & (1.1)22 & (1.1)17 & (1.1)16 \\ (1.1)23 & (1.1)30 & (1.1)24 & (1.1)19 \end{bmatrix}$$

$$= \begin{bmatrix} 18.7 & 24.2 & 18.7 & 17.6 \\ 25.3 & 33 & 26.4 & 20.9 \end{bmatrix}$$

The operation of multiplying a matrix A by a real number c is called **scalar multiplication**. (In matrix algebra any real number is called a **scalar**.) The new matrix, cA, is the result of multiplying each element in A by c.

Example 1 If $M = \begin{bmatrix} 3 & 1 & 5 \\ 4 & 0 & -2 \end{bmatrix}$, find M^t, $2M$, and $-3M$.

Solution $M^t = \begin{bmatrix} 3 & 4 \\ 1 & 0 \\ 5 & -2 \end{bmatrix}$ ← Interchange the rows and columns of M to get the transpose of M.

$$2M = \begin{bmatrix} 2(3) & 2(1) & 2(5) \\ 2(4) & 2(0) & 2(-2) \end{bmatrix} = \begin{bmatrix} 6 & 2 & 10 \\ 8 & 0 & -4 \end{bmatrix}$$

$$-3M = \begin{bmatrix} -3(3) & -3(1) & -3(5) \\ -3(4) & -3(0) & -3(-2) \end{bmatrix} = \begin{bmatrix} -9 & -3 & -15 \\ -12 & 0 & 6 \end{bmatrix}$$

Two matrices are *equal* if and only if they have the same dimensions, and the elements in all corresponding positions (same row, same column) are equal. Matrices of different dimensions can never be equal.

Two matrices having the same dimensions can be added to produce a new matrix by finding the sums of the corresponding elements of the matrices. This operation is called **matrix addition**. Since corresponding elements are added, two matrices cannot be added if they have different dimensions.

Matrix subtraction is similar to real number subtraction: to subtract a matrix, we add the *additive inverse* of the matrix. (The additive inverse of a matrix is defined in Section 14-3.) More simply, two matrices can be subtracted by finding the differences of the corresponding elements of the matrices.

Example 2 Let $A = \begin{bmatrix} 3 & 8 & 1 \\ 4 & 0 & -3 \\ -2 & 1 & 5 \end{bmatrix}$ and $B = \begin{bmatrix} 2 & 0 & 9 \\ 4 & -6 & -5 \\ 0 & 7 & 2 \end{bmatrix}$.

Find $A + B$ and $A - B$.

Solution
$$A + B = \begin{bmatrix} 3+2 & 8+0 & 1+9 \\ 4+4 & 0+(-6) & -3+(-5) \\ -2+0 & 1+7 & 5+2 \end{bmatrix} = \begin{bmatrix} 5 & 8 & 10 \\ 8 & -6 & -8 \\ -2 & 8 & 7 \end{bmatrix}$$

$$A - B = \begin{bmatrix} 3-2 & 8-0 & 1-9 \\ 4-4 & 0-(-6) & -3-(-5) \\ -2-0 & 1-7 & 5-2 \end{bmatrix} = \begin{bmatrix} 1 & 8 & -8 \\ 0 & 6 & 2 \\ -2 & -6 & 3 \end{bmatrix}$$

If two matrices are used to model a real-world application, their sum or difference may not always give a meaningful matrix. For example, the following matrices give baseball statistics for 1981, a year in which the season was divided in half because of a players' strike. Matrices A and B give the games won and lost and the winning percentage, expressed as a decimal, for the top three teams in the National League East Division. The teams were the Philadelphia Phillies, the St. Louis Cardinals, and the Montreal Expos.

First half of the season

	W	L	PCT
Phillies	34	21	.618
Cardinals	30	20	.600
Expos	30	25	.545

$= A$

Second half of the season

	W	L	PCT
Phillies	25	27	.481
Cardinals	29	23	.558
Expos	30	23	.566

$= B$

Notice what happens when A and B are added:

	W	L	PCT
Phillies	59	48	1.099
Cardinals	59	43	1.158
Expos	60	48	1.111

$= A + B$

Elements in the first two columns make sense because they give the number of wins and losses for the whole season.

The elements in the third column do not make sense because the winning percentage for the whole season is *not* the sum of the percentages for each half of the season. Clearly, no team can win more than 100% of its games.

Error Analysis

Encourage students always to note the dimensions of a matrix before doing a problem. This will prevent errors or difficulties in adding and subtracting matrices, and later in matrix multiplication.

CLASS EXERCISES

For Exercises 1–3, use matrices A, B, C, and D.

6. **b.** $\begin{bmatrix} 5 & 3 & 0 \\ -1 & -4 & 6 \end{bmatrix}$

$$A = \begin{bmatrix} 2 \\ 5 \end{bmatrix} \quad B = [2 \quad 5] \quad C = [5 \quad 2] \quad D = \begin{bmatrix} 2 & \frac{10}{2} \end{bmatrix}$$

1. Give the dimensions of each matrix. 2×1, 1×2, 1×2, 1×2
2. Which matrices are transposes of each other? *A and B, A and D*
3. Which matrices are equal to each other? *B and D*

For Exercises 4–6, use matrices A, B, C, and D.

$$A = \begin{bmatrix} 5 & -1 \\ 3 & -4 \\ 0 & 6 \end{bmatrix} \quad B = \begin{bmatrix} 3 & 7 \\ 2 & 9 \\ -3 & 8 \end{bmatrix} \quad C = \begin{bmatrix} 1 & 2 & 0 \\ 5 & 3 & 1 \\ 4 & 0 & 7 \end{bmatrix} \quad D = \begin{bmatrix} 1 & 5 & 9 \\ -2 & 3 & 4 \end{bmatrix}$$

4. Give the dimensions of each matrix. 3×2, 3×2, 3×3, 2×3
5. Which matrices can be added? Find their sum. *A and B;* $\begin{bmatrix} 8 & 6 \\ 5 & 5 \\ -3 & 14 \end{bmatrix}$
6. **a.** Give the dimensions of A^t and B^t. 2×3
   b. Find A^t. See above.
7. *Discussion* The matrix below gives nutritional information for single servings of tomato juice and orange juice. All elements are given in grams.

$$\begin{array}{r} \\ \text{Tomato juice} \\ \text{Orange juice} \end{array} \begin{array}{cccc} \text{Protein} & \text{Carbohydrate} & \text{Fat} & \text{Sodium} \\ \begin{bmatrix} 2.2 & 10.4 & 0.2 & 0.486 \\ 1.5 & 24.7 & 0.5 & 0.002 \end{bmatrix} \end{array} = N$$

   a. Explain the meaning of the number 10.4 in matrix N.
   b. Explain the meaning of the number 0.002 in matrix N.
   c. Give a real-world meaning for the matrix 2N.
   d. Matrix N could be called a "juice-type by nutrient" matrix. How could you form a "nutrient by juice-type" matrix?
8. *Discussion* Explain how a teacher's grade list could be considered a matrix. *It is a "student by grade" matrix.*

WRITTEN EXERCISES

Simplify.

A 1. $\begin{bmatrix} -3 \\ 0 \end{bmatrix} + \begin{bmatrix} 1 \\ 2 \end{bmatrix} \begin{bmatrix} -2 \\ 2 \end{bmatrix}$

2. $[12 \quad 7] + [3 \quad -4] [15 \quad 3]$

3. $\begin{bmatrix} 8 & 1 \\ -1 & 5 \end{bmatrix} - \begin{bmatrix} 3 & -2 \\ 4 & -1 \end{bmatrix} \begin{bmatrix} 5 & 3 \\ -5 & 6 \end{bmatrix}$

4. $[1 \quad -1 \quad 0] - [6 \quad \begin{matrix} [-5 & 8 & 1] \\ -9 & -1] \end{matrix}$

520   *Chapter Fourteen*

5. $\begin{bmatrix} 8 & 2 & -2 \\ -3 & 1 & 14 \end{bmatrix} + \begin{bmatrix} 12 & 3 & 10 \\ 0 & 0 & -6 \end{bmatrix}$ See below.

6. $\begin{bmatrix} 1 & -5 \\ -3 & 1 \\ 4 & 2 \end{bmatrix} - \begin{bmatrix} 0 & -2 \\ 0 & 7 \\ -8 & 3 \end{bmatrix}\begin{bmatrix} 1 & -3 \\ -3 & -6 \\ 12 & -1 \end{bmatrix}$

7. $8\begin{bmatrix} 5 & -2 \\ 4 & 0 \end{bmatrix}\begin{bmatrix} 40 & -16 \\ 32 & 0 \end{bmatrix}$

8. $-4\begin{bmatrix} 0 & -1 \\ -6 & 5 \end{bmatrix}\begin{bmatrix} 0 & 4 \\ 24 & -20 \end{bmatrix}$

9. $2\begin{bmatrix} 3 & 0 \\ -4 & 1 \\ 0 & -1 \end{bmatrix} + \begin{bmatrix} -2 & -2 \\ 3 & 0 \\ 6 & 11 \end{bmatrix}\begin{bmatrix} 4 & -2 \\ -5 & 2 \\ 6 & 9 \end{bmatrix}$

10. $\begin{bmatrix} 18 & 12 & 5 \\ 2 & 0 & -3 \end{bmatrix} - 5\begin{bmatrix} 4 & -1 & 3 \\ -3 & 2 & 5 \end{bmatrix}$

$\begin{bmatrix} -2 & 17 & -10 \\ 17 & -10 & -28 \end{bmatrix}$

For Exercises 11–14, use matrices A, B, and C.

$A = \begin{bmatrix} 3 & 2 & 4 & 8 \\ 0 & 1 & 8 & -2 \\ 3 & 7 & 9 & 1 \end{bmatrix}$ $B = \begin{bmatrix} 8 & 1 & 4 \\ 3 & 0 & 7 \\ -2 & 5 & 11 \\ 6 & 4 & -3 \end{bmatrix}$ $C = \begin{bmatrix} 9 & 0 & 5 \\ 4 & 3 & 1 \\ 7 & 12 & 2 \\ 6 & 0 & 3 \end{bmatrix}$

5. $\begin{bmatrix} 20 & 5 & 8 \\ -3 & 1 & 8 \end{bmatrix}$

11. a. Give the dimensions of each matrix.
   b. Which sum is defined, $A + B$ or $B + C$? Find this sum.

12. a. Give the dimensions of A^t.
   b. Find A^t, $A^t + B$, and $C - A^t$.

13. Find $2B$ and $2B + C$.

14. Find $3C$ and $3C - B$.

B **15. Business** The Handy Hardware Company has two locations, one downtown and one at the mall. During April, the downtown store sold 31 of the lowest-priced lawn mowers, 42 of the medium-priced mowers, and 18 of the highest-priced mowers. Also during April, the mall store sold 22 of the lowest-priced mowers, 25 of the medium-priced mowers, and 11 of the highest-priced mowers.

   a. Represent this information in an April sales matrix A.
   b. Do the dimensions of your matrix A represent a "mower-type by location" matrix or a "location by mower-type" matrix?
   c. Suppose that during May, the Handy downtown store sold 28 of the lowest-priced mowers, 29 of the medium-priced ones and 20 of the highest-priced ones, and that the mall store sold 20 lowest-priced ones, 18 medium-priced ones, and 9 highest-priced ones. Represent this information in a May matrix M that has the same dimensions as A.
   d. *Writing* Find $A + M$ and describe what this matrix sum tells you.
   e. If the manager of the Handy stores expects next year's lawn mower sales to rise about 8%, about how many highest-priced lawn mowers does the manager expect to sell at the downtown store next April? If you were the manager, would you round your calculations up or down? What scalar multiple of matrix A would assist you in planning next April's sales?
   f. Suppose the manager estimates that the total lawn mower sales for next April and May will be given by the matrix $1.09A + 1.15M$. What sales increases is the manager expecting for next April? for next May?

11. a. 3×4, 4×3, 4×3
   b. $B + C =$
$\begin{bmatrix} 17 & 1 & 9 \\ 7 & 3 & 8 \\ 5 & 17 & 13 \\ 12 & 4 & 0 \end{bmatrix}$

12. a. 4×3
   b. $A^t = \begin{bmatrix} 3 & 0 & 3 \\ 2 & 1 & 7 \\ 4 & 8 & 9 \\ 8 & -2 & 1 \end{bmatrix}$,

$A^t + B = \begin{bmatrix} 11 & 1 & 7 \\ 5 & 1 & 14 \\ 2 & 13 & 20 \\ 14 & 2 & -2 \end{bmatrix}$,

$C - A^t = \begin{bmatrix} 6 & 0 & 2 \\ 2 & 2 & -6 \\ 3 & 4 & -7 \\ -2 & 2 & 2 \end{bmatrix}$

13. $2B = \begin{bmatrix} 16 & 2 & 8 \\ 6 & 0 & 14 \\ -4 & 10 & 22 \\ 12 & 8 & -6 \end{bmatrix}$,

$2B + C = \begin{bmatrix} 25 & 2 & 13 \\ 10 & 3 & 15 \\ 3 & 22 & 24 \\ 18 & 8 & -3 \end{bmatrix}$

14. $3C = \begin{bmatrix} 27 & 0 & 15 \\ 12 & 9 & 3 \\ 21 & 36 & 6 \\ 18 & 0 & 9 \end{bmatrix}$,

$3C - B = \begin{bmatrix} 19 & -1 & 11 \\ 9 & 9 & -4 \\ 23 & 31 & -5 \\ 12 & -4 & 12 \end{bmatrix}$

Written Exercises 20 and 21 could be extended for more ambitious or able students. This would be an appropriate place to state and prove some simple theorems about matrices, such as:

$$k(A + B) = kA + kB$$
$$(kA)^t = kA^t$$
$$(A + B)^t = A^t + B^t$$

where k is a scalar and A and B are matrices. Start with specific 2×2 matrices. Later you may be able to generalize to $m \times n$ matrices using subscripted variables.

16. **Manufacturing** The matrix S gives this week's shipping order for various sofas, chairs, coffee tables, and end tables to be delivered from a warehouse to its store location. Each furniture piece is available in three styles: colonial, traditional, and modern. Matrix T gives the total living room furniture inventory of the warehouse.

$$
\begin{array}{c}
\text{Styles} \\
\begin{array}{ccc}
\text{C} & \text{T} & \text{M}
\end{array} \\
\begin{array}{c}
\text{s} \\ \text{c} \\ \text{ct} \\ \text{et}
\end{array}
\begin{bmatrix}
2 & 0 & 1 \\
10 & 2 & 3 \\
2 & 4 & 3 \\
6 & 8 & 2
\end{bmatrix} = S
\end{array}
\qquad
\begin{array}{c}
\text{Styles} \\
\begin{array}{ccc}
\text{C} & \text{T} & \text{M}
\end{array} \\
\begin{array}{c}
\text{s} \\ \text{c} \\ \text{ct} \\ \text{et}
\end{array}
\begin{bmatrix}
12 & 10 & 15 \\
40 & 15 & 17 \\
17 & 42 & 18 \\
24 & 24 & 24
\end{bmatrix} = T
\end{array}
$$

c. $T - S = \begin{bmatrix} 10 & 10 & 14 \\ 30 & 13 & 14 \\ 15 & 38 & 15 \\ 18 & 16 & 22 \end{bmatrix}$ inventory of warehouse after shipment

10 colonial chairs to be shipped; 40 colonial chairs in warehouse

a. What is the meaning of the 10 in matrix S? of the 40 in matrix T?

b. Give the dimensions of S and T. $4 \times 3, 4 \times 3$

c. Calculate $T - S$ and give its real-world significance. See above.

d. Because of an upcoming anniversary sale, the number of furniture pieces sold for each model type is expected to be 50% higher next week than this week. Express in terms of S and T a matrix that will give the warehouse inventory after next week's order is shipped. $T - 2.5S$

Find the values of the variable for which the given statement is true.

17. $\begin{bmatrix} x & 3 \\ -7 & 0 \end{bmatrix} = \begin{bmatrix} 2 & y \\ -7 & z+1 \end{bmatrix}$ $\begin{array}{l} x = 2, \\ y = 3, \\ z = -1 \end{array}$
  18. $\begin{bmatrix} a-3 & 5 \\ 2-c & 4b \end{bmatrix} = \begin{bmatrix} -1 & 2x+1 \\ -8 & 12 \end{bmatrix}$ $\begin{array}{l} a = 2, \\ b = 3, \\ c = 10, \\ x = 2 \end{array}$

19. For any matrix A, give a simpler name for the matrix $(A^t)^t$. A

20. Is it true or false that $(2A)^t = 2A^t$ for any matrix A? Give an example to support your answer. True

21. Is it true that $(A + B)^t = A^t + B^t$ for any matrices A and B that have the same dimensions? Give an example to support your answer. Yes

22. *Writing* Give a real-world example of two matrices having the same dimensions and whose sum makes no sense. Write a sentence explaining why the resulting matrix is not meaningful.

23. *Research* Refer to a world almanac. Write a matrix that gives the number of stories and the heights in meters of the five tallest buildings in the United States.

24. *Research* In a newspaper or magazine, find a table of numbers that can be expressed as a matrix.

14-2 Matrix Multiplication

Objective *To find the product of two matrices.*

Suppose a teacher calculates your test average for the term by using a formula that counts, or weighs, each of your five tests a certain percentage of your grade, as shown in matrix W below. Notice that the weights must add up to 100%.

$$\begin{array}{ccccc} & & \text{(midterm)} & & \text{(final exam)} \\ \text{Test 1} & \text{Test 2} & \text{Test 3} & \text{Test 4} & \text{Test 5} \end{array}$$
$$\text{weight} \quad [\; 15\% \quad 15\% \quad 25\% \quad 15\% \quad 30\% \;] = W$$

How would you use these weights to find the weighted score, or *weighted average*, of students A, B, and C, whose test scores are given in matrix T at the right.

$$\begin{array}{c} \\ \text{Test 1} \\ \text{Test 2} \\ \text{Test 3} \\ \text{Test 4} \\ \text{Test 5} \end{array} \begin{array}{ccc} A & B & C \\ \begin{bmatrix} 82 & 92 & 74 \\ 85 & 88 & 68 \\ 78 & 95 & 73 \\ 75 & 85 & 82 \\ 84 & 94 & 81 \end{bmatrix} \end{array} = T$$

To calculate a student's weighted score (or weighted average), you must multiply each test score by its weight and then add the products.

A's weighted score $= 0.15(82) + 0.15(85) + 0.25(78) + 0.15(75) + 0.30(84)$
$= 81$

B's weighted score $= 0.15(92) + 0.15(88) + 0.25(95) + 0.15(85) + 0.30(94)$
$= 91.70 \approx 92$

C's weighted score $= 0.15(74) + 0.15(68) + 0.25(73) + 0.15(82) + 0.30(81)$
$= 76.15 \approx 76$

The students' weighted scores for the term are shown in matrix S.

$$\begin{array}{ccc} A & B & C \end{array}$$
$$\text{weighted score} \; [\; 81 \quad 92 \quad 76 \;] = S$$

Matrix S can be thought of as the *matrix product* of matrices W and T.

$$\begin{array}{ccc} W & \cdot & T & = S \end{array}$$

$$[0.15 \quad 0.15 \quad 0.25 \quad 0.15 \quad 0.30] \cdot \begin{bmatrix} 82 & 92 & 74 \\ 85 & 88 & 68 \\ 78 & 95 & 73 \\ 75 & 85 & 82 \\ 84 & 94 & 81 \end{bmatrix} = [81 \quad 92 \quad 76]$$

Teaching Notes, p. 516B

Warm-Up Exercises

1. Complete: If an operation $*$ is commutative, then $a * b = \underline{\;?\;}$. $b * a$

2. If $b * h = \frac{1}{2} bh$, is $*$ a commutative operation?
   Yes; $\frac{1}{2} bh = \frac{1}{2} hb$.

3. If $r * h = \frac{1}{3} \pi r^2 h$, is $*$ a commutative operation?
   No; $\frac{1}{3} \pi r^2 h \neq \frac{1}{3} \pi h^2 r$.

4. Find the dot product $(1, 4, -3) \cdot (-7, -2, -1)$.
   -12

5. Is finding the dot product of two vectors a commutative operation?
   Yes; by property 1 on p. 442, $\mathbf{u} \cdot \mathbf{v} = \mathbf{v} \cdot \mathbf{u}$.

Motivating the Section

Ask students whether anyone has ever heard of arithmetic in which sums or products are not commutative. Commutativity is so ingrained that some development may be necessary before students even consider this a sensible question. For example, let $x * y$ be defined by $x * y = 3xy + 1$ and $x \otimes y$ be defined as $x \otimes y = 2x + y$. Several computations should make it clear that $*$ is commutative but $\otimes$ is not. Matrix multiplication is an operation that is not generally commutative.

1. Let $A = \begin{bmatrix} 2 & -1 & 3 \end{bmatrix}$ and

$B = \begin{bmatrix} 5 & -1 \\ 0 & -4 \\ -2 & 7 \end{bmatrix}$.

a. Is BA defined? Explain.
b. Show that $(AB)^t = B^t A^t$.

a. No; B has 2 columns and A has 1 row. Since $2 \neq 1$, the product BA is not defined.

b. $AB =$

$\begin{bmatrix} 2 & -1 & 3 \end{bmatrix} \begin{bmatrix} 5 & -1 \\ 0 & -4 \\ -2 & 7 \end{bmatrix} =$

$\begin{bmatrix} 4 & 23 \end{bmatrix}$

$(AB)^t = \begin{bmatrix} 4 \\ 23 \end{bmatrix}$

$B^t A^t =$

$\begin{bmatrix} 5 & 0 & -2 \\ -1 & -4 & 7 \end{bmatrix} \begin{bmatrix} 2 \\ -1 \\ 3 \end{bmatrix} =$

$\begin{bmatrix} 10 + 0 - 6 \\ -2 + 4 + 21 \end{bmatrix} = \begin{bmatrix} 4 \\ 23 \end{bmatrix}$

In the example on the previous page, notice how the dimensions of the matrices agree: W has five *columns* and T has five *rows*. Also, notice that the column labels of W match the row labels of T.

$$W_{1 \times 5} \cdot T_{5 \times 3} = S_{1 \times 3}$$

weight of test • test score by student = weighted score by student

In general, the **matrix product** of $A_{m \times n}$ and $B_{n \times p}$ is the matrix whose dimensions are $m \times p$ and whose element in the ith row and jth column is the sum of the products of the corresponding elements of the ith row of A and the jth column of B. From this definition we can see that matrices A and B can be multiplied only if the number of columns of A equals the number of rows of B.

Example 1   Find the matrix product $\begin{bmatrix} 1 & 2 & 0 \\ 3 & -5 & 2 \end{bmatrix} \begin{bmatrix} -3 & 2 \\ 0 & 4 \\ 1 & -1 \end{bmatrix}$.

Solution   First determine if the product is defined. The dimensions are:

$$2 \times 3 \text{ and } 3 \times 2.$$

Since the number of columns in the first matrix equals the number of rows in the second matrix, the two matrices can be multiplied. The matrix product has dimensions 2×2.

$$\begin{bmatrix} 1 & 2 & 0 \\ 3 & -5 & 2 \end{bmatrix} \begin{bmatrix} -3 & 2 \\ 0 & 4 \\ 1 & -1 \end{bmatrix} = \begin{bmatrix} 1(-3) + 2(0) + 0(1) & 1(2) + 2(4) + 0(-1) \end{bmatrix}$$

$$\begin{bmatrix} 1 & 2 & 0 \\ 3 & -5 & 2 \end{bmatrix} \begin{bmatrix} -3 & 2 \\ 0 & 4 \\ 1 & -1 \end{bmatrix} = \begin{bmatrix} 3(-3) + (-5)(0) + 2(1) & 3(2) + (-5)(4) + 2(-1) \end{bmatrix}$$

$$\begin{bmatrix} 1 & 2 & 0 \\ 3 & -5 & 2 \end{bmatrix} \begin{bmatrix} -3 & 2 \\ 0 & 4 \\ 1 & -1 \end{bmatrix} = \begin{bmatrix} -3 & 10 \\ -7 & -16 \end{bmatrix}$$

Example 2   Machine I and Machine II produce items X, Y, and Z at the hourly rate given in matrix H. Matrix D gives the number of hours each machine runs during the week.

$$\begin{array}{c} \\ X \\ Y \\ Z \end{array} \begin{array}{cc} \text{I} & \text{II} \\ \begin{bmatrix} 3 & 2 \\ 5 & 4 \\ 1 & 2 \end{bmatrix} \end{array} = H \qquad \begin{array}{c} \\ \text{I} \\ \text{II} \end{array} \begin{array}{ccccc} \text{M} & \text{T} & \text{W} & \text{Th} & \text{F} \\ \begin{bmatrix} 8 & 8 & 8 & 7 & 7 \\ 6 & 10 & 12 & 11 & 9 \end{bmatrix} \end{array} = D$$

---

a. Give the dimensions of H, D, and HD.

b. Find HD. What information does HD give?

c. How many Y items are produced on Monday? How many Z items are produced on Thursday?

Solution

a. H is a 3×2 matrix and D is a 2×5 matrix. So HD is a 3×5 matrix.

b.

$$HD = \begin{bmatrix} 3 & 2 \\ 5 & 4 \\ 1 & 2 \end{bmatrix} \begin{bmatrix} 8 & 8 & 8 & 7 & 7 \\ 6 & 10 & 12 & 11 & 9 \end{bmatrix} = \begin{array}{c} X \\ Y \\ Z \end{array} \begin{array}{ccccc} M & T & W & Th & F \\ \end{array} \begin{bmatrix} 36 & 44 & 48 & 43 & 39 \\ 64 & 80 & 88 & 79 & 71 \\ 20 & 28 & 32 & 29 & 25 \end{bmatrix}$$

HD gives a meaningful result since the column labels of matrix H match the row labels of matrix D. HD is an "item by day" matrix, and so each element gives the quantity of a particular item that is produced on a particular day.

c. From the matrix product HD above, you can see that 64 Y items are produced on Monday and 29 Z items are produced on Thursday.

Notice that in Example 2 the product $H_{3 \times 2} \cdot D_{2 \times 5}$ is a 3×5 matrix. However, the product $D_{2 \times 5} \cdot H_{3 \times 2}$ is not defined because there are more elements in each row of D than there are in each column of H.

CLASS EXERCISES

In Exercises 1–6, you are given the dimensions of matrices A and B. Is AB defined? Is BA defined? Give the dimensions of each possible product.

1. $A_{3 \times 5}$ Yes; no; $AB_{3 \times 2}$
   $B_{5 \times 2}$

2. $A_{7 \times 6}$ No; yes; $BA_{2 \times 6}$
   $B_{2 \times 7}$

3. $A_{1 \times 4}$ No; yes;
   $B_{3 \times 1}$ $BA_{3 \times 4}$

4. $A_{5 \times 7}$ Yes; no; $AB_{5 \times 3}$
   $B_{7 \times 3}$

5. $A_{2 \times 4}$ No; no
   $B_{3 \times 4}$

6. $A_{3 \times 4}$ Yes; yes;
   $B_{4 \times 3}$ $AB_{3 \times 3}$;
   $BA_{4 \times 4}$

Find each matrix product.

7. $\begin{bmatrix} 1 & 1 \\ 2 & 3 \end{bmatrix} \begin{bmatrix} 4 & 2 \\ 5 & 6 \end{bmatrix}$ $\begin{bmatrix} 9 & 8 \\ 23 & 22 \end{bmatrix}$

8. $\begin{bmatrix} 1 & 3 & 1 \\ 2 & 0 & 4 \end{bmatrix} \begin{bmatrix} 2 & 1 \\ 1 & 3 \\ -1 & 0 \end{bmatrix}$ $\begin{bmatrix} 4 & 10 \\ 0 & 2 \end{bmatrix}$

9. Matrix M gives the number of different models sold per month by a certain car dealership for a three month period. Matrix C gives the costs per model. Which matrix is defined, MC or CM? CM

10. **Discussion** A and B are both 2×3 matrices. Can you multiply them? Can you multiply A^t and B? How might this be useful? No; yes

Additional Examples cont.

2. The juniors at Adams High School held a two-day bake sale to raise money for a class trip. Cupcakes, cookies, and pies were sold for $0.40, $0.25, and $4.00 respectively. The first day, 84 cupcakes, 210 cookies, and 27 pies were sold. The second day, 95 cupcakes, 184 cookies, and 17 pies were sold. Display this information in matrix form. Then use matrix multiplication to find the amount of money raised each day and altogether.

$$\begin{bmatrix} 84 & 210 & 27 \\ 95 & 184 & 17 \end{bmatrix} \begin{bmatrix} 0.40 \\ 0.25 \\ 4.00 \end{bmatrix} =$$
$$\begin{bmatrix} 194.10 \\ 152.00 \end{bmatrix}$$

$194.10 was raised the first day, $152.00 the second day, and $346.10 in all.

Suggested Assignments

Discrete Math
526/1–19 odd

Supplementary Materials

Alternative Assessment, 43–44
Using Technology, 44–45

Incorrect:

$$\begin{bmatrix} 2 & 3 \\ 1 & 2 \end{bmatrix}\begin{bmatrix} 1 & 2 & 1 \\ -2 & 1 & 3 \end{bmatrix} = \begin{bmatrix} 2 & 4 & 2 \\ -2 & 1 & 3 \end{bmatrix}$$

Correct:

$$\begin{bmatrix} 2 & 3 \\ 1 & 2 \end{bmatrix}\begin{bmatrix} 1 & 2 & 1 \\ -2 & 1 & 3 \end{bmatrix} = \begin{bmatrix} 2 + (-6) & 4 + 3 & 2 + 9 \\ 1 + (-4) & 2 + 2 & 1 + 6 \end{bmatrix}$$

Additional Answers Written Exercises

11. a. Let $A = \begin{bmatrix} 1 & 0 \\ 0 & 2 \end{bmatrix}$ and

$B = \begin{bmatrix} 3 & -1 \\ -2 & 1 \end{bmatrix}$;

$AB = \begin{bmatrix} 3 & -1 \\ -4 & 2 \end{bmatrix}$ and

$BA = \begin{bmatrix} 3 & -2 \\ -2 & 2 \end{bmatrix}$.

WRITTEN EXERCISES

Find each matrix product, if it is defined.

A **1.** $\begin{bmatrix} 4 & 3 \\ -1 & -2 \end{bmatrix}\begin{bmatrix} 5 \\ 1 \end{bmatrix}$ $\begin{bmatrix} 23 \\ -7 \end{bmatrix}$

2. $\begin{bmatrix} -6 \\ 2 \end{bmatrix}\begin{bmatrix} -1 & 12 \\ 0 & -4 \end{bmatrix}$ Not defined

3. $\begin{bmatrix} 1 & -5 \\ 2 & 3 \end{bmatrix}\begin{bmatrix} 4 & -4 \\ 0 & 1 \end{bmatrix}$ $\begin{bmatrix} 4 & -9 \\ 8 & -5 \end{bmatrix}$

4. $\begin{bmatrix} -2 & 3 \\ 4 & 2 \end{bmatrix}\begin{bmatrix} 0 & 3 \\ -6 & 5 \end{bmatrix}$ $\begin{bmatrix} -18 & 9 \\ -12 & 22 \end{bmatrix}$

5. $\begin{bmatrix} 8 & -10 \\ 0 & 3 \\ -6 & 4 \end{bmatrix}\begin{bmatrix} -2 \\ -9 \\ 1 \end{bmatrix}$ Not defined

6. $[7 \quad 1 \quad -3 \quad 4]\begin{bmatrix} 4 & 1 \\ -3 & 8 \\ 9 & 5 \\ -2 & 6 \end{bmatrix}$ $[-10 \quad 24]$

7. $\begin{bmatrix} 9 & -4 & 4 \\ 2 & -1 & -6 \end{bmatrix}\begin{bmatrix} 2 & -1 & 0 \\ 0 & 1 & -3 \\ 3 & 5 & 2 \end{bmatrix}$ $\begin{bmatrix} 30 & 7 & 20 \\ -14 & -33 & -9 \end{bmatrix}$

8. $\begin{bmatrix} 0 \\ -2 \end{bmatrix}\begin{bmatrix} 4 \\ 1 \end{bmatrix}$ Not defined

9. $\begin{bmatrix} 1 & 0 & 0 \\ 0 & 1 & 0 \\ 0 & 0 & 1 \end{bmatrix}\begin{bmatrix} a & b & c \\ d & e & f \\ g & h & i \end{bmatrix}$ $\begin{bmatrix} a & b & c \\ d & e & f \\ g & h & i \end{bmatrix}$

10. $\begin{bmatrix} 9 & 4 \\ 3 & 1 \\ 2 & 8 \\ 1 & 5 \end{bmatrix}\begin{bmatrix} 4 & 2 & 1 \\ 3 & 0 & 2 \end{bmatrix}$ $\begin{bmatrix} 48 & 18 & 17 \\ 15 & 6 & 5 \\ 32 & 4 & 18 \\ 19 & 2 & 11 \end{bmatrix}$

B **11. a.** Multiplication of real numbers is commutative. That is, $ab = ba$ for all real numbers a and b. However, multiplication of matrices is *not* commutative. Give an example of two 2×2 matrices A and B for which $AB \neq BA$.

b. *Investigation* Is matrix addition commutative? Give examples and write a convincing argument to explain your answer. Yes

12. In Exercise 11, you found matrices A and B such that $AB \neq BA$. However, it is possible to find other matrices C and D such that $CD = DC$. Verify this for

$$C = \begin{bmatrix} 5 & 2 \\ 7 & 3 \end{bmatrix} \text{ and } D = \begin{bmatrix} 3 & -2 \\ -7 & 5 \end{bmatrix}. \quad CD = DC = \begin{bmatrix} 1 & 0 \\ 0 & 1 \end{bmatrix}$$

13. a. Multiplication of real numbers is associative. That is, $a(bc) = (ab)c$ for all real numbers a, b, and c. Choose three 2×2 matrices A, B, C and check to see whether $A(BC) = (AB)C$.

b. Compare your results in part(a) with those of your classmates. Does it appear that matrix multiplication is associative? Yes

c. *Investigation* Decide whether matrix addition is associative. Give examples to support your answer. Yes

14. a. Multiplication of real numbers is distributive over addition. That is, $a(b + c) = ab + ac$ for all real numbers a, b, and c. Choose three 2×2 matrices A, B, C and check to see whether $A(B + C) = AB + AC$.

b. Does $(B + C)A = BA + CA$? Yes

526 *Chapter Fourteen*

15. Business Matrix S gives the number of three types of cars sold in March by two car dealers, and matrix P gives the profit for each type of car sold.

$$\begin{array}{c} \text{dealer} \\ \begin{array}{cc} 1 & 2 \end{array} \end{array}$$

$$\begin{array}{c} \text{compact} \\ \text{mid-size} \\ \text{full-size} \end{array} \begin{bmatrix} 18 & 15 \\ 24 & 17 \\ 16 & 20 \end{bmatrix} = S$$

$$\begin{array}{ccc} \text{compact} & \text{mid-size} & \text{full-size} \end{array}$$

$$\text{profit } [\ \ \$400 \qquad \$650 \qquad \$900\ \ ] = P$$

Which matrix is defined, SP or PS? Find this matrix and interpret its elements. [37,200 35,050] represents the profit for each dealer in the month of March.

16. Animal Science A dog breeder finds that certain brands of dog food contain different amounts of three main nutrients, measured in milligrams per serving, as shown in the matrix N. The dog breeder decides to mix the brands in order to give the healthiest feeding mixture possible. Matrix P gives the portion of the mixture for each brand.

$$\begin{array}{c} \text{brands} \\ \begin{array}{cccc} \text{W} & \text{X} & \text{Y} & \text{Z} \end{array} \end{array}$$

$$\begin{array}{c} \text{nutrient 1} \\ \text{nutrient 2} \\ \text{nutrient 3} \end{array} \begin{bmatrix} 250 & 480 & 360 & 200 \\ 320 & 510 & 475 & 315 \\ 180 & 200 & 230 & 155 \end{bmatrix} = N$$

$$\begin{array}{c} \text{part of mixture} \\ \begin{array}{c} \text{W} \\ \text{X} \\ \text{Y} \\ \text{Z} \end{array} \begin{bmatrix} 40\% \\ 10\% \\ 15\% \\ 35\% \end{bmatrix} = P \end{array}$$

$$\begin{bmatrix} 272 \\ 360.5 \\ 180.75 \end{bmatrix} = NP$$

a. Which matrix is defined, NP or PN? Find this matrix. NP

b. How many milligrams of nutrient 2 are in a serving of the mixture? 360.5 mg

17. Business Matrix P gives the monthly production schedule for three models of calculators. Matrix M gives the number of components needed to construct each model. Matrix R gives the number of relays needed for each component.

$$\begin{array}{c} \qquad\quad \text{Jan.} \quad \text{Feb.} \quad \text{Mar.} \\ \begin{array}{c} \text{scientific} \\ \text{business} \\ \text{graphing} \end{array} \begin{bmatrix} 500 & 600 & 600 \\ 200 & 200 & 200 \\ 100 & 300 & 400 \end{bmatrix} = P \end{array}$$

$$\begin{array}{c} \text{components} \\ \begin{array}{ccc} \text{A} & \text{B} & \text{C} \end{array} \end{array}$$

$$\begin{array}{c} \text{scientific} \\ \text{business} \\ \text{graphing} \end{array} \begin{bmatrix} 2 & 1 & 2 \\ 2 & 1 & 4 \\ 2 & 5 & 6 \end{bmatrix} = M$$

$$\begin{array}{c} \text{components} \\ \begin{array}{ccc} \text{A} & \text{B} & \text{C} \end{array} \end{array}$$

$$\begin{array}{c} \text{relay x} \\ \text{relay y} \end{array} \begin{bmatrix} 5 & 3 & 2 \\ 6 & 4 & 3 \end{bmatrix} = R$$

a. Explain why the product PM is defined, but not meaningful.

b. Find $M^t P$. Use your answer to tell how many A components will be needed in March.

c. Explain why MR is not defined.

d. Find RM^t. What information does this product give?

e. Find $RM^t P$. What information does this product give?

18. a. cost of each menu item
per day, cost per day of
each menu item

b. $CA = \begin{bmatrix} 120 \\ 100 \\ 144 \\ 124 \end{bmatrix}$; cost of

one serving of each of
the menus

c. $BD =$
[222 229 229 235 213];
number of meals served
per day

18. Nutrition Patients in a large hospital have a choice of four menus, each with meat, potato, vegetable, salad, and dessert. For example, Menu 1 has baked ham, sweet potatoes, peas, green salad, and baked apples. Matrix C gives the costs per serving (in cents) for each item in these menus. Matrix D gives the number of menus ordered each day last week.

	meat	potato	vegetable	salad	dessert
Menu 1	74	8	15	11	12
Menu 2	48	10	16	12	14
Menu 3	87	9	12	16	20
Menu 4	62	7	18	15	22

$= C$

	M	Tu	W	Th	F
Menu 1	50	48	44	52	38
Menu 2	63	60	55	55	53
Menu 3	71	79	83	87	78
Menu 4	38	42	47	41	44

$= D$

a. Explain the meaning of C^tD and D^tC.
b. Let A be the 5×1 matrix whose elements are all 1. Find CA. What information does CA give?
c. Let B be the 1×4 matrix whose elements are all 1. Find BD. What information does BD give?

19. Industrial Design At a regional competition for industrial design students, each project submitted is given a score from 1 to 10 in each of three categories: precision, weight = 30%; surface finish, weight = 20%; artistic design, weight = 50%. The total score for a project is found by adding the products of each category score and its weight.

a. Stephanie has a precision score of 8, a surface finish score of 7, and an artistic design score of 9. What is her total project score? 8.3
b. The six finalists have the scores given in the table below. Use matrix multiplication to determine the first, second, and third place winners. 1st: Joyce, 2nd: Eduardo, 3rd: Stephanie

	precision	finish	design
Stephanie	8	7	9
Joyce	8	6	10
Eduardo	6	8	10
Frank	9	10	7
Asabi	10	10	6
Dara	8	7	8

20. Tickets to the Senior Class Play cost $2.00 for students, $5.00 for adults, and $4.00 for senior citizens. At Friday night's performance, there were 121 students, 164 adults, and 32 senior citizens. At Saturday night's performance, there were 183 students, 140 adults, and 25 senior citizens. Display this information in matrix form. Then use matrix multiplication to find the ticket sales income for Friday and Saturday nights' performances.

21. a. Business A Chicago company wants to send some of its key personnel to a convention in London. In the company's Research and Development Division, five people plan to fly first class, three people plan to fly business class, and two people plan to fly coach class. In the Sales Division, four people plan to fly business class, and eight people coach class. Display this information in a 2×3 travel matrix T.

b. Round-trip prices for four different airlines are as follows: Airline A charges $1,280 for first class, $922 for business class, and $676 for coach. Airline B charges $1,400 for first class, $1,024 for business class, and $728 for coach. Airline C charges $1,320 for first class, $905 for business class, and $654 for coach. Finally, Airline D charges $1,450 for first class, $1,050 for business class, and $734 for coach. Display this information in a price matrix P that can be multiplied with matrix T to give the travel costs for each company division per airline. Then find this product.

c. How much will it cost to fly the Sales Division on Airline D?

d. Which airline will cost the Research and Development Division the least?

C **22. Business** Consider the price matrix P, shown below, which gives the costs for flying first class, business class, and coach class on four different airlines, A, B, C, and D.

$$
\begin{array}{c}
 \\
\text{first} \\
\text{business} \\
\text{coach}
\end{array}
\begin{array}{cccc}
A & B & C & D \\
\left[\begin{array}{cccc}
a_1 & b_1 & c_1 & d_1 \\
a_2 & b_2 & c_2 & d_2 \\
a_3 & b_3 & c_3 & d_3
\end{array}\right] = P
\end{array}
$$

Suppose that the prices in matrix P are all going to increase as follows:

Airline A: increase all prices by 5%;
Airline B: increase all prices by 4%;
Airline C: increase all prices by 6%;
Airline D: increase all prices by 2%.

Write a 4×4 increase matrix N such that the product PN is the new price matrix; that is, each element of PN gives the new price for each class on a particular airline. Show the matrix PN.

**Additional Answers
Written Exercises**

$$
\begin{array}{c}
 \\
20. \ \text{cost}
\end{array}
\begin{array}{ccc}
\text{s} & \text{a} & \text{s.c.} \\
[2 & 5 & 4],
\end{array}
$$

$$
\begin{array}{c}
 \\
\text{s} \\
\text{a} \\
\text{s.c.}
\end{array}
\begin{array}{cc}
\text{Fri.} & \text{Sat.} \\
\left[\begin{array}{cc}
121 & 183 \\
164 & 140 \\
32 & 25
\end{array}\right];
\end{array}
$$

Fri.: $1190, Sat.: $1166

Matrices **529**

Warm-Up Exercises

For Exercises 1-3, let $A = \begin{bmatrix} 3 & -1 \\ -4 & 1 \end{bmatrix}$ and $B = \begin{bmatrix} 2 & 0 \\ 1 & 3 \end{bmatrix}$.

1. Find $A \cdot A$, $B \cdot B$, and $A \cdot A - B \cdot B$.

   $\begin{bmatrix} 13 & -4 \\ -16 & 5 \end{bmatrix}, \begin{bmatrix} 4 & 0 \\ 5 & 9 \end{bmatrix},$
   $\begin{bmatrix} 9 & -4 \\ -21 & -4 \end{bmatrix}$

2. Find $A + B$, $A - B$, and $(A + B)(A - B)$.

   $\begin{bmatrix} 5 & -1 \\ -3 & 4 \end{bmatrix}; \begin{bmatrix} 1 & -1 \\ -5 & -2 \end{bmatrix};$
   $\begin{bmatrix} 10 & -3 \\ -23 & -5 \end{bmatrix}$

3. Are $A \cdot A - B \cdot B$ and $(A + B)(A - B)$ equal? Explain why or why not.

   No; $(A + B)(A - B) = A \cdot A - AB + BA - B \cdot B$; since $AB \neq BA$, $(A + B)(A - B) \neq A \cdot A - B \cdot B$.

Motivating the Section

After defining an identity matrix, write the matrix $A = \begin{bmatrix} 1 & 1 \\ 0 & 1 \end{bmatrix}$ on the board.

Ask students if they can give another matrix B such that $AB = I_{2 \times 2}$. They will probably be successful after a bit of trial and error.

$\left(B = \begin{bmatrix} 1 & -1 \\ 0 & 1 \end{bmatrix} \right)$

Ask if $BA = I_{2 \times 2}$ is also true. Matrices A and B are inverses. A major goal of this section is finding and using inverses of matrices.

Objective   *To find the inverse of a 2 × 2 matrix and to solve linear systems using matrices.*

In the previous sections, you may have discovered that matrix addition and matrix multiplication have many of the properties of ordinary addition and multiplication.

	Properties of Real Numbers	Properties of Matrices
Property	Let a, b, and c be real numbers.	Let A, B, and C be $m \times n$ matrices.
Commutative	$a + b = b + a$ $ab = ba$	$A + B = B + A$ $AB \neq BA$
Associative	$(a + b) + c = a + (b + c)$ $(ab)c = a(bc)$	$(A + B) + C = A + (B + C)$ $(AB)C = A(BC)$
Distributive	$a(b + c) = ab + ac$ $= ba + ca$ $= (b + c)a$	$A(B + C) = AB + AC$ $(B + C)A = BA + CA$

An important exception to the similarity of these properties is that matrix multiplication is *not* commutative: in general, $AB \neq BA$. Since products do not commute, *left-multiplication* can give a different product from *right-multiplication*.

Before discussing other properties, we first need to identify some important matrices. Any $m \times n$ matrix whose elements are all zero is called a **zero matrix**, denoted by $O_{m \times n}$. The following matrices are zero matrices.

$$O_{1 \times 3} = [0 \quad 0 \quad 0] \qquad O_{2 \times 1} = \begin{bmatrix} 0 \\ 0 \end{bmatrix} \qquad O_{2 \times 2} = \begin{bmatrix} 0 & 0 \\ 0 & 0 \end{bmatrix} \qquad O_{3 \times 2} = \begin{bmatrix} 0 & 0 \\ 0 & 0 \\ 0 & 0 \end{bmatrix}$$

A **square matrix** is any matrix having the same number of rows as columns. The *main diagonal* of a square matrix is the diagonal that extends from upper left to lower right. Any $n \times n$ matrix whose main diagonal elements are 1 and whose other elements are 0 is called an **identity matrix**, denoted by $I_{n \times n}$. The following matrices are identity matrices.

$$I_{2 \times 2} = \begin{bmatrix} 1 & 0 \\ 0 & 1 \end{bmatrix} \qquad\qquad I_{3 \times 3} = \begin{bmatrix} 1 & 0 & 0 \\ 0 & 1 & 0 \\ 0 & 0 & 1 \end{bmatrix}$$

Powers of square matrices are defined just like powers of real numbers:

$$A^n = \underbrace{A \cdot A \cdot A \cdot \cdots \cdot A}_{n \text{ factors}}$$

As with real numbers, matrix addition and multiplication both have identity properties. It is also true that matrix addition and multiplication both have inverse properties. The real numbers 2 and -2 are called additive inverses since $2 + (-2) = 0$. Similarly, a matrix can have an additive inverse. The **additive inverse** of a matrix A, denoted by $-A$, is the matrix in which each element is the opposite of its corresponding element in A. For example,

$$\text{if } A = \begin{bmatrix} 2 & -3 & 5 \\ 0 & 1 & -8 \end{bmatrix}, \text{ then } -A = \begin{bmatrix} -2 & 3 & -5 \\ 0 & -1 & 8 \end{bmatrix}.$$

	Properties of Real Numbers	Properties of Matrices
Properties for Addition	Let a be any real number.	Let A be any $m \times n$ matrix. Let O be the $m \times n$ zero matrix.
Identity	$a + 0 = 0 + a = a$	$A + O = O + A = A$
Inverse	$a + (-a) = -a + a = 0$	$A + (-A) = -A + A = O$

You should verify that the matrices A and $-A$ given above satisfy the inverse property for addition. The real numbers 2 and 2^{-1} are called multiplicative inverses since $2 \cdot 2^{-1} = 1$. Similarly, a matrix can have a *multiplicative inverse*. The examples below illustrate the identity and inverse properties for multiplication.

$$\begin{bmatrix} 2 & 4 \\ -3 & 1 \end{bmatrix} \begin{bmatrix} 1 & 0 \\ 0 & 1 \end{bmatrix} = \begin{bmatrix} 1 & 0 \\ 0 & 1 \end{bmatrix} \begin{bmatrix} 2 & 4 \\ -3 & 1 \end{bmatrix} = \begin{bmatrix} 2 & 4 \\ -3 & 1 \end{bmatrix}$$

$$\begin{bmatrix} 2 & 1 \\ 5 & 3 \end{bmatrix} \begin{bmatrix} 3 & -1 \\ -5 & 2 \end{bmatrix} = \begin{bmatrix} 3 & -1 \\ -5 & 2 \end{bmatrix} \begin{bmatrix} 2 & 1 \\ 5 & 3 \end{bmatrix} = \begin{bmatrix} 1 & 0 \\ 0 & 1 \end{bmatrix}$$

Finding the multiplicative inverse of a square matrix can involve a lot of computations. The easiest inverse to find is that of a 2×2 matrix. In Written Exercise 29, you are asked to derive the following formula.

Inverse of a 2 × 2 Matrix

If $A = \begin{bmatrix} a & b \\ c & d \end{bmatrix}$, then $A^{-1} = \dfrac{1}{|A|} \begin{bmatrix} d & -b \\ -c & a \end{bmatrix}$, where $|A| = ad - bc$.

Recall that $|A|$ is called the *determinant* of A (see page 458). The formula above tells you that if $|A| \neq 0$ then A^{-1} exists. Thus, the multiplicative inverse property of matrices is not quite the same as the corresponding property of real numbers. Whereas 0 is the only real number that has no inverse, there are infinitely many matrices that have no inverse. The properties for multiplication are summarized on the next page.

Students learned about systems of linear equations in earlier courses. Help them see the connection to matrices by doing the following example in class.

Let $A = \begin{bmatrix} 1 & -2 \\ 3 & 2 \end{bmatrix}$, $X = \begin{bmatrix} x \\ y \end{bmatrix}$, and $C = \begin{bmatrix} 7 \\ 5 \end{bmatrix}$. Use matrix multiplication to show that if $AX = C$, then:

$$\begin{bmatrix} 1 & -2 \\ 3 & 2 \end{bmatrix} \begin{bmatrix} x \\ y \end{bmatrix} = \begin{bmatrix} 7 \\ 5 \end{bmatrix}$$

$$\begin{bmatrix} x - 2y \\ 3x + 2y \end{bmatrix} = \begin{bmatrix} 7 \\ 5 \end{bmatrix}$$

Thus, $x - 2y = 7$
$\qquad 3x + 2y = 5$.

This sets the stage for students to understand how systems of equations can be represented by matrix equations.

In Example 1, you may want to discuss the fact that A^{-1} can be written two ways, as

$$\frac{1}{7}\begin{bmatrix} 3 & -4 \\ -2 & 5 \end{bmatrix} \text{ or}$$

as $\begin{bmatrix} \frac{3}{7} & -\frac{4}{7} \\ -\frac{2}{7} & \frac{5}{7} \end{bmatrix}$. The preferred

form will depend on what is to be done with this matrix. For example,

$$\frac{1}{7}\begin{bmatrix} 3 & -4 \\ -2 & 5 \end{bmatrix} \cdot 14\begin{bmatrix} 2 & -1 \\ 0 & 1 \end{bmatrix}$$

is easier to compute than

$$\begin{bmatrix} \frac{3}{7} & -\frac{4}{7} \\ -\frac{2}{7} & \frac{5}{7} \end{bmatrix}\begin{bmatrix} 28 & -14 \\ 0 & 14 \end{bmatrix} \text{ for}$$

most students.

Additional Examples

1. Solve the matrix equation for X.

$$\begin{bmatrix} 7 & -2 \\ -4 & 1 \end{bmatrix} X - \begin{bmatrix} 3 & -5 \\ 0 & 4 \end{bmatrix}$$

$$= \begin{bmatrix} -1 & 5 \\ 2 & -3 \end{bmatrix}$$

$$\begin{bmatrix} 7 & -2 \\ -4 & 1 \end{bmatrix} X = \begin{bmatrix} 2 & 0 \\ 2 & 1 \end{bmatrix}$$

If $A = \begin{bmatrix} 7 & -2 \\ -4 & 1 \end{bmatrix}$, then

$|A| = 7(1) - (-4)(-2) = -1$ and $A^{-1} =$

$$\frac{1}{-1}\begin{bmatrix} 1 & 2 \\ 4 & 7 \end{bmatrix} = \begin{bmatrix} -1 & -2 \\ -4 & -7 \end{bmatrix}.$$

Left-multiply each side of the equation by A^{-1}:

$$X = \begin{bmatrix} -1 & -2 \\ -4 & -7 \end{bmatrix}\begin{bmatrix} 2 & 0 \\ 2 & 1 \end{bmatrix} =$$

$$\begin{bmatrix} -6 & -2 \\ -22 & -7 \end{bmatrix}$$

You can check the solution by substitution.

	Properties of Real Numbers	Properties of Matrices
Properties for Multiplication	Let a be any real number.	Let A be any $n \times n$ matrix. Let I be the $n \times n$ identity matrix and O be the $n \times n$ zero matrix.
Identity	$a \cdot 1 = 1 \cdot a = a$	$A \cdot I = I \cdot A = A$
Inverse	$a \cdot a^{-1} = a^{-1} \cdot a = 1$ $(a \neq 0)$	$A \cdot A^{-1} = A^{-1} \cdot A = I$ $(|A| \neq 0)$
Multiplicative Property of Zero	$a \cdot 0 = 0 \cdot a = 0$	$A \cdot O = O \cdot A = O$

Example 1   If $A = \begin{bmatrix} 5 & 4 \\ 2 & 3 \end{bmatrix}$, find A^{-1}. Check your answer by finding $A \cdot A^{-1}$.

Solution $|A| = 5 \cdot 3 - 4 \cdot 2 = 7$. Since $|A| \neq 0$, A^{-1} exists.

$$A^{-1} = \frac{1}{7}\begin{bmatrix} 3 & -4 \\ -2 & 5 \end{bmatrix} = \begin{bmatrix} \frac{3}{7} & -\frac{4}{7} \\ -\frac{2}{7} & \frac{5}{7} \end{bmatrix}$$

Check: $A \cdot A^{-1} = \begin{bmatrix} 5 & 4 \\ 2 & 3 \end{bmatrix}\begin{bmatrix} \frac{3}{7} & -\frac{4}{7} \\ -\frac{2}{7} & \frac{5}{7} \end{bmatrix} = \begin{bmatrix} 1 & 0 \\ 0 & 1 \end{bmatrix} = I$

When solving the linear equation $3x + 7 = 20$, you add -7 to both sides of the equation and then multiply both sides by 3^{-1}, or $\frac{1}{3}$. Any matrix equation in the form $AX + B = C$ can be solved in a similar way.

Example 2   Let A, B, C, and X be $n \times n$ matrices, where A^{-1} exists. Solve $AX + B = C$ for X.

Solution

$AX + B = C$	
$AX + B + (-B) = C + (-B)$	Add the additive inverse of B.
$AX + O = C - B$	Additive Inverse Property
$AX = C - B$	Additive Identity Property
$A^{-1}(AX) = A^{-1}(C - B)$	Left-multiply by A^{-1}.
$(A^{-1}A)X = A^{-1}(C - B)$	Associative Property
$IX = A^{-1}(C - B)$	Multiplicative Inverse Property
$X = A^{-1}(C - B)$	Multiplicative Identity Property

Example 3

Use matrices to solve the given system of equations:

$$5x + 4y = -2$$
$$2x + 3y = -5$$

Solution

This system of two equations in two unknowns can be replaced by a single matrix equation:

$$\begin{array}{ccc} A & \cdot\, X & = & C \end{array}$$
$$\begin{bmatrix} 5 & 4 \\ 2 & 3 \end{bmatrix}\begin{bmatrix} x \\ y \end{bmatrix} = \begin{bmatrix} -2 \\ -5 \end{bmatrix}$$

Notice that the *coefficient matrix A* is the matrix whose inverse was found in Example 1.

$$A^{-1} = \frac{1}{7}\begin{bmatrix} 3 & -4 \\ -2 & 5 \end{bmatrix}$$

If you left-multiply both sides of the matrix equation by A^{-1}, you get:

$$\frac{1}{7}\begin{bmatrix} 3 & -4 \\ -2 & 5 \end{bmatrix}\begin{bmatrix} 5 & 4 \\ 2 & 3 \end{bmatrix}\begin{bmatrix} x \\ y \end{bmatrix} = \frac{1}{7}\begin{bmatrix} 3 & -4 \\ -2 & 5 \end{bmatrix}\begin{bmatrix} -2 \\ -5 \end{bmatrix}$$

$$\frac{1}{7}\begin{bmatrix} 7 & 0 \\ 0 & 7 \end{bmatrix}\begin{bmatrix} x \\ y \end{bmatrix} = \frac{1}{7}\begin{bmatrix} 14 \\ -21 \end{bmatrix}$$

$$\begin{bmatrix} 1 & 0 \\ 0 & 1 \end{bmatrix}\begin{bmatrix} x \\ y \end{bmatrix} = \begin{bmatrix} 2 \\ -3 \end{bmatrix}$$

$$\begin{bmatrix} x \\ y \end{bmatrix} = \begin{bmatrix} 2 \\ -3 \end{bmatrix}$$

Thus, the solution to the system is $x = 2$ and $y = -3$.

Example 4

Rewrite the following system as a single matrix equation.

$$3x_1 - 5x_2 + 4x_3 + 7x_4 = 12$$
$$6x_1 + 7x_2 \qquad\quad - 8x_4 = 29$$
$$5x_1 + 9x_2 - 7x_3 - 5x_4 = -52$$
$$x_2 - 5x_3 + 9x_4 = 41$$

Solution

$$\begin{bmatrix} 3 & -5 & 4 & 7 \\ 6 & 7 & 0 & -8 \\ 5 & 9 & -7 & -5 \\ 0 & 1 & -5 & 9 \end{bmatrix}\begin{bmatrix} x_1 \\ x_2 \\ x_3 \\ x_4 \end{bmatrix} = \begin{bmatrix} 12 \\ 29 \\ -52 \\ 41 \end{bmatrix}\quad\begin{bmatrix} -17.26137 \\ 40.35170 \\ 33.59641 \\ 18.73671 \end{bmatrix}$$

The matrix equation in Example 4 has the form $AX = C$, whose solution is $X = A^{-1}C$, provided that A^{-1} exists. This textbook will not show how to find the inverse of matrices larger than 2×2 because the calculations can be tedious. However, many computer software programs and graphing calculators can calculate $A^{-1}C$ efficiently when given specific matrices for A and C. If you have access to a graphing calculator or computer program, you should find the solution to the system in Example 4. See solution above.

Additional Examples cont.

2. a. Write
$$2x - y + z = 3$$
$$3x + 2y - 4z = 23$$
$$x - 3y - 2z = 14$$
as a single matrix equation.

b. If the inverse of the coefficient matrix is
$$\begin{bmatrix} \frac{16}{45} & \frac{5}{45} & -\frac{2}{45} \\ -\frac{2}{45} & \frac{5}{45} & -\frac{11}{45} \\ \frac{11}{45} & -\frac{5}{45} & -\frac{7}{45} \end{bmatrix},$$
find the solution of the system.

a. $\begin{bmatrix} 2 & -1 & 1 \\ 3 & 2 & -4 \\ 1 & -3 & -2 \end{bmatrix}\begin{bmatrix} x \\ y \\ z \end{bmatrix} = \begin{bmatrix} 3 \\ 23 \\ 14 \end{bmatrix}$

b. If the matrix equation in part (a) is represented by $AX = C$, then $X = A^{-1}C$. Thus,

$$\begin{bmatrix} x \\ y \\ z \end{bmatrix} =$$

$$\begin{bmatrix} \frac{16}{45} & \frac{5}{45} & -\frac{2}{45} \\ -\frac{2}{45} & \frac{5}{45} & -\frac{11}{45} \\ \frac{11}{45} & -\frac{5}{45} & -\frac{7}{45} \end{bmatrix}\begin{bmatrix} 3 \\ 23 \\ 14 \end{bmatrix} =$$

$$\begin{bmatrix} 3 \\ -1 \\ -4 \end{bmatrix}.$$

The solution is $x = 3$, $y = -1$, and $z = -4$.

1. $-B = \begin{bmatrix} -7 & 1 \\ -6 & -2 \end{bmatrix}$,

$-C = \begin{bmatrix} -3 & -2 & -5 \\ -4 & -3 & 0 \end{bmatrix}$

2. $A^{-1} = \begin{bmatrix} 2 & 3 \\ 5 & 8 \end{bmatrix}$,

$B^{-1} = \begin{bmatrix} \dfrac{1}{10} & \dfrac{1}{20} \\ -\dfrac{3}{10} & \dfrac{7}{20} \end{bmatrix}$

4. $\begin{bmatrix} 79 & -30 \\ -50 & 19 \end{bmatrix}$; either

6. $3X = \begin{bmatrix} -9 & -6 \\ -3 & -15 \end{bmatrix}$

$X = \begin{bmatrix} -3 & -2 \\ -1 & -5 \end{bmatrix}$

1. a. $\begin{bmatrix} -\dfrac{7}{3} & \dfrac{4}{3} \\ 2 & -1 \end{bmatrix}$

b. $\begin{bmatrix} 1 & 0 \\ 0 & 1 \end{bmatrix}$

2. a. $\begin{bmatrix} \dfrac{7}{2} & \dfrac{11}{2} \\ -\dfrac{3}{2} & -\dfrac{5}{2} \end{bmatrix}$

b. $\begin{bmatrix} 1 & 0 \\ 0 & 1 \end{bmatrix}$

3. a. $\begin{bmatrix} -9 & -3 \\ -12 & -4 \end{bmatrix}$

b. not defined

4. a. not defined

b. $\begin{bmatrix} -2 & 0 & 8 & -1 \\ -5 & -7 & -3 & -9 \end{bmatrix}$

5. a. $\begin{bmatrix} -3 & -1 \\ -4 & -\dfrac{4}{3} \end{bmatrix}$

b. not defined

6. a. not defined

b. $\begin{bmatrix} 26 & 66 \\ -18 & -46 \end{bmatrix}$

CLASS EXERCISES

For Exercises 1–4, use matrices A, B, C, and D.

$$A = \begin{bmatrix} 8 & -3 \\ -5 & 2 \end{bmatrix} \qquad B = \begin{bmatrix} 7 & -1 \\ 6 & 2 \end{bmatrix} \qquad C = \begin{bmatrix} 3 & 2 & 5 \\ 4 & 3 & 0 \end{bmatrix} \qquad D = \begin{bmatrix} 3 & 6 \\ 4 & 8 \end{bmatrix}$$

1. Find the additive inverse of B and of C.

2. Find the multiplicative inverse of A and of B.   C is not a square matrix,

3. Explain why C and D do not have multiplicative inverses. and $|D| = 0$.

4. **Discussion** Find A^2. To find A^3, do you calculate $A^2 \cdot A$ or $A \cdot A^2$? Explain.

5. **Discussion** Refer to the solution to Example 2 on page 532. If the equation $AX = C - B$ is right-multiplied by A^{-1} instead of left-multiplied, then you would get $(AX)A^{-1}$ on the left side of the equation. Explain why $(AX)A^{-1}$ cannot be simplified. Mult. of matrices is not comm.

6. The matrix equation below contains the unknown matrix X. The first two steps of the solution are shown.

$$2\begin{bmatrix} 5 & 1 \\ 3 & 4 \end{bmatrix} + 3X = \begin{bmatrix} 1 & -4 \\ 3 & -7 \end{bmatrix}$$

Simplify by using $\left.\right\}$ $\begin{bmatrix} 10 & 2 \\ 6 & 8 \end{bmatrix} + 3X = \begin{bmatrix} 1 & -4 \\ 3 & -7 \end{bmatrix}$
scalar multiplication.

Subtract the first matrix $\left.\right\}$ $3X = \begin{bmatrix} 1 & -4 \\ 3 & -7 \end{bmatrix} - \begin{bmatrix} 10 & 2 \\ 6 & 8 \end{bmatrix}$
from both sides.

Complete the solution, giving reasons for each step.

WRITTEN EXERCISES

For Exercises 1–6, use the following matrices to find the specified matrices. If a matrix does not exist, write *not defined*.

$$M = \begin{bmatrix} 3 & 4 \\ 6 & 7 \end{bmatrix} \qquad N = \begin{bmatrix} 5 & 11 \\ -3 & -7 \end{bmatrix} \qquad C = \begin{bmatrix} 9 & 3 \\ 12 & 4 \end{bmatrix} \qquad D = \begin{bmatrix} 2 & 0 & -8 & 1 \\ 5 & 7 & 3 & 9 \end{bmatrix}$$

A

1. **a.** M^{-1} **b.** MM^{-1} 2. **a.** N^{-1} **b.** $N^{-1}N$

3. **a.** $-C$ **b.** C^{-1} 4. **a.** D^{-1} **b.** $-D$

5. **a.** $-\frac{1}{3}C$ **b.** $-2D + C$ 6. **a.** D^2 **b.** N^3

Solve each matrix equation for X. Assume any required inverse matrix exists.

7. **a.** $DX = E \;\; X = D^{-1}E$ 8. **a.** $TX - S = R \;\; X = T^{-1}(R + S)$
   b. $XD = E \;\; X = ED^{-1}$ **b.** $XT - S = R \;\; X = (R + S)T^{-1}$

Solve each matrix equation for X.

9. $X + 3\begin{bmatrix} 2 & 5 \\ 1 & 4 \end{bmatrix} = \begin{bmatrix} 12 & 7 \\ 0 & -8 \end{bmatrix}\begin{bmatrix} 6 & -8 \\ -3 & -20 \end{bmatrix}$ **10.** $2X - \begin{bmatrix} 1 & 5 \\ 9 & 2 \end{bmatrix} = \begin{bmatrix} 7 & 11 \\ -1 & 12 \end{bmatrix}$ See below.

11. $\begin{bmatrix} 3 & 2 \\ 7 & 5 \end{bmatrix}X = \begin{bmatrix} 9 & 3 \\ 9 & 1 \end{bmatrix}\begin{bmatrix} 27 & 13 \\ -36 & -18 \end{bmatrix}$ **12.** $\begin{bmatrix} 5 & 1 \\ 7 & 2 \end{bmatrix}X = \begin{bmatrix} 3 & 0 & 5 \\ 1 & 4 & 2 \end{bmatrix}$

13. $\begin{bmatrix} 4 & 2 \\ 1 & 1 \end{bmatrix}X + \begin{bmatrix} 2 & 7 \\ 6 & 3 \end{bmatrix} = \begin{bmatrix} 2 & 3 \\ 0 & 1 \end{bmatrix}\begin{bmatrix} 6 & 0 \\ -12 & -2 \end{bmatrix}$ **14.** $\begin{bmatrix} 5 & 3 \\ 3 & 2 \end{bmatrix}X - \begin{bmatrix} 1 & 3 \\ 2 & 1 \end{bmatrix} = \begin{bmatrix} 1 & 2 \\ -1 & 3 \end{bmatrix}$

Write each system of equations as a single matrix equation. Solve the system using the method shown in Example 3.

15. a. $8x - 2y = -14$ $x = -\frac{1}{2}, y = 5$
     $12x + 3y = 9$

   b. $8x - 2y = 38$ $x = 4, y = -3$
     $12x + 3y = 39$

16. a. $12x - 4y = 3$ $x = \frac{17}{8}, y = \frac{45}{8}$
     $17x - 5y = 8$

   b. $12x - 4y = 0$ $x = -8, y = -24$
     $17x - 5y = -16$

B **17. a.** Solve the system using matrices.

     $12x + 20y = 40$
     $9x + 15y = 45$

10. $\begin{bmatrix} 4 & 8 \\ 4 & 7 \end{bmatrix}$ **12.** $\begin{bmatrix} \frac{5}{3} & -\frac{4}{3} & \frac{8}{3} \\ -\frac{16}{3} & \frac{20}{3} & -\frac{25}{3} \end{bmatrix}$

   b. Sketch the graphs of the two equations on the same set of axes. What do the graphs tell you about the solution of this system? **14.** $\begin{bmatrix} 1 & -2 \\ -1 & 5 \end{bmatrix}$

18. a. Solve the system using matrices.

     $14x - 21y = 56$
     $x - \frac{3}{2}y = 4$

   b. Sketch the graphs of the two equations on the same set of axes. What do the graphs tell you about the solution of this system?

19. *Writing* Write a paragraph explaining why the identity and the inverse properties for multiplication apply only to square matrices.

20. a. Using the distributive property, factor the following expression. Then simplify.

$$\begin{bmatrix} 1 & 2 \\ 3 & 4 \end{bmatrix}\begin{bmatrix} 0 & -5 \\ 7 & 9 \end{bmatrix} + \begin{bmatrix} 1 & 2 \\ 3 & 4 \end{bmatrix}\begin{bmatrix} 4 & 1 \\ -1 & -3 \end{bmatrix}\begin{bmatrix} 16 & 8 \\ 36 & 12 \end{bmatrix}$$

   b. Solve the following equation:

$$X\begin{bmatrix} 5 & 0 \\ 2 & 3 \end{bmatrix} - X\begin{bmatrix} 4 & -7 \\ 2 & 2 \end{bmatrix} = \begin{bmatrix} 2 & 3 \\ -5 & 7 \end{bmatrix}\begin{bmatrix} 2 & -11 \\ -5 & 42 \end{bmatrix}$$

21. Solve the matrix equation $X - AX = D$ for X. (*Hint*: Rewrite X either as XI or as IX. Which way is helpful?) $IX; X = (I - A)^{-1}D$

22. Solve the matrix equation $X - XA = D$ for X. $X = D(I - A)^{-1}$

Review Note

Remind students that the lines determined by the equations in Exercise 17 have the same slope but different x-intercepts. The equations in Exercise 18, by contrast, determine the same line.

Additional Answers
Written Exercises

17. a. It cannot be done because the coefficient matrix has no inverse.
   b. There is no solution because the lines are parallel.

18. a. It cannot be done because the coefficient matrix has no inverse.
   b. There are infinitely many solutions.

Suggested Assignments

Discrete Math
Day 1: 534/1, 9, 13, 15, 20, 21
Day 2: 536/23, 25–27

Supplementary Materials

Alternative Assessment, 44
Activities Book, 38–40
Using Technology, 35–36
Precalculus Plotter Plus,
93–94, 97–98
Student Resource Guide,
130–133

For Exercises 23–27, first write a matrix equation, and then solve the equation. You will need to use computer software or a calculator that computes matrix operations in order to solve the equation.

23. $3x + 5y - 9z = 26$
$4x + 7y + 2z = -7$
$6x - 9y - 8z = 3$
$x = -2, \; y = 1, \; z = -3$

24. $3a + 4b - 5c - 11d = 5$
$5a - 2b - 7c + 8d = 6$
$2a - 4b - 8c + 16d = -3$
$3b + 2c + 8d = 4$
$a = \frac{5}{2}, \; b = \frac{3}{5},$
$c = \frac{5}{6}, \; d = \frac{1}{15}$

25. Nutrition A dietician wants to combine four foods (I, II, III, and IV) to make a meal having 78 units of vitamin A, 67 units of vitamin B, 146 units of vitamin C, and 153 units of vitamin D. The matrix below gives the vitamin content per ounce of each food. How many ounces of each food should be included in the meal? Begin by writing the unknowns in a 4×1 matrix. Set up a matrix equation and then solve, giving answers to the nearest ounce.

$$\begin{array}{c} \\ A \\ B \\ C \\ D \end{array} \begin{array}{cccc} \text{I} & \text{II} & \text{III} & \text{IV} \\ \begin{bmatrix} 3 & 2 & 2 & 6 \\ 2 & 3 & 5 & 0 \\ 8 & 6 & 4 & 7 \\ 5 & 5 & 8 & 6 \end{bmatrix} \end{array}$$
I: 4 oz,
II: 6 oz,
III: 8 oz,
IV: 6 oz

26. a. Manufacturing A company makes oak tables, chairs, and desks. Each item requires labor time in minutes, as given in the matrix below. The amount of time available for labor each week is 20,250 min for carpentry, 12,070 min for assembly, and 17,000 min for finishing. If the production manager wants to use all of the available labor, how many tables, chairs, and desks should the manager schedule for production each week?

$$\begin{array}{c} \\ \text{carpentry} \\ \text{assembly} \\ \text{finishing} \end{array} \begin{array}{ccc} \text{tables} & \text{chairs} & \text{desks} \\ \begin{bmatrix} 120 & 105 & 125 \\ 40 & 65 & 110 \\ 80 & 90 & 125 \end{bmatrix} \end{array}$$
25 tables,
150 chairs,
12 desks

b. Suppose that because of vacation schedules the amount of labor available is less for the coming week. The amount of labor available is 14,960 min for carpentry, 8,970 min for assembly, and 12,590 min for finishing. How many tables, chairs, and desks should the manager schedule for this week?
18 tables, 110 chairs, 10 desks

27. Suppose the points $(4, 11)$, $(-6, -9)$, and $(8, 61)$ are known to lie on the parabola $y = ax^2 + bx + c$. Write three equations in terms of a, b, and c. Express this system as a single matrix equation, and then solve to find an equation of the parabola. $y = \frac{3}{4}x^2 + \frac{7}{2}x - 15$

28. No. For example, let $A = \begin{bmatrix} 1 & 0 \\ 0 & 0 \end{bmatrix}$ and $B = \begin{bmatrix} 0 & 0 \\ 0 & 1 \end{bmatrix}$. Then $AB = O$, but $A \neq O$ and $B \neq O$.

C 28. **Investigation** If a and b are real numbers and $ab = 0$, then either $a = 0$ or $b = 0$. If A and B are $n \times n$ matrices, does the fact that $AB = O_{n \times n}$ imply that either A or B must be a zero matrix? Explain, giving examples for A and B.

29. The purpose of this exercise is to derive the formula for the inverse of a 2×2 matrix. Let the matrix $A = \begin{bmatrix} a & b \\ c & d \end{bmatrix}$. Suppose that the inverse of A is represented by $A^{-1} = \begin{bmatrix} w & x \\ y & z \end{bmatrix}$. Use the fact that $AA^{-1} = \begin{bmatrix} 1 & 0 \\ 0 & 1 \end{bmatrix}$ to solve for w, x, y, and z. $w = \dfrac{d}{ad - bc}$, $x = -\dfrac{b}{ad - bc}$, $y = -\dfrac{c}{ad - bc}$, $z = \dfrac{a}{ad - bc}$

Olga Taussky-Todd (1906–)

Born in what is now the Czech Republic, Olga Taussky-Todd studied number theory in Vienna. During World War II, she worked outside London studying the stability of matrices and applying her results to the oscillations of airplane wings at supersonic speeds. Later in her career, she said that matrix theory emerged as her major subject.

After working ten years in the United States as a consultant for the National Bureau of Standards, Taussky-Todd began teaching at the California Institute of Technology. In 1970 she received a Ford Prize for a paper on the sums of squares.

■ Applications of Matrices

14-4 Communication Matrices

> **Objective** To solve communication network problems using matrices.

The points A, B, C, and D in the diagram at the right represent four ships at sea that are within communication range. The arrows indicate the direction of radio transmissions. For example, Ship B can send a message to Ship A, but it cannot send a message to Ship C. Although Ship D cannot communicate directly with Ship A, it can communicate indirectly by using Ship C as a relay.

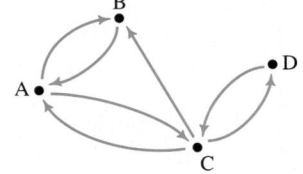

Teaching Notes, p. 516C

Warm-Up Exercises

In the matrix below, an element "1" indicates that direct transmission from the row computer to the column computer is possible. An element "0" indicates that direct transmission is not possible. For example, the first row shows that data can be sent directly from computer A to computer C, but not to computers B or D. Use the matrix for Exercises 1-6.

TO Computer

		A	B	C	D
FROM Computer	A	0	0	1	0
	B	1	0	1	1
	C	1	0	0	1
	D	1	1	0	0

1. Which computers can send data directly *to* computer C? A and B

2. Which computers can receive data directly *from* computer C? A and D

3. Which computer can send data directly *to* the most computers? B

4. Which computer can receive data directly *from* the most computers? A

5. Which two *pairs* of computers have the capacity for two-way direct transmission of data? A and C, B and D

6. Let the matrix above be matrix A. Find A^2.

$$\begin{bmatrix} 1 & 0 & 0 & 1 \\ 2 & 1 & 1 & 1 \\ 1 & 1 & 1 & 0 \\ 1 & 0 & 2 & 1 \end{bmatrix}$$

Motivating the Section

Open with a discussion of networks in the real world. Ask students whether they know of any organizations or structures labeled as networks. Everyone will know about television, radio, and telephone networks. Other examples that might arise are computer information networks, bank automatic teller networks, and the interpersonal systems people with common interests develop. Matrices provide the analytic tool for understanding and investigating these and other networks.

◫ Using Technology

Have students use either computer software or a calculator with a matrix feature to find M^3. Then relate M^3 and $M + M^2 + M^3$ to the diagram on page 537.

Additional Examples

1. Draw a communication network that is described by the given communication matrix.

$$
\begin{array}{c}
 & \begin{array}{ccccc} A & B & C & D & E \end{array} \\
\begin{array}{c} A \\ B \\ C \\ D \\ E \end{array} &
\left[\begin{array}{ccccc}
0 & 1 & 0 & 1 & 0 \\
0 & 0 & 0 & 1 & 1 \\
0 & 1 & 0 & 1 & 0 \\
1 & 0 & 1 & 0 & 1 \\
1 & 1 & 0 & 0 & 0
\end{array}\right]
\end{array}
$$

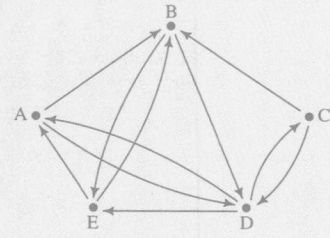

Although the diagram for the communication network of just four ships is easy to understand, a communication diagram can become very complicated when many ships are involved. Using matrices gives a way to keep track of who can communicate with whom. For example, the *communication matrix M* shown below models the communication between the four ships shown in the diagram on page 537. In such a matrix, a ''1'' in any row indicates that direct communication from one ship to another is possible. A ''0'' indicates that direct communication is not possible.

$$
\text{from ship} \quad
\begin{array}{c}
 & \begin{array}{c} \text{to ship} \\ \begin{array}{cccc} A & B & C & D \end{array} \end{array} \\
\begin{array}{c} A \\ B \\ C \\ D \end{array} &
\left[\begin{array}{cccc}
0 & 1 & 1 & 0 \\
1 & 0 & 0 & 0 \\
1 & 1 & 0 & 1 \\
0 & 0 & 1 & 0
\end{array}\right] = M
\end{array}
$$

Notice that in matrix M, the main diagonal elements are zeros because we are assuming that no ship will send a message directly to itself. Make sure that you understand what the elements in M represent.

While M gives the direct communication links between ships, M^2 gives the indirect two-step paths that use one ship as a relay. For example, A can communicate with D using C as a relay. This gives a two-step path from A to D, namely, A–C–D.

$$
M^2 =
\begin{bmatrix}
0 & 1 & 1 & 0 \\
1 & 0 & 0 & 0 \\
1 & 1 & 0 & 1 \\
0 & 0 & 1 & 0
\end{bmatrix}^2
=
\begin{bmatrix}
2 & 1 & 0 & 1 \\
0 & 1 & 1 & 0 \\
1 & 1 & 2 & 0 \\
1 & 1 & 0 & 1
\end{bmatrix}
$$

This means there are 2 two-step paths from A to itself. (A–B–A and A–C–A)

This means there is exactly 1 two-step path from D to B. (D–C–B)

If M^2 gives the number of two-step paths and M gives the number of one-step paths, then $M + M^2$ gives the number of paths that take either one or two steps.

$$
M + M^2 =
\begin{bmatrix}
0 & 1 & 1 & 0 \\
1 & 0 & 0 & 0 \\
1 & 1 & 0 & 1 \\
0 & 0 & 1 & 0
\end{bmatrix}
+
\begin{bmatrix}
2 & 1 & 0 & 1 \\
0 & 1 & 1 & 0 \\
1 & 1 & 2 & 0 \\
1 & 1 & 0 & 1
\end{bmatrix}
=
\begin{bmatrix}
2 & 2 & 1 & 1 \\
1 & 1 & 1 & 0 \\
2 & 2 & 2 & 1 \\
1 & 1 & 1 & 1
\end{bmatrix}
$$

This means there are two paths from C to B that take no more than 2 steps.

Similarly, it can be proved that the matrix M^3 gives the number of three-step paths between any two ships. The matrix $M + M^2 + M^3$ gives the number of communication paths that take no more than three steps.

Example

The matrix M below models the communication between Computers P, Q, R, S, and T.

$$
\begin{array}{c}
\\
\text{from}
\end{array}
\begin{array}{c}
\phantom{\text{to}}\ \ \text{to} \\
\text{P} \\
\text{Q} \\
\text{R} \\
\text{S} \\
\text{T}
\end{array}
\begin{array}{c}
\begin{array}{ccccc}
\text{P} & \text{Q} & \text{R} & \text{S} & \text{T}
\end{array} \\
\left[
\begin{array}{ccccc}
0 & 1 & 0 & 1 & 1 \\
1 & 0 & 1 & 1 & 0 \\
0 & 1 & 0 & 0 & 1 \\
1 & 0 & 0 & 0 & 1 \\
1 & 0 & 0 & 0 & 0
\end{array}
\right] = M
\end{array}
$$

a. Draw a diagram that illustrates this communication network.

b. Name the computer(s) that can send data along the greatest number of routes via one relay.

Solution

a.

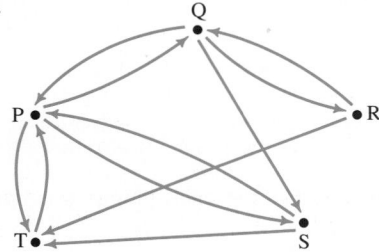

b. M^2 shows the number of ways data can be sent from one computer to another using one relay.

$$
\begin{array}{c}
\text{P} \\
\text{Q} \\
\text{R} \\
\text{S} \\
\text{T}
\end{array}
\begin{array}{c}
\begin{array}{ccccc}
\text{P} & \text{Q} & \text{R} & \text{S} & \text{T}
\end{array} \\
\left[
\begin{array}{ccccc}
3 & 0 & 1 & 1 & 1 \\
1 & 2 & 0 & 1 & 3 \\
2 & 0 & 1 & 1 & 0 \\
1 & 1 & 0 & 1 & 1 \\
0 & 1 & 0 & 1 & 1
\end{array}
\right] = M^2
\end{array}
$$

The row sums of M^2 for Computers P, Q, R, S, and T are 6, 7, 4, 4, and 3, respectively. Thus, Computer Q, with a total of 7, sends data along the greatest number of two-step paths. You can verify this by checking the diagram in part (a).

Using a computer network, students share files.

---

2. For the given communication network, find the matrix that represents the number of ways messages can be sent from one point to another, using at most one relay. Verify the entries in the first row of the matrix.

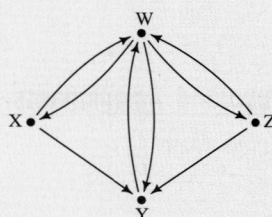

$$
M =
\begin{array}{c}
\text{W} \\
\text{X} \\
\text{Y} \\
\text{Z}
\end{array}
\begin{array}{c}
\begin{array}{cccc}
\text{W} & \text{X} & \text{Y} & \text{Z}
\end{array} \\
\left[
\begin{array}{cccc}
0 & 1 & 1 & 1 \\
1 & 0 & 1 & 0 \\
1 & 0 & 0 & 0 \\
1 & 0 & 1 & 0
\end{array}
\right]
\end{array}
$$

$$
M^2 =
\left[
\begin{array}{cccc}
3 & 0 & 2 & 0 \\
1 & 1 & 1 & 1 \\
0 & 1 & 1 & 1 \\
1 & 1 & 1 & 1
\end{array}
\right]
$$

$$
M + M^2 =
\left[
\begin{array}{cccc}
3 & 1 & 3 & 1 \\
2 & 1 & 2 & 1 \\
1 & 1 & 1 & 1 \\
2 & 1 & 2 & 1
\end{array}
\right]
$$

$M + M^2$ represents the number of ways messages can be sent using at most one relay.

3: W-X-W, W-Y-W, W-Z-W
1: W-X
3: W-Y, W-X-Y, W-Z-Y
1: W-Z

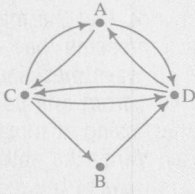
CLASS EXERCISES

Matrix M below describes a communication network for weather stations.

$$\text{from station}\quad \begin{array}{c} \\ A \\ B \\ C \\ D \end{array}\begin{array}{c} \text{to station} \\ \begin{array}{cccc} A & B & C & D \end{array} \\ \begin{bmatrix} 0 & 0 & 1 & 1 \\ 0 & 0 & 0 & 1 \\ 1 & 1 & 0 & 1 \\ 1 & 0 & 1 & 0 \end{bmatrix} \end{array} = M$$

1. To which stations can A send messages? C and D
2. From which stations can B receive messages? C
3. Draw a diagram that illustrates this communication network.
4. What information does M^2 give you? (Do not calculate M^2.)
5. **Discussion** Will the main diagonal of M^2 necessarily be all zeros? No; a station can send a message to itself using another station as a relay.

WRITTEN EXERCISES

Write the communication matrix for each communication network given.

A 1.   A

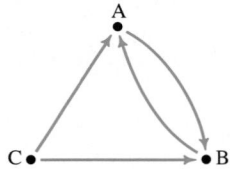

2.

3.

4.

Draw a communication network that is described by the given matrix.

5.
$$\begin{array}{c} \\ A \\ B \\ C \\ D \end{array}\begin{array}{c} \begin{array}{cccc} A & B & C & D \end{array} \\ \begin{bmatrix} 0 & 1 & 0 & 0 \\ 1 & 0 & 0 & 0 \\ 0 & 1 & 0 & 0 \\ 1 & 1 & 1 & 0 \end{bmatrix} \end{array}$$

6.
$$\begin{array}{c} \\ R \\ S \\ T \\ U \\ V \end{array}\begin{array}{c} \begin{array}{ccccc} R & S & T & U & V \end{array} \\ \begin{bmatrix} 0 & 0 & 1 & 0 & 1 \\ 1 & 0 & 0 & 1 & 0 \\ 0 & 0 & 0 & 0 & 0 \\ 0 & 1 & 1 & 0 & 1 \\ 0 & 0 & 0 & 1 & 0 \end{bmatrix} \end{array}$$

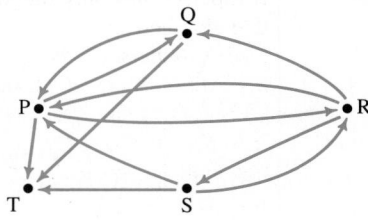

For some parts of the following exercises, you may wish to use computer software or a calculator that performs matrix operations.

7. a. Ship A sends messages to Ship B. Ship B sends and receives messages from Ships C and E. Ship D sends and receives messages from Ship C. Ship E receives messages from Ships A and C. Draw this network.
  b. Write the matrix M that models this communication network, labeling rows and columns alphabetically.
  c. Find M^2. Explain what the element in the fifth row, second column means.
  d. Find the matrix that represents the number of ways messages can be sent from one ship to another using at most one relay.
  e. Reason from the network diagram what the last element in row 4 of M^3 is. (You do not need to calculate M^3.)

8. a. Suppose radios of varying quality are used by forest rangers to communicate with each other in a large national park. Make a diagram showing five rangers, some pairs having two-way communication, some pairs having one-way communication, and some having no communication.
  b. Write a matrix M to model the communication between the forest rangers. Find $M + M^2$. What information does this matrix give you?

B **9.** The diagram below models a rumor network.

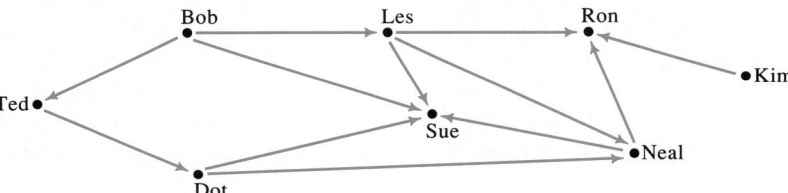

If a person has only outgoing arrows, then that person is a transmitter of the rumor. A person having only incoming arrows is a receiver of the rumor. A person having both outgoing and incoming arrows is a relay point for the spreading of the rumor.
  a. Identify the transmitters, receivers, and relays for this network.
  b. How can you identify a transmitter by looking at the corresponding rumor matrix? How can you identify a receiver?

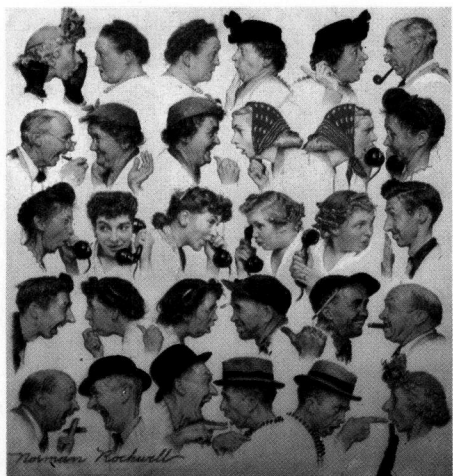

10. Let M be a matrix that illustrates communication among several Ham radio operators. If M^2 contains no zeros, what can you conclude?
Each operator can reach any other operator or itself by using 1 relay.

Matrices **541**

 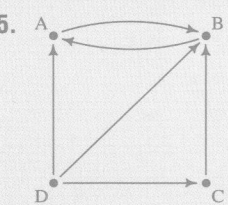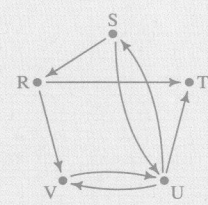
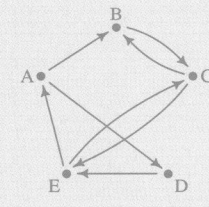

12. a.

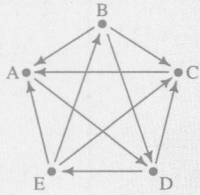

b. Bella and Enrico

c.
$$\begin{bmatrix} 0 & 0 & 1 & 0 & 1 \\ 1 & 0 & 1 & 1 & 1 \\ 0 & 0 & 0 & 1 & 0 \\ 2 & 1 & 1 & 0 & 0 \\ 2 & 0 & 1 & 2 & 0 \end{bmatrix}$$

d.
$$\begin{bmatrix} 0 & 0 & 1 & 1 & 1 \\ 2 & 0 & 2 & 2 & 1 \\ 1 & 0 & 0 & 1 & 0 \\ 2 & 1 & 2 & 0 & 1 \\ 3 & 1 & 2 & 2 & 0 \end{bmatrix};$$

Enrico, Bella, Denise

13. a.

$$\begin{array}{c c} & \begin{array}{c c c c c c} c & g & p & r & s & w \end{array} \\ \begin{array}{c} c \\ g \\ p \\ r \\ s \\ w \end{array} & \begin{bmatrix} 0 & 0 & 1 & 0 & 0 & 1 \\ 1 & 0 & 0 & 1 & 0 & 0 \\ 0 & 0 & 0 & 0 & 0 & 0 \\ 0 & 0 & 0 & 0 & 0 & 1 \\ 0 & 0 & 1 & 0 & 0 & 0 \\ 0 & 0 & 0 & 0 & 0 & 0 \end{bmatrix} \end{array}$$

b.
$$\begin{bmatrix} 0 & 0 & 0 & 0 & 0 & 0 \\ 0 & 0 & 1 & 0 & 0 & 2 \\ 0 & 0 & 0 & 0 & 0 & 0 \\ 0 & 0 & 0 & 0 & 0 & 0 \\ 0 & 0 & 0 & 0 & 0 & 0 \\ 0 & 0 & 0 & 0 & 0 & 0 \end{bmatrix};$$

wolves feed on two animals that feed on grass.

c. Put a "1" in the grass row, salmon column.

---

11. There are n ships at sea. Each ship in the network can communicate directly by radio with every ship (except itself). How many zeros are in the communications matrix? How many ones? n; $n^2 - n$

12. The *dominance matrix* shown below is a special kind of communication matrix. A **dominance relation** exists between members when, between any two members, one dominates the other. For example, the winner in the pair "dominates" the loser. In terms of a communication network, this means that, between any two members, only a one-way communication is possible.

a. Matrix M at the right represents the result of a chess tournament played by Ajani, Bella, Carmen, Denise, and Enrico. Make a diagram to illustrate this tournament.

$$\begin{array}{c} \\ A \\ B \\ C \\ D \\ E \end{array} \begin{array}{c} \begin{array}{c c c c c} A & B & C & D & E \end{array} \\ \begin{bmatrix} 0 & 0 & 0 & 1 & 0 \\ 1 & 0 & 1 & 1 & 0 \\ 1 & 0 & 0 & 0 & 0 \\ 0 & 0 & 1 & 0 & 1 \\ 1 & 1 & 1 & 0 & 0 \end{bmatrix} \end{array} = M$$

b. In the matrix above, the number of games won is indicated in the rows; the number of games lost is indicated in the columns. Which two players are tied with the best record for the number of wins and losses?

c. One method used to decide who comes in first, second, and third in a tournament is to look at the *second-stage dominances*. For example, Carmen has a second-stage dominance over Denise since Carmen beat Ajani and Ajani beat Denise. Find M^2, which gives the second-stage dominances.

d. The person who has the greatest total of direct and second-stage dominances is the winner of the tournament. Find the matrix that gives this information. Who won the tournament? Who finished second and third?

13. a. **Ecology** The diagram below models a possible food chain in Alaska.

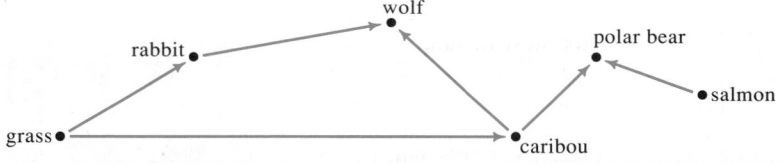

The arrow from the rabbit to the wolf indicates the rabbit is a source of food for the wolf. Any contamination of the rabbit's food, such as DDT poisoning, will also be harmful to the wolf. Write a matrix M that represents this food chain. Label the rows and columns alphabetically.

b. Find M^2. Tell the meaning of the element in the grass row, wolf column.

c. Although grass is not food for salmon, any insecticides on grasslands can wash into rivers, possibly killing salmon or affecting their reproduction. How would you change matrix M to reflect this fact?

542 *Chapter Fourteen*

14. *Research* Draw a food chain diagram for an ecological area of interest to you (for example, a swamp, a field, a pond, or a tropical region). Model this food chain with a matrix M and calculate M^2, M^3, and $M + M^2 + M^3$. Write a paragraph interpreting what the matrix $M + M^2 + M^3$ tells you.

C **15. *Writing*** Write a paragraph explaining why M^3 gives the number of three-step paths between points in a communication network.

14-5 Transition Matrices

Objective   *To make predictions using powers of matrices.*

The manufacturers of Highlight shampoo, the best-selling shampoo in the nation, have done a market analysis to predict the buying trends of customers in the marketplace over the long run. Their research tells them two things:

(1) the current market shares
(2) the percentages of people who change from brand to brand each month

The percentages of people who change brands each month can be written as a *transition matrix*, as illustrated in Example 1.

Example 1   Of all people who buy Highlight one month, 90% will buy Highlight the next month, 5% will change to Silky Shine the next month, and 5% will change to another brand. Of those people who buy Silky Shine one month, 10% will change to Highlight the next month, 70% will stay with Silky Shine, and 20% will change to another brand. Of the users of other brands, 20% will change to Highlight, and 20% will change to Silky Shine. Write this information as a transition matrix T.

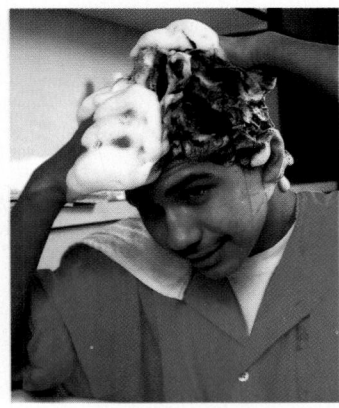

Solution

$$\begin{array}{c c} & \begin{array}{c c c} \text{Highlight} & \text{Silky Shine} & \text{Other} \end{array} \\ \begin{array}{c} \text{Highlight} \\ \text{Silky Shine} \\ \text{Other} \end{array} & \left[\begin{array}{c c c} 0.9 & 0.05 & 0.05 \\ 0.1 & 0.7 & 0.2 \\ 0.2 & 0.2 & 0.6 \end{array} \right] = T \end{array}$$

The rows in matrix T above tell what the users of each product will use next month. Notice that the percentages have been expressed as decimals and that the elements in each row add up to 1.

Matrices   **543**

Teaching Notes, p. 516D

Warm-Up Exercises

For Exercises 1 and 2, let $A =$ [0.8   0.2] and $T = \begin{bmatrix} 0.9 & 0.1 \\ 0.3 & 0.7 \end{bmatrix}$.

1. Find $B = AT$, then $C = BT$, and finally $D = CT$.
[0.78   0.22];
[0.768   0.232];
[0.7608   0.2392]

2. Compare corresponding elements of matrices B, C, and D found in Exercise 1. Describe what appears to be happening to the elements in each successive matrix? The values seem to be approaching limits.

3. Solve for x: $[x \quad 1 - x]$ · $\begin{bmatrix} 0.9 & 0.1 \\ 0.3 & 0.7 \end{bmatrix} = [x \quad 1 - x]$.
$x = 0.75$

4. Examine Exercise 3 and your answer. What is the significance of the matrix [0.75   0.25] as it relates to matrix T above? When T is multiplied from the left by this matrix, the product is this matrix.

Motivating the Section

Transition matrices provide a mathematical technique for analyzing, describing, and predicting change.

Additional Examples

1. Each year at Central High School, 30% of the students studying a foreign language will not continue their studies. Also, 20% of the students not studying a foreign language will begin a foreign language course. Suppose 60% of the students are now studying a foreign language.
   a. Give the transition matrix T and the current population matrix P_0.
   b. Find the percent of students studying a foreign language next year and two years from now.
   c. Show that $S = [0.4 \quad 0.6]$ is the steady-state matrix for T.

   a. $T = \begin{bmatrix} 0.7 & 0.3 \\ 0.2 & 0.8 \end{bmatrix}$; $P_0 = [0.6 \quad 0.4]$.
   b. $P_1 = P_0T =$
   $[0.6 \quad 0.4]\begin{bmatrix} 0.7 & 0.3 \\ 0.2 & 0.8 \end{bmatrix} =$
   $[0.5 \quad 0.5]$; next year, 50% of the students will be studying a foreign language. $P_2 = P_1T =$
   $[0.5 \quad 0.5]\begin{bmatrix} 0.7 & 0.3 \\ 0.2 & 0.8 \end{bmatrix} =$
   $[0.45 \quad 0.55]$; in two years, 45% of the students will be studying a foreign language.
   c. $ST =$
   $[0.4 \quad 0.6]\begin{bmatrix} 0.7 & 0.3 \\ 0.2 & 0.8 \end{bmatrix} =$
   $[0.4 \quad 0.6] = S$, so S is the steady-state matrix.

2. Find the steady-state matrix S for the transition matrix
   $T = \begin{bmatrix} 0.6 & 0.4 & 0 \\ 0.2 & 0.3 & 0.5 \\ 0 & 1 & 0 \end{bmatrix}$.

Suppose the market survey also indicates that the current market shares are Highlight users, 40%; Silky Shine users, 30%; and other brands, 30%. We can write this information as a current market share matrix M_0.

$$\text{market share } [\quad \overset{\text{Highlight}}{0.4} \quad \overset{\text{Silky Shine}}{0.3} \quad \overset{\text{Other}}{0.3} \quad] = M_0$$

Example 2 Use the transition matrix T and the current market share matrix M_0 to calculate the market shares for the next month.

Solution The matrix M_1, which gives the market shares one month from now, is the product M_0T.

$$M_1 = M_0T$$
$$= [0.4 \quad 0.3 \quad 0.3]\begin{bmatrix} 0.9 & 0.05 & 0.05 \\ 0.1 & 0.7 & 0.2 \\ 0.2 & 0.2 & 0.6 \end{bmatrix}$$
$$= [0.45 \quad 0.29 \quad 0.26]$$

To see why M_0T gives the next month's market share, let us consider how the first entry in the product M_0T is formed.

$$(0.4)(0.9) \quad + \quad (0.3)(0.1) \quad + \quad (0.3)(0.2) \quad = \quad 0.45$$

| 40% use Highlight, and 90% of these will buy it next month. | 30% use Silky Shine, and 10% of these will change to Highlight. | 30% use another brand, and 20% of these will change to Highlight. | 45% will use Highlight next month. |

To find a matrix M_2 that gives the market shares two months from now, multiply the next month's market share matrix M_1, which you found in Example 1, by the transition matrix T.

$$M_2 = M_1T$$
$$= (M_0T)T \leftarrow \text{substituting } M_1 = M_0T$$
$$= M_0T^2$$

Likewise, the market share matrix three months from now is:

$$M_3 = M_2T$$
$$= (M_0T^2)T$$
$$= M_0T^3$$

Generalizing, we see that M_0T^n gives the market share n months from now. Using computer software or a calculator that performs computations of matrices makes finding the calculations of M_0T^n easy. Some of these calculations are shown on the next page.

544 *Chapter Fourteen*

---

544 Chapter 14

$$M_1 = [0.45 \quad 0.29 \quad 0.26]$$
$$M_2 = [0.4860 \quad 0.2775 \quad 0.2365]$$
$$M_3 = [0.51245 \quad 0.26585 \quad 0.22170]$$
$$M_4 = [0.532130 \quad 0.256058 \quad 0.211813]$$
$$M_5 = [0.546885 \quad 0.248209 \quad 0.204906]$$

.
.
.

$$M_{10} = [0.581189 \quad 0.228840 \quad 0.189971]$$

.
.
.

$$M_{20} = [0.591878 \quad 0.222638 \quad 0.185485]$$

.
.
.

$$M_{30} = [0.592548 \quad 0.222248 \quad 0.185204]$$

The situation we have been discussing is an example of a **Markov chain**, a sequence of observations or predictions $M_0, M_1, M_2, \ldots$, each of which depends on the immediately preceding one, and only on it. The matrices M_n give the predicted market shares n months from now. Notice that in Example 2, there is little difference between M_{20} and M_{30}. If you were to continue calculating M_n for larger values of n, you would see that the market share values eventually *stabilize* at about 59%, 22%, and 19%. We say that the market shares have reached a *steady state*. Of course, achieving a steady state assumes that the transition matrix T remains the same. However, T *could* change due to an advertising campaign or a price change for one of the shampoos. If this happens, the Markov chain based on T terminates and a new chain based on a new transition matrix could begin.

The *steady-state* matrix [0.59 0.22 0.19] was calculated by assuming not only a constant transition matrix T but also by assuming a current market share $M_0 = [0.4 \ 0.3 \ 0.3]$. If the current market share were different, how would the steady-state matrix be affected, if at all? The answers to this and other interesting prediction problems can be found in the exercises.

CLASS EXERCISES

1. *Discussion* For the product $M_0T = M_1 = [0.45 \ 0.29 \ 0.26]$, the calculation of the number 0.45 is explained in the text. Give similar explanations for the numbers 0.29 and 0.26.

2. In the shampoo example, we saw that $M_2 = M_1T = (M_0T)T$. We then said that $(M_0T)T = M_0T^2$. What property of matrix multiplication justifies this last equation? Associative

Additional Examples cont.

Since the sum of the entries of S must be 1, S must have the form $[x \quad y \quad 1 - x - y]$. Thus,
$$[x \quad y \quad 1 - x - y] \cdot$$
$$\begin{bmatrix} 0.6 & 0.4 & 0 \\ 0.2 & 0.3 & 0.5 \\ 0 & 1 & 0 \end{bmatrix} =$$
$$[x \quad y \quad 1 - x - y]$$
1. $0.6x + 0.2y + 0 = x$
2. $0.4x + 0.3y + 1 - x - y = y$
3. $0 + 0.5y + 0 = 1 - x - y$

From (1), $y = 2x$. From (3), $x + 1.5y = 1$.
$x + 1.5(2x) = 1;$
$x = 0.25;$
$y = 2(0.25) = 0.5$
$S = [0.25 \quad 0.5 \quad 0.25]$
You can verify that $ST = S$.

◫ Using Technology

As indicated in the directions preceding Written Exercise 5, computations of higher powers of transition matrices are simply not practicable without the use of technology. Using a computer or calculator becomes a necessity, not a convenience.

Suggested Assignments

Discrete Math
Day 1: 546/1–7 odd
Day 2: 548/9, 11, 13, 14

Supplementary Materials

Alternative Assessment, 45
Precalculus Plotter Plus, 95–96

3. a. Geography Suppose that census takers have found that from one year to the next, 10% of those living in a certain city will move to its suburbs while 3% of those living in the suburbs will move into the city. Write a transition matrix T that describes this situation.

    b. If 60% of the metropolitan population is now in the city and 40% is in the suburbs, give the current population as a 1×2 matrix M_0.

    c. Calculate M_0T and interpret its elements.

    d. What do the elements of M_0T^2 tell you?

    e. How could one use a computer to determine if the population distribution reaches a steady state?

WRITTEN EXERCISES

A **1. Transportation** Commuters to a city commute either by car or by public transportation. Suppose each year 18% of the car commuters change to public transportation and 5% of those using public transportation change to cars.

    a. Give the transition matrix T.

    b. If 60% of the commuters now use public transportation and 40% use cars, write a 1×2 matrix M_0 to represent this information.

    c. Calculate $M_1 = M_0T$ and explain the meaning of each element.

    d. Calculate $M_2 = M_1T$ and explain the meaning of each element.

2. Geography Over the years, it has been observed that each year, 3% of those living in a certain city's urban area move to the suburbs and 2% of those in the suburbs move into the urban area.

    a. Give the transition matrix T.

    b. Suppose one third of the city's population now lives in the suburbs. Give a 1×2 matrix M_0 that describes the current population.

    c. Find the percentages of the population living in the urban area and in the suburbs one year from now and also two years from now.

    d. With a computer one could determine that $M_0T^{80} = [0.4 \ 0.6]$, so in 80 years, 40% will live in the urban area and 60% in the suburbs. Find the population percentages for 81 years from now without using a computer.

    e. Use your answer to part (d) to give the steady-state matrix.

3. Business Dixie's and Sargent's are two competing restaurants located next door to each other. A survey of those who eat regularly at one place or the other shows that customer loyalty is quite high. Only 10% of those who eat at Dixie's on one day will switch to Sargent's the next day and only 20% of those who eat at Sargent's will switch to Dixie's the next day. Suppose that of those who always eat lunch at one place or the other, 60% ate at Dixie's on Monday.

    a. What percentage ate at Dixie's on Tuesday? on Wednesday? 62%; 63%

    b. Suppose that two months from now, two thirds of the regular lunch crowd ate at Dixie's and one third ate at Sargent's. What fractional amount of the lunch crowd will be at each place on the next day? See below.

    c. Use your answer to part (b) to give the steady-state matrix. $\begin{bmatrix} \dfrac{2}{3} & \dfrac{1}{3} \end{bmatrix}$

    b. $\dfrac{2}{3}$ at Dixie's, $\dfrac{1}{3}$ at Sargent's

4. Psychology In an experiment to see how well mice can learn, mice are placed in the maze shown. A mouse that enters room 1 is rewarded with cheese, but there is no reward in room 2. Studies show that 70% of mice who go to room 1 will go to room 1 to be rewarded the next time. Those who go to room 2 one time are as likely to go to room 1 as to room 2 the next time.

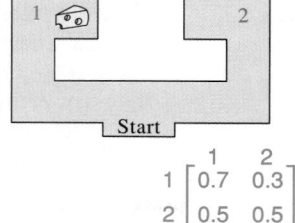

$$\begin{array}{c} \\ 1 \\ 2 \end{array} \begin{array}{cc} 1 & 2 \\ \left[\begin{array}{cc} 0.7 & 0.3 \\ 0.5 & 0.5 \end{array}\right] \end{array}$$

a. Write a transition matrix.
b. If on the twentieth trial, 62.5% of the mice go to room 1 to get the cheese, what percentage will do this on the twenty-first trial? 62.5%
c. What appears to be the steady-state matrix? [0.625 0.375]

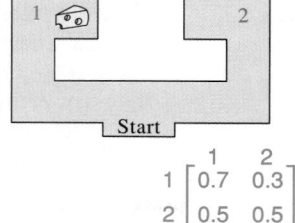 **For the parts of the exercises designated in green, you will need to use computer software or a calculator that performs matrix operations.**

5. Statistics In the land of Oz, the weather is a little unpredictable. However, the Wizard of Oz, who is an amateur weatherman, has made the following observations through the centuries: 80% of all sunny days are followed by another sunny day, 10% of sunny days are followed by cloudy ones, and 10% are followed by rainy ones. A cloudy day has a 40% chance of being followed by another cloudy day and a 60% chance of being followed by a rainy day. A rainy day is always followed by a sunny one.

a. Write the matrix T that represents the day-to-day weather transitions.
b. Calculate T^2. Give the chance that if Monday is sunny, then Wednesday will be rainy.
c. Calculate T^3. Give the chance that Thursday will be cloudy if Monday is sunny.
d. Calculate T^{30}. What is the chance that 30 days from now will be sunny if today is sunny? if today is cloudy? if today is rainy?
e. Calculate T^{40}. Then answer the three questions in part (d).
f. Over a long period, what percentage of the days are sunny? cloudy? rainy?

6. Business Every year, each of several rent-a-car companies will buy all new cars from the "big three" automobile manufacturing companies (A, B, and C). The three companies keep track of how these orders change from year to year. They find that 50% of A cars will be replaced by new A cars the next year, while the other 50% of A cars will be replaced by equal numbers of B and C cars. Sixty percent of B cars will be replaced by new B cars while 30% will be replaced by A cars and 10% by C cars. Thirty percent of C cars will be replaced by B cars and 25% will be replaced by A cars.

a. Write a transition matrix. Label the rows and columns.
b. The 1990 fleet of rental cars was distributed as follows: 42% A cars, 37% B cars, and 21% C cars. Find the following:

　1. the 1991 percentages for the three auto manufacturers.
　2. the percentage of B cars in the year 2000.
　3. the steady-state matrix.

---

5. a.
$$\begin{array}{c} \\ s \\ c \\ r \end{array} \begin{array}{ccc} s & c & r \\ \left[\begin{array}{ccc} 0.8 & 0.1 & 0.1 \\ 0 & 0.4 & 0.6 \\ 1 & 0 & 0 \end{array}\right] \end{array}$$

b. $T^2 =$
$$\begin{array}{c} \\ s \\ c \\ r \end{array} \begin{array}{ccc} s & c & r \\ \left[\begin{array}{ccc} 0.74 & 0.12 & 0.14 \\ 0.6 & 0.16 & 0.24 \\ 0.8 & 0.1 & 0.1 \end{array}\right] \end{array};$$
there is a 14% chance that Wed. will be rainy if Mon. is sunny.

c. $T^3 =$
$$\begin{array}{c} \\ s \\ c \\ r \end{array} \begin{array}{ccc} s & c & r \\ \left[\begin{array}{ccc} 0.732 & 0.122 & 0.146 \\ 0.72 & 0.124 & 0.156 \\ 0.74 & 0.12 & 0.14 \end{array}\right] \end{array};$$
there is a 12.2% chance that Thurs. will be cloudy if Mon. is sunny.

d. There is a 73% chance of a sunny day 30 days from now if today is sunny; 73% chance sunny if cloudy; 73% chance sunny if rainy.

e. same answers as in (d)

f. sunny: 73%
cloudy: 12%
rainy: 15%

6. a.
$$\begin{array}{c} \\ A \\ B \\ C \end{array} \begin{array}{ccc} A & B & C \\ \left[\begin{array}{ccc} 0.5 & 0.25 & 0.25 \\ 0.3 & 0.6 & 0.1 \\ 0.25 & 0.3 & 0.45 \end{array}\right] \end{array}$$

b. (1) [0.42 0.37 0.21] ·
$$\left[\begin{array}{ccc} 0.5 & 0.25 & 0.25 \\ 0.3 & 0.6 & 0.1 \\ 0.25 & 0.3 & 0.45 \end{array}\right] =$$
[0.3735 0.39 0.2365]
A: 37.35%; B: 39%; C: 23.65%
(2) 40.28%

7. **Statistics** Last winter a certain strain of flu was studied by medical researchers. They determined that from one five-day period to the next, a well person had a 20% chance of getting the flu. Of those persons who had the flu during a five-day period, 50% still had it during the next five-day period, while the other 50% recovered and became immune.
   a. Write a transition matrix T for the three conditions: well, flu, and immune.
   b. Suppose that after Christmas vacation, 5% of the students at a college returned with the flu, 10% were immune, and 85% were well but not immune. After 5 days, what percentage of the students had the flu? What percentage were immune?
   c. Find the percentages of students who were well, who had the flu, and who were immune after 10 days, after 30 days, and after 50 days.

8. Consider the transition matrix $T = \begin{bmatrix} \frac{3}{5} & \frac{2}{5} \\ \frac{1}{5} & \frac{4}{5} \end{bmatrix}$ and the matrix $S = \begin{bmatrix} \frac{1}{3} & \frac{2}{3} \end{bmatrix}$.
   a. Show that $ST = S$.
   b. *Writing* Write a sentence or two explaining how part (a) shows that matrix S is the steady-state matrix.

B 9. The purpose of this exercise is to find without a computer the steady-state matrix S for a given transition matrix. Let $T = \begin{bmatrix} 0.3 & 0.7 \\ 0.1 & 0.9 \end{bmatrix}$. By definition, S has the form $S = [x \quad 1 - x]$. Since ST must equal S, you can solve for x. Do so, and then write S.

10. Use the technique of Exercise 9 to find the steady-state matrix S for the given transition matrix T.
   a. $T = \begin{bmatrix} 0.25 & 0.75 \\ 0.45 & 0.55 \end{bmatrix}$ [0.375 0.625]   **b.** $T = \begin{bmatrix} \frac{1}{3} & \frac{2}{3} \\ \frac{4}{5} & \frac{1}{5} \end{bmatrix}$ $\begin{bmatrix} \frac{6}{11} & \frac{5}{11} \end{bmatrix}$

11. If T is a 3×3 transition matrix, then the steady-state matrix must have the form $S = [x \quad y \quad 1 - x - y]$. Find S if:
$$T = \begin{bmatrix} 0.4 & 0.3 & 0.3 \\ 0.5 & 0.5 & 0 \\ 0.3 & 0.3 & 0.4 \end{bmatrix} \quad \begin{bmatrix} \frac{5}{12} & \frac{3}{8} & \frac{5}{24} \end{bmatrix}$$

12. **Business** Refer to the shampoo problem on page 543. Recall that when the initial market share matrix was $M_0 = [0.4 \ 0.3 \ 0.3]$, the market share matrix for n months in the future stabilized at $M_0 T^n = [0.59 \ 0.22 \ 0.19]$ when n was large. This matrix described the steady state of the market shares. What would this matrix be if the initial market share had been different? Consider the following two possibilities for the initial market share.
   a. $M_0 = [0.6 \ 0.3 \ 0.1]$ **b.** $M_0 = [0.8 \ 0.1 \ 0.1]$
   [0.59 0.22 0.19]   [0.59 0.22 0.19]
   The matrix is independent of M_0.

548   *Chapter Fourteen*

13. The diagram at the right shows a five-room maze for mice. During a one-minute time period each mouse moves to an adjoining room. Thus from room 1, a mouse moves to either room 2, room 4, or room 5. If we assume that any of these possibilities is as likely as another, then the chances that a mouse moves from room 1 to room 2 , from room 1 to room 4, or from room 1 to room 5 are all $\frac{1}{3}$. This means that the first row of the transition matrix is $\left[0 \ \frac{1}{3} \ 0 \ \frac{1}{3} \ \frac{1}{3} \right]$.

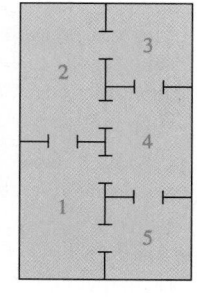

a. Complete the transition matrix.
b. If a mouse is placed in room 1, what are the chances that after one hour it will be in room 1? room 2? room 3? room 4? room 5?
c. Answer part (b) if the mouse begins in room 4.
d. Does the answer to part (b) seem to depend on where the mouse begins?

14. **Biology** Consider a species of insects that lives for four years subject to the following conditions:
1. 40% of those in their first year survive to the second year.
2. 50% of those in their second year survive to the third year.
3. 30% of those in their third year survive to the fourth year.
4. A female alive in the fourth year lays many eggs of which only an average of 40 will hatch the next spring. Assume that each male mates with exactly one female and that there are equal numbers of males and females. Thus, for each insect, male or female, alive in the fourth year, 20 new insects are produced the next year.

a. The above information has been incorporated into the matrix T. The 0.4 in the first row means that 40% of those in their first year survive to the second. Since it is impossible to go directly from year 1 to year 3, a zero appears in the third column of row 1. Explain the other elements in T.

$$\begin{array}{c} \\ 1 \\ 2 \\ 3 \\ 4 \end{array} \begin{array}{cccc} 1 & 2 & 3 & 4 \\ \left[\begin{array}{cccc} 0 & 0.4 & 0 & 0 \\ 0 & 0 & 0.5 & 0 \\ 0 & 0 & 0 & 0.3 \\ 20 & 0 & 0 & 0 \end{array} \right] \end{array} = T$$

b. Suppose that in a garden area the number of insects alive in their first, second, third, and fourth years is given by the population matrix $M_0 = [800 \ 300 \ 200 \ 100]$. The population one year later is $M_1 = M_0 T$. Calculate this product and explain the meaning of each element. Assume that the insects and their offspring continue to live in the same garden area.
c. The population two years later is $M_2 = M_1 T = M_0 T^2$. Calculate M_2.
d. Find the insect population ten years later.

15. **Biology** Suppose that the insects in Exercise 14 are able to reproduce in their third year as well as in their fourth year. Suppose also that the birthrate is the same for the three-year-olds as it is for the four-year-olds: 20 new insects for each older one. Change one element of T to account for this new information. Then answer parts (b), (c), and (d) of Exercise 14.

C 16. **Biology** Consider the following population cycle for beavers that are subdivided into five age categories:

A_1 (0–2 years), A_2 (2–4 years), A_3 (4–6 years), A_4 (6–8 years), and A_5 (8–10 years)

The numbers $S_1, S_2, S_3, S_4,$ and S_5 are the survival rates from one age group to the next; for example, S_2 gives the percentage that survives from age group A_2 to A_3. Since beavers do not live more than 10 years, $S_5 = 0$. The numbers $B_1, B_2, B_3, B_4,$ and B_5 are the birthrates for each age group; for example, if $B_3 = 1.6$, then for each beaver in age group A_3, an average of 1.6 beavers will be born into age group A_1 during the next two-year period. Since very young beavers cannot reproduce, $B_1 = 0$. The following diagram illustrates this population cycle.

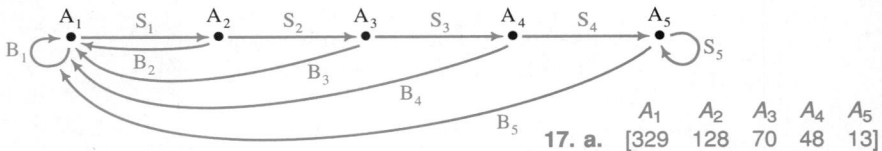

				A_1	A_2	A_3	A_4	A_5
17. a.				[329	128	70	48	13]

Write a 5 × 5 matrix T that gives the population rates for moving from one age group to another over a two-year period. (*Hint:* The first row of T is $[B_1 \ S_1 \ 0 \ 0 \ 0]$.)

17. **Biology** Use the matrix found in Exercise 16 to solve the following problem: Let the birthrates $B_1 = 0$, $B_2 = 1.2$, $B_3 = 1.5$, $B_4 = 2.0$, $B_5 = 0.2$ and the survival rates $S_1 = 0.4$, $S_2 = 0.6$, $S_3 = 0.7$, $S_4 = 0.3$, $S_5 = 0$. Suppose that in a certain locale, the population for the five age groups is $M_0 = [58 \ 68 \ 70 \ 48 \ 30]$.

  a. Find the population for the various age groups 20 years from now. Note that 20 years represents 10 two-year periods. See above.

  b. When will the population reach 1000 or more? (To find the total of the elements in a 1 × 5 population matrix, multiply it on the right by a 5 × 1 matrix whose elements are all 1.) In 38 years

14-6 Transformation Matrices

Objective   *To find the images of points under different types of transformations using matrices.*

Many movies and TV programs have introductions in which computer graphics are used to enlarge or rotate an image or printed message. Often the image or message seems to be transformed into three dimensions even though you are looking at a two-dimensional screen. Geometric transformations in two dimensions can be represented by 2×2 matrices and three-dimensional transformations can be represented by 3×3 matrices.

In Chapter 4 you learned that a function is a correspondence or rule that assigns to every element in one set of objects (the domain) exactly one element in a second set of objects (the range). For most of the functions studied in this book, the domain and range have been sets of numbers. If the domain and range are sets of points, then the function is usually called a **transformation of the plane**.

Consider the two-dimensional transformation T that maps each point $P(x, y)$ to its image point $P'(x + 2y, 3x + y)$. Under this *linear transformation*, line segments are mapped onto line segments, so the effect of T on an n-sided figure is easily found by finding just the images of the vertices of the figure. For example, consider $\triangle ABC$ shown below. To find its image, $\triangle A'B'C'$, we find the images of $A(0, 2)$, $B(3, 1)$, and $C(2, -3)$ using $T: (x, y) \rightarrow (x + 2y, 3x + y)$ as follows:

Since $A = (0, 2)$, $T: (0, 2) \rightarrow (0 + 2(2), 3(0) + 2)$,   so $A' = (4, 2)$.
Since $B = (3, 1)$, $T: (3, 1) \rightarrow (3 + 2(1), 3(3) + 1)$,   so $B' = (5, 10)$.
Since $C = (2, -3)$, $T: (2, -3) \rightarrow (2 + 2(-3), 3(2) + -3)$,   so $C' = (-4, 3)$.

Plot $A'(4, 2)$, $B'(5, 10)$, and $C'(-4, 3)$ and draw $\triangle A'B'C'$. Notice that $\triangle A'B'C'$ has the reverse orientation to that of $\triangle ABC$: reading the vertices in alphabetical order, one must move clockwise around $\triangle ABC$ and counterclockwise around $\triangle A'B'C'$.

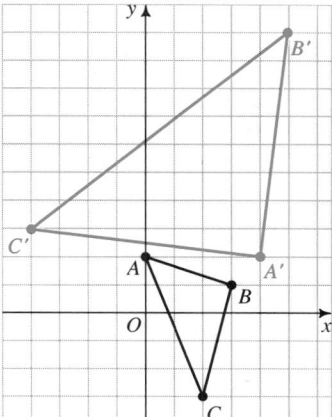

In the transformation T above, if the image of $P(x, y)$ is $P'(x', y')$, then:

$$x + 2y = x'$$
$$3x + y = y'$$

These equations can be simplified by writing a single matrix equation.

$$\begin{bmatrix} 1 & 2 \\ 3 & 1 \end{bmatrix} \begin{bmatrix} x \\ y \end{bmatrix} = \begin{bmatrix} x' \\ y' \end{bmatrix}$$

Teaching Notes, p. 516D

Warm-Up Exercises

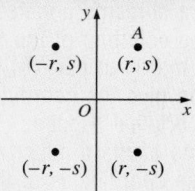

In the figure above, point A has coordinates (r, s). Give the coordinates of the point obtained by:

1. reflecting A in the x-axis. $(r, -s)$

2. reflecting A in the y-axis. $(-r, s)$

3. translating A 2 units to the right and 3 units down. $(r + 2, s - 3)$

4. rotating A 180° around point O. $(-r, -s)$

5. reflecting A in the x-axis and then translating the resulting point 1 unit up; translating A 1 unit up and then reflecting the resulting point in the x-axis. $(r, -s + 1)$, $(r, -s - 1)$

Motivating the Section

Tell students that computer aided design (CAD) and manufacturing (CAM) have revolutionized industrial processes worldwide. CAD and CAM depend on being able to represent and manipulate shapes and structures. Geometric transformations in matrix form provide an effective method for accomplishing this kind of manipulation.

It is customary to use the letter T to represent both the transformation and its corresponding matrix.

$$T = \begin{bmatrix} 1 & 2 \\ 3 & 1 \end{bmatrix} \quad \text{represents} \quad T: (x, y) \rightarrow (x + 2y, 3x + y).$$

Although we have already found the images of A, B, and C using the transformation $T: (x, y) \rightarrow (x + 2y, 3x + y)$, a more efficient method is to multiply the transformation matrix T by a matrix whose columns give the x- and y-coordinates of A, B, and C.

$$\overset{T}{\begin{bmatrix} 1 & 2 \\ 3 & 1 \end{bmatrix}} \overset{A \quad B \quad C}{\begin{bmatrix} 0 & 3 & 2 \\ 2 & 1 & -3 \end{bmatrix}} = \overset{A' \quad B' \quad C'}{\begin{bmatrix} 4 & 5 & -4 \\ 2 & 10 & 3 \end{bmatrix}}$$

Example 1 Consider the transformation

$$T: (x, y) \rightarrow (2x, -y),$$

and the "fish" figure F shown at the right.
  a. Find the transformation matrix T.
  b. Using matrices, find the images of the nine points determining F.
  c. Plot the image points and draw F'. Compare the orientations of F and F'.

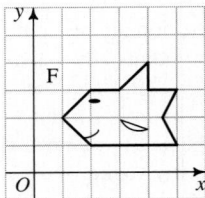

Solution

a. $T = \begin{bmatrix} 2 & 0 \\ 0 & -1 \end{bmatrix}$

b. $\begin{bmatrix} 2 & 0 \\ 0 & -1 \end{bmatrix} \begin{bmatrix} 1 & 2 & 2 & 3 & 4 & 4 & \frac{9}{2} & 5 & 5 \\ 2 & 1 & 3 & 3 & 3 & 4 & 2 & 1 & 3 \end{bmatrix}$

$= \begin{bmatrix} 2 & 4 & 4 & 6 & 8 & 8 & 9 & 10 & 10 \\ -2 & -1 & -3 & -3 & -3 & -4 & -2 & -1 & -3 \end{bmatrix}$

c.

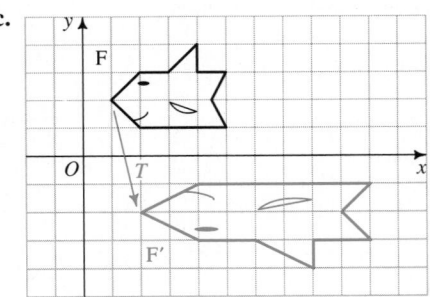

F and F' have opposite orientations.

One advantage of using matrices is that the determinant of the transformation matrix gives two useful pieces of information, as you will discover by completing the following activity.

Activity

a. By counting squares, approximate the areas of F and F′ in the diagram given in the solution to Example 1(c). 7 sq. units, 14 sq. units

b. Find the ratio of the area of F′ to the area of F. 2

c. Find the determinant of T. −2

d. Compare your answers to parts (b) and (c). They are opposites.

Depending on how good your approximations are in the previous activity, you can see that the following theorem is true for the transformation in Example 1.

Theorem 1

Let $T = \begin{bmatrix} a & b \\ c & d \end{bmatrix}$ represent a linear transformation that maps a region R to an image region R′. Then:

1. $\dfrac{\text{area of R}'}{\text{area of R}}$ = absolute value of the determinant of T.

2. If the determinant of T is positive, then R′ and R have the same orientation.

   If the determinant of T is negative, then R′ and R have opposite orientations.

   If the determinant of T is zero, then R′ lies on a line through the origin.

The reason behind the first statement of Theorem 1 can be seen by examining the unit square under the transformation $T: (x, y) \rightarrow (ax + by, cx + dy)$. Its image is a parallelogram, whose area is $|ad - bc|$ times the area of the original figure. You are asked to prove this fact in Written Exercise 30.

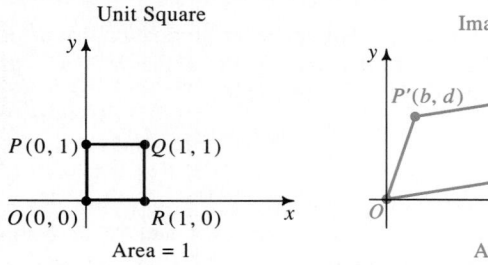

Unit Square

$P(0, 1)$ $Q(1, 1)$

$O(0, 0)$ $R(1, 0)$

Area = 1

Image of Unit Square

$P'(b, d)$ $Q'(a+b, c+d)$

$R'(a, c)$

Area = $|ad - bc|$

Theorem 2

The columns of a transformation matrix are the images of the points $(1, 0)$ and $(0, 1)$, respectively.

Proof: Let $T = \begin{bmatrix} a & b \\ c & d \end{bmatrix}$. Then

$$\begin{bmatrix} a & b \\ c & d \end{bmatrix} \begin{bmatrix} 1 \\ 0 \end{bmatrix} = \begin{bmatrix} a \\ c \end{bmatrix} \quad \text{and}$$

$$\begin{bmatrix} a & b \\ c & d \end{bmatrix} \begin{bmatrix} 0 \\ 1 \end{bmatrix} = \begin{bmatrix} b \\ d \end{bmatrix}.$$

Therefore $T: (1, 0) \to (a, c)$ and $T: (0, 1) \to (b, d)$.

Theorem 2 gives a method for finding the corresponding matrix of a transformation described in words. Example 2 illustrates this method.

Example 2 Let S be a reflection in the y-axis. Find the matrix S.

Solution Visualize what happens to the points $(1, 0)$ and $(0, 1)$ when each is reflected in the y-axis. Then use Theorem 2 to write the matrix S.

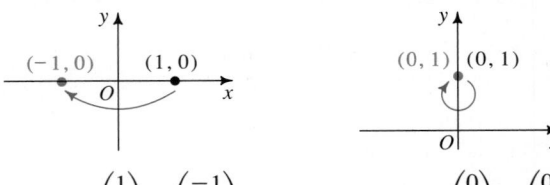

$$\text{image of } \begin{pmatrix} 1 \\ 0 \end{pmatrix} = \begin{pmatrix} -1 \\ 0 \end{pmatrix} \qquad \text{image of } \begin{pmatrix} 0 \\ 1 \end{pmatrix} = \begin{pmatrix} 0 \\ 1 \end{pmatrix}$$

$\therefore S = \begin{bmatrix} -1 & 0 \\ 0 & 1 \end{bmatrix}$ represents the reflection in the y-axis.

Suppose $T: \text{F} \to \text{F}'$ and $S: \text{F}' \to \text{F}''$. The single transformation that maps F to F'' is called the composite of S and T, written as $S \circ T$. Recall that with composition of functions, the rightmost function is performed first.

$$(g \circ f)(x) = g(f(x)) \qquad (S \circ T)(\text{F}) = S(T(\text{F}))$$
$$= S(\text{F}')$$
$$= \text{F}''$$

If T and S are the transformations given in Examples 1 and 2, the composite mapping of fish F is shown on the next page.

554 *Chapter Fourteen*

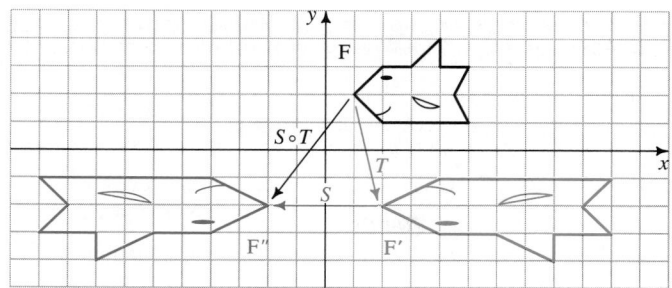

The next theorem, illustrated in Written Exercise 5, gives a nice connection between composition of functions and multiplication of matrices.

Theorem 3

Let S and T be transformations with corresponding matrices

$$S = \begin{bmatrix} a & b \\ c & d \end{bmatrix} \text{ and } T = \begin{bmatrix} e & f \\ g & h \end{bmatrix}.$$

Then the matrix representing the transformation $S \circ T$ is the matrix ST, where

$$ST = \begin{bmatrix} a & b \\ c & d \end{bmatrix} \begin{bmatrix} e & f \\ g & h \end{bmatrix}.$$

Example 3   Consider the transformations given in Examples 1 and 2. Find the corresponding matrix for the transformation $S \circ T$.

Solution   Refer to $S = \begin{bmatrix} -1 & 0 \\ 0 & 1 \end{bmatrix}$ and $T = \begin{bmatrix} 2 & 0 \\ 0 & -1 \end{bmatrix}$ given in Examples 2 and 1.

Then by Theorem 2, the matrix representing the transformation $S \circ T$ is

$$ST = \begin{bmatrix} -1 & 0 \\ 0 & 1 \end{bmatrix} \begin{bmatrix} 2 & 0 \\ 0 & -1 \end{bmatrix} = \begin{bmatrix} -2 & 0 \\ 0 & -1 \end{bmatrix}$$

You will see that the transformation matrices given in the exercises correspond to the transformations you studied in Chapter 4: reflections, translations, and expansions (stretching) or contractions (shrinking). In the Written Exercises you will explore another basic transformation that is not as familiar, namely, a rotation. New transformations can be created by the composition of these four basic transformations in any combination. For example, $T: (x, y) \to (2x, -y)$, given in Example 1, is the composite of an expansion and a reflection.

The graphs shown for Exercises 7–9 represent similarity transformations, since the shapes of the triangles are preserved.

Additional Answers
Class Exercises

4. $T = \begin{bmatrix} 3 & -2 \\ 4 & 1 \end{bmatrix}$, $|T| = 11$

CLASS EXERCISES

1. Let $g: 5 \rightarrow 3$ and $f: -1 \rightarrow 5$. Find $(g \circ f)(-1)$. 3
2. Let $g: x \rightarrow 2x - 4$ and $f: x \rightarrow -x + 1$. Find $(g \circ f)(x)$. $-2x - 2$

For Exercises 3–6, use $T: (x, y) \rightarrow (x', y')$, where $\begin{aligned} 3x - 2y &= x' \\ 4x + y &= y'. \end{aligned}$ $\begin{bmatrix} 3 & -2 \\ 4 & 1 \end{bmatrix} \begin{bmatrix} x \\ y \end{bmatrix} = \begin{bmatrix} x' \\ y' \end{bmatrix}$

3. Express the given two equations as a single matrix equation.
4. State the transformation matrix T and find its determinant.
5. If T maps $\triangle ABC$ to $\triangle A'B'C'$, how do the areas compare? $\dfrac{\text{area } \triangle A'B'C'}{\text{area } \triangle ABC} = 11$
6. Do the triangles have the same or opposite orientation?
   Same

The graphs below illustrate four basic transformations. (An expansion and a contraction are considered the same type of transformation: a *dilation*.) Compare $\triangle ABC$ and $\triangle A'B'C'$ in each graph. Then complete the table.

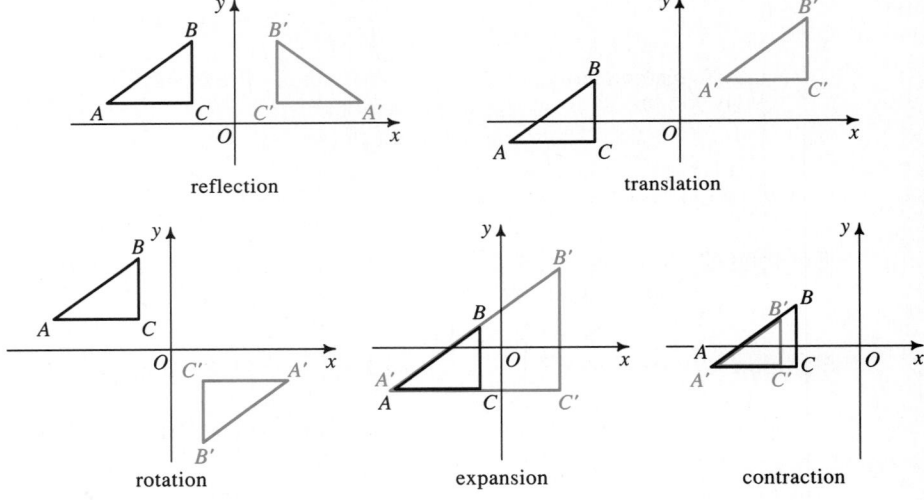

reflection   translation

rotation   expansion   contraction

	Transformation	Changes orientation	Changes size	Changes shape
	Reflection	yes	no	no
7.	Translation	? No	? No	? No
8.	Rotation	? No	? No	? No
9.	Dilation	? No	? Yes	? No

A **1.** Consider the transformation $T: (x, y) \rightarrow (2x, 3y)$.

    a. Plot points $A(2, 1)$, $B(4, 0)$, and $C(1, -3)$ and their images A', B', and C'.

    b. Write the transformation matrix T. Find the determinant of this matrix.

    c. Find the ratio of the area of $\triangle A'B'C'$ to the area of $\triangle ABC$.

    d. Compare the orientations of $\triangle ABC$ and $\triangle A'B'C'$.

2. Consider the transformation matrix $T = \begin{bmatrix} 1 & 1 \\ 3 & -1 \end{bmatrix}$.

    a. Draw $\triangle ABC$ with vertices $A(0, 3)$, $B(2, 4)$, and $C(5, 0)$.

    b. Write a matrix M that gives the coordinates of points A, B, and C. Calculate the product TM to find the coordinates of A', B', and C'.

    c. Draw $\triangle A'B'C'$ on the same set of axes used in part (a).

    d. Find the determinant of T. What two things does this determinant tell you about $\triangle ABC$ and $\triangle A'B'C'$?

3. a. Draw $\triangle PQR$ with vertices $P(-1, 0)$, $Q(0, 3)$, and $R(2, -3)$.

    b. Draw $\triangle P'Q'R'$, the image of $\triangle PQR$ under the following transformation: $T: (x, y) \rightarrow (2x + y, 3x + y)$.

    c. Find the transformation matrix T and its determinant. Then compare the areas and the orientations of the two triangles.

4. a. Draw $\triangle XYZ$ with vertices $X(1, 0)$, $Y(-1, 2)$, and $Z(0, 2)$.

    b. Draw the image of $\triangle XYZ$ under the following transformation: $T: (x, y) \rightarrow (4x - 6y, 2x - 3y)$.

    c. Find the transformation matrix T and its determinant. Explain what this tells you about the images of X, Y, and Z.

5. Consider the point $P(1, 2)$ and the transformation matrices

$$S = \begin{bmatrix} 1 & 3 \\ 4 & -2 \end{bmatrix} \text{ and } T = \begin{bmatrix} 4 & 1 \\ 0 & 5 \end{bmatrix}.$$

    a. Find $(S \circ T)(P)$ by first finding $T(P) = P'$ and then $S(P') = P''$. (36, 4)

    b. Find $(S \circ T)(P)$ in a different way by first finding the transformation matrix ST and then multiplying ST by the column matrix for point P. $\begin{bmatrix} 36 \\ 4 \end{bmatrix}$

6. Consider the point $P(x, y)$ and the transformation matrices

$$S = \begin{bmatrix} 3 & 7 \\ 1 & 4 \end{bmatrix} \text{ and } T = \begin{bmatrix} 1 & 2 \\ 2 & 5 \end{bmatrix}. \begin{bmatrix} 17x + 41y \\ 9x + 22y \end{bmatrix}$$

Find $(S \circ T)(P)$.

7. *Investigation* The transformation $T: (x, y) \rightarrow (4x + 2y, 2x + y)$ maps every point of the plane onto a line L. By experimenting with different points and their images, find the equation of L. Also find the matrix of the transformation and its determinant.

8. *Investigation* Repeat Exercise 7 for the following transformation:

$$T: (x, y) \rightarrow (5x - 15y, -x + 3y)$$

Matrices **557**

Suggested Assignments

Discrete Math
557/1, 3, 5, 13, 15, 19, 25

Supplementary Materials

Alternative Assessment, 46
Precalculus Plotter Plus, 67
Student Resource Guide,
134–137

The *translation* $G: (x, y) \rightarrow (x + h, y + k)$ moves every point P horizontally h units and vertically k units to its image point P'. This can be expressed by either a system of equations or a single matrix equation:

$$x + h = x'$$
$$y + k = y'$$

$$\begin{bmatrix} x \\ y \end{bmatrix} + \begin{bmatrix} h \\ k \end{bmatrix} = \begin{bmatrix} x' \\ y' \end{bmatrix}$$

Describe the motion of each translation.

9. $G: (3, -1) \rightarrow (8, -2)$
To the right 5, down 1

10. $x + 6 = x'$
$y - 1 = y'$
To the right 6, down 1

11. $\begin{bmatrix} x \\ y \end{bmatrix} + \begin{bmatrix} 0 \\ -2 \end{bmatrix} = \begin{bmatrix} x' \\ y' \end{bmatrix}$
Down 2

Express the translation $G: P(x, y) \rightarrow P'(x', y')$ by a matrix equation.

12. $P(2, -4)$; $P'(3, 1)$ 13. $P(5, 3)$; $P'(5, -1)$ 14. $P(-2, -2)$; $P'(4, -2)$

Use the method shown in Example 2 to find the matrix that represents each transformation described.

B 15. X, a reflection in the x-axis   16. L, a reflection in the line $y = x$
17. M, a reflection in the line $y = -x$ 18. R_{180}, a half-turn rotation about the origin O
19. R_{90}, a 90° counterclockwise rotation about the origin O 20. R_{-90}, a 90° clockwise rotation about the origin O

$R_{90} = \begin{bmatrix} 0 & -1 \\ 1 & 0 \end{bmatrix}$ $R_{-90} = \begin{bmatrix} 0 & 1 \\ -1 & 0 \end{bmatrix}$

Let $Y = \begin{bmatrix} -1 & 0 \\ 0 & 1 \end{bmatrix}$ represent a reflection in the y-axis. Use this matrix as well as the matrices found in Exercises 15–20 to do Exercises 21–26.

21. Find XY and show that the result equals R_{180}.
22. Find LX and show that the result equals R_{90}.
23. Find $R_{90}R_{-90}$. What special name is given to this matrix?
24. Find LM and show that the result equals R_{180}.
25. Find $(R_{90})^2$. Which matrix equals this product?
26. Find $(R_{-90})^{51}$ by considering the effect of performing R_{-90} fifty-one times.
27. *Writing* Find X^2, L^2, and M^2. Write a paragraph to explain what conclusion you can draw.

28. Consider R_α, the counterclockwise rotation of α degrees about the origin. The diagram shows that the image of $(1, 0)$ under this transformation is $(\cos \alpha, \sin \alpha)$.
 a. Find the image of $(0, 1)$.
 b. Find the transformation matrix R_α.
 c. Find the determinant of the transformation matrix. What does this tell you about the area of any figure F and the area of its image F'?

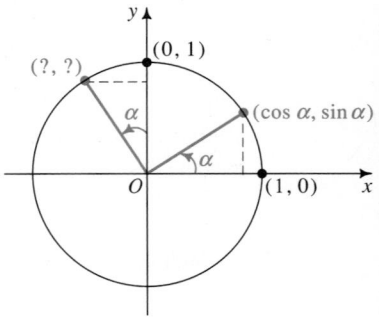

C **29.** Let R_α be the counterclockwise rotation of α degrees about the origin, as shown in Exercise 28. Let R_β be the counterclockwise rotation of β degrees about the origin.
  a. What is the transformation matrix R_β?
  b. *Writing* Find $(R_\alpha)^2$ and the product $R_\alpha R_\beta$, and simplify your answers. Write a paragraph explaining why your answers make sense.

30. Consider the transformation $T: (x, y) \rightarrow (ax + by, cx + dy)$, which maps a unit square to a parallelogram.
  a. Refer to the unit square and its image on page 553, and verify that the coordinates of the parallelogram are correct.
  b. Show that the area of the image is $|ad - bc|$. (*Hint:* Draw line segments from point Q' perpendicular to both axes.)

31. Consider the transformation T that maps a point $P(x, y, z)$ in a three-dimensional coordinate system to the point $P'(2x, 3y, y + 4z)$.
  a. Give a 3×3 matrix for T. See below.
  b. The diagrams below show a unit cube and its image, which is a parallelepiped. A parallelepiped is a polyhedron, all of whose faces are parallelograms. Guess what the determinant of T tells you about the cube and the parallelepiped. Ratio of volumes and orientation of points   **31. a.** $\begin{bmatrix} 2 & 0 & 0 \\ 0 & 3 & 0 \\ 0 & 1 & 4 \end{bmatrix}$

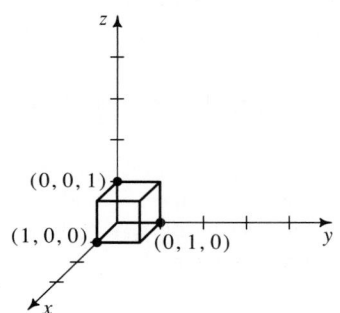

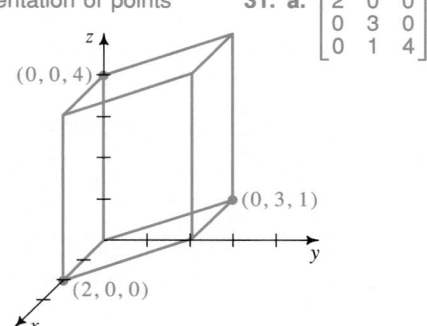

  c. How are the columns of T related to the three points labeled in the diagram at the left above? They are the coords. of the images of (1, 0, 0), (0, 1, 0), and (0, 0, 1).

32. a. Find the transformation M for a 90° counterclockwise rotation about the z-axis. (*Hint:* See Exercise 31(c).)
  b. Find the transformation matrix N for a 180° rotation about the y-axis.
  c. Find the product NM. Tell what transformation this product represents.
  d. *Investigation* Do you think that $NM = MN$? Explain your answer. Then find each product to see if you are correct.

33. Consider the transformations $Y: (x, y) \rightarrow (-x, y)$ and $G: (x, y) \rightarrow (x + h, y + k)$. Suppose that $G \circ Y: P(x, y) \rightarrow P'(x', y')$.
  a. Give transformation matrices Y and G and column matrices P and P'.
  b. Which of the following matrix equations is correct?
       (1) $P' = GYP$ (2) $P' = YP + G$

29. a. $\begin{bmatrix} \cos\beta & -\sin\beta \\ \sin\beta & \cos\beta \end{bmatrix}$
  b. $(R_\alpha)^2 =$
  $\begin{bmatrix} \cos 2\alpha & -\sin 2\alpha \\ \sin 2\alpha & \cos 2\alpha \end{bmatrix}$,
  $R_\alpha R_\beta =$
  $\begin{bmatrix} \cos(\alpha+\beta) & -\sin(\alpha+\beta) \\ \sin(\alpha+\beta) & \cos(\alpha+\beta) \end{bmatrix}$

30. a. T: $(0, 0) \rightarrow (0, 0)$;
  T: $(1, 0) \rightarrow (a, c)$;
  T: $(1, 1) \rightarrow$
  $(a + b, c + d)$;
  T: $(0, 1) \rightarrow (b, d)$

32. a. $\begin{bmatrix} 0 & -1 & 0 \\ 1 & 0 & 0 \\ 0 & 0 & 1 \end{bmatrix}$
  b. $\begin{bmatrix} -1 & 0 & 0 \\ 0 & 1 & 0 \\ 0 & 0 & -1 \end{bmatrix}$
  c. $\begin{bmatrix} 0 & 1 & 0 \\ 1 & 0 & 0 \\ 0 & 0 & -1 \end{bmatrix}$;
  a 90° counterclockwise rotation about the z-axis followed by a 180° rotation about the y-axis.
  d. No; consider (1, 0, 0);
  NM: (1, 0, 0) $\rightarrow$ (0, 1, 0) and MN: (1, 0, 0) $\rightarrow$ (0, −1, 0);
  $MN = \begin{bmatrix} 0 & -1 & 0 \\ -1 & 0 & 0 \\ 0 & 0 & -1 \end{bmatrix}$

33. a. $Y = \begin{bmatrix} -1 & 0 \\ 0 & 1 \end{bmatrix}$,
  $G = \begin{bmatrix} h \\ k \end{bmatrix}$,
  $P = \begin{bmatrix} x \\ y \end{bmatrix}$,
  $P' = \begin{bmatrix} -x + h \\ y + k \end{bmatrix}$
  b. (2)

Additional Answers
Chapter Test

1. **a.**
$$M \begin{matrix} s & m & l \end{matrix}$$
$$\text{1. a.} \ P \begin{bmatrix} 7 & 14 & 10 \\ 8 & 20 & 9 \\ 15 & 20 & 12 \end{bmatrix}$$

b.
$$1.5S = \begin{bmatrix} 10.5 & 21 & 15 \\ 12 & 30 & 13.5 \\ 22.5 & 30 & 18 \end{bmatrix}$$

3. Let $A = \begin{bmatrix} 1 & 2 \\ 3 & -1 \end{bmatrix}$,

$B = \begin{bmatrix} 0 & -1 \\ 1 & 2 \end{bmatrix}$;

$AB = \begin{bmatrix} 2 & 3 \\ -1 & -5 \end{bmatrix}$,

$BA = \begin{bmatrix} -3 & 1 \\ 7 & 0 \end{bmatrix}$

Chapter Summary

1. A *matrix* is a rectangular array of numbers. These numbers are the *elements* of the matrix. A matrix with m rows and n columns has *dimension* $m \times n$. If you interchange the rows and columns of a matrix, you get the *transpose* of the given matrix. The matrix cA is the result of multiplying each element in a matrix A by a real number c. This operation is called *scalar multiplication*.

2. Two matrices having the same dimensions can be *matrix added* (or *subtracted*) by finding the sums (or differences) of the corresponding elements.

3. The *matrix product* of $A_{m \times n}$ and $B_{n \times p}$ is an $m \times p$ matrix whose element in the ith row and jth column is the sum of the products of the corresponding elements of the ith row of A and the jth column of B. Matrices A and B can be multiplied only if the number of columns of A equals the number of rows of B.

4. Matrix properties (pages 530–532) can be used to solve matrix equations. Matrices can also be used to solve systems of linear equations (see page 533).

5. A^{-1} is called the *multiplicative inverse* of A. If $A = \begin{bmatrix} a & b \\ c & d \end{bmatrix}$, then $A^{-1} = \frac{1}{|A|} \begin{bmatrix} d & -b \\ -c & a \end{bmatrix}$, where $|A| = ad - bc$. If $|A| \neq 0$, then A^{-1} exists.

6. Matrices can be used to model a variety of problems. Applications include communication networks, prediction problems, and motion of graphics.

Key vocabulary and ideas

matrix, elements of a matrix (p. 517)
dimension of a matrix (p. 517)
transpose of a matrix (p. 517)
scalar multiplication (p. 518)
matrix addition and subtraction (p. 518)

matrix product (p. 524)
multiplicative inverse (p. 531)
communication matrix (p. 538)
transition matrix (p. 543)
transformation matrix (p. 552)

Chapter Test

1. A store sells Mama, Papa, and Baby Bears, each in small, medium, and large sizes. One week, the store sold 7 small, 14 medium, and 10 large Mama Bears. It also sold 8 small, 20 medium, and 9 large Papa Bears and 15 small, 20 medium, and 12 large Baby Bears.   14-1
   a. Express this information in a sales matrix S.
   b. Next week, the store hopes to sell 50% more of each type of bear. Write a sales matrix that reflects the projected increase in sales.

2. If $A = \begin{bmatrix} 1 & 2 & 0 \\ -4 & 1 & 5 \end{bmatrix}$ and $B = \begin{bmatrix} -3 & 4 & 6 \\ -2 & -1 & 2 \end{bmatrix}$, find $2A - B$. $\begin{bmatrix} 5 & 0 & -6 \\ -6 & 3 & 8 \end{bmatrix}$

3. **Writing** Give two 2×2 matrices A and B such that $AB \neq BA$. Use A 14-2
   and B to explain why matrix multiplication is not commutative.

4. Buddy's Nut Shop sells two varieties of mixed nuts. The mixes contain different amounts of peanuts, pecans, cashews, and walnuts. Matrix P below gives the portion of the total weight of each mix for each type of nut. The owner plans to change the mixes soon, and so decides to combine what is left of the two mixes and sell this combined mixture. Matrix N gives the current amounts for each mix.

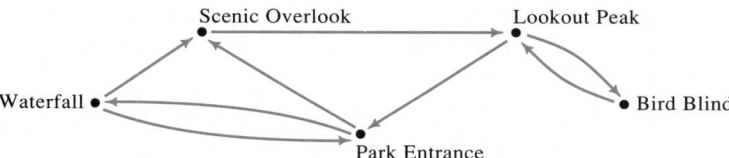

$$\begin{array}{c} \\ \text{Regular Mix} \\ \text{Deluxe Mix} \end{array} \begin{array}{cccc} \text{pt} & \text{pn} & \text{c} & \text{w} \\ \begin{bmatrix} 40\% & 25\% & 15\% & 20\% \\ 25\% & 25\% & 25\% & 25\% \end{bmatrix} \end{array} = P$$

$$\begin{array}{c} \text{Regular} \quad \text{Deluxe} \\ \text{weight in lbs } [\quad 12 \qquad 4 \quad] = N \end{array}$$

a. Which matrix is defined, PN or NP? Find this matrix. *NP; NP* = [5.8 4 2.8 3.4]
b. What is the amount of cashews in the combined mixture? 2.8 lbs

5. Solve this system using matrices: $3x + 5y = -10$ **14-3**
$\qquad\qquad\qquad\qquad\qquad 2x - 7y = 45$ *x* = 5, *y* = −5

6. The diagram below indicates how various sites in a park are connected by roads. **14-4**

Scenic Overlook • Lookout Peak •

Waterfall • • Bird Blind

• Park Entrance

a. Write the matrix M that models this network. Label the rows and columns in alphabetical order.
b. Find M^2. Which site has the greatest number of two-step paths to the other sites?

7. a. A survey done by a certain city has found that from one year to the next, 8% of those living in the city will move to its suburbs while 3% of those living in its suburbs will move into the city. Write a transition matrix T that describes this situation. **14-5**

$$\begin{array}{c} \\ \text{c} \\ \text{s} \end{array} \begin{array}{cc} \text{c} & \text{s} \\ \begin{bmatrix} 0.92 & 0.08 \\ 0.03 & 0.97 \end{bmatrix} \end{array}$$

b. If 60% of the population is now in the city and 40% is in the suburbs, write the current population matrix M_0. [0.6 0.4]
c. Find the percentages of the population living in the city and in the suburbs one year from now. 56.4% in the city, 43.6% in the suburbs

8. Consider the transformation $T: (x, y) \rightarrow (2x + y, x - y)$. **14-6**
a. Draw $\triangle PQR$ with vertices $P(0, -1)$, $Q(1, 3)$, and $R(-2, 2)$.
b. Draw $\triangle P'Q'R'$, the image of $\triangle PQR$ under the transformation T.
c. Find the transformation matrix T and its determinant. Explain what this value tells you about the relationship between $\triangle PQR$ and $\triangle P'Q'R'$.

6. a.
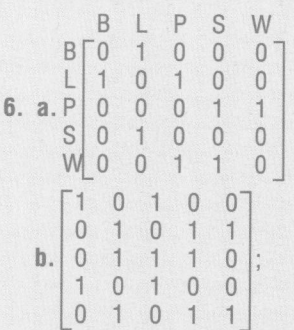

$$\begin{array}{c} \\ \text{B} \\ \text{L} \\ \text{P} \\ \text{S} \\ \text{W} \end{array} \begin{array}{ccccc} \text{B} & \text{L} & \text{P} & \text{S} & \text{W} \\ \begin{bmatrix} 0 & 1 & 0 & 0 & 0 \\ 1 & 0 & 1 & 0 & 0 \\ 0 & 0 & 0 & 1 & 1 \\ 0 & 1 & 0 & 0 & 0 \\ 0 & 0 & 1 & 1 & 0 \end{bmatrix} \end{array}$$

b.
$$\begin{bmatrix} 1 & 0 & 1 & 0 & 0 \\ 0 & 1 & 0 & 1 & 1 \\ 0 & 1 & 1 & 1 & 0 \\ 1 & 0 & 1 & 0 & 0 \\ 0 & 1 & 0 & 1 & 1 \end{bmatrix};$$

Lookout Peak, Park Entrance, Scenic Overlook, and Waterfall each have two two-step paths to other sites.

8. a, b.

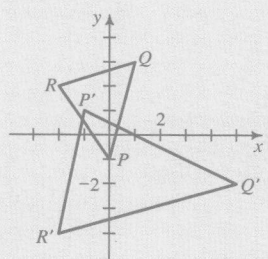

c. $\begin{bmatrix} 2 & 1 \\ 1 & -1 \end{bmatrix}$, -3; area of $\triangle P'Q'R'$ is 3 times the area of $\triangle PQR$, and the triangles have opposite orientations.

1. a.

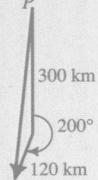

300 km

200°

120 km

2. a. $(-6, 8)$ **b.** $(3, 1)$ **c.** $5\sqrt{2}$

1. A ship travels 300 km south from port and then 120 km on a course of 200°.
   a. Make a diagram that illustrates the result of the ship's sailing.
   b. Find the ship's current distance and course from port. 414.8 km, 185.7°

2. If $\mathbf{u} = (2, -1)$ and $\mathbf{v} = (-3, 4)$, find **(a)** $2\mathbf{v}$, **(b)** $3\mathbf{u} + \mathbf{v}$, and **(c)** $|\mathbf{u} - \mathbf{v}|$. See above.

3. **a.** Find a vector equation of the line through the point $(-2, -4)$ and parallel to the line $(x, y) = (1, 3) + t(-2, 5)$. $(x, y) = (-2, -4) + t(-2, 5)$
   b. What are the corresponding parametric equations for this line? See below.

4. At time t, the position of an object with constant velocity is given by the parametric equations $x = 1 + 2t$ and $y = -1 - 4t$.
   a. What are the velocity and speed of the object? $(2, -4), 2\sqrt{5}$
   b. When and where does it cross the line $2x - y = 11$? $t = 1, (3, -5)$

5. Given the points $A(1, 3)$, $B(-2, -5)$, and $C(6, -4)$, find the measure of $\angle B$ to the nearest tenth of a degree. 62.3°

   $(x, y, z) = (2, 0, 5) + t(-5, -1, -1)$

6. Write vector and parametric equations for the line containing the points $A(2, 0, 5)$ and $B(-3, -1, 4)$. $x = 2 - 5t, y = -t, z = 5 - t$

7. Given the points $A(1, 7, 3)$ and $B(-3, 5, 5)$, find a Cartesian equation of the plane perpendicular to $\overline{AB}$ at its midpoint, M. $2x + y - z = 0$

8. Evaluate: $\begin{vmatrix} 10 & 4 & 3 \\ 0 & 2 & 5 \\ -1 & -7 & 0 \end{vmatrix}$ 336 **9.** Use Cramer's rule to solve: $\begin{aligned} 4x - 5y &= 2 \\ 5x - 7y &= 1 \end{aligned}$ (3, 2)

10. Find a vector perpendicular to the plane containing the points $P(1, 0, 5)$, $Q(-4, 3, 2)$, and $R(1, -1, -4)$. Then find an equation of this plane. See below.

11. State whether each sequence is arithmetic, geometric, or neither, and find a formula for t_n. $t_n = -\dfrac{4}{9}\left(-\dfrac{3}{2}\right)^n$ $t_n = \dfrac{2n}{2n + 1}$ $t_n = n - \dfrac{1}{3}$
    a. $\dfrac{2}{3}, -1, \dfrac{3}{2}, -\dfrac{9}{4}, \ldots$ **b.** $\dfrac{2}{3}, \dfrac{4}{5}, \dfrac{6}{7}, \dfrac{8}{9}, \ldots$ **c.** $\dfrac{2}{3}, \dfrac{5}{3}, \dfrac{8}{3}, \dfrac{11}{3}, \ldots$ Arith.
    Geom. Neither

12. A sequence is defined recursively by $t_1 = -2$ and $t_n = 4 + 2t_{n-1}$.
    a. Find the first five terms of the sequence. $-2, 0, 4, 12, 28$
    b. Find an explicit definition for the sequence. $t_n = 2^n - 4$
    c. Find t_{10}. 1020

13. Find S_{20} for the arithmetic series with $t_1 = 10$ and $t_5 = 2$. -180

14. Find S_6 for the geometric series with $r = \dfrac{1}{3}$ and $t_1 = 81$. $\dfrac{364}{3}$

15. Find the limit of each sequence. If the limit does not exist, say so.
    a. $\dfrac{2}{3}, \dfrac{3}{4}, \dfrac{4}{5}, \dfrac{5}{6}, \ldots$ 1 **b.** $\dfrac{1}{2}, -1, 2, -4, 8, -16, \ldots$
    Does not exist.

16. Evaluate the given limit or state that the limit does not exist. If the sequence approaches ∞ or $-\infty$, say so.
    Does not exist.
    a. $\lim\limits_{n \to \infty} \dfrac{3n^2 + 2n + 7}{n^2 - 9}$ 3 **b.** $\lim\limits_{n \to \infty} \dfrac{5n^3 - 2n}{n^2 + 1}$ ∞ **c.** $\lim\limits_{n \to \infty} [1 + (-1)^n]$

562 *Cumulative Review* **3. b.** $x = -2 - 2t$ **10.** $(6, 9, -1); 6x + 9y - z = 1$
    $y = -4 + 5t$

17. Find the interval of convergence for the geometric series $2 - \dfrac{x}{2} + \dfrac{x^2}{8} - \cdots$, and express the sum of this series in terms of x. See below.

18. Express the repeating decimal $0.345345345 \ldots$ as a rational number. $\dfrac{115}{333}$

19. Express the series $7 + 12 + 17 + \cdots + 102$ using sigma notation, and then evaluate. $\displaystyle\sum_{n=1}^{20} (5n + 2)$; 1090   17. $-4 < x < 4$; $\dfrac{8}{x + 4}$

20. Use mathematical induction to prove that $\displaystyle\sum_{i=1}^{n} \dfrac{1}{(2i - 1)(2i + 1)} = \dfrac{n}{2n + 1}$.

21. Let $A = \begin{bmatrix} -1 & 6 & -2 \\ 3 & 4 & 0 \end{bmatrix}$ and $B = \begin{bmatrix} 2 & -1 \\ -3 & 0 \\ 5 & 4 \end{bmatrix}$. 21. a. $\begin{bmatrix} -4 & 7 \\ 15 & 8 \\ -9 & -4 \end{bmatrix}$

    a. Find $2A^t - B$.

    b. Find AB and BA. Is the matrix A^tB defined? See below.

22. Solve this matrix equation for X: $\begin{bmatrix} 3 & 4 \\ -3 & -3 \end{bmatrix} X - \begin{bmatrix} 2 & -5 \\ 0 & 1 \end{bmatrix} = \begin{bmatrix} 0 & 4 \\ -2 & 1 \end{bmatrix}$

23. Write the system of equations shown at the right as a single matrix equation, and then solve the system by solving the matrix equation.

    $3x + 5y = 4$ $\begin{bmatrix} -2 \\ 2 \end{bmatrix}$
    $x + 6y = 10$

24. a. Write the communication matrix for the communication network shown at the right.

    b. Find the matrix that represents the number of ways messages can be sent from one point to another point using at most one relay.

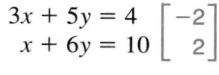

25. A store sells three different brands of granola cereal. Brand A accounts for 20% of the granola purchases, Brand B for 50%, and Brand C for the remaining 30%. The store plans to include a discount coupon for Brand A in its advertising circular for the next week. In the past, the coupon has caused 35% of Brand B buyers and 45% of Brand C buyers to switch to Brand A, with all other purchasers staying with their original brand.

    a. Write a transition matrix that describes this situation.

    b. What will be the market share for each brand after the coupon is run?

    c. If the coupon were to run for a second week, what would be the market share of each brand after the coupon is run again?

26. a. Draw $\triangle XYZ$ with vertices $X(-1, 0)$, $Y(3, 5)$, and $Z(9, 1)$.

    b. Draw $\triangle X'Y'Z'$, the image of $\triangle XYZ$ under the following transformation:
        $T: (x, y) \rightarrow (2x - 3y, x + y)$

    c. Find the transformation matrix T and its determinant. Explain what this value tells you about the relationship between $\triangle XYZ$ and $\triangle X'Y'Z'$.

21. b. $\begin{bmatrix} -30 & -7 \\ -6 & -3 \end{bmatrix}$, $\begin{bmatrix} -5 & 8 & -4 \\ 3 & -18 & 6 \\ 7 & 46 & -10 \end{bmatrix}$; no

22. $\begin{bmatrix} 2 & -\dfrac{5}{3} \\ \dfrac{2}{3} & 1 \\ 0 & 1 \end{bmatrix}$

24. a.

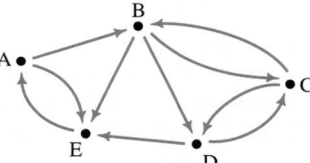

	A	B	C	D	E
A	0	1	0	0	1
B	0	0	1	1	1
C	0	1	0	1	0
D	0	0	1	0	1
E	1	0	0	0	0

    b. $\begin{bmatrix} 1 & 1 & 1 & 1 & 2 \\ 1 & 1 & 2 & 2 & 2 \\ 0 & 1 & 2 & 2 & 2 \\ 1 & 1 & 1 & 1 & 1 \\ 1 & 1 & 0 & 0 & 1 \end{bmatrix}$

25. a.

	A	B	C
A	1	0	0
B	0.35	0.65	0
C	0.45	0	0.55

    b. 51%, 32.5%, 16.5%

    c. 69.8%, 21.1%, 9.1%

26. a, b.

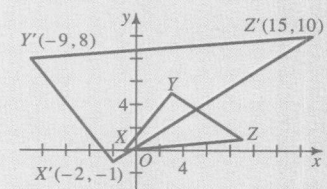

    c. $T = \begin{bmatrix} 2 & -3 \\ 1 & 1 \end{bmatrix}$; $|T| = 5$; the area of $\triangle X'Y'Z'$ is 5 times the area of $\triangle XYZ$. Since $|T| > 0$, $\triangle XYZ$ and $\triangle X'Y'Z'$ have the same orientation.

Chapter 15 Combinatorics

Overview

The ability to calculate combinations and permutations of objects provides a foundation useful in the study of probability, statistics, computer science, and many other fields. This chapter carefully builds the concepts needed to answer the question "In how many ways can you ...?" in a variety of contexts.

The first section introduces Venn diagrams, which students use to determine the number of elements in the union and intersection of finite sets. The next three sections present various principles (the multiplication, addition, and complement principles) and formulas (for combinations and permutations, including permutations with repetition and circular permutations) that students use to solve counting problems. The last section introduces Pascal's triangle and the binomial theorem, which students use to find coefficients when expanding a binomial raised to a whole-number power.

Objectives

15-1 To use Venn diagrams to illustrate intersections and unions of sets and to use the inclusion-exclusion principle of solve counting problems involving intersections and unions of sets.

15-2 To use the multiplication, addition, and complement principles to solve counting problems.

15-3 To solve problems involving permutations and combinations.

15-4 To solve counting problems that involve permutations with repetition and circular permutations.

15-5 To use the binomial theorem and Pascal's triangle.

Supplementary Resources

Tests, pp. 48–49
Alt. Assess., pp. 46–48, 70
Activities, pp. 41–43
Student Res. Guide, pp. 140–147

Pacing Guide

Section	15-1	15-2	15-3	15-4	15-5	Review	Test	Total
Discrete Math (days)	1	2	2	3	2	1	1	12

Teaching Notes

Presenting the Section

You might wish to introduce this section by developing a Venn diagram showing the numbers of students in the class with two different attributes. Using the class as the universal set, draw a single circle to represent the girls in the class and enter the appropriate numbers for this set and its complement. It would be useful to point out that a complement, in this case "non-girls," may have its own label, in this case "boys." Next consider a second attribute, such as taking a science course, and draw a second circle that overlaps the first. Point out that the numbers used in the original Venn diagram must now be "broken down"; by shows of hands, students can indicate to which subset they belong, and you can add the numbers to the diagram.

A second example that leads to an understanding of the inclusion-exclusion principle might be to ask, "How many positive integers less than 200 are divisible by 3 or 5?" A student might propose that the 66 multiples of 3 be added to the 39 multiples of 5, forgetting that this counts some numbers (the thirteen multiples of 15) twice. This example also demonstrates that "brute force" (listing all the numbers), though tempting, is usually an inefficient way to solve problems.

Review

Students do not always understand that the use of an overbar for complementation implies grouping. You should show students that $\overline{A \cup B}$ means the complement of $(A \cup B)$. Also illustrate the difference between $\overline{A \cap B}$ and $\overline{A} \cap \overline{B}$ by drawing a shaded Venn diagram for each.

Cooperative Learning

Ask students, working in groups of four, to define two sets and construct a Venn diagram showing the sets so that each of the four regions of the diagram contains exactly one member of the group. (Example: Set A might be students with an older brother, and set B might be students with a pet cat.)

Presenting the Section

The following example may help motivate a discussion of the concepts in this section:

In 1990 the government of Hungary became concerned that the country was running out of possible automobile license-plate numbers. Hungarian license plates were therefore changed from 2 letters followed by a four-digit number to 3 letters followed by a three-digit number. How many more plates are available under the new scheme? (An increase from 6,760,000 to 17,576,000 plates)

To illustrate the rate at which the number of ways of arranging items grows as the number of items grows, ask two students to come to the front of the class and to line up in as many ways as possible. Next ask a third student to join them and have them repeat the activity. If you wish, repeat with a fourth student. Next ask students to estimate how long it would take the entire class to line up in every possible arrangement. (*Note*: Even if the arrangement of students could be changed every second, it would take over a year to arrange 11 students. A class of 25 would take 5×10^{17} years!)

Application

Problems using the multiplication principle can sometimes be clarified by writing a "decision algorithm." For example, consider the following problem:

Four different car companies each manufacture a sedan, a station wagon, and a sports car. Mr. and Mrs. Jackson each have a different car, but both are manufactured by the same company. How many possibilities are there?

In this case, the decision algorithm involves three decisions:

1. Select the company: 4 ways.
2. Select one of the cars: 3 ways.
3. Select the other car: 2 ways.

This gives $4 \times 3 \times 2 = 24$ possibilities.

Presenting the Section

Central to this section is the distinction between a permutation of items and a combination of items. Students should be encouraged to write down one specific arrangement for a given problem, form a new arrangement using the same items as the first, and then decide whether the new arrangement should be considered different from the first one. (The decision, of course, will be based on whether the *order* of the items is important.)

Using Technology

Have students evaluate $_{100}P_3$ by using a factorial key on a calculator. They will most likely find that evaluating $\frac{100!}{97!}$ results in an error message. Point out that they should approach a problem like this by rewriting it as follows:

$$\frac{100!}{97!} = \frac{100 \cdot 99 \cdot 98 \cdot 97!}{97!} = 100 \cdot 99 \cdot 98$$

Assessment

Divide the class into small groups and distribute a list of situations which they are to identify as permutations or combinations. Do not hesitate to include ambiguous situations. Some possibilities are:

1. Selecting three toppings for ice cream.
2. Selecting the starting players for a basketball game.
3. Selecting three students to call on for answers.
4. Selecting your two favorite musical groups.
5. Selecting four toppings for a hot dog.
6. Selecting four ingredients for a soup.
7. Selecting three officers for a club.
8. Selecting three colors for tie-dyeing fabric.
9. Selecting three equation transformations to be used to solve $4x^2 + 16 = 64$.
10. Selecting three students to speak at graduation.

Using Technology

The BASIC program given at the top of the next column prints all permutations of the first n positive integers taken two at a time. Have students enter and run this program on a computer or programmable calculator.

```
10   INPUT "ENTER A POSITIVE INTEGER: ";
     N
20   FOR A = 1 TO N
30   FOR B = 1 TO N
40   IF A = B THEN GOTO 60
50   PRINT A, B
60   NEXT B
70   NEXT A
80   END
```

To convert the program to one that gives combinations rather than permutations, have students delete line 40 and replace line 30 with the following:

```
30   FOR B = 1 TO A − 1
```

Students interested in programming might be asked to modify the two programs to give permutations and combinations of the first n positive integers taken three at a time.

Presenting the Section

You might begin this section with a reminder that deciding whether we are counting permutations or combinations depends on what we consider to be "different" results.

To illustrate the effect of repetitions, present the following problem:

Cash prizes are contained in a red envelope, a blue envelope, and a yellow envelope. In how many ways can the envelopes be distributed to four people? ($_4P_3 = 24$)

Now tell students to suppose that the red envelope contains $100, while each of the other two contains $20. We can therefore consider the solution "red to first person, blue to second, and yellow to third" to be the same as "red to first person, yellow to second, and blue to third." Ask students how many different possibilities there are now. (12)

As a second example, have students list on the board all permutations of the letters of BIG. Now ask for a way to easily get all permutations of the letters of BIT. Students should suggest simply replacing all G's with T's. Do so. Now consider permutations of BIB. Write next to each permutation of BIT the result of replacing all T's with B's and point out that pairs of permutations are duplicates. Finally, consider the permutations of BBB by replacing I's with B's and note that there is just one permutation that occurs 6 times. This should lead to the rule for permutations with repetition.

For circular permutations you might consider this problem:

By using the repeat button on my carousel CD player, I can play three CD's in order over and over. Since I will start my CD player well before my party begins, it doesn't matter which CD plays first; only the order of the CD's matters. What different arrangements are possible for the three CD's X, Y, and Z?

As students suggest answers, write them on the board as long repeated strings; for example, the suggestion "XYZ" should be written as:

"$\ldots X Y Z X Y Z X Y \ldots$"

This should help students see that YZX and ZXY are not new arrangements.

Using Technology

Many computer programs include loops that repeat the same steps over and over. Consider the following lines from a program designed to generate a table of factorials:

a: PRINT N; "! = "; F
b: LET N = N + 1
c: LET F = F * N

Suppose these 3 lines are to be inserted into the following program:

```
10    LET F = 1
20    LET N = 1
30    REM BEGIN LOOP
40    ?
50    ?
60    ?
70    IF N < 6 THEN GOTO 30
80    END
```

Have students use a computer or programmable calculator to complete the tasks listed at the top of the next column.

1. Show the results of all 6 possible permutations of statements a, b, c.

a-b-c	a-c-b	b-a-c
$1! = 1$	$1! = 1$	$2! = 1$
$2! = 2$	$2! = 1$	$3! = 2$
$3! = 6$	$3! = 2$	$4! = 6$
$4! = 24$	$4! = 6$	$5! = 24$
$5! = 120$	$5! = 24$	$6! = 120$

b-c-a	c-a-b	c-b-a
$2! = 2$	$1! = 1$	$2! = 1$
$3! = 6$	$2! = 2$	$3! = 2$
$4! = 24$	$3! = 6$	$4! = 6$
$5! = 120$	$4! = 24$	$5! = 24$
$6! = 720$	$5! = 120$	$6! = 120$

2. Identify which permutations produce correct results. How are the permutations related? (The permutations a-b-c, b-c-a, and c-a-b produce correct results; they represent the same circular permutation (as do a-c-b, b-a-c, and c-b-a, which produce incorrect results).)

15-5   pages 590–594

Presenting the Section

This section introduces two methods for finding the coefficients when expanding a power of a binomial: Pascal's triangle and calculation of the number of combinations possible to obtain each term. A good introductory exercise for this section would be to ask students to multiply three or four binomials of the form $(x + a)$ together without recording any intermediate results. Earlier work with the sum and product of roots (see Section 2-7) can be used to obtain the coefficients of the second and last terms, but what about the others? For example, ask students to consider the product $(x + 2)(x + 3)(x + 1)$ and to determine the coefficient of x^2 ($2 + 3 + 1 = 6$) and the coefficient of x ($2 \cdot 3 + 2 \cdot 1 + 3 \cdot 1 = 11$).

Application

A solid understanding of the binomial theorem and its derivation will prove invaluable to students when they use the theorem to find binomial probabilities in the next chapter.

564

15-1 Venn Diagrams

Objective | *To use Venn diagrams to illustrate intersections and unions of sets and to use the inclusion-exclusion principle to solve counting problems involving intersections and unions of sets.*

In this chapter we investigate the theory of counting, more formally known as **combinatorics**. Although counting may seem to be a very simple activity—one that merely involves the pairing of the natural numbers 1, 2, 3, . . . with the objects in some set—the task can be challenging when the objects are numerous. Shortcuts in counting are therefore an important part of combinatorics.

Before we can get started counting, we must know precisely what we are counting. We can use sets to separate the objects to be counted from all others, and we can use a *Venn diagram* to illustrate those sets.

When drawing a Venn diagram, we begin with a rectangle, which represents a *universal set U*. Inside the rectangle we draw circles to represent *subsets* of the universal set. For example, consider the Venn diagram at the right. Here we have a universal set of 1000 typical Americans. The set *A* represents those who have a certain protein called antigen A on the surface of their red blood cells. Likewise, the set *B* represents those who have antigen B. Of our 1000 typical Americans, a total of 450 are in set *A* and a total of 140 are in set *B*.

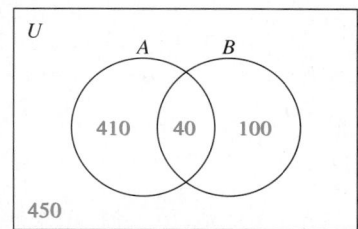

Notice that sets *A* and *B* overlap. This represents the fact that 40 of our 1000 typical Americans have *both* antigen A and antigen B on the surface of their red blood cells. We call the set of elements that any two sets *A* and *B* have in common the **intersection** of *A* and *B*, which we denote by $A \cap B$. It is the shaded region in the Venn diagram at the left below.

The set of people who have *either* antigen A or antigen B represents the **union** of sets *A* and *B*, which we denote by $A \cup B$. It is the shaded region in the Venn diagram at the right below.

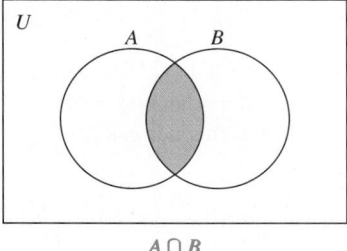

$A \cap B$

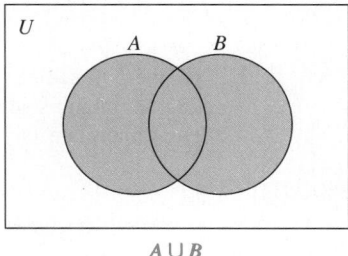

$A \cup B$

◀ How many license plates can be made using three letters followed by three digits? In this chapter you will learn how to find the answer—17,576,000.

Combinatorics **565**

Warm-Up Exercises

For Exercises 1-5, consider the integers from 1 to 10 inclusive.

1. List all the integers that are:
   a. prime. 2, 3, 5, 7
   b. odd. 1, 3, 5, 7, 9
   c. factors of 6. 1, 2, 3, 6

2. List each integer that is both prime and odd. 3, 5, and 7

3. List each integer that is prime, odd, and a factor of 6. 3

4. List each integer that is *not* prime, *not* odd, and *not* a factor of 6. 4, 8, 10

5. List each integer that is both odd and a factor of 6, but is not prime. 1

Motivating the Section

A consumer-advocate group in Wisconsin wants to direct a mailing to homeowners in the state who heat their homes with natural gas. The shaded region in the diagram below indicates the portion of the state's population to whom the mailing will be sent. In the diagram, *R* represents residents of Wisconsin, *H* represents homeowners in Wisconsin, and *N* represents people in Wisconsin whose homes or apartments are heated by natural gas.

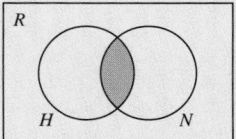

When discussing the Venn diagrams on the preceding page, be sure students understand that sets *A* and *B* are defined in terms of *antigens*, not *blood types*. Thus, there is a total of 450 people who have antigen A, and of these, 410 have "type A" blood. Ask students to make a similar statement regarding antigen B and "type B" blood. (A total of 140 people have antigen B, and of these, 100 have "type B" blood.)

Making Connections

Venn diagrams can also be used to illustrate logical statements (see Lewis Caroll's *Symbolic Logic*). Ask students to draw a Venn diagram to illustrate each of the following statements:

1. All A's are B's. (Circle for A is completely inside circle for B.)

2. No A's are B's. (Circle for A is completely outside circle for B.)

3. Some A's are B's. (Circle for A partially overlaps circle for B.)

Assessment Note

In discussing the inclusion-exclusion principle, ask students for $n(A \cup B)$ if A and B are disjoint sets, that is, if $A \cap B = \emptyset$. ($n(A \cup B) = n(A) + n(B)$ since $n(A \cap B) = 0$.)

The set of all elements *not* in a set A is called the **complement** of A and is denoted $\overline{A}$. In the union diagram on the right at the bottom of the preceding page, the unshaded region inside the rectangle represents $\overline{A \cup B}$, the complement of $A \cup B$. This complement consists of people who have *neither* antigen A nor antigen B on the surface of their red blood cells. (Such people are said to have "type O" blood. Moreover, those with antigen A but not antigen B have "type A" blood, those with antigen B but not antigen A have "type B" blood, and those with both antigen A and antigen B have "type AB" blood.)

Look again at the first Venn diagram shown on the preceding page. If we let $n(A \cup B)$ designate the number of elements in the union of sets A and B, we see that $n(A \cup B) = 550$. Notice that

$$n(A \cup B) \neq n(A) + n(B)$$

(that is, $550 \neq 450 + 140$), because the number of people in $A \cap B$ is counted twice when $n(A)$ and $n(B)$ are added. To compensate, we must subtract $n(A \cap B)$ once from the sum:

$$n(A \cup B) = n(A) + n(B) - n(A \cap B)$$
$$550 = 450 + 140 - 40$$

This gives us the following counting principle.

The Inclusion-Exclusion Principle

For any sets A and B, $n(A \cup B) = n(A) + n(B) - n(A \cap B)$.

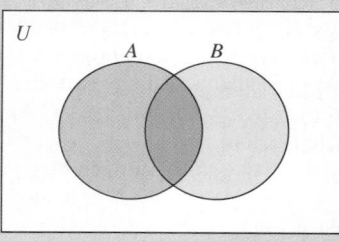

Example 1 Of the 540 seniors at Central High School, 335 are taking mathematics, 287 are taking science, and 220 are taking both mathematics and science. How many are taking neither mathematics nor science?

Solution Let U = the universal set of seniors at Central High School,

M = the subset of seniors taking mathematics,

and S = the subset of seniors taking science.

These sets and the number of people in each are shown in the Venn diagram at the top of the next page.

Using the inclusion-exclusion principle, we have:

$$n(M \cup S) = n(M) + n(S) - n(M \cap S)$$
$$= 335 + 287 - 220$$
$$= 402$$

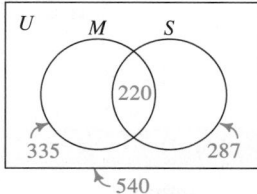

Thus, the number of seniors *not* taking mathematics or science is:

$$n(\overline{M \cup S}) = 540 - 402 = 138$$

Example 2

In a survey, 113 business executives were asked if they regularly read the *Wall Street Journal*, *Business Week* magazine, and *Time* magazine. The results of the survey are as follows:

88 read the *Journal*.   6 read only the *Journal*.
76 read *Business Week*.   5 read only *Business Week*.
85 read *Time*.   8 read only *Time*.
42 read all three.

How many read none of the three publications?

Solution

Letting *J*, *B*, and *T* represent the sets of *Journal*, *Business Week*, and *Time* readers, we draw a Venn diagram and place the numbers given above in the diagram, as shown at the right. (Be sure you understand the placement of each number in the diagram.) To find the numbers for the remaining regions of the diagram, we let *a* represent the number of executives who read just the *Journal* and *Business Week*, *b* represent the number who read just *Business Week* and *Time*, and *c* represent the number who read just the *Journal* and *Time*. We can then write the following system of equations:

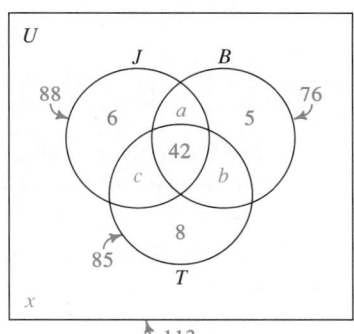

$$88 = 6 + 42 + a + c, \quad \text{or} \quad 40 = a + c$$
$$76 = 5 + 42 + a + b, \quad \text{or} \quad 29 = a + b$$
$$85 = 8 + 42 + b + c, \quad \text{or} \quad 35 = b + c$$

Solving this system gives $a = 17$, $b = 12$, and $c = 23$. The number x, representing those who read none of the three publications, can be found by adding the numbers within the three circles and subtracting the sum from 113. This gives $x = 0$, which means that every one of the business executives surveyed reads at least one of the three publications.

Example Note

Students may need help reading the numbers in a Venn diagram: A number by itself inside a region that has *not* been subdivided indicates how many objects are in that region only (for example, the "42" in Example 2 refers to $n(J \cap B \cap T)$); a number with an arrow pointing to a region that *has* been subdivided indicates the total number of objects in all of the subdivisions of the region (for example, the "88" in Example 2 refers to $n(J)$).

Suggested Assignments

Discrete Math
568/1, 7, 9, 13–19 odd

Supplementary Materials

Alternative Assessment, 46

Additional Examples

1. For each of the following, copy the Venn diagram and shade the specified set.

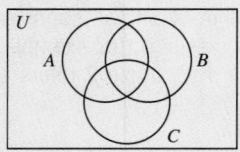

a. $\overline{A \cap B}$
b. $A \cap (B \cup C)$
c. $B \cup (\overline{A} \cap \overline{C})$

a.
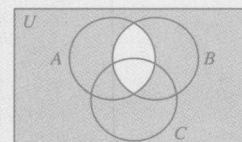

b.
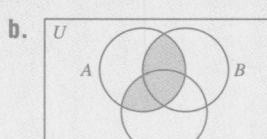

c.
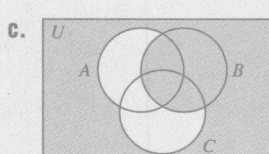

In Example 2 we found that the set $\overline{J \cup B \cup T}$ has no elements in it. Such a set is called the **empty set** and is denoted by the Greek letter ϕ (phi). Thus, $\overline{J \cup B \cup T} = \phi$.

CLASS EXERCISES

A certain small college has 1000 students. Let $F =$ the set of college freshmen, and let $M =$ the set of music majors. These sets are shown in the Venn diagram at the right. Describe each of the following sets in words and tell how many members it has.

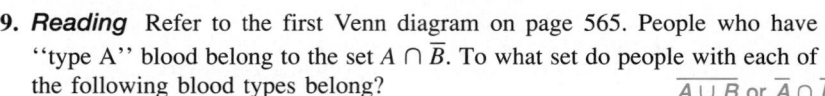

1. $F \cap M$ 15
2. $F \cup M$ 350
3. $\overline{F}$ 700
4. $\overline{M}$ 935
5. $\overline{F} \cap M$ 50
6. $F \cap \overline{M}$ 285
7. $\overline{F \cup M}$ 650
8. $F \cup \overline{M}$ 950

9. **Reading** Refer to the first Venn diagram on page 565. People who have "type A" blood belong to the set $A \cap \overline{B}$. To what set do people with each of the following blood types belong?
   a. B $B \cap \overline{A}$
   b. AB $A \cap B$
   c. O $\overline{A \cup B}$ or $\overline{A} \cap \overline{B}$

10. If A is any subset of a universal set U, complete the following.
    a. $A \cup \phi = \underline{\ ?\ }$ A
    b. $A \cap \phi = \underline{\ ?\ }$ ϕ
    c. $A \cup \overline{A} = \underline{\ ?\ }$ U
    d. $A \cap \overline{A} = \underline{\ ?\ }$ ϕ
    e. $A \cup U = \underline{\ ?\ }$ U
    f. $A \cap U = \underline{\ ?\ }$ A

WRITTEN EXERCISES

For Exercises 1–4, draw a Venn diagram and shade the region representing the set given in part (a). Then draw a separate Venn diagram and shade the region representing the set given in part (b).

 1. **a.** $P \cap Q$
   b. $P \cup Q$
2. **a.** $\overline{P} \cap Q$
   b. $P \cap \overline{Q}$
3. **a.** $\overline{P \cup Q}$
   b. $\overline{P} \cap \overline{Q}$
4. **a.** $\overline{P \cap Q}$
   b. $\overline{P} \cup \overline{Q}$

Let $U =$ the universal set of all teachers in your school. Let the subsets of mathematics teachers, biology teachers, physics teachers, and chemistry teachers be represented by M, B, P, and C, respectively. Describe in words each of the following sets, and name a teacher belonging to each set if such a teacher exists in your school.

5. **a.** $M \cup P$ **b.** $M \cap \overline{P}$
6. **a.** $P \cap C$ **b.** $\overline{P \cup C}$
7. **a.** $B \cup (P \cap C)$ **b.** $(B \cup P) \cap (B \cap C)$
8. **a.** $(B \cup C) \cap M$ **b.** $(B \cap M) \cup (C \cap M)$

9. a. $\overline{M \cup P \cup C}$ **b.** $\overline{M} \cap \overline{P} \cap \overline{C}$
10. a. $\overline{M \cap B \cap P \cap C}$ **b.** $\overline{M \cup B \cup P \cup C}$

In Exercises 11–14, draw a Venn diagram to illustrate each situation described. Then use the diagram to answer the question asked.

11. In an election-day survey of 100 voters leaving the polls, 52 said they voted for Proposition 1, and 38 said they voted for Proposition 2. If 18 said they voted for both, how many voted for neither? 28

12. Although the weather was perfect for the beach party, 17 of the 30 people attending got a sunburn and 25 people were bitten by mosquitoes. If 12 people were both bitten and sunburned, how many had neither affliction? 0

13. In a survey of 48 high school students, 20 liked classical music and 16 liked bluegrass music. Twenty students said they didn't like either. How many liked classical but not bluegrass? 12

14. Of the 52 teachers at Roosevelt High School, 27 said they like to go sailing, 25 said they like to go fishing, and 12 said they don't enjoy either recreation. How many enjoy fishing but not sailing? 13

B **15. Astronomy** Consider the sets defined below. (You may need to consult an encyclopedia to determine the elements of each set.)

> Let U = the universal set of planets in our solar system,
> S = the subset of planets smaller than Earth,
> and F = the subset of planets farther from the Sun than Earth.

a. Draw a Venn diagram with overlapping circles representing S and F inside a rectangular region U. Inside each of the four regions of your diagram, list the planets described by that region.

b. The smallest set to which the planet Venus belongs is $S \cap \overline{F}$. What is the smallest set to which the planet Uranus belongs? $\overline{S} \cap F$

16. Geography Consider the sets defined below.

> Let U = the universal set of states in the United States,
> P = the subset of states bordering the Pacific Ocean,
> and M = the subset of states bordering Mexico.

a. Name all the states in the set $\overline{P} \cap M$. Arizona, New Mexico, Texas
b. What is the only state in the set $P \cap M$? California
c. How many states are in the set $\overline{P \cup M}$? 42

---

2. Of the first 100 positive integers, 25 are prime, 9 are factors of 100, and 68 are neither prime nor a factor of 100. How many are:

a. both prime and a factor of 100?

b. a factor of 100 but not prime?

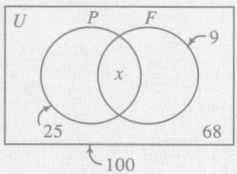

$n(P \cup F) =$
$n(P) + n(F) -$
$n(P \cap F) =$
$25 + 9 - x;$
$(34 - x) + 68 = 100;$
$x = 2$

a. Two numbers (2 and 5) are both prime and factors of 100.

b. Seven (9 − 2) numbers are factors of 100 but not prime.

Additional Answers Written Exercises

15. a. $\overline{S} \cap \overline{F}$ = {Earth}
$S \cap \overline{F}$ ={Mercury, Venus}
$S \cap F$ = {Mars, Pluto}
$\overline{S} \cap F$ = {Jupiter, Saturn, Uranus, Neptune}

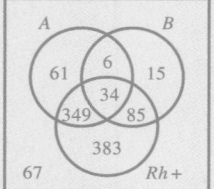
17. Geography Consider the sets defined below. (You may need to consult an encyclopedia or atlas to determine the elements of each set.)

Let U = the universal set of 13 countries in South America,
A = the subset of South American countries bordering the Atlantic Ocean or the Caribbean Sea,
P = the subset of South American countries bordering the Pacific Ocean,
and R = the subset of countries through which the Amazon River runs.

a. Name two countries that belong to $\overline{A} \cap \overline{P} \cap \overline{R}$. Bolivia, Paraguay

b. Draw a Venn diagram with overlapping circles representing A, P, and R all inside a rectangular region U. Inside each of the eight regions of your diagram, list the South American countries described by that region.

18. Medical Science In addition to antigens A and B, a red blood cell may have a protein called the Rhesus factor on its surface. If the protein is present, the blood is said to be "Rh positive"; if not, the blood is "Rh negative." About 85% of the American population is "Rh positive." Using this information, redraw the first Venn diagram on page 565 and introduce a third set, $Rh+$, that overlaps sets A and B. Then indicate the approximate number of people who belong to each of the eight regions of the new Venn diagram.

19. In a parking lot containing 85 cars, there are 45 cars with automatic transmissions, 43 cars with rear-wheel drive, and 46 cars with four-cylinder engines. Of the cars with automatic transmissions, 26 also have rear-wheel drive. Of the cars with rear-wheel drive, 29 also have four-cylinder engines. Of the cars with four-cylinder engines, 27 also have automatic transmissions. There are 21 cars with all three features.

a. How many cars do not have automatic transmissions and rear-wheel drive but do have four-cylinder engines? 11

b. How many cars do not have any of the three features? 12

20. Of the 415 girls at Gorham High School last year, 100 played fall sports, 98 played winter sports, and 96 played spring sports. Twenty-two girls played sports all three seasons while 40 played only in fall, 47 only in winter, and 33 only in spring.

a. How many girls played fall and winter sports but not a spring sport? 13

b. How many girls did not play sports in any of the three seasons? 219

570 *Chapter Fifteen*

21. For a universal set U, what is $\overline{U}$? ϕ

22. If A is a subset of a universal set U, find $\overline{\overline{A}}$. A

C **23.** When we say that multiplication is distributive over addition, we mean that for all real numbers a, b, and c,

$$a(b + c) = ab + ac.$$

Similarly, when we say that intersection is distributive over union, we mean that for all sets A, B, and C,

$$A \cap (B \cup C) = (A \cap B) \cup (A \cap C).$$

Show that this equation is true by drawing two three-circle Venn diagrams. In one diagram shade the region corresponding to $A \cap (B \cup C)$, and in the other shade the region corresponding to $(A \cap B) \cup (A \cap C)$.

24. Use the method of Exercise 23 to determine whether union distributes over intersection. Yes, union distributes over intersection.

25. a. One of De Morgan's laws states that $\overline{A \cup B} = \overline{A} \cap \overline{B}$. Use the method of Exercise 23 to verify this law.

b. Another of De Morgan's laws writes $\overline{A \cap B}$ in terms of $\overline{A}$ and $\overline{B}$. Use Venn diagrams to determine what this law is. $\overline{A \cap B} = \overline{A} \cup \overline{B}$

c. Use parts (a) and (b) to write $\overline{A \cap (B \cup C)}$ in terms of $\overline{A}$, $\overline{B}$, and $\overline{C}$. $\overline{A} \cup (\overline{B} \cap \overline{C})$

26. Simplify each of the following by using the properties of intersection, union, and complementation from Exercises 23–25.

a. $A \cap (\overline{A} \cup B)$ $A \cap B$ **b.** $A \cup (\overline{A} \cup B)$ $A \cup \overline{B}$

27. Extend the inclusion-exclusion principle to three sets. (*Hint*: Even though $n(A \cup B \cup C) \neq n(A) + n(B) + n(C)$, you can alter the right side of this inequality to obtain equality.) $n(A \cup B \cup C) =$
$n(A) + n(B) + n(C) - n(B \cap C) - n(A \cap B) - n(A \cap C) + n(A \cap B \cap C)$

15-2 The Multiplication, Addition, and Complement Principles

Objective   *To use the multiplication, addition, and complement principles to solve counting problems.*

In Section 15-1 we saw that Venn diagrams and the inclusion-exclusion principle are helpful in certain counting problems. In this section we consider three other counting principles.

The Multiplication Principle

If an action can be performed in n_1 ways, and for each of these ways another action can be performed in n_2 ways, then the two actions can be performed together in $n_1 n_2$ ways.

Teaching Notes, p. 564B

Warm-Up Exercises

Seven sculptures were included in an art display. Alex, Beth, Chen-Fa, and Donna were the North High School artists, and José, Kim, and Lois were the South High School artists.

1. In how many ways can a best-of-show prize be awarded? 7

2. a. List the different ways in which exactly one student from each school could be awarded a prize. *AJ, AK, AL, BJ, BK, BL, CJ, CK, CL, DJ, DK, DL*

b. Describe a way to determine the number of possibilities found in Ex. 2. Multiply the number of students from North H.S., 4, by the number from South H.S., 3.

3. If the three sculptures created by the South High School students are lined up, list all the possible arrangements. *JKL, JLK, KJL, KLJ, LJK, LKJ*

Motivating the Section

Ask students if they can predict which offers more choices for license plates:

1. a plate with three different letters of the alphabet in any order

2. a plate with four different nonzero digits in any order

(The plate with three different letters offers 15,600 choices; the plate with four different digits offers 3024 choices.)

Students sometimes object to the use of the multiplication principle in Example 2, for they argue that the selection of the first person in line influences the choices available for the next position. For example, if Tom is first in line, he can't be second; while if Jill is first, then Tom *can* be second. Be sure students understand that the multiplication principle relies on each choice leaving the same *number* of choices for the next step, not necessarily the same specific choices.

Mathematical Note

You may wish to introduce an alternate notation for *n*!:

$$\text{For } n \geq 1, \; n! = \prod_{i=1}^{n} i$$

Note that Π (pi) stands for "product" just as Σ (sigma) stands for "sum."

⌧ Using Technology

Many calculators have a factorial key. Students may gain an appreciation for the rate at which *n*! grows by finding various factorials on their calculators. Ask them:

1. What is the largest value of *n* for which the calculator can evaluate *n*!?

2. What is the largest value of *n* for which the calculator displays the *exact* value of *n*!?

3. Using the answer for question 2, calculate $\frac{(n + 1)!}{n + 1}$. Is the answer exact?

Example 1 If you have 4 sweaters and 2 pairs of jeans, how many different sweater-and-jeans outfits can you make?

Solution There are two actions to perform:

(1) choosing a sweater   and   (2) choosing a pair of jeans

Since there are 4 sweaters from which to choose and 2 pairs of jeans from which to choose, there are $4 \times 2 = 8$ ways to choose a sweater-and-jeans outfit. If we call the sweaters A, B, C, and D, and the pairs of jeans 1 and 2, the so-called *tree diagram* at the right shows the 8 possible outfits.

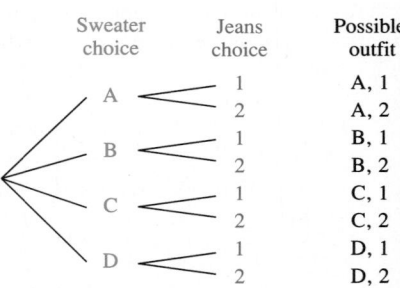

The multiplication principle can be extended to three or more actions that are performed together, as the next two examples show.

Example 2 In how many ways can 8 people line up in a cafeteria line?

Solution The diagram below represents the eight places in the cafeteria line.

The first place in line can be filled by any of the 8 people. Then the second place in line can be filled by any of the remaining 7 people, the third place by any of the remaining 6 people, and so on. The diagram below illustrates this reasoning.

| 8 | 7 | 6 | 5 | 4 | 3 | 2 | 1 |

By the multiplication principle, the answer is:

$$8 \cdot 7 \cdot 6 \cdot 5 \cdot 4 \cdot 3 \cdot 2 \cdot 1 = 40{,}320$$

In Example 2, the product $8 \cdot 7 \cdot 6 \cdot 5 \cdot 4 \cdot 3 \cdot 2 \cdot 1$ can be abbreviated as 8! (read "8 factorial"). The definition of *n*!, where *n* is a nonnegative integer, is:

$$n! = n \cdot (n - 1) \cdot (n - 2) \cdots 3 \cdot 2 \cdot 1$$
$$0! = 1$$

Notice in Example 2 that no person can occupy more than one space in the cafeteria line. Thus, the spaces of the cafeteria line are filled *without repetition*. This is in contrast with the next example, where repetition does occur.

Example 3 How many license plates can be made using 2 letters followed by 3 digits?

Solution The diagram below represents the five spaces of a license plate.

Each of the first two spaces can be filled with any of the 26 letters of the alphabet. Each of the last three spaces can be filled with any of the 10 digits. The diagram below illustrates this reasoning.

Thus, by the multiplication principle, the number of license plates is:

$$26 \cdot 26 \cdot 10 \cdot 10 \cdot 10 = 676,000$$

In Example 3, suppose that each of the five spaces of a license plate can be filled with *either* a letter *or* a digit. For any given space, then, there are 26 ways of choosing a letter, 10 ways of choosing a digit, and $26 + 10 = 36$ ways of filling the space. Note that we *add* the number of ways of performing the actions of choosing a letter and choosing a digit because the actions are **mutually exclusive**, that is, they cannot be performed together. This leads us to the following principle.

> ### The Addition Principle
>
> If two actions are mutually exclusive, and the first can be done in n_1 ways and the second in n_2 ways, then one action *or* the other can be done in $n_1 + n_2$ ways.

Example 4 In Morse code, the letters of the alphabet are represented by sequences of dots (·) and dashes (−). For example, · − represents the letter A and − − · represents the letter G. Show that sequences of no more than 4 symbols (dots and dashes) are needed to represent all of the alphabet.

Solution We can think of a 1-symbol sequence as a box to be filled with either a dot or a dash; a 2-symbol sequence as two boxes to be filled, each with a dot or a dash; and so on. Using the multiplication principle and then the addition principle, we have:

$n_1 = $ number of 1-symbol sequences = 2   = 2
$n_2 = $ number of 2-symbol sequences = 2 × 2   = 4
$n_3 = $ number of 3-symbol sequences = 2 × 2 × 2   = 8
$n_4 = $ number of 4-symbol sequences = 2 × 2 × 2 × 2   = 16

Total number of sequences with no more than 4 symbols = 30

Thus, the 30 possible sequences are obviously enough to accommodate the 26 letters of the alphabet.

Combinatorics **573**

Communication Note

Emphasize that multiplication is used for situations in which you select from set *A and* from set *B*; addition is used when you select from set *A or* from set *B*. A nice contrast between the addition and multiplication principles can be made by returning to Example 1 and assuming that the student agrees to loan a sibling either a sweater *or* a pair of jeans. A tree diagram for this situation might be:

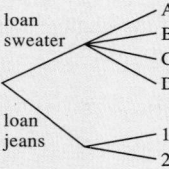

Note that 6 different loans can be made.

Example Note

Point out that Example 4 uses both the multiplication and addition principles. You might ask a student to obtain the complete Morse code and give a brief report on its history.

When counting the number of elements in a set, you may find it easier to count the number in the complement of the set and then use the following principle, which is illustrated in Example 5.

1. Classify as true or false:
$(3 \cdot 4)! = 3! \cdot 4!$
$(3 \cdot 4)! = 12! =$
$12 \cdot 11 \cdot 10 \cdot \ldots \cdot 3 \cdot 2 \cdot 1$
$3! \cdot 4! =$
$3 \cdot 2 \cdot 1 \cdot 4 \cdot 3 \cdot 2 \cdot 1$
Thus, $(3 \cdot 4)! \neq 3! \cdot 4!$

The Complement Principle

If A is a subset of a universal set U, then:
$$n(A) = n(U) - n(\overline{A})$$

2. Given the digits 0, 1, 2, 3, 4, and 5, how many four-digit numbers can be formed if:
a. digits can be repeated?
b. digits cannot be repeated?
c. digits cannot be repeated and the number must be even?
In each case, the first digit cannot be zero.
a. $\boxed{5} \cdot \boxed{6} \cdot \boxed{6} \cdot \boxed{6} =$
1080 numbers
b. Choices for first digit: 1, 2, 3, 4, or 5. Choices for second digit: 0 and the 4 digits remaining after the first choice.
$\boxed{5} \cdot \boxed{5} \cdot \boxed{4} \cdot \boxed{3} =$
300 numbers
c. The last digit must be 0, 2, or 4. When the last digit is 0, we have
$\boxed{5} \cdot \boxed{4} \cdot \boxed{3} \cdot \boxed{1} =$
60 choices. When the last digit is 2, we have
$\boxed{4} \cdot \boxed{4} \cdot \boxed{3} \cdot \boxed{1} =$
48 choices. When the last digit is 4, we have
$\boxed{4} \cdot \boxed{4} \cdot \boxed{3} \cdot \boxed{1} =$
48 choices. By the addition principle, there are
$60 + 48 + 48 = 156$
even 4-digit numbers without repetition of digits.

Example 5   Find the number of 4-digit numbers containing at least one digit 5.

Solution   Let U = the universal set of all 4-digit numbers,
   A = the set of 4-digit numbers containing at least one digit 5,
and $\overline{A}$ = the set of 4-digit numbers containing no 5's.

Then: $n(A) = \quad n(U) \quad - \quad n(\overline{A})$
$= 9 \cdot 10 \cdot 10 \cdot 10 - 8 \cdot 9 \cdot 9 \cdot 9$
$= 9000 - 5832 = 3168$

CLASS EXERCISES

1. Evaluate: **a.** 2! 2 **b.** 3! 6 **c.** 4! 24
2. If 9! = 362,880, what is 10!? 3,628,800
3. *Reading* In Example 3, suppose the first of the three digits in a license plate cannot be 0. How many such license plates are possible? 608,400
4. **a.** If a girl has 6 different skirts and 10 different blouses, how many different skirt-and-blouse outfits are possible? 60
   b. If she also has 3 different sweaters, how many skirt-blouse-and-sweater outfits are possible? 180
5. A boy has 2 sports coats and 4 sweaters.
   a. How many coat-and-sweater outfits can he wear? 8
   b. Suppose he decides to wear either a sports coat or a sweater, but not both. How many choices does he have? 6
6. If 10 runners compete in a race, in how many different ways can prizes be awarded for first, second, and third places? 720
7. The ''home row'' of a standard typewriter gives one arrangement of the letters A, S, D, F, G, H, J, K, L.
   a. How many *other* arrangements of these letters are possible? 362,879
   b. If *any* 9 letters of the alphabet could be placed on the ''home row'' of a typewriter, how many arrangements of the letters would be possible? See below.
8. *Reading* Give the conclusion of the complement principle in words rather than symbols. Then use a Venn diagram to illustrate this principle.
   7. b. 1,133,836,704,000

WRITTEN EXERCISES

A

1. Evaluate: **a.** 5! 120 **b.** 6! 720 **c.** 7! 5040 **d.** 0! 1

2. Evaluate: **a.** $\dfrac{10!}{9!}$ 10 **b.** $\dfrac{20!}{18!}$ 380 **c.** $\dfrac{n!}{(n-1)!}$ n **d.** $\dfrac{(n+1)!}{(n-1)!}$ $n^2 + n$

3. In how many different orders can you arrange 5 books on a shelf? 120

4. In how many different orders can 9 people stand in a line? 362,880

5. In how many different ways can you answer 10 true-false questions? 1024

6. In how many different ways can you answer 10 multiple-choice questions if each question has 5 choices? $5^{10} = 9,765,625$

7. Many radio stations have 4-letter call signs beginning with K. How many such call signs are possible if letters **(a)** can be repeated? **(b)** cannot be repeated? See below.

8. How many 3-digit numbers can be formed using the digits 4, 5, 6, 7, 8 if the digits **(a)** can be repeated? **(b)** cannot be repeated? 125; 60

9. In how many ways can 4 people be seated in a row of 12 chairs? 11,880

10. In how many ways can 4 different prizes be given to any 4 of 10 people if no person receives more than 1 prize? 5040

11. The tree diagram shows the four possible outcomes when a coin is tossed twice. If H and T represent "heads" and "tails," respectively, then the four outcomes are HH, HT, TH, and TT.

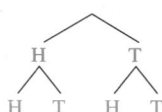

   a. Make a tree diagram showing the outcomes if a coin is tossed 3 times.
   b. The solid lines of the tree diagram are called *branches* and the elements of the bottom row (H, T, H, and T) are called *leaves*. How many branches and leaves are in your diagram for part (a)? 14 branches, 8 leaves
   c. How many branches and leaves would there be in a tree diagram showing the toss of a coin 10 times? 2046 branches, 1024 leaves

12. Four cards numbered 1 through 4 are shuffled and 3 different cards are chosen one at a time. Make a tree diagram showing the various possible outcomes.

B

13. **Sports** A high school coach must decide on the batting order for a baseball team of 9 players.
   a. The coach has how many different batting orders from which to choose? 362,880
   b. How many different batting orders are possible if the pitcher bats last? 40,320
   c. How many different batting orders are possible if the pitcher bats last and the team's best hitter bats third? 5040

---

7. **a.** 17,576 **b.** 13,800

Combinatorics **575**

Assessment Note

The first letter of the call signs for radio stations throughout the United States is limited to K and W. (Generally speaking, K is used west of the Mississippi River, and W is used east.) Ask students how many four-letter call signs are available nationwide.
$(2 \times 26 \times 26 \times 26 = 35,152)$

Suggested Assignments

Discrete Math
Day 1: 575/1–15 odd
Day 2: 576/17–31 odd

Supplementary Materials

Alternative Assessment, 46–47
Student Resource Guide, 140–142

14. **Sports** A track coach must choose a 4-person 400 m relay team and a 4-person 800 m relay team from a squad of 7 sprinters, any of whom can run on either team. If the fastest sprinter runs last in both races, in how many ways can the coach form the two teams if each of the 6 remaining sprinters runs only once and each different order is counted as a different team? 720

15. How many numbers consisting of 1, 2, or 3 digits (without repetitions) can be formed using the digits 1, 2, 3, 4, 5, 6? 156

16. If you have 5 signal flags and can send messages by hoisting one or more flags on a flagpole, how many messages can you send? 325

17. In some states license plates consist of 3 letters followed by 2 or 3 digits (for example, RRK-54 or ABC-055). How many such possibilities are there for those plates with 2 digits? for those with 3 digits? In all, how many license plates are possible? 1,757,600; 17,576,000; 19,333,600

18. How many possibilities are there for a license plate with 2 letters and 3 or 4 nonzero digits? 4,928,040

19. **a.** How many 3-digit numbers contain no 7's? 648
    b. How many 3-digit numbers contain at least one 7? 252

20. **a.** How many 4-digit numbers contain no 8's or 9's? 3584
    b. How many 4-digit numbers contain at least one 8 or 9? 5416

21. How many numbers from 5000 to 6999 contain at least one 3? 542

22. Many license plates in the U.S. consist of 3 letters followed by a 3-digit number from 100 to 999. How many of these contain at least one of the vowels A, E, I, O, and U? 7,483,500

23. Telephone numbers in the U.S. and Canada have 10 digits as follows:

    3-digit area code number: first digit is *not* 0 or 1;
                                     second digit *must be* 0 or 1
    3-digit exchange number: first and second digits are *not* 0 or 1
    4-digit line number: *not* all zeros

    a. How many possible area codes are there? 160
    b. The area code for Chicago is 312. Within this area code how many exchange numbers are possible? 640
    c. One of the exchange numbers for Chicago is 472. Within this exchange, how many line numbers are possible? 9999
    d. How many 7-digit phone numbers are possible in the 312 area code? 6,399,360
    e. How many 10-digit phone numbers are possible in the U.S. and Canada? 1,023,897,600

31,996,800

24. Suppose a state has 5 telephone area codes. Refer to Exercise 23 and tell how many phone numbers there could be in the state without adding any more area codes.

25. **a.** How many 9-letter "words" can be formed using the letters of the word FISHERMAN? (*Note*: We allow any arrangement of letters, such as "HAMERSNIF," to count as a "word." We also assume each letter is used exactly once.) 362,880

    b. How many 9-letter "words" begin and end with a vowel? (*Hint*: There are 3 choices for the first letter and 2 for the last letter, as shown below.) 30,240

    $$\boxed{3}\;\boxed{?}\;\boxed{?}\;\boxed{?}\;\boxed{?}\;\boxed{?}\;\boxed{?}\;\boxed{?}\;\boxed{2}$$

26. Suppose the letters of VERMONT are used to form "words."

    a. How many 7-letter "words" can be formed? 5040

    b. How many 6-letter "words" can be formed? How does your answer here compare with your answer in part (a)? 5040; =

    c. How many 5-letter "words" begin with a vowel and end with a consonant? 600

27. A school has 677 students. Explain why at least two students must have the same pair of initials. $26^2 = 676 < 677$

28. **a.** Ohio State University has about 40,000 students. Explain why at least two students must have the same first, middle, and last initials. $26^3 < 40{,}000$

    b. What is the minimum number of students needed to be *certain* that at least two students have the same three initials? 17,577

29. *Writing* Exercises 27 and 28 involve the *pigeonhole principle*, which states:

    If you are putting x pigeons in y pigeonholes, and $x > y$, then __?__.

    Write a paragraph in which you complete the statement of the pigeonhole principle and explain its use in Exercises 27 and 28. Be sure to tell what corresponds to pigeons and what corresponds to pigeonholes in those exercises.

30. *Research* In a computer, a *bit* stores the smallest unit of information. Bits are usually arranged in groups of 8, called *bytes*, in order to store larger pieces of information. Find out what a bit is and from this determine how many different pieces of information can be stored in a byte. Then find out how "byte-sized" pieces of information are part of the ASCII (pronounced "ask′ ee") code used by many computers for converting text characters to numbers.

31. Use your calculator to evaluate $\log_{10} 9!$. Then, without using your calculator, find the value of $\log_{10} 10!$ Check your answer with a calculator. See below.

32. Use the properties of logarithms to evaluate $\log_{10} 100! - \log_{10} 99!$ 2

33. **a.** Show that $10 \cdot 9 \cdot 8 \cdot 7 = \dfrac{10!}{6!}$. $\dfrac{10!}{6!} = \dfrac{10 \cdot 9 \cdot 8 \cdot 7 \cdot 6!}{6!} = 10 \cdot 9 \cdot 8 \cdot 7$

    b. For $1 \le r \le n$, show that:

    $$n \cdot (n - 1) \cdot (n - 2) \cdots (n - r + 1) = \frac{n!}{(n - r)!}$$

C 34. In how many zeros does the number 100! end? 24

31. $\log_{10} 9! \approx 5.56$; $\log_{10} 10! \approx 6.56$

Exercise Note

Emphasize that the exercises in this section and in following sections consider a "word" to be any arrangement of letters, even meaningless arrangements such as *XZGJ*.

Additional Answers Written Exercises

29. If you are putting x pigeons in y pigeonholes, and $x > y$, then at least one pigeonhole will contain more than one pigeon. In Exercises 27 and 28, the students correspond to pigeons and the initials correspond to pigeonholes. Because our alphabet has 26 letters, there are 26^2 sets of initials with 2 letters and 26^3 sets of initials with 3 letters. If a given number of people is greater than the number of sets of possible initials, then there will be people who have the same initials.

30. We can think of a bit as storing either a 0 or a 1. A byte is composed of 8 bits and thus can store 2^8 or 256 numbers in binary code. The ASCII code associates with each keyboard character a number between 0 and 255. Thus, when a key is pressed on the keyboard, the computer stores not the character itself but a number corresponding to that character. For example, the ASCII code for the character H is 72, or 01001000 in binary code.

Warm-Up Exercises

Evaluate each expression.

1. $7!$ 5040 **2.** $\frac{7!}{4!}$ 210

3. $\frac{7!}{4!3!}$ 35

Tell whether the order in which the following choices are made is important or *not*.

4. Four students are chosen from a class of 87 students to be student council representatives. not imp.

5. Four digits are chosen for a bank-card identification code. imp.

Motivating the Section

Ideas from combinatorics have long been used in encoding and decoding information. One recent application arose with the appearance of "cash stations," where people can make machine-assisted withdrawals from or deposits to their accounts. Suppose an automatic teller machine requires a customer to enter a 4-digit code in order to initiate a transaction. How many different 4-digit codes can be generated using the digits 0 through 9? (10,000)

Communication Note

Students will need help in reading the symbols for permutations and combinations out loud; $_nP_r$ is usually read as "the number of permutations of n things taken r at a time."

15-3 Permutations and Combinations

Objective   *To solve problems involving permutations and combinations.*

In some situations involving choices, the order in which the choices are made is important, whereas in others it is not. Suppose that a club with 12 members wishes to choose a president, a vice president, and a treasurer. In this case, the order of the choices is important: for instance, the order A, B, and C for president, vice president, and treasurer, respectively, is different from the order B, A, and C for the 3 offices. The number of ways of filling the 3 offices is:

$$12 \cdot 11 \cdot 10 = 1320$$

Suppose, on the other hand, that the club merely wants to choose a governing council of 3. In this case, the order of selection is *not* important, since a selection of A, B, and C is the same as a selection of B, A, and C. Now for each governing council of 3 that can be chosen (for example, A, B, C), there are 3! different slates of officers (ABC, ACB, BAC, BCA, CAB, CBA). Thus:

$$n(\text{governing councils}) \times 3! = n(\text{slates of officers})$$

$$n(\text{governing councils}) = \frac{n(\text{slates of officers})}{3!}$$

$$= \frac{12 \cdot 11 \cdot 10}{1 \cdot 2 \cdot 3} = 220$$

In the first situation, where the order of selection is important, each selection is called a **permutation** of 3 people from a set of 12; the number of such selections is denoted $_{12}P_3$. In the second situation, where the order of selection is not important, each selection is called a **combination** of 3 people from a set of 12; the number of such selections is denoted $_{12}C_3$.

In general, for $0 \le r \le n$, the symbols $_nP_r$ and $_nC_r$ denote the number of permutations and the number of combinations, respectively, of r things chosen from n things. (Sometimes $P(n, r)$ and $C(n, r)$ are used instead of $_nP_r$ and $_nC_r$ in order to emphasize that permutations and combinations are *functions* of n and r.)

	Formula	Example
Permutation (order important)	$_nP_r = n(n-1)(n-2)\cdots(n-r+1)$ $= \dfrac{n!}{(n-r)!}$ $\left\{\begin{array}{l}\text{Formula derived} \\ \text{in Exercise 33} \\ \text{on page 577}\end{array}\right.$	$_{12}P_3 = 12 \cdot 11 \cdot 10$ $= \dfrac{12!}{(12-3)!}$ $= 1320$
Combination (order not important)	$_nC_r = \dfrac{n(n-1)(n-2)\cdots(n-r+1)}{1 \cdot 2 \cdot 3 \cdots r}$ $= \dfrac{n!}{(n-r)!\,r!}$ $\left\{\begin{array}{l}\text{Formula derived} \\ \text{in Exercise 28} \\ \text{on page 582}\end{array}\right.$	$_{12}C_3 = \dfrac{12 \cdot 11 \cdot 10}{1 \cdot 2 \cdot 3}$ $= \dfrac{12!}{(12-3)!\,3!}$ $= 220$

578   *Chapter Fifteen*

Example 1 A company advertises two job openings, one for a copywriter and one for an artist. If 10 people who are qualified for either position apply, in how many ways can the openings be filled?

Solution Since the jobs are different, the order of selecting people matters: X as copywriter and Y as artist is different from Y as copywriter and X as artist. Thus, the solution is found by counting permutations:

$$_{10}P_2 = 10 \cdot 9 = 90$$

Example 2 A company advertises two job openings for computer programmers, both with the same salary and job description. In how many ways can the openings be filled if 10 people apply?

Solution Since the two jobs are identical, the order of selecting people is not important: Choosing X and Y for the positions is the same as choosing Y and X. Thus, the solution is found by counting combinations:

$$_{10}C_2 = \frac{10 \cdot 9}{1 \cdot 2} = 45$$

If you have a calculator with permutation and combination keys (or just a factorial key), you can use it to help solve counting problems that involve large numbers, as in the next example.

Example 3 As shown below, a standard deck of playing cards consists of 52 cards, with 13 cards in each of four *suits* (clubs, spades, diamonds, and hearts). Clubs and spades are black cards, and diamonds and hearts are red cards. Also, jacks, queens, and kings are called *face cards*.

Clubs (♣): ace, 2, 3, 4, 5, 6, 7, 8, 9, 10, jack, queen, king
Spades (♠): ace, 2, 3, 4, 5, 6, 7, 8, 9, 10, jack, queen, king
Diamonds (♦): ace, 2, 3, 4, 5, 6, 7, 8, 9, 10, jack, queen, king
Hearts (♥): ace, 2, 3, 4, 5, 6, 7, 8, 9, 10, jack, queen, king

How many ways are there to deal 13 cards from a standard deck if the order in which the cards are dealt is **(a)** important? **(b)** not important?

Solution **a.** $_{52}P_{13} = \frac{52!}{39!} \approx 3.95 \times 10^{21}$ **b.** $_{52}C_{13} = \frac{52!}{39!13!} \approx 6.35 \times 10^{11}$

Example Note

Point out that the answer to Example 1 is twice that of Example 2. Ask what would happen if there were 3 jobs instead of 2. (The answer to Example 1 would be 3!, or 6, times the answer to Example 2.)

Additional Examples

1. For a certain raffle, 845 tickets are sold.
   a. In how many ways can four $50 gift certificates be awarded?
   b. In how many ways can a $100, a $50, a $20, and a $10 gift certificate be awarded?
   a. Since the awards are the same, order is unimportant. $_{845}C_4 = \frac{845!}{841!4!} = \frac{845 \cdot 844 \cdot 843 \cdot 842}{4 \cdot 3 \cdot 2 \cdot 1} \approx 2.11 \times 10^{10}$
   b. Since the awards are different, the order is important. $_{845}P_4 = \frac{845!}{841!} = 845 \cdot 844 \cdot 843 \cdot 842 \approx 5.06 \times 10^{11}$

2. At the Red Lion Diner, an omelet can be ordered plain or with any or all of the following fillings: cheese, onions, peppers. How many different kinds of omelets are possible?
   A customer can choose 0, 1, 2, or 3 fillings.
   $_3C_0 + {_3C_1} + {_3C_2} + {_3C_3} = \frac{3!}{3!0!} + \frac{3!}{2!1!} + \frac{3!}{1!2!} + \frac{3!}{0!3!} = 1 + 3 + 3 + 1 = 8$

CLASS EXERCISES

In Exercises 1–4, find the value of each expression.

1. a. $_5P_2$ 20   b. $_5C_2$ 10   2. a. $_6P_3$ 120   b. $_6C_3$ 20

3. a. $_{10}P_3$ 720   b. $_{10}C_3$ 120   4. a. $_4P_4$ 24   b. $_4C_4$ 1

5. In how many ways can a club with 10 members choose a president, a vice president, and a treasurer? 720

6. In how many ways can a club with 10 members choose a 3-person governing council? 120

7. A lock on a safe has a dial with 50 numbers on it. To open it, you must turn the dial left, then right, and then left to 3 different numbers. Such a lock is usually called a combination lock, but a more accurate name would be permutation lock. Explain. Order of the 3 numbers is important.

8. a. Four people (A, B, C, and D) apply for three jobs (clerk, secretary, and receptionist). If each person is qualified for each job, make a list of the ways the jobs can be filled. For example, ABC means that A is clerk, B is secretary, and C is receptionist. This is different from BAC and CBA.
   b. Your list should contain $_4P_3$ entries. How many entries involve persons A, B, and C? How many entries involve persons A, C, and D? 6; 6
   c. If A, B, C, and D apply for three identical job openings as a clerk, make a list of the number of ways the openings can be filled. ABC, ABD, ACD, BCD

WRITTEN EXERCISES

A 1. a. In how many ways can a club with 20 members choose a president and a vice president? 380
   b. In how many ways can the club choose a 2-person governing council? 190

2. a. In how many ways can a club with 13 members choose 4 different officers? 17,160
   b. In how many ways can the club choose a 4-person governing council? 715

3. a. In how many ways can a host-couple choose 4 couples to invite for dinner from a group of 10 couples? 210
   b. Ten students each submit a woodworking project in an industrial arts competition. There are to be first-, second-, and third-place prizes plus an honorable mention. In how many ways can these awards be made? 5040

4. A teacher has a collection of 20 true-false questions and wishes to choose 5 of them for a quiz. How many quizzes can be made if the order of the questions is considered (a) important? (b) unimportant? 1,860,480; 15,504

5. Each of the 200 students attending a school dance has a ticket with a number for a door prize. If 3 different numbers are selected, how many ways are there to award the prizes, given that the 3 prizes are (a) identical? (b) different? See below

6. Suppose you bought 4 books and gave one to each of 4 friends. In how many ways can the books be given if they are (a) all different? (b) all identical? 24; 1

580 *Chapter Fifteen* **5. a.** 1,313,400 **b.** 7,880,400

7. Eight people apply for 3 job positions. In how many different ways can the 3 positions be filled if the positions are (a) all different? (b) all the same? 336; 56

8. How many different ways are there to deal a hand of 5 cards from a standard deck of 52 cards if the order in which the cards are dealt is (a) important? (b) not important? 311,875,200; 2,598,960

9. In how many ways can 6 hockey players be chosen from a group of 12 if the playing positions are (a) considered? (b) not considered? 665,280; 924

10. Of the 12 players on a school's basketball team, the coach must choose 5 players to be in the starting lineup. In how many ways can this be done if the playing positions are (a) considered? (b) not considered? 95,040; 792

11. a. A hiker would like to invite 7 friends to go on a trip but has room for only 4 of them. In how many ways can they be chosen? 35
    b. If there were room for only 3 friends, in how many ways could they be chosen? How is your answer related to the answer for part (a)? Why? 35

12. a. In how many ways can you choose 3 letters from the word LOGARITHM if the order of letters is unimportant? 84
    b. In how many ways can you choose 6 letters from LOGARITHM if the order of letters is unimportant? Compare with part (a). 84; =

13. Show that $_{100}C_2 = _{100}C_{98}$ in the following ways: (a) by using the formula for $_nC_r$; (b) by explaining how choosing 2 out of 100 is related to choosing 98 out of 100.

14. a. Show that $_{11}C_3 = _{11}C_8$.
    b. Study part (a) and Exercise 13. Then make a generalization and prove it.

15. a. Evaluate $_5C_0$ to find how many ways you can select no objects from a group of 5. 1
    b. Evaluate $_5C_5$ to find how many ways you can select 5 objects from a group of 5. 1

16. How does the formula for $_nC_n$ suggest the definition $0! = 1$?

B 17. A certain chain of ice cream stores sells 28 different flavors, and a customer can order a single-, double-, or triple-scoop cone. Suppose on a multiple-scoop cone that the order of the flavors is important and that the flavors can be repeated. How many possible cones are there? 22,764

18. Refer to Exercise 17 and find the number of double-scoop ice cream cones that are possible if repetition of flavors is allowed but the order of flavors is unimportant. 406

Combinatorics **581**

11. b. The answer is the same as that in part (a). Choosing 3 out of 7 to go is the same as choosing 4 out of 7 to stay.

13. a. $_{100}C_2 = \dfrac{100!}{98!2!}$; $_{100}C_{98} = \dfrac{100!}{2!98!}$. Therefore, $_{100}C_2 = _{100}C_{98}$.

14. a. $_{11}C_3 = \dfrac{11!}{8!3!}$ and $_{11}C_8 = \dfrac{11!}{3!8!}$; therefore, $_{11}C_3 = _{11}C_8$.

    b. $_nC_r = _nC_{n-r}$. $_nC_r = \dfrac{n!}{(n-r)!r!}$ and $_nC_{n-r} = \dfrac{n!}{[n-(n-r)]!(n-r)!} = \dfrac{n!}{r!(n-r)!}$; thus, $_nC_r = _nC_{n-r}$.

16. $_nC_n = \dfrac{n!}{0!n!} = \dfrac{1}{0!}$. There is only one way to select n objects from a group of n; therefore, $\dfrac{1}{0!} = 1$ and $0! = 1$.

21. a. 792   **b.** 658,008   **c.** 1,940,952

19. Three couples go to the movies and sit together in a row of six seats. In how many ways can these people arrange themselves if each couple sits together? 48

20. A mathematics teacher uses 4 algebra books, 2 geometry books, and 3 precalculus books for reference. In how many ways can the teacher arrange the books on a shelf if books covering the same subject matter are kept together? 1728

21. From a standard deck of 52 cards, 5 cards are dealt and the order of the cards is unimportant. In how many ways can you receive **(a)** all face cards? **(b)** no face cards? **(c)** at least one face card? See above.

22. Answer part (c) of Exercise 21 if the order of receiving the 5 cards is important. 232,914,240

23. a. How many line segments can be drawn joining the six points A, B, C, D, E, and F shown? 15

b. If the six points were positioned differently, the answer to part (a) could be different. Explain.

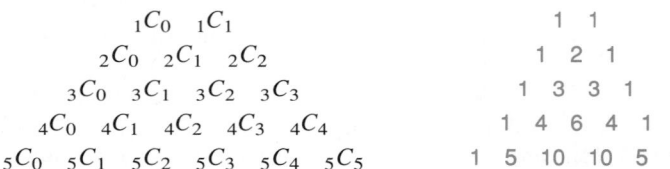

24. A convex polygon has n vertices. Find the number of diagonals by calculating the total number of ways of connecting two vertices and then subtracting the number of sides. (You may want to compare your answer with that for Exercise 25 on page 482 where difference equations were used.) $\dfrac{n(n-3)}{2}$

25. Solve for n: $_nC_2 = 45$ 10

26. Solve for n: $_nC_2 = _{n-1}P_2$ 4

27. *Investigation* Give the values of the combinations shown in the triangular array below. What patterns can you discover in this array?

$$
\begin{array}{ccccccc}
 & & _1C_0 & _1C_1 & & & \\
 & & _2C_0 & _2C_1 & _2C_2 & & \\
 & _3C_0 & _3C_1 & _3C_2 & _3C_3 & & \\
 & _4C_0 & _4C_1 & _4C_2 & _4C_3 & _4C_4 & \\
_5C_0 & _5C_1 & _5C_2 & _5C_3 & _5C_4 & _5C_5 &
\end{array}
$$

$$
\begin{array}{ccccccc}
 & & 1 & 1 & & & \\
 & & 1 & 2 & 1 & & \\
 & 1 & 3 & 3 & 1 & & \\
 1 & 4 & 6 & 4 & 1 & & \\
1 & 5 & 10 & 10 & 5 & 1 &
\end{array}
$$

28. For $1 \le r \le n$, show that $\dfrac{n(n-1)(n-2)\cdots(n-r+1)}{1 \cdot 2 \cdot 3 \cdot \cdots r} = \dfrac{n!}{(n-r)!\,r!}$.

29. If the 52 cards of a standard deck are dealt to four people, 13 cards at a time, the first person can receive $_{52}C_{13}$ possible hands. Then the second person can receive 13 of the remaining 39 cards in $_{39}C_{13}$ possible ways. The third person can receive 13 of the remaining 26 cards in $_{26}C_{13}$ ways, and the fourth person can receive 13 of the remaining 13 cards in $_{13}C_{13}$ ways. Thus, the total number of ways of distributing the 52 cards into four 13-card hands is:

$$_{52}C_{13} \cdot {}_{39}C_{13} \cdot {}_{26}C_{13} \cdot {}_{13}C_{13}$$

Without using a calculator, show that the above product simplifies to $\dfrac{52!}{(13!)^4}$.

Then use a calculator to find this number. 5.36×10^{28}

582   *Chapter Fifteen*

Write a program that outputs the values of $_nP_r$ and $_nC_r$ for input values of n and r where $0 \le r \le n$.

15-4 Permutations with Repetition; Circular Permutations

Objective | *To solve counting problems that involve permutations with repetition and circular permutations.*

Activity

a. Write down all possible arrangements of the letters of the word MOP.
b. Write down all *distinguishable* arrangements of the letters of the word MOM.
c. What accounts for the fact that part (b) gives fewer permutations than part (a)?

As the preceding activity shows, fewer permutations result when some of the objects being rearranged are the same. For example, consider the two 6-letter words MEXICO and CANADA. Since the letters in the word MEXICO are all different, there are 6! = 720 possible arrangements of the letters. On the other hand, the word CANADA has fewer distinguishable arrangements of its letters because of the 3 identical A's. To see why, suppose we distinguish the A's by color, as follows:

$$\text{CANADA}$$

Since the six letters are now all different, there are 6! arrangements of the letters, including:

NADACA	NADACA	NADACA
NADACA	NADACA	NADACA

As soon as the 3 different A's are changed back to the same A, however, the 6 different arrangements above become the *same* arrangement NADACA. Because the 3 different A's have 3! different arrangements, we have:

$$n(\text{arrangements of CANADA}) = \frac{n(\text{arrangements of CANADA})}{3!}$$

$$= \frac{6!}{3!} = 120$$

In generalizing this discussion, let us say that CANADA has 6 letters of several different "types": 3 A's, 1 C, 1 N, and 1 D. The number of permutations is:

$$\frac{6!}{3!\ 1!\ 1!\ 1!}$$

The general form of this result is given at the top of the next page.

Teaching Notes, p. 564C

Warm-Up Exercises

1. List all the possible arrangements of the digits in the number 1234.

1234	1243	1324	1342
1423	1432	2134	2143
2314	2341	2413	2431
3124	3142	3214	3241
3412	3421	4123	4132
4213	4231	4312	4321

2. List all the distinguishable (different) arrangements of the digits in the number 1134.

1134	1314	1341	1143
1413	1431	3114	3141
3411	4113	4131	4311

3. How many possible arrangements of two 1's are possible? of r 1's?
1; 1

4. Compare the number of arrangements in Exercise 1 with the number in Exercise 2. Then guess how to determine the number of arrangements of n objects if r of them are identical.
In Ex. 1, there are 4! = 24 arrangements; in Ex. 2, there are $\frac{4!}{2!} = 12$. In general, there are $\frac{n!}{r!}$ arrangements of n objects if r of them are identical.

Motivating the Section

In this section, students will extend their understanding of permutations to cover more complex situations.

Point out that the formula given at the top of the page assumes that

$$n_1 + n_2 + n_3 + \ldots + n_k = n.$$

When discussing circular permutations (that is, arrangements of objects in a circle), stress that the important consideration is the position of the objects *relative to each other*. For example, in the circular arrangements shown at the bottom of the page, A can serve as a reference point for B, C, and D: In all four arrangements, B is always on A's left, C is always opposite A, and D is always on A's right; thus, the four arrangements are the same.

Example Note

In discussing Example 1, point out that the 1! factors in the denominator do not affect the calculation and thus can be omitted.

The Number of Permutations of Things Not All Different

Let S be a set of n elements of k different types. Let $n_1 =$ the number of elements of type 1, $n_2 =$ the number of elements of type 2, ..., $n_k =$ the number of elements of type k. Then the number of distinguishable permutations of the n elements is:

$$\frac{n!}{n_1! \, n_2! \, n_3! \cdots n_k!}$$

Example 1   How many permutations are there of the letters of MASSACHUSETTS?

Solution   Of the 13 letters of MASSACHUSETTS, there are 4 S's, 2 A's, 2 T's, 1 M, 1 C, 1 H, 1 U, and 1 E. Thus, the number of permutations is:

$$\frac{13!}{4! \, 2! \, 2! \, 1! \, 1! \, 1! \, 1! \, 1!} = 64{,}864{,}800$$

Example 2   The grid shown at the right represents the streets of a city. A person at point X is going to walk to point Y by always traveling south or east. How many routes from X to Y are possible?

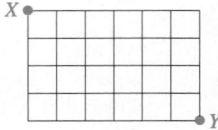

Solution   **Method 1**   To get from X to Y, the person must travel 4 blocks south (S) and 6 blocks east (E). One possible route can be symbolized by the "word" SSESEEESEE. Other routes can be symbolized by other 10-letter words having 4 S's and 6 E's. The number of these "words" is $\dfrac{10!}{4! \, 6!} = 210$.

Method 2   Every route from X to Y covers 10 blocks of which 4 must be south. The number of ways to choose which 4 of the 10 blocks are to be south is $_{10}C_4 = \dfrac{10!}{6! \, 4!} = 210$.

Thus far we have considered only *linear* permutations, but permutations may also be *circular* (or *cyclic*). For example, the diagrams below show that seating four people around a circular table is different from seating them in a row.

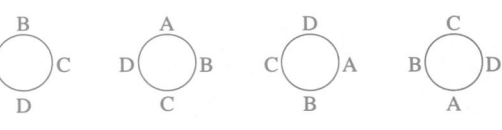

These circular permutations are the *same*, because in each one, A is to the right of B, who is to the right of C, who is to the right of D.

ABCD   DABC   CDAB   BCDA

These linear permutations are *different*.

Example 3 How many circular permutations are possible when seating four people around a table?

Solution The diagrams at the bottom of the preceding page show that 1 circular permutation corresponds to 4 linear permutations. Thus:

$$\text{number of circular permutations} = \frac{1}{4} \text{ (number of linear permutations)}$$

$$= \frac{1}{4} \cdot 4!$$

$$= 3! = 6$$

CLASS EXERCISES

In Exercises 1–4, find the number of permutations of the letters of each word.

1. a. MALE 24   **b.** MALL 12   **2. a.** MOUSE 120   **b.** MOOSE 60

3. a. VERIFY   **b.** VIVIFY   **4. a.** SHAKEUPDOWN   **b.** SHAKESPEARE
720   180   39,916,800   1,663,200

5. Find the number of permutations of the letters of: Answers will vary.

  a. your last name   **b.** your full name   Answers will vary.

6. *Reading* Of the two solutions given for Example 2, which do you prefer? Why?

7. Refer to the diagram for Example 2 and suppose that the street corner just north of Y is Z. How many ways are there to walk from X to Z if you always travel south or east? 84

8. All but one of the following linear permutations correspond to the same circular permutation. Which one does not? BCDAE

  CDEAB   EABCD   BCDAE   DEABC

WRITTEN EXERCISES

In Exercises 1–4, find the number of permutations of the letters of the given word.

A

1. MISSOURI 10,080

2. PENNSYLVANIA 39,916,800

3. MISSISSIPPI 34,650

4. CONNECTICUT 1,663,200

5. Each of the 10 finalists in the state spelling bee contest receives a prize. The prizes are six $25 bonds, three $50 bonds, and one $100 bond. In how many ways can these prizes be given? 840

6. The 12 workers in a cafeteria crew rotate among three kinds of jobs. In how many ways can the crew be assigned the jobs of 2 cooks, 7 servers, and 3 dishwashers? 7920

Mathematical Note

You may wish to help students generalize the formula for circular permutations of *n* objects. Since each circular permutation may be written as a linear permutation in *n* ways, there are $\frac{n!}{n} = (n-1)!$ circular permutations (see Exercise 12, page 586).

Additional Examples

1. How many different integers can be formed using all the digits of 253,225?

  There are 6 digits, of which 3 are two's and 2 are fives. The number of integers is
  $\frac{6!}{3!2!} = \frac{6 \cdot 5 \cdot 4}{2 \cdot 1} = 60.$

2. A person walks from *A* to *B*, and then from *B* to *C*, always traveling south or east on the streets shown. How many paths are possible?

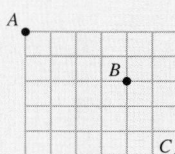

  The route from *A* to *B* consists of 6 blocks, of which 4 are eastward and 2 are southward. There are
  $\frac{6!}{4!2!} = 15$ such routes.

  Similarly, there are $\frac{5!}{2!3!} =$ 10 routes from *B* to *C*. Then, by the multiplication principle, there are
  15 · 10 = 150 routes from *A* to *C* via *B*.

Additional Answers
Written Exercises

15. a. Pair the position of a name from the grade-book list with the letter in the same position in the 25 letter "word."

7. In the English language, *antidisestablishmentarianism* is one of the longest words. If a computer could print out every permutation of this word at the rate of one word per second, how long would it take? 9.71×10^{14} years

8. Which of the 50 states has the least number of permutations of the letters of its name? OHIO

In Exercises 9 and 10, a person bicycles along the city streets shown at the right by always traveling south and east.

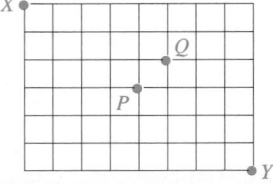

9. Find the number of possible routes from:
   a. X to Y 3003   b. X to Q 21
   c. Q to Y 35   d. X to Y via Q 735

10. To the nearest tenth of a percent, what percent of the routes from X to Y pass through P? 40.8%

11. How many circular permutations are possible when seating 5 people around a circular table? 24

12. The photo shows the mythical Round Table of King Arthur. Note that the table indicates where King Arthur sat as well as where 24 knights sat. In how many ways can n people be seated around a similar table if:
    a. a head of the table is designated?
    b. *no* head of the table is designated?
    a. $n!$ b. $(n-1)!$

B 13. **Meteorology** A typical January in Boston has 3 days of snow, 12 days of rain, and 16 days without precipitation. In how many ways can such weather be distributed throughout the month? [*Hint*: Think of a weather distribution pattern as a 31-letter word made up of S's (snow), R's (rain), and N's (no precipitation).] 1.37×10^{11}

14. In the card game of bridge, 52 cards are dealt to 4 people who are called North, East, South, and West. A 52-letter ''word'' consisting of 13 N's, 13 E's, 13 S's, and 13 W's will tell you who has what cards. For example, if a word begins NEEWS . . . , then North has the ace of spades, East has the 2 and 3 of spades, West has the 4 of spades, South has the 5 of spades, and so on. How many such arrangements are there? (Compare your answer with the method shown in Exercise 29 of Section 15-3.) 5.36×10^{28}

15. A teacher's grade book alphabetically lists 25 students in a class. This teacher, who does not give pluses and minuses with letter grades, announces that on the last test there were 5 A's, 10 B's, 6 C's, 3 D's, and 1 E.
    a. Explain how a 25-letter ''word'' could tell who got what grade.
    b. In how many ways can these grades be distributed to the 25 students? 8.25×10^{12}

586   *Chapter Fifteen*

16. A group of 4 people orders from a restaurant menu that lists 3 main courses: fish, chicken, and steak. How many combinations of main-course orders are possible in this group? (*Hint:* You can form "words" by using 4 x's to tally the orders and 2 bars to separate the three types of orders. For example, the "word" xx |x| x represents 2 orders of fish, 1 order of chicken, and 1 order of steak. Similarly, the "word" xxx | | x represents 3 orders of fish, 0 orders of chicken, and 1 order of steak.) 15

17. A group of 12 friends goes to a cinema complex that is showing 6 different movies. If the group splits up into subgroups based on movie preferences, how many subgroup combinations are possible? (See the hint for Exercise 16.) 6188

C 18. The word ABSTEMIOUS contains the 5 vowels A, E, I, O, and U in alphabetical order. How many permutations of this word have the vowels in alphabetical order? 15,120

MIXED COMBINATORICS EXERCISES

The following exercises are counting problems like those from the last four sections. They test your ability to choose the appropriate counting technique.

A 1. Three identical door prizes are to be given to three lucky people in a crowd of 100. In how many ways can this be done? 161,700

2. The license plates in a certain state consist of 3 letters followed by 3 nonzero digits. How many such license plates are possible? 12,812,904

3. How many 4-digit numbers (**a**) contain no 0's? (**b**) contain no 1's? (**c**) begin with an even digit and end with an odd digit? 6561; 5832; 2000

4. A student must take four final exams, scheduled by computer, during the morning and afternoon testing periods on Monday through Friday of exam week. If the order of the student's four exams is important, in how many ways can a computer schedule the exams? 5040

5. **a.** A railway has 30 stations. On each ticket, the departure station and the destination station are printed. How many different tickets are possible? 870
   b. If a ticket can be used in either direction between two stations, how many different tickets are needed? 435

6. In how many ways can the letters of each of the following words be arranged?
   a. RADISH 720
   b. SQUASH 360
   c. TOMATO 180

7. There are 3 roads from town A to town B, 5 roads from town B to town C, and 4 roads from town C to town D. How many ways are there to go from A to D via B and C? How many different round trips are possible? 60; 3600

8. A teacher must pick 3 high school students from a class of 30 to prepare and serve food at the junior high school picnic. How many choices are possible? 4060

9. All students at John Jay High School must take at least one of the school's three science courses: biology, chemistry, and physics. In order to project the future enrollment in these courses, the principal sent questionnaires to 297 junior high school students and got these results:

> 132 intend to take biology and chemistry.
> 107 intend to take chemistry and physics.
> 88 intend to take biology and physics.
> 43 intend to take only biology.
> 55 intend to take only chemistry.
> 38 intend to take only physics.

How many students intend to take all three science courses? 83

10. A radio show plays the top 10 musical hits of the previous week. During the first full week of last January, these 10 hits were chosen from 70 possibilities, and they were played in order of increasing popularity. How many possible orders were there? *Note*: Since calculators cannot evaluate large factorials, use the approximation given by the formula $n! \approx \sqrt{2\pi n} \left(\dfrac{n}{e}\right)^n$. 1.44×10^{18}

B 11. The locks on the gymnasium lockers have dials with numbers from 0 to 39. Each locker combination consists of 3 numbers.
   a. Pythagoras remembers that his numbers are 8, 15, and 17, but he can't remember in which order they appear. If he can try one possibility every 10 s, what is the maximum amount of time that it would take him to find the right combination? 1 min
   b. Hypatia cannot remember any of her combination numbers, but she does remember that exactly two of them are the same. If she can try one possibility every 10 s, what is the maximum amount of time that it would take her to find the right combination? 13 h

12. In how many ways can 8 jackets of different styles be hung:
   a. on a straight bar? 40,320   b. on a circular rack? 5040

13. A town council consists of 8 members including the mayor.
   a. How many different committees of 4 can be chosen from this council? 70
   b. How many of these committees include the mayor? 35
   c. How many do not include the mayor? 35
   d. Verify that the answer to part (a) is the sum of the answers to part (b) and part (c). 35 + 35 = 70

14. Repeat Exercise 13 if the council has 9 members including the mayor. See below.

15. If you have a $1 bill, a $5 bill, a $10 bill, and a $20 bill, how many different sums of money can you make using one or more of these bills? 15

14. a. 126   b. 56   c. 70   d. 56 + 70 = 126

16. **Communications** In Morse code, letters, digits, and various punctuation marks are represented by a sequence of dots and dashes. Sequences can be from 1 unit to 6 units in length. How many such sequences are possible? 126

17. The Pizza Place offers pepperoni, mushrooms, sausages, onions, anchovies, and peppers as toppings for their regular plain pizza. How many different pizzas can be made? 64

18. **a.** How many 4-letter "words" can be formed by using the 8 letters of the word TRIANGLE? 1680
   b. How many of the "words" formed in part (a) have no vowels? 120
   c. How many of the "words" formed in part (a) have at least one vowel? 1560

19. Five boys and five girls stand in a line. How many arrangements are possible **(a)** if all of the boys stand in succession? **(b)** if the boys and girls stand alternately? 86,400; 28,800

20. Answer Exercise 19 if the 5 boys and 5 girls stand in a circle instead of a line. 14,400; 2880

21. How many 5-digit numbers contain at least one 3? 37,512

22. In the diagram at the right, all paths go from X toward Y and consist only of south and east steps. For example, there are 3 paths that go from X to the point marked with a 3. Similarly, there are 4 paths that go from X to the point marked with a 4.

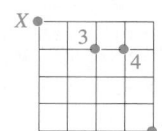

   a. Explain how the numbers 3 and 4 are obtained.
   b. Copy the grid and label each point with the number of paths that go from X to that point.

C 23. **Sports** In the World Series, two teams, A and B, play each other until one team has won 4 games. For example, the "word" ABBAAA represents a 6-game series in which team A wins games 1, 4, 5, and 6.
   a. Explain why the number of different 6-game series won by team A is $\dfrac{5!}{3!\,2!}$.
   b. Without actually listing the various series between team A and team B, show that there are 70 different sequences of games possible.

24. **Sports** In the tennis championship between players A and B at Wimbledon, the first player to win 3 sets is champion. Find the number of different ways for player A to win the championship. 10

Combinatorics **589**

22. **a.** Of the 3 blocks that must be traveled, 1 must be south; $_3C_1 = 3$. Of the 4 blocks that must be traveled, 1 must be south; $_4C_1 = 4$.

   b.

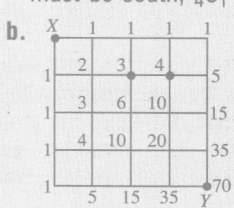

23. **a.** The team winning the series must win the last game of the series. In a six-game series, of the remaining 5 games, A wins 3. Which 3 of the 5 games is unimportant. Therefore, there are $_5C_3 = \dfrac{5!}{2!3!}$ different series.

   b. A wins 4 games: $_3C_3 = 1$ way;
   A wins 5 games: $_4C_3 = 4$ ways;
   A wins 6 games: $_5C_3 = 10$ ways;
   A wins 7 games: $_6C_3 = 20$ ways.
   A parallel analysis can be made for B's winning the series. Thus, there are $2(1 + 4 + 10 + 20) = 70$ different sequences of games possible.

1. Expand $(a + b)^1$, $(a + b)^2$, $(a + b)^3$, and $(a + b)^4$.

$a + b$; $a^2 + 2ab + b^2$;
$a^2 + 3a^2b + 3ab^2 + b^3$;
$a^4 + 4a^3b + 6a^2b^2 + 4ab^3 + b^4$

2. Replace each combination in the array below with its value. Compare the results with your answers to Exercise 1. What do you notice?

$$_1C_0 \quad _1C_1$$
$$_2C_0 \quad _2C_1 \quad _2C_2$$
$$_3C_0 \quad _3C_1 \quad _3C_2 \quad _3C_3$$
$$_4C_0 \quad _4C_1 \quad _4C_2 \quad _4C_3 \quad _4C_4$$

$$1 \quad 1$$
$$1 \quad 2 \quad 1$$
$$1 \quad 3 \quad 3 \quad 1$$
$$1 \quad 4 \quad 6 \quad 4 \quad 1$$

The values are the coefficients of the terms in the expansions of Exercise 1.

3. Use the pattern in Exercise 2 to give the next row of the array and then evaluate each entry.

$_5C_0 = 1$; $_5C_1 = 5$; $_5C_2 = 10$; $_5C_3 = 10$; $_5C_4 = 5$; $_5C_5 = 1$

4. Use your answers from Exercise 3 to write the expansion of $(a + b)^5$.

$a^5 + 5a^4b + 10a^3b^2 + 10a^2b^3 + 5ab^4 + b^5$

15-5 The Binomial Theorem; Pascal's Triangle

Objective *To use the binomial theorem and Pascal's triangle.*

Our goal in this section is to derive a formula for expanding $(a + b)^n$ for positive integers n. For small values of n we get the following expansions:

$$(a + b)^1 = 1a + 1b$$
$$(a + b)^2 = 1a^2 + 2ab + 1b^2$$
$$(a + b)^3 = 1a^3 + 3a^2b + 3ab^2 + 1b^3$$
$$(a + b)^4 = 1a^4 + 4a^3b + 6a^2b^2 + 4ab^3 + 1b^4$$
$$(a + b)^5 = 1a^5 + 5a^4b + 10a^3b^2 + 10a^2b^3 + 5ab^4 + 1b^5$$

From these examples we can see that the first term in the expansion of $(a + b)^n$ is always $1a^nb^0$. In successive terms the exponents of a decrease by 1 and the exponents of b increase by 1, so that the sum of the two exponents in a term is always n. The coefficients of the terms also have a pattern, which can be seen by studying the array of numbers below. The first five rows of this array are like the array above except that the a's, b's, and plus signs have been omitted.

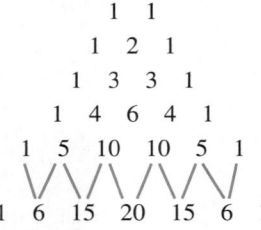

$$1 \quad 1$$
$$1 \quad 2 \quad 1$$
$$1 \quad 3 \quad 3 \quad 1$$
$$1 \quad 4 \quad 6 \quad 4 \quad 1$$
$$1 \quad 5 \quad 10 \quad 10 \quad 5 \quad 1$$
$$1 \quad 6 \quad 15 \quad 20 \quad 15 \quad 6 \quad 1$$

This array is called **Pascal's triangle**, named for the French mathematician Blaise Pascal (1623–1662). Because the numbers in the array are the coefficients of the terms in the expansion of $(a + b)^n$, they are called **binomial coefficients**. Notice that (except for the 1's) each number is the sum of the two numbers just above it. Hence, from the fifth row of the triangle, we can quickly form the sixth row, as shown above. You should now be able to write the expansion of $(a + b)^6$.

So far, it might seem that to get the numbers in the sixth row of Pascal's triangle you must first know the numbers in the fifth row, but this is not necessary. As the computations at the top of the next page show, each number in the sixth row can be calculated directly by formula.

Blaise Pascal

$$\text{6th row}\begin{cases} \begin{array}{ccccccc} {}_6C_0 & {}_6C_1 & {}_6C_2 & {}_6C_3 & {}_6C_4 & {}_6C_5 & {}_6C_6 \\ 1 & \dfrac{6}{1} & \dfrac{6\cdot5}{1\cdot2} & \dfrac{6\cdot5\cdot4}{1\cdot2\cdot3} & \dfrac{6\cdot5\cdot4\cdot3}{1\cdot2\cdot3\cdot4} & \dfrac{6\cdot5\cdot4\cdot3\cdot2}{1\cdot2\cdot3\cdot4\cdot5} & \dfrac{6!}{6!} \\ 1 & 6 & 15 & 20 & 15 & 6 & 1 \end{array} \end{cases}$$

You may be wondering why combinations have anything to do with the sixth row of Pascal's triangle and the expansion of $(a + b)^6$. To see why, consider $(a + b)^6$ in factored form:

$$(a + b)^6 = (a + b)(a + b)(a + b)(a + b)(a + b)(a + b)$$

The term a^6 in the expansion of $(a + b)^6$ is obtained by multiplying the a's in the 6 factors. The term containing a^5b is obtained by multiplying an a from 5 of the factors and a b from 1 of them. Since there are $_6C_1 = 6$ ways to choose the factor that contributes the b, there are 6 of these a^5b terms. Likewise, the term containing a^4b^2 is obtained by multiplying an a from 4 of the factors and a b from 2 of the factors. Since there are $_6C_2 = 15$ ways to choose the 2 factors that contribute the b's, there are 15 of these a^4b^2 terms. A similar argument can be used to prove the following theorem.

The Binomial Theorem

If n is a positive integer, then:

$$(a + b)^n = {}_nC_0a^nb^0 + {}_nC_1a^{n-1}b^1 + {}_nC_2a^{n-2}b^2 + {}_nC_3a^{n-3}b^3 + \cdots + {}_nC_na^0b^n$$

Equivalently:

$$(a + b)^n = 1a^nb^0 + \frac{n}{1}a^{n-1}b^1 + \frac{n(n-1)}{1\cdot2}a^{n-2}b^2 + \frac{n(n-1)(n-2)}{1\cdot2\cdot3}a^{n-3}b^3 + \cdots + 1a^0b^n$$

Example

Give the first four terms in the expansion of $(x - 2y)^{10}$ in simplified form.

Solution

First, we use the binomial theorem to find the first four terms in the expansion of $(a + b)^{10}$:

$$a^{10} + 10a^9b + \frac{10\cdot9}{1\cdot2}a^8b^2 + \frac{10\cdot9\cdot8}{1\cdot2\cdot3}a^7b^3, \text{ or}$$
$$a^{10} + 10a^9b + 45a^8b^2 + 120a^7b^3$$

Then, substituting x for a and $-2y$ for b in the expression above, we obtain the first four terms in the expansion of $(x - 2y)^{10}$:

$$x^{10} + 10x^9(-2y) + 45x^8(-2y)^2 + 120x^7(-2y)^3, \text{ or}$$
$$x^{10} - 20x^9y + 180x^8y^2 - 960x^7y^3$$

Combinatorics **591**

Students are frequently confused by references to the "fourth" term of $(a + b)^n$ as $_nC_3a^{n-3}b^3$, since there are no 4's in the expression. Help them by pointing out that there are $n + 1$ terms in all, that the exponents of b will run from 0 to n (which accounts for the $n + 1$ terms), and that the exponent of b in the kth term is $k - 1$. Some students may want to have an expression for the kth term; namely:

$$_nC_{k-1}a^{n-(k-1)}b^{k-1}$$

Additional Answers
Class Exercises

1. The numbers in Pascal's triangle have symmetry in a vertical line drawn through the center of the triangle.

◩ Using Technology

Students may wish to write a program to print the first n rows of Pascal's triangle. One possible program in BASIC would be:

```
10 INPUT "NUMBER OF
   ROWS: "; N
20 FOR R =1 TO N
30 LET C = 1
40 FOR T = 1 TO R + 1
50 PRINT C; " ";
60 LET C = C *
   (R − T + 1)/T
70 NEXT T
80 PRINT
90 NEXT R
100 END
```

A program that uses arrays can also be written. Such programs will help students complete Exercises 27 and 28.

CLASS EXERCISES

1. *Visual Thinking* Study Pascal's triangle (page 590). Describe the symmetry that the numbers in the triangle have.

2. Give the seventh row of Pascal's triangle. 1 7 21 35 35 21 7 1

3. **a.** Find the first 4 numbers in the eighth row of Pascal's triangle. 1 8 28 56
   b. State the first 4 terms of the expansion of $(x + y)^8$. x^8, $8x^7y$, $28x^6y^2$, $56x^5y^3$
   c. State the first 4 terms of the expansion of $(x - y)^8$. x^8, $-8x^7y$, $28x^6y^2$, $-56x^5y^3$

In Exercises 4–6, use Pascal's triangle to give the expansion of each binomial.

4. $(a - b)^3$ $a^3 - 3a^2b + 3ab^2 - b^3$ **5.** $(a + b)^4$ See below.   **6.** $(a - b)^4$ See below.

7. Find **(a)** the third term in the expansion of $(x + y)^6$, and **(b)** the fourth term in the expansion of $(x + y)^9$. $15x^4y^2$; $84x^6y^3$

5. $a^4 + 4a^3b + 6a^2b^2 + 4ab^3 + b^4$ 6. $a^4 - 4a^3b + 6a^2b^2 - 4ab^3 + b^4$

WRITTEN EXERCISES

In Exercises 1–12, give the expansion of each binomial. Simplify your answers.

A

1. a. $(a + b)^3$	**b.** $(20 + 1)^3$	**c.** $(20 - 1)^3$
2. a. $(x + y)^4$	**b.** $(10 + 1)^4$	**c.** $(10 - 1)^4$
3. a. $(a + b)^5$	**b.** $(a - b)^5$	**c.** $(2a + 1)^5$
4. a. $(p + q)^6$	**b.** $(p - q)^6$	**c.** $(3p - 2)^6$
5. a. $(x + y)^7$	**b.** $(x - y)^7$	**c.** $(x^2 - 2y)^7$
6. a. $(a + b)^8$	**b.** $(a - b)^8$	**c.** $(2a - b^2)^8$
7. $(x^2 - y^2)^3$	**8.** $(2x^2 - 1)^4$	**9.** $(x^2 + 1)^5$
10. $\left(1 + \dfrac{x}{2}\right)^3$	**11.** $\left(x + \dfrac{1}{x}\right)^6$	**12.** $\left(2x - \dfrac{1}{x}\right)^8$

In Exercises 13–16, find the first four terms of the expansion of the given expression. Do *not* simplify your answers.

13. $(a^2 - b)^{100}$ 14. $(3p + 2q)^{20}$

15. $(\sin x + \sin y)^{10}$ 16. $(\sin x - \cos y)^{30}$

17. Find the value of $(1.01)^5$ to the nearest hundredth by considering the expansion of $(1 + 0.01)^5$. 1.05

18. Find the value of $(0.99)^5$ to the nearest hundredth by considering the expansion of $(1 - 0.01)^5$. 0.95

B 19. In the expansion of $(a + b)^{12}$, what is the coefficient **(a)** of a^8b^4? **(b)** of a^4b^8? 495 495

20. In the expansion of $(a + b)^{20}$, what is the coefficient **(a)** of $a^{17}b^3$? **(b)** of a^3b^{17}? 1140; 1140

21. In the expansion of $\left(x^2 + \dfrac{2}{x}\right)^{12}$, find and simplify the constant term. 126,720

22. In the expansion of $(a^3 - 2)^{10}$, find the term containing a^{18}. $3360a^{18}$

23. A municipal council consists of a mayor and n councilors. The council needs to choose a committee of k members.
   a. How many committees are possible if any council member can be chosen? $_{n+1}C_k$
   b. How many committees are possible if the mayor must be included? $_nC_{k-1}$
   c. How many committees are possible if the mayor must not be included? $_nC_k$
   d. Use parts (a), (b), and (c) to complete the following equation:
   $$_{n+1}C_k = {}_?C_{k-1} + {}_?C_k \quad {}_{n+1}C_k = {}_nC_{k-1} + {}_nC_k$$
   e. Use algebra to prove that the equation in part (d) is correct.

24. Rewrite the binomial theorem using sigma (Σ) notation. $\displaystyle\sum_{k=0}^{n} {}_nC_k a^{n-k}b^k$

For Exercises 25 and 26, use the alternate form of Pascal's triangle shown at the right. Note the single 1 at the top of the triangle. This is sometimes called the "zeroth row" because it represents the expansion of $(a + b)^0$.

```
        1
      1   1
    1   2   1
  1   3   3   1
1   4   6   4   1
```

25. In what column of Pascal's triangle do the triangular numbers (see page 491) occur? Explain why they occur. 3

26. a. What special sequence of numbers results from summing the numbers in each row of Pascal's triangle? The nonnegative integral powers of 2
   b. Consider the diagonals that originate at a 1 on the left side of Pascal's triangle and have a slope of 1. What special sequence of numbers results from summing the numbers along each diagonal? Fibonacci sequence

27. *Investigation* Copy Pascal's triangle as given on page 590. Use a highlighting marker to color all the even numbers in the triangle. What patterns do you notice? (*Note*: You may need to add quite a few rows to the triangle before any patterns emerge.)

28. *Investigation* Repeat Exercise 27, but this time color all the numbers that are multiples of 3. What patterns do you notice?

29. a. Show that the expansion of $(\cos\theta + i\sin\theta)^3$ simplifies to:
   $$(\cos^3\theta - 3\cos\theta\sin^2\theta) + i(3\sin\theta\cos^2\theta - \sin^3\theta)$$
   b. Use De Moivre's theorem (see page 408) to find $(\cos\theta + i\sin\theta)^3$.
   c. Use parts (a) and (b) to show that:
   $$\cos 3\theta = \cos^3\theta - 3\cos\theta\sin^2\theta$$
   $$\text{and } \sin 3\theta = 3\sin\theta\cos^2\theta - \sin^3\theta$$

30. Use De Moivre's theorem and the expansion of $(\cos\theta + i\sin\theta)^4$ to show that:
   $$\cos 4\theta = \cos^4\theta - 6\cos^2\theta\sin^2\theta + \sin^4\theta$$
   $$\text{and } \sin 4\theta = 4\cos^3\theta\sin\theta - 4\cos\theta\sin^3\theta$$

Combinatorics **593**

Suggested Assignments

Discrete Math
Day 1: 592/1–13 odd
Day 2: 592/17–23

Supplementary Materials

Alternative Assessment, 48
Student Resource Guide,
143–145

**Additional Answers
Written Exercises**

27. Pascal's triangle for which even numbers are highlighted shows a pattern of inverted equilateral triangles of various sizes. Specifically, in each row n, where $n = 2^k$ and $k = 1$, 2, 3, . . . , an inverted triangle having $n - 1$ numbers on a side begins. Moreover, to the left and to the right of each inverted triangle is the same pattern of triangles that appears immediately above the inverted triangle.

28. Pascal's triangle for which numbers divisible by three are highlighted shows a pattern of inverted equilateral triangles of various sizes. Specifically, at each row $n = 3^k$ for $k = 1$, 2, 3, . . . , there begins a trio of equilateral triangles having $n - 1$ numbers on a side. (The triangles in each trio are themselves arranged in a triangular pattern.) Moreover, in the regions between the triangles of each trio is the same pattern of triangles that appears immediately above the uppermost triangle in the trio.

In the binomial theorem, if we substitute 1 for a and x for b, we get:

$$(1 + x)^n = 1 + nx + \frac{n(n-1)}{1 \cdot 2} x^2 + \frac{n(n-1)(n-2)}{1 \cdot 2 \cdot 3} x^3 + \cdots$$

If n is not a positive integer, the right side is an infinite series with sum $(1 + x)^n$ when $|x| < 1$. (A proof requires calculus.) Use this fact to do Exercises 31–34.

C **31.** Show that $(1 + x)^{-1} = 1 - x + x^2 - x^3 + \cdots$. (As a check on your work, notice that the right side of the equation is an infinite geometric series with first term 1 and ratio $-x$. Thus, the series has sum $\dfrac{1}{1 - (-x)} = (1 + x)^{-1}$ when $|x| < 1$.)

32. Show that $(1 + x)^{-2} = 1 - 2x + 3x^2 - 4x^3 + \cdots$.

33. Show that $(1 + x)^{1/2} \approx 1 + \frac{1}{2}x$ when $|x|$ is small. Use the result to approximate $\sqrt{1.04}$ and $\sqrt{0.98}$. 1.02; 0.99

34. Show that $(1 + x)^{1/3} \approx 1 + \frac{1}{3}x$ when $|x|$ is small. Use the result to approximate $\sqrt[3]{1.12}$ and $\sqrt[3]{67}$. $\left(Hint: \sqrt[3]{67} = \sqrt[3]{64\left(1 + \frac{3}{64}\right)}\right)$ 1.04; 4.06

Chapter Summary

1. If sets A and B are subsets of a universal set U, then:

(1) the *intersection* of A and B, denoted $A \cap B$, consists of those elements in *both* A and B,

(2) the *union* of A and B, denoted $A \cup B$, consists of those elements in *either* A or B,

(3) the *complement* of A, denoted $\overline{A}$, consists of those elements of U that are *not* in A.

Venn diagrams are used to illustrate a universal set and its subsets, along with their intersections, unions, and complements.

2. Four basic counting principles are presented.

a. *The inclusion-exclusion principle*: For any sets A and B,

$$n(A \cup B) = n(A) + n(B) - n(A \cap B).$$

b. *The multiplication principle*: If an action can be performed in n_1 ways, and for each of these ways another action can be performed in n_2 ways, then the two actions can be performed together in $n_1 n_2$ ways.

c. *The addition principle*: If two actions are mutually exclusive, and the first can be done in n_1 ways and the second in n_2 ways, then one action *or* the other can be done in $n_1 + n_2$ ways.

d. *The complement principle*: If A is a subset of a universal set U, then

$$n(A) = n(U) - n(\overline{A}).$$

3. The choice of r objects from a set of n objects is called a *permutation* if the order of choosing is important and a *combination* if the order is unimportant. The number of possible choices in each case is given by:

$$_nP_r = \frac{n!}{(n-r)!} \quad \text{and} \quad _nC_r = \frac{n!}{(n-r)!\,r!}$$

4. Let S be a set of n elements of k different types. Let n_1 = the number of elements of type 1, n_2 = the number of elements of type 2, . . . , n_k = the number of elements of type k. Then the number of distinguishable permutations of the n elements is $\dfrac{n!}{n_1!\,n_2!\,n_3! \cdots n_k!}$.

5. For a positive integer n, the *binomial theorem* gives the expansion of $(a + b)^n$:

$$(a + b)^n = {}_nC_0 a^n b^0 + {}_nC_1 a^{n-1} b^1 + {}_nC_2 a^{n-2} b^2 + \cdots + {}_nC_n a^0 b^n$$

Pascal's triangle can be used to find the *binomial coefficients*, $_nC_k$.

Key vocabulary and ideas

Venn diagram (p. 565)
intersection (p. 565)
union (p. 565)
complement (p. 566)
inclusion-exclusion principle (p. 566)
multiplication principle (p. 571)
addition principle (p. 573)

complement principle (p. 574)
permutation, combination (p. 578)
permutations with repetition (p. 584)
linear and circular permutations (p. 584)
Pascal's triangle (p. 590)
binomial coefficient (p. 590)
binomial theorem (p. 591)

Chapter Test

1. Of the 320 children at a movie, 190 buy popcorn and 245 buy something to drink. If 151 children buy both, how many buy neither? 36   15-1

2. A basketball team has 4 guards, 5 forwards, and 3 centers on its roster. In how many ways can a team consisting of 1 left guard, 1 right guard, 1 left forward, 1 right forward, and 1 center be formed? 720   15-2

3. A bank customer selects a password consisting of 4 different letters or 4 different digits. How many different passwords are possible? 363,840

4. A bookshelf has space for 5 books. If 7 different books are available, how many different arrangements can be made on the shelf? 2520   15-3

5. Six people meet at a party, and each pair of people shakes hands. How many handshakes are there? 15

6. A boat has 3 red, 3 blue, and 2 yellow flags with which to signal other boats. All 8 flags are flown in various sequences to denote different messages. How many such sequences are possible? 560   15-4

7. **Writing** Write a paragraph contrasting linear and circular permutations. Illustrate the difference between them with a specific example.

8. In the expansion of $(x - 2)^{15}$, find the coefficients of x^{13} and x^{12}.   15-5
   420; −3640

Combinatorics   **595**

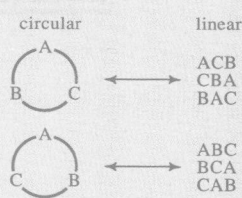

Chapter 16 Probability

Overview

Probability is a field of mathematics of steadily growing importance. Students interested in the biological sciences, business, and the social sciences may find their study of probability to be extremely useful.

The first two sections introduce the basic concepts of probability, which students use to find the probability of a single event, of either of two events, and of two events occurring together. The third section introduces the binomial probability theorem, which students use to find or approximate the multiple occurrences of an event on repeated trials of a binomial experiment. The fourth section gives students an alternative approach (based on combinations) to calculating probabilities. The last two sections introduce students to applications of probability; students solve problems of prediction and payoff from medicine, insurance, game theory, and so on.

Objectives

16-1 To find a sample space of an experiment and the probability of an event or either of two events.

16-2 To find the probability of events occurring together and to determine whether two events are independent.

16-3 To use the binomial probability theorem to find the probability of a given outcome on repeated independent trials of a binomial experiment and to approximate the probability when the trials are not independent.

16-4 To use combinations to solve probability problems.

16-5 To solve problems involving conditional probability.

16-6 To find expected value in situations involving gains and losses and to determine whether a game is fair.

Supplementary Resources

Tests, pp. 50–51
Alt. Assess., pp. 49–52, 70
Activities, pp. 44–45
Student Res. Guide, pp. 148–156

Pacing Guide

Section	16-1	16-2	16-3	16-4	16-5	16-6	Review	Test	Total
Discrete Math (days)	1	2	1	2	2	2	1	1	12

Teaching Notes

16-1 pages 597–605

Presenting the Section

The activity suggested in the text is an easy way to introduce many of the important concepts of probability. Although there are two distinct outcomes for the tossing of a thumbtack, they are *not* equally likely. As each pair of students reports results, you might want to compute a "running probability" based on the accumulated number of tacks landing point up and the accumulated number of tosses. For example, if the first group reports that 13 out of 20 tosses were point up, then the "running probability" is $\frac{13}{20} = 0.65$; if the second group reports that 11 out of 20 tosses were point up, then the "running probability" becomes $\frac{13 + 11}{20 + 20} = 0.60$; and so on. Although the "running probabilities" may vary at first, they should eventually stabilize.

To help students see the difference between empirical and theoretical probability, you might ask students how they would determine the probability of getting "heads" on the flip of a coin if:

1. they assume that the coin is *fair*. (Here theoretical probability can be employed: Since "heads" and "tails" are equally likely, the probability of getting "heads" is taken to be $\frac{1}{2}$.)
2. they assume that the coin is *biased*. (Here empirical probability must be employed: By flipping the coin, say, 100 times and getting "heads" 45 of those times, the probability of getting "heads" is taken to be 0.45.)

Communication

Note that the text does not require that the outcomes listed in a sample space be equally likely but only that (1) they are mutually exclusive, and (2) together they cover all possible results. Emphasize, however, that calculation of theoretical probabilities *does* require a list of equally likely outcomes.

Other ways of expressing the basic concept of the probability of an event are:

$$P = \frac{\text{number of "favorable" outcomes}}{\text{number of outcomes}}$$

and

$$P = \frac{\text{number of "successes"}}{\text{number of trials}}$$

Moreover, when a Venn diagram is used to picture a sample space S and a subset A, then the probability of A can be defined as:

$$P(A) = \frac{n(A)}{n(S)}$$

You may also wish to point out to students that the "odds in favor" of an event A can be expressed as the ratio $P(A):(1 - P(A))$, while the "odds against" an event can be expressed as the inverse ratio $(1 - P(A)):P(A)$.

Using Technology

A computer simulation for determining theoretical probability is suggested in the side-column notes on page 601. The program makes use of the computer's capacity to generate random numbers.

16-2 pages 605–613

Presenting the Section

To introduce the concept of independent events, you might return to the activity from the first section and again distribute thumbtacks to your students. This time, however, prepare the tacks in advance so that one third of the tacks have shafts bent to make a 60° angle with the heads. Have student pairs record the results of 20 tack tosses, then gather the results into a chart like this one:

	No. of tosses	No. landing point up
Bent tacks		
Normal tacks		
All tacks		

Discuss the overall probability for tacks landing point up and compare it to the probabilities for normal tacks only and bent tacks only to determine whether the way the tack lands is independent of whether it is bent.

When discussing the rules for the probability of two events occurring together, you might want to point out that Rule 2 is actually biconditional. That is, events A and B are independent *if and only if* $P(A \text{ and } B) = P(A) \cdot P(B)$. (Students may prove this by using the definition of independent events, page 607, and Rule 1, page 608.) We therefore have an alternate means of checking for independent events. For example, to answer the question in part (c) of Example 4, page 609, we can see that $P($student studies chemistry and French$) = \frac{7}{30}$ does not equal the product $P($student studies chemistry$) \cdot P($student studies French$) = \frac{4}{15} \cdot \frac{7}{15} = \frac{28}{225}$, so the events "student studies chemistry" and "student studies French" are not independent.

Communication

You may wish to point out that the mathematical definition of independent events requires exact equality of the probabilities $P(B)$ and $P(B \mid A)$ and that this will also result in $P(B \mid \text{not } A) = P(B) = P(B \mid A)$. In small sets, such equality may be impossible! For example, consider a class of 14 girls and 12 boys. Suppose we assume that the sex of a student is independent of whether he or she receives honors grades. If 7 girls receive honors grades, how many boys should receive honors grades to satisfy the mathematical definition of independent events? (6) What if 8 girls receive honors? (6.86 boys—an obvious impossibility!) Such a discussion may motivate an interest in the study of confidence intervals in the next chapter.

16-3   pages 613–618

Presenting the Section

The binomial probability theorem is a useful mathematical tool for determining the probability of k occurrences of an event A in n repeated independent trials: $P(k A\text{'s}) = {}_nC_k \, p^k(1 - p)^{n-k}$ where p and $1 - p$ are the probabilities of A's occurrence and nonoccurrence, respectively, on any trial. If the sample space is sufficiently large, the trials need not be independent; the theorem will give a close approximation of the actual probability.

You may want to tell students that any experiment having exactly two mutually exclusive outcomes (such as "heads" or "tails," six or non-six) is called a *Bernoulli trial* after the Swiss mathematician Jakob Bernoulli (1654–1705). One outcome of a Bernoulli trial is usually considered "success," while the other is "failure."

When discussing the binomial probability theorem, be sure to stress that four conditions must be met:

1. There is a fixed number n of repeated trials.
2. Each trial results in success or failure (that is, each trial is a Bernoulli trial).
3. There is independence between trials.
4. The probability p of success (and therefore the probability $1 - p$ of failure) remains constant from trial to trial.

Review

In order to complete the work of this section, students need to be thoroughly familiar with the formula for ${}_nC_r$ from Section 15-3.

Using Technology

Many scientific calculators have a combinations key that students can use for the exercises in this and subsequent sections.

16-4   pages 619–623

Presenting the Section

The technique for solving probability problems in this section makes use of a variety of skills students already have. Their knowledge of combinations, conditional probability, and the multiplication principle are all brought into play to solve problems that cannot easily be solved by using the rules for conditional probability alone. The crucial skill for students to develop here is the ability to analyze the problems properly so that they can make use of what they already know.

Detailed class discussion of the examples and of the Class Exercises should help students develop the analytical skill required for solving the types of problems given in the Written Exercises. Encourage them to refer back to the examples as they solve the problems.

Cooperative Learning

The B- and C-level exercises could be assigned to groups of three or four students, with each group being assigned one or two problems. Solutions could then be shared with the entire class.

Communication

The Communication feature on page 623 asks students to consider the meanings of words that are shared between mathematics and other disciplines and between mathematics and everyday discourse.

16-5   pages 624–629

Presenting the Section

Notice that this section introduces no new mathematics, but rather gives students an opportunity to apply what they have learned in the previous sections to new kinds of problems, including some with real-world applications.

The flu example may be more clearly explained using a table like the one below with two vertical columns for "flu" and "no flu" and two horizontal rows for "high temperature" and "no high temperature." For students who find the analysis difficult, you may wish to consider a hypothetical school with 1000 students. Enter $0.35(1000) = 350$ at the bottom of the first column and $1000 - 350 = 650$ at the bottom of the second. Then separate the 350 "flu" cases into $0.9(350) = 315$ with high temperatures and $350 - 315 = 35$ without. Separate the 650 "no flu cases" into $0.12(650) = 78$ with high temperatures and $650 - 78 = 572$ without. Finally, place the number of "high temperature" cases, 393, to the right of the first row and the total number of "no high temperature" cases, 607, to the right of the second row. (These are often called "marginal" probabilities.) Encourage students to use the integral values in the table to find the probabilities asked for in parts (b) and (c) of the example.

	Flu	No flu	Total
High temperature	315	78	393
No high temperature	35	572	607
Total	350	650	1000

Applications

Written Exercises 11 and 12 discuss an important application of conditional probability to medicine. Students need to realize that most medical tests are not foolproof.

16-6   pages 630–635

Presenting the Section

The basic concepts of expected value and fair game from game theory are introduced here. Students should find of interest the applications to games of chance, test taking, insurance, consumer economics, and business.

You should stress that the expected value of a game is the *average* gain or loss if the game is played many times. For instance, a person who plays the lottery described in Example 2, page 631, throws away an average of $0.47 each time he or she buys a lottery ticket. That is, while the person may win once in a very great while, the accumulated costs of the $1 tickets overwhelm the winnings.

Cooperative Learning

Have groups of students invent fair games and/or games that will make a profit for the group "running" the game. Have groups exchange games and write critiques of each other's efforts.

16 *Probability*

596

Finding Probabilities

16-1 Introduction to Probability

Objective *To find a sample space of an experiment and the probability of an event or either of two events.*

Probability theory is the branch of mathematics that deals with uncertainty. Although the foundations of probability theory were laid in the 1500s and 1600s by mathematicians who were interested in questions about gambling, the subject has since become associated with many other fields, such as meteorology and genetics.

At its most basic level, probability theory assigns a number (between 0 and 1, inclusive) to an *event* as a means of indicating the *probability*, or likelihood, of the occurrence of the event. For example, if a weather report states that the probability of precipitation tomorrow is 60%, then in the past, whenever weather conditions similar to the current conditions existed, it rained 3 out of 5 times.

The probability of an event is determined either *empirically* or *theoretically*. The precipitation probability mentioned above is an example of an empirical probability because it is based on previous observations of the weather. Another example of empirical probability is found in the following activity.

Activity

For this activity, you should work with a partner. You and your partner will need a cup and a thumbtack with a flat head. (All the thumbtacks used by the class should be alike.)

a. Repeat the following at least 20 times: Place the thumbtack in the cup, shake it, and "pour" it onto your desk. Your partner should then record whether the tack lands "point up" or "point down," as shown at the right.

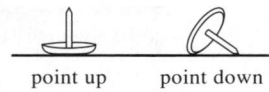
point up   point down

b. Based on your record of results from part (a), what would you say is the probability that the tack lands "point up"?

c. Combine the results from part (a) for all the pairs of partners in the class. Based on the combined results, what would you say is the probability that the tack lands "point up"?

d. Of the two probabilities from parts (b) and (c), which do you think is a better predictor of the tack's "behavior"? Explain your reasoning.

e. Describe how changing the shape of the tack would affect the probability that it lands "point up."

◀ As Mark Twain reportedly said, "Everybody talks about the weather, but nobody ever does anything about it!" *Probability theory,* however, and records of similar conditions allow us to at least make reasonable weather predictions.

Probability   **597**

Warm-Up Exercises

1. You toss a dime twice. Copy and complete the table below to show all possible outcomes. You can get "heads" or "tails" on each toss.

First toss	Second toss
H	H
H	T
T	H
T	T

2. Which principle explains the number of possible outcomes? The mult. principle; there are 2 outcomes (H or T) for each toss, so there are $2 \cdot 2 = 4$ outcomes in all.

3. Find the probability that you get the same outcome on both tosses? $\frac{2}{4} = \frac{1}{2}$

4. Find the probability that you get at least one "heads." $\frac{3}{4}$

5. Find the probability that you get "heads" on the first toss. $\frac{1}{2}$

Motivating the Section

Ask students to identify some situations in which probabilities are used. (Two such situations are weather forecasts (precipitation probabilities) and sports contests (odds of winning).)

Emphasize the phrase "exactly one" in the definition of a sample space and ask students to give examples of the two different errors possible if that phrase is ignored: a set that has no entry corresponding to a possible outcome and a set that has more than one entry for a possible outcome. For the experiment of selecting a random coin from the set of all United States coins, the first error is represented by {penny, nickel, dime, quarter}, which omits the half dollar and the silver dollar. The second error is represented by {copper coin, silver coin, coin from Philadelphia mint, coin from Denver mint} since both mints produce copper and silver coins.

Although empirical probabilities are unavoidable in many applied areas of probability theory, in this chapter we will usually consider events whose probabilities can be determined simply by reasoning about the events. For example, by assuming that the mass of a coin is evenly distributed so that each of the two sides (which are called "heads" and "tails") has an equal chance of turning up when the coin is flipped, we can say that the theoretical probability of getting "heads" is $\frac{1}{2}$. Likewise, the shape and composition of a die lead us to believe that each of the six faces has an equal chance of turning up, so the theoretical probability of rolling, say, a "3" is $\frac{1}{6}$.

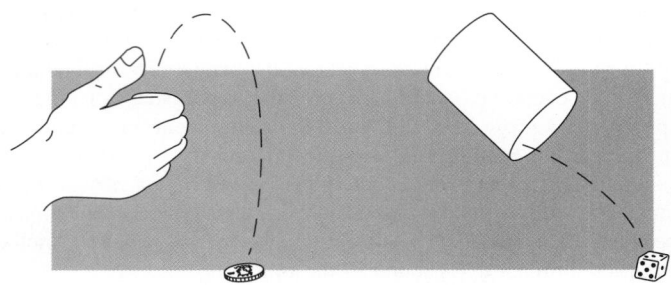

At this point, let us generalize our discussion of theoretical probability. We refer to an action having various outcomes that occur unpredictably (such as tossing a coin or rolling a die) as an **experiment**. A **sample space** of an experiment is a set S such that each outcome of the experiment corresponds to exactly one element of S. An **event** is any subset of a sample space. If a sample space of an experiment contains n equally likely outcomes and if m of the n outcomes correspond to some event A, then the **probability** of event A, denoted $P(A)$, is:

$$P(A) = \frac{m}{n}$$

If all n outcomes correspond to event A, then the event is certain to occur and $P(A) = \frac{n}{n} = 1$. Similarly, if no outcomes correspond to event A, then the event is certain *not* to occur and $P(A) = \frac{0}{n} = 0$.

Example 1 Suppose a die is rolled. Give a sample space for this experiment. Then find the probability of rolling a prime number.

Solution A sample space for the die-rolling experiment is {1, 2, 3, 4, 5, 6}. Of the six equally likely outcomes in the sample space, three outcomes (2, 3, and 5) correspond to the event "rolling a prime number." Thus:

$$P(\text{prime number}) = \frac{3}{6} = \frac{1}{2}$$

598 *Chapter Sixteen*

To determine the probability of the occurrence of either one of two different events, we can use a Venn diagram (introduced in Section 15-1). For example, in the roll of a die, consider the events "rolling a number divisible by 2" and "rolling a number divisible by 3." The Venn diagram at the right shows these two events. Notice that one number, 6, is common to both events, so that:

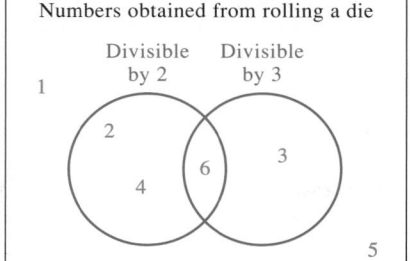

Numbers obtained from rolling a die

$$P\left(\begin{matrix} \text{divisible} \\ \text{by 2 or 3} \end{matrix}\right) = P\left(\begin{matrix} \text{divisible} \\ \text{by 2} \end{matrix}\right) + P\left(\begin{matrix} \text{divisible} \\ \text{by 3} \end{matrix}\right) - P\left(\begin{matrix} \text{divisible} \\ \text{by 2 and 3} \end{matrix}\right)$$

$$= \quad \frac{3}{6} \quad + \quad \frac{2}{6} \quad - \quad \frac{1}{6} \quad = \frac{2}{3}$$

Note that this result is based on the inclusion-exclusion principle (page 566).

Probability of Either of Two Events

For any two events A and B,
$$P(A \text{ or } B) = P(A) + P(B) - P(A \text{ and } B).$$

Two events A and B that cannot occur simultaneously are called **mutually exclusive** events. For example, in the roll of a die, the events "rolling a 3" and "rolling an even number" are mutually exclusive. Moreover:

$$P(3 \text{ or even number}) = P(3) + P(\text{even number})$$

$$= \quad \frac{1}{6} \quad + \quad \frac{3}{6} \quad = \frac{2}{3}$$

This result, generalized below, is a special case of the result stated above.

Probability of Either of Two Mutually Exclusive Events

If events A and B are mutually exclusive, then
$$P(A \text{ or } B) = P(A) + P(B).$$

The event "not A" occurs when event A does not. Since events A and "not A" are mutually exclusive and since one or the other is certain to occur, we have:

$$P(A \text{ or not } A) = 1$$
$$P(A) + P(\text{not } A) = 1$$
$$P(\text{not } A) = 1 - P(A)$$

1. Suppose you toss a coin four times.
   a. How many different equally likely outcomes are possible?
   b. Find the probability of obtaining no heads.
   c. Find the probability of obtaining at least one head.
   d. Find the probability of obtaining exactly one head.

   a. On each toss, there are two outcomes, heads and tails. By the multiplication principle, there are $2 \cdot 2 \cdot 2 \cdot 2 = 16$ possible outcomes, each of which is equally likely to occur.
   b. $P(TTTT) = \frac{1}{16}$
   c. The event of obtaining at least one head is the complement of the event in part (b), so $P(\text{at least one head}) = 1 - \frac{1}{16} = \frac{15}{16}$.
   d. $P(\text{exactly one head}) = P(HTTT, THTT, TTHT, TTTH) = \frac{4}{16} = \frac{1}{4}$

2. Each of five cards is labeled with a letter A, B, C, D, or E. Two cards are chosen without the first card being replaced.
   a. List all the possible outcomes.
   b. Find the probability that both letters chosen are consonants.

   a. $S = \{AB, AC, AD, AE, BA, BC, BD, BE, CA, CB, CD, CE, DA, DB, DC, DE, EA, EB, EC, ED\}$

Example 2 As shown below, a standard deck of playing cards consists of 52 cards, with 13 cards in each of four *suits* (clubs, spades, diamonds, and hearts). Clubs and spades are black cards, and diamonds and hearts are red cards. Also, jacks, queens, and kings are called *face cards*.

Clubs (♣): ace, 2, 3, 4, 5, 6, 7, 8, 9, 10, jack, queen, king
Spades (♠): ace, 2, 3, 4, 5, 6, 7, 8, 9, 10, jack, queen, king
Diamonds (♦): ace, 2, 3, 4, 5, 6, 7, 8, 9, 10, jack, queen, king
Hearts (♥): ace, 2, 3, 4, 5, 6, 7, 8, 9, 10, jack, queen, king

If the deck is well-shuffled, what is the probability that the top card is:
a. a black ace?   b. not a black ace?   c. a diamond face card?

Solution a. Since the cards are well-shuffled, any one of them has the same chance of being the top card. With 2 black aces in the deck,
$$P(\text{black ace}) = \frac{2}{52} = \frac{1}{26}.$$

b. $P(\text{not a black ace}) = 1 - P(\text{black ace})$
$$= 1 - \frac{1}{26} = \frac{25}{26}$$

c. With 3 diamond face cards in the deck,
$$P(\text{diamond face card}) = \frac{3}{52}.$$

In Example 2, notice that the card-choosing experiment has such sample spaces as:

$S_1 =$ the set of 52 cards listed in Example 2
$S_2 = \{$club, spade, diamond, heart$\}$
$S_3 = \{$black card, red card$\}$
$S_4 = \{$spade, non-spade$\}$

We often refer to the elements of a sample space as *sample points*. For example, the outcome of choosing, say, the 7 of clubs corresponds to the sample point "club" in the sample space S_2, to the sample point "black card" in the sample space S_3, and to the sample point "non-spade" in the sample space S_4.

If you were asked for the probability of drawing a spade from a standard deck, sample space S_4 would not be helpful, because its two sample points "spade" and "non-spade" are not equally likely. (Thus, it would be incorrect to say that the probability of a spade is $\frac{1}{2}$.) On the other hand, the four sample points of sample space S_2 *are* equally likely, so we can say $P(\text{spade}) = \frac{1}{4}$. We can also use the 52 equally likely sample points of S_1 and say $P(\text{spade}) = \frac{13}{52} = \frac{1}{4}$.

Example 3 Suppose a red die and a white die are rolled. What is the probability that the sum of the numbers showing on the dice is 9 or 10?

Solution The table below gives a sample space consisting of 36 ordered pairs (r, w) where r and w are the numbers showing on the red die and the white die.

White die

		1	2	3	4	5	6
	1	1, 1	1, 2	1, 3	1, 4	1, 5	1, 6
	2	2, 1	2, 2	2, 3	2, 4	2, 5	2, 6
Red	3	3, 1	3, 2	3, 3	3, 4	3, 5	3, 6
die	4	4, 1	4, 2	4, 3	4, 4	4, 5	4, 6
	5	5, 1	5, 2	5, 3	5, 4	5, 5	5, 6
	6	6, 1	6, 2	6, 3	6, 4	6, 5	6, 6

Sum = 9   Sum = 10

Since four of the equally likely sample points give a sum of 9 and three give a sum of 10, $P(\text{sum is } 9 \text{ or } 10) = \dfrac{4}{36} + \dfrac{3}{36} = \dfrac{7}{36}$.

Sometimes it is convenient to use a tree diagram to obtain a sample space, as shown in the next example.

Example 4 Suppose you toss a coin three times. What is the probability that exactly two of the tosses result in "heads"?

Solution The tree diagram at the right illustrates the experiment of tossing a coin three times and obtaining a "head" (H) or a "tail" (T) on each toss. By following each path of the tree, you get:

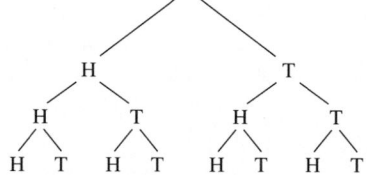

Sample space = {HHH, HHT, HTH, HTT, THH, THT, TTH, TTT}

Thus, $P(2 \text{ "heads"}) = P(\text{HHT or HTH or THH}) = \dfrac{3}{8}$.

The chances of winning a sporting event are often given as *odds* instead of as a probability. If the odds in favor of a team winning are 6 to 5, this means that the team's winning chances to its losing chances are in a 6 to 5 ratio. In other words, the probability of winning is $\dfrac{6}{6 + 5} = \dfrac{6}{11}$, and the probability of losing is $\dfrac{5}{11}$.

---

b. $P(BC, BD, CB, CD, DB, DC) = \dfrac{6}{20} = \dfrac{3}{10}$. (Note that there are three consonants, so there are $3 \cdot 2 = 6$ ways of choosing two without replacement.)

📉 **Using Technology**

A computer or programmable calculator can be used to simulate any of the situations described in the examples of this section. For instance, the following program simulates Example 4:

```
10  PRINT "HOW MANY
    TRIALS";
20  INPUT N
30  LET S = 0
40  FOR I = 1 TO N
50  LET H = 0
60  FOR T = 1 TO 3
70  IF RND(1) < 0.5 THEN
    LET H = H + 1
80  NEXT T
90  IF H = 2 THEN LET S =
    S + 1
100 NEXT I
110 PRINT S;" OUT OF ";N;
    "TRIALS RESULTED IN
    EXACTLY TWO HEADS."
120 PRINT "PROBABILITY OF
    GETTING TWO HEADS IN
    THREE TOSSES OF A
    COIN IS ABOUT "; S/N
130 END
```

Of course, this program gives an *empirical* probability. Students who use this program should see that the empirical probability approximates the theoretical probability better as the number of trials increases.

Students may wish to write simulation programs for the other examples in this section.

Have students translate into odds the probabilities of getting 0, 1, 2, or 3 heads when three coins are tossed. (The odds in favor are 1:7 for 0 heads and 3 heads, and 3:5 for 1 head and 2 heads.)

Additional Answers
Class Exercises

9. *S* is not a sample space because some outcomes, such as queen of hearts, correspond to more than one element in *S*.

1. As in Example 1, suppose a die is rolled. Find each probability.
   a. P(perfect square) $\frac{1}{3}$ **b.** P(factor of 60) 1   **c.** P(negative number) 0

2. If a card is drawn at random from a standard deck of 52 cards, what is the probability of getting:
   a. the queen of hearts? $\frac{1}{52}$ **b.** a heart? $\frac{1}{4}$ $\frac{3}{13}$ **c.** a queen? $\frac{1}{13}$ $\frac{3}{26}$
   d. a red card? $\frac{1}{2}$ **e.** a face card? $\frac{3}{13}$ **f.** a red face card? $\frac{3}{26}$

3. If two dice are rolled, what is the probability that both show the same number? $\frac{1}{6}$

4. If two dice are rolled, find the probability of getting:
   a. a sum of 3 $\frac{1}{18}$
   b. a sum of 4 $\frac{1}{12}$
   c. a sum of 3 or 4 $\frac{5}{36}$

5. **Meteorology** If the probability of rain tomorrow is 40%, what is the probability of no rain tomorrow? 60%

6. **Occupational Safety** If the probability of no accidents in a manufacturing plant during one month is 0.82, what is the probability of at least one accident? 0.18

7. *Discussion* In the solution of Example 3, another possible sample space is the set of the 11 possible sums for the two dice:

$$S = \{2, 3, 4, 5, 6, 7, 8, 9, 10, 11, 12\}$$

Since two of these 11 possibilities correspond to a sum of 9 or 10, some people might be tempted to say that the probability of a sum of 9 or 10 is $\frac{2}{11}$. Discuss why this reasoning is incorrect. The possible sums are not equally likely.

8. A penny, a nickel, and a dime are each tossed.
   a. Does the set $S = \{0, 1, 2, 3\}$, which gives the four different numbers of "heads" that could come up, satisfy the definition of a sample space? Yes
   b. What is wrong with reasoning that since one of the four sample points in part (a) corresponds to 2 "heads," then $P(2 \text{ "heads"}) = \frac{1}{4}$? The sample points are not equally likely.

9. A card is picked at random from a standard deck. Explain why the set $S = \{$club, spade, red card, face card$\}$ is *not* a sample space.

10. Comment on the following reasoning:
    There are 3 states whose names begin with the letter C (California, Colorado, and Connecticut); call them "C-states." Likewise, there are 3 "O-states" (Ohio, Oklahoma, and Oregon). Thus, if a person is chosen at random from the U.S. population, that person has the same probability of being from a "C-state" as from an "O-state." This is incorrect because population is not evenly distributed among the states.

602   *Chapter Sixteen*

For Exercises 1–4, suppose a card is drawn from a well-shuffled standard deck of 52 cards. Find the probability of drawing each of the following.

A 1. **a.** a black card $\frac{1}{2}$ **b.** a spade $\frac{1}{4}$ **c.** not a spade $\frac{3}{4}$

2. **a.** a black face card $\frac{3}{26}$ **b.** a black jack $\frac{1}{26}$ **c.** not a black jack $\frac{25}{26}$

3. **a.** a red diamond $\frac{1}{4}$ **b.** a black diamond 0 **c.** not a black diamond 1

4. **a.** a jack $\frac{1}{13}$ **b.** a jack or king $\frac{2}{13}$ **c.** neither jack nor king $\frac{11}{13}$

5. Mr. and Mrs. Smith each bought 10 raffle tickets. Each of their three children bought 4 tickets. If 4280 tickets were sold in all, what is the probability that the grand prize winner is:
   a. Mr. or Mrs. Smith? $\frac{1}{214}$ **b.** one of the 5 Smiths? $\frac{4}{535}$ **c.** none of the Smiths? $\frac{531}{535}$

6. One of the integers between 11 and 20, inclusive, is picked at random. What is the probability that the integer is:
   a. even? $\frac{1}{2}$ **b.** divisible by 3? $\frac{3}{10}$ **c.** a prime? $\frac{2}{5}$

7. New York City is divided into five boroughs: Manhattan, Queens, the Bronx, Brooklyn, and Staten Island. Suppose that a New York City telephone number is randomly chosen.
   a. Explain why the probability that it is a Manhattan telephone number is *not* $\frac{1}{5}$.
   b. What do you need to know in order to find the correct probability in part (a)?

8. Suppose that a member of the U.S. Senate and a member of the U.S. House of Representatives are randomly chosen to be photographed with the President. Explain why $\frac{1}{50}$ is the probability that the senator is from Iowa, and why $\frac{1}{50}$ is *not* the probability that the representative is from Iowa.

In Exercises 9–12, use the table on page 601, which gives the 36 equally likely outcomes when two dice are rolled. Find the probability of each event.

9. **a.** Sum is 6. $\frac{5}{36}$ **b.** Sum is 7. $\frac{1}{6}$ **c.** Sum is 8. $\frac{5}{36}$ $\frac{35}{36}$

10. **a.** Sum is even. $\frac{1}{2}$ **b.** Sum is 12. $\frac{1}{36}$ **c.** Sum is less than 12.

11. The two dice show different numbers. $\frac{5}{6}$

12. The red die shows a greater number than the white die. $\frac{5}{12}$

Additional Answers
Written Exercises

7. **a.** There are not an equal number of phone numbers in each borough.
    b. The number of phone numbers in Manhattan and the number of phone numbers in New York City

8. Each state has an equal number of senators, but the number of representatives varies according to the population of the state.

Suggested Assignments

Discrete Math
603/1, 5, 7, 9, 15, 17, 25, 27, 29

Supplementary Materials

Alternative Assessment, 49
Activities Book, 44–45

13. b. $\frac{1}{4}$

14. b. $\frac{1}{3}$

19. a. {{Alvin, Bob}, {Alvin, Carol}, {Alvin, Donna}, {Bob, Carol}, {Bob, Donna}, {Carol, Donna}}

b. $\frac{1}{6}$

c. $\frac{1}{2}$

20. a. {(black, black), (black, red), (red, black), (red, red)}

b. After a red card is drawn, a black card is more likely to be drawn because there are 26 black cards left and only 25 red cards.

21. The inclusion-exclusion principle states that "For any sets A and B, $n(A \cup B) = n(A) + n(B) - n(A \cap B)$." If A and B are events in a sample space of s equally likely outcomes, then dividing both sides of the above equation by s will turn each term into a probability: $\frac{n(A \cup B)}{s} =$ $\frac{n(A)}{s} + \frac{n(B)}{s} - \frac{n(A \cap B)}{s}$ becomes $P(A \text{ or } B) = P(A) + P(B) - P(A \text{ and } B)$, which is the equation for the probability of either of two events.

---

13. A die is rolled and a coin is tossed.
   a. Make a tree diagram showing the 12 possible outcomes of this experiment.
   b. Find the probability that the die's number is even and the coin is "heads."

14. The numbers 1, 2, 3, and 4 are written on separate slips of paper and placed in a hat. Two slips of paper are then randomly drawn, one after the other and without replacement.
   a. Make a tree diagram showing the 12 possible outcomes of this experiment.
   b. Find the probability that the sum of the numbers picked is 6 or more.

15. Suppose that the odds in favor of the National League's winning the All-Star Game are 4 to 3.
   a. What is the probability that the National League wins? $\frac{4}{7}$
   b. What is the probability that the American League wins? $\frac{3}{7}$

16. Suppose that the odds in favor of the incumbent's winning an election are 2 to 3.
   a. What is the probability that the incumbent wins? $\frac{2}{5}$
   b. What is the probability that the incumbent's challenger wins? $\frac{3}{5}$

B **17.** Suppose you roll two dice, each of which is a regular octahedron with faces numbered 1 to 8.
   a. What is the probability that the sum of the numbers showing is 2? $\frac{1}{64}$
   b. What is the probability that the sum is 3? $\frac{1}{32}$
   c. What sum is most likely to appear? 9

18. Suppose you roll two dice, each of which is a regular dodecahedron with faces numbered 1 to 12.
   a. What is the probability that the sum of the numbers showing is 24? $\frac{1}{144}$
   b. What is the probability that the sum is 23? $\frac{1}{72}$
   c. What sum is most likely to appear? 13

19. From a group consisting of Alvin, Bob, Carol, and Donna, two people are to be randomly selected to serve on a committee.
   a. Give a sample space for this experiment.
   b. Find the probability that Bob and Carol are selected.
   c. Find the probability that Carol is not selected.

20. From a standard deck of cards, two cards are randomly drawn, one after the other and without replacement. The color of each card is noted.
   a. Give a sample space for this experiment.
   b. Explain why two red cards are less likely to be drawn than a red card and then a black card.

21. *Reading* Explain how the inclusion-exclusion principle (see page 566) can be used to find the probability of either of two events (see page 599).

---

604 *Chapter Sixteen*

22. Of the 1260 households in a small town, 632 have dogs, 568 have cats, and 114 have both types of pet. If a household is chosen at random, what is the probability that the household has either type of pet? $\frac{181}{210}$

23. The letters of the word TEXAS are arranged in a random order. What is the probability that the letters spell TAXES? $\frac{1}{120}$

24. The letters of the word COSINE are arranged in a random order. What is the probability that the letters spell SONICE? $\frac{1}{720}$

25. A number k is randomly chosen from $\{-3, -2, -1, 0, 1, 2, 3, 4\}$. What is the probability that the expression $x^2 - 2x + k$ can be written as a product of two linear factors, each with integral coefficients? $\frac{3}{8}$

26. A number c is randomly chosen from $\{1, 2, 3, 4, 5, 6\}$. What is the probability that the graph of $y = x^2 - 4x + c$ intersects the x-axis? $\frac{2}{3}$

27. If a 3-letter ''word'' is formed by randomly choosing 3 letters from the word OCEAN, what is the probability that it is composed only of vowels? $\frac{1}{10}$

28. If a 3-letter ''word'' is formed by randomly choosing 3 letters from the word PAINTED, what is the probability that it is composed only of vowels? $\frac{1}{35}$

29. A number between 100 and 999, inclusive, is chosen at random. What is the probability that it contains (a) no 0's? (b) at least one 0? a. $\frac{81}{100}$ b. $\frac{19}{100}$

30. A number between 1000 and 9999, inclusive, is chosen at random. What is the probability that it contains (a) no 9's? (b) at least one 9? a. $\frac{81}{125}$ b. $\frac{44}{125}$

16-2 Probability of Events Occurring Together

Objective   *To find the probability of events occurring together and to determine whether two events are independent.*

If an experiment involves two or more events occurring together (whether successively or simultaneously), a tree diagram is a useful way to calculate the probabilities of the possible outcomes.

Activity

The tree diagram at the right illustrates an experiment in which a coin is tossed twice. (In the diagram, H and T represent ''heads'' and ''tails,'' respectively.)

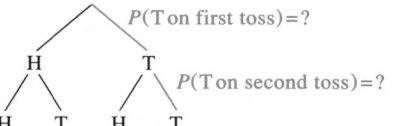

$P(\text{T on first toss}) = ?$

$P(\text{T on second toss}) = ?$

a. Use the tree diagram to give a sample space for the experiment. $\{HH, HT, TH, TT\}$

b. Use your answer to part (a) to find $P(TT)$. $\frac{1}{4}$

c. What are the probabilities indicated in red in the tree diagram? $\frac{1}{2}; \frac{1}{2}$

d. How could you obtain your answer to part (b) using your answers to part (c)? Multiply.

e. If the coin is tossed a third time, use the method from part (d) to find $P(TTT)$. $\frac{1}{8}$

Probability   **605**

A pile of cards consists of three red cards, two blue cards, and one white card.

1. You draw one card from the pile without looking. What is the probability that the card is:

   a. red? $\frac{1}{2}$ **b.** blue? $\frac{1}{3}$

   c. white? $\frac{1}{6}$

2. Assume that on your first drawing the card you selected was white and that it *was not* placed back in the pile. You then draw another card (without looking) from the remaining cards in the pile. What is the probability that this card is:

   a. red? $\frac{3}{5}$ **b.** blue? $\frac{2}{5}$

   c. white? 0

3. Assume again that the card you selected on the first drawing was white, but that it *was* placed back in the pile before you draw a second time. What is the probability that this second card is:

   a. red? $\frac{1}{2}$ **b.** blue? $\frac{1}{3}$

   c. white? $\frac{1}{6}$

4. Compare your answers for Exercises 2 and 3 to those for Exercise 1. Does replacement of the card selected in the first drawing affect the probabilities for the color of the card selected in the second drawing? yes

Motivating the Section

Pose the following question: If a person in this school building is chosen at random, which of the following events has the *least* probability: (a) the person is a teacher, (b) the person wears glasses, or (c) the person is a teacher who wears glasses? Students should choose event (c), because both events (a) and (b) must occur for event (c) to occur. The probability of events occurring together is the topic of this section.

Example Note

Before discussing Example 2, ask students if they think that not replacing the balls will increase the probability, decrease it, or leave it unchanged.

As the activity on the preceding page shows, if you know the probabilities that correspond to the branches of a tree diagram for some experiment, then you can find the probability of a particular outcome by multiplying the probabilities along the path leading to that outcome. We use this approach in the following examples.

Example 1   Two yellow balls and three green balls are placed in a jar. One ball is randomly chosen, its color is noted, and the ball is put back in the jar. This procedure is repeated for a second ball. Find the probability that both balls are the same color.

Solution   The probability of a yellow (Y) first ball is $\frac{2}{5}$, and the probability of a green (G) first ball is $\frac{3}{5}$. Because the first ball is put back in the jar before the second ball is chosen, the second ball has the same probabilities of being yellow and of being green as the first ball. These probabilities are shown next to the branches of the tree diagram at the right. Thus, the probability of getting two yellow balls is:

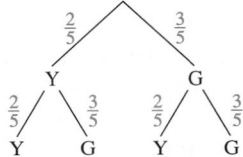

$$P(YY) = \frac{2}{5} \cdot \frac{2}{5} = \frac{4}{25}$$

Similarly:

$$P(YG) = \frac{2}{5} \cdot \frac{3}{5} = \frac{6}{25}$$

$$P(GY) = \frac{3}{5} \cdot \frac{2}{5} = \frac{6}{25}$$

$$P(GG) = \frac{3}{5} \cdot \frac{3}{5} = \frac{9}{25}$$

Note that the sum of these probabilities is 1.

From the above calculations, we find that

$$P(\text{both same color}) = P(YY) + P(GG)$$
$$= \frac{4}{25} + \frac{9}{25} = \frac{13}{25}.$$

Example 2   Rework Example 1 if the first ball is *not* put back in the jar before the second ball is chosen.

Solution   In this situation, the probability of a yellow (Y) first ball is still $\frac{2}{5}$, and the probability of a green (G) first ball is still $\frac{3}{5}$. The probabilities for the second ball, however, are different this time.

606   *Chapter Sixteen*

If the first ball chosen is yellow, then there are 1 yellow and 3 green balls left, so that the probability of a yellow second ball is $\frac{1}{4}$ and the probability of a green second ball is $\frac{3}{4}$. If the first ball is green, however, then there are 2 yellow and 2 green balls left, so that the probability of a yellow second ball is $\frac{2}{4}$ and the probability of a green second ball is also $\frac{2}{4}$. Thus:

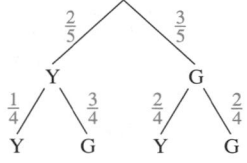

$$P(\text{both same color}) = P(\text{YY}) + P(\text{GG})$$

$$= \frac{2}{5} \cdot \frac{1}{4} + \frac{3}{5} \cdot \frac{2}{4} = \frac{8}{20} = \frac{2}{5}$$

In Examples 1 and 2, we denote the probability that the second ball is yellow given that the first ball is yellow by:

$$P(\text{second is Y} \mid \text{first is Y})$$

This probability is called a **conditional probability**, because it gives the probability that the second ball is yellow on the condition that the first ball is yellow. Other conditional probabilities are considered below.

In Example 1: $P(\text{second is G} \mid \text{first is G}) = \frac{3}{5}$

Note that these are equal.

$\qquad\qquad\qquad P(\text{second is G} \mid \text{first is Y}) = \frac{3}{5}$

In Example 2: $P(\text{second is G} \mid \text{first is G}) = \frac{2}{4}$

Note that these are different.

$\qquad\qquad\qquad P(\text{second is G} \mid \text{first is Y}) = \frac{3}{4}$

In Example 1, as the calculations on the preceding page show, the probability that the second ball is green is $\frac{3}{5}$ whether or not the first ball is green. Since the probability that the second ball is green does *not* depend on the color of the first ball, the two events ''second ball is green'' and ''first ball is green'' are called *independent events*. This is not the case, however, in Example 2 where the probability that the second ball is green *does* depend on whether or not the first ball is green.

In general, two events A and B are **independent** if and only if the occurrence of A does not affect the probability of the occurrence of B. In other words, events A and B are independent if and only if

$$P(B \mid A) = P(B).$$

Probability   **607**

In discussing Example 3, you
should point out that the independence of the two events
has nothing to do with how
many actions are performed.
In this case, a single action—
drawing a card—is performed
and two events—"jack" and
"spade"—are defined for the
action.

Additional Examples

1. Two cards are drawn from
the top of a well-shuffled
deck of cards. If A represents an ace and N represents a "non-ace," find
$P(AA)$, $P(AN)$, $P(NA)$, and
$P(NN)$ in the following
cases:

a. the first card is replaced
and the cards are shuf-
fled before the second
card is drawn.

b. the cards are drawn
consecutively without
replacement.

a.

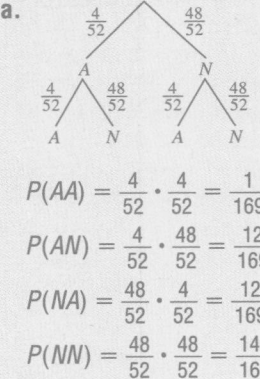

$$P(AA) = \frac{4}{52} \cdot \frac{4}{52} = \frac{1}{169}$$

$$P(AN) = \frac{4}{52} \cdot \frac{48}{52} = \frac{12}{169}$$

$$P(NA) = \frac{48}{52} \cdot \frac{4}{52} = \frac{12}{169}$$

$$P(NN) = \frac{48}{52} \cdot \frac{48}{52} = \frac{144}{169}$$

Examples 1 and 2 suggest the following rules for events occurring together.

Probability of Events Occurring Together

Rule 1. For any two events A and B,

$$P(A \text{ and } B) = P(A) \cdot P(B \mid A).$$

Rule 2. If events A and B are independent, then

$$P(A \text{ and } B) = P(A) \cdot P(B).$$

Example 3   A card is randomly drawn from a standard deck.

a. Show that the events "jack" and "spade" are independent.

b. Show how Rule 2 given above can be used to find the probability of
drawing the jack of spades.

Solution   **a.** Since 13 of the 52 cards are spades, $P(\text{spade}) = \frac{13}{52}$.

Since 1 of the 4 jacks is a spade, $P(\text{spade} \mid \text{jack}) = \frac{1}{4}$.

Thus, since $P(\text{spade} \mid \text{jack}) = P(\text{spade})$, the events "jack" and
"spade" satisfy the condition for independence stated at the bottom of
the preceding page.

b. If we think of drawing the jack of spades as the simultaneous occurrence of the events "jack" and "spade," Rule 2 gives:

$$P(\text{jack and spade}) = P(\text{jack}) \cdot P(\text{spade}) = \frac{1}{13} \cdot \frac{1}{4} = \frac{1}{52}$$

We can write Rule 1 given above as:

$$P(B \mid A) = \frac{P(A \text{ and } B)}{P(A)}$$

In this form, the rule gives us a means of calculating conditional probability. Another way to
calculate $P(B \mid A)$ can be seen by picturing
events A and B as sets in a Venn diagram, as
shown at the right. Since event A is known to
have happened, we can ignore everything outside circle A in the diagram. The probability of B
given A is just the fraction of A's sample points
that are in B. We therefore have the rule

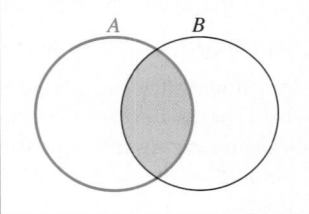

$$P(B \mid A) = \frac{n(A \cap B)}{n(A)}$$

which is equivalent to the one above.

Example 4 Each student in a class of 30 studies one foreign language and one science. The students' choices are shown in the table below.

	Chemistry (C)	Physics (P)	Biology (B)	Totals
French (F)	7	4	3	14
Spanish (S)	1	6	9	16
Totals	8	10	12	30

a. Find the probability that a randomly chosen student studies chemistry.

b. Find the probability that a randomly chosen student studies chemistry given that the student studies French.

c. Are the events "student studies chemistry" and "student studies French" independent?

Solution

a. Since 8 of the 30 students study chemistry, $P(C) = \frac{8}{30} = \frac{4}{15}$.

b. Of the 14 students who study French, 7 study chemistry. Thus:
$$P(C \mid F) = \frac{n(C \cap F)}{n(F)} = \frac{7}{14} = \frac{1}{2}$$

c. Parts (a) and (b) show that $P(C \mid F) \neq P(C)$, so the events "student studies chemistry" and "student studies French" are *not* independent.

CLASS EXERCISES

1. *Reading* Find an instance where each of the two rules given at the top of the preceding page is used in Examples 1 or 2.

2. In Example 3, are the events "jack" and "face card" independent? No

3. Consider the class of 30 students described in Example 4.
   a. Find the probability that a randomly chosen student studies Spanish. $\frac{8}{15}$
   b. Find the probability that a randomly chosen student studies biology given that the student studies Spanish. $\frac{9}{16}$
   c. Are the events "student studies Spanish" and "student studies biology" independent? No

WRITTEN EXERCISES

 1. From a box containing 3 red balls and 5 green balls, 2 balls are randomly picked, one after the other and without replacement.
   a. Draw a tree diagram showing the probabilities of each branch of the tree.
   b. Are the events "second ball is red" and "first ball is red" independent? No
   c. Find the probability that both balls are the same color. c. $\frac{13}{28}$ d. $\frac{15}{28}$
   d. Find the probability that one ball is red and one is green.

b.

$P(AA) = P(A \mid A) =$
$\frac{4}{52} \cdot \frac{3}{51} = \frac{1}{221}$
$P(AN) = P(N \mid A) =$
$\frac{4}{52} \cdot \frac{48}{51} = \frac{16}{221}$
$P(NA) = P(A \mid N) =$
$\frac{48}{52} \cdot \frac{4}{51} = \frac{16}{221}$
$P(NN) = P(N \mid N) =$
$\frac{48}{52} \cdot \frac{47}{51} = \frac{188}{221}$

2. Two dice are rolled. Show that the events "both numbers are the same" and "sum is 6" are *not* independent events. Refer to Example 3 in Section 16-1.

P(numbers are the same) $= \frac{6}{36} = \frac{1}{6}$.

{sum is 6} = {(1, 5), (2, 4), (3, 3), (4, 2), (5, 1)}, so

P(numbers are the same | sum is 6) $= \frac{1}{5}$.

Since P(numbers are the same) $\neq P$(numbers are the same | sum is 6), the events are *not* independent.

Additional Answers Class Exercises

1. Rule 1 is used in Example 2 when finding $P(YY)$ and $P(GG)$. Rule 2 is used in Example 1 when finding $P(YY)$ and $P(GG)$.

Additional Answers
Written Exercises

2. b. No

   c. $\frac{1}{2}$

   d. $\frac{1}{2}$

3. b. Yes

   c. $\frac{17}{32}$

   d. $\frac{15}{32}$

4. b. Yes

   c. $\frac{5}{8}$

   d. $\frac{3}{8}$

7. a. $P(F) = P(F|D) = \frac{3}{13}$

   b. $P(R) = \frac{1}{2}$, $P(R|D) = 1$;
   $P(R) \neq P(R|D)$

10. b. $\frac{1}{216}$, $\frac{5}{72}$, $\frac{25}{72}$, $\frac{125}{216}$

Suggested Assignments

Discrete Math
Day 1: 609/1, 5, 7, 11, 13–18
Day 2: 612/21–25, 27, 29

Supplementary Materials

Alternative Assessment, 50

---

2. Repeat Exercise 1 if the box contains 1 red ball and 3 green balls.

3. Repeat Exercise 1 if the first ball chosen is put back in the box and mixed with the other balls before the second ball is picked.

4. Repeat Exercise 2 if the box contains 1 red ball and 3 green balls and if the first ball chosen is put back in the box and mixed with the other balls before the second ball is picked.

5. Two cards are dealt from the top of a well-shuffled standard deck of cards.
   a. Draw a tree diagram showing the probabilities of a heart (H) and non-heart (N) for each of the two cards.
   b. Find $P(HH)$, $P(HN)$, $P(NH)$, and $P(NN)$. $\frac{1}{17}$, $\frac{13}{68}$, $\frac{13}{68}$, $\frac{19}{34}$

6. A coin is slightly bent so that the probability of getting "heads" on a toss is 0.55 instead of 0.50.
   a. Draw a tree diagram showing the probabilities for two tosses of the coin. Use H for "heads" and T for "tails."
   b. Find $P(HH)$, $P(HT)$, $P(TH)$, and $P(TT)$. 0.3025, 0.2475, 0.2475, 0.2025

7. A card is randomly chosen from a standard deck.
   a. Show that the events "face card" and "diamond" are independent.
   b. Show that the events "red card" and "diamond" are *not* independent.

8. The following 6 cards are placed face down on a table: ace, 2, and 3 of hearts; ace and 2 of clubs; and ace of spades. One card is randomly chosen.
   a. Are the events "ace" and "heart" independent? No
   b. Are the events "ace" and "club" independent? Yes

9. A penny, a nickel, and a dime are tossed one after the other.
   a. Draw a tree diagram showing the probabilities of a "head" (H) and "tail" (T) for each toss.
   b. Find $P(3 \text{ "heads"})$, $P(\text{exactly 2 "heads"})$, $P(\text{exactly 1 "head"})$, and $P(\text{no "heads"})$. $\frac{1}{8}$, $\frac{3}{8}$, $\frac{3}{8}$, $\frac{1}{8}$

10. A die is rolled 3 times.
    a. Draw a tree diagram showing the probabilities of a six (S) and non-six (N) for each roll.
    b. Find $P(3 \text{ sixes})$, $P(\text{exactly 2 sixes})$, $P(\text{exactly 1 six})$, and $P(\text{no sixes})$.

11. The letters of the word COMPUTER are written on separate slips of paper and placed in a hat. Three letters are then randomly drawn from the hat, one after the other and without replacement. To find the probability that all 3 letters are vowels, two methods can be used.

    Method 1: $P(3 \text{ vowels}) = \dfrac{\text{number of 3-letter words having 3 vowels}}{\text{number of 3-letter words}}$

    Method 2: $P(3 \text{ vowels}) = P(\text{first is vowel}) \times$
    $P(\text{second is vowel} \mid \text{first is vowel}) \times$
    $P(\text{third is vowel} \mid \text{first and second are vowels})$

    a. Use both methods to find $P(3 \text{ vowels})$. $\frac{1}{56}$
    b. Using whichever method you prefer, find the probability that all 3 letters are consonants. $\frac{5}{28}$

12. In a math class with 13 girls and 11 boys, the teacher randomly selects four students to put homework problems on the chalkboard. What is the probability that all are girls? (*Hint*: Use one of the methods described in Exercise 11.) $\frac{65}{966}$

In Exercises 13–18, use the table below, which gives the areas of concentration of the 400 students at a small college.

Area of concentration	Class				
	Freshman	Sophomore	Junior	Senior	Totals
Natural sciences	50	35	33	29	147
Social sciences	20	25	28	24	97
Humanities	40	40	39	37	156
Totals	110	100	100	90	400

13. If a student is selected at random, what is the probability that:
   a. the student's area of concentration is the natural sciences? $\frac{147}{400}$
   b. the student is a freshman in the social sciences? $\frac{1}{20}$

14. If a sophomore is selected at random, what is the probability that his or her area of concentration is the humanities? $\frac{2}{5}$

15. a. What is the probability that a senior's area of concentration is the natural sciences? $\frac{29}{90}$
   b. What is the probability that a student in the natural sciences is a senior? $\frac{29}{147}$

16. a. What is the probability that a junior's area of concentration is the humanities? $\frac{39}{100}$
   b. What is the probability that a student in the humanities is a junior? $\frac{39}{156}$

17. A student is selected at random. Are the events "student is a junior" and "student is in the humanities" independent? Yes

18. A student is selected at random. Are the events "student is a junior" and "student is in the natural sciences" independent? No

B **19.** *Reading* Given a sample space S with subsets A and B, use $n(S)$ to show that the rule $P(B \mid A) = \frac{n(A \cap B)}{n(A)}$ is equivalent to the rule $P(B \mid A) = \frac{P(A \text{ and } B)}{P(A)}$.

20. *Reading* Read again the two rules given at the top of page 608. Explain why Rule 2 is a special case of Rule 1.
If A and B are independent, $P(B \mid A) = P(B)$.

Probability **611**

21. Three cards are dealt from a well-shuffled standard deck.

    a. Complete the calculation: $P(3 \text{ clubs}) = \frac{13}{52} \cdot \frac{12}{51} \cdot \frac{?}{?} = ?$

    b. Find the probability that all 3 cards are red.

    c. Find the probability that all 3 cards are aces.

    d. Find the probability that none of the cards is an ace.

    e. Find the probability that at least one of the cards is an ace.

    f. Find the probability that either one or two of the cards is an ace.

22. Five cards are dealt from a well-shuffled standard deck.

    a. Find the probability that all 5 cards are spades. (*Hint:* See Exercise 21(a).)

    b. Find the probability that all 5 cards are of the same suit.

    c. Find the probability that none of the cards is an ace.

    d. Find the probability that at least one card is an ace.

23. Repeat Exercise 21 if each card is replaced before the next is drawn.

24. Repeat Exercise 22 if each card is replaced before the next is drawn.

25. Three people are randomly chosen. Find the probability of each event.

    a. All were born on different days of the week.   **a.** $\frac{30}{49}$ **b.** $\frac{19}{49}$

    b. At least two people were born on the same day of the week.

26. A die is rolled 3 times. Find the probability of each event.

    a. All 3 numbers are different. **a.** $\frac{5}{9}$ **b.** $\frac{4}{9}$

    b. At least 2 of the numbers are the same.

27. A bag contains 3 red marbles and 2 black marbles. Two persons, A and B, take turns drawing a marble from the bag without replacing any of the marbles drawn. The first person to draw a black marble wins. What is the probability that person A, who draws first, wins? $\frac{3}{5}$

28. Repeat Exercise 27 if the bag contains 4 red marbles and 1 black marble. $\frac{3}{5}$

29. **Sports** The championship series of the National Basketball Association consists of a series of at most 7 games between two teams X and Y. The first team to win 4 games is the champion and the series is over. At any time before or after a game, the status of the series can be recorded as a point (x, y) on the grid shown at the right. The point $A(3, 1)$, for example, means that team X

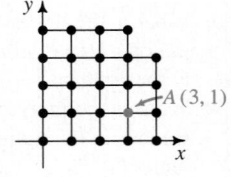

has won 3 games and team Y has won 1 game. From point A, the series can end in a championship for team X in 3 ways (X, YX, YYX). If you assume that team X has a probability of 0.6 of winning each and every remaining game, then the probability that team X becomes champion from point A is $P(X) + P(YX) + P(YYX) = 0.6 + (0.4)(0.6) + (0.4)(0.4)(0.6) = 0.936$.

    a. Find the probability that team Y becomes champion from point A. 0.064

    b. If team X has won 1 game and team Y has won 3 games, find the probability that team Y becomes champion. 0.784

    c. If team X has won 2 games and team Y has won 1 game, find the probability that team X becomes champion. 0.8208

30. Sports Refer to Exercise 29 and suppose that the *home* team has a probability
p of winning each game and a probability $q = 1 - p$ of losing.
  a. If team X is the home team for games 1, 2, 6, and 7, what is the probability
  that team X wins the championship? Give the answer in terms of p only.
  b. Use a computer or graphing calculator to find the approximate value of p
  that makes team X's probability of winning the championship 0.5.

16-3 The Binomial Probability Theorem

Objective To use the binomial probability theorem to find the probability of a given outcome on repeated independent trials of
a binomial experiment and to approximate the probability
when the trials are not independent.

Consider the experiment in which a die is
rolled three times. On each roll, the probability of getting a six (S) is $p = \frac{1}{6}$ and
the probability of getting a non-six (N) is
$1 - p = \frac{5}{6}$. These probabilities are
shown in the tree diagram at the right.
We can use the tree diagram to create the
table below, which summarizes the results of the experiment.

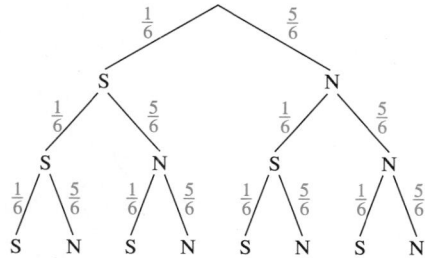

	3 sixes	2 sixes			1 six			0 sixes
Outcome	SSS	SSN	SNS	NSS	SNN	NSN	NNS	NNN
Probability	$\frac{1}{6}\cdot\frac{1}{6}\cdot\frac{1}{6}$	$\frac{1}{6}\cdot\frac{1}{6}\cdot\frac{5}{6}$	$\frac{1}{6}\cdot\frac{5}{6}\cdot\frac{1}{6}$	$\frac{5}{6}\cdot\frac{1}{6}\cdot\frac{1}{6}$	$\frac{1}{6}\cdot\frac{5}{6}\cdot\frac{5}{6}$	$\frac{5}{6}\cdot\frac{1}{6}\cdot\frac{5}{6}$	$\frac{5}{6}\cdot\frac{5}{6}\cdot\frac{1}{6}$	$\frac{5}{6}\cdot\frac{5}{6}\cdot\frac{5}{6}$
	$\left(\frac{1}{6}\right)^3$	$3\left(\frac{1}{6}\right)^2\left(\frac{5}{6}\right)$			$3\left(\frac{1}{6}\right)\left(\frac{5}{6}\right)^2$			$\left(\frac{5}{6}\right)^3$

Now compare:

$$[p + (1-p)]^3 = p^3 + 3p^2(1-p) + 3p(1-p)^2 + (1-p)^3$$

If the die were rolled 4 times instead of 3, we would have the following
distribution of probabilities:

Outcome	4 sixes	3 sixes	2 sixes	1 six	0 sixes
Probability	$\left(\frac{1}{6}\right)^4$	$4\left(\frac{1}{6}\right)^3\left(\frac{5}{6}\right)$	$6\left(\frac{1}{6}\right)^2\left(\frac{5}{6}\right)^2$	$4\left(\frac{1}{6}\right)\left(\frac{5}{6}\right)^3$	$\left(\frac{5}{6}\right)^4$

$$[p + (1-p)]^4 = p^4 + 4p^3(1-p) + 6p^2(1-p)^2 + 4p(1-p)^3 + (1-p)^4$$

Probability **613**

Warm-Up Exercises

Write the expansion of each
binomial, simplifying your an-
swer. Then substitute $\frac{2}{3}$ for x

and $\frac{1}{3}$ for y in the expansion
and evaluate the sum.

1. $(x + y)^2$ $x^2 + 2xy + y^2$;
$\frac{4}{9} + \frac{4}{9} + \frac{1}{9} = 1$

2. $(x + y)^3$ $x^3 + 3x^2y +$
$3xy^2 + y^3$; $\frac{8}{27} + \frac{12}{27} +$
$\frac{6}{27} + \frac{1}{27} = 1$

3. $(x + y)^4$ $x^4 + 4x^3y +$
$6x^2y^2 + 4xy^3 + y^4$; $\frac{16}{81} +$
$\frac{32}{81} + \frac{24}{81} + \frac{8}{81} + \frac{1}{81} = 1$

Evaluate each expression for
$p = \frac{1}{4}$ and $q = \frac{3}{4}$.

4. $_7C_5 \cdot p^5 \cdot q^2$ $\frac{189}{16,384}$

5. $_6C_3 \cdot p^3 \cdot q^3$ $\frac{135}{1024}$

6. $_5C_3 \cdot p^3 \cdot q^2$ $\frac{45}{512}$

Motivating the Section

A good introduction to this
section might be to consider
the problem of jury selection.
Suppose that the general population eligible for jury duty is
52% women. In an 8-person
jury we would expect four
women; three women would
be less likely, and one woman
even more unusual. But how
unusual? What is the probability of that happening?

The two illustrations on the preceding page can be generalized to give the following theorem, which is based on the binomial theorem (page 591).

The Binomial Probability Theorem

Suppose an experiment consists of a sequence of n repeated independent trials, each trial having two possible outcomes, A or not A. If on each trial, $P(A) = p$ and $P(\text{not } A) = 1 - p$, then the binomial expansion of $[p + (1 - p)]^n$,

$$_nC_np^n + \cdots + {_nC_k}p^k(1 - p)^{n - k} + \cdots + {_nC_0}(1 - p)^n,$$

gives the following probabilities for the occurrences of A:

Outcome	n A's	$\cdots$	k A's	$\cdots$	0 A's
Probability	$_nC_np^n$	$\cdots$	$_nC_kp^k(1 - p)^{n - k}$	$\cdots$	$_nC_0(1 - p)^n$

Example 1   A coin is tossed 10 times. What is the probability of exactly 4 "heads"?

Solution   Substitute $n = 10$, $k = 4$, and $p = 1 - p = \frac{1}{2}$ in the theorem above.

$$P(4 \text{ out of 10 tosses result in "heads"}) = {_{10}C_4}\left(\frac{1}{2}\right)^4\left(\frac{1}{2}\right)^6$$

$$= 210\left(\frac{1}{2}\right)^{10} = \frac{210}{1024} \approx 0.205$$

Example 2   In a survey of the 1000 students in a school, 950 indicated that they were right-handed. Find the probability that at least one of four randomly chosen students from the school is left-handed.

Solution   Technically speaking, this is not a binomial experiment because the sampling of the four students is done without replacement. In this experiment the probability of the first student's being right-handed is $\frac{950}{1000} = 0.95$.

The probability of the second student's being right-handed is either $\frac{949}{999}$ or $\frac{950}{999}$ depending on whether the first student is right-handed. These two probabilities (as well as the probabilities of the third and fourth student's being right-handed) are so close to 0.95 that we will use the binomial probability theorem even though the theorem assumes a constant 0.95 probability. Thus:

$$P(\text{at least one left-handed}) = 1 - P(\text{all right-handed})$$
$$\approx 1 - {_4C_4}(0.95)^4$$
$$\approx 1 - 0.815 = 0.185$$

Example 3

Of the computer operators who work for a large temporary employment agency, 60% have a car. If 5 of the computer operators are randomly chosen to work on a job that requires car transportation, what is the probability that at least 4 have cars?

Solution

As was the case in Example 2, this is not a binomial experiment because the sampling of the 5 computer operators is done without replacement. Nevertheless, if there is a large number of computer operators, the probability of choosing each of the 5 is close enough to 0.6 that we can use the binomial probability theorem to obtain a good approximation of the probability that at least 4 have cars. Thus:

$$P(\text{at least 4 have cars}) = P(\text{exactly 4 have cars}) + P(5 \text{ have cars})$$
$$\approx {}_5C_4(0.6)^4(0.4)^1 + {}_5C_5(0.6)^5$$
$$\approx 0.259 + 0.078 = 0.337$$

CLASS EXERCISES

1. In the expansion of $\left(\frac{1}{2} + \frac{1}{2}\right)^4$, shown below, the first term represents the probability of getting 4 "heads" in 4 tosses of a coin. What probabilities do the other terms represent?

$$\left(\frac{1}{2}\right)^4 + 4\left(\frac{1}{2}\right)^3\left(\frac{1}{2}\right) + 6\left(\frac{1}{2}\right)^2\left(\frac{1}{2}\right)^2 + 4\left(\frac{1}{2}\right)\left(\frac{1}{2}\right)^3 + \left(\frac{1}{2}\right)^4$$

2. **Discussion** In your own words, tell how to find the probability of getting exactly 3 fives when 4 dice are rolled.

3. A coin is tossed 9 times and comes up "heads" each time. What is the probability that it will come up "heads" on the next toss? $\frac{1}{2}$

4. **a.** A jar contains 50 red marbles and 50 white marbles. Two marbles are randomly drawn, one after the other and without replacement. What is the probability of drawing 2 red marbles?
   b. Explain why the binomial probability theorem does not apply in part (a).
   c. What does the binomial probability theorem give as an *approximation* of the probability in part (a)?

a. $P(1 \text{ blue}) =$
$${}_3C_1 \cdot \frac{3}{5} \cdot \left(\frac{2}{5}\right)^2 =$$
$$\frac{36}{125} = 0.288$$
$P(2 \text{ blue}) =$
$${}_3C_2 \cdot \left(\frac{3}{5}\right)^2 \cdot \frac{2}{5} = \frac{54}{125} =$$
0.432
$P(3 \text{ blue}) =$
$${}_3C_3 \cdot \left(\frac{3}{5}\right)^3 = \frac{27}{125} =$$
0.216

b. $P(1 \text{ blue}) = P(BRR) + P(RBR) + P(RRB) =$
$$\frac{3}{5} \cdot \frac{2}{4} \cdot \frac{1}{3} + \frac{2}{5} \cdot \frac{3}{4} \cdot \frac{1}{3} + \frac{2}{5} \cdot \frac{1}{4} \cdot \frac{3}{3} = \frac{3}{10} = 0.3$$
$P(2 \text{ blue}) = P(BBR) + P(BRB) + P(RBB) =$
$$\frac{3}{5} \cdot \frac{2}{4} \cdot \frac{2}{3} + \frac{3}{5} \cdot \frac{2}{4} \cdot \frac{2}{3} + \frac{2}{5} \cdot \frac{3}{4} \cdot \frac{2}{3} = \frac{3}{5} = 0.6$$
$P(3 \text{ blue}) = \frac{3}{5} \cdot \frac{2}{4} \cdot \frac{1}{3} =$
$$\frac{1}{10} = 0.1$$

Additional Answers
Class Exercises

1. 3 heads and 1 tail, 2 heads and 2 tails, 1 head and 3 tails, 4 tails

2. Find the second term in $\left(\frac{1}{6} + \frac{5}{6}\right)^4$.

4. **a.** $\frac{49}{198}$
   b. The first and second drawings are not independent trials.
   c. $\frac{1}{4}$

Error Analysis

Students often believe that the probability of 2 successes in 4 trials is the same as that of 3 successes in 6 trials since each represents half the trials being successful. You can help students see the error in this reasoning by first calculating the probability of getting exactly one head in two tosses of a coin (0.5) and then asking if they feel the probability is the same for getting *exactly* 10 heads in 20 tosses.

Additional Answers
Written Exercises

7. a. $\dfrac{1}{4}$

   b. $\dfrac{9}{16}, \dfrac{3}{8}, \dfrac{1}{16}$

Suggested Assignments

Discrete Math
616/7–19 odd, 20

Supplementary Materials

Alternative Assessment, 50–51

WRITTEN EXERCISES

A

1. Make a table, like the ones on page 613, showing the 8 different ways in which "heads" (H) and "tails" (T) can occur if a coin is tossed 3 times. Then find the probability of getting:
   a. 3 "heads" $\dfrac{1}{8}$ b. 2 "heads" $\dfrac{3}{8}$ c. 1 "head" $\dfrac{3}{8}$ d. 0 "heads" $\dfrac{1}{8}$

2. Make a table, like the ones on page 613, showing the 16 different ways in which sixes (S) and non-sixes (N) can occur when a die is rolled 4 times. Then find the probability of getting:
   a. 4 sixes $\dfrac{1}{1296}$ b. 3 sixes $\dfrac{5}{324}$ c. 2 sixes $\dfrac{25}{216}$ d. 1 six $\dfrac{125}{324}$ e. 0 sixes $\dfrac{625}{1296}$

3. Consider the set of families with exactly 4 children. If P(child is a boy) $= \dfrac{1}{2}$, find the probability that one of these families, picked at random, has:
   a. 4 boys $\dfrac{1}{16}$ b. 3 boys $\dfrac{1}{4}$ c. 2 boys $\dfrac{3}{8}$ d. 1 boy $\dfrac{1}{4}$ e. 0 boys $\dfrac{1}{16}$

4. Suppose a coin is bent so that the probability of its coming up "heads" on any toss is $\dfrac{2}{5}$. If the coin is tossed 3 times, find the probability of getting:
   a. 3 "heads" $\dfrac{8}{125}$ b. 2 "heads" $\dfrac{36}{125}$ c. 1 "head" $\dfrac{54}{125}$ d. 0 "heads" $\dfrac{27}{125}$

5. What is the probability of getting exactly 2 fives in 4 rolls of a die? ≈ 0.116

6. What is the probability of getting exactly 1 three in 7 rolls of a die? ≈ 0.391

7. a. If one card is drawn from a well-shuffled standard deck, what is the probability of drawing a spade?
   b. If one card is drawn from each of two well-shuffled standard decks, what is the probability of drawing 0 spades? 1 spade? 2 spades?

8. If one card is drawn from each of three well-shuffled standard decks, find the probability of drawing:
   a. 3 spades $\dfrac{1}{64}$ b. 2 spades $\dfrac{9}{64}$ c. 1 spade $\dfrac{27}{64}$ d. 0 spades $\dfrac{27}{64}$

9. A quiz has 6 multiple-choice questions, each with 4 choices. If you guess at every question, what is the probability of getting:
   a. all 6 questions right? $\dfrac{1}{4096}$ b. 5 out of 6 questions right? $\dfrac{9}{2048}$

10. A jar contains 4 red balls and 3 white balls, all the same size. Suppose you pull out a ball and note its color, put it back, and mix up the contents of the jar. If you do this twice more, find the probability of getting:
    a. 0 red balls $\dfrac{27}{343}$ b. 1 red ball $\dfrac{108}{343}$ c. 2 red balls $\dfrac{144}{343}$ d. 3 red balls $\dfrac{64}{343}$

B

11. Eight out of every ten nutritionists recommend Brand X. If nutritionists A, B, and C are asked their opinions on Brand X, what is the probability that:
    a. all three recommend Brand X? See below.   b. none recommends Brand X? $\dfrac{1}{125}$
    c. at least one recommends Brand X? $\dfrac{124}{125}$

12. In a certain high school, one third of the senior boys are at least 6 ft tall. In a randomly selected group of 7 senior boys, what is the probability that:
    a. all are less than 6 ft tall? See below.   b. none are less than 6 ft tall? $\dfrac{1}{2187}$
    c. all but one are less than 6 ft tall? $\dfrac{448}{2187}$

11. a. $\dfrac{64}{125}$ 12. a. $\dfrac{128}{2187}$

616   *Chapter Sixteen*

13. In the *random-number table* shown below, the probability that each digit, 0–9, occurs in any given position is 0.1. We can use the table to perform *Monte Carlo simulations* in order to find certain probabilities empirically.

```
49487  52802    28667  62058    87822  14704    18519  17889    45869  14454
29480  91539    46317  84803    86056  62812    33584  70391    77749  64906
25252  97738    23901  11106    86864  55808    22557  23214    15021  54268
02431  42193    96960  19620    29188  05863    92900  06836    13433  21709
69414  89353    70724  67893    23218  72452    03095  68333    13751  37260
```

For example, to use Monte Carlo simulations to find the probability of getting 0 "heads" in 3 tosses of a coin, first associate the digits 0–4 with the outcome "heads" and the digits 5–9 with the outcome "tails." Then the first row of the random-number table represents 16 simulations of 3 tosses of a coin.

```
494|87 5|280|2 28|667| 620|58 8|782|2 14|704 185|19 1|788|9 45|869| 144|54
HTH|TT T|HTH|H HT|TTT| THH|TT T|TTT|H HH|THH HTT|HT H|TTT|T HT|TTT| HHH|
```

a. The empirical probability (0.3125) is greater than the theoretical probability (0.125).

a. Of the 16 simulations, 5 produce 0 "heads." Thus, based on 16 simulations, $P(0 \text{ "heads"}) = 0.3125$. Compare this probability to the theoretical probability of getting 0 "heads" in 3 tosses of a coin (see Exercise 1(d)).

b. Begin with the last 2 digits of the first row and perform 14 more simulations. Using all 30 simulations, find the probability of getting 0 "heads." $0.1\overline{6}$

c. How many simulations can be done using the whole table? Based on these simulations, what is the probability of getting 0 "heads"? 83; ≈ 0.13253

d. *Discussion* When you use Monte Carlo simulations to approximate the theoretical probability of an event, will increasing the number of simulations always improve the accuracy of the approximation? Explain.

14. a. *Discussion* In order to use Monte Carlo simulations (see Exercise 13) to find the probabilities asked for in Exercises 2–12, how should the digits 0–9 be assigned to the possible outcomes for each exercise? (*Hint:* It may be helpful to eliminate some digits. Cross off all occurrences of those digits in the table before actually doing a simulation.)

b. Perform Monte Carlo simulations for Exercise 2 and Exercise 7. Compare the probabilities you find to the theoretical probabilities found using the binomial probability theorem.

15. a. **Sports** A basketball player's free-throw percent is .750. What is the probability that she scores on exactly 4 of her next 5 free throws?

b. *Discussion* Using the binomial probability theorem to complete part (a) assumes that the probability of a successful free throw is always .750. Is this assumption valid? Explain.

a. 0.396 **b.** No; the player's free-throw percent changes each time she takes a free throw.

Probability **617**

Cooperative Learning

Exercises 13 and 14 are perhaps best done in small groups.

Exercise Note

Exercises 13 and 14 can lead to a discussion of the difference between theoretical and empirical probability. To increase the variety of empirical probabilities calculated, students can use different strings of random numbers for the Monte Carlo simulations in Exercise 14. To create a new string of random numbers, begin *anywhere* in the table and move in *any* direction from that number. Alternatively, students can use a calculator or computer with a random-number generator to create a random-number table.

N   Using Technology

Students also can write a computer program to conduct Monte Carlo simulations (see the Computer Exercise on the next page).

16. **Investigation** Let $n = 6$ and $P(A) = 0.5$ in the binomial probability theorem. Plot the points $(k, P(k \text{ occurrences of } A))$ for $k = 0, 1, \ldots, 6$, and describe the shape of a smooth curve that can be drawn through the points. What happens to the shape of this curve for values of $P(A)$ other than 0.5?

17. In a package of tomato seeds, 9 seeds out of 10 sprout on the average. What is the probability that of the first 10 seeds planted, 1 does not sprout? 0.387

18. One out of every 5 boxes of Rice Toasties has a secret message decoder ring. You buy 5 boxes hoping to get at least 2 rings. What are your chances? ≈ 0.263

 **Parts of Exercises 19 and 20 require the use of a calculator or computer. If you use a computer, you may need to write your own short programs.**

19. **a.** If n people are randomly chosen, for what value of n is the probability of at least two having the same birthday equal to 1? 366 (ignoring leap year)

   b. Suppose you guess that the probability in part (a) is 0.5 if $n = 10$. To test the guess, use the probability that no two of 10 people share a birthday:

$$P(\text{no match}) = P\begin{pmatrix} \text{2nd person} \\ \text{does not} \\ \text{match first} \end{pmatrix} \times P\begin{pmatrix} \text{3rd person does} \\ \text{not match either} \\ \text{of first two} \end{pmatrix} \times \cdots \times P\begin{pmatrix} \text{10th person does} \\ \text{not match any} \\ \text{of first nine} \end{pmatrix}$$

$$= \frac{364}{365} \times \frac{363}{365} \times \frac{362}{365} \times \cdots \times \frac{356}{365} = \frac{364 P_9}{365^9}$$

   Thus, $P(\text{at least one match}) = 1 - P(\text{no match}) = 1 - \frac{364 P_9}{365^9}$.

   Use a calculator to evaluate the above expression for $P(\text{at least one match})$. 0.117

   c. The probability that of 10 randomly chosen people at least two have the same birthday is less than 0.5. Use the method of part (b) to determine how many people are needed for the probability to be 0.5. 23

20. **a.** $1 - \left(\frac{364}{365}\right)^n$

20. **a.** What is the probability that in a group of n randomly chosen people, at least one has the same birthday as you?

   b. Find the value of n that makes the probability in part (a) approximately 0.5. 253

//// COMPUTER EXERCISE

Write a program that does Monte Carlo simulations (see Exercise 13) of the binomial probability theorem. Use the computer's random-number generator to simulate N independent trials with probability P that a given outcome occurs on each trial. Count the number of times T that the outcome occurs. Repeat the simulation of N trials many times and count how often $T = 0$, $T = 1$, $\ldots$, $T = N$. Convert these tallies to approximate probabilities by dividing by the number of simulations.

16-4 Probability Problems Solved with Combinations

Objective *To use combinations to solve probability problems.*

In Example 1 we show two methods of finding the probability that five cards randomly drawn from a standard deck are all hearts. Method 1 uses conditional probability. Method 2, which uses combinations, is presented because it can be used to solve many problems that are not readily solved using conditional probability.

Example 1 Five cards are drawn at random from a standard deck. Find the probability that all 5 cards are hearts.

Solution **Method 1** Since there are 13 hearts in the 52-card deck, the probability that the first card drawn is a heart is $\frac{13}{52}$. With 12 hearts among the 51 cards remaining, the probability that the second card drawn is a heart is $\frac{12}{51}$. Continuing in this way, we have:

$$P(\text{all hearts}) = \frac{13}{52} \cdot \frac{12}{51} \cdot \frac{11}{50} \cdot \frac{10}{49} \cdot \frac{9}{48}$$

$$= \frac{33}{66,640} \approx 0.000495$$

Method 2 The experiment of choosing 5 cards from 52 has $_{52}C_5$ equally likely outcomes. Of these, there are $_{13}C_5$ combinations that contain 5 (of the 13) hearts. Thus:

$$P(\text{all hearts}) = \frac{_{13}C_5}{_{52}C_5} = \frac{1287}{2,598,960} \approx 0.000495$$

Example 2 Five cards are drawn at random from a standard deck. Find the probability that exactly 2 are hearts.

Solution In this example it is impractical to use a card-by-card approach based on conditional probability, as in Method 1 of Example 1. The difficulty is that we do not know whether the hearts will appear as the first and second cards, as the first and third cards, or as some other two cards. Instead, we determine the number of 5-card combinations that contain exactly 2 hearts. There are $_{13}C_2$ choices for the 2 hearts, and there are $_{39}C_3$ choices for the other 3 cards. By the multiplication principle (page 571), the number of 5-card combinations containing exactly 2 hearts is $_{13}C_2 \cdot {}_{39}C_3$. Since there are $_{52}C_5$ possible 5-card combinations, we have:

$$P\binom{2 \text{ hearts and}}{3 \text{ non-hearts}} = \frac{_{13}C_2 \cdot {}_{39}C_3}{_{52}C_5} = \frac{78 \cdot 9139}{2,598,960} \approx 0.274$$

Probability **619**

Warm-Up Exercises

Three cards are drawn from a well-shuffled standard deck of 52 cards, one after the other and without replacement.

1. Find the probability of drawing:

   a. all clubs. $\frac{11}{850}$

   b. no clubs. $\frac{703}{1700}$

   c. exactly one club. (Hint: The club can occur on the first, second, or third drawing.) $\frac{741}{1700}$

2. Evaluate:

   a. $\frac{_{13}C_3}{_{52}C_3}$ $\frac{11}{850}$

   b. $\frac{_{39}C_3}{_{52}C_3}$ $\frac{703}{1700}$

   c. $\frac{_{13}C_1 \cdot {}_{39}C_2}{_{52}C_3}$ $\frac{741}{1700}$

3. Compare your answers to Exercises 1 and 2. What do you notice? 1(a) and 2(a) are the same, as are 1(b) and 2(b), and 1(c) and 2(c).

4. Use the results of Exercises 1-3 to write and evaluate an expression using combinations to find the probability of getting exactly two clubs.
   $\frac{_{13}C_2 \cdot {}_{39}C_1}{_{52}C_3} = \frac{117}{850}$

Motivating the Section

You may wish to introduce this section by doing an example whose probability can be determined by listing all possibilities, such as selecting a committee of 3 from 5 students.

In discussing Example 3, you may wish to point out that most calculators can give only an approximation to $_{52}C_{13}$. To obtain the exact value, students may need to write out the factors and do as much canceling as possible before multiplying.

Additional Examples

1. Four balls are chosen at random from a bag containing 4 white balls, 3 green balls, and 5 black balls.
   a. Find the probability of choosing 0, 1, 2, 3, or 4 black balls and show that the sum of the probabilities is 1.
   b. Find the probability of choosing 2 white balls and 2 black balls.

   a. $P(0 \text{ black}) = \frac{_7C_4}{_{12}C_4} =$
   $\frac{35}{495} \approx 0.071$

   $P(1 \text{ black}) = \frac{_7C_3 \cdot _5C_1}{_{12}C_4} =$
   $\frac{35 \cdot 5}{495} \approx 0.354$

   $P(2 \text{ black}) = \frac{_7C_2 \cdot _5C_2}{_{12}C_4} =$
   $\frac{21 \cdot 10}{495} \approx 0.424$

   $P(3 \text{ black}) = \frac{_7C_1 \cdot _5C_3}{_{12}C_4} =$
   $\frac{7 \cdot 10}{495} \approx 0.141$

   $P(4 \text{ black}) =$
   $\frac{_5C_4}{_{12}C_4} = \frac{5}{495} \approx 0.010$

   $\frac{35 + 175 + 210 + 70 + 5}{495} = 1$

   b. $P(2 \text{ white}, 2 \text{ black}) =$
   $\frac{_4C_2 \cdot _5C_2}{_{12}C_4} = \frac{6 \cdot 10}{495} \approx 0.121$

Example 3 Thirteen cards are dealt from a well-shuffled standard deck. What is the probability that the 13 cards contain exactly 4 aces and exactly 3 kings?

Solution
1. There are $_4C_4$ choices for the aces.
2. There are $_4C_3$ choices for the kings.
3. Since the remaining 6 cards must be chosen from the 44 cards that are neither aces nor kings, there are $_{44}C_6$ choices for the remaining cards.
4. There are $_{52}C_{13}$ choices for the 13 cards.
5. Using these results and the multiplication principle, we have:

$$P\left(\begin{array}{c}13 \text{ cards with exactly} \\ 4 \text{ aces and exactly 3 kings}\end{array}\right) = \frac{_4C_4 \cdot _4C_3 \cdot _{44}C_6}{_{52}C_{13}}$$

$$= \frac{1 \cdot 4 \cdot 7,059,052}{635,013,559,600} \approx 0.0000445$$

CLASS EXERCISES

1. Three marbles are picked at random from a bag containing 4 red marbles and 5 white marbles. Match each event with its probability.

   Events: *Probabilities:*

   a. All 3 marbles are red. $\frac{_4C_1 \cdot _5C_2}{_9C_3}$ c

   b. Exactly 2 marbles are red. $\frac{_4C_2 \cdot _5C_1}{_9C_3}$ b

   c. Exactly 1 marble is red. $\frac{_4C_3 \cdot _5C_0}{_9C_3}$ a

   d. No marble is red. $\frac{_4C_0 \cdot _5C_3}{_9C_3}$ d

2. Five cards are drawn at random from a standard deck. Match each event with its probability.

   Events: *Probabilities:*

   a. All 4 aces are chosen. $\frac{_4C_0 \cdot _{48}C_5}{_{52}C_5}$ b

   b. No aces are chosen. $\frac{_4C_4 \cdot _{48}C_1}{_{52}C_5}$ a

   c. Exactly 4 diamonds are chosen. $\frac{_4C_4 \cdot _4C_1}{_{52}C_5}$ d

   d. Four aces and one jack are chosen. $\frac{_{13}C_4 \cdot _{39}C_1}{_{52}C_5}$ c

---

3. Five cards are drawn at random from a standard deck.

    a. You can find the probability of getting at least 1 ace by calculating the sum:

$$P\left(\begin{array}{c}\text{exactly}\\ \text{1 ace}\end{array}\right) + P\left(\begin{array}{c}\text{exactly}\\ \text{2 aces}\end{array}\right) + P\left(\begin{array}{c}\text{exactly}\\ \text{3 aces}\end{array}\right) + P(4 \text{ aces})$$

    It is far easier, however, to find $1 - P(\underline{\ ?\ })$. No aces

    b. Find the probability of getting at least 1 ace using the method suggested in part (a). 0.3412

WRITTEN EXERCISES

In Exercises 1 and 2, leave your answers in terms of factorials unless directed otherwise by your teacher.

A **1.** Five cards are dealt from a well-shuffled standard deck. What is the probability of getting:
    See below.
    a. all hearts? $\dfrac{13!\ 47!}{8!\ 52!}$ **b.** no hearts? $\dfrac{39!\ 47!}{34!\ 52!}$ **c.** at least one heart?

  2. Thirteen cards are dealt from a well-shuffled standard deck. What is the probability of getting:
    See below.
    a. all clubs? $\dfrac{13!\ 39!}{52!}$ **b.** no clubs? $\dfrac{39!\ 39!}{26!\ 52!}$ **c.** at least one club?

  3. A bag contains 5 red marbles and 3 white marbles. If 2 marbles are randomly drawn, one after the other and without replacement, what is the probability that the number of red marbles is 0? 1? 2? (Check to see that the sum of the probabilities is 1.) $\dfrac{3}{28}, \dfrac{15}{28}, \dfrac{5}{14}$

  4. Repeat Exercise 3 if the bag contains 6 red marbles and 2 white marbles; that is, find the probabilities of drawing 0, 1, and 2 red marbles.

  5. Free concert tickets are distributed to 4 students chosen at random from 8 juniors and 12 seniors in the school orchestra. What is the probability that free tickets are received by:
    a. 4 seniors?
    b. exactly 3 seniors?
    c. exactly 2 seniors?
    d. exactly 1 senior?
    e. no seniors?

  6. A town council consists of 8 Democrats, 7 Republicans, and 5 Independents. A committee of 3 is chosen by randomly pulling names from a hat. What is the probability that the committee has:
    a. 2 Democrats and 1 Republican?   **1. c.** $1 - \dfrac{39!\ 47!}{34!\ 52!}$ **2. c.** $1 - \dfrac{39!\ 39!}{26!\ 52!}$
    b. 3 Independents?
    c. no Independents?
    d. 1 Democrat, 1 Republican, and 1 Independent?

Probability   **621**

2. A carton contains 200 batteries, of which 5 are defective. If a random sample of 5 batteries is chosen, what is the probability that at least one of them is defective?

$P(\text{at least one defective}) = 1 - P(\text{none defective}) =$

$1 - \dfrac{_{195}C_5}{_{200}C_5} =$

$1 -$

$\dfrac{(195)(194)(193)(192)(191)}{(200)(199)(198)(197)(196)} \approx$

0.12

Exercise Note

The technique suggested in Class Exercise 3 is one that students should keep in mind and apply whenever they are asked to find probabilities in "at least one" situations.

Additional Answers Written Exercises

4. $\dfrac{1}{28}, \dfrac{3}{7}, \dfrac{15}{28}$

5. a. $\dfrac{33}{323} \approx 0.102$

  b. $\dfrac{352}{969} \approx 0.363$

  c. $\dfrac{616}{1615} \approx 0.381$

  d. $\dfrac{224}{1615} \approx 0.139$

  e. $\dfrac{14}{969} \approx 0.014$

6. a. $\dfrac{49}{285} \approx 0.172$

  b. $\dfrac{1}{114} \approx 0.009$

  c. $\dfrac{91}{228} \approx 0.399$

  d. $\dfrac{14}{57} \approx 0.246$

Suggested Assignments

Discrete Math
Day 1: 621/1–12
Day 2: 622/13–20

Supplementary Materials

Alternative Assessment, 51
Student Resource Guide,
148–151

Cooperative Learning

Consider using the B-level
exercises for small-group
work.

**Additional Answers
Written Exercises**

16. a. 6.30×10^{-12}
   b. 2.21×10^{-4}
   c. 6.30×10^{-11}
   d. 0.987

A committee of 3 people is to be randomly selected from the 6 people Archibald (A), Beatrix (B), Charlene (C), Denise (D), Eloise (E), and Fernando (F). The sample space below lists the 20 possible committees.

{A, B, C}	{A, C, E}	{B, C, D}	{B, E, F}
{A, B, D}	{A, C, F}	{B, C, E}	{C, D, E}
{A, B, E}	{A, D, E}	{B, C, F}	{C, D, F}
{A, B, F}	{A, D, F}	{B, D, E}	{C, E, F}
{A, C, D}	{A, E, F}	{B, D, F}	{D, E, F}

In Exercises 7–14, find the probability in each of two ways:

> Method 1: Use the sample space given above.
> Method 2: Use combinations.

Sample Find the probability that Archibald and Beatrix are on the committee.

Solution **Method 1** Since 4 of the 20 sample points include A and B,

$$P(\text{A and B}) = \frac{4}{20} = \frac{1}{5}.$$

Method 2 $P(\text{A and B}) = \frac{{}_2C_2 \cdot {}_4C_1}{{}_6C_3} = \frac{1 \cdot 4}{20} = \frac{1}{5}$

7. Find the probability that Eloise is on the committee. $\frac{1}{2}$
8. Find the probability that Eloise and Fernando are on the committee. $\frac{1}{5}$
9. Find the probability that either Eloise or Fernando is on the committee. $\frac{4}{5}$
10. Find the probability that neither Eloise nor Fernando is on the committee. $\frac{1}{5}$
11. Find the probability that Archibald is on the committee and Beatrix is not. $\frac{3}{10}$
12. Find the probability that Archibald and Beatrix are on the committee but Charlene is not. $\frac{3}{20}$
13. Find the probability that Denise is on the committee given that Archibald is. $\frac{2}{5}$
14. Find the probability that Denise is on the committee given that neither Archibald nor Beatrix is. $\frac{3}{4}$

B 15. Thirteen cards are dealt from a well-shuffled standard deck. What is the probability of getting:
   a. all red cards? 1.64×10^{-5} b. 7 diamonds and 6 hearts? 4.64×10^{-6}
   c. at least 1 face card? 0.981 d. all face cards? 0

16. Thirteen cards are dealt from a well-shuffled standard deck. What is the probability of getting:
   a. all cards from the same suit?   b. 7 spades, 3 hearts, and 3 clubs?
   c. all of the 12 face cards?   d. at least one diamond?

17. A committee of 4 is chosen at random from a group of 5 married couples. What is the probability that the committee includes no two people who are married to each other? $\frac{8}{21}$

18. The letters of the word ABRACA-DABRA are written on separate slips of paper and placed in a hat. Five slips of paper are then randomly drawn, one after the other and without replacement. What is the probability of getting:
   a. all A's?   **a.** $\frac{1}{462}$ **b.** $\frac{2}{11}$
   b. both R's?
   c. at least one B?   **c.** $\frac{8}{11}$ **d.** $\frac{10}{231}$
   d. one of each letter?

19. **Manufacturing** A quality control inspector randomly inspects 4 microchips in every lot of 100. If one or more microchips are defective, the entire lot is rejected for shipment. Suppose a lot contains 10 defective microchips and 90 acceptable ones. What is the probability that the lot is rejected?

20. **Manufacturing** A lot of 20 television sets consists of 6 defective sets and 14 good ones. If a sample of 3 sets is chosen, what is the probability that the sample contains:
   a. all defective sets? $\frac{1}{57}$
   b. at least one defective set? $\frac{194}{285}$

C 21. A standard deck of 52 cards is shuffled and the cards are dealt face up one at a time until an ace appears.
   a. What is the probability that the first ace appears on the third card?
   b. Show that the probability of getting the first ace on or before the ninth card is greater than 50%.

COMMUNICATION: Reading

Vocabulary in Mathematics

Like most disciplines, mathematics has a special vocabulary. Many of the terms found in a book like this are used only in mathematics. A few examples are *polynomial, logarithm, asymptote,* and *cosine.* Other terms, however, are used in other disciplines or in everyday discourse with meanings different from their mathematical meanings.

1. Consider the word *experiment.* Explain the difference between its meaning in probability and its meaning in science.

2. Select one of the words *degree, function, radical, polar, series,* and look it up in a dictionary. Write down two or three meanings for the word, including its mathematical meaning. For each meaning, write a sentence that uses the word with that meaning.

Probability   **623**

Application

In connection with Exercises 19 and 20, you might want to discuss defective pacemakers and defective space telescopes to point up how levels of acceptability for defective products can vary greatly.

Additional Answers Written Exercises

19. $\frac{8279}{23,765} \approx 0.348$

21. a. 0.0681
   b. P(1st ace on or before 9th card) = P(ace on 1st) + P(ace on 2nd) + $\cdots$ + P(ace on 9th) = 0.0769 + 0.0724 + 0.0681 + 0.0639 + 0.0599 + 0.0561 + 0.0524 + 0.0489 + 0.0456 = 0.5442 > 50%

For Exercises 1-6, use the tree diagram shown. Find each probability.

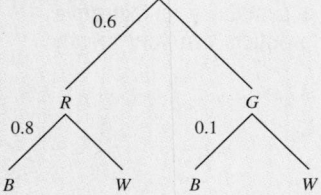

1. $P(G)$ 0.4
2. $P(W \mid R)$ 0.2
3. $P(W \mid G)$ 0.9
4. $P(B)$ (Hint: $P(B) = P(RB) + P(GB)$) 0.52
5. $P(W)$ 0.48
6. Show that $P(RB) + P(RW) + P(GB) + P(GW) = 1$.
   $0.48 + 0.12 + 0.04 + 0.36 = 1$

Motivating the Section

Use a simple example, such as life expectancy, to introduce this section. Suppose, for example, that in a given population, 60% of the females live to the age of 75 and beyond, while only 42% of the males live 75+ years. If the population is 52% female and 48% male, then conditional probability can be used to determine the probability that a 75-year-old person is female. ($P(75) = P(F \text{ and } 75) + P(M \text{ and } 75) \approx 0.51$. Then $P(F \mid 75) \approx 0.61$.)

■ Applications of Probability

16-5 Working with Conditional Probability

Objective   *To solve problems involving conditional probability.*

In this section we will consider the probability of a certain cause when a certain effect is observed. For example, a flu can cause symptoms such as high fever and sore throat, but these can be symptoms of other disorders besides flu. Example 1 examines the probability that a person who has a fever (an effect) also has the flu (a possible cause of the fever).

Example   During a flu epidemic, 35% of a school's students have the flu. Of those with the flu, 90% have high temperatures. However, a high temperature is also possible for people without the flu; in fact, the school nurse estimates that 12% of those without the flu have high temperatures.
a. Incorporate the facts given above into a tree diagram.
b. About what percent of the student body have a high temperature?
c. If a student has a high temperature, what is the probability that the student has the flu?

Solution   a. In the tree diagram at the right, F and F' represent the events "has the flu" and "does not have the flu," and T and T' represent "has a high temperature" and "does not have a high temperature." The probabilities in red 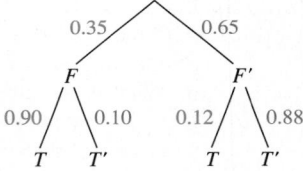 come directly from the description in the example. The probabilities in blue are deduced from the fact that all the branches from any given point of a tree must have probabilities that total 1.

b. To find $P(T)$, the probability that a student has a high temperature, we add the probabilities of the two paths leading to a T:

$$P(T) = P(F \text{ and } T) + P(F' \text{ and } T)$$
$$= 0.35 \times 0.90 + 0.65 \times 0.12$$
$$= 0.315 + 0.078 = 0.393$$

Thus, 39.3% of the student body have a high temperature.

c. Since a high temperature already exists, we consider only the *portion* of the students who have high temperatures. Part (b) has shown that this portion, 0.393, is the sum of 0.315 (those with high temperatures and the flu) and 0.078 (those with high temperatures and no flu). Thus:

$$P(F \mid T) = \frac{P(F \text{ and } T)}{P(T)} = \frac{0.315}{0.393} \approx 0.802$$

624   *Chapter Sixteen*

The technique used in the preceding example can be generalized. Suppose we have a situation in which there is a cause C and an effect E. Also suppose we know the probability of the occurrence of C and the probabilities of the occurrence of E given the occurrence or nonoccurrence of C. (These three probabilities are shown in red in the tree diagram above.) The problem is to find the probability that C has occurred given that E has.

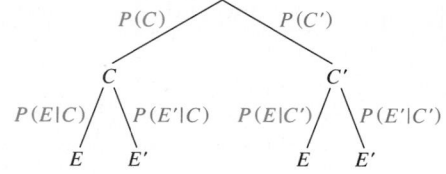

Let us first symbolically list what we know and can immediately deduce:

1. $P(C)$ and therefore $P(C') = 1 - P(C)$
2. $P(E|C)$ and therefore $P(E'|C) = 1 - P(E|C)$
3. $P(E|C')$ and therefore $P(E'|C') = 1 - P(E|C')$

Then, to find $P(C|E)$, we take the following two steps:

Step 1. Find $P(E) = P(C \text{ and } E) + P(C' \text{ and } E)$
$$= P(C) \cdot P(E|C) + P(C') \cdot P(E|C')$$

Step 2. Find $P(C|E) = \dfrac{P(C \text{ and } E)}{P(E)}$.   Each of these is calculated in Step 1.

CLASS EXERCISES

1. **a.** Copy the tree diagram at the right and find the missing probabilities.
   b. The question mark at the lower left refers to $P(Y'|X)$. To what probabilities do the other two question marks refer? $P(Y|X')$, $P(X')$
   c. Find $P(X \text{ and } Y)$. 0.18
   d. Find $P(X' \text{ and } Y)$. 0.14
   e. Find $P(Y)$. 0.32
   f. Find $P(X|Y)$. 0.5625

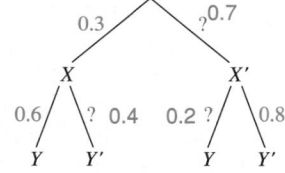

2. Use the tree diagram at the right to find each probability.
   a. $P(A|R)$ 0.2
   b. $P(A|S)$ 0
   c. $P(R \text{ and } C)$ 0.05
   d. $P(S \text{ and } C)$ 0.36
   e. $P(C)$ 0.41
   f. $P(R|C)$ 0.122

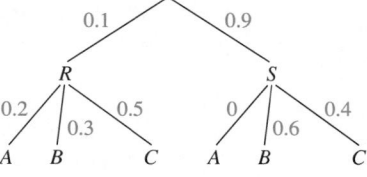

3. **Reading** Study the example in this section. Given the results of parts (b) and (c) of the example, show two different ways to find the probability that a student who has a high temperature does *not* have the flu.

1. Use the probability tree to find the specified probabilities.

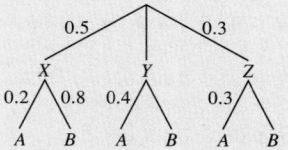

a. $P(Y)$	**b.** $P(B|Y)$
c. $P(B|Z)$	**d.** $P(X \cap A)$
e. $P(Y \cap A)$	**f.** $P(Z \cap A)$
g. $P(A)$	**h.** $P(B)$
i. $P(X|A)$	**j.** $P(Z|A)$

a. $P(Y) = 1 - 0.5 - 0.3 = 0.2$
b. $P(B|Y) = 1 - 0.4 = 0.6$
c. $P(B|Z) = 1 - 0.3 = 0.7$
d. $P(X \cap A) = P(X) \cdot P(A|X) = 0.5(0.2) = 0.1$
e. $P(Y \cap A) = P(Y) \cdot P(A|Y) = 0.2(0.4) = 0.08$
f. $P(Z \cap A) = P(Z) \cdot P(A|Z) = 0.3(0.3) = 0.09$
g. $P(A) = P(X \cap A) + P(Y \cap A) + P(Z \cap A) = 0.1 + 0.08 + 0.09 = 0.27$
h. $P(B) = 1 - P(A) = 1 - 0.27 = 0.73$
i. $P(X|A) = \dfrac{P(X \cap A)}{P(A)} = \dfrac{0.1}{0.27} \approx 0.37$
j. $P(Z|A) = \dfrac{P(Z \cap A)}{P(A)} = \dfrac{0.09}{0.27} \approx 0.33$

(*Continues on next page*)

2. Jar A contains 1 red, 2 white, and 3 green balls. Jar B contains 4 red and 5 white balls. A die is rolled. If a "1" or a "6" comes up, a ball is randomly picked from Jar A. Otherwise, a ball is randomly picked from Jar B.
   a. Draw a tree diagram that shows the given facts.
   b. Find the probability of picking a white ball.
   c. If a white ball is picked, what is the probability that it came from Jar A?

a.

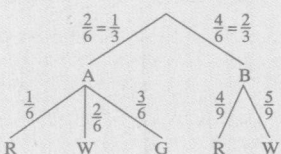

b. $P(W) = P(A \cap W) + P(B \cap W) =$
   $P(A) \cdot P(W|A) + P(B) \cdot P(W|B) =$
   $\frac{1}{3} \cdot \frac{2}{6} + \frac{2}{3} \cdot \frac{5}{9} = \frac{13}{27}$

c. $P(A|W) = \dfrac{P(A \cap W)}{P(W)} =$
   $\frac{1}{9} \div \frac{13}{27} = \frac{3}{13}$
   (Notice that $P(A|G) = 1$ and $P(B|G) = 0$.)

WRITTEN EXERCISES

A 1. Use the tree diagram at the left below to find each probability.
   a. $P(X')$ 0.6   b. $P(Y'|X)$ 0.7   c. $P(Y'|X')$ 0.5   d. $P(X \text{ and } Y)$ 0.12
   e. $P(X' \text{ and } Y)$ 0.3   f. $P(Y)$ 0.42   g. $P(X|Y)$ 0.286   h. $P(X'|Y)$ 0.714

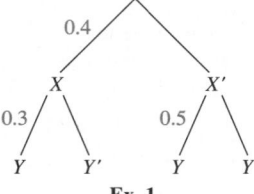 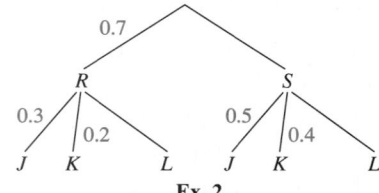

Ex. 1   **Ex. 2**

2. Use the tree diagram at the right above to find each probability.
   a. $P(S)$ 0.3   b. $P(L|R)$ 0.5   c. $P(L|S)$ 0.1   d. $P(J \text{ and } R)$ 0.21
   e. $P(J \text{ and } S)$ 0.15   f. $P(J)$ 0.36   g. $P(R|J)$ 0.583   h. $P(S|J)$ 0.417

3. Use the tree diagram at the right to find each sum of probabilities.
   a. $P(A \text{ and } D) + P(A \text{ and } E)$ 0.5
   b. $P(B \text{ and } D) + P(B \text{ and } E)$ 0.4
   c. $P(C \text{ and } D) + P(C \text{ and } E)$ 0.1

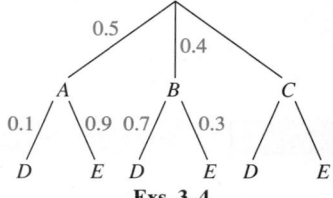

4. Refer to the tree in Exercise 3 and suppose that event D cannot possibly happen if event C happens. Find $P(A|D)$, $P(B|D)$, and $P(C|D)$.

Exs. 3, 4

5. Jar A contains 2 red balls and 3 white balls. Jar B contains 4 red balls and 1 white ball. A coin is tossed. If it shows "heads," a ball is randomly picked from Jar A; if it shows "tails," a ball is randomly picked from Jar B.

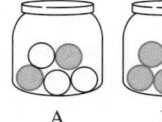

A   B

   a. Draw a tree diagram showing the probabilities of each jar and then the probabilities of picking a red ball or a white ball.
   b. Find the probability of picking a red ball. 0.6
   c. If a red ball is picked, find probability that it came from Jar A. 0.333

6. Jars A, B, and C contain red and white balls as shown. A die is rolled. If an even number comes up, a ball is randomly picked from Jar A. If a "1" or a "3" comes up, a ball is randomly picked from Jar B. If a "5" comes up, a ball is randomly picked from Jar C.

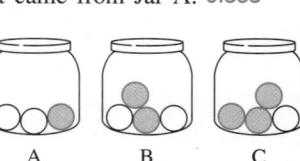

A   B   C

   a. Incorporate the facts given above into a tree diagram.
   b. Find the probability of picking a red ball. $\frac{11}{24} \approx 0.458$
   c. If a red ball is picked, what is the probability that it came from Jar A? from Jar B? from Jar C? $\frac{4}{11} \approx 0.364$; $\frac{4}{11} \approx 0.364$; $\frac{3}{11} \approx 0.273$

7. Manufacturing Machine A produces 60% of the ball bearings manufactured by a factory and Machine B produces the rest. Five percent of Machine A's bearings fail to have the required precision, and two percent of Machine B's bearings fail.

  a. Incorporate the facts given above into a tree diagram.

  b. What percent of the bearings fail to have the required precision? 3.8%

  c. If a bearing is inspected and fails to have the required precision, what is the probability that it was produced by Machine A? 0.789

8. Manufacturing Five percent of the welds on an automobile assembly line are defective. The defective welds are found using an X-ray machine. The machine correctly rejects 92% of the defective welds and correctly accepts all of the good welds.

  a. Incorporate the facts given above into a tree diagram.

  b. What percent of the welds are accepted by the machine? 95.4%

  c. Find the probability that an accepted weld is defective. 0.004

9. Insurance An auto insurance company charges younger drivers a higher premium than it does older drivers because younger drivers as a group tend to have more accidents. The company has 3 age groups: Group A includes those under 25 years old, 22% of all its policyholders. Group B includes those 25–39 years old, 43% of all of its policyholders. Group C includes those 40 years old or older. Company records show that in any given one-year period, 11% of its Group A policyholders have an accident. The percentages for groups B and C are 3% and 2%, respectively.

  a. What percent of the company's policyholders are expected to have an accident during the next 12 months? 4.41%

  b. Suppose Mr. X has just had a car accident. If he is one of the company's policyholders, what is the probability that he is under 25? 0.549

10. Insurance Suppose the insurance company of Exercise 9 not only classifies drivers by age, but in the case of drivers under 25 years old, it also notes whether they have had a driver's education course. One quarter of its policyholders under 25 have had driver's education and 5% of these have an accident in a one-year period. Of those under 25 who have not had driver's education, 13% have an accident within a one year period. A 20-year-old woman takes out a policy with this company and within one year she has an accident. What is the probability that she did *not* have a driver's education course? 0.886

Problem Solving

As an extension of Exercise 8, have students suppose that each week the assembly line makes 1000 welds and suppose further that a defective weld that is undetected costs the company $75.00. Then ask students to answer the following questions.

1. How much will the company lose on bad welds each week? ($300)

2. Suppose, to save money, the company decides not to use the X-ray machine. Now how much will the company lose on bad welds each week? ($3750)

3. Finally, suppose each X-ray costs $C. For what values of C is the use of X-rays a wise financial decision? (Use the machine if each X-ray costs less than $3.45.)

Suggest that students make tree diagrams to help them solve Exercises 13–16.

Additional Answers
Written Exercises

11. a. Answers may vary. A false negative can be serious because someone with the disease will not receive treatment. On the other hand, a false positive can be serious because someone who is healthy might receive expensive, potentially damaging treatment that isn't needed.
b.

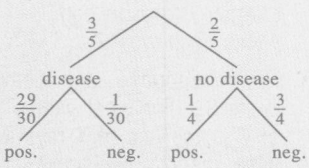

c. 0.853
d. 0.938

B **11. Medicine** A medical research lab proposes a screening test for a disease. In order to try out this test, it is given to 100 people, 60 of whom are known to have the disease and 40 of whom are known not to have the disease. A positive test indicates the disease and a negative test indicates no disease. Unfortunately, such medical tests can produce two kinds of errors:

(1) A *false negative* test: For the 60 people who do have the disease, this screening test indicates that 2 do *not* have it.
(2) A *false positive* test: For the 40 people who do not have the disease, this screening test indicates that 10 *do* have it.

a. Which of the false tests do you think is more serious? Why?
b. Incorporate the facts given above into a tree diagram. (Be sure to convert the given integers into probabilities.)
c. Suppose the test is given to a person not in the original group of 100 people. It is not known whether this person has the disease, but the test result is positive. What is the probability that the person really does have the disease?
d. Suppose the test is given to a person whose disease status is unknown. If the test result is negative, what is the probability that the person does *not* have the disease?

12. Medicine Part (c) of Exercise 11 indicates that about 85% of those who test positive really do have the disease, so that 15% of those who test positive do not have it. This 15% error may seem high, but people with a positive screening test are usually given a more thorough diagnostic test. Even the diagnostic test can yield errors but they are much less likely than the screening test, as the diagram shows.

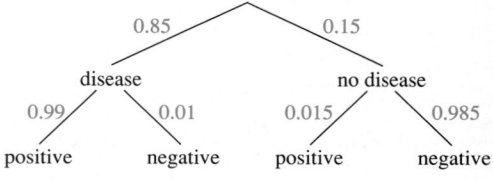

a. What is the probability that the diagnostic test gives:
(1) a false negative result? 0.0085
(2) a false positive result? 0.00225
b. What is the probability that:
(1) the diagnostic test gives the correct result? 0.989
(2) a person with a positive diagnostic test has the disease? 0.997
(3) a person with a negative diagnostic test does not have the disease? 0.946

13. The children of a math professor play two games that use dice. In one game, two dice are rolled and the sum of the numbers on the dice is called out. In the other game, a single die is rolled and its number is called out. The professor hears the children in another room call out the number 2, and knowing that they play the two games about equally often, the professor is able to calculate the probability they are playing the two-dice game. What is this probability?

14. Solve Exercise 13 if the children call out: **13.** $\frac{1}{7} \approx 0.143$
a. the number 4 $\frac{1}{3}$ **b.** the number 7 1   **c.** the number 1 0

628   *Chapter Sixteen*

15. Suppose you have two pairs of dice. One pair is fair, but each die of the other pair is weighted so that a six comes up with probability $\frac{1}{4}$ instead of the usual $\frac{1}{6}$. If you randomly choose one pair of dice, roll them, and obtain two sixes, what is the probability that you rolled the weighted dice? $\frac{9}{13} \approx 0.692$

16. Suppose you have 2 coins. One of them is fair, but the other has two "heads." You choose one coin at random, flip it n times, and get "heads" each time.
   a. Find the probability that the coin is two-headed for $n = 1$, $n = 2$, and $n = 10$.
   b. In terms of n, find the probability that the coin is two-headed.

C 17. Refer to the two jars pictured in Exercise 5. Suppose that someone randomly picks one of the jars, but you don't know which one. Before guessing which jar the "mystery jar" is, you have your choice of doing either of the following experiments.

   Experiment 1: Pick 2 balls from the "mystery jar" and note their colors. Replace the first ball before choosing the second.

   Experiment 2: Pick 2 balls from the "mystery jar" and note their colors. Do *not* replace the first ball before choosing the second.

   The tree diagram for Experiment 1 looks like this:

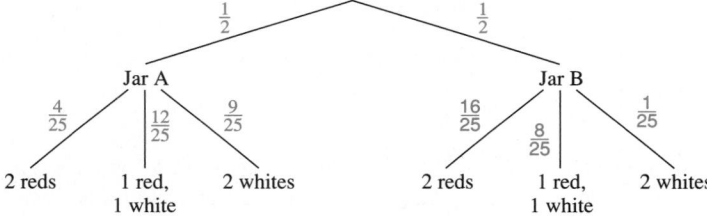

   a. Copy and complete the diagram.
   b. Suppose you performed Experiment 1 and got 2 red balls. Then you would no doubt guess that they came from Jar B. Which jar would you guess if you got 2 white balls? 1 red and 1 white ball? What is the probability that you would guess the correct jar if you performed the first experiment?
   c. Make a tree diagram for Experiment 2.
   d. If you performed Experiment 2, which jar would you guess if you got 2 red balls? 2 white balls? 1 red and 1 white ball? What is the probability that you would guess correctly if you performed the second experiment?
   e. Which experiment gives you the better chance of guessing correctly?
   f. Suppose a third experiment allows you to pick just one ball and then guess the jar. How likely are you to guess correctly in this experiment? Compare your answer with the probability of being correct in the first or second experiment. Are you surprised?

In Exercises 1 and 2, find
u · v.

1. **u** = (3, 5, −2) and **v** =
$\left(\frac{2}{5}, \frac{1}{5}, \frac{2}{5}\right)$ $\frac{7}{5}$

2. **u** = (20, 10, −5) and **v** =
(0.1, 0.3, 0.6) 2.0

In Exercises 3 and 4, find the
probabilities.

3. Two dice are rolled and the
sum of the numbers is
given below.

$P(6)$	?	$\frac{5}{36}$
$P(7)$	?	$\frac{6}{36}$
$P(8)$	?	$\frac{5}{36}$
$P(9)$	?	$\frac{4}{36}$
$P(10)$	?	$\frac{3}{36}$
$P(11)$	?	$\frac{2}{36}$
$P(12)$	?	$\frac{1}{36}$

4. A jar contains 6 red marbles and 2 white marbles.
Two marbles are drawn
without replacement.

$P(2 \text{ red})$	?	$\frac{15}{28}$
$P(2 \text{ white})$	?	$\frac{1}{28}$
$P(1 \text{ red and } 1 \text{ white})$	?	$\frac{3}{7}$

16-6 Expected Value

Objective *To find expected value in situations involving gains and losses and to determine whether a game is fair.*

In this section we will consider probabilistic situations involving gains or losses. For example:

1. What can you expect to win or lose in various games of chance?
2. What is the value of a $1 ticket in a $1 million lottery?
3. Should someone who is 18 years old pay $160 for collision damage insurance on his or her $1500 car?

Let us begin with the first question. Consider a simple game in which a die is rolled and you win points from, or lose points to, another player as follows.

Event	Die shows 1, 2, or 3	Die shows 4 or 5	Die shows 6
Gain or loss	+10 points	−13 points	−1 point
Probability	$\frac{3}{6}$	$\frac{2}{6}$	$\frac{1}{6}$

To calculate the *expected value* of this game, you must multiply each gain or loss by its probability and then add the products.

$$\text{Expected value} = 10 \cdot \frac{3}{6} + (-13) \cdot \frac{2}{6} + (-1) \cdot \frac{1}{6}$$
$$= +\frac{3}{6} = 0.5 \text{ point}$$

The expected value of 0.5 point means that if you were to play this game many times, your *average gain per game* would be 0.5 point. The other player would expect to lose 0.5 point per game.

In general, if a given situation involves various payoffs (gains or losses of points, money, time, and so on), then its **expected value** is calculated as follows.

Payoff	x_1	x_2	x_3	$\cdots$	x_n
Probability	$P(x_1)$	$P(x_2)$	$P(x_3)$	$\cdots$	$P(x_n)$

$$\text{Expected value} = x_1 \cdot P(x_1) + x_2 \cdot P(x_2) + x_3 \cdot P(x_3) + \cdots + x_n \cdot P(x_n)$$

If the expected value of a game is 0, then the game is called a **fair game**. The dice game previously discussed is not a fair game because the expected value is 0.5 point and not 0. An example of a fair game is a coin toss in which you win a point if the coin comes up ''heads'' and you lose a point if the coin comes up ''tails.'' The expected value is:

$$(1 \text{ point}) \times \frac{1}{2} + (-1 \text{ point}) \times \frac{1}{2} = 0$$

630 *Chapter Sixteen*

Example 1

If the sum of two rolled dice is 8 or more, you win $2; if not, you lose $1.
a. Show that this is not a fair game.
b. To have a fair game, the $2 winnings should instead be what amount?

Solution

Make a table of payoffs. For the probabilities, use the table on page 601.

Event	Sum $\geq$ 8	Sum $<$ 8
Payoff	$2	$-$1
Probability	$\frac{15}{36}$	$\frac{21}{36}$

a. Expected value of game $= 2\left(\frac{15}{36}\right) + (-1)\left(\frac{21}{36}\right) = +\frac{1}{4} = \$.25$

b. To have a fair game, let x be the winnings for a sum greater than or equal to 8 and find the value of x that produces an expected value of 0.

$$\text{Expected value} = x\left(\frac{15}{36}\right) + (-1)\left(\frac{21}{36}\right) = 0$$

$$15x - 21 = 0$$

$$x = \$1.40$$

Example 2

In a certain state's lottery, six numbers are randomly chosen without repetition from the numbers 1 to 40. If you correctly pick all 6 numbers, only 5 of the 6, or only 4 of the 6, then you win $1 million, $1000, or $100, respectively. What is the value of a $1 lottery ticket?

Solution

First we must find the probabilities of winning:

$$P(\text{all 6 correct}) = \frac{_6C_6}{_{40}C_6} \approx 0.00000026$$

$$P(\text{5 of 6 correct}) = \frac{_6C_5 \cdot _{34}C_1}{_{40}C_6} \approx 0.000053$$

$$P(\text{4 of 6 correct}) = \frac{_6C_4 \cdot _{34}C_2}{_{40}C_6} \approx 0.0022$$

Then we multiply these probabilities by their associated gains, temporarily ignoring the cost of the $1 ticket:

Expected gain
$\approx 1,000,000(0.00000026) + 1000(0.000053) + 100(0.0022)$
$= 0.26 + 0.053 + 0.22 \approx \$.53$

When we consider the $1 cost of the ticket, the expected value of the ticket is about $\$0.53 - \$1 = -\$0.47$.

Motivating the Section

To introduce this section, sketch on the board a spinner divided into quarters. Mark two of the regions "Win $1," mark one of them "Win $5," and mark the last region "Lose $15." Tell students that the spinner will be spun 40 times; ask how often they expect to land on "Win $5" (10 times) and then write $10 \times \$5 = \$50$ on the board. Ask how often they expect to land on "Win $1" (20 times) and write $20 \times \$1 = \$20$ on the board. Finally, ask how often they expect to land on "Lose $15" (10 times) and write $10 \times (-\$15) = -\$150$ on the board. Add the expected winnings and losses to obtain $-\$80$. Explain that this is an average loss of $80 in 40 spins, or $80/40 = \$2.00$ per spin. Distributing the division as shown below should help students see the reasoning behind multiplying payoffs by the corresponding probability:

$$\frac{\$5 \times 10 + \$1 \times 20 + (-\$15) \times 10}{40}$$

$$= \$5 \times \frac{10}{40} + \$1 \times \frac{20}{40} +$$

$$(-\$15) \times \frac{10}{40} =$$

$$\$5(0.25) + \$1(0.50) +$$
$$(-\$15)(0.25) = -\$2.00$$

Example Note

After discussing Example 2, you might wish to ask students why they think anyone would buy a ticket worth $-\$0.47$. Common replies might be "Someone has to win" or "It's entertaining."

1. Two coins are tossed. If both land heads up, then player A wins $4 from player B. If exactly one coin lands heads up, then B wins $1 from A. If both land tails up, then B wins $2 from A. Show that this is a fair game.

$P(HH) = \frac{1}{2} \cdot \frac{1}{2} = \frac{1}{4}$

$P(TT) = \frac{1}{2} \cdot \frac{1}{2} = \frac{1}{4}$

$P(\text{one head}) =$
$1 - \frac{1}{4} - \frac{1}{4} = \frac{1}{2}$

Heads	2	1	0
Payoff to A	$4	-$1	-$2
Probability	$\frac{1}{4}$	$\frac{1}{2}$	$\frac{1}{4}$

Expected value of game =
$\$4\left(\frac{1}{4}\right) + \left(-\$1\right)\left(\frac{1}{2}\right) +$
$(-\$2)\left(\frac{1}{4}\right) = 0$
Thus, the game is fair.

2. Three cards are drawn at random, without replacement, from a standard deck. Find the expected value for the occurrence of hearts.

Hearts	Probability
0	$\frac{_{39}C_3}{_{52}C_3} = \frac{703}{1700}$
1	$\frac{_{39}C_2 \cdot _{13}C_1}{_{52}C_3} = \frac{741}{1700}$
2	$\frac{_{39}C_1 \cdot _{13}C_2}{_{52}C_3} = \frac{234}{1700}$
3	$\frac{_{13}C_3}{_{52}C_3} = \frac{22}{1700}$

Example 3 An 18-year-old student must decide whether to spend $160 for one year's car collision damage insurance. The insurance carries a $100 deductible, which means that when the student files a damage claim, the student must pay $100 of the damage amount, with the insurance company paying the rest (up to the value of the car). Because the car is only worth $1500, the student consults with an insurance agent who draws up a table of possible damage amounts and their probabilities based on the driving records for 18-year-olds in the region.

Event	Accident costing $1500	Accident costing $1000	Accident costing $500	No accident
Payoff	$1400	$900	$400	$0
Probability	0.05	0.02	0.03	0.90

What is the expected value of this insurance?

Solution We temporarily ignore the $160 cost and calculate the expected payoff from the table above.

$$\text{Expected payoff} = 1400(0.05) + 900(0.02) + 400(0.03) + 0(0.90)$$
$$= 70 + 18 + 12 + 0$$
$$= \$100$$

Now we consider the $160 cost and find that:

$$\text{Expected value} = 100 - 160 = -\$60$$

While an insurance policy with an expected value of −$60 may not seem like a "fair game" between the student and the insurance company, the company uses the $60 to pay for operating expenses, salaries, and profit.

CLASS EXERCISES

Find the expected payoff.

1.
Payoff	5	10
Probability	0.6	0.4
7

2.
Payoff	3	1	−1
Probability	0.5	0.1	0.4
1.2

3. Players A and B play a game in which a die is rolled and A wins 2 points from B if a 5 or 6 appears. Otherwise, B wins 1 point from A. Decide if this is a fair game. Yes

4. *Reading* The game in Example 1 can be made fair by changing the winnings from $2 to $1.40 (see part (b)). Find a different way to make the game fair.

5. Suppose you toss 3 coins and win payoffs as shown in the table below. Complete the table and find your expected payoff.

Number of "heads"	3	2	1	0
Payoff	$5	$3	$1	−$9
Probability	?	?	?	?

$\frac{1}{8}; \frac{3}{8}; \frac{3}{8}; \frac{1}{8}; 1

6. **Discussion** Discuss why the student in Example 3 may want the car collision damage insurance even though its expected value to the student is negative.

WRITTEN EXERCISES

Find the expected payoff.

A

1.
Payoff	9	7	−5
Probability	0.1	0.3	0.6

0

2.
Payoff	6	3	−5
Probability	0.2	0.1	0.7

−2

3.
Payoff	60	52	50
Probability	0.4	0.5	0.1

55

4.
Payoff	13	−7	−12
Probability	0.4	0.2	0.4

−1

For Exercises 5–8, decide if each game is a fair game. If not, state which player has the advantage.

5. A die is rolled. If the number that shows is odd, player A wins $1 from player B. If it is a 6, A wins $2 from B. Otherwise B wins $3 from A. Not fair; B

6. A box contains 2 red balls and 1 white ball. Two balls are randomly chosen without replacement. If both are red, player A wins $5 from player B. Otherwise B wins $2 from A. Not fair; A

7. Two dice are rolled. If the sum is 6, 7, or 8, player A wins $5 from player B. Otherwise B wins $4 from A. Fair

8. Two dice are rolled. If the sum of the numbers showing on the dice is odd, player A wins $1 from player B. If both dice show the same number, A wins $3 from B. Otherwise B wins $3 from A. Fair

9. Suppose you play a game in which you make a bet and then draw a card from a well-shuffled deck that includes the standard 52 cards as well as 2 jokers. If you draw a joker, you keep your bet and win $5; if you draw a face card, you keep your bet and win $2; and if you draw any other card, you lose your bet. What is your expected gain or loss on this game if your bet is $1? An 11¢ loss

10. Suppose you have $10 to bet on the game described in Exercise 9. Is your expected gain or loss any different if you bet the whole $10 on one game rather than betting $1 at a time on 10 successive games?

Probability **633**

Additional Examples cont.

The expected value of hearts occurring is:

$0\left(\frac{703}{1700}\right) + 1\left(\frac{741}{1700}\right) +$
$2\left(\frac{234}{1700}\right) + 3\left(\frac{22}{1700}\right) =$
$\frac{1275}{1700} = 0.75$

Additional Answers
Class Exercises

4. Answers may vary. For example, if the sum is greater than 7, win $1; if the sum equals 7, no money is exchanged; and if the sum is less than 7, lose $1.

6. The student may not have $1500 available to replace the car. If the student has $160 for insurance, he or she would be able to replace the car.

Additional Answers
Written Exercises

10. The expected loss for one $10 game is $6.78. The expected loss for ten $1 games is $1.10.

Suggested Assignments

Discrete Math
Day 1: 633/1–13 odd
Day 2: 634/14, 15–23 odd

Supplementary Materials

Alternative Assessment, 52
Student Resource Guide, 152–154

Review Note

Students may need to review notation from Chapters 12 and 13 in order to answer Exercises 15 and 16.

Additional Answers
Written Exercises

15. $\sum_{i=1}^{n} x_i\, P(x_i)$

16. Let $\mathbf{u} = (x_1,\, x_2,\, \ldots,\, x_n)$ and $\mathbf{v} = (P(x_1),\, P(x_2),\, \ldots,\, P(x_n))$; then the expected value equals $\mathbf{u} \cdot \mathbf{v}$.

B **11. Test Taking** On a multiple-choice test, a student is given five possible answers for each question. The student receives 1 point for a correct answer and loses $\frac{1}{4}$ point for an incorrect answer. If the student has no idea of the correct answer for a particular question and merely guesses, what is the student's expected gain or loss on the question? 0

12. Test Taking Suppose you are taking the multiple-choice test described in Exercise 11. Suppose also that on one of the questions you can eliminate two of the five answers as being wrong. If you guess at one of the remaining three answers, what is your expected gain or loss on the question? $\frac{1}{6}$

13. A box contains 3 red balls and 2 green balls. Two balls are randomly chosen without replacement. If both are green, you win $2. If just one is green, you win $1. Otherwise you lose $1. What is your expected gain or loss? 50¢ gain

14. In the carnival game "chuck-a-luck," you pick a number from 1 to 6 and roll 3 dice in succession. If your number comes up all 3 times, you win $3; if your number comes up twice, you win $2; if it comes up once, you win $1; otherwise you lose $1. What is your expected gain or loss? (*Hint*: Make a table like the one in Class Exercise 5.) 8¢ loss

15. Rewrite the definition of expected value (page 630) using sigma notation.

16. Rewrite the definition of expected value (page 630) as a dot product of two vectors.

17. Farming A dairy farmer estimates that next year the farm's cows will produce about 25,000 gallons of milk. Because of variation in the market price of milk and the cost of feeding the cows, the profit per gallon may vary with the probabilities given in the table below. Estimate the profit on the 25,000 gallons. $20,850

Gain per gallon	$1.10	$.90	$.70	$.40	$.00	−$.10
Probability	0.30	0.38	0.20	0.06	0.04	0.02

18. Insurance At many airports, a person can pay only $1 for a $100,000 life insurance policy covering the duration of the flight. In other words, the insurance company pays $100,000 if the insured person dies from a possible flight crash; otherwise the company gains $1 (before expenses). Suppose that past records indicate 0.45 deaths per million passengers. How much can the company expect to gain on one policy? on 100,000 policies? 95.5¢; $95,500

19. **Business** A construction company wants to submit a bid for remodeling a school. The research and planning needed to make the bid cost $4000. If the bid is accepted, the company would make $26,000. Would you advise the company to spend the $4000 if the bid has only a 20% probability of being accepted? Explain your reasoning.

20. **Consumer Economics** Suppose the warranty period for your family's new television is about to expire and you are debating about whether to buy a one-year maintenance contract for $35. If you buy the contract, all repairs for one year are free. Consumer information shows that 12% of the televisions like yours require an annual repair that costs $140 on the average. Would you advise buying the maintenance contract? Explain your reasoning.

21. A lottery has one $1000 prize, five $100 prizes, and twenty $10 prizes. What is the expected gain from buying one of the 2000 tickets sold for $1 each? 15¢ loss

22. In a state lottery, five numbers are randomly chosen from the numbers 1 to 30. If you pick all 5 numbers, you win $100,000; and if you pick 4 of the 5 numbers, you win $100. What is the value of a $1 lottery ticket? −21¢

23. Players A and B are playing a game in which A wins a point every time a coin lands "heads" and B wins a point every time the coin lands "tails." (No points are lost by either player.) The first person to reach 3 points wins $100. If A currently has 2 points and B has 1 point, what is A's expected gain? (*Hint*: Make a tree diagram showing the ways in which the game can be finished. From the diagram, determine the probability that A wins.) $75

24. Suppose you play a game in which you make a bet, toss a coin, and either win an amount equal to your bet or lose your bet depending on whether you correctly call "heads" or "tails." Also suppose you begin with a $1 bet and double your bet on each toss until you win once and leave the game or until you have lost $15. What can you expect to win with this betting strategy? $0

C 25. A die is rolled repeatedly until a "1" appears.
   a. Complete the table.

Number of rolls until "1" appears	1	2	3	4	5	⋯	n	⋯
Probability	$\frac{1}{6}$	$\frac{5}{6} \cdot \frac{1}{6}$	$\left(\frac{5}{6}\right)^2 \cdot \frac{1}{6}$	?	?	⋯	?	⋯

   b. Express the expected number of rolls as the sum of an infinite series. Then factor $\frac{1}{6}$ from the sum.

   c. It can be proved (most easily with calculus) that

$$\sum_{n=1}^{\infty} nx^{n-1} = \frac{1}{(1-x)^2}.$$

   Use this to show that the expected number of rolls until a "1" appears is 6.

19. Answers may vary. The expected value is positive, but the company stands an 80% chance of losing $4000. Perhaps the company should bid on other jobs with lower costs of bidding and higher probabilities of being accepted.

20. Answers may vary. No, because after considering the cost of the policy, the expected value is negative.

25. a. $\left(\frac{5}{6}\right)^3 \cdot \frac{1}{6}; \left(\frac{5}{6}\right)^4 \cdot \frac{1}{6};$ $\left(\frac{5}{6}\right)^{n-1} \cdot \frac{1}{6}$

   b. $\sum_{n=1}^{\infty} n\left(\frac{5}{6}\right)^{n-1} \cdot \frac{1}{6} = \frac{1}{6} \sum_{n=1}^{\infty} n\left(\frac{5}{6}\right)^{n-1}$

   c. $\frac{1}{6} \sum_{n=1}^{\infty} n\left(\frac{5}{6}\right)^{n-1} = \frac{1}{6} \frac{1}{\left(1-\frac{5}{6}\right)^2} = \frac{1}{6} \cdot \frac{1}{\left(\frac{1}{6}\right)^2} = 6$

Additional Answers
Chapter Test

1. Empirical probability is determined by observing what previously happened. Theoretical probability is determined by reasoning about the events.

Chapter Summary

1. The probability of an event is a number (between 0 and 1, inclusive) that indicates the likelihood of the event's occurrence. Probabilities can be determined either empirically or theoretically.

2. An *experiment* is any action having various outcomes that occur unpredictably. A set S is a *sample space* of an experiment if each outcome of the experiment corresponds to exactly one element of S. Any subset of S is an *event*.

3. If the sample space of an experiment consists of n equally likely outcomes, m of which correspond to event A, then the probability of A is $P(A) = \frac{m}{n}$. The event "not A" occurs whenever event A does not, and $P(\text{not } A) = 1 - P(A)$.

4. Events A and B are *mutually exclusive* if they cannot occur simultaneously.

5. For any two events A and B, $P(A \text{ or } B) = P(A) + P(B) - P(A \text{ and } B)$, or, if A and B are mutually exclusive, $P(A \text{ or } B) = P(A) + P(B)$.

6. For two events A and B, the probability that B occurs on the condition that A occurs is denoted $P(B|A)$ and is called a *conditional probability*.

7. Events A and B are *independent* if the occurrence of one does not affect the probability of the other's occurrence, that is, $P(B|A) = P(B)$.

8. For any two events A and B, $P(A \text{ and } B) = P(A) \cdot P(B|A)$, or, if A and B are independent, $P(A \text{ and } B) = P(A) \cdot P(B)$.

9. The binomial probability theorem gives the probabilities of 0, 1, 2, . . . , n occurrences of the event A in n repeated independent trials.

10. If a given situation involves payoffs $x_1, x_2, \ldots, x_n$ with corresponding probabilities $P(x_1), P(x_2), \ldots, P(x_n)$, then the situation's *expected value* is:

$$x_1 \cdot P(x_1) + x_2 \cdot P(x_2) + x_3 \cdot P(x_3) + \cdots + x_n \cdot P(x_n)$$

Key vocabulary and ideas

probability (p. 597)
empirical probability (p. 597)
theoretical probability (p. 597, p. 598)
experiment, sample space, event (p. 598)
probability of event A, or $P(A)$ (p. 598)
$P(A \text{ or } B)$ (p. 599)
mutually exclusive events (p. 599)
$P(\text{not } A)$ (p. 599)

odds (p. 601)
conditional probability (p. 607)
independent events (p. 607)
$P(A \text{ and } B)$ (p. 608)
binomial probability theorem (p. 614)
Monte Carlo simulation (p. 617)
expected value (p. 630)
fair game (p. 630)

Chapter Test

1. ***Writing*** Write a paragraph in which you discuss the difference between, and give examples of, empirical and theoretical probability.   16-1

---

636   *Chapter Sixteen*

2. An experiment consists of randomly drawing a card from a standard deck and rolling a die. Find the probability that:
   a. the card is a spade and the die shows a 5 $\frac{1}{24}$
   b. the card is *not* a spade and the die shows a 5 $\frac{1}{8}$
   c. the card is *not* a spade and the die does *not* show a 5 $\frac{5}{8}$

3. A fish bowl contains slips of paper numbered 1 through 9. Two slips are drawn, one after the other and without replacement. Find the probability of the each of the following events.   16-2
   a. Both numbers are odd. $\frac{5}{18}$
   b. The first number drawn is 6. $\frac{1}{9}$
   c. The first number drawn is greater than 6 and the second number is less than 4. $\frac{1}{8}$

4. For a certain brand of marigold seeds, the seeds sprout on average 9 out of 10 times. If 5 seeds are selected at random and planted, find the probability that: 16-3
   a. all 5 seeds sprout 0.59
   b. at least 3 of the 5 seeds sprout 0.991

5. A Central School PTA committee is to consist of 3 teachers and 3 parents, who are to be chosen at random from the 24 teachers and 145 parents involved in the PTA. If half of the teachers and 99 parents are women, find the probability that the committee has: 16-4
   a. only female members 0.034
   b. only 1 parent and only 1 teacher who are women 0.081

6. Ninety-five percent of the sneakers manufactured by a shoe company have no defects. In order to find the 5% that do have defects, inspectors carefully look over every pair of sneakers. Still, the inspectors sometimes make mistakes because 8% of the defective pairs pass inspection and 1% of the good pairs fail the inspection test.   16-5
   a. Incorporate the facts given above into a tree diagram.
   b. What percent of the pairs of sneakers pass inspection? 94.5%
   c. If a pair of sneakers passes inspection, what is the probability that it has a defect? 0.004

7. Suppose you play a game in which 5 coins are tossed simultaneously. If 1, 2, 3, or 4 "heads" occur, you win $1 for each "head." If all "heads" or all "tails" occur, however, you lose $20.   16-6
   a. Copy and complete the following table.

Number of "heads"	0	1	2	3	4	5
Payoff	−$20	?	?	?	?	?
Probability	$\frac{1}{32}$	?	?	?	?	?

$1 $2 $3 $4 −$20

$\frac{5}{32}$ $\frac{10}{32}$ $\frac{10}{32}$ $\frac{5}{32}$ $\frac{1}{32}$

   b. What is the game's expected payoff? $1.09

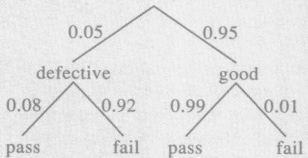

Chapter 17 Statistics

Overview

In this chapter, students learn to represent data graphically, to summarize data using statistics, and to analyze samples. In the first three sections, students graph sets of data in a variety of ways, including stem-and-leaf plots, histograms, frequency polygons, and box-and-whisker plots. Students also find various measures of central tendency and measures of dispersion for sets of data. The fourth section introduces normal distributions, for which students use the standard normal table (page 664) to find the percent of data in a given interval. The last two sections involve sampling. Students first look for errors in sampling procedures and then use their knowledge of normal distributions to find confidence intervals for the results of surveys and polls.

Objectives

17-1 To display a set of data using various statistical graphs and to find the mean, median, and mode.

17-2 To draw a box-and-whisker plot for a set of data and to use box-and-whisker and stem-and-leaf plots to compare sets of data.

17-3 To find the variance and standard deviation of a set of data and to convert data to standard values.

17-4 To recognize various types of distributions, to determine for a normal distribution the percent of data within a given interval, and to find percentiles.

17-5 To recognize different types of sampling procedures, to identify their limitations, and to estimate population characteristics based on samples.

17-6 To use a sample proportion to find a confidence interval for a corresponding population proportion.

Supplementary Resources

Tests, pp. 52–54
Alt. Assess., pp. 53–55, 71
Activities, pp. 46–47
Student Res. Guide, pp. 157–166

Software

Precalculus Plotter Plus
Activities, pp. 78–79, 80–81
Data Analyzer, Sampling Experiment

Pacing Guide

Section	17-1	17-2	17-3	17-4	17-5	17-6	Review	Test	Total
Discrete Math (days)	2	1	2	2	1	2	1	1	12

Teaching Notes

17-1 | pages 639–648

Presenting the Section

To introduce this chapter, you may want to tell students a little about the origins of statistics: The branch of mathematics called statistics has its origins in the need for governments to know certain facts about the governed. The ancient Romans, for example, took a regular census of people and property for the purpose of taxation. (Thus, we get the word *statistics* from the Latin word *status*, meaning state or condition.) Likewise, the United States government uses a census, conducted every 10 years, to determine the number of each state's representatives in the U.S. House of Representatives.

Students may have been introduced to the concepts of mean, median, and mode in previous mathematics courses. On the other hand, the stem-and-leaf plot and the cumulative frequency polygon (also known as an *ogive*) may be new to most students.

One way to introduce the stem-and-leaf plot is to begin with a set of test scores, such as 76, 84, 65, 68, 92, 77, 77, 63, 88, 82, 93, 95. Write these scores on the board in a location that can be later covered with a pull-down screen. On a different section of the board, construct a horizontal histogram for the set, using intervals of 60–69, 70–79, and so on. Cover the original data and ask students what the second lowest score is. Point out that the convenience of a histogram is balanced by the loss of information about individual items of data. Then return to the original data and construct a stem-and-leaf plot. Point out the similarity between the stem-and-leaf plot and the histogram, but be sure to note that (unlike a histogram) individual items of data are ''recoverable'' from a stem-and-leaf plot.

Communication

Throughout this chapter you should give students an opportunity to discuss the advantages and disadvantages of various statistical measures and techniques. In this section, for example, you could have students weigh the usefulness of statistical graphs and averages against the loss of information about individual items of data.

Using Technology

You may wish to have students write a computer program that calculates the mean, median, and mode of a set of data. Since finding the median requires an ordered set of data, you may wish to discuss one or more sorting algorithms (such as bubble sort and insertion sort) before students attempt to write the program. Students can test the program using the data in Written Exercises 1–4, page 643.

Cooperative Learning

Consider assigning the program described above to groups of three students. Each student in a group can be responsible for writing and testing a subroutine that calculates one of the three averages. The three students can then jointly write and test the main routine that calls the three subroutines.

17-2 | pages 648–653

Presenting the Section

To construct box-and-whisker plots (sometimes simply called *boxplots*), students need to be able to identify quartiles. Students may find a diagram like the one below helpful in understanding quartiles and medians. In the diagram, a vertical bar represents a quartile or median that is not an item of data but is instead the mean of the two data items on either side; an asterisk represents a quartile or median that is an actual item of data. (Of course, the numbers that appear between the vertical bars and asterisks indicate how many data items are in each quarter.)

For 40 data items: 10 | 10 | 10 | 10
For 41 data items: 10 | 10 * 10 | 10
For 42 data items: 10 * 10 | 10 * 10
For 43 data items: 10 * 10 * 10 * 10

Communication

Point out that in finding three medians—the median of all data items, the median of data items less than the median of all data items, and the median of data items greater than the median of all data items—the set of data is separated into four equal subsets. This process explains the use of the word *quartile*.

Assessment

You can reinforce the idea of quartiles by asking students to give the median and quartiles for various sets of consecutive integers. Here are four examples that you can use:

Set	Median	Lower quartile	Upper quartile
$\{1, 2, \ldots, 15\}$	8	4	12
$\{1, 2, \ldots, 14\}$	7.5	4	11
$\{1, 2, \ldots, 13\}$	7	3.5	10.5
$\{1, 2, \ldots, 12\}$	6.5	3.5	9.5

17-3 | pages 653–660

Presenting the Section

The focus of this section is on measures of dispersion, which describe the spread of data about the center. The section's opening discussion points out the fact that very different sets of data can have the same mean and even the same range. This shows the need for some other statistics that will make distinctions between the sets of data clear. The statistics discussed here are variance and standard deviation.

At this point in their study of statistics, students will probably enjoy discussing (and arguing about) the various methods of grading that teachers use. Point out that there are two basic types of grading, "absolute" and "relative." In absolute grading, objectives are established and students' performance is measured against them. (For example, a student who demonstrates mastery of, say, at least 90% of the objectives receives a grade of A.) In relative grading, the performance of one student is compared with the performance of the others. (For example, a student who performs better than, say, 80% of the class receives a grade of A. Alternately, an "A" student can be defined as one whose score is at least, say, 1.5 standard deviations above the mean.)

A discussion of relative grading naturally leads to a discussion of standardized tests, the scores for which are often interpreted relative to some base group. This is done through the use of standard values, which are presented at the end of the section.

Using Technology

Most scientific and programmable calculators will compute standard deviation. Make students aware of the fact that such calculators may have *two* keys for standard deviation, one that uses n in the denominator and one that uses $n - 1$. The appropriate use of these keys usually depends on whether the standard deviation is being calculated for the full set of data (use n) or for some sample (or subset) of the full set (use $n - 1$). In this text, n is always used.

17-4 | pages 660–669

Presenting the Section

The bell-shaped curve that signifies normal distribution is introduced gradually. Students first look at histograms depicting uniform, skewed, and normal distributions. Next, the histogram showing a normal distribution is subdivided into rectangles of lesser width, which then suggests the drawing of a smooth curve across the tops of the rectangles.

It is important for students to be comfortable with the basic properties of a normal distribution in order to understand the material presented in the rest of the chapter. You may need to supplement with other examples like Example 1 on page 663. Also, you may want to take the opportunity to discuss "grading on a curve" (where "curve" refers to a normal curve). Many students mistakenly believe that this procedure guarantees them a better grade. Point out, however, that if test scores are normally distributed and an "average grade" of C is to be assigned to any test score within 1 standard deviation of the mean, then the bulk of the students (in fact, about two thirds of them) will receive a grade of C. (This is good news for students who perform below average on the test, but bad news for those who perform above average.)

Students will need help in learning to use the standard normal table given on page 664. Be sure they can read the table both forward (for example, to find $P(2.1)$) and backward (for example, to find z such that $P(z) = 0.2743$).

Remind students of the formula from the previous section for converting "raw" data values to standard values.

Assessment

Use examples like Class Exercises 9a and 10a to check students' understanding of normal distributions before you discuss the standard normal distribution.

17-5 **pages 669–674**

Presenting the Section

This section has obvious relevance to many aspects of your students' lives. An awareness of different sampling procedures and their limitations will make students better-informed citizens. Because this section contains a lot of information, you may want students to read and then outline the section before discussing it in class.

Applications

Newspapers and magazines are handy sources of information related to sampling. This would be an opportune time to discuss recent polls concerning public opinion about the President's popularity, various political issues, and so on. Students may wish to conduct their own polls on school-related issues.

Cooperative Learning

The section lends itself well to group projects. One possibility would be to divide the class into small groups and have each group obtain poll data using the group itself as a sample of the entire class. The discussion may be more interesting if you can form the groups so that there is some discernible bias (all members of one group being of the same sex, or all being varsity athletes, or all being honors students, for example). Some questions that might be used for the intragroup polling are:

1. Should this school have a stricter dress code?
2. Should the school day be interrupted for pep rallies?
3. Should the school year be lengthened?
4. Should graduation requirements be made more demanding?

When group results are reported to the class, encourage students to discuss any glaring differences. They should consider not only bias in the composition of the groups, but also bias in the poll questions.

17-6 **pages 674–680**

Presenting the Section

Determining a confidence interval for a sample gives a more precise mathematical meaning to some of the ideas discussed in Section 17-5. As with that section, the material in this section is highly relevant to many situations in which students—and people in general—are called upon to make decisions.

In discussing the section, be sure students understand that different samples will give different estimates of a population proportion having a given characteristic. But because these sample proportions are normally distributed, we can determine an interval centered on any one of them such that there is a high probability that the interval actually contains the population proportion.

Cooperative Learning

You can help students understand the relationship between the mean of a sample and the mean of an entire population by having them consider all possible samples of a given size from a finite population. For example, have them consider $\{1, 2, 3, 4, 5, 6\}$, which has a mean of 3.5. Ask pairs of students to list the 15 possible samples of two items from the set and to calculate the mean of each sample, as shown.

Sample	Mean	Sample	Mean
{1, 2}	1.5	{2, 6}	4
{1, 3}	2	{3, 4}	3.5
{1, 4}	2.5	{3, 5}	4
{1, 5}	3	{3, 6}	4.5
{1, 6}	3.5	{4, 5}	4.5
{2, 3}	2.5	{4, 6}	5
{2, 4}	3	{5, 6}	5.5
{2, 5}	3.5		

Each pair of students should then draw a histogram of the sample means and note the familiar bell shape of a normal distribution. Have students find the mean of the sample means. (3.5)

17 Statistics

Descriptive Statistics

17-1 Tables, Graphs, and Averages

Objective *To display a set of data using a stem-and-leaf plot, a histogram, a frequency polygon, or a cumulative frequency polygon and to find the mean, median, and mode.*

The subject of statistics can be divided into two parts, *descriptive statistics* and *inferential statistics*. **Descriptive statistics** involves collecting, organizing, and summarizing numerical facts, called *data,* about some group. **Inferential statistics** involves making inferences or decisions based on the data. The first and second parts of this chapter deal with descriptive and inferential statistics, respectively.

Tables and Graphs

Shown below are the mathematics scores for the 50 seniors in a Midwestern high school who took a standardized achievement test.

480	520	670	580	500	570	700	540	690	500
520	450	570	540	620	450	460	790	580	610
550	630	460	470	510	630	510	720	490	540
690	510	500	710	520	480	750	450	670	610
320	680	530	510	350	400	400	640	430	480

To make some sense out of these scores, we can organize them into a **stem-and-leaf plot**. The first digit of each score is called the *stem* and the second digit is called the *leaf*. The result of recording the first column of scores (480, 520, 550, 690, and 320) is shown at the right. (Notice that all the leaves for a given stem are simply written one after the other.) Two complete stem-and-leaf plots are shown below. The ordered plot (on the right) is useful when calculating certain statistics, as you will see.

stem	leaf
3	2
4	8
5	2 5
6	9
7	

3|2 represents a score of 320.

3	2 5
4	8 5 6 7 5 8 0 6 0 5 9 3 8
5	2 5 2 1 7 0 3 8 4 1 0 1 2 7 1 4 8 0 4
6	9 3 8 7 2 3 4 9 7 1 1
7	1 0 5 9 2

3	2 5
4	0 0 3 5 5 5 6 6 7 8 8 8 9
5	0 0 0 1 1 1 1 2 2 2 3 4 4 4 5 7 7 8 8
6	1 1 2 3 3 4 7 7 8 9 9
7	0 1 2 5 9

Scores recorded by column   Scores rearranged in increasing order

◄ One can record the weights and heights of zebras in a small random sample, then use statistics to make a confident prediction about the average weights and heights in the population as a whole.

Teaching Notes, p. 638B

Warm-Up Exercises

1. Find the mean (or average) of the following numbers: 2, 37, 40, 44, 45. 33.6

2. Does the mean found in Exercise 1 represent the data well? Explain your answer. No; the mean does not fall in the interval 37–45 in which most of the numbers lie.

3. Is the mean of a group of numbers *always, sometimes,* or *never* a number in the group? sometimes

4. Describe how to find the mean of *n* numbers. Add the numbers and divide the sum by *n*.

Motivating the Section

Give students a set of data, such as the heights (in inches) of the players on a basketball team. Ask students what they think the "average" height is. Although some students may calculate the mean height, point out that there are different kinds of averages, to be discussed in this section.

Communication Note

Draw students' attention to the "decoding key" just below the partial stem-and-leaf plot on this page. Without this key, we do not know whether 3|2 represents 32, 320, 0.32, and so on. Require that students always include such a key with each stem-and-leaf plot.

Additional Examples

1. The 12 National League baseball teams had shutouts for one season as shown in the table below.

Team	Shutouts
Atlanta	4
Chicago	10
Cincinnati	13
Houston	15
Los Angeles	24
Montreal	12
New York	22
Philadelphia	6
Pittsburgh	11
St. Louis	14
San Diego	9
San Francisco	13

a. Summarize the data in a stem-and-leaf plot.
b. Find the mean, median, and mode.

a. 0 | 4 6 9
1 | 0 1 2 3 3 4 5
2 | 2 4
0 | 4 means 4 shutouts.

b. The sum of the data is 153, so the mean is $\frac{153}{12} = 12.75$;
median = mean of 6th and 7th numbers = $\frac{1}{2}(12 + 13) = 12.5$;
mode = 13

The test score data can also be organized and displayed in a **histogram** with either horizontal or vertical bars, as shown below. Notice that the histogram with horizontal bars looks very much like the stem-and-leaf plots at the bottom of the preceding page. Also notice that a frequency axis is not necessary when the frequency is noted on or near each bar.

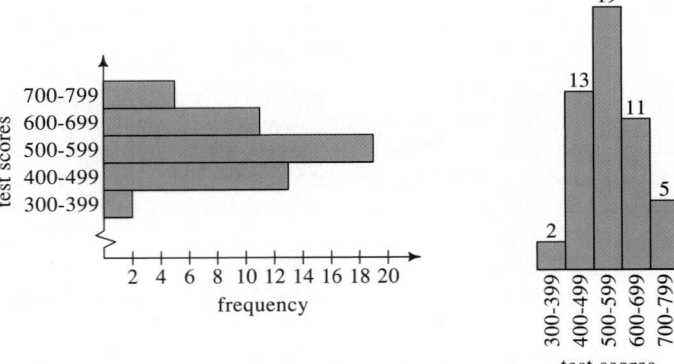

The test score data can also be given in a **frequency table,** as shown at the left below. Notice that the column labeled "frequency" gives the numbers of scores in each *class,* or group, of scores. The column labeled "cumulative frequency," on the other hand, gives the number of scores (from the entire set of scores) less than or equal to the largest score in each class. For example, the frequency for the class 400–499 is 13 (because there are 13 scores in the 400's), but the cumulative frequency for the same class is 2 + 13 = 15 (because there are a total of 15 scores less than or equal to 499).

Frequency Table		
Class	Frequency	Cumulative frequency
300–399	2	2
400–499	13	15
500–599	19	34
600–699	11	45
700–799	5	50
Total = 50		

Relative Frequency Table		
Class	Relative frequency	Relative cumulative frequency
300–399	4%	4%
400–499	26%	30%
500–599	38%	68%
600–699	22%	90%
700–799	10%	100%
Total = 100%		

The table on the right at the bottom of the preceding page is like the one on the left except that it gives *relative frequencies* (expressed as percents). For example, while the left table shows that 13 out of 50 scores are in the class 400–499, the right table records this as $\frac{13}{50} = 26\%$. Likewise, while the left table shows that the cumulative frequency for the class 400–499 is 15, the right table records this as a relative cumulative frequency of $\frac{15}{50} = 30\%$.

Using the frequency tables, we can construct the **frequency polygon** shown at the left below. Its vertical axis can be labeled either as integer frequencies, as shown, or as relative frequencies in percents. Either way, the horizontal axis of the graph shows the middle score for each class. Note that a frequency polygon is usually drawn so that it begins and ends on the horizontal axis.

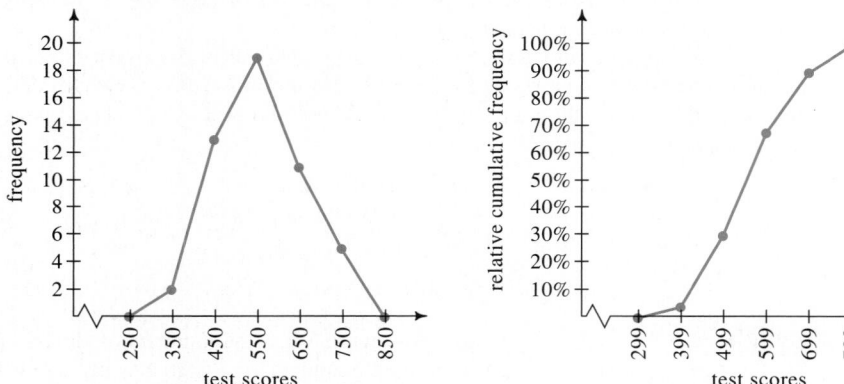

Using the frequency tables, we can also construct the **cumulative frequency polygon** shown at the right above. Although its vertical axis can be labeled either in integers or in percents, the use of percents is most common. The horizontal axis shows the largest score for each class.

Averages

Sometimes we use a single number, called an *average,* to represent an entire set of data. In statistics the three most common averages are the *mean,* the *median,* and the *mode.* Because each average points to some "center" of the data, each is called a *measure of central tendency.*

The **mean** of a set of data is the sum of the data divided by the number of items of data. The mean is the familiar average that you may use when talking about your "test average." If the data are given in a list $x_1, x_2, x_3, \ldots, x_n$, then the mean, which is denoted $\bar{x}$ (read "x bar"), is given by:

$$\bar{x} = \frac{x_1 + x_2 + x_3 + \cdots + x_n}{n} = \frac{\sum_{i=1}^{n} x_i}{n}$$

Statistics   **641**

2. Given below are the sales-tax rates, to the nearest whole percent, for the 50 states in a recent year.

% rate	No. of states
0	5
1	0
2	0
3	3
4	15
5	14
6	11
7	1
8	1

a. Draw a frequency polygon and a cumulative frequency polygon.

b. Find the mean, median, and mode(s).

a.

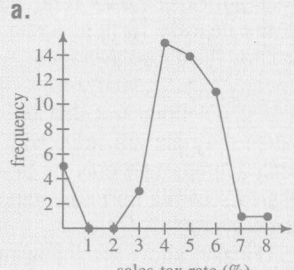

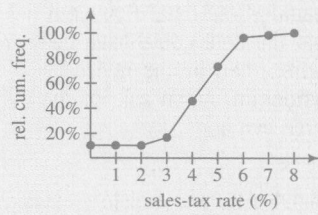

b. Sum of data is 220, so the mean is $\frac{220}{50} = 4.4\%$; median = mean of 25th and 26th rates = $\frac{1}{2}(5 + 5) = 5\%$; mode = 4%

When discussing the three types of average, present the family-size histogram shown below. Ask students to calculate the three averages for the data. (mean: 3.52, median: 4, mode: 4)

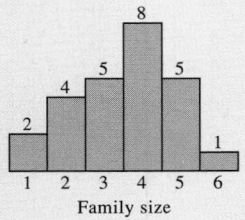

Family size

Now relabel the histogram as "Minutes to read a history essay" and renumber it as 10, 20, 30, 40, 50, and 60. Do not change the frequencies, however. Point out that the histogram gives data rounded to the nearest 10 min. Again ask for the three averages. (mean: 35.2, median: 40, mode: 40) Point out that because the data are rounded, each average can only be an *estimate* of the corresponding "true average."

Finally, show the following stem-and-leaf plot of the reading-time data. Point out that the data could have been represented by the revised histogram. Again ask for the three averages. (mean: 36.04, median: 38, mode: 20)

Minutes to read a history essay

```
1 | 2 2
2 | 0 0 0 0 8
3 | 0 2 2 3 8 8 9
4 | 0 0 2 2 4 8
5 | 2 2 4 4 9
```

1 | 2 represents 12 min.

The formula for the mean given at the bottom of the preceding page can be used to find the mean of the 50 test scores given on page 639. Since the sum of the scores is 27,480, the mean is $\frac{27,480}{50} = 549.6$.

The **median** of a set of data is found by arranging the data in increasing or decreasing order. If the number of items of data is odd, then the median is the middle number in the ordered set. For example, the median of the data 1, 3, 3, 5, 6, 8, 9 is 5. If the number of items of data is even, then the median is the mean of the two middle numbers. For example, the median of the data 2, 3, 3, 5, 8, 9 is 4, which is the mean of 3 and 5.

If a set of data is given in a stem-and-leaf plot with ordered leaves, then the median of the data can be found easily. For example, the median of the 50 test scores in the ordered stem-and-leaf plot at the bottom of page 639 is halfway between the 25th and 26th scores; it is 525.

The **mode** of a set of data is the one item of data that occurs most often. If there are two (or three or more) items of data that occur most often, the set of data is called *bimodal* (or *trimodal* or *multimodal*). If no item of data is repeated, then we say that there is no mode. For example:

1. The data 1, 3, 3, 5, 8, 9 have mode 3.
2. The data 1, 3, 3, 5, 8, 8 have modes 3 and 8.
3. The data 1, 2, 3, 5, 8, 9 have no mode.
4. The mode of the 50 test scores on page 639 is 510.

Example The frequency table below summarizes the numbers of siblings (brothers and sisters) for each of the 25 students in a statistics class. For the data, find **(a)** the mean, **(b)** the median, and **(c)** the mode.

Number of siblings	0	1	2	3	4	5
Number of students	2	10	8	4	0	1

Solution **a.** Mean = $\dfrac{\text{total number of siblings}}{25}$

$= \dfrac{0 \cdot 2 + 1 \cdot 10 + 2 \cdot 8 + 3 \cdot 4 + 4 \cdot 0 + 5 \cdot 1}{25}$

$= \dfrac{43}{25} = 1.72$

b. Since the data consist of 25 numbers, the middle one is the thirteenth. Reading from the bottom row of the table, we see that twelve (2 + 10) students have either 0 or 1 siblings. Thus, the thirteenth student has 2 siblings, which is the median number.

c. The mode of the given data is 1 because this number occurs most frequently (10 times).

Part (a) of the preceding example suggests an alternate way of calculating the mean of a set of data. If the data are given in a frequency table like this:

Item of data	x_1	x_2	$\cdots$	x_n
Frequency	f_1	f_2	$\cdots$	f_n

then the mean is given by:

$$\bar{x} = \frac{x_1 f_1 + x_2 f_2 + \cdots + x_n f_n}{f_1 + f_2 + \cdots + f_n} = \frac{\displaystyle\sum_{i=1}^{n} x_i f_i}{\displaystyle\sum_{i=1}^{n} f_i}$$

CLASS EXERCISES

1. Give the mean, median, and mode of each set of data.
   a. 1, 2, 3, 5, 5   b. 3, 4, 4, 7, 8, 8, 10, 11   c. 1, 2, 3
   3.2; 3; 5   6.875; 7.5; 4 and 8   2; 2; none

   d.
Observation	0	1	2	3	4
Frequency	3	2	5	8	2

   2.2; 2.5; 3

2. **Investigation** Record the number of siblings that each member of your class has. Tabulate the data and draw a histogram. Then calculate the mean, median, and mode of the data.

3. a. Is it possible for the mean of a set of integers not to be an integer? Yes
   b. Is it possible for the median of a set of integers not to be an integer? Yes
   c. Is it possible for the mode of a set of integers not to be an integer? No

4. In a set of data, suppose the smallest number (not a mode) is replaced with an even smaller number. How are the mean, median, and mode affected?
   The mean decreases. The median and mode do not change.

WRITTEN EXERCISES

In Exercises 1–4, find the mean, median, and mode for each set of data.

A 1. The numbers of mice born in 9 different litters were:

   5, 7, 6, 3, 8, 6, 4, 6, 4 5.4; 6; 6

2. The test scores for a statistics student were:

   85, 74, 92, 87, 84, 78, 90 84.3; 85; none

3. The numbers of items produced in 1 h by 12 different machines were:

   0, 1, 3, 5, 5, 5, 7, 9, 9, 11, 15, 99 14.1; 6; 5

4. The heights in centimeters of the members of a college basketball team are:

   185, 189, 191, 193, 193, 195, 196, 198, 198, 200 193.8; 194; 193 and 198

Additional Answers Written Exercises

6. a.

5	2 3 3 5 5 5 6 7 8 9
6	0 0 1 2 2 3 4 7 8
7	0 3 4 4 5 6 8
8	0 1 2 3

5 | 2 = 52°

b.

5	2 3 3
•	5 5 5 6 7 8 9
6	0 0 1 2 2 3 4
•	7 8
7	0 3 4 4
•	5 6 8
8	0 1 2 3

5 | 2 = 52°

c. 62.5, 55

7. a. 24, 25, 26, 27

b.

24	2 5
25	2 6 7 9
26	0 1 1 3 9
27	1 6 7

24 | 2 = .242

c. .2605

8. a. 9, 10, 11, 12, 13, 14

b.

9	4
10	1 8
11	6 7
12	2 6 7 7 9
13	0 4
14	2 5

9 | 4 = 94

c. 126.5

13.5; 9, 13, and 16

5. a. The number of trials required by 20 different puppies to learn a trick is recorded in the stem-and-leaf plot at the left below. Rearrange the data in increasing order, and then find the median and the mode of the data.

b. The stem-and-leaf plot at the left below has just three stems (0, 1, and 2). In order to spread out the data more, you can draw a plot with 6 stems, as shown at the right below. The stem 2 is used for data from 20 to 24, while the dot below the 2 is the stem for data from 25 to 29. The other dots have similar meanings. Complete this plot.

0	9 7 9 4 8
1	8 2 3 0 9 1 6 4 5 6 3
2	1 2 8 5

Stem-and-leaf plot with 3 stems

0	4
•	7 8 9 9
1	0 1 2 3 3 4
•	5 6 6 8 9
2	1 2
•	5 8

Stem-and-leaf plot with 6 stems

6. Meteorology The maximum Fahrenheit temperatures at the municipal airport for each of the 30 days of September were:

82°, 75°, 78°, 83°, 81°, 74°, 80°, 74°, 73°, 62°, 57°, 64°, 68°, 70°, 76°, 67°, 63°, 60°, 55°, 58°, 62°, 59°, 55°, 53°, 53°, 56°, 61°, 55°, 52°, 60°

a. Summarize the data in a stem-and-leaf plot having stems 5, 6, 7, and 8.

b. Summarize the data in a stem-and-leaf plot with 8 stems. (See part (b) of Exercise 5.)

c. Find the median and mode of these 30 temperatures.

7. Sports The 14 American League baseball teams had team batting averages for one season as shown in the table at the right.

a. To summarize the batting-average data in a stem-and-leaf plot, what stems would you use?

b. Using your answer to part (a), draw a stem-and-leaf plot.

c. What is the median of the data?

8. Sports The 14 American League baseball teams had total home runs for one season as shown in the table at the right.

a. To summarize the home-run data in a stem-and-leaf plot, what stems would you use?

b. Using your answer to part (a), draw a stem-and-leaf plot.

c. What is the median of the data?

American League team	Batting average	Home runs
Baltimore	.252	129
Boston	.277	108
California	.256	145
Chicago	.271	94
Cleveland	.245	127
Detroit	.242	116
Kansas City	.261	101
Milwaukee	.259	126
Minnesota	.276	117
New York	.269	130
Oakland	.261	127
Seattle	.257	134
Texas	.263	122
Toronto	.260	142

9. A 10-question true-false quiz was given to 50 students and the number of questions correct ranged from 3 to 10, as shown in the cumulative frequency polygon at the right.

percent of students with score $\leq x$

score = x

a. What percent of the students had scores less than or equal to 7? greater than 8? 60%; 20%

b. What is the median score? 6

10. Draw a frequency polygon and a cumulative frequency polygon for the following data: 20, 21, 22, 24, 24, 25, 25, 25, 27, 28.

11. Physiology The table below gives the results of a driver-education experiment. The experiment measures the time between the appearance of a stimulus on a screen and a student's reaction of depressing a brake pedal.

Time (to the nearest 0.1 s)	0.1	0.2	0.3	0.4	0.5	0.6	0.7
Frequency	1	4	5	4	3	1	2

a. Draw a histogram for the data.

b. Find the mean, median, and mode.

c. Draw a cumulative frequency polygon for the data.

12. The table below gives the number of questions answered correctly by 30 students on the 20-question written portion of a driver education test.

Number correct	12	13	14	15	16	17	18	19	20
Frequency	1	0	3	3	3	5	6	5	4

a. Draw a histogram for the data.

b. Find the mean, median, and mode.

c. Draw a cumulative frequency polygon for the data.

13. Five workers on an assembly line have hourly wages of $8.00, $8.00, $8.50, $10.50, and $12.00.

a. Find the mean, median, and mode.

b. If the hourly wage of the highest paid worker is raised to $20.00 per hour, how are the mean, median, and mode affected?

14. A teacher told her principal that her class of 10 astronomy students had done well on the last test by getting a modal score of 88. The students, however, thought that the test was difficult because the mean score was 68. Make up 10 scores that fit this description and find their median.

15. Writing For some set of data, suppose that you are given only a frequency polygon (with integer labels on the frequency axis). Write a paragraph in which you explain how to estimate the mean of the data from the graph. Then apply your method to the frequency polygon on page 641.

Statistics **645**

Additional Answers Written Exercises

11. b. 0.375, 0.35, 0.3

12. b. 17.2, 17.5, 18

13. a. $9.40, $8.50, $8.00

b. The mean is raised to $11.00 per hour. The median and mode are unchanged.

14. Answers may vary. For example: 45, 50, 55, 60, 65, 70, 75, 84, 88, 88; median: 67.5

15. The x- and y-coordinates of each vertex of a frequency polygon are the middle value and the frequency of each class, respectively. Thus, we can approximate the mean of the data by summing the products of the x- and y-coordinates of each vertex and then dividing by the sum of the y-coordinates. For the frequency polygon on page 641, the sum of the products of the x- and y-coordinates is $2(350) + 13(450) + 19(550) + 11(650) + 5(750) = 27,900$; the sum of the y-coordinates is $2 + 13 + 19 + 11 + 5 = 50$; so the mean of the data is about $\frac{27,900}{50} = 558$.

16. In architecture, an ogive is the rib of a Gothic vault. Because a cumulative frequency polygon tends to rise sharply and then level off, its shape suggests an ogive.

17. b. $4.98, $5.00, $4.50

18. b. 7.36 h, 7 h, 7 h

21. Since the sum of the number of children is divided by 100, at most two places to the right of the decimal point would be needed to express the mean. The median must be an integer or end in .5. The mode must be an integer.

16. *Research* A cumulative frequency polygon is sometimes called an *ogive* (pronounced ''oh-jive''). This word is also used in architecture. Find out what the architectural meaning of ogive is and explain how this meaning relates to the mathematical graph of the same name.

B **17. Consumer Economics** A consumer organization collected last week's prices for a chef's salad at 100 restaurants. The prices are given below.

Price in cents	Number of restaurants		Price in cents	Number of restaurants
325–374	8		575–624	10
375–424	10		625–674	6
425–474	30		675–724	2
475–524	20		725–774	1
525–574	12		775–824	1

a. Draw a frequency polygon for the data.
b. Calculate the mean, median, and mode using the middle price in each class as the representative price for the class. (That is, use 350 (cents), 400, 450, 500, and so on.)

18. Each of the 50 students in two mathematics classes was asked to keep track of the number of hours spent doing mathematics homework during one week. The results are given below.

Hours spent on mathematics homework	0–2	3–5	6–8	9–11	12–14
Number of students	3	11	17	15	4

a. Draw a frequency polygon for the data.
b. Calculate the mean, median, and mode using the middle number of hours for each homework category (1, 4, 7, 10, and 13).

19. *Research* Find a frequency table in a newspaper or magazine. Calculate the mean, median, and any modes.

20. Suppose you told a friend your ''average'' weekly earnings for the 12 weeks of summer. Could your friend calculate your total summer earnings if your ''average'' was **(a)** a mean? **(b)** a median? **(c)** a mode? yes; no; no

21. A pollster reports that for 100 families interviewed, the mean number of children per family was 2.038, the median was 1.9, and the mode was 1.82. Explain why each of these figures must be wrong.

22. Geography The table below gives data on the percent of the U.S. population ever married.

Age group	15–19	20–24	25–29	30–34	35–44	45–54
Percent of females ever married	6	39	71	84	92	95
Percent of males ever married	2	22	57	75	89	94

a. At the right is part of a comparative histogram showing two side-by-side bars for each of the age groups. Complete the histogram.

b. *Writing* Write a paragraph explaining what the data show.

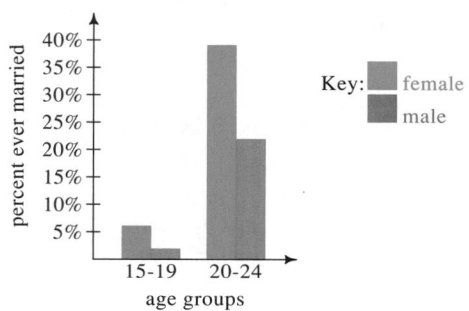

23. Nutrition The table below gives data on the recommended energy intake, in kilocalories (Cal), for average females and males in various age groups.

Age group	11–14	15–18	19–22	23–50	51–75	≥76
Energy needs (in Cal) for females	2200	2100	2100	2000	1800	1600
Energy needs (in Cal) for males	2700	2800	2900	2700	2400	2050

a. Draw a comparative histogram showing two side-by-side bars for each of the age groups. (See Exercise 22 for an example of such a histogram.)

b. *Writing* Write a paragraph explaining what the data show.

24. For a group of 10 teenagers, the mean age is 17.1, the median is 16.5, and the mode is 16. If a 21-year-old joins the group, give the mean, median, and mode for the ages of the 11 people. 17.5; 17; 16

The following data are the average low monthly temperatures (°F) for Berlin, Germany:

J	F	M	A	M	J
26	27	32	38	46	51
J	A	S	O	N	D
55	54	48	41	33	29

1. Find the median temperature. 39.5

2. Find the median of those temperatures that are less than the median temperature. 30.5

3. Find the median of those temperatures that are greater than the median temperature. 49.5

4. Suppose the data were given in degrees Celsius rather than degrees Fahrenheit. What differences would you expect in the answers to Exercises 1-3? Temperatures would shift down; all three values would be lower.

Give students a set of data by writing 12 numbers, in random order, on the board. Ask students to arrange the data in increasing order and to find the median. Then ask students to find the median of the smallest 6 numbers as well as the median of the largest 6 numbers. Point out that the three medians divide the data into quarters, for which a box-and-whisker plot, the subject of this section, can be drawn.

25. **Economics** The distribution of salaries in a company is shown below. The median salary is $14,500 and the mean salary is $17,800.

Job	5 executives	15 supervisors	80 production workers
Salary range	$40,200–$87,500	$15,800–$25,000	$9200–$18,700

a. If the 15 supervisors are each given a $1000 raise and no one else gets a raise, what are the new mean and median? $17,950; $14,500

b. If the 80 production workers each get a $1000 raise and no one else gets a raise, what are the new mean and median? $18,600; $15,500

17-2 Box-and-Whisker Plots

Objective To draw a box-and-whisker plot for a set of data and to use box-and-whisker and stem-and-leaf plots to compare sets of data.

Drawing Box-and-Whisker Plots

The stem-and-leaf plot below shows the ages of winners of the Academy Award for Best Actress from 1928 to 1989. Marlee Matlin was 21 at the time that she received an Oscar for her 1986 role in *Children of a Lesser God*. Since 21 is the youngest age of an Oscar-winning actress, it is called the *lower extreme* of the data. The *upper extreme* of the data is 80, Jessica Tandy's age when she won an Oscar for her 1989 role in *Driving Miss Daisy*. You might want to update the data with the winners of the Academy Award for Best Actress since 1989. You can find the information in an almanac.

Ages of Academy Award Winners, 1928–1989

Best Actresses

```
2 | 1 2 4 4 4 4 4
• | 5 6 6 6 6 6 6 6 7 7 7 8 8 9 9 9
3 | 0 0 0 0 1 1 2 3 3 3 4 4 4 4 4 4
• | 5 5 6 7 7 8 8 8
4 | 0 1 1 1 1 1 2
• | 5 8 9
5 |
• |
6 | 0 1 1 2
• |
7 | 4
• |
8 | 0
```

2|1 represents an age of 21.

The stem-and-leaf plot shows 63 ages (one more than expected because there were two winners in 1968). If these 63 ages are arranged from left to right in increasing order, the median is the 32nd age in the list. The median of the ages to the left of the median position is called the *lower quartile,* and the median of the ages to the right of the median position is called the *upper quartile.*

	Lower extreme	Lower quartile	Median	Upper quartile	Upper extreme
Position in list	1	16	32	48	63
Age	21	27	33	40	80

Note that the positions of the median, quartiles, and upper extreme shift as new data are added to the list. For example, if the age of the 1990 winner is included, the median would be halfway between the 32nd and 33rd ages, the lower quartile would be halfway between the 16th and 17th ages, the upper quartile would be halfway between the 48th and 49th ages, and the upper extreme would be at the 64th age.

We can display the data above in a **box-and-whisker plot.** To do so, we first plot the median, quartiles, and extremes not on, but just below, a number line.

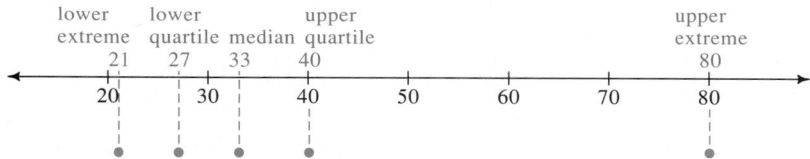

We then draw a *box* extending from the lower quartile to the upper quartile and *whiskers* extending from the quartiles to the extremes. We also draw a line through the box at the median as shown.

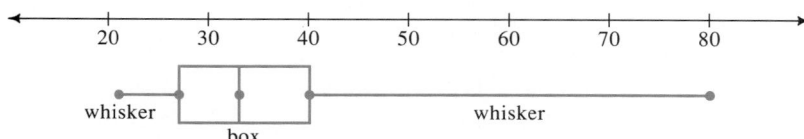

The difference between the two extremes is called the **range;** in this case, the range is $80 - 21 = 59$. About 50% of the data lie in the interval of the box, from the lower quartile 27 to the upper quartile 40. The length of this interval, which is $40 - 27 = 13$, is called the *interquartile range.*

If you refer to the stem-and-leaf plot on the preceding page, you will see that the 6 oldest ages are separated from the cluster of the remaining data. These 6 ages (60, 61, 61, 62, 74, and 80) are called *outliers.*

Error Analysis

Point out that the height of the box in a horizontal box-and-whisker plot is really irrelevant; only the length of the box is significant.

Additional Examples

1. The stem-and-leaf plot shows the ages of the first 41 U.S. Presidents at the time each took office.

4	·	5	·	6	·
2	6	0	5	0	5
3	7	1	5	1	8
	8	1	5	1	9
	9	1	5	1	
	9	1	6	2	
		1	6	4	
		2	6	4	
		2	7		
		4	7		
		4	7		
		4	7		
		4	8		

a. Find the extremes and the range.
b. Find the median and lower and upper quartiles.
c. Find the interquartile range.
d. Are there any outliers?
e. Draw a box-and-whisker plot.

a. Extremes: 42 and 69; range: $69 - 42 = 27$
b. Median: 21st age, or 55; lower quartile: mean of 10th and 11th ages, or 51; upper quartile: mean of 31st and 32nd ages, or 59
c. Interquartile range: $59 - 51 = 8$

(Continues on next page)

d. $1.5(8) = 12$
$51 - 12 = 39$
$59 + 12 = 71$
Since all the ages are between 39 and 71, there are no outliers.

e.

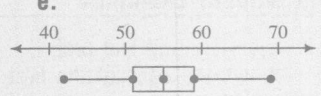

2. Compare the box-and-whisker plot that you drew in Additional Example 1 with the one for best actresses on this page. What do you notice?

The median age of presidents, 55, is considerably higher than the median age of best actresses, 33. The range of the presidents' ages, 27, is considerably smaller than the range of the best actresses' ages, 59; but almost equal to the restricted range for best actresses' ages, 28, when the outliers are excluded.

Cooperative Learning

Divide students into small groups and give ten thumbtacks to each group of students. For a given group, the tacks should be alike; but between groups, the tacks should be different shapes and sizes. Have students record data on the number of "point up" tacks when all ten tacks are flipped. Each group should perform 25 trials and then draw box-and-whisker plots of the results. Compare the plots from the various groups and ask whether any differences appear to be significant.

Although mathematicians do not always agree on how far out an item of data must be in order to be considered an outlier, we will define an outlier to be any item of data whose distance to the nearer quartile exceeds 1.5 times the interquartile range. For the age data, then, $1.5 \times$ interquartile range $= 1.5 \times 13 = 19.5$, so that any age more than 19.5 from the box in the box-and-whisker plot on the preceding page is an outlier. We therefore modify the plot by letting the whiskers include only the data that are not outliers and by using small x's to show the outliers.

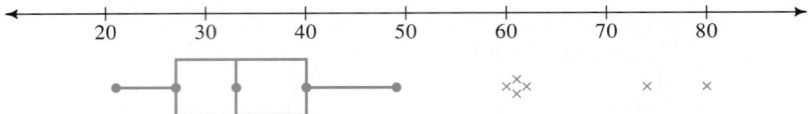

Comparing Sets of Data

How do the ages of winners of the Academy Award for Best Actor compare with the ages of Oscar-winning actresses? One way to make this comparison is with the back-to-back stem-and-leaf plot shown below. The youngest age for a best actor is 30 (Marlon Brando, *On the Waterfront*, 1954) and the oldest age is 76 (Henry Fonda, *On Golden Pond*, 1981).

Ages of Academy Award Winners, 1928–1989

Best Actors		Best Actresses
	2	1 2 4 4 4 4 4
	•	5 6 6 6 6 6 6 7 7 7 8 8 9 9 9
4 4 3 3 2 2 2 1 0	3	0 0 0 0 1 1 2 3 3 3 4 4 4 4 4 4
9 9 8 8 8 8 8 7 7 5 5 5 5	•	5 5 6 7 7 8 8 8
4 4 3 3 3 3 3 2 2 1 1 1 1 0 0 0 0	4	0 1 1 1 1 1 2
9 9 9 8 8 8 7 7 6 6 5	•	5 8 9
3 2 1 1	5	
6 6 6 5	•	
2 1 0	6	0 1 1 2
	•	
	7	4
6	•	
	8	0

$1|3|0$ represents an actor age of 31 and an actress age of 30.

	Best Actors	Best Actresses
Lower quartile:	38	27
Median:	42	33
Upper quartile:	48	40

The stem-and-leaf plot above makes it appear that best actors tend to be older than best actresses. Another way to make this comparison is with side-by-side box-and-whisker plots, as shown at the top of the next page.

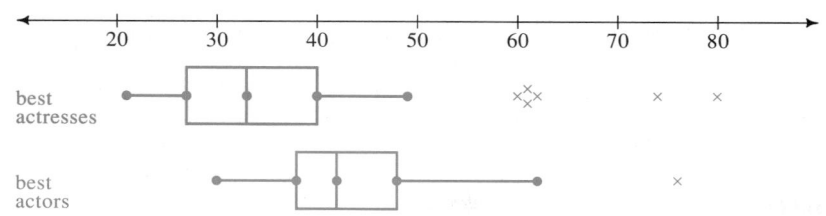

The apparent difference in the ages of best actors and best actresses can be shown by statistical methods to be *statistically significant*. This means that it is highly unlikely for the difference in ages to be due merely to chance.

CLASS EXERCISES

For Exercises 1–4, use the box-and-whisker plot shown below. It gives the weights in pounds of the 42 players on a high school football team.

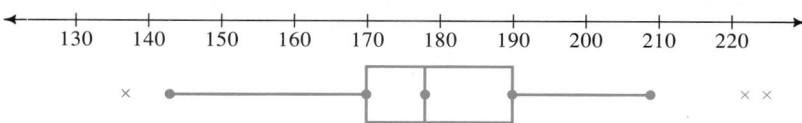

1. What are the extremes? What is the range? 137, 225; 88
2. What are the lower and upper quartiles? What is the interquartile range? 170, 190; 20
3. What is the median? 178
4. Explain why the weight of 225 lb is an outlier. What are the other outliers?

WRITTEN EXERCISES

A

1. **a.** Recently, several states with small populations, such as Vermont, had only one representative in the U.S. House of Representatives. On the other hand, California had 52 representatives, more than any other state. The median number of representatives was 6, the lower quartile was 2, and the upper quartile was 10. With these facts, draw a box-and-whisker plot. Show whiskers extending to the extremes of the data.
   b. Would Vermont's number of representatives be an outlier? What about California?

2. **a.** Several U.S. presidents, including George Washington, had no children. John Tyler, the tenth U.S. president, had 15 children (from two marriages), more than any other president. The median number of children of a president is 3, and the lower and upper quartiles are 2 and 5, respectively. With these facts, draw a box-and-whisker plot. Show whiskers extending to the extremes of the data.
   b. Would the number of Tyler's children be an outlier? Explain.

Statistics   **651**

3. In general, the students have better mathematics scores than verbal scores. The median mathematics score is higher than the median verbal score. Although the upper quartiles are the same, the distance to the median of the mathematics scores is much smaller, indicating more mathematics than verbal scores clustered in the upper 500 range. The verbal scores extended over a wider range than the mathematics scores.

4. In general, the students grew. The median height increased about 5 cm. The interquartile range and the range are greater, indicating that the heights are less clustered. There is an outlier now, indicating that one student grew considerably taller than the others.

Cooperative Learning

Separate the class into groups of 6 and assign each member of a group one of Exercises 9–14. Group members should share their work and make presentations of their graphs to the entire class.

Supplementary Materials

Alternative Assessment, 53

3. **Writing** The mathematics and verbal SAT scores for seniors at Van Buren High School are shown in the box-and-whisker plots below. Write a paragraph explaining what the plots tell you.

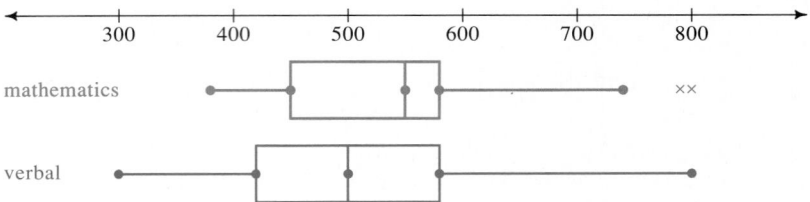

4. **Writing** Box-and-whisker plots can be displayed on a vertical scale as well as on a horizontal one. The two plots shown at the right compare the heights in centimeters of this year's senior class with their heights three years ago when they were freshmen. Write a paragraph explaining what the plots tell you.

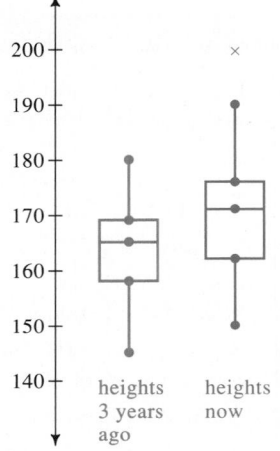

Nutrition For Exercises 5–8 at the top of the next page, use the table shown below, which gives nutritional information about some of the menu items at a restaurant.

Menu item	Calories per serving	Protein (in grams)	Carbohydrates (in grams)	Fat (in grams)
Regular salad	110	7.1	6.2	6.6
Chef salad	230	20.5	7.5	13.3
Fish sandwich	440	13.8	37.9	26.1
Chicken sandwich	490	19.2	39.8	28.6
Regular hamburger	260	12.3	30.6	9.5
Regular cheeseburger	310	15.0	31.2	13.8
Quarter-pound hamburger	410	23.1	34.0	20.7
Quarter-pound cheeseburger	520	28.5	35.1	29.2
Deluxe hamburger	560	25.2	42.5	32.4

5. Draw a stem-and-leaf plot and a box-and-whisker plot for the number of calories in one serving of each menu item.

6. Draw a stem-and-leaf plot and a box-and-whisker plot for the grams of fat (rounded to the nearest gram) in one serving of each menu item.

7. Draw a back-to-back stem-and-leaf plot and a side-by-side box-and-whisker plot comparing grams of protein and grams of carbohydrates (each rounded to the nearest gram).

8. The point (6.6, 110), plotted in the graph at the right, indicates that a regular salad has 6.6 grams of fat and 110 calories. Copy the graph and plot a point for each of the other menu items. Does the graph suggest any trends to you? (*Note:* The graph is an example of a *scatter plot,* which will be discussed in Section 18-1.)

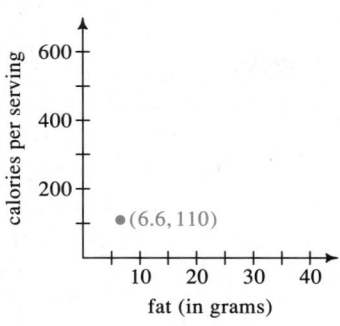

Research In Exercises 9–14, use an almanac to find data for each of the 50 states. Then draw stem-and-leaf and box-and-whisker plots.

9. Per pupil expenditure for education

10. Percent of eligible voters who actually voted in the last presidential election

11. Mean family income

12. Salary of governor

13. Comparison of land areas of states east and west of the Mississippi River (Consider Louisiana and Minnesota to be west of the Mississippi, since most of each state's area is west of the river.)

14. Comparison of populations of states east and west of the Mississippi River

15. *Research* From an almanac, select data that you find interesting. Then draw stem-and-leaf and box-and-whisker plots of the data.

17-3 Variability

| **Objective** | To find the variance and standard deviation of a set of data and to convert data to standard values. |

A **statistic** is a number that describes some characteristic of a set of data. For example, the mean, median, and mode (see Section 17-1) are statistics used to describe the center of a set of data, and the range and interquartile range (see Section 17-2) are statistics used to describe the spread of data about the center.

In this section we will discuss other statistics that are used to describe the spread of data about the center. These statistics are called *measures of dispersion,* because they indicate the amount of variability in the data.

Statistics **653**

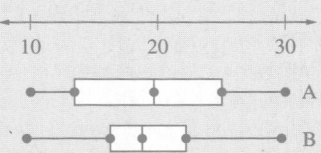

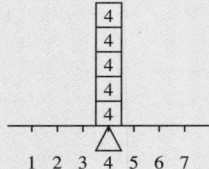
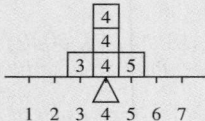
Consider the algebra test scores, given below, for two classes of 25 students.

Class I: 50, 64, 68, 70, 70, 72, 73, 75, 75, 76, 78, 78, 79, 80, 81, 82, 82, 82, 85, 88, 88, 89, 90, 92, 93

Class II: 56, 57, 58, 60, 61, 65, 67, 68, 70, 71, 74, 75, 77, 79, 82, 87, 88, 89, 92, 95, 96, 97, 98, 99, 99

Both classes have the same mean score, 78.4. Moreover, both classes have the same range, since $93 - 50 = 99 - 56 = 43$. Even though the ranges are the same, the histograms below show that the scores for Class I are packed more closely to the mean than the scores for Class II. It seems that the students in Class I are more alike—at least with respect to algebra—than the students in Class II.

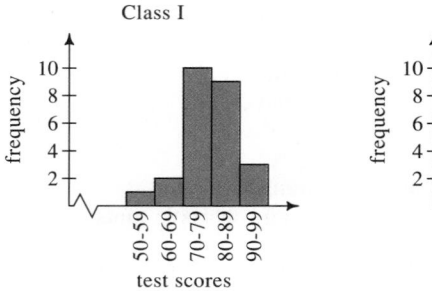
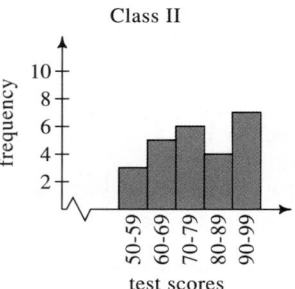

Two statistics used to describe the spread of data about the mean are called the *variance* and the *standard deviation*. The **variance**, denoted s^2 or σ^2 (sigma squared), is the mean of the squares of the deviations from $\bar{x}$.

$$s^2 = \frac{(x_1 - \bar{x})^2 + (x_2 - \bar{x})^2 + \cdots + (x_n - \bar{x})^2}{n} = \frac{\sum_{i=1}^{n}(x_i - \bar{x})^2}{n} \qquad (1)$$

The **standard deviation**, denoted s or σ, is the positive square root of the variance.

$$s = \sqrt{\text{variance}} = \sqrt{\frac{\sum_{i=1}^{n}(x_i - \bar{x})^2}{n}} \qquad (2)$$

The formulas for variance and standard deviation are sometimes written with a denominator of $n - 1$ instead of n. The reasons for this are best left to a full course in statistics. If you have a calculator that gives you these statistics, check to see whether it uses the formulas given above by reading its instruction booklet or by doing a calculation (as in Example 1 on the next page).

Formulas (1) and (2) can be expressed in equivalent forms, given at the top of the next page, that are often easier to use if you need to do the calculations for variance and standard deviation by hand.

$$\text{variance:} \quad s^2 = \frac{\sum_{i=1}^{n} x_i^2}{n} - \bar{x}^2 = \text{mean of squares} - \text{square of mean} \qquad (1a)$$

$$\text{standard deviation:} \quad s = \sqrt{\frac{\sum_{i=1}^{n} x_i^2}{n} - \bar{x}^2} \qquad (2a)$$

You can check for yourself that the standard deviations of the algebra test scores given on the preceding page are:

Class I: standard deviation ≈ 9.6
Class II: standard deviation ≈ 14.5

The smaller standard deviation for Class I indicates that the test scores for this class are packed more closely to the mean.

Example 1 Compute the variance and standard deviation of the data 3, 5, 6, 7, 9.

Solution We first find the mean:

$$\bar{x} = \frac{3 + 5 + 6 + 7 + 9}{5} = 6$$

Since $\bar{x}$ is an integer, formulas (1) and (1a) are equally easy to use. We will illustrate the use of both in finding the variance and standard deviation of the data.

Using formula (1):

x_i	$x_i - \bar{x}$	$(x_i - \bar{x})^2$
3	−3	9
5	−1	1
6	0	0
7	1	1
9	3	9
		20 = sum

$$\text{variance} = s^2 = \frac{\sum_{i=1}^{n} (x_i - \bar{x})^2}{n} = \frac{20}{5} = 4$$

$$\text{standard deviation} = s = \sqrt{4} = 2$$

Using formula (1a):

x_i	x_i^2
3	9
5	25
6	36
7	49
9	81
	200 = sum

$$\text{variance} = s^2 = \frac{\sum_{i=1}^{n} x_i^2}{n} - \bar{x}^2 = \frac{200}{5} - 6^2 = 4$$

$$\text{standard deviation} = s = \sqrt{4} = 2$$

Cooperative Learning

Have students discuss, in small groups, the strengths and weaknesses of each measure of dispersion. (For example, the range is easy to calculate but not a very good indicator of variability since it depends on only two items of data (the extremes). Variance and standard deviation, on the other hand, are more difficult to calculate but they are also sensitive to all the data.)

Mathematical Note

Encourage students to learn and apply both formulas for the variance. It may help to show the derivation of formula (1a) from formula (1). (See the solution to Exercise 18.)

Using Technology

If a program is to be written for calculating the mean and standard deviation of a set of data, then formula (2a) offers a better algorithm than formula (2): Formula (2) requires *two* passes through the data, first to sum the data (in order to calculate the mean) and then to sum the squares of the deviations from the mean (in order to calculate the standard deviation); formula (2a), however, requires *only one* pass through the data to sum both the data and the squares of the data. Encourage students with programming experience to write a program based on each of the two formulas and to compare the programs.

If data appear in a frequency table instead of a list, then we need to modify formula (1a). Suppose n items of data are given as r distinct numbers $x_1, \ldots, x_r$, and each x_i occurs with frequency f_i. Formula (1a) then becomes:

$$\text{variance: } s^2 = \frac{\sum\limits_{i=1}^{r} x_i^2 \cdot f_i}{n} - \overline{x}^2 \qquad (1b)$$

Example 2 Find the mean and standard deviation for the data in the table below.

x_i = item of data	3	6	7
f_i = frequency	2	5	3

Solution The number of distinct data items is $r = 3$.
The total number of data items is $n = 2 + 5 + 3 = 10$.
Therefore the mean is:

$$\overline{x} = \frac{\sum\limits_{i=1}^{r} x_i f_i}{n} = \frac{3 \cdot 2 + 6 \cdot 5 + 7 \cdot 3}{10} = \frac{57}{10} = 5.7$$

Using formula (1b) to find the variance, we have:

$$s^2 = \frac{\sum\limits_{i=1}^{r} x_i^2 \cdot f_i}{n} - \overline{x}^2 = \frac{3^2 \cdot 2 + 6^2 \cdot 5 + 7^2 \cdot 3}{10} - 5.7^2$$

$$= \frac{18 + 180 + 147}{10} - 5.7^2$$

$$= 34.5 - 32.49$$

$$= 2.01$$

Therefore the standard deviation is $s = \sqrt{2.01} \approx 1.42$.

Standard Value

As you will see in the next section, it is useful to convert a set of data into a related set having a mean of 0 and a standard deviation of 1. Each item of data x is converted to a **standard value** z by the formula:

$$z = \frac{x - \overline{x}}{s}$$

Written in words, the conversion formula is

$$\text{standard value} = \frac{\text{item of data} - \text{mean}}{\text{standard deviation}},$$

from which you can see that the standard value gives the number of standard deviations between an item of data and the mean.

Example 3   The mean and standard deviations of the algebra test scores on page 654 are as follows:

$$\text{Class I:} \quad \text{mean} = 78.4, \text{ standard deviation} \approx 9.6$$
$$\text{Class II:} \quad \text{mean} = 78.4, \text{ standard deviation} \approx 14.5$$

For each class, give the standard value of the scores 92 and 68.

Solution   For a score of 92:

Class I: $z \approx \dfrac{92 - 78.4}{9.6} \approx 1.4$

Class II: $z \approx \dfrac{92 - 78.4}{14.5} \approx 0.9$

> A score of 92 is 1.4 and 0.9 standard deviations *above* the mean for Class I and Class II, respectively.

For a score of 68:

Class I: $z \approx \dfrac{68 - 78.4}{9.6} \approx -1.1$

Class II: $z \approx \dfrac{68 - 78.4}{14.5} = -0.7$

> A score of 68 is 1.1 and 0.7 standard deviations *below* the mean for Class I and Class II, respectively.

CLASS EXERCISES

1. a. What does the following information tell you about two different basketball teams? Team A's heights are more closely clustered around 192 than Team B's.

Team A: mean height = 192 cm, standard deviation = 4 cm
Team B: mean height = 192 cm, standard deviation = 8 cm

b. Which team do you think has a better chance of getting rebounds? Team B

2. Complete the table at the left below and then the calculations at the right below in order to find the standard deviation of the data 1, 7, 9, 15.

x_i	x_i^2
1	? 1
7	? 49
9	? 81
15	? 225
sum = ? 32	? = sum 356

mean = $\bar{x}$ = ? 8

variance = $s^2 = \dfrac{\sum\limits_{i=1}^{n} x_i^2}{n} - \bar{x}^2$ = ? 25

standard deviation = $s = \sqrt{?} \approx$? 25; 5

3. If each of the x_i in Exercise 2 were decreased by 1, tell how the mean and standard deviation would change.

4. The mean and standard deviation of a set of data are $\bar{x} = 10$ and $s = 4$. Give the standard values of the following numbers from the set.
a. 14 1   **b.** 18 2   **c.** 20 2.5   **d.** 8 −0.5   **e.** 10 0

Statistics   **657**

1. Suppose the frequency of each data item in Example 2 is doubled. What is the effect, if any, on the mean and standard deviation of the data?

x_i	3	6	7
f_i	4	10	6

$n = 4 + 10 + 6 = 20$

$\bar{x} = \dfrac{\sum\limits_{i=1}^{3} x_i \cdot f_i}{n} =$

$\dfrac{3(4) + 6(10) + 7(6)}{20} = 5.7$

$s^2 = \dfrac{\sum\limits_{i=1}^{3} x_i^2 \cdot f_i}{n} - \bar{x}^2 =$

$\dfrac{3^2 \cdot 4 + 6^2 \cdot 10 + 7^2 \cdot 6}{20} -$

$(5.7)^2 = \dfrac{690}{20} - 32.49 =$

$2.01; s = \sqrt{2.01} \approx 1.42$

The mean and standard deviation are unchanged.

2. Shamea's algebra class had a mean score of 71 and a standard deviation of 8 on a chapter test. Carlos' algebra class had a mean score of 74 and a standard deviation of 6 on the same test. If Shamea's and Carlos' scores were 75 and 76, respectively, which student had the better relative score?

Shamea's standard value:

$\dfrac{x - \bar{x}}{s} = \dfrac{75 - 71}{8} = 0.5$

Carlos' standard value:

$\dfrac{y - \bar{y}}{s} = \dfrac{76 - 74}{6} \approx 0.3$

Therefore, Shamea had the better relative score.

Additional Answers
Written Exercises

3. a. Row A: 3, 1.41
Row B: 13; 1.41
The means differ by 10 and the standard deviations are equal.
b. 23, 1.41

4. a. Row A: 7, 1.73
Row B: 28, 6.93
Both the mean and the standard deviation of B are four times as large as those of A.
b. 70, 17.3

Suggested Assignments

Discrete Math
Day 1: 658/1–8
Day 2: 659/9–15

Supplementary Materials

Alternative Assessment, 54

---

5. *Reading* In Example 1, suppose the mean had not been an integer. To compute the variance, which formula, (1) or (1a), would have been easier to use? 1a

6. *Discussion* Is it possible to have a variance of zero? Explain.
Yes, if all of the data are equal.

WRITTEN EXERCISES

In Exercises 1 and 2, compute the mean and standard deviation for each set of data.

A 1. The number of grammatical errors on five daily French quizzes:
10, 8, 7, 5, 5 7; 1.90

2. The number of fish caught on each day of vacation: 3, 8, 7, 0, 4, 7, 13 6; 3.85

3. **a.** Each number in row B is 10 more than the corresponding number in row A. Compare the mean and standard deviation of each row of numbers.

Row A	1	2	3	4	5
Row B	11	12	13	14	15

   b. Estimate the mean and standard deviation of the numbers 21, 22, 23, 24, 25.

4. **a.** Each number in row B is four times the corresponding number in row A. Compare the mean and standard deviation of each row of numbers.

Row A	5	5	7	7	8	10
Row B	20	20	28	28	32	40

   b. Estimate the mean and standard deviation of the numbers 50, 50, 70, 70, 80, 100.

5. A teacher gives a test in which the average of the test scores is 68 and the standard deviation is 14. Realizing that the test was probably too difficult, the teacher decides to scale the tests by adding 10 points to all scores. What is the mean and standard deviation of the scaled test scores? 78; 14

6. A group of students has a mean height of 66 in. with a standard deviation of 4 in. If the heights had been measured in centimeters instead of inches, what 165 cm; would be the mean and standard deviation? Use the fact that 1 in. $\approx$ 2.5 cm. 10 cm

7. A set of test scores has a mean of 78 and a standard deviation of 6. Give the standard value for each of the following test scores:
   a. 90 2   **b.** 75 −0.5   **c.** 78 0   **d.** 70 −1.3

8. For all of the apartments in a large building, the number of kilowatt hours of electricity used last month had a mean of 760 and a standard deviation of 80. Give the standard value for each of the following usages:
   a. 720 −0.5   **b.** 860 1.25   **c.** 600 −2   **d.** 776 0.2

658 *Chapter Seventeen*

9. Suppose the means for the mathematics and verbal scores on a standardized achievement test are 500 and 504, respectively, and the standard deviations are 100 and 98, respectively. If a student has a mathematics score of 610 and a verbal score of 580, give the standard values of the student's scores. 1.1; 0.8

10. Sue's scores on her first and second physics tests were 82 and 88, respectively. On which test did she do better relative to the rest of the class? Use the following information: Test I

Test I: mean = 72, standard deviation = 5
Test II: mean = 80, standard deviation = 6

11. The mathematics achievement scores of 45 physics students are compared with the mathematics achievement scores of 45 students selected at random. Which group of students do you think has the greater mean score? the greater standard deviation in its scores? Give your reasons. See above.

12. The typing rates (in words per minute) of 25 professional typists are compared with the typing rates of 25 randomly selected people who own typewriters. Which group do you think has the greater standard deviation of typing rates? Give your reasons. Randomly selected people

13. Four people each think of an integer from 1 to 9. What are the smallest and largest possible standard deviations of the four numbers? 0; 4

B **14. Consumer Economics** Thirty samples of milk purchased at various stores were tested for bacteria. The bacteria count per milliliter in these 30 samples is given in the table below. Find the mean and standard deviation of the data. 6; 1.29

x_i = bacteria count	3	4	5	6	7	8
f_i = frequency	1	3	6	9	7	4

15. Twenty puppies were taught to sit and stay on command. Find the mean and standard deviation of the number of trials required before they learned to do this. 10; 1.45

Number of trials	Number of puppies
7	1
8	2
9	5
10	4
11	4
12	4

C **16.** Prove that adding a constant c to each item in a set of data does not change the standard deviation.

17. Prove that multiplying each item in a set of data by a positive constant c makes the standard deviation c times as large.

18. Derive formula (1a) on page 655 from formula (1) on page 654.

Gertrude Mary Cox (1900–1978)

Born and educated in Iowa, Gertrude Mary Cox studied psychological statistics and later coordinated the teaching of statistical theory and methodology at the University of North Carolina and North Carolina State College.

Her 1950 book *Experimental Designs* is a classic textbook on the design and analysis of replicated experiments.

Cox organized conferences in the South on plant and animal science, agricultural economics, quality control, and taste testing. She also helped develop statistical programs for Egypt and Thailand.

17-4 The Normal Distribution

Objective | *To recognize uniform, skewed, and normal distributions, to determine for a normal distribution the percent of data within a given interval, and to find percentiles.*

As we have seen in previous sections, histograms have various shapes according to the distribution of data. The distribution of the last digits of 500 different telephone numbers shown in figure (a) below is an example of a *uniform distribution*. Each digit occurs about 10% of the time.

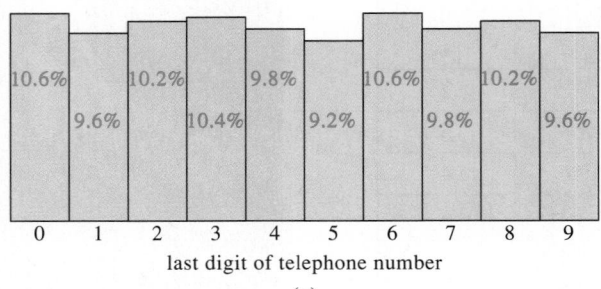

last digit of telephone number

(a)

Figure (b) at the right shows the distribution of precipitation for one year in a Midwestern city. On 242 days of the year the precipitation was less than 1 cm, on 99 days it was greater than or equal to 1 cm but less than 2 cm, and so on. This distribution is an example of a *skewed distribution* with a hump at one end and a long tail at the other.

In this section, we will consider a third type of distribution, called the *normal distribution*. The normal distribution is extremely important in that it occurs in a wide variety of data, including heights and weights of people, dimensions of manufactured goods, test scores, blood cholesterol levels, and times for marathon races.

Consider the following experiment: A psychologist asks 100 children to solve a certain puzzle and records their solution times. The results are shown in figure (c) below. This histogram gives the percent of children who completed the puzzle in the indicated time interval. Each interval includes its left endpoint but not its right endpoint. Thus, 1% of the children solved the puzzle in a time greater than or equal to 25 s, but less than 35 s.

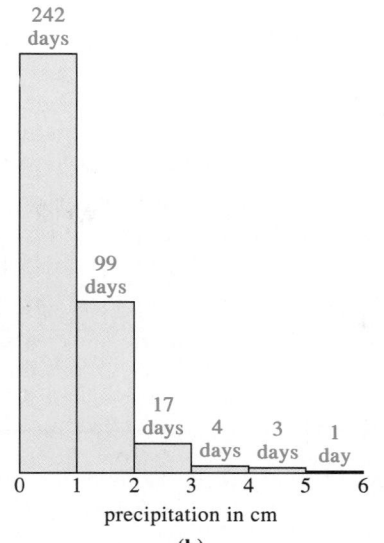

precipitation in cm

(b)

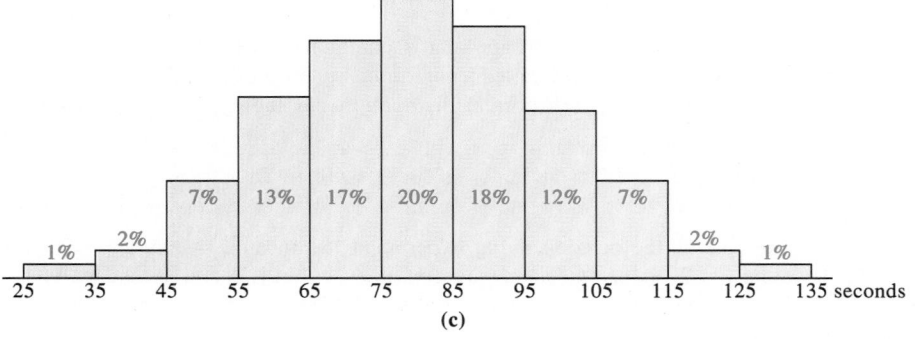

(c)

Each of the class intervals in figure (c) is 10 units wide. If the psychologist had observed a much larger population, the data would be more specific, and it would make sense to use smaller class intervals, as shown in figure (d) below.

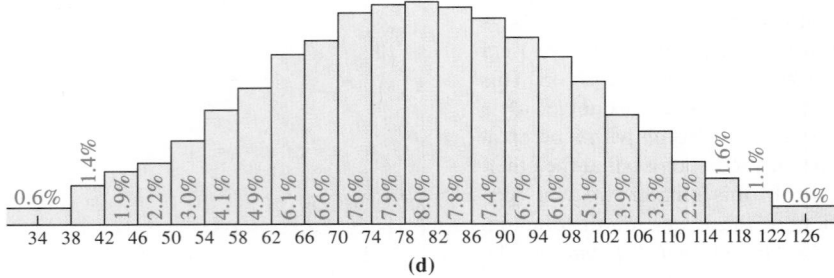

(d)

Activity

Use figure (d) to answer the following questions.

a. What appears to be the mean time for the puzzle-solving experiment? 80 s

b. If the standard deviation of the experiment is $s = 20$ s, approximately what percent of the data is within 1 standard deviation of the mean? 69%

If a smooth curve is drawn through the tops of the rectangles in figure (d), we get figure (e) below.

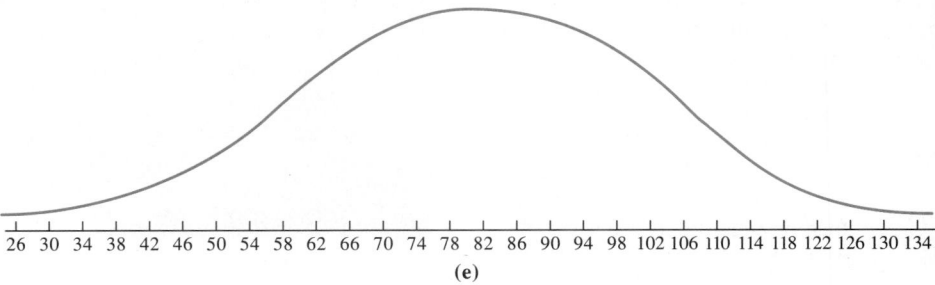

(e)

Figure (e) has the bell shape that is common to the graphs of all **normal distributions**. Such a graph, called a *normal curve,* reaches its maximum height at the mean. Moreover, every normal distribution has these important properties:

About 68% of the distribution is within 1 standard deviation of the mean.
About 95% of the distribution is within 2 standard deviations of the mean.
About 99% of the distribution is within 3 standard deviations of the mean.

For example, if the puzzle-solving experiment has mean $\bar{x} = 80$ s and standard deviation $s = 20$ s, then the percentages of times within 1, 2, and 3 standard deviations of the mean are shown in figure (f) at the top of the next page.

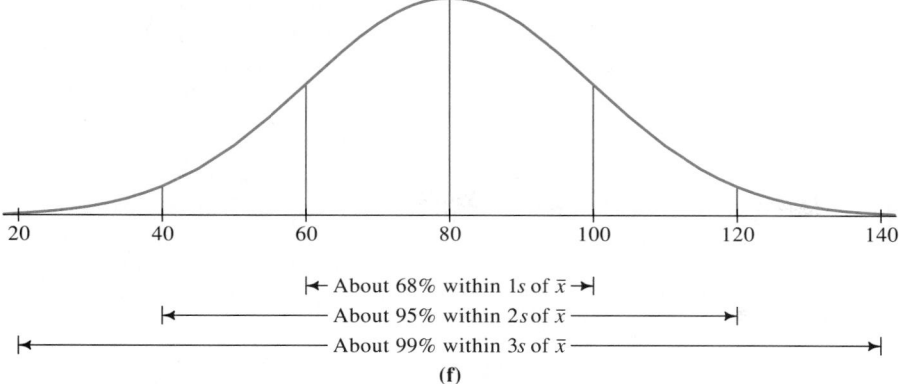

|←— About 68% within 1s of $\bar{x}$ —→|

|←———————— About 95% within 2s of $\bar{x}$ ————————→|

|←———————————— About 99% within 3s of $\bar{x}$ ————————————→|

(f)

Example 1   Over the last 10 years, the mean weight $\bar{x}$ of newborn babies in a large metropolitan hospital has been 3.4 kg and the standard deviation s has been 0.4 kg. Use this information to sketch a normal curve showing the weights at 1, 2, and 3 standard deviations from the mean. Also show the percents of the weights within 1, 2, and 3 standard deviations of the mean.

Solution

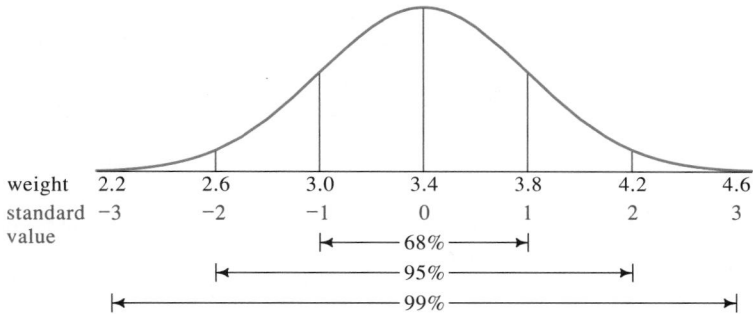

weight	2.2	2.6	3.0	3.4	3.8	4.2	4.6
standard value	−3	−2	−1	0	1	2	3

|←——— 68% ———→|

|←———— 95% ————→|

|←———— 99% ————→|

The **standard normal distribution** is the normal distribution having a mean of 0 and a standard deviation of 1. The standard normal distribution is particularly important because any normal distribution can be related to it through the use of standard values.

The graph of the standard normal distribution, called the *standard normal curve,* is shown at the right. Its equation is:

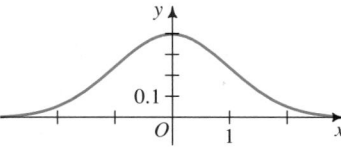

$$y = \frac{1}{\sqrt{2\pi}} e^{-x^2/2}$$

◩ Using Technology

To help students better appreciate the equation of the standard normal curve, you might have them use computers or graphing calculators to obtain the graphs of the following sequence of equations.

1. $y = e^{-x}$ for $x \geq 0$

2. $y = e^{-|x|}$ for all real x

3. $y = e^{-x^2}$ for all real x

4. $y = e^{-x^2/2}$ for all real x

5. $y = \frac{1}{\sqrt{2\pi}} e^{-x^2/2}$ for all real x

Point out that only in the last case is the area under the curve (that is, between the graph and the x-axis) equal to 1.

Mathematical Note

Note that the "hump" on the standard normal curve shown at the bottom of this page is exaggerated, since the scales on the x- and y-axes are not the same.

Statistics   **663**

The standard normal curve has the important property that the total area under the curve (and above the x-axis) is 1. Also, the area under the curve to the left of a number z is the proportion of the data having standard values less than z, as indicated in the figure at the left below. These proportions are given in the standard normal table below.

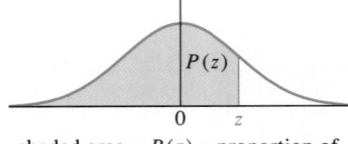

shaded area = $P(z)$ = proportion of data less than z

Examples:
$P(-2.0) = 0.0228$
$P(-0.3) = 0.3821$
$P(0.5) = 0.6915$
$P(1.7) = 0.9554$

z	.0	.1	.2	.3	.4	.5	.6	.7	.8	.9
−3	.0013	.0010	.0007	.0005	.0003	.0002	.0002	.0001	.0001	.0000+
−2	.0228	.0179	.0139	.0107	.0082	.0062	.0047	.0035	.0026	.0019
−1	.1587	.1357	.1151	.0968	.0808	.0668	.0548	.0446	.0359	.0287
−0	.5000	.4602	.4207	.3821	.3446	.3085	.2743	.2420	.2119	.1841
0	.5000	.5398	.5793	.6179	.6554	.6915	.7257	.7580	.7881	.8159
1	.8413	.8643	.8849	.9032	.9192	.9332	.9452	.9554	.9641	.9713
2	.9772	.9821	.9861	.9893	.9918	.9938	.9953	.9965	.9974	.9981
3	.9987	.9990	.9993	.9995	.9997	.9998	.9998	.9999	.9999	1.0000−

Standard Normal Table

Example 2 Use the data of Example 1 to find the approximate percent of newborn babies having weights:
a. less than 4.0 kg   b. more than 4.0 kg   c. between 3.0 kg and 4.0 kg

Solution We must first convert each weight to its standard value so that the standard normal distribution can be used.

For 4.0 kg:
$$z = \frac{4.0 - 3.4}{0.4} = 1.5$$

For 3.0 kg:
$$z = \frac{3.0 - 3.4}{0.4} = -1.0$$

The graph at the right shows these weights and their standard values.

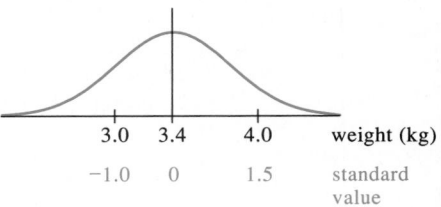

a. From the standard normal table, $P(1.5) = 0.9332$. Thus, about 93% of the babies weigh less than 4.0 kg.
b. About $100\% - 93\% = 7\%$ of the babies weigh more than 4.0 kg.
c. The proportion of weights between 3.0 kg and 4.0 kg is the same as the proportion of the standard normal values between -1.0 and 1.5:

$$P(1.5) - P(-1.0) = 0.9332 - 0.1587 = 0.7745$$

Thus, about 77% of the babies weigh between 3.0 kg and 4.0 kg.

Percentiles

Sometimes a set of data is arranged in ascending order and divided into 100 equal parts. The 99 points that divide the data are called **percentiles.** Percentiles are often used in reporting scores on a standardized test. If you score at the 70th percentile, for instance, you know that 70% of all people taking the test had a score less than or equal to yours. Similarly, the 25th percentile separates the bottom 25% of the test scores. The 25th percentile, 50th percentile (the median), and 75th percentile are also called the first quartile, the second quartile, and the third quartile. As the diagram below shows, the quartiles divide the population into quarters.

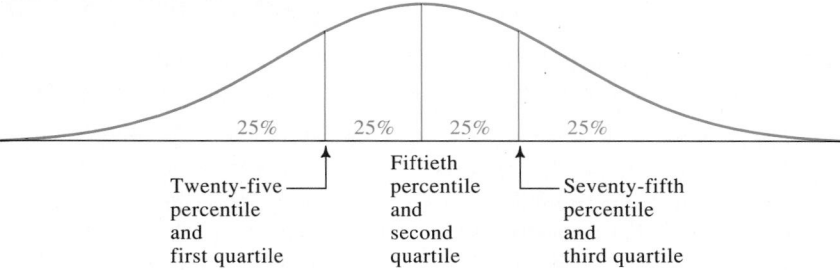

The next example shows that we can use the standard normal table to find various percentiles for any set of data having a normal distribution.

Example 3 The scores on a standardized test are normally distributed with mean $\bar{x} = 50$ and standard deviation $s = 10$. Find the 90th percentile.

Solution When we look at the standard normal table to find a number close to $90\% = 0.90$, we find 0.9032, which corresponds to the standard value $z = 1.3$. Using the formula for standard value, we have:

$$z = \frac{x - \bar{x}}{s}$$

$$1.3 = \frac{x - 50}{10}$$

$$x = 63$$

Thus, a score of 63 is approximately the 90th percentile.

Additional Examples cont.

2. In a normal distribution, about what percent of the data:
   a. fall within 1.5 standard deviations of the mean?
   b. do not fall within 1.5 standard deviations of the mean?

   a. $P(1.5) - P(-1.5) = 0.9332 - 0.0668 = 0.8664$. Thus, about 86.6% of the data fall within 1.5 standard deviations of the mean.
   b. $1 - 0.8664 = 0.1336$; about 13.4% of the data do not fall within 1.5 standard deviations of the mean.

Mathematical Note

Point out that percentiles can be used with *any* distribution of data, not just normal distributions.

Example Note

Students often encounter percentiles when receiving the results of standardized tests. Since a person's score on a standardized test is expected to vary (due to such variables as the person's degree of anxiety and ability to concentrate), a test result is often reported not as a single score, but as a range of scores. Have students suppose that a person's score on the standardized test described in Example 3 is reported as 63 ± 4 (that is, a range of scores from 59 to 67). Ask students what the corresponding percentile range is. (82nd to 96th percentile)

CLASS EXERCISES

1. **Discussion** For each of the following sets of data, discuss whether you expect the distribution to be uniform, skewed, or normal. Then collect the data to check your expectation. (*Note:* If your class size is small, you may not get the type of distribution that you expect.)

   a. The number of times in a day that students in this class use their lockers
   b. The heights of females in this class
   c. The last digit of the home street addresses of students in this class
   d. The ages, in months, of students in this class
   e. The shoe sizes of males in this class
   f. The number of books that students have brought with them to this class.

For Exercises 2–7, use the standard normal table to find the percent of the data in the standard normal distribution satisfying the given inequality. 7. 68.26%

2. $z < 1.2$ 88.49% | 3. $z < -1.8$ 3.59% | 4. $z > 0.6$ 27.43%
5. $z > -1.3$ 90.32% | 6. $0 < z < 1$ 34.13% | 7. $-1 < z < 1$

8. Explain how the standard normal table can be used to show that about 95% of the data is within 2 standard deviations of the mean.

9. In a certain large school district, the set of all standardized mathematics scores is normally distributed with mean $\bar{x} = 500$ and standard deviation $s = 100$.
   a. Make a sketch of the distribution showing the scores at 1, 2, and 3 standard deviations from the mean. Also show the percents of the scores within 1, 2, and 3 standard deviations of the mean.
   b. Find the percent of scores that are:
      (1) less than 300   (2) greater than 700   (3) between 400 and 600
   c. Find the score at the 90th percentile.
   d. Find the score at the first quartile.

10. The reaction times of all people in a psychology experiment were normally distributed with a mean $\bar{x} = 2$ s and a standard deviation $s = 0.5$ s.
    a. Make a sketch of the distribution showing the reaction times at 1, 2, and 3 standard deviations from the mean. Also show the percents of the times within 1, 2, and 3 standard deviations of the mean.
    b. Find the percent of reaction times that are:
       (1) less than 0.5 s   (2) greater than 3 s   (3) between 1.5 s and 2.5 s
    c. Find the reaction time at the 20th percentile.
    d. Find the reaction time at the third quartile.

When necessary to use the standard normal table, round standard values to the nearest tenth.

A

1. On a standardized aptitude test, scores are normally distributed with mean $\bar{x} = 100$ and standard deviation $s = 10$. Find the percent of scores that are:
   a. less than 95 30.85%
   b. greater than 115 6.68%
   c. between 90 and 110 68.26%
   d. between 105 and 125 30.23%

2. **Horticulture** At a tree nursery, the heights of the trees are normally distributed with a mean of 200 cm and a standard deviation of 20 cm. Find the percent of the trees with heights that are:
   a. less than 150 cm 0.62%
   b. greater than 220 cm 15.87%
   c. between 170 cm and 230 cm 86.64%
   d. between 160 cm and 190 cm 28.57%

3. The standardized verbal scores of students entering a large university are normally distributed with mean $\bar{x} = 600$ and standard deviation $s = 80$. Find the percent of scores that are:
   a. less than 500 9.68%
   b. greater than 640 30.85%
   c. between 600 and 720 43.32%
   d. between 760 and 800 1.66%

4. **Operations Research** The number of letters handled daily by a post office is normally distributed with a mean of 20,000 letters and a standard deviation of 100 letters. Find the percent of the days on which the post office handles:
   a. more than 20,200 letters 2.28%
   b. fewer than 19,900 letters 15.87%
   c. between 19,950 and 20,050 letters 38.3%
   d. between 20,100 and 20,150 letters 9.19%

5. **Operations Research** At a very large factory, employees must report to work at 8 A.M. The arrival times of the employees are normally distributed with mean 7:52 A.M. and standard deviation 4 min.
   a. On a typical day, what percent of the employees are late? 2.28%
   b. What percent of the employees arrive between 7:59 and 8:00 A.M.? 1.31%
   c. What percent of the employees arrive before 7:45 A.M.? 3.59%

6. **Manufacturing** A light bulb manufacturing company claims that a 100 watt bulb has a mean life of 750 h with a standard deviation of 70 h. Find the percent of light bulbs that have a life:
   d. 23.58%
   a. less than 610 h 2.28%
   b. more than 940 h 0.35%
   c. between 700 h and 800 h 51.6%
   d. between 730 h and 770 h

7. **Manufacturing** A manufacturer makes ball bearings that must have diameters between 17.0 mm and 18.0 mm. Quality control procedures indicate that the bearings produced have a mean diameter of 17.6 mm and a standard deviation of 0.3 mm. What percent of these ball bearings fail to meet the specifications? 11.96%

8. **Transportation** An airline has room for 262 passengers on its New York to Los Angeles flight. Because some people who book the flight do not show up, the airline books 290 people on the flight. If the mean number of "no-shows" is 45 people and the standard deviation is 10 people, what percent of these flights will be unable to seat all the passengers who do show up? 4.46%

To determine students' understanding of normal distributions, ask them to tell what information they need to evaluate statements like the following:

1. A score of 90 on the test deserves an A.

2. This Shetland sheepdog is 28 in. tall, which is exceptionally tall for the breed.

3. Today's high temperature was far above the normal high for this date.

4. Her body temperature had fallen dangerously below normal.

5. The car does not give nearly the gas mileage that it is supposed to give.

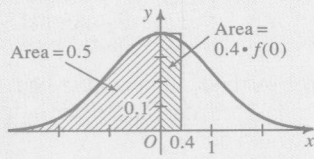
9. Suppose that in a large school district, the standardized mathematics test scores have a mean of 500 and a standard deviation of 100. Find the score at:
  a. the 80th percentile 580   **b.** the first quartile 430

10. At a certain university, the combined mathematics plus verbal standardized test scores of incoming freshmen had a mean of 1250 and a standard deviation of 70. Find the combined score at:
  a. the 90th percentile 1341   **b.** the third quartile 1299

B **11.** The professor teaching a large freshman course decides to give a grade of A to the top 10% of the final exam scores and a grade of B to the next 30%. If the mean score was 72 and the standard deviation was 9, what were the minimum scores needed for an A and a B? 84; 75

12. Refer to Exercise 11 and suppose the professor decided that 5% will fail and 10% will get a grade of D. Find **(a)** the minimum passing grade and **(b)** the minimum grade needed to get a C. 58; 63

13. **Manufacturing** A 2 L beverage bottle is filled from a machine that fills the bottle with a mean of 2.01 L and a standard deviation of 0.002 L.
  a. What percent of the bottles actually contain more than 2 L of beverage? 100%
  b. Ten percent of the bottles contain less than ___?___ L. 2.0074
  c. Eighty percent of the bottles contain more than ___?___ L. 2.0084

14. **Physiology** The weights of all students in a certain grade in a school district are normally distributed with $\bar{x} = 65$ kg and $s = 5$ kg.
  a. The "middle half" have weights between ___?___ kg and ___?___ kg. 61.5; 68.5
  b. The lightest 10% have weights less than ___?___ kg. 58.5
  c. The heaviest 5% have weights greater than ___?___ kg. 73.5

15. **Physiology** The reaction times for a particular task are normally distributed with a mean of 0.92 s and a standard deviation of 0.24 s. A reaction time in excess of 1.2 s is considered too slow. What percent of the people tested had reaction times that were too slow? 11.51%

C **16.** **Sports** Those wanting to participate in a state marathon race must first qualify by running in a regional marathon. The times of the 3750 regional runners are normally distributed with a mean of 198 min 36 s and a standard deviation of 23 min 14 s. If there are to be only 600 runners in the state marathon, what is the slowest time that will qualify a regional runner for the state race? 175 min 22 s

17. Let $f(x) = \dfrac{1}{\sqrt{2\pi}} e^{-x^2/2}$. The

graph of f is the standard normal curve, shown at the right.

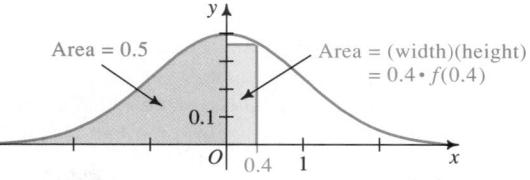

Area = 0.5   Area = (width)(height) = $0.4 \cdot f(0.4)$

0.1

O 0.4   1 x

a. **Visual Thinking** Explain how the shape of the curve implies that the region in red has an area of 0.5.

b. Use a calculator to find the area of the region in blue.

c. Use parts (a) and (b) to approximate $P(0.4)$.

d. **Visual Thinking** Examine the graph to determine whether the approximation found in part (c) is too large or too small. Then compare the approximation with the value of $P(0.4)$ given in the standard normal table.

e. Suppose the region in blue has a height of $f(0)$ instead of $f(0.4)$. Draw a graph to illustrate this. Then repeat parts (b), (c), and (d).

f. **Investigation** Find and use another method of approximating $P(0.4)$ that improves on the approximations found so far.

■ Inferential Statistics

17-5 Sampling

| Objective | To recognize different types of sampling procedures, to identify their limitations, and to estimate population characteristics based on samples. |

In statistics, the word *population* does not necessarily refer to people. It may, for example, refer to a set of stereo receivers sold or to a set of airline luggage claim tickets issued. A **population** is the entire set of individuals or objects in which we are interested, whether they be stereo receivers whose prices are recorded or claim tickets whose numbers are crosschecked. A subset of the population is called a **sample,** and the number of objects in a sample is called the **sample size.**

The process of selecting a sample that is representative of the total population is called **sampling. Sampling theory** is the branch of statistics that deals with the questions that arise when a sample is taken. How should the sample be selected? How large should the sample size be? How reliable are the conclusions drawn from the sample? The answers to such questions often depend upon the circumstances under which the sampling is done, the resources available to do the sampling, and the results desired from the sampling.

Teaching Notes, p. 638D

Warm-Up Exercises

1. Suppose you are the manager at a factory that manufacturers nails. Do you think that every nail should be inspected for flaws? Explain your answer. Answers may vary. Students are likely to say that inspecting every nail is not necessary; doing so would be expensive and time-consuming. Using a representative sample for inspection would probably be the best way to ensure quality while limiting the related expense.

2. Can you think of a situation in which it might be advisable to inspect every manufactured item? Answers may vary. For example, a very expensive machine, such as a car should be inspected since a defect is likely to cause considerable ill-will. Also, a health- or safety-related item such as a knee-replacement device or a smoke detector should be inspected individually.

We can distinguish between *probability sampling* and *nonprobability sampling*. In **nonprobability sampling,** the procedures for selecting a sample are not based on random processes. Examples of nonprobability sampling are *convenience sampling, judgment sampling,* and *sampling by questionnaire.* Such sampling procedures may produce a sample that is biased; that is, the sample and the total population may differ in important characteristics.

Consider, for example, a quality control inspector who inspects the top few apples in several crates to determine the percent of the fruit that is bruised. This is called **convenience sampling** because it is more convenient to sample the top apples than the bottom ones. The resulting sample is not representative of all the apples because apples near the bottom are more likely to be bruised.

In **judgment sampling,** one or more experts select a representative sample using their subjective judgment. It is quite likely that no two experts will agree on what is truly ''representative.''

One problem that occurs with **sampling by questionnaire** is that conclusions are based on voluntary responses. This is also true of other voluntary responses, such as phoned responses to questions that are broadcast on live television or radio. Usually only those who feel strongly about an issue will respond. Another problem with questionnaires is that the questions may be biased. For example, consider whether you agree with statements (a) and (b) below.

    (a) It is best to buy the least expensive brand.
    (b) It is best to buy the cheapest brand.

Some people might agree with statement (a) but disagree with (b) because ''cheapest'' also may mean ''poorest quality.''

A famous example of a biased poll occurred in 1936. A popular magazine mailed straw-poll ballots to telephone and automobile owners asking whether they favored Roosevelt or Landon for president. The poll showed that Landon would be victorious by a landslide, yet Roosevelt won the actual election. The problem, of course, was that the magazine's telephone and automobile owners were not representative of the population in 1936. The sample was biased toward voters who preferred Landon. Can you see why this is an example of convenience sampling?

In **probability sampling,** the procedures for selecting a sample are based on random processes. The most basic type of probability sampling, called **simple random sampling,** is a procedure in which every element of the population has an equal and independent chance of being chosen. This can be done by any chance method, such as flipping a coin or drawing lots, but random numbers generated by a computer or taken from a table of random digits are most frequently used.

For example, suppose a school has 960 students and the school newspaper wants to poll 60 randomly selected students. Although the 960 student names could be written on slips of paper and put in a box so that 60 names could be chosen at random, that would be tedious. It would be easier to have a computer generate 60 three-digit numbers from 001 to 960 and then take the 60 corresponding students from an alphabetized list.

Another type of probability sampling is **stratified random sampling,** in which the entire population is broken into groups, called *strata,* chosen because the individuals in each *stratum* have some common property. For example, you might have two strata (male and female) or four strata (freshmen, sophomores, juniors, and seniors). The only requirement for choosing strata is that you know what percent of the entire population each stratum represents. You then take a simple random sample of each stratum and combine the results.

Example

Southbrook High School has 960 students and 40 teachers. The school newspaper interviews 60 students and 12 teachers to see whether they favor changing a school policy. The results are given in the table below.

Stratum	Population size	Sample size	Number favoring change
Students	960	60	32
Teachers	40	12	4
Total	1000	72	36

Estimate the percent of the school population favoring change.

Solution

First note that 50% of the sample (36 out of 72) are in favor of change. Nevertheless, it would be a mistake to estimate that 50% of the school population favors change because the teachers are disproportionately represented in the sample. Instead, we take the following steps.

Step 1 For each stratum in the sample, we find the proportion of people favoring change:

$$\overline{p}_1 = \text{sample proportion of students in favor} = \frac{32}{60} = \frac{8}{15}$$

$$\overline{p}_2 = \text{sample proportion of teachers in favor} = \frac{4}{12} = \frac{1}{3}$$

Step 2 To estimate p, the proportion of the whole population favoring change, we take a weighted average of $\overline{p}_1$ and $\overline{p}_2$ by assigning weights according to the number in each stratum:

$$\overline{p} = \text{estimate of } p = \frac{960}{1000}\overline{p}_1 + \frac{40}{1000}\overline{p}_2$$

$$= \frac{96}{100} \cdot \frac{8}{15} + \frac{4}{100} \cdot \frac{1}{3} = \frac{788}{1500} \approx 52.5\%$$

Thus, about 52.5% of the students and teachers favor change.

1. The subscribers of the television magazine may have
opinions that are different
from those of the general
public. In fact, the magazine may have influenced
the subscribers' opinions.

2. The consultant must use
his or her own judgment to
define "successful" and
then choose 50 successful
managers. Another consultant might define and
choose differently.

3. The location of the market
or the time of day may
make people's answers
unrepresentative of those
of the general public.

4. There are no apparent limitations.

5. Each question is biased.
The first gives a favorable
impression of the store,
while the second gives an
unfavorable impression.

6. People who still use their
cars may be more likely to
respond than those who do
not.

CLASS EXERCISES

**In Exercises 1–6, state whether the sampling procedure described is an exam-
ple of (a) convenience sampling, (b) judgment sampling, (c) sampling by questionnaire, (d) simple random sampling, or (e) stratified random sampling.
Mention any limitations that the sampling procedure might have.**

1. A television rating organization mails questionnaires to subscribers of a television magazine. On the basis of responses to this questionnaire, it names the
most popular shows. c

2. A popular financial magazine would like to survey the techniques of successful managers. A managerial consultant is asked to select a group of 50 successful managers to participate in the survey. b

3. A manufacturing company would like to determine the approximate market
share of a certain product. A representative of the company is asked to stand in
front of a certain grocery store and ask the first 100 people who go into the
store whether they use the product. a

4. A television cable company with 20,000 subscribers would like to determine
its most popular services. To do so, the company's computer is programmed to
select 1000 subscribers at random. What these subscribers watch is then elec-
tronically monitored. d

5. In order to determine the public response to a proposed change in regu-
lations concerning small businesses,
a questionnaire is being developed.
Two alternative questions differing
by just one word are suggested for
the survey. c
   (a) Do you regret the disappearance
   of the friendly corner grocery
   store?
   (b) Do you regret the disappearance
   of the old corner grocery store?

6. An automobile manufacturer sends
out questionnaires to all people who
purchased one of its cars 10 years
ago. Of those who responded, 90%
reported that they still use the cars.
The manufacturer cites this result as
evidence of vehicle durability. c

7. Senior citizens are 20% of Littleton's voting population. In a poll of 100
citizens, half of whom were senior citizens, 30 senior citizens voted yes and 20
nonsenior citizens voted yes. Estimate the percent of the whole voting population who would vote yes. 44%

672   *Chapter Seventeen*

In Exercises 1–4, state any errors that you think might occur in the sampling situation discussed.

A 1. A senator explains a vote in favor of a new bill by stating that 60% of the mail received favored the bill.

2. A city newspaper asks readers to mail in a form on which they can choose one of three ways to finance improvements to the zoo. Based on the responses, the newspaper reports that financing the improvements by raising the admission charge is the first choice of the majority of the citizens.

3. A radio talk show host invites listeners to telephone the station and talk about their feelings on a proposed highway to be built in their county.

4. A newspaper reporter randomly stops people going in a grocery store and asks, "Does your family use newspaper coupons?"

5. In Farmington High School, there are 360 ninth- or tenth-grade students and 320 eleventh- or twelfth-grade students. A poll shows that 12 out of 40 ninth- or tenth-grade students intend to vote for Lahey as student council president. It also shows that 24 out of 40 eleventh- or twelfth-grade students intend to vote for Lahey. Estimate the percent of the student body favoring Lahey. 44%

6. Twenty-five percent of a city's employees live in the city and 75% live in the surrounding suburbs. The mayor conducts a small survey asking 10 employees who live in the city and 10 who live in the suburbs if they would prefer increases in pay or increases in benefits. Eight of the 10 city dwellers and four of the 10 suburbanites preferred an increase in pay. Estimate the percent of city employees preferring an increase in pay to an increase in benefits. 50%

B 7. Estimate the total number of years of service of 200 factory workers based on a sample of size $n = 5$ where the years of service reported are 15, 8, 20, 5, 12. 2400 yr

8. **Manufacturing** In one hour, a factory packaged 150 boxes of light bulbs. Estimate the total number of defective light bulbs packaged that hour based on a sample of 8 boxes containing the following numbers of defective bulbs: 5, 3, 5, 1, 2, 9, 4, 3. 600

9. **Ecology** The State Conservation Department wants to estimate the number of deer in a mountainous area. It captures and tags 80 deer and then releases them. Later it captures 156 deer and finds that 12 are tagged. Estimate the number of deer in the area. 1040

Statistics **673**

Cooperative Learning

Have students work in small groups to pool their ideas for Exercises 1–4. Then have the groups present their results to the entire class for discussion.

Additional Answers
Written Exercises

1–4. Answers may vary. Examples are given.

1. One special interest group may have been responsible for most of the favorable mail.

2. People who do not want their taxes raised to support the zoo may have been more likely to respond in favor of an increased admission charge in order to put the financing burden on those who use the zoo.

3. It is likely that people who call will have strong feelings about the highway. These feelings may not reflect those of the general population.

4. People may not respond truthfully. Busy people may not take the time to answer. The people who shop at this store may be more or less affluent than the general population.

13. During the 1948 presidential campaign, most public opinion polls predicted that the challenger, Thomas E. Dewey, would win a landslide victory over the incumbent, Harry S. Truman. Truman, however, won the election. The polls were incorrect for two basic reasons: (1) they were conducted too far in advance of the election (so that people had time to change their minds, and they did), and (2) they were based on quota samples that did not accurately reflect the people who voted.

Exercise Note

The Computer Exercise should help students see that increasing the sample size decreases the dispersion of the sample means. Thus, a sample mean's likelihood of being close to the true mean increases as the sample size increases.

Teaching Notes, p. 638D

Warm-Up Exercises

Suppose that 6 batteries out of a sample of 500 taken from a shipment were defective.

1. Estimate the percent of defective batteries in the entire shipment. 1.2%

10. Ecology The number of fish in a pond is unknown; call it x. Initially, c fish are caught, marked, and released. Later on, d fish are caught, and m of them are found to be marked. Estimate the value of x in terms of c, d, and m. $\frac{cd}{m}$

For Exercises 11 and 12, use the table below, which gives both the distribution (by class) of students at a small college and the responses of 25 randomly selected members of each class to two survey questions, one about meals and one about the required English course.

Stratum	Stratum size	Sample size	Pleased with meals	Pleased with English course
Freshmen	300	25	15	20
Sophomores	250	25	12	14
Juniors	250	25	10	18
Seniors	200	25	9	16

11. Estimate the percent of the student body pleased with the meals at the college. 47%

12. Estimate the percent of the student body pleased with the English course. 69%

13. *Research* Read about the 1948 presidential election, for which most poll-based predictions were incorrect. Report on what happened and why.

//// COMPUTER EXERCISE

Write a program that randomly selects a specified number of samples of a specified size from the set of 50 test scores on page 639. For each sample selected, have the program calculate the mean and display it in a histogram of sample means. Then, holding the number of samples constant, run the program for samples of size 5, 10, 15, and 20. (If possible, avoid having to enter the test score data on each run. In BASIC, for example, you can use READ and DATA statements.) Knowing that the mean of all 50 test scores is 549.6, write a paragraph in which you state any conclusions that you draw from comparing the histograms from the four runs.

17-6 Confidence Intervals for Surveys and Polls

Objective To use a sample proportion to find a confidence interval for the corresponding population proportion.

When you know all the relevant information about a population, you can use probability theory to make deductions about a sample. For example, knowing the contents of a standard 52-card deck, you can make probability predictions about a 5-card sample that you are about to deal. On the other hand, there are many situations in which you know all about a sample but not about the population from which it comes. In this case, you can use statistics to make inferences about the population based on the sample.

In everyday situations, samples often take the form of a survey or poll. For example, suppose that in a recent state poll of 100 randomly selected voters, 56 indicated that they were going to vote in favor of Proposition 1. Now, 100 voters are not very many compared with an entire state's voting population. If 560 out of 1000 had favored Proposition 1 instead of just 56 out of 100, we could be more confident of the poll's result. Suppose also that a different poll showed that as many as 60 out of 100 voters favor Proposition 1, and another showed that only 49 out of 100 favor it. The question is, how close is the sample proportion favoring Proposition 1 to the population proportion favoring it? At this point, we need to introduce some notation:

Let p = population proportion having a given characteristic.
Let $\bar{p}$ = sample proportion having a given characteristic.

For example, if a poll of 15 doctors shows that 12 recommend daily exercise, then $\bar{p} = \frac{12}{15} = 0.8$. Likewise, if 76 out of 200 voters surveyed say that they favor candidate Smith, then $\bar{p} = \frac{76}{200} = 0.38$.

You can see that $\bar{p}$ will vary from sample to sample and will depend partly on the sample size n. Moreover, an advanced theorem in statistics states three things about the values of $\bar{p}$:

1. The values of $\bar{p}$ are normally distributed.
2. The mean of this distribution is p.
3. The standard deviation of this distribution is approximately $\sqrt{\dfrac{\bar{p}(1-\bar{p})}{n}}$.

Since $\bar{p}$ is normally distributed, we can apply the 68%-95%-99% normal rule given at the bottom of page 662. For instance, 95% of the time, $\bar{p}$ is within 2 standard deviations of its mean, p. Since we use $\bar{p}$ to estimate p, the following equivalent statement is more useful: 95% of the time, p is within 2 standard deviations of $\bar{p}$. This interval in which p lies 95% of the time is called a **95% confidence interval** and is given by:

$$\bar{p} - 2 \times (\text{standard deviation}) < p < \bar{p} + 2 \times (\text{standard deviation})$$

$$\bar{p} - 2\sqrt{\frac{\bar{p}(1-\bar{p})}{n}} < p < \bar{p} + 2\sqrt{\frac{\bar{p}(1-\bar{p})}{n}}$$

Statistics   **675**

Warm-Up Exercises cont.

2. Can you be certain that your estimate in Ex. 1 is accurate? Explain. No; the sample may not be representative of the entire shipment.

3. You know that about 68% of a distribution is within 1 standard deviation of the mean. Let your estimate in Ex. 1 represent the mean and use 0.5% as the standard deviation. Complete: There is a 68% likelihood that the percent of defective batteries is more than _?_ and less than _?_. 0.7%; 1.7%

4. If you increase the number of batteries in a sample, will your estimate likely become more accurate, less accurate, or equally accurate? More accurate

Motivating the Section

If students prepared the report suggested in Exercise 13 of the previous section, you might use one of the reports as motivation for studying confidence intervals. The fact that predictions based on samples can be wrong leads naturally to considering how much confidence one can have in such predictions.

Communication Note

In many statistics texts, the term *parameter* is used for measures of an entire population, while the term *statistic* is reserved for measures of a sample. You may wish to adopt this vocabulary when discussing this section.

1. Of 400 households surveyed, 144 reported watching the Channel 7 news at least twice a week. Find a 99% confidence interval for p, the unknown proportion of households that watch the Channel 7 news at least twice a week.

$\bar{p} = \frac{144}{400} = 0.36$

Standard deviation of $\bar{p}$ is:

$\sqrt{\frac{\bar{p}(1 - \bar{p})}{n}} =$

$\sqrt{\frac{(0.36)(0.64)}{400}} = 0.024$

The 99% confidence interval for p is:

$0.36 - 3(0.024) < p <$
$0.36 + 3(0.024)$, or
$0.288 < p < 0.432$

2. Refer to Additional Example 1. Suppose that the general manager of Channel 5 claims that less than 30% of the households watch the Channel 7 news at least twice a week. Use a 95% confidence interval to show that it is unlikely that the manager is correct.

A 95% confidence interval is:

$0.36 - 2(0.024) < p <$
$0.36 + 2(0.024)$, or
$0.312 < p < 0.408$

Since 0.3 is not within the 95% confidence interval, it is unlikely that the manager is correct.

Example 1 Suppose that 56 out of 100 voters polled favor Proposition 1. What is the 95% confidence interval for p, the unknown proportion of the population favoring Proposition 1?

Solution **Step 1** $\bar{p} = \frac{56}{100} = 0.56$

Step 2 The standard deviation of $\bar{p}$ is:

$$\sqrt{\frac{\bar{p}(1 - \bar{p})}{n}} = \sqrt{\frac{(0.56)(0.44)}{100}} \approx 0.05$$

Step 3 The 95% confidence interval for p is:

$$0.56 - 2(0.05) < p < 0.56 + 2(0.05)$$
$$0.46 < p < 0.66$$

In Example 1, other samples of 100 voters will yield other 95% confidence intervals. Most of these confidence intervals (about 95% of them) will contain p, but a few of them (about 5%) will not. For this reason, we can be 95% confident that the unknown value of p is between 0.46 and 0.66. Even so, the 95% confidence interval from 0.46 to 0.66 is rather wide and does not pinpoint p to any great extent. (In fact, we cannot tell whether a majority favor Proposition 1.) Example 2 shows that we can obtain a narrower confidence interval by taking a larger sample.

Example 2 Suppose that 560 out of 1000 voters polled favor Proposition 1. Find a 95% confidence interval for p.

Solution **Step 1** $\bar{p} = \frac{560}{1000} = 0.56$

Step 2 The standard deviation of $\bar{p}$ is:

$$\sqrt{\frac{\bar{p}(1 - \bar{p})}{n}} = \sqrt{\frac{(0.56)(0.44)}{1000}} \approx 0.016$$

Step 3 The 95% confidence interval for p is:

$$0.56 - 2(0.016) < p < 0.56 + 2(0.016)$$
$$0.528 < p < 0.592$$

The results of Examples 1 and 2 are summarized in the table at the right. Notice that the sample of size 1000 gives a much narrower confidence interval than the sample of size 100. In fact, with the larger sample, we can be quite confident that a majority of the voters favors Proposition 1, since the smaller endpoint of the sample's 95% confidence interval, 0.528, is greater than 0.5. Bear in mind, however, that the larger sample may be more costly and time consuming to obtain than the smaller one.

Sample size	95% confidence interval
100	$0.46 < p < 0.66$
1000	$0.528 < p < 0.592$

Sometimes a 95% confidence interval is not enough. In testing new medical drugs or procedures, a *99% confidence interval* may be required before the new drug or procedure is approved for general use. A **99% confidence interval** is based on p being within 3 standard deviations of $\bar{p}$:

$$\bar{p} - 3 \times (\text{standard deviation}) < p < \bar{p} + 3 \times (\text{standard deviation})$$

$$\bar{p} - 3\sqrt{\frac{\bar{p}(1 - \bar{p})}{n}} < p < \bar{p} + 3\sqrt{\frac{\bar{p}(1 - \bar{p})}{n}}$$

Example 3

A new, medically approved wheelchair was judged to be an improvement over a standard wheelchair by 320 out of 400 patients in a survey.

a. Find a 99% confidence interval for p, the proportion of the wheelchair-using population for whom the new wheelchair would be an improvement.

b. If 72% of the wheelchair-using population has found the standard wheelchair to be acceptable, does it appear that manufacturing the new wheelchair is a good idea?

Solution

a. Step 1 $\bar{p} = \dfrac{320}{400} = 0.80$

Step 2 The standard deviation of $\bar{p}$ is:

$$\sqrt{\frac{\bar{p}(1 - \bar{p})}{n}} = \sqrt{\frac{(0.80)(0.20)}{400}} = 0.02$$

Step 3 The 99% confidence interval for p is:

$$0.80 - 3(0.02) < p < 0.80 + 3(0.02)$$
$$0.74 < p < 0.86$$

b. Since 0.72 lies outside the 99% confidence interval, the new wheelchair is probably a little better than the standard wheelchair. However, what the new one will cost, how widely available it will be, and other related questions must be considered in deciding whether to proceed with its manufacture.

CLASS EXERCISES

1. b. $0.2 < p < 0.6$; $0.1 < p < 0.7$
1. a. If $\bar{p} = 0.4$ and $n = 24$, find the standard deviation of $\bar{p}$. 0.1
  b. Find both a 95% and a 99% confidence interval for p. See above.
2. a. If $\bar{p} = 0.5$ and $n = 25$, find the standard deviation of $\bar{p}$. 0.1
  b. Find both a 95% and a 99% confidence interval for p. $0.3 < p < 0.7$; $0.2 < p < 0.8$

Example Note

When discussing Example 3, point out that we can expect the percentage of people who find the new wheelchair an improvement to be between 74% and 86%, but that percentages in this range are *not* equally likely. Also, you may need to help students understand the conclusion in part (b): We can be 99% confident that there will be more people who like the new wheelchair than there are people who like the standard one. Even so, the new wheelchair may not be manufactured if costs and other concerns outweigh benefits.

Additional Answers
Written Exercises

1. 0.7, 0.145;
   $0.41 < p < 0.99$

2. 0.6, 0.155;
   $0.29 < p < 0.91$

3. 0.36, 0.048;
   $0.264 < p < 0.456$

4. 0.64, 0.096;
   $0.448 < p < 0.832$

5. 0.0625, 0.012;
   $0.0385 < p < 0.0865$

6. 0.8, 0.04; $0.72 < p < 0.88$

7. $0.015 < p < 0.045$

8. $0.77 < p < 0.83$

9. $0.004 < p < 0.176$

10. $0.01 < p < 0.19$

11. $0.724 < p < 0.876$

12. $0.658 < p < 0.842$

Suggested Assignments

Discrete Math
Day 1: 678/1–15 odd
Day 2: 679/17, 19, 21–23

Supplementary Materials

Alternative Assessment, 55
Student Resource Guide,
162–164

3. **a.** If 50 out of 100 people surveyed favor brand X, find $\bar{p}$ and the standard deviation of $\bar{p}$. 0.5, 0.05

   b. Find a 95% confidence interval for p, the proportion of the population favoring brand X. $0.4 < p < 0.6$

4. *Reading* Does the formula for the standard deviation of $\bar{p}$ depend on the sample size or the population size? Sample size

5. Consider the two polls described below. Is there any difference in the 95% confidence intervals of these two polls? No

   Poll A: 23 out of 100 randomly chosen students in Central High School favor having an additional gym class in place of a study period.

   Poll B: 23 out of 100 randomly chosen high school students from around the state favor having an additional gym class in place of a study period.

WRITTEN EXERCISES

In Exercises 1–6, find $\bar{p}$ and the standard deviation of $\bar{p}$. Also, give a 95% confidence interval for p, the proportion of the entire population having the indicated characteristic.

A 1. Seven out of 10 dentists surveyed recommend Smile toothpaste.

2. Six out of 10 doctors polled recommend eating apples.

3. Thirty-six out of 100 voters surveyed favor Proposition 4.

4. Sixteen out of 25 students questioned prefer summer to winter.

5. Twenty-five out of 400 flashbulbs tested were defective.

6. Eighty out of 100 students have after-school jobs.

In Exercises 7–12, find a 99% confidence interval for p.

7. Of 1000 adults randomly selected for reading tests, 30 were illiterate.

8. Of 1600 eligible voters who were polled, 1280 favored the proposal.

9. Of 100 workers surveyed, 9 have been promoted.

10. In a television poll, 10 out of 100 television sets were tuned to the concert.

11. Of 250 students interviewed, 200 prefer tennis to racquetball.

12. In a survey, 150 out of 200 people preferred brand A to other brands.

13. If you wish to halve the standard deviation of $\bar{p}$, you must multiply the sample size by __?__. 4

14. If you wish to divide the standard deviation of $\bar{p}$ by 10, you must multiply the sample size by __?__. 100

B 15. *Polling* After conducting a poll, a national magazine reported in an article that 64% of the 1000 voters surveyed approved a proposed law. The article also reported a "margin of error of plus or minus 3 percentage points." Is this 3% margin of error associated with a 95% or a 99% confidence interval? 95%

16. **Polling** A city newspaper reported in an article that 55% of the city's registered voters intend to vote for Connelly for mayor. The article noted that this figure was based on a telephone survey of 380 voters and that there was a 5% margin of error in this estimate. Does this margin of error correspond to a 95% or a 99% confidence interval? 95%

17. In a poll of 240 college freshmen, 144 said that they exercise regularly. Find both a 95% and a 99% confidence interval for p, the proportion of the population of college freshmen who exercise regularly.

18. An audit of 840 randomly selected tax returns shows that 30% require payment of additional taxes. Find both a 95% and a 99% confidence interval for p, the proportion of tax returns requiring additional payments.

19. **a.** Graph $f(\overline{p}) = \overline{p}(1 - \overline{p})$.
    b. Find the maximum value of f.
    c. The standard deviation of $\overline{p}$ is always less than or equal to $\sqrt{\dfrac{1}{4n}}$. Explain.

20. Use Exercise 19 to prove that the widest possible 95% confidence interval is:

$$\overline{p} - \sqrt{\dfrac{1}{n}} < p < \overline{p} + \sqrt{\dfrac{1}{n}}$$

For Exercises 21–23, use the following sample.

Sample   A researcher wants to ensure that 95% of the time $\overline{p}$ is within 0.1 of p. How large a sample is needed?

Solution   The 95% confidence interval is to be $\overline{p} - 0.1 < p < \overline{p} + 0.1$.

$$\sqrt{\dfrac{1}{n}} < 0.1 \quad \longleftarrow \quad \text{See Exercise 20.}$$

$$\dfrac{1}{n} < 0.01$$

$$n > 100$$

Thus, a sample size greater than 100 is needed.

21. **Polling** A pollster wants a 95% confidence interval in which p is within 0.01 of $\overline{p}$. How large a sample is needed? 10,000

22. **Polling** Seeking to find the proportion of the population of voters who favor an increased budget, a pollster wants a 95% confidence interval of the form $\overline{p} - 0.03 < p < \overline{p} + 0.03$. How large a sample is needed? 1112

Statistics   **679**

Exercise Note

Exercises 21–23 suggest a method for deciding upon the size of sample needed to achieve a specific confidence interval. You might have students find the answers to these exercises for a 99% confidence interval.

**Additional Answers
Written Exercises**

17. $0.537 < p < 0.663$,
    $0.505 < p < 0.695$

18. $0.268 < p < 0.332$,
    $0.253 < p < 0.347$

19. a.

b. $\dfrac{1}{4}$

c. $s = \sqrt{\dfrac{f(\overline{p})}{n}} \leq \sqrt{\dfrac{\frac{1}{4}}{n}} = \sqrt{\dfrac{1}{4n}}$

20. 95% confidence interval is $\overline{p} - 2s < p < \overline{p} + 2s$. The maximum value for $2s$ is $2\sqrt{\dfrac{1}{4n}} = \sqrt{\dfrac{1}{n}}$. Thus, the widest possible 95% confidence interval is $\overline{p} - \sqrt{\dfrac{1}{n}} < p < \overline{p} + \sqrt{\dfrac{1}{n}}$.

23. **Manufacturing** A quality control inspector wishes to estimate the percent p of computer chips that are defective. Suppose that $|p - \bar{p}| < 0.05$ is a requirement. How large a sample is needed for a 95% confidence interval? 400

24. **Polling** If only 18 out of 50 voters surveyed favor Article VII, find an 80% confidence interval for the proportion of all voters favoring it.
$$0.272 < p < 0.448$$

Chapter Summary

1. We can display a set of data in a *histogram, frequency polygon, cumulative frequency polygon, stem-and-leaf plot,* or *box-and-whisker plot.*

2. The *mean, median,* and *mode* are three measures of central tendency. For the data $x_1, x_2, \ldots, x_n$:
   a. The *mean,* denoted $\bar{x}$, is the sum of the data divided by the number of items of data.
   b. If the data are arranged in increasing or decreasing order, the *median* is the middle item of data if n is odd and is the mean of the two middle items of data if n is even.
   c. The *mode* is the one item of data that occurs most often. A set of data can have more than one mode or no mode at all.

3. The *range, interquartile range, variance,* and *standard deviation* are all measures of dispersion. Definitions are given on pages 649 and 654.

4. A set of data $x_1, x_2, \ldots, x_n$ having mean $\bar{x}$ and standard deviation s can be converted to *standard values* $z_1, z_2, \ldots, z_n$ by the formula $z_i = \frac{x_i - \bar{x}}{s}$.

5. A *normal distribution* has a bell-shaped graph called a *normal curve.* In a normal distribution, about 68%, 95%, and 99% of the data are within 1, 2, and 3 standard deviations of the mean, respectively. The *standard normal distribution* has mean 0 and standard deviation 1. Any normal distribution can be related to the standard normal distribution through standard values.

6. A set of data arranged in ascending order can be divided into 100 equal parts by *percentiles.* The median is the 50th percentile, the *first quartile* is the 25th percentile, and the *third quartile* is the 75th percentile.

7. A sample should be chosen so that it is representative of the whole population. Methods of sampling include *nonprobability sampling,* such as *convenience sampling, judgment sampling,* and *sampling by questionnaire,* and *probability sampling,* such as *simple random sampling* and *stratified random sampling.*

8. A *confidence interval* indicates how well a sample proportion $\bar{p}$ estimates the true population proportion p. Confidence intervals have the form
$$\bar{p} - k \cdot \sqrt{\frac{\bar{p}(1 - \bar{p})}{n}} < p < \bar{p} + k \cdot \sqrt{\frac{\bar{p}(1 - \bar{p})}{n}}$$
where the value of k is the number of standard deviations related to the confidence level selected. See Examples 1–3 on pages 676 and 677.

Chapter Test

1. **Writing** In a paragraph, discuss *descriptive* and *inferential* statistics.   **17-1**

2. **a.** Draw a histogram for the real-estate sales data shown in the table.

Price (×$1000)	70–89.9	90–109.9	110–129.9	130–149.9
Number of sales	4	9	11	6

   b. Using the middle price (80, 100, and so on) as the representative price for each class, find the mean, median, and mode.

3. The stem-and-leaf plot shown at   **17-2**
   the right gives the page lengths of
   a book's 24 chapters.

   ```
   1 | 8 9 9
   2 | 0 1 2 2 3 3 5 5 5 7 8 8 8 8 9
   3 | 1 1 3 4 4 6
   ```

   Find **(a)** the median, lower quartile, and upper quartile, and **(b)** the range and interquartile range. **(c)** Draw a box-and-whisker plot.

4. Zina scored an 80 on both an English test and a math test. The mean   **17-3**
   score in the English class was 70, with a standard deviation of 5. In
   the math class, the mean score was 65, with a standard deviation of
   10. For which test did Zina rank higher with respect to her class? English

5. Calculate the standard deviation of: 4, 7, 10, 11, 12, 14, 15, 21, 32. 7.83

6. At a large university, the mathematics-plus-verbal standardized test   **17-4**
   scores of incoming students are normally distributed with a mean
   score of 1200 and a standard deviation of 100. Find the percent of
   incoming students with scores **(a)** greater than 1300 and **(b)** less than
   1000. Find **(c)** the 90th percentile and **(d)** the third quartile of the scores.

7. At a certain college, 62% of the students live on campus. Of 50 ran-   **17-5**
   domly selected on-campus students, 27 favor a change in housing
   rules. Of 50 randomly selected off-campus students, 14 favor the
   change. Estimate the percent of all students favoring the change. 44%

8. In a survey of 600 school children, 486 said that their favorite sand-   **17-6**
   wich filling is peanut butter. Find 95% and 99% confidence intervals
   for p, the true proportion of school children who prefer peanut butter.

Statistics **681**

Additional Answers
Chapter Test

1. Descriptive statistics involves collecting, organizing, and summarizing data. Inferential statistics involves drawing conclusions about a population based on a sample.

2. **a.**

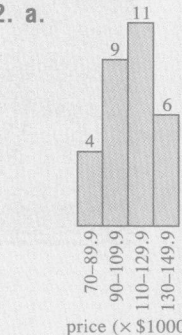

   price (× $1000)

   b. $112,667; $120,000; $120,000

3. **a.** 26, 22, 30
   b. 18, 8
   c.

   15 20 25 30 35 40

6. **a.** 15.87%
   b. 2.28%
   c. 1330
   d. 1270

8. $0.778 < p < 0.842$,
   $0.762 < p < 0.858$

Chapter 18 Curve Fitting and Models

Overview

In this chapter, students learn to create models for sets of paired data by fitting curves to the data. The first section introduces the least-squares line and the correlation coefficient, which students use throughout the chapter: In the first section, students fit a least-squares line to the data (x, y), and a high correlation allows students to write y as a linear function of x. In the second section, students fit a least-squares line to the data $(x, \log y)$, and a high correlation allows students to write y as an exponential function of x. In the third section, students fit a least-squares line to the data $(\log x, \log y)$, and a high correlation allows students to write y as a power function of x. In the fourth section, students must choose the best model for a given set of data.

Objectives

18-1 To find the line of best fit for a set of data and to find the correlation coefficient for a set of data.

18-2 To find an exponential function that models certain sets of data.

18-3 To fit a power curve to a set of data.

18-4 To choose a model that best describes a set of data.

Supplementary Resources

Tests, pp. 55–57, 58–60
Alt. Assess., pp. 56–57, 71
Activities, pp. 48–50
Using Tech., pp. 37–38, 42–43
Student Res. Guide, pp. 167–173

Software

Precalculus Plotter Plus
Activities, pp. 82, 83–84, 85–86
Data Analyzer

Pacing Guide

Section	18-1	18-2	18-3	18-4	Review	Test	Total
Discrete Math (days)	2	2	2	2	1	1	10

Teaching Notes

18-1 | pages 683–691

Presenting the Section

The idea of mathematical modeling is central to all applications of mathematics. Students have already worked with models, possibly without realizing it. Note that a goal in selecting or creating a model is inclusion of the basic, essential characteristics of the phenomenon under consideration. Point out, however, that data often contain "noise" attributable to inaccuracies of measurement, the influences of variables not considered or accounted for, and so on. While even good models may not fit the data precisely, they should fit the data well enough to permit reasonably accurate predictions.

The least-squares line is one way to fit a line to a set of data points. (See Appendix 2 for another way.) Point out that the least-squares line is the best linear model possible when given the requirement of minimizing the squares of the vertical distances of the data points from the line.

The correlation coefficient is a measure of the "relatedness" of two variables. Although a correlation coefficient near 1 or -1 does not necessarily imply a cause-and-effect relationship, it may very well suggest that the strongly correlated data have a common cause. Help students to see that $|r| \approx 1$ means that the data points fit very nearly onto a line; an equation of the line can be determined using the least-squares technique.

Review

Take the opportunity to review graphs in general and graphs of linear equations in particular. Review the various standard forms of a line, with special attention to the point-slope form. Students should be pleased to see an important application of that form. It may be useful to give the equation of a least-squares line in point-slope form:

$$y - \bar{y} = m(x - \bar{x}), \text{ where } m = \frac{\overline{xy} - \bar{x} \cdot \bar{y}}{s_x^2}$$

Cooperative Learning

Before formally introducing least-squares lines, consider making a "game" out of curve fitting. Plot a set of data points on a grid. Make several overhead transparencies of the graph. Ask pairs of students to draw a line to fit the data. Have students superimpose their graphs one by one, explaining as they go. Conclude by showing the actual least-squares line, the "best possible" line.

Using Technology

Graphing calculators and computers are indispensable tools for this chapter. Be sure to spend some class time familiarizing students with data entry, data graphing, and data analysis (that is, obtaining the correlation coefficient and the slope and y-intercept of a least-squares line). Students will then feel more confident using technology to complete the exercises.

18-2 | pages 691–697

Presenting the Section

In this section students are introduced to nonlinear curves that can be fitted to data. Exponential curves are widely used in modeling; for example, students saw exponential growth and decay models in Chapter 5.

Pay extra attention to the explanation on page 692 of how to recognize data suitable for an exponential model. Remind students that whenever variables w and v are related by an equation of the form $w = av + b$, with a and b constants, then w is a linear function of v. You may want to use a substitution of variables to underscore this point:

$$y = ab^x$$
$$\log y = \log a + x \log b$$
$$\log y = (\log b)x + \log a$$
Let $z = \log y$, $m = \log b$, and $k = \log a$.
Then $z = mx + k$.

Another way to amplify the idea that $\log y$ is a linear function of x when $y = ab^x$ is to make a chart, as shown below.

x	0	1	2
y	a	ab	ab^2
$\log y$	$\log a$	$\log a + \log b$	$\log a + 2 \log b$

Point out that every unit change in x produces a constant change of log b in log y. This is the fundamental idea of a linear function.

Applications

This section itself represents a real application of earlier work. In particular, students see a useful application of their work with linear functions in Chapter 1 and of properties of logarithms in Chapter 5. This section also constitutes an extension and a major application of the previous section, since least-squares lines are the linear functions relating x and log y and generating, in turn, exponential functions.

Assessment

Have students work in groups of three. Student A completes the third row of the chart below with any arithmetic sequence having positive terms.

x	0	1	2	3
y				
log y				

Student B checks that log y is a linear function of x. This will probably involve determining whether successive differences in log y are constant. Then Student C completes the second row of the chart. Student C will probably use trial and error, substituting for y and checking that log y is as given; or Student C may compute values of y directly by evaluating $10^{\log y}$ for the given values of log y. Finish by having all three students work together to find the explicit equation, $y = ab^x$, that relates x and y. They will need to find the constants a and b. Point out that $a = ab^0$, and $b = \dfrac{ab^1}{ab^0} = \dfrac{ab}{a}$.

18-3 | pages 698–704

Presenting the Section

Build on students' work in Chapter 2 by pointing out that power curves are actually polynomial graphs. Discuss some of the fundamental properties of these curves. For example, point out that the graph of $y = ax^m$ ($a > 0$, $m > 0$) always contains the origin; moreover, the greater the exponent m, the steeper the curve.

You may need to clarify the difference between power curves and exponential curves. The curves sometimes look alike (the graphs of $y = x^2$ and $y = 2^x$ are nearly identical for $0 \le x \le 4$, for example), but there is no power curve equivalent to an exponential curve and vice versa. Remind students that the graph of $y = ab^x$ (where $a > 0$ and $b > 1$) eventually rises much faster than the graph of $y = ax^m$ (where $a > 0$ and $m > 0$) for $x \ge 0$.

Using Technology

The use of a computer or graphing calculator can be especially instructive with the data and functions of this section. Use the data given in Exercise 7 or 8, page 701, for introductory work. Have students successively graph (x, y), $(\log x, y)$, $(x, \log y)$, and $(\log x, \log y)$. Only the last plot will give an unequivocally linear graph. To emphasize the linearity of this graph, have students compute

$$\frac{\log y_1 - \log y_2}{\log x_1 - \log x_2}$$

for any values of $(\log x_1, \log y_1)$ and $(\log x_2, \log y_2)$. Students appreciate the kind of concrete validation provided when these slopes are all found to be the same.

As a historical note on technology, you could show students slide rules with log scales, allowing multiplication and division through adding and subtracting lengths. Also, they may find logarithmic graph paper interesting and still practical, since this kind of graph paper readily reveals exponential and power functions. Point out that ''semilog'' paper is used for exponential functions and ''log-log'' paper for power functions.

Communication

A power curve can be thought of as a variation. If $y = ax^m$, ask students to explain the relationship between x and y for both positive and negative exponents. Have them express these relationships in terms of direct and inverse variation. For example, if $m = 3$, then y varies directly as the cube of x; if $m = -2$, then y varies inversely as the square of x; and if $m = \frac{1}{2}$, then y varies directly as the square root of x.

Presenting the Section

The chapter concludes with a section that should be challenging and stimulating for students. They must "pull together" all that they have learned in this chapter in order to recognize and apply relevant models for given problem situations.

Briefly review the models developed earlier in the chapter. Some problems may point to a particular model because of the nature of the problem. Others may suggest the model once a graph is made:

Graph of data	Model
(x, y) linear	Least-squares line
$(x, \log y)$ linear	Exponential curve
$(\log x, \log y)$ linear	Power curve

Students must learn to be flexible, however. This means that they must be ready to accept more than one model as "correct" for a given problem.

Applications

The entire exercise set for this section consists of application problems. Scan the exercises' subject headings with students, and emphasize that models are used in a wide variety of fields, not just in science and technology.

Cooperative Learning

The exercises in this section are diverse and substantial. They could be assigned singly to small groups, commensurate with students' abilities. Each group member can graph one of the following: (x, y), $(x, \log y)$, and $(\log x, \log y)$. Then the students can work as a whole group to compare results and choose a best model. One or two members could be responsible for presenting the problem and its solution to the class.

18 *Curve Fitting and Models*

682

18-1 Introduction to Curve Fitting; the Least-Squares Line

Objective *To find the line of best fit for a set of data and to find the correlation coefficient for a set of data.*

Fitting curves to a set of data points is a major goal of this chapter. The process of finding an equation of a curve that fits observed data is called *curve fitting,* and the equation is said to *model* the data. We will begin our study by learning to fit a line to a set of data points.

For example, ten 25-year-olds are surveyed to find their years of formal education (*x*) and their annual incomes (*y*). The data are given in the table below and the ten points (*x*, *y*) are graphed in a *scatter plot,* as shown at the left below.

Years of education	8	10	12	12	13	14	16	16	18	19
Income (in thousands)	13	14	17	17.5	20.6	21	24	25	30	31

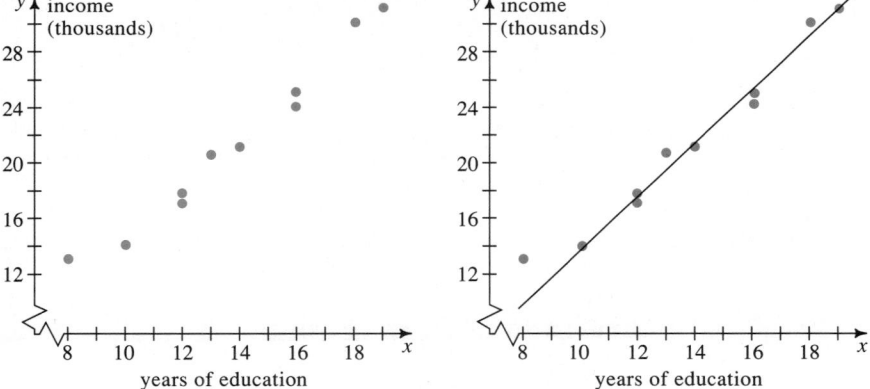

It is possible to fit a line to the scatter plot at the left above by using just your eye and a piece of black thread. Holding the thread taut, move the thread until most of the points lie on or close to the thread. One such line is shown in black at the right above. Some of the points lie on this line, while others are above it and still others are below it. Associated with every point is its vertical distance to the line. The sum of all the distances is a measure of how well the line fits the points. The smaller the sum of the distances, the better the fit.

◀ Will the Leaning Tower of Pisa ever fall down? A structural engineer might wish to collect and analyze data on the past shifting of the tower. If the data conformed to some mathematical model, the engineer might even predict, with some precision, future shifts.

Curve Fitting and Models **683**

Warm-Up Exercises

In Exercises 1-2, describe any trend, or pattern, that you notice in the data. Each set of data refers to the Olympic gold medalists in the specified event.

1. Women's 200-meter Dash

Year	Time(s)	
1972	22.4	Each gold medalist ran faster than the previous one.
1976	22.37	
1980	22.03	
1984	21.81	
1988	21.34	

2. Men's Platform Diving

Year	Points
1972	504.12
1976	600.51
1980	835.65
1984	710.91
1988	638.61

No trend

Motivating the Section

Ask students if they have ever wondered how scientists and other researchers determine links among various factors. Suggest that one way is through a mathematical analysis of the correlations among these factors. A strong correlation between paired data reveals a linear relationship in the data, although there may not be a cause-and-effect connection.

To find the line that best fits the data, statisticians prefer to work with the sum of the *squares* of these distances, rather than the vertical distances themselves. The line that minimizes the sum of these squares is called the *least-squares line*. It is shown in blue in the figure at the right.

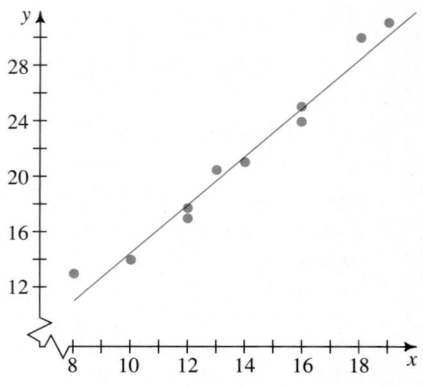

Properties of the Least-Squares Line

1. The line contains the point $(\bar{x}, \bar{y})$.
2. The line has slope $= \dfrac{\overline{xy} - \bar{x} \cdot \bar{y}}{s_x^{\,2}}$

$$m = \frac{\text{(mean of the products)} - \text{(product of the means)}}{\text{(standard deviation of } x\text{'s)}^2}.$$

An equation of the least-squares line is computed routinely by many computer software packages and scientific calculators. Example 1 illustrates how this can be done by hand. In this example, just four points are used in order to make the calculations easier to follow.

Example 1 Find an equation of the least-squares line for the data $(x, y) = (1, 2)$, $(2, 3)$, $(3, 5)$, and $(4, 5)$.

Solution Organize the data in a table.

x	y	xy	x^2
1	2	2	1
2	3	6	4
3	5	15	9
4	5	20	16
Sums 10	15	43	30

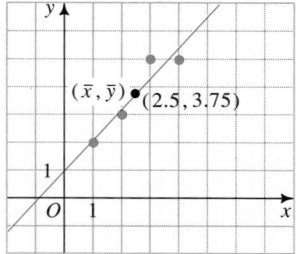

To find an equation of the least-squares line, we must compute the means of the x- and y-coordinates and the mean of the product of the x- and y-coordinates. Then we compute the standard deviation of the x-coordinates.

684 *Chapter Eighteen*

Since $n = 4$: $\bar{x} = \dfrac{10}{4} = 2.5$ $\bar{y} = \dfrac{15}{4} = 3.75$ $\overline{xy} = \dfrac{43}{4} = 10.75$

$$s_x = \sqrt{\dfrac{\displaystyle\sum_{i=1}^{n} x_i^2 - n(\bar{x})^2}{n}} = \sqrt{\dfrac{30 - 4(2.5)^2}{4}} = \sqrt{1.25} \approx 1.118$$

The slope of the least-squares line is:

$$\dfrac{\overline{xy} - \bar{x} \cdot \bar{y}}{s_x^2} = \dfrac{10.75 - (2.5)(3.75)}{1.25} = 1.1$$

Since the line contains $(\bar{x}, \bar{y}) = (2.5, 3.75)$, an equation of the least-squares line is:

$$y - 3.75 = 1.1(x - 2.5), \text{ or } y = 1.1x + 1$$

The Correlation Coefficient

In order to measure how closely points tend to cluster about the least-squares line, statisticians use a *correlation coefficient,* denoted r. If the data fit perfectly on a line with positive slope, the correlation coefficient is said to be $+1$. If the data fit perfectly on a line with negative slope, the correlation coefficient is said to be -1. If the points tend not to lie on any line, then the correlation coefficient is close to 0. Some calculators will compute the correlation coefficient for you.

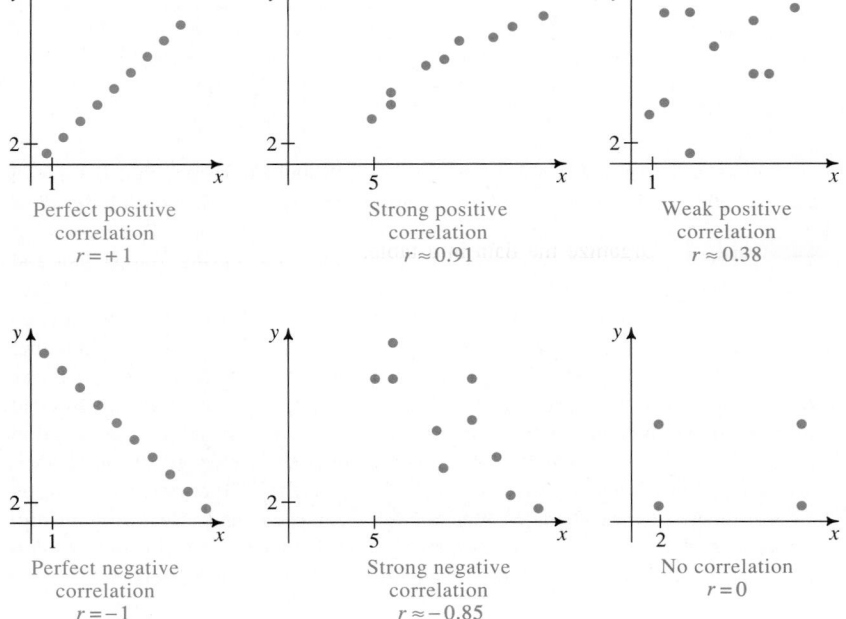

| Perfect positive correlation $r = +1$ | Strong positive correlation $r \approx 0.91$ | Weak positive correlation $r \approx 0.38$ |
| Perfect negative correlation $r = -1$ | Strong negative correlation $r \approx -0.85$ | No correlation $r = 0$ |

a. Find the correlation coefficient.

b. Find an equation of the least-squares line.

c. Bobby Hull played 1063 games. Use the least-squares line to predict the number of goals he might have scored.

a. $n = 4$

$\bar{x} = \dfrac{5056}{4} = 1264$

$\bar{y} = \dfrac{2825}{4} = 706.25$

$\overline{xy} = \dfrac{3{,}689{,}493}{4} =$
$922{,}373.25$

$s_x = \sqrt{\dfrac{\displaystyle\sum_{i=1}^{4} x_i^2 - 4(\bar{x})^2}{4}} =$

$\sqrt{\dfrac{6{,}968{,}950 - 4(1264)^2}{4}} =$

$\sqrt{144{,}541.5} \approx 380.19$

$s_y = \sqrt{\dfrac{\displaystyle\sum_{i=1}^{4} y_i^2 - 4(\bar{y})^2}{4}} =$

$\sqrt{\dfrac{2{,}019{,}755 - 4(706.25)^2}{4}} =$

$\sqrt{6149.6875} \approx 78.42$

$r = \dfrac{\overline{xy} - \bar{x} \cdot \bar{y}}{s_x s_y} \approx$
$\dfrac{922{,}373.25 - 1264(706.25)}{(380.19)(78.42)}$
≈ 0.995

b. slope $= \dfrac{\overline{xy} - \bar{x} \cdot \bar{y}}{s_x^2} =$
$\dfrac{922{,}373.25 - (1264)(706.25)}{144{,}541.5}$
≈ 0.205

The line contains
$(\bar{x}, \bar{y}) = (1264, 706.25)$.
Thus, $y - 706.25 =$
$0.205(x - 1264)$ is an equation of the line.

c. $y \approx$
$0.205(1063 - 1264) +$
$706.25 \approx 665$ goals

Mathematical Note

You might show students that multiplying the correlation coefficient r by $\frac{s_y}{s_x}$ gives the slope m of the least-squares line.

The Correlation Coefficient

The correlation coefficient for a set of data (x_1, y_1), (x_2, y_2), . . . , (x_n, y_n) is given by:

$$r = \frac{\overline{xy} - \overline{x} \cdot \overline{y}}{s_x s_y}$$

$$= \frac{\text{(mean of the products)} - \text{(product of the means)}}{\text{(product of the standard deviations)}}$$

Example 2   Find the correlation coefficient for the data in Example 1.

Solution   Most of the needed calculations can be found in the solution to Example 1. We need only calculate $\sum\limits_{i=1}^{4} y_i^2$ and s_y in order to find the correlation coefficient.

$$\sum_{i=1}^{4} y_i^2 = 2^2 + 3^2 + 5^2 + 5^2 = 63$$

$$s_y = \sqrt{\frac{\sum\limits_{i=1}^{n} y_i^2 - n(\overline{y})^2}{n}} = \sqrt{\frac{63 - 4(3.75)^2}{4}} = \sqrt{1.6875} \approx 1.299$$

Therefore:

$$r = \frac{\overline{xy} - \overline{x} \cdot \overline{y}}{s_x s_y} \approx \frac{10.75 - (2.5)(3.75)}{(1.118)(1.299)} \approx 0.95$$

If two variables are highly correlated, you should not infer that there is a cause-and-effect relationship between them. For example, the reading level of children aged four to ten years is highly correlated to shoe size. The reason, of course, is not that bigger feet help children to read better but that bigger feet and increased reading level are both related to age.

As another example, consider the following situation. A certain state university administers a test to each of its incoming freshmen. The university finds that there is a 0.80 correlation coefficient between the test scores and first-year grade averages. The positive correlation indicates that higher test scores generally yield higher first-year averages, but good test scores do not themselves cause good averages. Rather, the scores and the averages are both influenced by other variables such as knowledge, ability, and drive. In short, the 0.80 correlation between test scores and first-year averages does not indicate a cause-and-effect relationship. Nevertheless, it does indicate a relationship and is therefore useful, along with other data, in predicting a student's academic success at the university.

To show how such a prediction might be done, consider the scatter plot below. A relatively strong correlation, such as 0.80, indicates that the points tend to cluster around the least-squares line. With a computer or calculator, we find that an equation of the least-squares line is $y = 0.00263x - 0.541$. Then we can use it to predict y (a student's first-year grade average) from x (the student's test score). For example, if a freshman at the university has a total test score of 1200, then the predicted first-year average is about 2.6. This predicted value is approximate and should be regarded as the center of a range of values. Advanced statistics books present formulas for finding the range of values. For our purposes it is sufficient to be aware that this prediction and those predictions called for in Exercises 13–16 are approximate.

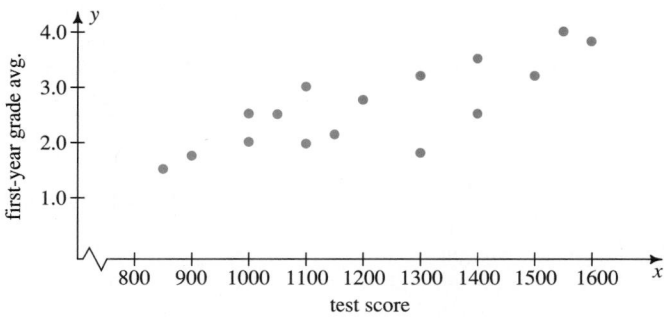

CLASS EXERCISES

The shoe sizes of eight boys are plotted versus their ages.

1. Use your eye to fit a line to the data.
   a. What is the slope of this line?
   b. What is the y-intercept of the line?
   c. What is an equation of the line?

2. a. Use the coordinates of the eight points (x, y) shown at the right to find $\bar{x}$, $\bar{y}$, $\overline{xy}$, s_x, and s_y.
   b. Name one point that the least-squares line contains.
   c. Find the slope of the least-squares line.
   d. Find an equation of the least-squares line.
   e. Find the value of r, the correlation coefficient.

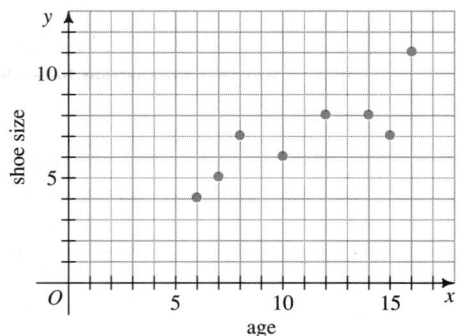

3. **Discussion** How do you think the *least-squares line* got its name?

1. Answers may vary. Examples are given.
   a. $\frac{1}{2}$
   b. 2
   c. $y = \frac{1}{2}x + 2$

2. a. $\bar{x} = 11$; $\bar{y} = 7$; $\overline{xy} = 83$; $s_x = 3.571$; $s_y = 2$
   b. $(11, 7)$
   c. 0.471
   d. $y = 0.471x + 1.82$
   e. 0.84

3. It is the line for which the sum of the squares of the vertical distances from the data points to the line is least or minimum.

Additional Answers
Class Exercises

7. No; both the number of firefighters and the number of students are related to the population of Chesterfield.

Additional Answers
Written Exercises

1. Positive; generally, the more time spent studying, the better one knows the subject and the better one does on a test.

2. Negative; the more absences in the course, the more poorly the class is likely to perform on examinations.

Suggested Assignments

Discrete Math
Day 1: 688/1–13 odd
Day 2: 690/18–21

Supplementary Materials

Alternative Assessment, 56
Using Technology, 42–43
Precalculus Plotter Plus, 82

4. *Visual Thinking* Six scatter plots are shown below. Match each scatter plot to one of the following estimated values of r: 0.8, 0.6, 0.2, 0, −0.2, −0.6, −0.8.

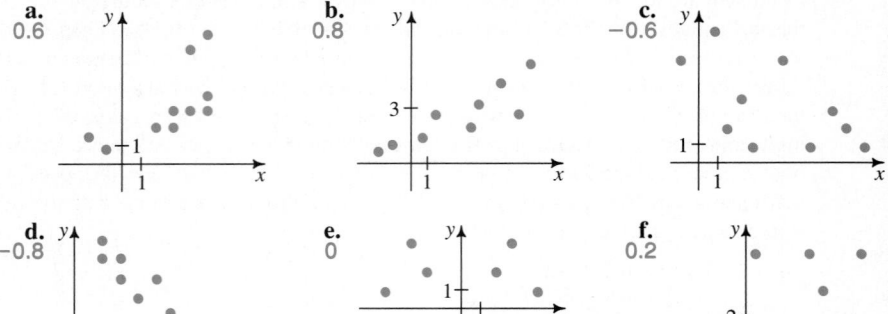

5. *Visual Thinking* For each pair of variables, tell whether you think the correlation will be positive, negative, or approximately zero.
a. The height and weight of a person Positive
b. The height of a person and the average height of his or her parents Positive
c. The value of an automobile and its age Negative
d. The incidence of flu and the outside temperature Negative
e. The height of a person and the number of years of formal education that person has completed Zero

6. *Reading* Refer to the least-squares line for predicting annual income at age 25 from the years of formal education (page 684).
a. Estimate the slope of this line. (Note that the axes have different scales.) About 1.75
b. In what units is the slope measured? Thousands of dollars per year
c. For every year of formal education, about how much does annual income increase? About $1750

7. *Discussion* The correlation between the number of students in the Chesterfield High School and the number of firefighters in the Chesterfield Fire Department is 0.9. Can you conclude that a change in the student population will cause a change in the number of firefighters? Give your reasons.

WRITTEN EXERCISES

For each pair of variables, tell whether you think the correlation will be positive, negative, or approximately zero. Briefly give your reasons.

A **1.** The number of hours spent studying statistics and the grade on a statistics test
2. The total number of absences in a chemistry course and the average of all the grades in that course

3. The weight of a car and the car's fuel economy

4. The blood pressure of an adult and his or her weight

5. The blood pressure of an adult and his or her age

6. Standardized test mathematics scores and standardized test verbal scores

7. The supply of an item and the demand for that item

8. The number of hours of sleep and the ability to memorize

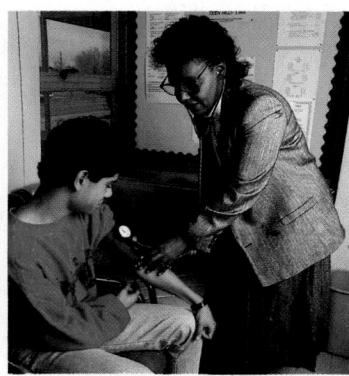

Display each set of data in a scatter plot and find the correlation coefficient. Also, find an equation of the least squares line and draw the line on your scatter plot. You may find a calculator helpful.

9.

x	y
1	1
1	2
3	3
3	4

10.

x	y
1	4
2	3
2	1
3	0

11.

x	y
1	6
2	6
3	5
4	3
5	0

12.

x	y
-2	-1
-1	-1
0	0
2	3
6	4

13. In a ten-week course, the final grades A, B, C, D, and F were given the scores $y = 4, 3, 2, 1,$ and 0, respectively. The least-squares line for predicting scores from x, the number of absences, was $y = 3.5 - 0.5x$. What letter grade can you predict for a student with **(a)** 1 absence? **(b)** 3 absences? **(c)** 8 absences? B; C; F

14. When the sales volume in hundreds of units is plotted against x, the money spent on advertising in thousands of dollars, researchers obtained a least-squares line with the equation $y = 14x + 0.7$. What average sales volumes (in hundreds of units) can you predict for the following amounts (in thousands of dollars) spent on advertising? **a.** 10 140.7 **b.** 5 70.7 **c.** 7 98.7

Find an equation of the least-squares line, then use it to complete the table.

B **15.**

x = amount of nitrogen (pounds per acre)	111	169	230	287	335	?
y = amount of alfalfa (tons per acre)	2	3	4	5	6	7

16.

x = standardized test math score	400	500	500	600	700	800	550
y = college first-year math avg.	50	65	70	80	85	95	?

15. $y = 0.018x + 0.007$; 388.5 **16.** $y = 0.105x + 12.7$; 70

Communication Note

You might want to point out that the process used in completing the table in Exercise 15 is called *extrapolation* ("beyond the data"), while the process used in Exercise 16 is called *interpolation* ("between the data").

**Additional Answers
Written Exercises**

3. Negative; the lighter a car, the better its fuel economy is likely to be.

4. Positive; obesity can strain the heart and increase blood pressure.

5. Positive; blood pressure generally increases with age.

6. Approximately zero; ability in one area is not generally likely to be associated with ability in the other area.

7. Negative; demand is generally high when supply is low.

8. Positive; the better rested one is, the better one can concentrate and think clearly.

9. $r = 0.89$, $y = x + 0.5$

10. $r = -0.89$, $y = -2x + 6$

11. $r = -0.93$,
$y = -1.5x + 8.5$

12. $r = 0.94$, $y = 0.7x + 0.3$

17. *Writing* In each of Exercises 15 and 16 you made a prediction. Which of these predictions do you think has a better chance of being correct? Write a few sentences explaining your reasoning.

The table shows statistics for the ten players of a college basketball team in its opening game of the season. Use the data to complete Exercises 18–20.

minutes played	32	30	25	21	21	18	16	15	12	10
points scored	22	15	11	8	6	4	8	10	4	6
fouls committed	4	3	5	2	3	2	2	1	0	1

 Use computer software to make a scatter plot, draw the least-squares line, find its equation, and find the correlation coefficient for each set of data.

	x	*y*
18.	minutes	points
19.	minutes	fouls
20.	fouls	points

21. *Writing* Refer to the results of Exercise 20. Does the positive correlation between fouls committed and points scored suggest a cause-and-effect relationship between fouls and points? As a coach, would you advise playing more defensively and thus committing more fouls as a way of increasing your players' points scored? Write a short paragraph explaining your reasoning.

C 22. The diagram at the right shows a scatter plot of a set of data and a line with slope m that contains the point $(\bar{x}, \bar{y})$.

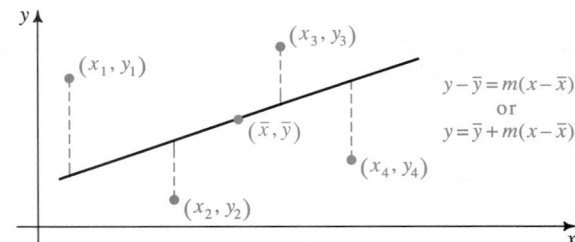

$$y - \bar{y} = m(x - \bar{x})$$
or
$$y = \bar{y} + m(x - \bar{x})$$

a. For each point (x_i, y_i), let d_i represent the vertical distance between the point and the line, that is, $d_i = |y_i - y|$. Show that:

$$d_i = |(y_i - \bar{y}) - m(x_i - \bar{x})|$$

b. Let $D = \sum_{i=1}^{n} d_i^2$. Show that:

$$D = m^2 \sum_{i=1}^{n} (x_i - \bar{x})^2 - 2m \sum_{i=1}^{n} (y_i - \bar{y})(x_i - \bar{x}) + \sum_{i=1}^{n} (y_i - \bar{y})^2$$

c. Note that $\bar{x}$, $\bar{y}$, and each x_i, y_i are known, so that D is a quadratic function of m, the slope of the line. Show that D is minimized when

$$m = \frac{\sum_{i=1}^{n} (y_i - \bar{y})(x_i - \bar{x})}{\sum_{i=1}^{n} (x_i - \bar{x})^2}.$$

d. In part (c), divide the numerator and denominator by n, then expand the product $(y_i - \bar{y})(x_i - \bar{x})$, and show that the value of m that minimizes D is:

$$\frac{\overline{xy} - \bar{x} \cdot \bar{y}}{s_x^2}$$

Hint: Remember that $\bar{x}$ and $\bar{y}$ are constants. Thus,

$$\frac{\sum_{i=1}^{n} \bar{x} y_i}{n} = \frac{\bar{x} \sum_{i=1}^{n} y_i}{n} = \bar{x} \cdot \bar{y}.$$

18-2 Fitting Exponential Curves

Objective *To find an exponential function that models certain sets of data.*

In Section 18-1, we fitted *lines* to data. In this section and the next, we will fit *curves* to data. Knowing what equations and curves are good candidates for fitting a set of data takes experience. However, in this section, we will consider only the exponential curve, $y = ab^x$, which you studied in Chapter 5. This curve and a translation of it can be used to model exponential growth ($b > 1$) and exponential decay ($0 < b < 1$), including the following phenomena.

Exponential Models, $y = ab^x$

1. Increasing costs due to inflation (See Example 1.)
2. Decreasing values due to depreciation (See Exercise 9.)
3. Population growth (See Exercises 10 and 11.)
4. Learning experiments (See Exercise 16.)
5. Performance in contests (See Exercise 17.)

Teaching Notes, p. 682B

Warm-Up Exercises

1. If $\log y = 3x + 2$, show that $y = 100 \cdot (1000)^x$.

If $\log y = 3x + 2$, then $y = 10^{3x+2} = 10^{3x} \cdot 10^2 = (10^3)^x \cdot 100 = 100 \cdot (1000)^x$.

2. The table below shows the population P of a bacteria colony h hours after an experiment began. Do you think that a linear model is appropriate for the data? Justify your answer.

h	P
0	2
2	19
4	185
6	1775
8	17,058
10	163,926

No; reasons will vary. Students could use a graph and show that there is no line that fits closely to the five data points. Another method is to compute the correlation coefficient, which in this case is about 0.7.

Motivating the Section

Ask students what the following have in common.

1. The value 20 years from now of a dollar invested today at a given interest rate.

2. The population of a given nation in the next century.

3. The time required for radioactive waste to decompose to safe levels.

4. The cost of a college education for a baby born today.

These are all questions that are best answered by using an exponential model.

Mathematical Note

The connection between an exponential curve with points (x, y) and a line with points $(x, \log y)$ is independent of the logarithm base. If $y = ab^x$, $\log_n y = \log_n a + x \log_n b$. Regardless of the value of n, $\log_n y$ is a linear function of x. In fact, the change of base formula for logarithms can be used, giving:

$$\log_n y = \frac{\log a}{\log n} + x \frac{\log b}{\log n}$$

$$= \frac{1}{\log n}(\log a + x \log b)$$

Note that the graph of $\log_n y$ as a function of x is simply a vertical stretch (or shrink) of the graph of $\log y$ as a function of x, with a scale factor of $\frac{1}{\log n}$.

One way to see if an exponential equation fits a set of data (x, y) is to plot the points $(x, \log y)$. If the points $(x, \log y)$ lie on a line, then the original data points (x, y) lie on an exponential curve. The reasoning behind this method follows.

If $y = ab^x$, then

$$\log y = \log a + x \log b$$
$$\log y = (\log b)x + \log a$$
$$\downarrow \qquad \downarrow \qquad \downarrow$$
$$\log y = \quad mx \quad + \quad k$$

Thus, $\log y$ is a *linear* function of x.

Since the steps above are reversible, we have proved the following:

Test for an Exponential Model

The points (x, y) lie on an exponential curve, $y = ab^x$, if and only if the points $(x, \log y)$ lie on a line.

Example 1 The average price of an adult ticket to the movies in the United States has increased, as the entries in the table below show.

$x = $ year	$y = $ price
1948	$0.36
1958	$0.68
1967	$1.22
1978	$2.34
1988	$4.11

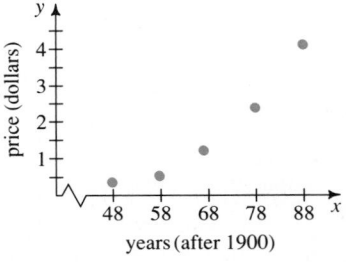

a. Use the movie price data above to plot the log of the price y versus the year x. Represent the years 1948, 1958, . . . as $x = 0, 10, \ldots$.

b. Fit a line to your plot in part (a) and find an equation of the line.

c. Find an exponential model for the price data.

d. Use the equation found in part (c) to predict the price of a movie ticket in the year 2000.

Solution   **a.** Find the logarithm of each price.

x	0	10	19	30	40
y	0.36	0.68	1.22	2.34	4.11
$\log y$	-0.4437	-0.1675	0.0864	0.3692	0.6138

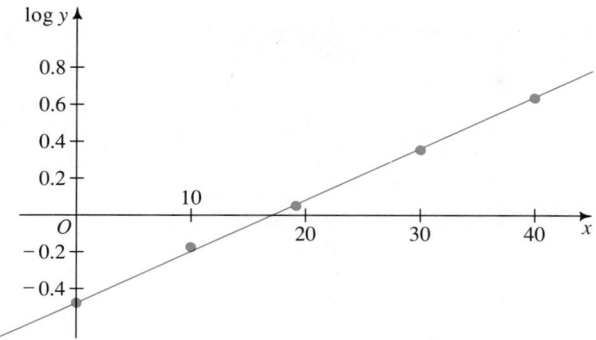

b. Applying the least-squares method from Section 18-1, we obtain the following values for the slope and the vertical intercept of the least-squares line.

$$\text{slope} \approx 0.0265$$
$$\text{vertical intercept} \approx -0.4331$$

Therefore, $\log y \approx 0.0265x - 0.4331$.

c. If $\log y \approx 0.0265x - 0.4331$, then:

$$y \approx 10^{0.0265x - 0.4331} = 10^{-0.4331} \cdot (10^{0.0265})^x$$
$$y \approx (0.369)1.063^x$$

d. For the year 2000, $x = 52$ and $y \approx (0.369)1.063^{52} \approx \$8.85$.
In the year 2000, a movie ticket will cost about \$9.00.

Fitting an Exponential Curve $y = ab^x$ to a Set of Data

1. Compute $(x, \log y)$ for each data point.
2. Fit a least-squares line to the set of data points in Step 1.
3. Convert the equation in Step 2 to give y as a function of x.

There are calculators and computer software that can give you the least-squares line needed in Step 2. Some will also carry out Step 3. If you do not have such a calculator or software available, proceed as shown in the next example.

Curve Fitting and Models   **693**

Additional Examples

1. Express y in terms of x if $\log y = 0.7x - 1.2$.

Since $\log y$ means $\log_{10} y$, $\log_{10} y = 0.7x - 1.2$.
In exponential form this is $y = 10^{0.7x - 1.2}$.
$y = (10^{0.7})^x \cdot 10^{-1.2}$
$10^{-1.2} \approx 0.0631$; $10^{0.7} \approx 5.012$
Thus, $y \approx 0.0631(5.012)^x$.

2. a. Use substitution to fit an exponential curve to the data in Written Exercise 8.

x	2	2.5	4	5.5
y	3.63	3.81	4.39	5.07

b. Name an advantage and a disadvantage that this method has as compared to the method described on this page.

a. Using the data pairs $(x, y) = (2, 3.63)$ and $(4, 4.39)$, we substitute into $y = ab^x$:
$3.63 = ab^2$; $4.39 = ab^4$
$\dfrac{3.63}{b^2} = a$; $\dfrac{4.39}{b^4} = a$
$\dfrac{3.63}{b^2} = \dfrac{4.39}{b^4}$
$b^2 = \dfrac{4.39}{3.63} \approx 1.209$
$b \approx 1.1$
$a \approx \dfrac{3.63}{1.209} \approx 3$
$y = ab^x$, so $y \approx 3(1.1)^x$.

(*Continues on next page*)

b. Answers may vary. An advantage is that this method is simpler, involving fewer computations than the method in the text. A disadvantage is that the data pairs chosen for substitution may produce a curve that varies substantially from the curve near which most of the data pairs lie, thereby giving an equation that models the data poorly; the least-squares method finds the "best" equation.

Suggested Assignments

Discrete Math
Day 1: 694/1–9 odd, 10
Day 2: 696/13–17

Supplementary Materials

Alternative Assessment, 56
Using Technology, 37–38
Precalculus Plotter Plus,
85–86

Example 2 Find y in terms of x if $\log y = 2x + 3$.

Solution If $\log y = 2x + 3$, then $y = 10^{2x+3}$. Therefore:
$$y = 10^{2x} \cdot 10^3$$
$$y = (10^2)^x \cdot 10^3$$
$$y = (1000) \cdot 100^x$$

CLASS EXERCISES

1. **a.** Suppose $y = 2 \cdot 3^x$. Express $\log y$ in terms of x. $\log y = (\log 3)x + \log 2$
   b. If $\log y$ is plotted versus x, what is the slope of the resulting line? $\log 3$
   c. What does the number $\log 2$ tell you about the graph? The vert.-int. is $\log 2$.

In Exercises 2–5, express y in terms of x.

2. $\log y = 1.3x + 0.8$ $y \approx (6.31)19.95^x$

3. $\log y = \frac{1}{2}x + 2$ $y \approx (100)3.162^x$

4. $\log y = -1.4x + 2.5$ $y \approx (316.2)0.0398^x$

5. $\ln y = 1.5x + 4$ $y \approx (54.6)4.482^x$

WRITTEN EXERCISES

In Exercises 1–6, express y in terms of x.

4. $y \approx (0.012)0.0891^x$

2. $y \approx (5.623)257.0^x$

A 1. $\log y = 1.6x + 1.8$ $y \approx (63.1)39.81^x$ 2. $\log y = 2.41x + 0.75$

3. $\log y = -0.4x + 2.3$ $y \approx (199.5)0.3981^x$ 4. $\log y = -1.05x - 1.92$ See above.

5. $\ln y = 0.5x + 3.2$ $y \approx (24.53)1.649^x$ 6. $\ln y = 4.5 - 2x$
$y \approx (90.02)0.1353^x$

In Exercises 7 and 8, fit an exponential curve to the data. See below.

7.
x	1	3	4	5
y	3.00	6.75	10.13	15.19

8.
x	2	2.5	4	5.5
y	3.63	3.81	4.39	5.07

9. **Consumer Economics** Each year after he bought his new car, Mr. Brown kept track of the market value of the car.

x = year after purchase	1	2	3	4	5	6
y = market value (in dollars)	12,000	9600	7700	6200	4900	3900

   a. Fit an exponential curve to the data. $y = (15,070)0.7989^x$
   b. Predict the value of the car when it is 10 years old. \$1596
   c. Estimate the amount that Mr. Brown paid for the car. about \$15,000

7. $y = (2)1.5^x$ 8. $y = (3)1.1^x$

694 *Chapter Eighteen*

The tables give the population P in millions and the population density D in people per square mile for the United States.

year	P	D
1790	3.9	4.5
1800	5.3	6.1
1810	7.2	4.3
1820	9.6	5.5
1830	12.9	7.4
1840	17.1	9.8
1850	23.2	7.9

year	P	D
1860	31.4	10.6
1870	39.8	13.4
1880	50.2	16.9
1890	62.9	21.2
1900	76.0	25.6
1910	92.0	31.0
1920	105.7	35.6

year	P	D
1930	122.8	41.2
1940	131.7	44.2
1950	150.7	50.7
1960	179.3	50.6
1970	203.3	57.4
1980	226.5	64.0
1990	248.7	70.3

10. a. Plot $\log P$ versus time. Let 1800 correspond to $x = 0$, and let 1810 and 1820 be $x = 10$ and $x = 20$, respectively. The year 1790 will correspond to $x = -10$.
   b. Fit a line to your plot in part (a) and find its equation.
   c. Geography Use your equation in part (b) to find an equation giving population as an exponential function of x, the number of years since 1800. Use your equation to predict the United States population in the year 2000.

11. a. *Research* Even though the population of the United States increased between 1800 and 1810, the population density decreased because the United States acquired a vast land area when it made the Louisiana Purchase in 1803. How do you account for the other two decreases in population density shown in the table?
   b. *Discussion* Assuming that the future population grows exponentially, can you assume that population density will also grow exponentially? Explain your answer.

12. Astronomy The observed brightness of stars is classified by magnitude. The table shows how the difference d in the magnitudes of two stars is related to the ratio of their brightness r. Comparing a first magnitude star with a sixth magnitude star, we have the magnitude difference $d = 6 - 1 = 5$, and we can see from the table that the first magnitude star appears 100 times as bright.

d	0	0.5	0.75	1.0	1.5	2	3	4	5	6	10	15
r	1	1.6	2	2.5	4	6.3	16	40	100	251	10^4	10^6

   a. By plotting $\log r$ versus d, show that $d \approx 2.5 \log r$.
   b. Express r in terms of d.
   c. The difference in magnitudes of the brightness of the sun and the full moon is 14. What is their approximate brightness ratio?

**Additional Answers
Written Exercises**

10. b. $\log P = 0.009x + 0.8563$
   c. $P = (7.183)1.021^x$; 458.6 million

11. a. Between 1840 and 1850, Florida, Iowa, Wisconsin, and Texas became states. Between 1950 and 1960, Alaska and Hawaii became states.
   b. Yes, given that $D = P/A$ and that A is likely to remain constant, exponential growth in P leads to exponential growth in D.

12. b. $r = 2.512^d$
   c. 400,000

13. **Electronics** An electrical circuit containing a capacitor has voltage 1.52 V. When the electrical circuit is broken, the capacitor gradually discharges. At various times t (in minutes) after the circuit is broken, the voltage V (in volts) is read on a voltmeter. The table below shows the data obtained.

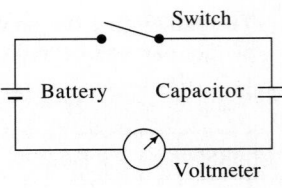

t	0	0.5	1.0	1.5	2.0	2.5	3.0	3.5
V	1.52	1.25	1.01	0.82	0.67	0.54	0.45	0.37

a. Plot (t, log V). Find an equation of the resulting straight line in the form
$$\log V = mt + k. \quad \log V = -0.1766t + 0.1811$$
b. Solve the equation from part (a) for V in terms of t. $V = (1.517)0.6659^t$

Exercises 14–17 concern exponential decay for which the graph has a horizontal asymptote $y = c$. An equation of such a curve has the form

$y - c = ab^x$, where $0 < b < 1$.

To find values for a, b, and c, proceed as follows:

Step 1 Plot the points (x, y) and visually find the value of c.

Step 2 Plot $(x, \log (y - c))$. Express $\log (y - c)$ as a linear function of x.

Step 3 Solve the equation in Step 2 for y.

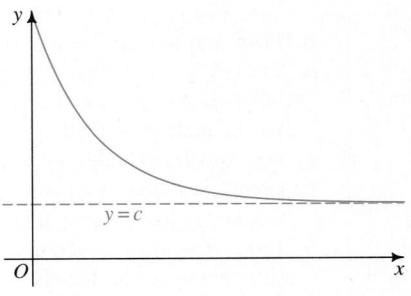

B 14. The scatter plot below suggests exponential decay with a horizontal asymptote $y = 6$.

x	y
0	20.0
1	17.2
2	15.0
4	11.7
7	9.0
10	7.5
20	6.2

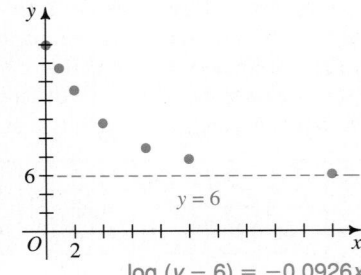

$$\log (y - 6) = -0.0926x + 1.1331$$

a. Plot log $(y - 6)$ versus x and find an equation of the resulting line.
b. Find y as a function of x. $y = (13.59)0.808^x + 6$

---

696 *Chapter Eighteen*

15. **Thermodynamics** A cup of hot tea just poured at 158°F slowly cools over time t (in minutes) and its temperature T is recorded, as shown in the table.

t	0	10	30	50	70	90	110	120	125	130
T	158	132.8	105.8	92.3	84.2	79.2	76.1	75	74.7	74.5

   a. Estimate the temperature of the room. 74°F
   b. If the points (t, T) were plotted, what would be the equation of the horizontal asymptote? $T = 74$
   c. Find an equation that models the data. $T = (103.3)0.9627^t + 74$

16. Each student in a typing class is tested at various times in the course and the average number of errors for the class is recorded.

t = time of testing (days)	2	10	14	21	30	45	63	70	91
y = avg. no. of errors	45.2	36.1	30.2	23.1	18.7	11.0	5.6	4.3	2.4

   a. Plot these points and connect them to make a "learning curve." What is the equation of the horizontal asymptote? $y = 0$
   b. Find an equation that models the data. $y = (48.89)0.9667^x$

17. **Sports** The scatter plot at the left below shows the winning times y (in seconds) for the Olympic men's 800-meter run. (Due to World War I and World War II, there were no Olympics in 1916, 1940, and 1944.) It suggests exponential decay with a horizontal asymptote at $y = 100$. The scatter plot at the right below shows the ordered pairs $(x, \log (y - 100))$.
   a. Imagine a line fitted to the data in the scatter plot on the left. What impossible event does the x-intercept suggest?
   b. Fit a line by eye to the plot at the right below. Write an equation of this line. Then write an exponential equation that expresses y in terms of x.

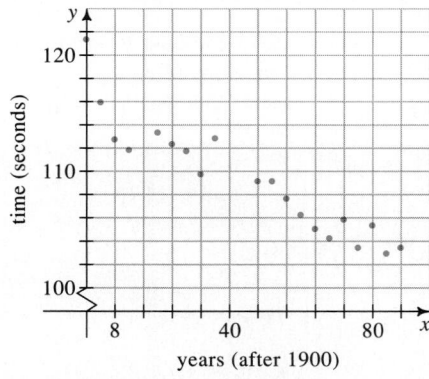

years (after 1900)

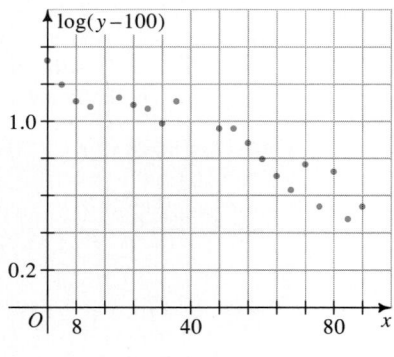

18. **Visual Thinking** Suppose that $y = ab^x$ and the points $(x, \log y)$ are plotted. If a line through these points has negative slope, what can you say about b? $0 < b < 1$

18-3 Fitting Power Curves

| Objective | To fit a power curve to a set of data. |

Sometimes data points (x, y) collected experimentally do not appear to lie on a line or an exponential curve. In such a case, the points may lie along a power curve, $y = ax^m$. Power curves occur often in geometric and scientific applications. Here are three illustrations.

1. The quadratic function $y = ax^2$ is often used when y is the area of a figure, or a price based on the area, and x is a linear dimension of the figure. For example, the cost y of paving a square parking area x meters on a side is modeled by $y = ax^2$ and we say that y varies as the square of x.

2. The cubic function $y = ax^3$ is often used when y is the volume or weight of some object and x is a linear dimension of the object. In this case, we say that y varies as the cube of x.

3. The power function $y = ax^{1.5}$ is used to model the periods of orbiting satellites and planets. (See Exercises 16 and 17.)

To see if a set of points (x, y) lies on a power curve, you can plot the points $(\log x, \log y)$ and see whether they lie on a line. The following reasoning shows why the technique works.

$$\text{If } y = ax^m, \text{ then } \log y = \log a + m \log x.$$

Since $(\log y) = k + m(\log x)$, it follows that $(\log y)$ is a linear function of $(\log x)$.

Example 1 Two biologists hope to find an equation that models how an alligator's weight depends on its length. After measuring and weighing 25 alligators, they plot the points $(x, y) = $ (length, weight). (See figure (a) at the top of the next page.) When they take natural logarithms of length and weight, the points $(\ln x, \ln y)$ seem to have a linear relationship. (See figure (b) at the top of the next page.)

a. Find an equation of a line that fits the data in figure (b).

b. Use the equation to express y in terms of x.

c. Estimate the weight of an alligator that is 10 ft long.

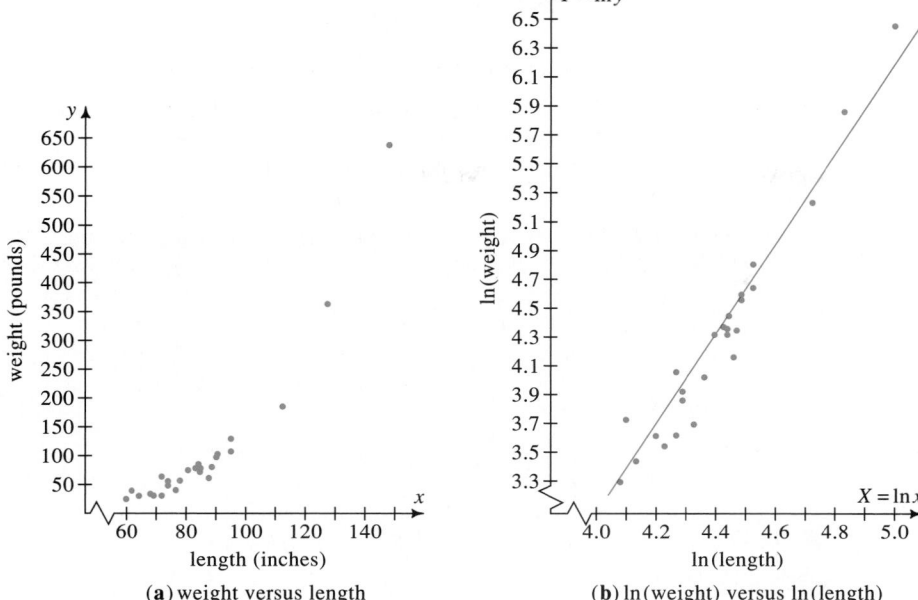

(a) weight versus length

(b) ln(weight) versus ln(length)

Solution

a. First notice that the axes of the scatter plot in figure (b) do *not* intersect at the origin $(0, 0)$. The $X = (\ln x)$-axis begins at 4.0 and the $Y = (\ln y)$-axis begins at 3.3.

Fit a line by sight to the data. One such line is shown in blue. To find the slope of the line, use two points that are far apart. For example, the line appears to go through the points $(4.1, 3.4)$ and $(5.0, 6.2)$, so we can approximate the slope as follows:

$$m \approx \frac{6.2 - 3.4}{5.0 - 4.1} = \frac{2.8}{0.9} \approx 3.1$$

Using the slope 3.1 and the point $(5.0, 6.2)$, we can write:

$$\frac{Y - 6.2}{X - 5.0} = 3.1$$

$$\text{or} \quad Y = 3.1X - 9.3$$

$$\text{or } \ln y = 3.1 \ln x - 9.3$$

b. If $\ln y = 3.1 \ln x - 9.3$, then $y = e^{3.1 \ln x - 9.3}$.

$$y = e^{3.1 \ln x} \cdot e^{-9.3}$$

$$= e^{\ln x^{3.1}} \cdot e^{-9.3} \qquad \longleftarrow \quad 3.1 \ln x = \ln x^{3.1}$$

$$\approx x^{3.1} \cdot (0.000091) \qquad \longleftarrow \quad e^{\ln N} = N$$

(Solution continues on the next page.)

Curve Fitting and Models **699**

Additional Examples

1. a. Use substitution to fit a power curve to the data in Written Exercise 7.

x	2.0	3.0	5.0	6.0
y	4.0	13.5	62.5	108.0

b. Name an advantage and a disadvantage that this method has compared to the method described on page 700.

a. Using the data pairs $(2.0, 4.0)$ and $(6.0, 108.0)$, we substitute into $y = ax^m$:

$$4.0 = a(2.0)^m;$$
$$108.0 = a(6.0)^m$$

$$\frac{4}{2^m} = a; \ \frac{108}{6^m} = a$$

$$\frac{4}{2^m} = \frac{108}{6^m}$$

$$27 = \frac{6^m}{2^m} = \left(\frac{6}{2}\right)^m = 3^m$$

$$m = 3$$

$$a = \frac{4}{2^3} = \frac{4}{8} = \frac{1}{2}$$

$$y = \frac{1}{2}x^3$$

b. Answers may vary. An advantage is that the method is simpler, involving fewer computations than the method in the text. A disadvantage is that the data pairs chosen for substitution may produce a curve that varies substantially from the curve near which most of the data pairs lie, thereby giving an equation that models the data poorly; the least-squares method finds the "best" equation.

(*Continues on next page*)

2. Each of the following data pairs (x, y) can be modeled by a power function $y = ax^m$. Using what you know about geometry, conjecture what the value of m is in each case. Then write a few sentences justifying your answer.
   a. $x = $ amount of paint needed to cover a wooden toy cube; $y = $ the height of the toy cube
   b. $x = $ the diameter of a balloon; $y = $ the amount of air enclosed by the balloon

   a. The amount of paint depends on the surface area of the cube, and the surface area varies as the square of a linear dimension, such as the height of the toy cube. Thus, $x = ky^2$, so $y = a\sqrt{x} = ax^{1/2}$; $m = \frac{1}{2}$.
   b. The amount of air enclosed by the balloon depends on the volume of the balloon, and the volume varies as the cube of the diameter. Thus, $y = ax^3$; $m = 3$.

c. A 10 ft alligator is 120 in. long. Therefore, substitute $x = 120$ into the equation of part (b).

$$y = 0.000091x^{3.1} = 0.000091(120)^{3.1} \approx 254 \text{ lb}$$

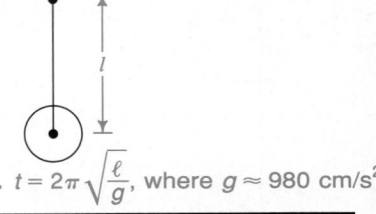 There are calculators and computer software that will find the power curve that best fits a set of data. To understand fully the process of fitting a power curve, it is recommended that you proceed as described below. You may wish to use a calculator or computer to check your answers.

Fitting a Power Curve $y = ax^m$ to a Set of Data

1. Plot the points $(X, Y) = (\log x, \log y)$ for each data point.
2. Fit a line to the set of data points obtained in Step 1.
3. Find an equation of this line:
$$Y = mX + k$$
$$\log y = m \log x + k$$
4. Convert the equation in Step 3 to give y as a function of x.

Example 2   Find y in terms of x if $\log y = 2.1 \log x + 1.5$.

Solution
$$\log y = 2.1 \log x + 1.5$$
$$y = 10^{2.1(\log x)+1.5}$$
$$= 10^{2.1(\log x)} \cdot 10^{1.5}$$
$$= 10^{\log x^{2.1}} \cdot 10^{1.5}$$
$$\approx (x^{2.1})(31.6)$$
$$y = 31.6x^{2.1}$$

Activity

The time t required for a pendulum to swing back and forth once is called the *period* of the pendulum.

1. Make two pendulums by hanging different weights from two strings of the same length. With a stopwatch, time ten back-and-forth swings of the pendulum with the lighter weight, and then divide the sum of the times by ten to get the period of the pendulum. Do the same for the other pendulum. Does it seem that a pendulum's period depends on the weight of the pendulum? No
2. Vary the length l of a pendulum's string, measuring the length from the pivot point to the center of the pendulum weight. Find the period t of the pendulum for various values of l and make a table of ordered pairs (l, t). Ans. may vary.
3. Plot $\log t$ versus $\log l$, and use the resulting line to find an equation giving t in terms of l.

3. $t = 2\pi\sqrt{\dfrac{\ell}{g}}$, where $g \approx 980$ cm/s^2

700   *Chapter Eighteen*

CLASS EXERCISES

1. **a.** Suppose $y = 4x^3$. Express $\log y$ in terms of $\log x$. $\log y = \log 4 + 3 \log x$
   b. If $\log y$ is plotted versus $\log x$, what is the slope of the resulting line? 3

Express y in terms of x.

2. $\log y = 5 \log x + 2$ $y = 100x^5$

3. $\log y = 2.5 \log x + 3.5$ $y = 3162x^{2.5}$

4. $\ln y = -3 \ln x + 1$ $y = ex^{-3}$

5. $\ln y = 2 \ln x + 0.5$ $y = 1.649x^2$

6. **Discussion** The cost y of a cheese pizza depends on the diameter x of the pizza. If $y = ax^m$, what do you think the value of m is? Explain.

7. **Discussion** The weight y of a great blue whale depends on its length x. If $y = ax^m$, what do you think the value of m is? Explain.

WRITTEN EXERCISES

Express y in terms of x.

A

1. $\log y = 4 \log x + 1$ $y = 10x^4$

2. $\log y = 2.5 \log x + 1.3$ $y = 19.95x^{2.5}$

3. $\log y = -\log x + 0.7$ $y = \dfrac{5.012}{x}$

4. $\log y = -2.6 \log x - 0.7$ $y = 0.1995x^{-2.6}$

5. $\ln y = 1.6 \ln x + 3.2$ $y = 24.53x^{1.6}$

6. $\ln y = -0.4 \ln x + 4.2$ $y = 66.69x^{-0.4}$

Fit a power curve to the data in Exercises 7 and 8. Answers may vary.

7.

x	2.0	3.0	5.0	6.0
y	4.0	13.5	62.5	108.0

$y = \frac{1}{2}x^3$

8.

x	0.9	1.5	2.4	3.0
y	4.94	1.78	0.70	0.44

$y = 4x^{-2}$

Writing **Each of the following sets of data (x, y) can be modeled by a power function $y = ax^m$. Using what you know about geometry, conjecture what the value of m is in each case. Then write a few sentences justifying your answer.**

x	y
9. **a.** length of bear footprint	area of the footprint 2
b. length of bear footprint	weight of the bear 3
10. **a.** diameter of a spherical tank	capacity of the tank 3
b. diameter of a spherical tank	amount of paint needed to paint it 2
11. diagonal of a television screen	area of the television screen 2
12. weight of a wooden cube	length of the cube's diagonal $\frac{1}{3}$
13. cost of a circular rug	diameter of the rug $\frac{1}{2}$
14. surface area of a sphere	volume of the sphere $\frac{3}{2}$
15. weight of an elephant	surface area of the elephant's ear $\frac{2}{3}$

Curve Fitting and Models **701**

Making Connections

In Written Exercises 9–15, students use their knowledge of formulas for surface area and volume for various solids. You should relate these exercises back to the conclusions reached for Class Exercises 6 and 7.

Additional Answers
Class Exercises

6. 2; the cost of a pizza depends on its area; the area varies as the square of x.

7. 3; the weight of a whale depends on its volume; the volume varies as the cube of x.

Suggested Assignments

Discrete Math
Day 1: 701/1–15 odd
Day 2: 703/17, 19, 20, 21

Supplementary Materials

Alternative Assessment, 57
Precalculus Plotter Plus, 83–84

Exercise 16 involves Kepler's third law of planetary motion: The square of a planet's period of revolution is proportional to the cube of its mean distance from the sun. To see that the equation derived in the exercise is equivalent to this statement, square both sides to eliminate the nonintegral exponent:

$$y = 0.1995x^{1.5}$$
$$y^2 = 0.0398x^3$$

Recall that whenever $z = kw$, with k a constant, we say that z is proportional to, or varies directly as, w. Observing that y^2 is the square of a planet's period of revolution and x^3 is the cube of the distance from the sun makes the equivalence clear.

For Exs. 16–21, answers may vary.

B **16. Astronomy** Johannes Kepler (1571–1630) spent years trying to discover a relationship between the time for a planet to revolve around the sun and the distance of the planet from the sun. A table and graph of the data for the nine planets are shown below. (Uranus, Neptune, and Pluto were not known in Kepler's time.) Also included is a graph of the logarithms of the data, which appears to be a straight line.

Planet	x = avg. dist. from sun (millions of kilometers)	y = time of orbit (days)
Mercury	57.9	88
Venus	108.2	225
Earth	149.6	365
Mars	227.9	687
Jupiter	778.3	4333
Saturn	1427	10,759
Uranus	2871	30,685
Neptune	4497	60,189
Pluto	5913	90,800

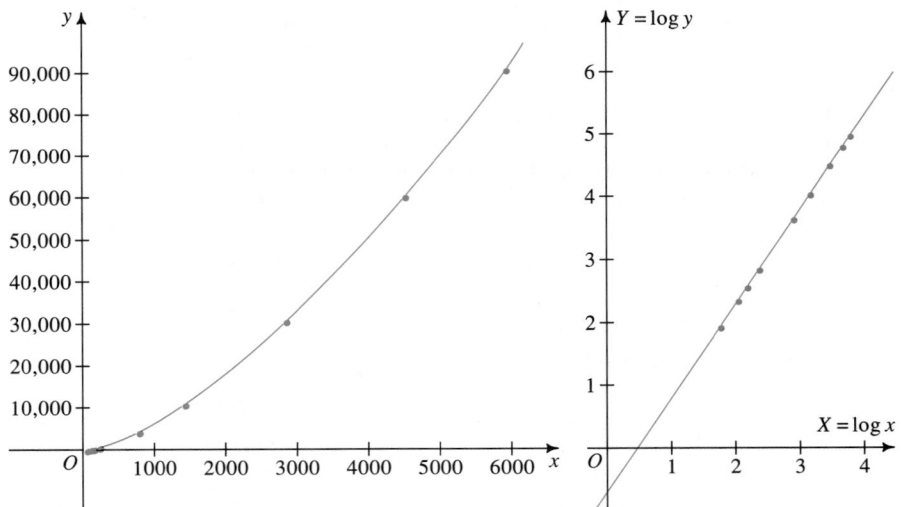

16.c. About 116,800 days

   a. Use the graph at the right on the preceding page to find a linear equation for Y in terms of X. $Y = 1.5X - 0.7$ $y = 0.1995x^{1.5}$

   b. Use your answer to part (a) to find an equation giving y in terms of x.

   c. Suppose that some time in the future a new planet is discovered and its average distance from the sun is estimated at 7000 million kilometers. Use your equation in part (b) to estimate its time of orbit around the sun. See above.

17. Astronomy In 1610, Galileo discovered how the time, T, required for each of Jupiter's satellites to revolve about Jupiter is related to the average distance, a, of the satellite from Jupiter. By plotting $\log T$ versus $\log a$, see if you can discover the relationship. See below.

Satellite	a (kilometers)	T (hours)
Io	422,000	42.5
Europa	671,000	85.2
Ganymede	1,070,000	171.7
Callisto	1,883,000	400.5

18. Sports The winning times (in seconds) for various men's races (in meters) in the 1988 Olympics are given below.

x = distance (m)	100	200	400	800	1500	5000	10,000
y = time (s)	9.92	19.75	43.87	103.45	215.96	791.70	1641.46

$\log y = 1.1253 \log x - 1.2677$

   a. Plot $\log y$ versus $\log x$. Find an equation of the resulting line.

   b. Find an equation that gives y in terms of x. $y = (0.054)x^{1.125}$

   c. If there had been a 1000-meter race in these Olympics, what would you predict the winning time would have been? About 128 s

19. Forestry Forest managers keep records giving the number of board feet B (in hundreds of board feet) of lumber produced by harvested white pine trees with diameter d (in inches). Consider the following data for nine white pine.

d	15	16	17	18	19	20	21	22	23
B	18.68	22.85	27.61	33.00	39.06	45.84	53.38	61.72	70.90

   a. Plot $\log B$ versus $\log d$ and show that you get a line with slope about 3.

   b. *Writing* Explain why part (a) shows that volume varies approximately with the cube of the diameter.

   c. *Writing* Explain why the result of part (b) is not surprising if you think of a white pine as a cylinder-like object whose height h is proportional to the radius r.

17. $T = (1.553 \times 10^{-7})a^{3/2}$

Curve Fitting and Models   **703**

Cooperative Learning

Exercise 23 could be assigned to small groups of students, with each group investigating the costs for different shops or chains. The entire class could then compare results from the different groups to decide which is the best buy.

Additional Answers
Written Exercises

20. b. log P versus log D;
$P = (154.3)D^{-0.6583}$
   c. About 13% of such shots succeed.
   d. No; very few shots are attempted at such great distances and few succeed. The model is based on data that are mostly at or below 30 ft. It would be impossible to tell whether the model would apply beyond the experimental range.

22. a. The cost of a pizza depends on its area. Since the area varies as the square of the diameter, cost varies as the square of the diameter.
   b. $y = 0.024x^2 + 1.6$
   c. About $5.06

20. Sports A statistician for a college basketball team keeps track of every shot the team's players attempt during games, and then computes for each player statistics that show how well the player shoots at various distances from the basket. The figures below are for the team's leading scorer, a 6'3" guard.

D = dist. from basket (ft)	1–5	6–10	11–15	16–20	21–25	26–30	over 30
P = % of shots made	70	43	30	23	19	17	15

   a. Make two plots: (1) log P versus D and (2) log P versus log D. Use the middle value of each distance interval. Thus when D is in the interval from 1 to 5 ft, use $D = 3$ and $P = 70$. For D over 30 ft, use $D = 33$.
   b. Which of your plots seems to fit the data better? Use the better model to find an equation that gives P in terms of D.
   c. Suppose the player whose statistics are given above attempts a shot from half-court (45 ft). How likely is it that the player will make the shot?
   d. *Discussion* Is it appropriate to use your model for half-court and full-court distances? Why or why not?

21. The table gives data on some famous roller coasters. Find a power function $y = ax^m$ that gives the roller coaster's maximum speed (y) as a function of its greatest single vertical drop (x). $y = (3.065)x^{0.6141}$

Roller coaster		x = vertical drop (ft)	y = max. speed (mi/h)
Texas Giant	Arlington, TX	137	62
The Beast	Cincinnati, OH	141	65
American Eagle	Gurnee, IL	147	66
Hercules	Allentown, PA	157	68

22. Consumer Economics The cost of making a plain pizza of diameter x can be modeled by an equation of the form $y = ax^2 + k$, where k is a constant that accounts for fixed costs, such as electricity for the ovens, rent, and so forth.
   a. Explain why the term ax^2 makes sense.
   b. Use the data below to write two equations in the unknowns a and k. Then solve for a and k. Give an equation for y in terms of x.

x = diameter (in.)	10	15
y = cost (dollars)	4	7

   c. Use the equation from part (b) to predict the cost of a 12 in. pizza.

23. *Research* Go to a pizza shop and find the costs of various sizes of plain pizza. Decide whether your data can be modeled by an equation like the one in Exercise 22(b).

704   *Chapter Eighteen*

18-4 Choosing the Best Model

Objective *To choose the model that best describes a set of data.*

The purpose of this section is to give you the opportunity to choose from among several models the one that best describes a set of data. Some of the most common models are the following:

1. Linear Function

Least-squares line $y - \bar{y} = m(x - \bar{x})$
This line is discussed on page 684.

2. Exponential Function

$y = ab^x$
A list of exponential growth and decay phenomena is given on page 691.

3. Power Function

$y = ax^m$
Situations involving $m = 2$, 3, and 1.5 are discussed on page 698.

4. Functions with Asymptotic Behavior

Exponential: $y - c = ab^x$

If, after having analyzed the problem, you're still not sure what model best fits a set of data, try one of the following strategies:
1. Graph the data and see what type of curve the graph suggests.
 2. Graph (x, y), $(x, \log y)$, and $(\log x, \log y)$ with a graphing calculator or computer and determine which looks most linear.

The equations above are not the only types of models used to describe real world situations. Difference equations and equations using matrices and calculus are also important. In many situations, a system of equations instead of a single equation must be applied. You should also know that there are many phenomena that at present seem to have no mathematical model at all. Where such phenomena seem to be chaotic in their behavior, some modeling progress has recently been made by using iteration of functions, a topic to be discussed in Chapter 19.

MIXED EXERCISES

For Exs. 1–14, answers may vary.

Model the data in each exercise. To do this, you must choose an appropriate model and then write an equation.

In Exercises 1–4, write an equation that models the data in part (a), then use your equation to answer part (b).

$y = (8.385 \times 10^{-4})x^{2.69}$

A 1. a. **Biology** A scientist caught, measured, weighed, and then released eight Maine landlocked salmon. The data were then recorded in a table.

x = length (in.)	5.5	10.6	15	17	19.6	22	25	28
y = weight (lb)	0.1	0.4	1.0	1.6	2.5	3.5	5.4	7.4

b. Predict the weight of a salmon that is 12 in. long. About 0.7 lb

Curve Fitting and Models **705**

Model the data. Then predict the unknown *y*-value.

x	*y*
0.3	8.011
1.5	1.591
2	0.8112
5	0.01426
1.4	?

Try fitting a line to the points (x, y), $(x, \log y)$, or $(\log x, \log y)$. The correlation coefficient for the data pairs $(x, \log y)$ is approximately equal to -1, so choose $\log y \approx mx + k$. The least-squares line is $\log y \approx -0.585x + 1.08$. Thus: $y \approx 10^{-0.585x + 1.08}$
$y \approx (10^{1.08})(10^{-0.585})^x$
$y \approx 12(0.26)^x$
When $x = 1.4$,
$y \approx 12(0.26)^{1.4} \approx 1.82$.

Additional Answers
Mixed Exercises

5. **a.** $W = aC^3$; weight depends on volume; the volume varies with the cube of the radius, and the radius varies with the circumference.
   b. $W = 0.394C^3$
   c. About 41 lb
   d. The equation used was based on one item of data.

Suggested Assignments

Discrete Math
Day 1: 705/1–5
Day 2: 707/6, 9–11

---

2. **a.** **Economics** The table below shows the average salaries (in thousands of dollars) of major league baseball players for selected years from 1967 to 1989. $y = (17.03)1.177^x$, x = no. of years after 1967

year	1967	1976	1979	1982	1984	1986	1989
salary (×$1000)	19.0	51.5	114	241	329	413	497

b. Estimate the average salary in the year 1970. $27,770

3. **a.** **Physics** The boiling point of water increases with atmospheric pressure. At sea level, where the atmospheric pressure is about 760 mm of Hg (mercury), water boils at 100°C. The table below shows the boiling point of water for different values of atmospheric pressure. $T = (13.13)P^{0.3061}$

P = pressure (mm of Hg)	149	234	355	526	760	1075
T = boiling point (°C)	60	70	80	90	100	110

b. When the atmospheric pressure is 620 mm of Hg, at what temperature does water boil? About 94°C

4. **a.** **Oceanography** Various depths (in meters) below the water surface and corresponding water pressures (in pounds per square inch) at those depths are given in the table below. $P = 1.47d$

d = depth (m)	10	20	30	40	50	60
P = pressure (lb/in.^{2})	14.7	29.4	44.1	58.8	73.5	88.2

b. Predict the water pressure at a depth of 13 m. 19.11 lb/in.^{2}

5. *The Guinness Book of World Records* (1990 edition) says that the world's largest pumpkin has a circumference of 11 ft $11\frac{1}{4}$ in. and weighs 671 lb.

   a. *Reading* Which function would you choose to model a pumpkin's weight as a function of its circumference? Explain your answer.
   b. Use the given data to find an equation that gives pumpkin weight as a function of circumference.
   c. Predict the weight of a pumpkin whose *diameter* is 18 in.
   d. *Writing* In one or two sentences, explain why your prediction in part (c) might be a poor estimate of the actual weight of a pumpkin with diameter 18 in.

B **6. Economics** The Consumer Price Index (CPI), published by the United States Bureau of Labor Statistics, compares the present cost of goods and services with their cost at another time. The period 1982–1984 is considered the base period and is assigned an index of 100. All other costs are expressed as a percent of the costs in the 1982–1984 period.

year	1945	1955	1965	1975	1985
CPI	18.0	26.8	31.5	53.8	107.6

a. What model do you think will best fit the data? Explain.

b. Find an equation that models the data above.

c. *Discussion* Use your equation to predict the CPI in the year 2000. What factors might affect the accuracy of your prediction?

7. Economics The gross national product (GNP) is the total sum spent on goods and services by a country in any year. The table below gives the GNP in billions of dollars for the United States for various years.

year	1900	1920	1940	1960	1980	1988
GNP	19.0	92.0	100	507	2732	4881

a. *Research* The GNP declined sharply in the early 1930s and rose dramatically in the early 1940s. To what historical events would you attribute these changes?

b. Find an equation that models the data above.

c. Use your equation to predict the GNP in the year 2000.

8. Medicine A doctor measures the pulse rate of a patient who has just completed vigorous exercise. The results shown in the graph at the right and the table below suggest that the graph of pulse rate versus time has a horizontal asymptote at $r = 70$.

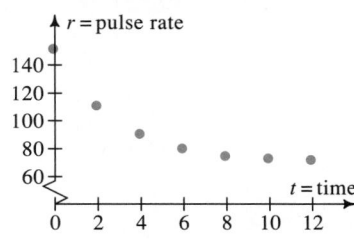

t = time after stopping (min)	0	2	4	6	8	10	12
r = pulse rate (beats/min)	150	110	90	78	73	72	71

Find an equation that gives pulse rate in terms of the time elapsed since the exercise ended. $r = (80.46)0.6869^t + 70$

Supplementary Materials

Alternative Assessment, 57
Activities Book, 48–50
Student Resource Guide, 167–171

Additional Answers
Mixed Exercises

6. a. Exponential growth because it is the best model for increasing costs due to inflation.

b. $C = (16.51)1.044^x$ where x is the number of years since 1945

c. About 176.3; unexpected occurrences such as war, a new inexpensive fuel source, changes in political alliances, or changes in demographics

7. a. The Great Depression and World War II

b. $G = (17.15)1.063^x$ where x is the number of years since 1900

c. About 7720

c. $\log(19 - T) = -0.0166t + 1.4037$

9. a. Thermodynamics The table shows the temperatures of a cold glass of water left to warm in a room. Study the data and give the approximate temperature of the room. About 19°C

t = time (min)	0	20	40	60	80	100	110	120
T = temperature (°C)	3.5	8.1	12.2	15.4	17.0	18.2	18.6	18.9

b. Plot T versus t and find an equation of the horizontal asymptote. $T = 19$
c. Plot $\log(19 - T)$ versus t and find an equation of the resulting line. See above.
d. Use your answer to part (c) to find T in terms of t. $T = 19 - (25.33)0.9625^{t}$

10. Manufacturing Last year, a chair manufacturer marketed a new style of chair. As the manufacturer produced more and more chairs, the production cost per chair decreased as shown below.

x = no. of chairs	50	300	800	2000	3000
y = cost per chair	$517	$387	$250	$162	$152

a. Study the data. What seems to be the limit of the manufacturing cost per chair as the number of chairs becomes very large? $150
b. Find an equation that models the data. $y = (404.9)0.9982^{x} + 150$
c. Predict the cost per chair when 500 chairs have been produced. About $314

11. Sports The table gives the average speed of a runner for various distances.

x = distance (m)	y = speed (m/s)
100	8.7
200	8.6
400	7.7
800	6.6
1500	6.1
3000	5.7

If the data were plotted, the line $y = 5$ would be an approximate asymptote. Find an equation that models the data. See below.

12. Research Use an almanac or other reference book to research two data sets that you think might be linearly related. Make a scatter plot and fit a line to the data. Finally, use the line to explore a question that the data might suggest.

11. $y = (3.371)0.9995^{x} + 5$

13. An English teacher returns students' papers with many comments, but never a letter grade. The data shown in the scatter plot correlate the grade each student *thought* he or she had received with the actual grade that the teacher put in the grade book. The numbers from 0 to 12 represent letter grades from E to A⁺. A (2) or (3) above a point indicates that 2 or 3 students, respectively, have the same data point.

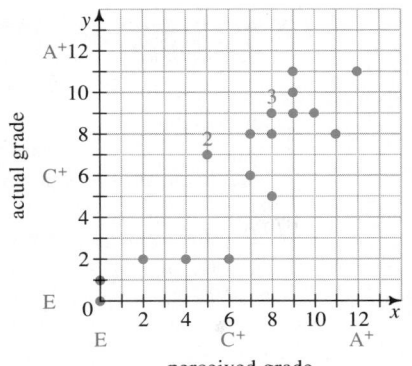

perceived grade

a. According to the scatter plot, how many students *thought* they got a B on their papers? Find the average (mean) grade that these students *actually* received. 5; 8 = B

b. Find an equation of the least-squares line that fits the data. Use this equation to predict an actual grade for a student who thinks she got a B. Compare this with your answer to the second question of part (a). $y = 0.9054x + 0.4932$

about 7.7

14. *Research* From reference books, magazines, or newspapers, collect data that can be represented as ordered pairs. Plot the data, fit a curve to the data, and write a report that explains what you have done and summarizes your findings.
Ans. will vary. Examples: weight vs. daily calorie intake, income vs. taxes

Chapter Summary

1. The process of finding an equation of a curve that fits observed data is called *curve fitting*. The equation is said to *model* the data.

2. A *scatter plot* provides a visual picture of any relationship between two variables. The line that best fits a set of data points is called the *least-squares line*. The *correlation coefficient* is a number between −1 and 1 that measures how closely the observed points tend to cluster about the least-squares line.

3. The points (x, y) lie on an *exponential curve*, $y = ab^x$, if and only if the points $(x, \log y)$ lie on a line. Exponential growth and decay phenomena are modeled by exponential functions. An exponential model with a horizontal asymptote has the form $y - c = ab^x$.

4. The points (x, y) lie on a *power curve*, $y = ax^m$, if and only if the points $(\log x, \log y)$ lie on a line. Power curves occur often in geometric and scientific applications.

Key vocabulary and ideas

curve fitting (p. 683)	exponential curve $y = ab^x$ (p. 691)
model (p. 683)	fitting an exponential curve (p. 693)
scatter plot (p. 683)	exponential model $y - c = ab^x$ (p. 696)
least-squares line (p. 684)	power curve $y = ax^m$ (p. 698)
correlation coefficient (pp. 685–686)	fitting a power curve (p. 700)

Assessment Note

Exercise 14 suggests an excellent means of checking students' understanding of the concepts of this chapter.

Chapter Test

1. The weather bureau for a certain city correlated the average annual temperature (°F) and annual precipitation (in.) for the last 40 years. Letting x represent temperature and y represent precipitation, the statisticians found the following: $\bar{x} = 55$, $\bar{y} = 40$, $\overline{xy} = 2240$, $s_x = 7$, and $s_y = 8$.
   a. Calculate the correlation coefficient. ≈ 0.71
   b. Find an equation of the least-squares line. $y = 0.82x - 4.9$
   c. *Writing* Evaluate the usefulness of the least-squares line in predicting the annual precipitation of the city for a given temperature.   18-1

2. Let x represent a student's score on the SAT portion of the College Entrance Exams and let y represent that person's score on the Achievement portion of the Exams. If x and y have a high correlation, can you infer that a high score on the SAT's causes a high score on the Achievements? Explain.

3. Express y in terms of x if $\log y = 3x + 2$. $y = (100)1000^x$ 18-2

4. The table gives the per capita personal income y, in dollars, for various years, x.

x	1935	1945	1955	1965	1975	1985
y	474	1223	1881	2773	5851	13,896

   a. Fit an exponential curve to the data. $y = (57.86)1.065^x$, $x =$ no. of years after 1900
   b. Predict the per capita personal income in 1995. $22,940

5. Express y in terms of x if $\log y = 1.7 \log x - 2$. $y = (0.01)x^{1.7}$ 18-3

6. *Writing* Let x represent the length of a diagonal of a square rug and let y represent the cost of the rug. The data (x, y) can be modeled by a power function $y = ax^m$. Using what you know about geometry, conjecture what the value of m is. Then write a few sentences justifying your answer. $m = 2$

7. The following table gives the length of the year, y, in Earth-days for three planets, and the mean distance, d, in miles, of these planets from the sun.   18-4

	d (in miles)	y (in Earth-days)
Mercury	36,000,000	88
Venus	67,250,000	225
Mars	141,700,000	687

   a. Find an equation that gives y in terms of d. $y = (4.099 \times 10^{-10})d^{1.5}$
   b. Use your equation to predict the distance of Earth from the sun. About 92,560,000 mi

710   *Chapter Eighteen*

1. An analysis of two test items revealed that 62 of 100 students responded correctly to the first item, while 48 responded correctly to the second. If 37 students got both items correct, how many students missed both questions? 27

2. How many different sequences can be formed using one consonant followed by one vowel (a, e, i, o, u) followed by a different consonant? 2100

3. To get an A on her report, Kyoko knows her teacher expects her to use at least 4 sources. In how many ways can she choose 4 books from a list of 9 books? 126

4. In how many ways can a judge choose a first-place winner, a second-place winner, and a third-place winner from 9 entries submitted in an art show? 504

5. How many permutations are there of the letters of the word BANANA? 60

6. Give the binomial expansion of $\left(2x - \frac{1}{2}\right)^4$. $16x^4 - 16x^3 + 6x^2 - x + \frac{1}{16}$

7. Suppose a card is drawn from a well-shuffled standard deck of 52 cards. Find the probability of drawing each of the following:
   a. a diamond $\frac{1}{4}$
   b. not a diamond $\frac{3}{4}$
   c. a red club 0
   d. not a red club 1
   e. a 10 $\frac{1}{13}$
   f. an ace or a king $\frac{2}{13}$

8. Three cards are randomly chosen from a well-shuffled standard deck.
   a. What is the probability that all 3 cards are spades?
   b. Consider the events "at least one ace is drawn," "all cards are black," and "none of the cards is the ace of spades." Which events are independent?

9. What is the probability of getting exactly 2 threes in 5 rolls of a die? $\frac{625}{3888}$

10. Given that the first card randomly drawn from a standard deck is a diamond, what is the probability that the second card randomly drawn is also a diamond? a spade?

11. In a manufacturing plant, Assembly Line A produces 60% of the plant's output and Assembly Line B produces the rest. Four percent of Line A's output is defective, and seven percent of Line B's output is defective.
    a. What percent of the plant's output is defective? 5.2%
    b. What is the probability that a randomly selected item found to be defective is from Line A? $\frac{6}{13}$

12. On a game show, a contestant spins a giant wheel that is divided into 10 equal segments, 4 marked $50, 3 marked $100, 2 marked $500, and 1 marked $1000. What is the expected payoff of a spin? $250

13. The table below shows the number of customers of Sheffield Village (the frequency) making telephone payments on different days of the month (the items of data). Draw a histogram for this data. Then find the mean, median, and mode.

Day of month	1	2	3	4	5	6	7	8	9	10
No. of customers	9	3	0	2	5	9	5	9	8	15

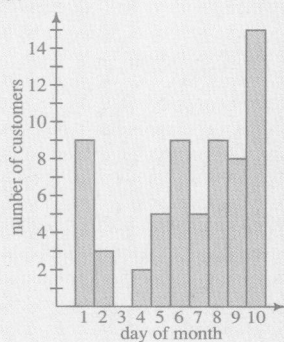

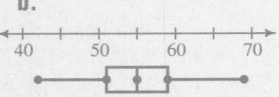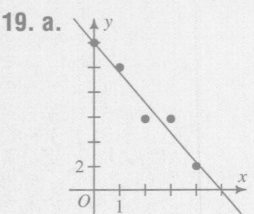
14. The age at which each of the United States presidents from George Washing-
ton to George Bush began his term is listed below. Draw **(a)** a stem-and-leaf
plot and **(b)** a box-and-whisker plot of the data.

57, 61, 57, 57, 58, 57, 61, 54, 68, 51, 49, 64, 50, 48, 65, 52, 56, 46, 54, 49, 51,
47, 55, 55, 54, 42, 51, 56, 55, 51, 54, 51, 60, 62, 43, 55, 56, 61, 52, 69, 64

15. Compute the mean and standard deviation for the following set of data: 1, 4, 9,
16, 25, 36, 49, 64, 81, 100 38.5; 32.4

16. The number of chocolate chips in each cookie sampled from a bag of cookies
is approximately normally distributed with a mean of 25 and a standard deviation of 4. Find the approximate percent of cookies with the number of chocolate chips being:
a. fewer than 18 4%   **b.** at least 20 90%   **c.** between 22 and 27 48%

17. A polling organization is trying to determine the level of support for a tax
measure. Describe any errors that might occur when the organization randomly calls 100 private homes during business hours.

18. Of 1000 university students surveyed, 430 said they would use the library for
studying during final exam week. Find a 99% confidence interval for p, the
true proportion of all university students who would use the library. $0.383 < p < 0.477$

19. Consider this set of data: $(x, y) = (0, 12), (1, 10), (2, 6), (3, 6)$, and $(4, 2)$
a. Draw a scatter plot of the data, and find the correlation coefficient.
b. Find an equation of the least squares line, and draw the line on your scatter
plot. You may find a calculator helpful.

20. Tell whether the relationships between the two quantities would most likely be
modeled by a linear function, an exponential function, a power function, or a
function with asymptotic behavior. Give reasons for your answers.
a. The diameter of a pipe and the volume of the pipe. Power
b. The cost of a calculator and the sales tax on the calculator. Linear
c. The amount of time a tea kettle has been heated and the temperature of the
water in the tea kettle. Asymptotic

21. Fit a power curve to the data in the following table.

x	1.5	2.0	2.5	3.0
y	6.7	15.9	31.2	54.2

$y = 1.97x^{3.016}$

22. The following table gives the population size, P, of bacteria at various regular
time intervals, t.

t	0	1	2	3	4	5	6	7
P	1	2	4	8	12	21	37	61

a. What model do you think will best fit the data? Explain. Exponential
b. Find an equation that models the data above. $P = (1.156)1.786^t$
c. Predict the population size when $t = 10$. About 382

712   *Cumulative Review*

Careers in Genetics and Statistics

The twentieth century has seen unprecedented progress in our ability to understand, control, and cure disease. Many fields of study contribute to medical research, and the methods of data analysis play an important role.

GENETICS

Genetics, the study of heredity, encompasses a number of areas of research. Examples include population genetics, which uses statistical analysis to find patterns in gene distribution and the occurrence of genetic diseases; microbial genetics, which explores the process of mutation by studying rapidly multiplying microorganisms; and cytogenetics, which looks at hereditary activity at the cellular level.

Agnes Stroud-Lee, a University of Chicago Ph.D., has done research in cytogenetics at the Los Alamos National Laboratory. By observing the effects of radiation and toxic chemicals on chromosomes, she sought clues to abnormalities in genetic material. Her results have improved our understanding of birth defects and cancer. "I have made inroads," she says, " . . . other scientists can use my work to further their pursuit of science." Ms. Stroud-Lee retired in 1979.

Agnes Stroud-Lee

STATISTICS

Statisticians are experts in the design of experiments, surveys, and questionnaires used to generate accurate and useful data, and in the meaningful analysis of data. Their skills are indispensable to medical and health research.

The National Center for Health Statistics is an important source of information for researchers and policymakers. NCHS analysis of data reveals connections between such varied health factors as socioeconomic status, type of health care used, nutrition, exposure to risks, and age.

Joe Fred Gonzalez, Jr., a mathematical statistician for NCHS whose work includes developing estimation procedures and national sample survey designs, always found math "challenging and intellectually stimulating." His desire to understand applications of mathematics led him to an M.S. degree in statistics from the George Washington University. Believing in the value of education, Mr. Gonzalez is also a part-time university and college professor.

Joe Fred Gonzalez, Jr.

Careers in Genetics and Statistics **713**

Preference Testing

French fries or baked? Milk or cream? Country-western or rock-and-roll? Vanilla or chocolate? Most everyone at some time has been asked to state a preference. In fact, the study of people's preferences plays an important part in decision-making. Marketing strategies, for instance, often rely heavily on determining consumer likes and dislikes. To get an idea of how this might be done, let's explore one of the simplest techniques for evaluating preferences.

Materials:

a random number generator   a graphing calculator or computer software

The Sign Test

How can you quantify an opinion? If you have ever participated in a survey, you probably have seen a scale like the one shown below.

strongly dislike = 1   dislike = 2   neutral = 3   like = 4   strongly like = 5

To determine whether most people prefer one item over another, you need a way of comparing people's responses. Here is one method. Suppose that *everyone* in a given population rated two items of interest using the scale above. A small portion of the results might look like this:

Person	Item #1 rating	Item #2 rating	Item #1 – Item #2	Sign of the difference
1	4	3	1	+
2	4	5	−1	−
3	3	3	0	omit data
4	5	2	3	+

Notice that when there is a tie, we omit the piece of data. Now, let p denote the population proportion of + signs out of the total number of signs. If $p > 0.5$, then the population must have rated Item #1 higher than Item #2. Why? However, as you learned in Chapter 17, using a random sample to represent a population is often far more reasonable than using the entire population. For this reason, statisticians have developed a rule known as the *sign test*. We can use a simplified version of the sign test to determine whether people significantly prefer Item #1. First, find the *critical value z* using the following formula.

$$z = \frac{2x - n}{\sqrt{n}},$$

where n is the number of pairs left in the sample *after discarding ties*, and x is the number of + signs in the sample. For $n \geq 10$, we can say that the following is true.

If $z > -1.645$, then $p > 0.5$. Thus, the population probably prefers Item #1. If $z \leq -1.645$, then $p \leq 0.5$. Thus, the population probably does not prefer Item #1.

If $n < 10$, then the sample size is too small for the test to be valid. Are you ready to give the sign test a try?

Conduct Your Survey.

■ An essential first step is to prepare a well-designed study. Think of two ratable things that you'd like to compare, and label them Item #1 and Item #2. The items needn't necessarily be commodities. They could be, for example, two viewpoints, such as attitudes towards an issue before and after a debate. Order your rating scale carefully according to your choice of topics.

■ Next, describe the population that you wish to survey. If at all possible, use one of the methods suggested in Section 17-5 to select a simple random sample. Otherwise, choose a sample that you believe to be random. Instruct the people in your sample to rate each of the two items with the scale that you've developed. Be sure that each person responds independently. After you've collected all of your data, calculate the value of n. You may need to add a few more people to your sample so that $n \geq 10$.

■ Illustrate your data by graphing on the same set of axes the ordered pairs

$$(x, y) = (\text{Item \#1 rating, Item \#2 rating})$$

and the line $y = x$. Notice that data points below the line $y = x$ indicate a preference for Item #1.

■ Use the sign test to determine whether your population significantly prefers Item #1. Based on your choice of a population, why do you think your results turned out as they did? Compare the outcome with the actual sample proportion $\frac{x}{n}$. Does there seem to be any conflicting information? Remember that even though a preference may exist in a sample, it may not be statistically significant enough to represent the population.

Report Your Results.

Any well-written statistical report should contain a section of summary statistics. For example, supply as much demographic information (mean age, number of males and females, years of education completed, and so forth) as possible about your sample. You should also provide a tally of ratings along with a table listing each item's rating and the corresponding sign of the difference between the ratings. Don't forget to include your graph as part of your data description. Explain why your choice of a scale fits the items you choose to rate. Justify your method of sample selection. State your values of x, n, and z, and cite the sign test rule you followed to obtain your conclusion. Finally, use your demographic data to help interpret your results.

Chapter 19 Limits, Series, and Iterated Functions

Overview

This chapter revisits and expands on a number of topics presented in Chapter 13. Once students have seen how the concept of limit can be extended from sequences (see Section 13-4) to functions, students use limits to determine the continuity of functions (Section 19-1) and to graph rational functions (Section 19-2). Students then use the concept of infinite series (see Section 13-5) to approximate the area under a curve (Section 19-3) and to find power series for certain functions (Section 19-4). Finally, once students have seen the relationship between recursive definitions of sequences (see Section 13-2) and iterated functions, students analyze orbits for iterated functions (Section 19-5) and use iterated functions to examine money and population growth (Section 19-6).

Objectives

19-1 To find the limit of a function or the quotient of two functions and to determine whether a function is continuous.

19-2 To sketch the graph of a rational function.

19-3 To use technology to approximate the area under a curve.

19-4 To use the power series of a given function to find an infinite series for a functional value or for a related function.

19-5 To analyze orbits for iterations of a given function.

19-6 To use iterated functions to model money and population growth.

Supplementary Resources

Tests, pp. 61–63
Alt. Assess., pp. 57–60, 72
Activities, pp. 51–52
Student Res. Guide, pp. 174–182

Software

Precalculus Plotter Plus
Activities, pp. 50–51, 76–77, 87–88
Function Plotter

Pacing Guide

Section	19-1	19-2	19-3	19-4	19-5	19-6	Review	Test	Total
Discrete Math (days)	2	2	–	–	–	–	–	1	5

Teaching Notes

19-1 pages 717–725

Presenting the Section

In this section students are reintroduced to the concept of limit. New avenues are pursued:

1. Expanding the domain from positive integers to real numbers.
2. The limit of $f(x)$ as x approaches $-\infty$ as well as ∞.
3. The limit of $f(x)$ as x approaches a constant value c.
4. An introduction to continuous functions.
5. The limit of a quotient of two functions.

Keep in mind that the limit concept and the definition of continuity are fundamental to a student's understanding of calculus. Time spent here solidifying students' confidence in mastering the material will not be wasted. Start by reinforcing their intuitive understanding of the limit of a sequence. Do this first by referring back to Chapter 13. Emphasize that when determining the limit of $f(x)$ as x approaches $-\infty$, students must take care in observing the sign of x.

Students' understanding of the section will be facilitated by maintaining a clear distinction between the domain and range of the function being considered. Stress that a limit is not the value of the function at a given x-value, but the value that the function approaches as x approaches the given x-value.

Using Technology

Here is an ideal place to use graphing calculators or computers. When illustrating the limit of a function at a point, you can deliberately adjust the viewing rectangle so that the point lies just off-screen. Students can then "predict" the value of the function at the point.

Cooperative Learning

Students can work in groups of two or three on the Class Exercises. You may then arbitrarily select students from each group to demonstrate and explain their solutions to the rest of the class.

Communication

If you want to stress the formal definitions of limit, consider having students play the "epsilon-delta" game in which they challenge each other to find a delta for a given epsilon for a specified limit problem. You can easily pick limit problems from the exercises or even have students make up their own. For instance, in Written Exercise 8, the limit is 6. The challenger may say, "Find a neighborhood of 3 on the x-axis that gives function values within a neighborhood of $\frac{1}{100}$ of 6 (on the y-axis)." (If x is within $\frac{1}{200}$ of 3, then $f(x)$ will be within $\frac{1}{100}$ of 6.)

Assessment

Upon completion of this section, it would be beneficial to have students submit a detailed written response to the following questions.

1. Explain what it means for a function $f(x)$ to have a limit as x approaches a certain value.
2. What concepts in Section 19-1 do you feel that you still do not understand?

Naturally, you cannot expect "textbook responses" to the first question. However, students' replies should clearly reflect their grasp of the basic ideas presented in the section.

19-2 pages 726–729

Presenting the Section

In this section students apply the knowledge of limits and continuity that they gained in the previous section to graphing rational functions. The general procedure is outlined and a couple of examples are given.

In discussing step 5 of Example 1, you may wish to write the first limit as $\lim_{x \to \infty} \frac{x-3}{x^2-1} = 0^+$ to emphasize the fact that the value of $\frac{x-3}{x^2-1}$ is *greater than* 0 but *approaching* 0 as x approaches infinity. Also, you can write the second limit as $\lim_{x \to -\infty} \frac{x-3}{x^2-1} = 0^-$ to emphasize the fact that the value of $\frac{x-3}{x^2-1}$ is *less than* 0 but *approaching* 0 as x approaches negative infinity.

The aforementioned limits imply that the x-axis is a horizontal asymptote for the graph of $y = \frac{x - 3}{x^2 - 1}$. Point out to students that a rational function's graph may cross a horizontal asymptote (as this one does at $x = 3$), but it can never cross a vertical asymptote.

Various properties of rational functions and their graphs will manifest themselves through the exercises. A discussion of Class Exercise 7 should prove interesting after students have considered a few examples. It would be useful to expose them to as many different rational functions as possible so that they do not make incorrect generalizations.

Using Technology

A graphing calculator is ideal as a tool for verifying graphs. It can also be useful for testing students' conjectures about properties of rational functions. For instance, how does the repeated root in the denominator in Exercise 15 affect the graph? Take advantage of the power of technology to streamline the investigation of this and other questions.

Assessment

A worthwhile activity would be to present students with a series of graphs and ask them to obtain a rational function having the given graph. This clearly illustrates the connection between the function and the graph. Just be mindful that the less detail you provide, the greater the number of functions that can describe the given graph.

Communication

Have students organize and share their own summaries of the characteristics of rational functions.

19-3 | pages 729–733

Presenting the Section

In this section students are introduced to the notion of the area under a curve. This involves finding a Riemann sum, a technique which gives students their first taste of integral calculus.

A BASIC computer program for finding the area

under a curve is given on page 730, and certain modifications are suggested to obtain better approximations.

The question arises of what yields the "best" approximation. Explore this question to help students see how limits can be utilized in such problems. This realization, and a similar connection made to the slope of a curve in the next chapter, form a solid foundation for the study of calculus.

You may wish to take a more general approach than that taken in the text to approximating the area under the curve $y = \cos x$ from $x = 0$ to $x = \frac{\pi}{2}$ as follows:

Begin by constructing n rectangles of constant width and variable height, as shown below. To do so, divide the interval $0 \le x \le \frac{\pi}{2}$ into n subintervals, each having width:

$$\frac{\frac{\pi}{2} - 0}{n} = \frac{\pi}{2n}$$

Thus, the first subinterval is $0 \cdot \frac{\pi}{2n} \le x \le 1 \cdot \frac{\pi}{2n}$, the second subinterval is $1 \cdot \frac{\pi}{2n} \le x \le 2 \cdot \frac{\pi}{2n}$, and the kth subinterval is $(k - 1) \cdot \frac{\pi}{2n} \le x \le k \cdot \frac{\pi}{2n}$. The rectangle constructed on each subinterval has a height equal to the cosine of the right-hand end of the subinterval. Thus, on the kth subinterval, the height of the rectangle is $\cos \frac{k\pi}{2n}$.

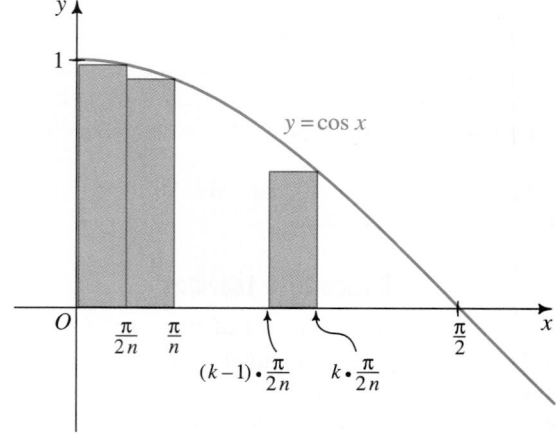

Since the kth rectangle has width $\frac{\pi}{2n}$ and height $\cos\frac{k\pi}{2n}$, the area of the rectangle is $\frac{\pi}{2n}\cdot\cos\frac{k\pi}{2n}$. The sum of the areas of the n rectangles should give a reasonable approximation of the area of R:

$$\text{Area of } R \approx \sum_{k=1}^{n} \frac{\pi}{2n}\cdot\cos\frac{k\pi}{2n}$$

The approximation will be less than the area of R because the rectangles are completely contained in R without covering all of R. On the other hand, the approximation improves as the number of rectangles increases. In fact, the area of R can be defined using limit notation as follows:

$$\text{Area of } R = \lim_{n\to\infty} \sum_{k=1}^{n} \frac{\pi}{2n}\cdot\cos\frac{k\pi}{2n}$$

We can use the following BASIC program for approximating the area of R. Notice that the loop in lines 50–70 keeps a running sum of the areas of the n rectangles used to approximate the area of R.

```
10   PRINT "NUMBER OF RECTANGLES";
20   INPUT N
30   LET W = 3.14159 / (2 * N)
40   LET A = 0
50   FOR K = 1 TO N
60   LET A = A + W * COS(K * W)
70   NEXT K
80   PRINT "AREA IS APPROXIMATELY "; A
90   END
```

Running the program for increasing values of n produces this output:

```
NUMBER OF RECTANGLES?10
AREA IS APPROXIMATELY .919403344

NUMBER OF RECTANGLES?100
AREA IS APPROXIMATELY .992125474

NUMBER OF RECTANGLES?1000
AREA IS APPROXIMATELY .999214398
```

The limit of the sequence of approximations seems to be 1. Using calculus, one can in fact show that the area of R is exactly 1.

Cooperative Learning

Since some students will have experience with computer programming, it may be worthwhile to have such students explain programs to their classmates. For instance, the day before the lesson, ask for a student who would like to volunteer to teach a portion of the lesson. This gives the volunteer a chance to prepare a well-thought-out explanation of the program. On the day of the lesson, you can introduce the topic and explain the concepts involved and then let the "guest speaker" take over and explain the program.

If you plan to have students modify the programs in the text, it is clearly helpful to have them work in pairs. In this way students with limited or no programming experience can be teamed up with those who have some expertise. Allow them to experiment with different functions. Be sure to emphasize that they choose functions carefully. If they do get strange results (because the functions are discontinuous or have negative values on the given interval), you might have them investigate why and share their findings with the rest of the class.

Applications

Many problems in physics require the use of the definite integral (the formalized way of finding the area under a curve). For example, *work* is the product of the magnitude of a force applied to an object and the distance the object is moved by the force. If the force is not constant, but varies over the distance, we must divide the distance into small subintervals so that we can treat the force as being constant on each subinterval. The sum of the products of these constant forces and small distances gives an approximation of the work done in moving the object.

Assessment

If you want to evaluate whether students truly understand the process involved in approximating the area under a curve by using rectangles, ask them to perform the same procedure but with rectangles whose height is determined by the left-hand endpoint of each subinterval. In the case of the cosine curve, this will always yield an approximation slightly larger than the actual area. Note that the limits of the two areas found by using left- and right-hand endpoints must be the same (see Class Exercise 3).

Presenting the Section

In this section the concept of an infinite geometric series is generalized through the presentation of a few special power series. The intervals of convergence are also discussed. The section concludes with a presentation of the p-series and the special case where $p = 1$, the harmonic series.

You may want to spend some time discussing the fact that the harmonic series diverges (see Exercise 19). Since the nth term, $\frac{1}{n}$, approaches 0 as n approaches infinity, many students have difficulty accepting the fact that the series does not converge. Consider having students write a computer program which demonstrates that the sum of the series must get larger than any fixed number.

Using Technology

To show students how the power series of a function approximates the function, you might have students use a calculator and the power series for e^x to approximate e^2 as follows:

One term of the power series:
$$e^2 \approx 1$$

Two terms of the power series:
$$e^2 \approx 1 + \frac{2}{1!} = 3$$

Three terms of the power series:
$$e^2 \approx 1 + \frac{2}{1!} + \frac{2^2}{2!} = 5$$

Four terms of the power series:
$$e^2 \approx 1 + \frac{2}{1!} + \frac{2^2}{2!} + \frac{2^3}{3!} = 6.\overline{3}$$

Five terms of the power series:
$$e^2 \approx 1 + \frac{2}{1!} + \frac{2^2}{2!} + \frac{2^3}{3!} + \frac{2^4}{4!} = 7$$

Six terms of the power series:
$$e^2 \approx 1 + \frac{2}{1!} + \frac{2^2}{2!} + \frac{2^3}{3!} + \frac{2^4}{4!} + \frac{2^5}{5!} = 7.2\overline{6}$$

Point out that in the case of e^x, using a finite number of terms always gives an approximation that is smaller than the actual value.

You might also have students use this technique to see what happens when an x-value is chosen outside the interval of convergence for a power series (for example, using the power series for $\ln (1 + x)$ to approximate $\ln 3$).

Review

In discussing the power series for $\sin x$ and $\cos x$, you may want to remind students about even and odd functions (see Exercises 31–36 on page 137). You can then ask students how the power series for $\sin x$ and $\cos x$ support the fact that sine is an odd function and cosine is an even function. (You may also ask students to examine the power series for $\mathrm{Tan}^{-1} x$ to determine whether the inverse tangent function is an even or odd function.)

Presenting the Section

An understanding of iterated functions, as for many other advanced topics, is best gained by using a hands-on approach. It would be beneficial to have students iterate a function by using a calculator and by drawing the accompanying web diagram.

For this section and the next—as well as the project at the end of the chapter—you may wish to do some background reading. An excellent resource is Robert Devaney's *Chaos, Fractals, and Dynamics: Computer Experiments in Mathematics* (Menlo Park, CA: Addison Wesley, 1990).

Assessment

Students may not understand the purpose of the line $y = x$ in a web diagram. Consider asking them to explain why it is included. The verbalization of this concept may help their understanding.

Applications

There are many applications of iterated functions, both practical and esthetic. In the next section money and population growth are discussed. Any such recursive problem involves iteration.

Using Technology

A computer or graphing calculator can be programmed to perform iterations and display web diagrams. Given below is a program for the TI-81 graphing calculator. Before running it, go to the "Y = " menu and enter the function to be iterated as Y_1 and the function $y = x$ as Y_2; be sure that the other two functions, Y_3 and Y_4, are inactive.

```
Prgm:ITERATE
:ClrDraw
:Connected
:Disp "Xmin"
:Input F
:Disp "Xmax"
:Input G
:Disp "Ymin"
:Input H
:Disp "Ymax"
:Input I
:F → Xmin
:G → Xmax
:H → Ymin
:I → Ymax
:Disp "ITERATIONS = "
:Input N
:Disp "SEED = "
:Input S
:ClrHome
:DispGraph
:0 → Y
:S → X
:1 → M
:Lbl 1
:Pause
:Y₁ → Z
:Disp Z
:Pause
:Line(X,Y,X,Z)
:Z → Y
:Line(X,Y,Z,Y)
:Z → X
:M + 1 → M
:If M ≤ N
:Goto 1
:Pause
:DispHome
```

Presenting the Section

This section demonstrates the power of iterated-function techniques in solving realistic problems. Stress that even though finance problems and population problems may look different on the surface, they have the same mathematical structure. This is an extension of the concept of modeling from Chapter 18.

Applications

You may want to have students acquire data from local banks about interest rates and then create and solve money-growth problems of their own. They can also examine population data for some species in a given environment and construct their own logistic model of changes in population over time.

Cooperative Learning

While investigating the logistic function, it is helpful to have students develop conjectures on their own about the behavior of the function for various values of c. This is the intention of Written Exercises 5–10. Students are likely to learn more from these exercises when working in small groups than when working individually.

Review

Exponential growth and decay were discussed in Section 5-7. Before beginning this section, it may be a good idea to review the concepts developed there. Also, Section 14-5, on transition matrices and Markov chains, is closely connected to the population models illustrated here. It would be valuable to point out the connection to students.

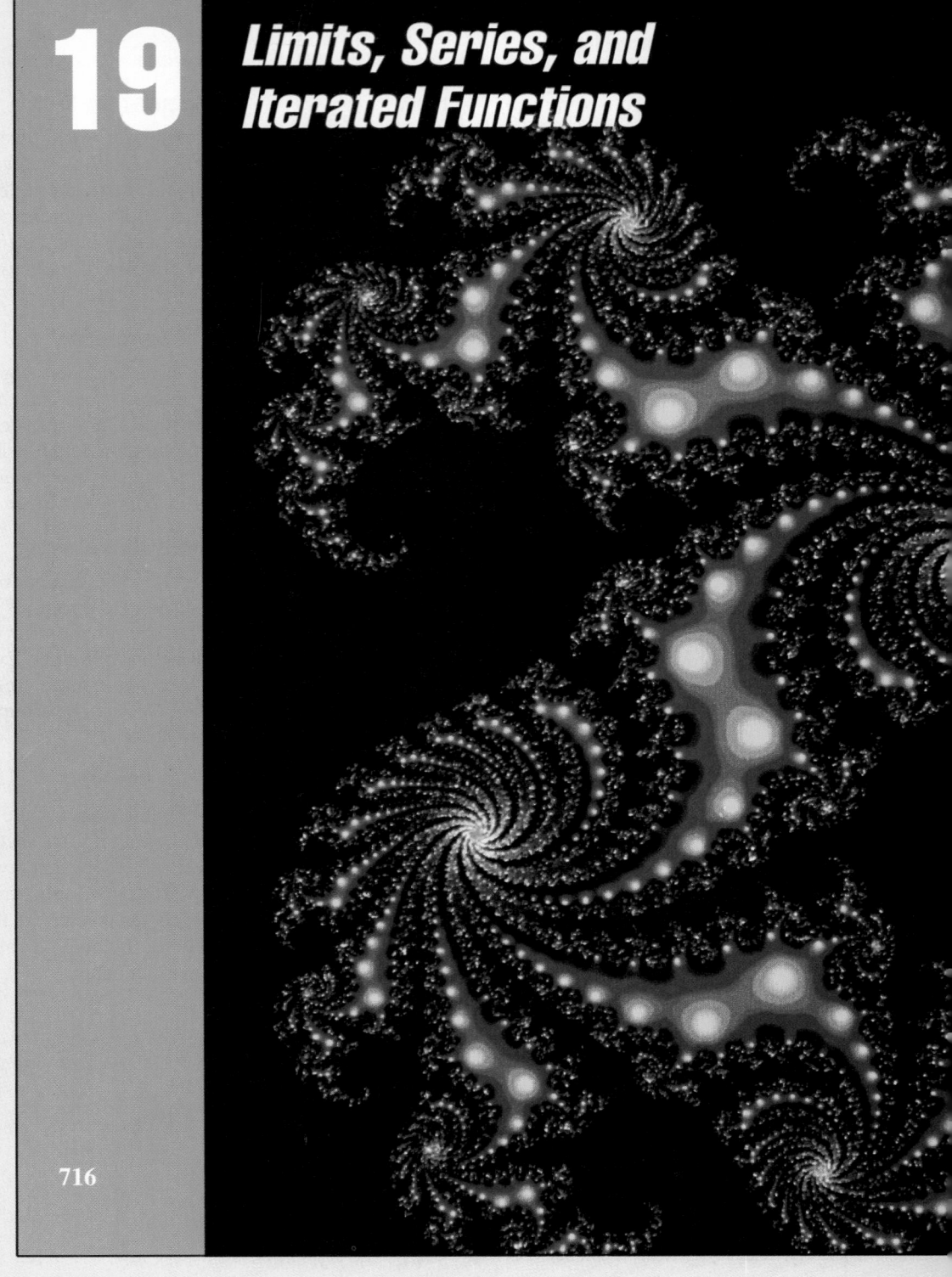

19 Limits, Series, and Iterated Functions

Limits

19-1 Limits of Functions

Objective | *To find the limit of a function or the quotient of two functions and to determine whether a function is continuous.*

The Limit as *x* Approaches ∞ or −∞

In Chapter 13 we discussed limits of infinite sequences and saw, for example, that the sequence

$$\frac{2}{1}, \frac{3}{2}, \frac{4}{3}, \ldots, \frac{n+1}{n}, \ldots$$

has limit 1. We expressed this fact by writing:

(1) $$\lim_{n \to \infty} \frac{n+1}{n} = 1$$

Likewise, for the function $f(x) = \frac{x+1}{x}$, we write

(2) $$\lim_{x \to \infty} \frac{x+1}{x} = 1$$

and say ''the limit of $\frac{x+1}{x}$ as x approaches infinity is 1.'' The only difference between statements (1) and (2) is that the sequence $t_n = \frac{n+1}{n}$ is defined only for positive integers n, whereas the function $f(x) = \frac{x+1}{x}$ is defined for all real numbers $x \neq 0$.

Because $f(x)$ is defined for negative as well as positive numbers, we can also talk about ''the limit of $f(x)$ as x approaches negative infinity.'' For the case in which $f(x) = \frac{x+1}{x}$, this limit is again 1, and we write:

$$\lim_{x \to -\infty} \frac{x+1}{x} = 1$$

These symbols mean that $\frac{x+1}{x}$ can be made as close to 1 as we like just by considering negative values of x with large enough absolute value. For example, when $x = -1,000,000$,

$$\frac{x+1}{x} = \frac{-999,999}{-1,000,000} \approx 1.$$

◀ The *Julia set* shown at the left is produced by iterating the complex polynomial function $f(z) = z^2 + 0.32 + 0.043i$. In a Julia set, as in all *fractals*, the larger patterns repeat at smaller levels of scale.

Limits, Series, and Iterated Functions   **717**

Teaching Notes, p. 716B

Warm-Up Exercises

In Exercises 1-3, evaluate the given limit or state that the limit does not exist.

1. $\lim_{n \to \infty} \frac{2n^3 - n^2}{3n^3 + 5n}$ $\frac{2}{3}$

2. $\lim_{n \to \infty} \frac{5n - 7}{8\sqrt{n} + 2}$ does not exist

3. $\lim_{n \to \infty} \cos\left(\frac{1}{n}\right)$ 1

4. Describe a numerical method for estimating the limit of an infinite sequence. Substitute very large values for n in the formula for the nth term and use a calculator to evaluate the expression.

Motivating the Section

The introduction to limits presented in this section will be most helpful to students who plan to take calculus, since the concept of limit is the cornerstone of the study of calculus. The ideas of this section are also essential to success in understanding iterated functions, which students will see applied later in the chapter to the dynamics of money growth and population growth.

Review Note

You may find that it is beneficial to review Section 13-4, on limits of infinite sequences, before discussing limits of functions with real-number domains.

Mathematical Note

Remind students that ∞ and $-\infty$ are not numbers. To convince students of this fact, have them assume that ∞ is a number and write the obviously true statement $\infty + \infty = \infty$; then show students that this statement implies $\infty = 0$, which is obviously false.

Example Note

When discussing Example 2, point out that $\lim\limits_{x \to \infty} (x \sin x)$ does not exist, because as x becomes arbitrarily large, $x \sin x$ neither approaches a finite number nor becomes infinitely positive or infinitely negative.

Ⓝ **Using Technology**

When discussing Example 2, you may wish to have students use graphing calculators to obtain the graph of $y = x \sin x$ in order to observe the behavior of the graph for large values of x.

A function f for which $|f(x)|$ becomes arbitrarily large as x approaches ∞ or $-\infty$ has no finite limit. It instead has an infinite limit, either ∞ or $-\infty$. Although we say that an infinite limit exists, we recognize that ∞ and $-\infty$ are not numbers, and so we must define finite and infinite limits differently. Formal definitions of these limits are given on pages 722 and 723. The following examples, however, should help to clarify the concept of an infinite limit.

Example 1   Evaluate: **a.** $\lim\limits_{x \to \infty} x^{1/3}$ **b.** $\lim\limits_{x \to -\infty} x^{1/3}$

Solution   **a.** Since the value of $x^{1/3}$ becomes arbitrarily large as x becomes arbitrarily large, $\lim\limits_{x \to \infty} x^{1/3} = \infty$.

b. When x is negative, the value of $x^{1/3}$ is negative. As x approaches negative infinity, $|x^{1/3}|$ becomes arbitrarily large. Thus, $\lim\limits_{x \to -\infty} x^{1/3} = -\infty$.

Example 2   Explain why $\lim\limits_{x \to \infty} x \sin x \neq \infty$.

Solution   Since the value of the sine function oscillates between 1 and -1, the graph of $y = x \sin x$ oscillates between the lines $y = x$ and $y = -x$ (see below). In particular, as x approaches infinity, the graph of $y = x \sin x$ crosses the x-axis infinitely often, so there is no value of x beyond which the value of $x \sin x$ becomes—and *stays*—arbitrarily large. Thus, $\lim\limits_{x \to \infty} x \sin x \neq \infty$.

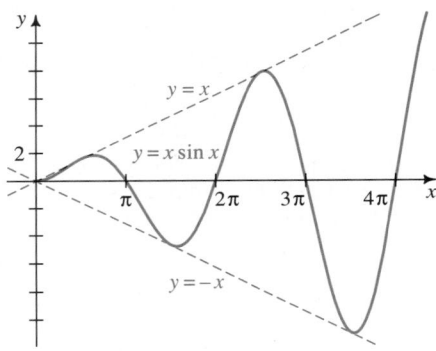

The Limit as x Approaches a Real Number c

To determine the behavior of a function $f(x)$ as x approaches a real number c, we consider the following two limits:

(1) $\lim\limits_{x \to c^+} f(x)$, read ''the limit of $f(x)$ as x approaches c from the right''

(2) $\lim\limits_{x \to c^-} f(x)$, read ''the limit of $f(x)$ as x approaches c from the left''

The meaning of limits of this type is discussed in the next two examples.

718   *Chapter Nineteen*

Example 3   Using the graph of $f(x)$ shown below, find $\lim\limits_{x \to 2^+} f(x)$ and $\lim\limits_{x \to 2^-} f(x)$.

Solution   The fact that $f(2) = 3$ has nothing to do with the solution. To evaluate $\lim\limits_{x \to 2^+} f(x)$, we consider the value of $f(x)$ for x near, but greater than, 2. Since the value of $f(x)$ gets closer and closer to -2 as x approaches 2 from the right,

$$\lim\limits_{x \to 2^+} f(x) = -2.$$

Also, since the value of $f(x)$ gets closer and closer to 5 as x approaches 2 from the left,

$$\lim\limits_{x \to 2^-} f(x) = 5.$$

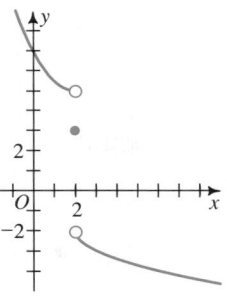

Example 4   If $f(x) = \dfrac{x^2 - 4}{x - 2}$, describe the behavior of $f(x)$ near $x = 2$.

Solution   The fact that $f(2)$ is undefined has nothing to do with the solution. We wish to determine what happens to $f(x)$ as x gets closer to 2. Substituting values of x near 2 gives the values of $f(x)$ shown below. These values suggest that $f(x)$ approaches 4 as x approaches 2 from the right or the left.

$f(2.1) = 4.1$	$f(1.9) = 3.9$
$f(2.01) = 4.01$	$f(1.99) = 3.99$
$f(2.001) = 4.001$	$f(1.999) = 3.999$
$\downarrow$	$\downarrow$
$\lim\limits_{x \to 2^+} f(x) = 4$	$\lim\limits_{x \to 2^-} f(x) = 4$

We can confirm these limits by examining the graph of $f(x)$ near the point where $x = 2$. The graph is easier to sketch if we first reduce $f(x)$:

$$f(x) = \frac{x^2 - 4}{x - 2} = \frac{(x + 2)(x - 2)}{x - 2} = x + 2$$

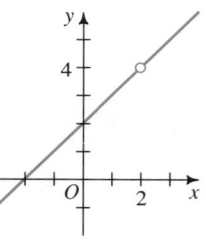

The graph of $f(x)$ is a line; however, since $x = 2$ is not in the domain of f, the graph has a "hole" at the point $(2, 4)$. Be aware that a graphing calculator display might not show the hole.

Notice in Example 4 that $\lim\limits_{x \to 2^+} f(x) = \lim\limits_{x \to 2^-} f(x) = 4$. In this case we can speak of "*the* limit of $f(x)$ as x approaches 2" and write $\lim\limits_{x \to 2} f(x) = 4$. In Example 3, however, $\lim\limits_{x \to 2^+} f(x) = -2$ and $\lim\limits_{x \to 2^-} f(x) = 5$. Since the right-hand and left-hand limits are different, $\lim\limits_{x \to 2} f(x)$ does not exist. In general:

$\lim\limits_{x \to c} f(x)$ exists if and only if $\lim\limits_{x \to c^+} f(x)$ and $\lim\limits_{x \to c^-} f(x)$ exist and agree.

Limits, Series, and Iterated Functions   **719**

Example Note

When discussing Example 3, emphasize that $f(2) = 3$ has nothing to do with the solution. In fact, the solution is the same no matter what the value of $f(2)$ is (or even whether $f(2)$ is defined).

◩ **Using Technology**

Example 4 provides a good opportunity to emphasize that important features of a graph (such as the hole in the graph of $f(x) = \dfrac{x^2 - 4}{x - 2}$) may not show up on a graph displayed on a graphing calculator. Students must use their own mathematical knowledge to refine graphs generated by technology.

Continuous Functions

The easiest limits to evaluate often involve a *continuous function*. Roughly speaking, a function is continuous if you can draw its graph without lifting your pencil from the paper. For example, any polynomial function, such as the cubic function shown at the right, is continuous.

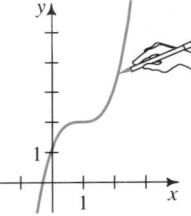

Formally speaking, a function $f(x)$ is **continuous** at a real number c if:

$$\lim\limits_{x \to c} f(x) = f(c)$$

In this definition, note that there are *three* conditions for continuity at $x = c$:

1. $\lim\limits_{x \to c} f(x)$ must exist.
2. $f(c)$ must exist.
3. (1) and (2) must be equal.

A function can fail to be continuous at $x = c$ in different ways. The function of Example 3 on page 719 is discontinuous at $x = 2$ because $\lim\limits_{x \to 2} f(x)$ does not exist. The function of Example 4 on page 719 is discontinuous at $x = 2$ because $f(2)$ does not exist. The function graphed at the right is discontinuous at $x = 2$ because, although both $\lim\limits_{x \to 2} f(x)$ and $f(2)$ exist, $\lim\limits_{x \to 2} f(x) \neq f(2)$.

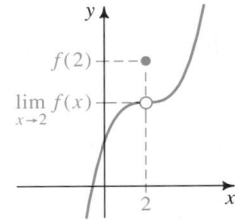

The Limit of a Quotient of Two Functions

The following theorem is useful for limit problems that involve a quotient of two functions. (A proof is difficult and will not be given.) Since the theorem applies whether we consider the limit as x approaches ∞, $-\infty$, or a real number c, the theorem uses the general notation "lim" instead of $\lim\limits_{x \to \infty}$, $\lim\limits_{x \to -\infty}$, or $\lim\limits_{x \to c}$.

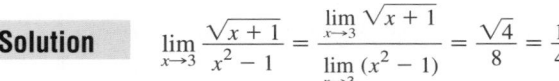

Quotient Theorem for Limits

If $\lim f(x)$ and $\lim g(x)$ both exist, and $\lim g(x) \neq 0$, then:

$$\lim \frac{f(x)}{g(x)} = \frac{\lim f(x)}{\lim g(x)}$$

Example 5   Evaluate $\lim\limits_{x \to 3} \dfrac{\sqrt{x+1}}{x^2 - 1}.$

Solution $\lim\limits_{x \to 3} \dfrac{\sqrt{x+1}}{x^2 - 1} = \dfrac{\lim\limits_{x \to 3} \sqrt{x+1}}{\lim\limits_{x \to 3} (x^2 - 1)} = \dfrac{\sqrt{4}}{8} = \dfrac{1}{4}$

720   *Chapter Nineteen*

Evaluating $\lim \dfrac{f(x)}{g(x)}$ is not always as simple as Example 5 suggests. Sometimes we need to use special techniques, described below.

Techniques for Evaluating $\lim \dfrac{f(x)}{g(x)}$

1. If possible, use the quotient theorem for limits.

2. If $\lim f(x) = 0$ and $\lim g(x) = 0$, try the following techniques.

   a. Factor $g(x)$ and $f(x)$ and reduce $\dfrac{f(x)}{g(x)}$ to lowest terms.

   b. If $f(x)$ or $g(x)$ involves a square root, try multiplying both $f(x)$ and $g(x)$ by the conjugate of the square root expression.

3. If $\lim f(x) \neq 0$ and $\lim g(x) = 0$, then either statement (a) or (b) below is true.

   a. $\lim \dfrac{f(x)}{g(x)}$ does not exist.

   b. $\lim \dfrac{f(x)}{g(x)} = \infty$ or $\lim \dfrac{f(x)}{g(x)} = -\infty$.

4. If x is approaching infinity or negative infinity, divide the numerator and the denominator by the highest power of x in the denominator.

5. If all else fails, you can guess $\lim\limits_{x \to \infty} \dfrac{f(x)}{g(x)}$ by evaluating $\dfrac{f(x)}{g(x)}$ for very large values of x, and you can guess $\lim\limits_{x \to c} \dfrac{f(x)}{g(x)}$ by evaluating $\dfrac{f(x)}{g(x)}$ for x-values very near $x = c$. These limits can also be guessed by using a graphing calculator to examine the graph of $y = \dfrac{f(x)}{g(x)}$ for very large values of x, or for x-values very near $x = c$. (Remember: A graphing calculator might not show points of discontinuity.)

Example 6 Evaluate: **a.** $\lim\limits_{x \to 1} \dfrac{x^2 + 2x - 3}{x^2 - 1}$ **b.** $\lim\limits_{x \to 0} \dfrac{1 - \sqrt{1 + x}}{x}$

Solution

a. Using technique (2a), we have:
$$\lim_{x \to 1} \frac{x^2 + 2x - 3}{x^2 - 1} = \lim_{x \to 1} \frac{(x - 1)(x + 3)}{(x - 1)(x + 1)} = \lim_{x \to 1} \frac{x + 3}{x + 1} = \frac{4}{2} = 2$$

b. Using technique (2b), we have:
$$\lim_{x \to 0} \frac{1 - \sqrt{1 + x}}{x} = \lim_{x \to 0} \frac{1 - \sqrt{1 + x}}{x} \cdot \frac{1 + \sqrt{1 + x}}{1 + \sqrt{1 + x}}$$
$$= \lim_{x \to 0} \frac{-x}{x(1 + \sqrt{1 + x})} = \lim_{x \to 0} \frac{-1}{1 + \sqrt{1 + x}} = -\frac{1}{2}$$

Example Note

In part (a) of Example 6, some students may feel that the limit should be 0 because the numerator is 0 upon substitution of 1. Remind students that $\dfrac{0}{0}$ is undefined.

◩ Using Technology

Many students have difficulty with the limit concept. Using a graphing calculator or computer as suggested in part (5) of the chart is an excellent way for students to gain an intuitive feel for limits.

Have students confirm the limits found in Example 6 by using a graphing calculator to graph the functions. As in Example 4 on page 719, each of these graphs has a hole in it—emphasize again that a calculator's display may not reveal these points of discontinuity.

Limits, Series, and Iterated Functions   **721**

In the formal definition of $\lim\limits_{x \to c} f(x) = L$, the inequality $|f(x) - L| < \epsilon$ implies $L - \epsilon < f(x) < L + \epsilon$, and the inequality $0 < |x - c| < \delta$ implies $c - \delta < x < c + \delta$ where $x \neq c$.

Communication Note

The formal definitions of limit given on the next page can be skipped. If the formal definitions are to be covered, consider stating them in less formal terms to help students understand them.

For example, the definition of $\lim\limits_{x \to c} f(x) = L$ basically says that for any interval (no matter how small) that is centered on L (on the y-axis), there can be found an interval that is centered on c (on the x-axis) such that any x-value in the horizontal interval (except possibly $x = c$) has an $f(x)$-value in the corresponding vertical interval.

Thus, the definition can be thought of as a "game" between a person trying to show that $\lim\limits_{x \to c} f(x) = L$ and some skeptic:

First, the skeptic declares how close the $f(x)$-values must be to L, and then the person trying to show that the limit is L must find an interval centered on c such that all x-values in the interval (except possibly $x = c$) have $f(x)$-values that are at least as close to L as the skeptic wants.

 The techniques for evaluating limits described in part (5) of the chart on page 721 are also useful for *confirming* limits found using the algebraic methods described in parts (1)–(4). In the examples that follow, a graphing calculator is used both to help evaluate a limit (Example 7(b)), and to confirm a limit found algebraically (Example 8).

Example 7   Evaluate: **a.** $\lim\limits_{x \to 1} \dfrac{1}{(x - 1)^2}$ **b.** $\lim\limits_{x \to 1} \dfrac{1}{x - 1}$

Solution   In each case we recognize that technique (3) applies. In addition, we can use the methods suggested in technique (5) to help evaluate these limits.

a. As x approaches 1, the values of $(x - 1)^2$ are positive and approach 0 so that the reciprocals of these values approach infinity. For example:

When $x = 0.999$, we have $\dfrac{1}{(x - 1)^2} = \dfrac{1}{(-0.001)^2} = 1{,}000{,}000$.

When $x = 1.001$, we have $\dfrac{1}{(x - 1)^2} = \dfrac{1}{(0.001)^2} = 1{,}000{,}000$.

Thus, $\qquad\qquad \lim\limits_{x \to 1} \dfrac{1}{(x - 1)^2} = \infty$.

b. A graph of the equation $y = \dfrac{1}{x - 1}$ shows that $\lim\limits_{x \to 1} \dfrac{1}{x - 1}$ does not exist because:

$$\lim\limits_{x \to 1^+} \dfrac{1}{x - 1} = \infty \quad \text{and} \quad \lim\limits_{x \to 1^-} \dfrac{1}{x - 1} = -\infty$$

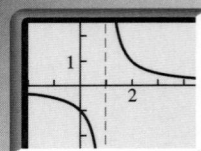

Example 8   Evaluate $\lim\limits_{x \to \infty} \dfrac{x^3 - 4x}{2x^4 + 5}$.

Solution   Using technique (4), we divide numerator and denominator by x^4 and get:

$$\lim\limits_{x \to \infty} \dfrac{x^3 - 4x}{2x^4 + 5} = \lim\limits_{x \to \infty} \dfrac{\dfrac{1}{x} - \dfrac{4}{x^3}}{2 + \dfrac{5}{x^4}} = \dfrac{0 - 0}{2 + 0} = 0$$

A graph of $y = \dfrac{x^3 - 4x}{2x^4 + 5}$ confirms this limit.

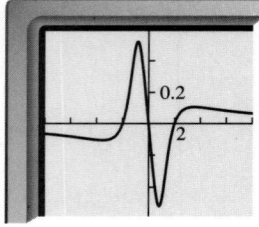

Formal Definitions of Limit

Although the idea of a limit is fairly easy to understand on an intuitive level, it is quite difficult to express formally. Therefore, read carefully the three formal definitions on page 723. The first two, in which the limit L is a real number, define finite limits. The last one defines an infinite limit.

1. $\lim\limits_{x \to c} f(x) = L$ means that for any small positive number ϵ (Greek epsilon), there is a positive number δ (Greek delta) such that

$$|f(x) - L| < \epsilon$$

whenever x is in the domain of f and $0 < |x - c| < \delta$.

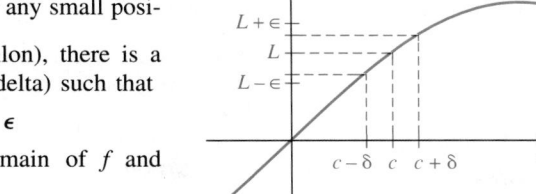

2. $\lim\limits_{x \to \infty} f(x) = L$ means that for any small positive number ϵ, there is a value of x, call it x_1, such that

$$|f(x) - L| < \epsilon$$

whenever x is in the domain of f and $x > x_1$. ($\lim\limits_{x \to -\infty} f(x) = L$ is similarly defined.)

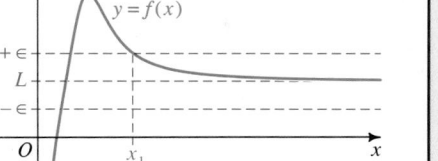

3. $\lim\limits_{x \to \infty} f(x) = \infty$ means that for any large positive number M, there is a value of x, call it x_1, such that

$$f(x) > M$$

whenever x is in the domain of f and $x > x_1$. (Other statements, such as $\lim\limits_{x \to -\infty} f(x) = \infty$, are similarly defined.)

CLASS EXERCISES

1. Read each equation aloud and explain what it means. **4. b.** $\lim\limits_{x \to 0^-} f(x) \neq \lim\limits_{x \to 0^+} f(x)$

   a. $\lim\limits_{x \to \infty} \dfrac{2x^2}{x^2 + 1} = 2$ **b.** $\lim\limits_{x \to -\infty} \dfrac{3x + 1}{2x - 5} = \dfrac{3}{2}$ **c.** $\lim\limits_{x \to 2^+} f(x) = 3$ **d.** $\lim\limits_{x \to 2^-} g(x) = \infty$

In Exercises 2–4, refer to the graph of $y = f(x)$ shown at the right below.

2. Evaluate $\lim\limits_{x \to \infty} f(x)$, $\lim\limits_{x \to -\infty} f(x)$, $\lim\limits_{x \to 2} f(x)$, and $f(2)$. 0, -1, 4, 1

3. Explain why $f(x)$ is discontinuous at $x = 2$. $\lim\limits_{x \to 2} f(x) \neq f(2)$

4. **a.** Evaluate $\lim\limits_{x \to 0^+} f(x)$ and $\lim\limits_{x \to 0^-} f(x)$. 0, -3

   b. Explain why $\lim\limits_{x \to 0} f(x)$ does not exist. See above.

5. **Reading** Express the quotient theorem for limits in words. The limit of a quotient of two functions equals the quotient of the limits of the functions.

6. Evaluate each limit.

   a. $\lim\limits_{x \to \infty} \dfrac{5x^2 - 3x + 1}{7x^2 + 9}$ $\dfrac{5}{7}$ **b.** $\lim\limits_{x \to 1^+} \dfrac{1}{x - 1}$ ∞ **c.** $\lim\limits_{x \to 1^-} \dfrac{1}{x - 1}$ $-\infty$ **d.** $\lim\limits_{x \to 4} \dfrac{x^2 - 16}{x - 4}$ 8

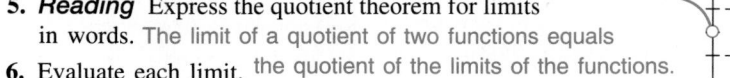

Exercise Note

Remind students that even though the function in Class Exercise 6d has a limit of 8 as x approaches 4, the function is not defined at $x = 4$. Students often overlook the fact that $f(x)$ does not need to be defined at $x = c$ in order for $\lim\limits_{x \to c} f(x)$ to exist.

Assessment Note

Upon completion of the Class Exercises, have students prepare a summary of the section by describing methods for determining limits and how the concept of limit relates to that of continuity.

Additional Answers
Class Exercises

1. **a.** For large values of x, the value of $\dfrac{2x^2}{x^2 + 1}$ is close to 2.

   b. For negative values of x where $|x|$ is large, the value of $\dfrac{3x + 1}{2x - 5}$ is close to $\dfrac{3}{2}$.

   c. For values of x near 2, but greater than 2, the value of $f(x)$ is close to 3.

   d. For values of x near 2, but less than 2, the value of $g(x)$ is arbitrarily large.

Review Note

Some students continue to have trouble with the greatest integer function (see Exercise 22). Consider reviewing the function with the entire class rather than having students review it on their own.

Exercise Note

Students who are confused by part (c) of Exercises 19–22 should be directed to the last paragraph on page 719.

Making Connections

As a prelude to Section 19-2, you may want to discuss horizontal and vertical asymptotes. Point out that a horizontal asymptote indicates the behavior of a function's graph as $|x|$ becomes arbitrarily large. This discussion can arise out of Exercises 1–22 if students are using technology to check their answers, or out of Exercises 23 and 24.

Suggested Assignments

Discrete Math
Day 1: 724/1–27 odd
Day 2: 725/29–47 odd

Supplementary Materials

Alternative Assessment, 57–58

WRITTEN EXERCISES

DNE = does not exist

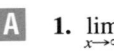 **Evaluate the given limit or state that it does not exist. You may wish to use a computer or a graphing calculator to confirm your answers.**

A 1. $\lim\limits_{x\to\infty} \dfrac{3x - 5}{4x + 9}$ $\dfrac{3}{4}$

2. $\lim\limits_{x\to\infty} \dfrac{2x^2 - 7x}{3x^2 + 5}$ $\dfrac{2}{3}$

3. $\lim\limits_{x\to-\infty} \dfrac{8x^2 - 7x + 5}{4x^2 + 9}$ 2

4. $\lim\limits_{x\to-\infty} \dfrac{5x^3}{7x^3 + 8x^2}$ $\dfrac{5}{7}$

5. $\lim\limits_{x\to\infty} \dfrac{(x^2 + 1)(x^2 - 1)}{2x^4}$ $\dfrac{1}{2}$

6. $\lim\limits_{x\to\infty} \dfrac{x^2 \cos\frac{1}{x}}{2x^2 - 1}$ $\dfrac{1}{2}$

7. $\lim\limits_{x\to1} \dfrac{x^2 - 1}{x - 1}$ 2

8. $\lim\limits_{x\to3} \dfrac{2x^2 - 6x}{x - 3}$ 6

9. $\lim\limits_{x\to2} \dfrac{x^2 + x - 6}{2x - 4}$ $\dfrac{5}{2}$

10. $\lim\limits_{x\to4} \dfrac{x - 4}{x^2 - x - 12}$ $\dfrac{1}{7}$

11. $\lim\limits_{x\to0^+} \dfrac{x + 1}{x}$ ∞

12. $\lim\limits_{x\to-1^+} \dfrac{x}{x + 1}$ $-\infty$

13. $\lim\limits_{x\to3^+} \dfrac{x - 4}{x - 3}$ $-\infty$

14. $\lim\limits_{x\to3^-} \dfrac{2x - 6}{x^2 - 3x}$ $\dfrac{2}{3}$

15. $\lim\limits_{x\to-2^-} \dfrac{x^2 + 4x + 4}{x^2 + 3x + 1}$ 0

16. $\lim\limits_{x\to-2^+} \dfrac{3x + 6}{2x + 4}$ $\dfrac{3}{2}$

17. $\lim\limits_{x\to0} \dfrac{x}{1 - \sqrt{1 - x}}$ 2

18. $\lim\limits_{x\to0} \dfrac{2 - \sqrt{4 - x}}{x}$ $\dfrac{1}{4}$

19. a. $\lim\limits_{x\to0^+} \dfrac{|x|}{x}$ 1

b. $\lim\limits_{x\to0^-} \dfrac{|x|}{x}$ -1

c. $\lim\limits_{x\to0} \dfrac{|x|}{x}$ DNE

20. a. $\lim\limits_{x\to1^+} \dfrac{x - 3}{x^2 - 1}$ $-\infty$

b. $\lim\limits_{x\to1^-} \dfrac{x - 3}{x^2 - 1}$ ∞

c. $\lim\limits_{x\to1} \dfrac{x - 3}{x^2 - 1}$ DNE

21. a. $\lim\limits_{x\to2^+} \dfrac{x - 2}{\sqrt{x^2 - 4}}$ 0

b. $\lim\limits_{x\to2^-} \dfrac{x - 2}{\sqrt{x^2 - 4}}$ DNE

c. $\lim\limits_{x\to2} \dfrac{x - 2}{\sqrt{x^2 - 4}}$ DNE

22. a. $\lim\limits_{x\to3^+} \lfloor x \rfloor$ 3

b. $\lim\limits_{x\to3^-} \lfloor x \rfloor$ 2

c. $\lim\limits_{x\to3} \lfloor x \rfloor$ DNE

(*Note:* $\lfloor x \rfloor$ is the greatest integer function. See Exercise 21 on page 124.)

Visual Thinking Use the given function's graph to evaluate each limit.

a. $\lim\limits_{x\to\infty} f(x)$ b. $\lim\limits_{x\to-\infty} f(x)$ c. $\lim\limits_{x\to2^+} f(x)$ d. $\lim\limits_{x\to2^-} f(x)$

B 23.

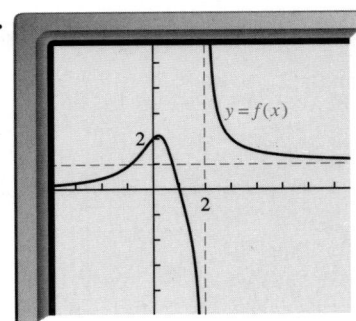

a. 1 b. 0 c. ∞ d. $-\infty$

24.

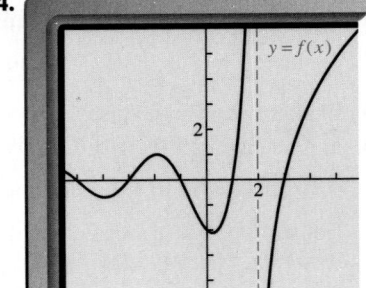

a. ∞ b. 0 c. $-\infty$ d. ∞

In Exercises 25–30, evaluate the given limit.

25. $\lim\limits_{x \to 0} \dfrac{1 - \sqrt{x^2 + 1}}{x^2} \quad -\dfrac{1}{2}$

26. $\lim\limits_{x \to -\infty} (\sqrt{4x^2 - 4x} - 2x) \quad \infty$

27. $\lim\limits_{x \to \infty} (\sqrt{x + 1} - \sqrt{x}) \quad 0$

28. $\lim\limits_{x \to \infty} (\sqrt{x^2 + 2x} - x) \quad 1$

29. a. $\lim\limits_{h \to 0} \dfrac{(1 + h)^2 - 1}{h} \quad 2$
   b. $\lim\limits_{h \to 0} \dfrac{(2 + h)^2 - 4}{h} \quad 4$
   c. $\lim\limits_{h \to 0} \dfrac{(x + h)^2 - x^2}{h} \quad 2x$

30. a. $\lim\limits_{h \to 0} \dfrac{(1 + h)^3 - 1}{h} \quad 3$
   b. $\lim\limits_{h \to 0} \dfrac{(2 + h)^3 - 8}{h} \quad 12$
   c. $\lim\limits_{h \to 0} \dfrac{(x + h)^3 - x^3}{h} \quad 3x^2$

Determine whether each function is continuous. If it is discontinuous, state where any discontinuities occur.

31. $f(x) = \begin{cases} 2 - x^2 & \text{if } x \le 1 \\ x & \text{if } x > 1 \end{cases}$ Cont.

32. $f(x) = \begin{cases} x^2 + 1 & \text{if } x < 0 \\ x^2 & \text{if } x \ge 0 \end{cases}$ Discont. at $x = 0$

33. $f(x) = \begin{cases} \dfrac{x^2 - 4}{x - 2} & \text{if } x \ne 2 \\ 4 & \text{if } x = 2 \end{cases}$ Cont.

34. $f(x) = \dfrac{x - 1}{1 - x}$ Discont. at $x = 1$

Determine values for a and b so that each function is continuous.

35. $f(x) = \begin{cases} x^2 & \text{if } x \le -1 \\ ax + b & \text{if } -1 < x < 1 \\ -x^2 & \text{if } x \ge 1 \end{cases}$ $a = -1$, $b = 0$

36. $f(x) = \begin{cases} 1 - 2x & \text{if } x \le -2 \\ ax + b & \text{if } -2 < x < 1 \\ 3x + 2 & \text{if } x \ge 1 \end{cases}$ $a = 0$, $b = 5$

Evaluate each limit or state that it does not exist.

37. $\lim\limits_{x \to \infty} (43{,}987)^{1/x} \quad 1$

38. $\lim\limits_{x \to 0} 2^{1/x} \quad$ DNE

39. $\lim\limits_{x \to 0} \dfrac{3^x - 3^{-x}}{3^x + 3^{-x}} \quad 0$

40. $\lim\limits_{x \to -\infty} \dfrac{3^x - 3^{-x}}{3^x + 3^{-x}} \quad -1$

41. $\lim\limits_{x \to 0^+} \sin \dfrac{1}{x} \quad$ DNE

42. $\lim\limits_{x \to \infty} \sin \dfrac{1}{x} \quad 0$

43. a. In the diagram at the right, explain why arc PQ has length θ, if θ is in radians.
   b. **Visual Thinking** Use the diagram to explain why

$$\lim\limits_{\theta \to 0} \dfrac{\sin \theta}{\theta} = 1.$$

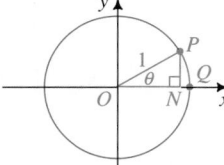

Evaluate each limit. The result of part (b) of Exercise 43 may be helpful.

44. $\lim\limits_{\theta \to 0} \dfrac{\tan \theta}{\theta} \quad 1$

45. $\lim\limits_{\theta \to 0} \dfrac{\sin 2\theta}{\theta} \quad 2$

46. $\lim\limits_{t \to 0} \dfrac{1 - \cos t}{t^2} \quad \dfrac{1}{2}$

$\left(\textit{Hint}: \text{Multiply by } \dfrac{1 + \cos t}{1 + \cos t}.\right)$

47. $\lim\limits_{x \to \infty} x \sin \dfrac{1}{x} \quad 1$

$\left(\textit{Hint}: \text{Let } t = \dfrac{1}{x}. \text{ As } x \to \infty, t \to 0.\right)$

Limits, Series, and Iterated Functions **725**

Using Technology

After students have determined the behavior of the functions in Exercises 31–34, have them try using graphing calculators or computers to obtain the graphs of the functions. If a piecewise function like the one in Exercise 31 can be graphed at all, its rule probably needs to be entered in the form $(2 - x^2)(x \le 1) + x(x > 1)$. The calculator or computer gives each inequality a value of 1 or 0 depending on whether the current value of x does or does not satisfy the inequality. For example, when finding the value of $(2 - x^2)(x \le 1) + x(x > 1)$ for $x = -3$, the calculator or computer produces $(2 - (-3)^2)(1) + (-3)(0) = -7$; for $x = 3$, the calculator or computer produces $(2 - 3^2)(0) + (3)(1) = 3$.

Additional Answers Written Exercises

43. a. The length of $\overgroup{PQ}$ is $r\theta$, which equals θ when $r = 1$.

   b. $\lim\limits_{\theta \to 0} \dfrac{\sin \theta}{\theta} =$

   $\lim\limits_{\theta \to 0} \dfrac{PN}{\text{length of } \overgroup{PQ}} = 1,$

   since as $\theta \to 0$, the length of $\overgroup{PQ}$ approaches PN.

For each of the following functions, name the x- and y-intercepts and all values for which the function is undefined. Then tell whether the graph of the function is symmetric about the x-axis, about the y-axis, or neither.

1. $y = \dfrac{x}{x-1}$ x-int.: 0; y-int.: 0; $x = 1$; neither

2. $y = \dfrac{1}{(x+1)^2}$ no x-int.; y-int.: 1; $x = -1$; neither

3. $y = \dfrac{4}{x^2+1}$ no x-int.; y-int.: 4; none; symm. about y-axis

4. $y = \dfrac{x+3}{x^2-4}$ x-int.: -3; y-int.: $-\dfrac{3}{4}$; $x = \pm 2$; neither

Motivating the Section

This section presents a useful application of the knowledge of limits gained in Section 19-1. Here students will see why graphs of rational functions look the way they do and how the graphs can be sketched quickly.

19-2 Graphs of Rational Functions

Objective   *To sketch the graph of a rational function.*

A **rational function** has the form $y = \dfrac{f(x)}{g(x)}$ where $f(x)$ and $g(x)$ are polynomials and $g(x) \neq 0$. In the following Activity, you will use a computer or a graphing calculator to learn about several features of the graph of a rational function.

Activity

c. To find the x-intercepts, solve $x^2 - 9 = 0$. To find the the y-intercept, evaluate $y = \dfrac{x^2-9}{x^2-4}$ when $x = 0$.

 a. Use a computer or a graphing calculator to graph $y = \dfrac{x^2-9}{x^2-4}$. $x = \pm 3$, $y = 2.25$

b. Use the TRACE feature to find the x- and y-intercepts of the graph.

c. How can you find the graph's x- and y-intercepts algebraically? See above.

d. Give equations of all lines that appear to be asymptotes of the graph. $x = \pm 2$, $y = 1$

e. How can you find the equations of the asymptotes algebraically?

e. To find the equations of the vertical asymptotes. solve $x^2 - 4 = 0$. To find the equation of the horizontal asymptote, evaluate $y = \lim_{x \to \pm\infty} \dfrac{x^2-9}{x^2-4}$

Procedure for Graphing the Rational Function $y = \dfrac{f(x)}{g(x)}$

1. Locate the zeros of $f(x)$. A zero of $f(x)$ will be an x-intercept of the graph of $y = \dfrac{f(x)}{g(x)}$ unless it is also a zero of $g(x)$.

2. Locate the zeros of $g(x)$. These values, of course, are not in the domain of $y = \dfrac{f(x)}{g(x)}$. Usually, vertical lines through these values are asymptotes. If $f(x)$ and $g(x)$ have a zero in common, however, reduce $\dfrac{f(x)}{g(x)}$ to lowest terms before determining vertical asymptotes.

3. Using the techniques of Section 2-3, perform a sign analysis of $\dfrac{f(x)}{g(x)}$, which should be in lowest terms. (Essentially, the sign changes at each zero of $f(x)$ or $g(x)$ unless the zero is a double zero, in which case the sign remains the same.)

4. If $x = c$ is a vertical asymptote, then the graph will approach infinity as it nears the asymptote in a region where the values of the function are positive. The graph will approach negative infinity as it nears the asymptote in a region where the values of the function are negative.

5. You can find nonvertical asymptotes by considering $\lim_{x \to \infty} \dfrac{f(x)}{g(x)}$ and $\lim_{x \to -\infty} \dfrac{f(x)}{g(x)}$. If $\lim_{x \to \pm\infty} \dfrac{f(x)}{g(x)} = c$, then $y = c$ is a horizontal asymptote.

726   *Chapter Nineteen*

Example 1   Sketch the graph of $y = \dfrac{x-3}{x^2-1}$.

Solution   We use the procedure described on the preceding page, as follows.

Step 1   The numerator $x - 3$ has a zero at $x = 3$, which is not a zero of the denominator. Hence 3 is an x-intercept of the graph.

Step 2   Since $x^2 - 1 = (x + 1)(x - 1)$, the denominator has zeros at $x = -1$ and $x = 1$. Hence the graph has vertical asymptotes $x = -1$ and $x = 1$.

Step 3   A sign analysis of $\dfrac{x-3}{(x+1)(x-1)}$ shows that the values of the function are negative for $x < -1$ and $1 < x < 3$, and that they are positive for $-1 < x < 1$ and $x > 3$.

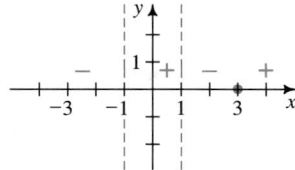

Step 4   Since the values of the function are negative in the region immediately to the right of the asymptote $x = 1$, the graph approaches negative infinity as it nears $x = 1$ from the right. Since the values of the function are positive in the region to the left of the asymptote, the graph approaches infinity as it nears $x = 1$ from the left. Similar reasoning shows that the graph approaches infinity as it nears $x = -1$ from the right, and that the graph approaches negative infinity as it nears $x = -1$ from the left.

Step 5   Since
$$\lim_{x \to \infty} \frac{x-3}{x^2-1} = 0$$
and
$$\lim_{x \to -\infty} \frac{x-3}{x^2-1} = 0,$$
the graph has the x-axis as a horizontal asymptote at the left (where the graph will approach from below) and at the right (where the graph will approach from above). The graph is at the right.

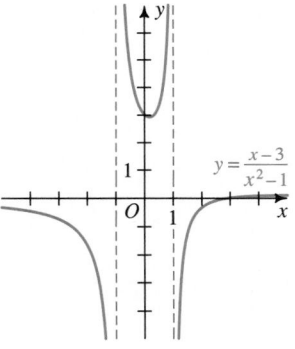

$y = \dfrac{x-3}{x^2-1}$

Example 2   Sketch the graph of $y = \dfrac{x^2 - 2x - 3}{x^2 - 4}$.

Solution   This time we will shorten the steps taken in Example 1.

Steps 1–3 $\dfrac{x^2 - 2x - 3}{x^2 - 4} = \dfrac{(x-3)(x+1)}{(x-2)(x+2)}$

(1) x-intercepts: -1, 3
(2) vertical asymptotes: $x = \pm 2$
(3) sign analysis at the right

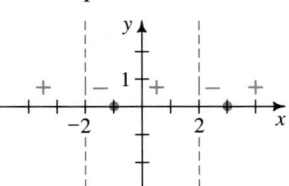

(Solution continues on the next page.)

---

1. Sketch the graph of $y = \dfrac{x+2}{x}$. Show vertical and horizontal asymptotes and x-intercepts.

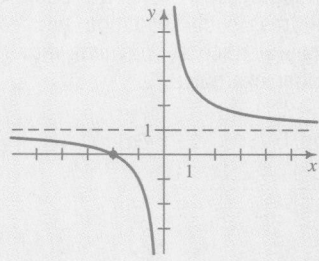

2. Sketch the graph of $y = \dfrac{x^2}{x-1}$. Show asymptotes and x-intercepts.

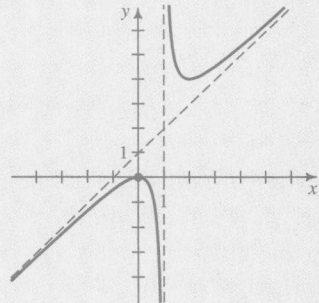

Error Analysis

To avoid giving students the misconception that the graph of a rational function must head in opposite directions on either side of a vertical asymptote, have them consider the graph of $y = \dfrac{1}{x^2}$.

The examples of this section do not include any situations in which there is a "hole" in the graph. (See Examples 4 and 6 in Section 19-1.) Although Class Exercise 7 does address such situations, you may want to supplement with other examples.

Additional Answers
Class Exercises

1. x-intercept: 2; vertical asymptotes: $x = 1$, $x = 3$; horizontal asymptote: $y = 0$

2. x-intercept: 0; vertical asymptotes: $x = \pm 2$; horizontal asymptote: $y = 0$

3. x-intercepts: ± 2; vertical asymptote: $x = 0$

4. x-intercepts: ± 3; horizontal asymptote: $y = 1$

5. x-intercept: 0; vertical asymptotes: $x = \pm\sqrt{6}$; horizontal asymptote: $y = 2$

6. x-intercept: 0; vertical asymptotes: $x = \pm 1$

7. If c is a zero of greater multiplicity for g than for f, there is a vertical asymptote at $x = c$. Otherwise, there is a "hole" in the graph at $x = c$.

Suggested Assignments

Discrete Math
Day 1: 728/1–11 odd
Day 2: 728/4, 12–17

Supplementary Materials

Alternative Assessment, 58
Precalculus Plotter Plus, 50–51, 87–88
Student Resource Guide, 174–176

Steps 4–5

(4)
$$\lim_{x \to -2^-} \frac{x^2 - 2x - 3}{x^2 - 4} = \infty$$
$$\lim_{x \to -2^+} \frac{x^2 - 2x - 3}{x^2 - 4} = -\infty$$
$$\lim_{x \to 2^-} \frac{x^2 - 2x - 3}{x^2 - 4} = \infty$$
$$\lim_{x \to 2^+} \frac{x^2 - 2x - 3}{x^2 - 4} = -\infty$$

(5)
$$\lim_{x \to \infty} \frac{x^2 - 2x - 3}{x^2 - 4} = 1$$
$$\lim_{x \to -\infty} \frac{x^2 - 2x - 3}{x^2 - 4} = 1$$

horizontal asymptote: $y = 1$

The graph is shown at the right.

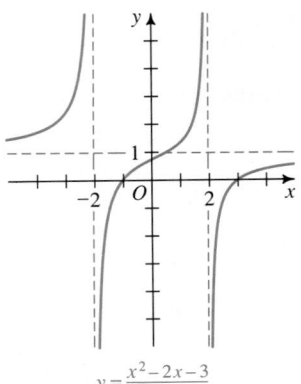

$$y = \frac{x^2 - 2x - 3}{x^2 - 4}$$

CLASS EXERCISES

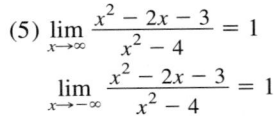

For each function, give the x-intercepts and the equations of any vertical or horizontal asymptotes. You may wish to use a computer or a graphing calculator to confirm your answers.

1. $y = \dfrac{x - 2}{(x - 1)(x - 3)}$

2. $y = \dfrac{x}{x^2 - 4}$

3. $y = \dfrac{x^2 - 4}{x}$

4. $y = \dfrac{x^2 - 9}{x^2 + 9}$

5. $y = \dfrac{2x^2}{x^2 - 6}$

6. $y = \dfrac{x^3}{x^2 - 1}$

7. **Discussion** Let $f(x)$ and $g(x)$ be polynomial functions. Suppose that for $x = c$, $f(c) = g(c) = 0$. At $x = c$, what happens to the graph of $y = \dfrac{f(x)}{g(x)}$?

WRITTEN EXERCISES

In Exercises 1–15, sketch the graph of each function. Show vertical and horizontal asymptotes and x-intercepts. You may wish to use a computer or a graphing calculator to confirm your answers.

A
1. $y = \dfrac{x - 1}{x + 1}$

2. $y = \dfrac{x}{x - 2}$

3. $y = \dfrac{x}{x^2 - 1}$

4. $y = \dfrac{2x^2}{x^2 - 9}$

5. $y = \dfrac{x^2 + 4}{x^2 - 4}$

6. $y = \dfrac{x^2 - 4}{x^2 + 4}$

7. $y = \dfrac{x - 4}{(x - 1)(x + 2)}$

8. $y = \dfrac{x + 3}{(x + 1)(x - 3)}$

9. $y = \dfrac{12}{x^2 + 2}$

10. $y = \dfrac{12x}{x^2 + 2}$

11. $y = \dfrac{12x^2}{x^2 + 2}$

12. $y = \dfrac{3x^2}{x^2 - 3x}$

B 13. $y = \dfrac{x^2 - 2x}{x^2 - 4}$ 14. $y = \dfrac{x^2 - 16}{x^2 - 6x + 8}$ 15. $y = \dfrac{x^2 - x - 2}{x^2 - 2x + 1}$

16. **Writing** Write a paragraph in which you explain why the line $y = x$ is an asymptote for the graph of $y = \dfrac{x^3}{x^2 + 1}$. Then sketch the graph.

17. **Writing** Write a paragraph in which you explain why the line $y = -x$ is an asymptote for the graph of $y = \dfrac{1 - x^2}{x}$. Then sketch the graph.

C 18. Consider the function $f(x) = \dfrac{\sin x}{x}$, where x is in radians.
   a. What is the domain of f? b. What are the zeros of f?
   c. Evaluate $\lim\limits_{x \to \infty} f(x)$ and $\lim\limits_{x \to 0} f(x)$. d. Sketch the graph of $y = f(x)$.

■ Series

19-3 Using Technology to Approximate the Area Under a Curve

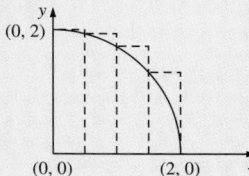 **Objective** *To use technology to approximate the area under a curve.*

In this section we will use technology to approximate the areas of regions in a coordinate plane. If the region is bounded by the graph of a function $y = f(x)$, the x-axis, and a pair of vertical lines $x = a$ and $x = b$, then we will speak of finding the *area under the curve* $y = f(x)$ from $x = a$ to $x = b$. For instance, consider the region R shown at the right. In this case we want to approximate the area under the cosine curve from $x = 0$ to $x = \dfrac{\pi}{2}$.

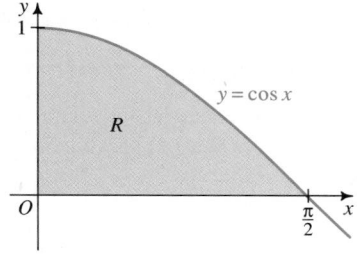

The area of region R can be approximated by finding the areas of rectangles having base vertices at $0, 0.1, 0.2, \ldots, 1.5$, as shown at the right. Since the height of the leftmost rectangle is $\cos(0.1)$, its area is $0.1 \cos(0.1)$. Likewise, the next rectangle has a height equal to $\cos(0.2)$ and an area equal to $0.1 \cos(0.2)$, the one after has a height equal to $\cos(0.3)$ and an area equal to $0.1 \cos(0.3)$, and so on. Note that the sum of the areas of all the rectangles is less than the area of region R because each rectangle is completely contained in R.

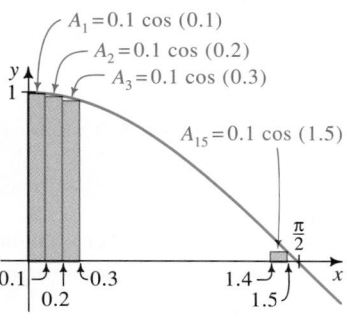

Limits, Series, and Iterated Functions **729**

18. a. Nonzero real numbers
   b. $x = n\pi$ where n is a nonzero integer
   c. 0, 1
   d.

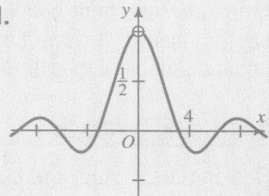

Teaching Notes, p. 716C

Warm-Up Exercises

The figure below shows the first-quadrant portion of the circle $x^2 + y^2 = 4$. The dotted lines mark the points at which $x = 0.5, 1.0, 1.5,$ and 2.

1. Estimate the area of the quarter-circle by finding the sum of the areas of the four rectangles shown. Round your answer to two decimal places. 3.50

2. Use the formula for the area of a circle to find the area of the quarter-circle shown to two decimal places. 3.14

3. Is your estimate in Exercise 1 larger or smaller than the actual area? Why does this occur? Larger; each rectangle includes a region that is not inside the circle; this overestimates the area of the quarter-circle.

The material in this section is important for two reasons. First, it is the basis for an intuitive understanding of integral calculus. Second, it illustrates how the limit concept can be used in a powerful yet understandable application.

Additional Example

Find the area under the curve $y = 1 + \cos x$ from $x = 0$ to $x = \frac{\pi}{2}$. Use the fact that the area under the curve $y = \cos x$ from $x = 0$ to $x = \frac{\pi}{2}$ is 1. Also give a geometric argument to defend your answer.

Area $= \frac{\pi}{2} + 1$. The graph of $y = 1 + \cos x$ is a vertical translation of the graph of $y = \cos x$. The area under the curve $y = 1 + \cos x$ from $x = 0$ to $x = \frac{\pi}{2}$ therefore consists of the area of a rectangle $\frac{\pi}{2}$ units wide and 1 unit high added to the area under the curve $y = \cos x$ from $x = 0$ to $x = \frac{\pi}{2}$.

◪ Using Technology

When running the program shown on this page, be aware that some computers may occasionally exit the FOR-NEXT loop after the second-to-last rectangle, and thus not add in the area of the last rectangle. This problem, which will affect the answer slightly, is a result of a computer round-off error in line 20.

The calculations involved in approximating the area of region R are ideally suited for a computer or programmable calculator. The program below, written in BASIC, approximates the area of R by keeping a running sum of the areas of the rectangles we've discussed.

```
10 LET A = 0
20 FOR X = .1 TO 1.5 STEP .1
30 LET Y = COS (X)
40 LET A = A + .1 * Y
50 NEXT X
60 PRINT ''THE AREA IS APPROXIMATELY ''; A
70 END
```

When the program is run, we get the following result.

```
THE AREA IS APPROXIMATELY .950200462
```

A better approximation can be obtained if we use rectangles having base vertices 0, 0.01, 0.02, . . . , 1.57, instead of 0, 0.1, 0.2, . . . , 1.5. All that we need to do is to change lines 20 and 40 as shown below.

```
20 FOR X = .01 TO 1.57 STEP .01
40 LET A = A + .01 * Y
```

These changes give the following result.

```
THE AREA IS APPROXIMATELY .994995334
```

Activity

Using rectangles of width 0.001, modify and run the program given above to obtain a third approximation of the area of region R. .999499996

The results that we have obtained suggest that the area approximations are approaching 1. In calculus, it is possible to show that the area of region R is *exactly* 1 without using approximations.

Example   Using rectangles of width 0.01, modify and run the program given above to approximate the area under the curve $y = 2^x$ from $x = 0$ to $x = 1$.

Solution   First, we change lines 20–40 as shown below.

```
20 FOR X = .01 TO 1 STEP .01
30 LET Y = 2 ^ X
40 LET A = A + .01 * Y
```

Then, running the modified program, we get the following result.

```
THE AREA IS APPROXIMATELY 1.44770082
```

CLASS EXERCISES

For Exercises 1 and 2, tell how you would modify the computer program on the preceding page to approximate each area.

1. The area under the curve $y = \sin x$ from $x = 0$ to $x = \pi$

2. The area between the curves $y = \sin x$ and $y = -\sin x$ from $x = 0$ to $x = \pi$

3. In the program on the preceding page, suppose we change line 20 as follows.

```
20 FOR X = 0 TO 1.5 STEP .1
```

   a. Draw a graph, like the one at the bottom of page 729, to illustrate how this program change affects the rectangles used to approximate the area of R.

   b. *Visual Thinking* Will the approximation from part (a) be less than or greater than the actual area? Explain.

WRITTEN EXERCISES

 Many of these Written Exercises require the use of a computer or programmable calculator.

For Exercises 1–4, modify and run the program on the preceding page to approximate each area. Use rectangles of width 0.01.

A

1. The area under the curve $y = \tan x$ from $x = 0$ to $x = \frac{\pi}{4}$ 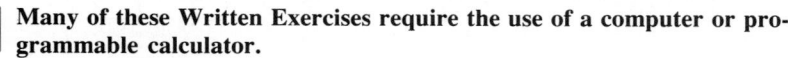 0.346158926

2. The area under the curve $y = \sin x$ from $x = 0$ to $x = \pi$ 1.99999006

3. The area under the curve $y = x^2$ from $x = 0$ to $x = 1$ 0.338350001

4. The area under the curve $y = x^3$ from $x = 0$ to $x = 1$ 0.255025001

5. Consider the area under the curve $y = x^n$ from $x = 0$ to $x = 1$.
   a. When $n = 1$, you know from geometry that the area is __?__. $\frac{1}{2}$

   b. When $n = 2$, Exercise 3 suggests that the area is $\frac{1}{3}$. When $n = 3$, Exercise 4 suggests that the area is $\frac{1}{4}$. Make a conjecture for $n = 4$. Then use a computer or programmable calculator to check your guess. $\frac{1}{5}$; 0.205033334

6. Consider the regions A, B, and C, each of which is bounded by the graph of $y = \frac{1}{x}$, the x-axis, and a pair of vertical lines. These pairs of vertical lines are $x = 1$ and $x = 2$ for A, $x = 1$ and $x = 3$ for B, and $x = 1$ and $x = 6$ for C.
   a. Use a computer or programmable calculator to approximate the areas of regions A, B, and C. Compare your answers with $\ln 2$, $\ln 3$, and $\ln 6$.

   b. What relationship exists among the areas of regions A, B, and C? (*Hint:* $\ln 2 + \ln 3 = \ln$ __?__.) What conclusion can you draw about areas under the curve $y = \frac{1}{x}$?

Limits, Series, and Iterated Functions   **731**

Chapter 19   **731**

Application

According to Hooke's law, the force $f(x)$ required to hold a spring stretched x inches beyond its length at rest is $f(x) = cx$ for some constant c. Suppose we have a spring that is 6 in. long at rest and we know that 20 lb of force is required to stretch it 2 in. How much force will be required to stretch the spring to a length of 10 in., and how much work is done? ($20 = c(2)$, so $c = 10$. By the definition of work (work = force × distance), $W = 10x \cdot x = 10x^2$, so the amount of work done is the area under the curve $W = 10x^2$ from $x = 0$ to $x = 4$.)

Cooperative Learning

Have students work in pairs to explore the areas under curves of their own choosing.

Supplementary Materials

Alternative Assessment, 59

Point out that the answer to Exercise 7 is approximately *e*. You might also tell students that in calculus they will learn that the area under the curve $y = \frac{1}{x}$ from $x = 1$ to $x = a$ is ln *a*. Thus, when $a = e$, the area is ln $e = 1$.

Communication Note

After students have completed the exercises and explored various functions, have them compile a summary of their results and share their generalizations with the class.

Additional Answers
Written Exercises

13. $y = \cos x$ is symmetric in the *y*-axis.

14. $y = \cos x$ and $y = -\cos x$ are symmetric with respect to each other in the *x*-axis.

15. The horizontal values remain the same while the vertical values are doubled. Thus, the area is doubled.

16. The area under $y = \sin x$ from $x = 0$ to $x = \pi$ is the same as the area under $y = \cos x$ from $x = -\frac{\pi}{2}$ to $x = \frac{\pi}{2}$.

17. $y = |\cos x|$ is symmetric in the vertical line $x = \frac{\pi}{2}$. Thus, the area from $x = 0$ to $x = \frac{\pi}{2}$ equals the area from $x = \frac{\pi}{2}$ to $x = \pi$.

B **7.** In Exercise 6, you found that the area under the curve $y = \frac{1}{x}$ from $x = 1$ to $x = 2$ was less than 1, and that the area from $x = 1$ to $x = 3$ was greater than 1. Thus, there must be a number *x* between 2 and 3 so that the area is exactly 1. Use a computer or programmable calculator to approximate *x*. 2.72

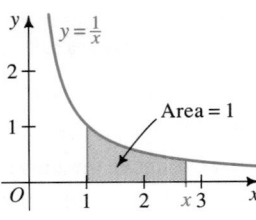

8. The area under the curve $y = \dfrac{4}{1 + x^2}$ from $x = 0$ to $x = 1$ is a special number.

Use a computer or programmable calculator to find this number. π

The area $A(x)$ of each shaded region depends only on the right-hand boundary x, which can vary from 0 to 5. Sketch the graph of $A(x)$ for $0 \le x \le 5$.

9.
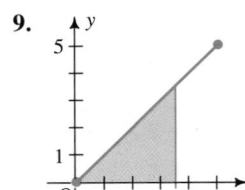

10.
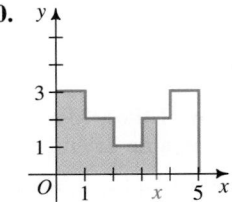

11.
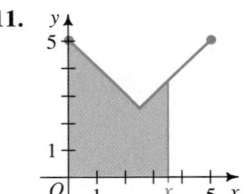

12.
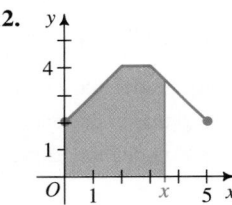

Visual Thinking **Find each area using the fact that the area under the curve $y = \cos x$ from $x = 0$ to $x = \frac{\pi}{2}$ is 1. Also give a geometric argument to defend your answer.**

13. The area under the curve $y = \cos x$ from $x = -\frac{\pi}{2}$ to $x = \frac{\pi}{2}$ $A = 2$

14. The area between the curves $y = \cos x$ and $y = -\cos x$ from $x = -\frac{\pi}{2}$ to $x = \frac{\pi}{2}$ $A = 4$

15. The area under the curve $y = 2 \cos x$ from $x = 0$ to $x = \frac{\pi}{2}$ $A = 2$

16. The area under the curve $y = \sin x$ from $x = 0$ to $x = \pi$ $A = 2$

17. The area under the curve $y = |\cos x|$ from $x = 0$ to $x = \pi$ $A = 2$

18. a. What do you think is the exact area between the x-axis and the curve $y = \cos x$ from $x = 0$ to $x = \pi$?

  b. Modify and run the program on page 730 to approximate the area described in part (a).

  c. Does the result of part (b) agree with your answer to part (a)? If not, use a geometric argument to explain why.

19-4 Power Series

Objective *To use the power series of a given function to find an infinite series for a functional value or for a related function.*

In Section 13-5, we discussed the infinite geometric series

$$a + ax + ax^2 + ax^3 + \cdots + ax^n + \cdots.$$

It is a special case of a **power series**

$$a_0 + a_1 x + a_2 x^2 + a_3 x^3 + \cdots + a_n x^n + \cdots.$$

Note that the coefficients of a power series need not all be the same.

Some extremely useful power series are given below. The *interval of convergence* for each series is also given. Note that several of the series involve $n!$, where $n! = 1 \times 2 \times 3 \times \cdots \times n$ for positive integral values of n and $0! = 1$.

Power series	Interval of convergence
(1) $e^x = 1 + \dfrac{x}{1!} + \dfrac{x^2}{2!} + \dfrac{x^3}{3!} + \cdots = \displaystyle\sum_{n=0}^{\infty} \dfrac{x^n}{n!}$	all real x
(2) $\cos x = 1 - \dfrac{x^2}{2!} + \dfrac{x^4}{4!} - \dfrac{x^6}{6!} + \cdots = \displaystyle\sum_{n=0}^{\infty} \dfrac{(-1)^n x^{2n}}{(2n)!}$	all real x
(3) $\sin x = x - \dfrac{x^3}{3!} + \dfrac{x^5}{5!} - \dfrac{x^7}{7!} + \cdots = \displaystyle\sum_{n=0}^{\infty} \dfrac{(-1)^n x^{2n+1}}{(2n+1)!}$	all real x
(4) $\mathrm{Tan}^{-1} x = x - \dfrac{x^3}{3} + \dfrac{x^5}{5} - \dfrac{x^7}{7} + \cdots = \displaystyle\sum_{n=0}^{\infty} \dfrac{(-1)^n x^{2n+1}}{2n+1}$	$-1 \le x \le 1$
(5) $\ln(1 + x) = x - \dfrac{x^2}{2} + \dfrac{x^3}{3} - \dfrac{x^4}{4} + \cdots = \displaystyle\sum_{n=1}^{\infty} \dfrac{(-1)^{n+1} x^n}{n}$	$-1 < x \le 1$

Limits, Series, and Iterated Functions **733**

Additional Answers
Written Exercises

18. a. 2
  b. 0
  c. No, because the program calculates the area from $x = \dfrac{\pi}{2}$ to $x = \pi$ as a negative number that when added to the area from $x = 0$ to $x = \dfrac{\pi}{2}$ produces an area of 0.

Teaching Notes, p. 716E

Warm-Up Exercises

In Exercises 1–3, use your calculator to find the sum, to two decimal places, of the first six terms of the given series. Tell what number appears to be the sum of the series.

1. $1.57 - \dfrac{(1.57)^3}{3!} + \dfrac{(1.57)^5}{5!} - \dfrac{(1.57)^7}{7!} + \cdots$ 1.00; 1

2. $4.71 - \dfrac{(4.71)^3}{3!} + \dfrac{(4.71)^5}{5!} - \dfrac{(4.71)^7}{7!} + \cdots$ −1.08; −1

3. $1 + \dfrac{1}{1!} + \dfrac{1}{2!} + \dfrac{1}{3!} + \cdots$ about 2.72; e

Motivating the Section

Ask students to state ways of approximating the value of sin 1. They might suggest that an approximate value can be obtained by reading it from a graph of the sine function, by referring to a table of trigonometric values, or by pressing keys on a calculator.

Chapter 19 **733**

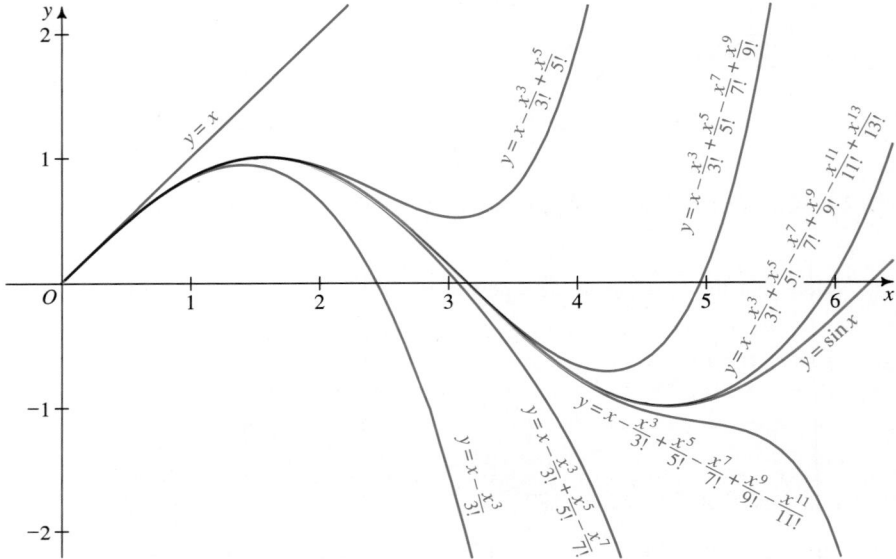

Using Technology

Additional Examples

1. **a.** Find a power series for e^{2x}. State the interval of convergence.
   b. Approximate e^2 by using the series from part (a) and a calculator.

   a. Substitute $2x$ for x in series (1).
   $$e^{2x} = 1 + \frac{2x}{1!} + \frac{(2x)^2}{2!} +$$
   $$\frac{(2x)^3}{3!} + \cdots = \sum_{n=0}^{\infty} \frac{(2x)^n}{n!}$$
   for all real x.

   b. When $x = 1$, $e^2 =$
   $$\sum_{n=0}^{\infty} \frac{2^n}{n!} = 1 + \frac{2}{1!} +$$
   $$\frac{4}{2!} + \frac{8}{3!} + \cdots \approx 7.39;$$
   a calculator gives $e^2 \approx$
   $(2.7182818)^2 \approx 7.39$.

2. Use the equation $e^{i\theta} = \cos \theta + i \sin \theta$ (proved in Exercise 14) to show that $e^{i\theta_1} \cdot e^{i\theta_2} = e^{i(\theta_1 + \theta_2)}$.

   $e^{i\theta_1} \cdot e^{i\theta_2} =$
   $(\cos \theta_1 + i \sin \theta_1) \cdot$
   $(\cos \theta_2 + i \sin \theta_2) =$
   $(\cos \theta_1 \cos \theta_2 - \sin \theta_1 \cdot$
   $\sin \theta_2) + i(\cos \theta_1 \sin \theta_2 +$
   $\sin \theta_1 \cos \theta_2) =$
   $\cos (\theta_1 + \theta_2) +$
   $i \sin (\theta_1 + \theta_2) = e^{i(\theta_1 + \theta_2)}$

Supplementary Materials

Power series are used by some calculators and computers to give approximations. For example, a computer may have a subroutine that evaluates several terms of the power series for sin x. If the computer encounters the number SIN (2) in a program, it will substitute $x = 2$ into the series. The computer is programmed to evaluate enough terms so that the result will be accurate to a predetermined number of decimal places. The diagram below shows that sin 2 is quite closely approximated by just three terms of the sine series. To evaluate sin 4 or sin 5, however, more terms are needed.

The following example shows how series (1) through (5) on the preceding page can be used to find other series.

Example   Find an infinite series for: **a.** sin 1   **b.** sin $(-x)$

Solution   **a.** In series (3), we substitute 1 for x and obtain:

$$\sin 1 = 1 - \frac{1}{3!} + \frac{1}{5!} - \frac{1}{7!} + \cdots$$

b. In series (3), we substitute $-x$ for x and obtain:

$$\sin (-x) = -x - \frac{(-x)^3}{3!} + \frac{(-x)^5}{5!} - \frac{(-x)^7}{7!} + \cdots$$
$$= -x + \frac{x^3}{3!} - \frac{x^5}{5!} + \frac{x^7}{7!} - \cdots$$

Notice that the series for sin $(-x)$ is the opposite of the series for sin x.

In addition to power series, the infinite *p-series* (so called because the *n*th term, $\frac{1}{n^p}$, involves the power p) is important in calculus. The following theorem, which Exercises 19–21 ask you to prove, gives the values of p for which the *p*-series converges and diverges.

Convergence and Divergence of the *p*-series

The *p*-series $\displaystyle\sum_{n=1}^{\infty} \frac{1}{n^p}$ converges if $p > 1$ and diverges if $p \leq 1$.

The *p*-series in which $p = 1$, $\displaystyle\sum_{n=1}^{\infty} \frac{1}{n}$, is called the **harmonic series**.

CLASS EXERCISES

1. *Visual Thinking* Suppose you want to use the power series for sin x to obtain a reasonable approximation of sin 4. What minimum number of terms in the power series would you use? Explain your answer by referring to the graph on page 734.
2. *Reading* The result of part (b) of the example on the preceding page agrees with the fact that sin $(-x) =$ __?__ for all x. $-\sin x$

WRITTEN EXERCISES

For Exercises 1–18, refer to series (1)–(5) on page 733.

A 1. Write infinite series in expanded form for $\cos \frac{\pi}{4}$ and $\sin \left(-\frac{\pi}{4}\right)$.

2. Write infinite series in expanded form for e and e^{-1}.
3. **a.** Substitute $x = 1$ in the series for ln $(1 + x)$ to get an infinite series for ln 2.
   b. Why can't you substitute $x = 2$ to get an infinite series for ln 3?
4. Substitute $x = 1$ in the series for $\text{Tan}^{-1} x$ to get an infinite series for a number involving π. $\dfrac{\pi}{4} = 1 - \dfrac{1}{3} + \dfrac{1}{5} - \dfrac{1}{7} + \cdots$

In Exercises 5–10, find a power series for each function. State the interval of convergence in each case.

5. e^{-x} 6. e^{-x^2} 7. $\text{Tan}^{-1} 2x$

8. ln $(1 - x)$ 9. $\sin x^2$ 10. $\cos 2x$

Challenge students to write a program that determines the number of terms needed before the sum of the harmonic series exceeds a given value greater than 1.

```
10 INPUT "VALUE (>1) TO
   BE EXCEEDED BY SUM-
   MING THE TERMS OF THE
   HARMONIC SERIES: "; M
20 LET S = 0
30 LET N = 1
40 LET S = S + 1/N
50 IF S > M THEN GOTO 80
60 LET N = N + 1
70 GOTO 40
80 PRINT "NUMBER OF
   TERMS NEEDED IS "; N
90 END
```

Students may be surprised by the number of terms needed for the sum to exceed, say, 9. Point out that divergence is based not on how many terms are needed for the sum to exceed a given value, but on the fact that the sum eventually exceeds any given value.

B 11. Use the sine series to show that $\lim\limits_{x \to 0} \dfrac{\sin x}{x} = 1$.

12. Use the cosine series to find $\lim\limits_{x \to 0} \dfrac{1 - \cos x}{x^2}$. $\dfrac{1}{2}$

13. Use series (5) to find $\lim\limits_{x \to 0} \dfrac{\ln(1 + x)}{x}$. 1

14. Use series (1), (2), and (3) to prove that $e^{i\theta} = \cos\theta + i\sin\theta$.

15. Use Exercise 14 to show that $e^{i\pi} = -1$ and $e^{2i\pi} = 1$. (*Note:* These identities were discovered by the Swiss mathematician Leonhard Euler. Notice the combination of arithmetic (the number -1), algebra (the number i), geometry (the number π), and analysis (the number e).)

16. Use Exercise 14 to show that $\cos\theta = \dfrac{e^{i\theta} + e^{-i\theta}}{2}$.

17. Use Exercise 14 to show that $\sin\theta = \dfrac{e^{i\theta} - e^{-i\theta}}{2i}$.

18. Show that the expressions for $\sin\theta$ and $\cos\theta$ given in Exercises 16 and 17 satisfy the equation $(\sin\theta)^2 + (\cos\theta)^2 = 1$.

The following exercises constitute a proof of the theorem on page 735.

C 19. Show that the harmonic series diverges by rewriting the series as

$$1 + \tfrac{1}{2} + \left(\tfrac{1}{3} + \tfrac{1}{4}\right) + \left(\tfrac{1}{5} + \tfrac{1}{6} + \tfrac{1}{7} + \tfrac{1}{8}\right) + \left(\tfrac{1}{9} + \cdots + \tfrac{1}{16}\right) + \cdots$$

and showing that the terms within each pair of parentheses total more than $\tfrac{1}{2}$.

20. Using a technique like the one in Exercise 19, show that the *p*-series converges if $p > 1$.

21. Show that the *p*-series diverges if $p < 1$.

Benoit Mandelbrot (1924–)

Born in Poland and educated in France, Benoit Mandelbrot is best known for developing the geometry of *fractals,* objects having similar features at many levels of scale.

Mandelbrot's work with fractals has application to many fields, helping to describe erratic stock market fluctuations, weather patterns, turbulence in fluid flow, disturbances in heart rhythms, and the irregular structures of coastlines, clouds, and galaxies.

Fractals have even been applied to film-making, where computer-generated fractals are used to create alien landscapes that look real.

736   *Chapter Nineteen*

■ Iterated Functions

19-5 Analyzing Orbits

Objective   *To analyze orbits for iterations of a given function.*

In this section and the next, we will discuss a branch of mathematics called *dynamical systems*. Generally speaking, a dynamical system is some situation that is undergoing change, whether the change is as orderly as the motion of a planet or as erratic as the fluctuations of a stock market. The goal is to predict the behavior of a dynamical system.

We can easily create a dynamical system by starting with an initial value, called a *seed,* and repeatedly applying a function to it. That is, given a seed x_0 and a function f, we will examine the sequence

$$x_0$$
$$x_1 = f(x_0)$$
$$x_2 = f(x_1) = f(f(x_0))$$
$$x_3 = f(x_2) = f(f(f(x_0)))$$
$$\vdots$$

This sequence is called the *orbit* of x_0. Notice that the terms of the sequence involve successive *iterations* of f. To identify iterations more easily, we will use superscripts: $f^2(x_0)$ will mean $f(f(x_0))$, $f^3(x_0)$ will mean $f(f(f(x_0)))$, and so on. Thus, the $(n + 1)$st term in the orbit of x_0 is:

$$x_n = f^n(x_0)$$

Note that n indicates the number of times that f has been applied to x_0; it does *not* indicate a power of f.

Activity 1

a. Using a calculator, obtain decimal approximations for the numbers in the sequence $10, \sqrt{10}, \sqrt{\sqrt{10}}, \sqrt{\sqrt{\sqrt{10}}}, \ldots$, which is the orbit of $x_0 = 10$ for iterations of $f(x) = \sqrt{x}$. What seems to be the limit of the orbit? 1

b. Repeat part (a), this time using $x_0 = 0.1$. 1

Limits, Series, and Iterated Functions   **737**

Teaching Notes, p. 716E

Warm-Up Exercises

For Exercises 1-4, let $f(x) = 1 - \dfrac{1}{x}$.

1. Find $f(0.5)$, $f(f(0.5))$, and $f(f(f(0.5)))$. -1; 2; 0.5

2. Find the next three terms in the sequence begun in Exercise 1. -1; 2; 0.5

3. Describe the pattern found in the answers to Exercises 1 and 2. a repeating pattern of three numbers: -1, 2, and 0.5

4. Repeat the process in Exercises 1 and 2 using $f(0.8)$ as the first step. What are your results? a repeating pattern of three numbers: -0.25, 5, and 0.8

Motivating the Section

Have students look at the photo on page 716. Point out that this intricate picture is the result of iterating a rather simple function having the set of complex numbers as its domain. Analyzing iterations of functions with *real* domains is the topic of this section. (Functions with complex domains are discussed in the Project on pages 750–755.)

Ⓝ **Using Technology**

Because of the repetitious nature of iteration, it is natural to program a calculator or computer to illustrate the process. Encourage students to do this on their own.

Although Activity 1 shows that we can examine an orbit by looking at *numbers*, we can also examine an orbit by looking at a *graph*. For example, we can use the graph of $y = \sqrt{x}$, shown below, to examine the orbit of $x_0 = 10$ for iterations of the square root function. Substituting x_0 into $y = \sqrt{x}$, we obtain the corresponding y-value, $y_0 \approx 3.16$. Then, letting $x_1 = y_0$ and substituting x_1 into $y = \sqrt{x}$, we obtain $y_1 \approx 1.78$. Letting $x_2 = y_1$ and substituting x_2 into $y = \sqrt{x}$, we obtain $y_2 \approx 1.33$. Continuing in this way, we see that the orbit of $x_0 = 10$ tends toward 1. We can write this observation as $\lim\limits_{n \to \infty} f^n(10) = 1$ where $f(x) = \sqrt{x}$.

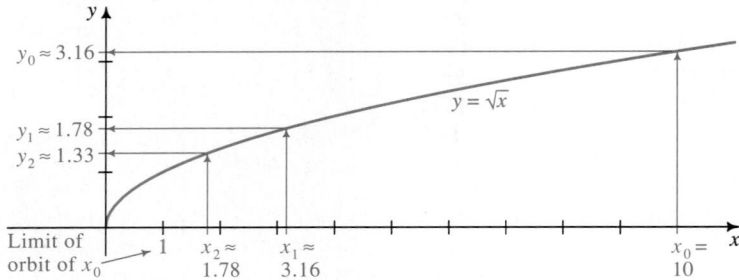

Since each y-value that is obtained becomes the next x-value in the iteration of a function, we can use the line $y = x$ to simplify the graphical analysis of a given seed's orbit. Consider the diagram below, where we start with $x_0 = 10$ on the x-axis and move vertically to the curve, horizontally to the line, vertically to the curve, horizontally to the line, and so on. A diagram like this, with the dotted lines omitted, is called a *web diagram* (because it looks like a cobweb).

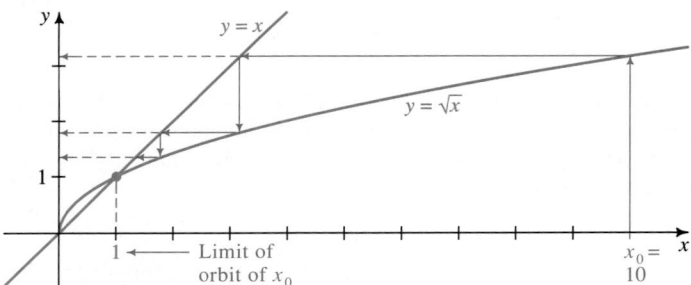

Example 1 Use **(a)** a calculator and **(b)** a web diagram to analyze the orbit of $x_0 = 1$ for iterations of the function $f(x) = 2 \sin x$.

Solution **a.** With your calculator in radian mode, enter the seed and iterate the function to produce the sequence at the top of the next page.

738 *Chapter Nineteen*

$$x_0 = 1$$
$$x_1 = 2 \sin x_0 = 2 \sin 1 \approx 1.683$$
$$x_2 = 2 \sin x_1 \approx 2 \sin 1.683 \approx 1.987$$
$$x_3 = 2 \sin x_2 \approx 2 \sin 1.987 \approx 1.829$$
$$\vdots$$

You should find a limiting value of approximately 1.895.

b. The web diagram below shows that $\lim\limits_{n \to \infty} f^n(1) \approx 1.895$. Notice that this limit is the x-coordinate of the point of intersection of the graphs of $y = 2 \sin x$ and $y = x$.

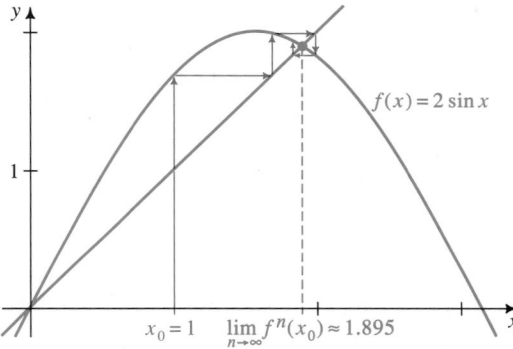

$f(x) = 2 \sin x$

$x_0 = 1 \qquad \lim\limits_{n \to \infty} f^n(x_0) \approx 1.895$

Activity 2

Use a web diagram to analyze the orbit in part (b) of Activity 1 on page 737. Does using some other seed between 0 and 1 affect the analysis? $\lim\limits_{n \to \infty} f^n(0.1) = 1$; no

---

Activity 2 indicates that for any seed x_0 such that $0 < x_0 < 1$, the orbit of x_0 for iterations of $f(x) = \sqrt{x}$ tends toward 1. The same is true for any seed $x_0 > 1$, as the web diagram at the right indicates. We have thus accounted for all seeds except $x_0 = 0$ and $x_0 = 1$. Since $\sqrt{0} = 0$ and $\sqrt{1} = 1$, the orbit of each seed is constant (that is, the orbit of 0 is 0, 0, 0, . . . , and the orbit of 1 is 1, 1, 1, . . .). The numbers 0 and 1 are therefore called *fixed points* of f. In general, a fixed point of a function f is any value of x satisfying the equation

$$f(x) = x.$$

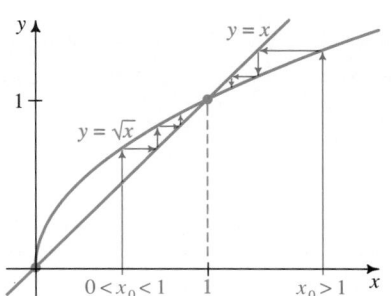

$y = x$

$y = \sqrt{x}$

$0 < x_0 < 1 \qquad 1 \qquad x_0 > 1$

2. To the nearest hundredth, find the value of the continued fraction

$$\cfrac{1}{2 + \cfrac{1}{2 + \cfrac{1}{2 + \cdots}}}$$

by iterating the function $f(x) = \dfrac{1}{2 + x}$ for the seed $x_0 = 1$.

$f(1) = \dfrac{1}{2 + 1} \approx 0.333$

$f^2(1) = \dfrac{1}{2 + f(1)} \approx 0.429$

$f^3(1) = \dfrac{1}{2 + f^2(1)} \approx 0.412$

$f^4(1) = \dfrac{1}{2 + f^3(1)} \approx 0.415$

$f^5(1) = \dfrac{1}{2 + f^4(1)} \approx 0.414$

Thus, the continued fraction has a value of 0.41 to the nearest hundredth. (Note that solving $x = \dfrac{1}{2 + x}$ gives $x = -1 \pm \sqrt{2}$, and $-1 + \sqrt{2} \approx 0.41$.)

◪ Using Technology

Because accurate web diagrams are difficult to draw by hand, you might have students use graphing calculators that are programmed to draw web diagrams. See the Teaching Notes for this section.

Although 0 and 1 are both fixed points, they differ in one important aspect. The orbit of any seed near 1 moves toward 1, while the orbit of any seed near 0 moves away from 0. Consequently, 1 is called an *attracting* fixed point, and 0 is called a *repelling* fixed point.

Example 2 Analyze all orbits for iterations of the function $f(x) = \frac{1}{2}x^2 - 1$.

Solution

Step 1 We first find the fixed points of f by solving the equation $f(x) = x$:

$$\frac{1}{2}x^2 - 1 = x$$
$$x^2 - 2x - 2 = 0$$
$$x = 1 \pm \sqrt{3}$$

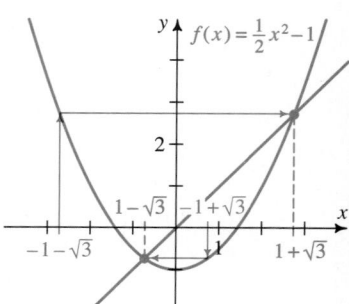

The fixed points of f are therefore $x_0 = 1 + \sqrt{3}$ and $x_0 = 1 - \sqrt{3}$.

Moreover, since $f(-x) = f(x)$ for all x, we can see that $x_0 = -1 - \sqrt{3}$ and $x_0 = -1 + \sqrt{3}$ each become fixed after one iteration of f. (See the web diagram above.) They are therefore called *eventually fixed* points.

Step 2 We now draw web diagrams, like the ones below, to examine the orbits of seeds other than the fixed and eventually fixed points. (Note that different scales are used for the two web diagrams.) The web diagram at the left below indicates that if $|x_0| > 1 + \sqrt{3}$, the orbit of x_0 tends toward infinity. On the other hand, the web diagram at the right below indicates that if $|x_0| < 1 + \sqrt{3}$, the orbit of x_0 tends to $1 - \sqrt{3}$.

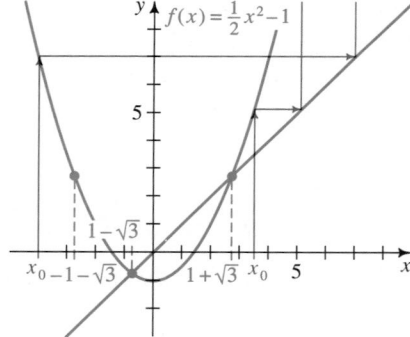

If $|x_0| > 1 + \sqrt{3}$, then $\lim_{n \to \infty} f^n(x_0) = \infty$.

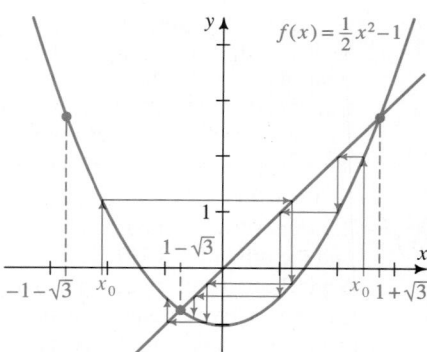

If $|x_0| < 1 + \sqrt{3}$, then $\lim_{n \to \infty} f^n(x_0) = 1 - \sqrt{3}$.

740   *Chapter Nineteen*

In Exercises 1–4, find the first four terms in the orbit of x_0 for iterations of the given function f.

1. $f(x) = x + 2$, $x_0 = 5$ 5, 7, 9, 11

2. $f(x) = 1 - x$, $x_0 = 0$ 0, 1, 0, 1

3. $f(x) = 3x + 2$, $x_0 = -1$ $-1, -1, -1, -1$

4. $f(x) = \frac{1}{10}x + 1$, $x_0 = 1$
   1, 1.1, 1.11, 1.111

5. Give an example showing the difference between $f^2(x_0)$ and $[f(x_0)]^2$.

6. a. The web diagram below shows the orbit of $x_0 = 2$ for iterations of the sine function. What is the limit of this orbit? 0

   b. If you set your calculator in radian mode and compute the orbit in part (a), does it seem that the limit is approached rapidly or slowly? Slowly

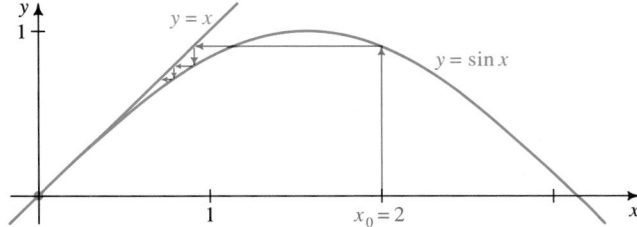

7. **Visual Thinking** Describe where the fixed points of a function f occur on the graph of f. At the graph's intersection with the graph of $y = x$

8. **Reading** Tell whether each of the two fixed points in Example 2 is attracting or repelling. $1 - \sqrt{3}$ is attracting; $1 + \sqrt{3}$ is repelling.

9. Consider the function $f(x) = x^2$.
   a. What are the fixed points of f? 0, 1
   b. Does f have any eventually fixed points? If so, what are they? Yes; -1
   c. Using web diagrams, find $\lim_{n \to \infty} f^n(x_0)$ if $|x_0| > 1$ and if $|x_0| < 1$. ∞, 0
   d. Based on your answers to parts (a) and (c), tell whether each fixed point of f is attracting or repelling. 0 is attracting; 1 is repelling.

10. Repeat Exercise 9 for the function $f(x) = x^3$.

In Exercises 1–4, find the first four terms in the orbit of x_0 for iterations of the given function f.

0, 1, 1.41, 1.55

A 1. $f(x) = 2x - 1$, $x_0 = 5$ 5, 9, 17, 33

2. $f(x) = \sqrt{x + 1}$, $x_0 = 0$

3. $f(x) = \frac{1}{x}$, $x_0 = 2.5$ 2.5, 0.4, 2.5, 0.4

4. $f(x) = \frac{x}{2} - 2$, $x_0 = 20$ 20, 8, 2, -1

Limits, Series, and Iterated Functions **741**

Assessment Note

After completing Example 2, consider giving a quick quiz on terminology. Ask students to define the following:

1. seed
2. orbit
3. fixed point
4. attracting fixed point
5. repelling fixed point
6. eventually fixed point

Additional Answers
Class Exercises

5. Answers may vary. For example, let $f(x) = x$. Then $f^2(x_0) = x_0$, and $(f(x_0))^2 = x_0^2$.

10. a. $-1, 0, 1$
    b. No
    c. If $x_0 > 1$,
       $\lim_{n \to \infty} f^n(x_0) = \infty$;
       if $x_0 < -1$,
       $\lim_{n \to \infty} f^n(x_0) = -\infty$;
       if $|x_0| < 1$,
       $\lim_{n \to \infty} f^n(x_0) = 0$.
    d. 0 is attracting; -1 and 1 are repelling.

Supplementary Materials

Alternative Assessment, 60

In Exercises 5–10, determine the fixed points of the given function f.

5. $f(x) = 5x - 16$ 4

6. $f(x) = 3 - \dfrac{1}{2}x$ 2

7. $f(x) = x^2 + \dfrac{1}{4}$ $\dfrac{1}{2}$

8. $f(x) = x^2 - 1$ $\dfrac{1 \pm \sqrt{5}}{2}$

9. $f(x) = 2 + \sqrt{x}$ 4

10. $f(x) = \dfrac{1}{2}\left(x + \dfrac{4}{x}\right)$ $-2, 2$

11. Use a calculator to examine the orbit of $x_0 = 235$ for iterations of the function $f(x) = \log_{10} x$. Describe what happens and explain why.

12. Use a calculator to examine the orbit of $x_0 = 101$ for iterations of the function $f(x) = \sqrt{x - 1}$. Describe what happens and explain why.

13. a. Draw a web diagram to analyze the orbit of $x_0 = 6$ for iterations of the function $f(x) = \dfrac{1}{2}x + 1$. Then find $\lim\limits_{n\to\infty} f^n(x_0)$.

   b. If you change x_0, how does the limit change?

14. Repeat Exercise 13 if $f(x) = 1 - \dfrac{2}{3}x$.

15. a. Using a calculator set in radian mode, examine the orbit of $x_0 = 1$ for iterations of the function $f(x) = \cos x$. Approximate $\lim\limits_{n\to\infty} f^n(x_0)$.

   b. Using web diagrams, convince yourself that the limit found in part (a) applies to *any* seed x_0.

16. Repeat Exercise 15 if $f(x) = \dfrac{1}{2}\cos x$.

B   **17.** Consider the function $f(x) = \ln(x + 1)$.
   a. What is $\lim\limits_{n\to\infty} f^n(x_0)$ for any positive seed x_0? 0
   b. What is $\lim\limits_{n\to\infty} f^n(x_0)$ for any seed x_0 such that $-1 < x_0 < 0$? Lim. not defined

18. Consider the function $f(x) = x - x^2$.
   a. What is $\lim\limits_{n\to\infty} f^n(x_0)$ for any seed x_0 such that $0 < x_0 < 1$? 0
   b. What is $\lim\limits_{n\to\infty} f^n(x_0)$ for any seed x_0 such that $x_0 < 0$ or $x_0 > 1$? $-\infty$

19. Analyze all orbits for iterations of the function $f(x) = 2x - x^2$.

20. Analyze all orbits for iterations of the function $f(x) = 2 \sin x$. (*Hint:* The orbit given in Example 1 on page 738 tends toward one of the fixed points of f. There are two other fixed points.)

21. Consider the set of linear functions $f(x) = mx + k$. For what values of m do iterations of f produce orbits that have a finite limit? (*Note:* Exercises 13 and 14 are two cases.) $|m| < 1$ and $m = 1$ when $k = 0$.

22. Consider the set of quadratic functions $f(x) = x^2 + c$.
   a. For what values of c does f have one or more fixed points? $c \le \dfrac{1}{4}$
   b. Given any value of c for which f has *no* fixed points, find $\lim\limits_{n\to\infty} f^n(x_0)$ for any seed x_0. ∞

23. The expression below is an example of a *continued fraction*.

$$1 + \cfrac{1}{1 + \cfrac{1}{1 + \cfrac{1}{1 + \cdots}}}$$

a. By iterating the function $f(x) = 1 + \dfrac{1}{x}$ for the seed $x_0 = 1$, you can find the value of the continued fraction. Approximate this value to the nearest hundredth. 1.62

b. Try seeds other than $x_0 = 1$. Does changing the seed affect $\lim\limits_{n \to \infty} f^n(x_0)$?

Explain. No. In a web diagram, 1.62 is an attracting pt.

24. To the nearest hundredth, find the value of the continued fraction

$$2 - \cfrac{1}{2 - \cfrac{1}{2 - \cdots}}$$

by iterating the function $f(x) = 2 - \dfrac{1}{x}$ for the seed $x_0 = 2$. 1.00

25. The "divide and average" algorithm for finding the square root of a positive number N is as follows:

1. Let the seed x_0 be any approximation value of $\sqrt{N}$.

2. Compute the orbit of x_0 by iterating the function $f(x) = \dfrac{x + \dfrac{N}{x}}{2}$.

a. Use the algorithm to find the first four approximations for $\sqrt{10}$. Begin with $x_0 = 3$.

b. Show that a fixed point of f is $\sqrt{N}$. Using web diagrams, also show that this is an attracting fixed point for any positive seed x_0.

26. *Investigation* Consider the function f defined for positive integers as shown below:

$$f(x) = \begin{cases} \dfrac{x}{2} & \text{if } x \text{ is even} \\ 3x + 1 & \text{if } x \text{ is odd} \end{cases}$$

The orbit of $x_0 = 4$ for iterations of f is:

$$4, 2, 1, 4, 2, 1, 4, 2, 1, \ldots$$

Since the orbit repeats the same three numbers over and over, the orbit is called *periodic* with *fundamental period* 3. Similarly, the orbit of $x_0 = 5$ is:

$$5, 16, 8, 4, 2, 1, 4, 2, 1, 4, 2, 1, \ldots$$

This orbit is called *eventually periodic*. Examine and describe the orbits of other seeds. (You may wish to write a computer program that generates a given seed's orbit. If you do, be sure to try $x_0 = 27$.)

Additional Answers Written Exercises

25. a. 3.17, 3.16, 3.16, 3.16

b. $f(\sqrt{N}) = \dfrac{\sqrt{N} + \dfrac{N}{\sqrt{N}}}{2} = \dfrac{\sqrt{N} + \sqrt{N}}{2} = \sqrt{N}$

26. The seeds $x_0 = 1$, $x_0 = 2$, and $x_0 = 4$ all have periodic orbits. All other integral seeds less than 27 have orbits that are eventually periodic, settling down to the same cycle of 4, 2, 1, 4, 2, 1, . . . within the first twenty iterations. The seed $x_0 = 27$, however, requires over 100 iterations before its orbit becomes periodic. One might conjecture that all seeds other than 1, 2, and 4 have orbits that are eventually periodic.

Exercise Note

When discussing Exercise 23, point out that the orbit of the seed $x_0 = 1$ is the sequence $\dfrac{1}{1}, \dfrac{2}{1}, \dfrac{3}{2}, \dfrac{5}{3}, \dfrac{8}{5}, \ldots$, which consists of ratios of successive terms in the Fibonacci sequence. Furthermore, the limit of the orbit is 1.618 . . . , which is the golden ratio.

Problem Solving

As an extension of Exercise 26, have students modify the definition of *f*. For example:

$$f(x) = \begin{cases} \dfrac{x}{2} & \text{if } x \text{ is even} \\ 3x - 1 & \text{if } x \text{ is odd} \end{cases}$$

Students can then investigate the orbits of seeds for iterations of the modified function.

For Exercises 1-3, let $f(x) = 2.1x(1 - x)$.

1. Use a calculator to examine the orbit of $x_0 = 0.3$ for iterations of f. Describe your findings. There is a limiting value of about 0.524.

2. Repeat Exercise 1 using $x_0 = 0.1$. Again, there is a limiting value of about 0.524.

3. Write a recursive definition for the terms in the orbit of $x_0 = 0.5$ for iterations of f. $x_0 = 0.5$; $x_n = 2.1x_{n-1}(1 - x_{n-1})$

Motivating the Section

Situations involving growth over time can be represented as iterated functions. The ability to utilize these functions gives students the power to handle problems in a variety of fields, including finance and biology.

Additional Examples

1. Use the function $f(x) = 2.5x(1 - x)$ to find the next generation of a deer population in a wooded area where the current population is 50% of capacity.

   $f(0.50) = (2.5)(0.50)(0.50) = 0.625$, or 62.5% of capacity.

19-6 Applications of Iterated Functions

Objective To use iterated functions to model money and population growth.

The iteration of a function is closely related to recursion formulas, which we discussed in Section 13-2. Example 1 compares these ideas.

Example 1 An investor initially deposits $2000 in an account that pays 6% annual interest, and each year thereafter the investor deposits $1000 in the account. Define the sequence of yearly account balances (a) by using a recursion formula and (b) by using an iterated function.

Solution It is helpful to see the first few dollar amounts in the sequence.

$$x_0 = 2000$$
$$x_1 = 1.06(2000) + 1000 = 3120$$
$$x_2 = 1.06(3120) + 1000 = 4307.20$$
$$\vdots$$

a. The sequence can be defined recursively as follows:

$$x_0 = 2000 \qquad \longleftarrow \text{ initial amount}$$
$$x_n = 1.06(x_{n-1}) + 1000 \qquad \longleftarrow \text{ recursion formula}$$

b. The sequence can also be defined as the orbit of $x_0 = 2000$ for iterations of the function $f(x) = 1.06x + 1000$.

An investment program like that in Example 1 is called an *annuity*. To see how changes in the interest rate or in the deposited amounts affect an annuity, a computer program like the one on page 481 is very helpful.

Example 1 dealt with the dynamics of money growth. Let us now look at the dynamics of population growth. Consider a population in some environment, such as bass in a lake, lions on a savanna, or owls in a forest. From one generation to the next, the population typically changes. Moveover, the environment in which the population exists can usually sustain the population up to some maximum number, or *capacity*. If we define the capacity to be 1, then we can always express the population as some fraction of the capacity. For example, to say that a population is 0.62 means that the population is at 62% of capacity.

When a population is low (that is, near 0% of capacity), the next generation tends to be larger because ample food is usually available, disease is not easily spread, and so on. However, when a population is high (that is, near 100% of capacity), the next generation tends to be smaller because food becomes scarce, disease is more likely, and so on. Therefore, a reasonable mathematical model for predicting the size of the next generation of a population would involve the following quantities:

> (1) the current population, x (where $0 < x < 1$)
>
> (2) the room available for growth, $1 - x$

One of the simplest such growth models assumes that the next generation is jointly proportional with x and $1 - x$. For example, if the current population is 0.62 and the room for growth is $1 - 0.62 = 0.38$, then the model predicts that the next generation is some positive proportionality constant times the product of 0.38 and 0.62. Thus, the model is based on the function

$$f(x) = cx(1 - x), \text{ where } c \text{ is a positive proportionality constant.}$$

The function $f(x) = cx(1 - x)$ is called a *logistic* function. It has become very important not only to biologists studying population growth but also to mathematicians studying dynamical systems.

 Using a computer or programmable calculator and the program given below, we can iterate $f(x) = cx(1 - x)$ for various proportionality constants c and various seeds x_0, as Example 2 shows.

```
10   PRINT ''WHAT IS THE PROPORTIONALITY CONSTANT (C > 0)''
20   INPUT C
30   PRINT ''WHAT IS THE INITIAL POPULATION (0 < X < 1)''
40   INPUT X
50   PRINT ''GENERATION'', ''POPULATION''
60   FOR G = 1 TO 100
70   LET X = C * X * (1 - X)
80   PRINT G, X
90   NEXT G
100  END
```

Example 2 Using the program above and the seed $x_0 = 0.2$, iterate each function.
a. $f(x) = 2.9x(1 - x)$ **b.** $f(x) = 3.1x(1 - x)$

Solution Although the program gives 100 generations of a population, only the last 6 generations are shown for each function.

a. GENERATION	POPULATION	b. GENERATION	POPULATION
95	.655169176	95	.558014125
96	.655175328	96	.76456652
97	.655169792	97	.558014125
98	.655174774	98	.76456652
99	.65517029	99	.558014125
100	.655174325	100	.76456652

Part (a) of Example 2 shows that the orbit of $x_0 = 0.2$ for iterations of the logistic function $f(x) = 2.9x(1 - x)$ seems to have a limit of approximately 0.655 (that is, about 65.5% of capacity). We therefore say that 0.655 is an *equilibrium point* for the population.

Part (b) of Example 2 shows that the orbit of $x_0 = 0.2$ for iterations of the logistic function $f(x) = 3.1x(1 - x)$ seems to cycle between approximately 0.558 and 0.765 (that is, about 55.8% and 76.5% of capacity). We therefore say that the population has a *period-2 cycle*.

Although iterating either of the functions in Example 2 using seeds other than $x_0 = 0.2$ produces essentially the same orbit each time, not all logistic functions behave so well. There are values of c for which iterating $f(x) = cx(1 - x)$ using seeds that are nearly identical produces orbits that are wildly different. (For example, see Written Exercise 10.) Such functions are said to behave *chaotically*.

CLASS EXERCISES

2. 0.32, 0.435, 0.492, 0.500; 0.500 is equilibrium pt.

1. A loan of $100 at 1% monthly interest is paid back at the rate of $20.60 per month. The loan's monthly outstanding balances form a sequence. Define this sequence **(a)** by using a recursion formula and **(b)** by using an iterated function.

2. Using a calculator and the seed $x_0 = 0.2$, find the first four iterations of the logistic function $f(x) = 2x(1 - x)$. (*Note:* The "memory in" and "memory recall" keys on your calculator are quite useful in iterating f.) What do you observe about the resulting sequence of population values? See above.

3. Repeat Exercise 2, this time using the function $f(x) = 0.5x(1 - x)$.
  0.08, 0.037, 0.018, 0.009; 0 is equilibrium pt.

WRITTEN EXERCISES

[N] **Many of these Written Exercises require the use of a computer or programmable calculator.**

In Exercises 1–4, use (a) a recursion formula and (b) an iterated function to describe each situation. In each case, be sure to state the initial value as well as the formula or function.

[A] 1. **Personal Finance** A savings account with an annual interest rate of 6%, an opening balance of $100, and no deposits or withdrawals

2. **Personal Finance** An annuity with an annual interest rate of 8%, an initial deposit of $5000, and annual deposits of $3000

3. **Personal Finance** A retirement account with an annual interest rate of 6.5%, a balance of $400,000 at retirement, and annual withdrawals of $30,000 after retirement

4. **Personal Finance** A loan of $400 with a monthly interest rate of 1.5% and monthly payments of $23.30

Writing For Exercises 5–10, use the program on page 745 to iterate the logistic function $f(x) = cx(1 - x)$ for the given value of c and the three given values of x_0. Write a sentence or two to describe your findings.

5. $c = 1.5$; $x_0 = 0.9$, $x_0 = 0.75$, $x_0 = 0.001$

6. $c = 3.2$; $x_0 = 0.4$, $x_0 = 0.8$, $x_0 = 0.99$

7. $c = 3.5$; $x_0 = 0.162$, $x_0 = 0.437$, $x_0 = 0.85$

8. $c = 3.55$; $x_0 = 0.2$, $x_0 = 0.73$, $x_0 = 0.825$

9. $c = 3.83$; $x_0 = 0.15$, $x_0 = 0.48$, $x_0 = 0.625$

10. $c = 4$; $x_0 = 0.5$, $x_0 = 0.50001$, $x_0 = 0.49999$

B **11.** If you win a certain state's million dollar lottery, you are given payments of $50,000 each year for 20 years. Use a computer or programmable calculator to complete the following.

  a. How much does it cost the state to pay you off if the state finances the payoffs by opening a savings account that pays 10% per year and has a balance of $0 after 20 annual withdrawals of $50,000 have been made?

  b. Explain why you would be much better off at the end of the 20 years if the state had given you $1,000,000 outright and you had deposited it in an account paying 10% annual interest and then withdrew $50,000 per year.

12. Finance Suppose an account with annual interest rate r has an opening balance of x_0 dollars, and no deposits or withdrawals are ever made. To obtain the sequence of yearly account balances, we would iterate the function

$$f(x) = (1 + r)x$$
$$(1 + r)^n x_0$$

using the seed x_0. If the function is iterated n times, then $f^n(x_0) = $ _?_ . This shows that iteration leads to what type of growth? Exponential

13. The function given in Exercise 12 is based on a *constant* rate of growth, r. Although iterating the function accurately models the growth of money in a bank account, it is inadequate for modeling the growth of a population (except under ideal circumstances), because the rule for the function does not take into account the room-available-for-growth factor discussed on page 745. Suppose the rate of growth, r, is proportional to the room available for growth, $1 - x$. That is, $r = c(1 - x)$ for some positive proportionality constant c. Then:

$$\begin{array}{ll} \text{population of} \\ \text{next generation} \end{array} = \begin{array}{ll} \text{population of} \\ \text{current generation} \end{array} + \begin{array}{ll} \text{rate of} \\ \text{growth} \end{array} \times \begin{array}{ll} \text{population of} \\ \text{current generation} \end{array}$$

$$f(x) = x + rx$$
$$= (1 + r)x$$
$$= (1 + c(1 - x))x$$

  a. *Writing* Suppose $c = 0.8$ in $f(x) = (1 + c(1 - x))x$. Using a computer or programmable calculator, iterate the function for various seeds x_0 such that $0 < x_0 < 1$. Write a sentence or two to describe the results.

  b. *Investigation* Using other values of c, repeat part (a). Can you find any c-values that produce periodic orbits? chaotic orbits?

Limits, Series, and Iterated Functions **747**

Cooperative Learning

Because of the investigatory nature of the exercises, it would be helpful to have students work in groups of 2 or 3 on Exercises 5–10. You may also consider assigning some of the B-level exercises to these groups and asking them to share their findings with the entire class.

Additional Answers
Written Exercises

5. Equilibrium point at 0.333

6. Period-2 cycle: 0.513, 0.799

7. Period-4 cycle: 0.501, 0.875, 0.383, 0.827

8. Period-8 cycle: 0.506, 0.887, 0.355, 0.813, 0.540, 0.882, 0.370, 0.828

9. Period-3 cycle: 0.505, 0.957, 0.156

10. For $x_0 = 0.5$, the function has an equilibrium point at 0. For $x_0 = 0.50001$ or 0.49999, the function fluctuates chaotically.

11. a. $425,678.19
   b. Twenty iterations of the function $f(x) = 1.1x - 50,000$ for the seed $x_0 = 1,000,000$ result in 20 annual withdrawals of $50,000 and an account balance of $3,863,749.98.

13. a. Equilibrium point at 1, that is, 100% of capacity
   b. Answers may vary. If $c = 2.5$ and $x_0 = 0.6$, a period-2 cycle is produced. If $c = 2.7$ and $x_0 = 0.5$, the function behaves chaotically.

14. For a given population, if a fraction h of the capacity (which is defined to be 1) is added or removed each year, then the function in Exercise 13 becomes:

$$f(x) = (1 + c(1 - x))x + h$$

For example, given a population of trees in a forest, if 10% of the forest's capacity is harvested each year, then $h = -0.1$ in the function above.

a. Writing Suppose $c = 0.8$ and $h = -0.1$ in the function above. Using a computer or programmable calculator, iterate the function for various seeds x_0 such that $0 < x_0 < 1$. Write a sentence or two to describe the results.

b. Investigation Using other values of c and h, repeat part (a). Can you find any c- and h-values that produce periodic orbits? chaotic orbits?

15. In Exercise 14, suppose a fraction h of the *current population* (instead of the capacity) is added or removed each year. Give a rule for the modified function. Then repeat parts (a) and (b) of Exercise 14.

Chapter Summary

1. The *limit* of a function $f(x)$ as x approaches a real number c, infinity, or negative infinity (denoted $\lim\limits_{x\to c} f(x)$, $\lim\limits_{x\to\infty} f(x)$, and $\lim\limits_{x\to -\infty} f(x)$, respectively) is discussed in Section 19-1. (Formal definitions are given on pages 722 and 723.) For example, $\lim\limits_{x\to\infty} \dfrac{1}{x} = 0$ means that the value of $\dfrac{1}{x}$ gets arbitrarily close to 0 as x becomes arbitrarily large.

2. A function $f(x)$ is *continuous* at $x = c$ if $\lim\limits_{x\to c} f(x) = f(c)$.

3. Various techniques for evaluating the limit of the quotient of two functions are given on page 721.

4. A *rational* function has the form $y = \dfrac{f(x)}{g(x)}$ where $f(x)$ and $g(x)$ are polynomial functions and $g(x) \neq 0$. The procedure for graphing a rational function is given on page 726.

5. The area of a region bounded by the graph of the function $y = f(x)$, the x-axis, and the vertical lines $x = a$ and $x = b$ can be approximated by dividing the region into rectangles of equal width and summing the rectangles' areas.

6. The *power series* for several important functions are given on page 733.

7. The infinite *p-series* converges if $p > 1$ and diverges if $p \leq 1$.

8. One way to create a *dynamical system* is to *iterate* a function $f(x)$ for a given *seed* x_0. The *orbit* of x_0 is the sequence $x_0, f(x_0), f^2(x_0), f^3(x_0), \ldots$, which can be analyzed graphically using a *web diagram*.

9. If x is a *fixed point* of a function f, then $f(x) = x$. Orbits of nearby seeds move toward an *attracting* fixed point and away from a *repelling* fixed point.

10. Iterated functions model both money and population growth. The *logistic function* $f(x) = cx(1 - x)$ models population growth. Iterating $f(x)$ for a given seed may result in an *equilibrium point*, a *periodic cycle*, or a *chaotic orbit*.

Key vocabulary and ideas

$\lim_{x \to \infty} f(x)$, $\lim_{x \to -\infty} f(x)$ (pp. 717–718)

$\lim_{x \to c} f(x)$ (pp. 718–719)

continuous function (p. 720)
rational function (p. 726)
area under a curve (p. 729)
power series (p. 733)
p-series (p. 735)

dynamical system (p. 737)

seed, orbit, iteration (p. 737)

web diagram (p. 738)
fixed point (p. 739)
attracting and repelling fixed points (p. 740)
logistic function (p. 745)
equilibrium point, periodic cycle (p. 746)

Chapter Test

1. Evaluate each limit or explain why the limit does not exist.   **19-1**

   a. $\lim\limits_{x \to \infty} \dfrac{x^2 - x}{x^2 - 1}$ 1   **b.** $\lim\limits_{x \to 1} \dfrac{x^2 - x}{x^2 - 1}$ $\frac{1}{2}$ **c.** $\lim\limits_{x \to 9} \dfrac{\sqrt{x} - 3}{x - 9}$ $\frac{1}{6}$ **d.** $\lim\limits_{x \to 0} \dfrac{|x|}{x}$

2. Graph each rational function. Show all asymptotes and x-intercepts.   **19-2**

   a. $y = \dfrac{x}{x - 4}$ **4. a.** $1 - \dfrac{1}{2!} + \dfrac{1}{4!} - \dfrac{1}{6!} + \cdots$ **b.** $y = \dfrac{x^2 + 1}{x^2 - 1}$

 Part (a) of Exercise 3 requires the use of a computer or programmable calculator.   **4. b.** $x^2 - \dfrac{(x^2)^2}{2} + \dfrac{(x^2)^3}{3} - \dfrac{(x^2)^4}{4} + \cdots, \; -1 \le x \le 1$

3. a. Modify and run the program on page 730 to approximate the area   **19-3** under the curve $y = 4 - x^2$ from $x = 0$ to $x = 2$. 5.31330004

   b. *Visual Thinking* Using part (a) and the symmetry of the curve $y = 4 - x^2$, give an approximation of the area under the curve from $x = -2$ to $x = 2$. 10.62660008

4. Find an infinite series for **(a)** $\cos(-1)$ and **(b)** $\ln(1 + x^2)$. See above **19-4**

5. If $f(x) = e^x - 2$, use a calculator to approximate $\lim\limits_{n \to \infty} f^n(1)$. -1.841 **19-5**

6. Analyze all orbits for iterations of the function $f(x) = x^3 + x$.

7. *Writing* In the logistic function $f(x) = cx(1 - x)$, the constant c re-   **19-6** flects the environmental factors that affect a population's growth. If $0 < c \le 1$, find $\lim\limits_{n \to \infty} f^n(x_0)$ for any seed x_0. Describe the significance of this result for any population to which the model is applied.

Additional Answers
Chapter Test

1. d. $\lim\limits_{x \to 0^-} \dfrac{|x|}{x} = -1$, but

$\lim\limits_{x \to 0^+} \dfrac{|x|}{x} = 1$; thus, $\lim\limits_{x \to 0} \dfrac{|x|}{x}$ does not exist.

2. a.

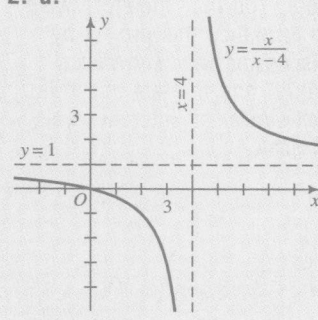

b.

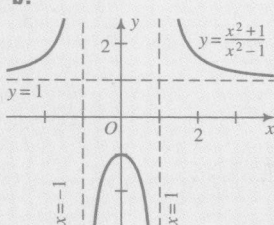

6. 0 is a repelling fixed point; $\lim\limits_{n \to \infty} f^n(x_0) = -\infty$ if $x_0 < 0$; $\lim\limits_{n \to \infty} f^n(x_0) = \infty$ if $x_0 > 0$.

7. $\lim\limits_{n \to \infty} f^n(x_0) = 0$. The population will not survive regardless of the initial population.

In Exercises 47 and 48 of Section 13-5, students were introduced to the "snowflake curve." This fractal is much easier to describe and illustrate than the Julia sets or the Mandelbrot set discussed here. You may therefore wish to introduce this project by discussing the "snowflake curve" and the fact that it is self-similar at different levels of scale.

PROJECT

Chaos in the Complex Plane

If you've explored some of the dynamical systems presented in Sections 19-5 and 19-6, then you're probably well aware that iterating a function—no matter how simple its rule—can produce unexpectedly complicated orbits for certain seeds. The results are even more unusual and exciting when a function's domain and range, which until now were sets of real numbers, are allowed to be sets of complex numbers instead. In fact, iterating a function with a complex domain and range can produce exceptionally intricate and beautiful graphic images, like the one shown.

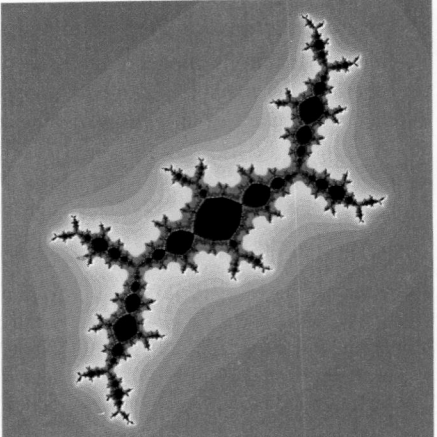

Filled-in Julia set of $f(z) = z^2 - 0.112 + 0.861i$

Materials:

a computer or programmable calculator with graphics mode (preferably color graphics)

a printer that can print a graphics screen

Complex Orbits

Suppose we have a function $f(z)$ where z is a complex number. Iterating $f(z)$ using some complex seed z_0 gives the orbit of z_0:

$$z_0$$
$$z_1 = f(z_0)$$
$$z_2 = f(z_1) = f^2(z_0)$$
$$z_3 = f(z_2) = f^3(z_0)$$
$$\vdots$$

⟵ { Recall from Section 19-5 that $f^2(z_0)$ means $f(f(z_0))$.

For example, the orbit of $z_0 = 1 + 2i$ for iterations of the squaring function $f(z) = z^2$ is:

$$z_0 = 1 + 2i$$
$$z_1 = (1 + 2i)^2 = -3 + 4i$$
$$z_2 = (-3 + 4i)^2 = -7 - 24i$$
$$z_3 = (-7 - 24i)^2 = -527 + 336i$$
$$\vdots$$

⟵ { Recall from Section 1-5 that $(x + yi)^2 = x^2 - y^2 + 2xyi$.

The calculation of such orbits is best left to a computer or programmable calculator. At the top of the next page is a BASIC program that performs 20 iterations of $f(z) = z^2$ for any seed z_0.

750 *Project*

```
10   PRINT ''REAL PART OF SEED'';
20   INPUT X0
30   PRINT ''IMAGINARY PART OF SEED'';
40   INPUT Y0
50   PRINT ''ITERATION'', ''ORBIT''
60   FOR I = 0 TO 20
70   PRINT I, X0;'' + '';Y0;''I''
80   LET X1 = X0 * X0 - Y0 * Y0
90   LET Y1 = 2 * X0 * Y0
100   LET X0 = X1
110   LET Y0 = Y1
120   NEXT I
130   END
```

■ Run the program for several values of $z_0 = x_0 + y_0 i$ such that
$$|z_0| = \sqrt{x_0^2 + y_0^2} < 1.$$
(Note that in the complex plane, these values of z_0 all lie inside the unit circle.) You should find that all orbits tend to zero. The reason is based on these facts:

1. Since $z_1 = z_0^2$, $|z_1| = |z_0^2| = |z_0|^2$.
2. Since $|z_0| < 1$, $|z_0|^2 < |z_0|$.
3. Thus, $|z_1| = |z_0|^2 < |z_0|$, which means that z_1 is closer to the origin of the complex plane than z_0 is.
4. Likewise, z_2 is closer to the origin than z_1, z_3 is closer than z_2, and so on.

■ Choose several values of z_0 such that $|z_0| > 1$. (Where are these values in the complex plane?) This time you should find that all orbits escape to infinity. (How does your computer or calculator indicate this?) Use an argument similar to the one for $|z_0| < 1$ to explain why.

■ Now let's consider the orbit of z_0 when $|z_0| = 1$. In this case, note that
$$|z_1| = |z_0^2| = |z_0|^2 = 1^2 = 1.$$

Likewise, $|z_2| = 1$, $|z_3| = 1$, and so on. In other words, the orbit of a point on the unit circle in the complex plane always stays on the unit circle. Test this result by running the program for several values of z_0 such that $|z_0| = 1$. (Be sure to try values other than $z_0 = \pm 1$ and $z_0 = \pm i$. For example, try $z_0 = 0.8 - 0.6i$.) You should find that some of the orbits fall off the unit circle. This happens because the orbits are very sensitive to slight round-off errors made by the computer or calculator. The orbits exhibit what is called *sensitive dependence on initial conditions,* a hallmark of *chaos.*

Julia Sets

For a complex polynomial function (such as $f(z) = z^2$), the set of seeds on the boundary between those whose orbits converge or cycle in some way and those whose orbits escape to infinity is called a *Julia set,* named after the French mathematician Gaston Julia. In the case of $f(z) = z^2$, the Julia set consists of all points on the unit circle.

So what do the Julia sets of other functions, such as

$$f(z) = z^2 + c$$

where c is some complex constant, look like? This is by no means an easy question to answer. In fact, Gaston Julia and others first asked this question in the early 1900s, but they lacked the technological tools that are used today to answer it. Not until the 1970s did mathematicians have computers that could quickly and accurately generate and display Julia sets. Thus, your own investigation of Julia sets puts you at the forefront of a very exciting branch of modern mathematics.

To obtain the graph of the Julia set of $f(z) = z^2 + c$ for a particular complex constant c, it seems that we would need to look at the orbits of *all* seeds z_0 in the complex plane. Fortunately, the following theorem limits the scope of our work.

Escaping Theorem

Consider the orbit of z_0 for iterations of the function $f(z) = z^2 + c$ for some complex constant c. If $|f^n(z_0)| > 2$ for some integer $n \geq 0$, then the orbit of z_0 escapes to infinity.

The escaping theorem tells us that we don't need to examine the orbit of any seed z_0 whose distance from the origin of the complex plane is more than 2 units. (Why?) Thus, if $z_0 = x_0 + y_0 i$, we can impose the restrictions $-2 \leq x_0 \leq 2$ and $-2 \leq y_0 \leq 2$ on the real and imaginary parts of z_0. These restrictions give us a square in the complex plane that we can directly associate with some square grid, say 100 dots by 100 dots, on a computer or calculator's graphics screen.

On a typical graphics screen, the point in the upper left corner is $(0, 0)$ in the computer or calculator's memory, and the point that is 100 dots over and 100 dots down from $(0, 0)$ is $(100, 100)$. We want these two graphics screen points to correspond respectively to the points $(-2, 2)$ and $(2, -2)$ in the complex plane. The transformation that converts graphics screen coordinates (a, b) to complex plane coordinates (x, y) is given by the pair of equations:

$$x = -2 + \frac{a}{25} \quad \text{and} \quad y = 2 - \frac{b}{25}$$

(You should be able to derive these equations for yourself. Try it.)

The BASIC program on the next page uses the transformation equations above to convert each graphics screen point (a, b) to a complex plane point (x_0, y_0). This point represents the seed $z_0 = x_0 + y_0 i$ whose orbit for iterations of $f(z) = z^2 + c$ (for an input value of c) is then examined. If the orbit of z_0 escapes to infinity within 20 iterations of f, then the point (a, b) is *skipped* (so that its color remains the background color of the screen). If the orbit of z_0 does *not* escape to infinity within 20 iterations of f, then the point (a, b) is *plotted* (so that its color is different from the background color of the screen). Notice in line 130 of the program that the test for an orbit escaping to infinity is based on $|f^n(z_0)|^2 > 4$ rather than $|f^n(z_0)| > 2$. (Why is this more efficient?)

752   *Project*

■ Lines 50 and 170 of the program below have been left blank, because different versions of BASIC differ in their commands for obtaining a blank graphics screen and for plotting points on the screen. Determine the proper commands for your version of BASIC and then complete the program.

```
10   PRINT ''REAL PART OF COMPLEX CONSTANT'';
20   INPUT CX
30   PRINT ''IMAGINARY PART OF COMPLEX CONSTANT'';
40   INPUT CY
50   ←——
60   FOR A = 0 TO 100
70   FOR B = 0 TO 100
80   LET X0 = -2 + A / 25
90   LET Y0 = 2 - B / 25
100  FOR I = 1 TO 20
110  LET X1 = X0 * X0 - Y0 * Y0 + CX
120  LET Y1 = 2 * X0 * Y0 + CY
130  IF X1 * X1 + Y1 * Y1 > 4 THEN GOTO 180
140  LET X0 = X1
150  LET Y0 = Y1
160  NEXT I
170  ←——
180  NEXT B
190  NEXT A
200  END
```

For the version of BASIC that you're using, insert the command(s) for obtaining a blank graphics screen. (*Note:* If the screen does not have a minimum width and depth of 100 dots each, then you will need to adjust lines 60–90.)

For the version of BASIC that you're using, insert the command(s) for plotting the point (A, B).

■ Run the program for $c = 0 + 0i$. (It may take several minutes before any points appear on the screen, so be patient.) You should get a disk of radius 1. Of course, the *edge* of the disk (that is, the unit circle) is the Julia set of $f(z) = z^2$. The disk itself is called the *filled-in* Julia set of f.

■ Run the program for various values of c, such as $c = -1$, $c = -0.1 + 0.8i$, and $c = 0.3 - 0.5i$. You should get some fascinating filled-in Julia sets. (Remember that the edge of each graph is the actual Julia set.) If possible, obtain printouts of your graphs.

■ If your graphics screen can display multiple colors, you might modify the program so that different colors are used to indicate the "speeds" with which orbits escape to infinity outside the filled-in Julia sets. For example, one color can be given seeds that escape in 10 or fewer iterations, while another color can be given seeds that escape in more than 10 iterations. Using more colors will produce an even more striking work of art, like the one shown.

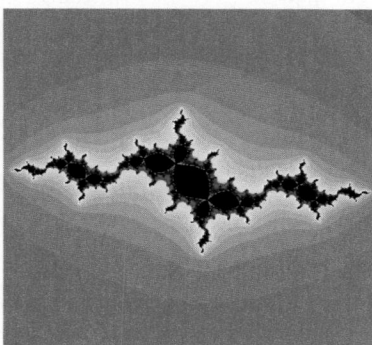

Filled-in Julia set of
$f(z) = z^2 - 1.139 + 0.238i$

To complete the programs on this page and page 755, here are the commands that you would most likely need if you have an IBM or Apple computer:

To obtain a blank screen:

IBM: SCREEN 1:CLS
Apple: HGR

To plot the point (A, B):

IBM: PSET(A, B)
Apple: HCOLOR = 3:
        HPLOT A, B

The program on this page plots the points whose orbits do not escape to infinity and skips the points whose orbits do escape to infinity. Consequently, quite a few minutes may pass before anything appears on the screen. To avoid this, you can reverse the plotting procedure (by plotting points whose orbits escape and skipping points whose orbits do not escape). Just change line 130 to:

130 IF X1 * X1 + Y1 * Y1 > 4
     THEN (plot (A, B)):
     GOTO 180

and delete line 170. A similar change can be made to the program on page 755.

Also, since the Julia set of $f(z) = z^2 + c$ has symmetry in the origin, the speed of the program can be greatly increased by plotting the points (A, B) and (100 − A, 100 − B) at the same time. If this is done, line 60 should be changed to:

60 FOR A = 0 TO 50

Fractals

One of the most unusual properties of Julia sets is *self-similarity at different levels of scale.* That is, if some portion of a Julia set is magnified, the magnified portion will resemble the original set. (See the pair of graphic images below.) Moreover, magnified portions of portions still resemble the original set. This amazing property is what typifies objects known as *fractals.*

To investigate the fractal quality of Julia sets, we need to alter the transformation equations discussed on page 752. If we want a complex plane point (x, y) in the square region defined by $m \leq x \leq m + s$ and $n \leq y \leq n + s$ to correspond to the graphics screen point (a, b) in the square region defined by $0 \leq a \leq 100$ and $0 \leq b \leq 100$, then we would use the transformation equations:

$$x = m + \frac{s \cdot a}{100} \quad \text{and} \quad y = n + s - \frac{s \cdot b}{100}$$

For example, if we want the square region defined by $-1.25 \leq x \leq -0.75$ and $-1 \leq y \leq -0.5$ to fill the screen grid of 100 dots by 100 dots, then we would modify the program on the preceding page by changing lines 80 and 90 as follows:

```
80 LET X0 = -1.25 + A / 200
90 LET Y0 = -0.5 - B / 200
```

- If you have previously obtained the filled-in Julia set of $f(z) = z^2 + c$ for $c = 0.3 - 0.5i$, then try running the program with the modifications shown above to enlarge a portion of the graph. What do you notice?
- Experiment by enlarging portions of other filled-in Julia sets whose graphs you've already obtained.

Filled-in Julia set of $f(z) = z^2 + 0.36 + 0.1i$ Magnified portion of the Julia set

The Mandelbrot Set

As we have seen, obtaining the Julia set of $f(z) = z^2 + c$ involves fixing the value of c and then iterating $f(z)$ for various seeds z_0. Suppose we turn the tables, however. What happens if we fix the value of z_0 and iterate $f(z)$ for various values of c? The result is a famous figure known as the *Mandelbrot set,* first discovered in the 1970s by Benoit Mandelbrot (see page 736).

The program below generates the Mandelbrot set. Note that the seed z_0 is given the value $0 + 0i$ in lines 60 and 70. (For reasons that we will not discuss here, the orbit of $z_0 = 0 + 0i$ is called the *critical orbit* for iterations of $f(z)$.) The program tests c-values to determine whether the orbit of $z_0 = 0 + 0i$ escapes to infinity. If the orbit does *not* escape for a particular c-value, then the c-value belongs to the Mandelbrot set (and is therefore plotted).

- Complete lines 10 and 150 of the program below and then run it. You should get a figure like the one shown in black at the right.

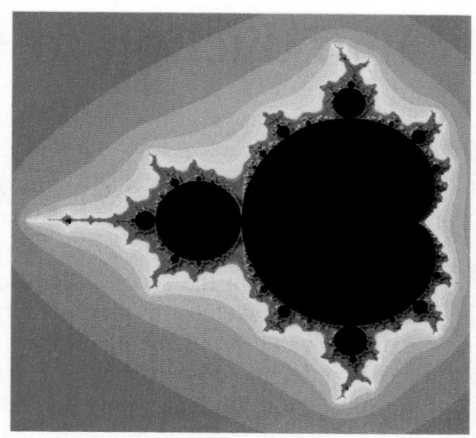

Mandelbrot set (in black)

```
10
20    FOR A = 0 TO 100          ←—— For the version of BASIC that you're
30    LET CX = - 2 + A / 25          using, insert the command(s) for obtain-
40    FOR B = 0 TO 100              ing a blank graphics screen. (Note: If the
50    LET CY = 2 - B / 25            screen does not have a minimum width
60    LET X0 = 0                     and depth of 100 dots each, then you will
70    LET Y0 = 0                     need to adjust lines 20–50.)
80    FOR I = 1 TO 20
90    LET X1 = X0 * X0 - Y0 * Y0 + CX
100   LET Y1 = 2 * X0 * Y0 + CY
110   IF X1 * X1 + Y1 * Y1 > 4 THEN GOTO 160
120   LET X0 = X1
130   LET Y0 = Y1
140   NEXT I
150                            ←—— For the version of BASIC that you're
160   NEXT B                        using, insert the command(s) for plotting
170   NEXT A                        the point (A, B).
180   END
```

- If color graphics are available, try modifying and running the program to produce an image similar to the one shown above, where color is used outside the Mandelbrot set to indicate the ''speeds'' of escaping orbits.
- Also try modifying and running the program so that one of the small ''bulbs'' on the Mandelbrot set is magnified. What do you notice?

Preparing a Report

Now that you've investigated some Julia sets as well as the Mandelbrot set, it is time to compile your work and summarize your findings. If you've obtained print-outs, be sure to indicate on them both the function that was iterated and the portion of the complex plane that was examined. Also point out any unusual features or unexpected results that you observe.

Chapter 20 Introduction to Calculus

Overview

This chapter introduces students to some of the concepts and techniques of differential calculus. In the first section, students learn to find the derivatives of certain functions and to interpret a derivative as the slope of the line tangent to the graph of a function. In the second section, students use the geometric interpretation of derivative to sketch the graphs of functions. In the third section, students solve extreme value problems again (see Section 2-4), this time using derivatives (instead of technology) to obtain exact (rather than approximate) answers. In the fourth section, students apply derivatives to physics in order to find the instantaneous velocities and accelerations of moving objects.

Objectives

20-1 To find derivatives of functions.
20-2 To sketch the graphs of functions using derivatives.
20-3 To solve extreme value problems using derivatives.
20-4 To find instantaneous velocities and accelerations.

Supplementary Resources

Tests, pp. 64–65, 66–67, 68–75
Alt. Assess., pp. 61–62, 72
Activities, pp. 53–54
Student Res. Guide, pp. 183–188

Software

Precalculus Plotter Plus
Activities, pp. 89–90, 91, 92
Function Plotter

Pacing Guide

Section	20-1	20-2	20-3	20-4	Review	Test	C. Rev.	Total
Optional (days)	2	2	2	2	1	1	1	11

Teaching Notes

20-1 pages 757–763

Presenting the Section

Emphasize that the slope of a curve at any point is the same as the slope of the tangent to the curve at that point. The tangent line can be thought of as the limiting position of secant lines "anchored" at the point, as shown in the diagram below.

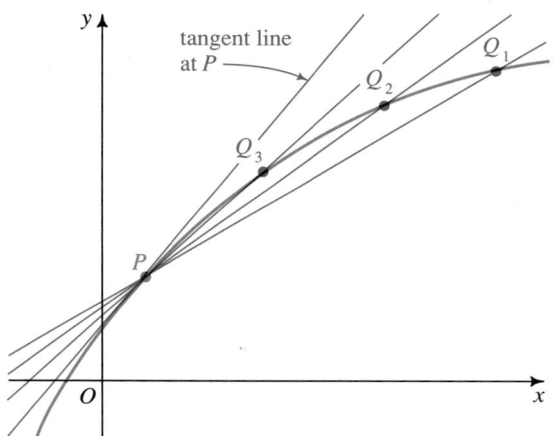

As point Q moves along the curve toward P, the secant lines approach a limiting position, which is the tangent line.

In discussing the definition of $f'(x)$, you may write $\lim_{h\to 0} \frac{f(x+h)-f(x)}{(x+h)-x}$ instead of $\lim_{h\to 0} \frac{f(x+h)-f(x)}{h}$ to emphasize that it is the limit of the slope between the points $(x, f(x))$ and $(x+h, f(x+h))$ on the curve $y = f(x)$.

Using Technology

There is calculus software available commercially that can be used to find the derivative of a function. Some software offers tutorial capabilities, while other software offers drill and practice. An example of such software is *Calculus*, published by Brøderbund.

Cooperative Learning

Working in a group may be beneficial for those students who have difficulty understanding either the notion of limit or the theorems about derivatives given in this section. Have the groups go through the proofs of the theorems by discussing them intuitively and then formally. Encourage students to demonstrate examples of the theorems as well.

Applications

Finding the slope of a curve at a specific point on the curve has many applications in physics. For example, if the curve corresponds to the position of an object with respect to time, then the slope of the curve at a given point corresponds to the instantaneous velocity of the object at a given moment (see Section 20-4).

20-2 pages 764–769

Presenting the Section

You may want to introduce this section by reviewing how to find the vertex of a parabola from Chapter 2. Then show how the newly defined derivative function helps to find the vertex.

Demonstrate how the derivative function can be used to help sketch the graph of a function. Point out that if the derivative is positive in an interval, then the graph of the function rises as x increases. If the derivative is negative in an interval, then the graph of the function falls as x increases. If the derivative is 0, then either the graph has reached a turning point (that is, a local maximum or minimum point) or it has "flattened out." These situations are distinguished by whether, on either side of the point where the derivative is 0, the derivative changes sign (a turning point) or has the same sign (a "flattening out").

Using Technology

You can use a calculator or computer to graph both a polynomial function and its derivative function. Students should see that whether the graph of the derivative function is above, on, or below the x-axis indicates whether the graph of the polynomial function is rising, horizontal, or falling.

Cooperative Learning

You might want to prepare a worksheet showing the graphs of a few functions. After dividing the class into small groups, have students within each group divide up the graphs and analyze them, noting the behavior of the graphed function and the behavior of its derivative. Then have students discuss their results with their groups. Other members of the group can point out additional features that they see, make corrections, and offer suggestions.

Applications

Determining the local maximum or minimum points on curves is an important application of calculus and can be used to model problems from economics, business, and physics. In business, for example, the maximum value of a function can represent the maximum gross profit based on the selling price of an item and the demand for it.

Assessment

You might want to ask students to summarize the relationship between the graph of a polynomial $f(x)$ and its derivative $f'(x)$. Their responses should agree with the following statements.

1. If $f'(c) = 0$, then the graph of $f(x)$ has a horizontal tangent at $x = c$.
2. The point $(c, f(c))$ is a local maximum point if $f'(c) = 0$, $f'(x) > 0$ for $x < c$, and $f'(x) < 0$ for $x > c$.
3. The point $(c, f(c))$ is a local minimum point if $f'(c) = 0$, $f'(x) < 0$ for $x < c$, and $f'(x) > 0$ for $x > c$.

20-3 | pages 769–774

Presenting the Section

You may wish to begin this section by asking students to look back at Section 2-4, where they solved extreme-value problems using technology. Point out that the use of technology can give only *approximate* solutions. The use of calculus, as in this section, can give *exact* solutions.

To help students see the importance of checking the endpoints of an interval when maximizing or minimizing a function, you might have them consider the following example: Suppose a person is at point A on a north-south road and wishes to walk to point B on an east-west road, as shown in the diagram below. The person can walk 1 mi north to the intersection of the roads and then 1 mi east to B, or the person can walk x mi ($0 \leq x < 1$) north and then cut diagonally to B. As long as the person stays on a road, the person's walking speed is 5 mi/h; if the person leaves the road, the person's walking speed is 3 mi/h (due to poor terrain). Question: What should the person do to minimize the time that it takes to get from A to B?

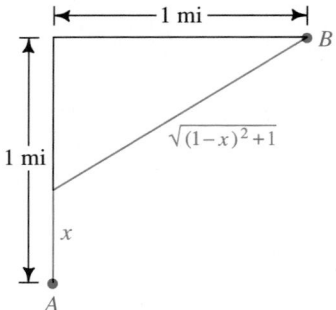

To answer the question, you can use the fact that time $= \frac{\text{distance}}{\text{rate}}$ to set up the time function:

$$t(x) = \frac{x}{5} + \frac{\sqrt{(1-x)^2 + 1}}{3} \text{ for } 0 \leq x < 1$$

Although students do not know enough calculus at this point to find the derivative of $t(x)$, you can ask them to accept the fact that

$$t'(x) = \frac{1}{5} + \frac{x-1}{3\sqrt{(1-x)^2 + 1}},$$

from which they can find that $t'(x) = 0$ when $x = \frac{1}{4}$.

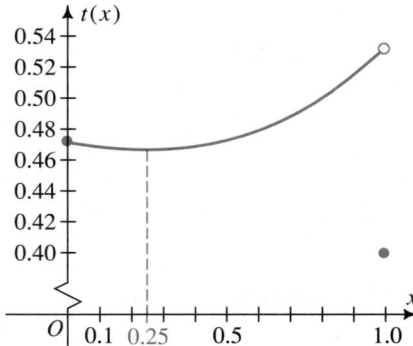

As the graph of $t(x)$ shows, $t\left(\frac{1}{4}\right) = \frac{7}{15}$, or 28 min, is a minimum time on the interval $0 \le x < 1$. But now consider the case where $x = 1$: The person walks 2 mi at 5 mi/h in $\frac{2}{5}$ h, or 24 min. Thus, the *global* minimum for $0 \le x \le 1$ is achieved at an endpoint, namely $x = 1$.

Cooperative Learning

For a class project, have students simulate the manufacturing concerns for producing a box. Divide students into groups of five. Give each group a different size of a large sheet of paper or cardboard. The task for each group is to generate all the data necessary to answer the question, "What is the box of largest volume that can be made from your sheet of paper?" Let students come up with their own ways of solving the problem, but you can encourage them to: (1) decide how the box is to fold, that is, if there is to be a top, a double bottom, and so on; (2) find the original measurements of the sheet; (3) determine the volume function; (4) find some experimental data values for the volume; (5) graph the volume function (using a graphing calculator or computer); and (6) find the maximum volume using the method presented in this section.

20-4 | pages 774–781

Presenting the Section

Encourage students to discuss the distance-versus-time graph on page 774 in detail. Ask questions such as, "How can you determine the direction of the car by looking at the position graph?" and "What does the maximum point of the graph represent?"

You can explain the notion of velocity by referring to a trip taken by car. If the car travels 100 mi in 2 h, its average speed is 50 mi/h. The car probably did not maintain that exact speed for the entire trip, however. Since the speedometer reading at any given moment represents the magnitude of the car's instantaneous velocity, the instantaneous velocity can be higher or lower than the average velocity. The instantaneous velocity corresponds to the derivative of the position function. Evaluating the derivative of the position function for any given time gives the instantaneous velocity of the car.

In discussing Example 1 on page 777, you may want to ask students for the position, velocity, and acceleration of the object at time $t = 0$ s. (In this case, $s(0) = 0$, which means that the object is initially at the origin; $s'(0) = 5$, which means that its initial velocity is 5 m/s to the right; and $s''(0) = -6$, which means that its initial acceleration is 6 m/s^2 to the left. These numbers mean that as the object leaves the origin, it moves to the right but is slowing down.)

Using Technology

It might be useful for students to try to approximate the instantaneous velocity at time $t = 1$ s for the thrown ball described on page 776. Have them use the position function with very small values of Δt, such as $\Delta t = 1$ s, $\Delta t = 0.1$ s, and so on. (A calculator will be helpful.)They should continue to decrease the size of the time interval until the calculations approach the expected limit. (20)

Cooperative Learning

Give each group of three students a set of position functions that have first and second derivatives (corresponding to velocity and acceleration functions). One student in each group should graph the position function, another student should graph the first-derivative function, and the third student should graph the second-derivative function. Have each student explain the nature of his or her curve to the other members of the group by answering questions such as "Why may some parts of the curve be horizontal?" and "Why are other parts rising or falling?" If students can describe and interpret their graphs, they will understand more about the graphs' physical meanings.

Applications

There are numerous applications of the derivative as described in this section. Other applications that involve a rate of change are unemployment rates, the rate of water flow, production rates, and population growth rates.

20 An Introduction to Calculus

20-1 The Slope of a Curve

| **Objective** | *To find derivatives of functions.* |

The *slope of a curve* at a point P is a measure of the steepness of the curve. If Q is a point on the curve near P, then the slope of the curve at P is approximately the slope of segment PQ. The approximation becomes better as Q moves along the curve toward P. We therefore define the **slope of the curve** at P to be the limit of the slope of $\overline{PQ}$ as Q approaches P along the curve:

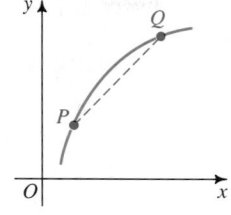

$$\text{slope of curve at } P = \lim_{Q \to P} (\text{slope of } \overline{PQ})$$

Suppose we want to find the slope of the curve $y = x^2$ at $P(1, 1)$. We can let Q be a point on the curve near P by letting the x-coordinate of Q be $1 + h$, which differs from the x-coordinate of P by h units. The y-coordinate of Q is $(1 + h)^2$, as shown in the diagram below. We now calculate the slope of $\overline{PQ}$:

$$\text{slope of } \overline{PQ} = \frac{(1 + h)^2 - 1}{h}$$

$$= \frac{(1 + 2h + h^2) - 1}{h}$$

$$= \frac{h(2 + h)}{h} = 2 + h$$

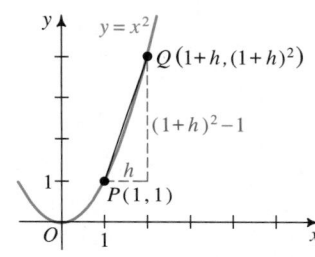

As Q approaches P, h approaches 0. Thus:

$$\text{slope of curve at } (1, 1) = \lim_{Q \to P} (\text{slope of } \overline{PQ})$$

$$= \lim_{h \to 0} (2 + h) = 2$$

Therefore, the slope of the curve $y = x^2$ at $P(1, 1)$ is 2.

To find the slope of the curve $y = x^2$ at $P(2, 4)$, we proceed similarly:

$$\text{slope of } \overline{PQ} = \frac{(2 + h)^2 - 4}{h}$$

$$= \frac{4 + 4h + h^2 - 4}{h}$$

$$= \frac{4h + h^2}{h} = 4 + h$$

$$\lim_{Q \to P} (\text{slope of } \overline{PQ}) = \lim_{h \to 0} (4 + h) = 4$$

Therefore, the slope of the curve $y = x^2$ at $P(2, 4)$ is 4.

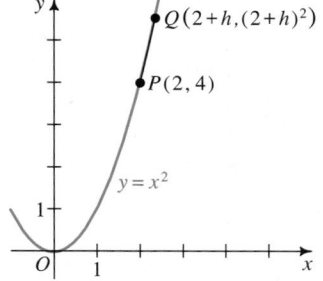

◀ The strobe photo at the left shows how a *continuous* motion can be seen as an infinite sequence of *discrete* steps. Calculus accomplishes this mathematically, allowing us to determine, for example, the *instantaneous velocity* of a drumstick at any given instant.

An Introduction to Calculus **757**

We have seen that the slope of the curve $y = x^2$ is 2 when $x = 1$ and that the slope is 4 when $x = 2$. Example 1 generalizes these results by finding the slope of the curve $y = x^2$ at an arbitrary point (x, x^2) on the curve.

Example 1   Find the slope of the curve $y = x^2$ at the point $P(x, x^2)$.

Solution   Let Q be any other point on the graph, as shown at the right. Then:

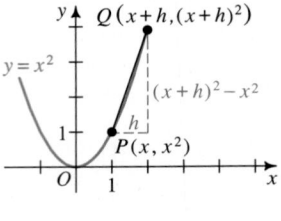

$$\text{slope of } \overline{PQ} = \frac{(x + h)^2 - x^2}{h}$$

$$= \frac{x^2 + 2xh + h^2 - x^2}{h}$$

$$= \frac{2xh + h^2}{h} = 2x + h$$

$$\lim_{Q \to P} (\text{slope of } \overline{PQ}) = \lim_{h \to 0} (2x + h) = 2x$$

Thus, the slope of the curve $y = x^2$ at $P(x, x^2)$ is $2x$.

In Example 1, notice that the slope, $2x$, is a function of x. In fact, the slope for *any* curve $y = f(x)$ at an arbitrary point $(x, f(x))$ on the curve is a function of x. This slope function, called the **derivative of $f(x)$** and denoted $f'(x)$, is defined as:

$$f'(x) = \lim_{h \to 0} \frac{f(x + h) - f(x)}{h}$$

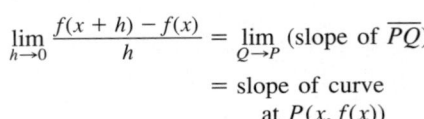

For a given value of x, the value of $f'(x)$ is the slope of the curve at $(x, f(x))$ for the following reason:

$$\lim_{h \to 0} \frac{f(x + h) - f(x)}{h} = \lim_{Q \to P} (\text{slope of } \overline{PQ})$$

$$= \text{slope of curve}$$
$$\text{at } P(x, f(x))$$

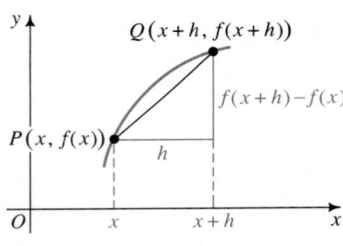

Example 2   **a.** If $f(x) = x^3$, find $f'(x)$.
  b. Find the slope of the curve $y = f(x)$ at the point $(2, 8)$.

Solution   **a.** $f'(x) = \lim\limits_{h \to 0} \dfrac{f(x + h) - f(x)}{h} = \lim\limits_{h \to 0} \dfrac{(x + h)^3 - x^3}{h}$

$$= \lim_{h \to 0} \frac{x^3 + 3x^2h + 3xh^2 + h^3 - x^3}{h}$$

$$= \lim_{h \to 0} \frac{3x^2h + 3xh^2 + h^3}{h} = \lim_{h \to 0} (3x^2 + 3xh + h^2)$$

$$= 3x^2$$

  b. Since $f'(2) = 3 \cdot 2^2 = 12$, the slope at $(2, 8)$ is 12.

The next example shows that the derivative of a function is not always defined at a point where the function itself is defined.

Example 3 If $f(x) = |x|$, show that $f'(0)$ is undefined.

Solution For $f'(0)$ to be defined, $\lim\limits_{h \to 0} \dfrac{f(0 + h) - f(0)}{h}$ must exist. Consider the right- and left-hand limits, however:

$$\lim_{h \to 0^+} \frac{f(0 + h) - f(0)}{h} = \lim_{h \to 0^+} \frac{|h| - |0|}{h}$$

$$= \lim_{h \to 0^+} \frac{h - 0}{h} \quad \longleftarrow \quad \begin{cases} \text{Since } h > 0, \\ |h| = h. \end{cases}$$

$$= \lim_{h \to 0^+} 1 = 1$$

$$\lim_{h \to 0^-} \frac{f(0 + h) - f(0)}{h} = \lim_{h \to 0^-} \frac{|h| - |0|}{h}$$

$$= \lim_{h \to 0^-} \frac{-h - 0}{h} \quad \longleftarrow \quad \begin{cases} \text{Since } h < 0, \\ |h| = -h. \end{cases}$$

$$= \lim_{h \to 0^-} -1 = -1$$

Since the right- and left-hand limits are not equal, $\lim\limits_{h \to 0} \dfrac{f(0 + h) - f(0)}{h}$ does not exist (see page 719), and therefore $f'(0)$ is undefined.

The following four theorems enable us to find the derivative of every polynomial function, and some other functions as well. Although the proof of Theorem 1 is given, Written Exercises 37–39 ask you to prove Theorems 2–4.

Theorem 1

If $f(x)$ is a constant function (that is, $f(x) = c$ for all x), then $f'(x) = 0$ for all x.

Proof $f'(x) = \lim\limits_{h \to 0} \dfrac{f(x + h) - f(x)}{h}$

$$= \lim_{h \to 0} \frac{c - c}{h}$$

$$= \lim_{h \to 0} \frac{0}{h}$$

$$= \lim_{h \to 0} 0$$

$$= 0$$

An Introduction to Calculus **759**

Mathematical Note

Emphasize that some curves may not have a slope at every point, as illustrated by Example 3. You might also point out that continuity (see Section 19-1) does not guarantee differentiability: $f(x) = |x|$ is continuous at $(0, 0)$ but it is not differentiable there.

Assessment Note

As an extension of Example 3, ask students to find $f'(x)$ for any $x > 0$ and for any $x < 0$. ($f'(x) = 1$ for $x > 0$; $f'(x) = -1$ for $x < 0$)

Additional Examples

1. **a.** Find the derivative of
   $$f(x) = \frac{1}{x} + 4x^2.$$
   b. For what value(s) of x does $f'(x) = 0$?
   a. $f(x) = x^{-1} + 4x^2$
   $f'(x) = -1 \cdot x^{-1-1} + 4 \cdot 2x^{2-1} =$
   $-x^{-2} + 8x = 8x - \dfrac{1}{x^2}$
   b. $f'(x) = 0$ when
   $8x - \dfrac{1}{x^2} = 0.$
   $8x^3 - 1 = 0$
   $x^3 = \dfrac{1}{8}$
   $x = \dfrac{1}{2}$

2. Find a function $g(x)$ such that $g'(x) = 5x^{3/2}$.
   By Theorem 3, $g(x)$ must have the form $cx^{5/2}$. Since the derivative of $cx^{5/2}$ is $\dfrac{5}{2}cx^{3/2}$, $\dfrac{5}{2}cx^{3/2} = 5x^{3/2}$; $\dfrac{1}{2}c = 1$; $c = 2$. Thus, $g(x) = 2x^{5/2}$.

Theorem 2

If n is any nonzero real number and $f(x) = x^n$, then $f'(x) = nx^{n-1}$.

Note that the results of Examples 1 and 2, where we saw that the derivative of $f(x) = x^2$ is $f'(x) = 2x$ and the derivative of $f(x) = x^3$ is $f'(x) = 3x^2$, agree with Theorem 2.

Theorem 3

If c and n are any nonzero real numbers and $f(x) = cx^n$, then $f'(x) = cnx^{n-1}$.

Theorem 4

If $f(x) = p(x) + q(x)$, then $f'(x) = p'(x) + q'(x)$.

Example 4 Find the derivative of each function.
  a. $f(x) = x^5$ **b.** $f(x) = 3x^5$
  c. $f(x) = 12x^2 - 4x + 7$ **d.** $f(x) = 8\sqrt{x} - \dfrac{7}{x^2}$

Solution **a.** To find the derivative of $f(x) = x^5$, we use Theorem 2 with $n = 5$:
$$f'(x) = 5x^4$$

  b. To find the derivative of $f(x) = 3x^5$, we use Theorem 3 with $c = 3$ and $n = 5$:
$$f'(x) = 3 \cdot 5x^4 = 15x^4$$

  c. We regard $f(x) = 12x^2 - 4x + 7$ as the sum of three functions defined by its terms and use Theorems 1, 3, and 4:
$$f'(x) = 12 \cdot 2x^1 - 4 \cdot 1x^0 + 0$$
$$= 24x - 4$$

  d. We rewrite $f(x) = 8\sqrt{x} - \dfrac{7}{x^2}$ as $f(x) = 8x^{1/2} - 7x^{-2}$ and proceed as in part (c) of this example.
$$f'(x) = 8 \cdot \frac{1}{2}x^{(1/2)-1} - 7(-2)x^{-2-1}$$
$$= 4x^{-1/2} + 14x^{-3} = \frac{4\sqrt{x}}{x} + \frac{14}{x^3}$$

760 *Chapter Twenty*

The line **tangent** to a curve at a point on the curve is defined to be the line passing through the point and having a slope equal to the slope of the curve at that point.

Example 5 Find an equation of the line that is tangent to the curve $y = \sqrt{x}$ at the point $(9, 3)$.

Solution Since $f(x) = \sqrt{x} = x^{1/2}$, $f'(x) = \frac{1}{2}x^{-1/2} = \frac{1}{2\sqrt{x}}$. Therefore:

$$\text{slope of tangent at } (9, 3) = \text{slope of curve at } (9, 3)$$
$$= f'(9)$$
$$= \frac{1}{2\sqrt{9}} = \frac{1}{6}$$

The tangent line has slope $\frac{1}{6}$ and contains the point $(9, 3)$. Thus, an equation of the line is

$$\frac{y - 3}{x - 9} = \frac{1}{6},$$

or $y = \frac{1}{6}x + \frac{3}{2}$.

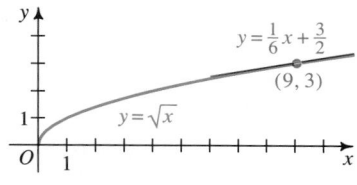

CLASS EXERCISES

In Exercises 1–10, find the derivative of each function. Express the derivative without using fractional or negative exponents.

1. $f(x) = x^8$ $8x^7$

2. $f(x) = x^{12}$ $12x^{11}$

3. $f(x) = x^{-4}$ **3.** $-\frac{4}{x^5}$ **5.** $\frac{3}{2}\sqrt{x}$

4. $g(x) = 3x^7$ $21x^6$

5. $h(x) = x^{3/2}$

6. $f(x) = 4x^{-3/2}$

7. $g(x) = \sqrt[3]{x}$ $\frac{\sqrt[3]{x}}{3x}$

8. $h(x) = \frac{1}{\sqrt[3]{x}}$ **6.** $-\frac{6\sqrt{x}}{x^3}$ **8.** $-\frac{\sqrt[3]{x^2}}{3x^2}$

9. $f(x) = 8x^3 - 7x^2 + 3x - 2$ See below.

10. $g(x) = 4x^{-5} - 2x^{-3} + 9x^{-1} + 5$

11. a. If $f(x) = 4x^2$, then $f'(x) = $ __?__ and $f'(-1) = $ __?__. $8x$; -8

$-\frac{20}{x^6} + \frac{6}{x^4} - \frac{9}{x^2}$

   b. What is the slope of $y = 4x^2$ at the point $(-1, 4)$? -8

12. a. If $f(x) = mx + k$ where $m \neq 0$, find $f'(1)$, $f'(2)$, and $f'(3)$.

   b. *Visual Thinking* Give a geometric interpretation of the results of part (a).

13. *Discussion* The graph of $f(x) = |x|$, shown at the right, has a "corner" at $x = 0$. Using part (b) of Exercise 12, discuss why $f'(0)$ is undefined. (See Example 3 on page 759.)

9. $24x^2 - 14x + 3$

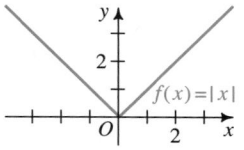

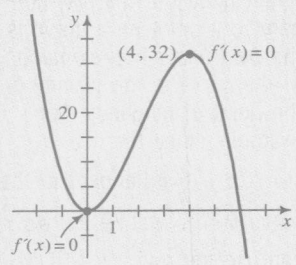
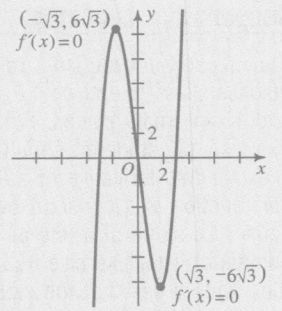
WRITTEN EXERCISES

A 1. $P(1, 1)$ is a point on the graph of $y = x^3$. Find the slope of $\overline{PQ}$ if Q is: 3.31
  a. $(2, 8)$ 7   b. $(1.5, 3.375)$ 4.75   c. $(1.1, 1.331)$
  d. If $f(x) = x^3$, find $f'(x)$ and $f'(1)$. $3x^2$; 3

2. $P\left(2, \dfrac{1}{2}\right)$ is a point on the graph of $y = \dfrac{1}{x}$. Find the slope of $\overline{PQ}$ if Q is:
  a. $(3, f(3))$ $-\dfrac{1}{6}$ b. $(2.5, f(2.5))$ $-\dfrac{1}{5}$ c. $(2.1, f(2.1))$ $-\dfrac{5}{21}$
  d. If $f(x) = \dfrac{1}{x}$, find $f'(x)$ and $f'(2)$. $-\dfrac{1}{x^2}$; $-\dfrac{1}{4}$

Find the derivative of the given function. Express the derivative without using fractional or negative exponents.

3. $f(x) = 2x^5$ $10x^4$ 4. $f(x) = 3x^6$ $18x^5$ 5. $f(x) = -2x^3$ $-6x^2$ 6. $f(x) = 8x^{3/4}$ $\dfrac{6\sqrt[4]{x^3}}{x}$

7. $f(x) = \dfrac{3}{x^2} - \dfrac{6}{x^3}$ 8. $g(x) = \dfrac{4}{\sqrt{x}} - \dfrac{2\sqrt{x}}{x^2}$ 9. $f(x) = \sqrt{5x}$ $\dfrac{\sqrt{5x}}{2x}$ 10. $f(x) = \dfrac{1}{4x^4} - \dfrac{1}{x^5}$

11. $g(x) = \dfrac{7}{2}x^2 - 5x + 3 - \dfrac{1}{x}$ $7x - 5 + \dfrac{1}{x^2}$ 12. $f(x) = \dfrac{4}{3}x^3 - \dfrac{2}{x} - \dfrac{5}{x^2} + \pi$

$$4x^2 + \dfrac{2}{x^2} + \dfrac{10}{x^3}$$

Find the slope of each curve at the given point P.

13. $y = x^3$; $P(-1, -1)$ 3   14. $y = \dfrac{x^4}{2}$; $P\left(-1, \dfrac{1}{2}\right)$ -2

15. $y = 3x^2 - 2x + 1$; $P(0, 1)$ -2 16. $y = 2\sqrt{x}$; $P(9, 6)$ $\dfrac{1}{3}$

17. $y = 8x^4 - 7x^2 + 5x + 6$; $P(-1, 2)$ 18. $y = x^3 - 5x^2 + 4x + 2$; $P(2, -2)$ -4
                       -13

B 19. $y = \dfrac{2}{x} - \dfrac{4}{x^2}$; $P(2, 0)$ $\dfrac{1}{2}$ 20. $y = \sqrt[3]{x}$; $P(-1, -1)$ $\dfrac{1}{3}$

21. a. If $f(x) = ax^2 + bx + c$ where $a \neq 0$, find $f'(x)$. $2ax + b$
  b. **Visual Thinking** Evaluate $f'\left(-\dfrac{b}{2a}\right)$ and explain what the result says about the tangent to the parabola $y = f(x)$ at the vertex. 0; horizontal tangent

22. **Visual Thinking** Give a geometric argument to explain why two functions whose rules differ only by a constant have the same derivative.

Find an equation of the line tangent to the given curve at the given point P.

23. $y = x^3$; $P(-2, -8)$ $y = 12x + 16$ 24. $y = 3x^2 - 5x$; $P(2, 2)$ See below.

25. $y = \dfrac{1}{x^2}$; $P(-1, 1)$ $y = 2x + 3$ 26. $y = \dfrac{1}{\sqrt[3]{x}}$; $P\left(8, \dfrac{1}{2}\right)$

27. a. Sketch the graph of the function $f(x) = 6x^2 - x^3$.
  b. For what values of x does $f'(x) = 0$?
  c. On your graph of $f(x)$, indicate the two points where $f'(x) = 0$.

28. Repeat Exercise 27 using the function $f(x) = x^3 - 9x$.

24. $y = 7x - 12$ 26. $y = -\dfrac{1}{48}x + \dfrac{2}{3}$

29. a. If $f(x) = x^{2/3}$, find $f'(x)$.

    b. For what values of x is $f'(x)$ undefined?

    c. Sketch the graph of $f(x)$. (If you use a computer or graphing calculator, you may need to enter the rule for the function as $(x^2)^{1/3}$ to obtain the complete graph.) Explain how the graph of $f(x)$ supports the result of part (b).

30. Repeat Exercise 29 using the function $f(x) = x^{1/3}$.

In Exercises 31–36, find a function that has the given derivative. Answers may vary by a constant.

31. $f'(x) = 4x^3 \quad x^4$ **32.** $f'(x) = 5x^4 \quad x^5$ **33.** $g'(x) = 3x^5 \quad \frac{1}{2}x^6$

34. $g'(x) = x^7 \quad \frac{1}{8}x^8$ **35.** $h'(x) = 3\sqrt{x} \quad 2x^{3/2}$ **36.** $h'(x) = \sqrt[3]{x} \quad \frac{3}{4}x^{4/3}$

C **37. a.** Use the binomial theorem (page 591) to write the first three terms and the last term in the expansion of $(x + h)^n$.

    b. Use part (a) and the definition of $f'(x)$ to prove Theorem 2 on page 760 for positive integral values of n.

    (*Note:* The binomial theorem can be generalized to apply to any nonzero real number n. This form of the binomial theorem can then be used to give a general proof of Theorem 2.)

38. Prove Theorem 3 on page 760. (*Hint:* Use the definition of $f'(x)$.)

39. Prove Theorem 4 on page 760. (*Hint:* Use the definition of $f'(x)$.)

COMMUNICATION: *Visual Thinking*

The saying "a picture is worth a thousand words" is often true in mathematics. One well-drawn diagram may be a very convincing argument on its own, without being a formal proof. Diagrams and sketches can offer evidence that a conjecture may be true, and they can offer clues that lead to interesting discoveries.

For example, consider the diagram at the right, where the lengths a and b are unequal positive numbers.

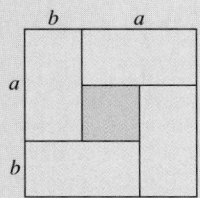

1. Explain how the areas of the squares and rectangles in the diagram allow you to conclude that $(a + b)^2 = (a - b)^2 + 4ab$.

2. Explain how the saying "the whole is greater than its parts" allows you to conclude from the diagram that $(a + b)^2 > 4ab$.

3. By taking the square root of each side of the inequality above, what relationship do you discover between the arithmetic mean, $\dfrac{a + b}{2}$, and the geometric mean, $\sqrt{ab}$, of a and b?

20-2 Using Derivatives in Curve Sketching

Objective *To sketch the graphs of functions using derivatives.*

In Section 2-3, we discussed techniques for sketching the graph of a polynomial function. In this section, we will see how the derivative of a polynomial function can be used in sketching the graph.

Consider the function $f(x) = 6x - x^2$, whose derivative is $f'(x) = 6 - 2x$. Values of x, $f(x)$, and $f'(x)$ are given in the following table.

x	0	1	2	3	4	5	6
$f(x) = 6x - x^2$	0	5	8	9	8	5	0
$f'(x) = 6 - 2x$	6	4	2	0	-2	-4	-6

Much can be learned about the graph of $f(x)$ by plotting each point $(x, f(x))$ from the table and then drawing the tangent to the curve at each point plotted. The tangent in each case is the line through $(x, f(x))$ that has slope $f'(x)$. Figure (a) below shows these points and tangent lines. Figure (b) shows a smooth curve drawn through the points and having the corresponding tangent lines. The smooth curve is the graph of $f(x)$.

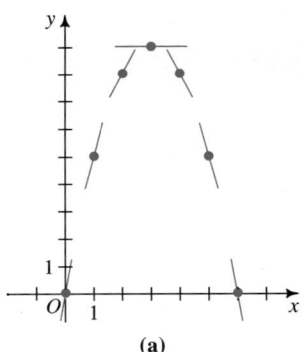
(a)

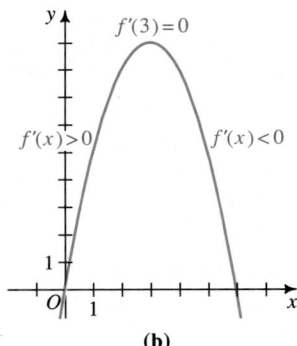
(b)

In figure (b), notice that as x increases, the curve rises whenever $f'(x) > 0$, and the curve falls whenever $f'(x) < 0$. When $f'(x) = 0$, the curve has a horizontal tangent and $f(x)$ attains its maximum value.

We can perform a similar analysis for the graph of
$$f(x) = x^3 - 9x^2 + 24x - 14.$$

We have:
$$f'(x) = 3x^2 - 18x + 24$$
$$= 3(x^2 - 6x + 8)$$
$$= 3(x - 2)(x - 4)$$

764 *Chapter Twenty*

The signs of $f'(x) = 3(x - 2)(x - 4)$ are given in the following table.

	$x < 2$	$x = 2$	$2 < x < 4$	$x = 4$	$x > 4$
$f'(x)$	$+$	0	$-$	0	$+$

From the graph of $f(x)$ shown at the right, we see that the curve is rising whenever $f'(x) > 0$, the curve is falling whenever $f'(x) < 0$, and the curve has a horizontal tangent when $f'(x) = 0$. A point such as $(2, 6)$, where $f'(x) = 0$ and the curve changes from rising to falling as x increases, is called a *local maximum point,* meaning that there is no nearby point for which $f(x) > 6$. Similarly, a point such as $(4, 2)$, where $f'(x) = 0$ and the curve changes from falling to rising as x increases, is called a *local minimum point,* meaning that there is no nearby point for which $f(x) < 2$.

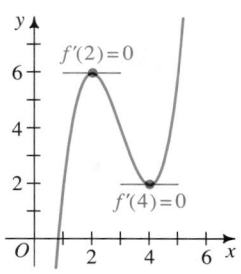

If the graph of a function $f(x)$ has a local maximum (or minimum) point, then the y-coordinate of the point is a local maximum (or minimum) *value* of the function. Thus, $f(x) = x^3 - 9x^2 + 24x - 14$ has a local maximum value of 6 at $x = 2$ and a local minimum value of 2 at $x = 4$.

The observations that we have so far made for polynomial functions are valid for any function $f(x)$ that has a derivative on an interval I:

1. If $f'(x) > 0$ on an interval I, then the graph of $f(x)$ rises as x increases. The function $f(x)$ is said to be *increasing on the interval I.*

2. If $f'(x) < 0$ on an interval I, then the graph of $f(x)$ falls as x increases. The function $f(x)$ is said to be *decreasing on the interval I.*

3. If $f'(c) = 0$, then the graph of $f(x)$ has a horizontal tangent at $x = c$. The function may have a local maximum or minimum value, or neither, as shown below.

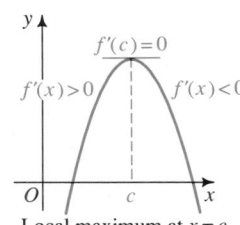

Local maximum at $x = c$

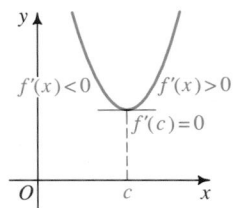

Local minimum at $x = c$

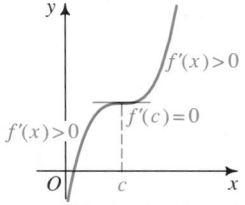

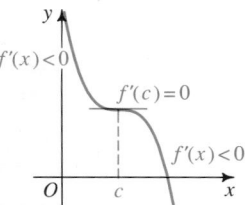

Neither local maximum nor local minimum at $x = c$; $f'(x)$ has the same sign on both sides of $x = c$.

An Introduction to Calculus **765**

Be sure students understand that if $(c, f(c))$ is a local minimum or maximum point on the graph of $f(x)$, then we say that $f(x)$ has a local minimum or maximum *value* of $f(c)$ at $x = c$; that is, the y-coordinate of $(c, f(c))$ gives the extreme value of $f(x)$ and the x-coordinate tells where it occurs.

Have students use the graphs at the bottom of the page to formulate what is known in calculus as the *first-derivative test:* $f(c)$ is a local *maximum value* of f if $f'(c) = 0$ and $f'(x)$ changes from being positive for $x < c$ to being negative for $x > c$; $f(c)$ is a local *minimum value* of f if $f'(c) = 0$ and $f'(x)$ changes from being negative for $x < c$ to being positive for $x > c$.

Additional Examples

1. Use the derivative to find the vertex of the parabola $f(x) = \frac{1}{2}x^2 - 4x + 2$.

$f'(x) = \frac{1}{2} \cdot 2x - 4 = x - 4$
$f'(x) < 0$ when $x < 4$.
$f'(x) = 0$ when $x = 4$.
$f'(x) > 0$ when $x > 4$.
Thus, the vertex occurs at $(4, f(4)) = (4, -6)$.

(*Continues on next page*)

2. Graph the function $f(x) = 2x + x^{-2}$. Identify local maximum and minimum points and asymptotes. $f'(x) = 2 - 2x^{-3}$. $f'(x)$ is undefined when $x = 0$, and $f'(x) = 0$ when $x = 1$. Since $f'(x) < 0$ for $0 < x < 1$, and $f'(x) > 0$ for $x > 1$, a local minimum point is $(1, f(1)) = (1, 3)$. When $|x|$ is large, $f(x) \approx 2x$, so $y = 2x$ is an asymptote. Also, since $\lim_{x \to 0^-} f(x) = \infty$ and $\lim_{x \to 0^+} f(x) = \infty$, the y-axis is an asymptote. The graph of f is shown below.

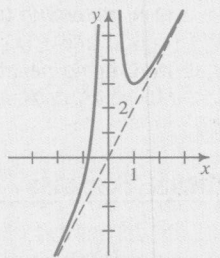

Mathematical Note

Emphasize that $f'(c) = 0$ does not necessarily imply that a local minimum or maximum value of $f(x)$ occurs at $x = c$.

Example 1 Graph $f(x) = x^4 - 4x^3 + 15$. Identify any local maximum and minimum points.

Solution To determine where local maximum and minimum points may occur, we find $f'(x)$ and solve $f'(x) = 0$.

$$f(x) = x^4 - 4x^3 + 15$$
$$f'(x) = 4x^3 - 12x^2$$
$$= 4x^2(x - 3)$$
$$f'(x) = 0 \text{ when } x = 0 \text{ or } x = 3.$$

	$f'(x)$
$x < 0$	$-$
$x = 0$	0
$0 < x < 3$	$-$
$x = 3$	0
$x > 3$	$+$

Note that the graph of $f(x)$ does *not* have a local maximum or minimum point at $x = 0$, since $f'(x)$ is negative on both sides of $x = 0$. The graph of $f(x)$ *does* have a local minimum point at $x = 3$, since $f'(3) = 0$ and the curve changes from falling to rising. The local minimum point is $(3, f(3)) = (3, -12)$.

Example 2 Graph $f(x) = x + x^{-1}$. Identify any local maximum and minimum points.

Solution To determine where local maximum and minimum points may occur, we find $f'(x)$ and solve $f'(x) = 0$.

$$f(x) = x + x^{-1}$$
$$f'(x) = 1 - x^{-2} = 1 - \frac{1}{x^2}$$
$$f'(x) = 0 \text{ when } x = -1 \text{ or } x = 1.$$

	$x < -1$	$x = -1$	$-1 < x < 1,$ $x \neq 0$	$x = 1$	$x > 1$
$f'(x)$	$+$	0	$-$	0	$+$

Neither $f(x)$ nor $f'(x)$ is defined at $x = 0$. A local maximum point is $(-1, -2)$. A local minimum point is $(1, 2)$. Observe that for large $|x|$, $y = x + x^{-1}$ is approximately the same as $y = x$. This indicates that $y = x$ is an asymptote. Also, since

$$\lim_{x \to 0^+} f(x) = \infty$$

and

$$\lim_{x \to 0^-} f(x) = -\infty,$$

the y-axis is an asymptote.

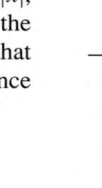

CLASS EXERCISES

Visual Thinking **Study the graph of each polynomial function. Then give the values of x for which $f'(x)$ is (a) equal to 0, (b) positive, and (c) negative.**

1.

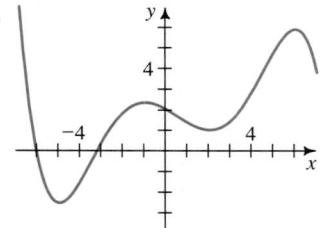

2.

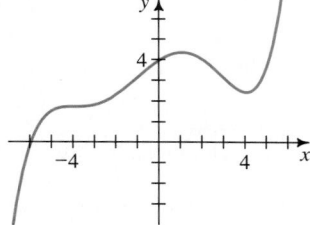

3. *Visual Thinking* Is it possible for a function to have a local maximum value that is less than a local minimum value? Sketch a graph to illustrate.

4. Suppose $f(x)$ is a second-degree polynomial function such that $f'(1) = 4$, $f'(2) = 0$, and $f'(3) = -4$. At $x = 2$, does $f(x)$ have a local maximum value, a local minimum value, or neither? Local maximum

5. The derivative of a polynomial function $f(x)$ has values as given in the table below. Give the values of x for which $f(x)$ has local maximum and local minimum values. Local maximum at $x = -1$; local minimum at $x = 2$

	$x < -1$	$x = -1$	$-1 < x < 0$	$x = 0$	$0 < x < 2$	$x = 2$	$x > 2$
$f'(x)$	+	0	−	0	−	0	+

6. Give the x-coordinates of the local maximum and minimum points on the graph of the function $f(x) = \frac{1}{3}x^3 - 4x$. Local maximum at $x = -2$; Local minimum at $x = 2$

Exercise Note

In Class Exercises 1 and 2, students will need to visualize tangent lines at various points along the curve in order to determine where the slopes are positive, negative, or zero. In Written Exercises 17 and 18, students will need to go one step further and attach (estimated) numerical values to the slopes of the tangents in order to draw approximate graphs of $f'(x)$.

Additional Answers
Class Exercises

1. a. $x = -5$, $x = -1$, $x = 2$, $x = 6$
  b. $-5 < x < -1$, $2 < x < 6$
  c. $x < -5$, $-1 < x < 2$, $x > 6$

2. a. $x = -4$, $x = 1$, $x = 4$
  b. $x < -4$, $-4 < x < 1$, $x > 4$
  c. $1 < x < 4$

3. Yes

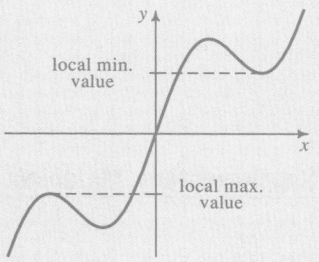

local min. value > local max. value

Additional Answers
Written Exercises

1. Max: $(-2, 16)$; min: $(2, -16)$

2. Max: $(0, 0)$; min: $(2, -4)$

3. Max: $(\sqrt{3}, 6\sqrt{3})$; min: $(-\sqrt{3}, -6\sqrt{3})$

4. Max: $(-3, 6)$; min: $\left(\frac{1}{3}, -\frac{338}{27}\right)$

5. Max: $(-1, 1)$, $(1, 1)$; min: $(0, 0)$

6. Max: $\left(\frac{2}{3}, \frac{1156}{27}\right)$; min: $(-4, -8)$

7. Max: $(0, 8)$; min: $(-\sqrt{3}, -1)$, $(\sqrt{3}, -1)$

8. Max: $(0, -9)$; min: $(-2, -25)$, $(2, -25)$

9. Max: $(0, 0)$; min: $(-1, -5)$, $(2, -32)$

10. Max: $(2, 16)$; no min

11. Max: $(1, 1)$; min: $(0, 0)$

12. Max: $(-1, 32)$; min: $(3, 0)$

Supplementary Materials

Alternative Assessment, 61
Precalculus Plotter Plus, 91, 92

WRITTEN EXERCISES

Graph each function. Identify any local maximum and minimum points.

A
1. $f(x) = x^3 - 12x$
2. $f(x) = x^3 - 3x^2$
3. $f(x) = 9x - x^3$
4. $f(x) = x^3 + 4x^2 - 3x - 12$
5. $g(x) = 2x^2 - x^4$
6. $h(x) = 40 + 8x - 5x^2 - x^3$
7. $g(x) = x^4 - 6x^2 + 8$
8. $f(x) = x^4 - 8x^2 - 9$
9. $f(t) = 3t^4 - 4t^3 - 12t^2$
10. $g(t) = 8t^3 - 3t^4$
11. $h(x) = 5x^4 - 4x^5$
12. $f(x) = x^3 - 3x^2 - 9x + 27$

In Exercises 13–16, use the derivative to find the vertex of each parabola.

13. $y = x^2 - 6x + 4$ $(3, -5)$
14. $y = 5 + 8x - 4x^2$ $(1, 9)$

B
15. $y = (x - a)^2$ $(a, 0)$
16. $y = ax^2 + bx + c$ $\left(-\dfrac{b}{2a}, \dfrac{4ac - b^2}{4a}\right)$

17. Refer to Class Exercise 1 and draw an approximate graph of $f'(x)$.

18. Refer to Class Exercise 2 and draw an approximate graph of $f'(x)$.

19. Show that the function $f(x) = x^3 - 3x^2 + 3x - 5$ has neither a local maximum value nor a local minimum value.

20. The symmetry point of the graph of the equation $y = ax^3 + bx^2 + cx + d$ occurs at $x = -\dfrac{b}{3a}$.
    a. Find the symmetry point S of the graph of $y = x^3 - 6x^2 + 9x - 5$.
    b. Find the coordinates of the local maximum point P and the local minimum point Q of the graph of $y = x^3 - 6x^2 + 9x - 5$.
    c. Show that P, S, and Q are collinear.
    d. Show that S is the midpoint of $\overline{PQ}$.

21. a. Evaluate $\displaystyle\lim_{h \to 0} \frac{\sin(0 + h) - \sin 0}{h}$ to determine the derivative of $\sin x$ when $x = 0$. (*Hint:* See Exercise 43 on page 725.) 1
    b. **Visual Thinking** Make a sketch of the curve $y = \sin x$, where x is in radians and the same scale is used on the x- and y-axes. Then, using your sketch and your answer to part (a), complete the table below.

At the point on the sine curve having x-coordinate:	0	$\dfrac{\pi}{2}$	π	$\dfrac{3\pi}{2}$	2π
The slope of the tangent to the sine curve is:	1 ?	0 ?	−1 ?	0 ?	1 ?

    c. Using the table in part (b), try to guess what function is the derivative of $\sin x$. cos x

---

22. Given that the derivative of $f(x) = \cos x$ is $f'(x) = -\sin x$, show that the function $g(x) = x + 2 \cos x$ has infinitely many local maximum and minimum values.

Graph each function. Identify local maximum and minimum points and asymptotes.

23. $y = x + 2x^{-1}$

24. $y = x^2 + x^{-2}$

C 25. $y = x^2 + 3x + \dfrac{1}{x}$

26. $y = x^2 + \dfrac{1}{x}$

20-3 Extreme Value Problems

Objective   *To solve extreme value problems using derivatives.*

Shown at the right is the graph of the function $f(x) = x^3 - 3x^2 - 9x + 11$ on the interval $-2 \le x \le 6$. Notice that the function has local *extreme* values (that is, local maximum or minimum values) not only at $x = -1$ and $x = 3$, where $f'(x) = 0$, but also at $x = -2$ and $x = 6$, which are the *endpoints* of the interval.

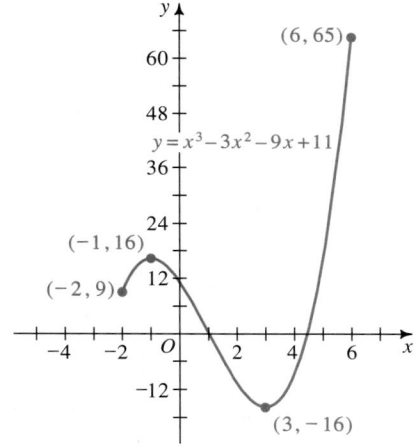

Of the two local minimum values, $f(-2) = 9$ and $f(3) = -16$, we call $f(3) = -16$ the *global minimum value* of $f(x)$ on the interval $-2 \le x \le 6$ because there is no other x-value in the interval for which $f(x)$ is smaller. Likewise, $f(6) = 65$ is the *global maximum value* of $f(x)$ on the interval.

Activity

Use the graph of $f(x) = x^3 - 3x^2 - 9x + 11$ shown above to answer the following questions.   **a.** 16, 9; 11, −16; 11, 0

a. What are the global maximum and minimum values of $f(x)$ on the interval $-2 \le x \le 0$? on the interval $0 \le x \le 4$? on the interval $0 \le x \le 1$?

b. Does $f(x)$ have either a global maximum value or a global minimum value on the interval $-2 < x < 6$? Explain. Global min. exists; global max. doesn't.

c. Consider the function $g(x) = |f(x)|$. What are the global maximum and minimum values of $g(x)$ on the interval $-2 \le x \le 6$? 65, 0

d. In part (c), what is true about $g'(x)$ at those x-values for which $g(x)$ has its global minimum value on the given interval? $g'(x)$ is undefined.

Teaching Notes, p. 756C

Warm-Up Exercises

Identify any local maximum and minimum points of the given function.

1. $f(x) = x(200 - 2x)$
   (50, 5000) is a local max.

2. $g(x) = x(4 - 2x)(4 - 2x)$
   local max. at $\left(\dfrac{2}{3}, 4\dfrac{20}{27}\right)$;
   local min. at (2, 0)

3. $h(x) = x(6 - x^2)$ local min. at $(-\sqrt{2}, -4\sqrt{2})$;
   local max. at $(\sqrt{2}, 4\sqrt{2})$

4. $k(x) = x^3 - x^4$ local max. at $\left(\dfrac{3}{4}, \dfrac{27}{256}\right)$

Motivating the Section

Remind students that in Chapter 2 they used technology to approximate the extreme values of polynomial functions. In this section they will see how derivatives can be used not to approximate, but to find exactly, the extreme values of polynomial and other functions.

Activity Note

The point of the activity on this page is to show students that a continuous function always has global minimum and maximum values on a *closed* interval (that is, an interval for which the endpoints are included). Moreover, the global extreme values occur either at an endpoint of the interval or at a point where the derivative of the function is 0 or undefined.

The preceding discussion and activity suggest that a continuous function $f(x)$ always has a global maximum value and a global minimum value on the interval $a \le x \le b$. Moreover, these global extreme values will occur:

1. at the endpoints $x = a$ or $x = b$ of the interval,
2. at x-values between a and b where $f'(x) = 0$, or
3. at x-values between a and b where $f'(x)$ is undefined.

When we find the global maximum or minimum value of a function on a given interval, we say that we are *maximizing* or *minimizing* the function.

Example 1   Congruent squares are cut from the corners of a 1 m square piece of tin, and the edges are then turned up to make an open rectangular box. How large should the squares cut from the corners be in order to maximize the volume of the box?

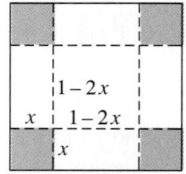

Solution   Letting x represent the length of a side of the square cut from a corner, we can express the volume V of the box as a function of x:

$$V(x) = (1 - 2x)(1 - 2x)x$$
$$= x - 4x^2 + 4x^3$$

The domain of $V(x)$ is $0 \le x \le \frac{1}{2}$. Since $V'(x) = 1 - 8x + 12x^2$ is always defined, the global maximum value of $V(x)$ must occur at an endpoint of the domain or where $V'(x) = 0$. Since $V(0) = V\left(\frac{1}{2}\right) = 0$, we can eliminate these values. We next solve $V'(x) = 0$:

$$1 - 8x + 12x^2 = 0$$
$$(1 - 6x)(1 - 2x) = 0$$
$$x = \frac{1}{6} \quad \text{or} \quad x = \frac{1}{2}$$

Since we have already eliminated $x = \frac{1}{2}$, we check $x = \frac{1}{6}$:

	$0 < x < \frac{1}{6}$	$x = \frac{1}{6}$	$\frac{1}{6} < x < \frac{1}{2}$
$V'(x) = (1 - 6x)(1 - 2x)$	$+$	0	$-$

Thus, the global maximum value of $V(x)$ occurs at $x = \frac{1}{6}$. The squares cut from each corner should therefore be $\frac{1}{6}$ m by $\frac{1}{6}$ m for the box to have its greatest volume. Since $V\left(\frac{1}{6}\right) = \frac{2}{27}$, the maximum volume is $\frac{2}{27}$ m^3.

Example 2   A manufacturer produces cardboard boxes with square bases. The top of each box is a double flap that opens as shown. The bottom of the box has a double layer of cardboard for strength. If each box must have a volume of 12 ft^3, what dimensions will minimize the amount of cardboard used?

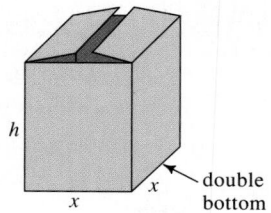

Solution   **Step 1**   Let the dimensions of the box be x, x, and h, as indicated above. Express the amount of cardboard C as a function of x and h:

$$C(x, h) = \text{double bottom} + \text{top} + 4 \text{ sides}$$
$$= 2x^2 + x^2 + 4xh$$
$$= 3x^2 + 4xh$$

Step 2   Express C as a function of one variable, x, by using the fact that the volume is fixed at 12 ft^3:

$$\text{Volume} = x^2 h = 12$$
$$h = \frac{12}{x^2}$$

Therefore, $C(x) = 3x^2 + 4x\left(\dfrac{12}{x^2}\right) = 3x^2 + \dfrac{48}{x}$, where $x > 0$.

Step 3   Find $C'(x)$ and solve $C'(x) = 0$:

$$C'(x) = 6x - 48x^{-2} = 0$$
$$6x = \frac{48}{x^2}$$
$$6x^3 = 48$$
$$x = 2$$

Note that when $x = 2$, $h = \dfrac{12}{x^2} = 3$.

Step 4   Check to see whether $x = 2$ gives a minimum or maximum value of $C(x)$:

	$0 < x < 2$	$x = 2$	$x > 2$
$C'(x) = 6x - 48x^{-2}$ $= \dfrac{6x^3 - 48}{x^2}$ $= \dfrac{6(x^3 - 8)}{x^2}$	$-$	0	$+$

Thus, $C(x)$ has a minimum value when $x = 2$. The dimensions of the box should be 2 ft by 2 ft by 3 ft.

An Introduction to Calculus   **771**

CLASS EXERCISES

For each function defined on the given interval, determine where the global extreme values occur and what they are.

1. $f(x) = 2x - 1$; $0 \le x \le 3$

2. $f(x) = -x + 3$; $1 \le x \le 4$

3. $g(x) = x^2 + 2$; $-1 \le x \le 2$

4. $g(x) = 2 - x^2$; $-2 \le x \le 1$

5. $h(x) = x^3 - 3x$; $0 \le x \le 2$

6. $h(x) = x^3 + 3x$; $-2 \le x \le 2$

7. $f(x) = 2 \sin x$; $0 \le x \le \pi$

8. $f(x) = \cos \frac{x}{2}$; $0 \le x \le \pi$

WRITTEN EXERCISES

A 1. The base of a rectangle is on the x-axis and the upper two vertices are on the parabola $y = 3 - x^2$, as shown at the left below.
   a. Show that the area of the rectangle is $A(x) = 6x - 2x^3$, $x \ge 0$.
   b. Show that the rectangle has the greatest area when $x = 1$.

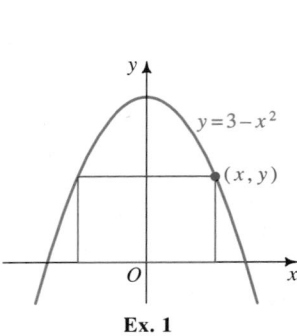

Ex. 1

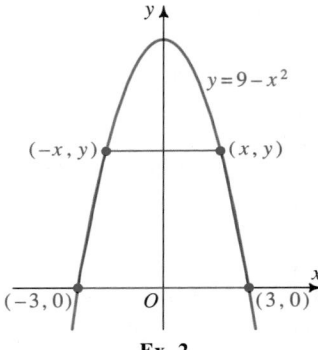

Ex. 2

2. An isosceles trapezoid is inscribed in the parabola $y = 9 - x^2$ with the base of the trapezoid on the x-axis, as shown at the right above.

   a. Show that the area of the trapezoid is $A(x) = \frac{1}{2}(9 - x^2)(6 + 2x)$, $x \ge 0$.

   b. Show that the trapezoid has its greatest area when $x = 1$.

3. A rectangle has one vertex at the origin, another on the positive x-axis, another on the positive y-axis, and the fourth vertex on the line $y = 8 - 2x$. What is the greatest area the rectangle can have? 8 square units

4. A rectangle has area A, where A is a constant. The length of the rectangle is x.
   a. Express the width of the rectangle in terms of A and x.
   b. Express the perimeter in terms of A and x.
   c. Find the length and width that minimize the perimeter.
   d. Of all rectangles with a given area, which one has the least perimeter?

B **5. Manufacturing** Congruent squares are cut from the corners of a rectangular sheet of metal that is 8 cm wide and 15 cm long. The edges are then turned up to make an open box. Let x be the length of a side of the square cut from each corner. $V(x) = x(8 - 2x)(15 - 2x)$, $0 \le x \le 4$
   a. Express the volume V of the box as a function of x.
   b. What value of x maximizes the volume? $\frac{5}{3}$ cm

6. Manufacturing A 1 m by 2 m sheet of metal is cut and folded, as shown, to make a box with a top.
   a. Express the volume V of the box as a function of x.
   b. What value of x maximizes the volume?

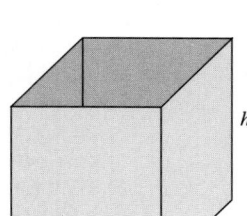
Ex. 6

7. The graphs of $y = x$ and $y = x^3$ intersect three times.
   a. Find the coordinates of the points of intersection. $(-1, -1)$, $(0, 0)$, $(1, 1)$
   b. For what positive value of x between the intersection points is the vertical distance between the graphs the greatest? $\frac{\sqrt{3}}{3}$

8. Manufacturing A manufacturer produces a metal box with a square base, no top, and a volume of 4000 cm³.
   a. If the base edges are x cm long, express the height h in terms of x.
   b. Show that the area of the metal used in the box is given by the function

   $$A(x) = x^2 + 16{,}000x^{-1}.$$

   c. What values of x and h minimize the amount of metal used to make the box?

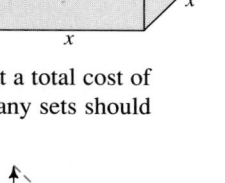

9. Business A toy manufacturer can make x toy tool sets a day at a total cost of $(300 + 12x + 0.2x^{3/2})$ dollars. Each set sells for \$18. How many sets should be made each day in order to maximize the profit? 400

10. Point B is 100 km east of point A. At noon, a truck leaves A and travels north at 60 km/h, while a car leaves B and travels west at 80 km/h.
   a. Write an expression for the distance d between the car and the truck after t hours of travel. See below.
   b. When will the distance between the vehicles be a minimum? (*Hint:* Minimize the *square* of the distance.) 12:48 P.M.
   c. What is the minimum distance between the vehicles? 60 km

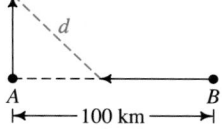
100 km

11. Manufacturing A box company produces a box with a square base and no top that has a volume of 8 ft³. Material for the bottom costs \$6/ft² and the material for the sides costs \$3/ft². Find the dimensions of the box that minimize the cost. 2 ft by 2 ft by 2 ft

C **12.** A cone has height 12 and radius 6. A cylinder is inscribed in the cone, as shown.
   a. Show that $h = 12 - 2r$ for $0 \le r \le 6$.
   b. Find the maximum volume of the inscribed cylinder.

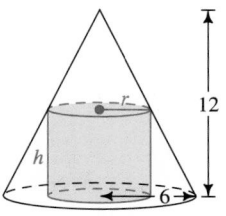
Ex. 12

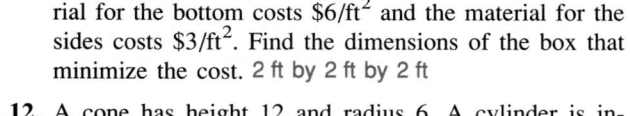

10. a. $d(t) = \sqrt{(60t)^2 + (100 - 80t)^2}$, $t \ge 0$ *An Introduction to Calculus*   **773**

Error Analysis

You may want to discuss the domain restrictions for a problem like Exercise 5. Here the value of x must be less than 4 cm because the sheet of metal is only 8 cm wide. Cutting away two 4-cm squares would use the entire width of the sheet. Similar arguments can be made for the other exercises.

Exercise Note

Suggest that students draw a cross-sectional diagram for Exercise 12 by using a plane through the vertex and perpendicular to the base. Students should then use similar triangles to solve for h in terms of r.

Additional Answers
Written Exercises

4. a. Width = $\frac{A}{x}$
   b. Perimeter = $2x + \frac{2A}{x}$
   c. Length = width = $\sqrt{A}$
   d. The square with the given area

6. a. $V(x) = x(1 - x)(1 - 2x)$, $0 \le x \le \frac{1}{2}$
   b. $x = \frac{3 - \sqrt{3}}{6} \approx 0.21$ m

8. a. $h = \frac{4000}{x^2}$
   b. $A(x) = x^2 + 4xh =$ $x^2 + 4x\left(\frac{4000}{x^2}\right) =$ $x^2 + \frac{16{,}000}{x} =$ $x^2 + 16{,}000x^{-1}$
   c. $x = 20$ cm and $h = 10$ cm

Additional Answers
Written Exercises

13. b. $\frac{32\pi}{3}$ cubic units

14. $\frac{4000\pi\sqrt{3}}{9}$ cubic units

15. a. $6.60; $6.52
   b. 60 mi/h
   c. A speed of 60 mi/h may be over the local speed limit or it may be an unsafe speed given the truck design and load.

Teaching Notes, p. 756D

Warm-Up Exercises

A ball is thrown upward from the top of a 45-ft hill with a velocity of 16 ft/s. The ball's height in feet above the ground t seconds after being thrown is given by $h(t) = 45 + 16t - 16t^2$.

1. How many seconds after being thrown does the ball reach its maximum height? 0.5 s

2. What is the maximum height reached by the ball? 49 ft

3. How many seconds after being thrown does the ball hit the ground? 2.25 s

4. For what values of t is the ball falling? $0.5 < t < 2.25$

5. Give the domain and the range of h (from the time the ball is thrown until it first hits the ground).
domain: $\{t | 0 \le t \le 2.25\}$;
range: $\{h(t) | 0 \le h(t) \le 49\}$

13. A cone is inscribed in a sphere with radius 3, as shown at the right.
   a. Show that the volume of the cone is

$$V(x) = \frac{1}{3}\pi(9 - x^2)(3 + x) \text{ for } 0 \le x \le 3.$$

   b. Find the maximum volume of the cone.

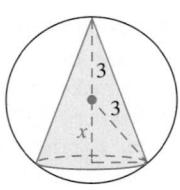

Ex. 13

14. Find the maximum volume of a cylinder inscribed in a sphere with radius 10.

15. **Business** It costs a construction company $(6 + 0.004x)$ dollars per mile to operate a truck at x miles per hour. In addition, it costs $14.40 per hour to pay the driver.
   a. What is the total cost per mile if the truck is driven at 30 mi/h? at 40 mi/h?
   b. At what speed should the truck be driven to minimize the total cost per mile?
   c. Why might the company decide to *not* require the use of the speed found in part (b)?

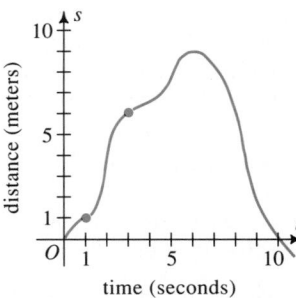

20-4 Velocity and Acceleration

Objective   *To find instantaneous velocities and accelerations.*

Imagine a toy car that can move forward or backward along a long straight track. If s is the directed distance of the car from its starting point, then s is a function of time t. This *position function* is illustrated in the graph below.

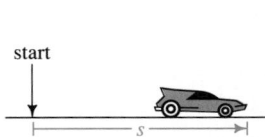

We can tell a great deal about the motion of the car from the graph. For example, the car apparently slowed and nearly stopped at time $t = 4$ s. At time $t = 6$ s, the car stopped moving forward and began to move backward. At time

$t = 10$ s, the car returned to its starting point. After 10 s, s is negative, indicating that the car is to the left of its starting point.

We can also learn about the car's velocity from the graph. For example, if we consider the time interval from $t = 1$ s to $t = 3$ s, we see that the car has moved from 1 m to the right of its starting point to 6 m to the right, a distance of 5 m in 2 s. Thus, the *average velocity* of the car in that 2-second interval is $\frac{5}{2} = 2.5$ m/s.

In general, suppose the positions of the car are s_1 and s_2 at times t_1 and t_2, respectively. If we use the Greek letter Δ (pronounced "delta") to denote "change in," then the car's change in position is $\Delta s = s_2 - s_1$, and the change in time is $\Delta t = t_2 - t_1$. Thus, the **average velocity** of the car is:

$$\text{average velocity} = \frac{\Delta s}{\Delta t} = \frac{s_2 - s_1}{t_2 - t_1}$$
$$= \text{slope of line joining } (t_1, s_1) \text{ and } (t_2, s_2)$$

Notice that the average velocity can be interpreted geometrically as the slope of the line joining points (t_1, s_1) and (t_2, s_2) on the graph of the position function. When the slope is positive, the velocity is positive and the car is moving forward. When the slope is negative (as shown in figure (a) below), the velocity is negative and the car is moving backward.

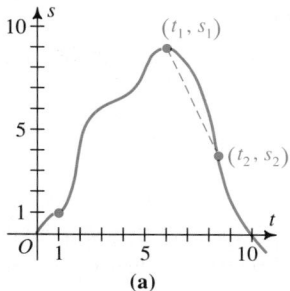

(a)

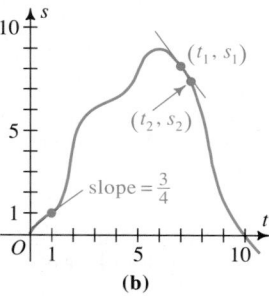

(b)

When the time interval between t_1 and t_2 is very small, the line joining (t_1, s_1) and (t_2, s_2) is almost the same as the line tangent to the graph at (t_1, s_1), as shown in figure (b) above. The slope of this tangent line is called the *instantaneous velocity* of the car at the time t_1 to distinguish it from the average velocity of the car over the time interval containing t_1. For example, figure (b) shows that the instantaneous velocity of the car at time $t = 1$ s is $\frac{3}{4}$ m/s.

In general, if the position of an object is given as a function $s(t)$ of time, then the **instantaneous velocity** at time t is defined as follows:

$$v(t) = \text{instantaneous velocity} = \lim_{\Delta t \to 0} \frac{s(t + \Delta t) - s(t)}{\Delta t} = s'(t)$$

Note that the instantaneous velocity is the derivative of the position function.

Motivating the Section

Ask students to imagine a driver who has just pulled onto a highway. At first, the driver depresses the gas pedal in order to increase the car's speed (which is the magnitude of the car's velocity); because the rate of change in the speed is positive, the car is said to be accelerating. When the speed limit is reached, the driver holds the gas pedal steady to maintain the car's speed; the acceleration is zero. The velocity and acceleration of a moving object are the subject of this section.

Communication Note

Point out that *average velocity* indicates the rate of change in position with respect to an interval of time, while *instantaneous velocity* indicates the rate of change in position at an instant of time.

Mathematical Note

If the graph of a position function is not a straight line, the average velocity depends on the time interval used. Instantaneous velocity is defined by a limiting process involving smaller and smaller intervals of time. As smaller time intervals are considered, Δs and Δt both approach 0, but their ratio $\frac{\Delta s}{\Delta t}$ approaches a limit, namely the derivative of the position function at a given instant. Graphically, the instantaneous velocity is the slope of the tangent to the position curve.

Communication Note

In common language, accelerating means "picking up speed." The technical definition from physics involves the rate at which the velocity changes. For linear motion, increases in velocity imply positive acceleration, decreases in velocity imply negative acceleration (or deceleration), and constant velocity implies zero acceleration.

Mathematical Note

Point out that a "freely falling" object near the surface of Earth experiences a constant acceleration of approximately -32 ft/s$^2 \approx -9.8$ m/s^2 due to gravity. In the case of the thrown ball described on this page, $a(t) = h''(t) = -10$.

Additional Examples

1. A ball is thrown upward from the top of an 85 m building with a velocity of 80 m/s. Its height in meters above the ground after t seconds is approximately $h(t) = 85 + 80t - 5t^2$.
   a. Find the average velocity from $t = 0$ s to $t = 3$ s.
   b. Find the instantaneous velocity at time t and at time $t = 3$ s.
   c. Find the maximum height reached by the ball.
   d. For what values of t is the ball falling?
   e. Show that the acceleration at time t is constant.

   a. $\dfrac{h(3) - h(0)}{3 - 0} = 65$ m/s
   b. $h'(t) = 80 - 10t$;
   $h'(3) = 80 - 10(3) = 50$ m/s

Consider now a ball that is thrown vertically upward with an initial speed of 30 m/s. Its height in meters t seconds later is approximately:

$$h(t) = 30t - 5t^2$$

The average velocity of the ball between times $t = 1$ s and $t = 2$ s is:

$$\text{average velocity} = \frac{h(2) - h(1)}{2 - 1}$$

$$= \frac{40 - 25}{2 - 1} = 15 \text{ m/s}$$

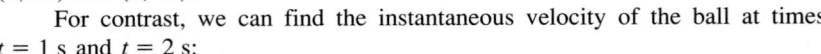

In the graph of $h(t)$ at the right, notice that the average velocity is the slope of the line joining the points $(1, 25)$ and $(2, 40)$.

For contrast, we can find the instantaneous velocity of the ball at times $t = 1$ s and $t = 2$ s:

$$\text{instantaneous velocity} = h'(t)$$
$$h(t) = 30t - 5t^2$$
$$h'(t) = 30 - 10t$$
$$h'(1) = 20 \quad \text{and} \quad h'(2) = 10$$

The instantaneous velocity of the ball is 20 m/s at $t = 1$ s and 10 m/s at $t = 2$ s. These values are, of course, the slopes of the tangents to the graph of the function at the points where $t = 1$ and $t = 2$. The velocity of the ball is 0 m/s when $t = 3$ s. At this time the ball has reached its maximum height of $h(3) = 45$ m. The ball "stops" for an instant, changing its direction of motion from upward ($h'(t) > 0$) to downward ($h'(t) < 0$).

Acceleration is a measure of the rate of change of the velocity of an object. If the position function of the object is $s(t)$, then the velocity function is $v(t) = s'(t)$. The **average acceleration** from time t_1 to t_2 is:

$$\text{average acceleration} = \frac{v(t_2) - v(t_1)}{t_2 - t_1} = \frac{s'(t_2) - s'(t_1)}{t_2 - t_1}$$

The **instantaneous acceleration** at time t is defined as:

$$a(t) = \text{instantaneous acceleration} = \lim_{\Delta t \to 0} \frac{v(t + \Delta t) - v(t)}{\Delta t} = v'(t)$$

Acceleration is the derivative of velocity, and velocity is the derivative of the position function. Acceleration is called the *second derivative* of the position function, and we write:

$$a(t) = s''(t)$$

Since acceleration is the rate of change of velocity over time, it is given in units such as meters per second per second, (m/s)/s, or feet per second per second, (ft/s)/s. These are often abbreviated as m/s^2 and ft/s^2, respectively.

776 *Chapter Twenty*

Example 1 | An object moves along a straight line so that the object's distance in meters to the right of its starting point after t seconds is given by the function $s(t) = t^3 - 3t^2 + 5t$. Find the velocity and acceleration of the object at the end of the first two seconds of motion.

Solution

Position function: $\qquad\qquad s(t) = t^3 - 3t^2 + 5t$

Velocity function: $\qquad v(t) = s'(t) = 3t^2 - 6t + 5$

Acceleration function: $\quad a(t) = s''(t) = 6t - 6$ ⟵ derivative of $s'(t)$

Thus, when $t = 2$ s, the velocity of the object is $s'(2) = 5$, or 5 m/s to the right, and the acceleration is $s''(2) = 6$, or 6 m/s^2 to the right.

Example 2 | Suppose a particle travels along a number line. The graph at the right shows the particle's distance s to the right of the origin as a function of the time t. Sketch approximate graphs of **(a)** the velocity function $v(t)$, and **(b)** the acceleration function $a(t)$.

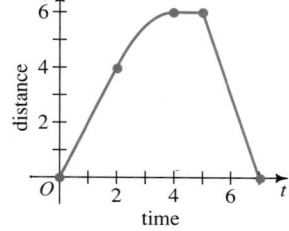

Solution

a. The velocity function $v(t)$ is the derivative of the distance function $s(t)$. This means that $v(t)$ gives the slope of the graph of $s(t)$. Thus:

For $0 \le t \le 2$, $v(t) = 2$.

For $2 \le t \le 4$, $v(t)$ decreases to 0.

For $4 \le t < 5$, $v(t) = 0$.

For $5 < t \le 7$, $v(t) = -3$.

Note that $v(t)$ is not defined for $t = 5$ because the graph of $s(t)$ has a "corner" there.

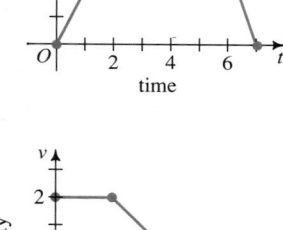

b. The acceleration function $a(t)$ is the derivative of the velocity function $v(t)$. This means that $a(t)$ gives the slope of the graph of $v(t)$. Thus:

For $0 \le t < 2$, $a(t) = 0$.

For $2 < t < 4$, $a(t) = -1$.

For $4 < t < 5$, $a(t) = 0$.

For $5 < t \le 7$, $a(t) = 0$.

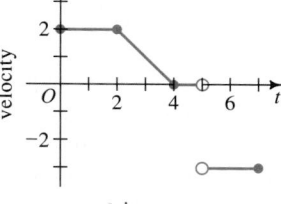

Note that $a(t)$ is not defined for $t = 2$ and $t = 4$ because the graph of $v(t)$ has "corners" there. Also, $a(t)$ is not defined at $t = 5$ because $v(t)$ is not defined there.

An Introduction to Calculus **777**

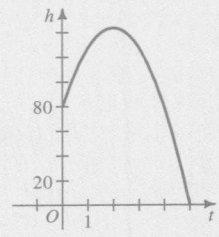
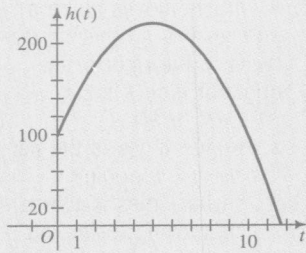
CLASS EXERCISES

1. A ball is thrown upward so that its height in feet above the ground after t seconds is $h(t) = 48t - 16t^2$.
   a. What is the velocity at time t? at time $t = 1$ s? $(48 - 32t)$ ft/s; 16 ft/s
   b. What is the acceleration at time t? -32 ft/s^2

2. *Visual Thinking* A distance-time graph for a toy car is shown at the right.
   a. How far does the toy car travel in the first two seconds? 2 m
   b. What is its average velocity during the first two seconds? 1 m/s
   c. When does the car move to the right? to the left? $0 \le t < 2$; $t > 2$
   d. When does the car return to its starting point? $t = 4$ s **2. e.** 1 m/s; 0 m/s; -1 m/s
   e. What is the instantaneous velocity of the car at $t = 1$ s? $t = 2$ s? $t = 3$ s?
   f. At what times does the car move fastest to the right? to the left? $t = 0$ s; $t = 4$ s
   g. What is the farthest the car moves to the right during the first 6 seconds? 2 m

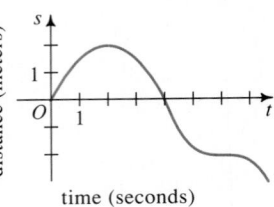

3. A particle moves along a line so that its distance in meters to the right of its starting point after t seconds is $s(t) = t^2 - 5t$.
   a. Does the particle begin moving left or right from the origin? Left
   b. Where is the particle after 1 s? 4 m to the left of the origin
   c. How far does the particle move in the first two seconds? 6 m
   d. What is its average velocity over the first two seconds? -3 m/s
   e. When does the particle return to the origin? $t = 5$ s
   f. What is the acceleration of the particle at time $t = 1$ s? at $t = 2$ s? 2 m/s^2; 2 m/s^2

4. Suppose a particle moves along a number line and its position to the right of the origin is given by the function $s(t)$, where t represents time.
   a. If $s(t) < 0$ at some time t, then the particle is __?__ at that time.
   b. If $s'(t) > 0$ at some time t, then the particle is __?__ at that time.
   c. If $s''(t) < 0$ at some time t, then the particle is __?__ at that time.
   4. a. to the left of the origin **b.** moving to the right **c.** decelerating

WRITTEN EXERCISES

A 1. **Physics** A ball is thrown upward from the top of an 80 ft building so that its height in feet above the ground after t seconds is $h(t) = 80 + 64t - 16t^2$.
   a. Find the average velocity over the interval from $t = 0$ s to $t = 2$ s. 32 ft/s
   b. What is the instantaneous velocity at time t? at time $t = 1$ s? $(64 - 32t)$ ft/s; 32 ft/s
   c. When is the velocity equal to 0? $t = 2$ s
   d. What is the ball's maximum height above the ground? 144 ft
   e. When does the ball hit the ground? $t = 5$ s
   f. For what values of t is the ball falling? $2 < t < 5$
   g. What is the acceleration at time t? -32 ft/s^2
   h. Graph the function $h(t)$.

2. Physics A helicopter climbs vertically from the top of a 98 m building so that its height in meters above the ground after t seconds is:

$$h(t) = 98 + 49t - 4.9t^2$$

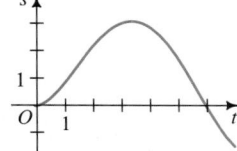

a. Find the average velocity over the interval from $t = 0$ s to $t = 2$ s.

b. What is the instantaneous velocity at time t? at time $t = 1$ s?

c. When is the velocity equal to 0?

d. What is the helicopter's maximum height above the ground?

e. When does the helicopter reach the ground?

f. For what values of t is the helicopter descending?

g. What is the acceleration at time t?

h. Graph the function $h(t)$.

Visual Thinking **In Exercises 3 and 4, answer the questions of Class Exercise 2 for the graphs shown below. In each graph, s is the distance (in meters) to the right of the starting point and t is the time (in seconds).**

3.

4.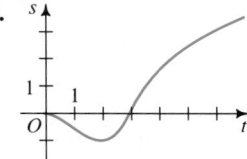

5. *Visual Thinking* Study the graph of $s(t)$ in Exercise 3. Then sketch an approximate graph of the velocity function $v(t)$.

6. *Visual Thinking* Study the graph of $s(t)$ in Exercise 4. Then sketch an approximate graph of the velocity function $v(t)$.

Visual Thinking **In Exercises 7 and 8, copy the given graph of $s(t)$. Then, on separate sets of axes aligned vertically with the copied graph, sketch the graphs of $v(t)$ and $a(t)$.**

B **7.**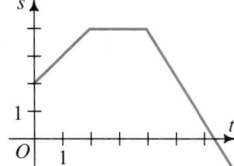

8.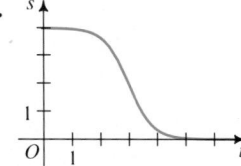

Exercise Note

In Written Exercises 7 and 8, students should sketch the graph of $v(t)$ by estimating the slopes of tangents to the graph of $s(t)$. Likewise, they should sketch the graph of $a(t)$ by estimating the slopes of tangents to the graph of $v(t)$. (Of course, since the graph of $a(t)$ is based on estimates of estimates, you should not expect very accurate acceleration graphs.)

Additional Answers Written Exercises

3. a. 2 m
   b. 1 m/s
   c. $0 \le t < 3.5$; $t > 3.5$
   d. $t = 6$ s
   e. 1 m/s; 1 m/s; 0.5 m/s
   f. $t = 1.5$ s; $t = 5$ s
   g. 3 m

4. a. 1 m
   b. -0.5 m/s
   c. $t > 2$; $0 \le t < 2$
   d. $t = 3$ s
   e. -0.5 m/s; 0 m/s; 2 m/s
   f. $t = 3$ s; $t = 1$ s
   g. 3 m

5.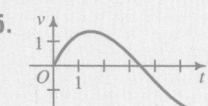

6.

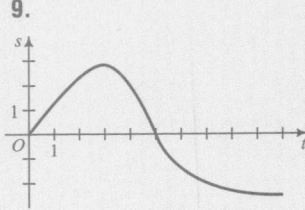
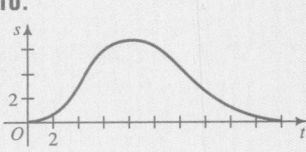
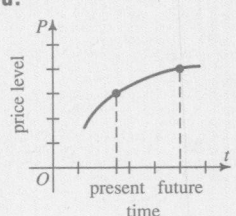
Visual Thinking In Exercises 9 and 10, read the given description of the motion of a particle along a number line. Then make an approximate graph (like those in Exercises 3 and 4) showing the distance of the particle to the right of the origin as a function of time.

9. A particle leaves the origin, traveling at a velocity of 1 m/s. Then it gradually slows until it reverses its direction at time $t = 3$ s. It returns to the origin at $t = 5$ s, traveling at a velocity of -2 m/s and then gradually slows until it stops to the left of the origin at time $t = 10$ s.

10. A particle leaves the origin, traveling right very slowly, and gradually increasing its velocity until it reaches 2 m/s at time $t = 4$ s. Then the velocity begins to decrease and finally becomes zero at $t = 8$ s. At this time, the particle begins to move slowly left, and at $t = 20$ s it finally returns to its starting point and stops.

11. **Economics** If $P(t)$ represents the general price level of goods and services in an economy at a given time t, then during an inflationary period, $P(t)$ increases rapidly. Suppose that our economy is in such an inflationary period, but that the president's economic advisers find that the rate of inflation is slowing. In a subsequent address to the nation, the president announces that "while inflation is still with us, it has been brought under control" and predicts that "prices will soon stabilize."
   a. Use a derivative to describe why "inflation is still with us."
   b. Use a derivative to describe why the president believes that inflation "has been brought under control."
   c. Use a derivative to describe the president's prediction that "prices will soon stabilize."
   d. *Visual Thinking* Assuming that the president is correct, sketch a graph of $P(t)$ from the present to the time when prices stabilize.

12. **Biology** At the right is the graph of $P(t)$, which shows the growth over time in the population of a protected endangered species.
   a. Compare the signs of $P'(t_1)$ and $P''(t_1)$ with $P'(t_2)$ and $P''(t_2)$ at times t_1 and t_2.
   b. *Writing* What conclusions might you expect biologists who are studying the endangered species to make at times t_1 and t_2?

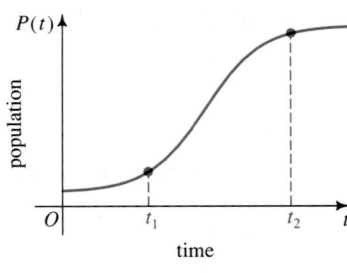

13. **Physics** A particle moves along a number line so that its distance to the right of the origin at time t is $s(t) = 2t^3 - 6t^2 + 8$.
   a. At what times is the particle at the origin? $t = 2$ s
   b. At what times is the particle not moving? $t = 0$ s and $t = 2$ s
   c. At what time is the velocity of the particle neither increasing nor decreasing; that is, at what time is the particle neither accelerating nor decelerating? $t = 1$ s

14. a. 7.5 mi/h eastward **b.** 15 mi west of Rockford; 1:00 P.M.

14. Physics Suppose a car is traveling so that its distance, in miles, west of Rockford at a time t hours after noon is $s(t) = 10t^{3/2} - 15t + 20$.

   a. How fast is the car going, and in which direction, at 12:15 P.M.?
   b. Where is the car, and what is the time, when its velocity is zero?

C 15. Physics A ball is thrown horizontally from the top of a 144 ft building with an initial speed of 80 ft/s. Unlike Exercise 1 where the motion of the ball is along a line, here the motion is in a plane and the position (x, y) of the ball at time t seconds is a vector $\mathbf{s}(t) = (x(t), y(t)) = (80t, 144 - 16t^2)$.

   a. Make a sketch showing the path of the ball and its position at times $t = 0$ s, $t = 1$ s, $t = 2$ s, and $t = 3$ s.
   b. Show that the distance between the ball's positions at times $t = 1$ s and $t = 2$ s is approximately 93.3 ft.
   c. From $t = 1$ s to $t = 2$ s, the ball actually traveled along a curved path which is a little longer than 93.3 ft. Thus, its average speed over the time interval is somewhat more than 93.3 ft/s. Use similar reasoning to find an approximate average speed of the ball over the time interval from $t = 1.9$ s to $t = 2.0$ s.
   d. Using calculus, one can show that the instantaneous velocity of the ball is the vector $\mathbf{v}(t) = (x'(t), y'(t))$. In other words, the components of the velocity vector are the derivatives of the components of the position vector. Moreover, the speed of the ball at time t is $|\mathbf{v}(t)| = \sqrt{(x'(t))^2 + (y'(t))^2}$. Use this information to find the velocity vector and speed of the ball when $t = 2$ s and at the instant the ball hits the ground.

Chapter Summary

1. The *derivative* of a function $f(x)$ is denoted $f'(x)$ and gives the slope of the graph of $f(x)$ at the point $(x, f(x))$. By definition,

$$f'(x) = \lim_{h \to 0} \frac{f(x + h) - f(x)}{h}.$$

2. Theorems 1–4 on pages 759 and 760 offer short cuts for calculating derivatives.

3. The graph of a function $f(x)$ rises where $f'(x) > 0$ and falls where $f'(x) < 0$.

4. A point $(c, f(c))$ is a *local maximum point* or a *local minimum point* on the graph of a function $f(x)$ if $f'(c) = 0$ and $f'(x)$ changes sign at $x = c$. The y-coordinate of a local maximum or minimum point is the local maximum or minimum *value* of the function.

5. If $s(t)$ is the position function of an object at time t, then the *instantaneous velocity* of the object at time t is:

$$v(t) = \lim_{\Delta t \to 0} \frac{s(t + \Delta t) - s(t)}{\Delta t} = s'(t)$$

The velocity function is the derivative of the position function.

An Introduction to Calculus **781**

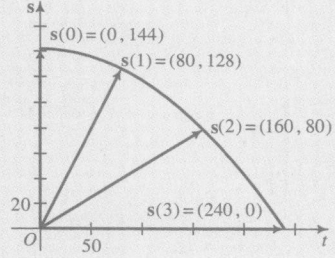

6. If $s(t)$ is the position function and $v(t)$ is the instantaneous velocity function of an object at time t, then the object's *instantaneous acceleration* at time t is:

$$a(t) = \lim_{\Delta t \to 0} \frac{v(t + \Delta t) - v(t)}{\Delta t} = v'(t) = s''(t)$$

The function $a(t)$ is the *first* derivative of $v(t)$ and the *second* derivative of $s(t)$.

Key vocabulary and ideas

derivative (p. 758)
curve sketching (pp. 764–765)
local maximum (or minimum) point (p. 765)
local maximum (or minimum) value (p. 765)

position function (p. 774)
instantaneous velocity (p. 775)
instantaneous acceleration (p. 776)
second derivative (p. 776)

Chapter Test

1. Find the derivative of each function.   20-1

   a. $f(x) = 5x^3 - 3x^2 + x - 1$ **b.** $f(x) = \dfrac{3}{x^2} - \dfrac{2}{x} + 4$

   c. $f(x) = \sqrt[3]{3x^2} - \sqrt[3]{\dfrac{1}{x}}$ **d.** $f(x) = \dfrac{2}{x^6} + 3x^6$

2. *Writing* Describe how to graph a function using its derivative.   20-2

3. Graph the function $f(x) = 12x - x^3$. Identify any local maximum and local minimum points.

4. Given a function $f(x)$, if $f'(2) = 0$, $f'(x) > 0$ for $x < 2$, and $f'(x) < 0$ 20-3
for $x > 2$, what can you say about $f(2)$?

5. a. A rectangular solid has a square base, as shown. Express its surface area A as a function of x and h.

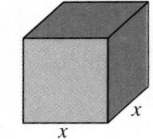

   b. If the volume of the solid is 1000 cubic units, express h in terms of x.

   c. Express A as a function of x alone.

   d. Show that the surface area is minimized when the solid is a cube.

6. Physics A particle is moving along a number line so that its position   20-4
to the right of the origin at time t is $s(t) = 3t^2 - 5t + 4$. Find the position, velocity, and acceleration of the particle at time $t = 5$.

7. Physics An arrow is shot upward from the bottom of a canyon that is 336 ft below the edge of a cliff. If the initial velocity of the arrow is 160 ft/s, then $s(t) = 160t - 16t^2$ gives the arrow's height, in feet, above the bottom of the canyon t seconds after the arrow is shot.

   a. In how many seconds does the arrow first pass the cliff's edge? 3 s

   b. How high does the arrow travel before it starts to descend? 400 ft

   c. If it lands on the cliff's edge, how long was the arrow in the air? 7 s

   d. At what velocity does the arrow hit the cliff's edge? -64 ft/s

You will need to use a computer or programmable calculator to complete Exercises 3 and 7.

1. Evaluate each limit or state that it does not exist.

   a. $\lim\limits_{x \to 1} \dfrac{2x^2 - x - 1}{2x^2 - 3x + 1}$ 3
   b. $\lim\limits_{x \to \infty} \dfrac{3x^2 + 2}{2x^2 + x}$ $\dfrac{3}{2}$
   c. $\lim\limits_{x \to 2^+} \dfrac{x + 1}{2 - x}$ $-\infty$

2. Sketch the graph of each function. Show vertical and horizontal asymptotes and x-intercepts.

   a. $y = \dfrac{x + 2}{1 - x}$
   b. $y = \dfrac{x}{x^2 - 4}$
   c. $y = \dfrac{x + 1}{x^2 + 1}$

3. **a.** Using rectangles of width 0.01, modify and run the program on page 730 to approximate the area under the curve $y = \log x$ from $x = 1$ to $x = 2$.
   b. *Visual Thinking* Using your answer to part (a), give an approximation of the area under the curve $y = 1 + \log x$ from $x = 1$ to $x = 2$. Also give a geometric argument to defend your answer.

4. **a.** Use series (4) on page 733 to find a power series for $\mathrm{Tan}^{-1}\dfrac{1}{x}$.
   b. For what values of x is the series in part (a) convergent?

5. Consider the function $f(x) = 0.2x - 1$.
   a. Find the first four terms in the orbit of $x_0 = 10$ for iterations of $f(x)$. 10, 1, −0.8, −1.16
   b. Find $\lim\limits_{n \to \infty} f^n(10)$. −1.25

6. Analyze all orbits for iterations of the function $f(x) = x^2 + \dfrac{1}{4}$.

7. *Writing* Modify and run the program on page 745 to iterate the logistic function $f(x) = 2.5x(1 - x)$ for the seed $x_0 = 0.3$. Write a sentence or two to describe your findings.

8. Find the derivative of each function. Express the derivative without using fractional or negative exponents.

   a. $f(x) = 3x^2 - x + \dfrac{4}{x} - \dfrac{1}{2x^2}$
   b. $f(x) = \sqrt{2x} - \dfrac{2}{\sqrt{x}}$

9. Find the slope of the curve $y = 1 - 2x^2$ at the point $P(-1, -1)$. 4

10. Use the derivative of $f(x) = 3x^4 + 4x^3 - 12x^2 + 6$ to sketch the graph of $f(x)$. Identify any local maximum and minimum points.

11. From a square piece of cardboard 10 in. on a side, a box with a square base and an open top is to be constructed by cutting out four squares at the corners and folding up the sides. What should the dimensions of the box be in order to maximize the volume of the box? $6\frac{2}{3}$ in. by $6\frac{2}{3}$ in. by $1\frac{2}{3}$ in.

12. A particle moves along a line so that its distance in feet to the right of its starting point after t seconds is $s(t) = 8t - t^2$. Find the times during which the particle **(a)** moves to the right and **(b)** moves to the left. **a.** $0 \le t < 4$ **b.** $t > 4$

Additional Answers
Cumulative Review

2. a. Vert. asymptote: $x = 1$; hor. asymptote: $y = -1$; x-int.: -2
   b. Vert asymptotes: $x = \pm 2$; hor. asymptote: $y = 0$; x-int.: 0
   c. No vert. asymptotes; hor. asymptote: $y = 0$; x-int.: -1

3. a. Approx. 0.169
   b. Approx. 1.169

4. a. $\mathrm{Tan}^{-1}\dfrac{1}{x} = \dfrac{1}{x} - \dfrac{1}{3x^3} + \dfrac{1}{5x^5} - \dfrac{1}{7x^7} + \cdots$
   b. $x \le -1$ or $x \ge 1$

6. $x_0 = \dfrac{1}{2}$ is a fixed point, and $x_0 = -\dfrac{1}{2}$ is an eventually fixed point. If $|x_0| < \dfrac{1}{2}$, then the orbit of x_0 tends to $\dfrac{1}{2}$. If $|x_0| > \dfrac{1}{2}$, then the orbit of x_0 tends to infinity.

7. The population has an equilibrium point of 0.6 (that is, 60% of capacity).

8. a. $f'(x) = 6x - 1 - \dfrac{4}{x^2} + \dfrac{1}{x^3}$
   b. $f'(x) = \dfrac{\sqrt{2x}}{2x} + \dfrac{\sqrt{x}}{x^2}$

10. Local max: (0, 6); local min: (−2, −26), (1, 1)

1. Conditions for congruent triangles

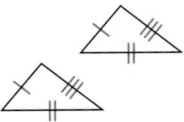 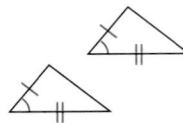 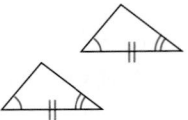

SSS: 3 pairs of corresponding sides congruent

SAS: 2 pairs of corresponding sides and included angles congruent

ASA: 2 pairs of corresponding angles and included sides congruent

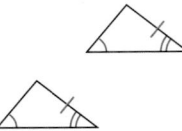 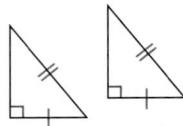

AAS: 2 pairs of corresponding angles and a pair of corresponding non-included sides congruent

HL: for right triangles, one pair of corresponding legs and hypotenuses congruent

2. Conditions for similar triangles

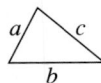 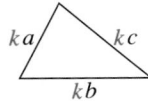 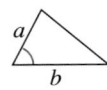 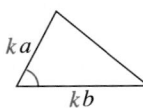 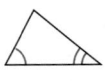 

SSS: 3 pairs of corresponding sides proportional

SAS: 2 pairs of corresponding sides proportional, and included angles congruent

AA: 2 pairs of corresponding angles congruent

3. Special triangle relationships

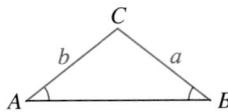 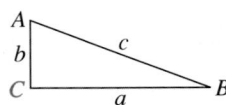

$a = b$ if and only if $\angle A \cong \angle B$.
$\angle A + \angle B + \angle C = 180°$.

$c^2 = a^2 + b^2$ if and only if $\angle C = 90°$; otherwise, $c^2 = a^2 + b^2 - 2ab \cos C$.

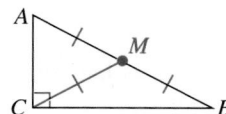

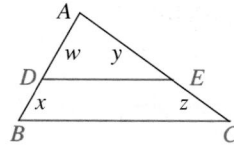

The midpoint of the hypotenuse of a right triangle is equidistant from the 3 vertices.

$\overline{DE} \parallel \overline{BC}$ if and only if

(1) $\dfrac{w}{x} = \dfrac{y}{z}$, or

(2) $\dfrac{w}{w + x} = \dfrac{y}{y + z} = \dfrac{DE}{BC}$.

If $w = x$ and $y = z$, then $DE = \frac{1}{2}BC$.

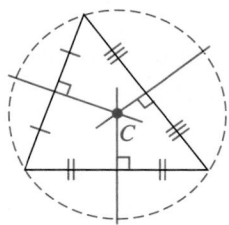

Perpendicular bisectors meet at circumcenter C.

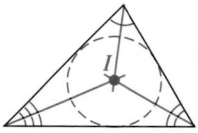

Angle bisectors meet at incenter I.

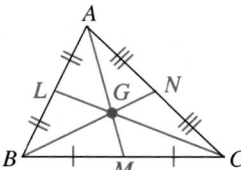

Medians meet at centroid G;
$AG = \frac{2}{3}AM,\ BG = \frac{2}{3}BN,$
$CG = \frac{2}{3}CL.$

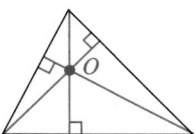

Altitudes meet at orthocenter O.

4. Parallel lines

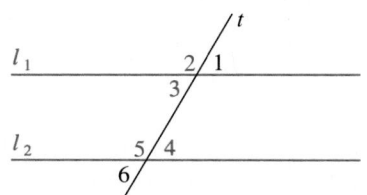

$l_1 \parallel l_2$ if and only if
(1) $\angle 2 \cong \angle 5$, or
(2) $\angle 3 \cong \angle 4$, or
(3) $\angle 3$ and $\angle 5$ are supplements.

5. Special quadrilaterals

Parallelogram Properties

   Opposite sides $\parallel$ and $\cong$.
   Opposite angles $\cong$.
   Diagonals bisect each other.
   If a rhombus: Diagonals $\perp$.
   If a rectangle: Diagonals $\cong$.

Trapezoid Properties

   Length of median is half the sum of base
     lengths.
   Median $\parallel$ bases.
   If isosceles: Base angles $\cong$.

6. Area formulas

Triangle:	$A = \frac{1}{2}bh$	Circle:	$A = \pi r^2$
Parallelogram:	$A = bh$	Sphere:	$A = 4\pi r^2$
Trapezoid:	$A = \frac{1}{2}h(b_1 + b_2)$	Cylinder:	$A = 2\pi r(r + h)$

Cone: $A = \pi r(r + s)$ (s = slant height)

Regular Pyramid: $A = B + \frac{1}{2}nas$ (B = area of base, n = number of base sides,
                                     a = length of one base side, s = slant height)

7. Volume formulas

Prism:	$V = Bh$ (B = area of base)	Sphere:	$V = \frac{4}{3}\pi r^3$
Cone:	$V = \frac{1}{3}\pi r^2 h$	Cylinder:	$V = \pi r^2 h$
Pyramid:	$V = \frac{1}{3}Bh$ (B = area of base)		

8. Circles and angle relationships

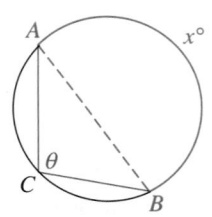

(1) $\theta = \frac{1}{2}x$
(2) $\theta = 90°$ if $\overline{AB}$ is a diameter.

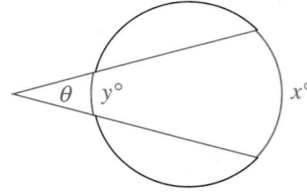

$\theta = \frac{1}{2}(x - y)$

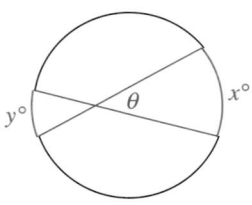

$\theta = \frac{1}{2}(x + y)$

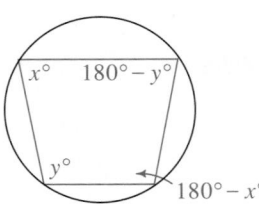

Opposite angles of an inscribed
quadrilateral are supplementary.

9. Circles and chord, secant, and tangent relationships

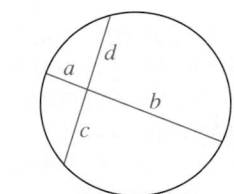

$$a \cdot b = c \cdot d$$

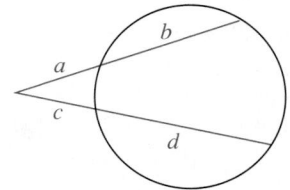

$$a \cdot (a + b) = c \cdot (c + d)$$

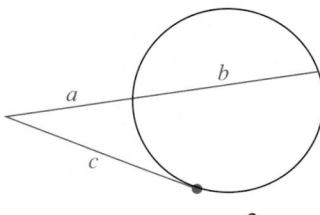

$$a \cdot (a + b) = c^2$$

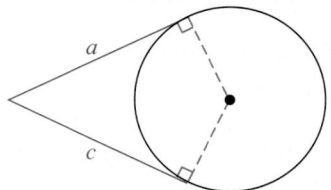

(1) $a = c$
(2) Radius $\perp$ tangent.

Properties of the Real Number System

Axioms, or **postulates,** for the real numbers are statements that are accepted as true without proof. The axioms for the real numbers are the basis for computation in arithmetic and algebra. A set of real numbers, together with the operations of addition and multiplication, that satisfies all the axioms of equality, addition, and multiplication, as well as the distributive axiom, is an example of what mathematicians call a **field.** In the field of real numbers, the following axioms of equality are assumed to be true.

Axioms of Equality

For all real numbers a, b, and c:

Reflexive Property $a = a$
Symmetric Property   If $a = b$, then $b = a$.
Transitive Property   If $a = b$ and $b = c$, then $a = c$.

The following *Substitution Property* is also useful.

Substitution Property   For all real numbers a and b, if $a = b$, then a can be replaced by b in any expression without changing the value of the expression.

Throughout our discussion we will use the symbol $\mathcal{R}$ to denote the set of real numbers. The symbol $\in$ means "is a member of" or "belongs to." Thus, $a \in \mathcal{R}$ means "a belongs to the set of real numbers $\mathcal{R}$." The following axioms are true for all a, b, and $c \in \mathcal{R}$ unless noted otherwise.

Real Number Axioms

Axiom Name	Addition	Multiplication
Closure	$a + b$ is a unique real number.	ab is a unique real number.
Commutative	$a + b = b + a$	$ab = ba$
Associative	$(a + b) + c = a + (b + c)$	$(ab)c = a(bc)$
Identity	There is a unique real number 0 such that $a + 0 = 0 + a = a$	There is a unique real number 1 such that $a \cdot 1 = 1 \cdot a = a.$
Inverse	For every $a \in \mathcal{R}$ there is a unique $-a \in \mathcal{R}$ such that $a + (-a) = (-a) + a = 0.$	For every nonzero $a \in \mathcal{R}$ there is a unique $\frac{1}{a} \in \mathcal{R}$ such that $a\left(\frac{1}{a}\right) = \left(\frac{1}{a}\right)a = 1.$
Distributive	$a(b + c) = ab + ac$	

Trigonometry Review

Definitions

$$\tan \theta = \frac{\sin \theta}{\cos \theta}$$

$$\cot \theta = \frac{1}{\tan \theta} = \frac{\cos \theta}{\sin \theta}$$

$$\sec \theta = \frac{1}{\cos \theta}$$

$$\csc \theta = \frac{1}{\sin \theta}$$

Pythagorean Formulas

$$\sin^2 \theta + \cos^2 \theta = 1$$
$$1 + \tan^2 \theta = \sec^2 \theta$$
$$1 + \cot^2 \theta = \csc^2 \theta$$

Addition Formulas

$$\sin (\alpha + \beta) = \sin \alpha \cos \beta + \cos \alpha \sin \beta$$
$$\sin (\alpha - \beta) = \sin \alpha \cos \beta - \cos \alpha \sin \beta$$
$$\cos (\alpha + \beta) = \cos \alpha \cos \beta - \sin \alpha \sin \beta$$
$$\cos (\alpha - \beta) = \cos \alpha \cos \beta + \sin \alpha \sin \beta$$

$$\tan (\alpha + \beta) = \frac{\tan \alpha + \tan \beta}{1 - \tan \alpha \tan \beta}$$

$$\tan (\alpha - \beta) = \frac{\tan \alpha - \tan \beta}{1 + \tan \alpha \tan \beta}$$

Double Angle Formulas

$$\sin 2\alpha = 2 \sin \alpha \cos \alpha$$
$$\cos 2\alpha = \cos^2 \alpha - \sin^2 \alpha$$
$$= 2 \cos^2 \alpha - 1$$
$$= 1 - 2 \sin^2 \alpha$$
$$\tan 2\alpha = \frac{2 \tan \alpha}{1 - \tan^2 \alpha}$$

Half-Angle Formulas

$$\sin \frac{\alpha}{2} = \pm \sqrt{\frac{1 - \cos \alpha}{2}}$$

$$\cos \frac{\alpha}{2} = \pm \sqrt{\frac{1 + \cos \alpha}{2}}$$

$$\tan \frac{\alpha}{2} = \pm \sqrt{\frac{1 - \cos \alpha}{1 + \cos \alpha}} = \frac{\sin \alpha}{1 + \cos \alpha}$$
$$= \frac{1 - \cos \alpha}{\sin \alpha}$$

Area of Triangle $K = \frac{1}{2} ab \sin C$

Law of Sines $\dfrac{\sin A}{a} = \dfrac{\sin B}{b} = \dfrac{\sin C}{c}$

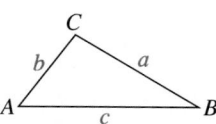

Law of Cosines $c^2 = a^2 + b^2 - 2ab \cos C, \ \cos C = \dfrac{a^2 + b^2 - c^2}{2ab}$

$$\cos C = \frac{\overrightarrow{CA} \cdot \overrightarrow{CB}}{|\overrightarrow{CA}| \, |\overrightarrow{CB}|} \quad (\textit{Vector form of law})$$

Sectors

	θ in radians	θ in degrees
Arc length	$s = r\theta$	$s = \dfrac{\theta}{360} \cdot 2\pi r$
Area	$K = \dfrac{1}{2} r^2 \theta = \dfrac{1}{2} rs$	$K = \dfrac{\theta}{360} \cdot \pi r^2$

College Entrance Examinations

The Scholastic Aptitude Test (SAT)

There are two 30-minute mathematical sections in the SAT. The questions (a total of 60) in these sections are designed to measure fundamental quantitative abilities closely related to college-level work.

In some questions you will be asked to apply graphic, spatial, numerical, symbolic, and logical techniques to problems that may be similar to exercises in your textbooks. Other questions may require you to do some original thinking.

The mathematical content is restricted to what is typically taught through first-year algebra and some geometry. The *arithmetic* includes simple addition, subtraction, multiplication, and division; percent; average; properties of odd and even numbers; prime numbers; and divisibility of numbers. The *algebra* includes negative numbers; simplifying algebraic expressions; factoring; linear equations; inequalities; simple quadratic equations; positive integer exponents; and roots. The *geometry* includes area, perimeter, circumference, and volume of a box and cube; special properties of isosceles, equilateral, and right triangles; 30°-60°-90° and 45°-45°-90° triangles; properties of parallel and perpendicular lines; and locating points on a coordinate grid.

Two kinds of multiple-choice questions are in the mathematical sections:
(1) Standard multiple-choice questions (approximately $\frac{2}{3}$ of the test)
(2) Quantitative comparison questions (approximately $\frac{1}{3}$ of the test)

Some sample questions of both types are given on the following pages. Answers to these questions appear at the end of the book.

Note: The directions given for both types of questions, as well as the examples given for the quantitative comparison questions, are printed as they would appear in an actual test booklet. Please disregard references made to using extra space for scratchwork and filling in answers on answer sheets.

The questions and accompanying text are reprinted, with permission, from copyrighted publications of the College Entrance Examination Board and Educational Testing Service, the copyright owner of the sample questions. Permission to reprint SAT Test and Achievement Test material does not constitute review or endorsement by Educational Testing Service or the College Board of this publication.

Standard Multiple-Choice Questions

Directions: In this section solve each problem, using any available space on the page for scratchwork. Then decide which is the best of the choices given and fill in the corresponding oval on the answer sheet.

The following information is for your reference in solving some of the problems:

Circle of radius r:
Area $= \pi r^2$; Circumference $= 2\pi r$
The number of degrees of arc in a circle is 360. The measure in degrees of a straight angle is 180.

Triangle: The sum of the measures in degrees of the angles of a triangle is 180.

If $\angle CDA$ is a right angle, then

(1) area of $\triangle ABC = \dfrac{AB \times CD}{2}$

(2) $AC^2 = AD^2 + DC^2$

Definitions of symbols:
$=$ is equal to $\neq$ is unequal to
$<$ is less than $\leq$ is less than or equal to
$>$ is greater than $\geq$ is greater than or equal to
$\parallel$ is parallel to $\perp$ is perpendicular to

Note: Figures that accompany problems in this test are intended to provide information useful in solving the problems. They are drawn as accurately as possible EXCEPT when it is stated in a specific problem that its figure is not drawn to scale. All figures lie in a plane unless otherwise indicated. All numbers used are real numbers.

1. If $2a + b = 5$, then $4a + 2b =$
(A) $\frac{5}{4}$ (B) $\frac{5}{2}$ (C) 10 (D) 20 (E) 25

2. If $16 \cdot 16 \cdot 16 = 8 \cdot 8 \cdot P$, then $P =$
(A) 4 (B) 8 (C) 32 (D) 48 (E) 64

3. The town of Mason is located on Eagle Lake. Canton is west of Mason. Sinclair is east of Canton, but west of Mason. Dexter is east of Richmond, but west of Sinclair and Canton. If these towns are in the United States, which town is farthest west?
(A) Mason (B) Dexter (C) Canton
(D) Sinclair (E) Richmond

4. If the average of seven x's is 7, what is the average of fourteen x's?
(A) $\frac{1}{7}$ (B) $\frac{1}{2}$ (C) 1 (D) 7 (E) 14

5. If an asterisk (*) between two expressions indicates that the expression on the right exceeds the expression on the left by 1, which of the following is (are) true for all real numbers x?
I. $x(x + 2)*(x + 1)^2$ II. $x^2*(x + 1)^2$
III. $\frac{x}{y} * \frac{x + 1}{y + 1}$
(A) None (B) I only (C) II only
(D) III only (E) I and III

6. If a car travels x kilometers of a trip in h hours, in how many hours can it travel the next y kilometers at this rate?
(A) $\frac{xy}{h}$ (B) $\frac{hy}{x}$ (C) $\frac{hx}{y}$ (D) $\frac{h + y}{x}$
(E) $\frac{x + y}{h}$

7. Which of the following fractions is greater than $\frac{2}{3}$ and less than $\frac{3}{4}$?
(A) $\frac{5}{8}$ (B) $\frac{4}{5}$ (C) $\frac{7}{12}$ (D) $\frac{9}{16}$ (E) $\frac{7}{10}$

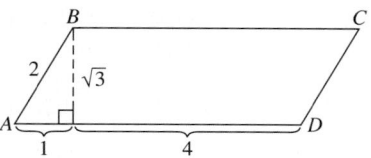

8. The figure above shows a piece of paper in the shape of a parallelogram with measurements as indicated. If the paper is tacked at its center to a flat surface and then rotated about its center, the points covered by the paper will be a circular region of diameter
(A) $\sqrt{3}$ (B) 2 (C) 5 (D) $\sqrt{28}$
(E) $\sqrt{39}$

9. An accurate 12-hour clock shows 3 o'clock at a certain instant. Exactly 11,999,999,995 hours later, what time does the clock show?
(A) 5 o'clock (B) 7 o'clock
(C) 8 o'clock (D) 9 o'clock
(E) 10 o'clock

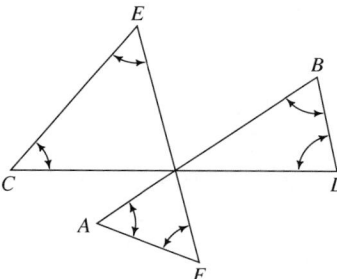

10. In the triangles above, if AB, CD, and EF are line segments, what is the sum of the measures of the six marked angles?
(A) 180° (B) 360° (C) 540° (D) 720°
(E) It cannot be determined from the information given.

11. If the area of a square is 16 and the coordinates of one corner are $(3, 2)$, which of the following could be the coordinates of the diagonally opposite corner?
(A) $(7, 2)$ (B) $(6, 7)$ (C) $(3, 6)$
(D) $(-1, 2)$ (E) $(-1, 6)$

Additional Answers
SAT

12. B

13. D

14. B

15. E

16. B

17. C

18. B

19. D

12. A box with a square base is filled with grass seed. How many cubic feet of seed will it contain when full, if the height of the box is 2 feet and one side of the base measures 18 *inches*?

(A) 3 ft^3 (B) 4$\frac{1}{2}$ ft^3 (C) 6 ft^3 (D) 36 ft^3
(E) 648 ft^3

13. The houses on the east side of a street are numbered with the consecutive even integers from 256 to 834 inclusive. How many houses are there on the east side of the street?

(A) 287   (B) 288   (C) 289   (D) 290
(E) 291

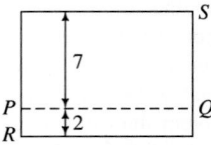

14. If the 9-inch by 12-inch piece of paper shown in the figure above were folded flat along the line PQ, then R would be how many inches closer to S?

(A) 4   (B) 2   (C) $\sqrt{193} - \sqrt{153}$
(D) $15 - \sqrt{193}$ (E) $\sqrt{193} - 13$

15. If 90 percent of p is 30 percent of q, then q is what percent of p?

(A) 3%   (B) 27%   (C) 30%   (D) 270%
(E) 300%

Quantitative Comparison Questions

Quantitative comparison questions emphasize the concepts of equalities, inequalities, and estimation. They generally involve less reading, take less time to answer, and require less computation than regular multiple-choice questions.

Directions: Each of the following questions consists of two quantities, one in Column A and one in Column B. You are to compare the two quantities and on the answer sheet fill in the oval

A   if the quantity in *Column A* is greater;
B   if the quantity in *Column B* is greater;
C   if the two quantities are equal;
D   if the relationship cannot be determined from the information given.

(*Reminder:* Since Questions 16–23 which follow are merely samples, no answer spaces are provided in this text.)

Notes: (1) In certain questions, information concerning one or both of the quantities to be compared is centered above the two columns.

(2) A symbol that appears in both columns represents the same thing in *Column A* as it does in *Column B*.

(3) Letters such as x, n, and k stand for real numbers.

EXAMPLES

	Column A	Column B	Answers
E1.	2×6	$2 + 6$	● ⓑ ⓒ ⓓ

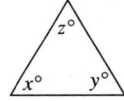

	Column A	Column B	Answers
E2.	$x + y$	$180 - z$	ⓐ ⓑ ● ⓓ
E3.	$p - q$	$q - p$	ⓐ ⓑ ⓒ ●

Column A	Column B
	$a < 0$
	$b < 0$

16. $a + b$ $\qquad$ $a - b$

17. The average of $(30 - 4)$ and $(30 + 4)$ $\qquad$ The average of 28, 29, 30, 31, and 32

18. $\sqrt{2} - 1$ $\qquad$ $\sqrt{3} - 1$

$0 < m < 6$
$0 < n < 8$

19. m $\qquad$ n

Column A	Column B		Column A	Column B

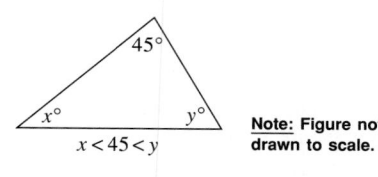

Note: Figure not drawn to scale.

$x < 45 < y$

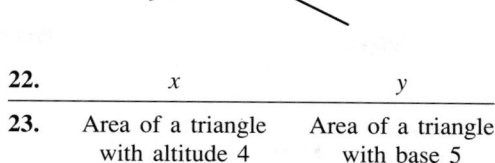

	Column A	Column B
20.	y	90
21.	$x + 1$	$2x + 1$
22.	x	y
23.	Area of a triangle with altitude 4	Area of a triangle with base 5

How the Tests Are Scored

The SAT Test, and also the Achievement Tests described next, are scored as follows: You get one point for each right answer, and you lose a fraction of a point for each wrong answer. You neither gain nor lose points for questions you omit. If you mark more than one answer for a question, it is scored like an omitted question.

For questions with 4 answer-choices (questions 16–23), your raw score =
$$\text{number right} - \tfrac{1}{3}(\text{number wrong}).$$
For questions with 5 answer-choices (questions 1–15), your raw score =
$$\text{number right} - \tfrac{1}{4}(\text{number wrong}).$$

Your total raw score is the sum of these raw scores. The College Board converts your raw score for the entire SAT test or achievement test to a scaled score between 200 and 800.

Because of the way the test is scored, haphazard or random guessing on questions you know nothing about is not recommended. When you know that one or more choices can be eliminated, however, guessing from the remaining choices should be to your advantage.

The Achievement Tests

The College Board offers three Achievement Tests in mathematics, Mathematics Level I, Level II, and Level IIC.

Mathematics Level I is a broad survey test that covers content typical of three years of college-preparatory mathematics (two years of algebra and a year of geometry). The Level I test contains questions in algebra, geometry, basic trigonometry, algebraic functions, and miscellaneous topics such as elementary statistics, counting problems, data interpretation, logic, elementary number theory, and arithmetic and geometric sequences.

Mathematics Level II is intended for students who have taken college-preparatory mathematics for more than three years—two years of algebra and a year of geometry, elementary functions (precalculus), and/or trigonometry. The Level II test contains questions in algebra, geometry, trigonometry, elementary functions, and miscellaneous topics such as probability, permutations and combinations, statistics, logic and proof, elementary number theory, and sequences and limits.

Mathematics Level IIC requires the use of a scientific calculator. The mathematical preparation required and topics tested are the same as for the Level II test. Students taking IIC must use one of the calculators specified by The College Board.

The questions on the Level IIC test fall into three categories with respect to calculator use:

Category (1) Calculator inactive—there is no advantage (perhaps even a disadvantage) to using a calculator. See sample questions 1 and 11.

Category (2) Calculator neutral—these problems could be solved without a calculator, but a calculator may be useful. See sample questions 12 and 16.

Category (3) Calculator active—a calculator is necessary to solve these problems. See sample questions 4 and 9.

Approximately 40% of the questions are in category 1 and 60% are in categories 2 and 3. Skill in knowing both when and how to use a scientific calculator is required. If you have not used a scientific calculator on a regular basis, you should take the Level II test instead.

Comparisons among the Tests: Level I, Level II, and Level IIC

Both Level I and Level II tests consist of approximately 50 multiple-choice questions with a 1 hour time limit. The content of the Level I test overlaps somewhat with that of the Level II and Level IIC tests, especially in questions on elementary algebra, coordinate geometry, statistics, and basic trigonometry. Consequently, some questions may be appropriate for any of these tests. However, the emphasis in the Level II and Level IIC tests is on more advanced content, with a greater percentage of questions devoted to trigonometry, elementary functions, and precalculus topics. Further, a significant percentage of questions on Level I is devoted to plane geometry, a topic not tested directly in Level II or IIC; the geometry questions on Levels II and IIC cover topics such as coordinate geometry in two or three dimensions, transformations, and solid geometry. The trigonometry questions on Level I are primarily limited to right triangle trigonometry and the fundamental relationships among the trigonometric ratios, whereas the Level II and IIC tests place more emphasis on the properties and graphs of the trigonometric functions, the inverse trigonometric functions, trigonometric equations and identities, and the laws of sines and cosines.

If you have had preparation in trigonometry and elementary functions and have attained grades of B or better in your mathematics courses, you should select either the Level II or Level IIC test. Students who are sufficiently prepared to take the Level II or IIC test, but who elect to take the Level I test in hopes of receiving high scores, sometimes do not do as well as they expect. No student taking Level I, Level II, or Level IIC is necessarily expected to have studied every topic on the test.

The chart below shows approximately how the questions in each test are distributed among the major curriculum areas. Comparison of the percentages for the levels should help you decide which test you are better prepared to take.

Topics Covered	Approximate Percentage of Test	
	Level I	Levels II and IIC
Algebra	30	18
Geometry		
Plane Geometry	20	—
Solid Geometry	6	8
Coordinate Geometry	12	12
Trigonometry	8	20
Functions	12	24
*Miscellaneous	12	18

*In Level 1, one-half of these questions include elementary statistics, counting problems, and data interpretation. In Level II and Level IIC, one-third of these questions include probability, permutations and combinations, and statistics.

As a general rule, do not try to work a problem by testing each of the answer choices. Such a method is time-consuming and, in many instances, will not work. However, you should look at the answer choices while studying the question. The form of the answer choices may help to clarify the problem.

The following sample questions are reprinted, with permission, from copyrighted publications of the College Board and Educational Testing Service, the copyright owner of the questions.

Mathematics Level I

Directions: For each of the following problems, decide which is the best of the choices given. Then fill in the corresponding oval on the answer sheet.

Notes: (1) Figures that accompany problems in this test are intended to provide information useful in solving the problems. They are drawn as accurately as possible EXCEPT when it is stated in a specific problem that its figure is not drawn to scale. All figures lie in a plane unless otherwise indicated.

(2) Unless otherwise specified, the domain of a function f is assumed to be the set of all real numbers x for which $f(x)$ is a real number.

MATHEMATICS LEVEL I TEST–SAMPLE QUESTIONS

1. If $\dfrac{3^x}{3^2} = 3^8$, then $x =$

   (A) 4   (B) 6   (C) 10   (D) 16   (E) 64

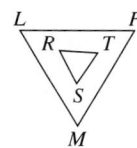

2. The figure above consists of two equilateral triangles, LMP and RST. If $RS = \frac{1}{3}LM$, then $\dfrac{\text{area } \triangle RST}{\text{area } \triangle LMP} =$

   (A) $\frac{1}{9}$ (B) $\frac{1}{6}$ (C) $\frac{1}{4}$ (D) $\frac{1}{3}$ (E) $\frac{1}{\sqrt{3}}$

3. The positive difference between the mean and the median of the numbers 27, 27, 29, 32, and 35 is

   (A) 0   (B) 1   (C) 2   (D) 3   (E) 8

4. If $i = \sqrt{-1}$, which of the following is an expression for $\sqrt{-49} - \sqrt{-25}$ in the form $a + bi$, where a and b are real numbers?

   (A) $0 + (\sqrt{24})i$ (B) $0 + (\sqrt{74})i$
   (C) $0 + 24i$ (D) $0 + 2i$ (E) $2 + 0i$

5. In the figure, $\sin \angle P =$

   (A) $\frac{2}{5}$ (B) $\frac{2}{3}$ (C) $\frac{\sqrt{5}}{3}$

   (D) $\frac{\sqrt{5}}{2}$ (E) $\frac{3}{2}$

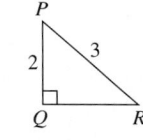

6. If $x + 2 = y$, what is the value of $|x - y| + |y - x|$?
   (A) -4 (B) 0   (C) 2   (D) 4
   (E) It cannot be determined.

7. Two rectangular solids have dimensions 4, 6, h, and 8, 2, $(2h - 1)$, respectively. Their volumes are equal when $h =$
   (A) $\frac{1}{8}$ (B) $\frac{4}{5}$ (C) 1   (D) 2   (E) 4

8. If $f(x) = 5x + 6$, then $f(x) < 16$ if and only if
   (A) $x < 2$ (B) $x > 2$ (C) $x = 2$
   (D) $-2 < x < 2$ (E) x is any real number

9. If x is the measure of an acute angle such that $\tan x = \dfrac{k}{3}$, then $\sin x =$

   (A) $\dfrac{k}{3 + k}$ (B) $\dfrac{3}{\sqrt{9 - k^2}}$ (C) $\dfrac{k}{\sqrt{9 - k^2}}$

   (D) $\dfrac{3}{\sqrt{9 + k^2}}$ (E) $\dfrac{k}{\sqrt{9 + k^2}}$

College Entrance Examinations **795**

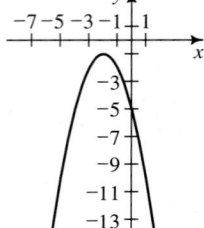

10. The circle in the figure above has center at O. If PQ and QR are secants and if $x = 40$, what is y?

(A) 10　(B) 20　(C) 30　(D) 40
(E) It cannot be determined.

11. On the curve shown in the figure, determine the y-coordinate of each point at which $y = 2x$.

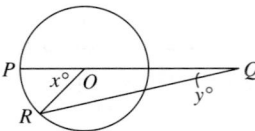

(A) There is no such point.
(B) -1
(C) -2 only
(D) -5
(E) -2 and -10

12. In how many points do the graphs of $x^2 + y^2 = 9$ and $x^2 = 8y$ intersect?

(A) One　(B) Two　(C) Three　(D) Four
(E) More than four

13. If $f(x) = 2x + 1$ and $g(x) = 3x - 1$, then $f(g(x)) =$

(A) $6x - 1$　(B) $6x + 2$　(C) $x - 2$
(D) $5x$　(E) $6x^2 + x - 1$

14. If a and b are positive integers and if 9 is a factor of their product, which of the following must be true?

(A) 9 is a factor of both a and b.
(B) 9 is a factor of one number but not of the other.
(C) 9 is a factor of at least one of the numbers a, b.
(D) 3 is a factor of both a and b.
(E) 3 is a factor of at least one of the numbers a, b.

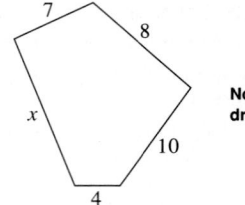

Note: Figure not drawn to scale.

15. In the figure above, a pentagon has four of its sides with lengths 4, 7, 8, and 10. If the length x of the fifth side is also an integer, then the greatest possible value of x is

(A) 16　(B) 22　(C) 26　(D) 28　(E) 30

16. If one angle of a rhombus is $60°$, then the ratio of the length of its longer diagonal to the length of its shorter diagonal is

(A) 2　(B) $\sqrt{3}$　(C) $\sqrt{2}$　(D) $\dfrac{\sqrt{3}}{2}$

(E) $\dfrac{\sqrt{2}}{2}$

17. The solution set of $2x^2 - 4x = 30$ is $\{-3, 5\}$. What is the solution set of $2(x - 4)^2 - 4(x - 4) = 30$?

(A) $\{-12, 20\}$　(B) $\{-7, 1\}$
(C) $\left\{-\dfrac{3}{4}, \dfrac{5}{4}\right\}$　(D) $\{1, 9\}$　(E) $\{12, -20\}$

18. Which quadrants of the plane contain points of the graph $2x - y > 4$?

(A) First, second, and third only
(B) First, second, and fourth only
(C) First, third, and fourth only
(D) Second, third, and fourth only
(E) First, second, third, and fourth

19. If $x = bc$ and $y = bd$, then $y - x =$

(A) $bc(1 - d)$　(B) $bd(1 - c)$
(C) $d(b - c)$　(D) $b(d - c)$
(E) $c(b - d)$

20. The graph of a certain linear function has a negative slope. If its x-intercept is negative, then its y-intercept must be

(A) positive　(B) negative　(C) nonnegative
(D) less than its x-intercept
(E) greater than its x-intercept

Mathematics Level II

Directions: For each of the following problems, decide which is the best of the choices given. Then fill in the corresponding oval on the answer sheet.

Notes: (1) Figures that accompany problems in this test are intended to provide information useful in solving the problems. They are drawn as accurately as possible EXCEPT when it is stated in a specific problem that its figure is not drawn to scale. All figures lie in a plane unless otherwise indicated.

(2) Unless otherwise specified, the domain of a function f is assumed to be the set of all real numbers x for which $f(x)$ is a real number.

MATHEMATICS LEVEL II TEST − SAMPLE QUESTIONS

1. If $f(x) = x^2 - x^3$, then $f(-1) =$
   (A) 2   (B) 1   (C) 0   (D) −1   (E) −2

2. The base of an equilateral triangle lies on the x-axis. What is the sum of the slopes of the three sides?
   (A) − 1   (B) 0   (C) 1   (D) $2\sqrt{3}$
   (E) $1 + 2\sqrt{3}$

3. For what real numbers x is $y = 2^{-x}$ a negative number?
   (A) All real x (B) $x > 0$ only
   (C) $x \geq 0$ only   (D) $x < 0$ only
   (E) No real x

4. If $f(x) = 2x$ and $f(g(x)) = -x$, then $g(x) =$
   (A) $-3x$ (B) $-\frac{x}{2}$ (C) $\frac{x}{2}$ (D) $2 - \frac{x}{2}$
   (E) x

5. $\lim\limits_{n \to \infty} \dfrac{3n + 12}{2n - 12} =$
   (A) − 6   (B) $-\frac{3}{2}$ (C) −1   (D) $\frac{2}{3}$ (E) $\frac{3}{2}$

6. If $0 < y < x < \frac{\pi}{2}$, which of the following are true?
         I. $\sin y < \sin x$
         II. $\cos y < \cos x$
         III. $\tan y < \tan x$
   (A) None   (B) I and II only
   (C) I and III only   (D) II and III only
   (E) I, II, and III

7. $\sin\left(\text{Arcsin } \frac{1}{10}\right) =$
   (A) 0   (B) $\frac{1}{10}$ (C) $\frac{1}{9}$ (D) $\frac{9}{10}$ (E) 1

8. If $f(x) = \log_{10}(x + 3)$ and f^{-1} denotes the inverse of f, then $f^{-1}(2) =$
   (A) $\frac{1}{10}$ (B) $\frac{1}{2}$ (C) 97   (D) 103
   (E) 100,000

9. The graph of $\begin{cases} x = 4t - 2 \\ y = 4t^2 \end{cases}$ in the xy-plane is
   (A) a circle   (B) a parabola
   (C) an ellipse that is not a circle
   (D) a hyperbola   (E) a straight line

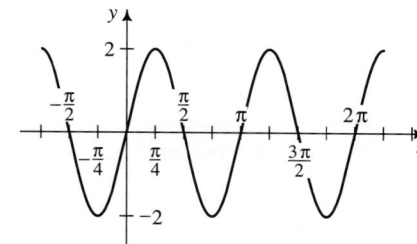

10. Which of the following equations has the graph shown in the figure above?
   (A) $y = \sin \frac{x}{2} + 1$ (B) $y = \sin 2x$
   (C) $y = 2 \sin \frac{x}{2}$ (D) $y = 2 \sin x$
   (E) $y = 2 \sin 2x$

11. All real numbers x and y that satisfy the equation $x + iy = \dfrac{1}{x - iy}$ also satisfy which of the following equations?
   (A) $x^2 - y^2 = 0$ (B) $x^2 - y^2 = 1$
   (C) $x^2 + y^2 = 0$ (D) $x^2 + y^2 = 1$
   (E) $xy = 1$

Additional Answers Level II Test

1. A
2. B
3. E
4. B
5. E
6. C
7. B
8. C
9. B
10. E
11. D

12. Lines l_1 and l_2 both intersect line l_3 at right angles, but do not necessarily lie in the same plane. Which of the following statements about l_1 and l_2 is correct?
    (A) They are necessarily parallel to each other.
    (B) They cannot be parallel to each other.
    (C) They are necessarily either parallel or perpendicular to each other.
    (D) They are necessarily perpendicular to each other.
    (E) They are not necessarily parallel or perpendicular to each other.

13. The probability that R hits a certain target is $\frac{3}{5}$ and the probability that T hits it is $\frac{5}{7}$. What is the probability that R hits the target and T misses it?
    (A) $\frac{4}{35}$ (B) $\frac{6}{35}$ (C) $\frac{3}{7}$ (D) $\frac{21}{25}$ (E) $\frac{31}{35}$

14. $\dfrac{(n+1)!}{n!} - n =$
    (A) 0   (B) 1   (C) n (D) $n+1$ (E) $n!$

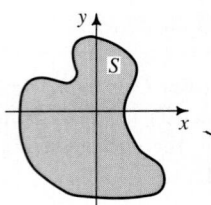

15. In the figure, shaded region S has area 10. What is the area of region T, which consists of all points $(x-1, y+4)$ where the point (x, y) is in S?
    (A) 40   (B) 30   (C) 20   (D) $10\sqrt{2}$ (E) 10

16. The graph of $\begin{cases} x = 2\cos\theta \\ y = 4\sin\theta \end{cases}$ in the xy-plane is
    (A) a line   (B) a circle of radius 2
    (C) a circle of radius 4
    (D) an ellipse with x-intercepts ± 4
    (E) an ellipse with y-intercepts ± 4

Mathematics Level IIC

Directions: For each of the following problems, look for your solution among the choices given. **In some cases, the choices given are approximate.** Decide which is the best of the choices by rounding your solution when appropriate. Then fill in the corresponding oval on the answer sheet.

Notes: (1) A calculator may be necessary for answering some (but not all) of the questions in this test. For each question, you will have to decide whether you should use a calculator. The calculator you use should be a scientific calculator; programmable calculators and calculators that can display graphs are NOT permitted.

(2) For some questions in this test you may have to decide whether your calculator should be in the radian mode or the degree mode.

(3) Figures that accompany problems in this test are intended to provide information useful in solving the problems. They are drawn as accurately as possible EXCEPT when it is stated in a specific problem that its figure is not drawn to scale. All figures lie in a plane unless otherwise indicated.

(4) Unless otherwise specified, the domain of a function f is assumed to be the set of all real numbers x for which $f(x)$ is a real number.

MATHEMATICS LEVEL IIC TEST – SAMPLE QUESTIONS

1. If $f(x) = 2x$ and $f(g(x)) = -x$, then $g(x) =$
   (A) $-3x$ (B) $-\frac{x}{2}$ (C) $\frac{x}{2}$
   (D) $2 - \frac{x}{2}$ (E) x

2. If $x - 2$ is a factor of $x^3 + kx^2 + 12x - 8$, then $k =$
   (A) -6 (B) -3 (C) 2   (D) 3   (E) 6

798   *College Entrance Examinations*

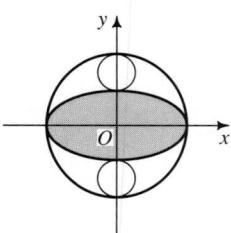

3. The formula for the area enclosed by an ellipse is $A = \pi ab$, where a and b are one-half the respective lengths of the major and minor axes. In the figure, both small circles have diameter 2, and the equation of the larger circle is $x^2 + y^2 = 15$. What is the area enclosed by the ellipse?
(A) 20   (B) 23   (C) 26   (D) 29   (E) 32

4. If $2x^2 + 4x = 3$, then, to the nearest tenth, what is the positive value of x?
(A) 0.6   (B) 0.8   (C) 1.2   (D) 1.6
(E) There is no positive value of x that satisfies the equation.

5. If $f(x) = (x + 3)^2 + 1$, what is the minimum value of the function f?
(A) -3 (B) 0   (C) 1   (D) 3   (E) 4

6. What is the area of a right triangle with an angle of 28° and with longer leg of length 13?
(A) 40   (B) 45   (C) 75   (D) 90   (E) 159

7. If $\log_2 3 = x$, then $x =$
(A) 0.4   (B) 0.6   (C) 1.6   (D) 2.5
(E) 3.0

8. $\lim\limits_{n \to \infty} \dfrac{3n + 12}{2n - 12} =$
(A) -6 (B) $-\dfrac{3}{2}$ (C) -1
(D) $\dfrac{2}{3}$ (E) $\dfrac{3}{2}$

9. The diameter and height of a right circular cylinder are equal. If the volume of the cylinder is 2, what is the height of the cylinder?
(A) 1.37   (B) 1.08   (C) 0.86   (D) 0.80
(E) 0.68

10. If $f(x) = \sqrt[3]{x^3 + 1}$, what is $f^{-1}(1.5)$?
(A) 3.4   (B) 2.4   (C) 1.6   (D) 1.5   (E) 1.3

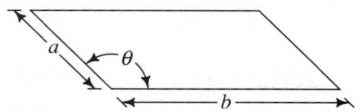

11. The area of the parallelogram in the figure is
(A) ab (B) $ab \cos \theta$ (C) $ab \sin \theta$
(D) $ab \tan \theta$ (E) $a^2 + b^2 - 2ab \cos \theta$

12. A club has 14 members, consisting of 6 men and 8 women. How many slates of 3 officers—president, vice-president, and secretary—can be formed if the president must be a woman and the vice-president must be a man?
(A) 2744   (B) 2184   (C) 672   (D) 576
(E) 336

13. What is the range of the function defined by
$f(x) = \dfrac{1}{x} + 2$?
(A) All real numbers
(B) All real numbers except $-\dfrac{1}{2}$
(C) All real numbers except 0
(D) All real numbers except 2
(E) All real numbers between 2 and 3

14. The probability that R hits a certain target is $\dfrac{3}{5}$ and the probability that T hits it is $\dfrac{5}{7}$. What is the probability that R hits the target and T misses it?
(A) $\dfrac{4}{35}$ (B) $\dfrac{6}{35}$ (C) $\dfrac{3}{7}$ (D) $\dfrac{21}{25}$ (E) $\dfrac{31}{35}$

15. If x is a real number between 0 and 2π, and $\sin x = 5 \cos x$, what is the value of x?
(A) 0.197   (B) 0.201   (C) 1.373
(D) 11.310   (E) 78.690

16. $\sin (\text{Arccos} (- 0.6)) =$
(A) 2.2   (B) 0.8   (C) 0.4   (D) -0.8
(E) -0.4

Table 1.   Example: $\sqrt{830} = 10\sqrt{8.3} \approx 28.81$

Tables 2 and 3.   Example **a.** $\log 4.72 = 0.6739$

Example **b.** If $\log N = 0.4843$, then $N = 3.05$

Example **c.** $\log 718 = \log (100 \cdot 7.18) = \log 10^2 + \log 7.18$
$$= 2 + 0.8561 = 2.8561$$
We call 2 the characteristic and 0.8561 the mantissa.

Tables 4 and 5.   To evaluate a trigonometric function of θ when $0° \leq \theta \leq 45°$, locate θ in the left-hand column and use the trigonometric-function headings at the top of the table. When $45° < \theta \leq 90°$, locate θ in the right-hand column and use the headings at the bottom of the table.

Examples: **a.** $\cos 14°10' = 0.9696$ **b.** $\tan 51.9° = 1.275$

Table 6.   Examples: **a.** $\sin 0.95 = 0.8134$ **b.** $\csc 1.51 = 1.002$

Table 7.   Example   **a.** $e^{2.6} = e^{2.5} \cdot e^{0.1} \approx (12.182)(1.1052) \approx 13.464$

Example   **b.** $e^{-3.21} = e^{-3} \cdot e^{-0.21} \approx (0.0498)(0.8106) \approx 0.0404$

Interpolation.   Examples: Use interpolation to find:

a. $\log 8.217$ **b.** x if $\log x = 0.5725$
c. $\tan 68°34'$ **d.** θ if $\sin \theta = 0.5065$

a. From Table 3:

N	$\log N$
8.220	.9149
8.217	x
8.210	.9143

$.10$; $.07$; $x - .9143$; $.0006$

$$\frac{.07}{.10} = \frac{x - .9143}{.0006}$$

$$.10x = .000042 + .09143$$

$$x \approx 0.91472$$

b. From Table 3:

N	$\log N$
3.740	.5729
x	.5725
3.730	.5717

$.010$; $x - 3.730$; $.0008$; $.0012$

$$\frac{x - 3.730}{.010} = \frac{.0008}{.0012}$$

$$x = 3.730 + \frac{8}{12}(.010)$$

$$\approx 3.737$$

c. From Table 4:

θ	$\tan \theta$
68°40'	2.560
68°34'	x
68°30'	2.539

$10'$; $4'$; $x - 2.539$; $.021$

$$\frac{4}{10} = \frac{x - 2.539}{.021}$$

$$x = 2.539 + \frac{4}{10}(.021)$$

$$\approx 2.547$$

d. From Table 4:

θ	$\sin \theta$
30°30'	.5075
30°x'	.5065
30°20'	.5050

$10'$; $x - 20'$; $.0015$; $.0025$

$$\frac{x - 20}{10} = \frac{.0015}{.0025}$$

$$x = 20 + \frac{3}{5}(10) = 26$$

$$\theta \approx 30°26'$$

Table 1   Squares and Square Roots

N	N^2	$\sqrt{N}$	$\sqrt{10N}$	N	N^2	$\sqrt{N}$	$\sqrt{10N}$
1.0	1.00	1.000	3.162	5.5	30.25	2.345	7.416
1.1	1.21	1.049	3.317	5.6	31.36	2.366	7.483
1.2	1.44	1.095	3.464	5.7	32.49	2.387	7.550
1.3	1.69	1.140	3.606	5.8	33.64	2.408	7.616
1.4	1.96	1.183	3.742	5.9	34.81	2.429	7.681
1.5	2.25	1.225	3.873	6.0	36.00	2.449	7.746
1.6	2.56	1.265	4.000	6.1	37.21	2.470	7.810
1.7	2.89	1.304	4.123	6.2	38.44	2.490	7.874
1.8	3.24	1.342	4.243	6.3	39.69	2.510	7.937
1.9	3.61	1.378	4.359	6.4	40.96	2.530	8.000
2.0	4.00	1.414	4.472	6.5	42.25	2.550	8.062
2.1	4.41	1.449	4.583	6.6	43.56	2.569	8.124
2.2	4.84	1.483	4.690	6.7	44.89	2.588	8.185
2.3	5.29	1.517	4.796	6.8	46.24	2.608	8.246
2.4	5.76	1.549	4.899	6.9	47.61	2.627	8.307
2.5	6.25	1.581	5.000	7.0	49.00	2.646	8.367
2.6	6.76	1.612	5.099	7.1	50.41	2.665	8.426
2.7	7.29	1.643	5.196	7.2	51.84	2.683	8.485
2.8	7.84	1.673	5.292	7.3	53.29	2.702	8.544
2.9	8.41	1.703	5.385	7.4	54.76	2.720	8.602
3.0	9.00	1.732	5.477	7.5	56.25	2.739	8.660
3.1	9.61	1.761	5.568	7.6	57.76	2.757	8.718
3.2	10.24	1.789	5.657	7.7	59.29	2.775	8.775
3.3	10.89	1.817	5.745	7.8	60.84	2.793	8.832
3.4	11.56	1.844	5.831	7.9	62.41	2.811	8.888
3.5	12.25	1.871	5.916	8.0	64.00	2.828	8.944
3.6	12.96	1.897	6.000	8.1	65.61	2.846	9.000
3.7	13.69	1.924	6.083	8.2	67.24	2.864	9.055
3.8	14.44	1.949	6.164	8.3	68.89	2.881	9.110
3.9	15.21	1.975	6.245	8.4	70.56	2.898	9.165
4.0	16.00	2.000	6.325	8.5	72.25	2.915	9.220
4.1	16.81	2.025	6.403	8.6	73.96	2.933	9.274
4.2	17.64	2.049	6.481	8.7	75.69	2.950	9.327
4.3	18.49	2.074	6.557	8.8	77.44	2.966	9.381
4.4	19.36	2.098	6.633	8.9	79.21	2.983	9.434
4.5	20.25	2.121	6.708	9.0	81.00	3.000	9.487
4.6	21.16	2.145	6.782	9.1	82.81	3.017	9.539
4.7	22.09	2.168	6.856	9.2	84.64	3.033	9.592
4.8	23.04	2.191	6.928	9.3	86.49	3.050	9.644
4.9	24.01	2.214	7.000	9.4	88.36	3.066	9.695
5.0	25.00	2.236	7.071	9.5	90.25	3.082	9.747
5.1	26.01	2.258	7.141	9.6	92.16	3.098	9.798
5.2	27.04	2.280	7.211	9.7	94.09	3.114	9.849
5.3	28.09	2.302	7.280	9.8	96.04	3.130	9.899
5.4	29.16	2.324	7.348	9.9	98.01	3.146	9.950
5.5	30.25	2.345	7.416	10	100.00	3.162	10.000

Table 2 **Natural Logarithms**

x	$\ln x$	x	$\ln x$	x	$\ln x$
		4.5	1.5041	9.0	2.1972
0.1	-2.3026	4.6	1.5261	9.1	2.2083
0.2	-1.6094	4.7	1.5476	9.2	2.2192
0.3	-1.2040	4.8	1.5686	9.3	2.2300
0.4	-0.9163	4.9	1.5892	9.4	2.2407
0.5	-0.6931	5.0	1.6094	9.5	2.2513
0.6	-0.5108	5.1	1.6292	9.6	2.2618
0.7	-0.3567	5.2	1.6487	9.7	2.2721
0.8	-0.2231	5.3	1.6677	9.8	2.2824
0.9	-0.1054	5.4	1.6864	9.9	2.2925
1.0	0.0000	5.5	1.7047	10	2.3026
1.1	0.0953	5.6	1.7228	11	2.3979
1.2	0.1823	5.7	1.7405	12	2.4849
1.3	0.2624	5.8	1.7579	13	2.5649
1.4	0.3365	5.9	1.7750	14	2.6391
1.5	0.4055	6.0	1.7918	15	2.7081
1.6	0.4700	6.1	1.8083	16	2.7726
1.7	0.5306	6.2	1.8245	17	2.8332
1.8	0.5878	6.3	1.8405	18	2.8904
1.9	0.6419	6.4	1.8563	19	2.9444
2.0	0.6931	6.5	1.8718	20	2.9957
2.1	0.7419	6.6	1.8871	25	3.2189
2.2	0.7885	6.7	1.9021	30	3.4012
2.3	0.8329	6.8	1.9169	35	3.5553
2.4	0.8755	6.9	1.9315	40	3.6889
2.5	0.9163	7.0	1.9459	45	3.8067
2.6	0.9555	7.1	1.9601	50	3.9120
2.7	0.9933	7.2	1.9741	55	4.0073
2.8	1.0296	7.3	1.9879	60	4.0943
2.9	1.0647	7.4	2.0015	65	4.1744
3.0	1.0986	7.5	2.0149	70	4.2485
3.1	1.1314	7.6	2.0281	75	4.3175
3.2	1.1632	7.7	2.0412	80	4.3820
3.3	1.1939	7.8	2.0541	85	4.4427
3.4	1.2238	7.9	2.0669	90	4.4998
3.5	1.2528	8.0	2.0794	100	4.6052
3.6	1.2809	8.1	2.0919	110	4.7005
3.7	1.3083	8.2	2.1041	120	4.7875
3.8	1.3350	8.3	2.1163	130	4.8676
3.9	1.3610	8.4	2.1282	140	4.9416
4.0	1.3863	8.5	2.1401	150	5.0106
4.1	1.4110	8.6	2.1518	160	5.0752
4.2	1.4351	8.7	2.1633	170	5.1358
4.3	1.4586	8.8	2.1748	180	5.1930
4.4	1.4816	8.9	2.1861	190	5.2470

Table 3 **Common Logarithms of Numbers***

N	0	1	2	3	4	5	6	7	8	9
1.0	0000	0043	0086	0128	0170	0212	0253	0294	0334	0374
1.1	0414	0453	0492	0531	0569	0607	0645	0682	0719	0755
1.2	0792	0828	0864	0899	0934	0969	1004	1038	1072	1106
1.3	1139	1173	1206	1239	1271	1303	1335	1367	1399	1430
1.4	1461	1492	1523	1553	1584	1614	1644	1673	1703	1732
1.5	1761	1790	1818	1847	1875	1903	1931	1959	1987	2014
1.6	2041	2068	2095	2122	2148	2175	2201	2227	2253	2279
1.7	2304	2330	2355	2380	2405	2430	2455	2480	2504	2529
1.8	2553	2577	2601	2625	2648	2672	2695	2718	2742	2765
1.9	2788	2810	2833	2856	2878	2900	2923	2945	2967	2989
2.0	3010	3032	3054	3075	3096	3118	3139	3160	3181	3201
2.1	3222	3243	3263	3284	3304	3324	3345	3365	3385	3404
2.2	3424	3444	3464	3483	3502	3522	3541	3560	3579	3598
2.3	3617	3636	3655	3674	3692	3711	3729	3747	3766	3784
2.4	3802	3820	3838	3856	3874	3892	3909	3927	3945	3962
2.5	3979	3997	4014	4031	4048	4065	4082	4099	4116	4133
2.6	4150	4166	4183	4200	4216	4232	4249	4265	4281	4298
2.7	4314	4330	4346	4362	4378	4393	4409	4425	4440	4456
2.8	4472	4487	4502	4518	4533	4548	4564	4579	4594	4609
2.9	4624	4639	4654	4669	4683	4698	4713	4728	4742	4757
3.0	4771	4786	4800	4814	4829	4843	4857	4871	4886	4900
3.1	4914	4928	4942	4955	4969	4983	4997	5011	5024	5038
3.2	5051	5065	5079	5092	5105	5119	5132	5145	5159	5172
3.3	5185	5198	5211	5224	5237	5250	5263	5276	5289	5302
3.4	5315	5328	5340	5353	5366	5378	5391	5403	5416	5428
3.5	5441	5453	5465	5478	5490	5502	5514	5527	5539	5551
3.6	5563	5575	5587	5599	5611	5623	5635	5647	5658	5670
3.7	5682	5694	5705	5717	5729	5740	5752	5763	5775	5786
3.8	5798	5809	5821	5832	5843	5855	5866	5877	5888	5899
3.9	5911	5922	5933	5944	5955	5966	5977	5988	5999	6010
4.0	6021	6031	6042	6053	6064	6075	6085	6096	6107	6117
4.1	6128	6138	6149	6160	6170	6180	6191	6201	6212	6222
4.2	6232	6243	6253	6263	6274	6284	6294	6304	6314	6325
4.3	6335	6345	6355	6365	6375	6385	6395	6405	6415	6425
4.4	6435	6444	6454	6464	6474	6484	6493	6503	6513	6522
4.5	6532	6542	6551	6561	6571	6580	6590	6599	6609	6618
4.6	6628	6637	6646	6656	6665	6675	6684	6693	6702	6712
4.7	6721	6730	6739	6749	6758	6767	6776	6785	6794	6803
4.8	6812	6821	6830	6839	6848	6857	6866	6875	6884	6893
4.9	6902	6911	6920	6928	6937	6946	6955	6964	6972	6981
5.0	6990	6998	7007	7016	7024	7033	7042	7050	7059	7067
5.1	7076	7084	7093	7101	7110	7118	7126	7135	7143	7152
5.2	7160	7168	7177	7185	7193	7202	7210	7218	7226	7235
5.3	7243	7251	7259	7267	7275	7284	7292	7300	7308	7316
5.4	7324	7332	7340	7348	7356	7364	7372	7380	7388	7396

*Decimal points omitted.

Table 3 | Common Logarithms of Numbers

N	0	1	2	3	4	5	6	7	8	9
5.5	7404	7412	7419	7427	7435	7443	7451	7459	7466	7474
5.6	7482	7490	7497	7505	7513	7520	7528	7536	7543	7551
5.7	7559	7566	7574	7582	7589	7597	7604	7612	7619	7627
5.8	7634	7642	7649	7657	7664	7672	7679	7686	7694	7701
5.9	7709	7716	7723	7731	7738	7745	7752	7760	7767	7774
6.0	7782	7789	7796	7803	7810	7818	7825	7832	7839	7846
6.1	7853	7860	7868	7875	7882	7889	7896	7903	7910	7917
6.2	7924	7931	7938	7945	7952	7959	7966	7973	7980	7987
6.3	7993	8000	8007	8014	8021	8028	8035	8041	8048	8055
6.4	8062	8069	8075	8082	8089	8096	8102	8109	8116	8122
6.5	8129	8136	8142	8149	8156	8162	8169	8176	8182	8189
6.6	8195	8202	8209	8215	8222	8228	8235	8241	8248	8254
6.7	8261	8267	8274	8280	8287	8293	8299	8306	8312	8319
6.8	8325	8331	8338	8344	8351	8357	8363	8370	8376	8382
6.9	8388	8395	8401	8407	8414	8420	8426	8432	8439	8445
7.0	8451	8457	8463	8470	8476	8482	8488	8494	8500	8506
7.1	8513	8519	8525	8531	8537	8543	8549	8555	8561	8567
7.2	8573	8579	8585	8591	8597	8603	8609	8615	8621	8627
7.3	8633	8639	8645	8651	8657	8663	8669	8675	8681	8686
7.4	8692	8698	8704	8710	8716	8722	8727	8733	8739	8745
7.5	8751	8756	8762	8768	8774	8779	8785	8791	8797	8802
7.6	8808	8814	8820	8825	8831	8837	8842	8848	8854	8859
7.7	8865	8871	8876	8882	8887	8893	8899	8904	8910	8915
7.8	8921	8927	8932	8938	8943	8949	8954	8960	8965	8971
7.9	8976	8982	8987	8993	8998	9004	9009	9015	9020	9025
8.0	9031	9036	9042	9047	9053	9058	9063	9069	9074	9079
8.1	9085	9090	9096	9101	9106	9112	9117	9122	9128	9133
8.2	9138	9143	9149	9154	9159	9165	9170	9175	9180	9186
8.3	9191	9196	9201	9206	9212	9217	9222	9227	9232	9238
8.4	9243	9248	9253	9258	9263	9269	9274	9279	9284	9289
8.5	9294	9299	9304	9309	9315	9320	9325	9330	9335	9340
8.6	9345	9350	9355	9360	9365	9370	9375	9380	9385	9390
8.7	9395	9400	9405	9410	9415	9420	9425	9430	9435	9440
8.8	9445	9450	9455	9460	9465	9469	9474	9479	9484	9489
8.9	9494	9499	9504	9509	9513	9518	9523	9528	9533	9538
9.0	9542	9547	9552	9557	9562	9566	9571	9576	9581	9586
9.1	9590	9595	9600	9605	9609	9614	9619	9624	9628	9633
9.2	9638	9643	9647	9652	9657	9661	9666	9671	9675	9680
9.3	9685	9689	9694	9699	9703	9708	9713	9717	9722	9727
9.4	9731	9736	9741	9745	9750	9754	9759	9763	9768	9773
9.5	9777	9782	9786	9791	9795	9800	9805	9809	9814	9818
9.6	9823	9827	9832	9836	9841	9845	9850	9854	9859	9863
9.7	9868	9872	9877	9881	9886	9890	9894	9899	9903	9908
9.8	9912	9917	9921	9926	9930	9934	9939	9943	9948	9952
9.9	9956	9961	9965	9969	9974	9978	9983	9987	9991	9996

θ Degrees	θ Radians	$\sin \theta$	$\cos \theta$	$\tan \theta$	$\cot \theta$	$\sec \theta$	$\csc \theta$		
0° 00′	.0000	.0000	1.0000	.0000	Undefined	1.000	Undefined	1.5708	90° 00′
10′	.0029	.0029	1.0000	.0029	343.8	1.000	343.8	1.5679	50′
20′	.0058	.0058	1.0000	.0058	171.9	1.000	171.9	1.5650	40′
30′	.0087	.0087	1.0000	.0087	114.6	1.000	114.6	1.5621	30′
40′	.0116	.0116	.9999	.0116	85.94	1.000	85.95	1.5592	20′
50′	.0145	.0145	.9999	.0145	68.75	1.000	68.76	1.5563	10′
1° 00′	.0175	.0175	.9998	.0175	57.29	1.000	57.30	1.5533	89° 00′
10′	.0204	.0204	.9998	.0204	49.10	1.000	49.11	1.5504	50′
20′	.0233	.0233	.9997	.0233	42.96	1.000	42.98	1.5475	40′
30′	.0262	.0262	.9997	.0262	38.19	1.000	38.20	1.5446	30′
40′	.0291	.0291	.9996	.0291	34.37	1.000	34.38	1.5417	20′
50′	.0320	.0320	.9995	.0320	31.24	1.001	31.26	1.5388	10′
2° 00′	.0349	.0349	.9994	.0349	28.64	1.001	28.65	1.5359	88° 00′
10′	.0378	.0378	.9993	.0378	26.43	1.001	26.45	1.5330	50′
20′	.0407	.0407	.9992	.0407	24.54	1.001	24.56	1.5301	40′
30′	.0436	.0436	.9990	.0437	22.90	1.001	22.93	1.5272	30′
40′	.0465	.0465	.9989	.0466	21.47	1.001	21.49	1.5243	20′
50′	.0495	.0494	.9988	.0495	20.21	1.001	20.23	1.5213	10′
3° 00′	.0524	.0523	.9986	.0524	19.08	1.001	19.11	1.5184	87° 00′
10′	.0553	.0552	.9985	.0553	18.07	1.002	18.10	1.5155	50′
20′	.0582	.0581	.9983	.0582	17.17	1.002	17.20	1.5126	40′
30′	.0611	.0610	.9981	.0612	16.35	1.002	16.38	1.5097	30′
40′	.0640	.0640	.9980	.0641	15.60	1.002	15.64	1.5068	20′
50′	.0669	.0669	.9978	.0670	14.92	1.002	14.96	1.5039	10′
4° 00′	.0698	.0698	.9976	.0699	14.30	1.002	14.34	1.5010	86° 00′
10′	.0727	.0727	.9974	.0729	13.73	1.003	13.76	1.4981	50′
20′	.0756	.0756	.9971	.0758	13.20	1.003	13.23	1.4952	40′
30′	.0785	.0785	.9969	.0787	12.71	1.003	12.75	1.4923	30′
40′	.0814	.0814	.9967	.0816	12.25	1.003	12.29	1.4893	20′
50′	.0844	.0843	.9964	.0846	11.83	1.004	11.87	1.4864	10′
5° 00′	.0873	.0872	.9962	.0875	11.43	1.004	11.47	1.4835	85° 00′
10′	.0902	.0901	.9959	.0904	11.06	1.004	11.10	1.4806	50′
20′	.0931	.0929	.9957	.0934	10.71	1.004	10.76	1.4777	40′
30′	.0960	.0958	.9954	.0963	10.39	1.005	10.43	1.4748	30′
40′	.0989	.0987	.9951	.0992	10.08	1.005	10.13	1.4719	20′
50′	.1018	.1016	.9948	.1022	9.788	1.005	9.839	1.4690	10′
6° 00′	.1047	.1045	.9945	.1051	9.514	1.006	9.567	1.4661	84° 00′
10′	.1076	.1074	.9942	.1080	9.255	1.006	9.309	1.4632	50′
20′	.1105	.1103	.9939	.1110	9.010	1.006	9.065	1.4603	40′
30′	.1134	.1132	.9936	.1139	8.777	1.006	8.834	1.4573	30′
40′	.1164	.1161	.9932	.1169	8.556	1.007	8.614	1.4544	20′
50′	.1193	.1190	.9929	.1198	8.345	1.007	8.405	1.4515	10′
7° 00′	.1222	.1219	.9925	.1228	8.144	1.008	8.206	1.4486	83° 00′
10′	.1251	.1248	.9922	.1257	7.953	1.008	8.016	1.4457	50′
20′	.1280	.1276	.9918	.1287	7.770	1.008	7.834	1.4428	40′
30′	.1309	.1305	.9914	.1317	7.596	1.009	7.661	1.4399	30′
40′	.1338	.1334	.9911	.1346	7.429	1.009	7.496	1.4370	20′
50′	.1367	.1363	.9907	.1376	7.269	1.009	7.337	1.4341	10′
8° 00′	.1396	.1392	.9903	.1405	7.115	1.010	7.185	1.4312	82° 00′
10′	.1425	.1421	.9899	.1435	6.968	1.010	7.040	1.4283	50′
20′	.1454	.1449	.9894	.1465	6.827	1.011	6.900	1.4254	40′
30′	.1484	.1478	.9890	.1495	6.691	1.011	6.765	1.4224	30′
40′	.1513	.1507	.9886	.1524	6.561	1.012	6.636	1.4195	20′
50′	.1542	.1536	.9881	.1554	6.435	1.012	6.512	1.4166	10′
9° 00′	.1571	.1564	.9877	.1584	6.314	1.012	6.392	1.4137	81° 00′
		$\cos \theta$	$\sin \theta$	$\cot \theta$	$\tan \theta$	$\csc \theta$	$\sec \theta$	θ Radians	θ Degrees

θ Degrees	θ Radians	sin θ	cos θ	tan θ	cot θ	sec θ	csc θ		
9° 00′	.1571	.1564	.9877	.1584	6.314	1.012	6.392	1.4137	81° 00′
10′	.1600	.1593	.9872	.1614	6.197	1.013	6.277	1.4108	50′
20′	.1629	.1622	.9868	.1644	6.084	1.013	6.166	1.4079	40′
30′	.1658	.1650	.9863	.1673	5.976	1.014	6.059	1.4050	30′
40′	.1687	.1679	.9858	.1703	5.871	1.014	5.955	1.4021	20′
50′	.1716	.1708	.9853	.1733	5.769	1.015	5.855	1.3992	10′
10° 00′	.1745	.1736	.9848	.1763	5.671	1.015	5.759	1.3963	80° 00′
10′	.1774	.1765	.9843	.1793	5.576	1.016	5.665	1.3934	50′
20′	.1804	.1794	.9838	.1823	5.485	1.016	5.575	1.3904	40′
30′	.1833	.1822	.9833	.1853	5.396	1.017	5.487	1.3875	30′
40′	.1862	.1851	.9827	.1883	5.309	1.018	5.403	1.3846	20′
50′	.1891	.1880	.9822	.1914	5.226	1.018	5.320	1.3817	10′
11° 00′	.1920	.1908	.9816	.1944	5.145	1.019	5.241	1.3788	79° 00′
10′	.1949	.1937	.9811	.1974	5.066	1.019	5.164	1.3759	50′
20′	.1978	.1965	.9805	.2004	4.989	1.020	5.089	1.3730	40′
30′	.2007	.1994	.9799	.2035	4.915	1.020	5.016	1.3701	30′
40′	.2036	.2022	.9793	.2065	4.843	1.021	4.945	1.3672	20′
50′	.2065	.2051	.9787	.2095	4.773	1.022	4.876	1.3643	10′
12° 00′	.2094	.2079	.9781	.2126	4.705	1.022	4.810	1.3614	78° 00′
10′	.2123	.2108	.9775	.2156	4.638	1.023	4.745	1.3584	50′
20′	.2153	.2136	.9769	.2186	4.574	1.024	4.682	1.3555	40′
30′	.2182	.2164	.9763	.2217	4.511	1.024	4.620	1.3526	30′
40′	.2211	.2193	.9757	.2247	4.449	1.025	4.560	1.3497	20′
50′	.2240	.2221	.9750	.2278	4.390	1.026	4.502	1.3468	10′
13° 00′	.2269	.2250	.9744	.2309	4.331	1.026	4.445	1.3439	77° 00′
10′	.2298	.2278	.9737	.2339	4.275	1.027	4.390	1.3410	50′
20′	.2327	.2306	.9730	.2370	4.219	1.028	4.336	1.3381	40′
30′	.2356	.2334	.9724	.2401	4.165	1.028	4.284	1.3352	30′
40′	.2385	.2363	.9717	.2432	4.113	1.029	4.232	1.3323	20′
50′	.2414	.2391	.9710	.2462	4.061	1.030	4.182	1.3294	10′
14° 00′	.2443	.2419	.9703	.2493	4.011	1.031	4.134	1.3265	76° 00′
10′	.2473	.2447	.9696	.2524	3.962	1.031	4.086	1.3235	50′
20′	.2502	.2476	.9689	.2555	3.914	1.032	4.039	1.3206	40′
30′	.2531	.2504	.9681	.2586	3.867	1.033	3.994	1.3177	30′
40′	.2560	.2532	.9674	.2617	3.821	1.034	3.950	1.3148	20′
50′	.2589	.2560	.9667	.2648	3.776	1.034	3.906	1.3119	10′
15° 00′	.2618	.2588	.9659	.2679	3.732	1.035	3.864	1.3090	75° 00′
10′	.2647	.2616	.9652	.2711	3.689	1.036	3.822	1.3061	50′
20′	.2676	.2644	.9644	.2742	3.647	1.037	3.782	1.3032	40′
30′	.2705	.2672	.9636	.2773	3.606	1.038	3.742	1.3003	30′
40′	.2734	.2700	.9628	.2805	3.566	1.039	3.703	1.2974	20′
50′	.2763	.2728	.9621	.2836	3.526	1.039	3.665	1.2945	10′
16° 00′	.2793	.2756	.9613	.2867	3.487	1.040	3.628	1.2915	74° 00′
10′	.2822	.2784	.9605	.2899	3.450	1.041	3.592	1.2886	50′
20′	.2851	.2812	.9596	.2931	3.412	1.042	3.556	1.2857	40′
30′	.2880	.2840	.9588	.2962	3.376	1.043	3.521	1.2828	30′
40′	.2909	.2868	.9580	.2994	3.340	1.044	3.487	1.2799	20′
50′	.2938	.2896	.9572	.3026	3.305	1.045	3.453	1.2770	10′
17° 00′	.2967	.2924	.9563	.3057	3.271	1.046	3.420	1.2741	73° 00′
10′	.2996	.2952	.9555	.3089	3.237	1.047	3.388	1.2712	50′
20′	.3025	.2979	.9546	.3121	3.204	1.048	3.357	1.2683	40′
30′	.3054	.3007	.9537	.3153	3.172	1.049	3.326	1.2654	30′
40′	.3083	.3035	.9528	.3185	3.140	1.049	3.295	1.2625	20′
50′	.3113	.3062	.9520	.3217	3.108	1.050	3.265	1.2595	10′
18° 00′	.3142	.3090	.9511	.3249	3.078	1.051	3.236	1.2566	72° 00′
		cos θ	sin θ	cot θ	tan θ	csc θ	sec θ	θ Radians	θ Degrees

θ Degrees	θ Radians	$\sin \theta$	$\cos \theta$	$\tan \theta$	$\cot \theta$	$\sec \theta$	$\csc \theta$		
18° 00′	.3142	.3090	.9511	.3249	3.078	1.051	3.236	1.2566	**72° 00′**
10′	.3171	.3118	.9502	.3281	3.047	1.052	3.207	1.2537	50′
20′	.3200	.3145	.9492	.3314	3.018	1.053	3.179	1.2508	40′
30′	.3229	.3173	.9483	.3346	2.989	1.054	3.152	1.2479	30′
40′	.3258	.3201	.9474	.3378	2.960	1.056	3.124	1.2450	20′
50′	.3287	.3228	.9465	.3411	2.932	1.057	3.098	1.2421	10′
19° 00′	.3316	.3256	.9455	.3443	2.904	1.058	3.072	1.2392	**71° 00′**
10′	.3345	.3283	.9446	.3476	2.877	1.059	3.046	1.2363	50′
20′	.3374	.3311	.9436	.3508	2.850	1.060	3.021	1.2334	40′
30′	.3403	.3338	.9426	.3541	2.824	1.061	2.996	1.2305	30′
40′	.3432	.3365	.9417	.3574	2.798	1.062	2.971	1.2275	20′
50′	.3462	.3393	.9407	.3607	2.773	1.063	2.947	1.2246	10′
20° 00′	.3491	.3420	.9397	.3640	2.747	1.064	2.924	1.2217	**70° 00′**
10′	.3520	.3448	.9387	.3673	2.723	1.065	2.901	1.2188	50′
20′	.3549	.3475	.9377	.3706	2.699	1.066	2.878	1.2159	40′
30′	.3578	.3502	.9367	.3739	2.675	1.068	2.855	1.2130	30′
40′	.3607	.3529	.9356	.3772	2.651	1.069	2.833	1.2101	20′
50′	.3636	.3557	.9346	.3805	2.628	1.070	2.812	1.2072	10′
21° 00′	.3665	.3584	.9336	.3839	2.605	1.071	2.790	1.2043	**69° 00′**
10′	.3694	.3611	.9325	.3872	2.583	1.072	2.769	1.2014	50′
20′	.3723	.3638	.9315	.3906	2.560	1.074	2.749	1.1985	40′
30′	.3752	.3665	.9304	.3939	2.539	1.075	2.729	1.1956	30′
40′	.3782	.3692	.9293	.3973	2.517	1.076	2.709	1.1926	20′
50′	.3811	.3719	.9283	.4006	2.496	1.077	2.689	1.1897	10′
22° 00′	.3840	.3746	.9272	.4040	2.475	1.079	2.669	1.1868	**68° 00′**
10′	.3869	.3773	.9261	.4074	2.455	1.080	2.650	1.1839	50′
20′	.3898	.3800	.9250	.4108	2.434	1.081	2.632	1.1810	40′
30′	.3927	.3827	.9239	.4142	2.414	1.082	2.613	1.1781	30′
40′	.3956	.3854	.9228	.4176	2.394	1.084	2.595	1.1752	20′
50′	.3985	.3881	.9216	.4210	2.375	1.085	2.577	1.1723	10′
23° 00′	.4014	.3907	.9205	.4245	2.356	1.086	2.559	1.1694	**67° 00′**
10′	.4043	.3934	.9194	.4279	2.337	1.088	2.542	1.1665	50′
20′	.4072	.3961	.9182	.4314	2.318	1.089	2.525	1.1636	40′
30′	.4102	.3987	.9171	.4348	2.300	1.090	2.508	1.1606	30′
40′	.4131	.4014	.9159	.4383	2.282	1.092	2.491	1.1577	20′
50′	.4160	.4041	.9147	.4417	2.264	1.093	2.475	1.1548	10′
24° 00′	.4189	.4067	.9135	.4452	2.246	1.095	2.459	1.1519	**66° 00′**
10′	.4218	.4094	.9124	.4487	2.229	1.096	2.443	1.1490	50′
20′	.4247	.4120	.9112	.4522	2.211	1.097	2.427	1.1461	40′
30′	.4276	.4147	.9100	.4557	2.194	1.099	2.411	1.1432	30′
40′	.4305	.4173	.9088	.4592	2.177	1.100	2.396	1.1403	20′
50′	.4334	.4200	.9075	.4628	2.161	1.102	2.381	1.1374	10′
25° 00′	.4363	.4226	.9063	.4663	2.145	1.103	2.366	1.1345	**65° 00′**
10′	.4392	.4253	.9051	.4699	2.128	1.105	2.352	1.1316	50′
20′	.4422	.4279	.9038	.4734	2.112	1.106	2.337	1.1286	40′
30′	.4451	.4305	.9026	.4770	2.097	1.108	2.323	1.1257	30′
40′	.4480	.4331	.9013	.4806	2.081	1.109	2.309	1.1228	20′
50′	.4509	.4358	.9001	.4841	2.066	1.111	2.295	1.1199	10′
26° 00′	.4538	.4384	.8988	.4877	2.050	1.113	2.281	1.1170	**64° 00′**
10′	.4567	.4410	.8975	.4913	2.035	1.114	2.268	1.1141	50′
20′	.4596	.4436	.8962	.4950	2.020	1.116	2.254	1.1112	40′
30′	.4625	.4462	.8949	.4986	2.006	1.117	2.241	1.1083	30′
40′	.4654	.4488	.8936	.5022	1.991	1.119	2.228	1.1054	20′
50′	.4683	.4514	.8923	.5059	1.977	1.121	2.215	1.1025	10′
27° 00′	.4712	.4540	.8910	.5095	1.963	1.122	2.203	1.0996	**63° 00′**
		$\cos \theta$	$\sin \theta$	$\cot \theta$	$\tan \theta$	$\csc \theta$	$\sec \theta$	θ Radians	θ Degrees

Table 4

Trigonometric Functions of θ (θ in degrees and minutes)

θ Degrees	θ Radians	sin θ	cos θ	tan θ	cot θ	sec θ	csc θ		θ Degrees
27° 00′	.4712	.4540	.8910	.5095	1.963	1.122	2.203	1.0996	63° 00′
10′	.4741	.4566	.8897	.5132	1.949	1.124	2.190	1.0966	50′
20′	.4771	.4592	.8884	.5169	1.935	1.126	2.178	1.0937	40′
30′	.4800	.4617	.8870	.5206	1.921	1.127	2.166	1.0908	30′
40′	.4829	.4643	.8857	.5243	1.907	1.129	2.154	1.0879	20′
50′	.4858	.4669	.8843	.5280	1.894	1.131	2.142	1.0850	10′
28° 00′	.4887	.4695	.8829	.5317	1.881	1.133	2.130	1.0821	62° 00′
10′	.4916	.4720	.8816	.5354	1.868	1.134	2.118	1.0792	50′
20′	.4945	.4746	.8802	.5392	1.855	1.136	2.107	1.0763	40′
30′	.4974	.4772	.8788	.5430	1.842	1.138	2.096	1.0734	30′
40′	.5003	.4797	.8774	.5467	1.829	1.140	2.085	1.0705	20′
50′	.5032	.4823	.8760	.5505	1.816	1.142	2.074	1.0676	10′
29° 00′	.5061	.4848	.8746	.5543	1.804	1.143	2.063	1.0647	61° 00′
10′	.5091	.4874	.8732	.5581	1.792	1.145	2.052	1.0617	50′
20′	.5120	.4899	.8718	.5619	1.780	1.147	2.041	1.0588	40′
30′	.5149	.4924	.8704	.5658	1.767	1.149	2.031	1.0559	30′
40′	.5178	.4950	.8689	.5696	1.756	1.151	2.020	1.0530	20′
50′	.5207	.4975	.8675	.5735	1.744	1.153	2.010	1.0501	10′
30° 00′	.5236	.5000	.8660	.5774	1.732	1.155	2.000	1.0472	60° 00′
10′	.5265	.5025	.8646	.5812	1.720	1.157	1.990	1.0443	50′
20′	.5294	.5050	.8631	.5851	1.709	1.159	1.980	1.0414	40′
30′	.5323	.5075	.8616	.5890	1.698	1.161	1.970	1.0385	30′
40′	.5352	.5100	.8601	.5930	1.686	1.163	1.961	1.0356	20′
50′	.5381	.5125	.8587	.5969	1.675	1.165	1.951	1.0327	10′
31° 00′	.5411	.5150	.8572	.6009	1.664	1.167	1.942	1.0297	59° 00′
10′	.5440	.5175	.8557	.6048	1.653	1.169	1.932	1.0268	50′
20′	.5469	.5200	.8542	.6088	1.643	1.171	1.923	1.0239	40′
30′	.5498	.5225	.8526	.6128	1.632	1.173	1.914	1.0210	30′
40′	.5527	.5250	.8511	.6168	1.621	1.175	1.905	1.0181	20′
50′	.5556	.5275	.8496	.6208	1.611	1.177	1.896	1.0152	10′
32° 00′	.5585	.5299	.8480	.6249	1.600	1.179	1.887	1.0123	58° 00′
10′	.5614	.5324	.8465	.6289	1.590	1.181	1.878	1.0094	50′
20′	.5643	.5348	.8450	.6330	1.580	1.184	1.870	1.0065	40′
30′	.5672	.5373	.8434	.6371	1.570	1.186	1.861	1.0036	30′
40′	.5701	.5398	.8418	.6412	1.560	1.188	1.853	1.0007	20′
50′	.5730	.5422	.8403	.6453	1.550	1.190	1.844	.9977	10′
33° 00′	.5760	.5446	.8387	.6494	1.540	1.192	1.836	.9948	57° 00′
10′	.5789	.5471	.8371	.6536	1.530	1.195	1.828	.9919	50′
20′	.5818	.5495	.8355	.6577	1.520	1.197	1.820	.9890	40′
30′	.5847	.5519	.8339	.6619	1.511	1.199	1.812	.9861	30′
40′	.5876	.5544	.8323	.6661	1.501	1.202	1.804	.9832	20′
50′	.5905	.5568	.8307	.6703	1.492	1.204	1.796	.9803	10′
34° 00′	.5934	.5592	.8290	.6745	1.483	1.206	1.788	.9774	56° 00′
10′	.5963	.5616	.8274	.6787	1.473	1.209	1.781	.9745	50′
20′	.5992	.5640	.8258	.6830	1.464	1.211	1.773	.9716	40′
30′	.6021	.5664	.8241	.6873	1.455	1.213	1.766	.9687	30′
40′	.6050	.5688	.8225	.6916	1.446	1.216	1.758	.9657	20′
50′	.6080	.5712	.8208	.6959	1.437	1.218	1.751	.9628	10′
35° 00′	.6109	.5736	.8192	.7002	1.428	1.221	1.743	.9599	55° 00′
10′	.6138	.5760	.8175	.7046	1.419	1.223	1.736	.9570	50′
20′	.6167	.5783	.8158	.7089	1.411	1.226	1.729	.9541	40′
30′	.6196	.5807	.8141	.7133	1.402	1.228	1.722	.9512	30′
40′	.6225	.5831	.8124	.7177	1.393	1.231	1.715	.9483	20′
50′	.6254	.5854	.8107	.7221	1.385	1.233	1.708	.9454	10′
36° 00′	.6283	.5878	.8090	.7265	1.376	1.236	1.701	.9425	54° 00′
		cos θ	sin θ	cot θ	tan θ	csc θ	sec θ	θ Radians	θ Degrees

θ Degrees	θ Radians	sin θ	cos θ	tan θ	cot θ	sec θ	csc θ		
36° 00′	.6283	.5878	.8090	.7265	1.376	1.236	1.701	.9425	**54° 00′**
10′	.6312	.5901	.8073	.7310	1.368	1.239	1.695	.9396	50′
20′	.6341	.5925	.8056	.7355	1.360	1.241	1.688	.9367	40′
30′	.6370	.5948	.8039	.7400	1.351	1.244	1.681	.9338	30′
40′	.6400	.5972	.8021	.7445	1.343	1.247	1.675	.9308	20′
50′	.6429	.5995	.8004	.7490	1.335	1.249	1.668	.9279	10′
37° 00′	.6458	.6018	.7986	.7536	1.327	1.252	1.662	.9250	**53° 00′**
10′	.6487	.6041	.7969	.7581	1.319	1.255	1.655	.9221	50′
20′	.6516	.6065	.7951	.7627	1.311	1.258	1.649	.9192	40′
30′	.6545	.6088	.7934	.7673	1.303	1.260	1.643	.9163	30′
40′	.6574	.6111	.7916	.7720	1.295	1.263	1.636	.9134	20′
50′	.6603	.6134	.7898	.7766	1.288	1.266	1.630	.9105	10′
38° 00′	.6632	.6157	.7880	.7813	1.280	1.269	1.624	.9076	**52° 00′**
10′	.6661	.6180	.7862	.7860	1.272	1.272	1.618	.9047	50′
20′	.6690	.6202	.7844	.7907	1.265	1.275	1.612	.9018	40′
30′	.6720	.6225	.7826	.7954	1.257	1.278	1.606	.8988	30′
40′	.6749	.6248	.7808	.8002	1.250	1.281	1.601	.8959	20′
50′	.6778	.6271	.7790	.8050	1.242	1.284	1.595	.8930	10′
39° 00′	.6807	.6293	.7771	.8098	1.235	1.287	1.589	.8901	**51° 00′**
10′	.6836	.6316	.7753	.8146	1.228	1.290	1.583	.8872	50′
20′	.6865	.6338	.7735	.8195	1.220	1.293	1.578	.8843	40′
30′	.6894	.6361	.7716	.8243	1.213	1.296	1.572	.8814	30′
40′	.6923	.6383	.7698	.8292	1.206	1.299	1.567	.8785	20′
50′	.6952	.6406	.7679	.8342	1.199	1.302	1.561	.8756	10′
40° 00′	.6981	.6428	.7660	.8391	1.192	1.305	1.556	.8727	**50° 00′**
10′	.7010	.6450	.7642	.8441	1.185	1.309	1.550	.8698	50′
20′	.7039	.6472	.7623	.8491	1.178	1.312	1.545	.8668	40′
30′	.7069	.6494	.7604	.8541	1.171	1.315	1.540	.8639	30′
40′	.7098	.6517	.7585	.8591	1.164	1.318	1.535	.8610	20′
50′	.7127	.6539	.7566	.8642	1.157	1.322	1.529	.8581	10′
41° 00′	.7156	.6561	.7547	.8693	1.150	1.325	1.524	.8552	**49° 00′**
10′	.7185	.6583	.7528	.8744	1.144	1.328	1.519	.8523	50′
20′	.7214	.6604	.7509	.8796	1.137	1.332	1.514	.8494	40′
30′	.7243	.6626	.7490	.8847	1.130	1.335	1.509	.8465	30′
40′	.7272	.6648	.7470	.8899	1.124	1.339	1.504	.8436	20′
50′	.7301	.6670	.7451	.8952	1.117	1.342	1.499	.8407	10′
42° 00′	.7330	.6691	.7431	.9004	1.111	1.346	1.494	.8378	**48° 00′**
10′	.7359	.6713	.7412	.9057	1.104	1.349	1.490	.8348	50′
20′	.7389	.6734	.7392	.9110	1.098	1.353	1.485	.8319	40′
30′	.7418	.6756	.7373	.9163	1.091	1.356	1.480	.8290	30′
40′	.7447	.6777	.7353	.9217	1.085	1.360	1.476	.8261	20′
50′	.7476	.6799	.7333	.9271	1.079	1.364	1.471	.8232	10′
43° 00′	.7505	.6820	.7314	.9325	1.072	1.367	1.466	.8203	**47° 00′**
10′	.7534	.6841	.7294	.9380	1.066	1.371	1.462	.8174	50′
20′	.7563	.6862	.7274	.9435	1.060	1.375	1.457	.8145	40′
30′	.7592	.6884	.7254	.9490	1.054	1.379	1.453	.8116	30′
40′	.7621	.6905	.7234	.9545	1.048	1.382	1.448	.8087	20′
50′	.7650	.6926	.7214	.9601	1.042	1.386	1.444	.8058	10′
44° 00′	.7679	.6947	.7193	.9657	1.036	1.390	1.440	.8029	**46° 00′**
10′	.7709	.6967	.7173	.9713	1.030	1.394	1.435	.7999	50′
20′	.7738	.6988	.7153	.9770	1.024	1.398	1.431	.7970	40′
30′	.7767	.7009	.7133	.9827	1.018	1.402	1.427	.7941	30′
40′	.7796	.7030	.7112	.9884	1.012	1.406	1.423	.7912	20′
50′	.7825	.7050	.7092	.9942	1.006	1.410	1.418	.7883	10′
45° 00′	.7854	.7071	.7071	1.000	1.000	1.414	1.414	.7854	**45° 00′**
		cos θ	sin θ	cot θ	tan θ	csc θ	sec θ	θ Radians	θ Degrees

Table 5

Trigonometric Functions of θ (θ in decimal degrees)

θ Degrees	θ Radians	$\sin\theta$	$\cos\theta$	$\tan\theta$	$\cot\theta$	$\sec\theta$	$\csc\theta$		
0.0	.0000	.0000	1.0000	.0000	undefined	1.000	undefined	1.5708	**90.0**
0.1	.0017	.0017	1.0000	.0017	573.0	1.000	573.0	1.5691	89.9
0.2	.0035	.0035	1.0000	.0035	286.5	1.000	286.5	1.5673	89.8
0.3	.0052	.0052	1.0000	.0052	191.0	1.000	191.0	1.5656	89.7
0.4	.0070	.0070	1.0000	.0070	143.2	1.000	143.2	1.5638	89.6
0.5	.0087	.0087	1.0000	.0087	114.6	1.000	114.6	1.5621	89.5
0.6	.0105	.0105	.9999	.0105	95.49	1.000	95.49	1.5603	89.4
0.7	.0122	.0122	.9999	.0122	81.85	1.000	81.85	1.5586	89.3
0.8	.0140	.0140	.9999	.0140	71.62	1.000	71.62	1.5568	89.2
0.9	.0157	.0157	.9999	.0157	63.66	1.000	63.66	1.5551	89.1
1.0	.0175	.0175	.9998	.0175	57.29	1.000	57.30	1.5533	**89.0**
1.1	.0192	.0192	.9998	.0192	52.08	1.000	52.09	1.5516	88.9
1.2	.0209	.0209	.9998	.0209	47.74	1.000	47.75	1.5499	88.8
1.3	.0227	.0227	.9997	.0227	44.07	1.000	44.08	1.5481	88.7
1.4	.0244	.0244	.9997	.0244	40.92	1.000	40.93	1.5464	88.6
1.5	.0262	.0262	.9997	.0262	38.19	1.000	38.20	1.5446	88.5
1.6	.0279	.0279	.9996	.0279	35.80	1.000	35.81	1.5429	88.4
1.7	.0297	.0297	.9996	.0297	33.69	1.000	33.71	1.5411	88.3
1.8	.0314	.0314	.9995	.0314	31.82	1.000	31.84	1.5394	88.2
1.9	.0332	.0332	.9995	.0332	30.14	1.001	30.16	1.5376	88.1
2.0	.0349	.0349	.9994	.0349	28.64	1.001	28.65	1.5359	**88.0**
2.1	.0367	.0366	.9993	.0367	27.27	1.001	27.29	1.5341	87.9
2.2	.0384	.0384	.9993	.0384	26.03	1.001	26.05	1.5324	87.8
2.3	.0401	.0401	.9992	.0402	24.90	1.001	24.92	1.5307	87.7
2.4	.0419	.0419	.9991	.0419	23.86	1.001	23.88	1.5289	87.6
2.5	.0436	.0436	.9990	.0437	22.90	1.001	22.93	1.5272	87.5
2.6	.0454	.0454	.9990	.0454	22.02	1.001	22.04	1.5254	87.4
2.7	.0471	.0471	.9989	.0472	21.20	1.001	21.23	1.5237	87.3
2.8	.0489	.0488	.9988	.0489	20.45	1.001	20.47	1.5219	87.2
2.9	.0506	.0506	.9987	.0507	19.74	1.001	19.77	1.5202	87.1
3.0	.0524	.0523	.9986	.0524	19.08	1.001	19.11	1.5184	**87.0**
3.1	.0541	.0541	.9985	.0542	18.46	1.001	18.49	1.5167	86.9
3.2	.0559	.0558	.9984	.0559	17.89	1.002	17.91	1.5149	86.8
3.3	.0576	.0576	.9983	.0577	17.34	1.002	17.37	1.5132	86.7
3.4	.0593	.0593	.9982	.0594	16.83	1.002	16.86	1.5115	86.6
3.5	.0611	.0610	.9981	.0612	16.35	1.002	16.38	1.5097	86.5
3.6	.0628	.0628	.9980	.0629	15.89	1.002	15.93	1.5080	86.4
3.7	.0646	.0645	.9979	.0647	15.46	1.002	15.50	1.5062	86.3
3.8	.0663	.0663	.9978	.0664	15.06	1.002	15.09	1.5045	86.2
3.9	.0681	.0680	.9977	.0682	14.67	1.002	14.70	1.5027	86.1
4.0	.0698	.0698	.9976	.0699	14.30	1.002	14.34	1.5010	**86.0**
4.1	.0716	.0715	.9974	.0717	13.95	1.003	13.99	1.4992	85.9
4.2	.0733	.0732	.9973	.0734	13.62	1.003	13.65	1.4975	85.8
4.3	.0750	.0750	.9972	.0752	13.30	1.003	13.34	1.4957	85.7
4.4	.0768	.0767	.9971	.0769	13.00	1.003	13.03	1.4940	85.6
4.5	.0785	.0785	.9969	.0787	12.71	1.003	12.75	1.4923	85.5
4.6	.0803	.0802	.9968	.0805	12.43	1.003	12.47	1.4905	85.4
4.7	.0820	.0819	.9966	.0822	12.16	1.003	12.20	1.4888	85.3
4.8	.0838	.0837	.9965	.0840	11.91	1.004	11.95	1.4870	85.2
4.9	.0855	.0854	.9963	.0857	11.66	1.004	11.71	1.4853	85.1
5.0	.0873	.0872	.9962	.0875	11.43	1.004	11.47	1.4835	**85.0**
5.1	.0890	.0889	.9960	.0892	11.20	1.004	11.25	1.4818	84.9
5.2	.0908	.0906	.9959	.0910	10.99	1.004	11.03	1.4800	84.8
5.3	.0925	.0924	.9957	.0928	10.78	1.004	10.83	1.4783	84.7
5.4	.0942	.0941	.9956	.0945	10.58	1.004	10.63	1.4765	84.6
5.5	.0960	.0958	.9954	.0963	10.39	1.005	10.43	1.4748	84.5
5.6	.0977	.0976	.9952	.0981	10.20	1.005	10.25	1.4731	84.4
5.7	.0995	.0993	.9951	.0998	10.02	1.005	10.07	1.4713	84.3
5.8	.1012	.1011	.9949	.1016	9.845	1.005	9.895	1.4696	84.2
5.9	.1030	.1028	.9947	.1033	9.677	1.005	9.728	1.4678	84.1
6.0	.1047	.1045	.9945	.1051	9.514	1.006	9.567	1.4661	**84.0**
		$\cos\theta$	$\sin\theta$	$\cot\theta$	$\tan\theta$	$\csc\theta$	$\sec\theta$	θ Radians	θ Degrees

θ Degrees	θ Radians	$\sin \theta$	$\cos \theta$	$\tan \theta$	$\cot \theta$	$\sec \theta$	$\csc \theta$		
6.0	.1047	.1045	.9945	.1051	9.514	1.006	9.567	1.4661	**84.0**
6.1	.1065	.1063	.9943	.1069	9.357	1.006	9.411	1.4643	83.9
6.2	.1082	.1080	.9942	.1086	9.205	1.006	9.259	1.4626	83.8
6.3	.1100	.1097	.9940	.1104	9.058	1.006	9.113	1.4608	83.7
6.4	.1117	.1115	.9938	.1122	8.915	1.006	8.971	1.4591	83.6
6.5	.1134	.1132	.9936	.1139	8.777	1.006	8.834	1.4574	83.5
6.6	.1152	.1149	.9934	.1157	8.643	1.007	8.700	1.4556	83.4
6.7	.1169	.1167	.9932	.1175	8.513	1.007	8.571	1.4539	83.3
6.8	.1187	.1184	.9930	.1192	8.386	1.007	8.446	1.4521	83.2
6.9	.1204	.1201	.9928	.1210	8.264	1.007	8.324	1.4504	83.1
7.0	.1222	.1219	.9925	.1228	8.144	1.008	8.206	1.4486	**83.0**
7.1	.1239	.1236	.9923	.1246	8.028	1.008	8.091	1.4469	82.9
7.2	.1257	.1253	.9921	.1263	7.916	1.008	7.979	1.4451	82.8
7.3	.1274	.1271	.9919	.1281	7.806	1.008	7.870	1.4434	82.7
7.4	.1292	.1288	.9917	.1299	7.700	1.008	7.764	1.4416	82.6
7.5	.1309	.1305	.9914	.1317	7.596	1.009	7.661	1.4399	82.5
7.6	.1326	.1323	.9912	.1334	7.495	1.009	7.561	1.4382	82.4
7.7	.1344	.1340	.9910	.1352	7.396	1.009	7.463	1.4364	82.3
7.8	.1361	.1357	.9907	.1370	7.300	1.009	7.368	1.4347	82.2
7.9	.1379	.1374	.9905	.1388	7.207	1.010	7.276	1.4329	82.1
8.0	.1396	.1392	.9903	.1405	7.115	1.010	7.185	1.4312	**82.0**
8.1	.1414	.1409	.9900	.1423	7.026	1.010	7.097	1.4294	81.9
8.2	.1431	.1426	.9898	.1441	6.940	1.010	7.011	1.4277	81.8
8.3	.1449	.1444	.9895	.1459	6.855	1.011	6.927	1.4259	81.7
8.4	.1466	.1461	.9893	.1477	6.772	1.011	6.845	1.4242	81.6
8.5	.1484	.1478	.9890	.1495	6.691	1.011	6.765	1.4224	81.5
8.6	.1501	.1495	.9888	.1512	6.612	1.011	6.687	1.4207	81.4
8.7	.1518	.1513	.9885	.1530	6.535	1.012	6.611	1.4190	81.3
8.8	.1536	.1530	.9882	.1548	6.460	1.012	6.537	1.4172	81.2
8.9	.1553	.1547	.9880	.1566	6.386	1.012	6.464	1.4155	81.1
9.0	.1571	.1564	.9877	.1584	6.314	1.012	6.392	1.4137	**81.0**
9.1	.1588	.1582	.9874	.1602	6.243	1.013	6.323	1.4120	80.9
9.2	.1606	.1599	.9871	.1620	6.174	1.013	6.255	1.4102	80.8
9.3	.1623	.1616	.9869	.1638	6.107	1.013	6.188	1.4085	80.7
9.4	.1641	.1633	.9866	.1655	6.041	1.014	6.123	1.4067	80.6
9.5	.1658	.1650	.9863	.1673	5.976	1.014	6.059	1.4050	80.5
9.6	.1676	.1668	.9860	.1691	5.912	1.014	5.996	1.4032	80.4
9.7	.1693	.1685	.9857	.1709	5.850	1.015	5.935	1.4015	80.3
9.8	.1710	.1702	.9854	.1727	5.789	1.015	5.875	1.3998	80.2
9.9	.1728	.1719	.9851	.1745	5.730	1.015	5.816	1.3980	80.1
10.0	.1745	.1736	.9848	.1763	5.671	1.015	5.759	1.3963	**80.0**
10.1	.1763	.1754	.9845	.1781	5.614	1.016	5.702	1.3945	79.9
10.2	.1780	.1771	.9842	.1799	5.558	1.016	5.647	1.3928	79.8
10.3	.1798	.1788	.9839	.1817	5.503	1.016	5.593	1.3910	79.7
10.4	.1815	.1805	.9836	.1835	5.449	1.017	5.540	1.3893	79.6
10.5	.1833	.1822	.9833	.1853	5.396	1.017	5.487	1.3875	79.5
10.6	.1850	.1840	.9829	.1871	5.343	1.017	5.436	1.3858	79.4
10.7	.1868	.1857	.9826	.1890	5.292	1.018	5.386	1.3840	79.3
10.8	.1885	.1874	.9823	.1908	5.242	1.018	5.337	1.3823	79.2
10.9	.1902	.1891	.9820	.1926	5.193	1.018	5.288	1.3806	79.1
11.0	.1920	.1908	.9816	.1944	5.145	1.019	5.241	1.3788	**79.0**
11.1	.1937	.1925	.9813	.1962	5.097	1.019	5.194	1.3771	78.9
11.2	.1955	.1942	.9810	.1980	5.050	1.019	5.148	1.3753	78.8
11.3	.1972	.1959	.9806	.1998	5.005	1.020	5.103	1.3736	78.7
11.4	.1990	.1977	.9803	.2016	4.959	1.020	5.059	1.3718	78.6
11.5	.2007	.1994	.9799	.2035	4.915	1.020	5.016	1.3701	78.5
11.6	.2025	.2011	.9796	.2053	4.872	1.021	4.973	1.3683	78.4
11.7	.2042	.2028	.9792	.2071	4.829	1.021	4.931	1.3666	78.3
11.8	.2059	.2045	.9789	.2089	4.787	1.022	4.890	1.3648	78.2
11.9	.2077	.2062	.9785	.2107	4.745	1.022	4.850	1.3631	78.1
12.0	.2094	.2079	.9781	.2126	4.705	1.022	4.810	1.3614	**78.0**
		$\cos \theta$	$\sin \theta$	$\cot \theta$	$\tan \theta$	$\csc \theta$	$\sec \theta$	θ Radians	θ Degrees

Table 5

Trigonometric Functions of θ (θ in decimal degrees)

θ Degrees	θ Radians	$\sin \theta$	$\cos \theta$	$\tan \theta$	$\cot \theta$	$\sec \theta$	$\csc \theta$		
12.0	.2094	.2079	.9781	.2126	4.705	1.022	4.810	1.3614	**78.0**
12.1	.2112	.2096	.9778	.2144	4.665	1.023	4.771	1.3596	77.9
12.2	.2129	.2113	.9774	.2162	4.625	1.023	4.732	1.3579	77.8
12.3	.2147	.2130	.9770	.2180	4.586	1.023	4.694	1.3561	77.7
12.4	.2164	.2147	.9767	.2199	4.548	1.024	4.657	1.3544	77.6
12.5	.2182	.2164	.9763	.2217	4.511	1.024	4.620	1.3526	77.5
12.6	.2199	.2181	.9759	.2235	4.474	1.025	4.584	1.3509	77.4
12.7	.2217	.2198	.9755	.2254	4.437	1.025	4.549	1.3491	77.3
12.8	.2234	.2215	.9751	.2272	4.402	1.025	4.514	1.3474	77.2
12.9	.2251	.2233	.9748	.2290	4.366	1.026	4.479	1.3456	77.1
13.0	.2269	.2250	.9744	.2309	4.331	1.026	4.445	1.3439	**77.0**
13.1	.2286	.2267	.9740	.2327	4.297	1.027	4.412	1.3422	76.9
13.2	.2304	.2284	.9736	.2345	4.264	1.027	4.379	1.3404	76.8
13.3	.2321	.2300	.9732	.2364	4.230	1.028	4.347	1.3387	76.7
13.4	.2339	.2317	.9728	.2382	4.198	1.028	4.315	1.3369	76.6
13.5	.2356	.2334	.9724	.2401	4.165	1.028	4.284	1.3352	76.5
13.6	.2374	.2351	.9720	.2419	4.134	1.029	4.253	1.3334	76.4
13.7	.2391	.2368	.9715	.2438	4.102	1.029	4.222	1.3317	76.3
13.8	.2409	.2385	.9711	.2456	4.071	1.030	4.192	1.3299	76.2
13.9	.2426	.2402	.9707	.2475	4.041	1.030	4.163	1.3282	76.1
14.0	.2443	.2419	.9703	.2493	4.011	1.031	4.134	1.3265	**76.0**
14.1	.2461	.2436	.9699	.2512	3.981	1.031	4.105	1.3247	75.9
14.2	.2478	.2453	.9694	.2530	3.952	1.032	4.077	1.3230	75.8
14.3	.2496	.2470	.9690	.2549	3.923	1.032	4.049	1.3212	75.7
14.4	.2513	.2487	.9686	.2568	3.895	1.032	4.021	1.3195	75.6
14.5	.2531	.2504	.9681	.2586	3.867	1.033	3.994	1.3177	75.5
14.6	.2548	.2521	.9677	.2605	3.839	1.033	3.967	1.3160	75.4
14.7	.2566	.2538	.9673	.2623	3.812	1.034	3.941	1.3142	75.3
14.8	.2583	.2554	.9668	.2642	3.785	1.034	3.915	1.3125	75.2
14.9	.2601	.2571	.9664	.2661	3.758	1.035	3.889	1.3107	75.1
15.0	.2618	.2588	.9659	.2679	3.732	1.035	3.864	1.3090	**75.0**
15.1	.2635	.2605	.9655	.2698	3.706	1.036	3.839	1.3073	74.9
15.2	.2653	.2622	.9650	.2717	3.681	1.036	3.814	1.3055	74.8
15.3	.2670	.2639	.9646	.2736	3.655	1.037	3.790	1.3038	74.7
15.4	.2688	.2656	.9641	.2754	3.630	1.037	3.766	1.3020	74.6
15.5	.2705	.2672	.9636	.2773	3.606	1.038	3.742	1.3003	74.5
15.6	.2723	.2689	.9632	.2792	3.582	1.038	3.719	1.2985	74.4
15.7	.2740	.2706	.9627	.2811	3.558	1.039	3.695	1.2968	74.3
15.8	.2758	.2723	.9622	.2830	3.534	1.039	3.673	1.2950	74.2
15.9	.2775	.2740	.9617	.2849	3.511	1.040	3.650	1.2933	74.1
16.0	.2793	.2756	.9613	.2867	3.487	1.040	3.628	1.2915	**74.0**
16.1	.2810	.2773	.9608	.2886	3.465	1.041	3.606	1.2898	73.9
16.2	.2827	.2790	.9603	.2905	3.442	1.041	3.584	1.2881	73.8
16.3	.2845	.2807	.9598	.2924	3.420	1.042	3.563	1.2863	73.7
16.4	.2862	.2823	.9593	.2943	3.398	1.042	3.542	1.2846	73.6
16.5	.2880	.2840	.9588	.2962	3.376	1.043	3.521	1.2828	73.5
16.6	.2897	.2857	.9583	.2981	3.354	1.043	3.500	1.2811	73.4
16.7	.2915	.2874	.9578	.3000	3.333	1.044	3.480	1.2793	73.3
16.8	.2932	.2890	.9573	.3019	3.312	1.045	3.460	1.2776	73.2
16.9	.2950	.2907	.9568	.3038	3.291	1.045	3.440	1.2758	73.1
17.0	.2967	.2924	.9563	.3057	3.271	1.046	3.420	1.2741	**73.0**
17.1	.2985	.2940	.9558	.3076	3.251	1.046	3.401	1.2723	72.9
17.2	.3002	.2957	.9553	.3096	3.230	1.047	3.382	1.2706	72.8
17.3	.3019	.2974	.9548	.3115	3.211	1.047	3.363	1.2689	72.7
17.4	.3037	.2990	.9542	.3134	3.191	1.048	3.344	1.2671	72.6
17.5	.3054	.3007	.9537	.3153	3.172	1.049	3.326	1.2654	72.5
17.6	.3072	.3024	.9532	.3172	3.152	1.049	3.307	1.2636	72.4
17.7	.3089	.3040	.9527	.3191	3.133	1.050	3.289	1.2619	72.3
17.8	.3107	.3057	.9521	.3211	3.115	1.050	3.271	1.2601	72.2
17.9	.3124	.3074	.9516	.3230	3.096	1.051	3.254	1.2584	72.1
18.0	.3142	.3090	.9511	.3249	3.078	1.051	3.236	1.2566	**72.0**
		$\cos \theta$	$\sin \theta$	$\cot \theta$	$\tan \theta$	$\csc \theta$	$\sec \theta$	θ Radians	θ Degrees

θ Degrees	θ Radians	$\sin\theta$	$\cos\theta$	$\tan\theta$	$\cot\theta$	$\sec\theta$	$\csc\theta$		
18.0	.3142	.3090	.9511	.3249	3.078	1.051	3.236	1.2566	**72.0**
18.1	.3159	.3107	.9505	.3269	3.060	1.052	3.219	1.2549	71.9
18.2	.3177	.3123	.9500	.3288	3.042	1.053	3.202	1.2531	71.8
18.3	.3194	.3140	.9494	.3307	3.024	1.053	3.185	1.2514	71.7
18.4	.3211	.3156	.9489	.3327	3.006	1.054	3.168	1.2497	71.6
18.5	.3229	.3173	.9483	.3346	2.989	1.054	3.152	1.2479	71.5
18.6	.3246	.3190	.9478	.3365	2.971	1.055	3.135	1.2462	71.4
18.7	.3264	.3206	.9472	.3385	2.954	1.056	3.119	1.2444	71.3
18.8	.3281	.3223	.9466	.3404	2.937	1.056	3.103	1.2427	71.2
18.9	.3299	.3239	.9461	.3424	2.921	1.057	3.087	1.2409	71.1
19.0	.3316	.3256	.9455	.3443	2.904	1.058	3.072	1.2392	**71.0**
19.1	.3334	.3272	.9449	.3463	2.888	1.058	3.056	1.2374	70.9
19.2	.3351	.3289	.9444	.3482	2.872	1.059	3.041	1.2357	70.8
19.3	.3368	.3305	.9438	.3502	2.856	1.060	3.026	1.2339	70.7
19.4	.3386	.3322	.9432	.3522	2.840	1.060	3.011	1.2322	70.6
19.5	.3403	.3338	.9426	.3541	2.824	1.061	2.996	1.2305	70.5
19.6	.3421	.3355	.9421	.3561	2.808	1.062	2.981	1.2287	70.4
19.7	.3438	.3371	.9415	.3581	2.793	1.062	2.967	1.2270	70.3
19.8	.3456	.3387	.9409	.3600	2.778	1.063	2.952	1.2252	70.2
19.9	.3473	.3404	.9403	.3620	2.762	1.064	2.938	1.2235	70.1
20.0	.3491	.3420	.9397	.3640	2.747	1.064	2.924	1.2217	**70.0**
20.1	.3508	.3437	.9391	.3659	2.733	1.065	2.910	1.2200	69.9
20.2	.3526	.3453	.9385	.3679	2.718	1.066	2.896	1.2182	69.8
20.3	.3543	.3469	.9379	.3699	2.703	1.066	2.882	1.2165	69.7
20.4	.3560	.3486	.9373	.3719	2.689	1.067	2.869	1.2147	69.6
20.5	.3578	.3502	.9367	.3739	2.675	1.068	2.855	1.2130	69.5
20.6	.3595	.3518	.9361	.3759	2.660	1.068	2.842	1.2113	69.4
20.7	.3613	.3535	.9354	.3779	2.646	1.069	2.829	1.2095	69.3
20.8	.3630	.3551	.9348	.3799	2.633	1.070	2.816	1.2078	69.2
20.9	.3648	.3567	.9342	.3819	2.619	1.070	2.803	1.2060	69.1
21.0	.3665	.3584	.9336	.3839	2.605	1.071	2.790	1.2043	**69.0**
21.1	.3683	.3600	.9330	.3859	2.592	1.072	2.778	1.2025	68.9
21.2	.3700	.3616	.9323	.3879	2.578	1.073	2.765	1.2008	68.8
21.3	.3718	.3633	.9317	.3899	2.565	1.073	2.753	1.1991	68.7
21.4	.3735	.3649	.9311	.3919	2.552	1.074	2.741	1.1973	68.6
21.5	.3752	.3665	.9304	.3939	2.539	1.075	2.729	1.1956	68.5
21.6	.3770	.3681	.9298	.3959	2.526	1.076	2.716	1.1938	68.4
21.7	.3787	.3697	.9291	.3979	2.513	1.076	2.705	1.1921	68.3
21.8	.3805	.3714	.9285	.4000	2.500	1.077	2.693	1.1903	68.2
21.9	.3822	.3730	.9278	.4020	2.488	1.078	2.681	1.1886	68.1
22.0	.3840	.3746	.9272	.4040	2.475	1.079	2.669	1.1868	**68.0**
22.1	.3857	.3762	.9265	.4061	2.463	1.079	2.658	1.1851	67.9
22.2	.3875	.3778	.9259	.4081	2.450	1.080	2.647	1.1833	67.8
22.3	.3892	.3795	.9252	.4101	2.438	1.081	2.635	1.1816	67.7
22.4	.3910	.3811	.9245	.4122	2.426	1.082	2.624	1.1798	67.6
22.5	.3927	.3827	.9239	.4142	2.414	1.082	2.613	1.1781	67.5
22.6	.3944	.3843	.9232	.4163	2.402	1.083	2.602	1.1764	67.4
22.7	.3962	.3859	.9225	.4183	2.391	1.084	2.591	1.1746	67.3
22.8	.3979	.3875	.9219	.4204	2.379	1.085	2.581	1.1729	67.2
22.9	.3997	.3891	.9212	.4224	2.367	1.086	2.570	1.1711	67.1
23.0	.4014	.3907	.9205	.4245	2.356	1.086	2.559	1.1694	**67.0**
23.1	.4032	.3923	.9198	.4265	2.344	1.087	2.549	1.1676	66.9
23.2	.4049	.3939	.9191	.4286	2.333	1.088	2.538	1.1659	66.8
23.3	.4067	.3955	.9184	.4307	2.322	1.089	2.528	1.1641	66.7
23.4	.4084	.3971	.9178	.4327	2.311	1.090	2.518	1.1624	66.6
23.5	.4102	.3987	.9171	.4348	2.300	1.090	2.508	1.1606	66.5
23.6	.4119	.4003	.9164	.4369	2.289	1.091	2.498	1.1589	66.4
23.7	.4136	.4019	.9157	.4390	2.278	1.092	2.488	1.1572	66.3
23.8	.4154	.4035	.9150	.4411	2.267	1.093	2.478	1.1554	66.2
23.9	.4171	.4051	.9143	.4431	2.257	1.094	2.468	1.1537	66.1
24.0	.4189	.4067	.9135	.4452	2.246	1.095	2.459	1.1519	**66.0**
		$\cos\theta$	$\sin\theta$	$\cot\theta$	$\tan\theta$	$\csc\theta$	$\sec\theta$	θ Radians	θ Degrees

θ Degrees	θ Radians	$\sin \theta$	$\cos \theta$	$\tan \theta$	$\cot \theta$	$\sec \theta$	$\csc \theta$		
24.0	.4189	.4067	.9135	.4452	2.246	1.095	2.459	1.1519	**66.0**
24.1	.4206	.4083	.9128	.4473	2.236	1.095	2.449	1.1502	65.9
24.2	.4224	.4099	.9121	.4494	2.225	1.096	2.439	1.1484	65.8
24.3	.4241	.4115	.9114	.4515	2.215	1.097	2.430	1.1467	65.7
24.4	.4259	.4131	.9107	.4536	2.204	1.098	2.421	1.1449	65.6
24.5	.4276	.4147	.9100	.4557	2.194	1.099	2.411	1.1432	65.5
24.6	.4294	.4163	.9092	.4578	2.184	1.100	2.402	1.1414	65.4
24.7	.4311	.4179	.9085	.4599	2.174	1.101	2.393	1.1397	65.3
24.8	.4328	.4195	.9078	.4621	2.164	1.102	2.384	1.1380	65.2
24.9	.4346	.4210	.9070	.4642	2.154	1.102	2.375	1.1362	65.1
25.0	.4363	.4226	.9063	.4663	2.145	1.103	2.366	1.1345	**65.0**
25.1	.4381	.4242	.9056	.4684	2.135	1.104	2.357	1.1327	64.9
25.2	.4398	.4258	.9048	.4706	2.125	1.105	2.349	1.1310	64.8
25.3	.4416	.4274	.9041	.4727	2.116	1.106	2.340	1.1292	64.7
25.4	.4433	.4289	.9033	.4748	2.106	1.107	2.331	1.1275	64.6
25.5	.4451	.4305	.9026	.4770	2.097	1.108	2.323	1.1257	64.5
25.6	.4468	.4321	.9018	.4791	2.087	1.109	2.314	1.1240	64.4
25.7	.4485	.4337	.9011	.4813	2.078	1.110	2.306	1.1222	64.3
25.8	.4503	.4352	.9003	.4834	2.069	1.111	2.298	1.1205	64.2
25.9	.4520	.4368	.8996	.4856	2.059	1.112	2.289	1.1188	64.1
26.0	.4538	.4384	.8988	.4877	2.050	1.113	2.281	1.1170	**64.0**
26.1	.4555	.4399	.8980	.4899	2.041	1.114	2.273	1.1153	63.9
26.2	.4573	.4415	.8973	.4921	2.032	1.115	2.265	1.1135	63.8
26.3	.4590	.4431	.8965	.4942	2.023	1.115	2.257	1.1118	63.7
26.4	.4608	.4446	.8957	.4964	2.014	1.116	2.249	1.1100	63.6
26.5	.4625	.4462	.8949	.4986	2.006	1.117	2.241	1.1083	63.5
26.6	.4643	.4478	.8942	.5008	1.997	1.118	2.233	1.1065	63.4
26.7	.4660	.4493	.8934	.5029	1.988	1.119	2.226	1.1048	63.3
26.8	.4677	.4509	.8926	.5051	1.980	1.120	2.218	1.1030	63.2
26.9	.4695	.4524	.8918	.5073	1.971	1.121	2.210	1.1013	63.1
27.0	.4712	.4540	.8910	.5095	1.963	1.122	2.203	1.0996	**63.0**
27.1	.4730	.4555	.8902	.5117	1.954	1.123	2.195	1.0978	62.9
27.2	.4747	.4571	.8894	.5139	1.946	1.124	2.188	1.0961	62.8
27.3	.4765	.4586	.8886	.5161	1.937	1.125	2.180	1.0943	62.7
27.4	.4782	.4602	.8878	.5184	1.929	1.126	2.173	1.0926	62.6
27.5	.4800	.4617	.8870	.5206	1.921	1.127	2.166	1.0908	62.5
27.6	.4817	.4633	.8862	.5228	1.913	1.128	2.158	1.0891	62.4
27.7	.4835	.4648	.8854	.5250	1.905	1.129	2.151	1.0873	62.3
27.8	.4852	.4664	.8846	.5272	1.897	1.130	2.144	1.0856	62.2
27.9	.4869	.4679	.8838	.5295	1.889	1.132	2.137	1.0838	62.1
28.0	.4887	.4695	.8829	.5317	1.881	1.133	2.130	1.0821	**62.0**
28.1	.4904	.4710	.8821	.5340	1.873	1.134	2.123	1.0804	61.9
28.2	.4922	.4726	.8813	.5362	1.865	1.135	2.116	1.0786	61.8
28.3	.4939	.4741	.8805	.5384	1.857	1.136	2.109	1.0769	61.7
28.4	.4957	.4756	.8796	.5407	1.849	1.137	2.103	1.0751	61.6
28.5	.4974	.4772	.8788	.5430	1.842	1.138	2.096	1.0734	61.5
28.6	.4992	.4787	.8780	.5452	1.834	1.139	2.089	1.0716	61.4
28.7	.5009	.4802	.8771	.5475	1.827	1.140	2.082	1.0699	61.3
28.8	.5027	.4818	.8763	.5498	1.819	1.141	2.076	1.0681	61.2
28.9	.5044	.4833	.8755	.5520	1.811	1.142	2.069	1.0664	61.1
29.0	.5061	.4848	.8746	.5543	1.804	1.143	2.063	1.0647	**61.0**
29.1	.5079	.4863	.8738	.5566	1.797	1.144	2.056	1.0629	60.9
29.2	.5096	.4879	.8729	.5589	1.789	1.146	2.050	1.0612	60.8
29.3	.5114	.4894	.8721	.5612	1.782	1.147	2.043	1.0594	60.7
29.4	.5131	.4909	.8712	.5635	1.775	1.148	2.037	1.0577	60.6
29.5	.5149	.4924	.8704	.5658	1.767	1.149	2.031	1.0559	60.5
29.6	.5166	.4939	.8695	.5681	1.760	1.150	2.025	1.0542	60.4
29.7	.5184	.4955	.8686	.5704	1.753	1.151	2.018	1.0524	60.3
29.8	.5201	.4970	.8678	.5727	1.746	1.152	2.012	1.0507	60.2
29.9	.5219	.4985	.8669	.5750	1.739	1.154	2.006	1.0489	60.1
30.0	.5236	.5000	.8660	.5774	1.732	1.155	2.000	1.0472	**60.0**
		$\cos \theta$	$\sin \theta$	$\cot \theta$	$\tan \theta$	$\csc \theta$	$\sec \theta$	θ Radians	θ Degrees

θ Degrees	θ Radians	$\sin\theta$	$\cos\theta$	$\tan\theta$	$\cot\theta$	$\sec\theta$	$\csc\theta$		
30.0	.5236	.5000	.8660	.5774	1.732	1.155	2.000	1.0472	**60.0**
30.1	.5253	.5015	.8652	.5797	1.725	1.156	1.994	1.0455	59.9
30.2	.5271	.5030	.8643	.5820	1.718	1.157	1.988	1.0437	59.8
30.3	.5288	.5045	.8634	.5844	1.711	1.158	1.982	1.0420	59.7
30.4	.5306	.5060	.8625	.5867	1.704	1.159	1.976	1.0402	59.6
30.5	.5323	.5075	.8616	.5890	1.698	1.161	1.970	1.0385	59.5
30.6	.5341	.5090	.8607	.5914	1.691	1.162	1.964	1.0367	59.4
30.7	.5358	.5105	.8599	.5938	1.684	1.163	1.959	1.0350	59.3
30.8	.5376	.5120	.8590	.5961	1.678	1.164	1.953	1.0332	59.2
30.9	.5393	.5135	.8581	.5985	1.671	1.165	1.947	1.0315	59.1
31.0	.5411	.5150	.8572	.6009	1.664	1.167	1.942	1.0297	**59.0**
31.1	.5428	.5165	.8563	.6032	1.658	1.168	1.936	1.0280	58.9
31.2	.5445	.5180	.8554	.6056	1.651	1.169	1.930	1.0263	58.8
31.3	.5463	.5195	.8545	.6080	1.645	1.170	1.925	1.0245	58.7
31.4	.5480	.5210	.8535	.6104	1.638	1.172	1.919	1.0228	58.6
31.5	.5498	.5225	.8526	.6128	1.632	1.173	1.914	1.0210	58.5
31.6	.5515	.5240	.8517	.6152	1.625	1.174	1.908	1.0193	58.4
31.7	.5533	.5255	.8508	.6176	1.619	1.175	1.903	1.0175	58.3
31.8	.5550	.5270	.8499	.6200	1.613	1.177	1.898	1.0158	58.2
31.9	.5568	.5284	.8490	.6224	1.607	1.178	1.892	1.0140	58.1
32.0	.5585	.5299	.8480	.6249	1.600	1.179	1.887	1.0123	**58.0**
32.1	.5603	.5314	.8471	.6273	1.594	1.180	1.882	1.0105	57.9
32.2	.5620	.5329	.8462	.6297	1.588	1.182	1.877	1.0088	57.8
32.3	.5637	.5344	.8453	.6322	1.582	1.183	1.871	1.0071	57.7
32.4	.5655	.5358	.8443	.6346	1.576	1.184	1.866	1.0053	57.6
32.5	.5672	.5373	.8434	.6371	1.570	1.186	1.861	1.0036	57.5
32.6	.5690	.5388	.8425	.6395	1.564	1.187	1.856	1.0018	57.4
32.7	.5707	.5402	.8415	.6420	1.558	1.188	1.851	1.0001	57.3
32.8	.5725	.5417	.8406	.6445	1.552	1.190	1.846	.9983	57.2
32.9	.5742	.5432	.8396	.6469	1.546	1.191	1.841	.9966	57.1
33.0	.5760	.5446	.8387	.6494	1.540	1.192	1.836	.9948	**57.0**
33.1	.5777	.5461	.8377	.6519	1.534	1.194	1.831	.9931	56.9
33.2	.5794	.5476	.8368	.6544	1.528	1.195	1.826	.9913	56.8
33.3	.5812	.5490	.8358	.6569	1.522	1.196	1.821	.9896	56.7
33.4	.5829	.5505	.8348	.6594	1.517	1.198	1.817	.9879	56.6
33.5	.5847	.5519	.8339	.6619	1.511	1.199	1.812	.9861	56.5
33.6	.5864	.5534	.8329	.6644	1.505	1.201	1.807	.9844	56.4
33.7	.5882	.5548	.8320	.6669	1.499	1.202	1.802	.9826	56.3
33.8	.5899	.5563	.8310	.6694	1.494	1.203	1.798	.9809	56.2
33.9	.5917	.5577	.8300	.6720	1.488	1.205	1.793	.9791	56.1
34.0	.5934	.5592	.8290	.6745	1.483	1.206	1.788	.9774	**56.0**
34.1	.5952	.5606	.8281	.6771	1.477	1.208	1.784	.9756	55.9
34.2	.5969	.5621	.8271	.6796	1.471	1.209	1.779	.9739	55.8
34.3	.5986	.5635	.8261	.6822	1.466	1.211	1.775	.9721	55.7
34.4	.6004	.5650	.8251	.6847	1.460	1.212	1.770	.9704	55.6
34.5	.6021	.5664	.8241	.6873	1.455	1.213	1.766	.9687	55.5
34.6	.6039	.5678	.8231	.6899	1.450	1.215	1.761	.9669	55.4
34.7	.6056	.5693	.8221	.6924	1.444	1.216	1.757	.9652	55.3
34.8	.6074	.5707	.8211	.6950	1.439	1.218	1.752	.9634	55.2
34.9	.6091	.5721	.8202	.6976	1.433	1.219	1.748	.9617	55.1
35.0	.6109	.5736	.8192	.7002	1.428	1.221	1.743	.9599	**55.0**
35.1	.6126	.5750	.8181	.7028	1.423	1.222	1.739	.9582	54.9
35.2	.6144	.5764	.8171	.7054	1.418	1.224	1.735	.9564	54.8
35.3	.6161	.5779	.8161	.7080	1.412	1.225	1.731	.9547	54.7
35.4	.6178	.5793	.8151	.7107	1.407	1.227	1.726	.9529	54.6
35.5	.6196	.5807	.8141	.7133	1.402	1.228	1.722	.9512	54.5
35.6	.6213	.5821	.8131	.7159	1.397	1.230	1.718	.9495	54.4
35.7	.6231	.5835	.8121	.7186	1.392	1.231	1.714	.9477	54.3
35.8	.6248	.5850	.8111	.7212	1.387	1.233	1.710	.9460	54.2
35.9	.6266	.5864	.8100	.7239	1.381	1.235	1.705	.9442	54.1
36.0	.6283	.5878	.8090	.7265	1.376	1.236	1.701	.9425	**54.0**
		$\cos\theta$	$\sin\theta$	$\cot\theta$	$\tan\theta$	$\csc\theta$	$\sec\theta$	θ Radians	θ Degrees

θ Degrees	θ Radians	$\sin \theta$	$\cos \theta$	$\tan \theta$	$\cot \theta$	$\sec \theta$	$\csc \theta$		
36.0	.6283	.5878	.8090	.7265	1.376	1.236	1.701	.9425	**54.0**
36.1	.6301	.5892	.8080	.7292	1.371	1.238	1.697	.9407	53.9
36.2	.6318	.5906	.8070	.7319	1.366	1.239	1.693	.9390	53.8
36.3	.6336	.5920	.8059	.7346	1.361	1.241	1.689	.9372	53.7
36.4	.6353	.5934	.8049	.7373	1.356	1.242	1.685	.9355	53.6
36.5	.6370	.5948	.8039	.7400	1.351	1.244	1.681	.9338	53.5
36.6	.6388	.5962	.8028	.7427	1.347	1.246	1.677	.9320	53.4
36.7	.6405	.5976	.8018	.7454	1.342	1.247	1.673	.9303	53.3
36.8	.6423	.5990	.8007	.7481	1.337	1.249	1.669	.9285	53.2
36.9	.6440	.6004	.7997	.7508	1.332	1.250	1.666	.9268	53.1
37.0	.6458	.6018	.7986	.7536	1.327	1.252	1.662	.9250	**53.0**
37.1	.6475	.6032	.7976	.7563	1.322	1.254	1.658	.9233	52.9
37.2	.6493	.6046	.7965	.7590	1.317	1.255	1.654	.9215	52.8
37.3	.6510	.6060	.7955	.7618	1.313	1.257	1.650	.9198	52.7
37.4	.6528	.6074	.7944	.7646	1.308	1.259	1.646	.9180	52.6
37.5	.6545	.6088	.7934	.7673	1.303	1.260	1.643	.9163	52.5
37.6	.6562	.6101	.7923	.7701	1.299	1.262	1.639	.9146	52.4
37.7	.6580	.6115	.7912	.7729	1.294	1.264	1.635	.9128	52.3
37.8	.6597	.6129	.7902	.7757	1.289	1.266	1.632	.9111	52.2
37.9	.6615	.6143	.7891	.7785	1.285	1.267	1.628	.9093	52.1
38.0	.6632	.6157	.7880	.7813	1.280	1.269	1.624	.9076	**52.0**
38.1	.6650	.6170	.7869	.7841	1.275	1.271	1.621	.9058	51.9
38.2	.6667	.6184	.7859	.7869	1.271	1.272	1.617	.9041	51.8
38.3	.6685	.6198	.7848	.7898	1.266	1.274	1.613	.9023	51.7
38.4	.6702	.6211	.7837	.7926	1.262	1.276	1.610	.9006	51.6
38.5	.6720	.6225	.7826	.7954	1.257	1.278	1.606	.8988	51.5
38.6	.6737	.6239	.7815	.7983	1.253	1.280	1.603	.8971	51.4
38.7	.6754	.6252	.7804	.8012	1.248	1.281	1.599	.8954	51.3
38.8	.6772	.6266	.7793	.8040	1.244	1.283	1.596	.8936	51.2
38.9	.6789	.6280	.7782	.8069	1.239	1.285	1.592	.8919	51.1
39.0	.6807	.6293	.7771	.8098	1.235	1.287	1.589	.8901	**51.0**
39.1	.6824	.6307	.7760	.8127	1.230	1.289	1.586	.8884	50.9
39.2	.6842	.6320	.7749	.8156	1.226	1.290	1.582	.8866	50.8
39.3	.6859	.6334	.7738	.8185	1.222	1.292	1.579	.8849	50.7
39.4	.6877	.6347	.7727	.8214	1.217	1.294	1.575	.8831	50.6
39.5	.6894	.6361	.7716	.8243	1.213	1.296	1.572	.8814	50.5
39.6	.6912	.6374	.7705	.8273	1.209	1.298	1.569	.8796	50.4
39.7	.6929	.6388	.7694	.8302	1.205	1.300	1.566	.8779	50.3
39.8	.6946	.6401	.7683	.8332	1.200	1.302	1.562	.8762	50.2
39.9	.6964	.6414	.7672	.8361	1.196	1.304	1.559	.8744	50.1
40.0	.6981	.6428	.7660	.8391	1.192	1.305	1.556	.8727	**50.0**
40.1	.6999	.6441	.7649	.8421	1.188	1.307	1.552	.8709	49.9
40.2	.7016	.6455	.7638	.8451	1.183	1.309	1.549	.8692	49.8
40.3	.7034	.6468	.7627	.8481	1.179	1.311	1.546	.8674	49.7
40.4	.7051	.6481	.7615	.8511	1.175	1.313	1.543	.8657	49.6
40.5	.7069	.6494	.7604	.8541	1.171	1.315	1.540	.8639	49.5
40.6	.7086	.6508	.7593	.8571	1.167	1.317	1.537	.8622	49.4
40.7	.7103	.6521	.7581	.8601	1.163	1.319	1.534	.8604	49.3
40.8	.7121	.6534	.7570	.8632	1.159	1.321	1.530	.8587	49.2
40.9	.7138	.6547	.7559	.8662	1.154	1.323	1.527	.8570	49.1
41.0	.7156	.6561	.7547	.8693	1.150	1.325	1.524	.8552	**49.0**
41.1	.7173	.6574	.7536	.8724	1.146	1.327	1.521	.8535	48.9
41.2	.7191	.6587	.7524	.8754	1.142	1.329	1.518	.8517	48.8
41.3	.7208	.6600	.7513	.8785	1.138	1.331	1.515	.8500	48.7
41.4	.7226	.6613	.7501	.8816	1.134	1.333	1.512	.8482	48.6
41.5	.7243	.6626	.7490	.8847	1.130	1.335	1.509	.8465	48.5
41.6	.7261	.6639	.7478	.8878	1.126	1.337	1.506	.8447	48.4
41.7	.7278	.6652	.7466	.8910	1.122	1.339	1.503	.8430	48.3
41.8	.7295	.6665	.7455	.8941	1.118	1.341	1.500	.8412	48.2
41.9	.7313	.6678	.7443	.8972	1.115	1.344	1.497	.8395	48.1
42.0	.7330	.6691	.7431	.9004	1.111	1.346	1.494	.8378	**48.0**
		$\cos \theta$	$\sin \theta$	$\cot \theta$	$\tan \theta$	$\csc \theta$	$\sec \theta$	θ Radians	θ Degrees

θ Degrees	θ Radians	$\sin\theta$	$\cos\theta$	$\tan\theta$	$\cot\theta$	$\sec\theta$	$\csc\theta$		
42.0	.7330	.6691	.7431	.9004	1.111	1.346	1.494	.8378	**48.0**
42.1	.7348	.6704	.7420	.9036	1.107	1.348	1.492	.8360	47.9
42.2	.7365	.6717	.7408	.9067	1.103	1.350	1.489	.8343	47.8
42.3	.7383	.6730	.7396	.9099	1.099	1.352	1.486	.8325	47.7
42.4	.7400	.6743	.7385	.9131	1.095	1.354	1.483	.8308	47.6
42.5	.7418	.6756	.7373	.9163	1.091	1.356	1.480	.8290	47.5
42.6	.7435	.6769	.7361	.9195	1.087	1.359	1.477	.8273	47.4
42.7	.7453	.6782	.7349	.9228	1.084	1.361	1.475	.8255	47.3
42.8	.7470	.6794	.7337	.9260	1.080	1.363	1.472	.8238	47.2
42.9	.7487	.6807	.7325	.9293	1.076	1.365	1.469	.8221	47.1
43.0	.7505	.6820	.7314	.9325	1.072	1.367	1.466	.8203	**47.0**
43.1	.7522	.6833	.7302	.9358	1.069	1.370	1.464	.8186	46.9
43.2	.7540	.6845	.7290	.9391	1.065	1.372	1.461	.8168	46.8
43.3	.7557	.6858	.7278	.9424	1.061	1.374	1.458	.8151	46.7
43.4	.7575	.6871	.7266	.9457	1.057	1.376	1.455	.8133	46.6
43.5	.7592	.6884	.7254	.9490	1.054	1.379	1.453	.8116	46.5
43.6	.7610	.6896	.7242	.9523	1.050	1.381	1.450	.8098	46.4
43.7	.7627	.6909	.7230	.9556	1.046	1.383	1.447	.8081	46.3
43.8	.7645	.6921	.7218	.9590	1.043	1.386	1.445	.8063	46.2
43.9	.7662	.6934	.7206	.9623	1.039	1.388	1.442	.8046	46.1
44.0	.7679	.6947	.7193	.9657	1.036	1.390	1.440	.8029	**46.0**
44.1	.7697	.6959	.7181	.9691	1.032	1.393	1.437	.8011	45.9
44.2	.7714	.6972	.7169	.9725	1.028	1.395	1.434	.7994	45.8
44.3	.7732	.6984	.7157	.9759	1.025	1.397	1.432	.7976	45.7
44.4	.7749	.6997	.7145	.9793	1.021	1.400	1.429	.7959	45.6
44.5	.7767	.7009	.7133	.9827	1.018	1.402	1.427	.7941	45.5
44.6	.7784	.7022	.7120	.9861	1.014	1.404	1.424	.7924	45.4
44.7	.7802	.7034	.7108	.9896	1.011	1.407	1.422	.7906	45.3
44.8	.7819	.7046	.7096	.9930	1.007	1.409	1.419	.7889	45.2
44.9	.7837	.7059	.7083	.9965	1.003	1.412	1.417	.7871	45.1
45.0	.7854	.7071	.7071	1.0000	1.000	1.414	1.414	.7854	**45.0**
		$\cos\theta$	$\sin\theta$	$\cot\theta$	$\tan\theta$	$\csc\theta$	$\sec\theta$	θ Radians	θ Degrees

θ Radians	θ Degrees	sin θ	cos θ	tan θ	cot θ	sec θ	csc θ
0.00	0° 00′	0.0000	1.000	0.0000	Undefined	1.000	Undefined
.01	0° 34′	.0100	1.000	.0100	100.0	1.000	100.0
.02	1° 09′	.0200	0.9998	.0200	49.99	1.000	50.00
.03	1° 43′	.0300	0.9996	.0300	33.32	1.000	33.34
.04	2° 18′	.0400	0.9992	.0400	24.99	1.001	25.01
0.05	2° 52′	0.0500	0.9988	0.0500	19.98	1.001	20.01
.06	3° 26′	.0600	.9982	.0601	16.65	1.002	16.68
.07	4° 01′	.0699	.9976	.0701	14.26	1.002	14.30
.08	4° 35′	.0799	.9968	.0802	12.47	1.003	12.51
.09	5° 09′	.0899	.9960	.0902	11.08	1.004	11.13
0.10	5° 44′	0.0998	0.9950	0.1003	9.967	1.005	10.02
.11	6° 18′	.1098	.9940	.1104	9.054	1.006	9.109
.12	6° 53′	.1197	.9928	.1206	8.293	1.007	8.353
.13	7° 27′	.1296	.9916	.1307	7.649	1.009	7.714
.14	8° 01′	.1395	.9902	.1409	7.096	1.010	7.166
0.15	8° 36′	0.1494	0.9888	0.1511	6.617	1.011	6.692
.16	9° 10′	.1593	.9872	.1614	6.197	1.013	6.277
.17	9° 44′	.1692	.9856	.1717	5.826	1.015	5.911
.18	10° 19′	.1790	.9838	.1820	5.495	1.016	5.586
.19	10° 53′	.1889	.9820	.1923	5.200	1.018	5.295
0.20	11° 28′	0.1987	0.9801	0.2027	4.933	1.020	5.033
.21	12° 02′	.2085	.9780	.2131	4.692	1.022	4.797
.22	12° 36′	.2182	.9759	.2236	4.472	1.025	4.582
.23	13° 11′	.2280	.9737	.2341	4.271	1.027	4.386
.24	13° 45′	.2377	.9713	.2447	4.086	1.030	4.207
0.25	14° 19′	0.2474	0.9689	0.2553	3.916	1.032	4.042
.26	14° 54′	.2571	.9664	.2660	3.759	1.035	3.890
.27	15° 28′	.2667	.9638	.2768	3.613	1.038	3.749
.28	16° 03′	.2764	.9611	.2876	3.478	1.041	3.619
.29	16° 37′	.2860	.9582	.2984	3.351	1.044	3.497
0.30	17° 11′	0.2955	0.9553	0.3093	3.233	1.047	3.384
.31	17° 46′	.3051	.9523	.3203	3.122	1.050	3.278
.32	18° 20′	.3146	.9492	.3314	3.018	1.053	3.179
.33	18° 55′	.3240	.9460	.3425	2.920	1.057	3.086
.34	19° 29′	.3335	.9428	.3537	2.827	1.061	2.999
0.35	20° 03′	0.3429	0.9394	0.3650	2.740	1.065	2.916
.36	20° 38′	.3523	.9359	.3764	2.657	1.068	2.839
.37	21° 12′	.3616	.9323	.3879	2.578	1.073	2.765
.38	21° 46′	.3709	.9287	.3994	2.504	1.077	2.696
.39	22° 21′	.3802	.9249	.4111	2.433	1.081	2.630
0.40	22° 55′	0.3894	0.9211	0.4228	2.365	1.086	2.568
.41	23° 30′	.3986	.9171	.4346	2.301	1.090	2.509
.42	24° 04′	.4078	.9131	.4466	2.239	1.095	2.452
.43	24° 38′	.4169	.9090	.4586	2.180	1.100	2.399
.44	25° 13′	.4259	.9048	.4708	2.124	1.105	2.348
0.45	25° 47′	0.4350	0.9004	0.4831	2.070	1.111	2.299
.46	26° 21′	.4439	.8961	.4954	2.018	1.116	2.253
.47	26° 56′	.4529	.8916	.5080	1.969	1.122	2.208
.48	27° 30′	.4618	.8870	.5206	1.921	1.127	2.166
.49	28° 05′	.4706	.8823	.5334	1.875	1.133	2.125

Table 6 **Trigonometric Functions of** θ **(θ in radians)**

θ Radians	θ Degrees	$\sin \theta$	$\cos \theta$	$\tan \theta$	$\cot \theta$	$\sec \theta$	$\csc \theta$
0.50	28° 39′	0.4794	0.8776	0.5463	1.830	1.139	2.086
.51	29° 13′	.4882	.8727	.5594	1.788	1.146	2.048
.52	29° 48′	.4969	.8678	.5726	1.747	1.152	2.013
.53	30° 22′	.5055	.8628	.5859	1.707	1.159	1.978
.54	30° 56′	.5141	.8577	.5994	1.668	1.166	1.945
0.55	31° 31′	0.5227	0.8525	0.6131	1.631	1.173	1.913
.56	32° 05′	.5312	.8473	.6269	1.595	1.180	1.883
.57	32° 40′	.5396	.8419	.6410	1.560	1.188	1.853
.58	33° 14′	.5480	.8365	.6552	1.526	1.196	1.825
.59	33° 48′	.5564	.8309	.6696	1.494	1.203	1.797
0.60	34° 23′	0.5646	0.8253	0.6841	1.462	1.212	1.771
.61	34° 57′	.5729	.8196	.6989	1.431	1.220	1.746
.62	35° 31′	.5810	.8139	.7139	1.401	1.229	1.721
.63	36° 06′	.5891	.8080	.7291	1.372	1.238	1.697
.64	36° 40′	.5972	.8021	.7445	1.343	1.247	1.674
0.65	37° 15′	0.6052	0.7961	0.7602	1.315	1.256	1.652
.66	37° 49′	.6131	.7900	.7761	1.288	1.266	1.631
.67	38° 23′	.6210	.7838	.7923	1.262	1.276	1.610
.68	38° 58′	.6288	.7776	.8087	1.237	1.286	1.590
.69	39° 32′	.6365	.7712	.8253	1.212	1.297	1.571
0.70	40° 06′	0.6442	0.7648	0.8423	1.187	1.307	1.552
.71	40° 41′	.6518	.7584	.8595	1.163	1.319	1.534
.72	41° 15′	.6594	.7518	.8771	1.140	1.330	1.517
.73	41° 50′	.6669	.7452	.8949	1.117	1.342	1.500
.74	42° 24′	.6743	.7385	.9131	1.095	1.354	1.483
0.75	42° 58′	0.6816	0.7317	0.9316	1.073	1.367	1.467
.76	43° 33′	.6889	.7248	.9505	1.052	1.380	1.452
.77	44° 07′	.6961	.7179	.9697	1.031	1.393	1.437
.78	44° 41′	.7033	.7109	.9893	1.011	1.407	1.422
.79	45° 16′	.7104	.7038	1.009	.9908	1.421	1.408
0.80	45° 50′	0.7174	0.6967	1.030	0.9712	1.435	1.394
.81	46° 25′	.7243	.6895	1.050	.9520	1.450	1.381
.82	46° 59′	.7311	.6822	1.072	.9331	1.466	1.368
.83	47° 33′	.7379	.6749	1.093	.9146	1.482	1.355
.84	48° 08′	.7446	.6675	1.116	.8964	1.498	1.343
0.85	48° 42′	0.7513	0.6600	1.138	0.8785	1.515	1.331
.86	49° 17′	.7578	.6524	1.162	.8609	1.533	1.320
.87	49° 51′	.7643	.6448	1.185	.8437	1.551	1.308
.88	50° 25′	.7707	.6372	1.210	.8267	1.569	1.297
.89	51° 00′	.7771	.6294	1.235	.8100	1.589	1.287
0.90	51° 34′	0.7833	0.6216	1.260	0.7936	1.609	1.277
.91	52° 08′	.7895	.6137	1.286	.7774	1.629	1.267
.92	52° 43′	.7956	.6058	1.313	.7615	1.651	1.257
.93	53° 17′	.8016	.5978	1.341	.7458	1.673	1.247
.94	53° 52′	.8076	.5898	1.369	.7303	1.696	1.238
0.95	54° 26′	0.8134	0.5817	1.398	0.7151	1.719	1.229
.96	55° 00′	.8192	.5735	1.428	.7001	1.744	1.221
.97	55° 35′	.8249	.5653	1.459	.6853	1.769	1.212
.98	56° 09′	.8305	.5570	1.491	.6707	1.795	1.204
.99	56° 43′	.8360	.5487	1.524	.6563	1.823	1.196

$\dfrac{\pi}{6}$ → .53

$\dfrac{\pi}{4}$ → .79

Table 6 **Trigonometric Functions of θ** (θ in radians)

θ Radians	θ Degrees	sin θ	cos θ	tan θ	cot θ	sec θ	csc θ
1.00	57° 18′	0.8415	0.5403	1.557	0.6421	1.851	1.188
1.01	57° 52′	.8468	.5319	1.592	.6281	1.880	1.181
1.02	58° 27′	.8521	.5234	1.628	.6142	1.911	1.174
1.03	59° 01′	.8573	.5148	1.665	.6005	1.942	1.166
1.04	59° 35′	.8624	.5062	1.704	.5870	1.975	1.160
1.05	60° 10′	0.8674	0.4976	1.743	0.5736	2.010	1.153
1.06	60° 44′	.8724	.4889	1.784	.5604	2.046	1.146
1.07	61° 18′	.8772	.4801	1.827	.5473	2.083	1.140
1.08	61° 53′	.8820	.4713	1.871	.5344	2.122	1.134
1.09	62° 27′	.8866	.4625	1.917	.5216	2.162	1.128
1.10	63° 02′	0.8912	0.4536	1.965	0.5090	2.205	1.122
1.11	63° 36′	.8957	.4447	2.014	.4964	2.249	1.116
1.12	64° 10′	.9001	.4357	2.066	.4840	2.295	1.111
1.13	64° 45′	.9044	.4267	2.120	.4718	2.344	1.106
1.14	65° 19′	.9086	.4176	2.176	.4596	2.395	1.101
1.15	65° 53′	0.9128	0.4085	2.234	0.4475	2.448	1.096
1.16	66° 28′	.9168	.3993	2.296	.4356	2.504	1.091
1.17	67° 02′	.9208	.3902	2.360	.4237	2.563	1.086
1.18	67° 37′	.9246	.3809	2.428	.4120	2.625	1.082
1.19	68° 11′	.9284	.3717	2.498	.4003	2.691	1.077
1.20	68° 45′	0.9320	0.3624	2.572	0.3888	2.760	1.073
1.21	69° 20′	.9356	.3530	2.650	.3773	2.833	1.069
1.22	69° 54′	.9391	.3436	2.733	.3659	2.910	1.065
1.23	70° 28′	.9425	.3342	2.820	.3546	2.992	1.061
1.24	71° 03′	.9458	.3248	2.912	.3434	3.079	1.057
1.25	71° 37′	0.9490	0.3153	3.010	0.3323	3.171	1.054
1.26	72° 12′	.9521	.3058	3.113	.3212	3.270	1.050
1.27	72° 46′	.9551	.2963	3.224	.3102	3.375	1.047
1.28	73° 20′	.9580	.2867	3.341	.2993	3.488	1.044
1.29	73° 55′	.9608	.2771	3.467	.2884	3.609	1.041
1.30	74° 29′	0.9636	0.2675	3.602	0.2776	3.738	1.038
1.31	75° 03′	.9662	.2579	3.747	.2669	3.878	1.035
1.32	75° 38′	.9687	.2482	3.903	.2562	4.029	1.032
1.33	76° 12′	.9711	.2385	4.072	.2456	4.193	1.030
1.34	76° 47′	.9735	.2288	4.256	.2350	4.372	1.027
1.35	77° 21′	0.9757	0.2190	4.455	0.2245	4.566	1.025
1.36	77° 55′	.9779	.2092	4.673	.2140	4.779	1.023
1.37	78° 30′	.9799	.1994	4.913	.2035	5.014	1.021
1.38	79° 04′	.9819	.1896	5.177	.1931	5.273	1.018
1.39	79° 39′	.9837	.1798	5.471	.1828	5.561	1.017
1.40	80° 13′	0.9854	0.1700	5.798	0.1725	5.883	1.015
1.41	80° 47′	.9871	.1601	6.165	.1622	6.246	1.013
1.42	81° 22′	.9887	.1502	6.581	.1519	6.657	1.011
1.43	81° 56′	.9901	.1403	7.055	.1417	7.126	1.010
1.44	82° 30′	.9915	.1304	7.602	.1315	7.667	1.009
1.45	83° 05′	0.9927	0.1205	8.238	0.1214	8.299	1.007
1.46	83° 39′	.9939	.1106	8.989	.1113	9.044	1.006
1.47	84° 14′	.9949	.1006	9.887	.1011	9.938	1.005
1.48	84° 48′	.9959	.0907	10.98	.0911	11.03	1.004
1.49	85° 22′	.9967	.0807	12.35	.0810	12.39	1.003

$\dfrac{\pi}{3} \longrightarrow$

Table 6 — Trigonometric Functions of θ (θ in radians)

θ Radians	θ Degrees	$\sin \theta$	$\cos \theta$	$\tan \theta$	$\cot \theta$	$\sec \theta$	$\csc \theta$
1.50	85° 57′	0.9975	0.0707	14.10	0.0709	14.14	1.003
1.51	86° 31′	.9982	.0608	16.43	.0609	16.46	1.002
1.52	87° 05′	.9987	.0508	19.67	.0508	19.70	1.001
1.53	87° 40′	.9992	.0408	24.50	.0408	24.52	1.001
1.54	88° 14′	.9995	.0308	32.46	.0308	32.48	1.000
1.55	88° 49′	0.9998	0.0208	48.08	0.0208	48.09	1.000
1.56	89° 23′	.9999	.0108	92.62	.0108	92.63	1.000
1.57	89° 57′	1.000	.0008	1256	.0008	1256	1.000

$\dfrac{\pi}{2} \longrightarrow$ (1.57)

Table 7 — e^x and e^{-x}

x	e^x	e^{-x}	x	e^x	e^{-x}
0.00	1.0000	1.0000	0.36	1.4333	0.6977
0.01	1.0101	0.9900	0.37	1.4477	0.6907
0.02	1.0202	0.9802	0.38	1.4623	0.6839
0.03	1.0305	0.9704	0.39	1.4770	0.6771
0.04	1.0408	0.9608	0.40	1.4918	0.6703
0.05	1.0513	0.9512	0.41	1.5068	0.6637
0.06	1.0618	0.9418	0.42	1.5220	0.6570
0.07	1.0725	0.9324	0.43	1.5373	0.6505
0.08	1.0833	0.9231	0.44	1.5527	0.6440
0.09	1.0942	0.9139	0.45	1.5683	0.6376
0.10	1.1052	0.9048	0.46	1.5841	0.6313
0.11	1.1163	0.8958	0.47	1.6000	0.6250
0.12	1.1275	0.8869	0.48	1.6161	0.6188
0.13	1.1388	0.8781	0.49	1.6323	0.6126
0.14	1.1503	0.8694	0.50	1.6487	0.6065
0.15	1.1618	0.8607	1.0	2.7183	0.3679
0.16	1.1735	0.8521	1.5	4.4817	0.2231
0.17	1.1853	0.8437	2.0	7.3891	0.1353
0.18	1.1972	0.8353	2.5	12.182	0.0821
0.19	1.2092	0.8270	3.0	20.086	0.0498
0.20	1.2214	0.8187	3.5	33.115	0.0302
0.21	1.2337	0.8106	4.0	54.598	0.0183
0.22	1.2461	0.8025	4.5	90.017	0.0111
0.23	1.2586	0.7945	5.0	148.41	0.0067
0.24	1.2712	0.7866	5.5	244.69	0.0041
0.25	1.2840	0.7788	6.0	403.43	0.0025
0.26	1.2969	0.7711	6.5	665.14	0.0015
0.27	1.3100	0.7634	7.0	1096.6	0.0009
0.28	1.3231	0.7558	7.5	1808.0	0.0006
0.29	1.3364	0.7483	8.0	2981.0	0.0003
0.30	1.3499	0.7408	8.5	4914.8	0.0002
0.31	1.3634	0.7334	9.0	8103.1	0.0001
0.32	1.3771	0.7261	9.5	13359	0.0001
0.33	1.3910	0.7189	10.0	22026	.00005
0.34	1.4049	0.7118			
0.35	1.4191	0.7047			

1.

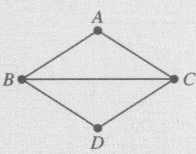

Vertex valences: *A*: 2; *B*: 3; *C*: 3, *D*: 2. No Euler circuit is possible.

2.

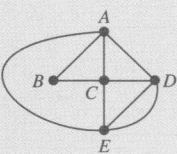

Vertex valences: *A*: 4; *B*: 2; *C*: 4; *D*: 4; *E*: 4. An Euler circuit is possible. Answers may vary. For example, *ABCDECAEDA*.

3. Answers may vary. For example, *BACBDC*.

4.

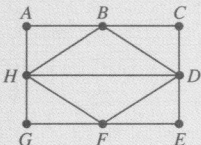

a. No; vertices *D* and *H* have valence 5.
b. Yes. For example, *HABCDEFGHBDFHD*.

Appendix 1

Graph Theory

A Famous Bridge Problem

In the 1700s, the European city of Königsberg (now Kaliningrad) had seven bridges connecting both sides of the Pregel River to two islands in the river, as shown below. As villagers strolled about the city, a problem was suggested: Is it possible to start at some point in the city, travel over each of the seven bridges exactly once, and return to the starting point? What do you think?

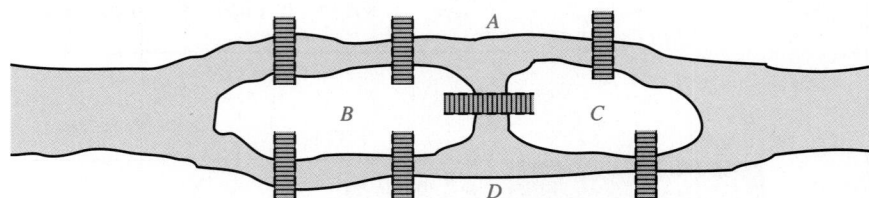

To analyze this problem, the Swiss mathematician Leonhard Euler (1707–1783) drew a *graph* like the one at the right. This graph, which is not like the graphs that you studied in algebra, is a diagram in which the four land masses of the city (*A*, *B*, *C*, and *D*) are pictured as points, or *vertices*, and the seven bridges are pictured as arcs, or *edges*. Next to each vertex is a number indicating how many edges are attached to the vertex. This number is called the *valence* of the vertex.

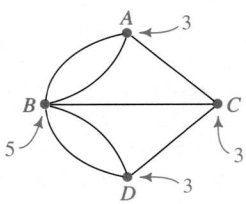

Euler reasoned that in order for a person to travel over every bridge once and return to the starting point, every vertex must have an even valence. This is because a person traveling *into* a vertex must also *leave* it; thus the edges must be paired: one "in" with one "out."

In the 7-bridge problem, none of the vertices has an even valence, so a circuit over all 7 bridges is impossible. However, if two more bridges are added (giving the 9 bridges shown at the right), then every vertex will have an even valence and a circuit over all 9 bridges will be possible. Such a circuit, which traces every edge of a graph exactly once and returns to its starting point, is called an *Euler circuit*. Can you find an Euler circuit for the graph at the right?

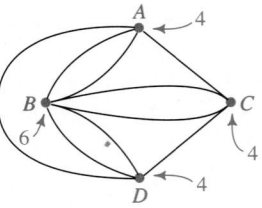

Draw a graph for each bridge problem pictured. Find the valence of each vertex and decide whether an Euler circuit is possible.

1.

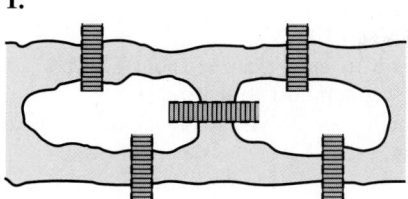

2.

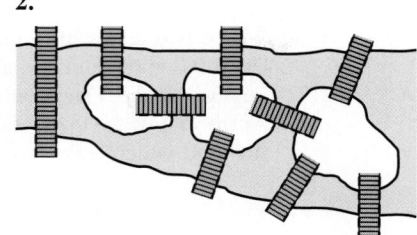

3. Although an Euler circuit is not possible in Exercise 1, it *is* possible to start at one point of the city, travel over each bridge once, and end at a part of the city different from where you started. Describe how to do this.

4. A network of city streets is shown at the right.
   a. If a member of the street department wants to inspect each street without traveling over any portion of a street twice, can the inspector do so by starting and ending at the same point?
   b. Can part (a) be done if the inspector starts and ends at different points?

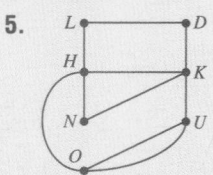

5. Consider a house having a floor plan as shown at the right.
   a. Can you enter the house, travel through each doorway once and only once, and return to your starting point? (*Hint:* Make a graph whose vertices represent the rooms and the outdoors.)
   b. Can you travel through each door exactly once and end at a different place from your starting point?

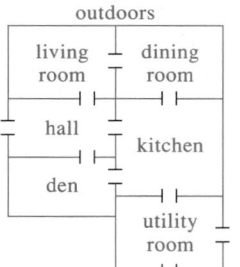

6. The graph at the right represents a network of one-way streets with the arrowheads indicating the direction of travel. Because 2 streets lead into vertex A, we say that the *invalence* of A is 2; likewise, because 2 streets lead out of A, we say that the *outvalence* of A is 2.
   a. Find the invalence and outvalence of each of the other vertices.
   b. Tell why an Euler circuit is not possible.
   c. Change the direction of just one arrowhead so that an Euler circuit is possible. Name the vertices in the order that they would be visited.

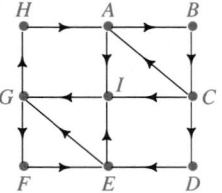

Graph Theory **823**

 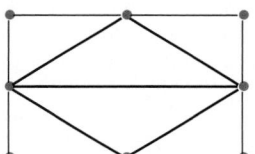

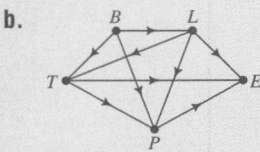
7. Five teams are involved in a tournament in which each team plays each of the other four exactly once. After a few games have been played, the results are shown as a graph in which an arrow from one team to another team indicates that the first team defeated the second.

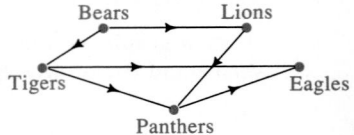

a. Which two teams have the most wins so far?

b. Copy the graph and add arrows to it to show that the Lions beat both the Eagles and the Tigers, while the Panthers lost to the Bears.

c. At the end of the competition, which team do you think should be ranked first and which team last? Explain.

A Traveler's Puzzle

A business traveler located in Denver must travel to each of the cities shown in the graph below. The airline on which the businessperson likes to travel connects certain pairs of cities, but not all pairs. Is it possible to make a circuit starting and ending in Denver and visiting each of the cities exactly once? This kind of circuit is called a *Hamilton circuit* in honor of William Rowan Hamilton (1805–1865), an Irish mathematician who worked on such circuits. Note that in a Hamilton circuit, each vertex is visited once, while in an Euler circuit, each edge is traveled once.

In the above graph, note that there are only 10 vertices (representing the 10 cities). Any other points where two edges intersect are not considered vertices. The graph *does* have a Hamilton circuit. Can you find it?

It is easy to tell whether a graph has an Euler circuit by investigating the valences of the vertices, but there is no known general rule for deciding when a Hamilton circuit is possible. For a small number of cities, you can use a trial-and-error method. For a larger number of cities, a computer can be programmed to use trial and error. But there are surprising limitations to what a computer can do, as the example at the top of the next page shows. The example presents a specific case of what is commonly called the *traveling salesman problem*.

Example

The graph at the right shows the distances between the 4 cities *A*, *B*, *C*, and *D*. A traveler wishes to leave *A*, visit the 3 other cities exactly once, and return to *A*. What is the shortest circuit the traveler can take?

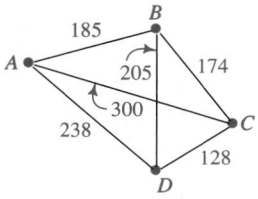

Solution

The traveler can visit any of the 3 cities first. From the first stop the traveler can go to either of the 2 remaining cities, and then the traveler must visit the 1 remaining city. Thus, there are $3 \times 2 \times 1 = 6$ possible circuits. The total distance for each circuit is shown below.

Circuit	Total distance
ABCDA	$185 + 174 + 128 + 238 = 725$
ABDCA	$185 + 205 + 128 + 300 = 818$
ACBDA	$300 + 174 + 205 + 238 = 917$
ADCBA	$238 + 128 + 174 + 185 = 725$
ACDBA	$300 + 128 + 205 + 185 = 818$
ADBCA	$238 + 205 + 174 + 300 = 917$

Thus, the shortest total distance is 725 mi for circuit *ABCDA* or its reverse circuit *ADCBA*.

The analysis used in the example for 4 cities can also be used for 10 cities, but now there are many more Hamilton circuits to analyze. In fact, the number of circuits is $9! = 9 \times 8 \times 7 \times 6 \times 5 \times 4 \times 3 \times 2 \times 1 = 362{,}880$. A reasonably fast computer can check all of these circuits and find the shortest one in less than a second. However, with 15 cities the computation time becomes about $1\frac{1}{2}$ minutes, and with 20 cities it becomes almost 4 years! Exercises 6–8 on pages 826 and 827 will help you understand why it takes even a fast computer so long to check all possible circuits.

Although the shortest circuit can be found only by testing all circuits, it is possible to get a very good circuit, although it might not be the shortest, by using the *nearest-neighbor algorithm*. With this method, you go from each city to the nearest city you haven't yet visited. Thus, for the graph shown above, you would go from *A* to *B* to *C* to *D* and back to *A*. This gives a total distance of 725 mi, which in this case is the shortest distance. Although the nearest-neighbor algorithm doesn't *always* give the shortest distance, it *is* easy and quick to use.

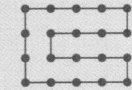
EXERCISES

Tell whether a Hamilton circuit is possible for each graph. If it is possible, name the vertices in the order visited. (More than one answer is possible.)

1.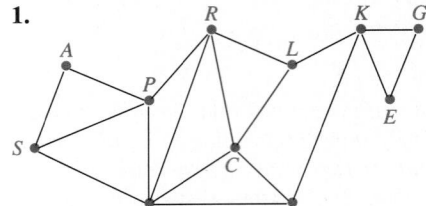

2.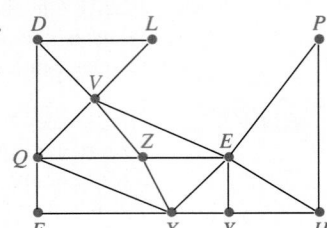

3. a. Is a Hamilton circuit possible for the 3 by 4 grid shown at the right?

b. Is a Hamilton circuit possible for a 4 by 6 grid?

c. Given an *x* by *y* grid, what must be true about *x* and *y* if a Hamilton circuit is possible?

4. Use the procedure of the example on page 825 to solve the traveler problem for the four cities shown in the graph at the right. Begin and end at vertex *F*.

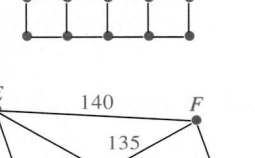

5. Use the nearest-neighbor algorithm for the graph in Exercise 4. Begin and end at vertex *F*. Does the algorithm give you the shortest circuit?

6. a. Suppose a traveler must leave city *A* and visit 14 other cities before returning to *A*. How many possible Hamilton circuits are there? (Give your answer as a factorial and then evaluate this factorial with a calculator.)

b. Since each circuit involves adding 15 numbers, there are 14 additions per circuit. Thus, the total number of additions to check all 14! circuits is 14 × 14!. If a computer can do 1 addition per nanosecond (one billionth of a second), how long will it take a computer to compute the distances for all circuits?

7. Suppose a traveler must leave city *A* and visit each of 20 cities before returning to *A*.

a. How many circuits are possible?

b. How many additions per circuit are there to compute the distance traveled?

c. What is the total number of additions required to find the distance traveled for each circuit?

d. How long would it take a computer to do all of the additions given in part (c)? Assume the computer can do 1 billion additions per second.

8. Repeat Exercise 7 if 25 cities are to be visited before returning to *A*.

9. The table below gives the cost of transportation between the 5 cities *A*, *B*, *C*, *D*, and *E*.

From \ To	A	B	C	D	E
A	—	$220	$150	$100	$130
B	$220	—	$160	$200	$240
C	$150	$160	—	$180	$110
D	$100	$200	$180	—	$190
E	$130	$240	$110	$190	—

a. Make a graph showing these costs.

b. Use the nearest-neighbor algorithm to find the circuit beginning and ending at *A*. (Note that "nearest neighbor" in this problem means "the neighbor it costs the least to reach.")

10. Many real-life problems can be modeled with Hamilton circuits. For example, a telephone company must decide on the most efficient route for a worker to collect money from pay phones. Give another example.

Minimizing the Cost of a Network

The graph at the left below gives the costs, in thousands of dollars, of joining several locations with roads. The total cost of all fourteen roads is $179,000. For far less money, just the roads colored in red at the right below could be built and people could still get from any location to any other. The cost of the red road network is $72,000, the sum of the costs for the 7 roads. Can you find another network of roads that costs less than this?

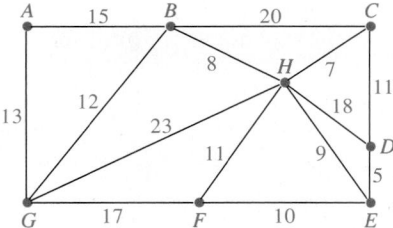 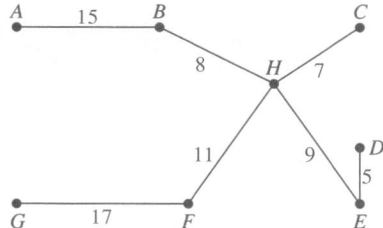

A network of least possible cost that allows one to travel from any vertex to any other vertex is called a *minimal spanning tree*. It is called "minimal" because the cost is least, it is called "spanning" because the network spans out to touch every vertex, and it is called a "tree" because it roughly resembles a tree with branches.

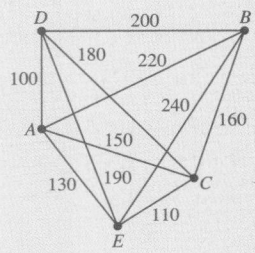

Algorithm for Finding a Minimal Spanning Tree

1. Build the least expensive road (edge) first.
2. Then build the road next lowest in cost.
3. At each stage, build the road that is next lowest in cost and *does not form a circuit*. Stop when all vertices have been spanned.

Example

Use the algorithm above to find the minimal spanning tree for the road network on the left at the bottom of the preceding page.

Solution

1. The least expensive road is *DE*.
2. The roads next lowest in cost are *HC, HB, HE,* and *EF*.
3. The road next lowest in cost is *HF*, but this road is not built because it forms the circuit *HEFH*. We therefore go to the road next lowest in cost, *BG*, and finally to *GA*.

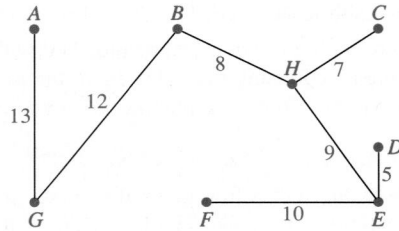

Thus, the total cost of the minimal spanning tree is

$5000 + $7000 + $8000 + $9000 + $10,000 + $12,000 + $13,000, or $64,000.

EXERCISES

Find a minimal spanning tree for each graph.

1.

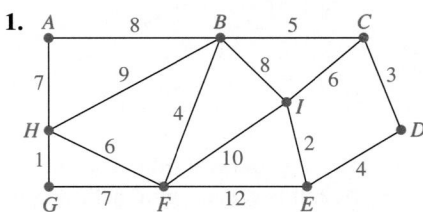

2.

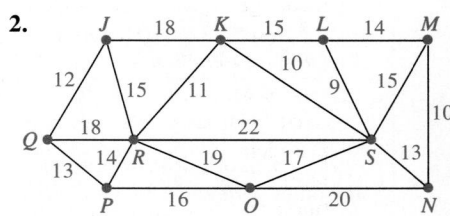

3. If a graph has *n* vertices, how many edges does a minimal spanning tree have?

4. The graph at the right shows the dollar costs required to connect various computers to form a computer network. What is the minimal cost of a network that links the six computers?

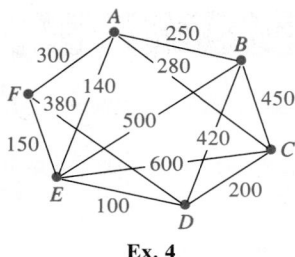

Ex. 4

5. a. Give an example of a situation in which a person or company might want to find a maximal spanning tree.
   b. Copy the graph in Exercise 1 and find a *maximal* spanning tree. Remember that a tree has no circuits.

6. A real estate developer wants to connect locations *A*, *B*, *C*, *D*, *E*, and *F* with roads while keeping costs to a minimum. The table below gives the estimated costs (in thousands of dollars) for building roads between pairs of locations. Make a graph and find the minimum cost of connecting the six locations.

	A	B	C	D	E	F
A	—	18	21	30	15	9
B		—	10	18	13	22
C			—	8	12	18
D				—	16	24
E					—	11
F						—

7. The graph at the right shows the costs (in hundreds of dollars) of building sidewalks between various locations on a college campus. Although a sidewalk between the administration building and the college president's home is the most expensive, the college president insists that it be built. What then is the minimum cost of joining the locations?

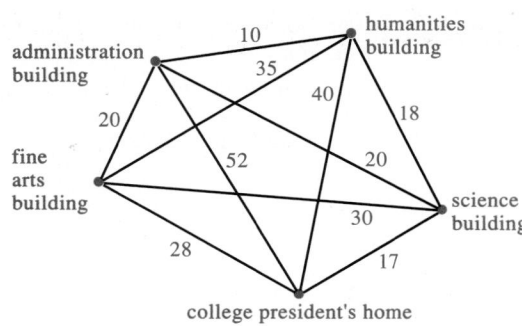

Additional Answers Exercises

3. $n - 1$

4. $840

5. a. Answers may vary. For example, agriculturalists might design a system of irrigation ditches to carry a maximum flow of water.
   b. Maximal spanning tree contains links *AB*, *BH*, *BI*, *IC*, *IF*, *FG*, *FE*, *ED*.

6. Minimal spanning tree contains links *AF*, *FE*, *EC*, *CB*, *CD*. Minimum cost is $9000 + $11,000 + $12,000 + $10,000 + $8000 = $50,000.

7. $9900

The Median-Median Line

In Section 18-1, we discussed how to fit a line, called the least-squares line, to a scatter plot. Another line that we can fit to a scatter plot is called the *median-median line*, and it has two advantages over the least-squares line. First, it is easier to use if you are not using a computer or calculator. Second, an outlying data point does not affect the median-median line very much, whereas it can have a great effect on the least-squares line.

When finding a median-median line, you must know how to find a *summary point* for a set of points (x_1, y_1), (x_2, y_2), . . . , (x_k, y_k).

summary point = (median of x-coordinates, median of y-coordinates)

To find the median of the x-coordinates of the points in a scatter plot, look at the points from left to right and draw a vertical line at the median x-position. To find the median of the y-coordinates, look at the points from the bottom to the top and draw a horizontal line at the median y-position. Two illustrations are shown below.

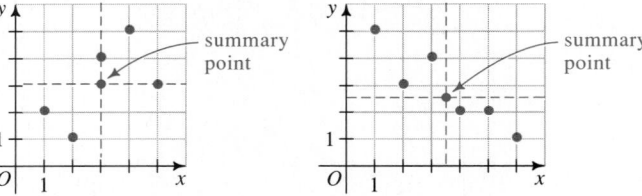

How to Find the Median-Median Line

Step 1 Draw two vertical lines that divide a scatter plot with n points into three parts as follows:

a. If n is divisible by 3, say $n = 3k$, then each part will have k points.

b. If $n = 3k + 1$, then the middle part will have one more data point than the two outside parts.

c. If $n = 3k + 2$, then the two outside parts will each have one more data point than the middle part.

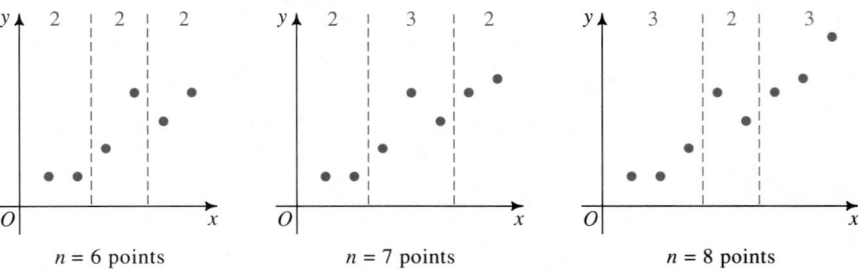

Step 2 **a.** Plot three summary points, one for each of the three subdivisions of the scatter plot. (See the graph at the right.)

b. Draw a line l_1 through the left and right summary points.

c. Draw a parallel line l_2 through the middle summary point.

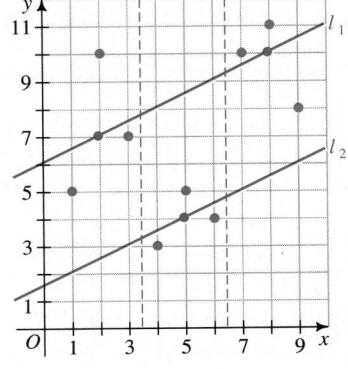

Step 3 The median-median line is drawn parallel to the two lines in Step 2 and such that its distance to line l_1 is half its distance to line l_2. This line is shown at the right and its equation is computed as follows:

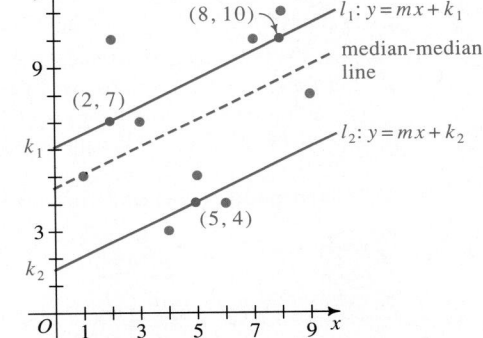

Equation of l_1:
$$y = mx + k_1$$

Equation of l_2:
$$y = mx + k_2$$

Equation of the median-median line:

$$y = mx + k, \text{ where } k = \frac{2}{3}k_1 + \frac{1}{3}k_2$$

Example Find an equation of the median-median line shown above.

Solution 1. Slope of l_1: $m = \dfrac{10 - 7}{8 - 2} = \dfrac{3}{6} = 0.5$

2. Since $(2, 7)$ and $(8, 10)$ lie on l_1, an equation of l_1 is:
$$y = 0.5x + 6$$

3. Since l_2 has slope 0.5 and $(5, 4)$ lies on l_2, an equation of l_2 is:
$$y = 0.5x + 1.5$$

4. Value of k: $k = \dfrac{2}{3}k_1 + \dfrac{1}{3}k_2 = \dfrac{2}{3}(6) + \dfrac{1}{3}(1.5) = 4.5$

An equation of the median-median line is $y = mx + k = 0.5x + 4.5$.

The Median-Median Line **831**

**Additional Answers
Exercises**

1. $(2.5, 3)$

2. $(4, 3)$

3. **a.** $8, 8, 8$
   b. $6, 5, 6$
   c. $20, 21, 20$

4. **a.** $l_1: y = \frac{1}{2}x + 2$

   $l_2: y = \frac{1}{2}x + \frac{7}{2}$

   b. $y = \frac{1}{2}x + \frac{5}{2}$

5. $y = \frac{1}{2}x + \frac{4}{3}$

6. $y = -\frac{3}{4}x + \frac{34}{3}$

7. $y = -\frac{1}{2}x + \frac{15}{4}$

8. $y = \frac{1}{2}x + \frac{17}{6}$

9. **a.** No change
   b. No change
   c. The point $(\bar{x}, \bar{y})$ is changed, and so is the least-squares line. Before replacement, $(\bar{x}, \bar{y}) = (3.6, 2)$. After replacement, $(\bar{x}, \bar{y}) = (3.6, 4)$.

EXERCISES

Find the summary point for each set of points.

1.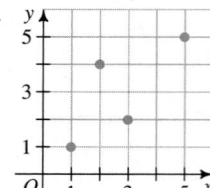

2.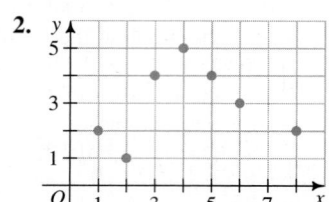

3. Refer to Step 1 in the instructions for finding the median-median line. How many points would you put into each of the three subdivisions of the data if the scatter plot has:
   a. 24 points?   **b.** 17 points?   **c.** 61 points?

4. Suppose the three summary points of a scatter plot are $(2, 3)$, $(5, 6)$, and $(8, 6)$.
   a. Plot these points and draw the lines l_1 and l_2 referred to in Step 2 on the preceding page. Find equations of these lines.
   b. Sketch the median-median line and find its equation.

Plot the given points (x, y) and find an equation of the median-median line.

5.

x	y
0	1
2	3
3	3
5	3
6	4
8	6

6.

x	y
2	9
6	7
7	8
8	5
9	6
10	3
14	1

7.

x	y
−3	5
−1	4
0	4
2	3
3	0
4	3
5	2
6	1
9	−1
11	−1

8.

x	y
1	0
2	4
3	4
4	5
6	5
6	7
8	7
8	9

9. Refer to the data in Exercise 7 and suppose that the point $(-3, 5)$ is replaced by $(-3, 25)$. Describe the effect, if any, that this replacement has on:
   a. the summary points for the three subdivisions of the data.
   b. the equation of the median-median line.
   c. the point $(\bar{x}, \bar{y})$, which lies on the least-squares line (see page 684).

10. ***Writing*** Fit a median-median line to the age-versus-shoe-size data whose scatter plot is given for Class Exercises 1 and 2 on page 687. Write a sentence or two comparing the median-median line with the least-squares line, which has the equation $y = 0.47x + 1.82$.

11. Parents of small children often wonder how tall their children will grow to be. In a study of 9 boys and 9 girls, their heights (in centimeters) were recorded on their second birthday and again on their eighteenth birthday, as shown below.

x = boy's height (age 2)	77	79	80	81	83	84	85	87	91
y = boy's height (age 18)	157	168	180	163	179	175	181	182	189

x = girl's height (age 2)	76	79	80	81	82	83	84	86	89
y = girl's height (age 18)	155	150	152	159	161	169	165	183	178

a. Using the data for either the boys or the girls, make a scatter plot of the set of points (x, y) and then fit a median-median line to the data.

b. Use the line in part (a) to predict how tall a two-year-old will be on his or her eighteenth birthday if he or she is now 78 cm tall.

12. During a promotion, an airline offers special fares to people flying from Boston, Massachusetts, to various cities in the United States. The tables below give the air distance (in miles) between Boston and each city as well as the one-way cost (based on a round-trip ticket) of each flight.

City	Distance	Cost
Cleveland, Ohio	551	$120.75
Denver, Colorado	1769	$194.25
Houston, Texas	1605	$183.25
Indianapolis, Indiana	807	$106.25
Kansas City, Missouri	1251	$119.25
Louisville, Kentucky	826	$141.25

City	Distance	Cost
New Orleans, Louisiana	1359	$119.25
Philadelphia, Pennsylvania	271	$72.75
Phoenix, Arizona	2300	$182.75
San Francisco, California	2699	$214.25
Seattle, Washington	2493	$214.75
Washington, D.C.	393	$89.00

a. Make a scatter plot of the set of points (x, y) = (distance, cost) and then fit a median-median line to the data.

b. If the air distance between Boston and Miami, Florida, is 1255 mi, what would you predict the one-way cost (based on a round-trip ticket) to be?

Additional Answers Exercises

10. The median-median line has equation $y = 0.38x + 2.54$; the least-squares line has equation $y = 0.47x + 1.82$. The median-median line has a greater y-intercept but a smaller slope. The median-median line predicts that a 20-year-old will wear a size 10 shoe; the least-squares line predicts a shoe size larger than 11.

11. a. Boys' median-median: $y = 1.75x + 29.75$. Girls' median-median: $y = 3.71x - 142.14$.
b. Boy at 18: $\approx$ 166 cm
Girl at 18: $\approx$ 147 cm

12. a. $y = 0.0554x + 66.9655$
b. $\approx$ $136.50

13. The table below gives the number of telephones and the number of television sets per 100 people for several countries.

Country	Phones	Televisions
Argentina	11.6	21.7
Australia	55.0	48.3
Canada	78.0	57.7
Costa Rica	14.7	7.9
Denmark	86.4	38.6
Egypt	2.8	8.3
Finland	61.7	37.4
France	60.8	33.3
India	0.6	0.7

Country	Phones	Televisions
Japan	55.5	58.7
Kenya	1.4	0.6
Malaysia	9.1	14.0
Mexico	9.6	12.0
Spain	39.6	36.8
Switzerland	85.6	40.5
U.K.	52.4	43.4
U.S.S.R.	11.3	31.4
U.S.A.	76.0	81.1

a. Make a scatter plot of the set of points (x, y) = (telephones, televisions) and then fit a median-median line to the data.
b. What nation's point has the greatest vertical distance to the line?
c. If you are told that Sweden has 89.0 telephones per 100 people, how many televisions per 100 people would you predict? (The actual number is 39.5 televisions per 100 people.)

14. The table below gives the life expectancy at birth (in years) for male and female United States citizens born in various years.

x = year of birth	1920	1930	1940	1950	1960	1970	1980	1990
y_m = life expectancy of males at birth	53.6	58.1	60.8	65.6	66.6	67.1	70.0	72.0
y_f = life expectancy of females at birth	54.6	61.6	65.2	71.1	73.1	74.7	77.4	78.8

a. Make separate scatter plots of the sets of points (x, y_m) and (x, y_f) and then fit a median-median line to each set of data.
b. *Writing* Write a paragraph in which you compare the two median-median lines in part (a).
c. What is the predicted life expectancy of a baby boy born in 2000? of a baby girl born in 2000?

15. *Research* Use an almanac or other reference book to research two data sets that you think might be linearly related. Make a scatter plot and fit a median-median line to the data. Then use the line to explore a question that the data might suggest.

Descartes' Rule of Signs and Bounds for Real Roots

The problem of determining the roots of polynomial equations occurs frequently in mathematics. Information that narrows down the set of *possible* roots of a polynomial equation greatly reduces the work involved in finding the roots. **Descartes' rule of signs** gives us information about the number of real roots of polynomial equations with real coefficients.

First consider a simplified polynomial $P(x)$ written in order of decreasing powers of x. Whenever the coefficients of two adjacent terms have opposite signs, we say that $P(x)$ has a *variation in sign*. Thus, the polynomial

$$\underset{1}{x^6 - 3x^4} - \underset{2}{2x^2 + 4x} - \underset{3}{5}$$

has three variations in sign. (Notice that you ignore terms having coefficient 0.)

Descartes' Rule of Signs

Let $P(x)$ be a polynomial with real coefficients; then
1. the number of positive roots of $P(x) = 0$ is either equal to the number of variations in sign of $P(x)$ or is less than this number by a positive even integer;
2. the number of negative roots of $P(x) = 0$ is either equal to the number of variations in sign of $P(-x)$ or is less than this number by a positive even integer.

Sometimes Descartes' rule gives complete information about the numbers of different types of roots. For example, $P(x) = x^4 + 11x - 6$ has *one* variation in sign and thus $P(x) = 0$ has *one positive root*. Since $P(-x) = (-x)^4 + 11(-x) - 6 = x^4 - 11x - 6$ has *one* variation in sign, $P(x) = 0$ has *one negative root*. Since $P(x)$ is of degree four, $P(x) = 0$ has exactly four roots. (See the fundamental theorem of algebra, page 85.) Since the equation has one positive real root and one negative real root, the remaining two roots must be imaginary.

More frequently, Descartes' rule leaves us with several possibilities.

Example 1 List the possibilities for the nature of the roots (positive, negative, and imaginary) of $P(x) = 0$ for the polynomial

$$P(x) = -2x^4 + x^3 - 7x^2 + 4x + 1.$$

Solution The given polynomial has three variations in sign, so the number of positive roots of $P(x) = 0$ is 3 or 1.

(Solution continues on the next page.)

Descartes' Rule of Signs and Bounds for Real Roots **835**

1.

pos.	neg.	imag.
1	1	0

2.

pos.	neg.	imag.
2	0	0
0	0	2

3.

pos.	neg.	imag.
2	1	0
0	1	2

4.

pos.	neg.	imag.
1	2	0
1	0	2

5.

pos.	neg.	imag.
2	2	0
2	0	2
0	2	2
0	0	4

6.

pos.	neg.	imag.
2	0	2
0	0	4

7.

pos.	neg.	imag.
3	1	0
1	1	2

8.

pos.	neg.	imag.
3	1	0
1	1	2

Examine $P(-x)$ to find the possible numbers of negative roots.

$$P(-x) = -2x^4 - x^3 - 7x^2 - 4x + 1$$

Since $P(-x)$ has one variation in sign, there is one negative root.

Since there are exactly four roots, we can make a chart to show the possible combinations of roots.

Number of positive real roots	Number of negative real roots	Number of imaginary roots
3	1	0
1	1	2

Descartes' rule of signs is used to determine the *number* of positive and negative roots of a polynomial equation $P(x) = 0$. The following rule can be used to narrow the region of possible real roots down to an interval on the number line, that is, to find *bounds* for the real roots of $P(x) = 0$.

Upper and Lower Bounds for Real Roots

Let $P(x)$ be a polynomial with real coefficients and a positive leading coefficient.

1. Let M be a nonnegative real number. If the coefficients of the quotient and remainder obtained by dividing $P(x)$ by $x - M$ are all nonnegative, then $P(x) = 0$ has no roots greater than M.

2. Let L be a nonpositive real number. If the coefficients of the quotient and remainder obtained by dividing $P(x)$ by $x - L$ are alternately nonnegative and nonpositive, then $P(x) = 0$ has no roots less than L.

The numbers M and L are called **upper** and **lower bounds,** respectively, for the real roots of $P(x) = 0$. To apply this rule, use synthetic division for different values of x until the last line of numbers satisfies the conditions for M or L.

Example 2 Find the least nonnegative integral upper bound and the greatest nonpositive integral lower bound for the roots of $3x^3 - 4x^2 - 4x + 4 = 0$.

Solution Use synthetic substitution with $x = \pm 1, \pm 2, \pm 3, \ldots$ until numbers that satisfy the respective conditions for upper and lower bounds are found. The results are shown in the table at the top of the next page.

x	coefficients of quotient and remainder			
1	3	−1	−5	−1
2	3	2	0	4	← all ≥ 0
−1	3	−7	3	1
−2	3	−10	16	−28	← alternately ≥ 0 and ≤ 0

Thus, $M = 2$ is the upper bound and $L = -2$ is the lower bound. If $3x^3 - 4x^2 - 4x + 4 = 0$ has any real roots, they must lie in the interval $-2 < x < 2$.

EXERCISES

Use Descartes' rule to make a chart summarizing the possible combinations of positive, negative, and imaginary roots of each polynomial equation.

1. $y^2 - y - 4 = 0$

2. $-y^2 + y - 3 = 0$

3. $y^3 - y^2 + 2y + 3 = 0$

4. $x^3 + 6x^2 - x - 12 = 0$

5. $2x^4 - 3x^3 - x^2 + x + 1 = 0$

6. $3y^4 - y^3 + 4y^2 + 8 = 0$

7. $5y^4 - y^3 + y^2 + 2y - 4 = 0$

8. $-4x^4 - 3x^3 + x^2 - x + 5 = 0$

For each polynomial equation (a) find the least nonnegative integral upper bound and the greatest nonpositive integral lower bound of the real roots, and (b) write an interval in which the real roots lie.

9. $x^3 - 2x^2 + 3x + 5 = 0$

10. $x^3 - 4x^2 + 8x - 5 = 0$

11. $2x^3 - 3x^2 - 3x + 3 = 0$

12. $2x^3 - 2x^2 + x + 8 = 0$

13. $y^4 + y^3 + 16 = 0$

14. $y^4 - 2y - 3 = 0$

15. $y^4 - 2y^3 - 3y^2 + 4y + 8 = 0$

16. $y^4 + 2y^3 - y^2 + 7y - 5 = 0$

Partial Fractions

In your algebra courses, you learned to simplify sums of rational expressions by using the least common denominator. For example,

$$\frac{-3}{x-2} + \frac{5}{x-1} = \frac{-3(x-1) + 5(x-2)}{(x-2)(x-1)} = \frac{2x-7}{(x-2)(x-1)}.$$

Some calculus problems involving rational expressions can be solved by reversing this process, that is, by rewriting a single rational expression as a sum of two or more rational expressions, called *partial fractions*:

$$\frac{2x-7}{(x-2)(x-1)} = \frac{-3}{x-2} + \frac{5}{x-1}$$

There may be several ways of rewriting a rational expression as the sum of other rational expressions. For example, both

$$\frac{2x}{(x-2)(x-1)} + \frac{-7}{(x-2)(x-1)} \quad \text{and} \quad \frac{x-5}{x-2} + \frac{6-x}{x-1}$$

are also equivalent to $\dfrac{2x-7}{(x-2)(x-1)}$. When writing a rational expression as the sum of partial fractions, however, the goal is to use partial fractions with different denominators and with numerators of the lowest degree possible. In the examples and in most of the exercises, the numerators of the partial fractions are constants.

Example 1 Write $\dfrac{3x+17}{(x-5)(x+3)}$ as a sum of partial fractions.

Solution Using the factors of the denominator as the denominators of the partial fractions, we want to find constants A and B such that

$$\frac{3x+17}{(x-5)(x+3)} = \frac{A}{x-5} + \frac{B}{x+3}.$$

Multiply both sides of the equation above by $(x-5)(x+3)$:

$$3x + 17 = A(x+3) + B(x-5) \qquad (1)$$

Expand the right side of equation (1) and combine like terms:

$$3x + 17 = (A + B)x + (3A - 5B) \qquad (2)$$

We can use either equation (1) or equation (2) to find A and B.

Method 1 In equation (2), A and B must satisfy the following system:

$$\begin{aligned} A + B &= 3 \qquad \longleftarrow \text{ The coefficients of } x \text{ must be equal.} \\ 3A - 5B &= 17 \qquad \longleftarrow \text{ The constant terms must be equal.} \end{aligned}$$

Solving this system of equations simultaneously gives the solution $A = 4$ and $B = -1$.

The values $A = 4$ and $B = -1$ make equation (2) true for *all* values of x. Then equation (1) must also be true for all values of x. Thus, even though the original rational expression is undefined for $x = 5$ and $x = -3$, we can legitimately substitute $x = 5$ and $x = -3$ into equation (1). This gives us a second method of finding A and B.

Method 2 We use equation (1):

$$3x + 17 = A(x + 3) + B(x - 5)$$

Substituting $x = 5$ causes the term $B(x - 5)$ to become 0:

$$3(5) + 17 = A(5 + 3) + 0$$
$$32 = 8A$$
$$A = 4$$

Substituting $x = -3$ causes the term $A(x + 3)$ to become 0:

$$3(-3) + 17 = 0 + B(-3 - 5)$$
$$8 = -8B$$
$$B = -1$$

Both methods give the solution $A = 4$ and $B = -1$. Thus.

$$\frac{3x + 17}{(x - 5)(x + 3)} = \frac{4}{x - 5} + \frac{-1}{x + 3}, \text{ or } \frac{4}{x - 5} - \frac{1}{x + 3}.$$

You can easily verify this answer by combining the partial fractions using the common denominator $(x - 5)(x + 3)$.

Example 2 shows how to deal with a rational expression in which the denominator, when factored, contains a repeated factor.

Example 2   Write $\dfrac{5x^2 - 27x + 21}{x^3 - 5x^2 - 8x + 48}$ as a sum of partial fractions.

Solution   Using the methods presented in Section 2-2 and Section 2-6, factor the denominator:

$$x^3 - 5x^2 - 8x + 48 = (x - 4)^2(x + 3)$$

Because the factor $(x - 4)$ is squared, the original rational expression *must* have a partial fraction with denominator $(x - 4)^2$, and it *may also* have a partial fraction with denominator $(x - 4)$. The factor $(x + 3)$ will be the denominator of a third partial fraction. Thus, we must find constants A, B, and C such that

$$\frac{5x^2 - 27x + 21}{(x - 4)^2(x + 3)} = \frac{A}{x - 4} + \frac{B}{(x - 4)^2} + \frac{C}{x + 3}.$$

Multiplying each side of the equation above by $(x - 4)^2(x + 3)$ produces equation (3), shown at the top of the next page.

(Solution continues on the next page.)

$$5x^2 - 27x + 21 = A(x - 4)(x + 3) + B(x + 3) + C(x - 4)^2 \qquad (3)$$

Expanding the right side of equation (3) would be time-consuming, so in this case we will use Method 2 presented in Example 1. Although we cannot find A directly by substituting a value for x in equation (3), we can find B and C by substituting $x = 4$ and $x = -3$, respectively.

When $x = 4$,

$$5(4)^2 - 27(4) + 21 = 0 + B(4 + 3) + 0$$
$$-7 = 7B$$
$$B = -1$$

When $x = -3$,

$$5(-3)^2 - 27(-3) + 21 = 0 + 0 + C(-3 - 4)^2$$
$$147 = 49C$$
$$C = 3$$

To find A, return to equation (3) and substitute -1 for B, 3 for C, and any convenient value, such as 0, for x:

$$5(0)^2 - 27(0) + 21 = A(0 - 4)(0 + 3) + (-1)(0 + 3) + (3)(0 - 4)^2$$
$$-24 = A(-12)$$
$$A = 2$$

Thus, $\quad \dfrac{5x^2 - 27x + 21}{x^3 - 5x^2 - 8x + 48} = \dfrac{2}{x - 4} - \dfrac{1}{(x - 4)^2} + \dfrac{3}{x + 3}.$

In Example 2, the denominator's repeated factor is squared. If the denominator of the original rational expression had contained a cubed factor, $(x - a)^3$, then we would have used each of $\dfrac{A}{x - a}$, $\dfrac{B}{(x - a)^2}$, and $\dfrac{C}{(x - a)^3}$ as partial fractions.

In both Example 1 and Example 2, the original rational expression has the following characteristics: (1) the degree of the numerator is less than the degree of the denominator, and (2) all factors of the denominator are linear. In the exercises, you will consider some other cases.

Exercises

In Exercises 1–15, write each rational expression as a sum of partial fractions.

1. $\dfrac{6x}{(x - 1)(x + 5)}$

2. $\dfrac{2x + 1}{(x - 4)(x - 1)}$ $\dfrac{3}{x - 4} - \dfrac{1}{x - 1}$

3. $\dfrac{7x + 6}{4x^2 + 9x - 9}$ $\dfrac{3}{4x - 3} + \dfrac{1}{x + 3}$

4. $\dfrac{4x + 1}{6x^2 - 7x + 2}$

5. $\dfrac{x^2 + 8}{x^3 + 3x^2 + 2x}$

6. $\dfrac{x - 6}{x^3 + x^2 - 6x}$

7. $\dfrac{x + 6}{x^2 - 4}$ $\dfrac{2}{x - 2} - \dfrac{1}{x + 2}$

8. $\dfrac{x^2 + x + 6}{x^3 + 2x^2 - 5x - 6}$

9. $\dfrac{3x - 1}{(x + 4)^2}$ $\dfrac{3}{x + 4} - \dfrac{13}{(x + 4)^2}$

10. $\dfrac{x^2 + x + 2}{(x - 2)^3}$ **11.** $\dfrac{x^2 + 1}{x^3 + 3x^2 + 3x + 1}$ **12.** $\dfrac{x + 2}{x^2 - 2x + 1}$ $\dfrac{1}{x - 1} + \dfrac{3}{(x - 1)^2}$

13. $\dfrac{2x^2}{x^3 + x^2 - x - 1}$ **14.** $\dfrac{6}{x^4 - x^2}$ $\dfrac{3}{x - 1} - \dfrac{3}{x + 1} - \dfrac{6}{x^2}$ **15.** $\dfrac{x^3 - x^2}{(x + 3)^4}$

16. Use Method 1 of Example 1 on page 838 to find A, B, and C in Example 2 on page 839. First show that A, B, and C satisfy the system of equations shown at the right. Then solve the system using substitution, determinants (see Section 12-8), or matrices (see Section 14-3). $A = 2$, $B = -1$, $C = 3$

$$A + C = 5$$
$$-A + B - 8C = -27$$
$$-12A + 3B + 16C = 21$$

17. a. Before trying to write the rational expression $\dfrac{2x^2 - 28}{x^2 - 6x - 16}$ as a sum of partial fractions, notice that its numerator and denominator are of the same degree. Explain why you cannot find constants A and B such that

$$\dfrac{2x^2 - 28}{x^2 - 6x - 16} = \dfrac{2x^2 - 28}{(x - 8)(x + 2)} = \dfrac{A}{x - 8} + \dfrac{B}{x + 2}.$$

    b. To write $\dfrac{2x^2 - 28}{x^2 - 6x - 16}$ as a sum of partial fractions, first use polynomial long division to divide the numerator by the denominator. Verify that

$$\dfrac{2x^2 - 28}{x^2 - 6x - 16} = 2 + \dfrac{12x + 4}{x^2 - 6x - 16}.$$

18. $2 + \dfrac{\frac{3}{5}}{x - 1} - \dfrac{\frac{21}{5}}{2x + 3}$

    c. Now write $\dfrac{12x + 4}{x^2 - 6x - 16}$ as a sum of partial fractions. Verify that

$$\dfrac{12x + 4}{x^2 - 6x - 16} = \dfrac{10}{x - 8} + \dfrac{2}{x + 2}.$$

19. $4x - 7 + \dfrac{1}{x} + \dfrac{8}{x + 1}$

    Thus, $\dfrac{2x^2 - 28}{x^2 - 6x - 16} = 2 + \dfrac{10}{x - 8} + \dfrac{2}{x + 2}.$

20. $x - 1 + \dfrac{1}{2x + 1} + \dfrac{2}{(2x + 1)^2}$

In Exercises 18–20, use the method presented in parts (b) and (c) of Exercise 17 to write the given rational expression as a sum of partial fractions.

18. $\dfrac{4x^2 - x}{2x^2 + x - 3}$ See above.   **19.** $\dfrac{4x^3 - 3x^2 + 2x + 1}{x^2 + x}$ See above.   **20.** $\dfrac{4x^3 - x + 2}{4x^2 + 4x + 1}$ See above.

21. In the rational expression $\dfrac{5x^2 - 2x - 1}{(x - 2)(x^2 + x - 1)}$, one factor of the denominator is a nonfactorable quadratic. Write the rational expression as a sum of partial fractions by finding constants A, B, and C such that

$$\dfrac{5x^2 - 2x - 1}{(x - 2)(x^2 + x - 1)} = \dfrac{A}{x - 2} + \dfrac{Bx + C}{x^2 + x - 1}. \quad \dfrac{3}{x - 2} + \dfrac{2x - 1}{x^2 + x - 1}$$

22. Write $\dfrac{x^4 + 8}{x^3 - 15x - 4}$ as a sum of partial fractions. $x + \dfrac{8}{x - 4} + \dfrac{7x}{x^2 + 4x + 1}$

Additional Answers Exercises

10. $\dfrac{1}{x - 2} + \dfrac{5}{(x - 2)^2} + \dfrac{8}{(x - 2)^3}$

11. $\dfrac{1}{x + 1} - \dfrac{2}{(x + 1)^2} + \dfrac{2}{(x + 1)^3}$

13. $\dfrac{\frac{1}{2}}{x - 1} + \dfrac{\frac{3}{2}}{x + 1} - \dfrac{1}{(x + 1)^2}$

15. $\dfrac{1}{x + 3} - \dfrac{10}{(x + 3)^2} + \dfrac{33}{(x + 3)^3} - \dfrac{36}{(x + 3)^4}$

17. a. For all values of A and B, the expression $A(x + 2) + B(x - 8)$ is linear and thus will not produce the quadratic term $2x^2$.

    b.
$$x^2 - 6x - 16 \overline{)\, 2x^2 + 0x - 28}$$
$$\underline{2x^2 - 12x - 32}$$
$$12x + 4$$

Thus, $\dfrac{2x^2 - 28}{x^2 - 6x - 16} =$

$2 + \dfrac{12x + 4}{x^2 - 6x - 16}$

    c. $\dfrac{10}{x - 8} + \dfrac{2}{x + 2} =$

$\dfrac{10(x + 2) + 2(x - 8)}{(x - 8)(x + 2)} =$

$\dfrac{12x + 4}{x^2 - 6x - 16}$

Tangents and Normals to Conic Sections

Families of Lines

A set of lines satisfying a given condition is called a *family of lines*. For example, the set of all nonvertical lines that contain the point (0, 0) is the family of nonvertical lines through the origin. Each nonvertical line has an equation in the form $y = mx$. The slope m is called the *parameter* of the family. Specifying a value for m determines the equation of a line in the family.

The diagram at the right shows a circle with center (5, 0) and radius $\sqrt{5}$, and also several lines from the family of nonvertical lines through the origin. Some of these lines intersect the circle in two points, some are tangent to the circle, and some do not intersect the circle. To classify such lines by the number of points of intersection, you can use your knowledge of the discriminant of a quadratic equation (see pages 31–32). The following four steps show how.

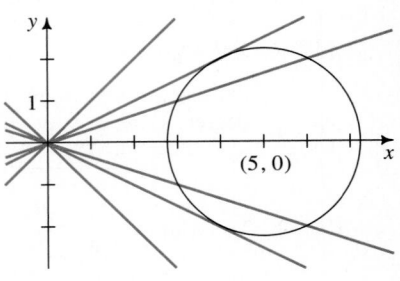

1. The equation $y = mx$ represents a nonvertical line through the origin. We wish to determine which values of m give equations of lines that intersect the circle.
2. We start as if we were solving the equations of the line and the circle simultaneously under the assumption that m is given.

$$\begin{aligned} \text{Line:} \quad & y = mx \\ \text{Circle:} \quad & (x - 5)^2 + y^2 = 5 \\ \text{Substituting:} \quad & (x - 5)^2 + (mx)^2 = 5 \\ & x^2 - 10x + 25 + m^2x^2 = 5 \\ & (1 + m^2)x^2 - 10x + 20 = 0 \quad (1) \end{aligned}$$

3. Instead of solving equation (1) we determine how many solutions it has. Equation (1) is a quadratic equation whose basic form is $ax^2 + bx + c = 0$. In this case $a = 1 + m^2$, $b = -10$, and $c = 20$. The discriminant of this quadratic equation is:

$$\begin{aligned} D = b^2 - 4ac &= (-10)^2 - 4(1 + m^2)(20) = 100 - 80 - 80m^2 \\ &= -80m^2 + 20 \end{aligned}$$

4. The graph of $D = -80m^2 + 20$ is a parabola that opens downward, crossing the horizontal axis when $m = -\frac{1}{2}$ and $m = \frac{1}{2}$. Using the graph to determine which values of m cause D to be positive, to be zero, and to be negative, we draw three conclusions.

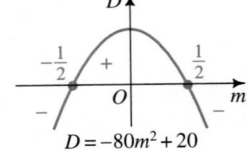

a. $D > 0$ when $-\frac{1}{2} < m < \frac{1}{2}$. In this case, equation (1) has two distinct roots, indicating that the line and the circle intersect twice. Thus, when $-\frac{1}{2} < m < \frac{1}{2}$, the line $y = mx$ intersects the circle in two distinct points.

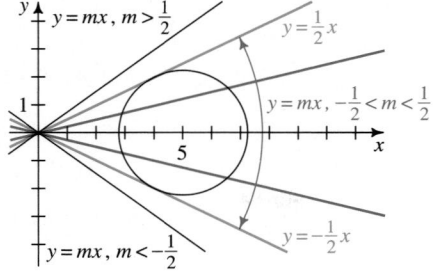

b. $D = 0$ when $m = \pm\frac{1}{2}$. In this case, equation (1) has one (double) root, indicating that the line and the circle intersect in exactly one point. Thus, when $m = -\frac{1}{2}$ or $m = \frac{1}{2}$, the line $y = mx$ is tangent to the circle.

c. $D < 0$ when $m < -\frac{1}{2}$ or $m > \frac{1}{2}$. In this case, equation (1) has no real roots, indicating that the line and the circle have no intersection points. Thus, when $m < -\frac{1}{2}$ or $m > \frac{1}{2}$, the line $y = mx$ does not intersect the circle.

Tangent and Normal Lines

The following example applies the method discussed above to finding the equation of a line tangent to a parabola. Notice that the lines of the family discussed in the example do not all pass through a common point; instead, these lines all have the same slope. The parameter for this family of lines is the y-intercept k.

Example 1 The diagram shows the parabola $y = x^2$ and several lines from the family of lines with slope 3. Find an equation of the line with slope 3 that is tangent to the parabola.

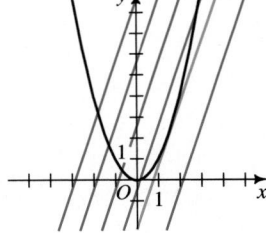

Solution The equation $y = 3x + k$ represents the family of lines with slope 3. We proceed as if we were solving the equation of the line and the parabola simultaneously under the assumption that k is given.

$$\text{Line:} \quad y = 3x + k$$
$$\text{Parabola:} \quad y = x^2$$
$$\text{Substituting:} \quad 3x + k = x^2$$
$$x^2 - 3x - k = 0 \quad (2)$$

Equation (2) is a quadratic equation with discriminant

$$D = b^2 - 4ac = (-3)^2 - 4(1)(-k) = 9 + 4k.$$

(Solution continues on the next page.)

We wish to find the value of k for which equation (2) has exactly one (double) root, indicating that the line intersects the parabola in a single point. This will happen when $D = 0$:

$$9 + 4k = 0$$

$$k = -\frac{9}{4}$$

Thus, the equation of the tangent line is $y = 3x - \frac{9}{4}$.

When a line is tangent to a curve, the point of intersection is called the point of tangency. A line that passes through the point of tangency and is perpendicular to the tangent line is called a **normal line.** Such a line is said to be *normal to the curve* at the point of tangency.

Example 2   As shown in Example 1, the line $y = 3x - \frac{9}{4}$ is tangent to the parabola $y = x^2$. Find **(a)** the coordinates of the point of tangency, and **(b)** the equation of the line normal to the parabola at the point of tangency.

Solution   **a.** We can take advantage of the fact that in Example 1 we began to solve the equations $y = 3x + k$ and $y = x^2$ simultaneously. We stopped when we reached equation (2):

$$x^2 - 3x - k = 0 \quad (2)$$

We then determined that when $k = -\frac{9}{4}$, equation (2) has exactly one root. This root is the x-coordinate of the only intersection point of the line $y = 3x - \frac{9}{4}$ and the parabola $y = x^2$. Thus, to find the x-coordinate of the point of tangency, substitute $k = -\frac{9}{4}$ in equation (2) and solve for x:

$$x^2 - 3x - \left(-\frac{9}{4}\right) = 0$$

$$x^2 - 3x + \frac{9}{4} = 0$$

$$\left(x - \frac{3}{2}\right)^2 = 0$$

$$x = \frac{3}{2}$$

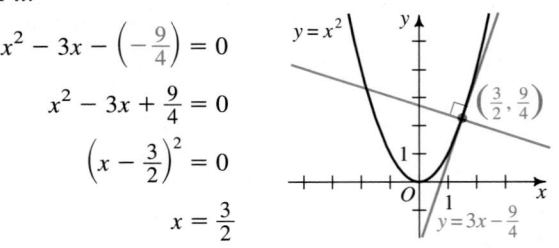

Substituting $x = \frac{3}{2}$ in the equation $y = x^2$ gives $y = \left(\frac{3}{2}\right)^2 = \frac{9}{4}$. Thus, the line and the parabola are tangent at the point $\left(\frac{3}{2}, \frac{9}{4}\right)$.

b. The slope of the tangent line $y = 3x - \dfrac{9}{4}$ is 3; therefore, the slope of the normal line is $-\dfrac{1}{3}$. Since the normal line passes through the point of tangency $\left(\dfrac{3}{2}, \dfrac{9}{4}\right)$, the equation of the normal line is

$$\frac{y - \dfrac{9}{4}}{x - \dfrac{3}{2}} = -\frac{1}{3}, \text{ or } y = -\frac{1}{3}x + \frac{11}{4}.$$

Exercises

2. $y = 3x + k$, where **(a)** $-10 < k < 10$, **(b)** $k = \pm 10$, **(c)** $k < -10$ or $k > 10$

1. Using a parameter, write an equation to represent each family of lines. **a.** $y = mx + 5$
   a. Nonvertical lines through the point $(0, 5)$ **b.** Lines with slope -2 $\;y = -2x + k$

2. From the family of lines with slope 3, which lines intersect the circle $x^2 + y^2 = 10$ in **(a)** 2 points? **(b)** 1 point? **(c)** 0 points? See above.

3. From the family of nonvertical lines through the origin, find an equation of each line that is tangent to the circle $x^2 + (y - 4)^2 = 8$. $\;y = x, \; y = -x$

4. Repeat Exercise 3 for the parabola $y = (x + 3)^2 - 5$. $\;y = 2x, \; y = 10x$

5. The diagram at the right shows the hyperbola $xy = 1$ and several lines from the family of lines with slope -1. Find an equation of each line with slope -1 that is tangent to the hyperbola $xy = 1$. $\;y = -x \pm 2$

6. From the family of lines with slope -3, find an equation of each line that is tangent to the ellipse $9x^2 + 4y^2 = 36$. Illustrate your solution with a sketch. $\;y = -3x \pm 3\sqrt{5}$

$xy = 1$

7. From the family of lines that contain the point $(0, -1)$, which lines are tangent to the ellipse $(x + 1)^2 + 4y^2 = 4$? Illustrate your solution with a sketch. $\;y = \dfrac{2}{3}x - 1, \; y = -1$

8. From the family of lines that contain the point $(-2, 0)$, which lines do not intersect the parabola $y = 6x - x^2$? $\;y = m(x + 2)$, where $2 < m < 18$

12. a. $y = 4x, \; y = -4x$ **b.** $(2, 8); (-2, 8)$ **c.** $y = -\dfrac{1}{4}x + \dfrac{17}{2}; \; y = \dfrac{1}{4}x + \dfrac{17}{2}$

In Exercises 9–12, do each of the following:

a. From the family of lines with the given condition, find an equation of each line that is tangent to the given curve. **9. a.** $y = -2x - 9$ **b.** $(-3, -3)$ **c.** $y = \dfrac{1}{2}x - \dfrac{3}{2}$

b. For each tangent line in part (a), find the coordinates of the point of tangency.   **10. a.** $y = 2x + 4, \; y = 2x - 4$ **b.** $(-1, 2); (1, -2)$ **c.** $y = -\dfrac{1}{2}x + \dfrac{3}{2};$

c. Find an equation of the line normal to the given curve at each point of tangency in part (b). Illustrate your solution with a sketch. $y = -\dfrac{1}{2}x - \dfrac{3}{2}$

9. $m = -2; \; y = x^2 + 4x$ See above.   **10.** $m = 2; \; xy = -2$ See above.

11. Contains $(0, 4); \; x^2 + y^2 = 8$ **12.** Contains $(0, 0); \; y = x^2 + 4$ See above.

11. a. $y = x + 4, \; y = -x + 4$ **b.** $(-2, 2); (2, 2)$ **c.** $y = -x; \; y = x$

Additional Answers
Exercises

6.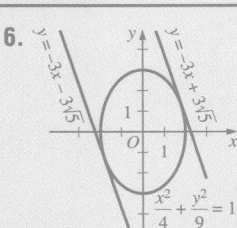

7.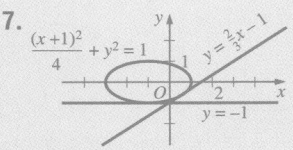

9. c.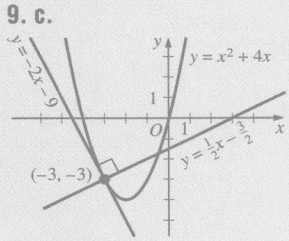

10. c.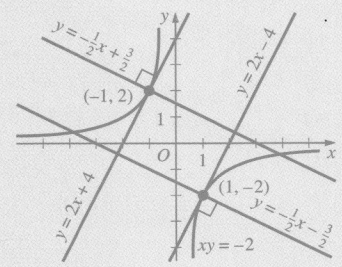

11. c.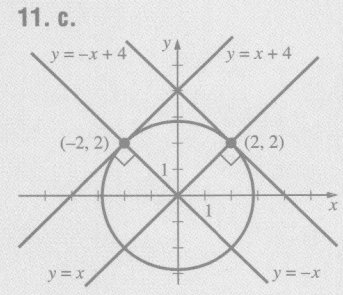

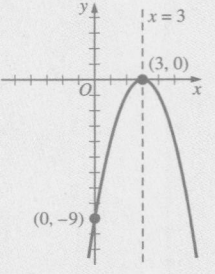
Refresher Exercises

Each set of Refresher Exercises provides review of prerequisite skills for the indicated chapter, as well as ongoing cumulative review of material covered in previous chapters.

Refresher Exercises
(for use before Chapter 1)

1. In what quadrant does the point $(-2, 5)$ lie? II
2. Is $(4, -1)$ a point on the graph of the equation $2x - 5y = 7$? No
3. Simplify: **a.** $3\sqrt{63}$ $9\sqrt{7}$ **b.** $\sqrt{10}(\sqrt{2} + 2)$ $2\sqrt{5} + 2\sqrt{10}$
4. The lengths of the legs of a right triangle are 3 cm and 6 cm. Find the length of the hypotenuse. $3\sqrt{5}$ cm
5. Let $3x + 4y = -9$. Find y when $x = -2$. $-\dfrac{3}{4}$
6. Solve for y: **a.** $5x - 8y = 16$ $y = \dfrac{5}{8}x - 2$ **b.** $\dfrac{y-4}{x+3} = \dfrac{3}{4}$ $y = \dfrac{3}{4}x + \dfrac{25}{4}$
7. Multiply: **a.** $(2x - 5)^2$ $4x^2 - 20x + 25$ **b.** $(\sqrt{3} + 2)(\sqrt{3} - 2)$ -1
8. Factor: **a.** $x^2 - 5x + 6$ $(x - 3)(x - 2)$ **b.** $x^2 - 25$ $(x - 5)(x + 5)$
9. Let $y = x^2 + 2x - 3$. Find x when $y = 0$. $-3, 1$
10. Solve for x by taking the square root of both sides: $(x - 2)^2 = 9$ $-1, 5$

Refresher Exercises
(for use before Chapter 2)

1. Classify each function as linear or quadratic.   quadratic   linear
   a. $f(x) = 5$ linear   **b.** $g(x) = 7x - 3x^2$ **c.** $h(x) = 2x - 15$
2. If $f(x) = 3x^2 - x - 1$, find $f(0)$, $f(-2)$, and $f(i)$. -1; 13; $-4 - i$
3. Is $x = -1$ a zero of the function $P(x) = x^3 - 3x^2 - x + 3$? Yes
4. Find all real and imaginary roots of each equation. $1, \pm 2i\sqrt{2}$
   a. $x^2 - 2x + 6 = 0$ $1 \pm i\sqrt{5}$ **b.** $6x - 4x^2 = 0$ $0, \dfrac{3}{2}$ **c.** $(x - 1)(x^2 + 8) = 0$
5. Factor completely: $5x^3 - 20x$ $5x(x - 2)(x + 2)$ **6.** $y = 2x^2 + 4x - 6$
6. Find an equation of the parabola graphed at the right.
7. **a.** Sketch the graph of $f(x) = -x^2 + 6x - 9$. Label the vertex, the axis of symmetry, and the intercepts.
   b. What is the maximum value of f? 0
8. The sides of a rectangle are x and $3 - 2x$.
   a. Express the rectangle's area as a function of x. $A(x) = x(3 - 2x)$
   b. Express the rectangle's perimeter as a function of x.
   c. Explain why x cannot equal 2. $P(x) = 6 - 2x$
9. The height and the diameter of a cylinder are equal. Express the volume of the cylinder as a function of its radius. $V(r) = 2\pi r^3$
10. Find the sum and the product of $1 - 3i$ and $1 + 3i$.   2; 10

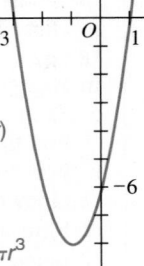

Ex. 6

Refresher Exercises *(for use before Chapter 3)*

1. Graph the inequality $-3 \leq x < 2$ on a number line.

2. Solve the inequality $5 - 4x \leq -7$ for x. $x \geq 3$

3. Give the values of x that make each statement true.
   a. $|x| = 5$ $5, -5$ **b.** $|x| < 5$ $-5 < x < 5$ **c.** $|x| > 5$ $x < -5$ or $x > 5$

4. Graph $P(x) = x(x - 3)(x + 1)$. Then tell if the graph of $y = P(x)$ is above or below the x-axis for each given set of x-values.
   a. $x < -1$ below   **b.** $-1 < x < 0$ above   **c.** $0 < x < 3$ below   **d.** $x > 3$ above

5. Sketch the graphs of $y = x^2 - 4x + 3$ and $x - 2y = -6$ on the same set of axes. Find the coordinates of each intersection point. $(0, 3), \left(\frac{9}{2}, \frac{21}{4}\right)$

6. Let $P = 10x + 8y$. In parts (a)–(e), evaluate P for the ordered pair (x, y).
   a. $(0, 0)$ 0   **b.** $(0, 5)$ 40   **c.** $(3, 3)$ 54   **d.** $(5, 2)$ 66   **e.** $(6, 0)$ 60
   f. Which ordered pair in parts (a)–(e) gives the largest value for P? $(5, 2)$

7. Factor: $a^3 - 2a^2 - a + 2 = 0$ $(a - 2)(a + 1)(a - 1)$

8. Solve for x: $4x^4 - 21x^2 + 27 = 0$ $\pm\frac{3}{2}, \pm\sqrt{3}$

Refresher Exercises *(for use before Chapter 4)*

1. For what value of x is the function $g(x) = \dfrac{2x + 1}{x + 7}$ undefined? -7

2. Write an inequality that describes the set of y-values on the graph of the function $f(x) = (x - 2)^2 - 1$. $y \geq -1$

3. Let $k(x) = 3x + 2$. Find $k(a)$, $k(2a)$, and $k(a + 1)$. $3a + 2$; $6a + 2$; $3a + 5$

In Exercises 4–6, refer to the graph of $y = f(x)$ shown at the right below. Draw the graph obtained by changing the graph of f in the way specified.

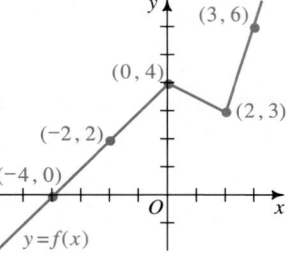

4. Shift each point of the graph of f to the right 3 units. For example, instead of plotting the point $(2, 3)$, plot the point $(5, 3)$.

5. Divide the y-coordinate of each point of the graph of f by 2. For example, instead of plotting the point $(2, 3)$, plot the point $(2, 1.5)$.

6. Reverse the x- and y-coordinates of each point of the graph of f. For example, instead of plotting the point $(2, 3)$, plot the point $(3, 2)$.

7. Solve $x = y^3 - 4$ for y in terms of x. $y = \sqrt[3]{x + 4}$

8. Find the vertex and the axis of symmetry for the parabola $y = 2x^2 + 8x + 5$. $(-2, -3); x = -2$

9. Give the dimensions of three different rectangles with area 6 cm^2. See below.

10. Each leg of an isosceles triangle is twice as long as its base. Express the perimeter of the triangle in terms of the length b of the base. $p = 5b$

9. For example, $2 \text{ cm} \times 3 \text{ cm}$, $6 \text{ cm} \times 1 \text{ cm}$, $\sqrt{2} \text{ cm} \times 3\sqrt{2} \text{ cm}$

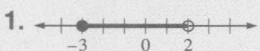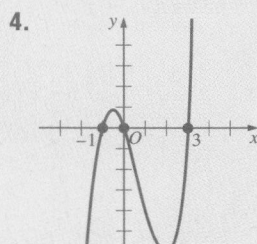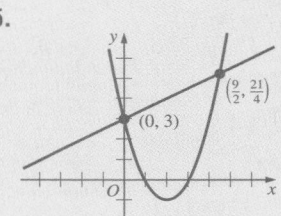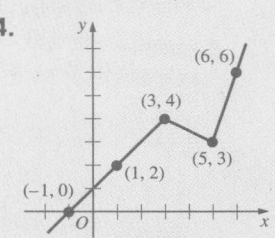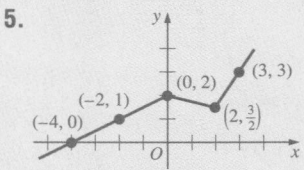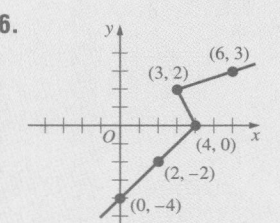

9.

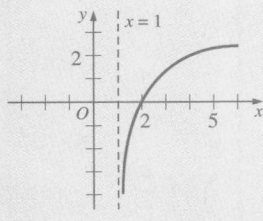

2.

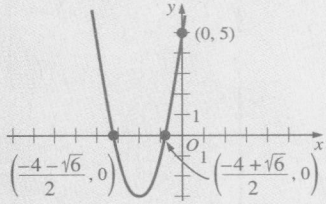

6. b. $y = -\frac{1}{2}x + 5$ contains

M, the midpoint of $\overline{AB}$.
The slope of $\overline{AB}$ is

$\frac{8-0}{4-0} = 2$.

The slope of $y = -\frac{1}{2}x + 5$

is $-\frac{1}{2}$; therefore, $y =$

$-\frac{1}{2}x + 5$ is perpendicu-

lar to $\overline{AB}$.

Thus, $y = -\frac{1}{2}x + 5$ is

the perpendicular bisec-
tor of $\overline{AB}$.

Refresher Exercises   *(for use before Chapter 5)*

1. Simplify: **a.** $x^3 \cdot x^2(4x^3)^2$ $16x^{11}$ **b.** $\dfrac{(3x)^3 y^4}{(6xy^2)^2}$ $\dfrac{3x}{4}$

2. Use a calculator to evaluate each expression to the nearest thousandth.

   a. $2(1.23)^3$ 3.722   **b.** $\left(1 + \dfrac{1}{n}\right)^n$ when $n = 9999$ 2.718

3. Use a calculator to evaluate $9^{0.5}$, $100^{0.5}$, and $16^{0.5}$. Use your answers to give the common meaning of $x^{0.5}$.   3, 10, 4; $\sqrt{x},\, x \ge 0$

In Exercises 4–7, solve for x.

4. $2^x = 64$ 6   5. $10^x = 1000$ 3   6. $\left(\dfrac{x}{2}\right)^3 = 125$ 10   7. $x^5 = 0.00001$ 0.1

8. This year a radio costs \$89. The cost is expected to increase by 7% next year. Find the expected cost of the radio next year.   \$95.23

9. Sketch the graph of the inverse of the function $y = f(x)$ whose graph is shown at the right.

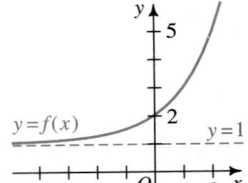

10. Let f be a linear function such that $f(2) = 5$ and $f(6) = -1$. Find an equation for $f(x)$. $f(x) = -\frac{3}{2}x + 8$

Refresher Exercises   *(for use before Chapter 6)*

For Exercises 1–3, let $f(x) = 2x^2 + 8x + 5$. $f(x) = 2(x + 2)^2 - 3$

1. **a.** Rewrite the equation for $f(x)$ in the form $f(x) = a(x - h)^2 + k$.
   b. The graph of $y = f(x)$ is a parabola. Give the coordinates of its vertex. $(-2, -3)$

2. Sketch the graph of $y = f(x)$. Label the x- and y-intercepts.

3. Suppose the graph of $y = f(x)$ is shifted 2 units to the left and 1 unit up. Write an equation for the new graph. $y = 2(x + 4)^2 - 2$

In Exercises 4–7, $\triangle ABC$ has vertices $A(0, 0)$, $B(4, 8)$, and $C(10, 0)$.

4. Find the coordinates of M, the midpoint of $\overline{AB}$. $(2, 4)$

5. Find an equation of the line that contains M and is parallel to $\overline{BC}$. $4x + 3y = 20$

6. **a.** Find an equation of the line through points C and M. $y = -\frac{1}{2}x + 5$
   b. Show that the line in part (a) is the perpendicular bisector of $\overline{AB}$.

7. What type of triangle is $\triangle ABC$? Justify your answer.   isosceles; $AC = BC = 10$

8. Find the domain and range of each function.   **a.** $-3 \le x \le 3;\ 0 \le y \le 3$
   a. $f(x) = \sqrt{9 - x^2}$ **b.** $g(x) = -\sqrt{x - 3}$ $x \ge 3;\ y \le 0$

9. Find the coordinates of any points of intersection of the graphs of $2x - y = 1$ and $y = 5 - 5x - 3x^2$. $(-3, -7)$, $\left(\dfrac{2}{3}, \dfrac{1}{3}\right)$

10. Test the equation $x^2 + 4y^2 = 1$ for symmetry in **(a)** the x-axis, **(b)** the y-axis, **(c)** the line $y = x$, and **(d)** the origin.   **a.** yes   **b.** yes   **c.** no   **d.** yes

848   *Refresher Exercises*

Refresher Exercises

(for use before Chapter 7)

1. a. Sketch the graph of the circle $x^2 + (y - 2)^2 = 25$.
   b. Find the circumference and the area of the circle in part (a). ≈ 31.4; ≈ 78.5

2. In the diagram at the right, point $P(x, 4)$ is located 6 units from the origin. Find the value of x. $-2\sqrt{5}$

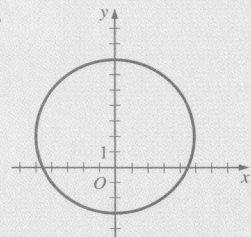

3. Find the height of an equilateral triangle with sides of length 10. $5\sqrt{3}$

4. The legs of a right isosceles triangle are 5 cm long. Find the length of the triangle's hypotenuse. $5\sqrt{2}$ cm

5. Give the equations of all asymptotes of the graph of the given equation.
   a. $y = \log(x + 2)$ $x = -2$
   b. $y = 2^x - 3$ $y = -3$
   c. $xy = 1$ $x = 0, y = 0$
   d. $36 = x^2 - 9y^2$ $y = \pm\frac{1}{3}x$

In Exercises 6–8, let $f(x) = x^2 + 2$.

6. Explain why the function f does not have an inverse. f is not one-to-one. For example, $f(-1) = f(1) = 3$.

7. Restrict the domain of f so that f has an inverse. Then find a rule for $f^{-1}(x)$.

8. Show that $f(f^{-1}(x)) = f^{-1}(f(x)) = x$.   **7.** $x \geq 0$; $f^{-1}(x) = \sqrt{x - 2}$

Refresher Exercises

(for use before Chapter 8)

1. Graph $y = \sin x$ for $-2\pi \leq x \leq 2\pi$.

2. If $\tan \theta = \frac{5}{2}$ and $180° < \theta < 270°$, find each of the following.
   a. $\csc \theta$ $-\frac{\sqrt{29}}{5}$
   b. $\sec \theta$ $-\frac{\sqrt{29}}{2}$
   c. $\cot \theta$ $\frac{2}{5}$

3. Find the exact value of each expression in radians.
   a. $\text{Cos}^{-1} 0$ $\frac{\pi}{2}$
   b. $\text{Sin}^{-1}\left(-\frac{1}{2}\right)$ $-\frac{\pi}{6}$
   c. $\text{Tan}^{-1} 1$ $\frac{\pi}{4}$

For Exercises 4 and 5, refer to the graph of the function f shown below.

4. Find the fundamental period and the amplitude of f.   4; 1

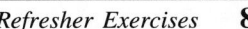

5. Sketch the graph of each function.
   a. $y = f(2x)$ **b.** $y = f(x) + 2$ **c.** $y = f(-x)$
   d. $y = 2f(x)$ **e.** $y = f(x - 1)$ **f.** $y = -f(x)$

In Exercises 6–8, solve for the variable.

6. $2y = 2y(y - 1)$ $0, 2$
7. $2t^2 - t = 2 - \frac{1}{t}$ $-1, \frac{1}{2}, 1$
8. $2\sqrt{a} = a - 3$ 9

9. Simplify the complex fraction: $\dfrac{\dfrac{a}{b} - \dfrac{b}{a}}{a - b}$ $\dfrac{a + b}{ab}$

10. Solve the system: $\begin{aligned} y^2 &= 1 - x^2 \\ y^2 &= x^2 - 3x + 2 \end{aligned}$ $(1, 0), \left(\frac{1}{2}, \frac{\sqrt{3}}{2}\right), \left(\frac{1}{2}, -\frac{\sqrt{3}}{2}\right)$

Additional Answers
Refresher Exercises

(before Chapter 7)

1.

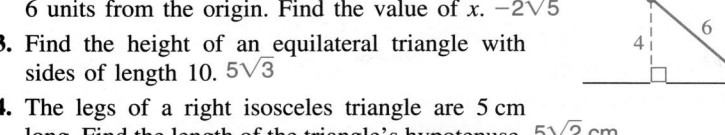

8. $f(f^{-1}(x)) = (\sqrt{x - 2})^2 + 2$
   $= x - 2 + 2 = x,\ x \geq 2$;
   $f^{-1}(f(x)) = \sqrt{(x^2 + 2) - 2}$
   $= \sqrt{x^2} = x,\ x \geq 0$

(before Chapter 8)

5. a.

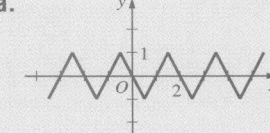

b.

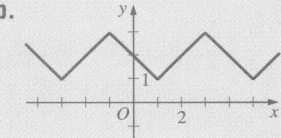

c.

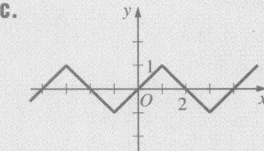

d.

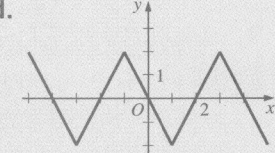

(before Chapter 9)

4.

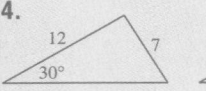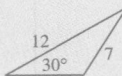

(before Chapter 10)

1. $\cos x = -\dfrac{8}{17}$, $\tan x = \dfrac{15}{8}$,

$\sec x = -\dfrac{17}{8}$, $\csc x =$

$-\dfrac{17}{15}$, $\cot x = \dfrac{8}{15}$

5. $\tan(-\theta) = \dfrac{\sin(-\theta)}{\cos(-\theta)} =$

$\dfrac{-\sin \theta}{\cos \theta} = -\dfrac{\sin \theta}{\cos \theta} = -\tan \theta$

9.

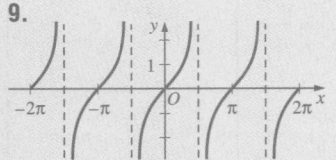

Domain: $\{x : x \neq \dfrac{\pi}{2} + n\pi\}$,
range: {all real numbers}

Refresher Exercises   *(for use before Chapter 9)*

In Exercises 1 and 2, find the value of x to three significant digits.

1. $\dfrac{x}{10} = \sin 40°$ 6.43   **2.** $\dfrac{3}{x} = \tan 27°$ 5.89

3. Find the area of the triangle shown in the diagram at the right. 20

4. Sketch two noncongruent triangles such that each triangle has one side of length 12, one side of length 7, and one angle that measures 30°.

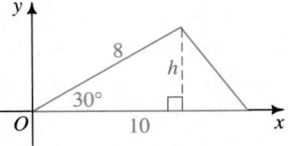

In Exercises 5 and 6, solve for $0° < \theta < 180°$. Give your answers to the nearest tenth of a degree.

5. $9 = 16 + 36 - 2(4)(6)\cos \theta$ 26.4°   **6.** $\dfrac{\sin \theta}{4} = \dfrac{\sin 27.2°}{5}$ 21.4°

7. In the diagram at the right, the blue arrows indicate the direction of true north. A ship sails from point A to point B along a *course* of 230°, as shown. The ship then changes its direction, sailing to point C along a course of 120°. Find the measure of $\angle ABC$. 70°

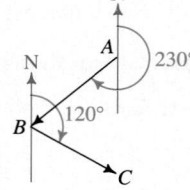

8. Graph $y = \sin x$ and $y = \ln x$ on the same set of axes. Between what two integers does the equation $\sin x = \ln x$ have a solution? 2 and 3

Refresher Exercises   *(for use before Chapter 10)*

1. If $\sin x = -\dfrac{15}{17}$ and $\dfrac{\pi}{2} < x < \dfrac{3\pi}{2}$, find the exact values of the other five trigonometric functions.

2. Find the exact value of $\sin\left(\text{Cos}^{-1}\left(-\dfrac{1}{3}\right)\right)$. $\dfrac{2\sqrt{2}}{3}$

3. Use a calculator to evaluate $\sin 149° \cos 121° + \cos 149° \sin 121°$. -1

4. Without using a calculator, evaluate $\dfrac{\tan 210° + \tan 120°}{1 - (\tan 210°)(\tan 120°)}$. $-\dfrac{\sqrt{3}}{3}$

5. Prove $\tan(-\theta) = -\tan \theta$.

6. Find the slope and the inclination of the line with equation $3x - 5y = 15$. $\dfrac{3}{5}$, 31°

7. Simplify $\sin \theta \cdot \sec^2 (90° - \theta)$. $\csc \theta$

8. Solve $2 + \cos^2 \theta = 3 \sin^2 \theta$ for $0° \leq \theta < 360°$. 60°, 120°, 240°, 300°

9. Sketch the graph of $y = \tan x$ for $-2\pi \leq x \leq 2\pi$. Then give the domain and range of the tangent function.

10. Graph $y = -2 \cos x$ and $y = \sqrt{x}$ on the same set of axes. How many solutions does the equation $-2 \cos x = \sqrt{x}$ have? 2

850   *Refresher Exercises*

Refresher Exercises (for use before Chapter 11)

1. Express $\cos 216°$ in terms of a reference angle. $-\cos 36°$

2. As shown in the diagram at the right, point P is 8 units from the origin on the terminal ray of a $135°$ angle in standard position. Find the x- and y-coordinates of P. $(-4\sqrt{2}, 4\sqrt{2})$

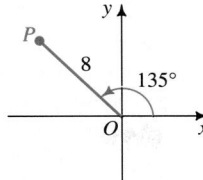

3. $\triangle ABC$ has vertices $A(0, 0)$, $B(4, 0)$, and $C(4, 7)$.
   a. Find AC to the nearest tenth. 8.1
   b. Find $\angle BAC$ to the nearest tenth of a degree. $60.3°$

In Exercises 4–7, express each complex number in the form $a + bi$.

4. $(1 + i)^3$ $-2 + 2i$

5. $(4 - i\sqrt{3})(2 + i\sqrt{3})$ $11 + 2i\sqrt{3}$

6. $(4 - i\sqrt{3}) + (2 + i\sqrt{3})$ 6

7. $\dfrac{2 - 4i}{1 + i}$ $-1 - 3i$

8. Show that $-\sqrt{3} + i$ is a square root of $2 - 2i\sqrt{3}$.

9. Express $\cos 4\theta$ in terms of $\cos \theta$. $8\cos^4 \theta - 8\cos^2 \theta + 1$

10. Solve $2x^2 + 2x + 1 = 0$. $\dfrac{-1 \pm i}{2}$

Refresher Exercises (for use before Chapter 12)

1. The diagram at the right shows $\triangle AOB$ with vertices $A(3, 4)$, $O(0, 0)$, and $B(7, 4)$.
   a. Find the length of each side of $\triangle AOB$.
   b. Find $\angle OAB$ to the nearest degree. $127°$

1. a. $AO = 5$, $AB = 4$, $OB = \sqrt{65}$

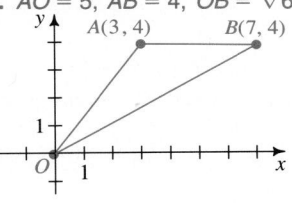

2. A plane flies 200 km north and then 250 km on a course of $110°$. How far is the plane from its starting point? Give your answer to the nearest kilometer. 261 km

3. Write an equation of the line that contains the points $(1, -5)$ and $(-2, 4)$. $y = -3x - 2$

4. Sketch the graph of the ellipse with equation $9(x - 1)^2 + 4(y + 2)^2 = 36$.

5. Given $A = (2, -3)$ and $B = (8, -11)$, find an equation of the circle with diameter AB. $(x - 5)^2 + (y + 7)^2 = 25$

6. Sketch the graphs of $2x + 3y = 6$ and $y = 2x^2 - 5x - 3$ to determine the number of intersection points. Then solve the equations simultaneously.

7. If $x = 2\cos^2 \theta$ and $y = \sin 2\theta$, show that $(x - 1)^2 + y^2 = 1$. **6.** 2; $(3, 0)$, $\left(-\dfrac{5}{6}, \dfrac{23}{9}\right)$

8. Write $8 \operatorname{cis} 240°$ in rectangular form. $-4 - 4i\sqrt{3}$

9. a. Let $2x + 3y - 4z = 10$. Find z when $x = 0$ and $y = 0$. $-\dfrac{5}{2}$
   b. Find two more *ordered triples* (x, y, z) that are solutions of the equation given in part (a). For example, $(5, 0, 0)$ and $(0, 2, -1)$

10. How many solutions does the system shown at the right have? What is the geometric relationship of the graphs of the two lines? 0; parallel

$9x - 6y = -12$
$y = 1.5x - 2$

Additional Answers
Refresher Exercises

(before Chapter 11)

8. $(-\sqrt{3} + i)^2 =$
$3 - 2i\sqrt{3} + i^2 =$
$3 - 2i\sqrt{3} - 1 =$
$2 - 2i\sqrt{3}$

(before Chapter 12)

4.

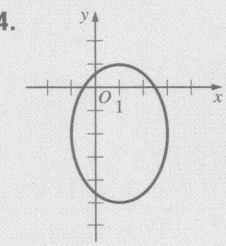

7. $(x - 1)^2 + y^2 =$
$(2\cos^2 \theta - 1)^2 + (\sin 2\theta)^2$
$= (\cos 2\theta)^2 + (\sin 2\theta)^2$
$= 1$

Additional Answers
Refresher Exercises

(before Chapter 13)

1. a.

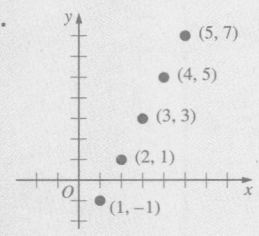

b.

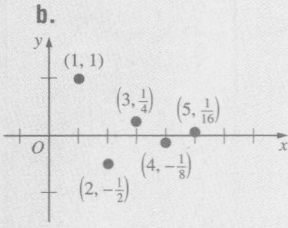

(before Chapter 14)

5. a. $(1, 0) \rightarrow (1, 0)$,
$\quad(0, 1) \rightarrow (0, -1)$;
b. $y = x - 1$

6. a. $(1, 0) \rightarrow (-1, 0)$,
$\quad(0, 1) \rightarrow (0, 1)$;
b. $y = x + 1$

7. a. $(1, 0) \rightarrow (0, 1)$,
$\quad(0, 1) \rightarrow (1, 0)$;
b. $y = -x + 1$

8. a. $(1, 0) \rightarrow (-1, 1)$,
$\quad(0, 1) \rightarrow (-2, 2)$;
b. $y = -x$

Refresher Exercises *(for use before Chapter 13)*

1. Plot the points $(x, f(x))$ for $x = 1, 2, 3, 4,$ and 5.

   a. $f(x) = 2x - 3$ **b.** $f(x) = \left(-\frac{1}{2}\right)^{x-1}$

2. Each leg of a right isosceles triangle has length 8. The midpoints of the sides of the triangle are joined to form another right isosceles triangle, and the midpoints of the sides of this triangle are joined to form still another right isosceles triangle, and so on.

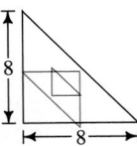

   a. Find the area of each of the first three triangles in this sequence of triangles. 32, 8, 2

   b. What is the area of the seventh triangle in the sequence? $\frac{1}{128}$

In Exercises 3–6, solve for n.

3. $4 + 3(n - 1) = 511$ 170   **4.** $5 \cdot 2^{n-1} = 20{,}480$ 13

5. $\dfrac{n(5 + 3n)}{2} = 175$ $10, -\dfrac{35}{3}$ **6.** $\dfrac{1}{1 - 4n} = 2$ $\dfrac{1}{8}$

7. Evaluate each expression to the nearest thousandth if $n = 1000$.

   a. $\dfrac{4n^2 + 3n - 5}{2n^2 - n}$ 2.002   **b.** $\dfrac{2 \sin n}{n}$ 0.002

8. Simplify and factor the expression $k^2 + k + 2(k + 1)$. $(k + 1)(k + 2)$

Refresher Exercises *(for use before Chapter 14)*

1. Solve each system. **3.** $(5, -8, 6)$; $(1, -13, -6)$; -13

   a. $2x + 5y = -11$ **b.** $x + 2y - z = -3$
       $3x - 7y = 27$ $(2, -3)$ $2x - y + 2z = 7$ $(1, -1, 2)$
                        $3y + 4z = 5$

2. Evaluate each determinant.

   a. $\begin{vmatrix} 3 & 2 \\ 1 & -5 \end{vmatrix}$ -17 **b.** $\begin{vmatrix} 1 & 2 & -3 \\ 4 & -4 & 1 \\ 2 & 1 & 0 \end{vmatrix}$ -33

3. Let $\mathbf{u} = (2, -7, 0)$ and $\mathbf{v} = (-3, 1, -6)$. Evaluate $\mathbf{u} - \mathbf{v}$, $2\mathbf{u} + \mathbf{v}$, and $\mathbf{u} \cdot \mathbf{v}$. See above.

4. Refer to the vectors $\mathbf{u}$ and $\mathbf{v}$ in Exercise 3. Show that $\mathbf{u} \times \mathbf{v} \neq \mathbf{v} \times \mathbf{u}$.
$\mathbf{u} \times \mathbf{v} = 42\mathbf{i} + 12\mathbf{j} - 19\mathbf{k}$, $\mathbf{v} \times \mathbf{u} = -42\mathbf{i} - 12\mathbf{j} + 19\mathbf{k}$

In Exercises 5–8, suppose that the graph of $y = -x + 1$, shown below, is changed in the way specified. (a) Tell what happens to points $(1, 0)$ and $(0, 1)$. (b) Give an equation of the new graph.

   5. The graph is reflected in the x-axis.

   6. The graph is reflected in the y-axis.

   7. The graph is reflected in the line $y = x$.

   8. The graph is shifted 2 units to the left and 1 unit up.

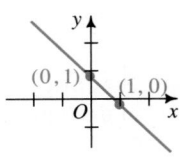

852   *Refresher Exercises*

1. From the State Park Visitors' Center, located at point V, you can choose one of three paths to follow to Pretty Pond, located at point P. There are 2 paths from the pond to an observation tower at point O. If you hike from the Visitors' Center to Pretty Pond and then go on to the tower, how many different routes can you follow? 6

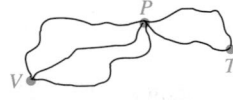

2. Make an organized list to determine the number of four-digit integers that can be formed using each of the digits 1, 2, 3, and 4 exactly once. 24

3. Make an organized list to determine the number of different ways you can choose two of the four digits 1, 2, 3, and 4. 6

4. Of the first ten counting numbers, how many are **(a)** prime? **(b)** odd? **(c)** both prime and odd? **(d)** neither prime nor odd? **a.** 4   **b.** 5   **c.** 3   **d.** 4

In Exercises 5 and 6, expand the given binomial.

5. $(x - y)^3$ $x^3 - 3x^2y + 3xy^2 - y^3$ 6. $(2x + 1)^3$ $8x^3 + 12x^2 + 6x + 1$

7. **a.** Express $2 + 2i$ in polar form. $2\sqrt{2}$ cis 45°
   b. Use De Moivre's Theorem to find $(2 + 2i)^4$. -64

8. Let $\mathbf{u} = (1, 4, -6)$ and $\mathbf{v} = (0.5, 0.25, 0.25)$.
   a. Find $\mathbf{u} \cdot \mathbf{v}$. 0
   b. Based on your answer to part (a), what can you conclude about $\mathbf{u}$ and $\mathbf{v}$? $\mathbf{u} \perp \mathbf{v}$

1. When one card is chosen at random from a standard deck of playing cards, there are 52 possible outcomes. What fraction of the possible outcomes are **(a)** clubs? **(b)** face cards? **(c)** aces? See below.

2. How many different 4-letter "words" can be formed using the letters of the given word?
   a. SINE 24   **b.** NEED 12   1. a. $\frac{1}{4}$ b. $\frac{3}{13}$ c. $\frac{1}{13}$

3. From a jar containing 3 green marbles and 2 yellow marbles, 3 marbles are randomly picked, one at a time and without being replaced. Draw a tree diagram showing all possible outcomes.

In Exercises 4–6, five cards are drawn at random from a well-shuffled standard deck of 52 cards. Find the number of different 5-card hands that contain the specified cards.

4. No diamonds 575,757   5. At least one diamond 2,023,203   6. All diamonds 1287

7. Evaluate the expression $\dfrac{_4C_1 \cdot _5C_2}{_9C_3}$. $\dfrac{10}{21}$

8. In the expansion of $(3x - 2y)^{10}$, find and simplify the middle term.
   $-1,959,552x^5y^5$

Additional Answers
Refresher Exercises

(before Chapter 16)

3.

```
            G           Y
          /   \       /   \
         G     Y     G     Y
        / \   / \   / \    |
       G  Y  G  Y  G  Y    G
```

4.

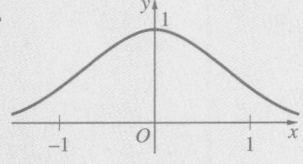

6. a.

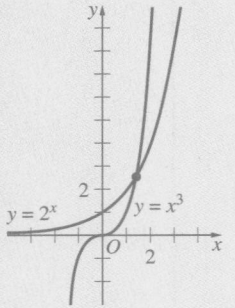

$y = 2^x$ $\quad y = x^3$

3. a.

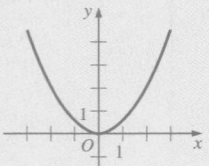

b.

$y = 2^{-x} + 2$

$y = 2$

c.

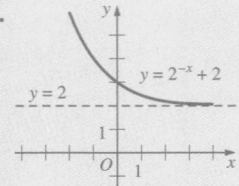

Refresher Exercises *(for use before Chapter 17)*

1. Twenty-eight percent of the 25 students in a precalculus class scored below 70 on a test. How many students in the class scored 70 or above on the test? 18

2. The table below shows the number of students whose English exam scores fell in each grade range. What percent of the students scored between 70 and 89? about 66%

Grade range	Below 60	60–69	70–79	80–89	90–100
Number of students	2	6	12	9	3

3. a. Let $x_1 = 6$, $x_2 = 8$, $x_3 = 9$, and $x_4 = 13$. Evaluate $m = \dfrac{\sum\limits_{i=1}^{4} x_i}{4}$. 9

b. Use the values of x_1, x_2, x_3, x_4, and m in part (a) to evaluate $\dfrac{\sum\limits_{i=1}^{4} (x_i - m)^2}{4}$. 6.5

4. Graph the function $f(x) = e^{-x^2}$.

5. Let $f(x) = x^2 - 4$ and $g(x) = \sqrt{x + 3}$. Find a rule for $(f \circ g)(x)$ and give the domain of the composite function. $f(g(x)) = x - 1$, $x \ge -3$

6. a. On a single set of axes, graph $y = x^3$ and $y = 2^x$.

b. Between what two integers is a solution of the equation $x^3 = 2^x$ located? 1 and 2

Refresher Exercises *(for use before Chapter 18)*

1. Four blocks of different sizes are cut from one piece of wood. The volume and weight of each block are found and then plotted as an ordered pair (volume, weight), as shown at the right. The four points lie on or near the line containing the points that represent the lightest and heaviest blocks. Find an equation of the line. $y = 0.659x + 0.237$

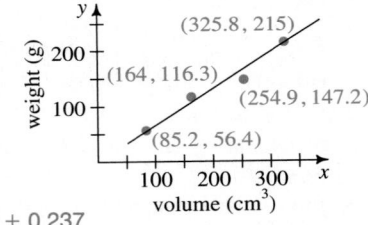

2. Find the standard deviation of the data 2.8, 3.7, 8.4, 9.6, 12.5. 3.65

3. Sketch the graph of each function.
  a. $y = 0.5x^2$ **b.** $y = 2^{-x} + 2$ **c.** $y = \sqrt{x}$

4. Find the value of y to three significant digits if $x = 6$.
  a. $y = 10 - 0.31x$ 8.14   **b.** $y = 50(1.06)^x$ 70.9   **c.** $y = 25.8x^{0.4}$ 52.8

5. Solve each equation. Give your answers to the nearest thousandth.
  a. $\log x = 0.72$ 5.248   **b.** $4^x = 3$ 0.792   **c.** $\ln x = 1.09$ 2.974

6. Express y in terms of x.
  a. $\log y = x + 2$ **b.** $\ln y = 2 \ln x$ $y = x^2$ **c.** $\log y = 4 \log x + 3$
  $y = (100)10^x$ $y = 1000x^4$

854 *Refresher Exercises*

Refresher Exercises

(for use before Chapter 19)

1. Evaluate each limit: **a.** $\lim\limits_{n \to +\infty} \sqrt{n}$ $+\infty$ **b.** $\lim\limits_{n \to \infty} \dfrac{2n^3 - n}{7n^3 - n^2}$ $\dfrac{2}{7}$

In Exercises 2 and 3, simplify the given expression.

2. $\dfrac{x^2 - 9}{x^2 + 4x + 3} \cdot \dfrac{x - 3}{x + 1}$

3. $\dfrac{x - 4}{\sqrt{x - 4}} \cdot \sqrt{x - 4}$

In Exercises 4 and 5, solve for x. **5.** $-3 < x < -1$ or $1 < x < 5$

4. $(x - 4)(x + 2)(x - 1)^2 \geq 0$
   $x \leq -2$ or $x = 1$ or $x \geq 4$

5. $\dfrac{(x - 5)(x + 3)}{(x - 1)(x + 1)} < 0$

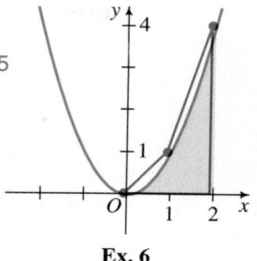

6. In the diagram at the right, the shaded region is bounded by the curve $y = x^2$, the x- and y-axes, and the vertical line $x = 2$. Approximate the area of the shaded region by finding the area of the convex quadrilateral shown in blue. 3

Ex. 6

7. Use your calculator to find the sum of the first six terms of the series $1 + \dfrac{1}{1!} + \dfrac{1}{2!} + \dfrac{1}{3!} + \cdots$. What number is approximated by the sum? $2.71\overline{6}$; e

8. Let $f(x) = \sqrt{x} + 1$. Find $f(1)$, $f(f(1))$, and $f(f(f(1)))$ to the nearest thousandth. 2; 2.414; 2.554

Refresher Exercises

(for use before Chapter 20)

$2x + 3y = -14$

1. Find an equation of the line with slope $-\dfrac{2}{3}$ that contains the point $(-1, -4)$.

2. Refer to the diagram at the right. **2. a.** $\dfrac{2^x - 1}{x}$
   a. Write an expression to represent the slope of a line that passes through the point $(0, 1)$ and through a second point $(x, 2^x)$ that lies on the graph of the function $f(x) = 2^x$. **b.** 1.5, 1, 0.828, 0.718
   b. Evaluate your expression in part (a) to the nearest thousandth for $x = 2$, $x = 1$, $x = 0.5$, and $x = 0.1$.

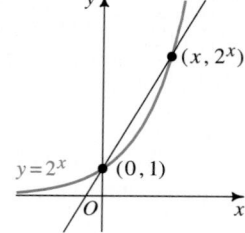

3. Simplify the expression $\dfrac{2(x + h)^2 + 1 - (2x^2 + 1)}{h}$. $4x + 2h$

4. Find the vertex of the parabola whose equation is $y = x^2 - 6x + 2$. $(3, -7)$

5. Sketch the graph of the equation $y = 4x^3 - 12x^2$.

6. The height of $\triangle ACE$ is 12, and base $AE = 20$. A rectangle is inscribed in $\triangle ACE$ as shown at the right.
   a. Express the rectangle's area as a function of x and y.
   b. Use the fact that $\triangle ACE \sim \triangle BCD$ to write y as a function of x.
   c. Express the rectangle's area as a function of x alone.
   d. Find the maximum possible area of the rectangle.

   a. $A = x(12 - y)$ **b.** $y = \dfrac{3}{5}x$ **c.** $A = -\dfrac{3}{5}x^2 + 12x$ **d.** 60

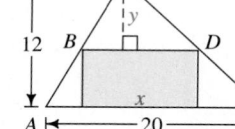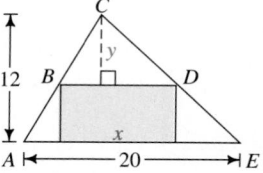

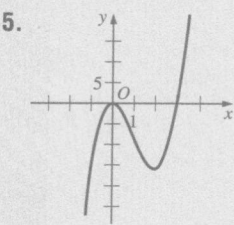
Refresher Exercises **855**

List of Symbols

Symbol	Meaning	Page
$P(x, y)$	point P with coordinates x and y	1
$\sqrt{}$	square root	4
$\overline{AB}$	line segment with endpoints A and B	4
AB	the distance from point A to point B	4
$\triangle$	triangle	5
$\angle$	angle	6
m	slope of a line	7
$\sim$	is similar to	12
$\cong$	is congruent to	12
$\parallel$	is parallel to	12
$\perp$	is perpendicular to	13
$f(x)$	f of x, or function f of x	19
i	$\sqrt{-1}$	26
$\bar{z}$	conjugate of the complex number z	27
$\overleftrightarrow{PB}$	the line through P and B	36
$|a|$	absolute value of a	37
x^n	the nth power of x	53
$\{(x, y)\,|\,y = f(x)\}$	the set of all ordered pairs (x, y) such that x is an element of the domain of f and y is the corresponding element of the range	119
$\lfloor x \rfloor$	the greatest integer less than or equal to x	124
$\lceil x \rceil$	the least integer greater than or equal to x	124
$(f \circ g)(x)$	$f(g(x))$, or f of g of x	126
$f^{-1}(x)$	f inverse of x	147
$P(l, w)$	P as a function of l and w	151
$\lim\limits_{n \to \infty}$	limit as n approaches infinity	186
e	irrational number that is approximately 2.718 . . .	187
$\log x$	common logarithm of x	191
$\log_b x$	logarithm to base b of x	193
$\ln x$	natural logarithm of x, or $\log_e x$	193
$25°20'6''$	25 degrees, 20 minutes, and 6 seconds	257
$P(r, \theta)$	point P with polar coordinates r and θ	395
$r\,\mathrm{cis}\,\theta$	$r(\cos \theta + i \sin \theta)$	403
$\mathbf{v}$	vector quantity	419
$\overrightarrow{AB}$	vector depicted by an arrow from point A to point B	419
$|\overrightarrow{AB}|$	magnitude of $\overrightarrow{AB}$	426
$\mathbf{v}_1 \cdot \mathbf{v}_2$	dot product of vectors $\mathbf{v}_1$ and $\mathbf{v}_2$	441
$\begin{vmatrix} a_1 & a_2 \\ b_1 & b_2 \end{vmatrix}$	2×2 determinant	458
$\mathbf{v}_1 \times \mathbf{v}_2$	cross product of vectors $\mathbf{v}_1$ and $\mathbf{v}_2$	465

Symbol	Meaning	Page
$\displaystyle\sum_{k=1}^{100} k^2$	sum of k^2 for values of k from 1 to 100	506
E^t	transpose of matrix E	517
A^{-1}	multiplicative inverse of matrix A	531
$|A|$	determinant of matrix A	531
$T: (x, y) \rightarrow$ $(x + 2y, 3x + y)$	transformation T that maps each point $P(x, y)$ to its image point $P'(x + 2y, 3x + y)$	551
$A \cap B$	intersection of sets A and B	565
$A \cup B$	union of sets A and B	565
$\overline{A}$	complement of set A	566
$n(A)$	number of elements in set A	566
$n!$	n factorial	572
$_nP_r$	number of permutations of r things chosen from n things	578
$_nC_r$	number of combinations of r things chosen from n things	578
$P(A)$	probability of event A	598
$P(B|A)$	probability of event B, given event A	607
$\overline{x}$	the mean of data values $x_1, x_2, x_3, \ldots, x_n$	641
s^2	variance	654
s	standard deviation	654
z	standard value	656
r	correlation coefficient	685
$f'(x)$	derivative of $f(x)$	758
Δs	the change in s values, or $s_2 - s_1$	775
$f''(x)$	second derivative of $f(x)$, or derivative of $f'(x)$	776

Metric Units

Time: second (s), minute (min)
Temperature:
  degree Celsius (°C)
Length: meter (m)
Capacity: liter (L)
Mass: kilogram (kg)*

Electrical charge: coulomb (C)
Force: newton (N)
Pressure: pascal (P)
Frequency: hertz (Hz)
Luminous intensity:
  candela (cd)

A prefix multiplies a unit by the factor given in the table. For example,

  1 MHz $= 10^6$ Hz $= 1{,}000{,}000$ Hz

Compound units may also be formed by multiplication or division. Examples are kilometers per hour (km/h) and cubic meters (m^3).

Factor	Prefix	Symbol
10^6	mega	M
10^3	kilo	k
10^{-2}	centi	c
10^{-3}	milli	m

*Although the kilogram is defined as the base unit, the gram (g) is used with prefixes to name other units of mass.

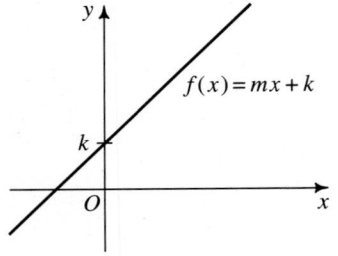

Linear

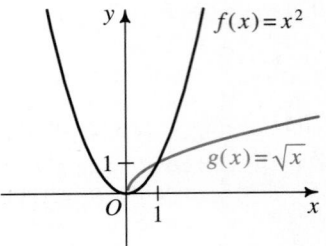

Quadratic; Square root

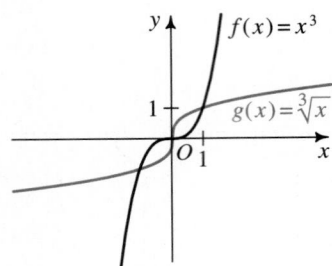

Cubic; Cube root

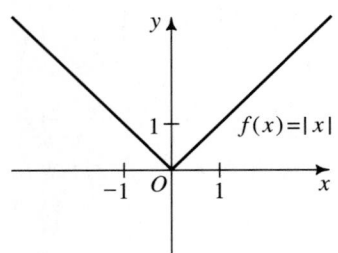

Absolute value

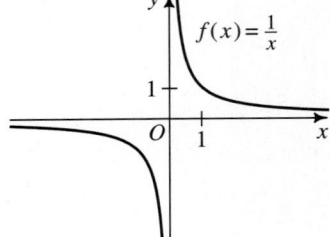

Reciprocal

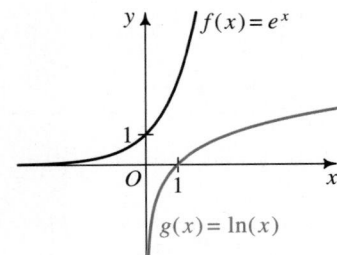

Exponential; Logarithmic

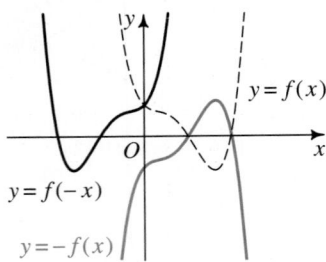

Reflection in the x- or y-axis

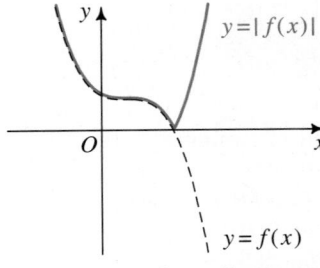

Absolute value reflection

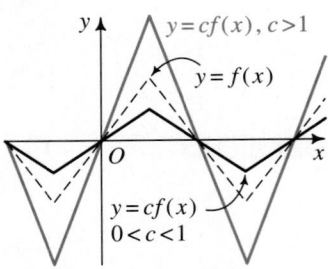

Vertical stretch or shrink

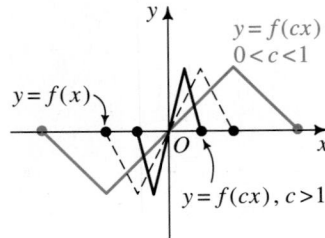

Horizontal stretch or shrink

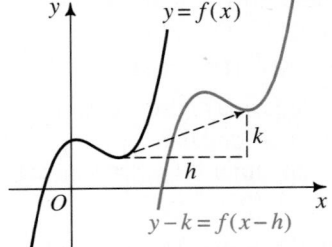

Translation

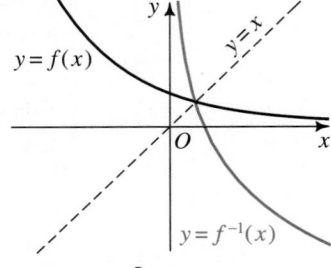

Inverse

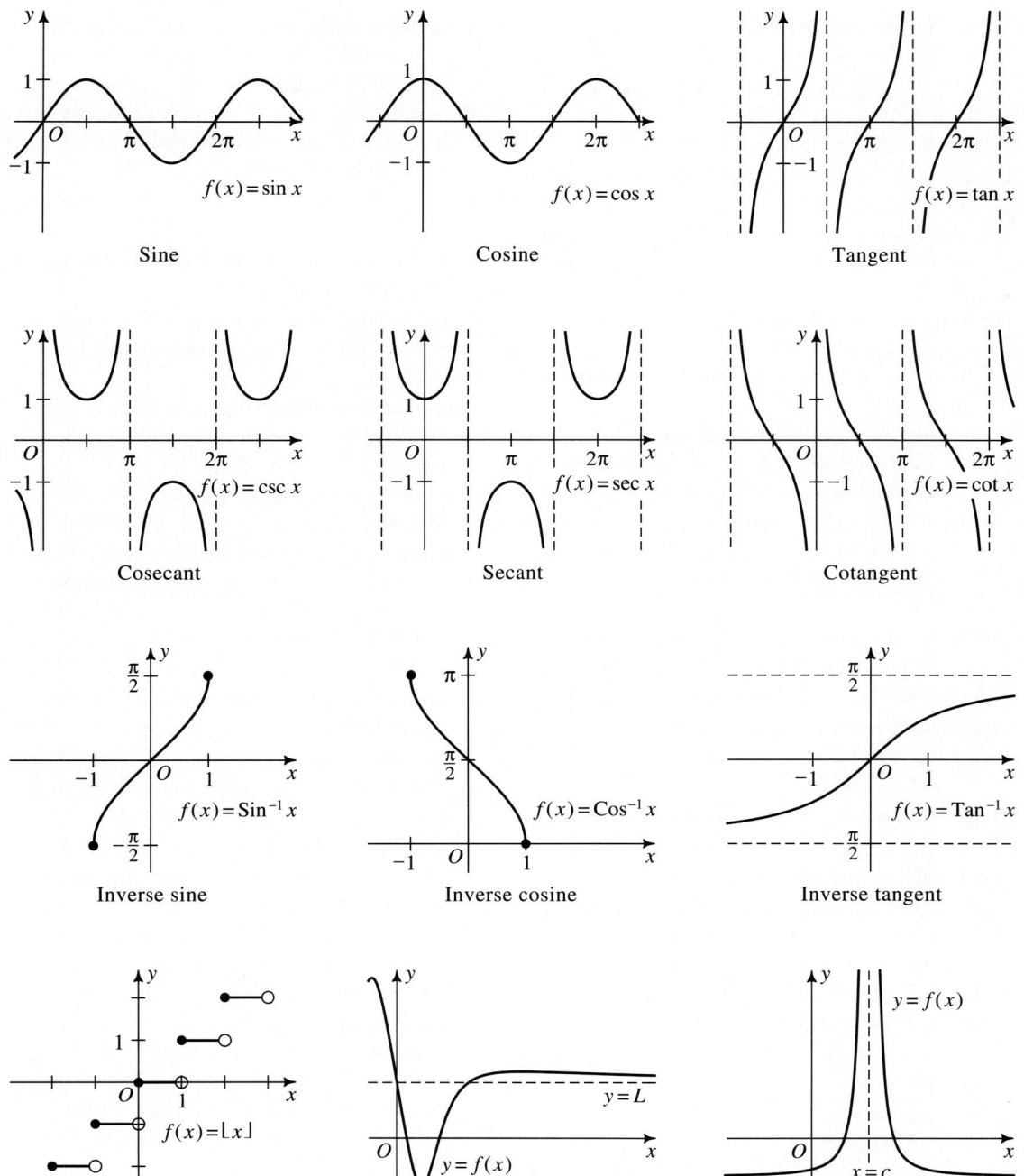

Sine

$f(x) = \sin x$

Cosine

$f(x) = \cos x$

Tangent

$f(x) = \tan x$

Cosecant

$f(x) = \csc x$

Secant

$f(x) = \sec x$

Cotangent

$f(x) = \cot x$

Inverse sine

$f(x) = \mathrm{Sin}^{-1} x$

Inverse cosine

$f(x) = \mathrm{Cos}^{-1} x$

Inverse tangent

$f(x) = \mathrm{Tan}^{-1} x$

Greatest integer

$f(x) = \lfloor x \rfloor$

$\lim_{x \to \infty} f(x) = L$

$y = f(x)$

$y = L$

$\lim_{x \to c} f(x) = \infty$

$y = f(x)$

$x = c$

Visual Glossary of Functions   **859**

Glossary

Absolute value of a complex number The length of the arrow representing the complex number. If $z = a + bi$, then $|z| = \sqrt{a^2 + b^2}$. (p. 403)

Absolute value of a number The distance from the number to zero on the number line. (p. 96)

Acceleration A measure of the rate of change of the velocity of an object over time. (p. 776)

Additive inverse of a matrix A (denoted $-A$) The matrix in which each element is the opposite of its corresponding element in A. (p. 531)

Amplitude of a periodic function One half the difference between the maximum and minimum values of the function. (p. 139)

Angle of depression Angle by which an observer's line of sight must be depressed from the horizontal to the point observed. (p. 335)

Angle of elevation Angle by which an observer's line of sight must be elevated from the horizontal to the point observed. (p. 332)

Apparent size The measure of an angle subtended at one's eye by an object being observed. (p. 264)

Argand diagram A diagram that represents complex numbers geometrically. (p. 403)

Arithmetic sequence A sequence of numbers such that the difference of any two consecutive terms is constant. (p. 473)

Asymptote A line that a curve approaches more and more closely. (p. 232)

Average A single number, such as the mean, median, and mode, used to represent an entire set of data. (p. 641)

Axis of symmetry A graph has a line called an axis of symmetry if it is possible to pair the points of the graph so that the line is the perpendicular bisector of the segment joining each pair. (p. 133)

Binomial coefficients The numbers in Pascal's triangle; the coefficients of the terms in the expansion of $(a + b)^n$. (p. 590)

Box-and-whisker plot A method for displaying the median, quartiles, and extremes of a set of data. (p. 649)

Cofunctions Sine, cosine; tangent, cotangent; secant, cosecant. (p. 318)

Combination An arrangement of a set of objects in which order is not important. (p. 578)

Common logarithm of a positive real number x The exponent when the number is written as a power of 10. (p. 191)

Complement of a set A (denoted $\overline{A}$) The set of all elements not in a set A. (p. 566)

Complex number Any number of the form $a + bi$, where a and b are real numbers and i is the imaginary unit. a is the *real part* and b is the *imaginary part* of $a + bi$. (p. 27)

Complex number plane The plane in which points represent complex numbers. (p. 403)

Composite of two functions $(f \circ g)(x) = f(g(x))$ where x is in the domain of g and $g(x)$ is in the domain of f. (p. 126)

Composition The operation that combines two functions to produce their composite. (p. 126)

Conditional probability The probability that an event will occur given that another event has occurred. (p. 607)

Conic sections Cross sections (circle, ellipse, hyperbola, parabola) resulting from slicing a double cone by a plane. When the result is a single point, a line, or a pair of lines, the cross section is called a *degenerate conic section*. (p. 213)

Constant A polynomial (or term of a polynomial) with degree 0. (p. 53)

Continuous function A function $f(x)$ is continuous at a real number c if $\lim_{x \to c} f(x) = f(c)$. (p. 720)

Coordinates An ordered pair of numbers associated with a point on the plane. (p. 1)

Correlation coefficient A number between -1 and 1, inclusive, that measures how closely plotted points tend to cluster about the least-squares line. (p. 685)

Cosecant $\csc \theta = \dfrac{r}{y}$, $y \neq 0$, where $P(x, y)$ is a point on circle O with equation $x^2 + y^2 = r^2$ and θ is an angle in standard position with terminal ray OP. Also, $\csc \theta = \dfrac{1}{\sin \theta}$, $\sin \theta \neq 0$. (p. 282)

Cosine $\cos \theta = \dfrac{x}{r}$ where $P(x, y)$ is a point on circle O with equation $x^2 + y^2 = r^2$ and θ is an angle in standard position with terminal ray OP. (p. 268)

Cotangent $\cot \theta = \dfrac{x}{y}$, $y \neq 0$, where $P(x, y)$ is a point on circle O with equation $x^2 + y^2 = r^2$ and θ is an angle in standard position with terminal ray OP. Also, $\cot \theta = \dfrac{\cos \theta}{\sin \theta}$ and $\cot \theta = \dfrac{1}{\tan \theta}$, $\sin \theta \neq 0$ and $\tan \theta \neq 0$. (p. 282)

Coterminal angles Two angles in standard position having the same terminal ray. (p. 260)

Cross product of two vectors If $\mathbf{v}_1 = (a_1, b_1, c_1)$ and $\mathbf{v}_2 = (a_2, b_2, c_2)$, then
$$\mathbf{v}_1 \times \mathbf{v}_2 = \begin{vmatrix} \mathbf{i} & \mathbf{j} & \mathbf{k} \\ a_1 & b_1 & c_1 \\ a_2 & b_2 & c_2 \end{vmatrix}. \text{ (p. 465)}$$

Cubic A polynomial (or term of a polynomial) with degree 3. (p. 53)

Degree A unit for measuring angles, $\frac{1}{360}$ of a revolution. (p. 257)

Degree of a polynomial in x The power of x contained in the leading term. (p. 53)

Derivative of $f(x)$ The slope function $f'(x)$. (p. 758)

Determinant A value associated with a square array of numbers. (pp. 458–459)

Dimensions of a matrix The number of rows and columns of the matrix. (p. 517)

Directrix of a parabola A fixed line such that all points of the parabola are equidistant from that line and a fixed point called the *focus*. (p. 238)

Discriminant The quantity $b^2 - 4ac$ that appears beneath the radical sign in the quadratic formula. (p. 31)

Domain of a function The set of values for which the function is defined. (pp. 20, 119)

Dot product of two vectors (also called *scalar product*) In two dimensions, $\mathbf{v}_1 \cdot \mathbf{v}_2 = x_1 x_2 + y_1 y_2$, where $\mathbf{v}_1 = (x_1, y_1)$ and $\mathbf{v}_2 = (x_2, y_2)$. (p. 441) In three dimensions, $\mathbf{u} \cdot \mathbf{v} = x_1 x_2 + y_1 y_2 + z_1 z_2$, where $\mathbf{u} = (x_1, y_1, z_1)$ and $\mathbf{v} = (x_2, y_2, z_2)$. (p. 448)

Doubling time Time required for a quantity growing exponentially to double. (p. 182)

e The irrational number, approximately equal to 2.718 . . . , which is the limit as n approaches infinity of $\left(1 + \dfrac{1}{n}\right)^n$. (p. 186)

Eccentricity of a conic section The positive constant e such that $PF{:}PD = e$, where F is a fixed point (*focus*) not on a fixed line d (*directrix*), and PD is the perpendicular distance from P to d. The curve is an ellipse if $0 < e < 1$, a parabola if $e = 1$, and a hyperbola if $e > 1$. (p. 247)

Element (or entry) of a matrix A number in a matrix. (p. 517)

Ellipse (1) The set of all points P in the plane such that $PF_1 + PF_2 = 2a$, where F_1 and F_2 are fixed points called *foci* and a is a positive constant. (p. 226)
(2) $\dfrac{x^2}{a^2} + \dfrac{y^2}{b^2} = 1$, where a and b are positive constants. (p. 227)

Even function f is an even function if $f(-x) = f(x)$. (p. 137)

Event Any subset of a sample space. (p. 598)

Expected value If a given situation has payoff x_1, with probability $P(x_1)$, payoff x_2, with probability $P(x_2)$, and so on, then the expected value is $x_1 \cdot P(x_1) + x_2 \cdot P(x_2) + \cdots + x_n \cdot P(x_n)$. (p. 630)

Exponential equation An equation that contains a variable in the exponent. (p. 203)

Exponential function with base b Any function of the form $f(x) = ab^x$, $a > 0$, $b > 0$, and $b \neq 1$. (p. 181)

Extreme values of a function Minimum and maximum values of the function. (p. 157)

Feasible solution Any point (x, y) that satisfies all the requirements of a problem. The set of these points is called the *feasible region*. (p. 109)

Frequency polygon A line graph used to display a set of data. (p. 641)

Frequency table A table that is used to organize and display data. (p. 640)

Function (1) A dependent relationship between quantities. (p. 19) (2) A correspondence or rule that assigns to every element in a set D (the *domain*) exactly one element in a set R (the *range*). (p. 119)

Geometric sequence A sequence such that the ratio (*common ratio*) of any two consecutive terms is constant. (p. 473)

Graph of an equation The set of all points in the coordinate plane corresponding to solutions of an equation. (p. 1)

Half-life The time required for half a quantity decaying exponentially to decay. (p. 182)

Histogram A bar graph used to display a set of data. (p. 640)

Hyperbola The set of all points P in the plane such that $|PF_1 - PF_2| = 2a$, where F_1 and F_2 are fixed points called *foci* and a is a positive constant. (p. 231)

Identity matrix (denoted $I_{n \times n}$) An $n \times n$ matrix whose main diagonal elements are 1 and whose other elements are 0. (p. 530)

Imaginary number A complex number $a + bi$ for which $b \neq 0$. (p. 27)

Imaginary unit i The number $\sqrt{-1}$ with the property $i^2 = -1$. (p. 26)

Inclination of a line The angle α, where $0° \leq \alpha < 180°$, that is measured from the positive x-axis to the line. (p. 296)

Independent events Two events A and B are independent if and only if the occurrence of A does not affect the probability of the occurrence of B. (p. 607)

Infinite sequence A sequence that does not have a last term. (p. 493)

Inverse functions Two functions f and g are inverse functions if $g(f(x)) = x$ for all x in the domain of f and $f(g(x)) = x$ for all x in the domain of g. f^{-1} represents the inverse of f. (p. 146)

Leading term of a polynomial in x The term containing the highest power of x. The coefficient of the leading term is called the *leading coefficient*. (p. 53)

Least-squares line The line that minimizes the sum of the squares of the distances of data points from the line. (p. 684)

Limit See pp. 722–723 for a formal definition.

Line of reflection A line that acts like a mirror and is located halfway between a point and its reflection. (p. 131)

Linear equation Any equation of the form $Ax + By = C$, where A and B are not both 0. (p. 1)

Linear function A function of the form $f(x) = mx + k$. (p. 19)

Linear programming A branch of mathematics that uses linear inequalities called *constraints* to solve certain decision-making problems where a linear expression is to be maximized or minimized. (p. 108)

Logarithm to base b of a positive number x
The exponent when the number is written as a power of b, where $b > 0$ and $b \neq 1$. (p. 193)

Logarithmic function with base b Inverse of exponential function with base b. (p. 193)

Markov chain A sequence of observations or predictions, each of which depends on the immediately preceding one, and only on it. (p. 545)

Mathematical induction A method of proof used to show that a statement is true for all positive integers. (p. 511)

Mathematical model One or more functions, graphs, tables, equations, or inequalities that describe a real-world situation. (p. 20)

Matrix (plural, *matrices*) A rectangular array of numbers. Each number is an *element* or *entry* of the matrix. (p. 517)

Mean (denoted $\bar{x}$) A statistical average found by dividing the sum of a set of data by the number of items of data. (p. 641)

Median A statistical average found by arranging a set of data in increasing or decreasing order and finding the middle number or the mean of the two middle numbers. (p. 642)

Mode A statistical average that is the one item of data that occurs most often. (p. 642)

Multiplicative inverse of a matrix A (denoted A^{-1}) The matrix A^{-1} such that $A \cdot A^{-1} = A^{-1} \cdot A = I$, where $|A| \neq 0$. (pp. 531–532)

Mutually exclusive events Events that cannot occur simultaneously. (p. 599)

Natural logarithm of a positive number x
$\log_e x$, or $\ln x$. (p. 193)

Normal distribution A distribution of data along a bell-shaped curve that reaches its maximum height at the mean. (p. 662)

Odd function f is an odd function if $f(-x) = -f(x)$. (p. 137)

One-to-one function A function that has an inverse. Each x-value corresponds to exactly one y-value, and each y-value corresponds to exactly one x-value. (p. 148)

Origin Point of intersection of x- and y-axes. (p. 1)

Parabola (1) The graph of a quadratic function. (p. 37) (2) The set of all points P in the plane that are equidistant from a fixed point F, called the *focus*, and a fixed line d, not containing F, called the *directrix*. (p. 238)

Pascal's triangle An array of the coefficients of the terms in the expansion of $(a + b)^n$. (p. 590)

Percentiles The 99 points that divide a set of data into 100 equal parts when the data are arranged in ascending order. (p. 665)

Periodic function A function f is periodic if there is a positive number p, called the *period* of f, such that $f(x + p) = f(x)$ for all x in the domain of f. (p. 138)

Permutation An arrangement of a set of objects in which order is important. (p. 578)

Point of symmetry A graph has a point of symmetry if it is possible to pair the points of the graph so that the point is the midpoint of the segment joining each pair. (p. 133)

Polar axis The reference ray, usually the nonnegative x-axis, in the polar coordinate system. (p. 395)

Polar coordinates The ordered pair (r, θ) that describes the position of a point P in the plane. r is the distance from the pole to P, and θ is the measure of an angle formed by ray OP and the polar axis. (p. 395)

Polar equation An equation of a curve given in terms of r and θ. (p. 396)

Pole The origin in the polar coordinate system. (p. 395)

Polynomial in x An expression that can be written in the form $a_n x^n + a_{n-1} x^{n-1} + \cdots + a_2 x^2 + a_1 x + a_0$ where n is a nonnegative integer. The numbers a_n, $a_{n-1}, \ldots, a_2, a_1$, and a_0 are called the *coefficients* of the polynomial. (p. 53)

Polynomial inequalities $P(x) < 0$ and $P(x) > 0$ where $P(x)$ is a polynomial. (p. 100)

Population   In statistics, the entire set of individuals or objects being studied. (p. 669)

Power series   An infinite series of the form $a_0 + a_1 x + a_2 x^2 + \cdots + a_n x^n + \cdots$. (p. 733)

Probability   (1) A number between 0 and 1, inclusive, that indicates the likelihood of the occurrence of an event. (p. 597) (2) If a sample space of an experiment contains n equally likely outcomes and if m of the n outcomes correspond to some event A, then $P(A) = \dfrac{m}{n}$. (p. 598)

Pure imaginary number   An imaginary number $a + bi$ for which $a = 0$ and $b \neq 0$. (p. 27)

Quadrantal angle   An angle in standard position with the terminal ray along an axis. (p. 259)

Quadrants   The four regions into which the coordinate plane is divided by the x- and y-axes. (p. 1)

Quadratic   A polynomial (or term of a polynomial) with degree 2. (p. 53)

Quadratic equation   Any equation that can be written in the form $x^2 + bx + c = 0$, where $a \neq 0$. (p. 30)

Quartic   A polynomial (or term of a polynomial) with degree 4. (p. 53)

Quintic   A polynomial (or term of a polynomial) with degree 5. (p. 53)

Radian   A unit for measuring angles. (p. 258)

Range of a function   (1) The set of output values of the function. (p. 20) (2) The set of elements to which the elements of the domain are assigned by the function. (p. 119)

Rational numbers   Numbers that are the ratio of two integers. (p. 25)

Real numbers   Zero and all positive and negative integers, rational numbers, and irrational numbers. (p. 26) All complex numbers $a + bi$ for which $b = 0$. (p. 27)

Reference angle   The acute positive angle formed by the terminal ray of an angle and the x-axis. (p. 275)

Resultant   Vector sum of two vectors. (p. 420)

Revolution   A complete circular motion. (p. 257)

Sample   A subset of the population. (p. 669)

Sample space of an experiment   A set S such that each outcome of the experiment corresponds to exactly one element of S. (p. 598)

Sampling   The process of selecting a sample representative of a total population. (p. 669)

Scalar   In work with vectors and matrices, any real number. (pp. 422, 518)

Secant $\sec \theta = \dfrac{r}{x}$, $x \neq 0$, where $P(x, y)$ is a point on a circle O with equation $x^2 + y^2 = r^2$ and θ is an angle in standard position with terminal ray OP. Also, $\sec \theta = \dfrac{1}{\cos \theta}$, $\cos \theta \neq 0$. (p. 282)

Sector of a circle   A region bounded by a central angle and the intercepted arc. (p. 263)

Segment of a circle   A region bounded by an arc of the circle and the chord connecting the endpoints of the arc. (p. 340)

Sequence   (1) A set of numbers, called *terms*, arranged in a particular order. (p. 473) (2) A function whose domain is the set of positive integers. (p. 474)

Series An indicated sum of the terms of a sequence. (p. 486)

Sign graph A number-line graph that shows when the value of a polynomial function is positive, negative, or zero. (p. 100)

Sine $\sin \theta = \frac{y}{r}$ where $P(x, y)$ is a point on circle O with equation $x^2 + y^2 = r^2$ and θ is an angle in standard position with terminal ray OP. (p. 268)

Slope of a curve at a point P The limit of the slope of $\overline{PQ}$ as Q approaches P along the curve, where Q is a point on the curve near P. (p. 758)

Slope of a nonvertical line (1) A number measuring the steepness of the line relative to the x-axis. (p. 7) (2) The tangent of the inclination of the line. (p. 296)

Slope function $f'(x)$, the derivative of $f(x)$. (p. 758)

Square matrix A matrix having the same number of rows as of columns. (p. 530)

Standard deviation The positive square root of the variance. (p. 654)

Standard position of an angle An angle shown in a coordinate plane with its vertex at the origin and its initial ray along the positive x-axis. (p. 259)

Stem-and-leaf plot A method for organizing and displaying data. (p. 639)

Synthetic substitution A method for evaluating a polynomial function $P(x)$ for any value of x. (p. 55)

Synthetic division A short-cut method of dividing a polynomial by a divisor of the form $x - a$. (p. 59)

Tangent $\tan \theta = \frac{y}{x}$, $x \neq 0$, where $P(x, y)$ is a point on circle O with equation $x^2 + y^2 = r^2$ and θ is an angle in standard position with terminal ray OP. Also $\tan \theta = \frac{\sin \theta}{\cos \theta}$, $\cos \theta \neq 0$. (p. 282)

Tangent to a curve at a point on the curve The line passing through the point and having a slope equal to the slope of the curve at that point. (p. 761)

Transformation of the plane A function whose domain and range are sets of points. (p. 551)

Transition matrix A matrix that specifies a transition from one observation to the next. (p. 543)

Transpose of a matrix The matrix obtained by interchanging the rows and columns of the matrix. (p. 517)

Trigonometric identity An equation that is true for all values of the variable for which each side of the equation is defined. (p. 318)

Unit circle The circle $x^2 + y^2 = 1$. (p. 269)

Variance A statistic used to describe the spread of data about the mean. (p. 654)

Vector (1) A quantity described by both direction and magnitude (p. 419). (2) A quantity named by an ordered pair (p. 426) or an ordered triple (p. 446) of numbers.

Venn diagram A diagram used to represent a relationship between sets. (p. 565)

x-axis and y-axis Perpendicular lines, usually horizontal and vertical, respectively, used in setting up a coordinate system. (p. 1)

x-intercept The x-coordinate of a point at which a graph intersects the x-axis. (p. 1)

y-intercept The y-coordinate of a point at which a graph intersects the y-axis. (p. 1)

Zero of a function If $f(a) = 0$, then a is a zero of the function f. (p. 19)

Zero of a polynomial function $P(x)$ Any value of x for which $P(x) = 0$. (p. 53)

Zero matrix (denoted $O_{m \times n}$) An $m \times n$ matrix whose elements are all zero. (p. 530)

Index

Electronics
  AM radio waves, 306
  FM radio waves, 307
  modeling voltage, 696
  power, 47
  voltage equations, 303–304
Engineering
  minimizing cost, 164
  volume of storage tank, 163, 344
Farming
  expected gain, 634
  fencing maximum area, 71
  grazing area, 266
Finance
  appreciation, 179
  compound interest, 189–190, 206
  doubling investment, 182, 184, 206
  effective annual yield, 189
  tripling investment, 205–206
  value of investment, 483, 747
Forestry
  lumber production, 703
Geography
  classifying countries, 570
  classifying states, 569
  distance on the globe, 281
  latitude and circumference, 277
  population distribution, 546, 647
  population growth, 184, 185, 205, 695
Geology
  geyser eruption, 107
  oil-well drilling, 353
  Richter scale, 202–203
Horticulture
  temperature range, 99
  tree heights, 667
Industrial design
  project-scoring matrix, 528
Insurance
  expected gain, 634
  probability, 627
Landscaping
  minimizing cost, 164

Linear programming, 108–111, 112, 113–114
Manufacturing
  ball-bearing diameter, 667
  extreme value, 773
  filling bottles, 668
  inventory matrix, 522
  labor-schedule matrix, 536
  light-bulb life, 667
  maximizing profit, 113–114
  maximizing volume, 72–73, 164
  minimizing cost, 162
  probability, 623, 627
  production costs, 156, 708
  sampling for defects, 673
  washer diameter, 108
Mechanics
  pistons, 282
Medicine
  drug half-life, 184
  probability, 628
  pulse rate, 707
  Rh factor, 570
Meteorology
  average temperatures, 644
  probability, 602
  temperature sine waves, 315
  weather distribution pattern, 586
  weather maps, 153
  wind chill, 155
Music
  note frequencies, 180
  vibration of piano string, 308
Navigation, 348, 357
  angle of depression, 335
  angle of elevation, 336
  great circles, 301
  setting a course, 424, 426, 430
  submarine dive, 356
Nutrition
  menu-cost matrix, 528
  menu information, 652–653
  recommended energy intake, 647

vitamin-content matrix, 536
Occupational safety
  probability, 602
Oceanography
  intensity of light in water, 207
  tides, 314
  water pressure, 706
Operations research
  employee arrival time, 667
  mail handling, 667
Optics
  apparent size, 266
  Snell's law, 300
Personal finance, 746
Physics
  boiling point of water, 706
  decibel levels, 192, 195
  displacement of spring, 307
  Earth's rotational speed, 281
  half-life, 182, 183, 206, 478
  height of rotating point, 316
  height of thrown object, 47, 71–72
  illumination, 131
  path of projectile, 438, 439
  pendulum motion, 308
  resultant force, 424, 429, 430
  rotational speed, 265
  sound and temperature, 130
  speed on an incline, 336
  stopping distance, 46
  velocity and acceleration, 778, 779, 780, 781, 782
  work, 445
Physiology
  calories burned, 155
  oxygen consumption, 129
  reaction time, 645, 668
  weight distribution, 668
Plumbing
  capacity of pipe, 48

Trigonometric functions, 256–293
  cofunction, 282, 318
  definitions, 268, 282
  identities, 317–326
  inverse, 286–288
  negatives, 317
  period and amplitude, 302
  Pythagorean relationships, 317
  reciprocal, 317
  signs in quadrants, 282
Trigonometric identities, 318–320
Trigonometric inverses, 286–288
  applications
    aviation, 338
    optics, 300
    space science, 338
  domain and range, 287, 288
  graphs, 287, 288
Trigonometry applications to navigation and surveying, 359–361

U

Union of sets, 565

V

Variability, 653–657
Variables, functions of two, 151–153
Variance, 654
Vector sum, 419
Vectors, 419–469
  absolute value, 419
  addition, 419–421
  algebraic representation, 426–428
  angle between nonzero, 442
  applications
    aviation, 424, 426
    navigation, 424, 426

    physics, 424, 425, 429, 438, 439, 445
    sports, 424
  component form, 426
  cross product, 465
  direction, 432
  dot product, 441, 442
  magnitude, 419
  multiples, 421–422
  negative of, 421
  operations, 427
  as ordered pairs, 426
  orthogonal, 441
  parallel, 441
  perpendicular, 441
  properties, 448, 465
  scalar product, 441
  subtraction, 421
  sum, 419
  in three dimensions, 446–449, 465–466
  zero, 421
Velocity
  average, 775
  instantaneous, 775
Venn diagram, 565
Vertex (vertices)
  ellipse, 227
  hyperbola, 233
  parabola, 37, 238
Vertical-line test, 121
Visual glossary of functions, 858–859
Visual thinking, 763
  exercises focused on, 45, 89, 128, 129, 135–136, 138, 153, 186, 244, 245, 261, 279, 281, 304, 322, 337, 343, 354, 361, 373, 378, 385, 390, 406, 407, 414, 430, 476, 483, 592, 669, 688, 697, 724, 725, 731, 732, 735, 741, 749, 761, 762, 767, 768, 778, 779–780, 783

W

Web diagram, 738
Whitman, Robert K., xxxvi
Wrapping function, 274 (Ex. 43)
Writing, exercises focused on,
  22, 24, 50, 78, 93, 99, 112, 115, 124, 167, 209, 223, 240, 253, 267, 274, 285, 290, 293, 306, 316, 322, 329, 365, 393, 407, 415, 430, 431, 436, 469, 477, 483, 503, 504, 515, 521, 522, 535, 543, 548, 558, 559, 560, 577, 595, 636, 645, 647, 652, 681, 690, 701, 703, 706, 710, 729, 747, 748, 749, 782, 783

X

x-axis, 1
x-coordinate, 1
x-intercept, 1
xy-coordinate plane, 446
xz-coordinate plane, 446

Y

Yamasaki, Minoru, 117
y-axis, 1
y-coordinate, 1
y-intercept, 1
yz-coordinate plane, 446

Z

Zero
  of a function, 19
  as polynomial, 53
Zero matrix, 530
Zuniga, Martha C., xxxiv

Try This, pages xxvi–xxxiii **2.b.** A sideways "S."
3.a. -2.3 **b.** $(-1.28, -2.57)$ **4.a.** -0.9
b. $-1.8, 2.8$ **5.** 2 **6.** 1.25 s

Class Exercises, pages 4–5 **1.** 10; $(4, 3)$ **2.** $2\sqrt{5}$;
$(5, 4)$ **3.** $6\sqrt{2}$; $(0, 1)$ **4.** 8; $(7, -5)$ **5.** $\overline{AM}$:
$(3, 4)$; $\overline{MB}$: $(5, 6)$ **6.** a, b, d **7.a.** For example,
$(0, 4), (1, 4), (2, 4)$ **b.** $y = 4$ **8.a.** For example,
$(8, -2), (8, 0), (8, 8)$ **b.** $x = 8$ **9.** $(0, -6)$,
$(4.5, 0)$ **10.a.** $(2, 3)$ **b.** $(2, 3)$ **c.** They are $=$.

Written Exercises, pages 5–7 **1.** 10; $(4, 4)$ **3.** 17;
$\left(-\frac{1}{2}, 1\right)$ **5.** $\frac{13}{2}$; $\left(-\frac{3}{4}, \frac{3}{2}\right)$ **7.** 5; $(4.8, -0.3)$
9. a, c **11.** Area of $\triangle OPQ = 3$ **13.** $(5, 3)$;
$y = 3, x = 5$ **15.** $(3, 0)$ **17.** $\left(-\frac{5}{4}, -\frac{7}{4}\right)$
19. Parallelogram **21.** Parallelogram
25.a. $PA = PB = 5$ **b.** -2 **27.** $(9, 0), (-15, 0)$
29. $(10, 3)$ **31.a.** $\angle C$ is a rt. $\angle$. **b.** $(0, 5)$
33. $A(2, -1), B(12, 7), C(8, 11)$

Class Exercises, page 11 **1.** $\frac{7}{5}$ **2.** 6 **3.** -1

4. $\frac{1}{2}$ **5.** 0 **6.** No slope **7.** $m = 3$; y-int. 4

8. $m = \frac{3}{5}$; y-int. -3 **9.** $m = -\frac{4}{3}$, y-int. 3

10. $m = 0$; y-int. -2 **11.** The change in x-values
is zero, so def. of slope would involve division by

zero. **12.** $\frac{1}{2}$ **13.a.** $-\frac{3}{2}$ **b.** $\frac{2}{3}$ **14.a.** Slope of $\overline{AB}$

and $\overline{CD} = \frac{4}{3}$; slope of $\overline{BC}$ and $\overline{DA} = -\frac{3}{4}$

b. $\left(\frac{4}{3}\right)\left(-\frac{3}{4}\right) = -1$

Written Exercises, pages 11–13 **1.** $\frac{3}{5}$ **3.** $-\frac{2}{3}$ **5.** 0

7. -4 **9.** -1 **11.** $m = 3$; y-int. 5 **13.** $m = 2$;
y-int. -4 **15.** $m = \frac{11}{3}$; y-int. 0 **17.** a‖b, a⊥c, b⊥c
19. Slope of line 1 = 1, slope of line 2 = -1.
21.a. 0 **b.** $\frac{17}{2}$ **23.a.** Slope of $\overline{AB}$ and $\overline{CD} = \frac{5}{3}$;

slope of $\overline{BC}$ and $\overline{AD} = \frac{1}{3}$ **b.** Both have midpt.

$(2, 0)$. **25.** Label the four points A, B, C, and D.

Slope of $\overline{BC}$ and $\overline{AD} = \frac{2}{3}$; slope of $\overline{AB} \neq$ slope of

$\overline{CD}$. $AB = CD = \sqrt{26}$. **29.a.** Slope of $l_1 = \frac{BC}{AC}$;

slope of $l_2 = \frac{EF}{DF}$ **31.a.** $OA = \sqrt{m_1^2 + 1}$,
$OB = \sqrt{m_2^2 + 1}$, $AB = |m_2 - m_1|$ **33.** $R(-3, 6)$,
$S(-1, 0)$ or $R(-9, 4), S(-7, -2)$

Calculator Exercises, page 13
1. Area $\triangle ABC = 6.37$ **2.** Midpt. M of $\overline{BC}$ is
$(1.0, 1.95)$. $AM = BM = CM = \sqrt{7.9625}$

Class Exercises, page 16 **1.** $y = \frac{5}{3}x - 2$

2. $\frac{y - 6}{x + 4} = 3$ **3.** $\frac{y - 2}{x - 1} = -\frac{1}{4}$ **4.** $-x + \frac{y}{6} = 1$

5. $\frac{y + 1}{x - 2} = \frac{2}{3}$ **6.a.** $y = mx$ **b.** $\frac{y}{x} = m$ **c.** No sol.

Written Exercises, pages 16–18 **Exs. 1–16:**
Ans. are given in general form. **1.** $2x + y = 8$
3. $2x - y = -4$ **5.** $2x - 3y = -14$ **7.** $y = -7$
9. $x = 2$ **11.** $0.25x - y = -1.8$ **13.** $x - 2y = 12$
15. $2x - y = -8$ **19.a.** $3x - 4y = -9$
b. $3x - 4y = -9$ **c.** Yes; yes **21.a.** $y = 3$,
$y = x + 1, y = -2x + 7$ **b.** $G(2, 3)$
23.a. $x = 6, x - y = 0, x + 2y = 18$ **b.** $O(6, 6)$
25. The first $\triangle$ **27.a.** $3\sqrt{10}$; $x + 3y = 3$
b. $3x - y = 19$ **c.** $(6, -1)$ **d.** $2\sqrt{10}$ **e.** 30
29.a. $bx - ay = bx_1 - ay_1$ **d.** 5
31.b. $x + y = -2, 7x - 7y = -10$

Class Exercises, pages 21–22 **1.a.** $f(3) = 5$;
$f(-3) = -25$ **b.** 2 **c.** $(0, -10), (2, 0)$
d. At the intersection with the horiz. axis **2.** a, b

3. $t(s) = -\frac{s}{2} + 4$

4.a, b.

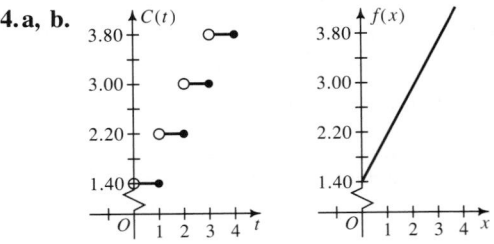

5. $f(a) - f(b) = m(a - b)$; $f(12.8) - f(12.3) = 25$

Written Exercises, pages 22–25 **1.a.** $f(2) = 1$;
$f(-2) = -2$ **b.** $\frac{2}{3}$ **3.** No **5.a.** 0

b. $g(x) = 0x + 2, h(x) = 0x - 1$ **7.a.** 1.5 **b.** $\frac{4}{3}$

c. x-int.: $\frac{4}{3}$, y-int.: -2 **9.** $f(n) = -2n + 12$

11. b. $f(x) = 2x + 3$ **13. a.** $D(t) = -0.16t + 4$
b. ≈ 25 min   **15. a.** $C(m) = 0.15m + 280$ **b.** 0.15
17. a. 51.8 mi/h; $d(t) = 185 + 51.8t$
b. $\{t | 0 \le t \le 4.25\}$ **c.** Slope is 51.8. **d.** They are =.
21. b. Answers may vary: $C(m) = 1.6m + 1.2$
23. $f(x) = \frac{25}{31}x + \frac{840}{31}$

Class Exercises, page 28   **1.** $7 + 2i$ **2.** $-1 - 5i$
3. 10   **4.** $28 - 6i$ **5.** 17   **6.** 34   **7.** 3
8. $a^2 + b^2$ **9.** $x = 2$, $y = -5$ **10.** If n is a pos.
multiple of 4, $i^n = 1$.
11. a. $\sqrt{4} \cdot \sqrt{9} = 2 \cdot 3 = 6 = \sqrt{36} = \sqrt{4 \cdot 9}$
b. $\sqrt{-4} \cdot \sqrt{9} = 2i \cdot 3 = 6i = \sqrt{-36} = \sqrt{-4 \cdot 9}$
c. $-6 \ne 6$ **d.** No

Written Exercises, pages 28–29   **1.** $7i$ **3.** -3
5. 2   **7.** $-2 + 5i$ **9.** $8 + 32i$ **11.** 37   **13.** 30
15. $31 - 34i$ **17.** $-9 - 40i$ **19.** $\frac{2}{29} - \frac{5}{29}i$
21. $\frac{12}{13} + \frac{5}{13}i$ **23.** $\frac{19}{51} + \frac{10\sqrt{2}}{51}i$ **25.** $-5i$ **27.** i
29. i **31.** i **33.** $x = 2$, $y = -3$
35. $(a + bi) + (a - bi) = (a + a) + (b - b)i = 2a$
37. a. Calculate 79^2 and compare with 6241.
b. Calculate $(3 - i)^2$ and compare with $8 - 6i$.
39. $\left[\frac{\sqrt{2}}{2}(1 + i)\right]^2 = \frac{1}{2}(2i) = i$ **41.** Real no.

Calculator Exercise, page 29
$42{,}660{,}913 + 30{,}082{,}416i$

Class Exercises, page 34   **1.** 16   **2.** 25   **3.** $\frac{49}{4}$
4. a^2 **5.** Complete square   **6.** Factor   **7.** Factor
8. Quad. formula   **9.** Quad. formula
10. Complete square   **11.** $\sqrt{b^2 - 4ac}$ is imag. if
$b^2 - 4ac < 0$.   **12.** $\sqrt{b^2 - 4ac} = 0$ if
$b^2 - 4ac = 0$, so (double) root is $-\frac{b}{2a}$.
13. $\sqrt{b^2 - 4ac}$ is a whole no. if $b^2 - 4ac$ is the
square of an integer.   **14.** No; $b = -\sqrt{5}$ isn't an
integer.   **15.** Yes; the derivation of the quad.
formula puts no restrictions (other than $a \ne 0$) on
the coefficients.   **16.** Dividing by $x + 4$ causes the
root -4 to be lost.   **17.** ± 6 **18.** $2, \frac{1}{4}$
19. Squaring introduces the extraneous root -2,
which does not satisfy the original equation.

Written Exercises, pages 35–36   **1.** $-1, \frac{7}{3}$
3. $2, -\frac{9}{2}$ **5.** $45, -35$ **7.** $34, -26$ **9.** $-3 \pm i$

11. $\frac{-1 \pm \sqrt{6}}{5}$ **13.** $2 \pm i$ **15.** 2, 8   **17.** $\frac{1}{2}, -\frac{7}{4}$
19. $-3, \frac{13}{3}$ **21.** $1, -\frac{5}{4}$ **23.** $0, \frac{1}{4}$ **25.** 0   **27.** 3
29. 16   **31.** 6   **33.** $\frac{-7 + \sqrt{1169}}{2}$ **35. a.** $64 - 16k$
b. $k = 4$ **c.** $k < 4$ **d.** $k > 4$ **e.** For example, $k = 0$,
$k = -12$, $k = -5$ **37.** $2\sqrt{2}, \frac{\sqrt{2}}{2}$ **39.** $3i$ **41.** $\frac{2}{9}$
43. (41) 0.22   (42) No sol.; No   **45. a.** 1, 2, 5, 7,
12 **b.** For example, 155, 222

Class Exercises, page 40   **1. a.** Upward
b. None   **2. a.** Downward   **b.** One
3. a. Downward   **b.** Two   **4.** $(5, 4)$ **5.** $(-2, -3)$
6. $(0, -10)$

Written Exercises, pages 41–43

1. a. 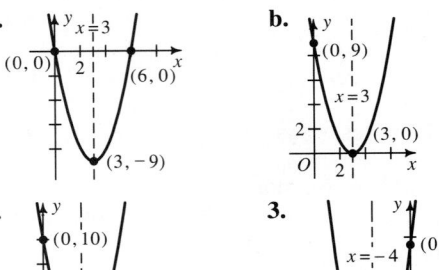   **b.**

c.   **3.**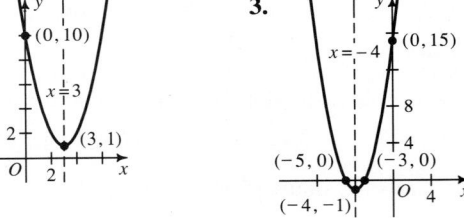

5.   **7.**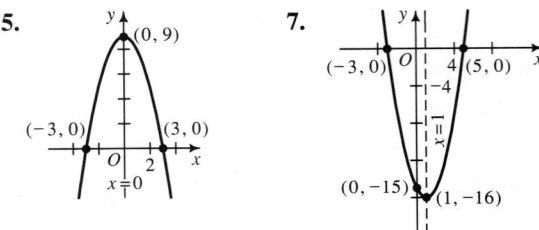

9. $V(1, -8)$; axis of sym. $x = 1$; x-ints. $1 \pm 2\sqrt{2}$;
y-int. -7 **11.** $V(1, -2)$; axis of sym. $x = 1$;
x-ints. $\frac{2 \pm \sqrt{2}}{2}$; y-int. 2   **13.** $V(-4, 0)$; axis of
sym. $x = -4$; x-int. -4; y-int. 8   **15.** $V(2, -1)$;
axis of sym. $x = 2$; x-ints. $\frac{3}{2}, \frac{5}{2}$; y-int. 15

17. $V(-1, 16)$; axis of sym. $x = -1$; no x-ints.; y-int. 17 **19.** $V(2, -9)$; axis of sym. $x = 2$; x-ints. $\frac{1}{2}, \frac{7}{2}$; y-int. 7 **21.** $(2, 0)$ **23.** $(-1, -5)$, $(-6, 0)$
25. $y = 2x - 5$ **27.** $f(x) = -3x^2 + 3x + 6$
29. $f(x) = -\frac{1}{2}x^2 + 4x$ **31.** $f(x) = \frac{7}{4}(x-3)^2 - 5$
33. $f(x) = \frac{3}{8}(x-4)^2$
35.a. $y = -\frac{29}{3600}(x-60)^2 + 30$ **b.** Yes
37.a. $f(1+k) = f(1-k) = 2k^2 + 5$ **41.a.** 2
b. $y = 2x - 1$

Class Exercises, page 45 **1.** Quad. **2.** Linear
3. Quad. **4.** Other

Written Exercises, pages 45–48
1. $f(x) = 2x^2 + 3x + 5$ **3.** $f(x) = \frac{1}{2}x^2 + 5x + 6$
5.b. Yes; \$8.75 **c.** No
7.a. $D(s) = -0.02s^2 + 1.6s + 2$ **b.** 21.5 mi
c. 55 mi/h **9.a.** 8 in. **b.** $f(x) = \frac{1}{2}x^2 - 5x + 20$
11.a. $h(t) = -4.9t^2 + 14t + 30$ **b.** $\frac{10}{7}$ s later
c. $\frac{30}{7}$ s later **13.** $\frac{10}{7}$ s
15.a. $R(x) = -100x^2 + 1400x + 24{,}000$ **b.** \$17
17.a. $f(t) = \pi L\left(\frac{D}{2} - t\right)^2$
b. $f\left(\frac{1}{4}D\right) = \frac{1}{4} \cdot f(0) = \frac{1}{16}\pi LD^2$

Chapter Test, pages 50–51 **1.a.** $2\sqrt{17}$
b. $(-3, -2)$ **2.** -3 **3.** $\left(-\frac{1}{2}, 1\right)$
4. $m = 2$, y-int.: $-\frac{7}{2}$ **5.** b$\|$c, a$\perp$b, a$\perp$c
6. $x + 3y = 0$ **7.** $4x + 3y = 35$
8. $5x + 3y = -15$ **9.** $x = 4$ **10.a.** Find the slope m and the midpt. M of the seg. Write an eq. of the line that contains M and has slope $-\frac{1}{m}$.
b. $3x - 4y = -4$ **11.a.** $f(t) = \frac{1}{9}t$
b. 207 min = 3 h 27 min **12.** $3i\sqrt{2}$
13. $-5 + 12i$ **14.** $\frac{2}{13} - \frac{3}{13}i$ **15.** $\frac{1}{2} + \frac{\sqrt{3}}{2}i$
16. $7 + 13i$ **17.** i **18.a.** $1, -\frac{2}{7}$ **b.** $2 \pm \sqrt{13}$
c. $-1 \pm i$ **19.** disc. = -20; $x = \frac{1 \pm i\sqrt{5}}{3}$

20.a. $V(1, 8)$;
   axis of sym. $x = 1$;
   x-ints. -1, 3;
   y-int. 6
b. $V(3, -4)$;
   axis of sym. $x = 3$
   x-ints. 1, 5
   y-int. 5

21.

22.a. $f(x) = 3x^2 - 11x + 35$ **b.** \$1.05

Class Exercises, page 55 **1.** Quad.; $5x^2$; 5; 2
2. Quartic; $5x^4$; 5; 4 **3.** Cubic; $-4x^3$; -4; 3
4. Quintic; $-2x^5$; -2; 5 **5.** 23 **6.** -13
7. $18n^2 + 5$ **8.** $2n^2 + 12n + 23$ **9.** $-5, 3$
10. $\pm 4, \pm 5i$ **11.** 3 **12.** $\frac{-b \pm \sqrt{b^2 - 4ac}}{2a}$
13. $0, -108$

Written Exercises, pages 56–57 **1.** Yes; $\frac{17}{3}$
3. Yes; none **5.** No; $-1, 4$ **7.** Yes; $-3, 0, 3$
9. Yes; $\frac{-q \pm \sqrt{q^2 - 4pr}}{2p}$ **11.** Yes; $-1, 0$
13.c. The first is undef. for $x = -2$; the second is never undef.
15.a. 13 **b.** $-2 - 10i$ **c.** $1 - i$ **d.** $18a^2 - 15a + 6$
17.a. $\frac{79\sqrt{2}}{27}$ **b.** $-12i\sqrt{3}$ **c.** $\frac{x^3}{27} - 3x$
d. $x^3 - 9x^2 + 18x$ **19.a.** 17 **b.** 75 **c.** -75
d. -183 **21.a.** 3 **b.** $-\frac{7}{3}$ **23.** -47.5 **25.** $\pm\sqrt{3}$
27.a. 1, 4 **b.** $f(x) = a(x-2)(x-3)$ for any real $a \neq 0$ **c.** $g(x) = ax(x-2)(x-3)$ for any real $a \neq 0$ **29.a.** 7 **b.** 7

Computer Exercises, page 58 **1.** Times may vary. The second program is faster. **2.** Faster than the first, but slower than the second

Class Exercises, page 60 **1.a.** 6 **b.** 4
2.a. $x^4 - 8x^2 + 5x - 1$; $x + 3$
b. $x^3 - 3x^2 + x + 2$; -7 **c.** -7
3.a. $2x^3 - 5x^2 + 8x - 4$; $x - \frac{1}{2}$ **b.** $2x^2 - 4x + 6$;
-1 **c.** -1 **4.** Yes; use the quadratic formula on the quotient of $P(x) \div (x + 1)$ **5.** If $P(a) = 0$, then by the remainder theorem, the remainder of $P(x) \div (x - a)$ is 0. Thus $(x - a)$ is a factor of $P(x)$. The converse is also true.

Written Exercises, page 61 **1.a.** -4 **b.** -2 **c.** 16
d. -16 **3.** $x^2 - x + 4$; 5

5. $x^3 - 3x^2 + 3x + 2$; 0
7. $x^4 + 3x^3 + 10x^2 + 30x + 91$; 273
9. $3x^2 - 8x + 21$; $-41x + 1$　**11.** $x - 1$ is; $x + 1$
is not.　**13.** c　**15.** $P(a) = a^n - a^n = 0$
17. $2x^3 - x^2 + 7x + 7$　**19.** $-1, \frac{1}{2}$　**21.** $\frac{1}{2}, 4$
23. $\pm i$

Class Exercises, page 65　**3.** (1) 2; double root
$x = 3$　(2) 3; double root $x = 3$　(3) 3; double roots
$x = 0$, $x = 3$　(4) 1; triple root $x = 2$　(5) 1; triple
root $x = 2$　(6) 2; triple root $x = 1$
5. a. $y = (x - 1)(x - 4)^2$　**b.** $y = -x(x + 1)(x - 3)$
c. $y = -(x - 1)^2$　**d.** $y = x(x - 2)^3$

Written Exercises, pages 66–68

1.
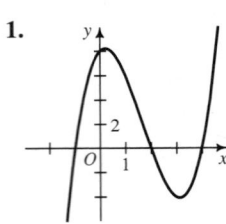

7.
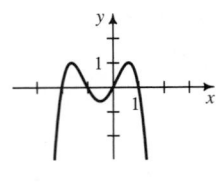

9.
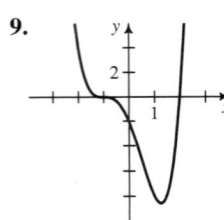

11.
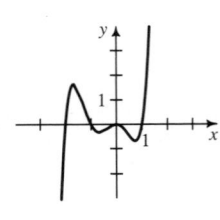

13. $x(x + 2)(x - 2)$　**15.** $x^2(x + 1)(x - 1)$
17. $(x + 1)(x - 1)^3$　**19. b.** $(-1, 1)$, $(0, 0)$, $(1, 1)$
21. $y = -(x + 3)(x + 1)(x - 1)$
23. $y = (x + 3)^2(x + 1)(x - 1)$　**25. b.** $P(2) = 24$,
$P(2.1) = 24.339$; highest point at $x \approx 2.3$
27. Intersect at $(-1, 3)$, $(0, 0)$, and $(1, -3)$
29. $y = -\frac{3}{8}(x + 4)(x - 2)$　**31.** $y = -\frac{1}{3}x(x + 2)^2$
33. d. Impossible　**35.** 0, 1, 2, 3, 4, 5, 6
37. $y = -2x^2(x - 3)$　**39.** $-(x + 3)(x + 1)(x - 2)$

Class Exercises, pages 70–71　**1. a.** Min.　**b.** 4
2. a. Max.　**b.** 2　**3. a.** Min.　**b.** $\frac{3}{2}$　**4. a.** Max.　**b.** $-\frac{2}{3}$
5. No; for example, $f(x) = x^3$.

7. Change line 20 to: 20 FOR X = 3 TO 5 STEP
0.01
Change line 40 to: 40 IF V > = M THEN 70
Change line 80 to: 80 PRINT "MINIMUM
VALUE IS APPROXIMATELY "; M; "AT
X = "; X1

Written Exercises, *Quadratic Functions*, pages 71–72
1. a. $A(x) = x(120 - 2x)$; Dom.: $\{x | 0 < x < 60\}$
b. 30　**3.** 433.5 m²　**5.** $\frac{x_1 + x_2}{2}$
9. a. 6 s　**b.** Dom.: $\{t | 0 \le t \le 6\}$　**c.** 45 m
11. $2.40　**13.** Width = 9, height = 6

Written Exercises, *Cubic Functions*, pages 72–74
1. b. Dom.: $\{x | 0 < x < 4\}$　**c.** $x \approx 1.64$ cm; max.
vol. ≈ 82.98 cm³　**3. b.** Dom.: $\{x | 0 < x < 2.5\}$
c. $x \approx 1.67$ ft; max. vol. ≈ 9.26 ft³
5. a. $V(x) = \frac{x^2(25 - x)}{4\pi}$; Dom.: $\{x | 0 < x < 25\}$
b. $x \approx 16.67$ cm; max. vol. ≈ 184.21 cm³
7 a. $V(x) = 2\pi x(25 - x^2)$; Dom.: $\{x | 0 < x < 5\}$
b. $x \approx 2.89$; max. vol. ≈ 302.30　**9.** $r \approx 3.33$;
$h \approx 3.33$; max. vol. ≈ 116.36

Calculator Exercises, page 74　**1.** 17.6 cm for
circle, 22.4 cm for square　**2.** Use all of the wire
for the circle.

Class Exercises, page 77　**1.** Between 1 and 2
2. Between 1 and 2　**3.** Between -1 and 0, or 2
and 3　**4.** Between -4 and -5, or -1 and 0, or 0
and 1　**5.** Ans. may vary. $x \approx -2.2$, 0.4, 2.8
6. There is no change in sign at a double root.

Written Exercises, pages 78–79　**1.** 1.5
3. $-1, -0.7, 2.7$　**5.** ± 2.2　**7.** 5.7　**9.** 0, 1.6
11. Quot.: $x^2 - 3x + 1$; $x = \frac{3 \pm \sqrt{5}}{2} \approx 0.38, 2.62$
15. 0.46

Class Exercises, page 83　**1.** $-4, \pm 3$　**2. a.** Let
$y = x^2$. Then the eq. becomes $y^2 - 5y + 4 = 0$.
b. Solve for y; substitute x^2 for y and solve for x.
3. a, b　**4. a.** Factor out x; factor (or use quad.
formula) on the remaining quad.　**b.** Cubic eqs.
with no constant term.　**5. a.** $\pm 4, \pm 2, \pm 1, \pm \frac{1}{2}$
b. $\pm 5, \pm 1, \pm \frac{5}{2}, \pm \frac{1}{2}, \pm \frac{5}{3}, \pm \frac{1}{3}, \pm \frac{5}{6}, \pm \frac{1}{6}$
6. No. There must be a real root, but it may be
irrational.　**7. a.** $P(-2) = -1 < 0$; $P(-1) = 1 > 0$
b. Irrational　**c.** $P(x) > 1$ for all $x > 0$

Written Exercises, pages 83–85 **1.** b; $\pm\sqrt{6}$, $\pm i\sqrt{2}$ **3.** a; ± 2, $\frac{16}{3}$ **5.** b; $\pm\frac{\sqrt{6}}{2}$, $\pm i\sqrt{5}$ **7.** a; -2, $\pm\sqrt{6}$ **9.** a; $\frac{3}{5}$, $\pm\frac{\sqrt{2}}{2}i$ **11.** $-1, 2$, $-1\pm i\sqrt{3}$, $\frac{1}{2}\pm\frac{\sqrt{3}}{2}i$ **13.** ± 1 **15.** ± 1, ± 3 **17.** -1, $\frac{1}{3}$, 2 **19.** -1, $\pm\sqrt{3}$ **21.** $\frac{2\pm\sqrt{2}}{2}$, -2 **23.** $f(x)=(x+3)(x+2)(x-3)$ **25.** $h(x)=(x+2)(x-2)(x+\sqrt{7})(x-\sqrt{7})$ **27.** $m(x)=(x-2)(2x+1)(2x+3)$ **31.** Intersect at $(2, 2)$; tang. at $(-1, 2)$ **33.** Tang. at $(\sqrt{2}, 4)$, $(-\sqrt{2}, 4)$ **35.** Intersect at $(2, 24)$; tang. at $(-3, 9)$ **39. b.** 2, $-1+\sqrt{6}$ **41.** $7\text{ cm}\times 10\text{ cm}\times 16\text{ cm}$

Class Exercises, page 89 **1.** $\sqrt{3}-2i$ **2.** $-3-\sqrt{2}$, $2i$ **3. a.** sum $=-\frac{5}{2}$; prod. $=-\frac{3}{2}$ **b.** sum $=\frac{1}{2}$; prod. $=\frac{3}{2}$ **c.** sum $=-2$; prod. $=0$ **d.** sum $=0$; prod. $=-32$ **4.** The graph must cross the x-axis at least once. **5.** The graph may lie entirely above or below the x-axis.

Written Exercises, pages 89–91 **1.** F **3.** F **5.** F **7.** T **9.** sum $=\frac{3}{4}$; prod. $=\frac{3}{2}$ **11.** sum $=-\frac{5}{3}$; prod. $=\frac{2}{3}$ **Exs. 13, 15:** Ans. may vary. **13.** $x^2-2x+2=0$ **15.** $x^2-6x+7=0$ **17.** -4, $2-i\sqrt{5}$ **Exs. 19–23:** Ans. may vary. **19.** $x^3-10x^2+33x-34=0$ **21.** $4x^3-12x^2+3x+19=0$ **23.** $x^4-10x^3+29x^2-10x+28=0$ **25.** $c=-6$, $d=-4$ **27.** -27 **39.** $-3-i$

Chapter Test, page 93 **1. a.** 20 **b.** 0 **c.** $-2-9i$
2. -2
3. $x^3+2x^2+2x+17$
4. -2 (double)
5. See graph at right.
6. $y=\frac{1}{2}(x+2)(x-1)(x-3)$
7. 800 m^2

8. a. Use the location principle: $P(1)=-2<0$ and $P(2)=7>0$. **b.** -0.6, 1.6 **9.** $\frac{3}{2}$, $\pm\sqrt{3}$
10. $-\frac{2}{3}$, 1, $\pm i\sqrt{2}$ **12.** sum $=\frac{5}{2}$; prod. $=-4$
13. Ans. may vary. $x^3+5x^2+9x+45=0$
14. If an imag. number is a double root, then so is the number's conjugate. This gives the cubic eq. at least four roots, when it can only have three.

Class Exercises, page 98 **1.** $x>-4$ **2.** $x<4$ **3.** $x>-5$ **4.** $-2<x<2$ **5.** $3<x<7$ **6.** $x<-4$ or $x>-2$ **7.** f **8.** a **9.** e **10.** c **11.** b **12.** d

Written Exercises, pages 98–99 **1.** $x<3$ **3.** $x<0$ **5.** $x\le 11$ **7.** $x>\frac{1}{5}$ **9.** $x<4$ **11.** $-3<x<3$ **13.** No sol. **15.** $1<x<7$ **17.** $x\le -10$ or $x\ge -4$ **19.** $x=4$ or $x=12$ **21.** $-\frac{1}{2}\le x\le\frac{9}{2}$ **23.** $-\frac{17}{4}\le x\le\frac{1}{4}$ **25.** $15\frac{5}{9}°\le C\le 18\frac{1}{3}°$ **27.** $1\le x\le 3$ or $5\le x\le 7$ **29.** $5<x<9$ **31.** $x<0$ or $x>\frac{1}{2}$ **33.** $0\le x\le 2$

Class Exercises, page 103 **1.** $-2<x<2$ or $x>4$ **2.** $x\le 1$ **3.** $1<x<2$ or $x>4$ **4.** $x<2$ or $4<x<7$ **5.** $x\le 5$ **6.** No sol.

Written Exercises, pages 103–104 **1.** $x<-4$ or $x>3$ **3.** $1<x<2$ or $x>4$ **5.** $x<4$, $x\ne 2$, $x\ne 3$ **7.** $-3<x<5$ **9.** $x\le -1$ or $x\ge 1.5$ **11.** $-3.5\le x\le 1$ **13.** $x<-\sqrt{5}$ or $x>\sqrt{5}$ **15.** $a>2$ **17.** $n<-3$ or $1<n<2$ **19.** $x<-2$ or $\frac{1}{2}<x<1$ **21.** $-2<y<2$ **23.** $x<-1$ or $x>1$ **25.** $x<3$ or $4<x<5$ **27.** $-4<x\le\frac{5}{2}$ or $x>7$ **29.** All real nos. **31.** $-2<x<-1.73$ or $x>1.73$ **33.** $-0.41\le x\le 0.50$ or $x\ge 2.41$ **35.** $x<-4$ or $x>1$ **37.** All real nos. **39.** $x\ge 0.70$ **41.** $-4.12\le x\le -3$ or $3\le x\le 4.12$ **43.** No sol.

Class Exercises, page 106 **1.** Above **2.** Below **3.** On **4.** On **5.** Above **6.** Below **7. a.** $x>0$, $y>0$ **b.** II: $x<0$, $y>0$; III: $x<0$, $y<0$; IV: $x>0$, $y<0$ **8.** $x\ge 0$, $y\ge 0$, $y\le -x+2$ **9.** $0\le y\le 3$, $x\ge 0$, $y\le 4-2x$ **10.** $y\ge x^2$, $y<x+2$

Written Exercises, pages 106–108 **1.** Pts. on or above the line $y=x$ **3.** Pts. below the line $3x+4y=12$ **5.** Pts. on or below the parabola $y=x^2$ **7.** Pts. below the parabola $y=2x^2-4x+1$ **9.** Pts. on or between the lines $x=0$ and $x=2$ **11.** Pts. below the line $y=-1$ or above the line $y=1$ **13.** Pts. between the lines $x=1$ and $x=5$ **15.** Pts. above the curve $y=x^3-9x$

17.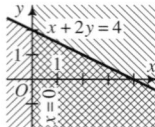
$x + 2y = 4$

19.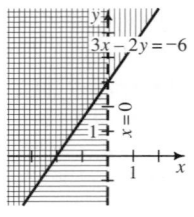
$3x - 2y = -6$
$x = 0$

21.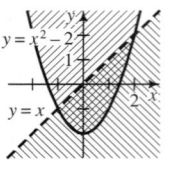
$y = x^2 - 2$
$y = x$

25.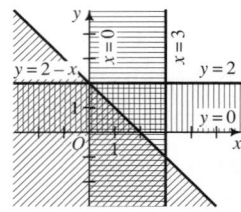
$y = 2 - x$
$x = 0$
$x = 3$
$y = 2$
$y = 0$

31. $1.5 \leq x \leq 5.5$
$10x + 30 \leq y \leq 10x + 50$

33.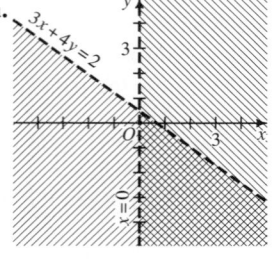
$(3, 8)$
$(8, 8)$
$\left(3, \frac{13}{2}\right)$
$\left(\frac{16}{3}, \frac{16}{3}\right)$

Written Exercises, pages 112–114 **1.a.** 5 consoles, 4 portables **3.** 5 consoles, 3 portables **5.** $(0, 2)$
7.a. $(7, 3)$ **b.** $(2, 8)$, $(7, 3)$ **9.** 300 tons alfalfa, 200 tons corn **11.a.** 12 racing skis, 30 free-style skis **b.** A max. profit of $960 occurs for 12 racing skis and 30 free-style skis, or for 20 racing skis and 18 free-style skis. **13.a.** 13 **b.** 32 **c.** 18

Chapter Test, page 115 **1.a.** $x > 3$ **b.** $x \leq 6$
2.a. $-2 < x < 8$ **b.** $x < -10$ or $x > -2$
c. $x = 0$ or $x = -\frac{8}{3}$ **d.** $1 \leq x \leq \frac{9}{5}$ **3.a.** $x \leq -4$ or
$x \geq \frac{1}{2}$ **b.** $-2 < x < \frac{3}{4}$ **c.** $-\frac{1}{2} < x < 0$ or $x > 1$
d. $-1 \leq x < 3$ or $3 < x \leq 6$ **4.** The pts. below the line $y = 2x - 5$ **5.a.** Pts. on or below the line $2 - 3y = 3x$ **b.** Pts. between the lines $x = -3$ and $x = 3$

6.a.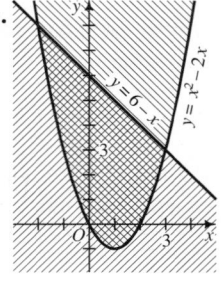
$3x + 4y = 2$
$x = 0$

b.
$y = 6 - x$
$y = x^2 - 2x$

7. $1 \leq x \leq 3$, $y \geq 2$, $y \leq -\frac{1}{2}x + \frac{9}{2}$ **8.** 20 trucks, 10 automobiles

Cumulative Review, page 116 **1.** $AB = \sqrt{58}$;
midpt. $= \left(-\frac{11}{2}, \frac{3}{2}\right)$ **2.a.** $x - y = 6$
b. $2x - 3y = -15$ **3.a.** $P(d) = 12.5d - 1027$
b. 123 **4.a.** $-2i$ **b.** $-12\sqrt{3}$ **c.** $\frac{2}{5} - \frac{3}{10}i$
5.a. $3 \pm 3i$ **b.** $\frac{3 \pm \sqrt{2}}{2}$ **c.** 4, 12
6. See graph below. **7.** $f(x) = -2x^2 + 20x - 42$
8. $f(x) = \frac{1}{4}x^2 + x - 6$ **9.a.** 15 **b.** 0 **c.** -6
d. $4 - 2i$ **10.a.** 2, -3 **b.** See graph below.

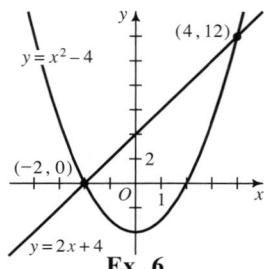
$y = x^2 - 4$
$(4, 12)$
$(-2, 0)$
$y = 2x + 4$
Ex. 6

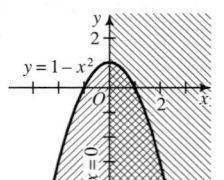
$y = x^3 + 2x^2 - 5x - 6$
Ex. 10.b.

11. $y = \frac{1}{6}(x + 2)^2(x - 3)$ **12.** 625 cm^2
13. -1.5, -0.3, 1.9 **14.** $\pm 2i$, 3
15. $3x^3 - 11x^2 + 20x + 8 = 0$, for example

16.a. $x < -7$ or $x > 17$ **17.**
b. $x \leq -3$ or $x \geq 1$

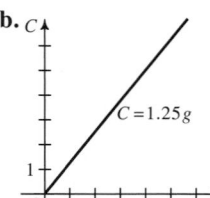
$y = 1 - x^2$
$y = 0$

Class Exercises, page 122 **1.a.** $g \geq 0$; $C \geq 0$
b.
$C = 1.25g$

2.a. $x \neq 2$ **b.** $t \geq 0$ **c.** $s \geq 4$
3.a. $f(2) = 3 \cdot 2^2 - 12 = 0$
b. -2 **c.** $f(x) \geq -12$
4. Dom.: all real nos.;
Range: $\{y | y \geq -2\}$;
zeros: -3, 0, 4

5. $x^2y = 4x - y$ passes; $y^2 = 4x^2$ fails.
6. $y^2 - x = 1$ **7.** Yes **8.** Yes **9.** No **10.** No
11. No; fails vert. line test **12.** Yes; passes vert. line test **13.** No; fails vert. line test

Written Exercises, pages 122–124   **1.** Yes; Dom.: $\{x|-2 \le x \le 5\}$; Range: $\{y|-1 \le y \le 2\}$ **3.** Yes; Dom.: $\{0, 2, 4, 6\}$; Range: $\{0, 1, 2, 3\}$ **5.** No **7.** $y = \pm\sqrt{1 - x^2}$, so there are two y-values for each x-value.   **9.a.** $x \ne 0$ **b.** $x \ne 9$ **c.** $x \ne \pm 2$ **11.a.** Dom.: all real nos.; Range: $\{y|y \ge 0\}$; zero: 0 **b.** Dom.: all real nos.; Range: $\{y|y \ge 0\}$; zero: 2 **c.** Dom.: all real nos.; Range: $\{y|y \ge -2\}$; zeros: ± 2 **13.** Range: $\{y|y \ge -1\}$; zeros: 2, 4 **15.** Range: all real nos.; zero: 2 **17.** Range: $\{h(u)| -2 < h(u) \le 4\}$; zeros: 0, 2 **19.a.** 36π **b.** Dom.: all solids; Range: nonnegative real nos.   **21.b.** Dom.: all real nos.; Range: all integers   **25.** d

Class Exercises, pages 127–128   **1.a.** 5 **b.** 8 **2.a.** ± 3 **b.** $-3 < x < 3$

4.a.
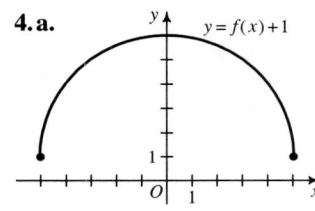

5. $x^2 + 2x + 1$
6. $x^2 - 1$
7. $x^3 + 2x^2 + x$
8. x, for $x \ne -1$
9.a. 12
  b. $x^2 + 3x + 2$
10.a. 7
  b. $x^2 + x + 1$

12. $F(M(X)); \; M(F(X))$ **13.** $B \circ F$

Written Exercises, pages 128–131

1.
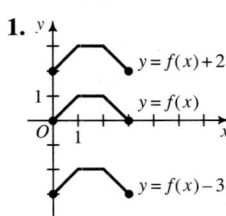

3.a. 1 **b.** Pos.: $0 < x < 4$; neg.: $-1 \le x < 0$ or $4 < x \le 6$; zero: $x = 0, 4$ **c.** 2   **5.** $x^3 + x - 2$ **7.** $x^4 - x^3 - x + 1$ **9.a.** 0 **b.** $x^3 - 3x^2 + 3x - 2$ **15.** 0.64

17.a. $f(g(x)) = g(f(x)) = x$ **b.** $f(h(x)) = 6x + 1$; $h(f(x)) = 6x - 7$ **19.a.** 3 **b.** $\sqrt{2x - 3}$ **21.a.** 0 **b.** $\dfrac{\sqrt{6x - 3}}{3}$ **23.** $j(f(h(x)))$ **25.** $f(j(h(x)))$ **27.a.** 1.5 m/s; 9 L/min **b.** 30 s **29.a.** $r = \dfrac{C}{2\pi}$ **b.** $A = \dfrac{C^2}{4\pi}$ **31.** $s(F) = 331 + \dfrac{1}{3}(F - 32)$ **33.** $(f \circ g)(x) = 2\sqrt{16 - x^2}$, Dom.: $\{x| -4 \le x \le 4\}$; $(g \circ f)(x) = 2\sqrt{4 - x^2}$, Dom.: $\{x|-2 \le x \le 2\}$

35. $(f \circ g)(x) = 1 - x$, Dom.: $\{x|x \le 1\}$; $(g \circ f)(x) = \sqrt{1 - x^2}$, Dom.: $\{x|-1 \le x \le 1\}$ **37.** 2.5, 2.75, 2.875   **39.a.** $E = \dfrac{130}{(t + 1)^2}$ **b.** 9 s

Class Exercises, pages 135–136   **2.a.** $x = 3$ **b.** $y = \pm x$; $(0, 0)$ **c.** $x = 0$, $y = 0$; $(0, 0)$ **3.a.** i, ii, iii, iv **b.** iv **c.** iv   **4.** In the origin **5.** No   **7.a.** $x = 4$ **b.** $x = 1$ **c.** $x = 0$ **8.a.** $(2, 1)$ **b.** $(-1, -5)$ **c.** $(0, 7)$

Written Exercises, pages 136–138

1.c.
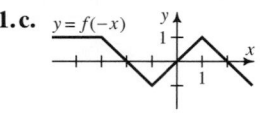

3.b.
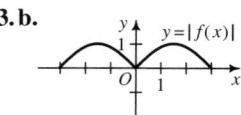

5.
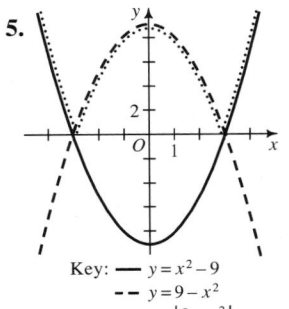
Key: — $y = x^2 - 9$
   -- $y = 9 - x^2$
   .... $y = |9 - x^2|$

7. $y = \dfrac{x + 4}{3}$

9.
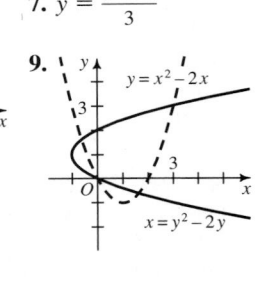

11. $y = \sqrt[3]{x}$ **13.** $x = |y| + 2$ or $y = \pm x \mp 2$ for $x \ge 2$ **15.a.** iv **b.** i, ii, iii, iv **c.** iv

19.
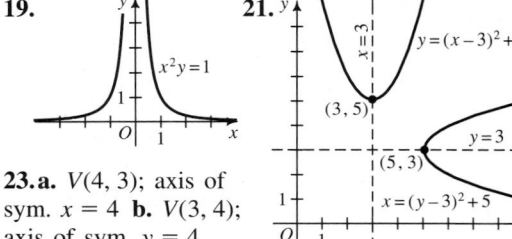

21.

23.a. $V(4, 3)$; axis of sym. $x = 4$ **b.** $V(3, 4)$; axis of sym. $y = 4$ **25.** $V(-1, -3)$; axis of sym. $y = -3$ **27.** $(-5, 11)$ **29.a.** $(0, 0)$ **b.** Pt. of sym. $(1, 2)$; local max. $(2, 4)$ **31.a.** Even **b.** Odd **c.** Neither **d.** Even **e.** Neither **f.** Odd   **33.a.** In the y-axis **b.** In the origin **37.a.** $m > 0$; $m < 0$; $m = 0$ **39.a.** dec. **b.** dec. **c.** inc.

Class Exercises, page 142   **1.a.** 4 **b.** $\dfrac{1}{2}$ **c.** $f(25) = 1$; $f(-25) = 0$ **2.** $p = 4$, $A = 1$; $p = 4$, $A = \dfrac{1}{4}$ **3.** $p = 2$, $A = \dfrac{1}{2}$; $p = 8$, $A = \dfrac{1}{2}$

4. $p = 4$, $A = \frac{1}{2}$; $p = 4$, $A = \frac{1}{2}$

5. If $(a, 0)$ is on $y = f(x)$, then $(a, 0)$ is on $y = c \cdot f(x)$, since $c \cdot f(a) = c \cdot 0 = 0$.

6. If $(0, a)$ is on $y = f(x)$, then $(0, a)$ is on $y = f(cx)$, since $f(c \cdot 0) = f(0) = a$.

Written Exercises, pages 143–146 **1.** Yes; $p = 6$, $A = 1$; $f(1000) = -1$, $f(-1000) = 1$ **3.** Yes; $p = 3$, $A = \frac{1}{2}$; $f(1000) = 2$, $f(-1000) = 3$

5.c.

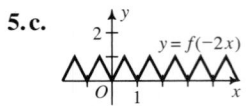

7.e.

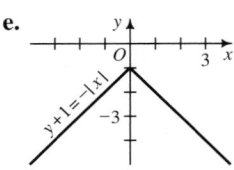

9.d.

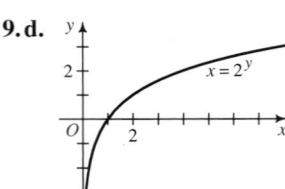

11. $(x - 8)^2 + (y - 4)^2 = 9$ **13.a.** x-ints. 0, 3; local max. (2, 32) **b.** x-ints. 0, 6; local max. (4, 64) **c.** x-ints. 2, 8; local max. (6, 32) **d.** x-ints. -2, 4; local max. (2, 32) **e.** $f(x) = -x^2(x - 6)$
15.a. $y = |x - 4| + 2$ **b.** $y = -|x| + 4$

c. $y = \frac{1}{2}|x| - 2$ **d.** $y = -2|x|$ **e.** $y = 2|x + 4|$

Class Exercises, pages 148–149 **1.** No; no **2.a.** 2
b. 3 **c.** 2 **3.a.** $g^{-1}(x) = \frac{x}{4}$ **b.** $g^{-1}(x) = \frac{x - 2}{3}$

c. $g^{-1}(x) = \frac{x + 1}{2}$ **d.** $g^{-1}(x) = \frac{4 - x}{5}$ **4.a.** (8, 2),
(1, 1), $(-1, -1)$, $(-8, -2)$ **b.** $f^{-1}(x) = \sqrt[3]{x}$
5.a. Yes; it passes the horiz. line test. **b.** No; $g(-1) = g(1) = 0$. **6.** Solve for the desired variable. **7.** g; g **8.** b; d **9.** No; $T(A) = T(B) = 2$, so T is not one-to-one.

Written Exercises, pages 149–150 **1.a.** 2 **b.** 3 **c.** 7
3. g is not one-to-one. **5.b.** $h^{-1}(x) = \frac{x + 3}{4}$

7. Yes **9.** No **11.** Yes; $f^{-1}(x) = \frac{x + 5}{3}$;

$f(f^{-1}(x)) = 3\left(\frac{x + 5}{3}\right) - 5 = x$; $f^{-1}(f(x)) =$

$\frac{(3x - 5) + 5}{3} = x$ **13.** Yes; $f^{-1}(x) = x^4$, $x \geq 0$
15. No **17.** No **19.** Yes; $f^{-1}(x) = \sqrt[3]{x^3 - 1}$

20.

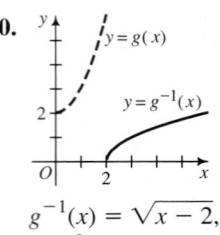

$g^{-1}(x) = \sqrt{x - 2}$,
$x \geq 2$

22. $g^{-1}(x) = 1 - \sqrt{x - 1}$, $x \geq 1$

24.

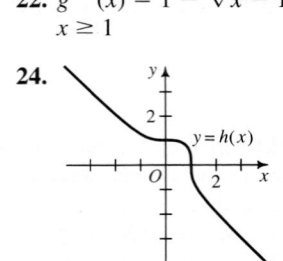

29. ii **31.** $P^{-1}(x) = \dfrac{-b - \sqrt{b^2 - 4a(c - x)}}{2a}$, $b^2 - 4a(c - x) \geq 0$

Calculator Exercises, page 150 **1.** f and g are inverse functions. **2.** It is. **3.** $x > 0$

Class Exercises, pages 153–154 **1.** Area of a rect. is a function of length and width. **2.** Dist. is a function of rate and time. **3.** Vol. of a cone is a function of base radius and ht. **4.** Force is a function of mass and acceleration. **5.** Density is a function of mass and vol. **6.** Surface area of a cylinder is a function of base radius and ht.
7. First-quadrant rays with vertices at the origin.
8. Ans. may vary. **a.** Tampa, San Francisco
b. Chihuahua, Ciudad Juarez, Phoenix **c.** Quebec, Montreal, Boston

10.a. 5; 5; 5 **b.**
11. For example:
$(-1, -1)$,
$(0, 0)$,
$(2, 2)$

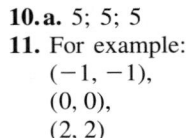

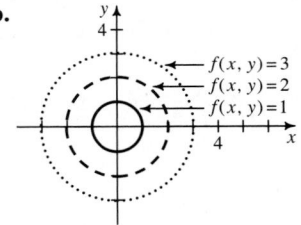

Written Exercises, pages 154–157 **1.a.** r, t
b. $A = 100(1.12)^t$, $A = 100(1.16)^t$ **c.** 9 yr; 6 yr; 5 yr **3.a.** Wind-chill temp. is a function of recorded temp. and wind speed. **b.** 16 mi/h **c.** The latter *feels* warmer. **d.** Between 5 mi/h and 10 mi/h
5.a. $A(8, 3) = 12$; $A(16, 6) = 48$ **b.** Area of a $\triangle$ is a function of base and ht. **c.** For example: (2, 5), (1, 10), (20, 0.5) **7.b.** If the base and ht. of a $\triangle$ are each tripled, then the area is multiplied by 9. **9.a.** $d(r, t) = rt$ **13.** $\{(x, y)|x \neq \pm y\}$
17.a. $A(4, 3, 5) = 94$; $A(6, 4, 7) = 188$ **b.** Surface area of a rect. prism is a function of length, width, and height. **21.** No; $R(37, 10, 8.5) \approx 5.85 > 5.5$

Class Exercises, pages 160–161 **1.a.** $d(s) = s\sqrt{2}$

b. $s(d) = \frac{d\sqrt{2}}{2}$ **c.** $A(d) = \frac{d^2}{2}$ **2.a.** $V(e) = e^3$

b. $V(d) = \frac{d^3\sqrt{3}}{9}$

3.a. $AP + PB = \sqrt{16 + x^2} + 8 - x$
b. $\{x|0 \le x \le 8\}$ **4.a.** $d(x, y) = \sqrt{(x - 2)^2 + y^2}$
b. $d(x) = \sqrt{(x - 2)^2 + x^4}$ **c.** With a computer or graphing calculator, find the y-value of the lowest point on the graph of $y = d(x)$. **5.** $d = 10t$

Written Exercises, pages 161–165 **1.** $A(h) = \frac{h^2\sqrt{3}}{8}$

3. $t(n) = \frac{11n}{36}$ **5.** $P(n) = 6n$ **7.** $A(h) = \frac{5}{8}\pi h^2$

9. $s(d) = 1.5d$ **11.** $V(w) = \frac{3w - 2w^3}{4}$

13.a. $C(t) = \pi t$ **b.** $A(t) = \frac{\pi}{4}t^2$

15.a. $C(w) = \frac{8w^3 + 192}{w}$ **b.** $\approx\$126$

17.b. 2:06 P.M. **c.** 4 km **19.a.** $V(h) = \frac{\pi}{48}h^3$

b. $h(t) = \sqrt[3]{\frac{240t}{\pi}}$

21.a. $d(x) = \sqrt{5x^2 - 40x + 100}$ **b.** Dom.: all real numbers; Range: $\{d(x)|d(x) \ge 2\sqrt{5}\}$
23.a. $P(x) = -2x^2 + 4x + 18$ **b.** $\{x|0 < x < 3\}$

c. 1 **25.a.** $t(x) = \sqrt{2500 + x^2} + \frac{1}{3}(100 - x)$

b. ≈ 80.5 s **27.a.** $C(x) = \frac{12x^2 + 480}{x}$ **b.** $\approx\$152$

29.a. $V(r) = \frac{2\pi r^4}{3(r^2 - 1)}$ **b.** $r \approx 1.41$, $h \approx 4.02$

Chapter Test, pages 166–167 **1.** Dom.:
$\{x| -3 \le x \le 3\}$; Range: $\{f(x)|0 \le f(x) \le 3\}$;
zeros: ± 3

2.a.

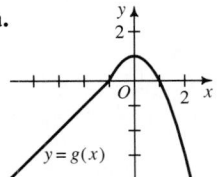

$y = g(x)$

b. Range: $\{g(x)|g(x) \le 1\}$;
zeros: ± 1
3.a. $x^2 + 3x + 2$
b. $x^2 + x - 2$
c. $x^3 + 4x^2 + 4x$
d. x, for $x \ne -2$

4.a. $x^2 + 6x + 8$
b. $x^2 + 2x + 2$
5. iv

6.a.

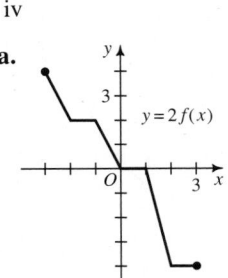

$y = 2f(x)$

b.

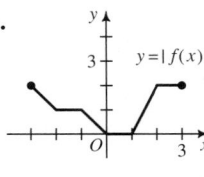

$y = |f(x)|$

c.

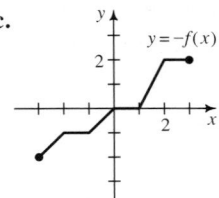

$y = -f(x)$

d.

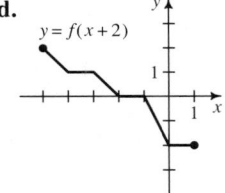

$y = f(x + 2)$

7.a. 4
b. max. 4; min. 1
c. $\frac{3}{2}$

8. The translated graph coincides with the original graph. Since for all x, $f(x + p) = f(x)$, the graphs of $y = f(x)$ and its horiz. translation $y = f(x + p)$ are identical. **9.a.** g; $g^{-1}(x) = x - 3$ **b.** f does not have an inverse because f is not one-to-one; for example, $f(-1) = f(1) = 2$.

10.a. $A(b, h) = \frac{1}{2}bh$ **b.** $A(3, 4) = 6$; $A(6, 5) = 15$

c.

h

11. $h(t) = \frac{5t}{2\pi}$

12.a. $A(x) = x(9 - x^2)$
b. $\{x|0 < x < 3\}$
c. ≈ 10.4

Class Exercises, page 172 **1.a.** 10.5% **b.** 1.03;
1.15; 1.046; 2.2 **2.a.** 32% **b.** 0.88; 0.925; 0.20; 0
3.a. $200(1.05)^t$ **b.** $20(1.08)^t$ **c.** 6%
4.a. $9800(0.8)^t$ **b.** $2200(0.85)^t$ **c.** 25% **5.a.** $\frac{1}{8}$

b. $\frac{1}{64}$ **6.a.** $\frac{3}{2}$ **b.** $\frac{9}{4}$ **7.a.** $\frac{4}{9}$ **b.** $\frac{1}{144}$ **8.a.** 8 **b.** $\frac{4}{3}$
9. 8 **10.** 6^n **11.** -2 **12.** $x^2 + 1$ **13.** $18a^{10}$
14. $3a^4 - 6a^7$ **15.** $5b^5 + 10b^8$ **16.** $50a^3b^8$

Written Exercises, pages 173–175 **1.a.** $\frac{1}{16}$ **b.** $-\frac{1}{16}$

3.a. $\frac{5}{8}$ **b.** $\frac{1}{1000}$ **5.a.** x^2 **b.** 1 **7.a.** $\frac{a}{3}$ **b.** $\frac{a}{3a + 1}$

9.a. $\frac{8}{3}$ **b.** 32 **11.a.** $\frac{ab}{b - a}$ **b.** ab **13.** $\$1214$;
$\$4910$ **15.** $\$1.97$; $\$3.87$ **17.** $\$27,422$; $\$11,569$
19. $\$0.83$; $\$0.69$ **21.** $\frac{81}{a}$ **23.** 3^6n^{12} **25.** $\frac{16}{a^4}$

27. $\frac{b^{10}}{a^5}$ **29.** $2x^2 - 4$ **31.** $2 + 3a^4$

33. 8000; 6400; 5100; 4100; 3300
35. $112.68   **37.a.** 3 **b.** 5 **c.** $\sqrt{25^{n+7}}$ **39.a.** b^n
b. b^{-2} **c.** b^{1-2n} **41.a.** $\frac{4}{}$ **b.** $\frac{1}{}$ **43.a.** $\frac{y+x}{}$
b. $\frac{1}{y+1}$ **45.** $\approx$55,000,000,000   **47.** $1, -1$
49. $-1, 1$

Class Exercises, page 177   1.a. 2 **b.** $\frac{1}{2}$ **2.a.** 8
b. $\frac{1}{8}$ **3.a.** -3 **b.** $-\frac{1}{3}$ **4.a.** 15 **b.** $8 + 2\sqrt{15}$
5. $\frac{5}{7}$ **6.** $\frac{8}{27}$ **7.** 2   **8.** 64   **9.** $\frac{8}{x}$ **10.** $\frac{5}{x^2}$ **11.** $\frac{x}{2}$
12. $8x$ **13.a.** $62 **b.** $2.5 **c.** -0.75 **14.** 6   **15.** $\frac{5}{2}$
16. 27   **17.** $\frac{1}{16}$

Written Exercises, pages 178–180   1.a. $\sqrt[3]{x^2}$
b. $\sqrt{x^3}$ **c.** $\sqrt{\frac{5}{x}}$ **d.** $\sqrt[3]{6x^2}$ **3.a.** $x^{5/2}$ **b.** $y^{2/3}$
c. $(2a)^{5/6}$ **d.** x **5.a.** $\frac{3}{5}$ **b.** $\frac{5}{3}$ **c.** $\frac{243}{3125}$ **d.** $\frac{125}{27}$ **7.** $\frac{1}{8}$
9. 36   **11.** $\frac{a^4}{4}$ **13.** $2x^2$ **15.a.** $182 **b.** $108
17. $a^2 - 2a$ **19.** $x^2 - 2x$ **21.** $x - 2$ **23.** $n - 2$
25. $3n - 4n^2$ **27.** $2\sqrt{a}$ **29.** 2   **31.** 2   **33.** $\frac{25}{3}$
35.a. $\frac{1}{32}$ **b.** $\frac{1}{2}$ **c.** $-\frac{31}{4}$ **37.** 12%   **39.** 4.3%
41. $\approx$7%; $24,400   **43.a.** $\frac{81}{4}$ **b.** 3   **45.** $-2, 3$
47. -2 **49.a.** $a^{1/2}b^{1/2}(a - b)$ **b.** $a^{1/2}b^{-1/2}(1 - ab)$
51.a. $(x^2 + 1)^{1/2}$ **b.** $2(x^2 + 2)^{-1/2}$
53. D $220 \cdot 2^{5/12}$; D$^{\#}$ $220 \cdot 2^{6/12}$; E $220 \cdot 2^{7/12}$;
F $220 \cdot 2^{8/12}$; F$^{\#}$ $220 \cdot 2^{9/12}$; G $220 \cdot 2^{10/12}$;
G$^{\#}$ $220 \cdot 2^{11/12}$; A 440 **55.** 27, 64   **57.** $\frac{1}{2}$

Class Exercises, pages 182–183   1. 7   **2.** $a = 5$;
$b = 3$ **3.** $b > 1$; $0 < b < 1$ **4.** Dom.: all real
numbers; Range: $\{y \mid y > 0\}$ **5.** 125;
$5\pi \approx 156.9925$; diff. ≈ 32 **6.** 8 yr; 1 yr
7.a. $1000 **b.** 10 yr   **8.a.** 90 **b.** 8 hr   **9.a.** 1000
b. 500 **c.** 4 days   **10.** Yes

Written Exercises, pages 183–186   1. $6\pi \approx 278.4$;
$\pi^6 \approx 961.4$ **3.** $f(x) = 3 \cdot 5^x$ **5.** $f(x) = 64\left(\frac{1}{4}\right)^x$
7.a. 1.6 kg **b.** 0.8 kg **c.** 0.1 kg **d.** $3.2\left(\frac{1}{2}\right)^{t/4}$ kg
9.a. 4 days **b.** 20 g **c.** $A(t) = 320\left(\frac{1}{2}\right)^{t/4}$
11.a. 15% **b.** 4 yr   **13.a.** $6727 **b.** 7 yr

15. 12 yr   **17.** $A_0\left(\frac{1}{2}\right)^{0.617}$ **19.b.** The graphs are
reflections of each other in the y-axis.   **21.b.** The
graph of $y = 2^{x-1}$ is the graph of $y = 2^x$ shifted
one unit to the right.   **23.a.** Ans. may vary.
$a = 18.924$; $b = 1.014$ **b.** In 2000, 305.2 million;
in 2050, 611.6 million.   **29.** For example:
$f(x) = 2^x$

Class Exercises, pages 188–189   1. 7.3891
2. 24.5325   **3.** 0.0183   **4.** 4.1133   **5.** 2.7183
6. e^π **7.a.** $\lim\limits_{n \to \infty}\left(1 + \frac{1}{n}\right)^n = e$ **b.** 2.718
8.a. $(1.02)^4$ **b.** $(1.0067)^{12}$ **c.** $e^{0.08}$
9. 5% compounded daily yields same interest as
5.13% annually.

Written Exercises, pages 189–190   1. 2.71801;
2.718282 **b.** $\approx$ equal   **3.** $e^{\sqrt{2}}$ **5.a.** $1.0614
b. $1.0617 **c.** $1.0618   **7.** 7.5%   **9.** Plan A
11. $1419.07   **13.b.** f is one-to-one.   **19.** 2.7083
23. e^2

Class Exercises, page 194   1.a. 1.4771 **b.** 2.7686
c. 3.8325 **d.** -0.5229 **e.** -3.5229 **2.** Error
message; the domain of the log function is pos.
reals.   **3.a.** 1.9031 **b.** 2.9031 **c.** 4.9031
d. -0.0969 **e.** -1.0969 **4.a.** 2^4 **b.** $10^{1.49}$
c. $10^{1.79}$ **5.a.** 2 **b.** 4 **c.** -3 **d.** -1 **e.** $\frac{1}{2}$
6.a. 0.6931 **b.** 1.0986 **c.** 0.9933 **d.** 1.0296 **e.** 1
7.a. 1.70 **b.** 3.91   **8.a.** 25 **b.** 36 **c.** 100 **d.** e^2
9.a. 11 **b.** 4 **c.** 2 **d.** 6

Written Exercises, pages 194–197   1. $4^2 = 16$
3. $6^{-2} = \frac{1}{36}$ **5.** $10^3 = 1000$ **7.** $e^{2.1} \approx 8$
9.a. $10^x = N$ **b.** 0.85; -0.25 **c.** $e^x = N$ **d.** 2.48;
-2.81 **11.a.** 2 **b.** 4 **c.** -2 **d.** -4 **13.a.** 2 **b.** 3
c. 5 **d.** 8   **15.a.** 3 **b.** -3 **c.** $\frac{1}{4}$ **d.** 0   **17.a.** 1 **b.** 2
c. -1 **d.** $\frac{1}{2}$ **19.a.** 2.6201 **b.** -0.3799 **c.** -1.3799
21.a. -2.3026 **b.** -4.6052 **c.** 4.6052
23.a. 68 dB **b.** 15 dB   **25.a.** 65 dB **b.** 67 dB

27. 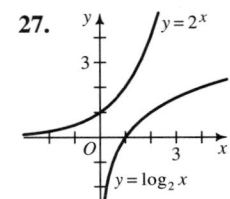   Domain $f = $ Range $f^{-1} = $
all real numbers
Range $f = $ Domain $f^{-1} = $
pos. real numbers

29.a. $(f \circ g)(x) = e^{\ln x} = x, \ x > 0$;
$(g \circ f)(x) = \ln(e^x) = x, \ x$ any real no.

31.a.

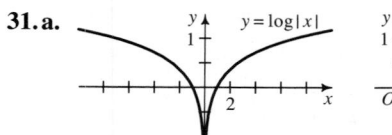 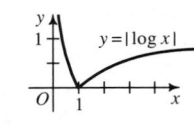

b. For $y = \log|x|$, dom. is $\{x \mid x \neq 0\}$; range is all real nos. For $y = |\log x|$, dom. is $\{x \mid x > 0\}$; range is $\{y \mid y \geq 0\}$.   **33.** For $y = \log x + 3$, dom. is $\{x \mid x > 0\}$; the range is all real nos.; zero is 0.001. For $y = \log(x + 3)$, dom. is $\{x \mid x > -3\}$; the range is all real nos.; zero is -2.
35.a. 1000   **b.** ± 1000 **c.** 1001, -999 **37.a.** 8
b. $e^{3/2}$ **c.** 1   **39.a.** 3   **b.** 10^{10} **c.** e^e **41.a.** 5.01
b. 5011.87   **c.** 0.50   **43.a.** 66.69   **b.** 0.22   **c.** 1.61
45. gastric juices: 2, acidic; acid rain: 4.52, acidic; pure water: 7, neutral; soil: 6.30, acidic; sea water: 8, alkaline   **47.a.** $\log_4 16 = 2$; $\log_{16} 4 = \frac{1}{2}$

b. $\log_9 27 = \frac{3}{2}$; $\log_{27} 9 = \frac{2}{3}$ **c.** $\log_a b = \dfrac{1}{\log_b a}$
49.b. $y \approx 10(3.16)^x$ **51.a.** (1) 145; (2) 72,382
b. 0.145; 0.072   **c.** Approaches zero

Class Exercises, page 199   **1.** $2 \log M + \log N$

2. $2 \log M - \log N$ **3.** $\frac{1}{2}(\log M - \log N)$

4. $\frac{1}{3}(\log M + \log N)$ **5.** $\log M + \frac{1}{2} \log N$

6. $2 \log M - 3 \log N$ **7.** $\log_5 6$ **8.** $\log_3 20$

9. $\log 4$ **10.** $\log 9$ **11.** $\ln 36$ **12.** $\ln \frac{5}{2}$

13. $\log MN^2$ **14.** $\log \left(\dfrac{P^2}{Q}\right)$ **15.** $\log_b MNP$

16. $\log_b \dfrac{MN}{P^3}$ **17.** $\ln \left(\dfrac{a}{b}\right)^{1/2}$ **18.** $\ln cd^{1/3}$ **19.** For example, $\log 10 + \log 1 = 1 + 0 = 1 \neq \log(11)$.
20. For example, $\dfrac{\log 1}{\log 10} = \dfrac{0}{1} = 0 \neq \log \dfrac{1}{10}$.

21. $\log \frac{1}{2} < 0$; dividing by neg. reverses the order of the inequality.

Written Exercises, pages 200–203

1. $2(\log M + \log N)$ **3.** $\frac{1}{3}(\log M - \log N)$

5. $2 \log M + \frac{1}{2} \log N$ **7.** $\log 24$ **9.** $\log_6 15$

11. $\ln 12$ **13.** $\log \left(\dfrac{M}{N^3}\right)$ **15.** $\log \dfrac{AB^2}{C^3}$

17. $\log_b \sqrt[3]{\dfrac{M^2}{NP}}$ **19.** $\log \pi r^2$ **21.** $\ln 4$ **23.a.** 2
b. 3   **c.** -1 **d.** $\frac{1}{2}$ **25.a.** x **b.** x **c.** x^2 **d.** $\frac{1}{x}$
27.a. 6   **b.** 36   **c.** 4000   **d.** $4e^3$ **29.a.** $y = x^2$
b. $y = 5x^3$ **31.a.** $y = \frac{1}{x}$ **b.** $y = 2x^2$
33.a. $y = 0.1(10^{1.2})^x$ **b.** $y = \frac{1}{e}(e^{1.2})^x$
35.a. Reflections in the x-axis, since
$\log_b \dfrac{1}{x} = -\log_b x$.   **37.b.** Shifting it up one unit
39. $y = 10^b(10^a)^x$; $\log y$ in terms of x is linear; y in terms of x is exponential.   **41.a.** $r + 2s$
b. $2r + 2s$ **c.** $r - 2s$ **d.** $r - 2$ **43.a.** $\frac{6}{5}$ **45.a.** 5
47.a. $0 < M < 1$ **b.** $1 < x < 3$ **c.** $M > 9$
d. $x < -2$ or $x > 2$ **e.** $0 < x < 5$ **49.a.** No sol.
51.a. ≈ 16 **b.** (1) ≈ 25; (2) ≈ 50 **53.** ≈ 1230 km

Class Exercises, page 205   **1.** $\sqrt[3]{81}$ **2.** 4   **3.** 3.17
4. $\pm \sqrt[4]{8}$ **5.** 0.48   **6.** 0.91   **7.** 2.41   **8.** 1.24

Written Exercises, pages 205–207   **1.** 2.26   **3.** 18.85
5. 2.89   **7.** 7.82   **9.a.** $\frac{9}{4}$ **b.** 2.16   **11.a.** 0   **b.** 0.22
13. 2007   **15.a.** $A(t) = 10,000(1.08)^t$
b. 14.3 yr   **19.a.** 177 g   **b.** 95.7 h   **23.b.** none
c. 2; 2   **d.** 0 or 2   **25.** $\log_2 3$ **27.** $\ln 2, \ln 3$
29. $\log_3 2, -1$ **31.a.** $m > n$ **b.** $m < n$
33.a. $N(t) = N_0\left(\frac{1}{2}\right)^{t/5700}$ **b.** 18,900 yr   **37.** 3
39. 1   **41.** $x > 3$ **43.** 49.5 cm; 279.4 cm

Chapter Test, page 209   **1.** $v(t) = 150,000(1.09)^t$
2.a. 8   **b.** $\frac{25}{26}$ **c.** 256   **d.** $\frac{4}{3}$
3.a. $\frac{2}{3}$ **b.** $\frac{5}{8}$ **4.** 5%
5. See graph at right. The graphs are reflections in the y-axis.

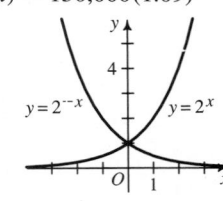

6. Ans. will vary. An initial amount P_0, growing at a rate of $r\%$, yields $P(t) = P_0(1 + r)^t$ at time t; decaying at a rate of $r\%$, $P(t) = P_0(1 - r)^t$. For example, \$1000 invested at 5% annual interest yields $1000(1 + 0.05)^2$ dollars after two yr. A \$10,000 car depreciating at 7% per yr will be worth $10,000(1 - 0.07)^2$ dollars after two yr.
7.a. \$1701.70   **b.** \$1703.98   **8.a.** 0.3979   **b.** 3.3979

c. -1.3979 **9.a.** 3 **b.** $\frac{1}{3}$ **c.** 8   **10.a.** $y = 4x^2$

b. $y = 4x$ **11.a.** $\frac{2}{3}\log_b M - \frac{1}{3}\log_b N$

b. $2\log_b M + 3\log_b N$ **12.** 5   **13.a.** 1 and 2

b. $\dfrac{\log 21}{\log 5} \approx 1.892$ **14.** 1.29

Class Exercises, pages 217–218   1. a, c   **2.** a, b, c

Class Exercises, page 222   1. $C(0, 0)$; $r = 4$

2. $C(2, 7)$; $r = 6$ **3.** $C(4, -7)$; $r = \sqrt{7}$
4. $C(0, -6)$; $r = 6$ **5.** $C(1, 3)$; $r = \sqrt{19}$
6. $C(0, 0)$; $r = 3$ **7.** $(x - 7)^2 + (y - 3)^2 = 36$
8. $(x + 5)^2 + (y - 4)^2 = 2$ **9.** $x^2 + y^2 = 169$
10.a. All pts. less than 5 units from $(0, 0)$. **b.** All
pts. more than 5 units from $(0, 0)$.   **12.** The line
and the circle do not intersect.

Written Exercises, pages 222–225
1. $(x - 4)^2 + (y - 3)^2 = 4$
3. $(x + 4)^2 + (y + 9)^2 = 9$ **5.** $(x - 6)^2 + y^2 = 15$
7. $(x - 2)^2 + (y - 3)^2 = 18$
9. $(x - 5)^2 + (y + 4)^2 = 16$
11. $(x - 4)^2 + (y + 5)^2 = 25$
13. $(x - 1)^2 + (y - 4)^2 = 1$; $C(1, 4)$, $r = 1$
15. $x^2 + (y - 6)^2 = 11$; $C(0, 6)$, $r = \sqrt{11}$
17. $\left(x - \frac{5}{2}\right)^2 + \left(y - \frac{9}{2}\right)^2 = 27$; $C\left(\frac{5}{2}, \frac{9}{2}\right)$, $r = 3\sqrt{3}$
27. $(8, 15)$, $(15, 8)$ **29.** Fail to intersect
31. $(-5, 12)$, $\left(\frac{135}{29}, -\frac{352}{29}\right)$ **33.** $(0, 0)$, $(2\sqrt{3}, 6)$
39.a. $x^2 + y^2 = 13$ **b.** $2x + 3y = 13$
41.a. $AC = BC = 13$ **b.** $M = \left(\frac{21}{2}, \frac{25}{2}\right)$; slope of

$\overline{CM} = 1$, slope of $\overline{AB} = -1$ **43.a.** The circle is
$x^2 + y^2 = 169$; $(-5)^2 + 12^2 = 169$ **b.** slope of

$\overline{PA} = -\frac{2}{3}$; slope of $\overline{PB} = \frac{3}{2}$

45. $(x - 6)^2 + y^2 = 36$; $C(6, 0)$, $r = 6$
47.a. $y = \sqrt{16 - x^2}$ **b.** $A(x) = 4\sqrt{16 - x^2}$
c. $\{x \mid -4 \le x \le 4\}$ **e.** 0 **f.** 16
49.a. $A(x) = (x + 4)\sqrt{16 - x^2}$ **b.** 2
51. $(x - 3)^2 + (y - 4)^2 = 25$
53. $(x - 2)^2 + y^2 = 50$ **55.** No pts.   **57.** The
x- and y-axes, and the circle centered at $(0, 0)$ with

$r = 1$ **59.b.** $\frac{8\pi}{3} - 2\sqrt{3}$

Class Exercises, page 228   1.a. Horiz.

b. $V(\pm 10, 0)$; $F(\pm 6, 0)$ **c.** 20 **d.** $\dfrac{x^2}{100} + \dfrac{y^2}{64} = 1$
2.a. Vert. **b.** $V(0, \pm 13)$; $F(0, \pm 12)$ **3.** The region

inside the ellipse $\dfrac{x^2}{9} + y^2 = 1$; the region outside

the ellipse $\dfrac{x^2}{9} + y^2 = 1$.   **4.a.** The surface of an
elliptical solid **b.** The interior of an elliptical solid

Written Exercises, pages 228–230   1. $V(\pm 6, 0)$;
$F(\pm 2\sqrt{5}, 0)$ **3.** $V(0, \pm 5)$; $F(0, \pm 3)$
5. $V(\pm 5, 0)$; $F(\pm 4, 0)$

7.a.

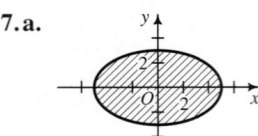

b.

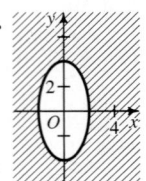

9. Dom.: $\{x \mid -6 \le x \le 6\}$; Range:
$\{y \mid -2 \le y \le 0\}$ **11.b.** Circle $= \pi$; ellipse $\approx 2\pi$
13. $\dfrac{x^2}{9} + \dfrac{y^2}{81} = 1$ **15.** $\dfrac{x^2}{289} + \dfrac{y^2}{225} = 1$
19. $(12, -8)$, $(16, -6)$ **21.** $(2, \sqrt{3})$, $(-2, \sqrt{3})$,
$(-2, -\sqrt{3})$, $(2, -\sqrt{3})$ **23.a.** The graph is shifted
down 6 units.   **25.** $C(-5, 4)$; $V_1(-10, 4)$, $V_2(0, 4)$
27. $C(-2, 5)$; $V_1(-2, 7)$, $V_2(-2, 3)$ **29.b.** $\frac{4}{3}\pi a^2 b$
31. $V_1(-6, 0)$, $V_2(-6, 8)$; $F_1(-6, 2)$, $F_2(-6, 6)$
33. $V_1(-4, -3)$, $V_2(2, -3)$; $F\left(-1 \pm \frac{3\sqrt{3}}{2}, -3\right)$
35. $V_1(-1, 6)$, $V_2(-1, 0)$; $F(-1, 3 \pm 2\sqrt{2})$
37.a. $y = \pm\frac{2}{3}\sqrt{36 - x^2}$ **b.** Yes

39. $\dfrac{4(x - 4)^2}{39} + \dfrac{(y + 1)^2}{16} = 1$

41. $\dfrac{(x - 5)^2}{25} + \dfrac{(y - 6)^2}{36} = 1$

Class Exercises, page 234   1.a. $V(\pm 2, 0)$;
$F(\pm\sqrt{13}, 0)$ **b.** Horiz. **c.** $\dfrac{x^2}{4} - \dfrac{y^2}{9} = 1$; $y = \pm\frac{3}{2}x$

2.a. $y = -\frac{2}{x}$ **b.** $y = \frac{2}{x}$ **3.a.** Vert. **b.** $V(0, \pm 5)$;

$F(0, \pm\sqrt{26})$ **c.** $y = \pm 5x$ **d.** $\dfrac{(y + 5)^2}{25} - \dfrac{(x - 6)^2}{1} = 1$

Written Exercises, pages 235–237   1. $V(\pm 3, 0)$;
asymptotes: $y = \pm\frac{2}{3}x$ **3.** $V(\pm 1, 0)$; asymptotes:

$y = \pm x$ **5.** $V(0, \pm 1)$; asymptotes: $y = \pm\frac{1}{2}x$
7. $V_1(2, 2)$, $V_2(-2, -2)$; asymptotes: $x = 0$, $y = 0$
9. Dom.: $\{x \mid |x| \ge 2\}$; Range: $\{y \mid y \ge 0\}$

12   *Answers to Selected Exercises*

11.a.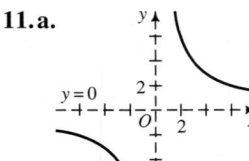

b.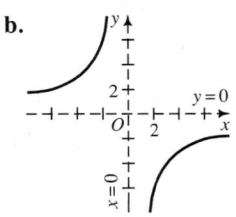

c.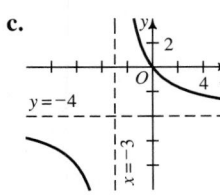

13.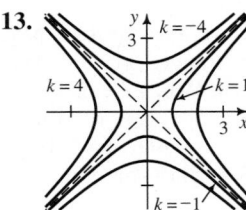

15. $(4.1, 0.9)$ **17.** $(-4, 5), (10, -2)$

19. $\frac{x^2}{36} - \frac{y^2}{64} = 1$ **21.** $\frac{y^2}{4} - \frac{x^2}{4} = 1$ **23.b.** 0.209,

$0.101, 0.010$ **27.c.** Equal

29.a. $\sqrt{x^2 + (y - 10)^2} - \sqrt{x^2 + (y + 10)^2} = \pm 12$

b. $\frac{y^2}{36} - \frac{x^2}{64} = 1$ **33.** $\frac{y^2}{36} - \frac{x^2}{9} = 1$; $y = \pm 2x$

35.a.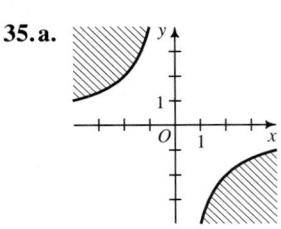

c.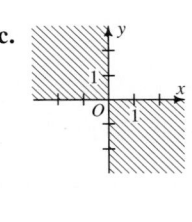

39. $V_1(12, 8), V_2(0, 8), F_1(16, 8), F_2(-4, 8)$;
asymptotes: $-4x + 3y = 0, 4x + 3y = 48$
41. $V_1(2, 2), V_2(2, 0); F(2, 1 \pm \sqrt{2})$; asymptotes:

$y = x - 1, y = -x + 3$ **43.** $\frac{(x - 5)^2}{16} - \frac{y^2}{9} = 1$

45. $a = b = \sqrt{2}$

Class Exercises, pages 239–240 **1.a.** $1; 1$ **b.** $2; 2$
c. $5; 5$ **d.** $\sqrt{x^2 + y^2}, y + 2$
2.a. (1) $\sqrt{(x + 4)^2 + y^2}$; (2) $|x|$ **b.** Set (1) and (2)
from Ex. 2a equal and simplify. **4.** $F(0, 1)$;
dir. $y = -1$; up **5.** $F(0, -1)$; dir. $y = 1$; down
6. $F\left(\frac{1}{4}, 0\right)$; dir. $x = -\frac{1}{4}$; right **7.** $F\left(-\frac{1}{4}, 0\right)$;

dir. $x = \frac{1}{4}$; left

Written Exercises, pages 240–241 **1.a.** $V(0, 0)$;
$F(0, 2)$; dir. $y = -2$ **b.** $V(0, 0); F(2, 0)$;

dir. $x = -2$ **3.a.** $V(0, 0); F\left(0, -\frac{1}{8}\right)$; dir. $y = \frac{1}{8}$

b. $V(0, 0); F\left(-\frac{1}{8}, 0\right)$; dir. $x = \frac{1}{8}$ **5.** $V(2, 1)$;

$F(2, 2)$; dir. $y = 0$ **7.** $V(4, 7); F\left(4\frac{1}{4}, 7\right)$;

dir. $y = 3\frac{3}{4}$ **9.** $x = \frac{1}{16}y^2$ **11.** $x = -\frac{1}{4}y^2$

13. $y = -x^2$ **15.** $x = -\frac{1}{16}y^2$ **17.** $y = \frac{1}{4}x^2 + 1$

19.a.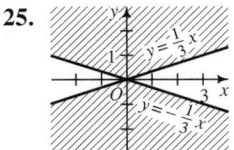

c.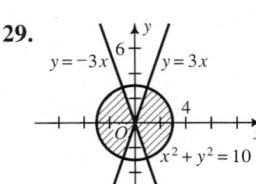

21.a. $\sqrt{x^2 + (y - 6)^2} = y$ **b.** $y - 3 = \frac{1}{12}(x - 0)^2$

23. $x = \frac{1}{4}y^2 + \frac{3}{2}y + \frac{5}{4}$ **25.** $y = \frac{1}{4}x^2 + \frac{7}{2}x + \frac{25}{4}$

27. $V(-1, 8); F\left(-1, \frac{95}{12}\right)$; dir. $y = \frac{97}{12}$

29. $V(4, -1); F(4, 0)$; dir. $y = -2$ **31.** $V(-1, 3)$;
$F(-5, 3)$; dir. $x = 3$

Class Exercises, page 244 **1.** Ans. may vary.
No sol.: $(x + 4)^2 + y^2 = 9; y^2 = x$.
1 sol.: $(x + 3)^2 + y^2 = 9; y^2 = x$.
2 sols.: $(x + 2)^2 + y^2 = 9; y^2 = x$.
3 sols.: $(x - 3)^2 + y^2 = 9; y^2 = x$.
4 sols.: $(x - 4)^2 + y^2 = 9; y^2 = x$. **2.** The graph
of $x^2 + y^2 = 9$ is a circle. The graph of
$(x + y)^2 = 9$ is two $\parallel$ lines. **3.** The graph of
$x^2 - y^2 = 4$ is a hyperbola. The graph of
$(x - y)^2 = 4$ is two $\parallel$ lines.

Written Exercises, pages 244–246 **1.** $(2, \pm 2\sqrt{3})$
3. $(4, 2), (0.8, -4.4)$ **5.** Tangent at $(2, -1)$
7.a. 4 **b.** $9x^2 + 9y^2 = 162$; circle **c.** An
intersection pt. must satisfy both equations, and $\therefore$
satisfies the eq. of their sum. **9.** $(0, \pm 2)$

11. $(0, -2), \left(\pm \frac{2\sqrt{17}}{3}, \frac{16}{9}\right)$ **13.** $(4, -3), (-4, 3)$,

$(3, -4), (-3, 4)$ **15.** No solution **17.** $(-1, -4)$
19. No solution **21.** $(-2, \pm\sqrt{5})$ **23.a.** $(4, 2)$;
$y = -2x + 10$ **b.** $y = -2x + 10$

25.

29.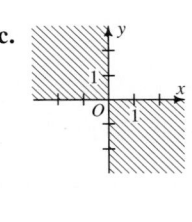

27. $(5, -3)$ **31.** ≈ 10.6 cm

Class Exercises, page 249 1.a. A: 4; B: 6; C: 8;
P: $|x - 3|$ **b.** A: 8; B: 10; C: 5; P: $|y + 6|$
2.a. $\sqrt{x^2 + y^2} = 2\sqrt{(x - 7)^2 + (y - 2)^2}$
b. $\sqrt{x^2 + y^2} = \frac{1}{2}\sqrt{(x + 3)^2 + (y - 4)^2}$
c. $\sqrt{(x + 5)^2 + (y + 1)^2} = |y + 6|$
d. $\sqrt{(x + 3)^2 + (y - 4)^2} = 2|y + 6|$
e. $\sqrt{(x - 7)^2 + (y - 2)^2} = \frac{1}{2}|y + 6|$

3. Hyperbola **4.a.** Hyperbola **b.** Ellipse
c. Parabola **d.** Circle **5.** The pt. $(0, 0)$ **6.** Two
lines, $x \pm 3y = 0$ **7.** No graph

Written Exercises, pages 250–251

1.a. $\frac{x^2}{4} + \frac{y^2}{3} = 1$ **b.** $(1, 0)$
c. $\sqrt{(x - 1)^2 + y^2} = 0.5|x - 4|$ simplifies to
$\frac{x^2}{4} + \frac{y^2}{3} = 1$ **3.** $\sqrt{(x + 6)^2 + y^2} = 2\left|x + \frac{3}{2}\right|$
simplifies to $\frac{x^2}{9} - \frac{y^2}{27} = 1$ **5.** Ellipse

7. Hyperbola **9.** Ellipse **11.** No graph
13. 0.016 **15.** $(\pm 3, 0)$, $(0, \pm 3)$ **17.** $(2\sqrt{5}, 2\sqrt{5})$,
$(-2\sqrt{5}, -2\sqrt{5})$, $(6, -2)$, $(-6, 2)$ **19.** $(\pm\sqrt{3}, 0)$,
$(2, -1)$, $(-2, 1)$

Chapter Test, page 253 1. Let the vertices of the
square be $A(a, 0)$, $B(a, a)$, $C(0, a)$ and $O(0, 0)$.
Slope of $\overline{AC} = -1$; slope of $\overline{OB} = 1$; $\therefore \overline{AC} \perp \overline{OB}$.
$AC = a\sqrt{2}$; $OB = a\sqrt{2}$; $\therefore AC = OB$. **2.** $(0, -2)$,
$(2, 0)$ **3.** $x^2 + (y - 2)^2 = 13$ **4.** $V(0, \pm 3)$;
$F(0, \pm 2)$ **5.** $(3, 0)$, $(0, 2)$ **6.** $\frac{x^2}{9} + \frac{y^2}{16} = 1$

7.a. $\frac{y^2}{9} - \frac{x^2}{4} = 1$

b. $\frac{x^2}{4} - \frac{y^2}{16} = 1$

8. See graph at right.

9. $V(0, 0)$; $F\left(0, \frac{3}{2}\right)$;

$y = -\frac{3}{2}$

10. $x = \frac{1}{12}y^2$

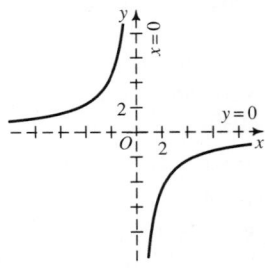
$y = 0$

11.

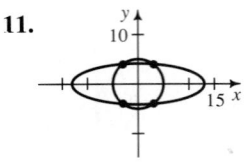

Intersect at $(3, \pm 4)$,
and $(-3, \pm 4)$

12.a. Circle
b. Ellipse
c. Parabola

13. $\sqrt{(x - 1)^2 + (y - 3)^2} = |y + 4|$;
$y + \frac{1}{2} = \frac{1}{14}(x - 1)^2$ **14.** If $B = C = 0$, the graph
is the y-axis, and if $A = B = 0$, the graph is the
x-axis. If $A = C = 0$, the graph is the x- and
y-axes. If only $A = 0$, the graph is two lines, $y = 0$
and $Bx + Cy = 0$; if only $C = 0$, the graph is two
lines, $x = 0$ and $Ax + By = 0$.

Cumulative Review, pages 254–255
1. Range: $\{y | y \geq -1\}$; zeros: 0, 2
2.a. 0 **b.** $2x - 2$ **c.** $-x^2 + 2x - 1$ **d.** $-1, x \neq 1$
e. $-x$ **f.** $2 - x$ **3.a.** iii **b.** ii **c.** i, ii, iv
4.a. $p = 6$; $A = 2$ **b.** $p = 4$; $A = 2$
5.a. Shift the graph of $y = f(x)$ left one unit.

b.
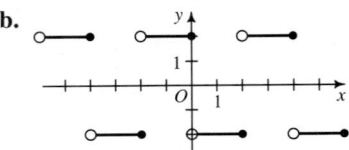

6.a. $f^{-1}(x) = \frac{8 - x}{2}$;
$f(f^{-1}(x)) = 8 - 2\left(\frac{8 - x}{2}\right) = 8 - 8 + x = x$;
$f^{-1}(f(x)) = \frac{8 - (8 - 2x)}{2} = \frac{2x}{2} = x$ **b.** No inverse
c. $f^{-1}(x) = \frac{1 + 2x}{x}$;
$f(f^{-1}(x)) = \frac{1}{\frac{1 + 2x}{x} - 2} = \frac{x}{1 + 2x - 2x} = x$;

$f^{-1}(f(x)) = \frac{1 + 2\left(\frac{1}{x - 2}\right)}{\frac{1}{x - 2}} = \frac{x - 2 + 2}{1} = x$

7.a. $S(x, y) = 6x + 4y$
b. $S(40, 25) = 340$;
$S(32, 48) = 384$
c. See graph at right.

8.a. $\frac{9}{16}$ **b.** $2y^{10}$ **c.** $\frac{x}{18}$
d. $\frac{8}{9}$ **e.** 9 **f.** $1 - a$

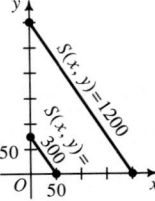

9.a. $\frac{3}{4}$ **b.** 1000 **c.** $\frac{6}{5}, \frac{4}{5}$ **10.** $f(x) = \frac{1}{2}(\sqrt{3})^x$
11.a. 3 h **b.** $N(t) = 150 \cdot 2^{t/3}$ **c.** 38,400
12.a. \$107.19 **b.** \$107.23 **c.** \$107.25 **13.** Plan A
14.a. -2 **b.** 4 **c.** $\frac{1}{3}$ **d.** -5 **e.** 9 **f.** 50

15.a. $2 \log M + 3 \log N$ **b.** $\frac{3}{2} \log M - \frac{1}{2} \log N$

c. $2 + \log M + \frac{1}{2} \log N$ **16.a.** $y = 100x$ **b.** $y = \frac{2}{x^3}$
c. $y = 10 \cdot 10^{2.5x}$ **17.a.** 2.08 **b.** 4.64 **c.** 17.67
18. See diagram below. *PQRS* is a rhombus.
$A = (a, b)$, $B = (-a, b)$, $C = (-a, -b)$,
$D = (a, -b)$. Slope of $\overline{AB} = 0$, slope of $\overline{BC}$ is
undef.; $\therefore \overline{AB} \perp \overline{BC}$. $AB = DC = 2a$,
$BC = AD = 2b$; $\therefore ABCD$ is ‖ogram. A ‖ogram
with 1 rt. $\angle$ is a rect.;
$\therefore ABCD$ is a rect.

19. $C(2, -7)$; $r = 5$
20. $C(0, 0)$;
$V(0, \pm 4)$; $F(0, \pm\sqrt{7})$
21. $\frac{x^2}{4} - y^2 = 1$
22. $V(-2, 1)$;
$F(-3, 1)$; dir. $x = -1$
23. $(\pm 3, 4)$
24. Ellipse

Ex. 18

Class Exercises, page 260 **1.a.** π **b.** $\frac{\pi}{2}$ **c.** $\frac{7\pi}{4}$
d. $\frac{\pi}{3}$ **e.** $\frac{2\pi}{3}$ **f.** $\frac{4\pi}{3}$ **g.** $\frac{\pi}{6}$ **h.** $\frac{\pi}{180}$ **2.a.** 360°
b. 180° **c.** 90° **d.** 45° **e.** 135° **f.** 300° **g.** 330°
h. 150° **3.** Ans. may vary. **a.** 370°, −350°
b. 460°, −260° **c.** 355°, −365° **d.** 40°, −320°
e. 3π, $-\pi$ **f.** $\frac{5\pi}{2}$, $-\frac{3\pi}{2}$ **g.** $\frac{5\pi}{3}$, $-\frac{7\pi}{3}$ **h.** $\pm 2\pi$
4.a. 60° **b.** $\theta = \frac{\pi}{3} + 2n\pi$ **5.a.** 1 **b.** 2 **c.** 0.75
6.a. 600° **b.** −990°

Written Exercises, pages 261–262 **1.a.** $\frac{7\pi}{4}$ **b.** $\frac{5\pi}{4}$
c. $\frac{\pi}{12}$ **d.** $-\frac{\pi}{4}$ **3.a.** $-\frac{2\pi}{3}$ **b.** $-\frac{4\pi}{3}$ **c.** $\frac{5\pi}{3}$ **d.** 2π
5.a. −90° **b.** 240° **c.** −135° **d.** −30° **7.a.** 180°
b. −270° **c.** 120° **d.** 210° **9.a.** 1.2 **b.** 0.75
11.a. 1.66 **b.** 1.92 **c.** 1.66 **d.** 2.08 **13.a.** 91°40′
or 91.7° **b.** 97°20′ or 97.4° **c.** 69°20′ or 69.3°
d. 75°40′ or 75.6° **15.** 0.5 **Exs. 17–21:** Ans. may
vary. **17. a.** 140°, −220° **b.** 300°, −420° **c.** $\frac{9\pi}{4}$,
$-\frac{7\pi}{4}$ **d.** $\frac{4\pi}{3}$, $-\frac{8\pi}{3}$ **19.a.** 388.5°, −331.5°
b. 476.3°, −243.7° **c.** 299.6°, −420.4° **d.** 44.7°,
−675.3° **21.a.** 0°30′, −359°30′ **b.** 269°20′,
−450°40′ **c.** 363°21′, −356°39′ **d.** 475°15′,
−244°45′ **23.** 29.7° + $n \cdot 360°$ **25.a.** 12,600°
b. 219.91 **27.a.** 900° **b.** 15.71 **29.a.** 5256°
b. 91.73 **33.a.** 69 mi

Class Exercises, page 264 **1.** $s = \pi$; area = 2π
2. $s = 4\pi$; area = 12π **3.** $s = 8$; area = 8
4.a. 100 m **b.** In a large circle, a small arc is very
nearly a straight line.

Written Exercises, pages 264–267 **1.** $s = 3$ cm;
area = 9 cm² **3.** $r = 5$ cm; area = 27.5 cm²
5. $r = 10$ cm; $s = 5$ cm **7.** 12 cm² **9.** 1.5 cm,
2 cm **11.** 402,000 km **13.a.** 72,000°; 1257
b. 7477 cm **c.** ≈125 cm/s **15.** 1.3×10^6 km
17. 1885 m² **19.** 350 m **21.** 4.1×10^{13} km
25.b. r; s

Class Exercises, page 271 **1.** $\sin \theta = \frac{\sqrt{10}}{10}$;
$\cos \theta = -\frac{3\sqrt{10}}{10}$ **2.** $\sin \theta = -\frac{2\sqrt{5}}{5}$; $\cos \theta = \frac{\sqrt{5}}{5}$
3. $\sin \theta = -\frac{4\sqrt{41}}{41}$; $\cos \theta = -\frac{5\sqrt{41}}{41}$ **4.a.** Pos.
b. Neg. **c.** Neg. **d.** Pos. **e.** Pos. **f.** Neg. **g.** Neg.
h. Pos. **i.** Pos. **j.** Neg. **k.** Neg. **l.** Neg. **5.a.** Dec.
b. Dec. **c.** Inc. **d.** Inc. **6.a.** Inc. **b.** Dec. **c.** Dec.
d. Inc. **7.** The terminal ray of θ intersects the
unit circle, $x^2 + y^2 = 1$, at $(\cos \theta, \sin \theta)$.
8. $\pm 90°$, $\pm 270°$ **9.a.** θ is coterminal with an
angle of 45°. **b.** $\theta = \frac{\pi}{4} + 2n\pi$

Written Exercises, pages 272–274 **1.a.** 0 **b.** −1
c. −1 **d.** 0 **3.a.** 0 **b.** −1 **c.** −1 **d.** 0 **5.a.** II
b. III **7.a.** $\frac{\pi}{2} + 2n\pi$ **b.** $\pi + 2n\pi$ **c.** $n\pi$
d. No sol. **9.a.** 0 **b.** Neg. **c.** Neg. **d.** Neg.
11.a. Pos. **b.** Neg. **c.** Pos. **d.** Pos. **13.a.** Neg.
b. Neg. **c.** 0 **d.** Pos. **15.a.** Pos. **b.** Neg. **c.** 0
d. Pos. **17.** $\sin \theta = \frac{4}{5}$; $\cos \theta = \frac{3}{5}$ **19.** $\sin \theta = \frac{12}{13}$;
$\cos \theta = -\frac{5}{13}$ **21.** $\frac{4}{5}$ **23.** $-\frac{7}{25}$ **25.** $-\frac{2\sqrt{6}}{5}$
27. $-\frac{\sqrt{7}}{4}$ **29.a.** $P\left(\frac{\sqrt{3}}{2}, \frac{1}{2}\right)$; $Q\left(-\frac{\sqrt{3}}{2}, \frac{1}{2}\right)$
31.c. $\sin z \approx z$ **33.** > **35.** = **37.** = **39.** >
41. sin 4, sin 3, sin 1, sin 2 **43.a.** $W(2) = (0, 1)$;
$W(3) = (0, 0)$; $W(4) = (1, 0)$; $W(5) = (1, 1)$
b. $p = 4$

Computer Exercises, page 274 **1.** For $x = 1$,
.84147; for $x = 2$, .90935; for $x = 1.5708$,
1.00000. SIN(1) = .84147; SIN(2) = .90930;
SIN(1.5708) = 1.00000. **2.** For $x = 1$, .54030; for
$x = 2$, −.41587; for $x = 3.1416$; −.97602
COS(1) = .54030; COS(2) = −.41615;
COS(3.1416) = −1

Class Exercises, pages 278–279 **1.a.** $10°$ **b.** $50°$
c. $25.1°$ **d.** $\pi - 3 \approx 0.14$ **2.** $110°$ **3.** $320°$
4.a. $\sin 10°$ **b.** $-\sin 30°$ **c.** $-\sin 15°$ **d.** $\sin 40°$
5.a. $-\cos 20°$ **b.** $-\cos 2°$ **c.** $-\cos 80°$ **d.** $\cos 5°$
6.a. -0.1392 **b.** -0.9397 **c.** 0.9848 **d.** -0.6428
7.a. 0.9842 **b.** 0.4787 **c.** -0.8161 **d.** 0.8471
8.a. 0.1411 **b.** -0.6536 **c.** -0.8415 **d.** -0.4161
9.a. -0.8011 **b.** 0.7452 **c.** 0.2555 **d.** 0.3802
10.a. $\frac{\sqrt{2}}{2}$ **b.** $\frac{\sqrt{2}}{2}$ **c.** $-\frac{\sqrt{2}}{2}$ **d.** $-\frac{\sqrt{2}}{2}$ **11.a.** $\frac{1}{2}$
b. $-\frac{1}{2}$ **c.** $\frac{1}{2}$ **d.** $\frac{1}{2}$ **12.a.** $\frac{1}{2}$ **b.** $-\frac{1}{2}$ **c.** $\frac{\sqrt{3}}{2}$ **d.** $\frac{\sqrt{3}}{2}$
13.a. $-\frac{1}{2}$ **b.** $\frac{\sqrt{3}}{2}$ **c.** $-\frac{1}{2}$ **d.** $-\frac{\sqrt{3}}{2}$ **14.a.** $\frac{\sqrt{2}}{2}$
b. $-\frac{\sqrt{3}}{2}$ **c.** $-\frac{\sqrt{3}}{2}$ **d.** $\frac{\sqrt{3}}{2}$ **15.** See graph below.

16. The graph of
$y = \cos \theta$ translated
$90°$ to the right
coincides with the
graph of $y = \sin \theta$.

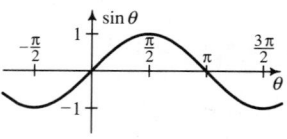

Ex. 15

17.a. Symmetry in the origin **b.** Symmetry in the
y-axis **18.** $\cos (90° - \theta) = \cos (-(\theta - 90°)) = \cos (\theta - 90°) = \sin \theta$

Written Exercises, pages 279–282 **1.a.** $\sin 52°$
b. $-\cos 52°$ **c.** $-\sin 37°$ **d.** $-\cos 40°$
3.a. $-\cos 44.5°$ **b.** $\cos 62°$ **c.** $\sin 34.3°$ **d.** $\sin 21°$
5.a. 0.4695 **b.** 0.7193 **c.** 0.4179 **d.** 0.4706
7.a. 0.3420 **b.** -0.9041 **c.** 0.5314 **d.** -0.4586
9.a. 0.9320 **b.** 0.8085 **c.** -0.9365 **d.** 0.2837
11.a. $\frac{\sqrt{2}}{2}$ **b.** $-\frac{\sqrt{2}}{2}$ **c.** $-\frac{1}{2}$ **d.** $-\frac{\sqrt{3}}{2}$ **13.a.** $\frac{1}{2}$
b. $-\frac{1}{2}$ **c.** $-\frac{\sqrt{2}}{2}$ **d.** $\frac{\sqrt{3}}{2}$ **15.a.** $\frac{\sqrt{3}}{2}$ **b.** $\frac{\sqrt{3}}{2}$ **c.** $-\frac{1}{2}$
d. $\frac{\sqrt{2}}{2}$ **17.a.** 1 **b.** $-\frac{1}{2}$ **c.** $-\frac{\sqrt{3}}{2}$ **d.** $-\frac{\sqrt{2}}{2}$
21.a. 3 **b.** $-1.90, 0, 1.90$ **23.a.** 2 **b.** $-0.82, 0.82$
25.a. 2 **b.** $45°, 225°$ **27.** 5395 mi **29.** 353 mi
31.a. 503 mi/h **b.** 955 mi/h **33.** 771 mi/h

Class Exercises, page 285 **1.a.** $90° + n \cdot 180°$
b. $90° + n \cdot 360°$ **c.** $135° + n \cdot 180°$ **d.** $n \cdot 180°$
2.a. 1.035 **b.** -1.035 **c.** -1.035 **d.** 1.035
3.a. -2.185 **b.** 11.43 **c.** 7.086 **d.** 1.079 **4.** III
5.a. $-\frac{3}{5}$ **b.** $-\frac{4}{3}$ **c.** $-\frac{3}{4}$ **d.** $-\frac{5}{3}$ **e.** $\frac{5}{4}$

Written Exercises, pages 285–286 **1.a.** -5.671
b. -0.1051 **c.** -1.043 **d.** -1.855 **3.a.** $-\sin 15°$
b. $\sec 80°$ **c.** $\tan 40°$ **d.** $-\sec (\pi - 2) \approx -\sec 1.14$
5.a. $-\tan 80°$ **b.** $\sec 70°$ **c.** $\cot 5°$
d. $-\csc (4 - \pi) \approx -\csc 0.86$ **7.a.** $n\pi$ **b.** None
c. $\frac{\pi}{2} + 2n\pi$ **d.** $\frac{3\pi}{2} + 2n\pi$

9.

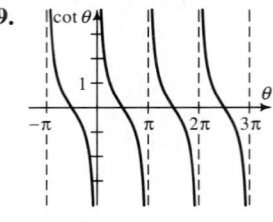

11. Infinitely many
13. $\cos x = -\frac{12}{13}$;

$\tan x = -\frac{5}{12}$;

$\csc x = \frac{13}{5}$;

$\sec x = -\frac{13}{12}$;

$\cot x = -\frac{12}{5}$

15. $\sin x = -\frac{3}{5}$; $\cos x = -\frac{4}{5}$; $\csc x = -\frac{5}{3}$;
$\sec x = -\frac{5}{4}$; $\cot x = \frac{4}{3}$ **17.** $\sin x = \frac{2\sqrt{2}}{3}$;
$\cos x = -\frac{1}{3}$; $\tan x = -2\sqrt{2}$; $\csc x = \frac{3\sqrt{2}}{4}$;
$\cot x = -\frac{\sqrt{2}}{4}$ **19.** $\cot x$ is undef. for $x = n\pi$, and
as x gets close to $n\pi$, $|\cot x|$ increases. **21.b.** All
real nos. except for odd multiples of $\frac{\pi}{2}$ **23.a.** $-\frac{1}{2}$
b. -2 **c.** $\frac{\sqrt{3}}{2}$ **d.** $-\sqrt{3}$ **25.a.** 1 **b.** -1 **c.** $\sqrt{3}$
d. Undef. **27.a.** Undef. **b.** $-\sqrt{3}$ **c.** 0 **d.** $-\frac{2\sqrt{3}}{3}$

Class Exercises, page 289 **1.** $50.2°$ **2.** $-17.5°$
3. $115.2°$ **4.** -1.24 **5.** 0.85 **6.** 1.51 **7.** An
error message; 1.7 is not in the domain of
$y = \mathrm{Sin}^{-1} x$. **8.a.** $\frac{3}{5}$ **b.** $\frac{4}{3}$ **c.** $\frac{3}{4}$

Written Exercises, pages 289–291 **1.a.** $64.2°$
b. $-64.2°$ **c.** $41.4°$ **d.** $138.6°$ **3.a.** 0.23 **b.** -0.23
c. 1.22 **d.** 1.92 **5.a.** 0 **b.** $\frac{\pi}{2}$ **c.** $\frac{\pi}{4}$ **d.** $-\frac{\pi}{4}$
7.a. $\frac{\pi}{6}$ **b.** $-\frac{\pi}{6}$ **c.** $\frac{\pi}{3}$ **d.** $\frac{2\pi}{3}$ **9.** 3 is not in the
domain of $y = \mathrm{Cos}^{-1} x$. Error message appears.
11.a. 0.42; $\frac{5}{12}$ **b.** -0.42; $-\frac{5}{12}$ **13.a.** 1.38; $\frac{29}{21}$
b. 1.15; $\frac{2\sqrt{3}}{3}$ **15.a.** $-1 \le x \le 1$; $-1 \le y \le 1$;
$y = x$, for $-1 \le x \le 1$ **c.** No **19.a.** True **b.** Not
true **21.a.** True **b.** Not true

23. Dom.: $\{x \mid x \le -1 \text{ or } x \ge 1\}$;

  Range: $\{y \mid 0 \le y < \frac{\pi}{2} \text{ or } \frac{\pi}{2} < y \le \pi\}$

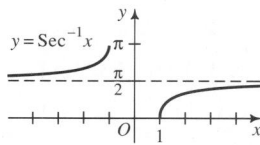

25. a. $\text{Sec}^{-1} 2 = \text{Cos}^{-1} \frac{1}{2} = \frac{\pi}{3}$

b. $\text{Sec}^{-1} x = \text{Cos}^{-1} \frac{1}{x}$ **27.** π

Chapter Test, page 293   1. a. $\frac{3\pi}{2}$ **b.** 3.35

2. a. 300° **b.** 143°10′ or 143.2° **3.** Ans. may vary.
a. 160°, −560° **b.** 673.2°, −46.8° **c.** $\frac{17\pi}{6}$, $-\frac{7\pi}{6}$
d. 502°10′, −217°50′   **4.** $s = 4.2$ cm;
area = 10.5 cm^2 **5.** 3480 km   **6. a.** $\sin\theta = \frac{\sqrt{2}}{2}$;
$\cos\theta = -\frac{\sqrt{2}}{2}$ **b.** $\sin\theta = -\frac{\sqrt{3}}{2}$; $\cos\theta = -\frac{1}{2}$
c. $\sin\theta = 1$; $\cos\theta = 0$ **7. a.** < **b.** > **c.** = **d.** <
8. a. $-\frac{\sqrt{2}}{2}$ **b.** 0 **c.** $\frac{1}{2}$ **d.** $\frac{\sqrt{3}}{2}$ **9. a.** −1 **b.** $-\frac{\sqrt{3}}{3}$
c. 2 **d.** Undef.   **10.** $\sin x = \frac{\sqrt{10}}{10}$; $\cos x = -\frac{3\sqrt{10}}{10}$;
$\csc x = \sqrt{10}$; $\sec x = -\frac{\sqrt{10}}{3}$; $\cot x = -3$
11. a. $\frac{\pi}{3}$ **b.** $\sqrt{3}$ **c.** $-\frac{5}{3}$ **12.** The restriction is
necessary to make $f(x) = \sin x$ one-to-one.

Class Exercises, page 299   1. 60°, 300°
2. No sol.   **3.** 30°, 150°   **4.** 135°, 315°
5. $\frac{5\pi}{6}$, $\frac{7\pi}{6}$ **6.** $\frac{\pi}{4}$, $\frac{5\pi}{4}$ **7.** $\frac{2\pi}{3}$, $\frac{5\pi}{3}$ **8.** No sol.
9. $(180 + 360n)°$ **10.** $(225 + 360n)°$,
$(315 + 360n)°$ **11.** $(45 + 180n)°$ **12.** −0.8391
13. 31°   **14.** 9°

Written Exercises, pages 299–301   1. 224.4°,
315.6°   **3.** 50.2°, 230.2°   **5.** 101.5°, 258.5°
7. 70.5°, 289.5°   **9.** 146.4°, 213.6°   **11.** 41.8°,
138.2°   **13.** 2.16, 5.30   **15.** 3.94, 5.49
17. 0.17, 3.31   **19.** 0.84, 5.44   **21.** 2.50, 5.64
23. $m = -\sqrt{3}$; $y = -\sqrt{3}x + 3 + 2\sqrt{3}$ **25.** 149°
27. 169°   **29.** 37°   **31.** $\frac{\pi}{2}$, $\frac{3\pi}{2}$ **33.** $\frac{\pi}{2}$ **35.** $\frac{\pi}{3}$, $\frac{4\pi}{3}$

37. Ellipse; $\frac{\pi}{4}$ **39.** Ellipse; $\frac{\pi}{4}$

41. Hyperbola; $\frac{\pi}{4}$ **43.** Parabola; $\frac{\pi}{4}$
45. a. $0° \le \alpha \le 90°$, $0° \le \beta \le 48.6°$ **b.** No

Class Exercises, page 304   1. $p = \pi$; $A = 4$
2. $p = 4\pi$; $A = 3$ **3.** $p = 7$; $A = 5$ **4.** $p = 3$;
$A = 6$ **5.** $p = 4$; $A = 3$; $y = 3\sin\frac{\pi}{2}x$ **6.** $p = 4\pi$;
$A = 2$; $y = -2\cos\frac{1}{2}x$ **7. a.** 1 **b.** 4 **c.** 3

Written Exercises, pages 305–308

1.

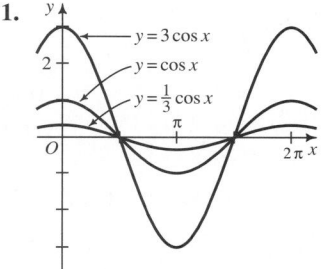

5. $A = 2$; $p = \frac{2\pi}{3}$

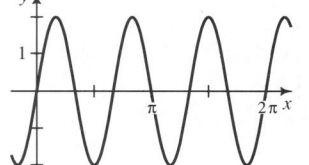

7. $A = 2$; $p = \pi$

9. $A = \frac{1}{2}$; $p = 1$
11. $A = 4$; $p = \pi$;
$y = 4\sin 2x$
13. $A = 3$; $p = 4\pi$;
$y = 3\cos\frac{1}{2}x$

15. $A = 3$; $p = 4$; $y = 3\sin\frac{\pi}{2}x$

17. $y = \pm 4\sin\frac{\pi}{6}x$ **19. a.** π or 3.14 **b.** $\frac{\pi}{2}$, $\frac{3\pi}{2}$ or

1.57, 4.71 **c.** $\frac{\pi}{3}$, π, $\frac{5\pi}{3}$ or 1.05, 3.14, 5.24
21. 0.72, 2.42, 3.86, 5.56   **23.** No sol.   **25.** 0.16,
1.84, 4.16, 5.84   **31. a.** 200 MHz
b. Max. 200.01 MHz; min. 199.99 MHz
33. b. ≈3263 **c.** 162 g

Class Exercises, page 312   1. 2   **2.** 4   **3.** $y = 5$
4. a. Horiz. 0; vert. 5 **b.** $y = 5 + 2\sin\frac{\pi}{2}x$
5. a. Horiz. 1; vert. 5 **b.** $y = 5 + 2\cos\frac{\pi}{2}(x - 1)$
6. 3: $p = \frac{2\pi}{3}$; 4: $A = 4$; 5: horiz. shift is left 5;
6: vert. shift is up 6.

Written Exercises, pages 313–316

1. $y = 2 + 2 \cos \frac{\pi}{2} x$ or $y = 2 + 2 \sin \frac{\pi}{2}(x - 3)$

3. $y = 2 + \cos \frac{\pi}{3}(x - 1)$ or $y = 2 + \sin \frac{\pi}{3}\left(x + \frac{1}{2}\right)$

5. $y = 3 \cos 2\left(x - \frac{5\pi}{12}\right)$ or $y = 3 \sin 2\left(x - \frac{\pi}{6}\right)$

7. $y = 3 + 2 \cos 8\left(x - \frac{\pi}{8}\right)$ or

$y = 3 + 2 \sin 8\left(x - \frac{\pi}{16}\right)$

13.

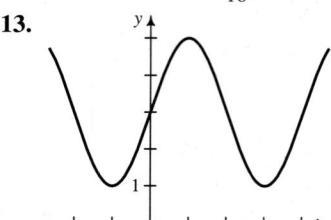

15. $y = 30 + 28 \cos\left(\frac{5\pi}{31} t\right)$ **17.a.** $p = 365$, $A = 3$,

$y = 12 + 3 \sin \frac{2\pi}{365}(x - 80)$ **b.** 9.1 h; 14.9 h

c. 123rd day (May 3) through 220th day (Aug. 8)

d. Greater   **19.a.** $T = 20 + 8 \cos \frac{2\pi}{365}(x - 197)$

b. 1st day (Jan. 1) through 66th day (Mar. 7),
328th day (Nov. 24) through 365th day (Dec. 31)
21.a. $y = -0.03 x + 18.38$

b. $y = 17.83 + 1.65 \cos \frac{2\pi}{365}(x - 172)$

25. $h = 7 + 6 \cos \frac{2\pi}{15}(t - 7.5)$

Class Exercises, page 320   **1.a.** $\cos^2 \theta$ **b.** $\sin^2 \theta$
2.a. $\tan^2 \theta$ **b.** 1   **3.a.** $\cot^2 \theta$ **b.** 1   **4.a.** $\sin \theta$
b. $\cot A$ **c.** $\sin x$ **5.a.** $\cos^2 x$ **b.** $-\cos^2 x$ **c.** $\tan^2 x$
6.a. 1 **b.** $\cos y$ **c.** -1 **7.a.** 1 **b.** 1 **c.** 1   **8.** Mult.
num. and den. by: **a.** t; $\tan A$ **c.** xy; $\cos \theta \sin \theta$
d. x; $\cos \theta$

Written Exercises, pages 321–322   **1.a.** 1 **b.** $\sin^2 \theta$
c. $-\cos^2 \theta$ **3.a.** $\csc^2 A$ **b.** $\cot^2 A$ **c.** 1   **5.a.** $\sin \theta$
b. 1 **c.** $\sin^2 \theta$ **7.** $\sec A$ **9.** 1   **11.** $\tan x$
13. $\cot x$ **15.** 2   **17.** $\csc \theta$ **19.** $\cos y$
21. $2 \sec \theta$ **23.** 1   **27.** Dom.: all real numbers;

$y = 1$ **41.** $\pm\dfrac{1}{\sqrt{1 - \sin^2 \theta}}$

Written Exercises, pages 326–327   **1.** 70.5°, 109.5°,
250.5°, 289.5°   **3.** 30°, 150°, 210°, 330°   **5.** 30°,
150°, 41.8°, 138.2°   **7.** 60°, 70.5°, 289.5°, 300°

9. $\frac{\pi}{4}, \frac{5\pi}{4}$ **11.** $0, \frac{\pi}{2}, \pi$ **13.** $\frac{\pi}{3}, \frac{\pi}{2}, \frac{3\pi}{2}, \frac{5\pi}{3}$

15. $\frac{3\pi}{4}, \frac{7\pi}{4}$ **17.** $0, \frac{\pi}{3}, \pi, \frac{5\pi}{3}$ **19.** $\frac{\pi}{6}, \frac{5\pi}{6}, \frac{7\pi}{6}, \frac{11\pi}{6}$

21. 0.67, 2.48   **23.** 0.34, 2.80   **25.** $0, \frac{\pi}{3}, \pi, \frac{5\pi}{3}$

27. 0, 5.36   **29.** 0.74   **31.** 0, 1.90   **33.** 1.52

Chapter Test, pages 328–329

1. 18°
2. 101.5°, 258.5°
3. 3.39, 6.03

4. $\frac{\pi}{12}, \frac{\pi}{4}, \frac{3\pi}{4},$

$\frac{11\pi}{12}, \frac{17\pi}{12}, \frac{19\pi}{12}$

5.

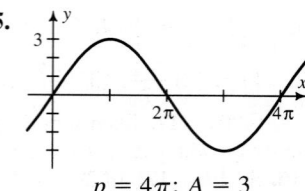

$p = 4\pi$; $A = 3$

6. $y = 1 + 2 \sin \frac{\pi}{2} x$

7.a.

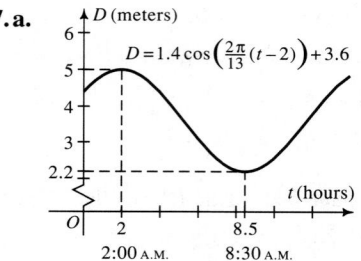

b. Ans. may vary. For example: The difference in
time between high and low tide is one-half the
period.   **8.a.** $\sin A$ **b.** $\csc \theta$ **c.** $\cos x$ **d.** $\tan \alpha$
9.a. 1.11, 4.25 **b.** $\frac{\pi}{2}, \frac{3\pi}{2}$ **10.a.** 30°, 90°, 150°
b. 26.6°, 90°, 206.6°, 270°

Class Exercises, page 333   **1.** $\sin = \dfrac{\text{opp.}}{\text{hyp.}}$;

$\cos = \dfrac{\text{adj.}}{\text{hyp.}}$; $\tan = \dfrac{\text{opp.}}{\text{adj.}}$; $\cot = \dfrac{\text{adj.}}{\text{opp.}}$; $\sec = \dfrac{\text{hyp.}}{\text{adj.}}$;

$\csc = \dfrac{\text{hyp.}}{\text{opp.}}$ **2.a.** $\sin \alpha = \frac{x}{z}$; $\cos \alpha = \frac{y}{z}$; $\tan \alpha = \frac{x}{y}$

b. $\sin \beta = \frac{y}{z}$; $\cos \beta = \frac{x}{z}$; $\tan \beta = \frac{y}{x}$ **c.** (1) T;
(2) T; (3) T   **3.** $\cot 28°$ **4.a.** 85.2 **b.** Use the
Pythagorean theorem   **5.a.** $\sin 50° = \frac{x}{10}$;

$\csc 50° = \frac{10}{x}$ **b.** $\cos 32° = \frac{5}{x}$; $\sec 32° = \frac{x}{5}$

c. $\tan 55° = \frac{x}{8}$; $\cot 55° = \frac{8}{x}$

d. $\cos 70° = \frac{3}{x}$; $\sec 70° = \frac{x}{3}$ **6.a.** $\tan \theta = \frac{3}{2}$;
$\cot \theta = \frac{2}{3}$ **b.** $\cos \theta = \frac{5}{8}$; $\sec \theta = \frac{8}{5}$ **c.** $\sin \theta = \frac{7}{12}$;
$\csc \theta = \frac{12}{7}$ **d.** $\cos \theta = \frac{7}{20}$; $\sec \theta = \frac{20}{7}$

Written Exercises, pages 334–338 **1.** $b = 7.61$;
$c = 16.3$ **3.** $d = 43.3$; $f = 42.3$ **5.a.** $\frac{5}{13}$ **b.** $\frac{5}{13}$
c. $\frac{5}{12}$ **d.** $\frac{5}{12}$ **e.** $\frac{13}{12}$ **f.** $\frac{13}{12}$ **7.** 36.9°, 53.1° **9.a.** $\sqrt{2}$
b. (1) 1; (2) $\frac{\sqrt{2}}{2}$; (3) $\frac{\sqrt{2}}{2}$ **c.** (1) 1; (2) 0.707;
(3) 0.707 **11.** Each length is the tan of the
corresponding $\angle$. **13.a.** 333,000 ft **b.** 335,000 ft
15. 38.2 ft **17.** 48.2°, 48.2°, 83.6°; area = 17.9
19. $PA = PB = 15.6$ **21.** 126.9° **23.** 21.5 m
25.a. 76.4 ft **b.** $AH = 196$ ft; $HC = 97.1$ ft
c. 8.3 ft/s **d.** 34.7 ft/s; too small
27.a. $\sec A = \frac{13}{12}$; $\csc A = \frac{13}{5}$ **b.** $\frac{1}{\sqrt{1-x^2}}$
31. $x = a \sin \beta \cot \alpha$; $y = a \sin \beta \csc \alpha$ **33.** 3 cm^2
37.a. For example, $\theta = \text{Tan}^{-1}\left(\frac{30000}{d}\right)$ **b.** 5.4°
39. 35.3° **41.** $A = nr^2 \sin \theta \cos \theta$

Class Exercises, page 341 **1.a.** 10 cm^2 **b.** 10 cm^2
2. 10 **3.** 10 **4.** $\frac{5\sqrt{2}}{2}$ **5.a.** 0.5 **b.** 30°, 150°
6. $\frac{\pi}{3} - 1 \approx 0.047$

Written Exercises, pages 342–344 **1.a.** 5 **b.** 5
3.a. $3\sqrt{2}$ **b.** $3\sqrt{2}$ **5.** $K = \frac{1}{2}ab$ **7.** 158 **9.** 30°,
150° **11.** $3200\sqrt{2}$ cm^2 **13.** 21 cm^2
15. $K = 6 \sin \theta$; Dom.: $\{\theta | 0 < \theta < \pi\}$; Range:
$\{K | 0 < K \le 6\}$ **17.b.** $2\frac{2}{3}$ **19.** 73 **21.** 13.6
23. 2.5 radians **27.** 22.1 **29.** 6.64 **31.a.** area
$\triangle BCD = \frac{\sqrt{3}}{4}ax$; area $\triangle ACD = \frac{\sqrt{3}}{4}bx$ **b.** $\frac{\sqrt{3}}{4}ab$
d. Yes **33.** 0.56 m from bottom **35.** 1.17

Class Exercises, pages 346–347 **1.** the ratio of the
sine of any $\angle$ to the length of the opp. side
2. They all $= \frac{1}{2}$. **3.** $\angle B < 90°$ **4.** $b > c$
5. $\angle A > \angle B$ **6.** $\frac{\sin 60°}{8} = \frac{\sin 25°}{x}$
7. $\frac{\sin 48°}{9} = \frac{\sin x°}{7}$ **8.** $\frac{\sin 112°}{7} = \frac{\sin 40°}{x}$

Written Exercises, pages 347–349 **1.a.** 0 **b.** 1 **c.** 2
d. 1 **3.** $\angle C = 75.0°$; $b = 7\sqrt{6}$; $c = 19.1$
5. $\angle C = 15.0°$; $a = 4\sqrt{2}$; $c = 2.07$
7. Either $\angle A = 115.7°$; $\angle B = 39.3°$; $a = 4.26$, or
$\angle A = 14.3°$; $\angle B = 140.7°$; $a = 1.17$
9. $\angle B = 18.9°$; $\angle C = 85.1°$; $c = 12.3$
11. $\angle A = 4.0°$; $\angle B = 88.0°$; $a = 0.489$
13. No sol. **15.** $\angle S = 28.8°$; $\angle T = 11.2°$
17. 15.5 km **19.** $\frac{25\sqrt{2}}{3}$ **21.a.** 1580 yd **b.** 1410 yd
c. 193 yd/min **d.** 5.71 knots
25.b. $K = \frac{1}{2}\left(\frac{\sin A \sin C}{\sin B}\right)b^2$; $K = \frac{1}{2}\left(\frac{\sin A \sin B}{\sin C}\right)c^2$

Class Exercises, page 352
1. $x^2 = 5^2 + 6^2 - 2 \cdot 5 \cdot 6 \cos 35°$
2. $x^2 = 5^2 + 10^2 - 2 \cdot 5 \cdot 10 \cos 115°$
3. $\cos x° = \frac{5^2 + 6^2 - 7^2}{2 \cdot 5 \cdot 6}$ **4.** 6.23 **5.** $\angle Y = 100.4°$;
$\angle X = 29.6°$ **6.** $\angle X < \angle Z < \angle Y$

Written Exercises, pages 352–354 **1.** $\angle A = 81.8°$;
$\angle B = 38.2°$; $c = 7$ **3.** $\angle X = 12.7°$; $\angle Y = 77.3°$;
$\angle Z = 90°$ **5.** $\angle P = 20.7°$; $\angle Q = 109.3°$; $r = 6.49$
7. $\angle A = 32.2°$; $\angle B = 27.8°$; $\angle C = 120.0°$
9. 7.07 **11.** $\sqrt{10}$ **13.** 9.74 cm, 13.3 cm **15.** 146
17. 15.8 mi **21.a.** 8; 12 **b.** $3\sqrt{6}$ **23.b.** $x = \frac{c}{2}$

Mixed Trigonometry Exercises, pages 355–358
1. Area $= 16.1$; $p = 5.57$ **3.** 90° **5.** $r \approx 3.35$;
$s = \sqrt{6} \approx 2.45$ **9.** $\sin A = 0.8$; $\tan A = -\frac{4}{3}$
11. 17.8 ft **13.** 32.7°, 147.3° **15.** 19.5°, 160.5°
17.a. $a = 14.0$; $e = 8.57$; $f = 14.3$ **b.** $49:100$
19. $6\sqrt{5} \approx 13.4$, $2\sqrt{13} \approx 7.21$ **21.** \$478.77
23. 1.15 nautical mi ≈ 1.32 mi **25.** $\sin A = \frac{\sqrt{2}}{2}$;
$\sin B = \frac{3}{5}$; $a = 15\sqrt{2}$ **27.** $\sin x = -\frac{\sqrt{26}}{26}$;
$\cos x = \frac{5\sqrt{26}}{26}$ **29.a.** $\angle A = 38.2°$; $\angle B = 81.8°$;
$\angle C = 60°$ **b.** $10\sqrt{3} \approx 17.3$ **c.** $\frac{5\sqrt{3}}{2} \approx 4.33$
d. $\sqrt{21} \approx 4.58$ **e.** $\frac{5\sqrt{7}}{3} \approx 4.41$ **f.** $\frac{7\sqrt{3}}{3} \approx 4.04$
g. $\sqrt{3} \approx 1.73$ **31.a.** $\angle A = 25.0°$; $\angle B = 28.1°$;
$\angle C = 126.9°$; $c = 17$ **b.** 36 **c.** 7.2
d. $4\sqrt{10} \approx 12.6$ **e.** $\frac{36\sqrt{17}}{13} \approx 11.4$ **f.** $\frac{85}{8} \approx 10.6$
g. 2 **33.a.** ≈ 2.80; ≈ 3.92 **43.** $\frac{6}{7}$

Class Exercises, pages 361–362 **2.** \$146 **3.** 040°
4. 115° **5.** 200° **6.** 300° **7.a.** 090° **b.** 270°

8.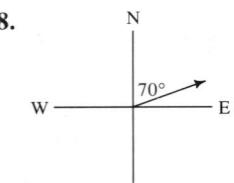

9.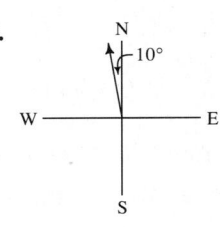

10.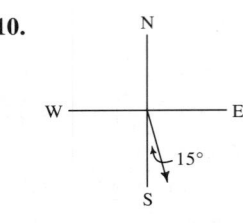

11.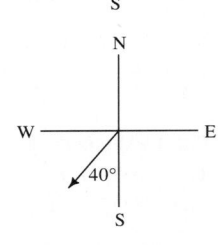

12. NE = 045°; SE = 135°; NW = 315°;
SW = 225° **13.** 337.5° **14.** 72°

Written Exercises, pages 362–364 **5.** 260°
7. 2255 km **9.** 030°; 17.3 km **11.** 18.6 nautical
mi **13.** 26 nautical mi; 358° **15.** 125,000 m^2
17. 29,600 m^2

Chapter Test, page 365 **1.** 75.5°; 75.5°; 29.0°
2. 53.2 m **3.** 38.0 in.^{2} **4.** 448 **5.a.** 0 **b.** 1
6. ≈32.0 km from A; ≈21.3 km from B **7.** 97.9°
8. 3.86 mi **9.** Refer to the chart at the top of
page 352. **10.** 332 km

Class Exercises, page 372 **Exs. 1–4:** Ans. may
vary. **1.** Let $x = 60°$, $y = 30°$. **2.** Let $a = 60°$,
$b = 0°$. **3.** Yes; let $b = 0°$. **4.** Yes; let $b = 0°$.
5. sin 3° **6.** sin 5° **7.** $\cos \frac{\pi}{2}$ **8.** cos 50°
9. $\sin(-\alpha) = \sin(0 - \alpha)$

Written Exercises, pages 373–374 **1.** 1 **3.** 0
5. sin x **11.** $y = \sin x$ **13.** $\frac{\sqrt{6} - \sqrt{2}}{4}$
15. $\frac{\sqrt{2} - \sqrt{6}}{4}$ **17.** $\frac{\sqrt{2} - \sqrt{6}}{4}$ **19.** $\frac{\sqrt{6} + \sqrt{2}}{4}$
21. cos θ **23.** cos x **25.** 0 **27.** $\frac{3}{5}$ **29.** $\frac{63}{65}$
31. Same graph as $y = \sin(x + 1)$ **33.** 2 tan α
35. $\sin(x + y)$ **37.** tan x + tan y **41.b.** $-\frac{\sqrt{2}}{2}$
43. 0

Class Exercises, page 377 **1.a.** −1 **b.** $-\frac{1}{7}$
2.a. 1 **b.** $\sqrt{3}$ **3.** The lines are ⊥.

Written Exercises, pages 377–379 **1.a.** $\frac{7}{4}$ **b.** $\frac{1}{8}$
3. 1 **5.** −1 **7.** 3 **9.** Use tan $(-\alpha)$ = tan $(0 - \alpha)$
11. $2 + \sqrt{3}$, $-2 + \sqrt{3}$ **13.** 26.6°, 153.4°
15. Graph of $y = \tan x$ shifted one unit left
17. tan $(\alpha + \beta) = 1$ **21.** Both are true for α or
$\beta = n\pi$, n an integer. **23.** $\frac{1}{8}$ **25.a.** $\frac{63}{65}$ **b.** $-\frac{16}{65}$
c. $-\frac{63}{16}$ **27.b.** $y = \frac{1 + \sqrt{10}}{3} x$

Class Exercises, page 383 **1.** sin 20° **2.** $\frac{\sqrt{3}}{2}$
3. cos 70° **4.** cos 50° **5.** tan 100° **6.** tan 80°
7. $\cos^2 x$ **8.** cos 2x **9.** sin 6α **10.** cos 10θ
11. Use the formula for $\cos \frac{\alpha}{2}$ with $\alpha = 70°$.

Written Exercises, pages 383–385 **1.** cos 20°
3. 2 tan 2β **5.** sin 70° **7.** tan 50° **9.** cos x
11. sin 40° **13.** $\frac{\sqrt{2}}{2}$ **15.** $\frac{\sqrt{3}}{2}$ **17.** $\frac{1}{4}$
19. $\sin 2A = \frac{120}{169}$; $\cos 2A = \frac{119}{169}$ **21.** $\sin 2A = \frac{24}{25}$;
$\sin 4A = \frac{336}{625}$ **23.** $\cos 2A = -\frac{23}{25}$; $\cos \frac{A}{2} = \frac{\sqrt{15}}{5}$
25.a. $\frac{\sqrt{2} - \sqrt{6}}{4}$ **b.** $-\frac{\sqrt{2 - \sqrt{3}}}{2}$ **27.** Range:
$\{y | -2 \leq y \leq 1.13\}$; $p = 2\pi$ **29.** Range:
$\{y | -6 \leq y \leq 4.13\}$; $p = \pi$ **39.** sin 2x **41.** cos 2x
43. 2 **47.** −1 **53.c.** 4

Class Exercises, page 389 **1.** Substitute
$2\cos^2 x - 1$ for cos 2x and solve algebraically for
x. **2.** Solve algebraically for x.
3. Rewrite as tan $x = 1$, cos $x \neq 0$, and solve for x.
4. Rewrite as tan 2$x = 1$, cos 2$x \neq 0$, and solve for
2x. **5.** Rewrite as tan 3$x = 1$, cos 3$x \neq 0$, and
solve for 3x. **6.** Solve for $x - 10°$. **7.** 2x **8.** 2x
9.b. $\frac{\pi}{4}$ **c.** $\frac{\pi}{4} < x \leq \frac{\pi}{2}$ **d.** $0 \leq x < \frac{\pi}{4}$

Written Exercises, pages 389–391 **1.** 0°, 120°, 240°
3. 45°, 225° **5.** 15°, 75°, 135°, 195°, 255°, 315°
7. 0°, 30°, 90°, 150°, 180°, 210°, 270°, 330° **9.b.** 0,
π, $\frac{\pi}{4}$, $\frac{3\pi}{4}$, $\frac{5\pi}{4}$, $\frac{7\pi}{4}$ **c.** $0 < x < \frac{\pi}{4}$, $\frac{\pi}{2} < x < \frac{3\pi}{4}$,
$\pi < x < \frac{5\pi}{4}$, $\frac{3\pi}{2} < x < \frac{7\pi}{4}$ **11.** $0 \leq x < 1.11$,
$4.25 < x \leq 6.28$ **13.** $0.52 < x < 1.57$,
$2.62 < x < 3.67$, $4.71 < x < 5.76$

15. $0.52 \le x \le 1.57$, $2.62 \le x \le 4.71$ **17.** $15°$, $255°$ **19.** $19.1°$, $199.1°$ **21.** $0°$, $60°$, $180°$, $300°$ **23.** $\frac{\pi}{4}$, $\frac{5\pi}{4}$ **25.** 0, π, $\frac{\pi}{6}$, $\frac{5\pi}{6}$, $\frac{7\pi}{6}$, $\frac{11\pi}{6}$ **27.** $\frac{\pi}{6}$, $\frac{5\pi}{6}$, $\frac{7\pi}{6}$, $\frac{11\pi}{6}$ **29.** $\frac{\pi}{6}$, $\frac{5\pi}{6}$ **31.** 0 **33.a.** No **c.** 3 **d.** 0.9 **35.** 5 **37.** ±5.41, ±3.80, ±2.14, 0 **39.** -0.42, 0.62, 2.40, 3.91, 5.50 **41.** $\frac{\pi}{4}$, $\frac{3\pi}{4}$, $\frac{5\pi}{4}$, $\frac{7\pi}{4}$, $\frac{\pi}{3}$, $\frac{5\pi}{3}$ **43.** $\frac{\sqrt{5}}{5}$

Chapter Test, page 393 **1.a.** $\frac{1}{2}$ **b.** 1 **c.** $\sqrt{3}\cos x$ **d.** $-\sqrt{2}\sin x$ **2.** $\frac{\sqrt{6}+\sqrt{2}}{4}$ **3.** 2 **4.** Use sum and diff. formulas to show they both $= \frac{1}{2}$. **5.** Ans. may vary. Use the slopes of the lines and the formula $\tan\theta = \frac{m_1-m_2}{1+m_1 m_2}$ to find one $\angle$. Find the second $\angle$ using the fact that the $\angle$s are supplementary. **6.a.** $\frac{3}{5}$ **b.** $\frac{7}{25}$ **c.** $\frac{24}{25}$ **d.** $\frac{336}{625}$

7.a. $\cot x$ **b.** $-2\tan^2 y$ **c.** $\tan\frac{t}{2}$ **d.** $\cos x$

8.a. $\frac{\sqrt{3}}{2}$ **b.** $\sqrt{3}$ **9.a.** $(1+\cot^2 x)(1-\cos 2x) = \csc^2 x(1-(1-2\sin^2 x)) = \csc^2 x(2\sin^2 x) = 2$
b. $\frac{\sin\theta\sec\theta}{\tan\theta+\cot\theta} = \frac{\sin\theta}{\cos\theta} \div \left(\frac{\sin\theta}{\cos\theta}+\frac{\cos\theta}{\sin\theta}\right) =$
$\frac{\sin\theta}{\cos\theta} \div \left(\frac{\sin^2\theta+\cos^2\theta}{\cos\theta\sin\theta}\right) = \frac{\sin\theta}{\cos\theta}\cdot\frac{\cos\theta\sin\theta}{1} =$
$\sin^2\theta = 1-\cos^2\theta = \cos^2\theta+\sin^2\theta-\cos^2\theta = \cos^2\theta-\cos 2\theta$

10.a. 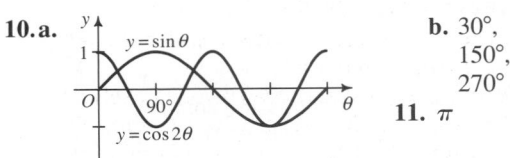 **b.** $30°$, $150°$, $270°$ **11.** π

Class Exercises, page 399 **Exs. 1–3.:** Ans. may vary.

1.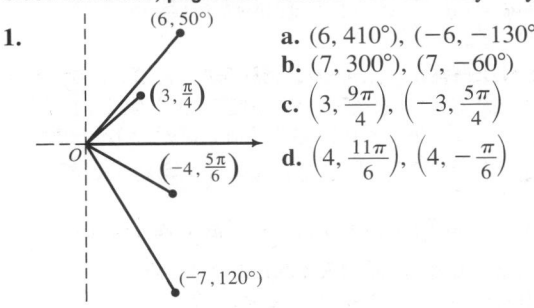
a. $(6, 410°)$, $(-6, -130°)$
b. $(7, 300°)$, $(7, -60°)$
c. $\left(3, \frac{9\pi}{4}\right)$, $\left(-3, \frac{5\pi}{4}\right)$
d. $\left(4, \frac{11\pi}{6}\right)$, $\left(4, -\frac{\pi}{6}\right)$

2.a. $(5, 315°)$, $(-5, 135°)$ **b.** $(2, 120°)$, $(-2, 300°)$
c. $\left(1, \frac{3\pi}{2}\right)$, $\left(-1, \frac{\pi}{2}\right)$ **d.** $\left(\frac{3}{2}, 0\right)$, $\left(-\frac{3}{2}, \pi\right)$
3. $(\sqrt{2}, 315°)$ **4.** $(-4, -4\sqrt{3})$
5. $(x-1)^2 + y^2 = 1$; a circle with $r = 1$, $C(1, 0)$; both graphs are the same.

Written Exercises, pages 400–402 **1.a.** $(2, 400°)$, $(-2, 220°)$ **b.** $(3, 260°)$, $(-3, 80°)$ **c.** $\left(5, -\frac{\pi}{2}\right)$, $\left(-5, \frac{\pi}{2}\right)$ **d.** $(4, 190°)$, $(-4, 370°)$

3.a. $(2\sqrt{2}, 135°)$ **b.** $(5, 0°)$ **c.** $(2, -45°)$
d. $(2, 150°)$ **5.a.** $\left(\sqrt{2}, \frac{5\pi}{4}\right)$ **b.** $\left(12, \frac{\pi}{2}\right)$ **c.** $\left(1, \frac{\pi}{3}\right)$
d. $(2, \pi)$ **7.a.** $(-2, 2\sqrt{3})$ **b.** $(0, -3)$
c. $\left(-\frac{\sqrt{3}}{2}, \frac{1}{2}\right)$ **d.** $(-\sqrt{2}, \sqrt{2})$ **9.a.** $(0.940, 0.342)$
b. $(1.879, 0.684)$ **c.** $(-0.416, 0.909)$
d. $(-0.416, -0.909)$ **11.a.** $(5, 53.1°)$
b. $(2.2, 63.4°)$ **c.** $(3.6, 123.7°)$ **d.** $(7.2, -56.3°)$

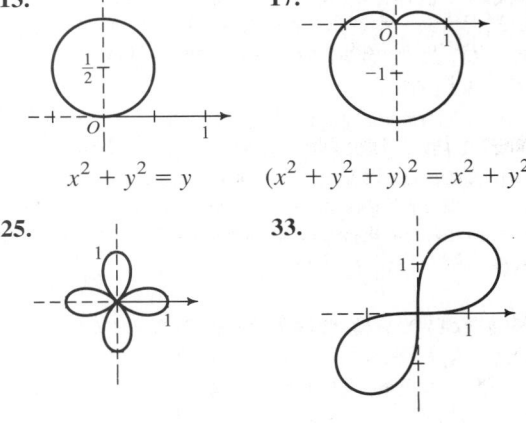

13.
$x^2 + y^2 = y$

17.
$(x^2 + y^2 + y)^2 = x^2 + y^2$

25.

33.

Computer Exercises, page 402 **1.b.** Roses with k leaves **d.** Roses with $2k$ leaves **e.** Roses with n leaves if n is odd and $2n$ leaves if n is even **3.** The graph will have a leaves and will cross the horiz. axis, $t = n\pi$, b times.

Class Exercises, page 405 **1.** $\sqrt{2}$ cis $45°$ **2.** cis $90°$ **3.** 3 cis $180°$ **4.** 2 cis $330°$ **5.** $2\left(\cos\frac{\pi}{4} + i\sin\frac{\pi}{4}\right)$ **6.** $5i$ **7.** -3 **8.** $2\sqrt{2} + 2i\sqrt{2}$ **9.** $3\sqrt{3} + 3i$ **10.** 24 cis $60°$ **11.** 10 cis π

Written Exercises, pages 406–407   **1.** $\sqrt{2}$ cis 135°
3. 2 cis 60°   **5.** 7 cis 180°   **7.** 5 cis 307°
9. $-1.04 + 5.91i$ **11.** $-\dfrac{9}{2} - \dfrac{9\sqrt{3}}{2}i$

13. 10 cis 90°, $10i$ **15.** 4 cis $\dfrac{5\pi}{3}$, $2 - 2i\sqrt{3}$

17.a. $4\sqrt{3} + 4i$ **b.** $z_1 = 4$ cis 60°, $z_2 = 2$ cis 330°,
$z_1 z_2 = 8$ cis 30° $= 4\sqrt{3} + 4i$ **19.a.** 8
b. $z_1 = 2\sqrt{2}$ cis 45°, $z_2 = 2\sqrt{2}$ cis 315°,
$z_1 z_2 = 8$ cis 0° $= 8$ **21.a.** $-22 + 32i$
b. $z_1 = 7.211$ cis 326.31°, $z_2 = 5.385$ cis 158.199°,
$z_1 z_2 = 38.831$ cis 484.509° $\approx -22 + 32i$ **23.** 45°
25. If $z = r$ cis α, $z^2 = r^2$ cis 2α.
27. $(\cos\theta + i\sin\theta)^3 = \cos 3\theta + i\sin 3\theta$

Class Exercises, page 409   **1.** 4 cis 90°
2. 8 cis 135°   **3.** 4 cis $(-72°)$ **4.** 1 cis 360°

5. 64 cis $\dfrac{\pi}{2}$ **6.** 27 cis 5π **7.a.** $z = \sqrt{2}$ cis 135°,

$z^6 = 8$ cis 90°   **b.** $8i$
8.a. 32 cis 3150° $= 32$ cis 270° $= -32i$
b. 32 cis $(-450°) = 32$ cis 270° $= -32i$ **c.** Part
(b) is easier; 3150° involves a larger multiple of
360° than $-450°$ does.

Written Exercises, pages 410–411   **1.** $z =$ cis 30°,
$z^2 =$ cis 60°, $z^3 =$ cis 90°   **3.a.** $z = \sqrt{2}$ cis 315°
b. $(\sqrt{2})^{12}$ cis $[12(-45°)] = 64$ cis $(-540°) = -64$
5.a,b. -8 **7.** Yes; z^2, z^4, z^5, z
9.a. $z^0 = r^0$ cis $(0 \cdot \alpha) = 1$ cis 0° $= 1$

b. $z^{-1} = \dfrac{1}{z} = \dfrac{1}{r}$ cis $(-\theta) = r^{-1}$ cis $(-1 \cdot \theta)$

11. Ans. may vary. $z = 1$ cis 90° or $z = -i$
13. $e^{i\pi} = 1$ cis $\pi = \cos\pi + i\sin\pi = -1$

Written Exercises, pages 413–414   **1.** $\dfrac{\sqrt{3}}{2} + \dfrac{1}{2}i$,

$-\dfrac{\sqrt{3}}{2} + \dfrac{1}{2}i$, $-i$ **3.** 2, $-1 + i\sqrt{3}$, $-1 - i\sqrt{3}$

5. 2, -2, $2i$, $-2i$ **7.b.** sum $= 0$, prod. $= 8i$
9. 2, $-1 + i\sqrt{3}$, $-1 - i\sqrt{3}$
11.a. $(z^2 + 2z + 2)(z^2 - 2z + 2)$ **b.** $1 + i$, $1 - i$,
$-1 + i$, $-1 - i$ **13.a.** Regular pentagon inscribed
in a circle with $C(0, 0)$ and $r = 10$ **b.** Regular
nonagon (9-gon) inscribed in a circle with $C(0, 0)$
and $r = 1$ **15.** $1.7895 + 0.3155i$,
$-1.1680 + 1.3920i$, $-0.6215 - 1.7075i$

17. $P(n) = 2n \sin\left(\dfrac{180°}{n}\right)$

Chapter Test, page 415

1.a. $(3\sqrt{2}, 45°)$ **2.a.**
b. $(0, 6)$
c. $(2, 270°)$
d. $(-8, 0)$
e. $(-1, -\sqrt{3})$
f. $(2, 120°)$

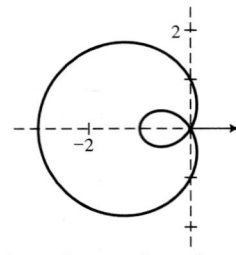

b. $(x^2 + y^2 + 2x)^2 = x^2 + y^2$

3.a,b,c.

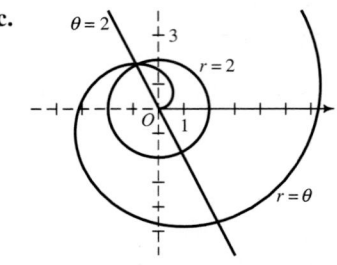

4.a. $z_1 = 2$ cis 330°; $z_2 = 4\sqrt{2}$ cis 45°;
$z_1 z_2 = 8\sqrt{2}$ cis 15°   **b.** See graph below.
5.a. 4 cis 240°; $-2 - 2i\sqrt{3}$

b. 4 cis 240° $= 4\left(-\dfrac{1}{2} - \dfrac{\sqrt{3}}{2}i\right) = -2 - 2i\sqrt{3}$

6.a. $\sqrt{2}$ cis 225°   **b.** See graph below.   **c.** $32i$

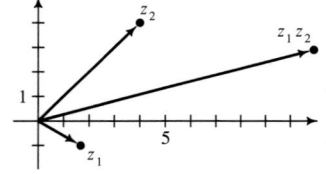
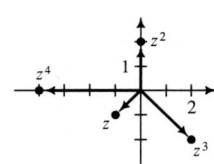

Ex. 4.b.   **Ex. 6.b.**

7. Ans. may vary. Express the complex
no. in polar form, $z = r$ cis θ. The n nth roots are
$\sqrt[n]{r}$ cis $\left(\dfrac{\theta}{n} + k \cdot \dfrac{360°}{n}\right)$ for $k = 0, 1, 2, \ldots, n - 1$.
Since $\sqrt[n]{z} = z^{1/n}$, by DeMoivre's Theorem,
$\sqrt[n]{z} = z^{1/n} = r^{1/n}$ cis $\left(\dfrac{1}{n} \cdot \theta\right) = \sqrt[n]{r}$ cis $\dfrac{\theta}{n}$.

8. $(\sqrt[6]{2}$ cis 135°$)^3 = \sqrt{2}$ cis 405° $= \sqrt{2}$ cis 45° $=$
$\sqrt{2}\left(\dfrac{\sqrt{2}}{2} + \dfrac{\sqrt{2}}{2}i\right) = 1 + i$; the other two cube
roots are $\sqrt[6]{2}$ cis 15° and $\sqrt[6]{2}$ cis 255°.

Cumulative Review, pages 416–417   **1.a.** 325°,
$-395°$ **b.** $\dfrac{17\pi}{6}$, $-\dfrac{7\pi}{6}$ **2.a.** $\dfrac{11\pi}{12}$ **b.** 3.63

3.a. $-210°$ **b.** $103°10'$ or $103.1°$ **4.** $r = 9$ cm;
$s = 1.8\pi$ cm **5.** $-\dfrac{\sqrt{5}}{3}$ **6.a.** $-\cos 56°$ **b.** $\sin 55°$
c. $-\sin 62°$

7.

	sin	cos	tan	csc	sec	cot
a. $\dfrac{\pi}{6}$	$\dfrac{1}{2}$	$\dfrac{\sqrt{3}}{2}$	$\dfrac{\sqrt{3}}{3}$	2	$\dfrac{2\sqrt{3}}{3}$	$\sqrt{3}$
b. $225°$	$-\dfrac{\sqrt{2}}{2}$	$-\dfrac{\sqrt{2}}{2}$	1	$-\sqrt{2}$	$-\sqrt{2}$	1
c. $-\dfrac{\pi}{2}$	-1	0	undef.	-1	undef.	0

8. $\sin x = -\dfrac{12}{13}$; $\cos x = -\dfrac{5}{13}$; $\tan x = \dfrac{12}{5}$;
$\csc x = -\dfrac{13}{12}$; $\sec x = -\dfrac{13}{5}$

9.a.
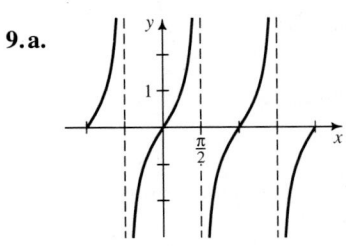

10.a. $\dfrac{\pi}{4}$ **b.** $\dfrac{\pi}{3}$
c. $\dfrac{5\pi}{6}$ **d.** $\dfrac{\pi}{3}$ **e.** 0
f. $\dfrac{4}{5}$ **g.** $\dfrac{13}{12}$ **h.** 0
11. 1.16, 5.12
12. $37°$

b. Dom.: $\left\{x \mid x \neq \dfrac{\pi}{2} + n\pi\right\}$;
Range: all real nos.; $p = \pi$

13.a.
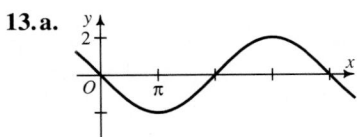

b.
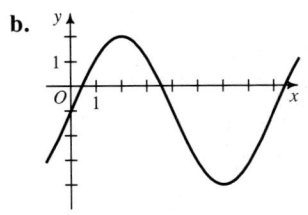

14. $-\dfrac{\cos\theta}{1 + \cos\theta}$ **15.** 0, $\dfrac{\pi}{4}$, π, $\dfrac{5\pi}{4}$ **16.** $a \approx 7.42$;
$b \approx 11.9$ **17.** 23.1 cm^2 **18.** 6.0 mi **19.** 10.5
20. 136 m **21.** $\dfrac{56}{65}$ **22.** $\dfrac{\sqrt{2} + \sqrt{6}}{4}$ **23.** -3
24. $\sin 2\alpha = \dfrac{4\sqrt{2}}{9}$; $\cos\dfrac{1}{2}\alpha = \dfrac{\sqrt{6}}{3}$ **25.** $(2, 120°)$;
$(2, -240°)$, $(2, 480°)$ **26.** $(-3\sqrt{3}, -3)$

27.a.
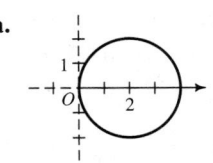

28.
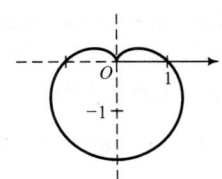

b. $(x - 2)^2 + y^2 = 4$ cardioid
29.a. $2 + 2i\sqrt{3}$ **b.** $z_1 = 2$ cis $300°$;
$z_2 = 2$ cis $120°$; $z_1 z_2 = 4$ cis $420°$
c. 4 cis $420° = 4$ cis $60° = 4\left(\dfrac{1}{2} + \dfrac{\sqrt{3}}{2}i\right) =$
$2 + 2i\sqrt{3}$ **30.** $-512i$ **31.** $\sqrt[6]{2}$ cis $105°$,
$\sqrt[6]{2}$ cis $225°$, $\sqrt[6]{2}$ cis $345°$

Class Exercises, pages 422–423 **1.** T **2.** F **3.** F
4. T **5.** T **6.** T **7.** T **8.** T **9.** Yes; no
10. $\overrightarrow{BC}$ **11.** 6 **12.** $\overrightarrow{BO}$ or $\overrightarrow{OD}$ **13.** $\overrightarrow{AC}$ **14.** $\overrightarrow{AC}$
15. $\overrightarrow{AB}$ **16.** $\overrightarrow{OD}$; $\overrightarrow{AD}$ **17.** $\overrightarrow{DB}$; $\overrightarrow{DC}$

Written Exercises, pages 423–426 **1.b.** 200 mi/h to
the NW **c.** 100 mi/h to the SE **3.a.** $\overrightarrow{PR}$ **b.** 5 **c.** 0

5.
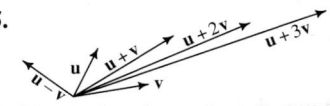

7. $2(\mathbf{u} + \mathbf{v}) = 2\mathbf{u} + 2\mathbf{v}$ **9.** $219.8°$ **11.** 436.8
knots; $85.4°$ **15.c.** $\sqrt{97}$ mi **17.a.** $\mathbf{a} + \mathbf{b}$
b. $2(\mathbf{a} + \mathbf{b}) = 2\mathbf{a} + 2\mathbf{b}$ **c.** The seg. joining the
midpts. of 2 sides of a $\triangle$ is $\parallel$ to the third side and
is half as long as the third side. **19.a.** $\dfrac{2}{3}\mathbf{v}$ **b.** $\dfrac{1}{3}\mathbf{v}$
c. $-\dfrac{1}{3}\mathbf{v}$ **21.a.** $\mathbf{u} + \mathbf{v}$ **b.** $\dfrac{1}{2}\mathbf{u} + \dfrac{1}{2}\mathbf{v}$ **c.** $\dfrac{1}{2}\mathbf{v} - \dfrac{1}{2}\mathbf{u}$
d. $\dfrac{1}{2}\mathbf{u} + \dfrac{7}{2}\mathbf{v}$ **23.a.** $\mathbf{w} - \mathbf{v}$ **b.** $\dfrac{1}{3}(\mathbf{w} - \mathbf{v})$
c. $\dfrac{2}{3}\mathbf{v} + \dfrac{1}{3}\mathbf{w}$ **d.** $\dfrac{4}{9}\mathbf{v} + \dfrac{2}{9}\mathbf{w}$ **e.** $\dfrac{2}{9}\mathbf{w} - \dfrac{5}{9}\mathbf{v}$
25. $158.9°$

Class Exercises, pages 428–429 **1.a.** $(1, 4)$;
$(4, -4)$ **b.** $|\overrightarrow{AB}| = \sqrt{17}$; $|\overrightarrow{CD}| = 4\sqrt{2}$ **c.** $\left(\dfrac{17}{4}, 1\right)$
d. $(0, -2)$ **2.a.** $(3, 6)$ **b.** $(4, 2)$ **c.** $(-2, 2)$
d. $(11, 4)$ **3.a.** $(-\sqrt{3}, 1)$ **b.** 2
4. $2\mathbf{v} = (6\cos 40°, 6\sin 40°)$;
$-\mathbf{v} = (-3\cos 40°, -3\sin 40°)$ **5.a.** $(7, 2)$
b. $(4, 9)$ **6.** Horiz.; $40\sqrt{3}$ lb ≈ 69.3 lb; yes

Written Exercises, pages 429–432 **1.** $(2, 0)$; 2
3. $(-2, 6)$; $2\sqrt{10}$ **5.** $(1.85, 5.71)$

7. $(-1, -1.73)$ **9.a.** $(-5, 5)$ **b.** $(11, -3)$
c. $(3, 1)$ **d.** $\sqrt{10}$ **11.a.** $\mathbf{0} = (0, 0)$ **b.** $|\mathbf{0}| = 0$
c. $2\sqrt{10}$ **13.** $\left(3, \frac{3}{2}\right)$ **15.** $(3, 6)$ **17.** $\left(-\frac{17}{5}, -\frac{11}{5}\right)$
21.a. $(20, 20\sqrt{3})$ **b.** Horiz.   **c.** More easily; the
horiz. component is greater with a $50°$ $\angle$ than with
a $60°$ $\angle$.   **23.** 4.6 knots S; 14.3 knots E

25.
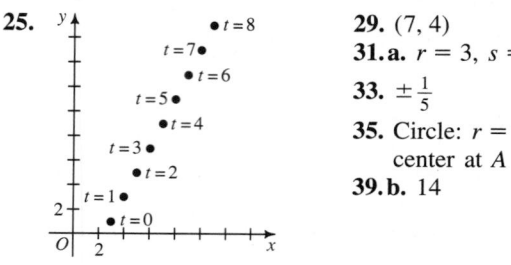

29. $(7, 4)$
31.a. $r = 3$, $s = 2$
33. $\pm\frac{1}{5}$
35. Circle: $r = 2$;
center at A
39.b. 14

Class Exercises, page 435   **1.a.** $(2, -5)$, $(3, -2)$,
$(4, 1)$ **b.** $(1, 3)$ **c.** $x = 2 + t$, $y = -5 + 3t$
2. $(x, y) = (-3, 1) + t(4, 3)$

Written Exercises, pages 435–440   Exs. 1–8: Ans.
may vary.
1. $(x, y) = (1, 5) + t(2, -1)$; $x = 1 + 2t$, $y = 5 - t$
3. $(x, y) = (1, 0) + t(2, -4)$; $x = 1 + 2t$,
$y = -4t$ **5.** $(x, y) = (-2, 3) + t(7, -2)$;
$x = -2 + 7t$, $y = 3 - 2t$
7. $(x, y) = (\pi, e) + t(1, 0)$; $x = \pi + t$, $y = e$

9.a.
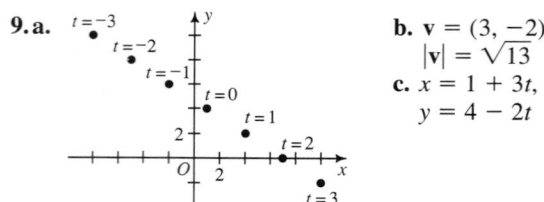

b. $\mathbf{v} = (3, -2)$;
$|\mathbf{v}| = \sqrt{13}$
c. $x = 1 + 3t$,
$y = 4 - 2t$

11. $(x, y) = (2, 3) + t(3, -1)$; $x = 2 + 3t$,
$y = 3 - t$ **13.** $x = 3 + 2t$, $y = 2 + 4t$; $2x - y = 4$
15.a. Vert. line through $(2, 0)$ **b.** $(0, 1)$ **c.** Slope is
undef.   **17.a.** 1.5   **b.** 1.5   **c.** Slopes are $=$.
d. $(x, y) = (7, 9) + t(2, 3)$

19.a. $\mathbf{v} = (-3, 2)$;
$|\mathbf{v}| = \sqrt{13}$
b. $t = -1$;
$(5, -3)$
21. $t = 1$, $(0, 2)$;
$t = -2$, $(3, -1)$
See graph at right.
23.a. Yes, at $(-1, 3)$
b. No   **c.** No

25. Circle: radius r; center $(0, 0)$
27.a. Ellipse; $V(0, \pm 5)$
b. $25x^2 + 4y^2 = 100$ **c.** $x = 3 \cos t$, $y = 6 \sin t$
29.a. $(x - 3)^2 + (y - 5)^2 = 36$ **b.** Before:
$x = 6 \cos t$, $y = 6 \sin t$; after: $x = 3 + 6 \cos t$,
$y = 5 + 6 \sin t$ **31.b.** (1) I (2) III (3) II (4) IV

35.
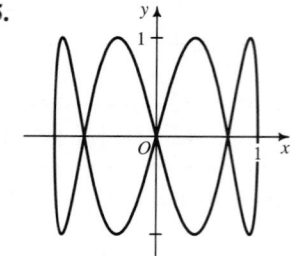

39.a. $y = \frac{\sqrt{3}}{3}x - \frac{4}{675}x^2$ **b.** No   **c.** For example,
$x = 90$, $y = 5t$, $0 \le t \le 2$ **d.** $36°$

Computer Exercise, page 440   **1.** 7.999672
2. 9.42271

Class Exercises, page 443   **1.** $\mathbf{u} \parallel \mathbf{v}$; $\mathbf{v} \perp \mathbf{w}$; $\mathbf{u} \perp \mathbf{w}$
2.a. 2   **b.** 4   **c.** 4   **d.** 4   **3.a.** 5   **b.** 25   **c.** 25
4.a. $5\sqrt{2}$ **b.** $5\sqrt{2}$ **c.** 40   **d.** $37°$ **5.** 0.8

Written Exercises, pages 444–446   **1.a.** -7 **b.** 1
3.a. 2   **b.** -4.5 **5.a.** 13   **b.** 13   **13.** $106.3°$
15. $53.1°$ **19.a.** $\cos C = 0.6$, $\sin C = 0.8$ **b.** 20
21. $73.7°$ **25.c.** 350

Class Exercises, page 449   **1.a.** $\sqrt{14}$; $\left(\frac{1}{2}, -1, \frac{3}{2}\right)$
b. $2\sqrt{14}$; $(4, 2, -1)$ **2.a.** $A = (4, 0, 0)$,
$B = (4, 5, 0)$, $C = (0, 5, 0)$, $D = (4, 0, 3)$,
$E = (0, 0, 3)$, $F = (0, 5, 3)$ **b.** $5\sqrt{2}$
3.a. $x^2 + y^2 + z^2 = 25$ **b.** $(x - 1)^2 +$
$(y - 2)^2 + (z - 3)^2 = 25$ **4.a.** $(5, 9, 4)$
b. 8   **5.** If $\mathbf{u} \cdot \mathbf{v} = 0$, $\mathbf{u} \perp \mathbf{v}$. If $\mathbf{u} = k\mathbf{v}$, $\mathbf{u} \parallel \mathbf{v}$.
6. Ans. may vary.   **a.** $(2, 5, 1)$, $(8, 12, 9)$
b. $x = 2 + 6t$, $y = 5 + 7t$, $z = 1 + 8t$ **c.** $(6, 7, 8)$
d. Dir. vectors are $=$.   **e.** Dot prod. of dir. vectors $= 0$.
7. Subst. parametric eqs. obtained from vector eq.
of line into eq. of sphere.   **8.** The vectors are $\parallel$.

Written Exercises, pages 450–452   **1.** $2\sqrt{6}$;
$(1, 4, -1)$ **3.** $2\sqrt{14}$; $(1, -2, 1)$ **5.** $A = (5, 0, 0)$,
$B = (5, 6, 0)$, $C = (0, 6, 0)$, $D = (5, 0, 4)$,
$E = (0, 0, 4)$, $F = (0, 6, 4)$; $|\overrightarrow{OG}| = \sqrt{77}$
7.a. $(11, 6, 2)$ **b.** -5 **c.** $\sqrt{35}$ **9.** Yes
11.a. $x^2 + y^2 + z^2 = 4$
b. $(x - 3)^2 + (y + 1)^2 + (z - 2)^2 = 4$

13. $(-1, 2, 3)$; 5   **15.** 70.5°   **17.b.** 9
19.a. $\cos A = \dfrac{\sqrt{3}}{2}$, $\sin A = \dfrac{1}{2}$ **b.** $\dfrac{\sqrt{3}}{2}$
21.a. $x = -2 + 4t$, $y = -t$, $z = 1 + t$
b. $(-2, 0, 1)$, $(2, -1, 2)$ **c.** $(-10, 2, -1)$
d. $(x, y, z) = (1, 2, 3) + t(4, -1, 1)$
23. $(x, y, z) = (4, 2, -1) + t(2, 1, 3)$; $x = 4 + 2t$,
$y = 2 + t$, $z = -1 + 3t$
27.a. $(x, y, z) = (1, 2, 3) + t(1, 4, -6)$; $x = 1 + t$,
$y = 2 + 4t$, $z = 3 - 6t$
b. $(x, y, z) = (5, 3, 0) + t(6, -3, -1)$; $x = 5 + 6t$,
$y = 3 - 3t$, $z = -t$ **31.a.** The $\perp$ bis. of $\overline{AB}$
b. The plane $\perp$ to $\overline{AB}$ and passing through its
midpt.   **33.a.** The circle with diameter $\overline{AB}$ **b.** The
sphere with diameter $\overline{AB}$ **35.a.** The ellipse with
foci at A and B and major axis 10   **b.** The ellipsoid
with foci at A and B and major axis 10
37.a. Ans. may vary. $4x + 6y + 12z = 49$
39.a. $x^2 + y^2 + z^2 = 9$ **b.** Both are spheres.
41. Circle: $C(1, 2, 0)$, $r = 4$;
$(x - 1)^2 + (y - 2)^2 = 16$ **43.b.** No   **45.b.** Ans.
may vary. $(x, y, z) = (3, -1, 5) + r(3, -10, 6)$

Class Exercises, pages 454–455   **1.** x-int. 2;
y-int. 6; z-int. 3   **2.a.** $z = 4$ **b.** $z = 0$ **c.** $x = 3$
d. $y = 6$ **3.a.** $(2, 3, 4)$ **b.** $(3, 0, -4)$ **c.** $(0, 0, 1)$
4. $x + y + z = 12$ **5.** $6x + 7y + 8z = 0$

Written Exercises, pages 455–458

1.   **3.**

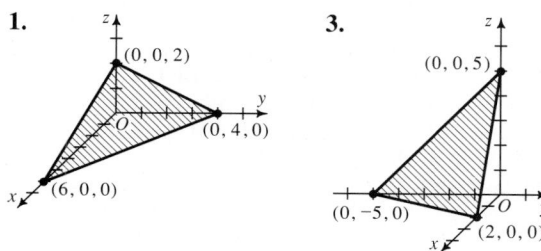

7.a. $z = 1$ **b.** $x = 4$ **c.** $y = 6$ **d.** $y = 0$
Exs. 9–13: Ans. may vary.   **9.** $(3, 4, 6)$
11. $(1, 1, 0)$ **13.** $(0, 0, 1)$ **15.** $2x + 3y + 5z = 44$
17. $z = 5$ **19.a.** $x + 2y + 3z = 26$
21.b. $2x + y + 2z = 9$ **23.** 16π **27.** 90°
29.a. $(2, 3, -1)$ **b.** Yes   **c.** Yes   **d.** Yes
31.a. $4x + y - 3z = 0$ **b.** $3x + 4z = 0$
33.a. $(4, -5, 2)$ **35.b.** $(2, 1, 0)$
37.a. $(x, y, z) = (3, 1, 5) + t(2, 2, 1)$
b. $(1, -1, 4)$ **c.** 3   **39.** $\dfrac{|ax_0 + by_0 + c|}{\sqrt{a^2 + b^2}}$

Class Exercises, page 460   **1.** -2 **2.** 12   **3.** 65
4. 31   **5.** 3; 2; 4

6. $4\begin{vmatrix} 5 & 7 \\ 8 & 6 \end{vmatrix} - 1\begin{vmatrix} 3 & 2 \\ 8 & 6 \end{vmatrix} + 9\begin{vmatrix} 3 & 2 \\ 5 & 7 \end{vmatrix}$ **7.** -3

Written Exercises, pages 460–461   **1.** -48
3. -3125 **5.** 435   **7.** 90   **9.** -30 **11.** No
13. 78

Class Exercises, page 463

1. $x = \dfrac{\begin{vmatrix} 7 & 4 \\ 8 & 6 \end{vmatrix}}{\begin{vmatrix} 3 & 4 \\ 5 & 6 \end{vmatrix}}$; $y = \dfrac{\begin{vmatrix} 3 & 7 \\ 5 & 8 \end{vmatrix}}{\begin{vmatrix} 3 & 4 \\ 5 & 6 \end{vmatrix}}$

2. $x = \dfrac{\begin{vmatrix} 4 & 1 & -2 \\ 1 & -1 & 4 \\ 3 & 2 & 7 \end{vmatrix}}{\begin{vmatrix} 3 & 1 & -2 \\ 2 & -1 & 4 \\ 1 & 2 & 7 \end{vmatrix}}$; $y = \dfrac{\begin{vmatrix} 3 & 4 & -2 \\ 2 & 1 & 4 \\ 1 & 3 & 7 \end{vmatrix}}{\begin{vmatrix} 3 & 1 & -2 \\ 2 & -1 & 4 \\ 1 & 2 & 7 \end{vmatrix}}$;

$z = \dfrac{\begin{vmatrix} 3 & 1 & 4 \\ 2 & -1 & 1 \\ 1 & 2 & 3 \end{vmatrix}}{\begin{vmatrix} 3 & 1 & -2 \\ 2 & -1 & 4 \\ 1 & 2 & 7 \end{vmatrix}}$ **3.** The den. $= 0$. There
is no sol. because
the lines are $\parallel$.
4. 2

Written Exercises, pages 463–464   **1.** $(1, 1)$
3. $(1, -2)$ **5.** $\left(\dfrac{1}{a + b}, \dfrac{1}{a + b}\right)$ **7.a.** No sol.
b. $\begin{vmatrix} a & b \\ d & e \end{vmatrix}$ **9.** 4   **11.** 8   **13.** 6   **15.** They are
collinear.   **17.** $(3, 2, 1)$ **19.** $(2, 0, 1)$ **21.** 5

Class Exercises, page 466

1.a. $\begin{vmatrix} i & j & k \\ 1 & 2 & 3 \\ 4 & 5 & 6 \end{vmatrix}$ **b.** $\begin{vmatrix} i & j & k \\ 4 & 5 & 6 \\ 7 & 8 & 9 \end{vmatrix}$

2.a. Vector   **b.** Scalar   **c.** Vector   **d.** Scalar
e. Vector   **f.** Scalar

Written Exercises, page 467   **1.** $(-1, -5, 4)$; $(1, 5, -4)$; yes   **5.** $\sqrt{42}$ **9.a.** $(-1, 4, 1)$
b. $-x + 4y + z = 3$ **c.** $\frac{3\sqrt{2}}{2}$ **11.a.** $(4, -4, -2)$
b. $2x - 2y - z = 2$ **c.** 3 **13.a.** $\frac{\sqrt{5}}{3}$ **b.** $\frac{2}{3}$ **15.** 73

Chapter Test, page 469   **1.** $112.6°$; 13 N
2.a. $-\mathbf{u}$ **b.** $\mathbf{v} - \mathbf{u}$ **c.** $\frac{1}{3}\mathbf{v} - \frac{1}{3}\mathbf{u}$ **d.** $\frac{2}{3}\mathbf{u} + \frac{1}{3}\mathbf{v}$
3. 367.7 mi/h N; 308.5 mi/h W   **4.** $(6, 6)$
5.a. $\mathbf{v} = (1, 4)$; $|\mathbf{v}| = \sqrt{17}$ **b.** $x = -2 + t$,
$y = -2 + 4t$ **c.** $t = \frac{3}{2}$, $\left(-\frac{1}{2}, 4\right)$; $t = 5$, $(3, 18)$
6. Since $\cos \theta = \dfrac{\mathbf{u} \cdot \mathbf{v}}{|\mathbf{u}|\,|\mathbf{v}|} = -1$, $\theta = 180°$ and $\therefore$ $\mathbf{u}$
and $\mathbf{v}$ are opp. in direction.   **7.** $136.8°$
8.a. $2\sqrt{62}$ **b.** $(-1, -1, 1)$
9.a. $(x, y, z) = (2, 1, 7) + t(4, -1, -3)$
b. $(-4, 0, 9)$ **10.** $2x - 2y - 5z = 20$ **11.a.** 22
b. 80 **12.** $\left(3, -\frac{1}{2}\right)$ **13.a.** Ans. may vary.
$(-16, 46, -24)$ **b.** $\dfrac{\sqrt{2948}}{2} \approx 27.1$

Class Exercises, page 476   **1.** Arith.; 5 **2.** Geom.; 2
3. Neither   **4.** Arith.; -6 **5.** Geom.; $-\frac{2}{3}$
6. Neither   **7.** $7, 12, 17, 22$; arith.   **8.** $\frac{2}{3}, \frac{3}{4}, \frac{4}{5}, \frac{5}{6}$;
neither   **9.** $3, 9, 27, 81$; geom.   **10.** $1, 8, 27, 64$;
neither   **11.a.** Yes; $d = 0$ **b.** Yes; $r = 1$ **12.** The
graph of the seq. is a set of discrete pts.; the graph
of the line is continuous.

Written Exercises, pages 476–479   **1.** $5, 7, 9, 11$;
arith.   **3.** $6, 12, 24, 48$; geom.   **5.** $2, \frac{5}{2}, \frac{10}{3}, \frac{17}{4}$;
neither   **7.** $-1, 2, -3, 4$; neither   **9.** $1, 0, -1, 0$;
neither   **11.** $t_n = -2 + 3n$ **13.** $t_n = 16(0.5)^n$
15. $t_n = 34 - 4n$ **17.** Arith.; $t_n = 13 + 4n$
19. Geom.; $t_n = \frac{16}{3}\left(\frac{3}{2}\right)^n$ **21.** Neither; $t_n = n^2$
23. Neither; $t_n = 10^n + 1$
25. Arith.; $t_n = (a - 3b) + n(a + b)$
27. Geom.; $t_n = 2^{-1/3}(2^n)$ **29.** 129 **31.** 149
33. 128 **35.** $-\frac{1}{2}$ **37.** 54 **39.** 38 **41.** 75 **43.** F
45. $t_n = \log a + (n - 1)\log r$ **47.a.** 3 **b.** 9
49. $x = 3$, $y = 2$ **51.** $\frac{1}{2}, \frac{1}{4}, \frac{1}{32}; \frac{1}{2^{d/2}}$
53. Ans. may vary. $1, 43, 85, 127, 169, \ldots$
55. $58, 135, 212, 289, 366$ **57.** $\triangle$ with sides $3n$,
$4n$, $5n$ **59.a.** Arith.: 6.5; geom.: 6 **b.** Arith.: 7.5;
geom.: $5\sqrt{2}$ **61.b.** y; c

Class Exercises, page 481   **1.** $5, 8, 11, 14$ **2.** $10,$
$12, 15, 19$ **3.** $3, 6, 12, 24$ **4.** $4, 7, 13, 25$ **5.** $5,$
$11, 24, 51$ **6.** $4, 16, 20, 36$

Written Exercises, pages 481–485   **1.** $14, 18, 22$
3. $5, 14, 41$ **5.** $9, 16, 25$ **7.** $6, 10, 16$ **9.** $9, 1,$
64 **11.** $t_n = 2 + 4n$ **13.** $t_1 = 9$; $t_n = t_{n-1} + 4$
15. $t_1 = 1$; $t_n = t_{n-1} + 2^{n-1}$ **17.** $t_1 = 1$;
$t_n = t_{n-1} + n$ **19.a.** $3, 5, 2, -3, -5, -2, 3, 5$
b. -3 **21.a.** $P_0 = 5,000,000$ and
$P_n = (1.02)P_{n-1} + 50,000$ **b.** $9,644,889$
23. $S_{n+1} = S_n + 2n + 1$ **25.a.** 5
b. $n - 2$ additional diagonals; $d_4 = 2$ and
$d_n = d_{n-1} + n - 2$
27.a. $6, 10$ **b.** $p_1 = 0$ and
$p_n = p_{n-1} + n - 1$ **29.a.** $A_0 = 5000$ and
$A_n = 1.1 A_{n-1} + 2000$ **b.** $\$118,997.93$
31.a. $M_2 = 3$; $M_3 = 7$ **c.** $M_4 = 15$; $M_5 = 31$;
$M_6 = 63$; $M_7 = 127$; $M_8 = 255$ **d.** $M_n = 2^n - 1$
e. 5.8×10^{11} yr   **33.a.** -0.05; neg. sign indicates
that the coffee is cooling.   **c.** After ≈ 7 min

Class Exercises, page 488   **1.** Ans. may vary. $1, 3,$
$5, \ldots$; $1 + 3 + 5 + \cdots$ **2.** Sum of Finite Geom.
Series; 2047 **3.a.** Neither   **b.** $S_0 = 0$,
$S_n = S_{n-1} + t_n$, where $t_n = 2^n + 1$ $(n \geq 1)$
c. 20 FOR N = 1 TO 15; 30 LET T = 2^N + 1

Written Exercises, pages 489–492   **1.** 210 **3.** 6375
5. 294 **7.** $500,500$ **9.a.** $\frac{255}{16}$ **b.** $\frac{85}{16}$ **17.** $t_n = 2n$;
$S_0 = 0$, $S_n = S_{n-1} + 2n$ **19.** $t_n = 8\left(\frac{3}{2}\right)^{n-1}$; $S_0 = 0$,
$S_n = S_{n-1} + 8\left(\frac{3}{2}\right)^{n-1}$ **21.** $25,333$ **23.** $33,835,000$
25.b. 20 FOR N = 1 TO 25; 30 LET T = 2^(−N)
c. 0.99999997 **d.** 1 **27.** $t_1 = 5$; $t_2 = 7$; $t_3 = 9$
29. $166,833$ **31.** $49,320$ **33.** 501
35. $305,175,780$ **37.a.** 2046 **b.** 19
39.a. $T_1 = 1$; $T_2 = 3$; $T_3 = 6$; $T_4 = 10$; $T_5 = 15$
b. $T_n + T_{n+1} = (n + 1)^2$
41. $M(n) = \dfrac{n(n^2 + 1)}{2}$
45.b. 9753.90; 97.54; 124.90
c. $A_n = 1.01 A_{n-1} - 222.44$

Class Exercises, page 496   **1.** 1 **2.** 1 **3.** $\frac{2}{3}$ **4.** $\frac{8}{5}$
5. 1 **6.** 0 **7.** 0 **8.** ∞ **9.** ∞ **10.** 0 **11.** No; the
terms do not "home in" on a single value.

Written Exercises, pages 496–498   **1.** 1 **3.** 1
5. $\frac{3}{8}$ **7.** 0 **9.** 0 **11.** 0 **13.** 0
15. Does not exist   **17.** Does not exist   **19.** ∞

21. $-\infty$ **23.** 0 **25.** 1 **27.** 0 **29.** 1 **31.b.** $\frac{3}{2}$

33. e^2 **35.b.** $\frac{1}{4}$ **c.** $\frac{1}{4}$

Calculator Exercises, pages 498–499 **1.a.** 0

b. $\dfrac{1}{\sqrt{n+1}+\sqrt{n}}$ **c.** The value of the expression

approaches zero. **2.a.** 1, 1.5, 1.4166667,

1.4142157, 1.4142136 **b.** $\sqrt{2}$ **3.a.** 1.618034

b. Same

Class Exercises, page 502 **1.** 1, $\frac{4}{3}$, $\frac{13}{9}$, $\frac{40}{27}$;

sum $=\frac{3}{2}$ **2.** $\frac{1}{2}$, $\frac{1}{4}$, $\frac{3}{8}$, $\frac{5}{16}$; sum $=\frac{1}{3}$ **3.** 1, 4, 13,

40; diverges **4.** 1, 1.1, 1.11, 1.111; sum $=\frac{10}{9}$

5.a. $-1<x<1$ **b.** $-\frac{1}{2}<x<\frac{1}{2}$ **6.a.** 0.3 **b.** 0.1

c. $\frac{1}{3}$ **8.** Diverges

Written Exercises, pages 502–505 **1.** 2 **3.** 16

5. $\frac{125}{24}$ **7.** 2 **9.** $\frac{1}{2}$ **11.** $-\frac{1}{3}$ **13.a.** $-1<x<1$

b. $\dfrac{1}{1-x^2}$ **15.a.** $2<x<4$ **b.** $\dfrac{1}{4-x}$

17.a. $x<-2$ or $x>2$ **b.** $\dfrac{x}{x+2}$

21. $|r|=1.5>1$; thus, series would diverge.

23. $\frac{7}{9}$ **25.** $\frac{400}{9}$ **27.** $\frac{1}{7}$ **29.** $\frac{1}{2}$, $\frac{2}{3}$, $\frac{3}{4}$, $\frac{4}{5}$;

$S_n=\dfrac{n}{n+1}$; $\lim\limits_{n\to\infty} S_n = 1$ **31.** $\frac{1}{4}$, $\frac{2}{7}$, $\frac{3}{10}$, $\frac{4}{13}$;

$S_n=\dfrac{n}{3n+1}$; $\lim\limits_{n\to\infty} S_n = \frac{1}{3}$ **33.** 56 m **35.a.** 288

b. $96+48\sqrt{2}$ **37.** 15 **39.** 10, 1, 0.1;

sum $=\frac{100}{9}$ s **41.** $5+5i$, $7+6i$, $9+7i$;

$t_{25}=49+27i$; $S_{25}=625+375i$ **43.a.** $\frac{4}{5}+\frac{2}{5}i$

b. $\frac{243}{10}-\frac{81}{10}i$ **45.** Odd: $\dfrac{1}{1+r}$;

even: $\dfrac{r}{1+r}$ **47.** $\dfrac{2\sqrt{3}}{5}$

Calculator Exercise, page 506 **a.** 6 **b.** 8 **c.** 11

Class Exercises, page 508

1. $5+10+15+20$ **2.** $9+16+25+36$

3. $1-1+1-1+1-1+1$

4. $1+\frac{1}{2}+\frac{1}{3}+\frac{1}{4}+\cdots$

5. $\sum\limits_{k=2}^{6} k^2$ **6.** $\sum\limits_{k=1}^{4}\dfrac{k}{k+1}$ **7.** $\sum\limits_{k=1}^{100} 3k$ **8.** $\sum\limits_{k=1}^{\infty} 3^{-k}$

Written Exercises, pages 508–510

1. $2+3+4+5+6$ **3.** $1+\frac{1}{2}+\frac{1}{3}+\frac{1}{4}+\frac{1}{5}$

5. $3+1+\frac{1}{3}+\frac{1}{9}+\cdots$ **7.** $\frac{1}{16}+\frac{1}{4}+1+4+16$

9. $\sum\limits_{k=1}^{5} 4k$ **11.** $\sum\limits_{k=1}^{25} 4k+1$ **13.** $\sum\limits_{k=1}^{\infty}\dfrac{1}{k^2}$

15. $\sum\limits_{k=1}^{\infty} \sin kx$ **19.** 0 **21.a.** 0 **b.** 8

25.a. $\sum\limits_{k=0}^{5}\left(-\frac{1}{2}\right)^k$ **b.** $\sum\limits_{k=0}^{5} -1\left(-\frac{1}{2}\right)^k$

27. $\sum\limits_{k=0}^{49} (-1)^k(2k+1)$ **33.** 343,400 **35.** 53,130

37.a. 204 **b.** $\dfrac{n(n+1)(2n+1)}{6}$

39. $\frac{1}{6}(n(n+1)(n+2))$

Computer Exercise, page 510 **a.** 11 **b.** 31
c. 12,367

Written Exercises, pages 513–514 **1.** For $n=1$,

$\dfrac{1(1+1)}{2}=\dfrac{2}{2}=1$; if $1+2+\cdots+k=\dfrac{k(k+1)}{2}$,

then $(1+2+\cdots+k)+(k+1)=$

$\dfrac{k(k+1)}{2}+\dfrac{2(k+1)}{2}=\dfrac{(k+1)(k+2)}{2}=$

$\dfrac{(k+1)[(k+1)+1]}{2}$. **11.** For $n=1$, $11^1-4^1=7$,

and 7 is a multiple of 7; if 11^k-4^k is a multiple

of 7, then $11^{k+1}-4^{k+1}=11\cdot 11^k-4\cdot 4^k=$

$(7+4)11^k-4\cdot 4^k=7\cdot 11^k+4(11^k-4^k)=$

the sum of 2 multiples of 7 = a multiple of 7.

13. $n\ge 3$ **19.** $\frac{1}{2}$, $\frac{1}{3}$, $\frac{1}{4}$, ...; $a_n=\dfrac{1}{n+1}$

21. $\frac{1}{4}$, $\frac{2}{7}$, $\frac{3}{10}$, $\frac{4}{13}$, ...; $a_n=\dfrac{n}{3n+1}$

23. 1, 5, 23, 119, ...; $S_n=(n+1)!-1$

Chapter Test, page 515 **1.a.** Arith.; $t_n=14-4n$

b. Neither; $t_n=n^2+2$ **c.** Geom.; $t_n=\frac{2}{3}(-3)^n$

2. 53.5 **3.** $\frac{9}{512}$ **4.** 128 **5.** $t_1=1$, $t_n=4t_{n-1}+1$

6.a. $t_n=94+10n$ **b.** $S_n=99n+5n^2$ **7.** $\frac{182}{9}$

8. Ans. will vary. An infinite seq. has a limit L if
the terms get arbitrarily close to L as n becomes
larger and larger.

9.a. $\frac{3}{4}$ **b.** Does not exist **c.** 0 **10.** $\sum_{n=1}^{\infty} 13(10)^{-2n}$;

$\frac{13}{99}$ **11.a.** $-\frac{2}{3} < x < \frac{2}{3}$ **b.** $\frac{2}{2 - 3x}$

12.a. $\sum_{n=0}^{4} \frac{(-1)^n}{(2n + 1)^2}$ **b.** $\sum_{n=1}^{7} (11 - 4n)$ **13.** 9870

14. For $n = 1$, $\frac{1(5 \cdot 1 - 1)}{2} = 2 = 5(1) - 3$; if

$2 + 7 + 12 + \cdots + (5k - 3) = \frac{k(5k - 1)}{2}$, then

$2 + 7 + 12 + \cdots + (5k - 3) + [5(k + 1) - 3] =$

$\frac{k(5k - 1)}{2} + \frac{2(5k + 2)}{2} = \frac{5k^2 + 9k + 4}{2} =$

$\frac{(k + 1)(5k + 4)}{2} = \frac{(k + 1)[5(k + 1) - 1]}{2}$

Class Exercises, page 520 **1.** $A_{2\times 1}$; $B_{1\times 2}$; $C_{1\times 2}$; $D_{1\times 2}$ **2.** A and B, A and D **3.** B and D
4. $A_{3\times 2}$; $B_{3\times 2}$; $C_{3\times 3}$; $D_{2\times 3}$

5. A and B; $\begin{bmatrix} 8 & 6 \\ 5 & 5 \\ -3 & 14 \end{bmatrix}$

6.a. 2×3 **b.** $\begin{bmatrix} 5 & 3 & 0 \\ -1 & -4 & 6 \end{bmatrix}$

7.a. 10.4 g of carbohydrate in a serving of tomato juice **b.** 0.002 g of sodium in a serving of orange juice **c.** Nutritional information for 2 servings **d.** N^t

Written Exercises, pages 520–522

1. $\begin{bmatrix} -2 \\ 2 \end{bmatrix}$ **3.** $\begin{bmatrix} 5 & 3 \\ -5 & 6 \end{bmatrix}$ **5.** $\begin{bmatrix} 20 & 5 & 8 \\ -3 & 1 & 8 \end{bmatrix}$

7. $\begin{bmatrix} 40 & -16 \\ 32 & 0 \end{bmatrix}$ **9.** $\begin{bmatrix} 4 & -2 \\ -5 & 2 \\ 6 & 9 \end{bmatrix}$

11.a. A: 3×4; B: 4×3; C: 4×3

b. $B + C$; $\begin{bmatrix} 17 & 1 & 9 \\ 7 & 3 & 8 \\ 5 & 17 & 13 \\ 12 & 4 & 0 \end{bmatrix}$

13. $\begin{bmatrix} 16 & 2 & 8 \\ 6 & 0 & 14 \\ -4 & 10 & 22 \\ 12 & 8 & -6 \end{bmatrix}$; $\begin{bmatrix} 25 & 2 & 13 \\ 10 & 3 & 15 \\ 3 & 22 & 24 \\ 18 & 8 & -3 \end{bmatrix}$

15. Ans. may vary. **a.** $\begin{array}{c} \\ D \\ M \end{array}\begin{matrix} l & m & h \\ \begin{bmatrix} 31 & 42 & 18 \\ 22 & 25 & 11 \end{bmatrix} \end{matrix}$

b. Location by mower-type. **c.** $\begin{array}{c} \\ D \\ M \end{array}\begin{matrix} l & m & h \\ \begin{bmatrix} 28 & 29 & 20 \\ 20 & 18 & 9 \end{bmatrix} \end{matrix}$

d. $\begin{array}{c} \\ D \\ M \end{array}\begin{matrix} l & m & h \\ \begin{bmatrix} 59 & 71 & 38 \\ 42 & 43 & 20 \end{bmatrix} \end{matrix}$; combined sales for April and May

e. 20; up; 1.08 **f.** 9%; 15% **17.** $x = 2$, $y = 3$, $z = -1$ **19.** A **21.** Yes

Class Exercises, page 525 **1.** Yes; no; $AB_{3\times 2}$
2. No; yes; $BA_{2\times 6}$ **3.** No; yes; $BA_{3\times 4}$ **4.** Yes; no; $AB_{5\times 3}$ **5.** No; no **6.** Yes; yes; $AB_{3\times 3}$;

$BA_{4\times 4}$ **7.** $\begin{bmatrix} 9 & 8 \\ 23 & 22 \end{bmatrix}$ **8.** $\begin{bmatrix} 4 & 10 \\ 0 & 2 \end{bmatrix}$ **9.** CM

10. No; yes

Written Exercises, pages 526–529

1. $\begin{bmatrix} 23 \\ -7 \end{bmatrix}$ **3.** $\begin{bmatrix} 4 & -9 \\ 8 & -5 \end{bmatrix}$ **5.** Not def.

7. $\begin{bmatrix} 30 & 7 & 20 \\ -14 & -33 & -9 \end{bmatrix}$ **9.** $\begin{bmatrix} a & b & c \\ d & e & f \\ g & h & i \end{bmatrix}$

11.b. Yes **13.b.** Yes **c.** Yes **15.** PS; [37,200   35,050]; profit for each dealer in March
17.a. No. of cols. of P = no. of rows of M, but cols. of P and rows of M do not represent the same things. **b.** 2400 **c.** There are 3 cols. in M and 2 rows in R.

d. $\begin{array}{c} \\ x \\ y \end{array}\begin{matrix} s & b & g \\ \begin{bmatrix} 17 & 21 & 37 \\ 22 & 28 & 50 \end{bmatrix} \end{matrix}$ **e.** $\begin{array}{c} \\ x \\ y \end{array}\begin{matrix} \text{Jan.} & \text{Feb.} & \text{Mar.} \\ \begin{bmatrix} 16,400 & 25,500 & 29,200 \\ 21,600 & 33,800 & 38,800 \end{bmatrix} \end{matrix}$

19.a. 8.3 **b.** 1st: Joyce; 2nd: Eduardo; 3rd: Stephanie **21.c.** $10,072 **d.** A

Class Exercises, page 534

1. $-B = \begin{bmatrix} -7 & 1 \\ -6 & -2 \end{bmatrix}$; $-C = \begin{bmatrix} -3 & -2 & -5 \\ -4 & -3 & 0 \end{bmatrix}$

2. $A^{-1} = \begin{bmatrix} 2 & 3 \\ 5 & 8 \end{bmatrix}$; $B^{-1} = \begin{bmatrix} \frac{1}{10} & \frac{1}{20} \\ -\frac{3}{10} & \frac{7}{20} \end{bmatrix}$

3. C is not a square matrix, and $|D| = 0$.

4. $\begin{bmatrix} 79 & -30 \\ -50 & 19 \end{bmatrix}$; both are correct because mult. of matrices is associative.

5. Mult. of matrices is not commutative.

6. $3X = \begin{bmatrix} -9 & -6 \\ -3 & -15 \end{bmatrix}$ (Perform the subtraction.)

$X = \begin{bmatrix} -3 & -2 \\ -1 & -5 \end{bmatrix}$ (Simplify using scalar div.)

Written Exercises, pages 534–537

1.a. $\begin{bmatrix} -\frac{7}{3} & \frac{4}{3} \\ 2 & -1 \end{bmatrix}$ **b.** $\begin{bmatrix} 1 & 0 \\ 0 & 1 \end{bmatrix}$ **3.a.** $\begin{bmatrix} -9 & -3 \\ -12 & -4 \end{bmatrix}$

b. Not def. **5.a.** $\begin{bmatrix} -3 & -1 \\ -4 & -\frac{4}{3} \end{bmatrix}$ **b.** Not def.

7.a. $X = D^{-1}E$ **b.** $X = ED^{-1}$ **9.** $\begin{bmatrix} 6 & -8 \\ -3 & -20 \end{bmatrix}$

11. $\begin{bmatrix} 27 & 13 \\ -36 & -18 \end{bmatrix}$ **13.** $\begin{bmatrix} 6 & 0 \\ -12 & -2 \end{bmatrix}$

15.a. $x = -\frac{1}{2}, y = 5$ **b.** $x = 4, y = -3$

17.a. Cannot be done; coef. matrix has no inverse. **b.** There is no sol. because the lines are $\parallel$.
21. IX; $X = (I - A)^{-1}D$ **23.** $x = -2, y = 1,$ $z = -3$ **25.** I: 4 oz; II: 6 oz; III: 8 oz; IV: 6 oz
27. $y = \frac{3}{4}x^2 + \frac{7}{2}x - 15$ **29.** $w = \dfrac{d}{ad - bc},$

$x = -\dfrac{b}{ad - bc}, y = -\dfrac{c}{ad - bc}, z = \dfrac{a}{ad - bc}$

Class Exercises, page 540 1. C and D 2. C

3.
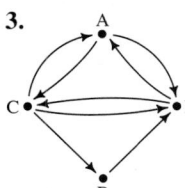

4. No. of ways a message can be sent using one relay
5. No; a station can send a message to itself using another station as a relay.

Written Exercises, pages 540–543

1. from $\begin{array}{c} \\ A \\ B \\ C \end{array} \begin{bmatrix} \text{to A} & \text{B} & \text{C} \\ 0 & 1 & 0 \\ 1 & 0 & 0 \\ 1 & 1 & 0 \end{bmatrix}$

3. from $\begin{array}{c} \\ A \\ B \\ C \\ D \\ E \end{array} \begin{bmatrix} \text{to A} & \text{B} & \text{C} & \text{D} & \text{E} \\ 0 & 0 & 1 & 0 & 1 \\ 1 & 0 & 0 & 0 & 0 \\ 1 & 0 & 0 & 1 & 0 \\ 0 & 1 & 0 & 0 & 0 \\ 1 & 1 & 0 & 0 & 0 \end{bmatrix}$

5. 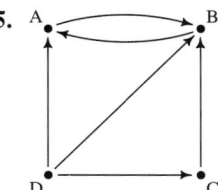 **7.a.**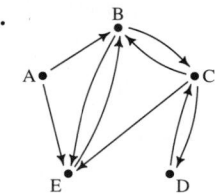

7.c. There is no route from E to B using one relay. **e.** 1 **9.a.** Transmitters: Kim, Bob; receivers: Ron, Sue; relays: Dot, Ted, Les, Neal
11. Zeros: n; ones: $n^2 - n$ **13.b.** Wolves feed on 2 animals that feed on grass. **c.** Put a "1" in the grass row, salmon col.

Class Exercises, pages 545–546 2. Associative

3.a. $\begin{array}{c} \\ c \\ s \end{array} \begin{bmatrix} c & s \\ 0.9 & 0.1 \\ 0.03 & 0.97 \end{bmatrix}$ **b.** $\begin{bmatrix} c & s \\ 0.6 & 0.4 \end{bmatrix}$ **c.** $\begin{bmatrix} c & s \\ 0.552 & 0.448 \end{bmatrix}$; population distribution in one year **d.** Population distribution in 2 years **e.** Calculate M_n for large values of n to see if its values stabilize.

Written Exercises, pages 546–550

1.a. $\begin{array}{c} \\ c \\ \text{p.t.} \end{array} \begin{bmatrix} c & \text{p.t.} \\ 0.82 & 0.18 \\ 0.05 & 0.95 \end{bmatrix}$ **b.** $\begin{bmatrix} c & \text{p.t.} \\ 0.4 & 0.6 \end{bmatrix}$

c. [0.358 0.642]; next year, 36% will commute by car and 64% by public transportation.
d. [0.326 0.674]; in 2 years, 33% will commute by car and 67% by public transportation.
3.a. 62%; 63% **b.** $\frac{2}{3}$ at Dixie's, $\frac{1}{3}$ at Sargent's

c. $\begin{bmatrix} \frac{2}{3} & \frac{1}{3} \end{bmatrix}$

5.a. $\begin{array}{c} \\ s \\ c \\ r \end{array} \begin{bmatrix} s & c & r \\ 0.8 & 0.1 & 0.1 \\ 0 & 0.4 & 0.6 \\ 1 & 0 & 0 \end{bmatrix}$ **b.** 14% **c.** 12.2%
d. 73%; 73%; 73% **e.** 73%; 73%; 73% **f.** 73%; 12%; 15%

7.a.
$$\begin{array}{c} & \begin{array}{ccc} w & f & i \end{array} \\ \begin{array}{c} w \\ f \\ i \end{array} & \left[\begin{array}{ccc} 0.8 & 0.2 & 0 \\ 0 & 0.5 & 0.5 \\ 0 & 0 & 1 \end{array}\right] \end{array}$$
b. 19.5%; 12.5%

c. 10 days: 54% well, 23% flu, 22% immune; 30 days: 22% well, 14% flu, 64% immune; 50 days: 9% well, 6% flu, 85% immune **9.** $x = 0.125$;

$S = [0.125 \quad 0.875]$ **11.** $\left[\frac{5}{12} \quad \frac{3}{8} \quad \frac{5}{24}\right]$

13.a.
$$\begin{array}{c} & \begin{array}{ccccc} 1 & 2 & 3 & 4 & 5 \end{array} \\ \begin{array}{c} 1 \\ 2 \\ 3 \\ 4 \\ 5 \end{array} & \left[\begin{array}{ccccc} 0 & \frac{1}{3} & 0 & \frac{1}{3} & \frac{1}{3} \\ \frac{1}{3} & 0 & \frac{1}{3} & \frac{1}{3} & 0 \\ 0 & \frac{1}{2} & 0 & \frac{1}{2} & 0 \\ \frac{1}{4} & \frac{1}{4} & \frac{1}{4} & 0 & \frac{1}{4} \\ \frac{1}{2} & 0 & 0 & \frac{1}{2} & 0 \end{array}\right] \end{array}$$
b. 21.4%; 21.4%; 14.3%; 28.6%; 14.3%
c. 21.4%; 21.4%; 14.3%; 28.6%; 14.3%
d. No

15. In the first col. of row 3 change 0 to 20.
b. [6000 320 150 60]
c. [4200 2400 160 45]
d. [444,768 45,824 25,190 6286]

17.a. $\begin{array}{ccccc} A_1 & A_2 & A_3 & A_4 & A_5 \end{array}$
[329 128 70 48 13] **b.** In 38 yr

Class Exercises, page 556 1. 3 **2.** $-2x - 2$

3. $\left[\begin{array}{cc} 3 & -2 \\ 4 & 1 \end{array}\right]\left[\begin{array}{c} x \\ y \end{array}\right] = \left[\begin{array}{c} x' \\ y' \end{array}\right]$ **4.** $T = \left[\begin{array}{cc} 3 & -2 \\ 4 & 1 \end{array}\right]$;

$|T| = 11$ **5.** $\frac{\text{area } \triangle A'B'C'}{\text{area } \triangle ABC} = 11$ **6.** Same

7. No; no; no **8.** No; no; no **9.** No; yes; no

Written Exercises, pages 557–559

1.b. $T = \left[\begin{array}{cc} 2 & 0 \\ 0 & 3 \end{array}\right]$; $|T| = 6$ **c.** 6 **d.** Same

3.c. $T = \left[\begin{array}{cc} 2 & 1 \\ 3 & 1 \end{array}\right]$; $|T| = -1$; the areas are $=$ and

the $\triangle$ have opp. orientations.

5.a. (36, 4) **b.** $\left[\begin{array}{c} 36 \\ 4 \end{array}\right]$ **7.** $y = \frac{1}{2}x$; $T = \left[\begin{array}{cc} 4 & 2 \\ 2 & 1 \end{array}\right]$,

$|T| = 0$ **9.** Right 5, down 1 **11.** Down 2

13. $\left[\begin{array}{c} x \\ y \end{array}\right] + \left[\begin{array}{c} 0 \\ -4 \end{array}\right] = \left[\begin{array}{c} x' \\ y' \end{array}\right]$ **15.** $\left[\begin{array}{cc} 1 & 0 \\ 0 & -1 \end{array}\right]$

17. $\left[\begin{array}{cc} 0 & -1 \\ -1 & 0 \end{array}\right]$ **19.** $\left[\begin{array}{cc} 0 & -1 \\ 1 & 0 \end{array}\right]$ **21.** $\left[\begin{array}{cc} -1 & 0 \\ 0 & -1 \end{array}\right]$

23. $\left[\begin{array}{cc} 1 & 0 \\ 0 & 1 \end{array}\right]$; $I_{2\times 2}$ **25.** $\left[\begin{array}{cc} -1 & 0 \\ 0 & -1 \end{array}\right]$; R_{180}

27. $X^2 = L^2 = M^2 = \left[\begin{array}{cc} 1 & 0 \\ 0 & 1 \end{array}\right]$

29.a. $\left[\begin{array}{cc} \cos\beta & -\sin\beta \\ \sin\beta & \cos\beta \end{array}\right]$ **31.a.** $\left[\begin{array}{ccc} 2 & 0 & 0 \\ 0 & 3 & 0 \\ 0 & 1 & 4 \end{array}\right]$

Chapter Test, pages 560–561

1.a.
$$\begin{array}{c} & \begin{array}{ccc} s & m & l \end{array} \\ \begin{array}{c} M \\ P \\ B \end{array} & \left[\begin{array}{ccc} 7 & 14 & 10 \\ 8 & 20 & 9 \\ 15 & 20 & 12 \end{array}\right] \end{array}$$

b. $1.5S = $
$$\begin{array}{c} & \begin{array}{ccc} s & m & l \end{array} \\ \begin{array}{c} M \\ P \\ B \end{array} & \left[\begin{array}{ccc} 10.5 & 21 & 15 \\ 12 & 30 & 13.5 \\ 22.5 & 30 & 18 \end{array}\right] \end{array}$$
2. $\left[\begin{array}{ccc} 5 & 0 & -6 \\ -6 & 3 & 8 \end{array}\right]$

3. Ans. may vary. $A = \left[\begin{array}{cc} 1 & 2 \\ 3 & -1 \end{array}\right]$, $B = \left[\begin{array}{cc} 0 & -1 \\ 1 & 2 \end{array}\right]$;

the 1st row, 1st col. entry of a product matrix XY comes from the product of the 1st row of X and the 1st col. of Y. These products are different for AB and BA. **4.a.** NP; $NP = [5.8 \quad 4 \quad 2.8 \quad 3.4]$
b. 2.8 lb **5.** $x = 5$, $y = -5$

6.a. from
$$\begin{array}{c} & \begin{array}{c} \text{to}\ \ B\ \ L\ \ P\ \ S\ \ W \end{array} \\ \begin{array}{c} B \\ L \\ P \\ S \\ W \end{array} & \left[\begin{array}{ccccc} 0 & 1 & 0 & 0 & 0 \\ 1 & 0 & 1 & 0 & 0 \\ 0 & 0 & 0 & 1 & 1 \\ 0 & 1 & 0 & 0 & 0 \\ 0 & 0 & 1 & 1 & 0 \end{array}\right] \end{array}$$

b. from
$$\begin{array}{c} & \begin{array}{c} \text{to}\ \ B\ \ L\ \ P\ \ S\ \ W \end{array} \\ \begin{array}{c} B \\ L \\ P \\ S \\ W \end{array} & \left[\begin{array}{ccccc} 1 & 0 & 1 & 0 & 0 \\ 0 & 1 & 0 & 1 & 1 \\ 0 & 1 & 1 & 1 & 0 \\ 1 & 0 & 1 & 0 & 0 \\ 0 & 1 & 0 & 1 & 1 \end{array}\right] \end{array}$$

Lookout Peak, Park Entrance, Scenic Overlook, and Waterfall have two 2-step paths to other sites.

7.a.
$\begin{array}{c} \;\;\text{c}\quad\;\; \text{s} \\ \begin{array}{c}\text{c}\\\text{s}\end{array}\begin{bmatrix} 0.92 & 0.08 \\ 0.03 & 0.97 \end{bmatrix}\end{array}$

b. [0.6 0.4]

c. City: 56.4%;
suburbs: 43.6%

8.a,b. See graph at right.

c. $T = \begin{bmatrix} 2 & 1 \\ 1 & -1 \end{bmatrix}$,

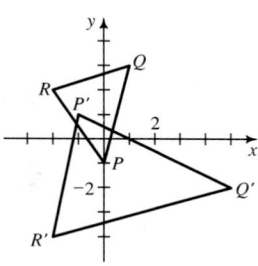

$|T| = -3$; $\dfrac{\text{area } \triangle P'Q'R'}{\text{area } \triangle PQR} = 3$ and the $\triangle$ have opp. orientations.

Cumulative Review, pages 562–563

1.a.
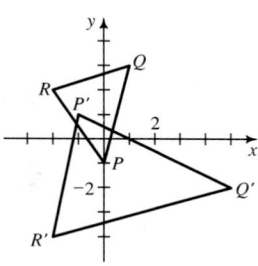

300 km
200°
120 km

b. 414.8 km; 185.7°

2.a. $(-6, 8)$ **b.** $(3, 1)$ **c.** $5\sqrt{2}$

3.a. $(x, y) = (-2, -4) + t(-2, 5)$

b. $x = -2 - 2t$, $y = -4 + 5t$

4.a. Vel. $= (2, -4)$, speed $= 2\sqrt{5}$

b. $t = 1$; $(3, -5)$

5. 62.3° **6.** $(x, y, z) = (2, 0, 5) + t(-5, -1, -1)$; $x = 2 - 5t$, $y = -t$, $z = 5 - t$ **7.** $2x + y - z = 0$

8. 336 **9.** $(3, 2)$ **10.** $(6, 9, -1)$;

$6x + 9y - z = 1$ **11.a.** Geom.; $t_n = -\frac{4}{9}\left(-\frac{3}{2}\right)^n$

b. Neither; $t_n = \dfrac{2n}{2n + 1}$ **c.** Arith.; $t_n = n - \frac{1}{3}$

12.a. $-2, 0, 4, 12, 28$; **b.** $t_n = 2^n - 4$ **c.** 1020

13. -180 **14.** $\frac{364}{3}$ **15.a.** 1 **b.** Does not exist.

16.a. 3 **b.** ∞ **c.** Does not exist. **17.** $-4 < x < 4$;

$\dfrac{8}{x + 4}$ **18.** $\dfrac{115}{333}$ **19.** $\displaystyle\sum_{n=1}^{20} (5n + 2)$; 1090

20. For $n = 1$, $\dfrac{1}{2(1) + 1} = \dfrac{1}{3} = \dfrac{1}{1 \cdot 3}$; if

$\dfrac{1}{1 \cdot 3} + \dfrac{1}{3 \cdot 5} + \cdots + \dfrac{1}{(2k - 1)(2k + 1)} = \dfrac{k}{2k + 1}$, then

$\dfrac{1}{1 \cdot 3} + \dfrac{1}{3 \cdot 5} + \cdots + \dfrac{1}{(2k - 1)(2k + 1)} +$

$\dfrac{1}{(2(k + 1) - 1)(2(k + 1) + 1)} =$

$\dfrac{k}{2k + 1} + \dfrac{1}{(2k + 1)(2k + 3)} =$

$\dfrac{k(2k + 3)}{(2k + 1)(2k + 3)} + \dfrac{1}{(2k + 1)(2k + 3)} =$

$\dfrac{2k^2 + 3k + 1}{(2k + 1)(2k + 3)} = \dfrac{(2k + 1)(k + 1)}{(2k + 1)(2k + 3)} = \dfrac{k + 1}{2(k + 1) + 1}$.

21.a. $\begin{bmatrix} -4 & 7 \\ 15 & 8 \\ -9 & -4 \end{bmatrix}$ **b.** $AB = \begin{bmatrix} -30 & -7 \\ -6 & -3 \end{bmatrix}$,

$BA = \begin{bmatrix} -5 & 8 & -4 \\ 3 & -18 & 6 \\ 7 & 46 & -10 \end{bmatrix}$; no **22.** $\begin{bmatrix} \frac{2}{3} & -\frac{5}{3} \\ 0 & 1 \end{bmatrix}$

23. $\begin{bmatrix} -2 \\ 2 \end{bmatrix}$ **24.a.**
$\begin{array}{c} \;\;\text{A}\;\;\text{B}\;\;\text{C}\;\;\text{D}\;\;\text{E} \\ \begin{array}{c}\text{A}\\\text{B}\\\text{C}\\\text{D}\\\text{E}\end{array}\begin{bmatrix} 0 & 1 & 0 & 0 & 1 \\ 0 & 0 & 1 & 1 & 1 \\ 0 & 1 & 0 & 1 & 0 \\ 0 & 0 & 1 & 0 & 1 \\ 1 & 0 & 0 & 0 & 0 \end{bmatrix}\end{array}$

b.
$\begin{array}{c} \;\;\text{A}\;\;\text{B}\;\;\text{C}\;\;\text{D}\;\;\text{E} \\ \begin{array}{c}\text{A}\\\text{B}\\\text{C}\\\text{D}\\\text{E}\end{array}\begin{bmatrix} 1 & 1 & 1 & 1 & 2 \\ 1 & 1 & 2 & 2 & 2 \\ 0 & 1 & 2 & 2 & 2 \\ 1 & 1 & 1 & 1 & 1 \\ 1 & 1 & 0 & 0 & 1 \end{bmatrix}\end{array}$

25.a.
$\begin{array}{c} \;\;\;\;\text{A}\;\;\;\;\;\text{B}\;\;\;\;\;\text{C} \\ \begin{array}{c}\text{A}\\\text{B}\\\text{C}\end{array}\begin{bmatrix} 1 & 0 & 0 \\ 0.35 & 0.65 & 0 \\ 0.45 & 0 & 0.55 \end{bmatrix}\end{array}$

b. A: 51%; B: 32.5%; C: 16.5% **c.** A: 69.8%; B: 21.1%; C: 9.1%

26.a,b.
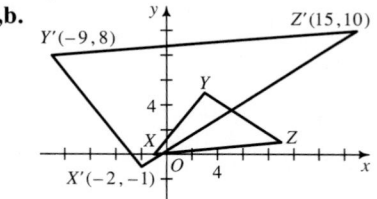

$Y'(-9, 8)$ $Z'(15, 10)$ $X'(-2, -1)$

c. $T = \begin{bmatrix} 2 & -3 \\ 1 & 1 \end{bmatrix}$, $|T| = 5$; $\dfrac{\text{area } \triangle X'Y'Z'}{\text{area } \triangle XYZ} = 5$ and the $\triangle$ have the same orientation.

Class Exercises, page 568 **1.** The set of freshmen music majors; 15 **2.** The set of students who are either freshmen or music majors; 350 **3.** The set of students who are not freshmen; 700 **4.** The set of students who are not music majors; 935 **5.** The set of music majors who are not freshmen; 50

6. The set of freshmen who are not music majors; 285 **7.** The set of students who are neither freshmen nor music majors; 650 **8.** The set of students who are either freshmen or are not music majors; 950 **9.a.** $B \cap \overline{A}$ **b.** $A \cap B$ **c.** $\overline{A \cup B}$ or $\overline{A} \cap \overline{B}$ **10.a.** A **b.** $\varnothing$ **c.** U **d.** $\varnothing$ **e.** U **f.** A

Written Exercises, pages 568–571

1.a.

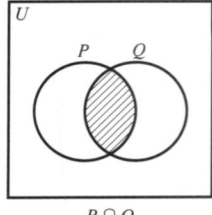

$P \cap Q$

3.a.

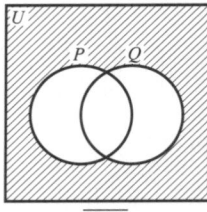

$\overline{P \cup Q}$

5.a. Teachers of either math or physics **b.** Teachers of math but not of physics **7.a.** Teachers of biology or of both physics and chemistry **b.** Teachers of both biology and chemistry **9.a.** Teachers of neither math nor physics nor chemistry **b.** Same as part a. **11.** 28 **13.** 12 **19.a.** 11 **b.** 12 **21.** $\varnothing$ **25.b.** $\overline{A \cap B} = \overline{A} \cup \overline{B}$ **c.** $\overline{A} \cup (\overline{B} \cap \overline{C})$ **27.** $n(A \cup B \cup C) = n(A) + n(B) + n(C) - n(B \cap C) - n(A \cap B) - n(A \cap C) + n(A \cap B \cap C)$

Class Exercises, page 574 1.a. 2 **b.** 6 **c.** 24 **2.** 3,628,800 **3.** 608,400 **4.a.** 60 **b.** 180 **5.a.** 8 **b.** 6 **6.** 720 **7.a.** 362,879 **b.** 1,133,836,704,000

Written Exercises, pages 575–577 1.a. 120 **b.** 720 **c.** 5040 **d.** 1 **3.** 120 **5.** 1024 **7.a.** 17,576 **b.** 13,800 **9.** 11,880 **11.b.** 14 branches, 8 leaves **c.** 2046 branches, 1024 leaves **13.a.** 362,880 **b.** 40,320 **c.** 5040 **15.** 156 **17.** 2 digits: 1,757,600; 3 digits: 17,576,000; total: 19,333,600 **19.a.** 648 **b.** 252 **21.** 542 **23.a.** 160 **b.** 640 **c.** 9999 **d.** 6,399,360 **e.** 1,023,897,600 **25.a.** 362,880 **b.** 30,240 **27.** $26^2 = 676 < 677$ **31.** $\log_{10} 9! \approx 5.56$; $\log_{10} 10! \approx 6.56$ **33.a.** $\dfrac{10!}{6!} = \dfrac{10 \cdot 9 \cdot 8 \cdot 7 \cdot 6!}{6!} = 10 \cdot 9 \cdot 8 \cdot 7$

Class Exercises, page 580 1.a. 20 **b.** 10 **2.a.** 120 **b.** 20 **3.a.** 720 **b.** 120 **4.a.** 24 **b.** 1 **5.** 720 **6.** 120 **7.** The order of the three nos. is important.

8.a.

ABC	ACB	BAC	BCA	CAB	CBA
ABD	ADB	BAD	BDA	DAB	DBA
ACD	ADC	CAD	CDA	DAC	DCA
BCD	BDC	CBD	CDB	DBC	DCB

b. A, B, and C: 6; A, C, and D: 6 **c.** ABC, ABD, ACD, BCD

Written Exercises, pages 580–582 1.a. 380 **b.** 190 **3.a.** 210 **b.** 5040 **5.a.** 1,313,400 **b.** 7,880,400 **7.a.** 336 **b.** 56 **9.a.** 665,280 **b.** 924 **11.a.** 35 **b.** 35 **13.a.** $_{100}C_2 = {}_{100}C_{98} = \dfrac{100!}{2! \, 98!}$ **15.a.** 1 **b.** 1 **17.** 22,764 **19.** 48 **21.a.** 792 **b.** 658,008 **c.** 1,940,952 **23.a.** 15 **b.** There will be fewer line segs. if three or more pts. are collinear. **25.** 10 **29.** 5.36×10^{28}

Class Exercises, page 585 1.a. 24 **b.** 12 **2.a.** 120 **b.** 60 **3.a.** 720 **b.** 180 **4.a.** 39,916,800 **b.** 1,663,200 **7.** 84 **8.** BCDAE

Written Exercises, pages 585–587 1. 10,080 **3.** 34,650 **5.** 840 **7.** 9.71×10^{14} years **9.a.** 3003 **b.** 21 **c.** 35 **d.** 735 **11.** 24 **13.** 1.37×10^{11} **15.b.** 8.25×10^{12} **17.** 6188

Mixed Combinatorics Exercises, pages 587–589 1. 161,700 **3.a.** 6561 **b.** 5832 **c.** 2000 **5.a.** 870 **b.** 435 **7.** 60 ways from A to D; 3600 round trips **9.** 83 **11.a.** 1 min **b.** 13 h **13.a.** 70 **b.** 35 **c.** 35 **15.** 15 **17.** 64 **19.a.** 86,400 **b.** 28,800 **21.** 37,512

Class Exercises, page 592 2. 1 7 21 35 35 21 7 1 **3.a.** 1 8 28 56 **b.** x^8, $8x^7y$, $28x^6y^2$, $56x^5y^3$ **c.** x^8, $-8x^7y$, $28x^6y^2$, $-56x^5y^3$ **4.** $a^3 - 3a^2b + 3ab^2 - b^3$ **5.** $a^4 + 4a^3b + 6a^2b^2 + 4ab^3 + b^4$ **6.** $a^4 - 4a^3b + 6a^2b^2 - 4ab^3 + b^4$ **7.a.** $15x^4y^2$ **b.** $84x^6y^3$

Written Exercises, pages 592–594 1.a. $a^3 + 3a^2b + 3ab^2 + b^3$ **b.** $8000 + 1200 + 60 + 1 = 9261$ **c.** $8000 - 1200 + 60 - 1 = 6859$ **3.a.** $a^5 + 5a^4b + 10a^3b^2 + 10a^2b^3 + 5ab^4 + b^5$ **b.** $a^5 - 5a^4b + 10a^3b^2 - 10a^2b^3 + 5ab^4 - b^5$ **c.** $32a^5 + 80a^4 + 80a^3 + 40a^2 + 10a + 1$ **5.a.** $x^7 + 7x^6y + 21x^5y^2 + 35x^4y^3 + 35x^3y^4 + 21x^2y^5 + 7xy^6 + y^7$ **b.** $x^7 - 7x^6y + 21x^5y^2 - 35x^4y^3 + 35x^3y^4 - 21x^2y^5 + 7xy^6 - y^7$

c. $x^{14} - 14x^{12}y + 84x^{10}y^2 - 280x^8y^3 +$
$560x^6y^4 - 672x^4y^5 + 448x^2y^6 - 128y^7$
7. $x^6 - 3x^4y^2 + 3x^2y^4 - y^6$
9. $x^{10} + 5x^8 + 10x^6 + 10x^4 + 5x^2 + 1$
11. $x^6 + 6x^4 + 15x^2 + 20 + 15x^{-2} + 6x^{-4} + x^{-6}$
13. $(a^2)^{100} + 100(a^2)^{99}(-b)^1 +$
$\frac{100 \cdot 99}{2}(a^2)^{98}(-b)^2 + \frac{100 \cdot 99 \cdot 98}{3 \cdot 2}(a^2)^{97}(-b)^3$
15. $(\sin x)^{10} + 10(\sin x)^9(\sin y) +$
$\frac{10 \cdot 9}{2}(\sin x)^8(\sin y)^2 + \frac{10 \cdot 9 \cdot 8}{3 \cdot 2}(\sin x)^7(\sin y)^3$
17. 1.05 19.a. 495 b. 495 21. 126,720
23.a. $_{n+1}C_k$ b. $_nC_{k-1}$ c. $_nC_k$
d. $_{n+1}C_k = {_nC_{k-1}} + {_nC_k}$ 25. Third
29.b. $\cos 3\theta + i \sin 3\theta$

Chapter Test, page 595 1. 36 **2.** 720 **3.** 363,840
4. 2520 **5.** 15 **6.** 560 **7.** Ans. may vary. A
linear permutation is like a football team of n
players seated on a bench. There is a first player,
then a second, etc., and the n positions can be
filled in $n!$ ways. A circular permutation is like
this same team in a huddle. There is no first
player, second player, etc. For each circular
arrangement there are n possible linear ones. Thus,
there are $\frac{n!}{n} = (n-1)!$ circular permutations.
8. x^{13}: 420; x^{12}: -3640

Class Exercises, page 602 1.a. $\frac{1}{3}$ **b.** 1 **c.** 0
2.a. $\frac{1}{52}$ **b.** $\frac{1}{4}$ **c.** $\frac{1}{13}$ **d.** $\frac{1}{2}$ **e.** $\frac{3}{13}$ **f.** $\frac{3}{26}$ **3.** $\frac{1}{6}$
4.a. $\frac{1}{18}$ **b.** $\frac{1}{12}$ **c.** $\frac{5}{36}$ **5.** 60% **6.** 0.18 **8.a.** Yes
b. The sample pts. are not equally likely.
9. Some outcomes, such as the queen of hearts,
correspond to more than one element of S.
10. This is incorrect because population is not
evenly distributed among the states.

Written Exercises, pages 603–605 1.a. $\frac{1}{2}$ **b.** $\frac{1}{4}$
c. $\frac{3}{4}$ **3.a.** $\frac{1}{4}$ **b.** 0 **c.** 1 **5.a.** $\frac{1}{214}$ **b.** $\frac{4}{535}$ **c.** $\frac{531}{535}$
7.a. There is not an equal number of phone nos.
in each borough. **b.** The no. of phone nos. in
Manhattan and the no. of phone nos. in all of New
York City **9.a.** $\frac{5}{36}$ **b.** $\frac{1}{6}$ **c.** $\frac{5}{36}$ **11.** $\frac{5}{6}$ **13.b.** $\frac{1}{4}$

15.a. $\frac{4}{7}$ **b.** $\frac{3}{7}$ **17.a.** $\frac{1}{64}$ **b.** $\frac{1}{32}$ **c.** 9 **19.b.** $\frac{1}{6}$ **c.** $\frac{1}{2}$
23. $\frac{1}{120}$ **25.** $\frac{3}{8}$ **27.** $\frac{1}{10}$ **29.a.** $\frac{81}{100}$ **b.** $\frac{19}{100}$

Class Exercises, page 609 2. No **3.a.** $\frac{8}{15}$ **b.** $\frac{9}{16}$
c. No

Written Exercises, pages 609–613

1.a.

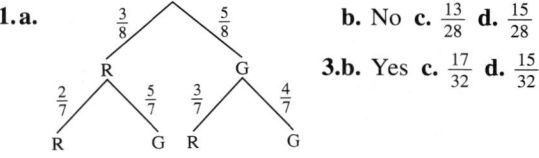

b. No **c.** $\frac{13}{28}$ **d.** $\frac{15}{28}$

3.b. Yes **c.** $\frac{17}{32}$ **d.** $\frac{15}{32}$

5.b. $P(HH) = \frac{1}{17}$; $P(HN) = \frac{13}{68}$; $P(NH) = \frac{13}{68}$;
$P(NN) = \frac{19}{34}$ **7.a.** $P(F) = P(F|D) = \frac{3}{13}$
b. $P(R) = \frac{1}{2}$, $P(R|D) = 1$; $P(R) \neq P(R|D)$
9.b. $P(3) = \frac{1}{8}$; $P(2) = \frac{3}{8}$; $P(1) = \frac{3}{8}$; $P(0) = \frac{1}{8}$
11.a. $\frac{1}{56}$ **b.** $\frac{5}{28}$ **13.a.** $\frac{147}{400}$ **b.** $\frac{1}{20}$ **15.a.** $\frac{29}{90}$ **b.** $\frac{29}{147}$
17. Yes **21.a.** $\frac{11}{50}$; $\frac{11}{850}$ **b.** $\frac{2}{17}$ **c.** $\frac{1}{5525}$ **d.** $\frac{4324}{5525}$
e. $\frac{1201}{5525}$ **f.** $\frac{48}{221}$ **25.a.** $\frac{30}{49}$ **b.** $\frac{19}{49}$ **27.** $\frac{3}{5}$
29.a. 0.064 **b.** 0.784 **c.** 0.8208

Class Exercises, page 615 1. $4\left(\frac{1}{2}\right)^3\left(\frac{1}{2}\right) =$
$P(3 \text{ H and } 1 \text{ T})$; $6\left(\frac{1}{2}\right)^2\left(\frac{1}{2}\right)^2 = P(2 \text{ H and } 2 \text{ T})$;
$4\left(\frac{1}{2}\right)\left(\frac{1}{2}\right)^3 = P(1 \text{ H and } 3 \text{ T})$; $\left(\frac{1}{2}\right)^4 = P(4 \text{ T})$ **3.** $\frac{1}{2}$
4.a. $\frac{49}{198}$ **b.** The 1st and 2nd drawings are not
independent trials. **c.** $\frac{1}{4}$

Written Exercises, pages 616–618 1.a. $\frac{1}{8}$ **b.** $\frac{3}{8}$ **c.** $\frac{3}{8}$
d. $\frac{1}{8}$ **3.a.** $\frac{1}{16}$ **b.** $\frac{1}{4}$ **c.** $\frac{3}{8}$ **d.** $\frac{1}{4}$ **e.** $\frac{1}{16}$ **5.** $\frac{25}{216}$
7.a. $\frac{1}{4}$ **b.** $P(0) = \frac{9}{16}$; $P(1) = \frac{3}{8}$; $P(2) = \frac{1}{16}$
9.a. $\frac{1}{4096}$ **b.** $\frac{9}{2048}$ **11.a.** $\frac{64}{125}$ **b.** $\frac{1}{125}$ **c.** $\frac{124}{125}$
13.a. Emp. prob. (0.3125) > theor. prob. (0.125)
b. 0.16 **c.** 83; $\approx$0.13253 **15.a.** 0.396 **17.** 0.387
19.a. 366 **b.** 0.117 **c.** 23

Class Exercises, pages 620–621 Exs. 1 and 2:
Probabilities in column match with events: **1.** c,
b, a, d **2.** b, a, d, c **3.a.** No aces **b.** 0.3412

Written Exercises, pages 621–623 1.a. $\frac{13!\,47!}{8!\,52!}$

b. $\frac{39!\,47!}{34!\,52!}$ **c.** $1 - \frac{39!\,47!}{34!\,52!}$ **3.** $P(0\ R) = \frac{3}{28}$;

$P(1\ R) = \frac{15}{28}$; $P(2\ R) = \frac{5}{14}$ **5.a.** $\frac{33}{323} \approx 0.102$

b. $\frac{352}{969} \approx 0.363$ **c.** $\frac{616}{1615} \approx 0.381$ **d.** $\frac{224}{1615} \approx 0.139$

e. $\frac{14}{969} \approx 0.014$ **7.** $\frac{1}{2}$ **9.** $\frac{4}{5}$ **11.** $\frac{3}{10}$ **13.** $\frac{2}{5}$
15.a. 1.64×10^{-5} **b.** 4.64×10^{-6} **c.** 0.981 **d.** 0

17. $\frac{8}{21}$ **19.** $\frac{8279}{23{,}765} \approx 0.348$ **21.a.** 0.0681

Class Exercises, page 625

1.a.
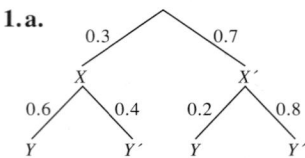
b. Lower: $P(Y|X')$;
upper: $P(X')$
c. 0.18
d. 0.14
e. 0.32
f. 0.5625

2.a. 0.2 **b.** 0 **c.** 0.05 **d.** 0.36 **e.** 0.41 **f.** 0.122

3. $P(F'|T) = 1 - P(F|T)$ or $\frac{P(F' \text{ and } T)}{P(T)}$

Written Exercises, pages 626–629 1.a. 0.6 **b.** 0.7
c. 0.5 **d.** 0.12 **e.** 0.3 **f.** 0.42 **g.** 0.286 **h.** 0.714
3.a. 0.5 **b.** 0.4 **c.** 0.1

5.a.
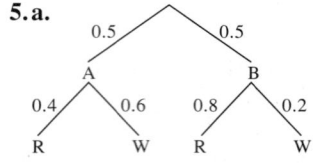
b. 0.6
c. 0.333
7.b. 3.8%
c. 0.789
9.a. 4.41%
b. 0.549

11.c. 0.853 **d.** 0.938 **13.** $\frac{1}{7} \approx 0.143$

15. $\frac{9}{13} \approx 0.692$ **17.b.** 0.74 **d.** 0.75 **e.** 2 **f.** 0.7

Class Exercises, pages 632–633 1. 7 **2.** 1.2
3. Yes **4.** Ans. may vary. **5.** $\frac{1}{8}, \frac{3}{8}, \frac{3}{8}, \frac{1}{8}$;
payoff = $1

Written Exercises, pages 633–635 1. 0 **3.** 55
5. Not fair; B **7.** Fair **9.** 11¢ loss **11.** 0
13. 50¢ gain **15.** Expected value $= \sum_{i=1}^{n} x_i P(x_i)$
17. $20,850 **21.** 15¢ loss **23.** $75

Chapter Test, pages 636–637 1. Empirical
probability is determined by observing what
previously happened. Theoretical probability is
determined by reasoning about the events.
2.a. $\frac{1}{24}$ **b.** $\frac{1}{8}$ **c.** $\frac{5}{8}$ **3.a.** $\frac{5}{18}$ **b.** $\frac{1}{9}$ **c.** $\frac{1}{8}$

4.a. 0.59 **b.** 0.991
5.a. 0.034 **b.** 0.081
6.a. See diagram
at right.
b. 94.5%
c. 0.004
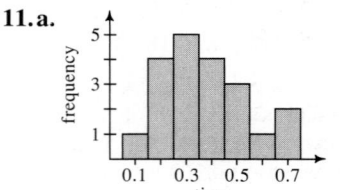
7.a. Payoffs: $1, $2, $3, $4, $−20;
probabilities: $\frac{5}{32}, \frac{10}{32}, \frac{10}{32}, \frac{5}{32}, \frac{1}{32}$ **b.** $1.09

Class Exercises, page 643 1.a. Mean = 3.2;
median = 3; mode = 5 **b.** Mean = 6.875;
median = 7.5; modes = 4 and 8 **c.** Mean = 2;
median = 2; no mode **d.** Mean = 2.2;
median = 2.5; mode = 3 **3.a.** Yes **b.** Yes **c.** No
4. Mean: decreases; median and mode: no change

Written Exercises, pages 643–648
1. Mean = 5.4; median = 6; mode = 6
3. Mean = 14.1; median = 6; mode = 5
5.a. Median = 13.5; modes = 9, 13, and 16
7.a. 24, 25, 26, 27 **b.**
c. .2605

24	2 5
25	2 6 7 9
26	0 1 1 3 9
27	1 6 7

24|2 represents .242
9.a. 60% of scores were ≤ 7; 20% were > 8. **b.** 6

11.a.

[histogram: frequency vs. time with values 0.1, 0.3, 0.5, 0.7]

b. Mean = 0.375;
median = 0.35;
mode = 0.3

c. See diagram at
right. Note that
frequencies can be
expressed as
integers, instead of
as percents.

[graph: rel. cum. freq. vs. time with 50% and 100% marks, time axis 0.1, 0.3, 0.5, 0.7]

13.a. Mean = $9.40; median = $8.50; mode = $8.00 **b.** The mean rises to $11.00; the median and mode do not change. **17.b.** Mean = $4.98; median = $5.00; mode = $4.50 **25.a.** Mean = $17,950; median = $14,500 **b.** Mean = $18,600; median = $15,500

Class Exercises, page 651 **1.** Extremes = 137, 225; range = 88 **2.** Lower = 170; upper = 190; iq range = 20 **3.** 178 **4.** $225 > 1.5(20) + 190$; other outliers are 137 and 222.

Written Exercises, pages 651–653 **1.b.** Vermont's no. is not; California's no. is.

5.
```
1 | 1
2 | 3 6
3 | 1
4 | 1 4 9
5 | 2 6
```
$5 \mid 2 = 520$ calories

(number line 0 to 600, box plot shown)

Class Exercises, pages 657–658 **1.a.** Team A's hts. are more closely clustered around 192 cm than Team B's. **b.** B **2.** x_i sum = 32; x_i^2 sum = 356; mean = 8; variance = 25; $s = 5$ **3.** The mean would dec. by 1; the standard deviation would not change. **4.a.** 1 **b.** 2 **c.** 2.5 **d.** -0.5 **e.** 0 **5.** 1a **6.** Yes, if all the data are equal.

Written Exercises, pages 658–660 **1.** Mean = 7; $s = 1.90$ **3.a.** Row A: $\bar{x} = 3$, $s = 1.41$; Row B: $\bar{x} = 13$, $s = 1.41$ **b.** $\bar{x} = 23$, $s = 1.41$ **5.** $\bar{x} = 78$, $s = 14$ **7.a.** 2 **b.** -0.5 **c.** 0 **d.** -1.3 **9.** Math: 1.1; verbal: 0.8 **13.** Smallest = 0; largest = 4 **15.** $\bar{x} = 10$, $s = 1.45$

Class Exercises, page 666 **2.** 88.49% **3.** 3.59% **4.** 27.43% **5.** 90.32% **6.** 34.13% **7.** 68.26% **8.** $P(2) - P(-2) = 0.9772 - 0.0228 = 0.9544 \approx 95\%$

9.a.

(normal curve with scores 200 300 400 500 600 700 800; standard value −3 −2 −1 0 1 2 3; 68%, 95%, 99%)

b. (1) 2.28%; (2) 2.28%; (3) 68.26% **c.** 630 **d.** 430

10.a.

(normal curve with reaction times (s) 0.5 1.0 1.5 2 2.5 3.0 3.5; standard value −3 −2 −1 0 1 2 3; 68%, 95%, 99%)

b. (1) 0.13%; (2) 2.28%; (3) 68.26% **c.** 1.6 s **d.** 2.4 s

Written Exercises, pages 667–669 **1.a.** 30.85% **b.** 6.68% **c.** 68.26% **d.** 30.23% **3.a.** 9.68% **b.** 30.85% **c.** 43.32% **d.** 1.66% **5.a.** 2.28% **b.** 1.31% **c.** 3.59% **7.** 11.96% **9.a.** 580 **b.** 430 **11.** A: 84; B: 75 **13.a.** 100% **b.** 2.0074 L **c.** 2.0084 L **15.** 11.51% **17.b.** 0.1473 **c.** 0.6473 **d.** Too small; $P(0.4) = 0.6554$

Class Exercises, page 672 **1.** c **2.** b **3.** a **4.** d **5.** c **6.** c **7.** 44%

Written Exercises, pages 673–674 **5.** 44% **7.** 2400 yr **9.** 1040 **11.** 47%

Class Exercises, pages 677–678 **1.a.** 0.1 **b.** 95%: $0.2 < p < 0.6$; 99%: $0.1 < p < 0.7$ **2.a.** 0.1 **b.** 95%: $0.3 < p < 0.7$; 99%: $0.2 < p < 0.8$ **3.a.** $\bar{p} = 0.5$, $s = 0.05$ **b.** $0.4 < p < 0.6$ **4.** Sample size **5.** No

Written Exercises, pages 678–680 **1.** $\bar{p} = 0.7$, $s = 0.145$; $0.41 < p < 0.99$ **3.** $\bar{p} = 0.36$, $s = 0.048$; $0.264 < p < 0.456$ **5.** $\bar{p} = 0.0625$, $s = 0.012$; $0.0385 < p < 0.0865$ **7.** $0.015 < p < 0.045$ **9.** $0.004 < p < 0.176$ **11.** $0.724 < p < 0.876$ **13.** 4 **15.** 95% **17.** 95%: $0.537 < p < 0.663$; 99%: $0.505 < p < 0.695$ **19.b.** $\frac{1}{4}$ **21.** 10,000 **23.** 400

Chapter Test, page 681
1. Descriptive statistics involves collecting, organizing, and summarizing data. Inferential statistics involves drawing conclusions about a population based on a sample.

2.a. See diagram at right.
  b. Mean = $112,667; median = $120,000; mode = $120,000

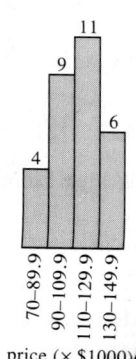

(bar graph; bars: 4, 9, 11, 6; price (× $1000); categories 70–89.9, 90–109.9, 110–129.9, 130–149.9)

3.a. Median = 26; **c.**

first = 22;

third = 30

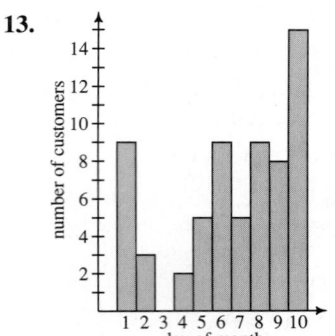

b. range = 18; iq range = 8
4. English **5.** 7.83 **6.a.** 15.87% **b.** 2.28%
c. 1330 **d.** 1270 **7.** 44% **8.** 95%:
$0.778 < p < 0.842$; 99%: $0.762 < p < 0.858$

Class Exercises, pages 687–688

1. Ans. may vary. **a.** $\frac{1}{2}$ **b.** 2 **c.** $y = \frac{1}{2}x + 2$

2.a. $\bar{x} = 11$; $\bar{y} = 7$; $\overline{xy} = 83$; $s_x = 3.571$; $s_y = 2$
b. (11, 7) **c.** 0.471 **d.** $y = 0.471x + 1.82$ **e.** 0.84
4.a. 0.6 **b.** 0.8 **c.** −0.6 **d.** −0.8 **e.** 0 **f.** 0.2
5.a. Pos. **b.** Pos. **c.** Neg. **d.** Neg. **e.** 0
6.a. ≈1.75 **b.** Thousands of dollars per yr
c. ≈$1750 **7.** No; the nos. of firefighters and students are both related to the population of Chesterfield.

Written Exercises, pages 688–691 **1.** Pos. **3.** Neg.
5. Pos. **7.** Neg. **9.** $r = 0.89$; $y = x + 0.5$
11. $r = -0.93$; $y = -1.5x + 8.5$ **13.a.** B **b.** C
c. F **15.** $y = 0.018x + 0.007$; 388.5
19. $y = 0.17x - 1.075$; $r = 0.8246$

Class Exercises, page 694
1.a. $\log y = (\log 3)x + \log 2$ **b.** log 3 **c.** Vert. int.
is log 2. **2.** $y \approx (6.31)19.95^x$ **3.** $y \approx (100)3.162^x$
4. $y \approx (316.2)0.0398^x$ **5.** $y \approx (54.6)4.482^x$

Written Exercises, pages 694–697
1. $y \approx (63.1)39.81^x$ **3.** $y \approx (199.5)0.3981^x$
5. $y \approx (24.53)1.649^x$ **7.** $y = (2)1.5^x$
9.a. $y = (15,070)0.7989^x$ **b.** $1596 **c.** ≈$15,000
13.a. $\log V = -0.1766t + 0.1811$
b. $V = (1.517)0.6659^t$ **15.a.** 74°F **b.** $T = 74$
c. $T = (103.3)0.9627^t + 74$ **17.b.** Ans. may vary.
$\log(y - 100) = -0.0075x + 1.225$;
$y = (16.79)0.9829^x + 100$

Class Exercises, page 701
1.a. $\log y = \log 4 + 3 \log x$ **b.** 3 **2.** $y = 100x^5$

$= 3162x^{2.5}$ **4.** $y = ex^{-3}$ **5.** $y = 1.649x^2$

7. 3

...cises, pages 701–704 **1.** $y = 10x^4$

... $y = 24.53x^{1.6}$ **7.** Ans. may vary.

... 11. 2 **13.** $\frac{1}{2}$ **15.** $\frac{2}{3}$

...ary.

$...^{5/2}$ **21.** $y = (3.065)x^{0.6141}$

Mixed Exercises, pages 705–709 Exs. 1–13: Ans.
may vary. **1.a.** $y = (8.385 \times 10^{-4})x^{2.69}$
b. ≈0.7 lb **3.a.** $T = (13.13)P^{0.3061}$ **b.** ≈94°C
5.a. $W = aC^3$ **b.** $W = 0.394C^3$ **c.** ≈41 lb
7.b. $G = (17.15)1.063^x$ **c.** ≈7720 **9.a.** ≈19°C
b. $T = 19$ **c.** $\log(19 - T) = -0.0166t + 1.4037$
d. $T = 19 - (25.33)0.9625^t$
11. $y = (3.371)0.9995^x + 5$ **13.a.** 5; 8 = B
b. $y = 0.9054x + 0.4932$; ≈7.7

Chapter Test, page 710 1.a. ≈0.71
b. $y = 0.82x - 4.9$ **c.** Since the corr. coeff. is only
0.71, the model probably has limited accuracy in
predicting precipitation. **2.** No; both scores are
probably related to other factors, perhaps I.Q.,
rather than to each other. **3.** $y = 100(1000)^x$
4.a. $y = (57.86)1.065^x$, $x =$ no. of yr after 1900
b. $22,940 **5.** $y = (0.01)x^{1.7}$ **6.** $m = 2$; cost
varies as area, which varies as the square of the
diagonal. **7.a.** $y = (4.099 \times 10^{-10})d^{1.5}$
b. ≈92,560,000 mi

Cumulative Review, pages 711–712 1. 27 **2.** 2100
3. 126 **4.** 504 **5.** 60
6. $16x^4 - 16x^3 + 6x^2 - x + \frac{1}{16}$ **7.a.** $\frac{1}{4}$ **b.** $\frac{3}{4}$ **c.** 0
d. 1 **e.** $\frac{1}{13}$ **f.** $\frac{2}{13}$ **8.a.** $\frac{11}{850}$ **b.** None **9.** $\frac{625}{3888}$
10. Diamond: $\frac{4}{17}$; spade: $\frac{13}{51}$ **11.a.** 5.2% **b.** $\frac{6}{13}$
12. $250

13.

mean ≈ 6.63
median = 7
mode = 10

number of customers (vertical axis: 2, 4, 6, 8, 10, 12, 14)

day of month (horizontal axis: 1 2 3 4 5 6 7 8 9 10)

14.a.

4	3 4
•	7 8 9
5	0 0 0 1 1 1 1 2 2 3 4
•	5 5 5 5 6 6 6 6 7 7 7 8 8 8 9
6	1 1 2 2 3 4
•	5 6 8
7	0

b.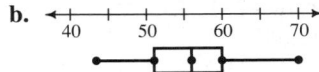

15. Mean = 38.5; $s = 32.4$ **16.a.** 4% **b.** 90%
c. 48%   **17.** Ans. may vary. The people polled are
not representative of the general population. The
factors which keep them at home may influence
their feelings about taxes. Opinions of those who
work during business hours would not be included.
18. $0.383 < p < 0.477$

19.a. 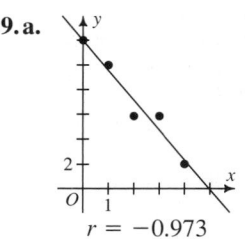   **b.** $y = -2.4x + 12$

$r = -0.973$

20.a. Power; vol. varies as the square of the
diameter. **b.** Linear; sales tax is a % of cost.
c. Asymptotic; the function's value does not go
beyond the boiling pt. of water.
21. $y = 1.97x^{3.016}$ **22.a.** Exponential; populations
tend to double at regular intervals.
b. $P = (1.156)1.786^t$ **c.** ≈ 382

Class Exercises, page 723   1. Meanings are given:
a. For large values of x, the value of $\dfrac{2x^2}{x^2 + 1}$ is
close to 2. **b.** For neg. values of x where $|x|$ is
large, the value of $\dfrac{3x + 1}{2x - 5}$ is close to $\dfrac{3}{2}$. **c.** For
values of x near 2, but greater than 2, the value of
$f(x)$ is close to 3. **d.** For values of x near 2, but
less than 2, the value of $g(x)$ is arbitrarily large.
2. 0, -1, 4, 1   **3.** $\lim\limits_{x \to 2} f(x) \neq f(2)$
4.a. $\lim\limits_{x \to 0^+} f(x) = 0$; $\lim\limits_{x \to 0^-} f(x) = -3$
b. $\lim\limits_{x \to 0^+} f(x) \neq \lim\limits_{x \to 0^-} f(x)$ **5.** The limit of the
quotient of two functions is the quotient of the
limits of the functions.   **6.a.** $\dfrac{5}{7}$ **b.** ∞ **c.** $-\infty$ **d.** 8

Written Exercises, pages 724–725   1. $\dfrac{3}{4}$ **3.** 2
5. $\dfrac{1}{2}$ **7.** 2   **9.** $\dfrac{5}{2}$ **11.** ∞ **13.** $-\infty$ **15.** 0   **17.** 2
19.a. 1 **b.** -1 **c.** Does not exist   **21.a.** 0 **b.** Does
not exist **c.** Does not exist   **23.a.** 1 **b.** 0 **c.** ∞

d. $-\infty$ **25.** $-\dfrac{1}{2}$ **27.** 0   **29.a.** 2 **b.** 4 **c.** $2x$
31. Cont.   **33.** Cont.   **35.** $a = -1$, $b = 0$ **37.** 1
39. 0   **41.** Does not exist   **45.** 2   **47.** 1

Class Exercises, page 728   1. x-int. 2; vert.
asymptotes: $x = 1$, $x = 3$; horiz. asymptote: $y = 0$
2. x-int. 0; vert. asymptotes: $x = \pm 2$;
horiz. asymptote: $y = 0$ **3.** x-ints. ± 2;
vert. asymptote: $x = 0$ **4.** x-ints. ± 3;
horiz. asymptote: $y = 1$ **5.** x-int. 0; vert.
asymptotes: $x = \pm\sqrt{6}$; horiz. asymptote: $y = 2$
6. x-int. 0; vert. asymptotes: $x = \pm 1$
7. If c is a zero of greater multiplicity for g than
for f, there is a vert. asymptote at $x = c$; otherwise,
there is a hole in the graph at $x = c$.

Written Exercises, pages 728–729

1.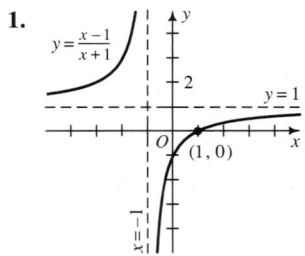

3.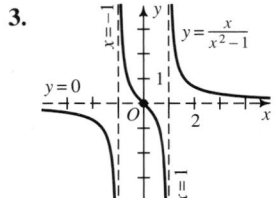

7.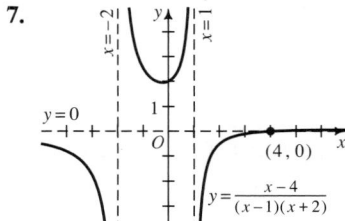

9.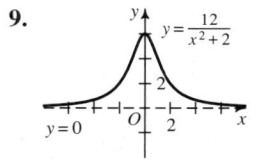

15.

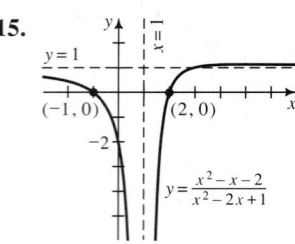

$$y = \frac{x^2 - x - 2}{x^2 - 2x + 1}$$

17.

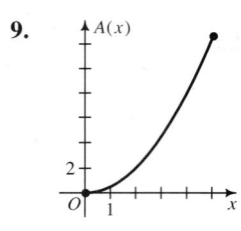

$$y = \frac{1 - x^2}{x}$$

Class Exercises, page 731

1. 20 FOR X = .1 TO 3.1 STEP .1
    30 LET Y = SIN(X)
2. 20 FOR X = .1 TO 3.1 STEP .1
    30 LET Y = 2 * SIN(X)
3.b. Greater

Written Exercises, pages 731–733   **1.** 0.346158926
3. 0.33835001   **5.a.** $\frac{1}{2}$ **b.** $\frac{1}{5}$; 0.20503334
7. 2.72

9.

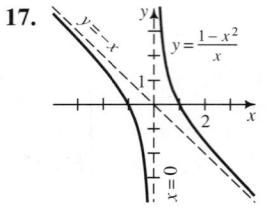

11.

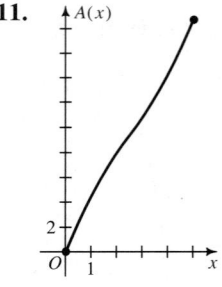

13. 2   **15.** 2   **17.** 2

Class Exercises, page 735   **1.** 6   **2.** $-\sin x$

Written Exercises, pages 735–736

1. $\cos \frac{\pi}{4} = 1 - \frac{\pi^2}{4^2 \cdot 2!} + \frac{\pi^4}{4^4 \cdot 4!} - \frac{\pi^6}{4^6 \cdot 6!} + \cdots$;

$\sin\left(-\frac{\pi}{4}\right) = -\frac{\pi}{4 \cdot 1!} + \frac{\pi^3}{4^3 \cdot 3!} - \frac{\pi^5}{4^5 \cdot 5!} +$

$\frac{\pi^7}{4^7 \cdot 7!} - \cdots$ **3.a.** $1 - \frac{1}{2} + \frac{1}{3} - \frac{1}{4} + \cdots$

b. 2 is not in the interval of convergence.

5. $1 - \frac{x}{1!} + \frac{x^2}{2!} - \frac{x^3}{3!} + \cdots$; all real x

7. $2x - \frac{(2x)^3}{3} + \frac{(2x)^5}{5} - \frac{(2x)^7}{7} + \cdots$; $-\frac{1}{2} \le x \le \frac{1}{2}$

9. $x^2 - \frac{(x^2)^3}{3!} + \frac{(x^2)^5}{5!} - \frac{(x^2)^7}{7!} + \cdots$; all real x

13. 1   **15.** $e^{i\pi} = \cos \pi + i \sin \pi = -1$;
$e^{2i\pi} = \cos 2\pi + i \sin 2\pi = 1$

Class Exercises, page 741   **1.** 5, 7, 9, 11   **2.** 0, 1,
0, 1   **3.** $-1, -1, -1, -1$ **4.** 1, 1.1, 1.11, 1.111
5. Ans. may vary. If $f(x) = x$, then $f^2(x_0) = x_0$;
but $(f(x_0))^2 = x_0^2$.   **6.a.** 0   **b.** Slowly   **7.** At the
graph's intersection with the graph of $y = x$
8. $1 - \sqrt{3}$ is attracting; $1 + \sqrt{3}$ is repelling.
9.a. 0, 1   **b.** Yes; -1 **c.** If $|x_0| > 1$,
$\lim_{n \to \infty} f^n(x_0) = \infty$; if $|x_0| < 1$, $\lim_{n \to \infty} f^n(x_0) = 0$.   **d.** 0
is attracting; 1 is repelling.   **10.a.** -1, 0, 1
b. No   **c.** If $x_0 > 1$, $\lim_{n \to \infty} f^n(x_0) = \infty$; if $x_0 < -1$,
$\lim_{n \to \infty} f^n(x_0) = -\infty$; if $|x_0| < 1$, $\lim_{n \to \infty} f^n(x_0) = 0$.
d. 0 is attracting; -1 and 1 are repelling.

Written Exercises, pages 741–743   **1.** 5, 9, 17, 33
3. 2.5, 0.4, 2.5, 0.4   **5.** 4   **7.** $\frac{1}{2}$ **9.** 4   **11.** x_4 is
not def. because $x_3 < 0$.   **13.a.** $\lim_{n \to \infty} f^n(x_0) = 2$
b. No change   **15.a.** 0.739   **17.a.** 0   **b.** Undef.
19. 0 and 1 are fixed pts.; 2 is an eventually fixed
pt.; $\lim_{n \to \infty} f^n(x_0) = \begin{cases} 1 & \text{if } 0 < x_0 < 2 \\ -\infty & \text{if } x_0 < 0 \text{ or } x_0 > 2 \end{cases}$
21. $|m| < 1$; $m = 1$ when $k = 0$.   **23.a.** 1.62
b. No; 1.62 is an attracting pt.   **25.a.** 3.17, 3.16,
3.16, 3.16

Class Exercises, page 746   **1.a.** $x_0 = 100$,
$x_n = 1.01(x_{n-1}) - 20.6$ **b.** The orbit of $x_0 = 100$
for iterations of $f(x) = 1.01x - 20.6$ **2.** 0.32,
0.435, 0.492, 0.500; 0.500 is equilibrium pt.
3. 0.08, 0.037, 0.018, 0.009; 0 is equilibrium pt.

Written Exercises, pages 746–748   **1.a.** $x_0 = 100$,
$x_n = 1.06(x_{n-1})$ **b.** The orbit of $x_0 = 100$ for
iterations of $f(x) = 1.06x$ **3.a.** $x_0 = 400{,}000$,
$x_n = 1.065(x_{n-1}) - 30{,}000$ **b.** The orbit of
$x_0 = 400{,}000$ for iterations of
$f(x) = 1.065x - 30{,}000$ **5.** Equilibrium pt. at
0.333   **7.** Period-4 cycle: 0.501, 0.875, 0.383,
0.827   **9.** Period-3 cycle: 0.505, 0.957, 0.156
11.a. \$425,678.19   **b.** Because you would also
receive the interest, you would have an additional
\$3,863,749.98.   **15.** $f(x) = (1 + c(1 - x))x + hx$

Chapter Test, page 749 **1.a.** 1 **b.** $\frac{1}{2}$ **c.** $\frac{1}{6}$ **d.** Does

not exist: $\lim\limits_{x \to 0^-} \dfrac{|x|}{x} = -1$, but $\lim\limits_{x \to 0^+} \dfrac{|x|}{x} = 1$.

2.a.

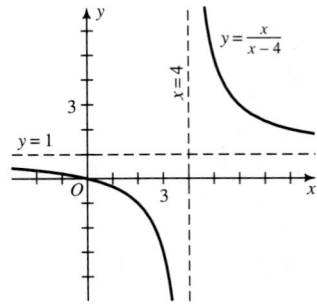

b.

$y = \dfrac{x^2+1}{x^2-1}$

3.a. 5.31330004
b. 10.62660008

4.a. $1 - \dfrac{1}{2!} + \dfrac{1}{4!} - \dfrac{1}{6!} + \cdots$

b. $x^2 - \dfrac{(x^2)^2}{2} + \dfrac{(x^2)^3}{3} - \dfrac{(x^2)^4}{4} + \cdots, \; -1 \le x \le 1$

5. -1.841 **6.** 0 is a repelling fixed pt.;

$\lim\limits_{n \to \infty} f^n(x_0) = \begin{cases} \infty & \text{if } x_0 > 0 \\ -\infty & \text{if } x_0 < 0 \end{cases}$

7. $\lim\limits_{n \to \infty} f^n(x_0) = 0$. The population will not survive

regardless of the initial population.

Class Exercises, page 761 **1.** $8x^7$ **2.** $12x^{11}$

3. $-\dfrac{4}{x^5}$ **4.** $21x^6$ **5.** $\dfrac{3}{2}\sqrt{x}$ **6.** $-\dfrac{6\sqrt{x}}{x^3}$ **7.** $\dfrac{\sqrt[3]{x}}{3x}$

8. $-\dfrac{\sqrt[3]{x^2}}{3x^2}$ **9.** $24x^2 - 14x + 3$

10. $-\dfrac{20}{x^6} + \dfrac{6}{x^4} - \dfrac{9}{x^2}$ **11.a.** $f'(x) = 8x; \; f'(1) = -8$

b. -8 **12.a.** $f'(1) = f'(2) = f'(3) = m$

Written Exercises, pages 762–763 **1.a.** 7 **b.** 4.75

c. 3.31 **d.** $f'(x) = 3x^2; \; f'(1) = 3$ **3.** $10x^4$

5. $-6x^2$ **7.** $-\dfrac{6}{x^3}$ **9.** $\dfrac{\sqrt{5x}}{2x}$ **11.** $7x - 5 + \dfrac{1}{x^2}$

13. 3 **15.** -2 **17.** -13 **19.** $\dfrac{1}{2}$ **21.a.** $2ax + b$

b. 0 **23.** $y = 12x + 16$ **25.** $y = 2x + 3$

27.a, c.

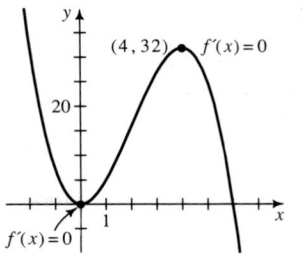

b. 0 or 4

29.a. $\dfrac{2}{3}x^{-1/3}$ **c.**

b. 0

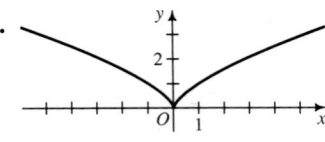

Exs. 31–35: Ans. may vary by a constant. **31.** x^4

33. $\dfrac{1}{2}x^6$ **35.** $2x^{3/2}$

37.a. $x^n + nx^{n-1}h + \dfrac{n(n-1)}{2}x^{n-2}h^2 + \cdots + h^n$

Class Exercises, page 767 **1.a.** $-5, -1, 2, 6$

b. $-5 < x < -1, \; 2 < x < 6$ **c.** $x < -5,$
$-1 < x < 2, \; x > 6$ **2.a.** $-4, 1, 4$ **b.** $x < -4,$
$-4 < x < 1, \; x > 4$ **c.** $1 < x < 4$ **3.** Yes
4. Local max. **5.** Local max. at $x = -1$; local
min. at $x = 2$ **6.** Local max. at $x = -2$; local
min. at $x = 2$

Written Exercises, pages 768–769

1.

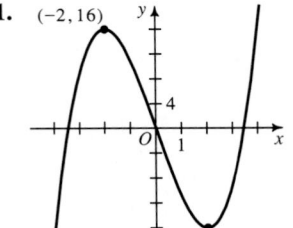

3.

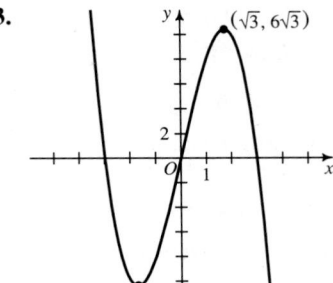

5. Local max.: $(-1, 1)$, $(1, 1)$; local min.: $(0, 0)$
7. Local max.: $(0, 8)$; local min.: $(-\sqrt{3}, -1)$, $(\sqrt{3}, -1)$ **9.** Local max.: $(0, 0)$; local min.: $(-1, -5)$, $(2, -32)$ **11.** Local max.: $(1, 1)$; local min.: $(0, 0)$ **13.** $(3, -5)$ **15.** $(a, 0)$

17.

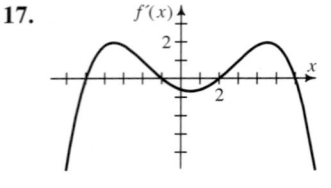

21.a. 1
b. 1; 0; −1; 0; 1
c. $\cos x$

23.

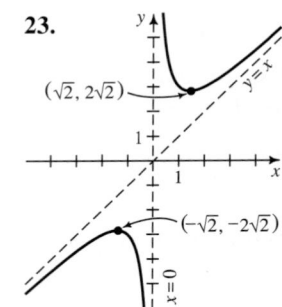

25. Local min.: $\left(\frac{1}{2}, 3\frac{3}{4}\right)$; asymptotes: $y = x^2 + 3x$; $x = 0$

Class Exercises, page 772   **1.** Global max. 5 at $x = 3$, global min. −1 at $x = 0$ **2.** Global max. 2 at $x = 1$, global min. −1 at $x = 4$ **3.** Global max. 6 at $x = 2$; global min. 2 at $x = 0$ **4.** Global max. 2 at $x = 0$; global min. −2 at $x = -2$ **5.** Global max. 2 at $x = 2$; global min. −2 at $x = 1$
6. Global max. 14 at $x = 2$; global min. −14 at $x = -2$ **7.** Global max. 2 at $x = \frac{\pi}{2}$; global min. 0 at $x = 0$, $x = \pi$ **8.** Global max. 1 at $x = 0$; global min. 0 at $x = \pi$

Written Exercises, pages 772–774   **3.** 8 sq. units
5.a. $V(x) = x(8 - 2x)(15 - 2x)$, $0 \le x \le 4$
b. $\frac{5}{3}$ cm   **7.a.** $(-1, -1)$, $(0, 0)$, $(1, 1)$ **b.** $\frac{\sqrt{3}}{3}$
9. 400   **11.** 2 ft by 2 ft by 2 ft   **13.b.** $\frac{32\pi}{3}$ cubic units   **15.a.** \$6.60; \$6.52   **b.** 60 mi/h

Class Exercises, page 778   **1.a.** $(48 - 32t)$ ft/s; 16 ft/s   **b.** -32 ft/s^2 **2.a.** 2 m   **b.** 1 m/s   **c.** Right: $0 \le t < 2$; left: $t > 2$ **d.** $t = 4$ s   **e.** 1 m/s; 0 m/s; -1 m/s   **f.** Right: $t = 0$ s; left: $t = 4$ s   **g.** 2 m

3.a. Left   **b.** 4 m to the left of the origin   **c.** 6 m
d. -3 m/s   **e.** $t = 5$ s   **f.** 2 m/s^2; 2 m/s^2 **4.a.** to the left of the origin   **b.** moving to the right   **c.** decelerating

Written Exercises, pages 778–781

1.a. 32 ft/s
b. $(64 - 32t)$ ft/s; 32 ft/s
c. $t = 2$ s
d. 144 ft
e. $t = 5$ s
f. $2 < t < 5$
g. -32 ft/s^2

h.

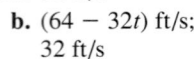

3.a. 2 m   **b.** 1 m/s   **c.** Right: $0 \le t < 3.5$; left: $t > 3.5$ **d.** $t = 6$ s   **e.** 1 m/s; 1 m/s; 0.5 m/s
f. Right: $t = 1.5$ s; left: $t = 5$ s   **g.** 3 m

5.

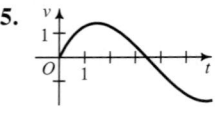

7.

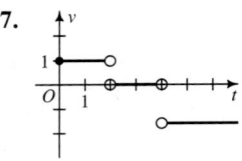

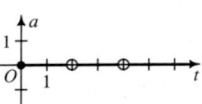

9.

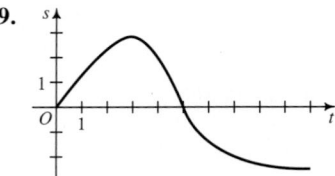

11.a. At the present time t, $P'(t) > 0$.   **b.** At the present time t, $P''(t) < 0$.   **c.** At some future time t, $P'(t) = 0$.   **13.a.** $t = 2$ s   **b.** $t = 0$ s, $t = 2$ s
c. $t = 1$ s   **15.c.** ≈ 101.5 ft/s   **d.** At $t = 2$ s, speed ≈ 102.4 ft/s; the ball hits the ground at $t = 3$ s at a speed of ≈ 125.0 ft/s.

Chapter Test, page 782   **1.a.** $15x^2 - 6x + 1$
b. $-\frac{6}{x^3} + \frac{2}{x^2}$ **c.** $\frac{2\sqrt[3]{3x^2}}{3x} + \frac{\sqrt[3]{x^2}}{3x^2}$ **d.** $-\frac{12}{x^7} + 18x^5$
2. Ans. may vary. Solve $f'(x) > 0$ and $f'(x) < 0$ to find the intervals on which the graph of $f(x)$ is rising and falling, respectively. At a pt. where

$f'(x) = 0$, the graph has a local max. or min. (if $f'(x)$ changes sign), or it flattens out (if $f'(x)$ doesn't change sign).

3.

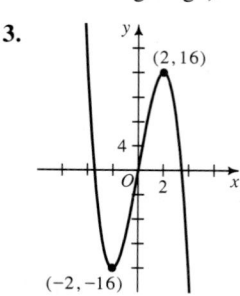

4. $f(2)$ is the max. value of the function.

5.a. $A(x, h) = 2x^2 + 4xh$ **b.** $h = \dfrac{1000}{x^2}$

c. $A(x) = 2x^2 + \dfrac{4000}{x},\ x > 0$

d. $A'(x) = 4x - \dfrac{4000}{x^2};\ 4x - \dfrac{4000}{x^2} = 0;\ x^3 = 1000;$

$x = 10;\ h = \dfrac{1000}{10^2} = 10.$ **6.** Position: 54 units to the right of the origin; velocity: 25 units/s; acceleration: 6 units/s² **7.a.** 3 s **b.** 400 ft **c.** 7 s **d.** −64 ft/s

Cumulative Review, page 783 **1.a.** 3 **b.** $\dfrac{3}{2}$ **c.** $-\infty$

2.a.

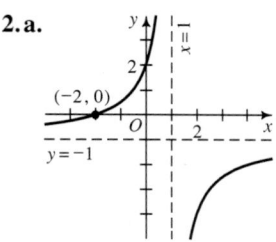

b.

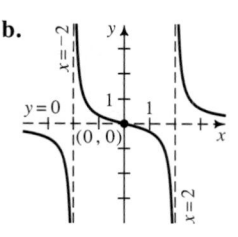

c.

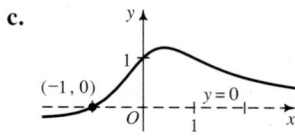

3.a. ≈ 0.169 **b.** ≈ 1.169; since the graph of $y = 1 + \log x$ is the graph of $y = \log x$ translated vertically 1 unit, the area between the two graphs from $x = 1$ to $x = 2$ is equivalent to the area of a rectangle 1 unit high and $2 - 1 = 1$ unit wide.

4.a. $\text{Tan}^{-1}\dfrac{1}{x} = \dfrac{1}{x} - \dfrac{1}{3x^3} + \dfrac{1}{5x^5} - \dfrac{1}{7x^7} + \cdots$

b. $x \le -1$ or $x \ge 1$ **5.a.** 10, 1, −0.8, −1.16 **b.** −1.25 **6.** $\dfrac{1}{2}$ is a fixed pt.; $-\dfrac{1}{2}$ is an eventually

fixed pt.; $\displaystyle\lim_{n\to\infty} f^n(x_0) = \begin{cases} \dfrac{1}{2} & \text{if } |x_0| < \dfrac{1}{2} \\ \infty & \text{if } |x_0| > \dfrac{1}{2} \end{cases}$

7. The population has an equilibrium pt. of 0.6.

8.a. $6x - 1 - \dfrac{4}{x^2} + \dfrac{1}{x^3}$ **b.** $\dfrac{\sqrt{2x}}{2x} + \dfrac{\sqrt{x}}{x^2}$ **9.** 4

10.

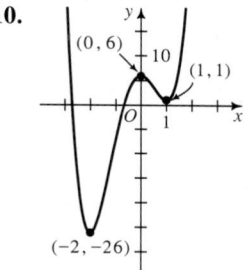

11. $6\dfrac{2}{3}$ in. by $6\dfrac{2}{3}$ in. by $1\dfrac{2}{3}$ in.

12.a. $0 \le t < 4$ **b.** $t > 4$

SAT Sample Questions, pages 790–793 **1.** C **2.** E
3. E **4.** D **5.** B **6.** B **7.** E **8.** E **9.** E
10. B **11.** E **12.** B **13.** D **14.** B **15.** E
16. B **17.** C **18.** B **19.** D **20.** A **21.** D
22. B **23.** D

Mathematics Level I Test Sample Questions, pages 795–796 **1.** C **2.** A **3.** B **4.** D **5.** C **6.** D
7. D **8.** A **9.** E **10.** E **11.** E **12.** B **13.** A
14. E **15.** D **16.** B **17.** D **18.** C **19.** D
20. B

Mathematics Level II Test Sample Questions, pages 797–798 **1.** A **2.** B **3.** E **4.** B **5.** E **6.** C
7. B **8.** C **9.** B **10.** E **11.** D **12.** E **13.** B
14. B **15.** E **16.** E

Mathematics Level IIC Test Sample Questions, pages 798–799 **1.** B **2.** A **3.** B **4.** A **5.** C **6.** B
7. C **8.** E **9.** A **10.** E **11.** C **12.** D **13.** D
14. B **15.** C **16.** B

Appendix 1 Exercises, pages 823–824

1.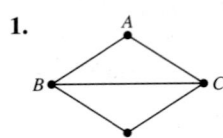

Valences: A: 2, B: 3
C: 3, D: 2
No Euler circuit

3. Ans. may vary. *BACBDC* **5.a.** No **b.** Yes

7.a. Bears and Tigers
c. First: Bears
last: Eagles

b.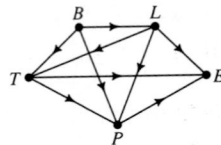

Appendix 1 Exercises, pages 826–827 **1.** No
3.a. Yes **b.** No **c.** At least one of the values x or y must be odd. **5.** *FGHEF*; yes

7.a. 20!
b. 20
c. 20 × 20!
d. ≈1542 yr
9.a. See graph at right.
b. *ADCEBA*

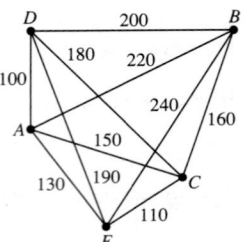

Appendix 1 Exercises, pages 828–829 **1.** *AH, HG, HF, FB, BC, CD, DE, EI* **3.** $n - 1$ **5. b.** *AB, BH, BI, IC, IF, FG, FE, ED* **7.** $9,900

Appendix 2 Exercises, pages 832–834 **1.** (2.5, 3)
3.a. 8; 8; 8 **b.** 6; 5; 6 **c.** 20; 21; 20
5. $y = \frac{1}{2}x + \frac{4}{3}$ **7.** $y = -\frac{1}{2}x + \frac{15}{4}$
9.a. No change **b.** No change **c.** Before replacement, $(\bar{x}, \bar{y}) = (3.6, 2)$; after replacement, $(\bar{x}, \bar{y}) = (3.6, 4)$. **11.a.** Boys: $y = 1.75x + 29.75$; girls: $y = 3.71x - 142.14$ **b.** Boy: ≈166 cm; girl: ≈147 cm **13.a.** $y = 0.4138x + 12.1471$
b. U.S.A. **c.** 49.0

Appendix 3 Exercises, page 837

1.

pos.	neg.	imag.
1	1	0

3.

pos.	neg.	imag.
2	1	0
0	1	2

5.

pos.	neg.	imag.
2	2	0
2	0	2
0	2	2
0	0	4

7.

pos.	neg.	imag.
3	1	0
1	1	2

9.a. $L = -1$; $M = 2$ **b.** $-1 < x < 2$
11.a. $L = -2$; $M = 3$ **b.** $-2 < x < 3$
13.a. $L = -1$; $M = 0$ **b.** $-1 < y < 0$
15.a. $L = -2$; $M = 3$ **b.** $-2 < y < 3$

Appendix 4 Exercises, pages 840–841

1. $\frac{1}{x-1} + \frac{5}{x+5}$ **3.** $\frac{3}{4x-3} + \frac{1}{x+3}$

5. $\frac{4}{x} - \frac{9}{x+1} + \frac{6}{x+2}$ **7.** $\frac{2}{x-2} - \frac{1}{x+2}$

9. $\frac{3}{x+4} - \frac{13}{(x+4)^2}$

11. $\frac{1}{x+1} - \frac{2}{(x+1)^2} + \frac{2}{(x+1)^3}$

13. $\frac{\frac{1}{2}}{x-1} + \frac{\frac{3}{2}}{x+1} - \frac{1}{(x+1)^2}$

15. $\frac{1}{x+3} - \frac{10}{(x+3)^2} + \frac{33}{(x+3)^3} - \frac{36}{(x+3)^4}$

19. $4x - 7 + \frac{1}{x} + \frac{8}{x+1}$ **21.** $\frac{3}{x-2} + \frac{2x-1}{x^2+x-1}$

Appendix 5 Exercises, page 845
1.a. $y = mx + 5$ **b.** $y = -2x + k$ **3.** $y = x$, $y = -x$ **5.** $y = -x \pm 2$ **7.** $y = \frac{2}{3}x - 1$, $y = -1$

9.a. $y = -2x - 9$ **b.** $(-3, 3)$ **c.** $y = \frac{1}{2}x - \frac{3}{2}$
11.a. $y = x + 4$, $y = -x + 4$ **b.** $(-2, 2)$; $(2, 2)$
c. $y = -x$; $y = x$

Refresher Exercises, Chapter 1, page 846
1. II **3.a.** $9\sqrt{7}$ **b.** $2\sqrt{5} + 2\sqrt{10}$ **5.** $-\frac{3}{4}$
7.a. $4x^2 - 20x + 25$ **b.** -1 **9.** $-3, 1$

Refresher Exercises, Chapter 2, page 846
1.a. Linear **b.** Quadratic **c.** Linear **3.** Yes
5. $5x(x - 2)(x + 2)$ **7.b.** 0 **9.** $V(r) = 2\pi r^3$

Refresher Exercises, Chapter 3, page 847
3.a. 5, -5 **b.** $-5 < x < 5$ **c.** $-5 < x$ or $x > 5$
5. $(0, 3)$, $\left(\frac{9}{2}, \frac{21}{4}\right)$ **7.** $(a - 2)(a + 1)(a - 1)$

Refresher Exercises, Chapter 4, page 847

1. -7 **3.** $3a + 2$; $6a + 2$; $3a + 5$

7. $y = \sqrt[3]{x + 4}$ **9.** Ans. may vary. For example, 2 cm $\times$ 3 cm, 6 cm $\times$ 1 cm, $\sqrt{2}$ cm $\times$ $3\sqrt{2}$ cm

Refresher Exercises, Chapter 5, page 848

1.a. $16x^{11}$ **b.** $\frac{3x}{4}$ **3.** 3, 10, 4; $\sqrt{x}$, $x \ge 0$ **5.** 3

7. 0.1

Refresher Exercises, Chapter 6, page 848

1.a. $f(x) = 2(x + 2)^2 - 3$ **b.** $(-2, -3)$

3. $y = 2(x + 4)^2 - 2$ **5.** $4x + 3y = 20$

7. Isosceles; $AC = BC = 10$ **9.** $(-3, -7)$, $\left(\frac{2}{3}, \frac{1}{3}\right)$

Refresher Exercises, Chapter 7, page 849

1.b. ≈ 31.4; ≈ 78.5 **3.** $5\sqrt{3}$ **5.a.** $x = -2$

b. $y = -3$ **c.** $x = 0$, $y = 0$ **d.** $y = \pm\frac{1}{3}x$

7. $x \ge 0$; $f^{-1}(x) = \sqrt{x - 2}$

Refresher Exercises, Chapter 8, page 849

3.a. $\frac{\pi}{2}$ **b.** $-\frac{\pi}{6}$ **c.** $\frac{\pi}{4}$ **7.** -1, $\frac{1}{2}$, 1 **9.** $\frac{a + b}{ab}$

Refresher Exercises, Chapter 9, page 850

1. 6.43   **3.** 20   **5.** 26.4°   **7.** 70°

Refresher Exercises, Chapter 10, page 850

1. $\cos x = -\frac{8}{17}$, $\tan x = \frac{15}{8}$, $\sec x = -\frac{17}{8}$,

$\csc x = -\frac{17}{15}$, $\cot x = \frac{8}{15}$ **3.** -1

5. $\tan(-\theta) = \dfrac{\sin(-\theta)}{\cos(-\theta)} = \dfrac{-\sin\theta}{\cos\theta} = -\dfrac{\sin\theta}{\cos\theta} = -\tan\theta$

7. $\csc\theta$ **9.** Domain: $\{x \mid x \neq \frac{\pi}{2} + n\pi\}$,

Range: {all real numbers}

Refresher Exercises, Chapter 11, page 851

1. $-\cos 36°$ **3.a.** 8.1   **b.** 60.3°   **5.** $11 + 2i\sqrt{3}$

7. $-1 - 3i$ **9.** $8\cos^4\theta - 8\cos^2\theta + 1$

Refresher Exercises, Chapter 12, page 851

1.a. $AO = 5$, $AB = 4$, $OB = \sqrt{65}$ **b.** 127°

3. $y = -3x - 2$ **5.** $(x - 5)^2 + (y + 7)^2 = 25$

7. $(x - 1)^2 + y^2 = (2\cos^2\theta - 1)^2 + (\sin 2\theta)^2 = (\cos 2\theta)^2 + (\sin 2\theta)^2 = 1$ **9.a.** $-\frac{5}{2}$

b. Ans. may vary. For example, $(5, 0, 0)$ and $(0, 2, -1)$

Refresher Exercises, Chapter 13, page 852

3. 170   **5.** 10, $-\frac{35}{3}$ **7.a.** 2.002   **b.** 0.002

Refresher Exercises, Chapter 14, page 852

1.a. $(2, -3)$ **b.** $(1, -1, 2)$ **3.** $(5, -8, 6)$; $(1, -13, -6)$; -13 **5.a.** $(1, 0) \rightarrow (1, 0)$; $(0, 1) \rightarrow (0, -1)$ **b.** $y = x - 1$

7.a. $(1, 0) \rightarrow (0, 1)$; $(0, 1) \rightarrow (1, 0)$ **b.** $y = -x + 1$

Refresher Exercises, Chapter 15, page 853

1. 6   **3.** 6   **5.** $x^3 - 3x^2y + 3xy^2 - y^3$

7.a. $2\sqrt{2}$ cis 45°   **b.** -64

Refresher Exercises, Chapter 16, page 853

1.a. $\frac{1}{4}$ **b.** $\frac{3}{13}$ **c.** $\frac{1}{13}$ **5.** 2,023,203   **7.** $\frac{10}{21}$

Refresher Exercises, Chapter 17, page 854

1. 18   **3.a.** 9   **b.** 6.5

5. $f(g(x)) = x - 1$, $x \ge -3$

Refresher Exercises, Chapter 18, page 854

1. $y = 0.659x + 0.237$ **5.a.** 5.248   **b.** 0.792   **c.** 2.974

Refresher Exercises, Chapter 19, page 855

1.a. $+\infty$ **b.** $\frac{2}{7}$ **3.** $\sqrt{x - 4}$ **5.** $-3 < x < -1$ or $1 < x < 5$ **7.** 2.716; e

Refresher Exercises, Chapter 20, page 855

1. $2x + 3y = -14$ **3.** $4x + 2h$

Credits

Cover: Concept by Ligature, Inc.; Photography by Ralph Brunke

Book Design: Morgan Cain & Associates

Technical art: Morgan Cain & Associates; **Illustrations:** Robert Grace: 237, 323, 375, 411, 499, 537, 660, 736

PHOTOGRAPHS

xxiv courtesy, Martha C. Zuniga. xxxv courtesy, Midori Kitagawa De Leon. xxxv NASA. xxxvi courtesy, American Indian Science and Engineering Society. xxxvi Martine Kempf. xxxvii NASA. xxxvii Wide World. xxxviii David J. Maenza/ The Image Bank. 6 John Lamb/Tony Stone Worldwide. 13 Susan Van Etten. 16 B. Gelburg/Sharpshooters. 19 Ken Straiton/The Stock Market. 20 Owen Franken/Stock Boston. 22 Malcolm Gilson/Tony Stone Worldwide. 23 Bill Gallery/Stock Boston. 29 © Susan Van Etten. 37 right, Ed Baverham Studio. 37 left, © Susan Van Etten. 42 © Bob Daemmrich. 45 Mauritius/Superstock. 46 Roy Morsch/The Stock Market. 48 Mark Segal/Tony Stone Worldwide. 52 Lew Merrim/ Monkmeyer Press Photos. 71 Bob Daemmrich/Stock Boston. 72 Spencer Grant/Stock Boston. 88 The Granger Collection. 94 © 1980, David Em. 99 © Susan Van Etten. 107 Timothy Eagan/Woodfin Camp & Associates. 108 Chris Jones/The Stock Market. 109 Michael L. Abramson/Woodfin Camp & Associates. 114 Bachman/Instock. 117 courtesy, Minoru Yamasaki Associates. 117 © Susan Van Etten. 118 NOAA/ NESOIS. 121 Cecile Brunswick/Peter Arnold, Inc. 126 Roger Dollarhide/Monkmeyer Press Photos. 138 C. Allan Morgan/ Peter Arnold, Inc. 139 Mimi Forsyth/Monkmeyer Press Photos. 146 Frank Siteman/The Picture Cube. 156 Derek Van Etten, Jr. 157 © Susan Van Etten. 165 Bob Daemmrich/Stock Boston. 168 Werner Forman Archive. 172 © Bob Daemmrich. 174 © Susan Van Etten. 175 Steve Elmore/Tony Stone Worldwide. 177 Gabe Palmer/The Stock Market. 179 Kennedy/ TexaStock. 183 Ted Horowitz/The Stock Market. 184 © Paul Conklin. 189 Jerry Howard/Stock Boston. 195 Al Satterwhite/ The Image Bank. 196 Phil Degginger/National Optical Astronomy Observatories. 201 Ellis Herwig/Stock Boston. 203 Kevin Schafer/Peter Arnold, Inc. 204 © Paul Conklin. 207 Roger Dollarhide/Monkmeyer Press Photos. 212 Kitt Peak National Observatory. 223 Grant Heilman/Grant Heilman Photography. 225 Kindra Clineff. 240 Richard Surman/Tony Stone Worldwide. 250 Richard Pasley/Stock Boston. 256 Butch Powell/Stockphotos, Inc. 262 Villafuerte/TexaStock. 265 H/O Photographers, Inc. 266 Moore & Moore Publishing/ Superstock. 271 Photri. 281 Scott Berner/The Picture Cube. 294 B. Christensen/Stock Boston. 300 George F. Riley/Stock Boston. 306 © Susan Van Etten. 307 Klaus Matwijow/Tony Stone Worldwide. 310 Leonard Harris/Stock Boston. 314 Steve Kaufman/Peter Arnold, Inc. 316 Peter Glass/Monkmeyer Press Photos. 327 Chris Luneski/Photo Researchers, Inc. 330 Joachim Messerschmidt/West Light. 332 Jim Anderson/ Woodfin Camp & Associates. 336 NASA. 348 Michael D. Sullivan/TexaStock. 356 © Bob Daemmrich. 362 Tony Freeman/Photo Edit. 362 Richard Pasley/Stock Boston. 363 Grant Le Duc/Monkmeyer Press Photos. 367 Steve Elmore/ Tony Stone Worldwide. 368 John Tesh/Instock. 369 David Joel/Tony Stone Worldwide. 380 Charles Gupton/Stock Boston. 386 Raymond G. Barnes/Tony Stone Worldwide. 390 C. Allan Morgan/Peter Arnold. 394 Kevin Vandivier/ TexaStock. 397 Kindra Clineff. 401 Photri.

409 Right, Manfred Kage/Peter Arnold, Inc. 409 Left, Ray Pfortner/Peter Arnold, Inc. 418 Rick Smolan/Stock Boston. 424 Faerder/Tony Stone Worldwide. 430 © Bob Daemmrich. 837 Herbert Lanks/Monkmeyer Press Photos. 439 Greg Meadows/Stock Boston. 451 Masa Vemura/Tony Stone Worldwide. 457 Peter Menzel/Stock Boston. 464 Doug Armand/Tony Stone Worldwide. 472 Jim Chandler. 476 Kindra Clineff. 483 John Running/Stock Boston. 484 Kindra Clineff. 509 Bill Anderson/Monkmeyer Press Photo. 510 © Susan Van Etten. 511 Runk/Schoenberger/Grant Heilman Photography, Inc. 516 Evans & Sutherland/Photo Edit. 517 © Susan Van Etten. 522 Charles Gupton/Uniphoto. 528 Peter Menzel/Stock Boston. 529 © Bob Daemmrich. 536 Jim Cambon/Tony Stone Worldwide. 538 Alex Langley/ DPI. 539 Hank Morgan/Rainbow. 541 Courtesy, Heirs of the Estate of Norman Rockwell, Norman Rockwell Museum, Sturbridge, MA. 542 Dr. E.R. Degginger. 543 © Bob Daemmrich. 549 Oxford Scientific Films/Animals, Animals. 550 Patty Murray/Animals, Animals. 551 © Orion Press/ West Light. 564 George Hall/Woodfin Camp and Associates. 569 Bob Daemmrich/Stock Boston. 570 Brian Parker/Tom Stack and Associates. 575 © Paul Conklin. 576 Larry Lawfer/The Picture Cube. 579 A & R Photo/TexaStock. 581 Steve Hansen/Stock Boston. 586 © Ronald Sheridan/Tony Stone Worldwide. 587 Takeshi Takahara/Photo Researchers. 589 Bruce Curtis/Peter Arnold, Inc. 590 The Granger Collection. 596 Dr. Vic Bradbury/Photo Researchers, Inc. 601 Susan Van Etten. 602 Bob Daemmrich/Stock Boston. 603 Photri. 604 Peter Menzel/Stock Boston. 611 Mike Malyszko/ Stock Boston. 615 Charles Feil/Uniphoto. 617 Bob Daemmrich/Stock Boston. 618 Ellis Herwig/Stock Boston. 621 © Bob Daemmrich. 623 P.A. Harrington/Peter Arnold, Inc. 627 Andrew Sacks/Tony Stone Worldwide. 634 © Robert Llewellyn. 638 Mitch Reardon/Tony Stone Worldwide. 647 Willie L. Hill/The Image Works. 648 Wide World. 659 Robert Pearcy/Animals, Animals. 661 Laura Dwight/Peter Arnold, Inc. 666 Larry Lawfer/The Picture Cube. 668 Martha Cooper/Peter Arnold, Inc. 669 Peter Gridley/FPG. 670 © Robert Llewellyn. 672 Barbara Filet/Tony Stone Worldwide. 673 Mimi Forsyth/Monkmeyer Press Photos. 675 ©Bob Daemmrich. 677 Boroff/TexaStock. 679 © Bob Daemmrich. 682 Steve Elmore/Tom Stack & Associates. 689 Bob Daemmrich/Stock Boston. 690 © Robert Llewellyn. 692 Michael Stuckey/Comstock. 698 Bob Daemmrich/Stock Boston. 703 Camerique/E.P. Jones. 706 Eric Carle/Stock Boston. 708 Zao Sulle/The Image Bank. 713 courtesy, Agnes Stroud-Lee. 713 courtesy, Joe Fred Gonzalez. 716 Homer Smith/Art Matrix. 737 Comstock. 744 Leonard Lee Rue, III/Stock Boston. 748 Wouterloot- Gregoire/Valan Photos. 750 Homer Smith/Art Matrix. 753 Homer Smith/Art Matrix. 754 both, Homer Smith/Art Matrix. 755 Homer Smith/Art Matrix. 756 Morgan Howarth/ Nawrocki Stock Photo. 774 J. Goell/The Picture Cube. 779 Photri.